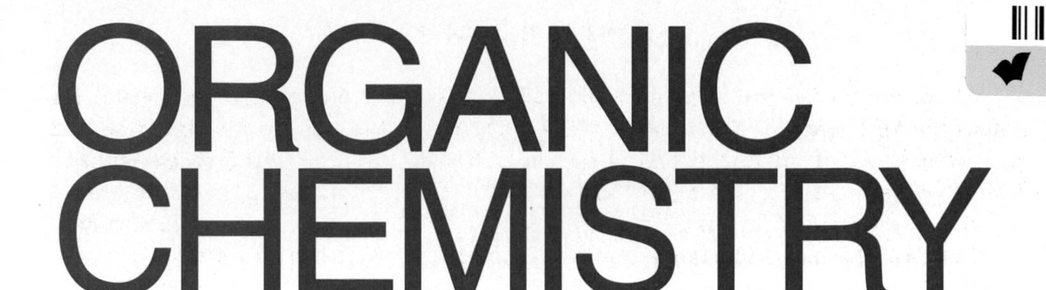

ORGANIC CHEMISTRY

THIRD EDITION

FRANCIS A. CAREY

Department of Chemistry
University of Virginia

THE McGRAW-HILL COMPANIES, INC.

New York St. Louis San Francisco Auckland
Bogotá Caracas Lisbon London Madrid
Mexico City Milan Montreal New Delhi San Juan
Singapore Sydney Tokyo Toronto

McGraw-Hill

ORGANIC CHEMISTRY

Acknowledgments appear on page ii and on this page by reference.

This book is printed on acid-free paper.

1 2 3 4 5 6 7 8 9 0 DOW DOW 9 0 9 8 7 6 5

ISBN 0-07-011212-6

This book was set in Times Roman by Progressive Information Technologies, Inc.
The editors were Karen Allanson and David A. Damstra;
the designers were Rafael Hernandez and Joan E. O'Connor;
the production supervisor was Leroy A. Young.
The photo editor was Kathy Bendo;
the photo researcher was Elyse Rieder.
New drawings were done by Fine Line Illustrations, Inc.
R. R. Donnelley & Sons Company was printer and binder.

Library of Congress Cataloging-in-Publication Data

Carey, Francis A., (date).
Organic chemistry / Francis A. Carey.—3rd ed.
p. cm.
Includes index.
ISBN 0-07-011212-6
1. Chemistry, Organic. I. Title.
QD251.2.C364 1996
547—dc20 95-25237

INTERNATIONAL EDITION

When ordering this title, use ISBN 0-07-114092-1

ABOUT THE AUTHOR

Francis A. Carey, a native of Pennsylvania, was educated in the public schools of Philadelphia, at Drexel University (B.S. in chemistry, 1959), and at Penn State (Ph.D. 1963). Following postdoctoral work at Harvard and military service, he was appointed to the chemistry faculty of the University of Virginia in 1966.

With his students, Professor Carey has published over forty research papers in synthetic and mechanistic organic chemistry. He is coauthor (with Richard J. Sundberg) of *Advanced Organic Chemistry,* a two-volume treatment designed for graduate students and advanced undergraduates, and (with Robert C. Atkins) of *Organic Chemistry: A Brief Course,* an introductory text for the one-semester organic course. He has recently written two articles on organic chemistry that appeared in the latest edition of the *Encyclopaedia Britannica.*

Since 1993 Professor Carey has been a member of the Committee of Examiners of the Graduate Record Examination in Chemistry. Not only does he contribute to the writing of the Chemistry GRE but he also participates in the annual working meetings, which provide a stimulating environment for sharing ideas about what should (and should not) be taught in college chemistry courses.

Professor Carey's main interest shifted from research to undergraduate education in the early 1980s. He regularly teaches both semesters of general chemistry and organic chemistry to classes of over 400 students. He enthusiastically embraces applications of electronic media to chemistry teaching and sees multimedia presentations as the wave of the present.

Frank and his wife Jill, who is a teacher/director of a preschool and a church organist, are the parents of three grown sons: Andy is an environmental chemist; Bob, an attorney; and Bill, a jazz guitarist.

This edition is dedicated to
all the students I have been privileged to teach
at the University of Virginia.

CONTENTS IN BRIEF

CONTENTS

CHAPTER 18
ENOLS AND ENOLATES 727

CHAPTER 19
CARBOXYLIC ACIDS 766

PREFACE

"So . . . what's new?"

In this edition, what's new is a new emphasis, new topics, new tools for learning, and a new look. What continues is a commitment to my students, past, present, and future: those I know personally and those I meet only through the words in the text.

CHANGES IN CONTENT AND ORGANIZATION

A New Emphasis . . .

Organic chemistry has a long tradition of relating the properties of a substance to structure and, more than anything else, the relationship between structure and properties is what *all of chemistry* is about. It is gratifying to see this idea beginning to influence the teaching of chemistry at all levels. New editions of mainstream general chemistry texts are packed with drawings of molecular models, the same kinds of drawings that have been a mainstay of organic textbooks for some time now. The familiarity with pictures of molecular models that students are bringing to their study of organic chemistry frees us to explore **molecular modeling** at a higher level. Two boxes ("Molecular Models" in Chapter 1 and "Molecular Mechanics Applied to Alkanes and Cycloalkanes" in Chapter 3) review the common types of molecular models and introduce the concept of strain-energy calculations. New end-of-chapter problems require the student to construct and analyze molecular models. To make these problems more realistic, we offer an option in which highly praised molecular modeling software can be purchased inexpensively with the text. This software incorporates some novel features, including tools to draw structural formulas suitable for illustrating writing assignments and laboratory reports, as well as the ability to translate a structural formula into a three-dimensional molecular model.

New Topics . . .

During the past several years I have participated in a number of workshops and symposia that assessed where organic chemistry teaching is now and where we should be going. At these meetings, there was a consistent call to reduce the content of the organic course by removing outdated, peripheral, and advanced material. The third edition fully embodies the spirit of this new trend. For example, qualitative tests for functional-group analysis such as the Lucas, Hinsberg, and Tollens test, which are almost never used in the modern practice of organic chemistry, are not included. The haloform reaction, outmoded as an analytical method and overrated as a synthetic one, was included in the first edition, dropped from the second, but reappears here in a much different, more modern form. "The Haloform Reaction and the Biosynthesis of Trihalomethanes" in Chapter 18 is a new boxed essay that connects this old warhorse to the currently controversial issue of halogenated organic compounds in the environment and how they get there. A second essay, "Economic and Environmental Factors in Organic Synthesis" in Chapter 15 traces its origin to another symposium in which I participated. It introduces the concept of *environmentally benign synthesis* to the beginning course in organic chemistry.

Another new essay, *Chiral Drugs,* connects the stereochemical principles of Chapter 7 to drug development in the pharmaceutical industry and explains how advances in organic synthesis provide opportunities to create better drugs. The essay, "Natural and 'Designed' Enediyne Antibiotics" in Chapter 9 shows how biologically active natural products of unusual structure can guide synthetic organic chemists in choosing promising target structures. There is a total of five new boxed essays in the third edition.

Functional-Group Transformations and Reaction Mechanisms . . .

Experience has shown that students learn organic chemistry most readily when they are presented with the small self-contained units that characterize the **functional-group approach.** As they learn more material, they begin to conceptualize it according to **reaction mechanisms.** As in earlier editions, the text is organized according to functional groups but stresses reaction mechanisms early and often. One of the most satisfying things that happens in my own course is that each spring, around Chapter 20, the better students are thinking almost entirely mechanistically and organizing the material in their minds that way. They have mastered the art of thinking like an organic chemist. To encourage this even more actively, the number of mechanistic problems has increased with each successive edition while, especially in this one, functional-group drill problems that seem unduly repetitious or unnecessary have been removed.

It has been brought to my attention that *the concepts and reactions covered in Chapter 4 easily permit the S_N1 and S_N2 mechanisms to be defined there* instead of their more customary place in Chapter 8. I have tried this out with my own students, found it to work well, and have revised Chapter 4 accordingly.

We often stress to students that a mechanism is our present best guess as to how a reaction takes place. It should come as no surprise to them or us that occasionally a "textbook mechanism" needs to be revised or replaced. It is in this spirit that the section describing the mechanism of addition of hydrogen halides to alkynes has been rewritten. The new version conforms to the currently held view that vinylic carbocations are not involved and that addition occurs via a termolecular transition state.

One of my goals has been to make the transition from general chemistry to organic chemistry smoother. Within reason, I have tried to use the same units and terminology that students have already learned. In this edition I have adopted the term "elementary step" to describe the individual steps in a reaction mechanism. This term, or one very similar to it, is used in most mainstream general chemistry texts.

A New Look . . .

All of the figures, previously hand-drawn, have been redone as **electronic art.** More than most fields of chemistry, organic chemistry is a visual discipline. We communicate not in the language of mathematics but in the imagery of structural formulas and molecular models. Computer-generated art is well suited to making our three-dimensional science actually look three-dimensional. On the other hand, the buckminsterfullerene cover, which was computer-generated in the earlier editions, is replaced by an artist's interpretation in this one. The drawing seems to me to be a lot like organic chemistry. Both appear complicated, even mysterious at first. But with the increasing familiarity that comes from regular contact, both become interesting, then intriguing, and even inspiring.

Inasmuch as **high-field nmr spectra** are now the standard in organic chemistry, we have replaced many of the 60 MHz spectra by 300 MHz versions. Many, but not all. Since extensive compilations of 60 MHz spectra are so common and promise to remain so for a long time, students need to be familiar with the somewhat different appearance of spectra run at 60 MHz versus 300 MHz.

The **writing style** in certain parts of the previous two editions was probably too formal. I have tried to remedy this and am pleased with the results. Making the language more direct has, I think, improved the clarity of the presentation and sharpened its focus.

STUDY AND LEARNING AIDS FOR THE STUDENT

Learning Aids:

We have not compromised with respect to the generous amount of space reserved for features designed to help students. The first edition included a number of features that were well received by teachers and students alike. These features, notable because of their novelty or prominence, included:

- annotated summary tables
- sample solutions to problems within the body of the text
- specific examples of reactions taken from the literature
- the routine naming of all of the compounds in an equation

All were refined and expanded in the second edition. There they were joined by additional features:

- a detailed glossary defining over 500 important terms
- boxed essays
- marginal notes

Both boxed essays and marginal notes were standard fare in general chemistry texts but neglected in organic texts prior to their appearance in the second edition. Again, users of the text found these features enhanced its value, and thus the number of boxed essays and marginal notes has been increased in this new edition.

SUPPLEMENTS

Study Guide and Solutions Manual by Francis A. Carey and Robert C. Atkins. This valuable supplement provides solutions to all problems in the text. More than simply providing answers, most solutions guide the student with the reasoning behind each problem. In addition, each chapter includes summaries headed *Important Terms and Concepts* or *Important Reactions.* These are intended to provide brief overviews of the major points and topics presented in each chapter of the text. Each chapter of the *Study Guide and Solutions Manual* concludes with a Self-Test designed to assess the student's mastery of the material.

Presenting Organic Chemistry by Francis A. Carey. Many schools are now, or soon will be, equipped with ''electronic classrooms'' that feature computer-driven projection systems. *Presenting Organic Chemistry* is a collection of more than 2,000 layered slides that I have prepared and tested for more than two years at the University of Virginia. These slides cover the core topics of a two-semester organic chemistry course and are available to adopters on a CD-ROM disk in both Macintosh and Windows versions.

Overhead Transparencies. 200 full color transparencies of illustrations from the text include reproductions of spectra, orbital diagrams, key tables, computer-generated molecular models, and step-by-step reaction mechanisms.

Electronic Overheads. *All* of the text figures and tables are available to adopters on one CD-ROM disk. They can be viewed on a monitor, projected onto a screen in an electronic classroom, or printed as hard copies or transparencies.

Test Bank. A collection of 1000 multiple-choice questions prepared by Professor Bruce Osterby of the University of Wisconsin-LaCrosse. It is available to adopters in print, Macintosh, or Windows format.

Molecular Modeling Software. A spectacular molecular modeling/structure drawing package now accompanies this text. The package, CS ChemOffice Ltd. 3.0, comprises two well-known and highly-regarded programs: ChemDraw and Chem3D. As mentioned earlier in this preface, this software package gives students the ability to draw structural formulas as well as the ability to translate a structural formula into a three-dimensional molecular model. In addition, it contains a library of assembled models cross-referenced to this text. McGraw-Hill is making it available to students with the text at an inexpensive price.

Please contact your McGraw-Hill representative for additional information concerning these supplements.

ACKNOWLEDGMENTS

Special thanks go to longtime teammates, Robert C. Atkins and David A. Damstra. Both have been on board since the first edition. Bob Atkins is much more than my coauthor on the *Solutions Manual* that accompanies this text. His sharp eye and careful attention to detail has kept me from committing more blunders than I care to admit. His enthusiasm for the project is exceeded only by his reliability. One couldn't ask for a better colleague. David Damstra, who supervised production at McGraw-Hill, is the ultimate professional—organized, dedicated, and willing to do the little bit extra that makes a difference.

It was a special pleasure working with Jennifer Speer, who guided the book through its early development. I would also like to thank Denise T. Schank, publisher of science and mathematics at McGraw-Hill, for her continuing support and encouragement.

This text has benefited from the comments offered by a large number of teachers of organic chemistry who reviewed it at various stages during its three editions. I appreciate their help and wish to acknowledge those who reviewed the text during the preparation of the third edition. They include:

Edward Alexander, San Diego Mesa College

Ronald Baumgarten, University of Illinois–Chicago

Barry Carpenter, Cornell University

John Cochran, Colgate University

I. G. Csizmadia, University of Toronto

Lorrain Dang, City College of San Francisco

Graham Darling, McGill University

Debra Dilner, U.S. Naval Academy

Charles Dougherty, Lehman College, CUNY

Fillmore Freeman, University of California–Irvine

Charles Garner, Baylor University

Rainer Glaser, University of Missouri–Columbia

Ron Gratz, Mary Washington College

Scott Gronert, San Francisco State University

Daniel Harvey, University of California–San Diego

John Henderson, Jackson Community College

Stephen Hixson, University of Massachusetts–Amherst

C. A. Kingsbury, University of Nebraska—Lincoln

Nicholas Leventis, University of Missouri–Rolla

Kwang-Ting Liu, National Taiwan University

Peter Livant, Auburn University

J. E. Mulvaney, University of Arizona

Marco Pagnotta, Barnard College

Michael Rathke, Michigan State University

Charles Rose, University of Nevada–Reno

Ronald Roth, George Mason University

Martin Saltzman, Providence College

Patricia Thorstenson, University of the District of Columbia

Marcus Tius, University of Hawaii at Manoa

Victoria Ukachukwu, Rutgers University

Thomas Waddell, University of Tennessee–Chattanooga

George Wahl, Jr., North Carolina State University

John Wasacz, Manhattan College

Comments, suggestions, and questions are welcome. The second edition produced a large number of e-mail messages from students. I found them very helpful and invite you to contact me at fac6q@virginia.edu.

Francis A. Carey

INTRODUCTION

At the root of all science is our own unquenchable curiosity about ourselves and our world. We marvel, as our ancestors did thousands of years ago, at the ability of a firefly to light up a summer evening. The colors and smells of nature bring subtle messages of infinite variety. Blindfolded, we know whether we are in a pine forest or near the seashore. We marvel. And we wonder. How does the firefly produce light? What are the substances that characterize the fragrance of the pine forest? What happens when the green leaves of summer are replaced by the red, orange, and gold of fall?

THE ORIGINS OF ORGANIC CHEMISTRY

As one of the tools that fostered an increased understanding of our world, the science of chemistry—the study of matter and the changes it undergoes—developed slowly until near the end of the eighteenth century. About that time, in connection with his studies of combustion the French nobleman Antoine-Laurent Lavoisier provided the clues that showed how chemical compositions could be determined by identifying and measuring the amounts of water, carbon dioxide, and other materials produced when various substances were burned in air. By the time of Lavoisier's studies, two branches of chemistry were becoming recognized. One branch was concerned with matter obtained from natural or living sources and was called *organic chemistry.* The other branch dealt with substances derived from nonliving matter—minerals and the like. It was called *inorganic chemistry.* Combustion analysis soon established that the compounds derived from natural sources contained carbon, and eventually a new definition of organic chemistry emerged: **organic chemistry is the study of carbon compounds.** This is the definition we still use today.

Lavoisier as portrayed on a 1943 French postage stamp.

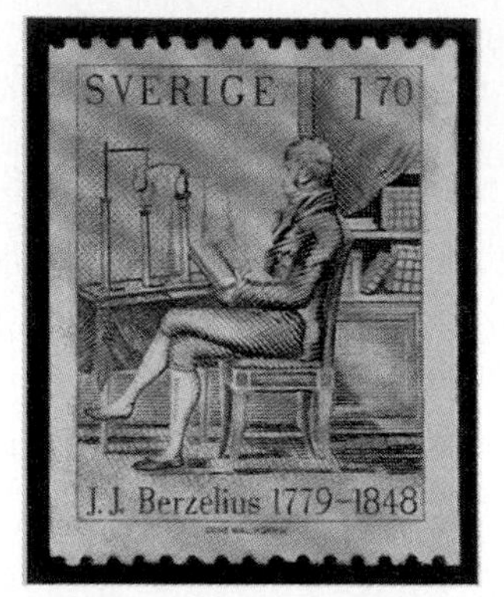

A 1979 Swedish stamp honoring Berzelius.

This German stamp depicts a molecular model of urea and was issued in 1982 to commemorate the 100th anniversary of Wöhler's death.

BERZELIUS, WÖHLER, AND VITALISM

As the eighteenth century gave way to the nineteenth, Jöns Jacob Berzelius emerged as one of the leading scientists of his generation. Berzelius, whose training was in medicine, had wide-ranging interests and made numerous contributions in diverse areas of chemistry. It was he who in 1807 coined the term *organic chemistry* for the study of compounds derived from natural sources. Berzelius, like almost everyone else at the time, subscribed to the doctrine known as **vitalism.** Vitalism held that living systems possessed a ''vital force'' which was absent in nonliving systems. Compounds derived from natural sources (organic) were thought to be fundamentally different from inorganic compounds; it was believed inorganic compounds could be synthesized in the laboratory, while organic compounds could not—at least not from inorganic materials.

In 1823 Friedrich Wöhler, fresh from completing his medical studies in Germany, traveled to Stockholm to study under Berzelius. A year later Wöhler accepted a position teaching chemistry and conducting research in Berlin. He went on to have a distinguished career, spending most of it at the University of Göttingen, but is best remembered for a brief paper he published in 1828. Wöhler noted that when he evaporated an aqueous solution of ammonium cyanate, he obtained ''colorless, clear crystals often more than an inch long,'' which were not ammonium cyanate but were instead urea.

$$\underset{\substack{\text{Ammonium cyanate} \\ \text{(an inorganic compound)}}}{NH_4^{+\,-}OCN} \longrightarrow \underset{\substack{\text{Urea} \\ \text{(an organic compound)}}}{O{=}C(NH_2)_2}$$

The transformation observed by Wöhler was one in which an *inorganic* salt, ammonium cyanate, was converted to urea, a known *organic* substance earlier isolated from urine. This experiment is now recognized as a scientific milestone, the first step toward overturning the philosophy of vitalism. Although Wöhler's synthesis of an organic compound in the laboratory from inorganic starting materials struck at the foundation of vitalist dogma, vitalism was not displaced overnight. Wöhler made no extravagant claims concerning the relationship of his discovery to vitalist theory, but the die was cast and over the next generation organic chemistry outgrew vitalism.

The article "Wöhler and the Vital Force" in the March 1957 issue of the *Journal of Chemical Education* (pp. 141–142) describes how Wöhler's experiment affected the doctrine of vitalism.

What particularly seemed to excite Wöhler and his mentor Berzelius about this experiment had very little to do with vitalism. Berzelius was interested in cases in which two clearly different materials had the same elemental composition and had invented the term **isomerism** to define it. The fact that an inorganic compound (ammonium cyanate) of molecular formula CH_4N_2O could be transformed into an organic compound (urea) of the same molecular formula had an important bearing on the concept of isomerism.

THE STRUCTURAL THEORY

It is from the concept of isomerism that we can trace the origins of the **structural theory**—the idea that a precise arrangement of atoms uniquely defines a substance. Ammonium cyanate and urea are different compounds because they have different structures. To some degree the structural theory was an idea whose time had come.

Three scientists stand out, however, in being credited with independently proposing the elements of the structural theory. They are August Kekulé, Archibald S. Couper, and Alexander M. Butlerov.

It is somehow fitting that August Kekulé's early training at the university in Giessen was as a student of architecture. Kekulé's contribution to chemistry lies in his description of the architecture of molecules. Two themes recur throughout Kekulé's work: critical evaluation of experimental information and a gift for visualizing molecules as particular assemblies of atoms. The essential features of Kekulé's theory, developed and presented while he taught at Heidelberg in 1858, were that carbon normally formed four bonds and had the capacity to bond to other carbons so as to form long chains. Isomers were possible because the same elemental composition (say, the CH_4N_2O molecular formula common to both ammonium cyanate and urea) accommodates more than one pattern of atoms and bonds.

A 1968 German stamp combines a drawing of the structure of benzene with a portrait of Kekulé.

Shortly thereafter, but independently of Kekulé, Archibald S. Couper, a Scot working in the laboratory of Charles-Adolphe Wurtz at the École de Medicine in Paris, and Alexander Butlerov, a Russian chemist at the University of Kazan, proposed similar theories.

The University of Kazan was home to a number of prominent nineteenth-century organic chemists. Their contributions are recognized in two articles published in the January and February 1994 issues of the *Journal of Chemical Education* (pp. 39–42 and 93–98).

ELECTRONIC THEORIES OF STRUCTURE AND REACTIVITY

In the late nineteenth and early twentieth centuries, major discoveries about the nature of atoms placed theories of molecular structure and bonding on a more secure foundation. Structural ideas progressed from simply identifying atomic connections to attempting to understand the bonding forces. In 1916 Gilbert N. Lewis of the University of California at Berkeley described covalent bonding in terms of shared electron pairs. Linus Pauling at the California Institute of Technology subsequently elaborated a more sophisticated bonding scheme based on Lewis' ideas and a concept called **resonance,** which he borrowed from the quantum mechanical treatments of theoretical physics.

Linus Pauling is portrayed on this 1977 Volta stamp. The chemical formulas depict the two resonance forms of benzene, and the explosion in the background symbolizes Pauling's efforts to limit the testing of nuclear weapons.

Once chemists gained an appreciation of the fundamental principles of bonding, a logical next step became the understanding of how chemical reactions occurred. Most notable among the early workers in this area were two British organic chemists, Sir Robert Robinson and Sir Christopher Ingold. Both held a number of teaching positions, with Robinson spending most of his career at Oxford while Ingold was at University College, London.

Robinson, who was primarily interested in the chemistry of natural products, had a keen mind and a penetrating grasp of theory. He was able to take the basic elements of Lewis' structural theories and apply them to chemical transformations by suggesting that chemical change can be understood by focusing on electrons. In effect, Robinson analyzed organic reactions by looking at the electrons and understood that atoms moved because they were carried along by the transfer of electrons. Ingold applied the quantitative methods of physical chemistry to the study of organic reactions so as to better understand the sequence of events, the **mechanism,** by which an organic substance is converted to a product under a given set of conditions.

Our current understanding of elementary reaction mechanisms is quite good. Most of the fundamental reactions of organic chemistry have been scrutinized to the degree that we have a relatively clear picture of the intermediates that occur during the passage of starting materials to products. Extension of the principles of mechanism to reactions that occur in living systems, on the other hand, is an area in which a large number of important questions remain to be answered.

THE IMPACT OF ORGANIC CHEMISTRY

Many organic compounds were known to and used by ancient cultures. Almost every known human society has manufactured and used beverages containing ethyl alcohol and has observed the formation of acetic acid when wine was transformed into vinegar. Early Chinese civilizations (2500 to 3000 B.C.) extensively used natural materials for treating illnesses and prepared a drug known as *Ma Huang* from herbal extracts. This drug was a stimulant and elevated blood pressure. We now know that it contains ephedrine, an organic compound similar in structure and physiological activity to adrenaline, a hormone secreted by the adrenal gland. Almost all drugs prescribed today for the treatment of disease are organic compounds—some are derived from natural sources; many others are the products of synthetic organic chemistry.

The discoverer of penicillin, Sir Alexander Fleming, has appeared on two stamps. This 1981 Hungarian issue includes both a likeness of Fleming and a structural formula for penicillin.

As early as 2500 B.C. in India, indigo was used to dye cloth a deep blue. The early Phoenicians discovered that a purple dye of great value, Tyrian purple, could be extracted from a Mediterranean sea snail. The beauty of the color and its scarcity made purple the color of royalty. The availability of dyestuffs underwent an abrupt change in 1856 when William Henry Perkin, an 18-year-old student, accidentally discovered a simple way to prepare a deep-purple dye, which he called *mauveine,* from extracts of coal tar. This led to a search for other synthetic dyes and forged a permanent link between industry and chemical research.

The synthetic fiber industry as we know it began in 1928 when E. I. Du Pont de Nemours & Company lured Professor Wallace H. Carothers from Harvard University to direct their research department. In a few years Carothers and his associates had produced *nylon,* the first synthetic fiber, and *neoprene,* a rubber substitute. Synthetic fibers and elastomers both represent important contemporary industries, with an economic impact far beyond anything imaginable in the middle 1920s.

COMPUTERS AND ORGANIC CHEMISTRY

A familiar arrangement of the sciences places chemistry between physics, which is highly mathematical, and biology, which is highly descriptive. Among chemistry's subdisciplines, organic chemistry is less mathematical than descriptive in that it emphasizes the qualitative aspects of molecular structure, reactions, and synthesis. The earliest applications of computers to chemistry took advantage of the ''number crunching'' power of mainframes to analyze data and to perform calculations concerned with the more quantitative aspects of bonding theory. More recently, organic chemists have found the graphics capabilities of minicomputers, workstations, and personal computers to be well-suited to visualizing a molecule as a three-dimensional object and assessing its ability to interact with another molecule. Given a biomolecule of known structure, a protein, for example, and a drug that acts on it, molecular-modeling software can evaluate the various ways in which the two may fit together. Such studies can provide information on the mechanism of drug action and guide the development of new drugs of greater efficacy.

The influence of computers on the practice of organic chemistry is a significant recent development and will be revisited numerous times in the chapters that follow.

CHALLENGES AND OPPORTUNITIES

A major contributor to the growth of organic chemistry during this century has been the accessibility of cheap starting materials. Petroleum and natural gas provide the building blocks for the construction of larger molecules. From petrochemicals comes a dazzling array of materials that enrich our lives: many drugs, plastics, synthetic fibers, films, and elastomers are made from the organic chemicals obtained from petroleum. As we enter an age of inadequate and shrinking supplies, the use to which we put petroleum looms large in determining the kind of society we will have. Alternative sources of energy, especially for transportation, will allow a greater fraction of the limited petroleum available to be converted to petrochemicals instead of being burned in automobile engines. At a more fundamental level, scientists in the chemical industry are trying to devise ways to use carbon dioxide as a carbon source in the production of building block molecules.

Many countries have celebrated their chemical industry on postage stamps. The stamp shown was issued in 1971 by Argentina.

Many of the most important processes in the chemical industry are carried out in the presence of **catalysts.** Catalysts increase the rate of a particular chemical reaction but are not consumed during it. In searching for new catalysts, we can learn a great deal from **biochemistry,** the study of the chemical reactions that take place in living organisms. All these fundamental reactions are catalyzed by enzymes. Rate enhancements of several millionfold are common when one compares an enzyme-catalyzed reaction with the same reaction performed in its absence. Many diseases are the result of specific enzyme deficiencies that interfere with normal metabolism. In the final analysis, effective treatment of diseases requires an understanding of biological processes at the molecular level—what the substrate is, what the product is, and the mechanism by which substrate is transformed to product. Enormous advances have been made in understanding biological processes. Because of the complexity of living systems, however, we have only scratched the surface of this fascinating field of study.

Spectacular strides have been made in genetics during the past few years. While generally considered a branch of biology, genetics is increasingly being studied at the molecular level by scientists trained as chemists. Gene-splicing techniques and methods for determining the precise molecular structure of DNA are just two of the tools driving the next scientific revolution.

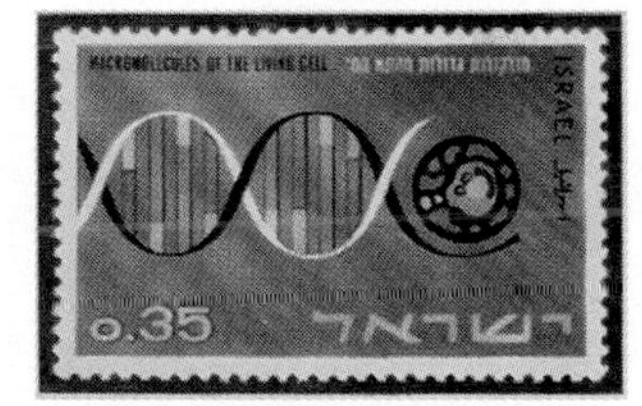

A DNA double helix as pictured on a 1964 postage stamp issued by Israel.

You are studying organic chemistry at a time of its greatest impact on our daily lives, at a time when it can be considered a mature science, and at a time when the challenging questions to which this knowledge can be applied have never been more important.

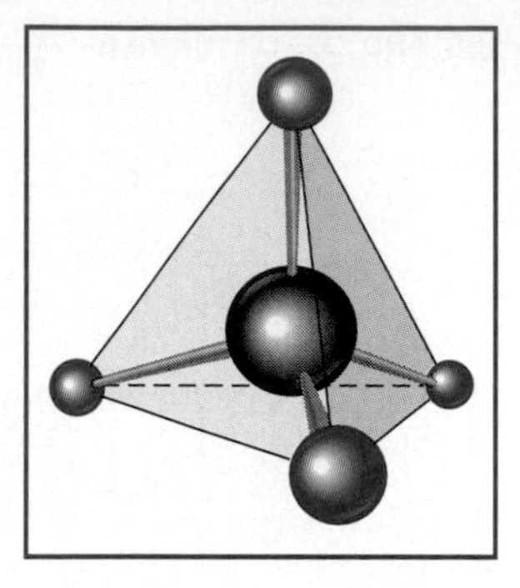

CHAPTER 1

CHEMICAL BONDING

Any attempt to understand the properties and reactions of organic compounds must begin with their **structure.** The way a substance behaves is directly related to the atoms it contains and the way these atoms are connected. This chapter reviews some fundamental principles of molecular structure and chemical **bonding** in organic compounds. By applying these principles you will learn to recognize bonding patterns that are more stable than others, and begin to develop skills in communicating chemical information by way of structural formulas that will be used throughout your study of organic chemistry.

1.1 ATOMS, ELECTRONS, AND ORBITALS

A glossary of important terms may be found immediately before the index at the back of the book.

Before discussing bonding principles, let us first review some fundamental relationships between atoms and electrons. Each element is characterized by a unique **atomic number *Z*,** which is equal to the number of protons in its nucleus. A neutral atom has equal numbers of protons, which are positively charged, and electrons, which are negatively charged.

Electrons were believed to be particles until 1924, when the French physicist Louis de Broglie suggested that they have wavelike properties as well. Two years later Erwin Schrödinger expressed the energy of an electron in a hydrogen atom in terms of a wave equation with a series of solutions called **wave functions** symbolized by the Greek letter ψ (psi). While the precise position of an electron can never be located with certainty, the probability of finding an electron at a particular point relative to the nucleus is given by the square of the wave function (ψ^2) at that point. Regions of space where there is a high probability of finding an electron are called **orbitals.**

Orbitals are described by specifying their size, shape, and spatial orientation. Those that are spherically symmetric are called *s orbitals* (Figure 1.1). The density of the color in the *s* orbital of Figure 1.1 relates to the probability of finding an electron at various positions. The number that describes the **energy level** of the orbital (1, 2, 3, etc.) is termed the **principal quantum number** and is given the symbol n. An electron in a 1*s* orbital is likely to be found closer to the nucleus, is lower in energy, and is more strongly held than an electron in a 2*s* orbital.

FIGURE 1.1 Probability distribution (ψ^2) for an electron in a 1*s* orbital.

Portions of a single orbital may be separated by **nodal surfaces** where the wave function changes its algebraic sign. The probability of finding an electron is zero at a nodal surface. A 1*s* orbital has no nodes, a 2*s* orbital one. A 1*s* and a 2*s* orbital are shown in cross section in Figure 1.2. The 2*s* wave function changes sign on passing through the node near the nucleus, as indicated by the plus (+) and minus (−) signs in Figure 1.2. *Do not confuse these signs with electric charges—they have nothing to do with electron or nuclear charge.*

When depicting orbitals, it is more customary to represent them by their **boundary surfaces,** as shown in Figure 1.3 for the 1*s* and 2*s* orbitals. The boundary surface encloses the region where the probability of finding an electron is high—on the order of 90 to 95 percent.

A hydrogen atom ($Z = 1$) has one electron; a helium atom ($Z = 2$) has two. The single electron of hydrogen occupies a 1*s* orbital, as do the two electrons of helium. The respective electron configurations are described as:

$$\text{Hydrogen: } 1s^1 \qquad \text{Helium: } 1s^2$$

In addition to being negatively charged, electrons possess the property of **spin.** The **spin quantum number** of an electron can have a value of either $+\frac{1}{2}$ or $-\frac{1}{2}$. According to the **Pauli exclusion principle,** two electrons may occupy the same orbital only when they have opposite, or "paired," spins. Since two electrons fill the 1*s* orbital, the third electron in lithium ($Z = 3$) must occupy an orbital of higher energy. After 1*s*, the next higher energy orbital is 2*s*. Therefore, the third electron in lithium occupies the 2*s* orbital, and the electron configuration of lithium is

$$\text{Lithium: } 1s^2, 2s^1$$

The **period** (or **row**) of the periodic table in which an element appears corresponds to the principal quantum number of the highest-numbered occupied orbital ($n = 1$ in the case of hydrogen and helium). Hydrogen and helium are first-row elements; lithium ($n = 2$) is a second-row element.

A complete periodic table of the elements is presented on the inside back cover.

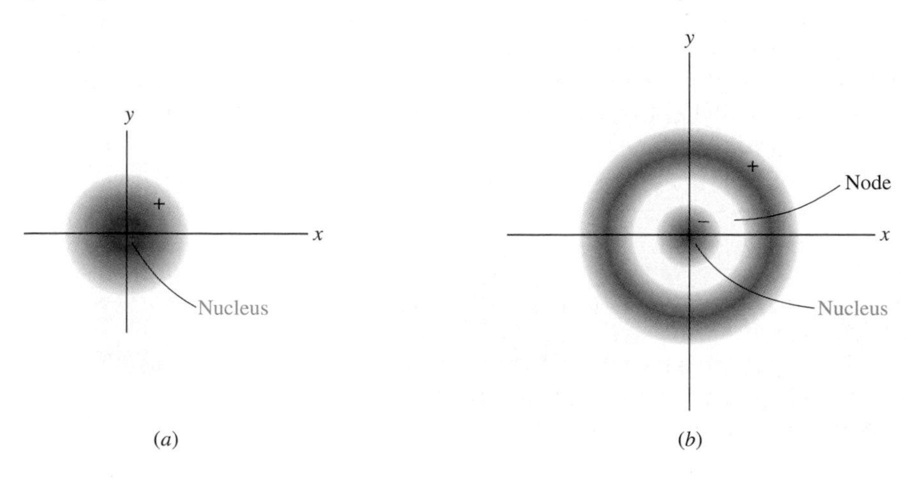

FIGURE 1.2 Cross sections of (*a*) a 1*s* orbital and (*b*) a 2*s* orbital. The wave function has the same sign over the entire 1*s* orbital. It is arbitrarily shown as +, but it could just as well have been designated as −. The 2*s* orbital has a spherical node where the wave function changes sign.

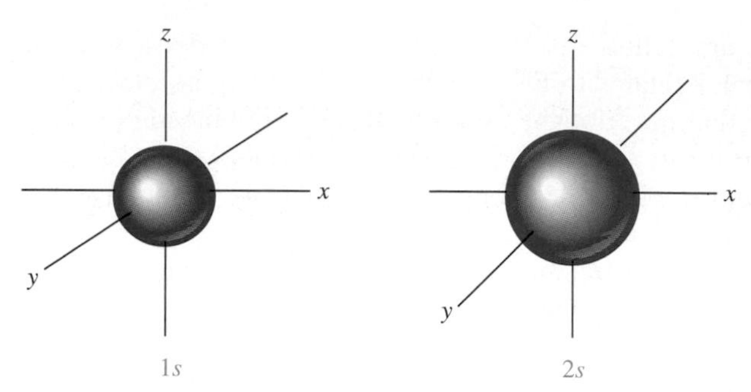

FIGURE 1.3 Representations of the boundary surfaces of a 1*s* orbital and a 2*s* orbital. The boundary surfaces enclose the volume where there is a 90 to 95 percent probability of finding an electron.

With beryllium ($Z = 4$), the 2*s* level becomes filled, and the next orbitals to be occupied in the remaining second-row elements are the $2p_x$, $2p_y$, and $2p_z$ orbitals. These orbitals, portrayed in Figure 1.4, have a boundary surface which is usually described as "dumbbell-shaped." Each orbital is composed of two "lobes," i.e., slightly flattened spheres that touch each other along a nodal plane passing through the nucleus. The two lobes of a single *p* orbital have wave functions of opposite sign. The $2p_x$, $2p_y$, and $2p_z$ orbitals are equal in energy and mutually perpendicular.

The electron configurations of the first 12 elements, hydrogen through magnesium, are given in Table 1.1. In filling the 2*p* orbitals, notice that each is singly occupied before any one is doubly occupied. This is a general principle for orbitals of equal energy known as **Hund's rule.** *Of particular importance in Table 1.1 are hydrogen, carbon, nitrogen, and oxygen.* Countless organic compounds contain nitrogen, oxygen, or both in addition to carbon, the essential element of organic chemistry. Most of them also contain hydrogen.

It is often convenient to speak of the **valence electrons** of an atom. These are the outermost electrons, the ones most likely to be involved in chemical bonding and reactions. For second-row elements these are the 2*s* and 2*p* electrons. Since four orbitals ($2s$, $2p_x$, $2p_y$, $2p_z$) are involved, the maximum number of electrons in the **valence shell** of any second-row element is 8. Neon, with all its 2*s* and 2*p* orbitals doubly occupied, has eight valence electrons and completes the second row of the periodic table.

Answers to all problems which appear within the body of a chapter are found in Appendix 2. A brief discussion of the problem and advice on how to do problems of the same type are offered in the Study Guide.

PROBLEM 1.1 How many valence electrons does carbon have?

Once the 2*s* and 2*p* orbitals are filled, the next level is the 3*s*, followed by the $3p_x$, $3p_y$, and $3p_z$ orbitals. Electrons in these orbitals are farther from the nucleus than those in the 2*s* and 2*p* orbitals and are of higher energy.

The material in this text can be understood without considering electrons in *d* or *f* orbitals, and so these higher-energy orbitals will not be discussed.

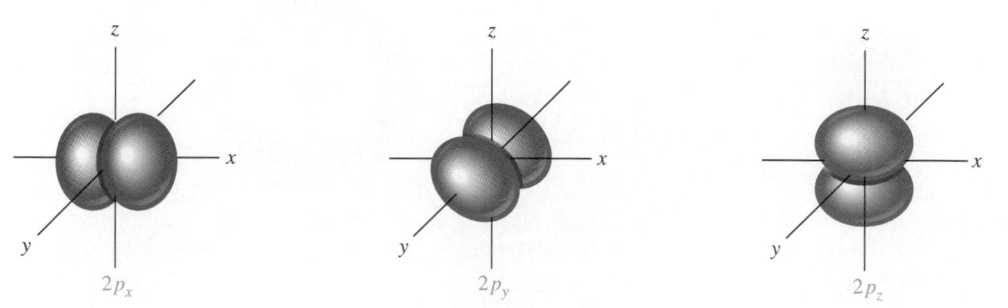

FIGURE 1.4 Representations of the boundary surfaces of the 2*p* orbitals.

TABLE 1.1
Electron Configurations of the First Twelve Elements of the Periodic Table

Element	Atomic number Z	Number of electrons in indicated orbital					
		$1s$	$2s$	$2p_x$	$2p_y$	$2p_z$	$3s$
Hydrogen	1	1					
Helium	2	2					
Lithium	3	2	1				
Beryllium	4	2	2				
Boron	5	2	2	1			
Carbon	6	2	2	1	1		
Nitrogen	7	2	2	1	1	1	
Oxygen	8	2	2	2	1	1	
Fluorine	9	2	2	2	2	1	
Neon	10	2	2	2	2	2	
Sodium	11	2	2	2	2	2	1
Magnesium	12	2	2	2	2	2	2

PROBLEM 1.2 Refer to the periodic table as needed and write electron configurations for all the elements in the third period.

SAMPLE SOLUTION The third period begins with sodium and ends with argon. The atomic number Z of sodium is 11, and so a sodium atom has 11 electrons. The maximum number of electrons in the $1s$, $2s$, and $2p$ orbitals is 10, and so the eleventh electron of sodium occupies a $3s$ orbital. The electron configuration of sodium is $1s^2$, $2s^2$, $2p_x^{\,2}$, $2p_y^{\,2}$, $2p_z^{\,2}$, $3s^1$.

In-chapter problems which contain multiple parts are accompanied by a sample solution to part (*a*). Answers to the other parts of the problem are found in Appendix 2, while detailed solutions are presented in the Study Guide.

Neon, in the second period, and argon, in the third, possess eight electrons in their valence shell; they are said to have a complete **octet** of electrons. Helium, neon, and argon belong to the class of elements known as **noble gases** or **rare gases.** The noble gases are characterized by an extremely stable "closed-shell" electron configuration and are very unreactive.

1.2 IONIC BONDS

Atoms combine with one another to give **compounds** having properties different from the atoms they contain. The attractive force between atoms in a compound is a **chemical bond.** When forming compounds, elements gain, lose, or share as many electrons as necessary to achieve the stable electron configuration of the nearest noble gas. Elements at the left of the periodic table tend to lose electrons to give the electron configuration of the preceding noble gas. Loss of an electron from sodium, for example, gives the species Na^+, which has the same closed-shell electron configuration as neon. Electrically charged species are called **ions,** and positively charged ions such as Na^+ are called **cations.**

$$\mathrm{Na}(g) \longrightarrow \mathrm{Na}^+(g) + e^-$$

Sodium atom $1s^22s^22p^63s^1$ — Sodium ion $1s^22s^22p^6$ — Electron

[The (g) indicates that the species is present in the gas phase.]

A large amount of energy, called the **ionization energy,** must be added to any atom in order to dislodge one of its electrons. The ionization energy of sodium, for example, is 496 kJ/mol (119 kcal/mol). Processes that absorb energy are said to be **endothermic.** Compared with other elements, sodium and its relatives in group IA have relatively low ionization energies. In general, ionization energy increases across a row in the periodic table.

The SI *(Système International d'Unites)* unit of energy is the *joule* (J). An older unit is the *calorie* (cal). Most organic chemists still express energy changes in units of kilocalories per mole (1 kcal/mol = 4.184 kJ/mol).

Elements at the right of the periodic table tend to gain electrons to reach the electron configuration of the next higher noble gas. Adding an electron to chlorine, for example, gives the species Cl^-, which has the same closed-shell electron configuration as the noble gas argon. Negatively charged ions such as Cl^- are called **anions.**

$$\mathrm{Cl}(g) + e^- \longrightarrow \mathrm{Cl}^-(g)$$

Chlorine atom $1s^22s^22p^63s^23p^5$ — Electron — Chloride ion $1s^22s^22p^63s^23p^6$

Energy is released when a chlorine atom captures an electron. Energy-releasing reactions are described as **exothermic,** and the energy change for an exothermic process has a negative sign. The energy change for addition of an electron to an atom is referred to as its **electron affinity** and is −349 kJ/mol (−83.4 kcal/mol) for chlorine.

PROBLEM 1.3 Which of the following ions possess a noble gas electron configuration?

(*a*) K^+
(*b*) He^+
(*c*) H^-
(*d*) O^-
(*e*) F^-
(*f*) Ca^{2+}

SAMPLE SOLUTION (*a*) Potassium has atomic number 19, and so a potassium atom has 19 electrons. The ion K^+, therefore, has 18 electrons, the same as the noble gas argon. The electron configurations of K^+ and Ar are the same: $1s^22s^22p^63s^23p^6$.

Transfer of an electron from a sodium atom to a chlorine atom yields a sodium cation and a chloride anion, both of which have a noble gas electron configuration:

$$\mathrm{Na}(g) + \mathrm{Cl}(g) \longrightarrow \mathrm{Na}^+\mathrm{Cl}^-(g)$$

Sodium atom — Chlorine atom — Sodium chloride

Were we to simply add the ionization energy of sodium (496 kJ/mol) and the electron affinity of chlorine (−349 kJ/mol), we would conclude that the overall process is endothermic with $\Delta H^\circ = +147$ kJ/mol. The energy liberated by adding an electron to chlorine is insufficient to override the energy required to remove an electron from sodium. This analysis, however, fails to consider the force of attraction between the oppositely charged ions Na^+ and Cl^-, which exceeds 500 kJ/mol and is more than sufficient to make the overall process exothermic. Attractive forces between oppositely charged particles are termed **electrostatic** or **coulombic** attractions and are what we mean by an **ionic bond** between two atoms.

Ionic bonding was proposed by the German physicist Walter Kossel in 1916, in order to explain the ability of substances such as sodium chloride to conduct an electric current.

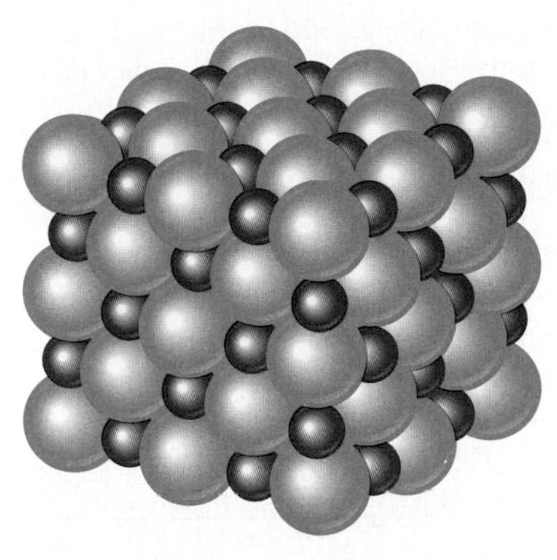

FIGURE 1.5 Lattice structure of crystalline sodium chloride. The small red spheres represent sodium ions; the large green spheres are chloride ions. Sodium chloride molecules, as such, do not exist in crystalline sodium chloride. Rather, each Na^+ cation is surrounded by six Cl^- anions and vice versa.

The preceding equation represents the formation of an Na^+Cl^- **ion pair** in the gas phase. Figure 1.5 shows the structure of solid sodium chloride, where it can be seen that individual NaCl units are not present but, instead, each sodium ion is surrounded by six chloride ions and vice versa. Such an ordered arrangement is called a *crystal lattice* and maximizes the electrostatic attractive forces. Compounds such as sodium chloride that are characterized by ionic bonding are called **ionic compounds.**

Ionic bonds are very common in inorganic compounds, but less common in organic chemistry. The ionization energy of carbon is much greater than that of sodium (1090 versus 496 kJ/mol) and its electron affinity is much less exothermic than chlorine's (−122 versus −349 kJ/mol). Carbon shows little tendency to act as the source of the cation or the anion in an ionic bond, and most of its compounds are characterized by *covalent bonding*. This topic is introduced in the following section.

PROBLEM 1.4 What is the electron configuration of C^+? Of C^-? Does either one of these ions have a noble gas (closed-shell) electron configuration?

1.3 COVALENT BONDS

The **covalent,** or **shared electron pair,** model of chemical bonding was first suggested by G. N. Lewis of the University of California in 1916. Lewis proposed that a *sharing* of two electrons by two hydrogen atoms permits each one to have a stable closed-shell electron configuration analogous to helium.

Gilbert Newton Lewis (born Weymouth, Massachusetts, 1875; died Berkeley, California, 1946) has been called the greatest American chemist. The January 1984 issue of the *Journal of Chemical Education* contains five articles describing Lewis' life and contributions to chemistry.

H· ·H	H:H
Two hydrogen atoms, each with a single electron	Hydrogen molecule: covalent bonding by way of a shared electron pair

Structural formulas of this type in which electrons are represented as dots are called **Lewis structures.**

The amount of energy required to dissociate a hydrogen molecule H_2 to two separate hydrogen atoms is called its **bond dissociation energy** (or **bond energy**) and is quite large, being equal to 435 kJ/mol (104 kcal/mol). A significant contributor to the strength of the covalent bond in H_2 is the increased binding force exerted on its two electrons. Each electron in H_2 "feels" the attractive force of two nuclei, rather than one as in an isolated hydrogen atom.

Covalent bonding in F_2 gives each fluorine eight electrons in its valence shell and a stable electron configuration equivalent to that of the noble gas neon:

:F̤̈· ·F̤̈:	:F̤̈:F̤̈:
Two fluorine atoms, each with seven electrons in its valence shell	Fluorine molecule: covalent bonding by way of a shared electron pair

---

PROBLEM 1.5 Hydrogen is bonded to fluorine in hydrogen fluoride by a covalent bond. Write a Lewis formula for hydrogen fluoride.

An important feature of the Lewis model is that the second-row elements (Li, Be, B, C, N, O, F, Ne) are limited to a total of eight electrons (shared plus unshared) in their valence shells. Hydrogen is limited to two. Most of the elements that we will encounter in this text obey the **octet rule:** *in forming compounds they gain, lose, or share electrons to give a stable electron configuration characterized by eight valence electrons.* When the octet rule is satisfied for carbon, nitrogen, oxygen, and fluorine, they have an electron configuration analogous to the noble gas neon.

The Lewis approach to the organic compounds methane and carbon tetrafluoride is illustrated as follows:

Combine ·Ċ· (with four single electrons) and four H· to write a Lewis structure for methane: H:C̤̈:H with H above and below C

Combine ·Ċ· (with four single electrons) and four ·F̤̈: to write a Lewis structure for carbon tetrafluoride: :F̤̈:C̤̈:F̤̈: with :F̤̈: above and below C

Carbon has eight electrons in its valence shell in both methane and carbon tetrafluoride. By forming covalent bonds to four other atoms, carbon achieves a stable electron configuration analogous to neon. Each covalent bond in methane and carbon tetrafluoride is quite strong—comparable to the bond between hydrogens in H_2 in bond dissociation energy.

PROBLEM 1.6 Given the information that it has a carbon-carbon bond, write a satisfactory Lewis structure for C_2H_6 (ethane).

It is customary to represent a two-electron covalent bond by a dash (—). Thus, the Lewis structures for hydrogen fluoride, fluorine, methane, and carbon tetrafluoride become:

H—F̤̈:	:F̤̈—F̤̈:	H—C—H (with H above and below C, joined by bonds)	:F̤̈—C—F̤̈: (with :F̤̈: above and below C, joined by bonds)
Hydrogen fluoride	Fluorine	Methane	Carbon tetrafluoride

1.4 MULTIPLE BONDING IN LEWIS STRUCTURES

An extension of the Lewis concept of shared electron pair bonds allows for four-electron **double bonds** and six-electron **triple bonds.** Carbon dioxide (CO_2) has two carbon-oxygen double bonds, and the octet rule is satisfied for both carbon and oxygen. Similarly, the most stable Lewis structure for hydrogen cyanide (HCN) has a carbon-nitrogen triple bond.

Carbon dioxide: :Ö::C::Ö: or :Ö=C=Ö:

Hydrogen cyanide: H:C:::N: or H—C≡N:

Multiple bonds are very common in organic chemistry. Ethylene (C_2H_4) contains a carbon-carbon double bond in its most stable Lewis structure, and each carbon has a completed octet. The most stable Lewis structure for acetylene (C_2H_2) contains a carbon-carbon triple bond. Here again, the octet rule is satisfied.

Ethylene: H₂C::CH₂ (H:C::C:H with two H on each carbon) or $H_2C=CH_2$

Acetylene: H:C:::C:H or H—C≡C—H

PROBLEM 1.7 Write the most stable Lewis structure for each of the following compounds:

(*a*) Formaldehyde, CH_2O. Both hydrogens are bonded to carbon. (A solution of formaldehyde in water is commonly used to preserve biological specimens.)
(*b*) Tetrafluoroethylene, C_2F_4. (The starting material for the preparation of *Teflon*.)
(*c*) Acrylonitrile, C_3H_3N. The atoms are connected in the order CCCN, and all hydrogens are bonded to carbon. (The starting material for the preparation of acrylic fibers such as *Orlon* and *Acrilan*.)

SAMPLE SOLUTION (*a*) Each hydrogen contributes 1 valence electron, carbon contributes 4, and oxygen 6 for a total of 12 valence electrons. We are told that both hydrogens are bonded to carbon. Since carbon forms four bonds in its stable compounds, join carbon and oxygen by a double bond. The partial structure so generated accounts for 8 of the 12 electrons. Add the remaining four electrons to oxygen as unshared pairs to complete the structure of formaldehyde.

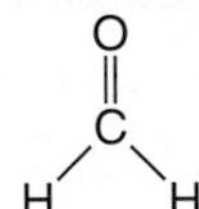

Partial structure showing covalent bonds

:O:
C
H H

Complete Lewis structure of formaldehyde

1.5 POLAR COVALENT BONDS AND ELECTRONEGATIVITY

Electrons in covalent bonds are not necessarily shared equally by the two atoms that they connect. If one atom has a greater tendency to attract electrons toward itself than the other, the electron distribution is said to be *polarized,* and the bond is referred to as a **polar covalent bond.** Hydrogen fluoride, for example, has a polar covalent bond. Fluorine attracts electrons more strongly than does hydrogen. The center of negative charge in the molecule is closer to fluorine, while the center of positive charge is closer to hydrogen. This polarization of electron density in the hydrogen-fluorine bond is represented in various ways.

$$^{\delta+}H—F^{\delta-} \qquad \overset{\longleftrightarrow}{H—F}$$

(The symbols $^{\delta+}$ and $^{\delta-}$ indicate partial positive and partial negative charge, respectively)

(The symbol $\longleftrightarrow$ represents the direction of polarization of electrons in the H—F bond)

Linus Pauling (1901–1994) was born in Portland, Oregon and was educated at Oregon State University and at the California Institute of Technology, where he earned a Ph.D. in chemistry in 1925. In addition to research in bonding theory, Pauling studied the structure of proteins and was awarded the Nobel Prize in chemistry for that work in 1954. Pauling won a second Nobel Prize (the Peace Prize) for his efforts to limit the testing of nuclear weapons. He was one of only four scientists to have won two Nobel Prizes. The first double winner was a woman. Can you name her?

The tendency of an atom to draw the electrons in a covalent bond toward itself is referred to as its **electronegativity.** An **electronegative** element attracts electrons; an **electropositive** one donates them. Electronegativity increases across a row in the periodic table. The most electronegative of the second-row elements is fluorine; the most electropositive is lithium. Electronegativity decreases in going down a column. Fluorine is more electronegative than chlorine. The most commonly cited electronegativity scale was devised by Linus Pauling and is presented in Table 1.2.

When centers of positive and negative charge are separated from each other, they constitute a *dipole.* The **dipole moment μ** of a molecule is the arithmetic product of the charge e (either the positive charge or the negative charge, since they must be equal) and the distance between them:

$$\mu = e \cdot d$$

Since the charge on an electron is 4.80×10^{-10} electrostatic units (esu) and distances within a molecule typically fall in the 10^{-8} cm range, molecular dipole moments are on the order of 10^{-18} esu·cm. The **debye, D** (named after the Dutch chemist, Peter J. W. Debye, who pioneered the systematic study of polar molecules), is defined as equal to

TABLE 1.2
Selected Values from the Pauling Electronegativity Scale

	Group number						
Period	**I**	**II**	**III**	**IV**	**V**	**VI**	**VII**
1	H 2.1						
2	Li 1.0	Be 1.5	B 2.0	C 2.5	N 3.0	O 3.5	F 4.0
3	Na 0.9	Mg 1.2	Al 1.5	Si 1.8	P 2.1	S 2.5	Cl 3.0
4	K 0.8	Ca 1.0					Br 2.8
5							I 2.5

TABLE 1.3
Selected Bond Dipole Moments

Bond*	Dipole moment, D	Bond*	Dipole moment, D
H—F	1.9	C—F	1.4
H—Cl	1.1	C—O	0.7
H—Br	0.8	C—N	0.2
H—I	0.4		
H—C	0.3		
H—N	1.3	C═O	2.4
H—O	1.5	C≡N	3.6

* The direction of the dipole moment is toward the more electronegative atom. In the above examples hydrogen and carbon are the positive ends of the dipoles. Carbon is the negative end of the dipole associated with the C—H bond.

1.0×10^{-18} esu·cm, a notational device that simplifies the reporting of molecular dipole moments. Thus, the experimentally determined dipole moment of hydrogen fluoride, 1.7×10^{-18} esu·cm, is stated as 1.7 D.

PROBLEM 1.8 The compounds FCl and ICl have dipole moments μ that are similar in magnitude (0.9 and 0.7 D, respectively) but opposite in direction. In one compound, chlorine is the positive end of the dipole; in the other it is the negative end. Specify the direction of the dipole moment in each compound and explain the reasoning behind your choice.

A summary of dipole moments associated with various bond types is given in Table 1.3. In particular, notice that the polarity of a carbon-hydrogen bond is relatively low; it is substantially less than that of carbon-oxygen and carbon-halogen bonds. As will be seen in later chapters, the kinds of reactions that an organic substance undergoes can often be related to the polarity of its bonds.

1.6 FORMAL CHARGE

Lewis structures frequently contain atoms that bear a positive or negative charge. If the molecule as a whole is neutral, the sum of its positive charges must equal the sum of its negative charges. An example is nitric acid, HNO_3:

$$\mathrm{H{-}\ddot{\underset{..}{O}}{-}\overset{+}{N}({=}\ddot{O}{:})({-}{:}\ddot{\underset{..}{O}}{:}^{-})}$$

As written, the structural formula for nitric acid depicts different bonding patterns for its three oxygens. One oxygen (red) is doubly bonded to nitrogen, another (black) is singly bonded to both nitrogen and hydrogen, and the third (green) has a single bond to nitrogen and a negative charge. Nitrogen (blue) is positively charged. The positive and negative charges are called **formal charges,** and the Lewis structure of nitric acid would be incomplete were they to be omitted.

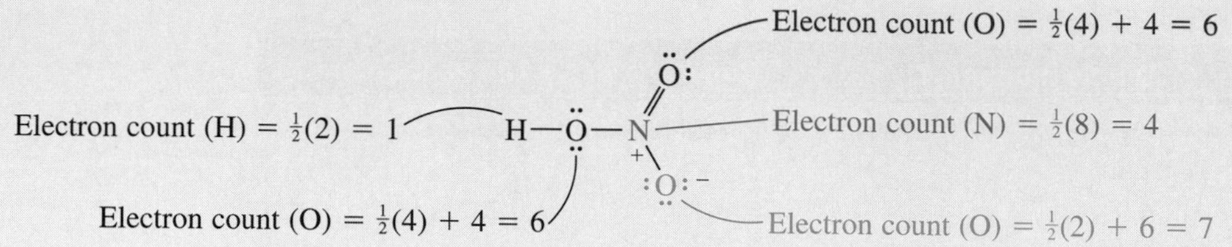

Electron count (O) = $\frac{1}{2}(4) + 4 = 6$

Electron count (H) = $\frac{1}{2}(2) = 1$ H—Ö—N Electron count (N) = $\frac{1}{2}(8) = 4$

Electron count (O) = $\frac{1}{2}(4) + 4 = 6$

Electron count (O) = $\frac{1}{2}(2) + 6 = 7$

FIGURE 1.6 Counting electrons in nitric acid. The electron count of each atom is equal to half the number of electrons it shares in covalent bonds plus the number of electrons in its own unshared pairs.

The number of valence electrons in an atom of a main-group element such as nitrogen is equal to its group number. In the case of nitrogen this is 5.

We calculate formal charges by counting the number of electrons ''owned'' by each atom in a Lewis structure and comparing this **electron count** with that of a neutral atom. Figure 1.6 illustrates how electrons are counted for each atom in nitric acid. Counting electrons for the purpose of computing the formal charge differs from counting electrons to see if the octet rule is satisfied. A second-row element has a filled valence shell if the sum of all the electrons, shared and unshared, is 8. Electrons that connect two atoms by a covalent bond count toward filling the valence shell of both atoms. When calculating the formal charge, however, only half the number of electrons in covalent bonds can be considered to be ''owned'' by an atom.

It will always be true that a covalently bonded hydrogen has no formal charge (formal charge = 0).

To illustrate, let us start with the hydrogen of nitric acid. As shown in Figure 1.6, hydrogen is associated with only two electrons—those in its covalent bond to oxygen. It shares those two electrons with oxygen, and so we say that the electron count of each hydrogen is $\frac{1}{2}(2) = 1$. Since this is the same as the number of electrons in a neutral hydrogen atom, the hydrogen in nitric acid has no formal charge.

It will always be true that a nitrogen with four covalent bonds has a formal charge of +1. (A nitrogen with four covalent bonds cannot have unshared pairs, because of the octet rule.)

Moving now to nitrogen, we see that it has four covalent bonds (two single bonds + one double bond), and so its electron count is $\frac{1}{2}(8) = 4$. A neutral nitrogen has five electrons in its valence shell. The electron count for nitrogen in nitric acid is 1 less than that of a neutral nitrogen atom. Since electrons have a charge of −1 and nitrogen has ''lost'' one electron, its formal charge is +1.

It will always be true that an oxygen with two covalent bonds and two unshared pairs has no formal charge.

While electrons in covalent bonds are counted as if they are shared equally by the atoms they connect, unshared electrons belong exclusively to a single atom. Thus the oxygen which is doubly bonded to nitrogen has an electron count of 6 (four electrons as two unshared pairs + two electrons from the double bond). Since this is the same as a neutral oxygen atom, its formal charge is 0. Similarly, the OH oxygen has two bonds plus two unshared electron pairs, giving it an electron count of 6 and no formal charge.

It will always be true that an oxygen with one covalent bond and three unshared pairs has a formal charge of −1.

The oxygen shown in green in Figure 1.6 owns three unshared pairs (six electrons) and shares two electrons with nitrogen to give it an electron count of 7. This is 1 more than the number of electrons in the valence shell of an oxygen atom, and so its formal charge is −1.

The method described for calculating formal charge has been one of reasoning through a series of logical steps. It can be reduced to the following equation:

$$\text{Formal charge} = \begin{matrix}\text{group number} \\ \text{in periodic table}\end{matrix} - \text{number of bonds} - \text{number of unshared electrons}$$

PROBLEM 1.9 Like nitric acid, each of the following inorganic compounds will be frequently encountered in this text. Calculate the formal charge on each of the atoms in the Lewis structures given.

(*a*) Thionyl chloride: :Cl—S(=O)—Cl: (Lewis structure: :Ö: above S; :C̈l—S̈—C̈l:)

(*b*) Phosphorus tribromide: :B̈r—P̈—B̈r: with :B̈r: below P

(*c*) Sulfuric acid: H—Ö—S—Ö—H with :Ö: above and below S

(*d*) Nitrous acid: H—Ö—N̈=Ö:

SAMPLE SOLUTION (*a*) The formal charge is the difference between the number of valence electrons in the neutral atom and the electron count in the Lewis structure. (The number of valence electrons is the same as the group number in the periodic table for the main-group elements.)

	Valence electrons of neutral atom	Electron count	Formal charge
Sulfur:	6	$\frac{1}{2}(6) + 2 = 5$	+1
Oxygen:	6	$\frac{1}{2}(2) + 6 = 7$	−1
Chlorine:	7	$\frac{1}{2}(2) + 6 = 7$	0

The formal charges are shown in the Lewis structure of thionyl chloride as :C̈l—S̈—C̈l: with :Ö:⁻ above S and + on S.

So far we have considered only neutral molecules—those in which the sums of the positive and negative formal charges were equal. With ions, of course, these sums will not be equal. Ammonium cation and borohydride anion, for example, are ions with net charges of +1 and −1, respectively. Nitrogen has a formal charge of +1 in ammonium ion, and boron has a formal charge of −1 in borohydride. None of the hydrogens in the Lewis structures shown for these ions bears a formal charge.

H—N⁺(H)(H)—H   H—B⁻(H)(H)—H

Ammonium ion   Borohydride ion

PROBLEM 1.10 Verify that the formal charges on nitrogen in ammonium ion and boron in borohydride ion are as shown.

Formal charges are based on Lewis structures in which electrons are considered to be shared equally between covalently bonded atoms. Actually, polarization of the N—H bonds in ammonium ion and B—H bonds in borohydride leads to some transfer of positive and negative charge, respectively, to the hydrogen substituents. Charge dispersal of this kind cannot be represented in a Lewis structure.

Determining formal charges on individual atoms of Lewis structures is an important element in good "electron bookkeeping." So much of organic chemistry can be made more understandable by keeping track of electrons that it is worth taking some time at the outset to develop a reasonable facility in the seemingly simple task of counting electrons.

1.7 WRITING STRUCTURAL FORMULAS OF ORGANIC MOLECULES

A structural formula attempts to show, in as much detail as necessary, the sequence of atomic connections in a molecule. The order in which the atoms are bonded defines the **constitution,** or **connectivity,** of a molecule. Organic chemists have devised a number of notational shortcuts to speed the writing of structural formulas. Representing covalent bonds by dashes is one of them. Another omits unshared electron pairs from Lewis structures. While omitting lone pairs from Lewis structures gives a less cluttered appearance, it requires that you be sufficiently proficient in electron bookkeeping that you can identify atoms that bear unshared electron pairs. Table 1.4 reviews the steps in the writing of Lewis structures.

Further simplification of Lewis structures results from deleting some or all of the covalent bonds and by indicating the number of identical groups attached to an atom by a subscript to give **condensed structural formulas.** These successive levels of simplification are illustrated as shown for isopropyl alcohol ("rubbing alcohol"):

```
   H  H  H
   |  |  |                        CH3CHCH3
H—C—C—C—H  written as            |          or condensed even further to
   |  |  |                           OH         (CH3)2CHOH
   H  O  H
      |
      H
```

PROBLEM 1.11 Expand the following condensed formulas so as to show all the bonds and unshared electron pairs.

(*a*) $HOCH_2CH_2NH_2$
(*b*) $(CH_3)_3CH$
(*c*) $ClCH_2CH_2Cl$
(*d*) CH_3CHCl_2
(*e*) $CH_3NHCH_2CH_3$
(*f*) $(CH_3)_2CHCH{=}O$

SAMPLE SOLUTION (*a*) The molecule contains two carbon atoms, which are bonded to each other. Both carbons bear two hydrogens. One carbon bears the group HO—; the other is attached to $—NH_2$.

```
       H  H  H
       |  |  |
H—O—C—C—N—H
       |  |
       H  H
```

When writing the constitution of a molecule, it is not necessary to concern yourself with the spatial orientation of the atoms. There are many other correct ways to represent the

TABLE 1.4
How to Write Lewis Structures

Step	Illustration
1. The molecular formula and the connectivity are determined experimentally and are included among the information given in the statement of the problem.	Methyl nitrite has the molecular formula CH_3NO_2. All hydrogens are bonded to carbon, and the order of atomic connections is CONO.
2. Count the number of valence electrons available. For a neutral molecule this is equal to the sum of the valence electrons of the constituent atoms.	Each hydrogen contributes 1 valence electron, carbon contributes 4, nitrogen contributes 5, and each oxygen contributes 6 for a total of 24 in CH_3NO_2.
3. Connect bonded atoms by a shared electron pair bond (:) represented by a dash (—).	For methyl nitrite we write the partial structure H | H—C—O—N—O | H
4. Count the number of electrons in shared electron pair bonds (twice the number of bonds) and subtract this from the total number of electrons to give the number of electrons to be added to complete the structure.	The partial structure in step 3 contains 6 bonds equivalent to 12 electrons. Since CH_3NO_2 contains 24 electrons, 12 more electrons need to be added.
5. Add electrons in pairs so that as many atoms as possible have 8 electrons. (Hydrogen is limited to 2 electrons.) When the number of electrons is insufficient to provide an octet for all atoms, assign electrons to atoms in order of decreasing electronegativity.	With 4 bonds, carbon already has 8 electrons. The remaining 12 electrons are added as indicated. Both oxygens have 8 electrons, but nitrogen (less electronegative than oxygen) has only 6. H | $H{-}C{-}\underset{..}{\ddot{O}}{-}\ddot{N}{-}\underset{..}{\ddot{O}}{:}$ | H
6. If one or more atoms have fewer than 8 electrons, use unshared pairs on an adjacent atom to form a double (or triple) bond to complete the octet.	An electron pair on the terminal oxygen is shared with nitrogen to give a double bond. H | $H{-}C{-}\underset{..}{\ddot{O}}{-}\ddot{N}{=}\ddot{O}{:}$ | H The structure shown is the best (most stable) Lewis structure for methyl nitrite. All atoms except hydrogen have 8 electrons (shared + unshared) in their valence shell.
7. Calculate formal charges.	None of the atoms in the Lewis structure shown in step 6 possesses a formal charge. An alternative Lewis structure for methyl nitrite, H | $H{-}C{-}\underset{..}{\overset{+}{O}}{=}\ddot{N}{-}\underset{..}{\ddot{O}}{:}^{-}$ | H while it satisfies the octet rule, is less stable than the one shown in step 6 because it has a separation of positive charge from negative charge.

compound shown. What is important is to show the sequence OCCN (or its equivalent NCCO) and to have the correct number of hydrogens present on each atom.

In order to locate unshared electron pairs, first count the total number of valence electrons brought to the molecule by its component atoms. Each hydrogen contributes 1, each carbon 4, nitrogen 5, and oxygen 6, for a total of 26. There are 10 bonds shown, accounting for 20 electrons; therefore 6 electrons must be contained in unshared pairs. Add pairs of electrons to oxygen and nitrogen so that their octets are complete, two unshared pairs to oxygen and one to nitrogen.

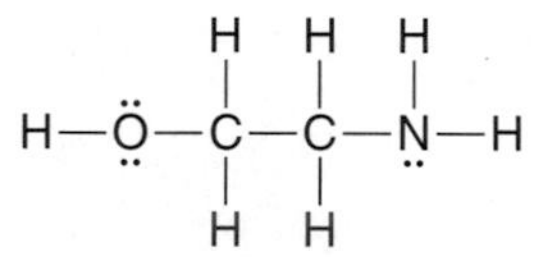

```
      H  H  H
      |  |  |
H—Ö—C—C—N—H
  ..  |  |  ..
      H  H
```

As you develop more practice, you will find it more convenient to remember patterns of electron distribution. A neutral oxygen with two bonds has two unshared electron pairs. A neutral nitrogen with three bonds has one unshared pair.

With practice, the writing of organic structures will soon become routine and may be simplified further. For example, by assuming that a carbon atom is present at both ends of a chain and at every bend in the chain, we can omit drawing individual carbons. The structures that result can be simplified even more by omitting the hydrogens attached to carbon.

$CH_3CH_2CH_2CH_3$ becomes [zigzag chain with H atoms] simplified to [zigzag line]

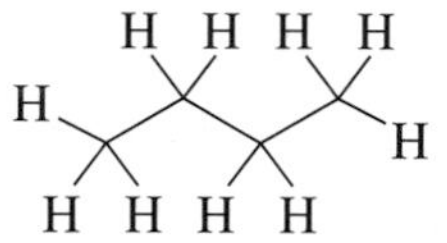

In these simplified representations, called **bond-line formulas** or **carbon skeleton diagrams,** the only atoms specifically written in are those that are neither carbon nor hydrogen bound to carbon. Hydrogens bound to these *heteroatoms* are shown, however.

$CH_3CH_2CH_2CH_2OH$ becomes [zigzag]OH

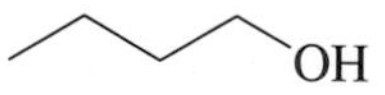

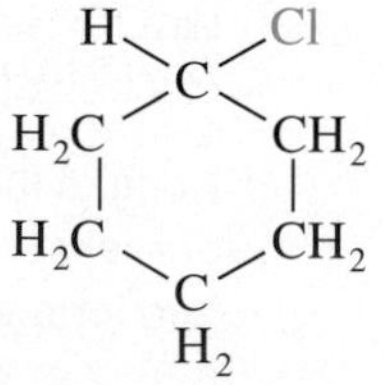

[cyclohexane ring: H, Cl on C; H_2C, CH_2, H_2C, CH_2, $C H_2$] becomes [cyclohexane ring with Cl]

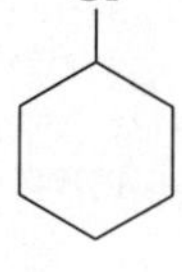

PROBLEM 1.12 Expand the following bond-line representations to show all the atoms including carbon and hydrogen.

(*a*) [zigzag chain] (*b*) [branched skeleton]

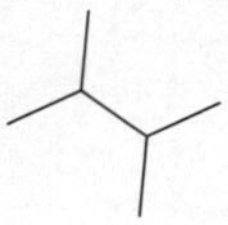

(*c*) 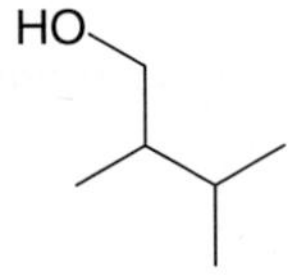

(*d*) 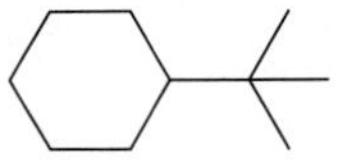

SAMPLE SOLUTION (*a*) There is a carbon at each bend in the chain and at the ends of the chain. Each of the 10 carbon atoms bears the appropriate number of hydrogen substituents so that it has four bonds.

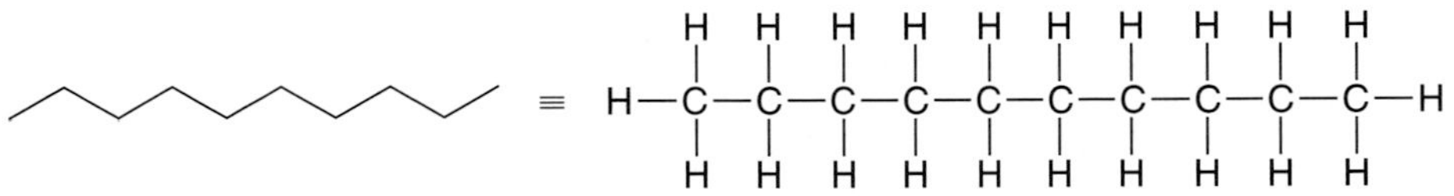

Alternatively, the structure could be written as $CH_3CH_2CH_2CH_2CH_2CH_2CH_2CH_2CH_2CH_3$ or in condensed form as $CH_3(CH_2)_8CH_3$.

1.8 CONSTITUTIONAL ISOMERS

In the introduction we noted that both Berzelius and Wöhler were fascinated by the fact that two different compounds with different properties, ammonium cyanate and urea, possessed exactly the same molecular formula, CH_4N_2O. Berzelius had studied examples of similar phenomena earlier and invented the word **isomer** to describe *different compounds that have the same molecular formula.*

We can illustrate the concept of isomerism by referring to two different compounds, *nitromethane* and *methyl nitrite,* both of which have the molecular formula CH_3NO_2.

The suffix *-mer* in the word *isomer* is derived from the Greek word *meros,* meaning "part," "share," or "portion." The prefix *iso-* is also from Greek (*isos,* "the same"). Thus isomers are different molecules that have the same parts (elemental composition).

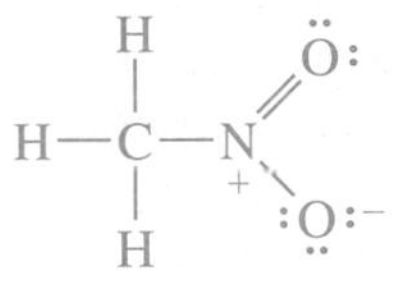

Nitromethane

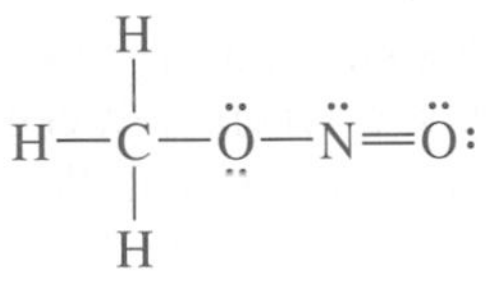

Methyl nitrite

Nitromethane, used as a high-energy fuel for race cars, is a liquid with a boiling point of 101°C. Methyl nitrite is a gas boiling at −12°C which when inhaled causes dilation of blood vessels. Isomers that differ in the order in which their atoms are bonded are often referred to as **structural isomers.** A more modern term is **constitutional isomer.** As noted in the previous section, the order of atomic connections that defines a molecule is termed its *constitution,* and we say that two compounds are *constitutionally isomeric* if they have the same molecular formula but differ in the order in which their atoms are connected.

PROBLEM 1.13 There are many more isomers of CH_3NO_2 other than nitromethane and methyl nitrite. Some, such as *carbamic acid,* an intermediate in the commercial preparation of urea for use as a fertilizer, are too unstable to isolate. Given the information that the nitrogen and both oxygens of carbamic acid are bonded to carbon and that one of the carbon-oxygen bonds is a double bond, write a Lewis structure for carbamic acid.

PROBLEM 1.14 Write structural formulas for all the constitutionally isomeric compounds having the given molecular formula.

(*a*) C_2H_6O (*b*) C_3H_8O (*c*) $C_4H_{10}O$

SAMPLE SOLUTION (*a*) Begin by considering the ways in which two carbons and one oxygen may be bonded. There are two possibilities: C—C—O and C—O—C. Add the six hydrogens so that each carbon has four bonds and each oxygen two. There are two constitutional isomers, ethyl alcohol and dimethyl ether.

```
    H  H                 H       H
    |  |                 |       |
 H—C—C—Ö—H          H—C—Ö—C—H
    |  |                 |       |
    H  H                 H       H
```

Ethyl alcohol Dimethyl ether

In Chapter 3 another type of isomerism, called **stereoisomerism,** will be introduced. Stereoisomers have the same constitution but differ in the three-dimensional arrangement of atoms in space.

1.9 RESONANCE

When we write a Lewis formula for a molecule, we restrict its electrons to certain well-defined locations, either between two nuclei linking them by a covalent bond, or as unshared electrons localized on a single atom. Sometimes more than one stable Lewis structure can be written for a molecule, especially in molecules that contain multiple bonds. An example of such a molecule often cited in introductory chemistry courses is ozone (O_3). Ozone is a form of oxygen that occurs naturally in large quantities in the upper atmosphere, where it screens the surface of the earth from much of the ultraviolet radiation present in sunlight. Were it not for this ozone layer, most forms of surface life on earth would be damaged or even destroyed by the rays of the sun. The following structural formula for ozone satisfies the octet rule; all three oxygens have eight electrons in their valence shell.

```
      +
      ..
      O
    //  \
  :O     O:
   ..    ..−
```

Bond distances in organic compounds are usually 1 to 2 Å (1 Å = 10^{-10} m). Since the angstrom (Å) is not an SI unit, we will express bond distances in picometers (1 pm = 10^{-12} m). Thus, 128 pm = 1.28 Å.

The Lewis formula shown, however, does not accurately portray the bonding in ozone, because it implies that the two terminal oxygens are bonded differently to the central oxygen. The central oxygen is depicted as doubly bonded to one and singly bonded to the other. Since it is generally true that double bonds are shorter than single bonds, we would expect ozone to exhibit two different O—O bond lengths, one of them characteristic of the O—O single bond distance (147 pm in hydrogen peroxide, H—O—O—H) and the other one characteristic of the O═O double bond distance (121 pm in O_2). Such is not the case. Both bond distances in ozone are exactly the same (128 pm)—somewhat shorter than the single bond distance and somewhat longer than the double bond distance. We conclude that *the central oxygen must be identically bonded to both terminal oxygens.*

In order to deal with circumstances such as the nature of the bonding in ozone, the notion of **resonance** between Lewis structures was developed. According to the resonance concept, when more than one Lewis structure may be written for a molecule, a single structure is not sufficient to describe it. Rather, the true structure has an electron distribution which is a "hybrid" of all the possible Lewis structures which may be written for the molecule. In the case of ozone, two equivalent Lewis structures may be written. We use a double-headed arrow to represent resonance between these two Lewis structures.

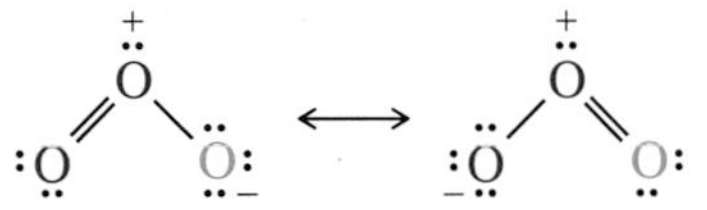

It is important to remember that the double-headed resonance arrow does not indicate a *process* in which the two Lewis structures interconvert. Ozone, for example, has a *single* structure; it does not oscillate back and forth between two Lewis structures. Its true structure is not adequately represented by any single Lewis structure.

Resonance attempts to correct a defect inherent in the nature of Lewis formulas. Lewis formulas depict electrons as being **localized;** they either are shared between two atoms in a covalent bond or are unshared electrons belonging to a single atom. In reality, electrons distribute themselves in the way that leads to their most stable arrangement. This sometimes means that a pair of electrons is **delocalized,** or shared by several nuclei. What we try to show by the resonance formulation of ozone is the delocalization of the lone-pair electrons of one oxygen and the electrons in the double bond over the three atoms of the molecule. Organic chemists often use curved arrows to show this electron delocalization. Alternatively, a superposition of two Lewis structures is sometimes represented using a dashed line to depict a "partial" bond. In the dashed-line notation the central oxygen is linked to the other two by bonds that are intermediate between a single bond and a double bond, and the terminal oxygens each bear one half of a unit negative charge.

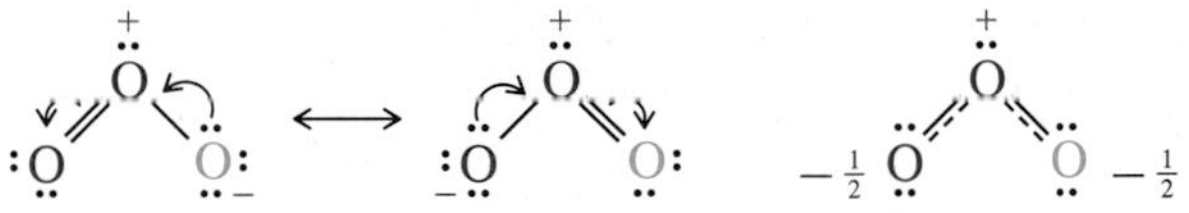

Curved arrow notation

Dashed-line notation

Electron delocalization in ozone

The rules to be followed when writing resonance structures to depict electron delocalization are summarized in Table 1.5.

PROBLEM 1.15 Electron delocalization can be important in ions as well as in neutral molecules. Using curved arrows, show how an equally stable resonance structure can be generated for each of the following anions:

(*a*)

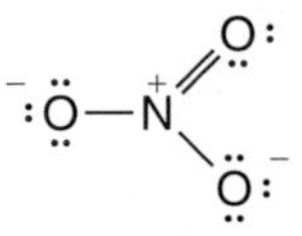

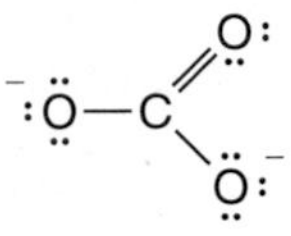

(*b*)

(*c*)

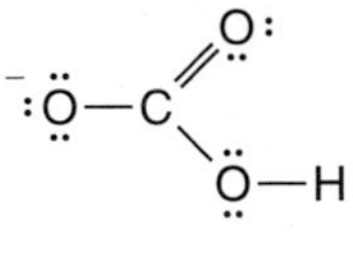

(*d*)

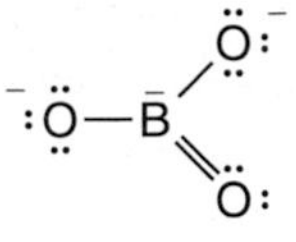

TABLE 1.5
Introduction to the Rules of Resonance*

Rule	Illustration
1. Atomic positions (connectivity) must be the same in all resonance structures; only the electron positions may vary among the various contributing structures.	The structural formulas 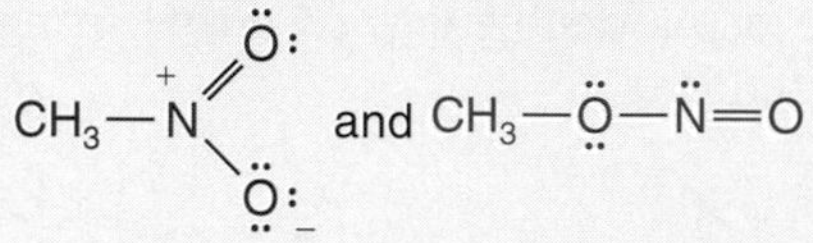 $CH_3-\overset{+}{N}(=\ddot{O}:)-\ddot{O}:^-$ (A) and $CH_3-\ddot{O}-\ddot{N}=O$ (B) represent different compounds, not different resonance forms of the same compound. A is a Lewis structure for *nitromethane;* B is *methyl nitrite.*
2. Lewis structures in which second-row elements own or share more than 8 valence electrons are especially unstable and make no contribution to the true structure. (The octet rule may be exceeded for elements beyond the second row.)	Structural formula C, 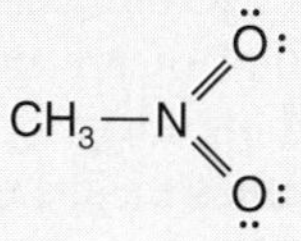 $CH_3-N(=\ddot{O}:)(=\ddot{O}:)$ (C) has 10 electrons around nitrogen. It is not a permissible Lewis structure for nitromethane and so cannot be a valid resonance form.
3. When two or more structures satisfy the octet rule, the most stable one is the one with the smallest separation of oppositely charged atoms.	The two Lewis structures D and E of methyl nitrite satisfy the octet rule: 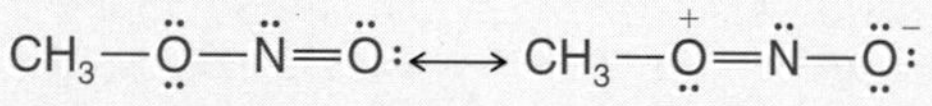 $CH_3-\ddot{O}-\ddot{N}=\ddot{O}: \longleftrightarrow CH_3-\overset{+}{O}=\ddot{N}-\ddot{O}:^-$ (D, E). Structure D has no separation of charge and is more stable than E, which does. The true structure of methyl nitrite is more like D than E.
4. Among structural formulas in which the octet rule is satisfied for all atoms and one or more of these atoms bears a formal charge, the most stable resonance form is the one in which negative charge resides on the most electronegative atom (or positive charge on the most electropositive one).	The most stable Lewis structure for cyanate ion is F because the negative charge is on its oxygen. 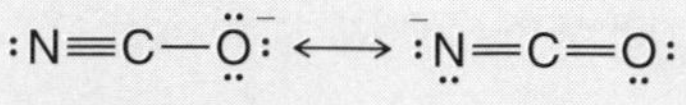 $:N\equiv C-\ddot{O}:^- \longleftrightarrow \ddot{N}^-=C=\ddot{O}:$ (F, G). In G the negative charge is on nitrogen. Oxygen is more electronegative than nitrogen and can better support a negative charge.

(Continued)

TABLE 1.5

Introduction to the Rules of Resonance* *(Continued)*

Rule	Illustration
5. Each contributing Lewis structure must have the same number of electrons and the same *net* charge, although the formal charges of individual atoms may vary among the various Lewis structures.	The Lewis structures $CH_3{-}\overset{+}{N}(=\ddot{O}:)(-\ddot{O}:^-)$ and $CH_3{-}\ddot{N}(-\ddot{O}:^-)(-\ddot{O}:)$ H I are *not* resonance forms of one another. Structure H has 24 valence electrons and a net charge of 0; I has 26 valence electrons and a net charge of − 2.
6. Each contributing Lewis structure must have the same number of *unpaired* electrons.	Structural formula J is a Lewis structure of nitromethane; K is not, even though it has the same atomic positions and the same number of electrons. $CH_3{-}\overset{+}{N}(=\ddot{O}:)(-\ddot{O}:^-)$ and $CH_3{-}\ddot{N}(-\dot{\ddot{O}}:)(-\dot{\ddot{O}}:)$ J K Structure K has 2 unpaired electrons. Structure J has all its electrons paired and is a more stable structure.
7. Electron delocalization stabilizes a molecule. A molecule in which electrons are delocalized is more stable than implied by any of the individual Lewis structures which may be written for it. The degree of stabilization is greatest when the contributing Lewis structures are of equal stability.	Nitromethane is stabilized by electron delocalization more than methyl nitrite is. *The two most stable resonance forms of nitromethane are equivalent to each other.* $CH_3{-}\overset{+}{N}(=\ddot{O}:)(-\ddot{O}:^-) \longleftrightarrow CH_3{-}\overset{+}{N}(-\ddot{O}:^-)(=\ddot{O}:)$ *The two most stable resonance forms of methyl nitrite are not equivalent.* $CH_3{-}\ddot{O}{-}\ddot{N}{=}\ddot{O}: \longleftrightarrow CH_3{-}\overset{+}{\ddot{O}}{=}\ddot{N}{-}\ddot{O}:^-$

*These are the most important rules to be concerned with at present. Additional aspects of electron delocalization, as well as additional rules for its depiction by way of resonance structures, will be developed as needed in subsequent chapters.

SAMPLE SOLUTION (*a*) When using curved arrows to represent the reorganization of electrons, begin at a site of high electron density, preferably an atom that is negatively charged. Move electron pairs until a proper Lewis structure results. For nitrate ion, this can be accomplished in two ways:

Three equally stable Lewis structures are possible for nitrate ion. The negative charge in nitrate is shared equally by all three oxygens.

It is good chemical practice to represent molecules by their most stable Lewis structure. The ability to write alternative resonance forms and to compare their relative stabilities, however, can provide insight into both molecular structure and chemical behavior. This will become particularly apparent in the last two-thirds of this text, where the resonance concept will be used regularly.

1.10 THE SHAPES OF SOME SIMPLE MOLECULES

Our concern to this point has emphasized ''electron bookkeeping.'' We now shift our emphasis to a consideration of the directional properties of bonds, understanding that the shape of a molecule is governed by the orientation of its bonds.

The *tetrahedron* is an important geometric shape in chemistry. Methane, for example, is described as a tetrahedral molecule because its four hydrogens occupy the corners of a tetrahedron which has carbon at its center (Figure 1.8, page 28). As an alternative to molecular models, chemists often show three-dimensionality in structural formulas by using a **solid wedge** (▬) to depict a bond projecting from the paper toward the reader, and a **dashed wedge** (┅) to depict one receding from the paper. A simple line (—) represents a bond that lies in the plane of the paper (Figure 1.9, page 28).

While reservations have been expressed concerning VSEPR as an *explanation* for molecular geometries, it remains a useful *tool* for predicting the shapes of organic compounds.

The tetrahedral geometry of methane is often explained in terms of the **valence-shell electron pair repulsion (VSEPR) model.** The VSEPR model rests on the premise that an electron pair, either a bonded pair or an unshared pair, associated with a particular atom will be as far away from the atom's other electron pairs as possible. Thus, a tetrahedral geometry permits the four bonds of methane to be maximally separated and is characterized by H—C—H angles of 109.5°, a value referred to as the **tetrahedral angle.**

Water, ammonia, and methane share the common feature of an approximately tetrahedral arrangement of four electron pairs. Since, however, we describe the shape of a molecule according to the positions of its atoms rather than the disposition of its electron pairs, water is said to be *bent,* and ammonia is *trigonal pyramidal* (Figure 1.10, page 28). The H—O—H angle in water (105°) and the H—N—H angle in ammonia (107°) are slightly less than the tetrahedral angle.

MOLECULAR MODELS

As early as the nineteenth century many chemists built scale models in order to better understand molecular structure. We can gain a heightened appreciation for features that affect structure and reactivity when we examine the three-dimensional shape of a molecular model. Three types of molecular models are encountered most often and are shown for methane in Figure 1.7. Probably the most familiar are ball-and-stick models (Figure 1.7*b*), which direct approximately equal attention to the atoms and the bonds that connect them. Framework models (Figure 1.7*a*) and space-filling models (Figure 1.7*c*) represent opposite extremes. Framework models emphasize the pattern of bonds of a molecule while ignoring the sizes of the atoms. Space-filling models emphasize the volume occupied by individual atoms at the cost of a clear depiction of the bonds; they are most useful in cases in which one wishes to examine the overall molecular shape and to assess how closely two nonbonded atoms approach each other.

The earliest ball-and-stick models were exactly that: wooden balls in which holes were drilled to accommodate dowels that connected the atoms. Plastic versions, including relatively inexpensive student sets, became available in the 1960s and proved to be a valuable learning aid. Precisely scaled stainless steel framework and plastic space-filling models, although relatively expensive, are standard equipment in most research laboratories.

Computer graphics–based representations are rapidly replacing classical molecular models. Indeed, the term *molecular modeling* as now used in organic chemistry implies computer generation of models. The methane models shown in Figure 1.7 were all drawn on a personal computer using software that possesses the feature of displaying and printing the same molecule in framework, ball-and-stick, and space-filling formats. In addition to permitting models to be constructed rapidly, even the simplest software allows the model to be turned and viewed from a variety of perspectives. More sophisticated programs not only draw molecular models but also incorporate computational tools that evaluate how structural changes affect the stability of a molecule. This facet of computer-assisted molecular modeling will be introduced in Chapter 3.

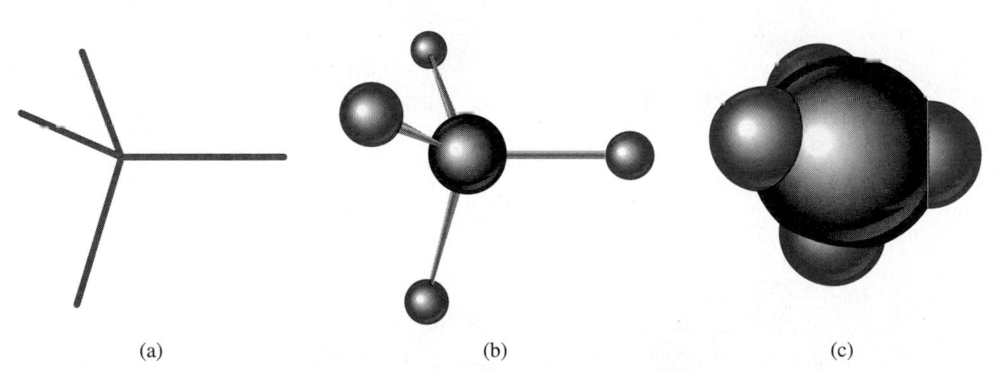

FIGURE 1.7 (*a*) A framework molecular model of methane (CH_4). A framework model shows the bonds connecting the atoms of a molecule, but not the atoms themselves. (*b*) A ball-and-stick model of methane. (*c*) A space-filling model of methane.

Boron trifluoride (BF_3; Figure 1.11) is a *trigonal planar* molecule. There are six electrons, two for each B—F bond, associated with the valence shell of boron. These three bonded pairs are farthest apart when they are coplanar, with F—B—F bond angles of 120°.

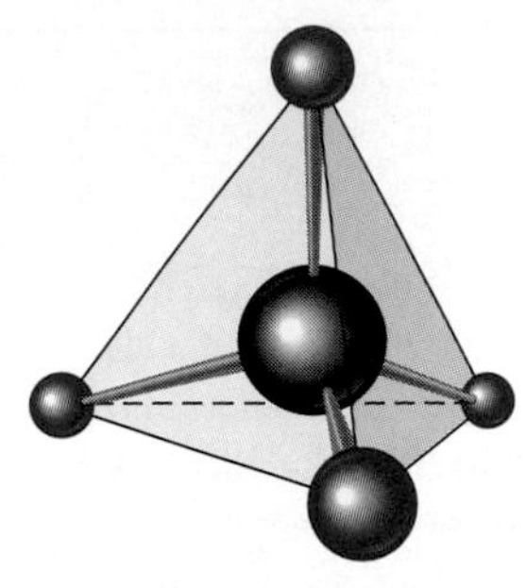

FIGURE 1.8 The tetrahedral geometry of methane: a methane molecule inscribed within a regular tetrahedron. Carbon is at the center and a hydrogen is at each of the corners.

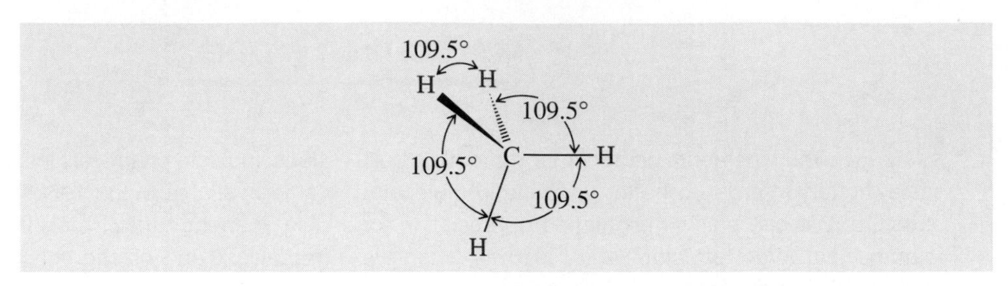

FIGURE 1.9 A wedge-and-dash depiction of the structure of methane. A solid wedge projects from the plane of the paper toward you; a dashed wedge projects away from you. A bond represented by a line drawn in the customary way lies in the plane of the paper.

PROBLEM 1.16 The salt sodium borohydride, $NaBH_4$, has an ionic bond between Na^+ and the anion BH_4^-. What are the H—B—H angles in the borohydride anion?

Multiple bonds are treated as a single unit in the VSEPR model. Formaldehyde (Figure 1.12) is a trigonal planar molecule in which the electrons of the double bond unit and those of the two single bonds are maximally separated. A linear arrangement of atoms in carbon dioxide (Figure 1.13) allows the electrons in one double bond to be as far away as possible from the electrons in the other double bond.

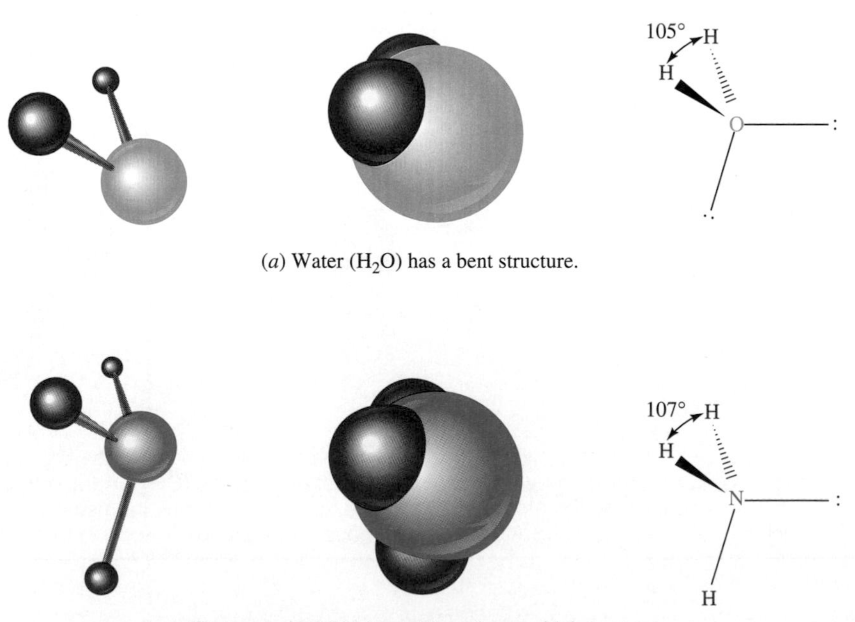

(*a*) Water (H_2O) has a bent structure.

(*b*) Ammonia (NH_3) has a trigonal pyramidal structure.

FIGURE 1.10 Ball-and-stick and space-filling models and wedge-and-dash depictions of (*a*) water and (*b*) ammonia. The shape of a molecule is specified according to the location of its atoms. An approximately tetrahedral arrangement of electron pairs translates into a bent geometry for water and a trigonal pyramidal geometry for ammonia.

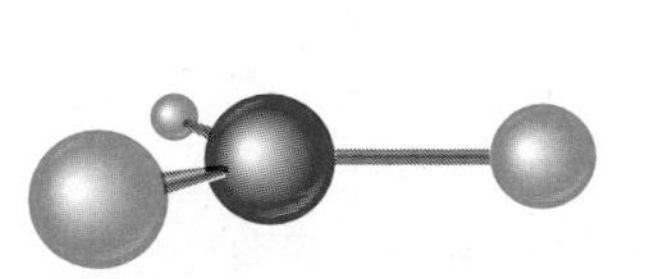

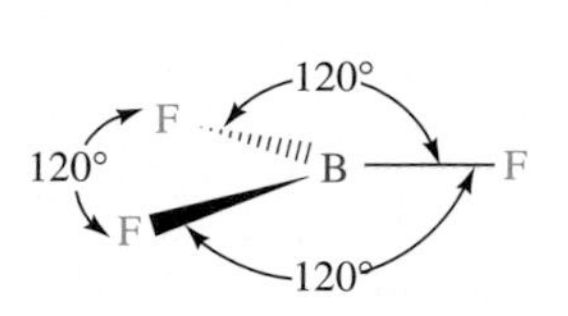

FIGURE 1.11 Representations of the trigonal planar geometry of boron trifluoride (BF_3). There are six electrons in the valence shell of boron, a pair for each covalent bond to fluorine. The three pairs of electrons are farthest apart when the F—B—F angle is 120°.

FIGURE 1.12 A space-filling model illustrating the trigonal planar structure of formaldehyde.

PROBLEM 1.17 Specify the geometry of the following:

(*a*) H—C≡N: (Hydrogen cyanide)
(*b*) H_4N^+ (Ammonium ion)
(*c*) $\overset{-}{:\ddot{N}}=\overset{+}{N}=\overset{-}{\ddot{N}:}$ (Azide ion)
(*d*) CO_3^{2-} (Carbonate ion)

SAMPLE SOLUTION (*a*) The structure shown accounts for all the electrons in hydrogen cyanide. There are no unshared electron pairs associated with carbon, and so the structure is determined by maximizing the separation between its single bond to hydrogen and the triple bond to nitrogen. Hydrogen cyanide is a *linear* molecule.

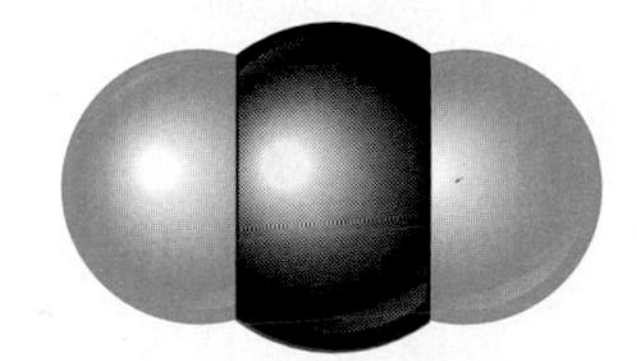

FIGURE 1.13 A space-filling model showing the linear geometry of carbon dioxide.

1.11 MOLECULAR DIPOLE MOMENTS

By combining a knowledge of molecular geometry with a feel for the polarity of chemical bonds (Section 1.5), it is possible to make reasonable predictions about the dipole moments of polyatomic molecules. Molecules, such as carbon dioxide, for example, can have polar bonds and yet, because of their shape, have no net dipole moment. The molecular dipole moment is the resultant, or vector sum, of all the individual bond dipole moments of a substance. In carbon dioxide, the linear arrangement of the atoms causes the two carbon-oxygen bond moments to cancel.

$$\overleftarrow{}\ \ \overrightarrow{}$$
$$:\ddot{O}=C=\ddot{O}: \qquad \text{Dipole moment} = 0\ \text{D}$$

Carbon dioxide

Carbon tetrachloride, with four polar C—Cl bonds and a tetrahedral shape, has no net dipole moment, because the resultant of the four bond dipoles, as shown in Figure 1.14, is zero. Dichloromethane, on the other hand, has a dipole moment of 1.62 D. The C—H bond dipoles reinforce the C—Cl bond dipoles.

PROBLEM 1.18 Which of the following compounds would you expect to have a dipole moment? If the molecule has a dipole moment, specify its direction.

(*a*) BF_3
(*b*) H_2O
(*c*) CH_4
(*d*) CH_3Cl
(*e*) CH_2O
(*f*) HCN

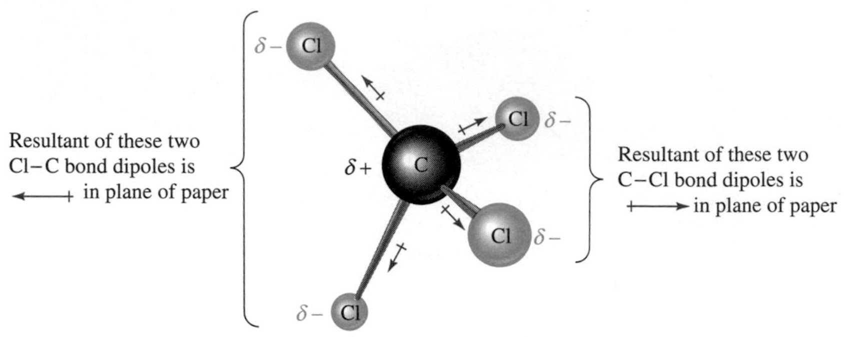

(*a*) There is a mutual cancellation of individual bond dipoles in carbon tetrachloride. It has no dipole moment.

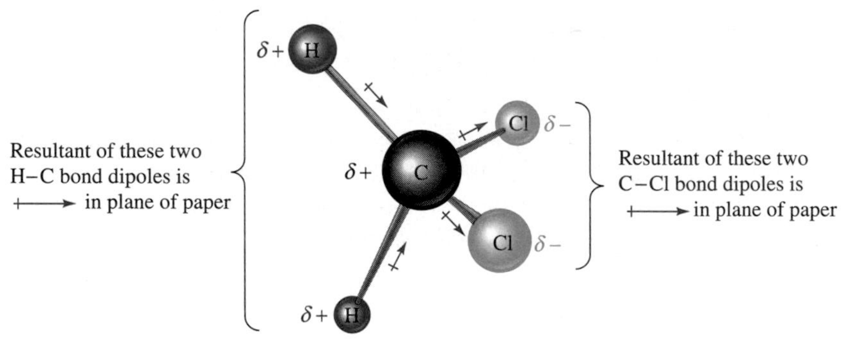

(*b*) The H–C bond dipoles reinforce the C–Cl bond moment in dichloromethane. The molecule has a dipole moment of 1.62 D.

FIGURE 1.14 Contribution of individual bond dipole moments to the molecular dipole moments of carbon tetrachloride and dichloromethane.

SAMPLE SOLUTION (*a*) Boron trifluoride is planar with 120° bond angles. Although each boron-fluorine bond is polar, their combined effects cancel and the molecule has no dipole moment.

$\mu = 0$ D

1.12 MOLECULAR ORBITALS OF THE HYDROGEN MOLECULE

The Lewis approach is a simple yet very useful conceptual model to describe chemical bonding. It is, however, not the only model. A second is the **molecular orbital** model. The molecular orbital approach rests on the idea that, as electrons in atoms occupy *atomic orbitals,* electrons in molecules occupy *molecular orbitals.* A molecular orbital may be associated with only one or two or a few of the atoms in a molecule, or it may encompass the entire molecule. Molecular orbitals, no matter how extensive, have many of the same properties that we associate with atomic orbitals. They are populated by electrons beginning with the orbital of lowest energy, and a molecular orbital is fully occupied when it contains two electrons of opposite spin.

Some of the fundamental principles of molecular orbital theory can be described by

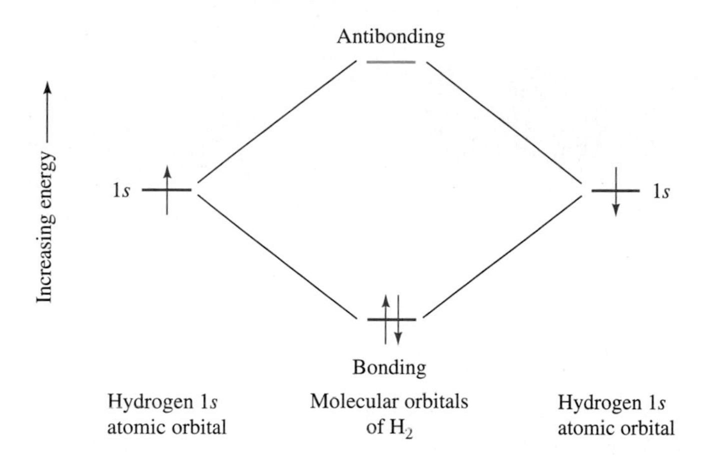

FIGURE 1.15 Two molecular orbitals are generated by combining two hydrogen 1*s* orbitals. One molecular orbital is a bonding molecular orbital and is lower in energy than either of the atomic orbitals that combine to produce it. The other molecular orbital is antibonding and is of higher energy than either atomic orbital. Each arrow indicates one electron; the electron spins are opposite in sign. The bonding orbital contains both electrons of H_2.

considering its application to the hydrogen molecule (H_2). Molecular orbitals of H_2 are generated by combining (or ''mixing'') the 1*s* atomic orbitals of two hydrogen atoms, as shown in Figure 1.15. The number of molecular orbitals so generated always equals the number of atomic orbitals that combine to produce them. Thus, two hydrogen 1*s* atomic orbitals (AOs) combine to produce two molecular orbitals (MOs) of an H_2 molecule. The lower-energy MO shown in Figure 1.15 arises by adding the two 1*s* wave functions, the higher-energy MO by subtracting them. The additive combination is analogous to two waves that overlap ''in phase'' to reinforce each other; the subtractive combination is analogous to the destructive interference between two waves that are ''out of phase.''

Figure 1.16 depicts the in-phase combination of the two 1*s* orbitals. The MO that results is lower in energy than either of the initial AOs and is called a **bonding molecular orbital.** Constructive reinforcement of the two waves increases the electron probability (electron density) in the region between the two nuclei and serves to bind them together. An electron in a bonding MO is held *more* strongly than on an isolated atom. The two electrons of H_2 occupy the bonding MO, making the H_2 molecule more stable than two independent hydrogen atoms.

Conversely, the out-of-phase combination depicted in Figure 1.17 results in a node between the two nuclei and gives an **antibonding molecular orbital.** This antibonding orbital is not only higher in energy than the bonding MO, but also higher in energy than either of the initial AOs. An electron in an antibonding MO is *less* strongly held than in one of the component AOs. Because H_2 has only two electrons and both occupy the bonding MO, the antibonding orbital of H_2 is vacant.

Both the bonding and the antibonding orbitals of H_2 have rotational symmetry around a line connecting the two atoms (the *internuclear axis*). A cross section of the orbital taken perpendicular to the internuclear axis is a circle. Orbitals that have this symmetry are classified as **σ (sigma) orbitals.** The antibonding orbital is designated by an asterisk and is referred to as a **σ^* (sigma star) orbital.**

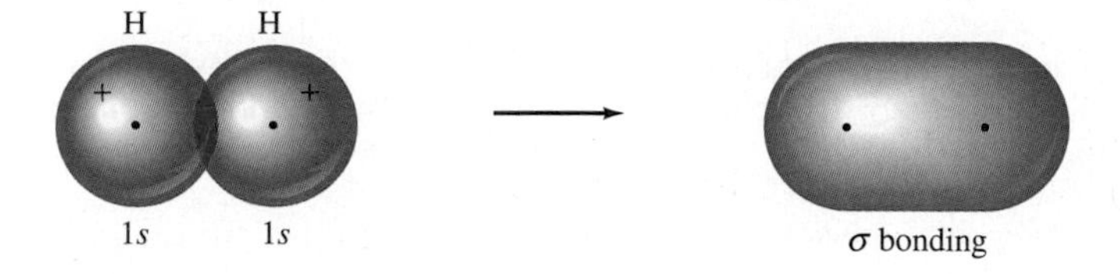

FIGURE 1.16 When two 1*s* atomic orbitals overlap so that the signs of their wave functions reinforce each other, a bonding molecular orbital results.

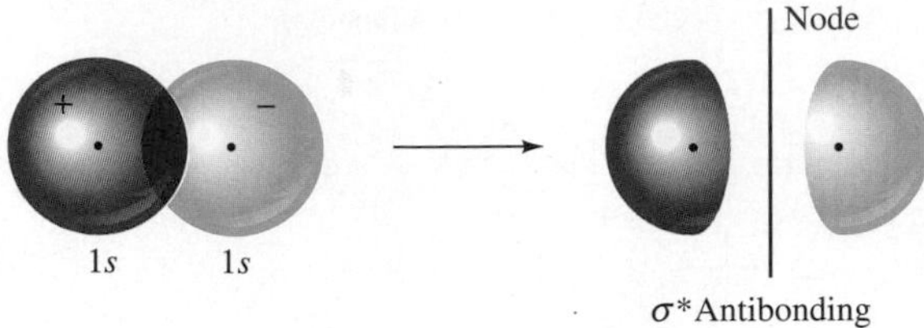

FIGURE 1.17 When two 1*s* atomic orbitals overlap so that the signs of their wave functions are opposite, an antibonding molecular orbital results.

1.13 sp^3 HYBRIDIZATION AND BONDING IN METHANE

While it is possible to construct molecular orbital diagrams of organic molecules, the information provided is not of the type that is useful to most chemists. Chemists, especially organic chemists, look at reactions in terms of changes that take place at individual bonds in a molecule, and the Lewis view of the covalent bond is more directly suited to most of their needs than the molecular orbital picture. A third approach, called **valence-bond theory,** has strongly influenced organic chemical thinking for more than half a century. According to valence-bond theory, the shared electron pair bond can be viewed as arising by overlap of a half-filled orbital of one atom with a half-filled orbital of another.

A second key element, introduced by Linus Pauling in the 1930s, is the idea of **orbital hybridization.** The classic example of this *orbital hybridization model* of bonding is methane. In order to form bonds to four hydrogens, valence-bond theory requires that carbon have four half-filled orbitals. However, as shown in Figure 1.18*a*, the stable electron configuration of a carbon atom contains only two such half-filled orbitals. Thus, it would seem that the electron configuration of carbon permits the compound CH_2, but not CH_4. We can get around the problem of an insufficient number of half-filled orbitals by simply "promoting" an electron from the 2*s* orbital to the vacant 2*p* orbital (Figure 1.18*b*). However, the resulting electron configuration, which has one half-filled 2*s* orbital and three half-filled 2*p* orbitals, is inconsistent

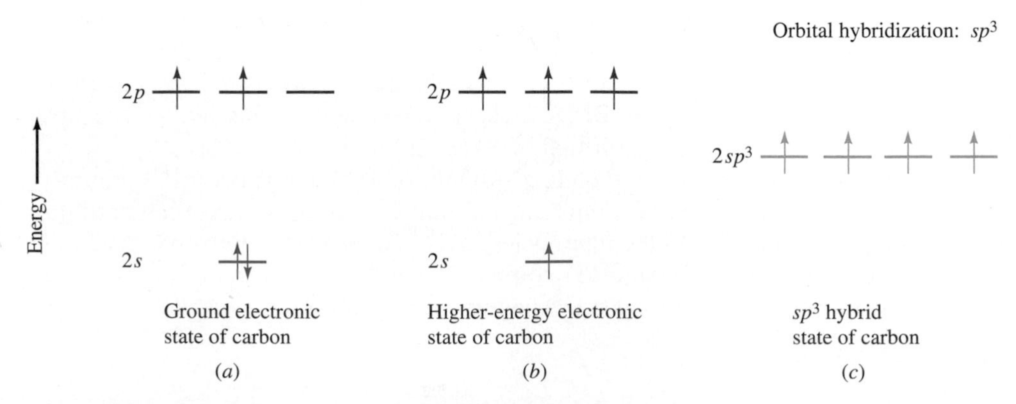

FIGURE 1.18 (*a*) Electron configuration of carbon in its most stable state. (*b*) An electron is "promoted" from the 2*s* orbital to the vacant 2*p* orbital. (*c*) The 2*s* orbital and the three 2*p* orbitals are combined to give a set of four equal-energy sp^3 hybridized orbitals, each of which contains one electron.

with the fact that all the C—H bonds of methane are equivalent. Pauling's idea was to mix together **(hybridize)** the four valence orbitals of carbon ($2s$, $2p_x$, $2p_y$, and $2p_z$) to give four half-filled orbitals of equal energy (Figure 1.18*c*). Conserving the "*s* character" brought by the $2s$ orbital and the total "*p* character" of the three $2p$ orbitals requires that each of the four new sp^3 **hybrid orbitals** contain 25 percent *s* character and 75 percent *p* character.

Figure 1.19 depicts this orbital hybridization in a form that emphasizes the shapes of the orbitals involved and of the sp^3 hybrid orbitals produced. Each sp^3 hybrid orbital has two lobes of unequal size, making the electron density greater on one side of the nucleus than the other. In a bond to hydrogen, it is the larger lobe of a carbon sp^3 orbital that overlaps with a hydrogen $1s$ orbital. The orbital overlaps corresponding to the four C—H bonds of methane are portrayed in Figure 1.20. Orbital overlap along the internuclear axis generates a bond with rotational symmetry—in this case a C($2sp^3$)—H($1s$) σ bond.

Our picture of the shape of sp^3 hybrid orbitals comes from mathematical treatments of orbital hybridization. These mathematical approaches also give the spatial orientation of the orbitals; the axes of the four sp^3 hybrid orbitals are directed toward the corners of a tetrahedron with carbon at the center. *A tetrahedral orientation of its σ bonds is characteristic of sp^3 hybridized carbon.*

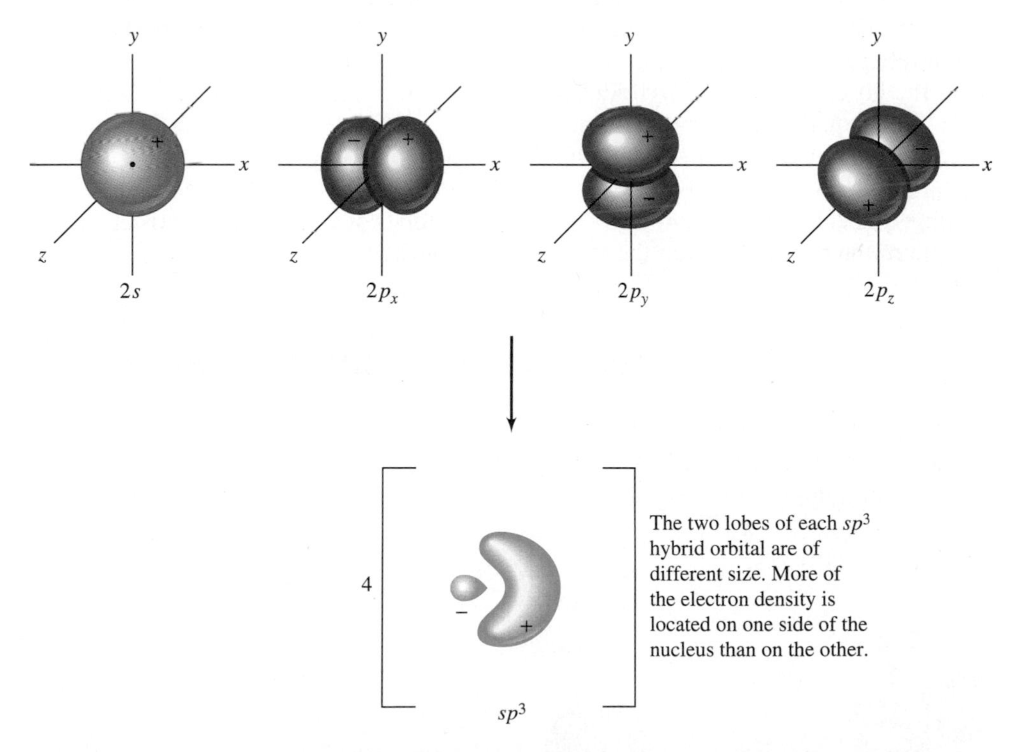
Combine one 2s and three 2p orbitals to give four equivalent sp3 hybrid orbitals:
y
x
z
+
−
2s
2px
2py
2pz
4
sp3
The two lobes of each sp3 hybrid orbital are of different size. More of the electron density is located on one side of the nucleus than on the other.

FIGURE 1.19 Representation of orbital mixing in sp^3 hybridization. Mixing of one *s* orbital with three *p* orbitals generates four sp^3 hybrid orbitals. Each sp^3 hybrid orbital has 25 percent *s* character and 75 percent *p* character. The four sp^3 hybrid orbitals have their major lobes directed toward the corners of a tetrahedron, which has the carbon atom at its center.

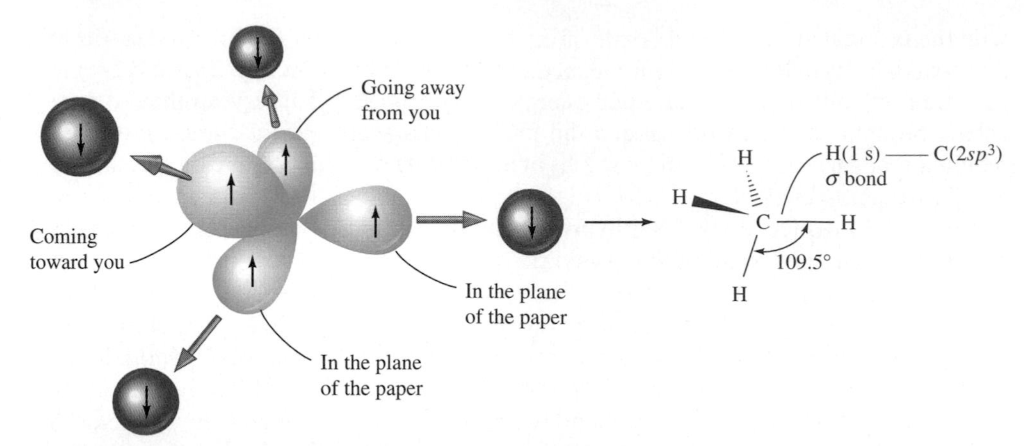

FIGURE 1.20 The sp^3 hybrid orbitals are arranged in a tetrahedral fashion around carbon. Each orbital contains one electron and can form a bond with a hydrogen atom to give a tetrahedral methane molecule. (*Note:* Only the major lobe of each sp^3 orbital is shown. As indicated in Figure 1.19, each orbital contains a smaller back lobe, which has been omitted for the sake of clarity.)

You should not think of orbital hybridization in methane as a sequential process with promotion, hybridization, and σ bond formation occurring as successive events. The electrons spontaneously adopt the configuration that provides the most stable structure. The sp^3 hybrid state permits carbon to form four σ bonds rather than two, and the shape of the sp^3 hybrid orbitals causes each σ bond to be stronger than it would be in the absence of orbital hybridization. An electron in an sp^3 orbital, with its two lobes of unequal size, spends almost all its time in the region between two atoms, where it can contribute to strong bonding. An electron in a spherically symmetric *s* orbital spends much of its time in regions where it contributes little or nothing to this bonding. An electron in a *p* orbital, which has two lobes of equal size, spends only half its time in the region between the two bonded atoms.

1.14 *sp*³ HYBRIDIZATION AND BONDING IN ETHANE

The C—H and C—C bond distances in ethane are 111 and 153 pm, respectively, and the bond angles are close to tetrahedral.

The orbital hybridization model of covalent bonding is readily extended to carbon-carbon bonds. As Figure 1.21 illustrates, ethane is described in terms of a carbon-carbon σ bond joining two CH_3 **(methyl)** groups. Each methyl group consists of an sp^3

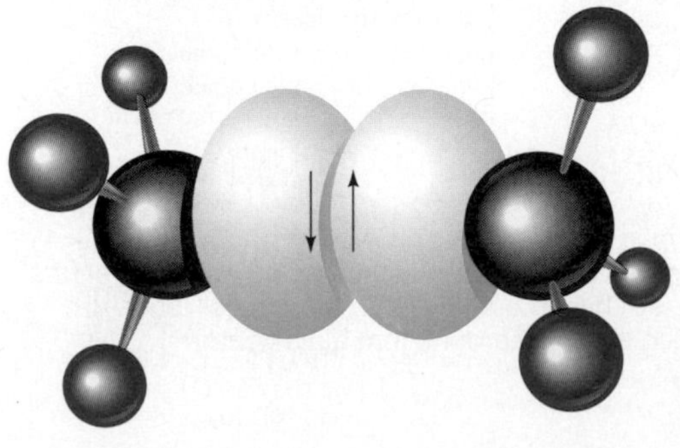

FIGURE 1.21 Orbital overlap description of an sp^3–sp^3 σ bond between two carbon atoms.

hybridized carbon attached to three hydrogens by sp^3-1s σ bonds. Overlap of the remaining half-filled orbital of one carbon with that of the other generates a σ bond between them. Here is a third kind of σ bond, one that has as its basis the overlap of two sp^3 hybridized orbitals. *In general, you can expect that carbon will be sp^3 hybridized when it is directly bonded to four atoms.*

The orbital hybridization model of bonding is not limited to compounds in which all the bonds are single bonds, but can be adapted to compounds with double and triple bonds, as described in the following two sections.

1.15 sp^2 HYBRIDIZATION AND BONDING IN ETHYLENE

Ethylene is a planar molecule, as the structural representations of Figure 1.22 indicate. Because sp^3 hybridization is associated with a tetrahedral geometry at carbon, it is not appropriate for ethylene, which has a trigonal planar geometry at both of its carbons. The hybridization scheme is determined by the number of atoms to which the carbon is directly attached. In ethane, four atoms are attached to carbon by σ bonds, and so four equivalent sp^3 hybrid orbitals are required. In ethylene, three atoms are attached to carbon, so that three equivalent hybrid orbitals are required for each carbon. As shown in Figure 1.23, these three orbitals are generated by mixing the carbon $2s$ orbital with two of the $2p$ orbitals and are called **sp^2 hybrid orbitals.** One of the $2p$ orbitals is left unhybridized.

Another name for ethylene is *ethene.*

Figure 1.24 illustrates the mixing of orbitals in sp^2 hybridization. The three sp^2 orbitals are of equal energy; each has one-third s character and two-thirds p character. Their axes are coplanar, and each has a shape much like that of an sp^3 orbital.

Each carbon of ethylene uses two of its sp^2 hybrid orbitals to form σ bonds to two hydrogen atoms, as illustrated in the first part of Figure 1.25. The remaining sp^2 orbitals, one on each carbon, overlap along the internuclear axis to give a σ bond connecting the two carbons.

As Figure 1.25 shows, each carbon atom still has, at this point, an unhybridized $2p$ orbital available for bonding. These two half-filled $2p$ orbitals have their axes perpendicular to the framework of σ bonds of the molecule and overlap in a side-by-side manner to give what is called a **π (pi) bond.** According to this analysis, the carbon-carbon double bond of ethylene is viewed as a combination of a σ bond plus a π bond. The additional increment of bonding makes a carbon-carbon double bond both stronger and shorter than a carbon-carbon single bond.

One measure of the strength of a bond is its *bond dissociation energy.* This topic will be introduced in Section 4.18 and applied to ethylene in Section 5.2.

Electrons in a π bond are called **π electrons.** The probability of finding a π electron is highest in the region above and below the plane of the molecule. The plane of the molecule corresponds to a nodal plane, where the probability of finding a π electron is zero.

In general, you can expect that carbon will be sp^2 hybridized when it is directly bonded to three atoms.

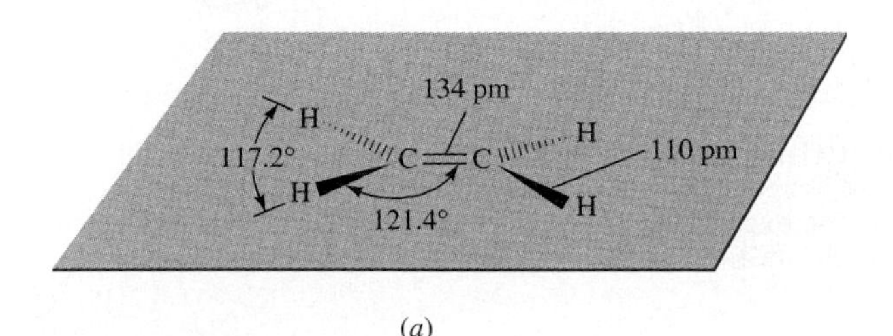

(*a*)

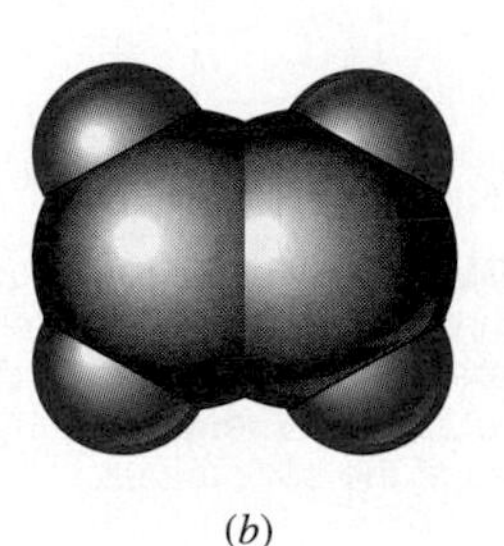

(*b*)

FIGURE 1.22 (*a*) All the atoms of ethylene lie in the same plane. Bond angles are close to 120°, and the carbon-carbon bond distance is significantly shorter than that of ethane. (*b*) A space-filling model of ethylene.

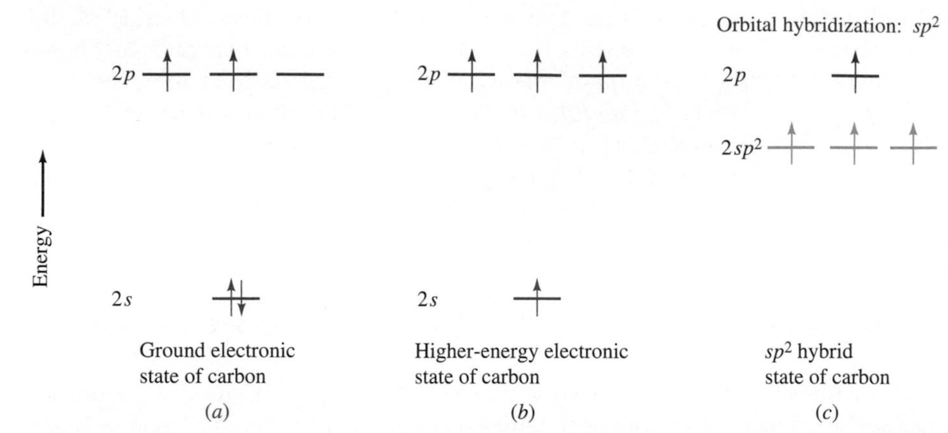

FIGURE 1.23 (*a*) Electron configuration of carbon in its most stable state. (*b*) An electron is "promoted" from the 2*s* orbital to the vacant 2*p* orbital. (*c*) The 2*s* orbital and two of the three 2*p* orbitals are combined to give a set of three equal-energy sp^2 hybridized orbitals. One of the 2*p* orbitals remains unchanged.

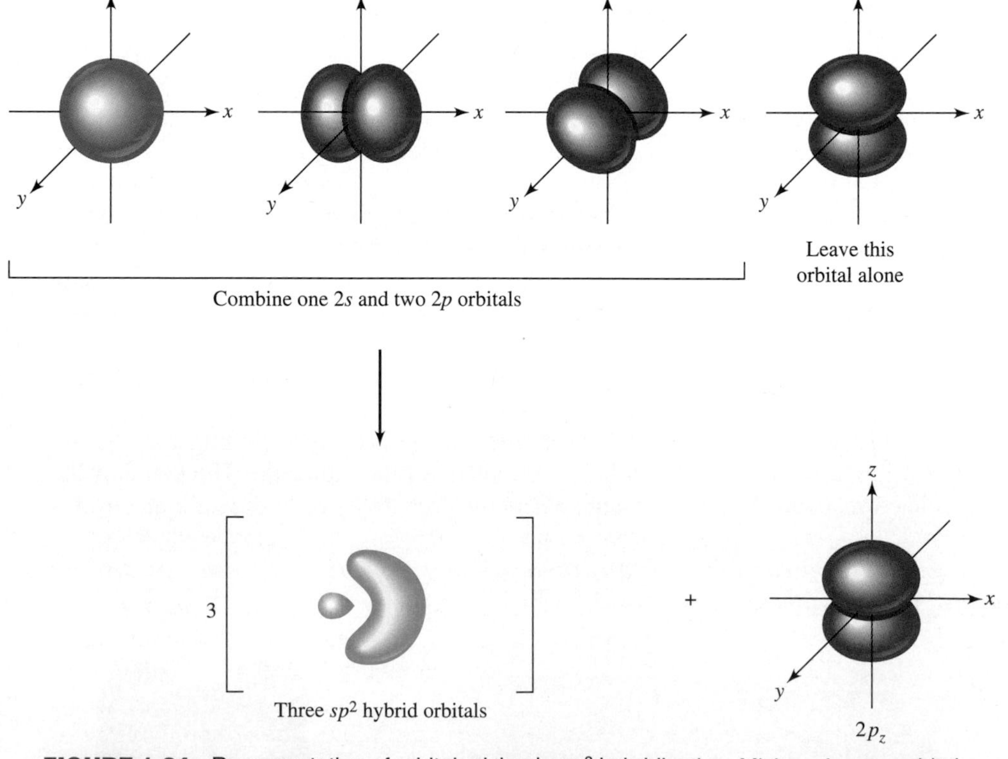

FIGURE 1.24 Representation of orbital mixing in sp^2 hybridization. Mixing of one *s* orbital with two *p* orbitals generates three sp^2 hybrid orbitals. Each sp^2 hybrid orbital has one-third *s* character and two-thirds *p* character. The axes of the three sp^2 hybrid orbitals are coplanar. One 2*p* orbital remains unhybridized, and its axis is perpendicular to the plane defined by the axes of the sp^2 orbitals.

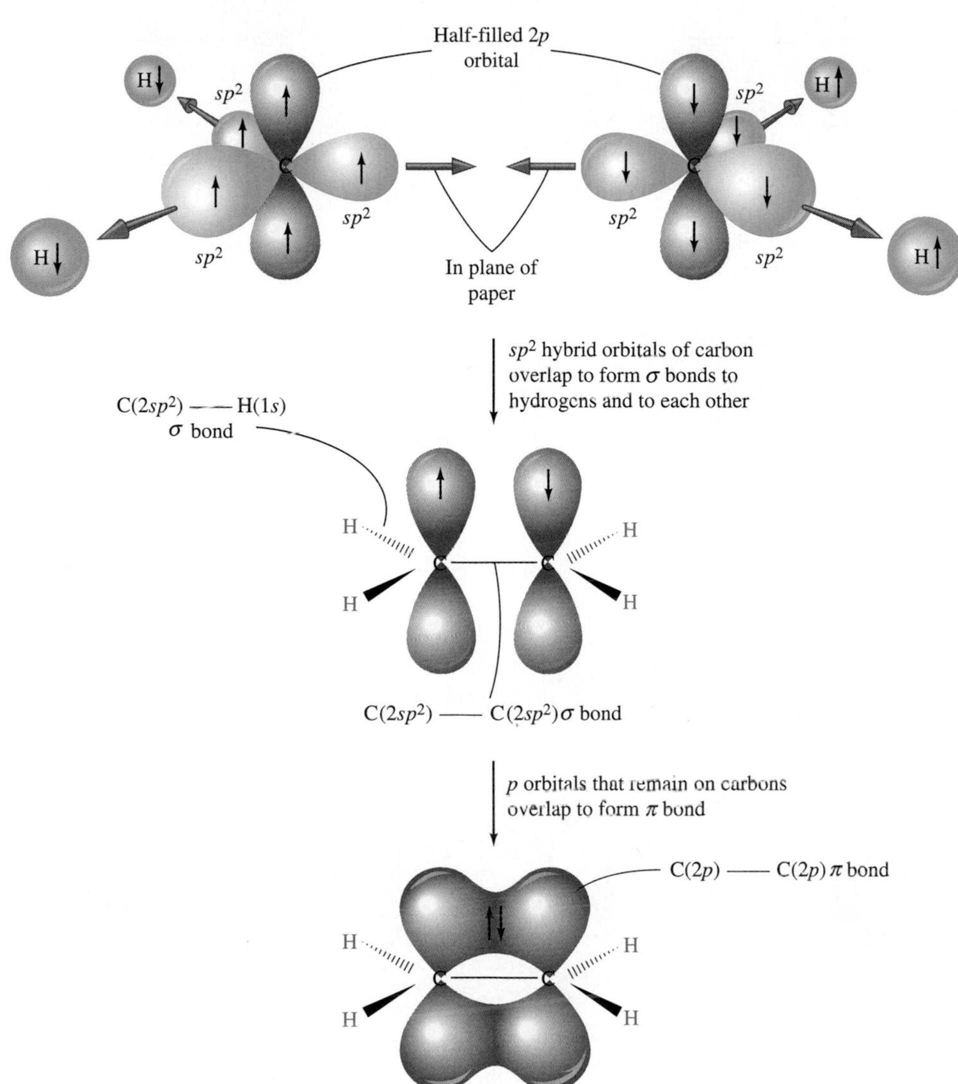

FIGURE 1.25 The carbon-carbon double bond in ethylene has a σ component and a π component. The σ component arises from overlap of sp^2 hybridized orbitals along the internuclear axis. The π component results from a side-by-side overlap of 2*p* orbitals.

1.16 *sp* HYBRIDIZATION AND BONDING IN ACETYLENE

Another name for acetylene is *ethyne.*

There is one more hybridization scheme that is important in organic chemistry. It is called ***sp* hybridization** and applies when carbon is directly bonded to two atoms, as it is in acetylene. The structure of acetylene is shown in Figure 1.26 along with its bond distances and bond angles.

Since each carbon in acetylene is bonded to two other atoms, the orbital hybridization model requires each carbon to have two equivalent orbitals available for the formation of σ bonds as outlined in Figures 1.27 and 1.28. According to this model the carbon 2*s* orbital and one of the 2*p* orbitals combine to generate a pair of two equivalent *sp* hybrid orbitals. Each *sp* hybrid orbital has 50 percent *s* character and 50 percent

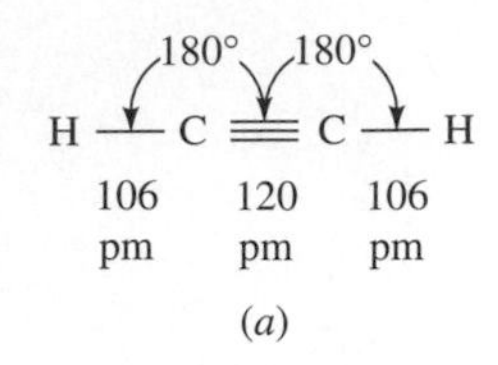

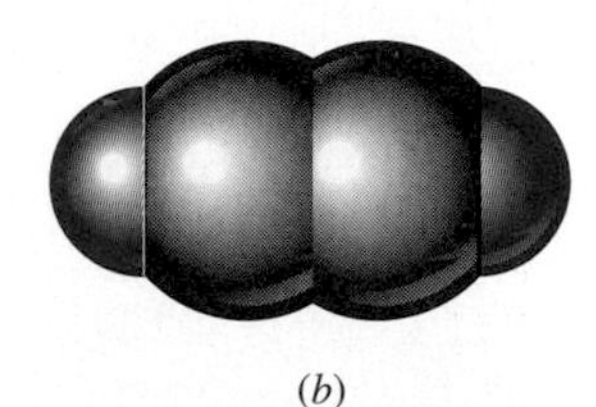

FIGURE 1.26 Acetylene is a linear molecule as indicated in the structural formula (*a*) and a space-filling model (*b*).

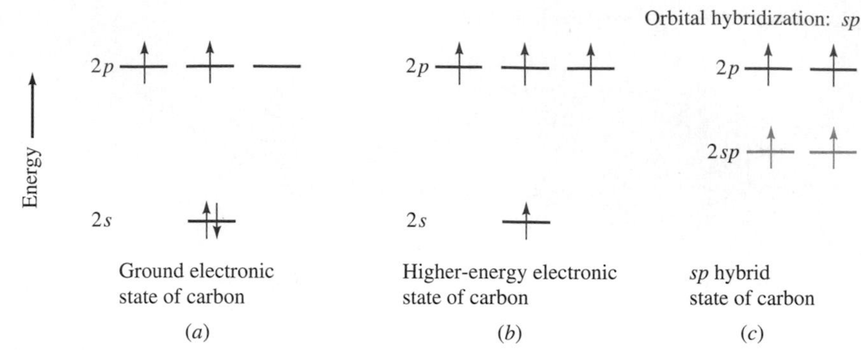

FIGURE 1.27 (*a*) Electron configuration of carbon in its most stable state. (*b*) An electron is "promoted" from the 2*s* orbital to the vacant 2*p* orbital. (*c*) The 2*s* orbital and one of the three 2*p* orbitals are combined to give a set of two equal-energy *sp* hybridized orbitals. Two of the 2*p* orbitals remain unchanged.

p character. These two *sp* orbitals share a common axis, but their major lobes are oriented at an angle of 180° to each other. Two of the original 2*p* orbitals remain unhybridized. Their axes are perpendicular to each other and to the common axis of the pair of *sp* hybrid orbitals.

As portrayed in Figure 1.29, the two carbons of acetylene are connected to each other by a 2*sp*–2*sp* σ bond, and each is attached to a hydrogen substituent by a 2*sp*–1*s* σ bond. The unhybridized 2*p* orbitals on one carbon overlap with their counterparts on

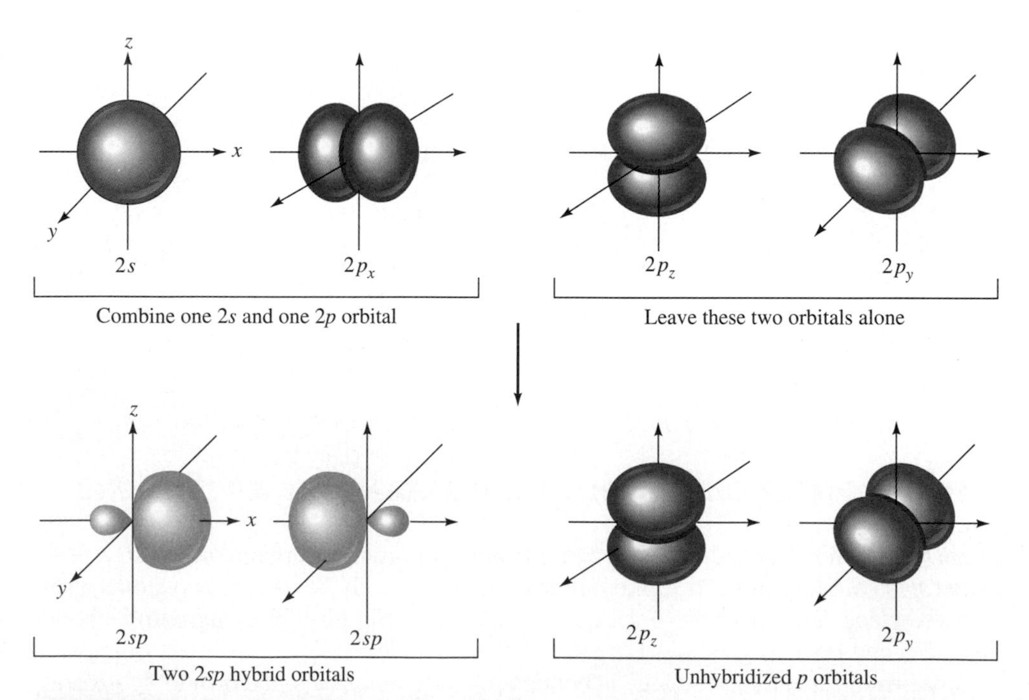

FIGURE 1.28 Representation of orbital mixing in *sp* hybridization. Mixing of the 2*s* orbital with one of the *p* orbitals generates two *sp* hybrid orbitals. Each *sp* hybrid orbital has 50 percent *s* character and 50 percent *p* character. The axes of the two *sp* hybrid orbitals are colinear. Two 2*p* orbitals remain unhybridized, and their axes are perpendicular to each other and to the long axis of the molecule.

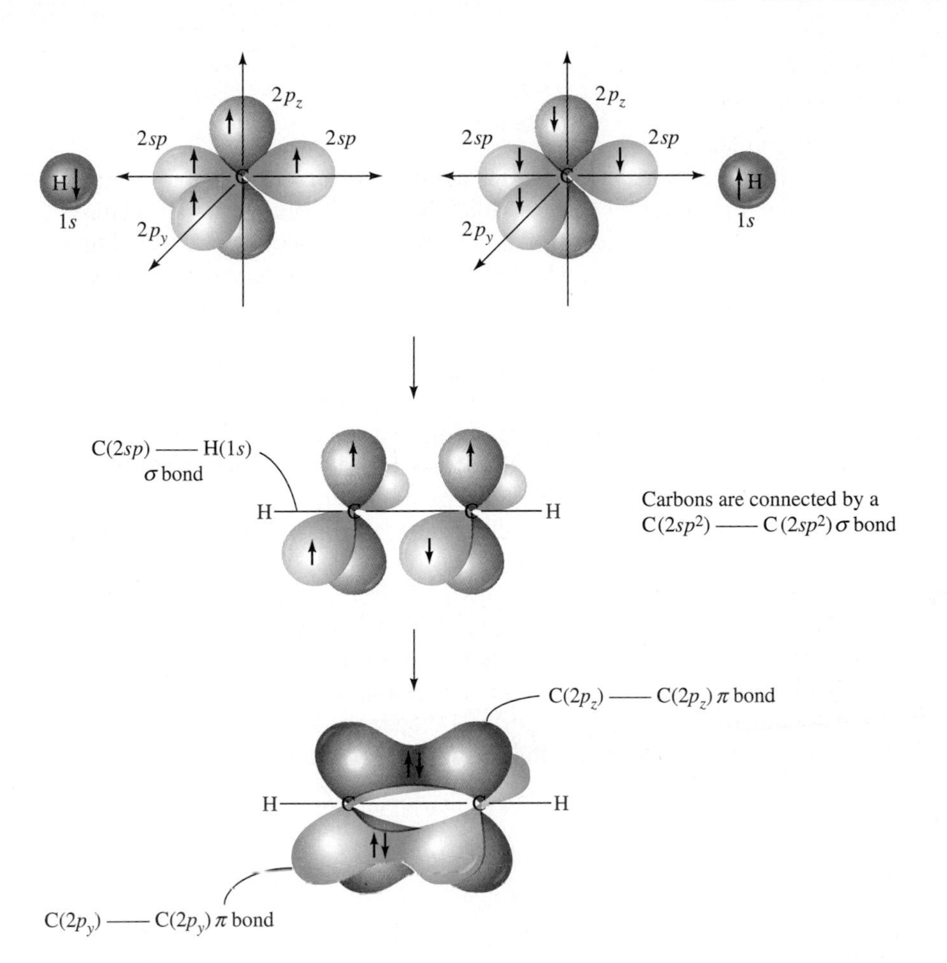

FIGURE 1.29 A description of bonding in acetylene based on *sp* hybridization of carbon. The carbon-carbon triple bond is viewed as consisting of one σ bond and two π bonds.

the other to form two π bonds. The carbon-carbon triple bond in acetylene, and in higher alkynes as well, is viewed as a multiple bond of the $\sigma + \pi + \pi$ type.

In general, you can expect that carbon will be sp hybridized when it is directly bonded to two atoms.

PROBLEM 1.19 Give the hybridization state of each carbon in the following compounds:

(*a*) Carbon dioxide ($O{=}C{=}O$)
(*b*) Formaldehyde ($H_2C{=}O$)
(*c*) Ketene ($H_2C{=}C{=}O$)
(*d*) Propene ($CH_3CH{=}CH_2$)
(*e*) Acetone [$(CH_3)_2C{=}O$]
(*f*) Acrylonitrile ($CH_2{=}CHC{\equiv}N$)

SAMPLE SOLUTION (*a*) Carbon in CO_2 is directly bonded to two other atoms. It is *sp* hybridized.

1.17 WHICH THEORY OF CHEMICAL BONDING IS BEST?

We have introduced three approaches to chemical bonding in this chapter:

1. The Lewis model

2. The molecular orbital model
3. The orbital hybridization model

Which one should you learn?

Generally speaking, the three models offer complementary information. Organic chemists use all three, emphasizing the one which best clarifies a particular feature of structure or reactivity. The Lewis and orbital hybridization models are used the most, the molecular orbital model the least. The Lewis rules are relatively straightforward, and you will find that your ability to write Lewis formulas increases rapidly with experience. (Get as much practice as you can early in the course. Success in organic chemistry depends on writing correct Lewis structures.) Orbital hybridization descriptions, since they too are based on the shared electron pair bond, enhance the information content of Lewis formulas by distinguishing among various types of atoms, electrons, and bonds in a molecule. As your familiarity with structural types increases, you will find that the term *sp^3 hybridized carbon* triggers a group of associations in your mind different from those of some other term, such as *sp^2 hybridized carbon,* for example.

The first applications of molecular orbital theory to chemical reactivity in this text appear in Chapter 10.

Molecular orbital theory can provide insights into structure and reactivity that the Lewis and orbital hybridization models cannot. It is the least intuitive of the three methods, however, and requires the most training before one can apply it in any meaningful way to the many-atom systems that are commonly encountered in organic chemistry. The degree to which molecular orbital theory is used in this text is modest and is designed to be more introductory than extensive.

1.18 SUMMARY

An understanding of organic chemistry depends on a sound understanding of the structure of carbon compounds, and this, in turn, requires a knowledge of how carbon bonds to other atoms.

The chapter begins with a review of **electron configurations** (Section 1.1). The number of electrons an atom has and the disposition of these electrons among various **orbitals** determines the kind of bonds it forms. Chemical bonds are classified as ionic or covalent. An **ionic bond** (Section 1.2) is the electrostatic attraction between two oppositely charged ions and occurs in substances such as sodium chloride. Atoms at the upper right of the periodic table, especially fluorine and oxygen, tend to gain electrons to form **anions.** Elements toward the left of the periodic table, especially metals such as sodium, tend to lose electrons to form **cations.**

Carbon forms neither anions nor cations easily and does not normally participate in ionic bonds; **covalent bonding** is observed instead. A covalent bond is the sharing of a pair of electrons between two atoms. **Lewis structures** for covalently bonded molecules are written on the basis of the **octet rule.** The most stable structures are those in which second-row elements are associated with eight electrons (shared + unshared) in their valence shells. In most of its compounds, carbon has **four bonds.** The bonds to carbon may be **single bonds** (two electrons), **double bonds** (four electrons), or **triple bonds** (six electrons) (Sections 1.3 and 1.4).

Covalent bonds between atoms of different electronegativity are **polarized** in the sense that the electrons in the bond are drawn closer to the more **electronegative** atom (Section 1.5).

$$\ce{>C^{\delta+}-X^{\delta-}}$$ Polarization of C—X bond; X is more electronegative than carbon

A particular atom in a Lewis structure may be neutral, positively charged, or negatively charged. We refer to the charge of an atom in a Lewis structure as its **formal charge,** and can calculate the formal charge by comparing the electron count of an atom in a molecule with that of the neutral atom itself. The procedure is described in Section 1.6.

Section 1.7 summarizes the procedure to be followed in writing Lewis structures for organic compounds. Table 1.4 describes notational conventions for presenting abbreviated structural formulas.

Isomers (Section 1.8) are different compounds that have the same molecular formula. They are different compounds because their structures are different. Isomers that differ in the order of their atomic connections are described as **constitutional isomers.**

Resonance between Lewis structures is a device used to describe electron delocalization in molecules (Section 1.9). Many molecules are not adequately described on the basis of a single Lewis structure because the Lewis rules restrict electrons to the region between only two nuclei. In those cases, the true structure is better understood as a hybrid of all possible structures that can be written which have the same atomic positions but differ only in respect to their electron distribution. The most fundamental rules for resonance are summarized in Table 1.5.

The **valence-shell electron pair repulsion method** (Section 1.10) predicts molecular geometries on the basis of repulsive interactions between the pairs of electrons that surround a central atom. A tetrahedral arrangement provides for the maximum separation of four electron pairs; a trigonal planar geometry is optimal for three electron pairs; and a linear arrangement is best for two electron pairs.

By combining knowledge concerning the shape of a molecule and the polarity of its individual bonds, the presence or absence of a **molecular dipole moment** and its direction may be determined (Section 1.11).

According to **molecular orbital theory** (Section 1.12), electrons in molecules are not localized in individual bonds but occupy regions in space—molecular orbitals—which may be so extensive as to include all the atoms of the molecule. Molecular orbitals are approximated as combinations of atomic orbitals; overlap of two atomic orbitals generates a *bonding* molecular orbital and an *antibonding* molecular orbital. Electrons in a bonding molecular orbital are more strongly bound than in an isolated atom; electrons in an antibonding molecular orbital are less strongly bound.

Orbitals characterized by rotational symmetry about the internuclear axis are called σ orbitals. The bond in H_2 is termed a σ bond; two electrons occupy the σ orbital and bind two hydrogen atoms together.

Bonding in organic compounds is often described according to an **orbital hybridization model** (Section 1.13). The ***sp*³ hybridization** state of carbon is derived by mixing its $2s$ and the three $2p$ orbitals to give a set of four equivalent orbitals that have their axes directed toward the corners of a tetrahedron. The four bonds in methane are C—H σ bonds generated by overlap of carbon sp^3 orbitals with hydrogen $1s$ orbitals.

The carbon-carbon bond in ethane (CH_3CH_3) is a σ bond generated by overlap of an sp^3 orbital of one carbon with the sp^3 orbital of the other (Section 1.14).

Carbon is ***sp*² hybridized** in ethylene, and the double bond is viewed as having a σ component and a π component. The sp^2 hybridization state is derived by mixing the $2s$ and two of the three $2p$ orbitals of carbon. Three equivalent sp^2 orbitals result, and the axes of these orbitals are coplanar. Overlap of an sp^2 orbital of one carbon with an sp^2 orbital of another produces a σ bond between them. Each carbon still has one unhybridized p orbital available for bonding, and ''side-by-side'' overlap of the p orbitals of adjacent carbons give a π bond between them (Section 1.15).

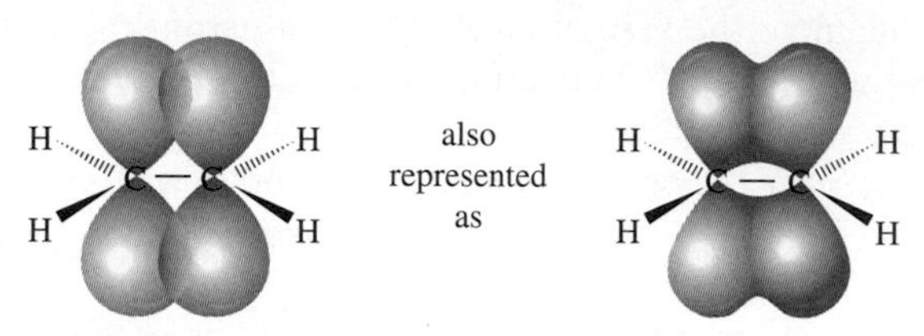

Representations of the π bond in ethylene

Carbon is ***sp* hybridized** in acetylene, and the triple bond is of the $\sigma + \pi + \pi$ type. The $2s$ orbital and one of the $2p$ orbitals are combined to give two equivalent sp orbitals that have their axes collinear. A σ bond between two carbons is supplemented by two π bonds formed by overlap of pairs of unhybridized p orbitals (Section 1.16).

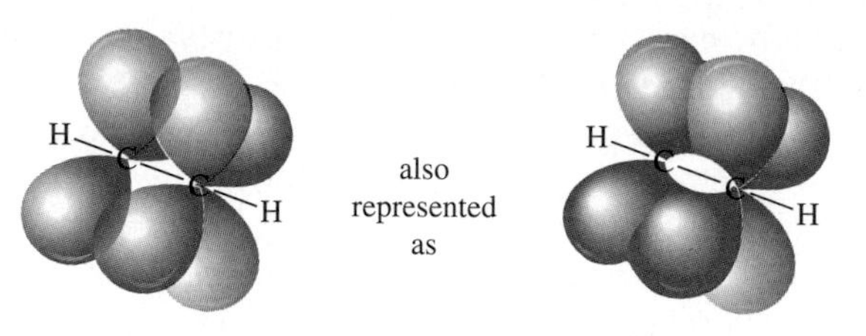

Representations of the π bond in acetylene

PROBLEMS

1.20 Each of the following species will be encountered at some point in this text. They all have the same number of electrons binding the same number of atoms and the same arrangement of bonds; i.e., they are *isoelectronic.* Specify which atoms, if any, bear a formal charge in the Lewis structure given and the net charge for each species.

(*a*) :N≡N:
(*b*) :C≡N:
(*c*) :C≡C:
(*d*) :N≡O:
(*e*) :C≡O:

1.21 You will meet all the following isoelectronic species in this text. Repeat the previous problem for these three structures.

(*a*) :Ö=C=Ö: (*b*) :N̈=N=N̈: (*c*) :Ö=N=Ö: (with an additional lone pair below the right-hand O)

1.22 All the following compounds are characterized by ionic bonding between a group I metal cation and a tetrahedral anion. Write an appropriate Lewis structure for each anion, remembering to specify formal charges where they exist.

(*a*) $NaBF_4$
(*b*) $LiAlH_4$
(*c*) K_2SO_4
(*d*) Na_3PO_4

1.23 Determine the formal charge at all the atoms in each of the following species and the net charge on the species as a whole.

(*a*) 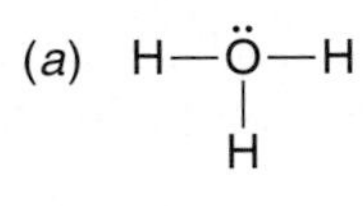

(*b*) H—C̈—H with H bonded below C

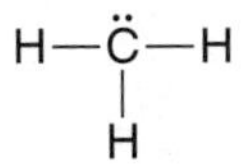

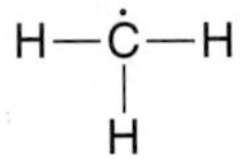

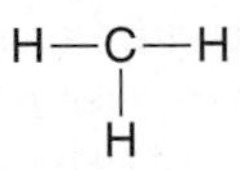

(*c*) H—Ċ—H with H bonded below C

(*d*) H—C—H with H bonded below C

(*e*) H—C̈—H

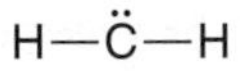

1.24 What is the formal charge of oxygen in each of the following Lewis structures?

(*a*) $CH_3\ddot{O}$: (*b*) $(CH_3)_2\ddot{O}$: (*c*) $(CH_3)_3O$:

1.25 Write a Lewis structure for each of the following organic molecules:

(*a*) C_2H_5Cl (ethyl chloride: sprayed from aerosol cans onto skin to relieve pain)
(*b*) C_2H_3Cl [vinyl chloride: starting material for the preparation of poly(vinyl chloride), or PVC, plastics]
(*c*) $C_2HBrClF_3$ (halothane: a nonflammable inhalation anesthetic; all three fluorines are bonded to the same carbon)
(*d*) $C_2Cl_2F_4$ (Freon 114: formerly used as a refrigerant and as an aerosol propellant; each carbon bears one chlorine)

1.26 Write a structural formula for the CH_3NO isomer characterized by the structural unit indicated. None of the atoms in the final structure should have a formal charge.

(*a*) C—N═O
(*b*) C═N—O
(*c*) O—C═N
(*d*) O═C—N

1.27 Consider structural formulas A, B, and C:

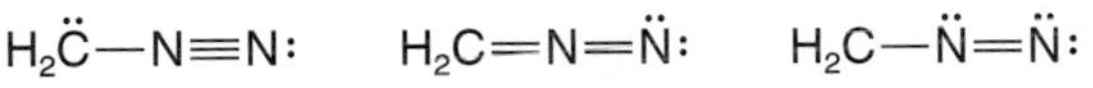

$H_2\ddot{C}$—N≡N: (A) H_2C═N═$\ddot{N}$: (B) H_2C—$\ddot{N}$═$\ddot{N}$: (C)

(*a*) Are A, B, and C constitutional isomers, or are they resonance forms?
(*b*) Which structures have a negatively charged carbon?
(*c*) Which structures have a positively charged carbon?
(*d*) Which structures have a positively charged nitrogen?
(*e*) Which structures have a negatively charged nitrogen?
(*f*) What is the net charge on each structure?
(*g*) Which is a more stable structure, A or B? Why?
(*h*) Which is a more stable structure, B or C? Why?

1.28 Consider structural formulas A, B, C, and D:

H—C̤═N═O̤: (A) H—C≡N—Ö̤: (B) H—C≡N═O̤: (C) H—C═N̈—Ö̤: (D)

(*a*) Which structures contain a positively charged carbon?
(*b*) Which structures contain a positively charged nitrogen?
(*c*) Which structures contain a positively charged oxygen?
(*d*) Which structures contain a negatively charged carbon?
(*e*) Which structures contain a negatively charged nitrogen?

(*f*) Which structures contain a negatively charged oxygen?
(*g*) Which structures are electrically neutral (contain equal numbers of positive and negative charges)? Are any of them cations? Anions?
(*h*) Which structure is the most stable?
(*i*) Which structure is the least stable?

1.29 In each of the following pairs, determine whether the two represent resonance forms of a single species or depict different substances. If two structures are not resonance forms, explain why.

(*a*) $:\ddot{\underset{..}{N}}-N\equiv N:$ and $:\underset{..}{N}=\underset{..}{N}=\underset{..}{N}:$

(*b*) $:\ddot{\underset{..}{N}}-N\equiv N:$ and $:\ddot{N}-N=\ddot{N}:$

(*c*) $:\ddot{\underset{..}{N}}-N\equiv N:$ and $:\ddot{\underset{..}{N}}-\ddot{\underset{..}{N}}-\ddot{\underset{..}{N}}:$

1.30 Among the following four structures, one is *not* a permissible resonance form. Identify the wrong structure. Why is it incorrect?

A: $\overset{+}{C}H_2-\ddot{N}(CH_3)-\ddot{\underset{..}{O}}:^-$

B: $CH_2=\overset{+}{N}(CH_3)-\ddot{\underset{..}{O}}:^-$

C: $CH_2=N(CH_3)=\underset{..}{O}:$

D: $:\overset{-}{C}H_2-\overset{+}{N}(CH_3)=\underset{..}{O}:$

1.31 Keeping the same atomic connections and moving only electrons, write a more stable Lewis structure for each of the following. Be sure to specify formal charges, if any, in the new structure.

(*a*) $H_3C-\ddot{N}=\overset{+}{N}:$ (C bearing three H)

(*b*) $H-C(-\ddot{\underset{..}{O}}:^-)(=\ddot{\overset{+}{O}}-H)$

(*c*) $H_2\overset{+}{C}-\overset{-}{\underset{..}{C}}H_2$

(*d*) $H_2\overset{+}{C}-CH=CH-\overset{-}{\underset{..}{C}}H_2$

(*e*) $H_2\overset{+}{C}-CH=CH-\ddot{\underset{..}{O}}:^-$

(*f*) $H_2\overset{-}{\underset{..}{C}}-CH=\ddot{O}:$

(*g*) $H-\overset{+}{C}=\ddot{O}:$

(*h*) $H_2\overset{+}{C}-\ddot{O}H$

(*i*) $H_2\overset{-}{C}:-\ddot{N}=\overset{+}{N}H_2$

1.32 Write structural formulas for all the constitutionally isomeric compounds having the molecular formula given.

(*a*) C_4H_{10}
(*b*) C_5H_{12}
(*c*) $C_2H_4Cl_2$
(*d*) C_4H_9Br
(*e*) C_3H_9N

1.33 Write structural formulas for all the constitutional isomers of

(*a*) C_3H_8 (*b*) C_3H_6 (*c*) C_3H_4

1.34 Write structural formulas for all the constitutional isomers of molecular formula C_3H_6O that contain

(*a*) Only single bonds (*b*) One double bond

1.35 For each of the following molecules that contain polar covalent bonds, indicate the positive and negative ends of the dipole, using the symbol $\leftrightarrow$. Refer to Table 1.2 as needed.

(*a*) HCl (*b*) ICl (*c*) HI (*d*) H_2O (*e*) HOCl

1.36 Which compound in each of the following pairs would you expect to have the greater dipole moment μ? Why?

(*a*) NaCl or HCl
(*b*) HF or HCl
(*c*) HF or BF_3
(*d*) $(CH_3)_3CH$ or $(CH_3)_3CCl$
(*e*) $CHCl_3$ or CCl_3F
(*f*) CH_3NH_2 or CH_3OH
(*g*) CH_3NH_2 or CH_3NO_2

1.37 Apply the VSEPR method to deduce the geometry around carbon in each of the following species:

(*a*) $:\overset{-}{C}H_3$ (*b*) $\overset{+}{C}H_3$ (*c*) $:CH_2$

1.38 Expand the following structural representations so as to more clearly show all the atoms and any unshared electron pairs.

(*a*) 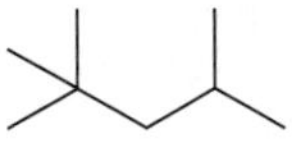 A component of high-octane gasoline

(*b*) 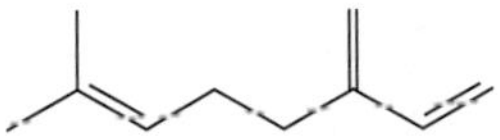 Occurs in bay and verbena oil

(*c*) 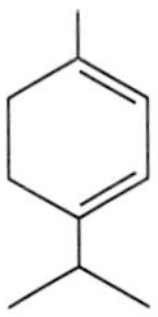 Pleasant-smelling substance found in marjoram oil

(*d*) OH — Present in oil of cloves

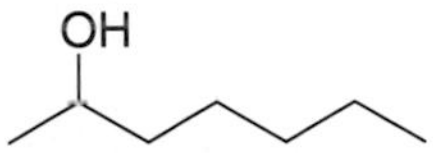

(*e*) O — Found in Roquefort cheese

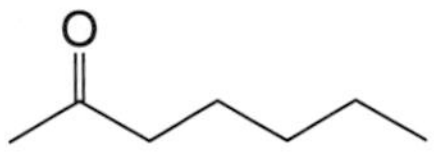

(*f*)  Benzene: parent compound of a large family of organic substances

(*g*) 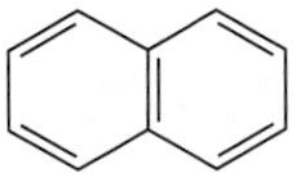 Naphthalene: sometimes used as a moth repellent

(*h*) $O\!\!=\!\!C(OCH_3)$ … $C(=O)OH$ — Aspirin

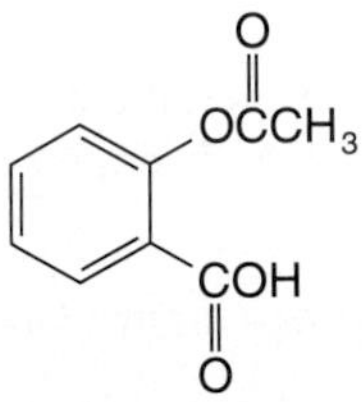

(*i*) Nicotine: a toxic substance present in tobacco

(*j*) Tyrian purple: a purple dye extracted from a species of Mediterranean sea snail

(*k*) Hexachlorophene: an antiseptic

1.39 Molecular formulas of organic compounds are customarily presented in the fashion $C_2H_5BrO_2$. The number of carbon and hydrogen atoms are presented first, followed by the other atoms in alphabetical order. Give the molecular formulas corresponding to each of the compounds in the preceding problem. Are any of them isomers?

1.40 Select the compounds in Problem 1.38 in which all the carbons are

(*a*) sp^3 hybridized (*b*) sp^2 hybridized

Do any of the compounds in Problem 1.38 contain an *sp* hybridized carbon?

1.41 Account for all the electrons in each of the following species, assuming sp^3 hybridization of the second-row element in each case. Which electrons are found in sp^3 hybridized orbitals? Which are found in σ bonds?

(*a*) Ammonia (NH_3)
(*b*) Water (H_2O)
(*c*) Hydrogen fluoride (HF)
(*d*) Ammonium ion (NH_4^+)
(*e*) Borohydride anion (BH_4^-)
(*f*) Amide anion ($:\ddot{N}H_2^-$)
(*g*) Methyl anion ($:CH_3^-$)

1.42 Of the orbital overlaps represented below, one is bonding, one is antibonding, and the other is nonbonding (neither bonding nor antibonding). Which pattern of orbital overlap corresponds to which interaction? Why?

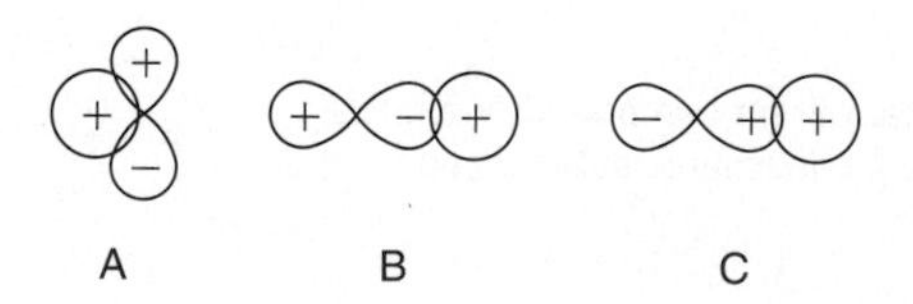

MOLECULAR MODELING EXERCISES

1.43 Practice working with your molecular modeling kit or software. Construct molecular models of methane, ethylene, and acetylene and verify that their molecular geometries are tetrahedral, planar, and linear, respectively.

1.44 Replace two of the hydrogens in your methane model by chlorine atoms. How many different structures (isomers) can you generate?

1.45 Replace two of the hydrogens in your methane model, one by a chlorine atom and the other by fluorine. How many different structures (isomers) can you generate?

1.46 Replace two of the hydrogens in your ethylene model by chlorine atoms. How many different structures (isomers) can you generate? Which, if any, have no dipole moment?

1.47 Construct molecular models for the three constitutional isomers of molecular formula C_3H_8O.

CHAPTER 2

ALKANES

Now that we have reviewed the Lewis and orbital hybridization models of covalent bonding, we are ready to begin examining organic compounds in respect to their *structure, reactions,* and *applications.* Estimates place the number of known organic compounds at more than 7 million. Were we to list the reactions available to each one separately, it would tax the capacity of the most powerful computers. Yet someone who is trained in organic chemistry can look at the structure of a substance and make reasonably confident predictions about its properties, including how it will behave in a chemical reaction. The basis for this predictive power rests in the fact that the more than 7 million known organic compounds belong to a relatively few structural types or families, and that there are even fewer reaction types than structural types.

Families of organic compounds are characterized by the presence of distinctive **functional groups.** A functional group is the structural unit responsible for a given molecule's chemical reactivity under a particular set of conditions. Time and experience have shown that by organizing the reactions of organic compounds according to functional groups, the task of associating structural type with reaction type is considerably simplified.

We begin the chapter with a survey of *hydrocarbon* structural types, proceed to introduce the various *functional group families,* and then return to hydrocarbons to discuss *alkanes* in some detail. Alkane names form the foundation upon which the most widely accepted system of *organic nomenclature* is based. The elements of this nomenclature system, the **IUPAC rules,** constitute one of the main topics of this chapter.

2.1 CLASSES OF HYDROCARBONS

Hydrocarbons are compounds that contain only carbon and hydrogen, and are divided into two main classes, **aliphatic hydrocarbons** and **aromatic hydrocarbons.** This classification dates from the nineteenth century, when organic chemistry was almost exclusively devoted to the study of materials from natural sources and terms were coined that reflected the origin of a substance. Two such sources were fats and oils, and the word *aliphatic* was derived from the Greek word *aleiphar* (''fat'') for these materials. Aromatic hydrocarbons, irrespective of their own odor, were typically obtained by chemical treatment of pleasant-smelling plant extracts.

Aliphatic hydrocarbons are further subdivided into three major groups, *alkanes, alkenes,* and *alkynes.* **Alkanes** are hydrocarbons in which all the bonds are single bonds, **alkenes** contain a carbon-carbon double bond, and **alkynes** contain a carbon-carbon triple bond. Examples of the three classes of aliphatic hydrocarbons are the two-carbon compounds *ethane, ethylene,* and *acetylene.* Both carbons in ethane are sp^3 hybridized, both are sp^2 hybridized in ethylene, and both are sp hybridized in acetylene.

Orbital hybridization and bonding in ethane, ethylene, and acetylene were discussed in Sections 1.14–1.16.

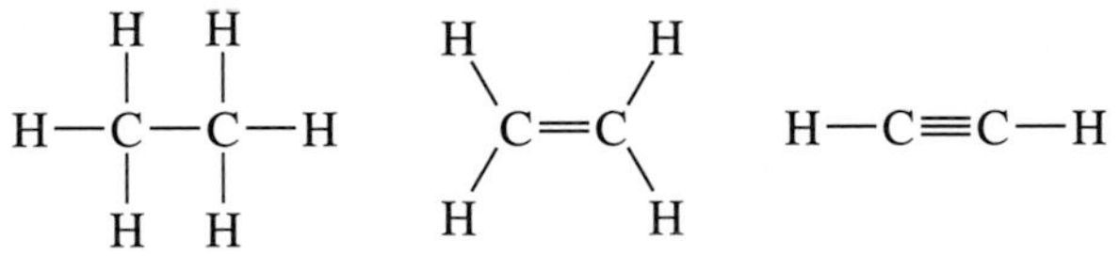

```
 H  H          H       H
 |  |           \     /
H—C—C—H          C=C          H—C≡C—H
 |  |           /     \
 H  H          H       H
```

Ethane (alkane) Ethylene (alkene) Acetylene (alkyne)

Another name for aromatic hydrocarbons is **arenes.** Arenes have properties that are much different from alkanes, alkenes, and alkynes. The most important aromatic hydrocarbon is *benzene.* All six carbons in benzene are sp^2 hybridized.

Orbital hybridization and bonding in benzene will be discussed in Section 11.7.

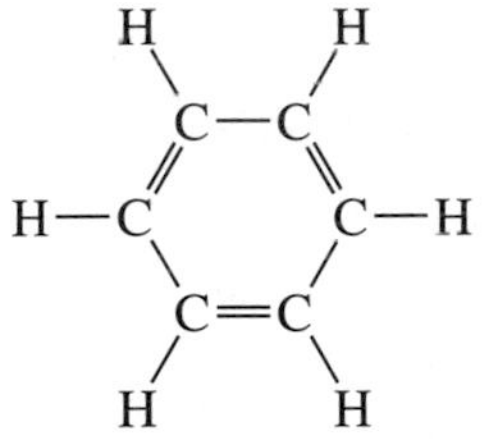

```
      H     H
       \   /
        C—C
       //   \\
   H—C       C—H
       \   /
        C=C
       /   \
      H     H
```

Benzene (arene)

Many of the principles of organic chemistry can be developed by examining the series of hydrocarbons in the order: alkanes, alkenes, alkynes, and arenes. Alkanes are introduced in the present chapter, alkenes will be discussed in Chapters 5 and 6, alkynes in Chapter 9, and arenes in Chapters 11 and 12.

2.2 FUNCTIONAL GROUPS IN HYDROCARBONS

A functional group may be as small as a single hydrogen atom, or it can encompass several atoms. The functional group of an alkane is any one of its hydrogen substituents. A reaction that we shall discuss in Chapter 4, illustrated here for the case of ethane, is one in which an alkane reacts with chlorine:

$$CH_3CH_3 + Cl_2 \longrightarrow CH_3CH_2Cl + HCl$$

Ethane Chlorine Chloroethane Hydrogen chloride

In this reaction one of the hydrogen atoms of ethane is replaced by a chlorine. This replacement of hydrogen by chlorine is a characteristic reaction of all alkanes and can be represented for the general case by the equation:

$$\underset{\text{Alkane}}{R{-}H} + \underset{\text{Chlorine}}{Cl_2} \longrightarrow \underset{\text{Alkyl chloride}}{R{-}Cl} + \underset{\text{Hydrogen chloride}}{HCl}$$

In the general equation the functional group (—H) is shown explicitly while the remainder of the alkane molecule is abbreviated as R. This is a commonly used notation which allows us to focus attention on the functional group transformation without being distracted by the parts of the molecule that remain unaffected. A hydrogen atom in one alkane is very much like the hydrogen atom of any other alkane in respect to its reactivity toward chlorine. Our ability to write general equations such as the one shown illustrates why the functional group approach is so useful in organic chemistry.

A hydrogen atom is a functional unit in alkenes and alkynes as well as in alkanes. However, these hydrocarbons contain a second functional group as well. The carbon-carbon double bond is a functional group in alkenes, and the carbon-carbon triple bond is a functional group in alkynes.

A hydrogen atom is a functional group in arenes, and we represent arenes as ArH to reflect this. What will become apparent when we discuss the reactions of arenes, however, is that their chemistry is much richer than that of alkanes and it is more appropriate to consider the ring in its entirety as the functional group.

2.3 FUNCTIONALLY SUBSTITUTED DERIVATIVES OF ALKANES

As a class, alkanes are not particularly reactive compounds, and the H in RH is not a particularly reactive functional group. Indeed, when a group other than hydrogen is present on an alkane framework, that group is almost always the functional group. Table 2.1 lists examples of some compounds of this type. All will be discussed at some point in this text.

TABLE 2.1
Functionally Substituted Alkanes

Class	Representative example	Name of example*	Generalized abbreviation
Alcohol	CH_3CH_2OH	Ethanol	ROH
Alkyl halide	CH_3CH_2Cl	Chloroethane	RCl
Amine†	$CH_3CH_2NH_2$	Ethanamine	RNH_2
Epoxide	$H_2C{-}CH_2$ (ring closed by O)	Oxirane	$R_2C{-}CR_2$ (ring closed by O)
Ether	$CH_3CH_2OCH_2CH_3$	Diethyl ether	ROR
Nitrile	$CH_3CH_2C{\equiv}N$	Propanenitrile	$RC{\equiv}N$
Nitroalkane	$CH_3CH_2NO_2$	Nitroethane	RNO_2
Thiol	CH_3CH_2SH	Ethanethiol	RSH

* Most compounds have more than one acceptable name.

† The example given is a *primary* amine (RNH_2). *Secondary* amines have the general structure R_2NH; *tertiary* amines are R_3N.

TABLE 2.2
Classes of Compounds That Contain a Carbonyl Group

Class	Representative example	Name of example	Generalized abbreviation
Aldehyde	$CH_3C(=O)H$	Ethanal	$RC(=O)H$
Ketone	$CH_3C(=O)CH_3$	2-Propanone	$RC(=O)R$
Carboxylic acid	$CH_3C(=O)OH$	Ethanoic acid	$RC(=O)OH$
Carboxylic acid derivatives:			
Acyl halide	$CH_3C(=O)Cl$	Ethanoyl chloride	$RC(=O)X$
Acid anhydride	$CH_3C(=O)OC(=O)CH_3$	Ethanoic anhydride	$RC(=O)OC(=O)R$
Ester	$CH_3C(=O)OCH_2CH_3$	Ethyl ethanoate	$RC(=O)OR$
Amide	$CH_3C(=O)NH_2$	Ethanamide	$RC(=O)NR_2$

2.4 CARBONYL-CONTAINING FUNCTIONAL GROUPS

Some of the most important families of organic compounds contain the **carbonyl group** ($>$C=O) and are listed in Table 2.2. Carbonyl-containing compounds are among the most abundant and biologically significant of naturally occurring substances. The reactions of the carbonyl group feature prominently in *organic synthesis* —that branch of organic chemistry concerned with preparing a desired compound of defined structure by an appropriate sequence of reactions.

Carbonyl-group chemistry is discussed in a block of five chapters (Chapters 17–21).

2.5 INTRODUCTION TO ALKANES: METHANE

Alkanes are characterized by the general molecular formula C_nH_{2n+2}. The simplest alkane, methane, comprises a significant portion of the atmospheres of Jupiter, Saturn, Uranus, and Neptune. The atmospheres of these planets are said to be *reducing*. They are rich in hydrogen, and both nitrogen and carbon are found in their reduced forms, ammonia (NH_3) and methane (CH_4), respectively. Earth's weaker gravity permitted its molecular hydrogen to escape long ago, and its present atmosphere is considered

FIGURE 2.1 Structure of methane. All H—C—H bond angles are 109.5°, and all C—H bond distances are 109 pm.

Methanogens ("methane makers") are one type of *archaeobacteria.* Archaeobacteria are among the oldest forms of life on earth. The prefix *archaeo-* is derived from the Greek word for "primitive," "original," or "ancient."

oxidizing. Nevertheless, large quantities of methane are present on Earth. It is formed when a class of bacteria called *methanogens* act on decaying plant and animal matter in the absence of oxygen. These bacteria live in aqueous environments—marshes, bogs, and the ocean, for example. Marsh gas is almost entirely methane. Natural gas, which contains 75 to 85 percent methane, usually occurs in association with petroleum deposits; these deposits are the legacy of marine plants that lived and died in inland seas millions of years ago. Not all of Earth's methane is ''old methane,'' however. It is estimated that methanogens that live in termites and in the digestive systems of plant-eating animals produce over 2 billion tons per year of methane! One source quotes a figure of 20 liters per day as the methane output of a large cow.

Methane is a colorless, odorless gas with a boiling point of −160°C at atmospheric pressure; its melting point is −182.5°C. The characteristic odor of natural gas used for cooking and heating our homes is not due to methane or any of the other hydrocarbons it contains. Natural gas is essentially odorless, and trace amounts of unpleasant-smelling sulfur-containing compounds such as ethanethiol (Table 2.1) are deliberately added to it in order to warn of any potentially dangerous leak.

Structurally, as was described in Section 1.10 and as summarized in Figure 2.1, methane is a tetrahedral molecule. Each H—C—H angle is 109.5° and each C—H bond distance is 109 pm. Carbon is sp^3 hybridized, and each C—H bond is a σ bond formed by overlap of an sp^3 hybridized orbital of carbon with a hydrogen 1s orbital.

2.6 ETHANE AND PROPANE

The next two members of the alkane family are **ethane** (C_2H_6) and **propane** (C_3H_8). Ethane has a much higher boiling point than methane, and propane a higher boiling point than ethane.

Boiling points cited in this text are at 1 atmosphere (760 mm of mercury) unless otherwise stated.

	CH_4	CH_3CH_3	$CH_3CH_2CH_3$
	Methane	Ethane	Propane
Boiling point:	−160°C	−89°C	−42°C

This will generally be true as we proceed to look at other alkanes; as the number of carbon atoms increases, so does the boiling point. All the alkanes with four carbons or fewer are gases at room temperature.

Ethane is, after methane, the second most abundant component of natural gas. The amount varies according to the source but is usually on the order of 5 to 10 percent. Ethane is an important starting material from which many industrial chemicals, especially ethylene, are made. Figure 2.2 shows the bond distances and bond angles in ethane.

Propane, like methane and ethane, is present in natural gas but only to the extent of a

FIGURE 2.2 The structure of ethane.

few percent. It is most familiar to us as the major component of *liquefied petroleum gas* (LPG), a clean-burning fuel in which propane and other low-boiling hydrocarbons are maintained in their liquid state under high pressure in steel containers.

Each carbon in propane ($CH_3CH_2CH_3$) is sp^3 hybridized, and all the bonds are σ bonds. Methyl (CH_3) groups form the ends of the three-carbon chain. The CH_2 group is referred to as a **methylene** group.

PROBLEM 2.1 How many σ bonds are there in propane? Identify the orbital overlaps that give rise to each σ bond.

2.7 ISOMERIC ALKANES: THE BUTANES

Methane is the only alkane of molecular formula CH_4, ethane the only one that is C_2H_6, and propane the only one that is C_3H_8. Beginning with C_4H_{10}, however, the possibility of constitutional isomerism (Section 1.8) arises; two alkanes have this particular molecular formula. In one, called ***n*-butane,** four carbons are joined in a continuous chain. The *n* is a notational device that stands for "normal" and means that the carbon chain is unbranched. The second isomer has a branched carbon chain and is called **isobutane.**

$$CH_3CH_2CH_2CH_3 \qquad \underset{\displaystyle CH_3}{CH_3\overset{}{C}HCH_3} \quad \text{or} \quad (CH_3)_3CH$$

	n-Butane	Isobutane
Boiling point:	−0.4°C	−10.2°C
Melting point:	−139°C	−160.9°C

As noted earlier, the group CH_3 is called *methyl* and CH_2 is called *methylene.* These are the only groups present in *n*-butane. Isobutane contains three methyl groups bonded to a CH unit. The group CH is called a **methine** group.

n-Butane and isobutane have the same molecular formula but differ in respect to the order in which their atoms are connected. They are constitutional isomers (Section 1.8) of each other. While both are gases at room temperature, *n*-butane has a boiling point almost 10° higher than that of isobutane, and a melting point that is over 20° higher. "Butane" lighters contain a mixture of *n*-butane (about 5 percent) and isobutane (about 95 percent), in a sealed container. The pressure produced by the two compounds is on the order of 3 to 4 atmospheres and is sufficient to maintain them in the liquid state until a small valve is opened to emit a fine stream of the vaporized mixture across a spark that ignites it.

The bonding in *n*-butane and isobutane is similar to that of ethane and propane. The bond angles are close to tetrahedral, each carbon atom is sp^3 hybridized, and all of the bonds are σ bonds. This generalization holds for all alkanes regardless of the number of carbons they possess.

2.8 HIGHER *n*-ALKANES

n-Alkanes are alkanes that have an unbranched carbon chain. ***n*-Pentane** and ***n*-hexane** are *n*-alkanes possessing five and six carbon atoms, respectively.

$CH_3CH_2CH_2CH_2CH_3$ — *n*-Pentane

$CH_3CH_2CH_2CH_2CH_2CH_3$ — *n*-Hexane

Their condensed structural formulas can be further abbreviated by indicating within parentheses the number of methylene groups in the chain. Thus, *n*-pentane may be written as $CH_3(CH_2)_3CH_3$ and *n*-hexane as $CH_3(CH_2)_4CH_3$. This shortcut is especially convenient with longer-chain alkanes. The laboratory synthesis of the "ultralong" alkane $CH_3(CH_2)_{388}CH_3$ was achieved in 1985; imagine trying to write a structural formula for this compound in anything other than an abbreviated way!

PROBLEM 2.2 An *n*-alkane of molecular formula $C_{28}H_{58}$ has been isolated from a certain fossil plant. Write a condensed structural formula for this alkane.

n-Alkanes have the general formula $CH_3(CH_2)_xCH_3$ and are said to belong to a **homologous series** of compounds. A homologous series is one in which successive members differ by a $—CH_2—$ group.

Unbranched alkanes are sometimes referred to as "straight-chain alkanes," but, as we shall see in Chapter 3, their chains are not straight but instead tend to adopt the "zigzag" shape portrayed in the bond-line formulas introduced in Section 1.7.

Bond-line formula of *n*-pentane

Bond-line formula of *n*-hexane

PROBLEM 2.3 Much of the communication between insects involves chemical messengers called *pheromones.* A species of cockroach secretes a substance from its mandibular glands that alerts other cockroaches to its presence and causes them to congregate. One of the principal components of this *aggregation pheromone* is the alkane shown in the bond-line formula that follows. Give the molecular formula of this substance and represent it by a condensed formula.

2.9 THE C_5H_{12} ISOMERS

There are three isomeric alkanes of molecular formula C_5H_{12}. The unbranched isomer is, as we have seen, *n*-pentane. The isomer with a single methyl branch is called

isopentane. The third isomer has a three-carbon chain with two methyl branches. It is called **neopentane.**

n-Pentane: $CH_3CH_2CH_2CH_2CH_3$ or $CH_3(CH_2)_3CH_3$ or

Isopentane: $CH_3CHCH_2CH_3$ (with CH_3 on the second carbon) or $(CH_3)_2CHCH_2CH_3$ or

Neopentane: CH_3CCH_3 (with CH_3 above and below the central carbon) or $(CH_3)_4C$ or

Are there any additional isomers of C_5H_{12}? Table 2.3 presents the number of possible isomers as a function of the number of carbon atoms in an alkane. According to the table, there are only three isomers of C_5H_{12}—the three indicated by the names *n*-pentane, isopentane, and neopentane. As the data in the table show, the number of isomers increases enormously with the number of carbon atoms and raises two important questions:

1. How can we tell when we have written all the possible isomers corresponding to a particular molecular formula?
2. How can we name alkanes so that each one has a unique name?

The answer to the first question is that you cannot easily calculate the number of isomers. The data in Table 2.3 were determined by a mathematician who concluded that there was no simple expression from which the number of isomers could be calculated. The best way to ensure that you have written all the isomers of a particular molecular formula is to work systematically, beginning with the unbranched chain and

TABLE 2.3
The Number of Constitutionally Isomeric Alkanes of Particular Molecular Formulas

Molecular formula	Number of constitutional isomers
CH_4	1
C_2H_6	1
C_3H_8	1
C_4H_{10}	2
C_5H_{12}	3
C_6H_{14}	5
C_7H_{16}	9
C_8H_{18}	18
C_9H_{20}	35
$C_{10}H_{22}$	75
$C_{15}H_{32}$	4,347
$C_{20}H_{42}$	366,319
$C_{40}H_{82}$	62,491,178,805,831

The number of C_nH_{2n+2} isomers has been calculated for values of n from 1 to 400 and the comment made that the number of isomers of $C_{167}H_{336}$ exceeds the number of particles in the known universe (10^{80}). These observations and the historical background of isomer calculation are described in a paper in the April 1989 issue of the *Journal of Chemical Education* (pp. 278–281).

then shortening it while adding branches one by one. It is essential that you be able to recognize when two different-looking structural formulas are actually the same molecule written in different ways. The key point is the *connectivity* of the carbon chain. For example, the following group of structural formulas do *not* represent different compounds; they are just a portion of the many ways we could write a structural formula for isopentane. Each one has a continuous chain of four carbons with a methyl branch located one carbon from the end of the chain.

$$\underset{\displaystyle CH_3}{CH_3CHCH_2CH_3} \qquad \overset{\displaystyle CH_3}{CH_3CHCH_2CH_3} \qquad \underset{\displaystyle CH_3}{CH_3CH_2CHCH_3}$$

$$\overset{\displaystyle CH_3}{CH_3CH_2CHCH_3} \qquad \underset{\displaystyle CH_3}{\overset{\displaystyle CH_3}{CHCH_2CH_3}}$$

PROBLEM 2.4 Write condensed and bond-line formulas for the five isomeric C_6H_{14} alkanes.

SAMPLE SOLUTION When writing isomeric alkanes, it is best to begin with the unbranched isomer.

$$CH_3CH_2CH_2CH_2CH_2CH_3 \quad \text{or}$$

Next, remove a carbon from the chain and use it as a one-carbon (methyl) branch at the carbon atom next to the end of the chain.

$$\underset{\displaystyle CH_3}{CH_3CHCH_2CH_2CH_3} \quad \text{or}$$

Now, write structural formulas for the remaining three isomers. Be sure that each one is a unique compound and not simply a different representation of one written previously.

The answer to the second question—how to provide a name that is unique to a particular structure—is presented in the following section. It is worth noting, however, that being able to name compounds in a *systematic* way is a great help in deciding whether two structural formulas represent isomeric substances or are the same compound represented in two different ways. By following a precise set of rules, one will always get the same systematic name for a compound, regardless of how it is written. Conversely, two different compounds will always have different names.

2.10 SYSTEMATIC IUPAC NOMENCLATURE OF UNBRANCHED ALKANES

Nomenclature in organic chemistry is of two types: **common** (or "trivial") and **systematic.** Some common names existed long before organic chemistry became an organized branch of chemical science and, indeed, before it was realized that different compounds had different structures. Methane, ethane, propane, *n*-butane, isobutane, *n*-pentane, isopentane, and neopentane are common names. One simply memorizes the name that goes with a particular compound in just the same way that one matches names with faces. So long as there are only a few names and a few compounds, the task is manageable. But there are 7 million organic compounds already known, and the list continues to grow! A system built on common names is not adequate to the task of communicating structural information. Beginning in 1892 chemists developed a set of rules for naming organic compounds based on their structures, which we now call the **IUPAC rules,** where *IUPAC* stands for the "International Union of Pure and Applied Chemistry." (See the accompanying box, "A Brief History of Systematic Organic Nomenclature.")

A more detailed account of the history of organic nomenclature may be found in the article "The Centennial of Systematic Organic Nomenclature" in the November 1992 issue of the *Journal of Chemical Education* (pp. 863–865).

The IUPAC rules assign names to unbranched alkanes as shown in Table 2.4. Methane, ethane, propane, and butane are retained for CH_4, CH_3CH_3, $CH_3CH_2CH_3$, and $CH_3CH_2CH_2CH_3$, respectively. Thereafter, the number of carbon atoms in the chain is specified by a Latin or Greek prefix preceding the suffix *-ane,* which identifies the compound as a member of the alkane family. Notice that the prefix *n*- is not part of the IUPAC system. The IUPAC name for $CH_3CH_2CH_2CH_3$ is butane, not *n*-butane.

PROBLEM 2.5 Refer to Table 2.4 as needed to answer the following questions:

(*a*) Beeswax contains 8 to 9 percent hentriacontane. Write a condensed structural formula for hentriacontane.

(*b*) Octacosane has been found to be present in a certain fossil plant. Write a condensed structural formula for octacosane.

(*c*) What is the IUPAC name of the alkane described in Problem 2.3 as a component of the cockroach aggregation pheromone?

SAMPLE SOLUTION (*a*) Note in Table 2.4 that hentriacontane has 31 carbon atoms. All the alkanes in Table 2.4 have unbranched carbon chains. Hentriacontane has the condensed structural formula $CH_3(CH_2)_{29}CH_3$.

TABLE 2.4
IUPAC Names of Unbranched Alkanes

Number of carbon atoms	Name	Number of carbon atoms	Name	Number of carbon atoms	Name
1	Methane	11	Undecane	21	Henicosane
2	Ethane	12	Dodecane	22	Docosane
3	Propane	13	Tridecane	23	Tricosane
4	Butane	14	Tetradecane	24	Tetracosane
5	Pentane	15	Pentadecane	30	Triacontane
6	Hexane	16	Hexadecane	31	Hentriacontane
7	Heptane	17	Heptadecane	32	Dotriacontane
8	Octane	18	Octadecane	40	Tetracontane
9	Nonane	19	Nonadecane	50	Pentacontane
10	Decane	20	Icosane*	100	Hectane

* Spelled "eicosane" prior to 1979 version of IUPAC rules.

A BRIEF HISTORY OF SYSTEMATIC ORGANIC NOMENCLATURE

The first successful formal system of chemical nomenclature was advanced in France in 1787 to replace the babel of common names which then plagued the science. Hydrogen (instead of "inflammable air") and oxygen (instead of "vital air") are just two of the substances that owe their modern names to the proposals described in the *Méthode de nomenclature chimique.* It was at this time that certain important compounds such as sulfuric, phosphoric, and carbonic acid were named and procedures for naming salts of these acids were outlined. The guidelines were more appropriate to inorganic compounds; it was not until the 1830s that names reflecting chemical composition, as opposed to origin or discoverer, began to appear in organic chemistry.

In 1889 a group with the imposing title of the International Commission for the Reform of Chemical Nomenclature was organized, and this group, in turn, sponsored a meeting of 34 prominent European chemists in Switzerland in 1892. Out of this meeting arose a system of organic nomenclature known as the **Geneva rules.** The principles upon which the Geneva rules were based are the forerunners of our present systems.

A second international conference was held in 1911, but the intrusion of World War I prevented any substantive revisions of the Geneva rules. The International Union of Chemistry was established in 1930 and undertook the necessary revision leading to publication in 1930 of what came to be known as the **Liège rules.**

After World War II, the International Union of Chemistry became the International Union of Pure and Applied Chemistry (known in the chemical community as the *IUPAC*). Since 1949 the IUPAC has issued reports on chemical nomenclature on a regular basis. The most recent **IUPAC rules** for organic chemistry were published in 1979. It should be pointed out that the IUPAC rules often offer several different ways to name a single compound. Thus while it is true that no two compounds can have the same name, it is incorrect to believe that there is only a single IUPAC name for a particular compound.

The IUPAC rules are not the only nomenclature system in use today. Chemical Abstracts Service surveys all the world's leading scientific journals that publish papers relating to chemistry and publishes brief abstracts of those papers. The publication *Chemical Abstracts* and its indexes are absolutely essential to the practice of chemistry. For many years *Chemical Abstracts* nomenclature was very similar to IUPAC nomenclature, but the tremendous explosion of chemical knowledge in recent years has required *Chemical Abstracts* to modify its nomenclature so that its indexes are better adapted to computerized searching. This means that whenever feasible, a compound has a single *Chemical Abstracts* name. Unfortunately, this *Chemical Abstracts* name may be different from any of the several IUPAC names. In general, it is easier to make the mental connection between a chemical structure and its IUPAC name than its *Chemical Abstracts* name.

It is worth noting that the **generic name** of a drug is not directly derived from systematic nomenclature. Furthermore, different pharmaceutical companies will call the same drug by their own trade name, which is different from its generic name. Generic names are invented on request (for a fee) by the U.S. Adopted Names Council, a private organization founded by the American Medical Association, the American Pharmaceutical Association, and the U.S. Pharmacopeial Convention.

In Problem 2.4 you were asked to write structural formulas for the five isomeric alkanes of molecular formula C_6H_{14}. In the next section you will see how the IUPAC rules generate a unique name for each isomer.

2.11 APPLYING THE IUPAC RULES: THE NAMES OF THE C_6H_{14} ISOMERS

We can present and illustrate the most important of the IUPAC rules for alkane nomenclature by naming the five C_6H_{14} isomers. By definition (Table 2.4), the unbranched C_6H_{14} isomer is hexane.

$$CH_3CH_2CH_2CH_2CH_2CH_3$$

IUPAC name: hexane
(common name: *n*-hexane)

The IUPAC rules name branched alkanes as *substituted derivatives* of the unbranched alkanes listed in Table 2.4. Consider the isomer represented by the structure

$$\begin{array}{l} CH_3\underset{|}{C}HCH_2CH_2CH_3 \\ \quad\;\; CH_3 \end{array}$$

Step 1

Identify the longest continuous carbon chain and find the IUPAC name in Table 2.4 that corresponds to the unbranched alkane having that number of carbons. This is the parent alkane from which the IUPAC name is to be derived.

In this case, the longest continuous chain has *five* carbon atoms; the compound is named as a derivative of pentane. The key word here is *continuous*. It does not matter whether the carbon skeleton is drawn in an extended straight-chain form or in one with many bends and turns. All that matters is the number of carbons linked together in an uninterrupted sequence.

Step 2

Identify the substituent groups attached to the parent chain.

The parent pentane chain bears a methyl (CH_3) group as a substituent.

Step 3

Number the longest continuous chain in the direction that gives the lowest number to the substituent groups at the first point of branching.

The numbering scheme

$$\begin{array}{l} \overset{1}{C}H_3\overset{2}{\underset{|}{C}}H\overset{3}{C}H_2\overset{4}{C}H_2\overset{5}{C}H_3 \\ \quad\;\; CH_3 \end{array} \quad \text{is equivalent to} \quad \begin{array}{l} CH_3\overset{2}{\underset{|}{C}}H\overset{3}{C}H_2\overset{4}{C}H_2\overset{5}{C}H_3 \\ \quad\;\; {}_1CH_3 \end{array}$$

Both schemes count five carbon atoms in their longest continuous chain and bear a methyl group as a substituent at the second carbon. An alternative numbering sequence, one that begins at the other end of the chain, is incorrect:

$$\begin{array}{l} \overset{5}{C}H_3\overset{4}{\underset{|}{C}}H\overset{3}{C}H_2\overset{2}{C}H_2\overset{1}{C}H_3 \\ \quad\;\; CH_3 \end{array} \quad \text{(methyl group attached to C-4)}$$

Step 4

Write the name of the compound. The parent alkane is the last part of the name and is preceded by the names of the substituent groups and their numerical locations (**locants**). Hyphens separate the locants from the names.

$$\begin{array}{l} CH_3\underset{|}{C}HCH_2CH_2CH_3 \\ \quad\;\; CH_3 \end{array}$$

IUPAC name: 2-methylpentane

The same sequence of four steps gives the IUPAC name for the isomer that has its methyl group attached to the middle carbon of the five-carbon chain.

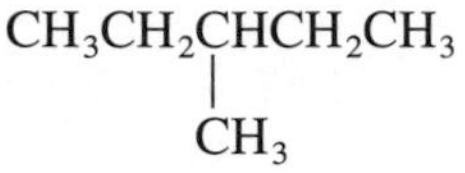

$$\begin{array}{l} CH_3CH_2\underset{\displaystyle CH_3}{\underset{|}{C}}HCH_2CH_3 \end{array}$$ **IUPAC name: 3-methylpentane**

Both remaining C_6H_{14} isomers have two methyl groups as substituents on a four-carbon chain. Thus the parent chain is butane. When the same substituent appears more than once, use the multiplying prefixes *di-*, *tri-*, *tetra-*, and so on. A separate locant is used for each substituent, and the locants are separated from each other by commas and from the words by hyphens.

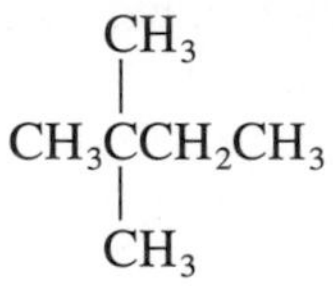

$$CH_3\overset{\displaystyle CH_3}{\overset{|}{\underset{\displaystyle CH_3}{\underset{|}{C}}}}CH_2CH_3 \qquad\qquad CH_3\overset{\displaystyle CH_3}{\overset{|}{C}}H\underset{\displaystyle CH_3}{\underset{|}{C}}HCH_3$$

IUPAC name: 2,2-dimethylbutane 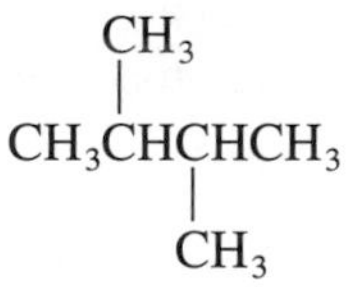 **IUPAC name: 2,3-dimethylbutane**

PROBLEM 2.6 Phytane is a naturally occurring alkane produced by the alga *Spirogyra* and is a constituent of petroleum. The IUPAC name for phytane is 2,6,10,14-tetramethylhexadecane. Write a structural formula for phytane.

PROBLEM 2.7 Derive the IUPAC names for

(*a*) The isomers of C_4H_{10}
(*b*) The isomers of C_5H_{12}
(*c*) $(CH_3)_3CCH_2CH(CH_3)_2$
(*d*) $(CH_3)_3CC(CH_3)_3$

SAMPLE SOLUTION (*a*) There are two C_4H_{10} isomers. Butane (Table 2.4) is the IUPAC name for the isomer that has an unbranched carbon chain. The other isomer has three carbons in its longest continuous chain with a methyl branch at the central carbon; its IUPAC name is 2-methylpropane.

$$CH_3CH_2CH_2CH_3 \qquad\qquad CH_3\underset{\displaystyle CH_3}{\underset{|}{C}}HCH_3$$

IUPAC name: butane (common name: *n*-butane)

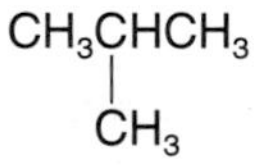

IUPAC name: 2-methylpropane (common name: isobutane)

Our discussion of the IUPAC rules for alkanes is not yet complete. Before proceeding further, however, it is necessary to consider substituent groups other than methyl that may be attached to the main carbon chain.

2.12 ALKYL GROUPS

Alkyl groups are structural units that lack one of the hydrogen substituents of an alkane. A methyl group (CH_3—) is an alkyl group derived from methane (CH_4). Unbranched alkyl groups in which the point of attachment is at the end of the chain are named in systematic nomenclature by replacing the *-ane* endings of Table 2.4 by *-yl*.

$CH_3CH_2—$ **Ethyl** group

$CH_3(CH_2)_5CH_2—$ **Heptyl** group

$CH_3(CH_2)_{16}CH_2—$ **Octadecyl** group

The dash at the end of the chain represents a potential point of attachment for some other atom or group.

Carbon atoms are classified according to their degree of substitution by other carbons. A **primary** carbon is one that is *directly* attached to one other carbon. Similarly, a **secondary** carbon is directly attached to two other carbons, a **tertiary** carbon to three, and a **quaternary** carbon to four. Alkyl groups are designated as primary, secondary, or tertiary according to the degree of substitution of the carbon that bears the potential point of attachment.

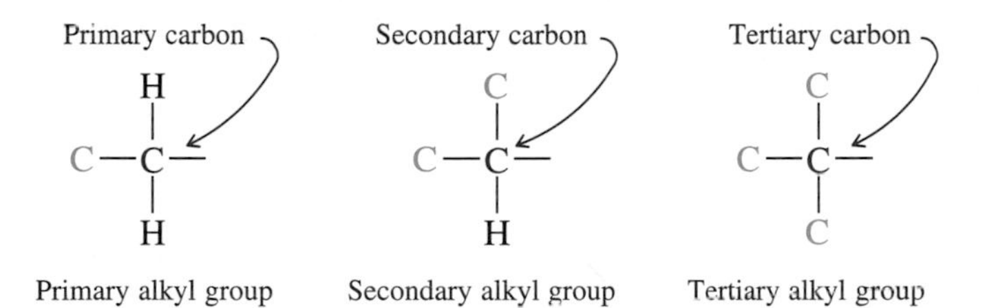

Ethyl ($CH_3CH_2—$), heptyl [$CH_3(CH_2)_5CH_2—$], and octadecyl [$CH_3(CH_2)_{16}CH_2—$] are examples of primary alkyl groups.

Branched alkyl groups are named by using the longest continuous chain that begins at the point of attachment as the base name. Thus, the systematic names of the two C_3H_7 alkyl groups are propyl and 1-methylethyl. Both are better known by their common names, *n*-propyl and isopropyl, respectively.

$CH_3CH_2CH_2—$

Propyl group
(common name: *n*-propyl)

$\underset{2}{CH_3}\overset{\overset{\displaystyle CH_3}{|}}{\underset{1}{CH}}—$ or $(CH_3)_2CH—$

1-Methylethyl group
(common name: isopropyl)

An isopropyl group is a *secondary* alkyl group. Its point of attachment is to a secondary carbon atom, one that is directly bonded to two other carbons.

The C_4H_9 alkyl groups may be derived either from the unbranched carbon skeleton of butane or from the branched carbon skeleton of isobutane. Those derived from butane are the butyl (*n*-butyl) group and the 1-methylpropyl (*sec*-butyl) group.

$CH_3CH_2CH_2CH_2—$

Butyl group
(common name: *n*-butyl)

$\underset{3}{CH_3}\underset{2}{CH_2}\overset{\overset{\displaystyle CH_3}{|}}{\underset{1}{CH}}—$

1-Methylpropyl group
(common name: *sec*-butyl)

Those derived from isobutane are the 2-methylpropyl (isobutyl) group and the 1,1-dimethylethyl (*tert*-butyl) group. Isobutyl is a primary alkyl group because its potential point of attachment is a primary carbon. *tert*-Butyl is a tertiary alkyl group because its potential point of attachment is a tertiary carbon.

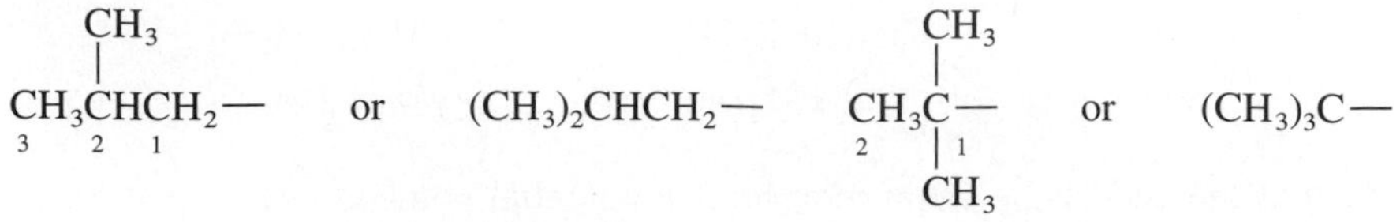

$$\underset{3\ \ 2\ \ 1}{CH_3\overset{\overset{\displaystyle CH_3}{|}}{C}HCH_2—} \quad \text{or} \quad (CH_3)_2CHCH_2—$$

2-Methylpropyl group
(common name: isobutyl)

$$CH_3\underset{\underset{\displaystyle CH_3}{|}}{\overset{\overset{\displaystyle CH_3}{|}}{C}}— \quad \text{or} \quad (CH_3)_3C—$$

(C-2 = CH_3, C-1 = C)

1,1-Dimethylethyl group
(common name: *tert*-butyl)

PROBLEM 2.8 Give the structures and IUPAC names of all the C_5H_{11} alkyl groups and identify them as primary, secondary, or tertiary alkyl groups, as appropriate.

SAMPLE SOLUTION Consider the alkyl group having the same carbon skeleton as $(CH_3)_4C$. All the hydrogens are equivalent, so that replacing any one of them by a potential point of attachment is the same as replacing any of the others.

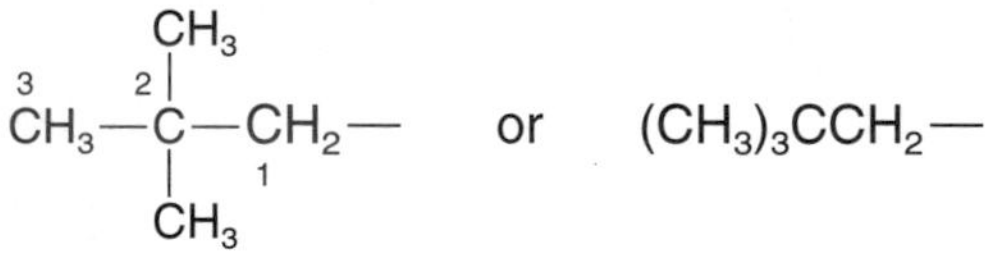

$$\overset{3}{CH_3}—\underset{\underset{\displaystyle CH_3}{|}}{\overset{\overset{\displaystyle CH_3}{2|}}{C}}—\underset{1}{CH_2}— \quad \text{or} \quad (CH_3)_3CCH_2—$$

Numbering always begins at the point of attachment and continues through the longest continuous chain. In this case the chain is three carbons and there are two methyl groups at C-2. The IUPAC name of this alkyl group is **2,2-dimethylpropyl.** (The common name for this group is *neopentyl*.) It is a *primary* alkyl group because the carbon that bears the potential point of attachment (C-1) is itself directly bonded to one other carbon.

The names and structures of the most frequently encountered alkyl groups are given on the inside back cover.

In addition to methyl and ethyl groups, we shall encounter *n*-propyl, isopropyl, *n*-butyl, *sec*-butyl, isobutyl, *tert*-butyl, and neopentyl groups many times throughout this text. Although these are common names, they have been integrated into the IUPAC system and are an acceptable adjunct to systematic nomenclature. You should be able to recognize these groups on sight and to give their structures when needed.

2.13 IUPAC NAMES OF HIGHLY BRANCHED ALKANES

By combining the fundamental principles of IUPAC notation with the names of the various alkyl groups, we can develop systematic names for highly branched alkanes. Let us start with a simple case, the alkane shown as follows, and increase its complexity by successively adding methyl groups at various positions.

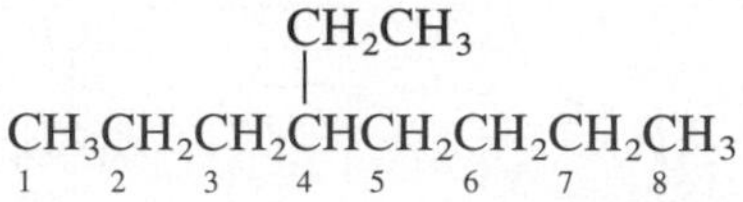

$$\underset{1}{CH_3}\underset{2}{CH_2}\underset{3}{CH_2}\underset{4}{\overset{\overset{\displaystyle CH_2CH_3}{|}}{C}H}\underset{5}{CH_2}\underset{6}{CH_2}\underset{7}{CH_2}\underset{8}{CH_3}$$

As numbered on the structural formula, the longest continuous chain contains eight carbons, and so the compound is named as a derivative of octane. Numbering begins at the end nearest the branch, and so the ethyl substituent is located at C-4 and the name of the alkane is **4-ethyloctane.**

What happens to the IUPAC name when another substituent, for example, a methyl group at C-3, is added to the structure?

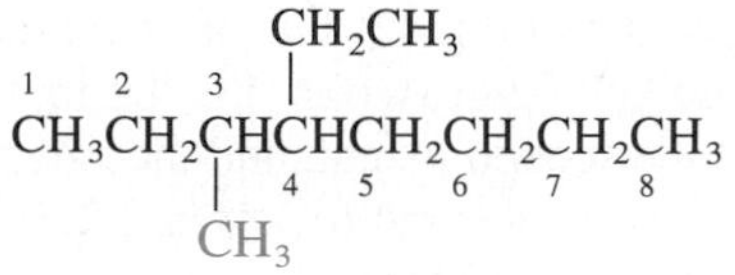

$$\begin{array}{c} \qquad\quad CH_2CH_3 \\ CH_3CH_2\underset{|}{C}H\overset{|}{C}HCH_2CH_2CH_2CH_3 \\ CH_3 \qquad\qquad\qquad \end{array}$$
(carbons numbered 1–8 along the main chain, from left to right)

The compound is named as an octane derivative that bears a C-3 methyl group and a C-4 ethyl group. *When two or more different substituents are present, they are listed in alphabetical order in the name.* The IUPAC name for this compound is **4-ethyl-3-methyloctane.**

Replicating prefixes such as *di-*, *tri-*, and *tetra-* (Section 2.11) are used as needed but are ignored when alphabetizing. Adding a second methyl group to the original structure, at C-5, for example, converts it to **4-ethyl-3,5-dimethyloctane.**

$$\begin{array}{c} \qquad\quad CH_2CH_3 \\ CH_3CH_2\underset{|}{C}H\overset{|}{C}H\underset{|}{C}HCH_2CH_2CH_3 \\ CH_3 \quad CH_3 \qquad\qquad \end{array}$$
(carbons numbered 1–8 along the main chain, from left to right)

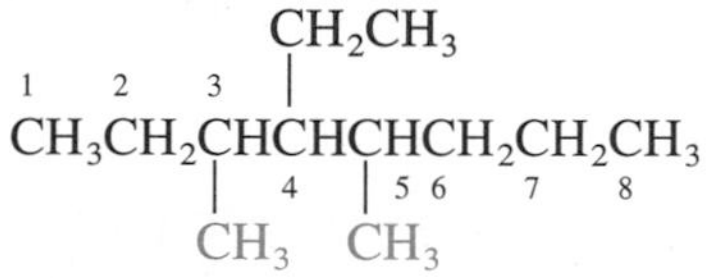

Italicized prefixes such as *sec-* and *tert-* are ignored when alphabetizing except when they are compared with each other. *tert*-Butyl precedes isobutyl, and *sec*-butyl precedes *tert*-butyl.

PROBLEM 2.9 Give an acceptable IUPAC name for each of the following alkanes:

(*a*)
$$\begin{array}{c} CH_2CH_3 \qquad\qquad \\ CH_3CH_2\underset{|}{C}H\overset{|}{C}H\underset{|}{C}HCH_2\underset{|}{C}HCH_3 \\ CH_3 \quad CH_3 \quad CH_3 \end{array}$$

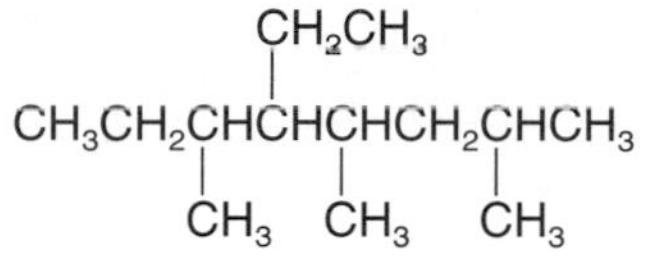

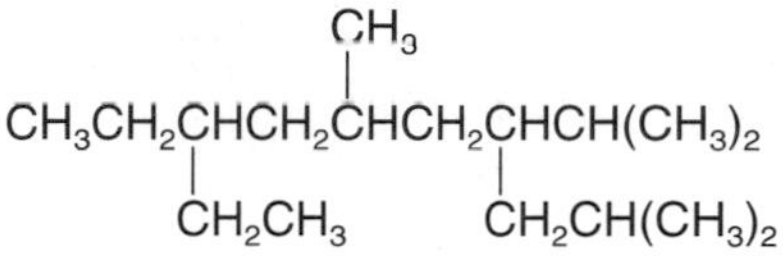

(*b*) $(CH_3CH_2)_2CHCH_2CH(CH_3)_2$

(*c*)
$$\begin{array}{c} CH_3 \qquad\qquad \\ CH_3CH_2\underset{|}{C}HCH_2\overset{|}{\underset{|}{C}}HCH_2CHCH(CH_3)_2 \\ CH_2CH_3 \qquad CH_2CH(CH_3)_2 \end{array}$$

SAMPLE SOLUTION (*a*) This problem extends the preceding discussion by adding a third methyl group to 4-ethyl-3,5-dimethyloctane, the compound just described. It is, theretore, an *ethyltrimethyloctane.* Notice, however, that the numbering sequence needs to be changed in order to adhere to the rule of numbering from the end of the chain nearest the first branch. When numbered properly, this compound has a methyl group at C-2 as its first-appearing substituent.

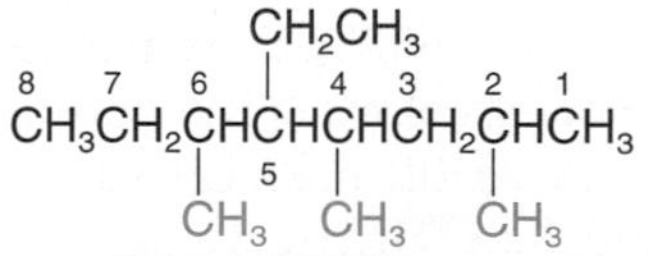

$$\begin{array}{c} CH_2CH_3 \qquad\qquad \\ CH_3CH_2\underset{|}{C}H\overset{|}{C}H\underset{|}{C}HCH_2\underset{|}{C}HCH_3 \\ CH_3 \quad CH_3 \quad CH_3 \end{array}$$
(carbons numbered 8–1 along the main chain, from left to right)

5-Ethyl-2,4,6-trimethyloctane

A pervasive element in IUPAC nomenclature as it pertains to the direction of numbering is the ''first point of difference'' rule. Consider the two directions in which the following alkane may be numbered:

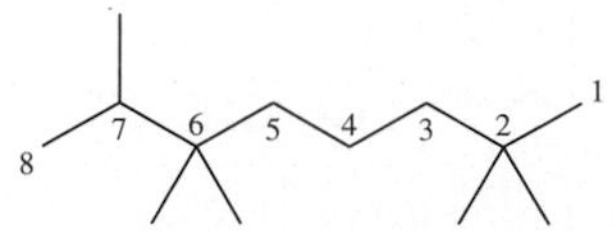

2,2,6,6,7-Pentamethyloctane (correct)

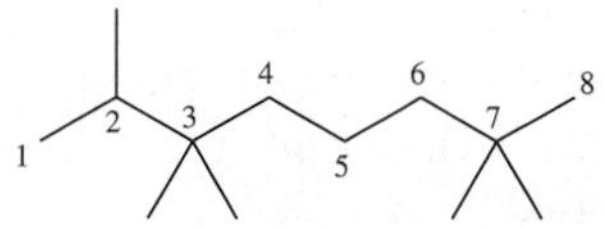

2,3,3,7,7-Pentamethyloctane (incorrect!)

When deciding on the proper direction, a point of difference occurs when one order gives a lower locant than another. Thus, while 2 is the first locant in both numbering schemes, the tie is broken at the second locant, and the rule favors 2,2,6,6,7, which has 2 as its second locant while 3 is the second locant in 2,3,3,7,7. Notice that locants are *not* added together, but examined one by one.

Finally, when equal locants are generated from two different numbering directions, the direction is chosen which gives the lower number to the substituent that appears first in the name. (Remember, substituents are listed alphabetically.)

A tabular summary of the IUPAC rules for alkane nomenclature appears on pages 76–77.

The IUPAC nomenclature system is inherently logical and incorporates healthy elements of common sense into its rules. Granted, some long, funny-looking, hard-to-pronounce names are generated. Once one knows the code (rules of grammar) though, it becomes a simple matter to convert those long names to unique structural formulas.

2.14 CYCLOALKANE NOMENCLATURE

Cycloalkanes are one class of *alicyclic* (*ali*phatic *cyclic*) hydrocarbons.

Cycloalkanes are alkanes that contain a ring of three or more carbons. They are frequently encountered in organic chemistry and are characterized by the molecular formula C_nH_{2n}. Some examples include:

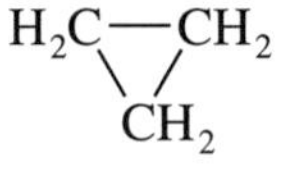

H_2C—CH_2, CH_2 usually represented as ▽

Cyclopropane

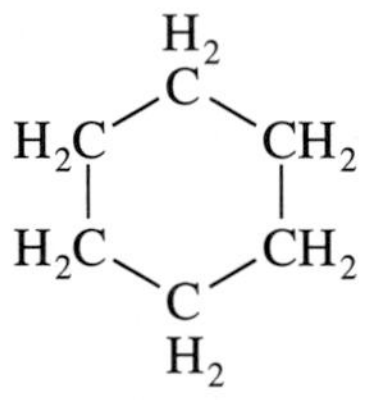

H_2C, H_2C, CH_2, CH_2, $C H_2$, $C H_2$ usually represented as ⬡

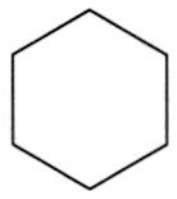

Cyclohexane

As you can see, cycloalkanes are named, under the IUPAC system, by adding the prefix *cyclo-* to the name of the unbranched alkane with the same number of carbons as the ring. Substituent groups are identified in the usual way. Their positions are specified by numbering the carbon atoms of the ring in the direction that gives the lowest number to the substituent groups at the first point of difference.

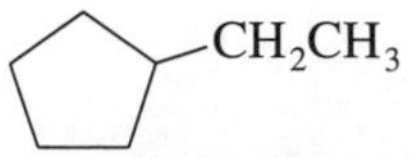

CH_2CH_3

Ethylcyclopentane

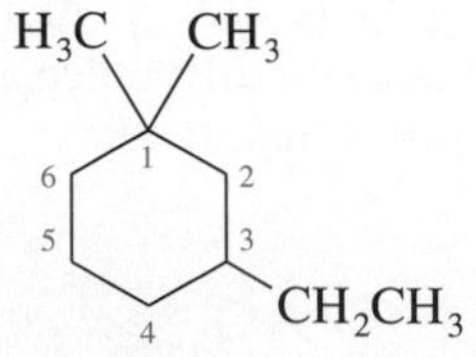

H_3C CH_3 1 2 3 4 5 6 CH_2CH_3

3-Ethyl-1,1-dimethylcyclohexane

(not 1-ethyl-3,3-dimethylcyclohexane, because first point of difference rule requires 1,1,3 substitution pattern rather than 1,3,3)

When the ring contains fewer carbon atoms than an alkyl group attached to it, the compound is named as an alkane and the ring is treated as a cycloalkyl substituent:

$$CH_3CH_2CHCH_2CH_3$$ (cyclobutyl on C-3)

3-Cyclobutylpentane

PROBLEM 2.10 Name each of the following compounds:

(*a*) [cyclononane]—$C(CH_3)_3$ (*b*) $(CH_3)_2CH$—[cyclododecane with H_3C, CH_3] (*c*) [cyclohexyl–cyclohexane]

SAMPLE SOLUTION (*a*) The molecule has a *tert*-butyl group bonded to a nine-membered cycloalkane. It is *tert*-butylcyclononane. Alternatively, the *tert*-butyl group could be named systematically as a 1,1-dimethylethyl group, and the compound would then be named (1,1-dimethylethyl)cyclononane. (Parentheses are used when necessary to avoid ambiguity. In this case the parentheses alert the reader that the locants 1,1 refer to substituents on the alkyl group and not to ring positions.)

2.15 SOURCES OF ALKANES AND CYCLOALKANES

As noted in earlier sections, natural gas is especially rich in methane and also contains ethane and propane, along with still smaller amounts of other low-molecular-weight alkanes. Natural gas is often found associated with petroleum deposits. Petroleum is a liquid mixture containing hundreds of substances, including approximately 150 hydrocarbons, roughly half of which are alkanes or cycloalkanes. Distillation of crude oil gives a number of fractions, which by custom are described by the names given in Figure 2.3. High-boiling fractions such as kerosene and gas oil find wide use as fuels for diesel engines and furnaces, and the nonvolatile residue can be processed to give lubricating oil, greases, petroleum jelly, paraffin wax, and asphalt.

The word *petroleum* is derived from the Latin words for "rock" (*petra*) and "oil" (*oleum*).

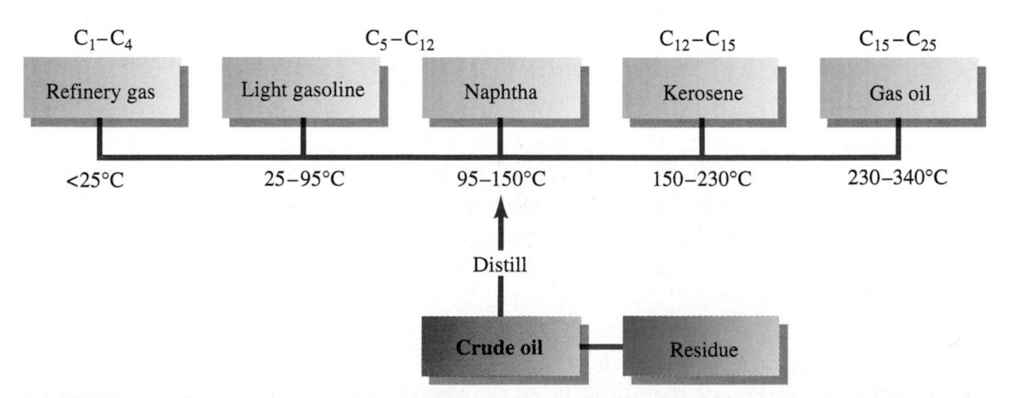

FIGURE 2.3 Distillation of crude oil yields a series of volatile fractions having the names indicated, along with a nonvolatile residue. The number of carbon atoms that characterize the hydrocarbons in each fraction is approximate.

Petroleum refining involves more than distillation, however, and includes two major additional operations:

1. **Cracking.** It is the more volatile, lower-molecular-weight hydrocarbons that are useful as automotive fuels and as a source of petrochemicals. Cracking increases the proportion of these hydrocarbons at the expense of higher-molecular-weight ones by processes that involve the cleavage of carbon-carbon bonds induced by heat (*thermal cracking*) or with the aid of certain catalysts (*catalytic cracking*).
2. **Reforming.** The physical properties of the crude oil fractions known as *light gasoline* and *naphtha* (Figure 2.3) are appropriate for use as a motor fuel, but their ignition characteristics in high-compression automobile engines are poor and give rise to preignition, or ''knocking.'' Reforming converts the hydrocarbons in petroleum to aromatic hydrocarbons and highly branched alkanes, both of which show less tendency to knock than unbranched alkanes and cycloalkanes.

The tendency of a gasoline to knock is given by its octane number. The lower the octane number, the greater the tendency to knock. The two standards are heptane (assigned a value of 0) and 2,2,4-trimethylpentane (assigned a value of 100). The octane number of a gasoline is equal to the percentage of 2,2,4-trimethylpentane in a mixture of 2,2,4-trimethylpentane and heptane that has the same tendency to knock as that sample of gasoline.

The leaves and fruit of many plants bear a waxy coating made up of alkanes that prevents loss of water. In addition to being present in beeswax (see Problem 2.5), hentriacontane, $CH_3(CH_2)_{29}CH_3$, is a component of the leaf wax of the tobacco plant.

Cyclopentane and cyclohexane are present in petroleum, but as a rule, unsubstituted cycloalkanes are rarely found in natural sources. Compounds based on cycloalkane structural units, however, are quite abundant. Such compounds include rings of various sizes that incorporate double bonds or bear substituent atoms or groups attached to the ring.

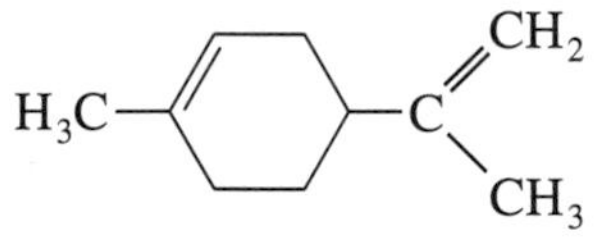

Limonene

(present in lemons and oranges)

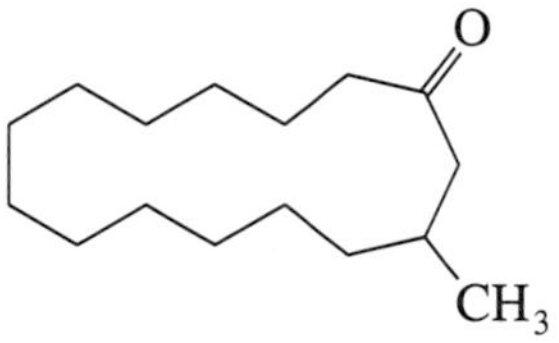

Muscone

(responsible for odor of musk; used in perfumery)

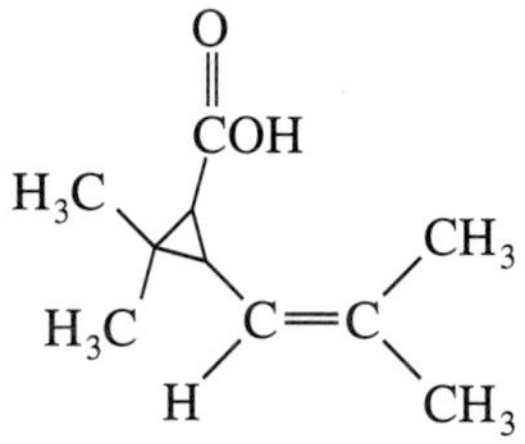

Chrysanthemic acid

(obtained from chrysanthemum flowers)

Many alkanes and cycloalkanes are the product of chemical synthesis.

2.16 PHYSICAL PROPERTIES OF ALKANES AND CYCLOALKANES

Appendix 1 lists selected physical properties for representative alkanes as well as members of other families of organic compounds.

As we have seen earlier in this chapter, methane, ethane, propane, and butane are gases at room temperature. The unbranched alkanes pentane (C_5H_{12}) through heptadecane ($C_{17}H_{36}$) are liquids, while higher homologs are solids. As shown in Figure 2.4, the boiling points of unbranched alkanes increase as the number of carbon atoms in the chain increases. Figure 2.4 also shows that the boiling points for 2-methyl-branched alkanes are lower than those of the unbranched isomer. By exploring at the molecular level the reasons for the increase in boiling point with the number of carbons and the reasons for the difference in boiling point between branched and unbranched alkanes, we can begin to develop some insights into the relationship between structure and properties.

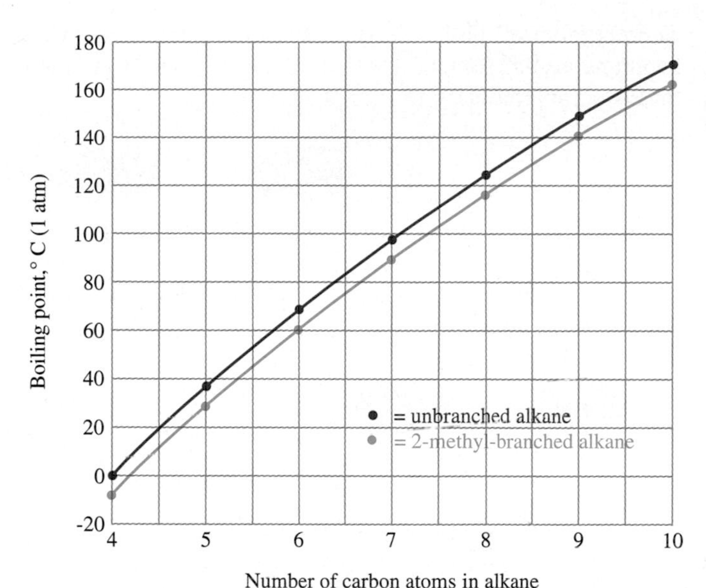

FIGURE 2.4 Boiling points of unbranched alkanes and their 2-methyl-branched isomers. (Temperatures in this text are expressed in degrees Celsius, °C. The SI unit of temperature is the kelvin, K. To convert degrees Celsius to kelvins add 273.15.)

A substance exists as a liquid rather than a gas because there are cohesive forces between molecules (**intermolecular attractive forces**) that are greater in the liquid state than in the gas phase. Attractive forces between neutral species (atoms or molecules, but not ions) are referred to as **van der Waals forces** and may be of three types:

1. dipole-dipole
2. dipole–induced dipole
3. induced dipole–induced dipole

These forces are electrical in nature, and in order to vaporize a substance, enough energy must be added to overcome them. Most alkanes have no measurable dipole moment, and therefore the only van der Waals force to be considered is the induced dipole–induced dipole attractive force.

Van der Waals forces of the induced dipole–induced dipole type are sometimes called *London forces,* or *dispersion forces.*

It might seem that two nearby molecules A and B of a nonpolar substance such as an alkane would be unaffected by each other.

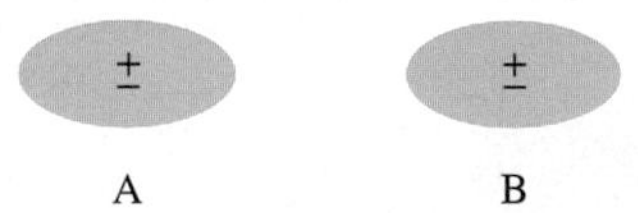

The electric field of a molecule, however, is not static but fluctuates rapidly. While, on average, the centers of positive and negative charge of an alkane nearly coincide, at any instant they may not, and molecule A can be considered to have a temporary dipole moment.

The neighboring molecule B ''feels'' the dipolar electric field of A and undergoes a spontaneous adjustment in its electron positions, giving it a temporary dipole moment that is complementary to that of A.

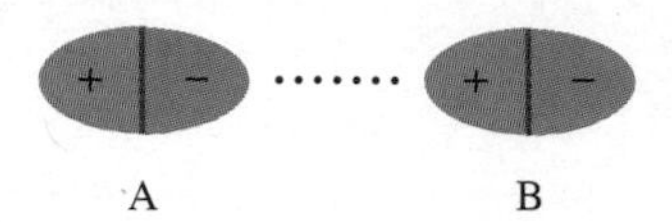

The electric fields of both A and B fluctuate, but always in a way that results in a weak attraction between them.

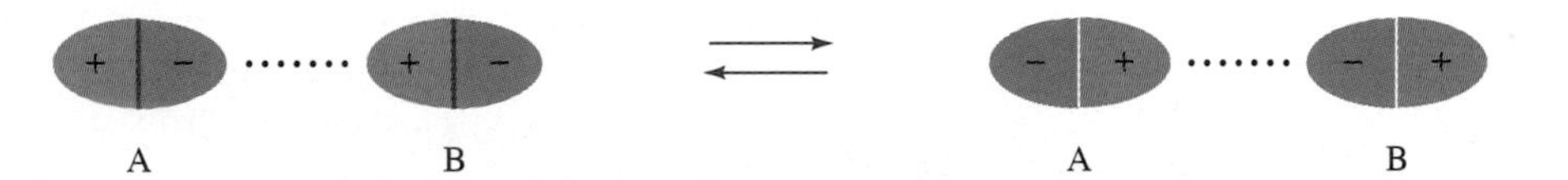

Extended assemblies of induced dipole–induced dipole attractions can accumulate to give substantial intermolecular attractive forces. An alkane with a higher molecular weight has more atoms and electrons and, therefore, more opportunities for intermolecular attractions and a higher boiling point than one with a lower molecular weight.

As noted earlier in this section, branched alkanes have lower boiling points than their unbranched isomers. Isomers have, of course, the same number of atoms and electrons, but a molecule of a branched alkane has a smaller surface area than an unbranched one. The extended shape of an unbranched alkane permits more points of contact for intermolecular associations. Compare the boiling points of pentane and its isomers:

$CH_3CH_2CH_2CH_2CH_3$	$CH_3\underset{\displaystyle CH_3}{\underset{\vert}{C}}HCH_2CH_3$	$CH_3\overset{\displaystyle CH_3}{\overset{\vert}{\underset{\displaystyle CH_3}{\underset{\vert}{C}}}}CH_3$
Pentane (bp 36°C)	2-Methylbutane (bp 28°C)	2,2-Dimethylpropane (bp 9°C)

The shapes of these isomers are clearly evident in the space-filling models depicted in Figure 2.5. Pentane has the most extended structure and the largest surface area available for ''sticking'' to other molecules by way of induced dipole–induced dipole

(*a*) Pentane: $CH_3CH_2CH_2CH_2CH_3$

(*b*) 2-Methylbutane: $(CH_3)_2CHCH_2CH_3$

(*c*) 2,2-Dimethylpropane: $(CH_3)_4C$

FIGURE 2.5 Space-filling models of (*a*) pentane, (*b*) 2-methylbutane, and (*c*) 2,2-dimethylpropane. The most branched isomer, 2,2-dimethylpropane, has the most compact, most spherical three-dimensional shape.

attractive forces; it has the highest boiling point. 2,2-Dimethylpropane has the most compact structure, engages in the fewest induced dipole–induced dipole attractions, and has the lowest boiling point.

Induced dipole–induced dipole attractions are very weak forces individually, but a typical organic substance can participate in so many of them that they are collectively the most important of all the contributors to intermolecular attraction in the liquid state. They are the only forces of attraction possible between nonpolar molecules such as alkanes.

PROBLEM 2.11 Match the boiling points with the appropriate alkanes.

Alkanes: octane, 2-methylheptane, 2,2,3,3-tetramethylbutane, nonane

Boiling points (*°C, 1 atm*): 106, 116, 126, 151

Solid alkanes are soft, generally low melting materials. The forces responsible for holding the crystal together are the same induced dipole–induced dipole interactions that operate between molecules in the liquid, but the degree of organization is greater in the solid phase. By measuring the distances between the atoms of one molecule and its neighbor in the crystal, it is possible to specify a distance of closest approach characteristic of an atom called its **van der Waals radius.** At distances close to the sum of their van der Waals radii, two nonbonded atoms experience their maximum mutual attraction. This weak attraction is replaced by a strong repulsion as the electron clouds of the two atoms penetrate each other at interatomic distances that are less than the sum of their van der Waals radii. The van der Waals radius for hydrogen is 120 pm. Thus, when two alkane molecules are brought together so that a hydrogen of one molecule is within 240 pm of a hydrogen of the other, the electrons of the two C—H bonds are forced into the same space and closer contact of the two molecules is strongly resisted.

A familiar physical property of alkanes is contained in the adage ''oil and water don't mix.'' Alkanes—indeed all hydrocarbons—are virtually insoluble in water. In order for an alkane to dissolve in water, the attractive forces between the alkane and water would have to be strong enough to replace the dipole-dipole attractive forces between water molecules. They are not. Alkanes, being nonpolar, interact only weakly with water molecules and, with densities in the 0.6–0.8 g/mL range, float on water (as the Alaskan oil spill of 1989 and the even larger Persian Gulf spill of 1991 remind us).

2.17 CHEMICAL PROPERTIES. COMBUSTION OF ALKANES

An older name for alkanes is **paraffin hydrocarbons.** *Paraffin* is derived from the Latin words *parum affinis* (''with little affinity'') and testifies to the low level of reactivity of alkanes. Like most other organic compounds, however, alkanes burn readily in air. This combination with oxygen is known as **combustion** and is quite exothermic. All hydrocarbons yield carbon dioxide and water as the products of their combustion.

Alkanes are so unreactive that George A. Olah of the University of Southern California was awarded the 1994 Nobel Prize in chemistry in part for developing novel substances that do react with alkanes.

$$\underset{\text{Methane}}{CH_4} + \underset{\text{Oxygen}}{2O_2} \longrightarrow \underset{\text{Carbon dioxide}}{CO_2} + \underset{\text{Water}}{2H_2O} \qquad \Delta H° = -890 \text{ kJ } (-212.8 \text{ kcal})$$

$$\underset{\text{2-Methylbutane}}{(CH_3)_2CHCH_2CH_3} + \underset{\text{Oxygen}}{8O_2} \longrightarrow \underset{\text{Carbon dioxide}}{5CO_2} + \underset{\text{Water}}{6H_2O} \qquad \Delta H° = -3529 \text{ kJ } (-843.4 \text{ kcal})$$

PROBLEM 2.12 Write a balanced chemical equation for the combustion of cyclohexane.

The heat released on complete combustion of one mole of a substance is called its **heat of combustion.** The heat of combustion is equal to $-\Delta H°$ for the reaction written in the direction shown. By convention

$$\Delta H° = H°_{products} - H°_{reactants}$$

where $H°$ is the heat content, or **enthalpy,** of a compound in its standard state, i.e., the gas, pure liquid, or crystalline solid at a pressure of 1 atm. In an exothermic process the heat content of the products is less than the heat content of the starting materials and $\Delta H°$ is a negative number.

Table 2.5 lists the heats of combustion of several alkanes. Unbranched alkanes have slightly higher heats of combustion than their 2-methyl-branched isomers, but the most important factor is the number of carbons. The unbranched alkanes and the 2-methyl-branched alkanes constitute two separate *homologous series* (Section 2.8) in which there is a regular increase of about 653 kJ/mol (156 kcal/mol) in the heat of combustion for each additional CH_2 group.

PROBLEM 2.13 Using the data in Table 2.5, estimate the heat of combustion of

(*a*) 2-Methylnonane (in kcal/mol) (*b*) Icosane (in kJ/mol)

SAMPLE SOLUTION (*a*) The last entry for the group of 2-methylalkanes in the table is 2-methylheptane. Its heat of combustion is 1306 kcal/mol. Since 2-methylnonane

TABLE 2.5
Heats of Combustion ($-\Delta H°$) of Representative Alkanes

		$-\Delta H°$	
Compound	**Formula**	**kJ/mol**	**kcal/mol**
Unbranched alkanes			
Hexane	$CH_3(CH_2)_4CH_3$	4,163	995.0
Heptane	$CH_3(CH_2)_5CH_3$	4,817	1151.3
Octane	$CH_3(CH_2)_6CH_3$	5,471	1307.5
Nonane	$CH_3(CH_2)_7CH_3$	6,125	1463.9
Decane	$CH_3(CH_2)_8CH_3$	6,778	1620.1
Undecane	$CH_3(CH_2)_9CH_3$	7,431	1776.1
Dodecane	$CH_3(CH_2)_{10}CH_3$	8,086	1932.7
Hexadecane	$CH_3(CH_2)_{14}CH_3$	10,701	2557.6
2-Methyl-branched alkanes			
2-Methylpentane	$(CH_3)_2CHCH_2CH_2CH_3$	4,157	993.6
2-Methylhexane	$(CH_3)_2CH(CH_2)_3CH_3$	4,812	1150.0
2-Methylheptane	$(CH_3)_2CH(CH_2)_4CH_3$	5,466	1306.3

has two more methylene groups than 2-methylheptane, its heat of combustion is 2 × 156 kcal/mol higher.

Heat of combustion of 2-methylnonane = 1306 + 2(156) = 1618 kcal/mol

---

One important use of heat of combustion data is in assessing the relative stability of isomeric hydrocarbons. Heats of combustion tell us not only which isomer is more stable than another, but by how much. Consider the group of alkanes:

$CH_3(CH_2)_6CH_3$ — Octane

$(CH_3)_2CHCH_2CH_2CH_2CH_2CH_3$ — 2-Methylheptane

$(CH_3)_3CCH_2CH_2CH_2CH_3$ — 2,2-Dimethylhexane

$(CH_3)_3CC(CH_3)_3$ — 2,2,3,3-Tetramethylbutane

Figure 2.6 compares the heats of combustion of these C_8H_{18} isomers on a *potential energy diagram.* **Potential energy** is comparable with enthalpy; it is the energy a molecule has exclusive of its kinetic energy of motion. A molecule with more potential energy is less stable than an isomeric molecule with less potential energy. Since these compounds are isomers and all undergo combustion to the same final state according to the equation

$$C_8H_{18} + \tfrac{25}{2}O_2 \longrightarrow 8CO_2 + 9H_2O$$

the differences in their heats of combustion translate directly to differences in their potential energies. *When comparing isomers, the one with the lowest potential energy*

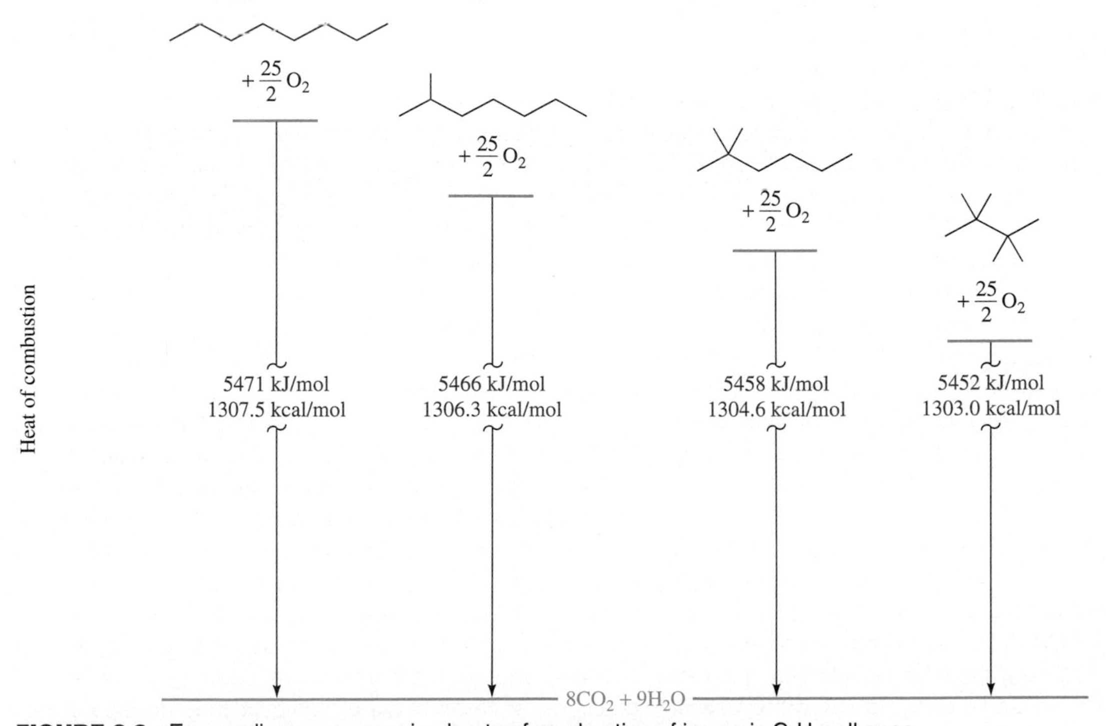

FIGURE 2.6 Energy diagram comparing heats of combustion of isomeric C_8H_{18} alkanes.

(in this case, the lowest heat of combustion) is the most stable. Among the C_8H_{18} alkanes, the most highly branched isomer, 2,2,3,3-tetramethylbutane, is the most stable, and the unbranched isomer octane is the least stable. It is generally true for alkanes that a more branched isomer is more stable than a less branched one.

THERMOCHEMISTRY

Thermochemistry is the study of the heat changes that accompany chemical processes. It has a long history dating back to the work of the French chemist Antoine-Laurent Lavoisier in the late eighteenth century. Thermochemistry provides quantitative information which complements the qualitative description of a chemical reaction and can help us understand why some reactions occur while others do not. It is of obvious importance when assessing the relative value of various materials as fuels, when comparing the stability of isomers, or when determining the practicality of a particular reaction. In the field of bioenergetics, thermochemical information is applied to the task of sorting out how living systems use chemical reactions to store and use the energy that originates in the sun.

By allowing compounds to react in a device called a **calorimeter,** it is possible to measure the heat evolved in an exothermic reaction or the heat absorbed in an endothermic reaction. Thousands of reactions have been studied to produce a rich library of thermochemical data. These data take the form of **heats of reaction** and correspond to the value of the enthalpy change $\Delta H°$ for a particular reaction of a particular substance.

In this section you have seen how heats of combustion can be used to determine relative stabilities of isomeric alkanes. In later sections we shall expand our scope to include the experimentally determined heats of certain other reactions, such as *bond dissociation energies* (Section 4.18) and *heats of hydrogenation* (Section 6.2), to see how $\Delta H°$ values from various sources can aid our understanding of structure and reactivity.

Heat of formation ($\Delta H_f°$), the enthalpy change for formation of a compound directly from the elements, is one type of heat of reaction. In cases such as the formation of CO_2 or H_2O from the combustion of carbon or hydrogen, respectively, the heat of formation of a substance can be measured directly. In most other cases, heats of formation are not measured experimentally but are calculated from the measured heats of other reactions. Consider, for example, the heat of formation of methane. The reaction that defines the formation of methane from the elements,

$$\underset{\text{Carbon}}{\text{C (graphite)}} + \underset{\text{Hydrogen}}{2\,H_2(g)} \longrightarrow \underset{\text{Methane}}{CH_4(g)}$$

can be expressed as the sum of three reactions:

$$\begin{array}{ll}
(1)\ \text{C (graphite)} + O_2(g) \longrightarrow CO_2(g) & \Delta H° = -393\text{ kJ} \\
(2)\ 2H_2(g) + O_2(g) \longrightarrow 2H_2O(l) & \Delta H° = -572\text{ kJ} \\
(3)\ CO_2(g) + 2H_2O(l) \longrightarrow CH_4(g) + 2O_2(g) & \Delta H° = +890\text{ kJ} \\
\hline
\text{C (graphite)} + 2H_2 \longrightarrow CH_4 & \Delta H° = -75\text{ kJ}
\end{array}$$

Equations (1) and (2) are the heats of formation of carbon dioxide and water, respectively. Equation (3) is the reverse of the combustion of methane, and so the heat of reaction is equal to the heat of combustion but opposite in sign. The **molar heat of formation** of a substance is the enthalpy change for formation of one mole of the substance from the elements. For methane $\Delta H_f° = -75$ kJ/mol.

The heats of formation of most organic compounds are derived from heats of reaction by arithmetic manipulations similar to that shown. Chemists find a table of $\Delta H_f°$ values to be convenient because it replaces many separate tables of $\Delta H°$ values for individual reaction types and permits $\Delta H°$ to be calculated for any reaction, real or imaginary, for which the heats of formation of reactants and products are available. It is more appropriate for our purposes, however, to connect thermochemical data to chemical processes as directly as possible, and therefore we will cite heats of particular reactions, such as heats of combustion and heats of hydrogenation, rather than heats of formation.

The small differences in stability between branched and unbranched alkanes depend on induced dipole–induced dipole attractive forces. We noted earlier that the more spherical shape of branched alkanes leads to *decreased* forces of attraction *between* molecules (intermolecular forces). The opposite holds true for induced dipole–induced dipole attractions *within* a molecule **(intramolecular forces).** Branching *increases* the number of intramolecular attractive forces, and these attractions stabilize a more branched alkane relative to a less branched isomer.

In the next chapter you will see what happens when atoms and groups within a molecule approach each other too closely and how molecules adjust to intramolecular repulsive forces.

PROBLEM 2.14 Without consulting Table 2.5 arrange the following compounds in order of decreasing heat of combustion: pentane, isopentane, neopentane, hexane.

The kind of reasoning that related heat of combustion data to the relative energies of molecules and after that to an explanation based on molecular structure is typical of that used in organic chemistry. You will encounter similar thought processes frequently throughout this text.

2.18 OXIDATION-REDUCTION IN ORGANIC CHEMISTRY

As we have just seen, the reaction of alkanes with oxygen to give carbon dioxide and water is called *combustion.* A more fundamental classification of reaction types places it in the *oxidation-reduction* category. To understand why this is so, let us review some principles of oxidation-reduction, beginning with the notion of **oxidation number** (also known as **oxidation state**).

There are a variety of methods for calculating oxidation numbers. In compounds that contain a single carbon, such as methane (CH_4) and carbon dioxide (CO_2), the oxidation number of carbon can be calculated from the molecular formula. Both molecules are neutral, and so the algebraic sum of all the oxidation numbers must equal zero. Assuming, as is customary, that the oxidation state of hydrogen is $+1$, the oxidation state of carbon in CH_4 is calculated to be -4. Similarly, assuming an oxidation state of -2 for oxygen, carbon is $+4$ in CO_2. This kind of calculation provides an easy way to develop a list of one-carbon compounds in order of increasing oxidation state, as shown in Table 2.6.

The carbon in methane has the lowest oxidation number (-4) of any of the compounds in Table 2.6. Methane contains carbon in its most *reduced* form. Carbon dioxide and carbonic acid have the highest oxidation numbers ($+4$) for carbon, corresponding to its most *oxidized* state. When methane or any alkane undergoes combustion to form carbon dioxide, carbon is oxidized because its oxidation number increases while oxygen is reduced from its elemental state (oxidation number $= 0$) to its -2 oxidation state.

A useful generalization from Table 2.6 is the following:

Oxidation of carbon corresponds to an increase in the number of bonds between carbon and oxygen and/or a decrease in the number of carbon-hydrogen bonds. Conversely, *reduction corresponds to an increase in the number of carbon-hydrogen bonds and/or a decrease in the number of carbon-oxygen bonds.* From Table 2.6 it can be seen that each successive increase in oxidation state increases the number of bonds between carbon and oxygen and decreases the number of carbon-hydrogen bonds. Methane has four C—H bonds and no C—O bonds; carbon dioxide has four C—O bonds and no C—H bonds.

TABLE 2.6
Oxidation Number of Carbon in One-Carbon Compounds

Compound	Structural formula	Molecular formula	Oxidation number
Methane	CH_4	CH_4	−4
Methanol	CH_3OH	CH_4O	−2
Formaldehyde	$H_2C{=}O$	CH_2O	0
Formic acid	$HC(=O)OH$	CH_2O_2	+2
Carbonic acid	$HOC(=O)OH$	H_2CO_3	+4
Carbon dioxide	$O{=}C{=}O$	CO_2	+4

Among the various classes of hydrocarbons, alkanes contain carbon in its most reduced state, and alkynes contain carbon in its most oxidized state.

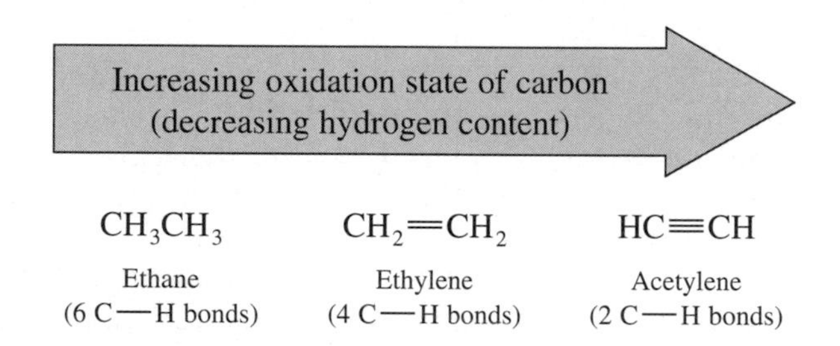

We can extend the generalization by recognizing that the pattern is not limited to increasing hydrogen and/or oxygen content. Any element *more electronegative* than carbon will have the same effect on oxidation number as oxygen. Thus, the oxidation numbers of carbon in CH_3Cl and in CH_3OH are the same (−2), and the reaction of methane with chlorine (to be discussed in Section 4.17) involves *oxidation* of carbon.

$$\underset{\text{Methane}}{CH_4} + \underset{\text{Chlorine}}{Cl_2} \longrightarrow \underset{\text{Chloromethane}}{CH_3Cl} + \underset{\text{Hydrogen chloride}}{HCl}$$

Any element *less electronegative* than carbon will have the same effect on oxidation number as hydrogen. Thus, the oxidation numbers of carbon in CH_3Li and in CH_4 are the same (−4), and the reaction of CH_3Cl with lithium (to be discussed in Section 14.3) involves *reduction* of carbon.

$$\underset{\text{Chloromethane}}{CH_3Cl} + \underset{\text{Lithium}}{2Li} \longrightarrow \underset{\text{Methyllithium}}{CH_3Li} + \underset{\text{Lithium chloride}}{LiCl}$$

The oxidation number of carbon *decreases* from −2 in CH_3Cl to −4 in CH_3Li.

The generalization can be expressed in terms broad enough to cover both the preceding reactions and many others as well, as follows: *Oxidation of carbon occurs when a bond between carbon and an atom which is less electronegative than carbon is replaced by a bond to an atom that is more electronegative than carbon. The reverse process is reduction.*

$$-\overset{|}{\underset{|}{C}}-X \underset{\text{reduction}}{\overset{\text{oxidation}}{\rightleftharpoons}} -\overset{|}{\underset{|}{C}}-Y$$

X is less electronegative than carbon — Y is more electronegative than carbon

Organic chemists are much more concerned with whether a particular reaction is an oxidation or a reduction of carbon than with determining the precise change in oxidation number. The generalizations described permit reactions to be examined in this way and eliminate the need for calculating oxidation numbers themselves.

Methods for calculating oxidation numbers in complex molecules are available. They are time-consuming to apply, however, and are rarely used in organic chemistry.

PROBLEM 2.15 The reactions shown will all be encountered in Chapter 6. Classify each according to whether it proceeds by oxidation of carbon, by reduction of carbon, or by a process other than oxidation-reduction.

(*a*) $CH_2{=}CH_2 + H_2O \longrightarrow CH_3CH_2OH$
(*b*) $CH_2{=}CH_2 + Br_2 \longrightarrow BrCH_2CH_2Br$
(*c*) $6CH_2{=}CH_2 + B_2H_6 \longrightarrow 2(CH_3CH_2)_3B$

SAMPLE SOLUTION (*a*) In this reaction one new C—H bond and one new C—O bond are formed. One carbon is reduced, the other is oxidized. Overall, there is no net change in oxidation state, and the reaction is not classified as an oxidation-reduction.

The ability to recognize when oxidation or reduction occurs is of value when deciding on the kind of reactant with which an organic molecule must be treated in order to convert it into some desired product. Many of the reactions to be discussed in subsequent chapters involve oxidation-reduction.

2.19 SUMMARY

The classes of hydrocarbons are **alkanes, alkenes, alkynes,** and **arenes** (Section 2.1). **Alkanes** are hydrocarbons in which all the bonds are *single* bonds and are characterized by the molecular formula C_nH_{2n+2}.

Functional groups are the structural units responsible for the characteristic reactions of a molecule. The functional groups in an alkane are its hydrogen substituents (Section 2.2). Other families of organic compounds, listed on the inside front cover and in Tables 2.1 and 2.2, bear more reactive functional groups, and the hydrocarbon chain to which they are attached can often be viewed as a supporting framework for the reactive function (Sections 2.3 and 2.4).

The simplest alkane is *methane,* CH_4 (Section 2.5); *ethane* is C_2H_6, and *propane* is C_3H_8 (Section 2.6). Constitutional isomers are possible for alkanes with four or more carbons. Thus there are two isomers of molecular formula C_4H_{10}. One of these has an unbranched carbon chain ($CH_3CH_2CH_2CH_3$) and is called n-*butane*; the other has a branched chain [$(CH_3)_3CH$)] and is called *isobutane* (Section 2.7). *n*-Butane and isobutane are **common names.** Unbranched alkanes are sometimes called *normal alkanes*

TABLE 2.7
Summary of IUPAC Nomenclature of Alkanes and Cycloalkanes

Rule	Example
A. Alkanes	
1. Find the longest continuous chain of carbon atoms and assign a basis name to the compound corresponding to the IUPAC name of the unbranched alkane having the same number of carbons.	The longest continuous chain in the alkane shown is six carbons (shown in red). 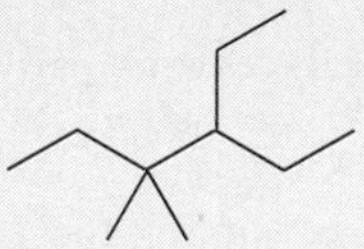 This alkane is named as a derivative of *hexane.*
2. List the substituents attached to the longest continuous chain in alphabetical order. Use the prefixes *di-, tri-, tetra-,* etc., when the same substituent appears more than once. Ignore these prefixes when alphabetizing.	The alkane bears two methyl groups and an ethyl group. It is an *ethyldimethylhexane.* 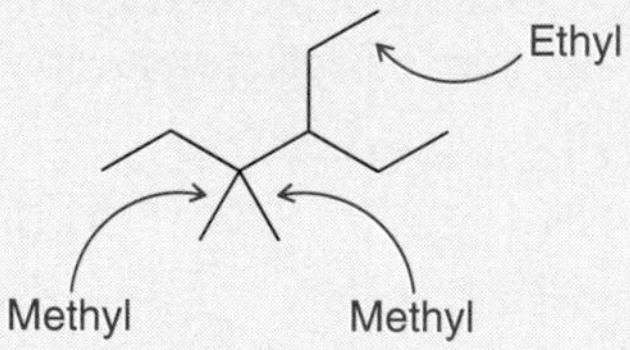 Ethyl Methyl Methyl
3. Number the chain in the direction that gives the lower locant to a substituent at the first point of difference.	When numbering from left to right, the substituents appear at carbons 3, 3, and 4. When numbering from right to left the locants are 3, 4, and 4. Therefore, number from left to right. 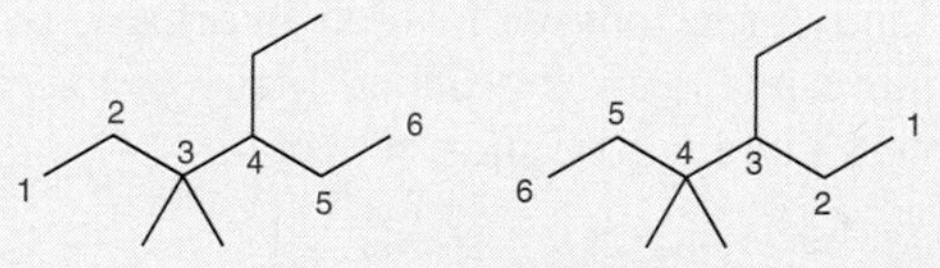 Correct Incorrect The correct name is *4-ethyl-3,3-dimethylhexane.*
4. When two different numbering schemes give equivalent sets of locants, choose the direction that gives the lower locant to the group that appears first in the name.	In the following example, the substituents are located at carbons 3 and 4 regardless of the direction in which the chain is numbered. 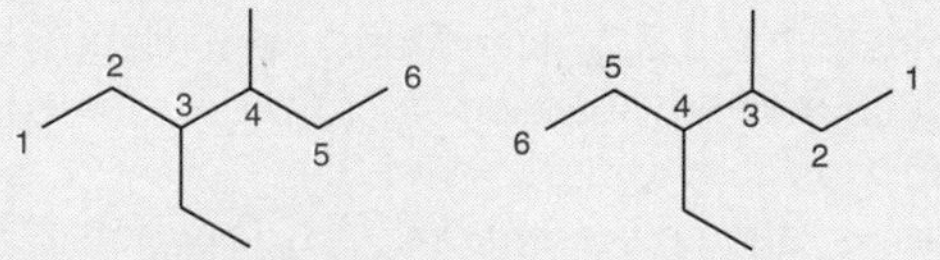 Correct Incorrect Ethyl precedes methyl in the name; therefore *3-ethyl-4-methylhexane* is correct.

(Continued)

TABLE 2.7
Summary of IUPAC Nomenclature of Alkanes and Cycloalkanes *(Continued)*

Rule	Example
5. When two chains are of equal length, choose the one with the greater number of substituents as the parent. (Although this requires naming more substituents, the substituents have simpler names.)	Two different chains contain five carbons in the alkane: 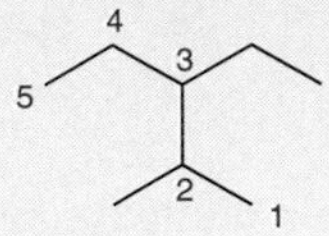 [structures labeled 1–5: Correct; Incorrect] The correct name is *3-ethyl-2-methylpentane* (disubstituted chain), rather than 3-isopropylpentane (monosubstituted chain). 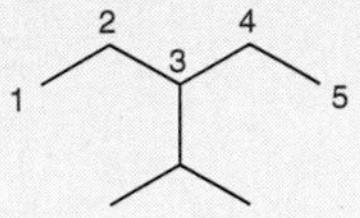
B. Cycloalkanes	
1. Count the number of carbons in the ring and assign a basis name to the cycloalkane corresponding to the IUPAC name of the unbranched alkane having the same number of carbons.	The compound shown contains five carbons in its ring. 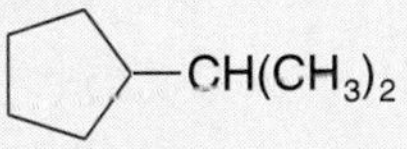 (cyclopentyl)$-CH(CH_3)_2$ It is named as a derivative of *cyclopentane.*
2. Name the alkyl group and append it as a prefix to the cycloalkane. No locant is needed if the compound is a monosubstituted cycloalkane. It is understood that the alkyl group is attached to C-1.	The compound shown above is *isopropylcyclopentane.* Alternatively, the alkyl group can be named according to the rules summarized in Table 2.8, whereupon the name becomes (*1-methylethyl*)*cyclopentane.* Parentheses are used to set off the name of the alkyl group as needed to avoid ambiguity.
3. When two or more different substituents are present, list them in alphabetical order and number the ring in the direction that gives the lower number at the first point of difference.	The compound shown is *1,1-diethyl-4-hexylcyclooctane.* 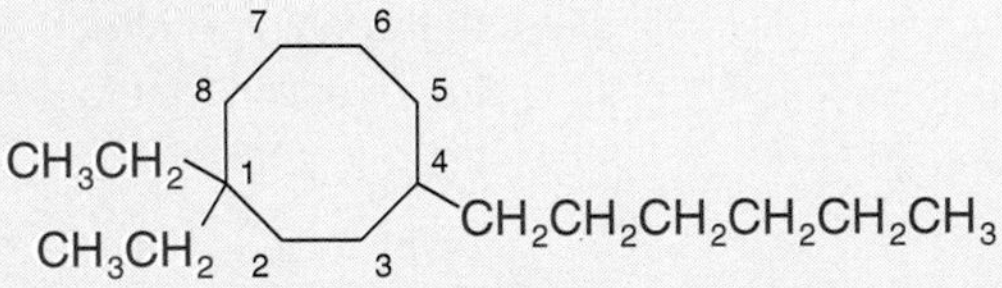 [cyclooctane ring numbered 1–8 with CH_3CH_2, CH_3CH_2 at C-1 and $CH_2CH_2CH_2CH_2CH_2CH_3$ at C-4]
4. Name the compound as a cycloalkyl-substituted alkane if the substituent has more carbons than the ring.	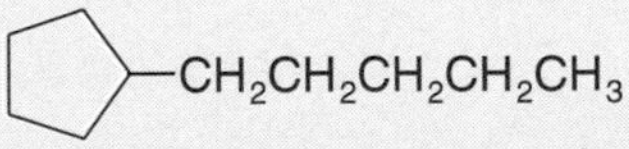 (cyclopentyl)$-CH_2CH_2CH_2CH_2CH_3$ is *pentylcyclopentane* but 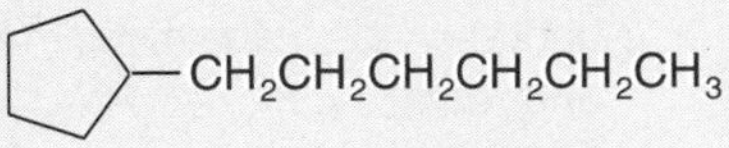 (cyclopentyl)$-CH_2CH_2CH_2CH_2CH_2CH_3$ is *1-cyclopentylhexane*

and are designated by the prefix *n*- in their common name (Section 2.8). The prefixes *n*- and "iso" are joined by "neo" in the common names of the three isomeric C_5H_{12} alkanes (Section 2.9):

$CH_3CH_2CH_2CH_2CH_3$	$(CH_3)_2CHCH_2CH_3$	$(CH_3)_4C$
n-Pentane	Isopentane	Neopentane

A single alkane may have different names; a name may be a common name, or it may be a *systematic* name developed by a well-defined set of rules. The system most widely used in chemistry is **IUPAC nomenclature** (Sections 2.10 through 2.13).

Cycloalkanes are alkanes in which a ring is present; they have the molecular formula C_nH_{2n} (Section 2.14). The IUPAC rules for alkanes and cycloalkanes are summarized in Table 2.7. The rules for alkyl groups are summarized in Table 2.8.

Natural gas is an abundant source of low-molecular-weight alkanes, while petroleum is a mixture of liquid hydrocarbons and includes many alkanes. Alkanes also occur naturally in the waxy coating of leaves and fruits (Section 2.15).

Alkanes and cycloalkanes are essentially nonpolar and are insoluble in water. The only forces of attraction between nonpolar molecules are relatively weak **induced dipole–induced dipole attractions.** Branched alkanes have lower boiling points than their unbranched isomers because their smaller surface area affords fewer points of contact between molecules (Section 2.16). There is a limit to how closely two molecules can approach each other, which is given by the sum of their **van der Waals radii.**

Alkanes and cycloalkanes burn in air to give carbon dioxide, water, and heat. This process is called **combustion** (Section 2.17).

TABLE 2.8
Summary of IUPAC Nomenclature of Alkyl Groups

Rule	Example
1. Number the carbon atoms beginning at the point of attachment, proceeding in the direction that follows the longest continuous chain.	The longest continuous chain that begins at the point of attachment in the group shown contains six carbons. $CH_3CH_2CH_2\overset{1|\,2}{C}(CH_3)CH_2\overset{3}{}\overset{4}{C}H(CH_3)\overset{5}{C}H_2\overset{6}{C}H_3$ (numbered 1–6 from the point of attachment; CH_3 groups on C-1 and C-3)
2. Assign a basis name according to the number of carbons in the corresponding unbranched alkane. Drop the ending *-ane* and replace it by *-yl*.	The alkyl group shown in step 1 is named as a substituted *hexyl* group.
3. List the substituents on the basis group in alphabetical order using replicating prefixes when necessary.	The alkyl group in step 1 is a *dimethylpropylhexyl* group.
4. Locate the substituents according to the numbering of the main chain described in step 1.	The alkyl group is a *1,3-dimethyl-1-propylhexyl* group.

$$\underset{\text{2-Methylbutane}}{(CH_3)_2CHCH_2CH_3} + \underset{\text{Oxygen}}{8O_2} \longrightarrow \underset{\text{Carbon dioxide}}{5CO_2} + \underset{\text{Water}}{6H_2O}$$

$$\Delta H^\circ = -3529 \text{ kJ } (-843.4 \text{ kcal})$$

The heat evolved on burning an alkane can be measured very accurately and increases with the number of carbon atoms. The relative stability of isomers may be determined by comparing their respective **heats of combustion.** The heat of combustion is the heat evolved in the reaction ($-\Delta H^\circ$) per mole of alkane. The more stable isomer has the lower heat of combustion.

Combustion of alkanes is an example of an **oxidation-reduction** reaction (Section 2.18). Carbon is oxidized; oxygen is reduced. While it is possible to calculate oxidation numbers of carbon in organic molecules, it is more convenient to regard oxidation of an organic substance as an increase in its oxygen content and/or a decrease in its hydrogen content.

PROBLEMS

2.16 Write structural formulas and give the IUPAC names for the nine alkanes that have the molecular formula C_7H_{16}.

2.17 From among the 18 constitutional isomers of C_8H_{18}, write structural formulas and give the IUPAC names for those that are named as derivatives of

(*a*) Heptane
(*b*) Hexane
(*c*) Pentane
(*d*) Butane

2.18 Write a structural formula for each of the following compounds:

(*a*) 6-Isopropyl-2,3-dimethylnonane
(*b*) 4-*tert*-Butyl-3-methylheptane
(*c*) 4-Isobutyl-1,1-dimethylcyclohexane
(*d*) *sec*-Butylcycloheptane
(*e*) Cyclobutylcyclopentane
(*f*) (2,2-Dimethylpropyl)cyclohexane
(*g*) Pentacosane
(*h*) 10-(1-methylpentyl)pentacosane

2.19 Give the IUPAC name for each of the following compounds:

(*a*) $CH_3(CH_2)_{25}CH_3$
(*b*) $(CH_3)_2CHCH_2(CH_2)_{14}CH_3$
(*c*) $(CH_3CH_2)_3CCH(CH_2CH_3)_2$
(*d*)
(*e*)
(*f*) 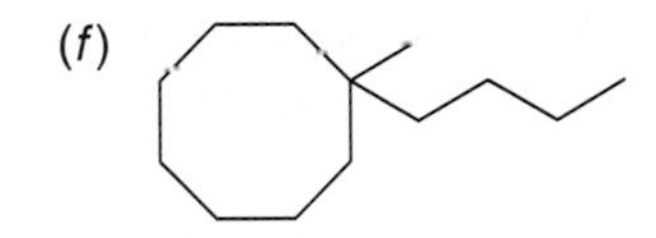
(*g*) 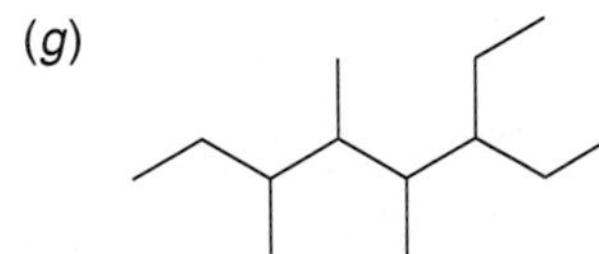

2.20 All the parts of this problem refer to the alkane having the carbon skeleton shown.

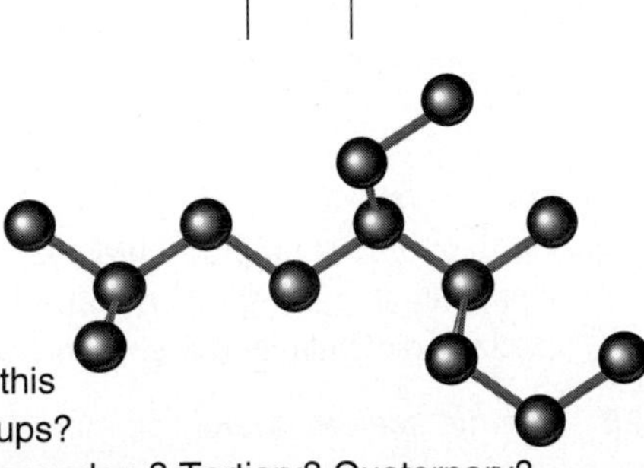

(*a*) What is the molecular formula of this alkane?
(*b*) What is its IUPAC name?
(*c*) How many methyl groups are present in this alkane? Methylene groups? Methine groups?
(*d*) How many carbon atoms are primary? Secondary? Tertiary? Quaternary?

2.21 Give the IUPAC name for each of the following alkyl groups and classify each one as primary, secondary, or tertiary:

(*a*) $CH_3(CH_2)_{10}CH_2—$

(*b*) $—CH_2CH_2CHCH_2CH_2CH_3$ (with CH_2CH_3 on the CH)

(*c*) $—C(CH_2CH_3)_3$

(*d*) $—CHCH_2CH_2CH_3$ (with a cyclopropyl ring on the CH)

(*e*) cyclohexyl$—CH_2CH_2—$

(*f*) cyclohexyl$—CH—$ (with CH_3 on the CH)

2.22 *Pristane* is an alkane that is present to the extent of about 14 percent in shark liver oil. Its IUPAC name is 2,6,10,14-tetramethylpentadecane. Write its structural formula.

2.23 How many σ bonds are there in pentane? In cyclopentane?

2.24 Hectane is the IUPAC name for the unbranched alkane that contains 100 carbon atoms.

(*a*) How many σ bonds are there in hectane?
(*b*) How many alkanes have names of the type *x*-methylhectane?
(*c*) How many alkanes have names of the type 2,*x*-dimethylhectane?
(*d*) Including those in the 1*s* level of carbon, how many electrons are there in hectane?

2.25 Which of the compounds in each of the following groups are isomers?

(*a*) Butane, cyclobutane, isobutane, 2-methylbutane
(*b*) Cyclopentane, neopentane, 2,2-dimethylpentane, 2,2,3-trimethylbutane
(*c*) Cyclohexane, hexane, methylcyclopentane, 1,1,2-trimethylcyclopropane
(*d*) Ethylcyclopropane, 1,1-dimethylcyclopropane, 1-cyclopropylpropane, cyclopentane
(*e*) 4-Methyltetradecane, 2,3,4,5-tetramethyldecane, pentadecane, 4-cyclobutyldecane

2.26 *Epichlorohydrin* is the common name of an industrial chemical used as a component in epoxy cement. The molecular formula of epichlorohydrin is C_3H_5ClO. Epichlorohydrin has an epoxide functional group; it does not have a methyl group. Write a structural formula for epichlorohydrin.

2.27

(*a*) Complete the structure of the pain-relieving drug *ibuprofen* on the basis of the fact that ibuprofen is a carboxylic acid, X is an isobutyl group, and Y is a methyl group.

X—(benzene ring)—CH—Z (with Y on the CH)

(*b*) *Mandelonitrile* may be obtained from peach flowers. Derive its structure from the template in part (*a*) given that X is hydrogen, Y is the functional group that characterizes alcohols, and Z characterizes nitriles.

2.28 *Isoamyl acetate* is the common name of the substance most responsible for the characteristic odor of bananas. Write a structural formula for isoamyl acetate given the information that it is an ester in which the carbonyl group bears a methyl substituent and there is a 3-methylbutyl group attached to one of the oxygens.

2.29 n-*Butyl mercaptan* is the common name of a foul-smelling substance obtained from skunk fluid. It is a thiol of the type RX, where R is an *n*-butyl group and X is the functional group that characterizes a thiol. Write a structural formula for this substance.

2.30 Write the structural formula of a compound of molecular formula $C_4H_8Cl_2$ in which

(*a*) All the carbons belong to methylene groups
(*b*) None of the carbons belong to methylene groups

2.31 Write the structural formula of a compound of molecular formula $C_5H_8Cl_4$ in which there are one quaternary carbon and four methylene groups.

2.32 A certain alkane isolated from a species of blue-green alga has a molecular weight of 240 and an unbranched carbon chain. Identify this alkane.

2.33 Female tiger moths signify their presence to male moths by giving off a sex attractant. The sex attractant has been isolated and found to be a 2-methyl-branched alkane having a molecular weight of 254. What is this material?

2.34 Write a balanced chemical equation for the combustion of each of the following compounds:

(*a*) Decane
(*b*) Cyclodecane
(*c*) Methylcyclononane
(*d*) Cyclopentylcyclopentane

2.35 The heats of combustion of methane and butane are 890 kJ/mol (212.8 kcal/mol) and 2876 kJ/mol (687.4 kcal/mol), respectively. When used as a fuel, would methane or butane generate more heat for the same mass of gas? Which would generate more heat for the same volume of gas?

2.36 In each of the following groups of compounds, identify the one with the largest heat of combustion and the one with the smallest. (Try to do this problem without consulting Table 2.5.)

(*a*) Hexane, heptane, octane
(*b*) Isobutane, pentane, isopentane
(*c*) Isopentane, 2-methylpentane, neopentane
(*d*) Pentane, 3-methylpentane, 3,3-dimethylpentane
(*e*) Ethylcyclopentane, ethylcyclohexane, ethylcycloheptane

2.37

(*a*) Given $\Delta H°$ for the reaction

$$H_2(g) + \tfrac{1}{2}O_2(g) \longrightarrow H_2O(l) \qquad \Delta H° = -286 \text{ kJ}$$

along with the information that the heat of combustion of ethane is 1560 kJ/mol and that of ethylene is 1410 kJ/mol, calculate $\Delta H°$ for the hydrogenation of ethylene:

$$CH_2{=}CH_2(g) + H_2(g) \longrightarrow CH_3CH_3(g)$$

(*b*) If the heat of combustion of acetylene is 1300 kJ/mol, what is the value of $\Delta H°$ for its hydrogenation to ethylene? To ethane?

(*c*) What is the value of $\Delta H°$ for the hypothetical reaction

$$2CH_2{=}CH_2(g) \longrightarrow CH_3CH_3(g) + HC{\equiv}CH(g)$$

2.38 Each of the following reactions will be encountered at some point in this text. Classify each one according to whether the organic substrate is oxidized or reduced in the process.

(*a*) $CH_3C{\equiv}CH + 2Na + 2NH_3 \longrightarrow CH_3CH{=}CH_2 + 2NaNH_2$

(*b*) $3\,CH_3CH(OH)CH_2CH_3 + Cr_2O_7^{2-} + 8H^+ \longrightarrow 3\,CH_3C(=O)CH_2CH_3 + 2Cr^{3+} + 7H_2O$

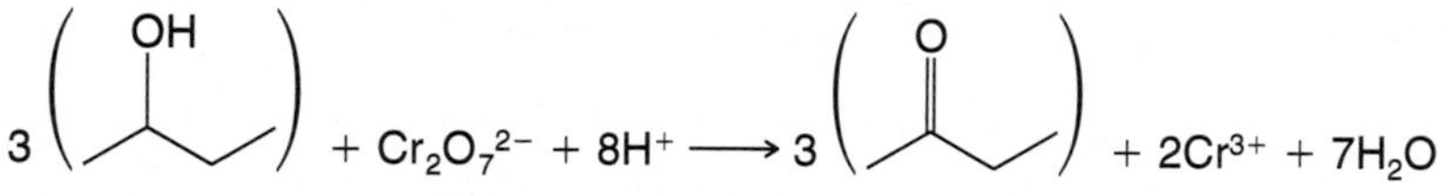

(*c*) $HOCH_2CH_2OH + HIO_4 \longrightarrow 2CH_2{=}O + HIO_3 + H_2O$

(*d*) $C_6H_5{-}NO_2 + 2Fe + 7H^+ \longrightarrow C_6H_5{-}\overset{+}{N}H_3 + 2Fe^{3+} + 2H_2O$

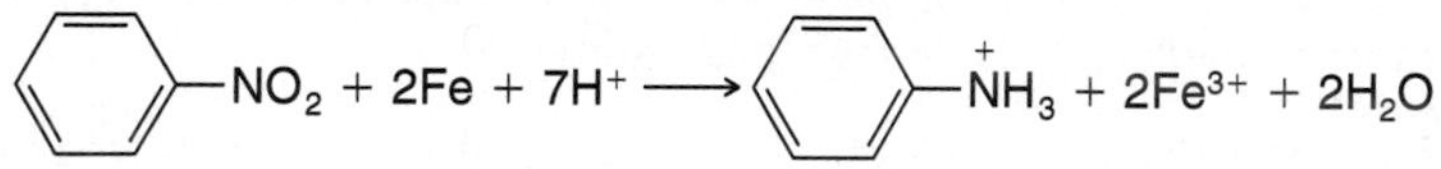

2.39 The reaction shown is important in the industrial preparation of dichlorodimethylsilane for eventual conversion to silicone polymers.

$$2CH_3Cl + Si \longrightarrow (CH_3)_2SiCl_2$$

Is carbon oxidized, or is it reduced in this reaction?

2.40 Compound A undergoes the following reactions:

$$CH_3\overset{O}{\overset{\|}{C}}C(CH_3)_3 \text{ (Compound A)} \longrightarrow CH_3\overset{O}{\overset{\|}{C}}OC(CH_3)_3$$

$$CH_3\overset{O}{\overset{\|}{C}}C(CH_3)_3 \longrightarrow CH_3CH_2C(CH_3)_3$$

$$CH_3\overset{O}{\overset{\|}{C}}C(CH_3)_3 \longrightarrow CH_3\underset{OH}{\underset{|}{C}H}C(CH_3)_3$$

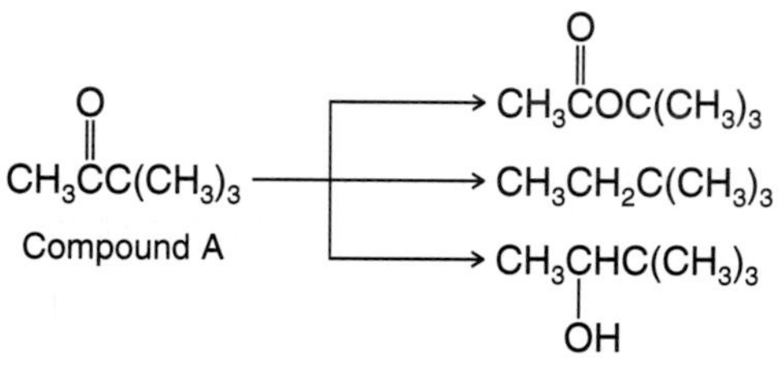

(*a*) To what class of compounds does compound A belong?
(*b*) Which of the reactions shown require(s) an oxidizing agent?
(*c*) Which of the reactions shown require(s) a reducing agent?
(*d*) Identify the class to which each of the reaction products belongs.

MOLECULAR MODELING EXERCISES

2.41 Construct molecular models for:

(*a*) two aldehydes
(*b*) a ketone

each of which has three sp^3 and one sp^2 hybridized carbon.

2.42 Verify, by a making a molecular model, that the carbon chain of an unbranched alkane is not "straight." What structural feature of sp^3-hybridized carbon is responsible for this fact?

What kind of hydrocarbon could have a "straight" carbon chain? Make a molecular model of such a hydrocarbon.

2.43 Construct molecular models for the three isomers of C_5H_{12}.

2.44 Construct a molecular model for *tert*-butylcyclopentane.

2.45 Construct a molecular model for an alcohol in which oxygen is bonded to a 1-methylbutyl group.

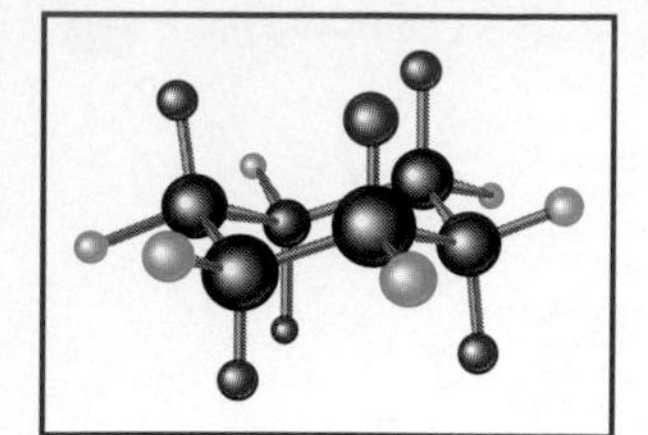

CHAPTER 3

CONFORMATIONS OF ALKANES AND CYCLOALKANES

Hydrogen peroxide is formed in the cells of plants and animals but is toxic to them. Consequently, living systems have developed mechanisms to rid themselves of hydrogen peroxide, usually by enzyme-catalyzed reduction to water. An understanding of how reactions take place, be they reactions in living systems or reactions in test tubes, begins with a thorough knowledge of the structure of the reactants, products, and catalysts. Even a simple molecule such as hydrogen peroxide may be structurally more complicated than you think. Suppose we wished to write the structural formula for H_2O_2 in sufficient detail to show the positions of the atoms relative to one another. We could write two different planar geometries A and B that differ by a 180° rotation about the O—O bond. We could also write an infinite number of nonplanar structures, of which C is but one example, that differ from one another by tiny increments of rotation about the O—O bond.

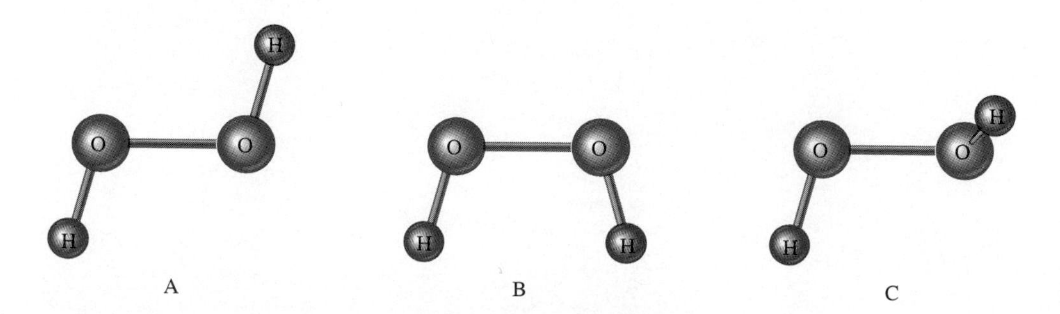

Structures A, B, and C represent different **conformations** of hydrogen peroxide. *Conformations are different spatial arrangements of a molecule that are generated by*

rotation about single bonds. Structural studies have determined that the most stable structure for hydrogen peroxide is C.

In this chapter we expand our structural perspective by examining the conformations that are adopted by individual molecules. We will focus most of our attention on three organic molecules: *ethane, butane,* and *cyclohexane.* Through a detailed discussion of these and related substances the fundamental principles of **conformational analysis** will be developed.

The particular conformation that a molecule adopts can exert a profound influence on its properties. Conformational analysis is a tool used not only by organic chemists but also by research workers in the life sciences as they attempt to develop a clearer picture of how molecules—from simple to very complex—interact with one another in living systems.

3.1 CONFORMATIONAL ANALYSIS OF ETHANE

Ethane is the simplest hydrocarbon that possesses distinct conformations. Two, called the **staggered conformation** and the **eclipsed conformation,** merit special attention and are illustrated in Figure 3.1. The C—H bonds in the staggered conformation are arranged so that each one bisects the angle defined by two C—H bonds on the adjacent carbon. In the eclipsed conformation each C—H bond is aligned with a C—H bond on the adjacent carbon. The staggered and eclipsed conformations are interconvertible by rotation of one carbon with respect to the other around the bond that connects them. Different conformations of the same molecule are sometimes referred to as **conformers** or **rotamers.**

Among the various ways in which the staggered and eclipsed forms are portrayed, wedge-and-dash, sawhorse, and Newman projection drawings are especially useful. These are shown for the staggered conformation of ethane in Figure 3.2 and for the eclipsed conformation in Figure 3.3.

We have used *wedge-and-dash* representations in earlier chapters, and so Figures 3.2*a* and 3.3*a* require no special explanation. A *sawhorse* drawing (Figures 3.2*b* and 3.3*b*) makes it possible to depict the conformation of a molecule without resorting to

You will find it helpful to construct molecular models of the staggered and eclipsed conformations of ethane for comparison purposes.

Staggered conformation of ethane:

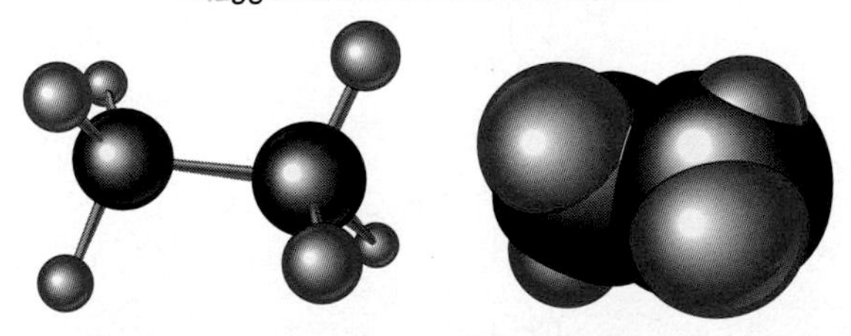

Eclipsed conformation of ethane:

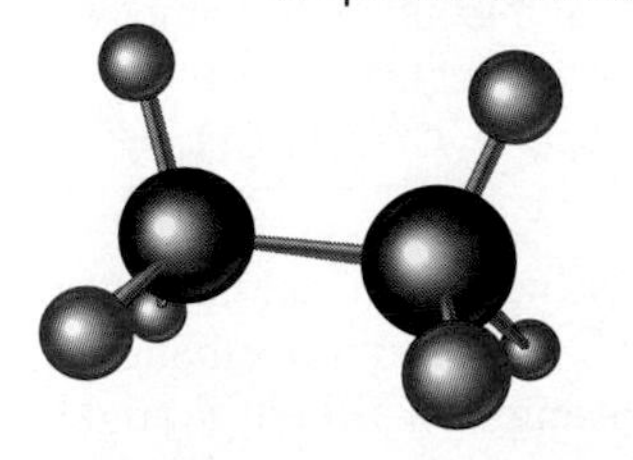

FIGURE 3.1 The staggered and eclipsed conformations of ethane shown as ball-and-stick models (left) and as space-filling models (right).

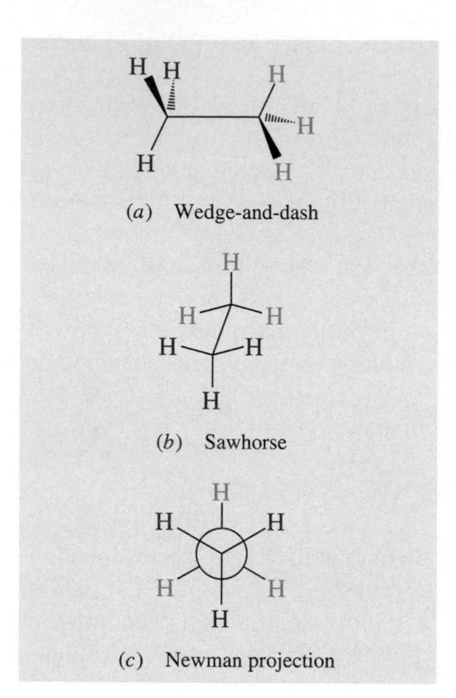

FIGURE 3.2 Some commonly used representations of the staggered conformation of ethane.

different styles of bonds. *Newman projection* formulas were devised by Professor M. S. Newman of Ohio State University. In a Newman projection of ethane (Figures 3.2*c* and 3.3*c*), we sight down the carbon-carbon bond, and represent the front carbon by a point and the back carbon by a circle. Each carbon has three substituents that are placed symmetrically around it.

PROBLEM 3.1 Identify the alkanes corresponding to each of the representations shown.

(*a*) Newman projection: CH_3, H, H, H, H, H

(*b*) Sawhorse: CH_3, H, H, H, H, CH_3

(*c*) Newman projection: CH_3, H, H, H, CH_3, CH_3

(*d*) Sawhorse: CH_3, H, CH_2CH_3, H, H, CH_2CH_3

SAMPLE SOLUTION (*a*) The Newman projection formula of this alkane resembles that of ethane except that one of the hydrogen substituents has been replaced by a methyl group. The drawing is a Newman projection formula of propane, $CH_3CH_2CH_3$.

The structural feature that Figures 3.2 and 3.3 illustrate is the spatial relationship between atoms on adjacent carbon atoms. Each H—C—C—H unit in ethane is characterized by a *torsion angle* or *dihedral angle.* Torsion angle or dihedral angle

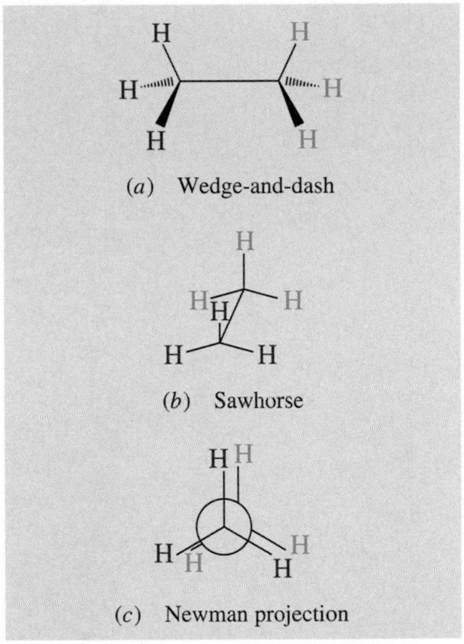

(a) Wedge-and-dash

(b) Sawhorse

(c) Newman projection

FIGURE 3.3 Some commonly used representations of the eclipsed conformation of ethane.

is the angle between the H—C—C plane and the C—C—H plane of an H—C—C—H unit. It is easily seen in a Newman projection of ethane as the angle between C—H bonds of adjacent carbons.

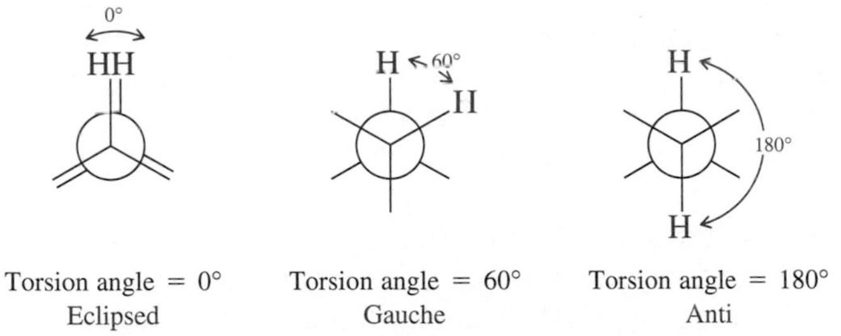

Torsion angle = 0°	Torsion angle = 60°	Torsion angle = 180°
Eclipsed	Gauche	Anti

Eclipsed bonds are characterized by a torsion angle of 0°. When the torsion angle is approximately 60°, we say that the spatial relationship is **gauche;** and when it is 180° we say that it is **anti**. Staggered conformations have only gauche or anti relationships between bonds on adjacent atoms.

In principle there are an infinite number of conformations of ethane, differing by only tiny increments in their torsion angles. This raises the question whether any one conformation is more stable than the others, and if so, which one. A related question has to do with interconversion of conformations. Rotation about the carbon-carbon bond of ethane converts a staggered conformation to an eclipsed one. How rapidly do these conformations interconvert?

3.2 INTERNAL ROTATION IN ETHANE

Rotation about the carbon-carbon bond in ethane, while very rapid, is not completely "free." Figure 3.4 shows how the potential energy of ethane changes as one of its methyl groups rotates through an angle of 360° around the carbon-carbon bond. Dia-

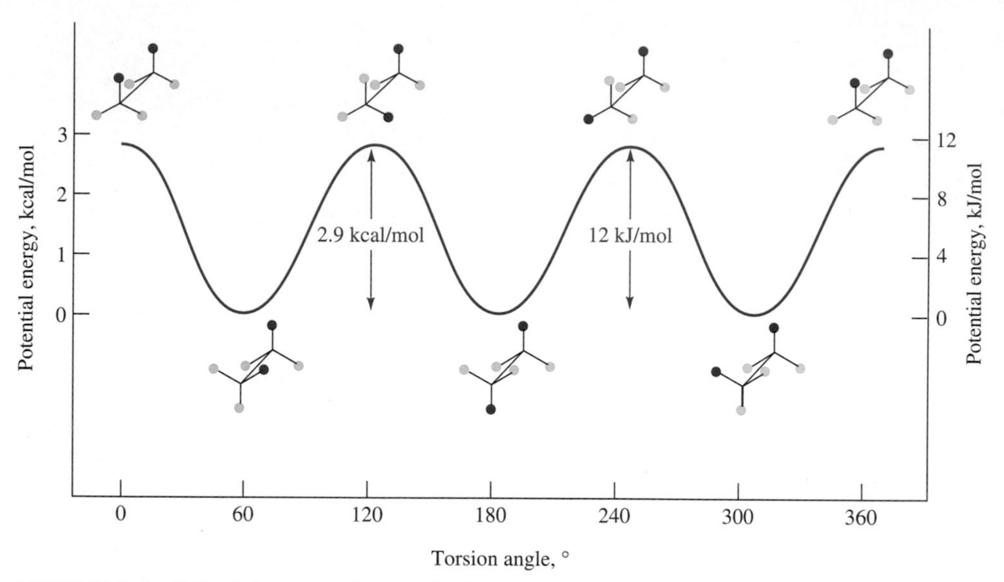

FIGURE 3.4 Potential energy diagram for internal rotation in ethane. Hydrogen substituents are represented by circles, two of which are shown in red so as to indicate more clearly the process of rotation about the carbon-carbon bond.

grams such as this, in which we plot the progress of a particular transformation as the x axis versus potential energy as the y axis, can aid the understanding of chemical and physical processes. As noted earlier in Section 2.17, potential energy is the energy a molecule possesses other than its kinetic energy. When comparing conformations, the one with the least potential energy is the most stable.

Figure 3.4 shows that the staggered conformations of ethane are 12 kJ/mol (2.9 kcal/mol) lower in energy than the eclipsed conformations. The three equivalent staggered conformations correspond to potential energy minima, while the three equivalent eclipsed conformations represent potential energy maxima.

What makes the staggered conformation of ethane more stable than the eclipsed conformation? A number of factors contribute, but the most important one seems to be that the staggered conformation allows for the maximum separation of bonded electron pairs. Electron pair repulsions are greatest when the bonds are eclipsed and least when the bonds are staggered. We call the destabilization associated with the eclipsing of bonds on adjacent atoms **torsional strain.** Since three pairs of eclipsed bonds produce 12 kJ/mol (2.9 kcal/mol) of torsional strain in ethane, it is reasonable to assign an "energy cost" of 4 kJ/mol (1 kcal/mol) to an alkane for each pair of eclipsed bonds it contains.

The difference in energy between the staggered and eclipsed conformation of ethane is the **energy of activation** (E_{act}) for the conversion of one staggered form to another staggered form. Every dynamic process in chemistry has an activation energy associated with it. Activation energy is the energy that a molecule must possess above the energy of its ground state in order to be transformed into some other species. Molecules must become energized in order to undergo chemical reaction or, as in this case, to undergo rotation around a carbon-carbon bond. Kinetic (thermal) energy is absorbed by a molecule from its surroundings and is transformed into potential energy. When the potential energy exceeds E_{act}, the unstable arrangement of atoms that exists at that instant can relax to a more stable structure, giving off its excess potential

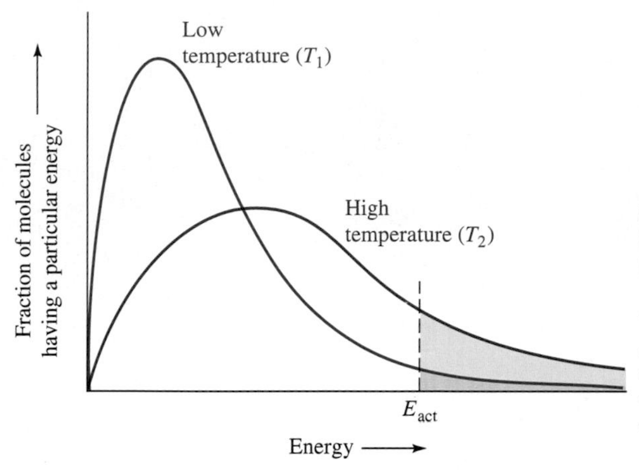

FIGURE 3.5 Distribution of molecular energies. (*a*) The number of molecules with energy greater than E_{act} at temperature T_1 is shown as the darker-green shaded area. (*b*) At some higher temperature T_2, the shape of the energy distribution curve is different, and more molecules have energies in excess of E_{act}.

energy as kinetic energy in collisions with other molecules or with the walls of a container. The point of maximum potential energy encountered by the reactants as they proceed to products is called the **transition state.** The eclipsed conformation is the transition state for the conversion of one staggered conformation of ethane to another.

The structure that exists at the transition state is sometimes referred to as the *transition structure* or the *activated complex.*

As with all chemical processes, the rate of rotation about a carbon-carbon bond increases with temperature. The reason for this can be seen by inspecting Figure 3.5. When examining the distribution of energies among large numbers of molecules, we find that the energies of most of them are clustered around some average value; some molecules are less energetic than the average, and others more energetic. Only molecules with a potential energy greater than E_{act}, however, are able to surmount the transition state for conversion to product. The number of these molecules is given by the shaded areas under the curve in Figure 3.5. The energy distribution curve flattens out at higher temperatures, and a greater proportion of molecules have energies in excess of E_{act} at T_2 than at T_1. The effect of temperature is quite pronounced; an increase of only 10°C produces a two- to threefold increase in the rate of a typical chemical process.

The 12 kJ/mol (2.9 kcal/mol) activation energy for rotation about the carbon-carbon bond in ethane is quite small. The thermal energy available from the surroundings is sufficient to cause staggered conformations of ethane to interconvert millions of times each second at room temperature. Internal rotation in ethane is not completely ‘‘free’’ but is an exceedingly rapid process, and conformational equilibrium is reached almost instantly.

3.3 CONFORMATIONAL ANALYSIS OF BUTANE

The next alkane that we will examine is butane. In particular, we will consider conformations related by rotation about the bond between carbon 2 and carbon 3. Unlike the case of ethane, where the staggered conformations are equivalent, there are two different staggered conformations of butane, as shown in Figure 3.6. One has a gauche relationship between methyl groups, the other an anti relationship. Since both are staggered, they are free of torsional strain, but two of the hydrogens of the gauche

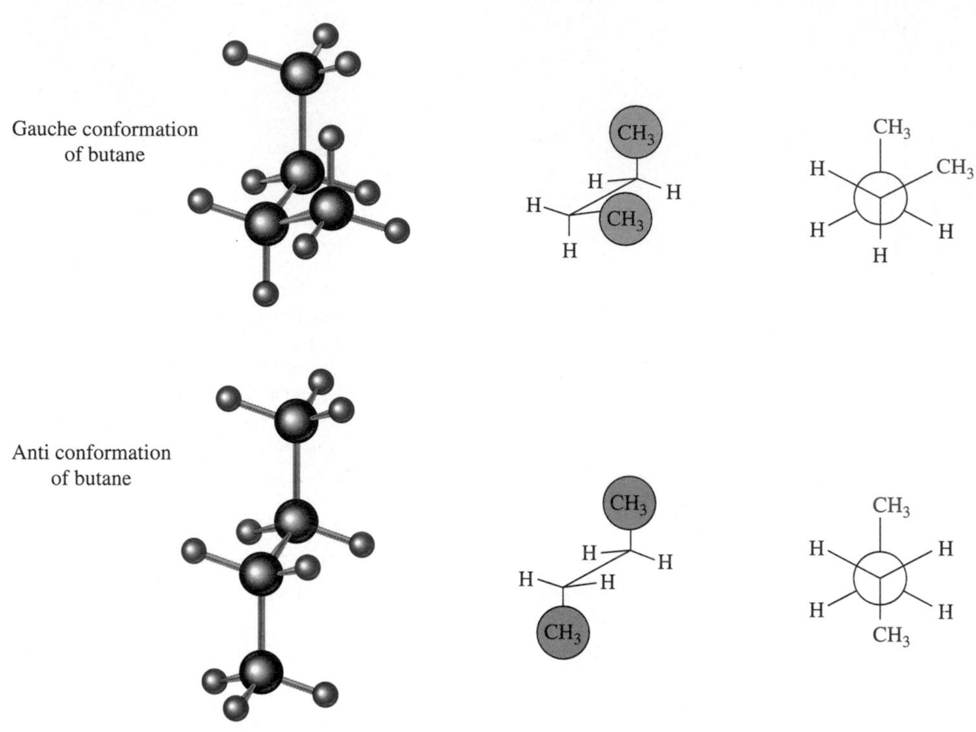

FIGURE 3.6 The gauche and anti conformations of butane shown as ball-and-stick models (left), sawhorse drawings (center), and Newman projections (right). The gauche conformation is less stable than the anti because of the van der Waals strain between the hydrogens shown in red (top left). These hydrogens are within 210 pm of each other, a distance less than the sum of their van der Waals radii (240 pm).

conformation (shown in red in Figure 3.6) are within 210 pm of each other. This distance is less than the sum of their van der Waals radii (240 pm), and there is a repulsive force between these two hydrogens. The destabilization of a molecule that results when two of its atoms are too close to each other is called **van der Waals strain, steric strain,** or **steric hindrance.** In the case of butane, van der Waals strain makes the gauche conformation approximately 3.2 kJ/mol (0.8 kcal/mol) less stable than the anti.

Figure 3.7 illustrates the potential energy relationships among the various conformations of butane. The staggered conformations are more stable than the eclipsed ones. At any instant, almost all the molecules exist in staggered conformations, and more are present in the anti conformation than in the gauche. The point of maximum potential energy lies some 25 kJ/mol (6.1 kcal/mol) above the anti conformation. The total strain in this structure is approximately equally divided between the torsional strain associated with three pairs of eclipsed bonds (12 kJ/mol; 2.9 kcal/mol) and the van der Waals strain between the methyl groups.

PROBLEM 3.2 Sketch a potential energy diagram for rotation around a carbon-carbon bond in propane. Clearly identify each potential energy maximum and minimum with a structural formula that shows the conformation of propane at that point. Does your diagram more closely resemble that of ethane or of butane? Would you expect the activation energy for internal rotation in propane to be more than or less than that of ethane? Of butane?

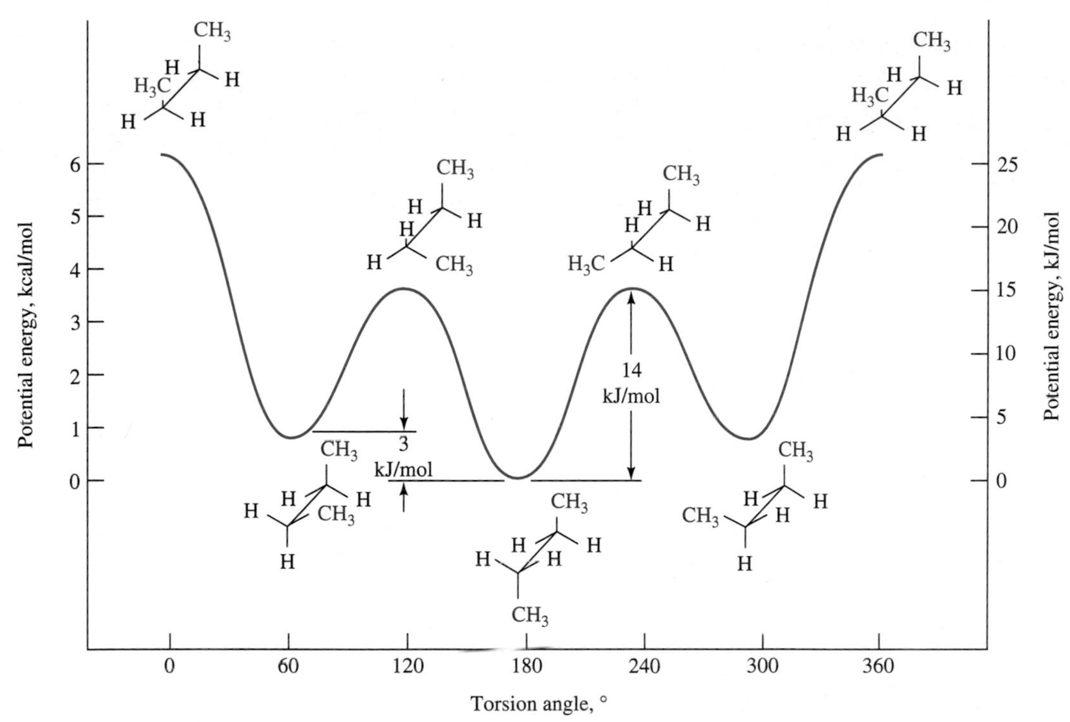

FIGURE 3.7 Potential energy diagram for rotation around the central carbon-carbon bond in butane.

MOLECULAR MECHANICS APPLIED TO ALKANES AND CYCLOALKANES

Of the numerous applications of computer technology to chemistry, one that has been enthusiastically embraced by organic chemists examines molecular structure from a perspective similar to that obtained by manipulating molecular models but with an additional quantitative dimension. *Molecular mechanics* is a computational method that allows us to assess the stability of a molecule by comparing selected features of its structure with those of ideal "unstrained" standards. Molecular mechanics makes no attempt to explain why the van der Waals radius of hydrogen is 120 pm, why the bond angles in methane are 109.5°, why the C—C bond distance in ethane is 153 pm, or why the staggered conformation of ethane is 12 kJ/mol more stable than the eclipsed, but uses these and related experimental observations as standards with which the corresponding features of other substances are compared.

If we assume that there are certain "ideal" values for bond angles, bond distances, etc., it follows that deviations from these ideal values will destabilize a particular structure and increase its potential energy. This increase in potential energy is referred to as the **strain energy** of the structure. Arithmetically, the total strain energy (E_s) of an alkane or cycloalkane can be considered as

$$E_s = E_{\text{bond stretching}} + E_{\text{angle bending}} + E_{\text{torsional}} + E_{\text{van der Waals}}$$

where

$E_{\text{bond stretching}}$ is the strain energy increment that results when C—C and C—H bond distances are distorted from their ideal values of 153 pm and 111 pm, respectively.

$E_{\text{angle bending}}$ is the strain that results from the expansion or contraction of bond angles from the normal values of 109.5° for sp^3 hybridized carbon.

$E_{\text{torsional}}$ is the portion of the total strain resulting from deviation of torsion angles from their stable staggered relationship.

$E_{\text{van der Waals}}$ is the energy associated with "nonbonded interactions."

Nonbonded interactions are the forces between atoms that are not bonded to one another; they may be either attractive or repulsive. It often happens that the overall molecular shape may cause two atoms to be close in space even though they are separated from each other by many bonds. Induced dipole–induced dipole interactions make van der Waals forces in alkanes weakly attractive at most distances, but when two atoms are closer to each other than the sum of their van der Waals radii, nuclear-nuclear and electron-electron repulsive forces between them dominate the $E_{\text{van der Waals}}$ term. The resulting destabilization is referred to as **van der Waals strain** or **steric strain.**

At its most basic level, separating the total strain of a structure into its components is a qualitative exercise. For example, a computer-drawn molecule of the eclipsed conformation of butane using ideal bond angles and bond distances reveals that the pairs of hydrogens shown in red in Figure 3.8 are separated by a distance of only 175 pm, a value considerably smaller than the sum of their van der Waals radii (2 × 120 pm = 240 pm). Thus, this conformation is destabilized not only by the torsional strain associated with its eclipsed bonds, but also by van der Waals strain.

At a higher level, molecular mechanics is applied quantitatively to strain energy calculations. Each component of strain is separately described by a mathematical expression developed and refined so that it gives solutions that match experimental observations for reference molecules. One then uses these empirically derived and tested expressions to calculate the most stable structure of a substance. The various structural features are interdependent; van der Waals strain, for example, might be decreased at the expense of introducing some angle strain, torsional strain, or both. The computer program is written so as to search for the combination of bond angles, distances, torsion angles, and nonbonded interactions that gives the molecule the lowest total strain energy. This procedure is called *strain energy minimization* and is based on the commonsense notion that the most stable structure is the one that has the least strain.

The most widely used molecular mechanics program was developed by Professor N. L. Allinger of the University of Georgia and is known in its latest version as *MM3.* MM3 has been refined to the extent that many structural features can be calculated more easily and more accurately than they can be measured experimentally.

Once requiring minicomputers and workstations, some molecular mechanics programs have been developed for personal computers. The information that strain energy calculations can provide is so helpful that molecular mechanics is no longer considered a novelty but rather as one more tool to be used by the practicing organic chemist.

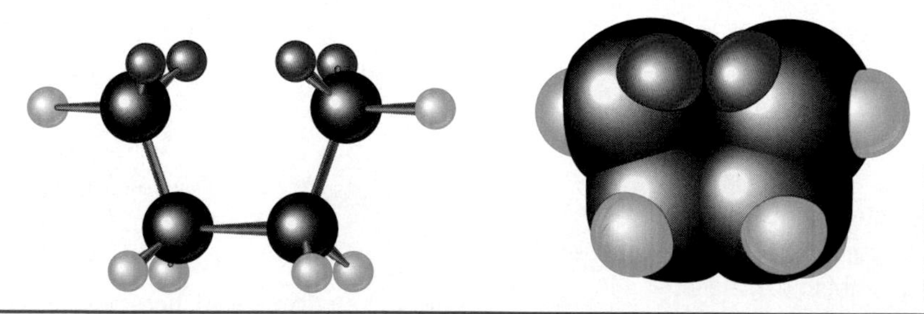

FIGURE 3.8 Ball-and-stick and space-filling models of the methyl-methyl eclipsed conformation of butane.

3.4 CONFORMATIONS OF HIGHER ALKANES

Higher alkanes having unbranched carbon chains are, like butane, most stable in their all-anti conformations. The potential energy difference between gauche and anti conformations, however, is similar to that of butane, and appreciable quantities of the gauche conformation are present in liquid alkanes at 25°C. In depicting the conformations of higher alkanes it is often more helpful to look at them from the side rather than end on as in a Newman projection. Viewed from this perspective, the most stable conformations of pentane and hexane have their carbon "backbones" arranged in a

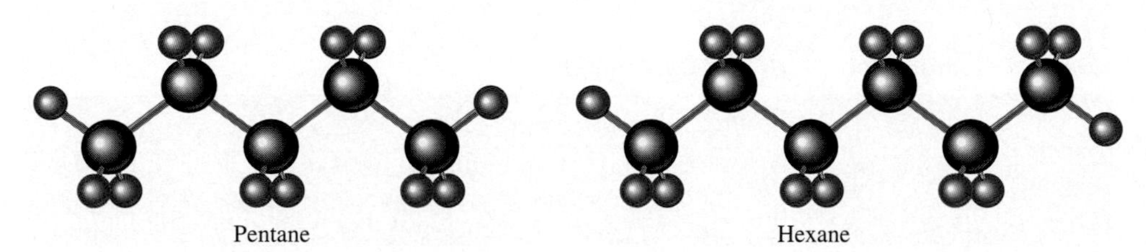

FIGURE 3.9 Ball-and-stick models of pentane and hexane in their all-anti (zigzag) conformations.

zigzag fashion, as shown in Figure 3.9. All the bonds are staggered, and the chains are characterized by anti arrangements of C—C—C—C units.

3.5 THE SHAPES OF CYCLOALKANES: PLANAR OR NONPLANAR?

During the nineteenth century it was widely believed—erroneously, as we shall see—that the carbon skeletons of cycloalkanes are planar. A prominent advocate of this point of view was the German chemist Adolf von Baeyer. Noting that rings containing fewer than five or more than six carbons seemed to be less abundant among naturally occurring materials as well as less stable than those related to cyclopentane and cyclohexane, Baeyer suggested that stability was related to how closely the angles of the corresponding planar regular polygons matched the tetrahedral value of 109.5°. Since the 60° bond angles required by the geometry of cyclopropane deviate by 49.5° from the tetrahedral value, Baeyer suggested that the *strain* in the three-membered ring of cyclopropane is responsible for its decreased stability relative to cyclopentane and cyclohexane. Similarly, each C—C—C angle of 90° in planar cyclobutane deviates by 19.5° from the ideal tetrahedral value. Cyclobutane, like cyclopropane, is said to be destabilized because of *angle strain*. **Angle strain** is the strain a molecule has because one or more of its bond angles deviate from the ideal value; in the case of alkanes the ideal value is 109.5°.

According to Baeyer, cyclopentane should be the most stable of all the cycloalkanes because the ring angles of a planar pentagon, 108°, are closer to the tetrahedral angle than those of any other cycloalkane. A prediction of the *Baeyer strain theory* is that the cycloalkanes beyond cyclopentane should become increasingly strained and correspondingly less stable. The angles of a regular hexagon are 120°, and the angles of larger polygons deviate more and more from the ideal tetrahedral angle.

While better known now for his incorrect theory that cycloalkanes were planar, Baeyer was responsible for notable advances in the chemistry of organic dyes such as indigo and was awarded the 1905 Nobel Prize in chemistry for his work in that area.

PROBLEM 3.3 The angles of a regular planar polygon of n sides are equal to

$$\frac{n-2}{n}(180°)$$

If cyclododecane had a planar carbon skeleton, how large would its C—C—C bond angles have to be?

As was done with isomeric alkanes in Section 2.17, we can employ heats of combustion to probe the relative energies of cycloalkanes. By so doing, some of the inconsistencies of the Baeyer strain theory will become evident. Table 3.1 lists the experimentally measured heats of combustion for a number of cycloalkanes. The most important column in the table is the heat of combustion per methylene (CH_2) group. Cyclopropane has the highest heat of combustion per methylene group, which is con-

TABLE 3.1
Heats of Combustion ($-\Delta H°$) of Cycloalkanes

Cycloalkane	Number of CH_2 groups	Heat of combustion		Heat of combustion per CH_2 group	
		kJ/mol	(kcal/mol)	kJ/mol	(kcal/mol)
Cyclopropane	3	2,091	(499.8)	697	(166.6)
Cyclobutane	4	2,721	(650.3)	681	(162.7)
Cyclopentane	5	3,291	(786.6)	658	(157.3)
Cyclohexane	6	3,920	(936.8)	653	(156.1)
Cycloheptane	7	4,599	(1099.2)	657	(157.0)
Cyclooctane	8	5,267	(1258.8)	658	(157.3)
Cyclononane	9	5,933	(1418.0)	659	(157.5)
Cyclodecane	10	6,587	(1574.3)	659	(157.5)
Cycloundecane	11	7,237	(1729.8)	658	(157.3)
Cyclododecane	12	7,845	(1875.1)	654	(156.2)
Cyclotetradecane	14	9,139	(2184.2)	653	(156.0)
Cyclohexadecane	16	10,466	(2501.4)	654	(156.3)

sistent with the idea that its internal energy is elevated because of angle strain. Cyclobutane has less angle strain at each of its carbon atoms and a lower heat of combustion per methylene group. Cyclopentane, as expected, has a lower value still. Notice, however, that contrary to the prediction of the Baeyer strain theory, cyclohexane has a smaller heat of combustion per methylene group than does cyclopentane. Each methylene group contributes a smaller increment to the internal energy of cyclohexane than the methylene units of cyclopentane. If bond angle distortion were greater in cyclohexane than in cyclopentane, the opposite would have been observed.

Furthermore, the heats of combustion per methylene group of the very large rings are all about the same and similar to that of cyclohexane. Rather than rising because of increasing angle strain in large rings, the heat of combustion per methylene group remains constant at approximately 653 kJ/mol (156 kcal/mol), the value cited in Section 2.17 as the difference between successive members of a homologous series of alkanes. Therefore, we conclude that the bond angles of large cycloalkanes are not much different from the bond angles of alkanes themselves. The prediction of the Baeyer strain theory that angle strain increases steadily with ring size is contradicted by experimental fact.

The Baeyer strain theory is useful to us in identifying angle strain as a destabilizing effect. It contains a fundamental flaw, however, in its assumption that the rings of cycloalkanes are planar. *With the exception of cyclopropane, cycloalkanes are nonplanar.* Sections 3.6 to 3.12 describe the shapes of cycloalkanes. Six-membered rings rank as the most important ring size among the organic compounds; thus let us begin with cyclohexane to examine the forces that determine the shapes of cycloalkanes.

3.6 CONFORMATIONS OF CYCLOHEXANE

Experimental evidence indicating that six-membered rings are nonplanar began to accumulate in the 1920s. Eventually, Odd Hassel of the University of Oslo established that the most stable conformation of cyclohexane has the shape shown in Figure 3.10.

Hassel shared the 1969 Nobel Prize in chemistry with Sir Derek Barton of Imperial College (London), now at Texas A&M University. Barton demonstrated how Hassel's structural results could be extended to an analysis of conformational effects on chemical reactivity.

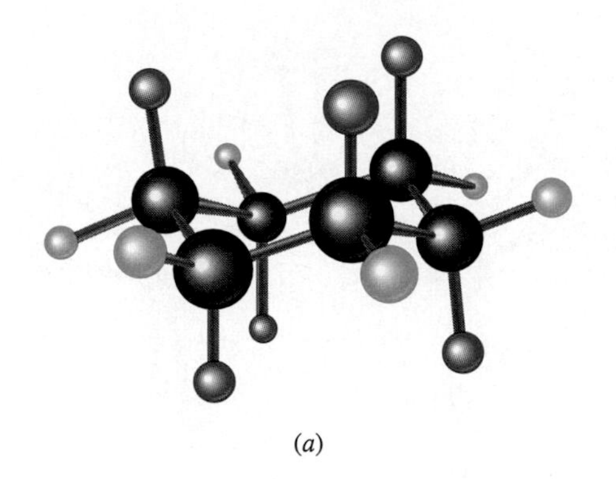

(*a*)

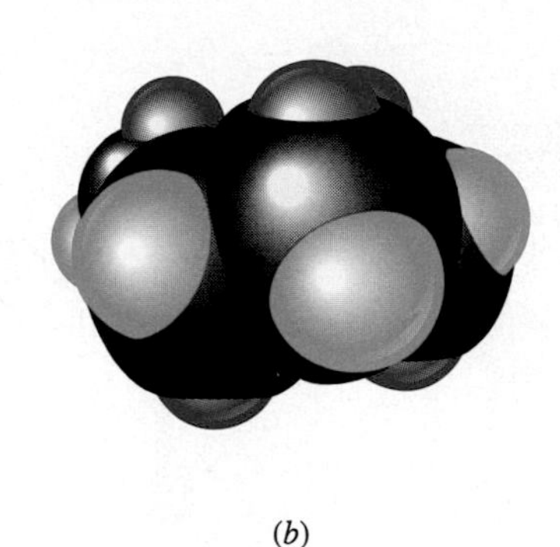

(*b*)

FIGURE 3.10 A ball-and-stick model (*a*) and a space-filling model (*b*) of the chair conformation of cyclohexane.

This is called the **chair** conformation. With C—C—C bond angles of 111°, the chair conformation is nearly free of angle strain. All its bonds are staggered, making it free of torsional strain as well. The staggered arrangement of bonds in the chair conformation of cyclohexane is apparent in a Newman-style projection.

Staggered arrangement of bonds in chair conformation of cyclohexane

A second, but much less stable, nonplanar conformation called the **boat** is shown in Figure 3.11. Like the chair, the boat conformation has bond angles that are approximately tetrahedral and is relatively free of angle strain. As noted in Figure 3.11, however, the boat is destabilized by van der Waals strain involving its two "flagpole" hydrogens, which are within 180 pm of each other. An even greater contribution to the estimated 27 kJ/mol (6.4 kcal/mol) energy difference between the chair and the boat is the torsional strain associated with eclipsed bonds on four of the carbons in the boat. Figure 3.12 depicts the eclipsed bonds and demonstrates how the associated torsional strain may be reduced by rotation about the carbon-carbon bonds to give the slightly more stable **twist boat,** or **skew boat,** conformation. The same bond rotations that reduce the torsional strain also reduce the van der Waals strain by increasing the distance between the two flagpole hydrogens from 180 pm to about 200 pm.

Recall from Section 3.3 that the sum of the van der Waals radii of two hydrogen atoms is 240 pm.

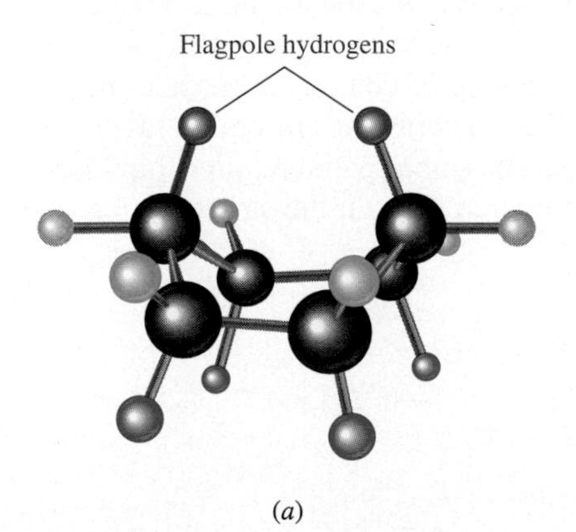
Flagpole hydrogens

(*a*)

Flagpole hydrogens are close enough
to touch each other

(*b*)

FIGURE 3.11 A ball-and-stick model (*a*) and a space-filling model (*b*) of the boat conformation of cyclohexane. The close approach of the two uppermost hydrogen substituents is clearly evident in the space-filling model.

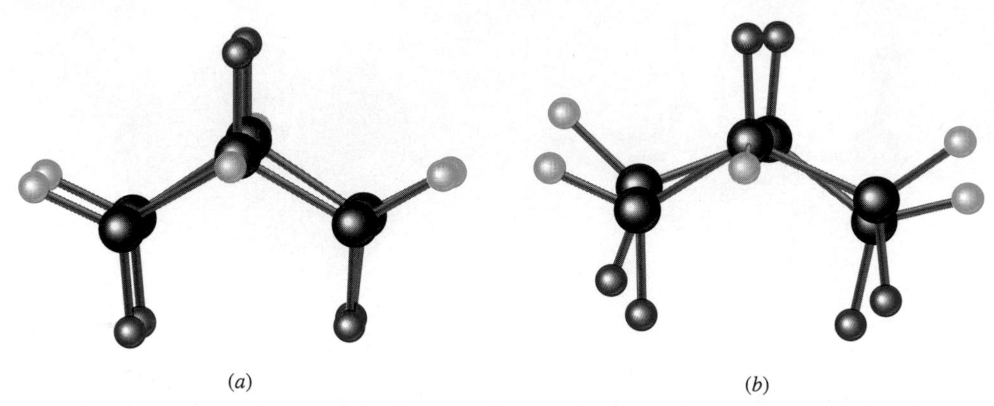

FIGURE 3.12 Relationship of boat (*a*) and skew boat (*b*) conformations of cyclohexane. A portion of the torsional strain in the boat is relieved by rotation about C—C bonds in going to skew boat. Bond rotation is accompanied by movement of flagpole hydrogens away from each other, which reduces van der Waals strain between them.

The various conformations of cyclohexane are in rapid equilibrium with one another, but at any moment almost all of the molecules exist in the chair conformation. Not more than one or two molecules per thousand are present in the higher-energy skew boat and boat conformations. Thus, the discussion of cyclohexane conformational analysis that follows focuses exclusively on the chair conformation.

3.7 AXIAL AND EQUATORIAL BONDS IN CYCLOHEXANE

One of the most significant findings to emerge from conformational studies of cyclohexane is that the spatial orientations of its 12 hydrogen atoms are not all identical but are divided into two groups, as shown in Figure 3.13. Six of the hydrogens, called **axial** hydrogens, have their bonds parallel to an axis of symmetry that passes through the ring's center. These axial bonds alternately are directed up and down on adjacent carbons. The second set of six hydrogens, called **equatorial** hydrogens, are located approximately along the equator of the molecule. Notice that the four bonds to each carbon are arranged tetrahedrally, consistent with an sp^3 hybridization of carbon.

The conformational features of six-membered rings are fundamental to organic chemistry, which makes it essential that you have a clear understanding of the directional properties of axial and equatorial bonds and be able to represent them accurately. Figure 3.14 offers some guidance on the drawing of chair cyclohexane rings.

It is no accident that sections of our chair cyclohexane drawings resemble sawhorse projections of staggered conformations of alkanes. The same spatial relationships seen in alkanes carry over to substituents on a six-membered ring. In the structure

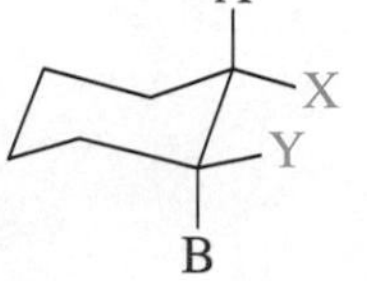

A X Y B

(The substituted carbons have the spatial arrangement shown)

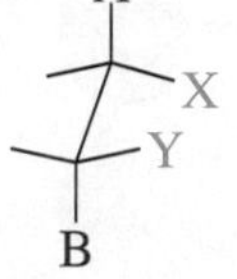

A X Y B

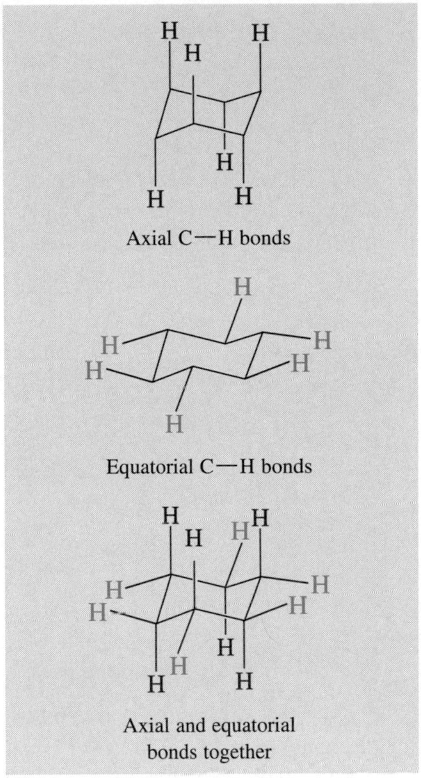

Axial C—H bonds

Equatorial C—H bonds

Axial and equatorial bonds together

FIGURE 3.13 Axial and equatorial bonds in cyclohexane.

substituents A and B are anti to each other, while the other relationships—A and Y, X and Y, and X and B—are gauche.

PROBLEM 3.4 Given the following partial structure, add a substituent X to C-1 so that it satisfies the indicated stereochemical requirement.

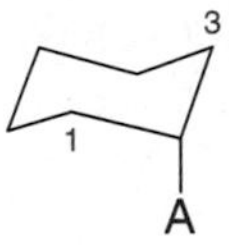

(*a*) Anti to A
(*b*) Gauche to A
(*c*) Anti to C-3
(*d*) Gauche to C-3

SAMPLE SOLUTION (*a*) In order to be anti to A, substituent X must be axial. The blue lines in the drawing show the A—C—C—X torsion angle to be 180°.

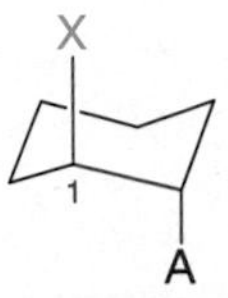

(1) Begin with the chair conformation of cyclohexane.

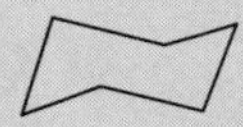

(2) Draw the axial bonds before the equatorial ones, alternating their direction on adjacent atoms. Always start by placing an axial bond "up" on the uppermost carbon or "down" on the lowest carbon.

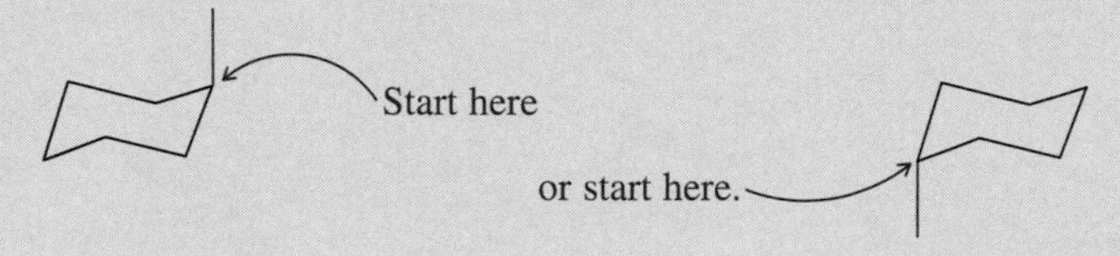

Then alternate to give

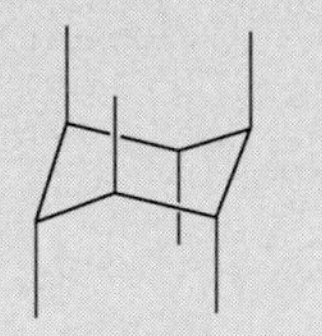

in which all the axial bonds are parallel to one another.

(3) Place the equatorial bonds so as to approximate a tetrahedral arrangement of the bonds to each carbon. The equatorial bond of each carbon should be parallel to the ring bonds of its two nearest-neighbor carbons.

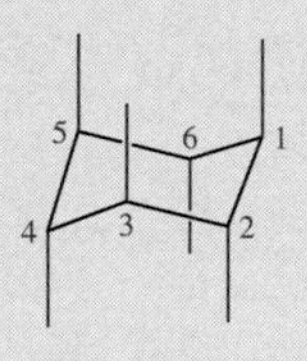

Place equatorial bond at C-1 so that it is parallel to the bonds between C-2 and C-3 and between C-5 and C-6.

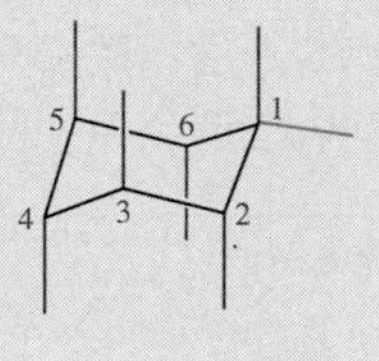

Following this pattern gives the complete set of equatorial bonds.

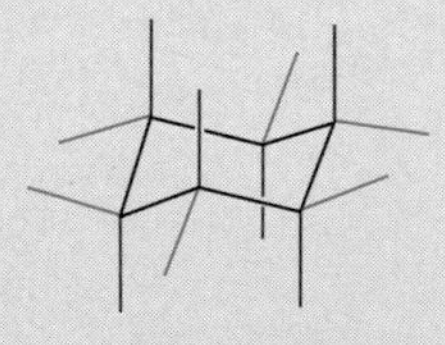

(4) Practice drawing cyclohexane chairs oriented in either direction.

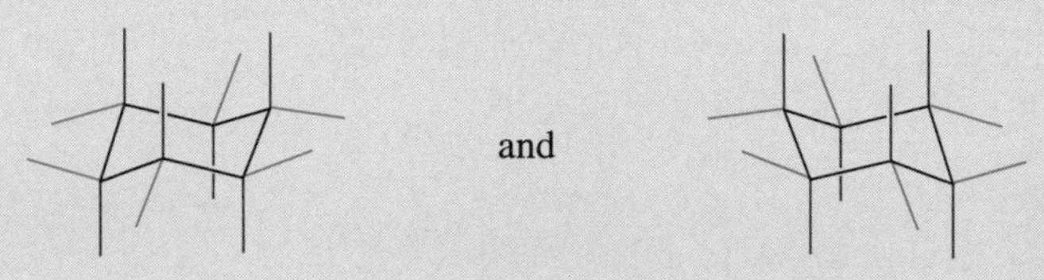

FIGURE 3.14 A guide to representing the orientations of the bonds in the chair conformation of cyclohexane.

3.8 CONFORMATIONAL INVERSION (RING FLIPPING) IN CYCLOHEXANE

We have seen that alkanes are not locked into a single conformation. Rotation around the central carbon-carbon bond in butane occurs rapidly, interconverting anti and gauche conformations. Cyclohexane, too, is conformationally mobile. Through a process known as **ring inversion, chair-chair interconversion,** or, more simply, **ring flipping,** one chair conformation is converted to another chair.

The activation energy for cyclohexane ring inversion is 45 kJ/mol (10.8 kcal/mol). It is a very rapid process; about half of all cyclohexane molecules undergo ring inversion every 10^{-5} s at 25°C. Figure 3.15 depicts a potential energy diagram for the process. In the first step the chair conformation is converted to a skew boat, which then proceeds to the inverted chair in the second step. The skew boat conformation is an *intermediate* in the process of ring inversion. Unlike a transition state, an **intermediate** is not a potential energy maximum but is a local minimum on the potential energy profile.

The most important result of ring inversion is that any substituent that is axial in the original chair conformation becomes equatorial in the ring-flipped form and vice versa.

very fast

X axial; Y equatorial

X equatorial; Y axial

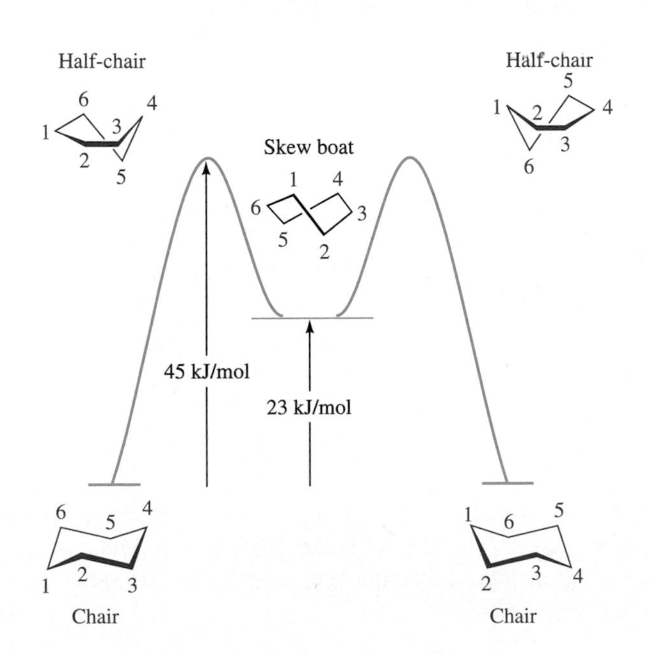

FIGURE 3.15 Energy diagram showing interconversion of various conformations of cyclohexane. In order to simplify the diagram, the boat conformation has been omitted. The boat is a transition state for the interconversion of skew boat conformations.

The consequences of this point are developed for a number of monosubstituted cyclohexane derivatives in the following section, beginning with methylcyclohexane.

3.9 CONFORMATIONAL ANALYSIS OF MONOSUBSTITUTED CYCLOHEXANES

Ring inversion in methylcyclohexane differs from that of cyclohexane in that the two chair conformations are not equivalent. In one chair the methyl group is axial; in the other it is equatorial. Structural studies have established that approximately 95 percent of the molecules of methylcyclohexane are in the chair conformation that has an equatorial methyl group while only 5 percent of the molecules have an axial methyl group at room temperature.

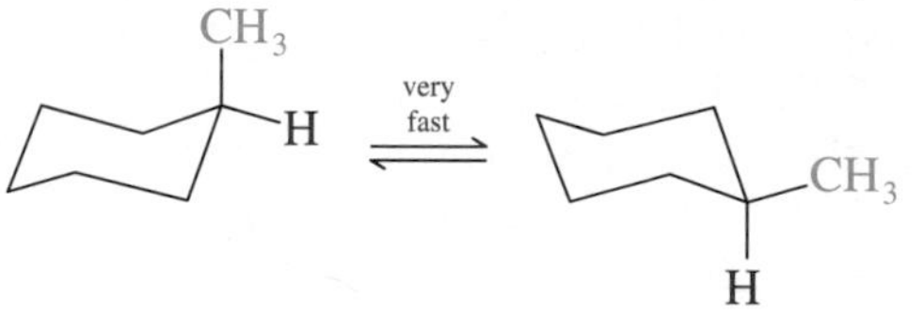

CH_3 H — very fast ⇌ — CH_3 H

5% 95%

See the box entitled "Enthalpy, Free Energy, and Equilibrium Constant" accompanying this section for a discussion of these relationships.

When two conformations of a molecule are in equilibrium with each other, the one with the lower free energy predominates. What then is the structural basis for the observation that equatorial methylcyclohexane is more stable than axial methylcyclohexane?

A methyl group is less crowded when it is equatorial than when it is axial. An axial methyl group in methylcyclohexane has one of its hydrogens within 190 to 200 pm of the axial hydrogens at C-3 and C-5. This distance is less than the sum of the van der Waals radii of two hydrogens (240 pm) and causes van der Waals strain in the axial conformation. When the methyl group is equatorial, it experiences no significant crowding.

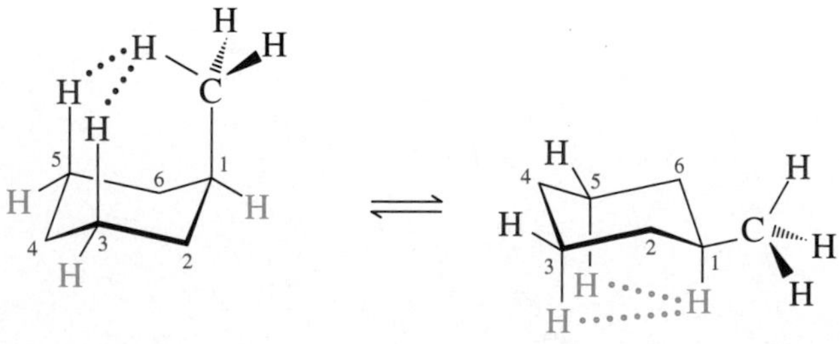

Van der Waals strain between hydrogen of axial CH_3 and axial hydrogens at C-3 and C-5

Smaller van der Waals strain between hydrogen at C-1 and axial hydrogens at C-3 and C-5

The greater stability of an equatorial methyl group, compared with an axial one, is another example of a *steric effect* (Section 3.3). An axial substituent is said to be crowded because of **1,3-diaxial repulsions** between itself and the other two axial substituents located on the same side of the ring.

PROBLEM 3.5 The following questions relate to a cyclohexane ring depicted in the chair conformation shown.

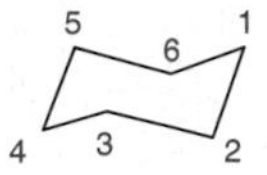

5 6 1
4 3 2

(*a*) Is a methyl group at C-6 that is "down" axial or equatorial?
(*b*) Is a methyl group which is "up" at C-1 more or less stable than a methyl group which is up at C-4?
(*c*) Place a methyl group at C-3 in its most stable orientation. Is it up or down?

SAMPLE SOLUTION (*a*) First indicate the directional properties of the bonds to the ring carbons. A substituent is down if it is below the other substituent on the same carbon atom. Therefore, a methyl group that is down at C-6 is axial.

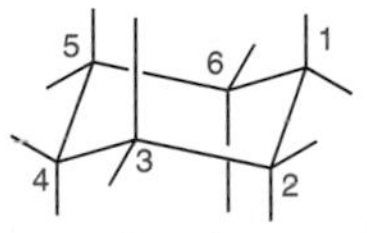

5 6 1 4 3 2

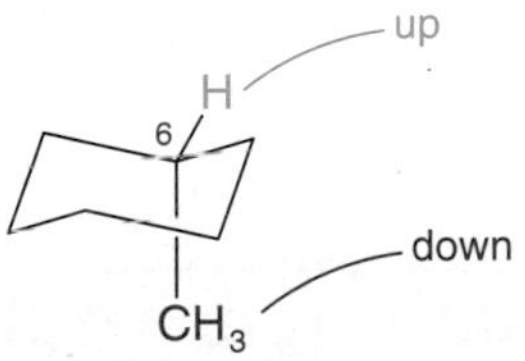

H up
6
CH_3 down

Other substituted cyclohexanes are similar to methylcyclohexane. Two chair conformations exist in rapid equilibrium, and the one in which the substituent is equatorial is more stable. The relative amounts of the two conformations depend on the effective size of the substituent. The size of a substituent, in the context of cyclohexane conformations, is related to the degree of branching at its point of connection to the ring. A single atom, such as a halogen substituent, does not take up much space, and its preference for an equatorial orientation is less pronounced than that of a methyl group.

The halogens F, Cl, Br, and I do not differ much in their preference for the equatorial position. As the atomic radius increases in the order F < Cl < Br < I, so does the carbon-halogen bond distance and the two effects tend to cancel.

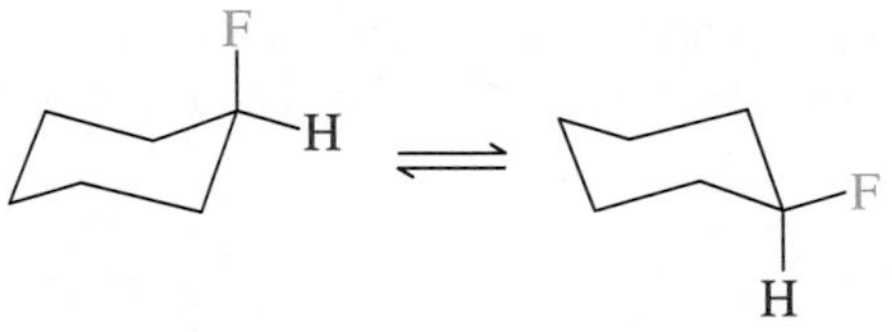

F H ⇌ F H

40 percent 60 percent

A branched alkyl group such as isopropyl exhibits a greater preference for the equatorial orientation than does methyl.

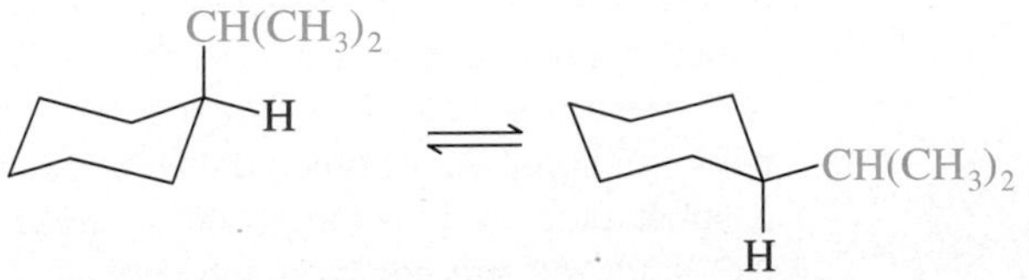

$CH(CH_3)_2$ H ⇌ $CH(CH_3)_2$ H

3 percent 97 percent

Highly branched groups such as *tert*-butyl are commonly described as "bulky."

A *tert*-butyl group is so large that *tert*-butylcyclohexane exists almost entirely in the conformation in which the *tert*-butyl group is equatorial. The amount of axial *tert*-butylcyclohexane present is too small to measure.

CH_3 CH_3 CH_3 C H H H ⇌ $C(CH_3)_3$ H

Less than 0.01 percent

(Serious 1,3-diaxial repulsions involving *tert*-butyl group)

Greater than 99.99 percent

(Decreased van der Waals strain)

PROBLEM 3.6 Draw the most stable conformation of 1-*tert*-butyl-1-methylcyclohexane.

ENTHALPY, FREE ENERGY, AND EQUILIBRIUM CONSTANT

One of the fundamental equations of thermodynamics concerns systems at equilibrium and relates the equilibrium constant K to the difference in **free energy** ($\Delta G°$) between the products and the reactants.

$$\Delta G° = G°_{products} - G°_{reactants} = -RT \ln K$$

where T is the absolute temperature in kelvins and R is a constant equal to 8.314 J/mol · K (1.99 cal/mol · K).

For the equilibrium between the axial and equatorial conformations of a monosubstituted cyclohexane,

X $\overset{K}{\rightleftharpoons}$ X

the equilibrium constant is given by the expression

$$K = \frac{[\text{products}]}{[\text{reactants}]}$$

Inserting the appropriate values for R, T (298 K), and K gives the values of $\Delta G°$ listed in the table (page 103) for the various substituents discussed in Section 3.9.

The relationship between $\Delta G°$ and K is presented in graphical form in Figure 3.16. We should note that the relationship represented in Figure 3.16 is a generally useful one and applies whenever two species are in equilibrium with each other, not just for the interconversion of conformations.

Reactions characterized by a negative $\Delta G°$ are described as *spontaneous* in the direction written. A larger value of K is associated with a more negative $\Delta G°$. Free energy and enthalpy are related by the expression

$$\Delta G° = \Delta H° - T\Delta S°$$

where $\Delta S°$ is the difference in *entropy* between the products and reactants. A positive $\Delta S°$ is accompanied by an increase in the disorder of a system. A positive $T\Delta S°$ term leads to a $\Delta G°$ that is more negative than $\Delta H°$ and a larger K than expected on the basis of enthalpy considerations alone. Conversely, a negative $\Delta S°$ gives a smaller K than expected. In the case of conformational equilibration between the chair forms of a substituted cyclohexane, $\Delta S°$ is close to zero and $\Delta G°$ and $\Delta H°$ are approximately equal.

Substituent X	Percent axial	Percent equatorial	K	$\Delta G^\circ_{298\,K}$ kJ/mol (kcal/mol)
—F	40	60	1.5	−1.0 (−0.24)
$—CH_3$	5	95	19	−7.3 (−1.7)
$—CH(CH_3)_2$	3	97	32.3	−8.6 (−2.1)
$—C(CH_3)_3$	<0.01	>99.99	>9999	−22.8 (−5.5)

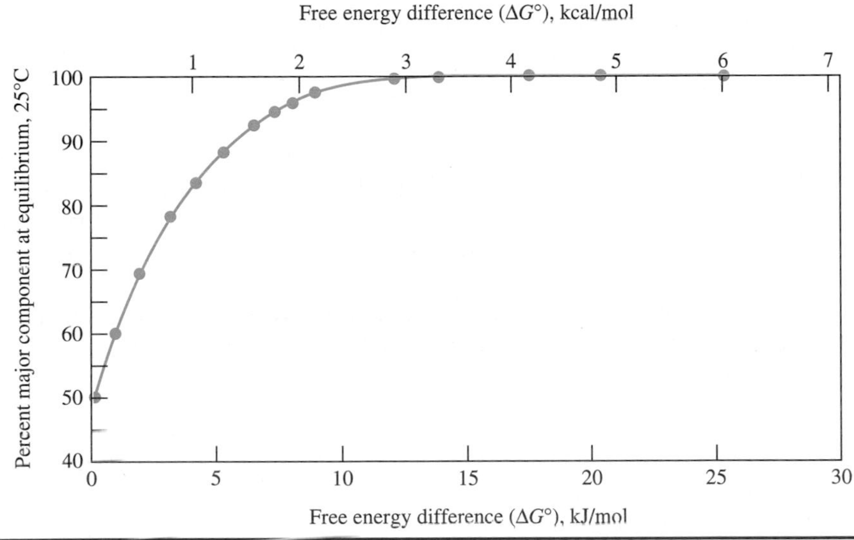

FIGURE 3.16 Distribution of two products at equilibrium plotted as a function of the difference in free energy (ΔG°) at 25°C between the two species.

3.10 SMALL RINGS: CYCLOPROPANE AND CYCLOBUTANE

Conformational analysis is far simpler in cyclopropane than in any other cycloalkane. Cyclopropane's three carbon atoms are, of geometric necessity, coplanar, and rotation about its carbon-carbon bonds is impossible. You saw in Section 3.5 how angle strain in cyclopropane leads to an abnormally large heat of combustion for a compound with only three methylene units. Let us now examine cyclopropane in more detail in order

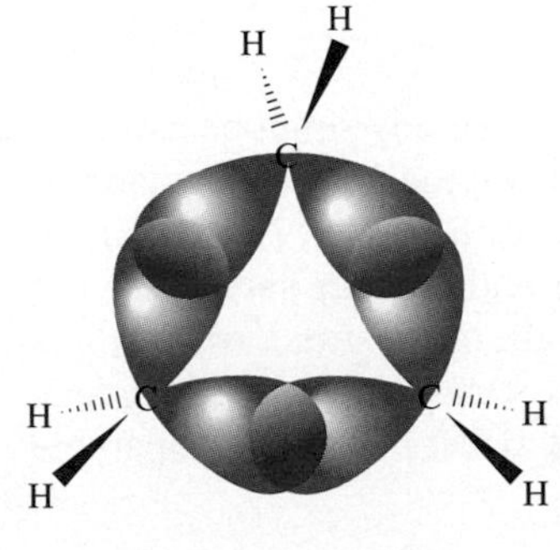

FIGURE 3.17 "Bent bonds" in cyclopropane. The orbitals involved in carbon-carbon bond formation overlap in a region that is displaced from the internuclear axis. Orbital overlap is less effective than in a normal carbon-carbon σ bond, and the carbon-carbon bond is weaker.

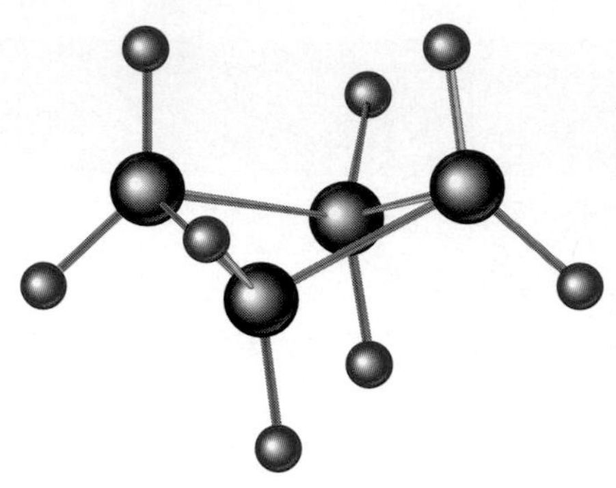

FIGURE 3.18 Nonplanar ("puckered") conformation of cyclobutane. The nonplanar conformation is more stable than the planar one because it avoids the eclipsing of bonds on adjacent carbons that characterizes the planar conformation.

to see how our orbital hybridization bonding model may be adapted to molecules of unusual geometry.

Strong sp^3–sp^3 σ bonds are not possible for cyclopropane, because the 60° bond angles of the ring do not permit the orbitals to be properly aligned for effective overlap (Figure 3.17). The less effective overlap that does occur leads to what chemists refer to as "bent" bonds. The electron density in the carbon-carbon bonds of cyclopropane does not lie along the internuclear axis but is distributed along an arc between the two carbon atoms. The ring bonds of cyclopropane are weaker than other carbon-carbon σ bonds.

In keeping with the "bent-bond" description of Figure 3.17, the carbon-carbon bond distance in cyclopropane (151 pm) is slightly shorter than that of ethane (153 pm) and cyclohexane (154 pm).

In addition to angle strain, cyclopropane is destabilized by torsional strain. Each C—H bond of cyclopropane is eclipsed with two others.

H H H

H H H

All adjacent pairs of bonds are eclipsed

Cyclobutane has less angle strain than cyclopropane and can reduce the torsional strain that goes with a planar geometry by adopting the nonplanar "puckered" conformation shown in Figure 3.18.

PROBLEM 3.7 The heats of combustion of ethylcyclopropane and methylcyclobutane have been measured as 3352 and 3384 kJ/mol (801.2 and 808.8 kcal/mol). Assign the correct heat of combustion to each isomer.

3.11 CYCLOPENTANE

Neighboring C—H bonds are eclipsed in any planar cycloalkane. Thus all planar conformations are destabilized by torsional strain.

Angle strain in the planar conformation of cyclopentane is relatively small because the 108° angles of a regular pentagon are not much different from the 109.5° bond angles that characterize sp^3 hybridized carbon. The torsional strain, however, is substantial, since five bonds are eclipsed on the top face of the ring and another set of five are eclipsed on the bottom face (Figure 3.19). Some, but not all, of this torsional strain is relieved in nonplanar conformations. Two nonplanar conformations of cyclopentane, the **envelope** (Figure 3.19*b*) and the **half-chair** (Figure 3.19*c*) are of comparable energy.

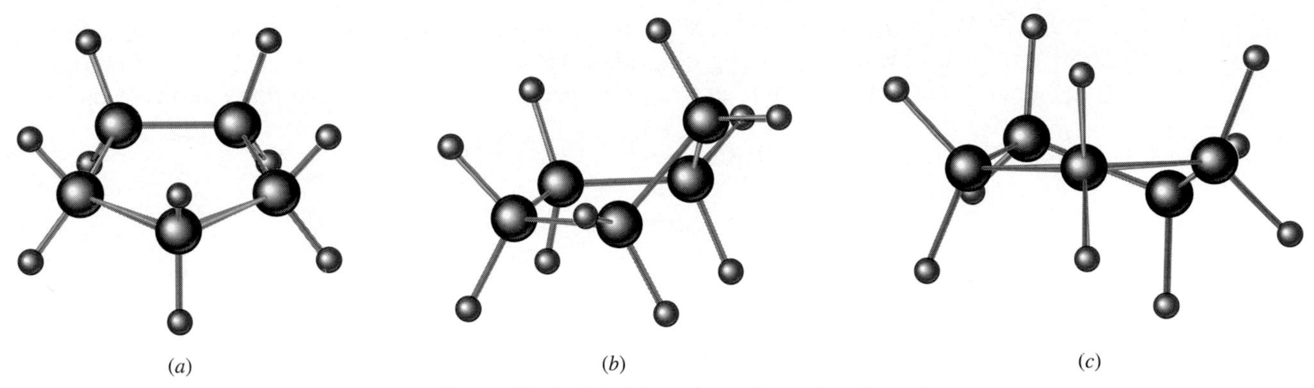

FIGURE 3.19 The planar (*a*), envelope (*b*), and half-chair (*c*) conformations of cyclopentane.

In the envelope conformation four of the carbon atoms are coplanar. The fifth carbon is out of the plane of the other four. There are three coplanar carbons in the half-chair conformation, with one carbon atom displaced above that plane and another below it. In both the envelope and the half-chair conformations, in-plane and out-of-plane carbons exchange positions rapidly. Equilibration between conformations of cyclopentane occurs at rates that are comparable with the rate of rotation about the carbon-carbon bond of ethane.

3.12 MEDIUM AND LARGE RINGS

Beginning with cycloheptane, which has four conformations of similar energy, conformational analysis of cycloalkanes becomes more complicated. The same fundamental principles apply to medium and large rings as apply to smaller ones—there are simply more atoms and more bonds to consider and more conformational possibilities. The following each depict one of the conformations available to the medium-ring cycloalkanes cyclooctane and cyclodecane, respectively.

In 1978 a German-Swiss team of organic chemists reported the synthesis of a cycloalkane with 96 carbons in its ring (cyclo-$C_{96}H_{192}$).

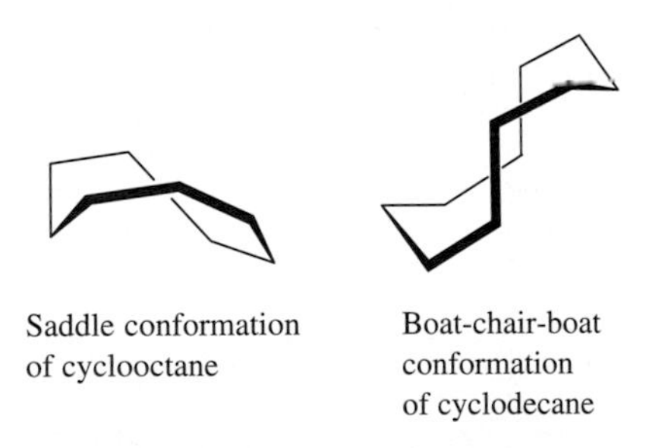

The shapes of alkanes, cycloalkanes, and monosubstituted cyclohexanes have served us well to this point as a vehicle to develop concepts related to the three-dimensional structure of molecules. In the next section we will examine disubstituted cycloalkanes to see how these principles may be extended to a yet higher level.

3.13 DISUBSTITUTED CYCLOALKANES. STEREOISOMERS

When a cycloalkane bears two substituents on different carbons—methyl groups, for example—these substituents may be on the same side or on opposite sides of the ring.

When the substituents are on the same side, we say they are **cis** to each other; when substituents are on opposite sides, we say they are **trans** to each other. Both terms come from the Latin, in which *cis* means "on this side" and *trans* means "across."

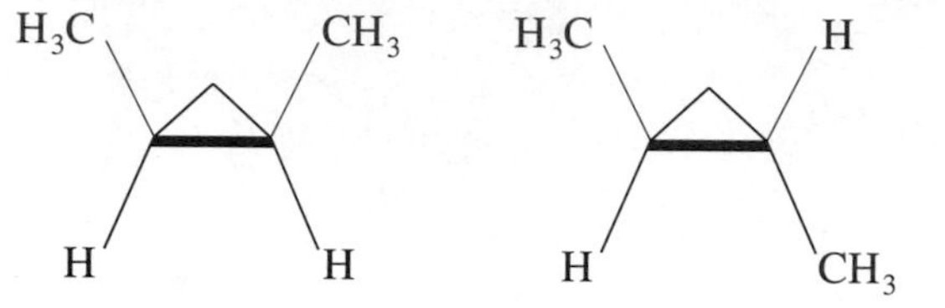

H_3C CH_3 H_3C H

H H H CH_3

cis-1,2-Dimethylcyclopropane *trans*-1,2-Dimethylcyclopropane

---

PROBLEM 3.8 Exclusive of compounds with double bonds, there are four hydrocarbons that are *constitutional* isomers of *cis*- and *trans*-1,2-dimethylcyclopropane. Identify these compounds.

The prefix *stereo-* is derived from the Greek word *stereos,* meaning "solid." *Stereochemistry* is the term applied to the three-dimensional aspects of molecular structure and reactivity.

The cis and trans forms of 1,2-dimethylcyclopropane are *stereoisomers.* **Stereoisomers** are isomers that have their atoms bonded in the same order—i.e., they have the same constitution, but they differ in the arrangement of atoms in space. Stereoisomers of the cis-trans type are sometimes referred to as *geometric isomers.* You learned in Section 2.17 that constitutional isomers could differ in stability. What about stereoisomers?

We can measure the energy difference between *cis*- and *trans*-1,2-dimethylcyclopropane by comparing their heats of combustion. As illustrated in Figure 3.20, the two compounds are isomers, and so the difference in their heats of combustion is a direct measure of the difference in their energies. Because the heat of combustion of *trans*-

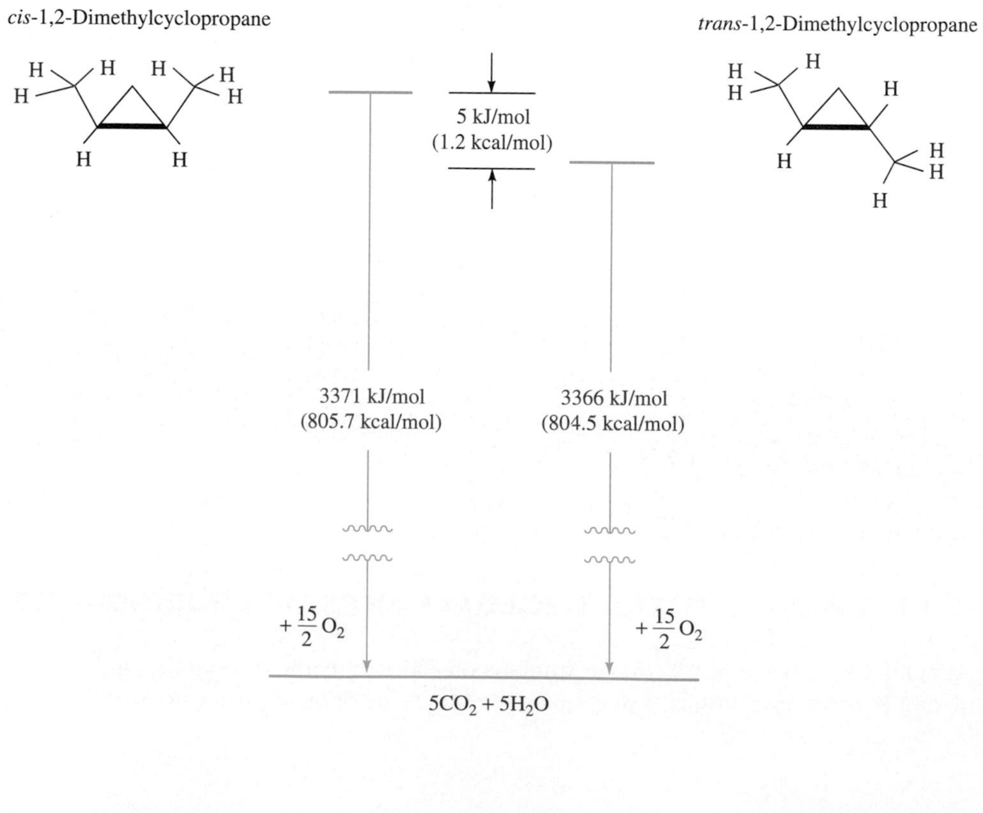

cis-1,2-Dimethylcyclopropane

trans-1,2-Dimethylcyclopropane

5 kJ/mol
(1.2 kcal/mol)

3371 kJ/mol
(805.7 kcal/mol)

3366 kJ/mol
(804.5 kcal/mol)

$+\frac{15}{2}O_2$ $+\frac{15}{2}O_2$

$5CO_2 + 5H_2O$

FIGURE 3.20 The enthalpy difference between *cis*- and *trans*-1,2-dimethylcyclopropane can be determined from their heats of combustion. Van der Waals strain between methyl groups on the same side of the ring make the cis isomer less stable than the trans isomer.

1,2-dimethylcyclopropane is 5 kJ/mol (1.2 kcal/mol) less than that of its cis stereoisomer, it follows that *trans*-1,2-dimethylcyclopropane is 5 kJ/mol (1.2 kcal/mol) more stable than *cis*-1,2-dimethylcyclopropane.

In this case, the relationship between stability and stereochemistry can be explained on the basis of van der Waals strain. The methyl groups on the same side of the ring in *cis*-1,2-dimethylcyclopropane crowd each other and increase the potential energy of this stereoisomer. Steric hindrance between methyl groups is absent in *trans*-1,2-dimethylcyclopropane.

Disubstituted cyclopropanes exemplify one of the simplest cases involving stability differences between stereoisomers. A three-membered ring has no conformational mobility, and there is no way the ring can adjust to reduce the van der Waals strain that exists between cis substituents on adjacent carbons. The situation is different, and more interesting, in disubstituted derivatives of cyclohexane.

3.14 CONFORMATIONAL ANALYSIS OF DISUBSTITUTED CYCLOHEXANES

Let us begin with *cis*- and *trans*-1,4-dimethylcyclohexane. A conventional method to represent cis and trans stereoisomers in cyclic systems uses wedge-and-dash descriptions as shown.

cis-1,4-Dimethylcyclohexane *trans*-1,4-Dimethylcyclohexane

Wedge-and-dash drawings fail to show conformation, and it is important to remember that the rings of *cis*- and *trans*-1,2-dimethylcyclohexane exist in a chair conformation. This fact must be taken into consideration when evaluating the relative stabilities of the stereoisomers.

Their respective heats of combustion (Table 3.2) indicate that *trans*-1,4-dimethyl cyclohexane is 7 kJ/mol (1.6 kcal/mol) more stable than the cis stereoisomer. It is

TABLE 3.2
Heats of Combustion of Isomeric Dimethylcyclohexanes

Compound	Orientation of methyl groups in most stable conformation	Heat of combustion kJ/mol (kcal/mol)	Difference in heat of combustion kJ/mol (kcal/mol)	More stable stereoisomer
cis-1,2-Dimethylcyclohexane	Axial-equatorial	5223 (1248.3)	6 (1.5)	trans
trans-1,2-Dimethylcyclohexane	Diequatorial	5217 (1246.8)		
cis-1,3-Dimethylcyclohexane	Diequatorial	5212 (1245.7)	7 (1.7)	cis
trans-1,3-Dimethylcyclohexane	Axial-equatorial	5219 (1247.4)		
cis-1,4-Dimethylcyclohexane	Axial-equatorial	5219 (1247.4)	7 (1.6)	trans
trans-1,4-Dimethylcyclohexane	Diequatorial	5212 (1245.8)		

unrealistic to invoke van der Waals strain between cis substituents as a destabilizing factor, because the methyl groups are too far away from each other. To understand why *trans*-1,4-dimethylcyclohexane is more stable than *cis*-1,4-dimethylcyclohexane, we need to examine each stereoisomer in its most stable conformation.

cis-1,4-Dimethylcyclohexane can adopt either of two equivalent chair conformations, *each having one axial methyl group and one equatorial methyl group.* The two are in rapid equilibrium with each other by ring flipping. The equatorial methyl group becomes axial and the axial methyl group becomes equatorial.

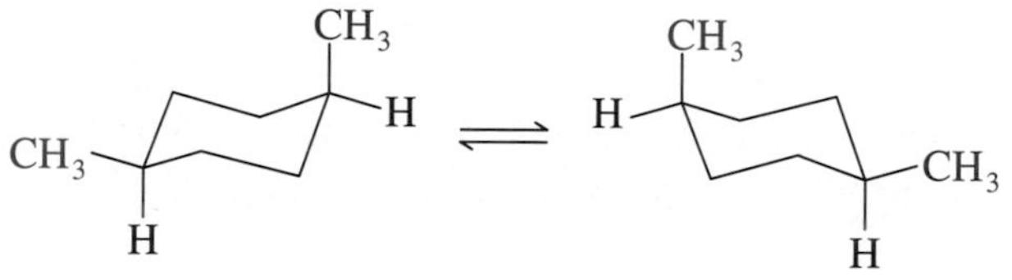

(One methyl group is axial, the other equatorial)

(One methyl group is axial, the other equatorial)

(Both methyl groups are up)

cis-1,4-Dimethylcyclohexane

The methyl groups are described as cis because both are up relative to the hydrogen substituents present at each carbon. If both methyl groups were down, they would still be cis to each other. Notice that ring flipping does not alter the cis relationship between the methyl groups. Nor does it alter their up versus down quality; substituents that are up in one conformation remain up in the ring-flipped form.

The most stable conformation of trans-*1,4-dimethylcyclohexane has both methyl groups in equatorial orientations.* The two chair conformations of *trans*-1,4-dimethylcyclohexane are not equivalent to each other. One has two equatorial methyl groups; the other, two axial methyl groups.

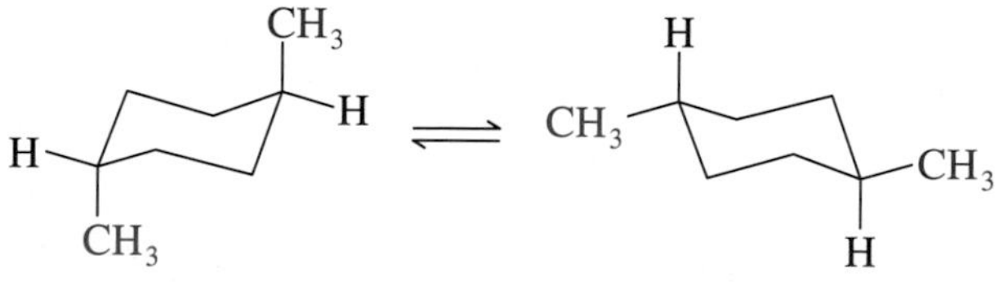

(Both methyl groups are axial: less stable chair conformation)

(Both methyl groups are equatorial: more stable chair conformation)

(One methyl group is up, the other down)

trans-1,4-Dimethylcyclohexane

The more stable chair—the one with both methyl groups equatorial—is the conformation adopted by most of the *trans*-1,4-dimethylcyclohexane molecules at equilibrium.

trans-1,4-Dimethylcyclohexane is more stable than *cis*-1,4-dimethylcyclohexane because both methyl groups are equatorial in the most stable conformation of *trans*-1,4-dimethylcyclohexane while one methyl group must be axial in the cis stereoisomer. Remember, it is a general rule that any substituent is more stable in an equatorial orientation than in an axial one. It is worth pointing out that the 7 kJ/mol (1.6 kcal/mol) energy difference between *cis*- and *trans*-1,4-dimethylcyclohexane is the same as the energy difference between the axial and equatorial conformations of methylcyclohexane. There is a simple reason for this: in both instances the less stable structure has one axial methyl group, and the 7 kJ/mol (1.6 kcal/mol) energy difference can be considered the ''energy cost'' of having a methyl group in an axial rather than an equatorial orientation.

Similarly, *trans*-1,2-dimethylcyclohexane has a lower heat of combustion (Table 3.2) and is more stable than *cis*-1,2-dimethylcyclohexane. The cis stereoisomer has two equivalent chair conformations, each containing one axial and one equatorial methyl group.

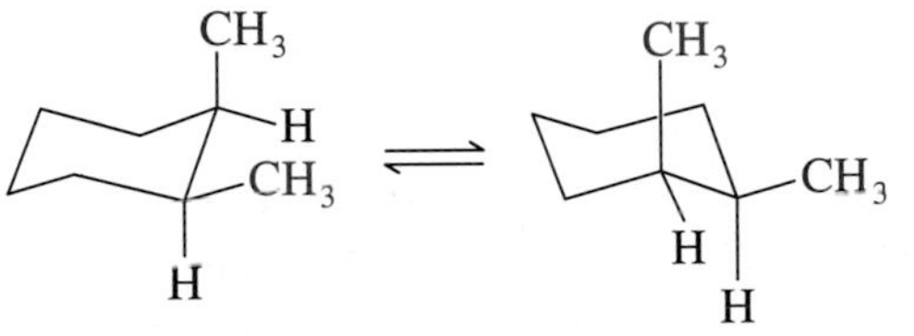

cis-1,2-Dimethylcyclohexane

Both methyl groups are equatorial in the most stable conformation of *trans*-1,2-dimethylcyclohexane.

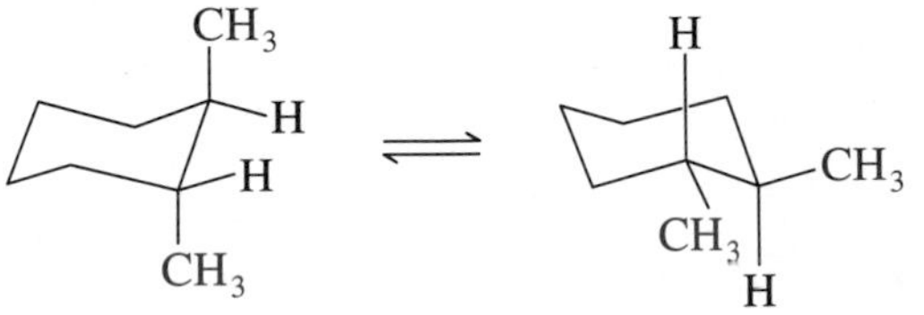

(Both methyl groups are axial: less stable chair conformation)

(Both methyl groups are equatorial: more stable chair conformation)

trans-1,2-Dimethylcyclohexane

As in the 1,4-dimethylcyclohexanes, the 6 kJ/mol (1.5 kcal/mol) energy difference between the more stable (trans) and the less stable (cis) stereoisomer is attributed to the strain associated with the presence of an axial methyl group in the cis isomer.

Probably the most interesting observation in Table 3.2 concerns the 1,3-dimethylcyclohexanes. Unlike the 1,2- and 1,4-dimethylcyclohexanes, where the trans stereoisomer is more stable than the cis, we find that *cis*-1,3-dimethylcyclohexane is 7 kJ/mol (1.7 kcal/mol) more stable than *trans*-1,3-dimethylcyclohexane. Why?

The most stable conformation of *cis*-1,3-dimethylcyclohexane has both methyl groups equatorial.

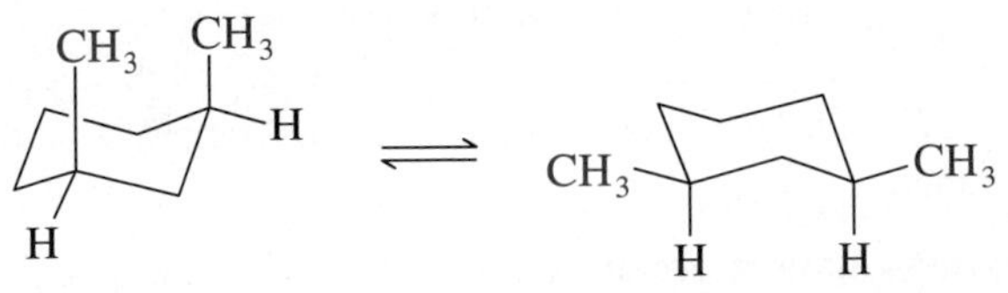

(Both methyl groups are axial: less stable chair conformation)

(Both methyl groups are equatorial: more stable chair conformation)

cis-1,3-Dimethylcyclohexane

The two chair conformations of *trans*-1,3-dimethylcyclohexane are equivalent to each other. Both contain one axial and one equatorial methyl group.

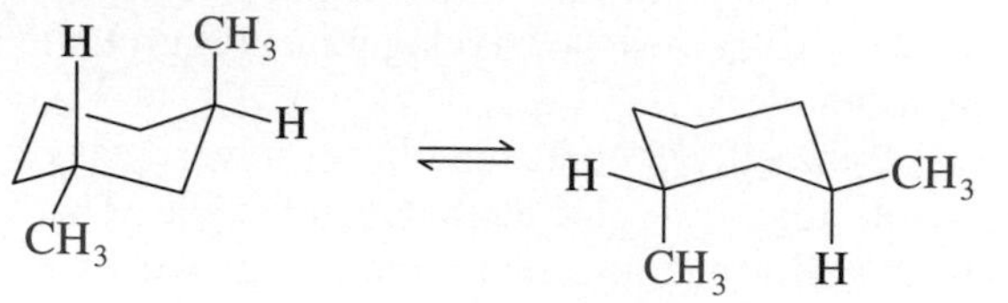

(One methyl group is axial, the other equatorial)

(One methyl group is axial, the other equatorial)

trans-1,3-Dimethylcyclohexane

Thus the trans stereoisomer, with one axial methyl group, is less stable than *cis*-1,3-dimethylcyclohexane where both methyl groups are equatorial.

PROBLEM 3.9 On the basis of what you know about disubstituted cyclohexanes, which of the following two stereoisomeric 1,3,5-trimethylcyclohexanes would you expect to be more stable?

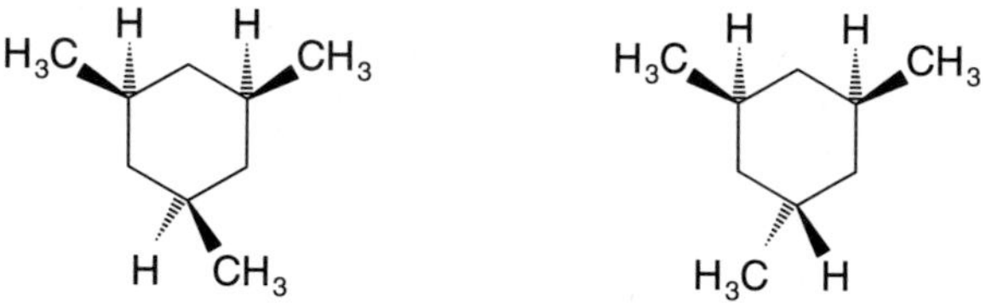

cis-1,3,5-Trimethylcyclohexane

trans-1,3,5-Trimethylcyclohexane

If in a disubstituted derivative of cyclohexane the two substituents are different, then the most stable conformation will be the chair that has the larger substituent in an equatorial orientation. This is most apparent when one of the substituents is a bulky group such as *tert*-butyl. Thus, the most stable conformation of *cis*-1-*tert*-butyl-2-methylcyclohexane has an equatorial *tert*-butyl group and an axial methyl group.

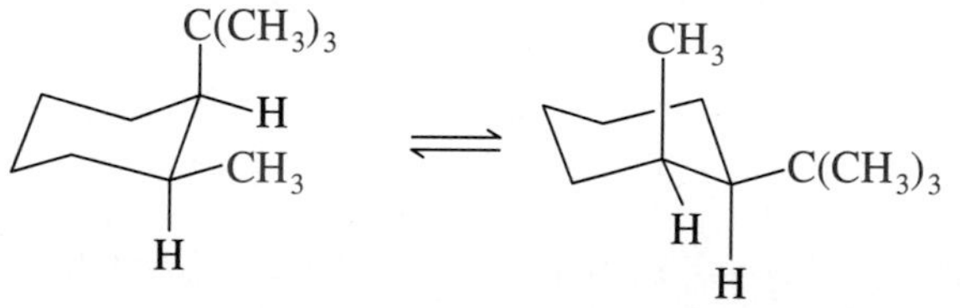

(Less stable conformation: larger group is axial)

(More stable conformation: larger group is equatorial)

cis-1-*tert*-Butyl-2-methylcyclohexane

PROBLEM 3.10 Write structural formulas for the most stable conformation of each of the following compounds:

(*a*) *trans*-1-*tert*-Butyl-3-methylcyclohexane
(*b*) *cis*-1-*tert*-Butyl-3-methylcyclohexane
(*c*) *trans*-1-*tert*-Butyl-4-methylcyclohexane
(*d*) *cis*-1-*tert*-Butyl-4-methylcyclohexane

SAMPLE SOLUTION (*a*) The most stable conformation is the one that has the larger substituent, the *tert*-butyl group, equatorial. Draw a chair conformation of cyclohexane, and place an equatorial *tert*-butyl group at one of its carbons. Add a methyl group at C-3 so that it is trans to the *tert*-butyl group.

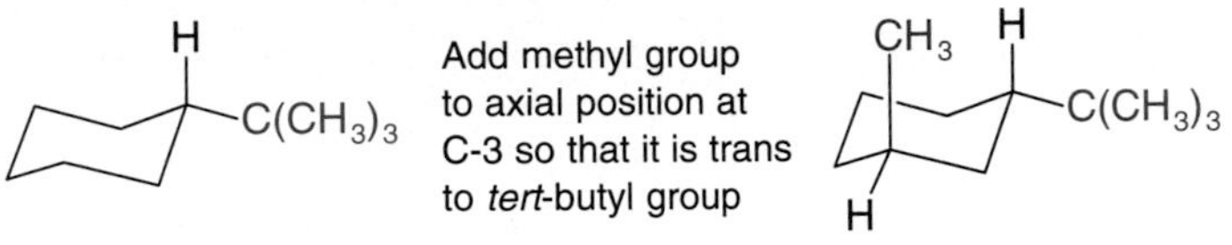

H
$C(CH_3)_3$

Add methyl group to axial position at C-3 so that it is trans to *tert*-butyl group

CH_3 H
$C(CH_3)_3$
H

tert-Butyl group equatorial on six-membered ring

trans-1-*tert*-Butyl-3-methylcyclohexane

Cyclohexane rings that bear *tert*-butyl substituents are examples of conformationally biased molecules. A *tert*-butyl group has such a pronounced preference for the equatorial orientation that it will strongly bias the equilibrium to favor such conformations. This does not mean that ring inversion does not occur, however. Ring inversion does occur, but at any instant only a tiny fraction of the molecules exist in conformations having axial *tert*-butyl groups. It is not strictly correct to say that *tert*-butylcyclohexane and its derivatives are "locked" into a single conformation; conformations related by ring flipping are in rapid equilibrium with one another, but the distribution between them strongly favors those in which the *tert*-butyl group is equatorial.

3.15 POLYCYCLIC RING SYSTEMS

Organic molecules in which *one* carbon atom is common to two rings are called **spirocyclic** compounds. The simplest spirocyclic hydrocarbon is *spiropentane,* a product of laboratory synthesis. More complicated spirocyclic hydrocarbons not only have been synthesized but also have been isolated from natural sources. *α-Alaskene,* for example, occurs in the fragrant oil given off by the needles of the Alaskan yellow cedar; one of its carbon atoms is common to both the six-membered ring and the five-membered ring.

CH_3 H CH_3
CH_3
CH_3

Spiropentane

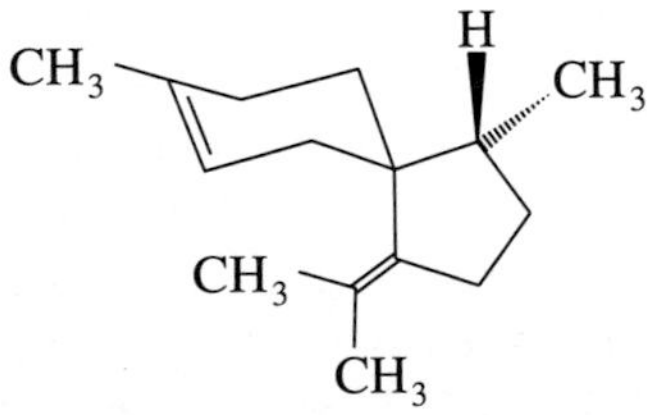

α-Alaskene

When *two* or more atoms are common to more than one ring, the compounds are called **polycyclic** ring systems. They are classified as *bicyclic, tricyclic, tetracyclic* (and so on), according to the number of bond cleavages required to generate a noncyclic structure. *Bicyclobutane* is the simplest bicyclic hydrocarbon; its four carbons form 2 three-membered rings that share a common side. *Camphene* is a naturally occurring bicyclic hydrocarbon obtained from pine oil. It is best regarded as a six-membered ring (indicated by blue bonds in the structure shown here) in which two of the carbons (designated by asterisks) are bridged by a CH_2 group.

CH_3 CH_3 CH_2

Bicyclobutane 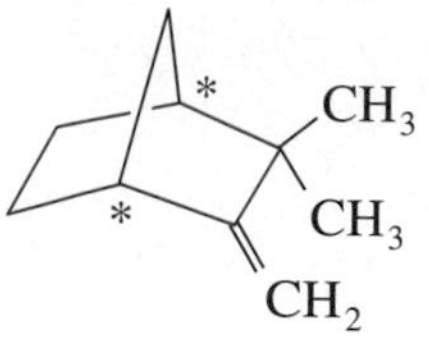   Camphene

---

PROBLEM 3.11 The conclusion that bicyclobutane is bicyclic can be drawn by inspection of its structural formula, which clearly shows the presence of two rings. The case of camphene is not so obvious. Use the bond-cleavage criterion to verify that camphene is bicyclic.

Bicyclic compounds are named in the IUPAC system by counting the number of carbons in the ring system, assigning to the structure the base name of the unbranched alkane having the same number of carbon atoms, and attaching the prefix bicyclo-. The number of atoms in each of the bridges connecting the common atoms is then placed, in descending order, within brackets.

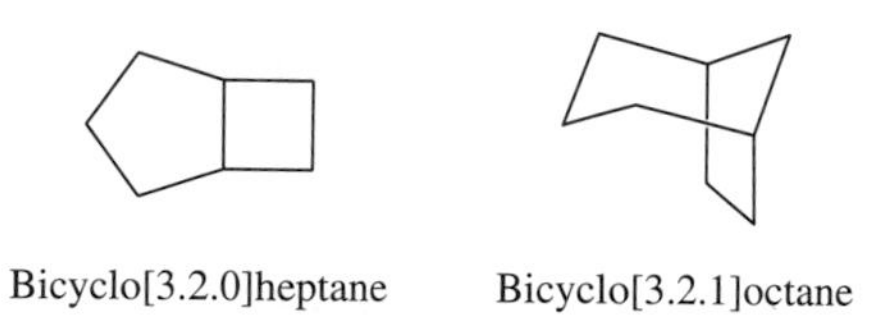

Bicyclo[3.2.0]heptane   Bicyclo[3.2.1]octane

PROBLEM 3.12 Write structural formulas for each of the following bicyclic hydrocarbons:

(*a*) Bicyclo[2.2.1]heptane
(*b*) Bicyclo[5.2.0]nonane
(*c*) Bicyclo[3.1.1]heptane
(*d*) Bicyclo[3.3.0]octane

SAMPLE SOLUTION (*a*) The bicyclo[2.2.1]heptane ring system is one of the most frequently encountered bicyclic structural types. It contains seven carbon atoms, as indicated by the suffix *-heptane.* The bridging groups contain two, two, and one carbon, respectively.

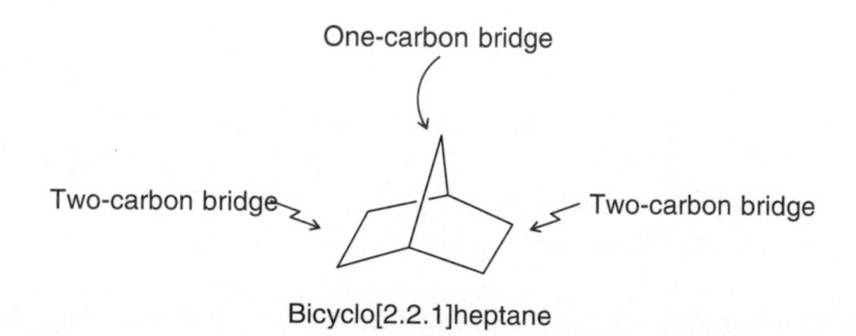

Bicyclo[2.2.1]heptane

Among the most important of the bicyclic hydrocarbons are the two stereoisomeric bicyclo[4.4.0]decanes, called *cis*- and *trans*-decalin. The hydrogen substituents at the ring junction positions are on the same side in *cis*-decalin and on opposite sides in *trans*-decalin. Both rings adopt the chair conformation in each stereoisomer.

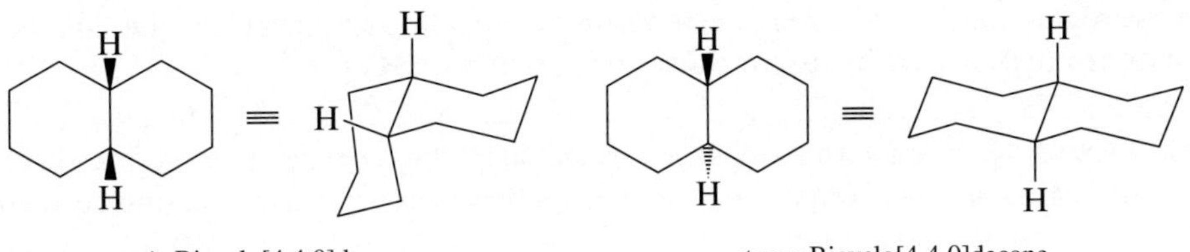

cis-Bicyclo[4.4.0]decane

(*cis*-decalin)

trans-Bicyclo[4.4.0]decane

(*trans*-decalin)

Decalin ring systems appear as structural units in a large number of naturally occurring substances, particularly the steroids. Cholic acid, for example, a steroid present in bile that promotes digestion, incorporates *cis*-decalin and *trans*-decalin units into a rather complex *tetracyclic* structure.

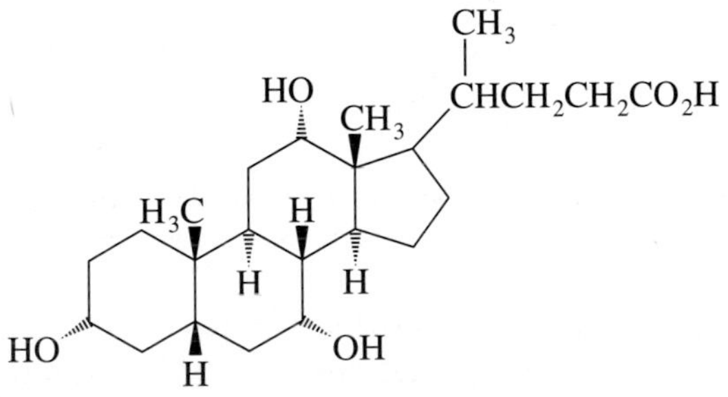

Cholic acid

3.16 HETEROCYCLIC COMPOUNDS

Not all cyclic compounds are hydrocarbons. Many substances include an atom other than carbon, called a *heteroatom* (Section 1.7), as part of a ring. A ring that contains at least one heteroatom is called a **heterocycle,** and a substance based on a heterocyclic ring is a **heterocyclic compound.** Each of the following heterocyclic ring systems will be encountered in this text:

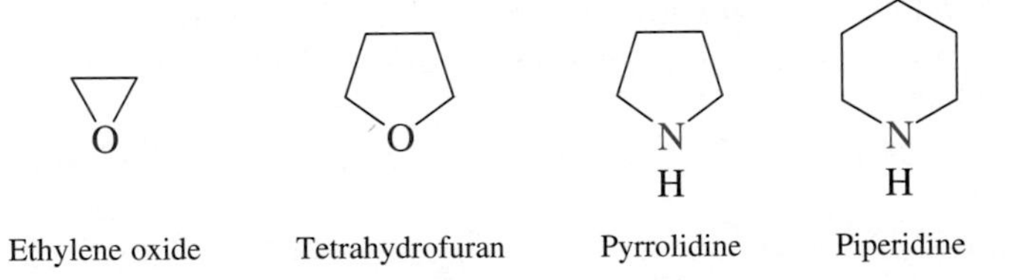

Ethylene oxide Tetrahydrofuran Pyrrolidine Piperidine

The names cited are common names, which have been in widespread use for a long time and are acceptable in IUPAC nomenclature. We will introduce the systematic nomenclature of these ring systems as needed in later chapters.

The shapes of heterocyclic rings are very much like those of their all-carbon analogs. Thus, six-membered heterocycles such as piperidine exist in a chair conformation analogous to cyclohexane.

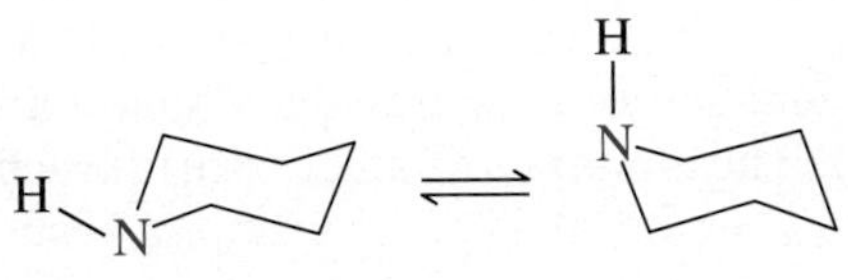

The hydrogen bonded to nitrogen can occupy an axial or an equatorial site, and both chair conformations are approximately equal in stability.

PROBLEM 3.13 Draw what would you expect to be the most stable conformation of the piperidine derivative in which the hydrogen bonded to nitrogen has been replaced by methyl. (This compound is called *N*-methylpiperidine. The N- in *N*-methylpiperidine is a locant signifying that the methyl group is attached to nitrogen.)

Sulfur-containing heterocycles are also common. Compounds in which sulfur is the heteroatom in three-, four-, five-, and six-membered rings, as well as larger rings, are all well known. Two interesting heterocyclic compounds that contain sulfur-sulfur bonds are *lipoic acid* and *lenthionine.*

S—S ring with side chain $CH_2CH_2CH_2CH_2\overset{O}{\overset{\|}{C}}OH$

Lipoic acid—a growth factor required by a variety of different organisms

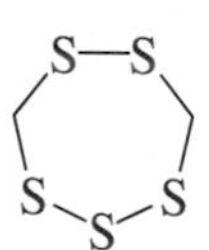

Lenthionine—contributes to the odor of Shiitake mushrooms

Many heterocyclic systems contain double bonds and are related to arenes. The most important representatives of this class are described in Sections 11.23 and 11.24.

3.17 SUMMARY

In this chapter we have explored the three-dimensional shapes of alkanes and cycloalkanes. The most important point to be taken from the chapter is that a molecule adopts the shape that minimizes its total **strain.** The sources of strain in alkanes and cycloalkanes are:

1. *Bond length distortion:* destabilization of a molecule that results when one or more of its bond distances are different from the normal values
2. *Angle strain:* destabilization that results from distortion of bond angles from their normal values
3. *Torsional strain:* destabilization that results from the eclipsing of bonds on adjacent atoms
4. *Van der Waals strain:* destabilization that results when atoms or groups on nonadjacent atoms are too close to one another

The various spatial arrangements available to a molecule by rotation about single bonds are called **conformations,** and **conformational analysis** is the study of the differences in stability and properties of the individual conformations. Rotation around carbon-carbon single bonds is normally very fast, occurring hundreds of thousands of times per second at room temperature. Molecules are rarely frozen into a single conformation but engage in rapid equilibration among the conformations that are energetically accessible.

The most stable conformation of ethane is the **staggered conformation** (Section 3.1). It is some 12 kJ/mol (3 kcal/mol) more stable than the **eclipsed,** which is the least stable conformation.

Staggered conformation of ethane (most stable conformation)

Eclipsed conformation of ethane (least stable conformation)

The difference in energy between the staggered and eclipsed forms is due almost entirely to the torsional strain in the eclipsed conformation. At any instant, almost all the molecules of ethane reside in the staggered conformation (Section 3.2).

The two staggered conformations of butane are not equivalent. The **anti** conformation is more stable than the **gauche** (Section 3.3).

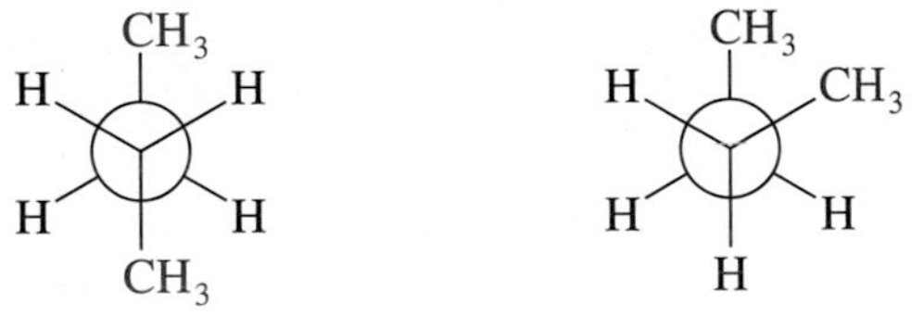

Anti conformation of butane

Gauche conformation of butane

Neither conformation suffers torsional strain, because each has a staggered arrangement of bonds. The gauche conformation is less stable because of van der Waals strain involving the methyl groups.

Higher alkanes adopt a zigzag conformation in which all the bonds are staggered (Section 3.4).

Three conformations of cyclohexane have approximately tetrahedral angles at carbon: the **chair,** the **boat,** and the **skew boat** (Section 3.6). The chair is by far the most stable; it is free of torsional strain while the boat and skew boat are not. When a cyclohexane ring is present as a structural unit in a compound, it almost always adopts a chair conformation. The C—H bonds in the chair conformation of cyclohexane are not all equivalent but are divided into two sets of six each, called **axial** and **equatorial** (Section 3.7). Cyclohexane undergoes a rapid conformational change referred to as **ring inversion** or **ring flipping** (Section 3.8). The process of ring inversion causes all axial bonds to become equatorial and vice versa.

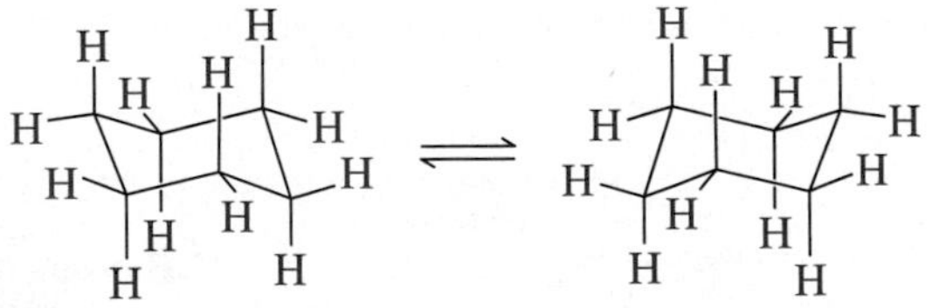

Substituents on a cyclohexane ring are more stable when they are equatorial than when they are axial (Section 3.9). Branched substituents, especially *tert*-butyl, have a pronounced preference for the equatorial position.

Cyclopropane (Section 3.10) is planar and strained (angle strain and torsional strain). Cyclobutane (Section 3.10) is nonplanar and less strained than cyclopropane. Cyclopentane (Section 3.11) has two nonplanar conformations that are of similar stability, the **envelope** and the **half-chair.**

Nonplanar conformation of cyclobutane

Envelope conformation of cyclopentane

Half-chair conformation of cyclopentane

Higher cycloalkanes have angles at carbon that are close to tetrahedral and are sufficiently flexible to be free of torsional strain. They tend to be populated by several different conformations of similar stability (Section 3.12).

Stereoisomers are isomers that have the same constitution but differ in the arrangement of atoms in space (Section 3.13). The relative stabilities of stereoisomeric disubstituted (and more highly substituted) cyclohexanes can be assessed by analyzing chair conformations for van der Waals strain involving axial substituents (Section 3.14).

Cyclic hydrocarbons can contain more than one ring (Section 3.15). **Spirocyclic** hydrocarbons are characterized by the presence of a single carbon that is common to two rings. **Bicyclic** alkanes contain two rings that share two or more atoms.

Substances that contain one or more atoms other than carbon as part of a ring are called **heterocyclic compounds** (Section 3.16). Rings in which the heteroatom is oxygen, nitrogen, or sulfur rank as both the most common and the most important.

PROBLEMS

3.14 Like hydrogen peroxide, the inorganic substances hydrazine (H_2NNH_2) and hydroxylamine (H_2NOH) possess conformational mobility. Write structural representations for two different staggered conformations of (*a*) hydrazine and (*b*) hydroxylamine.

3.15 Of the three conformations of propane shown in Figure 3.21, which one is the most stable? Which one is the least stable? Why?

3.16 Sight down the C-2—C-3 bond and draw Newman projection formulas for the

(*a*) Most stable conformation of 2,2-dimethylbutane
(*b*) Two most stable conformations of 2-methylbutane
(*c*) Two most stable conformations of 2,3-dimethylbutane

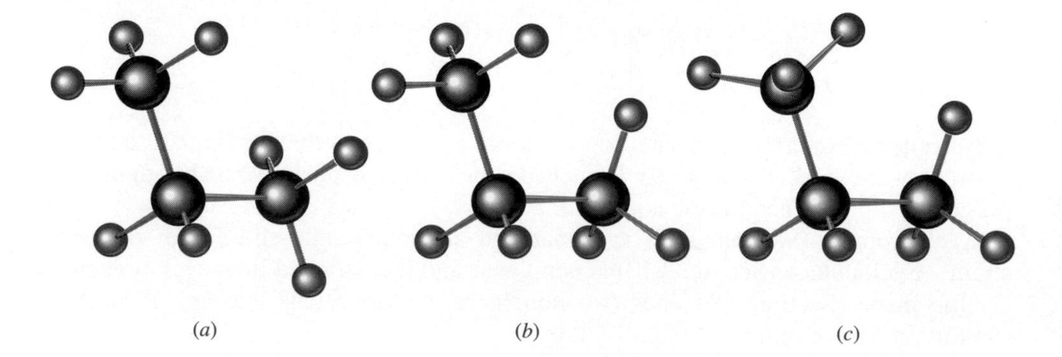

FIGURE 3.21 Conformations of propane (Problem 3.15).

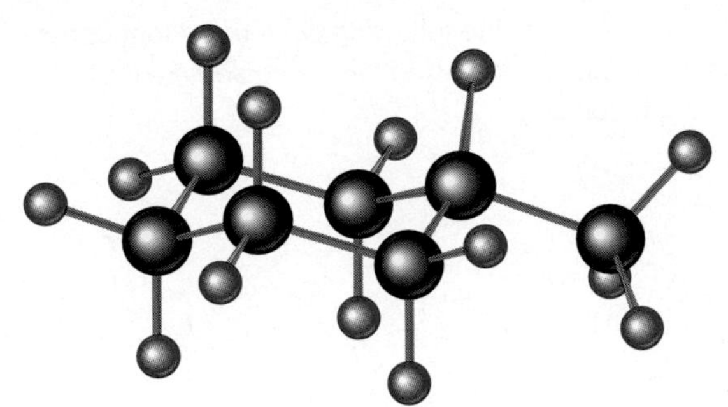

FIGURE 3.22 Conformation of equatorial methylcyclohexane to be considered in Problem 3.21.

3.17 One of the staggered conformations of 2-methylbutane in Problem 3.16*b* is more stable than the other. Which one is more stable? Why?

3.18 Sketch an approximate potential energy diagram similar to that shown in Figures 3.4 and 3.7 for rotation about the carbon-carbon bond in 2,2-dimethylpropane. Does the form of the potential energy curve of 2,2-dimethylpropane more closely resemble that of ethane or that of butane?

3.19 Repeat Problem 3.18 for the case of 2-methylbutane.

3.20 One of the C—C—C angles of 2,2,4,4-tetramethylpentane is very much larger than the others. Which angle? Why?

3.21 Even though the methyl group occupies an equatorial site, the conformation shown in Figure 3.22 is not the most stable one for methylcyclohexane. Explain.

3.22 Which of the structures shown in Figure 3.23 for the axial conformation of methylcyclohexane do you think is more stable, A or B? Why?

3.23 Which do you expect to be the more stable conformation of *cis*-1,3-dimethylcyclobutane, A or B? Why?

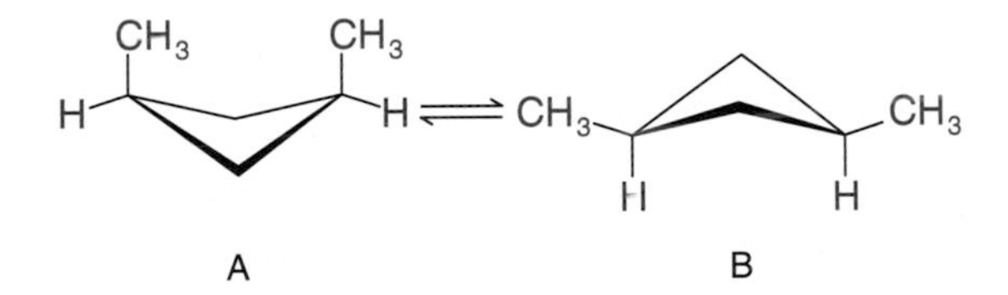

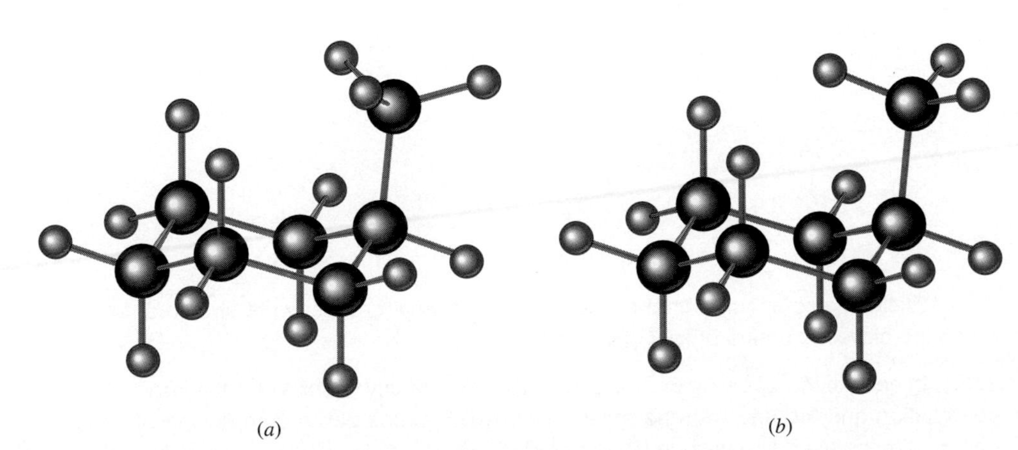

FIGURE 3.23 Nonequivalent conformations of axial methylcyclohexane (Problem 3.22).

3.24 Determine whether the two structures in each of the following pairs represent *constitutional isomers,* different *conformations* of the same compound, or *stereoisomers* that cannot be interconverted by rotation about single bonds.

(*a*)

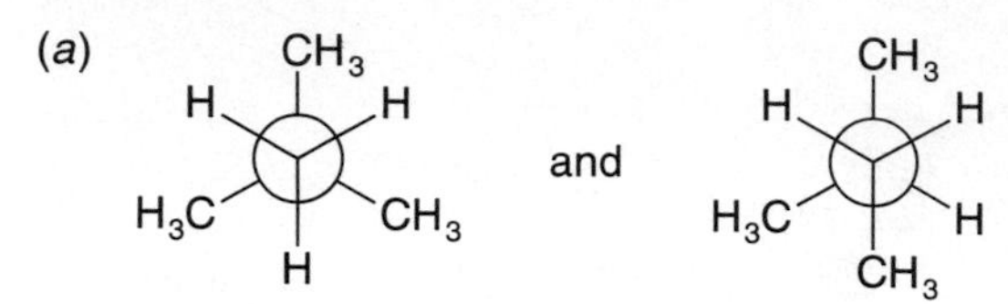

(*b*)

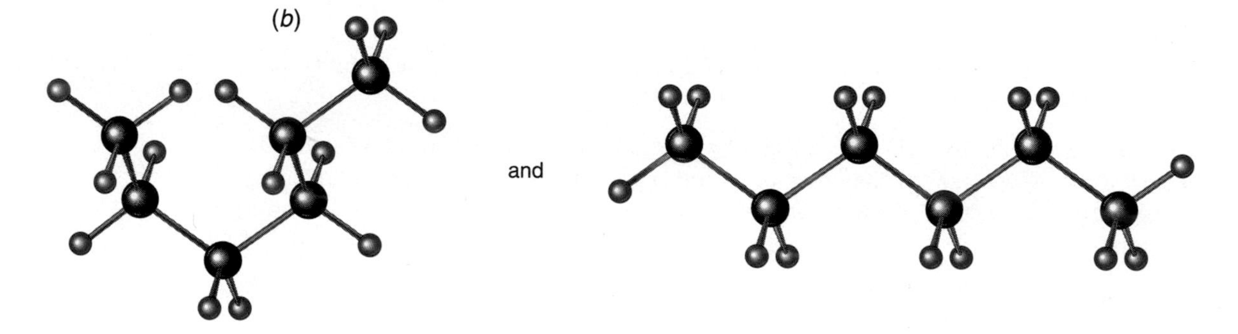

(*c*)

H, H_3C, CH_3, H_3C, CH_3, H and CH_3, H_3C, H, H_3C, H, CH_3

(*d*) *cis*-1,2-Dimethylcyclopentane and *trans*-1,3-dimethylcyclopentane

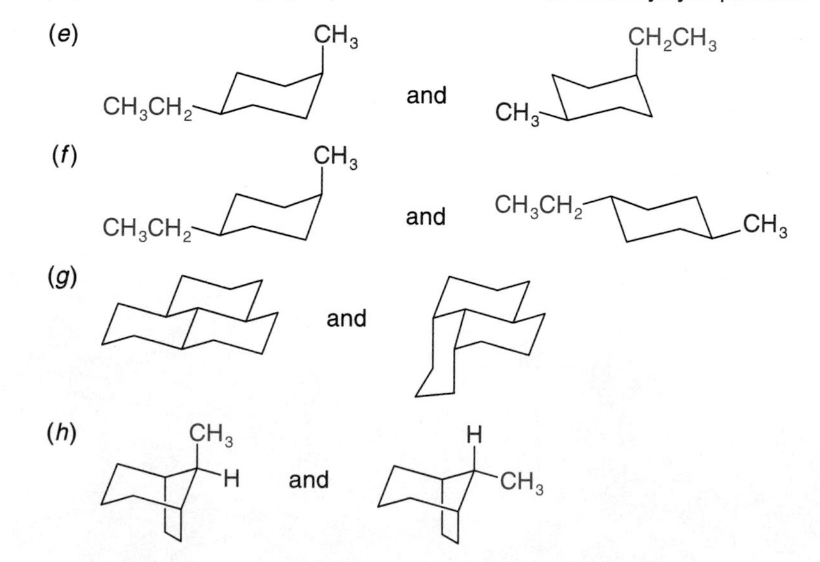

3.25 Excluding compounds that contain methyl or ethyl groups, write structural formulas for all the bicyclic isomers of (*a*) C_5H_8 and (*b*) C_6H_{10}.

3.26 In each of the following groups of compounds, identify the one with the largest heat of combustion and the one with the smallest. In which cases can a comparison of heats of combustion be used to assess relative stability?

(*a*) Cyclopropane, cyclobutane, cyclopentane

(*b*) *cis*-1,2-Dimethylcyclopentane, methylcyclohexane, 1,1,2,2-tetramethylcyclopropane

(*c*)

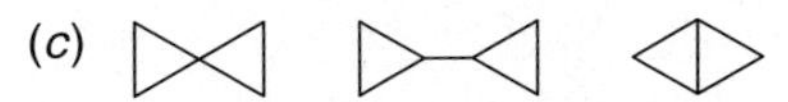

(*d*)

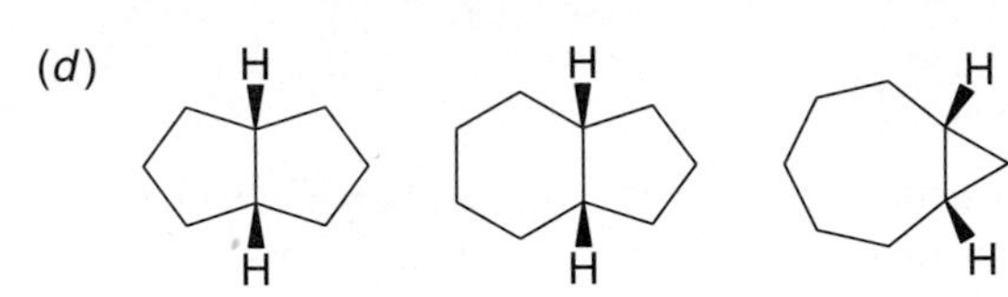

3.27 Write a structural formula for the most stable conformation of each of the following compounds:

(*a*) 2,2,5,5-Tetramethylhexane (Newman projection of conformation about C-3—C-4 bond)

(*b*) 2,2,5,5-Tetramethylhexane (zigzag conformation of entire molecule)

(*c*) *cis*-1-Isopropyl-3-methylcyclohexane

(*d*) *trans*-1-Isopropyl-3-methylcyclohexane

(*e*) *cis*-1-*tert*-Butyl-4-ethylcyclohexane

(*f*) *cis*-1,1,3,4-Tetramethylcyclohexane

(*g*)

3.28 Identify the more stable stereoisomer in each of the following pairs and give the reason for your choice:

(*a*) *cis*- or *trans*-1-Isopropyl-2-methylcyclohexane

(*b*) *cis*- or *trans*-1-Isopropyl-3-methylcyclohexane

(*c*) *cis*- or *trans*-1-Isopropyl-4-methylcyclohexane

(*d*) or

(*e*) or

(*f*) or

3.29 One stereoisomer of 1,1,3,5-tetramethylcyclohexane is 15 kJ/mol (3.7 kcal/mol) less stable than the other. Indicate which isomer is the less stable and identify the reason for its decreased stability.

3.30 One of the following two stereoisomers is 20 kJ/mol (4.9 kcal/mol) less stable than the other. Indicate which isomer is the less stable, and identify the reason for its decreased stability.

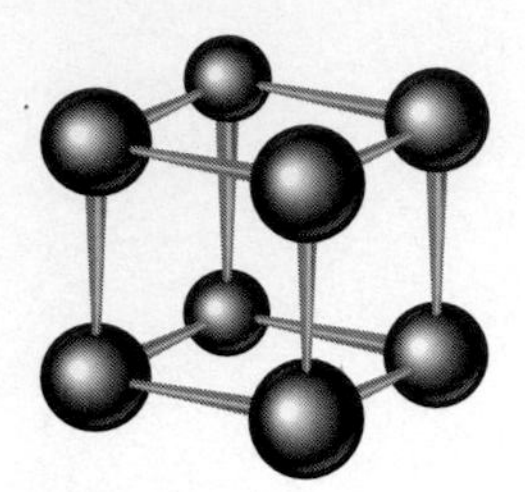

FIGURE 3.24 Carbon skeleton of cubane (Problem 3.31).

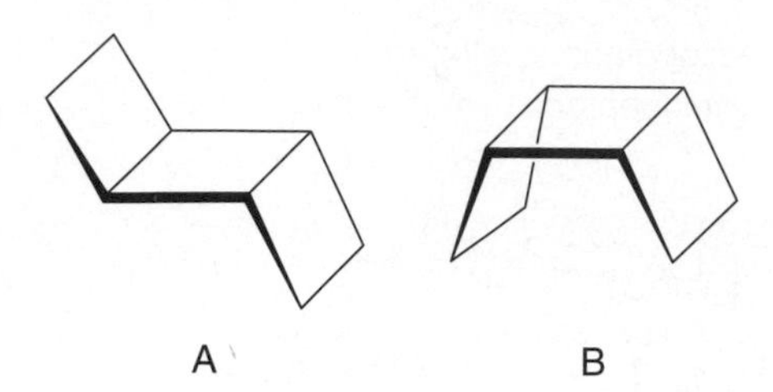

3.31 Cubane (C_8H_8) is the common name of a polycyclic hydrocarbon that was first synthesized in the early 1960s. As its name implies, its structure is that of a cube, as shown in Figure 3.24. How many rings are present in cubane?

3.32 The following are representations of two forms of glucose. The six-membered ring is known to exist in a chair conformation in each form. Draw clear representations of the most stable conformation of each. Are they two different conformations of the same molecule, or are they stereoisomers? Which substituents (if any) occupy axial sites?

3.33 A typical steroid skeleton is shown along with the numbering scheme used for this class of compounds. Specify in each case whether the designated substituent is axial or equatorial.

(*a*) Substituent at C-1 cis to the methyl groups
(*b*) Substituent at C-4 cis to the methyl groups
(*c*) Substituent at C-7 trans to the methyl groups
(*d*) Substituent at C-11 trans to the methyl groups
(*e*) Substituent at C-12 cis to the methyl groups

3.34 Repeat Problem 3.33 for the stereoisomeric steroid skeleton having a cis ring fusion between the first two rings.

3.35

(*a*) Write Newman projection formulas for the gauche and anti conformations of 1,2-dichloroethane ($ClCH_2CH_2Cl$).

(*b*) The measured dipole moment of $ClCH_2CH_2Cl$ is 1.12 D. Which one of the following statements about 1,2-dichloroethane is false?
(1) It may exist entirely in the anti conformation.
(2) It may exist entirely in the gauche conformation.
(3) It may exist as a mixture of anti and gauche conformations.

3.36
(*a*) Sketch the planar and the nonplanar conformations of *trans*-1,3-dibromocyclobutane.

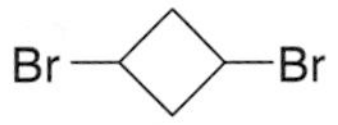

Br—◇—Br

1,3-Dibromocyclobutane

(*b*) The measured dipole moment of *trans*-1,3-dibromocyclobutane is 1.10 D. Which one of the following statements about this compound is false? Explain.
(1) It may exist entirely in the conformation in which the ring is planar.
(2) It may exist entirely in the conformation in which the ring is nonplanar.
(3) It may exist as a mixture of conformations.

MOLECULAR MODELING EXERCISES

3.37 Construct molecular models for the most important structures in this chapter.

(*a*) ethane (staggered and eclipsed conformations)
(*b*) butane (anti and gauche conformations)
(*c*) cyclohexane (chair conformation)
(*d*) methylcyclohexane (two nonequivalent chair conformations)

3.38 Construct a molecular model of the most stable conformation of *trans*-1-*tert*-butyl-3-methylcyclohexane.

3.39 Verify that an axial chlorine substituent at C-1 in chlorocyclohexane is closer to an axial hydrogen at C-3 than to an equatorial hydrogen at C-2.

3.40 Does your molecular models set allow you to make a model of cyclopropane? If not, why not?

3.41 Construct molecular models of the following hydrocarbons from Section 3.15.

(*a*) *cis*-decalin
(*b*) *trans*-decalin
(*c*) bicyclo[3.2.0]heptane
(*d*) bicyclo[3.2.1]octane

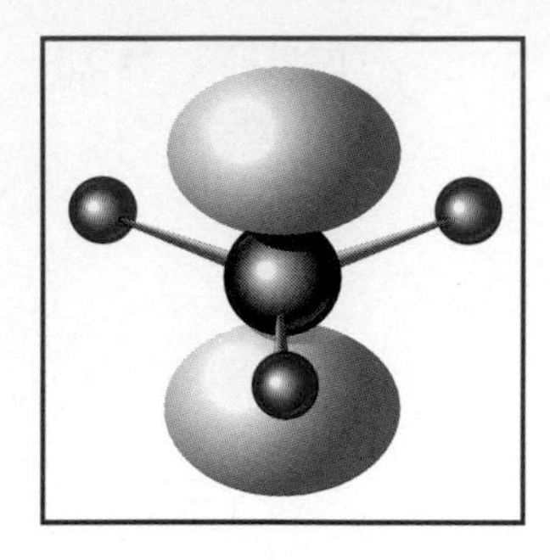

CHAPTER 4

ALCOHOLS AND ALKYL HALIDES

The first three chapters established some fundamental principles concerning the *structure* of organic molecules. In this chapter we begin our discussion of organic chemical *reactions* by directing attention to *alcohols* and *alkyl halides*. These two rank among the most useful classes of organic compounds because they often serve as starting materials for the preparation of numerous other families.

Two reactions that lead to alkyl halides will be described in this chapter. Both illustrate functional group transformations. In the first, the hydroxyl group of an alcohol is replaced by halogen on treatment with a hydrogen halide.

$$\underset{\text{Alcohol}}{R{-}OH} + \underset{\text{Hydrogen halide}}{H{-}X} \longrightarrow \underset{\text{Alkyl halide}}{R{-}X} + \underset{\text{Water}}{H{-}OH}$$

In the second, reaction with chlorine or bromine causes one of the hydrogen substituents of an alkane to be replaced by halogen.

$$\underset{\text{Alkane}}{R{-}H} + \underset{\text{Halogen}}{X_2} \longrightarrow \underset{\text{Alkyl halide}}{R{-}X} + \underset{\text{Hydrogen halide}}{H{-}X}$$

Both reactions are classified as *substitutions,* a term that describes the relationship between reactants and products—one functional group replaces another. In this chapter we go beyond the relationship of reactants and products and consider the *mechanism* of each reaction. A **mechanism** attempts to show *how* starting materials are converted into products in a chemical reaction.

While developing these themes of reaction and mechanism, we will also use alcohols and alkyl halides as vehicles to extend the principles of IUPAC nomenclature, continue to develop concepts of structure and bonding, and see how structure affects properties. A review of *acids and bases* comprises an important part of this chapter where a qualitative approach to proton-transfer equilibria will be developed that will be used throughout the remainder of the text.

4.1 IUPAC NOMENCLATURE OF ALKYL HALIDES

The IUPAC rules permit alkyl halides to be named in two different ways, called *radicofunctional* nomenclature and *substitutive* nomenclature. In **radicofunctional nomenclature** the alkyl group and the halide (*fluoride, chloride, bromide,* or *iodide*) are designated as separate words. The alkyl group is named on the basis of its longest continuous chain beginning at the carbon to which the halogen is attached.

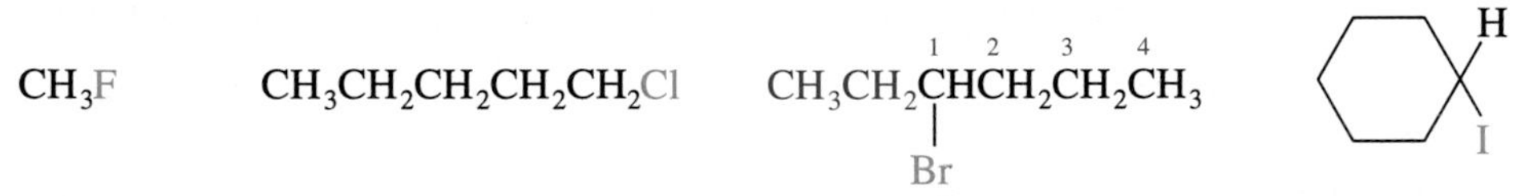

The IUPAC rules permit certain common alkyl group names to be used. These include *n*-propyl, isopropyl, *n*-butyl, *sec*-butyl, isobutyl, *tert*-butyl, and neopentyl (Section 2.12).

CH_3F — Methyl fluoride

$CH_3CH_2CH_2CH_2CH_2Cl$ — Pentyl chloride

$CH_3CH_2\overset{1}{C}H(Br)\overset{2}{C}H_2\overset{3}{C}H_2\overset{4}{C}H_3$ — 1-Ethylbutyl bromide

Cyclohexyl iodide (cyclohexane ring bearing H and I on one carbon)

Substitutive nomenclature of alkyl halides treats the halogen as a *halo-* (*fluoro-, chloro-, bromo-,* or *iodo-*) *substituent* on an alkane chain. The carbon chain is numbered in the direction that gives the substituted carbon the lower locant.

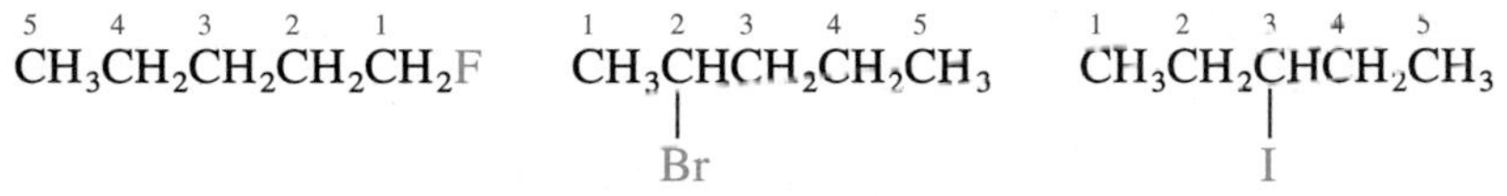

$\overset{5}{C}H_3\overset{4}{C}H_2\overset{3}{C}H_2\overset{2}{C}H_2\overset{1}{C}H_2F$ — 1-Fluoropentane

$\overset{1}{C}H_3\overset{2}{C}H(Br)\overset{3}{C}H_2\overset{4}{C}H_2\overset{5}{C}H_3$ — 2-Bromopentane

$\overset{1}{C}H_3\overset{2}{C}H_2\overset{3}{C}H(I)\overset{4}{C}H_2\overset{5}{C}H_3$ — 3-Iodopentane

When the carbon chain bears both a halogen and an alkyl substituent, the two substituents are considered of equal rank and the chain is numbered so as to give the lower number to the substituent nearer the end of the chain.

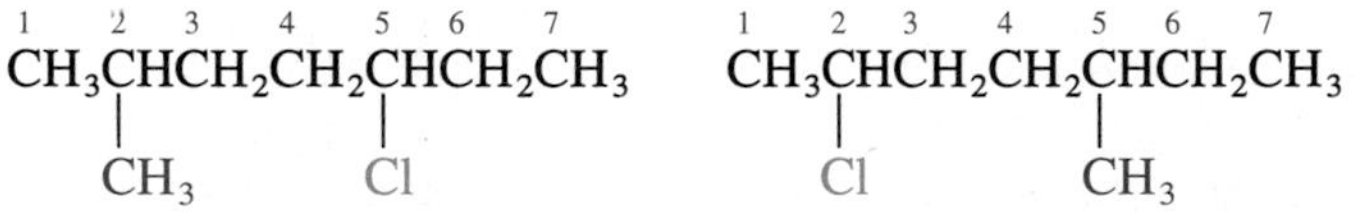

$\overset{1}{C}H_3\overset{2}{C}H(CH_3)\overset{3}{C}H_2\overset{4}{C}H_2\overset{5}{C}H(Cl)\overset{6}{C}H_2\overset{7}{C}H_3$ — 5-Chloro-2-methylheptane

$\overset{1}{C}H_3\overset{2}{C}H(Cl)\overset{3}{C}H_2\overset{4}{C}H_2\overset{5}{C}H(CH_3)\overset{6}{C}H_2\overset{7}{C}H_3$ — 2-Chloro-5-methylheptane

PROBLEM 4.1 Write structural formulas and give the radicofunctional and substitutive names of all the isomeric alkyl chlorides that have the molecular formula C_4H_9Cl.

Substitutive names are preferred, but radicofunctional names are sometimes more convenient or more familiar and are frequently encountered in organic chemistry.

4.2 IUPAC NOMENCLATURE OF ALCOHOLS

Radicofunctional names of alcohols are derived by naming the alkyl group that bears the hydroxyl substituent (—OH) and then adding *alcohol* as a separate word. The

Several alcohols are commonplace substances, well known by common names that reflect their origin (wood alcohol, grain alcohol) or use (rubbing alcohol). Wood alcohol is *methanol* (methyl alcohol, CH_3OH), grain alcohol is *ethanol* (ethyl alcohol, CH_3CH_2OH), and rubbing alcohol is *2-propanol* [isopropyl alcohol, $(CH_3)_2CHOH$].

chain is always numbered beginning at the carbon to which the hydroxyl group is attached.

Substitutive names of alcohols are developed by identifying the longest continuous chain that bears the hydroxyl group and replacing the *-e* ending of the corresponding alkane by the suffix *-ol*. The position of the hydroxyl group is indicated by number, choosing the sequence that assigns the lower locant to the carbon that bears the hydroxyl group.

	CH_3CH_2OH	$CH_3CH(OH)CH_2CH_2CH_2CH_3$	$CH_3C(CH_3)(OH)CH_2CH_2CH_3$
Radicofunctional name:	Ethyl alcohol	1-Methylpentyl alcohol	1,1-Dimethylbutyl alcohol
Substitutive name:	Ethanol	2-Hexanol	2-Methyl-2-pentanol

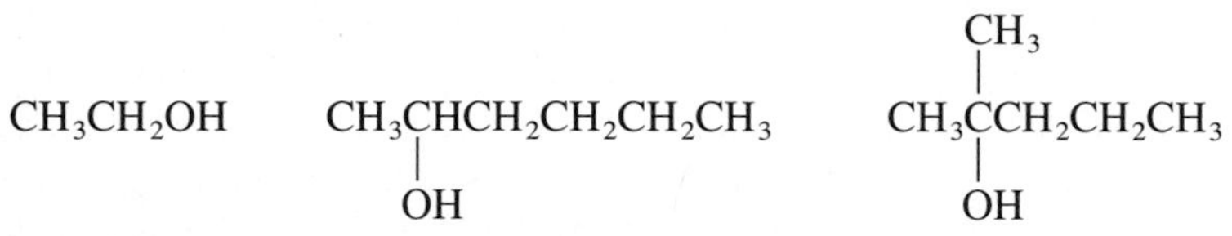

Hydroxyl groups take precedence over (''outrank'') alkyl groups and halogen substituents in determining the direction in which a carbon chain is numbered.

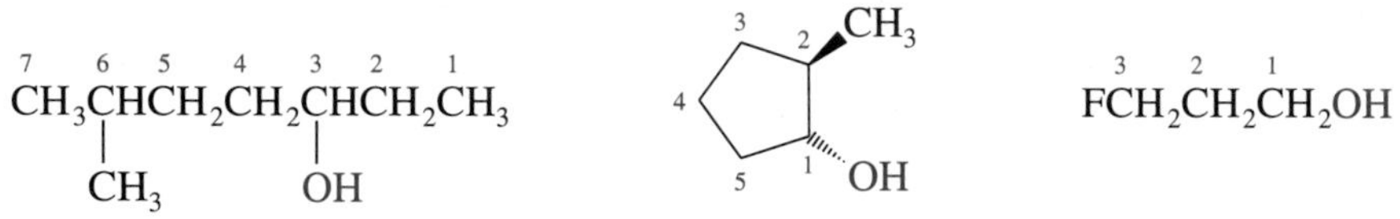

$\overset{7}{C}H_3\overset{6}{C}H(CH_3)\overset{5}{C}H_2\overset{4}{C}H_2\overset{3}{C}H(OH)\overset{2}{C}H_2\overset{1}{C}H_3$

6-Methyl-3-heptanol
(not 2-methyl-5-heptanol)

trans-2-Methylcyclopentanol

$F\overset{3}{C}H_2\overset{2}{C}H_2\overset{1}{C}H_2OH$

3-Fluoro-1-propanol

PROBLEM 4.2 Write structural formulas and give the radicofunctional and substitutive names of all the isomeric alcohols that have the molecular formula $C_4H_{10}O$.

4.3 CLASSES OF ALCOHOLS AND ALKYL HALIDES

Alcohols and alkyl halides are classified as primary, secondary, or tertiary according to the classification of the carbon that bears the functional group (Section 2.12). Thus, *primary alcohols* and *primary alkyl halides* are compounds of the type RCH_2G (where G is the functional group), *secondary alcohols* and *secondary alkyl halides* are compounds of the type R_2CHG, and *tertiary alcohols and tertiary alkyl halides* are compounds of the type R_3CG.

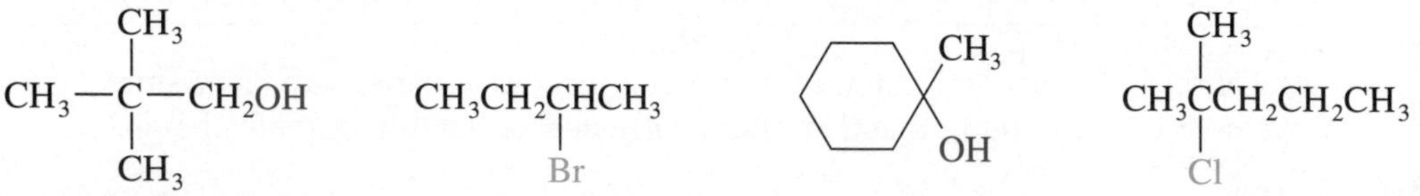

$CH_3—C(CH_3)_2—CH_2OH$

2,2-Dimethyl-1-propanol
(a primary alcohol)

$CH_3CH_2CH(Br)CH_3$

2-Bromobutane
(a secondary alkyl halide)

1-Methylcyclohexanol
(a tertiary alcohol)

$CH_3C(CH_3)(Cl)CH_2CH_2CH_3$

2-Chloro-2-methylpentane
(a tertiary alkyl halide)

PROBLEM 4.3 Classify the isomeric $C_4H_{10}O$ alcohols as primary, secondary, or tertiary.

Many of the properties of alcohols and alkyl halides are affected by whether their functional groups are attached to primary, secondary, or tertiary carbons. We will encounter numerous instances in which a functional group attached to a primary carbon is more reactive than one attached to a secondary or tertiary carbon, and numerous other instances in which the reverse is true.

4.4 BONDING IN ALCOHOLS AND ALKYL HALIDES

The carbon that bears the functional group is sp^3 hybridized in alcohols and alkyl halides. Figure 4.1 illustrates the orbital hybridization approach to bonding in alcohols for the specific example of methanol. The bond angles at carbon are approximately tetrahedral, as is the C—O—H angle. A similar orbital hybridization model applies to alkyl halides, with the halogen substituent connected to sp^3 hybridized carbon by a σ bond. Carbon-halogen bond distances in alkyl halides increase in the order C—F (140 pm) < C—Cl (179 pm) < C—Br (197 pm) < C—I (216 pm).

Carbon-oxygen and carbon-halogen bonds are polar covalent bonds, and carbon bears a partial positive charge in alcohols ($^{\delta+}$C—O$^{\delta-}$) and in alkyl halides ($^{\delta+}$C—X$^{\delta-}$). The presence of these polar bonds makes alcohols and alkyl halides such as methanol and chloromethane polar molecules. Their dipole moments are very similar to one another and to that of water.

H—O—H	H_3C—O—H	CH_3—Cl
Water (μ = 1.8 D)	Methanol (μ = 1.7 D)	Chloromethane (μ = 1.9 D)

PROBLEM 4.4 Bromine is less electronegative than chlorine, yet methyl bromide and methyl chloride have dipole moments that are very similar to each other. Can you think of a reason for this?

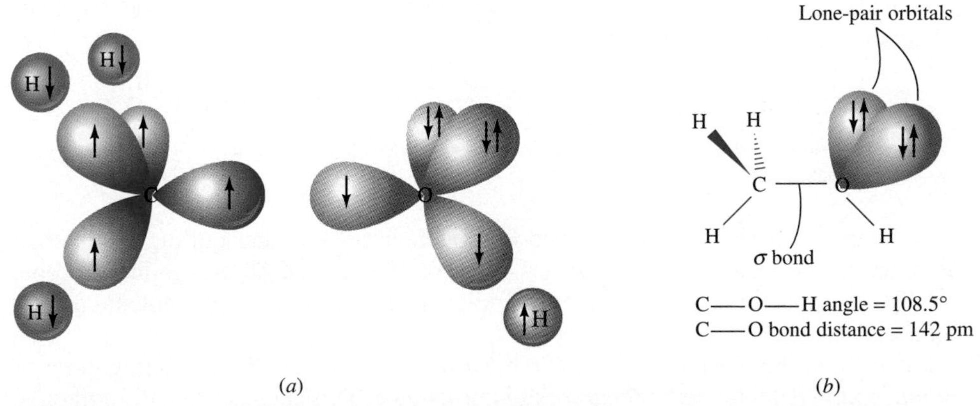

FIGURE 4.1 Orbital hybridization model of bonding in methanol. (*a*) The orbitals used in bonding are the 1*s* orbitals of hydrogen, sp^3 hybridized orbitals of carbon, and sp^3 hybridized orbitals of oxygen. (*b*) The bond angles at carbon and oxygen are close to tetrahedral, and the carbon-oxygen σ bond distance is about 10 pm shorter than a carbon-carbon single bond.

Relatively simple notions of attractive forces between opposite electrical charges are sufficient to account for many of the properties of chemical substances. You will find it helpful to keep the polarity of carbon-oxygen and carbon-halogen bonds in mind as the properties of alcohols and alkyl halides are developed in later sections.

4.5 PHYSICAL PROPERTIES OF ALCOHOLS AND ALKYL HALIDES. INTERMOLECULAR FORCES IN POLAR MOLECULES

Boiling Point. When describing the effect of alkane structure on boiling point in Section 2.16, we pointed out that the forces of attraction between molecules are of three types:

1. Induced dipole–induced dipole forces (also called *dispersion forces*)
2. Dipole–induced dipole forces
3. Dipole-dipole forces

Induced dipole–induced dipole forces are the only intermolecular attractive forces available to nonpolar molecules such as alkanes, and are the major forces of attraction in most substances. In addition to these, polar molecules also engage in attractive interactions of the dipole-dipole and dipole–induced dipole type. The **dipole-dipole** attractive force is easiest to visualize and is illustrated in Figure 4.2. Two molecules of a polar substance experience a mutual attraction between the positively polarized region of one molecule and the negatively polarized region of the other. As its name implies, the **dipole–induced dipole force** combines features of both the induced dipole–induced dipole and dipole-dipole attractive forces. A polar region of one molecule alters the electron distribution in a nonpolar region of another in a direction that produces an attractive force between them.

Because so many factors contribute to the net intermolecular attractive force, it is not always possible to predict which of two compounds will have the higher boiling point. We can, however, use the boiling point behavior of selected molecules to inform us of the relative importance of various intermolecular forces and the structural features that influence them.

Consider three compounds similar in size and shape: the alkane propane, the alcohol ethanol, and the alkyl halide fluoroethane.

$CH_3CH_2CH_3$	CH_3CH_2OH	CH_3CH_2F
Propane (μ = 0 D)	Ethanol (μ = 1.7 D)	Fluoroethane (μ = 1.9 D)
bp: −42°C	bp: 78°C	bp: −32°C

Both the polar compounds ethanol and fluoroethane have higher boiling points than the nonpolar propane. We attribute this to a combination of dipole–induced dipole and dipole-dipole attractive forces that stabilize the liquid states of ethanol and fluoroethane.

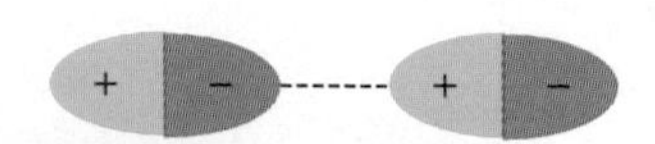

FIGURE 4.2 A dipole-dipole attractive force. Two molecules of a polar substance are oriented so that the positively polarized region of one and the negatively polarized region of the other attract each other.

The most striking aspect of the data, however, is the much higher boiling point of ethanol compared with both propane and fluoroethane. This suggests that the attractive forces in ethanol must be unusually strong. Figure 4.3 shows that this force results from a dipole-dipole attraction between the positively polarized proton of the —OH group of one ethanol molecule and the negatively polarized oxygen of another. The term **hydrogen bonding** is used to describe dipole-dipole attractive forces of this type.

The proton involved must be bonded to an electronegative element, usually oxygen or nitrogen. Protons in C—H bonds do not participate in hydrogen bonding. Thus fluoroethane, even though it is a polar molecule and engages in dipole-dipole attractions, does not form hydrogen bonds and, therefore, has a lower boiling point than ethanol.

Hydrogen bonding can be expected in molecules that have —OH or —NH groups. Individual hydrogen bonds are about 10–50 times weaker than typical covalent bonds, but their effects can be significant. More than other dipole-dipole attractive forces, intermolecular hydrogen bonds are strong enough to impose a relatively high degree of structural order on systems in which they are possible. As will be seen in Chapters 28 and 29, the three-dimensional structures adopted by proteins and nucleic acids, the organic molecules of life, are dictated by patterns of hydrogen bonds.

Hydrogen bonds between —OH groups are stronger than those between —NH groups, as a comparison of the boiling points of water (H_2O, 100°C) and ammonia (NH_3, −33°C) demonstrates.

PROBLEM 4.5 The constitutional isomer of ethanol, dimethyl ether (CH_3OCH_3), is a gas at room temperature. Suggest an explanation for this observation.

Table 4.1 lists the boiling points of some representative alkyl halides and alcohols. When comparing the boiling points of related compounds as a function of the *alkyl group,* we find that the boiling point increases with the number of carbon atoms as it does with alkanes. With respect to the *halogen* in a group of alkyl halides, the boiling point increases as one descends the periodic table; alkyl fluorides have the lowest boiling points, alkyl iodides the highest. This trend matches the order of increasing **polarizability** of the halogens. *Polarizability* is the ease with which the electron distribution around an atom is distorted by a nearby electric field and is a significant factor in determining the strength of induced dipole–induced dipole and dipole–induced dipole attractions. Forces that depend on induced dipoles are strongest when the halogen is a highly polarizable iodine, and weakest when the halogen is a nonpolarizable fluorine.

For a discussion concerning the boiling point behavior of alkyl halides, see the January 1988 issue of the *Journal of Chemical Education,* pp. 62–64.

The boiling points of the chlorinated derivatives of methane increase with the number of chlorine atoms because of an increase in the induced dipole–induced dipole attractive forces.

	CH_3Cl	CH_2Cl_2	$CHCl_3$	CCl_4
	Chloromethane (methyl chloride)	Dichloromethane (methylene dichloride)	Trichloromethane (chloroform)	Tetrachloromethane (carbon tetrachloride)
Boiling point:	−24°C	40°C	61°C	77°C

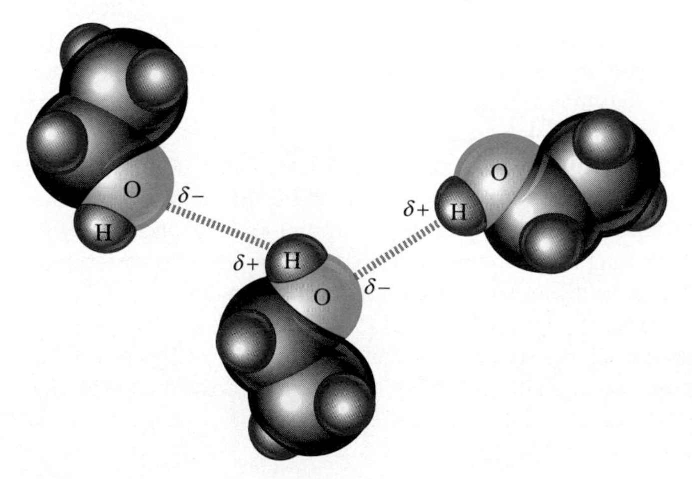

FIGURE 4.3 Hydrogen bonding in ethanol. The dipole-dipole attractive forces in ethanol involve the oxygen of one molecule and the proton of an —OH group of another. They are much stronger than most other types of dipole-dipole attractive forces.

TABLE 4.1
Boiling Points of Some Alkyl Halides and Alcohols

Name of alkyl group	Formula	Functional group X and boiling point, °C (1 atm)				
		X = F	X = Cl	X = Br	X = I	X = OH
Methyl	CH_3X	−78	−24	3	42	65
Ethyl	CH_3CH_2X	−32	12	38	72	78
Propyl	$CH_3CH_2CH_2X$	−3	47	71	103	97
Pentyl	$CH_3(CH_2)_3CH_2X$	65	108	129	157	138
Hexyl	$CH_3(CH_2)_4CH_2X$	92	134	155	180	157

Fluorine is unique among the halogens in that an accumulation of fluorine substituents does not give rise to successively higher and higher boiling points.

	CH_3CH_2F	CH_3CHF_2	CH_3CF_3	CF_3CF_3
	Fluoroethane	1,1-Difluoroethane	1,1,1-Trifluoroethane	Hexafluoroethane
Boiling point:	−32°C	−25°C	−47°C	−78°C

Thus, while the difluoride CH_3CHF_2 boils at a higher temperature than CH_3CH_2F, the trifluoride CH_3CF_3 boils at a lower temperature than either of them. Even more striking is the observation that the hexafluoride CF_3CF_3 is the lowest-boiling of any of the fluorinated derivatives of ethane. The boiling point of CF_3CF_3 is, in fact, only 11° higher than that of ethane itself. The reason for this behavior has to do with the very low polarizability of fluorine and a decrease in induced dipole–induced dipole forces that accompanies the incorporation of fluorine substituents into a molecule. Their weak intermolecular attractive forces give fluorinated hydrocarbons **(fluorocarbons)** certain desirable physical properties such as that found in the ''no stick'' *Teflon* coating of frying pans. Teflon is a *polymer* (Section 6.21) made up of long chains of $—CF_2CF_2—$ units.

Solubility in Water. Alkyl halides and alcohols differ markedly from one another in their solubility in water. All alkyl halides are insoluble in water, but low-molecular-weight alcohols (methyl, ethyl, *n*-propyl, and isopropyl) are soluble in water in all proportions. Their ability to participate in intermolecular hydrogen bonding not only affects the boiling points of alcohols, but also enhances their water solubility. Hydrogen-bonded networks of the type shown in Figure 4.4, in which alcohol and water molecules associate with one another, replace the alcohol-alcohol and water-water hydrogen-bonded networks present in the pure substances.

Higher alcohols become more ''hydrocarbonlike'' and less water-soluble. 1-Octanol, for example, dissolves to the extent of only 1 mL in 2000 mL of water. Attractive forces of the induced dipole–induced dipole type cause 1-octanol molecules to bind to each other through their long alkyl chains, and hydrogen bonding to water does not provide enough stabilization to dissociate these 1-octanol aggregates.

Density. Alkyl fluorides and chlorides are less dense than water; alkyl bromides and iodides are more dense than water. The following densities cited for the 1-halooctanes are representative.

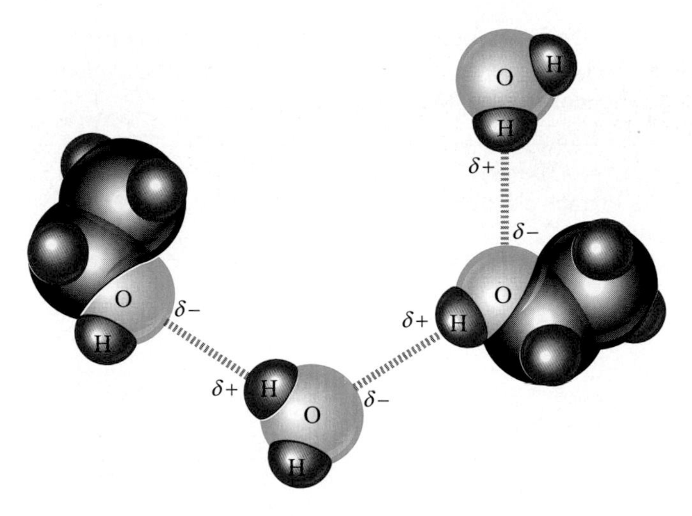

FIGURE 4.4 Hydrogen bonding between molecules of ethanol and water.

	$CH_3(CH_2)_6CH_2F$	$CH_3(CH_2)_6CH_2Cl$	$CH_3(CH_2)_6CH_2Br$	$CH_3(CH_2)_6CH_2I$
Density (20°C):	0.80 g/mL	0.89 g/mL	1.12 g/mL	1.34 g/mL

Since alkyl halides are insoluble in water, a mixture of an alkyl halide and water separates into two layers. When the alkyl halide is a fluoride or chloride, it is the upper layer and water is the lower. The situation is reversed when the alkyl halide is a bromide or an iodide. In these cases the alkyl halide is the lower layer. Polyhalogenation increases the density. The compounds CH_2Cl_2, $CHCl_3$, and CCl_4, for example, are all denser than water.

All liquid alcohols have densities of approximately 0.8 g/mL and are, therefore, less dense than water.

4.6 ACIDS AND BASES: GENERAL PRINCIPLES

A firm grasp of the theories of acidity and basicity is an exceedingly important element in understanding chemical reactivity. This and the following two sections review some principles and properties of acids and bases and examine how these principles apply to alcohols.

According to the theory proposed by Svante Arrhenius, a Swedish chemist and winner of the 1903 Nobel Prize in chemistry, an acid ionizes in aqueous solution to liberate protons (H^+, hydrogen ions), while bases ionize to liberate hydroxide ions (HO^-). A more general theory of acids and bases was devised independently by Johannes Brønsted (Denmark) and Thomas M. Lowry (England) in 1923. In the Brønsted-Lowry approach, an acid is a **proton donor,** and a base is a **proton acceptor.**

Curved arrow notation is used to show the electron pair of the base abstracting a proton from the acid. The pair of electrons in the H—A bond becomes an unshared pair in the anion $^-$:A. Curved arrows track **electron movement,** not atomic movement.

$$\underset{\text{Base}}{B:} + \underset{\text{Acid}}{H{-}A} \rightleftharpoons \underset{\text{Conjugate acid}}{\overset{+}{B}{-}H} + \underset{\text{Conjugate base}}{:A^-}$$

The Brønsted-Lowry definitions of acids and bases are widely used in organic chemistry. We will refer to proton donors and proton acceptors as *Brønsted acids* and

Brønsted bases, respectively. As noted in the preceding equation, the **conjugate acid** of a substance is formed when it accepts a proton from a suitable donor. Conversely, the proton donor is converted to its **conjugate base.**

PROBLEM 4.6 Write an equation for the reaction of ammonia ($:NH_3$) with hydrogen chloride (HCl). Use curved arrows to track electron movement, and identify the acid, base, conjugate acid, and conjugate base.

In aqueous solution, an acid transfers a proton to water. Water acts as a Brønsted base.

$$H_2\ddot{O}: + H{-}A \rightleftharpoons H_2\overset{+}{O}{-}H + :A^-$$

Water (base) | Acid | Conjugate acid of water | Conjugate base

The systematic name for the conjugate acid of water (H_3O^+) is **oxonium ion.** It is more commonly known as **hydronium ion.**

The strength of an acid in dilute aqueous solution is given by the **equilibrium constant** K for the reaction

$$HA + H_2O \overset{K}{\rightleftharpoons} H_3O^+ + A^-$$

where

$$K = \frac{[H_3O^+][A^-]}{[HA][H_2O]}$$

The concentration of water can be treated as a constant and incorporated into a new constant K_a called the **acid dissociation constant** or the **ionization constant.**

$$K_a = K[H_2O] = \frac{[H_3O^+][A^-]}{[HA]}$$

Table 4.2 lists a number of Brønsted acids and their acid dissociation constants. Strong acids are characterized by K_a values that are greater than that for hydronium ion (H_3O^+, $K_a = 55$). Essentially every molecule of a strong acid transfers a proton to water in dilute aqueous solution. Weak acids have K_a values less than that of H_3O^+; they are incompletely ionized in dilute aqueous solution.

A convenient way to express acid strength is through the use of pK_a, defined as follows:

$$pK_a = -\log_{10} K_a$$

Thus, water, with $K_a = 1.8 \times 10^{-16}$, has a pK_a of 15.7; ammonia, with $K_a = 10^{-36}$, has a pK_a of 36. The stronger the acid, the larger the value of its K_a and the smaller the value of pK_a. Water is a very weak acid, but is a far stronger acid than ammonia. Table 4.2 includes pK_a as well as K_a values for acids. Because both systems are frequently encountered as measures of acid strength, you should be proficient at converting one into the other.

TABLE 4.2
Acid Dissociation Constants K_a and pK_a Values for Some Brønsted Acids*

Acid	Formula†	Dissociation constant, K_a	pK_a	Conjugate base
Hydrogen iodide	HI	$\sim 10^{10}$	~ -10	I^-
Hydrogen bromide	HBr	$\sim 10^{9}$	~ -9	Br^-
Hydrogen chloride	HCl	$\sim 10^{7}$	~ -7	Cl^-
Sulfuric acid	$HOSO_2OH$	1.6×10^{5}	-4.8	$HOSO_2O^-$
Hydronium ion	$H\text{—}\overset{+}{O}H_2$	55	-1.7	H_2O
Hydrogen fluoride	HF	3.5×10^{-4}	3.5	F^-
Acetic acid	$CH_3\overset{O}{\overset{\|}{C}}OH$	1.8×10^{-5}	4.7	$CH_3\overset{O}{\overset{\|}{C}}O^-$
Ammonium ion	$H\text{—}\overset{+}{N}H_3$	5.6×10^{-10}	9.2	NH_3
Water	HOH	1.8×10^{-16}‡	15.7	HO^-
Methanol	CH_3OH	$\sim 10^{-16}$	~ 16	CH_3O^-
Ethanol	CH_3CH_2OH	$\sim 10^{-16}$	~ 16	$CH_3CH_2O^-$
Isopropyl alcohol	$(CH_3)_2CHOH$	$\sim 10^{-17}$	~ 17	$(CH_3)_2CHO^-$
tert-Butyl alcohol	$(CH_3)_3COH$	$\sim 10^{-18}$	~ 18	$(CH_3)_3CO^-$
Ammonia	H_2NH	$\sim 10^{-36}$	~ 36	H_2N^-
Dimethylamine	$(CH_3)_2NH$	$\sim 10^{-36}$	~ 36	$(CH_3)_2N^-$

* Acid strength decreases from top to bottom of the table. Strength of conjugate base increases from top to bottom of the table.

† The most acidic proton—the one that is lost on ionization—is indicated in red.

‡ The "true" K_a for water is 1×10^{-14}. Dividing this value by 55.5 (the number of moles of water in 1 L of water) gives a K_a of 1.8×10^{-16} and puts water on the same concentration basis as the other substances in the table. A paper in the May 1990 issue of the *Journal of Chemical Education* (p. 386) outlines the justification for this approach. For a dissenting view, see the March 1992 issue of the *Journal of Chemical Education* (p. 255).

PROBLEM 4.7 Hydrogen cyanide (HCN) has a pK_a of 9.1. What is its K_a? Is HCN a strong or a weak acid?

An important corollary of the Brønsted-Lowry view of acids and bases involves the relative strengths of an acid and its conjugate base. The stronger the acid, the weaker the conjugate base; the weaker the acid, the stronger the conjugate base. Ammonia (NH_3) is the second-weakest acid in Table 4.2. Therefore, its conjugate base, amide ion (H_2N^-), is the second-strongest base. Hydroxide (HO^-) is a moderately strong base, much stronger than the halide ions F^-, Cl^-, Br^-, and I^-, which are very weak bases. Fluoride is the strongest base of the halides but is 10^{12} times less basic than hydroxide ion on the basis of a comparison of their K_a's.

PROBLEM 4.8 As noted in Problem 4.7, hydrogen cyanide (HCN) has a pK_a of 9.1. Is cyanide ion (CN^-) a stronger base or a weaker base than hydroxide ion (HO^-)?

In any proton-transfer process the position of equilibrium favors formation of the weaker acid and the weaker base.

$$\text{Stronger acid} + \text{stronger base} \underset{}{\overset{K>1}{\rightleftharpoons}} \text{weaker acid} + \text{weaker base}$$

Table 4.2 is constructed so that the strongest acid is at the top of the acid column and the strongest base is at the bottom of the conjugate base column. An acid will transfer a proton to the conjugate base of any acid that lies below it in the table, and the equilibrium constant for the reaction will be greater than 1.

Table 4.2 contains both inorganic and organic compounds. Organic compounds are similar to inorganic ones when the functional groups responsible for their acid-base properties are the same. Thus, alcohols (ROH) are similar to water (HOH) in both their Brønsted acidity (ability to donate a proton *from oxygen*) and Brønsted basicity (ability to accept a proton *on oxygen*). Just as proton transfer to a water molecule gives oxonium ion (hydronium ion, H_3O^+), proton transfer to an alcohol gives an **alkyloxonium ion** (ROH_2^+).

$$\text{ROH (Alcohol)} + \text{H—A (Acid)} \rightleftharpoons \text{R}\overset{+}{\text{O}}\text{H}_2 \text{ (Alkyloxonium ion)} + :\text{A}^- \text{ (Conjugate base)}$$

We shall see that several important reactions of alcohols involve strong acids either as reagents or as catalysts to increase the rate of reaction. In all these reactions the first step is formation of an alkyloxonium ion by proton transfer from the acid to the oxygen of the alcohol.

PROBLEM 4.9 Write an equation for proton transfer from hydrogen chloride to *tert*-butyl alcohol. Use curved arrows to track electron movement, and identify the acid, base, conjugate acid, and conjugate base.

PROBLEM 4.10 Is the equilibrium constant for proton transfer from hydrogen chloride to *tert*-butyl alcohol greater than 1 or less than 1?

Alkyl halides are neither very acidic nor very basic and are absent from Table 4.2. In general, compounds, including alkyl halides, in which all the protons are bonded to carbon are exceedingly weak acids—too weak to be included in the table. They are also exceedingly weak bases. When exceptions are encountered, special mention will be made of that fact and the reasons for it explained.

4.7 ACID-BASE REACTIONS: A MECHANISM FOR PROTON TRANSFER

Potential energy diagrams of the type used in Chapter 3 to describe conformational processes can also help us understand more about chemical reactions. Consider the transfer of a proton from hydrogen bromide to a water molecule:

$$:\ddot{\text{Br}}\text{—H} + :\ddot{\text{O}}\text{H}_2 \rightleftharpoons :\ddot{\text{Br}}:^- + \text{H—}\overset{+}{\ddot{\text{O}}}\text{H}_2$$

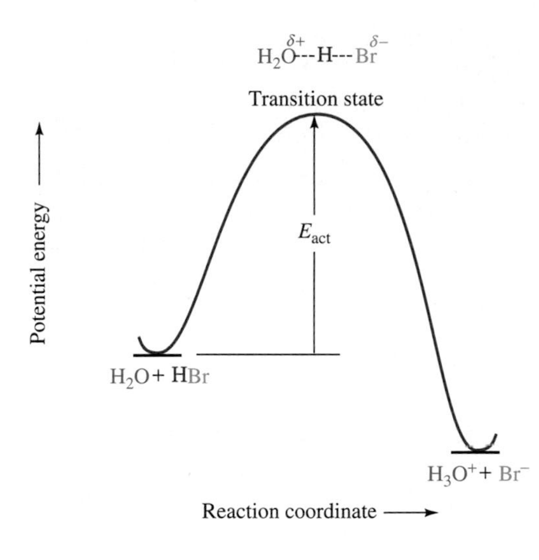

FIGURE 4.5 Energy diagram for concerted bimolecular proton transfer from hydrogen bromide to water.

A potential energy diagram for this reaction is shown in Figure 4.5. Because the transfer of a proton from hydrogen bromide to water is exothermic, the products are placed lower in energy than the reactants. The diagram depicts the reaction as occurring in a single **elementary step.** An elementary step is one that involves only one transition state. A reaction can proceed by way of a single elementary step, in which case it is described as a **concerted** reaction, or by a series of elementary steps. In the case of proton transfer from hydrogen bromide to water, breaking of the H—Br bond and making of the $H_2\overset{+}{O}$—H bond occur "in concert" with each other. The species present at the transition state is not a stable structure and cannot be isolated or examined directly. Its structure is assumed to be one in which the proton being transferred is partially bonded to both bromine and oxygen simultaneously, although not necessarily to the same extent.

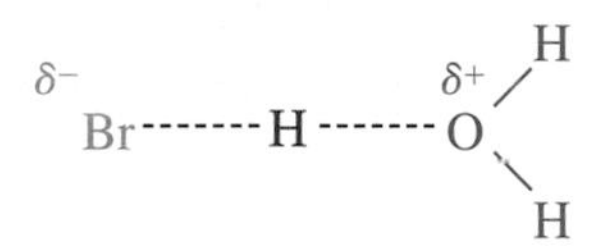

Dashed lines in transition-state structures represent *partial* bonds, i.e., bonds in the process of being made or broken.

The **molecularity** of an elementary step is given by the number of species that undergo a chemical change in that step. The elementary step

$$HBr + H_2O \rightleftharpoons Br^- + H_3O^+$$

is **bimolecular** because it involves one molecule of hydrogen bromide and one molecule of water.

PROBLEM 4.11 Represent the structure of the transition state for proton transfer from hydrogen chloride to *tert*-butyl alcohol.

The 1967 Nobel Prize in chemistry was awarded to Manfred Eigen, a German chemist who developed novel methods for measuring the rates of very fast reactions such as proton transfers.

Proton transfer from hydrogen bromide to water and alcohols is extremely fast. Measurements of the rate of proton transfers of this type indicate they occur at rates approaching those at which the molecules encounter one another in solution. Proton transfers from strong acids to water and alcohols rank among the most rapid chemical processes. Thus the height of the energy barrier separating reactants and products, the *activation energy* for the process, must be quite low.

The concerted nature of proton transfer is one reason for its rapid rate. In a concerted proton transfer, the energy cost of breaking the H—Br bond is partially offset by the energy released in making the $H_2\overset{+}{O}$—H bond. Thus, the activation energy is far less than it would be for a hypothetical stepwise process involving an initial, unassisted ionization of the H—Br bond, followed by a combination of the resulting H^+ with water.

4.8 ACIDITY OF ALCOHOLS. ALKOXIDE IONS

Alcohols are similar to one another in acidity and are slightly weaker acids than water. As Table 4.2 indicates, alcohols have K_a's in the 10^{-16} to 10^{-18} range (pK_a's = 16 to 18). The position of the Brønsted acid-base equilibrium for removal of a proton from the hydroxyl (—OH) group of an alcohol by hydroxide ion, therefore, lies to the left.

$$\underset{\text{Hydroxide ion}}{\text{H—}\ddot{\text{O}}\text{:}^-} + \underset{\text{Alcohol}}{\text{H—}\ddot{\text{O}}\text{R}} \overset{K<1}{\rightleftharpoons} \underset{\text{Water}}{\text{H—}\ddot{\text{O}}\text{—H}} + \underset{\text{Alkoxide ion}}{{}^-\text{:}\ddot{\text{O}}\text{R}}$$

In general, methyl and primary alcohols are slightly more acidic than secondary alcohols, and secondary alcohols are slightly more acidic than tertiary alcohols.

Alkoxide ions are also named as *alcoholates* according to the substitutive name of the alcohol. Thus $NaOCH_3$ may be called *sodium methanolate,* $NaOCH_2CH_3$ may be called *sodium ethanolate,* and so on.

The conjugate base of an alcohol is called an **alkoxide ion.** Alkoxide ions are strong bases, and the bonding in sodium and potassium alkoxides is ionic. As sodium hydroxide (NaOH) is a source of hydroxide ion, sodium methoxide ($NaOCH_3$) is a source of methoxide ion, sodium ethoxide ($NaOCH_2CH_3$) is a source of ethoxide ion, and so on.

Sodium and potassium alkoxides are among the most commonly used bases in organic chemistry. They are normally prepared by the reaction of the metal with the appropriate alcohol.

$$\underset{\text{Methyl alcohol}}{2CH_3OH} + \underset{\text{Sodium}}{2Na} \longrightarrow \underset{\text{Sodium methoxide}}{2NaOCH_3} + \underset{\text{Hydrogen}}{H_2}$$

$$\underset{\textit{tert}\text{-Butyl alcohol}}{2(CH_3)_3COH} + \underset{\text{Potassium}}{2K} \longrightarrow \underset{\text{Potassium }\textit{tert}\text{-butoxide}}{2KOC(CH_3)_3} + \underset{\text{Hydrogen}}{H_2}$$

Potassium is more reactive than sodium, and the order of alcohol reactivity is:

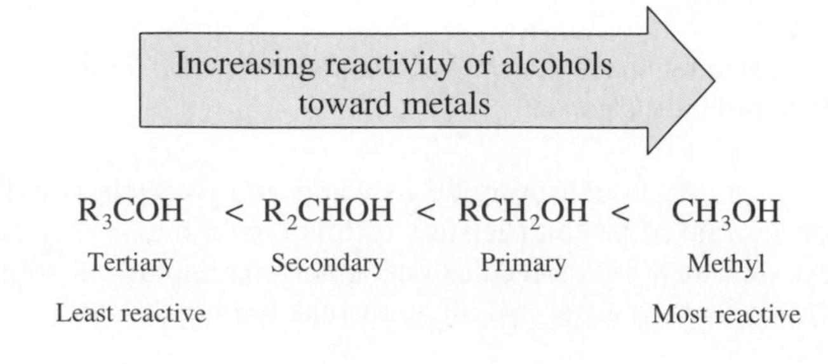

When the alcohol is primary or secondary, the reactivity of sodium is sufficient for relatively concentrated solutions of the sodium alkoxide in the alcohol to be easily prepared. When the alcohol is tertiary, the more reactive metal potassium is normally used.

Alternatively, sodium alkoxides are sometimes prepared by the reaction of an alcohol with sodium hydride:

$$\underset{\text{Alcohol}}{ROH} + \underset{\text{Sodium hydride}}{NaH} \longrightarrow \underset{\text{Sodium alkoxide}}{RONa} + \underset{\text{Hydrogen}}{H_2}$$

4.9 PREPARATION OF ALKYL HALIDES FROM ALCOHOLS AND HYDROGEN HALIDES

Much of what organic chemists do is directed toward practical goals. Chemists in the pharmaceutical industry, for example, synthesize new compounds as potential drugs for the treatment of disease. Agricultural chemicals designed to increase crop yields include organic compounds used for weed control, insecticides, and fungicides. Among the "building block" molecules used as starting materials to prepare new substances, alcohols and alkyl halides are especially valuable.

The procedures to be described in the remainder of this chapter use either an alkane or an alcohol as the starting material for preparing an alkyl halide. By knowing how to prepare alkyl halides, we can better appreciate the material in succeeding chapters, where alkyl halides figure prominently in key chemical transformations. The preparation of alkyl halides also serves as a focal point to develop the principles of reaction mechanisms. We begin with preparation of alkyl halides from alcohols by reaction with hydrogen halides according to the general equation:

$$\underset{\text{Alcohol}}{R{-}OH} + \underset{\text{Hydrogen halide}}{H{-}X} \longrightarrow \underset{\text{Alkyl halide}}{R{-}X} + \underset{\text{Water}}{H{-}OH}$$

The order of reactivity of the hydrogen halides parallels their acidity: HI > HBr > HCl ≫ HF. Hydrogen iodide is used infrequently, however, and the reaction of alcohols with hydrogen fluoride is not a useful method for the preparation of alkyl fluorides.

Among the various classes of alcohols, tertiary alcohols are observed to be the most reactive and primary alcohols the least reactive.

Increasing reactivity of alcohols toward hydrogen halides

$$\underset{\substack{\text{Methyl} \\ \text{Least reactive}}}{CH_3OH} < \underset{\text{Primary}}{RCH_2OH} < \underset{\text{Secondary}}{R_2CHOH} < \underset{\substack{\text{Tertiary} \\ \text{Most reactive}}}{R_3COH}$$

Tertiary alcohols are converted to alkyl chlorides in high yield within minutes on reaction with hydrogen chloride at room temperature and below.

$$(CH_3)_3COH + HCl \xrightarrow{25°C} (CH_3)_3CCl + H_2O$$

2-Methyl-2-propanol (*tert*-butyl alcohol) | Hydrogen chloride | 2-Chloro-2-methylpropane (*tert*-butyl chloride) (78–88%) | Water

The efficiency of a synthetic transformation is normally expressed as a **percent yield,** or percentage of the *theoretical yield.* Theoretical yield is the amount of product that could be formed if the reaction proceeded to completion and did not lead to any products other than those given in the equation.

Secondary and primary alcohols do not react with hydrogen chloride at rates fast enough to make the preparation of the corresponding alkyl chlorides a method of practical value. Therefore, the more reactive hydrogen halide HBr is used; even then, elevated temperatures are required in order to increase the rate of reaction.

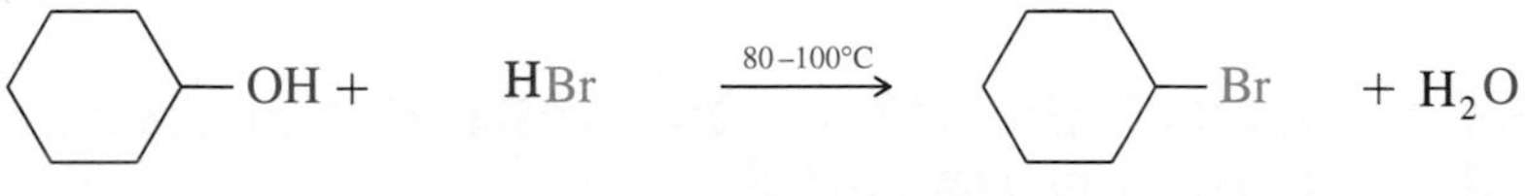

$$C_6H_{11}OH + HBr \xrightarrow{80-100°C} C_6H_{11}Br + H_2O$$

Cyclohexanol | Hydrogen bromide | Bromocyclohexane (73%) | Water

$$CH_3(CH_2)_5CH_2OH + HBr \xrightarrow{120°C} CH_3(CH_2)_5CH_2Br + H_2O$$

1-Heptanol | Hydrogen bromide | 1-Bromoheptane (87–90%) | Water

The same kind of transformation may be carried out more economically by heating an alcohol with sodium bromide and sulfuric acid.

$$CH_3CH_2CH_2CH_2OH \xrightarrow[\text{heat}]{NaBr,\ H_2SO_4} CH_3CH_2CH_2CH_2Br$$

1-Butanol (*n*-butyl alcohol) | 1-Bromobutane (70–83%) (*n*-butyl bromide)

Organic chemists often find it convenient to write chemical equations in the abbreviated form shown here, in which reagents, especially inorganic ones, are not included in the body of the equation but instead are indicated over the arrow. Inorganic products—in this case, water—are usually omitted. These simplifications focus our attention on the organic reactant and its functional group transformation.

PROBLEM 4.12 Write chemical equations for the reaction that takes place between each of the following pairs of reactants:

(*a*) 2-Butanol and hydrogen bromide
(*b*) 3-Ethyl-3-pentanol and hydrogen chloride
(*c*) 1-Tetradecanol and hydrogen bromide

SAMPLE SOLUTION (*a*) An alcohol and a hydrogen halide react to form an alkyl halide and water. In this case 2-bromobutane was isolated in 73 percent yield.

$$CH_3CH(OH)CH_2CH_3 + HBr \longrightarrow CH_3CH(Br)CH_2CH_3 + H_2O$$

2-Butanol | Hydrogen bromide | 2-Bromobutane | Water

4.10 MECHANISM OF THE REACTION OF ALCOHOLS WITH HYDROGEN HALIDES

The reaction of an alcohol with a hydrogen halide is a **substitution.** A halide, usually chloride or bromide, replaces a hydroxyl group as a substituent on carbon. Identifying the reaction as a substitution simply tells us the relationship between the organic reactant and its product but does not reveal the mechanism. In developing a mechanistic picture for a particular reaction some basic principles of chemical reactivity are combined with experimental observations to deduce the most likely sequence of elementary steps.

Consider the specific case of the reaction of *tert*-butyl alcohol with hydrogen chloride:

$$\underset{tert\text{-Butyl alcohol}}{(CH_3)_3COH} + \underset{\text{Hydrogen chloride}}{HCl} \longrightarrow \underset{tert\text{-Butyl chloride}}{(CH_3)_3CCl} + \underset{\text{Water}}{H_2O}$$

Overall Reaction:

$$\underset{tert\text{-Butyl alcohol}}{(CH_3)_3COH} + \underset{\text{Hydrogen chloride}}{HCl} \longrightarrow \underset{tert\text{-Butyl chloride}}{(CH_3)_3CCl} + \underset{\text{Water}}{HOH}$$

Step 1: Protonation of *tert*-butyl alcohol to give an oxonium ion:

$$\underset{tert\text{-Butyl alcohol}}{(CH_3)_3C{-}\ddot{O}{:}{-}H} + \underset{\text{Hydrogen chloride}}{H{-}\ddot{Cl}{:}} \rightleftharpoons \underset{tert\text{-Butyloxonium ion}}{(CH_3)_3C{-}\overset{+}{\ddot{O}}(H){-}H} + \underset{\text{Chloride ion}}{{:}\ddot{Cl}{:}^-}$$

Step 2: Dissociation of *tert*-butyloxonium ion to give a carbocation:

$$\underset{tert\text{-Butyloxonium ion}}{(CH_3)_3C{-}\overset{+}{\ddot{O}}(H){-}H} \rightleftharpoons \underset{tert\text{-Butyl cation}}{(CH_3)_3C^+} + \underset{\text{Water}}{{:}\ddot{O}(H){-}H}$$

Step 3: Capture of *tert*-butyl cation by chloride ion:

$$\underset{tert\text{-Butyl cation}}{(CH_3)_3C^+} + \underset{\text{Chloride ion}}{{:}\ddot{Cl}{:}^-} \longrightarrow \underset{tert\text{-Butyl chloride}}{(CH_3)_3C{-}\ddot{Cl}{:}}$$

FIGURE 4.6 The mechanism of formation of *tert*-butyl chloride from *tert*-butyl alcohol and hydrogen chloride.

The generally accepted mechanism for this reaction is presented as a series of three elementary steps in Figure 4.6. We say "generally accepted" because a reaction mechanism can never be proved to be correct. A generally accepted mechanism is our best present assessment of how a reaction proceeds and must account for all experimental observations. If new experimental data appear that are inconsistent with a mechanism, the mechanism must be modified to accommodate these new facts. If the new data are consistent with the proposed mechanism, our confidence grows that it is likely to be correct.

If you have not already written out the solutions to Problems 4.9 to 4.11, you should do so now.

We already know about step 1 of the mechanism outlined in Figure 4.6; it is an example of a Brønsted acid-base reaction of the type discussed in Section 4.6 and formed the basis of Problems 4.9 through 4.11.

Steps 2 and 3, however, are new to us. Step 2 involves dissociation of an alkyloxonium ion to a molecule of water and a **carbocation,** a species that contains a positively charged carbon. In step 3, this carbocation reacts with chloride ion to yield *tert*-butyl chloride. Both the alkyloxonium ion and the carbocation are **intermediates** in the reaction. They are not isolated, but are formed in one step and consumed in another during the passage of reactants to products. If we add the equations for steps 1 through 3 together, the equation for the overall process results. A valid reaction mechanism must account for the consumption of all reactants and the formation of all products, be they organic or inorganic. So that we may better understand the chemistry expressed in steps 2 and 3, we need to examine carbocations in more detail.

4.11 STRUCTURE, BONDING, AND STABILITY OF CARBOCATIONS

Carbocations are sometimes called *carbonium ions* or *carbenium ions.* An article in the November 1986 issue of the *Journal of Chemical Education,* pp. 930–933, traces the historical development of these and related terms.

Carbocations are classified as primary, secondary, or tertiary according to the number of carbons that are directly attached to the positively charged carbon. They are named by appending "cation" as a separate word after the IUPAC name of the appropriate alkyl group. The chain is numbered beginning with the positively charged carbon (the positive charge is always at C-1).

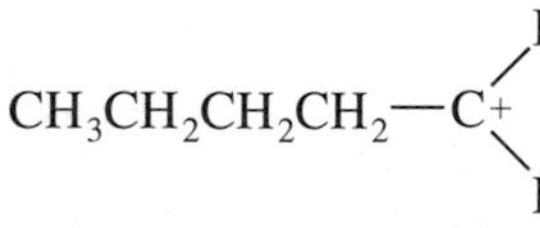

$CH_3CH_2CH_2CH_2—\overset{+}{C}H_2$

Pentyl cation
(primary carbocation)

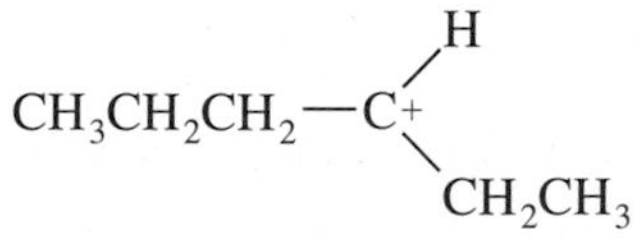

$CH_3CH_2CH_2—\overset{+}{C}H—CH_2CH_3$

1-Ethylbutyl cation
(secondary carbocation)

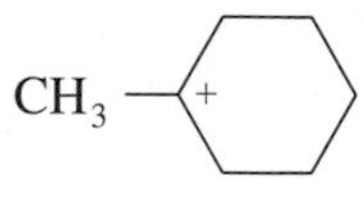

CH_3—(cyclohexyl, + at C-1)

1-Methylcyclohexyl cation
(tertiary carbocation)

Common names that have been incorporated into IUPAC nomenclature such as isopropyl, *sec*-butyl, and so on, are permitted. Thus 1,1-dimethylethyl cation $(CH_3)_3C^+$ may be called *tert*-butyl cation.

The properties of carbocations are intimately related to their structure, and so let us consider the bonding in methyl cation, CH_3^+. The positively charged carbon contributes three valence electrons and each hydrogen contributes one for a total of six electrons, which are used to form three 2-electron C—H σ bonds. As we saw in Section 1.15, carbon is sp^2 hybridized when it is bonded to three atoms or groups. Therefore, we choose the sp^2 hybridization model for bonding shown in Figure 4.7. Carbon forms σ bonds to three hydrogens by overlap of its sp^2 orbitals with hydrogen 1 *s* orbitals. The three σ bonds are coplanar. Remaining on carbon is an unhybridized

2*p* orbital that contains no electrons. The axis of this empty *p* orbital is perpendicular to the plane defined by the three σ bonds.

Evidence from a variety of sources convinces us that carbocations exist, but as relatively unstable, high-energy species. When carbocations are involved in chemical reactions, it is as reactive intermediates, formed in one step and consumed rapidly thereafter. Numerous studies of reactions involving carbocation intermediates have shown that the more stable a carbocation is, the faster it is formed. These studies also demonstrate that *alkyl groups directly attached to the positively charged carbon stabilize a carbocation.* Thus, the observed order of carbocation stability is

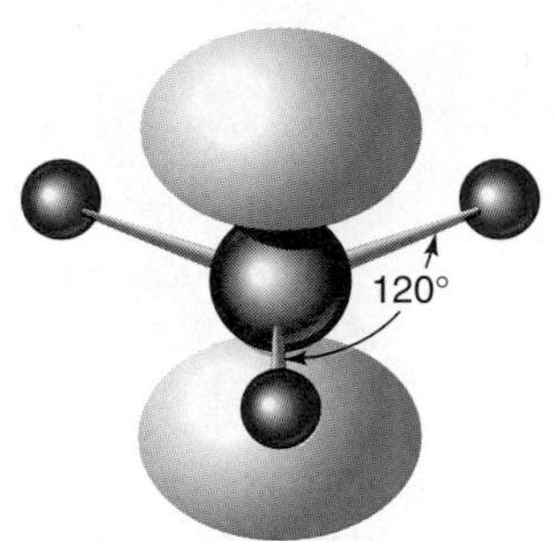

FIGURE 4.7 Structure of methyl cation CH_3^+. Carbon is sp^2 hybridized. Each hydrogen is attached to carbon by a σ bond formed by overlap of a hydrogen 1*s* orbital with an sp^2 hybrid orbital of carbon. All four atoms lie in the same plane. The unhybridized 2*p* orbital of carbon is unoccupied, and its axis is perpendicular to the plane of the atoms.

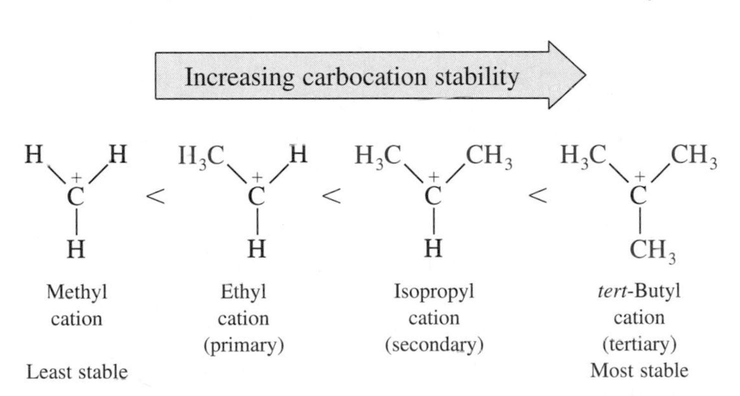

As carbocations go, CH_3^+ is particularly unstable, and its existence as an intermediate in chemical reactions has never been demonstrated. Primary carbocations, while more stable than CH_3^+, are still too unstable to be involved as intermediates in chemical reactions. The threshold of stability is reached with secondary carbocations. Many reactions, including the reaction of secondary alcohols with hydrogen halides, are believed to involve secondary carbocations. The evidence in support of tertiary carbocation intermediates is stronger yet.

PROBLEM 4.13 Of the isomeric $C_5H_{11}^+$ carbocations, which one is the most stable?

Because alkyl groups stabilize carbocations, we conclude that they release electrons to the positively charged carbon, dispersing the positive charge. They do this through a combination of effects. One involves polarization of the electron distribution in the σ bond that connects the alkyl group to the positively charged carbon. As illustrated for ethyl cation in Figure 4.8, the positively charged carbon draws the electrons in its σ bonds toward itself and away from the atoms attached to it. Electrons in a C—C σ bond are more polarizable than those in a C—H bond, so replacing hydrogens by alkyl substituents reduces the net charge on the sp^2 hybridized carbon. The electron-donating or electron-withdrawing effect of a group that is transmitted by the polarization of electrons in σ bonds is called an **inductive effect.**

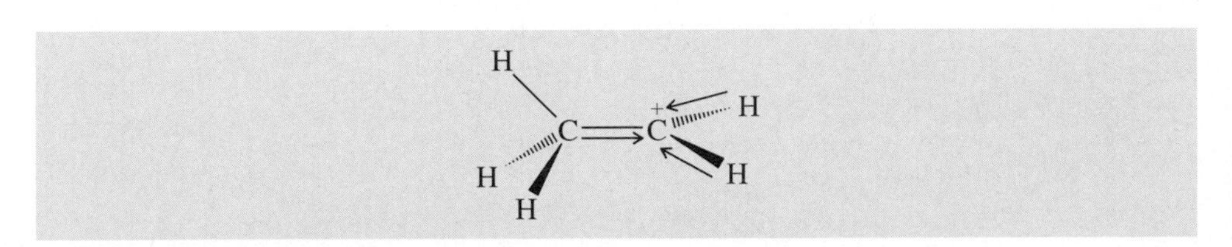

FIGURE 4.8 The charge in ethyl cation is stabilized by polarization of the electron distribution in the σ bonds between the positively charged carbon atom and its substituents. An alkyl substituent releases electrons better than does a hydrogen substituent.

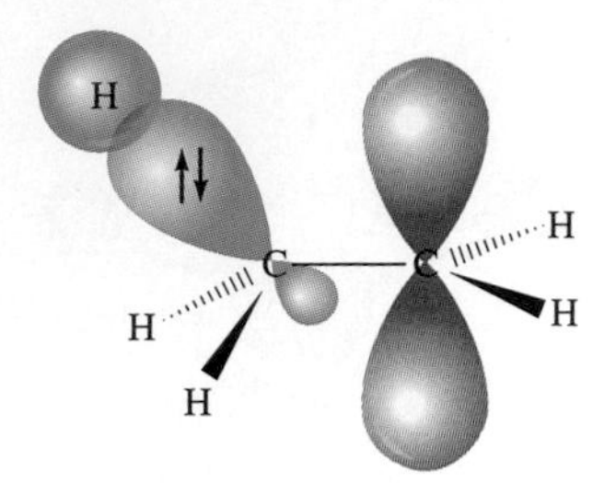

(*a*) Empty p orbital on positively charged carbon can overlap with σ orbital of methyl substituent.

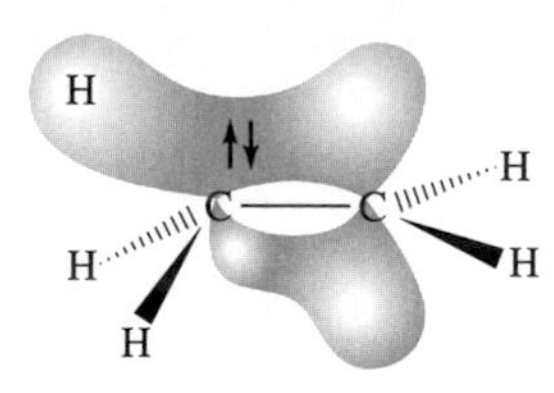

(*b*) Electrons in C——H σ bond are delocalized over a greater number of atoms, and the carbocation is stabilized.

FIGURE 4.9 Electrons in the C—H bonds of the methyl group are delocalized into the vacant 2*p* orbital of ethyl cation. Electron delocalization and dispersal of the positive charge stabilize the carbocation.

A second effect of comparable importance, called **hyperconjugation,** also calls on electrons in σ bonds to stabilize the carbocation. Again, consider ethyl cation. Begin at the positively charged carbon and label the bonds as α and β according to their increasing separation from this carbon.

$$\mathrm{H_3C}\overset{\alpha}{-}\overset{+}{\mathrm{C}}\mathrm{H_2}$$
(three C—H bonds of the CH_3 group labeled β; the two C—H bonds of the positively charged carbon and the C—C bond labeled α)

According to the hyperconjugation picture, *orbital overlap* provides a means by which the electrons in those σ bonds designated as β can be shared with the positively charged carbon to reduce its positive character. Figure 4.9 illustrates how the σ orbital of a β C—H bond overlaps with the vacant 2*p* orbital of the positively charged carbon. This overlap allows the electrons in the β bond to be delocalized, meaning that they are shared by both carbons. An increase in electron delocalization stabilizes the carbocation and assists in dispersing the positive charge. Only the electrons in the β bonds are involved; electrons in the α bonds cannot stabilize a carbocation by hyperconjugation. Thus, hyperconjugation is consistent with the increased stabilization that comes with an increasing number of alkyl substituents. Replacing any of the hydrogens attached to the positively charged carbon by alkyl groups brings more β bonds and their associated electrons, more charge dispersal, and greater stability.

The positive charge on carbon and the vacant *p* orbital combine to make carbocations strongly **electrophilic** (''electron-loving,'' or ''electron-seeking''). **Nucleophiles** have precisely complementary characteristics. A nucleophile is ''nucleus-seeking''; it has an unshared pair of electrons that it can use to form a covalent bond. Step 3 of the mechanism of the reaction of *tert*-butyl alcohol with hydrogen chloride is an example of a reaction between an electrophile and a nucleophile and is depicted from a structural perspective in Figure 4.10. The crucial electronic interaction is between an un-

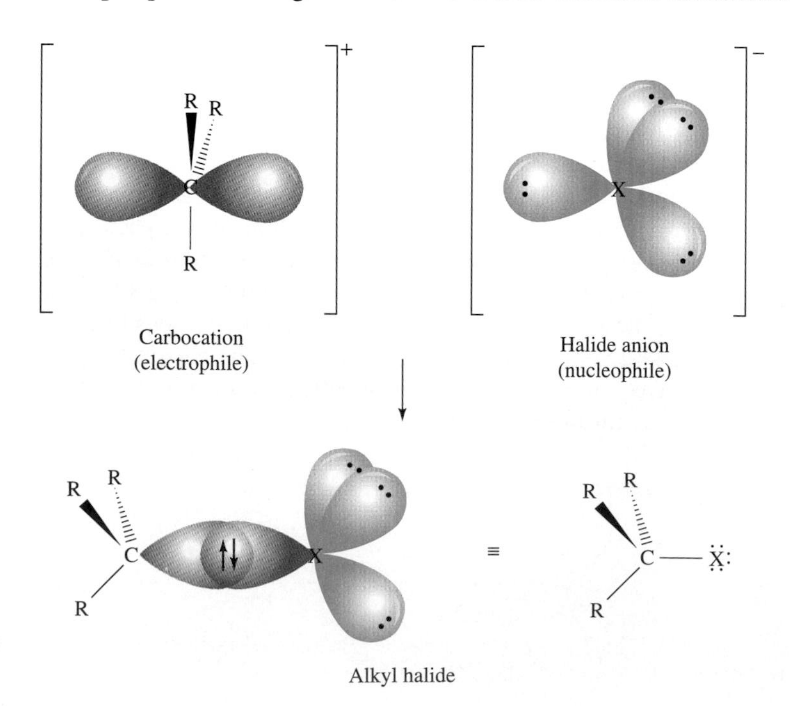

FIGURE 4.10 Combination of a carbocation and a halide anion to give an alkyl halide.

shared electron pair of the nucleophilic chloride anion and the vacant 2*p* orbital of the electrophilic carbocation.

A number of years ago G. N. Lewis (see page 3) extended our understanding of acid-base behavior to include reactions other than proton transfers. According to Lewis, *an acid is an electron pair acceptor* and *a base is an electron pair donor.* Thus, carbocations are electron pair acceptors and are **Lewis acids.** Halide anions are electron-pair donors and are **Lewis bases.** It is generally true that electrophiles are Lewis acids and nucleophiles are Lewis bases.

4.12 POTENTIAL ENERGY DIAGRAMS FOR MULTISTEP REACTIONS. THE S_N1 MECHANISM

The mechanism for the reaction of *tert*-butyl alcohol with hydrogen chloride presented in Figure 4.6 involves a sequence of three elementary steps. Therefore, it is not a concerted reaction. Remember from Section 4.7 that a concerted reaction proceeds in a single elementary step and only one transition state separates reactants from products. Each of the three steps in the mechanism for the reaction of *tert*-butyl alcohol with hydrogen chloride has its own transition state, and the potential energy diagram shown in Figure 4.11 for the overall process is a composite of the energy diagrams for the three successive steps.

Reading from left to right in Figure 4.11, the first maximum corresponds to the transition state for proton transfer from hydrogen chloride to *tert*-butyl alcohol. This step is **bimolecular.** The proton that is transferred is partially bonded to both chlorine and to the oxygen of the alcohol at the transition state.

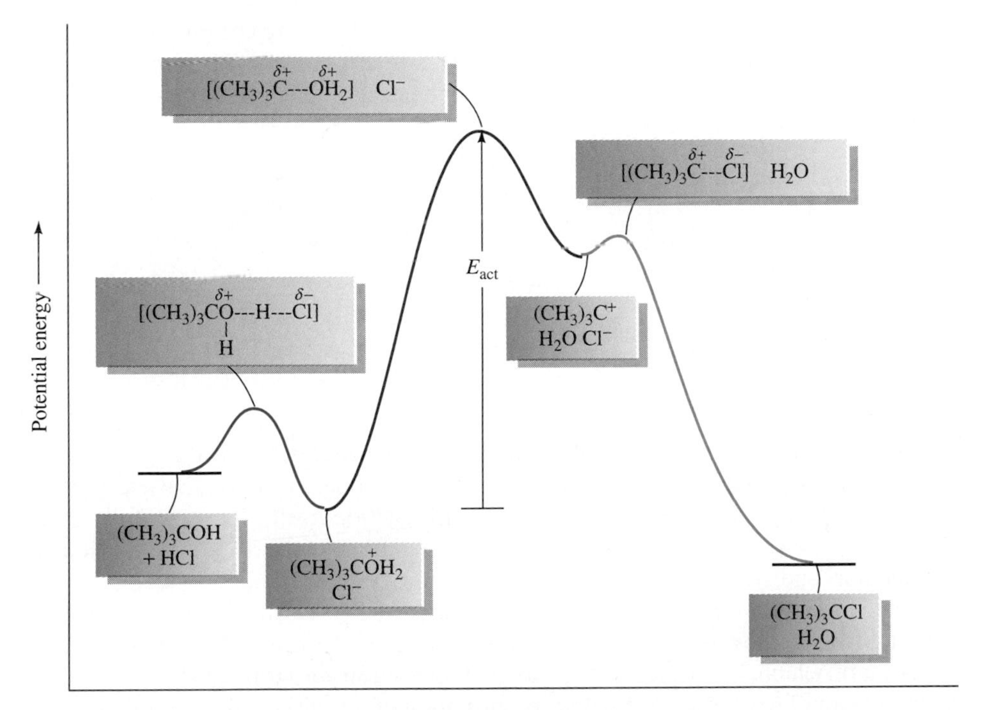

FIGURE 4.11 Energy diagram depicting the intermediates and transition states involved in the reaction of *tert*-butyl alcohol with hydrogen chloride.

$$(CH_3)_3C-\ddot{O}(H): + H-\ddot{Cl}: \longrightarrow (CH_3)_3C-\overset{\delta+}{\ddot{O}}(H)\text{---}H\text{---}\overset{\delta-}{\ddot{Cl}}: \longrightarrow (CH_3)_3C-\overset{+}{\ddot{O}}(H)-H + :\ddot{Cl}:^-$$

tert-Butyl alcohol | Hydrogen chloride | Transition state for proton transfer | *tert*-Butyloxonium ion | Chloride ion

This is a rapid process, and therefore the activation energy for the first step is relatively low.

Once formed, the alkyloxonium ion dissociates by cleavage of its carbon-oxygen bond, giving a carbocation.

$$(CH_3)_3C-\overset{+}{O}H_2 \longrightarrow (CH_3)_3\overset{\delta+}{C}\text{---}\overset{\delta+}{O}H_2 \longrightarrow (CH_3)_3C^+ + :\ddot{O}H_2$$

tert-Butyloxonium ion | Transition state for dissociation of alkyloxonium ion | *tert*-Butyl cation | Water

Only one species, the alkyloxonium ion, undergoes a chemical change in this step, making it **unimolecular**. Unlike the bimolecular proton transfer step that precedes it, in which formation of a new bond accompanies the cleavage of an old one, unimolecular dissociation of the alkyloxonium ion gives a carbocation without simultaneous formation of a new bond. Thus, the activation energy for carbocation formation is relatively high.

In the third step, the carbocation intermediate is captured by chloride ion, and the energy barrier for this cation-anion combination is relatively low. The transition state is characterized by partial bond formation between the nucleophile (chloride anion) and the electrophile (*tert*-butyl cation).

$$(CH_3)_3C^+ + :\ddot{Cl}:^- \longrightarrow (CH_3)_3\overset{\delta+}{C}\text{---}\overset{\delta-}{\ddot{Cl}}: \longrightarrow (CH_3)_3C-\ddot{Cl}:$$

tert-Butyl cation | Chloride anion | Transition state for cation-anion combination | *tert*-Butyl chloride

Two species, the carbocation and the anion, undergo chemical change in this step, making it **bimolecular.** Note that molecularity refers only to individual elementary steps in a multistep mechanism, not to the overall reaction itself. Step 1 of the mechanism (proton transfer) is bimolecular, step 2 (dissociation of the alkyloxonium ion) is unimolecular, and step 3 (cation-anion combination) is bimolecular.

Of the three steps in the reaction mechanism, step 2 has the highest activation energy and is the slowest step. A reaction can proceed no faster than the rate of its slowest step, which is referred to as the **rate-determining step.** In the reaction of *tert*-butyl alcohol with hydrogen chloride, formation of the carbocation by dissociation of the alkyloxonium ion is the rate-determining step.

Substitution reactions, of which the reaction of alcohols with hydrogen halides is but one example, will be discussed in more detail in Chapter 8. There, we will make extensive use of a shorthand notation for mechanism originally introduced by Sir Christopher Ingold in the 1930s. Ingold proposed the symbol S_N to stand for *substitution nucleophilic,* to be followed by the number *1* or *2* according to whether the rate-determining elementary step is unimolecular or bimolecular, respectively. The reaction of *tert*-butyl alcohol with hydrogen chloride, for example, is said to follow an S_N1 mechanism because its slow step is unimolecular (dissociation of *tert*-butyloxonium ion).

A brief synopsis of Sir Christopher Ingold's career appeared on page 3.

4.13 EFFECT OF ALCOHOL STRUCTURE ON REACTION RATE

We saw in Section 4.9 that the reactivity of alcohols with hydrogen halides increases in the order primary < secondary < tertiary. In order to be valid, the mechanism proposed in Figure 4.6 and represented by the energy diagram in Figure 4.11 must account for this order of relative reactivity. When considering rate effects, we focus on the slow step of a reaction mechanism and analyze how that step is influenced by changes in reactants or reaction conditions.

As mentioned, the slow step in the mechanism is the dissociation of the oxonium ion to the carbocation. The rate of this step is proportional to the concentration of the alkyloxonium ion:

$$\text{Rate} = k[\text{alkyloxonium ion}]$$

where k is a constant of proportionality called the *rate constant.* The value of k is related to the activation energy for alkyloxonium ion dissociation and is different for different alkyloxonium ions. The greater the value of k for a particular alkyloxonium ion, the more readily it dissociates to a carbocation. A low activation energy for carbocation formation implies a large value of k and a rapid rate of alkyloxonium ion dissociation. Conversely, a high activation energy is characterized by a low k for dissociation and a slow rate of reaction.

The rate of any chemical reaction increases with increasing temperature. Thus the value of k for a reaction is not constant, but increases as the temperature increases.

Consider what happens when the alkyloxonium ion dissociates to a carbocation and water. The positive charge resides mainly on oxygen in the alkyloxonium ion but is shared between oxygen and carbon at the transition state.

$$\text{C—}\overset{+}{\text{O}}\text{H}_2 \longrightarrow \overset{\delta+}{\text{C}}\text{---}\overset{\delta+}{\text{O}}\text{H}_2 \longrightarrow \text{C}^+ + :\ddot{\text{O}}\text{H}_2$$

Alkyloxonium ion | Transition state for dissociation of alkyloxonium ion | Carbocation | Water

The transition state for carbocation formation begins to resemble the carbocation. If we assume that structural features that stabilize carbocations also stabilize transition states that have carbocation character, it follows that alkyloxonium ions derived from tertiary alcohols have a lower energy of activation for dissociation and are converted to their corresponding carbocations faster than those derived from secondary and primary alcohols. Figure 4.12 depicts the effect of alkyloxonium ion structure on the activation

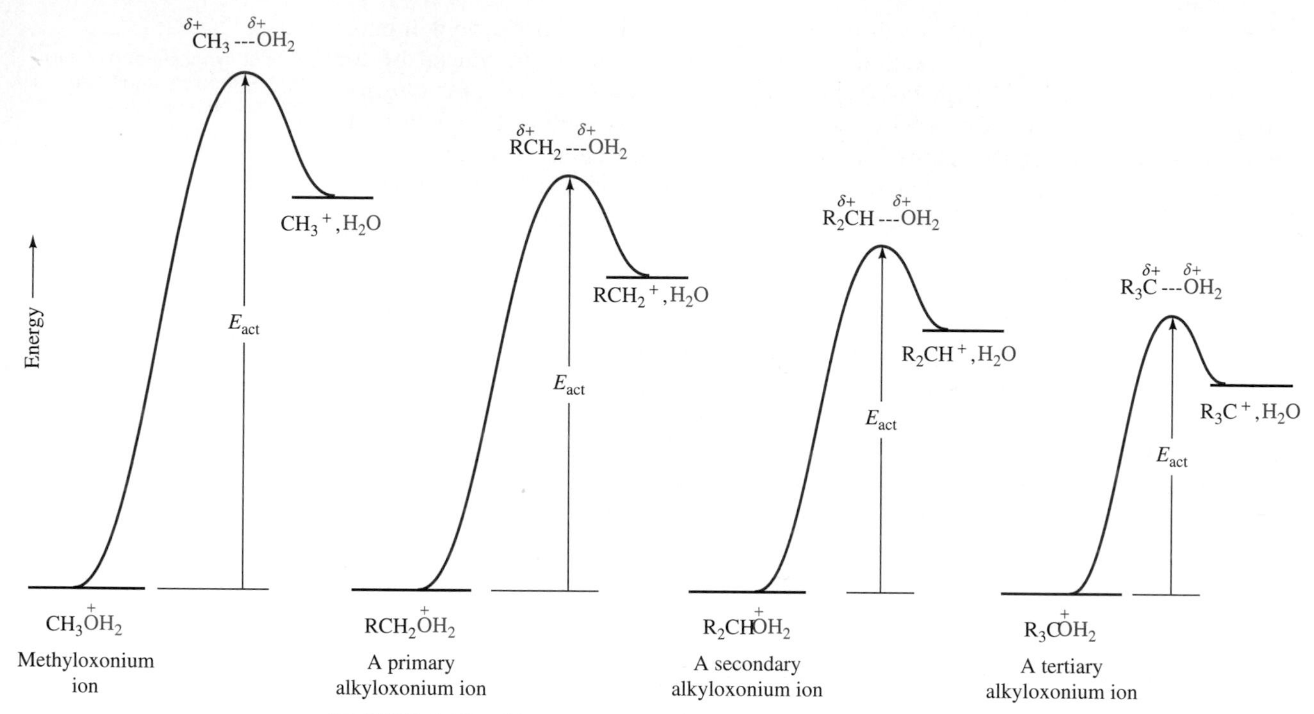

FIGURE 4.12 Diagrams comparing energies of activation for formation of carbocations from oxonium ions of methyl, primary, secondary, and tertiary alcohols.

energy for and thus the rate of carbocation formation. Once the carbocation is formed, it is rapidly captured by halide ion, so that the rate of alkyl halide formation is governed by the activation energy for dissociation of the alkyloxonium ion.

Inferring the structure of the transition state on the basis of what is known about the species that lead to it or may be formed by way of it is a practice with a long history in organic chemistry. A modern justification of this practice was advanced by George S. Hammond, who reasoned that *if two states, such as a transition state and an intermediate derived from it, are similar in energy, then they are similar in structure.* This rationale is known as **Hammond's postulate.** In the formation of a carbocation from an alcohol, the transition state is closer in energy to the carbocation than it is to the alcohol, and so its structure more closely resembles that of the carbocation and it responds in a similar way to the stabilizing effects of alkyl substituents.

4.14 REACTION OF PRIMARY ALCOHOLS WITH HYDROGEN HALIDES. THE S_N2 MECHANISM

Unlike tertiary and secondary carbocations, primary carbocations are of such high energy as to be inaccessible as intermediates in chemical reactions. Since primary alcohols are converted, albeit rather slowly, to alkyl halides on treatment with hydrogen halides, there must be some other mechanism for the formation of primary alkyl halides that avoids carbocation intermediates. This alternative mechanism is believed to be one in which the carbon-halogen bond begins to form before the carbon-oxygen bond of the oxonium ion is completely broken.

$$:\ddot{X}:^{-} + RCH_2\overset{+}{\ddot{O}}H_2 \longrightarrow \overset{\delta-}{:\ddot{X}}\text{---}\underset{}{\overset{R}{\overset{|}{C}}}H_2\text{---}\overset{\delta+}{\ddot{O}}H_2 \longrightarrow RCH_2\ddot{X}: + H_2\ddot{O}:$$

Halide ion | Primary alkyloxonium ion | Transition state | Primary alkyl halide | Water

In effect, the halide nucleophile helps to "push off" a water molecule from the oxonium ion. According to this mechanism, both the halide ion and the oxonium ion are involved in the same bimolecular elementary step. In Ingold's terminology, introduced in Section 4.12 and to be described in detail in Chapter 8, nucleophilic substitutions characterized by a bimolecular rate-determining step are given the mechanistic symbol $\mathbf{S_N2}$.

PROBLEM 4.14 1-Butanol and 2-butanol are converted to their corresponding bromides on being heated with hydrogen bromide. Write a suitable mechanism for each reaction, and assign each the appropriate symbol (S_N1 or S_N2).

4.15 OTHER METHODS FOR CONVERTING ALCOHOLS TO ALKYL HALIDES

Alkyl halides are such useful starting materials for the preparation of a large number of other functional group types that chemists have developed several different methods for converting alcohols to alkyl halides. Two methods, based on the inorganic reagents *thionyl chloride* and *phosphorus tribromide,* bear special mention.

Thionyl chloride reacts with alcohols to give alkyl chlorides. The inorganic byproducts in the reaction, sulfur dioxide and hydrogen chloride, are both gases at room temperature and are easily removed, making it an easy matter to isolate the alkyl chloride.

$$\underset{\text{Alcohol}}{ROH} + \underset{\text{Thionyl chloride}}{SOCl_2} \longrightarrow \underset{\text{Alkyl chloride}}{RCl} + \underset{\text{Sulfur dioxide}}{SO_2} + \underset{\text{Hydrogen chloride}}{HCl}$$

Since tertiary alcohols are so readily converted to chlorides with hydrogen chloride, thionyl chloride is used mainly to prepare primary and secondary alkyl chlorides. Reactions with thionyl chloride are normally carried out in the presence of potassium carbonate or the weak organic base pyridine.

$$\underset{\text{2-Octanol}}{CH_3\underset{\underset{OH}{|}}{C}H(CH_2)_5CH_3} \xrightarrow[K_2CO_3]{SOCl_2} \underset{\text{2-Chlorooctane (81\%)}}{CH_3\underset{\underset{Cl}{|}}{C}H(CH_2)_5CH_3}$$

$$\underset{\text{2-Ethyl-1-butanol}}{(CH_3CH_2)_2CHCH_2OH} \xrightarrow[\text{pyridine}]{SOCl_2} \underset{\text{1-Chloro-2-ethylbutane (82\%)}}{(CH_3CH_2)_2CHCH_2Cl}$$

Phosphorus tribromide reacts with alcohols to give alkyl bromides and phosphorous acid.

$$\underset{\text{Alcohol}}{3ROH} + \underset{\text{Phosphorus tribromide}}{PBr_3} \longrightarrow \underset{\text{Alkyl bromide}}{3RBr} + \underset{\text{Phosphorous acid}}{H_3PO_3}$$

Phosphorous acid is water-soluble and may be removed by washing the alkyl halide with water or with dilute aqueous base.

$$\underset{\text{Isobutyl alcohol}}{(CH_3)_2CHCH_2OH} \xrightarrow{PBr_3} \underset{\text{Isobutyl bromide (55–60\%)}}{(CH_3)_2CHCH_2Br}$$

$$\underset{\text{Cyclopentanol}}{C_5H_9OH} \xrightarrow{PBr_3} \underset{\text{Cyclopentyl bromide (78–84\%)}}{C_5H_9Br}$$

Thionyl chloride and phosphorus tribromide are specialized reagents used to bring about particular functional group transformations. For this reason, we will not discuss the mechanisms by which they convert alcohols to alkyl halides. Rather, we will limit our coverage to those mechanisms that have broad applicability and enhance our knowledge of fundamental principles. In those instances you will find that a mechanistic understanding is of immeasurable help in organizing the reaction types of organic chemistry.

4.16 HALOGENATION OF ALKANES

The remaining sections of this chapter describe the preparation of alkyl halides by a second type of substitution reaction, one that uses alkanes as reactants. It involves substitution of a halogen atom for one of the alkane's hydrogens and can be represented by the general equation:

$$\underset{\text{Alkane}}{R{-}H} + \underset{\text{Halogen}}{X_2} \longrightarrow \underset{\text{Alkyl halide}}{R{-}X} + \underset{\text{Hydrogen halide}}{HX}$$

The alkane is said to undergo *fluorination, chlorination, bromination,* or *iodination* according to whether X_2 is F_2, Cl_2, Br_2, or I_2, respectively. The general term is **halogenation. Chlorination** and **bromination** are the most widely used halogenation reactions and will be described in more detail in Sections 4.17 to 4.20.

Fluorine is an extremely aggressive oxidizing agent, and its reaction with alkanes is strongly exothermic and difficult to control. Direct fluorination of alkanes requires special equipment and techniques, is not a reaction of general applicability, and will not be discussed further.

Volume II of *Organic Reactions,* an annual series that reviews reactions of interest to organic chemists, contains the statement "Most organic compounds burn or explode when brought in contact with fluorine."

The reactivity of the halogens decreases in the order $F_2 > Cl_2 > Br_2 > I_2$. Chlorination of alkanes is less exothermic than fluorination, and bromination less exothermic than chlorination. *Iodine is unique among the halogens in that its reaction with alkanes is endothermic.* Endothermic reactions have equilibrium constants greater than 1 *only* when the $T\,\Delta S°$ term in the expression

$$\Delta G° = \Delta H° - T\,\Delta S°$$

exceeds that of the $\Delta H°$ term. Such is not the case here; $\Delta S°$ is close to zero for the iodination of an alkane. Thus, alkyl iodides are never prepared by iodination of alkanes. The reaction does not "go." Nor, as noted in Section 4.9, are alkyl iodides normally prepared from alcohols by reaction with hydrogen iodide. The customary method for the preparation of alkyl iodides will be described in Section 8.2.

4.17 CHLORINATION OF METHANE

The gas-phase chlorination of methane is a reaction of industrial importance and leads to a mixture of chloromethane (CH_3Cl), dichloromethane (CH_2Cl_2), trichloromethane ($CHCl_3$), and tetrachloromethane (CCl_4) by sequential substitution of hydrogens.

$$\underset{\text{Methane}}{CH_4} + \underset{\text{Chlorine}}{Cl_2} \xrightarrow{400-440°C} \underset{\substack{\text{Chloromethane}\\(\text{bp } -24°C)}}{CH_3Cl} + \underset{\substack{\text{Hydrogen}\\\text{chloride}}}{HCl}$$

$$\underset{\text{Chloromethane}}{CH_3Cl} + \underset{\text{Chlorine}}{Cl_2} \xrightarrow{400-440°C} \underset{\substack{\text{Dichloromethane}\\(\text{bp } 40°C)}}{CH_2Cl_2} + \underset{\substack{\text{Hydrogen}\\\text{chloride}}}{HCl}$$

$$\underset{\text{Dichloromethane}}{CH_2Cl_2} + \underset{\text{Chlorine}}{Cl_2} \xrightarrow{400-440°C} \underset{\substack{\text{Trichloromethane}\\(\text{bp } 61°C)}}{CHCl_3} + \underset{\substack{\text{Hydrogen}\\\text{chloride}}}{HCl}$$

$$\underset{\text{Trichloromethane}}{CHCl_3} + \underset{\text{Chlorine}}{Cl_2} \xrightarrow{400-440°C} \underset{\substack{\text{Tetrachloromethane}\\(\text{bp } 77°C)}}{CCl_4} + \underset{\substack{\text{Hydrogen}\\\text{chloride}}}{HCl}$$

Chlorination of methane provides approximately one-third of the annual U.S. production of chloromethane. The reaction of methanol with hydrogen chloride is the major synthetic method for the preparation of chloromethane.

One of the chief uses of chloromethane is as a starting material from which silicone polymers are made. Dichloromethane has been used to extract caffeine from coffee and as an aerosol propellant, and trichloromethane was once used as an inhalation anesthetic, but concerns about the toxicity of both are leading to their replacement by safer materials. Tetrachloromethane is the starting material for the preparation of several chlorofluorocarbons (CFCs) widely used as refrigerant gases. In 1987, most of the world's industrialized nations agreed to phase out all uses of CFCs by the year 2000 because these compounds have been implicated in atmospheric processes that degrade the earth's ozone layer.

The chlorination of methane is carried out at rather high temperatures (400 to 440°C), even though each substitution in the series is exothermic. The high temperature provides the energy to *initiate* the reaction. The term *initiation step* has a specific meaning in organic chemistry, one that is related to the mechanism of the reaction.

This mechanism, to be presented in Section 4.19, is fundamentally different from the mechanism by which alcohols react with hydrogen halides. Alcohols are converted to alkyl halides in reactions involving ionic (or ''polar'') intermediates—oxonium ions and carbocations. The intermediates in the chlorination of methane and other alkanes are quite different; they are neutral (''nonpolar'') species called **free radicals.**

4.18 STRUCTURE AND STABILITY OF FREE RADICALS

Free radicals are species that contain unpaired electrons. Alkyl radicals are classified as primary, secondary, or tertiary according to the number of carbon atoms directly attached to the carbon that bears the unpaired electron.

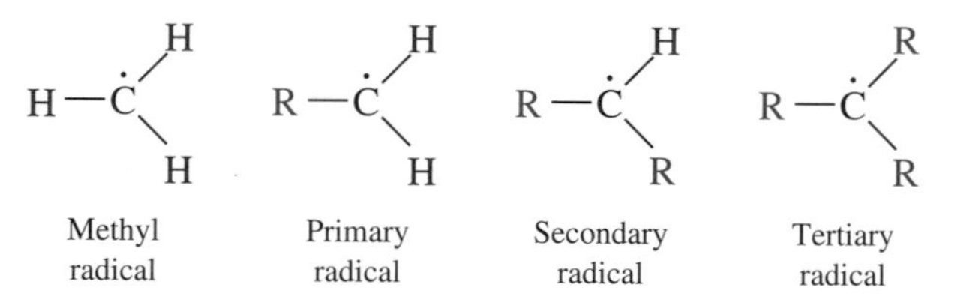

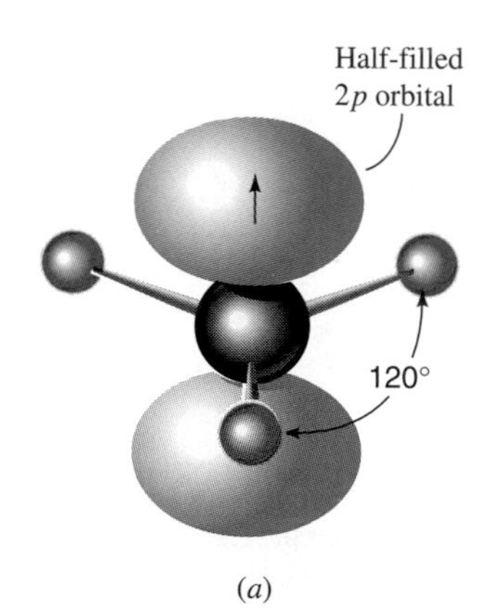

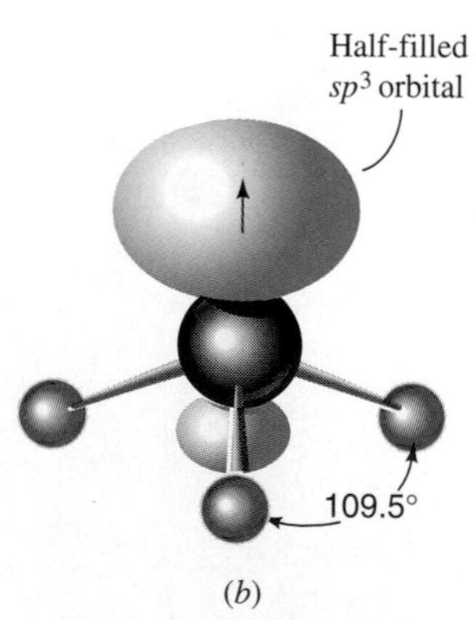

FIGURE 4.13 Orbital hybridization models of bonding in methyl radical. (*a*) If the structure of the CH_3 radical is planar, then carbon is sp^2 hybridized with an unpaired electron in a 2*p* orbital. (*b*) If CH_3 is pyramidal, carbon is sp^3 hybridized with an electron in an sp^3 orbital. Model (*a*) is more consistent with experimental observations.

An alkyl radical is neutral and has one more electron than the corresponding carbocation. Thus, bonding in methyl radical may be approximated by simply adding an electron to the vacant 2*p* orbital of sp^2 hybridized carbon in methyl cation (Figure 4.13a). A slightly different bonding model, shown in Figure 14.13*b*, holds the unpaired electron more strongly by placing it in an orbital with some *s* character. According to this model, carbon is sp^3 hybridized, and the unpaired electron occupies an orbital with 25 percent *s* character.

Of the two extremes, experimental studies indicate that the planar sp^2 model of Figure 4.13*a* describes the bonding in alkyl radicals better than the pyramidal sp^3 model of Figure 4.13*b*. Methyl radical is planar, and more highly substituted radicals such as *tert*-butyl radical are flattened pyramids closer in shape to that expected for sp^2 hybridized carbon than for sp^3.

Free radicals, like carbocations, have an unfilled 2*p* orbital and are stabilized by substituents, such as alkyl groups, that release electrons to the trivalent carbon. Consequently, the order of free-radical stability parallels that of carbocations.

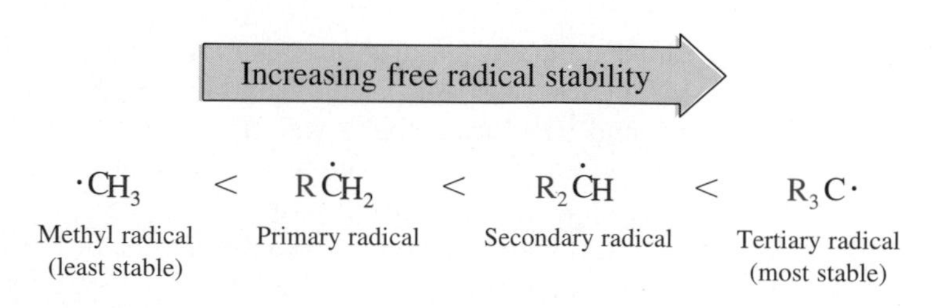

PROBLEM 4.15 Write a structural formula for the most stable of the free radicals having the formula C_5H_{11}.

Some of the evidence indicating that alkyl substituents stabilize free radicals comes from measurements of bond strengths. A **homolytic cleavage** is defined as the breaking of a covalent bond in a manner such that each of the initially bonded atoms retains one of the electrons in the bond.

A curved arrow shown as a single-barbed fishhook signifies the movement of *one* electron. "Normal" curved arrows track the movement of a *pair* of electrons.

$$X\!:\!Y \longrightarrow X\cdot + \cdot Y$$

Homolytic bond cleavage

In contrast, in a **heterolytic cleavage** one fragment retains both the electrons when the bond is broken.

$$X\!:\!Y \longrightarrow X^{+} + :Y^{-}$$

Heterolytic bond cleavage

We assess the relative stability of alkyl radicals by measuring the enthalpy change ($\Delta H°$) for the homolytic cleavage of a C—H bond in an alkane:

$$R\text{—}H \longrightarrow R\cdot + \cdot H$$

The more stable the radical, the lower the energy required to generate it by C—H bond homolysis.

The energy required for homolytic bond cleavage is called the **bond dissociation energy (BDE).** A list of some bond dissociation energies is given in Table 4.3.

As the table indicates, C—H bond dissociation energies in alkanes are approximately 375 to 435 kJ/mol (90 to 105 kcal/mol). Homolysis of the H—CH_3 bond in methane gives methyl radical and requires 435 kJ/mol (104 kcal/mol). The dissociation energy of the H—CH_2CH_3 bond in ethane, which gives a primary radical, is somewhat less (410 kJ/mol, or 98 kcal/mol) and is consistent with the notion that ethyl radical (primary) is more stable than methyl.

The dissociation energy of the terminal C—H bond in propane is exactly the same as that of ethane. The resulting free radical is primary ($R\dot{C}H_2$) in both cases.

$$CH_3CH_2CH_2\text{—}H \longrightarrow CH_3CH_2\dot{C}H_2 + H\cdot \qquad \Delta H° = +410 \text{ kJ/mol } (98 \text{ kcal/mol})$$

Propane; *n*-Propyl radical (primary); Hydrogen atom

Note, however, that Table 4.3 includes two entries for propane. The second entry corresponds to the cleavage of a bond to one of the hydrogens of the methylene (CH_2) group. It requires slightly less energy to break a C—H bond in the methylene group than in the methyl group.

$$CH_3CH(H)CH_3 \longrightarrow CH_3\dot{C}HCH_3 + H\cdot \qquad \Delta H° = +397 \text{ kJ/mol } (95 \text{ kcal/mol})$$

Propane; Isopropyl radical (secondary); Hydrogen atom

TABLE 4.3
Bond Dissociation Energies of Some Representative Compounds*

Bond	Bond dissociation energy kJ/mol (kcal/mol)	Bond	Bond dissociation energy kJ/mol (kcal/mol)
Diatomic molecules			
H—H	435 (104)		
F—F	159 (38)	H—F	568 (136)
Cl—Cl	242 (58)	H—Cl	431 (103)
Br—Br	192 (46)	H—Br	366 (87.5)
I—I	150 (36)	H—I	297 (71)
Alkanes			
CH_3—H	435 (104)	CH_3—CH_3	368 (88)
CH_3CH_2—H	410 (98)	CH_3CH_2—CH_3	355 (85)
$CH_3CH_2CH_2$—H	410 (98)		
$(CH_3)_2CH$—H	397 (95)		
$(CH_3)_2CHCH_2$—H	410 (98)	$(CH_3)_2CH$—CH_3	351 (84)
$(CH_3)_3C$—H	380 (91)	$(CH_3)_3C$—CH_3	334 (80)
Alkyl halides			
CH_3—F	451 (108)	$(CH_3)_2CH$—F	439 (105)
CH_3—Cl	349 (83.5)	$(CH_3)_2CH$—Cl	339 (81)
CH_3—Br	293 (70)	$(CH_3)_2CH$—Br	284 (68)
CH_3—I	234 (56)	$(CH_3)_3C$—Cl	330 (79)
CH_3CH_2—Cl	338 (81)	$(CH_3)_3C$—Br	263 (63)
$CH_3CH_2CH_2$—Cl	343 (82)		
Water and alcohols			
HO—H	497 (119)	CH_3CH_2—OH	380 (91)
CH_3O—H	426 (102)	$(CH_3)_2CH$—OH	385 (92)
CH_3—OH	380 (91)	$(CH_3)_3C$—OH	380 (91)

* Bond dissociation energies refer to bond indicated in structural formula for each substance.

Since the starting material (propane) and one of the products (hydrogen atom) are the same in both processes, the difference in bond dissociation energies is a measure of the energy difference between an *n*-propyl radical (primary) and an isopropyl radical (secondary). As depicted in Figure 4.14, the secondary radical is 13 kJ/mol (3 kcal/mol) more stable than the primary radical.

Similarly, comparing the bond dissociation energies of the two different types of C—H bonds in 2-methylpropane reveals a tertiary radical to be 30 kJ/mol (7 kcal/mol) more stable than a primary radical.

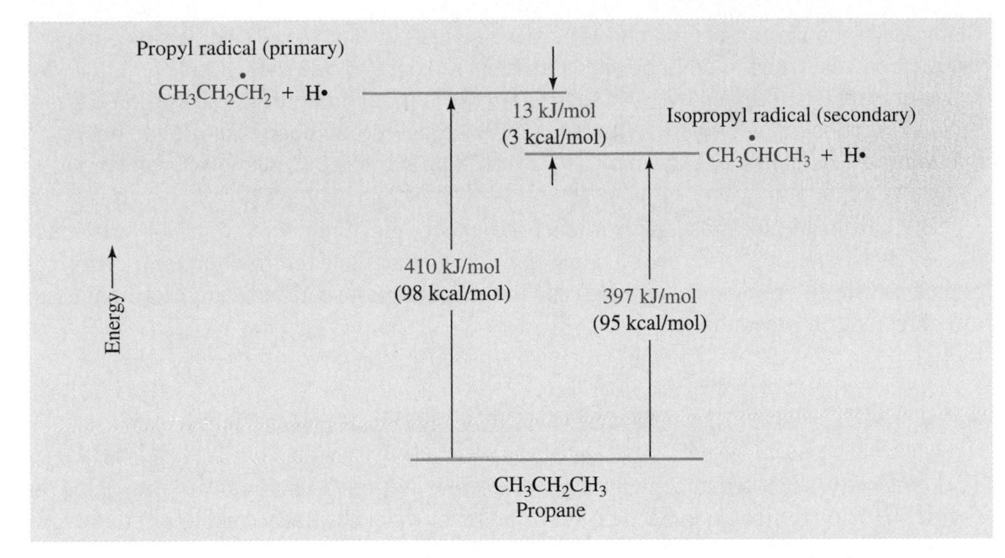

FIGURE 4.14 Diagram showing how bond dissociation energies of methylene and methyl C—H bonds in propane reveal a difference in stabilities between two isomeric free radicals. The secondary radical is more stable than the primary.

$$\underset{\text{2-Methylpropane}}{CH_3\underset{|}{C}HCH_2{-}H}\ (\text{with } CH_3 \text{ on C2}) \longrightarrow \underset{\text{Isobutyl radical (primary)}}{CH_3CH\dot{C}H_2}\ (\text{with } CH_3 \text{ on C2}) + \underset{\text{Hydrogen atom}}{H\cdot} \qquad \Delta H^\circ = +410 \text{ kJ/mol } (98 \text{ kcal/mol})$$

$$\underset{\text{2-Methylpropane}}{CH_3C(H)(CH_3)CH_3} \longrightarrow \underset{\textit{tert}\text{-Butyl radical (tertiary)}}{CH_3\dot{C}(CH_3)CH_3} + \underset{\text{Hydrogen atom}}{H\cdot} \qquad \Delta H^\circ = +380 \text{ kJ/mol } (91 \text{ kcal/mol})$$

PROBLEM 4.16 Carbon-carbon bond dissociation energies have been measured for alkanes. Without referring to Table 4.3, identify the alkane in each of the following pairs that has the lower carbon-carbon bond dissociation energy and explain the reason for your choice.

(*a*) Ethane or propane
(*b*) Propane or 2-methylpropane
(*c*) 2-Methylpropane or 2,2-dimethylpropane

SAMPLE SOLUTION (*a*) First write the equations that describe homolytic carbon-carbon bond cleavage in each alkane.

$$\underset{\text{Ethane}}{CH_3{-}CH_3} \longrightarrow \underset{\text{Two methyl radicals}}{\cdot CH_3 + \cdot CH_3}$$

$$\underset{\text{Propane}}{CH_3CH_2{-}CH_3} \longrightarrow \underset{\text{Ethyl radical}}{CH_3\dot{C}H_2} + \underset{\text{Methyl radical}}{\cdot CH_3}$$

Cleavage of the carbon-carbon bond in ethane yields two methyl radicals, while carbon-carbon bond cleavage in propane yields an ethyl radical and one methyl radical. Ethyl radical is more stable than methyl, and so less energy is required to break the carbon-carbon bond in propane than in ethane. The measured carbon-carbon bond dissociation energy in ethane is 368 kJ/mol (88 kcal/mol), while that in propane is 355 kJ/mol (85 kcal/mol).

Like carbocations, most free radicals are exceedingly reactive species—too reactive to be isolated but capable of being formed as transient intermediates in chemical reactions. Methyl radical, as we shall see in the following section, is an intermediate in the chlorination of methane.

4.19 MECHANISM OF METHANE CHLORINATION

The generally accepted mechanism for the chlorination of methane is presented in Figure 4.15. As indicated in Section 4.17, the reaction is normally carried out in the gas

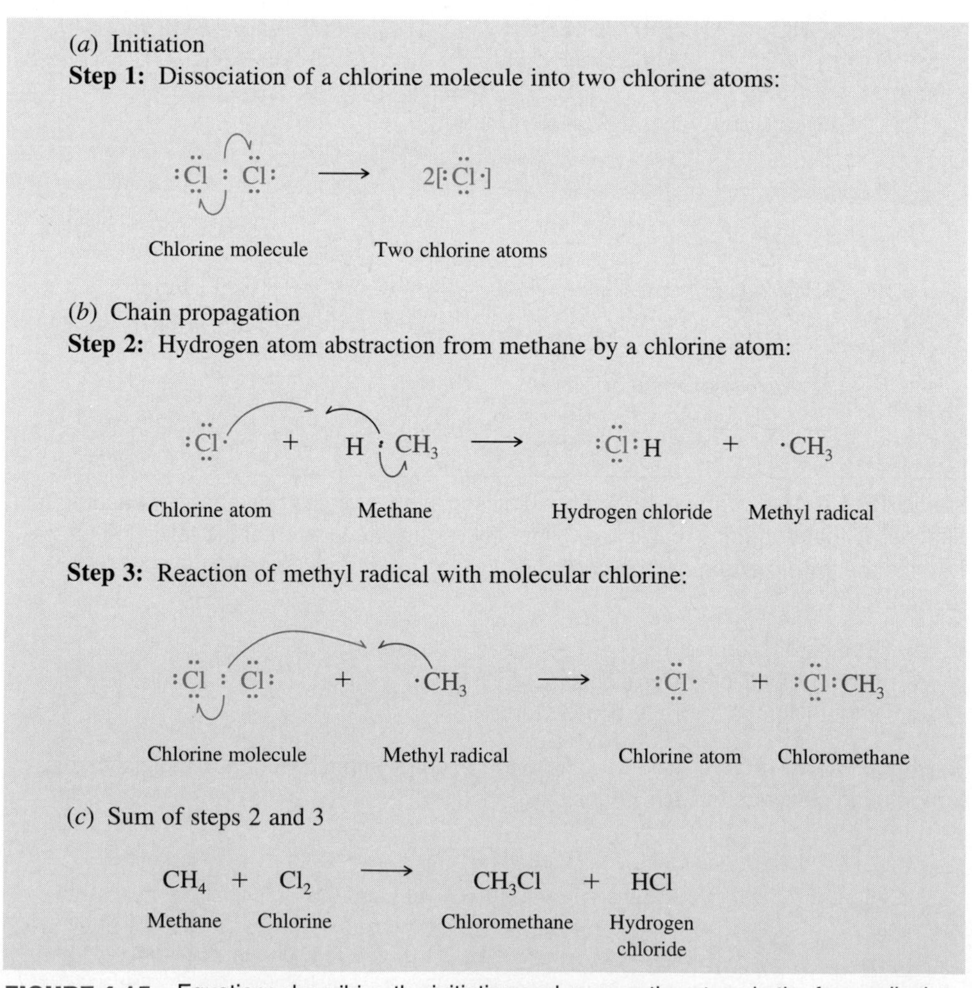

(*a*) Initiation
Step 1: Dissociation of a chlorine molecule into two chlorine atoms:
Chlorine molecule
Two chlorine atoms
(*b*) Chain propagation
Step 2: Hydrogen atom abstraction from methane by a chlorine atom:
Chlorine atom
Methane
Hydrogen chloride
Methyl radical
Step 3: Reaction of methyl radical with molecular chlorine:
Chlorine molecule
Methyl radical
Chlorine atom
Chloromethane
(*c*) Sum of steps 2 and 3
$CH_4 + Cl_2 \longrightarrow CH_3Cl + HCl$
Methane
Chlorine
Chloromethane
Hydrogen chloride

FIGURE 4.15 Equations describing the initiation and propagation steps in the free-radical mechanism for the chlorination of methane. Together the two propagation steps give the overall equation for the reaction.

phase at high temperature. The reaction itself is strongly exothermic, but energy must be put into the system in order to get the reaction going. This energy goes into breaking the weakest bond in the system, which, as we see from the bond dissociation energy data in Table 4.3, is the Cl—Cl bond with a bond dissociation energy of 242 kJ/mol (58 kcal/mol). The step in which Cl—Cl bond homolysis occurs is called the **initiation step.**

The bond dissociation energy of the other reactant, methane, is much higher. It is 435 kJ/mol (104 kcal/mol).

Each chlorine atom formed in the initiation step has seven valence electrons and is very reactive. Once formed, a chlorine atom abstracts a hydrogen atom from methane as shown in step 2 in Figure 4.15. Hydrogen chloride, one of the isolated products from the overall reaction, is formed in this step. A methyl radical is also formed, which then attacks a molecule of Cl_2 in step 3. Attack of methyl radical on Cl_2 gives chloromethane, the other product of the overall reaction, along with a chlorine atom which then cycles back to step 2, repeating the process. Steps 2 and 3 are called the **propagation steps** of the reaction and, when added together, give the overall equation for the reaction. Since one initiation step can result in a great many propagation cycles, the overall process is called a **free-radical chain reaction.**

In practice, side reactions intervene to reduce the efficiency of the propagation steps. The chain sequence is interrupted whenever two odd-electron species combine to give an even-electron product. Reactions of this type are called **chain-terminating steps.** Some commonly observed chain-terminating steps in the chlorination of methane are shown in the following equations.

Combination of a methyl radical with a chlorine atom:

$$\dot{C}H_3 \quad \cdot\ddot{\underset{..}{Cl}}: \longrightarrow CH_3\ddot{\underset{..}{Cl}}:$$

Methyl radical — Chlorine atom — Chloromethane

Combination of two methyl radicals:

$$\dot{C}H_3 \quad \dot{C}H_3 \longrightarrow CH_3CH_3$$

Two methyl radicals — Ethane

Combination of two chlorine atoms:

$$:\ddot{\underset{..}{Cl}}\cdot \quad \cdot\ddot{\underset{..}{Cl}}: \longrightarrow Cl_2$$

Two chlorine atoms — Chlorine molecule

Termination steps are, in general, less likely to occur than the propagation steps. Each of the termination steps requires two very reactive radicals to encounter each other in a medium that contains far greater quantities of other materials (methane and chlorine molecules, for example) with which they can react. While some chloromethane undoubtedly arises via direct combination of methyl radicals with chlorine atoms, most of it is formed by the propagation sequence shown in Figure 4.15.

FROM BOND ENERGIES TO HEATS OF REACTION

You have seen that measurements of heats of reaction, such as heats of combustion, can provide quantitative information concerning the relative stability of constitutional isomers (Section 2.17) and stereoisomers (Section 3.13). The box in Section 2.17 described how heats of reaction can be manipulated arithmetically to generate heats of formation (ΔH_f°) for many molecules. The following material shows how two different sources of thermochemical information, heats of formation and bond dissociation energies (Table 4.3), can reveal whether a particular reaction is exothermic or endothermic and by how much. The ability to make such predictions can be useful in deciding whether a reaction is feasible or not. Given the task of preparing a certain compound, an exothermic reaction would be a better candidate for further investigation than an endothermic one.

Consider the chlorination of methane to chloromethane. The heats of formation of the reactants and products appear beneath the equation. These heats of formation for the chemical compounds are taken from published tabulations; the heat of formation of chlorine, as it is for all elements, is zero.

	CH_4	+ Cl_2 ⟶	CH_3Cl	+ HCl
ΔH_f°: (kJ/mol)	−74.8	0	−81.9	−92.3

The overall heat of reaction is given by

$$\Delta H^\circ = \Sigma\text{ (heats of formation of products)} - \Sigma\text{ (heats of formation of reactants)}$$

$$\Delta H^\circ = (-81.9\text{ kJ/mol} - 92.3\text{ kJ/mol}) - (-74.8\text{ kJ/mol}) = -99.4\text{ kJ/mol}$$

Thus, the chlorination of methane is calculated to be an exothermic reaction on the basis of heat of formation data.

The same conclusion is reached using bond dissociation energies. The following equation shows the bond dissociation energies of the reactants and products taken from Table 4.3:

	CH_4	+ Cl_2 ⟶	CH_3Cl	+ HCl
BDE: (kJ/mol)	435	242	349	431

Since stronger bonds are formed at the expense of weaker ones, the reaction is exothermic and

$$\Delta H^\circ = \Sigma\text{ (BDE of bonds broken)} - \Sigma\text{ (BDE of bonds formed)}$$

$$\Delta H^\circ = (435\text{ kJ/mol} + 242\text{ kJ/mol}) - (349\text{ kJ/mol} + 431\text{ kJ/mol}) = -103\text{ kJ/mol}$$

This value is in good agreement with that obtained from heat of formation data.

Compare chlorination of methane with iodination. The relevant bond dissociation energies are given in the equation.

	CH_4	+ I_2 ⟶	CH_3I	+ HI
BDE: (kJ/mol)	435	150	234	297

$$\Delta H^\circ = \Sigma\text{ (BDE of bonds broken)} - \Sigma\text{ (BDE of bonds formed)}$$

$$\Delta H^\circ = (435\text{ kJ/mol} + 150\text{ kJ/mol}) - (234\text{ kJ/mol} + 297\text{ kJ/mol}) = +54\text{ kJ/mol}$$

A positive value for ΔH° signifies an **endothermic** reaction. The reactants are more stable than the products, and so iodination of alkanes is not a feasible reaction. You would not want to attempt the preparation of iodomethane by iodination of methane.

A similar analysis for fluorination of methane gives $\Delta H^\circ = -426$ kJ/mol for its heat of reaction. Fluorination of methane is 4 times as exothermic as chlorination. A reaction this exothermic, if it also occurs at a rapid rate, can proceed with explosive violence.

Bromination of methane is exothermic, but less exothermic than chlorination. The value calculated from bond dissociation energies is $\Delta H^\circ = -30$ kJ/mol. While bromination of methane is energetically feasible, economic considerations cause most of the methyl bromide prepared commercially to be made from methanol by reaction with hydrogen bromide.

4.20 HALOGENATION OF HIGHER ALKANES. REGIOSELECTIVITY

Like the chlorination of methane, chlorination of ethane is carried out on an industrial scale as a high-temperature gas-phase reaction.

$$CH_3CH_3 + Cl_2 \xrightarrow{420°C} CH_3CH_2Cl + HCl$$

Ethane | Chlorine | Chloroethane (78%) (ethyl chloride) | Hydrogen chloride

As in the chlorination of methane, it is often difficult to limit the reaction to monochlorination, and derivatives having more than one chlorine atom are also formed.

PROBLEM 4.17 Chlorination of ethane yields, in addition to ethyl chloride, a mixture of two isomeric dichlorides. What are the structures of these two dichlorides?

In the laboratory it is more convenient to use light, either visible or ultraviolet, as the source of energy to initiate the reaction. Reactions that occur when light energy is absorbed by a molecule are called **photochemical reactions.** Photochemical techniques permit the reaction of alkanes with chlorine to be performed at room temperature.

Photochemical energy is indicated by writing "light" or "$h\nu$" above the arrow. The symbol $h\nu$ is equal to the energy of a light photon and will be discussed in more detail in Section 13.1.

$$\text{Cyclobutane} + Cl_2 \xrightarrow{h\nu} \text{Cyclobutyl}-Cl + HCl$$

Cyclobutane | Chlorine | Chlorocyclobutane (73%) (cyclobutyl chloride) | Hydrogen chloride

Methane, ethane, and cyclobutane share the common feature that each one can give only a single monochloro derivative. All the hydrogens of cyclobutane, for example, are equivalent, and substitution of any one gives the same product as substitution of any other. Chlorination of alkanes in which all the hydrogens are not equivalent is more complicated in that a mixture of every possible monochloro derivative is formed, as the chlorination of butane illustrates:

$$CH_3CH_2CH_2CH_3 \xrightarrow[h\nu,\ 35°C]{Cl_2} CH_3CH_2CH_2CH_2Cl + CH_3\underset{\displaystyle Cl}{\underset{|}{C}}HCH_2CH_3$$

Butane | 1-Chlorobutane (28%) (*n*-butyl chloride) | 2-Chlorobutane (72%) (*sec*-butyl chloride)

The percentages cited in this equation reflect the composition of the monochloride fraction of the product mixture rather than the isolated yield of each component.

These two products arise because in one of the propagation steps a chlorine atom may abstract a hydrogen atom from either a methyl or a methylene group of butane.

$$CH_3CH_2CH_2CH_2{-}H + \cdot\ddot{\underset{..}{Cl}}\colon \longrightarrow CH_3CH_2CH_2\dot{C}H_2 + H\ddot{\underset{..}{Cl}}\colon$$

Butane | Butyl radical

$$CH_3\underset{\displaystyle H}{\underset{|}{C}}HCH_2CH_3 + \cdot\ddot{\underset{..}{Cl}}\colon \longrightarrow CH_3\dot{C}HCH_2CH_3 + H\ddot{\underset{..}{Cl}}\colon$$

Butane | *sec*-Butyl radical

The resulting free radicals react with chlorine in a succeeding propagation step to give the corresponding alkyl chlorides. Butyl radical gives only 1-chlorobutane; *sec*-butyl radical gives only 2-chlorobutane.

$$CH_3CH_2CH_2\dot{C}H_2 + Cl_2 \longrightarrow CH_3CH_2CH_2CH_2Cl + \cdot\ddot{\underset{..}{Cl}}:$$

Butyl radical — 1-Chlorobutane (*n*-butyl chloride)

$$CH_3\dot{C}HCH_2CH_3 + Cl_2 \longrightarrow CH_3\underset{\displaystyle Cl}{\underset{|}{C}}HCH_2CH_3 + \cdot\ddot{\underset{..}{Cl}}:$$

sec-Butyl radical — 2-Chlorobutane (*sec*-butyl chloride)

If every collision of a chlorine atom with a butane molecule resulted in hydrogen abstraction, the *n*-butyl/*sec*-butyl radical ratio and, therefore, the 1-chloro/2-chlorobutane ratio would be given by the relative numbers of hydrogens in the two equivalent methyl groups of $CH_3CH_2CH_2CH_3$ (six) compared with those in the two equivalent methylene groups (four). The product distribution expected on a *statistical* basis would be 60 percent 1-chlorobutane and 40 percent 2-chlorobutane. However, the *experimentally observed* product distribution is 28 percent 1-chlorobutane and 72 percent 2-chlorobutane. Therefore, *sec*-butyl radical is formed in greater amounts, and *n*-butyl radical in lesser amounts, than expected statistically.

The reason for this behavior stems from the greater stability of secondary compared with primary free radicals. The transition state for the step in which a chlorine atom abstracts a hydrogen from carbon has free-radical character at carbon.

$$CH_3CH_2CH_2\overset{\delta\cdot}{C}H_2 \cdots H \cdots \overset{\delta\cdot}{Cl}$$

Transition state for abstraction of a primary hydrogen

$$CH_3CH_2\overset{\delta\cdot}{C}HCH_3 \cdots H \cdots \overset{\delta\cdot}{Cl}$$

Transition state for abstraction of a secondary hydrogen

A secondary hydrogen is abstracted faster than a primary hydrogen because the transition state with secondary radical character is more stable than the one with primary radical character. The same factors that stabilize a secondary radical stabilize a transition state with secondary radical character more than one with primary radical character. Hydrogen atom abstraction from a CH_2 group occurs faster than from a CH_3 group. We can calculate how much faster a *single* secondary hydrogen is abstracted compared with a *single* primary hydrogen from the experimentally observed product distribution.

$$\frac{72\%\text{ 2-chlorobutane}}{28\%\text{ 1-chlorobutane}} = \frac{\text{rate of secondary H abstraction} \times 4 \text{ secondary hydrogens}}{\text{rate of primary H abstraction} \times 6 \text{ primary hydrogens}}$$

$$\frac{\text{Rate of secondary H abstraction}}{\text{Rate of primary H abstraction}} = \frac{72}{28} \times \frac{6}{4} = \frac{3.9}{1}$$

A single secondary hydrogen in butane is abstracted by a chlorine atom 3.9 times as fast as a single primary hydrogen.

PROBLEM 4.18 Assuming the relative rate of secondary to primary hydrogen atom abstraction to be the same in the chlorination of propane as it is in that of butane, calculate the relative amounts of propyl chloride and isopropyl chloride obtained in the free-radical chlorination of propane.

A similar study of the chlorination of 2-methylpropane established that a tertiary hydrogen is removed 5.2 times faster than each primary hydrogen.

$$CH_3\overset{H}{\underset{CH_3}{C}}CH_3 \xrightarrow[h\nu,\ 35°C]{Cl_2} CH_3\overset{H}{\underset{CH_3}{C}}CH_2Cl + CH_3\overset{Cl}{\underset{CH_3}{C}}CH_3$$

2-Methylpropane; 1-Chloro-2-methylpropane (63%) (isobutyl chloride); 2-Chloro-2-methylpropane (37%) (*tert*-butyl chloride)

Reactions in which more than one constitutional isomer can be formed from a single reactant, but where one isomer predominates, are said to be **regioselective.** The chlorination of alkanes is not very regioselective. Mixtures of isomers corresponding to substitution of each of the various hydrogens in the molecule are formed. The order of reactivity per hydrogen encompasses a factor of only 5 in rate.

The term *regioselective* was coined by Alfred Hassner of the University of Colorado in a paper published in the *Journal of Organic Chemistry* in 1968.

	R_3CH >	R_2CH_2 >	RCH_3
Relative rate (chlorination)	(tertiary) 5.2	(secondary) 3.9	(primary) 1

Bromine reacts with alkanes by a free-radical chain mechanism analogous to that of chlorine. There is an important difference between chlorination and bromination, however. Bromination is highly regioselective for substitution of *tertiary hydrogens.* The spread in reactivity among primary, secondary, and tertiary hydrogens is greater than 10^3.

	R_3CH >	R_2CH_2 >	RCH_3
Relative rate (bromination)	(tertiary) 1640	(secondary) 82	(primary) 1

In practice, this means that when an alkane contains primary, secondary, and tertiary hydrogens, it is usually only the tertiary hydrogen that is replaced by bromine.

$$CH_3\overset{H}{\underset{CH_3}{C}}CH_2CH_2CH_3 + Br_2 \xrightarrow[60°C]{h\nu} CH_3\overset{Br}{\underset{CH_3}{C}}CH_2CH_2CH_3 + HBr$$

2-Methylpentane; Bromine; 2-Bromo-2-methylpentane (76% isolated yield); Hydrogen bromide

The yield cited in this reaction is the isolated yield of purified product. Isomeric bromides constitute only a tiny fraction of the crude product.

PROBLEM 4.19 Give the structure of the principal organic product formed by free-radical bromination of each of the following:

(*a*) Methylcyclopentane
(*b*) 1-Isopropyl-1-methylcyclopentane
(*c*) 2,2,4-Trimethylpentane

SAMPLE SOLUTION (*a*) Write the structure of the starting hydrocarbon and identify any tertiary hydrogens that are present. The only tertiary hydrogen in methylcyclopentane is the one attached to C-1. This is the one replaced by bromine.

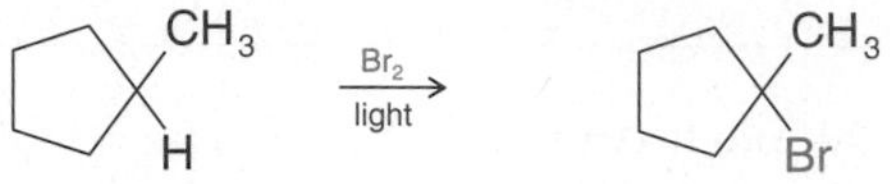

$$\text{Methylcyclopentane} \xrightarrow[\text{light}]{Br_2} \text{1-Bromo-1-methylcyclopentane}$$

This difference in regioselectivity between chlorination and bromination of alkanes has some practical consequences when one wishes to prepare an alkyl halide from an alkane:

1. Since chlorination of an alkane yields every possible monochloride, it is used only when all the hydrogens in an alkane are equivalent.
2. Bromination is normally used only to prepare tertiary alkyl bromides from alkanes.

Regioselectivity is not an issue in the conversion of alcohols to alkyl halides. Except for certain limitations to be discussed in Section 8.16, the location of the halogen substituent in the product corresponds to that of the hydroxyl group in the starting alcohol.

4.21 SUMMARY

Chemical reactivity and functional group transformations involving the formation of **alkyl halides** from **alcohols** and **alkanes** are the main themes of this chapter.

Alcohols and alkyl halides may be named using either **substitutive nomenclature** or **radicofunctional nomenclature** (Sections 4.1 and 4.2). In substitutive nomenclature alkyl halides are named as *halo-substituted* alkanes. The basis name is assigned according to the longest continuous chain that bears the functional group as a substituent, and the chain is numbered from the direction that gives the lowest number to the substituted carbon. The radicofunctional names of alkyl halides begin with the name of the alkyl group and end with identification of the halogen (as *halide*) as a separate word.

$CH_3CH(Br)CH_2CH_2CH_2CH_3$ — Substitutive name: 2-Bromohexane
Radicofunctional name: 1-Methylpentyl bromide

The substitutive names of alcohols are derived by replacing the *-e* ending of an alkane with *-ol*. Radicofunctional names of alcohols begin with the name of the alkyl group and end in the word *alcohol*.

$CH_3CH(OH)CH_2CH_2CH_2CH_3$ — Substitutive name: 2-Hexanol
Radicofunctional name: 1-Methylpentyl alcohol

Alcohols and alkyl halides are classified as primary, secondary, or tertiary according to the degree of substitution at the carbon that bears the functional group (Section 4.3).

The halogens (especially fluorine and chlorine) and oxygen are more electronegative than carbon, and the carbon-halogen bond in alkyl halides and the carbon-oxygen bond in alcohols are polar covalent bonds. Carbon is positively polarized and the halogen or oxygen is negatively polarized (Section 4.4).

Dipole–induced dipole and dipole-dipole attractive forces make alcohols higher-boiling than alkanes of similar molecular weight. The attractive force between —OH groups is called **hydrogen bonding.** Hydrogen bonding between the hydroxyl group and water molecules makes the water-solubility of alcohols greater than that of hydrocarbons. Low-molecular-weight alcohols (methanol, ethanol, 1-propanol, 2-propanol) are soluble in water in all proportions. Alkyl halides are insoluble in water (Section 4.5).

Brønsted acids are proton donors; **Brønsted bases** are proton acceptors (Sections 4.6 and 4.7). Strong acids transfer protons to alcohols to form **alkyloxonium ions.** An alkyloxonium ion is the **conjugate acid** of an alcohol. The **conjugate base** of an alcohol is an **alkoxide ion** (Section 4.8). Alkoxide ions are usually prepared by reaction of an alcohol with a group I metal such as sodium or potassium.

$$\underset{\text{Ethanol}}{2CH_3CH_2OH} + \underset{\text{Sodium}}{2Na} \longrightarrow \underset{\text{Sodium ethoxide}}{2NaOCH_2CH_3} + \underset{\text{Hydrogen}}{H_2}$$

Alkyl halides find many uses in organic chemistry, especially as starting materials for the preparation of various functional group classes. Table 4.4 summarizes the methods described in this chapter for the conversion of alcohols and alkanes to alkyl halides.

The conversion of an alcohol to an alkyl halide on reaction with a hydrogen halide is an example of a reaction type known as **nucleophilic substitution.** Nucleophilic substitutions are classified as S_N1 or S_N2 according to whether the rate-determining elementary step is unimolecular or bimolecular, respectively. Secondary and tertiary alcohols react with hydrogen halides by an S_N1 mechanism in which the rate-determining step is the unimolecular dissociation of an **alkyloxonium ion** to a **carbocation** (Sections 4.10 to 4.14).

$$(1)\quad \underset{\text{Alcohol}}{ROH} + \underset{\text{Hydrogen halide}}{HX} \xrightleftharpoons{\text{fast}} \underset{\text{Alkyloxonium ion}}{R\overset{+}{O}H_2} + \underset{\text{Halide anion}}{X^-}$$

$$(2)\quad \underset{\text{Alkyloxonium ion}}{R\overset{+}{O}H_2} \xrightarrow{\text{slow}} \underset{\text{Carbocation}}{R^+} + \underset{\text{Water}}{H_2O}$$

$$(3)\quad \underset{\text{Carbocation}}{R^+} + \underset{\text{Halide anion}}{X^-} \xrightarrow{\text{fast}} \underset{\text{Alkyl halide}}{RX}$$

Carbocations (Section 4.11) contain a positively charged carbon with only three atoms or groups bonded to it. Such a carbon atom is sp^2 hybridized, and has a vacant $2p$ orbital, the axis of which is perpendicular to the plane of the three bonds.

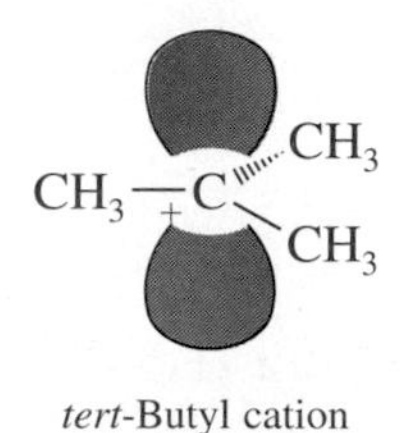

tert-Butyl cation

TABLE 4.4
Conversions of Alcohols and Alkanes to Alkyl Halides

Reaction (section) and comments	General equation and specific example(s)
Reactions of alcohols with hydrogen halides (Section 4.9) Alcohols react with hydrogen halides to yield alkyl halides. The reaction is useful as a synthesis of alkyl halides. The reactivity of hydrogen halides decreases in the order HI > HBr > HCl > HF. Alcohol reactivity decreases in the order tertiary > secondary > primary > methyl.	$ROH + HX \longrightarrow RX + H_2O$ Alcohol, Hydrogen halide, Alkyl halide, Water 1-Methylcyclopentanol $\xrightarrow{HCl}$ 1-Chloro-1-methylcyclopentane (96%)
Reaction of alcohols with thionyl chloride (Section 4.15) Thionyl chloride is a synthetic reagent used to convert alcohols to alkyl chlorides.	$ROH + SOCl_2 \longrightarrow RCl + SO_2 + HCl$ Alcohol, Thionyl chloride, Alkyl chloride, Sulfur dioxide, Hydrogen chloride $CH_3CH_2CH_2CH_2CH_2OH \xrightarrow[pyridine]{SOCl_2} CH_3CH_2CH_2CH_2CH_2Cl$ 1-Pentanol, 1-Chloropentane (80%)
Reaction of alcohols with phosphorus tribromide (Section 4.15) As an alternative to converting alcohols to alkyl bromides with hydrogen bromide, the inorganic reagent phosphorus tribromide is sometimes used.	$3ROH + PBr_3 \longrightarrow 3RBr + H_3PO_3$ Alcohol, Phosphorus tribromide, Alkyl bromide, Phosphorous acid $CH_3CH(OH)CH_2CH_2CH_3 \xrightarrow{PBr_3} CH_3CH(Br)CH_2CH_2CH_3$ 2-Pentanol, 2-Bromopentane (67%)
Free-radical halogenation of alkanes (Sections 4.16 through 4.20) Alkanes react with halogens by substitution of a halogen for a hydrogen on the alkane. The reactivity of the halogens decreases in the order $F_2 > Cl_2 > Br_2 > I_2$. The ease of replacing a hydrogen decreases in the order tertiary > secondary > primary > methyl. Chlorination is not very selective and so is used only when all the hydrogens of the alkane are equivalent. Bromination is highly selective, replacing tertiary hydrogens much more readily than secondary or primary ones.	$RH + X_2 \longrightarrow RX + HX$ Alkane, Halogen, Alkyl halide, Hydrogen halide Cyclodecane $\xrightarrow[h\nu]{Cl_2}$ Cyclodecyl chloride (64%) $(CH_3)_2CHC(CH_3)_3 \xrightarrow[h\nu]{Br_2} (CH_3)_2C(Br)C(CH_3)_3$ 2,2,3-Trimethylbutane, 2-Bromo-2,3,3-trimethylbutane (80%)

Carbocations are stabilized by alkyl substituents attached directly to the positively charged carbon. Alkyl groups are *electron-releasing* substituents. Stability increases in the order:

$$\text{(least stable)} \quad CH_3^+ < RCH_2^+ < R_2CH^+ < R_3C^+ \quad \text{(most stable)}$$

Carbocations are strongly **electrophilic.** They react with **nucleophiles,** substances with an unshared electron pair available for covalent bond formation.

Primary alcohols do not react with hydrogen halides by way of carbocation intermediates. The nucleophilic species (Br^-) attacks the alkyloxonium ion and "pushes off" a water molecule from carbon in a bimolecular step. This step is rate-determining, and the mechanism is S_N2 (Section 4.14).

A different kind of intermediate, a **free radical,** is involved in the chlorination of alkanes. The elementary steps (*a*) through (*c*) describe a **free-radical chain** mechanism for the reaction of an alkane with a halogen (Section 4.19).

(1) (initiation step)

$$X_2 \longrightarrow 2X\cdot$$

Halogen molecule → Two halogen atoms

(2) (propagation step)

$$RH + X\cdot \longrightarrow R\cdot + HX$$

Alkane + Halogen atom → Alkyl radical + Hydrogen halide

(3) (propagation step)

$$R\cdot + X_2 \longrightarrow RX + \cdot X$$

Alkyl radical + Halogen molecule → Alkyl halide + Halogen atom

Alkyl radicals are neutral species, characterized by the presence of an unpaired electron on a tricoordinate carbon.

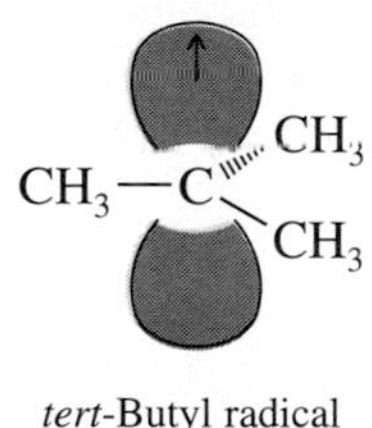

tert-Butyl radical

Like carbocations, free radicals are stabilized by alkyl substituents. The order of free-radical stability is

$$\text{(least stable)} \quad \cdot CH_3 < R\dot{C}H_2 < R_2\dot{C}H < R_3C\cdot \quad \text{(most stable)}$$

Multistep reaction mechanisms are commonplace in organic chemistry and will be encountered frequently as new reactions are introduced in the chapters that follow. Carbocations and free radicals are but two examples of reactive intermediates that occur during the conversion of starting materials to products. Both carbocations and free radicals are involved in reactions other than those described in this chapter and will reappear in a number of the functional group transformations that remain to be described.

PROBLEMS

4.20 Write structural formulas for each of the following alcohols and alkyl halides:

(*a*) Cyclobutanol
(*b*) *sec*-Butyl alcohol
(*c*) 3-Heptanol
(*d*) 1-Pentadecanol
(*e*) *trans*-2-Chlorocyclopentanol
(*f*) 4-Methyl-2-hexanol
(*g*) 2,6-Dichloro-4-methyl-4-octanol
(*h*) *trans*-4-*tert*-Butylcyclohexanol
(*i*) 1-Bromo-3-iodo-2-propanol
(*j*) 1-Cyclopropylethanol
(*k*) 2-Cyclopropylethanol
(*l*) Chlorotrifluoromethane

4.21 Name each of the following compounds according to substitutive IUPAC nomenclature:

(*a*) $CH_3(CH_2)_8I$
(*b*) $(CH_3)_2CHCH_2CH_2CH_2Br$
(*c*) $(CH_3)_2CHCH_2CH_2CH_2OH$
(*d*) Cl_3CCH_2Br
(*e*) $Cl_2CHCHBr$ (with Cl on the second carbon)
(*f*) CF_3CH_2OH
(*g*) F
(*h*) OH
(*i*) CH_3 OH CH_3
(*j*) Br
(*k*) OH

4.22 *Halothane* ($F_3CCHBrCl$) is a nonflammable, nonexplosive inhalation anesthetic. What is its substitutive IUPAC name?

4.23 Write structural formulas for all the constitutionally isomeric alcohols of molecular formula $C_5H_{12}O$. Assign a substitutive and a radicofunctional name to each one and specify whether it is a primary, secondary, or tertiary alcohol.

4.24 A hydroxyl group is a somewhat "smaller" substituent on a six-membered ring than is a methyl group. That is, the preference of a hydroxyl group for the equatorial orientation is less pronounced than that of a methyl group. Given this information, write structural formulas for all the isomeric methylcyclohexanols, showing each one in its most stable conformation. Give an acceptable IUPAC name for each isomer.

4.25 By assuming that the heat of combustion of the cis isomer was larger than the trans, structural assignments were made many years ago for the stereoisomeric 2-, 3-, and 4-methylcyclohexanols. This assumption is valid for two of the stereoisomeric pairs but is incorrect for the other. For which pair of stereoisomers is the assumption incorrect? Why?

4.26

(*a*) *Menthol,* used to flavor various foods and tobacco, is the most stable stereoisomer of 2-isopropyl-5-methylcyclohexanol. Draw its most stable conformation. Is the hydroxyl group cis or trans to the isopropyl group? To the methyl group?
(*b*) *Neomenthol* is a stereoisomer of menthol. That is, it has the same constitution but differs in the arrangement of its atoms in space. Neomenthol is the second-most-

stable stereoisomeric form of 2-isopropyl-5-methylcyclohexanol; it is less stable than menthol but more stable than any other stereoisomer. Write the structure of neomenthol in its most stable conformation.

4.27 Each of the following pairs of compounds undergoes a Brønsted acid-base reaction for which the equilibrium constant is greater than unity. Give the products of each reaction and identify the acid, the base, the conjugate acid, and the conjugate base. Use the acid dissociation constants in Table 4.2 as a guide.

(*a*) $HI + HO^- \rightleftharpoons$

(*b*) $CH_3CH_2O^- + CH_3\overset{O}{\overset{\|}{C}}OH \rightleftharpoons$

(*c*) $HF + H_2N^- \rightleftharpoons$

(*d*) $CH_3\overset{O}{\overset{\|}{C}}O^- + HCl \rightleftharpoons$

(*e*) $(CH_3)_3CO^- + H_2O \rightleftharpoons$

(*f*) $(CH_3)_2CHOH + H_2N^- \rightleftharpoons$

(*g*) $F^- + H_2SO_4 \rightleftharpoons$

4.28 Transition-state representations are shown for two acid-base reactions. For each one, write the equation for the reaction it represents in the direction for which the equilibrium constant is greater than 1. Label the acid, the base, the conjugate acid, and the conjugate base, and use curved arrows to show the flow of electrons.

(*a*) $CH_3—C(H)(CH_3)—\overset{\delta-}{O}\cdots\cdots H\cdots\cdots\overset{\delta-}{Br}$

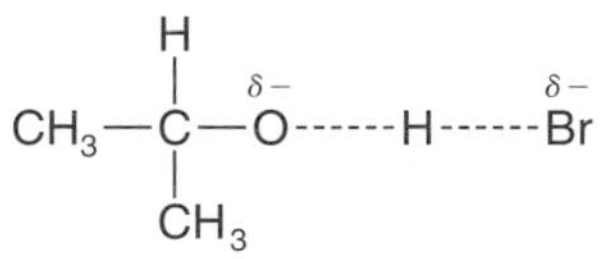

(*b*) $H—\overset{\delta-}{O}\cdots\cdots H\cdots\cdots\overset{\delta+}{O}(H)—CH_3$

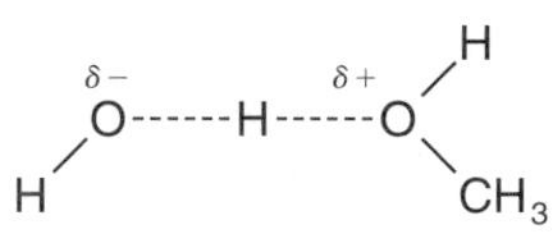

4.29 Calculate K_a for each of the following acids, given its pK_a. Rank the compounds in order of decreasing acidity.

(*a*) Aspirin: $pK_a = 3.48$
(*b*) Vitamin C (ascorbic acid): $pK_a = 4.17$
(*c*) Formic acid (present in sting of ants): $pK_a = 3.75$
(*d*) Oxalic acid (poisonous substance found in certain berries): $pK_a = 1.19$

4.30 The pK_a's of methanol (CH_3OH) and methanethiol (CH_3SH) are 16 and 11, respectively. Which is more basic, $KOCH_3$ or $KSCH_3$?

4.31 Write a chemical equation for the reaction of 1-butanol with each of the following:

(*a*) Sodium
(*b*) Sodium amide ($NaNH_2$)
(*c*) Hydrogen bromide, heat
(*d*) Sodium bromide, sulfuric acid, heat
(*e*) Phosphorus tribromide
(*f*) Thionyl chloride

4.32 Each of the following reactions has been described in the chemical literature and involves an organic starting material somewhat more complex than those we have encountered so far. Nevertheless, on the basis of the topics covered in this chapter, you should be able to write the structure of the principal organic product of each reaction.

(*a*) (cyclohexenyl)$—CH_2CH_2OH \xrightarrow[\text{pyridine}]{PBr_3}$

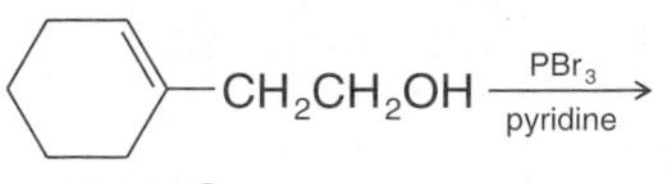

(*b*) (cyclopentane bearing CH_3, $\overset{O}{\overset{\|}{C}}OCH_2CH_3$, and OH) $\xrightarrow[\text{pyridine}]{SOCl_2}$

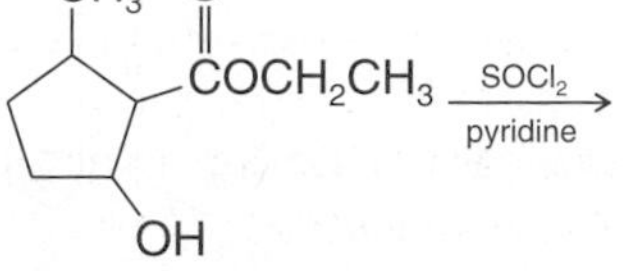

(*c*) Br CH_3 C—OH CH_3 $\xrightarrow{HCl}$

(*d*) $HOCH_2CH_2-C_6H_4-CH_2CH_2OH + 2HBr \xrightarrow{heat}$

(*e*) CH_3 CH_3 CH_3 OH $\xrightarrow{K}$

4.33 Select the compound in each of the following pairs that will be converted to the corresponding alkyl bromide more rapidly on being treated with hydrogen bromide. Explain the reason for your choice.

(*a*) 1-Butanol or 2-butanol
(*b*) 2-Methyl-1-butanol or 2-butanol
(*c*) 2-Methyl-2-butanol or 2-butanol
(*d*) 2-Methylbutane or 2-butanol
(*e*) 1-Methylcyclopentanol or cyclohexanol
(*f*) 1-Methylcyclopentanol or *trans*-2-methylcyclopentanol
(*g*) 1-Cyclopentylethanol or 1-ethylcyclopentanol

4.34 Assuming that the rate-determining elementary step in the reaction of cyclohexanol with hydrogen bromide to give cyclohexyl bromide is unimolecular, write an equation for this step. Use curved arrows to depict the flow of electrons.

4.35 Assuming that the rate-determining elementary step in the reaction of 1-hexanol with hydrogen bromide to give 1-bromohexane is an attack by a nucleophile on an oxonium ion, write an equation for this step. Use curved arrows to depict the flow of electrons.

4.36

(*a*) Evidence from a variety of sources indicates that cyclopropyl cation is much less stable than isopropyl cation. Why?

(*b*) The cation $CF_3\overset{+}{C}HCH_3$ is much less stable than isopropyl cation. Why?

4.37 The difference in energy between a secondary and a tertiary alkyl radical is about 12–17 kJ/mol (3–4 kcal/mol). The difference in energy between a secondary and tertiary carbocation is about 42–63 kJ/mol (10–15 kcal/mol). Can you think of a reason why carbocations are more sensitive to stabilization by substituents than free radicals are?

4.38 Basing your answers on the bond dissociation energy data in Table 4.3, calculate which of the following reactions are endothermic and which are exothermic:

(*a*) $(CH_3)_2CHOH + HF \longrightarrow (CH_3)_2CHF + H_2O$
(*b*) $(CH_3)_2CHOH + HCl \longrightarrow (CH_3)_2CHCl + H_2O$
(*c*) $CH_3CH_2CH_3 + HCl \longrightarrow (CH_3)_2CHCl + H_2$

4.39 By carrying out the reaction at −78°C it is possible to fluorinate 2,2-dimethylpropane to yield $(CF_3)_4C$. Write a balanced chemical equation for this reaction.

4.40 In a search for fluorocarbons having anesthetic properties, 1,2-dichloro-1,1-difluoropropane was subjected to photochemical chlorination. Two isomeric products were obtained, one of which was identified as 1,2,3-trichloro-1,1-difluoropropane. What is the structure of the second compound?

4.41 Among the isomeric alkanes of molecular formula C_5H_{12}, identify the one that on photochemical chlorination yields

(*a*) A single monochloride
(*b*) Three isomeric monochlorides
(*c*) Four isomeric monochlorides
(*d*) Two isomeric dichlorides

4.42 In both the following exercises, assume that all the methylene groups in the alkane are equally reactive as sites of free-radical chlorination.

(*a*) Photochemical chlorination of heptane gave a mixture of monochlorides containing 15 percent 1-chloroheptane. What other monochlorides are present? Estimate the percentage of each of these additional $C_7H_{15}Cl$ isomers in the monochloride fraction.
(*b*) Photochemical chlorination of dodecane gave a monochloride fraction containing 19 percent 2-chlorododecane. Estimate the percentage of 1-chlorododecane present in that fraction.

4.43 Photochemical chlorination of 2,2,4-trimethylpentane gives four isomeric monochlorides.

(*a*) Write structural formulas for these four isomers.
(*b*) The two primary chlorides make up 65 percent of the monochloride fraction. Assuming that all the primary hydrogens in 2,2,4-trimethylpentane are equally reactive, estimate the percentage of each of the two primary chlorides in the product mixture.

4.44 Photochemical chlorination of pentane gave a mixture of three isomeric monochlorides. The principal monochloride constituted 46 percent of the total, and the remaining 54 percent was approximately a 1 : 1 mixture of the other two isomers. Write structural formulas for the three monochloride isomers and specify which one was formed in greatest amount. (Recall that a secondary hydrogen is abstracted 3 times faster by a chlorine atom than a primary hydrogen.)

4.45 A certain substance contains only carbon and hydrogen and has a molecular weight of 70. Photochemical chlorination gave only one monochloride. Write the structure of the hydrocarbon and its monochloride derivative.

4.46 Two isomeric compounds A and B have the molecular formula C_3H_7Cl. Chlorination of A gave a mixture of two dichlorides of formula $C_3H_6Cl_2$. Chlorination of B gave three different compounds of formula $C_3H_6Cl_2$ (although these may not all be different from the dichlorides from A). What are the structural formulas of A and B and the dichlorides obtained from each?

4.47

(*a*) Write the structures of all the possible monobromides formed during the photochemical bromination of 2,2-dimethylbutane.
(*b*) Estimate the distribution of these monobromides in the product if the secondary/primary rate ratio for bromination is 82 : 1.

4.48 Deuterium oxide (D_2O) is water in which the protons (^1H) have been replaced by their heavier isotope deuterium (^2H). It is readily available and is used in a variety of mechanistic studies in organic chemistry and biochemistry. When D_2O is added to an alcohol (ROH), deuterium replaces the proton of the hydroxyl group.

$$ROH + D_2O \rightleftharpoons ROD + DOH$$

The reaction takes place extremely rapidly, and if D_2O is present in excess, all the alcohol is converted to ROD. This hydrogen-deuterium exchange can be catalyzed by either acids or bases. If D_3O^+ is the catalyst in acid solution and DO^- the catalyst in base, write reasonable reaction mechanisms for the conversion of ROH to ROD under conditions of (*a*) acid catalysis and (*b*) base catalysis.

MOLECULAR MODELING EXERCISES

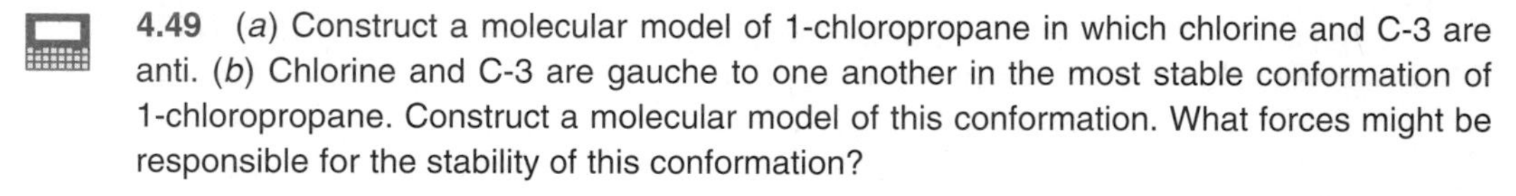

4.49 (*a*) Construct a molecular model of 1-chloropropane in which chlorine and C-3 are anti. (*b*) Chlorine and C-3 are gauche to one another in the most stable conformation of 1-chloropropane. Construct a molecular model of this conformation. What forces might be responsible for the stability of this conformation?

4.50 Two stereoisomers of 1-bromo-4-methylcyclohexane are formed when *trans*-4-methylcyclohexanol reacts with hydrogen bromide. Construct molecular models of:

(*a*) *trans*-4-methylcylohexanol
(*b*) The carbocation intermediate in this reaction
(*c*) The two stereoisomers of 1-bromo-4-methylcyclohexane

4.51 Make molecular models of ethyl cation ($CH_3CH_2^+$) in which the axis of the unfilled *p* orbital of the positively charged carbon is: (*a*) eclipsed with a C—H σ bond of the methyl group; and (*b*) perpendicular to a C—H σ bond of the methyl group. Which conformation best permits dispersal of the positive charge by hyperconjugation?

4.52 Make molecular models of methyl radical in which the carbon that bears the unshared electron is: (*a*) sp^3 hybridized, and (*b*) sp^2 hybridized.

4.53 Two inorganic reagents thionyl chloride ($SOCl_2$) and phosphorus tribromide (PBr_3) were introduced in this chapter. Make a molecular model of each of these compounds. (Oxygen and both chlorine atoms are attached to sulfur in $SOCl_2$, and all three bromines are attached to phosphorus in PBr_3.) Describe the geometry of each substance.

CHAPTER 5

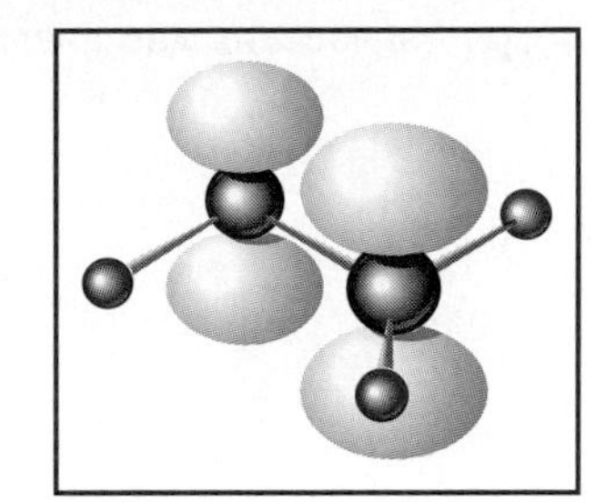

STRUCTURE AND PREPARATION OF ALKENES. ELIMINATION REACTIONS

Alkenes are hydrocarbons that contain a carbon-carbon double bond. A carbon-carbon double bond is both an important structural unit and an important functional group in organic chemistry. The shape of an organic molecule is influenced by the presence of a carbon-carbon double bond, and the double bond is the site of most of the chemical reactions that alkenes undergo. Some representative alkenes include *isobutylene* (an industrial chemical), *α-pinene* (a fragrant liquid obtained from pine trees), and *farnesene* (a naturally occurring alkene with three double bonds).

$(CH_3)_2C{=}CH_2$

Isobutylene
(used in the production of synthetic rubber)

CH_3 CH_3 H— CH_3

α-Pinene
(a major constituent of turpentine)

Farnesene
(present in the waxy coating found on apple skins)

This chapter is the first of two dealing with alkenes; it describes their structure, bonding, and preparation. Chapter 6 discusses their chemical reactions.

5.1 ALKENE NOMENCLATURE

Alkenes are named in the IUPAC system by replacing the *-ane* ending in the name of the corresponding alkane with *-ene*. The two simplest alkenes are **ethene** and **propene.** Both are also well known by their common names *ethylene* and *propylene.*

$CH_2{=}CH_2$ — IUPAC name: **ethene**; Common name: ethylene

$CH_3CH{=}CH_2$ — IUPAC name: **propene**; Common name: propylene

Ethylene is an acceptable synonym for *ethene* in the IUPAC system. *Propylene, isobutylene,* and other common names ending in *-ylene* are not acceptable IUPAC names.

The longest continuous chain that includes the double bond forms the base name of the alkene, and the chain is numbered in the direction that gives the doubly bonded carbons their lower numbers. The locant (or numerical position) of only one of the doubly bonded carbons is specified in the name; it is understood that the other doubly bonded carbon must follow in sequence.

$\overset{1}{C}H_2{=}\overset{2}{C}H\overset{3}{C}H_2\overset{4}{C}H_3$ — 1-Butene (not 1,2-butene)

$\overset{6}{C}H_3\overset{5}{C}H_2\overset{4}{C}H_2\overset{3}{C}H{=}\overset{2}{C}H\overset{1}{C}H_3$ — 2-Hexene (not 4-hexene)

Carbon-carbon double bonds take precedence over alkyl groups and halogen substituents in determining the main carbon chain and the direction in which it is numbered.

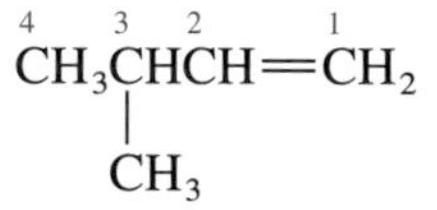

$\overset{4}{C}H_3\overset{3}{C}H(CH_3)\overset{2}{C}H{=}\overset{1}{C}H_2$ — 3-Methyl-1-butene (not 2-methyl-3-butene)

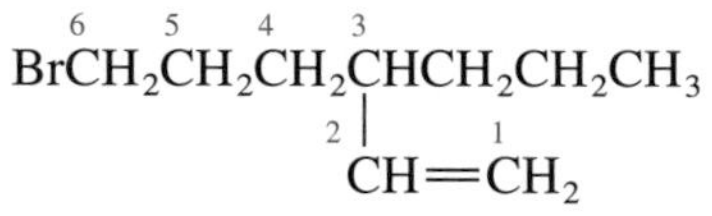

$Br\overset{6}{C}H_2\overset{5}{C}H_2\overset{4}{C}H_2\overset{3}{C}H(\overset{2}{C}H{=}\overset{1}{C}H_2)CH_2CH_2CH_3$ — 6-Bromo-3-propyl-1-hexene (longest chain that contains double bond is six carbons)

Hydroxyl groups, however, outrank the double bond. Compounds that contain both a double bond and a hydroxyl group use the combined suffix *-en* + *-ol* to signify that both structural units are present.

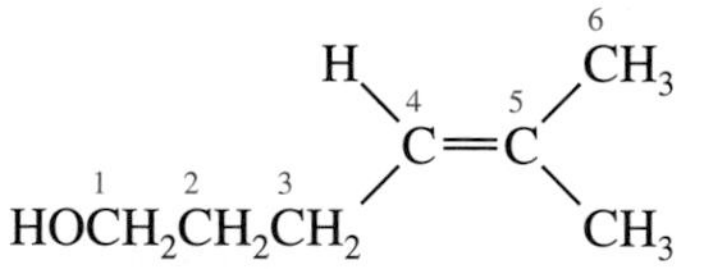

$HO\overset{1}{C}H_2\overset{2}{C}H_2\overset{3}{C}H_2(H)\overset{4}{C}{=}\overset{5}{C}(\overset{6}{C}H_3)CH_3$ — 5-Methyl-4-hexen-1-ol (not 2-methyl-2-hexen-6-ol)

PROBLEM 5.1 Name each of the following using systematic IUPAC nomenclature:

(*a*) $(CH_3)_2C{=}C(CH_3)_2$

(*b*) $(CH_3)_3CCH{=}CH_2$

(*c*) $(CH_3)_2C{=}CHCH_2CH_2CH_3$

(*d*) $CH_2{=}CHCH_2CH(Cl)CH_3$

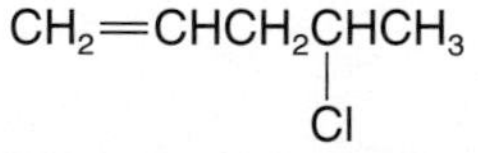

(*e*) $CH_2{=}CHCH_2CH(OH)CH_3$

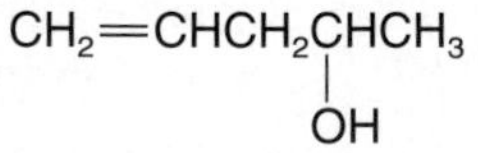

SAMPLE SOLUTION (*a*) The longest continuous chain in this alkene contains four carbon atoms. The double bond is between C-2 and C-3, and so it is named as a derivative of 2-butene.

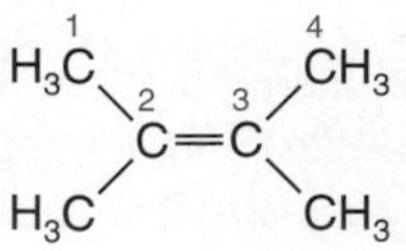

$(\overset{1}{H_3C})(H_3C)\overset{2}{C}{=}\overset{3}{C}(\overset{4}{C}H_3)CH_3$ — 2,3-Dimethyl-2-butene

Identifying the alkene as a derivative of 2-butene leaves two methyl groups to be accounted for as substituents attached to the main chain. This alkene is 2,3-dimethyl-2-butene. (It is sometimes called *tetramethylethylene,* but that is a common name, not a systematic IUPAC name.)

We noted earlier in Section 2.12 that the common names of certain frequently encountered *alkyl* groups, such as isopropyl and *tert*-butyl, are acceptable in the IUPAC system. Three *alkenyl* groups—**vinyl, allyl,** and **isopropenyl**—are treated the same way.

Vinyl chloride is an industrial chemical produced in large amounts (10^{10} lb/yr in the United States) and is used in the preparation of poly(vinyl chloride). Poly(vinyl chloride), often called simply *vinyl,* has many applications, including siding for houses, wall coverings, and PVC piping.

$CH_2{=}CH{-}$ as in $CH_2{=}CHCl$

Vinyl — Vinyl chloride

$CH_2{=}CHCH_2{-}$ as in $CH_2{=}CHCH_2OH$

Allyl — Allyl alcohol

$CH_2{=}C(CH_3){-}$ as in $CH_2{=}CCl(CH_3)$

Isopropenyl — Isopropenyl chloride

It frequently happens that a CH_2 group is doubly bonded to a ring. In such cases, the cycloalkane name corresponding to the ring is taken as the parent and the prefix *methylene* is added.

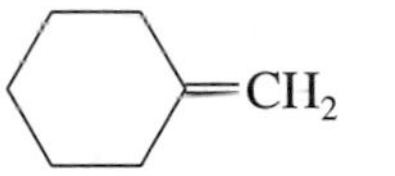

$=CH_2$

Methylenecyclohexane

Cycloalkenes and their derivatives are named by adapting cycloalkane terminology to the principles of alkene nomenclature.

Cyclopentene 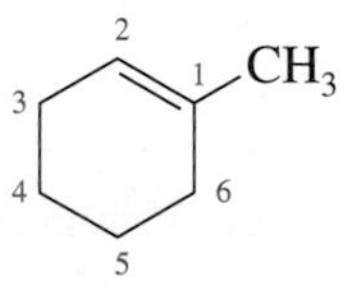 — 1-Methylcyclohexene 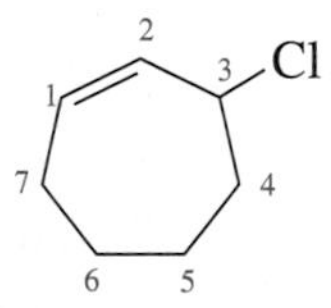 — 3-Chlorocycloheptene (not 1-chloro-2-cycloheptene)

No locants are needed in the absence of substituents; it is understood that the double bond connects C-1 and C-2. Substituted cycloalkenes are numbered beginning with the double bond, proceeding through it, and continuing in sequence around the ring. The direction of numbering is chosen so as to give the lower of two possible locants to the substituent group.

PROBLEM 5.2 Write structural formulas and give the IUPAC names of all the monochloro-substituted derivatives of cyclopentene.

ETHYLENE

Ethylene was known to chemists in the eighteenth century and was isolated in pure form in 1795. An early name for ethylene was *gaz oléfiant* (French for "oil-forming gas"), a term suggested to describe the fact that an oily liquid product is formed when two gases—ethylene and chlorine—react with each other.

$$\underset{\substack{\text{Ethylene} \\ (\text{bp: } -104^\circ\text{C})}}{CH_2{=}CH_2} + \underset{\substack{\text{Chlorine} \\ (\text{bp: } -34^\circ\text{C})}}{Cl_2} \longrightarrow \underset{\substack{\text{1,2-Dichloroethane} \\ (\text{bp: } 83^\circ\text{C})}}{ClCH_2CH_2Cl}$$

The term gaz oléfiant was the forerunner of the general term *olefin,* formerly used as the name of the class of compounds we now call *alkenes.*

Ethylene occurs naturally in small amounts as a plant hormone. Hormones are substances that act as messengers and play regulatory roles in biological processes. Ethylene is involved in the ripening of many fruits, where it is formed in a complex series of steps from a compound containing a cyclopropane ring:

$$\underset{\text{1-Aminocyclopropanecarboxylic acid}}{C_3H_4(\overset{+}{N}H_3)(CO_2^-)} \xrightarrow{\text{several steps}} \underset{\text{Ethylene}}{CH_2{=}CH_2} + \text{other products}$$

Even minute amounts of ethylene can stimulate ripening, and the rate of ripening increases with the concentration of ethylene. This property is used to advantage, for example, in the marketing of bananas. Bananas are picked green in the tropics, kept green by being stored with adequate ventilation to limit the amount of ethylene present, and then induced to ripen at their destination by passing ethylene over the fruit.*

Ethylene is the cornerstone substance of the world's mammoth petrochemical industry and is produced in vast quantities. In a typical year the amount of ethylene produced in the United States is approximately the same as the combined weight of all of its people (4×10^{10} lb). In one process, ethane from natural gas is heated to bring about its dissociation into ethylene and hydrogen:

$$\underset{\text{Ethane}}{CH_3CH_3} \xrightarrow{750^\circ C} \underset{\text{Ethylene}}{CH_2{=}CH_2} + \underset{\text{Hydrogen}}{H_2}$$

This reaction is known as **dehydrogenation** and is simultaneously both a source of ethylene and one of the methods by which hydrogen is prepared on an industrial scale. Most of the hydrogen so generated is subsequently used to reduce nitrogen to ammonia for the preparation of fertilizer.

Similarly, dehydrogenation of propane gives propene:

$$\underset{\text{Propane}}{CH_3CH_2CH_3} \xrightarrow{750^\circ C} \underset{\text{Propene}}{CH_3CH{=}CH_2} + \underset{\text{Hydrogen}}{H_2}$$

Propene is the second-most-important petrochemical and is produced on a scale about half that of ethylene.

Almost any hydrocarbon can serve as a starting material for production of ethylene and propene. Cracking of petroleum (Section 2.15) gives ethylene and propene by processes involving cleavage of carbon-carbon bonds of higher-molecular-weight hydrocarbons.

The major uses of ethylene and propene are as starting materials for the preparation of polyethylene and polypropylene plastics, fibers, and films. The chemical bases for these and other applications will be described in Chapter 6.

* For a review, see "Ethylene—An Unusual Plant Hormone" in the April 1992 issue of the *Journal of Chemical Education* (pp. 315–318).

5.2 STRUCTURE AND BONDING IN ALKENES

The structure of ethylene and the orbital hybridization model for the double bond were presented in Section 1.15. To review, Figure 5.1 depicts the planar structure of ethylene, its bond distances, and its bond angles. Each of the carbon atoms is sp^2 hybrid-

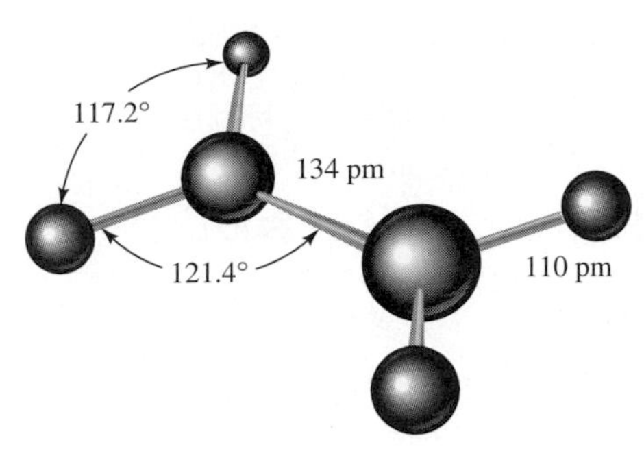

FIGURE 5.1 A ball-and-stick model of ethylene showing bond distances in picometers and bond angles in degrees. All six atoms are coplanar. The connections between atoms represent the framework of σ bonds in the molecule. The carbon-carbon bond is a double bond made up of the σ component shown and the π component illustrated in Figure 5.2.

ized, and the double bond possesses a σ component and a π component. The σ component results when an sp^2 orbital of one carbon, oriented so that its axis lies along the internuclear axis, overlaps with a similarly disposed sp^2 orbital of the other carbon. Each sp^2 orbital contains one electron, and the resulting σ bond contains two of the four electrons of the double bond. The π bond contributes the other two electrons and is formed by a ''side-by-side'' overlap of singly occupied p orbitals of the two carbons. The orientation of these p orbitals is depicted in Figure 5.2.

The double bond in ethylene is stronger than the C—C single bond in ethane, but it is not twice as strong. Thermochemical measurements indicate a C═C bond energy of 605 kJ/mol (144.5 kcal/mol) in ethylene versus a value of 368 kJ/mol (88 kcal/mol) for the carbon-carbon bond in ethane. Chemists do not agree on exactly how to apportion the total C═C bond energy between its σ and π components, but all agree that the π bond is weaker than the σ bond.

The simplest arithmetic approach subtracts the C—C σ bond energy of ethane (368 kJ/mol; 88 kcal/mol) from the C═C bond energy of ethylene (605 kJ/mol; 144.5 kcal/mol). This gives a value of 237 kJ/mol (56.5 kcal/mol) for the π bond energy.

Higher alkenes are related to ethylene by replacement of hydrogen substituents by alkyl groups. There are two different types of carbon-carbon bonds in propene, $CH_3CH{=}CH_2$. The double bond is of the $\sigma + \pi$ type, and the bond to the methyl group is a σ bond formed by sp^3–sp^2 overlap.

sp^3 hybridized carbon

C—C bond length = 150 pm
C═C bond length = 134 pm

sp^2 hybridized carbon

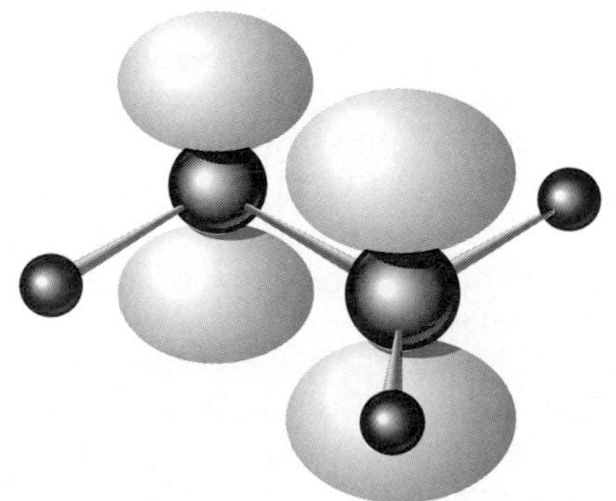

FIGURE 5.2 The p orbitals of ethylene represented by "dumbbell-shaped" lobes. The carbon atoms are sp^2 hybridized, and each is bonded to two hydrogens by sp^2–$1s$ σ bonds. The σ component of the carbon-carbon double bond arises by sp^2–sp^2 overlap and the π component by overlap of the p orbitals. An electron pair in the π bond is shared by the two carbons.

PROBLEM 5.3 We can use bond-line formulas to represent alkenes in much the same way that we use them to represent alkanes. Consider the following alkene:

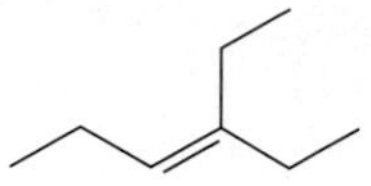

(*a*) What is the molecular formula of this alkene?
(*b*) What is its IUPAC name?
(*c*) How many carbon atoms are sp^2 hybridized in this alkene? How many are sp^3 hybridized?
(*d*) How many σ bonds are of the sp^2-sp^3 type? How many are of the sp^3-sp^3 type?

SAMPLE SOLUTION (*a*) Recall when writing bond-line formulas for hydrocarbons that there is a carbon at each end and at each bend in a carbon chain. The appropriate number of hydrogen substituents are attached so that each carbon has four bonds. Thus the compound shown is

$$CH_3CH_2CH{=}C(CH_2CH_3)_2$$

The general molecular formula for an alkene is C_nH_{2n}. Ethylene is C_2H_4; propene is C_3H_6. Counting the carbons and hydrogens of the compound shown (C_8H_{16}) reveals that it, too, corresponds to C_nH_{2n}.

5.3 ISOMERISM IN ALKENES

While ethylene is the only two-carbon alkene, and propene the only three-carbon alkene, there are *four* isomeric alkenes of molecular formula C_4H_8:

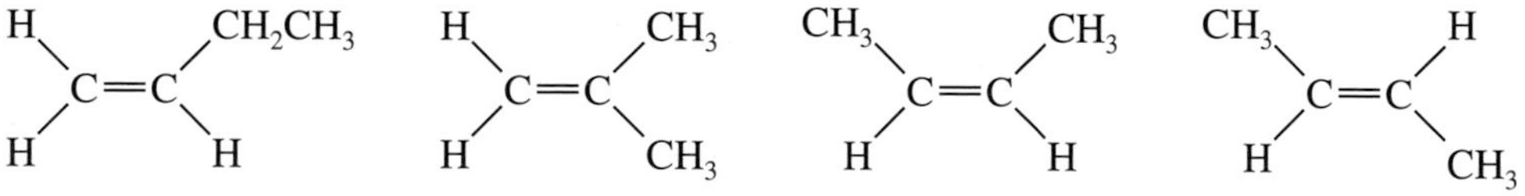

$H_2C{=}CHCH_2CH_3$ — 1-Butene
$(CH_3)_2C{=}CH_2$ — 2-Methylpropene
$CH_3CH{=}CHCH_3$ — *cis*-2-Butene
$CH_3CH{=}CHCH_3$ — *trans*-2-Butene

1-Butene has an unbranched carbon chain with a double bond between C-1 and C-2. It is a constitutional isomer of the other three. Similarly, 2-methylpropene, with a branched carbon chain, is a constitutional isomer of the other three.

The pair of isomers designated *cis*- and *trans*-2-butene have the same constitution; both have an unbranched carbon chain with a double bond connecting C-2 and C-3. They differ from each other, however, in that the cis isomer has both of its methyl groups on the same side of the double bond while the methyl groups in the trans isomer are on opposite sides of the double bond. Recall from Section 3.13 that isomers which have the same constitution but differ in the arrangement of their atoms in space are classified as *stereoisomers*. *cis*-2-Butene and *trans*-2-butene are stereoisomers, and the terms *cis* and *trans* specify the *configuration* of the double bond.

Stereoisomeric alkenes are sometimes referred to as *geometric isomers*.

Cis-trans stereoisomerism in alkenes is not possible when one of the doubly bonded carbons bears two identical substituents. Thus, neither 1-butene nor 2-methylpropene is capable of existing in stereoisomeric forms.

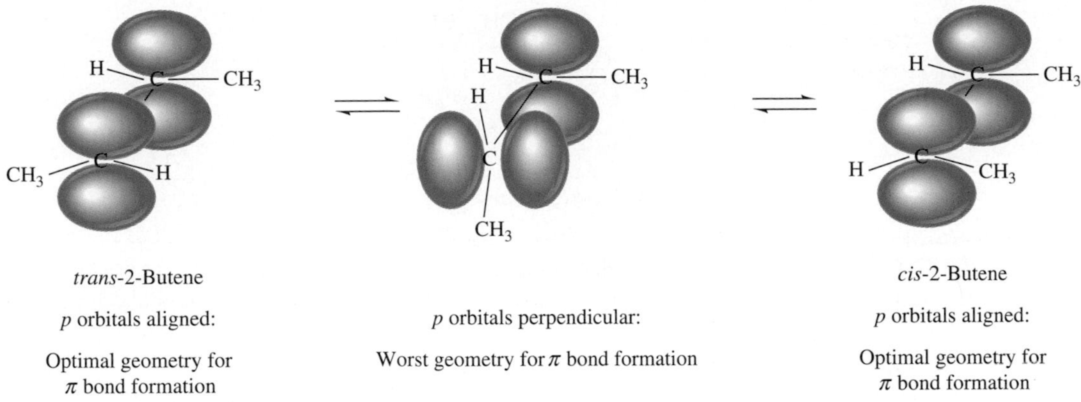

FIGURE 5.3 Interconversion of *cis*- and *trans*-2-butene proceeds by cleavage of the π component of the double bond.

Two identical substituents { H, H } $C{=}C$ (CH_2CH_3, H)

1-Butene
(no stereoisomers possible)

Two identical substituents { H, H } $C{=}C$ (CH_3, CH_3) } Two identical substituents

2-Methylpropene
(no stereoisomers possible)

PROBLEM 5.4 How many alkenes are there of molecular formula C_5H_{10}? Write their structures and give their IUPAC names. Specify the configuration of stereoisomers as cis or trans as appropriate.

In principle, *cis*-2-butene and *trans*-2-butene may be interconverted by rotation about the C-2═C-3 *double* bond. However, unlike rotation about the C-2—C-3 *single* bond in butane, which is quite fast, interconversion of the stereoisomeric 2-butenes does not occur under normal circumstances. It is sometimes said that rotation about a carbon-carbon double bond is *restricted,* but this is an understatement. Conventional laboratory sources of heat do not provide enough thermal energy for rotation about the double bond in alkenes to take place. As shown in Figure 5.3, rotation about a double bond requires the *p* orbitals of C-2 and C-3 to be twisted from their stable parallel alignment—in effect, the π component of the double bond must be broken at the transition state.

The activation energy for rotation about a typical carbon-carbon double bond is very high—on the order of 250 kJ/mol (about 60 kcal/mol). This quantity may be taken as a measure of the π bond contribution to the total C═C bond strength of 605 kJ/mol (144.5 kcal/mol) in ethylene and compares closely with the value estimated by manipulation of thermochemical data on page 171.

5.4 NAMING STEREOISOMERIC ALKENES BY THE *E-Z* NOTATIONAL SYSTEM

When the substituents on either end of a double bond are the same or are structurally similar to each other, it is a simple matter to describe the configuration of the double bond as cis or trans. Oleic acid, for example, a material that can be obtained from olive oil, has a cis double bond. Cinnamaldehyde, responsible for the characteristic odor of cinnamon, has a trans double bond.

$CH_3(CH_2)_6CH_2$ and $CH_2(CH_2)_6CO_2H$ on C=C, each carbon also bearing H (cis)

Oleic acid

C_6H_5 and H on one carbon, H and CH=O on the other (C=C)

Cinnamaldehyde

PROBLEM 5.5 Female houseflies attract males by sending a chemical signal known as a *pheromone.* The substance emitted by the female housefly that attracts the male has been identified as *cis*-9-tricosene, $C_{23}H_{46}$. Write a structural formula, including stereochemistry, for this compound.

However, *cis* and *trans* are ambiguous when it is not obvious which substituent on one carbon is ''similar'' or ''analogous'' to a reference substituent on the other. Fortunately, a completely unambiguous system for specifying double bond stereochemistry has been developed based on an *atomic number* criterion for ranking substituents on the doubly bonded carbons. When atoms of higher atomic number are on the *same* side of the double bond, we say that the double bond has the ***Z*** configuration. The *Z* descriptor stands for the German word ***zusammen,*** which means ''together.'' When atoms of higher atomic number are on *opposite* sides of the double bond, we say that the configuration is ***E.*** The symbol *E* stands for the German word ***entgegen,*** which means ''opposite.''

Higher → Cl, Br ← Higher; Lower → H, F ← Lower (C=C)

Z configuration

Higher-ranked substituents (Cl and Br) are on same side of double bond

Higher → Cl, F ← Lower; Lower → H, Br ← Higher (C=C)

E configuration

Higher-ranked substituents (Cl and Br) are on opposite sides of double bond

The priority rules were developed by R. S. Cahn and Sir Christopher Ingold (England) and Vladimir Prelog (Switzerland) in the context of a different aspect of organic stereochemistry; they will appear again in Chapter 7.

The substituent groups on the double bonds of most alkenes are, of course, more complicated than in this example. The rules for ranking substituents, especially alkyl groups, are described in Table 5.1.

PROBLEM 5.6 Determine the configuration of each of the following alkenes as *Z* or *E* as appropriate:

(*a*) H_3C and H on one carbon; CH_2OH and CH_3 on the other (C=C)

(*b*) H_3C and H on one carbon; CH_2CH_2F and $CH_2CH_2CH_2CH_3$ on the other (C=C)

(*c*) H_3C and H on one carbon; CH_2CH_2OH and $C(CH_3)_3$ on the other (C=C)

(*d*) cyclopropyl and CH_3CH_2 on one carbon; H and CH_3 on the other (C=C)

SAMPLE SOLUTION (*a*) A methyl group and a hydrogen occur as substituents on one of the doubly bonded carbons. According to the rules of Table 5.1, methyl outranks

TABLE 5.1
Cahn-Ingold-Prelog Priority Rules

Rule	Example
1. Higher atomic number takes precedence over lower. Bromine (atomic number 35) outranks chlorine (atomic number 17). Methyl (C, atomic number 6) outranks hydrogen (atomic number 1).	The compound Higher Br(Higher) CH_3 (Higher), Lower Cl, H (Lower): Br(Cl)C=C(CH_3)(H) has the *Z* configuration. Higher-ranked atoms (Br and C of CH_3) are on the same side of the double bond.
2. When two atoms directly attached to the double bond are identical, compare the atoms attached to these two on the basis of their atomic numbers. Precedence is determined at the first point of difference: Ethyl [—C(**C**,H,H)] outranks methyl [—C(**H**,H,H)] Similarly, *tert*-butyl outranks isopropyl, and isopropyl outranks ethyl: $—C(CH_3)_3 > —CH(CH_3)_2 > —CH_2CH_3$ —C(C,C,C) > —C(C,C,H) > —C(C,H,H)	The compound Higher Br, CH_3 Lower; Lower Cl, CH_2CH_3 Higher: Br(Cl)C=C(CH_3)(CH_2CH_3) has the *E* configuration.
3. Work outward from the point of attachment, comparing all the atoms attached to a particular atom before proceeding further along the chain: $—CH(CH_3)_2$ [—C(C,**C**,H)] outranks $—CH_2CH_2OH$ [—C(C,**H**,H)]	The compound Higher Br, CH_2CH_2OH Lower; Lower Cl, $CH(CH_3)_2$ Higher: Br(Cl)C=C(CH_2CH_2OH)($CH(CH_3)_2$) has the *E* configuration.
4. When working outward from the point of attachment, always evaluate substituent atoms one by one, never as a group. Since oxygen has a higher atomic number than carbon, $—CH_2OH$ [—C(**O**,H,H)] outranks $—C(CH_3)_3$ [—C(**C**,C,C)]	The compound Higher Br, CH_2OH Higher; Lower Cl, $C(CH_3)_3$ Lower: Br(Cl)C=C(CH_2OH)($C(CH_3)_3$) has the *Z* configuration.

(Continued)

TABLE 5.1
Cahn-Ingold-Prelog Priority Rules *(Continued)*

Rule	Example
5. An atom that is multiply bonded to another atom is considered to be replicated as a substituent on that atom: —CH(=O) is treated as if it were —C(O,O,H). The group —CH=O [—C(O,**O**,H)] outranks $—CH_2OH$ [—C(O,**H**,H)].	The compound (Higher) Br and (Lower) Cl on one carbon, CH_2OH (Lower) and CH=O (Higher) on the other, Br(Cl)C=C(CH_2OH)(CH=O), has the *E* configuration.
6. When it is necessary to consider substituents on a multiply bonded atom, the atom from which the multiple bond originates is counted: $—CH{=}CH_2$ is treated as if it were —C(H)(C)(C—C). A vinyl group $—CH{=}CH_2$ [—C(H)(C)(C—C)] outranks an isopropyl group $—CH(CH_3)_2$ [—C(H)(C(H,H,H))(C(H,H,H))].	The compound (Higher) Br and (Lower) Cl on one carbon, $CH{=}CH_2$ (Higher) and $CH(CH_3)_2$ (Lower) on the other, Br(Cl)C=C($CH{=}CH_2$)($CH(CH_3)_2$), has the *Z* configuration.

hydrogen. The other carbon atom of the double bond bears a methyl and a $—CH_2OH$ group. The $—CH_2OH$ group is of higher priority than methyl.

Higher (C) → H_3C ; CH_2OH ← Higher —C(O,H,H)

C=C

Lower (H) → H ; CH_3 ← Lower —C(H,H,H)

Higher-ranked substituents are on the same side of the double bond; the configuration is *Z*.

A table on the inside back cover (right page) lists some of the more frequently encountered substituent atoms and groups in organic chemistry in order of increasing precedence. You should not attempt to memorize this table, but should be able to derive the relative placement of one group versus another.

5.5 PHYSICAL PROPERTIES OF ALKENES

Alkenes resemble alkanes in most of their physical properties. The lower-molecular-weight alkenes through C_4H_8 are gases at room temperature and atmospheric pressure. The physical properties of a number of alkenes and cycloalkenes are collected in Appendix 1. Selected properties of the isomeric C_4H_8 alkenes are compared with the alkane butane in Table 5.2, where it is seen that the boiling points of the alkenes are similar to one another and to that of butane. As evidenced by the properties of *cis*- and *trans*-2-butene, it is apparent that stereoisomers, like constitutional isomers, can have different physical properties.

Table 5.2 notes that, like butane, the C_4H_8 alkenes are nonpolar and when measurable dipole moments are present, they are small. What neither the table nor the method by which dipole moments are measured reveals, however, is the *direction* of the small dipole moment in alkenes. This is deduced by examining substituent effects such as the following:

$H_2C{=}CH_2$	$H_2C{=}CHCl$	$CH_3CH{=}CH_2$	$CH_3CH{=}CHCl$ (trans)
Ethylene	Chloroethene	Propene	*trans*-1-Chloropropene
$\mu = 0$ D	$\mu = 1.4$ D	$\mu = 0.3$ D	$\mu = 1.7$ D

Using ethylene and chloroethene (vinyl chloride) as reference compounds, we note that ethylene, of course, has no dipole moment and that replacement of one of its hydrogens by an electronegative chlorine substituent gives rise to a substantial dipole moment (1.4 D in chloroethene). Replacement of a hydrogen by a methyl group has a much smaller effect: propene has a moment of 0.3 D. Does the methyl group act as an electron-attracting or as an electron-releasing substituent on the double bond? If methyl is electron-attracting, then the dipole moment of *trans*-$CH_3CH{=}CHCl$ will be smaller than that of $CH_2{=}CHCl$, because the effects of CH_3 and Cl will oppose each

TABLE 5.2
Selected Physical Properties of Butane and Four-Carbon Alkenes

Compound name	Structural formula	Boiling point, °C	Dipole moment μ, D
Butane	$CH_3CH_2CH_2CH_3$	0	0
1-Butene	$CH_2{=}CHCH_2CH_3$	−6	0.3
cis-2-Butene	$CH_3CH{=}CHCH_3$ (CH₃ groups cis)	4	0.3
trans-2-Butene	$CH_3CH{=}CHCH_3$ (CH₃ groups trans)	1	0
2-Methylpropene	$(CH_3)_2C{=}CH_2$	−7	0.5

other. The dipole moment of *trans*-$CH_3CH{=}CHCl$ is, in fact, larger than the dipole moment of $CH_2{=}CHCl$. The effect of the methyl group, therefore, reinforces that of chlorine, and so a methyl group must be an electron-releasing substituent.

A methyl group releases electrons to a double bond in much the same way that it releases electrons to the positively charged carbon of a carbocation—by an inductive effect and by hyperconjugation (Section 4.11). The increased *s* character of the sp^2 hybridized carbon of a carbon-carbon double bond makes it more electron-attracting than the sp^3 hybridized carbon of the methyl group; thus the electrons in the σ bond that connects the methyl group and the doubly bonded carbon are drawn closer to the doubly bonded carbon. Other alkyl groups behave similarly to methyl and act as electron-releasing substituents on a double bond.

5.6 RELATIVE STABILITIES OF ALKENES

We have described on earlier occasions (Sections 2.17, 3.13) how heats of combustion are used to evaluate relative stabilities of isomeric hydrocarbons. This technique has been applied to isomeric alkenes to reveal the ways in which alkene structure affects stability. Consider, for example, the heats of combustion of the isomeric C_4H_8 alkenes. All undergo combustion according to the equation

$$C_4H_8 + 6O_2 \longrightarrow 4CO_2 + 4H_2O$$

The heats of combustion of the various isomers are plotted on a common scale in Figure 5.4 and reveal their relative energies. The isomer of highest energy, i.e., the least stable one, is 1-butene, $CH_2{=}CHCH_2CH_3$. The isomer of lowest energy, the most stable one, is 2-methylpropene, $(CH_3)_2C{=}CH_2$.

By analyzing the heats of combustion of the C_4H_8 isomers, along with similar data for numerous other alkenes, we find two factors to be the most significant when comparing the stabilities of isomeric alkenes:

1. *Degree of substitution* (alkyl substituents stabilize a double bond)
2. *Van der Waals strain* (destabilizing when alkyl groups are cis to each other)

Degree of substitution. We classify double bonds as **monosubstituted, disubstituted, trisubstituted,** or **tetrasubstituted** according to the number of carbon atoms that are *directly* attached to the C=C structural unit.

Monosubstituted alkenes:

$RCH{=}CH_2$ as in $CH_3CH_2CH{=}CH_2$ (1-butene)

Disubstituted alkenes:
(R and R′ may be the same or different)

$RCH{=}CHR'$ as in $CH_3CH{=}CHCH_3$ (*cis*- or *trans*-2-butene)

$R(R')C{=}CH_2$ as in $(CH_3)_2C{=}CH_2$ (2-methylpropene)

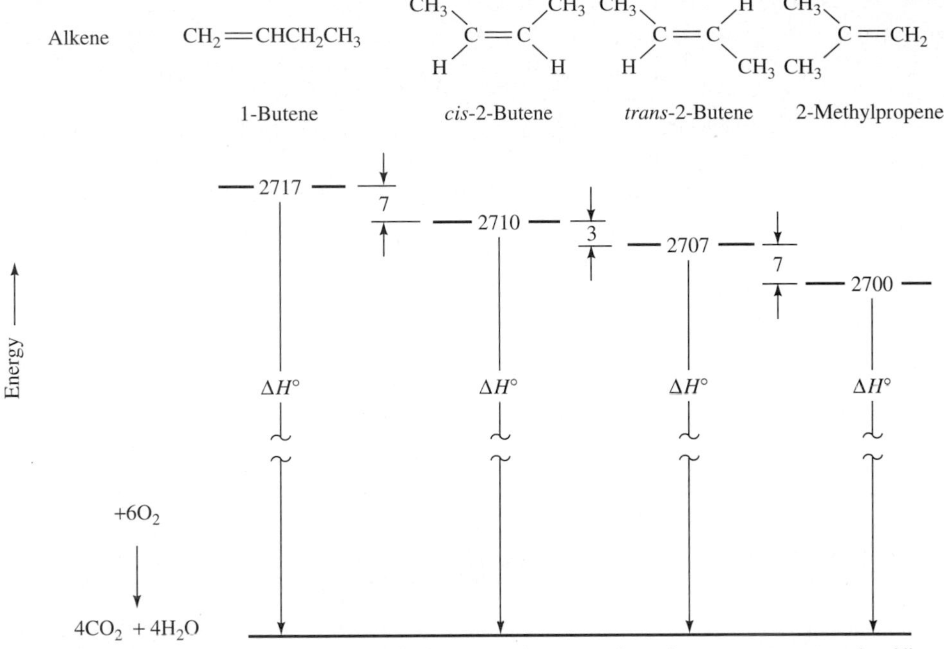

FIGURE 5.4 Heats of combustion of C_4H_8 alkene isomers plotted on a common scale. All energies are in kilojoules per mole. (An energy difference of 3 kJ/mole is equivalent to 0.7 kcal/mole; 7 kJ/mole is equivalent to 1.7 kcal/mol.)

Trisubstituted alkenes:
(R, R′, and R″ may be the same or different)

$R(R')C{=}C(R'')H$ as in $(CH_3)_2C{=}CHCH_2CH_3$ (2-methyl-2-pentene)

Tetrasubstituted alkenes:
(R, R′, R″, and R‴ may be the same or different)

$R(R')C{=}C(R'')R'''$ as in 1,2-dimethylcyclohexene (ring with two CH_3 groups on the double-bonded carbons; adjacent ring carbons marked *)

In the example shown, each of the ring carbons indicated by an asterisk counts as a unique substituent on the double bond.

PROBLEM 5.7 Write structural formulas and give the IUPAC names for all the alkenes of molecular formula C_6H_{12} that contain a trisubstituted double bond. (Don't forget to include stereoisomers.)

From the heats of combustion of the C_4H_8 alkenes in Figure 5.4 we see that each of the disubstituted alkenes

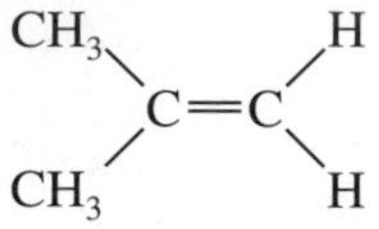

$(CH_3)_2C{=}CH_2$ 2-Methylpropene

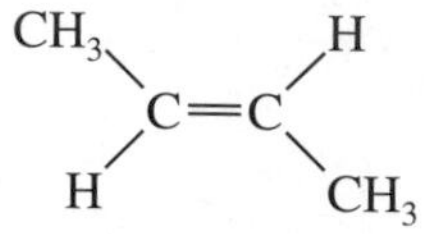

trans-$CH_3CH{=}CHCH_3$ *trans*-2-Butene

cis-$CH_3CH{=}CHCH_3$ *cis*-2-Butene

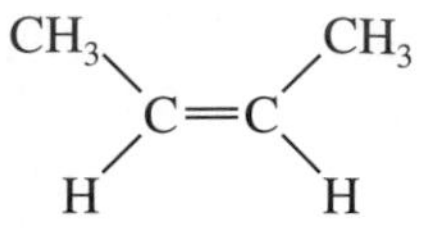

is more stable than the monosubstituted alkene

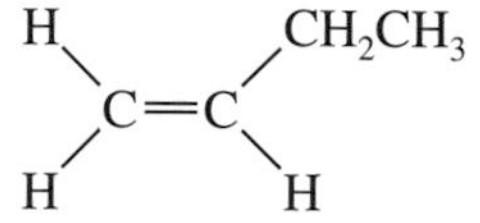

$H_2C{=}CHCH_2CH_3$

1-Butene

In general, alkenes with more highly substituted double bonds are more stable than isomers with less substituted double bonds.

PROBLEM 5.8 Give the structure of the most stable C_6H_{12} alkene.

Like the sp^2 hybridized carbons of carbocations and free radicals, the sp^2 hybridized carbons of double bonds are electron-attracting, and alkenes are stabilized by substituents that release electrons to these carbons. One mechanism by which this occurs is via a polarization of the electrons in the σ bond that connects the substituent to the C=C unit (Figure 5.5). As we saw in Section 4.11, alkyl groups are better electron-releasing substituents than hydrogen and are, therefore, better able to stabilize an alkene.

An effect that results when two or more atoms or groups interact so as to alter the electron distribution in a system is called an **electronic effect.** The greater stability of more highly substituted alkenes is an example of an electronic effect.

van der Waals strain. *Alkenes are more stable when large substituents are trans to each other than when they are cis.* As seen in Figure 5.4, *trans*-2-butene has a lower heat of combustion and is more stable than *cis*-2-butene. The energy difference between the two is 3 kJ/mol (0.7 kcal/mol). The source of this energy difference is illustrated in Figure 5.6, where it is seen that methyl groups approach each other very closely in *cis*-2-butene while the trans isomer is free of strain. An effect that results when two or more atoms are close enough in space that there is a van der Waals repulsion between them is one type of **steric effect.** The greater stability of trans alkenes compared with their cis counterparts is an example of a steric effect.

A similar steric effect was seen in Section 3.13, where van der Waals strain between methyl groups on the same side of the ring made *cis*-1,2-dimethylcyclopropane less stable than its trans stereoisomer.

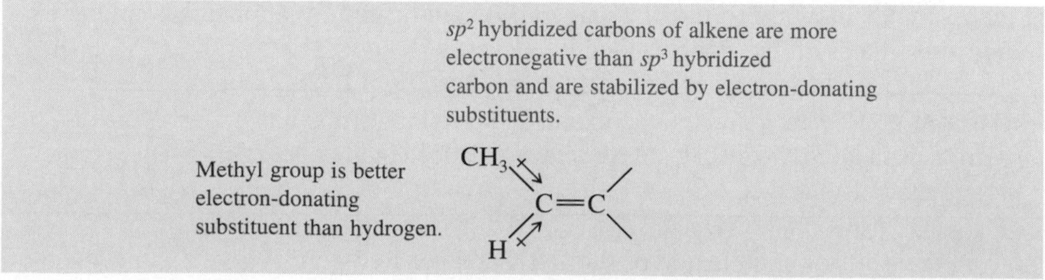

sp^2 hybridized carbons of alkene are more electronegative than sp^3 hybridized carbon and are stabilized by electron-donating substituents.

Methyl group is better electron-donating substituent than hydrogen.

CH_3 C=C H

FIGURE 5.5 Alkyl groups stabilize carbon-carbon double bonds by donating electrons to sp^2 hybridized carbons.

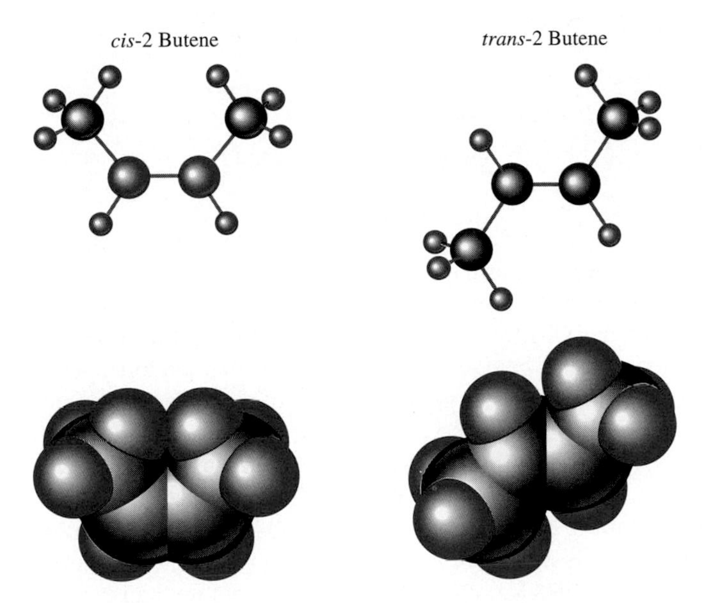

FIGURE 5.6 Ball-and-stick and space-filling molecular models of *cis*-2-butene and *trans*-2-butene. The space-filling model reveals the serious van der Waals strain between methyl groups in *cis*-2-butene. The molecule adjusts by expanding those bond angles that increase the separation between the crowded atoms in the region where the van der Waals strain is greatest. The combination of angle strain and van der Waals strain makes *cis*-2-butene less stable than *trans*-2-butene.

PROBLEM 5.9 Arrange the following alkenes in order of decreasing stability: 1-pentene; *(E)*-2-pentene; *(Z)*-2-pentene; 2-methyl-2-butene.

The difference in stability between stereoisomeric alkenes is even more pronounced with larger alkyl groups on the double bond. A particularly striking example compares *cis*- and *trans*-2,2,5,5-tetramethyl-3-hexene, where the heat of combustion of the cis stereoisomer is 44 kJ/mol (10.5 kcal/mol) higher than that of the trans. The cis isomer is destabilized by the large van der Waals strain between the bulky *tert*-butyl groups on the same side of the double bond.

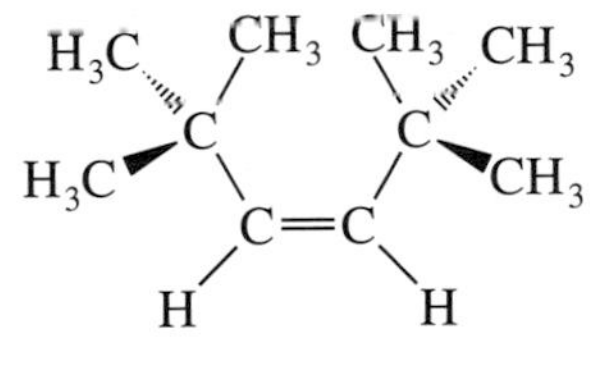

cis-2,2,5,5-Tetramethyl-3-hexene

Less stable

Energy difference = 44 kJ/mol (10.5 kcal/mol)

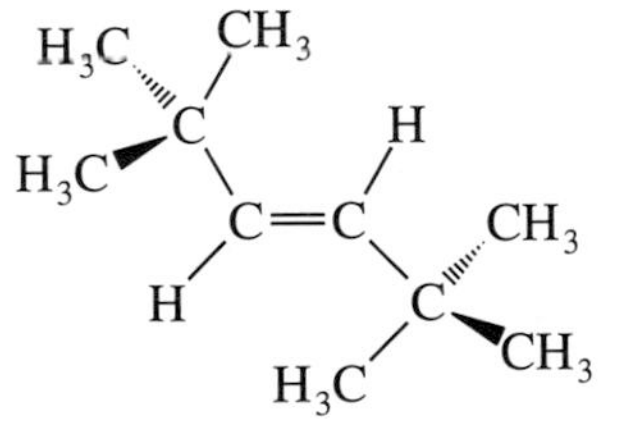

trans-2,2,5,5-Tetramethyl-3-hexene

More stable

The common names of these alkenes are *cis*- and *trans*-di-*tert*-butylethylene. In cases such as this the common names are somewhat more convenient than the IUPAC names because they are more readily associated with molecular structure.

5.7 SOURCES OF STRAIN IN CYCLOALKENES

Double bonds are accommodated by rings of all sizes. The simplest cycloalkene, cyclopropene, was first synthesized in 1922. A cyclopropene ring is present in sterculic acid, a substance derived from one of the components of the oil present in the seeds of a tree (*Sterculia foelida*) that grows in the Philippines and Indonesia.

Sterculic acid and related substances are the subject of an article in the July 1982 issue of *Journal of Chemical Education* (pp. 539–543).

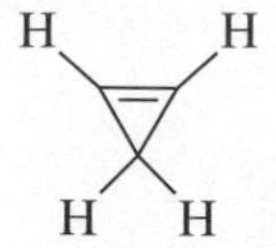

Cyclopropene

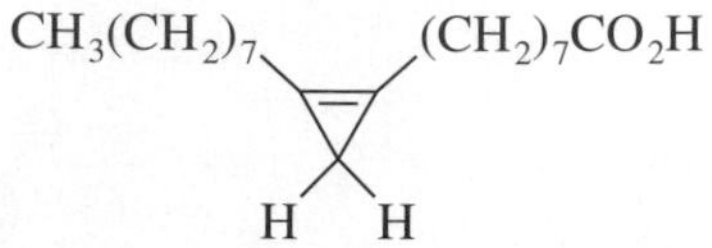

Sterculic acid

As we saw in Section 3.10, cyclopropane is destabilized by angle strain because its 60° bond angles are much smaller than the normal 109.5° angles associated with sp^3 hybridized carbon. Cyclopropene is even more strained because the deviation of the bond angles at its doubly bonded carbons from the normal sp^2 hybridization value of 120° is greater still. Cyclobutene has, of course, less angle strain than cyclopropene, and the angle strain of cyclopentene, cyclohexene, and higher cycloalkenes is negligible.

So far we have represented cycloalkenes by structural formulas in which the double bonds are of the cis configuration. If the ring is large enough, however, a trans stereoisomer is also possible. The smallest trans cycloalkene that is stable enough to be isolated and stored in a normal way is *trans*-cyclooctene. Thermochemical measurements reveal *trans*-cyclooctene to be 39 kJ/mol (9.2 kcal/mol) less stable than *cis*-cyclooctene.

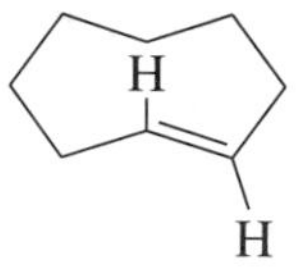

Energy difference = 39 kJ/mol (9.2 kcal/mol)

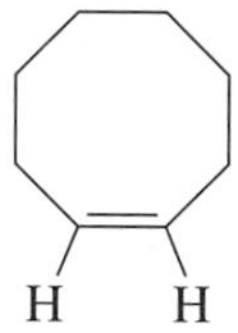

(*E*)-Cyclooctene (*trans*-cyclooctene) Less stable

(*Z*)-Cyclooctene (*cis*-cyclooctene) More stable

trans-Cycloheptene has been prepared and studied at low temperature (−90°C) but is too reactive to be isolated and stored at room temperature. Evidence has also been presented for the fleeting existence of the even more strained *trans*-cyclohexene as a reactive intermediate in certain reactions.

PROBLEM 5.10 Place a double bond in the carbon skeleton shown so as to represent

(*a*) (*Z*)-1-Methylcyclodecene
(*b*) (*E*)-1-Methylcyclodecene
(*c*) (*Z*)-3-Methylcyclodecene
(*d*) (*E*)-3-Methylcyclodecene
(*e*) (*Z*)-5-Methylcyclodecene
(*f*) (*E*)-5-Methylcyclodecene

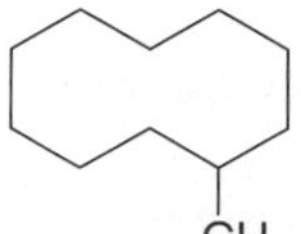

CH_3

SAMPLE SOLUTION (*a*) and (*b*) Since the methyl group must be at C-1, there are only two possible places to put the double bond:

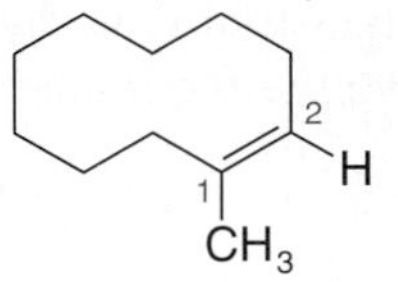

(Z)-1-Methylcyclodecene 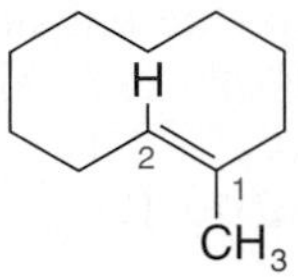 (E)-1-Methylcyclodecene

In the *Z* stereoisomer the two lower-priority substituents—the methyl group and the hydrogen—are on the same side of the double bond. In the *E* stereoisomer these substituents are on opposite sides of the double bond. The ring carbons are the higher-ranking substituents at each end of the double bond.

Because larger rings have more methylene groups with which to span the ends of a double bond, the strain associated with a trans cycloalkene decreases with increasing ring size. The strain eventually disappears when a 12-membered ring is reached:

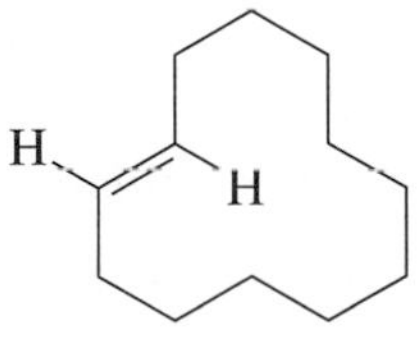

(*E*)-Cyclododecene (*trans*-cyclododecene) and (*Z*)-Cyclododecene (*cis*-cyclododecene)

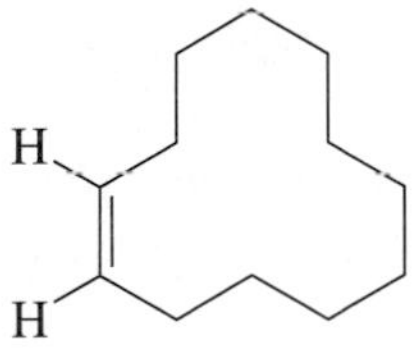

are of approximately equal stability.

When the rings are larger than 12-membered, trans cycloalkenes are more stable than cis. In these cases, the ring is large enough and flexible enough that it is energetically similar to a noncyclic alkene. As in noncyclic cis alkenes, van der Waals strain between carbons on the same side of the double bond destabilizes a cis cycloalkene.

5.8 PREPARATION OF ALKENES: ELIMINATION REACTIONS

The remaining portion of this chapter describes how alkenes are prepared by reactions of the type:

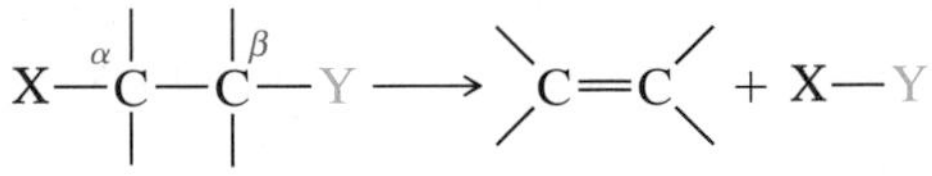

$$X-\overset{\alpha|}{\underset{|}{C}}-\overset{|\beta}{\underset{|}{C}}-Y \longrightarrow \rangle C=C\langle + X-Y$$

Alkene formation requires that X and Y be substituents on adjacent carbon atoms. By assigning X as the reference atom and identifying the carbon attached to it as the α carbon, we see that atom Y is a substituent on the β carbon. Carbons succeedingly more remote from the reference atom are designated γ, δ, and so on. Only β elimination reactions will be discussed in this chapter. [Beta (β) elimination reactions are also known as *1,2 eliminations*.]

α eliminations will be discussed in Section 14.13 and γ eliminations in Section 14.12. Neither α nor γ elimination reactions yield alkenes.

You are already familiar with one type of β elimination, having seen in Section 5.1

that ethylene and propene are prepared on an industrial scale by *dehydrogenation* of ethane and propane. Both these reactions involve the β elimination of H_2.

$$\underset{\text{Ethane}}{CH_3CH_3} \xrightarrow{750°C} \underset{\text{Ethylene}}{CH_2{=}CH_2} + \underset{\text{Hydrogen}}{H_2}$$

$$\underset{\text{Propane}}{CH_3CH_2CH_3} \xrightarrow{750°C} \underset{\text{Propene}}{CH_3CH{=}CH_2} + \underset{\text{Hydrogen}}{H_2}$$

Dehydrogenation of alkanes is not a practical *laboratory* synthesis for the vast majority of alkenes. The principal methods by which alkenes are prepared in the laboratory are two other β elimination processes, the **dehydration of alcohols** and the **dehydrohalogenation of alkyl halides.** A discussion of these two methods makes up the remainder of this chapter.

5.9 DEHYDRATION OF ALCOHOLS

In the dehydration of alcohols the elements of water are eliminated from adjacent carbons. An acid catalyst is necessary.

$$\underset{\text{Alcohol}}{H{-}\overset{|}{\underset{|}{C}}{-}\overset{|}{\underset{|}{C}}{-}OH} \xrightarrow{H^+} \underset{\text{Alkene}}{\gt C{=}C \lt} + \underset{\text{Water}}{H_2O}$$

Before dehydrogenation of ethane became the dominant method, ethylene was prepared by heating ethyl alcohol with sulfuric acid.

$$\underset{\text{Ethyl alcohol}}{CH_3CH_2OH} \xrightarrow[160°C]{H_2SO_4} \underset{\text{Ethylene}}{CH_2{=}CH_2} + \underset{\text{Water}}{H_2O}$$

Other alcohols behave similarly.

$$\underset{\text{Cyclohexanol}}{C_6H_{11}OH} \xrightarrow[140°C]{H_2SO_4} \underset{\text{Cyclohexene (79–87\%)}}{C_6H_{10}} + \underset{\text{Water}}{H_2O}$$

$$\underset{\text{2-Methyl-2-propanol}}{CH_3{-}\overset{CH_3}{\underset{OH}{C}}{-}CH_3} \xrightarrow[\text{heat}]{H_2SO_4} \underset{\text{2-Methylpropene (82\%)}}{(H_3C)_2C{=}CH_2} + \underset{\text{Water}}{H_2O}$$

Sulfuric acid (H_2SO_4) and phosphoric acid (H_3PO_4) are the acids most frequently used in alcohol dehydrations. Potassium hydrogen sulfate ($KHSO_4$) is also often used.

PROBLEM 5.11 Identify the alkene obtained on dehydration of each of the following alcohols:

(*a*) 3-Ethyl-3-pentanol
(*b*) 1-Propanol
(*c*) 2-Propanol
(*d*) 2,3,3-Trimethyl-2-butanol

SAMPLE SOLUTION (*a*) The hydrogen and the hydroxyl are lost from adjacent carbons in the dehydration of 3-ethyl-3-pentanol.

$$\underset{\text{3-Ethyl-3-pentanol}}{CH_3\overset{\beta}{C}H_2-\overset{\alpha}{C}(\overset{\beta}{C}H_2CH_3)(OH)-\overset{\beta}{C}H_2CH_3} \xrightarrow{H^+} \underset{\text{3-Ethyl-2-pentene}}{(CH_3CH_2)_2C=CHCH_3} + \underset{\text{Water}}{H_2O}$$

The hydroxyl group is lost from a carbon that bears three equivalent ethyl substituents. β elimination can occur in any one of three equivalent directions to give the same alkene, 3-ethyl-2-pentene.

5.10 REGIOSELECTIVITY IN ALCOHOL DEHYDRATION: THE ZAITSEV RULE

In the preceding examples, including those of Problem 5.11, only a single alkene product could be formed from each alcohol by β elimination. What about elimination in alcohols such as 2-methyl-2-butanol, where dehydration can occur in two different directions to give alkenes that are constitutional isomers? Here, a double bond can be generated between C-1 and C-2 or between C-2 and C-3. Both processes occur but not nearly to the same extent. Under the usual reaction conditions 2-methyl-2-butene is the major product, and 2-methyl-1-butene the minor one.

$$\underset{\text{2-Methyl-2-butanol}}{\overset{1}{C}H_3-\overset{2}{C}(OH)(\overset{1}{C}H_3)-\overset{3}{C}H_2\overset{4}{C}H_3} \xrightarrow[80°C]{H_2SO_4} \underset{\substack{\text{2-Methyl-1-butene}\\(10\%)}}{CH_2=C(CH_2CH_3)(CH_3)} + \underset{\substack{\text{2-Methyl-2-butene}\\(90\%)}}{(H_3C)_2C=CHCH_3}$$

As a second example, consider 2-methylcyclohexanol. It undergoes dehydration to yield a mixture of 1-methylcyclohexene (major) and 3-methylcyclohexene (minor).

$$\text{2-Methylcyclohexanol} \xrightarrow[\text{heat}]{H_3PO_4} \underset{(84\%)}{\text{1-Methylcyclohexene}} + \underset{(16\%)}{\text{3-Methylcyclohexene}}$$

For a previous example and a definition of regioselectivity, see Section 4.20.

Both dehydrations are *regioselective* in the sense that β elimination can occur in either of two directions to yield constitutionally isomeric alkenes, but one alkene is formed in greater amounts than the other.

Although Russian, Zaitsev published most of his work in German scientific journals, where his name was transliterated as *Saytzeff.* The spelling used here (*Zaitsev*) corresponds to the currently preferred style.

In 1875 Alexander M. Zaitsev of the University of Kazan (Russia) set forth a generalization describing the regioselectivity to be expected in β elimination reactions. **Zaitsev's rule** is an empirical one and summarizes the results of numerous experiments in which alkene mixtures were produced by β elimination. In its original form Zaitsev's rule stated that *the alkene formed in greatest amount is the one that corresponds to removal of the hydrogen from the β carbon having the fewest hydrogen substituents.*

$$R_2\overset{\beta}{C}H-\overset{\alpha}{C}(OH)(\overset{\beta}{C}H_3)-\overset{\beta}{C}H_2R \xrightarrow{-H_2O} R_2C=C(CH_2R)(CH_3)$$

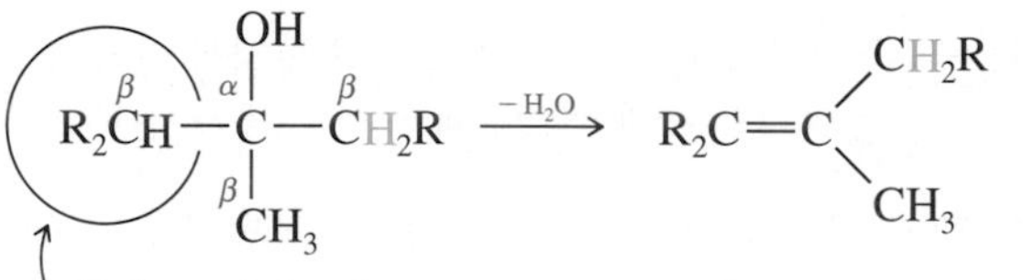

Hydrogen is lost from β carbon having the fewest hydrogen substituents

Alkene present in greatest amount in product

Zaitsev's rule as applied to the acid-catalyzed dehydration of alcohols is now more often expressed in a different way: *β elimination reactions of alcohols yield the most highly substituted alkene as the major product.* Since, as was discussed in Section 5.6, the most highly substituted alkene is also normally the most stable one, Zaitsev's rule is sometimes expressed as a preference for *predominant formation of the most stable alkene that could arise by β elimination.*

PROBLEM 5.12 Each of the following alcohols has been subjected to acid-catalyzed dehydration and yields a mixture of two isomeric alkenes. Identify the two alkenes in each case and predict which one is the major product on the basis of the Zaitsev rule.

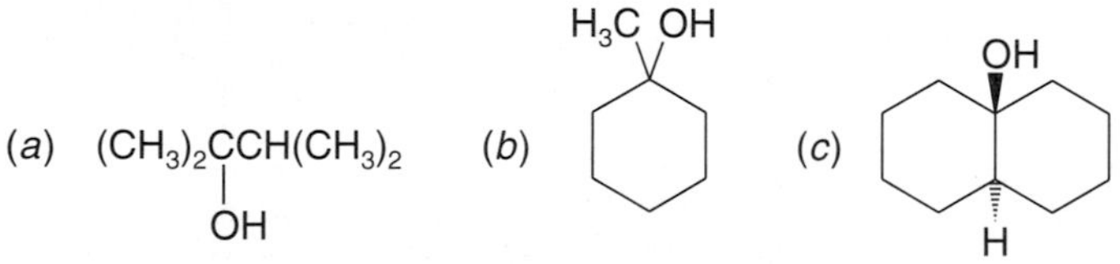

(*a*) $(CH_3)_2C(OH)CH(CH_3)_2$ (*b*) 1-methylcyclohexanol (H_3C, OH) (*c*) decalin with OH and H at ring fusion

SAMPLE SOLUTION (*a*) Dehydration of 2,3-dimethyl-2-butanol can lead to either 2,3-dimethyl-1-butene by removal of a C-1 hydrogen or to 2,3-dimethyl-2-butene by removal of a C-3 hydrogen.

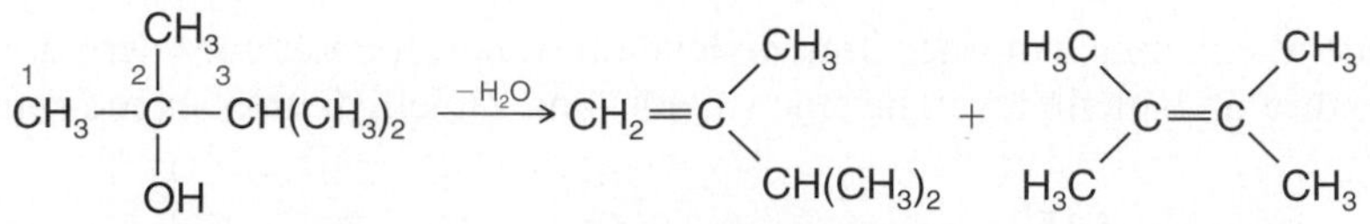

$$\overset{1}{C}H_3-\overset{2}{C}(CH_3)(OH)-\overset{3}{C}H(CH_3)_2 \xrightarrow{-H_2O} CH_2=C(CH_3)CH(CH_3)_2 + (H_3C)_2C=C(CH_3)_2$$

2,3-Dimethyl-2-butanol — 2,3-Dimethyl-1-butene (minor product) — 2,3-Dimethyl-2-butene (major product)

The major product is 2,3-dimethyl-2-butene. It has a tetrasubstituted double bond and is more stable than 2,3-dimethyl-1-butene, which has a disubstituted double bond. The major alkene product arises by loss of a hydrogen from the β carbon that has fewer hydrogen substituents (C-3) rather than from the β carbon that has the greater number of hydrogen substituents (C-1).

5.11 STEREOSELECTIVITY IN ALCOHOL DEHYDRATION

In addition to being regioselective, alcohol dehydration reactions are **stereoselective.** A stereoselective reaction is one in which a single starting material can yield two or more stereoisomeric products, but gives one of them in greater amounts than any other. Alcohol dehydrations tend to produce the more stable stereoisomeric form of an alkene. Dehydration of 3-pentanol, for example, yields a mixture of *trans*-2-pentene and *cis*-2-pentene in which the more stable trans stereoisomer predominates.

$$CH_3CH_2\underset{\underset{OH}{|}}{C}HCH_2CH_3 \xrightarrow[\text{heat}]{H_2SO_4} \underset{H}{\overset{H_3C}{}}C{=}C\underset{H}{\overset{CH_2CH_3}{}} + \underset{H}{\overset{H_3C}{}}C{=}C\underset{CH_2CH_3}{\overset{H}{}}$$

3-Pentanol — *cis*-2-Pentene (25%) (minor product) — *trans*-2-Pentene (75%) (major product)

PROBLEM 5.13 What three alkenes are formed in the acid-catalyzed dehydration of 2-pentanol?

5.12 THE MECHANISM OF ACID-CATALYZED DEHYDRATION OF ALCOHOLS

The dehydration of alcohols and the conversion of alcohols to alkyl halides by treatment with hydrogen halides are similar in two important ways:

1. Both reactions are promoted by acids.
2. The relative reactivity of alcohols decreases in the order tertiary > secondary > primary.

These common features suggest that carbocations are key intermediates in alcohol dehydration, just as they are in the conversion of alcohols to alkyl halides (Section 4.9). Figure 5.7 portrays a three-step mechanism for the sulfuric acid–catalyzed dehydration of *tert*-butyl alcohol. Steps 1 and 2 describe the generation of *tert*-butyl cation by a process similar to that which led to its formation as an intermediate in the reaction of *tert*-butyl alcohol with hydrogen chloride. Step 3 in Figure 5.7, however, is new to us and is the step in which the alkene product is formed from the carbocation intermediate.

Step 3 is an acid-base reaction. In this step the carbocation acts as a Brønsted acid, transferring a proton to a Brønsted base (water). This is the property of carbocations that is of the most significance to elimination reactions. Carbocations are strong acids; they are the conjugate acids of alkenes and readily lose a proton to form alkenes. Even weak bases such as water are sufficiently basic to abstract a proton from a carbocation.

Step 3 in Figure 5.7 shows water as the base which abstracts a proton from the carbocation. Other Brønsted bases present in the reaction mixture that can function in the same way include *tert*-butyl alcohol and hydrogen sulfate ion.

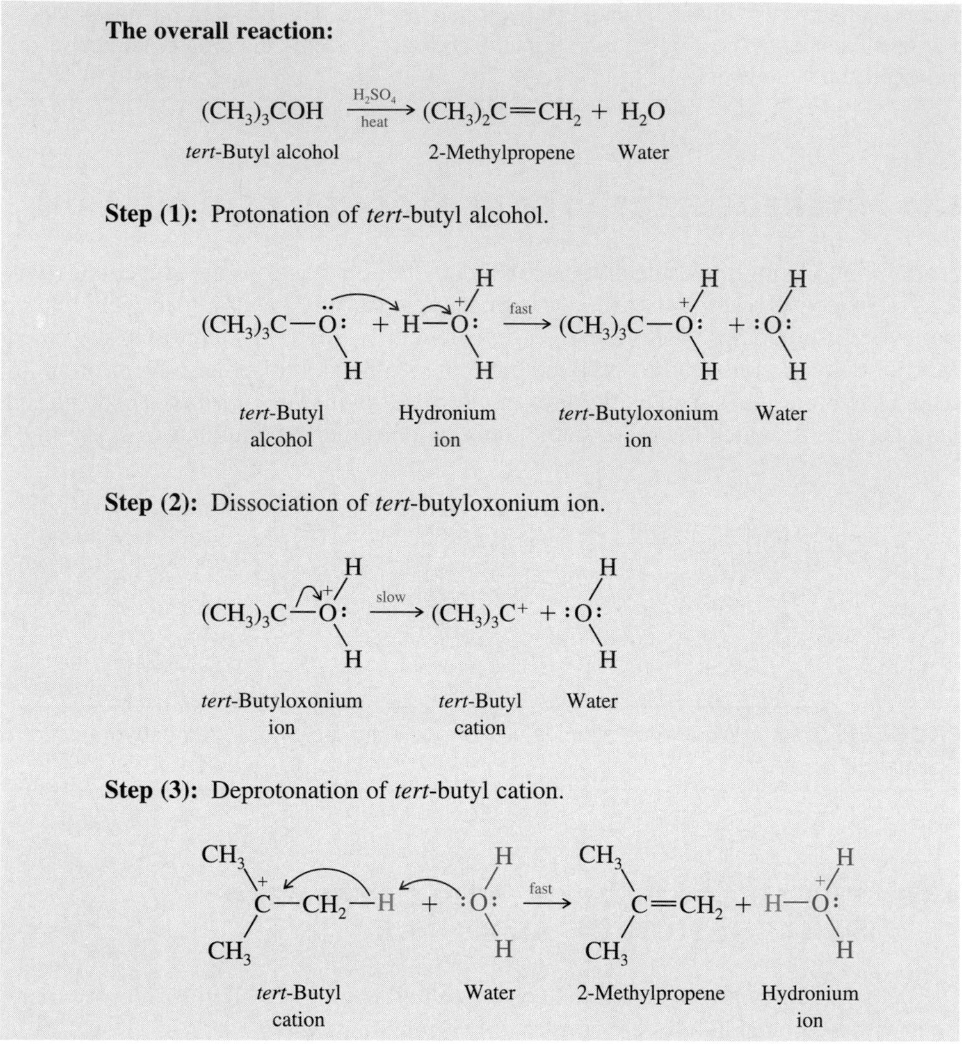

The overall reaction:

$$(CH_3)_3COH \xrightarrow[\text{heat}]{H_2SO_4} (CH_3)_2C{=}CH_2 + H_2O$$

tert-Butyl alcohol — 2-Methylpropene — Water

Step (1): Protonation of *tert*-butyl alcohol.

$$(CH_3)_3C{-}\ddot{O}{:}{-}H + H{-}\overset{+}{O}H_2 \xrightarrow{\text{fast}} (CH_3)_3C{-}\overset{+}{O}H_2 + {:}\ddot{O}H_2$$

tert-Butyl alcohol — Hydronium ion — *tert*-Butyloxonium ion — Water

Step (2): Dissociation of *tert*-butyloxonium ion.

$$(CH_3)_3C{-}\overset{+}{O}H_2 \xrightarrow{\text{slow}} (CH_3)_3C^+ + {:}\ddot{O}H_2$$

tert-Butyloxonium ion — *tert*-Butyl cation — Water

Step (3): Deprotonation of *tert*-butyl cation.

$$(CH_3)_2\overset{+}{C}{-}CH_2{-}H + {:}\ddot{O}H_2 \xrightarrow{\text{fast}} (CH_3)_2C{=}CH_2 + H{-}\overset{+}{O}H_2$$

tert-Butyl cation — Water — 2-Methylpropene — Hydronium ion

FIGURE 5.7 The mechanism for the acid-catalyzed dehydration of *tert*-butyl alcohol.

PROBLEM 5.14 Write a structural formula for the carbocation intermediate formed in the dehydration of each of the alcohols in Problem 5.12 (Section 5.10). Using curved arrow notation, show how each carbocation is deprotonated by water to give a mixture of alkenes.

SAMPLE SOLUTION (*a*) The carbon that bears the hydroxyl group in the starting alcohol is the one that becomes positively charged in the carbocation.

$$(CH_3)_2\underset{\displaystyle OH}{\underset{|}{C}}CH(CH_3)_2 \xrightarrow[-H_2O]{H^+} (CH_3)_2\overset{+}{C}CH(CH_3)_2$$

Water may remove a proton from either C-1 or C-3 of this carbocation. Loss of a proton from C-1 yields the minor product 2,3-dimethyl-1-butene. (This alkene has a disubstituted double bond.)

$$H_2\ddot{O}: + H-CH_2-\overset{+}{C}(CH_3)CH(CH_3)_2 \longrightarrow H_2\overset{+}{O}-H + CH_2=C(CH_3)CH(CH_3)_2$$

2,3-Dimethyl-1-butene

Loss of a proton from C-3 yields the major product 2,3-dimethyl-2-butene. (This alkene has a tetrasubstituted double bond.)

$$(H_3C)_2\overset{+}{C}-C(CH_3)_2H + :\ddot{O}H_2 \longrightarrow (H_3C)_2C=C(CH_3)_2 + H-\overset{+}{O}H_2$$

2,3-Dimethyl-2-butene

As noted earlier in Section 4.14 primary carbocations are too high in energy to be realistically invoked as intermediates in most chemical reactions. If primary alcohols do not form primary carbocations, then how do they undergo elimination? A modification of our general mechanism for alcohol dehydration offers a reasonable explanation. For primary alcohols it is believed that a proton is lost from the alkyloxonium ion in the same step in which carbon-oxygen bond cleavage takes place. For example, the rate-determining step in the sulfuric acid–catalyzed dehydration of ethanol may be represented as:

$$H_2\ddot{O}: + H-CH_2-CH_2-\overset{+}{O}H_2 \longrightarrow H_2\overset{+}{O}-H + CH_2=CH_2 + :\ddot{O}H_2$$

Water · Ethyloxonium ion · Hydronium ion · Ethylene · Water

Like their tertiary alcohol counterparts, secondary alcohols normally undergo dehydration to alkenes by way of carbocation intermediates.

In Chapter 4 you learned that carbocations could be captured by halide anions to give alkyl halides. In the present chapter, a second type of carbocation reaction has been introduced—a carbocation can lose a proton to form an alkene. In the next section a third aspect of carbocation behavior will be described, the *rearrangement* of one carbocation to another.

5.13 REARRANGEMENTS IN ALCOHOL DEHYDRATION

Some alcohols undergo dehydration to yield alkenes having carbon skeletons different from those of the starting alcohols. Not only has elimination taken place but the arrangement of atoms in the alkene is different from that in the alcohol. A **rearrangement** is said to have occurred. An example of an alcohol dehydration that is accompanied by rearrangement is the case of 3,3-dimethyl-2-butanol. This is one of many such experiments carried out by F. C. Whitmore and his students at Pennsylvania State University in the 1930s as part of a general study of rearrangement reactions.

$$CH_3-\underset{CH_3}{\overset{CH_3}{C}}-\underset{OH}{CH}-CH_3 \xrightarrow[\text{heat}]{H_3PO_4} CH_3-\underset{CH_3}{\overset{CH_3}{C}}-CH=CH_2 + (H_3C)_2C=C(CH_3)_2 + H_2C=\underset{H_3C}{C}-\underset{CH_3}{CH}-CH_3$$

3,3-Dimethyl-2-butanol; 3,3-Dimethyl-1-butene (3%); 2,3-Dimethyl-2-butene (64%); 2,3-Dimethyl-1-butene (33%)

A mixture of three alkenes was obtained in 80 percent yield, having the composition shown in the equation. The alkene having the same carbon skeleton as the starting alcohol, 3,3-dimethyl-1-butene, constituted only 3 percent of the alkene mixture. The two alkenes present in greatest amount, 2,3-dimethyl-2-butene and 2,3-dimethyl-1-butene, both have carbon skeletons different from that of the starting alcohol.

Whitmore proposed that the carbon skeleton rearrangement occurred in a separate step following carbocation formation. Once the alcohol was converted to the corresponding carbocation, that carbocation could either lose a proton to give an alkene having the same carbon skeleton or rearrange to a different carbocation, as shown in Figure 5.8. The rearranged alkenes arise by loss of a proton from the rearranged carbocation.

Why do carbocations rearrange? How do carbocations rearrange? The "why" is easy to understand. Estimates of their relative energies suggest that secondary carbocations are 40 to 60 kJ/mol (10 to 15 kcal/mol) less stable than tertiary carbocations. Consequently, rearrangement of a secondary to a tertiary carbocation decreases its potential energy and is an exothermic process. This is exactly what happens in the dehydration of 3,3-dimethyl-2-butanol. As Figure 5.8 depicts, the first-formed carbocation is secondary; the rearranged carbocation is tertiary. Almost all the alkene products come from the more stable tertiary carbocation.

The "how" of carbocation rearrangements requires that we examine the structural change that takes place at the transition state. Again referring to the initial (secondary) carbocation intermediate in the dehydration of 3,3-dimethyl-2-butanol, rearrangement

$$\underset{OH}{(CH_3)_3C\overset{3\ 2\ 1}{CHCH_3}} \xrightarrow[-H_2O]{H^+} CH_3-\underset{CH_3}{\overset{CH_3}{C}}-\overset{+}{C}HCH_3 \xrightarrow{\text{methyl shift from C-3 to C-2}} CH_3-\overset{CH_3}{\underset{+}{C}}-\underset{CH_3}{CHCH_3}$$

3,3-Dimethyl-2-butanol → 1,2,2-Trimethylpropyl cation (a secondary carbocation) → 1,1,2-Trimethylpropyl cation (a tertiary carbocation)

$\downarrow -H^+$ (from the secondary cation): $CH_3-\underset{CH_3}{\overset{CH_3}{C}}-CH=CH_2$ 3,3-Dimethyl-1-butene (3%)

$\downarrow -H^+$ (from the tertiary cation): $(CH_3)_2C=C(CH_3)_2$ 2,3-Dimethyl-2-butene (64%) + $CH_2=C(CH_3)CH(CH_3)_2$ 2,3-Dimethyl-1-butene (33%)

FIGURE 5.8 The first formed carbocation from 3,3-dimethyl-2-butanol is secondary and rearranges to a more stable tertiary carbocation by a methyl migration. The major portion of the alkene products is formed by way of the tertiary carbocation.

occurs when a methyl group shifts from C-3 to the positively charged carbon. The methyl group migrates with the pair of electrons that made up its original σ bond to C-3. In the curved arrow notation for this methyl migration, the arrow shows the movement of both the methyl group and the electrons in the σ bond.

To simplify the accompanying discussion, the carbons of the carbocation are numbered so as to correspond to their positions in the starting alcohol 3,3-dimethyl-2-butanol. These numbers are different from the locants in the IUPAC cation names, which are given under the structural formulas.

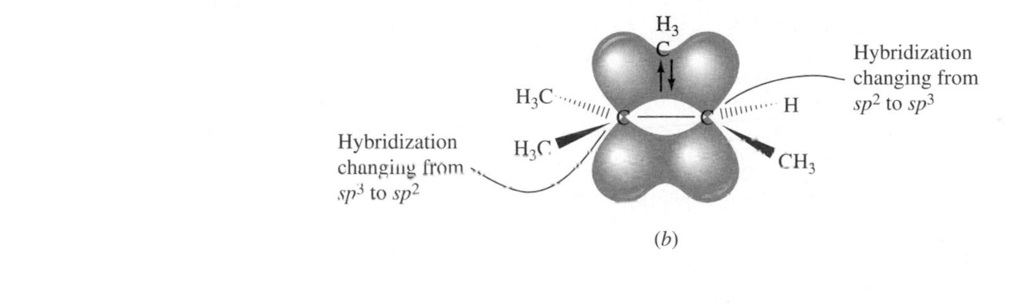

At the transition state for rearrangement, the methyl group is partially bonded both to its point of origin and to the carbon that will be its destination.

This rearrangement is depicted in orbital terms in Figure 5.9. The relevant orbitals of the secondary carbocation are shown in structure *a*, those of the transition state for rearrangement in *b*, and those of the tertiary carbocation in *c*. Delocalization of the

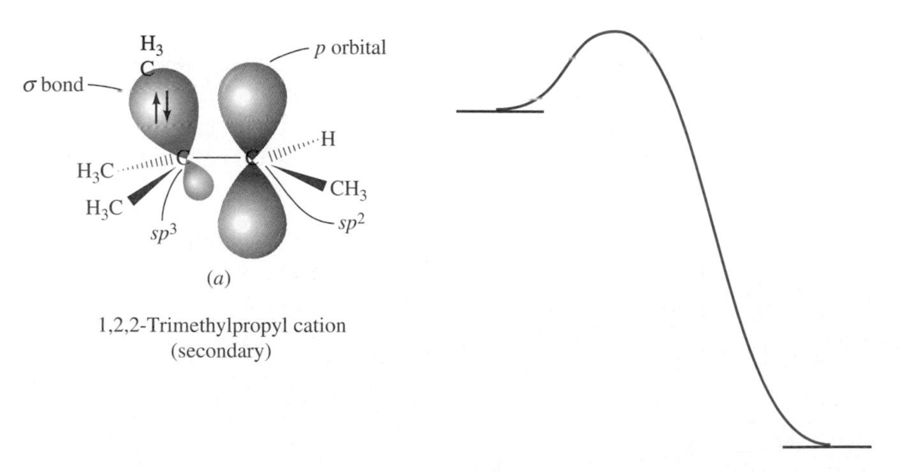

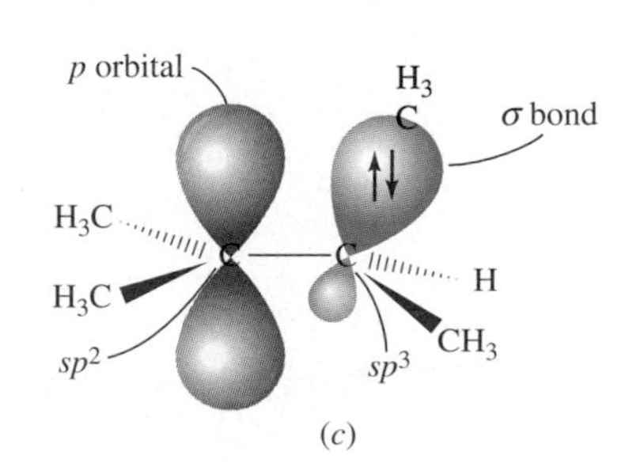

1,1,2-Trimethylpropyl cation (tertiary)

FIGURE 5.9 An orbital representation of methyl migration in 1,2,2,-trimethylpropyl cation. Structure *a* is the initial secondary carbocation, structure *b* is the transition state for methyl migration, and structure *c* is the final tertiary carbocation.

electrons of the C—CH_3 σ bond into the vacant *p* orbital of the positively charged carbon by hyperconjugation is present in both *a* and *c*, requires no activation energy, and stabilizes each carbocation. Migration of the *atoms* of the methyl group, however, occurs only when sufficient energy is absorbed by *a* to achieve the transition state *b*. The activation energy is modest, and carbocation rearrangements are normally quite rapid.

PROBLEM 5.15 The alkene mixture obtained on dehydration of 2,2-dimethylcyclohexanol contains appreciable amounts of 1,2-dimethylcyclohexene. Give a mechanistic explanation for the formation of this product.

Alkyl groups other than methyl are also capable of migrating to a positively charged carbon. Many carbocation rearrangements involve migration of a hydrogen substituent. These are called **hydride shifts.** The same requirements apply to hydride shifts as to alkyl group migrations; they proceed in the direction that leads to a more stable carbocation; the origin and destination of the migrating group are adjacent carbons, one of which must be positively charged; and the group migrates with a pair of electrons.

$$\text{A—C(H)(B)—}\overset{+}{\text{C}}\text{(Y)—X} \longrightarrow \text{A—}\overset{+}{\text{C}}\text{(B)—C(H)(Y)—X} \qquad \text{Hydride shift}$$

Hydride shifts often occur during the dehydration of primary alcohols. Thus, while 1-butene would be expected to be the only alkene formed on dehydration of 1-butanol, it is in fact only a minor product. The major product is a mixture of *cis*- and *trans*-2-butene.

$$CH_3CH_2CH_2CH_2OH \xrightarrow[140-170°C]{H_2SO_4} \underset{\text{1-Butene (12\%)}}{CH_3CH_2CH{=}CH_2} + \underset{\text{Mixture of cis-2-butene (32\%) and trans-2-butene (56\%)}}{CH_3CH{=}CHCH_3}$$

A mechanism consistent with the formation of these three alkenes is shown in Figure 5.10. Dissociation of the primary alkyloxonium ion is accompanied by a shift of

$$\underset{\text{1-Butanol}}{\overset{3\ \ 2\ \ 1}{CH_3CH_2CH_2CH_2OH}}$$

$$\downarrow H^+$$

$$\overset{3\ \ \ 2\ \ \ \ 1}{CH_3CH_2CH(H)—CH_2—\overset{+}{O}H_2{:}} \xrightarrow[-H_2O]{\text{hydride shift concerted with dissociation}} CH_3CH_2\overset{+}{C}HCH_3 \xrightarrow{-H^+} \underset{\text{1-Butene (12\%)}}{CH_3CH_2CH{=}CH_2} + \underset{\substack{\text{cis and trans-2-Butene}\\ \text{(32\% cis; 56\% trans)}}}{CH_3CH{=}CHCH_3}$$

FIGURE 5.10 Dehydration of 1-butanol is accompanied by a hydride shift from C-2 to C-1.

hydride from C-2 to C-1. This avoids the formation of a primary carbocation, leading instead to a secondary carbocation in which the positive charge is at C-2. Deprotonation of this carbocation yields the observed products. (Some 1-butene may also arise directly from the primary oxonium ion.)

This concludes discussion of our second functional group transformation involving *alcohols:* the first was the conversion of alcohols to alkyl halides (Chapter 4), and the second the conversion of alcohols to alkenes. In the remaining sections of the chapter the conversion of *alkyl halides* to alkenes by dehydrohalogenation reactions is described.

5.14 DEHYDROHALOGENATION OF ALKYL HALIDES

Dehydrohalogenation is the loss of the elements of a hydrogen halide (HX) from an alkyl halide. It is one of the most useful methods for preparing alkenes by β elimination.

$$H-\overset{|}{\underset{|}{C}}-\overset{|}{\underset{|}{C}}-X \longrightarrow \rangle C=C\langle + HX$$

Alkyl halide — Alkene — Hydrogen halide

When applied to the preparation of alkenes, the reaction is carried out in the presence of a strong base, such as sodium ethoxide in ethyl alcohol as solvent.

$$H-\overset{|}{\underset{|}{C}}-\overset{|}{\underset{|}{C}}-X + NaOCH_2CH_3 \longrightarrow \rangle C=C\langle + CH_3CH_2OH + NaX$$

Alkyl halide — Sodium ethoxide — Alkene — Ethyl alcohol — Sodium halide

$$\text{Cyclohexyl chloride} \xrightarrow[CH_3CH_2OH,\ 55^\circ C]{NaOCH_2CH_3} \text{Cyclohexene (100\%)}$$

Cyclohexyl chloride — Cyclohexene (100%)

Similarly, sodium methoxide is a suitable base and is used in methyl alcohol. Potassium hydroxide in ethyl alcohol is another base-solvent combination often employed in the dehydrohalogenation of alkyl halides. Potassium *tert*-butoxide is the preferred base when the alkyl halide is primary; it is used in either *tert*-butyl alcohol or dimethyl sulfoxide as solvent.

Dimethyl sulfoxide has the structure $(CH_3)_2\overset{+}{S}-\ddot{O}:^-$ and is commonly referred to as DMSO. It is a relatively inexpensive solvent, obtained as a by-product in paper manufacture.

$$CH_3(CH_2)_{15}CH_2CH_2Cl \xrightarrow[DMSO,\ 25^\circ C]{KOC(CH_3)_3} CH_3(CH_2)_{15}CH=CH_2$$

1-Chlorooctadecane — 1-Octadecene (86%)

The regioselectivity of dehydrohalogenation of alkyl halides follows the Zaitsev rule; β elimination predominates in the direction that leads to the more highly substituted alkene.

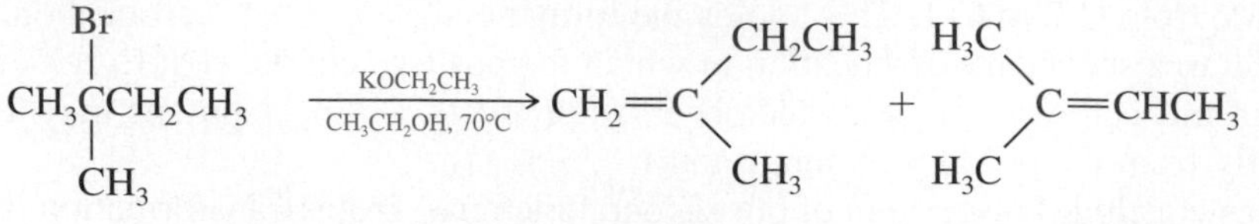

$$CH_3C(Br)(CH_3)CH_2CH_3 \xrightarrow[CH_3CH_2OH,\ 70°C]{KOCH_2CH_3} CH_2{=}C(CH_2CH_3)(CH_3) + (H_3C)_2C{=}CHCH_3$$

2-Bromo-2-methylbutane → 2-Methyl-1-butene (29%) + 2-Methyl-2-butene (71%)

PROBLEM 5.16 Write the structures of all the alkenes capable of being produced when each of the following alkyl halides undergoes dehydrohalogenation. Apply the Zaitsev rule to predict the alkene formed in greatest amount in each case.

(*a*) 2-Bromo-2,3-dimethylbutane
(*b*) *tert*-Butyl chloride
(*c*) 3-Bromo-3-ethylpentane
(*d*) 2-Bromo-3-methylbutane
(*e*) 1-Bromo-3-methylbutane
(*f*) 1-Iodo-1-methylcyclohexane

SAMPLE SOLUTION (*a*) First analyze the structure of 2-bromo-2,3-dimethylbutane with respect to the number of possible β elimination pathways.

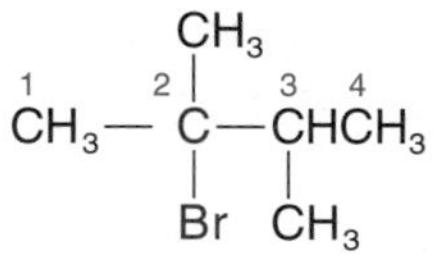

$$\overset{1}{C}H_3{-}\overset{2}{C}(CH_3)(Br){-}\overset{3}{C}H(CH_3)\overset{4}{C}H_3$$

Bromine must be lost from C-2; hydrogen may be lost from C-1 or from C-3

The two possible alkenes are

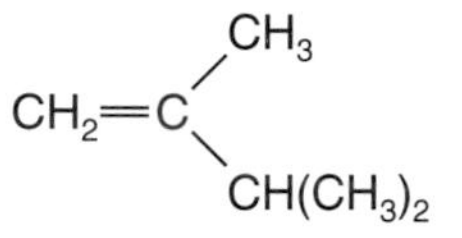

$$CH_2{=}C(CH_3)CH(CH_3)_2 \quad \text{and} \quad (H_3C)_2C{=}C(CH_3)_2$$

2,3-Dimethyl-1-butene (minor product) and 2,3-Dimethyl-2-butene (major product)

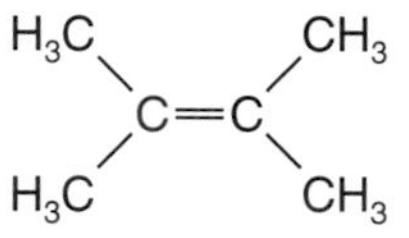

The major product, predicted on the basis of Zaitsev's rule, is 2,3-dimethyl-2-butene. It has a tetrasubstituted double bond. The minor alkene has a disubstituted double bond.

In addition to being regioselective, dehydrohalogenation of alkyl halides is stereoselective and favors formation of the more stable stereoisomer. Usually, as in the case of 5-bromononane, the trans (or *E*) alkene is formed in greater amounts than its cis (or *Z*) stereoisomer.

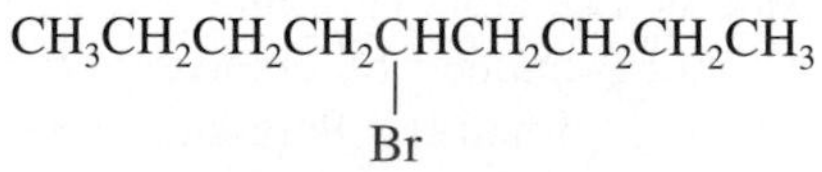

$$CH_3CH_2CH_2CH_2CH(Br)CH_2CH_2CH_2CH_3$$

5-Bromononane

$$\downarrow KOCH_2CH_3,\ CH_3CH_2OH$$

$$CH_3CH_2CH_2(H)C{=}C(H)CH_2CH_2CH_2CH_3 \ (cis) + CH_3CH_2CH_2(H)C{=}C(H)CH_2CH_2CH_2CH_3 \ (trans)$$

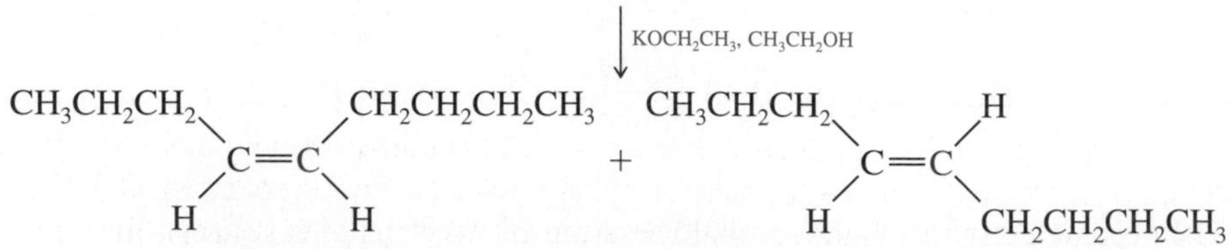

cis-4-Nonene (23%) + *trans*-4-Nonene (77%)

PROBLEM 5.17 Write structural formulas for all the alkenes capable of being produced in the reaction of 2-bromobutane with potassium ethoxide.

Elimination reactions of cycloalkyl halides lead exclusively to *cis* cycloalkenes when the ring has fewer than 10 carbons. As the ring becomes larger, it can accommodate either a cis or a trans double bond, and large-ring cycloalkyl halides give mixtures of cis and trans cycloalkenes.

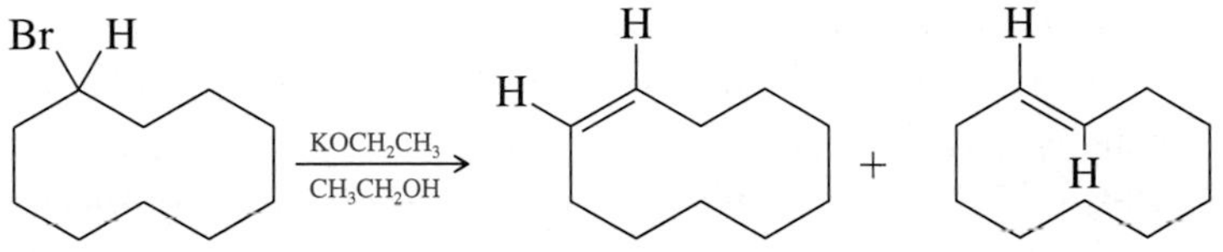

Bromocyclodecane $\xrightarrow[CH_3CH_2OH]{KOCH_2CH_3}$ *cis*-Cyclodecene [(*Z*)-cyclodecene] (85%) + *trans*-Cyclodecene [(*E*)-cyclodecene] (15%)

5.15 MECHANISM OF THE DEHYDROHALOGENATION OF ALKYL HALIDES: THE E2 MECHANISM

In the 1920s Sir Christopher Ingold proposed a mechanism for dehydrohalogenation that is still accepted as a valid description of how these reactions occur. Some of the information on which Ingold based his mechanism included these facts:

1. *The reaction exhibits second-order kinetics; it is first-order in alkyl halide and first-order in base.*

The rate equation may be written

$$\text{Rate} = k[\text{alkyl halide}][\text{base}]$$

Doubling the concentration of either the alkyl halide or the base doubles the reaction rate. Doubling the concentration of both reactants increases the rate by a factor of 4.

2. *The rate of elimination depends on the halogen, the reactivity of alkyl halides increasing with decreasing strength of the carbon-halogen bond.*

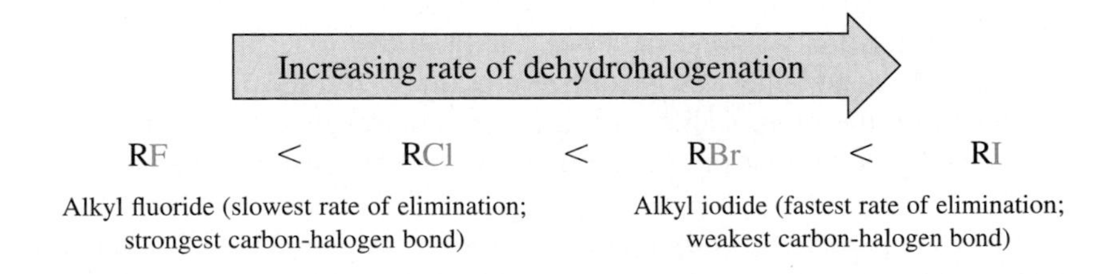

Cyclohexyl bromide, for example, is converted to cyclohexene by sodium ethoxide in ethanol at a rate that is over 60 times faster than that of cyclohexyl chloride. We say that iodide is the best **leaving group** in a dehydrohalogenation reaction, fluoride the poorest leaving group. Fluoride is such a poor leaving group that alkyl fluorides are rarely used as starting materials in the preparation of alkenes.

What are the implications of second-order kinetics? Ingold reasoned that second-order kinetics suggested a bimolecular rate-determining step involving both a molecule of the alkyl halide and a molecule of base. He concluded that proton removal from the β carbon by the base occurs during the rate-determining step rather than in a separate step following the rate-determining step.

What are the implications of the effects of the various halide leaving groups? Since it is the halogen with the weakest bond to carbon that reacts fastest, Ingold concluded that carbon-halogen bond cleavage must accompany carbon-hydrogen bond cleavage in the rate-determining step. The weaker the carbon-halogen bond, the lower is the activation energy for its cleavage.

On the basis of these observations, Ingold proposed a concerted mechanism for the dehydrohalogenation of alkyl halides in the presence of strong bases and coined the mechanistic symbol **E2** for it. The symbol stands for **elimination bimolecular.**

$$\bar{B}: \;\; H-C-C-X \longrightarrow \left[{}^{\delta-}B\text{---}H\text{---}C\text{=}\text{=}C\text{---}X_{\delta-} \right] \longrightarrow B-H + C=C + :X^-$$

Transition state for bimolecular elimination

In the E2 mechanism the three key elements

1. C—H bond breaking
2. C=C π bond formation
3. C—X bond breaking

all contribute to the transition state in a single-step transformation. In the E2 transition state, the carbon-hydrogen and carbon-halogen bonds are in the process of being broken, the base is becoming bonded to the hydrogen, a π bond is being formed, and the hybridization of carbon is changing from sp^3 to sp^2. An energy diagram depicting the E2 mechanism is presented in Figure 5.11.

PROBLEM 5.18 Use curved arrows to show electron movement in the dehydrohalogenation of *tert*-butyl chloride by sodium methoxide by the E2 mechanism.

The regioselectivity of elimination is accommodated in the E2 mechanism by noting that a partial double bond develops at the transition state. Since alkyl groups stabilize double bonds, they also stabilize a partially formed π bond in the transition state. Therefore, the more stable alkene requires a lower energy of activation for its formation and predominates in the product mixture because it is formed faster than a less stable one.

Ingold was a pioneer in applying quantitative measurements of reaction rates to the understanding of organic reaction mechanisms. Many of the reactions to be described in this text were studied by him and his students during the period of about 1920 to 1950. The facts disclosed by Ingold's experiments have been verified many times. His interpretations, although considerably refined during the decades that followed his original reports, still serve us well as a starting point for understanding how the funda-

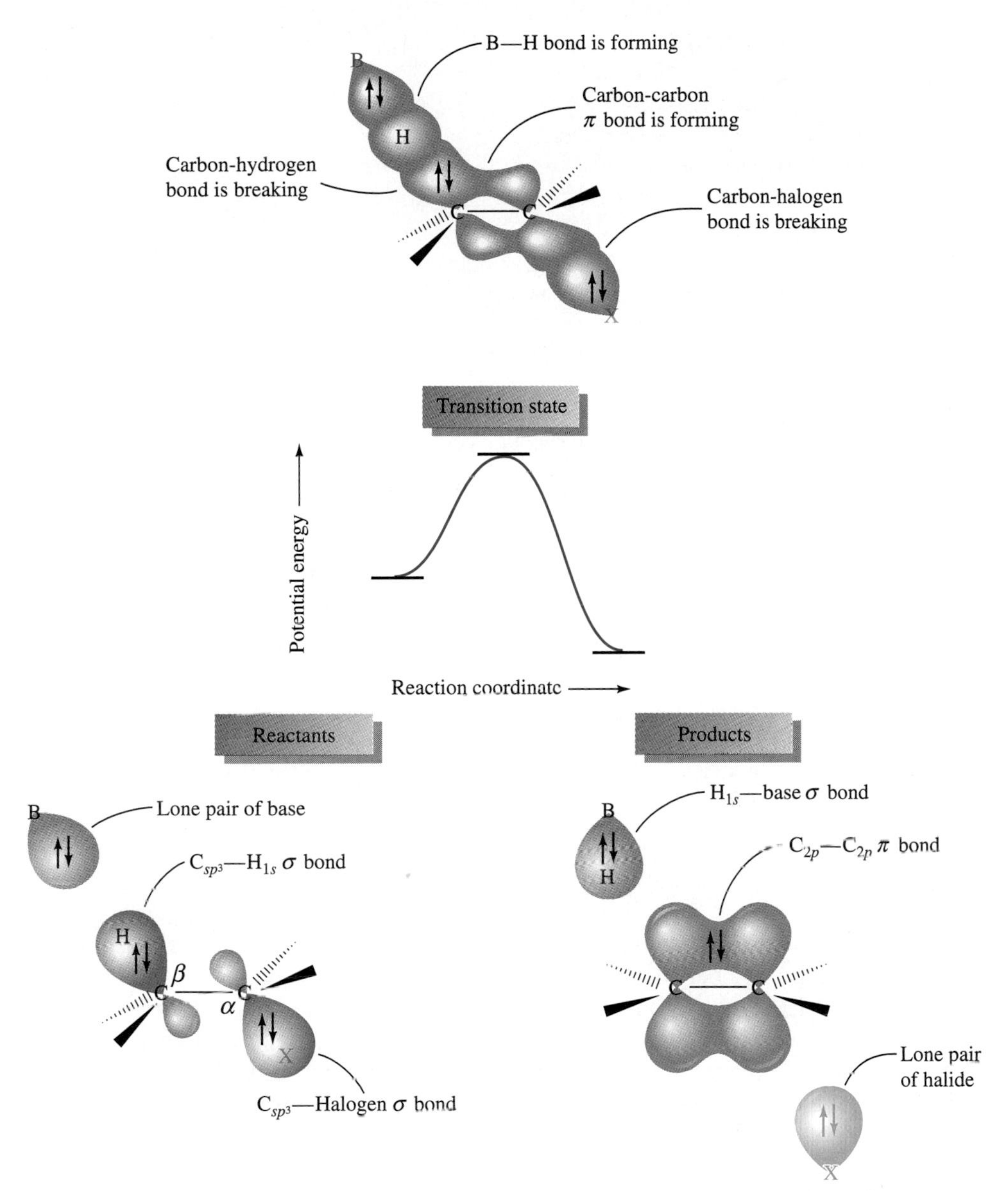

FIGURE 5.11 Potential energy diagram for concerted E2 elimination of an alkyl halide.

mental processes of organic chemistry take place. β elimination of alkyl halides by the E2 mechanism is one of those fundamental processes.

5.16 ANTI ELIMINATION IN E2 REACTIONS: STEREOELECTRONIC EFFECTS

Further insight into the mechanism of base-promoted elimination reactions of alkyl halides comes from stereochemical studies. In one such experiment the rates of elimination of the cis and trans isomers of 4-*tert*-butylcyclohexyl bromide were compared.

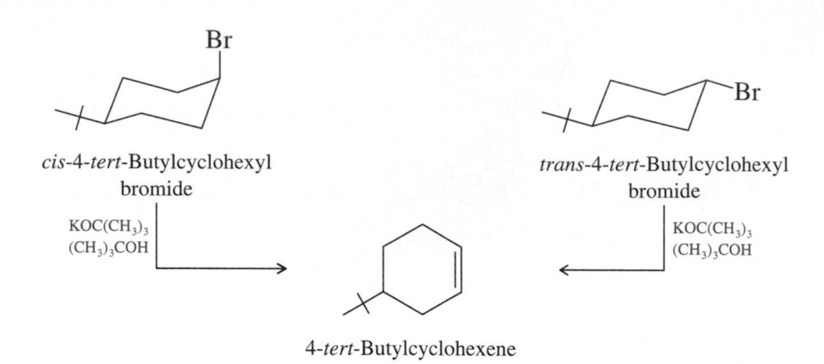

While both bromides yield 4-*tert*-butylcyclohexene as the only alkene, they do so at quite different rates. The cis isomer reacts over 500 times faster than the trans with potassium *tert*-butoxide in *tert*-butyl alcohol.

The disparity in reaction rate can be understood by considering π bond development in the E2 transition state. Since π overlap of *p* orbitals requires their axes to be parallel, it seems reasonable that π bond formation is best achieved when the four atoms of the H—C—C—X unit lie in the same plane at the transition state. The two conformations that permit this relationship are termed *syn periplanar* and *anti periplanar*.

The *peri-* in *periplanar* means "almost" or "nearly." While coplanarity of the *p* orbitals is the best geometry for the E2 process, modest deviations from this ideal can be tolerated.

Syn periplanar   Anti periplanar

Because adjacent bonds are eclipsed when the H—C—C—X assembly is syn periplanar, a transition state having this geometry is less stable than one that has an anti periplanar relationship between the proton and the leaving group.

As Figure 5.12 shows, bromine is axial in the most stable conformation of *cis*-4-*tert*-butylcyclohexyl bromide, while it is equatorial in the trans stereoisomer. An axial bromine is anti periplanar with respect to the axial hydrogens at C-2 and C-6, and so the proper geometry between the proton and the leaving group is already present in the

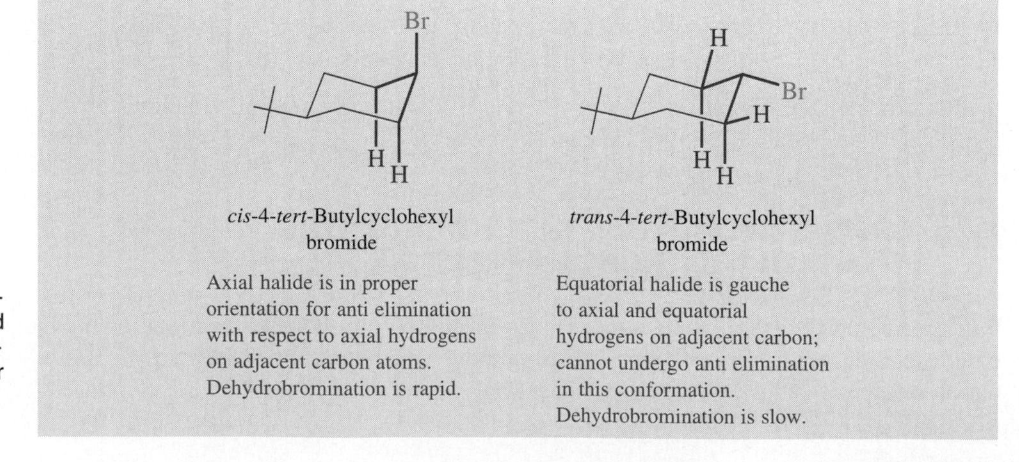

Br
H
H
H
Br
H
H
H
cis-4-*tert*-Butylcyclohexyl bromide
trans-4-*tert*-Butylcyclohexyl bromide
Axial halide is in proper orientation for anti elimination with respect to axial hydrogens on adjacent carbon atoms. Dehydrobromination is rapid.
Equatorial halide is gauche to axial and equatorial hydrogens on adjacent carbon; cannot undergo anti elimination in this conformation. Dehydrobromination is slow.

FIGURE 5.12 Conformations of *cis*- and *trans*-4-*tert*-butylcyclohexyl bromide and their relationship to the preference for an anti periplanar arrangement of proton and leaving group.

cis bromide, which undergoes E2 elimination rapidly. The less reactive stereoisomer, the trans bromide, has an equatorial bromine in its most stable conformation. An equatorial bromine is not anti periplanar with respect to any hydrogen substituents. The relationship between an equatorial leaving group and all the C-2 and C-6 hydrogens is gauche. In order to undergo E2 elimination, the trans bromide must adopt a geometry in which the ring is strained. Therefore, the transition state for its elimination is higher in energy, and reaction is slower.

PROBLEM 5.19 Use curved arrow notation to depict the bonding changes in the reaction of *cis*-4-*tert*-butylcyclohexyl bromide with potassium *tert*-butoxide. Be sure your drawing correctly represents the spatial relationship between the leaving group and the proton that is lost.

Effects that arise because one spatial arrangement of electrons (or orbitals or bonds) is more stable than another are called **stereoelectronic effects.** *There is a stereoelectronic preference for the anti periplanar arrangement of proton and leaving group in E2 reactions.*

5.17 A DIFFERENT MECHANISM FOR ALKYL HALIDE ELIMINATION: THE E1 MECHANISM

The E2 mechanism is a concerted process in which the carbon-hydrogen and carbon-halogen bonds both break in the same elementary step. What if these bonds break in separate steps?

One possibility is the two-step mechanism of Figure 5.13, in which the carbon-halogen bond breaks first to give a carbocation intermediate, followed by deprotonation of the carbocation in a second step.

The alkyl halide, in this case 2-bromo-2-methylbutane, ionizes to a carbocation and a halide anion by a heterolytic cleavage of the carbon-halogen bond. Like the dissociation of an oxonium ion to a carbocation, this step is rate-determining. Because the rate-determining step is unimolecular—it involves only the alkyl halide and not the base—this mechanism is known by the symbol **E1,** standing for **elimination unimolecular.** It exhibits first-order kinetics.

$$\text{Rate} = k[\text{alkyl halide}]$$

Typically, elimination by the E1 mechanism is important only for tertiary and some secondary alkyl halides, and then only when the base is weak or in low concentration. The reactivity order parallels the ease of carbocation formation.

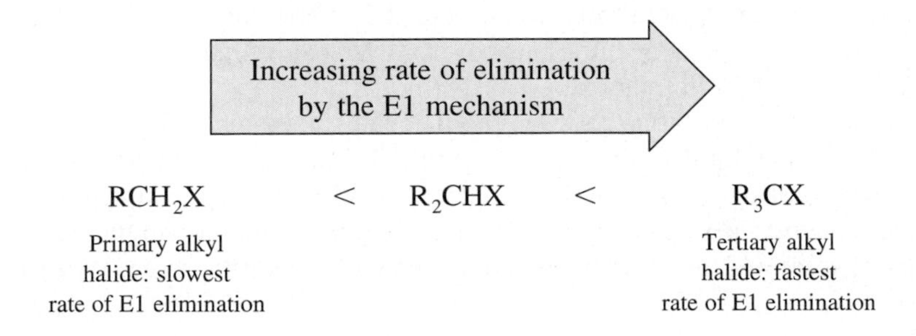

The reaction:

$$(CH_3)_2\underset{\displaystyle Br}{\underset{|}{C}}CH_2CH_3 \xrightarrow[\text{heat}]{CH_3CH_2OH} CH_2{=}\underset{\displaystyle CH_3}{\underset{|}{C}}CH_2CH_3 + (CH_3)_2C{=}CHCH_3$$

2-Bromo-2-methylbutane → 2-Methyl-1-butene (25%) + 2-Methyl-2-butene (75%)

The mechanism:

Step (1): Alkyl halide dissociates by heterolytic cleavage of carbon-halogen bond. (Ionization step)

$$CH_3\overset{\displaystyle CH_3}{\overset{|}{\underset{\displaystyle :\ddot{Br}:}{\underset{|}{C}}}}CH_2CH_3 \xrightarrow{\text{slow}} (CH_3)(CH_3CH_2)\overset{+}{C}CH_3 + :\ddot{Br}:^-$$

2-Bromo-2-methylbutane → 1,1-Dimethylpropyl cation + Bromide ion

Step (2): Ethanol acts as a base to remove a proton from the carbocation to give the alkene products. (Deprotonation step)

$$CH_3CH_2\ddot{O}H: + H{-}CH_2{-}\overset{+}{C}(CH_2CH_3)(CH_3) \xrightarrow{\text{fast}} CH_3CH_2\overset{+}{O}H_2 + CH_2{=}C(CH_2CH_3)(CH_3)$$

Ethanol + 1,1-Dimethylpropyl cation → Ethyloxonium ion + 2-Methyl-1-butene

$$CH_3CH_2\ddot{O}H: + H{-}\underset{}{\overset{\displaystyle CH_3}{\overset{|}{C}H}}{-}\overset{+}{C}(CH_3)(CH_3) \xrightarrow{\text{fast}} CH_3CH_2\overset{+}{O}H_2 + CH_3CH{=}C(CH_3)(CH_3)$$

Ethanol + 1,1-Dimethylpropyl cation → Ethyloxonium ion + 2-Methyl-2-butene

FIGURE 5.13 The E1 mechanism for the dehydrohalogenation of 2-bromo-2-methylbutane in ethanol.

Because the carbon-halogen bond breaks in the rate-determining step, the nature of the leaving group is important in terms of reaction rate. Alkyl iodides have the weakest carbon-halogen bond and are the most reactive; alkyl fluorides have the strongest carbon-halogen bond and are the least reactive.

The best examples of E1 eliminations are those carried out in the absence of added base. In the example cited in Figure 5.13, the base that abstracts the proton from the carbocation intermediate is a very weak one; it is a molecule of the solvent, ethyl alcohol. At even modest concentrations of strong base, elimination by the E2 mechanism is faster than E1 elimination.

There is a strong similarity between the mechanism shown in Figure 5.13 and the ones shown for alcohol dehydration in Figure 5.7. Indeed, we can describe the acid-catalyzed dehydration of alcohols as an E1 elimination of their conjugate acids. The main difference between the dehydration of 2-methyl-2-butanol and the dehydrohalogenation of 2-bromo-2-methylbutane is the source of the carbocation. When the alcohol is the substrate, it is the corresponding alkyloxonium ion that dissociates to form the carbocation. The alkyl halide ionizes directly to the carbocation.

$$CH_3\overset{CH_3}{\underset{\overset{+}{O}H_2}{C}}CH_2CH_3 \xrightarrow{-H_2O} \underset{H_3C\quad CH_2CH_3}{\overset{CH_3}{\overset{+}{C}}} \xleftarrow{-:\ddot{Br}:^-} CH_3\overset{CH_3}{\underset{:\ddot{Br}:}{C}}CH_2CH_3$$

Alkyloxonium ion — Carbocation — Alkyl halide

Like alcohol dehydrations, E1 reactions of alkyl halides can lead to alkenes with altered carbon skeletons resulting from carbocation rearrangements. Eliminations by the E2 mechanism, on the other hand, normally proceed without rearrangement. Consequently, if one wishes to prepare an alkene from an alkyl halide, conditions favorable to E2 elimination should be chosen. In practice this simply means carrying out the reaction in the presence of a strong base.

5.18 SUMMARY

Alkenes and cycloalkenes contain carbon-carbon double bonds. According to **IUPAC nomenclature** (Section 5.1), alkenes are named by substituting *-ene* for the *-ane* suffix of the alkane that has the same number of carbon atoms as the longest continuous chain that includes the double bond. The chain is numbered in the direction that gives the lower number to the first-appearing carbon of the double bond. The double bond takes precedence over alkyl groups and halogens in dictating the direction of numbering, but is outranked by the hydroxyl group.

$$\overset{1}{C}H_3(H)\overset{2}{C}{=}\overset{3}{C}(\overset{4}{C}H_2\overset{5}{C}H_3)CH_2CH_3 \qquad \text{3-Bromocyclopentene} \qquad \overset{4}{C}H_2{=}\overset{3}{C}H\overset{2}{C}H_2\overset{1}{C}H_2OH$$

3-Ethyl-2-pentene — 3-Bromocyclopentene — 3-Buten-1-ol

Ethylene is a planar molecule, and the carbon-carbon double bond with its four attached atoms is a planar structural unit in higher alkenes (Section 5.2). Bonding in alkenes is described according to an sp^2 orbital hybridization model. The double bond unites two sp^2 hybridized carbon atoms and is made up of a σ component and a π component. The σ bond arises by overlap of an sp^2 hybrid orbital on each carbon. The π bond is weaker than the σ bond and results from a side-by-side overlap of p orbitals.

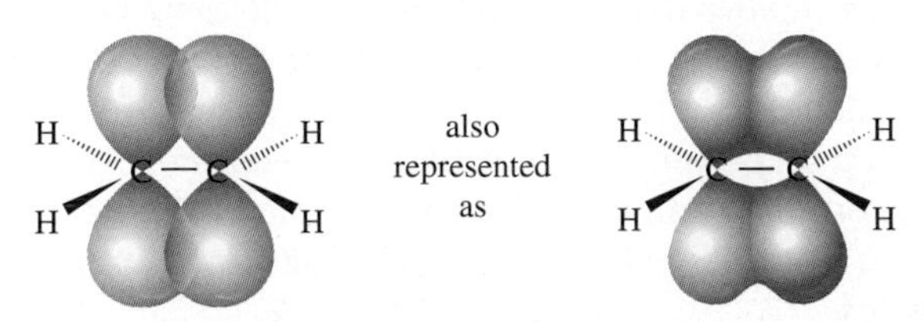

Representations of bonding in ethylene

Isomeric alkenes may be either **constitutional isomers** or **stereoisomers** (Section 5.3). There is a sizable barrier to rotation about a carbon-carbon double bond, which corresponds to the energy required to break the π component of the double bond. Stereoisomeric alkenes are configurationally stable under normal conditions. The **configurations** of stereoisomeric alkenes are described according to two notational systems. The simpler system adds the prefix *cis-* to the name of the alkene when similar substituents are on the same side of the double bond and the prefix *trans-* when they are on opposite sides. An alternative system (Section 5.4) ranks substituents according to a system of rules based on atomic number. The prefix *Z* is used for alkenes that have higher-ranked substituents on the same side of the double bond; the prefix *E* is used when higher-ranked substituents are on opposite sides.

$$(H_3C)(H)C{=}C(CH_2CH_3)(H) \qquad (H_3C)(H)C{=}C(H)(CH_2CH_3)$$

cis-2-Pentene [(*Z*)-2-pentene] *trans*-2-Pentene [(*E*)-2-pentene]

Alkenes, like alkanes, are relatively nonpolar (Section 5.5). Alkyl substituents donate electrons to an sp^2 hybridized carbon to which they are attached slightly better than do hydrogen substituents, and this electron release tends to stabilize a double bond (Section 5.6). The general order of alkene stability is as follows:

1. Tetrasubstituted alkenes ($R_2C{=}CR_2$) are the most stable.
2. Trisubstituted alkenes ($R_2C{=}CHR$) are next.
3. Among disubstituted alkenes, *trans*-RCH=CHR is normally more stable than *cis*-RCH=CHR. Exceptions are cycloalkenes, cis cycloalkenes being more stable than trans when the ring contains fewer than 11 carbons. Terminally disubstituted alkenes ($R_2C{=}CH_2$) may be slightly more or less stable than RCH=CHR, depending on their substituents.
4. Monosubstituted alkenes ($RCH{=}CH_2$) have a more stabilized double bond than ethylene (unsubstituted) but are less stable than disubstituted alkenes.

The greater stability of more highly substituted double bonds is an example of an **electronic effect.** The decreased stability that results from van der Waals strain between cis substituents is an example of a **steric effect.**

Cyclopropene and cyclobutene are destabilized by the angle strain that attends the incorporation of sp^2 hybridized carbons in small rings. Cycloalkenes that have trans double bonds in rings smaller than 12-membered are less stable than their cis stereoisomers. *trans*-Cyclooctene can be isolated and stored at room temperature, but *trans*-cycloheptene is not stable above $-30°C$ (Section 5.7).

Alkenes are formed by ***β* elimination reactions** (Section 5.8) of alcohols and alkyl halides. These reactions are summarized with examples in Table 5.3. In both cases, β elimination proceeds in the direction that yields the more highly substituted double bond **(Zaitsev's rule).**

Secondary and tertiary alcohols undergo **dehydration** by way of carbocation intermediates (Section 5.12).

TABLE 5.3
Preparation of Alkenes by Elimination Reactions of Alcohols and Alkyl Halides

Reaction (section) and comments	General equation and specific example
Dehydration of alcohols (Sections 5.9 through 5.13) Dehydration requires an acid catalyst; the order of reactivity of alcohols is tertiary > secondary > primary. Elimination is regioselective and proceeds in the direction that produces the most highly substituted double bond. When stereoisomeric alkenes are possible, the more stable one is formed in greater amounts. A carbocation intermediate is involved, and sometimes rearrangements take place during elimination.	$R_2CHCR'_2(OH) \xrightarrow{H^+} R_2C{=}CR'_2 + H_2O$ Alcohol; Alkene; Water 2-Methyl-2-hexanol $\xrightarrow{H_2SO_4,\ 80°C}$ 2-Methyl-1-hexene (19%) + 2-Methyl-2-hexene (81%)
Dehydrohalogenation of alkyl halides (Sections 5.14 through 5.17) Strong bases cause a proton and a halide to be lost from adjacent carbons of an alkyl halide to yield an alkene. Regioselectivity is in accord with the Zaitsev rule. The order of halide reactivity is I > Br > Cl > F and tertiary > secondary > primary. A concerted E2 reaction pathway is followed, carbocations are not involved, and rearrangements do not normally occur. An anti periplanar arrangement of the proton being removed and the halide being lost characterizes the transition state.	$R_2CHCR'_2(X) + :B^- \longrightarrow R_2C{=}CR'_2 + H{-}B + X^-$ Alkyl halide; Base; Alkene; Conjugate acid of base; Halide 1-Chloro-1-methylcyclohexane $\xrightarrow{KOCH_2CH_3,\ CH_3CH_2OH,\ 100°C}$ Methylenecyclohexane (6%) + 1-Methylcyclohexene (94%)

Step 1

$$R_2CH{-}CR'_2(\ddot{O}H) \xrightarrow{H^+} R_2CH{-}CR'_2(\overset{+}{O}H_2)$$

Alcohol → Alkyloxonium ion

Step 2

$$R_2CH{-}CR'_2(\overset{+}{O}H_2) \xrightarrow{-H_2\ddot{O}:} R_2CH{-}\overset{+}{C}R'_2$$

Alkyloxonium ion → Carbocation

Step 3

$$R_2C(H)-\overset{+}{C}R'_2 \xrightarrow{-H^+} R_2C=CR'_2$$

Carbocation   Alkene

Primary alcohols do not dehydrate as readily as secondary or tertiary alcohols, and their dehydration does not involve a carbocation. A proton is lost from the β carbon in the same step in which carbon-oxygen bond cleavage occurs (Section 5.12).

Alkene preparation using alcohol dehydration as the synthetic method is complicated by **carbocation rearrangements** (Section 5.13). A less stable carbocation can rearrange to a more stable one by an alkyl group migration or by a hydride shift, opening the possibility for alkene formation from two structurally distinct carbocation intermediates.

$$R-C(G)(R)-\overset{+}{C}(H)-R \longrightarrow R-\overset{+}{C}(R)-C(G)(H)-R$$

Secondary carbocation   Tertiary carbocation

(G is a migrating group; it may be either a hydrogen or an alkyl group)

Dehydrohalogenation of alkyl halides by alkoxide bases (Section 5.14) is not complicated by rearrangements, because carbocations are not intermediates. The **bimolecular (E2) mechanism** (Section 5.15) is a concerted process in which the base abstracts a proton from the β carbon while the bond between the halogen and the α carbon undergoes heterolytic cleavage.

$$\bar{B}: + H-{}_{\beta}C-C_{\alpha}-\ddot{X}: \longrightarrow B-H + C=C + :\ddot{X}:^-$$

Base   Alkyl halide   Conjugate acid of base   Alkene   Halide ion

The hydrogen abstracted and the halide lost must be in an **anti periplanar** relationship at the transition state in order to allow the developing p orbitals to overlap with each other to form a π bond (Section 5.16).

In the absence of a strong base, alkyl halides eliminate by the **unimolecular (E1) mechanism** (Section 5.17). The E1 mechanism involves rate-determining ionization of the alkyl halide to a carbocation, followed by deprotonation of the carbocation.

PROBLEMS

5.20 Write structural formulas for each of the following:

(*a*) 1-Heptene
(*b*) 3-Ethyl-2-pentene
(*c*) *cis*-3-Octene
(*d*) *trans*-1,4-Dichloro-2-butene
(*e*) (*Z*)-3-Methyl-2-hexene
(*f*) (*E*)-3-Chloro-2-hexene

(*g*) 1-Bromo-3-methylcyclohexene
(*h*) 1-Bromo-6-methylcyclohexene
(*i*) 4-Methyl-4-penten-2-ol
(*j*) Vinylcycloheptane
(*k*) 1,1-Diallylcyclopropane
(*l*) *trans*-1-Isopropenyl-3-methylcyclohexane

5.21 Write a structural formula and give a correct IUPAC name for each alkene of molecular formula C_7H_{14} that has a *tetrasubstituted* double bond.

5.22 Give the IUPAC names for each of the following compounds:

(*a*) $(CH_3CH_2)_2C{=}CHCH_3$
(*b*) $(CH_3CH_2)_2C{=}C(CH_2CH_3)_2$
(*c*) $(CH_3)_3CCH{=}CCl_2$
(*d*)
(*e*) H_3C H_3C H_3C
(*f*) H H
(*g*)

5.23

(*a*) A hydrocarbon isolated from fish oil and from plankton was identified as 2,6,10,14-tetramethyl-2-pentadecene. Write its structure.
(*b*) Alkyl isothiocyanates are compounds of the type RN=C=S. Write a structural formula for *allyl isothiocyanate,* a pungent-smelling compound isolated from mustard.

5.24

(*a*) The sex attractant of the Mediterranean fruit fly is (*E*)-6-nonen-1-ol. Write a structural formula for this compound, showing the stereochemistry of the double bond.
(*b*) Geraniol is a naturally occurring substance present in the fragrant oil of many plants. It has a pleasing, roselike odor. Geraniol is the *E* isomer of

$$(CH_3)_2C{=}CHCH_2CH_2\underset{\displaystyle CH_3}{\underset{|}{C}}{=}CHCH_2OH$$

Write a structural formula for geraniol, showing its stereochemistry.
(*c*) Nerol is a naturally occurring substance that is a stereoisomer of geraniol. Write its structure.
(*d*) The sex attractant of the codling moth is the 2*Z*, 6*E* stereoisomer of

$$CH_3CH_2CH_2\underset{\displaystyle CH_3}{\underset{|}{C}}{=}CHCH_2CH_2\underset{\displaystyle CH_2CH_3}{\underset{|}{C}}{=}CHCH_2OH$$

Write the structure of this substance in a way that clearly shows its stereochemistry.
(*e*) The sex pheromone of the honeybee is the *E* stereoisomer of the compound shown. Write a structural formula for this compound.

$$CH_3\overset{\displaystyle O}{\overset{\|}{C}}(CH_2)_4CH_2CH{=}CHCO_2H$$

(*f*) A growth hormone from the cecropia moth has the structure shown. Express the stereochemistry of the double bonds according to the *E-Z* system.

5.25 Match each alkene with the appropriate heat of combustion:

Heats of combustion (kJ/mol): 5293; 4658; 4650; 4638; 4632
Heats of combustion (kcal/mol): 1264.9; 1113.4; 1111.4; 1108.6; 1107.1

(*a*) 1-Heptene
(*b*) 2,4-Dimethyl-1-pentene
(*c*) 2,4-Dimethyl-2-pentene
(*d*) (*Z*)-4,4-Dimethyl-2-pentene
(*e*) 2,4,4-Trimethyl-2-pentene

5.26 Choose the more stable alkene in each of the following pairs. Explain your reasoning.

(*a*) 1-Methylcyclohexene or 3-methylcyclohexene
(*b*) Isopropenylcyclopentane or allylcyclopentane

(*c*)

Bicyclo[4.2.0]oct-7-ene   Bicyclo[4.2.0]oct-3-ene

(*d*) (*Z*)-Cyclononene or (*E*)-cyclononene
(*e*) (*Z*)-Cyclooctadecene or (*E*)-cyclooctadecene

5.27

(*a*) Suggest an explanation for the fact that 1-methylcyclopropene is some 42 kJ/mol (10 kcal/mol) less stable than methylenecyclopropane.

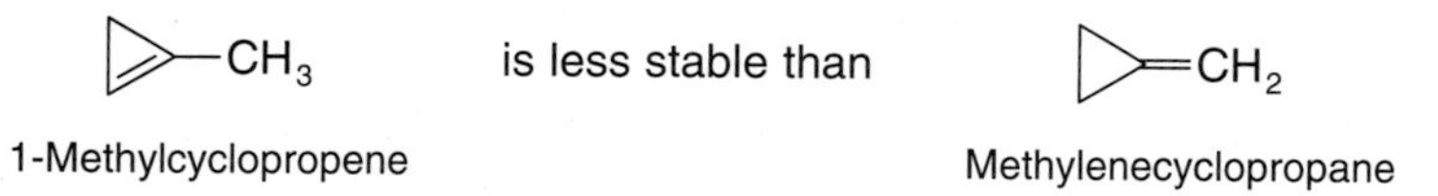

(*b*) On the basis of your answer to part (*a*), compare the expected stability of 3-methylcyclopropene with that of 1-methylcyclopropene and that of methylenecyclopropane.

5.28 How many alkenes would you expect to be formed from each of the following alkyl bromides under conditions of E2 elimination? Identify the alkenes in each case.

(*a*) 1-Bromohexane
(*b*) 2-Bromohexane
(*c*) 3-Bromohexane
(*d*) 2-Bromo-2-methylpentane
(*e*) 2-Bromo-3-methylpentane
(*f*) 3-Bromo-2-methylpentane
(*g*) 3-Bromo-3-methylpentane
(*h*) 3-Bromo-2,2-dimethylbutane

5.29 Write structural formulas for all the alkene products that could reasonably be formed from each of the following compounds under the indicated reaction conditions. Where more than one alkene is produced, specify the one that is the major product.

(*a*) 1-Bromo-3,3-dimethylbutane (potassium *tert*-butoxide, *tert*-butyl alcohol, 100°C)
(*b*) 1-Methylcyclopentyl chloride (sodium ethoxide, ethanol, 70°C)

(*c*) 3-Methyl-3-pentanol (sulfuric acid, 80°C)
(*d*) 2,3-Dimethyl-2-butanol (phosphoric acid, 120°C)
(*e*) 3-Iodo-2,4-dimethylpentane (sodium ethoxide, ethanol, 70°C)
(*f*) 2,4-Dimethyl-3-pentanol (sulfuric acid, 120°C)

5.30 Choose the compound of molecular formula $C_7H_{13}Br$ that gives each alkene shown as the exclusive product of E2 elimination.

(*a*) 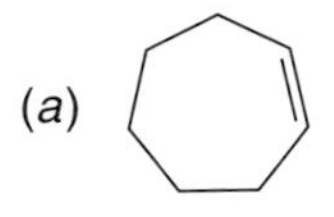

(*b*) 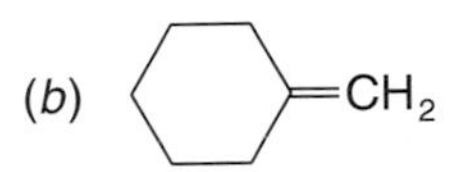

(*c*) —CH_3

(*d*) CH_3 CH_3

(*e*) —$CH{=}CH_2$

(*f*) $CH(CH_3)_2$

(*g*) —$C(CH_3)_3$

5.31 Give the structures of two different alkyl bromides both of which yield the indicated alkene as the exclusive product of E2 elimination.

(*a*) $CH_3CH{=}CH_2$
(*b*) $(CH_3)_2C{=}CH_2$
(*c*) $BrCH{=}CBr_2$
(*d*) CH_3 CH_3

5.32

(*a*) Write the structures of all the isomeric alkyl bromides having the molecular formula $C_5H_{11}Br$.
(*b*) Which one undergoes E1 elimination at the fastest rate?
(*c*) Which one is incapable of reacting by the E2 mechanism?
(*d*) Which ones can yield only a single alkene on E2 elimination?
(*e*) For which isomer is E2 elimination regiospecific while leading to the formation of two alkenes?
(*f*) Which one yields the most complex mixture of alkenes on E2 elimination?

5.33

(*a*) Write the structures of all the isomeric alcohols having the molecular formula $C_5H_{12}O$.
(*b*) Which one will undergo acid-catalyzed dehydration most readily?
(*c*) Write the structure of the most stable C_5H_{11} carbocation.
(*d*) Which alkenes may be derived from the carbocation in part (*c*)?
(*e*) Which alcohols can yield the carbocation in part (*c*) by a process involving a hydride shift?
(*f*) Which alcohols can yield the carbocation in part (*c*) by a process involving a methyl shift?

5.34 Identify the principal organic product of each of the following reactions. In spite of the structural complexity of some of the starting materials, the functional group transformations are all of the type described in this chapter.

(*a*) Br (meta-bromophenyl)–$\underset{\text{OH}}{\text{CHCH}_2\text{CH}_3}$ $\xrightarrow[\text{heat}]{\text{KHSO}_4}$

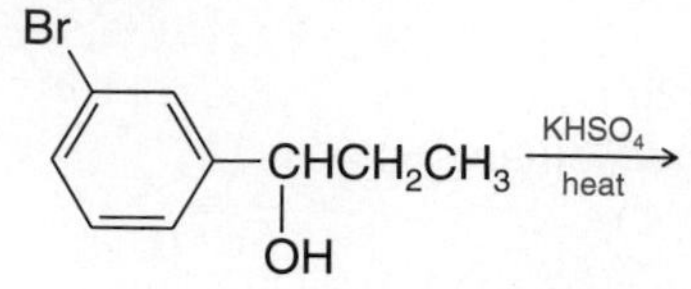

(*b*) $ICH_2CH(OCH_2CH_3)_2 \xrightarrow[(CH_3)_3COH,\ heat]{KOC(CH_3)_3}$

(*c*) (cyclohexane with H, CH_3 and H_3C, Br substituents) $\xrightarrow[CH_3CH_2OH,\ heat]{NaOCH_2CH_3}$

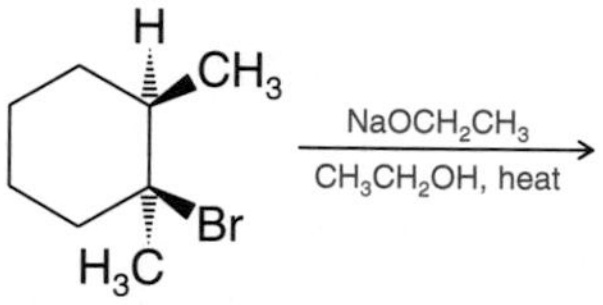

(*d*) (bicyclic compound with $(CH_3)_2CCl$ substituent) $\xrightarrow[(CH_3)_3COH,\ heat]{KOC(CH_3)_3}$

(*e*) (HO, CN; CH_3O-substituted tetrahydronaphthalene) $\xrightarrow[130–150°C]{KHSO_4}$ $(C_{12}H_{11}NO)$

(*f*) $HO\underset{CO_2H}{C}(CH_2CO_2H)_2 \xrightarrow[140–145°C]{H_2SO_4} (C_6H_6O_6)$

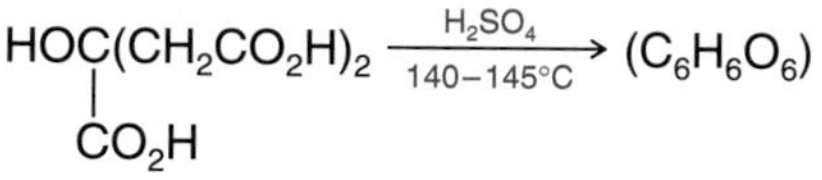

Citric acid

(*g*) (H_3C, CH_3, CH_3, Cl, Cl bicyclic compound) $\xrightarrow[DMSO,\ 70°C]{KOC(CH_3)_3}$ $(C_{10}H_{14})$

(*h*) Br–(tetracyclic bis-dioxolane compound)–Br $\xrightarrow[DMSO]{KOC(CH_3)_3}$ $(C_{14}H_{16}O_4)$

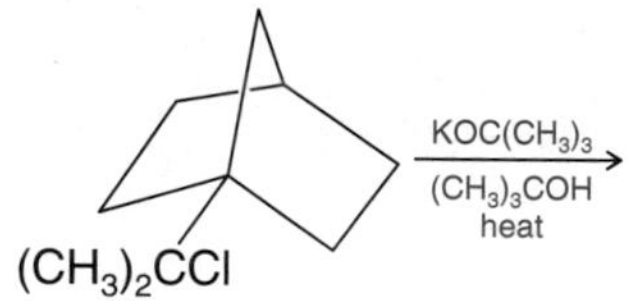

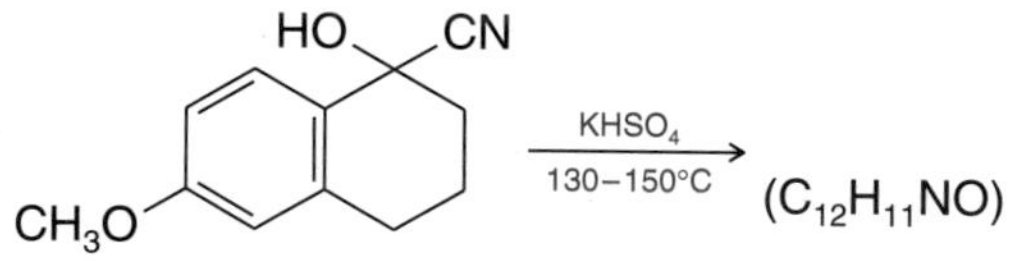

(*i*) (CH_3OCH_2, CH_3O, CH_3O, CH_3O, Br pyranose ring) $\xrightarrow[heat]{KOH}$ $(C_{10}H_{18}O_5)$

(*j*) (cyclohexane with Cl, CH_3 and $(CH_3)_3C$, H substituents) $\xrightarrow[CH_3OH,\ heat]{NaOCH_3}$

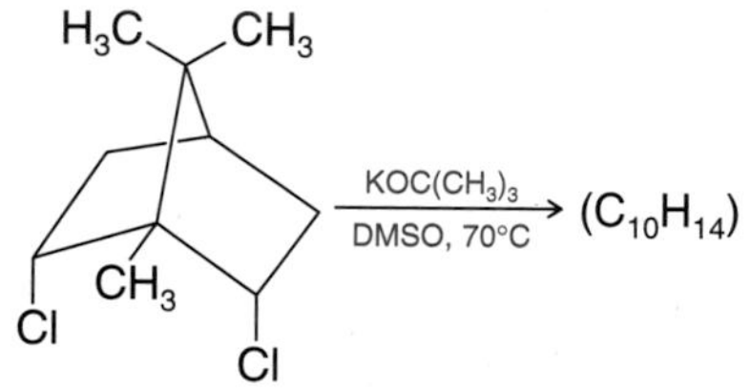

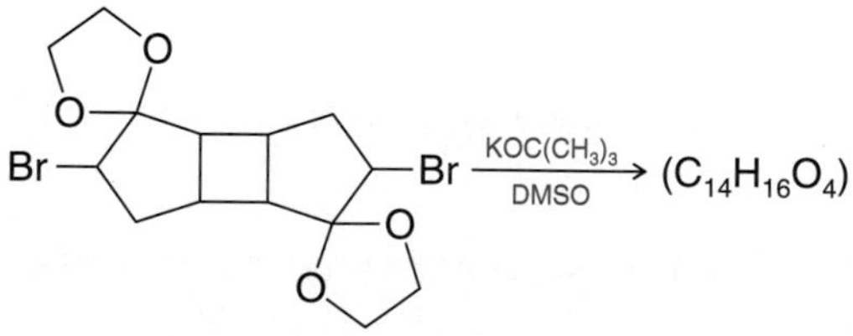

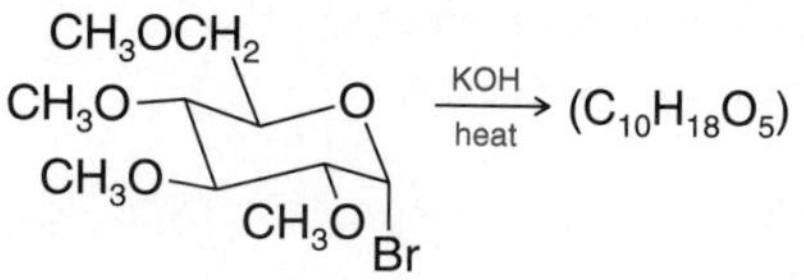

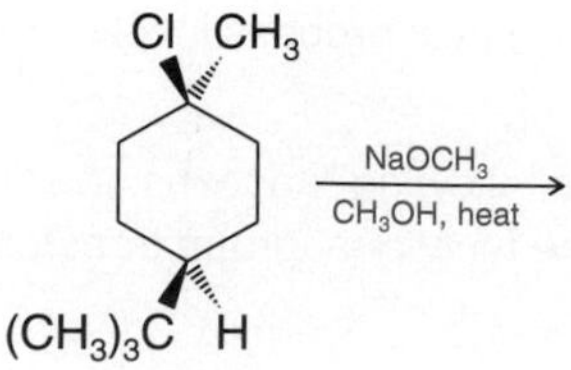

5.35 Evidence has been reported in the chemical literature that the reaction

$$(CH_3CH_2)_2CHCH_2Br + KNH_2 \longrightarrow (CH_3CH_2)_2C{=}CH_2 + NH_3 + KBr$$

proceeds by the E2 mechanism. Use curved arrow notation to represent the flow of electrons for this process.

5.36 The rate of the reaction

$$(CH_3)_3CCl + NaSCH_2CH_3 \longrightarrow (CH_3)_2C{=}CH_2 + CH_3CH_2SH + NaCl$$

is first-order in $(CH_3)_3CCl$ and first-order in $NaSCH_2CH_3$. Give the symbol (E1 or E2) for the most reasonable mechanism, and use curved arrow notation to represent the flow of electrons.

5.37 Menthyl chloride and neomenthyl chloride have the structures shown. One of these stereoisomers undergoes elimination on treatment with sodium ethoxide in ethanol much more readily than the other. Which reacts faster, menthyl chloride or neomenthyl chloride? Why?

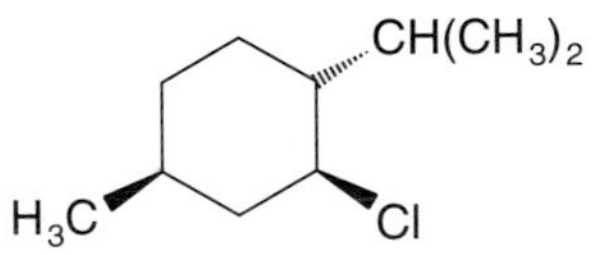

$CH(CH_3)_2$ H_3C Cl

Menthyl chloride

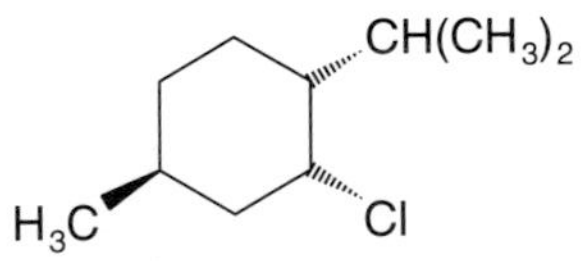

$CH(CH_3)_2$ H_3C Cl

Neomenthyl chloride

5.38 The stereoselectivity of elimination of 5-bromononane on treatment with potassium ethoxide was described in Section 5.14. Draw Newman projections of 5-bromononane showing the conformations that lead to *cis*-4-nonene and *trans*-4-nonene, respectively. Identify the proton that is lost in each case, and suggest a mechanistic explanation for the observed stereoselectivity.

5.39 In the acid-catalyzed dehydration of 2-methyl-1-propanol, what carbocation would be formed if a hydride shift accompanied cleavage of the carbon-oxygen bond in the oxonium ion? What ion would be formed as a result of a methyl shift? Which pathway do you think will predominate, a hydride shift or a methyl shift?

5.40 Each of the following carbocations has the potential to rearrange to a more stable one. Write the structure of the rearranged carbocation.

(*a*) $CH_3CH_2CH_2^+$

(*b*) $(CH_3)_2CH\overset{+}{C}HCH_3$

(*c*) $(CH_3)_3C\overset{+}{C}HCH_3$

(*d*) $(CH_3CH_2)_3CCH_2^+$

(*e*) (cyclopentyl cation, + on ring carbon) $-CH_3$

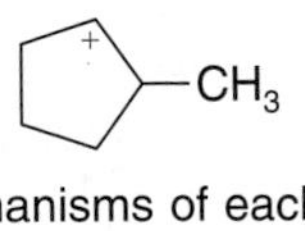

5.41 Write a sequence of steps depicting the mechanisms of each of the following reactions:

(*a*) $C(CH_3)_3$ OH $\xrightarrow{H^+}$ CH_3 C CH_2 CH_3

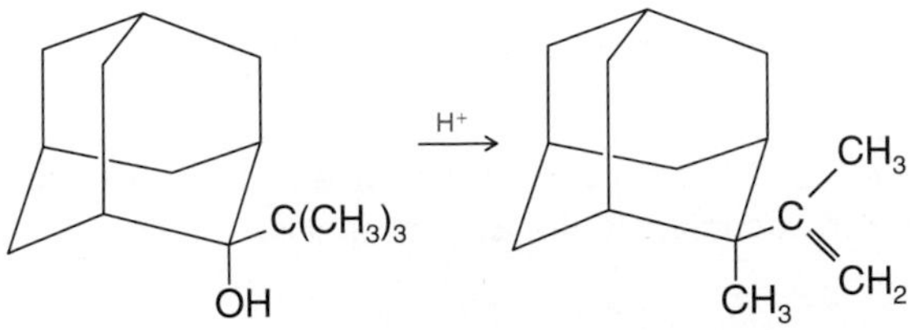

(*b*) 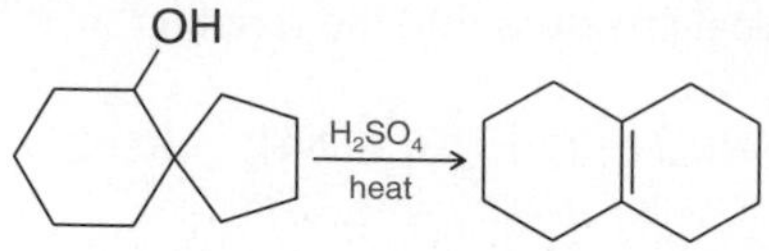 OH — H_2SO_4, heat →

(*c*) OH, H, CH_3, CH_3, CH_3 — $KHSO_4$, 170°C → CH_3, CH_3, CH_3

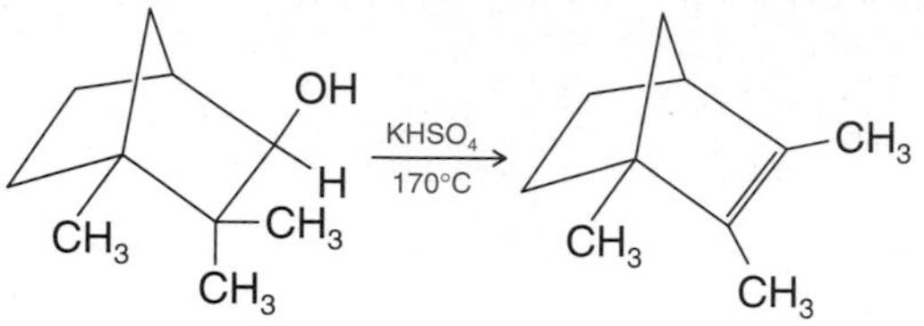

5.42 In Problem 5.15 (Section 5.13) we saw that acid-catalyzed dehydration of 2,2-dimethylcyclohexanol afforded 1,2-dimethylcyclohexene. To explain this product we must write a mechanism for the reaction in which a methyl shift transforms a secondary carbocation to a tertiary one. Another product of the dehydration of 2,2-dimethylcyclohexanol is isopropylidenecyclopentane. Write a mechanism to rationalize its formation.

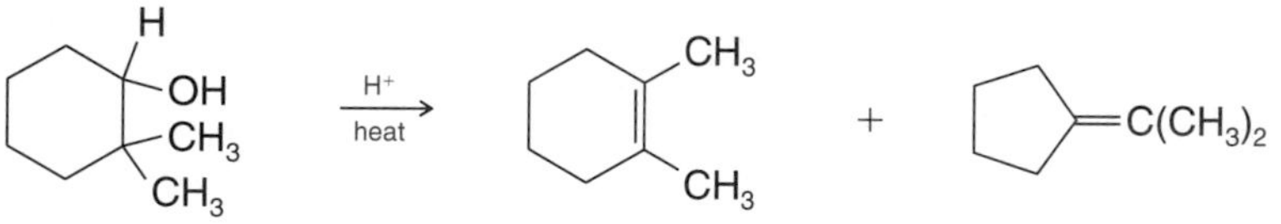

H, OH, CH_3, CH_3 — H^+, heat → CH_3, CH_3 + $=C(CH_3)_2$

2,2-Dimethylcyclohexanol 1,2-Dimethylcyclohexene Isopropylidenecyclopentane

5.43 Compound A (C_4H_{10}) gives two different monochlorides on photochemical chlorination. Treatment of either of these monochlorides with potassium *tert*-butoxide in dimethyl sulfoxide gives the same alkene B (C_4H_8) as the only product. What are the structures of compound A, the two monochlorides, and alkene B?

5.44 Compound A (C_6H_{14}) gives three different monochlorides on photochemical chlorination. One of these monochlorides is inert to E2 elimination. The other two monochlorides yield the same alkene B (C_6H_{12}) on being heated with potassium *tert*-butoxide in *tert*-butyl alcohol. Identify compound A, the three monochlorides, and alkene B.

MOLECULAR MODELING EXERCISES

5.45 Construct a molecular model of propene in which the vinyl hydrogen at C-2 is staggered with respect to the C—H bonds of the methyl group. (This conformation is more stable than one in which the C-2 hydrogen is eclipsed with the methyl group.)

5.46 Based on the most stable conformation of propene, described in the preceding problem, construct molecular models of *cis*-2-butene and *trans*-2-butene, each in its most stable conformation. Identify the van der Waals forces most responsible for destabilizing *cis*-2-butene compared to *trans*-2-butene.

5.47 Construct molecular models of *cis* and *trans*-cyclooctene.

5.48 Construct molecular models of the E and Z stereoisomers of 1-bromo-1,2-dichloro-2-fluoroethene.

5.49 Construct a molecular model of 1-bromo-2-methylpropane in the geometry it adopts at the transition state for elimination according to the E2 mechanism.

CHAPTER 6

REACTIONS OF ALKENES: ADDITION REACTIONS

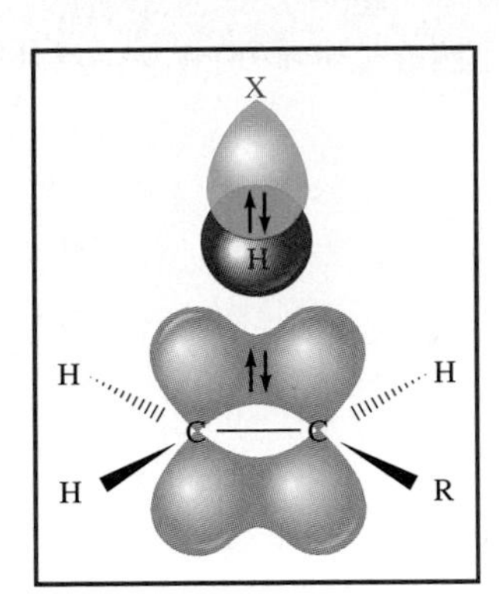

Now that we know something of the structure and preparation of alkenes, we are ready to look at their chemical reactions. The characteristic reaction of alkenes is **addition** to the double bond. The general form of addition to an alkene may be represented as

$$A{-}B + \;\rangle C{=}C\langle \;\longrightarrow A{-}\overset{|}{\underset{|}{C}}{-}\overset{|}{\underset{|}{C}}{-}B$$

The range of compounds represented as A—B in this equation is quite large, and addition reactions offer a wealth of opportunity for converting alkenes to a variety of other functional group types.

Addition and elimination are the reverse of each other. Alkenes are commonly described as **unsaturated** hydrocarbons because they have the capacity to react with substances which add to them. Alkanes, on the other hand, are said to be **saturated** hydrocarbons and are incapable of undergoing addition reactions.

6.1 HYDROGENATION OF ALKENES

In terms of the relationship between reactants and products, addition reactions are best exemplified by the *hydrogenation* of alkenes to yield alkanes. **Hydrogenation** is the addition of H_2 to a multiple bond. An example is the addition of hydrogen to the double bond of ethylene to form ethane.

$$\underset{\text{Ethylene}}{H_2C \overset{\pi}{=} CH_2} + \underset{\text{Hydrogen}}{H \overset{\sigma}{-} H} \xrightarrow{\text{Pt, Pd, Ni, or Rh}} \underset{\text{Ethane}}{H \overset{\sigma}{-} CH_2 - CH_2 \overset{\sigma}{-} H} \qquad \Delta H° = -136 \text{ kJ/mol } (-32.6 \text{ kcal/mol})$$

The bonds in the products are stronger than the bonds in the reactants; two C—H σ bonds of an alkane are formed at the expense of the H—H σ bond and the π component of the alkene's double bond. The overall reaction is *exothermic,* and the heat evolved on hydrogenation of an alkene is defined as its **heat of hydrogenation.** Heat of hydrogenation is a positive quantity equal to $-\Delta H°$ for the reaction.

The uncatalyzed addition of hydrogen to an alkene, although exothermic, is very slow. The rate of hydrogenation increases dramatically, however, in the presence of certain finely divided metal catalysts. *Platinum* is the hydrogenation catalyst most often used, although *palladium, nickel,* and *rhodium* are also effective. Metal-catalyzed addition of hydrogen is normally rapid at room temperature, and the alkane is produced in high yield, usually as the only product.

The French chemist Paul Sabatier received the 1912 Nobel Prize in chemistry for his discovery that finely divided nickel is an effective hydrogenation catalyst.

$$\underset{\text{2-Methyl-2-butene}}{(CH_3)_2C{=}CHCH_3} + \underset{\text{Hydrogen}}{H_2} \xrightarrow{\text{Pt}} \underset{\text{2-Methylbutane (100\%)}}{(CH_3)_2CHCH_2CH_3}$$

Step 1: Hydrogen molecules react with metal atoms at the catalyst surface (blue). The relatively strong hydrogen-hydrogen σ bond is broken and replaced by two weak metal-hydrogen bonds.

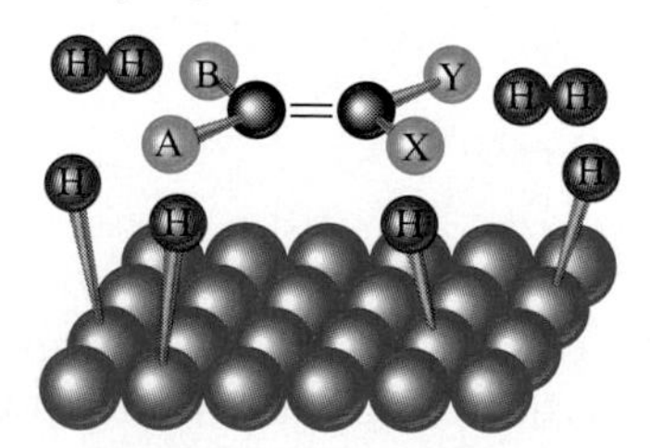

Step 2: The alkene reacts with the metal catalyst. The π component of the double bond between the two carbons is replaced by two relatively weak carbon-metal σ bonds.

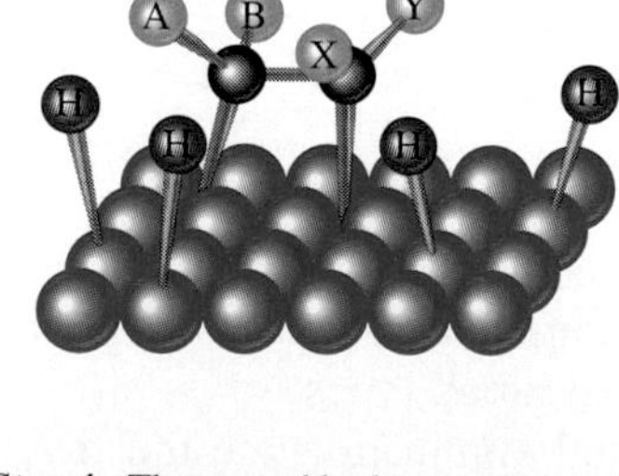

Step 3: A hydrogen atom is transferred from the catalyst surface to one of the carbons of the double bond.

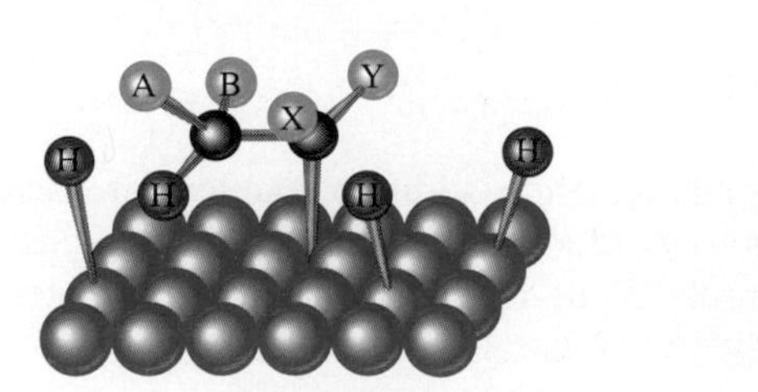

Step 4: The second hydrogen atom is transferred forming the alkane. The sites on the catalyst surface at which the reaction occurred are free to accept additional hydrogen and alkene molecules.

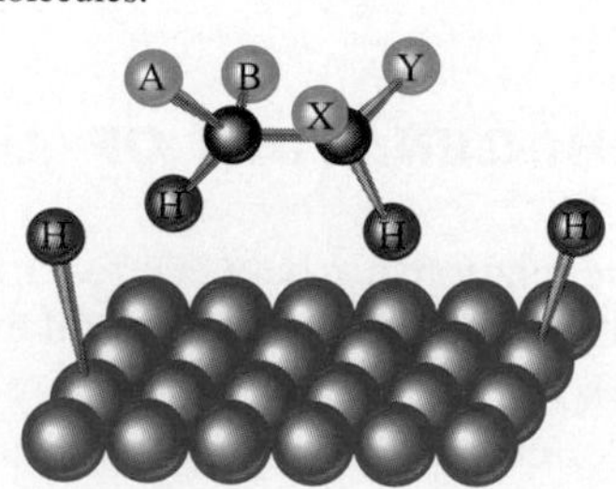

FIGURE 6.1 A mechanistic representation of heterogeneous catalysis in the hydrogenation of alkenes.

5,5-Dimethyl(methylene)cyclononane Hydrogen 1,1,5-Trimethylcyclononane (73%)

PROBLEM 6.1 What three alkenes yield 2-methylbutane on catalytic hydrogenation?

The solvent used in catalytic hydrogenation is chosen for its ability to dissolve the alkene and is typically ethanol, hexane, or acetic acid. The metal catalysts are insoluble in these solvents (or, indeed, in any solvent). Two phases, the solution and the metal, are present, and the reaction takes place at the interface between them. Reactions involving a substance in one phase with a different substance in a second phase are described as **heterogeneous reactions.**

Catalytic hydrogenation of an alkene is believed to proceed by the series of steps shown in Figure 6.1. As already noted, addition of hydrogen to the alkene is very slow in the absence of a metal catalyst, meaning that any uncatalyzed mechanism must have a very high activation energy. The metal catalyst accelerates the rate of hydrogenation by providing an alternative pathway that involves a sequence of several low-activation energy steps.

6.2 HEATS OF HYDROGENATION

Heats of hydrogenation have been used to assess the relative stabilities of alkenes in much the same way as was described for heats of combustion in Section 5.6. Catalytic hydrogenation of 1-butene, *cis*-2-butene, or *trans*-2-butene yields the same product—butane. As Figure 6.2 shows, the measured heats of hydrogenation reveal that *trans*-2-butene is 4 kJ/mol (1.0 kcal/mol) lower in energy than *cis*-2-butene and that *cis*-2-butene is 7 kJ/mol (1.7 kcal/mol) lower in energy than 1-butene.

Remember that a catalyst affects the rate of a reaction but not the energy relationships between reactants and products. Thus, the heat of hydrogenation of a particular alkene is the same irrespective of what catalyst is used.

The energy differences between the isomeric butenes as measured by their heats of hydrogenation are, within experimental error, equal to the differences in their heats of combustion. Both methods measure the differences in the energy of *isomers* by converting them to a product or products common to all.

Heats of hydrogenation can be used to *estimate* the stability of double bonds as structural units, even in alkenes that are not isomers. Table 6.1 lists the heats of hydrogenation for a representative collection of alkenes.

The pattern of alkene stability determined from heats of hydrogenation is

$CH_2{=}CH_2$	$RCH{=}CH_2$	$RCH{=}CHR$	$R_2C{=}CHR$	$R_2C{=}CR_2$
Ethylene	Monosubstituted	Disubstituted	Trisubstituted	Tetrasubstituted

and parallels exactly the pattern deduced from heats of combustion. Ethylene, which has no alkyl substituents to stabilize its double bond, has the highest heat of hydrogenation. Alkenes that are similar in structure to one another have similar heats of hydrogenation. For example, the heats of hydrogenation of the monosubstituted (terminal)

TABLE 6.1
Heats of Hydrogenation of Some Alkenes

Alkene	Structure	Heat of hydrogenation kJ/mol	kcal/mol
Ethylene	$CH_2{=}CH_2$	136	32.6
Monosubstituted alkenes			
Propene	$CH_2{=}CHCH_3$	125	29.9
1-Butene	$CH_2{=}CHCH_2CH_3$	126	30.1
1-Hexene	$CH_2{=}CHCH_2CH_2CH_2CH_3$	126	30.2
Cis disubstituted alkenes			
cis-2-Butene	$(H_3C)(H)C{=}C(CH_3)(H)$ (cis)	119	28.4
cis-2-Pentene	$(H_3C)(H)C{=}C(CH_2CH_3)(H)$ (cis)	117	28.1
Trans disubstituted alkenes			
trans-2-Butene	$(H_3C)(H)C{=}C(H)(CH_3)$ (trans)	115	27.4
trans-2-Pentene	$(H_3C)(H)C{=}C(H)(CH_2CH_3)$ (trans)	114	27.2
Terminally disubstituted alkenes			
2-Methylpropene	$CH_2{=}C(CH_3)_2$	117	28.1
2,3-Dimethyl-1-butene	$CH_2{=}C(CH_3)CH(CH_3)_2$	116	27.8
Trisubstituted alkenes			
2-Methyl-2-pentene	$(CH_3)_2C{=}CHCH_2CH_3$	112	26.7
Tetrasubstituted alkenes			
2,3-Dimethyl-2-butene	$(CH_3)_2C{=}C(CH_3)_2$	110	26.4

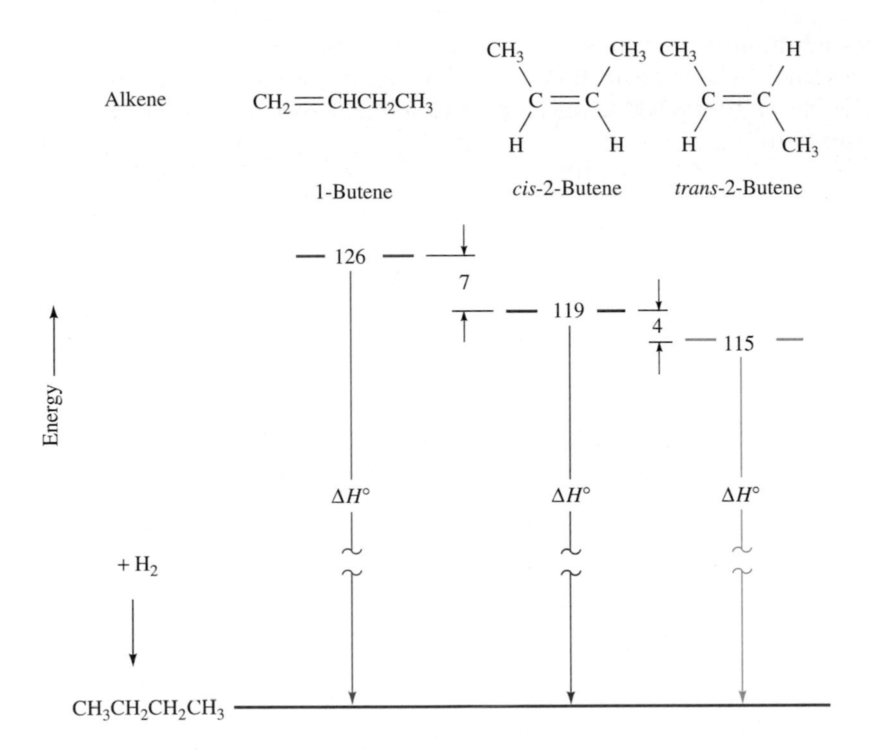

FIGURE 6.2 Heats of hydrogenation of butene isomers plotted on a common scale. All energies are in kilojoules per mole.

alkenes propene, 1-butene, and 1-hexene are almost identical. Cis disubstituted alkenes have lower heats of hydrogenation than monosubstituted alkenes but higher heats of hydrogenation than their more stable trans stereoisomers. Alkenes of the type $R_2C{=}CH_2$ may have heats of hydrogenation that are larger or smaller than either cis or trans disubstituted alkenes. Alkenes with trisubstituted double bonds have lower heats of hydrogenation than disubstituted alkenes, and tetrasubstituted alkenes have the lowest heats of hydrogenation.

PROBLEM 6.2 Match each alkene of Problem 6.1 with its correct heat of hydrogenation.

Heats of hydrogenation in kJ/mol (kcal/mol): 112 (26.7); 118 (28.2); 126 (30.2)

6.3 STEREOCHEMISTRY OF ALKENE HYDROGENATION

In our model for alkene hydrogenation shown in Figure 6.1, hydrogen atoms are transferred from the catalyst's surface to the alkene. Although the two hydrogens are not transferred simultaneously, both add to the same face of the double bond, as shown in the following example:

$$\text{Dimethyl cyclohexene-1,2-dicarboxylate} + H_2 \xrightarrow{Pt} \text{Dimethyl cyclohexane-}cis\text{-1,2-dicarboxylate (100\%)}$$

Dimethyl cyclohexene-1,2-dicarboxylate

Dimethyl cyclohexane-*cis*-1,2-dicarboxylate (100%)

The term **syn addition** is used to describe the stereochemistry of reactions such as catalytic hydrogenation where two atoms or groups add to the *same face* of a double bond. When atoms or groups add to *opposite faces* of the double bond, the process is called **anti addition.**

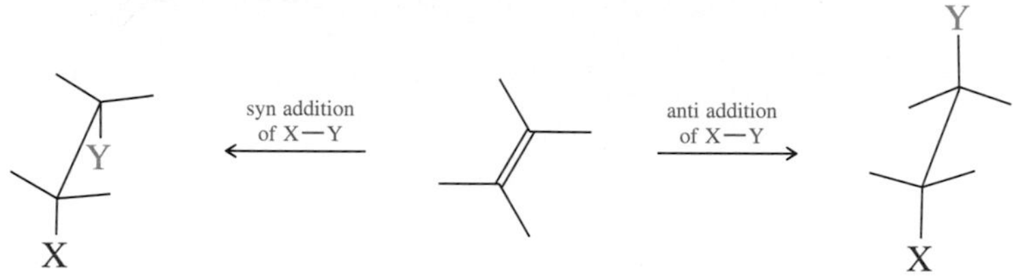

syn addition of X—Y ← → anti addition of X—Y

The concept of stereoselectivity was defined and introduced in connection with the formation of stereoisomeric alkenes in elimination reactions (Section 5.11).

A second stereochemical aspect of alkene hydrogenation concerns its **stereoselectivity.** A reaction in which a single starting material can give two or more stereoisomeric products but yields one of them in greater amounts than the other (or even to the exclusion of the other) is said to be **stereoselective.** The catalytic hydrogenation of α-pinene (a constituent of turpentine) is a stereoselective reaction for example. Syn addition of hydrogen can in principle lead to either *cis*-pinane or *trans*-pinane depending on which face of the double bond accepts the hydrogen atoms (shown in red in the equation).

cis-Pinane and *trans*-pinane are common names that denote the relationship between the pair of methyl groups on the bridge and the third methyl group.

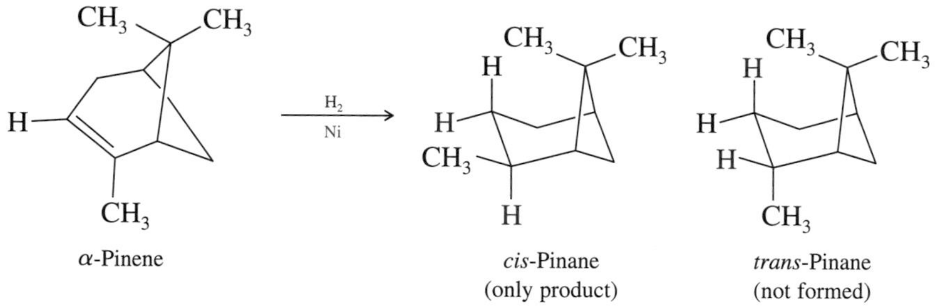

$\xrightarrow[\text{Ni}]{H_2}$

α-Pinene

cis-Pinane (only product)

trans-Pinane (not formed)

This hydrogenation is 100 percent stereoselective. The only product obtained is *cis*-pinane, none of the stereoisomeric *trans*-pinane being formed.

The observed stereoselectivity of this reaction depends on how the alkene approaches the catalyst surface. As Figure 6.3 shows, one of the methyl groups on the bridge carbon lies directly over the double bond and blocks that face from easy access

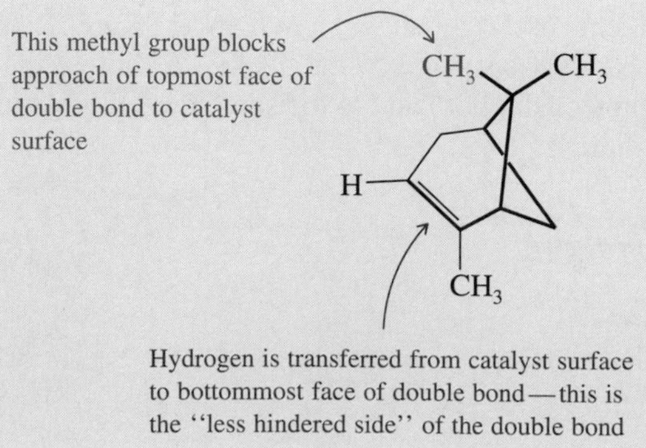

This methyl group blocks approach of topmost face of double bond to catalyst surface

Hydrogen is transferred from catalyst surface to bottommost face of double bond—this is the "less hindered side" of the double bond

FIGURE 6.3 The methyl group that lies over the double bond of α-pinene shields one face of it, preventing a close approach to the surface of the catalyst. Hydrogenation of α-pinene occurs preferentially from the bottom face of the double bond.

to the catalyst. The bottom face of the double bond is more exposed, and hydrogen is transferred from the catalyst to that face to yield *cis*-pinane.

Reactions such as catalytic hydrogenation that take place at the ''less hindered'' side of a reactant are common in organic chemistry and are examples of a steric effect (Section 5.6) on chemical reactivity.

6.4 ELECTROPHILIC ADDITION OF HYDROGEN HALIDES TO ALKENES

In many addition reactions the attacking reagent, unlike H_2, is a polar molecule or one that is easily polarizable. Hydrogen halides, which are polarized $^{\delta+}H{-}X^{\delta-}$, are among the simplest examples of polar substances that add to alkenes.

$$\mathrm{{>}C{=}C{<}} + \mathrm{HX} \longrightarrow \mathrm{H{-}\overset{|}{\underset{|}{C}}{-}\overset{|}{\underset{|}{C}}{-}X}$$

Alkene — Hydrogen halide — Alkyl halide

Addition occurs rapidly in a variety of solvents, including pentane, benzene, dichloromethane, chloroform, and acetic acid.

$$\mathrm{(CH_3CH_2)HC{=}CH(CH_2CH_3)} + \mathrm{HBr} \xrightarrow[\mathrm{CHCl_3}]{-30^\circ\mathrm{C}} \mathrm{CH_3CH_2CH_2\underset{\underset{\large Br}{|}}{C}HCH_2CH_3}$$

cis-3-Hexene — Hydrogen bromide — 3-Bromohexane (76%)

The reactivity of the hydrogen halides reflects their ability to donate a proton. Hydrogen iodide is the strongest acid of the hydrogen halides and reacts with alkenes at the fastest rate.

Increasing reactivity of hydrogen halides in addition to alkenes

$$\mathrm{HF} << \mathrm{HCl} < \mathrm{HBr} < \mathrm{HI}$$

Slowest rate of addition; least acidic — Fastest rate of addition; most acidic

We can gain a general understanding of the mechanism of hydrogen halide addition to alkenes by extending some of the principles of reaction mechanisms introduced earlier. In Section 5.12 we pointed out that carbocations are the conjugate acids of alkenes. Acid-base reactions are reversible processes. An alkene, therefore, can accept a proton from a hydrogen halide to form a carbocation.

$$\mathrm{R_2C{=}CR_2} + \mathrm{H{-}\ddot{\underset{..}{X}}{:}} \rightleftharpoons \mathrm{R_2\overset{+}{C}{-}\overset{\overset{\large H}{|}}{C}R_2} + \mathrm{{:}\ddot{\underset{..}{X}}{:}^-}$$

Alkene (base) — Hydrogen halide (acid) — Carbocation (conjugate acid) — Anion (conjugate base)

We have also seen (Section 4.10) that carbocations, when generated in the presence of halide anions, react with them to form alkyl halides.

$$R_2\overset{+}{C}-\underset{}{\overset{H}{|}}CR_2 \quad + \quad :\ddot{X}:^- \longrightarrow R_2\underset{:\ddot{X}:}{C}-\overset{H}{C}R_2$$

Carbocation (electrophile) Halide ion (nucleophile) Alkyl halide

Both steps in this general mechanism are based on precedent. It is called **electrophilic addition** because the reaction is triggered by the attack of an electrophile (an acid) on the π electrons of the carbon-carbon double bond. Alkenes are weak bases, and the site of their basicity is the π component of the double bond. Transfer of the two π electrons to an electrophile generates a carbocation as a reactive intermediate; normally this is the rate-determining step. This two-step general mechanism can be elaborated on by considering the regioselectivity of electrophilic addition.

6.5 REGIOSELECTIVITY OF HYDROGEN HALIDE ADDITION: MARKOVNIKOV'S RULE

In principle a hydrogen halide can add to an unsymmetrical alkene (an alkene in which the two carbons of the double bond are not equivalently substituted) in either of two ways. In practice, addition is so highly regioselective as to be considered regiospecific.

$$RCH{=}CH_2 + HX \longrightarrow \underset{X}{RCH}-\underset{H}{CH_2} \quad \text{rather than} \quad \underset{H}{RCH}-\underset{X}{CH_2}$$

$$R_2C{=}CH_2 + HX \longrightarrow \underset{X}{R_2C}-\underset{H}{CH_2} \quad \text{rather than} \quad \underset{H}{R_2C}-\underset{X}{CH_2}$$

$$R_2C{=}CHR + HX \longrightarrow \underset{X}{R_2C}-\underset{H}{CHR} \quad \text{rather than} \quad \underset{H}{R_2C}-\underset{X}{CHR}$$

An article in the December 1988 issue of the *Journal of Chemical Education* traces the historical development of Markovnikov's rule. In that article Markovnikov's name is spelled *Markownikoff,* which is the way it appeared in his original paper written in German.

In 1870 Vladimir Markovnikov, a colleague of Alexander Zaitsev at the University of Kazan, noting the pattern of hydrogen halide addition to alkenes, assembled his observations into a simple statement. **Markovnikov's rule** states that *when an unsymmetrically substituted alkene reacts with a hydrogen halide, the hydrogen adds to the carbon that has the greater number of hydrogen substituents, and the halogen adds to the carbon having fewer hydrogen substituents.* The preceding general equations illustrate regioselective addition in accordance with Markovnikov's rule, while the equations that follow provide some specific examples.

$$CH_3CH_2CH{=}CH_2 + HBr \xrightarrow{\text{acetic acid}} CH_3CH_2\underset{Br}{C}HCH_3$$

1-Butene Hydrogen bromide 2-Bromobutane (80%)

$$(CH_3)_2C{=}CH_2 + HBr \xrightarrow{\text{acetic acid}} CH_3{-}C(CH_3)_2{-}Br$$

2-Methylpropene | Hydrogen bromide | 2-Bromo-2-methylpropane (90%)

$$\text{1-Methylcyclopentene} + HCl \xrightarrow{0°C} \text{1-Chloro-1-methylcyclopentane}$$

1-Methylcyclopentene | Hydrogen chloride | 1-Chloro-1-methylcyclopentane (100%)

PROBLEM 6.3 Write the structure of the major organic product formed in the reaction of hydrogen chloride with each of the following:

(*a*) 2-Methyl-2-butene
(*b*) 2-Methyl-1-butene
(*c*) *cis*-2-butene
(*d*) $CH_3CH{=}$(cyclohexylidene)

SAMPLE SOLUTION (*a*) Hydrogen chloride adds to the double bond of 2-methyl-2-butene in accordance with Markovnikov's rule. The proton adds to the carbon that has one hydrogen substituent, chlorine to the carbon that has none.

$$(H_3C)_2C{=}C(H)CH_3$$

2-Methyl-2-butene

Chlorine becomes attached to this carbon

Hydrogen becomes attached to this carbon

$$CH_3{-}C(CH_3)(Cl){-}CH_2CH_3$$

2-Chloro-2-methylbutane

(major product from Markovnikov addition of hydrogen chloride to 2-methyl-2-butene)

Markovnikov's rule, like Zaitsev's, organizes experimental observations in a form suitable for predicting the major product of a reaction. The reasons why it works appear when we examine the mechanism of electrophilic addition in more detail.

6.6 MECHANISTIC BASIS FOR MARKOVNIKOV'S RULE

In the reaction of a hydrogen halide HX with an unsymmetrically substituted alkene $RCH{=}CH_2$, let us compare the carbocation intermediates for addition of HX according to Markovnikov's rule and opposite to Markovnikov's rule.

Addition according to Markovnikov's rule:

$$RCH{=}CH_2 + H{-}X \longrightarrow R\overset{+}{C}H{-}CH_3 + X^- \longrightarrow RCHCH_3 \text{ (with X on C-2)}$$

Secondary carbocation; Halide ion; Observed product

Addition opposite to Markovnikov's rule:

$$RCH{=}CH_2 + X{-}H \longrightarrow RCH(H){-}\overset{+}{C}H_2 + X^- \longrightarrow RCH_2CH_2X$$

Primary carbocation; Halide ion; Not formed

The transition state for proton transfer to the double bond has much of the character of a carbocation, and the activation energy for formation of the more stable carbocation (secondary) is less than that for formation of the less stable (primary) one. Figure 6.4 is a potential energy diagram illustrating these two competing modes of hydrogen halide addition to an unsymmetrical alkene. Both carbocations are rapidly captured by X^- to give an alkyl halide, with the major product derived from the carbocation that is formed faster. The energy difference between a primary carbocation and a secondary carbocation is so great and their rates of formation are so different that essentially all the product is derived from the secondary carbocation. Markovnikov's rule holds because addition of a proton to the doubly bonded carbon that already has the greater number of hydrogen substituents produces the more stable carbocation intermediate.

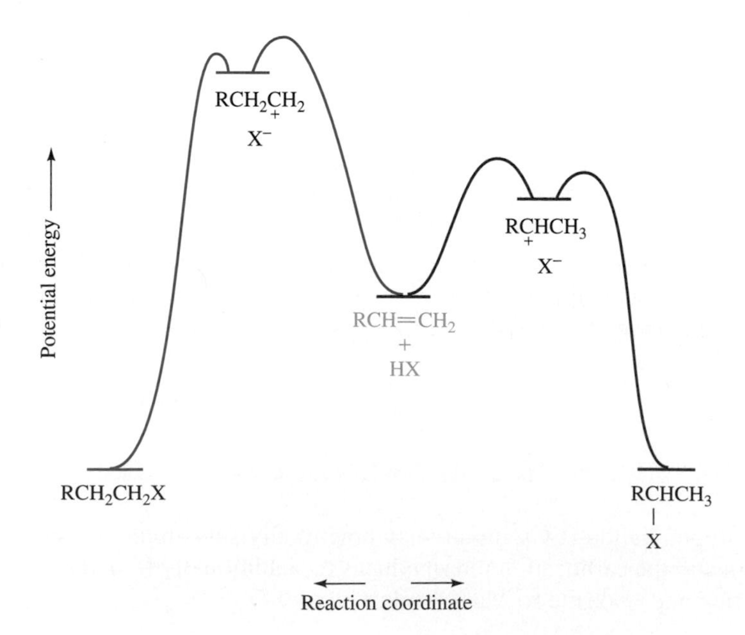

FIGURE 6.4 Energy diagram comparing addition of a hydrogen halide to an alkene in the direction corresponding to Markovnikov's rule and in the direction opposite to Markovnikov's rule. The alkene and hydrogen halide are shown in the center of the diagram. The lower-energy pathway that corresponds to Markovnikov's rule proceeds to the right and is shown in red; the higher-energy pathway proceeds to the left and is shown in blue.

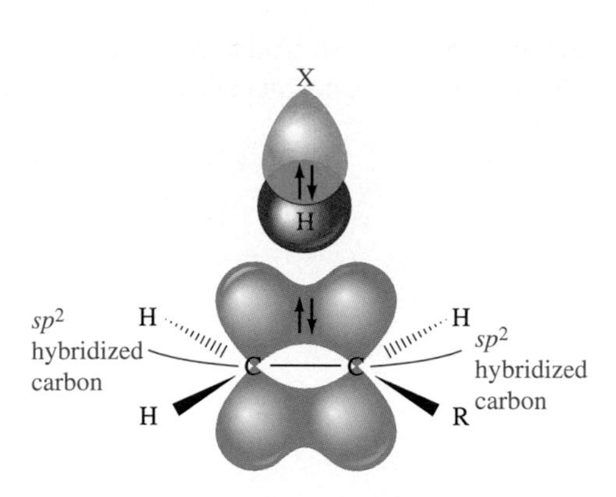

(a) The hydrogen halide (HX) and the alkene ($CH_2{=}CHR$) approach each other. The electrophile is the hydrogen halide, and the site of electrophilic attack is the orbital containing the π electrons of the double bond.

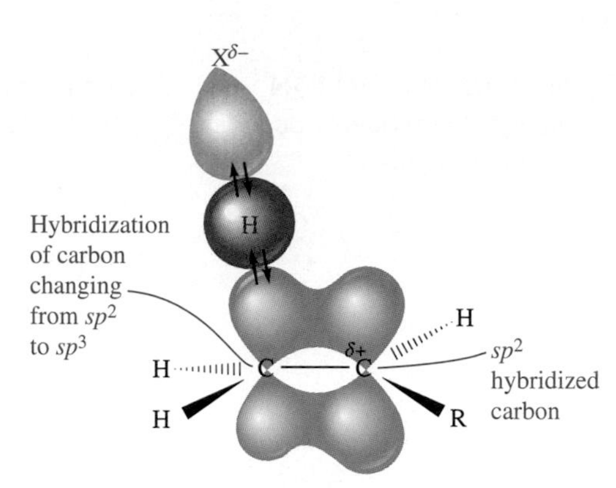

(b) Electrons flow from the π orbital of the alkene to the hydrogen halide. The π electrons flow in the direction that generates a partial positive charge on the carbon atom that bears the electron-releasing alkyl group (R). The hydrogen-halogen bond is partially broken and a C—H σ bond is partially formed at the transition state.

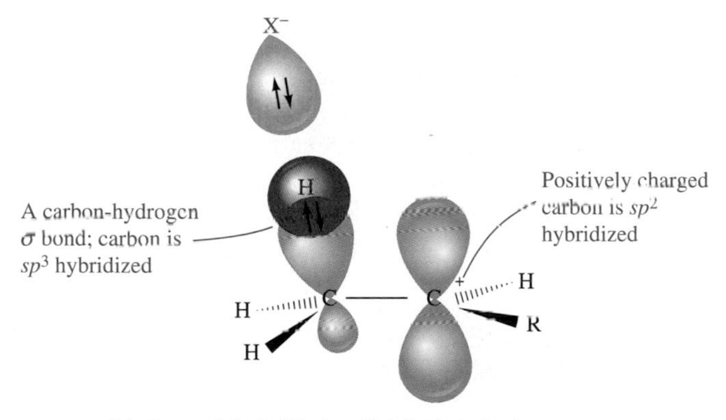

(c) Loss of the halide ion (X^-) from the hydrogen halide and C—H σ bond formation complete the formation of the more stable carbocation intermediate $CH_3\overset{+}{C}HR$.

FIGURE 6.5 Description of electron flow and orbital interactions in the transfer of a proton from a hydrogen halide to an alkene of the type $CH_2{=}CHR$.

Figure 6.5 focuses on the orbitals involved in proton transfer from a hydrogen halide to an alkene of the type $RCH{=}CH_2$, and shows how the π electrons of the alkene flow in the direction that generates the more stable of the two possible carbocations.

PROBLEM 6.4 Give a structural formula for the carbocation intermediate that leads to the major product in each of the reactions of Problem 6.3 (Section 6.5).

SAMPLE SOLUTION (*a*) Proton transfer to the carbon-carbon double bond of 2-methyl-2-butene can occur in a direction that leads to a tertiary carbocation or to a secondary carbocation.

$$\mathrm{(H_3C)_2\overset{2}{C}{=}\overset{3}{C}H\overset{4}{C}H_3}$$

2-Methyl-2-butene

Protonation of C-3 (faster) → Tertiary carbocation: $(H_3C)_2\overset{+}{C}-CH_2CH_3$

Protonation of C-2 (slower) → Secondary carbocation: $(CH_3)_2CH-\overset{+}{C}H-CH_3$

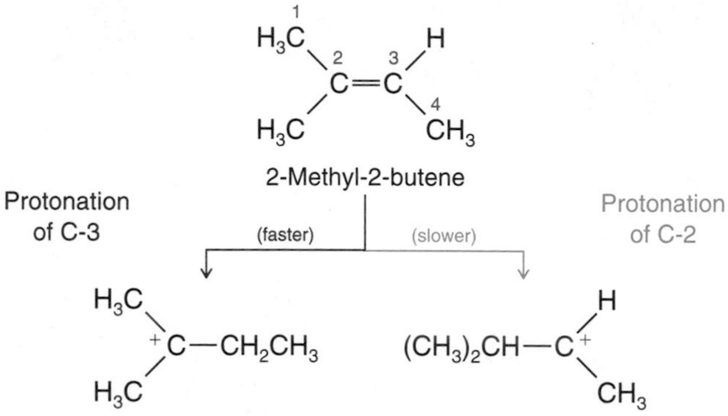

Tertiary carbocation | Secondary carbocation

The product of the reaction is derived from the more stable carbocation—in this case, it is a tertiary carbocation that is formed more rapidly than a secondary one.

In general, alkyl substituents on the double bond increase the reactivity of alkenes toward electrophilic addition. Alkyl groups are electron-releasing, and the more *electron-rich* a double bond is the better it can donate its π electrons to an electrophilic reagent. Along with the observed regioselectivity of addition, this supports the idea that carbocation formation, rather than carbocation capture, is rate-determining.

6.7 CARBOCATION REARRANGEMENTS IN HYDROGEN HALIDE ADDITION TO ALKENES

Our belief that carbocations are intermediates in the addition of hydrogen halides to alkenes is strengthened by the observation that rearrangements of the carbon skeleton analogous to those described in Section 5.13 sometimes occur. For example, the reaction of hydrogen chloride with 3-methyl-1-butene is expected to produce 2-chloro-3-methylbutane. Instead, a mixture of 2-chloro-3-methylbutane and 2-chloro-2-methylbutane results.

$$CH_2{=}CHCH(CH_3)_2 \xrightarrow[0^\circ C]{HCl} CH_3\underset{\large Cl}{\underset{|}{C}}HCH(CH_3)_2 + CH_3CH_2\underset{\large Cl}{\underset{|}{C}}(CH_3)_2$$

3-Methyl-1-butene — 2-Chloro-3-methylbutane (40%) — 2-Chloro-2-methylbutane (60%)

Addition begins in the usual way, by protonation of the double bond to give, in this case, a secondary carbocation. This carbocation can be captured by chloride to give 2-chloro-3-methylbutane (40%), or rearrange by way of a hydride shift to give a tertiary carbocation. The tertiary carbocation reacts with chloride ion to give 2-chloro-2-methylbutane (60%).

$$CH_3\overset{+}{C}H-\underset{|}{\overset{}{C}}(CH_3)_2 \xrightarrow{\text{hydride shift}} CH_3\underset{|}{C}H-\overset{+}{C}(CH_3)_2$$

(H on the carbon shown; 1,2-Dimethylpropyl cation (secondary) → 1,1-Dimethylpropyl cation (tertiary))

1,2-Dimethylpropyl cation (secondary) 1,1-Dimethylpropyl cation (tertiary)

The similar yields of the two alkyl chloride products indicate that the rate of attack by chloride on the secondary carbocation and the rate of rearrangement must be very similar.

PROBLEM 6.5 Addition of hydrogen chloride to 3,3-dimethyl-1-butene gives a mixture of two isomeric chlorides in approximately equal amounts. Suggest reasonable structures for these two compounds, and offer a mechanistic explanation for their formation.

6.8 FREE-RADICAL ADDITION OF HYDROGEN BROMIDE TO ALKENES

For a long time the regioselectivity of addition of hydrogen bromide to alkenes was unpredictable. Sometimes addition occurred according to Markovnikov's rule while at other times, seemingly under the same conditions, the opposite regioselectivity *(anti-Markovnikov addition)* was observed. In 1929 Morris S. Kharasch and his students at the University of Chicago began a systematic investigation of this puzzle. After hundreds of experiments, Kharasch concluded that anti-Markovnikov addition occurred when peroxides, i.e., organic compounds of the type ROOR, were present in the reaction mixture. He and his colleagues found, for example, that carefully purified 1-butene reacted with hydrogen bromide to give only 2-bromobutane—the product expected on the basis of Markovnikov's rule.

$$CH_2{=}CHCH_2CH_3 + HBr \xrightarrow[\text{}]{\text{no peroxides}} CH_3\underset{\underset{Br}{|}}{C}HCH_2CH_3$$

1-Butene Hydrogen bromide 2-Bromobutane (only product; 90% yield)

On the other hand, when the same reaction was performed in the presence of an added peroxide, only 1-bromobutane was formed.

$$CH_2{=}CHCH_2CH_3 + HBr \xrightarrow{\text{peroxides}} BrCH_2CH_2CH_2CH_3$$

1-Butene Hydrogen bromide 1-Bromobutane (only product; 95% yield)

Kharasch termed this phenomenon the **peroxide effect** and reasoned that it could occur even if peroxides were not deliberately added to the reaction mixture. Unless alkenes are protected from atmospheric oxygen, they become contaminated with small amounts of alkyl hydroperoxides, compounds of the type ROOH. These alkyl hydroperoxides act in the same way as deliberately added peroxides to promote addition in the direction opposite to that predicted by Markovnikov's rule.

PROBLEM 6.6 Kharasch's earliest studies in this area were carried out in collaboration with Frank R. Mayo. Mayo performed over 400 experiments in which allyl bromide (3-bromo-1-propene) was treated with hydrogen bromide under a variety of conditions, and determined the distribution of the "normal" and "abnormal" products formed during the reaction. What two products were formed? Which is the product of addition in accordance with Markovnikov's rule? Which one corresponds to addition opposite to the rule?

Kharasch proposed that hydrogen bromide can add to alkenes by two different mechanisms, both of which are, in modern terminology, regiospecific. The first mechanism is the one we discussed in the preceding section, electrophilic addition, and follows Markovnikov's rule. It is the mechanism followed when care is taken to ensure that no peroxides are present. The second mechanism is a free-radical chain process, initiated by homolytic cleavage of the weak oxygen-oxygen bond of a peroxide. It is presented in Figure 6.6.

Peroxides are *initiators* in this process; they are not incorporated into the product but act as a source of radicals necessary to get the chain reaction started. The oxygen-oxygen bond dissociation energies of most peroxides are in the 140 to 200 kJ/mol (35 to 50 kcal/mol) range, and the free-radical addition of hydrogen bromide to alkenes begins when a peroxide molecule undergoes homolytic cleavage to two alkoxy radicals. This is depicted in step 1 of Figure 6.6. A bromine atom is generated in step 2 when one of these alkoxy radicals abstracts a proton from hydrogen bromide. Once a bromine atom becomes available, the propagation phase of the chain reaction begins. In the propagation phase as shown in step 3, a bromine atom adds to the alkene in the direction that produces the more stable alkyl radical.

The addition of a bromine atom to C-1 gives a secondary alkyl radical.

$$\overset{4}{C}H_3\overset{3}{C}H_2\overset{2}{C}H{=}\overset{1}{C}H_2 + :\dot{\ddot{Br}}: \longrightarrow CH_3CH_2\dot{C}H{-}CH_2{-}\ddot{Br}:$$

Secondary alkyl radical

The addition of a bromine atom to C-2 gives a primary alkyl radical.

$$\overset{4}{C}H_3\overset{3}{C}H_2\overset{2}{C}H{=}\overset{1}{C}H_2 + :\dot{\ddot{Br}}: \longrightarrow CH_3CH_2CH(\ddot{Br}:){-}\dot{C}H_2$$

Primary alkyl radical

A secondary alkyl radical is more stable than a primary radical. Therefore, bromine adds to C-1 of 1-butene faster than it adds to C-2. Once the bromine atom has added to the double bond, the regioselectivity of addition is set. The alkyl radical then abstracts a hydrogen atom from hydrogen bromide to give the alkyl bromide product as shown in step 4 of Figure 6.6.

The regioselectivity of addition of hydrogen bromide to alkenes under normal (ionic addition) conditions is controlled by the tendency of a proton to add to the double bond so as to produce the more stable carbocation. Under free-radical conditions the regioselectivity is governed by addition of a bromine atom to give the more stable alkyl radical.

The overall reaction:

$$CH_3CH_2CH{=}CH_2 + HBr \xrightarrow[\text{light or heat}]{\text{ROOR}} CH_3CH_2CH_2CH_2Br$$

1-Butene, Hydrogen bromide, 1-Bromobutane

The mechanism:

(*a*) Initiation

Step 1: Dissociation of a peroxide into two alkoxy radicals:

$$R\ddot{O}:\ddot{O}R \xrightarrow[\text{heat}]{\text{light or}} 2R\ddot{O}\cdot$$

Peroxide, Two alkoxy radicals

Step 2: Hydrogen atom abstraction from hydrogen bromide by an alkoxy radical:

$$R\ddot{O}\cdot \quad H:\ddot{Br}: \longrightarrow R\ddot{O}:H + \cdot\ddot{Br}:$$

Alkoxy radical, Hydrogen bromide, Alcohol, Bromine atom

(*b*) Chain propagation

Step 3: Addition of a bromine atom to the alkene:

$$CH_3CH_2CH{=}CH_2 \quad \cdot\ddot{Br}: \longrightarrow CH_3CH_2\dot{C}H{-}CH_2:\ddot{Br}:$$

1-Butene, Bromine atom, (1-Bromomethyl)propyl radical

Step 4: Abstraction of a hydrogen atom from hydrogen bromide by the free radical formed in step 3:

$$CH_3CH_2\dot{C}H{-}CH_2Br \quad H:\ddot{Br}: \longrightarrow CH_3CH_2CH_2CH_2Br + \cdot\ddot{Br}:$$

(1-Bromomethyl)propyl radical, Hydrogen bromide, 1-Bromobutane, Bromine atom

FIGURE 6.6 The initiation and propagation steps in the free-radical addition of hydrogen bromide to 1-butene.

Another way that the free-radical chain reaction may be initiated is photochemically, either with or without added peroxides:

$$\text{(cyclopentane)}{=}CH_2 + HBr \xrightarrow{h\nu} \text{(cyclopentane)}(H)CH_2Br$$

Methylenecyclopentane, Hydrogen bromide, (Bromomethyl)cyclopentane (60%)

Among the hydrogen halides, only hydrogen bromide reacts with alkenes rapidly by both an ionic and a free-radical mechanism. Hydrogen iodide and hydrogen chloride always add to alkenes by an ionic mechanism in accordance with Markovnikov's rule. Hydrogen bromide normally reacts by the ionic mechanism, but if peroxides are present or if the reaction is initiated photochemically, the free-radical mechanism is followed.

PROBLEM 6.7 Give the major organic product formed when hydrogen bromide reacts with each of the alkenes in Problem 6.3 in the absence of peroxides and in their presence.

SAMPLE SOLUTION (*a*) The addition of hydrogen bromide in the absence of peroxides exhibits a regioselectivity just like that of hydrogen chloride addition; Markovnikov's rule is followed.

$$(H_3C)_2C{=}CHCH_3 \;+\; HBr \xrightarrow{\text{no peroxides}} CH_3{-}\underset{Br}{\overset{CH_3}{C}}{-}CH_2CH_3$$

2-Methyl-2-butene   Hydrogen bromide   2-Bromo-2-methylbutane

Under free-radical conditions in the presence of peroxides, addition takes place with a regioselectivity opposite to that of Markovnikov's rule.

$$(H_3C)_2C{=}CHCH_3 \;+\; HBr \xrightarrow{\text{peroxides}} CH_3{-}\underset{H}{\overset{CH_3}{C}}{-}\underset{Br}{C}HCH_3$$

2-Methyl-2-butene   Hydrogen bromide   2-Bromo-3-methylbutane

---

While the possibility of having two different reaction paths available to an alkene and hydrogen bromide may seem like a complication, it can be advantageous in organic synthesis. From a single alkene one may prepare either of two different alkyl bromides, with control of regioselectivity, simply by choosing reaction conditions that favor ionic addition or free-radical addition of hydrogen bromide.

6.9 ADDITION OF SULFURIC ACID TO ALKENES

Acids other than hydrogen halides also add to the carbon-carbon bond of alkenes. Concentrated sulfuric acid, for example, reacts with certain alkenes to form alkyl hydrogen sulfates.

$$\gt C{=}C\lt \;+\; H{-}OSO_2OH \longrightarrow H{-}\overset{|}{\underset{|}{C}}{-}\overset{|}{\underset{|}{C}}{-}OSO_2OH$$

Alkene   Sulfuric acid   Alkyl hydrogen sulfate

Notice that a proton adds to the carbon that has the greater number of hydrogen substituents and the hydrogen sulfate anion ($^-OSO_2OH$) adds to the carbon that has the fewer hydrogen substituents.

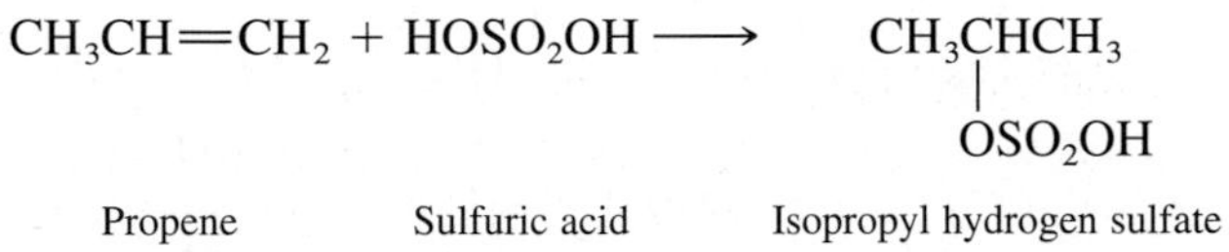

$$CH_3CH{=}CH_2 + HOSO_2OH \longrightarrow CH_3\underset{\substack{|\\ OSO_2OH}}{C}HCH_3$$

Propene | Sulfuric acid | Isopropyl hydrogen sulfate

Markovnikov's rule is obeyed because the mechanism of sulfuric acid addition to alkenes, illustrated for the case of propene in Figure 6.7, is analogous to that described earlier for the ionic addition of hydrogen halides.

Alkyl hydrogen sulfates are converted to alcohols by heating them with water or steam. This is called a **hydrolysis** reaction, because a bond is cleaved by reaction with water. (The suffix *-lysis* indicates cleavage.) It is the oxygen-sulfur bond that is broken when an alkyl hydrogen sulfate undergoes hydrolysis.

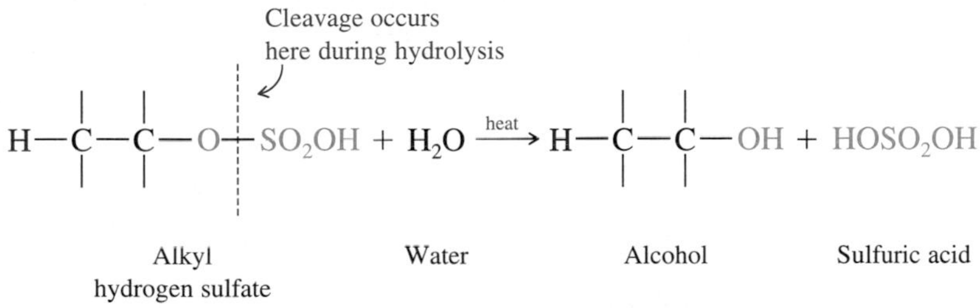

Cleavage occurs here during hydrolysis

$$H{-}\overset{|}{\underset{|}{C}}{-}\overset{|}{\underset{|}{C}}{-}O{-}SO_2OH + H_2O \xrightarrow{heat} H{-}\overset{|}{\underset{|}{C}}{-}\overset{|}{\underset{|}{C}}{-}OH + HOSO_2OH$$

Alkyl hydrogen sulfate | Water | Alcohol | Sulfuric acid

The combination of sulfuric acid addition to alkenes with hydrolysis of the resulting alkyl hydrogen sulfate is an important industrial process for preparation of ethyl alcohol and isopropyl alcohol.

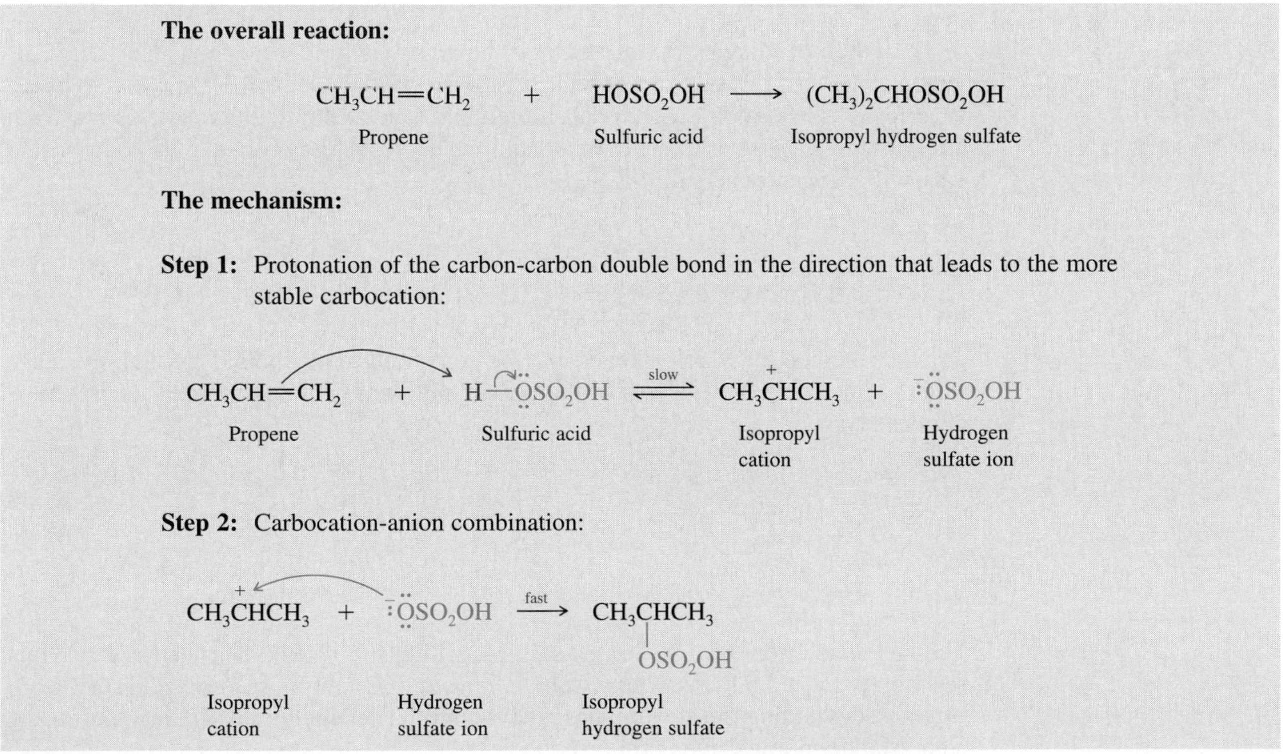

The overall reaction:

$$CH_3CH{=}CH_2 + HOSO_2OH \longrightarrow (CH_3)_2CHOSO_2OH$$

Propene | Sulfuric acid | Isopropyl hydrogen sulfate

The mechanism:

Step 1: Protonation of the carbon-carbon double bond in the direction that leads to the more stable carbocation:

$$CH_3CH{=}CH_2 + H{-}\ddot{O}SO_2OH \xrightleftharpoons{slow} CH_3\overset{+}{C}HCH_3 + {}^{-}{:}\ddot{O}SO_2OH$$

Propene | Sulfuric acid | Isopropyl cation | Hydrogen sulfate ion

Step 2: Carbocation-anion combination:

$$CH_3\overset{+}{C}HCH_3 + {}^{-}{:}\ddot{O}SO_2OH \xrightarrow{fast} CH_3\underset{\substack{|\\ OSO_2OH}}{C}HCH_3$$

Isopropyl cation | Hydrogen sulfate ion | Isopropyl hydrogen sulfate

FIGURE 6.7 Sequence of steps that describes the mechanism for addition of sulfuric acid to propene.

$$\underset{\text{Ethylene}}{CH_2{=}CH_2} \xrightarrow{H_2SO_4} \underset{\text{Ethyl hydrogen sulfate}}{CH_3CH_2OSO_2OH} \xrightarrow[\text{heat}]{H_2O} \underset{\text{Ethyl alcohol}}{CH_3CH_2OH}$$

$$\underset{\text{Propene}}{CH_3CH{=}CH_2} \xrightarrow{H_2SO_4} \underset{\text{Isopropyl hydrogen sulfate}}{CH_3\underset{OSO_2OH}{\underset{|}{C}}HCH_3} \xrightarrow[\text{heat}]{H_2O} \underset{\text{Isopropyl alcohol}}{CH_3\underset{OH}{\underset{|}{C}}HCH_3}$$

It is convenient in synthetic transformations involving more than one step simply to list all the reagents with a single arrow. Individual synthetic steps are indicated by number. Numbering the individual steps is essential so as to avoid the implication that everything is added to the reaction mixture at the same time.

We say that ethylene and propene have undergone **hydration.** In effect, a molecule of water has been added across the carbon-carbon double bond by the combination of reactions shown. In the same manner, cyclohexanol has been prepared from cyclohexene:

$$\text{Cyclohexene} \xrightarrow[\text{2. } H_2O\text{, heat}]{\text{1. } H_2SO_4} \text{Cyclohexanol (75\%)}$$

PROBLEM 6.8 Write a structural formula for the compound formed on electrophilic addition of sulfuric acid to cyclohexene (step 1 in the two-step transformation shown in the preceding equation).

Hydration of alkenes by this method, however, is limited to monosubstituted alkenes and disubstituted alkenes of the type $RCH{=}CHR$. Disubstituted alkenes of the type $R_2C{=}CH_2$, along with trisubstituted and tetrasubstituted alkenes, do not form alkyl hydrogen sulfates under these conditions but instead react in a more complicated way with concentrated sulfuric acid (to be discussed in Section 6.21).

6.10 ACID-CATALYZED HYDRATION OF ALKENES

Another method by which alkenes may be converted to alcohols is through the addition of a molecule of water across the carbon-carbon double bond under conditions of acid catalysis.

$$\underset{\text{Alkene}}{\gt C{=}C\lt} + \underset{\text{Water}}{HOH} \xrightarrow{H^+} \underset{\text{Alcohol}}{H{-}\overset{|}{\underset{|}{C}}{-}\overset{|}{\underset{|}{C}}{-}OH}$$

Unlike the addition of concentrated sulfuric acid to form alkyl hydrogen sulfates, this reaction is carried out in a *dilute acid* medium. A 50% water–sulfuric acid solution is often used, yielding the alcohol directly without the necessity of a separate hydrolysis step. Markovnikov's rule is followed:

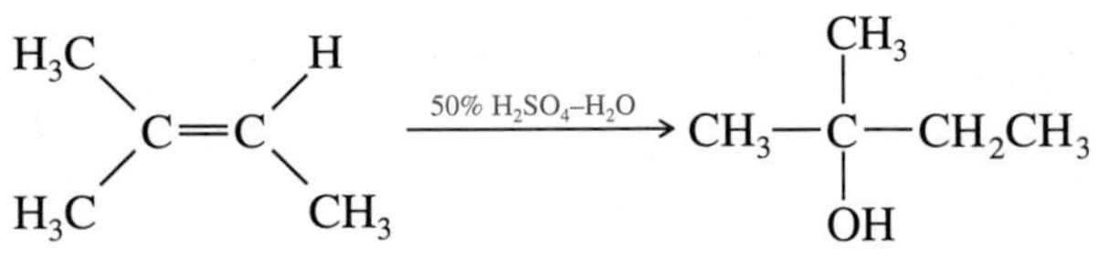

$$(CH_3)_2C{=}CHCH_3 \xrightarrow{50\%\ H_2SO_4\text{–}H_2O} CH_3{-}C(CH_3)(OH){-}CH_2CH_3$$

2-Methyl-2-butene → 2-Methyl-2-butanol (90%)

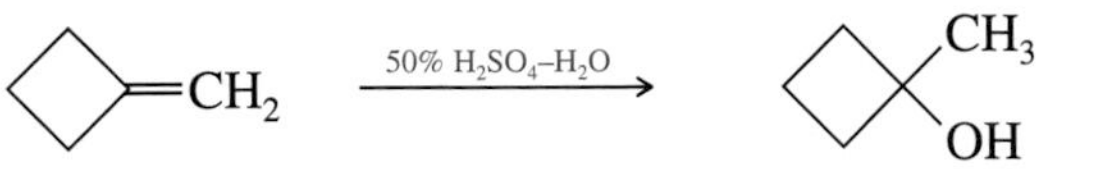

$$\text{Methylenecyclobutane} \xrightarrow{50\%\ H_2SO_4\text{–}H_2O} \text{1-Methylcyclobutanol (80\%)}$$

Methylenecyclobutane → 1-Methylcyclobutanol (80%)

Figure 6.8 extends the general principles of electrophilic addition to alkenes to their acid-catalyzed hydration. In the example cited in Figure 6.8, proton transfer to 2-methylpropene forms *tert*-butyl cation in the first step. This is followed in step 2 by reaction of the carbocation intermediate with a molecule of water acting as a nucleophile. The product of nucleophilic capture of the carbocation by water is an oxonium ion. The oxonium ion is simply the conjugate acid of *tert*-butyl alcohol. Deprotonation of the oxonium ion in step 3 yields the alcohol product and regenerates the acid catalyst.

PROBLEM 6.9 Instead of the three-step mechanism of Figure 6.8, the following two-step mechanism might be considered:

1. $(CH_3)_2C{=}CH_2 + H_3O^+ \xrightarrow{\text{slow}} (CH_3)_3C^+ + H_2O$
2. $(CH_3)_3C^+ + HO^- \xrightarrow{\text{fast}} (CH_3)_3COH$

This mechanism cannot be correct! What is its fundamental flaw?

The proposal that carbocation formation is rate-determining follows from observations of how the reaction rate is affected by the structure of the alkene. Table 6.2 cites some data showing that alkenes that yield relatively stable carbocations react faster than those that yield less stable carbocations. Protonation of ethylene, the least reactive alkene in the table, yields a primary carbocation; protonation of 2-methylpropene, the most reactive in the table, yields a tertiary carbocation. As we have seen on other occasions, the more stable the carbocation, the faster is its rate of formation.

PROBLEM 6.10 The rates of hydration of the two alkenes shown differ by a factor of over 7000 at 25°C. Which isomer is the more reactive? Why?

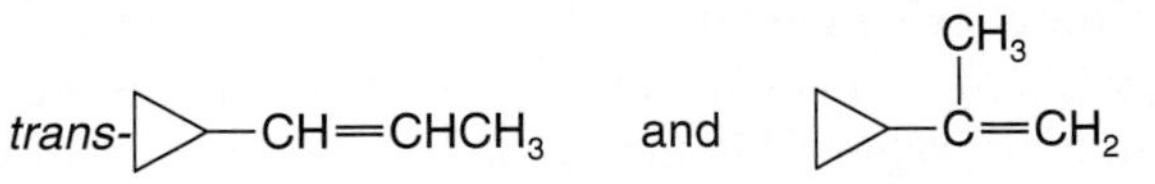

trans-(cyclopropyl)$-CH{=}CHCH_3$ and (cyclopropyl)$-C(CH_3){=}CH_2$

You may have noticed that the acid-catalyzed hydration of an alkene and the acid-catalyzed dehydration of an alcohol are the reverse of each other.

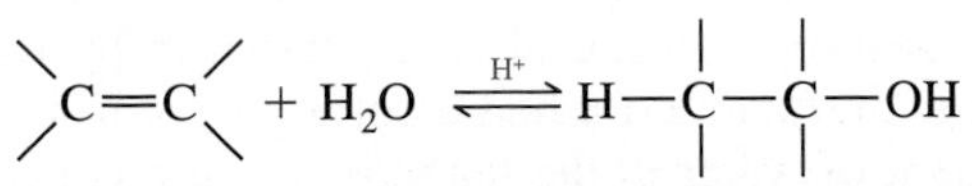

$$\text{C}{=}\text{C} + H_2O \underset{}{\overset{H^+}{\rightleftharpoons}} H{-}\text{C}{-}\text{C}{-}OH$$

Alkene + Water ⇌ Alcohol

The overall reaction:

$$(CH_3)_2C{=}CH_2 \quad + \quad H_2O \xrightarrow{H_3O^+} (CH_3)_3COH$$

2-Methylpropene — Water — *tert*-Butyl alcohol

The mechanism:

Step 1: Protonation of the carbon-carbon double bond in the direction that leads to the more stable carbocation:

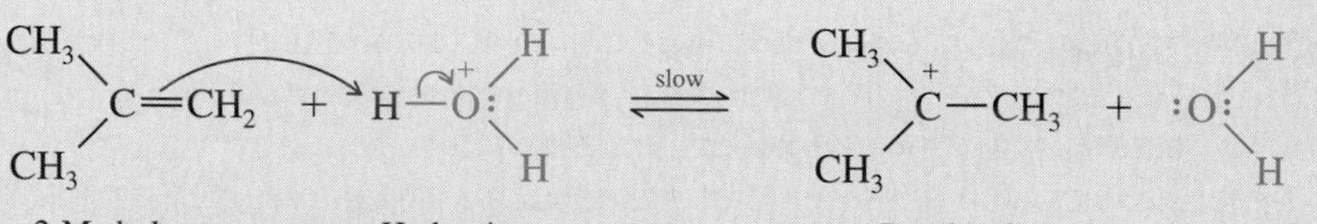

$$(CH_3)_2C{=}CH_2 + H{-}\overset{+}{O}H_2 \xrightleftharpoons{\text{slow}} (CH_3)_2\overset{+}{C}{-}CH_3 + :\ddot{O}H_2$$

2-Methylpropene — Hydronium ion — *tert*-Butyl cation — Water

Step 2: Water acts as a nucleophile to capture *tert*-butyl cation:

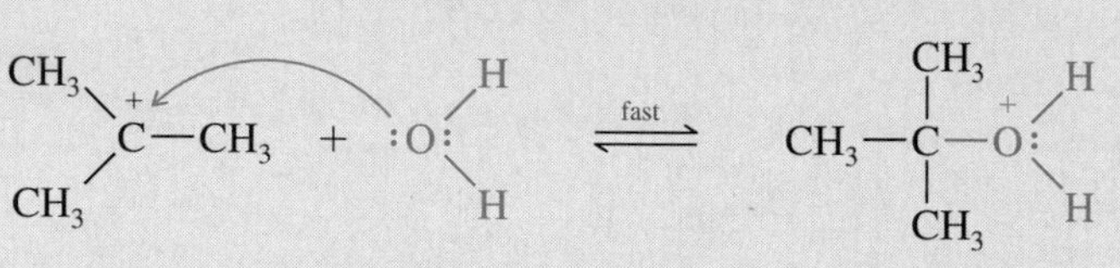

$$(CH_3)_2\overset{+}{C}{-}CH_3 + :\ddot{O}H_2 \xrightleftharpoons{\text{fast}} (CH_3)_3C{-}\overset{+}{O}H_2$$

tert-Butyl cation — Water — *tert*-Butyloxonium ion

Step 3: Deprotonation of *tert*-butyloxonium ion. Water acts as a Brønsted base:

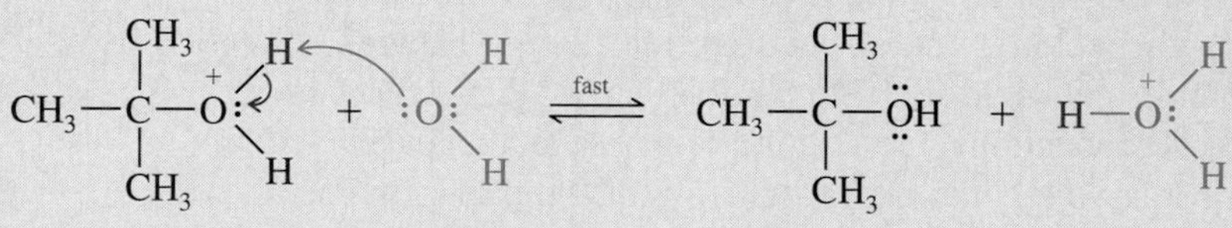

$$(CH_3)_3C{-}\overset{+}{O}H_2 + :\ddot{O}H_2 \xrightleftharpoons{\text{fast}} (CH_3)_3C{-}\ddot{O}H + H{-}\overset{+}{O}H_2$$

tert-Butyloxonium ion — Water — *tert*-Butyl alcohol — Hydronium ion

FIGURE 6.8 Sequence of steps describing the mechanism of acid-catalyzed hydration of 2-methylpropene.

According to **Le Châtelier's principle,** *a system at equilibrium adjusts so as to minimize any stress applied to it.* When the concentration of water is increased, the system responds by consuming water. This means that proportionally more alkene is converted to alcohol; the position of equilibrium shifts to the right. Thus, when we wish to prepare an alcohol from an alkene, we employ a reaction medium in which the molar concentration of water is high—dilute sulfuric acid, for example.

On the other hand, alkene formation is favored when the concentration of water is kept low. The system responds to the absence of water by causing more alcohol molecules to suffer dehydration, and when alcohol molecules dehydrate, they form more alkene. The amount of water in the reaction mixture is kept low by using concentrated strong acids as catalysts. Distilling the reaction mixture is an effective way of removing water as it is formed, causing the equilibrium to shift toward products. If the

TABLE 6.2
Relative Rates of Acid-Catalyzed Hydration of Some Representative Alkenes

Alkene	Structural formula	Relative rate of acid-catalyzed hydration*
Ethylene	$CH_2{=}CH_2$	1.0
Propene	$CH_3CH{=}CH_2$	1.6×10^6
2-Methylpropene	$(CH_3)_2C{=}CH_2$	2.5×10^{11}

* In water, 25°C.

alkene is low-boiling, it too can be removed by distillation. This offers the additional benefit of protecting the alkene from acid-catalyzed isomerization after it is formed.

In any equilibrium process, the sequence of intermediates and transition states encountered as reactants proceed to products in one direction must also be encountered, and in precisely the reverse order, in the opposite direction. This is called the **principle of microscopic reversibility.** Just as the reaction

$$\underset{\text{2-Methylpropene}}{(CH_3)_2C{=}CH_2} + \underset{\text{Water}}{H_2O} \xrightleftharpoons{H^+} \underset{\text{2-Methyl-2-propanol}}{(CH_3)_3COH}$$

is reversible with respect to reactants and products, so each tiny increment of progress along the reaction coordinate is considered reversible. Once we know the mechanism for the forward phase of a particular reaction, we also know what the intermediates and transition states must be for the reverse phase. In particular, the three-step mechanism for the acid-catalyzed hydration of 2-methylpropene in Figure 6.8 is the reverse of that for the acid-catalyzed dehydration of *tert*-butyl alcohol in Figure 5.7.

PROBLEM 6.11 Is the electrophilic addition of hydrogen chloride to 2-methylpropene the reverse of the E1 or the E2 elimination reaction of *tert*-butyl chloride?

6.11 HYDROBORATION-OXIDATION OF ALKENES

Acid-catalyzed hydration adds the elements of water to alkenes with Markovnikov-rule regioselectivity. Frequently, however, one needs an alcohol having a structure that corresponds to hydration of an alkene with a regioselectivity opposite to that of Markovnikov's rule **(anti-Markovnikov hydration).** The conversion of 1-decene to 1-decanol is an example of such a transformation.

$$\underset{\text{1-Decene}}{CH_3(CH_2)_7CH{=}CH_2} \longrightarrow \underset{\text{1-Decanol}}{CH_3(CH_2)_7CH_2CH_2OH}$$

The synthetic procedure that allows anti-Markovnikov hydration of alkenes is an indirect one, known as **hydroboration-oxidation.** It was developed by Professor Herbert C. Brown and his coworkers at Purdue University in the 1950s as part of a broad program designed to apply boron-containing reagents to organic chemical synthesis. The number of applications is so large (hydroboration-oxidation is just one of them) and the work so novel that Brown was a corecipient of the 1979 Nobel Prize in chemistry.

Hydroboration is a reaction in which a boron hydride, a compound of the type R_2BH, adds to a carbon-carbon bond. A new carbon-hydrogen bond and a carbon-boron bond result.

$$\mathrm{>C{=}C<} \quad + \quad R_2BH \longrightarrow \mathrm{H{-}\overset{|}{\underset{|}{C}}{-}\overset{|}{\underset{|}{C}}{-}BR_2}$$

Alkene   Boron hydride   Organoborane

Following hydroboration, the organoborane is oxidized by treatment with hydrogen peroxide in aqueous base. This is the **oxidation** stage of the sequence; hydrogen peroxide is the oxidizing agent, and the organoborane is converted to an alcohol.

With sodium hydroxide as the base, boron of the alkylborane is converted to the water-soluble and easily removed sodium salt of boric acid.

$$\mathrm{H{-}\overset{|}{\underset{|}{C}}{-}\overset{|}{\underset{|}{C}}{-}BR_2} + 3H_2O_2 + 4HO^- \longrightarrow$$

Organoborane   Hydrogen peroxide   Hydroxide ion

$$\mathrm{H{-}\overset{|}{\underset{|}{C}}{-}\overset{|}{\underset{|}{C}}{-}OH} + 2ROH + B(OH)_4^- + 3H_2O$$

Alcohol   Alcohol   Borate ion   Water

The combination of hydroboration and oxidation leads to the overall hydration of an alkene. Notice, however, that water is not a reactant. The hydrogen that becomes bonded to carbon is derived from the organoborane, and the hydroxyl group comes from hydrogen peroxide.

With this as introduction, let us now look at the individual steps in more detail for the specific case of hydroboration-oxidation of 1-decene. The boron hydride used most frequently in the hydroboration of alkenes is *diborane* (B_2H_6). Diborane adds to 1-decene to give tridecylborane according to the balanced equation:

Diglyme, shown in the equation above the arrow, is the customary solvent in which hydroboration reactions are carried out. Diglyme is an acronym for *di*ethylene *gly*col di*m*ethyl *e*ther, and its structure is $CH_3OCH_2CH_2OCH_2CH_2OCH_3$.

$$6CH_3(CH_2)_7CH{=}CH_2 + B_2H_6 \xrightarrow{\text{diglyme}} 2[CH_3(CH_2)_7CH_2CH_2]_3B$$

1-Decene   Diborane   Tridecylborane

There is a pronounced tendency for boron to become bonded to the less substituted carbon of the double bond. Thus, the hydrogen atoms of diborane add to C-2 of 1-decene, and boron to C-1. This is believed to be mainly a steric effect, but the regioselectivity of addition does correspond to Markovnikov's rule in the sense that hydrogen is the negatively polarized atom in a B—H bond and boron the positively polarized one. Oxidation of tridecylborane gives 1-decanol. The net result is the conversion of an alkene to an alcohol with a regioselectivity opposite to that obtainable by acid-catalyzed hydration.

$$[CH_3(CH_2)_7CH_2CH_2]_3B \xrightarrow[\text{NaOH}]{H_2O_2} CH_3(CH_2)_7CH_2CH_2OH$$

Tridecylborane   1-Decanol

It is customary to combine the two stages, hydroboration and oxidation, in a single equation with the operations numbered sequentially above and below the arrow.

$$CH_3(CH_2)_7CH{=}CH_2 \xrightarrow[\text{2. } H_2O_2,\ HO^-]{\text{1. } B_2H_6,\text{ diglyme}} CH_3(CH_2)_7CH_2CH_2OH$$

1-Decene → 1-Decanol (93%)

Another convenient hydroborating agent is the borane-tetrahydrofuran complex (BH_3–THF). It is very reactive, adding to alkenes within minutes at 0°C, and is used in tetrahydrofuran as the solvent.

$H_3\overset{-}{B}—\overset{+}{O}$: (tetrahydrofuran ring)

Borane-tetrahydrofuran complex

$$(CH_3)_2C{=}CHCH_3 \xrightarrow[\text{2. } H_2O_2,\ HO^-]{\text{1. } BH_3\text{–THF}} (CH_3)_2CHCHCH_3 \text{ (OH on C-2)}$$

2-Methyl-2-butene → 3-Mcthyl-2-butanol (98%)

Carbocation intermediates are not involved in hydroboration-oxidation. Hydration of double bonds takes place without rearrangement, even in alkenes as highly branched as the following:

$$\xrightarrow[\text{2. } H_2O_2,\ HO^-]{\text{1. } B_2H_6,\text{ diglyme}}$$

(*E*)-2,2,5,5-Tetramethyl-3-hexene → 2,2,5,5-Tetramethyl-3-hexanol (82%)

PROBLEM 6.12 Write the structure of the major organic product obtained by hydroboration-oxidation of each of the following alkenes:

(*a*) 2-Methylpropene
(*b*) *cis*-2-Butene
(*c*) (cyclobutane ring)$=CH_2$
(*d*) Cyclopentene
(*e*) 3-Ethyl-2-pentene
(*f*) 3-Ethyl-1-pentene

SAMPLE SOLUTION (*a*) In hydroboration-oxidation the elements of water (H and OH) are introduced with a regioselectivity opposite to that of Markovnikov's rule. In the case of 2-methylpropene, this leads to 2-methyl-1-propanol as the product.

$$(CH_3)_2C{=}CH_2 \xrightarrow[\text{2. oxidation}]{\text{1. hydroboration}} (CH_3)_2CH—CH_2OH$$

2-Methylpropene → 2-Methyl-1-propanol

Hydrogen becomes bonded to the carbon that has the fewer hydrogen substituents, hydroxyl to the carbon that has the greater number of hydrogen substituents. We say that hydroboration-oxidation leads to anti-Markovnikov hydration of alkenes.

6.12 STEREOCHEMISTRY OF HYDROBORATION-OXIDATION

A second aspect of hydroboration-oxidation concerns its stereochemistry. As illustrated for the case of 1-methylcyclopentene, the H and OH groups are added to the same face of the double bond.

1-Methylcyclopentene $\xrightarrow[2.\ H_2O_2,\ HO^-]{1.\ B_2H_6,\ \text{diglyme}}$ *trans*-2-Methylcyclopentanol (only product, 86% yield)

Overall, the reaction leads to syn addition of the elements of water to the double bond. This fact has an important bearing on the mechanism of the process.

PROBLEM 6.13 Hydroboration-oxidation of α-pinene (page 216), like catalytic hydrogenation, is stereoselective. Addition takes place at the less hindered face of the double bond, and a single alcohol is produced in high yield (89 percent). Suggest a reasonable structure for this alcohol.

6.13 MECHANISM OF HYDROBORATION-OXIDATION

The regioselectivity and syn stereochemistry of hydroboration-oxidation, coupled with a knowledge of the chemical properties of alkenes and boranes, contribute to our understanding of the reaction mechanism.

Borane (BH_3) does not exist as such under normal conditions of temperature and atmospheric pressure. Two molecules of BH_3 combine to give diborane (B_2H_6), which is the more stable form.

We can consider the hydroboration step as though it involved monomeric BH_3. It makes our mechanistic analysis easier to follow and is at variance with reality only in matters of detail. Borane is an electrophilic species; it has a vacant 2*p* orbital and can accept a pair of electrons into that orbital. The source of this electron pair is the π bond of an alkene. It is believed, as shown in Figure 6.9 for the example of the hydroboration of 1-methylcyclopentene, that the first step produces an unstable intermediate called a *π complex.* In this π complex boron and the two carbon atoms of the double bond are joined by a *three-center two-electron bond,* by which we mean that three atoms share two electrons. Three-center two-electron bonds are frequently encountered in boron chemistry. The π complex is formed by a transfer of electron density from the π orbital of the alkene to the 2*p* orbital of boron. This leaves each carbon of the complex with a small positive charge, while boron is slightly negative. The negative character of boron in this intermediate makes it easy for one of its hydrogen substituents to migrate with a pair of electrons (a hydride shift) from boron to carbon. The transition state for this process is shown in step 2(a) of Figure 6.9; completion of the migration in step 2(b) yields the alkylborane. According to this mechanism, the carbon-boron bond and the carbon-hydrogen bond are formed on the same side of the alkene. The hydroboration step is a syn addition process.

The regioselectivity of addition is consistent with the electron distribution in the π complex. Hydrogen is transferred with a pair of electrons to the carbon atom that can best support a positive charge, namely, the one that bears the methyl group.

Steric effects may be an even more important factor in controlling the regioselectivity of addition. Boron, with its attached substituents, is much larger than a hydrogen atom and becomes bonded to the less crowded carbon of the double bond, while hydrogen becomes bonded to the more crowded carbon.

A second aspect of the electrophilic character of boron is evident when we consider the oxidation of organoboranes. In the oxidation phase of the hydroboration-oxidation sequence, as presented in Figure 6.10, the anion of hydrogen peroxide attacks boron.

Step 1: A molecule of borane (BH_3) attacks the alkene. Electrons flow from the π orbital of the alkene to the $2p$ orbital of boron. A π complex is formed.

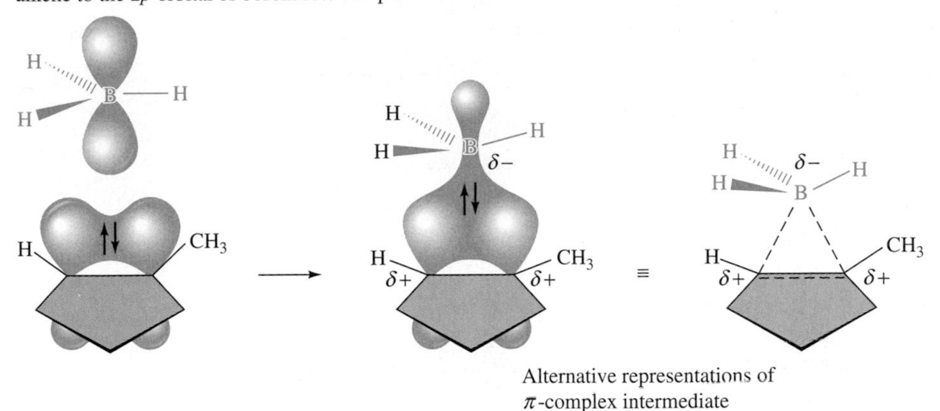

Step 2: The π complex rearranges to an organoborane. Hydrogen migrates from boron to carbon, carrying with it the two electrons in its bond to boron. Development of the transition state for this process is shown in 2(a), and its transformation to the organoborane is shown in 2(b).

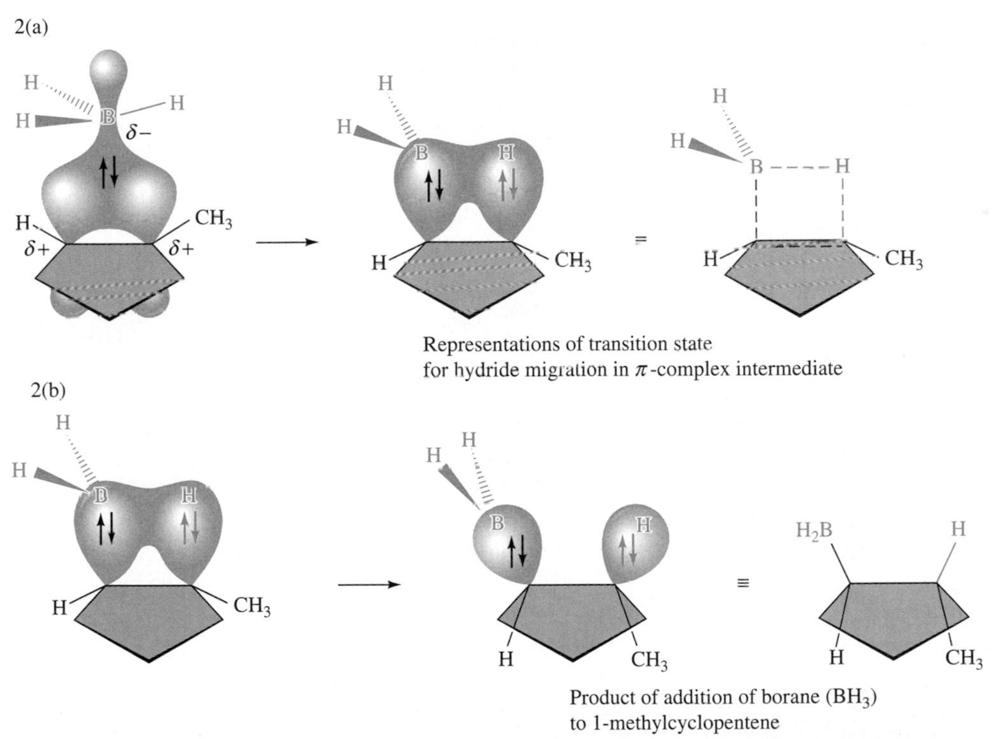

2(a)
Representations of transition state
for hydride migration in π-complex intermediate
2(b)
Product of addition of borane (BH3)
to 1-methylcyclopentene

FIGURE 6.9 Depiction of orbital interactions and electron redistribution in the hydroboration of 1-methylcyclopentene.

Hydroperoxide ion is formed in an acid-base reaction in step 1 and attacks boron in step 2. The empty $2p$ orbital of boron makes it electrophilic and permits nucleophilic reagents such as HOO^- to add to it.

The combination of a negative charge on boron and a weak oxygen-oxygen bond in its hydroperoxy substituent causes an alkyl group to migrate from boron to oxygen in

Step 1: Hydrogen peroxide is converted to its anion in basic solution:

$$H{-}O{-}O{-}H \; + \; {}^{-}OH \; \rightleftharpoons \; H{-}O{-}O^{-} \; + \; H{-}O{-}H$$

Hydrogen peroxide | Hydroxide ion | Hydroperoxide ion | Water

Step 2: Anion of hydrogen peroxide acts as a nucleophile, attacking boron and forming an oxygen-boron bond:

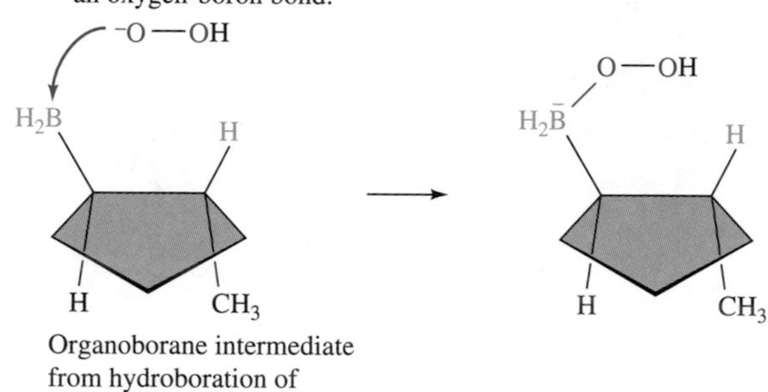

Organoborane intermediate from hydroboration of 1-methylcyclopentene

Step 3: Carbon migrates from boron to oxygen, displacing hydroxide ion. Carbon migrates with the pair of electrons in the carbon-boron bond; these become the electrons in the carbon-oxygen bond:

Representation of transition state for migration of carbon from boron to oxygen

Alkoxyborane

Step 4: Hydrolysis cleaves the boron-oxygen bond, yielding the alcohol:

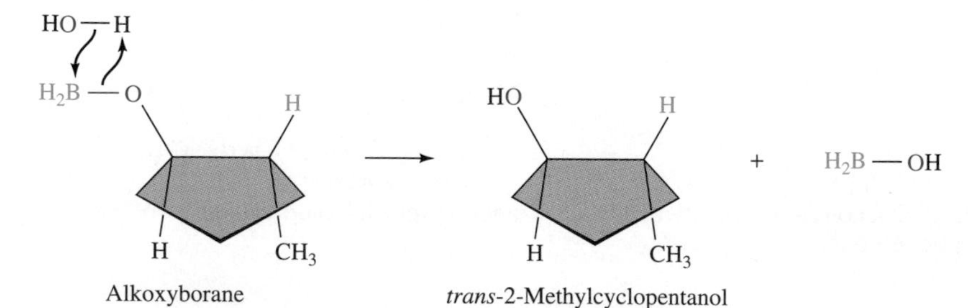

Alkoxyborane | *trans*-2-Methylcyclopentanol

FIGURE 6.10 Mechanistic description of the oxidation phase in the hydroboration-oxidation of 1-methylcyclopentene.

step 3 of the mechanism shown in Figure 6.10. This alkyl group migration occurs with loss of hydroxide ion and is the step in which the critical carbon-oxygen bond is formed. What is especially significant about this alkyl group migration is that the stereochemical orientation of the new carbon-oxygen bond is the same as that of the original carbon-boron bond. This is crucial to the overall syn stereochemistry of the hydroboration-oxidation sequence. Migration of the alkyl group from boron to oxygen is said to have occurred with *retention of configuration* at carbon. The alkoxyborane intermediate formed in step 3 undergoes subsequent base-promoted oxygen-boron bond cleavage in step 4 to give the alcohol product.

The mechanistic complexity of hydroboration-oxidation stands in contrast to the simplicity with which these reactions are carried out experimentally. Both the hydroboration and oxidation steps are extremely rapid reactions and are performed at room temperature with conventional laboratory equipment. Ease of operation, along with the fact that hydroboration-oxidation leads to syn hydration of alkenes and occurs with a regioselectivity contrary to Markovnikov's rule, makes this procedure one of great value to the synthetic chemist.

6.14 ADDITION OF HALOGENS TO ALKENES

In contrast to the free-radical substitution observed when halogens react with *alkanes,* halogens normally react with *alkenes* by electrophilic addition.

$$\mathrm{{>}C{=}C{<}} \;+\; \mathrm{X_2} \longrightarrow \mathrm{X{-}\overset{|}{\underset{|}{C}}{-}\overset{|}{\underset{|}{C}}{-}X}$$

Alkene Halogen Vicinal dihalide

The products of these reactions are called **vicinal** dihalides. Two substituents, in this case the halogen substituents, are vicinal if they are attached to adjacent carbons. The word is derived from the Latin *vicinalis,* which means "neighboring." The halogen is either chlorine (Cl_2) or bromine (Br_2), and addition takes place rapidly at room temperature and below in a variety of solvents, including acetic acid, carbon tetrachloride, chloroform, and dichloromethane.

$$\mathrm{CH_3CH{=}CHCH(CH_3)_2} + \mathrm{Br_2} \xrightarrow[0^\circ C]{CHCl_3} \mathrm{CH_3\underset{Br}{\underset{|}{C}}H{-}\underset{Br}{\underset{|}{C}}HCH(CH_3)_2}$$

4-Methyl-2-pentene Bromine 2,3-Dibromo-4-methylpentane (100%)

Rearrangements do not normally occur, which can mean either of two things. Either carbocations are not intermediates, or if they are, they are captured by a nucleophile faster than they rearrange. We shall see in Section 6.15 that the first of these is believed to be the case.

Fluorine addition to alkenes is a violent reaction, difficult to control, and accompanied by substitution of hydrogens by fluorine (Section 4.16). The addition of I_2 to alkenes is endothermic. Vicinal diiodides tend to lose I_2 and revert to alkenes. Thus, special conditions are normally required for their preparation, and their relative instability makes vicinal diiodides an infrequently encountered class of compounds.

6.15 STEREOCHEMISTRY OF HALOGEN ADDITION

The reaction of chlorine and bromine with cycloalkenes illustrates an important stereochemical feature of halogen addition. Anti addition is observed; the two bromine atoms of Br_2 or the two chlorines of Cl_2 add to opposite faces of the double bond.

Cyclopentene + Br_2 $\xrightarrow{CHCl_3}$ *trans*-1,2-Dibromocyclopentane

Cyclopentene Bromine *trans*-1,2-Dibromocyclopentane (80% yield; none of the cis isomer is formed)

Cyclooctene + Cl_2 $\xrightarrow[-60°C]{CHCl_3}$ *trans*-1,2-Dichlorocyclooctane

Cyclooctene Chlorine *trans*-1,2-Dichlorocyclooctane (73% yield; none of the cis isomer is formed)

These observations must be taken into account when considering the mechanism of halogen addition. They force the conclusion that a simple one-step "bond-switching" process of the following type cannot be correct. A process of this type requires syn addition; it is *not* consistent with the anti addition actually observed.

$$X{-}X \quad C{=}C \longrightarrow X{-}C{-}C{-}X$$

PROBLEM 6.14 The mass 82 isotope of bromine (^{82}Br) is radioactive and is used as a tracer to identify the origin and destination of individual atoms in chemical reactions and biological transformations. A sample of 1,1,2-tribromocyclohexane was prepared by adding $^{82}Br{-}^{82}Br$ to ordinary (nonradioactive) 1-bromocyclohexene. How many of the bromine atoms in the 1,1,2-tribromocyclohexane produced are radioactive? Which ones are they?

6.16 MECHANISM OF HALOGEN ADDITION TO ALKENES. HALONIUM IONS

Many of the features that characterize the generally accepted mechanism for the addition of halogens to alkenes can be introduced by referring to the reaction of ethylene with bromine:

$$CH_2{=}CH_2 + Br_2 \longrightarrow BrCH_2CH_2Br$$

Ethylene Bromine 1,2-Dibromoethane

Until it was banned in the United States in 1984, 1,2-dibromoethane (ethylene dibromide, or EDB) was produced on a large scale for use as a pesticide and soil fumigant.

Neither bromine nor ethylene is a polar molecule, but both are *polarizable,* and an induced dipole–induced dipole force causes them to be mutually attracted to each other. This induced dipole–induced dipole attraction sets the stage for Br_2 to act as an electrophile. Electrons flow from the π system of ethylene to Br_2, causing the weak bromine-bromine bond to break. By analogy to the customary mechanisms for electrophilic addition, we might represent this as the formation of a carbocation in a bimolecular elementary step.

$$CH_2{=}CH_2 + :\ddot{B}r{-}\ddot{B}r: \longrightarrow \overset{+}{C}H_2{-}CH_2{-}\ddot{B}r: + :\ddot{B}r:^-$$

Ethylene (nucleophile) | Bromine (electrophile) | 2-Bromoethyl cation | Bromide ion (leaving group)

A carbocation of the type shown, however, has been demonstrated to be less stable than an alternative structure called a **cyclic bromonium ion,** in which the positive charge resides on bromine, not carbon.

$$\underset{\diagdown\ \overset{+}{:Br:}\ \diagup}{CH_2{-}CH_2}$$

Ethylenebromonium ion

The chief reason why ethylenebromonium ion, in spite of its three-membered ring, is more stable than 2-bromoethyl cation is that all its atoms have octets of electrons, while carbon has only six electrons in the carbocation.

Thus, the mechanism for electrophilic addition of Br_2 to ethylene as presented in Figure 6.11 is characterized by the direct formation of a cyclic bromonium ion as its

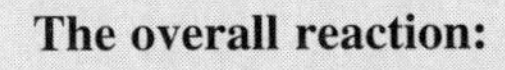

The overall reaction:

$$CH_2{=}CH_2 + Br_2 \longrightarrow BrCH_2CH_2Br$$

Ethylene | Bromine | 1,2-Dibromoethane

The mechanism:

Step 1: Reaction of ethylene and bromine to form a bromonium ion intermediate:

$$CH_2{=}CH_2 + :\ddot{B}r{-}\ddot{B}r: \longrightarrow \underset{\diagdown\ \underset{+}{:Br:}\ \diagup}{CH_2{-}CH_2} + :\ddot{B}r:^-$$

Ethylene | Bromine | Ethylenebromonium ion | Bromide ion

Step 2: Nucleophilic attack of bromide anion on the bromonium ion:

$$:\ddot{B}r:^- + \underset{\diagdown\ \underset{+}{:Br:}\ \diagup}{CH_2{-}CH_2} \longrightarrow :\ddot{B}r{-}CH_2{-}CH_2{-}\ddot{B}r:$$

Bromide ion | Ethylenebromonium ion | 1,2-Dibromoethane

FIGURE 6.11 Mechanistic description of electrophilic addition of bromine to ethylene.

first elementary step. Step 2 is the conversion of the bromonium ion to 1,2-dibromoethane by reaction with bromide ion (Br^-).

The effect of substituents on the rate of addition of bromine to alkenes (Table 6.3) is substantial and consistent with a rate-determining transition state in which electrons flow from the alkene to the halogen. Alkyl groups on the carbon-carbon double bond release electrons, stabilize the transition state for bromonium ion formation, and increase the reaction rate.

Representation of transition state for bromonium ion formation from an alkene and bromine

PROBLEM 6.15 Arrange the compounds 2-methyl-1-butene, 2-methyl-2-butene, and 3-methyl-1-butene in order of decreasing reactivity toward bromine.

Step 2 of the mechanism in Figure 6.11 is a nucleophilic attack by Br^- at one of the carbons of the cyclic bromonium ion. For reasons that will be explained in Chapter 8, reactions of this type normally take place via a transition state in which the nucleophile approaches carbon from the side opposite the bond that is to be broken. Recalling that the vicinal dibromide formed from cyclopentene is exclusively the trans stereoisomer, we see that attack by Br^- from the side opposite the C—Br bond of the bromonium ion intermediate can give only *trans*-1,2-dibromocyclopentane in accordance with the experimental observations.

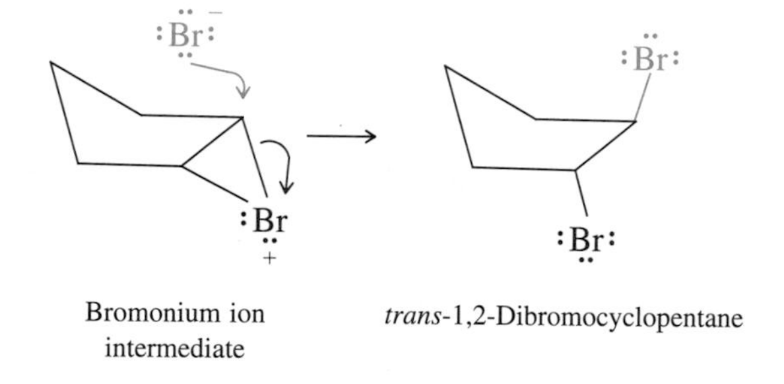

Bromonium ion intermediate

trans-1,2-Dibromocyclopentane

Some supporting evidence is described in the article "The Bromonium Ion," in the August 1963 issue of the *Journal of Chemical Education* (pp. 392–395).

That a cyclic bromonium ion could be an intermediate in bromine addition to alkenes was a novel concept at the time of its proposal in 1937. Much additional

TABLE 6.3
Relative Rates of Reaction of Some Representative Alkenes with Bromine

Alkene	Structural formula	Relative rate of reaction with bromine*
Ethylene	$CH_2{=}CH_2$	1.0
Propene	$CH_3CH{=}CH_2$	61
2-Methylpropene	$(CH_3)_2C{=}CH_2$	5,400
2,3-Dimethyl-2-butene	$(CH_3)_2C{=}C(CH_3)_2$	920,000

* In methanol, 25°C.

evidence, including the isolation and structure determination of a stable cyclic bromonium ion in 1969, has been obtained since then to reinforce the conclusion that the cyclic bromonium ion correctly describes the rate-determining intermediate in these reactions. Similarly, **cyclic chloronium ions** are believed to be involved in the addition of chlorine to alkenes. In the next section we shall see how cyclic chloronium and bromonium ions **(halonium ions)** are intermediates in a second reaction involving alkenes and halogens.

6.17 CONVERSION OF ALKENES TO VICINAL HALOHYDRINS

In *aqueous* solution chlorine, bromine, and iodine react with alkenes to form compounds known as **vicinal halohydrins,** which have a halogen and a hydroxyl group on adjacent carbons.

$$\text{C}=\text{C} + X_2 + H_2O \longrightarrow \text{HO}-\text{C}-\text{C}-\text{X} + \text{HX}$$

Alkene Halogen Water Halohydrin Hydrogen halide

$$CH_2{=}CH_2 + Br_2 \xrightarrow{H_2O} HOCH_2CH_2Br$$

Ethylene Bromine 2-Bromoethanol (70%)

Anti addition occurs. The halogen and the hydroxyl group add to opposite faces of the double bond.

$$\text{Cyclopentene} + Cl_2 \xrightarrow{H_2O} \textit{trans}\text{-2-Chlorocyclopentanol}$$

Cyclopentene Chlorine *trans*-2-Chlorocyclopentanol (52–56% yield; cis isomer not formed)

Halohydrin formation, as depicted in Figure 6.12, is mechanistically related to halogen addition to alkenes. A halonium ion intermediate is formed, which is attacked by water in aqueous solution.

It is convenient to think of the reaction as if it involved addition of a hypohalous acid (HOX) across the double bond when relating the regioselectivity of halohydrin formation to Markovnikov's rule.

$$(CH_3)_2C{=}CH_2 \xrightarrow[H_2O]{Br_2} (CH_3)_2C(OH){-}CH_2Br$$

2-Methylpropene 1-Bromo-2-methyl-2-propanol (77%)

Viewing the reaction as if it involved HOBr is simply a device to mentally connect halohydrin formation and other addition reactions of alkenes. While a small amount of HOBr is formed when bromine is dissolved in water, the principal electrophilic species that attacks the alkene in the first step is Br_2.

Oxygen is more electronegative than bromine (or chlorine, or iodine), and so the $\overset{\delta-}{\text{HO}}-\overset{\delta+}{\text{Br}}$ bond is polarized as indicated. The positively polarized bromine adds to the

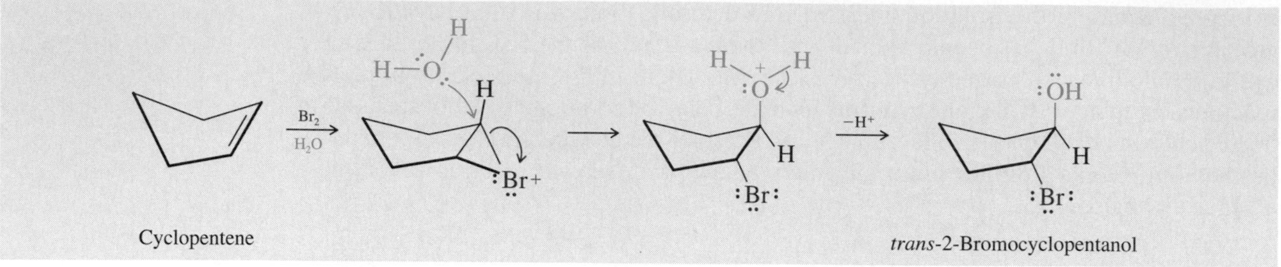

Cyclopentene $\xrightarrow[H_2O]{Br_2}$ ... $\xrightarrow{-H^+}$ *trans*-2-Bromocyclopentanol

FIGURE 6.12 Mechanistic description of bromohydrin formation from cyclopentene. A bridged bromonium ion is formed and is attacked by a water molecule from the side opposite the carbon-bromine bond. The bromine and the hydroxyl group are trans to each other in the product.

carbon atom of the double bond having the greater number of hydrogen substituents, and the negatively polarized oxygen adds to the carbon with the fewer.

PROBLEM 6.16 Give the structure of the product formed when each of the following alkenes reacts with bromine in water:

(*a*) 2-Methyl-1-butene
(*b*) 2-Methyl-2-butene
(*c*) 3-Methyl-1-butene
(*d*) 1-Methylcyclopentene

SAMPLE SOLUTION (*a*) First write the structural formula for 2-methyl-1-butene, and then apply the modified version of Markovnikov's rule just described. Bromine becomes attached to the CH_2 group of the double bond, and the hydroxyl group to C-2.

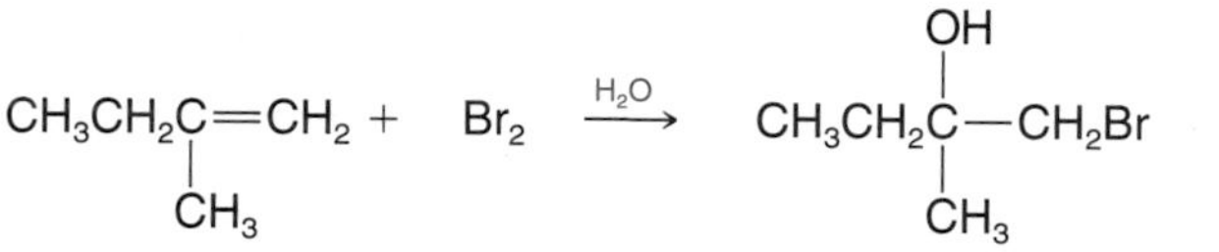

$$CH_3CH_2\underset{\underset{CH_3}{|}}{C}{=}CH_2 + Br_2 \xrightarrow{H_2O} CH_3CH_2\overset{\overset{OH}{|}}{\underset{\underset{CH_3}{|}}{C}}{-}CH_2Br$$

2-Methyl-1-butene Bromine 1-Bromo-2-methyl-2-butanol

A mechanistic explanation for the observed regioselectivity rests on the idea that one carbon-halogen bond in the halonium ion intermediate is weaker than the other. The bridging halogen is less strongly bonded to the carbon atom that can better support a positive charge, i.e., the one with the greater number of electron-releasing alkyl substituents. In resonance terms we say that the bromonium ion formed from 2-methylpropene has some of the character of the three resonance structures A, B, and C.

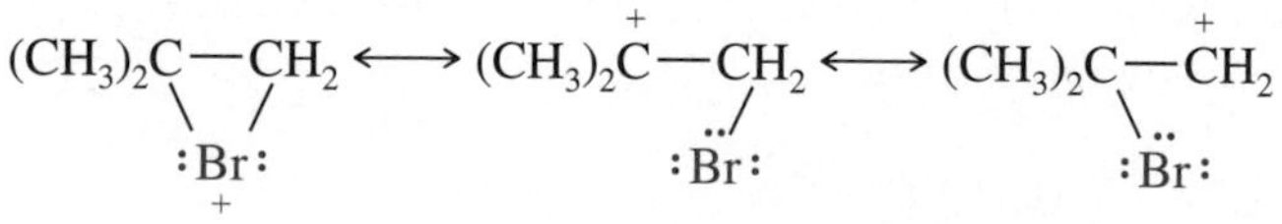

$$(CH_3)_2C{-}CH_2 \text{ (bridged by } :\overset{+}{Br}:) \longleftrightarrow (CH_3)_2\overset{+}{C}{-}CH_2{-}\ddot{Br}: \longleftrightarrow (CH_3)_2C(\ddot{Br}:){-}\overset{+}{C}H_2$$

A B C

Resonance form A is the most stable of the three and the best approximation of the true structure. Resonance form B is more stable than C because the positive charge resides on a tertiary carbon in B as opposed to a primary carbon in C. Water attacks the more highly substituted carbon because this carbon atom has a greater degree of carbocation character than does its less substituted counterpart.

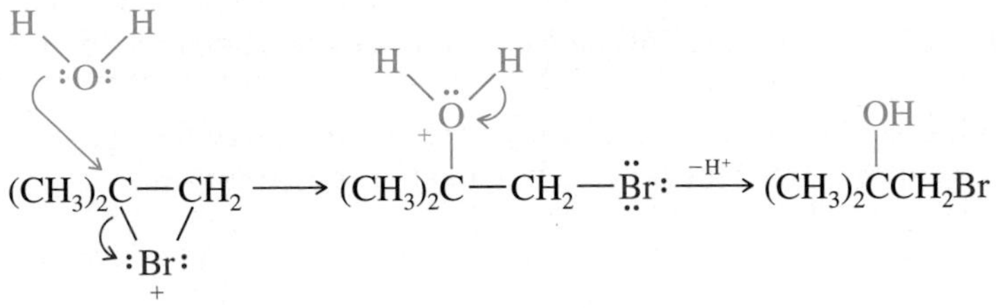

$$(CH_3)_2\overset{\frown}{C}{-}CH_2 \text{ (bromonium, attacked by } H_2\ddot{O}:) \longrightarrow (CH_3)_2\underset{|}{\overset{H\;\;H \atop {\overset{+}{\ddot{O}}}}{C}}{-}CH_2{-}\ddot{B}r: \xrightarrow{-H^+} (CH_3)_2\overset{OH}{C}CH_2Br$$

6.18 EPOXIDATION OF ALKENES

You have just seen that cyclic halonium ion intermediates are formed when sources of electrophilic halogen attack a double bond. Likewise, three-membered oxygen-containing rings are formed by the reaction of alkenes with sources of electrophilic oxygen.

Three-membered rings that contain oxygen are called *epoxides.* At one time, epoxides were named as oxides of alkenes. Ethylene oxide and propylene oxide, for example, are the common names of two industrially important epoxides.

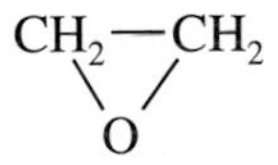

$CH_2{-}CH_2$ (O bridging) — Ethylene oxide

$CH_2{-}CHCH_3$ (O bridging) — Propylene oxide

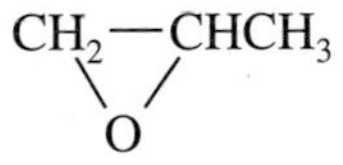

Substitutive IUPAC nomenclature names epoxides as *epoxy* derivatives of alkanes. According to this system, ethylene oxide becomes epoxyethane, and propylene oxide becomes 1,2-epoxypropane. The prefix *epoxy-* always immediately precedes the alkane ending; it is not listed in alphabetical order in the manner of other substituents.

A second method for naming epoxides in the IUPAC system is described in Section 16.1.

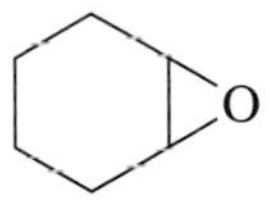

1,2-Epoxycyclohexane

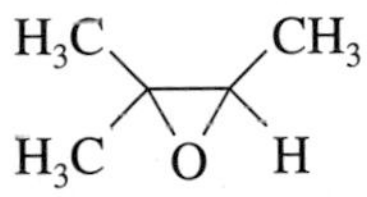

H_3C, CH_3, H_3C, O, H — 2-Methyl-2,3-epoxybutane

Functional group transformations of epoxide groups rank among the fundamental reactions of organic chemistry, and epoxides are commonplace natural products. The female gypsy moth, for example, attracts the male by emitting an epoxide known as *disparlure.* On detecting the presence of disparlure, the male follows the scent to its origin and mates with the female.

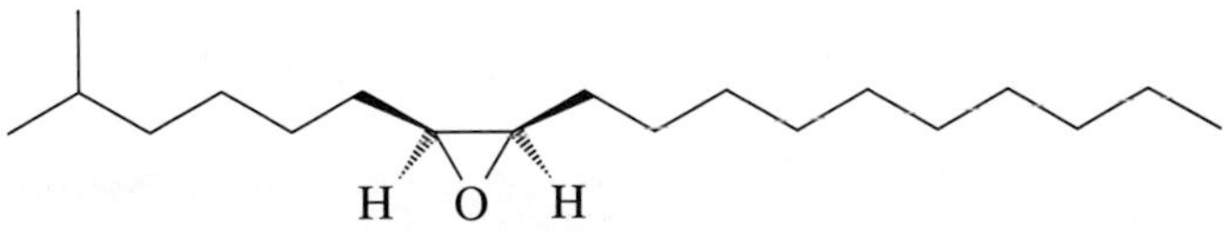

H O H

Disparlure

In one strategy designed to control the spread of the gypsy moth, infested areas are sprayed with synthetic disparlure. With the sex attractant everywhere, male gypsy moths become hopelessly confused as to the actual location of individual females. Many otherwise fertile female gypsy moths then live out their lives without producing hungry gypsy moth caterpillars.

PROBLEM 6.17 Give the IUPAC name, including stereochemistry, for disparlure.

Epoxides are very easy to prepare. They are the products of the reaction between an alkene and a peroxy acid. This process is known as **epoxidation.**

$$\text{>C=C<} + \text{RCOOH} \longrightarrow \text{>C—C<} \text{ (epoxide, bridged by O)} + \text{RCOH}$$

Alkene | Peroxy acid | Epoxide | Carboxylic acid

A commonly used peroxy acid is peroxyacetic acid (CH_3CO_2OH). Peroxyacetic acid is normally used in acetic acid as the solvent, but epoxidation reactions tolerate a variety of solvents and are often carried out in dichloromethane or chloroform.

$$CH_2{=}CH(CH_2)_9CH_3 + CH_3COOH \longrightarrow CH_2{-}CH(CH_2)_9CH_3 \text{ (bridged by O)} + CH_3COH$$

1-Dodecene | Peroxyacetic acid | 1,2-Epoxydodecane (52%) | Acetic acid

$$\text{Cyclooctene} + CH_3COOH \longrightarrow \text{1,2-Epoxycyclooctane} + CH_3COH$$

Cyclooctene | Peroxyacetic acid | 1,2-Epoxycyclooctane (86%) | Acetic acid

Epoxidation of alkenes with peroxy acids involves syn addition to the double bond. Substituents that are cis to each other in the alkene remain cis to each other in the epoxide; substituents that are trans in the alkene remain trans in the epoxide.

PROBLEM 6.18 Give the structure of the alkene, including stereochemistry, that you would choose as the starting material in a preparation of synthetic disparlure.

TABLE 6.4
Relative Rates of Epoxidation of Some Representative Alkenes with Peroxyacetic Acid

Alkene	Structural formula	Relative rate of epoxidation*
Ethylene	$CH_2{=}CH_2$	1.0
Propene	$CH_3CH{=}CH_2$	22
2-Methylpropene	$(CH_3)_2C{=}CH_2$	484
2-Methyl-2-butene	$(CH_3)_2C{=}CHCH_3$	6526

* In acetic acid, 26°C.

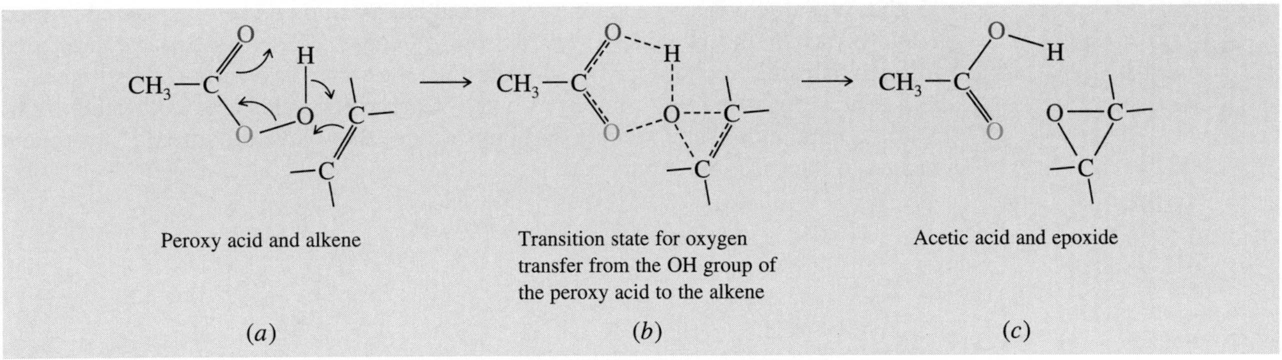

Peroxy acid and alkene

(a)

Transition state for oxygen transfer from the OH group of the peroxy acid to the alkene

(b)

Acetic acid and epoxide

(c)

FIGURE 6.13 A one-step mechanism for epoxidation of alkenes by peroxyacetic acid. In (*a*) the starting peroxy acid is shown in a conformation in which the proton of the OH group is hydrogen-bonded to the oxygen of the C=O group. (*b*) The weak O—O bond of the peroxy acid breaks, and both C—O bonds of the epoxide form in the same transition state leading to products (*c*).

As shown in Table 6.4, electron-releasing alkyl groups on the double bond increase the rate of epoxidation. This suggests that the peroxy acid acts as an electrophilic reagent toward the alkene.

The mechanism of alkene epoxidation is believed to be a concerted process involving a single bimolecular elementary step, as shown in Figure 6.13.

6.19 OZONOLYSIS OF ALKENES

Ozone (O_3) is the triatomic form of oxygen. It is a neutral but polar molecule that can be represented as a combination of its two most stable Lewis structures.

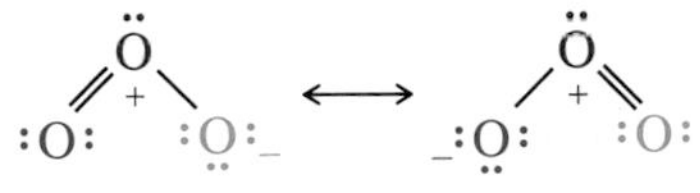

Ozone is a powerful electrophile and undergoes a remarkable reaction with alkenes in which both the σ and π components of the carbon-carbon double bond are cleaved. This reaction is known as **ozonation,** and its product is referred to as an **ozonide.** (The systematic IUPAC name for an ozonide is a 1,2,4-trioxolane.)

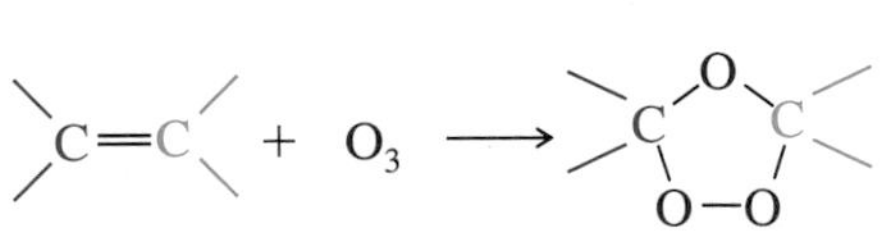

Alkene Ozone Ozonide

Ozonides undergo hydrolysis in water, giving carbonyl compounds.

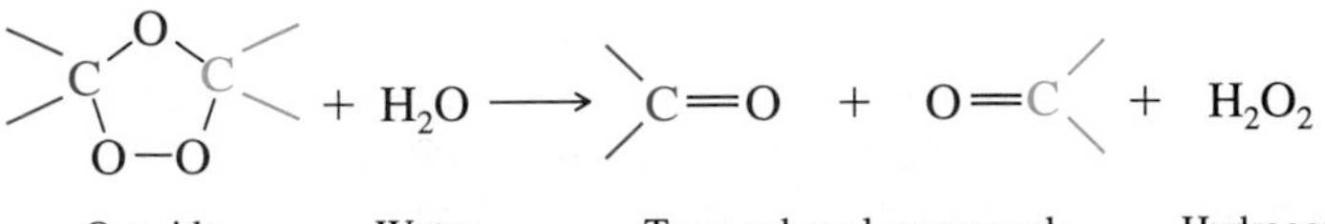

Ozonide Water Two carbonyl compounds Hydrogen Peroxide

Either two aldehydes, two ketones, or one aldehyde and one ketone may be formed on hydrolysis of the ozonides derived from alkenes. Aldehydes have at least one hydrogen substituent on the carbonyl group; ketones have two carbon substituents—alkyl groups, for example—on the carbonyl. Carboxylic acids are another family of organic compounds that contain a carbonyl group; they have a hydroxyl substituent attached to the carbonyl group.

$$\underset{\text{Formaldehyde}}{\mathrm{H{-}C({=}O){-}H}} \qquad \underset{\text{Aldehyde}}{\mathrm{R{-}C({=}O){-}H}} \qquad \underset{\text{Ketone}}{\mathrm{R{-}C({=}O){-}R'}} \qquad \underset{\text{Carboxylic acid}}{\mathrm{R{-}C({=}O){-}OH}}$$

Aldehydes are easily oxidized to carboxylic acids under conditions of ozonide hydrolysis. When one wishes to isolate the aldehyde itself, a reducing agent such as zinc is included during the hydrolysis step. Zinc reacts with the oxidants present (excess ozone and hydrogen peroxide), preventing them from oxidizing any aldehyde formed. An alternative, more modern technique follows ozonation of the alkene in methanol with reduction by dimethyl sulfide (CH_3SCH_3).

The two-stage reaction sequence is called **ozonolysis** and is represented by the general equation

$$\underset{\text{Alkene}}{\mathrm{(R)(H)C{=}C(R')(R'')}} \xrightarrow[\text{1. } O_3,\ CH_3OH;\ \text{2. } (CH_3)_2S]{\text{1. } O_3;\ \text{2. } H_2O,\ Zn \text{ or}} \underset{\text{Aldehyde}}{\mathrm{(R)(H)C{=}O}} + \underset{\text{Ketone}}{\mathrm{O{=}C(R')(R'')}}$$

Each carbon of the double bond becomes the carbon of a carbonyl group.

Ozonolysis has both synthetic and analytical applications in organic chemistry. In synthesis, ozonolysis of alkenes provides a method for the preparation of aldehydes and ketones.

$$\underset{\text{1-Octene}}{CH_3(CH_2)_5CH{=}CH_2} \xrightarrow[\text{2. } (CH_3)_2S]{\text{1. } O_3,\ CH_3OH} \underset{\text{Heptanal (75\%)}}{CH_3(CH_2)_5\overset{O}{\overset{\|}{C}}H} + \underset{\text{Formaldehyde}}{H\overset{O}{\overset{\|}{C}}H}$$

$$\underset{\text{2-Methyl-1-hexene}}{CH_3CH_2CH_2CH_2\underset{CH_3}{\underset{|}{C}}{=}CH_2} \xrightarrow[\text{2. } H_2O,\ Zn]{\text{1. } O_3} \underset{\text{2-Hexanone (60\%)}}{CH_3CH_2CH_2CH_2\overset{O}{\overset{\|}{C}}CH_3} + \underset{\text{Formaldehyde}}{H\overset{O}{\overset{\|}{C}}H}$$

When the objective is analytical, the products of ozonolysis are isolated and identified, thereby allowing the structure of the alkene to be deduced. In one such example, an alkene having the molecular formula C_8H_{16} was obtained from a chemical reaction and was then subjected to ozonolysis, giving acetone and 2,2-dimethylpropanal as the products.

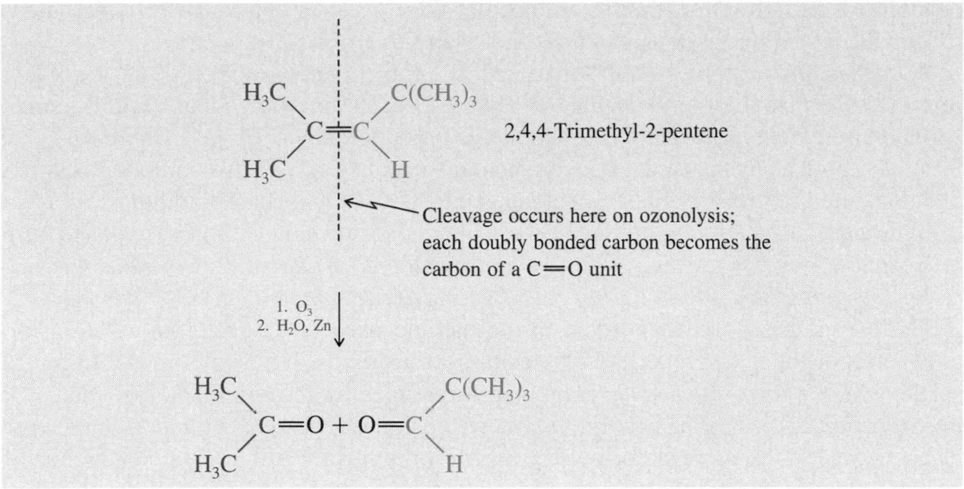

$$(H_3C)_2C{=}CH{-}C(CH_3)_3$$

2,4,4-Trimethyl-2-pentene

Cleavage occurs here on ozonolysis; each doubly bonded carbon becomes the carbon of a C=O unit

$$\xrightarrow[\text{2. } H_2O,\ Zn]{\text{1. } O_3}$$

$$(H_3C)_2C{=}O + O{=}CH{-}C(CH_3)_3$$

FIGURE 6.14 Ozonolysis of 2,4,4-trimethyl-2-pentene. On cleavage, each of the doubly bonded carbons becomes the carbon of a carbonyl (C=O) group.

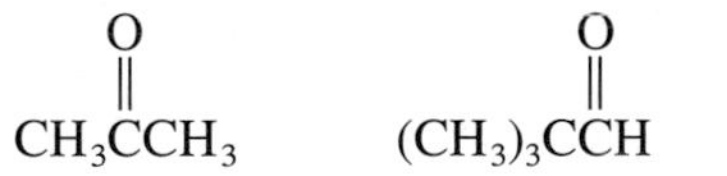

$$CH_3\overset{O}{\overset{\|}{C}}CH_3 \qquad (CH_3)_3C\overset{O}{\overset{\|}{C}}H$$

Acetone — 2,2-Dimethylpropanal

Together, these two products contain all eight carbons of the starting alkene. The two carbonyl carbons correspond to those that were doubly bonded in the original alkene. Therefore, one of the doubly bonded carbons bears two methyl substituents; the other bears a hydrogen and a *tert*-butyl group. The alkene is identified as 2,4,4-trimethyl-2-pentene, $(CH_3)_2C{=}CHC(CH_3)_3$, as shown in Figure 6.14.

PROBLEM 6.19 The same reaction that gave 2,4,4-trimethyl-2-pentene also yielded an isomeric alkene. This second alkene produced formaldehyde and 4,4-dimethyl-2-pentanone on ozonolysis. Identify this alkene.

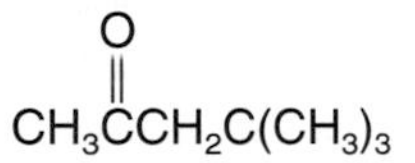

$$CH_3\overset{O}{\overset{\|}{C}}CH_2C(CH_3)_3$$

4,4-Dimethyl-2-pentanone

6.20 ANALYSIS OF ALKENES: MOLECULAR FORMULA AS A CLUE TO STRUCTURE

Chemists often confront the problem of determining the structure of an unknown compound. Sometimes the unknown can be shown to be a compound previously reported in the literature. On other occasions, the unknown may be truly that—a compound never before encountered by anyone. In such cases recourse is sometimes made to chemical reactions that convert the unknown compound to some known derivative. Ozonolysis, for example, can be used to identify the substituents on the double bond of an alkene, and catalytic hydrogenation to a known alkane serves to define the carbon skeleton. An arsenal of powerful instrumental techniques is also available, the most important of which will be described in Chapter 13. It should also be pointed out, however, that the molecular formula of a substance provides more information than might be apparent at first glance.

Consider, for example, a substance with the molecular formula C_7H_{16}. We know immediately that the compound is an alkane because its molecular formula corresponds to the general formula for that class of compounds, C_nH_{2n+2}, where $n = 7$.

What about a substance with the molecular formula C_7H_{14}? This compound cannot be an alkane but may be either a cycloalkane or an alkene, because both these classes of hydrocarbons correspond to the general molecular formula C_nH_{2n}. *Any time a ring or a double bond is present in an organic molecule, its molecular formula has two fewer hydrogen atoms than that of an alkane with the same number of carbons.*

A more detailed discussion can be found in the May 1995 issue of the *Journal of Chemical Education*, pp. 245–248.

Various names have been given to the method used to describe this relationship between molecular formulas and classes of hydrocarbons. It is sometimes referred to as the *index of hydrogen deficiency* and sometimes as the number of *elements of unsaturation* or *sites of unsaturation.* We will use the term *sum of double bonds and rings* because it is more descriptive than the others. In the interest of economy of space, however, the full term *sum of double bonds and rings* will be abbreviated as SODAR.

$$\text{Sum of double bonds and rings (SODAR)} = \tfrac{1}{2}(C_nH_{2n+2} - C_nH_x)$$

where C_nH_x is the molecular formula of the compound.

A molecule that has a molecular formula of C_7H_{14} has a SODAR of 1:

$$\text{SODAR} = \tfrac{1}{2}(C_7H_{16} - C_7H_{14})$$

$$\text{SODAR} = \tfrac{1}{2}(2) = 1$$

Thus, the structure has one ring or one double bond. A molecule of molecular formula C_7H_{12} has four fewer hydrogens than the corresponding alkane; it has a SODAR of 2 and can have two rings, two double bonds, one ring and one double bond, or one triple bond.

A halogen substituent, like hydrogen, is monovalent, and when present in a molecular formula is treated as if it were hydrogen for the purposes of computing the number of double bonds and rings. Oxygen atoms have no effect on the relationship between molecular formulas, double bonds, and rings. They are ignored when computing the SODAR of a substance.

How does one distinguish between rings and double bonds? This additional piece of structural information is revealed by catalytic hydrogenation experiments in which the amount of hydrogen that reacts is measured exactly. Each of a molecule's double bonds consumes one molar equivalent of hydrogen, while rings do not undergo hydrogenation. A substance with a SODAR of 5 that takes up 3 moles of hydrogen must therefore have two rings.

PROBLEM 6.20 How many rings are present in each of the following compounds? Each consumes 2 moles of hydrogen on catalytic hydrogenation.

(*a*) $C_{10}H_{18}$
(*b*) C_8H_8
(*c*) $C_8H_8Cl_2$
(*d*) C_8H_8O
(*e*) $C_8H_{10}O_2$
(*f*) C_8H_9ClO

SAMPLE SOLUTION (*a*) The molecular formula $C_{10}H_{18}$ contains four fewer hydrogens than the alkane having the same number of carbon atoms ($C_{10}H_{22}$). Therefore, the SODAR of this compound is 2. Since it consumes two molar equivalents of hydrogen on catalytic hydrogenation, it must have two double bonds and no rings.

6.21 REACTIONS OF ALKENES WITH ALKENES: POLYMERIZATION

While 2-methylpropene undergoes acid-catalyzed hydration in dilute sulfuric acid to form *tert*-butyl alcohol (Section 6.10 and Figure 6.8), an unusual reaction occurs in more concentrated solutions of sulfuric acid. Rather than form the expected alkyl hydrogen sulfate (Section 6.9), 2-methylpropene is converted to a mixture of two isomeric C_8H_{16} alkenes.

The structures of these two C_8H_{16} alkenes were determined by oxonolysis as described in Section 6.19.

$$2(CH_3)_2C{=}CH_2 \xrightarrow{65\%\ H_2SO_4} CH_2{=}\underset{\displaystyle CH_3}{\underset{|}{C}}CH_2C(CH_3)_3 + (CH_3)_2C{=}CHC(CH_3)_3$$

2-Methylpropene — 2,4,4-Trimethyl-1-pentene — 2,4,4-Trimethyl-2-pentene

With molecular formulas corresponding to twice that of the starting alkene, the products of this reaction are referred to as **dimers** of 2-methylpropene, which is, in turn, called the **monomer.** The suffix *-mer* is derived from the Greek *meros,* meaning ''part.'' Three monomeric units produce a **trimer,** four a **tetramer,** and so on. A high-molecular-weight material comprising a large number of monomeric subunits is called a **polymer.**

PROBLEM 6.21 The two dimers of 2-methylpropene shown in the equation can be converted to 2,2,4-trimethylpentane (known by its nonsystematic name *isooctane*) for use as a gasoline additive. Can you suggest a method for this conversion?

Alkene dimers are formed by the mechanism shown in Figure 6.15 for the case of 2-methylpropene. In step 1 protonation of the double bond generates a small amount of *tert*-butyl cation in equilibrium with the alkene. The carbocation is an electrophile and attacks a second molecule of 2-methylpropene in step 2, forming a new carbon-carbon bond and generating a C_8 carbocation. This new carbocation loses a proton in step 3 to form a mixture of 2,4,4-trimethyl-1-pentene and 2,4,4-trimethyl-2-pentene.

Dimerization in concentrated sulfuric acid occurs mainly with those alkenes that form tertiary carbocations. In some cases reaction conditions can be developed that favor the formation of higher-molecular-weight alkenes that are polymers of 2-methylpropene. Since these reactions proceed by way of carbocation intermediates, the process is referred to as **cationic polymerization.**

The uses to which ethylene and its relatives are put are summarized in an article entitled "Alkenes and Their Derivatives: The Alchemists' Dream Come True," in the August 1989 issue of the *Journal of Chemical Education* (pp. 670–672).

We made special mention in Section 5.1 of the enormous volume of ethylene and propene production in the petrochemical industry. The accompanying box summarizes the principal uses of these alkenes. Most of the ethylene is converted to **polyethylene,** a high-molecular-weight polymer of ethylene. Polyethylene cannot be prepared by

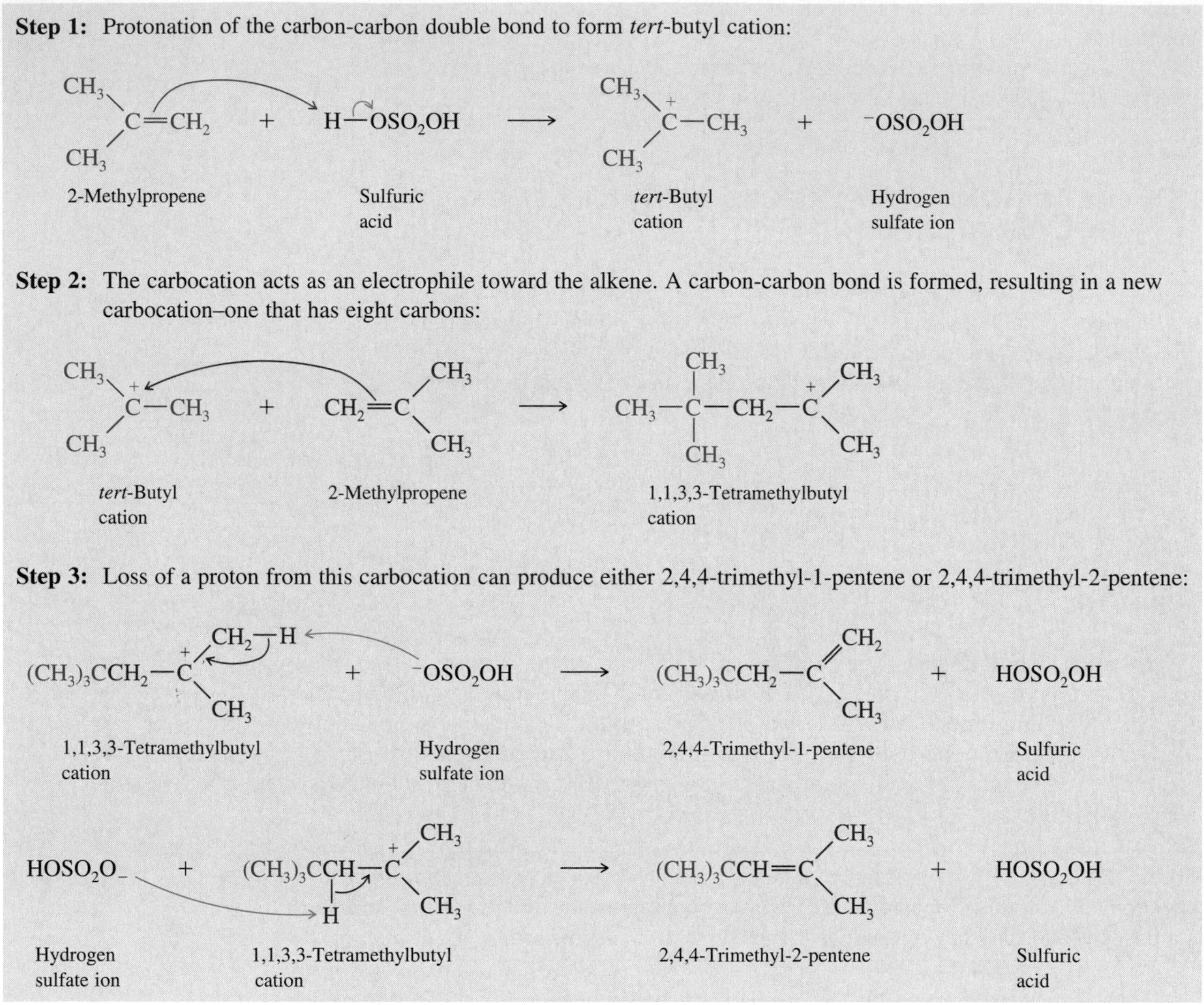

Step 1: Protonation of the carbon-carbon double bond to form *tert*-butyl cation:

$$(CH_3)_2C{=}CH_2 + H{-}OSO_2OH \longrightarrow (CH_3)_2\overset{+}{C}{-}CH_3 + {}^{-}OSO_2OH$$

2-Methylpropene · Sulfuric acid · *tert*-Butyl cation · Hydrogen sulfate ion

Step 2: The carbocation acts as an electrophile toward the alkene. A carbon-carbon bond is formed, resulting in a new carbocation–one that has eight carbons:

$$(CH_3)_2\overset{+}{C}{-}CH_3 + CH_2{=}C(CH_3)_2 \longrightarrow CH_3{-}C(CH_3)_2{-}CH_2{-}\overset{+}{C}(CH_3)_2$$

tert-Butyl cation · 2-Methylpropene · 1,1,3,3-Tetramethylbutyl cation

Step 3: Loss of a proton from this carbocation can produce either 2,4,4-trimethyl-1-pentene or 2,4,4-trimethyl-2-pentene:

$$(CH_3)_3CCH_2{-}\overset{+}{C}(CH_2{-}H)(CH_3) + {}^{-}OSO_2OH \longrightarrow (CH_3)_3CCH_2{-}C({=}CH_2)(CH_3) + HOSO_2OH$$

1,1,3,3-Tetramethylbutyl cation · Hydrogen sulfate ion · 2,4,4-Trimethyl-1-pentene · Sulfuric acid

$$HOSO_2O^{-} + (CH_3)_3CCH(H){-}\overset{+}{C}(CH_3)_2 \longrightarrow (CH_3)_3CCH{=}C(CH_3)_2 + HOSO_2OH$$

Hydrogen sulfate ion · 1,1,3,3-Tetramethylbutyl cation · 2,4,4-Trimethyl-2-pentene · Sulfuric acid

FIGURE 6.15 Sequence of steps that describes the mechanism of acid-catalyzed dimerization of 2-methylpropene.

cationic polymerization, but is the simplest example of a polymer that is produced on a large scale by **free-radical polymerization.**

In the free-radical polymerization of ethylene, ethylene is heated at high pressure in the presence of oxygen or a peroxide.

$$nCH_2{=}CH_2 \xrightarrow[O_2 \text{ or peroxides}]{200°C,\ 2000\text{ atm}} {-}CH_2{-}CH_2{-}(CH_2{-}CH_2)_{n-2}{-}CH_2{-}CH_2{-}$$

Ethylene · Polyethylene

In this reaction n can have a value of thousands.

An outline of the mechanism of free-radical polymerization of ethylene is shown in Figure 6.16. Dissociation of a peroxide initiates the process in step 1. The resulting peroxy radical adds to the carbon-carbon double bond in step 2, giving a new radical,

Step 1: Homolytic dissociation of a peroxide produces alkoxy radicals that serve as free-radical initiators:

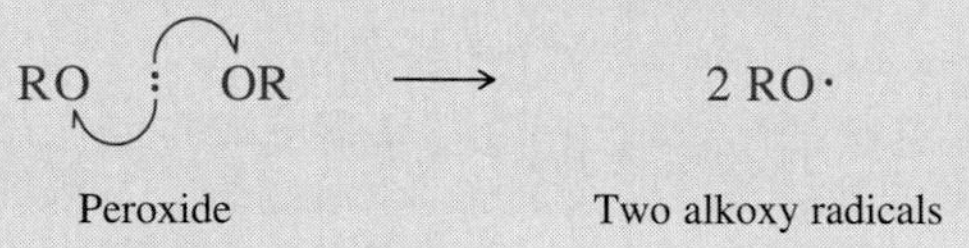

$$RO \vdots OR \longrightarrow 2\,RO\cdot$$

Peroxide — Two alkoxy radicals

Step 2: An alkoxy radical adds to the carbon-carbon double bond:

$$RO\cdot + CH_2{=}CH_2 \longrightarrow RO{-}CH_2{-}\dot{C}H_2$$

Alkoxy radical — Ethylene — 2-Alkoxyethyl radical

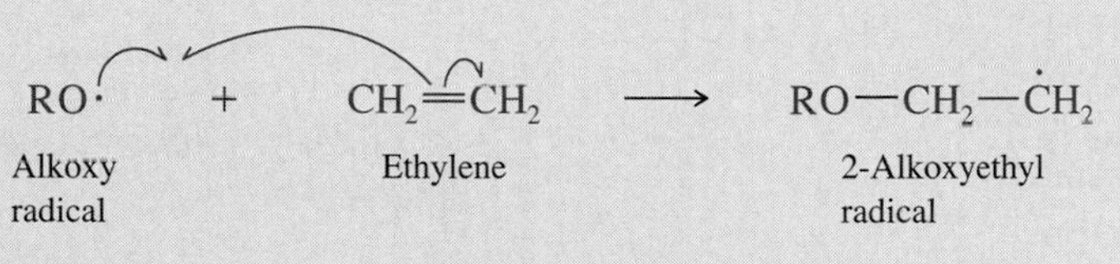

Step 3: The radical produced in step 2 adds to a second molecule of ethylene:

$$RO{-}CH_2{-}\dot{C}H_2 + CH_2{=}CH_2 \longrightarrow RO{-}CH_2{-}CH_2{-}CH_2{-}\dot{C}H_2$$

2-Alkoxyethyl radical — Ethylene — 4-Alkoxybutyl radical

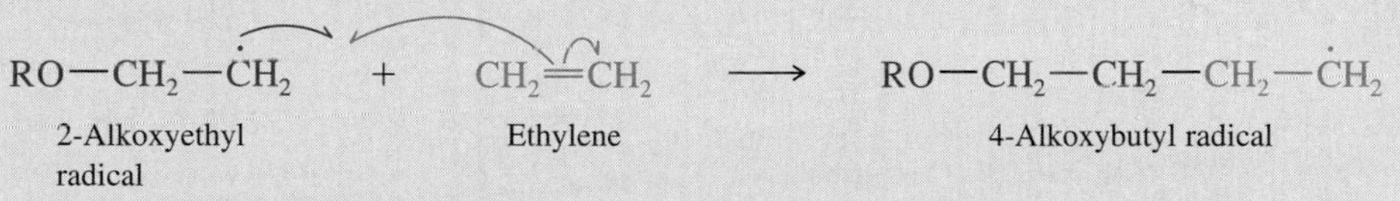

The radical formed in step 3 then adds to a third molecule of ethylene, and the process continues, forming a long chain of methylene groups.

FIGURE 6.16 Mechanistic description of peroxide-induced free-radical polymerization of ethylene.

which then adds to a second molecule of ethylene in step 3. The carbon-carbon bond-forming process in step 3 can be repeated thousands of times to give long carbon chains.

In spite of the *-ene* ending to its name, polyethylene is much more closely related to alkanes than to alkenes. It is simply a long chain of CH_2 groups bearing at its ends an alkoxy group (from the initiator) or a carbon-carbon double bond.

A large number of compounds with carbon-carbon double bonds have been polymerized to yield materials having useful properties. Some of the more important or familiar of these are listed in Table 6.5. Not all these monomers are effectively polymerized under free-radical conditions, and much research has been carried out to develop alternative polymerization techniques. One of these, **coordination polymerization,** employs a mixture of titanium tetrachloride, $TiCl_4$, and triethylaluminum, $(CH_3CH_2)_3Al$, as a catalyst. Polyethylene produced by coordination polymerization has a higher density than that produced by free-radical polymerization and somewhat different—in many applications, more desirable—properties. The catalyst system used in coordination polymerization was developed independently by Karl Ziegler in Germany and Giulio Natta in Italy in the early 1950s. They shared the Nobel Prize in chemistry in 1963 for this work. The Ziegler-Natta catalyst system also permits polymerization of propene to be achieved in a way that gives a form of **polypropylene** suitable for plastics and fibers. When propene is polymerized under free-radical conditions, the polypropylene has physical properties (such as a low melting point) that preclude its use in plastics and fibers.

TABLE 6.5
Some Compounds with Carbon-Carbon Double Bonds Used to Prepare Polymers

A. Alkenes of the type $CH_2{=}CH{-}X$ used to form polymers of the type $(-CH_2-\underset{\large X}{\underset{|}{CH}}-)_n$

Compound	Structure	—X in polymer	Application
Ethylene	$CH_2{=}CH_2$	—H	Polyethylene films as packaging material; "plastic" squeeze bottles are molded from high-density polyethylene.
Propene	$CH_2{=}CH{-}CH_3$	$-CH_3$	Polypropylene fibers for use in carpets and automobile tires; consumer items (luggage, appliances, etc.); packaging material.
Styrene	$CH_2{=}CH{-}C_6H_5$ (phenyl ring)	$-C_6H_5$ (phenyl ring)	Polystyrene packaging, housewares, luggage, radio and television cabinets.
Vinyl chloride	$CH_2{=}CH{-}Cl$	—Cl	Poly(vinyl chloride) (PVC) has replaced leather in many of its applications; PVC tubes and pipes are often used in place of copper.
Acrylonitrile	$CH_2{=}CH{-}C{\equiv}N$	$-C{\equiv}N$	Wool substitute in sweaters, blankets, etc.

B. Alkenes of the type $CH_2{=}CX_2$ used to form polymers of the type $(-CH_2-CX_2-)_n$

Compound	Structure	X in polymer	Application
1,1-Dichloroethene (vinylidene chloride)	$CH_2{=}CCl_2$	Cl	Saran used as air- and water-tight packaging film.
2-Methylpropene	$CH_2{=}C(CH_3)_2$	CH_3	Polyisobutene is component of "butyl rubber," one of earliest synthetic rubber substitutes.

C. Others

Compound	Structure	Polymer	Application
Tetrafluoroethene	$CF_2{=}CF_2$	$(-CF_2-CF_2-)_n$ (Teflon)	Nonstick coating for cooking utensils; bearings, gaskets, and fittings.
Methyl methacrylate	$CH_2{=}\underset{\large CH_3}{\underset{|}{C}}CO_2CH_3$	$(-CH_2-\overset{\large CO_2CH_3}{\overset{|}{\underset{\large CH_3}{\underset{|}{C}}}}-)_n$	When cast in sheets, is transparent; used as glass substitute (Lucite, Plexiglas).
2-Methyl-1,3-butadiene	$CH_2{=}\underset{\large CH_3}{\underset{|}{C}}CH{=}CH_2$	$(-CH_2\underset{\large CH_3}{\underset{|}{C}}{=}CH-CH_2-)_n$ (Polyisoprene)	Synthetic rubber.

Source: R. C. Atkins and F. A. Carey, *Organic Chemistry: A Brief Course,* McGraw-Hill, New York, 1990, p. 132.

ETHYLENE AND PROPENE: THE MOST IMPORTANT INDUSTRIAL ORGANIC CHEMICALS

Having examined the properties of alkenes and introduced the elements of polymers and polymerization, we narrow our focus to concentrate on some commercial applications of ethylene and propene, which, as the title states, are the two most important industrial organic chemicals.

ETHYLENE We discussed ethylene production in an earlier boxed essay (Section 5.1), where it was pointed out that the output of the U.S. petrochemical industry exceeds 4×10^{10} pounds per year. Approximately 90 percent of this material is used for the preparation of four compounds (polyethylene, ethylene oxide, vinyl chloride, and styrene), with polymerization to polyethylene accounting for half the total (see the accompanying summary diagram). Both vinyl chloride and styrene are polymerized to give poly(vinyl chloride) and polystyrene, respectively (Table 6.5). Ethylene oxide is a starting material for the preparation of ethylene glycol for use as an antifreeze in automobile radiators and in the production of polyester fibers (see the boxed essay "Condensation Polymers. Polyamides and Polyesters" in Chapter 20).

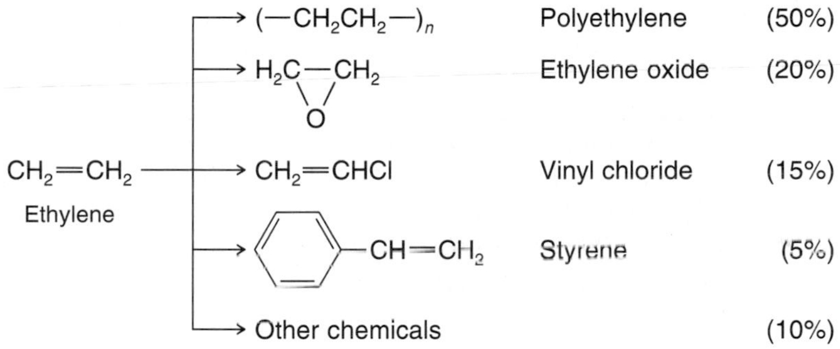

$CH_2{=}CH_2$ (Ethylene) →

Product	Name	Share
$(-CH_2CH_2-)_n$	Polyethylene	(50%)
H_2C-CH_2 (epoxide, bridged by O)	Ethylene oxide	(20%)
$CH_2{=}CHCl$	Vinyl chloride	(15%)
$C_6H_5-CH{=}CH_2$	Styrene	(5%)
Other chemicals		(10%)

Among the "other chemicals" prepared from ethylene are ethanol and acetaldehyde:

CH_3CH_2OH

Ethanol (industrial solvent; used in preparation of ethyl acetate; unleaded gasoline additive)

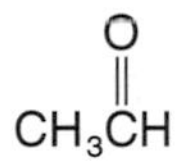

$CH_3\overset{\overset{O}{\|}}{C}H$

Acetaldehyde (used in preparation of acetic acid)

PROPENE The major use of propene is in the production of polypropylene (see the diagram, page 254). Two of the propene-derived organic chemicals shown, acrylonitrile and propylene oxide, are also starting materials for polymer synthesis. Acrylonitrile is used to make acrylic fibers (Table 6.5), while propylene oxide is one component in the preparation of *polyurethane* polymers. Cumene itself has no direct uses but rather serves as the starting material in a process which yields two valuable industrial chemicals, acetone and phenol.

We have not indicated the reagents employed in the reactions by which ethylene and propene are converted to the compounds shown. Because of patent requirements, different companies often use different processes. While the processes may be different, they share the common characteristic of being extremely efficient. The industrial chemist faces the challenge of producing valuable materials, at low cost. Thus, success in the industrial environment requires both an understanding of chemistry and an appreciation of the economics associated with alternative procedures. One measure of how successfully these challenges have been met can be seen in the fact that the United States maintains a positive trade balance in chemicals each year. In 1992 that surplus amounted to $16.2 billion in chemicals versus an overall trade deficit of $84.5 billion.

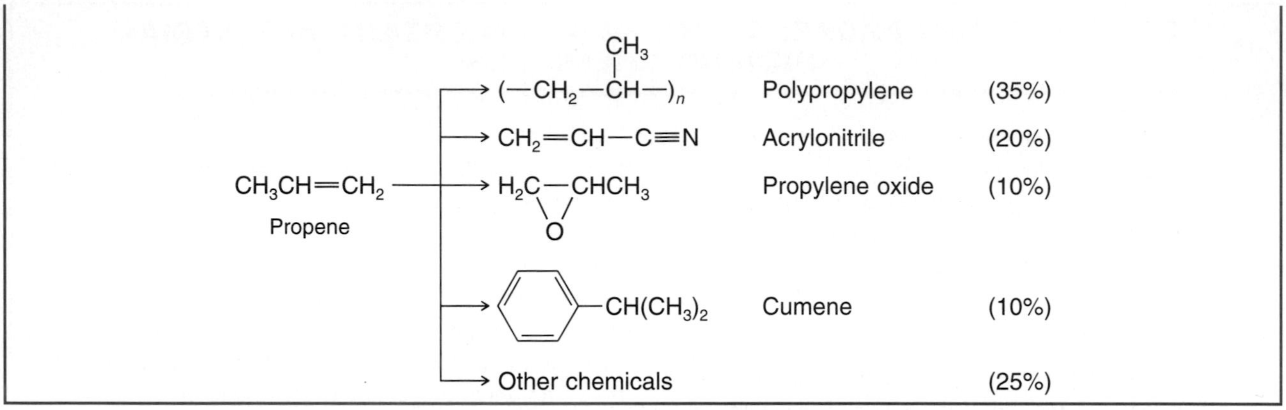

$CH_3CH{=}CH_2$ (Propene) →

Product	Name	Share
$(-CH_2-CH(CH_3)-)_n$	Polypropylene	(35%)
$CH_2{=}CH-C{\equiv}N$	Acrylonitrile	(20%)
$H_2C-CHCH_3$ (epoxide, O bridging)	Propylene oxide	(10%)
$C_6H_5-CH(CH_3)_2$	Cumene	(10%)
Other chemicals		(25%)

6.22 INTRODUCTION TO ORGANIC CHEMICAL SYNTHESIS

An important area of concern to chemists is synthesis, the task of preparing a particular compound in an economical way and with confidence that the synthetic method will lead precisely to the desired structure. Among the tools applied to synthetic objectives, functional group transformations are the most often encountered. In this section we will introduce the topic of synthesis, emphasizing the need for systematic planning in order to decide what is the best sequence of steps to convert a specified starting material to a desired product (the **target molecule**).

A critical feature of synthetic planning is *to reason backward from the target to the starting material.* A second is *always to use reactions that you know will work.*

Let us examine a simple example. Suppose you had to prepare cyclohexane, given cyclohexanol as the starting material. We have encountered no reactions to this point that permit us to carry out the indicated conversion in a single step.

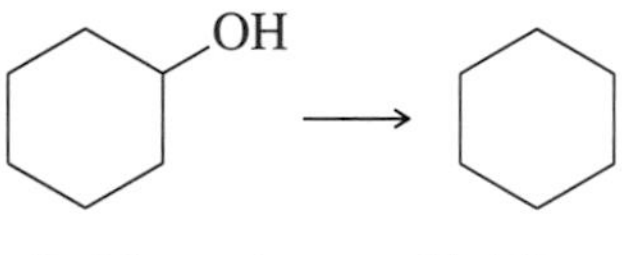

Cyclohexanol → Cyclohexane

Reasoning backward, however, we know that we can prepare cyclohexane by hydrogenation of cyclohexene. Therefore, let us use this reaction as the last step in our proposed synthesis.

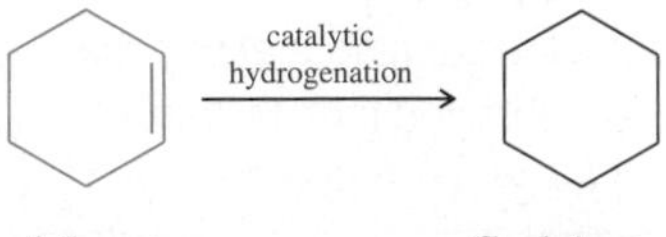

Cyclohexene —catalytic hydrogenation→ Cyclohexane

Recognizing that cyclohexene may be prepared by dehydration of cyclohexanol, a practical synthesis of cyclohexane from cyclohexanol becomes apparent.

$$\text{Cyclohexanol} \xrightarrow[\text{heat}]{H_2SO_4} \text{Cyclohexene} \xrightarrow[\text{Pt}]{H_2} \text{Cyclohexane}$$

Cyclohexanol Cyclohexene Cyclohexane

As a second example, consider the preparation of 1-bromo-2-methyl-2-propanol from *tert*-butyl alcohol.

$$(CH_3)_3COH \longrightarrow (CH_3)_2C(OH)CH_2Br$$

tert-Butyl alcohol 1-Bromo-2-methyl-2-propanol

Begin by asking the question, ''What kind of compound is the target molecule, and what methods are available for preparing that kind of compound?'' The desired product has a bromine and a hydroxyl on adjacent carbons; it is a *vicinal bromohydrin.* The only method we have learned so far for the preparation of vicinal bromohydrins involves the reaction of alkenes with Br_2 in water. Thus, a reasonable last step is:

$$(CH_3)_2C{=}CH_2 \xrightarrow[H_2O]{Br_2} (CH_3)_2C(OH)CH_2Br$$

2-Methylpropene 1-Bromo-2-methyl-2-propanol

We now have a new problem: Where does the necessary alkene come from? Alkenes are prepared from alcohols by acid-catalyzed dehydration (Section 5.9) or from alkyl halides by E2 elimination (Section 5.14). Since our designated starting material is *tert* butyl alcohol, we can combine its dehydration with bromohydrin formation to give the correct sequence of steps:

$$(CH_3)_3COH \xrightarrow[\text{heat}]{H_2SO_4} (CH_3)_2C{=}CH_2 \xrightarrow[H_2O]{Br_2} (CH_3)_2C(OH)CH_2Br$$

tert-Butyl alcohol 2-Methylpropene 1-Bromo-2-methyl-2-propanol

---

PROBLEM 6.22 Write a series of equations describing a synthesis of 1-bromo-2-methyl-2-propanol from *tert*-butyl bromide.

Often more than one synthetic route may be available to prepare a particular compound. Indeed, it is normal to find in the chemical literature that the same compound has been synthesized in a number of different ways. As we proceed through the text and develop a larger inventory of functional group transformations, our ability to evaluate alternative synthetic plans will increase. In most cases the best synthetic plan is the one with the fewest steps.

6.23 SUMMARY

Alkenes are **unsaturated hydrocarbons** and undergo reactions with substances that add to their multiple bonds. Representative addition reactions of alkenes are summarized in Table 6.6. Except for catalytic hydrogenation, these reactions proceed by **electrophilic** attack of the reagent on the π electrons of the double bond. Carbocations are intermediates in the electrophilic addition of hydrogen halides to alkenes.

$$\rangle C{=}C\langle \;+\; H{-}X \longrightarrow \;{}^{+}\!C{-}\overset{|}{\underset{|}{C}}{-}H \;+\; X^- \longrightarrow X{-}\overset{|}{\underset{|}{C}}{-}\overset{|}{\underset{|}{C}}{-}H$$

Alkene | Hydrogen halide | Carbocation | Halide ion | Alkyl halide

As described in the table, the **regioselectivity** of addition of hydrogen halides to alkenes and the hydration of alkenes can be predicted by applying **Markovnikov's rule.** Mechanistically, Markovnikov's rule results from protonation of the double bond in the direction that gives the more stable of two possible carbocations.

Hydrogen bromide is unique among the hydrogen halides in that it can add to alkenes either by an ionic mechanism or by a free-radical mechanism (Section 6.8). Under photochemical conditions or in the presence of peroxides, free-radical addition is observed and HBr adds to the double bond with a regioselectivity opposite to that of Markovnikov's rule.

$$\text{(cycloheptane)}{=}CH_2 \xrightarrow[h\nu]{HBr} \text{(cycloheptane)}\langle^{CH_2Br}_{H}$$

Methylenecycloheptane → (Bromomethyl)cycloheptane (61%)

Alkenes are cleaved to carbonyl compounds by **ozonolysis** (Section 6.19). This reaction is useful both for synthesis (preparation of aldehydes, ketones, or carboxylic acids) and analysis. When applied to analysis, the carbonyl compounds are isolated and identified, allowing the substituents attached to the double bond to be deduced.

$$CH_3CH{=}C(CH_2CH_3)_2 \xrightarrow[2.\ Zn,\ H_2O]{1.\ O_3} CH_3\overset{O}{\overset{\|}{C}}H + CH_3CH_2\overset{O}{\overset{\|}{C}}CH_2CH_3$$

3-Ethyl-2-pentene → Acetaldehyde (38%) + 3-Pentanone (57%)

Polymerization is the process whereby many alkene molecules react to give long hydrocarbon chains. Among the methods by which alkenes are polymerized are *cationic polymerization, free-radical polymerization,* and *coordination polymerization* (Section 6.21).

TABLE 6.6
Addition Reactions of Alkenes

Reaction (section) and comments	General equation and specific example
Catalytic hydrogenation (Sections 6.1–6.3) Alkenes react with hydrogen in the presence of a platinum, palladium, rhodium, or nickel catalyst to form the corresponding alkane.	$R_2C{=}CR_2 + H_2 \xrightarrow{\text{Pt, Pd, Rh, or Ni}} R_2CHCHR_2$ Alkene, Hydrogen → Alkane *cis*-Cyclododecene $\xrightarrow[\text{Pt}]{H_2}$ Cyclododecane (100%)
Addition of hydrogen halides (Sections 6.4–6.7) A proton and a halogen add to the double bond of an alkene to yield an alkyl halide. Addition proceeds in accordance with Markovnikov's rule; hydrogen adds to the carbon that has the greater number of hydrogen substituents, halide to the carbon that has the fewer hydrogen substituents.	$RCH{=}CR'_2 + HX \longrightarrow RCH_2{-}CR'_2(X)$ Alkene, Hydrogen halide → Alkyl halide Methylenecyclohexane ($=CH_2$) + HCl (Hydrogen chloride) $\longrightarrow$ 1-Chloro-1-methylcyclohexane (75–80%)
Addition of sulfuric acid (Section 6.9) Alkenes react with cold, concentrated sulfuric acid to form alkyl hydrogen sulfates. A proton and a hydrogen sulfate ion add across the double bond in accordance with Markovnikov's rule. Alkenes that yield tertiary carbocations on protonation tend to polymerize in concentrated sulfuric acid (Section 6.21).	$RCH{=}CR'_2 + HOSO_2OH \longrightarrow RCH_2{-}CR'_2(OSO_2OH)$ Alkene, Sulfuric acid → Alkyl hydrogen sulfate $CH_2{=}CHCH_2CH_3 + HOSO_2OH \longrightarrow CH_3{-}CH(OSO_2OH)CH_2CH_3$ 1-Butene, Sulfuric acid → *sec*-Butyl hydrogen sulfate
Acid-catalyzed hydration (Section 6.10) Addition of the elements of water to the double bond of an alkene takes place in aqueous acidic solution. Addition occurs according to Markovnikov's rule. A carbocation is an intermediate and is captured by a molecule of water acting as a nucleophile. For synthetic purposes this reaction ordinarily works well only for the preparation of tertiary alcohols.	$RCH{=}CR'_2 + H_2O \xrightarrow{H^+} RCH_2CR'_2(OH)$ Alkene, Water → Alcohol $CH_2{=}C(CH_3)_2 \xrightarrow{50\%\ H_2SO_4{-}H_2O} (CH_3)_3COH$ 2-Methylpropene → *tert*-Butyl alcohol (55–58%)

(Continued)

TABLE 6.6
Addition Reactions of Alkenes *(Continued)*

Reaction (section) and comments	General equation and specific example
Hydroboration-oxidation (Sections 6.11–6.13) This two-step sequence achieves hydration of alkenes in a stereospecific syn manner, with a regioselectivity opposite to that of Markovnikov's rule. An organoborane is formed by electrophilic addition of diborane to an alkene. Oxidation of the organoborane intermediate with hydrogen peroxide completes the process. Rearrangements do not occur.	$RCH{=}CR'_2 \xrightarrow[2.\ H_2O_2,\ HO^-]{1.\ B_2H_6,\ diglyme} RCH(OH)CHR'_2$ Alkene → Alcohol $(CH_3)_2CHCH_2CH{=}CH_2 \xrightarrow[2.\ H_2O_2,\ HO^-]{1.\ BH_3\text{–}THF} (CH_3)_2CHCH_2CH_2CH_2OH$ 4-Methyl-1-pentene → 4-Methyl-1-pentanol (80%)
Addition of halogens (Sections 6.14–6.16) Bromine and chlorine add across the carbon-carbon double bond of alkenes to form vicinal dihalides. A cyclic halonium ion is an intermediate. The reaction is stereospecific; anti addition is observed.	$R_2C{=}CR_2 + X_2 \longrightarrow X{-}CR_2{-}CR_2{-}X$ Alkene, Halogen → Vicinal dihalide $CH_2{=}CHCH_2CH_2CH_2CH_3 + Br_2 \longrightarrow BrCH_2{-}CH(Br)CH_2CH_2CH_2CH_3$ 1-Hexene, Bromine → 1,2-Dibromohexane (100%)
Halohydrin formation (Section 6.17) When treated with bromine or chlorine in aqueous solution, alkenes are converted to vicinal halohydrins. A halonium ion is an intermediate. The elements of hypobromous acid (HOBr) or hypochlorous acid (HOCl) add across the double bond in accordance with Markovnikov's rule. The positively polarized halogen adds to the carbon that has the greater number of hydrogen substituents.	$RCH{=}CR'_2 + X_2 + H_2O \longrightarrow X{-}CH(R){-}CR'_2{-}OH + HX$ Alkene, Halogen, Water → Vicinal halohydrin, Hydrogen halide Methylenecyclohexane $\xrightarrow[H_2O]{Br_2}$ (1-Bromomethyl)cyclohexanol (89%)
Epoxidation (Section 6.18) Peroxy acids transfer oxygen to the double bond of alkenes to yield epoxides. The reaction is a stereospecific syn addition.	$R_2C{=}CR_2 + R'COOH \longrightarrow R_2C{-}CR_2$ (epoxide, bridged by O) $+ R'COH$ Alkene, Peroxy acid → Epoxide, Carboxylic acid 1-Methylcycloheptene + CH_3COOH (Peroxyacetic acid) → 1-Methyl-1,2-epoxycycloheptane (65%) + CH_3COH (Acetic acid)

PROBLEMS

6.23 Write the structure of the principal organic product formed in the reaction of 1-pentene with each of the following:

(*a*) Hydrogen chloride
(*b*) Hydrogen bromide
(*c*) Hydrogen bromide in the presence of peroxides
(*d*) Hydrogen iodide
(*e*) Dilute sulfuric acid
(*f*) Diborane in diglyme, followed by basic hydrogen peroxide
(*g*) Bromine in carbon tetrachloride
(*h*) Bromine in water
(*i*) Peroxyacetic acid
(*j*) Ozone
(*k*) Product of part (*j*) treated with zinc and water

6.24 Repeat Problem 6.23 for 2-methyl-2-butene.

6.25 Repeat Problem 6.23 for 1-methylcyclohexene.

6.26 Match the following alkenes with the appropriate heats of hydrogenation:

(*a*) 1-Pentene
(*b*) (*E*)-4,4-Dimethyl-2-pentene
(*c*) (*Z*)-4-Methyl-2-pentene
(*d*) (*Z*)-2,2,5,5-Tetramethyl-3-hexene
(*e*) 2,4-Dimethyl-2-pentene

Heats of hydrogenation in kJ/mol (kcal/mol): 151(36.2); 122(29.3); 114(27.3); 111(26.5); 105(25.1).

6.27

(*a*) How many alkenes yield 2,2,3,4,4-pentamethylpentane on catalytic hydrogenation?
(*b*) How many yield 2,3-dimethylbutane?
(*c*) How many yield methylcyclobutane?
(*d*) Several alkenes undergo hydrogenation to yield a mixture of *cis*- and *trans*-1,4-dimethylcyclohexane. Only one substance, however, gives only *cis*-1,4-dimethylcyclohexane. What compound is this?

6.28 Specify reagents suitable for converting 3-ethyl-2-pentene to each of the following:

(*a*) 2,3-Dibromo-3-ethylpentane
(*b*) 3-Chloro-3-ethylpentane
(*c*) 2-Bromo-3-ethylpentane
(*d*) 3-Ethyl-3-pentanol
(*e*) 3-Ethyl-2-pentanol
(*f*) 3-Ethyl-2,3-epoxypentane
(*g*) 3-Ethylpentane

6.29

(*a*) Which primary alcohol of molecular formula $C_5H_{12}O$ cannot be prepared from an alkene? Why?

(*b*) Write equations describing the preparation of three isomeric primary alcohols of molecular formula $C_5H_{12}O$ from alkenes.

(*c*) Write equations describing the preparation of the tertiary alcohol of molecular formula $C_5H_{12}O$ from two different alkenes.

6.30 All the following reactions have been reported in the chemical literature. Give the structure of the principal organic product in each case.

(*a*) $CH_3CH_2CH{=}CHCH_2CH_3 + HBr \xrightarrow{\text{no peroxides}}$

(*b*) $(CH_3)_2CHCH_2CH_2CH_2CH{=}CH_2 \xrightarrow[\text{peroxides}]{\text{HBr}}$

(*c*) 2-*tert*-Butyl-3,3-dimethyl-1-butene $\xrightarrow[\text{2. } H_2O_2,\ HO^-]{\text{1. } B_2H_6}$

(*d*) 1,2-dimethylcyclohexene (CH_3, CH_3) $\xrightarrow[\text{2. } H_2O_2,\ HO^-]{\text{1. } B_2H_6}$

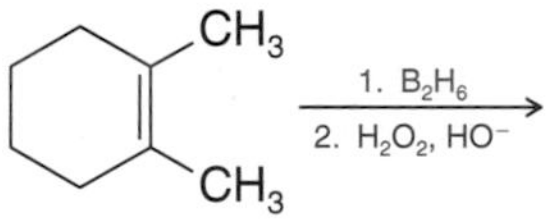

(*e*) $CH_2{=}C(CH_3)CH_2CH_2CH_3 + Br_2 \xrightarrow{CHCl_3}$

(*f*) $(CH_3)_2C{=}CHCH_3 + Br_2 \xrightarrow{H_2O}$

(*g*) 1-methylcyclopentene ($—CH_3$) $\xrightarrow[H_2O]{Cl_2}$

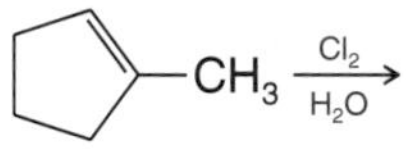

(*h*) $(CH_3)_2C{=}C(CH_3)_2 + CH_3\overset{O}{\overset{\|}{C}}OOH \longrightarrow$

(*i*) bicyclic alkene (octahydronaphthalene) $\xrightarrow[\text{2. } H_2O]{\text{1. } O_3}$

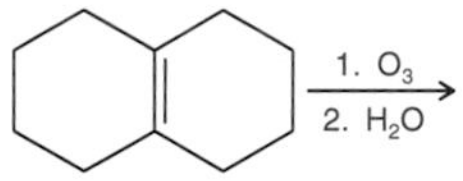

6.31 Suggest a sequence of reactions suitable for preparing each of the following compounds from the indicated starting material. You may use any necessary organic or inorganic reagents.

(*a*) 1-Propanol from 2-propanol
(*b*) 1-Bromopropane from 2-bromopropane
(*c*) 1,2-Dibromopropane from 2-bromopropane
(*d*) 1-Bromo-2-propanol from 2-propanol
(*e*) 1,2-Epoxypropane from 2-propanol
(*f*) *tert*-Butyl alcohol from isobutyl alcohol
(*g*) *tert*-Butyl iodide from isobutyl iodide
(*h*) *trans*-2-Chlorocyclohexanol from cyclohexyl chloride
(*i*) Cyclopentyl iodide from cyclopentane
(*j*) *trans*-1,2-Dichlorocyclopentane from cyclopentane
(*k*)  $H\overset{O}{\overset{\|}{C}}CH_2CH_2CH_2\overset{O}{\overset{\|}{C}}H$ from cyclopentanol

6.32 Two different compounds having the molecular formula $C_8H_{15}Br$ are formed when 1,6-dimethylcyclohexene reacts with hydrogen bromide in the dark and in the absence of peroxides. The same two compounds are formed from 1,2-dimethylcyclohexene. What are these two compounds?

6.33 Trimedlure is a powerful attractant for the Mediterranean fruit fly. It is a mixture of four isomers. In the synthesis of Trimedlure, the step in which four isomeric products are formed involves the reaction of hydrogen chloride with the compound shown.

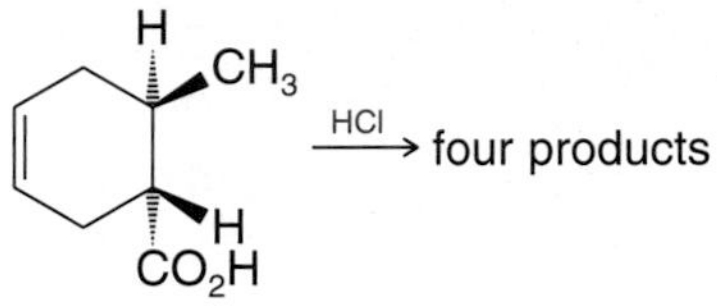

H, CH_3, H, CO_2H $\xrightarrow{HCl}$ four products

Write the structures of four isomeric products that could reasonably be formed under these reaction conditions.

6.34 On catalytic hydrogenation over a rhodium catalyst, the compound shown gave a mixture containing *cis*-1-*tert*-butyl-4-methylcyclohexane (88 percent) and *trans*-1-*tert*-butyl-4-methylcyclohexane (12 percent).

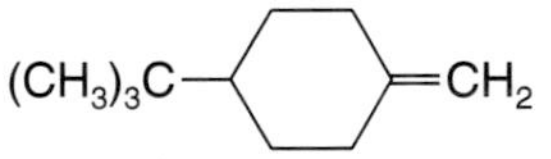

$(CH_3)_3C$—⟨cyclohexane⟩=CH_2 4-*tert*-Butyl(methylene)cyclohexane

(*a*) What two products are formed in the epoxidation of 4-*tert*-butyl(methylene)cyclohexane? Which one do you think will predominate?
(*b*) What two products are formed in the hydroboration-oxidation of 4-*tert*-butyl(methylene)cyclohexane? Which one do you think will predominate?

6.35 Compound A undergoes catalytic hydrogenation much faster than does compound B. Why?

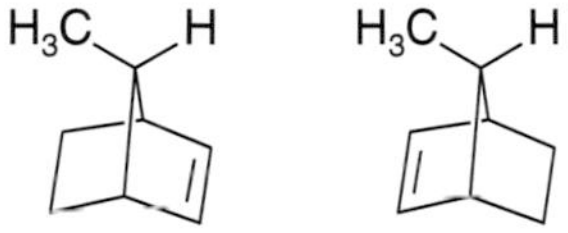

H_3C H H_3C H

Compound A Compound B

6.36 Catalytic hydrogenation of 1,4-dimethylcyclopentene yields a mixture of two products. Identify them. One of them is formed in much greater amounts than the other (observed ratio = 10 : 1). Which one is the major product?

6.37 There are two products that can be formed by syn addition of hydrogen to 2,3-dimethylbicyclo[2.2.1]-2-heptene. Write their structures.

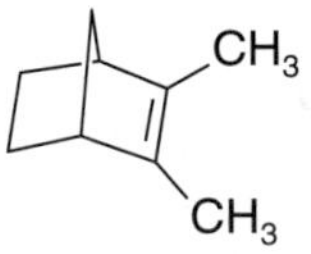

CH_3
CH_3 2,3-Dimethylbicyclo[2.2.1]-2-heptene

6.38 Hydrogenation of 3-carene is in principle capable of yielding two stereoisomeric products. Write their structures. Only one of them was actually obtained on catalytic hydrogenation over platinum. Which one do you think is formed? Explain your reasoning.

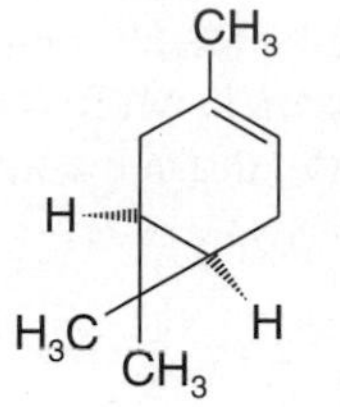

3-Carene

6.39 On the basis of the mechanism of acid-catalyzed hydration, can you suggest a reason why the reaction

$$CH_2{=}CHCH(CH_3)_2 \xrightarrow[H_2O]{H_2SO_4} CH_3\underset{\large OH}{\underset{|}{C}}HCH(CH_3)_2$$

would probably *not* be a good method for the synthesis of 3-methyl-2-butanol?

6.40 As a method for the preparation of alkenes, a weakness in the acid-catalyzed dehydration of alcohols is that the initially formed alkene (or mixture of alkenes) sometimes isomerizes under the conditions of its formation. Write a stepwise mechanism showing how 2-methyl-1-butene might isomerize to 2-methyl-2-butene in the presence of sulfuric acid.

6.41 When bromine is added to a solution of 1-hexene in methanol, the major products of the reaction are as shown:

$$\underset{\text{1-Hexene}}{CH_2{=}CHCH_2CH_2CH_2CH_3} \xrightarrow[CH_3OH]{Br_2} \underset{\text{1,2-Dibromohexane}}{BrCH_2\underset{Br}{\underset{|}{C}}HCH_2CH_2CH_2CH_3} + \underset{\text{1-Bromo-2-methoxyhexane}}{BrCH_2\underset{OCH_3}{\underset{|}{C}}HCH_2CH_2CH_2CH_3}$$

1,2-Dibromohexane is not converted to 1-bromo-2-methoxyhexane under the reaction conditions. Suggest a reasonable mechanism for the formation of 1-bromo-2-methoxyhexane.

6.42 The reaction of thiocyanogen ($N{\equiv}CS{-}SC{\equiv}N$) with *cis*-cyclooctene proceeds by anti addition.

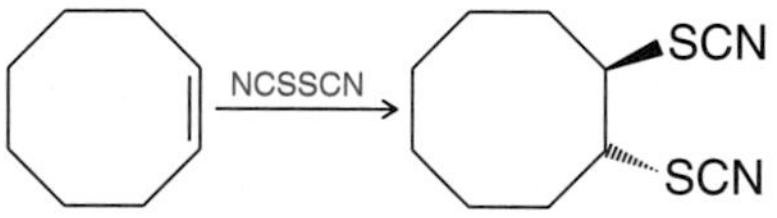

NCSSCN → SCN, SCN

A bridged *sulfonium ion* is presumed to be an intermediate. Write a stepwise mechanism for this reaction.

6.43 On the basis of the mechanism of cationic polymerization, predict the alkenes of molecular formula $C_{12}H_{24}$ that can most reasonably be formed when 2-methylpropene [$(CH_3)_2C{=}CH_2$] is treated with sulfuric acid.

6.44 On being heated with a solution of sodium ethoxide in ethanol, compound A ($C_7H_{15}Br$) yielded a mixture of two alkenes B and C, each having the molecular formula C_7H_{14}. Catalytic hydrogenation of the major isomer B or the minor isomer C gave only 3-ethylpentane. Suggest structures for compounds A, B, and C consistent with these observations.

6.45 Compound A ($C_7H_{15}Br$) is not a primary alkyl bromide. It yields a single alkene (compound B) on being heated with sodium ethoxide in ethanol. Hydrogenation of compound B yields 2,4-dimethylpentane. Identify compounds A and B.

6.46 Compounds A and B are isomers of molecular formula $C_9H_{19}Br$. Both yield the same alkene C as the exclusive product of elimination on being treated with potassium *tert*-butoxide in dimethyl sulfoxide. Hydrogenation of alkene C gives 2,3,3,4-tetramethylpentane. What are the structures of compounds A and B and alkene C?

6.47 Alcohol A ($C_{10}H_{18}O$) is converted to a mixture of alkenes B and C on being heated with potassium hydrogen sulfate ($KHSO_4$). Catalytic hydrogenation of B and C yields the same product. Assuming that dehydration of alcohol A proceeds without rearrangement, deduce the structures of alcohol A and alkene C.

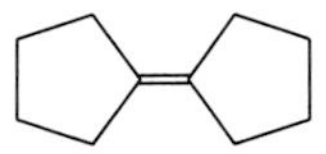

Compound B

6.48 Reaction of 3,3-dimethyl-1-butene with hydrogen iodide yields two compounds A and B, each having the molecular formula $C_6H_{13}I$, in the ratio A:B = 90:10. Compound A, on being heated with potassium hydroxide in *n*-propyl alcohol, gives only 3,3-dimethyl-1-butene. Compound B undergoes elimination under these conditions to give 2,3-dimethyl-2-butene as the major product. Suggest structures for compounds A and B, and write a reasonable mechanism for the formation of each.

6.49 Dehydration of 2,2,3,4,4-pentamethyl-3-pentanol gave two alkenes A and B. Ozonolysis of the lower-boiling alkene A gave formaldehyde ($CH_2{=}O$) and 2,2,4,4-tetramethyl-3-pentanone. Ozonolysis of B gave formaldehyde and 3,3,4,4-tetramethyl-2-pentanone. Identify A and B and suggest an explanation for the formation of B in the dehydration reaction.

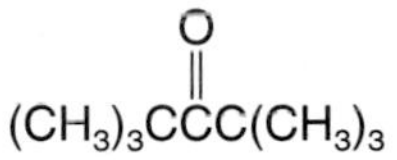

$(CH_3)_3CC(=O)C(CH_3)_3$

2,2,4,4-Tetramethyl-3-pentanone

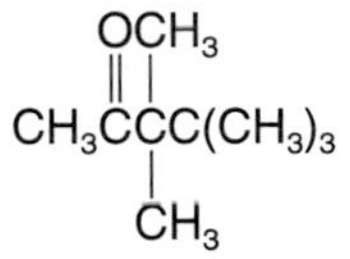

$CH_3C(=O)C(CH_3)(CH_3)C(CH_3)_3$

3,3,4,4-Tetramethyl-2-pentanone

6.50 Compound A ($C_7H_{13}Br$) is a tertiary bromide. On treatment with sodium ethoxide in ethanol, A is converted into B (C_7H_{12}). Ozonolysis of B gives C as the only product. Deduce the structures of A and B. What is the symbol for the reaction mechanism by which A is converted to B under the reaction conditions?

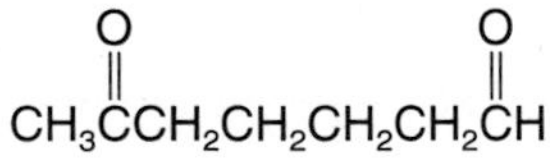

$CH_3C(=O)CH_2CH_2CH_2CH_2CH(=O)$

Compound C

6.51 East Indian sandalwood oil contains a hydrocarbon given the name *santene* (C_9H_{14}). Ozonation of santene followed by hydrolysis gives compound A. What is the structure of santene?

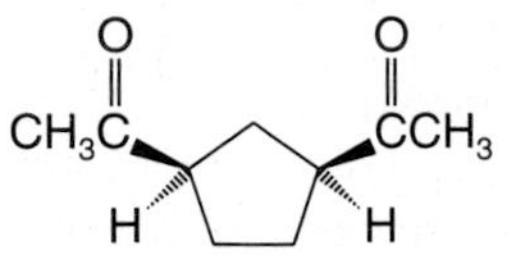

$CH_3C(=O)$ — $C(=O)CH_3$; H — H

Compound A

6.52 *Sabinene* and Δ^3-*carene* are isomeric natural products with the molecular formula $C_{10}H_{16}$. (*a*) Ozonolysis of sabinene followed by hydrolysis in the presence of zinc gives compound A. What is the structure of sabinene? What other compound is formed on ozonolysis? (*b*) Ozonolysis of Δ^3-carene followed by hydrolysis in the presence of zinc gives compound B. What is the structure of Δ^3-carene?

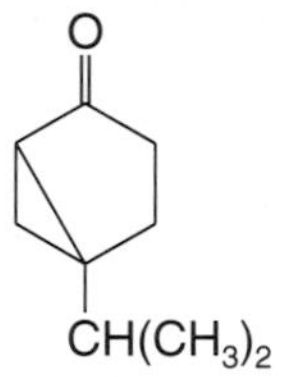

O

$CH(CH_3)_2$

Compound A

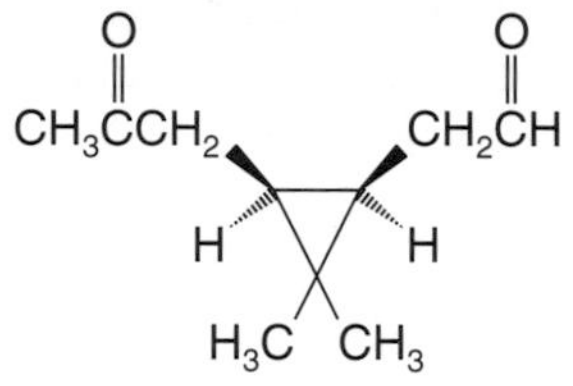

$CH_3\overset{O}{\overset{\|}{C}}CH_2$ $CH_2\overset{O}{\overset{\|}{C}}H$

H H

H_3C CH_3

Compound B

6.53 The sex attractant by which the female housefly attracts the male has the molecular formula $C_{23}H_{46}$. Catalytic hydrogenation yields an alkane of molecular formula $C_{23}H_{48}$. Ozonation followed by hydrolysis in the presence of zinc yields $CH_3(CH_2)_7\overset{O}{\overset{\|}{C}}H$ and $CH_3(CH_2)_{12}\overset{O}{\overset{\|}{C}}H$. What is the structure of the housefly sex attractant?

6.54 A certain compound of molecular formula $C_{19}H_{38}$ was isolated from fish oil and from plankton. On hydrogenation it gave 2,6,10,14-tetramethylpentadecane. Ozonation followed by hydrolysis in the presence of zinc gave $(CH_3)_2C{=}O$ and a 16-carbon aldehyde. What is the structure of the natural product? What is the structure of the aldehyde?

6.55 The sex attractant of the female arctiid moth contains, among other components, a compound of molecular formula $C_{21}H_{40}$ that yields $CH_3(CH_2)_{10}\overset{O}{\overset{\|}{C}}H$, $CH_3(CH_2)_4\overset{O}{\overset{\|}{C}}H$, and $H\overset{O}{\overset{\|}{C}}CH_2\overset{O}{\overset{\|}{C}}H$ on ozonolysis. What is the constitution of this material?

MOLECULAR MODELING EXERCISES

6.56 Construct a molecular model of the product formed by catalytic hydrogenation of 1,2-dimethylcyclohexene. Assume syn addition occurs.

6.57 Construct a molecular model of the product formed by anti addition of Br_2 to 1,2-dimethylcyclohexene.

6.58 Transform a molecular model of 1,2-dimethylcyclohexene to the product of its ozonolysis.

6.59 The epoxide formed by treating of methylenecyclohexane with peroxyacetic acid can exist in two nonequivalent chair conformations. Make a molecular model of each.

6.60 Construct a molecular model of α-pinene (page 216) and verify that the bottom face of the double bond is less sterically hindered than the top face.

CHAPTER 7

STEREOCHEMISTRY

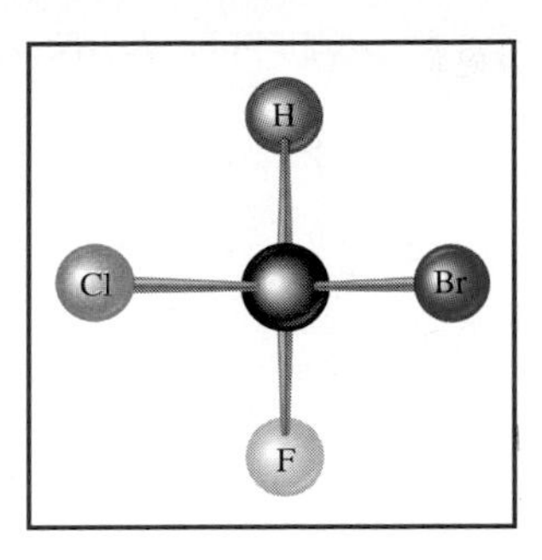

Stereos is a Greek word meaning "solid," and *stereochemistry* refers to chemistry in three dimensions. The foundations of organic stereochemistry were laid by Jacobus van't Hoff and Charles Le Bel in 1874. Independently of each other, van't Hoff and Le Bel proposed that the four bonds to carbon were directed toward the corners of a tetrahedron. One consequence of a tetrahedral arrangement of bonds to carbon is that two compounds may be different because the arrangement of their atoms in space is different. Isomers that have the same constitution but differ in the spatial arrangement of their atoms are called **stereoisomers.** We have already had considerable experience with certain types of stereoisomers—those involving cis and trans substitution patterns in alkenes and in cycloalkanes.

Van't Hoff was the recipient of the first Nobel Prize in chemistry in 1901 for his work in chemical dynamics and osmotic pressure—topics far removed from stereochemistry.

Our major objectives in this chapter are to develop a feeling for molecules as three-dimensional objects and to become familiar with stereochemical principles, terms, and notation. A full understanding of organic and biological chemistry requires an awareness of the spatial requirements for interactions between molecules; this chapter provides the basis for that understanding.

7.1 MOLECULAR CHIRALITY. ENANTIOMERS

Symmetry of form abounds in classical solid geometry. A sphere, a cube, a cone, and a tetrahedron are all identical with, and can be superposed point for point on, their mirror images. Mirror-image superposability also exists in many objects used every day. Cups and saucers, forks and spoons, chairs and beds, are all identical with their mirror images. Many other objects, however, cannot be superposed on their mirror images. Your left hand and your right hand, for example, are mirror images of each other but cannot be made to coincide point for point in three dimensions. In geometry, an object that is not superposable on its mirror image is said to be **dissymmetric,** where the prefix *dis-* signifies an opposite quality as in *discomfort* and *discontent.* In chemistry, the word that corresponds to *dissymmetric* is **chiral,** as in a *chiral molecule. Chiral* is derived from the Greek word *cheir,* meaning "hand," in that it refers to the "handedness" of molecules. The opposite of chiral is **achiral.** A molecule that *is* superposable on its mirror image is achiral.

Van't Hoff pointed out that a molecule is **asymmetric** (without symmetry) when four different groups are arranged in a tetrahedral fashion around one of its carbon atoms. All asymmetric structures are also dissymmetric and cannot be superposed on their mirror images. An example is bromochlorofluoromethane (BrClFCH), depicted in its two mirror-image forms in Figure 7.1. As demonstrated in the figure, these two mirror images cannot be brought to superposition upon each other. *Since mirror-image representations of bromochlorofluoromethane are nonsuperposable, BrClFCH is a chiral molecule.*

(*a*) Structures A and B are mirror-image representations of bromochlorofluoromethane (BrClFCH).

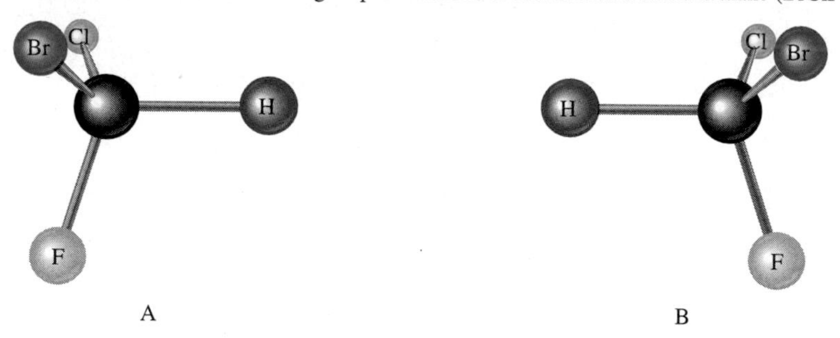

(*b*) To test for superposability, reorient B by turning it 180°.

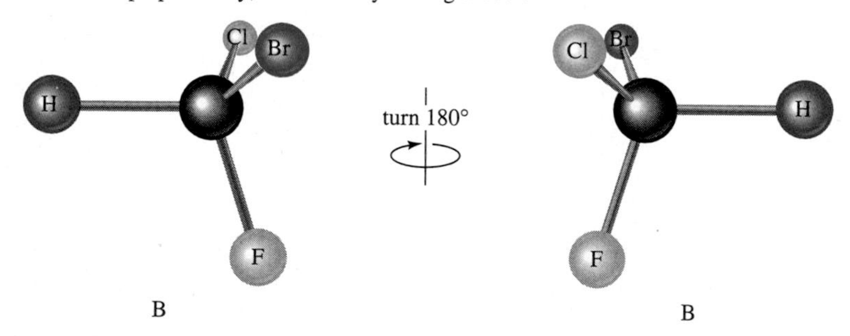

(*c*) Compare A and B. The two do not match. A and B cannot be superposed on each other. Therefore, bromochlorofluoromethane is a chiral molecule. The two mirror-image forms are enantiomers of each other.

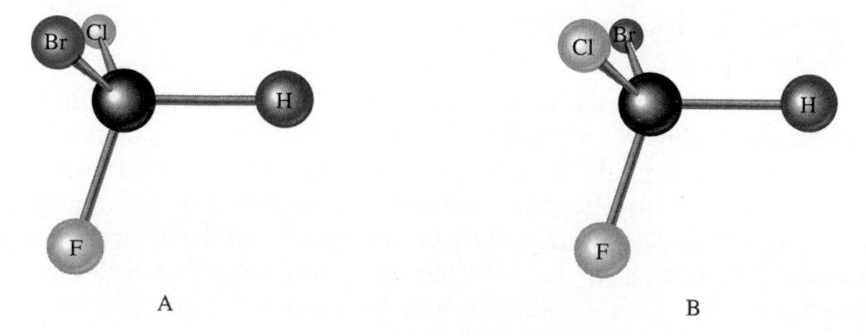

FIGURE 7.1 A molecule with four different groups attached to a single carbon is chiral. Its two mirror-image forms are not superposable.

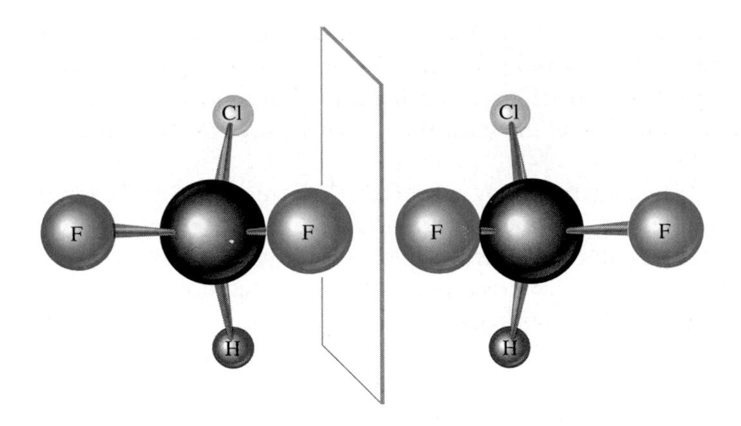

FIGURE 7.2 Mirror-image forms of chlorodifluoromethane are superposable on each other. Chlorodifluoromethane is achiral.

The two mirror-image representations of bromochlorofluoromethane have the same constitution. That is, the atoms are connected to each other in the same order. But the molecules differ in the arrangement of their atoms in space; they are **stereoisomers.** Stereoisomers that are related as an object and its nonsuperposable mirror image are classified as **enantiomers.** The word *enantiomer* describes a particular relationship between two objects. One cannot look at a single molecule in isolation and ask whether the molecule is an enantiomer any more than one can look at an individual human being and ask, ''Is that person a cousin?'' Furthermore, just as an object has one, and only one, mirror image, a chiral molecule can have one, and only one, enantiomer.

Bromochlorofluoromethane is a known compound, and samples selectively enriched in each enantiomer have been described in the chemical literature. In 1989 two chemists at Polytechnic University (Brooklyn, New York) described a method that they believe holds promise for the preparation of an enantiomerically homogeneous sample of BrClFCH.

Notice in Figure 7.1*c*, where the two enantiomers of bromochlorofluoromethane are similarly oriented, that the difference between them corresponds to an interchange of the positions of bromine and chlorine. It will generally be true for species of the type C(*w*, *x*, *y*, *z*), where *w*, *x*, *y*, and *z* are different atoms or groups, that an exchange of two atoms or groups converts a structure to its enantiomer while an exchange of three gives the original structure, albeit in a different orientation.

Consider next a molecule such as chlorodifluoromethane (ClF_2CH), in which two of the atoms attached to carbon are the same. Figure 7.2 shows two molecular models of ClF_2CH drawn so as to be related as an object and its mirror image. As is evident from these drawings, it is a simple matter to merge the two models so that all the atoms match. *Since mirror-image representations of chlorodifluoromethane are superposable on each other, ClF_2CH is an achiral molecule.*

The surest test for chirality is a careful examination of mirror-image forms for superposability. Working with models provides the best practice in dealing with molecules as three-dimensional objects and is strongly recommended.

7.2 THE STEREOGENIC CENTER

As we have just seen, molecules of the general type

x
w—C—y
z

An article in the December 1987 issue of the *Journal of Chemical Education* gives a thorough discussion of molecular chirality and examines some of its past and present terminology.

are chiral when *w, x, y,* and *z* are different substituents. A tetrahedral carbon atom that bears four different substituents is variously referred to as a *chiral center,* a *chiral carbon atom,* an *asymmetric center,* or an *asymmetric carbon atom.* A more modern term is **stereogenic center,** and that is the term that we shall use. (*Stereocenter* is synonymous with *stereogenic center.*)

Noting the presence of a stereogenic center in a given molecule is a simple, rapid way to determine that the molecule is chiral. For example, C-2 is a stereogenic center in 2-butanol; it bears a hydrogen atom and methyl, ethyl, and hydroxyl groups as its four different substituents. By way of contrast, none of the carbon atoms bear four different groups in the achiral alcohol 2-propanol.

$$CH_3-\underset{OH}{\overset{H}{C}}-CH_2CH_3 \qquad CH_3-\underset{OH}{\overset{H}{C}}-CH_3$$

2-Butanol: chiral; four different substituents at C-2

2-Propanol: achiral; two of the substituents at C-2 are the same

PROBLEM 7.1 Examine the following for stereogenic centers:

(*a*) 2-Bromopentane
(*b*) 3-Bromopentane
(*c*) 1-Bromo-2-methylbutane
(*d*) 2-Bromo-2-methylbutane

SAMPLE SOLUTION A stereogenic carbon has four different substituents. (*a*) In 2-bromopentane, C-2 satisfies this requirement. (*b*) None of the carbons in 3-bromopentane have four different substituents, and so none of its atoms are stereogenic centers.

$$CH_3-\underset{Br}{\overset{H}{C}}-CH_2CH_2CH_3 \qquad CH_3CH_2-\underset{Br}{\overset{H}{C}}-CH_2CH_3$$

2-Bromopentane

3-Bromopentane

Molecules with stereogenic centers are very common, both as naturally occurring substances and as the products of chemical synthesis. In the following examples, the stereogenic carbon is indicated by an asterisk. (Carbons that are part of a double bond or a triple bond cannot be stereogenic centers.)

$$CH_3CH_2CH_2-\overset{*}{\underset{CH_2CH_3}{\overset{CH_3}{C}}}-CH_2CH_2CH_2CH_3 \qquad (CH_3)_2C{=}CHCH_2CH_2-\overset{*}{\underset{OH}{\overset{CH_3}{C}}}-CH{=}CH_2$$

4-Ethyl-4-methyloctane (a chiral alkane)

Linalool (a pleasant-smelling oil obtained from orange flowers)

A carbon atom in a ring can be a stereogenic center if it bears two different substituents and the path traced around the ring from that carbon in one direction is different from that traced in the other. The carbon atom that bears the methyl group in 1,2-epoxypropane, for example, is a stereogenic center. The sequence of groups is CH_2—O as one proceeds clockwise around the ring from that atom, but is O—CH_2 in the anticlockwise direction. Similarly, C-4 is a stereogenic center in limonene.

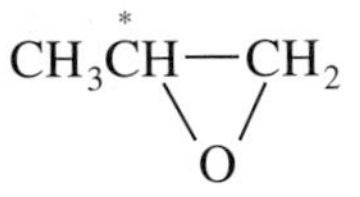

1-2-Epoxypropane (product of epoxidation of propene)

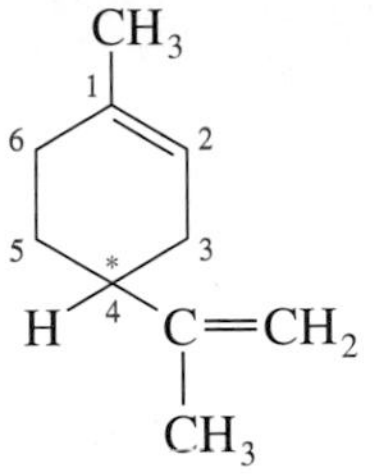

Limonene (a constituent of lemon oil)

PROBLEM 7.2 Identify the stereogenic centers, if any, in

(*a*) 2-Cyclopenten-1-ol
(*b*) 3-Cyclopenten-1-ol
(*c*) 1,1,2-Trimethylcyclobutane
(*d*) 1,1,3-Trimethylcyclobutane

SAMPLE SOLUTION (*a*) The hydroxyl-bearing carbon in 2-cyclopenten-1-ol is a stereogenic center.

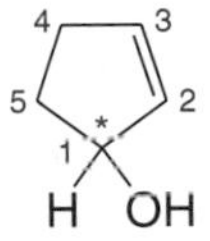

2-Cyclopenten-1-ol

(*b*) There is no stereogenic center in 3-cyclopenten-1-ol, since the sequence of atoms 1 → 2 → 3 → 4 → 5 is equivalent regardless of whether one proceeds clockwise or anticlockwise.

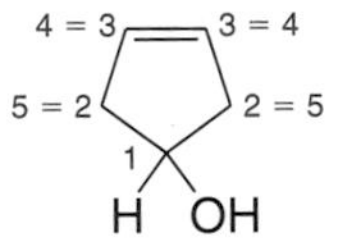

3-Cyclopenten-1-ol (does not have a stereogenic carbon)

Molecules with more than one stereogenic center may or may not be chiral; these will be discussed in Sections 7.10 through 7.13.

7.3 SYMMETRY IN ACHIRAL STRUCTURES

Certain structural features related to molecular symmetry can sometimes help us determine by inspection whether a molecule is chiral or achiral. For example, a molecule that has a *plane of symmetry* or a *center of symmetry* is superposable on its mirror image and is achiral.

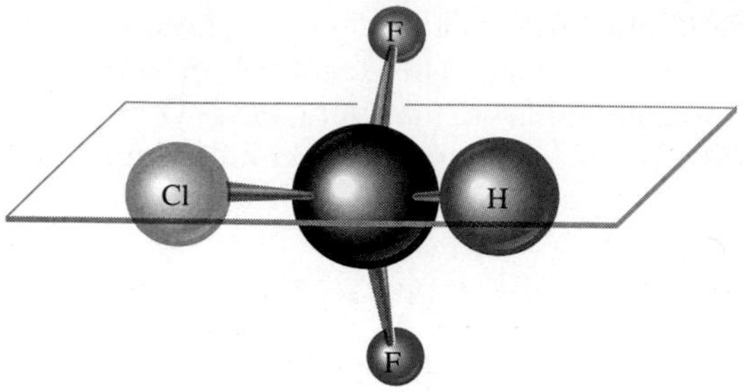

FIGURE 7.3 A plane of symmetry defined by the atoms H—C—Cl divides chlorodifluoromethane into two mirror-image halves.

A **plane of symmetry** bisects a molecule so that one half of the molecule is the mirror image of the other half. The achiral molecule chlorodifluoromethane, for example, has the plane of symmetry shown in Figure 7.3.

A point in a molecule is a **center of symmetry** if any line drawn from it to some element of the structure will, when extended an equal distance in the opposite direction, encounter an identical element. The cyclobutane derivative in Figure 7.4 lacks a plane of symmetry, yet is achiral because it possesses a center of symmetry.

PROBLEM 7.3 Locate any planes of symmetry or centers of symmetry in each of the following compounds. Which of the compounds are chiral? Which are achiral?

(*a*) (*E*)-1,2-Dichloroethene
(*b*) (*Z*)-1,2,Dichloroethene
(*c*) *cis*-1,2-Dichlorocyclopropane
(*d*) *trans*-1,2-Dichlorocyclopropane

SAMPLE SOLUTION (*a*) (*E*)-1,2-Dichloroethene is a planar molecule. The molecular plane is a plane of symmetry.

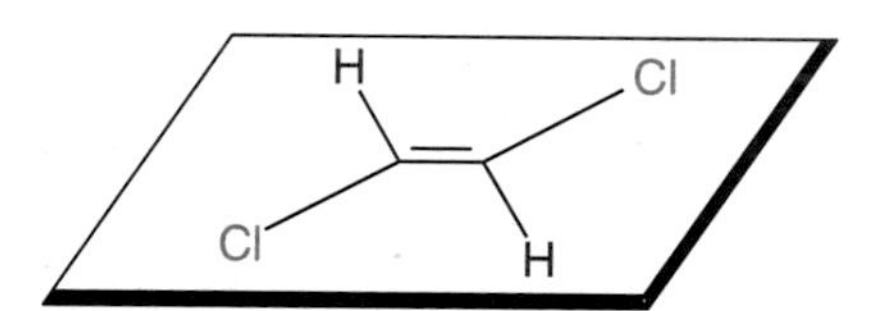

Further, (*E*)-1,2-dichloroethene has a center of symmetry located at the midpoint of the carbon-carbon double bond. It is an achiral molecule.

The preceding brief discussion covers most of the situations we shall encounter that involve symmetry and allows us to identify certain, but not all, structural features that lead to achiral molecules. Any molecule with a plane of symmetry or a center of symmetry is achiral, but the absence of these symmetry elements is not sufficient for a molecule to be chiral. A molecule lacking a center of symmetry or a plane of symmetry is *likely* to be chiral, but the superposability test should be applied to be certain.

FIGURE 7.4 (*a*) Structural formulas A and B are drawn as mirror images. (*b*) The two mirror images are superposable by rotating form B 180° about an axis passing through the center of the molecule. The center of the molecule is a center of symmetry.

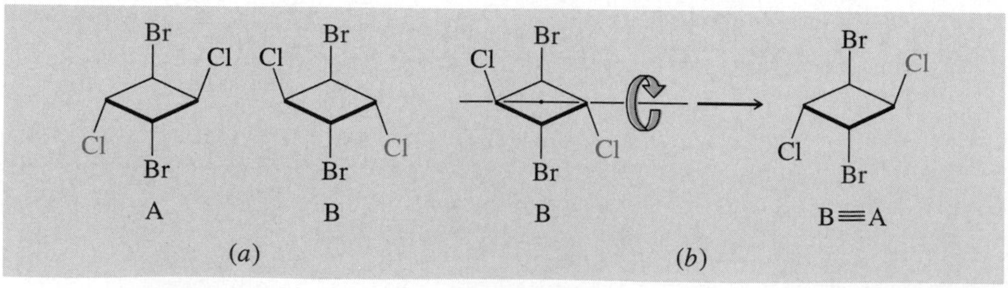

7.4 PROPERTIES OF CHIRAL MOLECULES: OPTICAL ACTIVITY

The phenomenon of optical activity was discovered by the French physicist Jean-Baptiste Biot in 1815.

The experimental facts that led van't Hoff and Le Bel to propose that molecules having the same constitution could differ in the arrangement of their atoms in space concerned a physical property called **optical activity**. Optical activity is the ability of a chiral substance to rotate the plane of **plane-polarized light** and is measured using an instrument called a **polarimeter.** Before describing the operation of a polarimeter, we need to examine some characteristics of light, especially plane-polarized light.

Electromagnetic radiation, including visible light, has a wave nature and has both an electric field and a magnetic field associated with it. We need consider only the electric field and can represent the wave property of light as a regular variation in the strength of its electric field, as shown in Figure 7.5*a*. The wavelength is the distance for one complete cycle and is 589 nm (589×10^{-9} m) for the yellow light produced by sodium lamps. This 589-nm light is called the *D line* of sodium and is the wavelength used most often when measuring optical activity. The amplitude of the electric field vector, represented by the arrows in Figure 7.5*a*, changes with time and, when viewed along the line of propagation, can be represented in composite form as in Figure 7.5*b*. Light such as this that vibrates in a single plane is plane-polarized light. However, the source emits waves in all possible planes, not just one. Thus the light emitted by the source is **unpolarized** (Figure 7.5*c*). One of the functions of a polarimeter is to extract plane-polarized light from the unpolarized light produced by the source.

Figure 7.6 outlines how optical activity is measured using a polarimeter. The unpolarized light from a sodium lamp is first passed through a *polarizer.* The polarizer is a device called a *Nicol prism,* which can be thought of as a grid that permits passage of only those components of the light beam that have the same plane of polarization. This light, now plane-polarized, passes through a polarimeter tube containing the substance to be examined, either in the liquid phase or as a solution in a suitable solvent (usually water, ethanol, or chloroform). The sample is ''optically active'' if it rotates the plane of polarized light. The occurrence of rotation, as well as its magnitude, is determined

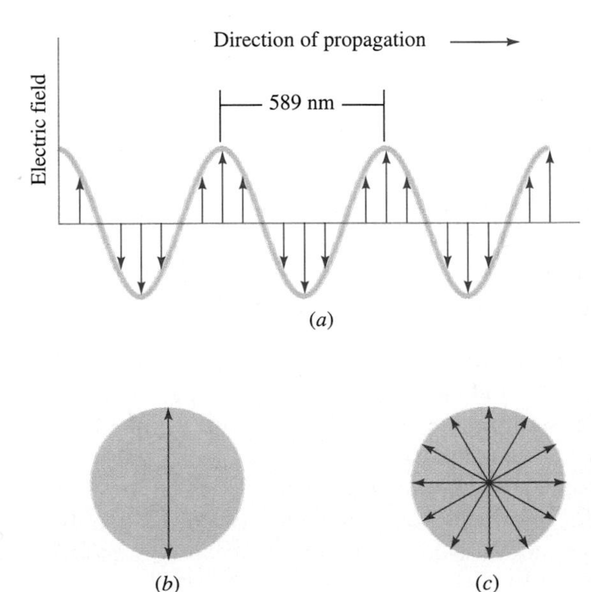

FIGURE 7.5 The yellow light from a sodium lamp that has a wavelength of 589 nm is the most common form of electromagnetic radiation used in measuring optical activity. Electromagnetic radiation undergoes a regular increase and decrease in the magnitude of its electric field vector as represented by the arrows in part *a*. A single wave is polarized in a single plane, and when viewed along the line of propagation, the increase and decrease in its amplitude can be represented as in *b*. A beam of light comprises many waves, and *c* represents the composite of the electric field vectors for six waves of unpolarized light that differ by increments of 30° from one another in their plane of polarization.

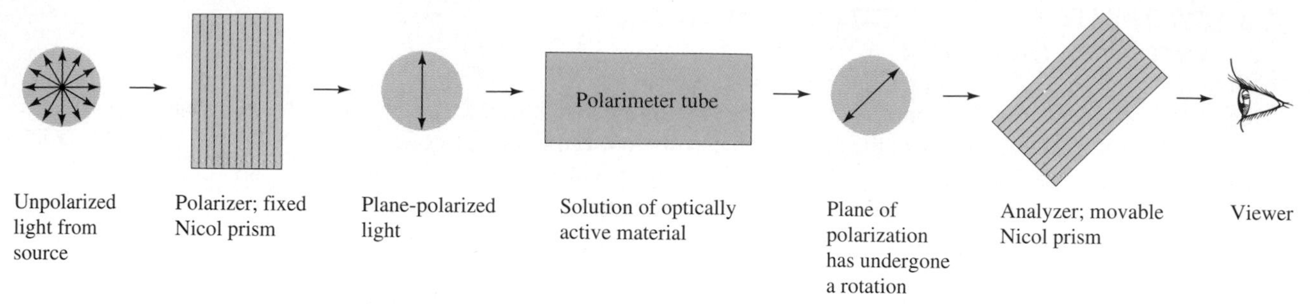

FIGURE 7.6 Measurement of optical activity using a polarimeter.

by using a second Nicol prism as an analyzer. If the plane of polarization undergoes a rotation, the second Nicol prism (the analyzer) must be oriented at an angle to the first (the polarizer) in order to transmit the light beam. Measuring this angle gives the observed rotation α of the sample.

In order to be optically active, the sample must contain a chiral substance and one enantiomer must be present in excess of the other. A substance that does not rotate the plane of polarized light is said to be optically inactive. *All achiral substances are optically inactive.*

What causes optical rotation? The plane of polarization of a light wave undergoes a minute rotation when it encounters a chiral molecule. Enantiomeric forms of a chiral molecule cause a rotation of the plane of polarization in exactly equal amounts but in opposite directions. Therefore a solution containing equal quantities of enantiomers exhibits no net rotation because all the tiny increments of clockwise rotation produced by molecules of one ''handedness'' are canceled by an equal number of increments of anticlockwise rotation produced by molecules of the opposite handedness.

Mixtures containing equal quantities of enantiomers are called **racemic mixtures.** Racemic mixtures are optically inactive. Conversely, when one enantiomer is present in excess, a net rotation of the plane of polarization is observed. At the limit, where all the molecules are of the same handedness, we say the substance is **optically pure.** Optical purity, or *percent enantiomeric excess,* is defined as follows:

$$\begin{aligned}\text{Optical purity} &= \text{percent enantiomeric excess}\\ &= \text{percent of one enantiomer} - \text{percent of other enantiomer}\end{aligned}$$

Thus, a material that is 50 percent optically pure contains 75 percent of one enantiomer and 25 percent of the other.

Rotation of the plane of polarized light in the clockwise sense is taken as positive (+), while rotation in the anticlockwise sense is taken as a negative (−) rotation. The classical terms for positive and negative rotations are *dextrorotatory* and *levorotatory,* from the Latin prefixes *dextro-* (''to the right'') and *levo-* (''to the left''), respectively. At one time, the symbols *d* and *l* were used to distinguish between enantiomeric forms of a substance. Thus the dextrorotatory enantiomer of 2-butanol was called *d*-2-butanol, and the levorotatory form *l*-2-butanol; a racemic mixture of the two was referred to as *dl*-2-butanol. Current custom favors using algebraic signs instead, as in (+)-2-butanol, (−)-2-butanol, and (±)-2-butanol, respectively.

The observed rotation α of an optically pure substance depends on how many molecules the light beam encounters. A filled polarimeter tube twice the length of another produces twice the observed rotation, as does a solution twice as concentrated.

In order to account for the effects of path length and concentration, chemists have defined the term **specific rotation,** given the symbol $[\alpha]$. Specific rotation is calculated from the observed rotation according to the expression

$$[\alpha] = \frac{100\alpha}{cl}$$

where c is the concentration of the sample in grams per 100 mL of solution and l is the length of the polarimeter tube in decimeters. (One decimeter is 10 cm.)

If concentration is expressed as g/mL of solution instead of g/100 mL, an equivalent expression is $[\alpha] = \frac{\alpha}{cl}$

Specific rotation is a physical property of a substance, just as melting point, boiling point, density, and solubility are. For example, the lactic acid obtained from milk is exclusively a single enantiomer. We cite its specific rotation in the form $[\alpha]_D^{25} = +3.8°$. The temperature in degrees Celsius and the wavelength of light at which the measurement was made are indicated as superscripts and subscripts, respectively.

PROBLEM 7.4 Cholesterol, when isolated from natural sources, is obtained as a single enantiomer. The observed rotation α of a 0.3-g sample of cholesterol in 15 mL of chloroform solution contained in a 10-cm polarimeter tube is $-0.78°$. Calculate the specific rotation of cholesterol.

PROBLEM 7.5 A sample of synthetic cholesterol was prepared consisting entirely of the enantiomer of natural cholesterol. A mixture of natural and synthetic cholesterol has a specific rotation $[\alpha]_D^{20}$ of $-13°$. What fraction of the mixture is natural cholesterol?

It is convenient to distinguish between enantiomers by prefixing the sign of rotation to the name of the substance. For example, we refer to one of the enantiomers of 2-butanol as (+)-2-butanol and the other as (−)-2-butanol. Optically pure (+)-2-butanol has a specific rotation $[\alpha]_D^{27}$ of $+13.5°$; optically pure (−)-2-butanol has an exactly opposite specific rotation $[\alpha]_D^{27}$ of $-13.5°$.

7.5 ABSOLUTE AND RELATIVE CONFIGURATION

The precise arrangement of substituents at a stereogenic center is its **absolute configuration.** Neither the sign nor the magnitude of rotation by itself provides any information concerning the absolute configuration of a substance. Thus, one of the following structures is (+)-2-butanol and the other is (−)-2-butanol, but in the absence of additional information we cannot tell which is which.

CH_3CH_2, H, C—OH, CH_3

H, CH_2CH_3, HO—C, CH_3

While no absolute configuration was known for any substance before 1951, organic chemists had experimentally determined the configurations of thousands of compounds relative to one another (their **relative configurations**) through chemical interconversion. To illustrate, consider (+)-3-buten-2-ol. Hydrogenation of this compound yields (+)-2-butanol.

$$\underset{\substack{\text{3-Buten-2-ol} \\ [\alpha]_D^{27}\ +33.2^\circ}}{\underset{\substack{| \\ OH}}{CH_3\overset{*}{C}HCH{=}CH_2}} + \underset{\text{Hydrogen}}{H_2} \xrightarrow{Pd} \underset{\substack{\text{2-Butanol} \\ [\alpha]_D^{27}\ +13.5^\circ}}{\underset{\substack{| \\ OH}}{CH_3\overset{*}{C}HCH_2CH_3}}$$

Since hydrogenation of the double bond does not involve any of the bonds to the stereogenic center, the spatial arrangement of substituents in (+)-3-buten-2-ol must be the same as that of the substituents in (+)-2-butanol. The fact that these two compounds have the same sign of rotation when they have the same relative configuration is established by the hydrogenation experiment; it could not have been predicted in advance of the experiment.

Sometimes compounds that have the same relative configuration have optical rotations of opposite sign. For example, treatment of (−)-2-methyl-1-butanol with hydrogen bromide converts it to (+)-1-bromo-2-methylbutane.

$$\underset{\substack{\text{2-Methyl-1-butanol} \\ [\alpha]_D^{25}\ -5.8^\circ}}{\underset{\substack{| \\ CH_3}}{CH_3CH_2\overset{*}{C}HCH_2OH}} + \underset{\substack{\text{Hydrogen} \\ \text{bromide}}}{HBr} \longrightarrow \underset{\substack{\text{1-Bromo-2-methylbutane} \\ [\alpha]_D^{25}\ +4.0^\circ}}{\underset{\substack{| \\ CH_3}}{CH_3CH_2\overset{*}{C}HCH_2Br}} + \underset{\text{Water}}{H_2O}$$

This reaction does not involve any of the bonds to the stereogenic center, and so both the starting alcohol and the product bromide have the same relative configuration.

An elaborate network connecting sign of rotation and relative configuration was developed that included the most important compounds of organic chemistry and biological chemistry. When, in 1951, the absolute configuration of a salt of (+)-tartaric acid was determined, the absolute configurations of all the compounds whose configurations had been related to (+)-tartaric acid stood revealed as well. Thus, returning to the pair of 2-butanol enantiomers that introduced this section, their absolute configurations are now known to be as shown.

CH_3CH_2, H, C—OH, CH_3	HO—C, H, CH_2CH_3, CH_3
(+)-2-Butanol	(−)-2-Butanol

7.6 THE CAHN-INGOLD-PRELOG *R-S* NOTATIONAL SYSTEM

Just as it makes sense to have a system of nomenclature that permits structural information to be communicated without the necessity of writing a constitutional formula for each compound, so also is it reasonable to have a notational system that simply and unambiguously describes the absolute configuration of a substance. We have already had experience with this principle when we distinguished between stereoisomers in cyclic compounds and in alkenes.

cis-4-Methylcyclohexanol

(*E*)-3-Isopropyl-2-hexene

In the case of alkenes, the system using the descriptors *E* and *Z* proved to be more versatile than the earlier cis and trans notation.

In the *E-Z* system, substituents are ranked according to a set of rules based on atomic number precedence (Section 5.4). Historically, the Cahn-Ingold-Prelog rules were first formulated to deal with the problem of the absolute configuration at a stereogenic center, and this is their major application. Table 7.1 shows how this system, called the **sequence rule** by its developers, is used to specify the absolute configuration at the stereogenic center in (+)-2-butanol.

TABLE 7.1
Absolute Configuration According to the Cahn-Ingold-Prelog Notational System

Step number	Example
Given that the absolute configuration of (+)-2-butanol is	(+)-2-Butanol
1. Identify the substituents at the stereogenic center and rank them in order of decreasing precedence according to the system described in Section 5.4. Precedence is determined by atomic number, working outward from the point of attachment at the stereogenic center.	In order of decreasing precedence, the four substituents attached to the stereogenic center of 2-butanol are $HO{-} > CH_3CH_2{-} > CH_3{-} > H{-}$ (highest) (lowest)
2. Orient the molecule so that the lowest-ranked substituent points away from you.	As represented in the wedge-and-dash drawing at the top of this table, the molecule is already appropriately oriented. Hydrogen is the lowest-ranked substituent attached to the stereogenic center, and points away from us.
3. Draw the three highest-ranked substituents as they appear to you when the molecule is oriented so that the lowest-ranked group points away from you.	
4. If the order of decreasing precedence of the three highest-ranked substituents appears in a clockwise sense, the absolute configuration is *R* (Latin *rectus,* "right," "correct"). If the order of decreasing precedence is anticlockwise, the absolute configuration is *S* (Latin *sinister,* "left").	The order of decreasing precedence is *anticlockwise.* The configuration at the stereogenic center is *S.* (highest) (second highest) (third highest)

The January 1994 issue of the *Journal of Chemical Education* contains an article that describes how to use your hands to assign *R* and *S* configurations.

As outlined in Table 7.1, (+)-2-butanol has the *S* configuration. Its mirror image is (−)-2-butanol, which has the *R* configuration.

CH_3CH_2, H, C—OH, CH_3 and HO—C, H, CH_2CH_3, CH_3

(*S*)-2-Butanol (*R*)-2-Butanol

Often, the *R* or *S* descriptor of absolute configuration and the sign of rotation are incorporated into the name of the compound, as in (*R*)-(−)-2-butanol and (*S*)-(+)-2-butanol. A racemic mixture can be denoted by combining *R* and *S*, as in (*R*)(*S*)-2-butanol.

PROBLEM 7.6 Assign absolute configurations as *R* or *S* to each of the following compounds:

(*a*) H_3C, H, C—CH_2OH, CH_3CH_2

(+)-2-Methyl-1-butanol

(*b*) H_3C, H, C—CH_2F, CH_3CH_2

(+)-1-Fluoro-2-methylbutane

(*c*) H, CH_3, C—CH_2Br, CH_3CH_2

(+)-1-Bromo-2-methylbutane

(*d*) H_3C, H, C—CH=CH_2, HO

(+)-3-Buten-2-ol

SAMPLE SOLUTION (*a*) The highest-ranking substituent at the stereogenic center of 2-methyl-1-butanol is CH_2OH; the lowest is H. Of the remaining two, ethyl outranks methyl.

Order of precedence: $CH_2OH > CH_3CH_2 > CH_3 > H$

The lowest-ranking substituent (hydrogen) points away from us, and so the molecule is oriented properly as drawn. The three highest-ranking substituents trace a clockwise path from $CH_2OH \rightarrow CH_3CH_2 \rightarrow CH_3$.

H_3C CH_2OH CH_3CH_2

Therefore, this compound has the *R* configuration. It is (*R*)-(+)-2-methyl-1-butanol.

The system is broadly applicable. Compounds in which a stereogenic center is part of a ring are handled in an analogous fashion. To determine, for example, whether the configuration of (+)-4-methylcyclohexene is *R* or *S*, treat the right- and left-hand paths around the ring as if they were independent substituents.

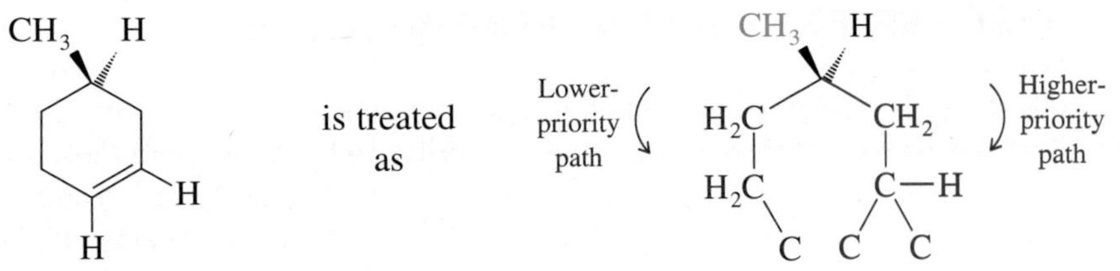

CH$_3$ H

H

H

(+)-4-Methylcyclohexene

is treated as

Lower-priority path

CH$_3$ H

H$_2$C CH$_2$

H$_2$C C—H

C C C

Higher-priority path

With the lowest-ranked substituent (hydrogen) directed away from us, we see that the order of decreasing sequence rule precedence is *clockwise*. The absolute configuration is *R*.

PROBLEM 7.7 Draw three-dimensional representations of

(*a*) The *R* enantiomer of

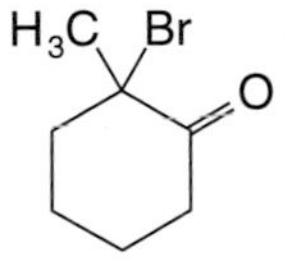

H$_3$C Br

O

(*b*) The *S* enantiomer of

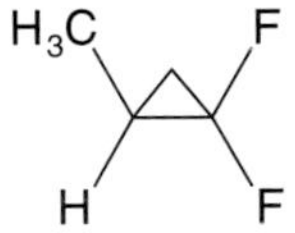

H$_3$C F

H F

SAMPLE SOLUTION (*a*) The stereogenic center is the one that bears the bromine. In order of decreasing precedence, the substituents attached to the stereogenic center are

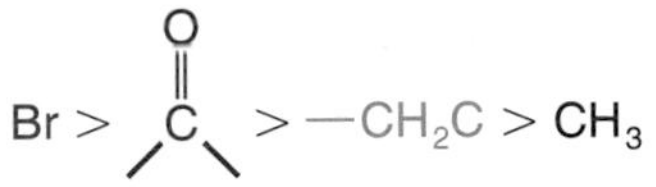

$$\text{Br} > \overset{\text{O}}{\overset{\|}{\text{C}}} > \text{—CH}_2\text{C} > \text{CH}_3$$

When the lowest-ranked substituent (the methyl group) is away from us, the order of decreasing precedence of the remaining groups must appear in a clockwise sense in the *R* enantiomer.

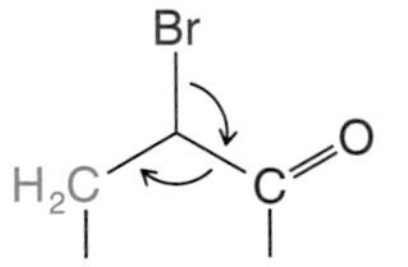

Br

H$_2$C C=O

Therefore, we can represent the *R* enantiomer as

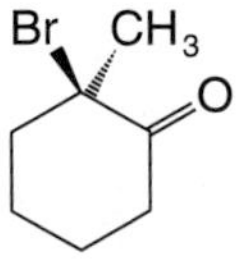

Br CH$_3$

O

(*R*)-2-Bromo-2-methylcyclohexanone

Since its inception in 1956, the Cahn-Ingold-Prelog system has received almost universal acceptance as the standard method of stereochemical notation.

7.7 FISCHER PROJECTION FORMULAS

Fischer was the foremost organic chemist of the late nineteenth century. He won the 1902 Nobel Prize in chemistry for his pioneering work in carbohydrate and protein chemistry.

Stereochemistry is concerned with the three-dimensional arrangement of a molecule's atoms, and we have attempted to show stereochemistry with wedge-and-dash drawings and computer-generated ball-and-stick models. It is possible, however, to convey stereochemical information in an abbreviated form using a method devised by the German chemist Emil Fischer.

As in Section 7.1, again consider bromochlorofluoromethane as a simple example of a chiral molecule. The two enantiomers of BrClFCH are shown as ball-and-stick models, as wedge-and-dash drawings, and as **Fischer projections** in Figure 7.7. Fischer projections are always generated the same way: the molecule is oriented so that the vertical bonds at the stereogenic center are directed away from you and the horizontal bonds point toward you. A projection of the bonds onto the page is a cross. The stereogenic carbon lies at the center of the cross but is not explicitly shown.

It is customary to orient molecules that have several carbons so that the carbon chain is vertical as shown for the Fischer projection of (*R*)-2-butanol.

The Fischer projection

CH_3
HO—+—H
CH_2CH_3

(*R*)-2-Butanol

corresponds to

CH_3
HO—C—H
CH_2CH_3

Assignment of *R* and *S* configurations to molecules on the basis of their Fischer projections is the subject of a brief note in the June 1989 issue of the *Journal of Chemical Education.*

When specifying a configuration as *R* or *S*, the safest course of action is to convert a Fischer projection to a three-dimensional representation, remembering that the horizontal bonds always point toward you.

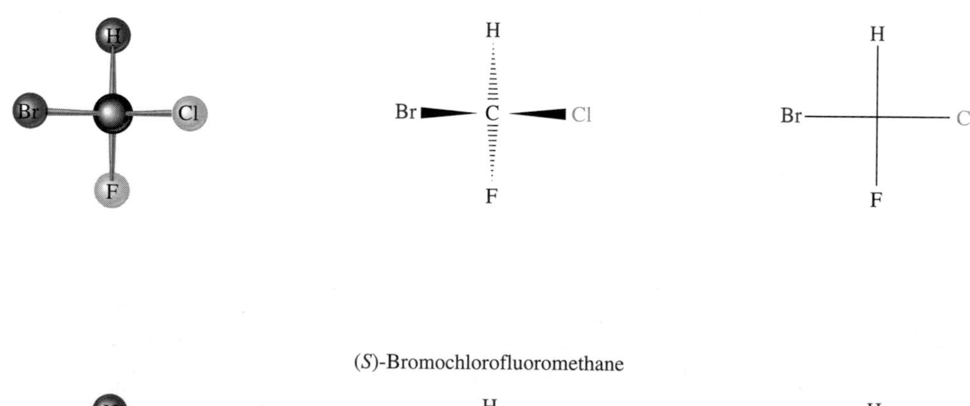

FIGURE 7.7 Ball-and-stick models (left), wedge-and-dash drawings (center), and Fischer projection formulas (right) of the *R* and *S* enantiomers of bromochlorofluoromethane.

PROBLEM 7.8 Write Fischer projections for each of the compounds of Problem 7.6.

SAMPLE SOLUTION (a) The structure of (*R*)-(+)-2-methyl-1-butanol is shown at the left. View the structural formula from a position chosen so that the $CH_3—C—CH_2CH_3$ segment is aligned vertically, with the methyl and the ethyl groups pointing away from you. Replace the wedge-and-dash bonds by lines to give the Fischer projection shown at the right.

CH_3, H, CH_3CH_2 on C—CH_2OH is the same as $HOCH_2$—C—H (with CH_3 above and CH_2CH_3 below) which becomes the Fischer projection $HOCH_2$—+—H (with CH_3 above and CH_2CH_3 below)

7.8 PHYSICAL PROPERTIES OF ENANTIOMERS

The usual physical properties of density, melting point, and boiling point are identical within experimental error for both enantiomers of a chiral compound.

Enantiomers can have striking differences, however, in properties that depend on the arrangement of atoms in space. Take, for example, the enantiomeric forms of carvone. (*R*)-(−)-Carvone is the principal component of spearmint oil. Its enantiomer, (*S*)-(+)-carvone, is the principal component of caraway seed oil. The two enantiomeric forms of carvone do not smell the same; each has its own characteristic odor.

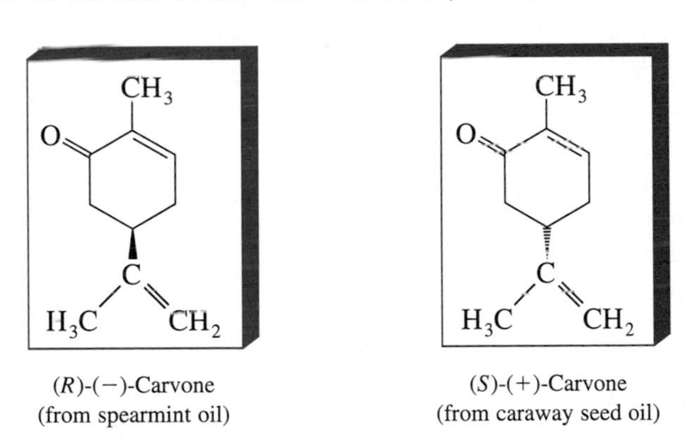

(*R*)-(−)-Carvone (from spearmint oil)   (*S*)-(+)-Carvone (from caraway seed oil)

The difference in odor between (*R*)- and (*S*)-carvone results from their different behavior toward receptor sites in the nose. It is believed that volatile molecules occupy only those odor receptors that have the proper shape to accommodate them. These receptor sites are themselves chiral, with the result that one enantiomer may fit one kind of receptor site while the other enantiomer fits a different kind of receptor. One analogy that can be drawn is to hands and gloves. Your left hand and your right hand are enantiomers. You can place your left hand into a left glove but not into a right one. The receptor site (the glove) can accommodate one enantiomer of a chiral object (your hand) but not the other.

The term *chiral recognition* refers to the process whereby some chiral receptor or reagent interacts selectively with one of the enantiomers of a chiral molecule. Very high levels of chiral recognition are common in biological processes. (−)-Nicotine, for example, is much more toxic than (+)-nicotine, and (+)-adrenaline is more active in the constriction of blood vessels than (−)-adrenaline. (−)-Thyroxine is an amino acid of the thyroid gland, which speeds up metabolism and causes nervousness and loss of weight. Its enantiomer, (+)-thyroxine, exhibits none of these effects but is sometimes given to heart patients to lower their cholesterol levels.

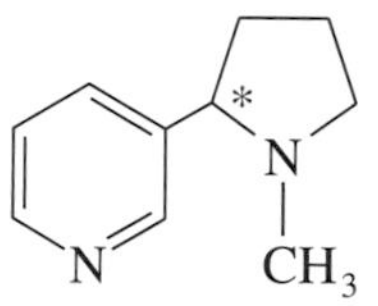

Nicotine

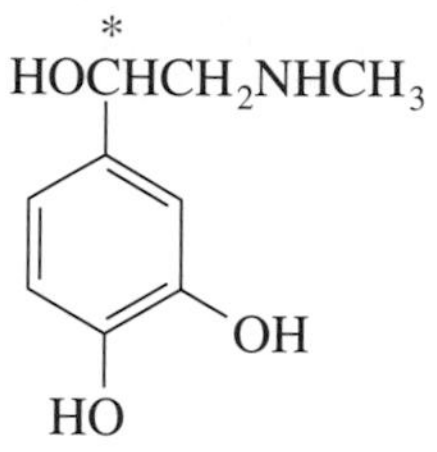

Adrenaline: $HO\overset{*}{C}HCH_2NHCH_3$

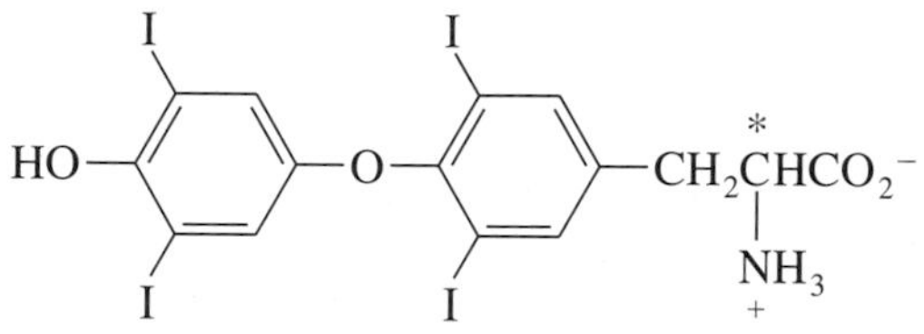

Thyroxine: $-CH_2\overset{*}{C}HCO_2^-$, NH_3^+

(Stereogenic centers identified by asterisks)

CHIRAL DRUGS

A recent estimate places the number of prescription and over-the-counter drugs marketed throughout the world at about 2000. Approximately one-third of these are either naturally occurring substances themselves or are prepared by chemical modification of natural products. Most of the drugs derived from natural sources are chiral and are almost always obtained as a single enantiomer rather than as a racemic mixture. Not so with the over 500 chiral substances represented among the more than 1300 drugs that are the products of synthetic organic chemistry. Until recently, such substances were, with few exceptions, prepared, sold, and administered as racemic mixtures even though the desired therapeutic activity resided in only one of the enantiomers. Spurred by a number of factors ranging from safety and efficacy to synthetic methodology and economics, this practice is undergoing rapid change as more and more chiral synthetic drugs become available in enantiomerically pure form.

Because of the high degree of chiral recognition inherent in most biological processes (Section 7.8), it is unlikely that both enantiomers of a chiral drug will exhibit the same level, or even the same kind, of effect. At one extreme, one enantiomer has the desired effect, while the other exhibits no biological activity at all. In this case, which is relatively rare, the racemic form is simply a drug that is 50 percent pure and contains 50 percent "inert ingredients." Real cases are more complicated. For example, it is the *S* enantiomer that is responsible for the pain-relieving properties of ibuprofen, normally sold as a racemic mixture. The 50 percent of racemic ibuprofen that is the *R* enantiomer is not completely wasted, however, because enzyme-catalyzed reactions in our body convert much of it to active (*S*)-ibuprofen.

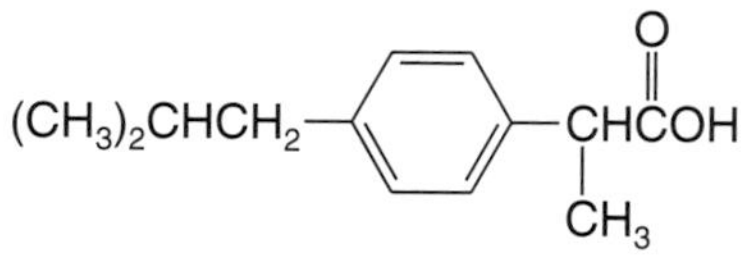

$(CH_3)_2CHCH_2-C_6H_4-CH(CH_3)COOH$

Ibuprofen

A much more serious drawback to using chiral drugs as racemic mixtures is illustrated by thalidomide, briefly employed as a sedative and antinausea drug in Europe and Great Britain during the period 1959–1962. The desired properties are those of (*R*)-thalidomide. (*S*)-Thalidomide, however, has a very different spectrum of biological activity and was shown to be responsible for over 2000 cases of serious birth defects in children born to women who took it while pregnant.

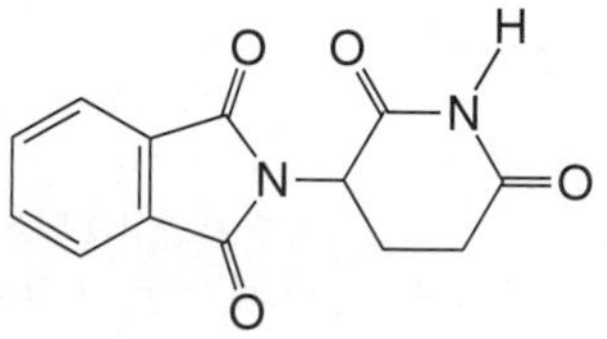

Thalidomide

Basic research directed toward understanding the factors that control the stereochemistry of chemical reactions has led to new synthetic methods that make it practical to prepare chiral molecules in enantiomerically pure form. Recognizing this, most major pharmaceutical companies are examining their existing drugs to see which ones are the best candidates for synthesis as single enantiomers and, when preparing a new drug, design its synthesis so as to provide only the desired enantiomer. In 1992 the United States Food and Drug Administration (FDA) issued guidelines that encouraged such an approach, but left open the door for approval of new drugs as racemic mixtures when special circumstances warrant. An incentive to developing enantiomerically pure versions of existing drugs is that through enhanced effectiveness or novel production methods, they are usually eligible for patent protection separate from that of the original drugs. Thus the temporary monopoly position that patent law views as essential to fostering innovation can be extended by transforming a successful chiral, but racemic, drug into an enantiomerically pure version.

7.9 REACTIONS THAT CREATE A STEREOGENIC CENTER

Many of the reactions we have already encountered can yield a chiral product from an achiral starting material. The epoxidation of propene, for example, illustrates the creation of a stereogenic center by addition to the double bond of an alkene.

$$CH_3CH{=}CH_2 \xrightarrow{CH_3CO_2OH} CH_3CH{-}CH_2 \text{ (with O bridging C1–C2)}$$

Propene (achiral) → 1,2-Epoxypropane (chiral)

In this, as in other reactions in which achiral reactants yield chiral products, the product is formed as a *racemic mixture* and is *optically inactive.* Remember, in order for a substance to be optically active, not only must it be chiral but one enantiomer must be present in excess of the other.

Figure 7.8 shows why equal amounts of (R)- and (S)-1,2-epoxypropane are formed when propene reacts with peroxyacetic acid. The figure illustrates that the peroxy acid is just as likely to transfer oxygen to one face of the double bond as the other. Enantiomeric transition states have the same energy. The rates of formation of the R and S enantiomers of the product are the same, and a racemic mixture of the two results.

It is often helpful, especially in a multistep reaction, to focus on the step that creates the stereogenic center. In the ionic addition of hydrogen bromide to 2-butene, for example, the stereogenic center is generated when bromide ion attacks *sec*-butyl cation.

$$CH_3CH{=}CHCH_3 \xrightarrow{HBr} CH_3\underset{Br}{\underset{|}{C}}HCH_2CH_3 \quad \text{via} \quad CH_3\overset{+}{C}HCH_2CH_3$$

(E)- or (Z)-2-butene (achiral) — 2-Bromobutane (chiral) — *sec*-Butyl cation (achiral)

As seen in Figure 7.9, the bonds to the positively charged carbon are coplanar and define a plane of symmetry in the carbocation, which is achiral. The rates at which bromide ion attacks the carbocation at its two mirror-image faces are equal, and the product, 2-bromobutane, although chiral, is optically inactive because it is formed as a racemic mixture.

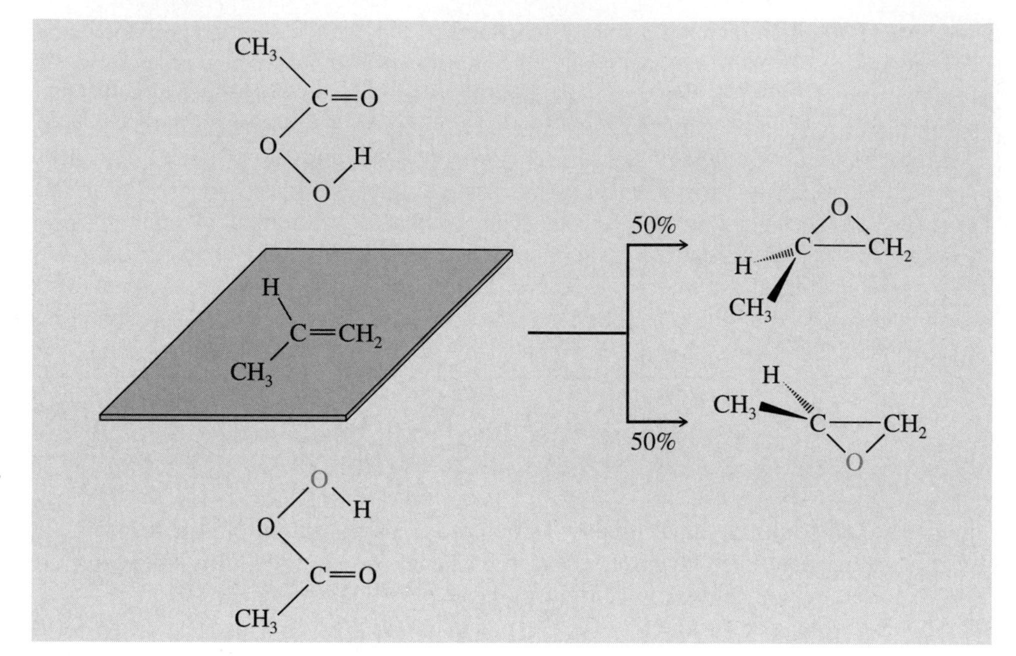

FIGURE 7.8 Epoxidation of propene produces a racemic mixture of (*R*)- and (*S*)-1,2-epoxypropane. These two epoxides are nonsuperposable mirror images (enantiomers) and are formed in equal amounts.

It is a general principle that optically active products cannot be formed when optically inactive substrates react with optically inactive reagents. This principle holds irrespective of whether the addition is syn or anti, concerted or stepwise. No matter how many steps are involved in a reaction, if the reactants are achiral, formation of one enantiomer is just as likely as the other and a racemic mixture results.

When a reactant is chiral but optically inactive because it is *racemic,* any products

$CH_3CH{=}CHCH_3$

$\downarrow H^+$

$[H, CH_3, C{-}CH_2CH_3]^+$

$\downarrow Br^-$

(50%) (*R*)-(−)-2-Bromobutane $[\alpha]_D$ −39° + (*S*)-(+)-2-Bromobutane $[\alpha]_D$ +39° (50%)

FIGURE 7.9 Ionic addition of hydrogen bromide to (*E*)- and (*Z*)-2-butene proceeds by way of an achiral carbocation, which leads to equal quantities of (*R*)- and (*S*)-2-bromobutane.

derived from its reactions with optically inactive reagents will be *optically inactive.* For example, 2-butanol is chiral and may be converted with hydrogen bromide to 2-bromobutane, which is also chiral. If racemic 2-butanol is used, each enantiomer will react at the same rate with the achiral reagent. Whatever happens to (*R*)-(−)-2-butanol is mirrored in a corresponding reaction of (*S*)-(+)-2-butanol, and a racemic, optically inactive product results.

$$(\pm)\text{-}CH_3\underset{\displaystyle OH}{\underset{|}{C}}HCH_2CH_3 \xrightarrow{HBr} (\pm)\text{-}CH_3\underset{\displaystyle Br}{\underset{|}{C}}HCH_2CH_3$$

2-Butanol (chiral but racemic) — 2-Bromobutane (chiral but racemic)

Optically inactive starting materials can give optically active products if they are treated with an optically active reagent or if the reaction is catalyzed by an optically active substance. The best examples are found in biochemical processes. Most biochemical reactions are catalyzed by enzymes. Enzymes are chiral and enantiomerically homogeneous; they provide an asymmetric environment in which chemical reaction can take place. Ordinarily, enzyme-catalyzed reactions occur with such a high level of stereoselectivity that one enantiomer of a substance is formed exclusively even when the substrate is achiral. The enzyme *fumarase,* for example, catalyzes the hydration of fumaric acid to malic acid in apples and other fruits. Only the *S* enantiomer of malic acid is formed in this reaction.

HO$_2$C(H)C=C(H)CO$_2$H + H_2O $\xrightleftharpoons{\text{fumarase}}$ HO$_2$C(H)C(OH)CH$_2$CO$_2$H

Fumaric acid — (*S*)-(−)-Malic acid

The reaction is reversible, and its stereochemical requirements are so pronounced that neither the cis isomer of fumaric acid (maleic acid) nor the *R* enantiomer of malic acid can serve as a substrate for the fumarase-catalyzed hydration-dehydration equilibrium.

PROBLEM 7.9 Biological reduction of pyruvic acid, catalyzed by the enzyme lactate dehydrogenase, gives (+)-lactic acid, represented by the Fischer projection shown. What is the configuration of (+)-lactic acid according to the Cahn-Ingold-Prelog *R-S* notational system?

$$CH_3\overset{\displaystyle O}{\overset{\|}{C}}CO_2H \xrightarrow{\text{biological reduction}} HO\text{—}\overset{\displaystyle CO_2H}{\underset{\displaystyle CH_3}{+}}\text{—}H$$

Pyruvic acid — (+)-Lactic acid

We shall return to the three-dimensional details of chemical reactions later in this chapter. Before doing that, however, we need to develop some additional stereochemical principles that apply to structures with more than one stereogenic center.

7.10 CHIRAL MOLECULES WITH TWO STEREOGENIC CENTERS

When a molecule contains two stereogenic centers, as does 2,3-dihydroxybutanoic acid, how many stereoisomers are possible?

$$\overset{4}{CH_3}\overset{*}{\underset{|}{\overset{3}{C}}}H\overset{*}{\underset{|}{\overset{2}{C}}}H\overset{1}{C}(=O)OH$$
$$\quad HO \quad OH$$

2,3-Dihydroxybutanoic acid

The answer can be determined to be 4 by employing a commonsense approach. The absolute configuration at C-2 may be *R* or *S*. Likewise, C-3 may have either the *R* or the *S* configuration. The four possible combinations of these two stereogenic centers are

(2*R*,3*R*) (stereoisomer I) (2*S*,3*S*) (stereoisomer II)
(2*R*,3*S*) (stereoisomer III) (2*S*,3*R*) (stereoisomer IV)

Compounds I through IV have the same constitution but differ in the arrangement of their atoms in space. They are stereoisomers. Figure 7.10 presents structural formulas for these four stereoisomers. Stereoisomers I and II are enantiomers of each other; the enantiomer of (*R*,*R*) is (*S*,*S*). Likewise stereoisomers III and IV are enantiomers of each other, the enantiomer of (*R*,*S*) being (*S*,*R*).

Stereoisomer I is not a mirror image of III or IV, and so it is not an enantiomer of either one. Stereoisomers that are not related as an object and its mirror image are called **diastereomers;** *diastereomers are stereoisomers that are not enantiomers.* Thus, stereoisomer I is a diastereomer of III and a diastereomer of IV. Similarly, II is a diastereomer of III and IV.

In order to convert a molecule with two stereogenic centers to its enantiomer, the configuration at both centers must be changed. Reversing the configuration at only one stereogenic center converts it to a diastereomeric structure.

Enantiomers must have equal and opposite specific rotations. Diastereomeric substances can have different rotations, with respect to both sign and magnitude. Thus, as Figure 7.10 shows, the (2*R*,3*R*) and (2*S*,3*S*) enantiomers (I and II) have specific rotations that are equal in magnitude but opposite in sign. The (2*R*,3*S*) and (2*S*,3*R*) enantiomers (III and IV) likewise have specific rotations that are equal to each other but opposite in sign. The magnitudes of rotation of I and II are different, however, from those of their diastereomers III and IV.

In writing Fischer projections of molecules with two stereogenic centers, the molecule is arranged in an eclipsed conformation for projection onto the page, as shown in Figure 7.11. Again, horizontal lines in the projection represent bonds coming toward you; vertical bonds point away.

Organic chemists use an informal nomenclature system based on Fischer projections to distinguish between diastereomers. When the carbon chain is vertical and like substituents are on the same side of the Fischer projection, the molecule is described as the **erythro** diastereomer. When like substituents are on opposite sides of the Fischer projection, the molecule is described as the **threo** diastereomer. Thus, as seen in the Fischer projections of the stereoisomeric 2,3-dihydroxybutanoic acids, compounds I and II are erythro stereoisomers and III and IV are threo.

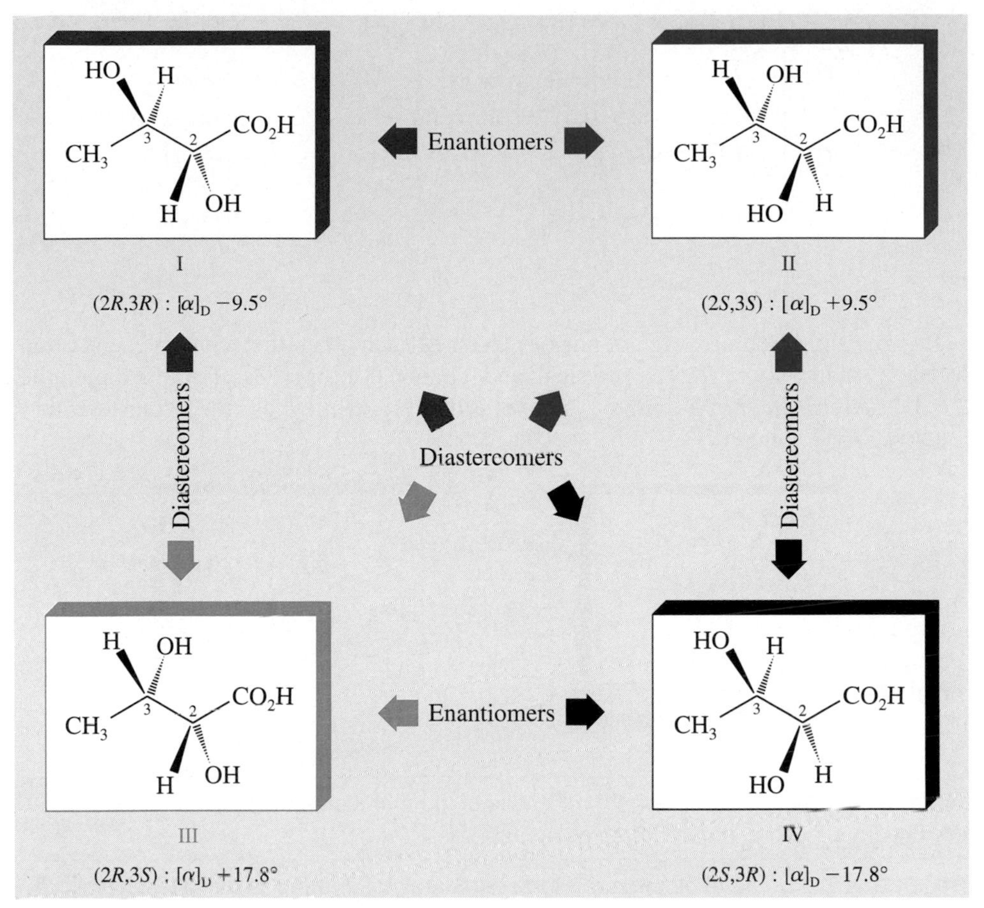

FIGURE 7.10 Stereoisomeric 2,3-dihydroxybutanoic acids. Stereoisomers I and II are enantiomers. Stereoisomers III and IV are enantiomers. All other relationships are diastereomeric (see text).

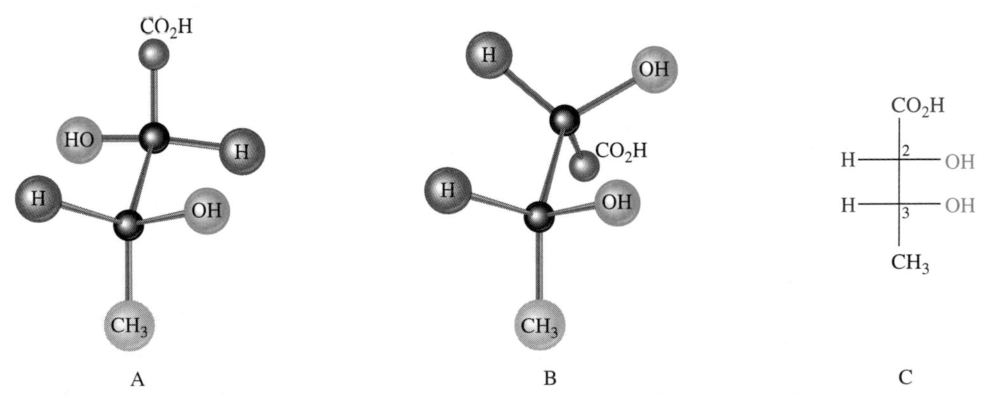

FIGURE 7.11 Representations of (2*R*,3*R*)-dihydroxybutanoic acid. The staggered conformation (A) is the most stable, but is not properly arranged to show stereochemistry according to the Fischer projection method. Rotation about C-2—C-3 bond produces the eclipsed conformation (B), and projection of the eclipsed conformation onto the page gives a correct Fischer projection (C).

Erythro and threo describe the *relative configuration* (Section 7.5) of two stereogenic centers within a single molecule.

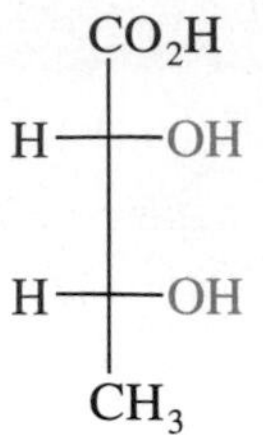

I	II	III	IV
CO_2H	CO_2H	CO_2H	CO_2H
H—OH	HO—H	H—OH	HO—H
H—OH	HO—H	HO—H	H—OH
CH_3	CH_3	CH_3	CH_3
erythro	erythro	threo	threo

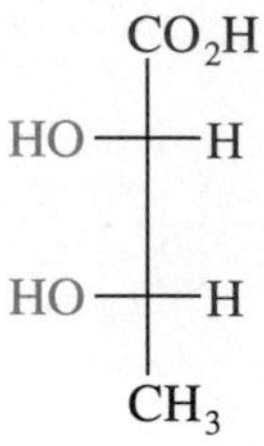

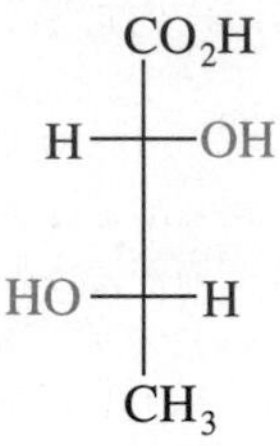

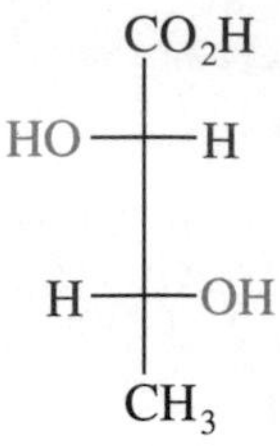

Because diastereomers are not mirror images of each other, they can have, and often do have, markedly different physical and chemical properties. For example, the (2*R*,3*R*) stereoisomer of 3-amino-2-butanol is a liquid, while the (2*R*,3*S*) diastereomer is a crystalline solid.

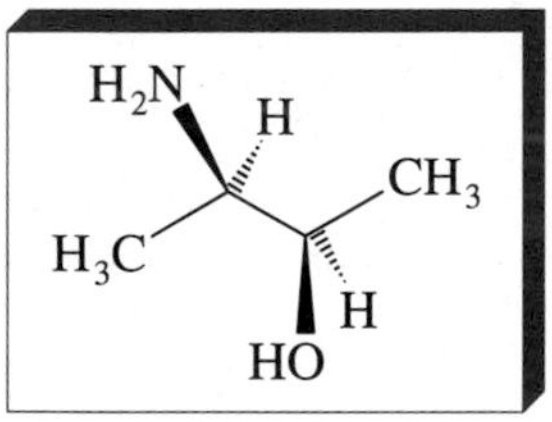

(2*R*,3*R*)-3-Amino-2-butanol
(liquid)

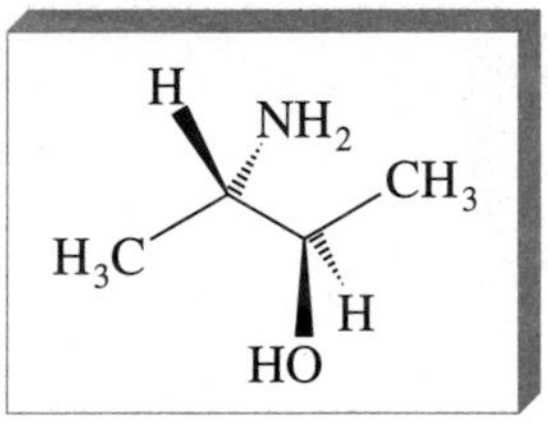

(2*R*,3*S*)-3-Amino-2-butanol
(solid, mp 49°C)

PROBLEM 7.10 Draw Fischer projections of the four stereoisomeric 3-amino-2-butanols and label each erythro or threo as appropriate.

PROBLEM 7.11 One other stereoisomer of 3-amino-2-butanol is a crystalline solid. Which one?

The situation is the same when the two stereogenic centers are present in a ring. There are four stereoisomeric 1-bromo-2-chlorocyclopropanes, a pair of enantiomers in which the halogens are trans and a pair in which they are cis. The cis compounds are diastereomers of the trans.

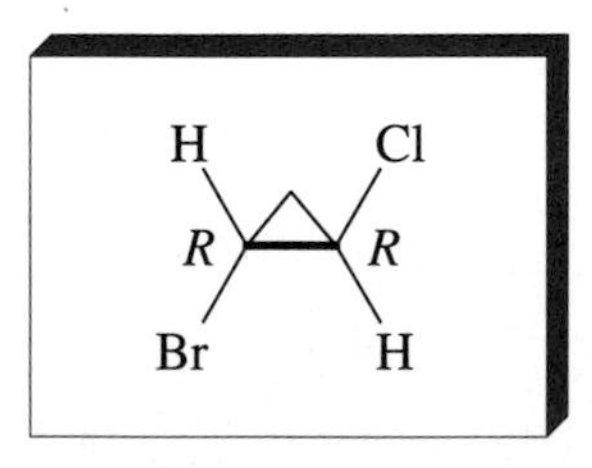

(1*R*,2*R*)-1-Bromo-2-chlorocyclopropane

Enantiomers

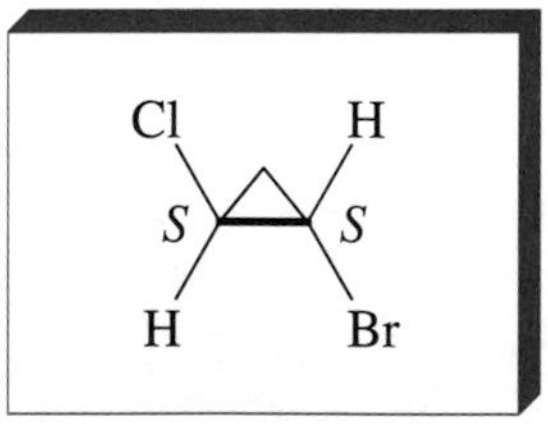

(1*S*,2*S*)-1-Bromo-2-chlorocyclopropane

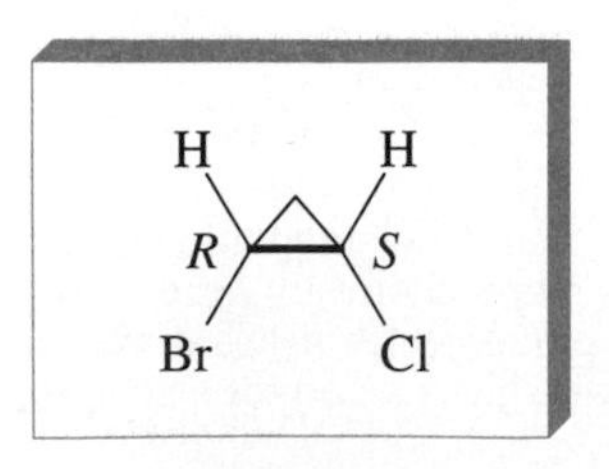

(1*R*,2*S*)-1-Bromo-2-chlorocyclopropane

Enantiomers

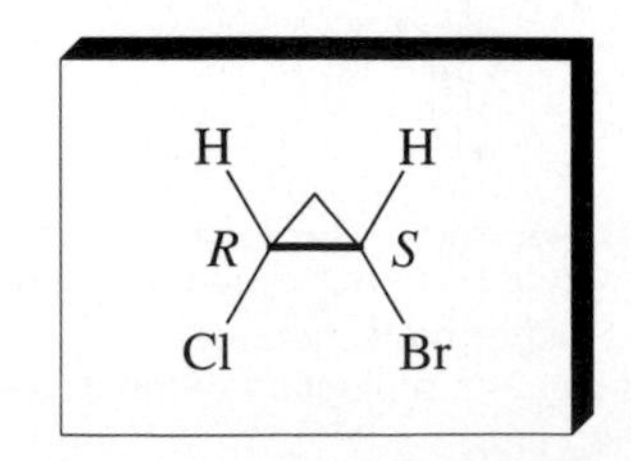

(1*S*,2*R*)-1-Bromo-2-chlorocyclopropane

7.11 ACHIRAL MOLECULES WITH TWO STEREOGENIC CENTERS

Now consider what happens when a molecule has two stereogenic centers that are equivalently substituted, as does 2,3-butanediol.

$$CH_3\overset{*}{C}H(OH)\overset{*}{C}H(OH)CH_3$$

2,3-Butanediol

Only *three,* not four, stereoisomeric 2,3-butanediols are possible. These three are shown in Figure 7.12. The (2*R*,3*R*) and (2*S*,3*S*) forms are enantiomers of each other and have equal and opposite optical rotations. A third combination of stereogenic centers, (2*R*,3*S*), however, gives an *achiral* structure that is superposable on its (2*S*,3*R*) mirror image. Because it is achiral, this third stereoisomer is *optically inactive.* We call achiral molecules that have stereogenic centers **meso forms.** The meso form in Figure 7.12 is known as *meso*-2,3-butanediol.

One way to demonstrate that *meso*-2,3-butanediol is achiral is to recognize that its eclipsed conformation has a plane of symmetry that passes through and is perpendicular to the C-2—C-3 bond, as illustrated in Figure 7.13*a*. The anti conformation is achiral as well. As Figure 7.13*b* shows, this conformation is characterized by a center of symmetry at the midpoint of the C-2—C-3 bond.

Fischer projection formulas are often helpful in identifying meso forms. Of the three stereoisomeric 2,3-butanediols, notice that only in the meso stereoisomer does a dashed line through the center of the Fischer projection divide the molecule into two mirror-image halves.

(2*R*,3*R*)-2,3-Butanediol	(2*S*,3*S*)-2,3-Butanediol	*meso*-2,3-Butanediol
CH_3	CH_3	CH_3
HO—+—H	H—+—OH	H—+—OH
		- - - - - - - - -
H—+—OH	HO—+—H	H—+—OH
CH_3	CH_3	CH_3

In the same way that a Fischer formula is a projection of the eclipsed conformation onto the page, the line drawn through its center is a projection of the plane of symmetry which is present in the eclipsed conformation of *meso*-2,3-butanediol.

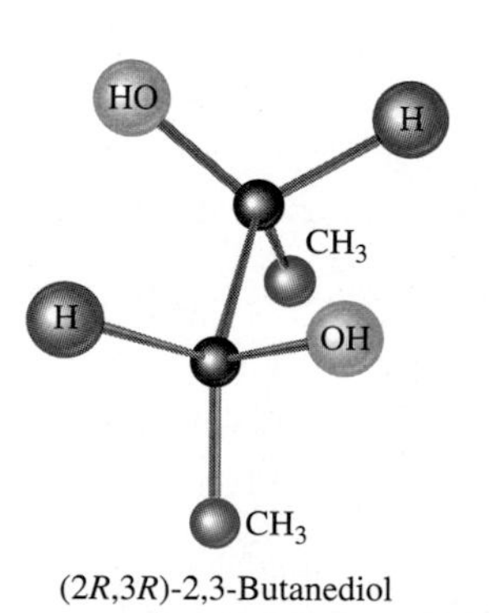

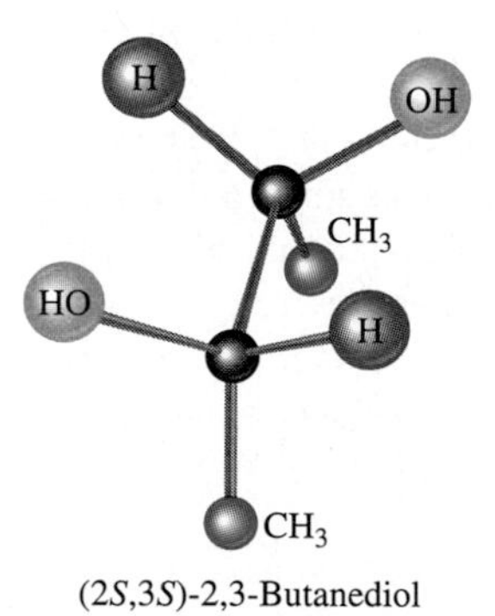

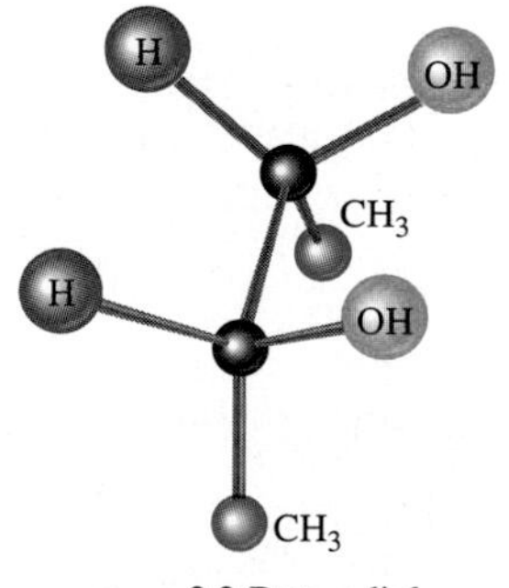

FIGURE 7.12 Stereoisomeric 2,3-butanediols shown in their eclipsed conformations for convenience. Stereoisomers A and B are enantiomers of each other. Structure C is a diastereomer of A and B and is achiral. It is called *meso*-2,3-butanediol.

When using Fischer projections for this purpose, however, be sure to remember what three-dimensional objects they stand for. One should not, for example, test for superposition of the two chiral stereoisomers by a procedure that involves moving any atoms out of the plane of the paper in any step.

CHIRALITY OF DISUBSTITUTED CYCLOHEXANES

Disubstituted cyclohexanes present us with a challenging exercise in stereochemistry. Consider the seven possible dichlorocyclohexanes: 1,1-; *cis*- and *trans*-1,2-; *cis*- and *trans*-1,3-; and *cis*- and *trans*-1,4-. Which are chiral? Which are achiral?

Four isomers—the ones that are achiral because they have a plane of symmetry—are relatively easy to identify:

Achiral dichlorocyclohexanes

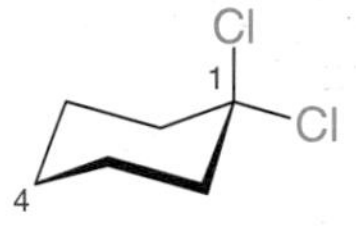

1,1

(plane of symmetry through C-1 and C-4)

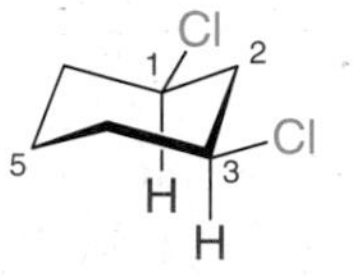

cis-1,3

(plane of symmetry through C-2 and C-5)

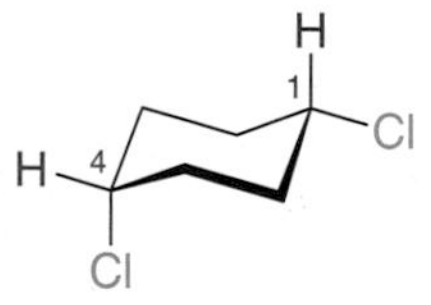

cis-1,4

(plane of symmetry through C-1 and C-4)

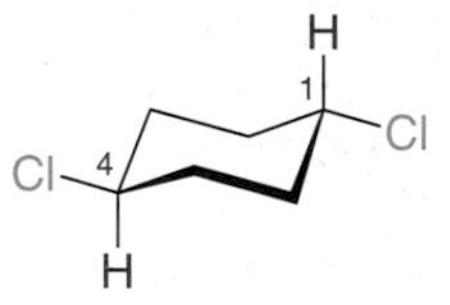

trans-1,4

(plane of symmetry through C-1 and C-4)

The remaining three isomers are chiral:

Chiral dichlorocyclohexanes

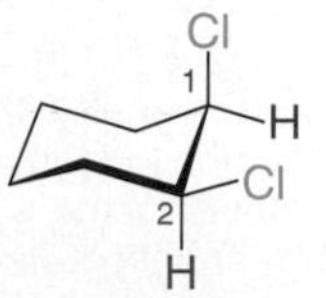

cis-1,2

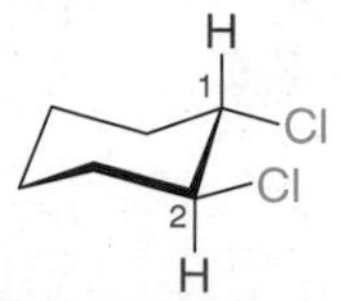

trans-1,2

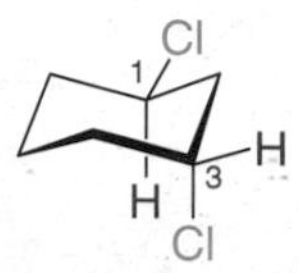

trans-1,3

Among all the isomers, *cis*-1,2-dichlorocyclohexane is unique in that the ring-flipping process typical of cyclohexane derivatives (Section 3.8) converts it to its enantiomer.

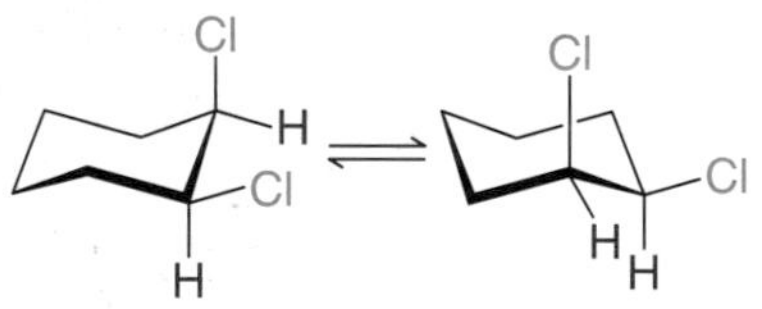

A ⇌ A′ which is equivalent to A′

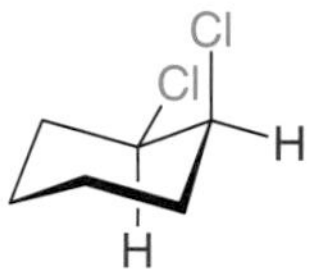

Structures A and A′ are nonsuperposable mirror images of each other. Thus while *cis*-1,2-dichlorocyclohexane is chiral, it is optically inactive under conditions where chair-chair interconversion occurs. Such interconversion is rapid at room temperature and converts optically active A to a racemic mixture of A and A′. Since A and A′ are enantiomers interconvertible by a conformational change, they are sometimes referred to as **conformational enantiomers.**

The same kind of spontaneous racemization occurs for any *cis*-1,2 disubstituted cyclohexane where both substituents are the same. Since such compounds are chiral, it is not correct to speak of them as meso compounds, which are achiral by definition. Rapid chair-chair interconversion, however, converts them to a 1 : 1 mixture of enantiomers, and this mixture is optically inactive.

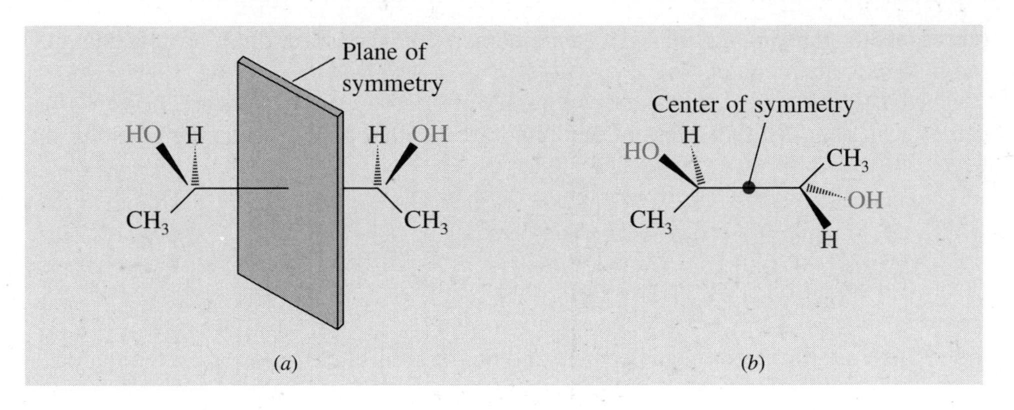

FIGURE 7.13 (*a*) The eclipsed conformation of *meso*-2,3-butanediol has a plane of symmetry. (*b*) The anti conformation of *meso*-2,3-butanediol has a center of symmetry.

PROBLEM 7.12 A meso stereoisomer is possible for one of the following compounds. Which one?

2,3-Dibromopentane; 2,4-dibromopentane; 3-bromo-2-pentanol; 4-bromo-2-pentanol

---

Turning to cyclic compounds, we see that there are three, not four, stereoisomeric 1,2-dibromocyclopropanes. Of these, two are enantiomeric *trans*-1,2-dibromocyclopropanes. The cis diastereomer is a meso form; it has a plane of symmetry.

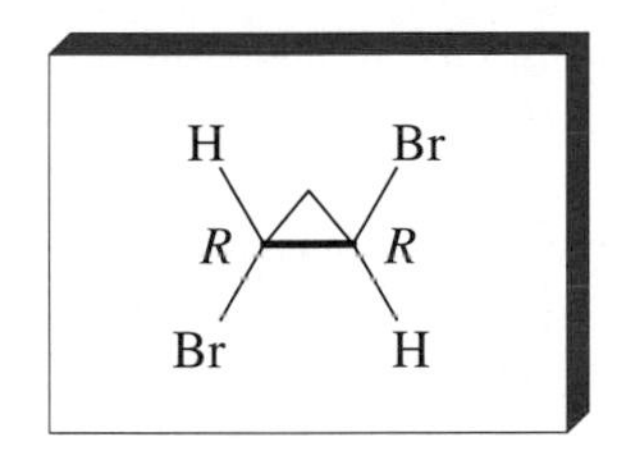

(1*R*,2*R*)-1,2-Dibromocyclopropane

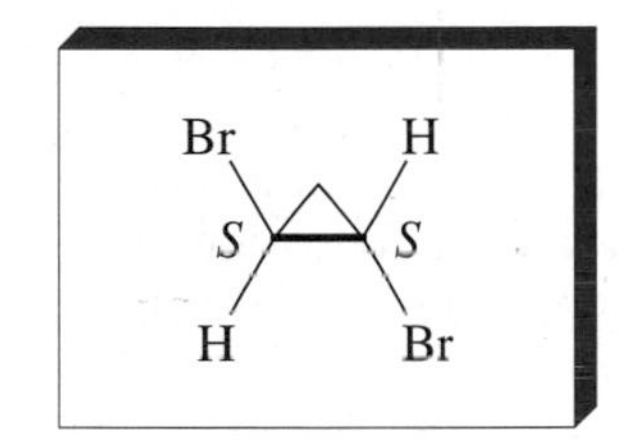

(1*S*,2*S*)-Dibromocyclopropane

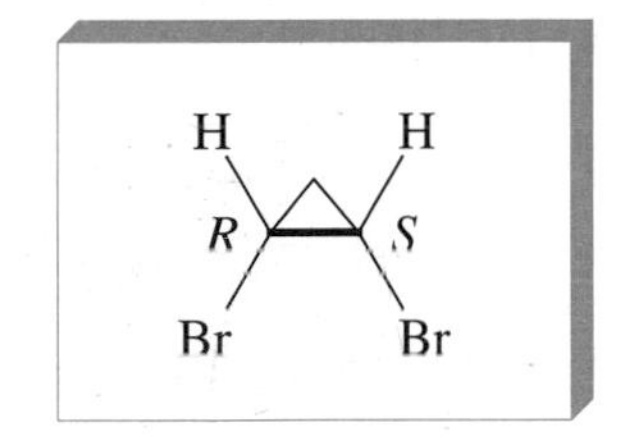

meso-1,2-Dibromocyclopropane

PROBLEM 7.13 One of the stereoisomers of 1,3-dimethylcyclohexane is a meso form. Which one?

7.12 MOLECULES WITH MULTIPLE STEREOGENIC CENTERS

Many naturally occurring compounds contain several stereogenic centers. By an analysis similar to that described for the case of two stereogenic centers, it can be shown that the maximum number of stereoisomers for a particular constitution is 2^n, where n is equal to the number of stereogenic centers.

PROBLEM 7.14 Using *R* and *S* descriptors, write all the possible combinations for a molecule with three stereogenic centers.

When two or more of a molecule's stereogenic centers are equivalently substituted, meso forms are possible, and the number of stereoisomers is then less than 2^n. Thus, 2^n

represents the *maximum* number of stereoisomers for a constitutional formula containing *n* stereogenic centers.

The best examples of substances with multiple stereogenic centers are the *carbohydrates* (Chapter 25). One class of carbohydrates, called *hexoses,* has the constitution

$$HOCH_2\overset{*}{C}H(OH)-\overset{*}{C}H(OH)-\overset{*}{C}H(OH)-\overset{*}{C}H(OH)-CH{=}O$$

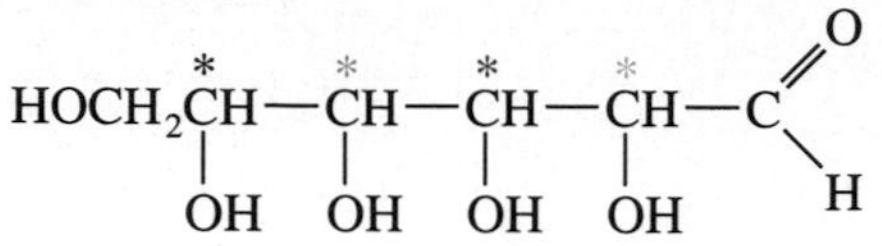

A hexose

Since there are four stereogenic centers and no possibility of meso forms, there are 2^4, or 16, stereoisomeric hexoses. All 16 are known, having been isolated either as natural products or as the products of chemical synthesis.

PROBLEM 7.15 A second category of six-carbon carbohydrates, called *2-hexuloses,* has the constitution shown. How many stereoisomeric 2-hexuloses are possible?

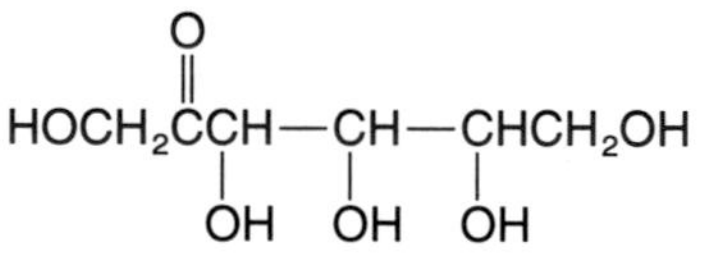

$$HOCH_2\overset{O}{\overset{\|}{C}}CH(OH)-CH(OH)-CH(OH)CH_2OH$$

A 2-hexulose

Steroids represent another class of natural products with multiple stereogenic centers. One such compound is *cholic acid,* which can be obtained from bile. Its structural formula is given in Figure 7.14. Cholic acid has 11 stereogenic centers, and so there are a total (including cholic acid) of 2^{11}, or 2048, stereoisomers that have this constitution. Of these 2048 stereoisomers, how many of them are diastereomers of cholic acid? Remember! Diastereomers are stereoisomers that are not enantiomers, and any object can have only one mirror image. Therefore, only one of the stereoisomers is an enantiomer of cholic acid, while all the rest are diastereomers. Of the 2048 stereoisomers, one is cholic acid, one is its enantiomer, and the other 2046 are diastereomers of cholic acid. Only a small fraction of these compounds are known, and (+)-cholic acid is the only one ever isolated from natural sources.

Eleven stereogenic centers may seem like a lot, but this number is nowhere close to a world record. *Palytoxin,* a poisonous polyhydroxylated substance produced by a Tahitian marine organism, has 64 stereogenic centers. Even 64 is a modest number when compared with the more than 100 stereogenic centers typical for most small proteins and the thousands of stereogenic centers that are present in nucleic acids.

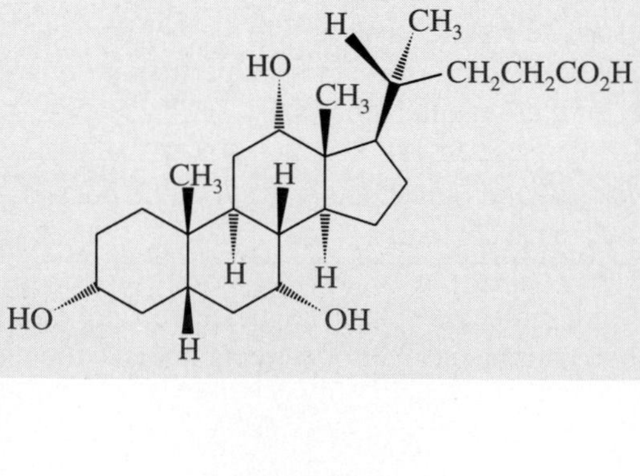

H CH3
HO CH2CH2CO2H
CH3
CH3 H
H H
HO OH
H

FIGURE 7.14 The structure of cholic acid. Its 11 stereogenic centers are those carbons at which stereochemistry is indicated in the diagram.

A molecule that contains both stereogenic centers and double bonds has additional opportunities for stereoisomerism. For example, the configuration of the stereogenic center in 3-penten-2-ol may be either *R* or *S*, and that of the double bond may be either *E* or *Z*. Therefore, there are four stereoisomers of 3-penten-2-ol even though it has only one stereogenic center.

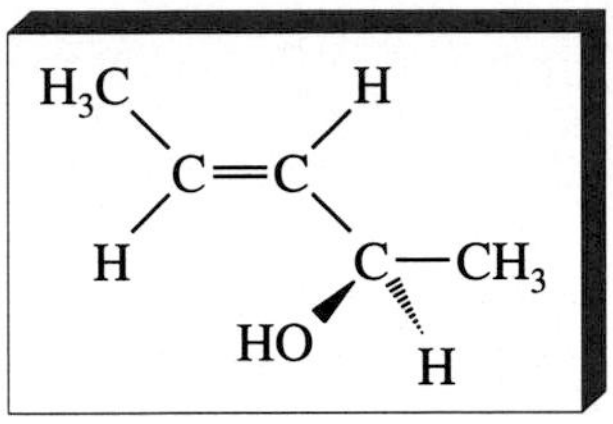

(2*R*,3*E*)-3-Penten-2-ol

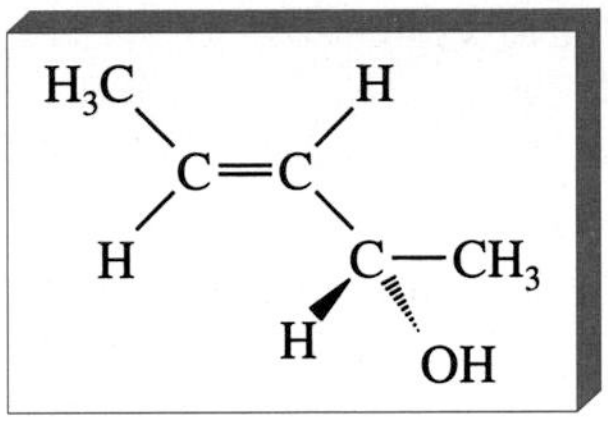

(2*S*,3*E*)-3-Penten-2-ol

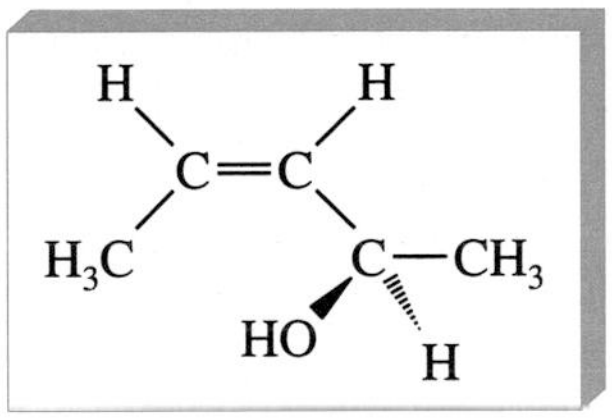

(2*R*,3*Z*)-3-Penten-2-ol

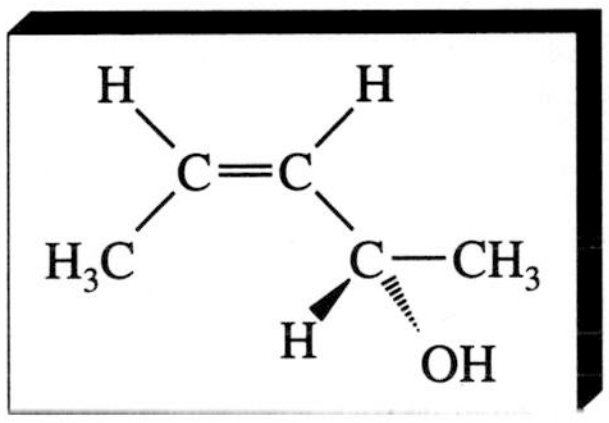

(2*S*,3*Z*)-3-Penten-2-ol

The relationship of the (2*R*,3*E*) stereoisomer to the others is that it is the enantiomer of (2*S*,3*E*)-3-penten-2-ol and is a diastereomer of the (2*R*,3*Z*) and (2*S*,3*Z*) isomers.

7.13 REACTIONS THAT PRODUCE DIASTEREOMERS

Once we grasp the idea of stereoisomerism in molecules with two or more stereogenic centers, we can explore further details of addition reactions of alkenes.

When bromine adds to (*Z*)- or (*E*)-2-butene, the product 2,3-dibromobutane contains two equivalently substituted stereogenic centers:

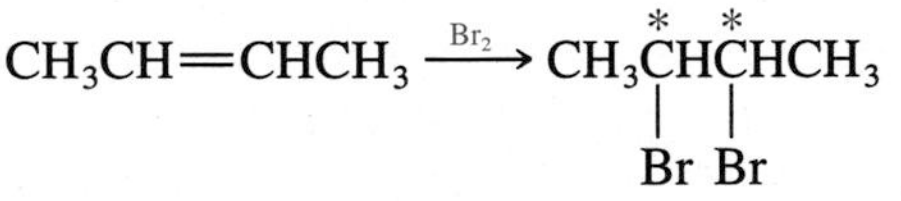

$$CH_3CH{=}CHCH_3 \xrightarrow{Br_2} CH_3\overset{*}{C}H(Br)\overset{*}{C}H(Br)CH_3$$

(*Z*)- or (*E*)-2-butene 2,3-Dibromobutane

Three stereoisomers are possible, a pair of enantiomers and a meso form.

Two factors combine to determine which stereoisomers are actually formed in the reaction.

1. The (*E*)- or (*Z*)-configuration of the starting alkene
2. The anti stereochemistry of addition

Figures 7.15 and 7.16 depict the stereochemical relationships associated with anti addition of bromine to (*E*)- and (*Z*)-2-butene, respectively. The trans alkene (*E*)-2-butene yields only *meso*-2,3-dibromobutane, while the cis alkene (*Z*)-2-butene gives a racemic mixture of (2*R*,3*R*)- and (2*S*,3*S*)-2,3-dibromobutane.

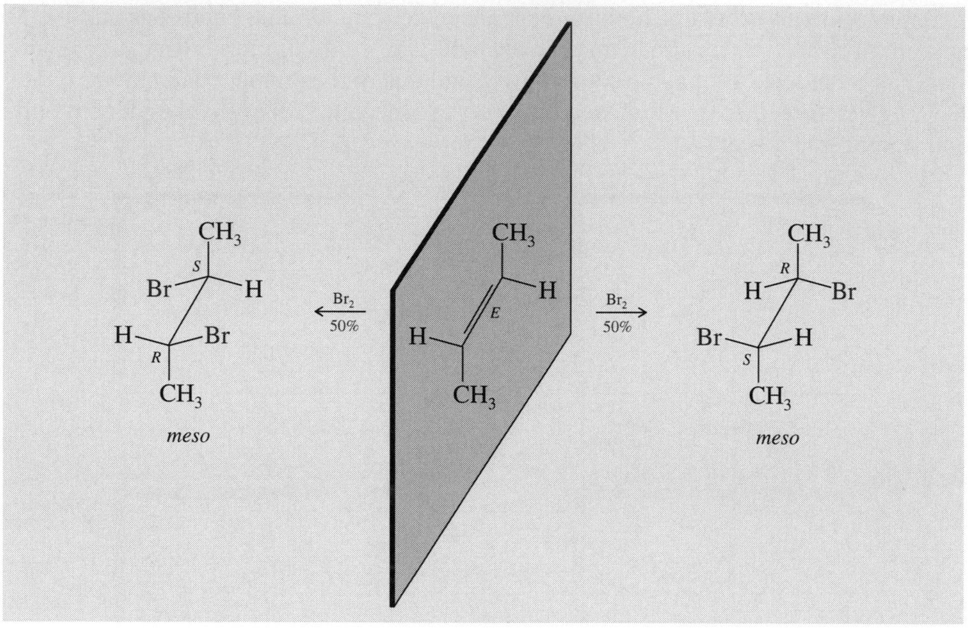

CH$_3$
S
Br
H
H
Br
R
CH$_3$
meso
Br$_2$
50%
CH$_3$
H
E
H
CH$_3$
Br$_2$
50%
CH$_3$
R
H
Br
Br
H
S
CH$_3$
meso

FIGURE 7.15 Anti addition of Br_2 to (*E*)-2-butene gives *meso*-2,3-dibromobutane.

Bromine addition to alkenes is an example of a **stereospecific reaction.** A stereospecific reaction is one in which stereoisomeric starting materials yield products that are stereoisomers of each other. In this case the starting materials, in separate reactions, are the *E* and *Z* stereoisomers of 2-butene. The chiral dibromides from (*Z*)-2-butene are stereoisomers (diastereomers) of the meso dibromide formed from (*E*)-2-butene.

Notice further that, consistent with the principle developed in Section 7.9, optically

H
R
Br
CH$_3$
H
Br
R
CH$_3$
2*R*, 3*R*
Br$_2$
50%
H
CH$_3$
Z
H
CH$_3$
Br$_2$
50%
H
S
H$_3$C
Br
Br
H
S
CH$_3$
2*S*, 3*S*

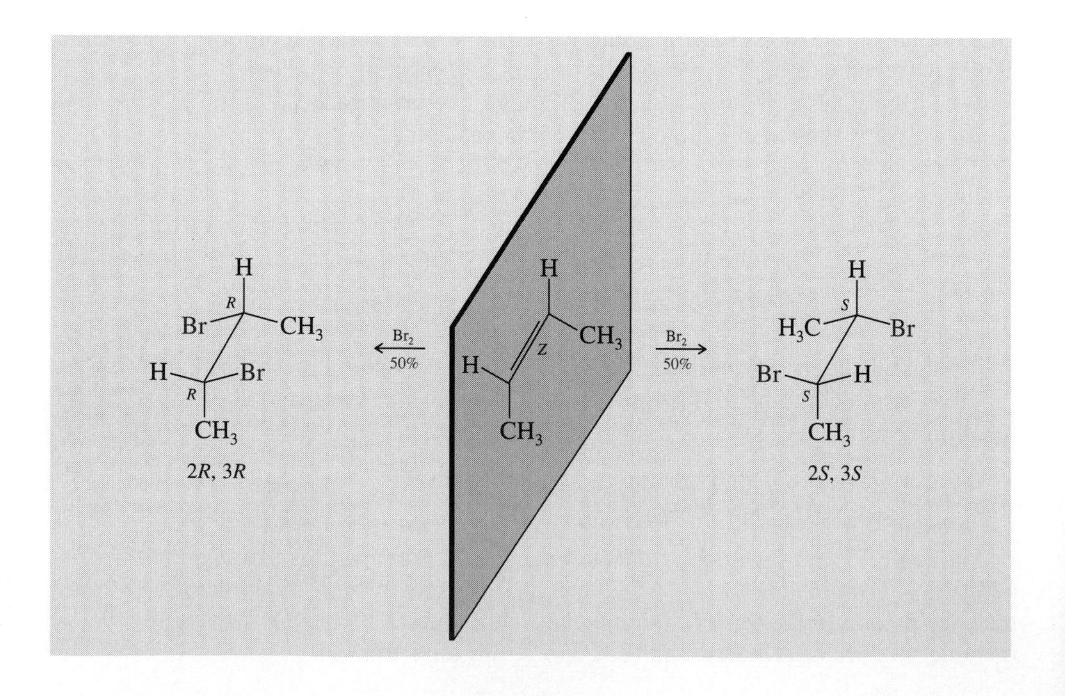

FIGURE 7.16 Anti addition of Br_2 to (*Z*)-2-butene gives a racemic mixture of (2*R*,3*R*)- and (2*S*,3*S*)-2,3-dibromobutane.

inactive starting materials (achiral alkenes and bromine) yield optically inactive products (a racemic mixture or a meso structure) in these reactions.

PROBLEM 7.16 Epoxidation of alkenes is a stereospecific syn addition. Which stereoisomer of 2-butene reacts with peroxyacetic acid to give *meso*-2,3-epoxybutane? Which one gives a racemic mixture of (2*R*,3*R*)- and (2*S*,3*S*)-2,3-epoxybutane?

A reaction that introduces a second stereogenic center into a starting material that already has one need not produce equal quantities of two possible diastereomers. Consider catalytic hydrogenation of 2-methyl(methylene)cyclohexane. As you might expect, both *cis*- and *trans*-1,2-dimethylcyclohexane are formed.

$\xrightarrow[\text{acetic acid}]{H_2,\ Pt}$

2-Methyl(methylene)cyclohexane → *cis*-1,2-Dimethylcyclohexane (68%) + *trans*-1,2-Dimethylcyclohexane (32%)

The relative amounts of the two products, however, are not equal; more *cis*-1,2-dimethylcyclohexane is formed than *trans*. The reason for this is that it is the less hindered face of the double bond that approaches the catalyst surface, and the face to which hydrogen is transferred. Hydrogenation of 2-methyl(methylene)cyclohexane occurs preferentially at the side of the double bond opposite that of the methyl group and leads to a faster rate of formation of the cis stereoisomer of the product.

PROBLEM 7.17 Could the fact that hydrogenation of 2-methyl(methylene)cyclohexane gives more *cis*-1,2-dimethylcyclohexane than *trans*- be explained on the basis of the relative stabilities of the two stereoisomeric products?

The hydrogenation of 2-methyl(methylene)cyclohexane is an example of a *stereoselective reaction,* meaning one in which stereoisomeric products are formed in unequal amounts from a single starting material (Section 5.11).

A common misconception is that a stereospecific reaction is simply one that is 100 percent stereoselective. The two terms have precise definitions that are independent of each other. A stereospecific reaction is defined as one which, when carried out with stereoisomeric starting materials, gives a product from one reactant that is a stereoisomer of the product from the other. A stereoselective reaction is one in which a single starting material gives a predominance of a single stereoisomer when two or more are possible. *Stereospecific* is more closely connected with features of the reaction itself than with the reactant. Thus terms such as syn *addition* and anti *elimination* describe the stereospecificity of reactions. *Stereoselective* is more closely connected with structural effects in the reactant as expressed in terms such as *addition to the less hindered side.* A stereospecific reaction can also be stereoselective. For example, syn addition describes stereospecificity in the catalytic hydrogenation of alkenes, while the preference for addition to the less hindered face of the double bond describes stereoselectivity.

Note that the terms *regioselective* and *regiospecific,* however, are defined in terms of each other. A regiospecific reaction is one that is 100 percent regioselective.

7.14 RESOLUTION OF ENANTIOMERS

The separation of a racemic mixture into its enantiomeric components is termed **resolution.** The first resolution, that of tartaric acid, was carried out by Louis Pasteur in 1848. Tartaric acid is a by-product of wine making and is almost always found as its dextrorotatory 2*R*,3*R* stereoisomer, shown here in a perspective drawing and in a Fischer projection.

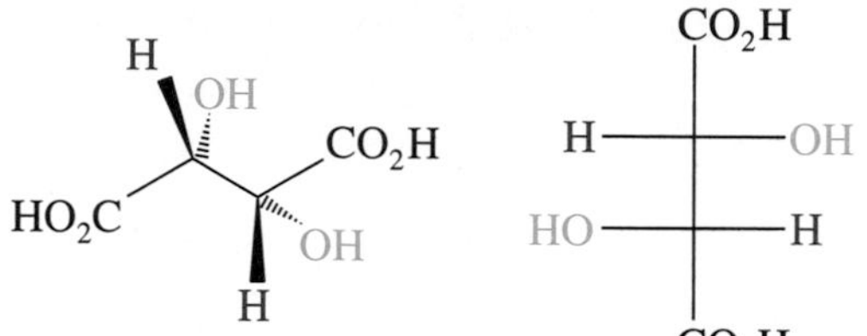

(2*R*,3*R*)-Tartaric acid (mp 170°C, $[\alpha]_D$ +12°)

PROBLEM 7.18 There are two other stereoisomeric tartaric acids. Write their Fischer projections and specify the configuration at their stereogenic centers.

Occasionally, an optically inactive sample of tartaric acid was obtained. Pasteur noticed that the sodium ammonium salt of optically inactive tartaric acid was a mixture of two mirror-image crystal forms. With microscope and tweezers, Pasteur carefully separated the two. He found that one kind of crystal (in aqueous solution) was dextrorotatory, while the mirror-image crystals rotated the plane of polarized light an equal amount but were levorotatory.

Although Pasteur was not able to provide a structural explanation—that had to wait for van't Hoff and Le Bel a quarter of a century later—he correctly deduced that the enantiomeric quality of the crystals of the sodium ammonium tartrates must be a consequence of enantiomeric molecules. The rare form of tartaric acid was optically inactive because it contained equal amounts of (+)-tartaric acid and (−)-tartaric acid. It had earlier been called *racemic acid* (from Latin *racemus,* "a bunch of grapes"), a name that subsequently gave rise to our present term for a mixture of enantiomers.

PROBLEM 7.19 Could the unusual, optically inactive form of tartaric acid studied by Pasteur have been *meso*-tartaric acid?

Pasteur's technique of separating enantiomers not only is laborious but requires that the crystal habits of enantiomers be distinguishable. This happens very rarely. Consequently, alternative and more general approaches to optical resolution of enantiomers have been developed. Most are based on a strategy of temporarily converting the enantiomers of a racemic mixture to diastereomeric derivatives, separating these diastereomers, then regenerating the enantiomeric starting materials.

Figure 7.17 illustrates this strategy. Let us say we have a mixture of enantiomers, which, for simplicity, we label as C(+) and C(−). Further, let us say that C(+) and C(−) bear some functional group that can combine with a reagent P to yield adducts C(+)—P and C(−)—P. Now, if reagent P is chiral, and if only a single enantiomer of P, say, P(+), is added to a racemic mixture of C(+) and C(−), as shown in step 1 of Figure 7.17, then the products of the reaction are C(+)—P(+) and C(−)—P(+).

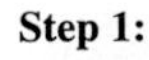

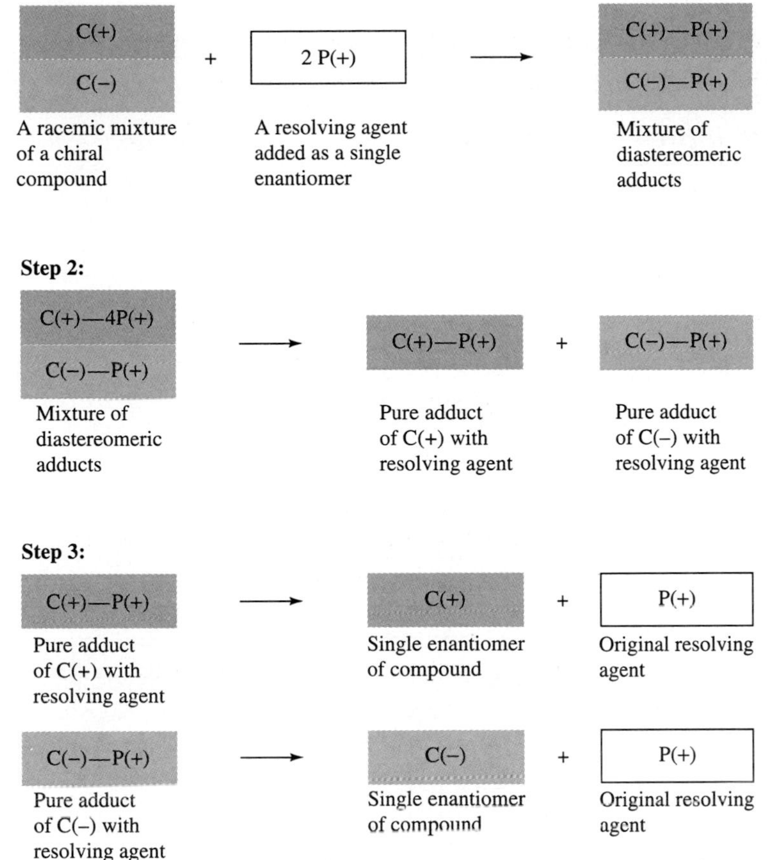

FIGURE 7.17 Diagram illustrating the general procedure followed in resolving a chiral substance into its enantiomers. You are invited to use a yellow highlighter to aid your understanding of the process. Highlight all the boxes that contain the symbol P(+) either by itself or in combination with C(+) or C(−). Thus, if you have used the highlighter, in step 1 the yellow resolving agent P(+) reacts with red C(+) to give orange C(+)—P(+), and with blue C(−) to give green C(−)—P(+). After separation of the orange and green diastereomers in step 2, orange C(+)—P(+) is converted to red C(+) and yellow P(+) in step 3. Similarly, green C(−)—P(+) is converted to blue C(−) and yellow P(+).

These products are not mirror images; they are diastereomers. Diastereomers can have different physical properties, and this difference in physical properties can serve as a means of separating them. In step 2, these diastereomers are separated, usually by recrystallization from a suitable solvent. In step 3, an appropriate chemical transformation is employed to remove the resolving agent from the separated adducts, thereby liberating the enantiomers and regenerating the resolving agent.

Whenever possible, the chemical reactions involved in the formation of diastereomers and their conversion to separate enantiomers are simple acid-base reactions. For example, naturally occurring (*S*)-(−)-malic acid is often used to resolve amines. One such amine that has been resolved in this way is 1-phenylethylamine. Amines are bases, and malic acid is an acid. Proton transfer from (*S*)-(−)-malic acid to a racemic mixture of (*R*)- and (*S*)-1-phenylethylamine gives a mixture of diastereomeric salts.

$$C_6H_5\overset{*}{C}(CH_3)HNH_2 + HO_2CCH_2\overset{*}{C}(OH)HCO_2H \longrightarrow C_6H_5\overset{*}{C}(CH_3)H\overset{+}{N}H_3 \quad {}^{-}O_2CCH_2\overset{*}{C}(OH)HCO_2H$$

1-Phenylethylamine (racemic mixture) | (*S*)-(−)-Malic acid (resolving agent) | 1-Phenylethylammonium (*S*)-malate (mixture of diastereomeric salts)

The diastereomeric salts are separated and the individual enantiomers of the amine liberated by treatment with a base:

$$\underbrace{\mathrm{C_6H_5\overset{*}{C}H\overset{+}{N}H_3 \quad {}^{-}O_2CCH_2\overset{*}{C}HCO_2H} \atop \mathrm{\quad | \qquad\qquad\qquad\qquad | } \atop \mathrm{CH_3 \qquad\qquad\qquad OH}}_{\text{1-Phenylethylammonium (S)-malate (a single diastereomer)}} + \underset{\text{Hydroxide}}{2OH^-} \longrightarrow$$

$$\underset{\text{1-Phenylethylamine (a single enantiomer)}}{\mathrm{C_6H_5\overset{*}{C}HNH_2 \atop |\;CH_3}} + \underset{\text{(S)-(−)-Malic acid (recovered resolving agent)}}{\mathrm{{}^{-}O_2CCH_2\overset{*}{C}HCO_2^{-} \atop |\;OH}} + \underset{\text{Water}}{2H_2O}$$

PROBLEM 7.20 In the resolution of 1-phenylethylamine using (−)-malic acid, the compound obtained by recrystallization of the mixture of diastereomeric salts is (*R*)-1-phenylethylammonium (*S*)-malate. The other component of the mixture is more soluble and remains in solution. What is the configuration of the more soluble salt?

This method is widely used for the resolution of chiral amines and carboxylic acids. Analogous methods based on the formation and separation of diastereomers have been developed for other functional groups; the precise approach depends on the kind of chemical reactivity associated with the functional groups present in the molecule.

7.15 STEREOREGULAR POLYMERS

Before the development of the Ziegler-Natta catalyst systems (Section 6.21), polymerization of propene was not a reaction of much value. The reason for this has a stereochemical basis. Consider a section of *polypropylene:*

$$\mathrm{-CH_2\underset{}{\overset{CH_3}{\overset{|}{C}}}HCH_2\overset{CH_3}{\overset{|}{C}}HCH_2\overset{CH_3}{\overset{|}{C}}HCH_2\overset{CH_3}{\overset{|}{C}}HCH_2\overset{CH_3}{\overset{|}{C}}HCH_2\overset{CH_3}{\overset{|}{C}}H-}$$

Representation of the polymer chain in an extended zigzag conformation, as shown in Figure 7.18, reveals several distinct structural possibilities differing with respect to the relative configurations of the carbons that bear the methyl groups.

One possible structure, represented in Figure 7.18*a*, has all the methyl groups oriented in the same direction with respect to the polymer chain. This stereochemical arrangement is said to be **isotactic.** Another form, shown in Figure 7.18*b*, has its methyl groups alternating in their stereochemical orientation along the chain. This arrangement is described as **syndiotactic.** Both the isotactic and the syndiotactic forms of polypropylene are known as **stereoregular polymers,** because each is characterized by a precise stereochemistry at the carbon atom that bears the methyl group. There is a third possibility, shown in Figure 7.18*c*, which is described as **atactic.** Atactic polypropylene has a random orientation of its methyl groups; it is not a stereoregular polymer.

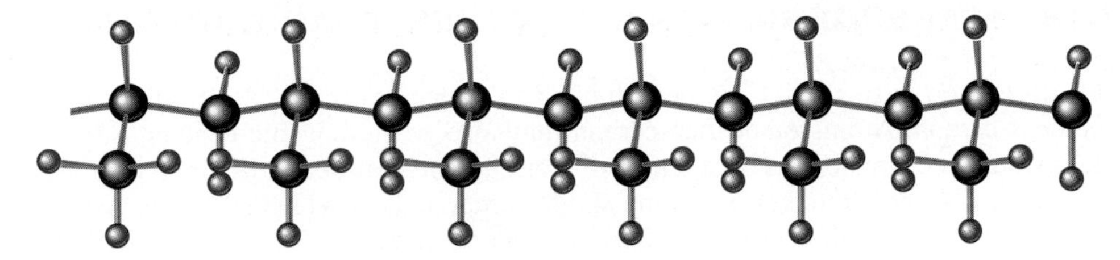

(*a*) Isotactic polypropylene

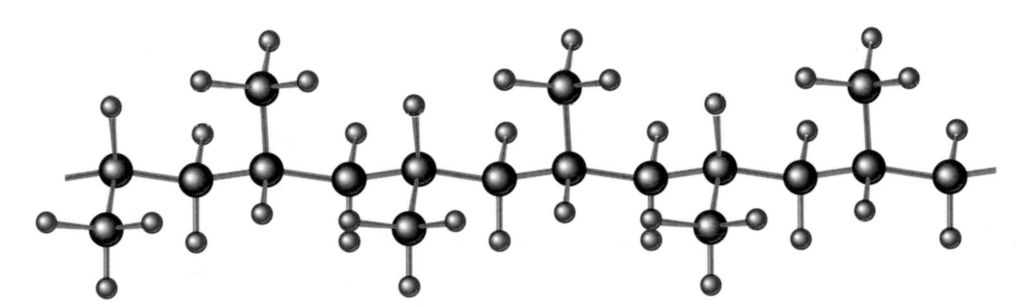

(*b*) Syndiotactic polypropylene

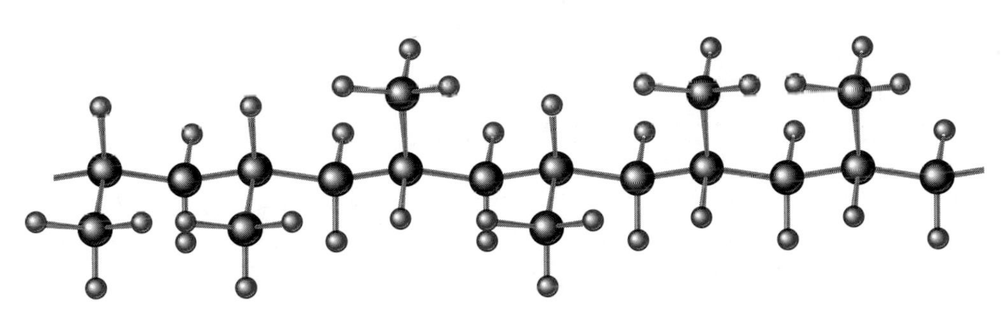

(*c*) Atactic polypropylene

FIGURE 7.18 Polymers of propene. The main chain is arranged in a zigzag conformation. Every other carbon bears a methyl substituent and is a stereogenic center. (*a*) All the methyl groups are on the same side of the carbon chain in isotactic polypropylene. (*b*) Methyl groups alternate from one side to the other in syndiotactic polypropylene. (*c*) The spatial orientation of the methyl groups is random in atactic polypropylene.

Polypropylene chains associate with each other because of attractive van der Waals forces. The extent of this association is relatively large for isotactic and syndiotactic polymers, because the stereoregularity of the polymer chains permits efficient packing. Atactic polypropylene, on the other hand, does not associate as strongly. It has a lower density and lower melting point than the stereoregular forms. The physical properties of stereoregular polypropylene are more useful for most purposes than those of atactic polypropylene.

When propene is polymerized under free-radical conditions, the polypropylene that results is atactic. Catalysts of the Ziegler-Natta type, however, permit the preparation of either isotactic or syndiotactic polypropylene. We see here an example of how proper choice of experimental conditions can affect the stereochemical course of a chemical reaction to the extent that entirely new materials with unique properties result.

7.16 STEREOGENIC CENTERS OTHER THAN CARBON

Our discussion to this point has been limited to molecules in which the stereogenic center is carbon. Atoms other than carbon may also be stereogenic centers. Silicon, like carbon, is characterized by a tetrahedral arrangement of bonds when it bears four substituents. A large number of organosilicon compounds in which silicon bears four different groups have been resolved into their enantiomers.

Trigonal pyramidal molecules are chiral if the central atom bears three different groups. If one is to resolve substances of this type, however, the pyramidal inversion that interconverts enantiomers must be slow at room temperature. In the case of amines, pyramidal inversion at nitrogen is very rapid (E_{act} = 24 to 40 kJ/mol, or 6 to 10 kcal/mol), and attempts at resolving amines that have no stereogenic centers other than nitrogen have been thwarted by their immediate racemization.

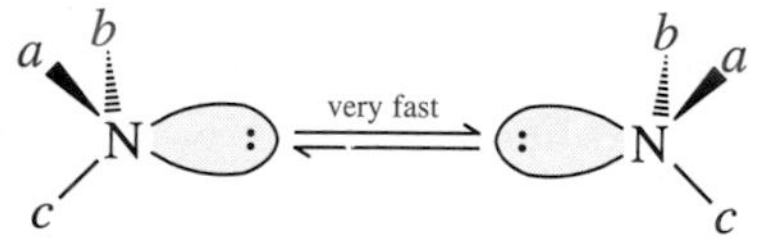

very fast

Phosphorus is in the same group of the periodic table as nitrogen, and tricoordinate phosphorus compounds (phosphines), like amines, are trigonal pyramidal. Phosphines, however, undergo pyramidal inversion much more slowly than amines (E_{act} = 120 to 140 kJ/mol, or 30 to 35 kcal/mol), and a number of optically active phosphines have been prepared. Similarly, tricoordinate sulfur compounds are chiral when sulfur bears three different substituents, because the rate of pyramidal inversion at sulfur is rather slow. Optically active sulfoxides, such as the one shown, are examples of compounds of this type.

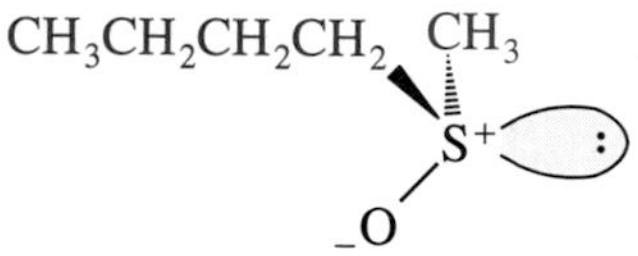

$CH_3CH_2CH_2CH_2$ CH_3 S^+ ^-O

(*S*)-(+)-Butyl methyl sulfoxide

The absolute configuration at sulfur is specified by the Cahn-Ingold-Prelog method with the provision that the unshared electron pair is considered to be the lowest-ranking substituent.

7.17 SUMMARY

Our concern in this chapter has been with the spatial arrangement of atoms and groups in molecules. Chemistry in three dimensions is known as **stereochemistry.** At its most fundamental level, stereochemistry deals with molecular structure; at another level, it is concerned with chemical reactivity. Table 7.2 summarizes some basic definitions relating to molecular structure and stereochemistry.

A detailed flowchart describing a more finely divided set of subcategories of isomers appears in the February 1990 issue of the *Journal of Chemical Education.*

A molecule is **chiral** if it cannot be superposed on its mirror image (Section 7.1). The most common kind of chiral molecule contains a carbon atom that bears four different substituents. Such a carbon is called a **stereogenic center** (Section 7.2). Table 7.2 shows the nonsuperposable mirror images of 2-chlorobutane; 2-chlorobutane is a chiral molecule and C-2 is a stereogenic center.

TABLE 7.2
Classification of Isomers*

Definition	Example
1. *Constitutional isomers* are isomers that differ in the order in which their atoms are connected.	There are three constitutionally isomeric compounds of molecular formula C_3H_8O: $CH_3CH_2CH_2OH$ (1-Propanol); $CH_3CH(OH)CH_3$ (2-Propanol); $CH_3CH_2OCH_3$ (Ethyl methyl ether)
2. *Stereoisomers* are isomers that have the same constitution but differ in the arrangement of their atoms in space.	
(*a*) *Enantiomers* are stereoisomers that are related as an object and its nonsuperposable mirror image.	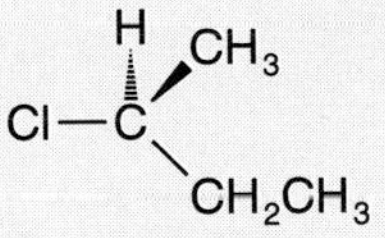 The two enantiomeric forms of 2-chlorobutane are 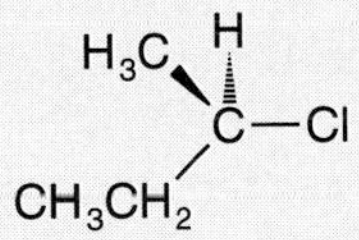 (*R*)-(−)-2-Chlorobutane and (*S*)-(+)-2-Chlorobutane
(*b*) *Diastereomers* are stereoisomers that are not enantiomers.	The cis and trans isomers of 4-methylcyclohexanol are stereoisomers, but they are not related as an object and its mirror image; they are diastereomers. 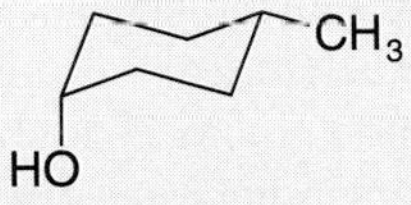 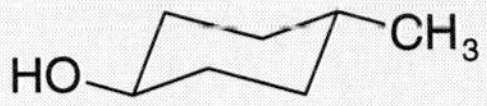 *cis*-4-Methylcyclohexanol; *trans*-4-Methylcyclohexanol

*Isomers are different compounds that have the same molecular formula. They may be either constitutional isomers or stereoisomers.

Achiral molecules are superposable on their mirror images. Any structure that has a plane of symmetry or a center of symmetry is achiral (Section 7.3). Both the cis and trans stereoisomers of 4-methylcyclohexanol shown in Table 7.2 are achiral. Each has a plane of symmetry that bisects the molecule into two mirror-image halves.

The **absolute configuration** of a molecule is a precise description of the arrangement of its atoms in space. **Relative configuration** compares the configuration of one compound with that of another (Section 7.5). We specify absolute configuration by the Cahn-Ingold-Prelog notational system using the descriptors *R* and *S* (Section 7.6). Table 7.2 identifies the enantiomers of 2-chlorobutane according to the Cahn-Ingold-Prelog system.

Fischer projection formulas (Section 7.7) are often used to show stereochemistry, especially when dealing with carbohydrates (Chapter 25) and amino acids (Chapter 27). A Fischer projection depicts how a tetrahedral molecule would appear if its bonds were projected onto a flat surface. Horizontal lines represent bonds coming toward you; vertical bonds point away from you. The projection is normally drawn so that the carbon chain is vertical.

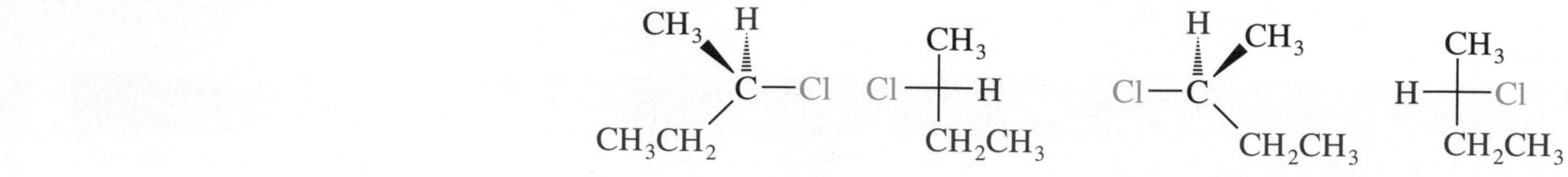

CH_3, H, C—Cl, CH_3CH_2 Cl—|—H CH_3 CH_2CH_3

(*R*)-2-Chlorobutane

H, CH_3, Cl—C, CH_2CH_3 H—|—Cl CH_3 CH_2CH_3

(*S*)-2-Chlorobutane

Optical activity, the capacity to rotate the plane of polarized light, is a physical property of chiral molecules (Section 7.4). Enantiomeric forms of the same molecule rotate the plane of polarization an equal amount but in opposite directions. The enantiomer that rotates the plane of polarization in the clockwise (or positive) sense is said to be *dextrorotatory;* the one that rotates plane-polarized light in the anticlockwise (or negative) sense is said to be *levorotatory.* In order to be optically active a substance must be chiral, and one enantiomer must be present in an amount greater than the other. A **racemic mixture** is optically inactive and contains equal quantities of enantiomers.

A chemical reaction can convert an achiral molecule such as an alkene to a chiral product. If the product contains a single stereogenic center, it is formed as a racemic mixture (Section 7.9). If addition to an alkene converts both the doubly bonded carbons to stereogenic centers, the stereochemistry of the product depends on the configuration (*E* or *Z*) of the alkene and whether the addition is syn or anti (Section 7.13). In all cases, if the reactants are optically inactive, the products will be too. Optically active products can be formed from optically inactive starting materials if some optically active agent is present. The best examples are biological processes in which enzymes catalyze the formation of a single enantiomer of a chiral molecule.

When a molecule contains two or more stereogenic centers, the maximum number of stereoisomers is 2^n, where n is equal to the number of structural units capable of stereochemical variation—usually this is the number of stereogenic centers, but it can include *E* and *Z* double bonds as well (Sections 7.10 through 7.12). The number of stereoisomers is reduced to less than 2^n when meso forms are possible. A meso form is an achiral molecule that contains stereogenic centers. Meso forms are optically inactive. Tartaric acid has two stereogenic centers, but only three stereoisomers are possible, because one of them is a meso form. These stereoisomers are shown here as Newman and as Fischer projections. The Newman projections are drawn as staggered conformations with the CO_2H groups anti to each other. The meso form has a center of symmetry in this conformation. The Fischer projections depict how the atoms are arranged in an eclipsed conformation. The meso form has a plane of symmetry in the eclipsed conformation (shown as a line that separates the top half of the Fischer projection from its mirror-image bottom half).

Newman projections of the tartaric acids:

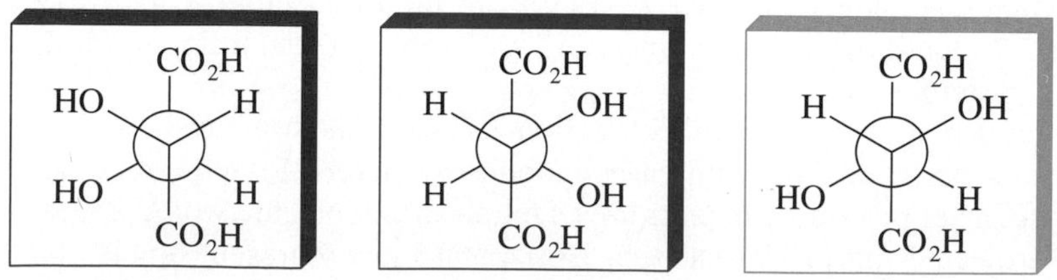

CO_2H, HO, H, HO, H, CO_2H

(2*R*,3*R*)

CO_2H, H, OH, H, OH, CO_2H

(2*S*,3*S*)

CO_2H, H, OH, HO, H, CO_2H

meso
(achiral)

Fischer projections of the tartaric acids:

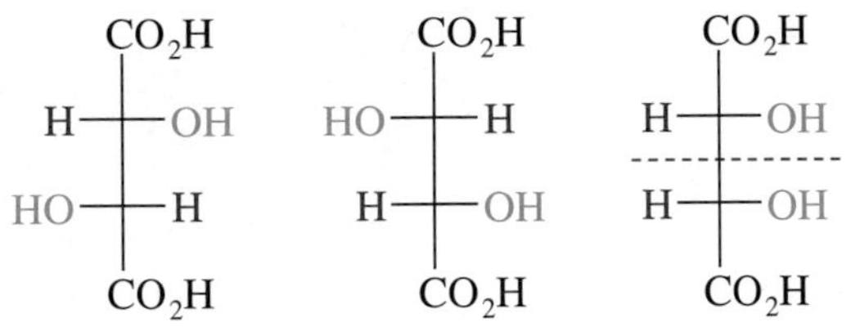

$$\begin{array}{c} CO_2H \\ H-\!\!\!-\!\!\!-OH \\ HO-\!\!\!-\!\!\!-H \\ CO_2H \\ (2R,3R) \end{array} \qquad \begin{array}{c} CO_2H \\ HO-\!\!\!-\!\!\!-H \\ H-\!\!\!-\!\!\!-OH \\ CO_2H \\ (2S,3S) \end{array} \qquad \begin{array}{c} CO_2H \\ H-\!\!\!-\!\!\!-OH \\ \hdashline H-\!\!\!-\!\!\!-OH \\ CO_2H \\ \text{meso} \end{array}$$

Resolution is the separation of a racemic mixture into its enantiomers (Section 7.14).

Certain polymers such as polypropylene contain stereogenic centers, and the relative configurations of these centers affect the physical properties of the polymers. Like substituents appear on the same side of a zigzag carbon chain in an **isotactic** polymer, alternate along the chain in a **syndiotactic** polymer, and appear in a random manner in an **atactic** polymer. Isotactic and syndiotactic polymers are referred to as **stereoregular** polymers.

Atoms other than carbon can be stereogenic centers. Examples cited in Section 7.16 included those based on tetracoordinate silicon and tricoordinate sulfur as the stereogenic atom. In principle, tricoordinate nitrogen can be a stereogenic center in compounds of the type N(x, y, z), where x, y, and z are different, but inversion of the nitrogen pyramid is so fast that racemization occurs virtually instantly at room temperature.

PROBLEMS

7.21 Which of the isomeric alcohols having the molecular formula $C_5H_{12}O$ are chiral? Which are achiral?

7.22 Write structural formulas for all the compounds that are trichloro derivatives of cyclopropane. (Don't forget to include stereoisomers.) Which are chiral? Which are achiral?

7.23 In each of the following pairs of compounds one is chiral and the other is achiral. Identify each compound as chiral or achiral, as appropriate.

(*a*) $\begin{array}{c} ClCH_2CHCH_2OH \\ \;\;| \\ \;\;OH \end{array}$ and $\begin{array}{c} HOCH_2CHCH_2OH \\ | \\ Cl \end{array}$

(*b*) $CH_3CH{=}CHCH_2Br$ and $\begin{array}{c} CH_3CHCH{=}CH_2 \\ | \\ Br \end{array}$

(*c*) $\begin{array}{c} CH_3 \\ H_2N-\!\!\!-\!\!\!-H \\ H-\!\!\!-\!\!\!-NH_2 \\ CH_3 \end{array}$ and $\begin{array}{c} CH_3 \\ H-\!\!\!-\!\!\!-NH_2 \\ H-\!\!\!-\!\!\!-NH_2 \\ CH_3 \end{array}$

(*d*) H Br; H_3C CH_3; H Br and H Br H Br; H_3C CH_3

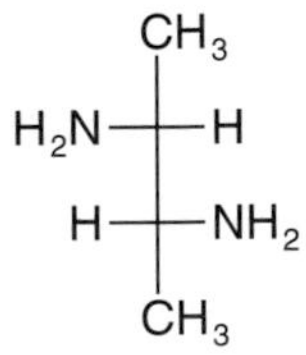

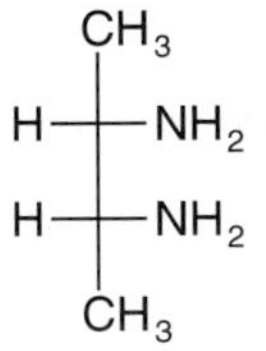

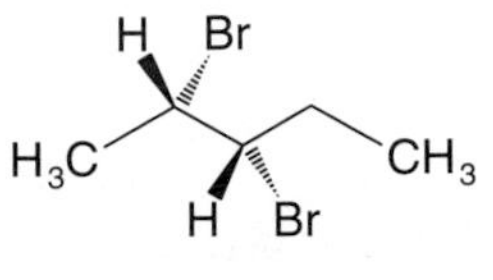

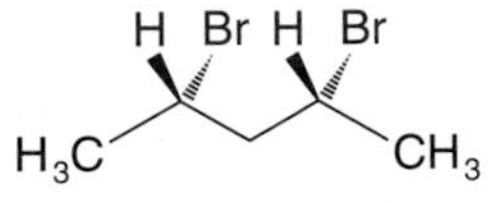

(*e*)

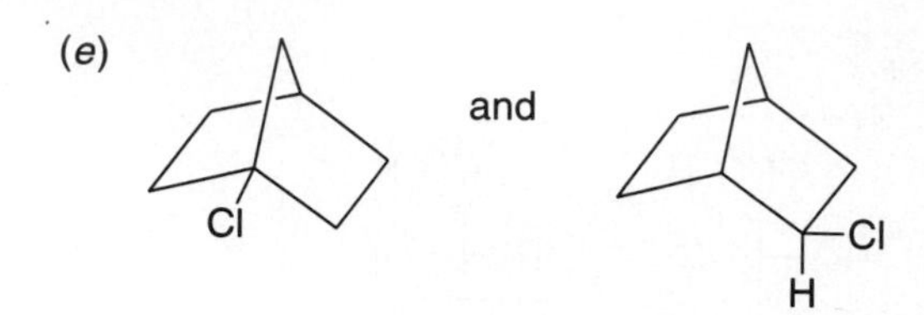

7.24 Compare 2,3-pentanediol and 2,4-pentanediol with respect to the number of stereoisomers possible for each constitution. Which stereoisomers are chiral? Which are achiral?

7.25 Specify the configuration as *R* or *S* in each of the following:

(*a*) (−)-2-Octanol

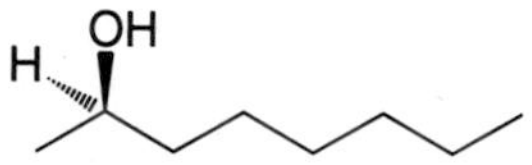

H OH

(*b*) Monosodium L-glutamate (only this stereoisomer is of any value as a flavor-enhancing agent)

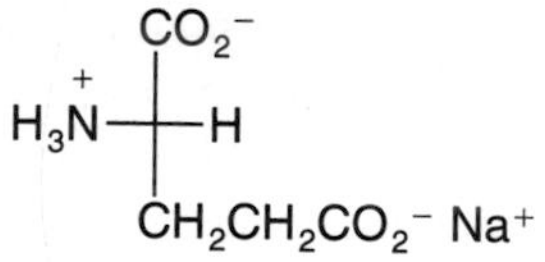

CO_2^-
$H_3\overset{+}{N}$—H
$CH_2CH_2CO_2^-$ Na^+

7.26 Identify the relationship in each of the following pairs. Do the drawings represent compounds which are constitutional isomers or stereoisomers, or are they identical? If they are stereoisomers, are they enantiomers or diastereomers? (Molecular models may prove useful in this problem.)

(*a*) H CH_3 C—CH_2Br HO and H_3C H C—CH_2OH Br

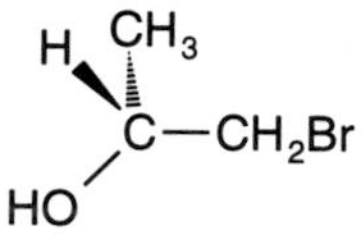

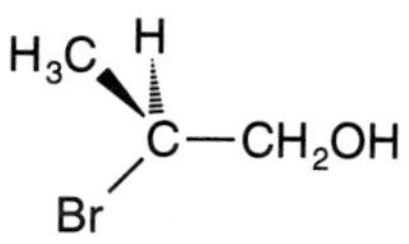

(*b*) H CH_3 C—Br CH_3CH_2 and H Br C—CH_3 CH_3CH_2

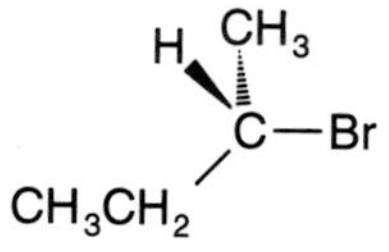

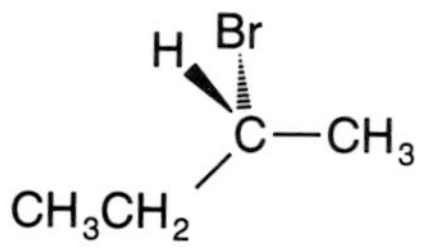

(*c*) H CH_3 C—Br CH_3CH_2 and H_3C H C—CH_2CH_3 Br

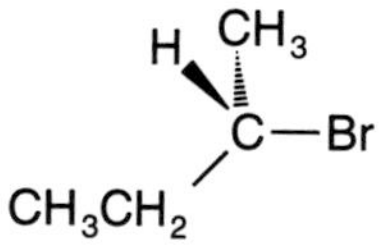

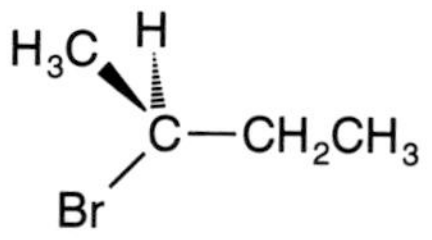

(*d*) H CH_3 C—Br CH_3CH_2 and CH_2CH_3 Br—CH_3 H

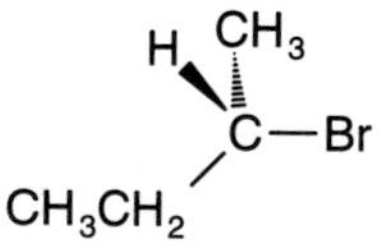

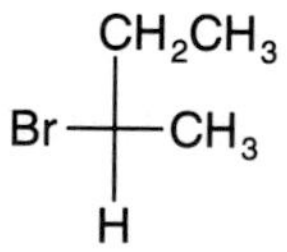

(*e*) CH_2OH H—OH H—H OH and CH_2OH HO—H H—H OH

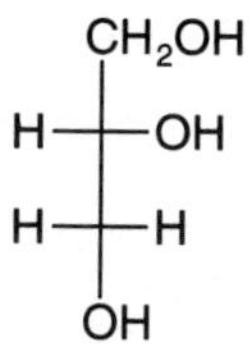

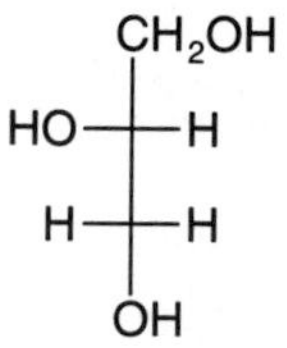

(*f*) H_3C H H Cl and H_3C Cl H H

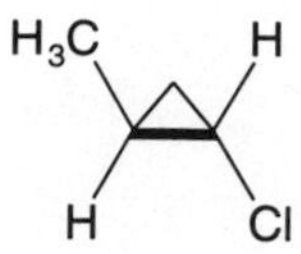

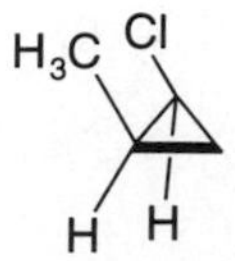

(*g*) 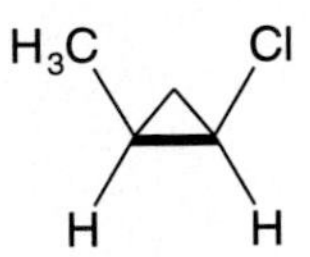 and 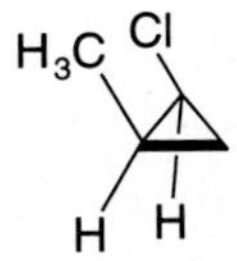

(*h*) 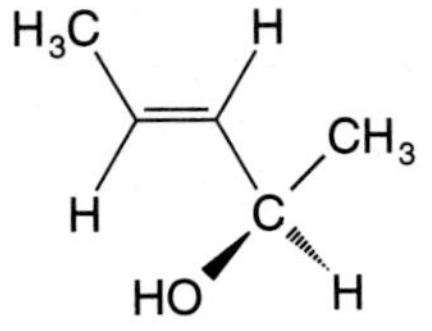 and 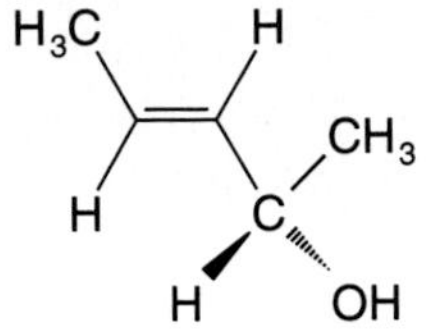

(*i*) 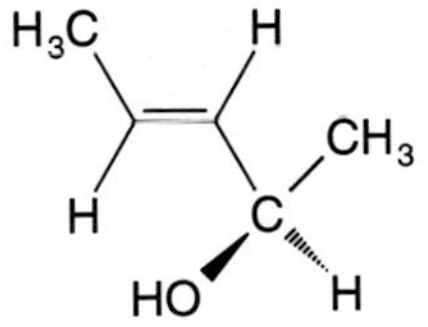 and 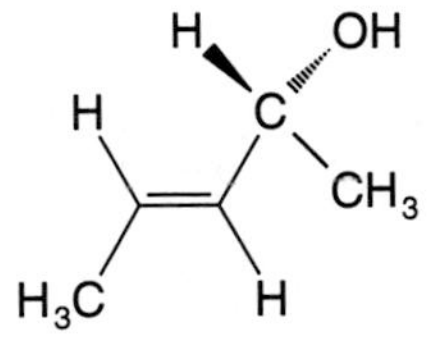

(*j*) 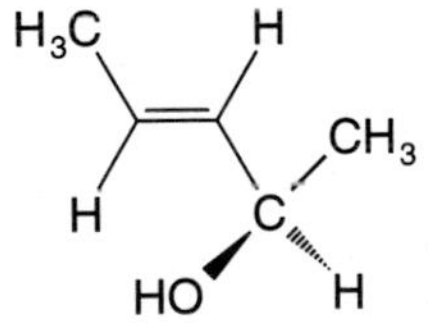 and 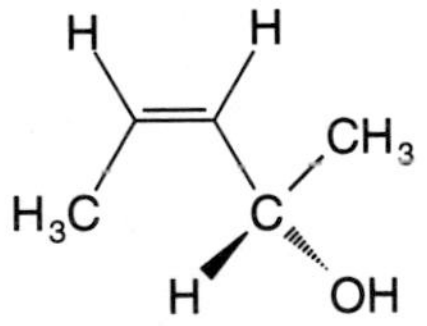

(*k*) 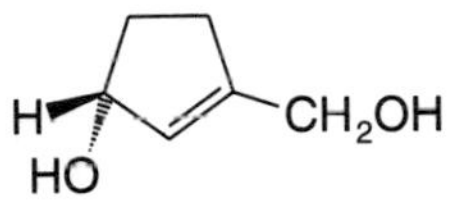 and 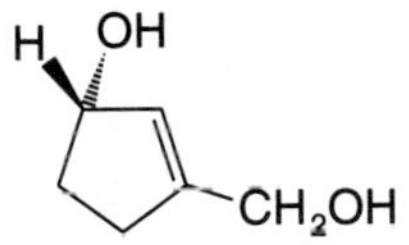

(*l*) 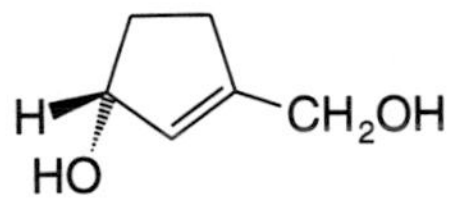 and 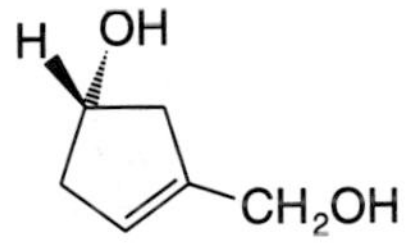

(*m*) 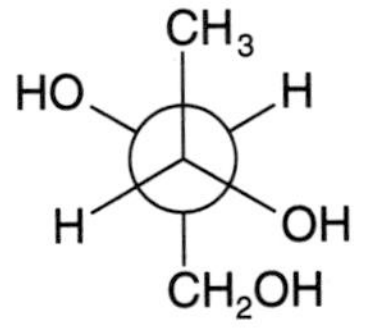 and 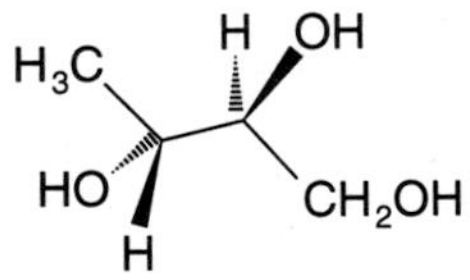

(*n*) 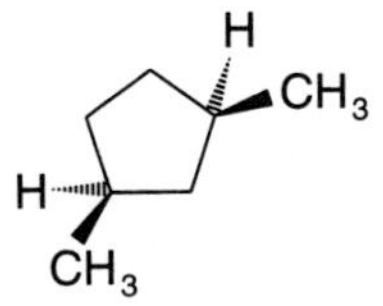 and 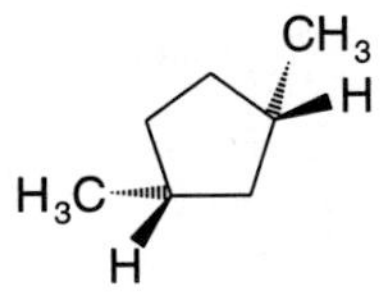

(*o*) 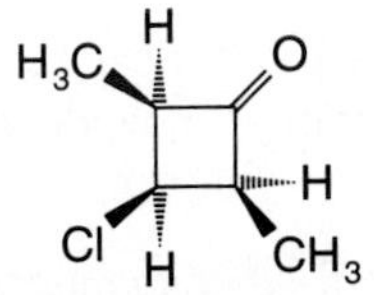 and 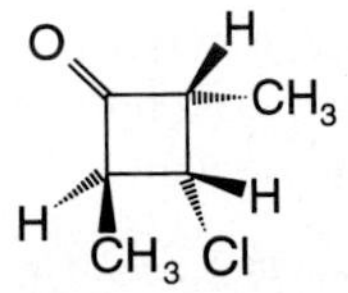

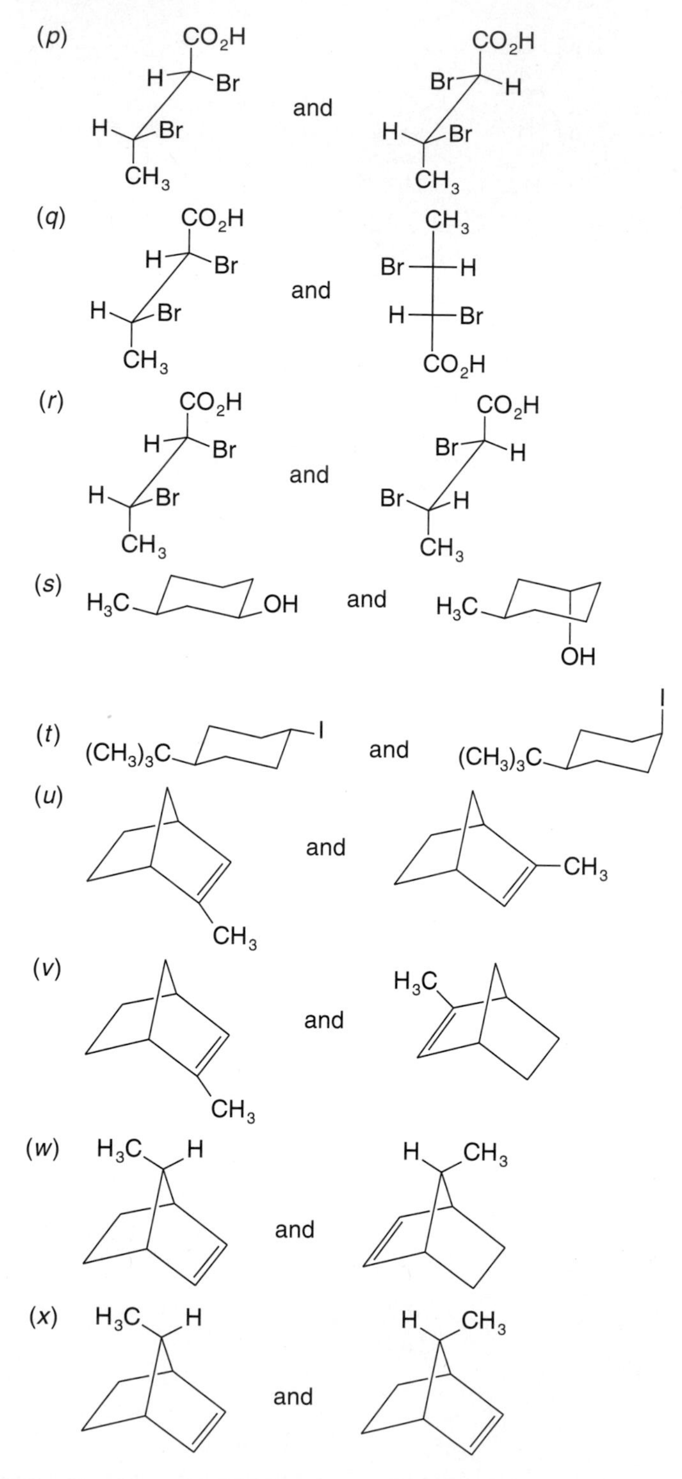

7.27 Chemical degradation of chlorophyll gives a number of substances including *phytol.* The constitution of phytol is given by the name 3,7,11,15-tetramethyl-2-hexadecen-1-ol. How many stereoisomers have this constitution?

7.28 *Muscarine* is a poisonous substance present in the mushroom *Amanita muscaria.* Its structure is represented by the constitution shown.

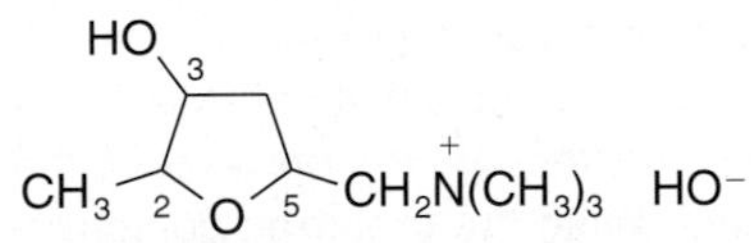

HO 3

CH_3 2 O 5 $CH_2\overset{+}{N}(CH_3)_3$ HO^-

(*a*) Including muscarine, how many stereoisomers have this constitution?

(*b*) One of the substituents on the ring of muscarine is trans to the other two. How many of the stereoisomers satisfy this requirement?

(*c*) Muscarine has the configuration 2*S*,3*R*,5*S*. Write a structural formula of muscarine showing its correct stereochemistry.

7.29 *Ectocarpene* is a volatile, sperm cell–attracting material released by the eggs of the seaweed *Ectocarpus siliculosus.* Its constitution is

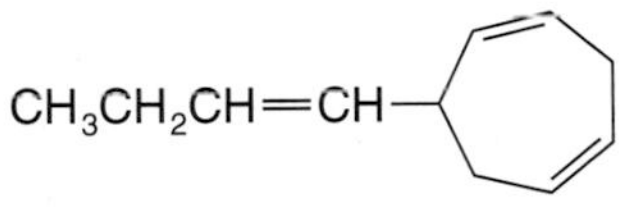

$$CH_3CH_2CH{=}CH-$$

All the double bonds are cis, and the absolute configuration of the stereogenic center is *S*. Write a stereochemically accurate representation of ectocarpene.

7.30 *Multifidene* is a sperm cell–attracting substance released by the female of a species of brown algae (*Cutleria multifida*). The constitution of multifidene is

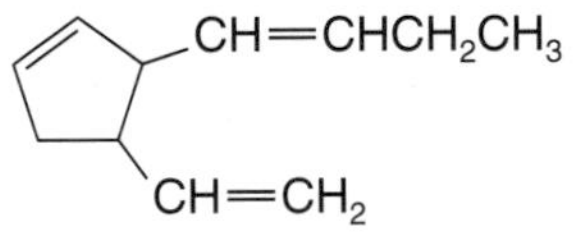

$CH{=}CHCH_2CH_3$

$CH{=}CH_2$

(*a*) How many stereoisomers are represented by this constitution?

(*b*) Multifidene has a cis relationship between its alkenyl substituents. Given this information, how many stereoisomers are possible?

(*c*) The butenyl side chain has the *Z* configuration of its double bond. On the basis of all the data, how many stereoisomers are possible?

(*d*) Draw stereochemically accurate representations of all the stereoisomers that satisfy the structural requirements of multifidene.

(*e*) How are these stereoisomeric multifidenes related (enantiomers or diastereomers)?

7.31 *Streptimidone* is an antibiotic and has the structure shown. How many diastereomers of streptimidone are possible? How many enantiomers? Using the *E*, *Z* and *R*, *S* descriptors, specify all essential elements of stereochemistry of streptimidone.

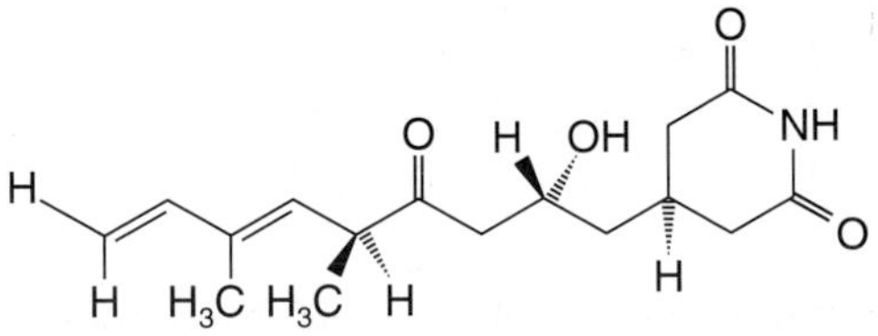

O H OH O NH

H O

H H_3C H_3C H H

7.32 In Problem 4.26 you were asked to draw the preferred conformation of menthol on the basis of the information that menthol is the most stable stereoisomer of 2-isopropyl-5-methylcyclohexanol. We can now completely describe (−)-menthol structurally by noting that it has the *R* configuration at the hydroxyl-substituted carbon.

(*a*) Write the preferred conformation of (−)-menthol in its correct configuration.

(*b*) (+)-Isomenthol has the same constitution as (−)-menthol. The configurations at C-1 and C-2 of (+)-isomenthol are the opposite of the corresponding stereogenic centers of (−)-menthol. Write the preferred conformation of (+)-isomenthol in its correct configuration.

7.33 A certain natural product having $[\alpha]_D$ +40.3° was isolated. Two structures have been independently proposed for this compound. Which one do you think is more likely to be correct? Why?

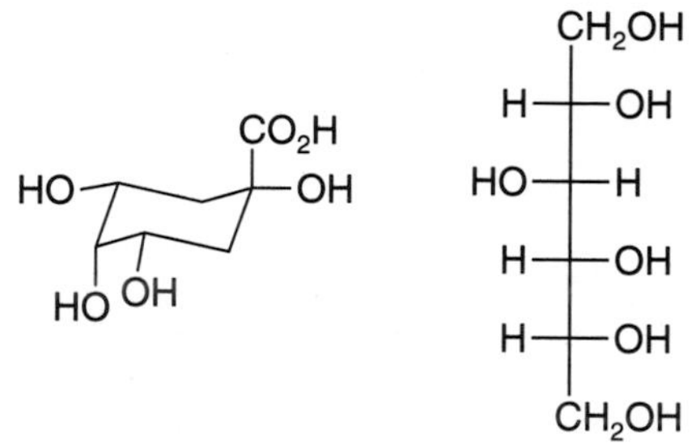

7.34 One of the principal substances obtained from primitive bacteria (*archaeobacteria:* see the marginal note on page 52) is derived from a 40-carbon diol. Given the fact that this diol is optically active, is it compound A or is it compound B?

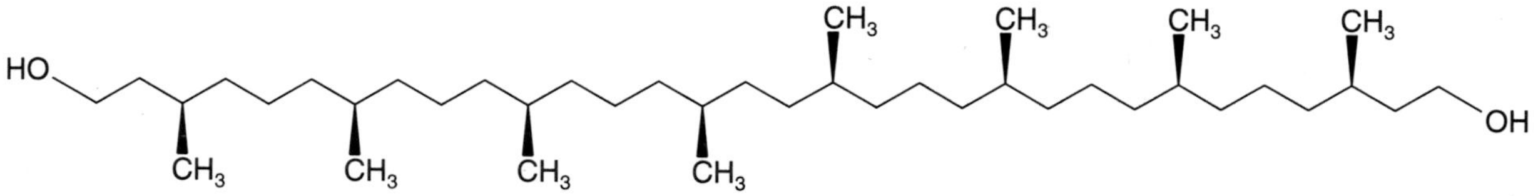

Compound A

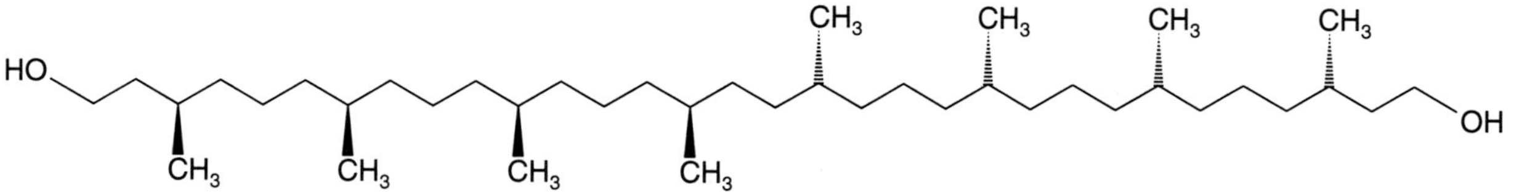

Compound B

7.35

(*a*) An aqueous solution containing 10 g of optically pure fructose was diluted to 500 mL with water and placed in a polarimeter tube 20 cm long. The measured rotation was −5.20°. Calculate the specific rotation of fructose.

(*b*) If this solution were mixed with 500 mL of a solution containing 5 g of racemic fructose, what would be the specific rotation of the resulting fructose mixture? What would be its optical purity?

7.36 Write the organic products of each of the following reactions. If two stereoisomers are formed, show both. Label all stereogenic centers *R* or *S* as appropriate.

(*a*) 1-Butene and hydrogen iodide

(*b*) (*E*)-2-Pentene and bromine in carbon tetrachloride

(*c*) (*Z*)-2-Pentene and bromine in carbon tetrachloride
(*d*) 1-Butene and peroxyacetic acid in dichloromethane
(*e*) (*Z*)-2-Pentene and peroxyacetic acid in dichloromethane
(*f*) 1,5,5-Trimethylcyclopentene and hydrogen in the presence of platinum
(*g*) 1,5,5-Trimethylcyclopentene and diborane in tetrahydrofuran followed by oxidation with hydrogen peroxide

7.37 The enzyme *aconitase* catalyzes the hydration of aconitic acid to two products, citric acid and isocitric acid. Isocitric acid is optically active; citric acid is not. What are the respective constitutions of citric acid and isocitric acid?

$$(HO_2CCH_2)(HO_2C)C{=}C(CO_2H)(H)$$

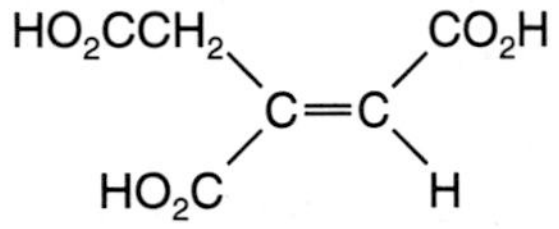

Aconitic acid

7.38 Consider the ozonolysis of *trans*-4,5-dimethylcyclohexene having the configuration shown.

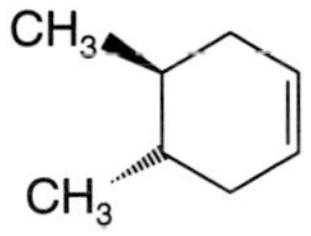

CH_3 / CH_3

Structures A, B, and C are three stereoisomeric forms of the reaction product.

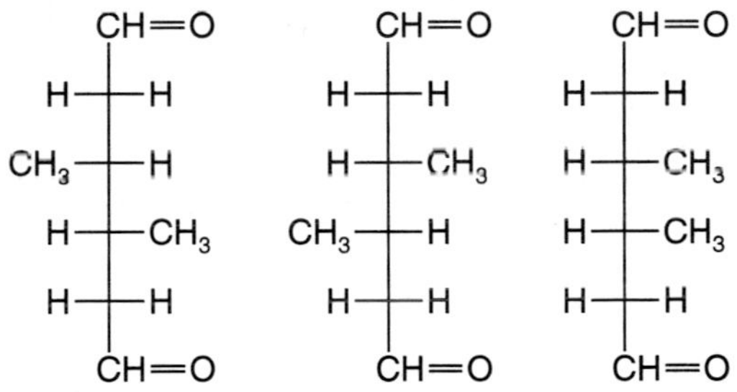

A	B	C
CH=O	CH=O	CH=O
H—H	H—H	H—H
CH_3—H	H—CH_3	H—CH_3
H—CH_3	CH_3—H	H—CH_3
H—H	H—H	H—H
CH=O	CH=O	CH=O

(*a*) Which, if any, of the compounds A, B, and C are chiral?
(*b*) What product is formed in the reaction?
(*c*) What product would be formed if the methyl groups were cis to each other in the starting alkene?

7.39

(*a*) On being heated with potassium ethoxide in ethanol (70°C), the deuterium-labeled alkyl bromide shown gave a mixture of 1-butene, *cis*-2-butene, and *trans*-2-butene. On the basis of your knowledge of the E2 mechanism, predict which alkene(s), if any, contained deuterium.

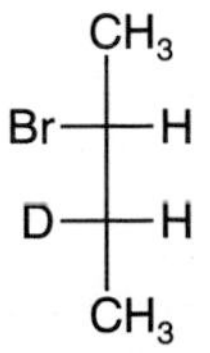

CH_3
Br—H
D—H
CH_3

(*b*) The bromide shown in part (*a*) is the erythro diastereomer. How would the deuterium content of the alkenes formed by dehydrohalogenation of the threo diastereomer differ from those produced in part (*a*)?

7.40 A compound (C_6H_{10}) contains a five-membered ring. When Br_2 adds to it, two diastereomeric dibromides are formed. Suggest reasonable structures for the compound and the two dibromides.

7.41 When optically pure 2,3-dimethyl-2-pentanol was subjected to dehydration, a mixture of two alkenes was obtained. Hydrogenation of this alkene mixture gave 2,3-dimethylpentane, which was 50 percent optically pure. What were the two alkenes formed in the elimination reaction, and what were the relative amounts of each?

7.42 When (*R*)-3-buten-2-ol is treated with a peroxy acid, two stereoisomeric epoxides are formed in a 60 : 40 ratio. The minor stereoisomer has the structure shown.

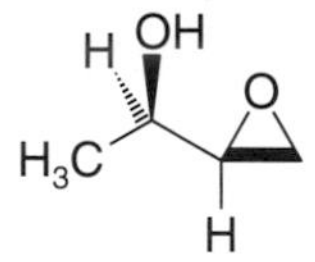

H OH
H_3C O
H

(*a*) Write the structure of the major stereoisomer.
(*b*) What is the relationship between the two epoxides? Are they enantiomers or diastereomers?
(*c*) What four stereoisomeric products are formed when racemic 3-buten-2-ol is epoxidized under the same conditions? How much of each stereoisomer is formed?

MOLECULAR MODELING EXERCISES

7.43 Verify that dibromochloromethane is achiral by superposing models of its two mirror image forms. In the same way, verify that bromochlorofluoromethane is chiral. (In general, attempted superposition of structures is easier with "real" molecular models than with molecular modeling software.)

7.44 Construct a molecular model of (*R*)-2-pentanol.

7.45 Construct a molecular model of (*S*)-3-chlorocyclopentene.

7.46 Construct a molecular model corresponding to the Fischer projection of *meso*-2,3-dibromobutane. Convert this molecular model to a staggered conformation in which the bromines are anti to one another. Are the methyl groups anti or gauche to one another in this staggered conformation?

7.47 The gauche conformation of butane is chiral. Verify this with the aid of molecular models. Butane, however, is not optically active. Why?

7.48 What alkene gives a racemic mixture of (2*R*,3*S*) and (2*S*,3*R*)-3-bromo-2-butanol on treatment with Br_2 in aqueous solution? (Hint: make a molecular model of one of the enantiomeric 3-bromo-2-butanols, arrange it in a conformation in which the Br and OH groups are anti to one another, then disconnect them.)

CHAPTER 8

NUCLEOPHILIC SUBSTITUTION

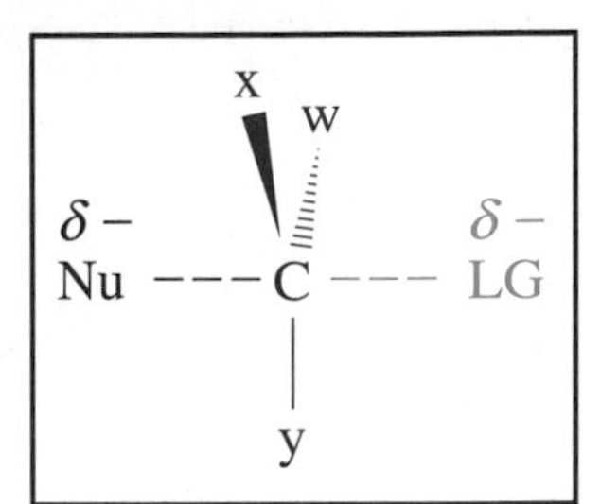

In our discussion of elimination reactions in Chapter 5, we learned that a Lewis base can react with an alkyl halide to form an alkene by dehydrohalogenation. In the present chapter, you shall find that the same kinds of reactants can also undergo a different reaction, one in which the Lewis base acts as a **nucleophile** to substitute for the halide substituent on carbon.

$$R{-}\ddot{\underset{..}{X}}: \; + \; Y:^- \longrightarrow RY \; + \; :\ddot{\underset{..}{X}}:^-$$

Alkyl halide	Lewis base	Product of nucleophilic substitution	Halide anion

We first encountered nucleophilic substitution in Chapter 4, in the reaction of alcohols with hydrogen halides to form alkyl halides. Now we will see how alkyl halides can themselves be converted to other classes of organic compounds by nucleophilic substitution.

This chapter has a mechanistic emphasis designed to achieve a practical result. By understanding the mechanisms by which alkyl halides are converted to products in substitution reactions, intelligent decisions can be made in choosing experimental conditions best suited to carry out a particular functional group transformation. The difference between a successful reaction that leads cleanly to a desired product and one that fails is often a subtle one. Mechanistic analysis helps us to appreciate these subtleties and use them to our advantage.

8.1 FUNCTIONAL GROUP TRANSFORMATION BY NUCLEOPHILIC SUBSTITUTION

Nucleophilic substitution reactions of alkyl halides are related to elimination reactions in that the halogen acts as a leaving group on carbon, and is lost as an anion. The carbon-halogen bond of the alkyl halide is broken **heterolytically:** the pair of electrons in that bond are lost with the leaving group.

The carbon-halogen bond in an alkyl halide is polar

$$\overset{\delta+}{R}-\overset{\delta-}{X} \qquad X = I, Br, Cl, F$$

and is cleaved on attack by a nucleophile so that the two electrons in the bond are retained by the halogen

$$^{-}Y\!:\ \ R-\ddot{X}\!: \longrightarrow R-Y + :\ddot{X}:^{-}$$

The most frequently encountered nucleophiles in functional group transformations are anions, which are used as their lithium, sodium, or potassium salts. If we use M to represent lithium, sodium, or potassium, some representative nucleophilic reagents are

MOR (a metal *alkoxide,* a source of the nucleophilic anion $R\ddot{O}:^{-}$)

MOCR (C=O) (a metal *carboxylate,* a source of the nucleophilic anion $R\overset{:O:}{\overset{\|}{C}}-\ddot{O}:^{-}$)

MSH (a metal *hydrogen sulfide,* a source of the nucleophilic anion $H\ddot{S}:^{-}$)

MCN (a metal *cyanide,* a source of the nucleophilic anion $:\overset{-}{C}\equiv N:$)

MN_3 (a metal *azide,* a source of the nucleophilic anion $:\overset{-}{N}=\overset{+}{N}=\overset{-}{N}:$)

Table 8.1 illustrates an application of each of these nucleophilic reagents to a functional group transformation. The anionic portion of the salt substitutes for the halogen of an alkyl halide. The metal cation portion becomes a lithium, sodium, or potassium halide.

$$M^{+}\ ^{-}Y: + R-\ddot{X}: \longrightarrow R-Y + M^{+}\ :\ddot{X}:^{-}$$

Nucleophilic reagent — Alkyl halide — Product of nucleophilic substitution — Metal halide

Alkenyl halides are also referred to as *vinylic halides.*

Notice that all the examples in Table 8.1 involve **alkyl halides,** that is, compounds in which the halogen is attached to an sp^3 hybridized carbon. **Alkenyl halides** and **aryl halides,** compounds in which the halogen is attached to sp^2 hybridized carbons, are essentially unreactive under these conditions, and the principles to be developed in this chapter do not apply to them.

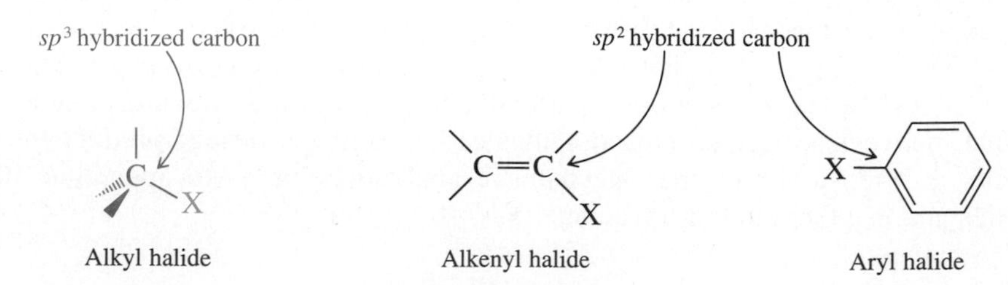

Alkyl halide Alkenyl halide Aryl halide

TABLE 8.1
Representative Functional Group Transformations by Nucleophilic Substitution Reactions of Alkyl Halides

Nucleophile and comments	General equation and specific example
Alkoxide ion ($R\ddot{O}:^-$) The oxygen atom of a metal alkoxide acts as a nucleophile to replace the halogen of an alkyl halide. The product is an *ether.*	$R'\ddot{O}:^- + R{-}X \longrightarrow R'\ddot{O}R + :\ddot{X}:^-$ Alkoxide ion, Alkyl halide, Ether, Halide ion $(CH_3)_2CHCH_2ONa + CH_3CH_2Br \xrightarrow{\text{isobutyl alcohol}} (CH_3)_2CHCH_2OCH_2CH_3 + NaBr$ Sodium isobutoxide, Ethyl bromide, Ethyl isobutyl ether (66%), Sodium bromide
Carboxylate ion ($RC(=O)-\ddot{O}:^-$) An *ester* is formed when the negatively charged oxygen of a carboxylate replaces the halogen of an alkyl halide.	$R'C(=O)\ddot{O}:^- + R{-}\ddot{X}: \longrightarrow R'C(=O)\ddot{O}R + :\ddot{X}:^-$ Carboxylate ion, Alkyl halide, Ester, Halide ion $KOC(=O)(CH_2)_{16}CH_3 + CH_3CH_2I \xrightarrow[\text{water}]{\text{acetone}} CH_3CH_2OC(=O)(CH_2)_{16}CH_3 + KI$ Potassium octadecanoate, Ethyl iodide, Ethyl octadecanoate (95%), Potassium iodide
Hydrogen sulfide ion ($H\ddot{S}:^-$) Use of hydrogen sulfide as a nucleophile permits the conversion of alkyl halides to compounds of the type RSH. These compounds are the sulfur analogs of alcohols and are known as *thiols.*	$H\ddot{S}:^- + R{-}\ddot{X}: \longrightarrow R\ddot{S}H + :\ddot{X}:^-$ Hydrogen sulfide ion, Alkyl halide, Thiol, Halide ion $KSH + CH_3CH(Br)(CH_2)_6CH_3 \xrightarrow[\text{water}]{\text{ethanol}} CH_3CH(SH)(CH_2)_6CH_3 + KBr$ Potassium hydrogen sulfide, 2-Bromononane, 2-Nonanethiol (74%), Potassium bromide
Cyanide ion ($:\bar{C}{\equiv}N:$) The negatively charged carbon atom of cyanide ion is usually the site of its nucleophilic character. Use of cyanide ion as a nucleophile permits the extension of a carbon chain by carbon-carbon bond formation. The product is an *alkyl cyanide,* or *nitrile.*	$:N{\equiv}\bar{C}: + R{-}\ddot{X}: \longrightarrow RC{\equiv}N: + :\ddot{X}:^-$ Cyanide ion, Alkyl halide, Alkyl cyanide, Halide ion $NaCN + C_5H_9{-}Cl \xrightarrow{\text{DMSO}} C_5H_9{-}CN + NaCl$ Sodium cyanide, Cyclopentyl chloride, Cyclopentyl cyanide (70%), Sodium chloride
Azide ion ($:\overset{-}{\ddot{N}}{=}\overset{+}{N}{=}\overset{-}{\ddot{N}}:$) Sodium azide is a reagent used for carbon-nitrogen bond formation. The product is an *alkyl azide.*	$:\overset{-}{\ddot{N}}{=}\overset{+}{N}{=}\overset{-}{\ddot{N}}: + R{-}\ddot{X}: \longrightarrow R\ddot{N}{=}\overset{+}{N}{=}\overset{-}{\ddot{N}}: + :\ddot{X}:^-$ Azide ion, Alkyl halide, Alkyl azide, Halide ion $NaN_3 + CH_3(CH_2)_4I \xrightarrow{\text{1-propanol-water}} CH_3(CH_2)_4N_3 + NaI$ Sodium azide, Pentyl iodide, Pentyl azide (52%), Sodium iodide

In order to ensure that reaction occurs in homogeneous solution, solvents are chosen that dissolve both the alkyl halide and the ionic salt. The alkyl halide substrates are soluble in organic solvents, but the salts often are not. Inorganic salts are soluble in water, but alkyl halides are not. Mixed solvents such as ethanol-water mixtures can often dissolve enough of both the substrate and the nucleophile to give fairly concentrated solutions and thus are frequently used. Many salts, as well as most alkyl halides, possess significant solubility in dimethyl sulfoxide (DMSO), which makes this a good medium for carrying out nucleophilic substitution reactions.

The use of DMSO as a solvent in *dehydrohalogenation* reactions was mentioned earlier, in Section 5.14.

PROBLEM 8.1 Write a structural formula for the principal organic product formed in the reaction of methyl bromide with each of the following compounds:

(*a*) NaOH (sodium hydroxide)
(*b*) $KOCH_2CH_3$ (potassium ethoxide)
(*c*) $NaOC(=O)C_6H_5$ (sodium benzoate)
(*d*) LiN_3 (lithium azide)
(*e*) KCN (potassium cyanide)
(*f*) NaSH (sodium hydrogen sulfide)

SAMPLE SOLUTION (*a*) The nucleophile in sodium hydroxide is the negatively charged hydroxide ion. The reaction that occurs is nucleophilic substitution of bromide by hydroxide. The product is methyl alcohol.

$$HO^- + CH_3Br \longrightarrow CH_3OH + Br^-$$

Hydroxide ion (nucleophile) | Methyl bromide (substrate) | Methyl alcohol (product) | Bromide ion (leaving group)

With this as background, you can begin to see how useful alkyl halides are in synthetic organic chemistry. Alkyl halides may be prepared from alcohols by nucleophilic substitution, from alkanes by free-radical halogenation, and from alkenes by addition of hydrogen halides. They then become available as starting materials for the preparation of other functionally substituted organic compounds by replacement of the halide leaving group with a nucleophile. The range of compounds that can be prepared by nucleophilic substitution reactions of alkyl halides is quite large; the examples shown in Table 8.1 illustrate only a few of them. Numerous other examples will be added to the list in this and subsequent chapters.

8.2 SUBSTITUTION OF ONE HALOGEN BY ANOTHER

In the reactions described in the preceding section, we saw examples of nucleophilic substitutions involving halide leaving groups. Halide ions may also act as nucleophiles. In a reaction known as **halide-halide exchange,** one halogen displaces another from an alkyl halide.

$$:\ddot{Y}:^- + R-\ddot{X}: \longrightarrow R-\ddot{Y}: + :\ddot{X}:^-$$

Halide ion (nucleophile) | Alkyl halide (substrate) | Alkyl halide (product) | Halide ion (leaving group)

Since the halide displaced is also a nucleophile, an equilibrium is established. Organic chemists have learned how to shift the position of equilibrium so as to make this reaction an effective one for the preparation of *alkyl fluorides* and *alkyl iodides*.

In the preparation of *alkyl fluorides,* an alkyl chloride, bromide, or iodide is heated with potassium fluoride in a high-boiling alcohol solvent such as ethylene glycol.

The boiling point of ethylene glycol is 198°C. It has the formula $HOCH_2CH_2OH$ and is made from ethylene. One of its uses is as a coolant and antifreeze for automobile radiators.

$$CH_3CH_2CH_2CH_2CH_2Br + KF \xrightarrow[120°C]{\text{ethylene glycol}} CH_3CH_2CH_2CH_2CH_2F + KBr$$

1-Bromopentane, bp 129°C; Potassium fluoride; 1-Fluoropentane, bp 65°C (50%); Potassium bromide

Alkyl fluorides have the lowest boiling points of all the alkyl halides and are removed from the reaction mixture by distillation as they are formed. In accordance with Le Châtelier's principle, the system responds by forming more alkyl fluoride at the expense of the original alkyl halide. Even if the alkyl fluoride were not removed by distillation, it would predominate at equilibrium because the reaction favors formation of the stronger C—F bond in place of weaker C—I, C—Br, or C—Cl bonds. Since reactions that lead to alkyl fluorides in good yield are relatively rare, this is a valuable synthetic method for these compounds.

Alkyl iodides may be prepared from alkyl chlorides and bromides by treatment with sodium iodide in acetone as the solvent.

Acetone, $(CH_3)_2C{=}O$, is a very good solvent for organic substances. One consumer use is as a nail-polish remover.

$$CH_2{=}CHCH_2Cl + NaI \xrightarrow{\text{acetone}} CH_2{=}CHCH_2I + NaCl \text{ (solid)}$$

3-Chloro-1-propene; Sodium iodide; 3-Iodo-1-propene (77%); Sodium chloride

$$CH_3\underset{\displaystyle Br}{\underset{|}{C}}HCH_3 + NaI \xrightarrow{\text{acetone}} CH_3\underset{\displaystyle I}{\underset{|}{C}}HCH_3 + NaBr \text{ (solid)}$$

2-Bromopropane; Sodium iodide; 2-Iodopropane (63%); Sodium bromide

Le Châtelier's principle is at work here as well. Sodium iodide is soluble in acetone, but sodium bromide and sodium chloride are not. In these reactions, sodium chloride and sodium bromide precipitate from the reaction mixture, causing the position of equilibrium to shift so as to favor formation of the alkyl iodides.

8.3 RELATIVE REACTIVITY OF HALIDE LEAVING GROUPS

Among alkyl halides, alkyl iodides undergo nucleophilic substitution at the fastest rate, alkyl fluorides at the slowest.

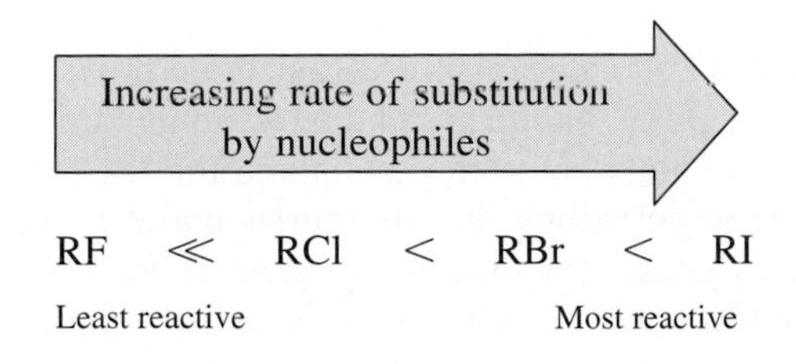

The order of alkyl halide reactivity in substitution reactions is the same as their order in elimination reactions. Iodine has the weakest bond to carbon, and iodide is the best leaving group. Alkyl iodides are several times more reactive than alkyl bromides and from 50 to 100 times more reactive than alkyl chlorides. Fluorine has the strongest bond to carbon, and fluoride is the poorest leaving group. Alkyl fluorides are rarely used as substrates in nucleophilic substitution because they are several thousand times less reactive than alkyl chlorides.

PROBLEM 8.2 A single organic product was obtained when 1-bromo-3-chloropropane was allowed to react with one molar equivalent of sodium cyanide in aqueous ethanol. What was this product?

Leaving-group ability is also related to basicity. A strongly basic anion is usually a poorer leaving group than a weakly basic one. Fluoride is the most basic and the poorest leaving group among the halide anions, iodide the least basic and the best leaving group.

8.4 THE BIMOLECULAR (S_N2) MECHANISM OF NUCLEOPHILIC SUBSTITUTION

The mechanisms by which nucleophilic substitution takes place have been the subject of much study. Extensive research by Sir Christopher Ingold and Edward D. Hughes and their associates at University College, London, during the 1930s emphasized kinetic and stereochemical measurements to probe the mechanisms of these reactions.

Recall that the term *kinetics* refers to the study of how the rate of a reaction varies with changes in concentration. Consider the nucleophilic substitution in which sodium hydroxide reacts with methyl bromide to form methyl alcohol and sodium bromide:

$$\underset{\text{Methyl bromide}}{CH_3Br} + \underset{\text{Hydroxide ion}}{HO^-} \longrightarrow \underset{\text{Methyl alcohol}}{CH_3OH} + \underset{\text{Bromide ion}}{Br^-}$$

The rate of this reaction is observed to be directly proportional to the concentration of both methyl bromide and sodium hydroxide. It is first-order in each reactant, or *second-order* overall.

$$\text{Rate} = k[CH_3Br][HO^-]$$

Hughes and Ingold interpreted second-order kinetic behavior to mean that the rate-determining step is *bimolecular,* i.e., that both hydroxide ion and methyl bromide are involved at the transition state. The symbol given to the detailed description of the mechanism that they developed is S_N2, standing for **substitution nucleophilic bimolecular.**

The S_N2 mechanism was introduced earlier in Section 4.14.

The Hughes and Ingold S_N2 mechanism is a concerted process, that is, a single-step reaction in which both the alkyl halide and the nucleophile are involved at the transition state. Cleavage of the bond between carbon and the leaving group is assisted by formation of a bond between carbon and the nucleophile. In effect, the nucleophile "pushes off" the leaving group from its point of attachment to carbon. For this reason, the S_N2 mechanism is sometimes referred to as a **direct displacement** process. The

S_N2 mechanism for the hydrolysis of methyl bromide may be represented by a single elementary step:

$$\underset{\text{Hydroxide ion}}{\ddot{\text{H}\ddot{\text{O}}}{:}^-} + \underset{\text{Methyl bromide}}{CH_3\ddot{\text{Br}}{:}} \longrightarrow \underset{\text{Transition state}}{\overset{\delta-}{H\ddot{O}}\text{---}CH_3\text{---}\overset{\delta-}{\ddot{Br}{:}}} \longrightarrow \underset{\text{Methyl alcohol}}{H\ddot{O}CH_3} + \underset{\text{Bromide ion}}{{:}\ddot{Br}{:}^-}$$

Carbon is partially bonded to both the incoming nucleophile and the departing leaving group at the transition state. Progress is made toward the transition state as the nucleophile begins to share a pair of its electrons with carbon and the halide ion leaves, taking with it the pair of electrons in its bond to carbon. The negative charge that develops on the leaving group is stabilized by hydrogen bonding of the leaving group to the solvent (water or an alcohol).

PROBLEM 8.3 Is the two-step sequence depicted in the following equations consistent with the second-order kinetic behavior observed for the hydrolysis of methyl bromide?

$$CH_3Br \xrightarrow{\text{slow}} CH_3^+ + Br^-$$

$$CH_3^+ + HO^- \xrightarrow{\text{fast}} CH_3OH$$

The S_N2 mechanism is believed to describe most substitutions in which simple primary and secondary alkyl halides react with anionic nucleophiles. All the examples cited in Sections 8.1 and 8.2 proceed by the S_N2 mechanism (or a mechanism very much like S_N2—remember, mechanisms can never be established with certainty but represent only our best present explanations of experimental observations). We will examine the S_N2 mechanism, particularly the structure of the transition state, in more detail in Section 8.6 after first looking at some stereochemical studies of nucleophilic substitution reactions.

8.5 STEREOCHEMISTRY OF S_N2 REACTIONS

Assuming that the transition state is bimolecular in reactions of primary and secondary alkyl halides with anionic nucleophiles, what is its structure? In particular, what is the spatial arrangement of the nucleophile in relation to the leaving group as reactants pass through the transition state on their way to products?

Two stereochemical possibilities present themselves. In the pathway shown in Figure 8.1*a*, the nucleophile simply assumes the position occupied by the leaving group. It attacks the substrate at the same face from which the leaving group departs. This is called "front-side displacement," or substitution with **retention of configuration.**

In a second possibility, illustrated in Figure 8.1*b*, the nucleophile attacks the substrate from the side opposite the bond to the leaving group. This is called "back-side displacement," or substitution with **inversion of configuration.**

Which of these two opposite stereochemical possibilities is the correct one was determined by carrying out nucleophilic substitution reactions with optically active alkyl halides. In one such experiment, Hughes and Ingold determined that hydrolysis

(*a*) Nucleophilic substitution with retention of configuration

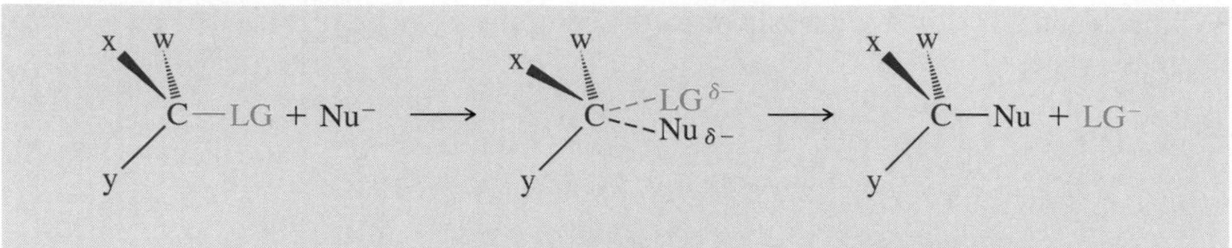

(*b*) Nucleophilic substitution with inversion of configuration

FIGURE 8.1 Two contrasting stereochemical pathways for substitution of a leaving group (LG) by a nucleophile (Nu^-). In (*a*) the nucleophile attacks carbon at the same side from which the leaving group departs. In (*b*) nucleophilic attack occurs at the side opposite the bond to the leaving group.

of 2-bromooctane in the presence of hydroxide ion gave 2-octanol having a configuration opposite that of the starting alkyl halide.

Although the alkyl halide and alcohol given in this example have opposite configurations when they have opposite signs of rotation, it cannot be assumed that this will be true for all alkyl halide–alcohol pairs. (See Section 7.5)

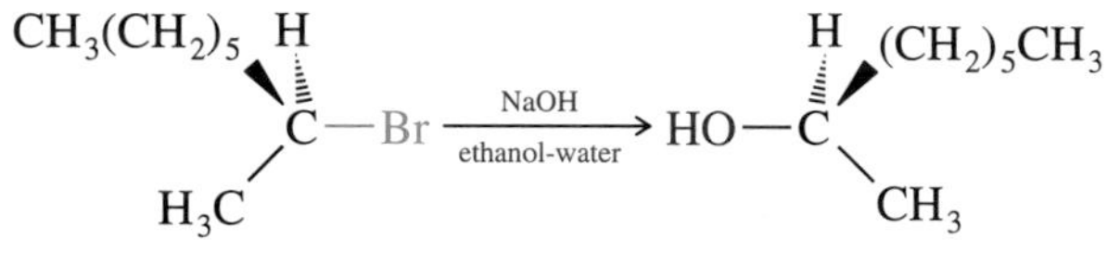

$$\text{(S)-(+)-2-Bromooctane} \xrightarrow[\text{ethanol-water}]{\text{NaOH}} \text{(R)-(−)-2-Octanol}$$

(*S*)-(+)-2-Bromooctane (*R*)-(−)-2-Octanol

Nucleophilic substitution in this case had occurred with inversion of configuration, consistent with the following transition state representation:

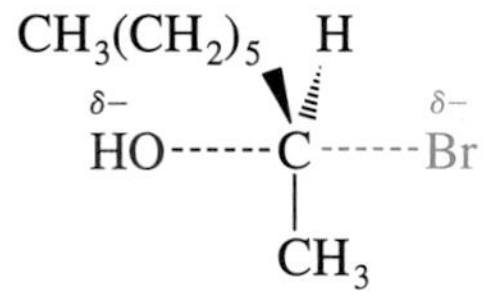

$$\overset{\delta-}{HO}\cdots C(H)(CH_3(CH_2)_5)(CH_3)\cdots\overset{\delta-}{Br}$$

PROBLEM 8.4 The Fischer projection formula for (+)-2-bromooctane is shown. Write the Fischer projection of the (−)-2-octanol formed from it by nucleophilic substitution with inversion of configuration.

CH_3 (top); H (left), Br (right); $CH_2(CH_2)_4CH_3$ (bottom)

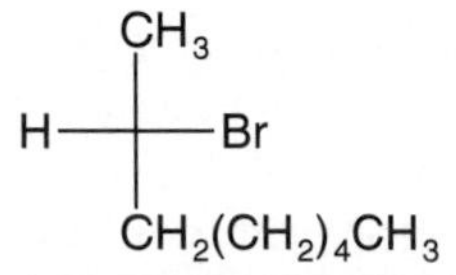

PROBLEM 8.5 Would you expect the 2-octanol formed by S_N2 hydrolysis of (−)-2-bromooctane to be optically active? If so, what will be its absolute configuration and sign of rotation? What about the 2-octanol formed by hydrolysis of racemic 2-bromooctane?

Numerous similar experiments have demonstrated the generality of this observation. Substitution reactions that proceed by the S_N2 mechanism are stereospecific and proceed with inversion of configuration at the carbon that bears the leaving group. *There is a stereoelectronic requirement for the nucleophile to approach carbon from the side opposite the bond to the leaving group.* Organic chemists often speak of this as a **Walden inversion,** after the German chemist Paul Walden, who described the earliest experiments in this area in the 1890s.

The first example of a stereoelectronic effect in this text concerned anti elimination in E2 reactions of alkyl halides (Section 5.15).

8.6 HOW S_N2 REACTIONS OCCUR

When we consider the overall reaction stereochemistry along with the kinetic data, a fairly complete picture of the bonding changes that take place during S_N2 reactions emerges. The potential energy diagram of Figure 8.2 for the hydrolysis of (*S*)-(+)-2-bromooctane is one that is consistent with the experimental observations.

Hydroxide ion acts as a nucleophile, using an unshared electron pair to attack carbon from the side opposite the bond to the leaving group. The hybridization of the carbon at which substitution occurs changes from sp^3 in the alkyl halide to sp^2 in the transition state. Both the nucleophile (hydroxide) and the leaving group (bromide) are partially bonded to this carbon in the transition state. We say that the S_N2 transition state is *pentacoordinate;* carbon is fully bonded to three substituents and partially bonded to both the leaving group and the incoming nucleophile. The bonds to the nucleophile and the leaving group are relatively long and weak at the transition state.

Once the reaction is past the transition state, the leaving group is expelled and carbon becomes tetracoordinate, its hybridization returning to sp^3.

During the passage of starting materials to products, three interdependent and synchronous changes take place:

1. Stretching, then breaking, of the bond to the leaving group
2. Formation of a bond to the nucleophile from the opposite side of the bond that is broken

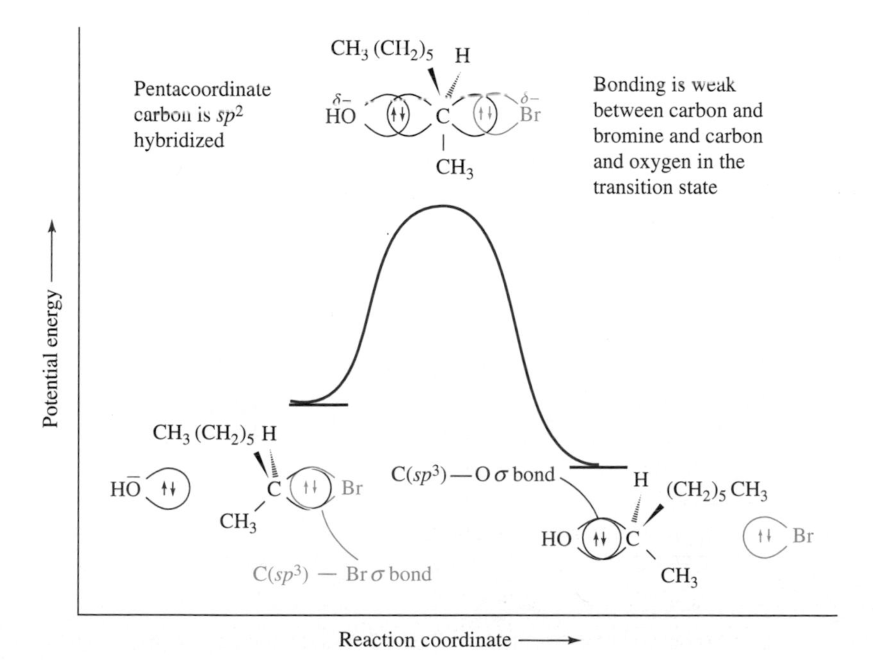

FIGURE 8.2 Hybrid orbital description of the bonding changes that take place at carbon during nucleophilic substitution by the S_N2 mechanism.

3. Stereochemical inversion of the tetrahedral arrangement of bonds to the carbon at which substitution occurs

Although this mechanistic picture developed from experiments involving optically active alkyl halides, it applies as well to cases not amenable to direct stereochemical study. Chemists even speak of methyl bromide as undergoing nucleophilic substitution with *inversion.* By this they mean that tetrahedral inversion of the bonds to carbon occurs as the reactant proceeds to product by way of a pentacoordinated transition state.

$$HO^{-} + CH_3Br \longrightarrow [HO^{\delta-}\cdots CH_3 \cdots Br^{\delta-}] \longrightarrow HOCH_3 + Br^{-}$$

Hydroxide ion | Methyl bromide | Transition state | Methyl alcohol | Bromide ion

We have noted earlier, in Section 8.3, that alkyl halides differ in reactivity according to the nature of their leaving group. The bond to the leaving group is partially broken in the S_N2 transition state, and alkyl iodides react faster than other alkyl halides with nucleophilic reagents because they have the weakest carbon-halogen bonds. In the next section we will see how reactivity in nucleophilic substitution reactions is affected by another aspect of alkyl halide structure.

8.7 STERIC EFFECTS IN S_N2 REACTIONS

There are very large differences in reactivity between alkyl halides, which depend on the degree of substitution at the carbon that bears the leaving group. As Table 8.2 shows for the reaction

$$\underset{\text{Alkyl bromide}}{RBr} + \underset{\text{Lithium iodide}}{LiI} \xrightarrow{\text{acetone}} \underset{\text{Alkyl iodide}}{RI} + \underset{\text{Lithium bromide}}{LiBr}$$

the rates of nucleophilic substitution of a series of alkyl bromides differ by a factor of over 10^6 when the most reactive member of the group (methyl bromide) and the least reactive member (*tert*-butyl bromide) are compared.

The large rate difference between methyl, ethyl, isopropyl, and *tert*-butyl bromides

TABLE 8.2

Reactivity of Some Alkyl Bromides toward Substitution by the S_N2 Mechanism*

Alkyl bromide	Structure	Class	Relative rate†
Methyl bromide	CH_3Br	Unsubstituted	221,000
Ethyl bromide	CH_3CH_2Br	Primary	1,350
Isopropyl bromide	$(CH_3)_2CHBr$	Secondary	1
tert-Butyl bromide	$(CH_3)_3CBr$	Tertiary	Too small to measure

* Substitution of bromide by lithium iodide in acetone.

† Ratio of second-order rate constant *k* for indicated alkyl bromide to *k* for isopropyl bromide at 25°C.

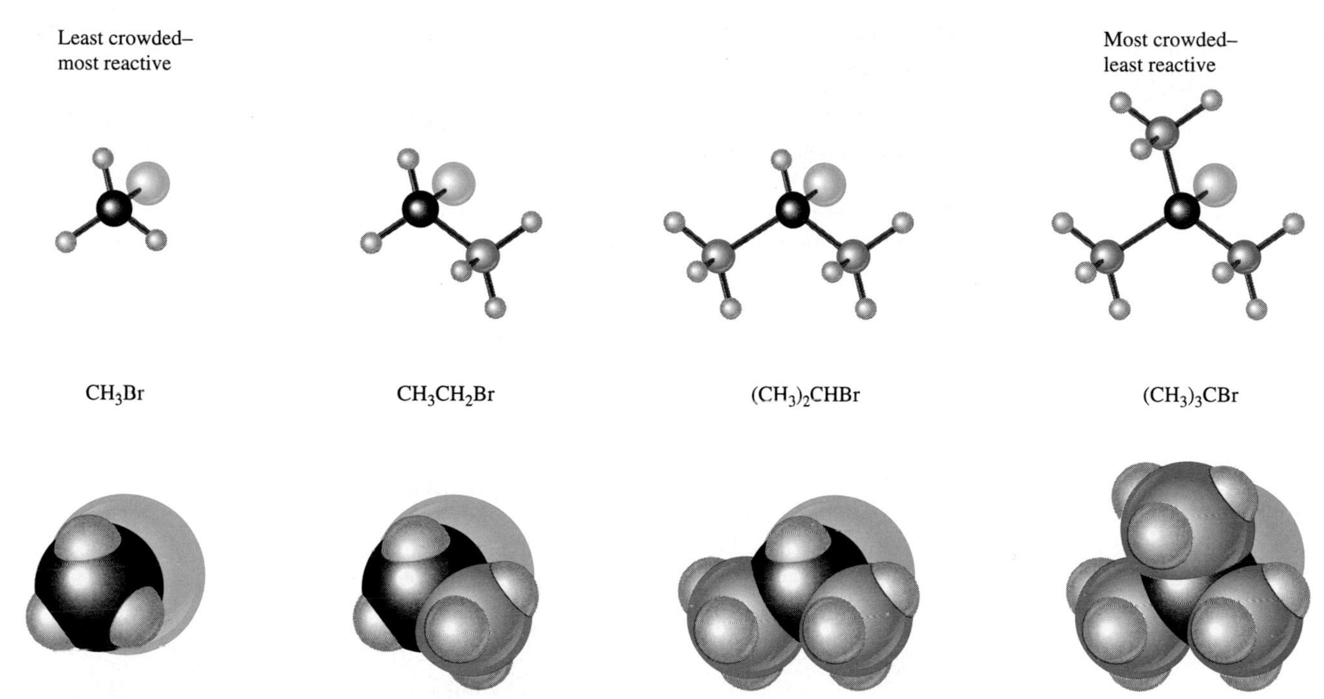

FIGURE 8.3 Ball-and-stick and space-filling models of alkyl bromides showing how substituents shield the carbon atom that bears the leaving group from attack by a nucleophile. The nucleophile must attack from the side opposite the bond to the leaving group.

rests on the degree of **steric hindrance** each offers to nucleophilic attack. The nucleophile must approach the alkyl halide from the side opposite the bond to the leaving group, and, as illustrated in Figure 8.3, this approach is hindered by alkyl substituents on the carbon that is being attacked. The three hydrogen substituents of methyl bromide offer little resistance to approach of the nucleophile, and a rapid reaction occurs. Replacing one of the hydrogens by a methyl group somewhat shields the carbon from attack by the nucleophile and causes ethyl bromide to be less reactive than methyl bromide. Replacing all three hydrogen substituents by methyl groups almost completely blocks back-side approach to the tertiary carbon of $(CH_3)_3CBr$ and shuts down bimolecular nucleophilic substitution.

In general, nucleophilic substitutions characterized by second-order kinetic behavior exhibit the following dependence of rate on substrate structure:

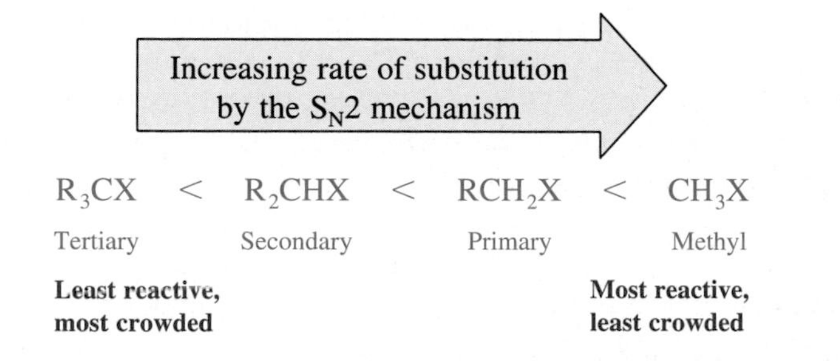

PROBLEM 8.6 Identify the compound in each of the following pairs that reacts with sodium iodide in acetone at the faster rate:

(*a*) 1-Chlorohexane or cyclohexyl chloride
(*b*) 1-Bromopentane or 3-bromopentane
(*c*) 2-Chloropentane or 2-fluoropentane
(*d*) 2-Bromo-2-methylhexane or 2-bromo-5-methylhexane
(*e*) 2-Bromopropane or 1-bromodecane

SAMPLE SOLUTION (*a*) Compare the structures of the two chlorides. 1-Chlorohexane is a primary alkyl chloride; cyclohexyl chloride is secondary. Primary alkyl halides are less crowded at the site of substitution than secondary ones and react faster in substitution by the S_N2 mechanism. 1-Chlorohexane is more reactive.

$CH_3CH_2CH_2CH_2CH_2CH_2Cl$

1-Chlorohexane
(primary, more reactive)

Cyclohexyl chloride
(secondary, less reactive)

Alkyl substituents at the carbon atom *adjacent* to the point of nucleophilic attack also decrease the rate of the S_N2 reaction. Compare the rates of nucleophilic substitution in the series of primary alkyl bromides shown in Table 8.3. Taking ethyl bromide as the standard and successively replacing its C-2 hydrogen substituents by methyl groups, we see that each additional methyl group decreases the rate of displacement of bromide by iodide. The effect is slightly smaller than that seen for alkyl substituents directly attached to the carbon that bears the leaving group, but is still substantial. When C-2 is completely substituted by methyl groups, as it is in neopentyl bromide [$(CH_3)_3CCH_2Br$], we see the unusual case of a primary alkyl halide that is practically inert to substitution by the S_N2 mechanism because of steric hindrance.

8.8 NUCLEOPHILES AND NUCLEOPHILICITY

The Lewis base that acts as the nucleophile in a nucleophilic substitution often is, but need not always be, an anion. Neutral Lewis bases can also serve as nucleophiles. Common examples of substitutions involving neutral nucleophiles include *solvolysis* reactions. **Solvolysis** reactions are substitutions in which the nucleophile is the solvent in which the reaction is carried out. Solvolysis in *water* converts an alkyl halide to an *alcohol.*

TABLE 8.3
Effect of Chain Branching on Reactivity of Primary Alkyl Bromides toward Substitution under S_N2 Conditions*

Alkyl bromide	Structure	Relative rate†
Ethyl bromide	CH_3CH_2Br	1.0
Propyl bromide	$CH_3CH_2CH_2Br$	0.8
Isobutyl bromide	$(CH_3)_2CHCH_2Br$	0.036
Neopentyl bromide	$(CH_3)_3CCH_2Br$	0.00002

* Substitution of bromide by lithium iodide in acetone.
† Ratio of second-order rate constant *k* for indicated alkyl bromide to *k* for ethyl bromide at 25°C.

$$H_2\ddot{O}: + R{-}X \xrightarrow{\text{slow}} H_2\overset{+}{O}{-}R + X^- \xrightarrow{\text{fast}} R\ddot{O}H + HX$$

Water | Alkyl halide | Alkyloxonium halide | Alcohol | Hydrogen halide

Solvolysis in *methyl alcohol* converts an alkyl halide to an *alkyl methyl ether.*

$$H_3C(H)\ddot{O}: + R{-}X \xrightarrow{\text{slow}} H_3C(H)\overset{+}{O}{-}R + X^- \xrightarrow{\text{fast}} R\ddot{O}CH_3 + HX$$

Methyl alcohol | Alkyl halide | Dialkyloxonium halide | Alkyl methyl ether | Hydrogen halide

In these and related solvolyses, nucleophilic substitution is the first step and is rate-determining. The proton transfer step that follows it is much faster.

Since, as we have seen, the nucleophile attacks the substrate in the rate-determining step of the S_N2 mechanism, it follows that the rate at which substitution occurs may vary from nucleophile to nucleophile. Just as some alkyl halides are more reactive than others, some nucleophiles are more reactive than others. Nucleophilic strength, or **nucleophilicity,** is a measure of how fast a Lewis base displaces a leaving group from a suitable substrate. By measuring the rate at which various Lewis bases react with methyl iodide in methanol, a list of their nucleophilicities relative to methanol as the standard nucleophile has been compiled. It is presented in Table 8.4.

Neutral Lewis bases such as water, alcohols, and carboxylic acids are much weaker nucleophiles than their conjugate bases. When comparing species that have the same nucleophilic atom, a negatively charged nucleophile is more reactive than a neutral one.

RO^- (Alkoxide ion) is more nucleophilic than ROH (Alcohol)

RCO^- with C=O (Carboxylate ion) is more nucleophilic than $RCOH$ with C=O (Carboxylic acid).

TABLE 8.4
Nucleophilicity of Some Common Nucleophiles

Reactivity class	Nucleophile	Relative reactivity*
Very good nucleophiles	I^-, HS^-, RS^-	$>10^5$
Good nucleophiles	Br^-, HO^-, RO^-, CN^-, N_3^-	10^4
Fair nucleophiles	NH_3, Cl^-, F^-, RCO_2^-	10^3
Weak nucleophiles	H_2O, ROH	1
Very weak nucleophiles	RCO_2H	10^{-2}

* Relative reactivity is *k*(nucleophile)/*k*(methanol) for typical S_N2 reactions and is approximate. Data pertain to methanol as the solvent.

So long as the nucleophilic atom is the same, the more basic the nucleophile, the more reactive it is. An alkoxide ion (RO^-) is more basic and more nucleophilic than a carboxylate ion (RCO_2^-).

RO^- is more nucleophilic than 

Stronger base	Weaker base
Conjugate acid is ROH:	Conjugate acid is RCO_2H:
$K_a = 10^{-16}$ ($pK_a = 16$)	$K_a = 10^{-5}$ ($pK_a = 5$)

The connection between basicity and nucleophilicity holds when comparing atoms in the *same row* of the periodic table. Thus, HO^- is more basic and more nucleophilic than F^-, and H_3N is more basic and more nucleophilic than H_2O. *It does not hold when proceeding down a column in the periodic table.* For example, I^- is the least basic of the halide ions but is the most nucleophilic. F^- is the most basic halide ion but the least nucleophilic. The factor that seems most responsible for the inverse relationship between basicity and nucleophilicity among the halide ions is the degree to which they are *solvated* by hydrogen bonds of the type illustrated in Figure 8.4. Smaller anions, because of their high charge-to-size ratio, are more strongly solvated than larger ones. In order to act as a nucleophile, the halide must shed some of the solvent molecules that surround it. Among the halide anions, F^- forms the strongest hydrogen bonds to water and alcohols, and I^- the weakest. Thus, the nucleophilicity of F^- is suppressed more than that of Cl^-, Cl^- more than Br^-, and Br^- more than I^-. Similarly, HO^- is smaller, more solvated, and less nucleophilic than HS^-.

A descriptive term applied to a highly polarizable species is *soft*. Iodide is a very soft nucleophile. Conversely, fluoride ion is not very polarizable and is said to be a *hard* nucleophile.

Nucleophilicity is also related to polarizability, or the ease of distortion of the electron "cloud" surrounding the nucleophile. The partial bond between the nucleophile and the alkyl halide that characterizes the S_N2 transition state is more fully developed at a longer distance when the nucleophile is very polarizable than when it is not. An increased degree of bonding to the nucleophile lowers the energy of the transition state and increases the rate of substitution. Among related atoms, polarizability increases with increasing size. Thus iodide is the most polarizable and most nucleophilic halide ion, fluoride the least.

PROBLEM 8.7 Sodium nitrite ($NaNO_2$) reacted with 2-iodooctane to give a mixture of two constitutionally isomeric compounds of molecular formula $C_8H_{17}NO_2$ in a combined yield of 88 percent. Suggest reasonable structures for these two isomers.

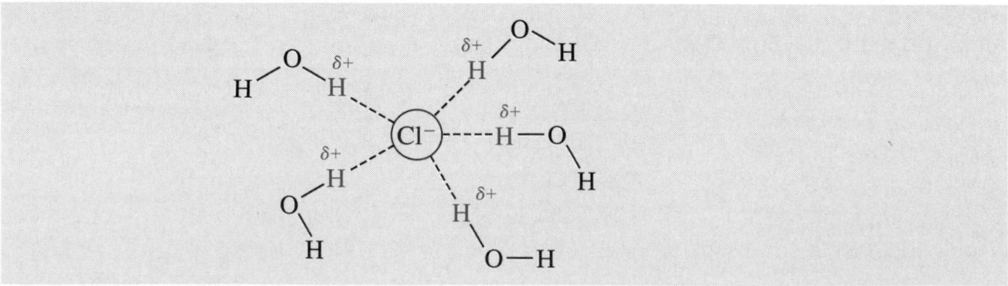

FIGURE 8.4 Solvation of a representative halide ion (chloride) by ion-dipole attractive forces with water. The negatively charged chloride ion interacts with the positively polarized hydrogens of water molecules to form hydrogen bonds.

8.9 THE UNIMOLECULAR (S_N1) MECHANISM OF NUCLEOPHILIC SUBSTITUTION

Recalling from Section 8.7 that tertiary alkyl halides are practically inert to substitution by the S_N2 mechanism because of steric hindrance, we might wonder whether they undergo nucleophilic substitution at all. We shall see in this section that they do, but by a mechanism different from S_N2.

Hughes and Ingold observed that the hydrolysis of *tert*-butyl bromide, which occurs readily, is characterized by a *first-order* rate law:

$$\underset{\textit{tert}\text{-Butyl bromide}}{(CH_3)_3CBr} + \underset{\text{Water}}{H_2O} \longrightarrow \underset{\textit{tert}\text{-Butyl alcohol}}{(CH_3)_3COH} + \underset{\text{Hydrogen bromide}}{HBr}$$

$$\text{Rate} = k[(CH_3)_3CBr]$$

They found that the rate of hydrolysis depends only on the concentration of *tert*-butyl bromide. Adding the stronger nucleophile hydroxide ion, moreover, causes no change

The overall reaction:

$$\underset{\textit{tert}\text{-Butyl bromide}}{(CH_3)_3CBr} + \underset{\text{Water}}{2H_2O} \longrightarrow \underset{\textit{tert}\text{-Butyl alcohol}}{(CH_3)_3COH} + \underset{\text{Hydronium ion}}{H_3O^+} + \underset{\text{Bromide ion}}{Br^-}$$

Step 1: The alkyl halide dissociates to a carbocation and a halide ion.

$$\underset{\textit{tert}\text{-Butyl bromide}}{(CH_3)_3C{-}\ddot{\underset{..}{Br}}:} \xrightarrow{\text{slow}} \underset{\textit{tert}\text{-Butyl cation}}{(CH_3)_3C^+} + \underset{\text{Bromide ion}}{:\ddot{\underset{..}{Br}}:^-}$$

Step 2: The carbocation formed in step 1 reacts rapidly with a water molecule. Water is a nucleophile. This step completes the nucleophilic substitution stage of the mechanism and yields an oxonium ion.

$$\underset{\textit{tert}\text{-Butyl cation}}{(CH_3)_3C^+} + \underset{\text{Water}}{:\ddot{O}H_2} \xrightarrow{\text{fast}} \underset{\textit{tert}\text{-Butyloxonium ion}}{(CH_3)_3C{-}\overset{+}{\ddot{O}}H_2}$$

Step 3: This step is a fast acid-base reaction that follows the nucleophilic substitution. Water acts as a base to remove a proton from the oxonium ion to give the observed product of the reaction, *tert*-butyl alcohol.

$$\underset{\textit{tert}\text{-Butyloxonium ion}}{(CH_3)_3C{-}\overset{+}{\ddot{O}}H_2} + \underset{\text{Water}}{:\ddot{O}H_2} \xrightarrow{\text{fast}} \underset{\textit{tert}\text{-Butyl alcohol}}{(CH_3)_3C{-}\ddot{O}H} + \underset{\text{Hydronium ion}}{H{-}\overset{+}{\ddot{O}}H_2}$$

FIGURE 8.5 Sequence of steps that describes the S_N1 mechanism for hydrolysis of *tert*-butyl bromide.

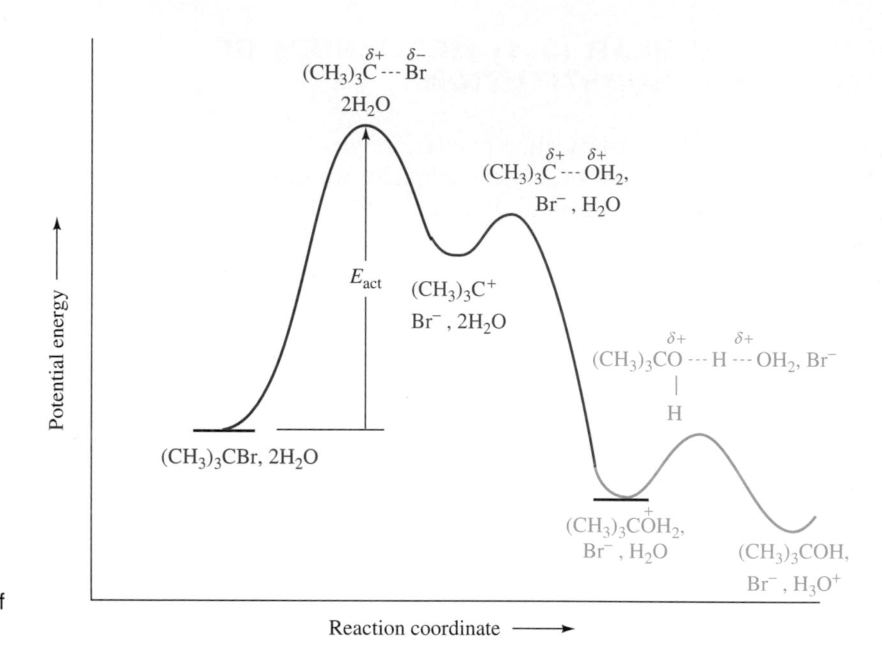

FIGURE 8.6 Energy diagram illustrating the S_N1 mechanism for hydrolysis of *tert*-butyl bromide.

in the rate of substitution, nor does this rate depend on the concentration of hydroxide. Just as second-order kinetics was interpreted as indicating a bimolecular rate-determining step, first-order kinetics was interpreted as evidence for a *unimolecular* rate-determining step—a step that involves only the alkyl halide.

The S_N1 mechanism was earlier introduced in Section 4.12.

The proposed mechanism is outlined in Figure 8.5 and is called S_N1, standing for **substitution nucleophilic unimolecular.** The first step, a unimolecular dissociation of the alkyl halide to form a carbocation as the key intermediate, is rate-determining. An energy diagram for the process is shown in Figure 8.6.

PROBLEM 8.8 Suggest a structure for the product of nucleophilic substitution obtained on solvolysis of *tert*-butyl bromide in methanol, and outline a reasonable mechanism for its formation.

The S_N1 mechanism is an *ionization* mechanism. The nucleophile does not participate until after the rate-determining step has taken place. Thus, the effects of nucleophile and substrate structure are expected to be different from those observed for reactions proceeding by the S_N2 pathway. How the structure of the alkyl halide affects the rate of S_N1 reactions is the topic of the next section.

8.10 CARBOCATION STABILITY AND THE RATE OF SUBSTITUTION BY THE S_N1 MECHANISM

Tertiary alkyl halides are good candidates for reaction by the S_N1 mechanism because they are too sterically hindered to react by the S_N2 mechanism and, since they form relatively stable carbocations, do not have prohibitively high activation energies for ionization. What about S_N1 reactions in other classes of alkyl halides?

In order to compare S_N1 substitution rates in a range of alkyl halides, experimental conditions are chosen in which competing substitution by the S_N2 route is very slow. One such set of conditions is solvolysis in aqueous formic acid (HCO_2H):

$$\underset{\text{Alkyl halide}}{RX} + \underset{\text{Water}}{H_2O} \xrightarrow{\text{formic acid}} \underset{\text{Alcohol}}{ROH} + \underset{\text{Hydrogen halide}}{HX}$$

Neither formic acid nor water is very nucleophilic, and so S_N2 substitution is suppressed. The relative rates of hydrolysis of a group of alkyl bromides under these conditions are presented in Table 8.5.

The relative rate order in S_N1 reactions is exactly the opposite of that seen in S_N2 reactions:

S_N1 reactivity: methyl < primary < secondary < tertiary
S_N2 reactivity: tertiary < secondary < primary < methyl

Clearly, the steric crowding that influences reaction rates in S_N2 processes plays no role in S_N1 reactions. The order of alkyl halide reactivity in S_N1 reactions is the same as the order of carbocation stability: the more stable the carbocation, the more reactive the alkyl halide. We have seen this situation before in the reaction of alcohols with hydrogen halides (Section 4.14), in the acid-catalyzed dehydration of alcohols (Section 5.9), and in the conversion of alkyl halides to alkenes by the E1 mechanism (Section 5.17). As in these other reactions, an electronic effect, specifically, the stabilization of the carbocation intermediate by alkyl substituents, is the decisive factor.

PROBLEM 8.9 Identify the compound in each of the following pairs that reacts at the faster rate in an S_N1 reaction:

(*a*) Isopropyl bromide or isobutyl bromide
(*b*) Cyclopentyl iodide or 1-methylcyclopentyl iodide
(*c*) Cyclopentyl bromide or 1-bromo-2,2-dimethylpropane
(*d*) *tert*-Butyl chloride or *tert*-butyl iodide

SAMPLE SOLUTION (*a*) Isopropyl bromide, $(CH_3)_2CHBr$, is a secondary alkyl halide, while isobutyl bromide, $(CH_3)_2CHCH_2Br$, is primary. Since the rate-determining step in an S_N1 reaction is carbocation formation and since secondary carbocations are more stable than primary carbocations, isopropyl bromide is more reactive than isobutyl bromide in nucleophilic substitution by the S_N1 mechanism.

TABLE 8.5
Reactivity of Some Alkyl Bromides toward Substitution by the S_N1 Mechanism*

Alkyl bromide	Structure	Class	Relative rate†
Methyl bromide	CH_3Br	Unsubstituted	1
Ethyl bromide	CH_3CH_2Br	Primary	2
Isopropyl bromide	$(CH_3)_2CHBr$	Secondary	43
tert-Butyl bromide	$(CH_3)_3CBr$	Tertiary	100,000,000

* Solvolysis in aqueous formic acid.
† Ratio of first-order rate constant *k* for indicated alkyl bromide to *k* for methyl bromide at 25°C.

Primary carbocations are so high in energy that their intermediacy in nucleophilic substitution reactions is unlikely. When ethyl bromide undergoes hydrolysis in aqueous formic acid, substitution probably takes place by a direct displacement of bromide by water in an S_N2-like process.

$$H_2\overset{\delta+}{O}\text{------}C(H)(H)(CH_3)\text{------}\overset{\delta+}{Br}$$

Representation of bimolecular transition state for hydrolysis of ethyl bromide

8.11 STEREOCHEMISTRY OF S_N1 REACTIONS

While nucleophilic substitutions that exhibit second-order kinetics are stereospecific and proceed with inversion of configuration at carbon, the situation is somewhat less clear-cut for reactions that have a first-order kinetic dependence. When the leaving group is attached to the stereogenic center of an optically active halide, ionization gives a carbocation intermediate that is achiral. It is achiral because the three bonds to the positively charged carbon lie in the same plane, and this plane is a plane of symmetry for the carbocation. As shown in Figure 8.7, such a carbocation should react with a nucleophile at the same rate at either of its two faces. We expect the product of substitution by the S_N1 mechanism to be formed as a racemic mixture and to be optically inactive. This outcome is rarely observed in practice. Normally, the product is formed with predominant, but not complete, inversion of configuration.

For example, the hydrolysis of 2-bromooctane in the absence of added base is a first-order reaction. When the starting alkyl halide is optically active, the resulting 2-octanol is formed with 66 percent inversion of configuration.

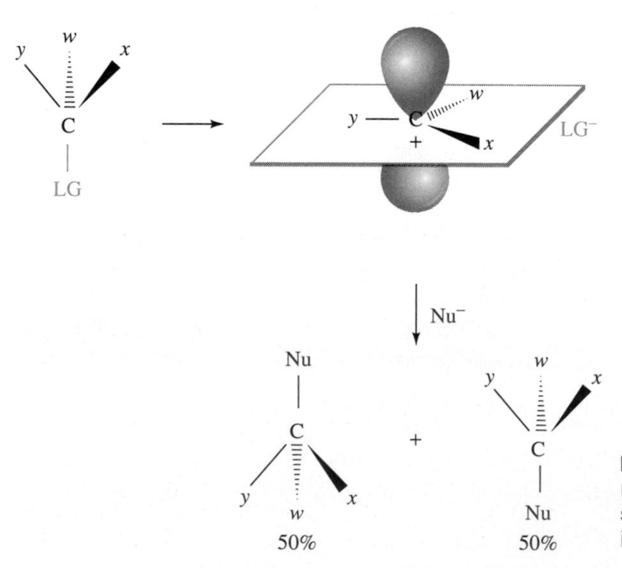

FIGURE 8.7 Formation of a racemic mixture by nucleophilic substitution via a carbocation intermediate.

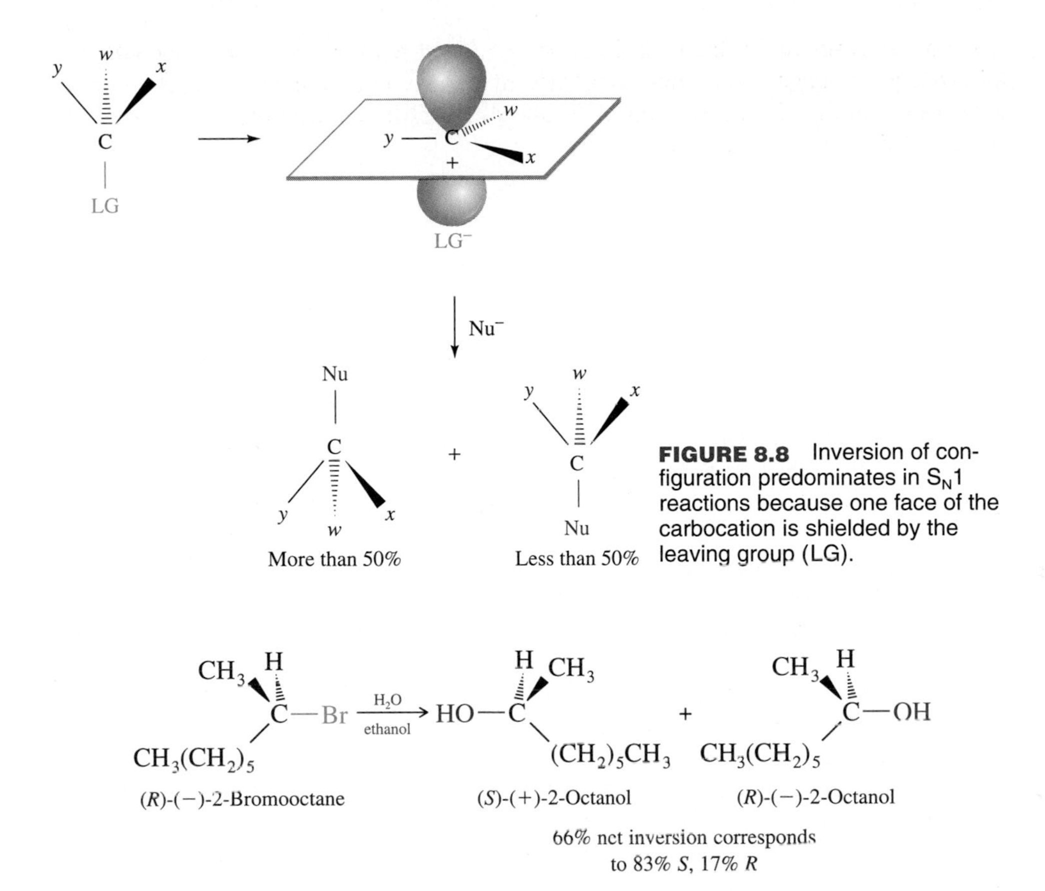

FIGURE 8.8 Inversion of configuration predominates in S_N1 reactions because one face of the carbocation is shielded by the leaving group (LG).

Partial but not complete loss of optical activity in S_N1 reactions is understood to mean that the carbocation is not free of its halide ion counterpart when it is attacked by the nucleophile. Ionization of the alkyl halide gives a carbocation–halide ion pair, as depicted in Figure 8.8. The anion of the leaving group shields one side of the carbocation, and the nucleophile captures the carbocation faster from the side opposite the leaving group. More product of inverted configuration is formed than product of retained configuration. In spite of the observation that the products of S_N1 reactions are only partially racemic, the fact that these reactions are not stereospecific is more consistent with the involvement of carbocation intermediates than with a concerted bimolecular mechanism.

PROBLEM 8.10 What two stereoisomeric substitution products would you expect to isolate from the hydrolysis of *cis*-1,4-dimethylcyclohexyl bromide? From hydrolysis of *trans*-1,4-dimethylcyclohexyl bromide?

8.12 CARBOCATION REARRANGEMENTS IN S_N1 REACTIONS

Additional evidence for the intermediacy of carbocations in certain nucleophilic substitutions comes from the observation that the products must in some cases have arisen by rearrangements of the kind customarily associated with carbocation intermediates. For example, hydrolysis of the secondary alkyl bromide 2-bromo-3-methylbutane yields the rearranged tertiary alcohol 2-methyl-2-butanol as the exclusive product of substitution.

$$\underset{\text{2-Bromo-3-methylbutane}}{CH_3CH(CH_3)CH(Br)CH_3} \xrightarrow{H_2O} \underset{\text{2-Methyl-2-butanol (93\%)}}{CH_3C(CH_3)(OH)CH_2CH_3}$$

A reasonable mechanism which explains this observation assumes rate-determining ionization of the substrate as the first step.

$$\underset{\text{2-Bromo-3-methylbutane}}{CH_3C(CH_3)(H)-CH(Br)CH_3} \xrightarrow{\text{slow}} \underset{\text{1,2-Dimethylpropyl cation (a secondary carbocation)}}{CH_3C(CH_3)(H)-\overset{+}{C}HCH_3} + Br^-$$

This is followed by a hydride shift which converts the secondary carbocation to a more stable tertiary one.

$$\underset{\text{1,2-Dimethylpropyl cation}}{CH_3C(CH_3)(H)-\overset{+}{C}HCH_3} \xrightarrow{\text{fast}} \underset{\text{1,1-Dimethylpropyl cation (a tertiary carbocation)}}{CH_3\overset{+}{C}(CH_3)CH(H)CH_3}$$

The tertiary carbocation then reacts with water to yield the observed product.

$$\underset{\text{1,1-Dimethylpropyl cation}}{CH_3\overset{+}{C}(CH_3)CH_2CH_3} \xrightarrow[\text{fast}]{H_2O} CH_3C(CH_3)(\overset{+}{O}H_2)CH_2CH_3 \xrightarrow{\text{fast}} \underset{\text{2-Methyl-2-butanol}}{CH_3C(CH_3)(OH)CH_2CH_3}$$

PROBLEM 8.11 Why does the carbocation intermediate in the hydrolysis of 2-bromo-3-methylbutane rearrange by way of a hydride shift rather than a methyl shift?

Rearrangements, when they do occur, are taken as evidence for carbocation intermediates and point to the S_N1 mechanism as the reaction pathway. Rearrangements are never observed in S_N2 reactions.

8.13 EFFECT OF SOLVENT ON THE RATE OF NUCLEOPHILIC SUBSTITUTION

The major effect of the solvent is on the *rate* of nucleophilic substitution. Thus we need to consider two related questions:

1. What are the properties of the *solvent* that influence the rate the most?
2. How does the activation energy of the rate-determining step of the *mechanism* respond to this property of the solvent?

Because the S_N1 and S_N2 mechanisms are so different from each other, we will examine each one separately.

Solvent Effects on the Rate of Substitution by the S_N1 Mechanism. Table 8.6 lists the relative rate constants for some typical S_N1 reactions—solvolyses of *tert*-butyl chloride—in several media in order of increasing **dielectric constant** (ϵ) of the solvent. Dielectric constant is a measure of the ability of a material, in this case the solvent, to moderate the force of attraction between oppositely charged particles compared with that of a standard. The standard dielectric is a vacuum, which is assigned a value ϵ of exactly 1. The higher the dielectric constant ϵ, the better the medium is able to support separated positively and negatively charged species. Solvents with high dielectric constants are classified as *polar solvents.* As Table 8.6 illustrates, the rate of solvolysis of *tert*-butyl chloride (which is equal to its rate of ionization) increases dramatically as the dielectric constant of the solvent increases.

According to the S_N1 mechanism, a neutral alkyl halide molecule ionizes to a positively charged carbocation and a negatively charged halide ion in the rate-determining step. As the alkyl halide approaches the transition state for this step, a partial positive charge develops on carbon and a partial negative charge on the halogen. Figure 8.9 contrasts the behavior of a nonpolar and a polar solvent on the transition state for ionization of an alkyl halide. Polar and nonpolar solvents are similar in their interaction with the starting alkyl halide, but differ markedly in how they affect the energy of the transition state. A solvent with a low dielectric constant has little effect on the energy of the transition state, while one with a high dielectric constant stabilizes the charge-separated transition state, lowers the activation energy, and increases the rate of reaction.

TABLE 8.6
Relative Rate of S_N1 Solvolysis of *tert*-Butyl Chloride as a Function of Solvent Polarity*

Solvent	Dielectric constant ϵ	Relative rate
Acetic acid	6	1
Methanol	33	4
Formic acid	58	5,000
Water	78	150,000

* Ratio of first-order rate constant for solvolysis in indicated solvent to that for solvolysis in acetic acid at 25°C.

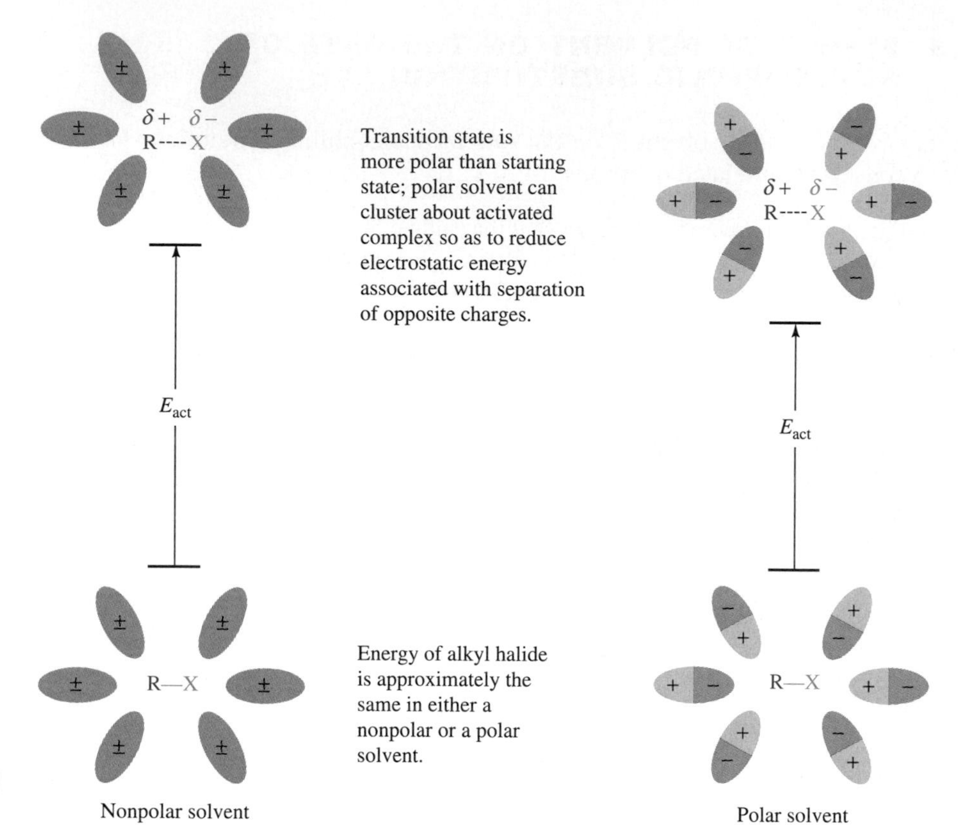
δ+ δ−
R----X
Transition state is more polar than starting state; polar solvent can cluster about activated complex so as to reduce electrostatic energy associated with separation of opposite charges.
δ+ δ−
R----X
E_{act}
E_{act}
R—X
Energy of alkyl halide is approximately the same in either a nonpolar or a polar solvent.
R—X
Nonpolar solvent
Polar solvent

FIGURE 8.9 A polar solvent stabilizes the transition state of an S_N1 reaction and increases its rate.

Solvent Effects on the Rate of Substitution by the S_N2 Mechanism. Polar solvents are required in typical bimolecular substitutions because ionic substances, such as the sodium and potassium salts cited earlier in Table 8.1, are not sufficiently soluble in nonpolar solvents to give a high enough concentration of the nucleophile to allow the reaction to occur at a rapid rate. Other than the requirement that the solvent be polar enough to dissolve ionic compounds, however, the effect of solvent polarity on the rate of S_N2 reactions is small. What is most important is whether or not the polar solvent is **protic** or **aprotic.**

Water (HOH), alcohols (ROH), and carboxylic acids (RCO_2H) are classified as *polar protic solvents;* they all have —OH groups that allow them to form hydrogen bonds to anionic nucleophiles as shown in Figure 8.10. Solvation forces such as these stabilize the anion and suppress its nucleophilicity. *Aprotic solvents,* on the other hand, lack —OH groups and do not solvate anions very strongly, leaving them much more able to express their nucleophilic character. Table 8.7 compares the second-order rate constants k for S_N2 substitution of 1-bromobutane by azide ion (a good nucleophile) in some common polar aprotic solvents with the corresponding k's for the much slower reactions observed in the polar protic solvents methanol and water.

$$CH_3CH_2CH_2CH_2Br + N_3^- \longrightarrow CH_3CH_2CH_2CH_2N_3 + Br^-$$

1-Bromobutane | Azide ion | 1-Azidobutane | Bromide ion

The large rate enhancements observed for bimolecular nucleophilic substitutions in

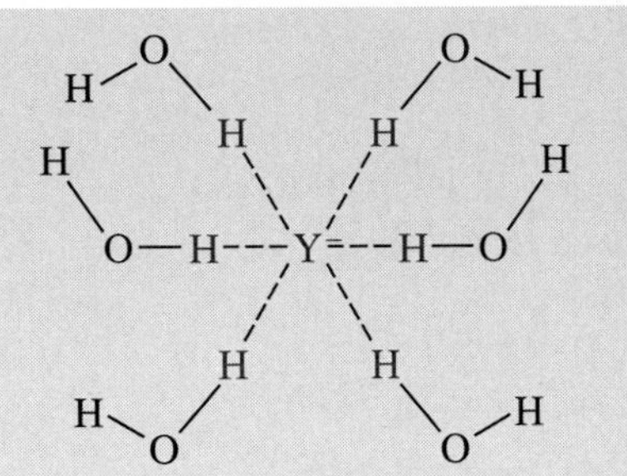

FIGURE 8.10 Hydrogen bonding of the solvent to the nucleophile stabilizes the nucleophile and makes it less reactive.

polar aprotic solvents are used to advantage in synthetic applications. An example can be seen in the preparation of alkyl cyanides (nitriles) by the reaction of sodium cyanide with alkyl halides:

$$\underset{\text{Hexyl halide}}{CH_3(CH_2)_4CH_2X} + \underset{\text{Sodium cyanide}}{NaCN} \longrightarrow \underset{\text{Hexyl cyanide}}{CH_3(CH_2)_4CH_2CN} + \underset{\text{Sodium halide}}{NaX}$$

When the reaction was carried out in aqueous methanol as the solvent, hexyl bromide was converted to hexyl cyanide in 71 percent yield by heating with sodium cyanide. While this is a perfectly acceptable synthetic reaction, a period of over *20 hours* was required. Changing the solvent to dimethyl sulfoxide brought about an increase in the reaction rate sufficient to allow the less reactive substrate hexyl chloride to be used instead, and the reaction was complete (91 percent yield) in only *20 minutes.*

The *rate* at which reactions occur can be important to the practice of organic chemistry in the laboratory, and understanding how solvents affect rate is of practical value. As we proceed through the text, however, and see how nucleophilic substitution applied to a variety of functional group transformations, be aware that it is the nature of the substrate and the nucleophile that, more than anything else, determines what *product* is formed.

TABLE 8.7
Relative Rate of S_N2 Displacement of 1-Bromobutane by Azide in Various Solvents*

Solvent	Structural formula	Dielectric constant ϵ	Type of solvent	Relative rate
Methanol	CH_3OH	32.6	Polar protic	1
Water	H_2O	78.5	Polar protic	7
Dimethyl sulfoxide	$(CH_3)_2S{=}O$	48.9	Polar aprotic	1300
N,N-Dimethylformamide	$(CH_3)_2NCH{=}O$	36.7	Polar aprotic	2800
Acetonitrile	$CH_3C{\equiv}N$	37.5	Polar aprotic	5000

* Ratio of second-order rate constant for substitution in indicated solvent to that for substitution in methanol at 25°C.

8.14 SUBSTITUTION AND ELIMINATION AS COMPETING REACTIONS

We have seen that an alkyl halide and a Lewis base can react together in either a substitution or an elimination reaction.

$$\text{H}-\overset{|}{\underset{|}{\text{C}}}-\overset{|}{\underset{\text{X}}{\text{C}}}- + \text{Y}^- \begin{cases} \xrightarrow{\beta \text{ elimination}} \text{C}=\text{C} + \text{HY} + \text{X}^- \\ \xrightarrow{\text{nucleophilic substitution}} \text{H}-\overset{|}{\underset{|}{\text{C}}}-\overset{|}{\underset{\text{Y}}{\text{C}}}- + \text{X}^- \end{cases}$$

Substitution can take place by the S_N1 or the S_N2 mechanism, elimination by E1 or E2.

How can we predict whether substitution or elimination will be the principal reaction observed with a particular combination of reactants? While many factors influence the competition between nucleophilic substitution and elimination, the two most important ones are the *structure of the alkyl halide* and the *basicity of the anion.* It is useful to approach the question from the premise that the characteristic reaction of alkyl halides with Lewis bases is *elimination,* and that substitution predominates only under certain special circumstances. Thus, a typical secondary alkyl halide such as isopropyl bromide reacts with a typical nucleophile such as sodium ethoxide mainly by elimination:

$$\underset{\text{Isopropyl bromide}}{CH_3\underset{\underset{Br}{|}}{C}HCH_3} \xrightarrow[CH_3CH_2OH,\ 55°C]{NaOCH_2CH_3} \underset{\text{Propene (87\%)}}{CH_3CH{=}CH_2} + \underset{\text{Ethyl isopropyl ether (13\%)}}{CH_3\underset{\underset{OCH_2CH_3}{|}}{C}HCH_3}$$

Figure 8.11 illustrates the close relationship between the E2 and S_N2 pathways for this combination of alkyl halide and Lewis base, and the results cited in the preceding equation tell us that E2 is faster than S_N2 for this case.

As crowding at the carbon that bears the leaving group decreases, the rate of nucleophilic attack by the Lewis base increases. A low level of steric hindrance to approach of the nucleophile is one of the special circumstances that permit substitution to pre-

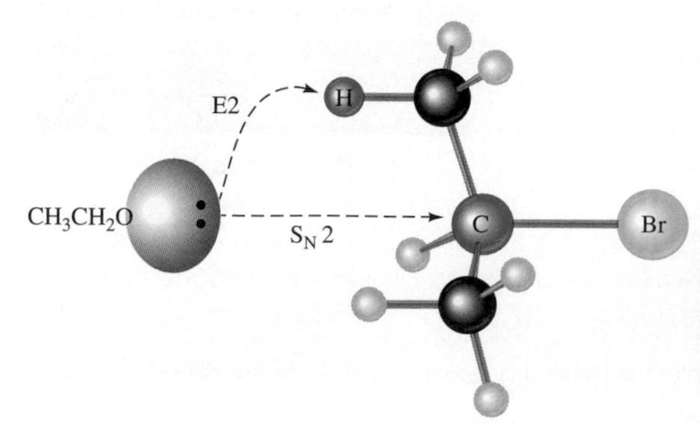

FIGURE 8.11 When a Lewis base reacts with an alkyl halide, either substitution or elimination can occur. Substitution (S_N2) occurs when the nucleophile attacks carbon to displace bromide. Elimination occurs when the Lewis base abstracts a proton from the β carbon. The alkyl halide shown is isopropyl bromide. The carbon atom that bears the leaving group is somewhat sterically hindered, and elimination (E2) predominates over substitution with alkoxide bases.

dominate, and primary alkyl halides react with alkoxide bases by an S_N2 mechanism in preference to E2:

$$\underset{\text{Propyl bromide}}{CH_3CH_2CH_2Br} \xrightarrow[CH_3CH_2OH,\ 55°C]{NaOCH_2CH_3} \underset{\text{Propene (9\%)}}{CH_3CH{=}CH_2} + \underset{\text{Ethyl propyl ether (91\%)}}{CH_3CH_2CH_2OCH_2CH_3}$$

However, if the base itself is a crowded one, such as potassium *tert*-butoxide, even primary alkyl halides undergo elimination rather than substitution:

$$\underset{\text{1-Bromooctadecane}}{CH_3(CH_2)_{15}CH_2CH_2Br} \xrightarrow[(CH_3)_3COH,\ 40°C]{KOC(CH_3)_3} \underset{\text{1-Octadecene (87\%)}}{CH_3(CH_2)_{15}CH{=}CH_2} + \underset{\textit{tert}\text{-Butyl octadecyl ether (13\%)}}{CH_3(CH_2)_{15}CH_2CH_2OC(CH_3)_3}$$

A second factor that can tip the balance in favor of substitution is weak basicity of the nucleophile. Nucleophiles that are less basic than hydroxide react with both primary and secondary alkyl halides to give the product of nucleophilic substitution in high yield. To illustrate, cyanide ion is about as nucleophilic as ethoxide (see Table 8.4) but is much less basic. Its reaction with 2-chlorooctane gives the corresponding alkyl cyanide as the principal product.

Cyanide is a weaker base than hydroxide because its conjugate acid HCN (pK_a 9.1) is a stronger acid than water (pK_a 15.7).

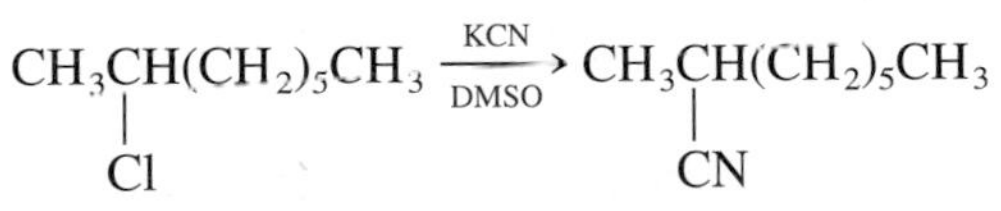

$$\underset{\text{2-Chlorooctane}}{CH_3\underset{\underset{Cl}{|}}{C}H(CH_2)_5CH_3} \xrightarrow[DMSO]{KCN} \underset{\text{2-Cyanooctane (70\%)}}{CH_3\underset{\underset{CN}{|}}{C}H(CH_2)_5CH_3}$$

Azide ion ($:\overset{-}{\ddot{N}}{=}\overset{+}{N}{=}\overset{-}{\ddot{N}}:$) is a good nucleophile and not strongly basic. It reacts with secondary alkyl halides mainly by substitution:

The conjugate acid of azide ion is called *hydrazoic acid* (HN_3). It has a pK_a of 4.6, and so is similar to acetic acid in its acidity.

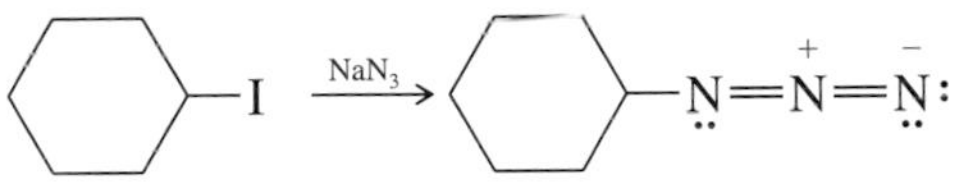

$$\underset{\text{Cyclohexyl iodide}}{C_6H_{11}{-}I} \xrightarrow{NaN_3} \underset{\text{Cyclohexyl azide (75\%)}}{C_6H_{11}{-}\ddot{N}{=}\overset{+}{N}{=}\overset{-}{\ddot{N}}:}$$

Hydrogen sulfide ion HS^-, and anions of the type RS^-, are substantially less basic than hydroxide ion and react with both primary and secondary alkyl halides to give mainly substitution products.

Hydrogen sulfide (pK_a 7.0) is a stronger acid than water (pK_a 15.7). Therefore HS^- is a much weaker base than HO^-.

Tertiary alkyl halides are so sterically hindered to nucleophilic attack that the presence of any anionic Lewis base leads to elimination as the major reaction path. Usually substitution predominates over elimination in tertiary alkyl halides only when the concentration of anionic Lewis bases is small. In the solvolysis of the tertiary bromide 2-bromo-2-methylbutane, for example, the ratio of substitution to elimination is 64 : 36 in pure ethanol but falls to 1 : 99 in the presence of 2 *M* sodium ethoxide.

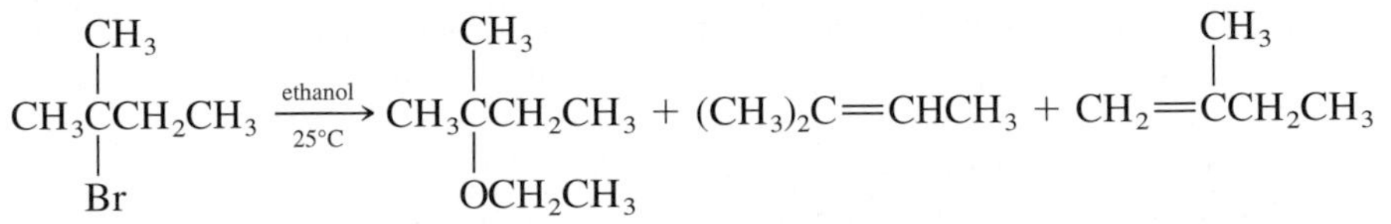

$$\underset{\text{2-Bromo-2-methylbutane}}{CH_3\overset{\overset{CH_3}{|}}{\underset{\underset{Br}{|}}{C}}CH_2CH_3} \xrightarrow[25°C]{ethanol} \underset{\text{2-Ethoxy-2-methylbutane}}{CH_3\overset{\overset{CH_3}{|}}{\underset{\underset{OCH_2CH_3}{|}}{C}}CH_2CH_3} + \underset{\text{2-Methyl-2-butene}}{(CH_3)_2C{=}CHCH_3} + \underset{\text{2-Methyl-1-butene}}{CH_2{=}\overset{\overset{CH_3}{|}}{C}CH_2CH_3}$$

(Major product in absence of sodium ethoxide)

(Alkene mixture is major product in presence of sodium ethoxide)

PROBLEM 8.12 Predict the major organic product of each of the following reactions:

(*a*) Cyclohexyl bromide and potassium ethoxide
(*b*) Ethyl bromide and potassium cyclohexanolate
(*c*) *sec*-Butyl bromide solvolysis in methanol
(*d*) *sec*-Butyl bromide solvolysis in methanol containing 2 *M* sodium methoxide

SAMPLE SOLUTION (*a*) Cyclohexyl bromide is a secondary halide and reacts with alkoxide bases by elimination rather than substitution. The major organic products are cyclohexene and ethanol.

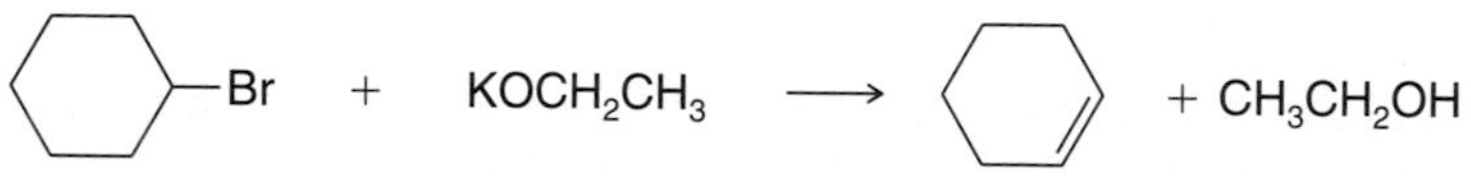

$$\text{C}_6\text{H}_{11}\text{—Br} + \text{KOCH}_2\text{CH}_3 \longrightarrow \text{C}_6\text{H}_{10} + \text{CH}_3\text{CH}_2\text{OH}$$

Cyclohexyl bromide   Potassium ethoxide   Cyclohexene   Ethanol

Regardless of the substrate, an increase in temperature causes both the rate of substitution and the rate of elimination to increase. The rate of elimination, however, usually increases faster than the rate of substitution, so that at higher temperatures the proportion of elimination products increases at the expense of substitution products.

As a practical matter, elimination can always be made to occur quantitatively. Strong bases, especially bulky ones such as *tert*-butoxide ion, react even with primary alkyl halides by an E2 process at elevated temperatures. The more difficult task is to find the set of conditions most conducive to substitution. In general, the best approach is to choose conditions that favor the S_N2 mechanism—an unhindered substrate, a good nucleophile that is not strongly basic, and the lowest practical temperature consistent with reasonable reaction rates.

Functional group transformations that rely on substitution by the S_N1 mechanism are not as generally applicable as those of the S_N2 type. Hindered substrates are prone to elimination by dehydrohalogenation, and the possibility of rearrangement reactions arises when carbocation intermediates are involved. Only in cases in which elimination is impossible are S_N1 reactions employed in functional group transformations of alkyl halides.

8.15 SULFONATE ESTERS AS SUBSTRATES IN NUCLEOPHILIC SUBSTITUTION

Two kinds of substrates have been examined in nucleophilic substitution reactions to this point. In Chapter 4 we saw how alcohols can be converted to alkyl halides by reaction with hydrogen halides, and it was pointed out that this process is a nucleophilic substitution taking place on the protonated form of the alcohol, with water serving as the leaving group. In the present chapter the substrates have been alkyl halides, and halide ions have been the leaving groups. There are a few other classes of organic compounds that undergo nucleophilic substitution reactions analogous to those of alkyl halides, the most important of these being alkyl esters of sulfonic acids.

Sulfonic acids are strong acids, comparable in acidity with sulfuric acid. Representative examples are methanesulfonic acid and *p*-toluenesulfonic acid.

$$CH_3-\overset{O}{\overset{\|}{\underset{\underset{O}{\|}}{S}}}-OH \qquad CH_3-C_6H_4-\overset{O}{\overset{\|}{\underset{\underset{O}{\|}}{S}}}-OH$$

Methanesulfonic acid — *p*-Toluenesulfonic acid

Alkyl sulfonate esters are derivatives of sulfonic acids in which the proton of the hydroxyl group is replaced by an alkyl group. They are prepared by treating an alcohol with the appropriate sulfonyl chloride.

$$R\ddot{O}H + R'SO_2Cl \longrightarrow R\ddot{O}-SO_2R' + HCl$$

Alcohol — Sulfonyl chloride — Sulfonate ester — Hydrogen chloride

These reactions are usually carried out in the presence of pyridine.

$$CH_3CH_2OH + CH_3-C_6H_4-SO_2Cl \xrightarrow{\text{pyridine}} CH_3CH_2OSO_2-C_6H_4-CH_3$$

Ethanol — *p*-Toluenesulfonyl chloride — Ethyl *p*-toluenesulfonate (72%)

Alkyl sulfonate esters resemble alkyl halides in their ability to undergo elimination and nucleophilic substitution.

$$Y:^- + R-OSO_2-C_6H_4-CH_3 \longrightarrow R-Y + {}^-OSO_2-C_6H_4-CH_3$$

Nucleophile — *p*-Toluenesulfonate ester — Product of nucleophilic substitution — *p*-Toluenesulfonate anion

The sulfonate esters used most frequently are the *p*-toluenesulfonates. They are commonly known as *tosylates* and given the abbreviated formula ROTs.

$$\text{(3-cyclopentenyl)}CH_2OTs \xrightarrow[\text{ethanol-water}]{\text{KCN}} \text{(3-cyclopentenyl)}CH_2CN$$

(3-Cyclopentenyl)methyl *p*-toluenesulfonate — 4-(Cyanomethyl)cyclopentene (86%)

p-Toluenesulfonate (TsO^-) is a very good leaving group. As Table 8.8 reveals, alkyl *p*-toluenesulfonates undergo nucleophilic substitution at rates that are even faster than those of alkyl iodides. A correlation of leaving-group abilities with carbon-halo-

TABLE 8.8
Approximate Relative Leaving-Group Abilities*

Leaving group	Relative rate	Conjugate acid of leaving group	K_a of conjugate acid	pK_a
F^-	10^{-5}	HF	3.5×10^{-4}	3.5
Cl^-	10^{0}	HCl	10^{7}	−7
Br^-	10^{1}	HBr	10^{9}	−9
I^-	10^{2}	HI	10^{10}	−10
H_2O	10^{1}	H_3O^+	55	−1.7
TsO^-	10^{5}	TsOH	6×10^{2}	−2.8
$CF_3SO_2O^-$	10^{8}	CF_3SO_2OH	10^{6}	−6

* Values are approximate and vary according to substrate. A greater spread in reactivity is observed for S_N1 reactions than for S_N2 reactions.

gen bond strengths was noted earlier, in Section 8.3. Note also the correlation with the basicity of the leaving group. Iodide is the weakest base among the halide anions and is the best leaving group, fluoride the strongest base and the poorest leaving group. A similar correlation with basicity is seen among oxygen-containing leaving groups. The weaker the base, the better the leaving group. Trifluoromethanesulfonic acid (CF_3SO_2OH) is a much stronger acid than *p*-toluenesulfonic acid, and therefore trifluoromethanesulfonate is a much weaker base than *p*-toluenesulfonate and a much better leaving group.

Trifluoromethanesulfonate esters are called *triflates.*

Notice too that strongly basic leaving groups are absent from Table 8.8. In general, any species that has a K_a less than 1 for its conjugate acid cannot be a leaving group in a nucleophilic substitution. Thus, hydroxide (HO^-) is far too strong a base to be displaced from an alcohol (ROH), and alcohols do not undergo nucleophilic substitution. In strongly acidic media, alcohols are protonated to give alkyloxonium ions, and these do undergo nucleophilic substitution, because the leaving group is a weakly basic water molecule.

Since halides are poorer leaving groups than *p*-toluenesulfonate, alkyl *p*-toluenesulfonates can be converted to alkyl halides by S_N2 reactions involving chloride, bromide, or iodide as the nucleophile.

$$\underset{\substack{\text{sec-Butyl}\\ p\text{-toluenesulfonate}}}{\underset{\displaystyle \text{OTs}}{CH_3\overset{}{C}HCH_2CH_3}} + \underset{\substack{\text{Sodium}\\ \text{bromide}}}{NaBr} \xrightarrow{\text{DMSO}} \underset{\substack{\text{sec-Butyl}\\ \text{bromide (82\%)}}}{\underset{\displaystyle \text{Br}}{CH_3CHCH_2CH_3}} + \underset{\substack{\text{Sodium}\\ p\text{-toluenesulfonate}}}{NaOTs}$$

PROBLEM 8.13 Write a chemical equation showing the preparation of octadecyl *p*-toluenesulfonate.

PROBLEM 8.14 Write equations showing the reaction of octadecyl *p*-toluenesulfonate with each of the following reagents:

(*a*) Potassium acetate ($KO\overset{O}{\overset{\|}{C}}CH_3$)
(*b*) Potassium iodide (KI)

(*c*) Potassium cyanide (KCN)
(*d*) Potassium hydrogen sulfide (KSH)
(*e*) Sodium butanethiolate ($NaSCH_2CH_2CH_2CH_3$)

SAMPLE SOLUTION All these reactions of octadecyl *p*-toluenesulfonate have been reported in the chemical literature, and all proceed in synthetically useful yield. You should begin by identifying the nucleophile in each of the parts to this problem. The nucleophile replaces the *p*-toluenesulfonate leaving group in an S_N2 reaction. In part (*a*) the nucleophile is acetate ion, and the product of nucleophilic substitution is octadecyl acetate.

$$CH_3\overset{O}{\overset{\|}{C}}O^- + \underset{(CH_2)_{16}CH_3}{\underset{|}{CH_2}}-OTs \longrightarrow CH_3\overset{O}{\overset{\|}{C}}OCH_2(CH_2)_{16}CH_3$$

Acetate ion Octadecyl tosylate Octadecyl acetate

Sulfonate esters are subject to the same limitations as alkyl halides in their use as substrates in nucleophilic substitution reactions. Competition from elimination needs to be considered when planning a functional group transformation that requires an anionic nucleophile, because tosylates undergo elimination reactions, just as alkyl halides do.

An advantage that sulfonate esters have over alkyl halides is that their preparation from alcohols does not involve any of the bonds to carbon. The alcohol oxygen becomes the oxygen that connects the alkyl group to the sulfonyl group. Thus, the configuration of a sulfonate ester is exactly the same as that of the alcohol from which it was prepared. If we wish to study the stereochemistry of nucleophilic substitution in an optically active substrate, for example, we know that a tosylate ester will have the same configuration and the same optical purity as the alcohol from which it was prepared.

$$CH_3(CH_2)_5(H)(H_3C)C-OH \xrightarrow[\text{pyridine}]{p\text{-Toluenesulfonyl chloride}} CH_3(CH_2)_5(H)(H_3C)C-O\overset{O}{\overset{\|}{\underset{\underset{O}{\|}}{S}}}-C_6H_4-CH_3$$

(*S*)-(+)-2-Octanol
$[\alpha]_D^{25}$ +9.9°
(optically pure)

(*S*)-(+)-1-Methylheptyl *p*-toluenesulfonate
$[\alpha]_D^{25}$ +7.9°
(optically pure)

The same cannot be said about reactions with alkyl halides as substrates. The conversion of optically active 2-octanol to the corresponding halide *does* involve a bond to the stereogenic center, and so the optical purity and absolute configuration of the alkyl halide need to be independently established.

The mechanisms by which sulfonate esters undergo nucleophilic substitution are the same as those of alkyl halides. Inversion of configuration is observed in S_N2 reactions of alkyl sulfonates and predominant inversion accompanied by racemization in S_N1 processes.

PROBLEM 8.15 The hydrolysis of sulfonate esters of 2-octanol is a stereospecific reaction and proceeds with complete inversion of configuration. Write a structural formula that shows the stereochemistry of the 2-octanol formed by hydrolysis of an optically pure sample of (*S*)-(+)-1-methylheptyl *p*-toluenesulfonate, identify the product as *R* or *S*, and deduce its specific rotation.

8.16 A RETROSPECTIVE LOOK: REACTIONS OF ALCOHOLS WITH HYDROGEN HALIDES

The principles developed in this chapter can be applied to a more detailed examination of the reaction of alcohols with hydrogen halides than was possible when this reaction was first introduced in Chapter 4.

$$\underset{\text{Alcohol}}{ROH} + \underset{\text{Hydrogen halide}}{HX} \longrightarrow \underset{\text{Alkyl halide}}{RX} + \underset{\text{Water}}{H_2O}$$

As pointed out in Chapter 4, the first step in the reaction is proton transfer to the alcohol from the hydrogen halide to yield an alkyloxonium ion. This is an acid-base reaction.

$$\underset{\substack{\text{Alcohol}\\\text{(base)}}}{R(H)\ddot{O}:} + \underset{\substack{\text{Hydrogen halide}\\\text{(acid)}}}{H-\ddot{X}:} \longrightarrow \underset{\substack{\text{Alkyloxonium ion}\\\text{(conjugate acid)}}}{R(H)\overset{+}{\ddot{O}}-H} + \underset{\substack{\text{Halide ion}\\\text{(conjugate base)}}}{:\ddot{X}:^-}$$

With primary alcohols, the next step is best described as an S_N2 reaction in which the halide ion, bromide, for example, displaces a molecule of water from the alkyloxonium ion.

$$\underset{\substack{\text{Bromide}\\\text{ion}}}{:\ddot{Br}:^-} + \underset{\substack{\text{Primary alkyl-}\\\text{oxonium ion}}}{RCH_2-\overset{+}{O}H_2} \longrightarrow \underset{\substack{\text{Representation of}\\S_N2\text{ transition state}}}{\overset{\delta-}{:\ddot{Br}}\cdots C(R)(H)(H)\cdots\overset{\delta+}{O}H_2} \longrightarrow \underset{\substack{\text{Primary}\\\text{alkyl bromide}}}{:\ddot{Br}-CH_2R} + \underset{\text{Water}}{H_2\ddot{O}:}$$

With secondary and tertiary alcohols, the step which follows protonation of the alcohol is best described as an S_N1 reaction in which the alkyloxonium ion undergoes rate-determining dissociation to a carbocation and water.

$$\underset{\substack{\text{Secondary}\\\text{alkyloxonium ion}}}{R_2CH-\overset{+}{O}H_2} \longrightarrow \underset{\substack{\text{Representation of}\\S_N1\text{ transition state}}}{\overset{\delta+}{R_2CH}\cdots\overset{\delta+}{O}H_2} \longrightarrow \underset{\substack{\text{Secondary}\\\text{carbocation}}}{R_2\overset{+}{C}H} + \underset{\text{Water}}{H_2\ddot{O}:}$$

Following its formation, the carbocation is captured by halide.

$$\underset{\substack{\text{Secondary}\\\text{carbocation}}}{R_2\overset{+}{C}H} + \underset{\substack{\text{Bromide}\\\text{ion}}}{:\ddot{Br}:^-} \xrightarrow{\text{fast}} \underset{\substack{\text{Secondary}\\\text{alkyl bromide}}}{R_2CH-\ddot{Br}:}$$

With optically active secondary alcohols the reaction proceeds with predominant, but incomplete, inversion of configuration.

$$(R)\text{-}(-)\text{-2-Butanol} \xrightarrow{HBr} (S)\text{-}(+)\text{-2-Bromobutane (87\%)} + (R)\text{-}(-)\text{-2-Bromobutane (13\%)}$$

The few studies that have been carried out with optically active tertiary alcohols indicate that almost complete racemization attends the preparation of tertiary alkyl halides by this method.

Rearrangement occurs with certain types of alcohols, and the desired alkyl halide is sometimes accompanied by an isomeric halide as a contaminant. An example is seen in the case of the secondary alcohol 2-octanol, which yields a mixture of 2- and 3-bromooctane:

$$\underset{\text{2-Octanol}}{CH_3CH(OH)CH_2(CH_2)_4CH_3} \xrightarrow{HBr} \underset{\text{1-Methylheptyl cation}}{CH_3\overset{+}{C}HCH_2(CH_2)_4CH_3} \longrightarrow \underset{\text{1-Ethylhexyl cation}}{CH_3CH_2\overset{+}{C}H(CH_2)_4CH_3}$$

$$\text{1-Methylheptyl cation} \xrightarrow{Br^-} \underset{\text{2-Bromooctane (93\%)}}{CH_3CH(Br)CH_2(CH_2)_4CH_3} \qquad \text{1-Ethylhexyl cation} \xrightarrow{Br^-} \underset{\text{3-Bromooctane (7\%)}}{CH_3CH_2CH(Br)(CH_2)_4CH_3}$$

PROBLEM 8.16 Treatment of 3-methyl-2-butanol with hydrogen chloride yielded only a trace of 2-chloro-3-methylbutane. An isomeric chloride was isolated in 97 percent yield. Suggest a reasonable structure for this product.

Unbranched primary alcohols and tertiary alcohols tend to react with hydrogen halides without rearrangement. The oxonium ions from primary alcohols react rapidly with bromide ion, for example, in an S_N2 process without significant development of positive charge at carbon. Tertiary alcohols give tertiary alkyl halides because tertiary carbocations are stable and show little tendency to rearrange.

When it is necessary to prepare secondary alkyl halides with assurance that no trace of rearrangement accompanies their formation, the corresponding alcohol is first converted to its *p*-toluenesulfonate ester and this ester is then allowed to react with sodium chloride, bromide, or iodide, as described in Section 8.15.

8.17 SUMMARY

Nucleophilic substitution plays a prominent role in functional group transformations (Section 8.1). Among the synthetically useful processes accomplished by nucleophilic substitution are:

1. Preparation of alkyl fluorides

$$RX + F^- \xrightarrow{\text{heat in high-boiling solvent}} RF + X^-$$

2. Preparation of alkyl iodides

$$RX + I^- \xrightarrow{\text{acetone}} RI + X^-$$

3. Preparation of alkyl cyanides

$$RX + CN^- \longrightarrow RCN + X^-$$

4. Preparation of ethers

$$RX + R'O^- \longrightarrow ROR' + X^-$$

5. Preparation of alkyl azides

$$RX + N_3^- \longrightarrow RN_3 + X^-$$

6. Preparation of thiols and thioethers

$$RX + R'S^- \longrightarrow RSR' + X^-$$

where R is a primary or secondary alkyl halide. Examples of these reactions are given in Sections 8.1 and 8.2.

We distinguish between two limiting mechanisms for nucleophilic substitution: S_N1 and S_N2, where S_N designates the reaction type as *substitution-nucleophilic* and 1 and 2 specify that the rate-determining step is either *unimolecular* or *bimolecular,* respectively. Table 8.9 compares various features of these two mechanisms.

Nucleophilicity, mentioned briefly in Table 8.9, warrants additional comment. As discussed in Section 8.8, nucleophilicity is a measure of how aggressively a particular Lewis base attacks an alkyl halide in a bimolecular substitution (S_N2). In general:

1. When the attacking atom is the same, nucleophilicity increases with basicity. For example, neutral nucleophiles such as water are less nucleophilic than their conjugate bases (HO^-).
2. When comparing Lewis bases, nucleophilicity decreases across a row of the periodic table. For example, HO^- is more nucleophilic than F^-.
3. Nucleophilicity increases down a column of the periodic table. For example, HS^- is more nucleophilic than HO^-.

Reactions in which the Lewis base is the solvent are said to take place under **solvolysis** conditions. (Solvolysis of alkyl halides normally yields products from both substitution and elimination pathways.)

When nucleophilic substitution is employed as a synthetic method, the competition between substitution and elimination must always be considered (Section 8.14). *The normal reaction of a secondary alkyl halide with a base as strong as or stronger than hydroxide is elimination by the E2 mechanism.* Substitution by the S_N2 mechanism predominates when the base is weaker than hydroxide or the alkyl halide is primary. Elimination predominates when tertiary alkyl halides react with any anion. Nucleophilic substitution by the S_N1 mechanism normally predominates only when secondary and tertiary alkyl halides undergo solvolysis in the absence of added base.

TABLE 8.9
Comparison of S_N1 and S_N2 Mechanisms of Nucleophilic Substitution in Alkyl Halides

	S_N1	S_N2
Characteristics of mechanism	Two elementary steps: Step 1: $R-\ddot{X}: \rightleftharpoons R^+ + :\ddot{X}:^-$ Step 2: $R^+ + :Nu^- \longrightarrow R-Nu$ Ionization of alkyl halide (step 1) is rate-determining. (Section 8.9)	Single step: $^-Nu: \; R-\ddot{X}: \longrightarrow Nu-R + :\ddot{X}:^-$ Nucleophile displaces leaving group; bonding to the incoming nucleophile accompanies cleavage of the bond to the leaving group. (Sections 8.4 and 8.6)
Rate-determining transition state	$^{\delta+}R\text{---}\ddot{X}:^{\delta-}$ (Section 8.9)	$^{\delta-}Nu\text{---}R\text{---}\ddot{X}:^{\delta-}$ (Sections 8.4 and 8.6)
Molecularity	Unimolecular (Section 8.9)	Bimolecular (Section 8.4)
Kinetics and rate law	First order: Rate = k[alkyl halide] (Section 8.9)	Second order: Rate = k[alkyl halide][nucleophile] (Section 8.4)
Relative reactivity of halide leaving groups	$RI > RBr > RCl \gg RF$ (Section 8.3)	$RI > RBr > RCl \gg RF$ (Section 8.3)
Effect of structure on rate	$R_3CX > R_2CHX > RCH_2X > CH_3X$ Rate is governed by stability of carbocation that is formed in ionization step. Tertiary alkyl halides can react only by the S_N1 mechanism; they never react by the S_N2 mechanism. (Section 8.10)	$CH_3X > RCH_2X > R_2CHX > R_3CX$ Rate is governed by steric effects (crowding in transition state). Methyl and primary alkyl halides can react only by the S_N2 mechanism; they never react by the S_N1 mechanism. (Section 8.7)
Effect of nucleophile on rate	Rate of substitution is independent of both concentration and nature of nucleophile. Nucleophile does not participate until after rate-determining step. (Section 8.9)	Rate depends on both nature of nucleophile and its concentration. (Section 8.4)
Effect of solvent on rate	Rate increases with increasing polarity of solvent as measured by its dielectric constant ϵ. (Section 8.13)	Polar aprotic solvents give fastest rates of substitution; solvation of $Nu:^-$ is minimal and nucleophilicity is greatest. (Section 8.13)
Stereochemistry	Not stereospecific: racemization accompanies inversion when leaving group is located at a stereogenic center. (Section 8.11)	Stereospecific: 100 percent inversion of configuration at reaction site. Nucleophile attacks carbon from side opposite bond to leaving group. (Section 8.5)
Potential for rearrangements	Carbocation intermediate capable of rearrangement (Section 8.12)	No carbocation intermediate; no rearrangement.

Nucleophilic substitution can occur with leaving groups other than halide. Alkyl *p*-toluenesulfonates (*tosylates*), which are prepared from alcohols by reaction with *p*-toluenesulfonyl chloride, are often used (Section 8.15).

$$\text{ROH} + \text{CH}_3\text{—C}_6\text{H}_4\text{—SO}_2\text{Cl} \xrightarrow{\text{pyridine}} \text{ROS(=O)}_2\text{—C}_6\text{H}_4\text{—CH}_3 \text{ (ROTs)}$$

Alcohol — *p*-Toluenesulfonyl chloride — Alkyl *p*-toluenesulfonate (alkyl tosylate)

In terms of its ability to act as a leaving group, *p*-toluenesulfonate is comparable to iodide.

$$\bar{\text{N}}\text{u}: \;\; \text{R—OTs} \longrightarrow \text{Nu—R} + {}^{-}\text{OTs}$$

Nucleophile — Alkyl *p*-toluenesulfonate — Substitution product — *p*-Toluenesulfonate ion

The reactions of alcohols with hydrogen halides to give alkyl halides (Chapter 4) are nucleophilic substitution reactions of oxonium ions in which water is the leaving group (Section 8.16). Primary alcohols react by an S_N2-like displacement of water from the oxonium ion by halide. Secondary and tertiary alcohols give oxonium ions which form carbocations in an S_N1-like process. Rearrangements are possible with secondary alcohols, and substitution takes place with predominant, but not complete, inversion of configuration.

PROBLEMS

8.17 Write the structure of the principal organic product to be expected from the reaction of 1-bromopropane with each of the following:

(*a*) Sodium iodide in acetone
(*b*) Sodium acetate ($CH_3\overset{O}{\overset{\|}{C}}ONa$) in acetic acid
(*c*) Sodium ethoxide in ethanol
(*d*) Sodium cyanide in dimethyl sulfoxide
(*e*) Sodium azide in aqueous ethanol
(*f*) Sodium hydrogen sulfide in ethanol
(*g*) Sodium methanethiolate ($NaSCH_3$) in ethanol

8.18 All the reactions of 1-bromopropane in the preceding problem give the product of nucleophilic substitution in high yield. High yields of substitution products are also obtained in all but one of the analogous reactions using 2-bromopropane as the substrate. In one case, however, 2-bromopropane is converted to propene, especially when the reaction is carried out at elevated temperature (about 55°C). Which reactant is most effective in converting 2-bromopropane to propene?

8.19 Each of the following nucleophilic substitution reactions has been reported in the chemical literature. Many of them involve reactants that are somewhat more complex than those we have dealt with to this point. Nevertheless, you should be able to predict the product by analogy to what you know about nucleophilic substitution in simple systems.

(*a*) $BrCH_2\overset{O}{\overset{\|}{C}}OCH_2CH_3 \xrightarrow[\text{acetone}]{\text{NaI}}$

(*b*) $O_2N-C_6H_4-CH_2Cl \xrightarrow[\text{acetic acid}]{CH_3\overset{O}{\overset{\|}{C}}ONa}$

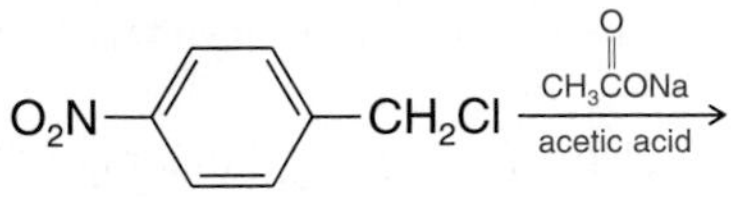

(*c*) $CH_3CH_2OCH_2CH_2Br \xrightarrow[\text{ethanol-water}]{\text{NaCN}}$

(*d*) $NC-C_6H_4-CH_2Cl \xrightarrow{H_2O,\ HO^-}$

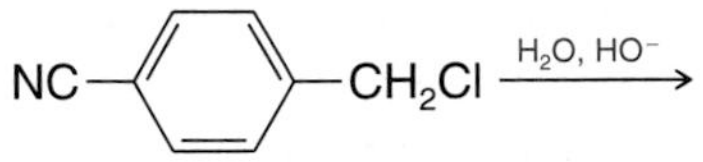

(*e*) $ClCH_2\overset{O}{\overset{\|}{C}}OC(CH_3)_3 \xrightarrow[\text{acetone-water}]{NaN_3}$

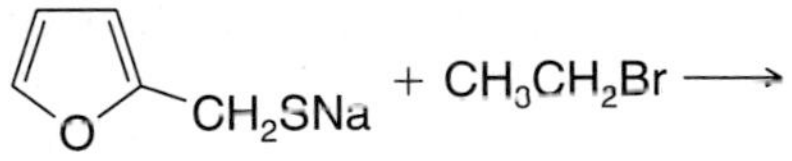

(*f*) 4-($TsOCH_2$)-2,2-dimethyl-1,3-dioxolane (CH_3, CH_3) $\xrightarrow[\text{acetone}]{\text{NaI}}$

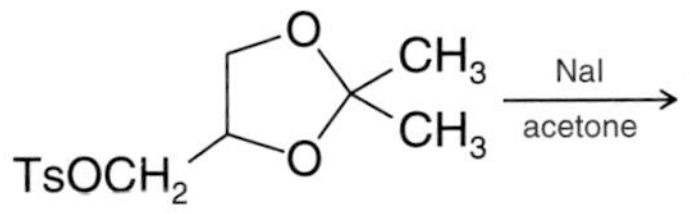

(*g*) 2-furyl-CH_2SNa + $CH_3CH_2Br \longrightarrow$

(*h*) 2,3,4-trimethoxyphenyl (OCH_3, CH_3O, CH_3O) $-CH_2CH_2CH_2CH_2OH \xrightarrow[\text{2. LiI, acetone}]{\text{1. TsCl, pyridine}}$

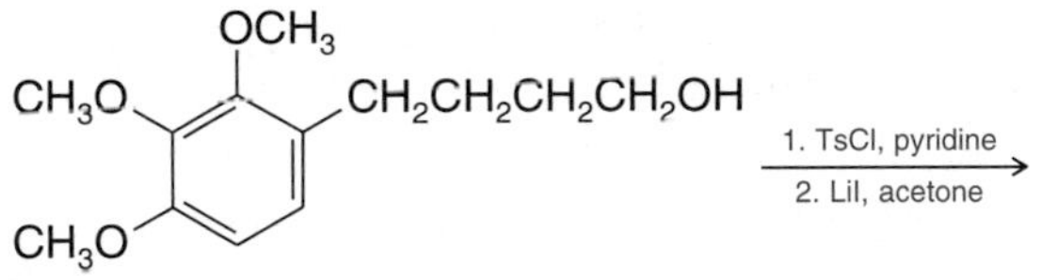

8.20 Each of the reactions shown involves nucleophilic substitution. The product of reaction (*a*) is an isomer of the product of reaction (*b*). What kind of isomer? By what mechanism does nucleophilic substitution occur? Write the structural formula of the product of each reaction.

(*a*) Cl-cyclohexane-$C(CH_3)_3$ (Cl equatorial) + $C_6H_5SNa \longrightarrow$

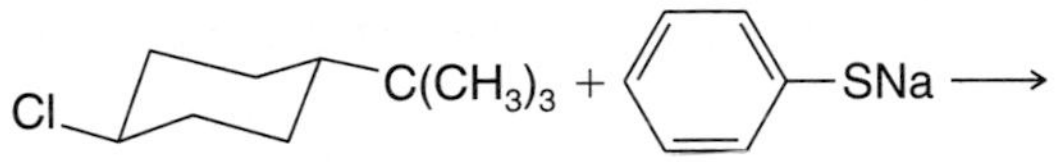

(*b*) Cl-cyclohexane-$C(CH_3)_3$ (Cl axial) + $C_6H_5SNa \longrightarrow$

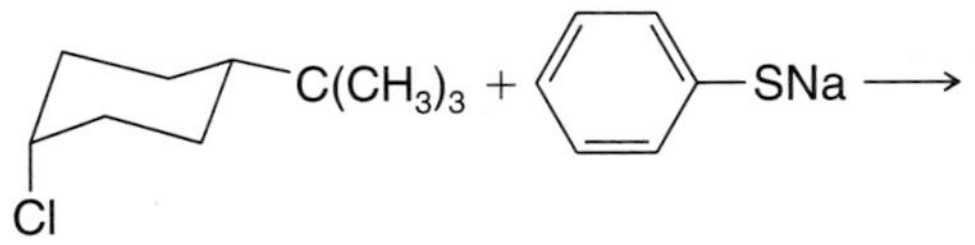

8.21 Arrange the isomers of molecular formula C_4H_9Cl in order of decreasing rate of reaction with sodium iodide in acetone.

8.22 There is an overall 29-fold difference in reactivity of 1-chlorohexane, 2-chlorohexane, and 3-chlorohexane toward potassium iodide in acetone.

(*a*) Which one is the most reactive? Why?
(*b*) Two of the isomers differ by only a factor of 2 in reactivity. Which two are these? Which one is the more reactive? Why?

8.23 In each of the following indicate which reaction will occur faster. Explain your reasoning.

(*a*) $CH_3CH_2CH_2CH_2Br$ or $CH_3CH_2CH_2CH_2I$ with sodium cyanide in dimethyl sulfoxide
(*b*) 1-Chloro-2-methylbutane or 1-chloropentane with sodium iodide in acetone

(*c*) Hexyl chloride or cyclohexyl chloride with sodium azide in aqueous ethanol
(*d*) Solvolysis of 1-bromo-2,2-dimethylpropane or *tert*-butyl bromide in ethanol
(*e*) Solvolysis of isobutyl bromide or *sec*-butyl bromide in aqueous formic acid
(*f*) Reaction of 1-chlorobutane with sodium acetate in acetic acid or with sodium methoxide in methanol
(*g*) Reaction of 1-chlorobutane with sodium azide or sodium *p*-toluenesulfonate in aqueous ethanol

8.24 Under conditions of photochemical chlorination, $(CH_3)_3CCH_2C(CH_3)_3$ gave a mixture of two monochlorides in a 4:1 ratio. The structures of these two products were assigned on the basis of their S_N1 hydrolysis rates in aqueous ethanol. The major product (compound A) underwent hydrolysis much more slowly than the minor one (compound B). Deduce the structures of compounds A and B.

8.25 The compound KSCN is a source of *thiocyanate* ion.

(*a*) Write the two most stable Lewis structures for thiocyanate ion and identify the atom in each that bears a formal charge of −1.
(*b*) Two constitutionally isomeric products of molecular formula C_5H_9NS were isolated in a combined yield of 87 percent in the reaction shown. (*DMF* stands for *N,N*-dimethylformamide, a polar aprotic solvent.) Suggest reasonable structures for these two compounds.

$$CH_3CH_2CH_2CH_2Br \xrightarrow[\text{DMF}]{\text{KSCN}}$$

(*c*) The major product of the reaction cited in (*b*) constituted 99 percent of the mixture of isomers. Its structure corresponds to attack by the most polarizable atom of thiocyanate ion on 1-bromobutane. What is this product?

8.26 Reaction of ethyl iodide with triethylamine $[(CH_3CH_2)_3N\!:]$ yields a crystalline compound $C_8H_{20}NI$ in high yield. This compound is soluble in polar solvents such as water but insoluble in nonpolar ones such as diethyl ether. It does not melt below about 200°C. Suggest a reasonable structure for this product.

8.27 Write an equation, clearly showing the stereochemistry of the starting material and the product, for the reaction of (*S*)-1-bromo-2-methylbutane with sodium iodide in acetone. What is the configuration (*R* or *S*) of the product?

8.28 Identify the product in each of the following reactions:

(*a*) $ClCH_2CH_2\underset{\displaystyle Cl}{\underset{|}{C}}HCH_2CH_3 \xrightarrow[\text{acetone}]{\text{NaI (1.0 equiv)}} C_5H_{10}ClI$

(*b*) $BrCH_2CH_2Br + NaSCH_2CH_2SNa \longrightarrow C_4H_8S_2$
(*c*) $ClCH_2CH_2CH_2CH_2Cl + Na_2S \longrightarrow C_4H_8S$

8.29 Give the mechanistic symbols (S_N1, S_N2, E1, E2) that are most consistent with each of the following statements:

(*a*) Methyl halides react with sodium ethoxide in ethanol only by this mechanism.
(*b*) Unhindered primary halides react with sodium ethoxide in ethanol mainly by this mechanism.
(*c*) When cyclohexyl bromide is treated with sodium ethoxide in ethanol, the major product is formed by this mechanism.

(*d*) The principal substitution product obtained by solvolysis of *tert*-butyl bromide in ethanol arises by this mechanism.
(*e*) In ethanol that contains sodium ethoxide, *tert*-butyl bromide reacts mainly by this mechanism.
(*f*) These reaction mechanisms represent concerted processes.
(*g*) Reactions proceeding by these mechanisms are stereospecific.
(*h*) These reaction mechanisms involve carbocation intermediates.
(*i*) These reaction mechanisms are the ones most likely to have been involved when the products are found to have a different carbon skeleton from the substrate.
(*j*) Alkyl iodides react faster than alkyl bromides in reactions that proceed by these mechanisms.

8.30 Outline an efficient synthesis of each of the following compounds from the indicated starting material and any necessary organic or inorganic reagents:

(*a*) Ethyl fluoride from ethyl alcohol
(*b*) Cyclopentyl cyanide from cyclopentane
(*c*) Cyclopentyl cyanide from cyclopentene
(*d*) Cyclopentyl cyanide from cyclopentanol
(*e*) $NCCH_2CH_2CN$ from ethyl alcohol
(*f*) Isobutyl iodide from isobutyl chloride
(*g*) Isobutyl iodide from *tert*-butyl chloride
(*h*) Isopropyl azide from isopropyl alcohol
(*i*) Isopropyl azide from 1-propanol
(*j*) (*S*)-*sec*-Butyl azide from (*R*)-*sec*-butyl alcohol
(*k*) (*S*)-$CH_3CH_2CHCH_3$ (with SH on the CH) from (*R*)-*sec*-butyl alcohol

8.31 Select the combination of alkyl bromide and potassium alkoxide that would be the most effective in the syntheses of the following ethers:

(*a*) $CH_3OC(CH_3)_3$ (*b*) cyclopentyl–OCH_3 (*c*) $(CH_3)_3CCH_2OCH_2CH_3$

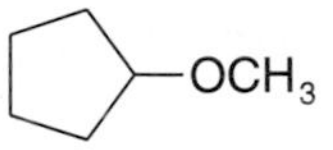

8.32 *(Note to the student: This problem previews an important aspect of Chapter 9 and is well worth attempting in order to get a head start on the material presented there.)*

Alkynes of the type RC≡CH may be prepared by nucleophilic substitution reactions in which one of the starting materials is sodium acetylide (Na^+ $:\bar{C}$≡CH).

(*a*) Devise a method for the preparation of CH_3CH_2C≡CH from sodium acetylide and any necessary organic or inorganic reagents.
(*b*) Given the information that K_a for acetylene (HC≡CH) is 10^{-26} (pK_a 26), comment on the scope of this preparative procedure with respect to R in RC≡CH. Could you prepare $(CH_3)_2CHC$≡CH or $(CH_3)_3CC$≡CH in good yield by this method?

8.33 Give the structures, including stereochemistry, of compounds A and B in the following sequence of reactions:

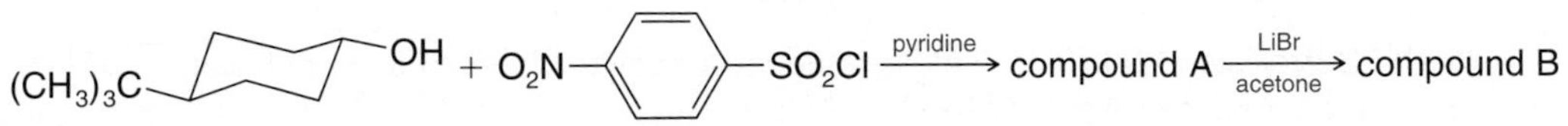

$(CH_3)_3C$–(cyclohexane)–OH + O_2N–C_6H_4–SO_2Cl $\xrightarrow{\text{pyridine}}$ compound A $\xrightarrow[\text{acetone}]{\text{LiBr}}$ compound B

8.34

(*a*) Suggest a reasonable series of synthetic transformations for converting *trans*-2-methylcyclopentanol to *cis*-2-methylcyclopentyl acetate.

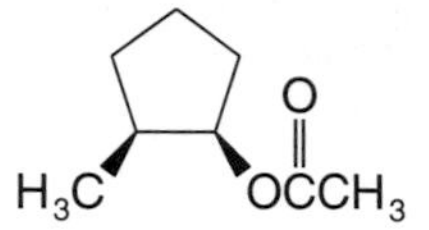

H_3C $OCCH_3$ (O) — *cis*-2-Methylcyclopentyl acetate

(*b*) How could you prepare *cis*-2-methylcyclopentyl acetate from 1-methylcyclopentanol?

8.35 Optically pure (*S*)-(+)-2-butanol was converted to its methanesulfonate ester according to the reaction shown.

CH_3 / H—C—OH / CH_2CH_3 $\xrightarrow[\text{pyridine}]{CH_3SO_2Cl}$ *sec*-butyl methanesulfonate

(*a*) Write the Fischer projection formula of the *sec*-butyl methanesulfonate formed in this reaction.

(*b*) The *sec*-butyl methanesulfonate in part (*a*) was treated with $NaSCH_2CH_3$ to give a product having an optical rotation α_D of $-25°$. Write the Fischer projection of this product. By what mechanism is it formed? What is its absolute configuration (*R* or *S*)?

(*c*) When treated with PBr_3, optically pure (*S*)-(+)-2-butanol gave 2-bromobutane having an optical rotation $\alpha_D = -38°$. This bromide was then allowed to react with $NaSCH_2CH_3$ to give a product having an optical rotation α_D of $+23°$. Write the Fischer projection for (−)-2-bromobutane and specify its configuration as *R* or *S*. Does the reaction of 2-butanol with PBr_3 proceed with predominant inversion or retention of configuration?

(*d*) What is the optical rotation of optically pure 2-bromobutane?

8.36 In one of the classic experiments of organic chemistry, Edward Hughes (a colleague of Ingold's at University College, London) studied the rate of racemization of 2-iodooctane by sodium iodide in acetone and compared it with the rate of incorporation of radioactive iodine into 2-iodooctane.

$$RI + [I^*]^- \longrightarrow RI^* + I^-$$

(I^* = radioactive iodine)

How will the rate of racemization compare with the rate of incorporation of radioactivity if

(*a*) Each act of exchange proceeds stereospecifically with retention of configuration?

(*b*) Each act of exchange proceeds stereospecifically with inversion of configuration?

(*c*) Each act of exchange proceeds in a stereorandom manner, i.e., retention and inversion of configuration are equally likely?

8.37 Solvolysis of 2-bromo-2-methylbutane in acetic acid containing potassium acetate gave three products. Identify them.

8.38 The ratio of elimination to substitution is exactly the same (26 percent elimination) for 2-bromo-2-methylbutane and 2-iodo-2-methylbutane in 80% ethanol/20% water at 25°C.

(*a*) By what mechanism does substitution most likely occur in these compounds under these conditions?
(*b*) By what mechanism does elimination most likely occur in these compounds under these conditions?
(*c*) Which substrate undergoes substitution faster?
(*d*) Which substrate undergoes elimination faster?
(*e*) What two substitution products are formed from each substrate?
(*f*) What two elimination products are formed from each substrate?
(*g*) Why do you suppose the ratio of elimination to substitution is the same for the two substrates?

8.39 Solvolysis of 1,2-dimethylpropyl *p*-toluenesulfonate in acetic acid (75°C) yields five different products. Three are alkenes and two are substitution products. Suggest reasonable structures for these five products.

8.40 Solution A was prepared by dissolving potassium acetate in methanol. Solution B was prepared by adding potassium methoxide to acetic acid. Reaction of methyl iodide either with solution A or with solution B gave the same major product. Why? What was this product?

8.41 If the temperature is not kept below 25°C during the reaction of primary alcohols with *p*-toluenesulfonyl chloride in pyridine, it is sometimes observed that the isolated product is not the desired alkyl *p*-toluenesulfonate but is instead the corresponding alkyl chloride. Suggest a mechanistic explanation for this observation.

8.42 The reaction of cyclopentyl bromide with sodium cyanide to give cyclopentyl cyanide

Cyclopentyl bromide $\xrightarrow[\text{ethanol-water}]{\text{NaCN}}$ Cyclopentyl cyanide

proceeds faster if a small amount of sodium iodide is added to the reaction mixture. Can you suggest a reasonable mechanism to explain the catalytic function of sodium iodide?

MOLECULAR MODELING EXERCISES

8.43 Illustrate that bimolecular nucleophilic substitution occurs with inversion of configuration by constructing molecular models of one of the enantiomers of a chiral alkyl bromide and the alcohol formed from it by hydrolysis under S_N2 conditions.

8.44 Illustrate the stereochemistry associated with unimolecular nucleophillic substitution by constructing molecular models of *cis*-4-*tert*-butylcyclohexyl bromide, its derived carbocation, and the alcohols formed from it by hydrolysis under S_N1 conditions.

8.45 Given the molecular formula $C_6H_{11}Br$, construct a molecular model of the isomer that is a primary alkyl bromide yet relatively unreactive toward bimolecular nucleophilic substitution.

8.46 Cyclohexyl bromide is less reactive than noncyclic secondary alkyl halides toward S_N2 substitution. Construct a molecular model of cyclohexyl bromide and suggest a reason for its low reactivity.

8.47 1-Bromobicyclo[2.2.1]heptane (the structure of which is shown) is exceedingly unreactive toward nucleophilic substitution by either the S_N1 or S_N2 mechanism. Use molecular models to help you understand why.

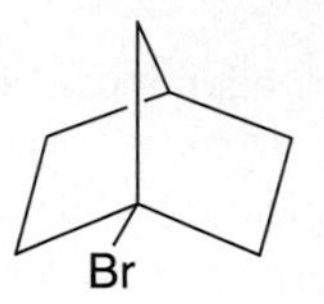

1-Bromobicyclo[2,2,1]heptane

CHAPTER 9

ALKYNES

H

H

Hydrocarbons characterized by the presence of a carbon-carbon triple bond are called **alkynes.** Noncyclic alkynes have the molecular formula C_nH_{2n-2}. *Acetylene* ($HC{\equiv}CH$) is the simplest alkyne. We call compounds that have their triple bond at the end of a carbon chain ($RC{\equiv}CH$) *monosubstituted,* or *terminal, alkynes.* Disubstituted alkynes ($RC{\equiv}CR'$) are said to have *internal* triple bonds. You will see in this chapter that a carbon-carbon triple bond is a functional group, reacting with many of the same reagents that react with alkenes.

A distinctive aspect of the chemistry of acetylene and terminal alkynes is their acidity. As a class, compounds of the type $RC{\equiv}CH$ are the most acidic of all simple hydrocarbons. The structural reasons for this property, as well as the ways in which the enhanced acidity of alkynes is used to advantage in chemical synthesis, comprise important elements of this chapter.

9.1 SOURCES OF ALKYNES

Acetylene was first characterized by the French chemist P. E. M. Berthelot in 1862 and did not command much attention until its large-scale preparation from calcium carbide in the last decade of the nineteenth century stimulated interest in industrial applications. In the first stage of that synthesis, limestone and coke are heated in an electric furnace to form calcium carbide.

Coke is a material rich in elemental carbon obtained from coal.

$$\underset{\text{(from limestone)}}{\underset{\text{Calcium oxide}}{CaO}} + \underset{\text{(from coke)}}{\underset{\text{Carbon}}{3C}} \xrightarrow{1800-2100°C} \underset{\text{Calcium carbide}}{CaC_2} + \underset{\text{Carbon monoxide}}{CO}$$

Calcium carbide is the calcium salt of the doubly negative carbide ion ($:\bar{C}\equiv\bar{C}:$). Carbide dianion is strongly basic and reacts with water to form acetylene:

$$Ca^{2+}\left[\begin{array}{c}:C\\|||\\C:\end{array}\right]^{2-} + 2H_2O \longrightarrow Ca(OH)_2 + HC\equiv CH$$

Calcium carbide   Water   Calcium hydroxide   Acetylene

PROBLEM 9.1 Use curved arrow notation to show how calcium carbide reacts with water to give acetylene.

Beginning in the middle of the twentieth century, alternative methods of acetylene production became practical. One of these is based on the thermal dehydrogenation of ethylene.

$$CH_2{=}CH_2 \rightleftharpoons HC\equiv CH + H_2$$

Ethylene   Acetylene   Hydrogen

The reaction is endothermic, and its position of equilibrium favors ethylene at low temperatures but shifts to favor acetylene above 1150°C. Indeed, at very high temperatures most hydrocarbons, even methane, are converted to acetylene. Higher alkynes are prepared from alkenes and from acetylene by methods to be described later in this chapter.

Natural products that contain carbon-carbon triple bonds are numerous. Two examples are *tariric acid,* from the seed fat of a Guatemalan plant, and *cicutoxin,* a poisonous substance isolated from water hemlock.

$$CH_3(CH_2)_{10}C\equiv C(CH_2)_4\overset{\overset{\large O}{\|}}{C}OH$$

Tariric acid

$$HOCH_2CH_2CH_2C\equiv C{-}C\equiv CCH{=}CHCH{=}CHCH{=}CH\underset{\underset{\large OH}{|}}{C}HCH_2CH_2CH_3$$

Cicutoxin

Diacetylene ($HC\equiv C{-}C\equiv CH$) has been identified as a component of the hydrocarbon-rich atmospheres of Uranus, Neptune, and Pluto. It is also present in the atmospheres of Titan and Triton, satellites of Saturn and Neptune, respectively.

9.2 NOMENCLATURE

In naming alkynes the usual IUPAC rules for hydrocarbons are followed, and the suffix *-ane* is replaced by *-yne*. Both acetylene and ethyne are acceptable IUPAC names for $HC\equiv CH$. The position of the triple bond along the chain is specified by number in a manner analogous to that used in alkene nomenclature.

$HC\equiv CCH_3$	$HC\equiv CCH_2CH_3$	$CH_3C\equiv CCH_3$	$(CH_3)_3CC\equiv CCH_3$
Propyne	1-Butyne	2-Butyne	4,4-Dimethyl-2-pentyne

PROBLEM 9.2 Write structural formulas and give the IUPAC names for all the alkynes of molecular formula C_5H_8.

A compound containing both a double bond and a triple bond is named as an *enyne.* Separate locants identify the position of each multiple bond, reflecting the direction of numbering that gives the lower locant to the one nearer the end of the chain. When the location of the double bond with respect to one end of the chain is the same as that of the triple bond with respect to the other end, the chain is numbered from the end nearer the double bond.

$CH_2{=}CHCH_2CH_2C{\equiv}CCH_3$	$(CH_3)_2C{=}CHCH_2CH_2C{\equiv}CH$	$CH_2{=}CHCH_2CH_2CH_2C{\equiv}CH$
1-Hepten-5-yne	6-Methyl-5-hepten-1-yne	1-Hepten-6-yne

In cases where the $-C{\equiv}CH$ group is named as a substituent, it is designated as an *ethynyl* group.

9.3 STRUCTURE AND BONDING IN ALKYNES. *sp* HYBRIDIZATION

Acetylene is a linear molecule with a carbon-carbon bond distance of 120 pm and carbon-hydrogen bond distances of 106 pm.

$$\underset{106\text{ pm}}{H{-}C}\underset{120\text{ pm}}{\equiv}\underset{106\text{ pm}}{C{-}H} \qquad \angle HCC = 180^\circ,\ \angle CCH = 180^\circ$$

There is a progressive shortening of the carbon-carbon bond distance in the series ethane (153 pm), ethylene (134 pm), and acetylene (120 pm). The carbon-hydrogen bond distances decrease in the series as well.

The orbital hybridization model for the triple bond in acetylene was presented in Section 1.17. To review, each carbon of acetylene is *sp* hybridized, and the triple bond contains a σ component and two π components. The σ component is formed when an *sp* orbital of one carbon, oriented so that its axis lies along the internuclear axis, overlaps with a similarly disposed *sp* orbital of the other carbon. Each *sp* orbital contains one electron, and the resulting σ bond contains two of the six electrons of the triple bond. The remaining four electrons reside in two π bonds, each formed by a "side-by-side" overlap of singly occupied *p* orbitals of the two carbons. The orientation of these π bonds is shown in Figure 9.1.

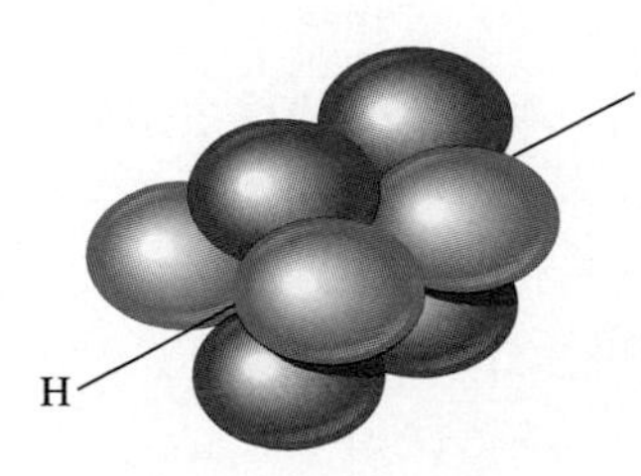

FIGURE 9.1 The carbon atoms of acetylene are connected by a $\sigma + \pi + \pi$ triple bond. The carbon atoms are *sp* hybridized, and each is bonded to a hydrogen by an *sp*–1*s* σ bond. The σ component of the triple bond arises by *sp*–*sp* overlap. Each carbon has two *p* orbitals, the axes of which are perpendicular to each other. One π bond is formed by overlap of the red *p* orbitals, the other by overlap of the blue *p* orbitals. Each π bond contains two electrons.

Table 9.1 compares some structural features in the two-carbon compounds ethane, ethylene, and acetylene. As we noted at the beginning of this section, the carbon-carbon bonds and carbon-hydrogen bonds become shorter as one proceeds from ethane to ethylene to acetylene. As the *s* character of the orbitals involved in σ bonds increases from 25 percent (sp^3) to $33\frac{1}{3}$ percent (sp^2) to 50 percent (sp), the electrons in those orbitals are, on the average, closer to the carbon nucleus, and this leads to a contraction in the internuclear distance. Further, the two π components of the carbon-carbon triple bond augment the σ bond and bring the two nuclei nearer each other.

Bond dissociation energies reveal a pattern of increasingly stronger bonds as the hybridization of carbon changes from sp^3 to sp^2 to sp. As the *s* character of the orbital that binds carbon to another atom increases, the pair of electrons in that orbital is more strongly held, and it requires more energy for homolytic cleavage of both the C—H and C—C bonds.

Table 9.1 also compares the acidities of ethane, ethylene, and acetylene. Acetylene and terminal alkynes are far more acidic than alkenes, and alkenes are more acidic than alkanes—important observations related to orbital hybridization, which will be addressed in greater detail in Section 9.6.

Higher alkynes combine structural features of alkynes and alkyl groups. Propyne has a linear arrangement of its H—C≡C—C unit, and its carbon-carbon and carbon-hydrogen bonds are shorter than those of propene, the analogous alkene.

H—C≡C—CH_3 (106 pm, 121 pm, 146 pm)

Propyne

$H_2C{=}CH{-}CH_3$ (C=C 134 pm, C—H 108 pm, C—CH_3 151 pm)

Propene

TABLE 9.1
Structural Features of Ethane, Ethylene, and Acetylene

Feature	Ethane	Ethylene	Acetylene
Systematic name	Ethane	Ethene	Ethyne
Molecular formula	C_2H_6	C_2H_4	C_2H_2
Structural formula	$H_3C{-}CH_3$	$H_2C{=}CH_2$	H—C≡C—H
C—C bond distance, pm	153	134	120
C—H bond distance, pm	111	110	106
H—C—C bond angles	111.0°	121.4°	180°
C—C bond dissociation energy, kJ/mol (kcal/mol)	368 (88)	611 (146)	820 (196)
C—H bond dissociation energy, kJ/mol (kcal/mol)	410 (98)	452 (108)	536 (128)
Hybridization of carbon	sp^3	sp^2	sp
s character in C—H bonds	25%	33%	50%
Approximate acidity as measured by K_a (pK_a)	10^{-62} (62)	10^{-45} (45)	10^{-26} (26)

Not only is the C(*sp*)—C(*sp*) triple bond distance shorter than the C(sp^2)—C(sp^2) double bond distance, but the C(*sp*)—C(sp^3) single bond distance in propyne is shorter than the C(sp^2)—C(sp^3) single bond distance in propene. This is another example of the general rule that an increase in the *s* character of the orbitals involved leads to a shorter bond.

PROBLEM 9.3 Which has a longer carbon-methyl bond, 1-butyne or 2-butyne? Why?

9.4 CYCLOALKYNES

Rings that are not large enough to accommodate a linear C—C≡C—C unit are destabilized by *angle strain.* Among the cycloalkynes, cyclooctyne is the smallest one stable enough to be isolated and studied at room temperature. It is relatively reactive, however, and polymerizes on standing. The next larger homolog, cyclononyne is more stable and can be stored indefinitely.

Cyclooctyne   Cyclononyne

Structural studies give values of 155° and 160° for the C(1)—C(2)—C(3) bond angles of cyclooctyne and cyclononyne, respectively, indicating substantial distortion from the ideal bond angles of 180° observed for C≡C—C structural units in noncyclic alkynes. Bending of the C—C≡C—C unit is clearly evident in the molecular model of cyclononyne displayed in Figure 9.2.

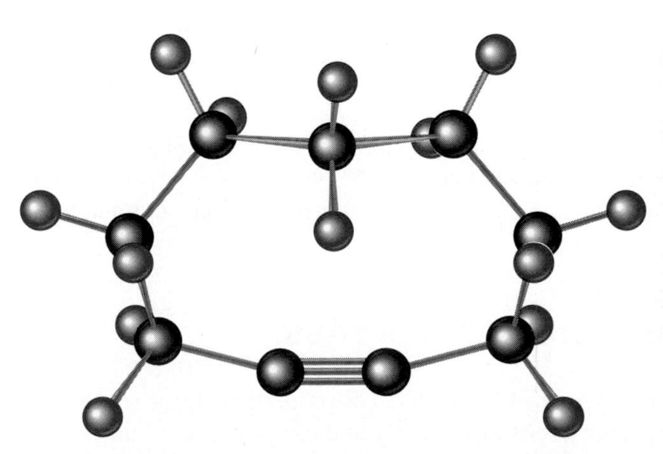

FIGURE 9.2 Molecular model of cyclononyne showing bending of bond angles associated with triply bonded carbons. This model represents the structure obtained when the strain energy is minimized according to molecular mechanics (MM2), and closely matches the structure determined experimentally by electron diffraction. Notice too the degree to which the staggering of bonds on adjacent atoms governs the overall shape of the ring.

NATURAL AND "DESIGNED" ENEDIYNE ANTIBIOTICS

Beginning in the 1980s, research directed toward the isolation of new drugs derived from natural sources identified a family of tumor-inhibitory antibiotic substances characterized by novel structures containing a C≡C—C=C—C≡C unit as part of a 9- or 10-membered ring. With one double bond and two triple bonds (*-ene* + *di-* + *-yne*), these compounds soon became known as *enediyne* antibiotics. The simplest member of the class is *dynemicin A;* most of the other enediynes have even more complicated structures.

Enediynes hold substantial promise as anticancer drugs because of their potency and selectivity. Not only do they inhibit cell growth, they have a greater tendency to kill cancer cells than they do normal cells. The mechanism by which enediynes act involves novel chemistry unique to the C≡C—C=C—C≡C unit, which leads to a species that cleaves DNA and halts tumor growth.

The history of drug development has long been based on naturally occurring substances. Often, however, compounds that might be effective drugs are produced by plants and microorganisms in such small amounts that their isolation from natural sources is not practical. If the structure is relatively simple, chemical synthesis provides an alternative source of the drug, making it more available at a lower price. Equally important, chemical synthesis, modification, or both can improve the effectiveness of a drug. Building on the enediyne core of dynemicin A, for example, Professor Kyriacos C. Nicolaou and his associates at the Scripps Research Institute and the University of California at San Diego have prepared a simpler analog that it is both more potent and more selective than dynemicin A. It is a "designed enediyne" in that its structure was conceived on the basis of chemical reasoning so as to carry out its biochemical task. The designed enediyne offers the additional advantage of being more amenable to large-scale synthesis.

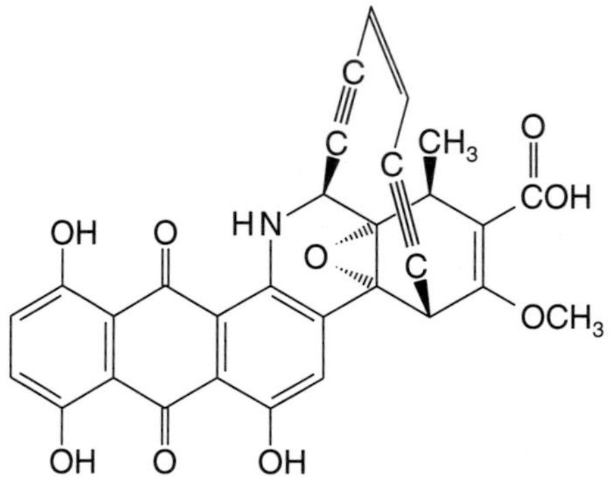

OH O HN CH$_3$ O COH OCH$_3$ OH O OH

Dynemicin A

$^-$O O$^-$ S ++ O O N O HOCH$_2$CH$_2$O

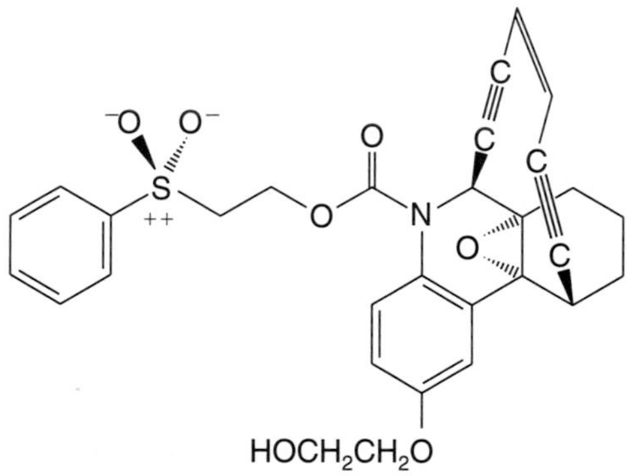

"Designed" enediyne

Cycloheptyne is too strained to be isolated as a stable species, forming a polymer soon after it is generated. Evidence suggests that cyclohexyne and even cyclopentyne may be formed as transitory intermediates in certain chemical reactions, but neither is stable enough to study.

9.5 PHYSICAL PROPERTIES OF ALKYNES

Examples of physical properties of alkynes are given in Appendix 1.

Alkynes resemble alkanes and alkenes in their physical properties. They share with these other hydrocarbons the properties of low density and low water-solubility. They are nonpolar and dissolve readily in typical organic solvents such as alkanes, diethyl ether, and chlorinated hydrocarbons. Alkynes generally have slightly higher boiling points than the corresponding alkanes and alkenes.

9.6 ACIDITY OF ACETYLENE AND TERMINAL ALKYNES

The C—H bonds of hydrocarbons show little tendency to ionize, and alkanes, alkenes, and alkynes are very weak acids. The ionization constant K_a for methane, for example, is too small to be measured directly but is estimated to be about 10^{-60} (pK_a 60).

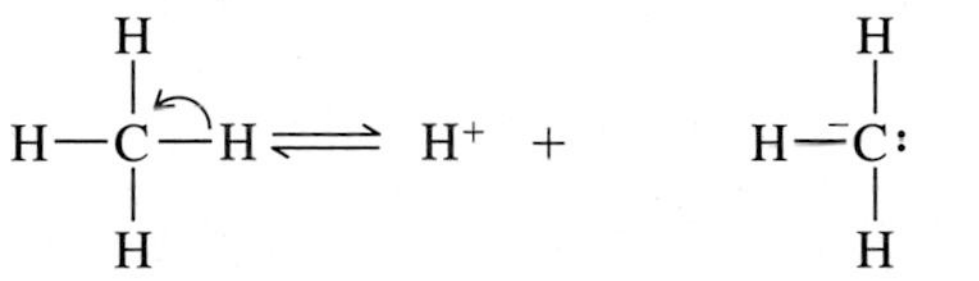

$$CH_4 \rightleftharpoons H^+ + {}^-:CH_3$$

Methane — Proton — Methide anion (a *carbanion*)

The conjugate base of a hydrocarbon is called a **carbanion.** It is an anion in which the negative charge is borne by carbon. Since it is derived from a very weak acid, a carbanion such as $^-:CH_3$ is an exceptionally strong base.

In general, the ability of an atom to bear a negative charge is related to its electronegativity. Both the electronegativity of an atom X and the acidity of H—X increase across a row in the periodic table.

CH_4	< NH_3	< H_2O	< HF
Methane	Ammonia	Water	Hydrogen fluoride
$K_a \sim 10^{-60}$	$\sim 10^{-36}$	1.8×10^{-16}	3.5×10^{-4}
p$K_a \sim 60$	~ 36	15.7	3.2
(weakest acid)			(strongest acid)

Conversely, the basicity of the conjugate base $:X^-$ decreases across a row.

$H_3C:^-$	> $H_2N:^-$	> $HO:^-$	> $:F:^-$
Methide	Amide	Hydroxide	Fluoride
(strongest base)			(weakest base)

As the electron-attracting power of the negatively charged atom becomes greater, the anion becomes less basic.

Alkynes are more acidic than alkenes, and alkenes are more acidic than alkanes.

HC≡CH	> CH_2=CH_2	> CH_3CH_3
Acetylene	Ethylene	Ethane
$K_a = 10^{-26}$	$\sim 10^{-45}$	$\sim 10^{-62}$
p$K_a = 26$	~ 45	~ 62
(strongest acid)		(weakest acid)

We can understand this order of acidity by examining how strongly each anion holds the unshared electron pair. The ionization of acetylene yields an anion in which the unshared electron pair occupies an *sp* hybridized orbital.

$$H{-}C{\equiv}C{-}H \rightleftharpoons H^+ + H{-}C{\equiv}C\!:^- \quad sp$$

Acetylene Proton Acetylide ion

Ionization of ethylene gives an anion in which the unshared electron pair occupies an sp^2 hybridized orbital.

$$H_2C{=}CH_2 \rightleftharpoons H^+ + H_2C{=}\overset{-}{C}H \quad sp^2$$

Ethylene Proton Vinyl anion

The equilibrium constant for ionization of acetylene is greater than that of ethylene because an electron pair in an orbital with 50 percent *s* character (*sp*) is more strongly bound than an electron pair in an orbital with $33\frac{1}{3}$ percent *s* character (sp^2). Acetylide ion holds its electron pair more strongly than vinyl anion and is less basic. According to the same reasoning, ethane is the weakest acid of the group because the unshared electron pair in ethyl anion ($CH_3\overset{-}{C}H_2\!:$) occupies an orbital with only 25 percent *s* character. An electron pair in ethyl anion is less strongly held than an electron pair in vinyl anion. Terminal alkynes ($RC{\equiv}CH$) are similar to acetylene in their acidity.

$$(CH_3)_3CC{\equiv}CH \qquad K_a = 3 \times 10^{-26}\ (pK_a = 25.5)$$

3,3-Dimethyl-1-butyne

While acetylene and terminal alkynes are far stronger acids than other hydrocarbons, it must be remembered that they are, nevertheless, very weak acids—much weaker than water and alcohols, for example. Hydroxide ion is too weak a base to convert acetylene to its anion in meaningful amounts. The position of the equilibrium described by the following equation lies overwhelmingly to the left:

$$H{-}C{\equiv}C{-}H + \,^{-}\!:\ddot{O}H \rightleftharpoons H{-}C{\equiv}C\!:^- + H{-}\ddot{O}H$$

Acetylene	Hydroxide ion	Acetylide ion	Water
(weaker acid)	(weaker base)	(stronger base)	(stronger acid)
$K_a = 10^{-26}$			$K_a = 1.8 \times 10^{-16}$
$pK_a = 26$			$pK_a = 15.7$

Because acetylene is a far weaker acid than water and alcohols, these substances are not suitable solvents for reactions involving acetylide ions. Acetylide is instantly converted to acetylene by proton transfer from compounds that contain hydroxyl groups.

Amide ion is a much stronger base than acetylide ion and converts acetylene to its conjugate base quantitatively.

$$H{-}C{\equiv}C{-}H + \,^{-}\!:\ddot{N}H_2 \rightleftharpoons H{-}C{\equiv}C\!:^- + H{-}\ddot{N}H_2$$

Acetylene	Amide ion	Acetylide ion	Ammonia
(stronger acid)	(stronger base)	(weaker base)	(weaker acid)
$K_a = 10^{-26}$			$K_a = 10^{-36}$
$pK_a = 26$			$pK_a = 36$

Solutions of sodium acetylide ($HC{\equiv}CNa$) may be prepared by adding *sodium amide* ($NaNH_2$) to acetylene in liquid ammonia as the solvent. Terminal alkynes react similarly to give species of the type $RC{\equiv}CNa$.

PROBLEM 9.4 Complete each of the following equations to show the conjugate acid and the conjugate base formed by proton transfer between the indicated species. For each reaction, specify whether the position of equilibrium lies to the side of reactants or products.

(*a*) $CH_3C{\equiv}CH + {}^-OCH_3 \rightleftharpoons$
(*b*) $HC{\equiv}CH + CH_3CH_2{}^- \rightleftharpoons$
(*c*) $CH_2{=}CH_2 + H_2N^- \rightleftharpoons$
(*d*) $CH_3C{\equiv}CCH_2OH + H_2N^- \rightleftharpoons$

SAMPLE SOLUTION (*a*) The equation representing the acid-base reaction between propyne and methoxide ion is:

$$CH_3C{\equiv}C{-}H + {}^-OCH_3 \rightleftharpoons CH_3C{\equiv}\ddot{C}{:}^- + HOCH_3$$

Propyne (weaker acid)	Methoxide ion (weaker base)	Propynide ion (stronger base)	Methanol (stronger acid)

Alcohols are stronger acids than acetylene, and so the position of equilibrium lies to the left. Methoxide ion is not a sufficiently strong base to abstract a proton from acetylene.

Anions of acetylene and terminal alkynes are nucleophilic and react with methyl and primary alkyl halides to form carbon-carbon bonds by nucleophilic substitution. Some useful applications of this reaction will be discussed in the following section.

9.7 PREPARATION OF ALKYNES BY ALKYLATION OF ACETYLENE AND TERMINAL ALKYNES

Organic synthesis makes use of two major reaction types:

1. Functional group transformations
2. Carbon-carbon bond–forming reactions

Both strategies are applied to the preparation of alkynes. In this section we shall see how to prepare alkynes by combining smaller structural units to build longer carbon chains. One of these structural units may be as simple as acetylene itself. By attaching alkyl groups to this unit, more complex alkynes can be prepared.

$$H{-}C{\equiv}C{-}H \longrightarrow R{-}C{\equiv}C{-}H \longrightarrow R{-}C{\equiv}C{-}R'$$

Acetylene	Monosubstituted or terminal alkyne	Disubstituted derivative of acetylene

Reactions that lead to attachment of alkyl groups to molecular fragments are called **alkylation** reactions. One way in which alkynes are prepared is by alkylation of acetylene.

Alkylation of acetylene is a synthetic process comprising two separate reactions carried out in sequence. In the first stage, acetylene is converted to its conjugate base by treatment with sodium amide.

$$\underset{\text{Acetylene}}{HC{\equiv}CH} + \underset{\text{Sodium amide}}{NaNH_2} \longrightarrow \underset{\text{Sodium acetylide}}{HC{\equiv}CNa} + \underset{\text{Ammonia}}{NH_3}$$

Next, an alkyl halide is added to the solution of sodium acetylide. Acetylide ion acts as a nucleophile, displacing halide from carbon and forming a new carbon-carbon bond. Substitution occurs by an S_N2 mechanism.

$$\underset{\text{Sodium acetylide}}{HC{\equiv}CNa} + \underset{\text{Alkyl halide}}{RX} \longrightarrow \underset{\text{Alkyne}}{HC{\equiv}CR} + \underset{\text{Sodium halide}}{NaX} \quad \text{via} \quad HC{\equiv}C{:}^{-} \; R{-}X$$

The synthetic sequence is usually carried out in liquid ammonia as the solvent. Alternatively, diethyl ether or tetrahydrofuran may be used.

$$\underset{\text{Sodium acetylide}}{HC{\equiv}CNa} + \underset{\text{1-Bromobutane}}{CH_3CH_2CH_2CH_2Br} \xrightarrow{NH_3} \underset{\text{1-Hexyne (70–77\%)}}{CH_3CH_2CH_2CH_2C{\equiv}CH}$$

An analogous sequence using terminal alkynes as starting materials yields alkynes of the type RC$\equiv$CR′.

$$\underset{\text{4-Methyl-1-pentyne}}{(CH_3)_2CHCH_2C{\equiv}CH} \xrightarrow[NH_3]{NaNH_2} (CH_3)_2CHCH_2C{\equiv}CNa \xrightarrow{CH_3Br} \underset{\text{5-Methyl-2-hexyne (81\%)}}{(CH_3)_2CHCH_2C{\equiv}CCH_3}$$

Dialkylation of acetylene can be achieved by carrying out the sequence twice.

$$\underset{\text{Acetylene}}{HC{\equiv}CH} \xrightarrow[\text{2. } CH_3CH_2Br]{\text{1. } NaNH_2,\ NH_3} \underset{\text{1-Butyne}}{HC{\equiv}CCH_2CH_3} \xrightarrow[\text{2. } CH_3Br]{\text{1. } NaNH_2,\ NH_3} \underset{\text{2-Pentyne (81\%)}}{CH_3C{\equiv}CCH_2CH_3}$$

As in other nucleophilic substitution reactions, alkyl *p*-toluenesulfonates may be used in place of alkyl halides.

PROBLEM 9.5 Outline efficient syntheses of each of the following alkynes from acetylene and any necessary organic or inorganic reagents:

(*a*) 1-Heptyne (*b*) 2-Heptyne (*c*) 3-Heptyne

SAMPLE SOLUTION (*a*) An examination of the structural formula of 1-heptyne reveals it to have a pentyl group attached to an acetylene unit. Alkylation of acetylene, by way of its anion, with a pentyl halide is a suitable synthetic route to 1-heptyne.

$$\underset{\text{Acetylene}}{HC{\equiv}CH} \xrightarrow[NH_3]{NaNH_2} \underset{\text{Sodium acetylide}}{HC{\equiv}CNa} \xrightarrow{CH_3CH_2CH_2CH_2CH_2Br} \underset{\text{1-Heptyne}}{HC{\equiv}CCH_2CH_2CH_2CH_2CH_3}$$

The most significant limitation to this reaction is that synthetically acceptable yields are obtained only with methyl halides and primary alkyl halides. Acetylide anions are very basic, much more basic than hydroxide, for example, and react with secondary and tertiary alkyl halides by elimination.

$$HC{\equiv}C{:}^- + H{-}CH_2{-}C(CH_3)_2{-}Br \xrightarrow{E2} HC{\equiv}CH + CH_2{=}C(CH_3)_2 + Br^-$$

Acetylide   *tert*-Butyl bromide   Acetylene   2-Methylpropene   Bromide

The desired S_N2 substitution pathway is observed only with methyl and primary alkyl halides. Alkyl p-toluenesulfonates are subject to the same limitations as alkyl halides.

PROBLEM 9.6 Refer to the various alkynes of molecular formula C_5H_8 of Problem 9.2 (Section 9.2) and select those that can be prepared in good yield by alkylation or dialkylation of acetylene. Explain why the preparation of the other isomers would not be practical.

A second strategy for alkyne synthesis, involving functional group transformation reactions, is described in the following section.

9.8 PREPARATION OF ALKYNES BY ELIMINATION REACTIONS

Just as it is possible to prepare alkenes by dehydrohalogenation of alkyl halides, so may alkynes be prepared by a *double dehydrohalogenation* of dihaloalkanes. The dihalide may be a **geminal dihalide,** one in which both halogens are substituents on the same carbon, or it may be a **vicinal dihalide,** one in which the halogens are substituents on adjacent carbons.

Double dehydrohalogenation of a geminal dihalide

$$R{-}CH_2{-}CX_2{-}R' + 2NaNH_2 \longrightarrow R{-}C{\equiv}C{-}R' + 2NH_3 + 2NaX$$

Geminal dihalide   Sodium amide   Alkyne   Ammonia   Sodium halide

Double dehydrohalogenation of a vicinal dihalide

$$R{-}CHX{-}CHX{-}R' + 2NaNH_2 \longrightarrow R{-}C{\equiv}C{-}R' + 2NH_3 + 2NaX$$

Vicinal dihalide   Sodium amide   Alkyne   Ammonia   Sodium halide

The most frequent applications of these procedures are in the preparation of terminal alkynes. Since the terminal alkyne product is a strong enough acid to transfer a proton to amide anion, one equivalent of base in excess of the two equivalents required for double dehydrohalogenation is required. Addition of water or acid after the reaction is complete converts the sodium salt to the corresponding alkyne.

Double dehydrohalogenation of a geminal dihalide

$$(CH_3)_3CCH_2CHCl_2 \xrightarrow[NH_3]{3NaNH_2} (CH_3)_3CC{\equiv}CNa \xrightarrow{H_2O} (CH_3)_3CC{\equiv}CH$$

1,1-Dichloro-3,3-dimethylbutane — Sodium salt of alkyne product (not isolated) — 3,3-Dimethyl-1-butyne (56–60%)

Double dehydrohalogenation of a vicinal dihalide

$$CH_3(CH_2)_7\underset{\displaystyle Br}{\underset{|}{C}}HCH_2Br \xrightarrow[NH_3]{3NaNH_2} CH_3(CH_2)_7C{\equiv}CNa \xrightarrow{H_2O} CH_3(CH_2)_7C{\equiv}CH$$

1,2-Dibromodecane — Sodium salt of alkyne product (not isolated) — 1-Decyne (54%)

Double dehydrohalogenation to form terminal alkynes may also be carried out by heating geminal and vicinal dihalides with potassium *tert*-butoxide in dimethyl sulfoxide.

PROBLEM 9.7 Give the structures of three isomeric dibromides that could be used as starting materials for the preparation of 3,3-dimethyl-1-butyne.

Since vicinal dihalides are prepared by addition of chlorine or bromine to alkenes (Section 6.14), we see that alkenes can serve as starting materials for the preparation of alkynes by the following sequence of functional group transformations:

$$RCH{=}CHR' + X_2 \longrightarrow R\underset{\displaystyle X}{\underset{|}{C}}H{-}\underset{\displaystyle X}{\underset{|}{C}}HR' \xrightarrow[2.\ H_2O]{1.\ NaNH_2,\ NH_3} RC{\equiv}CR'$$

Alkene — Chlorine or bromine — Vicinal dihalide — Alkyne

$$(CH_3)_2CHCH{=}CH_2 \xrightarrow{Br_2} (CH_3)_2CH\underset{\displaystyle Br}{\underset{|}{C}}HCH_2Br \xrightarrow[2.\ H_2O]{1.\ NaNH_2,\ NH_3} (CH_3)_2CHC{\equiv}CH$$

3-Methyl-1-butene — 1,2-Dibromo-3-methylbutane — 3-Methyl-1-butyne (52%)

PROBLEM 9.8 Show, by writing an appropriate series of equations, how you could prepare propyne from each of the following compounds as starting materials. You may use any necessary organic or inorganic reagents.

(*a*) 2-Propanol
(*b*) 1-Propanol
(*c*) Isopropyl bromide
(*d*) 1,1-Dichloroethane
(*e*) Ethyl alcohol

SAMPLE SOLUTION (*a*) Since we know that we can convert propene to propyne by the sequence of reactions

$$\underset{\text{Propene}}{CH_3CH{=}CH_2} \xrightarrow{Br_2} \underset{\text{1,2-Dibromopropane}}{\underset{\substack{|\\ Br}}{CH_3CHCH_2Br}} \xrightarrow[\text{2. } H_2O]{\text{1. } NaNH_2,\ NH_3} \underset{\text{Propyne}}{CH_3C{\equiv}CH}$$

all that remains in order to completely describe the synthesis is to show the preparation of propene from 2-propanol. Acid-catalyzed dehydration is suitable.

$$\underset{\text{2-Propanol}}{(CH_3)_2CHOH} \xrightarrow[\text{heat}]{H^+} \underset{\text{Propene}}{CH_3CH{=}CH_2}$$

The two dehydrohalogenation steps in the base-promoted formation of alkynes from vicinal and geminal dihalides occur sequentially. Alkenyl halides are intermediates.

$$\underset{\text{Geminal dihalide}}{RCH_2{-}\overset{\overset{X}{|}}{\underset{\underset{X}{|}}{C}}{-}R'} \quad \text{or} \quad \underset{\text{Vicinal dihalide}}{\underset{\substack{|\\X}}{RCH}{-}\underset{\substack{|\\X}}{CHR'}} \longrightarrow \underset{\text{Alkenyl halide}}{RCH{=}\underset{\substack{|\\X}}{CR'}} \longrightarrow \underset{\text{Alkyne}}{RC{\equiv}CR'}$$

The second dehydrohalogenation step is typically more difficult than the first, so rather strongly basic conditions or high temperatures are needed to convert dihalides to alkynes. When weaker bases and lower temperatures are used, the intermediate alkenyl halide may be isolated.

$$\underset{\text{3,4-Dichlorohexane}}{CH_3CH_2\underset{\substack{|\\Cl}}{CH}\underset{\substack{|\\Cl}}{CH}CH_2CH_3} \xrightarrow[\text{1-propanol, 90°C}]{KOH} \underset{\text{3-Chloro-3-hexene (90\%)}}{CH_3CH_2CH{=}\underset{\substack{|\\Cl}}{C}CH_2CH_3}$$

Subjecting alkenyl halides to more strongly basic conditions brings about their dehydrohalogenation to alkynes.

$$\underset{\text{2-Chloro-4-methyl-1-pentene}}{(CH_3)_2CHCH_2\underset{\substack{|\\Cl}}{C}{=}CH_2} \xrightarrow[\text{2. } H_2O]{\text{1. } 2NaNH_2} \underset{\text{4-Methyl-1-pentyne (80\%)}}{(CH_3)_2CHCH_2C{\equiv}CH}$$

9.9 REACTIONS OF ALKYNES

We have already discussed one important chemical property of alkynes, the acidity of acetylene and terminal alkynes. In the remaining sections of this chapter additional aspects of the reactions of alkynes will be explored. Most of the reactions to be encountered will be similar to those of alkenes. Like alkenes, alkynes undergo addition reactions. We shall begin with a reaction familiar to us from our study of alkenes, namely, catalytic hydrogenation.

9.10 HYDROGENATION OF ALKYNES

The conditions for hydrogenation of alkynes are similar to those employed for alkenes. In the presence of finely divided platinum, palladium, nickel, or rhodium, two molar equivalents of hydrogen add to the triple bond of an alkyne to yield an alkane.

$$\underset{\text{Alkyne}}{RC{\equiv}CR'} + \underset{\text{Hydrogen}}{2H_2} \xrightarrow{\text{Pt, Pd, Ni, or Rh}} \underset{\text{Alkane}}{RCH_2CH_2R'}$$

$$\underset{\text{4-Methyl-1-hexyne}}{CH_3CH_2\underset{\displaystyle CH_3}{\underset{|}{C}}HCH_2C{\equiv}CH} + \underset{\text{Hydrogen}}{2H_2} \xrightarrow{\text{Ni}} \underset{\text{3-Methylhexane (77\%)}}{CH_3CH_2\underset{\displaystyle CH_3}{\underset{|}{C}}HCH_2CH_2CH_3}$$

PROBLEM 9.9 Write a series of equations showing how you could prepare octane from acetylene and any necessary organic and inorganic reagents.

Substituents affect the heats of hydrogenation of alkynes in the same way they affect alkenes. Alkyl groups release electrons to *sp* hybridized carbon, stabilizing the alkyne and decreasing the heat of hydrogenation.

	$CH_3CH_2C{\equiv}CH$	$CH_3C{\equiv}CCH_3$
	1-Butyne	2-Butyne
$-\Delta H°$ (hydrogenation)	292 kJ/mol	275 kJ/mol
	(69.9 kcal/mol)	(65.6 kcal/mol)

Alkenes are intermediates in the hydrogenation of alkynes to alkanes.

$$\underset{\text{Alkyne}}{RC{\equiv}CR'} \xrightarrow[\text{catalyst}]{H_2} \underset{\text{Alkene}}{RCH{=}CHR'} \xrightarrow[\text{catalyst}]{H_2} \underset{\text{Alkane}}{RCH_2CH_2R'}$$

The high energy of acetylene is released when it is mixed with oxygen and burned in an *oxyacetylene torch.* The temperature of the flame so produced (about 3000°C) exceeds that obtainable with any other hydrocarbon fuel and is higher than the melting point of iron (1535°C).

The heats of hydrogenation of alkynes are typically somewhat greater than twice the heats of hydrogenation of analogous alkenes. Thus, as shown in Figure 9.3, the first hydrogenation step of 1-hexyne is more exothermic than the second. The carbon-carbon triple bond makes alkynes rather high in energy compared with other hydrocarbons.

Noting that alkenes are intermediates in the hydrogenation of alkynes leads us to consider the possibility of halting hydrogenation at the alkene stage. If partial hydrogenation (also called *semihydrogenation*) of an alkyne could be achieved, it would provide a useful synthesis of alkenes. In practice it is a simple matter to convert alkynes to alkenes by semihydrogenation in the presence of specially developed catalysts. The one most frequently used is the **Lindlar catalyst,** a palladium on calcium

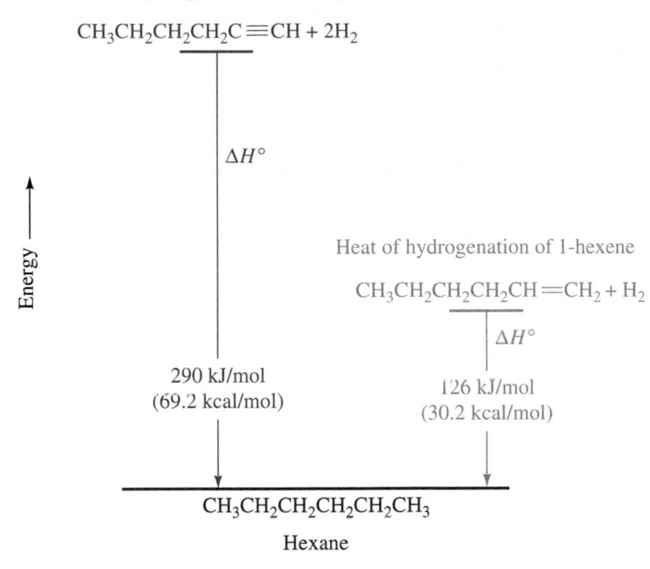

FIGURE 9.3 Diagram comparing heats of hydrogenation of 1-hexyne and 1-hexene. By subtracting the heat of hydrogenation of 1-hexene from that of 1-hexyne, we see that addition of the first molecule of H_2 liberates 164 kJ/mol (39.0 kcal/mol) of heat, the second only 126 kJ/mol (30.2 kcal/mol).

carbonate combination to which have been added lead acetate and quinoline. Lead acetate and quinoline partially deactivate (''poison'') the catalyst, making it a poor catalyst for alkene hydrogenation while retaining its ability to catalyze the addition of hydrogen to alkynes.

The structure of quinoline is shown on page 454.

1-Ethynylcyclohexanol + H_2 (Hydrogen) $\xrightarrow[\text{lead acetate, quinoline}]{Pd/CaCO_3}$ 1-Vinylcyclohexanol (90–95%)

In subsequent equations, we will not specify the components of the Lindlar palladium catalyst in detail but will simply write ''Lindlar Pd'' over the reaction arrow.

A number of other catalysts for semihydrogenation have been developed. These include palladium supported on barium sulfate, and a ''nickel boride'' catalyst prepared by reaction of nickel salts with sodium borohydride.

Hydrogenation of alkynes to alkenes is highly stereoselective and yields the cis (or Z) alkene by syn addition to the triple bond.

$$CH_3(CH_2)_3C{\equiv}C(CH_2)_3CH_3 \xrightarrow[\text{Lindlar Pd}]{H_2} CH_3(CH_2)_3CH{=}CH(CH_2)_3CH_3$$

5-Decyne → (Z)-5-Decene (87%)

PROBLEM 9.10 Oleic acid and stearic acid are naturally occurring compounds, which can be isolated from various fats and oils. In the laboratory, each can be prepared by hydrogenation of a compound known as *stearolic acid,* which has the formula

$CH_3(CH_2)_7C{\equiv}C(CH_2)_7CO_2H$. Oleic acid is obtained by hydrogenation of stearolic acid over Lindlar palladium; stearic acid is obtained by hydrogenation over platinum. What are the structures of oleic acid and stearic acid?

9.11 METAL-AMMONIA REDUCTION OF ALKYNES

A useful alternative to catalytic partial hydrogenation for converting alkynes to alkenes is reduction by a group I metal (lithium, sodium, or potassium) in liquid ammonia as the reaction medium. The unique feature of metal-ammonia reduction is that it converts alkynes to trans (or *E*) alkenes while catalytic semihydrogenation yields cis (or *Z*) alkenes. Thus, from the same alkyne one can prepare either a cis or a trans alkene by choosing the appropriate reaction conditions.

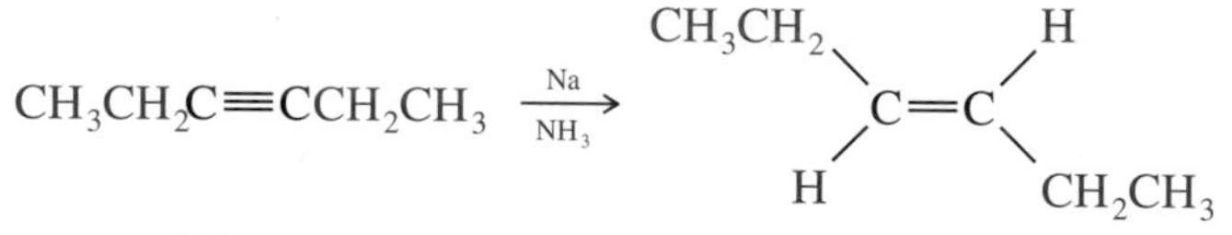

$$CH_3CH_2C{\equiv}CCH_2CH_3 \xrightarrow[NH_3]{Na} (E)\text{-}CH_3CH_2CH{=}CHCH_2CH_3$$

3-Hexyne → (*E*)-3-Hexene (82%)

PROBLEM 9.11 Sodium-ammonia reduction of stearolic acid (see Problem 9.10) yields a compound known as *elaidic acid.* What is the structure of elaidic acid?

PROBLEM 9.12 Suggest efficient syntheses of (*E*)- and (*Z*)-2-heptene from propyne and any necessary organic or inorganic reagents.

The stereochemistry of metal-ammonia reduction of alkynes differs from that of catalytic semihydrogenation because the mechanisms of the two reactions are different. The mechanism of semihydrogenation of alkynes is similar to that of catalytic hydrogenation of alkenes (Sections 6.1 and 6.3). A mechanism for metal-ammonia reduction of alkynes is outlined in Figure 9.4.

The mechanism includes two single-electron transfers (steps 1 and 3) and two proton transfers (steps 2 and 4). Experimental evidence indicates that step 2 is rate-determining, and it is believed that the observed trans stereoselectivity reflects the distribution of the two stereoisomeric alkenyl radical intermediates formed in this step.

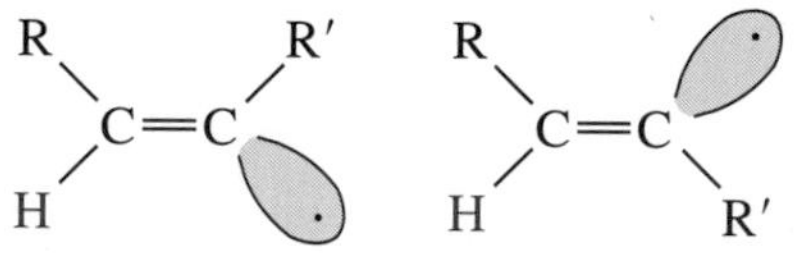

R(H)C=C(R′)• — (*Z*)-Alkenyl radical (less stable)

R(H)C=C(•)R′ — (*E*)-Alkenyl radical (more stable)

The more stable (*E*)-alkenyl radical, in which the alkyl groups R and R′ are trans to each other, is formed faster than its *Z* stereoisomer. Steps 3 and 4, which follow, are fast, and the product distribution is determined by the *E*/*Z* ratio of radicals produced in step 2.

Overall Reaction:

$$\underset{\text{Alkyne}}{RC{\equiv}CR'} + \underset{\text{Sodium}}{2Na} + \underset{\text{Ammonia}}{2NH_3} \longrightarrow \underset{\text{Trans alkene}}{RCH{=}CHR'} + \underset{\text{Sodium amide}}{2NaNH_2}$$

Step 1: Electron transfer from sodium to the alkyne. The product is an anion radical.

$$\underset{\text{Alkyne}}{RC{\equiv}CR'} + \underset{\text{Sodium}}{\cdot Na} \longrightarrow \underset{\text{Anion radical}}{R\dot{C}{=}\ddot{C}R'} + \underset{\text{Sodium ion}}{Na^+}$$

Step 2: The anion radical is a strong base and abstracts a proton from ammonia.

$$\underset{\text{Anion radical}}{R\dot{C}{=}\ddot{C}R'} + \underset{\text{Ammonia}}{H{-}\ddot{N}H_2} \longrightarrow \underset{\text{Alkenyl radical}}{R\dot{C}{=}CHR'} + \underset{\text{Amide ion}}{:\ddot{N}H_2}$$

Step 3: Electron transfer to the alkenyl radical.

$$\underset{\text{Alkenyl radical}}{R\dot{C}{=}CHR'} + \underset{\text{Sodium}}{\cdot Na} \longrightarrow \underset{\text{Alkenyl radical}}{R\ddot{C}{=}CHR'} + \underset{\text{Sodium ion}}{Na^+}$$

Step 4: Proton transfer from ammonia converts the alkenyl anion to an alkene.

$$\underset{\text{Ammonia}}{H_2\ddot{N}{-}H} + \underset{\text{Alkenyl anion}}{R\ddot{C}{=}CHR'} \longrightarrow \underset{\text{Alkene}}{RCH{=}CHR'} + \underset{\text{Amide ion}}{H_2\ddot{N}:}$$

FIGURE 9.4 Sequence of steps that describes the sodium-ammonia reduction of an alkyne.

9.12 ADDITION OF HYDROGEN HALIDES TO ALKYNES

Alkynes react with many of the same electrophilic reagents that add to the carbon-carbon double bond of alkenes. Hydrogen halides, for example, add to alkynes to form alkenyl halides.

$$\underset{\text{Alkyne}}{RC{\equiv}CR'} + \underset{\text{Hydrogen halide}}{HX} \longrightarrow \underset{\text{Alkenyl halide}}{RCH{=}\underset{X}{\underset{|}{C}}R'}$$

The regioselectivity of addition follows Markovnikov's rule. A proton adds to the carbon that has the greater number of hydrogen substituents, while halide adds to the carbon with the fewer hydrogen substituents.

$$\underset{\text{1-Hexyne}}{CH_3CH_2CH_2CH_2C{\equiv}CH} + \underset{\text{Hydrogen bromide}}{HBr} \longrightarrow \underset{\text{2-Bromo-1-hexene (60\%)}}{CH_3CH_2CH_2CH_2\underset{Br}{\underset{|}{C}}{=}CH_2}$$

When formulating a mechanism for the reaction of alkynes with hydrogen halides, we could propose a process analogous to that of electrophilic addition to alkenes in which the first step is formation of a carbocation and is rate-determining. The second step according to such a mechanism would be nucleophilic capture of the carbocation by a halide ion.

$$\underset{\text{Alkyne}}{RC{\equiv}CH} + \underset{\text{Hydrogen halide}}{H{-}\ddot{X}:} \xrightarrow{\text{slow}} \underset{\text{Alkenyl cation}}{R\overset{+}{C}{=}CH_2} + \underset{\text{Halide ion}}{:\ddot{X}:^-} \xrightarrow{\text{fast}} \underset{\text{Alkenyl halide}}{R\underset{:\ddot{X}:}{C}{=}CH_2}$$

Evidence from a variety of sources, however, indicates that alkenyl cations (also called *vinylic cations*) are much less stable than simple alkyl cations, and their involvement in these additions has been questioned. For example, although electrophilic addition of hydrogen halides to alkynes occurs more slowly than the corresponding additions to alkenes, the difference is not nearly as great as the difference in carbocation stabilities would suggest.

Further, kinetic studies reveal that electrophilic addition of hydrogen halides to alkynes follows a rate law that is third-order overall and second-order in hydrogen halide.

$$\text{Rate} = k[\text{alkyne}][\text{HX}]^2$$

This third-order rate dependence suggests a termolecular transition state, one that involves two molecules of the hydrogen halide. Figure 9.5 depicts such a termolecular process using curved arrow notation to show the flow of electrons, and dashed-line notation to indicate the bonds being made and broken at the transition state. This mechanism, called Ad_E3 for *addition-electrophilic-termolecular,* avoids the formation of a very unstable alkenyl cation intermediate by invoking nucleophilic participation by the halogen at an early stage. Nevertheless, since Markovnikov's rule is observed, it seems likely that some degree of positive character develops at carbon and controls the regioselectivity of addition.

For further discussion of this topic, see the article "The Electrophilic Addition to Alkynes" in the November 1993 edition of the *Journal of Chemical Education* (p. 873).

In the presence of excess hydrogen halide, geminal dihalides are formed by sequential addition of two molecules of hydrogen halide to the carbon-carbon triple bond.

$$\underset{\text{Alkyne}}{RC{\equiv}CR'} \xrightarrow{HX} \underset{\text{Alkenyl halide}}{RCH{=}\underset{X}{C}R'} \xrightarrow{HX} \underset{\text{Geminal dihalide}}{RCH_2\overset{X}{\underset{X}{C}}R'}$$

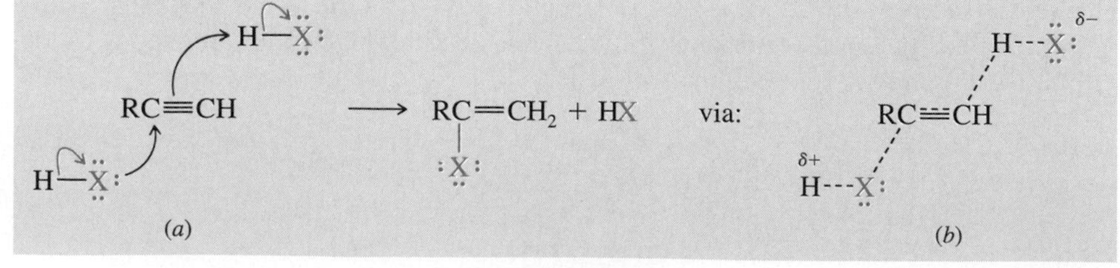

FIGURE 9.5 Curved arrow notation (*a*), and transition-state representation (*b*) for electrophilic addition of a hydrogen halide HX to an alkyne.

The hydrogen halide adds to the initially formed alkenyl halide in accordance with Markovnikov's rule. Overall, both protons become bonded to the same carbon and both halogens to the adjacent carbon.

$$\underset{\text{3-Hexyne}}{CH_3CH_2C{\equiv}CCH_2CH_3} + \underset{\text{Hydrogen fluoride}}{2HF} \longrightarrow \underset{\text{3,3-Difluorohexane (76\%)}}{CH_3CH_2CH_2CF_2CH_2CH_3}$$

PROBLEM 9.13 Write a series of equations showing how you could prepare 1,1-dichloroethane from

(*a*) Ethylene
(*b*) Vinyl chloride ($CH_2{=}CHCl$)
(*c*) Vinyl bromide ($CH_2{=}CHBr$)
(*d*) 1,1-Dibromoethane

SAMPLE SOLUTION (*a*) Reasoning backward, we recognize 1,1-dichloroethane as the product of addition of two molecules of hydrogen chloride to acetylene. Thus, the synthesis requires converting ethylene to acetylene as a key feature. As described in Section 9.8, this may be accomplished by conversion of ethylene to a vicinal dihalide, followed by double dehydrohalogenation. A suitable synthesis based on this analysis is as shown:

$$\underset{\text{Ethylene}}{CH_2{=}CH_2} \xrightarrow{Br_2} \underset{\text{1,2-Dibromoethane}}{BrCH_2CH_2Br} \xrightarrow[\text{2. } H_2O]{\text{1. } NaNH_2} \underset{\text{Acetylene}}{HC{\equiv}CH} \xrightarrow{2HCl} \underset{\text{1,1-Dichloroethane}}{CH_3CHCl_2}$$

Hydrogen bromide (but not hydrogen chloride or hydrogen iodide) adds to alkynes by a free-radical mechanism when peroxides are present in the reaction mixture. As in the free-radical addition of hydrogen bromide to alkenes (Section 6.8), a regioselectivity opposite to Markovnikov's rule is observed.

$$\underset{\text{1-Hexyne}}{CH_3CH_2CH_2CH_2C{\equiv}CH} + \underset{\text{Hydrogen bromide}}{HBr} \xrightarrow{\text{peroxides}} \underset{\text{1-Bromo-1-hexene (79\%)}}{CH_3CH_2CH_2CH_2CH{=}CHBr}$$

9.13 HYDRATION OF ALKYNES

By analogy to the hydration of alkenes, adding the elements of water to a triple bond is expected to yield an alcohol. The kind of alcohol produced by hydration of an alkyne, however, is of a special kind, one in which the hydroxyl group is a substituent on a carbon-carbon double bond. This type of alcohol is called an **enol** (the double bond suffix *-ene* plus the alcohol suffix *-ol*). An important property of enols is their rapid isomerization to aldehydes or ketones under the conditions of their formation.

$$\underset{\text{Alkyne}}{RC{\equiv}CR'} + \underset{\text{Water}}{H_2O} \xrightarrow{\text{slow}} \underset{\substack{\text{Enol}\\\text{(not isolated)}}}{RCH{=}C(OH)R'} \xrightarrow{\text{fast}} \underset{\substack{R' = \text{H, aldehyde}\\R' = \text{alkyl, ketone}}}{RCH_2C({=}O)R'}$$

The process by which enols are converted to aldehydes or ketones is called *keto-enol isomerism* (or *keto-enol tautomerism*) and proceeds by the sequence of proton transfers shown in Figure 9.6. Proton transfer to the double bond of an enol occurs readily because the carbocation that is produced is a very stable one. The positive charge on carbon is stabilized by electron release from oxygen and may be represented in resonance terms as

$$\underset{\text{A}}{RCH_2-\overset{:\ddot{O}H}{\overset{|}{\underset{+}{C}}}R'} \longleftrightarrow \underset{\text{B}}{RCH_2-\overset{+\ddot{O}H}{\overset{\|}{C}}R'}$$

Delocalization of an oxygen lone pair stabilizes the cation. All the atoms in B have octets of electrons, making it a more stable structure than A. Only six electrons are associated with the positively charged carbon in A.

PROBLEM 9.14 Give the structure of the enol formed by hydration of 2-butyne, and write a series of equations showing its conversion to its corresponding ketone isomer.

In general, ketones are more stable than their enol precursors and are the products actually isolated when alkynes undergo acid-catalyzed hydration. The standard method by which alkyne hydration is carried out employs aqueous sulfuric acid as the

Overall Reaction:

$$\underset{\text{Enol}}{RCH{=}\overset{OH}{\overset{|}{C}}R'} \longrightarrow \underset{\substack{\text{Ketone}\\ \text{(aldehyde if R}' = \text{H)}}}{RCH_2-\overset{O}{\overset{\|}{C}}R'}$$

Step 1: The enol is formed in aqueous acidic solution. The first step of its transformation to a ketone is proton transfer to the carbon-carbon double bond.

$$\underset{\text{Hydronium ion}}{H_2\overset{+}{\ddot{O}}{-}H} + \underset{\text{Enol}}{RCH{=}\overset{:\ddot{O}H}{\overset{|}{C}}R'} \rightleftharpoons \underset{\text{Water}}{H_2\ddot{O}:} + \underset{\text{Carbocation}}{\underset{H}{\underset{|}{RCH}}-\overset{:\ddot{O}H}{\overset{|}{\underset{+}{C}}}R'}$$

Step 2: The carbocation transfers a proton from oxygen to a water molecule, yielding a ketone.

$$\underset{\text{Carbocation}}{RCH_2-\overset{:\ddot{O}-H}{\overset{|}{\underset{+}{C}}}R'} + \underset{\text{Water}}{:\ddot{O}H_2} \longrightarrow \underset{\text{Ketone}}{RCH_2\overset{:\ddot{O}}{\overset{\|}{C}}R'} + \underset{\text{Hydronium ion}}{H{-}\overset{+}{\ddot{O}}H_2}$$

FIGURE 9.6 Conversion of an enol to a ketone takes place by way of two solvent-mediated proton transfers. A proton is transferred to carbon in the first step, then removed from oxygen in the second.

reaction medium and mercury(II) sulfate or mercury(II) oxide as a catalyst. Because alkynes possess only limited solubility in aqueous sulfuric acid, methanol or acetic acid is often added as a cosolvent.

Mercury(II) sulfate and mercury(II) oxide are also known as *mercuric* sulfate and oxide, respectively.

$$CH_3CH_2CH_2C{\equiv}CCH_2CH_2CH_3 + H_2O \xrightarrow{H^+,\ Hg^{2+}} CH_3CH_2CH_2CH_2\overset{O}{\overset{\|}{C}}CH_2CH_2CH_3$$

4-Octyne → 4-Octanone (89%)

Hydration of alkynes follows Markovnikov's rule; terminal alkynes yield methyl-substituted ketones.

$$HC{\equiv}CCH_2CH_2CH_2CH_2CH_2CH_3 + H_2O \xrightarrow[HgSO_4]{H_2SO_4} CH_3\overset{O}{\overset{\|}{C}}CH_2CH_2CH_2CH_2CH_2CH_3$$

1-Octyne → 2-Octanone (91%)

PROBLEM 9.15 Show by a series of equations how you could prepare 2-octanone from acetylene and any necessary organic or inorganic reagents. How could you prepare 4-octanone?

Because of the regioselectivity of alkyne hydration, acetylene is the only alkyne structurally capable of yielding an aldehyde under these conditions.

$$HC{\equiv}CH + H_2O \longrightarrow CH_2{=}CHOH \longrightarrow CH_3\overset{O}{\overset{\|}{C}}H$$

Acetylene, Water → Vinyl alcohol (not isolated) → Acetaldehyde

At one time acetaldehyde was prepared on an industrial scale by this method. More modern methods involve direct oxidation of ethylene and are more economical.

9.14 ADDITION OF HALOGENS TO ALKYNES

Alkynes react with chlorine and bromine to yield tetrahaloalkanes. Two molecules of the halogen add to the triple bond.

$$RC{\equiv}CR' + 2X_2 \longrightarrow R\overset{X}{\underset{X}{C}}-\overset{X}{\underset{X}{C}}R'$$

Alkyne + Halogen (chlorine or bromine) → Tetrahaloalkane

$$CH_3C{\equiv}CH + 2Cl_2 \longrightarrow CH_3\overset{Cl}{\underset{Cl}{C}}CHCl_2$$

Propyne + Chlorine → 1,1,2,2-Tetrachloropropane (63%)

A dihaloalkene is an intermediate and is the isolated product when the alkyne and the halogen are present in equimolar amounts. The stereochemistry of addition is anti.

$$CH_3CH_2C{\equiv}CCH_2CH_3 + Br_2 \longrightarrow (CH_3CH_2)(Br)C{=}C(Br)(CH_2CH_3)$$

3-Hexyne — Bromine — (*E*)-3,4-Dibromo-3-hexene (90%)

In contrast to the addition of hydrogen halides to alkynes and to their acid-catalyzed hydration, reactions that take place at rates only slightly slower than the corresponding reactions of alkenes, the addition of halogens to carbon-carbon triple bonds proceeds far more slowly than halogen addition to alkenes. This is presumed to be due to the less stable nature of cyclic halonium ions formed from alkynes compared with those formed from alkenes.

Cyclic halonium ion formed by addition to double bond (more stable)

Cyclic halonium ion formed by addition to triple bond (more strained, less stable)

9.15 OZONOLYSIS OF ALKYNES

Carboxylic acids are produced when alkynes are subjected to ozonolysis.

$$RC{\equiv}CR' \xrightarrow[2.\ H_2O]{1.\ O_3} RCOH + HOCR'$$

(each product containing C=O)

$$CH_3CH_2CH_2CH_2C{\equiv}CH \xrightarrow[2.\ H_2O]{1.\ O_3} CH_3CH_2CH_2CH_2CO_2H + HOCOH$$

1-Hexyne — Pentanoic acid (51%) — Carbonic acid

Recall that when carbonic acid is formed as a reaction product, it dissociates to carbon dioxide and water.

Ozonolysis is sometimes used as a tool in structure determination. By identifying the carboxylic acids produced, we can deduce the structure of the alkyne. As with many other chemical methods of structure determination, however, it has been superseded by spectroscopic methods.

PROBLEM 9.16 A certain hydrocarbon had the molecular formula $C_{16}H_{26}$ and contained two triple bonds. Ozonation followed by hydrolysis gave $CH_3(CH_2)_4CO_2H$ and $HO_2CCH_2CH_2CO_2H$ as the only products. Suggest a reasonable structure for this hydrocarbon.

9.16 SUMMARY

Alkynes are hydrocarbons that contain a carbon-carbon *triple bond.* Simple alkynes having no other functional groups or rings are represented by the general formula C_nH_{2n-2}. Acetylene ($HC{\equiv}CH$) is the simplest alkyne. It is prepared by heating hydrocarbons, especially methane or ethane, at temperatures of 1150°C or more (Section 9.1).

Alkynes are named in a manner analogous to alkenes, using the suffix *-yne* in place of *-ene* (Section 9.2).

Acetylene is a linear molecule, and alkynes have a linear geometry of their X—C≡C—Y units. The carbon-carbon triple bond in alkynes is composed of a σ

TABLE 9.2
Preparation of Alkynes

Reaction (section) and comments	General equation and specific example
Alkylation of acetylene and terminal alkynes (Section 9.7) The acidity of acetylene and terminal alkynes permits them to be converted to their conjugate bases on treatment with sodium amide. These anions are good nucleophiles and react with primary alkyl halides to form carbon-carbon bonds. Secondary and tertiary alkyl halides cannot be used, because they yield only elimination products under these conditions.	$RC{\equiv}CH + NaNH_2 \longrightarrow RC{\equiv}CNa + NH_3$ Alkyne; Sodium amide; Sodium alkynide; Ammonia $RC{\equiv}CNa + R'CH_2X \longrightarrow RC{\equiv}CCH_2R' + NaX$ Sodium alkynide; Primary alkyl halide; Alkyne; Sodium halide $(CH_3)_3CC{\equiv}CH \xrightarrow[2.\ CH_3I]{1.\ NaNH_2,\ NH_3} (CH_3)_3CC{\equiv}CCH_3$ 3,3-Dimethyl-1-butyne → 4,4-Dimethyl-2-pentyne (96%)
Double dehydrohalogenation of geminal dihalides (Section 9.8) An E2 elimination reaction of a geminal dihalide yields an alkenyl halide. If a strong enough base is used, sodium amide, for example, a second elimination step follows the first and the alkenyl halide is converted to an alkyne.	H X RC—CR' + $2NaNH_2 \longrightarrow RC{\equiv}CR' + 2NaX$ H X Geminal dihalide; Sodium amide; Alkyne; Sodium halide $(CH_3)_3CCH_2CHCl_2 \xrightarrow[2.\ H_2O]{1.\ 3NaNH_2,\ NH_3} (CH_3)_3CC{\equiv}CH$ 1,1-Dichloro-3,3-dimethylbutane → 3,3-Dimethyl-1-butyne (56–60%)
Double dehydrohalogenation of vicinal dihalides (Section 9.8) Dihalides in which the halogens are on adjacent carbons undergo two elimination processes analogous to those of geminal dihalides.	H H RC—CR' + $2NaNH_2 \longrightarrow RC{\equiv}CR' + 2NaX$ X X Vicinal dihalide; Sodium amide; Alkyne; Sodium halide $CH_3CH_2CHCH_2Br$ (Br on C-2) $\xrightarrow[2.\ H_2O]{1.\ 3NaNH_2,\ NH_3} CH_3CH_2C{\equiv}CH$ 1,2-Dibromobutane → 1-Butyne (78–85%)

and two π components (Section 9.3). The triply bonded carbons are *sp* hybridized. The σ component of the triple bond contains two electrons in an orbital generated by the overlap of *sp* hybridized orbitals on adjacent carbons. Each of these carbons also has two $2p$ orbitals, which overlap in pairs so as to give two π orbitals, each of which contains two electrons.

The linear geometry of the alkyne structural unit limits stable **cycloalkynes** to structures in which the ring has at least eight carbon atoms (Section 9.4).

Alkynes are similar to alkanes and alkenes in their physical properties (Section 9.5).

The order of electronegativity among the various hybridization states of carbon is:

$$sp^3 < sp^2 < sp$$

Least electronegative — Most electronegative

Since an *sp* hybridized carbon is more electronegative than an sp^2 or sp^3 hybridized carbon and can better bear a negative charge, acetylene and terminal alkynes are more *acidic* than other hydrocarbons (Section 9.6). They have K_a's for ionization of approximately 10^{-26}, compared with about 10^{-45} for alkenes and about 10^{-60} for alkanes.

Methods for the preparation of alkynes are summarized in Table 9.2. They include

1. Alkylation of acetylene and terminal alkynes (Section 9.7)
2. Double dehydrohalogenation of *geminal* dihalides (Section 9.8)
3. Double dehydrohalogenation of *vicinal* dihalides (Section 9.8)

Most reactions of alkynes (Section 9.9) are analogous to those of alkenes. They are summarized in Tables 9.3 and 9.4.

PROBLEMS

9.17 Write structural formulas and give the IUPAC names for all the alkynes of molecular formula C_6H_{10}.

9.18 Provide a systematic IUPAC name for each of the following alkynes:

(*a*) $CH_3CH_2CH_2C{\equiv}CH$

(*b*) $CH_3CH_2C{\equiv}CCH_3$

(*c*) $CH_3C{\equiv}CCHCH(CH_3)_2$ (with CH_3 on the CH carbon)

(*d*) cyclopropyl$-CH_2CH_2CH_2C{\equiv}CH$

(*e*) cycloalkyne ring: $CH_2C{\equiv}CCH_2$ within a ring

(*f*) $CH_3CH_2CH_2CH_2CHCH_2CH_2CH_2CH_2CH_3$ (with $C{\equiv}CCH_3$ on the CH carbon)

(*g*) $(CH_3)_3CC{\equiv}CC(CH_3)_3$

9.19 Write a structural formula corresponding to each of the following:

(*a*) 1-Octyne

(*b*) 2-Octyne

(*c*) 3-Octyne

(*d*) 4-Octyne

(*e*) 2,5-Dimethyl-3-hexyne

(*f*) 4-Ethyl-1-hexyne

(*g*) Ethynylcyclohexane

(*h*) 3-Ethyl-3-methyl-1-pentyne

9.20 All the compounds in Problem 9.19 are isomers except one. Which one?

9.21 Write structural formulas for all the alkynes of molecular formula C_8H_{14} that yield 3-ethylhexane on catalytic hydrogenation.

TABLE 9.3
Conversion of Alkynes to Alkenes and Alkanes

Reaction (section) and comments	General equation and specific example
Hydrogenation of alkynes to alkanes (Section 9.10) Alkynes are completely hydrogenated, yielding alkanes, in the presence of the customary metal hydrogenation catalysts.	$RC{\equiv}CR' + 2H_2 \xrightarrow{\text{metal catalyst}} RCH_2CH_2R'$ Alkyne, Hydrogen, Alkane 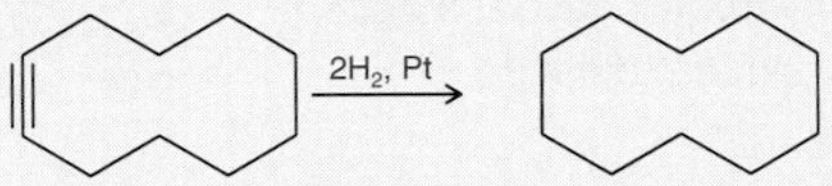 Cyclodecyne $\xrightarrow{2H_2,\ Pt}$ Cyclodecane (71%)
Semihydrogenation of alkynes to alkenes (Section 9.10) Hydrogenation of alkynes may be halted at the alkene stage by using special catalysts. Lindlar palladium is the metal catalyst employed most often. Hydrogenation occurs with syn selectivity and yields a cis alkene.	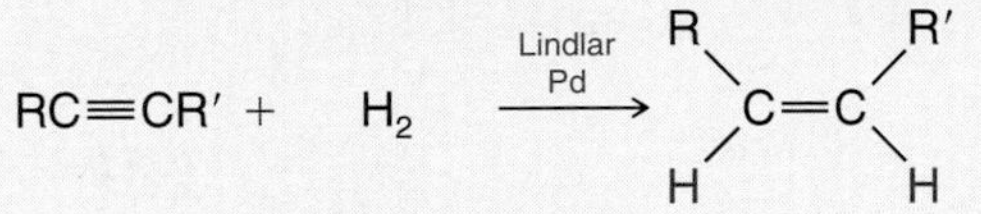 $RC{\equiv}CR' + H_2 \xrightarrow{\text{Lindlar Pd}}$ (R)(H)C=C(R')(H) Alkyne, Hydrogen, Cis alkene 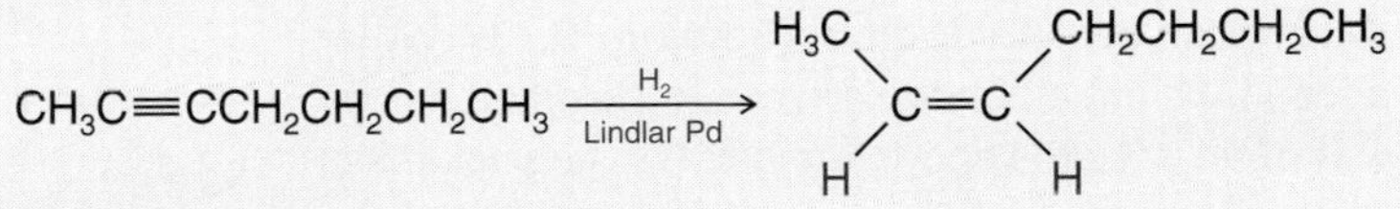 $CH_3C{\equiv}CCH_2CH_2CH_2CH_3 \xrightarrow[\text{Lindlar Pd}]{H_2}$ (H_3C)(H)C=C($CH_2CH_2CH_2CH_3$)(H) 2-Heptyne, *cis*-2-Heptene (59%)
Metal-ammonia reduction (Section 9.11) Group I metals—sodium is the one usually employed—in liquid ammonia as the solvent convert alkynes to trans alkenes. The reaction proceeds by a four-step sequence in which electron-transfer and proton-transfer steps alternate.	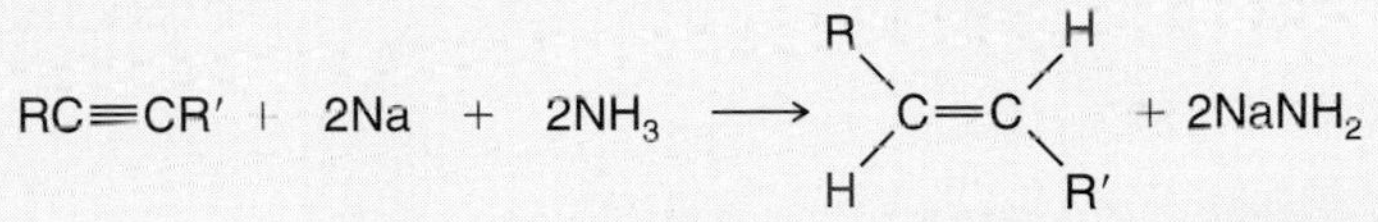 $RC{\equiv}CR' + 2Na + 2NH_3 \longrightarrow$ (R)(H)C=C(H)(R') $+ 2NaNH_2$ Alkyne, Sodium, Ammonia, Trans alkene, Sodium amide 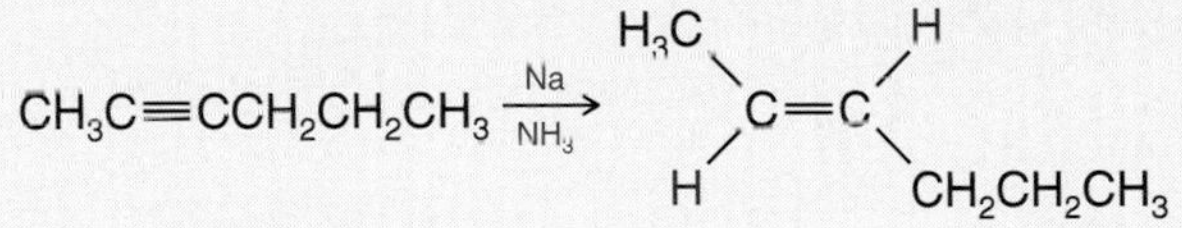 $CH_3C{\equiv}CCH_2CH_2CH_3 \xrightarrow[NH_3]{Na}$ (H_3C)(H)C=C(H)($CH_2CH_2CH_3$) 2-Hexyne, *trans*-2-Hexene (69%)

9.22 An unknown acetylenic amino acid obtained from the seed of a tropical fruit has the molecular formula $C_7H_{11}NO_2$. On catalytic hydrogenation over platinum this amino acid yielded homoleucine (an amino acid of known structure shown here) as the only product. What is the structure of the unknown amino acid?

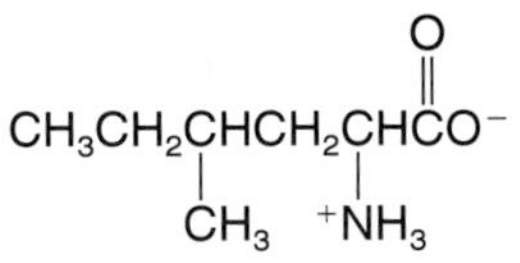

$$CH_3CH_2CH(CH_3)CH_2CH(\overset{+}{N}H_3)\overset{O}{\overset{\|}{C}}O^-$$ Homoleucine

9.23 Show by writing appropriate chemical equations how each of the following compounds could be converted to 1-hexyne:

(*a*) 1,1-Dichlorohexane
(*b*) 1-Hexene

TABLE 9.4
Electrophilic Addition to Alkynes

Reaction (section) and comments	General equation and specific example
Addition of hydrogen halides (Section 9.12) Hydrogen halides add to alkynes in accordance with Markovnikov's rule to give alkenyl halides. In the presence of 2 equivalents of hydrogen halide, a second addition occurs to give a geminal dihalide.	$RC{\equiv}CR' \xrightarrow{HX} RCH{=}CXR' \xrightarrow{HX} RCH_2CX_2R'$ Alkyne → Alkenyl halide → Geminal dihalide $CH_3C{\equiv}CH + 2HBr \longrightarrow CH_3CBr_2CH_3$ Propyne + Hydrogen bromide → 2,2-Dibromopropane (100%)
Acid-catalyzed hydration (Section 9.13) Water adds to the triple bond of alkynes to yield ketones by way of an unstable enol intermediate. The enol arises by Markovnikov hydration of the alkyne. Enol formation is followed by rapid isomerization of the enol to a ketone.	$RC{\equiv}CR' + H_2O \xrightarrow[Hg^{2+}]{H_2SO_4} RCH_2C({=}O)R'$ Alkyne + Water → Ketone $HC{\equiv}CCH_2CH_2CH_2CH_3 + H_2O \xrightarrow[HgSO_4]{H_2SO_4} CH_3C({=}O)CH_2CH_2CH_2CH_3$ 1-Hexyne + Water → 2-Hexanone (80%)
Halogenation (Section 9.14) Addition of 1 equivalent of chlorine or bromine to an alkyne yields a trans dihaloalkene. A tetrahalide is formed on addition of a second equivalent of the halogen.	$RC{\equiv}CR' \xrightarrow{X_2} (R)(X)C{=}C(X)(R')$ (trans) $\xrightarrow{X_2} RCX_2{-}CX_2R'$ Alkyne → Dihaloalkene → Tetrahaloalkane $CH_3C{\equiv}CH + 2Cl_2 \longrightarrow CH_3CCl_2CHCl_2$ Propyne + Chlorine → 1,1,2,2-Tetrachloropropane (63%)

(*c*) Acetylene
(*d*) 1-Iodohexane
(*e*) (*E*)-1-Bromo-1-hexene

9.24 Show by writing appropriate chemical equations how each of the following compounds could be converted to 3-hexyne:

(*a*) 1-Butene
(*b*) 1,1-Dichlorobutane
(*c*) 1-Chloro-1-butene
(*d*) Acetylene

9.25 When 1,2-dibromodecane was treated with potassium hydroxide in aqueous ethanol, it yielded a mixture of three isomeric compounds of molecular formula $C_{10}H_{19}Br$. Each of these compounds was converted to 1-decyne on reaction with sodium amide in dimethyl sulfoxide. Identify these three compounds.

9.26 Write the structure of the major organic product isolated from the reaction of 1-hexyne with

(*a*) Hydrogen (2 mol), platinum
(*b*) Hydrogen (1 mol), Lindlar palladium
(*c*) Lithium in liquid ammonia
(*d*) Sodium amide in liquid ammonia
(*e*) Product in part (*d*) treated with 1-bromobutane
(*f*) Product in part (*d*) treated with *tert*-butyl bromide
(*g*) Hydrogen chloride (1 mol)
(*h*) Hydrogen chloride (2 mol)
(*i*) Chlorine (1 mol)
(*j*) Chlorine (2 mol)
(*k*) Aqueous sulfuric acid, mercury(II) sulfate
(*l*) Ozone followed by hydrolysis

9.27 Write the structure of the major organic product isolated from the reaction of 3-hexyne with

(*a*) Hydrogen (2 mol), platinum
(*b*) Hydrogen (1 mol), Lindlar palladium
(*c*) Lithium in liquid ammonia
(*d*) Hydrogen chloride (1 mol)
(*e*) Hydrogen chloride (2 mol)
(*f*) Chlorine (1 mol)
(*g*) Chlorine (2 mol)
(*h*) Aqueous sulfuric acid, mercury(II) sulfate
(*i*) Ozone followed by hydrolysis

9.28 When 2-heptyne was treated with aqueous sulfuric acid containing mercury(II) sulfate, two products, each having the molecular formula $C_7H_{14}O$, were obtained in approximately equal amounts. What are these two compounds?

9.29 The alkane formed by hydrogenation of (*S*)-4-methyl-1-hexyne is optically active, while the one formed by hydrogenation of (*S*)-3-methyl-1-pentyne is not. Explain. Would you expect the products of semihydrogenation of these two compounds in the presence of Lindlar palladium to be optically active?

9.30 All the following reactions have been described in the chemical literature and proceed in good yield. In some cases the reactants are more complicated than those we have so far encountered. Nevertheless, on the basis of what you have already learned, you should be able to predict the principal product in each case.

(*a*) $NaC\equiv CH + ClCH_2CH_2CH_2CH_2CH_2CH_2I \longrightarrow$

(*b*) $BrCH_2\underset{\displaystyle Br}{\underset{|}{C}}HCH_2CH_2\underset{\displaystyle Br}{\underset{|}{C}}HCH_2Br \xrightarrow[\text{2. } H_2O]{\text{1. excess } NaNH_2,\ NH_3}$

(*c*) $\triangleright\!-\!C(Cl)_2CH_3 \xrightarrow[\text{heat}]{KOC(CH_3)_3,\ DMSO}$

(*d*) $C_6H_5C{\equiv}CNa + CH_3CH_2OS(=O)_2\!-\!C_6H_4\!-\!CH_3 \longrightarrow$

(*e*) Cyclodecyne $\xrightarrow[2.\ H_2O]{1.\ O_3}$

(*f*) 1-Ethynylcyclohexanol (cyclohexane bearing $C{\equiv}CH$ and OH on the same carbon) $\xrightarrow[2.\ H_2O]{1.\ O_3}$

(*g*) $CH_3CH(CH_3)CH_2C(OH)(CH_3)C{\equiv}CH \xrightarrow[HgO]{H_2O,\ H_2SO_4}$

(*h*) (*Z*)-$CH_3CH_2CH_2CH_2CH{=}CHCH_2(CH_2)_7C{\equiv}CCH_2CH_2OH \xrightarrow[2.\ H_2O]{1.\ Na,\ NH_3}$

(*i*) 2-[$O(CH_2)_8Cl$]-tetrahydropyran $+ NaC{\equiv}CCH_2CH_2CH_2CH_3 \longrightarrow$

(*j*) Product of part (*i*) $\xrightarrow[\text{Lindlar Pd}]{H_2}$

9.31 The ketone 2-heptanone has been identified as contributing to the odor of a number of dairy products, including condensed milk and cheddar cheese. Describe a synthesis of 2-heptanone from acetylene and any necessary organic or inorganic reagents.

$$CH_3C(=O)CH_2CH_2CH_2CH_2CH_3$$

2-Heptanone

9.32 (*Z*)-9-Tricosene [(*Z*)—$CH_3(CH_2)_7CH{=}CH(CH_2)_{12}CH_3$] is the sex pheromone of the female housefly. Synthetic (*Z*)-9-tricosene is used as bait to lure male flies to traps that contain insecticide. Using acetylene and alcohols of your choice as starting materials, along with any necessary inorganic reagents, show how you could prepare (*Z*)-9-tricosene.

9.33 Show by writing a suitable series of equations how you could prepare each of the following compounds from the designated starting materials and any necessary organic or inorganic reagents:

(*a*) 2,2-Dibromopropane from 1,1-dibromopropane
(*b*) 2,2-Dibromopropane from 1,2-dibromopropane
(*c*) 2-Chloropropene from 1-bromopropene
(*d*) 1,1,2,2-Tetrachloropropane from 1,2-dichloropropane
(*e*) 2,2-Diiodobutane from acetylene and ethyl bromide
(*f*) 1-Hexene from 1-butene and acetylene

(*g*) Decane from 1-butene and acetylene

(*h*) Cyclopentadecyne from cyclopentadecene

(*i*) 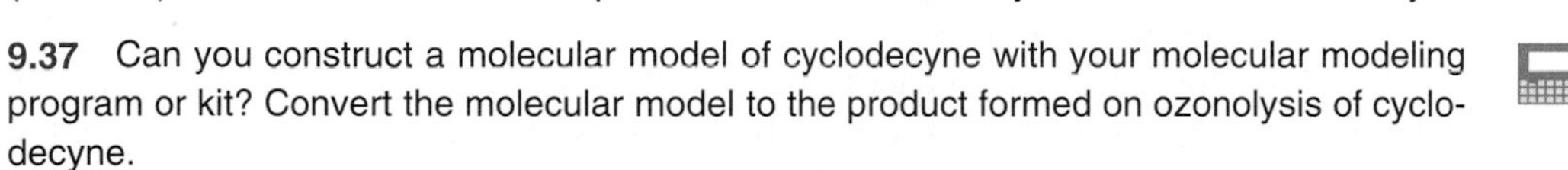

(*j*) *meso*-2,3-Dibromobutane from 2-butyne

9.34 Assume that you need to prepare 4-methyl-2-pentyne and discover that the only alkynes on hand are acetylene and propyne. You also have available methyl iodide, isopropyl bromide, and 1,1-dichloro-3-methylbutane. Which of these compounds would you choose in order to perform your synthesis, and how would you carry it out?

9.35 Compound A has the molecular formula $C_{14}H_{25}Br$ and was obtained by reaction of sodium acetylide with 1,12-dibromododecane. On treatment of compound A with sodium amide, it was converted to compound B ($C_{14}H_{24}$). Ozonolysis of compound B gave the diacid $HO_2C(CH_2)_{12}CO_2H$. Catalytic hydrogenation of compound B over Lindlar palladium gave compound C ($C_{14}H_{26}$), while hydrogenation over platinum gave compound D ($C_{14}H_{28}$). Sodium-ammonia reduction of compound B gave compound E ($C_{14}H_{26}$). Both C and E yielded $O{=}CH(CH_2)_{12}CH{=}O$ on ozonolysis. Assign structures to compounds A through E so as to be consistent with the observed transformations.

MOLECULAR MODELING EXERCISES

9.36 The melting point of 2-butyne (−32°C) is much higher than that of 1-butyne (−126°C). Based on the overall shape of each molecule, can you think of a reason why?

9.37 Can you construct a molecular model of cyclodecyne with your molecular modeling program or kit? Convert the molecular model to the product formed on ozonolysis of cyclodecyne.

9.38 Using molecular models compare the substitutents $—C{\equiv}CH$, $—CH{=}CH_2$, and $—CH_2CH_3$ with respect to their preference for an equatorial orientation when attached to a cyclohexane ring. One of these groups is very much different from the other two. Which one? Why?

9.39 Construct a model of a molecule that contains three carbons, only one of which is *sp* hybridized. What is its molecular formula? Is it an alkyne? What must be the hybridization state of the other two carbons? (You will learn more about compounds of this type in Chapter 10).

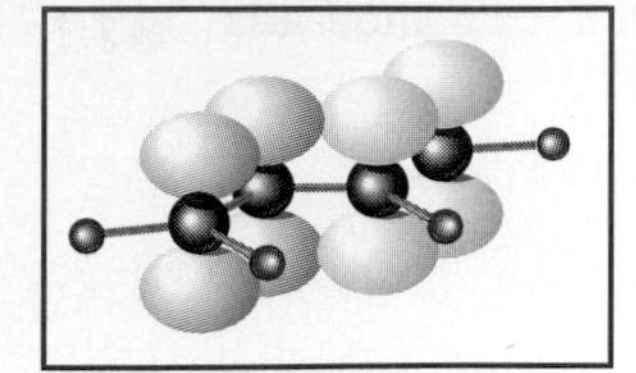

CHAPTER 10

CONJUGATION IN ALKADIENES AND ALLYLIC SYSTEMS

Not all the properties of alkenes can be understood by focusing exclusively on the functional group behavior of the double bond. A double bond can affect the properties of a second functional unit to which it is directly attached. It can be a substituent, for example, on a positively charged carbon in an **allylic carbocation,** or on a carbon that bears an unpaired electron in an **allylic free radical,** or it can be a substituent on a second double bond in a **conjugated diene.**

Allylic carbocation | Allylic free radical | Conjugated diene

Conjugare is a Latin verb meaning "to link or yoke together," and allylic carbocations, allylic free radicals, and conjugated dienes are all examples of **conjugated systems.** The concept of the functional group has served us well to this point as a device to organize patterns of organic chemical reactivity. In this chapter we consider how conjugation permits two functional units within a molecule to display a kind of reactivity that is qualitatively different from that of either unit alone.

Allyl is derived from the botanical name for garlic *(Allium sativum).* It was found in 1892 that the major component obtained by distilling garlic oil is $CH_2{=}CHCH_2SSCH_2CH{=}CH_2$, and the word *allyl* was coined for the $CH_2{=}CHCH_2—$ group on the basis of this origin.

10.1 THE ALLYL GROUP

The group $CH_2{=}CHCH_2—$ is known as **allyl,** which is both a common name and a permissible IUPAC name. It is most often encountered in functionally substituted derivatives, and the following compounds containing this group are much better known by their radicofunctional names than by their substitutive names:

$CH_2{=}CHCH_2OH$	$CH_2{=}CHCH_2Cl$	$CH_2{=}CHCH_2Br$
Allyl alcohol (2-propen-1-ol)	Allyl chloride (3-chloro-1-propene)	Allyl bromide (3-bromo-1-propene)

The adjective *allylic* denotes the structural unit C=C—C. The sp^3 hybridized carbon of an allylic unit is called the **allylic carbon,** and an **allylic substituent** is one that is attached to an allylic carbon. Conversely, the sp^2 hybridized carbons of a carbon-carbon double bond are called **vinylic carbons,** and substituents attached to either one of them are referred to as **vinylic substituents.**

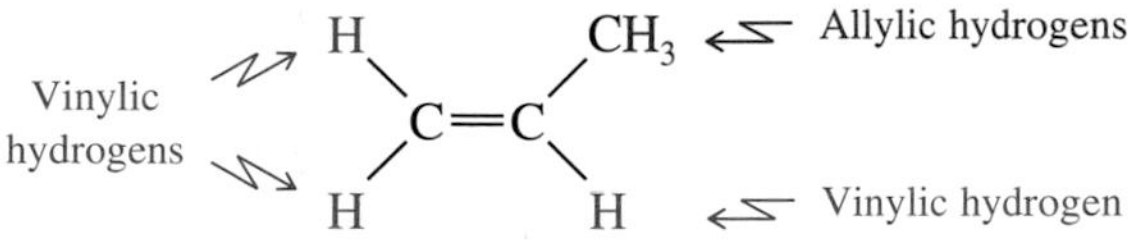

$$(H)(H)C{=}C(CH_3)(H)$$

Vinylic hydrogens → the two H on the left carbon; Allylic hydrogens → CH_3; Vinylic hydrogen → H on the right carbon

Allylic is often used as a generic term to refer to a molecule which bears a functional group at an allylic position. Thus, the following compounds represent an *allylic alcohol* and an *allylic chloride,* respectively. They are named according to the IUPAC rules of systematic nomenclature.

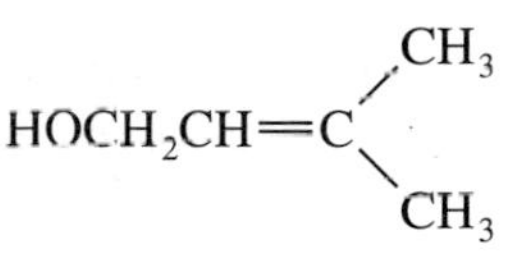

$$HOCH_2CH{=}C(CH_3)_2 \qquad CH_2{=}CHCCl(CH_3)_2$$

3-Methyl-2-buten-1-ol 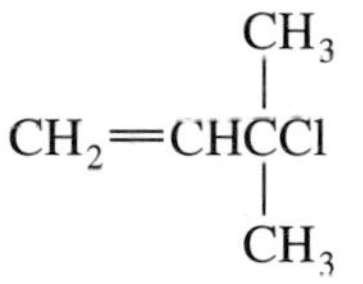 3-Chloro-3-methyl-1-butene

10.2 ALLYLIC CARBOCATIONS

Allylic carbocations are carbocations that have a vinyl group or substituted vinyl group as a substituent on their positively charged carbon. The allyl cation is the simplest allylic carbocation.

Representative allylic carbocations

$CH_2{=}CHCH_2^+$	$CH_3CH{=}CH\overset{+}{C}HCH_3$	(cyclopentene ring with + on ring carbon)
Allyl cation	1-Methyl-2-butenyl cation	2-Cyclopentenyl cation

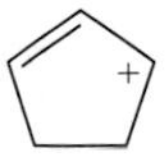

A substantial body of evidence indicates that allylic carbocations are more stable than simple alkyl cations. Compare, for example, the first-order rate constant k for the solvolysis of a typical tertiary alkyl chloride with that for solvolysis of a chloride that is both tertiary and allylic.

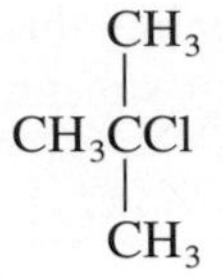

$$\underset{\underset{CH_3}{|}}{\overset{\overset{CH_3}{|}}{CH_3CCl}} \qquad \underset{\underset{CH_3}{|}}{\overset{\overset{CH_3}{|}}{CH_2{=}CHCCl}}$$

tert-Butyl chloride
Less reactive: *k*(rel) 1.0

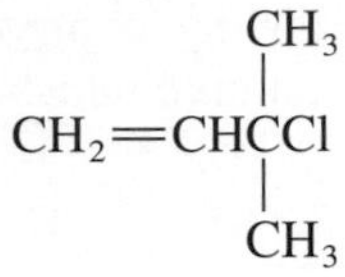

3-Chloro-3-methyl-1-butene
More reactive: *k*(rel) 123

The allylic chloride 3-chloro-3-methyl-1-butene is over 100 times more reactive toward ethanolysis (at 45°C) than is *tert*-butyl chloride. Both compounds undergo ethanolysis by an S_N1 mechanism, and their relative rates reflect their activation energies for carbocation formation. Since the allylic chloride is more reactive, we infer that it ionizes more rapidly because it forms a more stable carbocation. Structurally, the two carbocations differ in that the allylic carbocation has a vinyl substituent on its positively charged carbon in place of one of the methyl groups of *tert*-butyl cation.

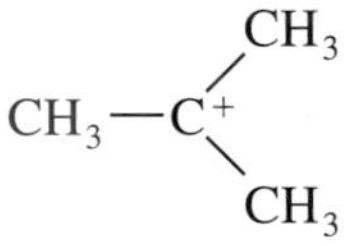

$$CH_3{-}\overset{+}{C}(CH_3)_2 \qquad CH_2{=}CH{-}\overset{+}{C}(CH_3)_2$$

tert-Butyl cation
(less stable)

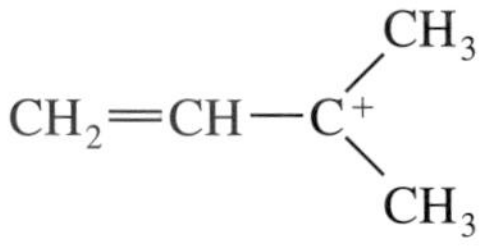

1,1-Dimethylallyl
cation
(more stable)

A rule of thumb is that a C=C substituent stabilizes a carbocation about as well as two alkyl groups. Although allyl cation ($CH_2{=}CHCH_2^+$) is a primary carbocation, it is about as stable as a typical secondary carbocation such as isopropyl cation, $(CH_3)_2CH^+$.

A vinyl group stabilizes a carbocation more than does a methyl group. Why?

A vinyl group is an extremely effective electron-releasing substituent. A resonance interaction of the type shown permits the π electrons of the double bond to be delocalized and disperses the positive charge.

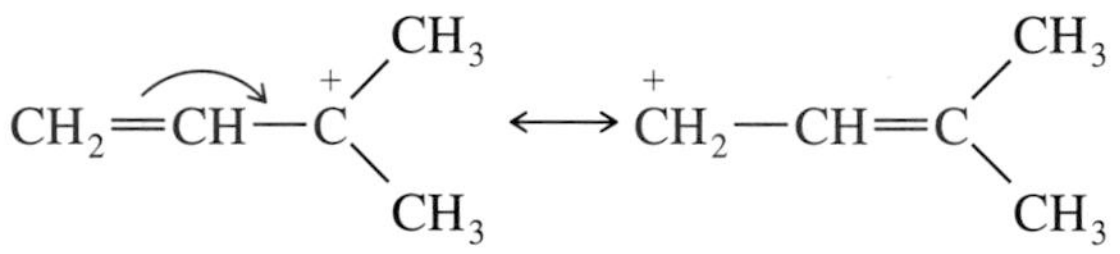

$$CH_2{=}CH{-}\overset{+}{C}(CH_3)_2 \longleftrightarrow \overset{+}{C}H_2{-}CH{=}C(CH_3)_2$$

PROBLEM 10.1 Write a second resonance structure for each of the following carbocations:

(*a*) $CH_3CH{=}CHCH_2^+$ (*b*) $CH_2{=}\underset{\underset{CH_3}{|}}{C}CH_2^+$ (*c*) cyclohexyl cation (+ on ring carbon) $=C(CH_3)_2$

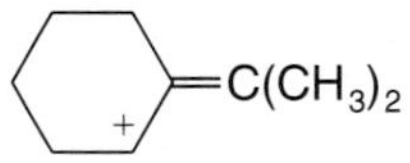

SAMPLE SOLUTION (*a*) When writing resonance forms of carbocations, electrons are moved in pairs from sites of high electron density toward the positively charged carbon.

$$CH_3CH{=}CH{-}\overset{+}{C}H_2 \longleftrightarrow CH_3\overset{+}{C}H{-}CH{=}CH_2$$

Electron delocalization in allylic carbocations is sometimes portrayed by using a dashed line to indicate sharing of a pair of π electrons by the three carbons. The structural formula is completed by appending a positive charge to the dashed-line symbol or by adding partial positive charges to the carbons at the end of the allylic system.

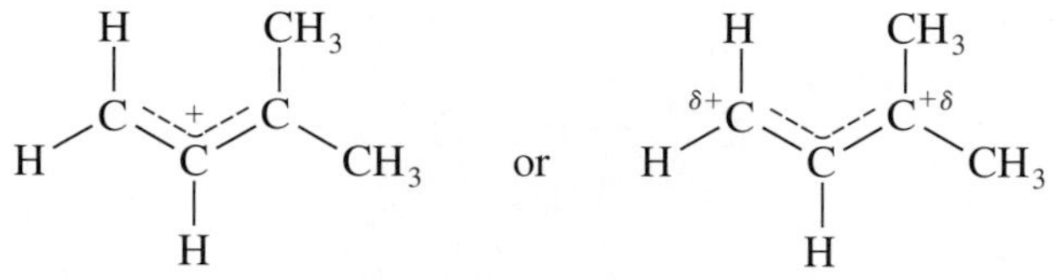

Two dashed-line representations of 1,1-dimethylallyl cation

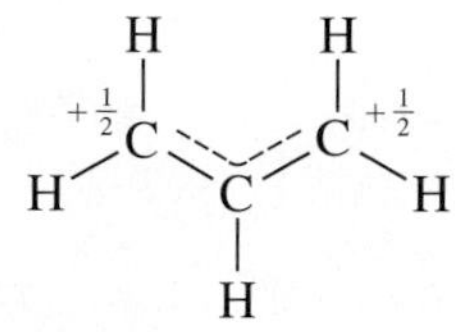

Allyl cation

In the case of the parent cation $CH_2{=}CH{-}CH_2^+$ (the rightmost representation) both the terminal carbons are equivalently substituted so that each bears exactly half of a positive charge. Notice that none of the positive charge is associated with the middle carbon of an allylic system.

An orbital overlap description of electron delocalization in 1,1-dimethylallyl cation $CH_2{=}CH{-}\overset{+}{C}(CH_3)_2$ is given in Figure 10.1. Figure 10.1*a* shows the π bond and the vacant *p* orbital as independent units. Figure 10.1*b* shows how the units can overlap to give an extended π orbital that encompasses all three carbons. This permits the two π electrons to be delocalized over three carbons and disperses the positive charge.

Since the positive charge in an allylic carbocation is shared by two carbons, there are two potential sites for attack by a nucleophile. Thus, hydrolysis of 3-chloro-3-methyl-1-butene gives a mixture of two allylic alcohols:

$$(CH_3)_2\underset{\displaystyle Cl}{C}CH{=}CH_2 \xrightarrow[Na_2CO_3]{H_2O} (CH_3)_2\underset{\displaystyle OH}{C}CH{=}CH_2 + (CH_3)_2C{=}CHCH_2OH$$

3-Chloro-3-methyl-1-butene; 2-Methyl-3-buten-2-ol (85%); 3-Methyl-2-buten-1-ol (15%)

Both alcohols are formed from the same carbocation. Water may react with the carbocation so as to give either a primary alcohol or a tertiary alcohol.

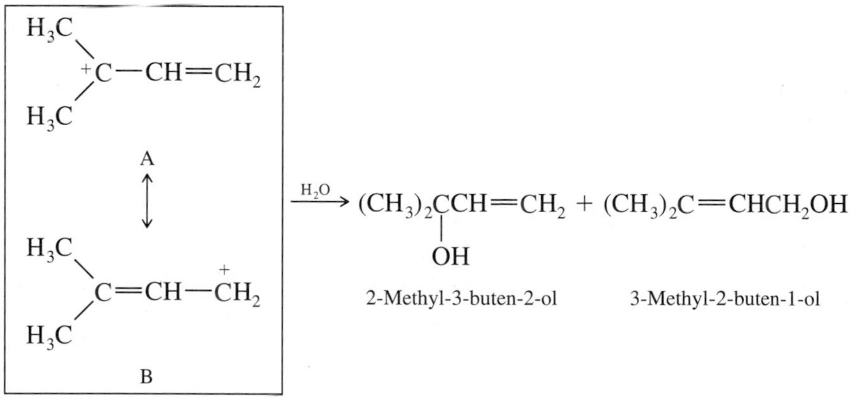

$$(H_3C)_2\overset{+}{C}{-}CH{=}CH_2 \text{ (A)} \longleftrightarrow (H_3C)_2C{=}CH{-}\overset{+}{C}H_2 \text{ (B)} \xrightarrow{H_2O} (CH_3)_2\underset{\displaystyle OH}{C}CH{=}CH_2 + (CH_3)_2C{=}CHCH_2OH$$

2-Methyl-3-buten-2-ol; 3-Methyl-2-buten-1-ol

It must be emphasized that we are not dealing with an equilibrium between two isomeric carbocations. *There is only one carbocation.* Its structure is not adequately represented by either of the individual resonance forms but is a hybrid having qualities of both of them. The carbocation has more of the character of A than B because resonance structure A is more stable than B. Water attacks faster at the tertiary carbon because it bears more of the positive charge.

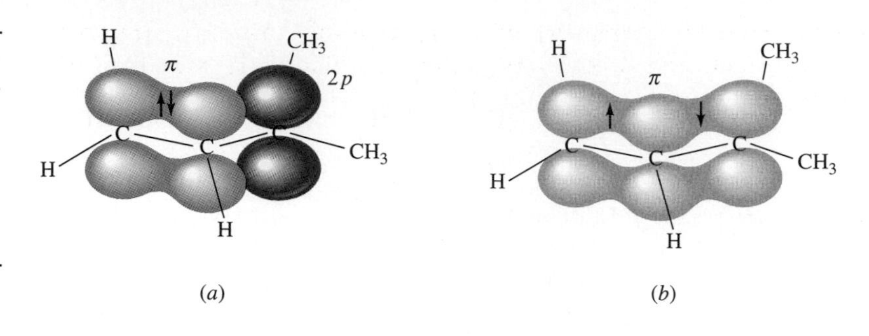

FIGURE 10.1 A representation of electron delocalization in an allylic carbocation. (*a*) The π orbital of the double bond, and the vacant 2*p* orbital of the positively charged carbon. (*b*) The overlap of the π orbital and the 2*p* orbital to give an extended π orbital that encompasses all three carbons. The two electrons in the π bond are delocalized over two carbons in (*a*) and over three carbons in (*b*).

The same two alcohols are formed in the hydrolysis of 1-chloro-3-methyl-2-butene:

$$\underset{\text{1-Chloro-3-methyl-2-butene}}{(CH_3)_2C{=}CHCH_2Cl} \xrightarrow[Na_2CO_3]{H_2O} \underset{\text{2-Methyl-3-buten-2-ol (85\%)}}{(CH_3)_2\underset{\displaystyle OH}{\underset{|}{C}}CH{=}CH_2} + \underset{\text{3-Methyl-2-buten-1-ol (15\%)}}{(CH_3)_2C{=}CHCH_2OH}$$

The carbocation formed on ionization of 1-chloro-3-methyl-2-butene is the same allylic carbocation as the one formed on ionization of 3-chloro-3-methyl-1-butene and gives the same mixture of products.

Reactions of allylic systems that yield products in which double-bond migration has occurred are said to have proceeded with **allylic rearrangement,** or by way of an **allylic shift.**

PROBLEM 10.2 From among the following compounds, choose the two that yield the same carbocation on ionization: 3-bromo-1-methylcyclohexene; 4-bromo-1-methylcyclohexene; 5-chloro-1-methylcyclohexene; 3-chloro-3-methylcyclohexene; 1-bromo-3-methylcyclohexene

---

Later in this chapter we will see how allylic carbocations are involved in electrophilic addition to dienes and how the principles developed in this section apply there as well.

10.3 ALLYLIC FREE RADICALS

Just as allyl cation is stabilized by electron delocalization, so is allyl radical:

$$H_2C{=}CH{-}\dot{C}H_2 \longleftrightarrow H_2\dot{C}{-}CH{=}CH_2 \quad \text{or} \quad H_2\overset{\frac{1}{2}\cdot}{C}{\cdots}CH{\cdots}\overset{\frac{1}{2}\cdot}{C}H_2$$

Allyl radical

Allyl radical is a conjugated system in which three electrons are delocalized over three carbons. The unpaired electron has an equal probability of being found at C-1 or C-3.

Reactions that generate allylic radicals occur more readily than those involving simple alkyl radicals. Compare the bond dissociation energies of the primary C—H bonds of propane and propene:

$$\underset{\text{Propane}}{CH_3CH_2CH_2{:}H} \longrightarrow \underset{\text{Propyl radical}}{CH_3CH_2\dot{C}H_2} + \underset{\text{Hydrogen atom}}{\cdot H} \qquad \Delta H° = +410 \text{ kJ/mol } (+98 \text{ kcal/mol})$$

$$\underset{\text{Propene}}{CH_2{=}CHCH_2{:}H} \longrightarrow \underset{\text{Allyl radical}}{CH_2{=}CH\dot{C}H_2} + \underset{\text{Hydrogen atom}}{\cdot H} \qquad \Delta H° = +368 \text{ kJ/mol } (+88 \text{ kcal/mol})$$

It requires less energy, by 42 kJ/mol (10 kcal/mol), to break a bond to a primary hydrogen atom in propene than in propane. The free radical produced from propene is allylic and stabilized by electron delocalization; the one from propane is not.

PROBLEM 10.3 Identify the allylic hydrogens in

(*a*) Cyclohexene
(*b*) 1-Methylcyclohexene
(*c*) 2,3,3-Trimethyl-1-butene
(*d*) 1-Octene

SAMPLE SOLUTION (*a*) Allylic hydrogens are bonded to an allylic carbon. An allylic carbon is an sp^3 hybridized carbon that is attached directly to an sp^2 hybridized carbon of an alkene. Cyclohexene has four allylic hydrogens.

These are allylic hydrogens

These are allylic hydrogens

These are vinylic hydrogens

10.4 ALLYLIC HALOGENATION

Of the reactions that involve carbon radicals, the most familiar are the chlorination and bromination of alkanes (Sections 4.16 through 4.20):

$$\underset{\text{Alkane}}{RH} + \underset{\text{Halogen}}{X_2} \xrightarrow[\text{light}]{\text{heat or}} \underset{\text{Alkyl halide}}{RX} + \underset{\text{Hydrogen halide}}{HX}$$

While alkenes typically react with chlorine and bromine by *addition* at room temperature and below (Section 6.14), *substitution* becomes competitive at higher temperatures, especially when the concentration of the halogen is low. When substitution does occur, it is highly regioselective for the allylic position. This forms the basis of an industrial preparation of allyl chloride:

$$\underset{\text{Propene}}{CH_2{=}CHCH_3} + \underset{\text{Chlorine}}{Cl_2} \xrightarrow{500°C} \underset{\substack{\text{Allyl chloride}\\(80-85\%)}}{CH_2{=}CHCH_2Cl} + \underset{\text{Hydrogen chloride}}{HCl}$$

The reaction proceeds by a free-radical chain mechanism, involving the following propagation steps:

$$\underset{\text{Propene}}{CH_2{=}CHCH_2{:}H} + \underset{\text{Chlorine atom}}{\cdot\ddot{\underset{..}{Cl}}:} \longrightarrow \underset{\text{Allyl radical}}{CH_2{=}CH\dot{C}H_2} + \underset{\text{Hydrogen chloride}}{H{:}\ddot{\underset{..}{Cl}}:}$$

$$\underset{\text{Allyl radical}}{CH_2{=}CH\dot{C}H_2} + \underset{\text{Chlorine}}{:\ddot{\underset{..}{Cl}}:\ddot{\underset{..}{Cl}}:} \longrightarrow \underset{\text{Allyl chloride}}{CH_2{=}CHCH_2\ddot{\underset{..}{Cl}}:} + \underset{\text{Chlorine atom}}{\cdot\ddot{\underset{..}{Cl}}:}$$

Allyl chloride is quite reactive toward nucleophilic substitutions, especially those that proceed by the S_N2 mechanism, and is used as a starting material in the synthesis of a variety of drugs and agricultural and industrial chemicals.

Allylic brominations are normally carried out by using one of a number of specialized reagents developed for that purpose. *N*-Bromosuccinimide (NBS) is the most frequently used of these reagents. An alkene is dissolved in carbon tetrachloride, *N*-bromosuccinimide is added, and the reaction mixture is heated, illuminated with a sunlamp, or both. The products are an allylic halide and succinimide.

N-Bromosuccinimide will be seen again as a reagent for regioselective bromination in Section 11.14.

$$\underset{\text{Cyclohexene}}{C_6H_{10}} + \underset{\substack{N\text{-Bromosuccinimide}\\(\text{NBS})}}{\text{NBr}} \xrightarrow[CCl_4]{\text{heat}} \underset{\substack{\text{3-Bromocyclohexene}\\(82-87\%)}}{C_6H_9Br} + \underset{\text{Succinimide}}{\text{NH}}$$

N-Bromosuccinimide provides a low concentration of molecular bromine, which reacts with alkenes by a mechanism analogous to that of other free-radical halogenations.

PROBLEM 10.4 Assume that *N*-bromosuccinimide serves as a source of Br_2, and write equations for the propagation steps in the formation of 3-bromocyclohexene by allylic bromination of cyclohexene.

While allylic brominations and chlorinations offer us a method to attach a reactive functional group to a hydrocarbon framework, we need to be aware of two important limitations. In order for allylic halogenation to be effective in a particular synthesis:

1. All the allylic hydrogens in the starting alkene must be equivalent.
2. Both resonance forms of the allylic radical must be equivalent.

In the two examples cited so far, the chlorination of propene and the bromination of cyclohexene, both criteria are met.

All the allylic hydrogens of propene (shown in red) are equivalent.

$$CH_2{=}CH{-}CH_3$$

The two resonance forms of allyl radical are equivalent.

$$CH_2{=}CH{-}\dot{C}H_2 \longleftrightarrow \dot{C}H_2{-}CH{=}CH_2$$

All the allylic hydrogens of cyclohexene (shown in red) are equivalent.

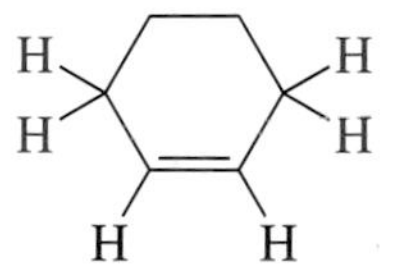

The two resonance forms of 2-cyclohexenyl radical are equivalent.

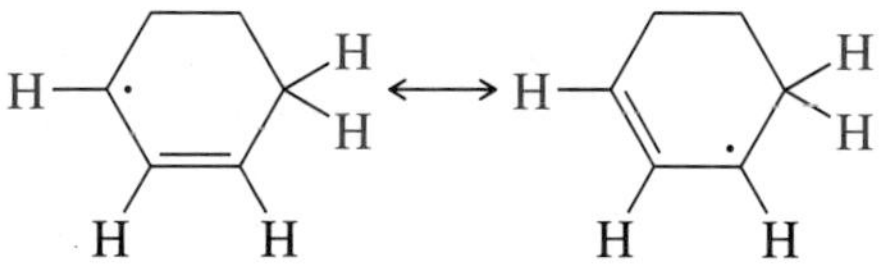

Unless both criteria are met, mixtures of constitutionally isomeric allylic halides result.

PROBLEM 10.5 The two alkenes 2,3,3-trimethyl-1-butene and 1-octene were each subjected to allylic halogenation with *N*-bromosuccinimide. One of these alkenes yielded a single allylic bromide, whereas the other gave a mixture of two constitutionally isomeric allylic bromides. Match the chemical behavior to the correct alkene and give the structure of the allylic bromide(s) formed from each.

10.5 CLASSES OF DIENES

Allylic carbocations and allylic radicals are conjugated systems that are involved as reactive intermediates in chemical reactions. The third type of conjugated system that we will examine, **conjugated dienes,** consists of stable molecules.

A hydrocarbon that contains two double bonds is called an **alkadiene,** and the relationship between the double bonds may be described as *isolated, conjugated,* or *cumulated.* **Isolated diene** units are those in which two carbon-carbon double bond units are separated from each other by one or more sp^3 hybridized carbon atoms. 1,4-Pentadiene and 1,5-cyclooctadiene have isolated double bonds:

$$CH_2{=}CHCH_2CH{=}CH_2$$

1,4-Pentadiene

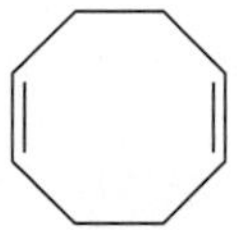

1,5-Cyclooctadiene

Conjugated dienes are those in which two carbon-carbon double bond units are directly connected to each other by a single bond. 1,3-Pentadiene and 1,3-cyclooctadiene are dienes that contain conjugated double bonds:

$$CH_2{=}CH{-}CH{=}CHCH_3$$

1,3-Pentadiene

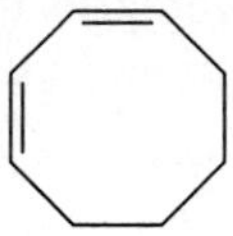

1,3-Cyclooctadiene

Cumulated dienes are those in which one carbon atom is common to two carbon-carbon double bonds. The simplest cumulated diene is 1,2-propadiene, also called *allene,* and compounds of this class are more usually called *allenes.*

$$CH_2{=}C{=}CH_2$$

1,2-Propadiene

(*Allene* is an acceptable IUPAC name for 1,2-propadiene.)

PROBLEM 10.6 Many naturally occurring substances contain several carbon-carbon double bonds: some isolated, some conjugated, and some cumulated. Identify the types of carbon-carbon double bonds found in each of the following substances:

(*a*) β-Springene (a scent substance obtained from the dorsal gland of springboks)

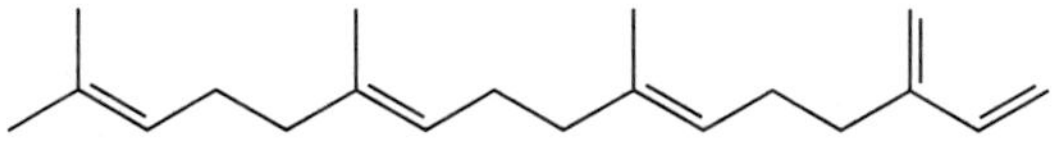

(*b*) Humulene (found in hops and oil of cloves)

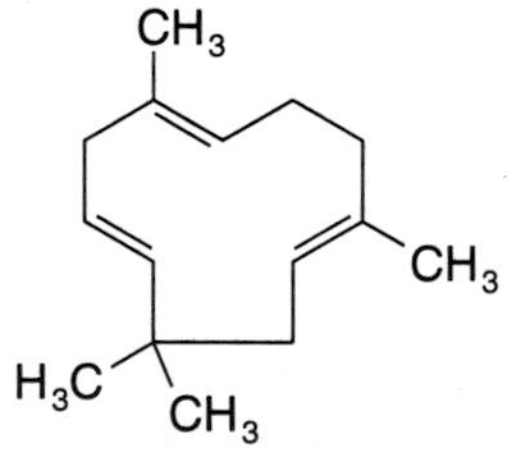

CH_3 CH_3 H_3C CH_3

(*c*) Cembrene (occurs in pine resin)

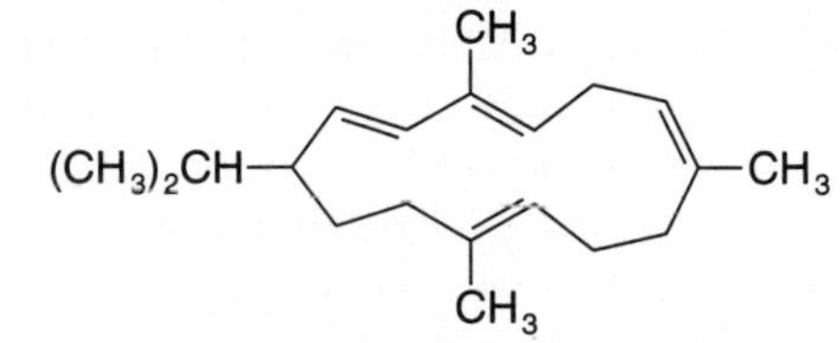

CH_3 $(CH_3)_2CH$ CH_3 CH_3

(*d*) The sex attractant of the male dried-bean beetle

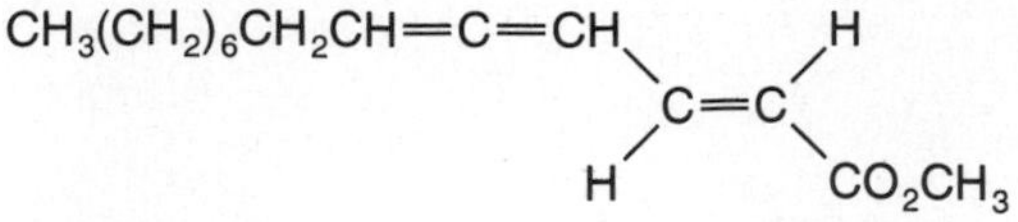

$$CH_3(CH_2)_6CH_2CH{=}C{=}CH\text{—}C(H){=}C(H)CO_2CH_3$$

SAMPLE SOLUTION (*a*) As indicated in the following structural formula, *β*-springene has three isolated double bonds and a pair of conjugated double bonds:

Conjugated

Isolated double bonds

Isolated double bonds are separated from other double bonds by at least one sp^3 hybridized carbon. Conjugated double bonds are joined by a single bond.

As may be apparent from some of the examples, alkadienes are named according to the IUPAC rules by replacing the *-ane* ending of an alkane with *-adiene* and locating the position of each double bond by number. In an analogous manner, compounds with three carbon-carbon double bonds are called *alkatrienes* and named accordingly, those with four double bonds are *alkatetraenes,* and so on.

10.6 RELATIVE STABILITIES OF ALKADIENES

Which is the most stable arrangement of double bonds in an alkadiene—isolated, conjugated, or cumulated?

As we have seen in Chapter 6, the stabilities of alkenes may be assessed by comparing their heats of hydrogenation. Figure 10.2 depicts the heats of hydrogenation of an isolated diene (1,4-pentadiene) and a conjugated diene (1,3-pentadiene), along with the alkenes 1-pentene and (*E*)-2-pentene. The figure shows that an isolated pair of double bonds behaves much like two independent alkene units. The measured heat of hydrogenation of the two double bonds in 1,4-pentadiene is 252 kJ/mol (60.2 kcal/

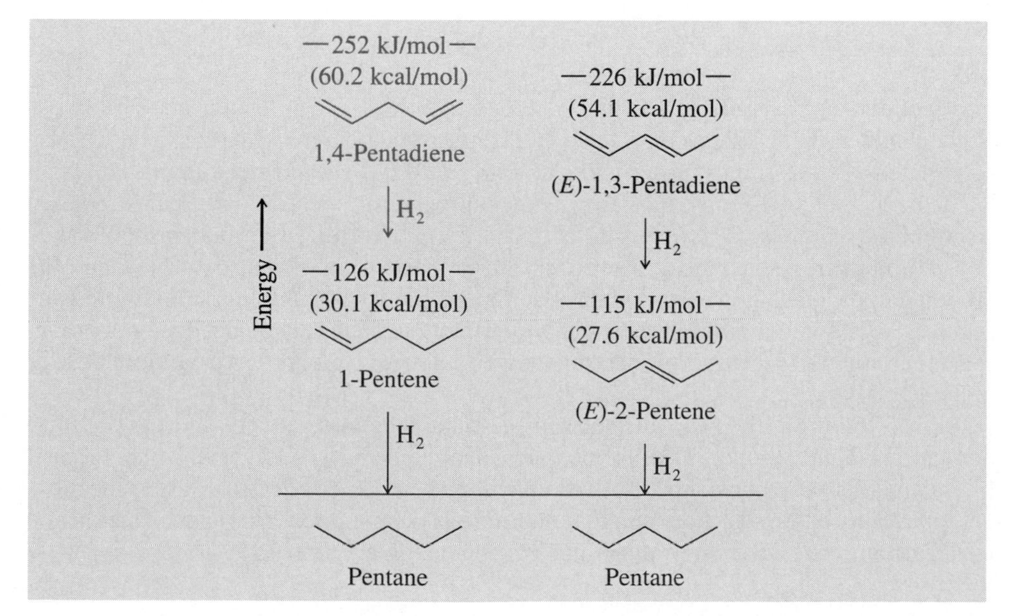

FIGURE 10.2 Heats of hydrogenation of some C_5H_{10} alkenes and C_5H_8 alkadienes.

mol), exactly twice the heat of hydrogenation of 1-pentene. Further, the heat evolved on hydrogenation of each double bond must be 126 kJ/mol (30.1 kcal/mol), since 1-pentene is an intermediate in the hydrogenation of 1,4-pentadiene to pentane.

By the same reasoning hydrogenation of the terminal double bond in the conjugated diene (E)-1,3-pentadiene releases only 111 kJ/mol (26.5 kcal/mol) when it is hydrogenated to (E)-2-pentene. Hydrogenation of the terminal double bond in the conjugated diene evolves 15 kJ/mol (3.6 kcal/mol) less heat than hydrogenation of a terminal double bond in the diene with isolated double bonds. *A conjugated double bond is thus 15 kJ/mol (3.6 kcal/mol) more stable than a simple double bond.* We call this increased stability due to conjugation the **delocalization energy, resonance energy,** or **conjugation energy.**

The cumulated double bonds of an allenic system are of relatively high energy. The heat of hydrogenation of allene is more than twice that of propene.

$$\underset{\text{Allene}}{CH_2{=}C{=}CH_2} + \underset{\text{Hydrogen}}{2H_2} \longrightarrow \underset{\text{Propane}}{CH_3CH_2CH_3} \qquad \Delta H^\circ = -295 \text{ kJ/mol } (-70.5 \text{ kcal/mol})$$

$$\underset{\text{Propene}}{CH_3CH{=}CH_2} + \underset{\text{Hydrogen}}{H_2} \longrightarrow \underset{\text{Propane}}{CH_3CH_2CH_3} \qquad \Delta H^\circ = -125 \text{ kJ/mol } (-29.9 \text{ kcal/mol})$$

PROBLEM 10.7 Another way in which energies of isomers may be compared is by their heats of combustion. Match the heat of combustion with the appropriate diene.

Dienes:	1,2-Pentadiene, (E)-1,3-pentadiene, 1,4-pentadiene
Heats of combustion:	3186 kJ/mol, 3217 kJ/mol, 3251 kJ/mol
	(761.6 kcal/mol, 768.9 kcal/mol, 777.1 kcal/mol)

Thus, the order of alkadiene stability decreases in the order: conjugated diene (most stable) → isolated diene → cumulated diene (least stable). In order to understand this ranking, we need to look at structure and bonding in alkadienes in more detail.

10.7 ELECTRON DELOCALIZATION IN CONJUGATED DIENES

The factor most responsible for the increased stability of conjugated double bonds is the greater delocalization of their π electrons compared with the π electrons of isolated double bonds. As shown in Figure 10.3a, the π electrons of an isolated diene system occupy, in pairs, two noninteracting π orbitals. Each of these π orbitals encompasses two carbon atoms. An sp^3 hybridized carbon insulates the two π orbitals from each other, preventing the exchange of electrons between them. In a conjugated diene, however, mutual overlap of the two π orbitals, represented in Figure 10.3b, gives an orbital system in which each π electron is delocalized over four carbon atoms. Delocalizing of electrons lowers their energy and gives a more stable molecule.

At 146 pm the C-2—C-3 distance in 1,3-butadiene is relatively short for a carbon-carbon single bond. This is most reasonably seen as a hybridization effect. In ethane both carbons are sp^3 hybridized and are separated by a distance of 153 pm. The carbon-carbon single bond in propene unites sp^3 and sp^2 hybridized carbons and is shorter than that of ethane. Both C-2 and C-3 are sp^2 hy-

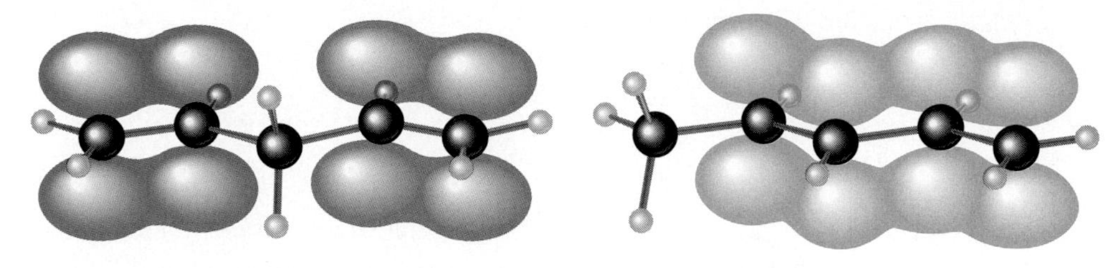

(*a*) Isolated double bonds (*b*) Conjugated double bonds

FIGURE 10.3 (*a*) Isolated double bonds are separated from each other by one or more sp^3 hybridized carbons and cannot overlap to give an extended π orbital. (*b*) In a conjugated diene, overlap of two π orbitals gives an extended π system encompassing four carbon atoms. 1,4-Pentadiene is presented as an example of an isolated diene; 1,3-pentadiene is the conjugated diene.

bridized in 1,3-butadiene, and a decrease in bond distance between them reflects the tendency of carbon to attract electrons more strongly as its *s* character increases.

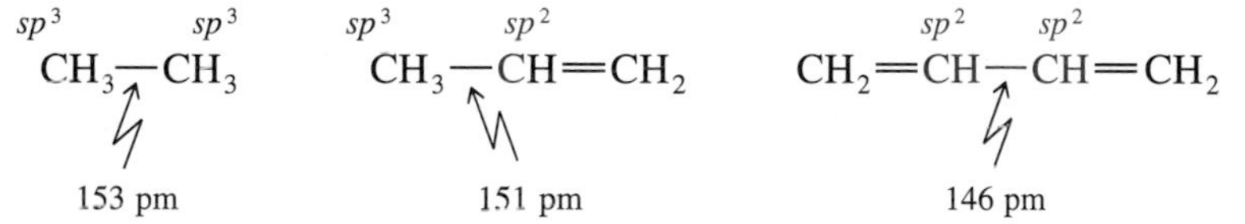

$$\overset{sp^3}{CH_3}-\overset{sp^3}{CH_3} \quad \overset{sp^3}{CH_3}-\overset{sp^2}{CH}=CH_2 \quad CH_2=\overset{sp^2}{CH}-\overset{sp^2}{CH}=CH_2$$

153 pm 151 pm 146 pm

Additional evidence for electron delocalization in 1,3-butadiene can be obtained by considering its conformations. Overlap of the two π electron systems is optimal when the four carbon atoms are coplanar. There are two conformations that allow this coplanarity: they are called the s-*cis* and s-*trans* conformations.

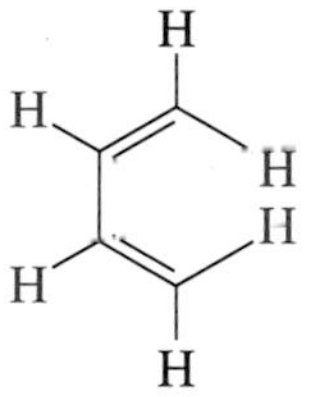

s-Cis conformation of 1,3-butadiene

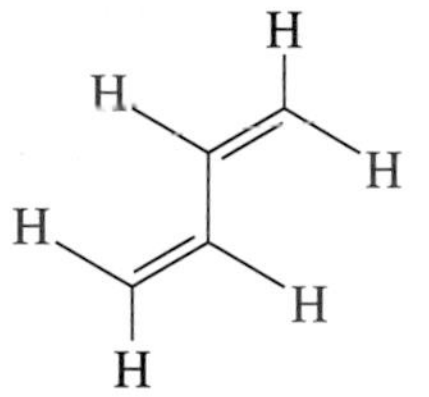

s-Trans conformation of 1,3-butadiene

The letter *s* in *s*-cis and *s*-trans refers to conformations around the C—C *single* bond in the diene. The *s*-trans conformation of 1,3-butadiene is 12 kJ/mol (2.8 kcal/mol) more stable than the *s*-cis conformation; the *s*-cis conformation contains an unfavorable van der Waals interaction between the C-1 and C-4 hydrogens. These two coplanar conformations interconvert by rotation around the C-2—C-3 bond, as illustrated in Figure 10.4. The conformation at the midpoint of this rotation, the *perpendicular conformation,* has its 2*p* orbitals in a geometry that precludes extended conjugation. It has localized double bonds. A significant contributor to the energy of activation for rotation about the single bond in 1,3-butadiene is the decrease in electron delocalization that attends conversion of the *s*-cis or *s*-trans conformation to the perpendicular conformation.

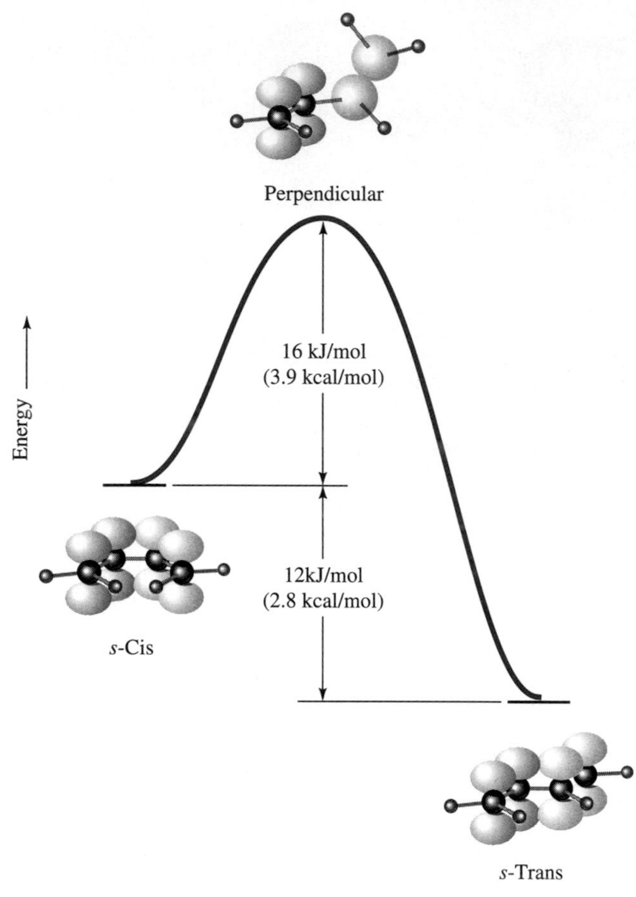

FIGURE 10.4 Conformations and electron delocalization in 1,3-butadiene. The *s*-cis and the *s*-trans conformations permit the axes of the 2*p* orbitals to be aligned parallel to one another for maximum π electron delocalization. The *s*-trans conformation is more stable than the *s*-cis. Stabilization resulting from π electron delocalization is least in the perpendicular conformation, which is a transition state for rotation about the C-2—C-3 single bond.

10.8 BONDING IN ALLENES

The three carbons of allene lie in a straight line, with relatively short carbon-carbon bond distances of 131 pm. The central carbon, since it bears only two substituents, is *sp* hybridized. The terminal carbons of allene are *sp*^{2} hybridized.

$$H_2C{=}C{=}CH_2$$

sp   *sp*^{2}
118.4°
108 pm   131 pm

Allene

Structural studies reveal allene to be nonplanar. As Figure 10.5 illustrates, the plane of one HCH unit is perpendicular to the plane of the other. Figure 10.5 also portrays the reason for the molecular geometry of allene. The 2*p* orbital of each of the terminal carbons overlaps with a different 2*p* orbital of the central carbon. Since the 2*p* orbitals of the central carbon are perpendicular to each other, the perpendicular nature of the two HCH units follows naturally.

(*a*) Planes defined by H(C-1)H and H(C-3)H are mutually perpendicular.

H····C=C=C(H)H (C-1 and C-3 labeled 1 and 3)

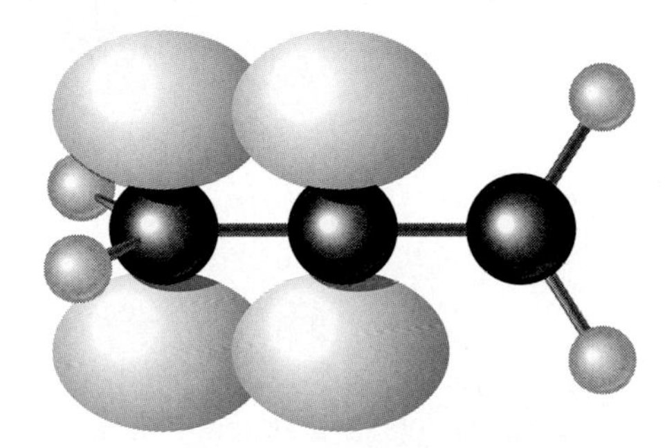

(*b*) The *p* orbital of C-1 and one of the *p* orbitals of C-2 can overlap so as to participate in π bonding.

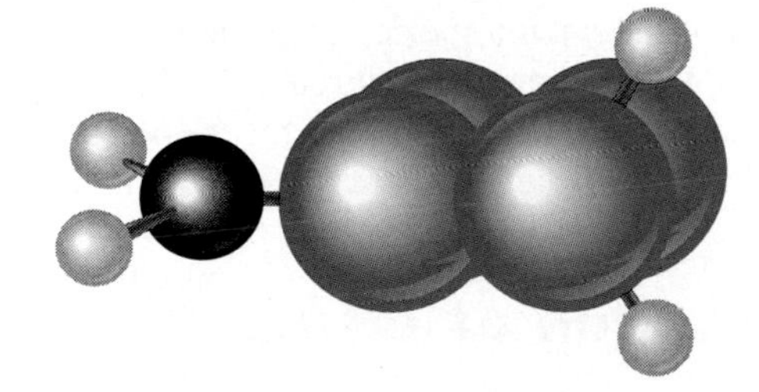

(*c*) The *p* orbital of C-3 and one of the *p* orbitals of C-2 can overlap so as to participate in a second π orbital perpendicular to the one in (*b*).

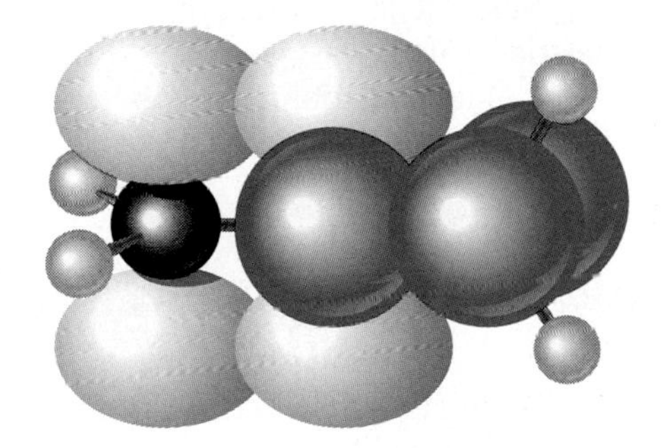

(*d*) Allene is a nonplanar molecule characterized by a linear carbon chain and two mutually perpendicular π bonds.

FIGURE 10.5 Bonding and geometry in 1,2-propadiene (allene).

The nonplanarity of the allene unit has an interesting stereochemical consequence. 1,3-Disubstituted allenes are not superposable on their mirror images; i.e., they are chiral. Even an allene as simple as 2,3-pentadiene has been obtained in optically active form.

H, CH_3 / C=C=C / H_3C, H

(+)-2,3-Pentadiene

H_3C, H / C=C=C / H_3C, H

(−)-2,3-Pentadiene

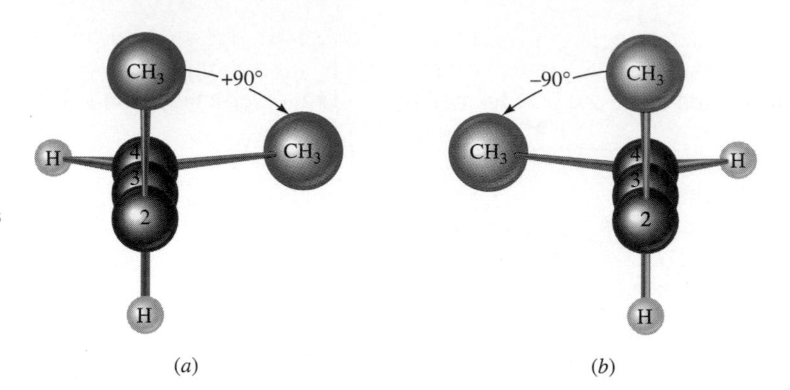

FIGURE 10.6 2,3-Pentadiene has as a stereogenic axis a line that passes through C-2, C-3, and C-4. A clockwise path around this axis is encountered when proceeding from the front methyl (C-1) to the back methyl (C-5) in (*a*) while an anticlockwise path is seen in (*b*).

The Cahn-Ingold-Prelog *R,S* notation has been extended to chiral allenes and other molecules that have a stereogenic axis. Such compounds are so infrequently encountered, however, we will not cover the rules for specifying their stereochemistry in this text.

Chiral allenes are examples of molecules that contain a **stereogenic axis,** also called a **chiral axis.** A mechanical analog of a stereogenic axis is a line passing lengthwise through the center of a screw. Figure 10.6 illustrates the right-handed and left-handed screwlike qualities of the enantiomeric 2,3-pentadienes.

Because of the linear geometry required of cumulated dienes, cyclic allenes, like cycloalkynes, are strained unless the rings are fairly large. 1,2-Cyclononadiene is the smallest cyclic allene that is sufficiently stable to be isolated and stored conveniently.

10.9 PREPARATION OF DIENES

The use of 1,3-butadiene in the preparation of synthetic rubber is discussed in the boxed essay "Diene Polymers" that appears later in this chapter.

The conjugated diene 1,3-butadiene is used in the manufacture of synthetic rubber and is prepared on an industrial scale in vast quantities. Production in the United States is currently 3.2×10^9 lb/yr. One industrial process is similar to that used for the preparation of ethylene: in the presence of a suitable catalyst, butane undergoes thermal dehydrogenation to yield 1,3-butadiene.

$$CH_3CH_2CH_2CH_3 \xrightarrow[\text{chromia-alumina}]{590-675°C} CH_2{=}CHCH{=}CH_2 + 2H_2$$

Laboratory syntheses of conjugated dienes can be achieved by elimination reactions of unsaturated alcohols and alkyl halides. Notice in the examples cited that the conjugated diene is produced in high yield even though an isolated diene is also possible:

$$CH_2{=}CHCH_2\underset{OH}{\overset{CH_3}{C}}CH_2CH_3 \xrightarrow{KHSO_4,\ \text{heat}} CH_2{=}CHCH{=}\overset{CH_3}{C}CH_2CH_3$$

3-Methyl-5-hexen-3-ol → 4-Methyl-1,3-hexadiene (88%)

$$CH_2{=}CHCH_2\underset{Br}{\overset{CH_3}{C}}CH_2CH_3 \xrightarrow{KOH,\ \text{heat}} CH_2{=}CHCH{=}\overset{CH_3}{C}CH_2CH_3$$

4-Bromo-4-methyl-l-hexene → 4-Methyl-1,3-hexadiene (78%)

As we have seen earlier, dehydrations and dehydrohalogenations are typically regioselective in the direction that leads to the most stable double bond. Conjugated dienes are more stable than isolated dienes and are formed faster via a lower-energy transition state.

PROBLEM 10.8 Give the constitutions of the dienes containing isolated double bonds capable of being formed, but not observed, according to the two preceding equations describing elimination in 3-methyl-5-hexen-3-ol and 4-bromo-4-methyl-1-hexene.

Dienes with isolated double bonds are formed when the structure of the substrate precludes the formation of a conjugated diene.

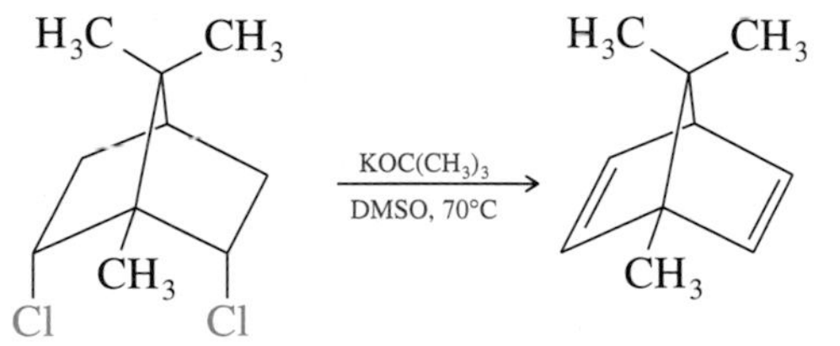

H_3C CH_3 / CH_3 / Cl Cl $\xrightarrow[\text{DMSO, 70°C}]{\text{KOC(CH}_3\text{)}_3}$ H_3C CH_3 / CH_3

2,6-Dichlorocamphane — Bornadiene (83%)

We will not discuss the preparation of cumulated dienes. They are prepared less readily than isolated or conjugated dienes and require special methods.

10.10 ADDITION OF HYDROGEN HALIDES TO CONJUGATED DIENES

Our discussion of chemical reactions of alkadienes will be limited to those of conjugated dienes. The reactions of isolated dienes are essentially the same as those of individual alkenes. The reactions of cumulated dienes are—like their preparation—so specialized that their treatment is better suited to an advanced course in organic chemistry.

Electrophilic addition is the characteristic chemical reaction of alkenes, and conjugated dienes undergo addition reactions with the same electrophiles that react with alkenes, and by similar mechanisms. As we saw in the reaction of hydrogen halides with alkenes (Section 6.5), the regioselectivity of electrophilic addition is governed by protonation of the double bond in the direction that gives the more stable of two possible carbocations. With conjugated dienes it is one of the terminal carbons that is protonated, because the species that results is an allylic carbocation which is stabilized by electron delocalization. Thus, when 1,3-cyclopentadiene reacts with hydrogen chloride, the product is 3-chlorocyclopentene.

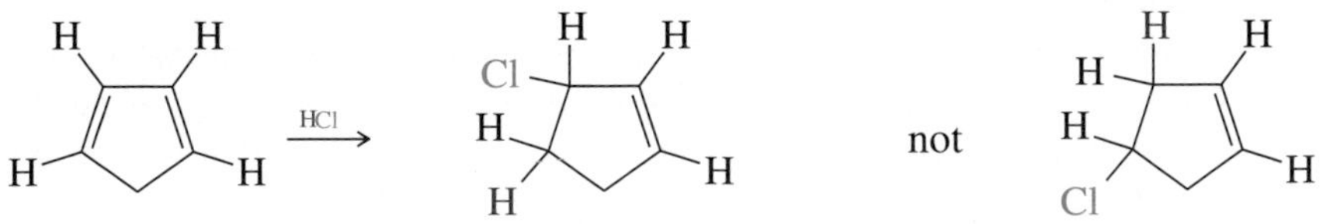

1,3-Cyclopentadiene $\xrightarrow{\text{HCl}}$ 3-Chlorocyclopentene (70–90%) not 4-Chlorocyclopentene

The carbocation that leads to the observed product is secondary and allylic; the other is secondary but not allylic.

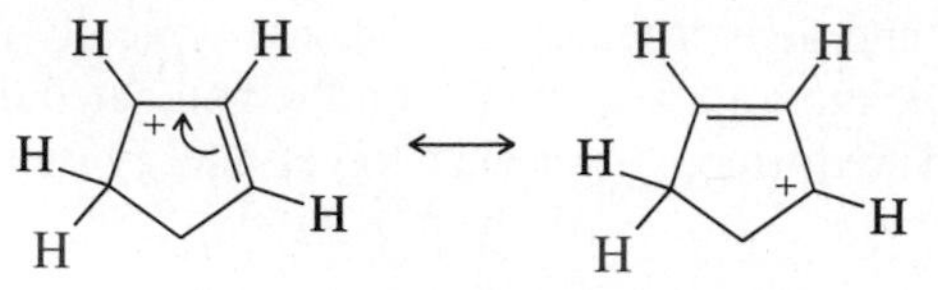

Protonation at end of diene unit gives a carbocation that is both secondary and allylic; product is formed from this carbocation.

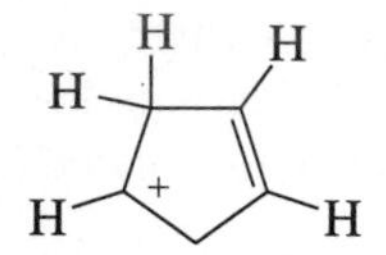

Protonation at C-2 gives a carbocation that is secondary but not allylic; less stable carbocation; not formed as rapidly.

Both resonance forms of the allylic carbocation from 1,3-cyclopentadiene are equivalent, and so attack at either of the carbons that share the positive charge gives the same product, 3-chlorocyclopentene. This is not the case with 1,3-butadiene, and so hydrogen halides add to 1,3-butadiene to give a mixture of two regioisomeric allylic halides. For the case of ionic addition of hydrogen bromide,

$$CH_2{=}CHCH{=}CH_2 \xrightarrow[\text{free-radical inhibitor}]{HBr,\ -80°C} CH_3\underset{Br}{\underset{|}{C}}HCH{=}CH_2 + CH_3CH{=}CHCH_2Br$$

1,3-Butadiene → 3-Bromo-1-butene (81%) + 1-Bromo-2-butene (19%)

The major product corresponds to addition of a proton at C-1 and bromide at C-2. This mode of addition is called **1,2 addition,** or **direct addition.** The minor product has its proton and bromide at C-1 and C-4, respectively, of the original diene system. This mode of addition is called **1,4 addition,** or **conjugate addition.** The double bond that was between C-3 and C-4 in the starting material remains there in the product from 1,2 addition but migrates to a position between C-2 and C-3 in the product from 1,4 addition.

Both the 1,2-addition product and the 1,4-addition product are derived from the same allylic carbocation.

$$\left[CH_3\overset{+}{C}HCH{=}CH_2 \longleftrightarrow CH_3CH{=}CH\overset{+}{C}H_2 \right] \xrightarrow{Br^-} CH_3\underset{Br}{\underset{|}{C}}HCH{=}CH_2 + CH_3CH{=}CHCH_2Br$$

3-Bromo-1-butene (major) + 1-Bromo-2-butene (minor)

The secondary carbon bears more of the positive charge than does the primary carbon, and attack by the nucleophilic bromide ion is faster there. Hence, the major product is the secondary bromide.

When the major product of a reaction is the one that is formed at the fastest rate, we say that the reaction is governed by **kinetic control** (or *rate control*). Most organic reactions fall into this category, and the electrophilic addition of hydrogen bromide to 1,3-butadiene at low temperature is a kinetically controlled reaction.

When, however, the ionic addition of hydrogen bromide to 1,3-butadiene is carried out at room temperature, the ratio of isomeric allylic bromides observed is different from that which is formed at −80°C. At room temperature, the 1,4-addition product comprises the major portion of the reaction product.

$$CH_2{=}CHCH{=}CH_2 \xrightarrow[\text{room temperature, free-radical inhibitor}]{HBr} CH_3\underset{\displaystyle Br}{\underset{|}{C}}HCH{=}CH_2 + CH_3CH{=}CHCH_2Br$$

1,3-Butadiene — 3-Bromo-1-butene (44%) — 1-Bromo-2-butene (56%)

Clearly, the temperature at which reaction occurs exerts a major influence on the product composition. In order to understand why, an important fact must be added. The 1,2- and 1,4-addition products *interconvert rapidly* by allylic rearrangement at elevated temperature in the presence of hydrogen bromide. Heating the product mixture to 45°C in the presence of hydrogen bromide leads to a mixture in which the ratio of 3-bromo-1-butene to 1-bromo-2-butene is 15:85.

$$CH_3\underset{\displaystyle Br}{\underset{|}{C}}HCH{=}CH_2 \underset{\text{cation-anion combination}}{\overset{\text{ionization}}{\rightleftharpoons}} CH_3CH\overset{\displaystyle H}{\overset{|}{C}}CH_2^{+}\ Br^- \underset{\text{ionization}}{\overset{\text{cation-anion combination}}{\rightleftharpoons}} CH_3CH{=}CHCH_2Br$$

3-Bromo-1-butene (less stable isomer) — Carbocation + bromide anion — 1-Bromo-2-butene (more stable isomer)

The product of 1,4 addition, 1-bromo-2-butene, contains an internal double bond and so is *more stable* than the product of 1,2 addition, 3-bromo-l-butene, which has a terminal double bond.

When addition occurs under conditions where the products can equilibrate, the composition of the reaction mixture no longer reflects the relative rates of formation of the products but tends to reflect their *relative stabilities*. Reactions of this type are said to be governed by **thermodynamic control** (or *equilibrium control*). A useful way to

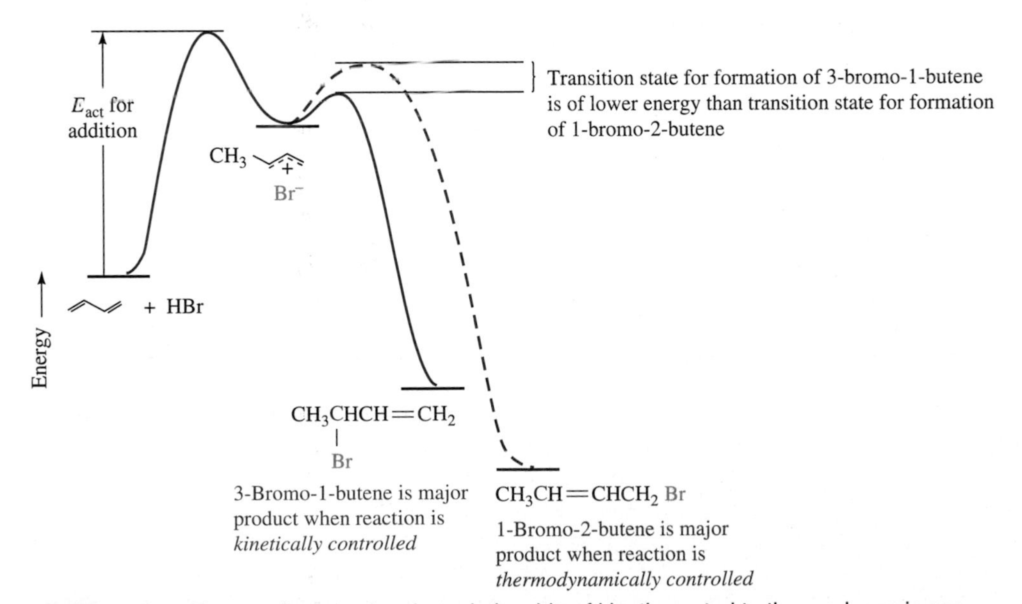

FIGURE 10.7 Energy diagram showing relationship of kinetic control to thermodynamic control in addition of hydrogen bromide to 1,3-butadiene.

illustrate kinetic and thermodynamic control in the addition of hydrogen bromide to 1,3-butadiene is by way of the energy diagram of Figure 10.7. At low temperature, addition takes place irreversibly. Isomerization is slow because insufficient thermal energy is available to permit the products to surmount the energy barrier for ionization. At higher temperatures isomerization is possible, and the more stable product predominates.

PROBLEM 10.9 Addition of hydrogen chloride to 2-methyl-1,3-butadiene is a kinetically controlled reaction and gives one product in much greater amounts than any isomers. What is this product?

10.11 HALOGEN ADDITION TO DIENES

Mixtures of 1,2- and 1,4-addition products are obtained when 1,3-butadiene reacts with chlorine or bromine.

$$CH_2{=}CHCH{=}CH_2 + Br_2 \xrightarrow{CHCl_3} BrCH_2\underset{\displaystyle Br}{\underset{|}{C}}HCH{=}CH_2 + \underset{H}{\overset{BrCH_2}{}}\!\!\!\!\!\!\!\!\!\!\!\!\!\! C{=}C \underset{CH_2Br}{\overset{H}{}}$$

1,3-Butadiene; Bromine; 3,4-Dibromo-1-butene (37%); (*E*)-1,4-Dibromo-2-butene (63%)

The tendency for conjugate addition is pronounced, and *E* double bonds are generated almost exclusively.

PROBLEM 10.10 Exclusive of stereoisomers, how many products are possible in the electrophilic addition of 1 equivalent of bromine to 2-methyl-1,3-butadiene?

DIENE POLYMERS

Some 500 years ago during Columbus's second voyage to what are now the Americas, he and his crew saw children playing with balls made from the latex of trees that grew there. Later, Joseph Priestley called this material "rubber" to describe its ability to erase pencil marks by rubbing, and in 1823 Charles Macintosh demonstrated how rubber could be used to make waterproof coats and shoes. Shortly thereafter Michael Faraday determined an empirical formula of C_5H_8 for rubber. It was eventually determined that rubber is a polymer of 2-methyl-1,3-butadiene.

$$CH_2{=}\underset{\displaystyle CH_3}{\underset{|}{C}}CH{=}CH_2$$

2-Methyl-1,3-butadiene (common name: *isoprene*)

The structure of rubber corresponds to 1,4 addition of several thousand isoprene units to one another:

All the double bonds in rubber have the Z (or cis) configuration. A different polymer of isoprene, called *gutta-percha,* has shorter polymer chains and E (or trans) double bonds. Gutta-percha is a tough, hornlike substance once used as a material for golf ball covers.*

In natural rubber the attractive forces between neighboring polymer chains are relatively weak, and there is little overall structural order. The chains slide easily past one another when stretched and return, in time, to their disordered state when the distorting force is removed. The ability of a substance to recover its original shape after distortion is defined as its *elasticity.* The elasticity of natural rubber is satisfactory only within a limited temperature range; it is too rigid when cold and too sticky when warm to be very useful. Rubber's elasticity is improved by *vulcanization,* a process discovered by Charles Goodyear in 1839. When natural rubber is heated with sulfur, a chemical reaction occurs in which neighboring polyisoprene chains become connected through covalent bonds to sulfur. While these sulfur "bridges" permit only limited movement of one chain with respect to another, their presence ensures that the rubber will snap back to its original shape once the distorting force is removed.

As the demand for rubber increased, so did the chemical industry's efforts to prepare a synthetic substitute. One of the first **elastomers** (a synthetic polymer that possesses elasticity) to find a commercial niche was *neoprene,* discovered by chemists at Du Pont in 1931. Neoprene is produced by free-radical polymerization of 2-chloro-1,3-butadiene and has the greatest variety of applications of any elastomer. Some uses include electrical insulation, conveyer belts, hoses, and weather balloons.

* A detailed discussion of the history, structure, and applications of natural rubber appears in the May 1990 issue of the *Journal of Chemical Education.*

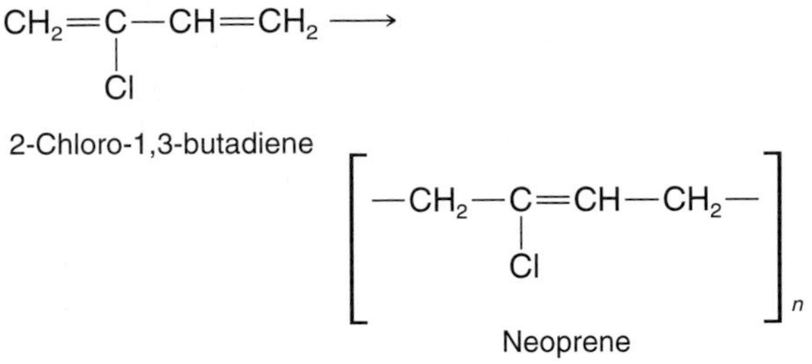

$$CH_2{=}C(Cl){-}CH{=}CH_2 \longrightarrow \left[{-}CH_2{-}C(Cl){=}CH{-}CH_2{-}\right]_n$$

2-Chloro-1,3-butadiene — Neoprene

The elastomer produced in greatest amount is *styrene-butadiene rubber* (SBR). Annually, just under 10^9 lb of SBR is produced in the United States, and almost all of this is used in automobile tires. As its name suggests, SBR is prepared from styrene and 1,3-butadiene. It is an example of a **copolymer,** a polymer assembled from two or more different monomers. Free-radical polymerization of a mixture of styrene and 1,3-butadiene gives SBR.

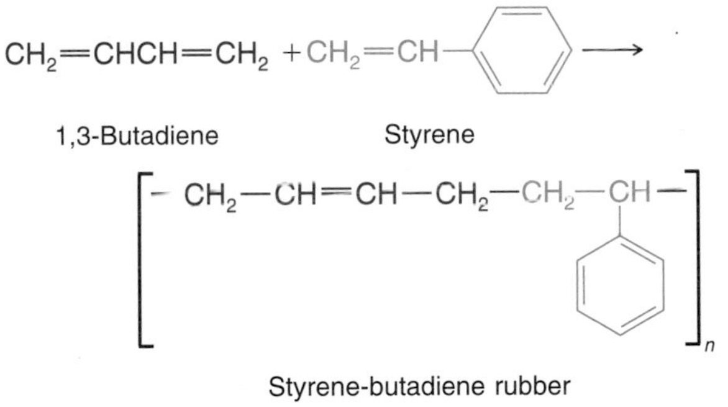

$$CH_2{=}CHCH{=}CH_2 + CH_2{=}CH{-}C_6H_5 \longrightarrow \left[{-}CH_2{-}CH{=}CH{-}CH_2{-}CH_2{-}CH(C_6H_5){-}\right]_n$$

1,3-Butadiene — Styrene — Styrene-butadiene rubber

Coordination polymerization of isoprene using Ziegler-Natta catalyst systems (Section 6.20) gives a material similar in properties to natural rubber, as does polymerization of 1,3-butadiene. Poly(1,3-butadiene) is produced in about two-thirds the quantity of SBR each year. It, too, finds its principal use in tires.

10.12 THE DIELS-ALDER REACTION

A particular kind of conjugate addition reaction earned the Nobel Prize in chemistry for Otto Diels and Kurt Alder of the University of Kiel (Germany) in 1950. The Diels-Alder reaction is the *conjugate addition of an alkene to a diene.* Using 1,3-butadiene as a typical diene, the Diels-Alder reaction may be represented by the general equation:

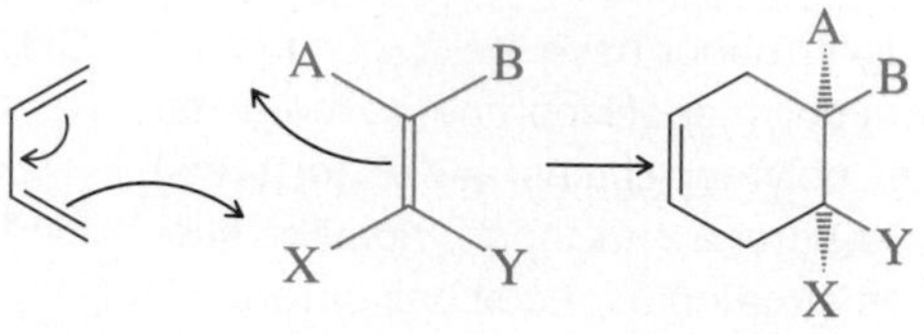

1,3-Butadiene   Dienophile   Diels-Alder adduct

The alkene that adds to the diene is called the **dienophile.** Because the Diels-Alder reaction leads to the formation of a ring, it is termed a **cycloaddition** reaction. The product contains a cyclohexene ring as a structural unit.

The Diels-Alder cycloaddition is one example of a class called **pericyclic reactions.** A pericyclic reaction is a one-step reaction that proceeds through a cyclic transition state. Bond formation occurs at both ends of the diene system, and the Diels-Alder transition state involves a cyclic array of six carbons and six π electrons. The diene must adopt the *s*-cis conformation in the transition state.

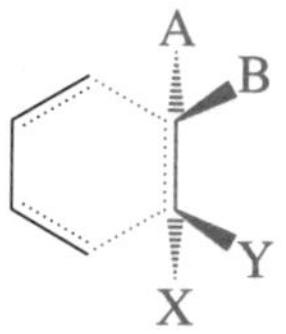

Representation of transition state for Diels-Alder cycloaddition

The simplest of all Diels-Alder reactions, cycloaddition of ethylene to 1,3-butadiene, does not proceed readily. It has a high activation energy and a low reaction rate. However, substituents such as C=O or C≡N, when *directly* attached to the double bond of the dienophile, increase its reactivity, and compounds of this type give high yields of Diels-Alder adducts at modest temperatures.

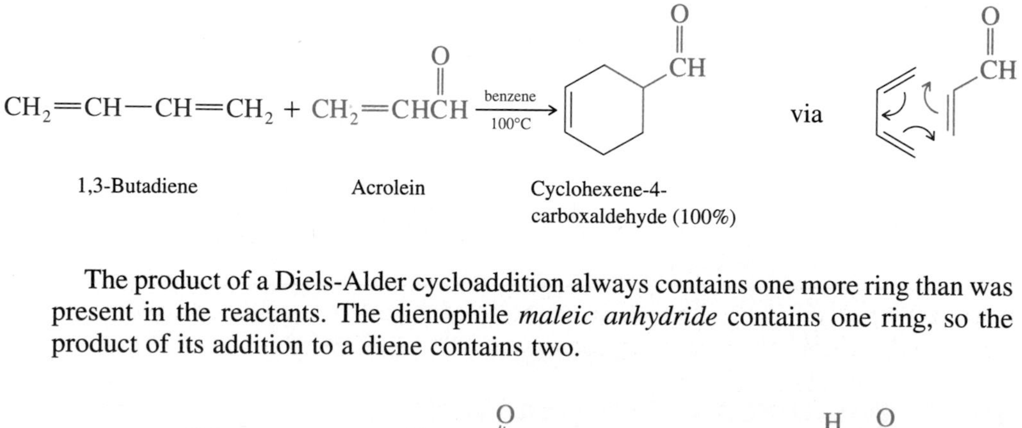

$$CH_2{=}CH{-}CH{=}CH_2 + CH_2{=}CH\overset{O}{\overset{\|}{C}}H \xrightarrow[100°C]{\text{benzene}}$$ (cyclohexene ring bearing CHO) via (s-cis diene + acrolein)

1,3-Butadiene   Acrolein   Cyclohexene-4-carboxaldehyde (100%)

The product of a Diels-Alder cycloaddition always contains one more ring than was present in the reactants. The dienophile *maleic anhydride* contains one ring, so the product of its addition to a diene contains two.

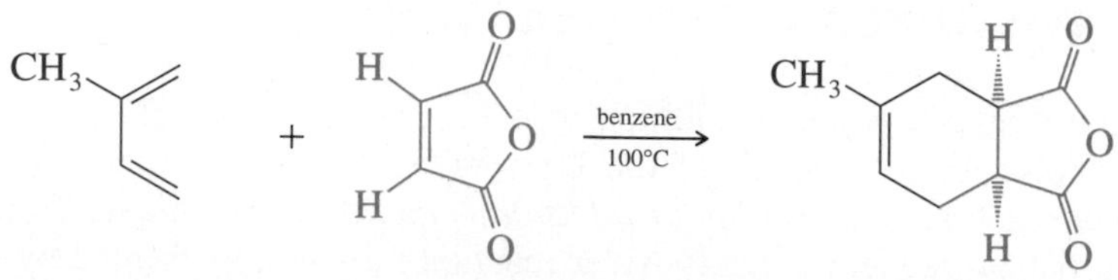

2-Methyl-1,3-butadiene + Maleic anhydride $\xrightarrow[100°C]{\text{benzene}}$ 1-Methylcyclohexene-4,5-dicarboxylic anhydride (100%)

PROBLEM 10.11 Benzoquinone is a very reactive dienophile. It reacts with 2-chloro-1,3-butadiene to give a single product, $C_{10}H_9ClO_2$, in 95 percent yield. Write a structural formula for this product.

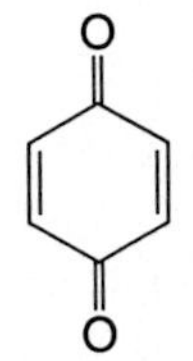

Benzoquinone

Acetylene, like ethylene, is a poor dienophile, but alkynes that bear C=O or C≡N substituents react readily with dienes. A cyclohexadiene derivative is the product.

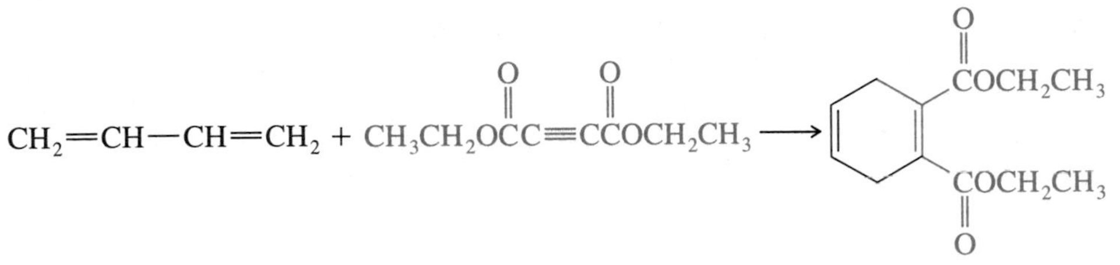

$$CH_2{=}CH{-}CH{=}CH_2 + CH_3CH_2O\overset{O}{\overset{\|}{C}}C{\equiv}C\overset{O}{\overset{\|}{C}}OCH_2CH_3 \longrightarrow$$

1,3-Butadiene | Diethyl acetylenedicarboxylate | Diethyl 1,4-cyclohexadiene-1,2-dicarboxylate (98%)

The Diels-Alder reaction is stereospecific. Substituents that are cis in the dienophile remain cis in the product; substituents that are trans in the dienophile remain trans in the product.

Recall from Section 7.13 that a stereospecific reaction is one in which each stereoisomer of a particular starting material yields a different stereoisomeric form of the reaction product. In the examples shown, the product from Diels-Alder cycloaddition of 1,3-butadiene to *cis*-cinnamic acid is a stereoisomer of the product from *trans*-cinnamic acid.

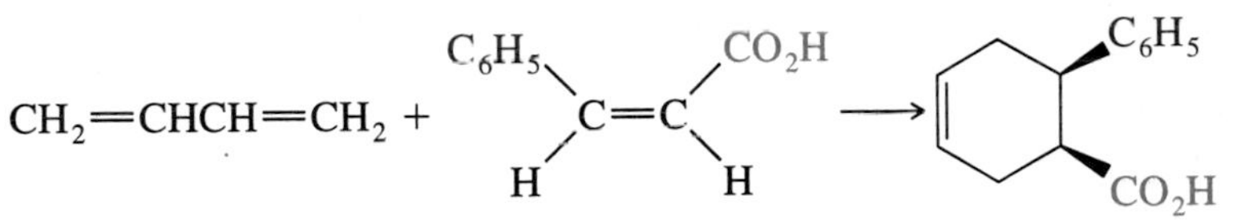

$$CH_2{=}CHCH{=}CH_2 + (C_6H_5)HC{=}CH(CO_2H) \longrightarrow$$

1,3-Butadiene | *cis*-Cinnamic acid | Only product

$$CH_2{=}CHCH{=}CH_2 + (C_6H_5)HC{=}CH(CO_2H) \longrightarrow$$

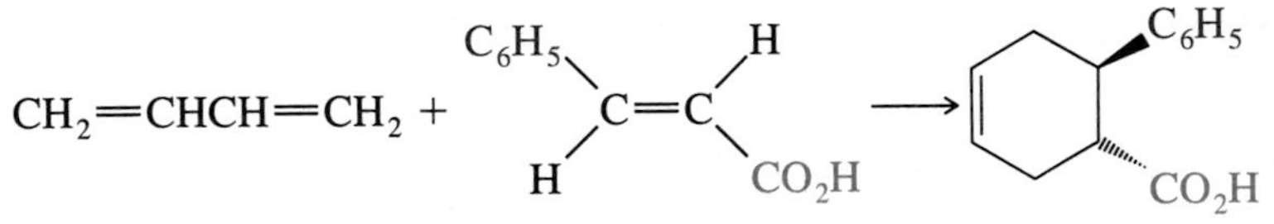

1,3-Butadiene | *trans*-Cinnamic acid | Only product

PROBLEM 10.12 What combination of diene and dienophile would you choose in order to prepare each of the following compounds?

(*a*) 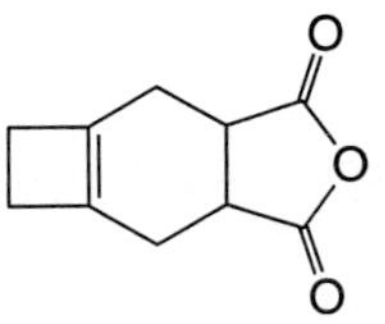 (*b*) C≡N C≡N (*c*) CH_3

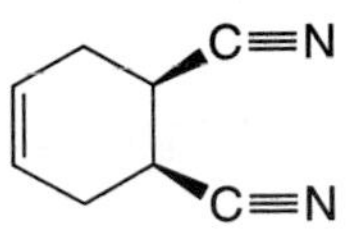

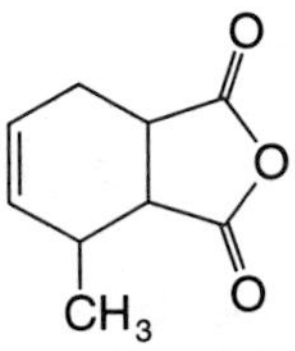

SAMPLE SOLUTION (*a*) We represent a Diels-Alder reaction according to the curved arrow formalism as

In order to deduce the identity of the diene and dienophile that lead to a particular Diels-Alder adduct, all we need to do is use curved arrows in the reverse fashion to "undo" the cyclohexene derivative. Start with the π component of the double bond in the six-membered ring, and move electrons in pairs.

Diels-Alder adduct is derived from Diene + Dienophile

Cyclic dienes yield bridged bicyclic Diels-Alder adducts.

1,3-Cyclopentadiene + Dimethyl fumarate → Dimethylbicyclo[2.2.1]hept-2-ene-*trans*-5,6-dicarboxylate

PROBLEM 10.13 Give the structure of the Diels-Alder adduct of 1,3-cyclohexadiene and dimethyl acetylenedicarboxylate ($CH_3O\overset{O}{\overset{\|}{C}}C{\equiv}C\overset{O}{\overset{\|}{C}}OCH_3$).

The importance of the Diels-Alder reaction is in synthesis. It gives us a method to form *two* new carbon-carbon bonds in a single operation and requires no reagents, such as acids or bases, that might affect other functional groups in the molecule.

10.13 STEREOSELECTIVITY OF DIELS-ALDER REACTIONS

Besides being stereospecific, Diels-Alder reactions are stereoselective, as the following example illustrates. Two stereoisomeric Diels-Alder adducts are possible and both are formed.

$$+ \ CH_2{=}CH\overset{O}{\overset{\|}{C}}OCH_3 \longrightarrow$$ Endo isomer + Exo isomer

1,3-Cyclopentadiene — Methyl acrylate — Endo isomer (75%) — Exo isomer (25%)

(Stereoisomeric forms of methyl bicyclo[2.2.1]hept-5-ene-2-carboxylate)

The stereoisomer formed in greater amount has its $\overset{O}{\overset{\|}{C}}OCH_3$ group syn to the CH=CH bridge and is called the *endo* isomer. The minor product has its $\overset{O}{\overset{\|}{C}}OCH_3$ group anti to the CH=CH bridge and is called the *exo* isomer. It has been shown that both isomers are of almost equal stability; therefore, we are not simply observing the formation of the more stable product in an equilibrium-controlled process. There is a kinetically controlled preference for the formation of the endo product.

Observations similar to this have been made many times in Diels-Alder reactions and have led to the formulation of the following empirical rule, known as the *Alder rule,* or the *rule of maximum accumulation of unsaturation:* In a Diels-Alder reaction, the major product is derived from the transition state in which unsaturated groups in the dienophile assume a syn rather than an anti orientation with respect to the diene.

The two transition-state geometries, one leading to the endo product and the other to the exo product, are shown in Figure 10.8. It is believed that attractive intermolecular forces between the π systems of the diene and the C=O group make the endo transition state more stable than the exo.

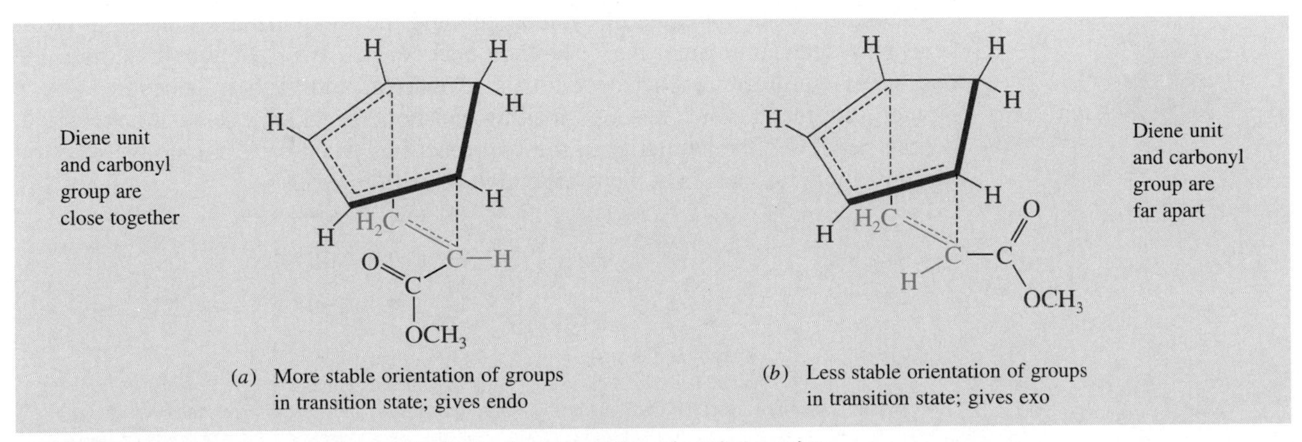

FIGURE 10.8 The observed stereoselectivity of cycloaddition of methyl acrylate ($CH_2{=}CHCO_2CH_3$) to 1,3-cyclopentadiene requires that the transition state represented by (*a*) be lower in energy than (*b*).

The stereoselective bias expressed in the Alder rule is, however, not sufficient to overcome the stereospecific requirement for syn addition. A dienophile in which the substituents are trans to one another on the double bond reacts with 1,3-cyclopentadiene to give a bicyclic Diels-Alder adduct in which one substituent is exo and the other endo.

PROBLEM 10.14 Can you find a reaction in Section 10.12 that illustrates the point of the preceding paragraph?

10.14 ELECTROCYCLIC REACTIONS OF POLYENES

While Diels-Alder cycloaddition has been known for many years and was in wide use as a synthetic method long before its mechanism became understood, other pericyclic reactions are of more recent vintage and of more mechanistic than synthetic interest. One of these is the ring opening to 1,3-butadiene that occurs when cyclobutene is heated:

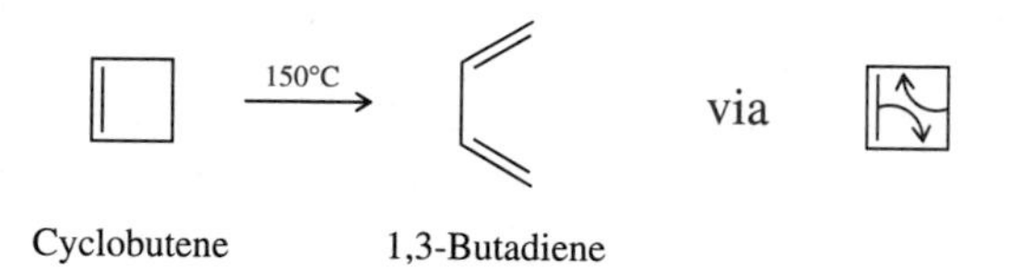

A second example is the thermal cyclization of *cis*-1,3,5-hexatriene to 1,3-cyclohexadiene:

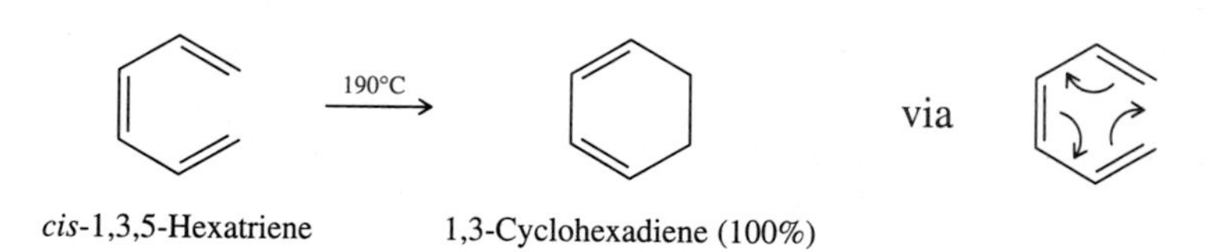

Although the ring opening of cyclobutene and the ring closure of *cis*-1,3,5-hexatriene may seem unrelated, the difference between the two reactions is simply the position of equilibrium. Each reaction is reversible, and each is believed to be a one-step process in which all bond-making and bond-breaking events are concerted. We can better see the similarity in the two reactions when we examine the electron reorganization that occurs at their respective transition states:

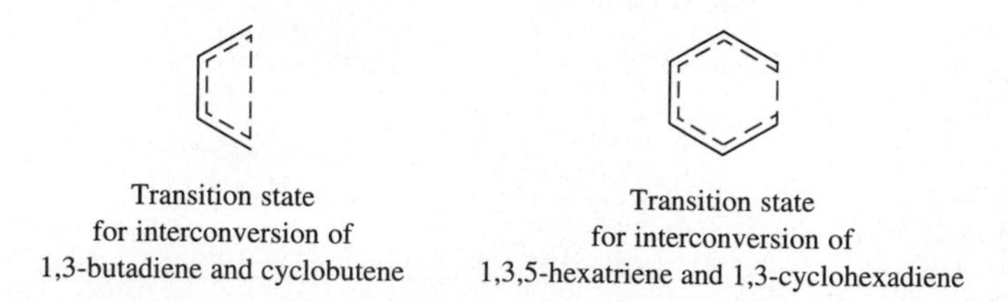

Transition state
for interconversion of
1,3-butadiene and cyclobutene
Transition state
for interconversion of
1,3,5-hexatriene and 1,3-cyclohexadiene

In terms of equilibria, *cis*-1,3,5-hexatriene closes to 1,3-cyclohexadiene because a stronger σ bond is formed at the expense of a weaker π bond. Cyclobutene opens to 1,3-butadiene because angle strain destabilizes the cycloalkene.

Reactions such as these that involve the formation of a σ bond between the ends of a conjugated π system, *or the reverse,* belong to the class of pericyclic transformations called **electrocyclic** reactions.

The most striking thing about electrocyclic reactions is their stereospecificity. As shown in Figure 10.9, for example, each stereoisomer of 3,4-dimethylcyclobutene gives a different stereoisomer of 2,4-hexadiene on being heated. The cyclobutene ring of each stereoisomer opens so that the sp^3 hybridized carbons (C-3 and C-4) rotate in the same direction as they become sp^2 hybridized in the diene product. We use the term **conrotation** to describe the stereochemical course of this ring opening and say that the thermal ring opening of cyclobutenes to butadiene derivatives is a **conrotatory** electrocyclic reaction.

The closure of conjugated hexatrienes to derivatives of 1,3-cyclohexadiene is also stereospecific, but in a sense opposite to the stereospecificity of the cyclobutene to 1,3-butadiene interconversion. Figure 10.10 shows that *trans,cis,trans-* and *cis,cis,trans*-2,4,6-octatriene cyclize to stereoisomeric 5,6-dimethyl-1,3-cyclohexadienes on heating. In this reaction, rotation of the substituents at one end of the triene system is clockwise, while at the other end it is anticlockwise. Such a process is described as a **disrotatory** electrocyclic reaction.

The difference in the stereochemical courses of the electrocyclic reactions of Figures 10.9 and 10.10 is unrelated to the fact that one reaction is a ring opening and the other a cyclization. The cyclobutene to conjugated diene interconversion is conrotatory, as the cyclohexadiene to conjugated triene interconversion is disrotatory irrespective of the positions of their equilibria.

PROBLEM 10.15 Demonstrate that both the forward and reverse steps of the electrocyclic reactions shown in Figure 10.9 must proceed with the same stereochemistry. What would you expect to observe if the forward reaction were conrotatory and the back reaction disrotatory? Repeat this exercise for Figure 10.10.

One of the most significant advances in mechanistic organic chemistry occurred during the 1960s when a coherent picture for pericyclic reactions was advanced on the

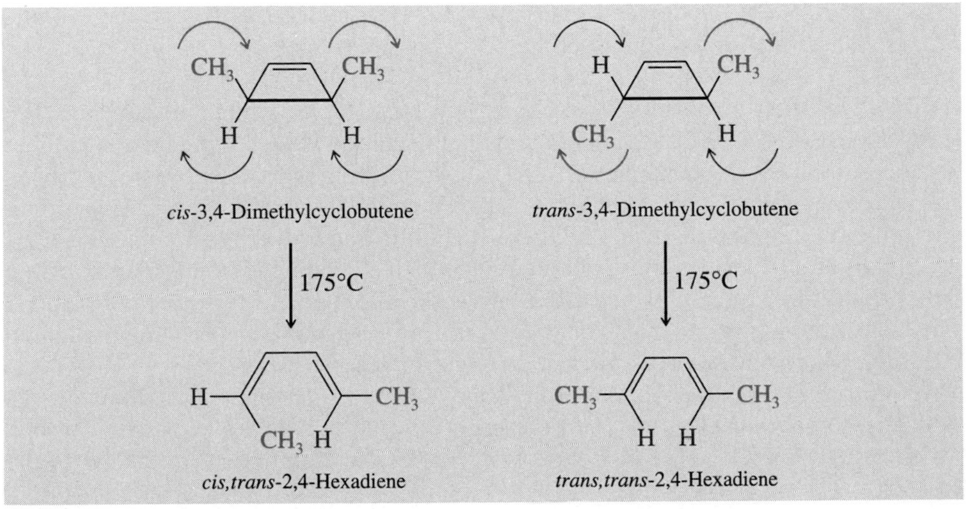

cis-3,4-Dimethylcyclobutene

175°C

cis,trans-2,4-Hexadiene

trans-3,4-Dimethylcyclobutene

175°C

trans,trans-2,4-Hexadiene

FIGURE 10.9 Conrotatory ring opening of *cis-* and *trans*-3,4-dimethylcyclobutene. Substituents at the termini of the developing diene system all rotate in the same direction during cleavage of the σ bond of the ring.

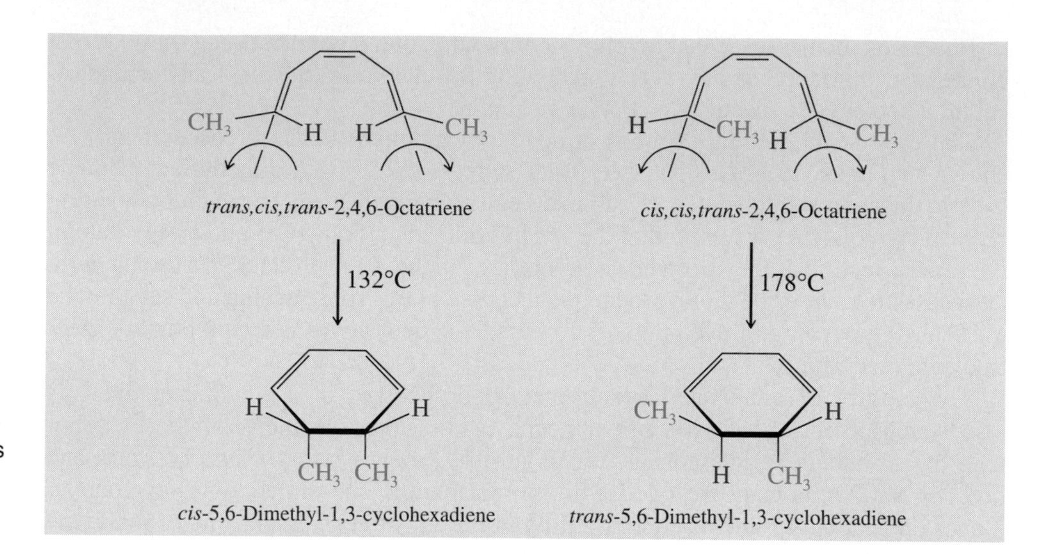

FIGURE 10.10 Disrotatory ring closure of stereoisomeric conjugated trienes. Substituents at the termini of the triene system rotate in opposite directions in the process of forming a σ bond between the two substituted carbons.

basis of molecular orbital theory. In order to understand that explanation we need to take a more detailed look at π bonding in alkenes and polyenes.

10.15 THE π MOLECULAR ORBITALS OF ALKENES AND CONJUGATED POLYENES

The valence-bond approach has served us well to this point as a tool to probe structure and reactivity in organic chemistry. An appreciation for the delocalization of π electrons through a system of overlapping *p* orbitals has given us insights into conjugated systems that are richer in detail than those obtained by examining Lewis formulas. An even deeper understanding can be gained by applying qualitative molecular orbital theory to these π electron systems. We shall see that useful information can be gained by directing attention to what are called the **frontier orbitals** of molecules. The frontier orbitals are the *highest occupied molecular orbital* (the *HOMO*) and the *lowest unoccupied molecular orbital* (the *LUMO*). When electrons are transferred *from* a molecule, it is the electrons in the HOMO that are involved, because they are the most weakly held. When electrons are transferred *to* a molecule, they go into the LUMO, because that is the lowest-energy orbital available.

Ethylene. Let us begin by examining the π molecular orbitals of ethylene. We learned in Section 1.13 that the number of σ molecular orbitals is equal to the number of atomic orbitals that combine to form them. Thus the 1*s* orbitals of two hydrogen atoms overlap to give both a bonding (σ) and an antibonding (σ^*) orbital. This principle applies to π orbitals as well. As Figure 10.11 illustrates for the case of ethylene, the 2*p* orbitals of adjacent carbons overlap to give both a bonding (π) and an antibonding (π^*) orbital. Notice that the σ electrons are not explicitly considered in Figure 10.11. These electrons are strongly held, and the collection of σ bonds can be thought of as an inert framework that supports the valence electrons of the π orbital.

As is true for all orbitals, a π orbital may contain a maximum of two electrons. Ethylene has two π electrons, and these occupy the bonding π molecular orbital, which is the HOMO. The antibonding π^* molecular orbital is vacant, and is the LUMO.

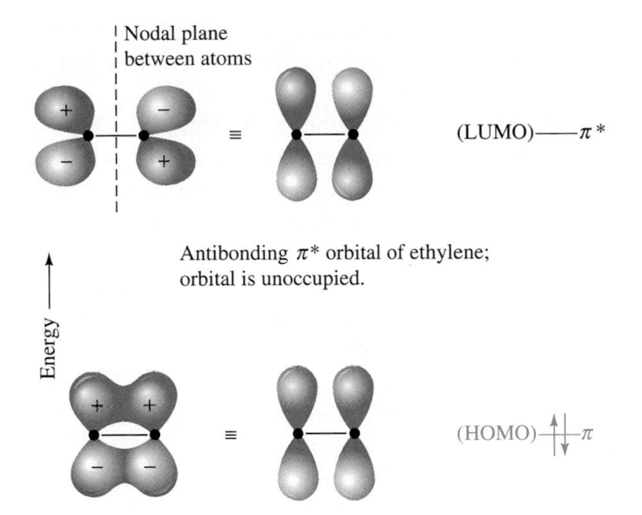

FIGURE 10.11 Representations of the bonding and antibonding π molecular orbitals of ethylene.

Both the π and π^* molecular orbitals of ethylene are *antisymmetric* with respect to the plane of the molecule. By this we mean that the wave function describing the orbital changes sign on passing through the molecular plane. We shall find it convenient to designate the signs of p orbital wave functions by shading one lobe of a p orbital in red and the other in blue instead of using plus (+) and minus (−) signs that might be confused with electronic charges. The plane of the molecule corresponds to a nodal plane where the probability of finding the π electrons is zero. The bonding π

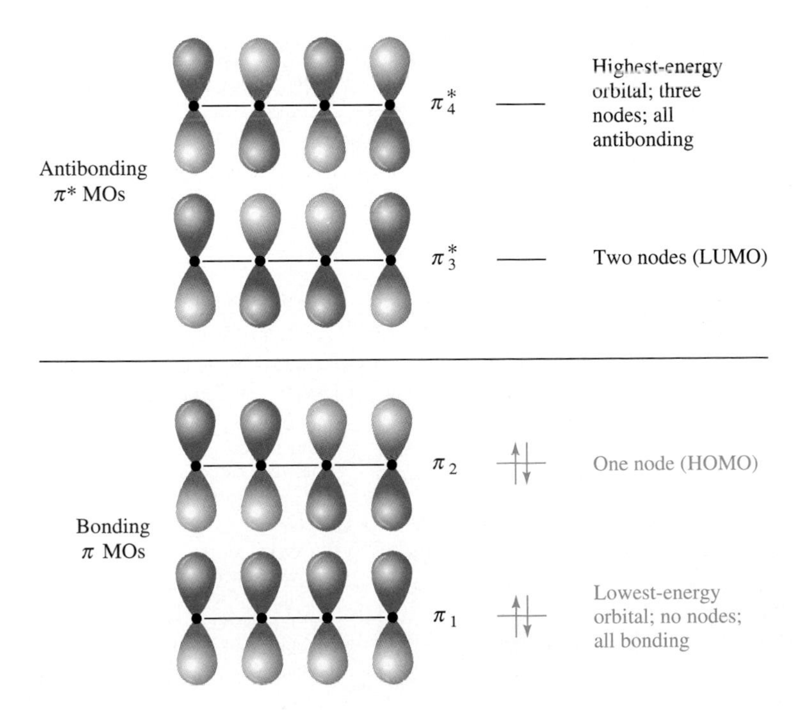

FIGURE 10.12 Electron distribution among the π molecular orbitals of 1,3-butadiene.

orbital has no nodes other than this plane, while the antibonding π^* orbital has a nodal plane between the two carbons. The more nodes an orbital has, the higher is its energy.

1,3-Butadiene. The π molecular orbitals of 1,3-butadiene are shown in Figure 10.12 (page 405). The four sp^2 hybridized carbons contribute four $2p$ atomic orbitals, and their overlap leads to four π molecular orbitals. Two are bonding (π_1 and π_2) and two are antibonding (π_3^* and π_4^*). Each π molecular orbital encompasses all four carbons of the diene. There are four π electrons, and these are distributed in pairs between the two orbitals of lowest energy (π_1 and π_2). Both bonding orbitals are occupied; π_2 is the HOMO. Both antibonding orbitals are vacant; π_3^* is the LUMO.

1,3,5-Hexatriene. Figure 10.13 depicts the six π molecular orbitals of 1,3,5-hexatriene. The six π electrons of 1,3,5-hexatriene occupy the three bonding orbitals (π_1, π_2, and π_3) while the antibonding orbitals (π_4^*, π_5^*, and π_6^*) are unoccupied. The lowest-energy orbital (π_1) has no nodes in the molecular plane, the next orbital (π_2) has one node, and the HOMO (π_3) has two.

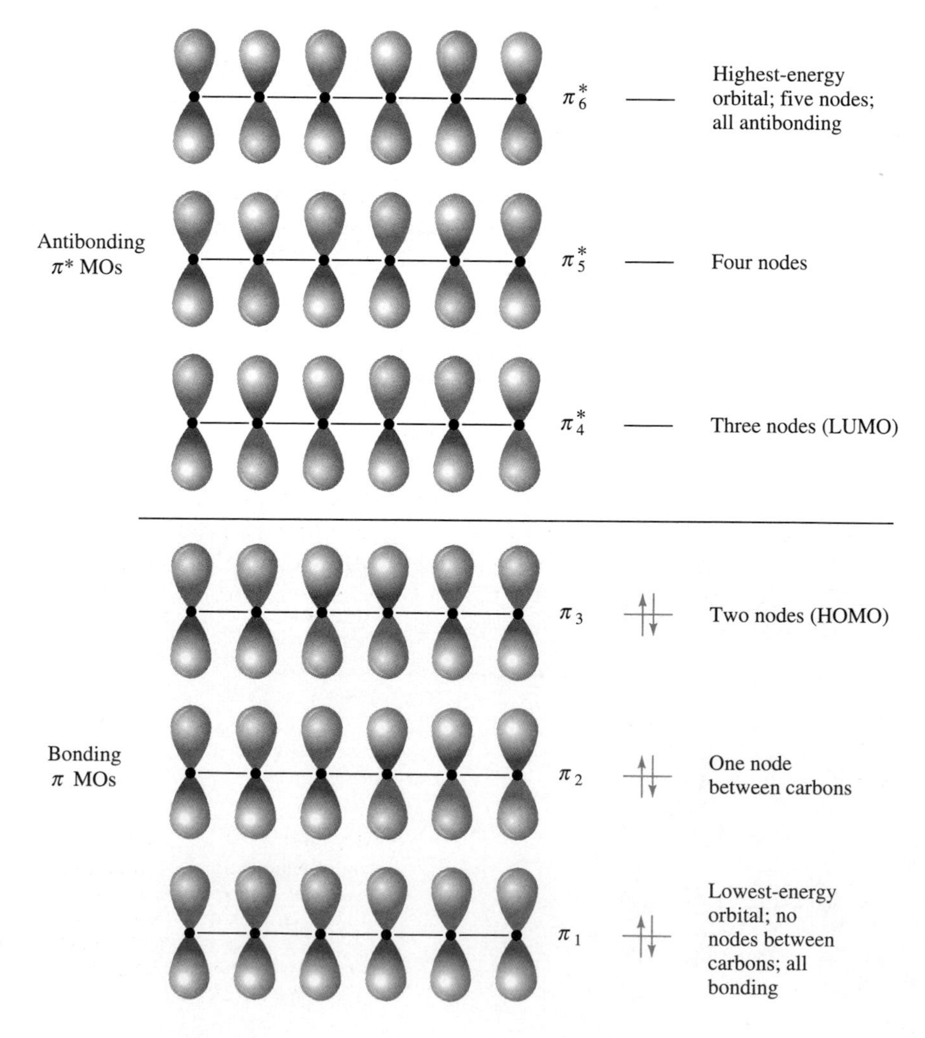

FIGURE 10.13 The π molecular orbitals of 1,3,5-hexatriene.

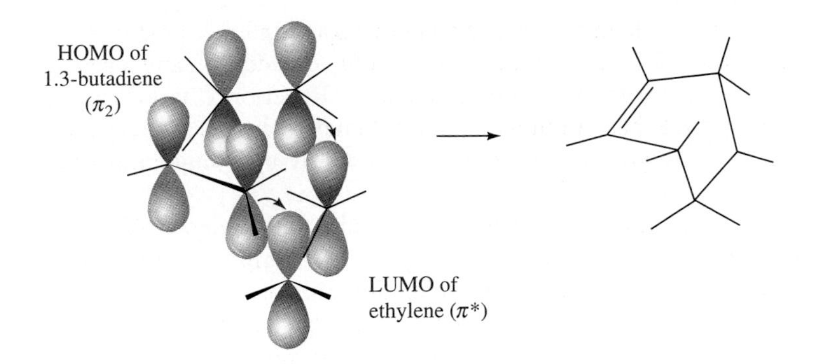

FIGURE 10.14 The HOMO of 1,3-butadiene and the LUMO of ethylene have the proper symmetry to allow σ bond formation to occur at both ends of the diene chain in the same transition state.

10.16 A π MOLECULAR ORBITAL ANALYSIS OF THE DIELS-ALDER REACTION

Let us now examine the Diels-Alder cycloaddition from a molecular orbital perspective. Chemical experience, such as the observation that the substituents that increase the reactivity of a dienophile tend to be those that attract electrons, suggests that electrons flow from the diene to the dienophile during the reaction. Thus, the orbitals to be considered are the HOMO of the diene and the LUMO of the dienophile. As shown in Figure 10.14 for the case of ethylene and 1,3-butadiene, the symmetry properties of the HOMO of the diene and the LUMO of the dienophile permit bond formation between the termini of the diene system and the two carbons of the dienophile double bond because the necessary orbitals overlap "in phase" with each other. Cycloaddition of a diene and an alkene is said to be a **symmetry-allowed** reaction.

Contrast the Diels-Alder reaction with a cycloaddition reaction that looks superficially similar, the combination of two alkenes to give a cyclobutane derivative.

Alkene Alkene Cyclobutane derivative

Reactions of this type are rather rare and seem to proceed in a stepwise fashion rather than by way of a concerted mechanism involving a single transition state.

Figure 10.15 shows the interaction between the HOMO of one alkene and the LUMO of another. In particular, notice that two of the carbons that are to become

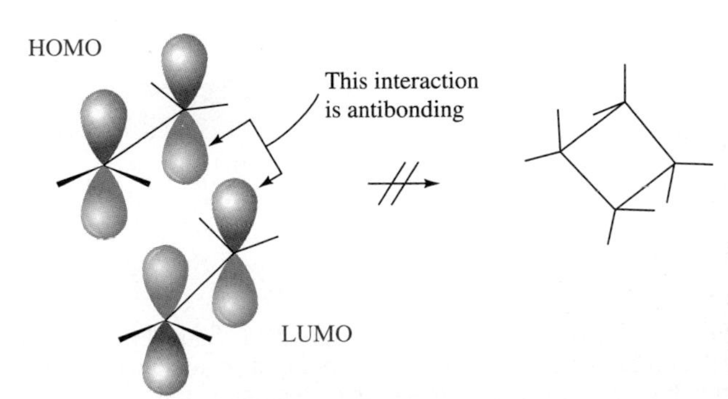

FIGURE 10.15 The HOMO of one ethylene molecule and the LUMO of another do not have the proper symmetry to permit two σ bonds to be formed in the same transition state for concerted cycloaddition.

σ-bonded to each other in the product experience an antibonding interaction during the cycloaddition process. This raises the activation energy for cycloaddition and leads the reaction to be classified as a **symmetry-forbidden** reaction. Reaction, if it does occur, takes place slowly and by a mechanism in which the two new σ bonds are formed in separate steps rather than by way of a concerted process involving a single transition state.

Frontier orbital analysis is a powerful technique that aids our understanding of a great number of organic reactions. Its early development is attributed to Professor Kenichi Fukui of Kyoto University, Japan. The application of frontier orbital methods to Diels-Alder reactions represents one part of what organic chemists refer to as the *Woodward-Hoffmann rules,* a beautifully simple analysis of organic reactions by Professor R. B. Woodward of Harvard University and Professor Roald Hoffmann of Cornell University. Professors Fukui and Hoffmann were corecipients of the 1981 Nobel Prize in chemistry for their work.

Woodward's death in 1979 prevented his being considered for a share of the 1981 prize with Fukui and Hoffmann. Woodward had earlier won a Nobel Prize (1965) for his achievements in organic synthesis.

In the following section we shall see how related reasoning can help us understand why some electrocyclic reactions are conrotatory and some are disrotatory.

10.17 A π MOLECULAR ORBITAL ANALYSIS OF ELECTROCYCLIC REACTIONS

Irrespective of whether the reaction is a ring opening or a ring closing, the Woodward-Hoffmann rules for electrocyclic reactions begin with an examination of the HOMO of the open-chain polyene. For 1,3-butadiene, the HOMO is π_2, for which the contributing *p* orbitals have the symmetry properties shown in Figure 10.16. Particular attention is paid to the symmetry properties of the *p* orbitals of C-1 and C-4, since these are the carbons that will be involved in σ bond formation. In order to form a σ bond the orbitals at C-1 and C-4 must overlap *in phase* with one another. This overlap can be accomplished only when ring closure is *conrotatory.* A disrotatory closure would force an out-of-phase overlap of orbitals and an antibonding interaction between C-1 and C-4. Bond formation stabilizes the conrotatory transition state for the one-step (concerted) 1,3-butadiene/cyclobutene equilibrium, and this is an *orbital symmetry–allowed* reaction. The disrotatory process, characterized by an antibonding interaction at the transition state, cannot occur in a concerted reaction and is said to be an *orbital symmetry–forbidden* reaction.

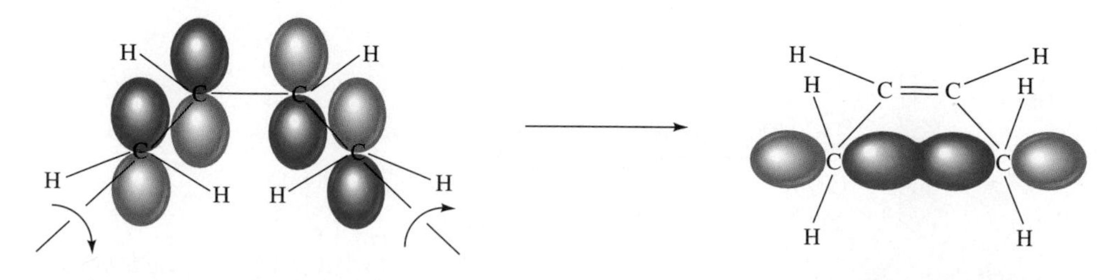
H
C
H
H
C
C
H
H
C
H
C=C
Conrotation permits orbitals at terminal carbons to overlap in phase with each other.
In-phase overlap gives σ bond of cyclobutene.

FIGURE 10.16 The highest occupied molecular orbital (π_2) of 1,3-butadiene can close to cyclobutene only when the process is conrotatory. Disrotation forces an out-of-phase overlap (antibonding) of the orbitals at the end of the diene system and is orbital symmetry–forbidden.

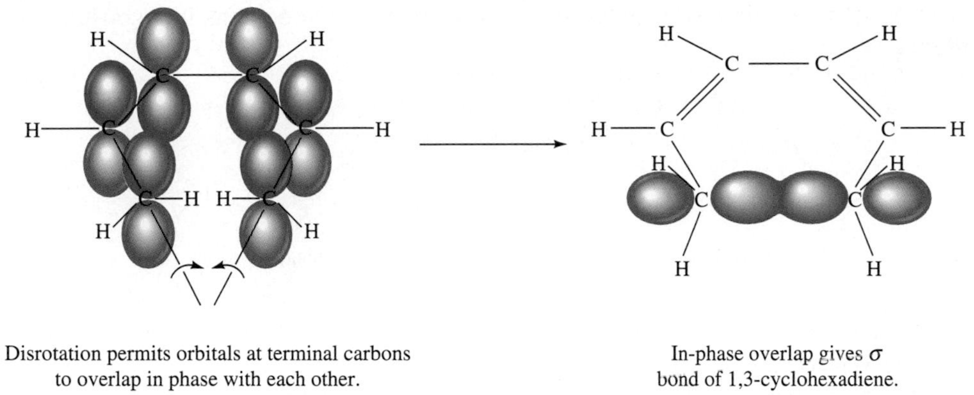

Disrotation permits orbitals at terminal carbons to overlap in phase with each other.

In-phase overlap gives σ bond of 1,3-cyclohexadiene.

FIGURE 10.17 The highest occupied molecular orbital (π_3) of 1,3,5-hexatriene can close to 1,3-cyclohexadiene only when the process is disrotatory. Conrotation forces an out-of-phase overlap (antibonding) of the orbitals at the end of the diene system and is orbital symmetry–forbidden.

Similar reasoning, as shown in Figure 10.17, leads to an opposite stereochemical requirement for the 1,3,5-hexatriene/1,3-cyclohexadiene equilibrium. The HOMO for 1,3,5-hexatriene is π_3, from which a bonding interaction between the terminal carbons (C-1 and C-6) is possible only with disrotation. Concerted disrotatory closure of 1,3,5-hexatriene is *orbital symmetry–allowed;* conrotatory closure is *orbital symmetry–forbidden.*

On the basis of the orbital symmetry patterns of polyenes, a general rule for electrocyclic reactions can be stated:

Thermal electrocyclic reactions are conrotatory when the number of π electrons in the conjugated polyene is equal to 4n (where n is an integer) and disrotatory when it is $4n + 2$.

While electrocyclic reactions were somewhat of a curiosity at the time, their analysis by molecular orbital methods in the 1960s stimulated a flurry of activity to test the theory. Thus, the thermal cyclization of the conjugated tetraene shown:

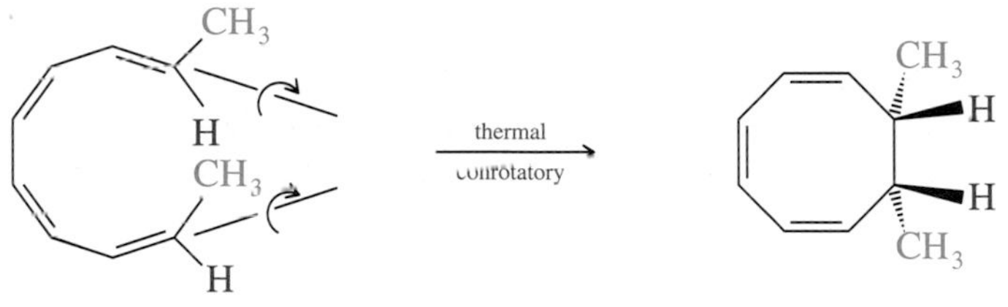

thermal
conrotatory

cis,cis,cis,trans-2,4,6,8-Decatetraene

cis-7,8-Dimethyl-1,3,5-cyclooctatriene

was found to conform to the rule which requires that the electrocyclic reaction of a conjugated polyene with eight π electrons (a $4n$ system where $n = 2$) should be conrotatory.

PROBLEM 10.16 What two stereoisomers of 2,4,6,8-decatetraene give *trans*-7,8-dimethyl-1,3,5-cyclooctatriene on cyclization?

10.18 SUMMARY

This chapter has focused on the effect of a carbon-carbon double bond as a substituent on a positively charged carbon in an **allylic carbocation,** on a carbon bearing an odd electron in an **allylic free radical,** and on a second double bond as in a **conjugated diene.**

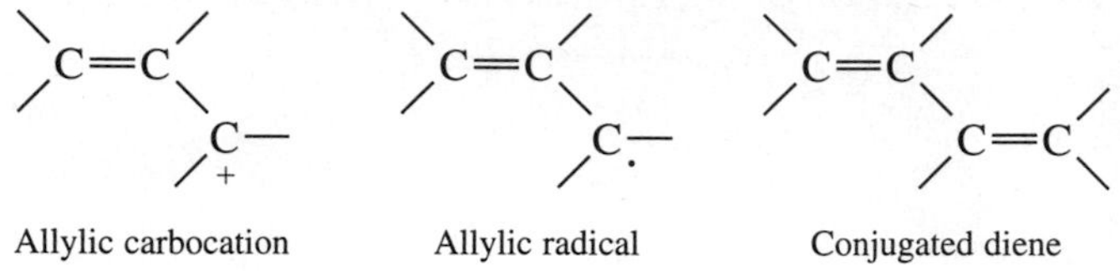

Allylic carbocation   Allylic radical   Conjugated diene

Allyl is the common name of the parent group $CH_2{=}CHCH_2{-}$ and is an acceptable name in IUPAC nomenclature (Section 10.1).

Allylic carbocations (Section 10.2) are stabilized by delocalization of the π electrons of the double bond. The positive charge is shared by the two carbons at the end of the three-carbon allylic system, which may be represented in resonance terms as:

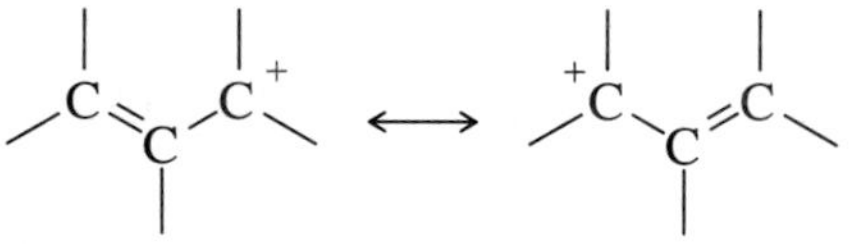

Allylic carbocations are more stable than simple alkyl cations, the activation energy for their formation is lower, and S_N1 reactions of allylic halides proceed at faster rates than the corresponding reactions of alkyl halides. Some reactions involving allylic carbocations are summarized in Table 10.1.

Allylic free radicals are more stable than structurally related alkyl radicals (Section 10.3). Free-radical halogenation of alkenes proceeds by substitution of an allylic hydrogen as illustrated by the example in Table 10.1.

Dienes are classified as having **isolated, conjugated,** or **cumulated** double bonds (Section 10.5). Conjugated dienes are slightly more stable than isolated dienes, and cumulated dienes are the least stable of the **alkadienes** (Section 10.6).

Conjugated dienes are stabilized by resonance to the extent of 12 to 16 kJ/mol (3 to 4 kcal/mol). Their two most stable conformations are designated as *s*-cis and *s*-trans.

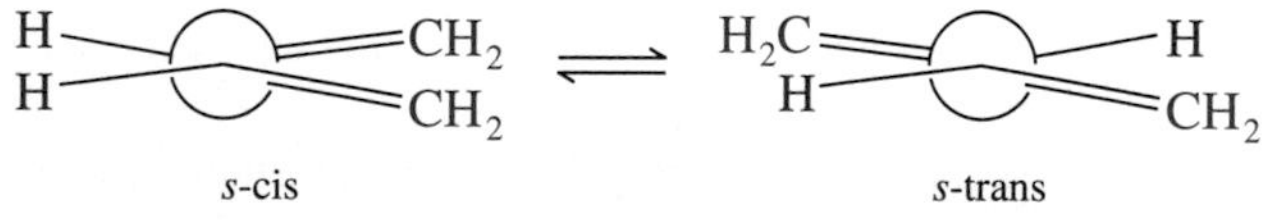

s-cis   *s*-trans

The *s*-trans conformation is normally more stable than the *s*-cis. Both conformations are planar; this aligns the four 2*p* orbitals so that they overlap to give an extended π system (Section 10.7).

1,2-Propadiene ($CH_2{=}C{=}CH_2$), also called **allene,** is the simplest cumulated diene. The two π bonds in an allene share an *sp* hybridized carbon and are at right angles to each other (Section 10.8). Certain allenes such as 2,3-pentadiene ($CH_3CH{=}C{=}CHCH_3$) possess a *stereogenic axis* and are chiral.

1,3-Butadiene is an industrial chemical and is prepared by dehydrogenation of butane. Elimination reactions such as dehydration and dehydrohalogenation are common routes to alkadienes.

TABLE 10.1
Reactions Involving Allylic Reactive Intermediates (Carbocations or Free Radicals)

Reaction (section) and comments	Example
Solvolysis of allylic halides (Section 10.2) The carbocations formed as intermediates when allylic halides undergo S_N1 reactions have their positive charge shared by the two end carbons of the allylic system and may be attacked by nucleophiles at either site. Products may be formed with the same pattern of bonds as the starting allylic halide or with *allylic rearrangement.*	$CH_3CH(Cl)CH{=}CH_2 \xrightarrow[H_2O]{Na_2CO_3} CH_3CH(OH)CH{=}CH_2 + CH_3CH{=}CHCH_2OH$ 3-Chloro-1-butene → 3-Buten-2-ol (65%) + 2-Buten-1-ol (35%) *via:* $CH_3\overset{+}{C}H{-}CH{=}CH_2 \longleftrightarrow CH_3CH{=}CH{-}\overset{+}{C}H_2$
Addition of hydrogen halides to conjugated dienes (Section 10.10) Protonation at the terminal carbon of a conjugated diene system gives an allylic carbocation that can be captured by the halide nucleophile at either of the two sites that share the positive charge. Nucleophilic attack at the carbon adjacent to the one that is protonated gives the product of *direct addition (1,2 addition).* Capture at the other site gives the product of *conjugate addition (1,4 addition).*	$CH_2{=}CHCH{=}CH_2 \xrightarrow{HCl} CH_3CH(Cl)CH{=}CH_2 + CH_3CH{=}CHCH_2Cl$ 1,3-Butadiene → 3-Chloro-1-butene (78%) + 1-Chloro-2-butene (22%) *via:* $CH_3\overset{+}{C}H{-}CH{=}CH_2 \longleftrightarrow CH_3CH{=}CH{-}\overset{+}{C}H_2$
Allylic halogenation of alkenes (Section 10.4) Alkenes react with *N*-bromosuccinimide (NBS) to give allylic bromides. NBS serves as a source of Br_2, and substitution occurs by a free-radical mechanism. The reaction is used for synthetic purposes only when the two resonance forms of the allylic radical are equivalent. Otherwise a mixture of isomeric allylic bromides is produced.	Cyclodecene $\xrightarrow[CCl_4,\ heat]{NBS}$ 3-Bromocyclodecene (56%) *via:* two equivalent resonance forms of the cyclodecenyl allylic radical (H, H, H labels)

$$\underset{\text{3-Methyl-5-hexen-3-ol}}{CH_2{=}CHCH_2C(CH_3)(OH)CH_2CH_3} \xrightarrow[\text{heat}]{KHSO_4} \underset{\text{4-Methyl-1,3-hexadiene (88\%)}}{CH_2{=}CHCH{=}C(CH_3)CH_2CH_3}$$

Elimination is typically regioselective and gives a conjugated diene rather than an isolated or cumulated diene system of double bonds (Section 10.9).

TABLE 10.2
Pericyclic Reactions Involving Conjugated Dienes and Polyenes

Reaction (section) and comments	Example
Diels-Alder cycloaddition (Sections 10.12–10.13) Conjugate addition of an alkene (the *dienophile*) to a conjugated diene gives a cyclohexene derivative. It is concerted and stereospecific; substituents that are cis to one another on the dienophile remain cis in the product. The reaction is used primarily for synthetic purposes.	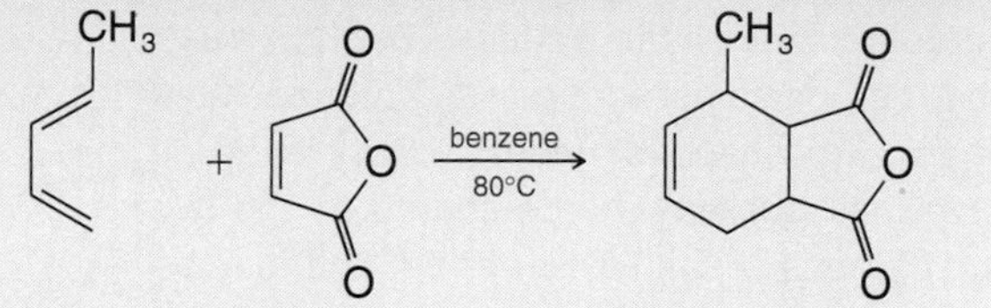 *trans*-1,3-Pentadiene + Maleic anhydride → (benzene, 80°C) 3-Methylcyclohexene-4,5-dicarboxylic anhydride (81%)
Electrocyclic reactions (Section 10.14) Electrocyclic processes include both ring-opening and ring-closing reactions. Their stereochemistry depends on the number of π electrons in the open-chain conjugated polyene component. When $4n$ π electrons are involved, the reaction is *conrotatory.* When the number of π electrons is $4n + 2$, the reaction is *disrotatory.* The example shown is a conrotatory closure of a conjugated tetraene (eight π electrons).	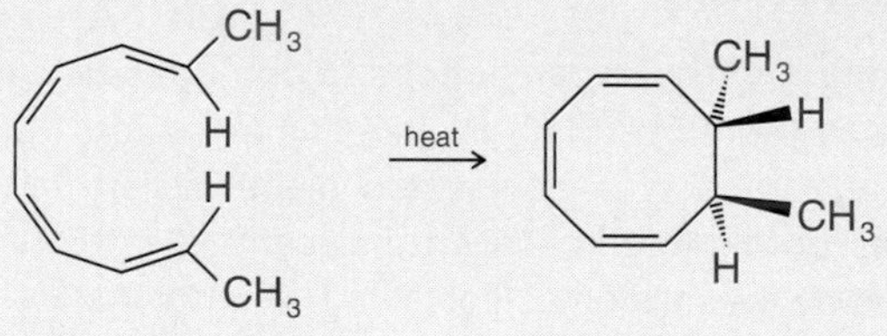 *trans,cis,cis,trans*-2,4,6,8-Decatetraene → (heat) *trans*-7,8-Dimethyl-1,3,5-cyclooctatriene

Electrophilic addition to a conjugated diene can proceed by either **direct addition** (1,2 addition) or **conjugate addition** (1,4 addition; Sections 10.10 and 10.11). Protonation of a conjugated diene by a hydrogen halide generates an allylic carbocation that can react with a nucleophile at either of two sites, as shown in Table 10.1. Usually the major product is the one that is formed faster **(kinetic control),** but when allylic rearrangement occurs under the reaction conditions, the more stable isomer predominates **(thermodynamic control).**

Pericyclic reactions are concerted reactions in which electron reorganization involves all the atoms that together constitute a cyclic transition state. Table 10.2 gives examples of two types of pericyclic reactions, **Diels-Alder cycloaddition** and **electrocyclic** reactions. The theoretical treatment of pericyclic reactions requires an understanding of the symmetry properties of the π orbitals of polyenes (Section 10.15) and rests on the idea that *p* orbitals in prescribed **frontier orbitals** must overlap in phase with each other in the transition state for bond formation (Sections 10.16 and 10.17).

PROBLEMS

10.17 Write structural formulas for each of the following:

(*a*) 3,4-Octadiene
(*b*) (*E,E*)-3,5-Octadiene
(*c*) (*Z,Z*)-1,3-Cyclooctadiene
(*d*) (*Z,Z*)-1,4-Cyclooctadiene
(*e*) (*E,E*)-1,5-Cyclooctadiene
(*f*) (2*E*,4*Z*,6*E*)-2,4,6-Octatriene
(*g*) 5-Allyl-1,3-cyclopentadiene
(*h*) *trans*-1,2-Divinylcyclopropane
(*i*) 2,4-Dimethyl-1,3-pentadiene

10.18 Give the IUPAC names for each of the following compounds:

(*a*) $CH_2{=}CH(CH_2)_5CH{=}CH_2$

(*b*) $(CH_3)_2C{=}C(CH_3)C(CH_3){=}C(CH_3)_2$

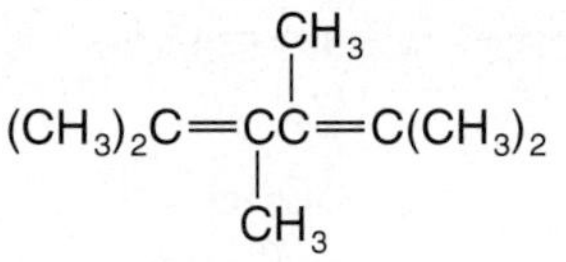

(*c*) $(CH_2{=}CH)_3CH$

(*d*)
CH$_3$
H C=CH$_2$

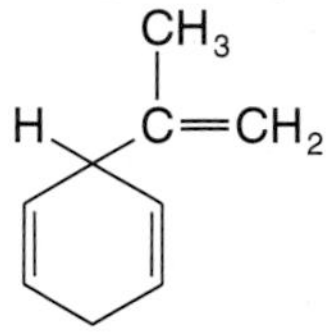

(*e*)
H H Cl
H
H
Cl H H

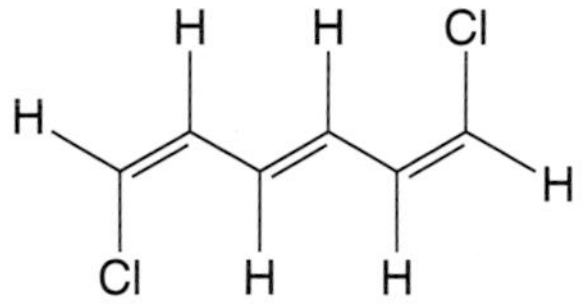

(*f*) $CH_2{=}C{=}CHCH{=}CHCH_3$

(*g*)

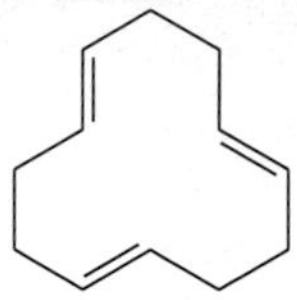

(*h*)
H_3C $CH{=}CH_2$
C=C
CH_3CH_2 CH_2CH_3

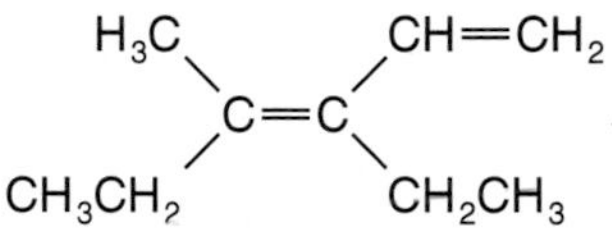

10.19

(*a*) What compound of molecular formula C_6H_{10} gives 2,3-dimethylbutane on catalytic hydrogenation over platinum?

(*b*) What two compounds of molecular formula $C_{11}H_{20}$ give 2,2,6,6-tetramethylheptane on catalytic hydrogenation over platinum?

10.20 Write structural formulas for all the

(*a*) Conjugated dienes (*b*) Isolated dienes (*c*) Cumulated dienes

that give 2,4-dimethylpentane on catalytic hydrogenation.

10.21 A certain species of grasshopper secretes an allenic substance of molecular formula $C_{13}H_{20}O_3$ that acts as an ant repellent. The carbon skeleton and location of various substituents in this substance are indicated in the partial structure shown. Complete the structure, adding double bonds where appropriate.

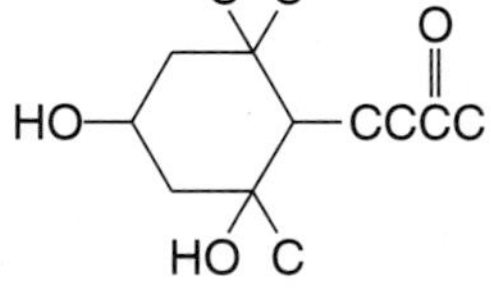

C C
O
HO— —CCCC
HO C

10.22 Show how you could prepare each of the following compounds from propene and any necessary organic or inorganic reagents:

(*a*) Allyl bromide
(*b*) 1,2-Dibromopropane
(*c*) 1,3-Dibromopropane
(*d*) 1-Bromo-2-chloropropane
(*e*) 1,2,3-Tribromopropane
(*f*) Allyl alcohol
(*g*) 1-Penten-4-yne ($CH_2{=}CHCH_2C{\equiv}CH$)
(*h*) 1,4-Pentadiene

10.23 Show, by writing a suitable sequence of chemical equations, how you could prepare each of the following compounds from cyclopentene and any necessary organic or inorganic reagents:

(*a*) 2-Cyclopenten-1-ol
(*b*) 3-Iodocyclopentene
(*c*) 3-Cyanocyclopentene
(*d*) 1,3-Cyclopentadiene
(*e*) [structure with two $\overset{O}{\|}COCH_3$ groups]

10.24 Give the structure, exclusive of stereochemistry, of the principal organic product formed on reaction of 2,3-dimethyl-1,3-butadiene with each of the following:

(*a*) 2 mol H_2, platinum catalyst
(*b*) 1 mol HCl (product of direct addition)
(*c*) 1 mol HCl (product of conjugate addition)
(*d*) 1 mol Br_2 (product of direct addition)
(*e*) 1 mol Br_2 (product of conjugate addition)
(*f*) 2 mol Br_2
(*g*) [structure]

10.25 Repeat the previous problem for the reactions of 1,3-cyclohexadiene.

10.26 Allene can be converted to a trimer (compound A) of molecular formula C_9H_{12}. Compound A reacts with dimethyl acetylenedicarboxylate to give compound B. Deduce the structure of compound A.

$$3CH_2{=}C{=}CH_2 \longrightarrow \text{compound A} \xrightarrow{CH_3O\overset{O}{\overset{\|}{C}}C{\equiv}C\overset{O}{\overset{\|}{C}}OCH_3}$$

H_2C ... $\overset{O}{\|}COCH_3$, $COCH_3$

Compound B

10.27 The reaction shown below gives only the product indicated. By what mechanism does this reaction most likely occur?

$$CH_3CH{=}CHCH_2Cl + C_6H_5{-}SNa \xrightarrow{\text{ethanol}} CH_3CH{=}CHCH_2S{-}C_6H_5$$

10.28 Suggest reasonable explanations for each of the following observations:

(*a*) The first-order rate constant for the solvolysis of $(CH_3)_2C{=}CHCH_2Cl$ in ethanol is over 6000 times greater than that of allyl chloride (25°C).
(*b*) After a solution of 3-buten-2-ol in aqueous sulfuric acid had been allowed to stand for 1 week, it was found to contain both 3-buten-2-ol and 2-buten-1-ol.
(*c*) Treatment of $CH_3CH{=}CHCH_2OH$ with hydrogen bromide gave a mixture of 1-bromo-2-butene and 3-bromo-1-butene.

(*d*) Treatment of 3-buten-2-ol with hydrogen bromide gave the same mixture of bromides as in part (*c*).
(*e*) The major product in parts (*c*) and (*d*) was 1-bromo-2-butene.

10.29 2-Chloro-1,3-butadiene (chloroprene) is the monomer from which the elastomer *neoprene* is prepared. 2-Chloro-1,3-butadiene is the thermodynamically controlled product formed by addition of hydrogen chloride to vinylacetylene ($CH_2{=}CHC{\equiv}CH$). The principal product under conditions of kinetic control is the allenic chloride 4-chloro-1,2-butadiene. Suggest a mechanism to account for the formation of each product.

10.30

(*a*) Write equations expressing the *s*-trans $\rightleftharpoons$ *s*-cis conformational equilibrium for (*E*)-1,3-pentadiene and for (*Z*)-1,3-pentadiene.
(*b*) For which stereoisomer will the equilibrium favor the *s*-trans conformation more strongly? Why?

10.31 The allene 2,3-pentadiene is chiral. Which of the following are chiral?

(*a*) 2-Methyl-2,3-pentadiene
(*b*) 2-Methyl-2,3-hexadiene
(*c*) 4-Methyl-2,3-hexadiene
(*d*) 2,4-Dimethyl-2,3-pentadiene

10.32

(*a*) Describe the molecular geometry expected for 1,2,3-butatriene ($CH_2{=}C{=}C{=}CH_2$).
(*b*) Two stereoisomers are expected for 2,3,4-hexatriene ($CH_3CH{=}C{=}C{=}CHCH_3$). What should be the relationship between these two stereoisomers?

10.33 Suggest reagents suitable for carrying out each step in the following synthetic sequence:

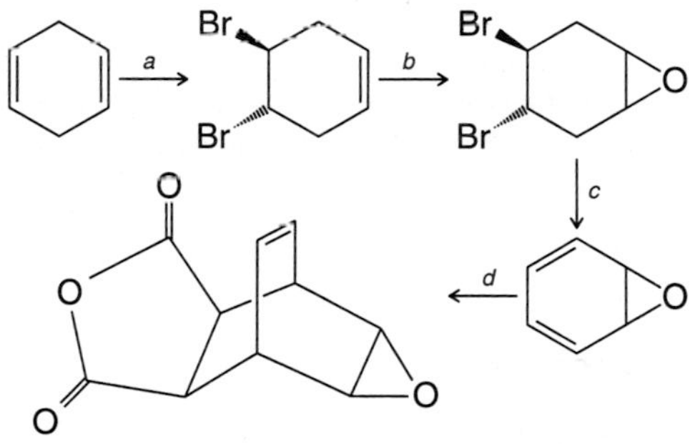

10.34 A very large number of Diels-Alder reactions are recorded in the chemical literature, many of which involve relatively complicated dienes, dienophiles, or both. On the basis of your knowledge of Diels-Alder reactions, predict the constitution of the Diels-Alder adduct that you would expect to be formed from the following combinations of dienes and dienophiles:

(*a*) 2,3-Dimethyl-1,3-butadiene + $C_6H_5SO_2CH{=}CH_2$

(*b*) OCH_3 / $(CH_3)_3SiO$ diene + $CH_3O_2CC{\equiv}CCO_2CH_3$

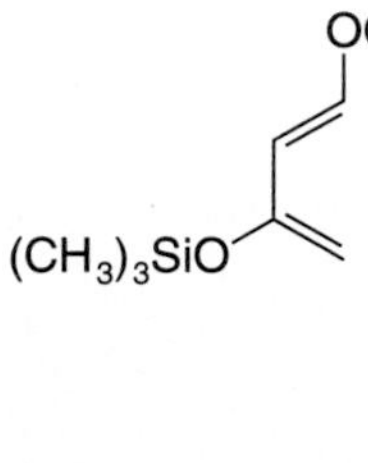

(*c*) [diene with two $OCCH_3$ (ester) groups] + $CH_2{=}CHCO_2CH_3$

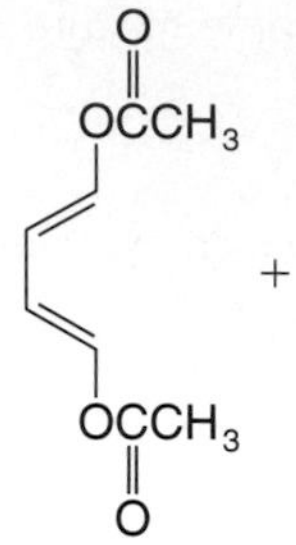

(*d*) [furan] + $CH_3O_2CC{\equiv}CCO_2CH_3$

(*e*) [5-(methoxymethyl)cyclopentadiene, CH_2OCH_3] + $CH_2{=}CHNO_2$

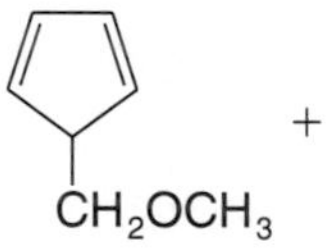

10.35 On standing, 1,3-cyclopentadiene is transformed into a new compound called *dicyclopentadiene,* having the molecular formula $C_{10}H_{12}$. Hydrogenation of dicyclopentadiene gives the compound shown. Suggest a structure for dicyclopentadiene. What kind of reaction is occurring in its formation?

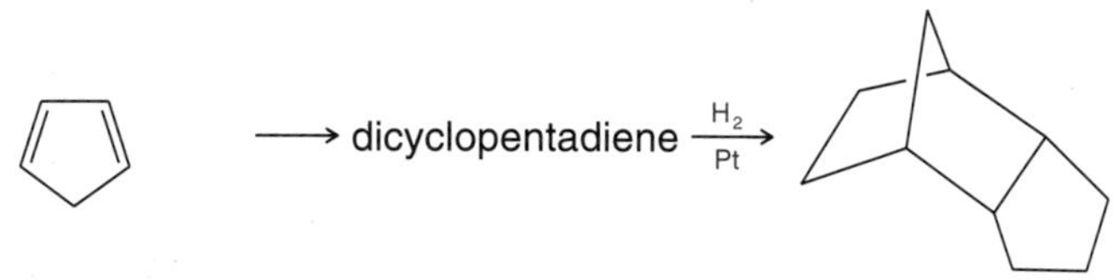

1,3-Cyclopentadiene $\longrightarrow$ dicyclopentadiene ($C_{10}H_{12}$) $\xrightarrow[Pt]{H_2}$ [tricyclic product] ($C_{10}H_{16}$)

10.36 When 2,3-disubstituted cyclobutenes are treated with bromine at room temperature, addition occurs in the usual manner to give the corresponding dibromides with the four-membered ring intact. If, however, the 2,3-disubstituted cyclobutene is first heated and then treated with bromine, the isolated dibromide is as shown in the following equation. Offer a mechanistic explanation for these observations.

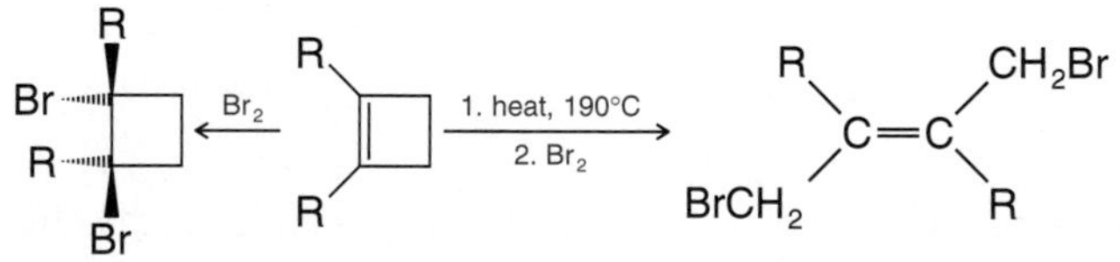

[dibromocyclobutane with R, Br, R, Br] $\xleftarrow{Br_2}$ [cyclobutene with two R groups] $\xrightarrow[2.\ Br_2]{1.\ heat,\ 190°C}$ $(R)(BrCH_2)C{=}C(CH_2Br)(R)$

10.37 Each of the following reactions has been reported in the chemical literature and gives the indicated product in high yield. Suggest a reasonable mechanism to account for each.

(*a*) [maleic anhydride] + [1,2-bis($COCH_3$)cyclobutene] $\xrightarrow{heat}$ [bicyclic anhydride product with two $COCH_3$ groups]

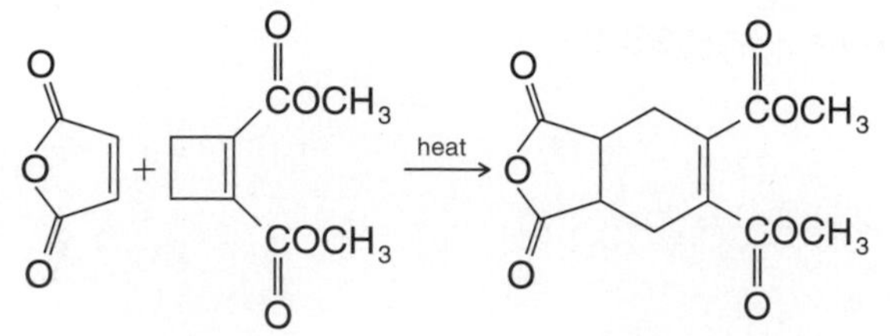

(a) O + $COCH_3$ / $COCH_3$ $\xrightarrow{\text{heat}}$ O / $COCH_3$ / $COCH_3$

10.38 Refer to the molecular orbital diagrams of allyl cation (Figure 10.18) and those presented earlier in this chapter for ethylene,1,3-butadiene, and 1,3,5-hexatriene (Figures 10.11 through 10.13) to decide which of the following cycloaddition reactions are allowed and which are forbidden according to the Woodward-Hoffmann rules.

(a) (b) (c) (d) (e)

10.39 Alkenes slowly undergo a reaction in air called *autoxidation* in which allylic hydroperoxides are formed.

Cyclohexene + O_2 ⟶ 3-Hydroperoxycyclohexene (—OOH)

Cyclohexene Oxygen 3-Hydroperoxycyclohexene

Keeping in mind that oxygen has two unpaired electrons ($\cdot\ddot{O}:\ddot{O}\cdot$), suggest a reasonable mechanism for this reaction.

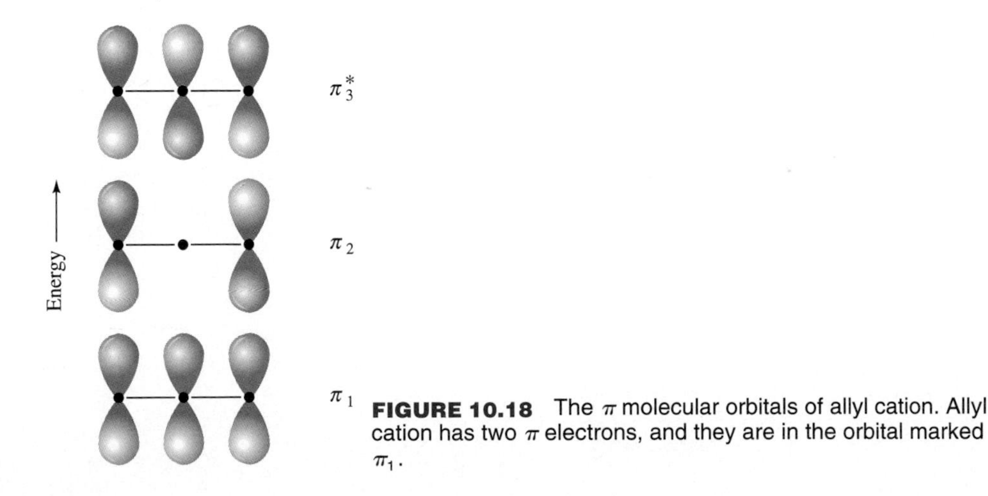

FIGURE 10.18 The π molecular orbitals of allyl cation. Allyl cation has two π electrons, and they are in the orbital marked π_1.

MOLECULAR MODELING EXERCISES

10.40 Construct molecular models of:

(*a*) 1,2-pentadiene (*b*) (*E*)-1,3-pentadiene (*c*) 1,4-pentadiene

In each of the models specify the hybridization state of each carbon and identify all allylic and vinylic hydrogens.

10.41 Construct molecular models of the *s*-cis and *s*-trans conformations of (*Z*)-1,3 pentadiene.

10.42 Show, using a molecular model, why 2,3-di-*tert*-butyl-1,3-butadiene is expected to be exceedingly unreactive toward Diels-Alder cycloaddition.

10.43 Construct molecular models of the two stereoisomeric forms of 1,3-dichloro-1,3-propadiene. What is the relationship between these two stereoisomers?

10.44 Construct a molecular model of 1,3-cyclohexadiene. Compare the preference of a substituent for the equatorial orientation of C-5 in 1,3-cyclohexadiene to its preference in cyclohexane.

CHAPTER 11

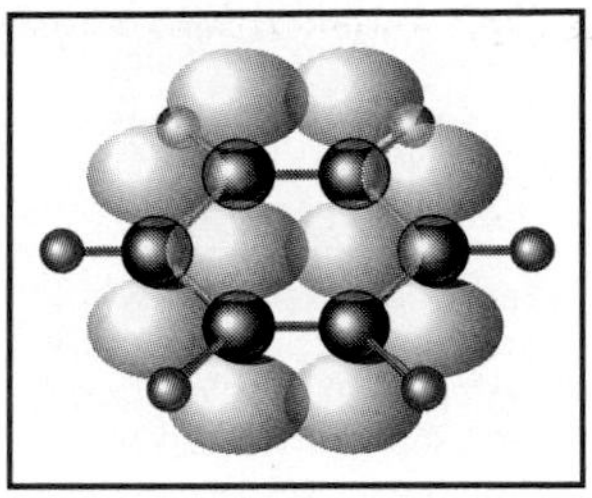

ARENES AND AROMATICITY

In this chapter and the next we extend our coverage of conjugated systems to include **arenes.** Arenes are hydrocarbons based on the benzene ring as a structural unit. Benzene, toluene, and naphthalene, for example, are arenes.

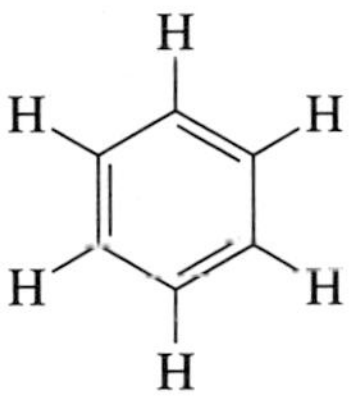

Benzene 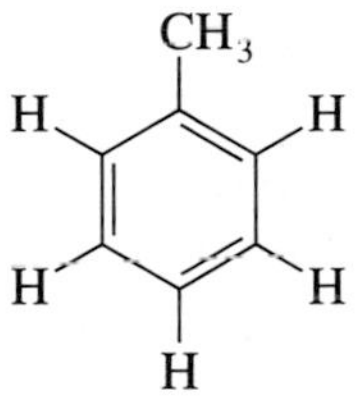   Toluene 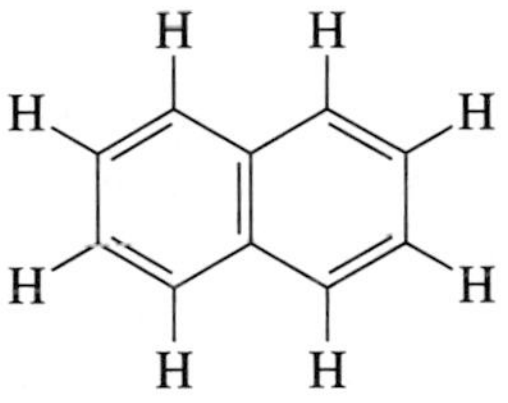   Naphthalene

One factor that makes conjugation in arenes distinctive is its cyclic nature. A conjugated system that closes upon itself can have properties that are much different from those of open-chain polyenes. Arenes are also referred to as **aromatic hydrocarbons.** Used in this sense, the word **aromatic** has nothing to do with odor but rather refers to a level of stability for arenes that is substantially greater than that expected on the basis of their formulation as conjugated trienes. Our main objective in this chapter will be to develop an appreciation for the concept of **aromaticity**—to see what are the properties of benzene and its derivatives that reflect its special stability, and to explore the reasons for it. This chapter develops the idea of the benzene ring as a fundamental structural unit and examines the effect of a benzene ring as a substituent. The chapter following this one describes reactions that involve the ring itself.

Aromaticity can be profitably introduced by tracing the history of benzene, its origin, and its structure. Many of the terms we use, including *aromaticity* itself, are of historical origin. Let us begin with the discovery of benzene.

11.1 BENZENE

Faraday is better known in chemistry for his laws of electrolysis and in physics for proposing the relationship between electric and magnetic fields and for demonstrating the principle of electromagnetic induction.

In 1825 Michael Faraday isolated a new hydrocarbon from illuminating gas, which he called "bicarburet of hydrogen." Nine years later Eilhardt Mitscherlich of the University of Berlin prepared the same substance by heating benzoic acid with lime and found it to be a hydrocarbon having the empirical formula C_nH_n.

$$\underset{\text{Benzoic acid}}{C_6H_5CO_2H} + \underset{\text{Calcium oxide}}{CaO} \xrightarrow{\text{heat}} \underset{\text{Benzene}}{C_6H_6} + \underset{\text{Calcium carbonate}}{CaCO_3}$$

Eventually, because of its relationship to benzoic acid, this hydrocarbon came to be named *benzin,* then later *benzene,* the name by which it is known today in the IUPAC nomenclature system.

Benzoic acid had been known for several hundred years by the time of Mitscherlich's experiment. Many trees exude resinous materials called *balsams* when cuts are made in their bark. Some of these balsams are very fragrant, which once made them highly prized articles of commerce, especially when the trees which produced them could be found only in exotic, faraway lands. *Gum benzoin* is one such substance and is obtained from a tree that grows in Java and Sumatra. *Benzoin* is a word derived from the French equivalent, *benjoin,* which in turn comes from the Arabic *luban jawi,* meaning "incense from Java." Benzoic acid is itself odorless but can easily be isolated from the mixture that makes up the material known as *gum benzoin.*

Compounds related to benzene were obtained from similar plant extracts. For example, a pleasant-smelling resin known as *tolu balsam* was obtained from the South American tolu tree. In the 1840s it was discovered that distillation of tolu balsam gave a methyl derivative of benzene, which, not surprisingly, came to be named *toluene.*

Although benzene and toluene are not particularly fragrant compounds themselves, their origins in aromatic plant extracts led them and compounds related to them to be classified as *aromatic hydrocarbons.* Alkanes, alkenes, and alkynes belong to another class, the **aliphatic hydrocarbons.** The word *aliphatic* comes from the Greek *aleiphar* (meaning "oil" or "unguent") and arose from the observation that hydrocarbons of this class could be obtained by the chemical degradation of fats.

Benzene was prepared from coal tar by August W. von Hofmann in 1845. This remained the primary source for the industrial production of benzene for many years, until petroleum-based technologies became competitive about 1950. Current production is about 6 million tons per year in the United States. A substantial portion of this benzene is converted to styrene for use in the preparation of polystyrene plastics and films.

Toluene is also an important organic chemical. Like benzene, its early production was from coal tar, but more recently petroleum sources have come to supply most of the toluene used for industrial purposes.

11.2 BENZENE REACTIVITY

The division of hydrocarbons into two groups designated as aliphatic and aromatic in the 1860s took place at a time when it was already apparent that there was something special about benzene, toluene, and their derivatives. The high carbon/hydrogen ratios evident in their molecular formulas (benzene is C_6H_6, toluene is C_7H_8) indicate that aromatic hydrocarbons are highly unsaturated and, like alkenes and alkynes, should undergo addition reactions. However, under conditions in which bromine adds rapidly to alkenes and alkynes, benzene proved to be inert. When bromination was carried out in the presence of catalysts such as Fe(III) bromide, the reaction that took place was not addition but substitution!

$$C_6H_6 + Br_2 \xrightarrow{CCl_4} \text{no observable reaction}$$

Benzene, Bromine

$$C_6H_6 + Br_2 \xrightarrow{FeBr_3} C_6H_5Br + HBr$$

Benzene, Bromine → Bromobenzene, Hydrogen bromide

Further, only one monobromination product of benzene was ever obtained. This implies that all the hydrogen atoms of benzene are equivalent. Substitution of one hydrogen by bromine gives the same product as substitution of any of the other hydrogens.

Organic chemists came to regard the six carbon atoms of benzene as a fundamental structural unit. Reactions could be carried out that altered its substituents, but the integrity of the benzene unit remained undisturbed. There must be something "special" about the structure of benzene that makes it inert to many of the reagents that add to alkenes and alkynes.

11.3 KEKULÉ'S FORMULATION OF THE BENZENE STRUCTURE

In 1866, only a few years after publishing his ideas concerning what we now recognize as the structural theory of organic chemistry, August Kekulé applied its principles to the structure of benzene. He based his reasoning on three premises:

1. Benzene is C_6H_6.
2. All the hydrogens of benzene are equivalent.
3. The structural theory requires that there be four bonds to each carbon.

Kekulé advanced the venturesome notion that the six carbon atoms of benzene were joined together in a ring. Four bonds to each carbon could be accommodated by a system of alternating single and double bonds with one hydrogen on each carbon.

In 1861 Johann Josef Loschmidt, who was later to become a professor at the University of Vienna, privately published a book containing a structural formula for benzene similar to that which Kekulé would propose five years later. Loschmidt's book reached few readers, and his ideas were not well known.

A flaw in the Kekulé formula for benzene was soon pointed out. According to the Kekulé formula, 1,2 disubstitution is different from 1,6 disubstitution.

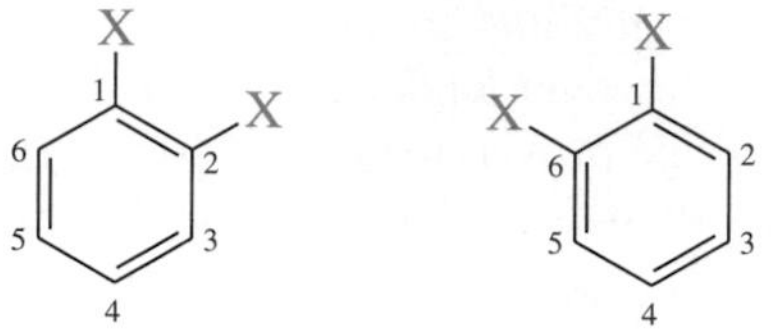

1,2-Disubstituted derivative of benzene

1,6-Disubstituted derivative of benzene

These two structural formulas differ in that the two substituted carbons are connected by a double bond in one but by a single bond in the other. Since all the available facts indicated that no such isomerism existed, Kekulé modified his hypothesis regarding benzene to one in which rapid bond migrations caused interconversion of the two structures.

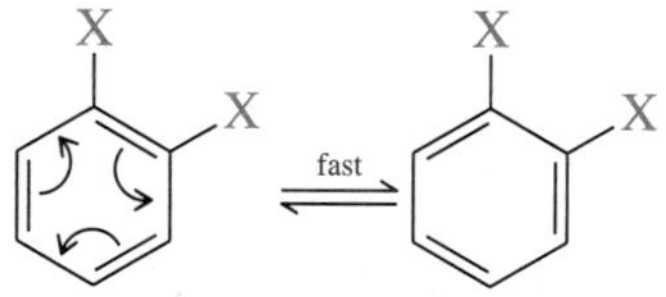

fast

BENZENE, DREAMS, AND CREATIVE THINKING

At ceremonies in Berlin in 1890 celebrating the twenty-fifth anniversary of his proposed structure of benzene, August Kekulé recalled the thinking that led him to it. He began by noting that the idea of the structural theory came to him during a daydream while on a bus in London. Kekulé went on to describe the origins of his view of the benzene structure.

> There I sat and wrote for my textbook; but things did not go well; my mind was occupied with other matters. I turned the chair towards the fireplace and began to doze. Once again the atoms danced before my eyes. This time smaller groups modestly remained in the background. My mental eye, sharpened by repeated apparitions of similar kind, now distinguished larger units of various shapes. Long rows, frequently joined more densely; everything in motion, twisting and turning like snakes. And behold, what was that? One of the snakes caught hold of its own tail and mockingly whirled round before my eyes. I awoke, as if by lightning; this time, too, I spent the rest of the night working out the consequences of this hypothesis.*

Concluding his remarks, Kekulé merged his advocacy of creative imagination with the rigorous standards of science by reminding his audience:

> Let us learn to dream, then perhaps we shall find the truth. But let us beware of publishing our dreams before they have been put to the proof by the waking understanding.

The imagery of a whirling circle of snakes evokes a vivid picture that engages one's attention when first exposed to Kekulé's model of the benzene structure. Recently, however, the opinion has been expressed that Kekulé might have engaged in some hyperbole during his speech. Professor John Wotiz of Southern Illinois University suggests that discoveries in science are the result of a disciplined analysis of a sufficient body of experimental observations to progress to a higher level of understanding. Wotiz' view that Kekulé's account is more fanciful than accurate has sparked a controversy with ramifications that go beyond the history of organic chemistry. How does creative thought originate? What can we do to become more creative? Because these are questions that have concerned psychologists for decades, the idea of a sleepy Kekulé being more creative than an alert Kekulé becomes more than simply a charming story he once told about himself.

* The Kekulé quotes are taken from the biographical article of K. Hafner published in *Angew. Chem. Internat. ed. Engl.* **18,** 641–651 (1979).

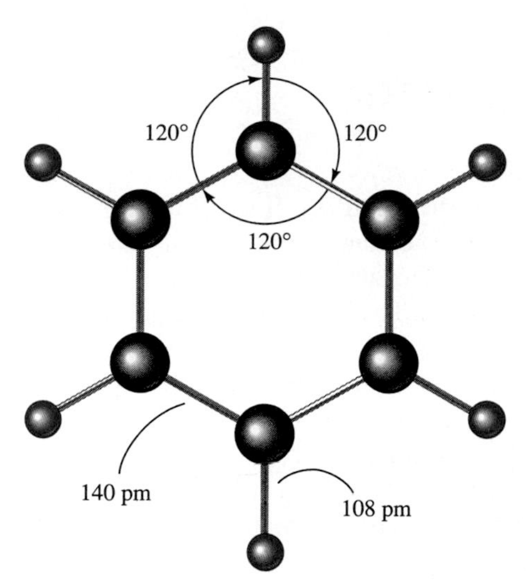

FIGURE 11.1 Bond distances and bond angles of benzene.

Kekulé's formulation addressed the *structure* of benzene but left unanswered important questions about its *reactivity*. Benzene does not behave at all the way we would expect cyclohexatriene to behave. *Benzene is not a static cyclohexatriene, nor is benzene a pair of rapidly equilibrating cyclohexatriene isomers.* It remained for twentieth-century electronic theories of organic chemistry to provide insight into why a benzene ring is such a stable structural unit. We will discuss modern interpretations of the stability of benzene shortly. First, however, let us examine what the results of a variety of experiments tell us about the details of its structure.

11.4 STRUCTURAL FEATURES OF BENZENE

Numerous studies leave no doubt that benzene is planar and that its carbon skeleton has the shape of a regular hexagon. There is no evidence to support structural formulations with alternating single and double bonds. As shown in Figure 11.1, all the carbon-carbon bonds are the same length (140 pm) and the 120° bond angles correspond to perfect sp^2 hybridization. Interestingly, the 140-pm bond distances in benzene are exactly midway between the typical sp^2–sp^2 single-bond distance of 146 pm and the sp^2–sp^2 double-bond distance of 134 pm. If bond distances are related to bond type, what kind of carbon-carbon bond is it that lies halfway between a single bond and a double bond in length?

11.5 A RESONANCE DESCRIPTION OF BONDING IN BENZENE

Twentieth-century descriptions of bonding in benzene include two major approaches, resonance and molecular orbital treatments, that taken together, provide a rather clear picture of aromaticity. We shall start with the resonance description of benzene.

The two Kekulé structures for benzene have the same arrangement of atoms, but differ in the placement of electrons. Thus they are resonance forms of each other, and neither one by itself correctly depicts the bonding in the actual molecule. Benzene is a hybrid of the two Kekulé structures. The hybrid structure is often represented by a hexagon containing an inscribed circle.

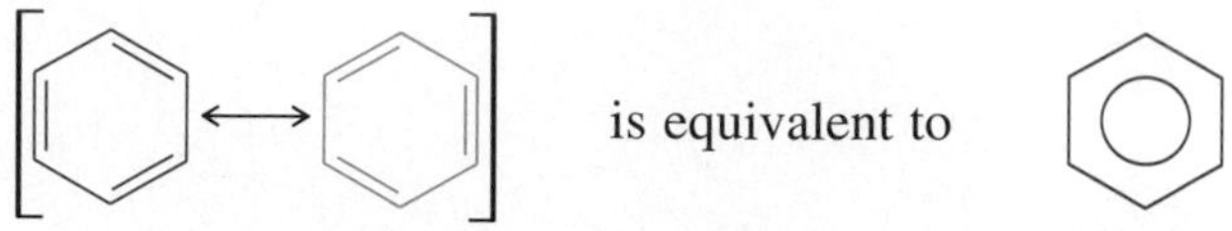

is equivalent to

The circle reminds us of the delocalized nature of the electrons. It was first suggested by the British chemist Sir Robert Robinson as a convenient symbol for the **aromatic sextet,** the six delocalized π electrons present in a benzene ring. In this text, we shall use both the Kekulé and the Robinson representations. The Robinson circle-in-a-hexagon symbol is a timesaving shorthand device, while Kekulé formulas are better for counting and keeping track of electrons, particularly when we study the chemical reactions of benzene and its derivatives.

PROBLEM 11.1 Write structural formulas for toluene ($C_6H_5CH_3$) and for benzoic acid ($C_6H_5CO_2H$) (*a*) as resonance hybrids of two Kekulé forms and (*b*) with the Robinson symbol.

Since the carbons that are singly bonded in one resonance form are doubly bonded in the other, the resonance description is consistent with the observed carbon-carbon bond distances in benzene. These distances not only are all identical but also are intermediate between typical single-bond and double-bond lengths.

We have come to associate electron delocalization, in conjugated dienes, for example, with increased stability. On that basis alone, benzene ought to be a stabilized molecule. It differs from other conjugated systems that we have seen, however, in that its π electrons are delocalized over a *cyclic conjugated* system. Both Kekulé structures of benzene are of equal energy, and one of the principles of resonance theory is that stabilization is greatest when the contributing structures are of similar energy. Cyclic conjugation in benzene, then, leads to a greater stabilization than is observed in noncyclic conjugated trienes. How much greater that stabilization is can be estimated from heats of hydrogenation.

11.6 THE STABILITY OF BENZENE

Hydrogenation of benzene and other arenes is more difficult than hydrogenation of alkenes and alkynes. Two of the more active catalysts are rhodium and platinum, and it is possible to hydrogenate arenes in the presence of these catalysts at room temperature and modest pressure. Benzene consumes three molar equivalents of hydrogen to give cyclohexane.

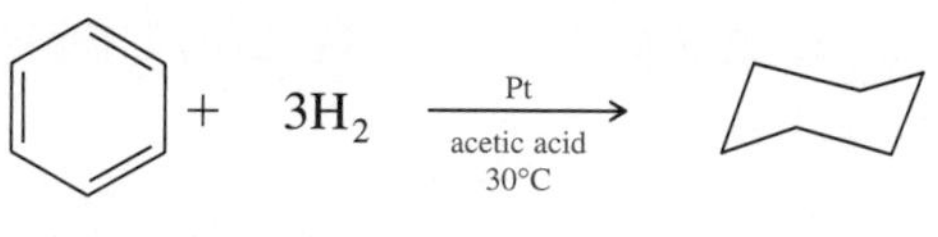

+ $3H_2$ $\xrightarrow[\text{acetic acid, 30°C}]{\text{Pt}}$

Benzene — Hydrogen (2–3 atm pressure) — Cyclohexane (100%)

Nickel catalysts, while less expensive than rhodium and platinum, are also less active. Hydrogenation of arenes in the presence of nickel requires high temperatures (100 to 200°C) and pressures (100 atm).

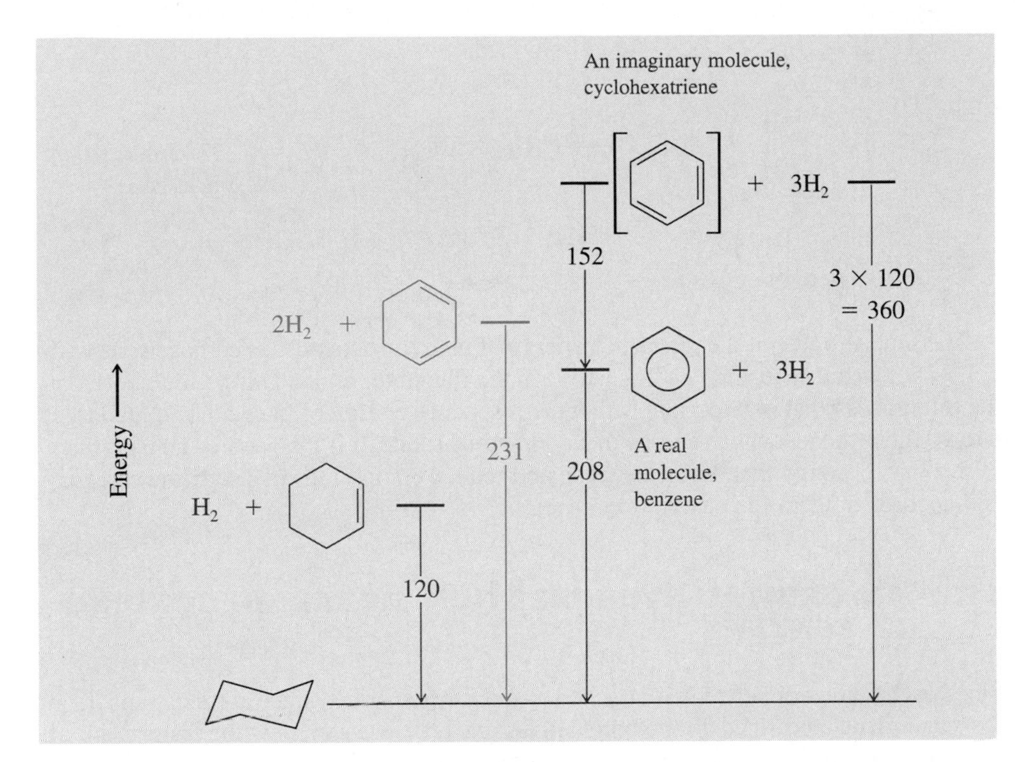

FIGURE 11.2 Heats of hydrogenation of cyclohexene, 1,3-cyclohexadiene, a hypothetical 1,3,5-cyclohexatriene, and benzene. All heats of hydrogenation are in kJ/mol.

The experimentally measured heat of hydrogenation of benzene to cyclohexane is, of course, the same regardless of the catalyst and is 208 kJ/mol (49.8 kcal/mol). To put this value into perspective, compare it with the heats of hydrogenation of cyclohexene and 1,3-cyclohexadiene, as shown in Figure 11.2.

The most striking feature of Figure 11.2 is that the heat of hydrogenation of benzene, with three "double bonds," is less than the heat of hydrogenation of the two double bonds of 1,3-cyclohexadiene. Our experience has been that some 125 kJ/mol (30 kcal/mol) is given off whenever a double bond is hydrogenated. When benzene combines with three molecules of hydrogen, the reaction is far less exothermic than we would expect it to be on the basis of a 1,3,5-cyclohexatriene structure for benzene.

How much less? Since 1,3,5-cyclohexatriene does not exist (if it did, it would instantly relax to benzene), we cannot measure its heat of hydrogenation in order to allow direct comparison with benzene. We can approximate the heat of hydrogenation of our hypothetical 1,3,5-cyclohexatriene as being equal to 3 times the heat of hydrogenation of cyclohexene, or a total of 360 kJ/mol (85.8 kcal/mol). The heat of hydrogenation of benzene is 152 kJ/mol (36 kcal/mol) *less* than expected for a hypothetical 1,3,5-cyclohexatriene with noninteracting double bonds. This is the **empirical resonance energy** of benzene. It is a measure of how much more stable benzene is than would be predicted on the basis of its formulation as a pair of rapidly interconverting 1,3,5-cyclohexatrienes.

We reach a similar conclusion when comparing benzene with the open-chain conjugated triene (*Z*)-1,3,5-hexatriene. Here we compare two real molecules, both conjugated trienes, but one is cyclic and the other is not. The heat of hydrogenation of (*Z*)-1,3,5-hexatriene is 337 kJ/mol (80.5 kcal/mol), a value which is 129 kJ/mol (30.7 kcal/mol) greater than that of benzene.

$$(Z)\text{-1,3,5-Hexatriene} + 3H_2 \longrightarrow CH_3(CH_2)_4CH_3 \qquad \Delta H^\circ = -337\ \text{kJ/mol}\ (-80.5\ \text{kcal/mol})$$

(Z)-1,3,5-Hexatriene Hydrogen Hexane

The precise value of the resonance energy of benzene depends, as comparisons with 1,3,5-cyclohexatriene and (*Z*)-1,3,5-hexatriene illustrate, on the compound chosen as the reference. What is important is that the resonance energy of benzene is quite large, 6 to 10 times the resonance energy of a conjugated triene. It is this very large increment of resonance energy that places benzene and related compounds in a separate category and accords to them the description *aromatic*.

11.7 AN ORBITAL HYBRIDIZATION MODEL OF BONDING IN BENZENE

Since each carbon of benzene is attached to three other atoms, since all the atoms lie in the same plane, and since all the bond angles are 120°, we can view the framework of carbon-carbon σ bonds as arising from the overlap of sp^2 hybrid orbitals. Figure 11.3 illustrates the σ framework of benzene.

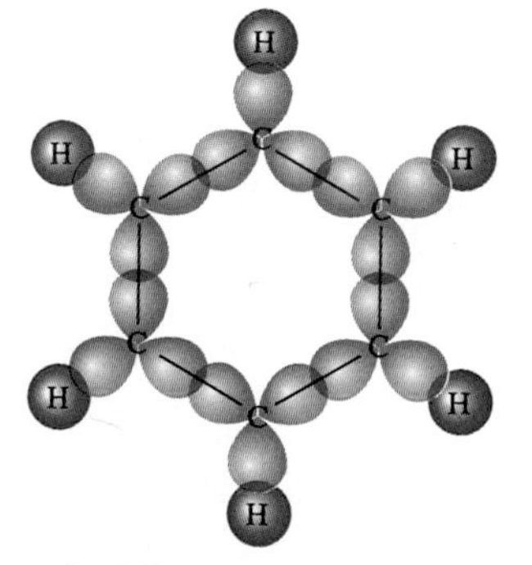

FIGURE 11.3 The assembly of σ bonds in benzene. Each carbon is sp^2 hybridized and forms σ bonds to two other carbons and to one hydrogen.

An unhybridized $2p$ orbital remains on each carbon. As shown in Figure 11.4, overlap of these $2p$ orbitals generates a continuous π system encompassing all the carbon atoms of the ring. The six π electrons are delocalized over all six carbons. Benzene is characterized by a cyclic conjugated π system containing six π electrons. The electron density associated with the π system of benzene is greatest in regions directly above and directly below the plane of the ring. The π electron density is zero in the plane of the ring.

11.8 THE π MOLECULAR ORBITALS OF BENZENE

The picture portrayed in Figure 11.4 is a useful model of electron distribution in benzene and emphasizes the delocalization of its π electrons. It is, of course, a superficial one, since six electrons cannot simultaneously occupy any one orbital, be it an

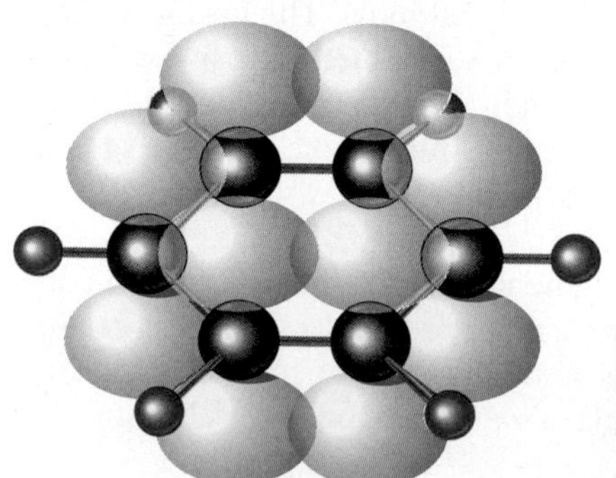

FIGURE 11.4 Each carbon of benzene contributes one $2p$ orbital. Overlap of these six $2p$ orbitals generates a π system encompassing the entire ring. There are regions of high π electron density above and below the plane of the ring.

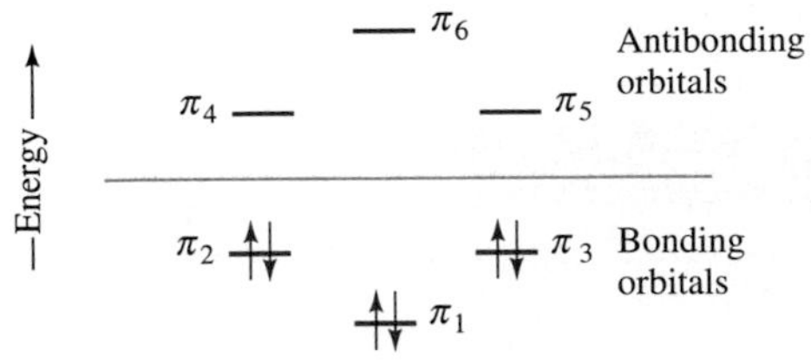

FIGURE 11.5 The π molecular orbitals of benzene arranged in order of increasing energy. The six π electrons of benzene occupy the three lowest-energy orbitals, all of which are bonding.

atomic orbital or a molecular orbital. A more rigorous molecular orbital analysis recognizes that overlap of the six $2p$ atomic orbitals of the ring carbons generates six π molecular orbitals. These six π molecular orbitals include three which are bonding and three which are antibonding. The relative energies of these orbitals and the distribution of the π electrons among them are illustrated in Figure 11.5. Benzene is said to have a **closed-shell** π electron configuration. All the bonding orbitals are filled, and there are no electrons in antibonding orbitals.

Other species said to have closed-shell electron configurations include the noble gases (helium, neon, argon, and so forth). Benzene and the noble gases share the common property of high stability and low reactivity.

Higher-level molecular orbital theory can provide quantitative information about orbital energies and how strongly a molecule holds it electrons. When one compares aromatic and nonaromatic species in this way, it is found that cyclic delocalization causes the π electrons of benzene to be more strongly bound (more stable) than they would be if restricted to a system with alternating single and double bonds.

We shall return to the molecular orbital description of benzene later in this chapter (Section 11.21) to see how other conjugated polyenes compare with benzene.

11.9 SUBSTITUTED DERIVATIVES OF BENZENE AND THEIR NOMENCLATURE

All compounds that contain a benzene ring are aromatic, and substituted derivatives of benzene make up the largest class of aromatic compounds. Many such compounds are named by attaching the name of the substituent as a prefix to *benzene.*

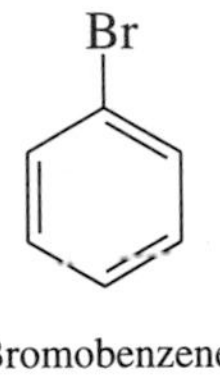

Bromobenzene 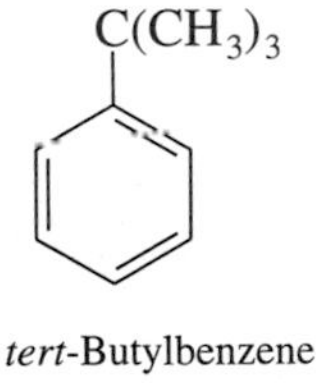   *tert*-Butylbenzene 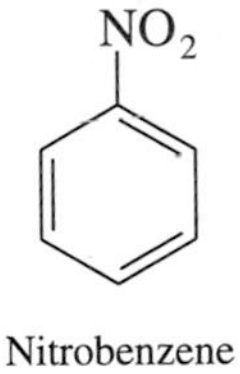   Nitrobenzene

Many simple monosubstituted derivatives of benzene have common names of long standing that have been retained in the IUPAC system. The common names benzaldehyde and benzoic acid, for example, are used far more frequently than their systematic counterparts benzenecarbaldehyde and benzenecarboxylic acid, respectively. Table 11.1 gives the common names of some others.

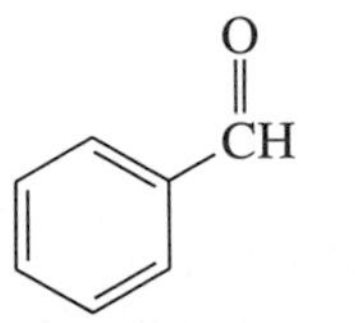

Benzenecarbaldehyde (benzaldehyde)   Benzenecarboxylic acid (benzoic acid)

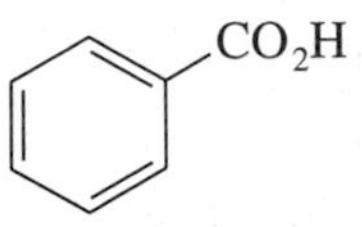

TABLE 11.1
Names of Some Common Benzene Derivatives

Structure	Name*
C_6H_5—$CH{=}CH_2$	Styrene
C_6H_5—$\overset{O}{\overset{\|\|}{C}}CH_3$	Acetophenone
C_6H_5—OH	Phenol
C_6H_5—OCH_3	Anisole
C_6H_5—NH_2	Aniline

* These common names are acceptable in IUPAC nomenclature.

Dimethyl derivatives of benzene are called *xylenes.* There are three xylene isomers, the *ortho* (*o*)-, *meta* (*m*)-, and *para* (*p*)-substituted derivatives.

CH_3 CH_3 | CH_3 CH_3 | CH_3 CH_3

o-Xylene (1,2-dimethylbenzene) | *m*-Xylene (1,3-dimethylbenzene) | *p*-Xylene (1,4-dimethylbenzene)

The prefix *ortho* signifies a 1,2-disubstituted benzene ring, *meta* signifies 1,3-disubstitution, and *para* signifies 1,4-disubstitution. The prefixes *o, m,* and *p* can be used when a substance is named as a benzene derivative or when a specific base name (such as acetophenone) is used. For example,

Cl Cl | NO_2 CH_3 | O C CH_3 F

o-Dichlorobenzene (1,2-dichlorobenzene) | *m*-Nitrotoluene (3-nitrotoluene) | *p*-Fluoroacetophenone (4-fluoroacetophenone)

PROBLEM 11.2 Write a structural formula for each of the following compounds:

(*a*) *o*-Ethylanisole (*b*) *m*-Chlorostyrene (*c*) *p*-Nitroaniline

SAMPLE SOLUTION (*a*) The parent compound in *o*-ethylanisole is anisole. Anisole, as shown in Table 11.1, has a methoxy (CH_3O—) substituent on the benzene ring. The ethyl group in *o*-ethylanisole is attached to the carbon adjacent to the carbon that bears the methoxy substituent.

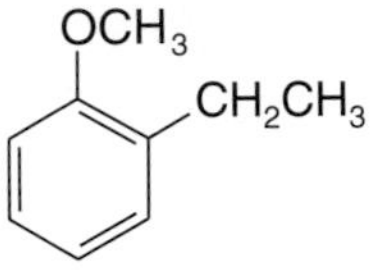

OCH_3 CH_2CH_3

o-Ethylanisole

The prefixes *o, m,* and *p* are *not* used when three or more substituents are present on benzene. Multiple substitution is described by identifying substituent positions on the ring.

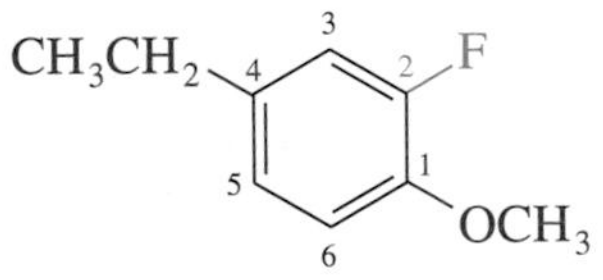

4-Ethyl-2-fluoroanisole

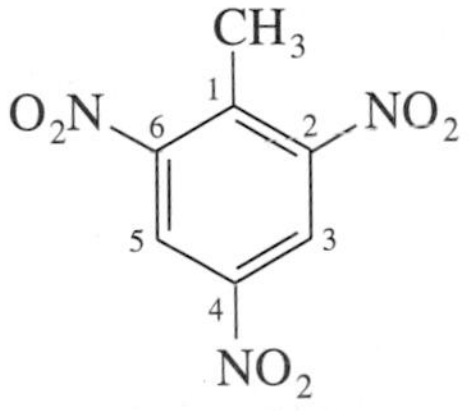

2,4,6-Trinitrotoluene

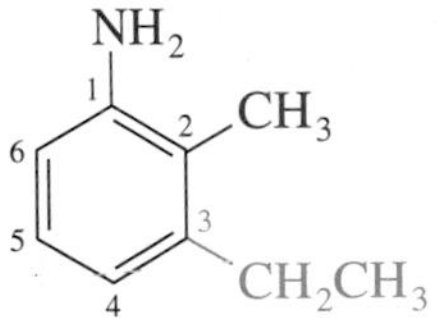

3-Ethyl-2-methylaniline

In these examples the numbering sequence is established by the base name of the benzene derivative: anisole has its methoxy group at C-1, toluene its methyl group at C-1, and aniline its amino group at C-1. The direction of numbering is chosen to give the next substituted position the lowest number irrespective of what substituent it bears. *The order of appearance of substituents in the name is alphabetical.* When no simple base name other than benzene is appropriate, positions are numbered so as to give the lowest locant at the first point of difference. Thus, each of the following examples is named as a 1,2,4-trisubstituted derivative of benzene rather than as a 1,3,4-derivative:

The "first point of difference" rule was introduced in Section 2.13.

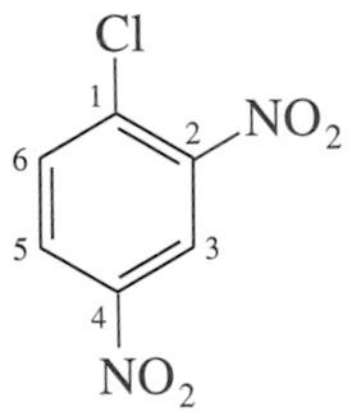

1-Chloro-2,4-dinitrobenzene

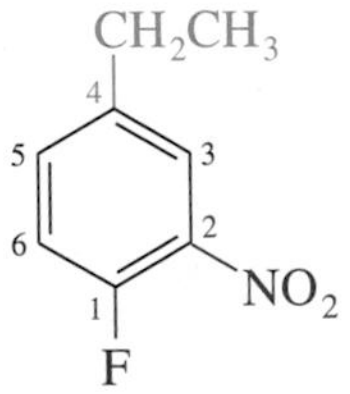

4-Ethyl-1-fluoro-2-nitrobenzene

In cases where the benzene ring is named as a substituent, the word *phenyl* is used to stand for C_6H_5—. Similarly, an arene named as a substituent is called an *aryl* group. A *benzyl* group is $C_6H_5CH_2$—.

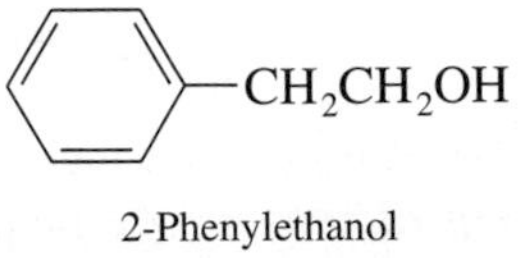

2-Phenylethanol

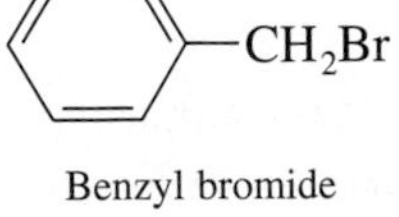

Benzyl bromide

Biphenyl is the accepted IUPAC name for the compound in which two benzene rings are connected by a single bond.

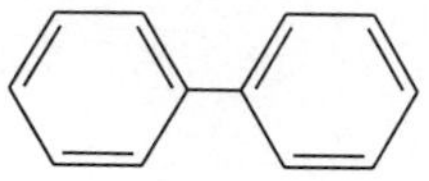

Biphenyl

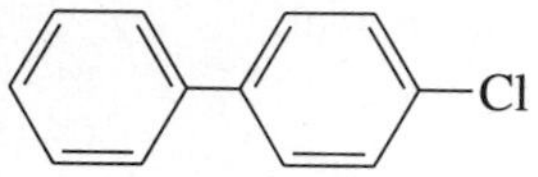

p-Chlorobiphenyl

11.10 POLYCYCLIC AROMATIC HYDROCARBONS

Members of a class of arenes called **polycyclic benzenoid aromatic hydrocarbons** possess substantial resonance energies because each is a collection of benzene rings fused together.

Naphthalene is a white crystalline solid melting at 80°C that sublimes readily. It has a characteristic odor and was formerly used as a moth repellent.

Naphthalene, anthracene, and phenanthrene are the three simplest members of this class. They are all present in **coal tar,** a mixture of organic substances formed when coal is converted to coke by heating at high temperatures (about 1000°C) in the absence of air. Naphthalene is **bicyclic** (has two rings), and its two benzene rings share a common side. Anthracene and phenanthrene are both **tricyclic** aromatic hydrocarbons. Anthracene has three rings fused in a "linear" fashion, while "angular" fusion characterizes phenanthrene. The structural formulas of naphthalene, anthracene, and phenanthrene are shown along with the numbering system used to name their substituted derivatives:

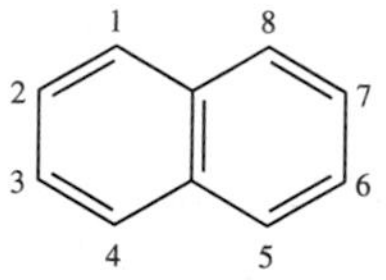

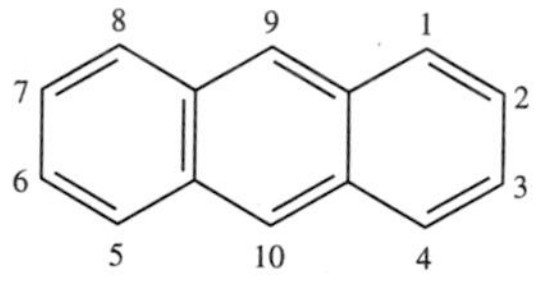

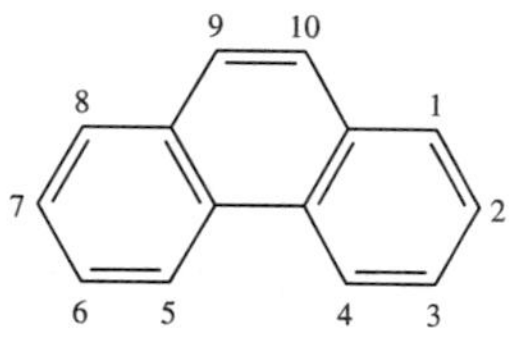

Arene:	Naphthalene	Anthracene	Phenanthrene
Resonance energy:	255 kJ/mol (61 kcal/mol)	347 kJ/mol (83 kcal/mol)	381 kJ/mol (91 kcal/mol)

In general, the most stable structural formula for a polycyclic aromatic hydrocarbon is the resonance form which has the greatest number of rings that correspond to Kekulé formulations of benzene. Naphthalene provides a fairly typical example:

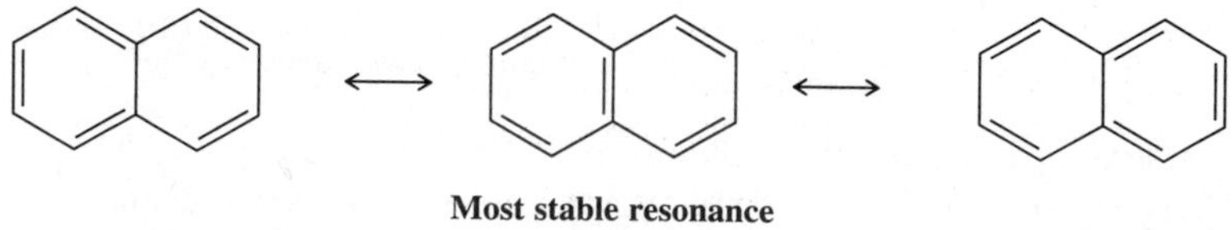

Only left ring corresponds to Kekulé benzene.

Most stable resonance form

Both rings correspond to Kekulé benzene.

Only right ring corresponds to Kekulé benzene.

Notice that anthracene cannot be represented by any single Lewis structure in which all three rings correspond to Kekulé formulations of benzene, while phenanthrene can.

PROBLEM 11.3 Chrysene is an aromatic hydrocarbon found in coal tar. The structure shown is not the most stable resonance form. Write the most stable resonance form for chrysene.

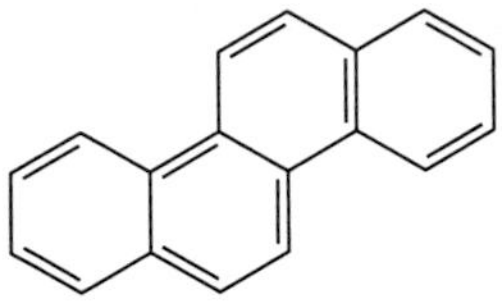

A large number of polycyclic benzenoid aromatic hydrocarbons are known. Many have been synthesized in the laboratory, and several of the others are products of combustion. Benzo[a]pyrene has been determined to be present in tobacco smoke. It contaminates food cooked on barbecue grills and collects in the soot of chimneys. Benzo[a]pyrene is a **carcinogen** (a cancer-causing substance). It is converted in the liver to an epoxy diol that can induce mutations leading to the uncontrolled growth of certain cells.

In 1775 the British surgeon Sir Percivall Pott suggested that scrotal cancer in chimney sweeps was caused by soot. This was the first proposal that cancer could be caused by chemicals present in the workplace.

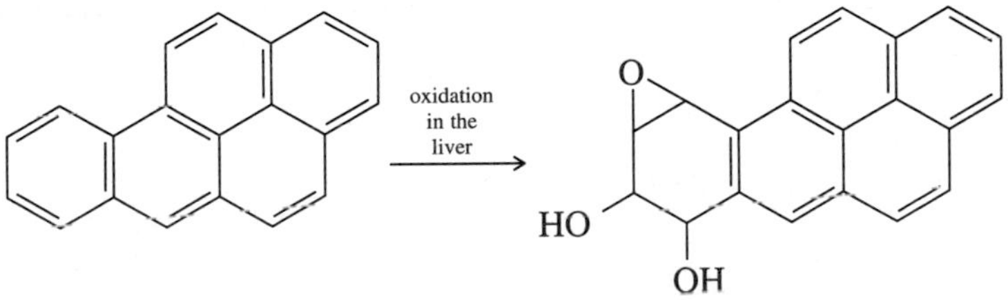

oxidation in the liver →

HO OH

Benzo[a]pyrene

7,8-Dihydroxy-9,10-epoxy-7,8,9,10-tetrahydrobenzo[a]pyrene

CARBON CLUSTERS AND FULLERENES

Harold W. Kroto is a physical chemist at the University of Sussex with interests in what seem to be two disparate areas—the structural chemistry of polyacetylenes and radio astronomy. These two interests merged in the 1970s when Kroto collaborated with David Walton, an organic chemistry colleague at Sussex, to prepare the compound HC≡C—C≡C—C≡N and, with radio astronomers at an observatory in Canada, to search for its presence in interstellar space. Kroto then wondered whether polyacetylenes of the type $HC{\equiv}C{-}(C{\equiv}C)_n{-}C{\equiv}CH$ might also occur as interstellar molecules. During a trip to Houston in the spring of 1984 he proposed an experiment to Richard Smalley of Rice University that led to some surprising and potentially far-reaching results.

Smalley had developed a method for the laser-induced evaporation of metals at very low pressure and was able to measure the molecular weights of the various clusters of atoms produced. Kroto suggested that by applying this technique to graphite (Figure 11.6) the gas-phase carbon species formed might be similar to those produced by certain stars.

Kroto returned to England, and before he and Smalley could arrange a mutually convenient time to do the experiment, one similar to it was carried out by a group of chemists at Exxon.

The Exxon group discovered that evaporation of carbon from a graphite surface gave a collection of many species containing as many as 100 carbons. They proposed that these species were small fragments of graphite composed of the same kinds of structural units as graphite itself. When Kroto, Smalley,

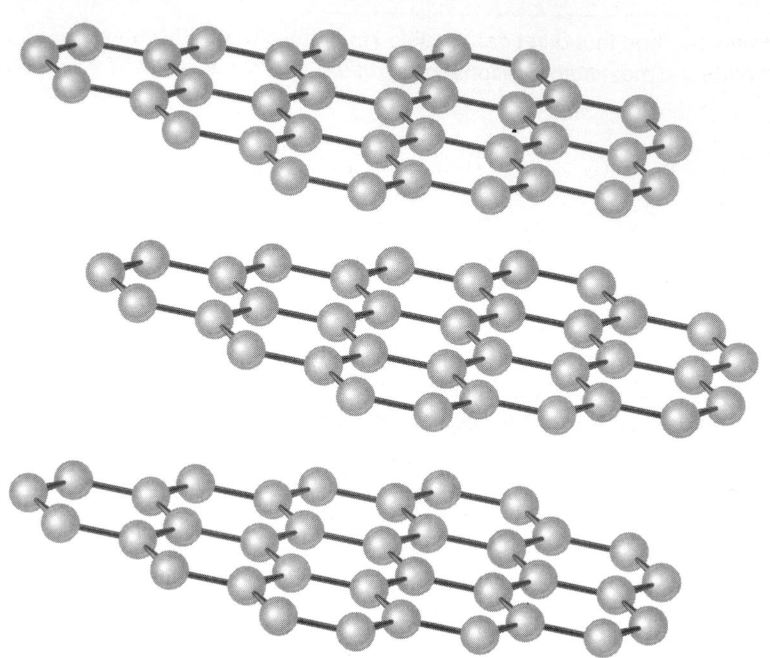

FIGURE 11.6 Graphite is a form of elemental carbon composed of parallel sheets of fused six-membered benzene-like rings.

and their coworkers repeated the experiment, they discovered that under certain conditions a species with a molecular formula of C_{60} was present in amounts much greater than all the others. If only graphite fragments are produced, there is no reason to believe that one fragment would be more abundant than any other fragment.

On speculating about what features would cause one species to be favored over all others, Kroto and Smalley concluded that C_{60} was formed preferentially because it was the most stable fragment and that its structure is the soccer ball–shaped cluster of carbon atoms shown in Figure 11.7*a*. Noting the similarity of the C_{60} cluster to the geodesic domes popularized by the American architect and inventor R. Buckminster Fuller, Kroto and Smalley suggested it be called *buckminsterfullerene* (an even less formal name is *buckyball*). They soon found that other carbon clusters, some

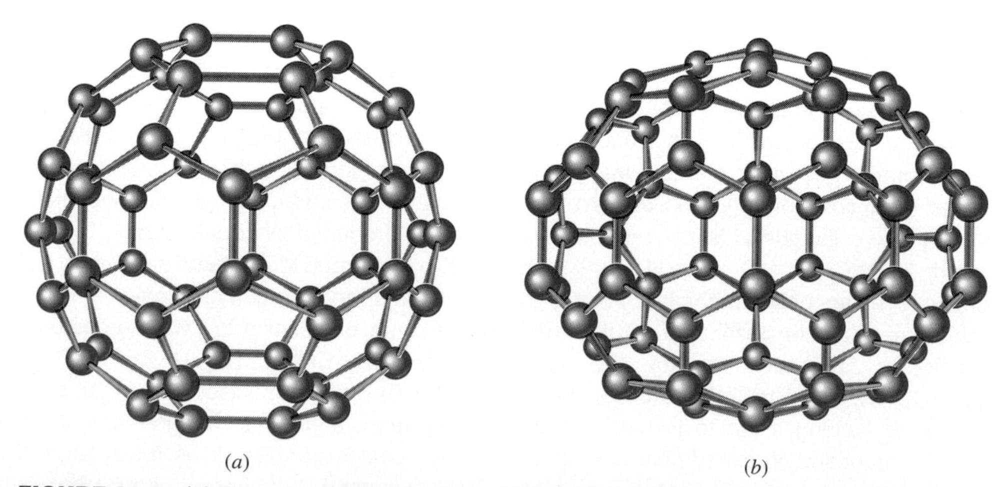

FIGURE 11.7 (*a*) Buckminsterfullerene (C_{60}). Note that all carbons are equivalent and that no five-membered rings are adjacent to one another. (*b*) The structure of C_{70} is more elongated than that of C_{60}, and not all the carbons are equivalent.

larger and some smaller than C_{60}, were formed by dissociation of graphite, and the general term *fullerene* is used to refer to spherical carbon clusters. Figure 11.7*b* depicts the structure of C_{70}.

All the carbons of C_{60} are equivalent and are sp^2 hybridized; each one simultaneously belongs to one five-membered ring and two benzenelike six-membered rings. The strain caused by distortion of the rings from coplanarity is equally distributed among all the carbons.

Confirmation of the Kroto-Smalley hypothesis for the structure of C_{60} required isolation of enough material to allow the arsenal of modern techniques of structure determination to be applied. A quantum leap in fullerene research came in 1990 when a team led by Wolfgang Krätschmer of the Max Planck Institute for Nuclear Physics in Heidelberg and Donald Huffman of the University of Arizona successfully prepared buckminsterfullerene in amounts sufficient for its isolation, purification and detailed study. Not only was the buckminsterfullerene structure shown to be correct, but academic and industrial scientists around the world seized the opportunity afforded by the availability of C_{60} in quantity to probe the properties of this novel form of carbon.

Speculation about the stability of C_{60} centered on the extent to which the aromaticity associated with its 20 benzene rings is degraded by their nonplanarity and the accompanying angle strain. It is now clear that C_{60} is a relatively reactive substance, reacting with many substances toward which benzene itself is inert. Many of these reactions are characterized by the addition of nucleophilic substances to buckminsterfullerene, converting sp^2 hybridized carbons to sp^3 hybridized ones and reducing the overall strain.

Thus far, the importance of carbon cluster chemistry has been in the discovery of new knowledge. Many scientists feel that the earliest industrial applications of the fullerenes will be based on their novel electrical properties. Buckminsterfullerene is an insulator, but has a high electron affinity and is a superconductor in its reduced form.

Although the question that began the fullerene story, the possibility that carbon clusters are formed in stars, still remains unanswered, the attempt to answer that question has opened the door to an entirely new area of chemistry.

11.11 PHYSICAL PROPERTIES OF ARENES

In general, arenes resemble other hydrocarbons in their physical properties. They are nonpolar materials, insoluble in water, and less dense than water. In the absence of polar substituent groups, intermolecular forces are weak and limited to van der Waals attractions of the induced dipole–induced dipole type.

Selected physical properties for a number of arenes are listed in Appendix 1.

Not long ago, and in spite of its flammability, benzene was widely used as a solvent. This use virtually disappeared once it was demonstrated that benzene is a carcinogen and statistical evidence revealed an increased incidence of leukemia among workers exposed to atmospheric levels of benzene as low as 1 ppm. Toluene has replaced benzene as an inexpensive organic solvent, because it has similar solvent properties but has not been determined to be carcinogenic in the cell systems and at the dose levels that benzene is.

11.12 REACTIONS OF ARENES: A PREVIEW

We shall examine the chemical properties of aromatic compounds from two different perspectives:

1. *One mode of chemical reactivity involves the ring itself as a functional group and includes*
   (*a*) Reduction
   (*b*) Electrophilic aromatic substitution

Reduction of arenes by catalytic hydrogenation was described in Section 11.6. A different method using group I metals as reducing agents, which gives 1,4-cyclohexadiene derivatives, will be presented in Section 11.13. **Electrophilic aromatic substitution** is the most important reaction type exhibited by benzene and its derivatives and comprises the entire subject matter of Chapter 12.

2. *The second family of reactions encompasses those in which the aryl group acts as a substituent and affects the reactivity of a functional unit to which it is attached.*

A carbon atom that is directly attached to a benzene ring is called a **benzylic** carbon (analogous to the allylic carbon of C=C—C). A phenyl group (C_6H_5—) is an even better conjugating substituent than a vinyl group (CH_2=CH—), and benzylic carbocations and radicals are more highly stabilized than their allylic counterparts. The double bond of an alkenylbenzene is stabilized to about the same extent as that of a conjugated diene.

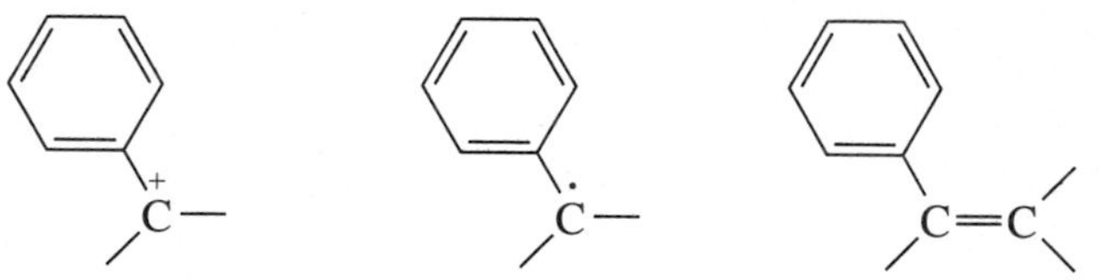

Benzylic carbocation Benzylic radical Alkenylbenzene

Reactions involving benzylic cations, benzylic radicals, and alkenylbenzenes will be discussed in Sections 11.14 through 11.20.

11.13 THE BIRCH REDUCTION

We saw in Section 9.11 that the combination of a group I metal and liquid ammonia is a powerful reducing system capable of reducing alkynes to trans alkenes. In the presence of an alcohol, this same combination reduces arenes to *nonconjugated dienes.* Thus, treatment of benzene with sodium and methanol or ethanol in liquid ammonia converts it to 1,4-cyclohexadiene.

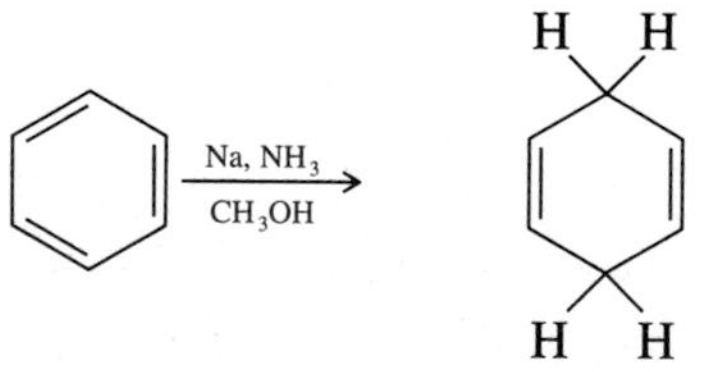

Na, NH_3 / CH_3OH

Benzene 1,4-Cyclohexadiene (80%)

Metal-ammonia-alcohol reductions of aromatic rings are known as **Birch reductions,** after the Australian chemist Arthur J. Birch, who demonstrated their usefulness beginning in the 1940s. The Birch reduction is one type of a more general class of reactions called *dissolving-metal reductions.*

The mechanism by which the Birch reduction of benzene takes place is analogous to the mechanism for the metal-ammonia reduction of alkynes (Figure 11.8). It involves a sequence of four steps in which steps 1 and 3 are single-electron transfers from the metal and steps 2 and 4 are proton transfers from the alcohol.

The overall reaction:

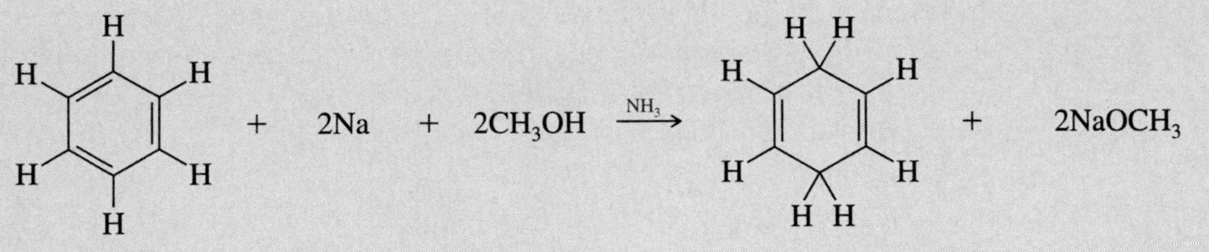

Benzene + 2Na (Sodium) + $2CH_3OH$ (Methanol) $\xrightarrow{NH_3}$ 1,4-Cyclohexadiene + $2NaOCH_3$ (Sodium methoxide)

The mechanism:

Step 1: An electron is transferred from sodium (the reducing agent) to the π system of the aromatic ring. The product is an anion radical.

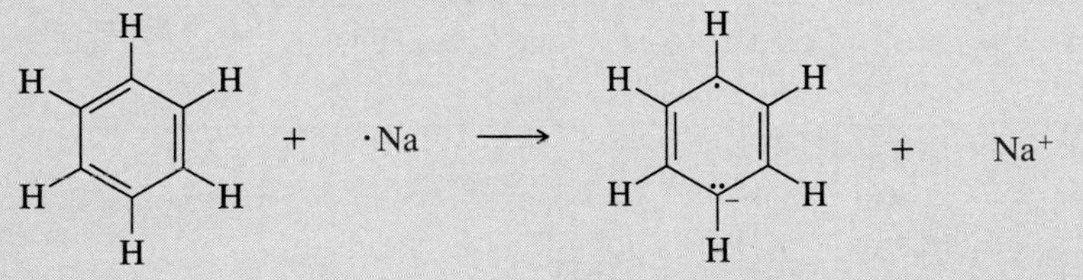

Benzene + ·Na (Sodium) ⟶ Benzene anion radical + Na^+ (Sodium ion)

Step 2: The anion radical is a strong base and abstracts a proton from methanol.

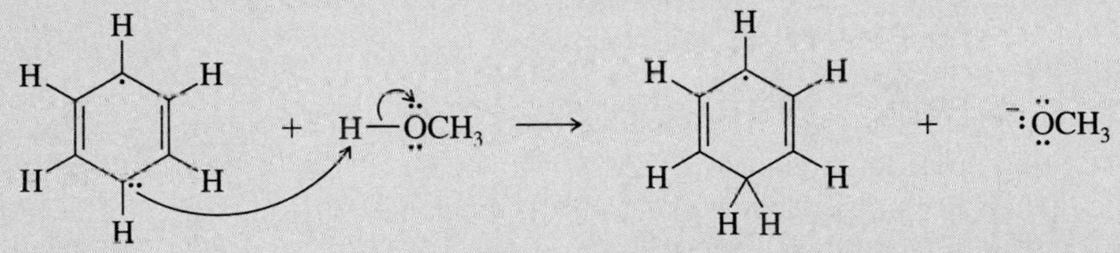

Benzene anion radical + $H—\ddot{O}CH_3$ (Methanol) ⟶ Cyclohexadienyl radical + $^-:\ddot{O}CH_3$ (Methoxide ion)

Step 3: The cyclohexadienyl radical produced in step 2 is converted to an anion by electron transfer from sodium.

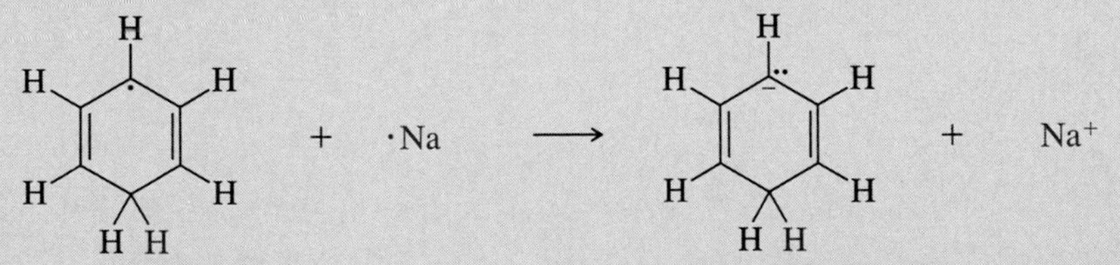

Cyclohexadienyl radical + ·Na (Sodium) ⟶ Cyclohexadienyl anion + Na^+ (Sodium ion)

Step 4: Proton transfer from methanol to the anion gives 1,4-cyclohexadiene.

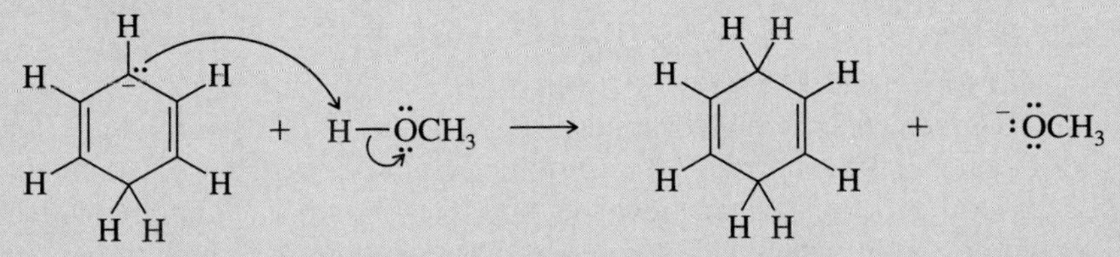

Cyclohexadienyl anion + $H—\ddot{O}CH_3$ ⟶ 1,4-Cyclohexadiene + $^-:\ddot{O}CH_3$ (Methoxide ion)

FIGURE 11.8 Mechanism of the Birch reduction.

The Birch reduction not only provides a method to prepare dienes from arenes, which cannot be accomplished by catalytic hydrogenation, but also gives a nonconjugated diene system rather than the more stable conjugated one. The reaction is kinetically controlled (Section 10.10) and, for reasons we need not go into, gives a 1,4-cyclohexadiene faster than a 1,3-cyclohexadiene.

Alkyl-substituted arenes give 1,4-cyclohexadienes in which the alkyl group is a substituent on the double bond.

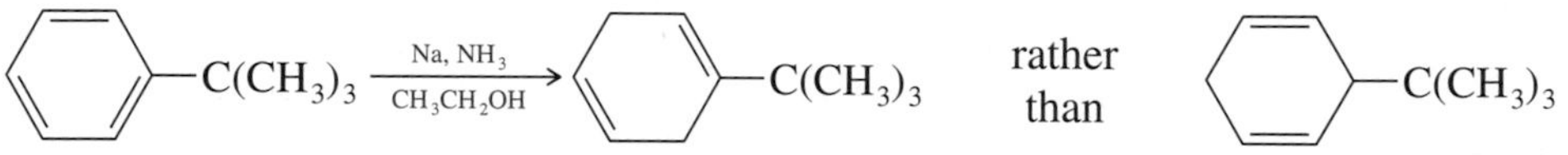

C_6H_5—$C(CH_3)_3$ $\xrightarrow[CH_3CH_2OH]{Na,\ NH_3}$ 1-tert-butyl-1,4-cyclohexadiene rather than 3-tert-butyl-1,4-cyclohexadiene

tert-Butylbenzene — 1-*tert*-Butyl-1,4-cyclohexadiene (86%) — 3-*tert*-Butyl-1,4-cyclohexadiene

---

PROBLEM 11.4 A single organic product was isolated after Birch reduction of *p*-xylene. Suggest a reasonable structure for this substance.

Substituents other than alkyl groups may also be present on the aromatic ring, but their reduction is beyond the scope of the present discussion.

11.14 FREE-RADICAL HALOGENATION OF ALKYLBENZENES

As noted in Section 11.12 the benzylic position in alkylbenzenes is analogous to the allylic position in alkenes. Thus a benzylic C—H bond, like an allylic one, is weaker than a C—H bond of an alkane, as the bond dissociation energies of toluene, propene, and 2-methylpropane attest:

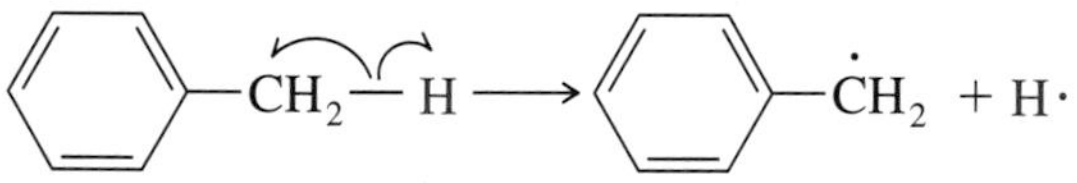

$$C_6H_5CH_2\text{—}H \longrightarrow C_6H_5\dot{C}H_2 + H\cdot \qquad \Delta H° = 356\text{ kJ/mol (85 kcal/mol)}$$

Toluene — Benzyl radical

$$CH_2{=}CHCH_2\text{—}H \longrightarrow CH_2{=}CH\dot{C}H_2 + H\cdot \qquad \Delta H° = 368\text{ kJ/mol (88 kcal/mol)}$$

Propene — Allyl radical

$$(CH_3)_3C\text{—}H \longrightarrow (CH_3)_3C\cdot + H\cdot \qquad \Delta H° = 397\text{ kJ/mol (95 kcal/mol)}$$

2-Methylpropane — *tert*-Butyl radical

We attributed the decreased bond dissociation energy in propene to stabilization of allyl radical by electron delocalization. Similarly, electron delocalization stabilizes benzyl radical and weakens the benzylic C—H bond. The unpaired electron is shared by the benzylic carbon and by the ring carbons that are ortho and para to it.

Most stable Lewis structure of benzyl radical

In orbital terms, as represented in Figure 11.9, benzyl radical is stabilized by delocalization of electrons throughout the extended π system formed by overlap of the p orbital of the benzylic carbon with the π system of the ring.

The comparative ease with which a benzylic hydrogen is abstracted leads to high regioselectivity in free-radical halogenations of alkylbenzenes. Thus, chlorination of toluene takes place exclusively at the benzylic carbon and is an industrial process for the preparation of the compounds shown.

$$\text{Toluene} \xrightarrow[\text{light or heat}]{Cl_2} \text{Benzyl chloride} \xrightarrow[\text{light or heat}]{Cl_2} \text{(Dichloromethyl)benzene} \xrightarrow[\text{light or heat}]{Cl_2} \text{(Trichloromethyl)benzene}$$

The common names of (dichloromethyl)benzene and (trichloromethyl)benzene are benzal chloride and benzotrichloride, respectively.

The propagation steps in the formation of benzyl chloride involve benzyl radical as an intermediate.

$$C_6H_5CH_3 + Cl\cdot \longrightarrow C_6H_5\dot{C}H_2 + HCl$$

Toluene, Chlorine atom, Benzyl radical, Hydrogen chloride

$$C_6H_5\dot{C}H_2 + Cl_2 \longrightarrow C_6H_5CH_2Cl + Cl\cdot$$

Benzyl radical, Chlorine, Benzyl chloride, Chlorine atom

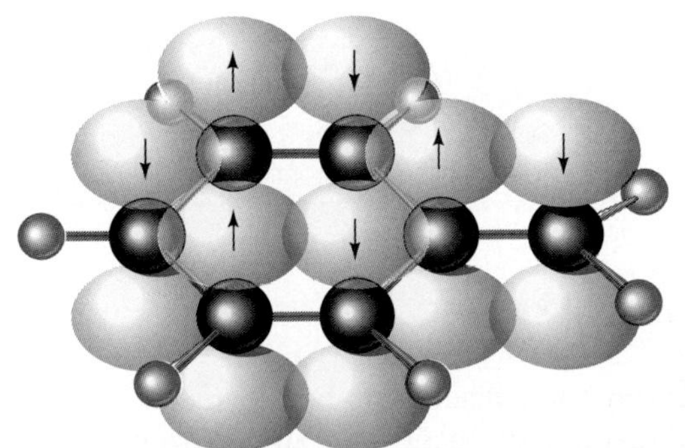

FIGURE 11.9 Benzyl radical is stabilized by overlap of its half-filled p orbital with the π system of the aromatic ring.

(Dichloromethyl)benzene and (trichloromethyl)benzene arise by further side-chain chlorination of benzyl chloride.

Benzylic bromination is a more commonly used laboratory procedure than chlorination and is typically carried out under conditions of photochemical initiation.

$$p\text{-}O_2NC_6H_4CH_3 + Br_2 \xrightarrow[\text{light}]{CCl_4,\ 80°C} p\text{-}O_2NC_6H_4CH_2Br + HBr$$

p-Nitrotoluene — Bromine — *p*-Nitrobenzyl bromide (71%) — Hydrogen bromide

As we saw when discussing allylic bromination in Section 10.4, *N*-bromosuccinimide is a convenient free-radical brominating agent. Benzylic brominations using this reagent are normally performed in carbon tetrachloride as the solvent in the presence of peroxides, which are added as initiators. As the example illustrates, free-radical bromination is selective for substitution of benzylic hydrogens

Benzoyl peroxide is a commonly used free-radical initiator. It has the formula $C_6H_5\overset{O}{\overset{\|}{C}}O O\overset{O}{\overset{\|}{C}}C_6H_5$.

$$C_6H_5CH_2CH_3 + \text{NBS} \xrightarrow[CCl_4,\ 80°C]{\text{benzoyl peroxide}} C_6H_5\underset{Br}{\underset{|}{C}}HCH_3 + \text{succinimide (NH)}$$

Ethylbenzene — *N*-Bromosuccinimide (NBS) — 1-Bromo-1-phenylethane (87%) — Succinimide

PROBLEM 11.5 The reaction of *N*-bromosuccinimide with the following compounds has been reported in the chemical literature. Each compound yields a single product in 95 percent yield. Identify the product formed from each starting material.

(*a*) *p-tert*-Butyltoluene (*b*) 4-Methyl-3-nitroanisole

SAMPLE SOLUTION (*a*) The only benzylic hydrogens in *p-tert*-butyltoluene are those of the methyl group that is attached directly to the ring. Substitution occurs there to give *p-tert*-butylbenzyl bromide.

$$(CH_3)_3C\text{–}C_6H_4\text{–}CH_3 \xrightarrow[\substack{CCl_4,\ 80°C \\ \text{free-radical} \\ \text{initiator}}]{NBS} (CH_3)_3C\text{–}C_6H_4\text{–}CH_2Br$$

p-tert-Butyltoluene — *p-tert*-Butylbenzyl bromide

11.15 OXIDATION OF ALKYLBENZENES

A striking example of the activating effect that a benzene ring has on reactions that take place at benzylic positions may be found in the reactions of alkylbenzenes with oxidizing agents. Chromic acid, for example, prepared by adding sulfuric acid to aqueous sodium dichromate, is a strong oxidizing agent but does not react either with benzene or with alkanes.

$$RCH_2CH_2R' \xrightarrow[H_2O,\ H_2SO_4,\ heat]{Na_2Cr_2O_7} \text{no reaction}$$

$$\text{Benzene} \xrightarrow[H_2O,\ H_2SO_4,\ heat]{Na_2Cr_2O_7} \text{no reaction}$$

An alternative oxidizing agent, similar to chromic acid in its reactions with organic compounds, is potassium permanganate ($KMnO_4$).

On the other hand, an alkyl side chain on a benzene ring is oxidized on being heated with chromic acid. The product is benzoic acid or a substituted derivative of benzoic acid.

$$C_6H_5CH_2R \text{ or } C_6H_5CHR_2 \xrightarrow[H_2O,\ H_2SO_4,\ heat]{Na_2Cr_2O_7} C_6H_5COH \ (\text{C=O})$$

Alkylbenzene — Benzoic acid

$$O_2N{-}C_6H_4{-}CH_3 \xrightarrow[H_2O,\ H_2SO_4]{Na_2Cr_2O_7} O_2N{-}C_6H_4{-}COH\ (\text{C=O})$$

p-Nitrotoluene — *p*-Nitrobenzoic acid (82–86%)

When two alkyl groups are present on the ring, both are oxidized.

$$CH_3{-}C_6H_4{-}CH(CH_3)_2 \xrightarrow[H_2O,\ H_2SO_4,\ heat]{Na_2Cr_2O_7} HOC{-}C_6H_4{-}COH\ (\text{both C=O})$$

p-Isopropyltoluene — *p*-Benzenedicarboxylic acid (45%)

Note that alkyl groups, regardless of their chain length, are converted to carboxyl groups ($—CO_2H$) attached directly to the ring. An exception is a *tert*-alkyl substituent. Because it lacks benzylic hydrogens, a *tert*-alkyl group is not susceptible to oxidation under these conditions.

PROBLEM 11.6 Chromic acid oxidation of 4-*tert*-butyl-1,2-dimethylbenzene yielded a single compound having the molecular formula $C_{12}H_{14}O_4$. What was this compound?

Side-chain oxidation of alkylbenzenes is important in certain metabolic processes. One way in which the body rids itself of foreign substances is by oxidation in the liver to compounds more easily excreted in the urine. Toluene, for example, is oxidized to benzoic acid by this process and is eliminated rather readily.

$$C_6H_5CH_3 \xrightarrow[\text{(an enzyme in the liver)}]{O_2,\ \text{cytochrome P-450}} C_6H_5COH\ (\text{C=O})$$

Toluene — Benzoic acid

Benzene, with no alkyl side chain, undergoes a different reaction in the presence of these enzymes, which convert it to a substance capable of inducing mutations in DNA (deoxyribonucleic acid). This difference in chemical behavior seems to be responsible for the fact that benzene is carcinogenic while toluene is not.

11.16 NUCLEOPHILIC SUBSTITUTION IN BENZYLIC HALIDES

Primary benzylic halides are ideal substrates for S_N2 reactions, since they are very reactive toward good nucleophiles and cannot undergo competing elimination.

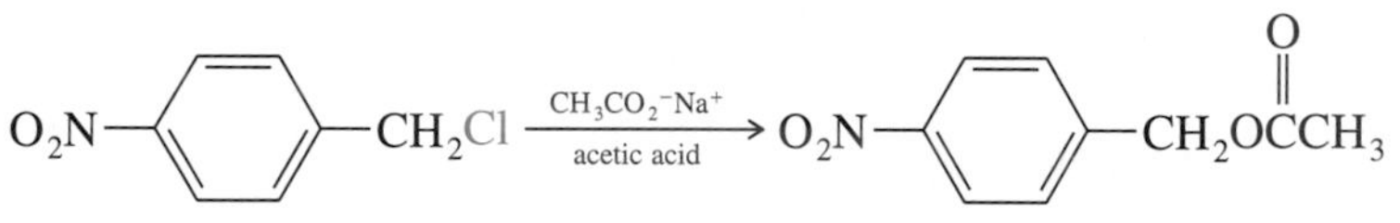

$$O_2N-C_6H_4-CH_2Cl \xrightarrow[\text{acetic acid}]{CH_3CO_2^-Na^+} O_2N-C_6H_4-CH_2OCCH_3 \;(C{=}O)$$

p-Nitrobenzyl chloride — *p*-Nitrobenzyl acetate (78–82%)

Benzylic halides that are secondary resemble secondary alkyl halides in that they undergo substitution only when the nucleophile is weakly basic. If the nucleophile is a strong base such as sodium ethoxide, elimination by the E2 mechanism is faster than substitution in secondary benzylic halides.

PROBLEM 11.7 Give the structure of the principal organic product formed on reaction of benzyl bromide with each of the following reagents:

(*a*) Sodium ethoxide
(*b*) Potassium *tert*-butoxide
(*c*) Sodium azide
(*d*) Sodium hydrogen sulfide
(*e*) Sodium iodide (in acetone)

SAMPLE SOLUTION (*a*) Benzyl bromide is a primary bromide and undergoes S_N2 reactions readily. It has no hydrogens β to the leaving group and so cannot undergo elimination. Ethoxide ion acts as a nucleophile, displacing bromide and forming benzyl ethyl ether.

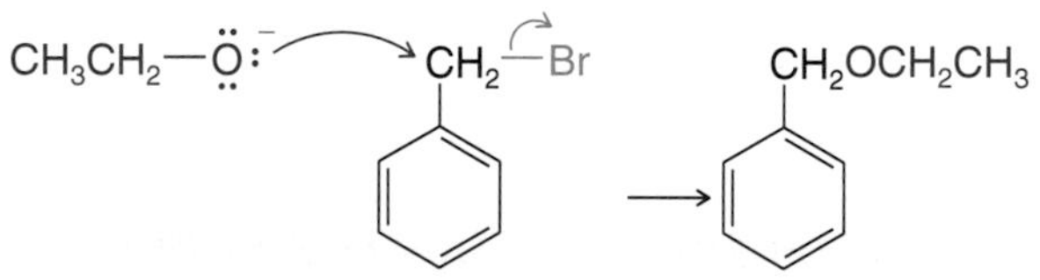

$$CH_3CH_2-\ddot{O}:^- + C_6H_5CH_2-Br \longrightarrow C_6H_5CH_2OCH_2CH_3$$

Ethoxide ion — Benzyl bromide — Benzyl ethyl ether

Benzylic halides resemble allylic halides in the readiness with which they form carbocations. On comparing the rate of S_N1 hydrolysis in aqueous acetone of the following two tertiary chlorides, we find that the benzylic chloride reacts over 600 times faster than does *tert*-butyl chloride.

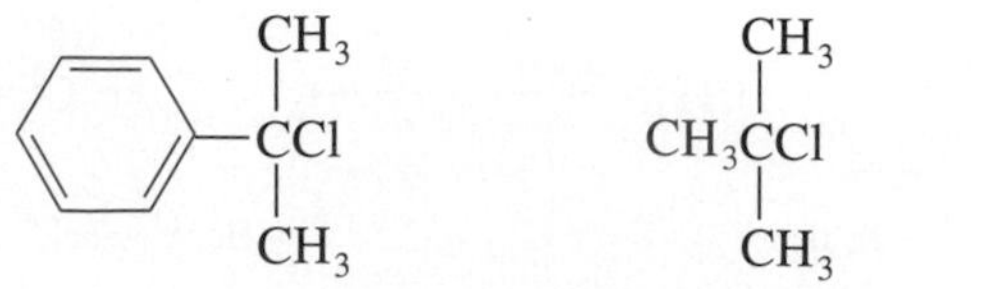

$$C_6H_5C(CH_3)_2Cl \qquad CH_3C(CH_3)_2Cl$$

2-Chloro-2-phenylpropane — 2-Chloro-2-methylpropane

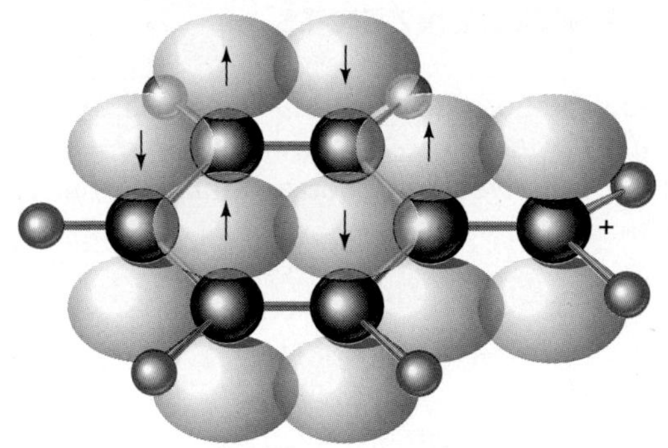

FIGURE 11.10 Benzyl cation is stabilized by overlap of its vacant *p* orbital with the π system of the aromatic ring.

Just as the odd electron in benzyl radical is shared by the carbons ortho and para to the benzylic carbon, the positive charge in benzyl cation is shared by these same positions.

Most stable Lewis structure of benzyl cation

Figure 11.10 depicts how the π system of a benzene ring stabilizes a benzylic carbocation by overlap with the vacant *p* orbital of the benzylic carbon.

Unlike the case with allylic carbocations, however, dispersal of the positive charge does not result in nucleophilic attack at more than one carbon. There is no ''benzylic rearrangement'' analogous to allylic rearrangement (Section 10.2), because the aromatic stabilization would be lost if the nucleophile became bonded to one of the ring carbons. Thus, when conditions are chosen that favor S_N1 substitution over E2 elimination (solvolysis, weakly basic nucleophile), benzylic halides give a single substitution product in high yield.

$$C_6H_5C(CH_3)_2Cl \xrightarrow{CH_3CH_2OH} C_6H_5C(CH_3)_2OCH_2CH_3 \quad \text{via} \quad C_6H_5\overset{+}{C}(CH_3)_2$$

2-Chloro-2-phenylpropane — 2-Ethoxy-2-phenylpropane (87%)

Additional phenyl substituents stabilize carbocations even more. Triphenylmethyl cation is particularly stable. Its perchlorate salt is ionic and stable enough to be isolated and stored indefinitely.

The triphenylmethyl group is often referred to as a *trityl* group.

$$(C_6H_5)_3\overset{+}{C} \quad [ClO_4]^-$$

Triphenylmethyl perchlorate

11.17 PREPARATION OF ALKENYLBENZENES

Alkenylbenzenes are prepared by the various methods described in Chapter 5 for the preparation of alkenes: *dehydrogenation, dehydration,* and *dehydrohalogenation.*

Dehydrogenation of alkylbenzenes is not a convenient laboratory method but is used industrially to convert ethylbenzene to styrene.

$$C_6H_5CH_2CH_3 \xrightarrow[ZnO]{630°C} C_6H_5CH{=}CH_2 + H_2$$

Ethylbenzene | Styrene | Hydrogen

Acid-catalyzed dehydration of benzylic alcohols is a useful route to alkenylbenzenes, as is dehydrohalogenation under E2 conditions.

$$m\text{-}ClC_6H_4CH(OH)CH_3 \xrightarrow[heat]{KHSO_4} m\text{-}ClC_6H_4CH{=}CH_2$$

1-(*m*-Chlorophenyl)ethanol | *m*-Chlorostyrene (80–82%)

$$p\text{-}H_3CC_6H_4CH_2CH(Br)CH_3 \xrightarrow[CH_3CH_2OH,\ 50°C]{NaOCH_2CH_3} p\text{-}H_3CC_6H_4CH{=}CHCH_3$$

2-Bromo-1-(*p*-methylphenyl)propane | 1-(*p*-Methylphenyl)propene (99%)

11.18 ADDITION REACTIONS OF ALKENYLBENZENES

Most of the reactions of alkenes that were discussed in Chapter 6 find a parallel in the reactions of alkenylbenzenes.

Hydrogenation of the side-chain double bond of an alkenylbenzene is much easier than hydrogenation of the aromatic ring and can be achieved with high selectivity, leaving the ring unaffected.

$$m\text{-}BrC_6H_4C(CH_3){=}CHCH_3 + H_2 \xrightarrow{Pt} m\text{-}BrC_6H_4CH(CH_3)CH_2CH_3$$

2-(*m*-Bromophenyl)-2-butene | Hydrogen | 2-(*m*-Bromophenyl)butane (92%)

PROBLEM 11.8 Both 1,2-dihydronaphthalene and 1,4-dihydronaphthalene may be selectively hydrogenated to 1,2,3,4-tetrahydronaphthalene.

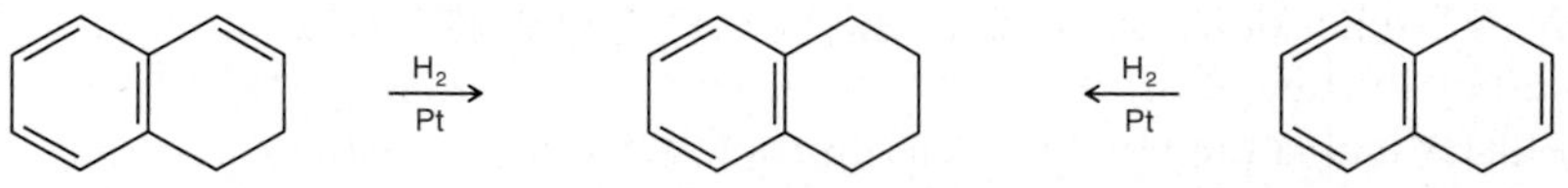

$\xrightarrow[Pt]{H_2}$ $\xleftarrow[Pt]{H_2}$

1,2-Dihydronaphthalene 1,2,3,4-Tetrahydronaphthalene 1,4-Dihydronaphthalene

One of these isomers has a heat of hydrogenation of 101 kJ/mol (24.1 kcal/mol), while the heat of hydrogenation of the other is 113 kJ/mol (27.1 kcal/mol). Match the heat of hydrogenation with the appropriate dihydronaphthalene.

The double bond in the alkenyl side chain undergoes addition reactions that are typical of alkenes when treated with electrophilic reagents.

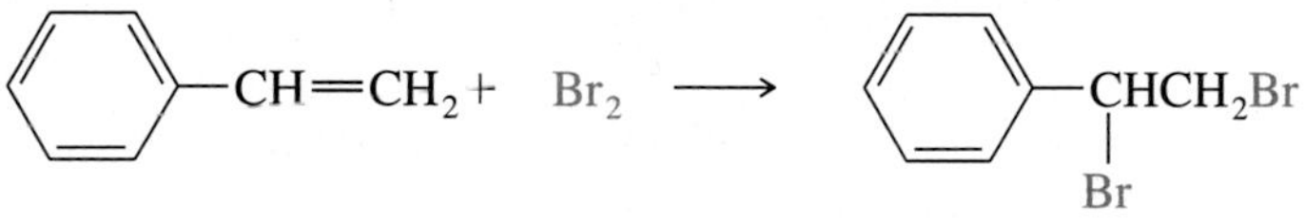

$$C_6H_5CH{=}CH_2 + Br_2 \longrightarrow C_6H_5CHBrCH_2Br$$

Styrene Bromine 1,2-Dibromo-1-phenylethane (82%)

The regioselectivity of electrophilic addition is governed by the ability of an aromatic ring to stabilize an adjacent carbocation. This is clearly seen in the addition of hydrogen chloride to indene. Only a single chloride is formed.

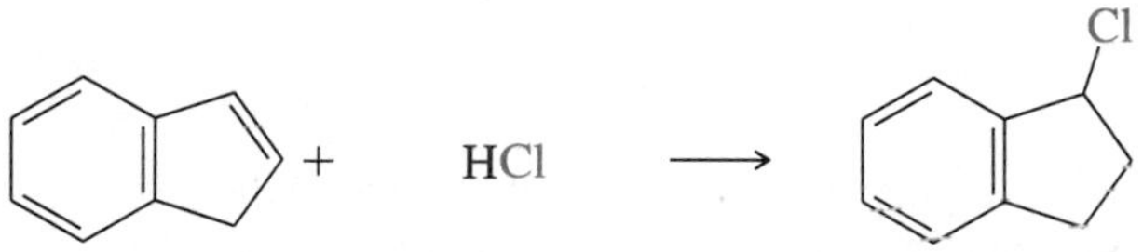

$$\text{Indene} + HCl \longrightarrow \text{1-Chloroindane}$$

Indene Hydrogen chloride 1-Chloroindane (75–84%)

Only the benzylic chloride is formed, because protonation of the double bond occurs in the direction that gives a secondary carbocation that is benzylic.

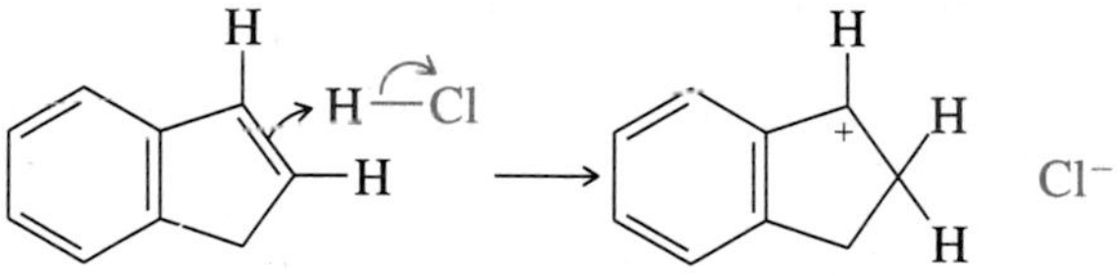

H H—Cl —H → H H H Cl⁻

Carbocation that leads to observed product

Protonation in the opposite direction also gives a secondary carbocation, but one that is not benzylic.

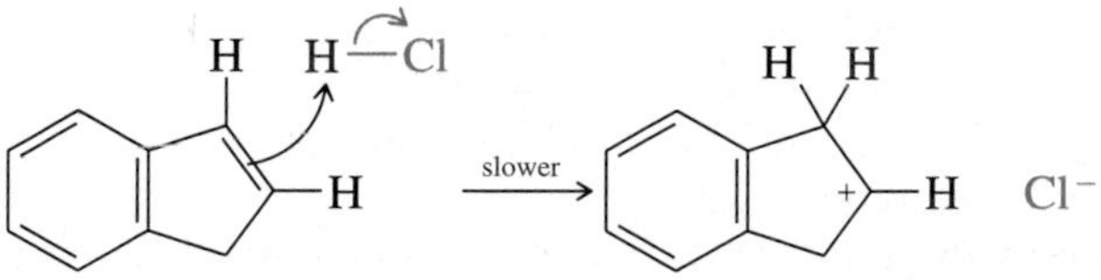

H H—Cl —H $\xrightarrow{\text{slower}}$ H H —H Cl⁻

Less stable carbocation

This alternative carbocation does not receive the extra increment of stabilization that its benzylic isomer does and so is formed more slowly. The orientation of addition is controlled by the rate of carbocation formation; the more stable benzylic carbocation is formed faster and is the one that determines the reaction product.

PROBLEM 11.9 Each of the following reactions has been reported in the chemical literature and gives a single organic product in high yield. Write the structure of the product for each reaction.

(*a*) 2-Phenylpropene + hydrogen chloride
(*b*) 2-Phenylpropene treated with diborane in tetrahydrofuran followed by oxidation with basic hydrogen peroxide
(*c*) Styrene + bromine in aqueous solution
(*d*) Styrene + peroxybenzoic acid (two organic products in this reaction; identify both)

SAMPLE SOLUTION (*a*) Addition of hydrogen chloride to the double bond takes place by way of a tertiary benzylic carbocation.

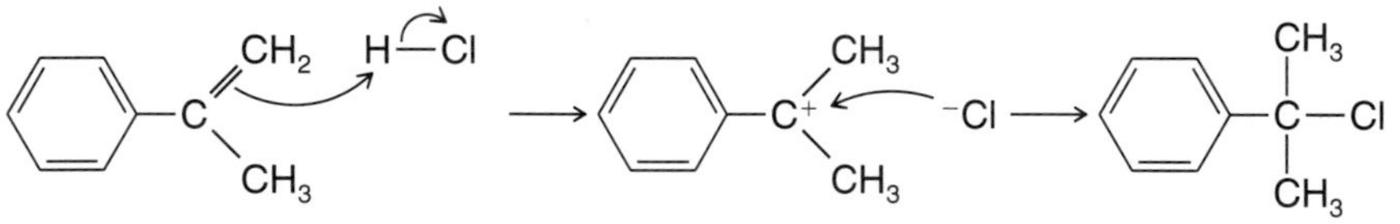

$$C_6H_5C(CH_3){=}CH_2 + H{-}Cl \longrightarrow C_6H_5\overset{+}{C}(CH_3)_2 \;\; ^{-}Cl \longrightarrow C_6H_5C(CH_3)_2Cl$$

2-Phenylpropene   Hydrogen chloride   2-Chloro-2-phenylpropane

---

In the presence of peroxides, hydrogen bromide adds to the double bond of styrene with a regioselectivity opposite to that of Markovnikov's rule. The reaction is a free-radical addition, and the regiochemistry is governed by preferential formation of the more stable radical.

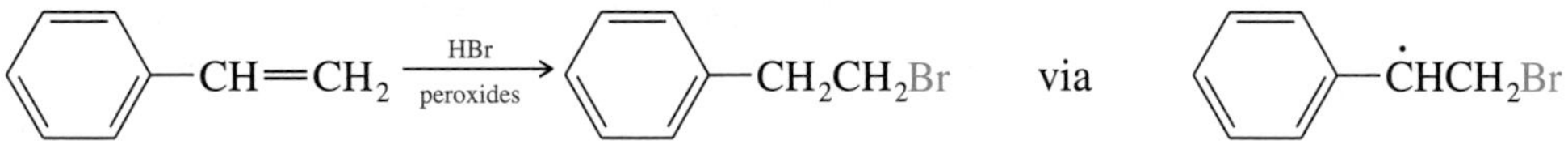

$$C_6H_5CH{=}CH_2 \xrightarrow[\text{peroxides}]{\text{HBr}} C_6H_5CH_2CH_2Br \quad \text{via} \quad C_6H_5\dot{C}HCH_2Br$$

Styrene   1-Bromo-2-phenylethane (major product)   2-Bromo-1-phenylethyl radical (secondary; benzylic)

11.19 POLYMERIZATION OF STYRENE

As described in the box "Diene Polymers" in Chapter 10, most synthetic rubber is a copolymer of styrene and 1,3-butadiene.

The annual production of styrene in the United States is on the order of 8×10^9 lb, with about 65 percent of this output used to prepare polystyrene plastics and films (see Table 6.5, page 252). Styrofoam coffee cups are made from polystyrene. Polystyrene can also be produced in a form that is very strong and impact-resistant and is used widely in luggage, television and radio cabinets, and furniture.

Polymerization of styrene is carried out under free-radical conditions, often with benzoyl peroxide as the initiator. Figure 11.11 illustrates a step in the growth of a polystyrene chain by a mechanism analogous to that of the polymerization of ethylene (Section 6.21).

$$\boxed{\text{Polymer}}-CH_2-\underset{\cdot}{\overset{C_6H_5}{\overset{|}{C}H}} + CH_2{=}CHC_6H_5 \longrightarrow \boxed{\text{Polymer}}-CH_2-\overset{C_6H_5}{\overset{|}{C}H}-CH_2-\underset{\cdot}{\overset{C_6H_5}{\overset{|}{C}H}}$$

FIGURE 11.11 Chain propagation step in polymerization of styrene. The growing polymer chain has a free radical site at the benzylic carbon. It adds to a molecule of styrene to extend the chain by one styrene unit. The new polymer chain is also a benzylic radical; it attacks another molecule of styrene, and the process repeats over and over again.

11.20 CYCLOBUTADIENE AND CYCLOOCTATETRAENE

During the course of our discussion of benzene and its derivatives, it may have occurred to you that cyclobutadiene and cyclooctatetraene offer possibilities for stabilization through delocalization of their π electrons in a manner analogous to that of benzene.

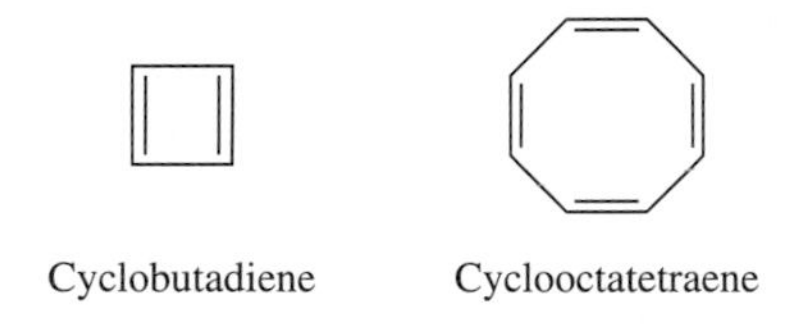

Cyclobutadiene   Cyclooctatetraene

The same thought occurred to early chemists. The complete absence of naturally occurring compounds based on cyclobutadiene and cyclooctatetraene contrasted starkly with the abundance of compounds based on the benzene nucleus. Attempts to synthesize cyclobutadiene and cyclooctatetraene met with failure and reinforced the growing conviction that these compounds would prove to be quite unlike benzene if, in fact, they could be isolated at all.

The first breakthrough came in 1911 when Richard Willstätter prepared cyclooctatetraene by a lengthy degradation of *pseudopelletierine,* a natural product obtained from the bark of the pomegranate tree. Nowadays, cyclooctatetraene is prepared from acetylene in a reaction catalyzed by nickel cyanide.

Willstätter's most important work, for which he won the 1915 Nobel Prize in chemistry, was directed toward determining the structure of chlorophyll.

$$4HC{\equiv}CH \xrightarrow[\text{heat, pressure}]{Ni(CN)_2} \text{Cyclooctatetraene}$$

Acetylene   Cyclooctatetraene (70%)

Analysis of thermochemical data gives a value of only about 20 kJ/mol (about 5 kcal/mol) for the resonance energy of cyclooctatetraene, indicating a degree of stabilization far less than the aromatic stabilization of benzene (150 kJ/mol; 36 kcal/mol).

PROBLEM 11.10 Both cyclooctatetraene and styrene have the molecular formula C_8H_8 and undergo combustion according to the equation

$$C_8H_8 + 10O_2 \longrightarrow 8CO_2 + 4H_2O$$

The measured heats of combustion are 4393 and 4543 kJ/mol (1050 and 1086 kcal/mol). Which heat of combustion belongs to which compound?

Structural studies confirm the absence of appreciable π electron delocalization in cyclooctatetraene. Its structure is as pictured in Figure 11.12—a *nonplanar* hydrocarbon with four short carbon-carbon bond distances and four longer carbon-carbon bond distances. Cyclooctatetraene is satisfactorily represented by a single Lewis structure having alternating single and double bonds in a tub-shaped eight-membered ring.

All the evidence indicates that cyclooctatetraene lacks the "special stability" of benzene, and is more appropriately considered as a conjugated polyene than as an aromatic hydrocarbon.

Cyclobutadiene escaped chemical characterization for more than 100 years. Despite numerous attempts, all synthetic efforts met with failure. It became apparent not only that cyclobutadiene was not aromatic but that it was exceedingly unstable. Beginning in the 1950s, a variety of novel techniques succeeded in generating cyclobutadiene as a transient, reactive intermediate.

PROBLEM 11.11 One of the chemical properties which make cyclobutadiene difficult to isolate is that it reacts readily with itself to give a dimer:

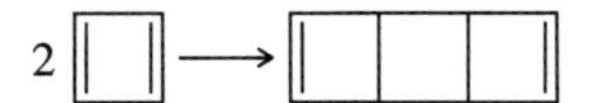

What reaction of dienes does this resemble?

Structural studies of cyclobutadiene and some of its derivatives reveal a pattern of alternating single and double bonds and a rectangular, rather than a square, shape. Bond distances in a stable, highly substituted derivative of cyclobutadiene illustrate this pattern of alternating short and long ring bonds.

$(CH_3)_3C$ $C(CH_3)_3$

138 pm

$(CH_3)_3C$ CO_2CH_3

151 pm

Methyl 2,3,4-tri-*tert*-butylcyclobutadiene-1-carboxylate

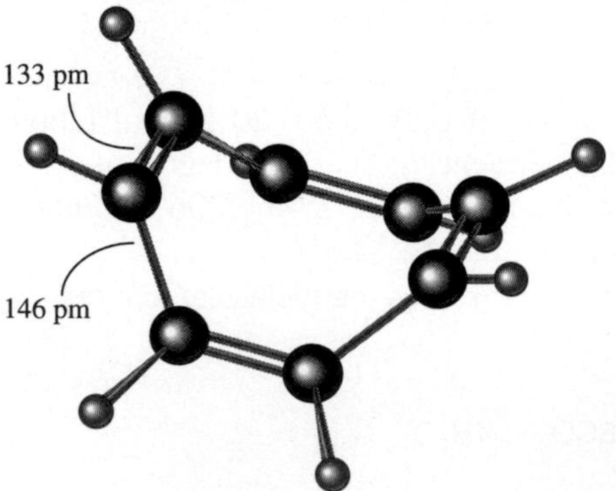

FIGURE 11.12 Molecular geometry of cyclooctatetraene. The ring is not planar, and the bond distances alternate between short double bonds and long single bonds. The structure shown was generated by minimization of strain energy using molecular mechanics and matches the experimentally determined structure.

Thus cyclobutadiene, like cyclooctatetraene, is not aromatic. *Cyclic conjugation, while necessary for aromaticity, is not sufficient for it.* Some other factor or factors must contribute to the special stability of benzene and its derivatives. To understand these factors, let us return to the molecular orbital description of benzene.

11.21 HÜCKEL'S RULE. ANNULENES

One of the early successes of molecular orbital theory occurred in 1931 when Erich Hückel discovered an interesting difference in the π orbital energy levels of benzene, compared with those of cyclobutadiene and cyclooctatetraene. By limiting his analysis to monocyclic conjugated polyenes and restricting the structures to planar geometries, Hückel found that these hydrocarbons are characterized by a set of π molecular orbitals in which one orbital is lowest in energy, another is highest in energy, and the rest are distributed in pairs between them.

Hückel was a German physical chemist. Before his theoretical studies of aromaticity, Hückel collaborated with Peter Debye (page 14) in developing what remains the most widely accepted theory of electrolyte solutions.

The arrangements of π orbitals for cyclobutadiene, benzene, and cyclooctatetraene as determined by Hückel are presented in Figure 11.13. Their interpretation can be summarized as follows:

Cyclobutadiene— According to the molecular orbital picture, square planar cyclobutadiene should be a diradical (have two unpaired electrons). The four π electrons are distributed so that two are in the lowest-energy orbital and, in accordance with Hund's rule, each of the two equal-energy nonbonding orbitals is half-filled. (Remember, Hund's rule tells us that when two orbitals have the same energy, each one is half-filled before either of them reaches its full complement of two electrons.)

Benzene— As seen earlier in Figure 11.5 (Section 11.8), the six π electrons of benzene are distributed in pairs among its three bonding orbitals. Benzene has a closed-shell electron configuration: all the bonding orbitals are occupied, and all the electron spins are paired.

Cyclooctatetraene— Six of the eight π electrons of cyclooctatetraene occupy three bonding orbitals. The remaining two π electrons occupy, one each, the two equal-energy nonbonding orbitals. Planar cyclooctatetraene should, like square cyclobutadiene, be a diradical.

As it turns out, neither cyclobutadiene nor cyclooctatetraene is a diradical in its most stable electron configuration. The Hückel approach treats them as planar regular polygons. Because the electron configurations associated with these geometries are not

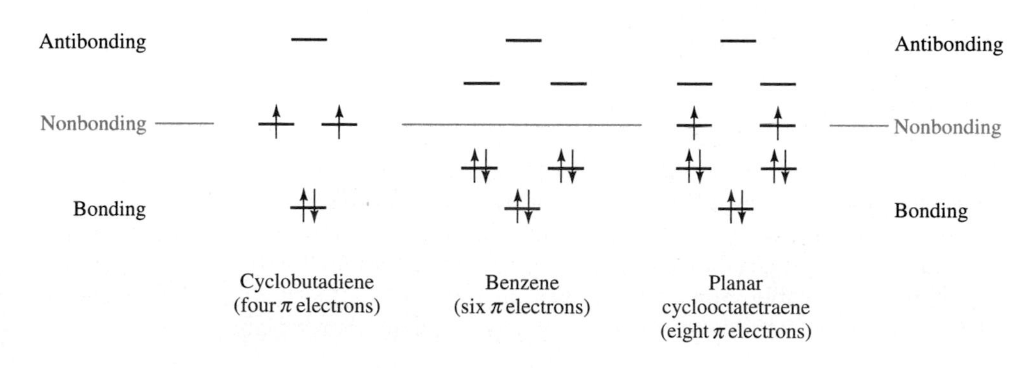

FIGURE 11.13 Distribution of π molecular orbitals and π electrons in cyclobutadiene, benzene, and planar cyclooctatetraene.

particularly stable, cyclobutadiene and cyclooctatetraene adopt structures other than planar regular polygons. Cyclobutadiene, rather than possessing a square shape with two unpaired electron spins, is a spin-paired rectangular molecule. Cyclooctatetraene is nonplanar, with all its π electrons paired in alternating single and double bonds.

On the basis of his analysis Hückel proposed that only certain numbers of π electrons could lead to aromatic stabilization. Only when the number of π electrons is 2, 6, 10, 14, and so on, can a closed-shell electron configuration be realized. These results are summarized in **Hückel's rule: Among planar, monocyclic, fully conjugated polyenes, only those possessing ($4n + 2$) π electrons, where n is an integer, will have special aromatic stability.**

Hückel's rule should not be applied to polycyclic aromatic hydrocarbons (Section 11.10). Hückel's analysis is limited to monocyclic systems.

The general term **annulene** has been coined to apply to completely conjugated monocyclic hydrocarbons. A numerical prefix specifies the number of carbon atoms. Cyclobutadiene is [4]-annulene, benzene is [6]-annulene, and cyclooctatetraene is [8]-annulene.

PROBLEM 11.12 Represent the π electron distribution among the π orbitals in

(*a*) [10]-Annulene (*b*) [12]-Annulene

SAMPLE SOLUTION (*a*) [10]-Annulene is cyclodecapentaene. It has 10 π orbitals and 10 π electrons. Like benzene, it should have a closed-shell electron configuration with all its bonding orbitals doubly occupied.

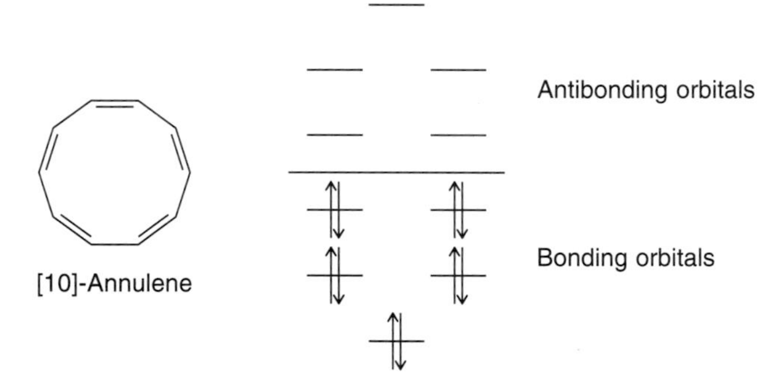

The prospect of observing aromatic character in conjugated polyenes having 10, 14, 18, etc., π electrons spurred efforts toward the synthesis of higher annulenes. A problem immediately arises in the case of the all-cis isomer of [10]-annulene, the structure of which is shown in the preceding problem. Geometry requires a 10-sided regular polygon to have 144° bond angles; sp^2 hybridization at carbon requires 120° bond angles. Therefore, aromatic stabilization due to conjugation in all-*cis*-[10]-annulene is opposed by the destabilizing effect of 24° of angle strain at each of its carbon atoms. All-*cis*-[10]-annulene has been prepared. It is not very stable and is highly reactive.

The size of each angle of a regular polygon is given by the expression

$$180° \times \frac{(\text{number of sides}) - 2}{(\text{number of sides})}$$

A second isomer of [10]-annulene, the cis, trans, cis, cis, trans stereoisomer, can have bond angles close to 120° but is destabilized by a close contact between two hydrogens directed toward the interior of the ring. In order to minimize the van der Waals strain between these hydrogens, the ring adopts a nonplanar geometry, which limits its ability to be stabilized by π electron delocalization. It, too, has been prepared and is not very stable. Similarly, the next-higher ($4n + 2$) system, [14]-annulene, is also somewhat destabilized by van der Waals strain and is nonplanar.

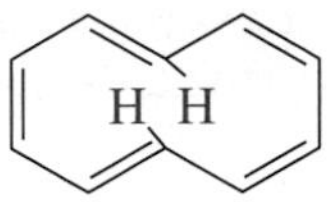

H H

cis,trans,cis,cis,trans-
[10]-Annulene

Planar geometry required for aromaticity destabilized by van der Waals repulsions between indicated hydrogens

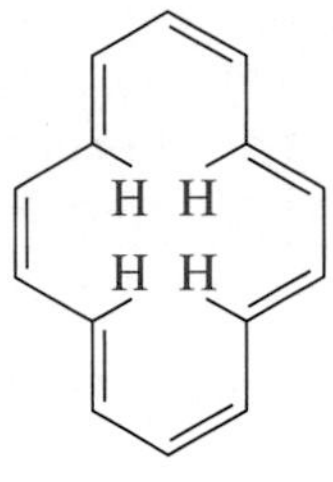

H H
H H

[14]-Annulene

When the ring contains 18 carbon atoms, it is large enough to be planar while still allowing its interior hydrogens to be far enough apart that they do not interfere with one another. The [18]-annulene shown is planar or nearly so and has all its carbon-carbon bond distances in the range 137 to 143 pm—very much like those of benzene. Its resonance energy is estimated to be about 418 kJ/mol (100 kcal/mol). While its structure and resonance energy attest to the validity of Hückel's rule, which predicts "special stability" for [18]-annulene, its chemical reactivity does not. [18]-Annulene behaves more like a polyene than like benzene in that it is hydrogenated readily, undergoes addition rather than substitution with bromine, and forms a Diels-Alder adduct with maleic anhydride.

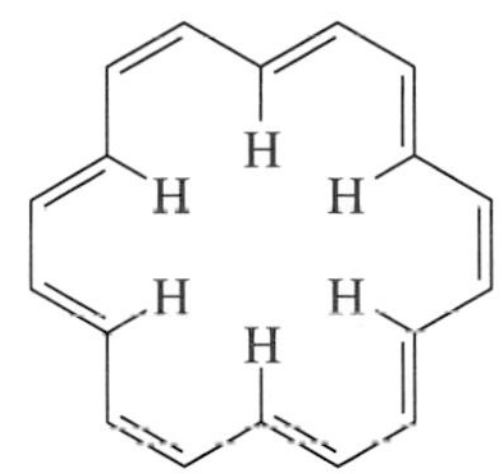

[18]-Annulene

No serious repulsions between six interior hydrogens; molecule is planar and aromatic.

According to Hückel's rule, annulenes with $4n$ π electrons are not aromatic. Cyclobutadiene and cyclooctatetraene are [$4n$]-annulenes, and, as we have seen, their properties are more in accord with their classification as cyclic polyenes than as aromatic hydrocarbons. Among higher [$4n$]-annulenes, [16]-annulene has been prepared. [16]-Annulene is not planar and shows a pattern of alternating short (average 134 pm) and long (average 146 pm) bonds typical of a nonaromatic cyclic polyene.

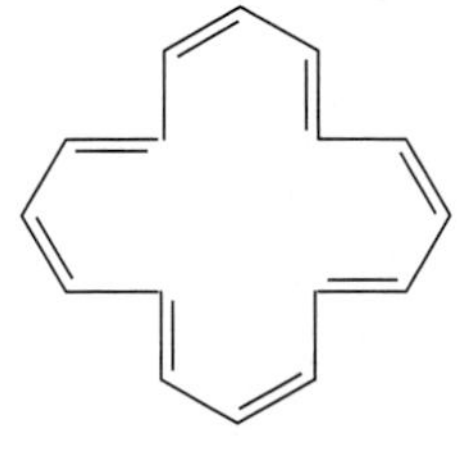

[16]-Annulene

PROBLEM 11.13 What does a comparison of the heats of combustion of benzene (3265 kJ/mol; 781 kcal/mol), cyclooctatetraene (4543 kJ/mol; 1086 kcal/mol), [16]-annulene (9121 kJ/mol; 2182 kcal/mol), and [18]-annulene (9806 kJ/mol; 2346 kcal/mol) reveal?

Most of the synthetic work directed toward the higher annulenes was carried out by Franz Sondheimer and his students, first at Israel's Weizmann Institute and later at the University of London. Sondheimer's research systematically explored the chemistry of these hydrocarbons and provided experimental verification of Hückel's rule.

11.22 AROMATIC IONS

Hückel realized that his molecular orbital analysis of conjugated systems could be extended beyond the realm of neutral hydrocarbons. He pointed out that cycloheptatrienyl cation contained a π system with a closed-shell electron configuration similar to that of benzene (Figure 11.14). Cycloheptatrienyl cation has a set of seven π molecular orbitals. Three of these are bonding and contain the six π electrons of the cation. These six π electrons are delocalized over seven carbon atoms, each of which contributes one $2p$ orbital to a planar, monocyclic, completely conjugated π system. Therefore, cycloheptatrienyl cation should be aromatic. It should be appreciably more stable than expected on the basis of any Lewis structure written for it.

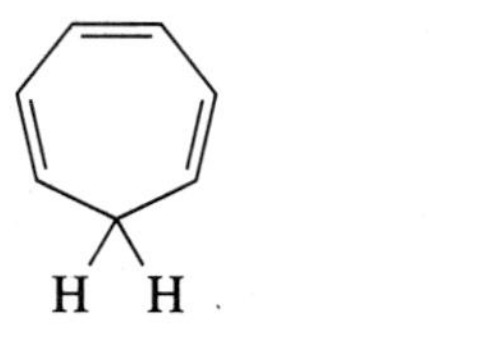

Cycloheptatriene

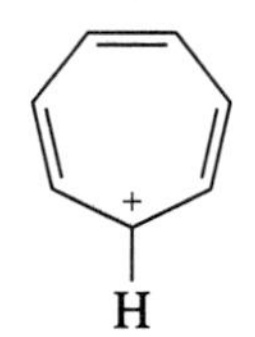

Cycloheptatrienyl cation (commonly referred to as *tropylium* cation)

It is important to recognize the difference between the hydrocarbon cycloheptatriene and cycloheptatrienyl (tropylium) cation. The carbocation, as we have just stated, is aromatic, whereas cycloheptatriene is not. Cycloheptatriene has six π electrons in a conjugated system, but its π system does not close upon itself. The ends of the triene system are joined by an sp^3 hybridized carbon, which precludes continuous electron delocalization. The ends of the triene system in the carbocation are joined by

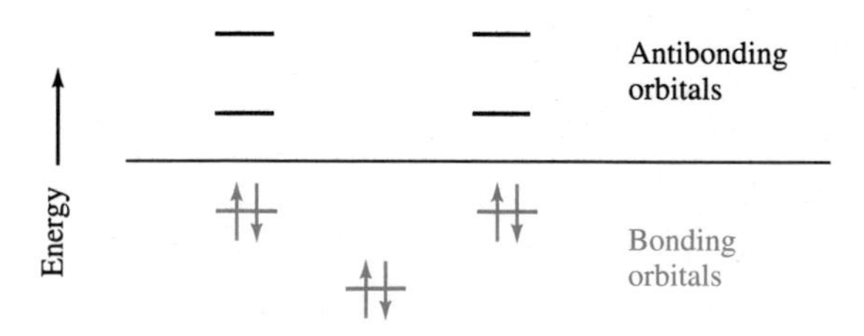

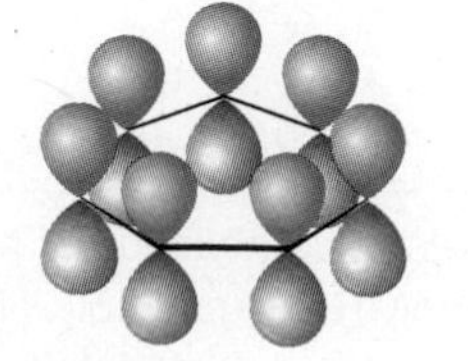

FIGURE 11.14 Electron distribution among the π molecular orbitals of cycloheptatrienyl (tropylium) cation.

an sp^2 hybridized carbon, which contributes an empty p orbital, thereby allowing continuous delocalization of the six π electrons. When we say cycloheptatriene is not aromatic but tropylium cation is, we are not comparing the stability of the two directly. Cycloheptatriene is a stable hydrocarbon but does not possess the *special stability* required to be called *aromatic*. Tropylium cation, while aromatic, is still a carbocation and reasonably reactive toward nucleophiles. Its special stability does not imply a rocklike passivity but rather a much greater ease of formation than expected on the basis of the Lewis structure drawn for it. A number of observations indicate that tropylium cation is far more stable than simple tertiary carbocations. To emphasize the aromatic nature of tropylium cation, it is sometimes written in the Robinson manner, representing the aromatic sextet with a circle in the ring and including a positive charge within the circle.

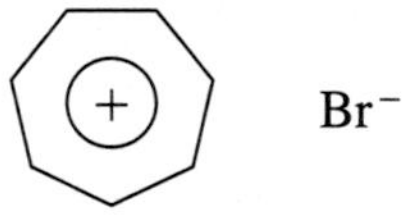

Br^-

Tropylium bromide

Tropylium bromide was first prepared, but not recognized as such, in 1891. The work was repeated in 1954, and the ionic properties of tropylium bromide were demonstrated. The ionic properties of tropylium bromide are apparent in its unusually high melting point (203°C), its solubility in water, and its complete lack of solubility in diethyl ether.

PROBLEM 11.14 Write resonance structures for tropylium cation sufficient to show the delocalization of the positive charge over all seven carbons.

Cyclopentadienide anion is an *aromatic anion*. It has six π electrons delocalized over a completely conjugated planar monocyclic array of five sp^2 hybridized carbon atoms.

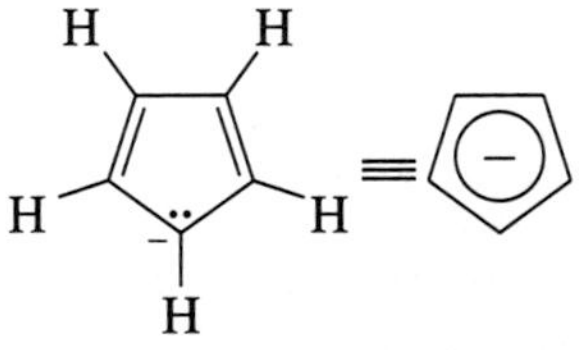

H H H H H ≡

Cyclopentadienide anion

PROBLEM 11.15 Write resonance structures for cyclopentadienide anion sufficient to show the delocalization of the negative charge over all five carbons.

Figure 11.15 presents Hückel's depiction of the molecular orbitals of cyclopentadienide anion. Like benzene, like tropylium cation, cyclopentadienide anion has a closed-shell configuration of six π electrons.

A convincing demonstration of the stability of cyclopentadienide anion can be found in the acidity of cyclopentadiene.

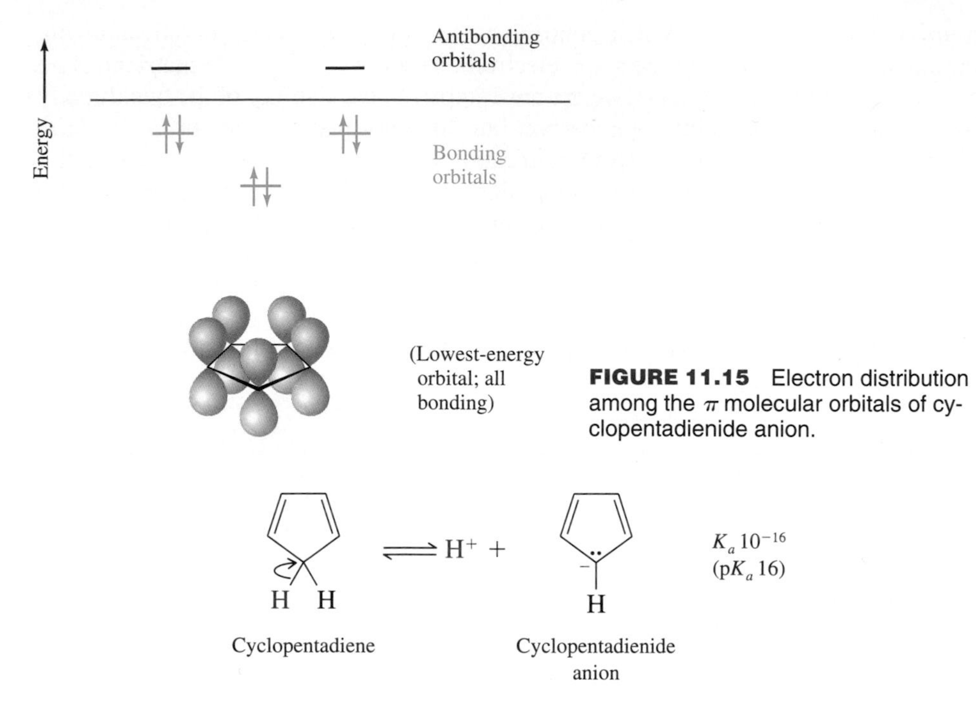

FIGURE 11.15 Electron distribution among the π molecular orbitals of cyclopentadienide anion.

Cyclopentadiene is only a slightly weaker acid than water. The position of equilibrium for its deprotonation is more favorable than for other hydrocarbons because cyclopentadienide anion possesses the special stability of aromaticity. The contrast is striking when we compare this equilibrium with that for loss of a proton from cycloheptatriene.

Cycloheptatriene $\rightleftharpoons$ H^+ + Cycloheptatrienide anion K_a 10^{-36} (pK_a 36)

While resonance structures can be written that show delocalization of the negative charge over all of its seven carbons, nevertheless, because cycloheptatrienide anion contains *eight π electrons,* it is not aromatic. The equilibrium constant for formation from the parent hydrocarbon is more favorable by 10^{20} (20 pK_a units) for the aromatic cyclopentadienide anion than for the nonaromatic cycloheptatrienide anion.

PROBLEM 11.16 A standard method for the preparation of sodium cyclopentadienide (C_5H_5Na) is by reaction of cyclopentadiene with a solution of sodium amide in liquid ammonia. Write a balanced equation for this reaction.

Hückel's rule is now taken to apply to planar, monocyclic, completely conjugated systems generally, not just to neutral hydrocarbons. **A planar, monocyclic, continuous system of *p* orbitals possesses aromatic stability when it contains $(4n + 2)$ π electrons.**

Other aromatic ions include cyclopropenyl cation (two π electrons) and cyclooctatetraene dianion (ten π electrons).

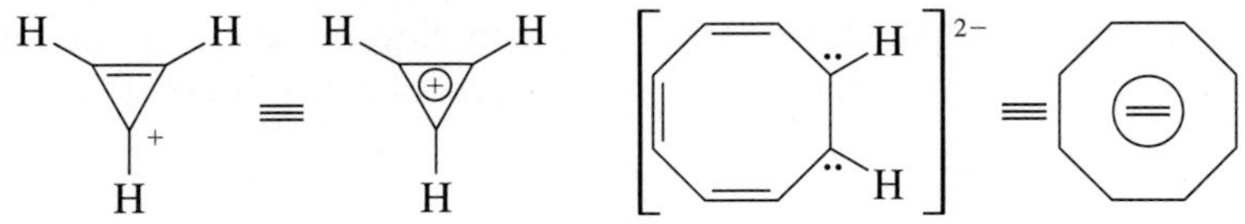

Cyclopropenyl cation

Cyclooctatetraene dianion

Here, liberties have been taken with the Robinson symbol. Instead of restricting its use to a sextet of electrons, organic chemists have come to adopt it as an all-purpose symbol for cyclic electron delocalization.

PROBLEM 11.17 Is either of the following ions aromatic?

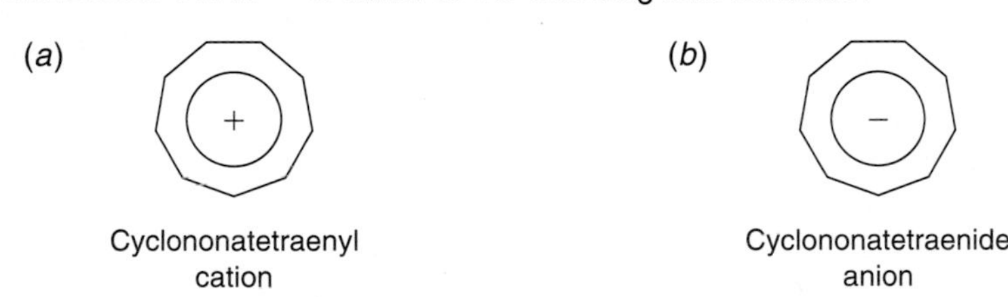

(*a*) Cyclononatetraenyl cation

(*b*) Cyclononatetraenide anion

SAMPLE SOLUTION (*a*) The crucial point is the number of π electrons. If there are $(4n + 2)$ π electrons, the ion is aromatic. Electron counting is facilitated by writing the ion as a single Lewis structure and remembering that each double bond contributes two π electrons, a negatively charged carbon contributes two, and a positively charged carbon contributes none.

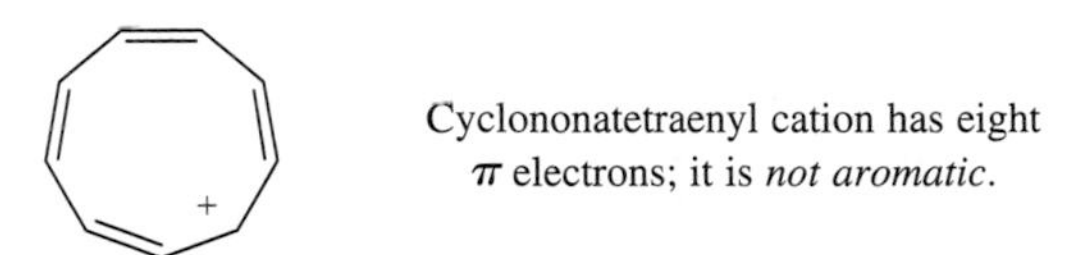

Cyclononatetraenyl cation has eight π electrons; it is *not aromatic.*

11.23 HETEROCYCLIC AROMATIC COMPOUNDS

Cyclic compounds that contain at least one atom other than carbon within their ring are called **heterocyclic compounds,** and those that possess aromatic stability are called **heterocyclic aromatic compounds.** Some representative heterocyclic aromatic compounds are *pyridine, pyrrole, furan,* and *thiophene.* The structures and the IUPAC numbering system used in naming their derivatives are shown. In their stability and chemical behavior, all these compounds resemble benzene more than they resemble alkenes.

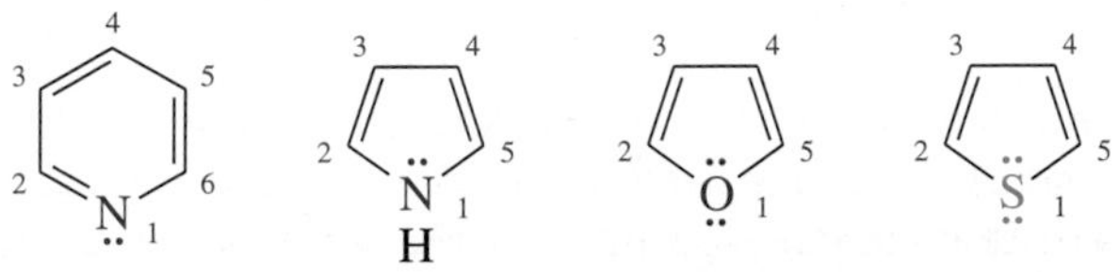

Pyridine Pyrrole Furan Thiophene

Pyridine, pyrrole, and thiophene, like benzene, are present in coal tar. Furan is prepared from a substance called *furfural* obtained from corncobs.

Heterocyclic aromatic compounds can be polycyclic as well. A benzene ring and a pyridine ring, for example, can share a common side in two different ways. One mode of fusion creates a compound called *quinoline;* the other gives *isoquinoline.*

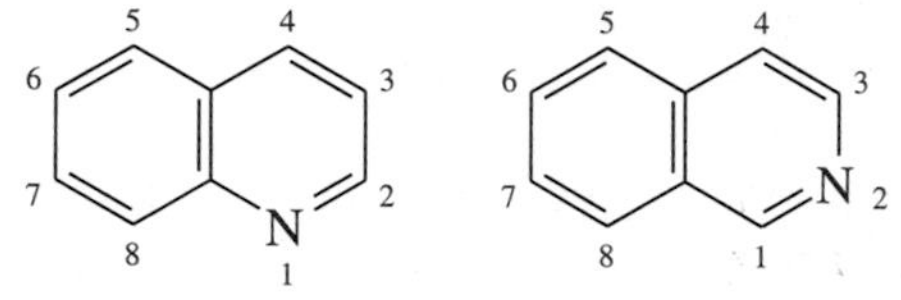

Quinoline Isoquinoline

Analogous compounds derived by fusion of a benzene ring to a pyrrole, furan, or thiophene nucleus are called *indole, benzofuran,* and *benzothiophene.*

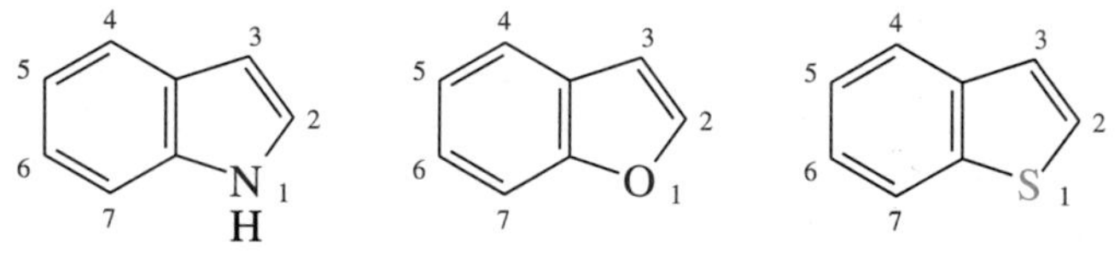

Indole Benzofuran Benzothiophene

PROBLEM 11.18 Unlike quinoline and isoquinoline, which are of comparable stability, the compounds indole and isoindole are quite different from each other. Which one is more stable? Explain the reason for your choice.

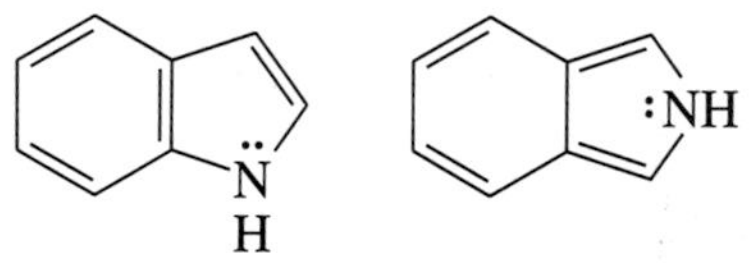

Indole Isoindole

A large group of heterocyclic aromatic compounds are related to pyrrole by replacement of one of the ring carbons β to nitrogen by a second heteroatom. Compounds of this type are called *azoles.*

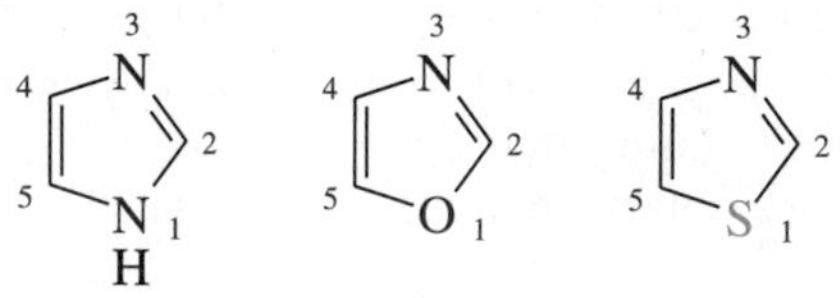

Imidazole Oxazole Thiazole

The most widely prescribed drug for the treatment of gastric ulcers has the generic name *cimetidine* and is a synthetic imidazole derivative. *Firefly luciferin* is a thiazole derivative that is the naturally occurring light-emitting substance present in fireflies.

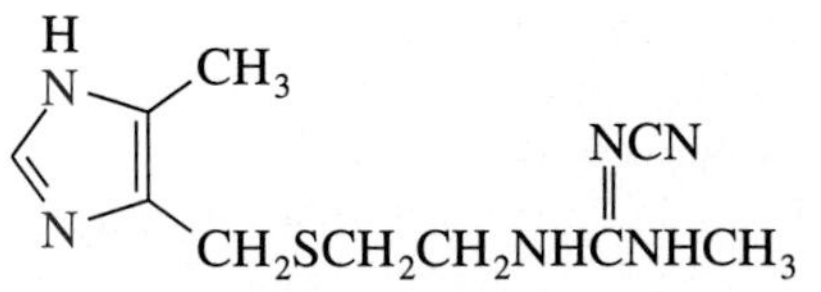

Cimetidine 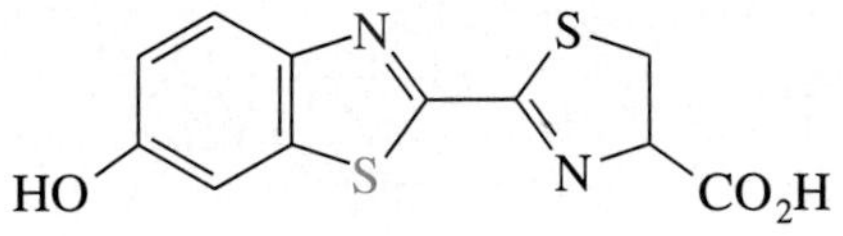 Firefly luciferin

Firefly luciferin is an example of an azole that contains a benzene ring fused to the five-membered ring. Such structures are fairly common. Another example is *benzimidazole,* present as a structural unit in vitamin B_{12} (page 593). Some compounds related to benzimidazole include *purine* and its amino-substituted derivative *adenine,* one of the so-called heterocyclic bases found in DNA and RNA (Chapter 27).

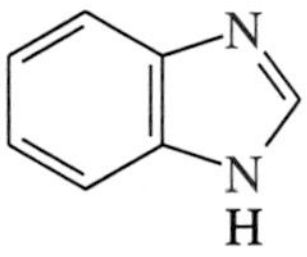

Benzimidazole 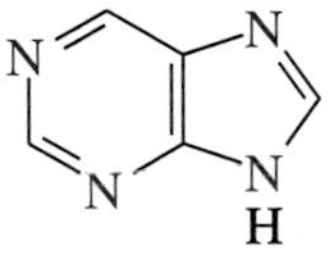 Purine 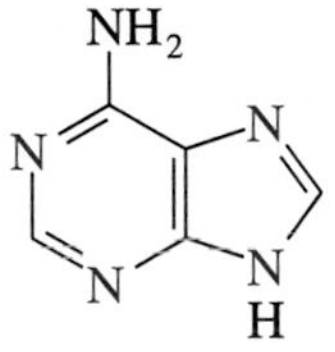 Adenine

---

PROBLEM 11.19 Can you deduce the structural formulas of *benzoxazole* and *benzothiazole?*

The structural types described in this section are but a tiny fraction of those possible. The chemistry of heterocyclic aromatic compounds is a rich and varied field with numerous applications.

11.24 HETEROCYCLIC AROMATIC COMPOUNDS AND HÜCKEL'S RULE

Hückel's rule can be extended to heterocyclic aromatic compounds. A single heteroatom can contribute either 0 or 2 of its lone-pair electrons as needed to the π system so as to satisfy the $(4n + 2)$ π electron requirement. The lone pair in pyridine, for example, is associated entirely with nitrogen and is not delocalized into the aromatic π system. As shown in Figure 11.16*a*, pyridine is simply a benzene ring in which a nitrogen atom has replaced a CH group. The nitrogen is sp^2 hybridized and the three double bonds of the ring contribute the necessary six π electrons to make pyridine a heterocyclic aromatic compound. The unshared electron pair of nitrogen occupies an sp^2 orbital in the plane of the ring, not a p orbital aligned with the π system.

In pyrrole, on the other hand, the unshared pair belonging to nitrogen must be added to the four π electrons of the two double bonds in order to meet the six-π-electron requirement. As shown in Figure 11.16*b*, the nitrogen of pyrrole is sp^2 hybridized and the pair of electrons occupies a p orbital where both electrons can participate in the aromatic π system.

Pyridine and pyrrole are both weak bases, but pyridine is much more basic than pyrrole. When pyridine is protonated, its unshared pair is used to bond to a proton and, since the unshared pair is not involved in the π system, the aromatic character of the ring is little affected. When pyrrole acts as a base, the two electrons used to form a

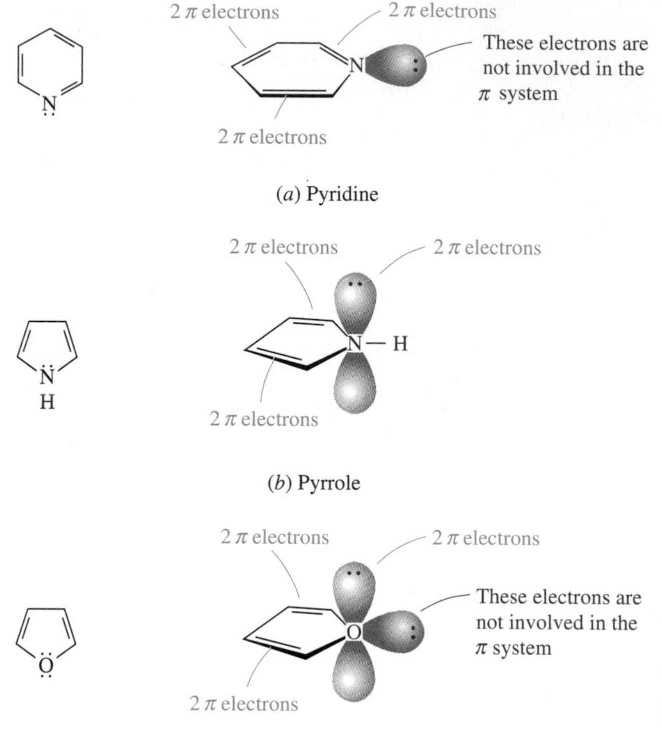

FIGURE 11.16 (*a*) Pyridine has six π electrons plus an unshared pair in a nitrogen sp^2 orbital. (*b*) Pyrrole has six π electrons. (*c*) Furan has six π electrons plus an unshared pair in an oxygen sp^2 orbital, which is perpendicular to the π system and does not interact with it.

bond to hydrogen must come from the π system and the aromaticity of the molecule is sacrificed on protonation.

PROBLEM 11.20 Imidazole is a much stronger base than pyrrole. Predict which nitrogen is protonated when imidazole reacts with an acid, and write a structural formula for the species formed.

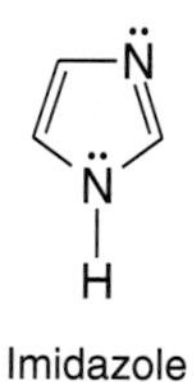

Imidazole

The oxygen in furan has two unshared electron pairs (Figure 11.16*c*). One pair is like the pair in pyrrole, occupying a *p* orbital and contributing two electrons to complete the six-π-electron requirement for aromatic stabilization. The other electron pair in furan is an ''extra'' pair, not needed to satisfy the $4n + 2$ rule for aromaticity, and occupies an sp^2 hybridized orbital like the unshared pair in pyridine.

The bonding in thiophene is similar to that of furan.

11.25 SUMMARY

Benzene (Section 11.1) is the parent of a class of hydrocarbons called **arenes,** or **aromatic hydrocarbons.** Benzene does not react with many of the reagents that react rapidly with alkenes (Section 11.2) and is substantially more stable than expected on the basis of structural formulas written for it. Neither of the two Kekulé formulas for benzene (Section 11.3) adequately describes its structure or properties. All the C—C bond distances in benzene, for example, are of equal length (Section 11.4), and benzene is said to be a **resonance hybrid** of the two Kekulé forms (Section 11.5).

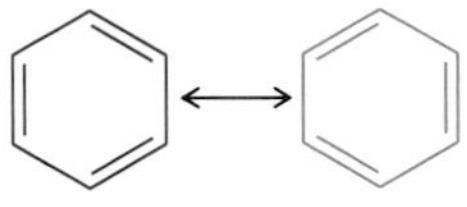

alternatively written as

Resonance between two Kekulé forms of benzene

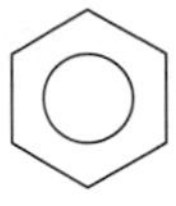

Robinson circle-in-a-hexagon symbol for benzene

The extent to which benzene is more stable than either of the Kekulé structures is its **resonance energy.** Experimental estimates of the resonance energy of benzene place it at 125 to 150 kJ/mol (30 to 36 kcal/mol; Section 11.6). The *special stability* of benzene has been attributed to delocalization of its six π electrons over the six carbons of the ring (Section 11.7). A molecular orbital description of benzene assigns the π electrons in pairs to three bonding orbitals in a closed-shell electron configuration analogous to that of the noble gases (Section 11.8).

Compounds that incorporate a benzene ring into their structure are also aromatic. Examples include substituted derivatives of benzene such as *toluene* (Section 11.9) and polycyclic aromatic hydrocarbons such as *naphthalene* (Section 11.10).

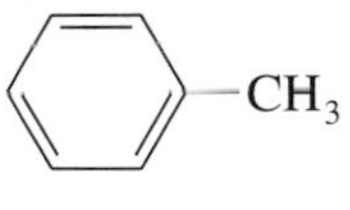

$—CH_3$

Toluene

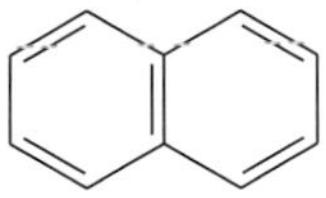

Naphthalene

The most important reaction of benzene and its derivatives involves replacement of a ring hydrogen by an electrophilic reagent *(electrophilic aromatic substitution),* which will be covered in Chapter 12.

An aromatic ring can be hydrogenated to a cyclohexane (Section 11.6) or reduced to a 1,4-cyclohexadiene derivative (the **Birch reduction,** Section 11.13). In the Birch reduction an arene is treated with a group I metal (usually sodium) in liquid ammonia in the presence of an alcohol.

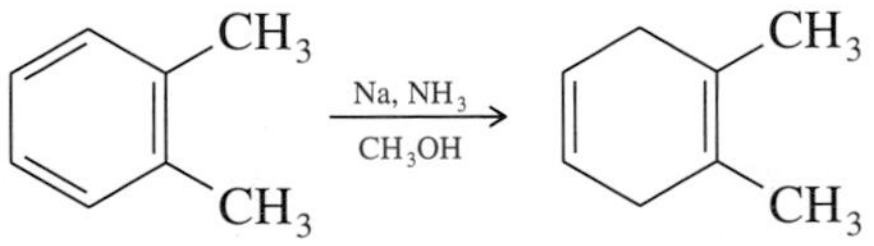

CH_3 CH_3 $\xrightarrow[CH_3OH]{Na,\ NH_3}$ CH_3 CH_3

o-Xylene

1,2-Dimethyl-1,4-cyclohexadiene (92%)

Table 11.2 summarizes some of the reactions that take place which involve hydrocarbon substituents of arenes. An aryl substituent stabilizes an adjacent radical, carbocation, or double bond by conjugation (Section 11.12). The **benzylic** position, the carbon adjacent to the ring, is the site at which free-radical halogenation (Section 11.14) and oxidation (Section 11.15) occur. Benzylic halides are more reactive toward nucleophilic substitution than simple alkyl halides (Section 11.16). Addition reactions to alkenylbenzenes occur at the double bond of the alkenyl substituent, and the regioselectivity of electrophilic addition is governed by carbocation formation at the benzylic carbon (Section 11.18).

TABLE 11.2
Reactions Involving Alkyl and Alkenyl Side Chains in Arenes and Arene Derivatives

Reaction (section) and comments	General equation and specific example
Halogenation (Section 11.14) Free-radical halogenation of alkylbenzenes is highly regioselective for substitution at the benzylic position. In the example shown, elemental bromine was used. Alternatively, *N*-bromosuccinimide is a convenient reagent for benzylic bromination.	$ArCHR_2 \xrightarrow[\text{benzoyl peroxide, } CCl_4,\ 80°C]{NBS} ArCR_2(Br)$ Arene → 1-Arylalkyl bromide $O_2N{-}C_6H_4{-}CH_2CH_3 \xrightarrow[CCl_4,\ \text{light}]{Br_2} O_2N{-}C_6H_4{-}CH(Br)CH_3$ *p*-Ethylnitrobenzene → 1-(*p*-Nitrophenyl)ethyl bromide (77%)
Oxidation (Section 11.15) Oxidation of alkylbenzenes occurs at the benzylic position of the alkyl group and gives a benzoic acid derivative. Oxidizing agents include sodium or potassium dichromate in aqueous sulfuric acid. Potassium permanganate ($KMnO_4$) is also an effective oxidant.	$ArCHR_2 \xrightarrow{\text{oxidize}} ArCO_2H$ Arene → Arenecarboxylic acid 2,4,6-$(O_2N)_3C_6H_2CH_3 \xrightarrow[H_2SO_4,\ H_2O]{Na_2Cr_2O_7}$ 2,4,6-$(O_2N)_3C_6H_2CO_2H$ 2,4,6-Trinitrotoluene → 2,4,6-Trinitrobenzoic acid (57–69%)
Hydrogenation (Section 11.18) Hydrogenation of aromatic rings is somewhat slower than hydrogenation of alkenes, and it is a simple matter to reduce the double bond of an unsaturated side chain in an arene while leaving the ring intact.	$ArCH{=}CR_2 + H_2 \xrightarrow{Pt} ArCH_2CHR_2$ Alkenylarene + Hydrogen → Alkylarene $m\text{-}BrC_6H_4{-}CH{=}CHCH_3 \xrightarrow[Pt]{H_2} m\text{-}BrC_6H_4{-}CH_2CH_2CH_3$ 1-(*m*-Bromophenyl)propene → *m*-Bromopropylbenzene (85%)
Electrophilic addition (Section 11.18) An aryl group stabilizes a benzylic carbocation and controls the regioselectivity of addition to a double bond involving the benzylic carbon. Markovnikov's rule is obeyed.	$ArCH{=}CH_2 \xrightarrow{\overset{\delta+}{E}{-}\overset{\delta-}{Y}} ArCH(Y){-}CH_2E$ Alkenylarene → Product of electrophilic addition $C_6H_5{-}CH{=}CH_2 \xrightarrow{HBr} C_6H_5{-}CH(Br)CH_3$ Styrene → 1-Phenylethyl bromide (85%)

While cyclic conjugation is a necessary requirement for aromaticity, this alone is not sufficient. If it were, cyclobutadiene and cyclooctatetraene would be aromatic. They are not (Section 11.20).

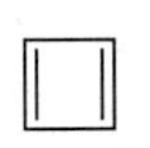

Cyclobutadiene
(four π electrons; not aromatic)

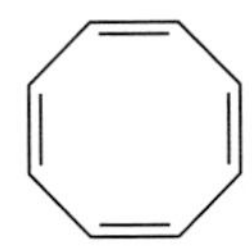

Cyclooctatetraene
(eight π electrons; not aromatic)

An additional requirement for aromaticity is that the number of π electrons in conjugated, planar, monocyclic species must be equal to $4n + 2$, where n is an integer. This is called **Hückel's rule** (Section 11.21). Benzene, with six π electrons, satisfies Hückel's rule for $n = 1$. Cyclobutadiene and cyclooctatetraene do not and so do not possess the closed-shell electron configuration required for aromatic stability. Planar, monocyclic, completely conjugated polyenes are called **annulenes** (Section 11.21).

Other than benzene, systems with six π electrons that are substantially more stable than expected include certain ionic species, such as *cyclopentadienide* anion and *cycloheptatrienyl* (tropylium) cation (Section 11.22).

Cyclopentadienide
anion

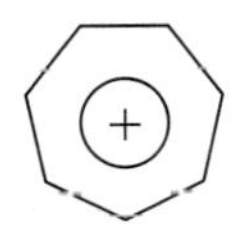

Cycloheptatrienyl
cation

Heterocyclic aromatic compounds are compounds that contain at least one atom other than carbon within an aromatic ring (Section 11.23). Usually the heteroatom is nitrogen, oxygen, or sulfur, and the ring is normally five-membered or six membered. Hückel's rule can be extended to heterocyclic aromatic compounds on the premise that unshared electron pairs of the heteroatom may be used as π electrons as necessary to satisfy the $4n + 2$ rule (Section 11.24).

PROBLEMS

11.21 Write structural formulas and give the IUPAC names for all the isomers of $C_6H_5C_4H_9$ that contain a monosubstituted benzene ring.

11.22 Write a structural formula corresponding to each of the following:

(*a*) Allylbenzene
(*b*) (*E*)-1-Phenyl-1-butene
(*c*) (*Z*)-2-Phenyl-2-butene
(*d*) (*R*)-1-Phenylethanol
(*e*) *o*-Chlorobenzyl alcohol
(*f*) *p*-Chlorophenol
(*g*) 2-Nitrobenzenecarboxylic acid
(*h*) *p*-Diisopropylbenzene
(*i*) 2,4,6-Tribromoaniline
(*j*) *m*-Nitroacetophenone
(*k*) 4-Bromo-3-ethylstyrene

11.23 Using numerical locants and the names in Table 11.1 as a guide, give an acceptable IUPAC name for each of the following compounds:

(*a*) *Estragole* (principal component of wormwood oil)

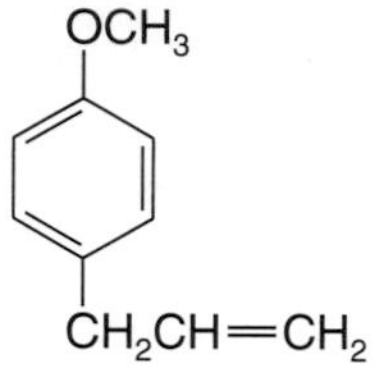

OCH_3

$CH_2CH{=}CH_2$

(*b*) *Diosphenol* (used in veterinary medicine to control parasites in animals)

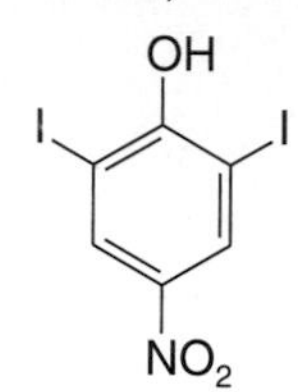

(*c*) *m*-Xylidine (intermediate in synthesis of lidocaine, a local anesthetic)

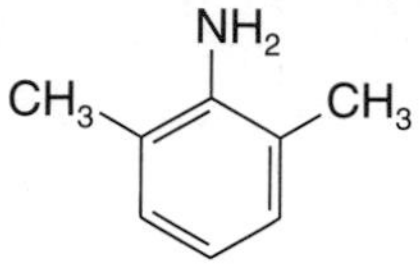

NH_2

CH_3 CH_3

11.24 Write structural formulas and give acceptable names for all the isomeric

(*a*) Nitrotoluenes
(*b*) Dichlorobenzoic acids
(*c*) Tribromophenols
(*d*) Tetrafluorobenzenes
(*e*) Naphthalenecarboxylic acids
(*f*) Bromoanthracenes

11.25 Which of the following are isomers of benzene? Do any of them have six equivalent carbons?

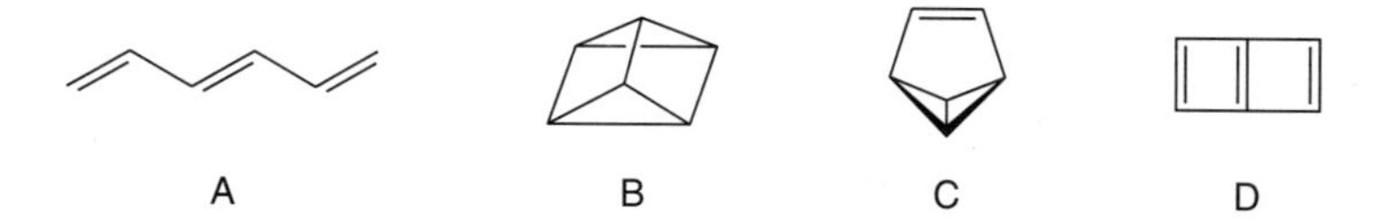

11.26 Mesitylene (1,3,5-trimethylbenzene) is the most stable of the trimethylbenzene isomers. Can you think of a reason why? Which isomer do you think is the least stable?

11.27 Which one of the dichlorobenzene isomers does not have a dipole moment?

11.28 Identify the longest and the shortest carbon-carbon bonds in styrene.

11.29 The resonance form shown is not the most stable one for the compound indicated. Write the most stable resonance form.

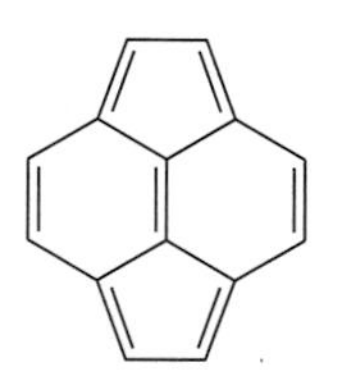

11.30 Each of the following may be represented by at least one alternative resonance structure in which all the six-membered rings correspond to Kekulé forms of benzene. Write such a resonance form for each.

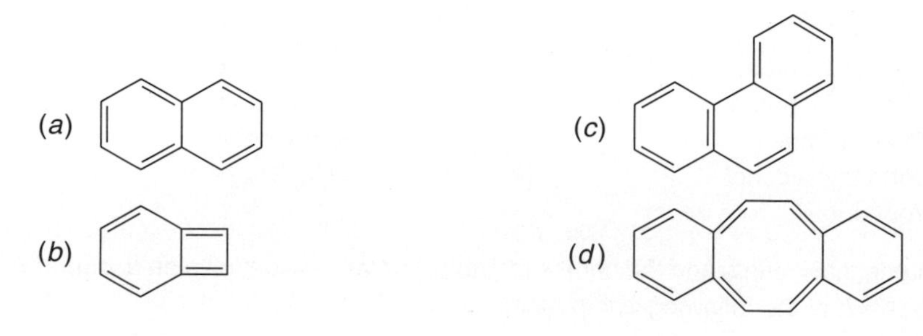

11.31 Give the structure of the expected product from the reaction of isopropylbenzene (*cumene*) with

(*a*) Hydrogen (3 moles), Pt
(*b*) Sodium and ethanol in liquid ammonia
(*c*) Sodium dichromate, water, sulfuric acid, heat
(*d*) *N*-Bromosuccinimide in CCl_4, heat, benzoyl peroxide
(*e*) The product of part (*d*) treated with sodium ethoxide in ethanol

11.32 Each of the following reactions has been described in the chemical literature and gives a single organic product in good yield. Identify the product of each reaction.

(*a*) 1-phenylcyclobutene (C_6H_5 on cyclobutene) $\xrightarrow[2.\ H_2O_2,\ HO^-]{1.\ B_2H_6,\ \text{diglyme}}$

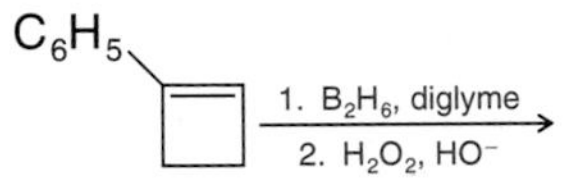

(*b*) 3-ethylindene (CH_2CH_3) + H_2 (1 mol) $\xrightarrow{Pt}$

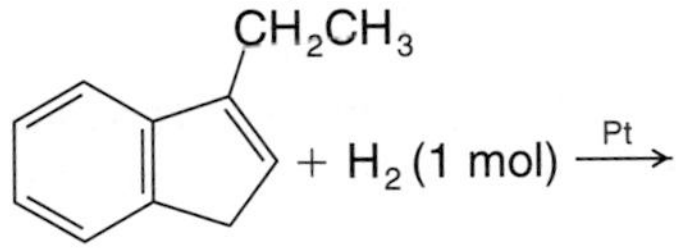

(*c*) $(C_6H_5)_2CH$—C₆H₄—CH_3 $\xrightarrow[CCl_4,\ \text{light}]{\text{excess } Cl_2}$ $C_{20}H_{14}Cl_4$

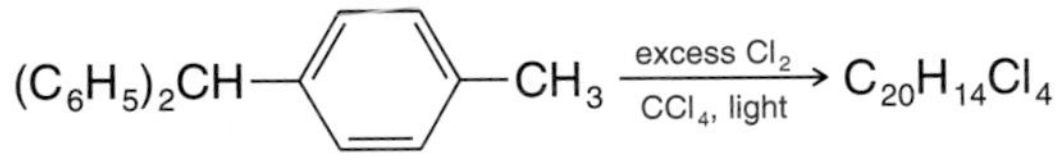

(*d*) (E)—$C_6H_5CH{=}CHC_6H_5$ $\xrightarrow[\text{acetic acid}]{CH_3CO_2OH}$

(*e*) H_3C–cyclohexane–C(OH)–phenyl $\xrightarrow[\text{acetic acid}]{H_2SO_4}$

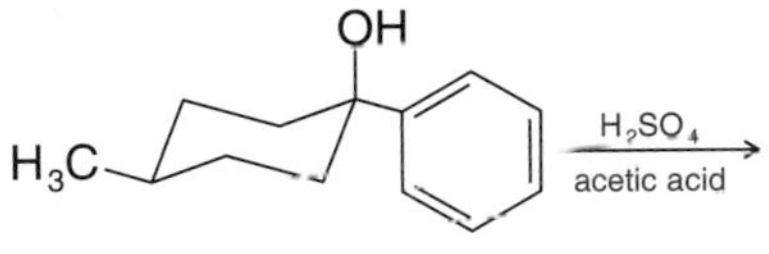

(*f*) p-$(CH_3)_2COH$–C₆H₄–$(CH_3)_2COH$ $\xrightarrow[\text{heat}]{KHSO_4}$ $C_{12}H_{14}$

(*g*) (Cl—C₆H₄—)$_2CHCCl_3$ $\xrightarrow[CH_3OH]{NaOCH_3}$ $C_{14}H_8Cl_4$

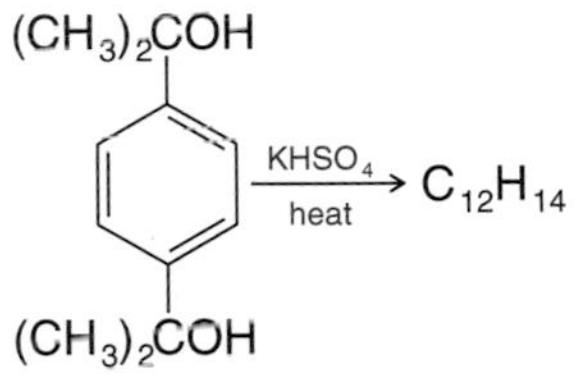

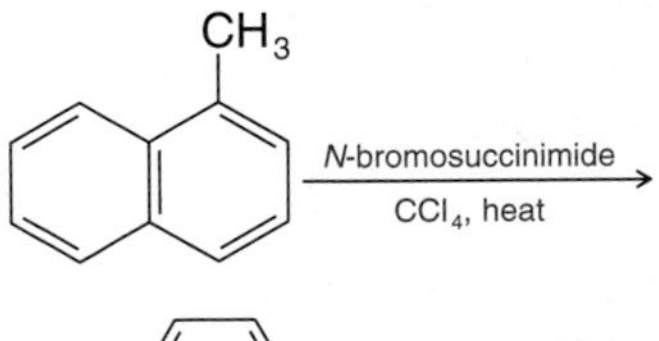

(DDT)

(*h*) 1-methylnaphthalene (CH_3) $\xrightarrow[CCl_4,\ \text{heat}]{N\text{-bromosuccinimide}}$

(*i*) NC—C₆H₄—CH_2Cl $\xrightarrow[\text{water}]{K_2CO_3}$ C_8H_7NO

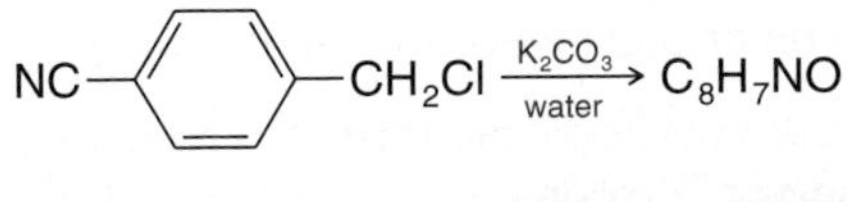

11.33 A certain compound A, when treated with *N*-bromosuccinimide and benzoyl peroxide under photochemical conditions in refluxing carbon tetrachloride, gave 3,4,5-tribromobenzyl bromide in excellent yield. Deduce the structure of compound A.

11.34 A compound was obtained from a natural product and had the molecular formula $C_{14}H_{20}O_3$. It contained three methoxy ($—OCH_3$) groups and a $—CH_2CH{=}C(CH_3)_2$ substituent. Oxidation with either chromic acid or potassium permanganate gave 2,3,5-trimethoxybenzoic acid. What is the structure of the compound?

11.35 Hydroboration-oxidation of (*E*)-2-(*p*-anisyl)-2-butene yielded an alcohol A, mp 60°C, in 72 percent yield. When the same reaction was performed on the *Z* alkene, an isomeric liquid alcohol B was obtained in 77 percent yield. Suggest reasonable structures for A and B, and describe the relationship between them.

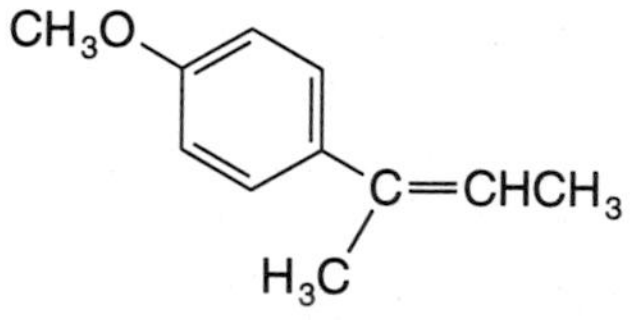

CH_3O–(p-C$_6$H$_4$)–$C(CH_3){=}CHCH_3$

2-(*p*-Anisyl)-2-butene

11.36 Dehydrohalogenation of the diastereomeric forms of 1-chloro-1,2-diphenylpropane is stereospecific. One diastereomer yields (*E*)-1,2-diphenylpropene, while the other yields the *Z* isomer. Which diastereomer yields which alkene? Why?

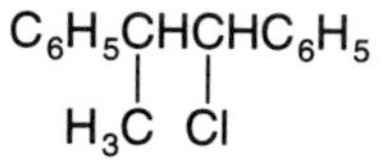

$C_6H_5CH(CH_3)CH(Cl)C_6H_5$ — 1-Chloro-1,2-diphenylpropane

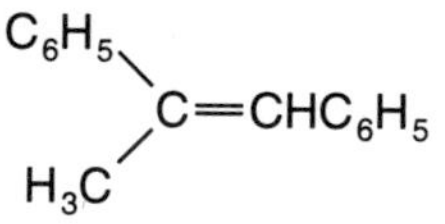

$(C_6H_5)(H_3C)C{=}CHC_6H_5$ — 1,2-Diphenylpropene

11.37 Suggest reagents suitable for carrying out each of the following conversions. In most cases more than one synthetic operation will be necessary.

(*a*) $C_6H_5CH_2CH_3 \longrightarrow C_6H_5CH(Br)CH_3$

(*b*) $C_6H_5CH(Br)CH_3 \longrightarrow C_6H_5CH(Br)CH_2Br$

(*c*) $C_6H_5CH{=}CH_2 \longrightarrow C_6H_5C{\equiv}CH$

(*d*) $C_6H_5C{\equiv}CH \longrightarrow C_6H_5CH_2CH_2CH_2CH_3$

(*e*) $C_6H_5CH_2CH_2OH \longrightarrow C_6H_5CH_2CH_2C{\equiv}CH$

(*f*) $C_6H_5CH_2CH_2Br \longrightarrow C_6H_5CH(OH)CH_2Br$

11.38 The relative rates of reaction of ethane, toluene, and ethylbenzene with bromine atoms have been measured. The most reactive hydrocarbon undergoes hydrogen atom abstraction 1 million times faster than does the least reactive one. Arrange these hydrocarbons in order of decreasing reactivity.

11.39 Write the principal resonance structures of *o*-methylbenzyl cation and *m*-methylbenzyl cation. Which one has a tertiary carbocation as a contributing resonance form?

11.40 The same anion is formed by loss of the most acidic proton from 1-methyl-1,3-cyclopentadiene as from 5-methyl-1,3-cyclopentadiene. Explain.

11.41 There are two different tetramethyl derivatives of cyclooctatetraene that have methyl groups on four adjacent carbon atoms. They are both completely conjugated and are not stereoisomeric. Write their structures.

11.42 Evaluate each of the following processes applied to cyclooctatetraene, and decide whether the species formed is aromatic or not.

(*a*) Addition of one more π electron, to give $C_8H_8^-$

(*b*) Addition of two more π electrons, to give $C_8H_8^{2-}$

(*c*) Removal of one π electron, to give $C_8H_8^+$

(*d*) Removal of two π electrons, to give $C_8H_8^{2+}$

11.43 Evaluate each of the following processes applied to cyclononatetraene and decide whether the species formed is aromatic or not:

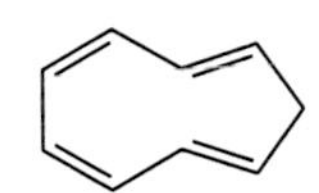

Cyclononatetraene

(*a*) Addition of one more π electron, to give $C_9H_{10}^-$

(*b*) Addition of two more π electrons, to give $C_9H_{10}^{2-}$

(*c*) Loss of H^+ from the sp^3 hybridized carbon

(*d*) Loss of H^+ from one of the sp^2 hybridized carbons

11.44 From among the molecules and ions shown, all of which are based on cycloundecapentaene, identify those which satisfy the criteria for aromaticity as prescribed by Hückel's rule.

(*a*) Cycloundecapentaene

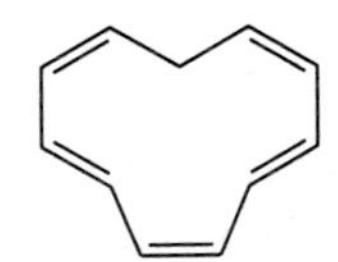

(*b*) Cycloundecapentaenyl radical

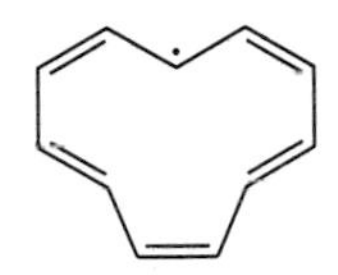

(*c*) Cycloundecapentaenyl cation

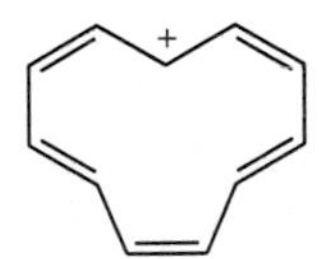

(*d*) Cycloundecapentaenide anion

11.45 (*a*) The molecule calicene, so called because it resembles a chalice (*calix* is the Latin word for "cup"), should be a fairly polar hydrocarbon. This prediction is based on the idea that one of the two dipolar structures A and B is expected to contribute appreciably to the resonance description of calicene. Which of these two resonance forms is more reasonable, A or B? Why?

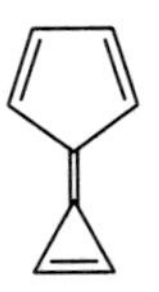

Calicene

A

B

(*b*) Which one of the following should be stabilized by resonance to a greater extent? (*Hint:* Consider the reasonableness of dipolar resonance forms.)

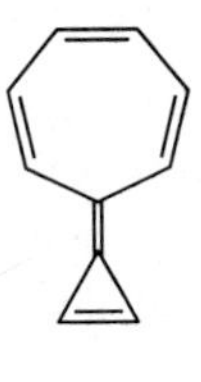

C

or

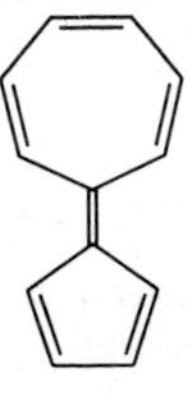

D

11.46 Classify each of the following heterocyclic molecules as aromatic or not, according to Hückel's rule:

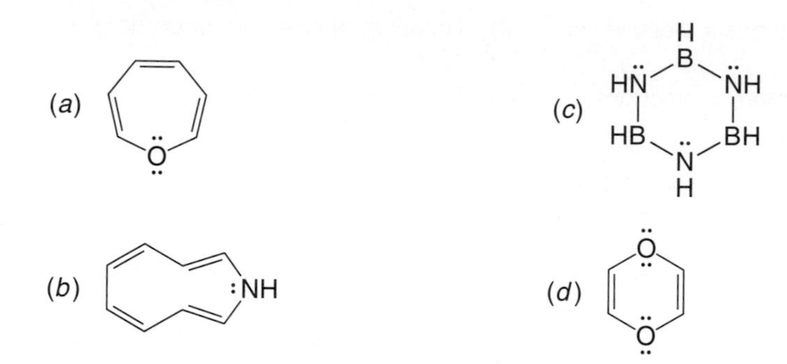

11.47 Pellagra is a disease caused by a deficiency of *niacin* ($C_6H_5NO_2$) in the diet. Niacin can be synthesized in the laboratory by the side-chain oxidation of 3-methylpyridine with chromic acid or potassium permanganate. Suggest a reasonable structure for niacin.

11.48 *Nitroxoline* is the generic name by which 5-nitro-8-hydroxyquinoline is sold as an antibacterial drug. Write its structural formula.

11.49 *Acridine* is a heterocyclic aromatic compound obtained from coal tar that is used in the synthesis of dyes. The molecular formula of acridine is $C_{13}H_9N$, and its ring system is analogous to that of anthracene except that one CH group has been replaced by N. The two most stable resonance structures of acridine are equivalent to each other, and both contain a pyridinelike structural unit. Write a structural formula for acridine.

MOLECULAR MODELING EXERCISES

11.50 Construct molecular models of benzene and cyclooctatetraene. Notice how the requirement of 120° bond angles for sp^2 hybridized carbon leads naturally to a planar geometry for benzene and a nonplanar one for cyclooctatetraene.

11.51 Make a molecular model of [10]-annulene having the double bond stereochemistry shown. Verify that this molecule is destabilized by van der Waals strain.

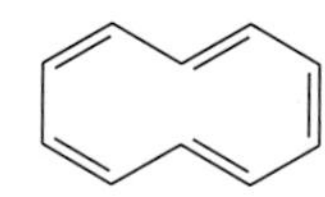

11.52 Construct a molecular model of benzyl cation ($C_6H_5CH_2^+$). Did you choose a planar or nonplanar geometry? Why?

11.53 Construct a molecular model of styrene. Did you choose a planar or nonplanar geometry? Why? How might the geometry of 2-phenylpropene differ from that of styrene?

11.54 Biphenyl is planar, but certain of its ortho-substituted derivatives are not and can possess a stereogenic axis. Construct a molecular model of a biphenyl derivative, using substituents W, X, Y, and Z of your choice, in a chiral conformation. What are the desirable qualities that characterize the substitutents you chose? Which of the substituents can be the same in order for the molecule to be chiral?

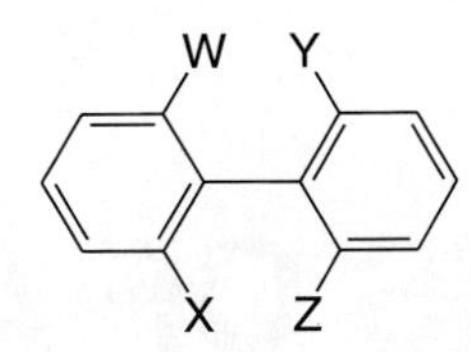

11.55 A phenyl substituent attached to cyclohexane has a less pronounced preference for the equatorial orientation than does a *tert*-butyl substituent. Construct a molecular model of 1-*tert*-butyl-1-phenylcyclohexane in its most stable conformation and suggest a reason for the smaller size of C_6H_5— relative to $(CH_3)_3C$—.

CHAPTER 12

REACTIONS OF ARENES. ELECTROPHILIC AROMATIC SUBSTITUTION

In the preceding chapter the *special stability* of benzene was described, along with reactions in which an aromatic ring was present as a substituent. In the present chapter we move from considering the aromatic ring as a substituent to studying it as a functional group. What kind of reactions are available to benzene and its derivatives? What sort of reagents react with arenes, and what products are formed in those reactions?

Characteristically, the reagents that react with the aromatic ring of benzene and its derivatives are *electrophiles.* We already have some experience with electrophilic reagents, particularly with respect to how they react with alkenes. Electrophilic reagents *add* to alkenes.

$$\text{>C=C<} + \overset{\delta+}{E}-\overset{\delta-}{Y} \longrightarrow E-\overset{|}{\underset{|}{C}}-\overset{|}{\underset{|}{C}}-Y$$

Alkene | Electrophilic reagent | Product of electrophilic addition

A different reaction takes place when electrophiles react with arenes. *Substitution is observed instead of addition.* If we represent an arene by the general formula ArH, where Ar stands for an aryl group, the electrophilic portion of the reagent replaces one of the hydrogens on the ring:

$$Ar-H + \overset{\delta+}{E}-\overset{\delta-}{Y} \longrightarrow Ar-E + H-Y$$

Arene | Electrophilic reagent | Product of electrophilic aromatic substitution

We call this reaction **electrophilic aromatic substitution;** it is one of the fundamental processes of organic chemistry.

12.1 REPRESENTATIVE ELECTROPHILIC AROMATIC SUBSTITUTION REACTIONS OF BENZENE

The scope of electrophilic aromatic substitution is quite large; both the arene and the electrophilic reagent are capable of wide variation. Indeed, it is this breadth of scope that makes electrophilic aromatic substitution so important. Electrophilic aromatic substitution is the principal method by which substituted derivatives of benzene are prepared both by industry and in the laboratory. We can gain a feeling for these reactions by examining a few typical examples in which benzene is the substrate. These reactions are listed in Table 12.1, and each will be discussed in more detail in

TABLE 12.1

Representative Electrophilic Aromatic Substitution Reactions of Benzene

Reaction and comments	Equation
1. Nitration Warming benzene with a mixture of nitric acid and sulfuric acid gives nitrobenzene. A nitro group ($-NO_2$) replaces one of the ring hydrogens.	$C_6H_5H + HNO_3 \xrightarrow[30-40°C]{H_2SO_4} C_6H_5NO_2 + H_2O$ Benzene, Nitric acid, Nitrobenzene (95%), Water
2. Sulfonation Treatment of benzene with hot concentrated sulfuric acid gives benzenesulfonic acid. A sulfonic acid group ($-SO_2OH$) replaces one of the ring hydrogens.	$C_6H_5H + HOSO_2OH \xrightarrow{heat} C_6H_5SO_2OH + H_2O$ Benzene, Sulfuric acid, Benzenesulfonic acid (100%), Water
3. Halogenation Bromine reacts with benzene in the presence of iron(III) bromide as a catalyst to give bromobenzene. Chlorine reacts similarly in the presence of iron(III) chloride to give chlorobenzene.	$C_6H_5H + Br_2 \xrightarrow{FeBr_3} C_6H_5Br + HBr$ Benzene, Bromine, Bromobenzene (65–75%), Hydrogen bromide
4. Friedel-Crafts alkylation Alkyl halides react with benzene in the presence of aluminum chloride to yield alkylbenzenes.	$C_6H_5H + (CH_3)_3CCl \xrightarrow[0°C]{AlCl_3} C_6H_5C(CH_3)_3 + HCl$ Benzene, *tert*-Butyl chloride, *tert*-Butylbenzene (60%), Hydrogen chloride
5. Friedel-Crafts acylation An analogous reaction occurs when acyl halides react with benzene in the presence of aluminum chloride. The products are acylbenzenes.	$C_6H_5H + CH_3CH_2\overset{O}{\overset{\Vert}{C}}Cl \xrightarrow[40°C]{AlCl_3} C_6H_5\overset{O}{\overset{\Vert}{C}}CH_2CH_3 + HCl$ Benzene, Propanoyl chloride, 1-Phenyl-1-propanone (88%), Hydrogen chloride

Sections 12.3 through 12.7. First, however, let us look at the general mechanism of electrophilic aromatic substitution.

12.2 MECHANISTIC PRINCIPLES OF ELECTROPHILIC AROMATIC SUBSTITUTION

Recall from Chapter 6 the general mechanism for electrophilic addition to alkenes:

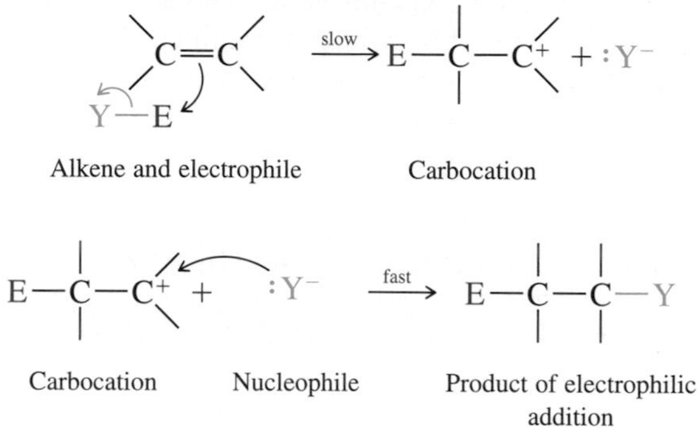

$$\text{C}=\text{C} + \text{Y}-\text{E} \xrightarrow{\text{slow}} \text{E}-\text{C}-\text{C}^{+} + :\text{Y}^{-}$$

Alkene and electrophile Carbocation

$$\text{E}-\text{C}-\text{C}^{+} + :\text{Y}^{-} \xrightarrow{\text{fast}} \text{E}-\text{C}-\text{C}-\text{Y}$$

Carbocation Nucleophile Product of electrophilic addition

The first step is rate-determining. It is the transfer of the pair of π electrons of the alkene to the electrophile to form a high-energy intermediate, a carbocation. Following its formation, the carbocation undergoes rapid capture by some Lewis base present in the medium.

The first step in the reaction of electrophilic reagents with benzene is similar. An electrophile accepts an electron pair from the π system of benzene to form a carbocation:

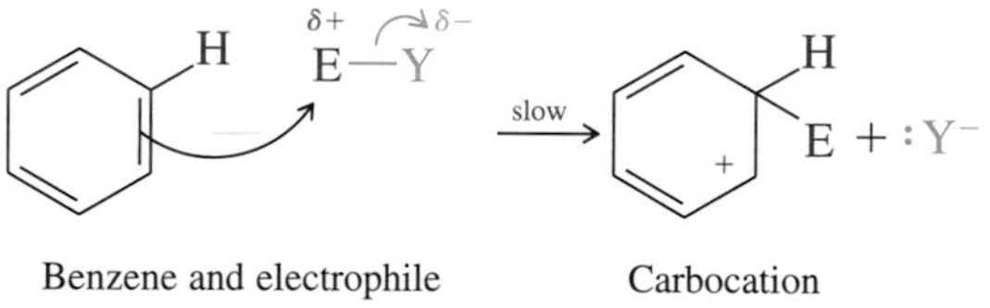

$$\text{C}_6\text{H}_6 + \overset{\delta+}{\text{E}}-\overset{\delta-}{\text{Y}} \xrightarrow{\text{slow}} \text{cyclohexadienyl cation (H, E)} + :\text{Y}^{-}$$

Benzene and electrophile Carbocation

This particular carbocation is a resonance-stabilized one of the allylic type. It is a **cyclohexadienyl cation** (often referred to as an **arenium ion**).

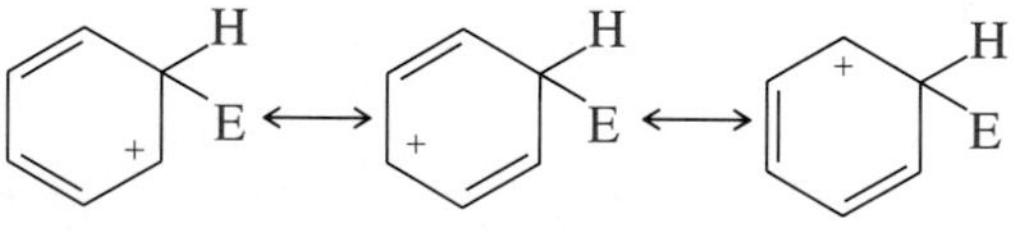

Resonance forms of a cyclohexadienyl cation

PROBLEM 12.1 In the simplest molecular orbital treatment of conjugated systems, it is assumed that the π system does not interact with the framework of σ bonds. When this MO method was used to calculate the charge distribution in cyclohexadienyl cation, it gave the results indicated. How does the charge at each carbon compare with that deduced by examining the most stable resonance structures for cyclohexadienyl cation?

(Structure: cyclohexadienyl cation with charges 0, +0.33, 0, +0.33, 0, +0.33)

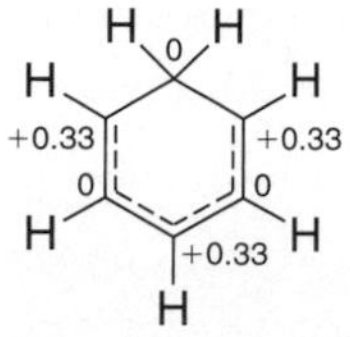

Most of the resonance stabilization of benzene is lost when it is converted to the cyclohexadienyl cation intermediate. In spite of being allyic, a cyclohexadienyl cation is *not* aromatic and possesses only a fraction of the resonance stabilization of benzene. Once formed, it rapidly loses a proton, restoring the aromaticity of the ring and giving the product of electrophilic aromatic substitution.

Cyclohexadienyl cation + $:Y^-$ —fast→ + HY

Observed product of electrophilic aromatic substitution

Not observed—not aromatic

If the Lewis base ($:Y^-$) had acted as a nucleophile and added to carbon, the product would have been a nonaromatic cyclohexadiene derivative. Addition and substitution products arise by alternative reaction paths of a cyclohexadienyl cation. Substitution occurs preferentially because there is a substantial driving force favoring rearomatization.

Figure 12.1 is a potential energy diagram describing the general mechanism of electrophilic aromatic substitution. In order for electrophilic aromatic substitution reactions to overcome the high activation energy that characterizes the first step, the electrophile must be a fairly reactive one. Many electrophilic reagents that react rap-

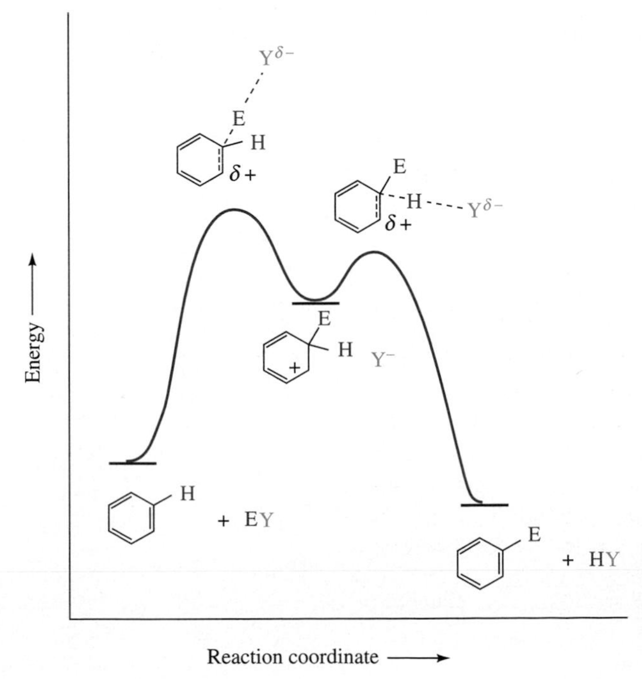

FIGURE 12.1 Energy diagram illustrating the energy changes associated with the two steps of electrophilic aromatic substitution.

idly with alkenes do not react at all with benzene. Peroxy acids and diborane, for example, fall into this category. Others, such as bromine, react with benzene only in the presence of catalysts that increase their electrophilicity. The low level of reactivity of benzene toward electrophiles stems from the substantial loss of resonance stabilization that accompanies transfer of a pair of its six π electrons to an electrophile.

With this as background, let us now examine each of the electrophilic aromatic substitution reactions presented in Table 12.1 in more detail, especially with respect to the electrophile that attacks benzene.

12.3 NITRATION OF BENZENE

On the basis of the general mechanism developed in the preceding section, we need only identify the specific electrophile in the nitration of benzene to have a fairly clear idea of how the reaction occurs. Figure 12.2 shows the application of those general principles to the reaction:

$$C_6H_5\text{—}H + HONO_2 \xrightarrow[30\text{–}40°C]{H_2SO_4} C_6H_5\text{—}NO_2 + H_2O$$

Benzene Nitric acid Nitrobenzene (95%) Water

In studies of nitration, Sir Christopher Ingold established that nitronium cation (NO_2^+) is the electrophile that attacks benzene in the rate-determining step. The concentration of NO_2^+ in nitric acid alone is too low to nitrate benzene at a convenient rate, but adding sulfuric acid to nitric acid causes the nitronium ion concentration to increase according to the equation

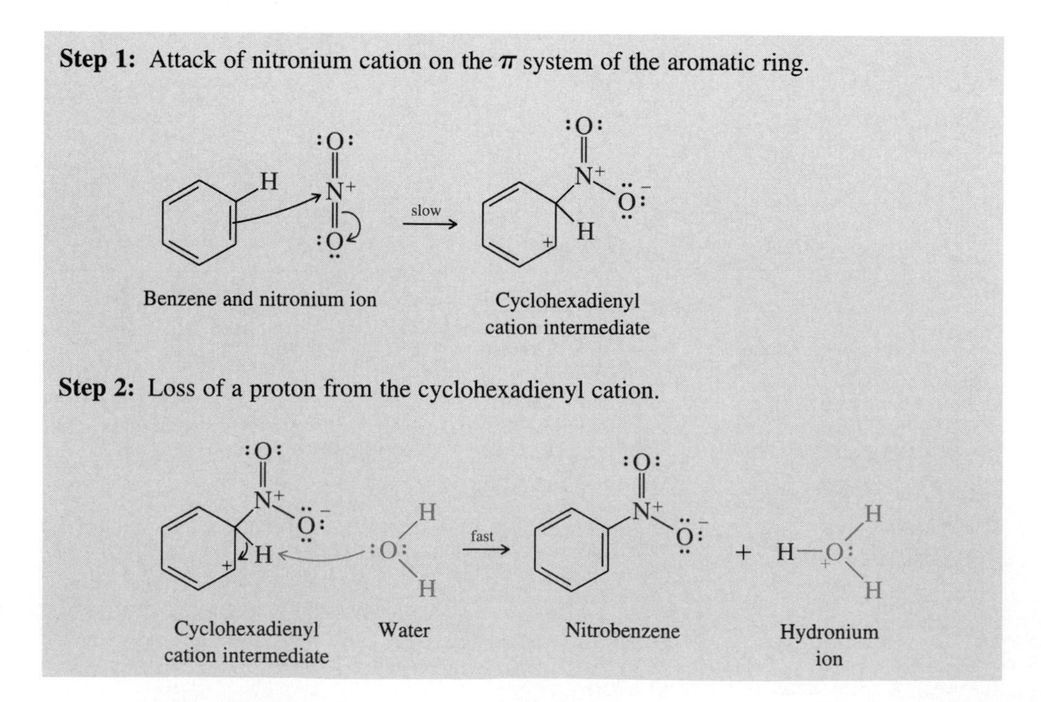

FIGURE 12.2 The mechanism of nitration of benzene.

$$\text{HO}-\overset{+}{\text{N}}(=\text{O})-\text{O}^- + 2\text{HOSO}_2\text{OH} \rightleftharpoons :\text{O}=\overset{+}{\text{N}}=\text{O}: + \text{H}_3\text{O}^+ + 2\text{HOSO}_2\text{O}^-$$

Nitric acid — Sulfuric acid — Nitronium ion — Hydronium ion — Hydrogen sulfate ion

Substitution of a nitro group for one of the ring hydrogens is a general reaction observed when benzene or one of its derivatives is treated with a mixture of nitric acid and sulfuric acid.

PROBLEM 12.2 Nitration of *p*-xylene gives a single product having the molecular formula $C_8H_9NO_2$ in high yield. Suggest a reasonable structure for this product. (If necessary, see Section 11.9 to remind yourself of the structure of *p*-xylene.)

12.4 SULFONATION OF BENZENE

The reaction of benzene with sulfuric acid to produce benzenesulfonic acid,

$$\text{C}_6\text{H}_5\text{H} + \text{HOSO}_2\text{OH} \overset{\text{heat}}{\rightleftharpoons} \text{C}_6\text{H}_5\text{SO}_2\text{OH} + \text{H}_2\text{O}$$

Benzene — Sulfuric acid — Benzenesulfonic acid — Water

is reversible but can be driven to completion by several techniques. Removing the water formed in the reaction, for example, allows benzenesulfonic acid to be obtained in virtually quantitative yield. When a solution of sulfur trioxide in sulfuric acid is used as the sulfonating agent, the rate of sulfonation is much faster and the equilibrium is displaced entirely to the side of products, according to the equation

$$\text{C}_6\text{H}_6 + \text{SO}_3 \xrightarrow{\text{H}_2\text{SO}_4} \text{C}_6\text{H}_5\text{SO}_2\text{OH}$$

Benzene — Sulfur trioxide — Benzenesulfonic acid

Among the variety of electrophilic species present in concentrated sulfuric acid, sulfur trioxide is probably the actual electrophile in aromatic sulfonation. We can represent the mechanism of sulfonation of benzene by sulfur trioxide by the sequence of steps shown in Figure 12.3.

PROBLEM 12.3 On being heated with sulfur trioxide in sulfuric acid, 1,2,4,5-tetramethylbenzene was converted to a product of molecular formula $C_{10}H_{14}O_3S$ in 94 percent yield. Suggest a reasonable structure for this product.

Step 1: Sulfur trioxide attacks benzene in the rate-determining step.

slow

Benzene and sulfur trioxide — Cyclohexadienyl cation intermediate

Step 2: A proton is lost from the sp^3 hybridized carbon of the intermediate to restore the aromaticity of the ring. The species shown that abstracts the proton is a hydrogen sulfate ion formed by ionization of sulfuric acid.

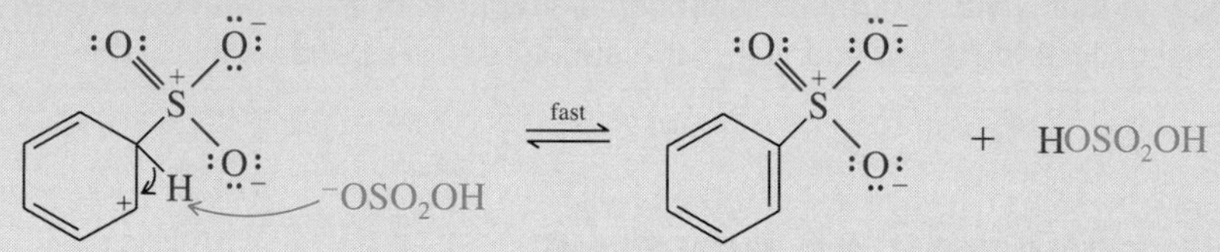

$^{-}OSO_2OH$ — fast — + $HOSO_2OH$

Cyclohexadienyl cation intermediate — Hydrogen sulfate ion — Benzenesulfonate ion — Sulfuric acid

Step 3: A rapid proton transfer from the oxygen of sulfuric acid to the oxygen of benzenesulfonate completes the process.

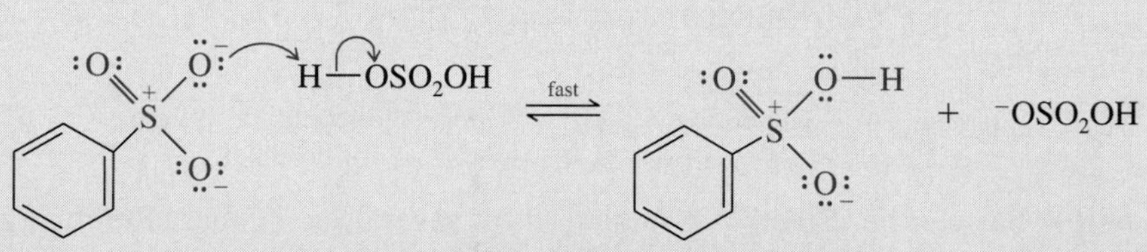

$H—OSO_2OH$ — fast — + $^{-}OSO_2OH$

Benzenesulfonate ion — Sulfuric acid — Benzenesulfonic acid — Hydrogen sulfate ion

FIGURE 12.3 The mechanism of sulfonation of benzene.

12.5 HALOGENATION OF BENZENE

According to the usual procedure for preparing bromobenzene, bromine is added to benzene in the presence of metallic iron (customarily a few carpet tacks) and the reaction mixture is heated.

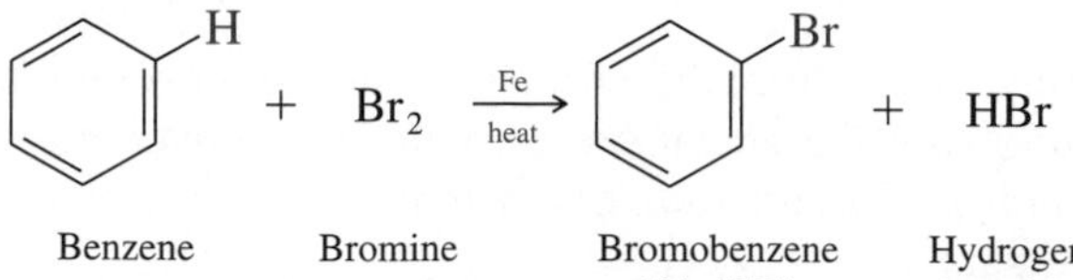

$$C_6H_5H + Br_2 \xrightarrow[\text{heat}]{\text{Fe}} C_6H_5Br + HBr$$

Benzene — Bromine — Bromobenzene (65–75%) — Hydrogen bromide

Bromine, while it adds rapidly to alkenes, is too weak an electrophile to react at an appreciable rate with benzene. A catalyst that increases the electrophilic properties of bromine must be present. Somehow carpet tacks can do this. How?

The active catalyst is not iron itself but iron(III) bromide, formed by reaction of iron and bromine.

Iron(III) bromide ($FeBr_3$) is also called *ferric bromide.*

$$2Fe + 3Br_2 \longrightarrow 2FeBr_3$$

Iron Bromine Iron(III) bromide

Iron(III) bromide is a weak Lewis acid. It combines with bromine to form a Lewis acid–Lewis base complex.

$$:\ddot{Br}-\ddot{Br}: + FeBr_3 \rightleftharpoons :\ddot{Br}-\overset{+}{\ddot{Br}}-\overset{-}{Fe}Br_3$$

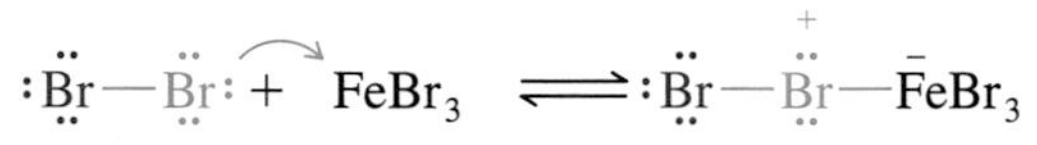

Lewis base Lewis acid Lewis acid–Lewis base complex

Complexation of bromine with iron(III) bromide makes bromine more electrophilic, and it attacks benzene to give a cyclohexadienyl intermediate as shown in step 1 of the mechanism depicted in Figure 12.4. In step 2, as in nitration and sulfonation, loss of a proton from the cyclohexadienyl cation is rapid and gives the product of electrophilic aromatic substitution.

Only small quantities of iron(III) bromide are required. It is a catalyst for the bromination and, as Figure 12.4 indicates, is regenerated in the course of the reaction. We shall see later in this chapter that some aromatic substrates are much more reactive than benzene and react rapidly with bromine even in the absence of a catalyst.

Chlorination is carried out in a manner similar to bromination and provides a ready route to chlorobenzene and related aryl chlorides. Fluorination and iodination of benzene and other arenes are rarely performed. Fluorine is so reactive that its reaction with benzene is difficult to control. Iodination is very slow and is characterized by an

Step 1: The bromine–iron(III) bromide complex is the active electrophile that attacks benzene. Two of the π electrons of benzene are used to form a bond to bromine and give a cyclohexadienyl cation intermediate.

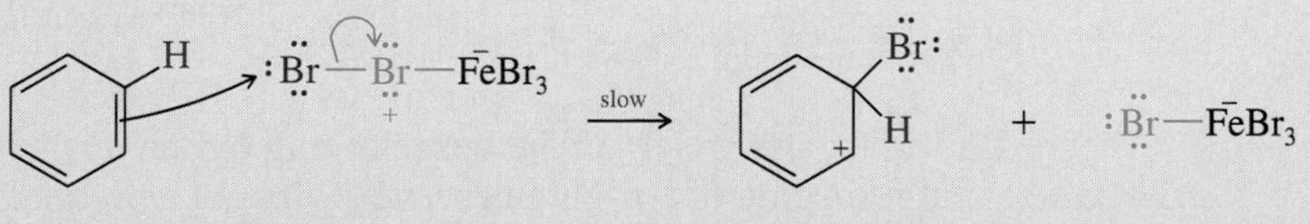

Benzene and bromine–iron(III) bromide complex — Cyclohexadienyl cation intermediate — Tetrabromoferrate ion

Step 2: Loss of a proton from the cyclohexadienyl cation yields bromobenzene.

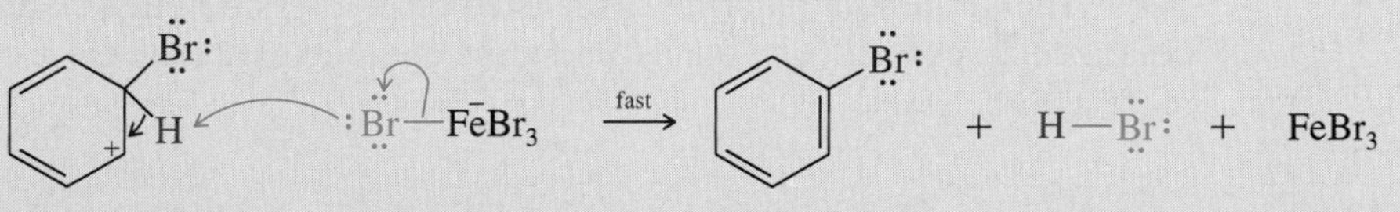

Cyclohexadienyl cation intermediate — Tetrabromoferrate ion — Bromobenzene — Hydrogen bromide — Iron(III) bromide

FIGURE 12.4 The mechanism of bromination of benzene.

unfavorable equilibrium constant. Syntheses of aryl fluorides and aryl iodides are normally carried out by way of functional group transformations of arylamines; these reactions will be described in Chapter 22.

12.6 FRIEDEL-CRAFTS ALKYLATION OF BENZENE

Alkyl halides react with benzene in the presence of aluminum chloride to yield alkylbenzenes.

$$C_6H_5\text{—}H + (CH_3)_3CCl \xrightarrow[0°C]{AlCl_3} C_6H_5\text{—}C(CH_3)_3 + HCl$$

Benzene; *tert*-Butyl chloride; *tert*-Butylbenzene (60%); Hydrogen chloride

Alkylation of benzene with alkyl halides in the presence of aluminum chloride was discovered by Charles Friedel and James M. Crafts in 1877. Crafts, who later became president of the Massachusetts Institute of Technology, collaborated with Friedel at the Sorbonne in Paris, and together they developed what we now call the **Friedel-Crafts reaction** into one of the most useful synthetic methods in organic chemistry.

Alkyl halides by themselves are insufficiently electrophilic to react with benzene. Aluminum chloride serves as a Lewis acid catalyst to enhance the electrophilicity of the alkylating agent. With tertiary and secondary alkyl halides, the addition of aluminum chloride leads to the formation of carbocations, which then attack the aromatic ring.

$$(CH_3)_3C\text{—}\ddot{\underset{..}{Cl}}: + AlCl_3 \longrightarrow (CH_3)_3C\text{—}\overset{+}{\ddot{\underset{..}{Cl}}}\text{—}\overset{-}{Al}Cl_3$$

tert-Butyl chloride; Aluminum chloride; Lewis acid–Lewis base complex

$$(CH_3)_3C\text{—}\overset{+}{\ddot{\underset{..}{Cl}}}\text{—}\overset{-}{Al}Cl_3 \longrightarrow (CH_3)_3C^+ + \overset{-}{Al}Cl_4$$

tert-Butyl chloride–aluminum chloride complex; *tert*-Butyl cation; Tetrachloroaluminate anion

Figure 12.5 illustrates attack on the benzene ring by the *tert*-butyl cation (step 1) and subsequent formation of *tert*-butylbenzene by loss of a proton from the cyclohexadienyl cation intermediate (step 2).

Secondary alkyl halides react by a similar mechanism involving attack on benzene by a secondary carbocation. Methyl and ethyl halides do not form carbocations when treated with aluminum chloride, but do alkylate benzene under Friedel-Crafts conditions. The aluminum chloride complexes of methyl and ethyl halides contain highly polarized carbon-halogen bonds, and these complexes are the electrophilic species that react with benzene.

$$CH_3\text{—}\overset{+}{\ddot{\underset{..}{X}}}\text{—}\overset{-}{Al}X_3 \qquad CH_3CH_2\text{—}\overset{+}{\ddot{\underset{..}{X}}}\text{—}\overset{-}{Al}X_3$$

Methyl halide–aluminum halide complex; Ethyl halide–aluminum halide complex

Step 1: Once generated by the reaction of *tert*-butyl chloride and aluminum chloride, *tert*-butyl cation attacks the π electrons of benzene, and a carbon-carbon bond is formed.

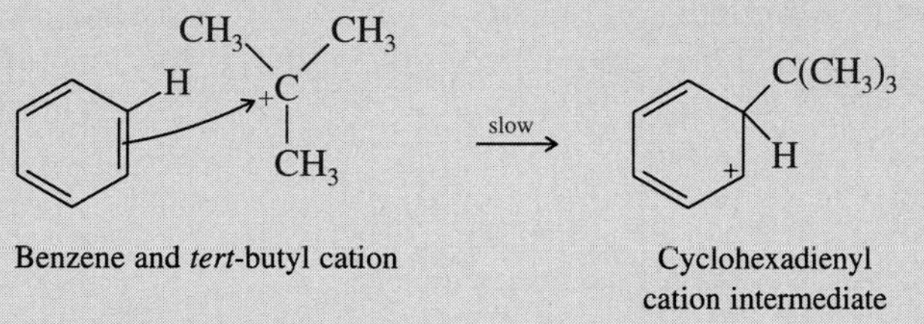

$$\text{C}_6\text{H}_6 + (CH_3)_3C^+ \xrightarrow{\text{slow}} \text{cyclohexadienyl cation (C(CH}_3)_3\text{, H)}$$

Benzene and *tert*-butyl cation — Cyclohexadienyl cation intermediate

Step 2: Loss of a proton from the cyclohexadienyl cation intermediate yields *tert*-butylbenzene.

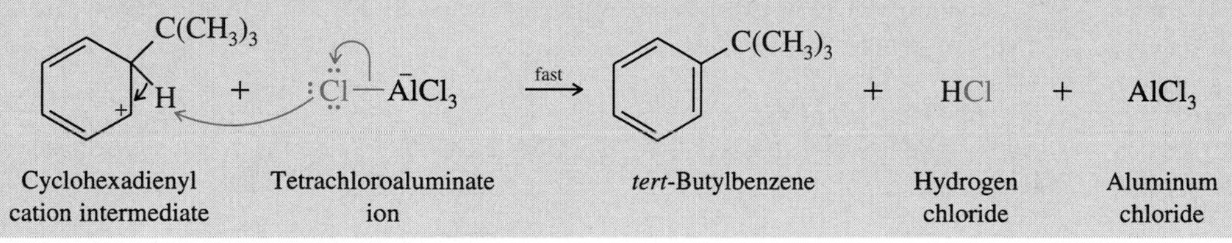

$$\text{Cyclohexadienyl cation} + :\ddot{Cl}-\bar{Al}Cl_3 \xrightarrow{\text{fast}} C_6H_5C(CH_3)_3 + HCl + AlCl_3$$

Cyclohexadienyl cation intermediate — Tetrachloroaluminate ion — *tert*-Butylbenzene — Hydrogen chloride — Aluminum chloride

FIGURE 12.5 The mechanism of Friedel-Crafts alkylation.

One drawback to Friedel-Crafts alkylation with primary alkyl halides is that secondary or tertiary carbocations can be formed from primary alkyl halides by rearrangement under the reaction conditions. This leads to the isolation of alkylbenzenes different from those to be expected in the absence of such rearrangements. For example, a Friedel-Crafts alkylation using isobutyl chloride (a primary alkyl halide) yields only *tert*-butylbenzene.

Other limitations to Friedel-Crafts reactions will be encountered in this chapter and are summarized in Table 12.4 (page 504).

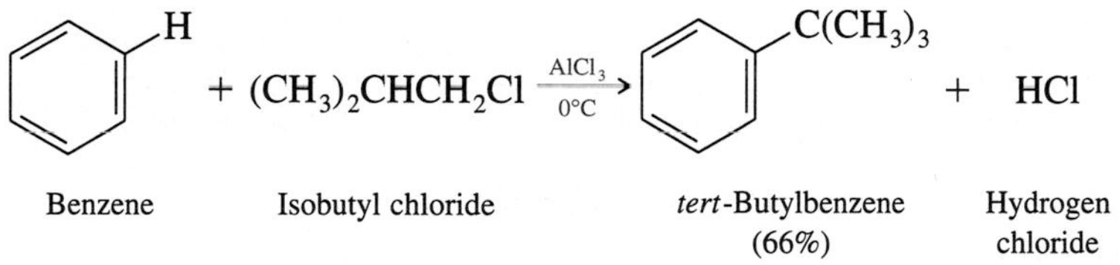

$$C_6H_6 + (CH_3)_2CHCH_2Cl \xrightarrow[0°C]{AlCl_3} C_6H_5C(CH_3)_3 + HCl$$

Benzene — Isobutyl chloride — *tert*-Butylbenzene (66%) — Hydrogen chloride

Here, the electrophile must be the *tert*-butyl cation formed by a hydride migration that accompanies ionization of the carbon-chlorine bond.

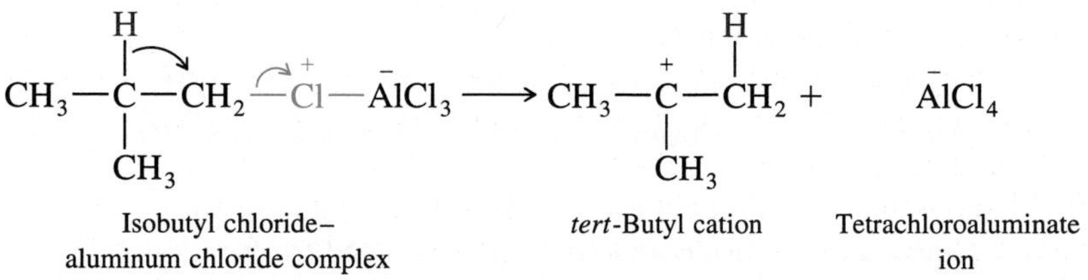

$$CH_3-\underset{CH_3}{\overset{H}{C}}-CH_2-\overset{+}{Cl}-\bar{Al}Cl_3 \longrightarrow CH_3-\underset{CH_3}{\overset{+}{C}}-\overset{H}{CH_2} + \bar{Al}Cl_4$$

Isobutyl chloride–aluminum chloride complex — *tert*-Butyl cation — Tetrachloroaluminate ion

PROBLEM 12.4 In an attempt to prepare propylbenzene, a chemist alkylated benzene with 1-chloropropane and aluminum chloride. To our chemist's chagrin, two isomeric hydrocarbons were obtained in a ratio of 2:1, the desired propylbenzene being the minor component. What do you think was the major component? How did it arise?

Other Lewis acid catalysts used in Friedel-Crafts reactions include iron(III) chloride ($FeCl_3$), zinc chloride ($ZnCl_2$), and boron trifluoride (BF_3).

Since electrophilic attack on benzene is simply another reaction available to a carbocation, alternative carbocation precursors can be used in place of alkyl halides. For example, alkenes, which are converted to carbocations by protonation, can serve to alkylate benzene.

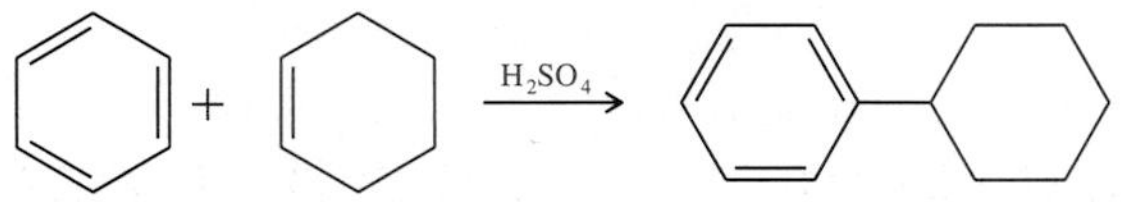

$$\text{Benzene} + \text{Cyclohexene} \xrightarrow{H_2SO_4} \text{Cyclohexylbenzene (65–68\%)}$$

Liquid hydrogen fluoride is sometimes used as both the proton donor and the solvent in these reactions.

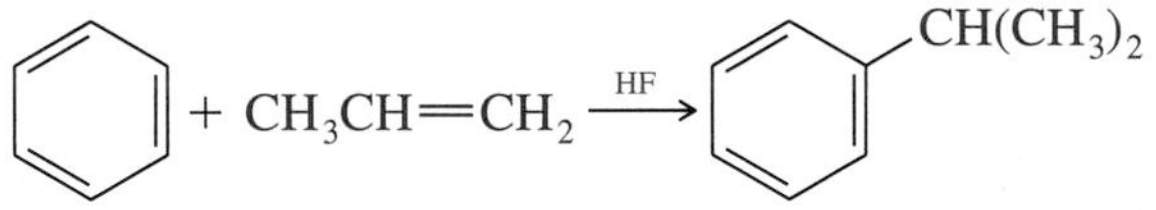

$$\text{Benzene} + CH_3CH{=}CH_2 \xrightarrow{HF} C_6H_5CH(CH_3)_2$$

Benzene, Propene, Isopropylbenzene (75%)

PROBLEM 12.5 Write a reasonable mechanism for the formation of isopropylbenzene from benzene and propene in liquid hydrogen fluoride.

Alcohols undergo a comparable reaction in acidic media or when treated with strong Lewis acids, leading to alkylation of benzene.

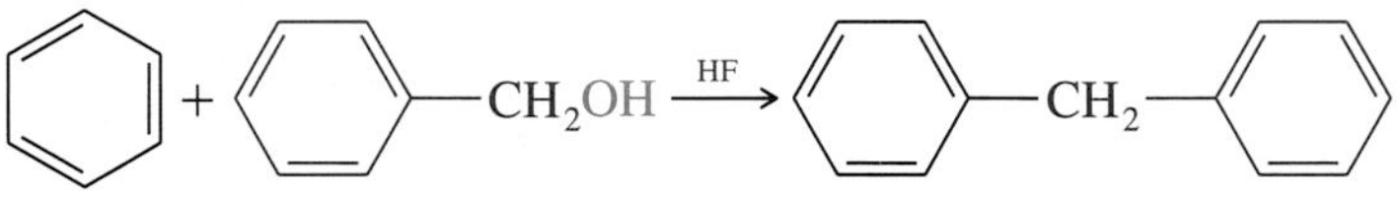

$$\text{Benzene} + C_6H_5CH_2OH \xrightarrow{HF} C_6H_5CH_2C_6H_5$$

Benzene, Benzyl alcohol, Diphenylmethane (65%)

Rearrangements can occur in these procedures, just as they do in Friedel-Crafts reactions of alkyl halides.

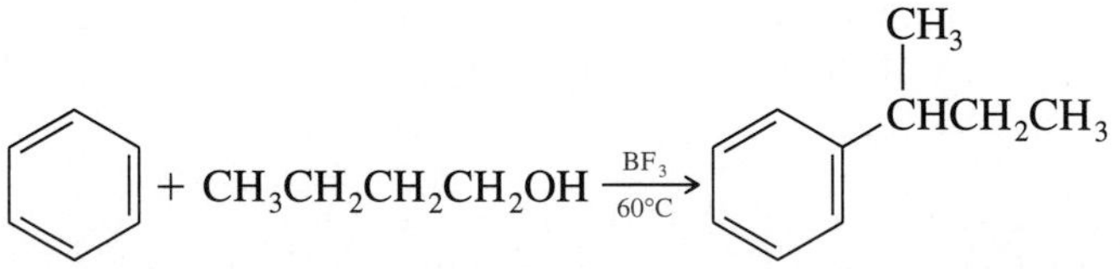

$$\text{Benzene} + CH_3CH_2CH_2CH_2OH \xrightarrow[60°C]{BF_3} C_6H_5CH(CH_3)CH_2CH_3$$

Benzene, 1-Butanol, *sec*-Butylbenzene (80%)

Alkenyl halides such as vinyl chloride ($CH_2{=}CHCl$) do *not* form carbocations on treatment with aluminum chloride and so cannot be used in Friedel-Crafts reactions. Thus, the industrial preparation of styrene from benzene and ethylene does not involve vinyl chloride but proceeds by way of ethylbenzene.

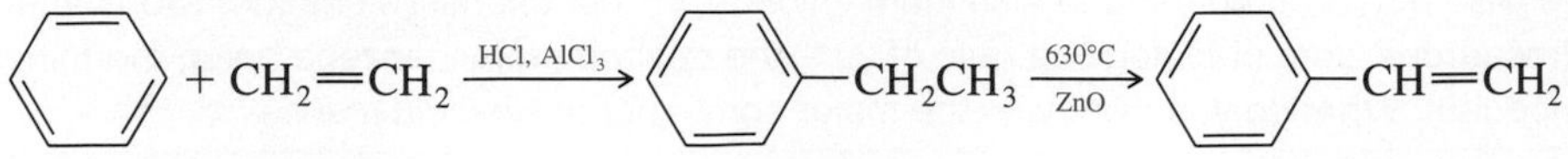

$$\text{Benzene} + CH_2{=}CH_2 \xrightarrow{HCl,\ AlCl_3} C_6H_5CH_2CH_3 \xrightarrow[ZnO]{630°C} C_6H_5CH{=}CH_2$$

Benzene, Ethylene, Ethylbenzene, Styrene

Dehydrogenation of alkylbenzenes, while useful in the industrial preparation of styrene, is not a general procedure and is not well-suited to the laboratory preparation of alkenylbenzenes. In such cases an alkylbenzene is subjected to benzylic bromination (Section 11.14), and the resulting benzylic bromide is treated with base to effect dehydrohalogenation.

PROBLEM 12.6 Outline a synthesis of 1-phenylcyclohexene from benzene and cyclohexene.

12.7 FRIEDEL-CRAFTS ACYLATION OF BENZENE

An acyl group has the general formula $RC(=O)—$.

Another version of the Friedel-Crafts reaction uses **acyl halides** and yields **acylbenzenes.**

$$C_6H_5H + CH_3CH_2CCl(=O) \xrightarrow[\text{carbon disulfide, 40°C}]{AlCl_3} C_6H_5CCH_2CH_3(=O) + HCl$$

Benzene; Propanoyl chloride; 1-Phenyl-1-propanone (88%); Hydrogen chloride

The electrophile in a Friedel-Crafts acylation reaction is an **acyl cation** (also referred to as an **acylium ion**). Acyl cations are stabilized by resonance. The acyl cation derived from propanoyl chloride is represented by the two resonance forms

$$CH_3CH_2\overset{+}{C}=\ddot{O}: \longleftrightarrow CH_3CH_2C\equiv\overset{+}{O}:$$

Most stable resonance form; oxygen and carbon have octets of electrons

Acyl cations form by coordination of an acyl chloride with aluminum chloride, followed by cleavage of the carbon-chlorine bond.

$$CH_3CH_2C(=\ddot{O})—\ddot{Cl}: + AlCl_3 \longrightarrow CH_3CH_2C(=\ddot{O})—\overset{+}{Cl}—\overset{-}{AlCl_3} \longrightarrow CH_3CH_2C\equiv\overset{+}{O}: + \overset{-}{AlCl_4}$$

Propanoyl chloride; Aluminum chloride; Lewis acid–Lewis base complex; Propanoyl cation; Tetrachloroaluminate ion

The electrophilic site of an acyl cation is its acyl carbon. The acyl cation reacts with benzene in a manner analogous to that of other electrophilic reagents as shown in Figure 12.6.

PROBLEM 12.7 The reaction shown gives a single product in 88 percent yield. What is that product?

$$1,2,3\text{-}(CH_3O)_3C_6H_3 \text{ (}CH_3O\text{, }OCH_3\text{, }OCH_3\text{)} + (CH_3)_2CHCH_2CCl(=O) \xrightarrow{AlCl_3}$$

Step 1: The acyl cation attacks benzene. A pair of π electrons of benzene is used to form a covalent bond to the carbon of the acyl cation.

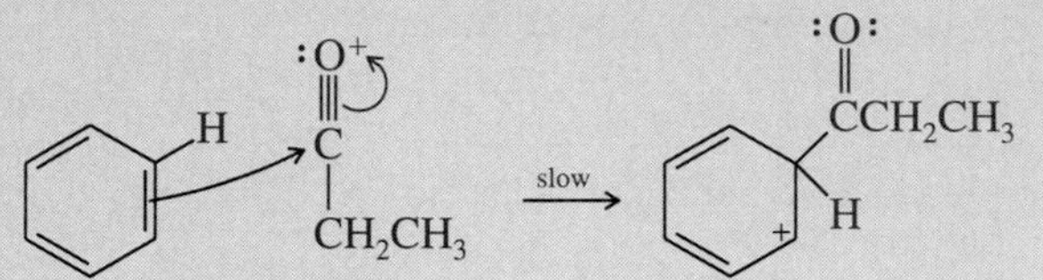

Benzene and propanoyl cation $\xrightarrow{\text{slow}}$ Cyclohexadienyl cation intermediate

Step 2: Aromaticity of the ring is restored when it loses a proton to give the acylbenzene.

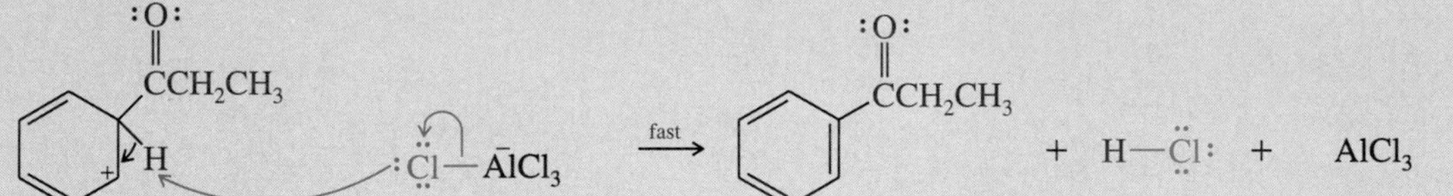

Cyclohexadienyl cation intermediate + Tetrachloroaluminate ion $\xrightarrow{\text{fast}}$ 1-Phenyl-1-propanone + Hydrogen chloride ($H-Cl$) + Aluminum chloride ($AlCl_3$)

FIGURE 12.6 The mechanism of Friedel-Crafts acylation.

Acyl chlorides are readily available. They are prepared by the reaction of a carboxylic acid with thionyl chloride.

$$\underset{\text{Carboxylic acid}}{RCOH} + \underset{\text{Thionyl chloride}}{SOCl_2} \longrightarrow \underset{\text{Acyl chloride}}{RCCl} + \underset{\text{Sulfur dioxide}}{SO_2} + \underset{\text{Hydrogen chloride}}{HCl}$$

Carboxylic acid anhydrides, compounds of the type RCOCR (with both carbonyl O's), can also serve as sources of acyl cations and, in the presence of aluminum chloride, acylate benzene. One acyl unit of an acid anhydride becomes attached to the benzene ring, while the other becomes part of a carboxylic acid.

Acetophenone is one of the commonly encountered benzene derivatives listed in Table 11.1 (page 428).

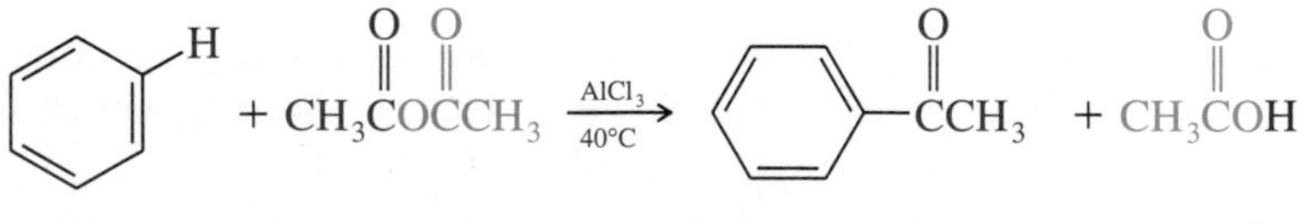

$$\underset{\text{Benzene}}{C_6H_5H} + \underset{\text{Acetic anhydride}}{CH_3COCCH_3} \xrightarrow[40°C]{AlCl_3} \underset{\text{Acetophenone (76–83\%)}}{C_6H_5CCH_3} + \underset{\text{Acetic acid}}{CH_3COH}$$

PROBLEM 12.8 *Succinic anhydride,* the structure of which is shown, is a cyclic anhydride often used in Friedel-Crafts acylations. Give the structure of the product obtained when benzene is acylated with succinic anhydride in the presence of aluminum chloride.

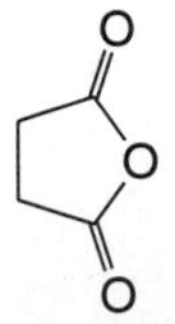

An important difference between Friedel-Crafts alkylations and acylations is that acylium ions do not rearrange. The acyl group of the acyl chloride or acid anhydride is transferred to the benzene ring unchanged. The reason for this is that an acylium ion is so strongly stabilized by resonance that it is more stable than any ion that could conceivably arise from it by a hydride or alkyl group shift.

$$-\overset{|}{\underset{\underset{R}{|}}{C}}-\overset{+}{C}\equiv O: \not\longrightarrow -\underset{+}{\overset{|}{C}}-C(=\ddot{O}:)R$$

More stable cation; all atoms have octets of electrons

Less stable cation; six electrons at carbon

12.8 SYNTHESIS OF ALKYLBENZENES BY ACYLATION-REDUCTION

Because acylation of an aromatic ring can be accomplished without rearrangement, it is frequently used as the first step in a procedure for the *alkylation* of aromatic compounds by *acylation-reduction.* As we saw in Section 12.6, Friedel-Crafts alkylation of benzene with primary alkyl halides normally yields products having rearranged alkyl groups as substituents. When a compound of the type $ArCH_2R$ is desired, a two-step transformation is used in which the first step is a Friedel-Crafts acylation using an acyl chloride or anhydride.

$$\text{C}_6\text{H}_6 \xrightarrow[\text{AlCl}_3]{\text{RCCl (O)}} \text{C}_6\text{H}_5\overset{O}{\overset{\|}{C}}R \xrightarrow{\text{reduction}} \text{C}_6\text{H}_5CH_2R$$

Benzene — Acylbenzene — Alkylbenzene

The second step is a reduction of the carbonyl group (C=O) to a methylene group (CH_2).

The most commonly used method for reducing an acylbenzene to an alkylbenzene employs a zinc-mercury amalgam in concentrated hydrochloric acid and is called the **Clemmensen reduction.**

The synthesis of butylbenzene illustrates the use of the acylation-reduction sequence.

$$\text{C}_6\text{H}_6 + CH_3CH_2CH_2\overset{O}{\overset{\|}{C}}Cl \xrightarrow{AlCl_3} \text{C}_6\text{H}_5\overset{O}{\overset{\|}{C}}CH_2CH_2CH_3 \xrightarrow[HCl]{Zn(Hg)} \text{C}_6\text{H}_5CH_2CH_2CH_2CH_3$$

Benzene — Butanoyl chloride — 1-Phenyl-1-butanone (86%) — Butylbenzene (73%)

Direct alkylation of benzene using 1-chlorobutane and aluminum chloride would yield *sec*-butylbenzene by rearrangement and so could not be used.

PROBLEM 12.9 Using benzene and any necessary organic or inorganic reagents, suggest efficient syntheses of

(*a*) Isobutylbenzene, $C_6H_5CH_2CH(CH_3)_2$
(*b*) Neopentylbenzene, $C_6H_5CH_2C(CH_3)_3$

SAMPLE SOLUTION (*a*) Friedel-Crafts alkylation of benzene with isobutyl chloride is not suitable, because it yields *tert*-butylbenzene by rearrangement.

$$C_6H_6 + (CH_3)_2CHCH_2Cl \xrightarrow{AlCl_3} C_6H_5C(CH_3)_3$$

Benzene   Isobutyl chloride   *tert*-Butylbenzene (66%)

The two-step acylation-reduction sequence is required. Acylation of benzene with $(CH_3)_2CH\overset{O}{\overset{\|}{C}}Cl$ puts the side chain on the ring with the correct carbon skeleton. Clemmensen reduction converts the carbonyl group to a methylene group.

$$C_6H_6 + (CH_3)_2CH\overset{O}{\overset{\|}{C}}Cl \xrightarrow{AlCl_3} C_6H_5\overset{O}{\overset{\|}{C}}CH(CH_3)_2 \xrightarrow[HCl]{Zn(Hg)} C_6H_5CH_2CH(CH_3)_2$$

Benzene   2-Methylpropanoyl chloride   2-Methyl-1-phenyl-1-propanone (84%)   Isobutylbenzene (80%)

An alternative method for reducing aldehyde and ketone carbonyl groups is the **Wolff-Kishner reduction.** Heating an aldehyde or a ketone with hydrazine (H_2NNH_2) and sodium or potassium hydroxide in a high-boiling alcohol such as diethylene glycol ($HOCH_2CH_2OCH_2CH_2OH$, bp 245°C) or triethylene glycol ($HOCH_2CH_2OCH_2CH_2OCH_2CH_2OH$, bp 287°C) converts the carbonyl to a CH_2 group.

$$C_6H_5\overset{O}{\overset{\|}{C}}CH_2CH_3 \xrightarrow[\text{triethylene glycol, 175°C}]{H_2NNH_2,\ KOH} C_6H_5CH_2CH_2CH_3$$

1-Phenyl-1-propanone   Propylbenzene (82%)

Both the Clemmensen and the Wolff-Kishner reductions are designed to carry out a specific functional group transformation, the reduction of an aldehyde or ketone carbonyl to a methylene group. Neither one will reduce the carbonyl group of a carboxylic acid, nor are carbon-carbon double or triple bonds affected by these methods. We will not discuss the mechanism of either the Clemmensen reduction or the Wolff-Kishner reduction, since both involve chemistry that is beyond the scope of what we have covered to this point.

12.9 RATE AND ORIENTATION IN ELECTROPHILIC AROMATIC SUBSTITUTION

So far we have been concerned only with electrophilic substitution on benzene. Two important questions arise when we turn to analogous substitutions on arenes that already bear at least one substituent:

1. What is the effect of a substituent on the *rate* of electrophilic aromatic substitution?
2. What is the effect of a substituent on the *regioselectivity* (orientation) of electrophilic aromatic substitution?

To illustrate substituent effects on rate, consider the nitration of benzene, toluene, and (trifluoromethyl)benzene.

CH_3 | | CF_3

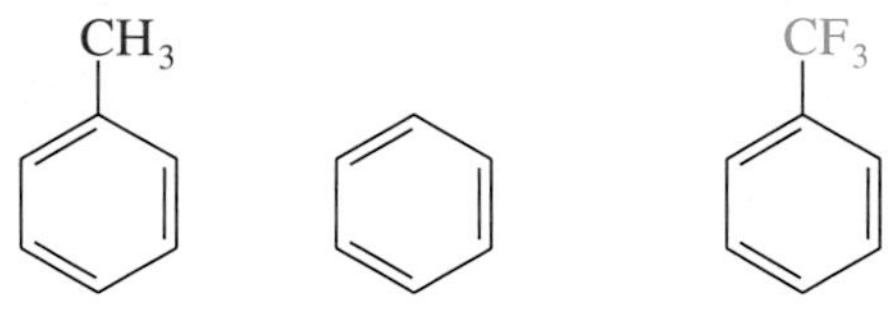

Toluene (most reactive) | Benzene | (Trifluoromethyl)benzene (least reactive)

Toluene undergoes nitration some 20 to 25 times faster than benzene. Because toluene is more reactive than benzene, we say that a methyl group *activates* the ring toward electrophilic aromatic substitution. (Trifluoromethyl)benzene, on the other hand, undergoes nitration about 40,000 times more slowly than benzene. We say that a trifluoromethyl group *deactivates* the ring toward electrophilic aromatic substitution.

Just as there is a marked difference in how methyl and trifluoromethyl substituents affect the rate of electrophilic aromatic substitution, so too there is a marked difference in how they affect its regioselectivity.

Three products are possible from nitration of toluene: *o*-nitrotoluene, *m*-nitrotoluene, and *p*-nitrotoluene. All are formed, but not in equal amounts. The meta isomer is formed to only a very small extent (3 percent). Together, the ortho- and para-substituted isomers make up 97 percent of the product mixture.

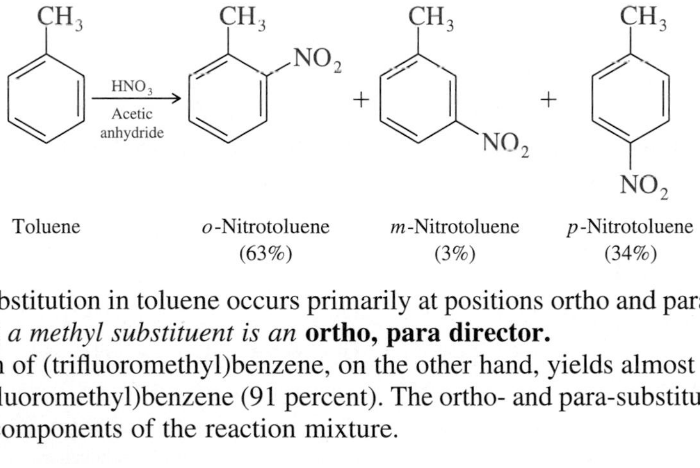

Toluene $\xrightarrow[\text{Acetic anhydride}]{HNO_3}$ *o*-Nitrotoluene (63%) + *m*-Nitrotoluene (3%) + *p*-Nitrotoluene (34%)

Because substitution in toluene occurs primarily at positions ortho and para to methyl, we say that *a methyl substituent is an* **ortho, para director.**

Nitration of (trifluoromethyl)benzene, on the other hand, yields almost exclusively *m*-nitro(trifluoromethyl)benzene (91 percent). The ortho- and para-substituted isomers are minor components of the reaction mixture.

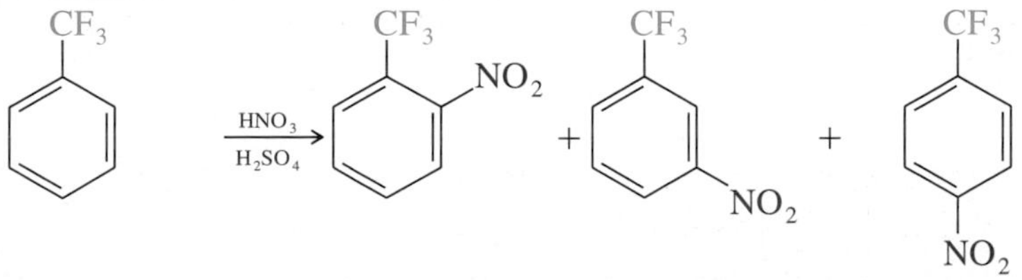

(Trifluoromethyl)benzene $\xrightarrow[H_2SO_4]{HNO_3}$ *o*-Nitro(trifluoromethyl)benzene (6%) + *m*-Nitro(trifluoromethyl)benzene (91%) + *p*-Nitro(trifluoromethyl)benzene (3%)

Because substitution in (trifluoromethyl)benzene occurs primarily at positions meta to the substituent, we say that *a trifluoromethyl group is a* **meta director.**

The regioselectivity of substitution, like the rate, is strongly affected by the substituent. In the following several sections we will examine the relationship between the structure of the substituent and its effect on rate and regioselectivity of electrophilic aromatic substitution.

12.10 RATE AND ORIENTATION IN THE NITRATION OF TOLUENE

Why is there such a marked difference between methyl and trifluoromethyl substituents in their influence on electrophilic aromatic substitution? Methyl is activating and ortho, para–directing; trifluoromethyl is deactivating and meta-directing. The first point to remember is that the regioselectivity of substitution is set once the cyclohexadienyl cation intermediate is formed. If we can explain why

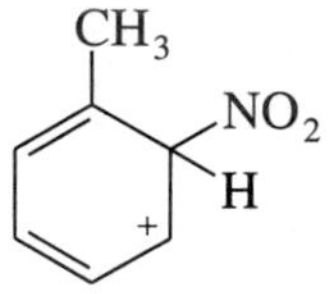

CH_3 NO_2 H and CH_3 H NO_2 are formed faster than CH_3 H NO_2

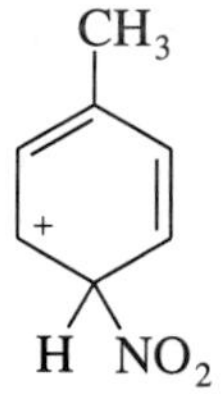

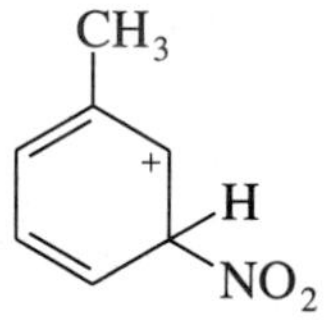

we will understand the reasons for the regioselectivity. A principle we have invoked previously serves us well here: a *more stable carbocation is formed faster than a less stable one.* The most likely reason for the directing effect of methyl must be that the cyclohexadienyl cation precursors to *o*- and *p*-nitrotoluene are more stable than the one leading to *m*-nitrotoluene.

One way to assess the relative stabilities of these various intermediates is to examine electron delocalization in them using a resonance description. The cyclohexadienyl cations leading to *o*- and *p*-nitrotoluene have tertiary carbocation character. Each has one principal resonance form in which the positive charge resides on the carbon that bears the methyl group.

Ortho attack

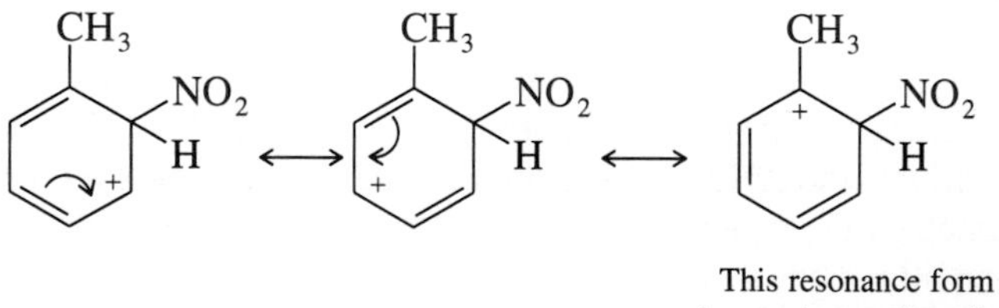

CH_3 NO_2 H ⟷ CH_3 NO_2 H ⟷ CH_3 NO_2 H

This resonance form is a tertiary carbocation

Para attack

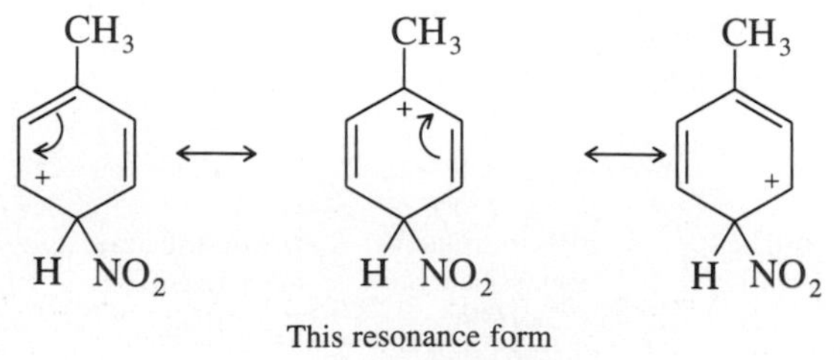

CH_3 H NO_2 ⟷ CH_3 H NO_2 ⟷ CH_3 H NO_2

This resonance form is a tertiary carbocation

The three resonance structures of the cyclohexadienyl cation intermediate leading to meta substitution are all secondary carbocations.

Meta attack

Because of their tertiary carbocation character the cyclohexadienyl cation intermediates leading to ortho and to para substitution are more stable and are formed faster than the secondary carbocation intermediate leading to meta substitution. They are also more stable than the secondary cyclohexadienyl cation intermediate formed during nitration of benzene. A methyl group is an activating substituent because it stabilizes the carbocation intermediate formed in the rate-determining step more than a hydrogen substituent does. It is ortho, para–directing because it stabilizes the carbocation formed by electrophilic attack at these positions more than it stabilizes the intermediate formed by attack at the meta position. Figure 12.7 compares the energies of activation for attack at the various positions of toluene.

A methyl group is an *electron-releasing* substituent and activates *all* of the ring carbons of toluene toward electrophilic attack. The ortho and para positions are activated more than the meta positions. The relative rates of attack at the various positions in toluene compared with a single position in benzene are as follows (for nitration at 25°C):

CH_3: 42, 42, 2.5, 2.5, 58 relative to 1, 1, 1, 1, 1, 1

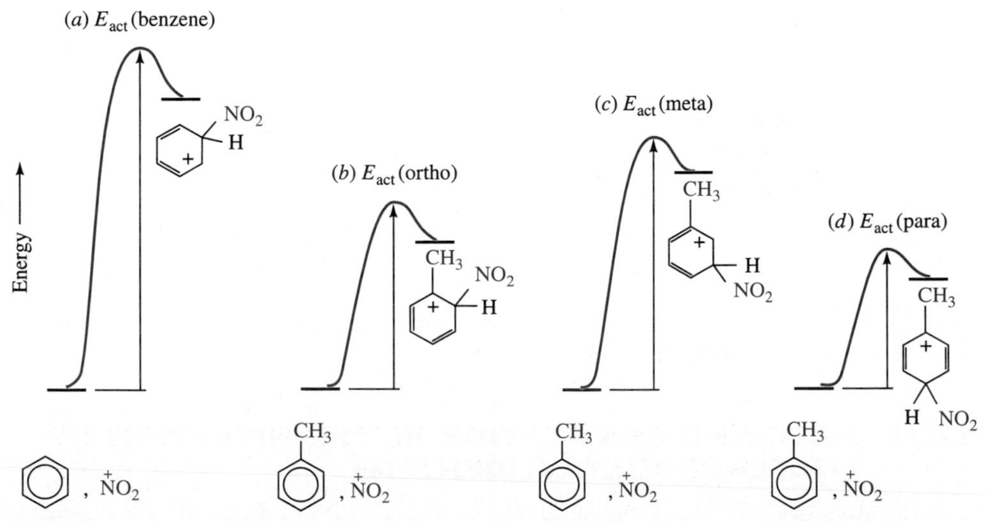

FIGURE 12.7 Comparative energy diagrams for nitronium ion attack (*a*) on benzene and at the (*b*) ortho, (*c*) meta, and (*d*) para positions of toluene. E_{act}(benzene) > E_{act}(meta) > E_{act}(ortho) > E_{act}(para).

These relative rate data per position are experimentally determined and are known as *partial rate factors.* They offer a convenient way to express substituent effects in electrophilic aromatic substitution reactions.

The major influence of the methyl group is *electronic.* The most important factor is relative carbocation stability. To a small extent, the methyl group sterically hinders the ortho positions, making attack slightly more likely at the para carbon than at a single ortho carbon. However, para substitution is at a statistical disadvantage, since there are two equivalent ortho positions but only one para position.

PROBLEM 12.10 The rates of nitration relative to a single position of benzene at the various positions of *tert*-butylbenzene are as shown.

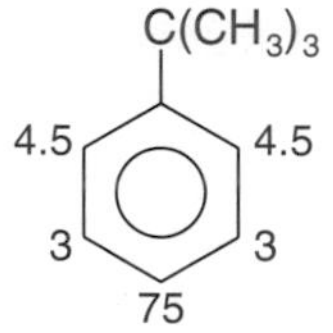

$C(CH_3)_3$

4.5 4.5

3 3

75

(*a*) How reactive is *tert*-butylbenzene toward nitration compared with benzene?

(*b*) How reactive is *tert*-butylbenzene toward nitration compared with toluene?

(*c*) Predict the distribution among the various mononitration products of *tert*-butylbenzene.

SAMPLE SOLUTION (*a*) Benzene has six equivalent sites at which nitration can occur. Summing the individual relative rates of attack at each position in *tert*-butylbenzene and benzene, we obtain

$$\frac{\textit{tert}\text{-Butylbenzene}}{\text{Benzene}} = \frac{2(4.5) + 2(3) + 75}{6(1)} = \frac{90}{6} = 15$$

tert-Butylbenzene undergoes nitration 15 times as fast as benzene.

All alkyl groups, not just methyl, are activating substituents and ortho, para directors. This is because any alkyl group, be it methyl, ethyl, isopropyl, *tert*-butyl, or any other, stabilizes a carbocation site to which it is directly attached. When R = alkyl,

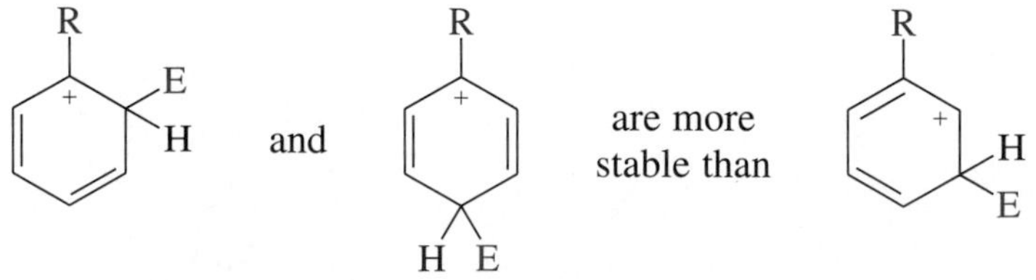

R, +, E, H and R, +, H E are more stable than R, +, H, E

where E is any electrophile. All three structures are more stable for R = alkyl than for R = H and are formed at faster rates.

12.11 RATE AND ORIENTATION IN THE NITRATION OF (TRIFLUOROMETHYL)BENZENE

Turning now to electrophilic aromatic substitution in (trifluoromethyl)benzene, we consider the electronic properties of a trifluoromethyl group. Because of their high electronegativity the three fluorine atoms polarize the electron distribution in their σ bonds to carbon, so that carbon bears a partial positive charge.

Recall from Section 4.11 that effects that are transmitted by the polarization of σ bonds are called *inductive effects.*

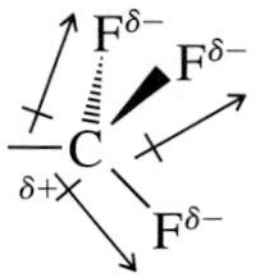

$F^{\delta-}$ $F^{\delta-}$ C $\delta+$ $F^{\delta-}$

Unlike a methyl group, which is slightly electron-releasing, a trifluoromethyl group is a powerful electron-withdrawing substituent. Consequently, a CF_3 group *destabilizes* a carbocation site to which it is attached.

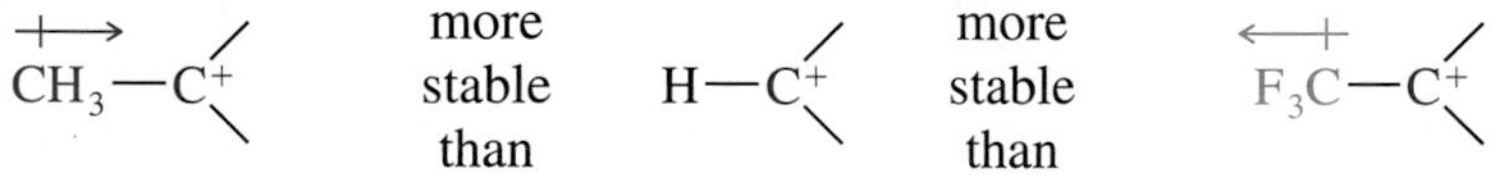

$CH_3—C^+$ more stable than $H—C^+$ more stable than $F_3C—C^+$

Methyl group releases electrons, stabilizes carbocation

Trifluoromethyl group withdraws electrons, destabilizes carbocation

When we examine the cyclohexadienyl cation intermediates involved in the nitration of (trifluoromethyl)benzene, we find that those leading to ortho and para substitution are strongly destabilized.

Ortho attack

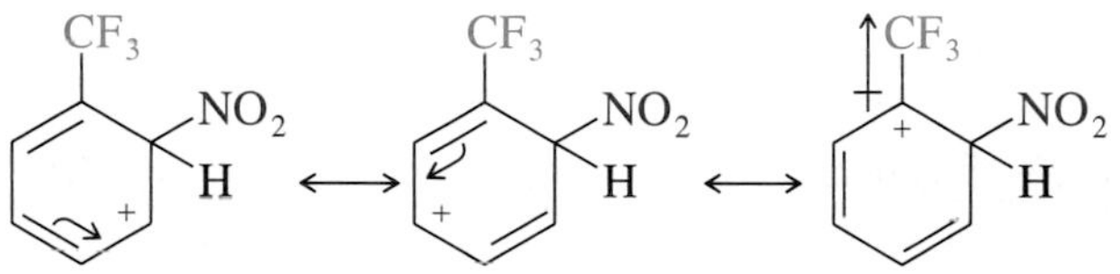

CF_3 NO_2 H ⟷ CF_3 NO_2 H ⟷ CF_3 NO_2 H

Positive charge on carbon bearing trifluoromethyl group; very unstable

Para attack

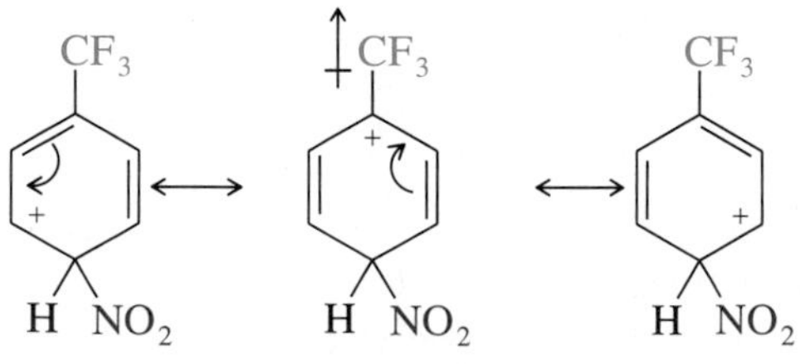

CF_3 H NO_2 ⟷ CF_3 H NO_2 ⟷ CF_3 H NO_2

Positive charge on carbon bearing trifluoromethyl group; very unstable

None of the three major resonance forms of the intermediate formed during attack at the meta position has a positive charge on the carbon bearing the trifluoromethyl substituent.

Meta attack

CF_3 ... H, NO_2 $\longleftrightarrow$ CF_3 ... H, NO_2 $\longleftrightarrow$ CF_3 ... H, NO_2

Attack at the meta position leads to a more stable intermediate than attack at either the ortho or the para position, and so meta substitution predominates. Even the intermediate corresponding to meta attack, however, is very unstable and is formed with difficulty. The trifluoromethyl group is only one bond farther removed from the positive charge here than it is in the ortho and para intermediates and so still exerts a significant, although somewhat diminished, destabilizing effect.

All the ring positions of (trifluoromethyl)benzene are deactivated as compared with benzene. The meta position is simply deactivated *less* than the ortho and para positions. We find that the relative rates of nitration of the various sites of (trifluoromethyl)benzene are

CF_3: ortho 4.5×10^{-6}, 4.5×10^{-6}; meta 67×10^{-6}, 67×10^{-6}; para 4.5×10^{-6} compared with benzene: 1, 1, 1, 1, 1, 1

Figure 12.8 compares the energy profile for nitronium ion attack at benzene with those

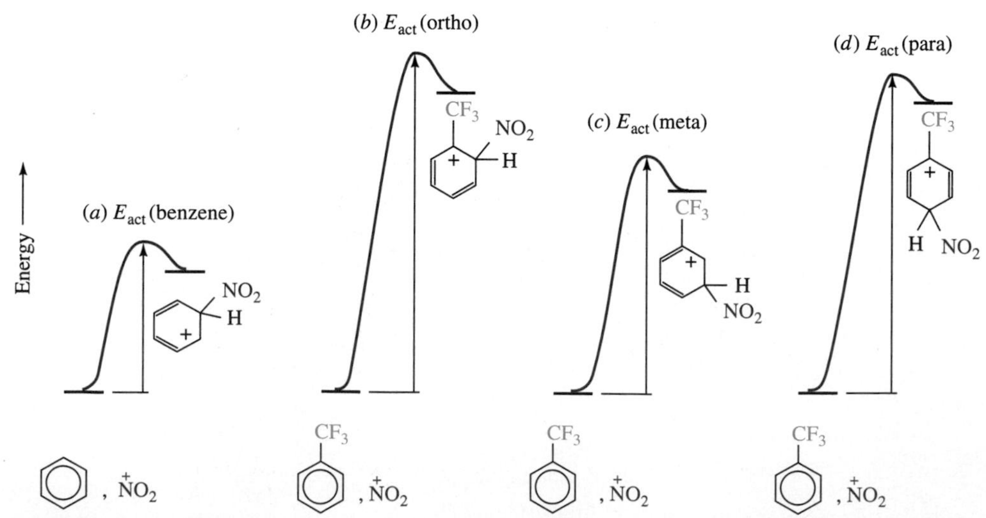

FIGURE 12.8 Comparative energy diagrams for nitronium ion attack (*a*) on benzene and at the (*b*) ortho, (*c*) meta, and (*d*) para positions of (trifluoromethyl)benzene. E_{act}(ortho) > E_{act}(para) > E_{act}(meta) > E_{act}(benzene).

for attack at the ortho, meta, and para positions of (trifluoromethyl)benzene. The presence of the electron-withdrawing trifluoromethyl group raises the activation energy for attack at all the ring positions, but the increase is least for attack at the meta position.

PROBLEM 12.11 The compounds benzyl chloride ($C_6H_5CH_2Cl$), (dichloromethyl)benzene ($C_6H_5CHCl_2$), and (trichloromethyl)benzene ($C_6H_5CCl_3$) all undergo nitration more slowly than benzene. The proportion of *m*-nitro-substituted product is 4 percent in one, 34 percent in another, and 64 percent in another. Classify the substituents $—CH_2Cl$, $—CHCl_2$, and $—CCl_3$ according to each one's effect on rate and orientation in electrophilic aromatic substitution.

12.12 SUBSTITUENT EFFECTS IN ELECTROPHILIC AROMATIC SUBSTITUTION: ACTIVATING SUBSTITUENTS

Our analysis of substituent effects has so far centered on two groups, methyl and trifluoromethyl. We have seen that a methyl substituent is activating and ortho, para–directing. A trifluoromethyl group is strongly deactivating and meta-directing. What about other substituents?

Table 12.2 summarizes orientation and rate effects in electrophilic aromatic substitution reactions for a variety of frequently encountered substituents. It is arranged in order of decreasing activating power: the most strongly activating substituents are at the top, the most strongly deactivating substituents are at the bottom. The main features of the table can be summarized as follows:

1. All activating substituents are ortho, para directors.
2. Halogen substituents are slightly deactivating but are ortho, para–directing.
3. Substituents more deactivating than halogen are meta directors.

Some of the most powerful *activating* substituents are those in which an oxygen atom is attached directly to the ring. These substituents include the hydroxyl group as well as alkoxy and acyloxy groups.

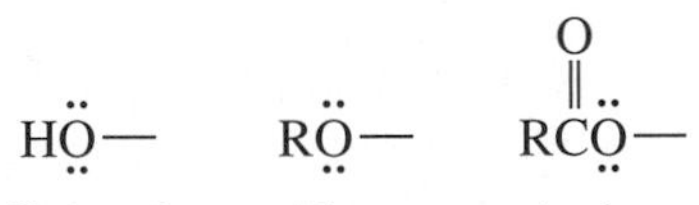

HÖ—	RÖ—	RC(=O)Ö—
Hydroxyl	Alkoxy	Acyloxy

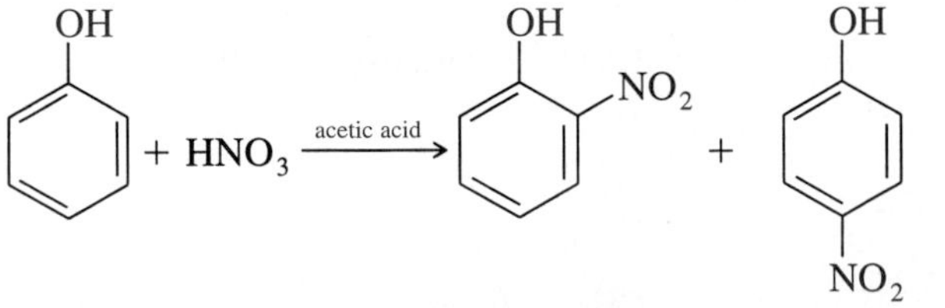

$$\text{Phenol} + HNO_3 \xrightarrow{\text{acetic acid}} o\text{-Nitrophenol (44\%)} + p\text{-Nitrophenol (56\%)}$$

Phenol | *o*-Nitrophenol (44%) | *p*-Nitrophenol (56%)

Phenol and *anisole* are among the commonly encountered benzene derivatives listed in Table 11.1 (page 428). Electrophilic aromatic substitution in phenol is discussed in more detail in Section 24.8.

Hydroxyl, alkoxy, and acyloxy groups activate the ring to such an extent that bromination occurs rapidly even in the absence of a catalyst.

OCH_3 → (Br_2, acetic acid) → OCH_3, Br

Anisole → *p*-Bromoanisole (90%)

TABLE 12.2
Classification of Substituents in Electrophilic Aromatic Substitution Reactions

Effect on rate	Substituent	Effect on orientation
Very strongly activating	$-\ddot{N}H_2$ (amino)	Ortho, para-directing
	$-\ddot{N}HR$ (alkylamino)	
	$-\ddot{N}R_2$ (dialkylamino)	
	$-\ddot{O}H$ (hydroxyl)	
Strongly activating	$-\ddot{N}H\overset{O}{\overset{\Vert}{C}}R$ (acylamino)	Ortho, para-directing
	$-\ddot{O}R$ (alkoxy)	
	$-\ddot{O}\overset{O}{\overset{\Vert}{C}}R$ (acyloxy)	
Activating	$-R$ (alkyl)	Ortho, para-directing
	$-Ar$ (aryl)	
	$-CH{=}CR_2$ (alkenyl)	
Standard of comparison	$-H$ (hydrogen)	
Deactivating	$-X$ (halogen) ($X{=}F, Cl, Br, I$)	Ortho, para-directing
	$-CH_2X$ (halomethyl)	
Strongly deactivating	$-\overset{O}{\overset{\Vert}{C}}H$ (formyl)	Meta-directing
	$-\overset{O}{\overset{\Vert}{C}}R$ (acyl)	
	$-\overset{O}{\overset{\Vert}{C}}OH$ (carboxylic acid)	
	$-\overset{O}{\overset{\Vert}{C}}OR$ (ester)	
	$-\overset{O}{\overset{\Vert}{C}}Cl$ (acyl chloride)	
	$-C{\equiv}N$ (cyano)	
	$-SO_3H$ (sulfonic acid)	
Very strongly deactivating	$-CF_3$ (trifluoromethyl)	Meta-directing
	$-NO_2$ (nitro)	

The inductive effect of hydroxyl and alkoxy groups, because of the electronegativity of oxygen, is to withdraw electrons and would seem to require that such substituents be deactivating. The electron-withdrawing inductive effect, however, is overcome by a much larger electron-releasing effect involving the unshared electron pairs of oxygen. Attack at positions ortho and para to a carbon that bears a substituent of the type —$\ddot{\text{O}}$R gives a cation stabilized by delocalization of an unshared electron pair of oxygen into the π system of the ring (a *resonance* or *conjugation* effect).

Ortho attack

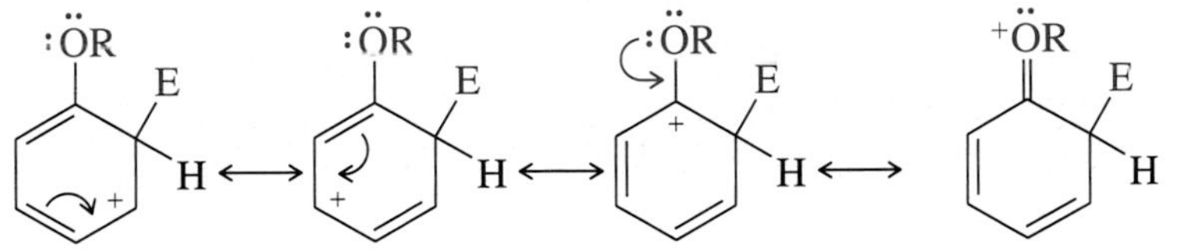

Most stable resonance form; oxygen and all carbons have octets of electrons

Para attack

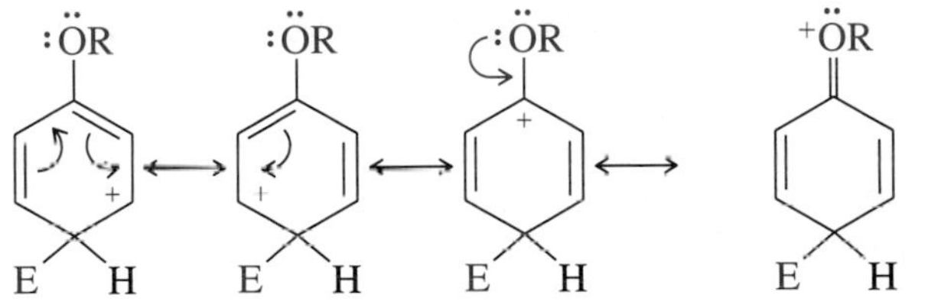

Most stable resonance form; oxygen and all carbons have octets of electrons

Oxygen-stabilized carbocations of this type are far more stable than tertiary carbocations. They are best represented by structures in which the positive charge is on oxygen because all the atoms have octets of electrons in such a structure. Their stability permits them to be formed rapidly, resulting in rates of electrophilic aromatic substitution that are much faster than that of benzene.

The lone pair on oxygen cannot be directly involved in carbocation stabilization when attack is meta to the substituent.

Meta attack

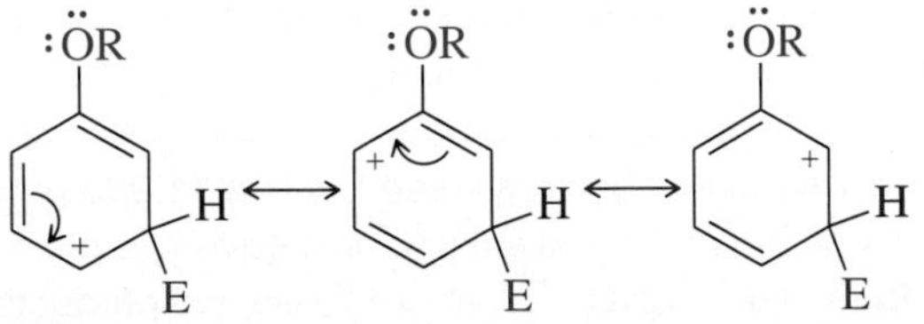

Oxygen lone pair cannot be used to stabilize positive charge in any of these structures; all have six electrons around positively charged carbon.

The greater stability of the carbocations arising from attack at the ortho and para positions compared with the carbocation formed by attack at the position meta to the oxygen substituent explains the ortho, para–directing property of hydroxyl, alkoxy, and acyloxy groups.

Nitrogen-containing substituents related to the amino group are even more strongly activating than the corresponding oxygen-containing substituents.

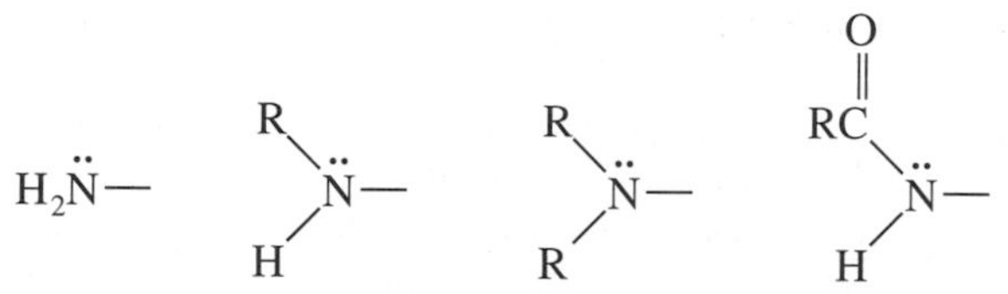

$H_2\ddot{N}-$ Amino; $R(H)\ddot{N}-$ Alkylamino; $R_2\ddot{N}-$ Dialkylamino; $RC(=O)(H)\ddot{N}-$ Acylamino

The nitrogen atom in each of these groups bears an electron pair that, like the unshared pair of an oxygen substituent, stabilizes a carbocation site to which it is attached. Since nitrogen is less electronegative than oxygen, it is a better electron pair donor and stabilizes the cyclohexadienyl cation intermediates in electrophilic aromatic substitution to an even greater degree.

Aniline and its derivatives are so reactive in electrophilic aromatic substitution that special strategies are usually necessary to carry out these reactions effectively. This topic is discussed in Section 22.15.

PROBLEM 12.12 Write structural formulas for the cyclohexadienyl cations formed from aniline ($C_6H_5NH_2$) during

(*a*) Ortho bromination (four resonance structures)
(*b*) Meta bromination (three resonance structures)
(*c*) Para bromination (four resonance structures)

SAMPLE SOLUTION (*a*) There are the customary three resonance structures for the cyclohexadienyl cation plus a resonance structure (the most stable one) derived by delocalization of the nitrogen lone pair into the ring.

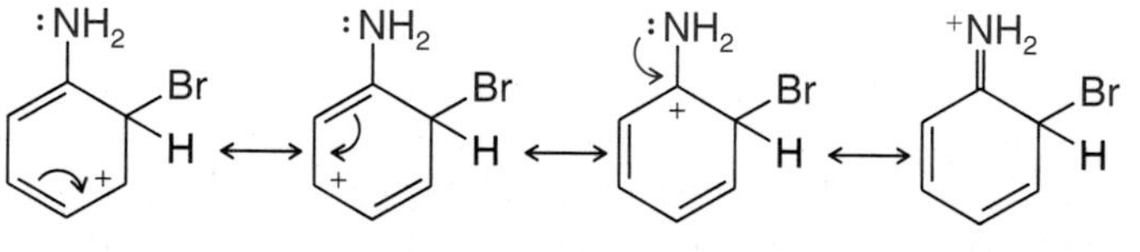

:NH₂, Br, H ⟷ :NH₂, Br, H ⟷ :NH₂, Br, H ⟷ ⁺NH₂, Br, H

Most stable resonance structure

Alkyl groups are, as we saw when we discussed the nitration of toluene in Section 12.10, activating and ortho, para–directing substituents. Aryl and alkenyl substituents resemble alkyl groups in this respect; they too are activating and ortho, para–directing.

PROBLEM 12.13 Treatment of biphenyl (see Section 11.9 to remind yourself of its structure) with a mixture of nitric acid and sulfuric acid gave two principal products both having the molecular formula $C_{12}H_9NO_2$. What are these two products?

The next group of substituents in Table 12.2 consists of the halogens. They are ortho, para–directing and slightly deactivating. We shall defer discussing the reasons for this behavior until Section 12.14. Instead, we shall turn our present attention to the substituents at the end of the table, those that are meta-directing and strongly deactivating.

12.13 SUBSTITUENT EFFECTS IN ELECTROPHILIC AROMATIC SUBSTITUTION: DEACTIVATING SUBSTITUENTS

As Table 12.2 indicates, there are a variety of substituent types that are *meta-directing and strongly deactivating*. We have already discussed one of these, the trifluoromethyl group. Several of the others are characterized by the presence of a carbonyl group ($\gt C{=}O$) at their point of attachment to the aromatic ring.

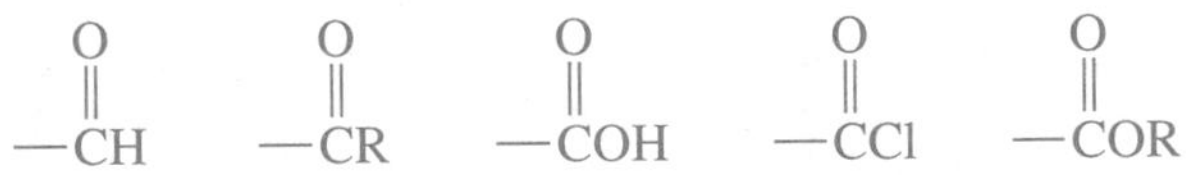

$-\overset{O}{\overset{\|}{C}}H$	$-\overset{O}{\overset{\|}{C}}R$	$-\overset{O}{\overset{\|}{C}}OH$	$-\overset{O}{\overset{\|}{C}}Cl$	$-\overset{O}{\overset{\|}{C}}OR$
Aldehyde	Ketone	Carboxylic acid	Acyl chloride	Ester

The behavior of the aldehyde group is typical of these carbonyl-containing substituents. Nitration of benzaldehyde takes place several thousand times more slowly than that of benzene and yields *m*-nitrobenzaldehyde as the major product.

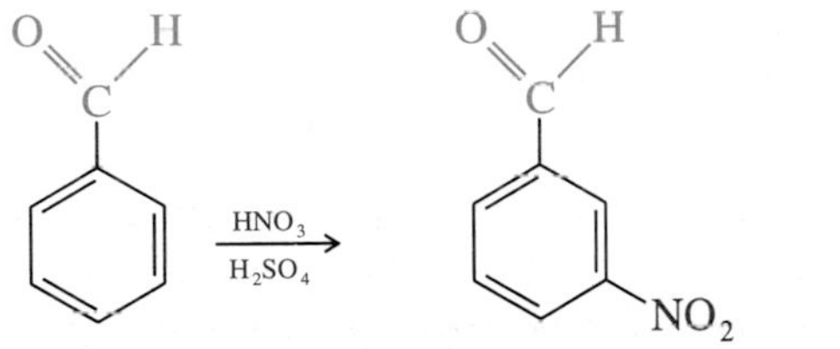

$$\text{Benzaldehyde} \xrightarrow[H_2SO_4]{HNO_3} m\text{-Nitrobenzaldehyde } (75\text{–}84\%)$$

Benzaldehyde   *m*-Nitrobenzaldehyde (75–84%)

To understand the effect of substituents in which there is a carbonyl group attached directly to the ring, consider the polarization of a carbon-oxygen double bond. The electrons in the carbon-oxygen double bond are drawn toward oxygen and away from carbon, leaving the carbon attached to the ring with a partial positive charge. Using benzaldehyde as an example,

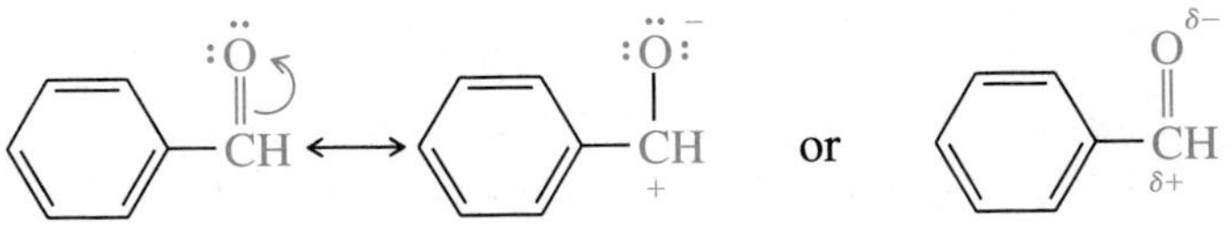

$$C_6H_5-\overset{:\ddot{O}}{\overset{\|}{C}}H \longleftrightarrow C_6H_5-\overset{:\ddot{O}:^-}{\overset{|}{\underset{+}{C}}}H \quad \text{or} \quad C_6H_5-\overset{O^{\delta-}}{\overset{\|}{\underset{\delta+}{C}}}H$$

Because the carbon atom attached to the ring is positively polarized, a carbonyl group behaves in much the same way as a trifluoromethyl group and *destabilizes* all the cyclohexadienyl cation intermediates in electrophilic aromatic substitution reactions. Attack at any ring position in benzaldehyde is slower than attack in benzene. The intermediates that lead to ortho and para substitution are particularly unstable because each is characterized by a resonance structure in which there is a positive charge on the carbon that bears the electron-withdrawing substituent. The intermediate leading to meta substitution avoids this unfavorable juxtaposition of positive charges, is not as unstable, and gives rise to most of the product. For the nitration of benzaldehyde:

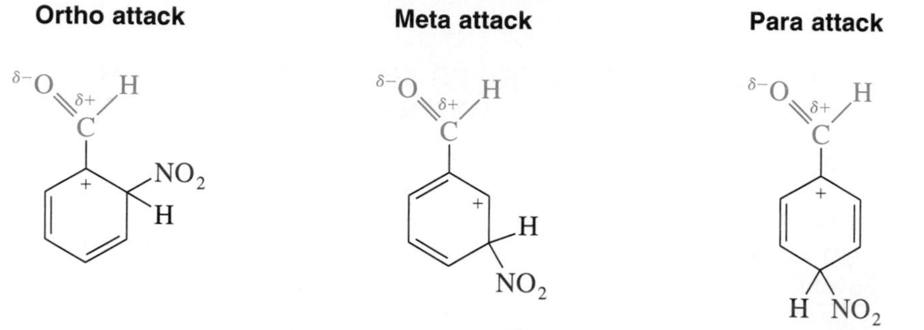

Ortho attack — Unstable because of adjacent positively polarized atoms

Meta attack — Positively polarized atoms not adjacent; most stable intermediate

Para attack — Unstable because of adjacent positively polarized atoms

---

PROBLEM 12.14 Each of the following reactions has been reported in the chemical literature, and the principal organic product has been isolated in good yield. Write a structural formula for the isolated product of each reaction.

(*a*) Treatment of benzoyl chloride ($C_6H_5\overset{O}{\overset{\|}{C}}Cl$) with chlorine and iron(III) chloride

(*b*) Treatment of methyl benzoate ($C_6H_5\overset{O}{\overset{\|}{C}}OCH_3$) with nitric acid and sulfuric acid

(*c*) Nitration of 1-phenyl-1-propanone ($C_6H_5\overset{O}{\overset{\|}{C}}CH_2CH_3$)

SAMPLE SOLUTION (*a*) Benzoyl chloride has a carbonyl group attached directly to the ring. A $—\overset{O}{\overset{\|}{C}}Cl$ substituent is meta-directing. The reaction conditions, namely, use of chlorine and iron(III) chloride, are those that introduce a chlorine onto the ring. The product is *m*-chlorobenzoyl chloride.

$$C_6H_5CCl \xrightarrow[FeCl_3]{Cl_2} m\text{-}ClC_6H_4CCl$$

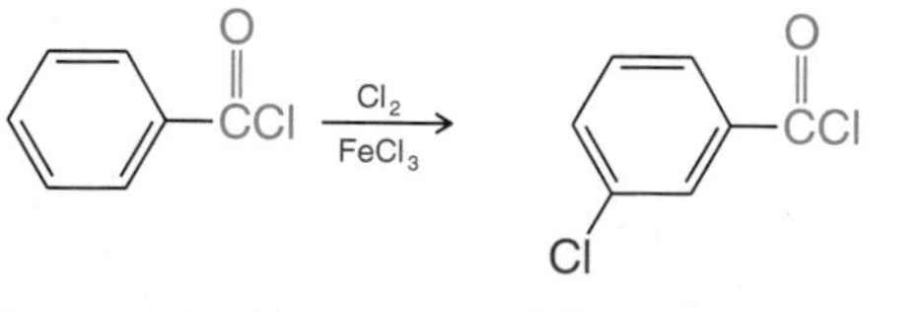

Benzoyl chloride → *m*-Chlorobenzoyl chloride (isolated in 62% yield)

A cyano substituent is similar to a carbonyl group for analogous reasons involving resonance of the type

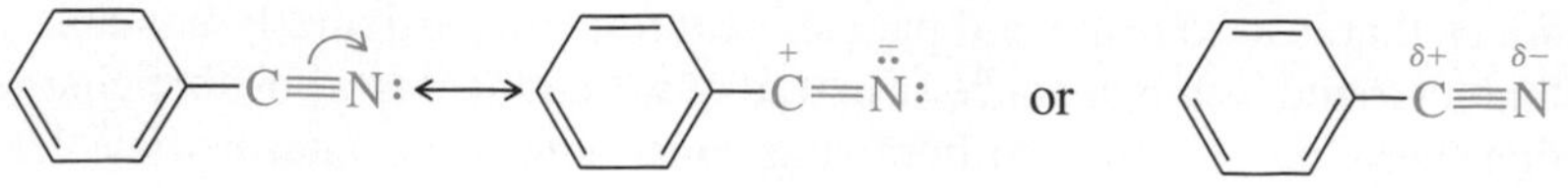

$$C_6H_5-C\equiv N: \longleftrightarrow C_6H_5-\overset{+}{C}=\ddot{N}:^- \quad \text{or} \quad C_6H_5-\overset{\delta+}{C}\equiv\overset{\delta-}{N}$$

Cyano groups are electron-withdrawing, deactivating, and meta-directing.

Sulfonic acid groups are electron-withdrawing because sulfur has a formal positive charge in several of the resonance forms of benzenesulfonic acid.

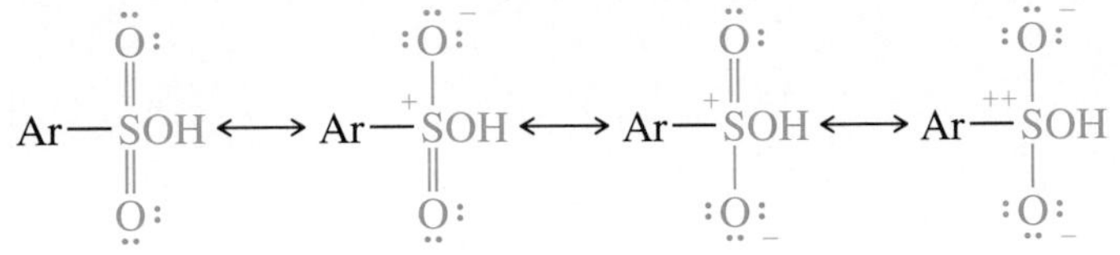

$$Ar{-}SO_2OH \longleftrightarrow Ar{-}\overset{+}{S}(O^-)(=O)OH \longleftrightarrow Ar{-}\overset{+}{S}(=O)(O^-)OH \longleftrightarrow Ar{-}\overset{++}{S}(O^-)_2OH$$

When benzene undergoes disulfonation, *m*-benzenedisulfonic acid is formed. The first sulfonic acid group to go on directs the second one meta to itself.

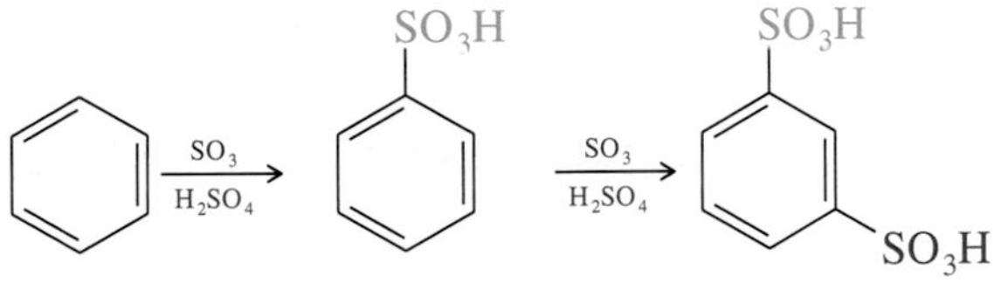

$$\text{Benzene} \xrightarrow[H_2SO_4]{SO_3} \text{Benzenesulfonic acid} \xrightarrow[H_2SO_4]{SO_3} m\text{-Benzenedisulfonic acid (90\%)}$$

Benzene — Benzenesulfonic acid — *m*-Benzenedisulfonic acid (90%)

The nitrogen atom of a nitro group bears a full positive charge in its two most stable Lewis structures.

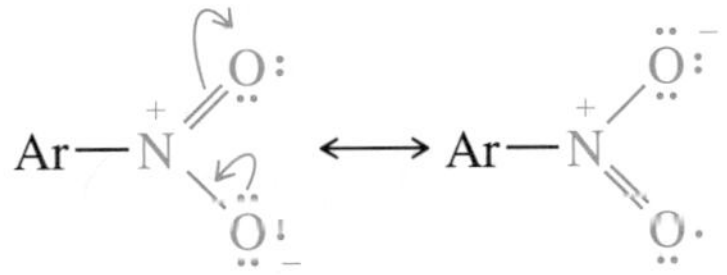

$$Ar{-}\overset{+}{N}(=O){-}O^- \longleftrightarrow Ar{-}\overset{+}{N}(-O^-){=}O$$

This makes it a powerful electron-withdrawing deactivating substituent and a meta director.

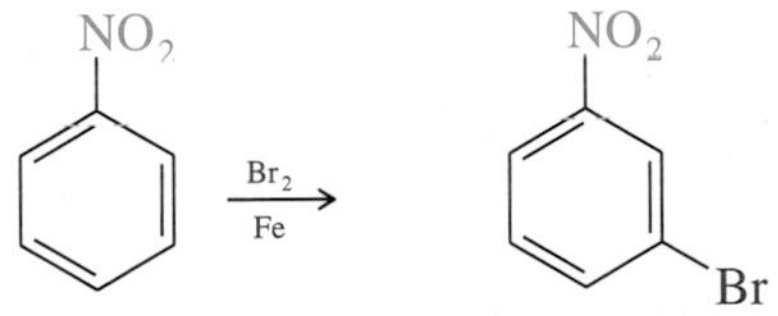

$$\text{Nitrobenzene} \xrightarrow[Fe]{Br_2} m\text{-Bromonitrobenzene (60–75\%)}$$

Nitrobenzene — *m*-Bromonitrobenzene (60–75%)

PROBLEM 12.15 Would you expect the substituent $—\overset{+}{N}(CH_3)_3$ to more closely resemble $—\ddot{N}(CH_3)_2$ or $—NO_2$ in its effect on rate and orientation in electrophilic aromatic substitution? Why?

12.14 SUBSTITUENT EFFECTS IN ELECTROPHILIC AROMATIC SUBSTITUTION: HALOGEN SUBSTITUENTS

Returning to Table 12.2, notice that *halogen substituents direct an incoming electrophile to the ortho and para positions but deactivate the ring toward substitution.*

Nitration of chlorobenzene is a typical example of electrophilic aromatic substitution in a halobenzene; it proceeds at a rate that is some 30 times slower than the corresponding nitration of benzene. The major products are *o*-chloronitrobenzene and *p*-chloronitrobenzene.

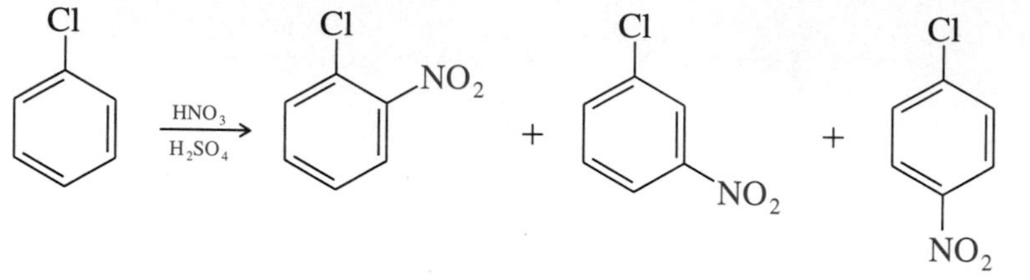

HNO_3 / H_2SO_4

Chlorobenzene → *o*-Chloronitrobenzene (30%) + *m*-Chloronitrobenzene (1%) + *p*-Chloronitrobenzene (69%)

PROBLEM 12.16 Reaction of chlorobenzene with 4-chlorobenzyl chloride and aluminum chloride gave a mixture of two products in good yield (76 percent). What were these two products?

Since we have come to associate activating substituents with ortho, para orientation and deactivating substituents with meta orientation, the properties of the halogen substituents appear on initial inspection to be unusual.

This seeming inconsistency between regioselectivity and rate can be understood by analyzing the two ways that a halogen substituent can affect the stability of a cyclohexadienyl cation. First, halogens are electronegative, and their inductive effect is to draw electrons away from the carbon to which they are bonded in the same way that a trifluoromethyl group does. Thus, all the cyclohexadienyl cation intermediates formed by electrophilic attack on a halobenzene are less stable than the corresponding cyclohexadienyl cation for benzene, and halobenzenes are less reactive than benzene.

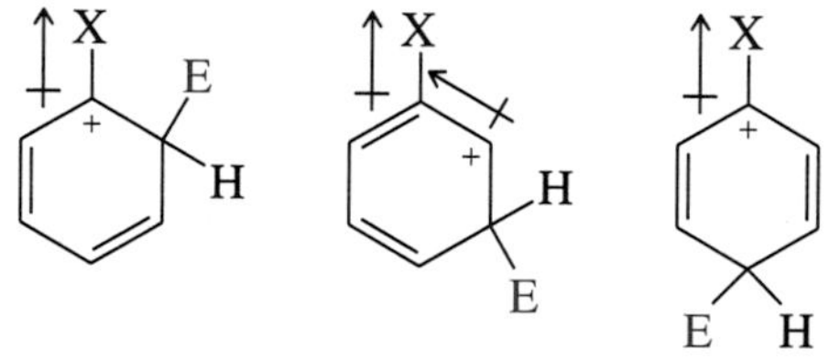

All these ions are less stable when X = F, Cl, Br, or I than when X = H

However, like hydroxyl groups and amino groups, halogen substituents possess unshared electron pairs that can be donated to a positively charged carbon. This electron donation into the π system stabilizes the intermediates derived from ortho and from para attack.

Ortho attack

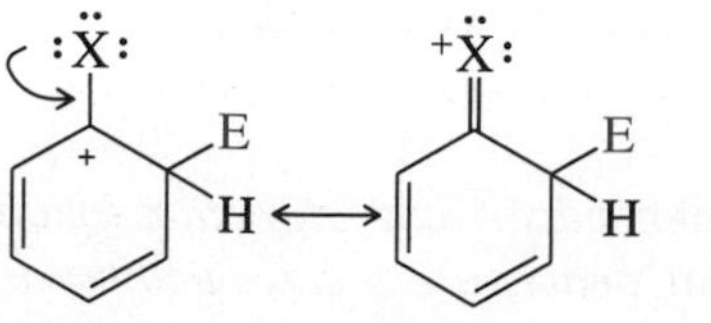

Para attack

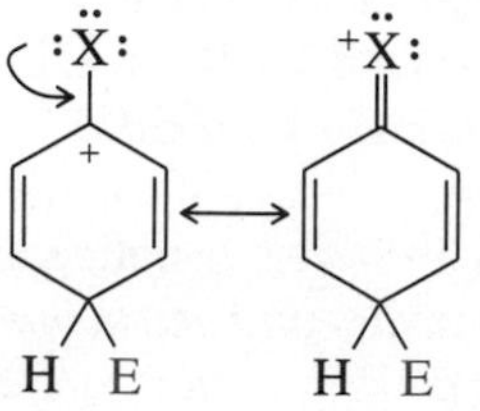

Comparable resonance stabilization of the intermediate leading to meta substitution is not possible. Thus, resonance involving halogen lone pairs causes electrophilic attack to be favored at the ortho and para positions but is weak and insufficient to overcome the electron-withdrawing inductive effect of the halogen, which deactivates all the ring positions. The experimentally observed partial rate factors for nitration of chlorobenzene result from this blend of inductive and resonance effects.

Cl
0.029 0.029
0.0009 0.0009
0.137

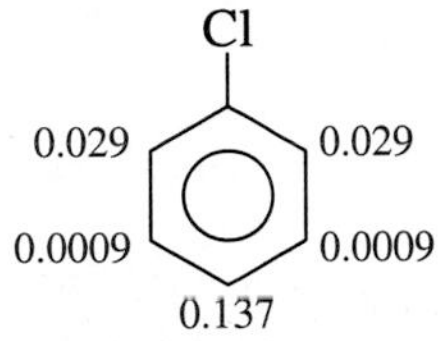

The mix of inductive and resonance effects varies from one halogen to another, but the net result is that fluorine, chlorine, bromine, and iodine are weakly deactivating, ortho, para–directing substituents.

12.15 MULTIPLE SUBSTITUENT EFFECTS

When a benzene ring bears two or more substituents, both its reactivity and the site of further substitution can usually be predicted from the cumulative effects of its substituents.

In the simplest cases all the available sites are equivalent and substitution at any one of them gives the same product.

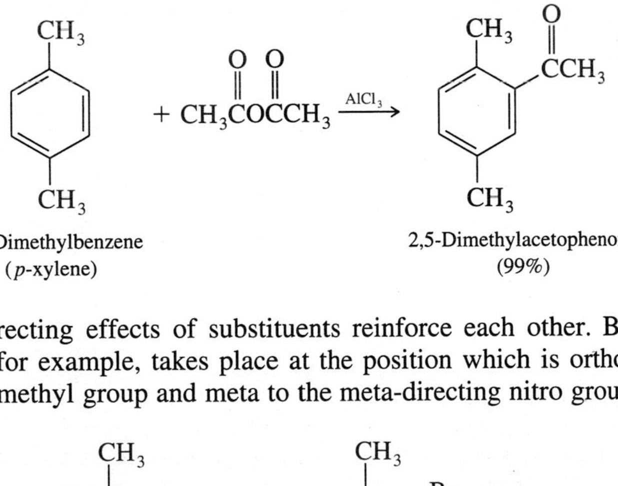

$$\text{1,4-Dimethylbenzene} + CH_3\overset{O}{\overset{\|}{C}}O\overset{O}{\overset{\|}{C}}CH_3 \xrightarrow{AlCl_3} \text{2,5-Dimethylacetophenone}$$

1,4-Dimethylbenzene (*p*-xylene) — 2,5-Dimethylacetophenone (99%)

Problems 12.2 (page 471), 12.3 (page 471), and 12.7 (page 477) offer additional examples of reactions in which only a single product of electrophilic aromatic substitution is possible.

Often the directing effects of substituents reinforce each other. Bromination of *p*-nitrotoluene, for example, takes place at the position which is ortho to the ortho, para–directing methyl group and meta to the meta-directing nitro group.

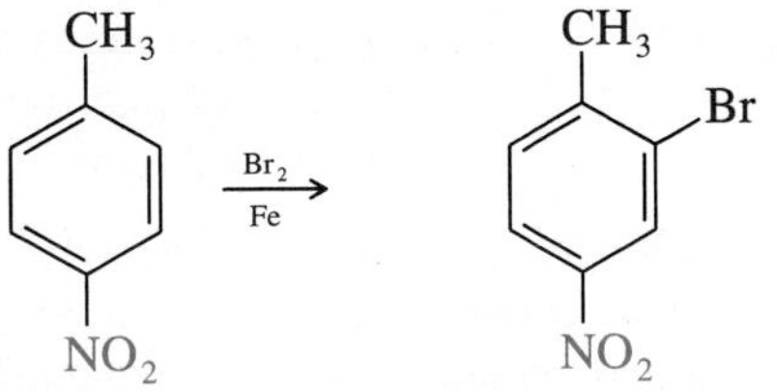

$$\text{p-Nitrotoluene} \xrightarrow[Fe]{Br_2} \text{2-Bromo-4-nitrotoluene}$$

p-Nitrotoluene — 2-Bromo-4-nitrotoluene (86–90%)

In almost all cases, including most of those in which the directing effects of individual substituents oppose each other, *it is the more activating substituent that controls the regioselectivity of electrophilic aromatic substitution.* Thus, bromination occurs ortho to the *N*-methylamino group in 4-chloro-*N*-methylaniline because this group is a very powerful activating substituent while the chlorine is weakly deactivating.

$$\text{4-Chloro-}N\text{-methylaniline} \xrightarrow[\text{acetic acid}]{Br_2} \text{2-Bromo-4-chloro-}N\text{-methylaniline (87\%)}$$

When two positions are comparably activated by alkyl groups, substitution usually occurs at the less hindered site. Nitration of *p-tert*-butyltoluene takes place at positions ortho to the methyl group in preference to those ortho to the larger *tert*-butyl group. This is an example of a *steric effect.*

$$p\text{-}tert\text{-Butyltoluene} \xrightarrow[H_2SO_4]{HNO_3} \text{4-}tert\text{-Butyl-2-nitrotoluene (88\%)}$$

Nitration of *m*-xylene is directed ortho to one methyl group and para to the other.

$$m\text{-Xylene} \xrightarrow[H_2SO_4]{HNO_3} \text{2,4-Dimethyl-1-nitrobenzene (98\%)}$$

The ortho position between the two methyl groups is less reactive because it is more sterically hindered.

PROBLEM 12.17 Write the structure of the principal organic product obtained on nitration of each of the following:

(*a*) *p*-Methylbenzoic acid
(*b*) *m*-Dichlorobenzene
(*c*) *m*-Dinitrobenzene
(*d*) *p*-Methoxyacetophenone
(*e*) *p*-Methylanisole
(*f*) 2,6-Dibromoanisole

SAMPLE SOLUTION (*a*) Of the two substituents in *p*-methylbenzoic acid, the methyl group is more activating and so controls the regioselectivity of electrophilic aromatic substitution. The position para to the ortho, para–directing methyl group already bears a substituent (the carboxyl group), and so substitution occurs ortho to the methyl group. This position is meta to the *m*-directing carboxyl group, so that the orienting properties of the two substituents reinforce each other. The product is 4-methyl-3-nitrobenzoic acid.

$$\text{p-Methylbenzoic acid} \xrightarrow[H_2SO_4]{HNO_3} \text{4-Methyl-3-nitrobenzoic acid}$$

p-Methylbenzoic acid 4-Methyl-3-nitrobenzoic acid

An exception to the rule that regioselectivity is controlled by the most activating substituent occurs in cases where the directing effects of alkyl groups and halogen substituents oppose each other. Since alkyl groups and halogen substituents are weakly activating and weakly deactivating, respectively, the difference between them is too small to allow a simple generalization.

Problem 12.38 (page 509) illustrates how partial rate factor data may be applied to such cases.

12.16 REGIOSELECTIVE SYNTHESIS OF DISUBSTITUTED AROMATIC COMPOUNDS

Since the position of electrophilic attack on an aromatic ring is controlled by the directing effects of substituents already present, the preparation of disubstituted aromatic compounds requires that careful thought be given to the order of introduction of the two groups.

Compare the independent preparations of *m*-bromoacetophenone and *p*-bromoacetophenone from benzene. Both syntheses require a Friedel-Crafts acylation step and a bromination step, but the major product is determined by the *order* in which the two steps are carried out. When the meta-directing acetyl group is introduced first, the final product is *m*-bromoacetophenone.

$$\text{Benzene} \xrightarrow[AlCl_3]{CH_3COCCH_3} \text{Acetophenone } (76\text{–}83\%) \xrightarrow[AlCl_3]{Br_2} m\text{-Bromoacetophenone } (59\%)$$

Benzene Acetophenone (76–83%) *m*-Bromoacetophenone (59%)

Aluminum chloride is a stronger Lewis acid than iron(III) bromide and has been used as a catalyst in electrophilic bromination when, as in the example shown, the aromatic ring bears a strongly deactivating substituent.

When the ortho, para–directing bromo substituent is introduced first, the major product is *p*-bromoacetophenone (along with some of its ortho isomer, from which it is separated by distillation).

$$\text{Benzene} \xrightarrow[Fe]{Br_2} \text{Bromobenzene } (65\text{–}75\%) \xrightarrow[AlCl_3]{CH_3COCCH_3} p\text{-Bromoacetophenone } (69\text{–}79\%)$$

Benzene Bromobenzene (65–75%) *p*-Bromoacetophenone (69–79%)

PROBLEM 12.18 Write chemical equations showing how you could prepare *m*-bromonitrobenzene as the principal organic product, starting with benzene and using any necessary organic or inorganic reagents. How could you prepare *p*-bromonitrobenzene?

A less obvious example of a situation in which the success of a synthesis depends on the order of introduction of substituents is illustrated by the preparation of *m*-nitroacetophenone. Here, even though both substituents are meta-directing, the only practical synthesis is the one in which Friedel-Crafts acylation is carried out first.

$$\text{Benzene} \xrightarrow[\text{AlCl}_3]{\text{CH}_3\overset{\text{O}}{\overset{\|}{\text{C}}}\text{O}\overset{\text{O}}{\overset{\|}{\text{C}}}\text{CH}_3} \text{Acetophenone (76–83\%)} \xrightarrow[\text{H}_2\text{SO}_4]{\text{HNO}_3} \textit{m}\text{-Nitroacetophenone (55\%)}$$

When the reverse order of steps is attempted, it is observed that the Friedel-Crafts acylation of nitrobenzene fails.

$$\text{Benzene} \xrightarrow[\text{H}_2\text{SO}_4]{\text{HNO}_3} \text{Nitrobenzene (95\%)} \xrightarrow[\text{AlCl}_3]{\text{CH}_3\overset{\text{O}}{\overset{\|}{\text{C}}}\text{O}\overset{\text{O}}{\overset{\|}{\text{C}}}\text{CH}_3} \text{no reaction}$$

Neither Friedel-Crafts acylation nor alkylation reactions can be carried out on nitrobenzene. The presence of a strongly deactivating substituent such as a nitro group on an aromatic ring so depresses its reactivity that Friedel-Crafts reactions do not take place. Nitrobenzene is so unreactive that it is sometimes used as a solvent in Friedel-Crafts reactions. The practical limit for Friedel-Crafts alkylation and acylation reactions is effectively a monohalobenzene. *An aromatic ring more deactivated than a monohalobenzene cannot be alkylated or acylated under Friedel-Crafts conditions.*

Sometimes the orientation of two substituents in an aromatic compound precludes its straightforward synthesis. *m*-Chloroethylbenzene, for example, has two ortho, para–directing groups in a meta relationship and so cannot be prepared either from chlorobenzene or from ethylbenzene. In cases such as this we couple electrophilic aromatic substitution with functional group manipulation to produce the desired compound. The key here is to recognize that an ethyl substituent can be introduced by Friedel-Crafts acylation followed by a Clemmensen or Wolff-Kishner reduction step later in the synthesis. If the chlorine is introduced prior to reduction, it will be directed meta to the acetyl group, giving the correct substitution pattern.

$$\text{Benzene} \xrightarrow[\text{AlCl}_3]{\text{CH}_3\overset{\text{O}}{\overset{\|}{\text{C}}}\text{O}\overset{\text{O}}{\overset{\|}{\text{C}}}\text{CH}_3} \text{Acetophenone} \xrightarrow[\text{AlCl}_3]{\text{Cl}_2} \textit{m}\text{-Chloroacetophenone} \xrightarrow[\text{HCl}]{\text{Zn(Hg)}} \textit{m}\text{-Chloroethylbenzene}$$

A related problem attends the synthesis of *p*-nitrobenzoic acid. Here, two meta-directing substituents are para to each other. This compound has been prepared from toluene according to the procedure shown:

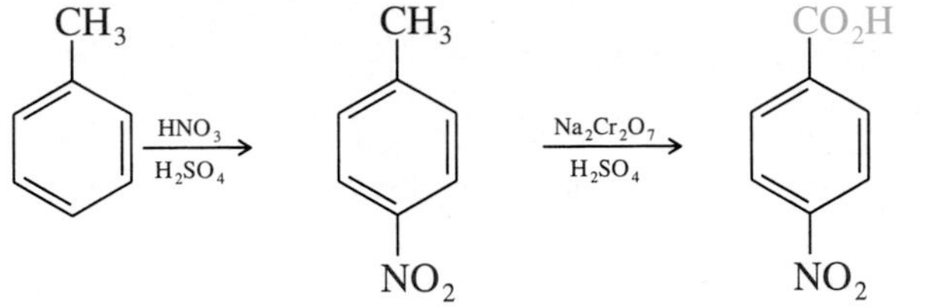

CH_3 $\xrightarrow[H_2SO_4]{HNO_3}$ CH_3 / NO_2 $\xrightarrow[H_2SO_4]{Na_2Cr_2O_7}$ CO_2H / NO_2

p-Nitrotoluene (separate from ortho isomer)

p-Nitrobenzoic acid (82–86%)

Since it may be oxidized to a carboxyl group (Section 11.15), a methyl group can be used to introduce the nitro substituent in the proper position.

PROBLEM 12.19 Suggest an efficient synthesis of *m*-nitrobenzoic acid from toluene.

12.17 SUBSTITUTION IN NAPHTHALENE

Polycyclic aromatic hydrocarbons undergo electrophilic aromatic substitution when treated with the same reagents that react with benzene. In general, polycyclic aromatic hydrocarbons are more reactive than benzene. Since, however, most lack the symmetry of benzene, mixtures of products may be formed even on monosubstitution. Among polycyclic aromatic hydrocarbons, we will discuss only naphthalene, and that only briefly.

Two sites are available for substitution in naphthalene, C-1 and C-2, C-1 being normally the preferred site of electrophilic attack.

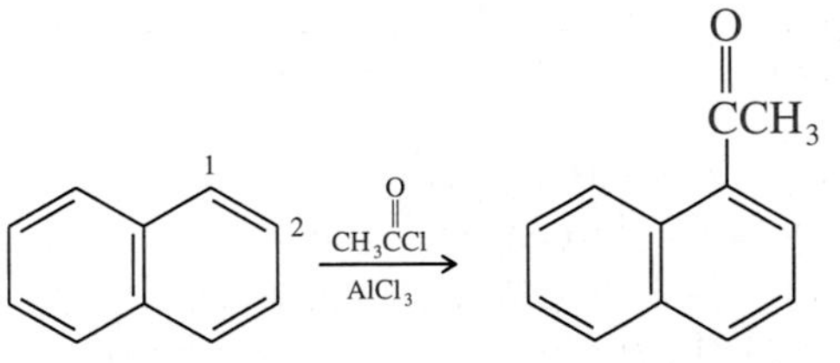

1 2 $\xrightarrow[AlCl_3]{CH_3\overset{O}{\overset{\|}{C}}Cl}$ $\overset{O}{\overset{\|}{C}}CH_3$

Naphthalene

1-Acetylnaphthalene (90%)

The C-1 position is the more reactive because the arenium ion formed by electrophilic attack there is a relatively stable one. Benzenoid character is retained in one ring while the positive charge is delocalized by allylic resonance.

Attack at C-1

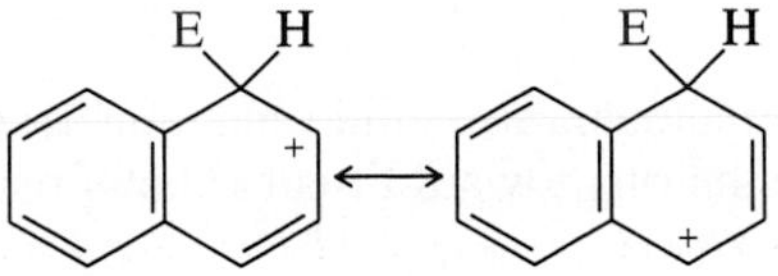

E H + $\longleftrightarrow$ E H +

Attack at C-2

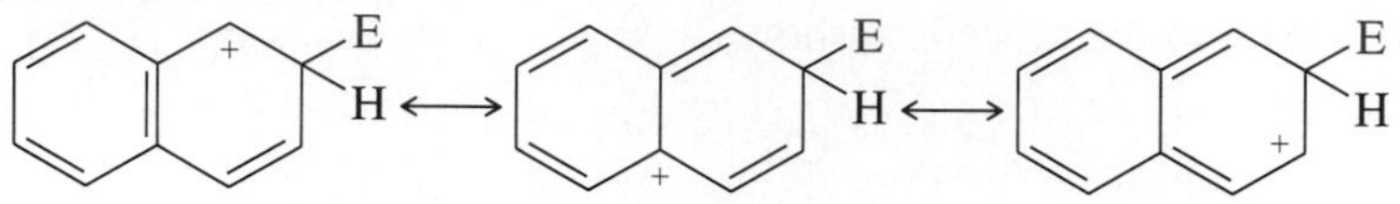

In order to involve allylic resonance in stabilizing the arenium ion formed during attack at C-2, the benzenoid character of the other ring is sacrificed.

PROBLEM 12.20 Sulfonation of naphthalene is reversible at elevated temperature. A different isomer of naphthalenesulfonic acid is the major product at 160°C than is the case at 0°C. Which isomer is the product of kinetic control? Which one is formed under conditions of thermodynamic control? Can you think of a reason why one isomer is more stable than the other?

12.18 SUBSTITUTION IN HETEROCYCLIC AROMATIC COMPOUNDS

The great variety of available structural types causes heterocyclic aromatic compounds to range from exceedingly reactive to practically inert toward electrophilic aromatic substitution.

Pyridine lies near one extreme in being far less reactive than benzene toward substitution by electrophilic reagents. In this respect it resembles strongly deactivated aromatic compounds such as nitrobenzene. It is incapable of being acylated or alkylated under Friedel-Crafts conditions, but can be sulfonated at high temperature. Electrophilic substitution in pyridine, when it does occur, takes place at C-3.

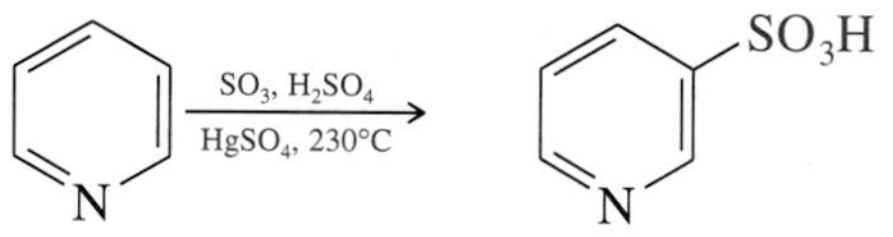

SO_3, H_2SO_4, $HgSO_4$, 230°C

Pyridine → Pyridine-3-sulfonic acid (71%)

One reason for the low reactivity of pyridine is that its nitrogen atom, since it is more electronegative than a CH in benzene, causes the π electrons to be held more strongly and raises the activation energy for their donation to an attacking electrophile. Another is that the nitrogen of pyridine is protonated in sulfuric acid and the resulting pyridinium ion is even more deactivated than pyridine itself.

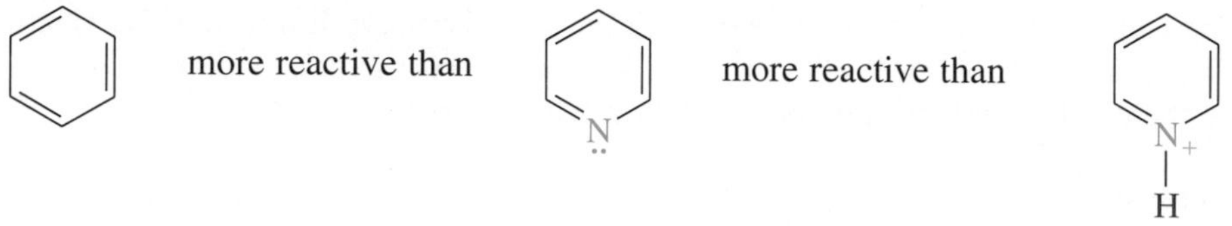

Benzene more reactive than Pyridine more reactive than Pyridinium ion

Lewis acid catalysts such as aluminum chloride and iron(III) halides also bond to nitrogen to strongly deactivate the ring toward Friedel-Crafts reactions and halogenation.

Pyrrole, furan, and thiophene, on the other hand, have electron-rich aromatic rings and are extremely reactive toward electrophilic aromatic substitution—more like phenol and aniline than benzene. Like benzene they have six π electrons, but these π electrons are delocalized over *five* atoms, not six, and are not held as strongly as those of benzene. Even when the ring atom is as electronegative as oxygen, substitution occurs readily.

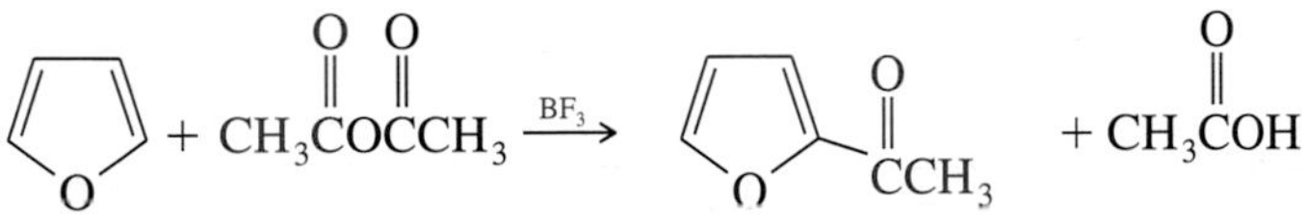

$$\text{Furan} + \underset{\text{Acetic anhydride}}{CH_3\overset{O}{\overset{\|}{C}}O\overset{O}{\overset{\|}{C}}CH_3} \xrightarrow{BF_3} \underset{\text{2-Acetylfuran (75–92\%)}}{\text{2-furyl}-\overset{O}{\overset{\|}{C}}CH_3} + \underset{\text{Acetic acid}}{CH_3\overset{O}{\overset{\|}{C}}OH}$$

The regioselectivity of substitution in furan is explained using a resonance description. When the electrophile attacks C-2, the positive charge is shared by three atoms, C-3, C-5, and O.

Attack at C-2: Carbocation *more stable;* positive charge shared by C-3, C-5, and O

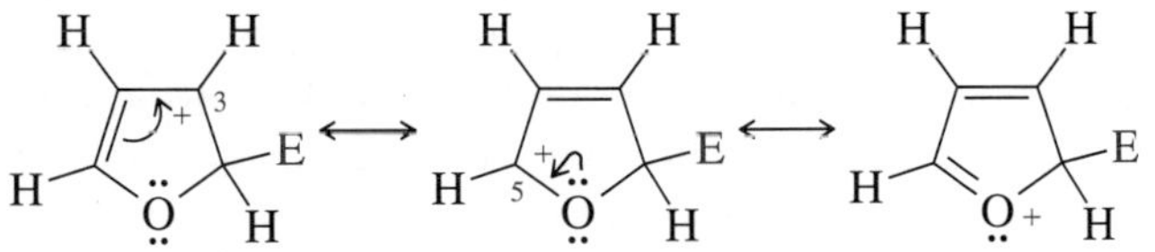

When the electrophile attacks at C-3, the positive charge is shared by only two atoms, C-2 and O, and the carbocation intermediate is less stable and formed more slowly.

Attack at C-3: Carbocation *less stable;* positive charge shared by C-2 and O

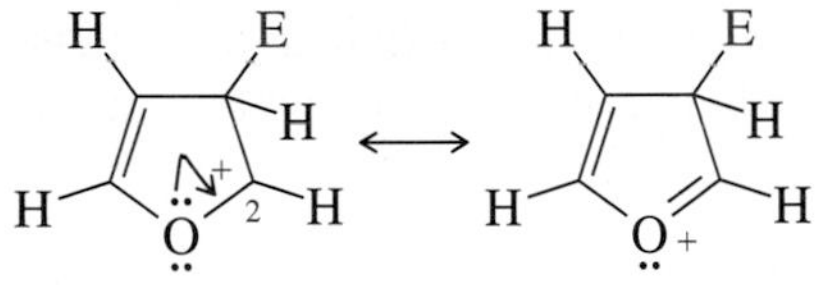

The regioselectivity of substitution in pyrrole and thiophene is like that of furan and for similar reasons.

PROBLEM 12.21 When benzene is prepared from coal tar, it is contaminated with thiophene, from which it cannot be separated by distillation because of very similar boiling points. Shaking a mixture of benzene and thiophene with sulfuric acid causes sulfonation of the thiophene ring but leaves benzene untouched. The sulfonation product of thiophene dissolves in the sulfuric acid layer, from which the benzene layer is separated; the benzene layer is then washed with water and distilled. Give the structure of the sulfonation product of thiophene.

12.19 SUMMARY

On reaction with electrophilic reagents, aromatic substances undergo substitution rather than addition.

$$\text{ArH} + \overset{\delta+}{\text{E}}-\overset{\delta-}{\text{Y}} \longrightarrow \text{ArE} + \text{HY}$$

Substitution occurs by attack of the electrophile on the π electrons of the aromatic ring in the rate-determining step, to form a cyclohexadienyl cation intermediate. Loss of a proton from this intermediate restores the aromaticity of the ring and yields the product of **electrophilic aromatic substitution** (Section 12.2).

$$\text{C}_6\text{H}_6 + \overset{\delta+}{\text{E}}-\overset{\delta-}{\text{Y}} \longrightarrow [\text{C}_6\text{H}_6\text{E}]^+ + \text{Y}^- \longrightarrow \text{C}_6\text{H}_5\text{E} + \text{HY}$$

Benzene | Electrophilic reagent | Cyclohexadienyl cation intermediate | Product of electrophilic aromatic substitution

Table 12.3 presents some typical examples of electrophilic aromatic substitution reactions; it illustrates conditions for carrying out the nitration, sulfonation, halogenation, and Friedel-Crafts alkylation and acylation of aromatic substances.

Friedel-Crafts reactions are among the most useful carbon-carbon bond–forming reactions in synthetic organic chemistry. They do, however, suffer from certain limitations, which are summarized in Table 12.4. As noted in the table, attempts to use primary alkyl halides in Friedel-Crafts alkylation reactions yield products resulting from attack by rearranged carbocations on the ring. Since acyl cations do not rearrange, the customary procedure for introducing a primary alkyl group onto an arene is by Friedel-Crafts acylation followed by Clemmensen or Wolff-Kishner reduction (Section 12.8).

Substituents on the ring can influence both the *rate* at which electrophilic aromatic substitution occurs and its *regioselectivity* (Section 12.9). Substituents are classified as *activating* or *deactivating* according to whether they cause the ring to react more rapidly or less rapidly than benzene. Substituents are arranged into three major categories:

1. **Activating and ortho, para–directing (Sections 12.10 and 12.12):** These substituents stabilize the cyclohexadienyl cation formed in the rate-determining step. They include $-\ddot{\text{N}}\text{R}_2$, $-\ddot{\text{O}}\text{R}$, —R, —Ar, and related species. The most strongly activating members of this group are bonded to the ring by a nitrogen or oxygen atom that bears an unshared pair of electrons.
2. **Deactivating and ortho, para–directing (Section 12.14):** The halogens are the most prominent members of this class. They withdraw electron density from all the ring positions by an inductive effect, making halobenzenes less reactive than benzene. Lone-pair electron donation stabilizes the cyclohexadienyl cations corresponding to attack at the ortho and para positions more than those formed by attack at the meta positions, giving rise to the observed regioselectivity.
3. **Deactivating and meta-directing (Sections 12.11 and 12.13):** These substituents are strongly electron-withdrawing and destabilize carbocations. They include $-\text{CF}_3$, $-\overset{\text{O}}{\overset{\|}{\text{C}}}\text{R}$, $-\text{C}\equiv\text{N}$, $-\text{NO}_2$, and related species. All the ring positions are deactivated, but since the *meta* positions are deactivated less than the ortho and para, meta substitution is favored.

TABLE 12.3

Representative Electrophilic Aromatic Substitution Reactions

Reaction (section) and comments	General equation and specific example
Nitration (Section 12.3) The active electrophile in the nitration of benzene and its derivatives is nitronium cation ($:\ddot{O}=\overset{+}{N}=\ddot{O}:$). It is generated by reaction of nitric acid and sulfuric acid. Very reactive arenes—those that bear strongly activating substituents—undergo nitration in nitric acid alone.	$ArH + HNO_3 \xrightarrow{H_2SO_4} ArNO_2 + H_2O$ Arene, Nitric acid, Nitroarene, Water Fluorobenzene $\xrightarrow[H_2SO_4]{HNO_3}$ *p*-Fluoronitrobenzene (80%)
Sulfonation (Section 12.4) Sulfonic acids are formed when aromatic compounds are treated with sources of sulfur trioxide. These sources can be concentrated sulfuric acid (for very reactive arenes) or solutions of sulfur trioxide in sulfuric acid (for benzene and arenes less reactive than benzene).	$ArH + SO_3 \longrightarrow ArSO_3H$ Arene, Sulfur trioxide, Arenesulfonic acid 1,2,4,5-Tetramethylbenzene $\xrightarrow[H_2SO_4]{SO_3}$ 2,3,5,6-Tetramethylbenzenesulfonic acid (94%)
Halogenation (Section 12.5) Chlorination and bromination of arenes are carried out by treatment with the appropriate halogen in the presence of a Lewis acid catalyst. Very reactive arenes undergo halogenation in the absence of a catalyst.	$ArH + X_2 \xrightarrow{FeX_3} ArX + HX$ Arene, Halogen, Aryl halide, Hydrogen halide Phenol $\xrightarrow[CS_2]{Br_2}$ *p*-Bromophenol (80–84%)
Friedel-Crafts alkylation (Section 12.6) Carbocations, usually generated from an alkyl halide and aluminum chloride, attack the aromatic ring to yield alkylbenzenes. The arene must be at least as reactive as a halobenzene. Carbocation rearrangements can occur, especially with primary alkyl halides.	$ArH + RX \xrightarrow{AlCl_3} ArR + HX$ Arene, Alkyl halide, Alkylarene, Hydrogen halide Benzene + Cyclopentyl bromide $\xrightarrow{AlCl_3}$ Cyclopentylbenzene (54%)
Friedel-Crafts acylation (Section 12.7) Acyl cations (acylium ions) generated by treating an acyl chloride or acid anhydride with aluminum chloride attack aromatic rings to yield ketones. The arene must be at least as reactive as a halobenzene. Acyl cations are relatively stable, and do not rearrange.	$ArH + RCCl(=O) \xrightarrow{AlCl_3} ArCR(=O) + HCl$ Arene, Acyl chloride, Ketone, Hydrogen chloride $ArH + RC(=O)OC(=O)R \xrightarrow{AlCl_3} ArC(=O)R + RC(=O)OH$ Arene, Acid anhydride, Ketone, Carboxylic acid Anisole $\xrightarrow[AlCl_3]{CH_3C(=O)OC(=O)CH_3}$ *p*-Methoxyacetophenone (90–94%)

TABLE 12.4
Limitations on Friedel-Crafts Reactions

1. The organic halide that reacts with the arene must be an alkyl halide (Section 12.6) or an acyl halide (Section 12.7).	*These will react with benzene under Friedel-Crafts conditions:* 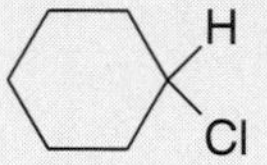 H, Cl; CH_2Cl; O, CCl — Alkyl halide, 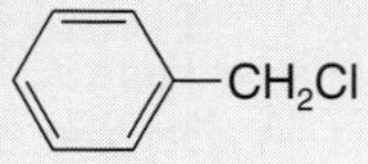 Benzylic halide, 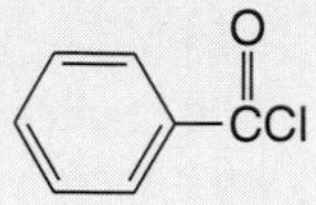 Acyl halide
Vinylic halides and aryl halides do not form carbocations under conditions of the Friedel-Crafts reaction and so cannot be used in place of an alkyl halide or an acyl halide.	*These will not react with benzene under Friedel-Crafts conditions:* 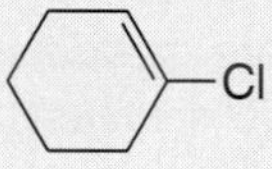 Cl; Cl — Vinylic halide, 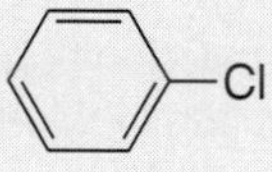 Aryl halide
2. Rearrangement of alkyl groups can occur (Section 12.6).	Rearrangement is especially prevalent with primary alkyl halides of the type RCH_2CH_2X and R_2CHCH_2X. Aluminum chloride induces ionization with rearrangement to give a more stable carbocation. Benzylic halides and acyl halides do not rearrange.
3. Strongly deactivated aromatic rings do not undergo Friedel-Crafts alkylation or acylation (Section 12.16). Friedel-Crafts alkylations and acylations fail when applied to compounds of the following type, where EWG is a strongly electron-withdrawing group: 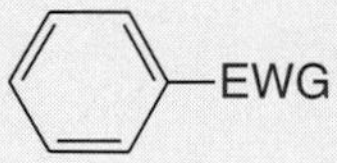 (benzene)—EWG	EWG: $—CF_3$, $—NO_2$, $—SO_3H$, —C≡N, —C(=O)H, —C(=O)R, —C(=O)OH, —C(=O)OR, —C(=O)Cl
4. It is sometimes difficult to limit Friedel-Crafts alkylation to monoalkylation.	The first alkyl group that goes on makes the ring more reactive toward further substitution because alkyl groups are activating substituents. Monoacylation is possible because the first acyl group to go on is strongly electron-withdrawing and deactivates the ring toward further substitution.

The regioselectivity of electrophilic aromatic substitution in arenes that bear two or more substituents is generally controlled by the directing effect of the more powerful *activating* substituent (Section 12.15).

When using electrophilic aromatic substitution reactions in multistep syntheses, careful thought needs to be given to the order of introduction of various substituents, since they can affect the observed regiochemistry. Regiochemistry can sometimes be controlled by using functional group manipulations to alter the directing properties of substituents (Section 12.16).

Polycyclic aromatic hydrocarbons (Section 12.17) and heterocyclic aromatic compounds (Section 12.18) are capable of undergoing electrophilic aromatic substitution. Polycyclic aromatic hydrocarbons tend to be more reactive than benzene. Heterocyclic aromatic compounds may be more reactive or less reactive than benzene. Pyridine is much less reactive than benzene, while pyrrole, furan, and thiophene are *electron-rich* and undergo electrophilic aromatic substitution at rapid rates.

PROBLEMS

12.22 Give reagents suitable for effecting each of the following reactions, and write the principal products. If an ortho, para mixture is expected, show both. If the meta isomer is the expected major product, write only that isomer.

(*a*) Nitration of benzene
(*b*) Nitration of the product of part (*a*)
(*c*) Bromination of toluene
(*d*) Bromination of (trifluoromethyl)benzene
(*e*) Sulfonation of anisole
(*f*) Sulfonation of acetanilide ($C_6H_5NH\overset{O}{\overset{\|}{C}}CH_3$)
(*g*) Chlorination of bromobenzene
(*h*) Friedel-Crafts alkylation of anisole with benzyl chloride
(*i*) Friedel-Crafts acylation of benzene with benzoyl chloride
(*j*) Nitration of the product from part (*i*)
(*k*) Clemmensen reduction of the product from part (*i*)
(*l*) Wolff-Kishner reduction of the product from part (*i*)

12.23 Write a structural formula for the most stable cyclohexadienyl cation intermediate formed in each of the following reactions. Is the cyclohexadienyl cation more or less stable than the corresponding intermediate formed by electrophilic attack on benzene?

(*a*) Bromination of *p*-xylene
(*b*) Chlorination of *m*-xylene
(*c*) Nitration of acetophenone
(*d*) Friedel-Crafts acylation of anisole with $CH_3\overset{O}{\overset{\|}{C}}Cl$
(*e*) Nitration of isopropylbenzene
(*f*) Bromination of nitrobenzene
(*g*) Sulfonation of furan
(*h*) Bromination of pyridine

12.24 In each of the following pairs of compounds choose which one will react faster with the indicated reagent and write a chemical equation for the faster reaction:

(*a*) Toluene or chlorobenzene with nitric acid and sulfuric acid
(*b*) Fluorobenzene or (trifluoromethyl)benzene with benzyl chloride and aluminum chloride
(*c*) Methyl benzoate ($C_6H_5\overset{O}{\overset{\|}{C}}OCH_3$) or phenyl acetate ($C_6H_5O\overset{O}{\overset{\|}{C}}CH_3$) with bromine in acetic acid
(*d*) Acetanilide ($C_6H_5NH\overset{O}{\overset{\|}{C}}CH_3$) or nitrobenzene with sulfur trioxide in sulfuric acid
(*e*) *p*-Dimethylbenzene (*p*-xylene) or *p*-di-*tert*-butylbenzene with acetyl chloride and aluminum chloride

(*f*) Benzophenone ($C_6H_5\overset{O}{\overset{\|}{C}}C_6H_5$) or biphenyl ($C_6H_5—C_6H_5$) with chlorine and iron(III) chloride

12.25 Arrange the following five compounds in order of decreasing rate of bromination: benzene, toluene, *o*-xylene, *m*-xylene, 1,3,5-trimethylbenzene (the relative rates are 2×10^7, 5×10^4, 5×10^2, 60, and 1).

12.26 Each of the following reactions has been carried out under conditions such that disubstitution or trisubstitution occurred. Identify the principal organic product in each case.

(*a*) Nitration of *p*-chlorobenzoic acid (dinitration)
(*b*) Bromination of aniline (tribromination)
(*c*) Bromination of *o*-aminoacetophenone (dibromination)
(*d*) Nitration of benzoic acid (dinitration)
(*e*) Bromination of *p*-nitrophenol (dibromination)
(*f*) Reaction of biphenyl with *tert*-butyl chloride and iron(III) chloride (dialkylation)
(*g*) Sulfonation of phenol (disulfonation)

12.27 Write equations showing how you could prepare each of the following from benzene or toluene and any necessary organic or inorganic reagents. If an ortho, para mixture is formed in any step of your synthesis, assume that you can separate the two isomers.

(*a*) Isopropylbenzene
(*b*) *p*-Isopropylbenzenesulfonic acid
(*c*) 2-Bromo-2-phenylpropane
(*d*) 4-*tert*-Butyl-2-nitrotoluene
(*e*) *m*-Chloroacetophenone
(*f*) *p*-Chloroacetophenone
(*g*) 3-Bromo-4-methylacetophenone
(*h*) 2-Bromo-4-ethyltoluene
(*i*) 1-Bromo-3-nitrobenzene
(*j*) 1-Bromo-2,4-dinitrobenzene
(*k*) 3-Bromo-5-nitrobenzoic acid
(*l*) 2-Bromo-4-nitrobenzoic acid
(*m*) Diphenylmethane
(*n*) 1-Phenyloctane
(*o*) 1-Phenyl-1-octene
(*p*) 1-Phenyl-1-octyne
(*q*) 1,4-di-*tert*-Butyl-1,4-cyclohexadiene

12.28 Write equations showing how you could prepare each of the following from anisole and any necessary organic or inorganic reagents. If an ortho, para mixture is formed in any step of your synthesis, assume that you can separate the two isomers.

(*a*) *p*-Methoxybenzenesulfonic acid
(*b*) 2-Bromo-4-nitroanisole
(*c*) 4-Bromo-2-nitroanisole
(*d*) *p*-Methoxystyrene

12.29 How many products are capable of being formed from toluene in each of the following reactions?

(*a*) Mononitration (HNO_3, H_2SO_4, 40°C).
(*b*) Dinitration (HNO_3, H_2SO_4, 80°C).
(*c*) Trinitration (HNO_3, H_2SO_4, 110°C). The explosive TNT (trinitrotoluene) is the major product obtained on trinitration of toluene. Which trinitrotoluene isomer is TNT?

12.30 Friedel-Crafts acylation of the individual isomers of xylene with acetyl chloride and aluminum chloride yields a single product, different for each xylene isomer, in high yield in each case. Write the structures of the products of acetylation of *o*-, *m*-, and *p*-xylene.

12.31 Reaction of benzanilide ($C_6H_5NH\overset{O}{\overset{\|}{C}}C_6H_5$) with chlorine in acetic acid yields a mixture of two monochloro derivatives formed by electrophilic aromatic substitution. Suggest reasonable structures for these two isomers.

12.32 Each of the following reactions has been reported in the chemical literature and gives a predominance of a single product in synthetically acceptable yield. Write the structure of the product. Only monosubstitution is involved in each case, unless otherwise indicated.

(*a*) 2-chloro-1,3-benzenedicarboxylic acid (CO_2H, Cl, CO_2H) $\xrightarrow[H_2SO_4,\ heat]{HNO_3}$

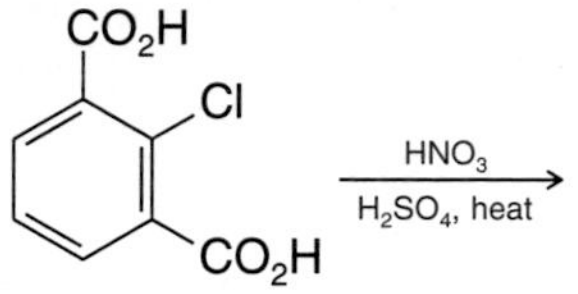

(*b*) O_2N–, CF_3, NH_2 substituted benzene $\xrightarrow[acetic\ acid]{Br_2}$

(*c*) 4-biphenyl–OH $\xrightarrow[CHCl_3]{Br_2}$

(*d*) $C(CH_3)_3$, $CH(CH_3)_2$ (meta) substituted benzene $\xrightarrow[acetic\ acid]{HNO_3}$

(*e*) benzene + $CH_2{=}CH(CH_2)_5CH_3$ $\xrightarrow[5–15°C]{H_2SO_4}$

(*f*) F-substituted benzene–OCH_3 + $CH_3\overset{O}{\overset{\|}{C}}O\overset{O}{\overset{\|}{C}}CH_3$ $\xrightarrow{AlCl_3}$

(*g*) $CH(CH_3)_2$, NO_2 (para) substituted benzene $\xrightarrow[H_2SO_4]{HNO_3}$

(*h*) OCH_3, CH_3 (para) substituted benzene + $(CH_3)_2C{=}CH_2$ $\xrightarrow{H_2SO_4}$

(*i*) benzene–CH_2–(benzene with H_3C, OH)–CH_3 $\xrightarrow[CHCl_3]{Br_2}$

(*j*) benzene–$\overset{O}{\overset{\|}{C}}$–benzene $\xrightarrow[triethylene\ glycol,\ 173°C]{H_2NNH_2,\ KOH}$

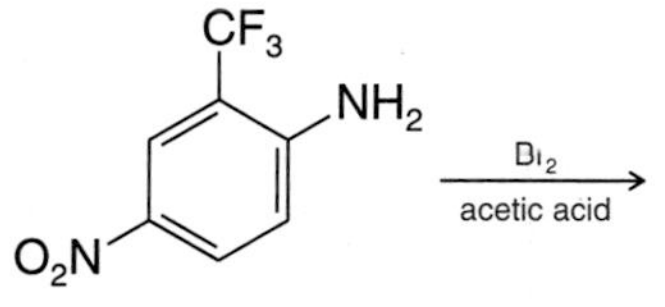

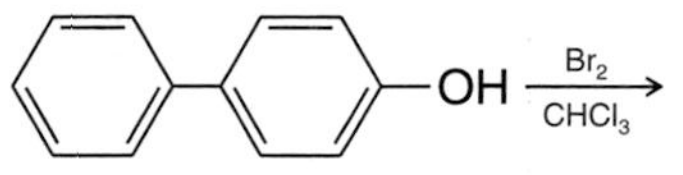

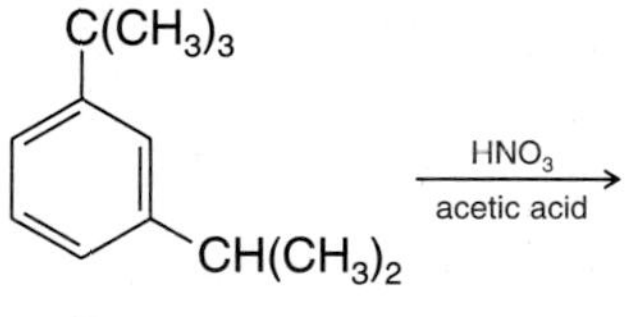

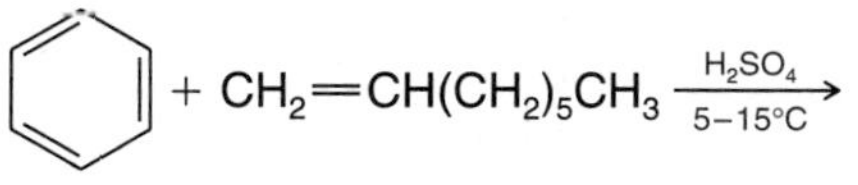

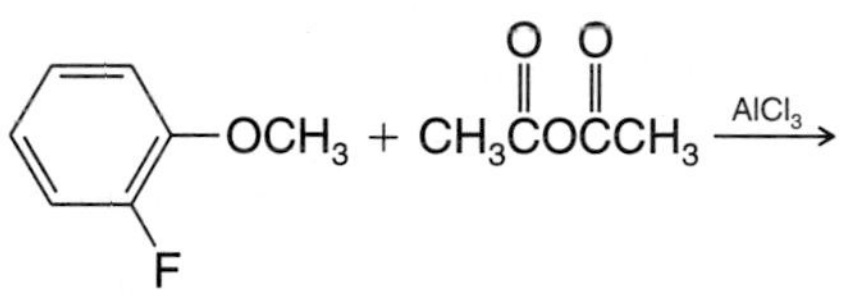

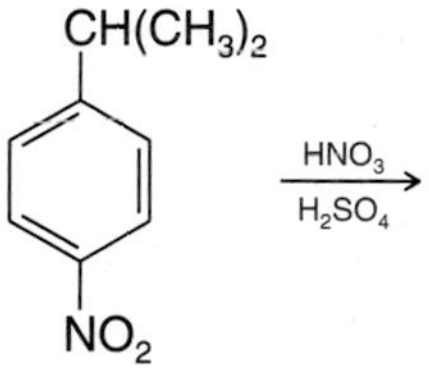

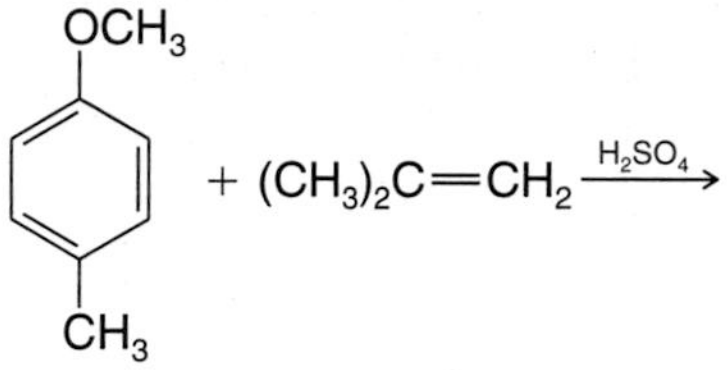

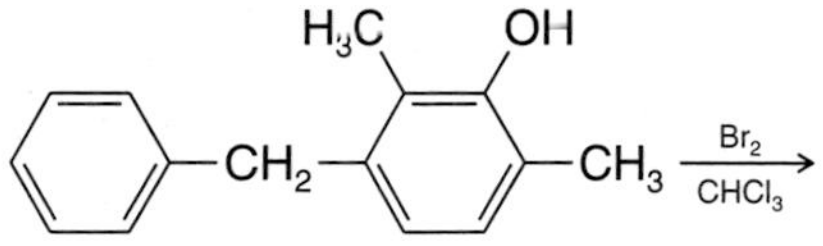

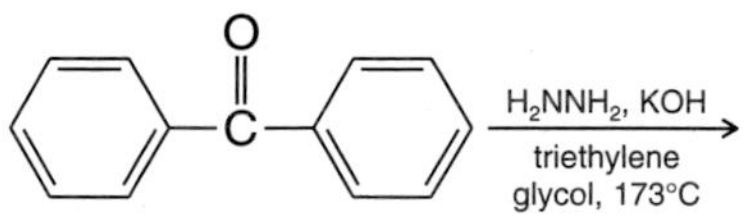

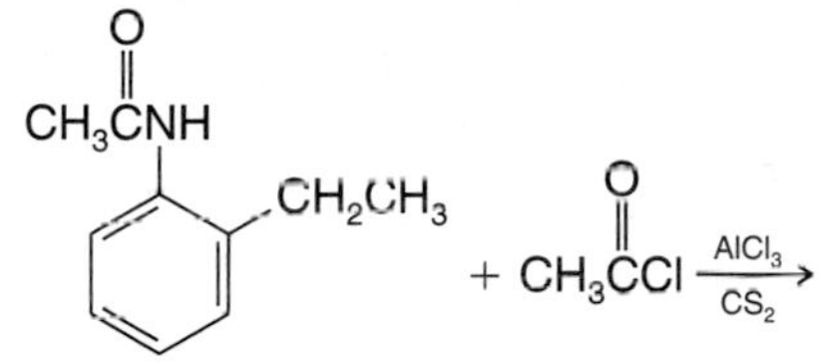

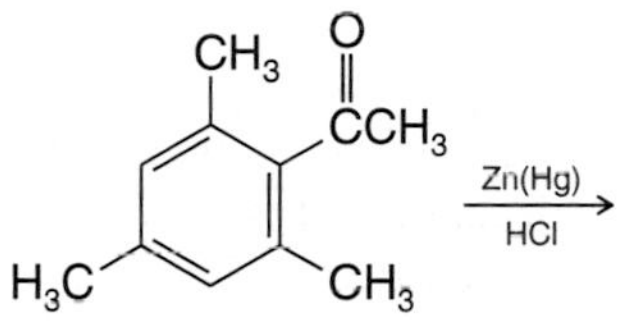

(*k*) benzene–F + benzene–CH_2Cl $\xrightarrow{AlCl_3}$

(*l*) $CH_3\overset{O}{\overset{\|}{C}}NH$, CH_2CH_3 (ortho) substituted benzene + $CH_3\overset{O}{\overset{\|}{C}}Cl$ $\xrightarrow[CS_2]{AlCl_3}$

(*m*) benzene with CH_3, $\overset{O}{\overset{\|}{C}}CH_3$, CH_3, H_3C $\xrightarrow[HCl]{Zn(Hg)}$

(*n*) thiophene-3-CO_2H (S) $\xrightarrow[acetic\ acid]{Br_2}$

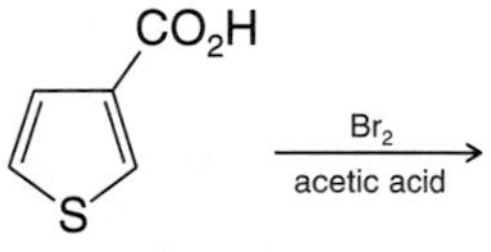

12.33 What combination of acyl chloride or acid anhydride and arene would you choose to prepare each of the following compounds by a Friedel-Crafts acylation reaction?

(*a*) $C_6H_5\overset{O}{\overset{\|}{C}}CH_2C_6H_5$

(*b*) H_3C–(benzene)–CH_3 with $\underset{O}{\underset{\|}{C}}CH_2CH_2CO_2H$

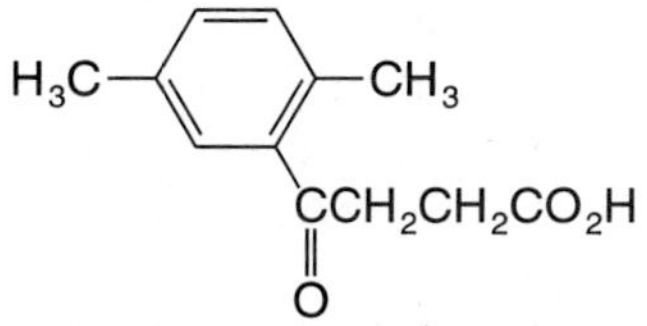

(*c*) O_2N–(benzene)–$\overset{O}{\overset{\|}{C}}$–(benzene)

(*d*) (H_3C)(H_3C)-benzene–$\overset{O}{\overset{\|}{C}}$–benzene

(*e*) H_3C–(benzene)–$\overset{O}{\overset{\|}{C}}$–(benzene with HO_2C)

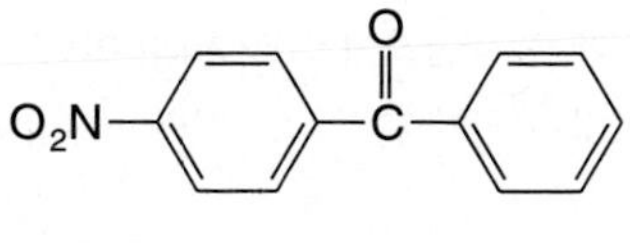

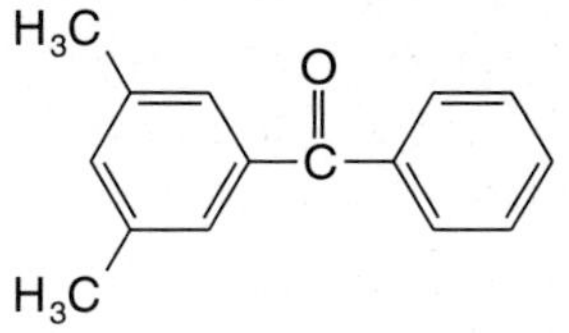

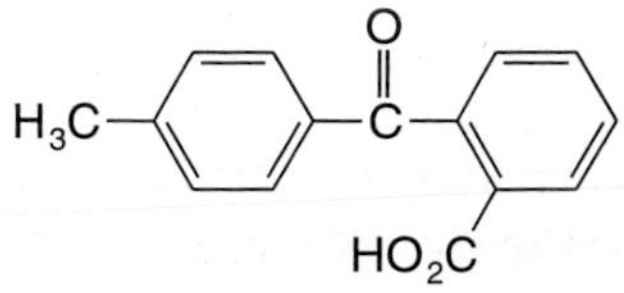

12.34 Suggest a suitable series of reactions for carrying out each of the following synthetic transformations:

(*a*) $CH(CH_3)_2$ to CO_2H SO_3H

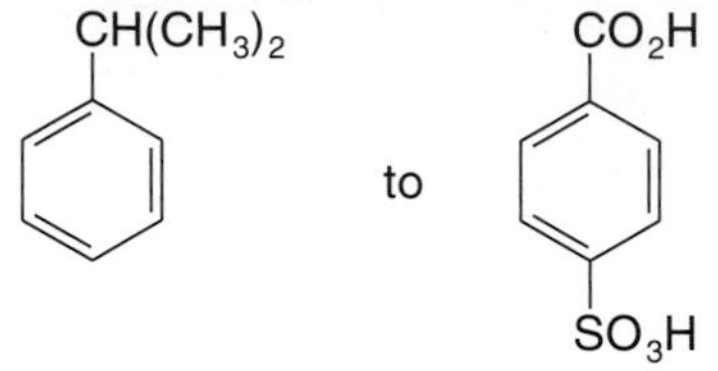

(*b*) CH_3 CH_3 to CO_2H CO_2H $C(CH_3)_3$

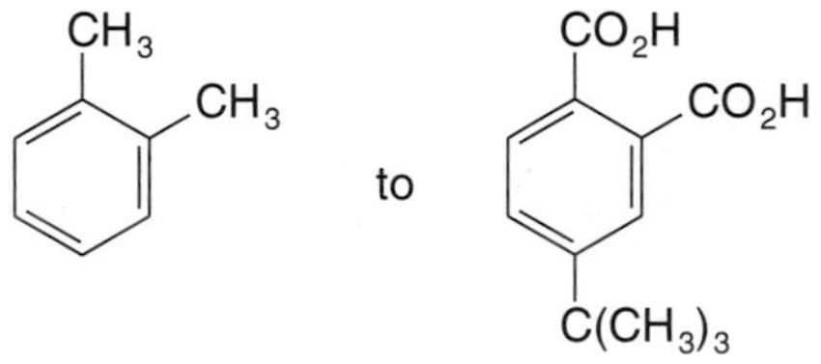

(*c*) O to O CH_3C

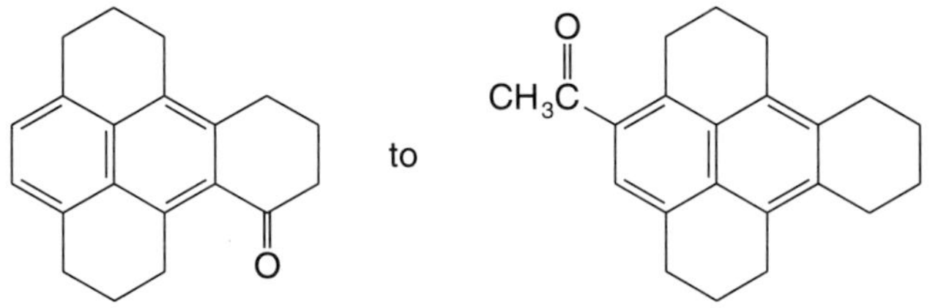

(*d*) OCH_3 OCH_3 to OCH_3 O_2N OCH_3 $C(CH_3)_3$

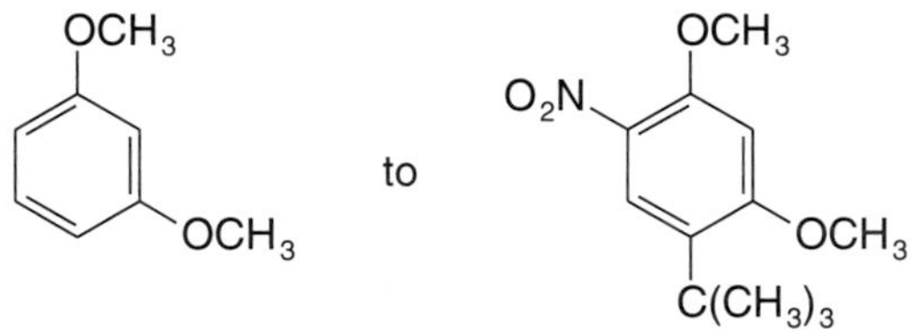

12.35 A standard synthetic sequence for building a six-membered cyclic ketone onto an existing aromatic ring is shown in outline as follows. Specify the reagents necessary for each step.

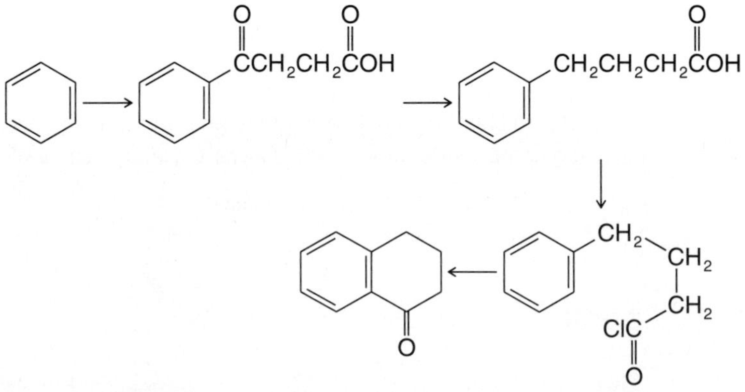

O O CCH_2CH_2COH → O $CH_2CH_2CH_2COH$ → CH_2 CH_2 CH_2 ClC O ← O

12.36 Each of the compounds indicated undergoes an intramolecular Friedel-Crafts acylation reaction to yield a cyclic ketone. Write the structure of the expected product in each case.

(*a*) 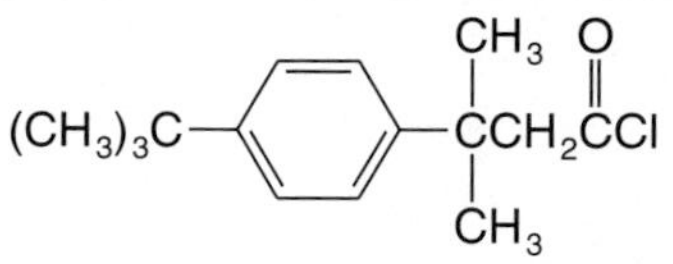 $(CH_3)_3C-C_6H_4-C(CH_3)_2CH_2C(=O)Cl$

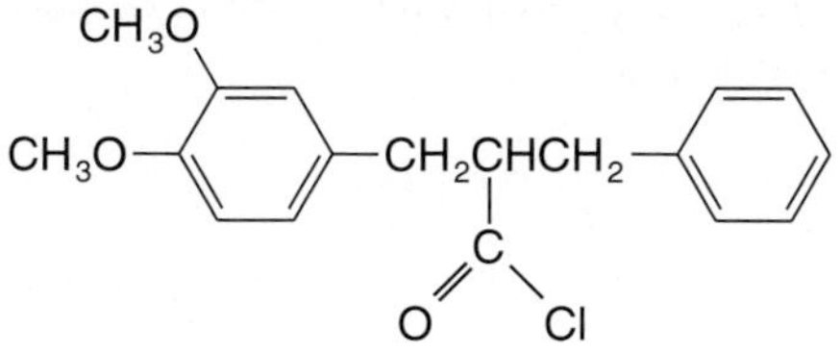

(*b*) 2-phenylcyclohexyl–$CH_2C(=O)Cl$

(*c*) $(CH_3O)_2C_6H_3-CH_2CH(C(=O)Cl)CH_2-C_6H_5$

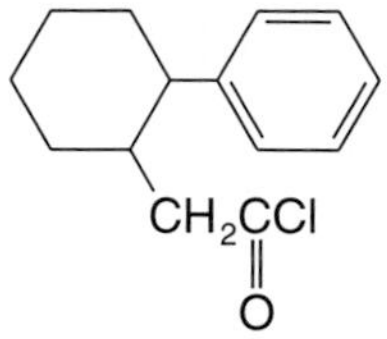

12.37 The relative rates of attack at the individual positions of biphenyl compared with that at a single position of benzene (taken as 1) in electrophilic chlorination are as shown.

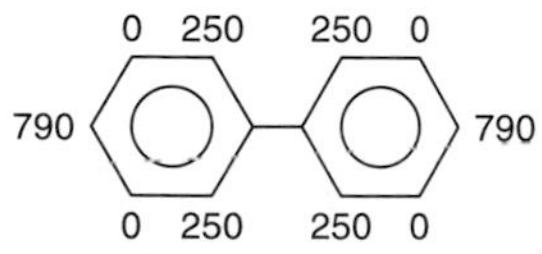

0 250 250 0
790 790
0 250 250 0

(*a*) What is the relative rate of chlorination of biphenyl compared with benzene?
(*b*) If, in a particular chlorination reaction, 10 g of *o*-chlorobiphenyl were formed, how much *p*-chlorobiphenyl would you expect to find?

12.38 Partial rate factors may be used to estimate product distributions in disubstituted benzene derivatives. The reactivity of a particular position in *o*-bromotoluene, for example, is given by the product of the partial rate factors for the corresponding position in toluene and bromobenzene. On the basis of the partial rate factor data given here for Friedel-Crafts acetylation, predict the major product of the reaction of *o*-bromotoluene with acetyl chloride and aluminum chloride.

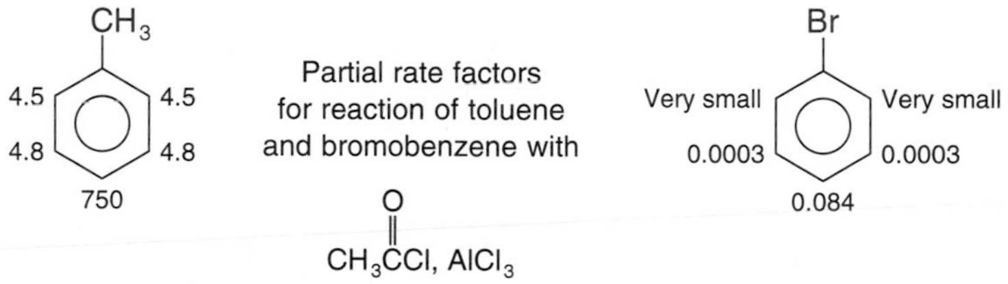

CH_3: 4.5, 4.5, 4.8, 4.8, 750

Partial rate factors for reaction of toluene and bromobenzene with CH_3CCl (O), $AlCl_3$

Br: Very small, Very small, 0.0003, 0.0003, 0.084

12.39 When 2-isopropyl-1,3,5-trimethylbenzene is heated with aluminum chloride (trace of HCl present) at 50°C, the major material present after 4 h is 1-isopropyl-2,4,5-trimethylbenzene. Suggest a reasonable mechanism for this isomerization.

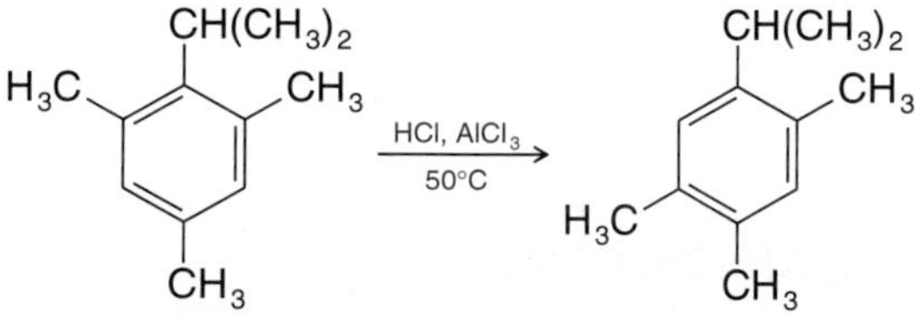

$CH(CH_3)_2$, H_3C, CH_3, CH_3 $\xrightarrow[50°C]{HCl,\ AlCl_3}$ $CH(CH_3)_2$, CH_3, H_3C, CH_3

12.40 When a dilute solution of 6-phenylhexanoyl chloride in carbon disulfide was slowly added (over a period of 8 days!) to a suspension of aluminum chloride in the same solvent, it yielded a product A ($C_{12}H_{14}O$) in 67 percent yield. Oxidation of A gave benzene-1,2-dicarboxylic acid.

$$C_6H_5(CH_2)_5\overset{O}{\overset{\|}{C}}Cl \xrightarrow[CS_2]{AlCl_3} C_{12}H_{14}O \xrightarrow[H_2SO_4,\ heat]{Na_2Cr_2O_7,\ H_2O} \text{benzene-1,2-}(CO_2H)_2$$

6-Phenylhexanoyl chloride — Compound A — Benzene-1,2-dicarboxylic acid

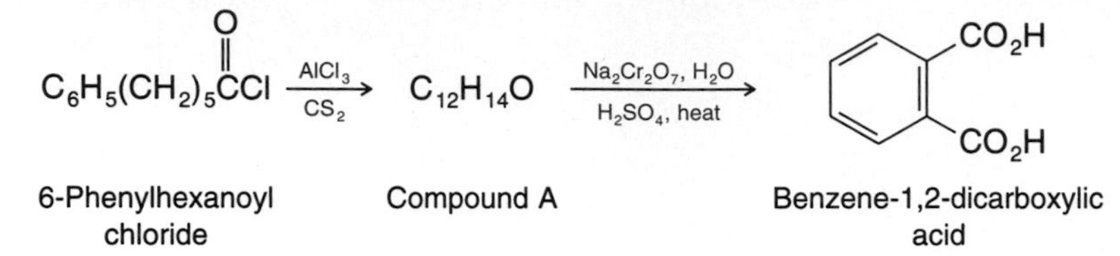

Formulate a reasonable structure for compound A.

12.41 Reaction of hexamethylbenzene with methyl chloride and aluminum chloride gave a salt A, which, on being treated with aqueous sodium bicarbonate solution, yielded compound B. Suggest a mechanism for the conversion of hexamethylbenzene to B by correctly inferring the structure of A.

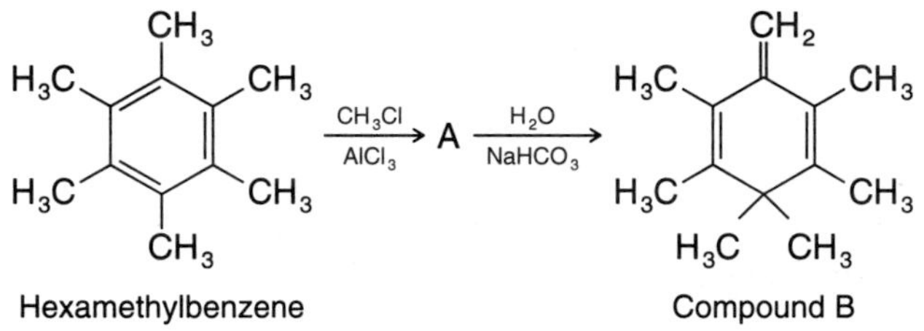

$$\text{Hexamethylbenzene} \xrightarrow[AlCl_3]{CH_3Cl} A \xrightarrow[NaHCO_3]{H_2O} \text{Compound B}$$

Hexamethylbenzene — Compound B

12.42 The synthesis of compound C was achieved by using compounds A and B as the sources of all carbon atoms. Suggest a synthetic sequence involving no more than three steps by which A and B may be converted to C.

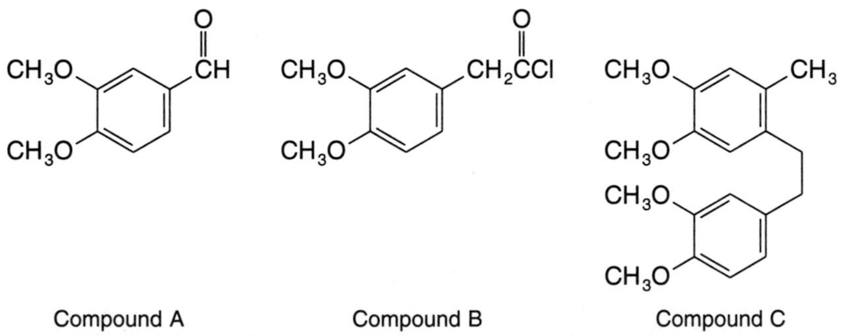

Compound A — Compound B — Compound C

12.43 When styrene is refluxed with aqueous sulfuric acid, two "styrene dimers" are formed as the major products. One of these styrene dimers is 1,3-diphenyl-1-butene; the other is 1-methyl-3-phenylindan. Suggest a reasonable mechanism for the formation of each of these compounds.

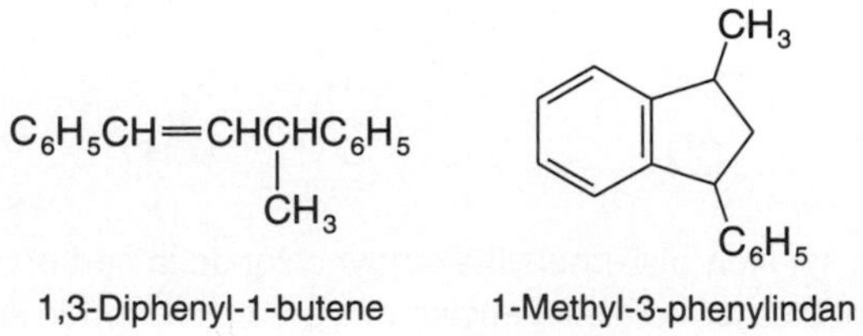

$$C_6H_5CH{=}CH\underset{CH_3}{\underset{|}{C}}HC_6H_5$$

1,3-Diphenyl-1-butene — 1-Methyl-3-phenylindan

12.44 Treatment of the alcohol whose structure is given below with sulfuric acid gave as the major organic product a tricyclic hydrocarbon of molecular formula $C_{16}H_{16}$. Suggest a reasonable structure for this hydrocarbon.

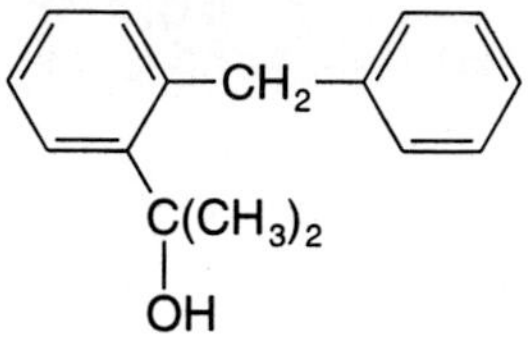

CH_2

$C(CH_3)_2$

OH

MOLECULAR MODELING EXERCISES

12.45 Nitronium ion ($^+NO_2$) is the active electrophile in most aromatic nitrations. Make a molecular model of nitronium ion. What is its geometry?

12.46 Using a molecular model of benzene, show the approximate direction taken by an incoming electrophile when it attacks the ring.

12.47 Construct a molecular model of the cyclohexadienyl cation intermediate in the nitration of benzene. Identify the hybridization state of each carbon.

12.48 Compare the crowding experienced by an incoming electrophile when it attacks C-2 versus C-4 of 1,3-dimethylbenzene.

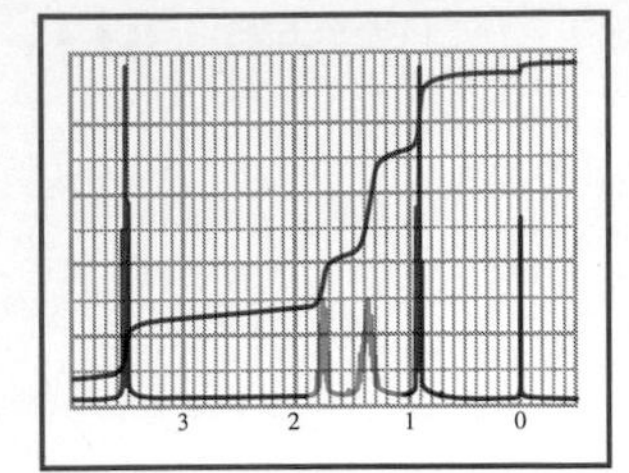

CHAPTER 13

SPECTROSCOPY

Until the second half of the twentieth century, the structure of a substance—a new natural product, for example—was determined using information obtained from chemical reactions. This information included the identification of functional groups by chemical tests, along with the results of degradation experiments in which the substance was broken down into smaller, more readily identified fragments. Typical of this approach is the demonstration of the presence of a double bond in an alkene by catalytic hydrogenation and subsequent determination of its location by ozonolysis. After considering all the available chemical evidence, the chemist proposed a candidate structure (or structures) consistent with the observations. Proof of structure was provided either by converting the substance to some already known compound or by an independent synthesis.

Qualitative tests and chemical degradation as structural probes have been supplemented and to a large degree superseded in present-day organic chemistry by instrumental methods of structure determination. The most prominent of these techniques are **nuclear magnetic resonance (nmr) spectroscopy, infrared (ir) spectroscopy, ultraviolet-visible (uv-vis) spectroscopy,** and **mass spectrometry (MS).** As diverse as these techniques are, all of them are based on the absorption of energy by a molecule, and all examine how a molecule responds to that absorption of energy. In describing these techniques our emphasis will be on their application to the task of structure determination. We begin with a brief discussion of the nature of electromagnetic radiation, a topic that is fundamental to understanding the physical bases on which molecular spectroscopy depends.

13.1 PRINCIPLES OF MOLECULAR SPECTROSCOPY: ELECTROMAGNETIC RADIATION

Electromagnetic radiation, of which visible light is but one example, is said to have a dual nature. It has the properties of both particles and waves. The particles are called

photons, and each possesses an amount of energy referred to as a **quantum.** In 1900 the German physicist Max Planck proposed that the energy of a photon (E) is directly proportional to its frequency (ν).

$$E = h\nu$$

The SI units of frequency are reciprocal seconds (s^{-1}), given the name *hertz* and the symbol Hz in honor of the nineteenth-century physicist Heinrich R. Hertz. The constant of proportionality h is called **Planck's constant** and has the value

$$h = 6.62 \times 10^{-27} \text{ erg} \cdot \text{s}$$

"Modern" physics dates from Planck's proposal that energy is quantized, which set the stage for the development of quantum mechanics. Planck received the 1918 Nobel Prize in physics.

Planck's equation gives the energy of a photon in ergs; 1 erg per molecule is equivalent to 6.0×10^{13} kJ/mol (1.44×10^{13} kcal/mol).

Electromagnetic radiation is propagated at the speed of light. The speed of light (c) is 3.0×10^8 m/s and is equal to the product of the frequency ν and the wavelength λ:

$$c = \nu\lambda$$

The range of photon energies is called the *electromagnetic spectrum* and is shown in Figure 13.1. Visible light occupies a very small region of the electromagnetic spectrum. It is characterized by wavelengths of 4×10^{-7} m (violet) to 8×10^{-7} m

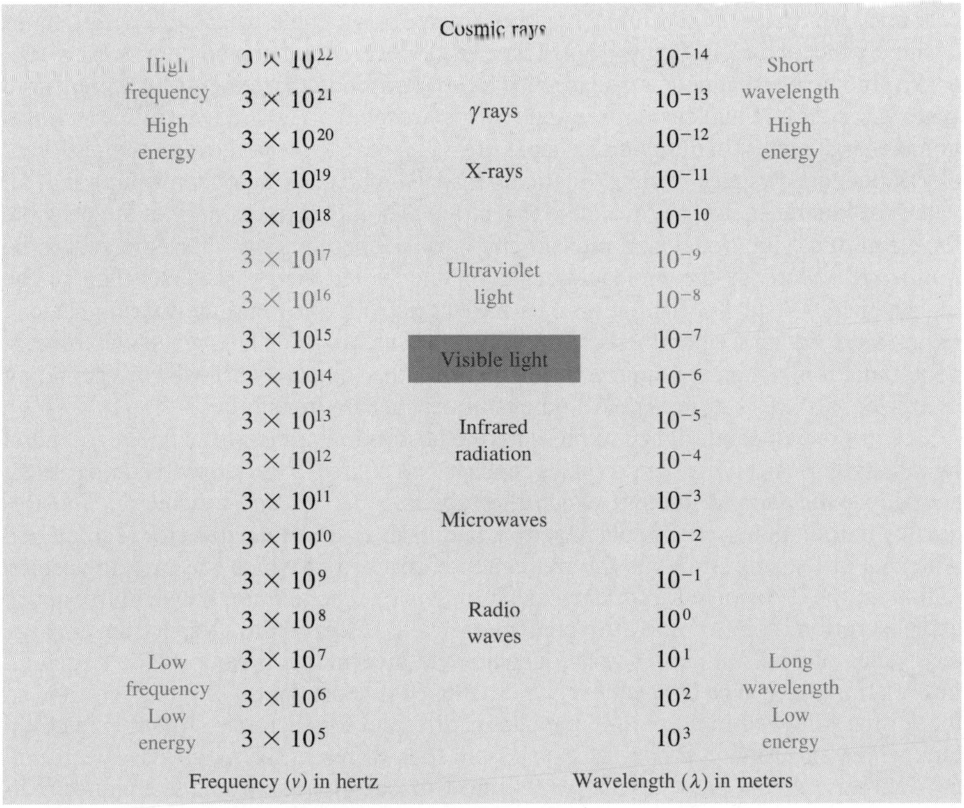

	Frequency (ν) in hertz		Wavelength (λ) in meters	
		Cosmic rays		
High frequency	3×10^{22}		10^{-14}	Short wavelength
	3×10^{21}	γ rays	10^{-13}	
High energy	3×10^{20}		10^{-12}	High energy
	3×10^{19}	X-rays	10^{-11}	
	3×10^{18}		10^{-10}	
	3×10^{17}	Ultraviolet light	10^{-9}	
	3×10^{16}		10^{-8}	
	3×10^{15}	Visible light	10^{-7}	
	3×10^{14}		10^{-6}	
	3×10^{13}	Infrared radiation	10^{-5}	
	3×10^{12}		10^{-4}	
	3×10^{11}	Microwaves	10^{-3}	
	3×10^{10}		10^{-2}	
	3×10^{9}		10^{-1}	
	3×10^{8}	Radio waves	10^{0}	
Low frequency	3×10^{7}		10^{1}	Long wavelength
	3×10^{6}		10^{2}	
Low energy	3×10^{5}		10^{3}	Low energy

FIGURE 13.1 The electromagnetic spectrum.

(red). When examining Figure 13.1 it is helpful to keep the following two relationships in mind:

1. *Frequency is inversely proportional to wavelength;* the greater the frequency, the shorter the wavelength.
2. *Energy is proportional to frequency;* electromagnetic radiation of higher frequency possesses more energy than radiation of lower frequency.

Depending on its source, a photon can have a vast amount of energy; cosmic rays and x-rays are streams of very high energy photons. Radio waves are of relatively low energy. Ultraviolet radiation is of higher energy than the violet end of visible light. Infrared radiation is of lower energy than the red end of visible light. When a molecule is exposed to electromagnetic radiation, it may absorb a photon, thereby increasing its energy by an amount equal to the energy of the photon. Molecules are highly selective with respect to the frequencies of radiation that they absorb. Only photons of certain specific frequencies are absorbed by a molecule. The particular photon energies absorbed by a molecule depend on molecular structure and can be measured with instruments called **spectrometers.** The data obtained are very sensitive indicators of molecular structure and have revolutionized the practice of chemical analysis.

13.2 PRINCIPLES OF MOLECULAR SPECTROSCOPY: QUANTIZED ENERGY STATES

What determines whether a photon is absorbed by a molecule or not? The most important condition is that the energy of the photon must match the energy difference between two states of the molecule. Consider, for example, the two energy states designated E_1 and E_2 in Figure 13.2. The energy difference between them is $E_2 - E_1$, or ΔE. In nuclear magnetic resonance (nmr) spectroscopy these are two different spin states of an atomic nucleus; in infrared (ir) spectroscopy, they are two different vibrational energy states; in ultraviolet-visible (uv-vis) spectroscopy, they are two different electronic energy states. Unlike kinetic energy, which is continuous, meaning that all values of kinetic energy are available to a molecule, only certain energies are possible for electronic, vibrational, and nuclear spin states. These energy states are said to be **quantized.** More of the molecules exist in the lower-energy state E_1 than in the higher-energy state E_2. Excitation of a molecule from a lower state to a higher one requires the addition of an increment of energy equal to ΔE. Thus, when electromagnetic radiation is incident upon a molecule, only the frequency whose corresponding energy equals ΔE is absorbed. All other frequencies are transmitted.

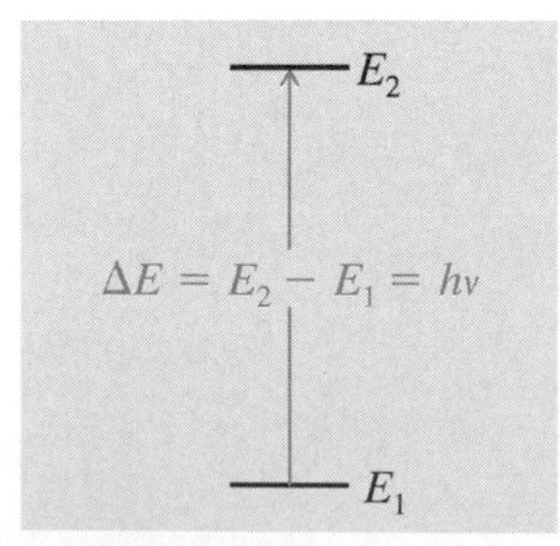

FIGURE 13.2 Two energy states of a molecule. Absorption of energy equal to $E_2 - E_1$ excites a molecule from its lower-energy state to the next-higher state.

Spectrometers are designed to measure the absorption of electromagnetic radiation by a sample. Basically, a spectrometer consists of a source of radiation, a compartment containing the sample through which the radiation passes, and a detector. The frequency of radiation is continuously varied, and its intensity at the detector is compared with that at the source. When the frequency is reached at which the sample absorbs radiation, the detector senses a decrease in intensity. The relation between frequency and absorption is plotted on a strip chart and is called a **spectrum.** A spectrum consists of a series of peaks at particular frequencies; its interpretation can provide structural information. Each type of spectroscopy developed independently of the others, and so the format followed in presenting the data is different for each one. An nmr spectrum looks different from an ir spectrum, and both look different from a uv-vis spectrum.

With this general background, we will now discuss spectroscopic techniques individually. Nmr, ir, and uv-vis spectroscopy provide complementary information, and all are useful. Among them, nmr provides the information most directly related to molecular structure and is the one we shall examine first.

13.3 PROTON MAGNETIC RESONANCE (^{1}H NMR) SPECTROSCOPY

As noted in the preceding section, nmr spectroscopy is based on transitions between **nuclear spin states.** What do we mean by nuclear spin, and how can it provide structural information?

An article in the May 1988 issue of the *Journal of Chemical Education* (pp. 426–433) introduces proton nmr spectroscopy at a level comparable to that described here.

A proton, like an electron, possesses the property of spin. Also, like an electron, a proton has two (nuclear) spin states with spin quantum numbers of $+\frac{1}{2}$ and $-\frac{1}{2}$. There is no difference in energy between these two nuclear spin states; a proton is just as likely to have a spin of $+\frac{1}{2}$ as $-\frac{1}{2}$. Since both spin states have the same energy, transitions between the two seem to offer no promise as the basis for a spectroscopic technique. There is a simple way, however, to render the energies of the two nuclear spin states unequal, and that is by placing the proton in a strong magnetic field.

A proton is a spinning charge and has associated with it a magnetic moment coinciding with the axis of rotation (Figure 13.3). In the presence of an external magnetic field H_0, the two nuclear spin states no longer have the same energy. The state in which the nuclear magnetic moment is aligned with the external field is lower in energy than the state in which it opposes the applied field. As depicted in Figure 13.4, the difference in energy between the two states is directly proportional to the strength of the applied field. Even in very powerful magnetic fields, the energy difference is quite small. At a field strength of 14,100 gauss (G) the energy difference between the two proton spin states is only 2.5×10^{-5} kJ/mol (6×10^{-6} kcal/mol). An energy difference of this magnitude corresponds to the radio-frequency region of the electromagnetic spectrum (see Figure 13.1).

Nuclear magnetic resonance of protons was first detected in 1946 by Edward Purcell (at Harvard) and by Felix Bloch (at Stanford). Purcell and Bloch shared the 1952 Nobel Prize in physics.

Figure 13.5 illustrates the essential features of a nuclear magnetic resonance spectrometer. At its heart is a powerful magnet, either a permanent magnet or an electromagnet. Consider the situation in which it is a permanent magnet of field strength 14,100 G. The sample is placed between the pole faces of the magnet, which causes the nuclear spins to align themselves either with the field or against the field. There is a small excess of nuclei that have their spins aligned with the field as compared with those aligned against it. For a sample of 1 million protons at 25°C and 14,100 G, approximately 500,005 are in the lower-energy state compared with 499,995 in the higher state. The sample cavity is surrounded by a radio-frequency source and the frequency is continuously varied. When the frequency of the source matches the energy difference between nuclear spin states, the two are said to be in *resonance* with each other. For protons at 14,100 G, this frequency is approximately 60×10^6 Hz, or

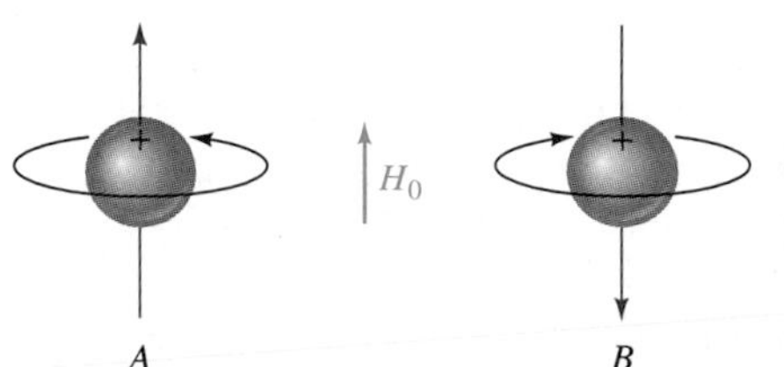

FIGURE 13.3 Nuclear spin states of a proton. Spin state *A*, in which the nuclear magnetic moment is parallel to the applied field H_0, is of lower energy than spin state *B*, in which the nuclear magnetic moment is antiparallel to the applied field. *(From Pine, Hendrickson, Cram, and Hammond, Organic Chemistry, McGraw-Hill, New York, 1980.)*

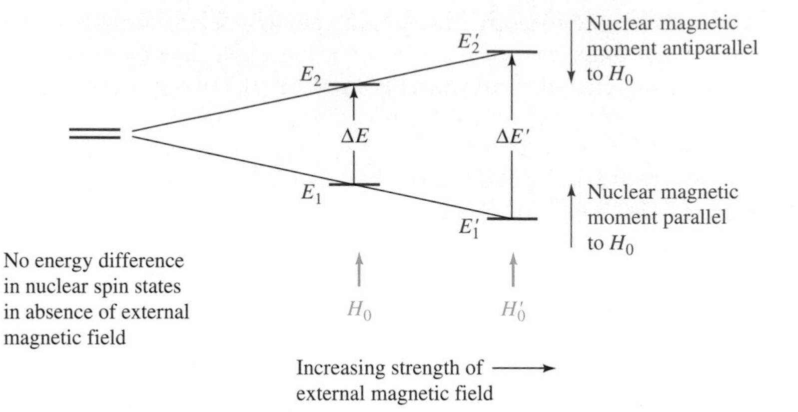

FIGURE 13.4 An external magnetic field causes the two nuclear spin states to have different energies. The difference in energy ΔE is proportional to the strength of the applied field.

60 MHz. Energy is absorbed and nuclei undergo a *spin flip* from the lower-energy orientation to the higher one. The absorption of energy is detected in a radio-frequency receiver and shown as a peak on the nmr spectrum.

Nuclear magnetic resonance spectrometers that use electromagnets operate in a complementary manner. The frequency of the source is maintained at a constant value, say, 60 MHz, and the magnetic field strength is varied until the energy gap between spin states matches that of the source. Both types of spectrometers are available, and the nmr spectra from permanent magnet instruments are identical with those from electromagnet spectrometers.

In 1994 the U.S. Department of Energy announced plans to construct a laboratory in Richland, Washington, that would house (in 1997) the world's first 1,000-MHz nmr spectrometer.

At one time, most nmr spectra were routinely obtained using 60-MHz instruments, and compilations containing thousands of such spectra are available in chemistry libraries. Spectrometers operating at magnetic field strengths greater than 60 MHz for protons (90 or 300 MHz, for example) give better-quality spectra and have, to a great degree, replaced 60-MHz instruments. Because both kinds of spectra are commonly encountered, this text includes examples recorded at 300 MHz as well as the more familiar 60-MHz type.

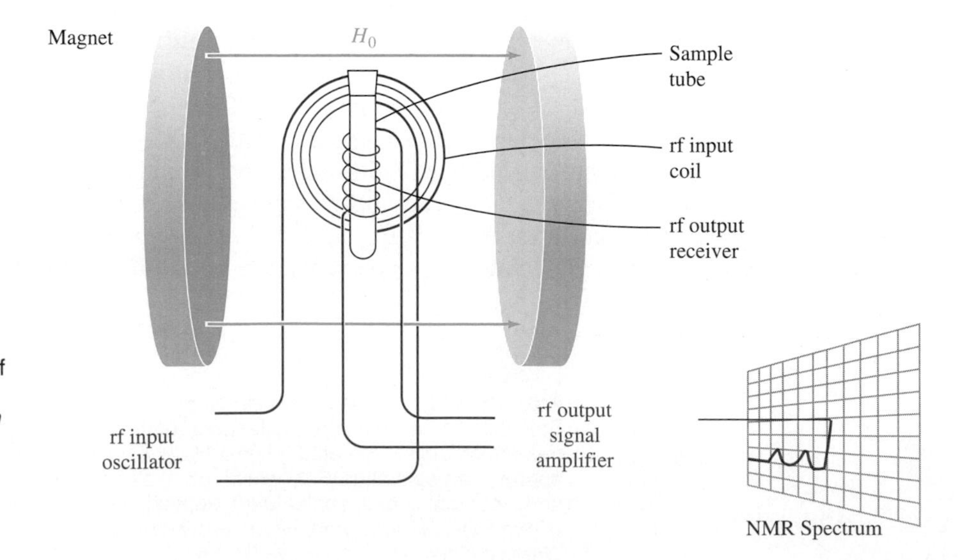

FIGURE 13.5 Diagram of a nuclear magnetic resonance spectrometer. (*From Pine, Hendrickson, Cram, and Hammond, Organic Chemistry, McGraw-Hill, New York, 1980, p. 136.*)

13.4 NUCLEAR SHIELDING AND CHEMICAL SHIFT

Our discussion to this point has centered on transitions between the two spin states of a "bare" proton, that is, an isolated hydrogen nucleus. In real molecules a proton is bonded to another atom by a two-electron covalent bond. These electrons, indeed all the electrons in a molecule, affect the magnetic environment of the proton. Since our concern is organic chemistry, let us examine the difference between a bare proton and a proton in some organic compound.

H^+ — Bare proton

H—C (with three further bonds) — Proton in an organic molecule

Assuming that we carry out the nmr experiment under conditions such that the radio-frequency source emits at a constant frequency of 60 MHz and the magnetic field strength is slowly increased until the resonance condition is realized, we find that a bare proton undergoes its spin flip at some value of the applied field H_0 close to 14,100 G. Figure 13.6*a* is an idealized representation of the spectrum obtained. Under the same conditions, as shown in Figure 13.6*b*, we find that a field strength slightly greater than H_0 is required to flip the spin of the proton in an organic molecule. We say that the nmr signal of the bound proton appears *at higher field* than the signal of the bare proton and call this difference between the two a **chemical shift.** *A chemical shift is the change in the resonance position of a nucleus that is brought about by its molecular environment.* (A peak at higher field than another is said to be **upfield**; a peak at lower field than another is said to be **downfield.**)

The reason that higher field strengths are required to achieve the same separation of energy levels of bound protons compared with bare protons is that the molecule's electrons *shield* the bound proton from the external field. Under the influence of an external field there is an induced magnetic field associated with the electrons that opposes the applied field. This is illustrated in Figure 13.7. Thus, the magnetic field "felt" by a proton is less than the actual field strength H_0. In order to achieve a separation between levels equal to the energy of the radio-frequency radiation, the strength of the applied field must be increased by an amount equal to the strength of

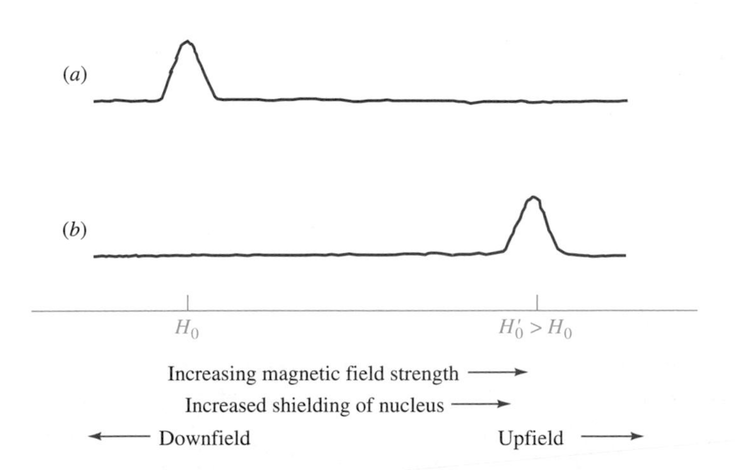

FIGURE 13.6 A bare proton (*a*) inverts its nuclear spin at a lower field strength than a proton in an organic molecule (*b*). The nmr signal for proton *b* is at higher field than that for proton *a*.

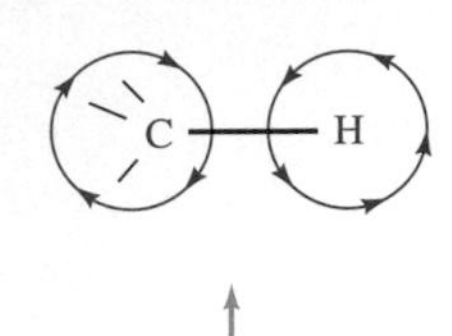

FIGURE 13.7 The induced magnetic field on the electrons in the carbon-hydrogen bond opposes the external magnetic field. The resultant magnetic field experienced by the proton is slightly less than H_0.

the opposing induced magnetic fields of the electrons. Increased shielding of a nucleus requires higher magnetic field strengths to achieve resonance.

13.5 HOW CHEMICAL SHIFT IS MEASURED

Chemists compare the shielding of protons in organic molecules by specifying their chemical shifts relative to a standard substance. This substance is *tetramethylsilane,* $(CH_3)_4Si$, abbreviated *TMS.* The protons of TMS are more shielded than those of almost all organic compounds. In a solution containing TMS, all the relevant signals appear to the left of the TMS peak. The orientation of the spectrum on the chart is adjusted electronically so that the TMS peak coincides with the zero grid line. Peak positions are measured in frequency units (hertz) downfield from the TMS peak. (Frequency units are more convenient to use than units of magnetic field strength. The two sets of units are directly proportional to each other.)

Figure 13.8 is the 60-MHz nmr spectrum of chloroform ($CHCl_3$) containing a few drops of TMS. The signal due to the proton in chloroform appears 437 Hz downfield from the TMS peak. By common agreement, chemical shifts (δ) are reported in parts per million (ppm) from the TMS peak.

$$\text{Chemical shift } (\delta) = \frac{\text{position of signal} - \text{position of TMS peak}}{\text{spectrometer frequency}} \times 10^6$$

Thus, the chemical shift for the proton in chloroform is:

$$\delta = \frac{437 \text{ Hz} - 0 \text{ Hz}}{60 \times 10^6 \text{ Hz}} \times 10^6 = 7.28 \text{ ppm}$$

Nuclear magnetic resonance spectra are recorded on chart paper that is calibrated in

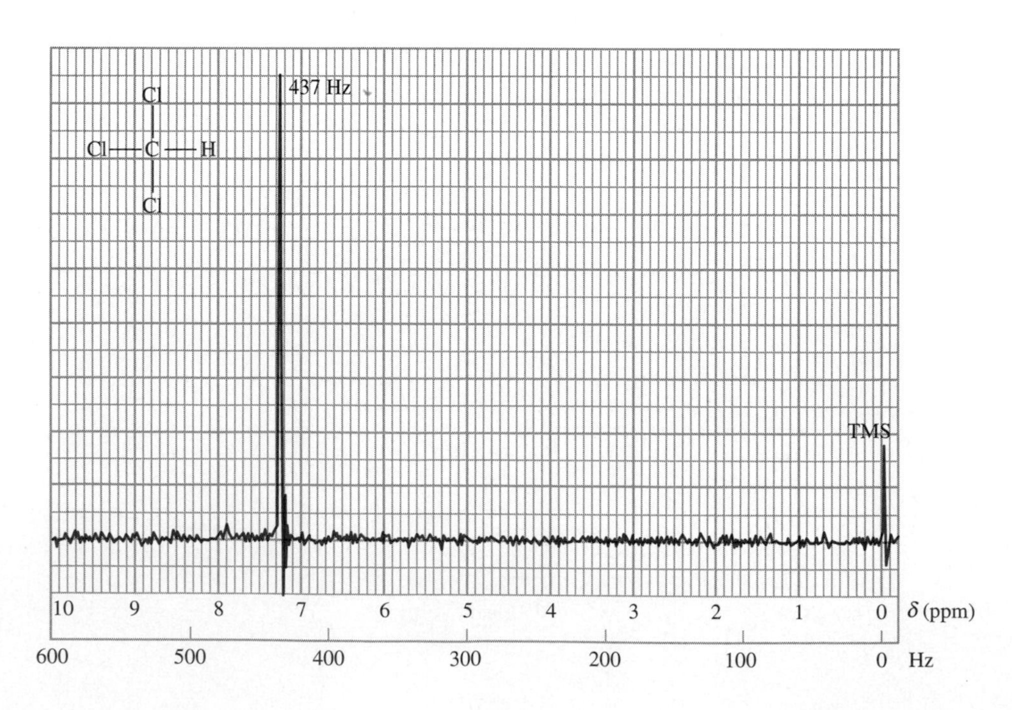

FIGURE 13.8 The proton magnetic resonance (¹H nmr) spectrum of chloroform ($CHCl_3$).

both parts per million (ppm) and hertz, and both refer to the TMS peak as the zero point. When reporting chemical shifts in frequency units (hertz), the field strength of the instrument must be specified. A 60-MHz nmr spectrometer separates the energy of nuclear spin states only 20 percent as much as does a 300-MHz spectrometer. Thus, a chemical shift of 437 Hz with a 60-MHz instrument appears at 2185 Hz with a 300-MHz spectrometer. By reporting chemical shifts in parts per million, this effect of field strength is taken into account; thus, irrespective of magnetic field strength, the signal due to the proton of chloroform appears at 7.28 ppm. All chemical shifts in this text are reported in parts per million.

PROBLEM 13.1 Calculate the chemical shift in parts per million for each of the following compounds, given their shifts in hertz as measured downfield from tetramethylsilane on a spectrometer of 60-MHz field strength.

(*a*) Bromoform ($CHBr_3$), 413 Hz
(*b*) Iodoform (CHI_3), 322 Hz
(*c*) Methyl chloride (CH_3Cl), 184 Hz

SAMPLE SOLUTION (*a*) The chemical shift in bromoform is calculated from the preceding equation:

$$\delta = \frac{413 \text{ Hz } (CHBr_3) - 0 \text{ Hz (TMS)}}{60 \times 10^6 \text{ Hz}} \times 10^6 = 6.88 \text{ ppm}$$

Although chloroform is a good solvent for organic compounds, it is not widely used as a medium for measuring nmr spectra; its own nmr signal could potentially obscure a signal in the sample. Chloroform-d ($CDCl_3$) is used instead. Because the magnetic properties of deuterium, the mass 2 isotope of hydrogen (D = ^{2}H), are different from those of protium (^{1}H), chloroform-d does not give an nmr signal under the conditions employed for measuring proton magnetic resonance spectra. Chloroform-d exhibits no peaks in the spectrum that would obscure those of an organic molecule being examined.

13.6 CHEMICAL SHIFT AND MOLECULAR STRUCTURE

What makes nuclear magnetic resonance spectroscopy such a powerful tool for structure determination is that *protons in different environments experience different degrees of shielding and have different chemical shifts.* In compounds of the type CH_3X, for example, the shielding of the methyl protons increases as X becomes less electronegative.

Increased shielding of methyl protons
Decreasing electronegativity of attached atom →

	CH_3F	CH_3OCH_3	$(CH_3)_3N$	CH_3CH_3
	Methyl fluoride	Dimethyl ether	Trimethylamine	Ethane
Chemical shift of methyl protons (δ), ppm:	4.3	3.2	2.2	0.9

A similar trend is seen in the methyl halides, where the protons in CH_3F are the least shielded (δ = 4.3 ppm) and those of CH_3I (δ = 2.2 ppm) the most.

An electronegative substituent decreases the electron density at the methyl carbon, and decreases the shielding of the methyl protons. The effects are cumulative, as the chemical shift data for the various chlorinated derivatives of methane indicate:

	$CHCl_3$	CH_2Cl_2	CH_3Cl
	Chloroform (trichloromethane)	Methylene chloride (dichloromethane)	Methyl chloride (chloromethane)
Chemical shift (δ), ppm:	7.3	5.3	3.1

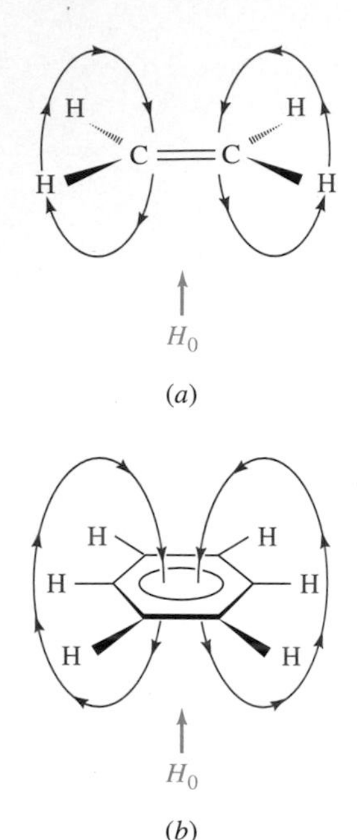

FIGURE 13.9 The induced magnetic field of the π electrons of (a) an alkene and (b) an arene reinforces the applied fields in the regions where vinyl and aryl protons are located.

Vinyl protons in alkenes and aryl protons in arenes are substantially less shielded than protons in alkanes:

	Benzene	Ethylene	Ethane (CH_3CH_3)
Chemical shift (δ), ppm:	7.3	5.3	0.9

One reason for the decreased shielding of vinyl and aryl protons is related to the directional properties of the induced magnetic field of the π electrons. As Figure 13.9 shows, the induced magnetic field due to the π electrons is just like that due to electrons in σ bonds; it opposes the applied magnetic field. However, all magnetic fields close upon themselves, and protons attached to a carbon-carbon double bond or an aromatic ring lie in a region where the induced field reinforces the applied field. This decreases the shielding of vinyl and aryl protons.

A similar, although much smaller, effect of π electron systems is seen in the chemical shifts of benzylic and allylic hydrogens. The methyl hydrogens in hexamethylbenzene and in 2,3-dimethyl-2-butene are less shielded than those in ethane.

	Hexamethylbenzene	2,3-Dimethyl-2-butene
Chemical shift (δ), ppm:	2.2	1.7

Table 13.1 collects chemical-shift information for protons of various types. Within each type, methyl (CH_3) protons are more shielded than methylene (CH_2) protons, and methylene protons are more shielded than methine (CH) protons. These differences are small—only about 0.7 ppm separates a methyl proton from a methine proton of the

TABLE 13.1
Chemical Shifts of Representative Types of Protons

Type of proton	Chemical shift (δ), ppm*	Type of proton	Chemical shift (δ), ppm*
H—C—R	0.9–1.8	H—C—NR	2.2–2.9
H—C—C=C	1.6–2.6	H—C—Cl	3.1–4.1
H—C—C(=O)—	2.1–2.5	H—C—Br	2.7–4.1
H—C≡C—	2.5	H—C—O	3.3–3.7
H—C—Ar	2.3–2.8	H—NR	1–3†
H—C=C<	4.5–6.5	H—OR	0.5–5†
H—Ar	6.5–8.5	H—OAr	6–8†
H—C(=O)—	9–10	H—OC(=O)—	10–13†

* Approximate values relative to tetramethylsilane; other groups within the molecule can cause a proton signal to appear outside of the range cited.

† The chemical shifts of protons bonded to nitrogen and oxygen are temperature- and concentration-dependent.

same type. Overall, proton chemical shifts among common organic compounds encompass a range of about 12 ppm. The protons in alkanes are the most shielded, while O—H protons of carboxylic acids are the least shielded.

The ability of an nmr spectrometer to separate signals that have similar chemical shifts is termed its *resolving power* and is directly related to the magnetic field strength of the instrument. Two closely spaced signals are resolved better when the spectrum is recorded using a 300-MHz spectrometer than a 60-MHz one. (Remember, though, that the chemical shift δ, cited in ppm, is independent of the field strength.)

13.7 INTERPRETING PROTON NMR SPECTRA

Analyzing an nmr spectrum in terms of a unique molecular structure makes use of the information contained in Table 13.1. By knowing the chemical shifts characteristic of various proton environments, the presence of a particular structural unit in an unknown compound may be inferred. An nmr spectrum also provides other information that aids the task of structure determination, including

1. The number of signals
2. The intensity of the signals, as measured by the area under each peak
3. The multiplicity, or *splitting,* of each signal

Protons that have different chemical shifts are said to be **chemical-shift–nonequivalent** (or **chemically nonequivalent**). A separate nmr signal is given for each chemical-shift–nonequivalent proton in a substance. The proton (^1H) nmr spectrum of chloromethyl methyl ether, $ClCH_2OCH_3$, shown in Figure 13.10, contains two peaks. One peak corresponds to the methylene protons, the other to the methyl protons. The methylene group bears two electronegative substituents, a chlorine and an oxygen, and so is less shielded than the methyl group, which bears only an oxygen. The signal that corresponds to the methylene protons appears at $\delta = 5.5$ ppm; the signal corresponding to the methyl protons is at $\delta = 3.5$ ppm.

Another way to assign the peaks in the nmr spectrum of chloromethyl methyl ether is by comparing their intensities. The three equivalent protons of the methyl group give rise to a more intense peak than the two equivalent protons of the methylene group. Intensities, as measured by peak areas, are proportional to the number of equivalent protons responsible for the signal. The area of the methyl signal in chloromethyl methyl ether is 50 percent greater than that of the methylene signal.

Peak areas are measured electronically. The spectrum is recorded in the usual manner and the nmr spectrometer is then switched from the normal mode to the integral mode. As the spectrum is scanned a second time, the instrument continuously adds the areas of all the peaks and superposes it as a series of steps on the original spectrum. The height of each step is proportional to the area under the peak and proportional to the number of protons responsible for the peak.

Integrated areas are relative; a 1.5 : 1 ratio of areas is just as consistent with a 6 : 4 ratio of protons as with a 3 : 2 ratio of protons.

Figure 13.11 shows the ^1H nmr spectrum of *p*-xylene. Like the spectrum of chloromethyl methyl ether (Figure 13.10), it exhibits two peaks in a 3 : 2 ratio. The six methyl protons of *p*-xylene are equivalent to one another and give rise to a single peak. Similarly, the four aryl protons are equivalent to one another.

Protons are equivalent to one another and have the same chemical shift when they are in equivalent environments. Often it is an easy matter to decide, simply by inspection of the chemical formula, when protons are equivalent to one another. In cases

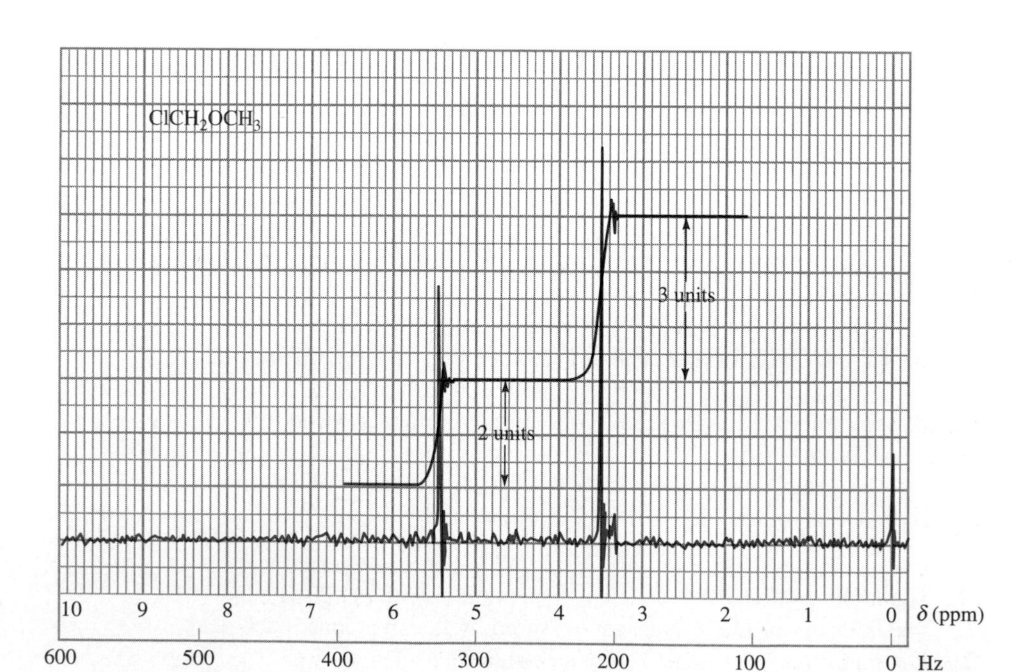

FIGURE 13.10 The 60-MHz ^1H nmr spectrum of chloromethyl methyl ether ($ClCH_2OCH_3$). Both the normal spectrum (blue) and its integral (red) are shown.

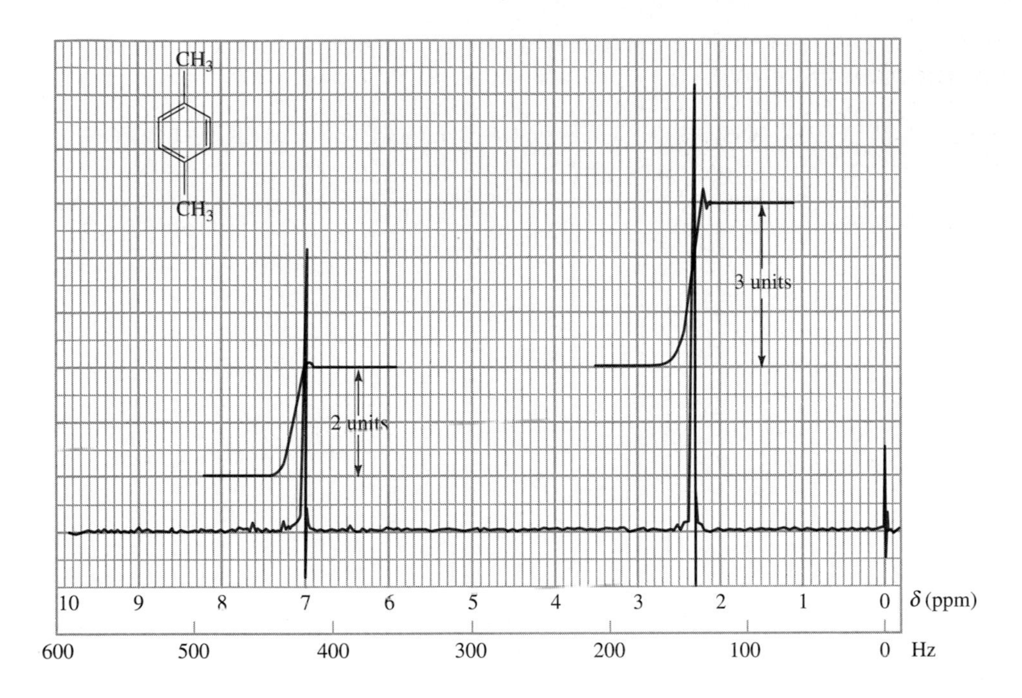

FIGURE 13.11 The 60-MHz ^1H nmr spectrum of *p*-xylene. Integrated areas are relative rather than absolute. The 6:4 ratio between methyl and aryl protons is reflected in a 1.5:1 ratio of peak areas.

where it is more difficult, a procedure involving mental replacement of a proton in a molecule by a ''test group'' is helpful. The procedure is illustrated here for the simple case of propane. To test for chemical equivalence, mentally replace one of the methyl protons of propane by chlorine to give 1-chloropropane. Replace a proton in the other methyl group by chlorine; the product is again 1-chloropropane.

$CH_3CH_2CH_3$	$ClCH_2CH_2CH_3$	$CH_3CH_2CH_2Cl$
Propane	1-Chloropropane	1-Chloropropane

If the two structures produced by mental replacement of two different hydrogens in a molecule by a test group are the same, the hydrogens are chemically equivalent. Thus, the six methyl protons of propane are all chemically equivalent to one another and have the same chemical shift.

Replacement of either one of the methylene protons of propane generates 2-chloropropane. Both methylene protons are equivalent. Neither of them is equivalent to any of the methyl protons.

The ^1H nmr spectrum of propane contains two signals, one for the six equivalent methyl protons, the other for the pair of equivalent methylene protons.

PROBLEM 13.2 How many signals would you expect to find in the ^1H nmr spectrum of each of the following compounds?

(*a*) 1-Bromobutane
(*b*) 1-Butanol
(*c*) Butane
(*d*) 1,4-Dibromobutane
(*e*) 2,2-Dibromobutane
(*f*) 2,2,3,3-Tetrabromobutane
(*g*) 1,1,4-Tribromobutane
(*h*) 1,1,1-Tribromobutane

SAMPLE SOLUTION (*a*) To test for chemical-shift equivalence, replace the protons at C-1, C-2, C-3, and C-4 of 1-bromobutane by some test group such as chlorine. Four constitutional isomers result:

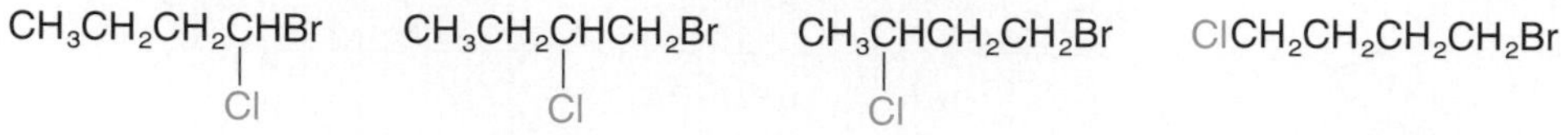

$CH_3CH_2CH_2CHBr$ (Cl on C-1)	$CH_3CH_2CHCH_2Br$ (Cl on C-2)	$CH_3CHCH_2CH_2Br$ (Cl on C-3)	$ClCH_2CH_2CH_2CH_2Br$
1-Bromo-1-chlorobutane	1-Bromo-2-chlorobutane	1-Bromo-3-chlorobutane	1-Bromo-4-chlorobutane

Thus, separate signals will be seen for the protons at C-1, C-2, C-3, and C-4. Barring any accidental overlap, we expect to find four signals in the nmr spectrum of 1-bromobutane.

Chemical-shift nonequivalence can occur when two environments are stereochemically different. The two vinyl protons of 2-bromopropene have different chemical shifts.

Br and H_3C on one carbon of C=C; H ($\delta = 5.3$ ppm, cis to Br) and H ($\delta = 5.5$ ppm, trans to Br) on the other.

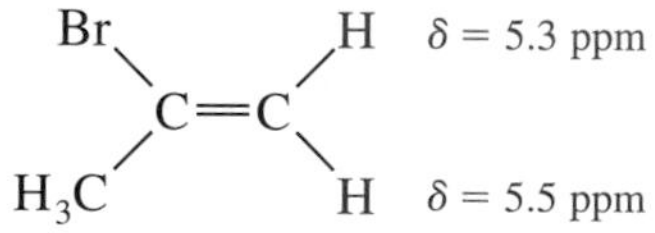

2-Bromopropene

One of the vinyl protons is cis to bromine; the other trans. Replacing one of the vinyl protons by some test group, say, chlorine, gives the *Z* isomer of 2-bromo-1-chloropropene; replacing the other gives the *E* stereoisomer. The *E* and *Z* forms of 2-bromo-1-chloropropene are stereoisomers that are not enantiomers; they are diastereomers. Protons that yield diastereomers on being replaced by some test group are described as **diastereotopic.** *The vinyl protons of 2-bromopropene are diastereotopic.* Diastereotopic protons can have different chemical shifts. Because their environments are similar, however, this difference in chemical shift is usually slight and it sometimes happens that two diastereotopic protons accidentally have the same chemical shift. Recording the spectrum on a higher-field nmr spectrometer is often helpful in resolving signals with similar chemical shifts.

PROBLEM 13.3 How many signals would you expect to find in the ^1H nmr spectrum of each of the following compounds?

(*a*) Vinyl bromide
(*b*) 1,1-Dibromoethene
(*c*) *cis*-1,2-Dibromoethene
(*d*) *trans*-1,2-Dibromoethene
(*e*) Allyl bromide
(*f*) 2-Methyl-2-butene

SAMPLE SOLUTION (*a*) Each of the protons of vinyl bromide is unique and has a chemical shift different from the other two. The least shielded proton is attached to the carbon that bears the bromine. The pair of protons at C-2 are diastereotopic with respect to each other; one is cis to bromine while the other is trans to bromine. There are three proton signals in the nmr spectrum of vinyl bromide. Their observed chemical shifts are as indicated.

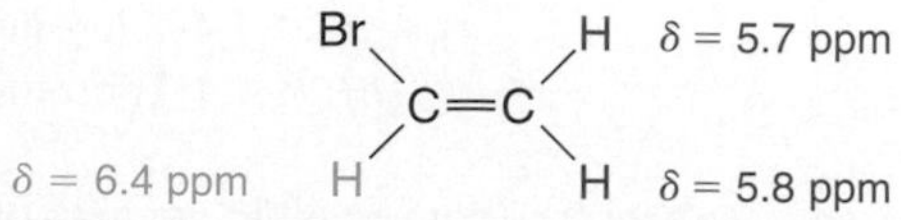

Br and H ($\delta = 6.4$ ppm) on one carbon of C=C; H ($\delta = 5.7$ ppm, cis to Br) and H ($\delta = 5.8$ ppm, trans to Br) on the other.

When enantiomers are generated by replacing first one proton and then another by a test group, the pair of protons are **enantiotopic** with respect to each other. The methylene protons at C-2 of 1-propanol, for example, are enantiotopic.

Enantiotopic protons

CH_3 — C — H, H, CH_2OH — 1-Propanol

CH_3 — C — Cl, H, CH_2OH — (*R*)-2-Chloro-1-propanol

CH_3 — C — H, Cl, CH_2OH — (*S*)-2-Chloro-1-propanol

Replacement of one of these protons by chlorine as a test group gives (*R*)-2 chloro-1-propanol; replacement of the other gives (*S*)-2-chloro-1-propanol. Enantiotopic protons have the same chemical shift, regardless of the field strength.

At the beginning of this section we noted that an nmr spectrum provides structural information based on chemical shift, the number of peaks, the intensities of the peaks as measured by their integrated areas, and the multiplicity, or splitting, of the peaks. We have discussed the first three of these features of ^1H nmr spectroscopy. Let us now direct our attention to peak splitting and see what kind of information it offers.

13.8 SPIN-SPIN SPLITTING IN NMR SPECTROSCOPY

Each signal in the nmr spectra of chloroform (Figure 13.8), chloromethyl methyl ether (Figure 13.10), and *p*-xylene (Figure 13.11) consists of a single sharp peak, referred to as a **singlet.** It is quite common to see nmr spectra, however, in which the signal due to a particular proton is not a singlet but instead appears as a collection of peaks. The signal may be split into two peaks (a **doublet**), three peaks (a **triplet**), four peaks (a **quartet**), and so on. Figure 13.12 shows the ^1H nmr spectrum of 1,1-dichloroethane.

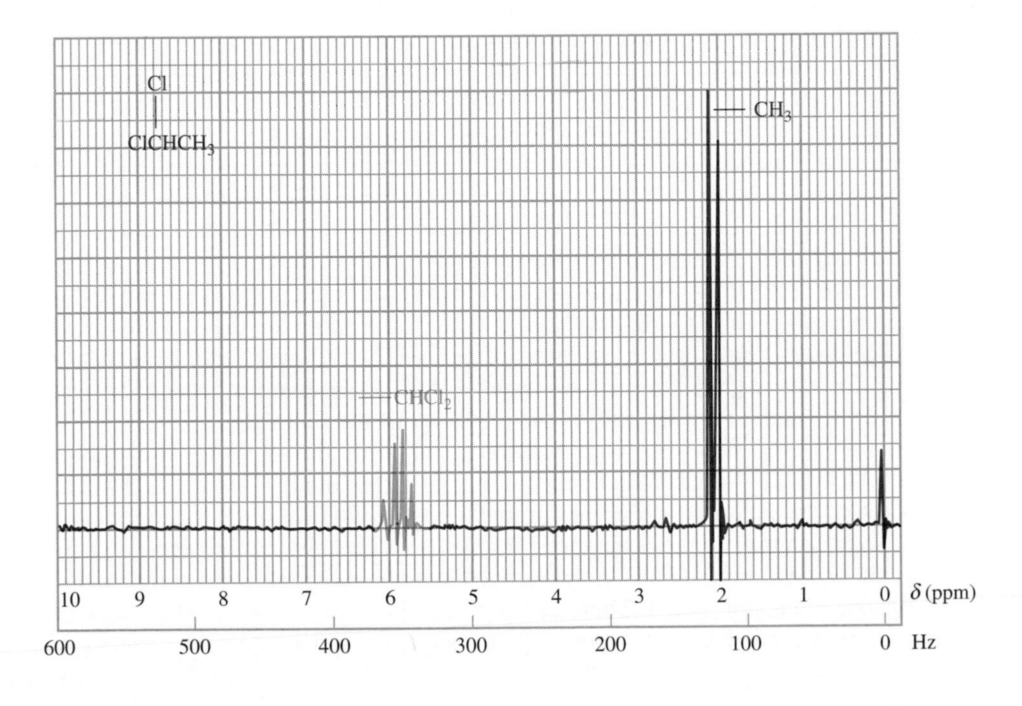

FIGURE 13.12 The 60-MHz ^1H nmr spectrum of 1,1-dichloroethane, showing the methine proton as a quartet and the methyl protons as a doublet.

The methyl protons appear as a doublet centered at $\delta = 2.0$ ppm, and the signal for the methine proton is a quartet at $\delta = 5.9$ ppm.

The number of peaks into which the signal for a particular proton is split is called its **multiplicity.** For simple cases the rule that allows us to predict splitting in ^{1}H nmr spectroscopy is

More complicated splitting patterns conform to an extension of the "$n + 1$" rule and will be discussed in Section 13.13.

$$\text{Multiplicity of signal for } H_a = n + 1$$

where n is equal to the number of protons that are vicinal to H_a. Two protons are vicinal to each other when they are bonded to adjacent atoms. Protons vicinal to H_a are separated from H_a by three bonds. The three methyl protons of 1,1-dichloroethane are vicinal to the methine proton and split its signal into a quartet. The single methine proton, in turn, splits the methyl protons' signal into a doublet.

This proton splits the signal for the methyl protons into a doublet.

$H-C(Cl)_2-CH_3$

These three protons split the signal for the methine proton into a quartet.

The physical basis for peak splitting can be described by examining the methyl doublet in the nmr spectrum of 1,1-dichloroethane. Splitting of the methyl signal into a doublet occurs because the chemical shift of the methyl protons is influenced by the spin of the methine proton on the adjacent carbon. (All the methyl protons in 1,1-dichloroethane are equivalent and all have the same chemical shift. Whatever we say about one of these methyl protons applies equally to the other two.)

Figure 13.13*a* recalls the fundamentals of an nmr experiment carried out at a constant frequency. There is a certain field strength H_0 at which the energy required to flip the spin of the methyl protons (designated H_a in the figure) is exactly equal to the energy of the radio-frequency source. The magnetic moment of the methine proton (designated H_b) can be aligned parallel (Figure 13.13*b*) or antiparallel (Figure 13.13*c*) to the applied field. The net field experienced by the methyl protons is the aggregate of the applied field, the induced fields due to the electrons, *and* the tiny magnetic field of the methine proton. When the magnetic moment of the methine proton is parallel to the applied field, as in Figure 13.13*b*, it reinforces it and slightly deshields the methyl protons. Thus, the magnetic field strength required to achieve resonance is less than H_0, and the signal due to the methyl protons appears at lower field. When the magnetic moment of the methine proton is antiparallel to the applied field, it opposes it, increasing the shielding of the methyl protons and causing a shift in their signal to higher field. Since both spin states are almost equally likely, the methyl protons of half the molecules undergo a shift to slightly higher field while those of the other half are deshielded by the methine proton and are shifted to slightly lower field. Instead of a single peak for the methyl protons, two peaks of equal intensity appear. The true chemical shift for the methyl protons is the midpoint of the doublet.

Turning now to the methine proton, its signal is split by the methyl protons into a quartet. The same kind of analysis applies here and is outlined in Figure 13.14. The methine proton "sees" eight different combinations of nuclear spins for the methyl protons. In one combination, the magnetic moments of all three methyl protons reinforce the applied field. At the other extreme, the magnetic moments of all three methyl protons oppose the applied field. There are three combinations in which the magnetic moments of two methyl protons reinforce the applied field while one opposes it.

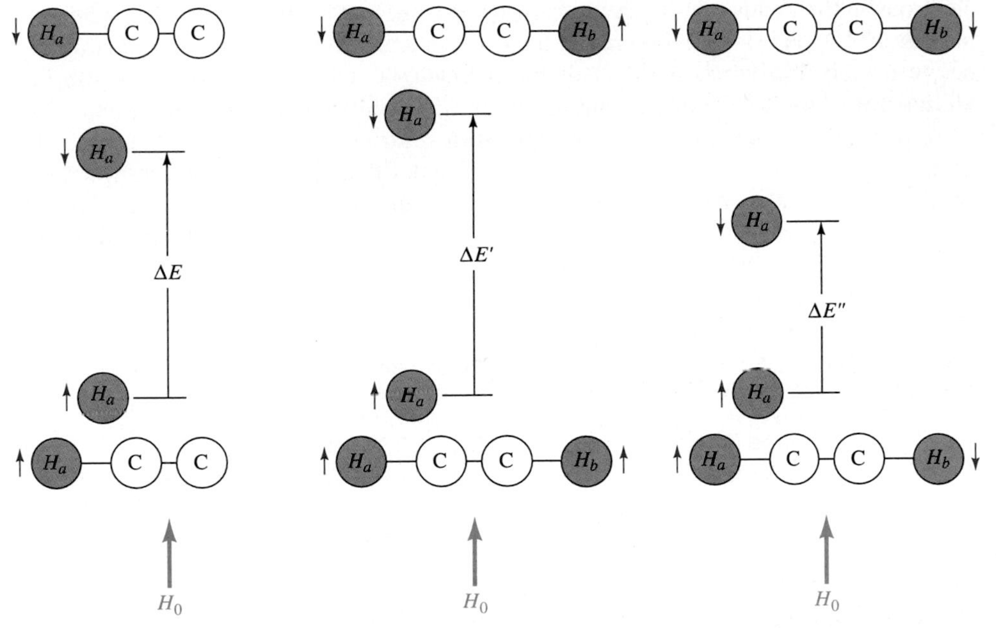

(a) Energy equal to ΔE is required to flip the spin of nucleus H_a. This corresponds to an external magnetic field strength of H_0.

(b) Magnetic moment of H_b is parallel to direction of external magnetic field and deshields nucleus H_a. A weaker external field is sufficient to make $\Delta E' = \Delta E$.

(c) Magnetic moment of H_b is antiparallel to direction of external magnetic field and shields nucleus H_a. A stronger external field is required in order to make $\Delta E'' = \Delta E$.

FIGURE 13.13 Diagram showing how two possible spin orientations of nucleus H_b split the signal of nucleus H_a into a doublet.

Finally, there are three combinations in which the magnetic moments of two methyl protons oppose the applied field and one reinforces it. These eight possible combinations give rise to four distinct peaks for the methine proton, with a ratio of intensities of 1:3:3:1.

We describe the observed splitting of nmr signals as **spin-spin splitting** and the physical basis for it as **spin-spin coupling.** It has its origin in the communication of nuclear spin information between nuclei. This information is transmitted by way of the

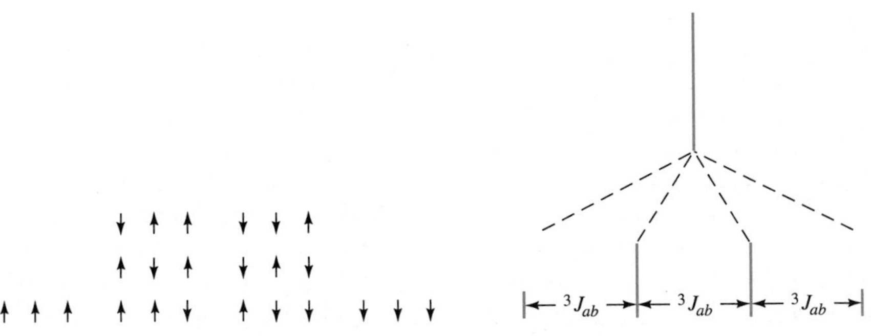

There are eight possible combinations of the nuclear spins of the three methyl protons in CH_3CHCl_2.

These eight combinations cause the signal of the $CHCl_2$ proton to be split into a quartet, in which the intensities of the peaks are in the ratio 1:3:3:1.

FIGURE 13.14 The methyl protons of 1,1-dichloroethane split the signal of the methine proton into a quartet.

electrons in the bonds that intervene between the nuclei. Its effect is greatest when the number of bonds is small. Vicinal protons are separated by three bonds, and coupling between vicinal protons, as in 1,1-dichloroethane, is called **three-bond coupling** or **vicinal coupling.** Four-bond couplings are weaker and not normally observable.

A very important characteristic of spin-spin splitting is that protons that have the same chemical shift do not split each other's signal. Ethane, for example, shows only a single sharp peak in its nmr spectrum. Even though there is a vicinal relationship between the protons of one methyl group and those of the other, they do not split each other's signal, because they are equivalent.

PROBLEM 13.4 Describe the appearance of the ^1H nmr spectrum of each of the following compounds. How many signals would you expect to find, and into how many peaks will each signal be split?

(*a*) 1,2-Dichloroethane
(*b*) 1,1,1-Trichloroethane
(*c*) 1,1,2-Trichloroethane
(*d*) 1,2,2-Trichloropropane
(*e*) 1,1,1,2-Tetrachloropropane

SAMPLE SOLUTION (*a*) All the protons of 1,2-dichloroethane ($ClCH_2CH_2Cl$) are chemically equivalent and have the same chemical shift. Protons that have the same chemical shift do not split each other's signal, and so the nmr spectrum of 1,2-dichloroethane consists of a single sharp peak.

Mutual coupling of nuclei requires that they split each other's signal equally. The separation in hertz between the two halves of the methyl doublet in 1,1-dichloroethane is equal to the separation between any two adjacent peaks of the methine quartet. The extent to which two nuclei are coupled is known as the **coupling constant *J*** and in simple cases is equal to the separation between adjacent lines of the signal of a particular proton. The three-bond coupling constant $^3J_{ab}$ in 1,1-dichloroethane has a value of 7 Hz. *The size of the coupling constant is independent of the field strength;* the separation between adjacent peaks in 1,1-dichloroethane is 7 Hz irrespective of whether the spectrum is recorded at 60 MHz or 300 MHz.

13.9 PATTERNS OF SPIN-SPIN SPLITTING: THE ETHYL GROUP

At first glance splitting may seem to complicate the interpretation of nmr spectra. In fact, it makes structure determination easier because it provides additional information. It tells us how many protons are vicinal to a proton responsible for a particular signal. With practice, we learn to pick out characteristic patterns of peaks, associating them with particular structural types. One of the most common of these patterns is that of the ethyl group, represented in the nmr spectrum of ethyl bromide in Figure 13.15.

In compounds of the type CH_3CH_2X, especially where X is an electronegative atom or group, such as bromine in ethyl bromide, the ethyl group appears as a *triplet-quartet pattern.* The methylene proton signal is split into a quartet by coupling with the methyl protons. The signal for the methyl protons is a triplet because of vicinal coupling to the two protons of the adjacent methylene group.

$$Br—CH_2—CH_3$$

These two protons split the methyl signal into a triplet.

These three protons split the methylene signal into a quartet.

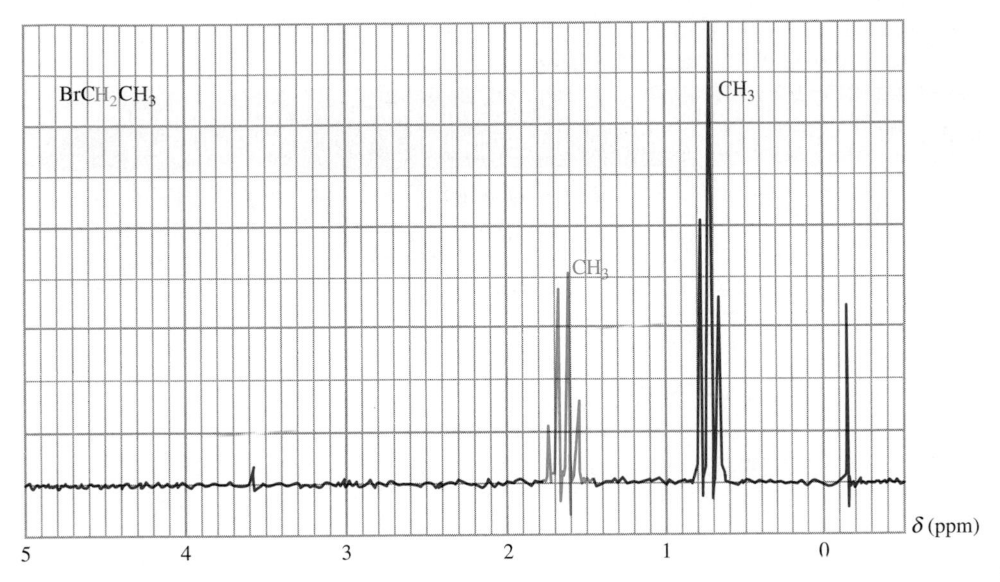

FIGURE 13.15 The 60-MHz ^1H nmr spectrum of ethyl bromide, showing the characteristic triplet-quartet pattern of an ethyl group.

We have discussed in the preceding section why methyl groups split the signals due to vicinal protons into a quartet. Splitting by a methylene group gives a triplet corresponding to the spin combinations shown in Figure 13.16 for ethyl bromide. The relative intensities of the peaks of this triplet are 1 : 2 : 1.

PROBLEM 13.5 Describe the appearance of the ^1H nmr spectrum of each of the following compounds. How many signals would you expect to find, and into how many peaks will each signal be split?

(*a*) $ClCH_2OCH_2CH_3$
(*b*) $CH_3CH_2OCH_3$
(*c*) $CH_3CH_2OCH_2CH_3$
(*d*) *p*-Diethylbenzene
(*e*) $ClCH_2CH_2OCH_2CH_3$

SAMPLE SOLUTION (*a*) Along with the triplet-quartet pattern of the ethyl group, the nmr spectrum of this compound will contain a singlet for the two protons of the chloromethyl group.

$ClCH_2—O—CH_2—CH_3$

Singlet; no protons vicinal to these; therefore, no splitting (→ $ClCH_2$)

Split into quartet by three protons of methyl group (→ CH_2)

Split into triplet by two protons of adjacent methylene group (→ CH_3)

Table 13.2 summarizes the splitting patterns and peak intensities expected for coupling to various numbers of protons. The intensities are equal to the coefficients of the binomial expansion.

The spectrum of ethyl bromide presented in Figure 13.15 illustrates a facet of splitting patterns that often occurs in nmr spectroscopy. The multiplets are not perfectly symmetrical but are skewed toward each other. The methylene quartet is not quite a 1 : 3 : 3 : 1 pattern, and the methyl triplet is not quite 1 : 2 : 1. The distortion becomes more pronounced as the chemical-shift difference between the coupled protons decreases, but decreases when the spectrum is recorded at higher field, 300 MHz rather than 60 MHz, for example.

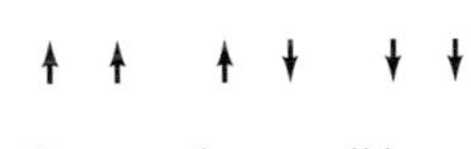

There are four possible combinations of the nuclear spins of the two methylene protons in CH_3CH_2Br.

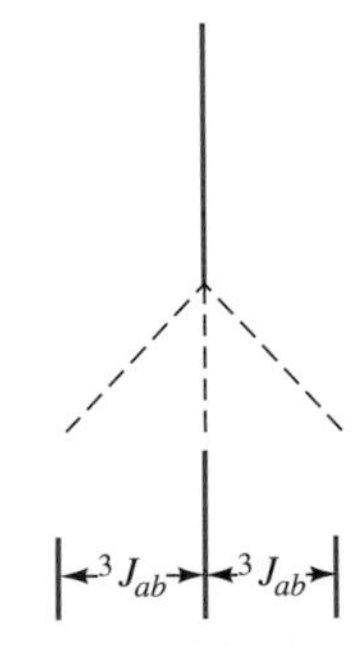

These four combinations cause the signal of the CH_3 protons to be split into a triplet, in which the intensities of the peaks are in the ratio 1:2:1.

FIGURE 13.16 The methylene protons of ethyl bromide split the signal of the methyl protons into a triplet.

TABLE 13.2
Splitting Patterns of Common Multiplets

Number of equivalent protons to which nucleus is coupled	Appearance of multiplet	Intensities of lines in multiplet
1	Doublet	1:1
2	Triplet	1:2:1
3	Quartet	1:3:3:1
4	Pentet	1:4:6:4:1
5	Hextet	1:5:10:10:5:1
6	Heptet	1:6:15:20:15:6:1

13.10 PATTERNS OF SPIN-SPIN SPLITTING: THE ISOPROPYL GROUP

The nmr spectrum of isopropyl chloride, shown in Figure 13.17, illustrates the appearance of an isopropyl group. The six equivalent methyl proton signals are split by the C-2 methine proton into a doublet at $\delta = 1.5$ ppm. The methine proton is equally coupled to all six methyl protons, and so its signal is split into a heptet. A *doublet-heptet* pattern is characteristic of an isopropyl group.

Sometimes, especially in spectra recorded at 60 MHz, the outermost lines of a heptet are too weak to be clearly seen.

13.11 PATTERNS OF SPIN-SPIN SPLITTING: PAIRS OF DOUBLETS

Another commonly encountered splitting pattern involves only two protons. Our simple splitting rules tell us that the signal for each proton will be split into a doublet because of coupling to the other. A representative example of this splitting pattern appears in the nmr spectrum of 2,3,4-trichloroanisole, shown in Figure 13.18.

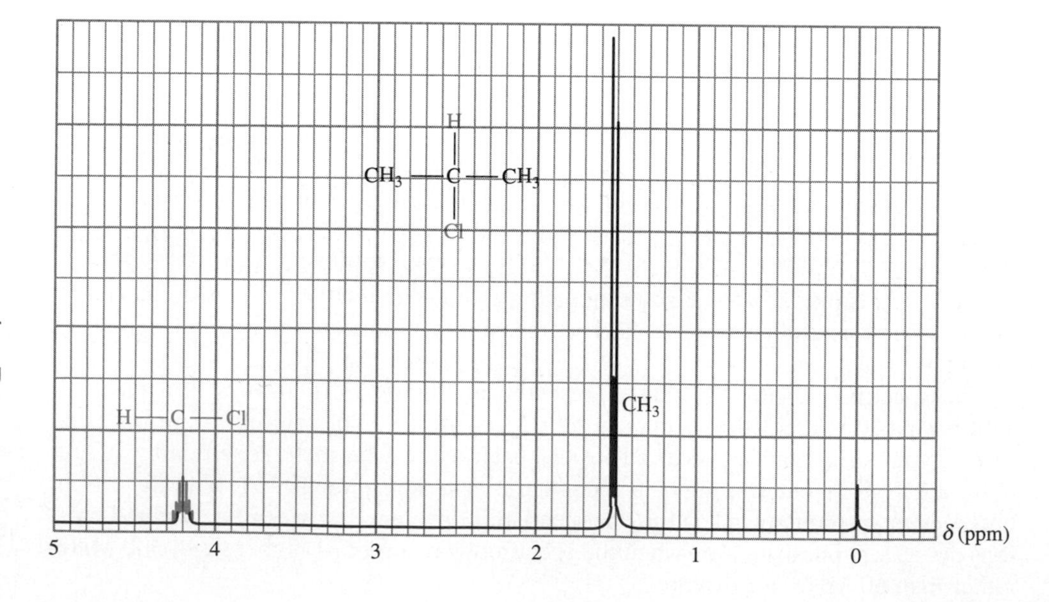

FIGURE 13.17 The 300-MHz ^1H nmr spectrum of isopropyl chloride, showing the doublet-heptet pattern of an isopropyl group. This is the first 300 MHz nmr spectrum depicted in this chapter. All of the earlier spectra were recorded at 60 MHz.

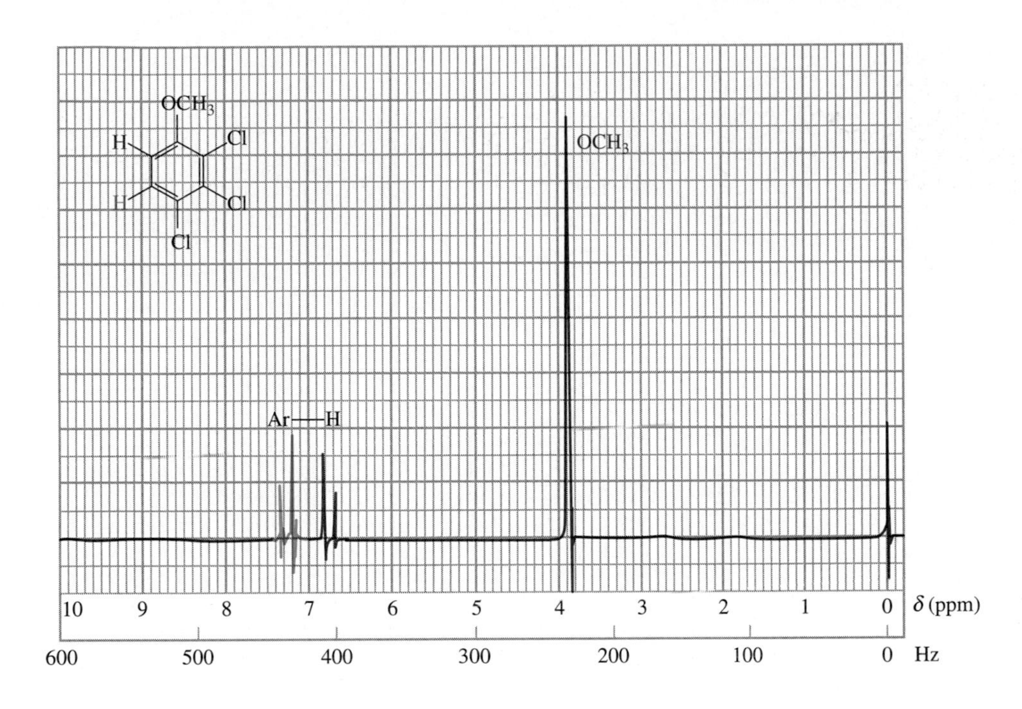

FIGURE 13.18 The 60-MHz ^{1}H nmr spectrum of 2,3,4-trichloroanisole, showing the splitting of the ring protons into a pair of doublets.

The singlet at $\delta = 3.9$ ppm corresponds to the protons of the —OCH_3 group. The two aryl protons are nonequivalent and mutually coupled; they give rise to two doublets at $\delta = 6.7$ ppm and $\delta = 7.25$ ppm. (The aryl proton that is ortho to the electron-releasing methoxyl group is slightly more shielded than the one meta to it.)

Doublet $\delta = 7.25$ ppm — H
Doublet $\delta = 6.7$ ppm — H
—OCH_3 Singlet $\delta = 3.9$ ppm
Cl, Cl, Cl

2,3,4-Trichloroanisole

In neither of the doublets is the intensity ratio of the peaks 1 : 1. Both doublets are skewed in the sense that the outermost two peaks are weaker than the innermost two. Skewing is minimal when the chemical-shift difference between two coupled protons is much larger than their coupling constant. Conversely, skewing is most pronounced when the chemical-shift difference is small.

PROBLEM 13.6 To which one of the following compounds does the nmr spectrum of Figure 13.19 correspond?

$ClCH_2C(Cl)(OCH_2CH_3)_2$ $\quad$ $Cl_2CHCH(OCH_2CH_3)_2$ $\quad$ $CH_3CH_2OCH(Cl)CH(Cl)OCH_2CH_3$

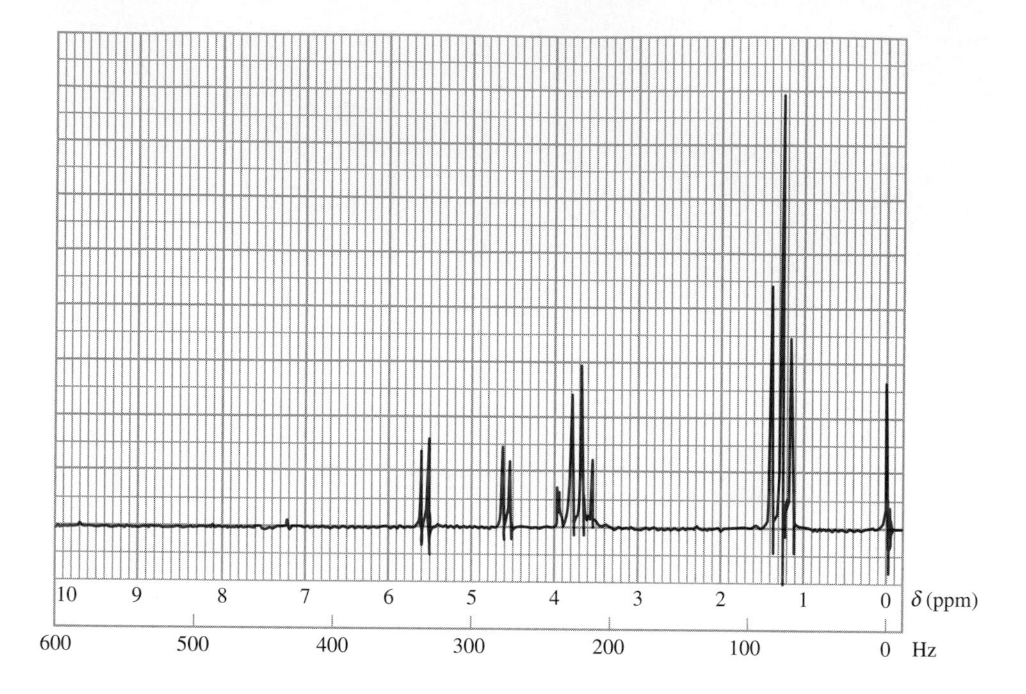

FIGURE 13.19 The 60-MHz ^1H nmr spectrum for Problem 13.6.

13.12 GEMINAL COUPLING

A pattern similar to that of vicinal coupling between two nonequivalent protons is seen in the mutual splitting of signals of *geminal* protons (protons bonded to the same carbon). Geminal protons are separated by two bonds, and geminal coupling is referred to as *two-bond coupling* (^2J) in the same way that vicinal coupling is referred to as *three-bond coupling* (^3J). An example of geminal coupling is provided by the compound 1-chloro-1-cyanoethene, where the two hydrogens appear as a pair of doublets. The splitting in each doublet is 2 Hz.

The protons in 1-chloro-1-cyanoethene are *diastereotopic* (Section 13.7). They are nonequivalent and have different chemical shifts. Remember, splitting can only occur between protons that have different chemical shifts.

Doublet

$^2J = 2$ Hz — $H_2C{=}C(Cl)(CN)$ — 1-Chloro-1-cyanoethene

Doublet

Splitting due to geminal coupling is seen only in CH_2 groups and only when the two protons are diastereotopic. All three protons of a methyl (CH_3) group are equivalent and cannot split one another's signal, and, of course, there are no protons geminal to a single methine (CH) proton.

13.13 COMPLEX SPLITTING PATTERNS

All the cases we have discussed to this point have involved splitting of a proton signal by coupling to other protons that were equivalent to one another. Indeed, we have stated the splitting rule in terms of the multiplicity of a signal as being equal to $n + 1$, where n is equal to the number of equivalent protons to which the proton that gives the signal is coupled. What about cases in which all the vicinal protons are not equivalent?

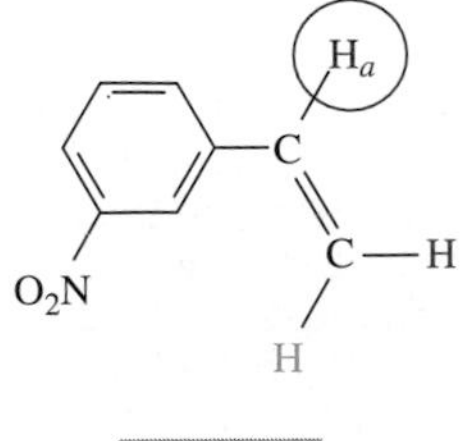

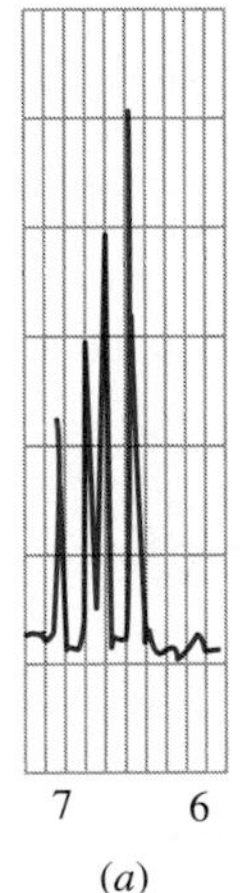

(a)

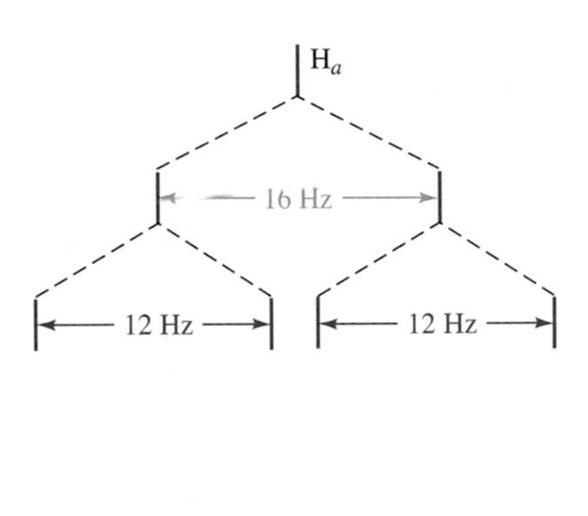

(b)

FIGURE 13.20 Splitting of a signal into a doublet of doublets by unequal coupling to two vicinal protons. (*a*) Appearance of the signal for the proton marked H_a in *m*-nitrostyrene as a set of four peaks. (*b*) Origin of these four peaks through successive splitting of the signal for H_a.

Figure 13.20*a* shows the signal for the proton marked $ArCH_a{=}CH_2$ in *m*-nitrostyrene, which appears as a set of four peaks in the range $\delta = 6.4$ to 6.9 ppm. These four peaks are in fact a "doublet of doublets." The proton in question is unequally coupled to the two protons at the end of the vinyl side chain. The size of the vicinal coupling constant between protons trans to each other on a double bond is normally larger than that between cis protons. In this case the trans coupling constant is 16 Hz and the cis coupling constant is 12 Hz. Thus, as shown in Figure 13.20*b*, the signal is split into a doublet with a spacing of 16 Hz by one vicinal proton, and each line of this doublet is then split into another doublet with a spacing of 12 Hz.

You will find it revealing to construct a splitting diagram similar to that of Figure 13.20 for the case in which the cis and trans **H**—C=C—**H** coupling constants are equal. Under those circumstances the four-line pattern simplifies to a triplet, as it should for a proton equally coupled to two vicinal protons.

PROBLEM 13.7 In addition to the proton marked H_a in *m*-nitrostyrene in Figure 13.20, there are two other vinylic protons. Assuming that the coupling constant between the two geminal protons in $ArCH{=}\mathbf{CH_2}$ is 2 Hz and the vicinal coupling constants are 12 Hz (cis) and 16 Hz (trans), describe the splitting pattern for each of these other two vinylic hydrogens.

The "$n + 1$ rule" should be amended to read: *When a proton* H_a *is coupled to* H_b, H_c, H_d, *etc., and* $J_{ab} \neq J_{ac}, \neq J_{ad}$, *etc., the original signal for* H_a *is split into* $n + 1$ *peaks by n* H_b *protons, each of these lines is further split into* $n + 1$ *peaks by n* H_c *protons, and each of these into* $n + 1$ *lines by n* H_d *protons, etc.*

PROBLEM 13.8 Describe the splitting pattern expected for the proton at

(*a*) C-2 in (*Z*)-1,3-dichloropropene

(*b*) C-2 in CH_3CHCH (with Br on C-2 and =O on the terminal C)

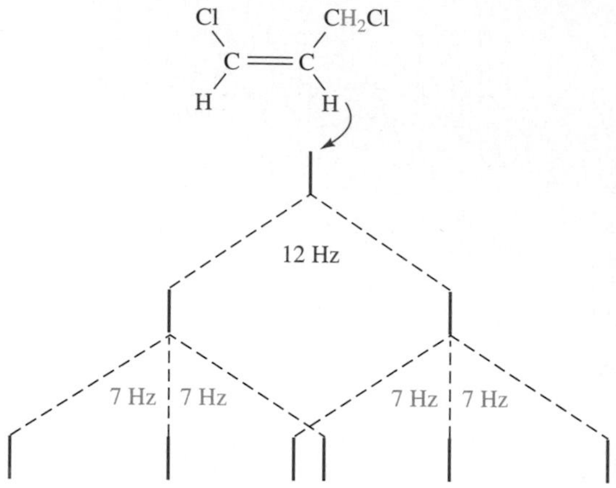

FIGURE 13.21 Splitting diagram for the signal corresponding to the proton at C-2 in (*Z*)-1,3-dichloropropene ($ClCH{=}\mathbf{CH}CH_2Cl$). The signal for this proton is split into a doublet by vicinal coupling to the proton cis to it on the double bond (^3J = 12 Hz). Each line of this doublet is split into a triplet by vicinal coupling to the two protons of the CH_2Cl group (^3J = 7 Hz). The signal appears as a set of six peaks; it is a triplet of doublets. Notice that the values of the coupling constants are such as to cause the two triplets to overlap.

SAMPLE SOLUTION (*a*) The signal of the proton at C-2 is split into a doublet by coupling to the proton cis to it on the double bond, and each line of this doublet is split into a triplet by the two protons of the CH_2Cl group.

This proton splits signal for proton at C-2 into a doublet.

Proton at C-2 appears as a doublet of triplets.

These protons split signal for proton at C-2 into a triplet.

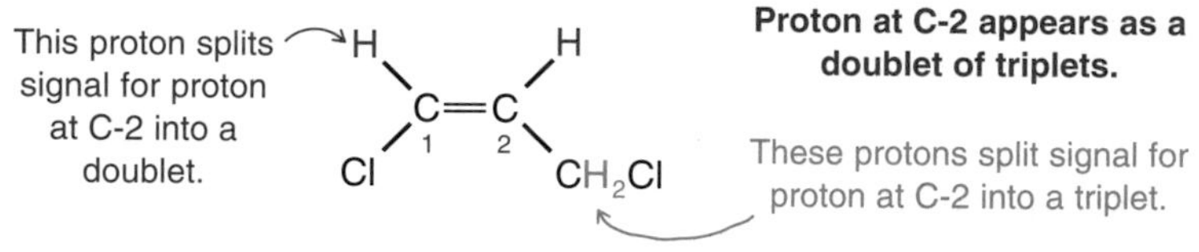

Figure 13.21 gives the splitting diagram for the C-2 proton based on the observed values of the coupling constants.

13.14 PROTON NMR SPECTRA OF ALCOHOLS

The hydroxyl proton of a primary alcohol RCH_2OH is vicinal to two protons, and its signal would be expected to be split into a triplet. Under certain conditions signal splitting of alcohol protons is observed, but usually it is not. Figure 13.22 presents the nmr spectrum of benzyl alcohol, showing the methylene and hydroxyl protons as singlets at δ = 4.7 and 1.8 ppm, respectively. (The aromatic protons also appear as a singlet, but that is because they all accidentally have the same chemical shift and so cannot split each other.)

The reason that splitting of the hydroxyl proton of an alcohol is not observed is that it is involved in rapid exchange reactions with other alcohol molecules. Transfer of the hydroxyl proton from an oxygen of one alcohol molecule to the oxygen of another is quite fast and effectively *decouples* it from other protons in the molecule. Factors that slow down this exchange of hydroxyl protons between molecules, such as diluting the solution, lowering the temperature, or increasing the crowding around the hydroxyl group, can cause splitting of hydroxyl resonances.

The chemical shift of the hydroxyl proton is variable, with a range of δ = 0.5 to 5 ppm, depending on the solvent, the temperature at which the spectrum is recorded, and the concentration of the solution. The alcohol proton shifts to lower field in more concentrated solutions.

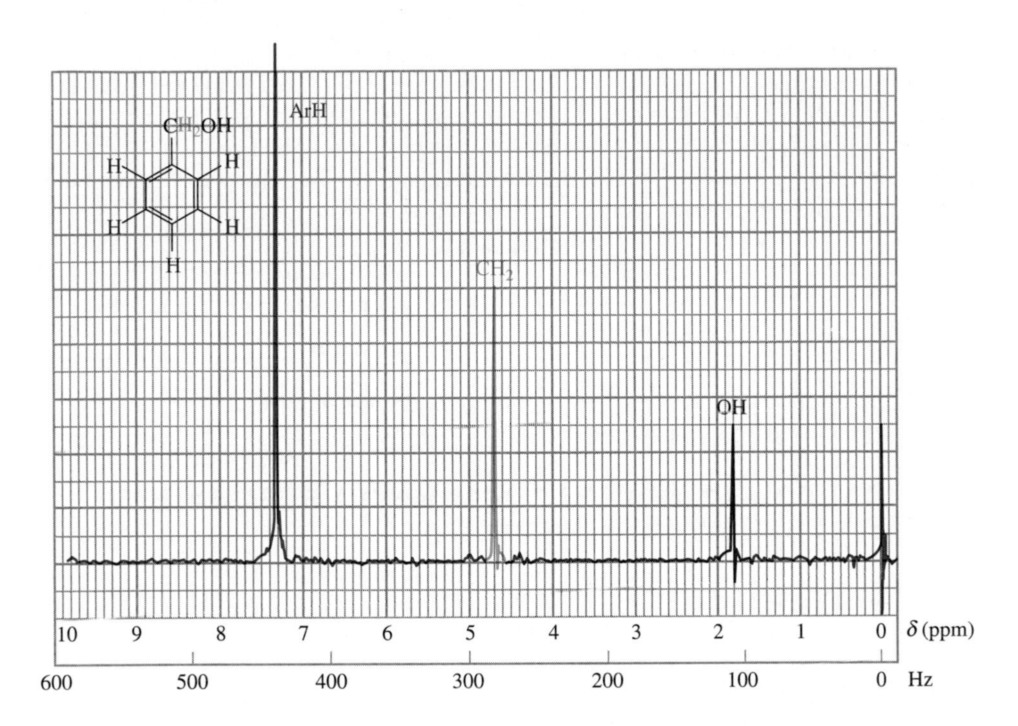

FIGURE 13.22 The 60-MHz ^{1}H nmr spectrum of benzyl alcohol. The hydroxyl proton and the methylene protons are vicinal but do not split each other because of the rapid intermolecular exchange of hydroxyl protons.

An easy way to verify that a particular signal belongs to a hydroxyl proton is to add D_2O. The hydroxyl proton is replaced by deuterium according to the equation:

$$RCH_2OH + D_2O \rightleftharpoons RCH_2OD + DOH$$

Deuterium does not give a signal under the conditions of proton nmr spectroscopy. Thus, exchange of a hydroxyl proton by deuterium leads to the disappearance of its nmr peak. Protons bonded to nitrogen and sulfur also undergo exchange with D_2O. Those bound to carbon normally do not, and so this technique is useful for assigning the proton resonances of —OH, >NH, and —SH groups.

13.15 NMR AND CONFORMATIONS

We know from Chapter 3 that the protons in cyclohexane exist in two different environments, axial and equatorial. The nmr spectrum of cyclohexane, however, shows only a single sharp peak at $\delta = 1.4$ ppm. All the protons of cyclohexane appear to be equivalent in the nmr spectrum. Why?

The answer is related to the very rapid rate of ring flipping in cyclohexane.

H_x H_y $\rightleftharpoons$ H_x H_y

One property of nmr spectroscopy is that it is too slow a technique to ''see'' the individual conformations of cyclohexane. What nmr sees is the *average* environment of the protons. Since chair-chair interconversion in cyclohexane converts each axial proton to an equatorial one and vice versa, the average environments of all the protons are the same. A single peak is observed that has a chemical shift midway between the true chemical shifts of the axial and the equatorial protons.

A similar effect occurs when measuring coupling constants. The magnitudes of vicinal coupling constants J depend on a variety of factors, including stereochemistry. Protons that are anti to each other have values of ^3J = 12 to 15 Hz, while those that are gauche have ^3J = 2 to 4 Hz. Most observed vicinal coupling constants, as measured from the splitting of peaks in the nmr spectrum, are on the order of 6 to 7 Hz, a value that reflects the weighted average of the coupling constants between two protons over all the conformations of the molecule.

13.16 CARBON-13 NUCLEAR MAGNETIC RESONANCE: THE SENSITIVITY PROBLEM

Magnetic resonance spectroscopy of nuclei other than protons is also possible. Two nuclei that have been extensively studied are ^{19}F and ^{31}P. Like ^1H, both ^{19}F and ^{31}P are the principal isotopes present in the ''natural'' state of each element, and both have nuclear spins of $\pm\frac{1}{2}$. What about carbon? When we use proton nmr as a tool for identifying organic compounds, we infer the structure of the carbon skeleton on the basis of the environments of the hydrogen substituents. Clearly, a spectroscopic method that examines the carbons directly would simplify the task of structure determination. The major isotopic form of carbon (^{12}C), however, has a nuclear spin of zero and is incapable of giving an nmr signal. The mass 13 isotope of carbon has a nuclear spin of $\pm\frac{1}{2}$ and so would be a suitable candidate for study by nmr were it not for a severe sensitivity problem. (By **sensitivity** we mean the ease with which signals may be detected under the conditions of measurement.) The nmr signal given by a ^{13}C nucleus is inherently much weaker than a proton signal, and in addition only 1.1 percent of all carbon atoms in a sample are ^{13}C. Thus, even with an nmr spectrometer properly tuned for ^{13}C magnetic resonance (15 MHz at 14,100 G), the peaks are too weak to be detected and are lost in the background noise. This hampered the development of ^{13}C nmr (cmr) spectroscopy as a routine technique for organic structure determination until a new generation of nmr spectrometers incorporating special sensitivity-enhancing features became available in the 1970s.

The strategy behind sensitivity enhancement rests on the fact that background noise is random but the signals of a particular sample, even though they may be weak, always appear at the same chemical shift regardless of how many times the spectrum is scanned. By scanning the spectrum of a compound hundreds of times, storing the spectra in a computer, and then adding all the spectra together, the spectrometer accumulates the nmr signals more than the background noise. An increase in the signal-to-noise ratio results.

Within the solution to the sensitivity problem lies another problem. All the 60-MHz ^1H nmr spectra that we have seen to this point have required several minutes for the instrument to scan from low field to high field. To record several hundred ^{13}C spectra under these conditions would take many hours or even days, and ^{13}C nmr spectroscopy would be a technique used only in special cases. Fortunately, there is another way to measure nmr spectra, a way in which a complete spectrum can be obtained in about 1 second and 1000 spectra can be obtained in 1000 seconds. This technique involves irradiating the sample with a brief but intense pulse of radio-frequency energy, excit-

Richard R. Ernst of the Swiss Federal Institute of Technology won the 1991 Nobel Prize in Chemistry for devising pulse-relaxation nmr techniques.

ing all the nuclei in the sample to their higher spin state. The excited nuclei then relax to their lower-energy state, and it is this process that is monitored. By a mathematical technique known as a *Fourier transform,* the data are converted to a spectrum that looks very much like one obtained in the usual manner. Instruments that perform in this way are known as **Fourier transform (FT) nmr spectrometers.** As a technique for sensitivity enhancement, FT nmr is practically essential for ^{13}C nmr. It is also used for most ^{1}H nmr spectra, and all the 300-MHz nmr spectra in this text were recorded on an FT instrument.

The raw data obtained from a ^{13}C nmr are analogous to the sound of an orchestra. Just as you can pick out the contribution of the violin section even when all the instruments in the orchestra are playing, Fourier transform analysis separates a collection of ^{13}C signals into peaks for individual carbons.

13.17 CARBON-13 NUCLEAR MAGNETIC RESONANCE: INTERPRETATION OF SPECTRA

Figure 13.23 shows three different nmr spectra for 1-chloropentane. Figure 13.23*a* and 13.23*b* are ^{1}H nmr spectra recorded at 60 and 300 MHz, respectively; Figure 13.23*c* is the ^{13}C spectrum. Comparing these three spectra will illustrate the effect of field strength on proton spectra and orient us in the appearance of ^{13}C spectra. Most importantly, it will demonstrate how much better ^{13}C nmr spectroscopy is than ^{1}H nmr in discriminating between nonequivalent nuclei.

First, notice the difference between the 60-MHz and 300-MHz ^{1}H nmr spectra. The only well-defined signal in the 60-MHz proton spectrum of 1-chloropentane (Figure 13.23*a*) is the triplet of the $—CH_2Cl$ group at $\delta = 3.5$ ppm. There is a badly skewed triplet at $\delta = 0.9$ ppm for the methyl group, while all the other proton signals appear as a broad multiplet at $\delta = 1$ to 2 ppm. The 300-MHz spectrum (Figure 13.23*b*) is better, but it is not completely resolved either. Both triplets are more clearly defined, and the cluster of peaks in the 1–2 ppm region has separated into two multiplets for the three nonequivalent methylene groups at C-2, C-3, and C-4 of $CH_3CH_2CH_2CH_2CH_2Cl$. The ^{13}C nmr spectrum in Figure 13.23*c*, on the other hand, consists of five well-defined peaks, one for each of the nonequivalent carbons of 1-chloropentane. Notice, too, the widely separated chemical shifts of these five peaks; they cover a range of over 30 ppm compared with less than 3 ppm for the proton signals of the same compound. In general, proton signals span a range of about 12 ppm; ^{13}C signals can span a range of over 200 ppm.

PROBLEM 13.9 How many signals would you expect to see in the ^{13}C nmr spectrum of each of the following compounds?

(*a*) Propylbenzene
(*b*) Isopropylbenzene
(*c*) 1,2,3-Trimethylbenzene
(*d*) 1,2,4-Trimethylbenzene
(*e*) 1,3,5-Trimethylbenzene

SAMPLE SOLUTION (*a*) The two ring carbons that are ortho to the propyl substituent are equivalent and so must have the same chemical shift. Similarly, the two ring carbons that are meta to the propyl group are equivalent to each other. The carbon atom para to the substituent is unique, as is the carbon that bears the substituent. Thus, there will be four signals for the ring carbons, designated *w, x, y,* and *z* in the structural formula. These four signals for the ring carbons added to those for the three nonequivalent carbons of the propyl group yield a total of *seven* signals.

x y
w z $—CH_2CH_2CH_3$ Propylbenzene
x y

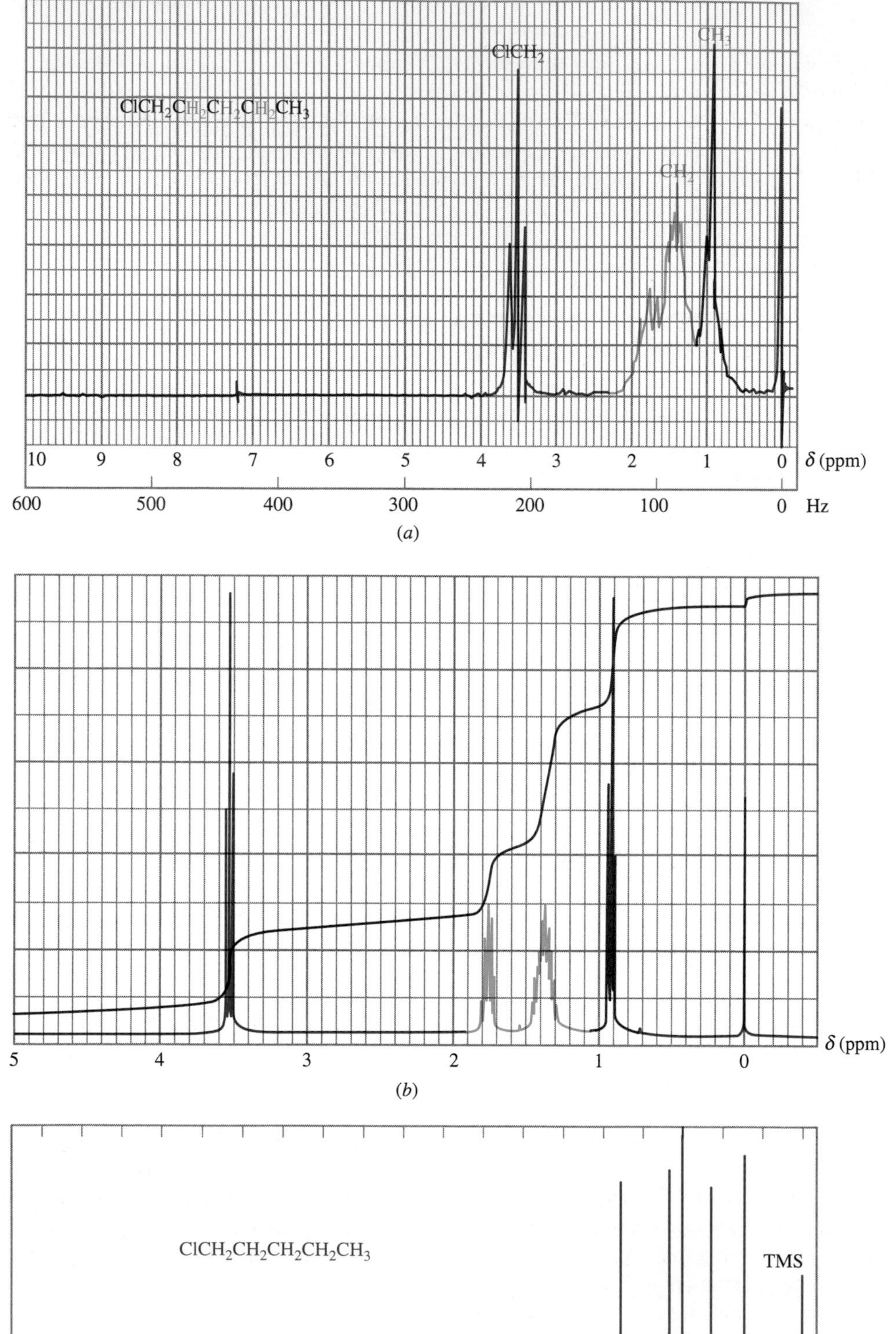

FIGURE 13.23 The (*a*) 60-MHz and (*b*) 300-MHz ^{1}H nmr spectra of 1-chloropentane. (*c*) The ^{13}C nmr spectrum of 1-chloropentane. *(The ^{13}C spectrum is taken from Carbon-13 NMR Spectra: A Collection of Assigned, Coded, and Indexed Spectra, by LeRoy F. Johnson and William A. Jankowski, Wiley-Interscience, New York, 1972. Reprinted by permission of John Wiley & Sons, Inc.)*

TABLE 13.3
Chemical Shifts of Representative Carbons

Type of carbon	Chemical shift (δ) ppm*	Type of carbon	Chemical shift (δ) ppm*
RCH_3	0–35	>C=C<	100–150
R_2CH_2	15–40		
RCH_2Br	20–40		
R_3CH	25–50	benzene ring	110–175
RCH_2Cl	25–50		
RCH_2NH_2	35–50	>C=O	190–220
RCH_2OH	50–65		
—C≡C—	65–90		

* Approximate values relative to tetramethylsilane.

Chemical shifts in ^1H nmr are measured relative to the *protons* of tetramethylsilane; chemical shifts in ^{13}C nmr are measured relative to the *carbons* of tetramethylsilane. Table 13.3 lists typical chemical-shift ranges for some representative types of carbon atoms.

A second aspect of the ^{13}C nmr spectrum of Figure 13.23*c* is that all the peaks are singlets. The reason for the lack of $^{13}C—^{13}C$ coupling is that only 1 percent of all the carbons in any given sample are ^{13}C. The probability that two ^{13}C atoms are present in the same molecule is quite small, and the probability that they are separated by only a few bonds is smaller still. The lack of splitting due to $^{13}C—H$ coupling, however, reflects a deliberate decision to measure the spectrum under conditions that suppress such splitting. A ^{13}C nucleus can couple strongly to protons directly attached to it as well as to protons separated from it by two, three, or even more bonds. The myriad of $^{13}C—H$ couplings that would result would make the spectrum too complicated for easy interpretation. A technique known as **broadband decoupling** removes all the proton-carbon couplings, producing the kind of spectrum shown in Figure 13.23*c*.

There is another technique, known as **off-resonance decoupling,** that removes all the carbon-proton couplings except those involving nuclei that are directly bonded. Thus, a methine (CH) carbon appears as a doublet, a methylene (CH_2) carbon as a triplet, and a methyl (CH_3) carbon as a quartet. Carbons that bear no hydrogen substituents are called *quaternary* carbons and appear as a singlet. Typically, a carbon spectrum is recorded in the broadband-decoupled mode, signals are tentatively assigned on the basis of their chemical shifts, and then an off-resonance–decoupled spectrum is taken to confirm the assignments. In the off-resonance–decoupled spectrum of 1-chloropentane, all the signals are triplets except for the peak at $\delta = 13.9$ ppm, which is a quartet and must correspond to the methyl group.

PROBLEM 13.10 Into how many peaks would each of the signals be split in the off-resonance–decoupled ^{13}C nmr spectrum of each of the following compounds?

(*a*) Propylbenzene
(*b*) Isopropylbenzene
(*c*) 1,2,3-Trimethylbenzene
(*d*) 1,2,4-Trimethylbenzene
(*e*) 1,3,5-Trimethylbenzene

SAMPLE SOLUTION (*a*) The number of peaks into which the signal for a particular carbon is split is 1 more than the number of protons bonded to it. The ring carbon that bears the propyl group has no hydrogen substituents and so appears as a singlet. All the other ring carbons bear one hydrogen, and each is split into a doublet. The two methylene groups of the side chain are each split into a triplet by their two hydrogens, and the methyl group appears as a quartet because of splitting by three hydrogens.

Propylbenzene: ring carbons *d*, *d*, *d*, *d*, *d*; ipso carbon *s*; $CH_2CH_2CH_3$ (*t*, *t*, *q*)

(*s* = singlet; *d* = doublet; *t* = triplet; *q* = quartet)

The last feature of ^{13}C nmr spectra to be pointed out concerns peak intensities. Spectra measured by the pulse relaxation technique used in ^{13}C FT nmr spectroscopy are subject to distortion of their signal intensities. This distortion weakens the signal for carbons that do not have hydrogen substituents, as the ^{13}C nmr spectrum of *m*-methylphenol (*m*-cresol) shown in Figure 13.24 illustrates. In addition to the signal for the methyl group at $\delta = 21$ ppm, there are six signals for the six aromatic carbons in the range $\delta = 112$ to 155 ppm. Two of these peaks are clearly smaller than the others and correspond to the ring carbons that bear the methyl group and the hydroxyl group.

PROBLEM 13.11 Which one of the compounds of Problems 13.9 and 13.10 corresponds to the broadband-decoupled ^{13}C nmr spectrum of Figure 13.25?

Both ^{13}C and ^{1}H nmr are powerful techniques for organic structure determination. They share the quality of being "nondestructive"; that is, the sample can be recovered in its entirety after the spectrum is taken. Thus, one never faces the problem of choosing whether to take a proton spectrum or a carbon spectrum—both can be recorded in successive experiments on the same sample. The two forms of magnetic resonance spectroscopy provide complementary information. In the next section, we will examine another nondestructive analytical method, infrared spectroscopy.

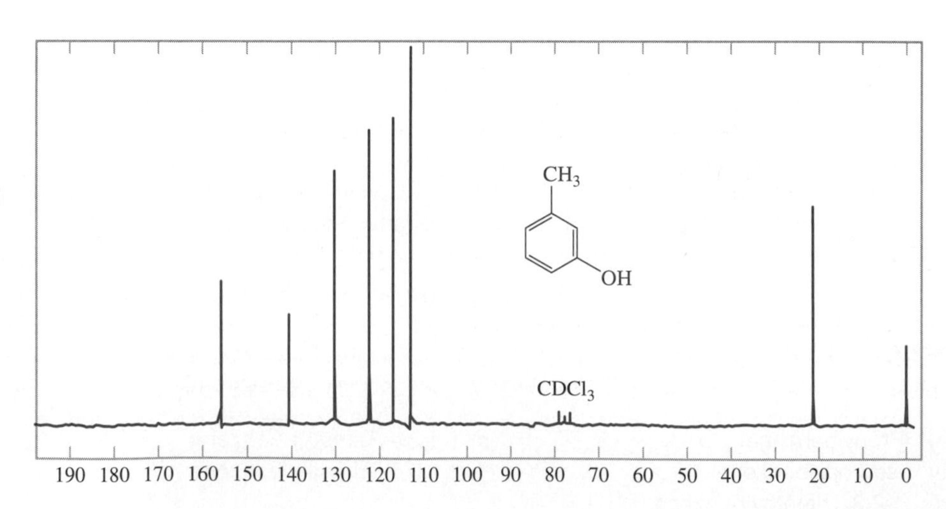

FIGURE 13.24 The ^{13}C nmr spectrum of *m*-cresol. The two weakest signals (at 140 and 157 ppm) correspond to the ring carbons that are not directly attached to a hydrogen. *(The ^{13}C spectrum is taken from Carbon-13 NMR Spectra: A Collection of Assigned, Coded, and Indexed Spectra, by LeRoy F. Johnson and William C. Jankowski, Wiley-Interscience, New York, 1972. Reprinted by permission of John Wiley & Sons, Inc.)*

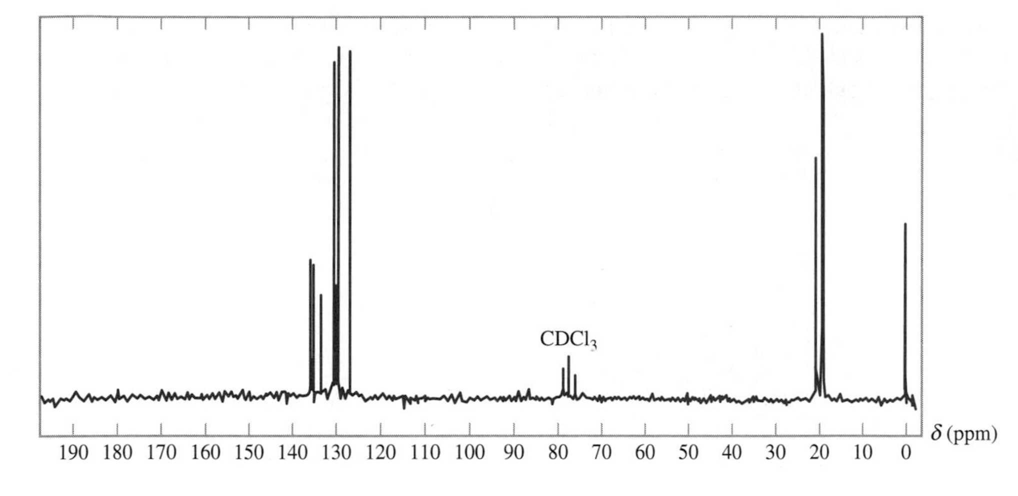

FIGURE 13.25 The ^{13}C nmr spectrum of the unknown compound of Problem 13.11. *(The ^{13}C spectrum is taken from Carbon-13 NMR Spectra: A Collection of Assigned, Coded, and Indexed Spectra, by LeRoy F. Johnson and William C. Jankowski, Wiley-Interscience, New York, 1972. Reprinted by permission of John Wiley & Sons, Inc.)*

MAGNETIC RESONANCE IMAGING

Like all photographs, a chest x-ray is a two-dimensional projection of a three-dimensional object. It is literally a collection of shadows produced by all the organs that lie between the source of the x-rays and the photographic plate. The clearest images in a chest x-ray are not the lungs (the customary reason for taking the x-ray in the first place) but rather the ribs and backbone. It would be desirable if we could limit x-ray absorption to two dimensions at a time rather than three. This is, in fact, what is accomplished by a technique known as *computerized axial tomography,* which yields its information in a form called a CT (or CAT) scan. With the aid of a computer, a CT scanner controls the movement of an x-ray source and detector with respect to the patient and to each other, stores the x-ray absorption pattern, and converts it to an image that is equivalent to an x-ray photograph of a thin section of tissue. It is a *noninvasive* diagnostic method, meaning that surgery is not involved nor are probes inserted into the patient's body.

As useful as the CT scan is, it has some drawbacks. Prolonged exposure to x-rays is harmful, and CT scans often require contrast agents to make certain organs more opaque to x-rays. Some patients are allergic to these contrast agents. An alternative technique was introduced in the 1980s that is not only safer but potentially more versatile than x-ray tomography. This technique is called *magnetic resonance imaging,* or MRI.

MRI is an application of nuclear magnetic resonance spectroscopy that makes it possible to examine the inside of the human body using radio-frequency radiation, which is lower in energy (Figure 13.1) and less damaging than x-rays and requires no imaging or contrast agents. By all rights MRI should be called NMRI, but the word *nuclear* was dropped from the name so as to avoid confusion with nuclear medicine, which involves radioactive isotopes.

While the technology of an MRI scanner is rather sophisticated, it does what we have seen other nmr spectrometers do; it detects protons. Thus, MRI is especially sensitive to biological materials such as water and lipids that are rich in hydrogen. Figure 13.26 shows an example of the use of MRI to detect a brain tumor. Regions of the image are lighter or darker according to the relative concentration of protons and to their environments.

MRI is such a new technique that using it as a substitute for x-ray tomography is only the first of what promises to be numerous medical applications. If, for example, the rate of data acquisition could be increased, then it would become possible to make the leap from the equivalent of still photographs to motion pictures. One could watch the inside of the body as it works—see the heart beat, see the lungs expand and contract—rather than merely examine the structure of an organ.

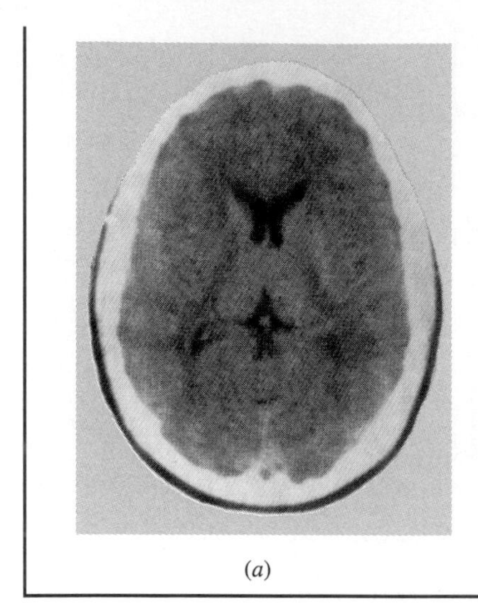

(a)

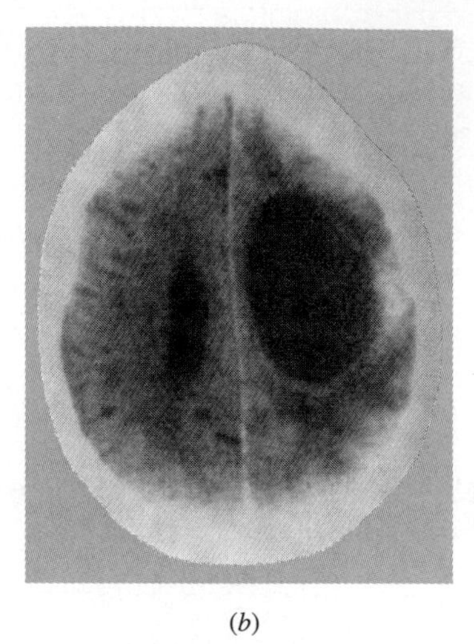

(b)

FIGURE 13.26 A magnetic resonance image of a two-dimensional cross section of (*a*) a normal brain and (*b*) a brain showing a tumor, indicated by the dark shadow. *(Photographs courtesy of NIH.)*

13.18 INFRARED SPECTROSCOPY

Before the advent of nmr spectroscopy, infrared (ir) spectroscopy was the instrumental method most often applied to structure determination of organic compounds. While nmr spectroscopy is, in general, more revealing of the structure of an unknown compound, ir still retains an important place in the chemist's inventory of spectroscopic methods because of its usefulness in identifying the presence of certain *functional groups* within a molecule.

Infrared radiation comprises the portion of the electromagnetic spectrum (Figure 13.1) between microwaves and visible light. The fraction of the infrared region of most use for structure determination lies between 2.5×10^{-6} m and 16×10^{-6} m in wavelength. Two derived units that are commonly employed in infrared spectroscopy are the *micrometer* and the *wave number.* One micrometer (μm) is 10^{-6} m, and infrared spectra record the region from 2.5 μm to 16 μm. Wave numbers are reciprocal centimeters (cm^{-1}), so that the region 2.5 to 16 μm corresponds to 4000 to 625 cm^{-1}. An advantage to using wave numbers is that they are directly proportional to energy. Thus, 4000 cm^{-1} is the high-energy end of the scale and 625 cm^{-1} is the low-energy end.

Electromagnetic radiation in the 4000 to 625 cm^{-1} region corresponds to the separation between adjacent **vibrational energy states** in organic molecules. Absorption of a photon of infrared radiation excites a molecule from its lowest, or *ground,* vibrational state to a higher one. These vibrations include stretching and bending modes of the type illustrated for a methylene group in Figure 13.27. A single molecule can have a large number of distinct vibrations available to it, and infrared spectra of different molecules, like fingerprints and snowflakes, are different. Superposability of their infrared spectra is commonly offered as proof that two compounds are the same.

Like nmr spectrometers, some ir spectrometers operate in a continuous-sweep mode, while others employ pulse–Fourier transform (FT-IR) technology. All the ir spectra in this text were obtained on an FT-IR instrument.

A typical infrared spectrum, such as that of hexane in Figure 13.28, appears as a series of absorption peaks of varying shape and intensity. Almost all organic compounds exhibit a peak or group of peaks near 3000 cm^{-1} because it is in this region that absorption due to carbon-hydrogen stretching vibrations occurs. The peaks at 1460, 1380, and 725 cm^{-1} are due to various bending vibrations.

Infrared spectra can be recorded on a sample regardless of its physical state—solid, liquid, gas, or dissolved in some solvent. The spectrum in Figure 13.28 was taken on

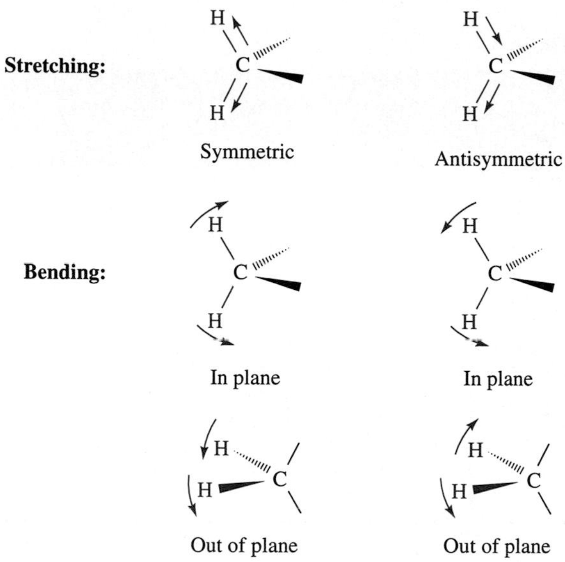

Stretching:

Symmetric

Antisymmetric

Bending:

In plane

In plane

Out of plane

Out of plane

FIGURE 13.27 Stretching and bending vibrations of a methylene unit.

the neat sample, meaning the pure liquid. A drop or two of hexane was placed between two sodium chloride disks, through which the infrared beam is passed. Solids may be dissolved in a suitable solvent such as carbon tetrachloride or chloroform. More commonly, though, a solid sample is mixed with potassium bromide and the mixture pressed into a thin wafer, which is placed in the path of the infrared beam.

In using infrared spectroscopy for structure determination, peaks in the range 1600 to 4000 cm^{-1} are usually emphasized because this is the region in which the vibrations characteristic of particular functional groups are found. The region 1300 to 625 cm^{-1} is known as the **fingerprint region;** it is here that the pattern of peaks varies most from compound to compound. Table 13.4 lists the frequencies (in wave numbers) associated with a variety of groups commonly found in organic compounds.

To illustrate how structural features affect infrared spectra, compare the spectrum of hexane (Figure 13.28) with that of 1-hexene (Figure 13.29). The two are quite

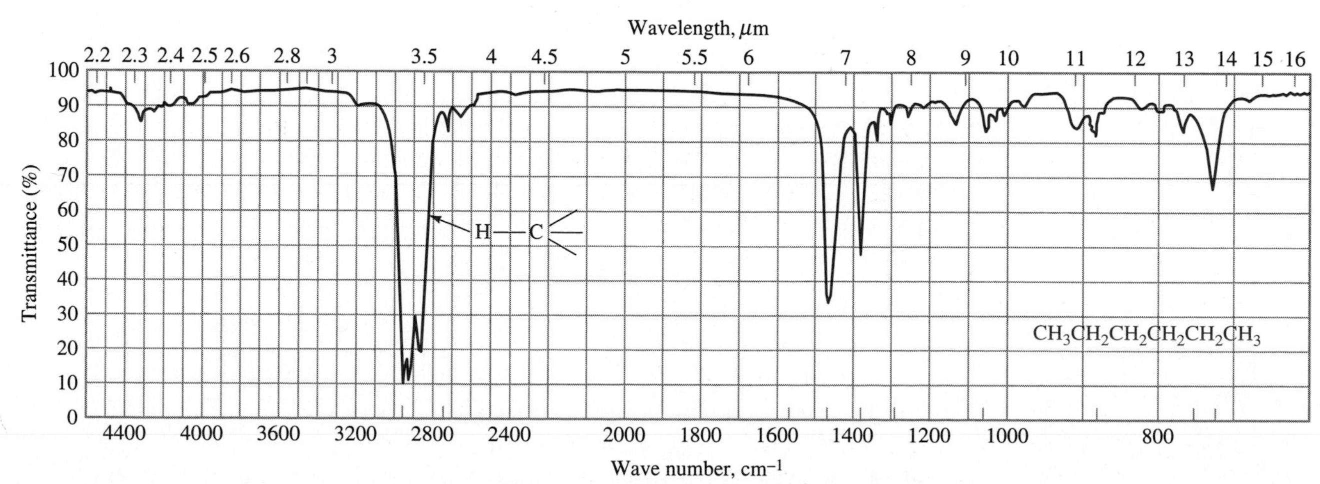

FIGURE 13.28 The infrared spectrum of hexane.

TABLE 13.4
Infrared Absorption Frequencies of Some Common Structural Units

Single bonds		Double bonds	
Structural unit	**Frequency, cm^{-1}**	**Structural unit**	**Frequency, cm^{-1}**
Stretching vibrations			
—O—H (alcohols)	3200–3600	>C=C<	1620–1680
—O—H (carboxylic acids)	2500–3600	>C=O	
>N—H	3350–3500	Aldehydes and ketones	1710–1750
sp C—H	3310–3320	Carboxylic acids	1700–1725
sp^2 C—H	3000–3100	Acid anhydrides	1800–1850 and 1740–1790
sp^3 C—H	2850–2950	Acyl halides	1770–1815
		Esters	1730–1750
sp^2 C—O	1200	Amides	1680–1700
sp^3 C—O	1025–1200		
		Triple bonds	
		—C≡C—	2100–2200
		—C≡N	2240–2280
Bending vibrations of diagnostic value			
Alkenes:		*Substituted derivatives of benzene:*	
$RCH{=}CH_2$	910, 990	Monosubstituted	730–770 and 690–710
$R_2C{=}CH_2$	890	Ortho-disubstituted	735–770
cis-RCH=CHR′	665–730	Meta-disubstituted	750–810 and 680–730
trans-RCH=CHR′	960–980	Para-disubstituted	790–840
$R_2C{=}CHR'$	790–840		

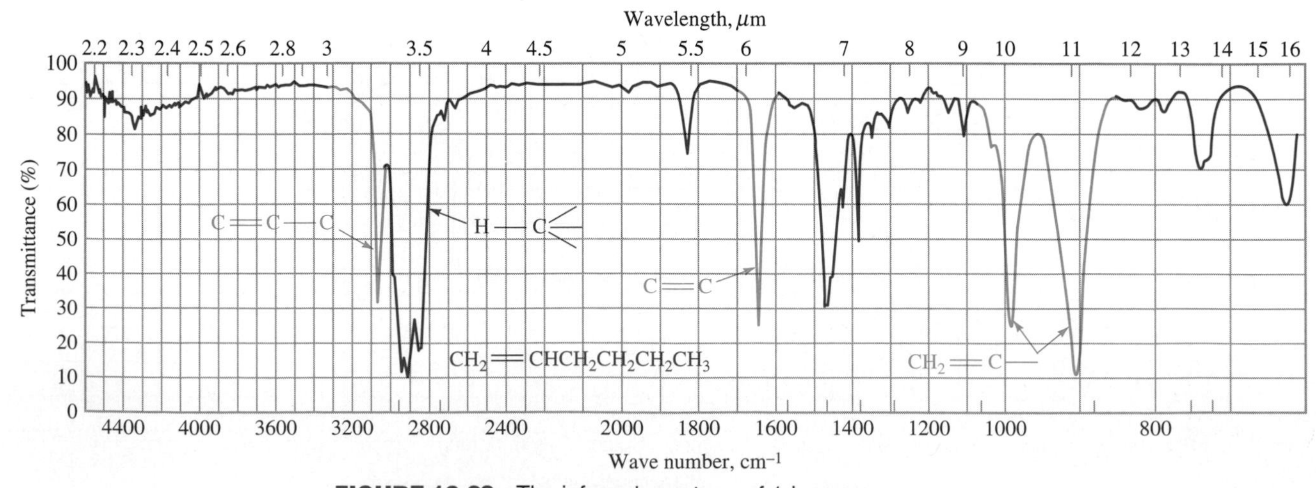

FIGURE 13.29 The infrared spectrum of 1-hexene.

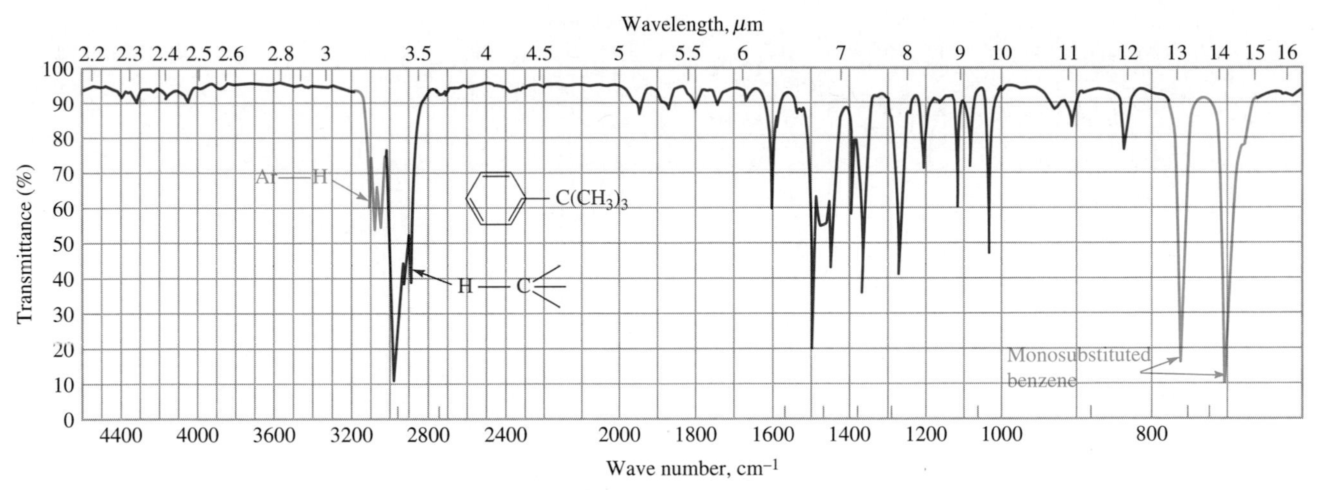

FIGURE 13.30 The infrared spectrum of *tert*-butylbenzene.

different. In the C—H stretching region of 1-hexene, there is a peak at 3095 cm^{-1}, while all the C—H stretching vibrations of hexane appear below 3000 cm^{-1}. A peak or peaks above 3000 cm^{-1} is characteristic of a hydrogen bonded to sp^2 hybridized carbon. The ir spectrum of 1-hexene also displays a peak at 1640 cm^{-1} corresponding to its C=C stretching vibration. The peaks near 1000 and 900 cm^{-1} in the spectrum of 1-hexene, absent in the spectrum of hexane, are bending vibrations involving the hydrogens of the doubly bonded carbons.

Carbon-hydrogen stretching vibrations with frequencies above 3000 cm^{-1} are also found in arenes such as *tert*-butylbenzene, as shown in Figure 13.30. This spectrum also contains two intense bands at 760 and 700 cm^{-1}, which are characteristic of monosubstituted benzene rings. Other substitution patterns, some of which are listed in Table 13.4, give different combinations of peaks.

In addition to sp^2 C—H stretching modes, there are other stretching vibrations that appear at frequencies above 3000 cm^{-1}. The most important of these is the O—H stretch of alcohols. Figure 13.31 shows the ir spectrum of 2-hexanol. It contains a

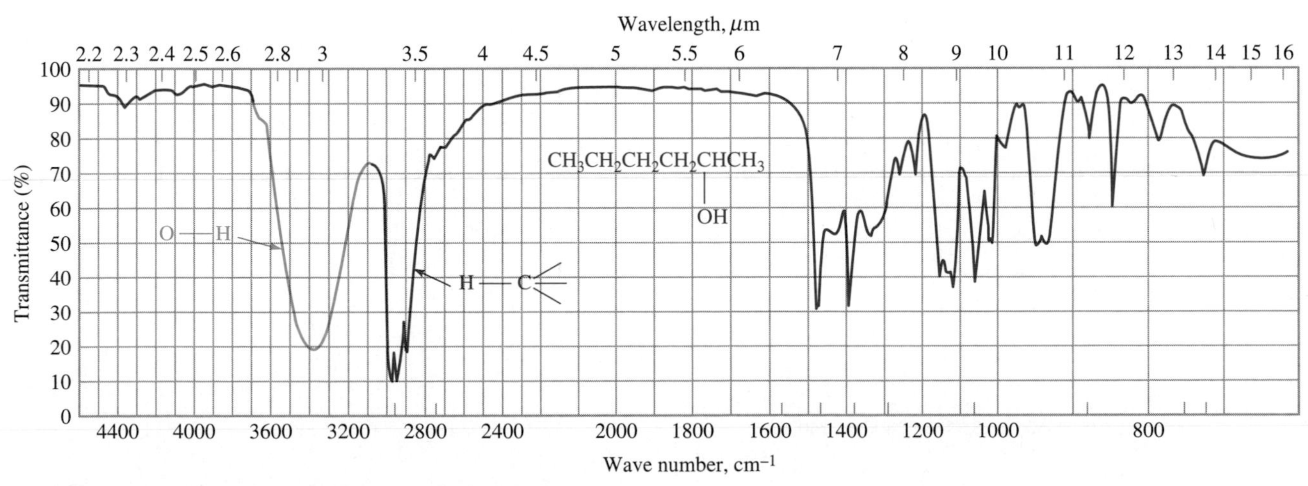

FIGURE 13.31 The infrared spectrum of 2-hexanol.

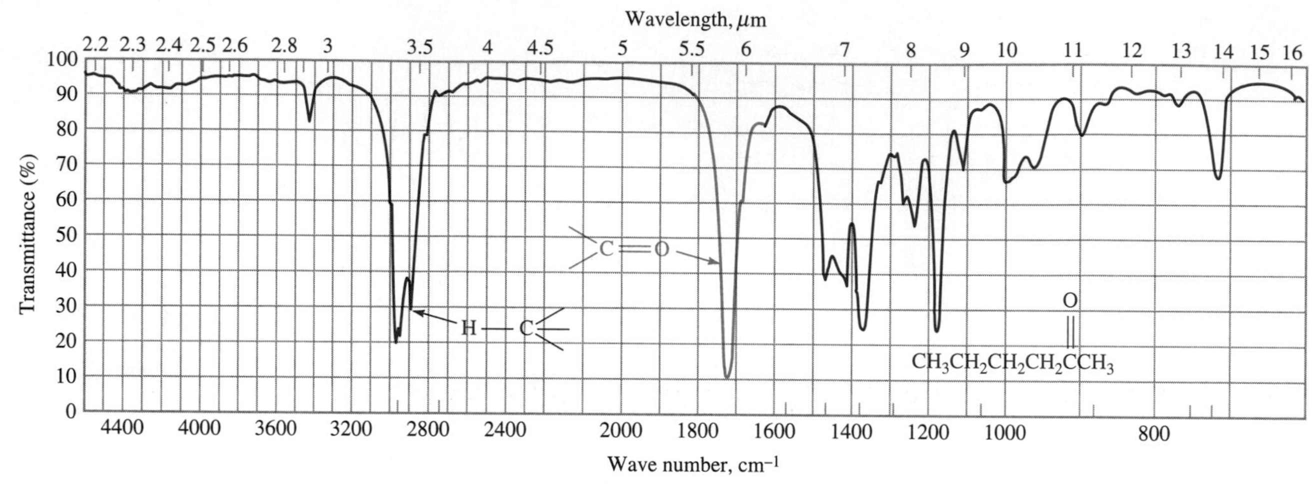

FIGURE 13.32 The infrared spectrum of 2-hexanone.

broad peak at 3300 cm^{-1} ascribable to O—H stretching of hydrogen-bonded alcohol groups. In dilute solution, where hydrogen bonding is less and individual alcohol molecules are present as well as hydrogen-bonded aggregates, an additional peak appears at approximately 3600 cm^{-1}.

Carbonyl groups rank among the structural units most readily revealed by ir spectroscopy. The carbon-oxygen double bond stretching mode gives rise to a very strong peak in the 1650 to 1800 cm^{-1} region. This peak is clearly evident in the spectrum of 2-hexanone, shown in Figure 13.32. The position of the carbonyl peak varies with the nature of the substituents on the carbonyl group. Thus, characteristic frequencies are associated with aldehydes and ketones, amides, esters, and so forth, as summarized in Table 13.4.

PROBLEM 13.12 Which one of the following compounds is most consistent with the infrared spectrum given in Figure 13.33? Explain your reasoning.

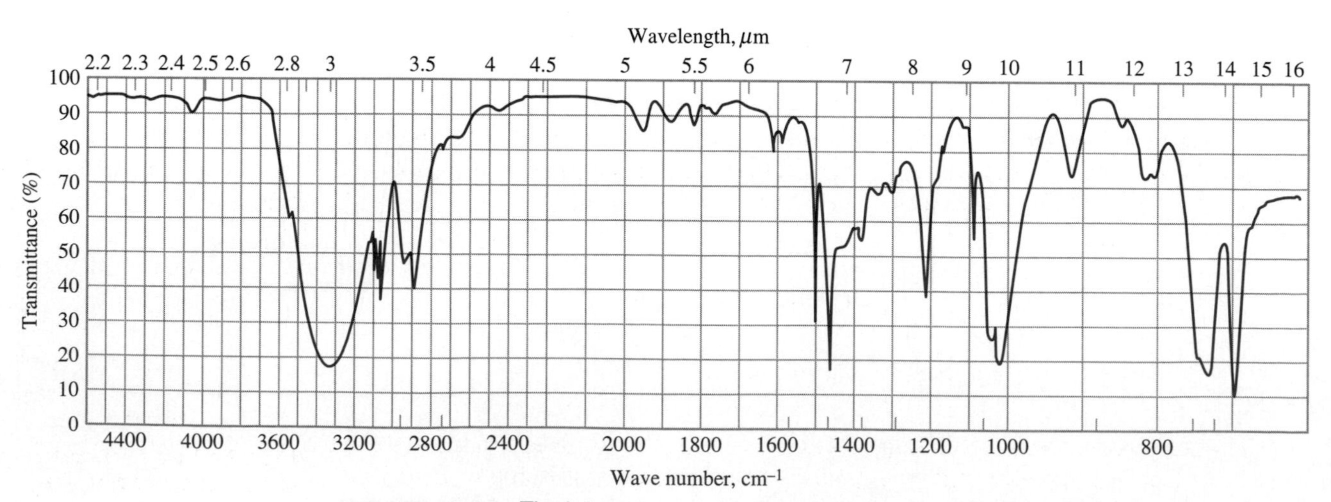

FIGURE 13.33 The infrared spectrum of the unknown compound in Problem 13.12.

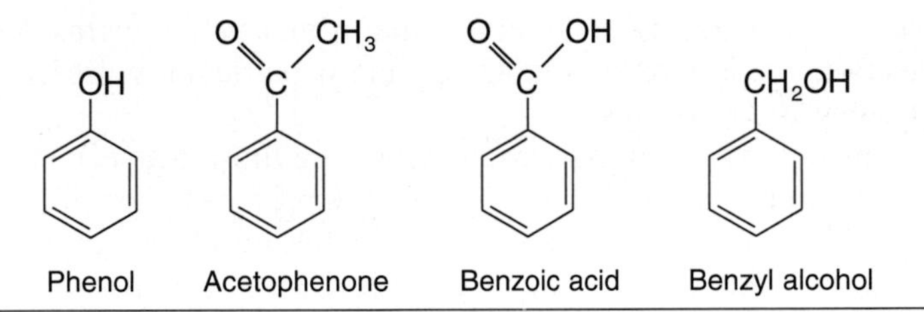

In later chapters, when families of compounds are discussed in detail, the infrared frequencies associated with each type of functional group will be described.

13.19 ULTRAVIOLET-VISIBLE (UV-VIS) SPECTROSCOPY

Visible light is the portion of the electromagnetic spectrum (Figure 13.1) that lies just beyond the infrared region. It occupies the relatively narrow region between 12,500 and 25,000 cm^{-1}. Wave numbers are directly proportional to energy, and so visible light is approximately 10 times more energetic than infrared radiation. Red light is the low-energy end of the visible region, violet light the high-energy end. Ultraviolet radiation lies beyond the violet end of the visible light spectrum; it encompasses the region from 25,000 to 50,000 cm^{-1}. Positions of absorption in visible spectroscopy are customarily expressed in units of 10^{-9} m, or *nanometers* (nm). Thus, the visible region corresponds to 800 to 400 nm, and the ultraviolet region to 400 nm to 200 nm.

Figure 13.34 shows the ultraviolet (uv) spectrum of a conjugated diene, *cis,trans*-1,3-cyclooctadiene, measured in ethanol as solvent in the range 200 to 280 nm. There are no additional absorption maxima beyond 280 nm, and so that portion of the spectrum has been omitted. As is typical of most uv spectra, the absorption peak is rather broad. The wavelength at which absorption is a maximum is referred to as the λ_{max} of the sample. In this case λ_{max} is 230 nm. The absorbance *A* of a sample is proportional to its concentration in solution and the path length through which the beam of ultraviolet radiation passes. To correct for concentration and path length, absorbance is converted to *molar absorptivity* ε by dividing it by the concentration *c* in moles per liter and the path length *l* in centimeters:

$$\varepsilon = \frac{A}{c \cdot l}$$

Molar absorptivity, when measured at λ_{max}, is cited as ε_{max}. It is normally expressed without units. Both λ_{max} and ε_{max} are affected by the solvent, which is therefore included when reporting uv-vis spectroscopic data. Thus, you might find a literature reference expressed in the form

cis, trans-1,3-Cyclooctadiene

$\lambda_{max}^{ethanol}$ 230 nm

$\varepsilon_{max}^{ethanol}$ 2630

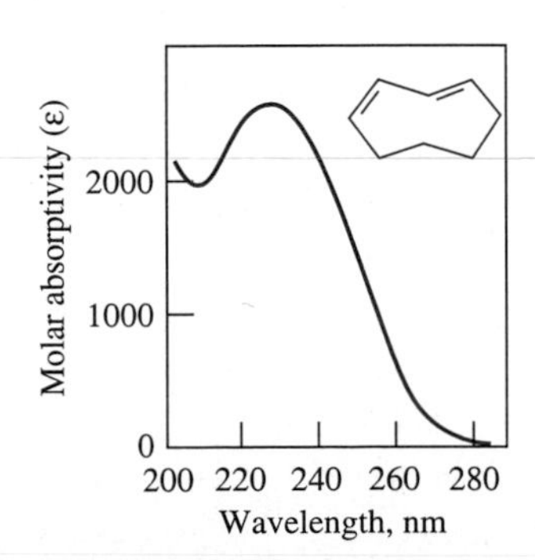

FIGURE 13.34 The ultraviolet spectrum of *cis,trans*-1,3-cyclooctadiene.

As might be expected from the physical appearance of a uv-vis spectrum or from the limited data contained in a literature citation, the structural information we derive from uv-vis spectroscopy is less than that derived from nmr and ir. Ultraviolet spectroscopy, however, does tell us some things that ir and nmr spectroscopy do not; it probes the electron distribution in a molecule and is particularly useful when conjugated π electron systems are present.

The transitions involved in uv-vis spectroscopy are those between electron energy levels. In the lowest-energy transition, an electron is promoted from the highest occupied molecular orbital (HOMO) of a molecule to the lowest unoccupied molecular orbital (LUMO). In alkenes and polyenes the HOMO is the highest-energy π orbital and the LUMO is the lowest-energy π^* orbital. Excitation of a π electron from a bonding to an antibonding orbital is referred to as a $\pi \rightarrow \pi^*$ transition. Figure 13.35 depicts the π electron configuration of 1,3-butadiene in its ground (most stable) electron configuration and in its excited state. The HOMO-LUMO energy gap is quite large (550 kJ/mol; 132 kcal/mol), and the $\pi \rightarrow \pi^*$ transition in 1,3-butadiene requires electromagnetic radiation of 217 nm.

Substituents, especially those that extend conjugation, decrease the HOMO-LUMO energy difference and cause a shift of λ_{max} to longer wavelengths. Thus, while λ_{max} is 217 nm for 1,3-butadiene, it is 258 nm for 1,3,5-hexatriene. A striking example of the effect of conjugation on the absorption of light can be found in the case of a compound known as *lycopene.* Lycopene is a red substance that contributes to the color of ripe tomatoes; it has a conjugated system of 11 double bonds and absorbs *visible light* (λ_{max} = 505 nm). By absorbing the blue-green fraction of visible light, lycopene appears red.

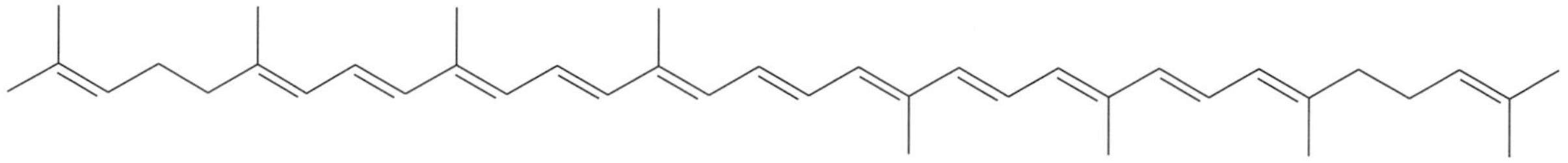

Lycopene

Many organic compounds such as lycopene are colored because their HOMO-LUMO energy gap is small enough that λ_{max} appears in the visible range of the spectrum. All that is required for a compound to be colored, however, is that it possess some absorption in the visible range. It often happens that a compound will have its

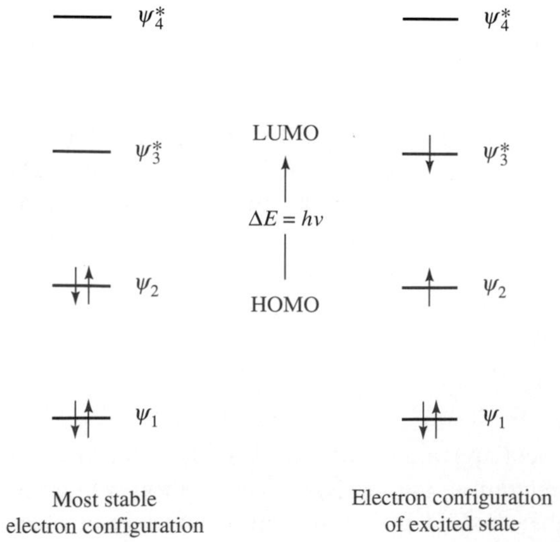

ψ_4^* ψ_3^* ψ_2 ψ_1

LUMO

$\Delta E = h\nu$

HOMO

Most stable electron configuration

Electron configuration of excited state

FIGURE 13.35 The nature of the $\pi \rightarrow \pi^*$ transition in 1,3-butadiene.

λ_{max} in the uv region but that the peak is broad and extends into the visible. Absorption of the blue to violet components of visible light occurs, and the compound appears yellow.

A second type of absorption that is important in uv-vis examination of organic compounds is the $n \rightarrow \pi^*$ transition of the carbonyl ($\diagdown C{=}O$ with $\diagup$) group. One of the electrons in a lone-pair orbital of oxygen is excited to an antibonding orbital of the carbonyl group. The n in $n \rightarrow \pi^*$ identifies the electron as one of the nonbonded electrons of oxygen. This transition gives rise to relatively weak absorption peaks ($\varepsilon_{max} < 100$) in the region 270 to 300 nm.

The structural unit associated with the electronic transition in uv-vis spectroscopy is called a **chromophore.** Chemists often refer to *model compounds* to help interpret uv-vis spectra. An appropriate model is a simple compound of known structure that incorporates the chromophore suspected of being present in the sample. Because remote substituents do not affect λ_{max} of the chromophore, a strong similarity between the spectrum of the model compound and that of the unknown can serve to identify the kind of π electron system present in the sample. There is a substantial body of data concerning the uv-vis spectral properties associated with a great many chromophores, as well as empirical correlations of substituent effects on λ_{max}. Such data are helpful when using uv-vis spectroscopy as a tool for structure determination.

13.20 MASS SPECTROMETRY

Mass spectrometry differs from the other instrumental methods discussed in this chapter in a fundamental way. It does not depend on the absorption of electromagnetic radiation but rather examines what happens when a molecule is bombarded with high-energy electrons. If an electron having an energy of about 10 electronvolts (10 eV = 230.5 kcal/mol) collides with an organic molecule, the energy transferred as a result of that collision is sufficient to dislodge one of the molecule's electrons.

$$\underset{\text{Molecule}}{A{:}B} + \underset{\text{Electron}}{e^-} \longrightarrow \underset{\text{Cation radical}}{A\overset{+}{\cdot}B} + \underset{\text{Two electrons}}{2e^-}$$

We say the molecule AB has been ionized by **electron impact.** The species that results, called the **molecular ion,** is positively charged and has an odd number of electrons—it is a **cation radical.** The molecular ion has the same mass (less the negligible mass of a single electron) as the molecule from which it is formed.

While energies of about 10 eV are required, energies of about 70 eV are used. Electrons this energetic not only cause ionization of a molecule but impart a large amount of energy to the molecular ion. The molecular ion dissipates this excess energy by dissociating into smaller fragments. Dissociation of a cation radical produces a neutral fragment and a positively charged fragment.

$$\underset{\text{Cation radical}}{A\overset{+}{\cdot}B} \longrightarrow \underset{\text{Cation}}{A^+} + \underset{\text{Radical}}{B\cdot}$$

Ionization and fragmentation produce a mixture of particles, some neutral and some positively charged. To understand what follows, we need to examine the design of an

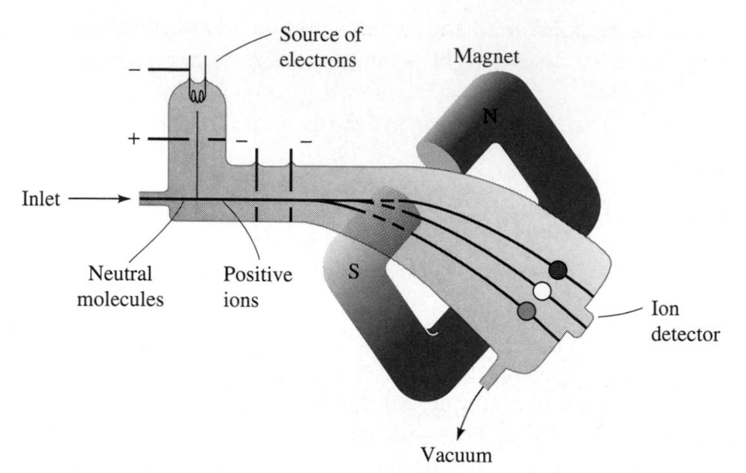

FIGURE 13.36 Schematic diagram of a mass spectrometer. Only positive ions are detected. The cation ● has the highest mass/charge ratio, and its path is deflected least by the magnet. The cation ● has the lowest mass/charge ratio, and its path is deflected most. The mass/charge ratio of cation ○ is such that its path carries it through a small slit to the detector. *(Adapted from John B. Russell, General Chemistry, McGraw-Hill, New York, 1980, p. 111.)*

electron-impact mass spectrometer, shown in a schematic diagram in Figure 13.36. The sample is bombarded with 70-eV electrons, and the resulting positively charged ions (the molecular ion as well as fragment ions) are directed into an analyzer tube surrounded by a magnet. This magnet deflects the ions from their original trajectory, causing them to adopt a circular path, the radius of which depends on their mass/charge ratio (m/z). Ions of small m/z are deflected more than those of larger m/z. By varying either the magnetic field strength or the degree to which the ions are accelerated on entering the analyzer, ions of a particular m/z can be selectively focused through a narrow slit onto a detector, where they are counted. Scanning all m/z values gives the distribution of positive ions, called a **mass spectrum,** characteristic of a particular compound.

Modern mass spectrometers are interfaced with computerized data-handling systems capable of displaying the mass spectrum according to a number of different formats. Bar graphs on which relative intensity is plotted versus m/z are the most common. Figure 13.37 shows the mass spectrum of benzene in bar graph form.

The mass spectrum of benzene is relatively simple and illustrates some of the information that mass spectrometry provides. The most intense peak in the mass spectrum is called the **base peak** and is assigned a relative intensity of 100. Ion abundances are proportional to peak intensities and are reported as intensities relative to the base peak. The base peak in the mass spectrum of benzene corresponds to the molecular ion (M^+) at $m/z = 78$.

$$C_6H_6 + e^- \longrightarrow C_6H_6^{+\cdot} + 2e^-$$

Benzene | Electron | Molecular ion of benzene | Two electrons

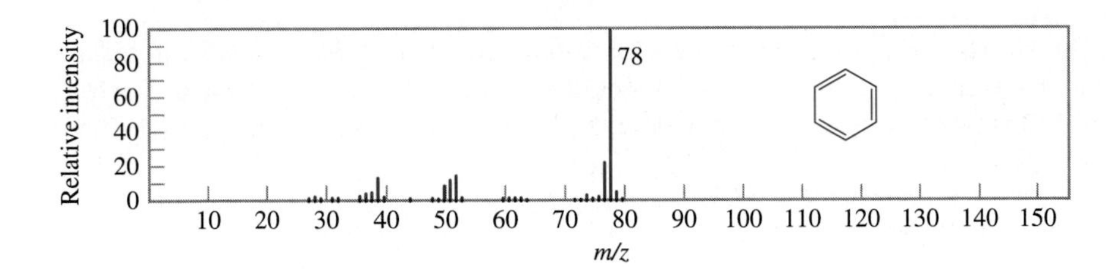

FIGURE 13.37 The mass spectrum of benzene. The peak at $m/z = 78$ corresponds to the C_6H_6 molecular ion.

Benzene does not undergo extensive fragmentation; none of the fragment ions in its mass spectrum are as abundant as the molecular ion.

There is a small peak one mass unit higher than M^+ in the mass spectrum of benzene. What is the origin of this peak? What we see in Figure 13.37 as a single mass spectrum is actually a superposition of the spectra of three isotopically distinct benzenes. Most of the benzene molecules contain only ^{12}C and ^1H and have a molecular mass of 78. Smaller proportions of benzene molecules contain ^{13}C in place of one of the ^{12}C atoms or ^2H in place of one of the protons. Both these species have a molecular mass of 79.

93.4% (all carbons are ^{12}C) Gives M^+ 78	6.5% (* = ^{13}C) Gives M^+ 79	0.1% (all carbons are ^{12}C) Gives M^+ 79

Not only the molecular ion peak but all the peaks in the mass spectrum of benzene are accompanied by a smaller peak one mass unit higher. Indeed, since all organic compounds contain carbon and most contain hydrogen, similar **isotopic clusters** will appear in the mass spectra of all organic compounds.

Isotopic clusters are especially apparent when atoms such as bromine and chlorine are present in an organic compound. The natural ratios of isotopes in these elements are

$$\frac{^{35}Cl}{^{37}Cl} = \frac{100}{32.7} \qquad \frac{^{79}Br}{^{81}Br} = \frac{100}{97.5}$$

Figure 13.38 presents the mass spectrum of chlorobenzene. There are two prominent molecular ion peaks, one at *m/z* 112 for $C_6H_5{}^{35}Cl$ and the other at *m/z* 114 for $C_6H_5{}^{37}Cl$. The peak at *m/z* 112 is three times as intense as the one at *m/z* 114.

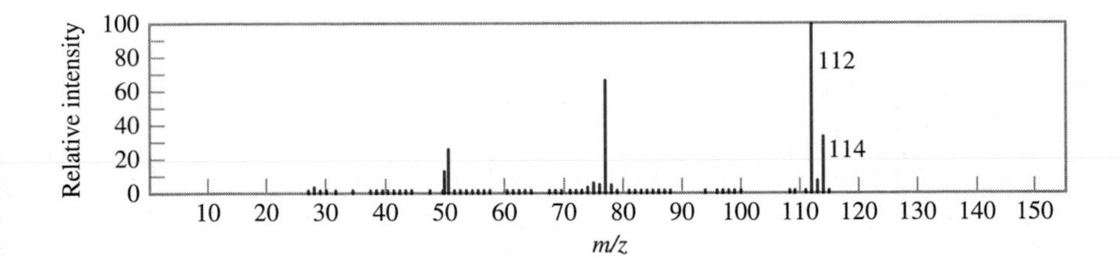

FIGURE 13.38 The mass spectrum of chlorobenzene.

PROBLEM 13.13 Knowing what to look for with respect to isotopic clusters can aid in interpreting mass spectra. How many peaks would you expect to see for the molecular ion in each of the following compounds? At what *m/z* values would these peaks appear? (Disregard the small peaks due to ^{13}C and ^{2}H.)

(*a*) *p*-Dichlorobenzene
(*b*) *o*-Dichlorobenzene
(*c*) *p*-Dibromobenzene
(*d*) *p*-Bromochlorobenzene

SAMPLE SOLUTION (*a*) The two isotopes of chlorine are ^{35}Cl and ^{37}Cl. There will be three isotopically different forms of *p*-dichlorobenzene present. They have the structures shown as follows. Each one will give an M^+ peak at a different value of *m/z*.

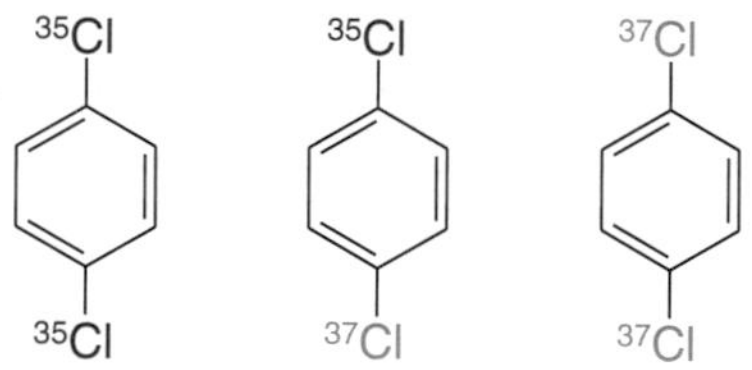

^{35}Cl–C_6H_4–^{35}Cl *m/z* 146
^{35}Cl–C_6H_4–^{37}Cl *m/z* 148
^{37}Cl–C_6H_4–^{37}Cl *m/z* 150

Unlike the case of benzene, in which ionization involves loss of a π electron from the ring, electron-impact–induced ionization of chlorobenzene involves loss of an electron from an unshared pair of chlorine. The molecular ion then fragments by carbon-chlorine bond cleavage.

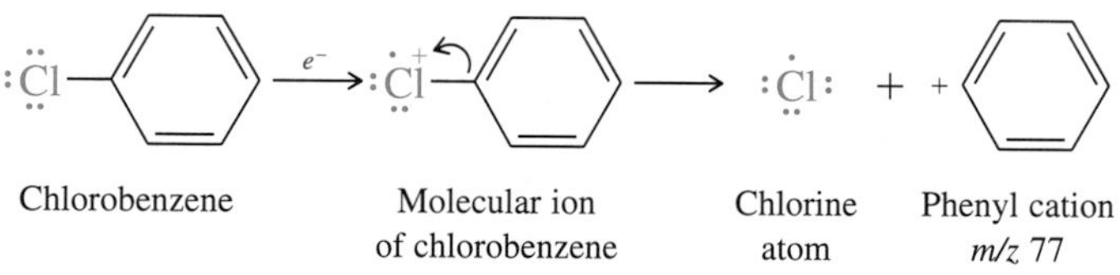

Chlorobenzene $\xrightarrow{e^-}$ Molecular ion of chlorobenzene $\longrightarrow$ Chlorine atom + Phenyl cation *m/z* 77

The peak at *m/z* 77 in the mass spectrum of chlorobenzene in Figure 13.38 is attributed to this fragmentation. Because there is no peak of significant intensity two atomic mass units higher, we know that the cation responsible for the peak at *m/z* 77 cannot contain chlorine.

Some classes of compounds are so prone to fragmentation that the molecular ion peak is very weak. The base peak in most unbranched alkanes, for example, is *m/z* 43, which is followed by peaks of decreasing intensity at *m/z* values of 57, 71, 85, and so on. These peaks correspond to cleavage of each possible carbon-carbon bond in the molecule. This pattern is evident in the mass spectrum of decane, depicted in Figure 13.39. The points of cleavage are indicated in the following diagram:

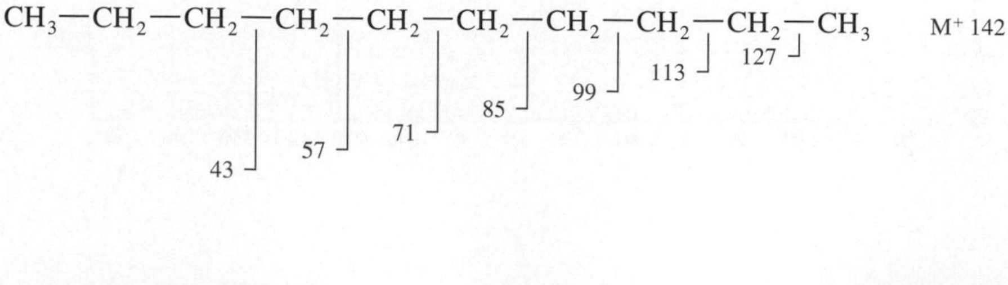

$$CH_3{-}CH_2{-}CH_2{-}CH_2{-}CH_2{-}CH_2{-}CH_2{-}CH_2{-}CH_2{-}CH_3 \quad M^+\ 142$$

Cleavage fragments: 43, 57, 71, 85, 99, 113, 127

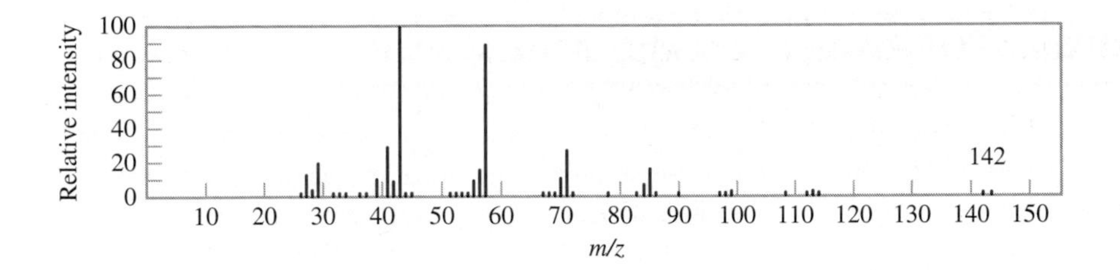

FIGURE 13.39 The mass spectrum of decane. The peak for the molecular ion is extremely small. The most prominent peaks arise by fragmentation.

Many fragmentations in mass spectrometry proceed so as to form a stable carbocation, and the principles that we have developed regarding carbocation stability apply. Alkylbenzenes of the type $C_6H_5CH_2R$ undergo cleavage of the bond to the benzylic carbon to give *m/z* 91 as the base peak. The mass spectrum in Figure 13.40 and the following fragmentation diagram illustrate this for propylbenzene.

C_6H_5—CH_2—CH_2—CH_3 M^+ 120

91

While this cleavage is probably driven by the stability of benzyl cation, evidence has been obtained suggesting that tropylium cation, formed by rearrangement of benzyl cation, is actually the species responsible for the peak.

The structure of tropylium cation is given in Section 11.22.

PROBLEM 13.14 The base peak appears at *m/z* 105 for one of the following compounds and at *m/z* 119 for the other two. Match the compounds with the appropriate *m/z* values for their base peaks.

CH_2CH_3 (ring with two CH_3) $CH_2CH_2CH_3$ (ring with CH_3) CH_3CHCH_3 (ring with CH_3)

Understanding how molecules fragment upon electron impact permits a mass spectrum to be analyzed in sufficient detail to deduce the structure of an unknown compound. Thousands of compounds of known structure have been examined by mass spectrometry, and the fragmentation patterns that characterize different classes are well documented. As various groups are covered in subsequent chapters, aspects of their fragmentation behavior under conditions of electron impact will be described.

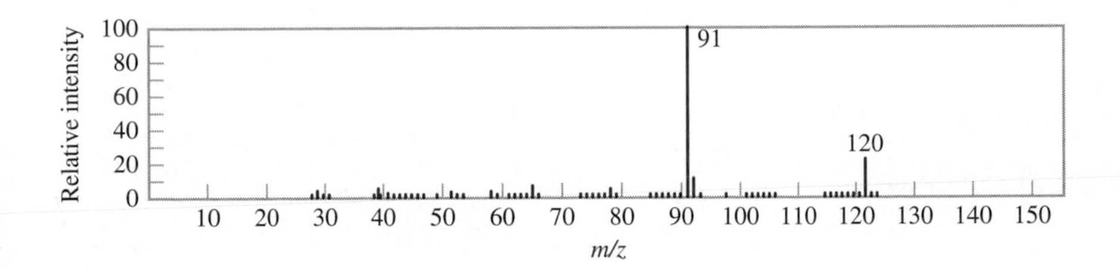

FIGURE 13.40 The mass spectrum of propylbenzene. The most intense peak is $C_7H_7^+$.

GAS CHROMATOGRAPHY, GC/MS, AND MS/MS

The spectra depicted in this chapter (^{1}H nmr, ^{13}C nmr, ir, uv-vis, and ms) were all obtained using pure substances. It is much more common, however, to encounter an organic substance, either formed as the product of a chemical reaction or isolated from natural sources, as but one component of a mixture. Just as the last half of the twentieth century has seen a revolution in the methods available for the *identification* of organic compounds, so too has it seen remarkable advances in methods for their *separation* and *purification*.

Classical methods for separation and purification include fractional distillation of liquids and recrystallization of solids, and these two methods are routinely included in the early portions of laboratory courses in organic chemistry. Because they are capable of being adapted to work on a large scale, fractional distillation and recrystallization are the preferred methods for purifying organic substances in the pharmaceutical and chemical industries.

Some other methods are more appropriate to the separation of small amounts of material in laboratory-scale work and are most often encountered there. Indeed, it is their capacity to deal with exceedingly small quantities that is the strength of a number of methods that together encompass the various forms of **chromatography.** The first step in all types of chromatography involves adsorbing the sample onto some material called the *stationary phase.* Next, a second phase (the *mobile phase*) is allowed to move across the stationary phase. Depending on the properties of the two phases and the components of the mixture, the mixture is separated into its components according to the rate at which each is removed from the stationary phase by the mobile phase.

In **gas chromatography** (GC) the stationary phase consists of beads of an inert solid support coated with a high-boiling liquid, and the mobile phase is a gas, usually helium. Figure 13.41 shows a typical gas chromatograph. The sample is injected by syringe onto a heated block where a stream of helium carries it onto a coiled column packed with the stationary phase. The components of the mixture move through the column at different rates. They are said to have different *retention times.* Gas chromatography is also referred to as *gas-liquid partition chromatography,* because the technique depends on how different substances partition themselves between the gas phase (dispersed in the helium carrier gas) and the liquid phase (dissolved in the coating on the beads of solid support).

Typically the effluent from a gas chromatograph is passed through a detector which feeds a signal to a recorder whenever a substance different from pure carrier gas leaves the column. Thus one determines the number of components in a mixture by counting the number of peaks on a strip chart. It is good practice to carry out the analysis under different conditions by varying the liquid phase, the temperature, and the flow rate of the carrier gas so as to ensure that two substances have not eluted together and given a single peak under the original conditions. Gas chromatogra-

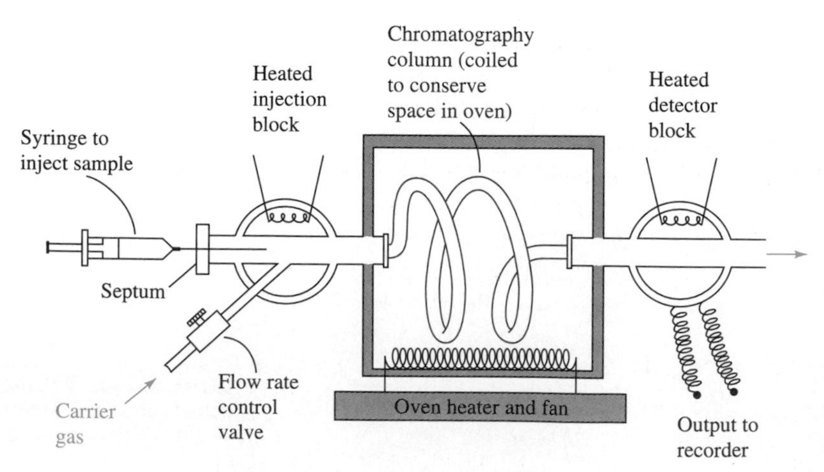

FIGURE 13.41 Diagram of a gas chromatograph. When connected to a mass spectrometer, as in GC/MS, the effluent is split into two streams as it leaves the column. One stream goes to the detector, the other to the mass spectrometer. *(Adapted from H. D. Durst and G. W. Gokel, Experimental Organic Chemistry, 2d ed., McGraw-Hill, New York, 1987.)*

phy can also be used to identify the components of a mixture by comparing their retention times with those of authentic samples.

In **gas chromatography–mass spectrometry** (GC/MS), the effluent from a gas chromatograph is passed into a mass spectrometer and a mass spectrum is taken for each peak. Thus gas chromatography is used to separate a mixture, and mass spectrometry used to analyze it. Gas chromatography–mass spectrometry is a very powerful analytical technique. One of its more visible applications involves the testing of athletes for steroids, stimulants, and other performance-enhancing drugs. These drugs are converted in the body to derivatives called *metabolites,* which are then excreted in the urine. When the urine is subjected to GC/MS analysis, the mass spectra of its organic components are identified by comparison with the mass spectra of known metabolites stored in the instrument's computer. Using a similar procedure, the urine of newborn infants is monitored by GC/MS for metabolite markers of genetic disorders that can be treated if detected early in life. Gas chromatography–mass spectrometry is also used for more routine analyses such as detecting and measuring the concentration of halogenated hydrocarbons in drinking water.

While GC/MS is the most widely used analytical method that combines a chromatographic separation with the identification power of mass spectrometry, it is not the only one. Chemists have coupled mass spectrometers to most of the instruments that are used to separate mixtures. Perhaps the ultimate is **mass spectrometry–mass spectrometry** (MS/MS), in which one mass spectrometer generates and separates the molecular ions of the components of a mixture and a second mass spectrometer examines their fragmentation patterns!

13.21 SUMMARY

Structure determination in modern-day organic chemistry relies on instrumental methods. Several of the most widely used methods are based on the absorption of electromagnetic radiation by a sample (Section 13.1). Absorption of electromagnetic radiation causes the molecule to be excited from its most stable state (the *ground state*) to a higher-energy state (an *excited state;* Section 13.2). **Nuclear magnetic resonance spectroscopy** depends on transitions between nuclear spin states, **infrared spectroscopy** on transitions between vibrational energy states, and **ultraviolet-visible spectroscopy** on transitions between electronic energy states. A fourth method, **mass spectrometry,** is not based on absorption of electromagnetic radiation, but instead monitors what happens when a substance is ionized on collision with a high-energy electron.

^{1}H Nuclear Magnetic Resonance Spectroscopy. In the presence of an external magnetic field, the $+\frac{1}{2}$ and $-\frac{1}{2}$ nuclear spin states of a proton have slightly different energies (Section 13.3). The energy required to ''flip'' the spin of a proton from the lower-energy spin state to the higher state depends on the extent to which a nucleus is shielded from the external magnetic field (Section 13.4).

Protons in different environments within a molecule have different **chemical shifts;** that is, the extent to which they are shielded is different, and they absorb radio-frequency radiation at different frequencies. Chemical shifts in ^{1}H nmr spectroscopy are measured in parts per million relative to the proton signal of tetramethylsilane as a standard (Section 13.5). Table 13.1 (Section 13.6) gives chemical-shift values for protons in various environments.

The *number of signals* in a ^{1}H nmr spectrum reveals the number of sets of different chemical-shift-equivalent protons in a molecule, and the *integrated areas* tell us their relative ratios (Section 13.7).

Spin-spin splitting of nmr signals results from coupling of the nuclear spin of the proton that gives the signal to protons separated from it by two bonds (in *geminal coupling*) or three bonds (in *vicinal coupling;* Section 13.8). *Protons that have the same chemical shift do not split each other's signal.*

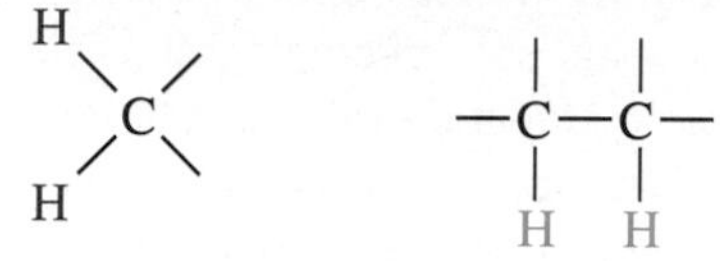

Geminal hydrogens   Vicinal hydrogens

In the simplest cases, the number of peaks into which a signal is split is equal to $n + 1$, where n is the number of protons to which the proton responsible for the signal is equally coupled. Some commonly encountered structural units are easily recognized by their characteristic splitting patterns. The methyl protons of an ethyl group appear as a *triplet* and the methylene protons as a *quartet* in compounds of the type **CH_3CH_2X** (Section 13.9). The methyl protons of an isopropyl group appear as a *doublet* and the methine proton as a *heptet* in compounds of the type **$(CH_3)_2CHX$** (Section 13.10). A pair of nonequivalent protons split each other's signal into a doublet giving a characteristic *doublet of doublets* pattern (Section 13.11). Vicinal coupling (^3J) is the most common, but geminal coupling (^2J) is also observed (Section 13.12). More complicated splitting patterns result when a proton is unequally coupled to two or more protons different from one another. These patterns are *multiplets of multiplets* (Section 13.13).

Splitting resulting from coupling to the O—H proton of alcohols is not normally observed, because the hydroxyl proton undergoes rapid intermolecular exchange with other alcohol molecules, which ''decouples'' it from other protons in the molecule (Section 13.14).

Many processes such as conformational changes take place faster than they can be detected by nmr, and nmr provides information about the *average* environment of a proton (Section 13.15). Thus, only a single peak appears for the 12 protons of cyclohexane even though, at any instant, six of them occupy axial sites and six are equatorial.

^{13}C Nuclear Magnetic Resonance Spectroscopy. The ^{13}C nucleus has a spin of $\pm\frac{1}{2}$ but is present to the extent of only 1.1 percent at natural abundance. By using special techniques for signal enhancement, high-quality ^{13}C nmr spectra may be obtained, and these provide a useful complement to proton spectra (Section 13.16).

Carbon-13 chemical shifts are summarized in Table 13.3 and cover a range of over 200 ppm. In many substances a separate signal is observed for each carbon atom, and the chemical shifts are characteristic of particular structural types. Carbon signals are normally presented as singlets, but through a technique known as *off-resonance decoupling* they appear as multiplets in which the number of peaks is 1 more than the number of hydrogens directly bonded to the carbon responsible for the signal (Section 13.17).

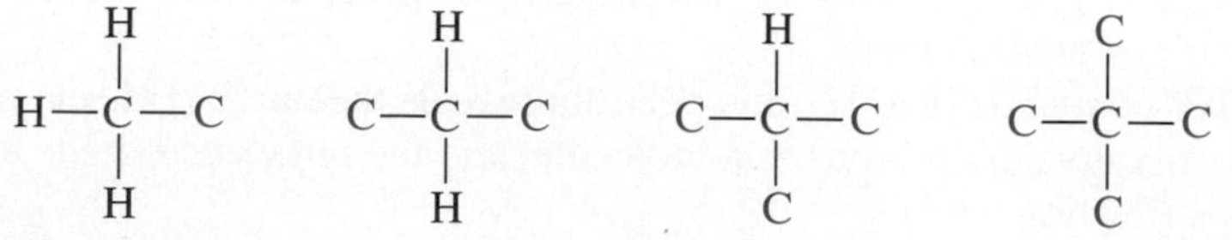

Primary carbon Quartet   Secondary carbon Triplet   Tertiary carbon Doublet   Quaternary carbon Singlet

Infrared Spectroscopy. This method probes molecular structure by examining transitions between vibrational energy levels using electromagnetic radiation in the 625 to 4000 cm^{-1} range. The presence or absence of a peak at a characteristic frequency in an ir spectrum tells us whether a certain *functional group* is present in a molecule (Section 13.18). Table 13.4 lists ir absorption frequencies for common structural units.

Ultraviolet-Visible Spectroscopy. Transitions between electronic energy levels involving electromagnetic radiation in the 200 nm to 800 nm range form the basis of uv-vis spectroscopy. The absorption peaks tend to be broad but are sometimes useful in indicating the presence of particular *π electron* systems within a molecule (Section 13.19).

Mass Spectrometry. Mass spectrometry exploits the information obtained when a molecule is ionized by electron impact and then dissociates to smaller fragments. Positive ions are separated and detected according to their mass/charge (m/z) ratio. By examining the fragments and by knowing how classes of molecules dissociate on electron impact, one can deduce the structure of a compound. Mass spectrometry is quite sensitive; as little as 10^{-9} g of compound is sufficient.

PROBLEMS

13.15 *Microwave spectroscopy* is used to probe transitions between rotational energy levels in molecules.

(*a*) A typical wavelength for microwaves is 10^{-2} m, compared with 10^{-5} m for infrared radiation. Is the energy separation between rotational energy levels in a molecule greater or less than the separation between vibrational energy levels?

(*b*) Microwave ovens cook food by heating the water in the food. Absorption of microwave radiation by the water excites it to a higher rotational energy state, and it gives off this excess energy as heat when it relaxes to its ground state. Why are vibrational and electronic energy states not involved in this process?

13.16 The peak in the uv-vis spectrum of acetone [$(CH_3)_2C{=}O$] corresponding to the $n \rightarrow \pi^*$ transition appears at 279 nm when hexane is the solvent, but shifts to 262 nm in water. Which is more polar, the ground electronic state or the excited state?

13.17 A particular vibration will give an absorption peak in the infrared spectrum only if the dipole moment of the molecule changes during the vibration. Which vibration of carbon dioxide, the symmetric stretch or the antisymmetric stretch, is "infrared-active"?

$$\overset{\leftarrow}{O}{=}C{=}\vec{O} \qquad \vec{O}{=}C{=}\vec{O}$$

Symmetric stretch Antisymmetric stretch

13.18 We have noted in Section 13.15 that an nmr spectrum is an average spectrum of the conformations populated by a molecule. From the following data, estimate the percentages of axial and equatorial bromide present in bromocyclohexane.

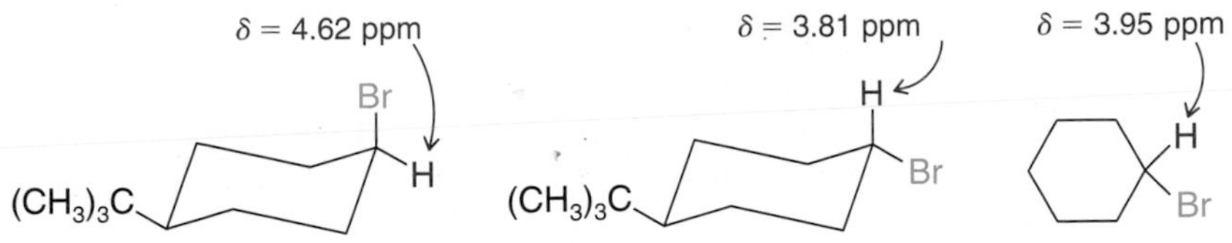

13.19 Infrared spectroscopy is an inherently "faster" method than nmr, and an ir spectrum is a superposition of the spectra of the various conformations, rather than an average of them. When 1,2-dichloroethane is cooled below its freezing point, the crystalline material gives an ir spectrum consistent with a single species that has a center of symmetry. At room temperature, the ir spectrum of liquid 1,2-dichloroethane retains the peaks present in the solid, but includes new peaks as well. Explain these observations.

13.20 Each of the following compounds is characterized by a ^{1}H nmr spectrum that consists of only a single peak having the chemical shift indicated. Identify each compound.

(*a*) C_8H_{18}; $\delta = 0.9$ ppm
(*b*) C_5H_{10}; $\delta = 1.5$ ppm
(*c*) C_8H_8; $\delta = 5.8$ ppm
(*d*) C_4H_9Br; $\delta = 1.8$ ppm
(*e*) $C_2H_4Cl_2$; $\delta = 3.7$ ppm
(*f*) $C_2H_3Cl_3$; $\delta = 2.7$ ppm
(*g*) $C_5H_8Cl_4$; $\delta = 3.7$ ppm
(*h*) $C_{12}H_{18}$; $\delta = 2.2$ ppm
(*i*) $C_3H_6Br_2$; $\delta = 2.6$ ppm

13.21 Each of the following compounds is characterized by a ^{1}H nmr spectrum that consists of two peaks, both singlets, having the chemical shift indicated. Identify each compound.

(*a*) C_6H_8; $\delta = 2.7$ ppm (4H) and 5.6 ppm (4H)
(*b*) $C_5H_{11}Br$; $\delta = 1.1$ ppm (9H) and 3.3 ppm (2H)
(*c*) $C_6H_{12}O$; $\delta = 1.1$ ppm (9H) and 2.1 ppm (3H)
(*d*) $C_6H_{10}O_2$; $\delta = 2.2$ ppm (6H) and 2.7 ppm (4H)

13.22 Deduce the structure of each of the following compounds on the basis of their ^{1}H nmr spectra and molecular formulas:

(*a*) C_8H_{10}; $\delta = 1.2$ ppm (triplet, 3H)
$\delta = 2.6$ ppm (quartet, 2H)
$\delta = 7.1$ ppm (broad singlet, 5H)

(*b*) $C_{10}H_{14}$; $\delta = 1.3$ ppm (singlet, 9H)
$\delta = 7.0$ to 7.5 ppm (multiplet, 5H)

(*c*) C_6H_{14}; $\delta = 0.8$ ppm (doublet, 12H)
$\delta = 1.4$ ppm (heptet, 2H)

(*d*) C_6H_{12}; $\delta = 0.9$ ppm (triplet, 3H)
$\delta = 1.6$ ppm (singlet, 3H)
$\delta = 1.7$ ppm (singlet, 3H)
$\delta = 2.0$ ppm (pentet, 2H)
$\delta = 5.1$ ppm (triplet, 1H)

(*e*) $C_4H_6Cl_4$; $\delta = 3.9$ ppm (doublet, 4H)
$\delta = 4.6$ ppm (triplet, 2H)

(*f*) $C_4H_6Cl_2$; $\delta = 2.2$ ppm (singlet, 3H)
$\delta = 4.1$ ppm (doublet, 2H)
$\delta = 5.7$ ppm (triplet, 1H)

(*g*) C_3H_7ClO; $\delta = 2.0$ ppm (pentet, 2H)
$\delta = 2.8$ ppm (singlet, 1H)
$\delta = 3.7$ ppm (triplet, 2H)
$\delta = 3.8$ ppm (triplet, 2H)

(*h*) $C_{14}H_{14}$; $\delta = 2.9$ ppm (singlet, 4H)
$\delta = 7.1$ ppm (broad singlet, 10H)

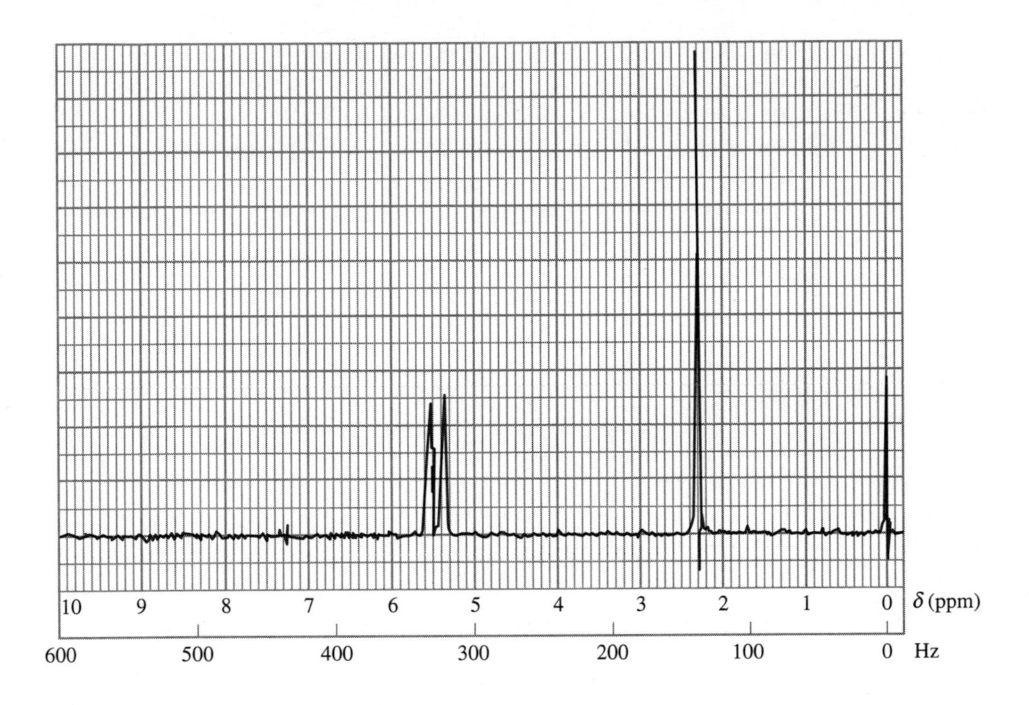

FIGURE 13.42 The 60-MHz ^1H nmr spectrum of isomer A of C_3H_5Br (Problem 13.24*a*).

13.23 From among the isomeric compounds of molecular formula C_4H_9Cl, choose the one having a ^1H nmr spectrum that

(*a*) Contains only a single peak
(*b*) Has several peaks including a doublet at $\delta = 3.4$ ppm
(*c*) Has several peaks including a triplet at $\delta = 3.5$ ppm
(*d*) Has several peaks including two distinct three-proton signals, one of them a triplet at $\delta = 1.0$ ppm and the other a doublet at $\delta = 1.5$ ppm

13.24 Identify the C_3H_5Br isomers on the basis of the following information:

(*a*) Isomer A has the ^1H nmr spectrum shown in Figure 13.42.
(*b*) Isomer B has three peaks in its ^{13}C nmr spectrum: $\delta = 32.6$ ppm (triplet); 118.8 ppm (triplet); and 134.2 ppm (doublet).
(*c*) Isomer C has two peaks in its ^{13}C nmr spectrum: $\delta = 12.0$ ppm (triplet) and 16.8 ppm (doublet). The peak at lower field is only half as intense as the one at higher field.

13.25 Identify each of the $C_4H_{10}O$ isomers on the basis of their ^{13}C nmr spectra:

(*a*) $\delta = 18.9$ ppm (quartet) (two carbons)
$\delta = 30.8$ ppm (doublet) (one carbon)
$\delta = 69.4$ ppm (triplet) (one carbon)
(*b*) $\delta = 10.0$ ppm (quartet)
$\delta = 22.7$ ppm (quartet)
$\delta = 32.0$ ppm (triplet)
$\delta = 69.2$ ppm (doublet)
(*c*) $\delta = 31.2$ ppm (quartet) (three carbons)
$\delta = 68.9$ ppm (singlet) (one carbon)

13.26 Identify the C_6H_{14} isomers on the basis of their ^{13}C nmr spectra:

(*a*) $\delta = 19.1$ ppm (quartet)
$\delta = 33.9$ ppm (doublet)
(*b*) $\delta = 13.7$ ppm (quartet)
$\delta = 22.8$ ppm (triplet)
$\delta = 31.9$ ppm (triplet)
(*c*) $\delta = 11.1$ ppm (quartet)
$\delta = 18.4$ ppm (quartet)
$\delta = 29.1$ ppm (triplet)
$\delta = 36.4$ ppm (doublet)

(*d*) $\delta = 8.5$ ppm (quartet)
$\delta = 28.7$ ppm (quartet)
$\delta = 30.2$ ppm (singlet)
$\delta = 36.5$ ppm (triplet)
(*e*) $\delta = 14.0$ ppm (quartet)
$\delta = 20.5$ ppm (triplet)
$\delta = 22.4$ ppm (quartet)
$\delta = 27.6$ ppm (doublet)
$\delta = 41.6$ ppm (triplet)

13.27 A compound (C_4H_6) has two signals of approximately equal intensity in its ^{13}C nmr spectrum; one is a triplet at $\delta = 30.2$ ppm, the other a doublet at $\delta = 136$ ppm. Identify the compound.

13.28 A compound ($C_3H_7ClO_2$) exhibited three peaks in its ^{13}C nmr spectrum at $\delta = 46.8$ (triplet), 63.5 (triplet), and 72.0 ppm (doublet). Excluding compounds that have Cl and OH on the same carbon, which are unstable, what is the most reasonable structure for this compound?

13.29 From among the compounds chlorobenzene, *o*-dichlorobenzene, and *p*-dichlorobenzene, choose the one that

(*a*) Gives the simplest ^{1}H nmr spectrum
(*b*) Gives the simplest ^{13}C nmr spectrum
(*c*) Has three peaks in its ^{13}C nmr spectrum
(*d*) Has four peaks in its ^{13}C nmr spectrum

13.30 Compounds A and B are isomers of molecular formula $C_{10}H_{14}$. Identify each one on the basis of the ^{1}H nmr spectra presented in Figure 13.43.

13.31 A compound ($C_8H_{10}O$) has the infrared and ^{1}H nmr spectra presented in Figure 13.44. What is its structure?

13.32 Deduce the structure of a compound having the mass spectrum and ^{1}H nmr spectrum presented in Figure 13.45.

13.33 Figure 13.46 presents several types of spectroscopic data (ir, ^{1}H nmr, ^{13}C nmr, and mass spectra) for a particular compound. What is it?

13.34 [18]-Annulene exhibits a ^{1}H nmr spectrum that is unusual in that in addition to a peak at $\delta = 8.8$ ppm, it contains a second peak having a chemical shift δ of -1.9 ppm. A negative value for the chemical shift δ indicates that the protons are *more* shielded than those of tetramethylsilane. This peak is 1.9 ppm *upfield* from the TMS peak. The high-field peak has half the area of the low-field peak. Suggest an explanation for these observations.

[18]-Annulene

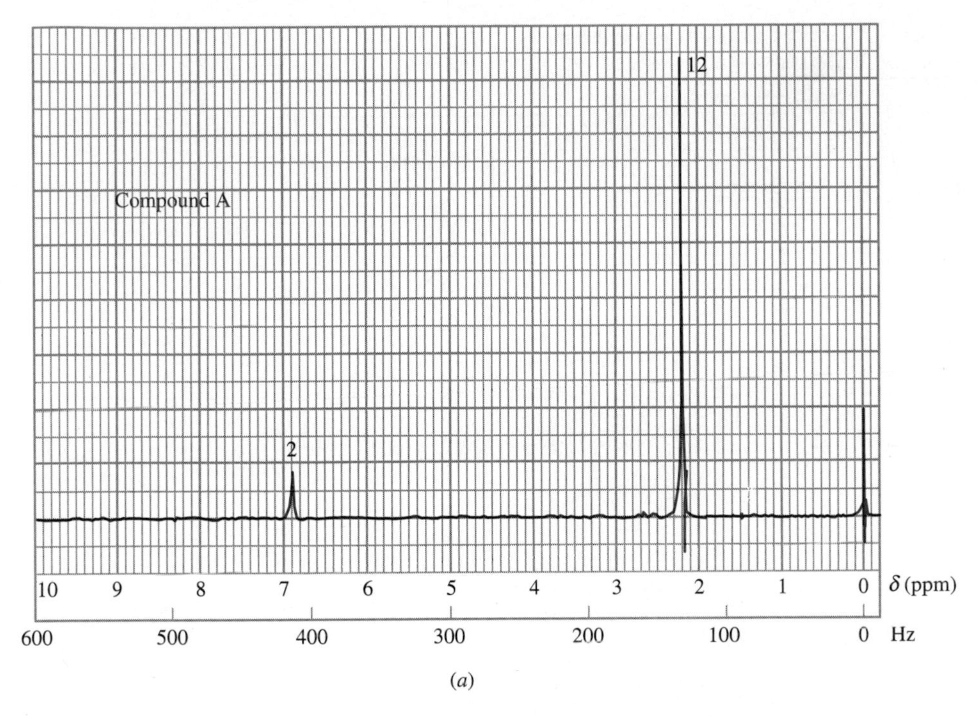

(*a*)

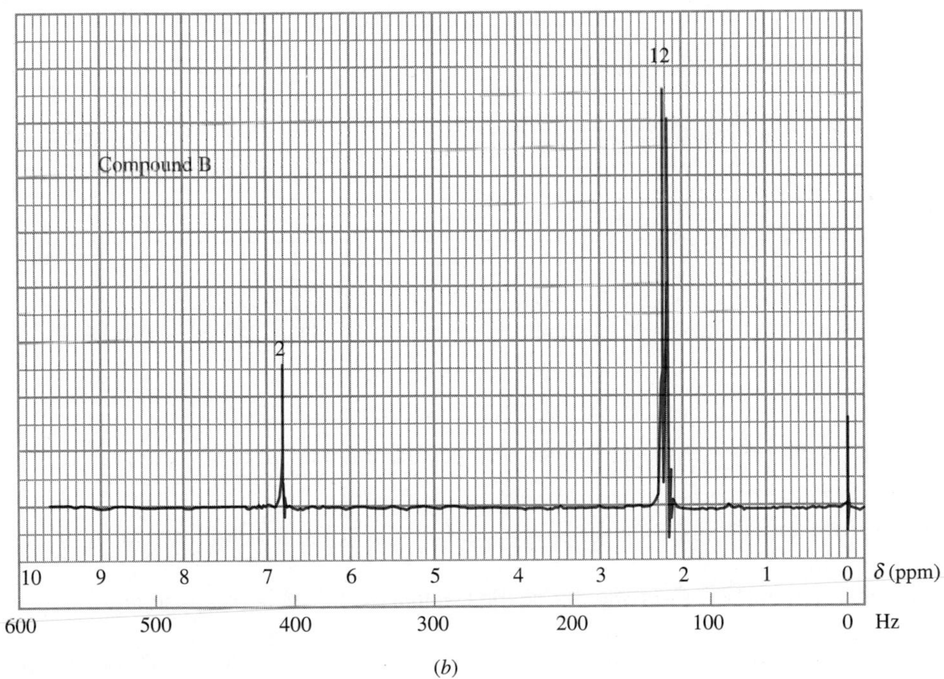

(*b*)

FIGURE 13.43 The 60-MHz ^1H nmr spectrum of (*a*) compound A and (*b*) compound B, isomers of $C_{10}H_{14}$ (Problem 13.30).

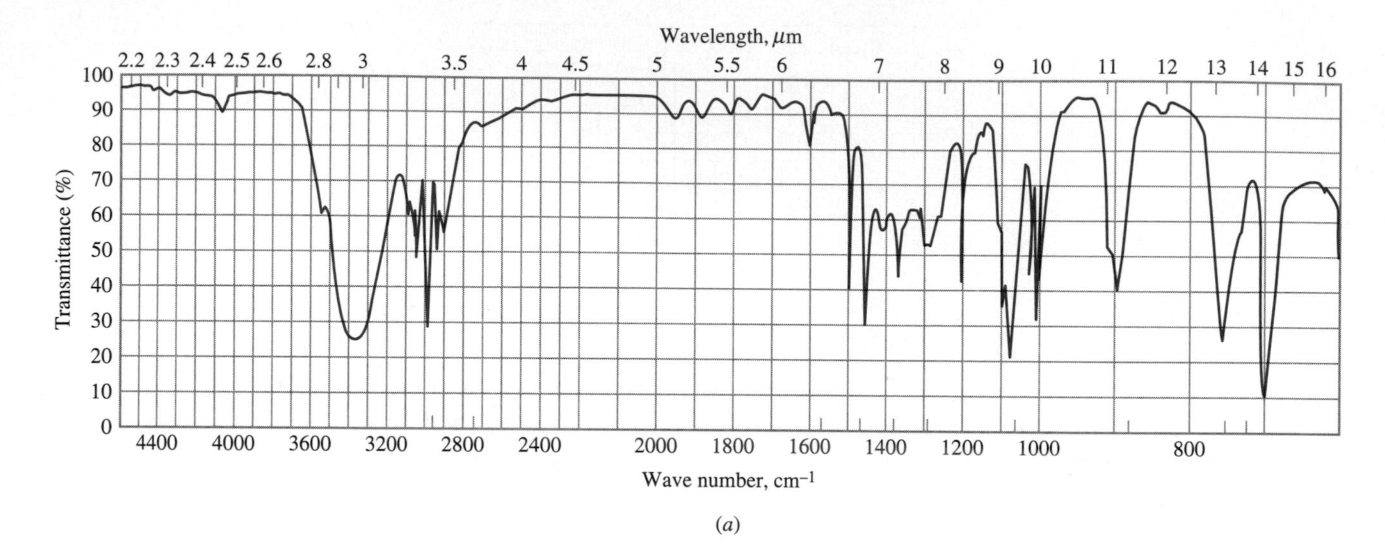

(a)

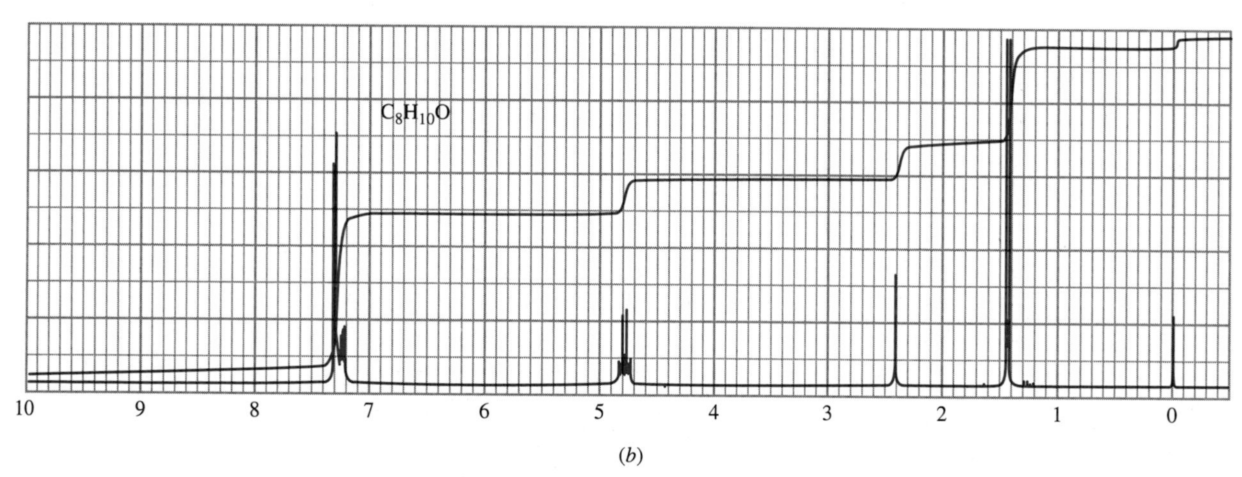

(b)

FIGURE 13.44 (*a*) Infrared and (*b*) 300-MHz ^{1}H nmr spectra of the compound $C_8H_{10}O$ (Problem 13.31).

13.35 ^{19}F is the only isotope of fluorine that occurs naturally, and it has a nuclear spin of $\pm\frac{1}{2}$.

(*a*) Into how many peaks will the proton signal in the ^{1}H nmr spectrum of methyl fluoride be split?

(*b*) Into how many peaks will the fluorine signal in the ^{19}F nmr spectrum of methyl fluoride be split?

(*c*) The chemical shift of the protons in methyl fluoride is $\delta = 4.3$ ppm. Given that the geminal ^{1}H—^{19}F coupling constant is 45 Hz, specify the δ values at which peaks are observed in the proton spectrum of this compound at 60 MHz.

13.36 Given the coupling constants of 1,1,1,2-tetrafluoroethane as shown in the diagram on the facing page, answer the following questions about the ^{1}H and ^{19}F nmr spectra of this compound:

(*a*) How many signals are there in the ^{1}H nmr spectrum of this compound? Into how many peaks are these signals split?

$^3J_{HF} = 8$ Hz

F–C(F)(F)–C(H)(F)–H

$^2J_{HF} = 45$ Hz

$^3J_{FF} = 15$ Hz

(*b*) How many signals are there in the ^{19}F nmr spectrum of this compound? Into how many peaks are these signals split?

13.37 ^{31}P is the only phosphorus isotope present at natural abundance and has a nuclear spin of $\pm\frac{1}{2}$. The ^{1}H nmr spectrum of trimethyl phosphite, $(CH_3O)_3P$, exhibits a doublet for the methyl protons with a splitting of 12 Hz.

(*a*) Into how many peaks is the ^{31}P signal split?

(*b*) What is the difference in chemical shift (in hertz) between the lowest- and highest-field peaks of the ^{31}P multiplet?

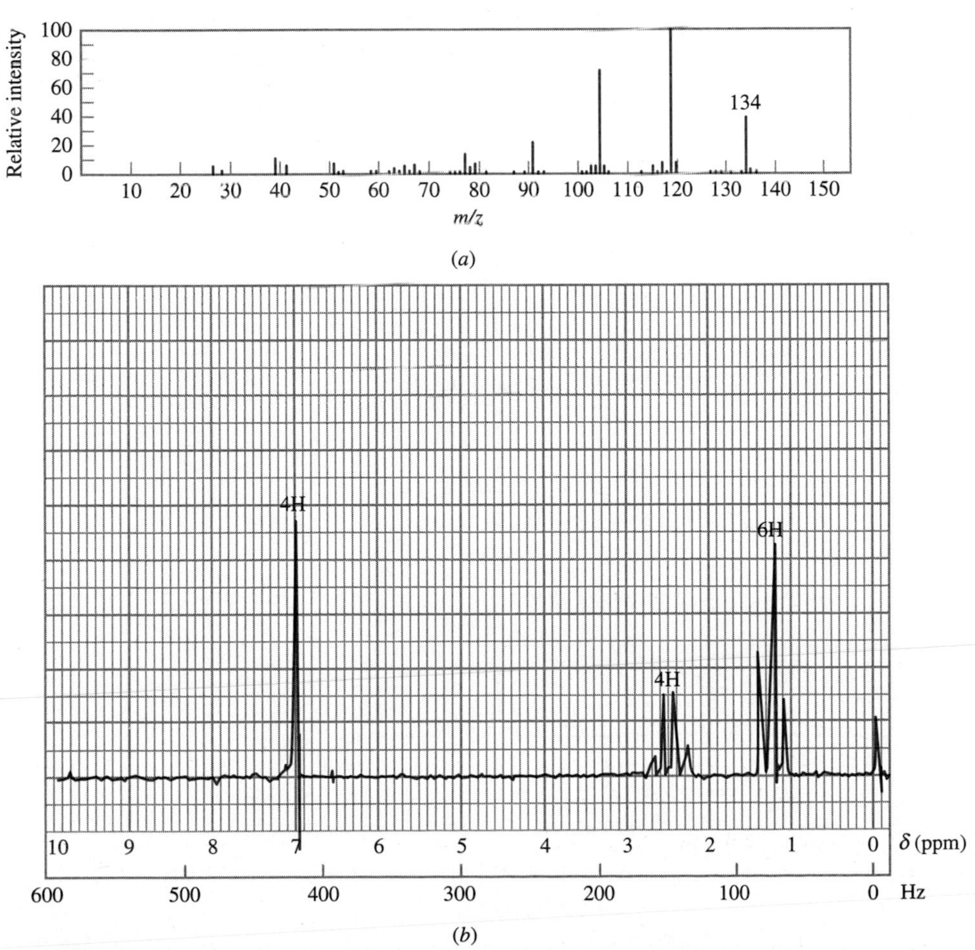

FIGURE 13.45 (*a*) Mass spectrum and (*b*) 60-MHz ^{1}H nmr spectrum of the compound in Problem 13.32.

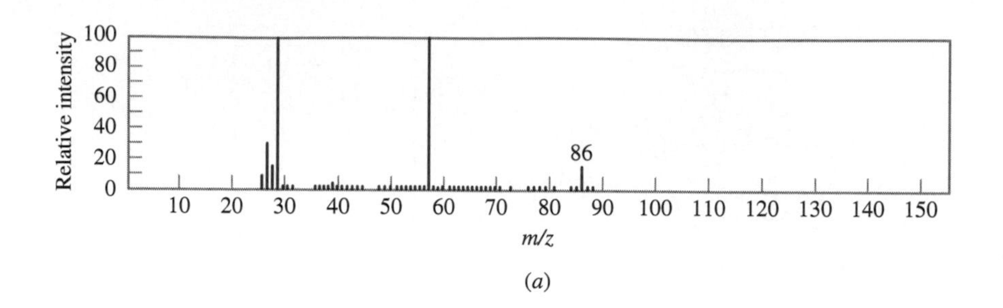

(a)

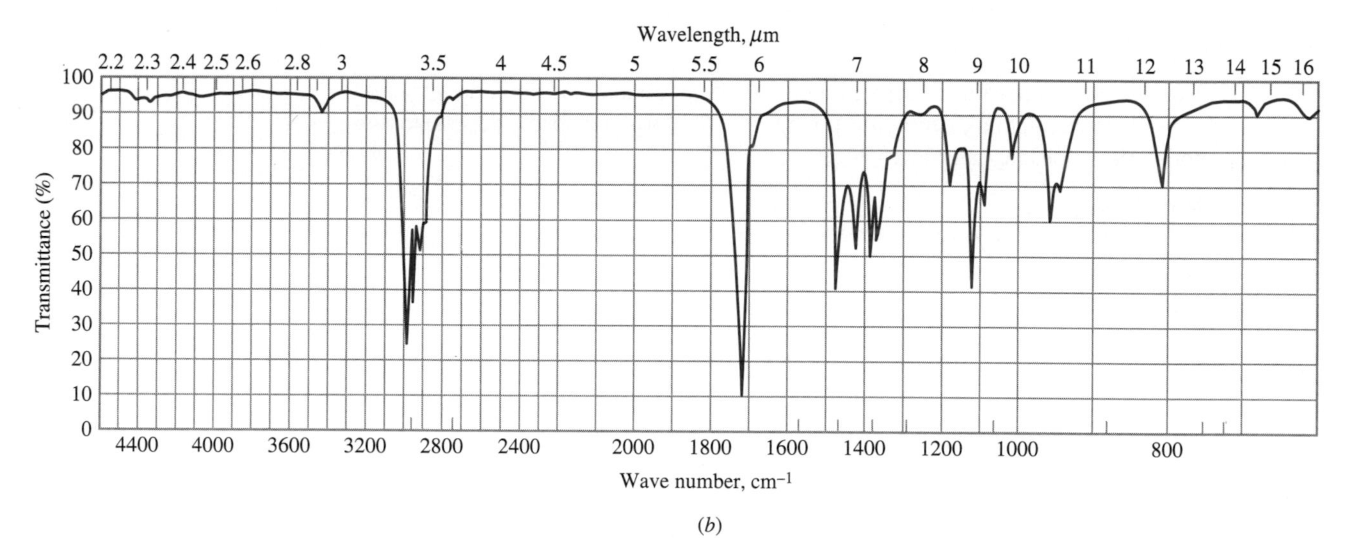

(b)

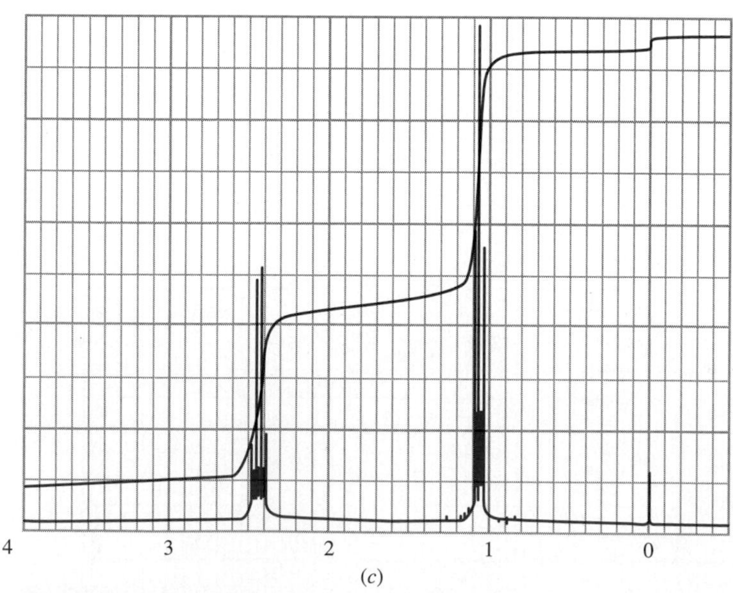

(c)

FIGURE 13.46 (*a*) Mass, (*b*) infrared, (*c*) 300-MHz ^{1}H nmr, and (*d*) ^{13}C nmr spectra for the compound of Problem 13.33.

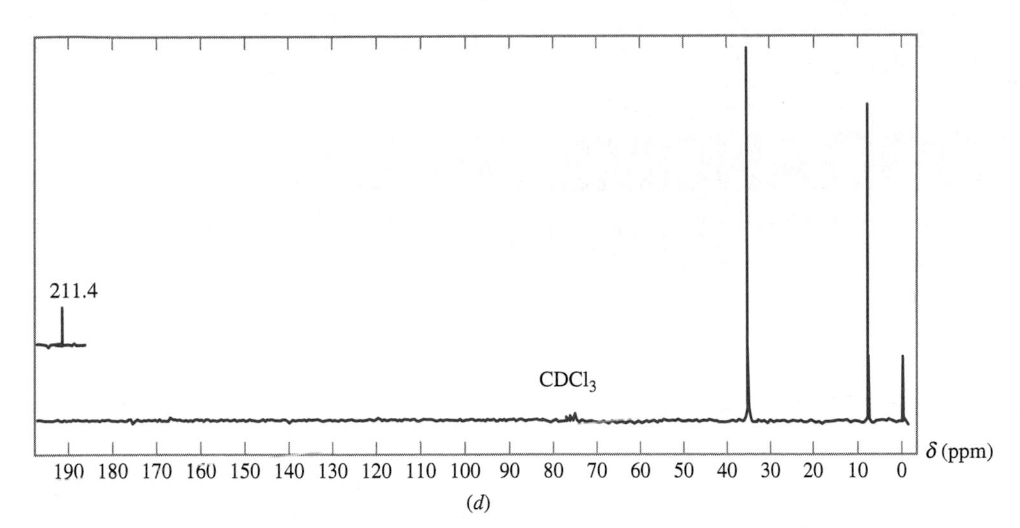

FIGURE 13.46
(Continued)

MOLECULAR MODELING EXERCISES

The dependence of ^3J on dihedral angle is referred to as the **Karplus relationship** after Martin Karplus (Harvard University) who offered the presently accepted theoretical treatment of it.

13.38 In general, the vicinal coupling constant between two protons varies with the angle between the C—H bonds of the H—C—C—H unit. The coupling constant is greatest when the protons are periplanar (dihedral angle = 0° or 180°) and smallest when the angle is approximately 90°. Compare, with the aid of appropriate molecular models, the splitting pattern expected for the proton at C-1 in *cis*-1-bromo-4-*tert*-butylcyclohexane with that of its trans stereoisomer.

13.39 Using the Karplus relationship (described in the previous problem), describe how you could distinguish between *cis*-1-bromo-2-chlorocyclopropane and its trans stereoisomer on the basis of their ^{1}H nmr spectra.

13.40 The π-π^* transition in the UV spectrum of *trans*-stilbene (*trans*-$C_6H_5CH{=}CHC_6H_5$) appears at 295 nm compared to 283 nm for the cis stereoisomer. The extinction coefficient ε_{max} is approximately twice as great for *trans*-stilbene as for *cis*-stilbene. Both facts are normally interpreted in terms of more effective conjugation of the π-electron system in *trans*-stilbene. Construct a molecular model of each stereoisomer and identify the reason for the decreased effectiveness of conjugation in *cis*-stilbene.

13.41 There are three vibrational modes available to a water molecule. Two are stretching modes, one is bending. Illustrate each of these modes with the aid of a molecular model.

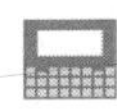

13.42 The protons in the methyl group shown in italics in the structure below are highly shielded and give a signal 0.38 ppm *upfield* from TMS. The other methyl group on the same carbon has a more normal chemical shift of 0.86 ppm downfield from TMS. Why is the indicated methyl group so highly shielded?

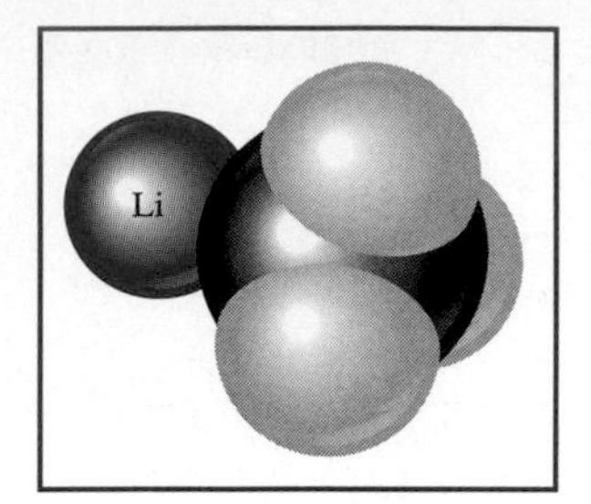

CHAPTER 14

ORGANOMETALLIC COMPOUNDS

Organometallic compounds occupy the interface at which organic and inorganic chemistry meet. *Organometallic compounds are compounds that contain a carbon-metal bond.* You are already familiar with at least one organometallic compound, sodium acetylide (NaC≡CH), which has an ionic bond between carbon and sodium. But simply having a metal and carbon in the same compound is not sufficient for classification as organometallic. Sodium methoxide ($NaOCH_3$), for example, is *not* an organometallic compound; sodium is bonded to oxygen, not to carbon.

Organometallic compounds have rather different properties from those of other classes of compounds that we have encountered so far. Most prominently, many organometallic compounds serve as powerful sources of nucleophilic carbon. It is this feature that makes them especially valuable to the synthetic organic chemist, a feature exploited in Chapter 9, where the preparation of alkynes by alkylation of sodium acetylide was described. Synthetic procedures employing organometallic reagents are the principal methods for carbon-carbon bond formation in organic chemistry. In this chapter you will learn how to prepare organic derivatives of lithium, magnesium, copper, zinc, and mercury and see how their novel properties are used to advantage in organic synthesis.

14.1 CARBON-METAL BONDS IN ORGANOMETALLIC COMPOUNDS

Carbon, a group IV element, is neither strongly electropositive nor strongly electronegative. When carbon is bonded to an atom more electronegative than itself, such as oxygen or chlorine, the electron distribution in the bond is polarized so that carbon is slightly positive and the more electronegative atom is slightly negative. Conversely, when carbon is bonded to an atom that is more electropositive, such as a metal, the electrons in the bond are more strongly attracted toward carbon.

$$\overset{\delta+}{C}-\overset{\delta-}{X} \qquad \overset{\delta-}{C}-\overset{\delta+}{M}$$

X is more electronegative than carbon — M is less electronegative than carbon

The electronegativity of carbon on the Pauling scale (Section 1.5) is 2.5. Table 14.1 compares the electronegativities of some metals with the familiar nonmetals nitrogen, oxygen, and fluorine. Hydrogen, with an electronegativity of 2.1, is only slightly more electropositive than carbon, and the carbon-hydrogen bond is not very polar. The group I and II metals in Table 14.1 are much more electropositive, and their bonds to carbon are relatively polar.

A species containing a negatively charged carbon is referred to as a **carbanion.** Covalently bonded organometallic compounds are said to have *carbanionic character.* As the metal becomes more electropositive, the ionic character of the carbon-metal bond becomes more pronounced. Organosodium and organopotassium compounds have ionic carbon-metal bonds; organolithium and organomagnesium compounds tend to have covalent, but rather polar, carbon-metal bonds with significant carbanionic character. *It is the carbanionic character of such compounds that is responsible for their usefulness as synthetic reagents.*

TABLE 14.1 Electronegativities of Some Representative Elements

Element	Electro-negativity
F	4.0
O	3.5
N	3.0
C	2.5
H	2.1
Cu	1.9
Hg	1.9
Zn	1.6
Mg	1.2
Li	1.0
Na	0.9
K	0.8

14.2 ORGANOMETALLIC NOMENCLATURE

Organometallic compounds are named as substituted derivatives of metals. The metal is the base name, and the attached alkyl groups are identified by the appropriate prefix.

Cyclopropyllithium — $CH_2{=}CHNa$ Vinylsodium — $(CH_3CH_2)_2Mg$ Diethylmagnesium

When the metal bears a substituent other than carbon, the substituent is treated as if it were an anion and named separately.

CH_3MgI Methylmagnesium iodide — $(CH_3CH_2)_2AlCl$ Diethylaluminum chloride

PROBLEM 14.1 Each of the following organometallic reagents will be encountered later in this chapter. Suggest a suitable name for each.

(*a*) $(CH_3)_3CLi$ (*b*) cyclohexyl(H)MgCl (*c*) ICH_2ZnI

SAMPLE SOLUTION (*a*) The metal lithium provides the base name for $(CH_3)_3CLi$. The alkyl group to which lithium is bonded is *tert*-butyl, and so the name of this organometallic compound is *tert*-butylithium. An alternative, equally correct name is 1,1-dimethylethyllithium.

An exception to this type of nomenclature is NaC≡CH, which is normally referred to as *sodium acetylide.* Both sodium acetylide and ethynylsodium are acceptable IUPAC names.

14.3 PREPARATION OF ORGANOLITHIUM COMPOUNDS

Before we describe the applications of organometallic reagents to organic synthesis, let us examine their preparation. Organolithium compounds and other group I organometallic compounds are prepared by the reaction of an alkyl halide with the appropriate metal.

The reaction of an alkyl halide with lithium was cited earlier (Section 2.18) as an example of an *oxidation-reduction.* Group I metals are powerful reducing agents.

$$\underset{\text{Alkyl halide}}{RX} + \underset{\text{Group I metal}}{2M} \longrightarrow \underset{\text{Group I organometallic compound}}{RM} + \underset{\text{Metal halide}}{M^+X^-}$$

$$\underset{\text{Butyl bromide}}{CH_3CH_2CH_2CH_2Br} + \underset{\text{Lithium}}{2Li} \xrightarrow[-10°C]{\text{diethyl ether}} \underset{\text{Butyllithium (80–90\%)}}{CH_3CH_2CH_2CH_2Li} + \underset{\text{Lithium bromide}}{LiBr}$$

$$\underset{\textit{tert}\text{-Butyl chloride}}{(CH_3)_3CCl} + \underset{\text{Lithium}}{2Li} \xrightarrow[-30°C]{\text{diethyl ether}} \underset{\textit{tert}\text{-Butyllithium (75\%)}}{(CH_3)_3CLi} + \underset{\text{Lithium chloride}}{LiCl}$$

Unlike elimination and nucleophilic substitution reactions, formation of organolithium compounds does not require that the halogen be bonded to sp^3 hybridized carbon. Compounds such as vinyl halides and aryl halides, in which the halogen is bonded to sp^2 hybridized carbon, react in the same way as alkyl halides, but somewhat less readily.

$$\underset{\text{Bromobenzene}}{C_6H_5Br} + \underset{\text{Lithium}}{2Li} \xrightarrow[35°C]{\text{diethyl ether}} \underset{\text{Phenyllithium (95–99\%)}}{C_6H_5Li} + \underset{\text{Lithium bromide}}{LiBr}$$

PROBLEM 14.2 Write an equation showing the formation of each of the following from the appropriate bromide:

(*a*) Isopropenyllithium (*b*) *sec*-Butyllithium (*c*) Benzylsodium

SAMPLE SOLUTION (*a*) In the preparation of organolithium compounds from organic halides, lithium becomes bonded to the carbon that bore the halogen. Therefore, isopropenyllithium must arise from isopropenyl bromide.

$$\underset{\text{Isopropenyl bromide}}{CH_2{=}C(Br)CH_3} + \underset{\text{Lithium}}{2Li} \longrightarrow \underset{\text{Isopropenyllithium}}{CH_2{=}C(Li)CH_3} + \underset{\text{Lithium bromide}}{LiBr}$$

Reaction with an alkyl halide takes place at the metal surface. In the first step, an electron is transferred from the metal to the alkyl halide.

$$\underset{\text{Alkyl halide}}{R{:}\ddot{\underset{..}{X}}{:}} + \underset{\text{Lithium}}{Li\cdot} \longrightarrow \underset{\text{Anion radical}}{[R{:}\ddot{\underset{..}{X}}{:}]^{\overset{\cdot}{-}}} + \underset{\text{Lithium cation}}{Li^+}$$

Since the alkyl halide has gained one electron, it is negatively charged and has an odd number of electrons. It is an *anion radical.* The extra electron occupies an antibonding orbital. This anion radical fragments to an alkyl radical and a halide anion.

$$\underset{\text{Anion radical}}{[R{:}\ddot{\underset{..}{X}}{:}]^{\overset{\cdot}{-}}} \longrightarrow \underset{\text{Alkyl radical}}{R\cdot} + \underset{\text{Halide anion}}{{:}\ddot{\underset{..}{X}}{:}^-}$$

Following fragmentation, the alkyl radical rapidly combines with a metal atom to form the organometallic compound.

$$\underset{\text{Alkyl radical}}{R\cdot} + \underset{\text{Lithium}}{Li\cdot} \longrightarrow \underset{\text{Alkyllithium}}{R{:}Li}$$

Among alkyl halides the order of reactivity is I > Br > Cl > F. Bromides combine high reactivity with ready availability and are used most often. Fluorides are relatively unreactive.

Organolithium compounds are sometimes prepared in hydrocarbon solvents such as pentane and hexane, but normally diethyl ether is used. *It is especially important that the solvent be anhydrous.* Even trace amounts of water or alcohols react with lithium to form insoluble hydroxide or alkoxide salts that coat the surface of the metal and prevent it from reacting with the alkyl halide. Furthermore, organolithium reagents are strong bases and react rapidly with even weak proton sources to form hydrocarbons. We shall discuss this property of organolithium reagents in Section 14.5.

14.4 PREPARATION OF ORGANOMAGNESIUM COMPOUNDS: GRIGNARD REAGENTS

The most important organometallic reagents in organic chemistry are organomagnesium compounds. They are called **Grignard reagents** after the French chemist Victor Grignard. Grignard developed efficient methods for the preparation of organic derivatives of magnesium and demonstrated their application in the synthesis of alcohols. For these achievements he was a corecipient of the 1912 Nobel Prize in chemistry.

Grignard shared the prize with Paul Sabatier, who, as was mentioned in Chapter 6, showed that finely divided nickel could be used to catalyze the hydrogenation of alkenes.

Grignard reagents are prepared directly from organic halides by reaction with magnesium, a group II metal.

$$\underset{\text{Organic halide}}{RX} + \underset{\text{Magnesium}}{Mg} \longrightarrow \underset{\text{Organomagnesium halide}}{RMgX}$$

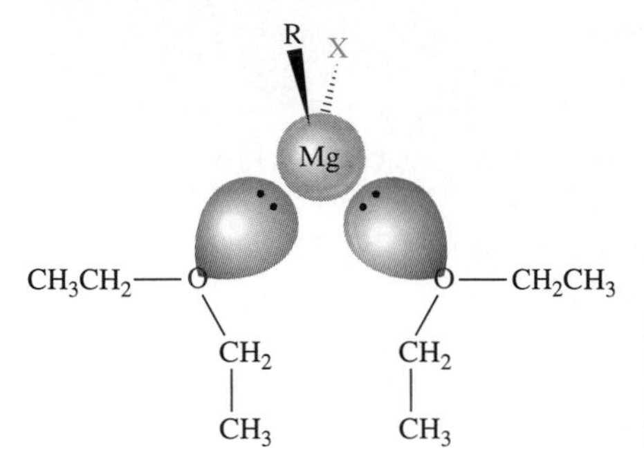

FIGURE 14.1 Solvation of an alkylmagnesium halide by diethyl ether. R, X, and the two diethyl ether molecules are arranged around magnesium in a tetrahedral geometry.

(R may be methyl or primary, secondary, or tertiary alkyl; it may also be a cycloalkyl, alkenyl, or aryl group.)

$$\text{Cyclohexyl chloride} + \text{Mg (Magnesium)} \xrightarrow[35°\text{C}]{\text{diethyl ether}} \text{Cyclohexylmagnesium chloride (96\%)}$$

$$\text{Bromobenzene} + \text{Mg (Magnesium)} \xrightarrow[35°\text{C}]{\text{diethyl ether}} \text{Phenylmagnesium bromide (95\%)}$$

Anhydrous diethyl ether is the customary solvent for the preparation of organomagnesium compounds. It stabilizes the Grignard reagent by forming a Lewis acid–Lewis base complex with it, as shown in Figure 14.1. Sometimes the reaction of an organic halide does not begin readily, but once started, it is exothermic and maintains the temperature of the reaction mixture at the boiling point of diethyl ether (35°C).

The order of halide reactivity is $I > Br > Cl > F$, and alkyl halides are more reactive than aryl and vinyl halides. Indeed, aryl and vinyl chlorides do not form Grignard reagents in diethyl ether. When more vigorous reaction conditions are required, tetrahydrofuran (THF) is used as the solvent. Its boiling point (65°C) is somewhat higher than that of diethyl ether and, more important, it forms a more stable complex with the Grignard reagent, which seems to increase the rate of its formation.

Recall the structure of tetrahydrofuran from Section 3.16:

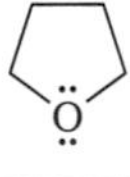

$$\underset{\text{Vinyl chloride}}{CH_2{=}CHCl} \xrightarrow[\text{THF, 60°C}]{\text{Mg}} \underset{\text{Vinylmagnesium chloride (92\%)}}{CH_2{=}CHMgCl}$$

PROBLEM 14.3 Write the structure of the alkylmagnesium halide or arylmagnesium halide formed from each of the following compounds on reaction with magnesium in diethyl ether:

(*a*) *p*-Bromofluorobenzene
(*b*) Allyl chloride
(*c*) Iodocyclobutane
(*d*) 1-Bromocyclohexene

SAMPLE SOLUTION (*a*) Of the two halogen substituents on the aromatic ring, bromine reacts much faster with magnesium than does fluorine. Therefore, fluorine is left intact on the ring, while the carbon-bromine bond is converted to a carbon-magnesium bond.

$$\text{F—C}_6\text{H}_4\text{—Br} + \text{Mg} \xrightarrow{\text{diethyl ether}} \text{F—C}_6\text{H}_4\text{—MgBr}$$

p-Bromofluorobenzene   Magnesium   p-Fluorophenylmagnesium bromide

---

The formation of a Grignard reagent can be viewed as taking place in a manner analogous to that of organolithium reagents except that each magnesium atom can participate in two separate one-electron transfer steps:

$$R:\ddot{\underset{..}{X}}: + \dot{Mg}\cdot \longrightarrow [R:\ddot{\underset{..}{X}}:]^{\overset{\cdot}{-}} + \overset{+}{Mg}\cdot$$

Alkyl halide   Magnesium   Anion radical

$$[R:\ddot{\underset{..}{X}}:]^{\overset{\cdot}{-}} \longrightarrow R\cdot + :\ddot{\underset{..}{X}}:^- \xrightarrow{\overset{+}{Mg}\cdot} R:Mg^+:\ddot{\underset{..}{X}}:^-$$

Anion radical   Alkyl radical   Halide ion   Alkylmagnesium halide

Organolithium and organomagnesium compounds find their chief use in the preparation of alcohols by reaction with aldehydes and ketones. Before discussing these reactions, let us first examine the reactions of these organometallic compounds with proton donors.

14.5 ORGANOLITHIUM AND ORGANOMAGNESIUM COMPOUNDS AS BRØNSTED BASES

Organolithium and organomagnesium compounds are stable species when prepared in suitable solvents such as diethyl ether. They are strongly basic, however, and react instantly with proton donors even as weakly acidic as water and alcohols. A proton is transferred from the hydroxyl group to the negatively polarized carbon of the organometallic compound to form a hydrocarbon.

$$\begin{matrix} {}^{\delta-}R & H^{\delta+} \\ | & | \\ {}_{\delta+}M & :\underset{..}{O}R' \\ & \delta- \end{matrix} \longrightarrow R\text{—}H + R'\ddot{\underset{..}{O}}:^- \quad M^+$$

$$CH_3CH_2CH_2CH_2Li + H_2O \longrightarrow CH_3CH_2CH_2CH_3 + LiOH$$

Butyllithium   Water   Butane (100%)   Lithium hydroxide

$$C_6H_5\text{—}MgBr + CH_3OH \longrightarrow C_6H_6 + CH_3OMgBr$$

Phenylmagnesium bromide   Methanol   Benzene (100%)   Methoxymagnesium bromide

Because of their basicity organolithium compounds and Grignard reagents cannot be prepared or used in the presence of any material that bears a hydroxyl group. Nor are these reagents compatible with —NH or —SH groups, which can also function as proton donors and convert an organolithium or organomagnesium compound to a hydrocarbon.

Organolithium and organomagnesium compounds possess a significant degree of carbanionic character in their carbon-metal bonds. Carbanions rank among the strongest bases that we will encounter in this text. Their conjugate acids are hydrocarbons—very weak acids indeed. The equilibrium constants K_a for ionization of hydrocarbons are much smaller than K_a for water and alcohols.

$$\text{C—H} \rightleftharpoons \text{H}^+ + \text{C}^-$$

Hydrocarbon — Proton — Carbanion

Table 14.2 presents some approximate data for the acid strengths of representative hydrocarbons.

Acidity increases in progressing from the top of Table 14.2 to the bottom. An acid will transfer a proton to the conjugate base of any acid above it in the table. Organolithium compounds and Grignard reagents function as carbanion equivalents and will abstract a proton from any species more acidic than a hydrocarbon. Thus, N—H groups and terminal alkynes (RC≡C—H) are converted to their conjugate bases by proton transfer to organolithium and organomagnesium compounds.

$$CH_3Li + NH_3 \longrightarrow CH_4 + LiNH_2$$

Methyllithium (stronger base); Ammonia (stronger acid: $K_a = 10^{-36}$); Methane (weaker acid: $K_a \cong 10^{-60}$); Lithium amide (weaker base)

$$CH_3CH_2MgBr + HC{\equiv}CH \longrightarrow CH_3CH_3 + HC{\equiv}CMgBr$$

Ethylmagnesium bromide (stronger base); Acetylene (stronger acid: $K_a \cong 10^{-26}$); Ethane (weaker acid: $K_a \cong 10^{-62}$); Ethynylmagnesium bromide (weaker base)

TABLE 14.2
Approximate Acidities of Some Hydrocarbons and Reference Materials

Compound	Formula*	K_a	pK_a	Conjugate base
2-Methylpropane	$(CH_3)_3C—H$	10^{-71}	71	$(CH_3)_3\bar{C}:$
Ethane	$CH_3CH_2—H$	10^{-62}	62	$CH_3\bar{C}H_2$
Methane	$CH_3—H$	10^{-60}	60	$H_3\bar{C}:$
Ethylene	$CH_2{=}CH—H$	10^{-45}	45	$CH_2{=}\bar{C}H$
Benzene	$C_6H_5—H$	10^{-43}	43	$C_6H_5^-$
Ammonia	$H_2N—H$	10^{-36}	36	$H_2\ddot{N}:^-$
Acetylene	$HC{\equiv}CH$	10^{-26}	26	$HC{\equiv}\bar{C}:$
Ethanol	$CH_3CH_2O—H$	10^{-16}	16	$CH_3CH_2O^-$
Water	$HO—H$	1.8×10^{-16}	15.7	HO^-

* The acidic proton in each compound is in red.

PROBLEM 14.4 Butyllithium is commercially available and is frequently used by organic chemists as a strong base. Show how you could use butyllithium to prepare solutions containing

(*a*) Lithium diethylamide, $(CH_3CH_2)_2NLi$
(*b*) Lithium 1-hexanolate, $CH_3(CH_2)_4CH_2OLi$
(*c*) Lithium benzenethiolate, C_6H_5SLi

SAMPLE SOLUTION When butyllithium is used as a base, it abstracts a proton, in this case a proton attached to nitrogen. The source of lithium diethylamide must be diethylamine.

$$\underset{\substack{\text{Diethylamine}\\ \text{(stronger acid)}}}{(CH_3CH_2)_2NH} + \underset{\substack{\text{Butyllithium}\\ \text{(stronger base)}}}{CH_3CH_2CH_2CH_2Li} \longrightarrow \underset{\substack{\text{Lithium}\\ \text{diethylamide}\\ \text{(weaker base)}}}{(CH_3CH_2)_2NLi} + \underset{\substack{\text{Butane}\\ \text{(weaker acid)}}}{CH_3CH_2CH_2CH_3}$$

While diethylamine is not specifically listed in Table 14.2, its strength as an acid ($K_a \cong 10^{-36}$) is, as might be expected, similar to that of ammonia.

It is sometimes necessary in a synthesis to reduce an alkyl halide to a hydrocarbon. In such cases converting the halide to a Grignard reagent and then adding water or an alcohol as a proton source is a satisfactory procedure. Adding D_2O to a Grignard reagent is a commonly used method for introducing deuterium into a molecule at a specific location.

Deuterium is the mass 2 isotope of hydrogen. Deuterium oxide (D_2O) is sometimes called "heavy water."

$$\underset{\text{1-Bromopropene}}{CH_3CH{=}CHBr} \xrightarrow[\text{THF}]{\text{Mg}} \underset{\text{Propenylmagnesium bromide}}{CH_3CH{=}CHMgBr} \xrightarrow{D_2O} \underset{\text{1-Deuteriopropene (70\%)}}{CH_3CH{=}CHD}$$

14.6 SYNTHESIS OF ALCOHOLS USING GRIGNARD REAGENTS

The main synthetic application of Grignard reagents is their reaction with certain carbonyl-containing compounds to produce alcohols. Carbon-carbon bond formation is rapid and exothermic when a Grignard reagent reacts with an aldehyde or ketone.

$$\overset{\delta+}{C}{=}\overset{\delta-}{\ddot{O}}: + \underset{\delta-}{R}{-}\underset{\delta+}{MgX} \longrightarrow -\underset{R}{\overset{|}{C}}-\ddot{\underset{..}{O}}:^{-}\ {}^{+}MgX \quad \text{normally written as} \quad -\underset{R}{\overset{|}{C}}OMgX$$

A carbonyl group is quite polar, and its carbon atom is electrophilic. Grignard reagents are nucleophilic and add to carbonyl groups, forming a new carbon-carbon bond. This addition step leads to formation of the alkoxymagnesium halide, which in the second stage of the synthesis is hydrolyzed to an alcohol.

TABLE 14.3
Reactions of Grignard Reagents with Aldehydes and Ketones

Reaction	General equation and specific example
Reaction with formaldehyde Grignard reagents react with formaldehyde ($CH_2{=}O$) to give *primary alcohols* having one more carbon than the Grignard reagent.	$\underset{\text{Grignard reagent}}{RMgX} + \underset{\text{Formaldehyde}}{HCH{=}O} \xrightarrow{\text{diethyl ether}} \underset{\text{Primary alkoxymagnesium halide}}{R{-}CH_2{-}OMgX} \xrightarrow{H_3O^+} \underset{\text{Primary alcohol}}{R{-}CH_2{-}OH}$ $\underset{\text{Cyclohexylmagnesium chloride}}{C_6H_{11}MgCl} + \underset{\text{Formaldehyde}}{HCH{=}O} \xrightarrow[\text{2. } H_3O^+]{\text{1. diethyl ether}} \underset{\text{Cyclohexylmethanol (64–69\%)}}{C_6H_{11}CH_2OH}$
Reaction with aldehydes Grignard reagents react with aldehydes ($RCH{=}O$) to give *secondary alcohols.*	$\underset{\text{Grignard reagent}}{RMgX} + \underset{\text{Aldehyde}}{R'CH{=}O} \xrightarrow{\text{diethyl ether}} \underset{\text{Secondary alkoxymagnesium halide}}{R{-}CH(R'){-}OMgX} \xrightarrow{H_3O^+} \underset{\text{Secondary alcohol}}{R{-}CH(R'){-}OH}$ $\underset{\text{Hexylmagnesium bromide}}{CH_3(CH_2)_4CH_2MgBr} + \underset{\text{Ethanal (acetaldehyde)}}{CH_3CH{=}O} \xrightarrow[\text{2. } H_3O^+]{\text{1. diethyl ether}} \underset{\text{2-Octanol (84\%)}}{CH_3(CH_2)_4CH_2CH(OH)CH_3}$
Reaction with ketones Grignard reagents react with ketones ($RC({=}O)R'$) to give *tertiary alcohols.*	$\underset{\text{Grignard reagent}}{RMgX} + \underset{\text{Ketone}}{R'C({=}O)R''} \xrightarrow{\text{diethyl ether}} \underset{\text{Tertiary alkoxymagnesium halide}}{R{-}C(R')(R''){-}OMgX} \xrightarrow{H_3O^+} \underset{\text{Tertiary alcohol}}{R{-}C(R')(R''){-}OH}$ $\underset{\text{Methylmagnesium chloride}}{CH_3MgCl} + \underset{\text{Cyclopentanone}}{C_5H_8O} \xrightarrow[\text{2. } H_3O^+]{\text{1. diethyl ether}} \underset{\text{1-Methylcyclopentanol (62\%)}}{1\text{-}CH_3\text{-}C_5H_8\text{-}1\text{-}OH}$

$$\underset{\text{Alkoxymagnesium halide}}{R-\overset{|}{\underset{|}{C}}-OMgX} + \underset{\text{Hydronium ion}}{H_3O^+} \longrightarrow \underset{\text{Alcohol}}{R-\overset{|}{\underset{|}{C}}-OH} + \underset{\text{Magnesium ion}}{Mg^{2+}} + \underset{\text{Halide ion}}{X^-} + \underset{\text{Water}}{H_2O}$$

The type of alcohol produced depends on the carbonyl-containing compound used. Substituents present on the carbonyl group of an aldehyde or ketone stay there—they become substituents on the carbon that bears the hydroxyl group in the product. Thus as shown in Table 14.3, formaldehyde reacts with Grignard reagents to yield primary alcohols, aldehydes yield secondary alcohols, and ketones yield tertiary alcohols.

PROBLEM 14.5 Write the structure of the product of the reaction of propylmagnesium bromide with each of the following. Assume that the reactions are worked up by the addition of dilute aqueous acid.

(*a*) Formaldehyde, $H\overset{O}{\overset{\|}{C}}H$

(*b*) Benzaldehyde, $C_6H_5\overset{O}{\overset{\|}{C}}H$

(*c*) Cyclohexanone, (cyclohexane ring)=O

(*d*) 2-Butanone, $CH_3\overset{O}{\overset{\|}{C}}CH_2CH_3$

SAMPLE SOLUTION (*a*) Grignard reagents react with formaldehyde to give primary alcohols having one more carbon atom than the alkyl halide from which the Grignard reagent was prepared. The product is 1-butanol.

$$\underset{\text{Propylmagnesium bromide + formaldehyde}}{CH_3CH_2CH_2-MgBr \quad H_2C{=}O} \xrightarrow{\text{diethyl ether}} CH_3CH_2\underset{\substack{|\\ H-C-OMgBr\\ |\\ H}}{CH_2} \xrightarrow{H_3O^+} \underset{\text{1-Butanol}}{CH_3CH_2CH_2CH_2OH}$$

The ability to form carbon-carbon bonds is fundamental to organic synthesis. The addition of Grignard reagents to aldehydes and ketones is one of the most frequently used reactions in synthetic organic chemistry. Not only does it permit the extension of carbon chains, but since the product is an alcohol, a wide variety of subsequent functional group transformations are possible.

14.7 SYNTHESIS OF ALCOHOLS USING ORGANOLITHIUM REAGENTS

Organolithium reagents react with carbonyl groups in the same way that Grignard reagents do. In their reactions with aldehydes and ketones, organolithium reagents are somewhat more reactive than Grignard reagents.

$$\underset{\text{Alkyllithium compound}}{RLi} + \underset{\text{Aldehyde or ketone}}{>C{=}O} \longrightarrow \underset{\text{Lithium alkoxide}}{R{-}\overset{|}{\underset{|}{C}}{-}OLi} \xrightarrow{H_3O^+} \underset{\text{Alcohol}}{R{-}\overset{|}{\underset{|}{C}}{-}OH}$$

$$\underset{\text{Vinyllithium}}{CH_2{=}CHLi} + \underset{\text{Benzaldehyde}}{C_6H_5\overset{O}{\overset{\|}{C}}H} \xrightarrow[\text{2. } H_3O^+]{\text{1. diethyl ether}} \underset{\text{1-Phenyl-2-propen-1-ol (76\%)}}{C_6H_5\underset{OH}{\underset{|}{C}}HCH{=}CH_2}$$

14.8 SYNTHESIS OF ACETYLENIC ALCOHOLS

The first organometallic compounds we encountered were compounds of the type $RC{\equiv}CNa$ obtained by treatment of terminal alkynes with sodium amide in liquid ammonia (Section 9.7):

$$\underset{\text{Terminal alkyne}}{RC{\equiv}CH} + \underset{\text{Sodium amide}}{NaNH_2} \xrightarrow[-33^\circ C]{NH_3} \underset{\text{Sodium alkynide}}{RC{\equiv}CNa} + \underset{\text{Ammonia}}{NH_3}$$

These compounds are sources of the nucleophilic anion $RC{\equiv}C{:}^-$, and their reaction with primary alkyl halides provides an effective synthesis of alkynes (Section 9.7). The nucleophilicity of acetylide anions is also evident in their reactions with aldehydes and ketones, which are entirely analogous to those of Grignard and organolithium reagents.

These reactions are normally carried out in liquid ammonia because that is the solvent in which the sodium salt of the alkyne is prepared.

$$\underset{\text{Sodium alkynide}}{RC{\equiv}CNa} + \underset{\text{Aldehyde or ketone}}{R'\overset{O}{\overset{\|}{C}}R''} \xrightarrow{NH_3} \underset{\text{Sodium salt of an alkynyl alcohol}}{RC{\equiv}C{-}\overset{R''}{\overset{|}{\underset{R'}{\underset{|}{C}}}}{-}ONa} \xrightarrow{H_3O^+} \underset{\text{Alkynyl alcohol}}{RC{\equiv}C\overset{R''}{\overset{|}{\underset{R'}{\underset{|}{C}}}}OH}$$

$$\underset{\text{Sodium acetylide}}{HC{\equiv}CNa} + \underset{\text{Cyclohexanone}}{\text{(cyclohexanone)}} \xrightarrow[\text{2. } H_3O^+]{\text{1. } NH_3} \underset{\text{1-Ethynylcyclohexanol (65–75\%)}}{\text{(1-HO, 1-C}{\equiv}\text{CH cyclohexane)}}$$

Acetylenic Grignard reagents of the type $RC{\equiv}CMgBr$ are prepared, not from an acetylenic halide, but by an acid-base reaction in which a Grignard reagent abstracts a proton from a terminal alkyne.

$$CH_3(CH_2)_3C{\equiv}CH + CH_3CH_2MgBr \xrightarrow{\text{diethyl ether}} CH_3(CH_2)_3C{\equiv}CMgBr + CH_3CH_3$$

1-Hexyne — Ethylmagnesium bromide — 1-Hexynylmagnesium bromide — Ethane

$$CH_3(CH_2)_3C{\equiv}CMgBr + \underset{\text{HCH}}{\overset{\text{O}}{\|}} \xrightarrow[\text{2. } H_3O^+]{\text{1. diethyl ether}} CH_3(CH_2)_3C{\equiv}CCH_2OH$$

1-Hexynylmagnesium bromide — Formaldehyde — 2-Heptyn-1-ol (82%)

14.9 RETROSYNTHETIC ANALYSIS

In earlier discussions of synthesis, the importance of reasoning backward from the target molecule to suitable starting materials has been stressed. A name for this process is *retrosynthetic analysis.* Organic chemists have employed this approach for many years, but the term was invented and a formal statement of its principles was set forth only relatively recently by E. J. Corey at Harvard University. Beginning in the 1960s, Corey began studies aimed at making the strategy of organic synthesis sufficiently systematic so that the power of electronic computers could be applied to assist synthetic planning.

Corey was honored with the 1990 Nobel Prize for his achievements in synthetic organic chemistry.

A symbol used to indicate a retrosynthetic step is an open arrow written from product to suitable precursors or fragments of those precursors.

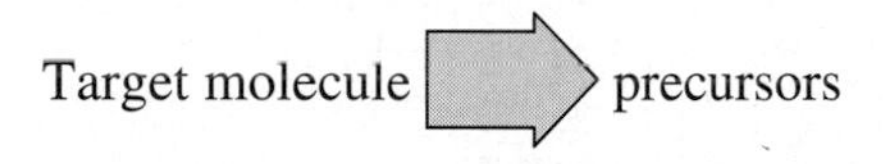

Often the precursor is not defined completely, but rather its chemical nature is emphasized by writing it as a species to which it is equivalent for synthetic purposes. Thus, a Grignard reagent or an organolithium reagent might be considered synthetically equivalent to a carbanion:

$$\text{RMgX or RLi} \quad \text{is synthetically equivalent to } R{:}^-$$

Figure 14.2 illustrates how retrosynthetic analysis can guide you in designing the synthesis of alcohols by identifying suitable Grignard reagent and carbonyl-containing precursors. In the first step, locate the carbon of the target alcohol that bears the hydroxyl group, remembering that this carbon originated in the C=O group. Next, as shown in Figure 14.2, step 2, mentally disconnect a bond between that carbon and one of its attached groups (other than hydrogen). The attached group is the group that is to be transferred from the Grignard reagent. Once you recognize these two structural fragments, the carbonyl partner and the carbanion that attacks it (Figure 14.2, step 3), you can readily determine the synthetic mode wherein a Grignard reagent is used as the synthetic equivalent of a carbanion (Figure 14.2, step 4).

Primary alcohols, by this analysis, are seen to be the products of Grignard addition to formaldehyde:

Disconnect this bond

$$\mathrm{R-CH_2-OH} \Rightarrow \mathrm{R:^-} \quad \mathrm{H_2C{=}O}$$

Secondary alcohols may be prepared by *two* different combinations of Grignard reagent and aldehyde:

Disconnect R—C   Disconnect R′—C

$$\mathrm{R:^-} \quad \mathrm{R'CH{=}O} \Leftarrow \mathrm{R-CH(R')-OH} \Rightarrow \mathrm{R':^-} \quad \mathrm{RCH{=}O}$$

Step 1: Locate the hydroxyl-bearing carbon.

$$\mathrm{X-C(R)(OH)-Y}$$

This carbon must have been part of the C=O group in the starting material

Step 2: Disconnect one of the organic substituents attached to the carbon that bears the hydroxyl group.

$$\mathrm{X-C(R)(OH)-Y}$$

Disconnect this bond

Step 3: Steps 1 and 2 reveal the carbonyl-containing substrate and the carbanionic fragment.

$$\mathrm{X-C(R)(OH)-Y} \Rightarrow \mathrm{R^-} \quad \mathrm{XC(=O)Y}$$

Step 4: Since a Grignard reagent may be considered as synthetically equivalent to a carbanion, this suggests the synthesis shown.

$$\mathrm{RMgBr} + \mathrm{XC(=O)Y} \xrightarrow[\text{2. } H_3O^+]{\text{1. diethyl ether}} \mathrm{R-C(X)(Y)-OH}$$

FIGURE 14.2 A retrosynthetic analysis of alcohol preparation by way of the addition of a Grignard reagent to an aldehyde or ketone.

Three combinations of Grignard reagent and ketone give rise to tertiary alcohols:

$$R{:}^- \quad R''(R')C{=}O \;\xleftarrow{\text{Disconnect } R{-}C}\; R{-}\overset{R''}{\underset{R'}{C}}{-}OH \;\xrightarrow{\text{Disconnect } R'{-}C}\; R'{:}^- \quad R''(R)C{=}O$$

$$R{-}\overset{R''}{\underset{R'}{C}}{-}OH \;\xrightarrow{\text{Disconnect } R''{-}C}\; R''{:}^- \quad R(R')C{=}O$$

Usually, there is little to choose among the various routes leading to a particular target alcohol. For example, all three of the following combinations have been used to prepare the tertiary alcohol 2-phenyl-2-butanol:

$$CH_3MgI + C_6H_5\overset{O}{\overset{\|}{C}}CH_2CH_3 \xrightarrow[2.\ H_3O^+]{1.\ \text{diethyl ether}} C_6H_5\overset{OH}{\underset{CH_3}{C}}CH_2CH_3$$

Methylmagnesium iodide; 1-Phenyl-1-propanone; 2-Phenyl-2-butanol

$$CH_3CH_2MgBr + C_6H_5\overset{O}{\overset{\|}{C}}CH_3 \xrightarrow[2.\ H_3O^+]{1.\ \text{diethyl ether}} C_6H_5\overset{OH}{\underset{CH_3}{C}}CH_2CH_3$$

Ethylmagnesium bromide; Acetophenone; 2-Phenyl-2-butanol

$$C_6H_5MgBr + CH_3\overset{O}{\overset{\|}{C}}CH_2CH_3 \xrightarrow[2.\ H_3O^+]{1.\ \text{diethyl ether}} C_6H_5\overset{OH}{\underset{CH_3}{C}}CH_2CH_3$$

Phenylmagnesium bromide; 2-Butanone; 2-Phenyl-2-butanol

PROBLEM 14.6 Suggest two ways in which each of the following alcohols might be prepared by using a Grignard reagent:

(*a*) 2-Hexanol, $CH_3\underset{OH}{\underset{|}{C}}HCH_2CH_2CH_2CH_3$ (*b*) 2-Phenyl-2-propanol, $C_6H_5\underset{OH}{\underset{|}{C}}(CH_3)_2$

SAMPLE SOLUTION (*a*) Since 2-hexanol is a secondary alcohol, we consider the reaction of a Grignard reagent with an aldehyde. Disconnection of bonds to the hydroxyl-bearing carbon generates two pairs of structural fragments:

$$\underset{\displaystyle OH}{CH_3\underset{|}{C}HCH_2CH_2CH_2CH_3} \Rightarrow :\bar{C}H_3 \qquad H\underset{\displaystyle \|\atop O}{C}CH_2CH_2CH_2CH_3$$

and

$$\underset{\displaystyle OH}{CH_3\underset{|}{C}HCH_2CH_2CH_2CH_3} \Rightarrow CH_3\underset{\displaystyle \|\atop O}{C}H \qquad :\bar{C}H_2CH_2CH_2CH_3$$

Therefore, one route involves the addition of a methyl Grignard reagent to a five-carbon aldehyde:

$$\underset{\text{Methylmagnesium iodide}}{CH_3MgI} + \underset{\text{Pentanal}}{CH_3CH_2CH_2CH_2\overset{O}{\overset{\|}{C}}H} \xrightarrow[2.\ H_3O^+]{1.\ \text{diethyl ether}} \underset{\text{2-Hexanol}}{CH_3CH_2CH_2CH_2\underset{OH}{\underset{|}{C}}HCH_3}$$

The other requires addition of a butylmagnesium halide to a two-carbon aldehyde:

$$\underset{\text{Butylmagnesium bromide}}{CH_3CH_2CH_2CH_2MgBr} + \underset{\text{Acetaldehyde}}{CH_3\overset{O}{\overset{\|}{C}}H} \xrightarrow[2.\ H_3O^+]{1.\ \text{diethyl ether}} \underset{\text{2-Hexanol}}{CH_3CH_2CH_2CH_2\underset{OH}{\underset{|}{C}}HCH_3}$$

All that has been said in this section applies with equal force to the use of organolithium reagents in the synthesis of alcohols. Grignard reagents are one source of nucleophilic carbon; organolithium reagents are another. Both have pronounced carbanionic character in their carbon-metal bonds and undergo the same kind of reaction with aldehydes and ketones.

14.10 PREPARATION OF TERTIARY ALCOHOLS FROM ESTERS AND GRIGNARD REAGENTS

Tertiary alcohols can be prepared by a variation of the Grignard synthesis that employs an ester as the carbonyl component. Methyl and ethyl esters are the most readily available esters and are the ones most often used. Two moles of a Grignard reagent are required per mole of ester; the first mole reacts with the ester, converting it to a ketone.

$$\underset{\text{Grignard reagent}}{RMgX} + \underset{\text{Methyl ester}}{R'\overset{O}{\overset{\|}{C}}OCH_3} \xrightarrow{\text{diethyl ether}} R'\overset{O-MgX}{\overset{|}{\underset{R}{\underset{|}{C}}}}-OCH_3 \longrightarrow \underset{\text{Ketone}}{R'\overset{O}{\overset{\|}{C}}R} + \underset{\text{Methoxymagnesium halide}}{CH_3OMgX}$$

The ketone is not isolated, but reacts rapidly with the Grignard reagent to give, after adding aqueous acid, a tertiary alcohol. Ketones are more reactive than esters toward Grignard reagents, and so it is not normally possible to interrupt the reaction at the ketone stage even if only one equivalent of the Grignard reagent is used.

$$\underset{\text{Ketone}}{R'\overset{O}{\overset{\|}{C}}R} + \underset{\text{Grignard reagent}}{RMgX} \xrightarrow[2.\ H_3O^+]{1.\ \text{diethyl ether}} \underset{\text{Tertiary alcohol}}{R'\overset{OH}{\overset{|}{\underset{R}{\underset{|}{C}}}}R}$$

Two of the groups bonded to the hydroxyl-bearing carbon of the alcohol are the same because they are derived from the Grignard reagent. For example,

$$\underset{\text{Methylmagnesium bromide}}{2CH_3MgBr} + \underset{\text{Methyl 2-methylpropanoate}}{(CH_3)_2CH\overset{O}{\overset{\|}{C}}OCH_3} \xrightarrow[2.\ H_3O^+]{1.\ \text{diethyl ether}} \underset{\text{2,3-Dimethyl-2-butanol (73\%)}}{(CH_3)_2CH\overset{OH}{\overset{|}{\underset{CH_3}{\underset{|}{C}}}}CH_3} + \underset{\text{Methanol}}{CH_3OH}$$

PROBLEM 14.7 What combination of ester and Grignard reagent could you use to prepare each of the following tertiary alcohols?

(*a*) $C_6H_5\underset{OH}{\underset{|}{C}}(CH_2CH_3)_2$ (*b*) $(C_6H_5)_2\underset{OH}{\underset{|}{C}}-$cyclopropyl

SAMPLE SOLUTION (*a*) To apply the principles of retrosynthetic analysis to this case, we disconnect both ethyl groups from the tertiary carbon and identify them as arising from the Grignard reagent. The phenyl group originates in an ester of the type $C_6H_5CO_2R$ (a benzoate ester).

$$C_6H_5\underset{OH}{\underset{|}{C}}(CH_2CH_3)_2 \Longrightarrow C_6H_5\underset{O}{\underset{\|}{C}}OR + 2\ CH_3CH_2MgX$$

An appropriate synthesis would be

$$\underset{\text{Ethylmagnesium bromide}}{2CH_3CH_2MgBr} + \underset{\text{Methyl benzoate}}{C_6H_5\overset{O}{\overset{\|}{C}}OCH_3} \xrightarrow[2.\ H_3O^+]{1.\ \text{diethyl ether}} \underset{\text{3-Phenyl-3-pentanol}}{C_6H_5\underset{OH}{\underset{|}{C}}(CH_2CH_3)_2}$$

14.11 ALKANE SYNTHESIS USING ORGANOCOPPER REAGENTS

Organometallic compounds of copper have been known for a long time, but their versatility as reagents in synthetic organic chemistry has only recently been recognized. The most useful organocopper reagents are the lithium dialkylcuprates, which result when a copper(I) halide reacts with two equivalents of an alkyllithium in diethyl ether or tetrahydrofuran.

Copper(I) salts are also known as *cuprous* salts.

$$\underset{\text{Alkyllithium}}{2RLi} + \underset{\substack{\text{Cu(I) halide} \\ (X = Cl, Br, I)}}{CuX} \xrightarrow[\text{THF}]{\text{diethyl ether or}} \underset{\substack{\text{Lithium} \\ \text{dialkylcuprate}}}{R_2CuLi} + \underset{\substack{\text{Lithium} \\ \text{halide}}}{LiX}$$

In the first stage of the preparation, one molar equivalent of alkyllithium displaces halide from copper to give an alkylcopper(I) species:

$$R{-}Li + Cu{-}I \longrightarrow \underset{\text{Alkylcopper}}{RCu} + \underset{\text{Lithium iodide}}{LiI}$$

The second molar equivalent of the alkyllithium adds to the alkylcopper to give a negatively charged dialkyl-substituted derivative of copper(I) called a *dialkylcuprate* anion. It is formed as its lithium salt, a lithium dialkylcuprate.

$$\underset{\text{Alkyllithium}}{Li{-}R} + \underset{\text{Alkylcopper}}{Cu{-}R} \longrightarrow \underset{\substack{\text{Lithium dialkylcuprate} \\ \text{(soluble in ether and in THF)}}}{[R{-}\overset{-}{Cu}{-}R]\,Li^+}$$

Lithium dialkylcuprates react with alkyl halides to produce alkanes by carbon-carbon bond formation between the alkyl group of the alkyl halide and the alkyl group of the dialkylcuprate:

$$\underset{\substack{\text{Lithium} \\ \text{dialkylcuprate}}}{R_2CuLi} + \underset{\text{Alkyl halide}}{R'X} \longrightarrow \underset{\text{Alkane}}{R{-}R'} + \underset{\text{Alkylcopper}}{RCu} + \underset{\substack{\text{Lithium} \\ \text{halide}}}{LiX}$$

Primary alkyl halides, especially iodides, are the best substrates. Elimination becomes a problem with secondary and tertiary alkyl halides:

$$\underset{\substack{\text{Lithium dimethyl-} \\ \text{cuprate}}}{(CH_3)_2CuLi} + \underset{\text{1-Iododecane}}{CH_3(CH_2)_8CH_2I} \xrightarrow[0°C]{\text{ether}} \underset{\text{Undecane (90\%)}}{CH_3(CH_2)_8CH_2CH_3}$$

Lithium diarylcuprates are prepared in the same way as lithium dialkylcuprates and undergo comparable reactions with primary alkyl halides:

$$(C_6H_5)_2CuLi + ICH_2(CH_2)_6CH_3 \xrightarrow[\text{ether}]{\text{diethyl}} C_6H_5CH_2(CH_2)_6CH_3$$

Lithium diphenylcuprate — 1-Iodooctane — 1-Phenyloctane (99%)

The most frequently used organocuprates are those in which the alkyl group is primary. Steric hindrance makes organocuprates that bear secondary and tertiary alkyl groups less reactive, and they tend to decompose before they react with the alkyl halide. The reaction of cuprate reagents with alkyl halides follows the usual S_N2 order: CH_3 > primary > secondary > tertiary, and I > Br > Cl > F. *p*-Toluenesulfonate esters are suitable substrates and are somewhat more reactive than halides. Because the alkyl halide and dialkylcuprate reagent should both be primary in order to produce satisfactory yields of coupled products, the reaction is limited to the formation of $RCH_2—CH_2R'$ and $RCH_2—CH_3$ bonds in alkanes.

A key step in the reaction mechanism appears to be nucleophilic attack on the alkyl halide by the negatively charged copper atom. The intermediate thus formed is unstable and fragments to the observed products:

$$R_2\bar{Cu} + R'—X \longrightarrow [R_2CuR'X^-] \longrightarrow RR' + RCu + X^-$$

Dialkylcuprate ion — Alkyl halide — Alkane — Alkylcopper — Halide ion

However, the intimate details of the reaction mechanism are not well understood. Indeed, there is probably more than one mechanism by which cuprates react with organic halogen compounds. Vinyl halides and aryl halides are known to be very unreactive toward nucleophilic attack, yet react smoothly with lithium dialkylcuprates:

$$(CH_3CH_2CH_2CH_2)_2CuLi + \text{(cyclohexenyl)}—Br \xrightarrow[\text{ether}]{\text{diethyl}} \text{(cyclohexenyl)}—CH_2CH_2CH_2CH_3$$

Lithium dibutylcuprate — 1-Bromocyclohexene — 1-Butylcyclohexene (80%)

$$(CH_3CH_2CH_2CH_2)_2CuLi + C_6H_5—I \xrightarrow[\text{ether}]{\text{diethyl}} C_6H_5—CH_2CH_2CH_2CH_3$$

Lithium dibutylcuprate — Iodobenzene — Butylbenzene (75%)

PROBLEM 14.8 Suggest a combination of organic halide and cuprate reagent appropriate for the preparation of each of the following compounds:

(*a*) 2-Methylbutane (*b*) 1,3,3-Trimethylcyclopentene

SAMPLE SOLUTION (*a*) First inspect the target molecule to see which bonds are capable of being formed by reaction of an alkyl halide and a cuprate, bearing in mind that neither the alkyl halide nor the alkyl group of the lithium dialkylcuprate should be secondary or tertiary.

$$CH_3—CH_2—\underset{}{\overset{CH_3}{\overset{|}{C}H}}—CH_3$$

A bond between a methyl group and a methylene group can be formed.

None of the bonds to the methine group can be formed efficiently.

There are two combinations, both acceptable, that give rise to the $CH_3—CH_2$ bond:

$$(CH_3)_2CuLi + BrCH_2CH(CH_3)_2 \longrightarrow CH_3CH_2CH(CH_3)_2$$

Lithium dimethylcuprate, 1-Bromo-2-methylpropane, 2-Methylbutane

$$CH_3I + LiCu[CH_2CH(CH_3)_2]_2 \longrightarrow CH_3CH_2CH(CH_3)_2$$

Iodomethane, Lithium diisobutylcuprate, 2-Methylbutane

14.12 AN ORGANOZINC REAGENT FOR CYCLOPROPANE SYNTHESIS

Zinc reacts with alkyl halides in a manner similar to that of magnesium.

$$RX + Zn \xrightarrow{\text{ether}} RZnX$$

Alkyl halide, Zinc, Alkylzinc halide

Victor Grignard was led to study organomagnesium compounds because of earlier work he performed with organic derivatives of zinc.

Organozinc reagents are not nearly so reactive toward aldehydes and ketones as Grignard reagents and organolithium compounds but are intermediates in a number of reactions of alkyl halides and dihalides which were once common synthetic procedures, but which have given way to more effective methods.

An organozinc compound that continues to occupy a special niche in organic synthesis is *iodomethylzinc iodide* (ICH_2ZnI), prepared by the reaction of zinc-copper couple [Zn(Cu), zinc that has had its surface activated with a little copper] with diiodomethane in ether.

Iodomethylzinc iodide is known as the *Simmons-Smith reagent,* after Howard E. Simmons and Ronald D. Smith of Du Pont, who first described its use in the preparation of cyclopropanes.

$$ICH_2I + Zn \xrightarrow[\text{Cu}]{\text{diethyl ether}} ICH_2ZnI$$

Diiodomethane, Zinc, Iodomethylzinc iodide

What makes iodomethylzinc iodide such a useful reagent is that it reacts with alkenes to give cyclopropanes.

$$CH_2{=}C(CH_2CH_3)(CH_3) \xrightarrow[\text{diethyl ether}]{CH_2I_2,\ Zn(Cu)} \text{1-ethyl-1-methylcyclopropane}$$

2-Methyl-1-butene, 1-Ethyl-1-methylcyclopropane (79%)

This reaction is called the *Simmons-Smith* reaction and is one of the few methods available for the synthesis of cyclopropanes. Mechanistically, the Simmons-Smith reaction seems to proceed by a single-step cycloaddition of a methylene (CH_2) unit from iodomethylzinc iodide to the alkene:

$$\gt C{=}C\lt \;\; ICH_2ZnI \longrightarrow [\text{transition state}] \longrightarrow \gt C{-}C\lt (CH_2) + ZnI_2$$

ICH$_2$ZnI, Transition state for methylene transfer

PROBLEM 14.9 What alkenes would you choose as starting materials in order to prepare each of the following cyclopropane derivatives by reaction with iodomethylzinc iodide?

(*a*) CH_3 (*b*)

SAMPLE SOLUTION (*a*) In a cyclopropane synthesis using the Simmons-Smith reagent, you should remember that a CH_2 unit is transferred. Therefore, retrosynthetically disconnect the bonds to a CH_2 group of a three-membered ring to identify the starting alkene.

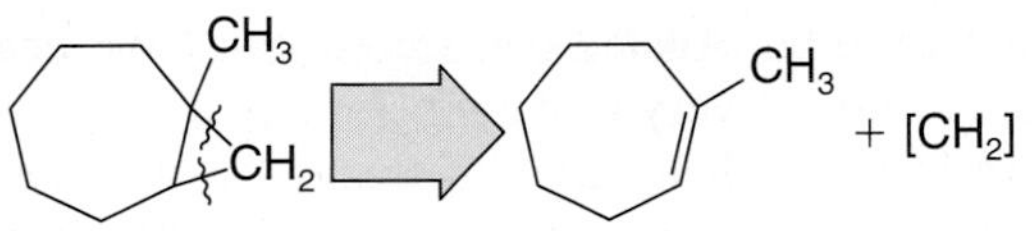

CH_3 CH_2 ⟹ CH_3 + $[CH_2]$

The complete synthesis is:

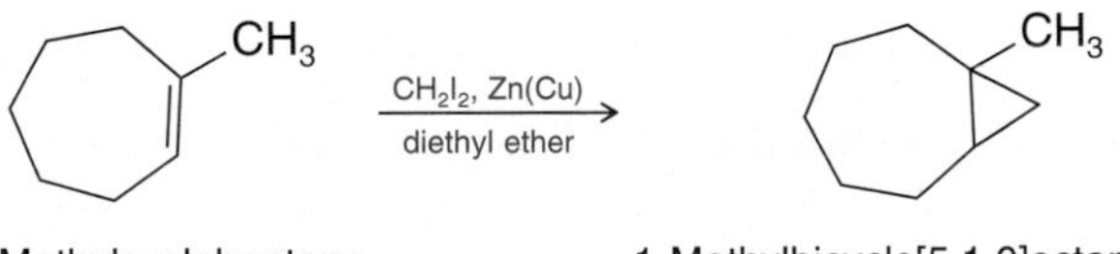

CH_3 $\xrightarrow[\text{diethyl ether}]{CH_2I_2,\ Zn(Cu)}$ CH_3

1-Methylcycloheptene 1-Methylbicyclo[5.1.0]octane (55%)

Methylene transfer from iodomethylzinc iodide is *stereospecific*. Substituents that were cis in the alkene remain cis in the cyclopropane.

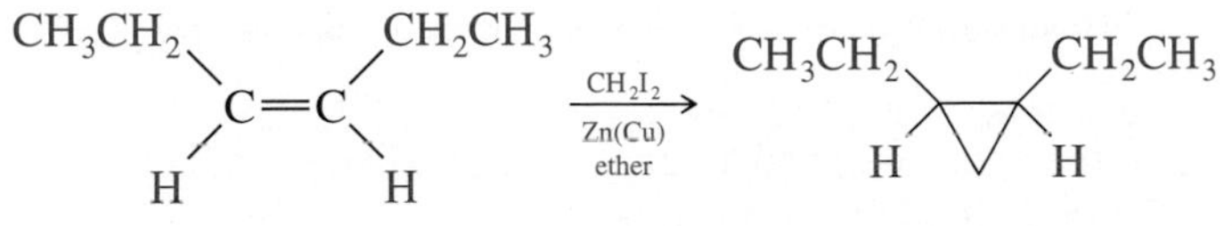

$CH_3CH_2(H)C{=}C(H)CH_2CH_3$ $\xrightarrow[\text{Zn(Cu)}\\ \text{ether}]{CH_2I_2}$ cyclopropane with CH_3CH_2, CH_2CH_3, H, H

(*Z*)-3-Hexene *cis*-1,2-Diethylcyclopropane (34%)

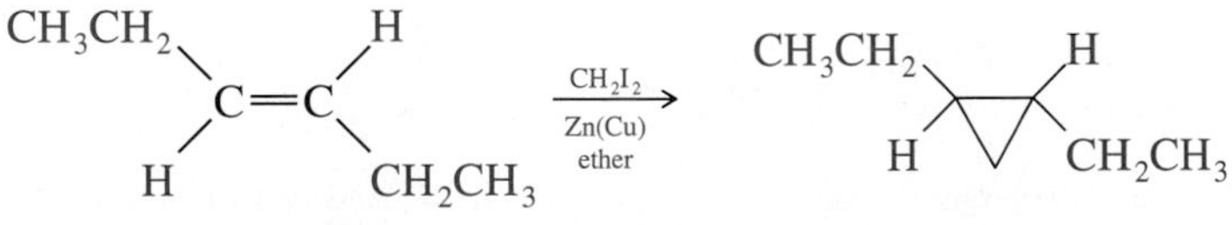

$CH_3CH_2(H)C{=}C(H)CH_2CH_3$ $\xrightarrow[\text{Zn(Cu)}\\ \text{ether}]{CH_2I_2}$ cyclopropane with CH_3CH_2, H, H, CH_2CH_3

(*E*)-3-Hexene *trans*-1,2-Diethylcyclopropane (15%)

Yields in Simmons-Smith reactions are sometimes low. Nevertheless, since it often provides the only feasible route to a particular cyclopropane derivative, it is a valuable addition to the organic chemist's store of synthetic methods.

14.13 CARBENES AND CARBENOIDS

Iodomethylzinc iodide is often referred to as a **carbenoid,** meaning that it resembles a **carbene** in its chemical reactions. Carbenes are neutral molecules in which one of the carbon atoms has six valence electrons. Such carbons are *divalent;* they are directly bonded to only two other atoms and have no multiple bonds. Iodomethylzinc iodide reacts as if it were a source of the carbene $H—\ddot{C}—H$.

It is clear that free $:CH_2$ is not involved in the Simmons-Smith reaction, but there is substantial evidence to indicate that carbenes are formed as intermediates in certain other reactions that convert alkenes to cyclopropanes. The most studied examples of these reactions involve dichlorocarbene and dibromocarbene.

$$Cl_2\ddot{C} \qquad Br_2\ddot{C}$$

Dichlorocarbene Dibromocarbene

Carbenes are too reactive to be isolated and stored, but have been trapped in frozen argon for spectroscopic study at very low temperatures.

Dihalocarbenes are formed when trihalomethanes are treated with a strong base, such as potassium *tert*-butoxide. The trihalomethyl anion produced on proton abstraction dissociates to a dihalocarbene and a halide anion:

$$Br_3C—H + :\ddot{O}^-C(CH_3)_3 \longrightarrow Br_3C:^- + H—\ddot{O}C(CH_3)_3$$

Tribromomethane *tert*-Butoxide ion Tribromomethide ion *tert*-Butyl alcohol

$$Br_3C:^- \longrightarrow Br_2C: + :\ddot{Br}:^-$$

Tribromomethide ion Dibromocarbene Bromide ion

When generated in the presence of an alkene, dihalocarbenes undergo cycloaddition to the double bond to give dihalocyclopropanes:

$$\text{Cyclohexene} + CHBr_3 \xrightarrow[(CH_3)_3COH]{KOC(CH_3)_3} \text{7,7-Dibromobicyclo[4.1.0]heptane (75\%)}$$

Cyclohexene Tribromomethane 7,7-Dibromobicyclo[4.1.0]heptane (75%)

The reaction of dihalocarbenes with alkenes is stereospecific, and syn addition is observed.

PROBLEM 14.10 The syn stereochemistry of dibromocarbene cycloaddition was demonstrated in experiments using *cis*- and *trans*-2-butene. Give the structure of the product obtained from addition of dibromocarbene to each alkene.

The process in which a dihalocarbene is formed from a trihalomethane corresponds to an elimination in which a proton and a halide are lost from the same carbon. It is an *α elimination* proceeding via the organometallic intermediate $K^+ [:CX_3]^-$.

14.14 ORGANIC DERIVATIVES OF MERCURY. OXYMERCURATION-DEMERCURATION OF ALKENES

Mercury is a relative of zinc in group IIB of the periodic table, but its organometallic chemistry is quite different. Mercury is not nearly so "metallic" as zinc and, for example, does not react with alkyl halides. The organomercury compounds that are the most useful in organic synthesis are formed when mercury(II) salts such as mercury(II) acetate react with alkenes. Mercury(II) acetate (also called *mercuric acetate*) has the formula

$$CH_3\overset{O}{\overset{\|}{C}}O-Hg-O\overset{O}{\overset{\|}{C}}CH_3 \qquad \text{also written as} \qquad Hg(O_2CCH_3)_2 \text{ or } Hg(OAc)_2$$

and reacts with alkenes in the presence of water to give compounds known as *β-hydroxyalkylmercury(II) acetates.*

$$\underset{\text{Alkene}}{>C=C<} + \underset{\text{Mercury(II) acetate}}{Hg(O_2CCH_3)_2} + \underset{\text{Water}}{H_2O} \longrightarrow \underset{\beta\text{-Hydroxyalkylmercury(II) acetate}}{HO-\overset{|}{\underset{|}{C}}-\overset{|}{\underset{|}{C}}-HgO_2CCH_3} + \underset{\text{Acetic acid}}{CH_3CO_2H}$$

This reaction is known as **oxymercuration.** It is best carried out in a mixture of water and tetrahydrofuran as the solvent. Oxymercuration is the first operation in a two-stage process called *oxymercuration-demercuration* by which alkenes are converted to alcohols.

The β-hydroxyalkylmercury(II) acetate formed by oxymercuration is not normally isolated but is treated directly with sodium borohydride ($NaBH_4$), which converts it to an alcohol.

$$\underset{\beta\text{-Hydroxyalkymercury(II) acetate}}{HO-\overset{|}{\underset{|}{C}}-\overset{|}{\underset{|}{C}}-HgO_2CCH_3} \xrightarrow[HO^-]{NaBH_4} \underset{\text{Alcohol}}{HO-\overset{|}{\underset{|}{C}}-\overset{|}{\underset{|}{C}}-H} + \underset{\text{Mercury}}{Hg} + \underset{\text{Acetate ion}}{CH_3CO_2^-}$$

This step is called **demercuration.** The carbon-mercury bond of the hydroxyalkylmercuric acetate is replaced by a carbon-hydrogen bond.

Experimentally, oxymercuration-demercuration reactions are very easy to carry out. The alkene is added to a solution of mercury(II) acetate in aqueous tetrahydrofuran. After about an hour the oxymercuration phase is complete. A basic solution of sodium borohydride is then added, whereupon metallic mercury separates from the reaction mixture and is removed and the alcohol is isolated.

$$\underset{\text{Cyclopentene}}{\text{Cyclopentene}} \xrightarrow[2.\ NaBH_4,\ HO^-]{1.\ Hg(OAc)_2,\ THF-H_2O} \underset{\text{Cyclopentanol (91\%)}}{\text{Cyclopentanol (OH)}}$$

The regioselectivity of oxymercuration-demercuration is identical with that of acid-catalyzed hydration in that Markovnikov's rule is followed. Hydrogen is introduced at the carbon that has the greater number of hydrogen substituents, and hydroxyl at the carbon with the fewer hydrogens. Rearrangements of the carbon skeleton do not occur.

$$(CH_3)_3CCH{=}CH_2 \xrightarrow[\text{2. NaBH}_4,\ HO^-]{\text{1. Hg(OAc)}_2,\ THF-H_2O} (CH_3)_3C\underset{\displaystyle OH}{\underset{|}{C}}HCH_3$$

3,3-Dimethyl-1-butene — 3,3-Dimethyl-2-butanol (94%)

PROBLEM 14.11 Oxymercuration-demercuration of *trans*-2-pentene gives comparable amounts of two isomeric alcohols in a combined yield of 91 percent. What are these two alcohols?

PROBLEM 14.12 Each of the following alcohols may be prepared in high yield by oxymercuration-demercuration of two different alkenes. Write the structures of the two alkenes that could produce

(*a*) 2-Methyl-2-butanol
(*b*) 1-Methylcyclopentanol
(*c*) 3-Hexanol

SAMPLE SOLUTION (*a*) The alcohol is formed by addition of a hydrogen to one carbon and a hydroxyl to the other carbon of a double bond. Examining the structure of the given alcohol reveals a tertiary hydroxyl group at C-2. Therefore, an alkene with its double bond either between C-1 and C-2 or between C-2 and C-3 could undergo hydration according to Markovnikov's rule to give 2-methyl-2-butanol:

2-Methyl-1-butene or 2-Methyl-2-butene $\xrightarrow[\text{2. NaBH}_4,\ HO^-]{\text{1. Hg(OAc)}_2,\ THF-H_2O}$ 2-Methyl-2-butanol

A mechanism for the oxymercuration stage of the sequence is outlined in Figure 14.3 and involves electrophilic addition to the double bond. The attacking electrophile is believed to be a mercury cation formed by dissociation of mercury(II) acetate.

$$CH_3\overset{O}{\overset{\|}{C}}O{-}Hg{-}O\overset{O}{\overset{\|}{C}}CH_3 \rightleftharpoons CH_3\overset{O}{\overset{\|}{C}}O{-}Hg^+ + {}^-O\overset{O}{\overset{\|}{C}}CH_3$$

This cation attacks the alkene (step 1 of Figure 14.3) to give a carbocation of a special type, one that has a carbon-mercury bond β to the positive charge. Carbocations of this type are relatively stable, are formed readily, and do not rearrange. Indeed, one school of thought holds that such ions are "bridged," meaning that mercury is simultaneously bonded to both the α and β carbons and bears a portion of the positive charge. A bridged ion of this type is called a **mercurinium ion.**

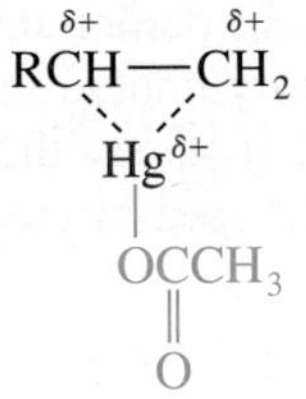

$$\underset{\delta+}{RCH}-\underset{\delta+}{CH_2}\ \text{(bridged to)}\ Hg^{\delta+}-OC(=O)CH_3$$

Like other carbocations, however, β-mercury-substituted carbocations are attacked by nucleophiles, and this is what happens in step 2, in which a molecule of water from the tetrahydrofuran-water mixture used as the solvent is the nucleophile. Loss of a proton in step 3 gives the β-hydroxyalkylmercury(II) acetate.

Step 1: A mercury cation acts as an electrophile, attacking the π electrons of the carbon-carbon double bond:

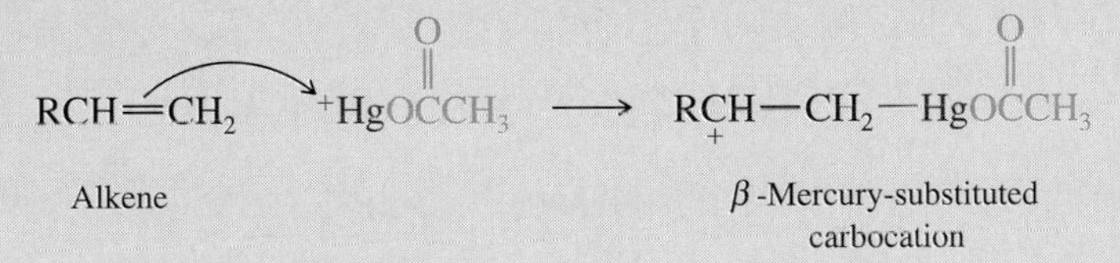

$$\underset{\text{Alkene}}{RCH{=}CH_2} \quad {}^{+}HgOCCH_3 \longrightarrow \underset{\beta\text{-Mercury-substituted carbocation}}{R\overset{+}{C}H-CH_2-HgOCCH_3}$$

(each $OCCH_3$ group is $OC(=O)CH_3$)

Step 2: Water acts as a nucleophile toward the mercury-substituted carbocation intermediate:

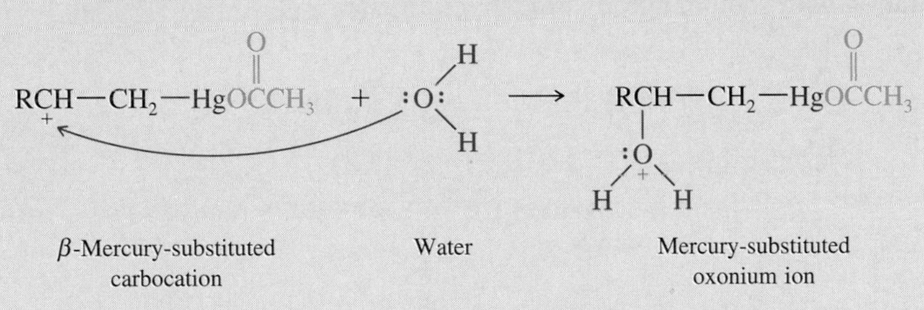

$$\underset{\beta\text{-Mercury-substituted carbocation}}{R\overset{+}{C}H-CH_2-HgOCCH_3} + \underset{\text{Water}}{H_2\ddot{O}:} \longrightarrow \underset{\text{Mercury-substituted oxonium ion}}{RCH(\overset{+}{O}H_2)-CH_2-HgOCCH_3}$$

Step 3: Loss of a proton from the oxonium ion yields the product of oxymercuration, a β-hydroxyalkylmercury(II) acetate:

$$\underset{\text{Mercury-substituted oxonium ion + acetate ion}}{RCH(\overset{+}{O}H_2)-CH_2-HgOCCH_3 \quad {}^{-}OCCH_3} \longrightarrow \underset{\beta\text{-Hydroxyalkylmercury(II) acetate}}{RCH(OH)-CH_2-HgOCCH_3} + \underset{\text{Acetic acid}}{HOCCH_3}$$

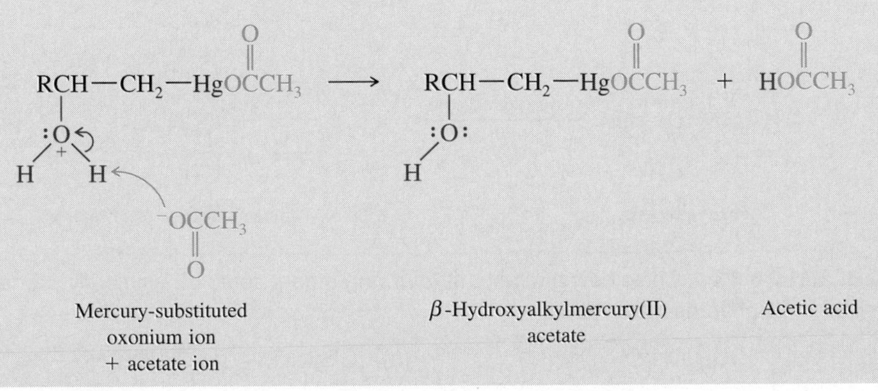

FIGURE 14.3 A mechanistic description of the oxymercuration of an alkene.

The second phase of the reaction, the demercuration phase, involves chemistry that is beyond the scope of an introductory course in organic chemistry and will not be discussed. The hydrogen atom that replaces the mercury substituent on carbon is derived from the sodium borohydride used in the demercuration reaction.

$$RCH{=}CH_2 \xrightarrow[2.\ NaBH_4,\ HO^-]{1.\ Hg(OAc)_2,\ THF{-}H_2O} \underset{\underset{OH}{|}}{RCH}{-}CH_2{-}H$$

OH: From water present in solvent ($THF{-}H_2O$)

H: From sodium borohydride ($NaBH_4$)

It is possible to modify the oxymercuration-demercuration procedure so that it is also applicable to the synthesis of *ethers*. As already noted, the hydroxyl group in the product alcohol is derived from water. If an alcohol is used as the solvent instead of aqueous tetrahydrofuran, then it captures the β-mercury-substituted carbocation to introduce an alkoxy group in the first stage of the reaction.

$$\underset{+}{RCH}CH_2HgOAc \xrightarrow{R'OH} \underset{\underset{R'\overset{+}{O}H}{|}}{RCH}CH_2HgOAc \xrightarrow{-H^+} \underset{\underset{R'O}{|}}{RCH}CH_2HgOAc$$

β-Mercury-substituted carbocation — β-Alkoxyalkylmercury(II) acetate

Demercuration with sodium borohydride gives an ether:

$$\underset{\underset{R'O}{|}}{RCH}CH_2HgOAc \xrightarrow[HO^-]{NaBH_4} \underset{\underset{R'O}{|}}{RCH}CH_3$$

β-Alkoxyalkylmercury(II) acetate — Ether

The term *solvomercuration-demercuration* has been applied to this general class of reactions. Ether synthesis by solvomercuration-demercuration, like alcohol synthesis by oxymercuration-demercuration, corresponds to addition to the alkene according to Markovnikov's rule.

$$C_6H_5\underset{\underset{CH_3}{|}}{C}{=}CH_2 \xrightarrow[2.\ NaBH_4,\ HO^-]{1.\ Hg(OAc)_2,\ CH_3OH} C_6H_5\underset{\underset{CH_3}{|}}{\overset{\overset{OCH_3}{|}}{C}}CH_3$$

2-Phenylpropene — 2-Methoxy-2-phenylpropane (100%)

PROBLEM 14.13 Show how each of the following ethers could be efficiently prepared by solvomercuration-demercuration:

(*a*) $C_6H_5\underset{\underset{CH_3}{|}}{C}HOCH_2CH_2CH_3$

(*b*) $CH_3\underset{\underset{OCH_2CH_3}{|}}{C}HCH_2CH_2CH_2CH_3$

SAMPLE SOLUTION (*a*) A decision must be made as to which one of the groups bonded to the ether oxygen is derived from an alkene and which is derived from the alcohol used as the solvent. Since the overall reaction follows Markovnikov's rule, a propyl group can originate only in the alcohol. Markovnikov addition to propene yields an isopropyl group. Therefore, the correct alkene is styrene and the correct alcohol is 1-propanol.

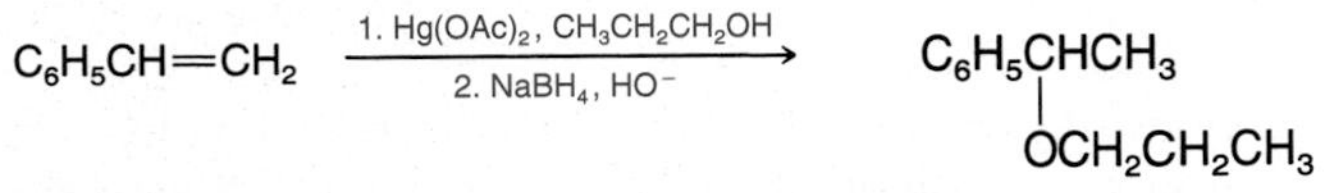

$$\underset{\text{Styrene}}{C_6H_5CH{=}CH_2} \xrightarrow[\text{2. NaBH}_4\text{, HO}^-]{\text{1. Hg(OAc)}_2\text{, CH}_3\text{CH}_2\text{CH}_2\text{OH}} \underset{\text{1-Phenylethyl propyl ether}}{C_6H_5CH(OCH_2CH_2CH_3)CH_3}$$

Mercury-based methods are often valuable in laboratory syntheses but present disposal problems when carried out on a larger scale. Mercury compounds are quite toxic to most organisms. For a long time, wastewater containing mercury salts from industrial operations was dumped in rivers and lakes in large amounts. There, inorganic mercury (Hg^{2+}) is converted by bacterial action to methylmercury (CH_3Hg^+) and dimethylmercury (CH_3HgCH_3), materials that accumulate in the tissues of fish. At even modest mercury levels fish populations are threatened. It is dangerous to eat fish caught in mercury-polluted waters, and hundreds of incidents of poisoning are known to have occurred because of this. While stricter controls on waste disposal have helped, so much mercury remains in the environment that it will take a long time for it to be reduced to a reasonable level in some riverbeds.

14.15 TRANSITION METAL ORGANOMETALLIC COMPOUNDS

A large number of organometallic compounds are based on transition metals. Examples include organic derivatives of iron, nickel, chromium, platinum, and rhodium. Many important industrial processes are catalyzed by transition metals or their complexes.

A noteworthy feature of many organic derivatives of transition metals is that the organic group is bonded to the metal through its π system rather than by a σ bond. The compound (benzene)tricarbonylchromium, for example, has three carbon monoxide ligands and a *benzene* ring—not a phenyl group—attached to chromium:

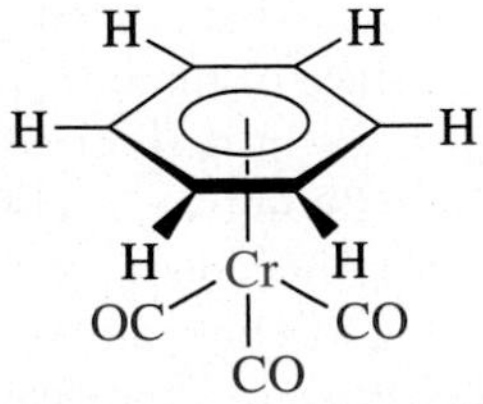

(Benzene)tricarbonylchromium

It is not uncommon to find stable transition metal organometallic compounds bearing as their organic ligand species that are known to be highly reactive in their free or uncoordinated state. One such organometallic compound is (cyclobutadiene)tricarbonyliron.

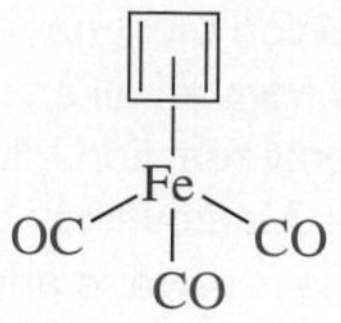

Fe
OC CO
CO

(Cyclobutadiene)tricarbonyliron

Many of the reactions of free cyclobutadiene have been studied by using (cyclobutadiene)tricarbonyliron as the source of cyclobutadiene. Oxidation causes dissociation of the complex, liberating free cyclobutadiene in solution.

One of the earliest and best examples of π-bonded organometallic compounds is *ferrocene.* It was expected that the σ-bonded species dicyclopentadienyliron could be prepared by adding iron(II) chloride (ferrous chloride) to cyclopentadienylsodium.

Cyclopentadienylsodium is ionic. Its anion is the cyclopentadienide ion, which contains six π electrons.

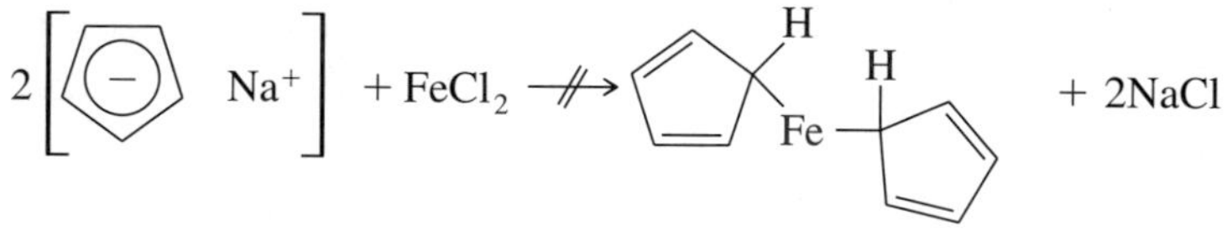

2 [Na$^+$] + $FeCl_2$ ⇸ H Fe H + 2NaCl

Cyclopentadienylsodium | Iron(II) chloride | (Not formed)

However, the isolated product, ferrocene, clearly was structurally much different from that expected. It was subsequently shown to have the doubly π-bonded *sandwich* structure shown:

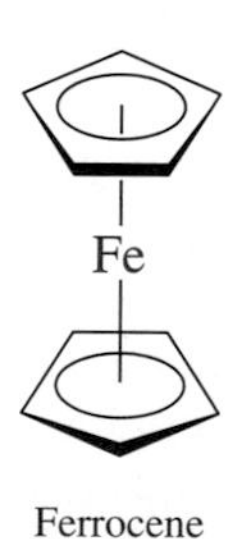

Ferrocene

Since then, numerous related molecules have been prepared—even some in which uranium is the metal! The precise nature of the bonding in such compounds served to foster development of bonding theories in both organic and inorganic chemistry and stimulate research at the point where these two disciplines merge. Two of the leading figures in the growth of organometallic chemistry of the transition metals, E. O. Fischer of the Technical University in Munich and Sir Geoffrey Wilkinson of Imperial College, London, shared the Nobel Prize in chemistry in 1973.

Naturally occurring compounds with carbon-metal bonds are very rare. The best example of such an organometallic compound is coenzyme B_{12}, which has a carbon-cobalt σ bond (Figure 14.4). Pernicious anemia results from a coenzyme B_{12} deficiency and can be treated by adding sources of cobalt to the diet. One source of cobalt is vitamin B_{12}, a compound structurally related to, but not identical with, coenzyme B_{12}.

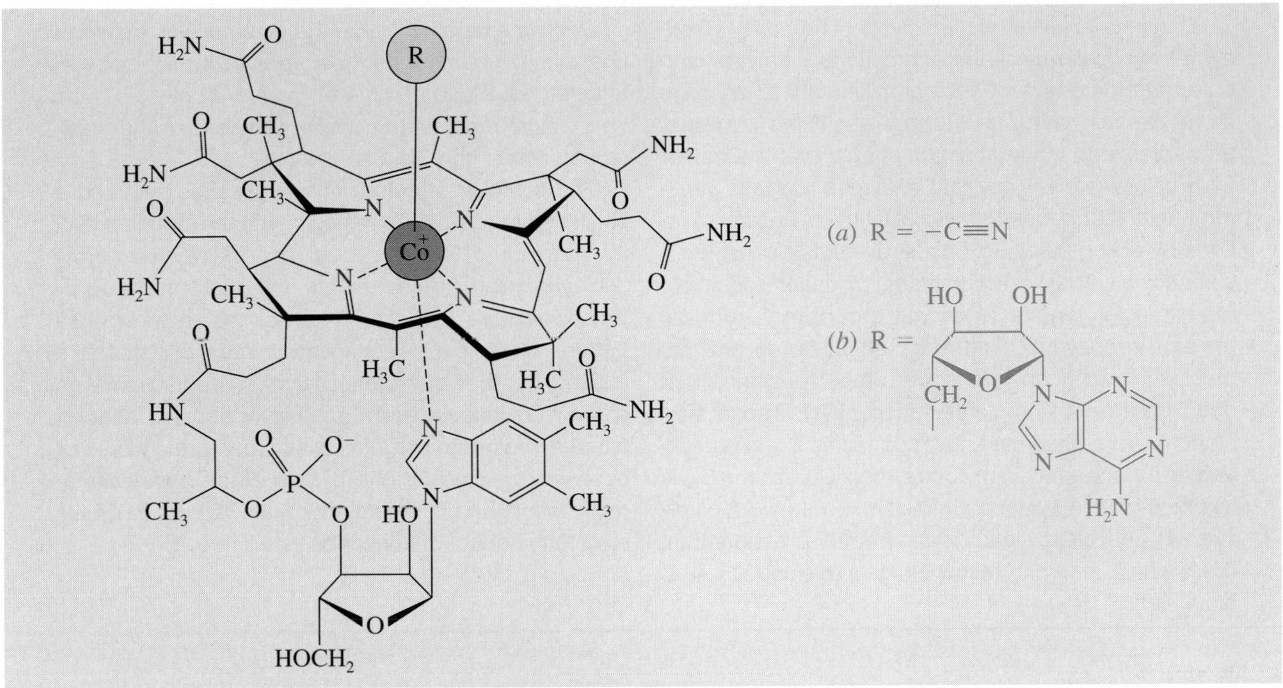

FIGURE 14.4 The structures of (*a*) vitamin B_{12} and (*b*) coenzyme B_{12}.

AN ORGANOMETALLIC COMPOUND THAT OCCURS NATURALLY: COENZYME B_{12}

Pernicious anemia is a disease characterized, as are all anemias, by a deficiency of red blood cells. Unlike ordinary anemia, pernicious anemia does not respond to treatment with sources of iron, and before effective treatments were developed early in this century, pernicious anemia was often fatal. Injection of liver extracts was one such treatment, and in 1948 chemists succeeded in isolating the "anti-pernicious anemia factor" from beef liver as a red crystalline compound, which they called **vitamin B_{12}**. This compound had the formula $C_{63}H_{88}CoN_{14}O_{14}P$. Its complexity precluded structure determination by classical degradation techniques, and spectroscopic methods were too primitive to be of much help. The structure was solved by Dorothy Crowfoot Hodgkin of Oxford University in 1955 using x-ray diffraction techniques and is shown in Figure 14.4*a*. Structure determination by x-ray crystallography can be superficially considered as taking a photograph of a molecule with x-rays. It is a demanding task and earned Hodgkin the 1964 Nobel Prize in chemistry. Modern structural studies by x-ray crystallography use computers to collect and analyze the diffraction data and take only a fraction of the time required years ago to solve the vitamin B_{12} structure.

The structure of vitamin B_{12} is interesting in that it contains a central cobalt atom that is surrounded by six atoms in an octahedral geometry. One substituent, the cyano (—CN) group, is what is known as an "artifact." It appears to be introduced into the molecule during the isolation process and leads to the synonym **cyanocobalamin** for vitamin B_{12} This is the material which is used to treat pernicious anemia, but is not the form in which it exerts its activity. The biologically active material is called **coenzyme B_{12}** and differs from vitamin B_{12} in the substituent attached to cobalt (Figure 14.4*b*). Coenzyme B_{12} is the only known naturally occurring substance that has a carbon-metal bond. Moreover, coenzyme B_{12} was discovered before any compound containing an alkyl group σ-bonded to cobalt had ever been isolated in the laboratory!

Enzymes are Nature's catalysts. The ability of enzymes to accelerate chemical reactions is truly remarkable, rate increases of several millionfold being common. We will have more to say about the structure, properties, and mode of action of enzymes in Chapter 27. For now we will say that enzymes possess structural units that permit them to catalyze biological reactions by acting as acids, bases, or relatively weak nucleophiles. Certain reaction types, oxidation-reduction, for example, require functional groups not normally present in enzymes, and to be effective, an enzyme must act in concert with some other substance that does possess the required functionality. These substances are **coenzymes,** and coenzyme B_{12} is one example. Most vitamins are coenzymes. Coenzyme B_{12} is involved in DNA synthesis in the bone marrow. A coenzyme B_{12} deficiency leads to the formation of abnormal DNA, which, in turn, is responsible for dysfunctional red blood cells.

Despite extensive study, the mechanisms by which coenzyme B_{12}–mediated reactions occur are not well understood. At least some of them are believed to be free-radical reactions in which cleavage of the weak carbon-cobalt bond is a key step.

The complex structure of vitamin B_{12} presented a challenge to chemical synthesis, and the completion of this task in 1972 was an especially noteworthy achievement. The leaders of the team were R. B. Woodward (Harvard) and Albert Eschenmoser (Swiss Federal Institute of Technology). Their success is a testimony to their own creativity, the skill and dedication of their coworkers, and the level of sophistication of modern synthetic methods. It also bears mention that observations made in the course of this synthesis directly influenced chemical theory in respect to orbital symmetry control of concerted reactions.

14.16 SUMMARY

Organometallic compounds contain a carbon-metal bond, polarized so that carbon bears a partial to complete negative charge and the metal bears a partial to complete positive charge (Section 14.1).

$$H_3C^{\delta-}\text{—}Li^{\delta+}$$

Methyllithium has a polar covalent carbon-lithium bond.

$$HC{\equiv}\bar{C}{:}\ Na^+$$

Sodium acetylide has an ionic bond between carbon and sodium.

A species that has a negatively charged carbon is called a **carbanion.** Organometallic substances with a significant degree of carbanionic character are strong bases and good nucleophiles. Many organometallic compounds are used in synthesis as reagents for carbon-carbon bond formation.

Organometallic compounds are named as alkyl (or aryl) derivatives of metals (Section 14.2). Butyllithium ($CH_3CH_2CH_2CH_2Li$) and phenylmagnesium bromide (C_6H_5MgBr) illustrate this nomenclature.

The most useful organometallic reagents are derived from magnesium and are called **Grignard reagents.** They are prepared by the reaction of magnesium and an organic halide, usually in diethyl ether. Alkyllithium reagents are prepared in an analogous way and undergo chemical reactions that are similar to those of Grignard reagents. The preparation of Grignard reagents and alkyllithium reagents is summarized in Table 14.4. The table also illustrates the preparation of *lithium dialkylcuprates* and

TABLE 14.4
Preparation of Organometallic Reagents Used in Synthesis

Type of organometallic reagent (section) and comments	General equation for preparation and specific example
Organolithium reagents (Section 14.3) Lithium metal reacts with organic halides to produce organolithium compounds. The organic halide may be alkyl, alkenyl, or aryl. Iodides react most and fluorides least readily; bromides are used most often. Suitable solvents include hexane, diethyl ether, and tetrahydrofuran.	$RX + 2Li \longrightarrow RLi + LiX$ Alkyl halide, Lithium, Alkyllithium, Lithium halide $CH_3CH_2CH_2Br \xrightarrow[\text{diethyl ether}]{\text{Li}} CH_3CH_2CH_2Li$ Propyl bromide → Propyllithium (78%)
Grignard reagents (Section 14.4) Grignard reagents are prepared in a manner similar to that used for organolithium compounds. Diethyl ether and tetrahydrofuran are appropriate solvents.	$RX + Mg \longrightarrow RMgX$ Alkyl halide, Magnesium, Alkylmagnesium halide (Grignard reagent) $C_6H_5CH_2Cl \xrightarrow[\text{diethyl ether}]{\text{Mg}} C_6H_5CH_2MgCl$ Benzyl chloride → Benzylmagnesium chloride (93%)
Lithium dialkylcuprates (Section 14.11) These reagents contain a negatively charged copper atom and are formed by the reaction of a copper(I) salt with two equivalents of an organolithium reagent.	$2RLi + CuX \longrightarrow R_2CuLi + LiX$ Alkyllithium, Copper(I) halide, Lithium dialkylcuprate, Lithium halide $2CH_3Li + CuI \xrightarrow{\text{diethyl ether}} (CH_3)_2CuLi + LiI$ Methyllithium, Copper(I) iodide, Lithium dimethylcuprate, Lithium iodide
Iodomethylzinc iodide (Section 14.12) This is the Simmons-Smith reagent. It is prepared by the reaction of zinc (usually in the presence of copper) with diiodomethane.	$CH_2I_2 + Zn \xrightarrow[\text{Cu}]{\text{diethyl ether}} ICH_2ZnI$ Diiodomethane, Zinc, Iodomethylzinc iodide

iodomethylzinc iodide, two other organometallic compounds that are synthetically useful.

Organolithium compounds and Grignard reagents are strong bases and react instantly with compounds that have —OH groups (Section 14.5):

$$R{-}M + H{-}O{-}R' \longrightarrow R{-}H + M^+\ {}^-O{-}R'$$

Therefore these organometallic compounds cannot be formed or used in solvents such as water and ethanol. The most commonly employed solvents are diethyl ether and tetrahydrofuran.

Table 14.5 shows how organometallic compounds based on magnesium, lithium, copper, and zinc are used in *carbon-carbon bond–forming reactions.* In particular, the

TABLE 14.5

Synthetic Applications of Organometallic Reagents

Reaction (section) and comments	General equation and specific example
Alcohol synthesis via the reaction of Grignard reagents with carbonyl compounds (Section 14.6) This is one of the most useful reactions in synthetic organic chemistry. Grignard reagents react with formaldehyde to yield primary alcohols, with aldehydes to give secondary alcohols, and with ketones to form tertiary alcohols.	$RMgX + R'\overset{O}{\overset{\|}{C}}R'' \xrightarrow[2.\ H_3O^+]{1.\ \text{diethyl ether}} R\overset{R'}{\overset{\mid}{\underset{R''}{\underset{\mid}{C}}}}OH$ Grignard reagent; Aldehyde or ketone; Alcohol $CH_3MgI + CH_3CH_2CH_2\overset{O}{\overset{\|}{C}}H \xrightarrow[2.\ H_3O^+]{1.\ \text{diethyl ether}} CH_3CH_2CH_2\underset{OH}{\underset{\mid}{C}}HCH_3$ Methylmagnesium iodide; Butanal; 2-Pentanol (82%)
Reaction of Grignard reagents with esters (Section 14.10) Tertiary alcohols in which two of the substituents on the hydroxyl carbon are the same may be prepared by the reaction of an ester with two equivalents of a Grignard reagent.	$2RMgX + R'\overset{O}{\overset{\|}{C}}OR'' \xrightarrow[2.\ H_3O^+]{1.\ \text{diethyl ether}} R\overset{R'}{\overset{\mid}{\underset{R}{\underset{\mid}{C}}}}OH$ Grignard reagent; Ester; Tertiary alcohol $2C_6H_5MgBr + C_6H_5\overset{O}{\overset{\|}{C}}OCH_2CH_3 \xrightarrow[2.\ H_3O^+]{1.\ \text{diethyl ether}} (C_6H_5)_3COH$ Phenylmagnesium bromide; Ethyl benzoate; Triphenylmethanol (89–93%)
Synthesis of alcohols using organolithium reagents (Section 14.7) Organolithium reagents react with aldehydes and ketones in a manner similar to that of Grignard reagents to produce alcohols.	$RLi + R'\overset{O}{\overset{\|}{C}}R'' \xrightarrow[2.\ H_3O^+]{1.\ \text{diethyl ether}} R\overset{R'}{\overset{\mid}{\underset{R''}{\underset{\mid}{C}}}}OH$ Alkyllithium; Aldehyde or ketone; Alcohol $\triangleright\!-Li + CH_3\overset{O}{\overset{\|}{C}}C(CH_3)_3 \xrightarrow[2.\ H_3O^+]{1.\ \text{diethyl ether}} \triangleright\!-\overset{OH}{\overset{\mid}{\underset{CH_3}{\underset{\mid}{C}}}}C(CH_3)_3$ Cyclopropyllithium; 3,3-Dimethyl-2-butanone; 2-Cyclopropyl-3,3-dimethyl-2-butanol (71%)
Preparation of alkanes using lithium dialkylcuprates (Section 14.11) Two alkyl groups may be coupled together to form an alkane by the reaction of an alkyl halide with a lithium dialkylcuprate. Both alkyl groups must be primary (or methyl). Aryl and vinyl halides may be used in place of alkyl halides.	$R_2CuLi + R'CH_2X \longrightarrow RCH_2R'$ Lithium dialkylcuprate; Primary alkyl halide; Alkane $(CH_3)_2CuLi + C_6H_5CH_2Cl \xrightarrow{\text{diethyl ether}} C_6H_5CH_2CH_3$ Lithium dimethylcuprate; Benzyl chloride; Ethylbenzene (80%)

TABLE 14.5
Synthetic Applications of Organometallic Reagents (*Continued*)

Reaction (section) and comments	General equation and specific example
The Simmons-Smith reaction (Section 14.12) Methylene transfer from iodomethylzinc iodide converts alkenes to cyclopropanes. The reaction is a stereospecific syn addition of a CH_2 group to the double bond.	$R_2C{=}CR_2$ (Alkene) + ICH_2ZnI (Iodomethylzinc iodide) $\xrightarrow{\text{diethyl ether}}$ Cyclopropane derivative + ZnI_2 (Zinc iodide) Cyclopentene $\xrightarrow[\text{diethyl ether}]{CH_2I_2,\ Zn(Cu)}$ Bicyclo[3.1.0]hexane (53%)
Oxymercuration-demercuration (Section 14.14) As an alternative to acid-catalyzed hydration, alkenes are treated with mercury(II) acetate in aqueous tetrahydrofuran, followed by sodium borohydride. The elements of water add across the double bond with a regioselectivity corresponding to Markovnikov's rule. Rearrangements do not occur. Ethers are formed when the reaction is carried out in an alcohol as the solvent.	$RCH{=}CH_2$ (Alkene) $\xrightarrow[\text{2. NaBH}_4,\ HO^-]{\text{1. Hg(OAc)}_2,\ THF-H_2O}$ $RCH(OH)CH_3$ (Alcohol) $CH_3(CH_2)_{15}CH{=}CH_2$ (1-Octadecene) $\xrightarrow[\text{2. NaBH}_4,\ HO^-]{\text{1. Hg(OAc)}_2,\ THF-H_2O}$ $CH_3(CH_2)_{15}CH(OH)CH_3$ (2-Octadecanol (93%))

reactions of organolithium compounds and Grignard reagents with aldehydes and ketones to produce alcohols rank among the most frequently employed of all synthetic procedures. Organometallic compounds are often intermediates in chemical reactions. Table 14.5 includes an example of one such process, the preparation of alcohols by oxymercuration-demercuration, in which an organomercury compound is an intermediate.

When planning the synthesis of a compound using an organometallic reagent, or indeed any synthesis, the best approach is to reason backward from the product. This method is called **retrosynthetic analysis** (Section 14.9).

A wide variety of organometallic compounds are known, and novel structural types continue to be discovered. Organometallic chemistry is one of the most active areas of contemporary scientific research and promises to be a fertile field for continued development.

PROBLEMS

14.14 Write structural formulas for each of the following compounds. Specify which compounds qualify as organometallic compounds.

(*a*) Cyclopentyllithium
(*b*) Ethoxymagnesium chloride
(*c*) 2-Phenylethylmagnesium iodide
(*d*) Lithium divinylcuprate
(*e*) Mercury(II) acetate
(*f*) Benzylpotassium

14.15 Suggest appropriate methods for the preparation of each of the following compounds from the starting material of your choice:

(a) $CH_3CH_2CH_2CH_2CH_2MgI$
(b) $CH_3CH_2C{\equiv}CMgI$
(c) $CH_3CH_2CH_2CH_2CH_2Li$
(d) $(CH_3CH_2CH_2CH_2CH_2)_2CuLi$
(e) $HOCH_2CH_2HgO\overset{O}{\overset{\|}{C}}CH_3$
(f) $CH_3OCH_2CH_2HgO\overset{O}{\overset{\|}{C}}CH_3$

14.16 Which compound in each of the following pairs would you expect to have the more polar carbon-metal bond?

(a) CH_3CH_2Li or $(CH_3CH_2)_2Hg$
(b) $(CH_3)_2Zn$ or $(CH_3)_2Mg$
(c) CH_3CH_2MgBr or $HC{\equiv}CMgBr$

14.17 Write the structure of the principal organic product of each of the following reactions:

(a) 1-Bromopropane with lithium in diethyl ether
(b) 1-Bromopropane with magnesium in diethyl ether
(c) 2-Iodopropane with lithium in diethyl ether
(d) 2-Iodopropane with magnesium in diethyl ether
(e) Product of part (a) with copper(I) iodide
(f) Product of part (e) with 1-bromobutane
(g) Product of part (e) with iodobenzene
(h) Product of part (b) with D_2O and DCl
(i) Product of part (c) with D_2O and DCl
(j) Product of part (a) with formaldehyde in ether, followed by dilute acid
(k) Product of part (b) with benzaldehyde in ether, followed by dilute acid
(l) Product of part (c) with cycloheptanone in ether, followed by dilute acid
(m) Product of part (d) with $CH_3\overset{O}{\overset{\|}{C}}CH_2CH_3$ in ether, followed by dilute acid
(n) Product of part (b) with $C_6H_5\overset{O}{\overset{\|}{C}}OCH_3$ in ether, followed by dilute acid
(o) 1-Octene with diiodomethane and zinc-copper couple in ether
(p) (E)-2-Decene with diiodomethane and zinc-copper couple in ether
(q) (Z)-3-Decene with diiodomethane and zinc-copper couple in ether
(r) 1-Pentene with mercury(II) acetate in aqueous tetrahydrofuran, followed by sodium borohydride
(s) 1-Pentene with mercury(II) acetate in ethanol, followed by sodium borohydride
(t) 1-Pentene with tribromomethane and potassium *tert*-butoxide in *tert*-butyl alcohol

14.18 Using 1-bromobutane and any necessary organic or inorganic reagents, suggest efficient syntheses of each of the following alcohols:

(a) 1-Pentanol
(b) 2-Hexanol
(c) 1-Phenyl-1-pentanol
(d) 3-Methyl-3-heptanol
(e) 1-Butylcyclobutanol

14.19 Using bromobenzene and any necessary organic or inorganic reagents, suggest efficient syntheses of each of the following:

(a) Benzyl alcohol
(b) 1-Phenyl-1-hexanol

(*c*) Bromodiphenylmethane
(*d*) 4-Phenyl-4-heptanol
(*e*) 1-Phenylcyclooctanol
(*f*) *trans*-2-Phenylcyclooctanol

14.20 Analyze the following structures so as to determine all the practical combinations of Grignard reagent and carbonyl compound that will give rise to each:

(*a*) $CH_3CH_2CHCH_2CH(CH_3)_2$ (with OH on C-3)

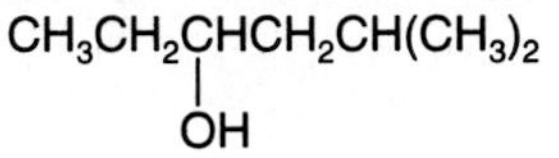

(*b*) cyclopropyl–CH(OH)–C$_6$H$_4$–OCH_3 (para)

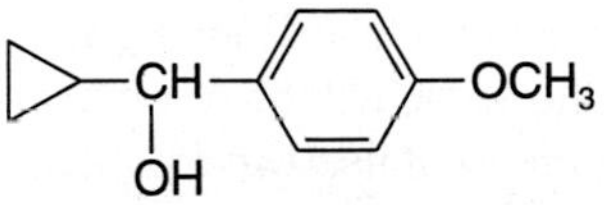

(*c*) $(CH_3)_3CCH_2OH$

(*d*) 6-Methyl-5-hepten-2-ol

(*e*) OH

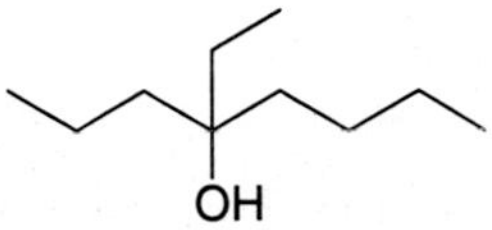

14.21 A number of drugs are prepared by reactions of the type described in this chapter. Indicate what you believe would be a reasonable last step in the synthesis of each of the following:

(*a*) $CH_3CH_2C(OH)(CH_3)C{\equiv}CH$ *Meparfynol,* a mild hypnotic or sleep-inducing agent

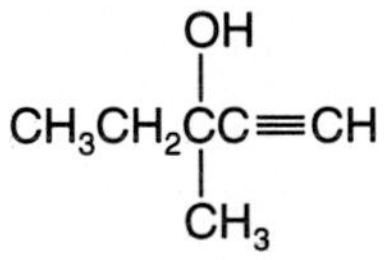

(*b*) $(C_6H_5)_2C(OH)CH(CH_3)$—N(piperidine) *Diphepanol,* an antitussive (cough suppressant)

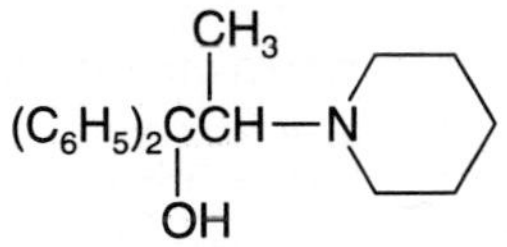

(*c*) CH_3, OH, $C{\equiv}CH$, CH_3O *Mestranol,* an estrogenic component of oral contraceptive drugs

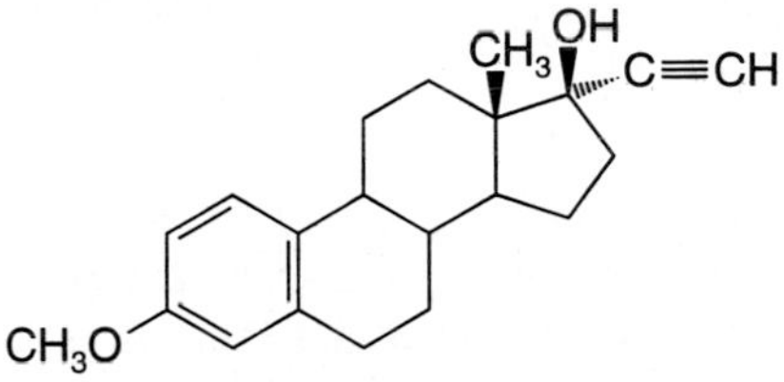

14.22 Predict the principal organic product of each of the following reactions:

(*a*) $C_6H_5COC_6H_5$ + $NaC{\equiv}CH$ $\xrightarrow[\text{2. } H_3O^+]{\text{1. liquid ammonia}}$

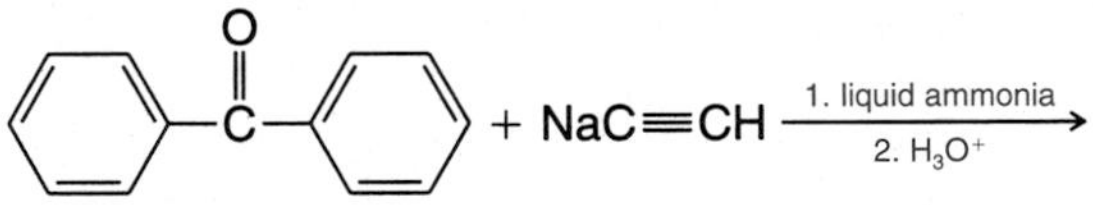

(*b*) (adamantanone, O) + CH_3CH_2Li $\xrightarrow[\text{2. } H_3O^+]{\text{1. diethyl ether}}$

(*c*) (1-bromocyclopentene) —Br $\xrightarrow{\text{1. Mg, THF; 2. HCHO; 3. } H_3O^+}$

(*d*) $C_6H_5CH_2CH{=}CH_2$ $\xrightarrow[\text{Zn(Cu), diethyl ether}]{CH_2I_2}$

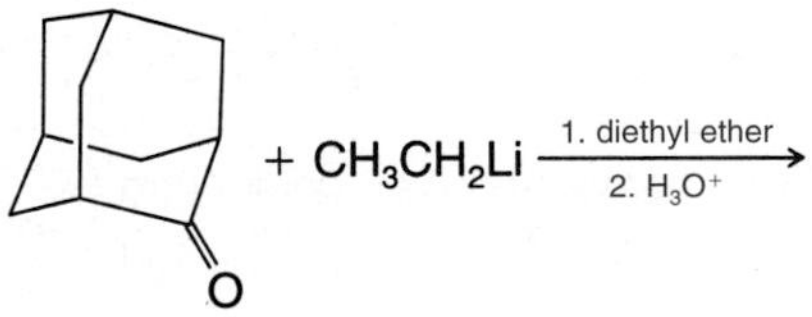

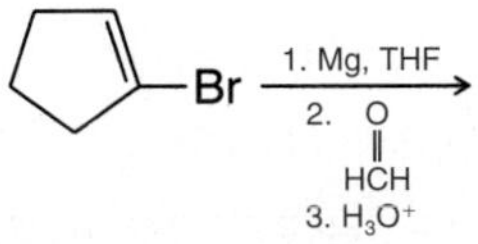

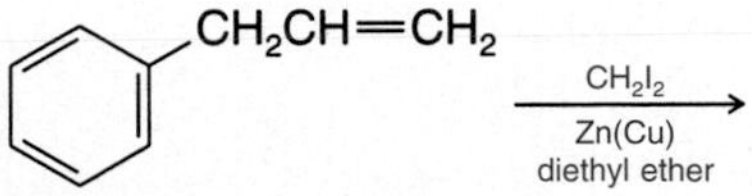

(*e*) $C_6H_5CH_2(H)C{=}C(H)CH_3$ $\xrightarrow[\text{Zn(Cu)}\ \text{ether}]{CH_2I_2}$

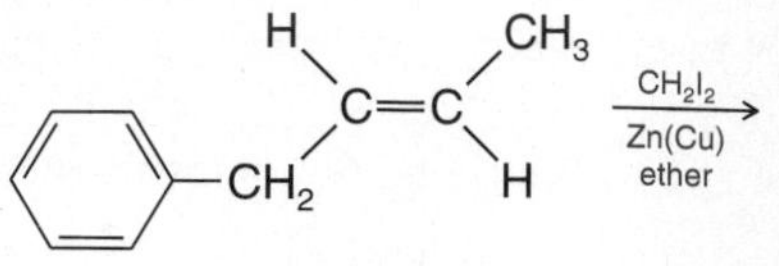

(*f*) [CH_3O-substituted benzonorbornadiene with I on the double bond] + $LiCu(CH_3)_2 \longrightarrow$

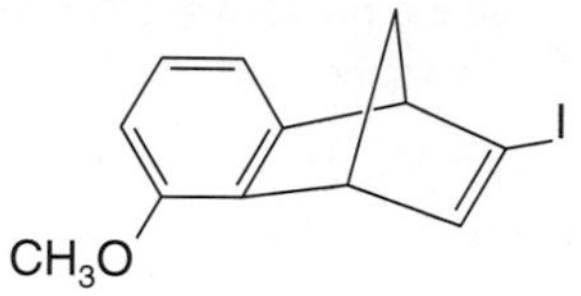

14.23 Addition of phenylmagnesium bromide to 4-*tert*-butylcyclohexanone gives two isomeric tertiary alcohols as products. Both alcohols yield the same alkene when subjected to acid-catalyzed dehydration. Suggest reasonable structures for these two alcohols.

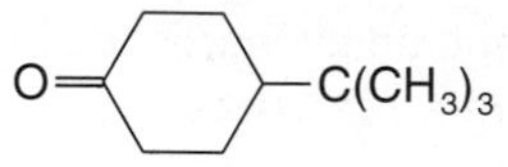

$O{=}C_6H_9{-}C(CH_3)_3$

4-*tert*-Butylcyclohexanone

14.24

(*a*) Unlike other esters, which react with Grignard reagents to give tertiary alcohols, ethyl formate ($H\overset{O}{\overset{\|}{C}}OCH_2CH_3$) yields a different class of alcohols on treatment with Grignard reagents. What kind of alcohol is formed in this case and why?

(*b*) Diethyl carbonate ($CH_3CH_2O\overset{O}{\overset{\|}{C}}OCH_2CH_3$) reacts with excess Grignard reagent to yield alcohols of a particular type. What is the structural feature that characterizes alcohols prepared in this way?

14.25 Reaction of lithium diphenylcuprate with optically active 2-bromobutane yields 2-phenylbutane, with high net inversion of configuration. When the 2-bromobutane used has the stereostructure shown, will the 2-phenylbutane formed have the *R* or the *S* configuration?

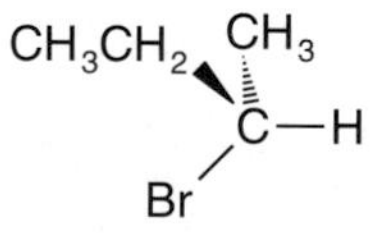

CH_3CH_2, CH_3, H, Br on C

14.26 Suggest reasonable structures for compounds A, B, and C in the following reactions:

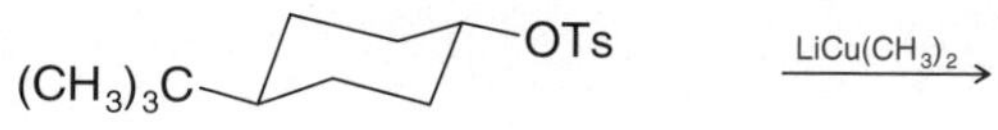

$(CH_3)_3C$–cyclohexyl–OTs (equatorial) $\xrightarrow{LiCu(CH_3)_2}$ compound A + compound B
($C_{11}H_{22}$) ($C_{10}H_{18}$)

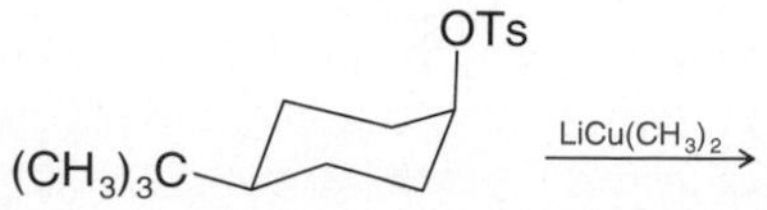

$(CH_3)_3C$–cyclohexyl–OTs (axial) $\xrightarrow{LiCu(CH_3)_2}$ compound B + compound C
($C_{11}H_{22}$)

Compound C is more stable than compound A.

14.27 The following conversion has been reported in the chemical literature. It was carried out in two steps, the first of which involved formation of a *p*-toluenesulfonate ester. Indicate the reagents for this step and show how you could convert the *p*-toluenesulfonate to the desired product.

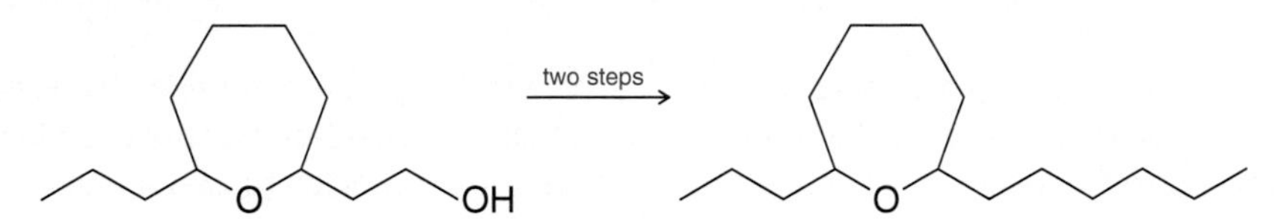

14.28 Oxymercuration-demercuration of *cis*-1,3,4-trimethylcyclopentene gives two different alcohols as products. Write their structures.

14.29 Sometimes the strongly basic properties of Grignard reagents can be turned to synthetic advantage. A chemist needed samples of butane specifically labeled with deuterium, the mass 2 isotope of hydrogen, as shown:

(*a*) $CH_3CH_2CH_2CH_2D$
(*b*) $CH_3CHDCH_2CH_3$

Suggest methods for the preparation of each of these using heavy water (D_2O) as the source of deuterium, butanols of your choice, and any necessary organic or inorganic reagents.

14.30 Diphenylmethane is significantly more acidic than benzene, and triphenylmethane is more acidic than either. Identify the most acidic proton in each compound, and suggest a reason for the trend in acidity.

C_6H_6	$(C_6H_5)_2CH_2$	$(C_6H_5)_3CH$
Benzene	Diphenylmethane	Triphenylmethane
$K_a \cong 10^{-45}$	$K_a \cong 10^{-34}$	$K_a \cong 10^{-32}$

MOLECULAR MODELING EXERCISES

14.31 Construct molecular models of methyllithium and vinyllithium. Assume each has a covalent carbon-metal bond.

14.32 Construct a molecular model of the major diastereomer formed in the addition of methylmagnesium bromide to 7,7-dimethylbicyclo[2.2.1]heptan-2-one.

14.33 The major diastereomer formed by addition of a Grignard reagent to an aldehyde or ketone can be predicted on the basis of *Cram's rule.* According to this rule, one orients the molecule as shown in the Newman projection on the following page, with the carbonyl oxygen anti to the largest group (L). The group transferred from the Grignard reagent is attached from the same side as the smaller (S) of the two remaining substituents. Use Cram's rule to predict the diastereomer formed in greatest amount in the reaction of methylmagnesium bromide with 2-phenylpropanal ($C_6H_5\underset{\substack{|\\CH_3}}{C}HCH{=}O$).

Donald Cram carried out the work that led to this empirical rule at U.C.L.A. in the 1950s. His later studies in molecular recognition led to a share of the 1987 Nobel Prize in chemistry.

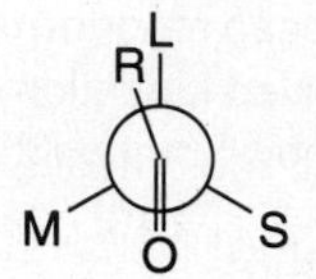

14.34 Verify, by making molecular models of each, that the products formed by addition of dichlorocarbene to the opposite faces of the double bond of *trans*-2-butene are enantiomers. Similarly, show that only a single stereoisomer is formed by addition to *cis*-2-butene.

14.35 The stereoselectivity of reaction of the Simmons-Smith reagent (ICH_2ZnI) with 3-cyclopentenol is believed to be governed by a Lewis acid–Lewis base interaction between Zn and the OH group. Construct a molecular model of 3-cyclopentenol, then add a CH_2 group to the double bond with the correct stereochemistry to give the observed product.

CHAPTER 15

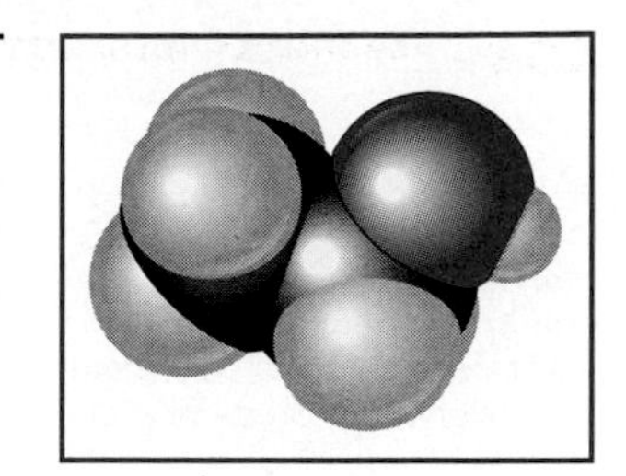

ALCOHOLS, DIOLS, AND THIOLS

The material in the next several chapters deals with the chemistry of various oxygen-containing functional groups. The interplay of these important classes of compounds—alcohols, ethers, aldehydes, ketones, carboxylic acids, and derivatives of carboxylic acids—is fundamental to the science of organic chemistry.

ROH	ROR′	RC(=O)H	RC(=O)R′	RC(=O)OH
Alcohol	Ether	Aldehyde	Ketone	Carboxylic acid

We begin by discussing in more detail a class of compounds already familiar to us, *alcohols.* Alcohols were introduced in Chapter 4 and have appeared regularly since then. With this chapter we extend our knowledge of alcohols, particularly with respect to their relationship to carbonyl-containing compounds. In the course of studying alcohols, we shall also look at some relatives. **Diols** are alcohols in which two hydroxyl groups (—OH) are present; **thiols** are compounds that contain an —SH group. **Phenols,** compounds of the type ArOH, share many properties in common with alcohols but are sufficiently different from them to warrant separate discussion in Chapter 24.

This chapter is a transitional one. It ties together much of the material encountered earlier and sets the stage for our study of other oxygen-containing functional groups in the chapters that follow.

15.1 SOURCES OF ALCOHOLS

Until the 1920s, the major source of *methanol* was as a by-product in the production of charcoal from wood—hence, the name *wood alcohol.* Now, most of the more than 10

Carbon monoxide is obtained from coal, and hydrogen is one of the products formed when natural gas is converted to ethylene and propene (Section 5.1).

billion pounds of methanol used annually in the United States is synthetic, prepared directly from carbon monoxide and hydrogen.

$$\underset{\text{Carbon monoxide}}{CO} + \underset{\text{Hydrogen}}{2H_2} \xrightarrow[400°C]{ZnO-Cr_2O_3} \underset{\text{Methanol}}{CH_3OH}$$

Almost half of this methanol is converted to formaldehyde as a starting material for various resins and plastics. Methanol is also used as a solvent, as an antifreeze, and as a convenient clean-burning liquid fuel. This last property makes it a candidate as a fuel for automobiles—methanol is already used to power Indianapolis-class race cars—but extensive emissions tests remain to be done before it can be approved as a gasoline substitute. Methanol is a colorless liquid, boiling at 65°C, and is miscible with water in all proportions. It is poisonous; drinking as little as 30 mL has been fatal. Ingestion of sublethal amounts can lead to blindness.

When vegetable matter ferments, its carbohydrates are converted to *ethanol* and carbon dioxide by enzymes present in yeast. Fermentation of barley produces beer; grapes give wine. The maximum ethanol content is on the order of 15 percent, because higher concentrations inactivate the enzymes, halting fermentation. Since ethanol boils at 78°C and water at 100°C, distillation of the fermentation broth can be used to give ''distilled spirits'' of increased ethanol content. Whiskey is the aged distillate of fermented grain and contains slightly less than 50 percent ethanol. Brandy and cognac are made by aging the distilled spirits from fermented grapes and other fruits. The characteristic flavors, odors, and colors of the various alcoholic beverages depend on both their origin and the way they are aged.

Synthetic ethanol is derived from petroleum by hydration of ethylene. In the United States, some 700 million pounds of synthetic ethanol is produced annually. It is relatively inexpensive and useful for industrial applications. In order to make it unfit for drinking, it is *denatured* by adding any of a number of noxious materials, a process that exempts it from the high taxes most governments impose on ethanol used in beverages.

Some of the substances used to denature ethanol include methanol, benzene, pyridine, castor oil, and gasoline.

Our bodies are reasonably well equipped to metabolize ethanol, making it less dangerous than methanol. Alcohol abuse and alcoholism have been and remain, however, persistent problems.

Isopropyl alcohol is prepared from petroleum by hydration of propene. With a boiling point of 82°C, isopropyl alcohol evaporates quickly from the skin, producing a cooling effect. Often containing dissolved oils and fragrances, it is the major component of rubbing alcohol. Isopropyl alcohol possesses weak antibacterial properties and is used to maintain medical instruments in a sterile condition and to clean the skin before minor surgery.

Methanol, ethanol, and isopropyl alcohol are included among the readily available starting materials commonly found in laboratories where organic synthesis is carried out. So, too, are many other alcohols. All alcohols of four carbons or less, as well as most of the five- and six-carbon alcohols and many higher alcohols, are commercially available at low cost. Some occur naturally; others are the products of efficient syntheses. Figure 15.1 presents the structures of a few naturally occurring alcohols. Table 15.1 summarizes the reactions encountered in earlier chapters that give alcohols and illustrates a thread that runs through the fabric of organic chemistry: *a reaction that is characteristic of one functional group often serves as a synthetic method for preparing another.*

As Table 15.1 indicates, reactions leading to alcohols are not in short supply.

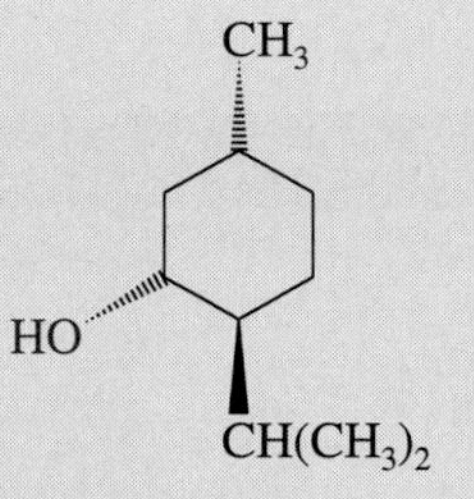

Menthol (obtained from oil of peppermint and used to flavor tobacco and food)

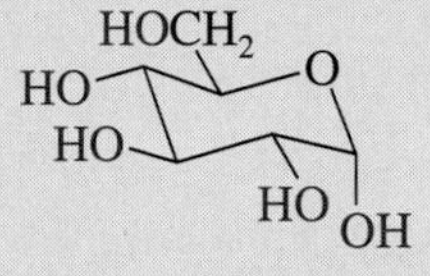

Glucose (a carbohydrate)

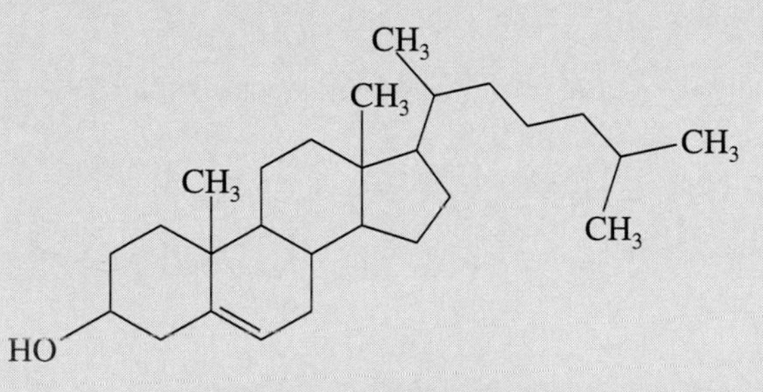

Cholesterol (principal constituent of gallstones and biosynthetic precursor of the steroid hormones)

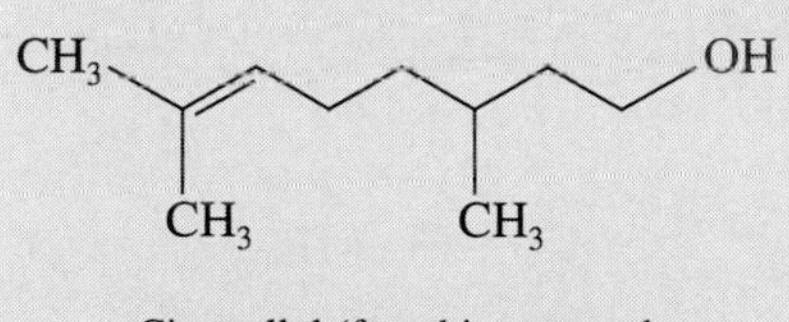

Citronellol (found in rose and geranium oil and used in perfumery)

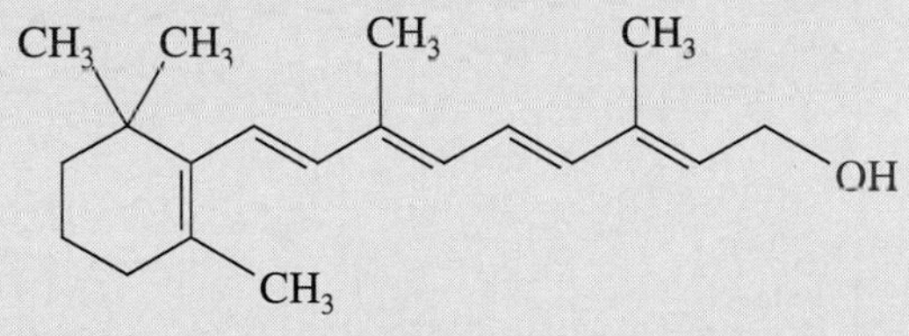

Retinol (vitamin A, an important substance in vision)

FIGURE 15.1 Some naturally occurring alcohols.

Nevertheless, several more will be added to the list in the present chapter—testimony to the importance of alcohols in synthetic organic chemistry. Some of these methods involve reduction of carbonyl groups:

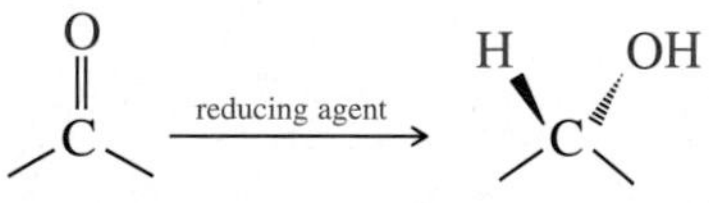

Recall from Section 2.18 that reduction corresponds to a decrease in the number of bonds between carbon and oxygen and/or an increase in the number of bonds between carbon and hydrogen.

$$\text{C=O} \xrightarrow{\text{reducing agent}} \text{H–C–OH}$$

We will begin with the reduction of aldehydes and ketones.

15.2 PREPARATION OF ALCOHOLS BY REDUCTION OF ALDEHYDES AND KETONES

The most obvious way to reduce an aldehyde or a ketone to an alcohol is by hydrogenation of the carbon-oxygen double bond. Like the hydrogenation of alkenes, the

TABLE 15.1
Summary of Reactions Discussed in Earlier Chapters That Yield Alcohols

Reaction (section) and comments	General equation and specific example
Acid-catalyzed hydration of alkenes (Section 6.10) The elements of water add to the double bond in accordance with Markovnikov's rule. This reaction is not used frequently for laboratory-scale synthesis of alcohols.	$R_2C{=}CR_2 + H_2O \xrightarrow{H^+} R_2CHCR_2(OH)$ Alkene, Water, Alcohol $(CH_3)_2C{=}CHCH_3 \xrightarrow[H_2SO_4]{H_2O} CH_3C(CH_3)(OH)CH_2CH_3$ 2-Methyl-2-butene → 2-Methyl-2-butanol (90%)
Oxymercuration-demercuration of alkenes (Section 14.14) Markovnikov addition of the elements of water to the double bond occurs. This is a useful synthetic reaction. Rearrangements do not occur.	$R_2C{=}CR_2 \xrightarrow[2.\ NaBH_4,\ HO^-]{1.\ Hg(O_2CCH_3)_2,\ THF-H_2O} R_2CHCR_2(OH)$ Alkene → Alcohol $C_6H_5C(CH_3){=}CH_2 \xrightarrow[2.\ NaBH_4,\ HO^-]{1.\ Hg(O_2CCH_3)_2,\ THF-H_2O} C_6H_5C(OH)(CH_3)CH_3$ 2-Phenylpropene → 2-Phenyl-2-propanol (95%)
Hydroboration-oxidation of alkenes (Section 6.11) The elements of water add to the double bond with regioselectivity opposite to that of Markovnikov's rule. This is a very good synthetic method; addition is syn, and no rearrangements are observed.	$R_2C{=}CR_2 \xrightarrow[2.\ H_2O_2,\ HO^-]{1.\ B_2H_6} R_2CHCR_2(OH)$ Alkene → Alcohol $CH_3(CH_2)_7CH{=}CH_2 \xrightarrow[2.\ H_2O_2,\ HO^-]{1.\ B_2H_6,\ diglyme} CH_3(CH_2)_7CH_2CH_2OH$ 1-Decene → 1-Decanol (93%)
Hydrolysis of alkyl halides (Section 8.1) A reaction useful only with substrates that do not undergo E2 elimination readily. It is rarely used for the synthesis of alcohols, since alkyl halides are normally prepared from alcohols.	$RX + HO^- \longrightarrow ROH + X^-$ Alkyl halide, Hydroxide ion, Alcohol, Halide ion $2,4,6\text{-}(CH_3)_3C_6H_2CH_2Cl \xrightarrow[heat]{H_2O,\ Ca(OH)_2} 2,4,6\text{-}(CH_3)_3C_6H_2CH_2OH$ 2,4,6-Trimethylbenzyl chloride → 2,4,6-Trimethylbenzyl alcohol (78%)

TABLE 15.1

Summary of Reactions Discussed in Earlier Chapters That Yield Alcohols *(Continued)*

Reaction (section) and comments	General equation and specific example
Reaction of Grignard reagents with aldehydes and ketones (Section 14.6) A method that allows for alcohol preparation with formation of new carbon-carbon bonds. Primary, secondary, and tertiary alcohols can all be prepared.	$RMgX + R'\overset{O}{\overset{\Vert}{C}}R'' \xrightarrow[2.\ H_3O^+]{1.\ \text{diethyl ether}} RC(R')(R'')OH$ Grignard reagent; Aldehyde or ketone; Alcohol Cyclopentylmagnesium bromide + $H\overset{O}{\overset{\Vert}{C}}H$ $\xrightarrow[2.\ H_3O^+]{1.\ \text{diethyl ether}}$ Cyclopentylmethanol ($C_5H_9CH_2OH$) Cyclopentylmagnesium bromide; Formaldehyde; Cyclopentylmethanol (62–64%)
Reaction of organolithium reagents with aldehydes and ketones (Section 14.7) Organolithium reagents react with aldehydes and ketones in a manner similar to that of Grignard reagents to form alcohols.	$RLi + R'\overset{O}{\overset{\Vert}{C}}R'' \xrightarrow[2.\ H_3O^+]{1.\ \text{diethyl ether}} RC(R')(R'')OH$ Organolithium reagent; Aldehyde or ketone; Alcohol $CH_3CH_2CH_2CH_2Li + C_6H_5\overset{O}{\overset{\Vert}{C}}CH_3 \xrightarrow[2.\ H_3O^+]{1.\ \text{diethyl ether}} CH_3CH_2CH_2CH_2{-}C(C_6H_5)(CH_3){-}OH$ Butyllithium; Acetophenone; 2-Phenyl-2-hexanol (67%)
Reaction of Grignard reagents with esters (Section 14.10) Produces tertiary alcohols in which two of the substituents on the hydroxyl-bearing carbon are derived from the Grignard reagent.	$2RMgX + R'\overset{O}{\overset{\Vert}{C}}OR'' \xrightarrow[2.\ H_3O^+]{1.\ \text{diethyl ether}} RC(R')(R)OH + R''OH$ $2CH_3CH_2CH_2CH_2CH_2MgBr + CH_3\overset{O}{\overset{\Vert}{C}}OCH_2CH_3 \xrightarrow[2.\ H_3O^+]{1.\ \text{diethyl ether}}$ $CH_3C(OH)(CH_2CH_2CH_2CH_2CH_3)CH_2CH_2CH_2CH_2CH_3$ Pentylmagnesium bromide; Ethyl acetate; 6-Methyl-6-undecanol (75%)

reaction is exothermic but exceedingly slow in the absence of a catalyst. Finely divided metals such as platinum, palladium, nickel, and ruthenium are effective catalysts for the hydrogenation of aldehydes and ketones. Aldehydes yield primary alcohols:

$$\underset{\text{Aldehyde}}{\overset{\displaystyle O}{\overset{\|}{RCH}}} + \underset{\text{Hydrogen}}{H_2} \xrightarrow{\text{Pt, Pd, Ni, or Ru}} \underset{\text{Primary alcohol}}{RCH_2OH}$$

$$CH_3O\text{—}C_6H_4\text{—}\overset{\displaystyle O}{\overset{\|}{CH}} \xrightarrow[\text{ethanol}]{H_2,\ Pt} CH_3O\text{—}C_6H_4\text{—}CH_2OH$$

p-Methoxybenzaldehyde → *p*-Methoxybenzyl alcohol (92%)

Ketones yield secondary alcohols:

$$\underset{\text{Ketone}}{\overset{\displaystyle O}{\overset{\|}{RCR'}}} + \underset{\text{Hydrogen}}{H_2} \xrightarrow{\text{Pt, Pd, Ni, or Ru}} \underset{\text{Secondary alcohol}}{\underset{\displaystyle OH}{\underset{|}{RCHR'}}}$$

$$\text{Cyclopentanone} \xrightarrow[\text{methanol}]{H_2,\ Pt} \text{Cyclopentanol (93–95\%)}$$

Cyclopentanone → Cyclopentanol (93–95%)

PROBLEM 15.1 Which of the isomeric $C_4H_{10}O$ alcohols can be prepared by hydrogenation of aldehydes? Which can be prepared by hydrogenation of ketones? Which cannot be prepared by hydrogenation of a carbonyl compound?

For most laboratory-scale reductions of aldehydes and ketones, catalytic hydrogenation has been replaced by methods based on metal hydride reducing agents. The two most common reagents are sodium borohydride and lithium aluminum hydride.

$$Na^+ \left[H\text{—}\overset{\displaystyle H}{\overset{|}{\underset{\displaystyle H}{\underset{|}{B}}}}{}^{-}\text{—}H \right] \qquad Li^+ \left[H\text{—}\overset{\displaystyle H}{\overset{|}{\underset{\displaystyle H}{\underset{|}{Al}}}}{}^{-}\text{—}H \right]$$

Sodium borohydride ($NaBH_4$) Lithium aluminum hydride ($LiAlH_4$)

Sodium borohydride is particularly easy to use, needing only to be added to an aqueous or alcoholic solution of an aldehyde or a ketone:

$$\underset{\text{Aldehyde}}{\mathrm{RCH}} \xrightarrow[\text{water, methanol, or ethanol}]{\mathrm{NaBH_4}} \underset{\text{Primary alcohol}}{\mathrm{RCH_2OH}}$$

$$m\text{-}O_2NC_6H_4CH{=}O \xrightarrow[\text{methanol}]{\mathrm{NaBH_4}} m\text{-}O_2NC_6H_4CH_2OH$$

m-Nitrobenzaldehyde → *m*-Nitrobenzyl alcohol (82%)

$$\underset{\text{Ketone}}{\mathrm{RCR'}} \xrightarrow[\text{water, methanol, or ethanol}]{\mathrm{NaBH_4}} \underset{\text{Secondary alcohol}}{\mathrm{RCHR'}\ (\mathrm{OH})}$$

$$\mathrm{CH_3CCH_2C(CH_3)_3}\ (\mathrm{C{=}O}) \xrightarrow[\text{ethanol}]{\mathrm{NaBH_4}} \mathrm{CH_3CHCH_2C(CH_3)_3}\ (\mathrm{OH})$$

4,4-Dimethyl-2-pentanone → 4,4-Dimethyl-2-pentanol (85%)

Lithium aluminum hydride reacts violently with water and alcohols, so it must be used in solvents such as anhydrous diethyl ether or tetrahydrofuran. Following reduction, a separate hydrolysis step is required to liberate the alcohol product:

$$\underset{\text{Aldehyde}}{\mathrm{RCH}} \xrightarrow[\text{2. } H_2O]{\text{1. } LiAlH_4\text{, diethyl ether}} \underset{\text{Primary alcohol}}{\mathrm{RCH_2OH}}$$

$$\underset{\text{Heptanal}}{\mathrm{CH_3(CH_2)_5CH}} \xrightarrow[\text{2. } H_2O]{\text{1. } LiAlH_4\text{, diethyl ether}} \underset{\text{1-Heptanol (86\%)}}{\mathrm{CH_3(CH_2)_5CH_2OH}}$$

$$\underset{\text{Ketone}}{\mathrm{RCR'}} \xrightarrow[\text{2. } H_2O]{\text{1. } LiAlH_4\text{, diethyl ether}} \underset{\text{Secondary alcohol}}{\mathrm{RCHR'}\ (\mathrm{OH})}$$

$$\mathrm{(C_6H_5)_2CHCCH_3}\ (\mathrm{C{=}O}) \xrightarrow[\text{2. } H_2O]{\text{1. } LiAlH_4\text{, diethyl ether}} \mathrm{(C_6H_5)_2CHCHCH_3}\ (\mathrm{OH})$$

1,1-Diphenyl-2-propanone → 1,1-Diphenyl-2-propanol (84%)

Sodium borohydride and lithium aluminum hydride react with carbonyl compounds in much the same way that Grignard reagents do, except that they function as *hydride donors* rather than as carbanion sources. Borohydride transfers a hydrogen with its pair of bonding electrons to the positively polarized carbon of a carbonyl group. The negatively polarized oxygen attacks boron. Ultimately, all four of the hydrogens of borohydride are transferred and a tetraalkoxyborate is formed.

$$H-\overline{B}H_3 \;+\; \underset{\delta+\;\;\;\delta-}{R_2C{=}O} \longrightarrow R_2\overset{H}{\overset{|}{C}}-\overset{\overline{B}H_3}{\overset{|}{O}} \xrightarrow{3R_2C{=}O} \underset{\text{Tetraalkoxyborate}}{(R_2CHO)_4\overline{B}}$$

Hydrolysis or alcoholysis converts the tetraalkoxyborate intermediate to the corresponding alcohol. The following equation illustrates the process for reactions carried out in water. An analogous process occurs in methanol or ethanol and yields the alcohol and $(CH_3O)_4B^-$ or $(CH_3CH_2O)_4B^-$.

$$R_2CHO-\overline{B}(OCHR_2)_3 \;+\; H-OH \longrightarrow R_2CHOH + HO\overline{B}(OCHR_2)_3 \xrightarrow{3H_2O} 3R_2CHOH + (HO)_4\overline{B}$$

A similar series of hydride transfers occurs when aldehydes and ketones are treated with lithium aluminum hydride.

$$H-\overline{Al}H_3 \;+\; \underset{\delta+\;\;\;\delta-}{R_2C{=}O} \longrightarrow R_2\overset{H}{\overset{|}{C}}-\overset{\overline{Al}H_3}{\overset{|}{O}} \xrightarrow{3R_2C{=}O} \underset{\text{Tetraalkoxyaluminate}}{(R_2CHO)_4\overline{Al}}$$

Addition of water converts the tetraalkoxyaluminate to the desired alcohol.

$$\underset{\text{Tetraalkoxyaluminate}}{(R_2CHO)_4\overline{Al}} + 4H_2O \longrightarrow \underset{\text{Alcohol}}{4R_2CHOH} + \overline{Al}(OH)_4$$

PROBLEM 15.2 Sodium borodeuteride ($NaBD_4$) and lithium aluminum deuteride ($LiAlD_4$) are convenient reagents for introducing deuterium, the mass 2 isotope of hydrogen, into organic compounds. Write the structure of the organic product of the following reactions, clearly showing the position of all the deuterium atoms in each:

(*a*) Reduction of $CH_3\overset{O}{\overset{\|}{C}}H$ (acetaldehyde) with $NaBD_4$ in H_2O

(*b*) Reduction of $CH_3\overset{O}{\overset{\|}{C}}CH_3$ (acetone) with $NaBD_4$ in CH_3OD

(*c*) Reduction of $C_6H_5\overset{O}{\overset{\|}{C}}H$ (benzaldehyde) with $NaBD_4$ in CD_3OH

(*d*) Reduction of $H\overset{O}{\overset{\|}{C}}H$ (formaldehyde) with $LiAlD_4$ in diethyl ether, followed by addition of D_2O

SAMPLE SOLUTION (*a*) Sodium borodeuteride transfers deuterium to the carbonyl group of acetaldehyde, forming a C—D bond.

$$D{-}\bar{B}D_3 \quad CH_3\underset{H}{C}{=}O \longrightarrow CH_3{-}\underset{H}{\overset{D}{C}}{-}\overset{\bar{B}D_3}{O} \xrightarrow{3CH_3\overset{O}{\overset{\|}{C}}H} (CH_3\overset{D}{C}HO)_4\bar{B}$$

Hydrolysis of $(CH_3CHDO)_4\bar{B}$ in H_2O leads to the formation of ethanol, retaining the C—D bond formed in the preceding step while forming of an O—H bond.

$$CH_3\overset{D}{C}H{-}O{-}\bar{B}(OCHDCH_3)_3 \;\; H{-}OH \longrightarrow \underset{\text{Ethanol-1-d}}{CH_3\overset{D}{C}H{-}OH} + \bar{B}(OCHDCH_3)_3{-}OH \xrightarrow{3H_2O} 3CH_3\overset{D}{C}HOH + \bar{B}(OH)_4$$

Neither sodium borohydride nor lithium aluminum hydride reduces isolated carbon-carbon double bonds. This makes possible the selective reduction of a carbonyl group in a molecule that contains both carbon-carbon and carbon-oxygen double bonds.

Catalytic hydrogenation would not be suitable for this transformation, because H_2 adds to carbon-carbon double bonds faster than it reduces carbonyl groups.

$$\underset{\text{6-Methyl-5-hepten-2-one}}{(CH_3)_2C{=}CHCH_2CH_2\overset{O}{\overset{\|}{C}}CH_3} \xrightarrow[\text{2. } H_2O]{\text{1. LiAlH}_4\text{, diethyl ether}} \underset{\text{6-Methyl-5-hepten-2-ol (90\%)}}{(CH_3)_2C{=}CHCH_2CH_2\overset{OH}{\overset{|}{C}}HCH_3}$$

15.3 PREPARATION OF ALCOHOLS BY REDUCTION OF CARBOXYLIC ACIDS AND ESTERS

Carboxylic acids are exceedingly difficult to reduce. Acetic acid, for example, is often used as a solvent in catalytic hydrogenations because it is inert under the reaction conditions. A very powerful reducing agent is required in order to convert a carboxylic acid to a primary alcohol. Lithium aluminum hydride is that reducing agent.

$$\underset{\text{Carboxylic acid}}{R\overset{O}{\overset{\|}{C}}OH} \xrightarrow[\text{2. } H_2O]{\text{1. LiAlH}_4\text{, diethyl ether}} \underset{\text{Primary alcohol}}{RCH_2OH}$$

$$\underset{\text{Cyclopropanecarboxylic acid}}{\triangleright{-}CO_2H} \xrightarrow[\text{2. } H_2O]{\text{1. LiAlH}_4\text{, diethyl ether}} \underset{\text{Cyclopropylmethanol (78\%)}}{\triangleright{-}CH_2OH}$$

Sodium borohydride is not nearly so potent a hydride donor as lithium aluminum hydride and does not reduce carboxylic acids.

Esters are more easily reduced than carboxylic acids. Two alcohols are formed from each ester molecule. The acyl group of the ester is cleaved, giving a primary alcohol.

$$\underset{\text{Ester}}{\overset{\overset{\displaystyle O}{\|}}{RCOR'}} \longrightarrow \underset{\text{Primary alcohol}}{RCH_2OH} + \underset{\text{Alcohol}}{R'OH}$$

Lithium aluminum hydride is the reagent of choice for reducing esters to alcohols.

$$\underset{\text{Ethyl benzoate}}{C_6H_5\overset{\overset{\displaystyle O}{\|}}{C}OCH_2CH_3} \xrightarrow[\text{2. } H_2O]{\text{1. } LiAlH_4\text{, diethyl ether}} \underset{\text{Benzyl alcohol (90\%)}}{C_6H_5CH_2OH} + \underset{\text{Ethanol}}{CH_3CH_2OH}$$

PROBLEM 15.3 Give the structure of an ester that will yield a mixture containing equimolar amounts of 1-propanol and 2-propanol on reduction with lithium aluminum hydride.

Sodium borohydride reduces esters, but the reaction is too slow to be useful. Hydrogenation of esters requires a special catalyst and extremely high pressures and temperatures; it is used in industrial settings but rarely in the laboratory.

15.4 PREPARATION OF ALCOHOLS FROM EPOXIDES

Although the chemical reactions of epoxides will not be covered in detail until the following chapter, we shall introduce their use in the synthesis of alcohols here.

Grignard reagents react with ethylene oxide to yield primary alcohols containing two more carbon atoms than the alkyl halide from which the organometallic compound was prepared.

$$\underset{\text{Grignard reagent}}{RMgX} + \underset{\text{Ethylene oxide}}{H_2C\overset{O}{—}CH_2} \xrightarrow[\text{2. } H_3O^+]{\text{1. diethyl ether}} \underset{\text{Primary alcohol}}{RCH_2CH_2OH}$$

$$\underset{\text{Hexylmagnesium bromide}}{CH_3(CH_2)_4CH_2MgBr} + \underset{\text{Ethylene oxide}}{H_2C\overset{O}{—}CH_2} \xrightarrow[\text{2. } H_3O^+]{\text{1. diethyl ether}} \underset{\text{1-Octanol (71\%)}}{CH_3(CH_2)_4CH_2CH_2CH_2OH}$$

Organolithium reagents react with epoxides in a similar manner.

PROBLEM 15.4 Each of the following alcohols has been prepared by reaction of a Grignard reagent with ethylene oxide. Select the appropriate Grignard reagent in each case.

(*a*) 2-methylphenyl–CH_2CH_2OH (benzene ring bearing CH_3 and CH_2CH_2OH on adjacent carbons)

(*b*) cyclohexyl–CH_2CH_2OH

SAMPLE SOLUTION (*a*) Reaction with ethylene oxide results in the addition of a $—CH_2CH_2OH$ unit to the Grignard reagent. The Grignard reagent derived from *o*-bromotoluene (or *o*-chlorotoluene or *o*-iodotoluene) is appropriate here.

$$o\text{-}CH_3C_6H_4MgBr + H_2C\underset{O}{—}CH_2 \xrightarrow[2.\ H_3O^+]{1.\ \text{diethyl ether}} o\text{-}CH_3C_6H_4CH_2CH_2OH$$

o-Methylphenylmagnesium bromide — Ethylene oxide — 2-(*o*-Methylphenyl)ethanol (66%)

Epoxide rings are readily opened with cleavage of the carbon-oxygen bond when attacked by nucleophiles. Grignard reagents and organolithium reagents react with ethylene oxide by serving as sources of nucleophilic carbon.

$$\overset{\delta-}{R}—\overset{\delta+}{MgX} + H_2C\underset{\ddot{O}}{—}CH_2 \longrightarrow R—CH_2—CH_2—\ddot{\underset{..}{O}}:^- \ ^+MgX \xrightarrow{H_3O^+} RCH_2CH_2OH$$

(may be written as RCH_2CH_2OMgX)

This kind of chemical reactivity of epoxides is rather general. Nucleophiles other than Grignard reagents react with epoxides, and epoxides more elaborate than ethylene oxide may be used. All these features of epoxide chemistry will be discussed in Sections 16.11 and 16.12.

15.5 PREPARATION OF DIOLS

Much of the chemistry of diols—compounds that bear two hydroxyl groups—is analogous to that of alcohols. Diols may be prepared, for example, from compounds that contain two carbonyl groups, using the same reducing agents employed in the preparation of alcohols. The following example shows the conversion of a dialdehyde to a diol by catalytic hydrogenation. Alternatively, the same transformation can be achieved by reduction with sodium borohydride or lithium aluminum hydride.

$$\overset{O}{\overset{\|}{HC}}CH_2\underset{CH_3}{\underset{|}{C}H}CH_2\overset{O}{\overset{\|}{C}H} \xrightarrow[\text{Ni, 125°C}]{H_2\ (100\ \text{atm})} HOCH_2CH_2\underset{CH_3}{\underset{|}{C}H}CH_2CH_2OH$$

3-Methylpentanedial — 3-Methyl-1,5-pentanediol (81–83%)

Diols are almost always given substitutive names. As the name of the product in the example indicates, the substitutive nomenclature of diols is similar to that of alcohols. The suffix *-diol* replaces *-ol*, and two locants, one for each hydroxyl group, are required. Note that the final *-e* of the alkane basis name is retained when the suffix begins with a consonant (*-diol*), but dropped when the suffix begins with a vowel (*-ol*).

PROBLEM 15.5 Write equations showing how 3-methyl-1,5-pentanediol could be prepared from a dicarboxylic acid or a diester.

Ethylene glycol and propylene glycol are prepared industrially from the corresponding alkenes by way of their epoxides. Some applications were given in the box in Section 6.21.

Vicinal diols are diols that have their hydroxyl groups on adjacent carbons. Two commonly encountered vicinal diols are 1,2-ethanediol and 1,2-propanediol.

$$HOCH_2CH_2OH \qquad CH_3\underset{\displaystyle OH}{\underset{|}{C}}HCH_2OH$$

1,2-Ethanediol (ethylene glycol) — 1,2-Propanediol (propylene glycol)

Ethylene glycol and *propylene glycol* are common names for these two diols and are acceptable IUPAC names. Aside from these two compounds, the IUPAC system does not use the word ''glycol'' for naming diols.

In the laboratory, vicinal diols are normally prepared from alkenes using the reagent *osmium tetraoxide* (OsO_4). Osmium tetraoxide reacts rapidly with alkenes to give cyclic osmate esters.

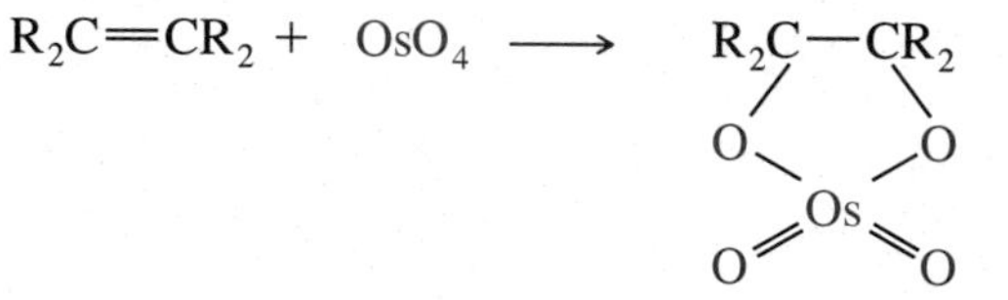

$$R_2C{=}CR_2 + OsO_4 \longrightarrow \text{(cyclic osmate ester: } R_2C{-}CR_2 \text{ with each C bonded to O, both O bonded to Os, which bears two =O)}$$

Alkene — Osmium tetraoxide — Cyclic osmate ester

Osmate esters are fairly stable but are readily cleaved in the presence of an oxidizing agent such as *tert*-butyl hydroperoxide.

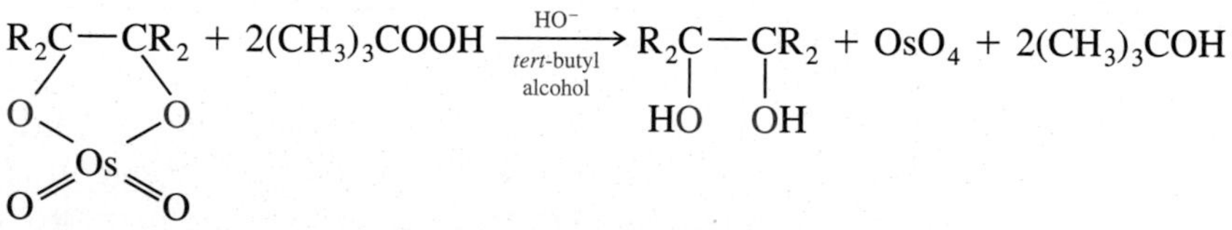

$$\text{Cyclic osmate ester} + 2(CH_3)_3COOH \xrightarrow[\textit{tert}\text{-butyl alcohol}]{HO^-} R_2\underset{\displaystyle HO}{\underset{|}{C}}{-}\underset{\displaystyle OH}{\underset{|}{C}}R_2 + OsO_4 + 2(CH_3)_3COH$$

tert-Butyl hydroperoxide — Vicinal diol — Osmium tetraoxide — *tert*-Butyl alcohol

Since osmium tetraoxide is regenerated in this step, alkenes can be converted to vicinal diols using only catalytic amounts of osmium tetraoxide. This is fortunate because osmium tetraoxide is both toxic and expensive. The entire process is performed in a single operation by simply allowing a solution of the alkene and *tert*-butyl hydroperoxide in *tert*-butyl alcohol containing a small amount of osmium tetraoxide and base to stand for several hours.

$$CH_3(CH_2)_7CH{=}CH_2 \xrightarrow[\textit{tert}\text{-butyl alcohol, } HO^-]{(CH_3)_3COOH,\ OsO_4(\text{cat})} CH_3(CH_2)_7\underset{\displaystyle OH}{\underset{|}{C}}HCH_2OH$$

1-Decene — 1,2-Decanediol (73%)

Overall, the reaction leads to addition of two hydroxyl groups to the double bond and is referred to as **hydroxylation.** Both oxygens of the diol come from osmium tetraoxide via the cyclic osmate ester. The reaction of OsO_4 with the alkene is a syn addition, and the conversion of the cyclic osmate to the diol involves cleavage of the bonds between oxygen and osmium. Thus, both hydroxyl groups of the diol become attached to the same face of the double bond; *syn hydroxylation of the alkene is observed.*

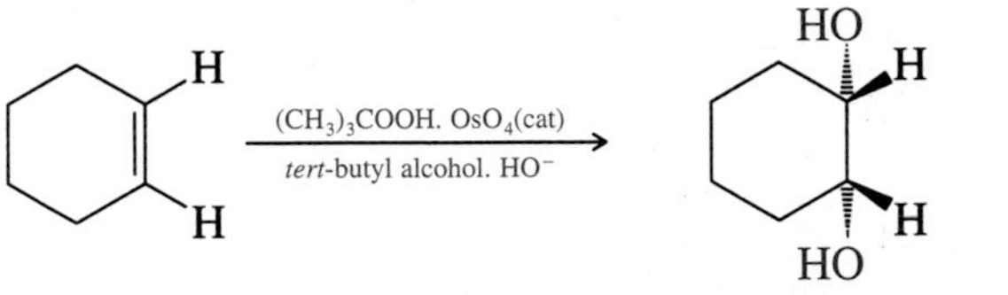

$(CH_3)_3COOH$. OsO_4(cat)
tert-butyl alcohol. HO^-

Cyclohexene → *cis*-1,2-Cyclohexanediol (62%)

PROBLEM 15.6 Give the structures, including stereochemistry, for the diols obtained by hydroxylation of *cis*-2-butene and *trans*-2-butene.

A complementary method, one that gives anti hydroxylation of alkenes by way of the hydrolysis of epoxides, will be described in Section 16.13.

15.6 REACTIONS OF ALCOHOLS: A REVIEW AND A PREVIEW

Alcohols are versatile starting materials for the preparation of a variety of organic functional groups. Several reactions of alcohols have already been seen in earlier chapters and are summarized in Table 15.2.

Alcohols can undergo reactions involving various combinations of the bonds to carbon and oxygen. The ionization of alcohols when they act as weak acids and the formation of alkoxides when they react with metals are reactions that take place at the O—H bond.

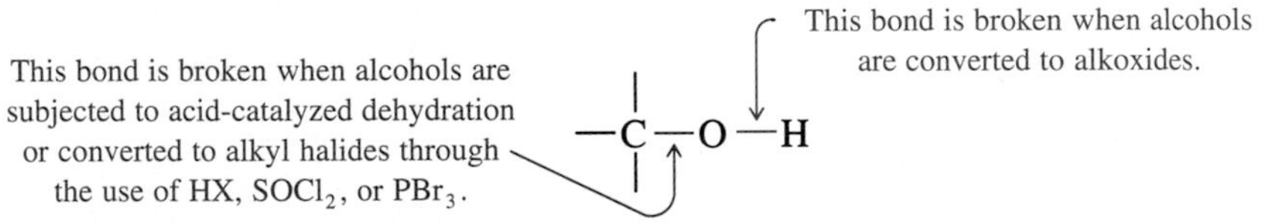

This bond is broken when alcohols are subjected to acid-catalyzed dehydration or converted to alkyl halides through the use of HX, $SOCl_2$, or PBr_3.

This bond is broken when alcohols are converted to alkoxides.

The carbon-oxygen bond of alcohols is cleaved when alcohols are converted to alkyl halides or undergo acid-catalyzed dehydration.

Some of the new reactions of alcohols that we will examine in this chapter occur by O—H bond breaking and some by C—O bond breaking. Additionally, primary and secondary alcohols can exhibit a third reaction type in which a carbon-oxygen double bond is formed by cleavage of both an O—H bond and a C—H bond:

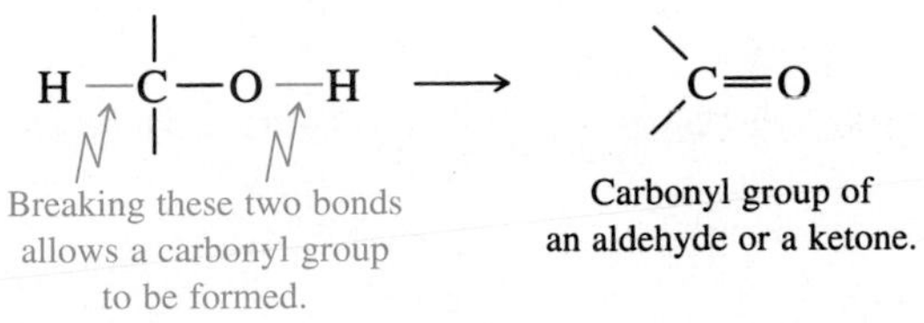

Breaking these two bonds allows a carbonyl group to be formed.

Carbonyl group of an aldehyde or a ketone.

TABLE 15.2
Summary of Reactions of Alcohols Discussed in Earlier Chapters

Reaction (section) and comments	General equation and specific example
Conversion to alkoxides (Section 4.8) Alcohols are slightly less acidic than water; alkoxides are slightly more basic than hydroxide. Alcohols react readily with group I metals to liberate hydrogen and form metal alkoxides.	$2ROH + 2M \longrightarrow 2RO^-M^+ + H_2$ Alcohol, Metal, Metal alkoxide, Hydrogen $CH_3CH(CH_3)OH \xrightarrow{Na} CH_3CH(CH_3)O^-Na^+$ 2-Propanol → Sodium 2-propanolate (100%)
Reaction with hydrogen halides (Section 4.9) The order of alcohol reactivity parallels the order of carbocation stability: $R_3C^+ > R_2CH^+ > RCH_2^+ > CH_3^+$. Benzylic alcohols react readily. The order of hydrogen halide reactivity is $HI > HBr > HCl \gg HF$.	$ROH + HX \longrightarrow RX + H_2O$ Alcohol, Hydrogen halide, Alkyl halide, Water $m\text{-}CH_3OC_6H_4CH_2OH \xrightarrow{HBr} m\text{-}CH_3OC_6H_4CH_2Br$ *m*-Methoxybenzyl alcohol → *m*-Methoxybenzyl bromide (98%)
Reaction with thionyl chloride (Section 4.15) Thionyl chloride converts alcohols to alkyl chlorides.	$ROH + SOCl_2 \longrightarrow RCl + SO_2 + HCl$ Alcohol, Thionyl chloride, Alkyl chloride, Sulfur dioxide, Hydrogen chloride $(CH_3)_2C{=}CHCH_2CH_2CH(OH)CH_3 \xrightarrow[\text{diethyl ether}]{SOCl_2,\ \text{pyridine}} (CH_3)_2C{=}CHCH_2CH_2CH(Cl)CH_3$ 6-Methyl-5-hepten-2-ol → 6-Chloro-2-methyl-2-heptene (67%)
Reaction with phosphorus trihalides (Section 4.15) Phosphorus trichloride and phosphorus tribromide convert alcohols to alkyl halides.	$3ROH + PX_3 \longrightarrow 3RX + H_3PO_3$ Alcohol, Phosphorus trihalide, Alkyl halide, Phosphorous acid $C_5H_9CH_2OH \xrightarrow{PBr_3} C_5H_9CH_2Br$ Cyclopentylmethanol → (Bromomethyl)cyclopentane (50%)
Acid-catalyzed dehydration (Section 5.9) This is a frequently used procedure for the preparation of alkenes. The order of alcohol reactivity parallels the order of carbocation stability: $R_3C^+ > R_2CH^+ > RCH_2^+$. Benzylic alcohols react readily. Rearrangements are sometimes observed.	$R_2C(OH)CHR_2 \xrightarrow[\text{heat}]{H^+} R_2C{=}CR_2 + H_2O$ Alcohol, Alkene, Water $m\text{-}BrC_6H_4CH(OH)CH_2CH_3 \xrightarrow[\text{heat}]{KHSO_4} m\text{-}BrC_6H_4CH{=}CHCH_3$ 1-(*m*-Bromophenyl)-1-propanol → 1-(*m*-Bromophenyl)propene (71%)

TABLE 15.2
Summary of Reactions of Alcohols Discussed in Earlier Chapters *(Continued)*

Reaction (section) and comments	General equation and specific example
Conversion to *p*-toluenesulfonate esters (Section 8.15) Alcohols react with *p*-toluenesulfonyl chloride to give *p*-toluenesulfonate esters. Sulfonate esters are reactive substrates for nucleophilic substitution and elimination reactions. The *p*-toluenesulfonate group is often abbreviated—OTs.	$ROH + H_3C-C_6H_4-SO_2Cl \longrightarrow ROS(=O)_2-C_6H_4-CH_3 + HCl$ Alcohol; *p*-Toluenesulfonyl chloride; Alkyl *p*-toluenesulfonate; Hydrogen chloride Cycloheptanol —OH $\xrightarrow[\text{pyridine}]{p\text{-toluenesulfonyl chloride}}$ —OTs Cycloheptanol; Cycloheptyl *p*-toluenesulfonate (83%)

Transformations such as this correspond to the loss of the elements of hydrogen from an alcohol and are sometimes termed *dehydrogenation* reactions. More normally, however, organic chemists classify the conversion of an alcohol to a carbonyl compound as an *oxidation* of the alcohol. These reactions are especially important in the practice of organic synthesis.

15.7 CONVERSION OF ALCOHOLS TO ETHERS

Primary alcohols are converted to ethers on heating in the presence of an acid catalyst, usually sulfuric acid.

$$2RCH_2OH \xrightarrow{H^+,\ \text{heat}} RCH_2OCH_2R + H_2O$$

Primary alcohol — Dialkyl ether — Water

This kind of reaction is said to be a **condensation.** A condensation is a reaction in which two molecules combine to form a larger one while liberating a small molecule. In this case two alcohol molecules combine to give an ether and water.

$$2CH_3CH_2CH_2CH_2OH \xrightarrow[130°C]{H_2SO_4} CH_3CH_2CH_2CH_2OCH_2CH_2CH_2CH_3 + H_2O$$

1-Butanol — Dibutyl ether (60%) — Water

When applied to the synthesis of ethers, the reaction is effective only with primary alcohols. Elimination to form alkenes predominates with secondary and tertiary alcohols.

The mechanism by which two molecules of a primary alcohol condense to give a dialkyl ether and water is outlined for the specific case of ethanol in Figure 15.2. The individual steps of this mechanism are analogous to those seen earlier. Nucleophilic attack on a protonated alcohol was encountered in the reaction of primary alcohols with hydrogen halides (Section 4.14), and the nucleophilic properties of alcohols were

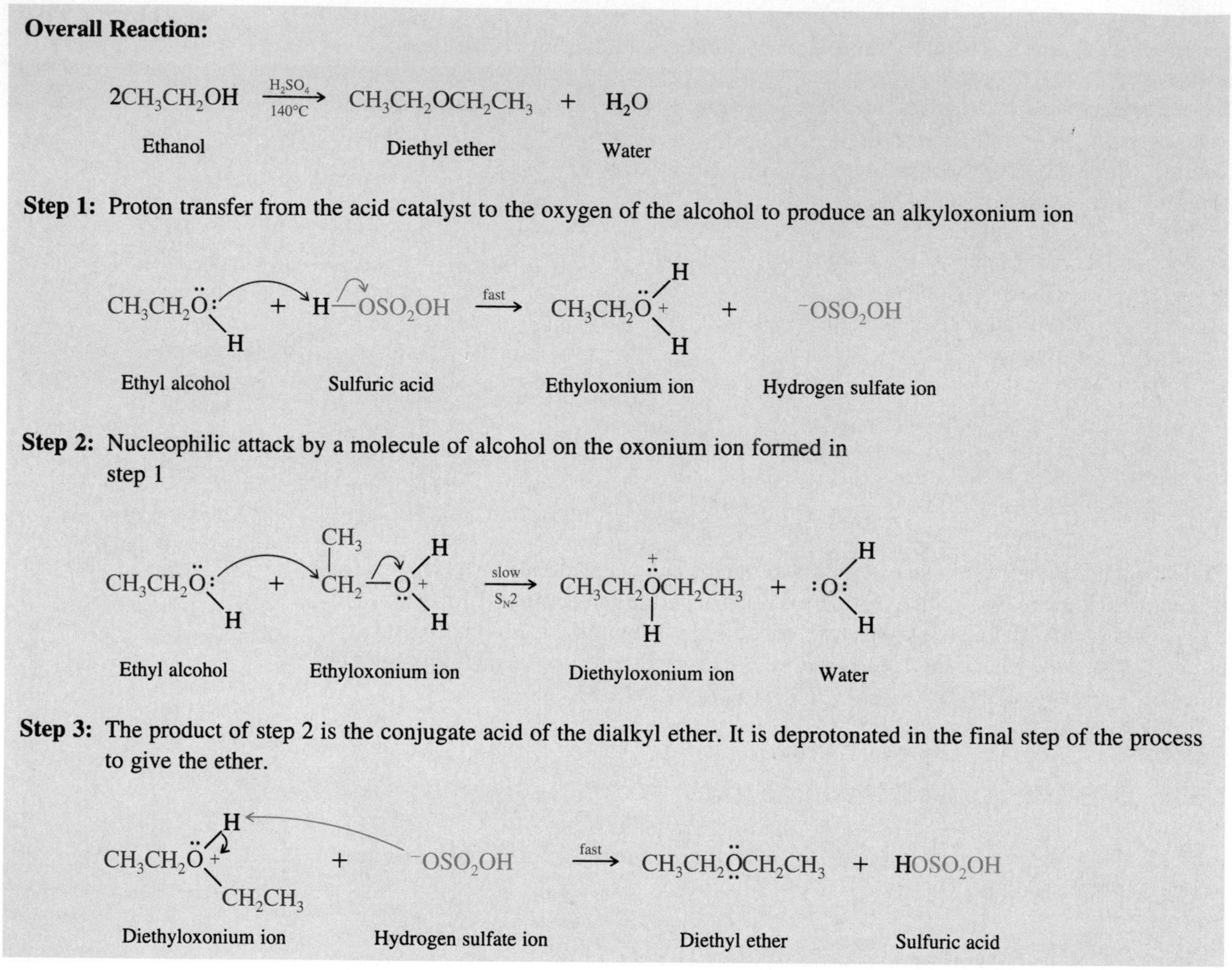

Overall Reaction:

$$2CH_3CH_2OH \xrightarrow[140°C]{H_2SO_4} CH_3CH_2OCH_2CH_3 + H_2O$$

Ethanol — Diethyl ether — Water

Step 1: Proton transfer from the acid catalyst to the oxygen of the alcohol to produce an alkyloxonium ion

$$CH_3CH_2\ddot{O}H + H{-}OSO_2OH \xrightarrow{fast} CH_3CH_2\overset{+}{O}H_2 + {}^-OSO_2OH$$

Ethyl alcohol — Sulfuric acid — Ethyloxonium ion — Hydrogen sulfate ion

Step 2: Nucleophilic attack by a molecule of alcohol on the oxonium ion formed in step 1

$$CH_3CH_2\ddot{O}H + CH_3CH_2{-}\overset{+}{O}H_2 \xrightarrow[S_N2]{slow} CH_3CH_2\overset{+}{O}(H)CH_2CH_3 + :\ddot{O}H_2$$

Ethyl alcohol — Ethyloxonium ion — Diethyloxonium ion — Water

Step 3: The product of step 2 is the conjugate acid of the dialkyl ether. It is deprotonated in the final step of the process to give the ether.

$$CH_3CH_2\overset{+}{O}(H)CH_2CH_3 + {}^-OSO_2OH \xrightarrow{fast} CH_3CH_2\ddot{O}CH_2CH_3 + HOSO_2OH$$

Diethyloxonium ion — Hydrogen sulfate ion — Diethyl ether — Sulfuric acid

FIGURE 15.2 The mechanism of acid-catalyzed formation of diethyl ether from ethyl alcohol. As an alternative in the third step, the base that abstracts the proton could be a molecule of the starting alcohol.

discussed in the context of solvolysis reactions (Section 8.8). Both the first and the last steps are proton-transfer reactions between oxygens.

Diethyl ether is prepared on an industrial scale by heating ethanol with sulfuric acid at 140°C. At higher temperatures elimination predominates, and ethylene is the major product.

Diols react intramolecularly to form cyclic ethers when a five-membered or six-membered ring can result.

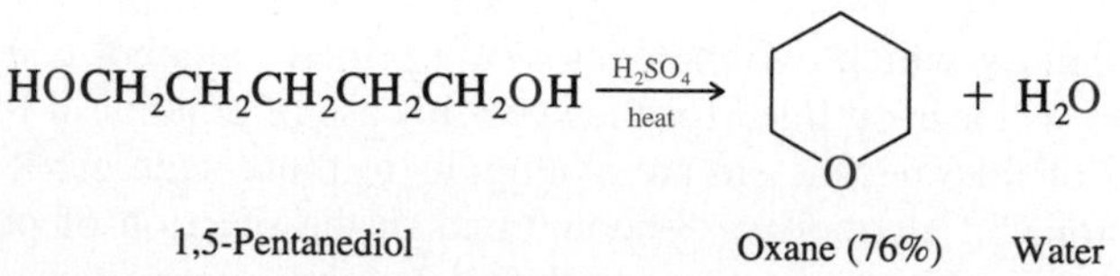

$$HOCH_2CH_2CH_2CH_2CH_2OH \xrightarrow[heat]{H_2SO_4} \text{Oxane} + H_2O$$

1,5-Pentanediol — Oxane (76%) — Water

In these intramolecular ether-forming reactions, the alcohol functions may be primary, secondary, or tertiary.

PROBLEM 15.7 On the basis of the mechanism for the acid-catalyzed formation of diethyl ether from ethanol in Figure 15.2, write a stepwise mechanism for the formation of oxane from 1,5-pentanediol (see the equation in the preceding paragraph).

15.8 ESTERIFICATION

Acid-catalyzed condensation of an alcohol and a carboxylic acid yields an ester and water and is known as the **Fischer esterification.**

$$\underset{\text{Alcohol}}{ROH} + \underset{\text{Carboxylic acid}}{R'\overset{O}{\overset{\|}{C}}OH} \overset{H^+}{\rightleftharpoons} \underset{\text{Ester}}{R'\overset{O}{\overset{\|}{C}}OR} + \underset{\text{Water}}{H_2O}$$

Fischer esterification is reversible, and the position of equilibrium lies slightly to the side of products when the reactants are simple alcohols and carboxylic acids. When the Fischer esterification is used for preparative purposes, the position of equilibrium can be made more favorable by using either the alcohol or the carboxylic acid in excess. In the following example, where an excess of the alcohol was employed, the yield indicated is based on the carboxylic acid as the limiting reagent.

$$\underset{\substack{\text{Methanol}\\(0.6\text{ mol})}}{CH_3OH} + \underset{\substack{\text{Benzoic acid}\\(0.1\text{ mol})}}{C_6H_5\overset{O}{\overset{\|}{C}}OH} \xrightarrow[\text{heat}]{H_2SO_4} \underset{\substack{\text{Methyl benzoate}\\\text{(isolated in 70\%}\\\text{yield based on}\\\text{benzoic acid)}}}{C_6H_5\overset{O}{\overset{\|}{C}}OCH_3} + \underset{\text{Water}}{H_2O}$$

Another way to shift the position of equilibrium to favor the formation of ester is to remove the other product (water) from the reaction mixture as it is formed. This can be accomplished by adding benzene as a cosolvent and distilling the azeotropic mixture of benzene and water.

An *azeotropic mixture* contains two or more substances that distill together at a constant boiling point. The benzene-water azeotrope contains 9% water and boils at 69°C.

$$\underset{\substack{\textit{sec}\text{-Butyl alcohol}\\(0.20\text{ mol})}}{CH_3\underset{OH}{\underset{|}{C}}HCH_2CH_3} + \underset{\substack{\text{Acetic acid}\\(0.25\text{ mol})}}{CH_3\overset{O}{\overset{\|}{C}}OH} \xrightarrow[\text{benzene, heat}]{H^+} \underset{\substack{\textit{sec}\text{-Butyl acetate}\\\text{(isolated in 71\%}\\\text{yield based on}\\\textit{sec}\text{-butyl alcohol)}}}{CH_3\overset{O}{\overset{\|}{C}}O\underset{CH_3}{\underset{|}{C}}HCH_2CH_3} + \underset{\substack{\text{Water}\\\text{(codistills}\\\text{with benzene)}}}{H_2O}$$

For steric reasons, the order of alcohol reactivity in the Fischer esterification is CH_3OH > primary > secondary > tertiary. Phenols are much less reactive than alcohols.

PROBLEM 15.8 Write the structure of the ester formed in each of the following reactions:

$$(a)\quad CH_3CH_2CH_2CH_2OH + CH_3CH_2\overset{O}{\overset{\|}{C}}OH \xrightarrow[\text{heat}]{H_2SO_4}$$

$$(b)\quad 2CH_3OH + HO\overset{O}{\overset{\|}{C}}-C_6H_4-\overset{O}{\overset{\|}{C}}OH \xrightarrow[\text{heat}]{H_2SO_4} (C_{10}H_{10}O_4)$$

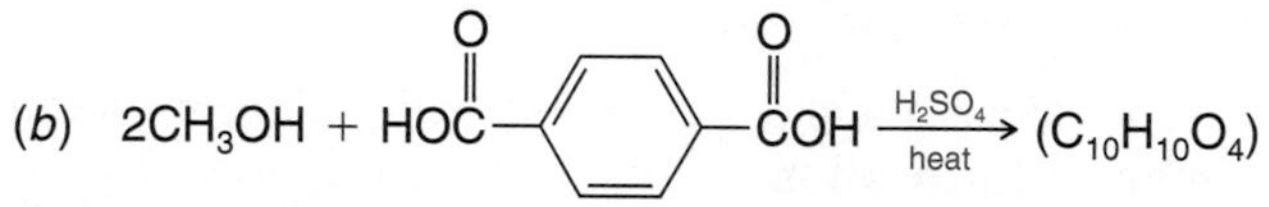

SAMPLE SOLUTION (*a*) By analogy to the general equation and to the examples cited in this section, we can write the equation

$$\underset{\text{1-Butanol}}{CH_3CH_2CH_2CH_2OH} + \underset{\text{Propanoic acid}}{CH_3CH_2\overset{O}{\overset{\|}{C}}OH} \xrightarrow[\text{heat}]{H_2SO_4} \underset{\text{Butyl propanoate}}{CH_3CH_2\overset{O}{\overset{\|}{C}}OCH_2CH_2CH_2CH_3} + \underset{\text{Water}}{H_2O}$$

As actually carried out in the laboratory, 3 moles of propanoic acid was used per mole of 1-butanol, and the desired ester was obtained in 78 percent yield.

Esters are also formed by the reaction of alcohols with acyl chlorides:

$$\underset{\text{Alcohol}}{ROH} + \underset{\text{Acyl chloride}}{R'\overset{O}{\overset{\|}{C}}Cl} \longrightarrow \underset{\text{Ester}}{R'\overset{O}{\overset{\|}{C}}OR} + \underset{\text{Hydrogen chloride}}{HCl}$$

This reaction is normally carried out in the presence of a weak base such as pyridine, which captures the hydrogen chloride that is formed.

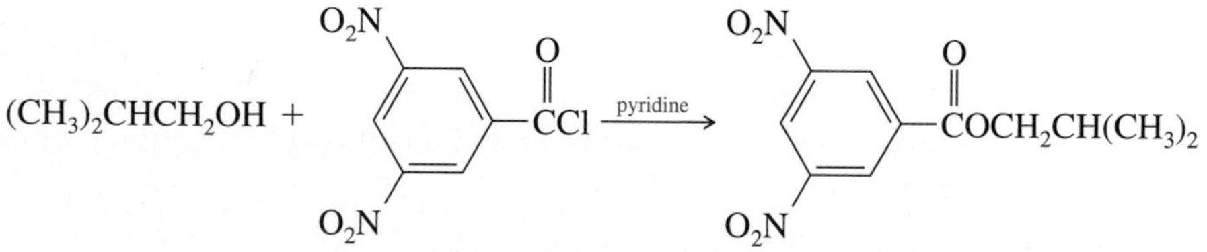

$$\underset{\text{Isobutyl alcohol}}{(CH_3)_2CHCH_2OH} + \underset{\text{3,5-Dinitrobenzoyl chloride}}{3,5\text{-}(O_2N)_2C_6H_3\overset{O}{\overset{\|}{C}}Cl} \xrightarrow{\text{pyridine}} \underset{\text{Isobutyl 3,5-dinitrobenzoate (86\%)}}{3,5\text{-}(O_2N)_2C_6H_3\overset{O}{\overset{\|}{C}}OCH_2CH(CH_3)_2}$$

Carboxylic acid anhydrides react similarly to acyl chlorides.

$$\underset{\text{Alcohol}}{ROH} + \underset{\text{Carboxylic acid anhydride}}{R'\overset{O}{\overset{\|}{C}}O\overset{O}{\overset{\|}{C}}R'} \longrightarrow \underset{\text{Ester}}{R'\overset{O}{\overset{\|}{C}}OR} + \underset{\text{Carboxylic acid}}{R'\overset{O}{\overset{\|}{C}}OH}$$

$$C_6H_5CH_2CH_2OH + CF_3\overset{O}{\overset{\|}{C}}O\overset{O}{\overset{\|}{C}}CF_3 \xrightarrow{\text{pyridine}} C_6H_5CH_2CH_2O\overset{O}{\overset{\|}{C}}CF_3 + CF_3\overset{O}{\overset{\|}{C}}OH$$

2-Phenylethanol | Trifluoroacetic anhydride | 2-Phenylethyl trifluoroacetate (83%) | Trifluoroacetic acid

The mechanisms of the Fischer esterification and the reactions of alcohols with acyl chlorides and acid anhydrides will be discussed in detail in Chapters 19 and 20 after some fundamental principles of carbonyl group reactivity have been developed. For the present, it is sufficient to point out that most of the reactions that convert alcohols to esters leave the C—O bond of the alcohol intact.

$$H—O—R \longrightarrow R'\overset{O}{\overset{\|}{C}}—O—R$$

This is the same oxygen that was attached to the group R in the starting alcohol.

The acyl group of the carboxylic acid, acyl chloride, or acid anhydride is transferred to the oxygen of the alcohol. This fact is most clearly evident in the esterification of chiral alcohols, where, since none of the bonds to the stereogenic center is broken in the process, *retention of configuration is observed.*

$$C_6H_5\overset{CH_3CH_2}{\underset{CH_3}{—\!\!\!\!\!|\!\!\!\!\!—}}OH + O_2N—C_6H_4—\overset{O}{\overset{\|}{C}}Cl \xrightarrow{\text{pyridine}} C_6H_5\overset{CH_3CH_2}{\underset{CH_3}{—\!\!\!\!\!|\!\!\!\!\!—}}O\overset{O}{\overset{\|}{C}}—C_6H_4—NO_2$$

(*R*)-(+)-2-Phenyl-2-butanol | *p*-Nitrobenzoyl chloride | (*R*)-(−)-1-Methyl-1-phenylpropyl *p*-nitrobenzoate (63% yield)

PROBLEM 15.9 A similar conclusion may be drawn by considering the reactions of the cis and trans isomers of 4-*tert*-butylcyclohexanol with acetic anhydride. On the basis of the information just presented, predict the product formed from each stereoisomer.

The reaction of alcohols with acyl chlorides is analogous to their reaction with *p*-toluenesulfonyl chloride described earlier (Section 8.15 and Table 15.2). In those reactions, a *p*-toluenesulfonate ester was formed by displacement of chloride from the sulfonyl group by the oxygen of the alcohol. Carboxylic esters arise by displacement of chloride from a carbonyl group by the alcohol oxygen.

15.9 ESTERS OF INORGANIC ACIDS

While the term *ester,* used without a modifier, is normally taken to mean an ester of a carboxylic acid, alcohols can react with inorganic acids in a process similar to the Fischer esterification. The products are esters of inorganic acids. For example, *alkyl nitrates* are esters formed by the reaction of alcohols with *nitric acid.*

$$ROH + HONO_2 \xrightarrow{H^+} RONO_2 + H_2O$$

Alcohol | Nitric acid | Alkyl nitrate | Water

$$\underset{\text{Methanol}}{CH_3OH} + \underset{\text{Nitric acid}}{HONO_2} \xrightarrow{H_2SO_4} \underset{\text{Methyl nitrate (66–80\%)}}{CH_3ONO_2} + \underset{\text{Water}}{H_2O}$$

PROBLEM 15.10 Alfred Nobel's fortune was based on his 1866 discovery that nitroglycerin, which is far too shock-sensitive to be transported or used safely, can be stabilized by adsorption onto a substance called *kieselguhr* to give what is familiar to us as *dynamite.* Nitroglycerin is the trinitrate of glycerol (1,2,3-propanetriol). Write a structural formula for nitroglycerin.

Dialkyl sulfates are esters of *sulfuric acid,* **trialkyl phosphites** are esters of *phosphorous acid* (H_3PO_3), and **trialkyl phosphates** are esters of *phosphoric acid* (H_3PO_4).

$$\underset{\text{Dimethyl sulfate}}{CH_3O\overset{O}{\overset{\|}{\underset{\underset{O}{\|}}{S}}}OCH_3} \qquad \underset{\text{Trimethyl phosphite}}{(CH_3O)_3P:} \qquad \underset{\text{Trimethyl phosphate}}{(CH_3O)_3\overset{+}{P}—\ddot{\underset{\cdot\cdot}{O}}:^{-}}$$

Some esters of inorganic acids, such as dimethyl sulfate, are used as reagents in synthetic organic chemistry. Certain naturally occurring alkyl phosphates play an important role in biological processes.

15.10 OXIDATION OF ALCOHOLS

As previewed in Section 15.6, oxidation of an alcohol yields a carbonyl compound. Whether the resulting carbonyl compound is an aldehyde, a ketone, or a carboxylic acid depends on the alcohol and on the oxidizing agent.

Primary alcohols may be oxidized either to an aldehyde or to a carboxylic acid:

$$\underset{\text{Primary alcohol}}{RCH_2OH} \xrightarrow{\text{oxidize}} \underset{\text{Aldehyde}}{R\overset{O}{\overset{\|}{C}}H} \xrightarrow{\text{oxidize}} \underset{\text{Carboxylic acid}}{R\overset{O}{\overset{\|}{C}}OH}$$

Vigorous oxidation leads to the formation of a carboxylic acid, but there are a number of methods that permit us to stop the oxidation at the intermediate aldehyde stage. The reagents that are most commonly used for oxidizing alcohols are based on high-oxidation-state transition metals, particularly chromium(VI).

Chromic acid (H_2CrO_4) is a good oxidizing agent and is formed when solutions containing chromate (CrO_4^{2-}) or dichromate ($Cr_2O_7^{2-}$) are acidified. Sometimes it is possible to obtain aldehydes in satisfactory yield before they are further oxidized, but in most cases carboxylic acids are the major products isolated on treatment of primary alcohols with chromic acid.

Potassium permanganate ($KMnO_4$) will also oxidize primary alcohols to carboxylic acids.

$$\underset{\text{3-Fluoro-1-propanol}}{FCH_2CH_2CH_2OH} \xrightarrow[H_2SO_4,\ H_2O]{K_2Cr_2O_7} \underset{\text{3-Fluoropropanoic acid (74\%)}}{FCH_2CH_2\overset{O}{\overset{\|}{C}}OH}$$

Conditions that do permit the easy isolation of aldehydes in good yield by oxidation of primary alcohols employ various Cr(VI) species as the oxidant in *anhydrous* media. Two such reagents are **pyridinium chlorochromate (PCC),** $C_5H_5NH^+\ ClCrO_3^-$, and **pyridinium dichromate (PDC),** $(C_5H_5NH)_2^{2+}\ Cr_2O_7^{2-}$; both are used in dichloromethane.

The pyridine complex of chromium trioxide $[(C_5H_5N)_2CrO_3]$ in CH_2Cl_2, called Collins' reagent, may also be used.

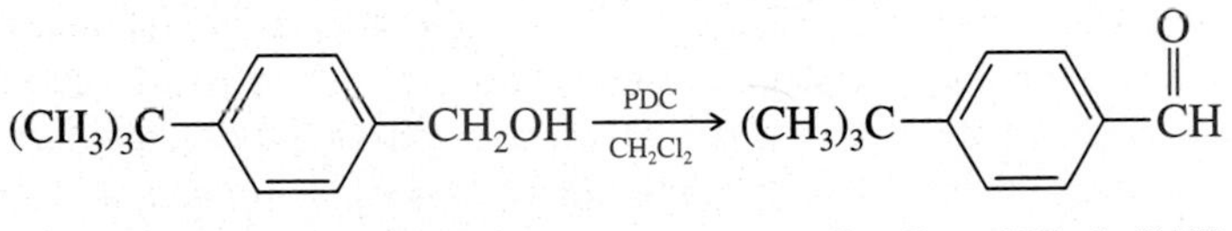

$$CH_3(CH_2)_5CH_2OH \xrightarrow[CH_2Cl_2]{PCC} CH_3(CH_2)_5\overset{O}{\overset{\|}{C}}H$$

1-Heptanol → Heptanal (78%)

$$(CH_3)_3C-C_6H_4-CH_2OH \xrightarrow[CH_2Cl_2]{PDC} (CH_3)_3C-C_6H_4-\overset{O}{\overset{\|}{C}}H$$

p-tert-Butylbenzyl alcohol → *p-tert*-Butylbenzaldehyde (94%)

Secondary alcohols are oxidized to ketones by the same reagents that oxidize primary alcohols:

$$R\overset{OH}{\overset{|}{C}}HR' \xrightarrow{oxidize} R\overset{O}{\overset{\|}{C}}R'$$

Secondary alcohol → Ketone

$$\text{Cyclohexanol} \xrightarrow[H_2SO_4,\ H_2O]{Na_2Cr_2O_7} \text{Cyclohexanone}$$

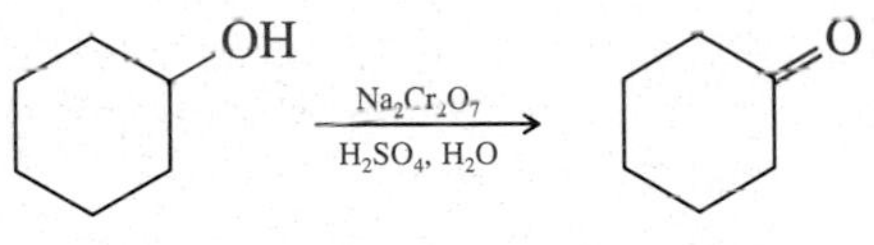

Cyclohexanol → Cyclohexanone (85%)

$$CH_2{=}CH\overset{OH}{\overset{|}{C}}HCH_2CH_2CH_2CH_2CH_3 \xrightarrow[CH_2Cl_2]{PDC} CH_2{=}CH\overset{O}{\overset{\|}{C}}CH_2CH_2CH_2CH_2CH_3$$

1-Octen-3-ol → 1-Octen-3-one (80%)

Tertiary alcohols have no hydrogen on their hydroxyl-bearing carbon and do not undergo oxidation readily:

$$R-\overset{R'}{\underset{R''}{\overset{|}{\underset{|}{C}}}}-OH \xrightarrow{oxidize} \text{no reaction except under forcing conditions}$$

In the presence of strong oxidizing agents at elevated temperatures, oxidation of tertiary alcohols leads to cleavage of the various carbon-carbon bonds at the hydroxyl-bearing carbon atom, and a complex mixture of products results.

ECONOMIC AND ENVIRONMENTAL FACTORS IN ORGANIC SYNTHESIS

Beyond the obvious difference in scale that is evident when one compares preparing tons of a compound versus preparing just a few grams of it, there are sharp distinctions between "industrial" and "laboratory" syntheses. On a laboratory scale, a chemist is normally concerned only with obtaining a modest amount of a substance. Sometimes making the compound is an end in itself, while on other occasions the compound is needed for some further study of its physical, chemical, or biological properties. Considerations such as the cost of reagents and solvents tend to play only a minor role when planning most laboratory syntheses. Faced with a choice between two synthetic routes to a particular compound, one based on the cost of chemicals and the other on the efficient use of a chemist's time, the decision is almost always made in favor of the latter.

Not so for synthesis in the chemical industry, where not only must a compound be prepared on a large scale, but it must be prepared at low cost. There is a pronounced bias toward reactants and reagents that are both abundant and inexpensive. The oxidizing agent of choice, for example, in the chemical industry is O_2, and extensive research has been devoted to developing catalysts for preparing various compounds by air oxidation of readily available starting materials. To illustrate, air and ethylene are the reactants for the industrial preparation of both acetaldehyde and ethylene oxide. Which of the two products is obtained depends on the catalyst employed.

$$CH_2{=}CH_2 + \tfrac{1}{2}O_2 \xrightarrow[H_2O]{PdCl_2,\ CuCl_2} CH_3\overset{O}{\overset{\|}{C}}H$$

Ethylene, Oxygen → Acetaldehyde

$$CH_2{=}CH_2 + \tfrac{1}{2}O_2 \xrightarrow[300^\circ C]{Ag} H_2C\overset{O}{—}CH_2$$

Ethylene, Oxygen → Ethylene oxide

Dating approximately from the creation of the U.S. Environmental Protection Agency (EPA) in 1970, dealing with the by-products of synthetic procedures has become an increasingly important consideration in designing a chemical synthesis. In terms of changing the strategy of synthetic planning, the chemical industry actually had a shorter road to travel than the pharmaceutical industry, academic laboratories, and research institutes. Simple business principles had long dictated that waste chemicals represented wasted opportunities. It made better sense for a chemical company to recover the solvent from a reaction and use it again than to throw it away and buy more. Similarly, it was far better to find a "value-added" use for a by-product from a reaction than to throw it away. By raising the cost of generating chemical waste, environmental regulations increased the economic incentive to design processes that produced less of it.

The term "environmentally benign" synthesis has been coined to refer to procedures explicitly designed to minimize the formation of by-products that present disposal problems. Both the National Science Foundation and the Environmental Protection Agency have allocated a portion of their grant budgets to encourage efforts in this vein.

The application of environmentally benign principles to laboratory-scale synthesis can be illustrated by revisiting the oxidation of alcohols. As noted in Section 15.10, the most widely used methods involve Cr(VI)-based oxidizing agents. Unfortunately, Cr(VI) compounds are carcinogenic and appear on the EPA list of compounds requiring special disposal methods. The best way to replace Cr(VI)-based oxidants would be to develop catalytic methods analogous to those used in industry. Another approach would be to use oxidizing agents that are less hazardous, such as sodium hypochlorite. Aqueous solutions of sodium hypochlorite are available as "swimming-pool chlorine," and procedures for their use in oxidizing secondary alcohols to ketones have been developed. One is described on page 71 of the January 1991 edition of the *Journal of Chemical Education.*

$$\underset{\text{2-Methyl-4-heptanol}}{(CH_3)_2CHCH_2\underset{\displaystyle OH}{\underset{|}{C}}HCH_2CH_2CH_3} \xrightarrow[\text{acetic acid–water}]{NaOCl} \underset{\text{2-Methyl-4-heptanone (77\%)}}{(CH_3)_2CHCH_2\overset{\displaystyle O}{\overset{\|}{C}}CH_2CH_2CH_3}$$

There is a curious irony in the nomination of hypochlorite as an environmentally benign oxidizing agent. It comes at a time of increasing pressure to eliminate chlorine and chlorine-containing compounds from the environment to as great a degree as possible. Any all-inclusive assault on chlorine needs to be carefully scrutinized, especially when one remembers that chlorination of the water supply has probably done more to extend human life than any other public health measure ever undertaken. (The role of chlorine in the formation of chlorinated hydrocarbons in water is discussed in Section 18.7.)

PROBLEM 15.11 Predict the principal organic product of each of the following reactions:

(*a*) $ClCH_2CH_2CH_2CH_2OH \xrightarrow[H_2SO_4,\ H_2O]{K_2Cr_2O_7}$

(*b*) $CH_3\underset{\displaystyle OH}{\underset{|}{C}}HCH_2CH_2CH_2CH_2CH_2CH_3 \xrightarrow[H_2SO_4,\ H_2O]{Na_2Cr_2O_7}$

(*c*) $CH_3CH_2CH_2CH_2CH_2CH_2CH_2OH \xrightarrow[CH_2Cl_2]{PCC}$

SAMPLE SOLUTION (*a*) The substrate is a primary alcohol and so can be oxidized either to an aldehyde or to a carboxylic acid. Aldehydes are the major products only when the oxidation is carried out in anhydrous media. Carboxylic acids are formed when water is present. The reaction shown produced 4-chlorobutanoic acid in 56 percent yield.

$$\underset{\text{4-Chloro-1-butanol}}{ClCH_2CH_2CH_2CH_2OH} \xrightarrow[H_2SO_4,\ H_2O]{K_2Cr_2O_7} \underset{\text{4-Chlorobutanoic acid}}{ClCH_2CH_2CH_2\overset{\displaystyle O}{\overset{\|}{C}}OH}$$

The mechanisms by which transition metal oxidizing agents convert alcohols to aldehydes and ketones are rather complicated and will not be dealt with in detail. In broad outline, chromic acid oxidation involves initial formation of an alkyl chromate:

$$\underset{\text{Alcohol}}{>C(H)OH} + \underset{\text{Chromic acid}}{HO\overset{\displaystyle O}{\overset{\|}{\underset{\displaystyle O}{\underset{\|}{Cr}}}}OH} \longrightarrow \underset{\text{Alkyl chromate}}{>C(H)O\overset{\displaystyle O}{\overset{\|}{\underset{\displaystyle O}{\underset{\|}{Cr}}}}OH} + H_2O$$

An alkyl chromate is an example of an ester of an inorganic acid (Section 15.9)

This alkyl chromate then undergoes an elimination reaction to form the carbon-oxygen double bond.

$$\text{Alkyl chromate} \longrightarrow \; \rangle C{=}O + H_3O^+ + HCrO_3^-$$

Alkyl chromate Aldehyde or ketone

In the elimination step, chromium is reduced from Cr(VI) to Cr(IV). Since the eventual product is Cr(III), further electron-transfer steps are also involved.

15.11 BIOLOGICAL OXIDATION OF ALCOHOLS

Many biological processes involve oxidation of alcohols to carbonyl compounds or the reverse process, reduction of carbonyl compounds to alcohols. Ethanol, for example, is metabolized in the liver to acetaldehyde. Such processes are catalyzed by enzymes; the enzyme that catalyzes the oxidation of ethanol is called *alcohol dehydrogenase.*

$$\underset{\text{Ethanol}}{CH_3CH_2OH} \xrightleftharpoons{\text{alcohol dehydrogenase}} \underset{\text{Acetaldehyde}}{CH_3\overset{O}{\overset{\|}{C}}H}$$

In addition to enzymes, biological oxidations require substances known as *coenzymes.* Coenzymes are organic molecules that, in concert with an enzyme, act upon a substrate to bring about chemical change. Most of the substances that we call vitamins are coenzymes. The coenzyme contains a functional group that is complementary to a functional group of the substrate; the enzyme catalyzes the interaction of these mutually complementary functional groups. If ethanol is oxidized, some other substance must be reduced. This other substance is the oxidized form of the coenzyme *nicotinamide adenine dinucleotide* (NAD). Chemists and biochemists abbreviate the oxidized form of this coenzyme as NAD^+ and its reduced form as NADH. More completely, the chemical equation for the biological oxidation of ethanol may be written:

$$\underset{\text{Ethanol}}{CH_3CH_2OH} + \underset{\text{Oxidized form of NAD coenzyme}}{NAD^+} \xrightleftharpoons{\text{alcohol dehydrogenase}} \underset{\text{Acetaldehyde}}{CH_3\overset{O}{\overset{\|}{C}}H} + \underset{\text{Reduced form of NAD coenzyme}}{NADH} + H^+$$

The structure of the oxidized form of nicotinamide adenine dinucleotide is shown in Figure 15.3. The only portion of the coenzyme that undergoes chemical change in an oxidation-reduction reaction is the substituted pyridine ring of the nicotinamide unit

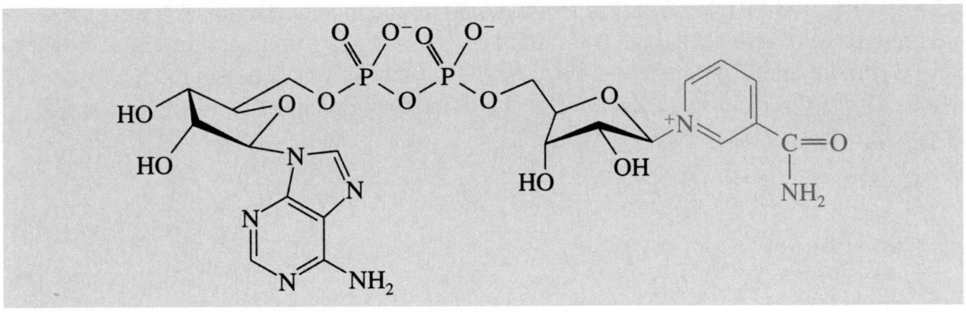

FIGURE 15.3 Structure of NAD^+, the oxidized form of the coenzyme nicotinamide adenine dinucleotide.

(shown in green in Figure 15.3). If the remainder of the coenzyme molecule is represented by R, its role as an oxidizing agent is shown in the equation

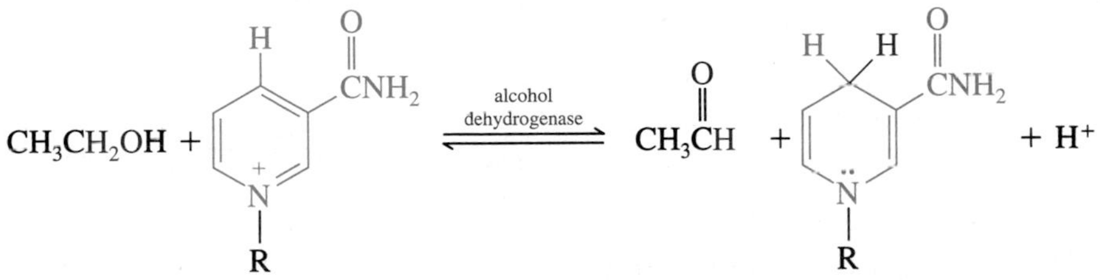

$$CH_3CH_2OH + NAD^+ \xrightleftharpoons{\text{alcohol dehydrogenase}} CH_3CH{=}O + NADH + H^+$$

Ethanol NAD+ Acetaldehyde NADH

According to one mechanistic interpretation, a hydrogen with a pair of electrons is transferred from ethanol to NAD^+, forming acetaldehyde and converting the positively charged pyridinium ring to a dihydropyridine:

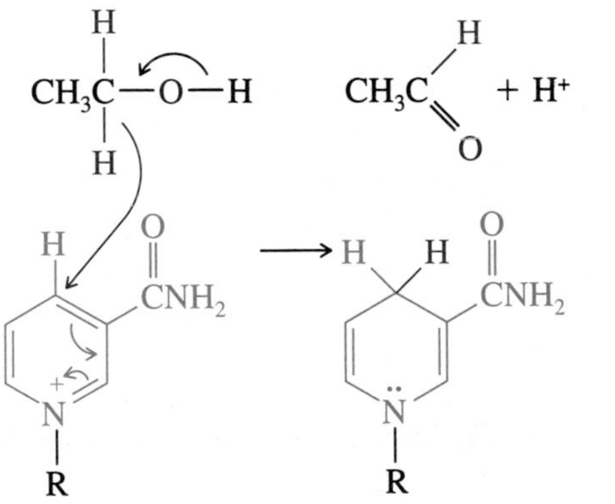

The pyridinium ring of NAD^+ serves as an acceptor of hydride ion in this picture of its role in biological oxidation.

PROBLEM 15.12 The mechanism of enzymatic oxidation has been studied by isotopic labeling with the aid of deuterated derivatives of ethanol. Specify the number of deuterium atoms that you would expect to find attached to the dihydropyridine ring of the reduced form of the nicotinamide adenine dinucleotide coenzyme following enzymatic oxidation of each of the alcohols given:

(*a*) CD_3CH_2OH (*b*) CH_3CD_2OH (*c*) CH_3CH_2OD

SAMPLE SOLUTION Examination of the proposed mechanism for biological oxidation of ethanol reveals that the hydrogen that is transferred to the coenzyme comes from C-1 of ethanol. Therefore, the dihydropyridine ring will bear no deuterium atoms when CD_3CH_2OH is oxidized, because all the deuterium atoms of the substrate are bound to C-2.

$$CD_3CH_2OH + NAD^+ \xrightleftharpoons{\text{alcohol dehydrogenase}} CD_3CH{=}O + NADH + H^+$$

2,2,2-Trideuterioethanol; NAD^+; 2,2,2-Trideuterioethanal; NADH

The reverse reaction is also observed to occur in living systems; NADH reduces acetaldehyde to ethanol in the presence of alcohol dehydrogenase. In this process, NADH serves as a hydride donor and is oxidized to NAD^+ while acetaldehyde is reduced.

The NAD^+–NADH coenzyme system is involved in a large number of biological oxidation-reduction reactions. Another reaction similar to the ethanol-acetaldehyde conversion is the oxidation of lactic acid to pyruvic acid by NAD^+ and the enzyme *lactic acid dehydrogenase:*

$$CH_3CH(OH)COOH + NAD^+ \xrightleftharpoons{\text{lactic acid dehydrogenase}} CH_3C(=O)COOH + NADH + H^+$$

Lactic acid; Pyruvic acid

We shall encounter other biological processes in which the $NAD^+ \rightleftharpoons NADH$ interconversion plays a prominent role as we proceed through our coverage of organic chemistry.

15.12 OXIDATIVE CLEAVAGE OF VICINAL DIOLS

A reaction characteristic of vicinal diols is their oxidative cleavage on treatment with periodic acid (HIO_4). The carbon-carbon bond of the vicinal diol unit is broken and two carbonyl groups result. Periodic acid is reduced to iodic acid (HIO_3).

$$R_2C(OH){-}C(OH)R'_2 + HIO_4 \longrightarrow R_2C{=}O + R'_2C{=}O + HIO_3 + H_2O$$

Vicinal diol; Periodic acid; Aldehyde or ketone; Aldehyde or ketone; Iodic acid; Water

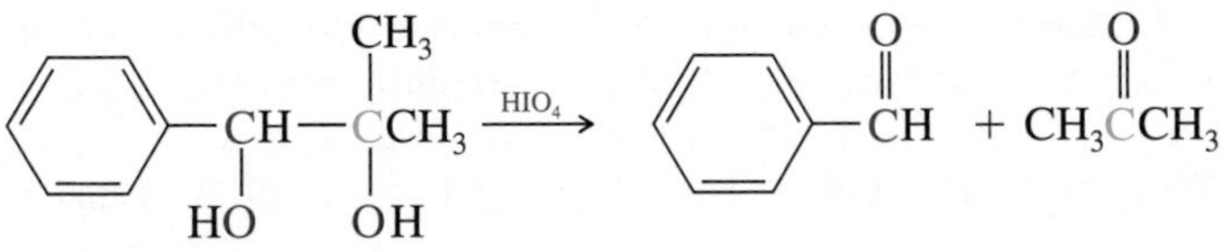

$$C_6H_5\underset{HO}{\underset{|}{C}}H-\overset{CH_3}{\overset{|}{\underset{OH}{\underset{|}{C}}}}CH_3 \xrightarrow{HIO_4} C_6H_5\overset{O}{\overset{\|}{C}}H + CH_3\overset{O}{\overset{\|}{C}}CH_3$$

2-Methyl-1-phenyl-1,2-propanediol — Benzaldehyde (83%) — Acetone

This reaction occurs only when the hydroxyl groups are on adjacent carbons.

PROBLEM 15.13 Predict the products formed on oxidation of each of the following with periodic acid:

(*a*) $HOCH_2CH_2OH$

(*b*) $(CH_3)_2CHCH_2\underset{HO}{\underset{|}{C}}H\underset{OH}{\underset{|}{C}}HCH_2C_6H_5$

(*c*) 1-(hydroxymethyl)cyclopentanol: cyclopentane ring bearing OH and CH_2OH on the same carbon

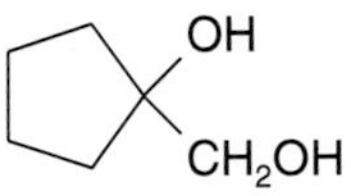

SAMPLE SOLUTION (*a*) The carbon-carbon bond of 1,2-ethanediol is cleaved by periodic acid to give two molecules of formaldehyde:

$$HOCH_2CH_2OH \xrightarrow{HIO_4} 2H\overset{O}{\overset{\|}{C}}H$$

1,2-Ethanediol — Formaldehyde

Cyclic diols give dicarbonyl compounds. The reactions are faster when the hydroxyl groups are cis than when they are trans, but both stereoisomers are oxidized by periodic acid.

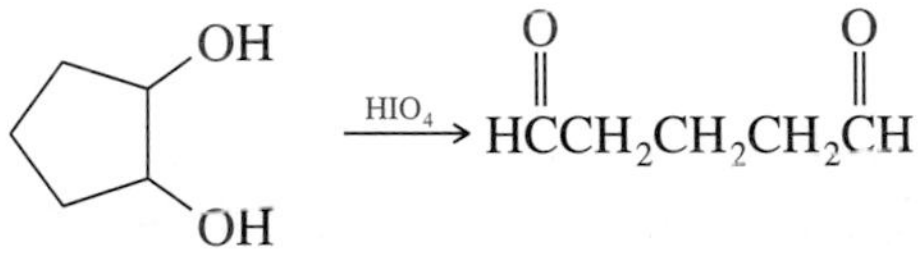

$$\text{1,2-Cyclopentanediol} \xrightarrow{HIO_4} H\overset{O}{\overset{\|}{C}}CH_2CH_2CH_2\overset{O}{\overset{\|}{C}}H$$

1,2-Cyclopentanediol (either stereoisomer) — Pentanedial

Periodic acid cleavage of vicinal diols is often used for analytical purposes as an aid in structure determination. By identifying the carbonyl compounds produced, the constitution of the starting diol may be deduced. This technique has found its widest application with carbohydrates and will be discussed more fully in Chapter 25.

15.13 PREPARATION OF THIOLS

Sulfur is the element most like oxygen in the periodic table, and many oxygen-containing organic compounds have sulfur analogs. The sulfur analogs of alcohols (ROH) are **thiols (RSH).** Thiols are given substitutive names by appending the suffix *-thiol* to the name of the corresponding alkane, numbering the chain in the direction that gives the lower locant to the carbon that bears the —SH group. As with diols (Section 15.5), the final *-e* of the alkane basis name is retained. When the —SH group is named as a substituent, it is called a *mercapto* group. It is also often referred to as a *sulfhydryl* group, but this is a generic term, not used in systematic nomenclature.

$(CH_3)_2CHCH_2CH_2SH$	$HSCH_2CH_2OH$	$HSCH_2CH_2CH_2SH$
3-Methyl-1-butanethiol	2-Mercaptoethanol	1,3-Propanedithiol

Thiols have a marked tendency to bond to mercury, and the word *mercaptan* comes from the Latin *mercurium captans,* which means "seizing mercury." The drug *dimercaprol* is used to treat mercury and lead poisoning; it is 2,3-dimercapto-1-propanol.

At one time thiols were given radicofunctional names as *mercaptans.* Thus, 3-methyl-1-butanethiol would have been called "3-methylbutyl mercaptan" according to this system. This mercaptan nomenclature was abandoned beginning with the 1965 revision of the IUPAC rules but is still sometimes encountered, especially in the older literature.

The preparation of thiols involves nucleophilic substitution of the S_N2 type on alkyl halides and uses the reagent *thiourea* as the source of sulfur. Reaction of the alkyl halide with thiourea gives a compound known as an *isothiouronium salt* in the first step. Hydrolysis of the isothiouronium salt in base gives the desired thiol (along with urea):

$$(H_2\ddot{N})_2C{=}\ddot{S}: + R{-}\ddot{X}: \longrightarrow H_2\overset{+}{N}{=}C(\ddot{N}H_2){-}\ddot{S}{-}R \;\; :\ddot{X}:^- \xrightarrow{HO^-} (H_2\ddot{N})_2C{=}\ddot{O}: + H\ddot{S}{-}R$$

Thiourea   Alkyl halide   Isothiouronium salt   Urea   Thiol

Both steps can be carried out sequentially in the same reaction vessel without isolation of the isothiouronium salt.

$$CH_3(CH_2)_4CH_2Br \xrightarrow[2.\ NaOH]{1.\ (H_2N)_2C{=}S} CH_3(CH_2)_4CH_2SH$$

1-Bromohexane   1-Hexanethiol (84%)

PROBLEM 15.14 Choose the correct enantiomer of 2-butanol that would permit you to prepare (*R*)-2-butanethiol by way of a *p*-toluenesulfonate ester.

15.14 PROPERTIES OF THIOLS

A historical account of the analysis of skunk scent and a modern determination of its composition appear in the March 1978 issue of the *Journal of Chemical Education.*

When one encounters a thiol for the first time, especially a low-molecular-weight thiol, its most obvious property is its foul odor. Ethanethiol is added to natural gas so that leaks can be detected without special equipment—your nose is so sensitive that it can detect less than one part of ethanethiol in 10,000,000,000 parts of air! 3-Methyl-1-butanethiol [$(CH_3)_2CHCH_2CH_2SH$] and *cis*- and *trans*-2-butene-1-thiol ($CH_3CH{=}CHCH_2SH$) are the principal components of the skunk's scent fluid. The odor of thiols weakens as the number of carbons increases, because both the volatility and the sulfur content decrease. 1-Dodecanethiol, for example, has only a faint odor.

Compare the boiling points of H_2S (−60°C) and H_2O (100°C).

The S—H bond is less polar than the O—H bond, and hydrogen bonding in thiols is much weaker than that of alcohols. Thus, methanethiol (CH_3SH) is a gas at room temperature (bp 6°C) while methanol (CH_3OH) is a liquid (bp 65°C).

Thiols are far more acidic than alcohols. We have seen that most alcohols have K_a values in the range 10^{-16} to 10^{-19} (pK_a = 16 to 19). The corresponding values for thiols are about $K_a = 10^{-10}$ ($pK_a = 10$). The significance of this difference is that a thiol can be quantitatively converted to its conjugate base (RS^-), called an **alkanethiolate** anion, by hydroxide:

$$\text{R}\ddot{\text{S}}\text{—H} + {}^{-}\!:\ddot{\text{O}}\text{H} \longrightarrow \text{R}\ddot{\text{S}}:^{-} + \text{H—}\ddot{\text{O}}\text{H}$$

Alkanethiol	Hydroxide ion	Alkanethiolate ion	Water
(stronger acid) ($pK_a = 10$)	(stronger base)	(weaker base)	(weaker acid) ($pK_a = 15.7$)

Thiols, therefore, dissolve in aqueous media when the pH is greater than 10.

Another difference between thiols and alcohols concerns their oxidation. We have seen earlier in this chapter that oxidation of alcohols gives compounds having carbonyl groups. Analogous oxidation of thiols to compounds with C=S functions does *not* occur. Only sulfur is oxidized, not carbon, and compounds containing sulfur in various oxidation states are possible. These include a series of acids classified as *sulfenic, sulfinic,* and *sulfonic* according to the number of oxygens attached to sulfur.

$$\text{RS—H} \longrightarrow \text{RS—OH} \longrightarrow \text{R}\overset{:\ddot{\text{O}}:^{-}}{\overset{|}{\text{S}^{+}}}\text{—OH} \longrightarrow \text{R}\overset{:\ddot{\text{O}}:^{-}}{\underset{:\ddot{\text{O}}:^{-}}{\text{S}^{2+}}}\text{—OH}$$

Thiol — Sulfenic acid — Sulfinic acid — Sulfonic acid

Of these the most important are the sulfonic acids. In general, however, sulfonic acids are not prepared by oxidation of thiols. Arenesulfonic acids ($ArSO_3H$), for example, are prepared by sulfonation of arenes (Section 12.4).

One of the most important oxidative processes, especially from a biochemical perspective, is the oxidation of thiols to **disulfides.**

$$\underset{\text{Thiol}}{2\text{RSH}} \underset{\text{Reduce}}{\overset{\text{Oxidize}}{\rightleftharpoons}} \underset{\text{Disulfide}}{\text{RSSR}}$$

While a variety of oxidizing agents are available for this transformation, it occurs so readily that thiols are slowly converted to disulfides by the oxygen in the air. Dithiols give cyclic disulfides by intramolecular sulfur-sulfur bond formation. An example of a cyclic disulfide is the coenzyme *α-lipoic acid.* The last step in the laboratory synthesis of α-lipoic acid is an iron(III)-catalyzed oxidation of the dithiol shown:

$$\underset{\text{6,8-Dimercaptooctanoic acid}}{\text{HSCH}_2\text{CH}_2\underset{}{\overset{\text{SH}}{\overset{|}{\text{C}}}}\text{H(CH}_2)_4\overset{\text{O}}{\overset{\|}{\text{C}}}\text{OH}} \xrightarrow{O_2,\ FeCl_3} \underset{\alpha\text{-Lipoic acid (78\%)}}{\text{(S—S five-membered ring)—(CH}_2)_4\overset{\text{O}}{\overset{\|}{\text{C}}}\text{OH}}$$

Rapid and reversible making and breaking of the sulfur-sulfur bond is essential to the biological function of α-lipoic acid.

15.15 SPECTROSCOPIC ANALYSIS OF ALCOHOLS

Characteristic features of the infrared spectra of alcohols were discussed in Section 13.18. The O—H stretching mode is particularly easy to identify, appearing in the 3200 to 3650 cm^{-1} region. As the infrared spectrum of cyclohexanol, presented in Figure 15.4, demonstrates, this peak is seen as a broad absorption of moderate intensity. The C—O bond stretching of alcohols gives rise to a moderate to strong absorbance between 1025 and 1200 cm^{-1}. It appears at 1070 cm^{-1} in cyclohexanol, a typical secondary alcohol, but is shifted to slightly higher energy in tertiary alcohols and slightly lower energy in primary alcohols.

Features that are helpful in identifying alcohols by nuclear magnetic resonance spectroscopy include the presence of a hydroxyl proton signal in the spectrum and the chemical shift of the proton of an H—C—O unit.

—C—O—H　　H—C—O—

δ = 0.5–5 ppm　　δ = 3.3–4.0 ppm

The chemical shift of the hydroxyl proton signal is variable, depending on solvent, temperature, and concentration. Its precise position is not particularly significant in structure determination. Because the signals due to hydroxyl protons are not usually split by other protons in the molecule and are often rather broad, they are often fairly easy to identify. To illustrate, Figure 15.5 shows the ^{1}H nmr spectrum of 2-phenylethanol, in which the hydroxyl proton signal appears as a singlet at $\delta = 2.2$ ppm. Of the two triplets in this spectrum, the one at lower field ($\delta = 3.8$ ppm) corresponds to the protons of the CH_2O unit. The higher-field triplet at $\delta = 2.8$ ppm arises from the benzylic CH_2 group. The assignment of a particular signal to the hydroxyl proton can be confirmed by adding D_2O. The hydroxyl proton is replaced by deuterium, and its ^{1}H nmr signal disappears.

In relation to the ^{13}C nmr spectrum of an alkane, the electronegative oxygen of an alcohol causes a pronounced downfield shift in the position of the carbon to which it is

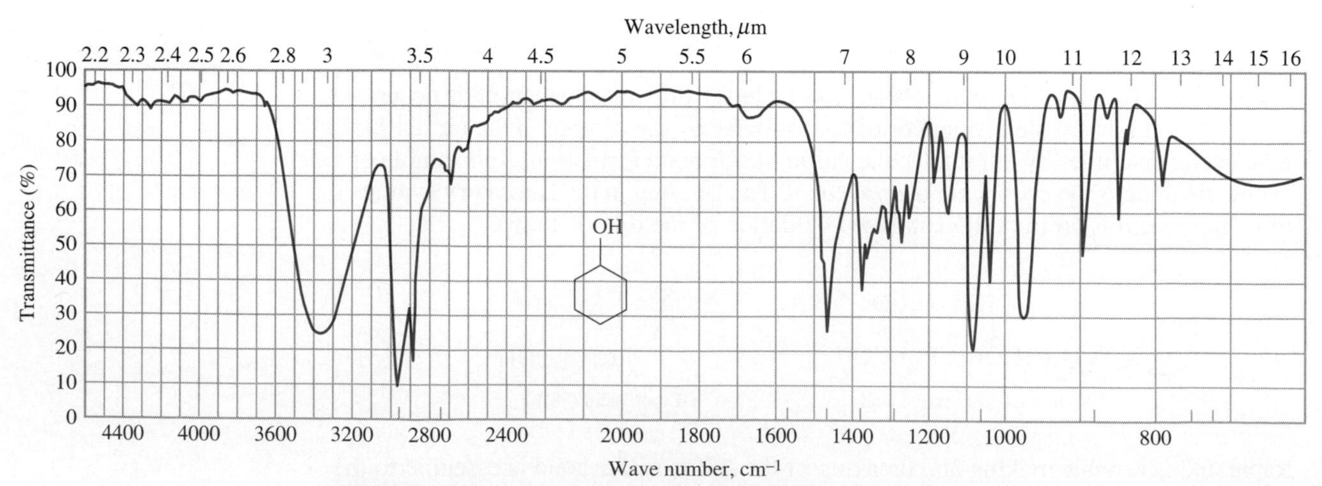

FIGURE 15.4 The ir spectrum of cyclohexanol.

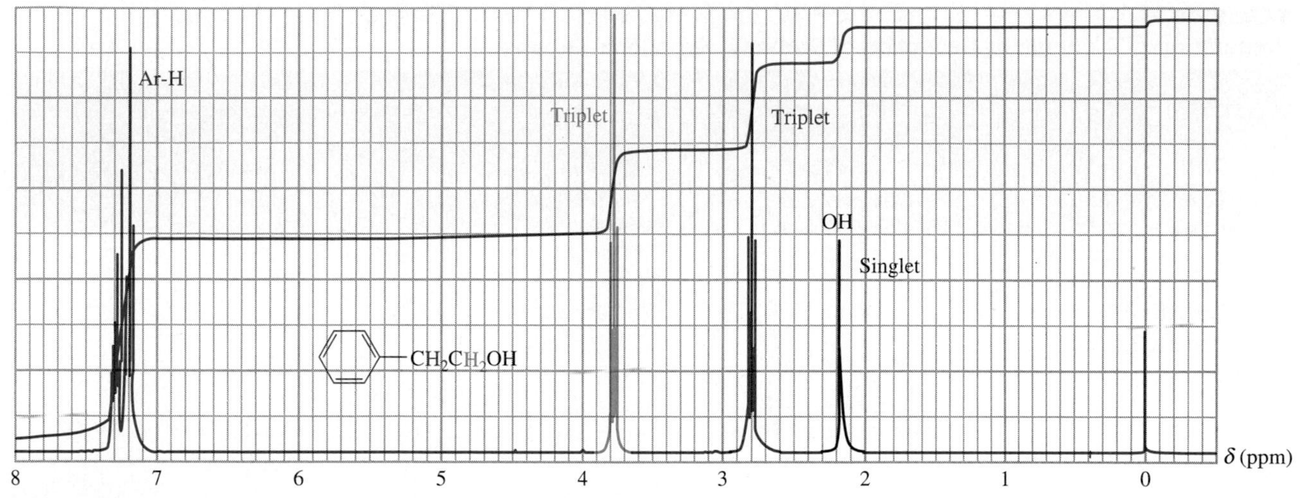

FIGURE 15.5 The 300 MHz ¹H nmr spectrum of 2-phenylethanol ($C_6H_5CH_2CH_2OH$).

attached. This can be seen by comparing chemical shifts for the carbons of ethyl alcohol and propane:

17.9 ppm	57.3 ppm	15.6 ppm	16.1 ppm

CH_3CH_2OH — Ethyl alcohol (17.9 ppm for CH_3, 57.3 ppm for CH_2)

$CH_3CH_2CH_3$ — Propane (15.6 ppm for CH_3, 16.1 ppm for CH_2)

15.16 MASS SPECTROMETRY OF ALCOHOLS

Peaks due to the molecular ion are normally quite small in the mass spectra of alcohols. A peak corresponding to loss of water from the molecular ion is often evident. Alcohols also fragment readily by a pathway in which the molecular ion loses an alkyl group from the hydroxyl-bearing carbon to form a stable cation. Thus, the mass spectra of most primary alcohols exhibit a prominent peak at *m/z* 31.

$$RCH_2\ddot{O}H \longrightarrow R{-}CH_2{-}\overset{\cdot +}{\ddot{O}}H \longrightarrow R\cdot + CH_2{=}\overset{+}{\ddot{O}}H$$

Primary alcohol — Molecular ion — Alkyl radical — Conjugate acid of formaldehyde, *m/z* 31

PROBLEM 15.15 Three of the most intense peaks in the mass spectrum of 2-methyl-2-butanol appear at *m/z* 59, 70, and 73. Explain the origin of these peaks.

15.17 SUMMARY

Functional group interconversions involving alcohols either as reactants or as products are the focus of this chapter. A few of the most important aspects of the chemistry of diols and thiols are introduced as well.

TABLE 15.3
Preparation of Alcohols by Reduction of Carbonyl Functional Groups

	Product of reduction of carbonyl compound by specified reducing agent		
Carbonyl compound	Lithium aluminum hydride ($LiAlH_4$)	Sodium borohydride ($NaBH_4$)	Hydrogen (in the presence of a catalyst)
Aldehyde RCH (with C=O) (Section 15.2)	Primary alcohol RCH_2OH	Primary alcohol RCH_2OH	Primary alcohol RCH_2OH
Ketone RCR' (with C=O) (Section 15.2)	Secondary alcohol $RCH(OH)R'$	Secondary alcohol $RCH(OH)R'$	Secondary alcohol $RCH(OH)R'$
Carboxylic acid $RCOH$ (with C=O) (Section 15.3)	Primary alcohol RCH_2OH	Not reduced	Not reduced
Carboxylic ester $RCOR'$ (with C=O) (Section 15.3)	Primary alcohol RCH_2OH plus $R'OH$	Reduced too slowly to be of practical value	Requires special catalyst, high pressures and temperatures

A number of reactions that produce alcohols had been encountered in earlier chapters and were summarized in Table 15.1 (Section 15.1).

Alcohols can be prepared from carbonyl compounds by reduction of aldehydes and ketones (Section 15.2), and carboxylic acids and esters (Section 15.3). Table 15.3 summarizes the type of alcohol produced as a function of the carbonyl compound and the reducing agent.

As we saw in Chapter 14 and as is evident in Table 15.1, Grignard reagents and organolithium reagents are among the most effective tools in alcohol synthesis. The present chapter described yet another way in which these organometallic reagents could be used. Treatment of a Grignard or organolithium reagent with ethylene oxide yields a primary alcohol (Section 15.4):

$$\underset{\text{Grignard reagent}}{RMgX} + \underset{\text{Ethylene oxide}}{H_2C\overset{O}{—}CH_2} \xrightarrow[\text{2. } H_3O^+]{\text{1. diethyl ether}} \underset{\text{Primary alcohol}}{RCH_2CH_2OH}$$

$$\underset{\text{Butylmagnesium bromide}}{CH_3CH_2CH_2CH_2MgBr} + \underset{\text{Ethylene oxide}}{H_2C\overset{O}{—}CH_2} \xrightarrow[\text{2. } H_3O^+]{\text{1. diethyl ether}} \underset{\text{1-Hexanol (60–62\%)}}{CH_3CH_2CH_2CH_2CH_2CH_2OH}$$

Many of the reactions used to prepare alcohols can be applied to the preparation of diols. A reaction unique to the preparation of **vicinal diols** employs osmium tetraoxide as the key reagent (Section 15.5):

$$C_6H_5C(CH_3){=}CH_2 \xrightarrow[tert\text{-butyl alcohol, } HO^-]{(CH_3)_3COOH,\ OsO_4(cat)} C_6H_5C(OH)(CH_3)CH_2OH$$

2-Phenylpropene → 2-Phenyl-1,2-propanediol (71%)

The reaction is called **hydroxylation** and proceeds by syn addition to the double bond.

Alcohols undergo a number of conversions that were introduced in earlier chapters as syntheses of other functional group families. These reactions were reviewed in Table 15.2 (Section 15.6). Table 15.4 lists the reactions discussed for the first time in the present chapter. The last entry in Table 15.4 points out that a general reaction of alcohols is oxidation to a carbonyl compound. The specifics of alcohol oxidation are summarized in Table 15.5. Biological oxidation of alcohols is described in Section 15.11.

Diols undergo many of the same reactions that simple alcohols do. A reaction characteristic of vicinal diols is their cleavage to two carbonyl compounds on treatment with periodic acid (Section 15.12):

$$R_2C(OH){-}C(OH)R_2 \xrightarrow{HIO_4} R_2C{=}O + O{=}CR_2$$

Diol → Two carbonyl-containing compounds

$$CH_3(CH_2)_7CH(OH){-}CH(OH)(CH_2)_7COOH \xrightarrow{HIO_4} CH_3(CH_2)_7CHO + HC(O)(CH_2)_7COOH$$

9,10-Dihydroxyoctadecanoic acid → Nonanal (89%) + 9-Oxononanoic acid (76%)

Thiols are prepared by the reaction of alkyl halides with thiourea. An intermediate isothiouronium salt is formed, which is subjected to basic hydrolysis in a separate step (Section 15.13).

$$RX \xrightarrow[2.\ NaOH]{1.\ (H_2N)_2C{=}S} RSH$$

Alkyl halide → Alkanethiol

$$CH_3(CH_2)_{11}Br \xrightarrow[2.\ NaOH]{1.\ (H_2N)_2C{=}S} CH_3(CH_2)_{11}SH$$

1-Bromododecane → 1-Dodecanethiol (79–83%)

TABLE 15.4
Summary of Reactions of Alcohols Presented in This Chapter

Reaction (section) and comments	General equation and specific example
Conversion to dialkyl ethers (Section 15.7) On being heated in the presence of an acid catalyst, two molecules of a primary alcohol combine to form an ether and water. Diols can undergo an intramolecular condensation if a five-membered or six-membered cyclic ether results.	$2RCH_2OH \xrightarrow[heat]{H^+} RCH_2OCH_2R + H_2O$ Alcohol; Dialkyl ether; Water $2(CH_3)_2CHCH_2CH_2OH \xrightarrow[150°C]{H_2SO_4} (CH_3)_2CHCH_2CH_2OCH_2CH_2CH(CH_3)_2$ 3-Methyl-1-butanol; Di-(3-methylbutyl) ether (27%)
Fischer esterification (Section 15.8) Alcohols and carboxylic acids yield an ester and water in the presence of an acid catalyst. The reaction is an equilibrium process that can be driven to completion by using either the alcohol or the acid in excess or by removing the water as it is formed.	$ROH + R'\overset{O}{\overset{\Vert}{C}}OH \xrightarrow{H^+} R'\overset{O}{\overset{\Vert}{C}}OR + H_2O$ Alcohol; Carboxylic acid; Ester; Water $CH_3CH_2CH_2CH_2CH_2OH + CH_3\overset{O}{\overset{\Vert}{C}}OH \xrightarrow{H^+} CH_3\overset{O}{\overset{\Vert}{C}}OCH_2CH_2CH_2CH_2CH_3$ 1-Pentanol; Acetic acid; Pentyl acetate (71%)
Esterification with acyl chlorides (Section 15.8) Acyl chlorides react with alcohols to give esters. The reaction is usually carried out in the presence of pyridine.	$ROH + R'\overset{O}{\overset{\Vert}{C}}Cl \longrightarrow R'\overset{O}{\overset{\Vert}{C}}OR + HCl$ Alcohol; Acyl chloride; Ester; Hydrogen chloride $(CH_3)_3COH + CH_3\overset{O}{\overset{\Vert}{C}}Cl \xrightarrow{pyridine} CH_3\overset{O}{\overset{\Vert}{C}}OC(CH_3)_3$ *tert*-Butyl alcohol; Acetyl chloride; *tert*-Butyl acetate (62%)
Esterification with carboxylic acid anhydrides (Section 15.8) Carboxylic acid anhydrides react with alcohols to form esters in the same way that acyl chlorides do.	$ROH + R'\overset{O}{\overset{\Vert}{C}}O\overset{O}{\overset{\Vert}{C}}R' \longrightarrow R'\overset{O}{\overset{\Vert}{C}}OR + R'\overset{O}{\overset{\Vert}{C}}OH$ Alcohol; Carboxylic acid anhydride; Ester; Carboxylic acid $m\text{-}CH_3OC_6H_4CH_2OH + CH_3\overset{O}{\overset{\Vert}{C}}O\overset{O}{\overset{\Vert}{C}}CH_3 \xrightarrow{pyridine} m\text{-}CH_3OC_6H_4CH_2O\overset{O}{\overset{\Vert}{C}}CH_3$ *m*-Methoxybenzyl alcohol; Acetic anhydride; *m*-Methoxybenzyl acetate (99%)
Formation of esters of inorganic acids (Section 15.9) Alkyl nitrates, dialkyl sulfates, trialkyl phosphites, and trialkyl phosphates are examples of alkyl esters of inorganic acids. In some cases, these compounds are prepared by the direct reaction of an alcohol and the inorganic acid.	$ROH + HONO_2 \xrightarrow{H^+} RONO_2 + H_2O$ Alcohol; Nitric acid; Alkyl nitrate; Water $C_5H_9OH \xrightarrow[H_2SO_4]{HNO_3} C_5H_9ONO_2$ Cyclopentanol; Cyclopentyl nitrate (69%)
Oxidation of alcohols (Section 15.10) See Table 15.5.	

TABLE 15.5
Oxidation of Alcohols

Class of alcohol	Desired product	Suitable oxidizing agent(s)
Primary, RCH_2OH	Aldehyde RCH (with C=O)	PCC* PDC
Primary, RCH_2OH	Carboxylic acid $RCOH$ (with C=O)	$Na_2Cr_2O_7$, H_2SO_4, H_2O H_2CrO_4
Secondary, $RCHR'$ (with OH on C)	Ketone RCR' (with C=O)	PCC PDC $Na_2Cr_2O_7$, H_2SO_4, H_2O H_2CrO_4

* PCC is pyridinium chlorochromate; PDC is pyridinium dichromate. Both are used in dichloromethane.

Thiols are more acidic than alcohols and are easily converted to their corresponding alkanethiolate ions on treatment with aqueous hydroxide. Compounds incorporating sulfur in higher oxidation states are formed when thiols are oxidized (Section 15.14) and include disulfides (RSSR), sulfenic acids (RSOH), sulfinic acids (RSO_2H), and sulfonic acids (RSO_3H).

PROBLEMS

15.16 Write chemical equations, showing all necessary reagents, for the preparation of 1-butanol by each of the following methods:

(*a*) Hydroboration-oxidation of an alkene
(*b*) Use of a Grignard reagent
(*c*) Use of a Grignard reagent in a way different from part (*b*)
(*d*) Reduction of a carboxylic acid
(*e*) Reduction of a methyl ester
(*f*) Reduction of a butyl ester
(*g*) Hydrogenation of an aldehyde
(*h*) Reduction with sodium borohydride

15.17 Write chemical equations, showing all necessary reagents, for the preparation of 2-butanol by each of the following methods:

(*a*) Hydroboration-oxidation of an alkene
(*b*) Oxymercuration-demercuration of an alkene
(*c*) Use of a Grignard reagent
(*d*) Use of a Grignard reagent different from that used in part *(c)*
(*e*)–(*g*) Three different methods for reducing a ketone

15.18 Write chemical equations, showing all necessary reagents, for the preparation of 2-methyl-2-propanol by each of the following methods:

(*a*) Oxymercuration-demercuration of an alkene
(*b*) Use of a Grignard reagent

(*c*) Use of the same Grignard reagent used in part (*b*) with a different carbonyl-containing compound

15.19 Which of the isomeric $C_5H_{12}O$ alcohols can be prepared by sodium borohydride reduction of a carbonyl compound?

15.20 Evaluate the feasibility of the route

$$RH \xrightarrow[\text{light or heat}]{Br_2} RBr \xrightarrow{KOH} ROH$$

as a method for preparing

(*a*) 1-Butanol from butane
(*b*) 2-Methyl-2-propanol from 2-methylpropane
(*c*) Benzyl alcohol from toluene
(*d*) (*R*)-1-Phenylethanol from ethylbenzene

15.21 Sorbitol is a sweetener often substituted for cane sugar, since it is better tolerated by diabetics. It is also an intermediate in the commercial synthesis of vitamin C. Sorbitol is prepared by high-pressure hydrogenation of glucose over a nickel catalyst. What is the structure (including stereochemistry) of sorbitol?

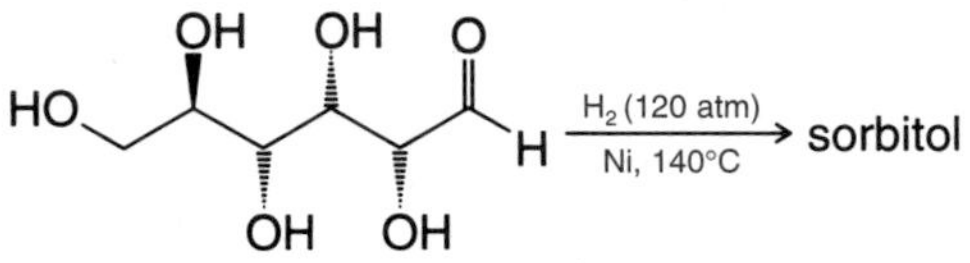

Glucose

15.22 Write equations showing how 1-phenylethanol ($C_6H_5\underset{OH}{\underset{|}{C}}HCH_3$) could be prepared from each of the following starting materials:

(*a*) Bromobenzene
(*b*) Benzaldehyde
(*c*) Benzyl alcohol
(*d*) Styrene
(*e*) Acetophenone
(*f*) Benzene

15.23 Write equations showing how 2-phenylethanol ($C_6H_5CH_2CH_2OH$) could be prepared from each of the following starting materials:

(*a*) Bromobenzene
(*b*) Styrene
(*c*) Phenylacetylene
(*d*) 2-Phenylethanal ($C_6H_5CH_2CHO$)
(*e*) Ethyl 2-phenylethanoate ($C_6H_5CH_2CO_2CH_2CH_3$)
(*f*) 2-Phenylethanoic acid ($C_6H_5CH_2CO_2H$)

15.24 Outline practical syntheses of each of the following compounds from alcohols containing no more than four carbon atoms and any necessary organic or inorganic reagents. In many cases the desired compound can be made from one prepared in an earlier part of the problem.

(*a*) 1-Fluorobutane
(*b*) 1-Butanethiol
(*c*) 1-Hexanol
(*d*) 2-Hexanol
(*e*) Hexanal, $CH_3CH_2CH_2CH_2CH_2CH{=}O$
(*f*) 2-Hexanone, $CH_3\overset{\overset{O}{\|}}{C}CH_2CH_2CH_2CH_3$
(*g*) Hexanoic acid, $CH_3(CH_2)_4CO_2H$

(*h*) Ethyl hexanoate, $CH_3(CH_2)_4\overset{O}{\overset{\|}{C}}OCH_2CH_3$

(*i*) 2-Methyl-1,2-propanediol

(*j*) 2,2-Dimethylpropanal, $(CH_3)_3\overset{O}{\overset{\|}{C}}H$

15.25 Outline practical syntheses of each of the following compounds from benzene, alcohols, and any necessary organic or inorganic reagents:

(*a*) 1-Chloro-2-phenylethane

(*b*) 2-Methyl-1-phenyl-1-propanone, $C_6H_5\overset{O}{\overset{\|}{C}}CH(CH_3)_2$

(*c*) Isobutylbenzene, $C_6H_5CH_2CH(CH_3)_2$

15.26 Show how each of the following compounds can be synthesized from cyclopentanol and any necessary organic or inorganic reagents. In many cases the desired compound can be made from one prepared in an earlier part of the problem.

(*a*) 1-Phenylcyclopentanol

(*b*) 1-Phenylcyclopentene

(*c*) *trans*-2-Phenylcyclopentanol

(*d*) C_6H_5 O

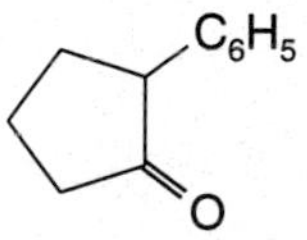

(*e*) OH C_6H_5 OH

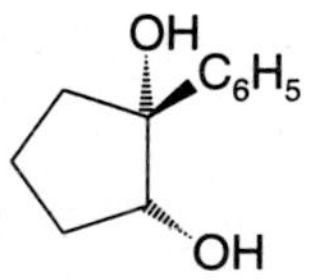

(*f*) $C_6H_5\overset{O}{\overset{\|}{C}}CH_2CH_2CH_2\overset{O}{\overset{\|}{C}}H$

(*g*) 1-Phenyl-1,5-pentanediol

15.27 Write the structure of the principal organic product formed in the reaction of 1-propanol with each of the following reagents:

(*a*) Sulfuric acid (catalytic amount), heat at 140°C

(*b*) Sulfuric acid (catalytic amount), heat at 200°C

(*c*) Nitric acid (H_2SO_4 catalyst)

(*d*) Pyridinium chlorochromate (PCC) in dichloromethane

(*e*) Potassium dichromate ($K_2Cr_2O_7$) in aqueous sulfuric acid, heat

(*f*) Metallic sodium

(*g*) Sodium amide ($NaNH_2$)

(*h*) Acetic acid ($CH_3\overset{O}{\overset{\|}{C}}OH$) in the presence of dissolved hydrogen chloride

(*i*) CH_3—⟨benzene ring⟩—SO_2Cl in the presence of pyridine

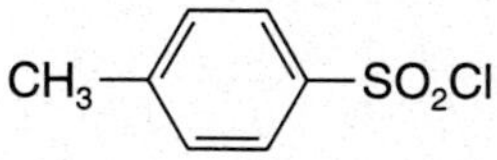

(*j*) CH_3O—⟨benzene ring⟩—$\overset{O}{\overset{\|}{C}}Cl$ in the presence of pyridine

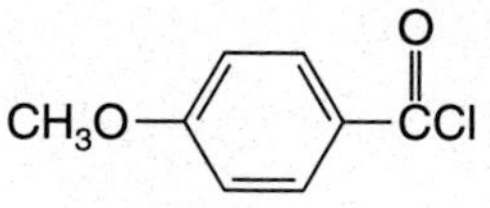

(*k*) $C_6H_5\overset{O}{\overset{\|}{C}}O\overset{O}{\overset{\|}{C}}C_6H_5$ in the presence of pyridine

(*l*) O O O in the presence of pyridine

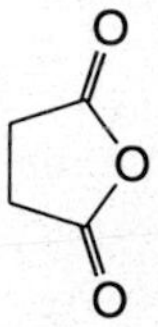

15.28 Each of the following reactions has been reported in the chemical literature. Predict the product in each case, showing stereochemistry where appropriate.

(*a*) CH_3–(cyclohexane)–C(OH)(C_6H_5) $\xrightarrow[\text{heat}]{H_2SO_4}$

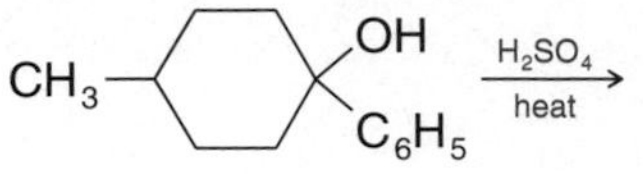

(*b*) $(CH_3)_2C{=}C(CH_3)_2 \xrightarrow[(CH_3)_3COH,\ HO^-]{(CH_3)_3COOH,\ OsO_4(\text{cat})}$

(*c*) 1-C_6H_5-cyclobutene $\xrightarrow[2.\ H_2O_2,\ HO^-]{1.\ B_2H_6,\ \text{diglyme}}$

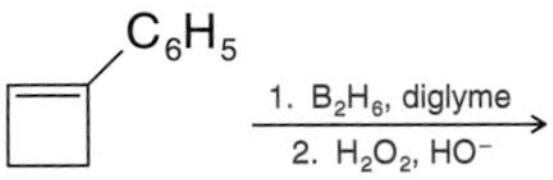

(*d*) cyclopentene–CO_2H $\xrightarrow[2.\ H_2O]{1.\ LiAlH_4,\ \text{diethyl ether}}$

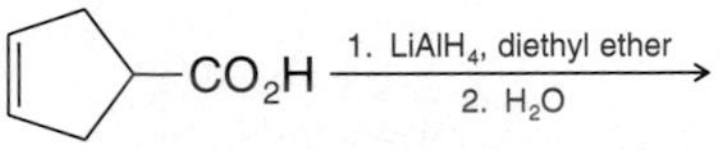

(*e*) $CH_3CH(OH)C{\equiv}C(CH_2)_3CH_3 \xrightarrow[H_2SO_4,\ H_2O,\ \text{acetone}]{H_2CrO_4}$

(*f*) $CH_3\overset{O}{\overset{\|}{C}}CH_2CH{=}CHCH_2\overset{O}{\overset{\|}{C}}CH_3 \xrightarrow[2.\ H_2O]{1.\ LiAlH_4,\ \text{diethyl ether}}$

(*g*) CH_3–(cyclohexane)–OH + 3,5-$(O_2N)_2C_6H_3\overset{O}{\overset{\|}{C}}Cl \xrightarrow{\text{pyridine}}$

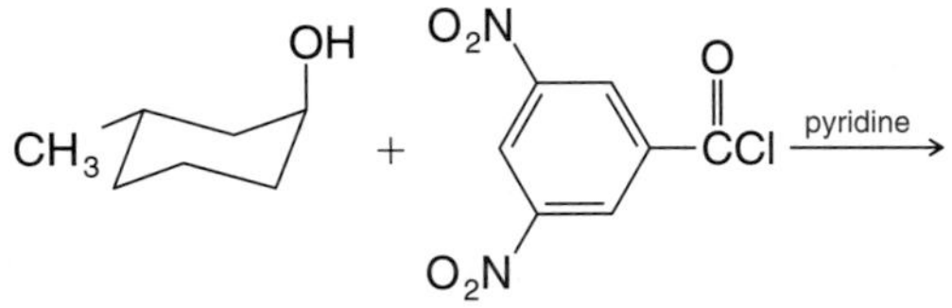

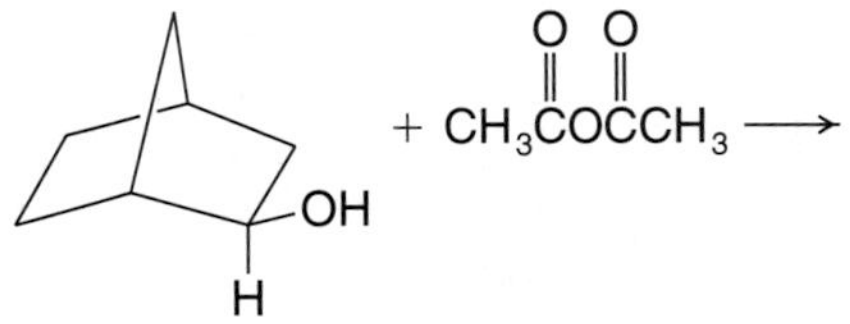

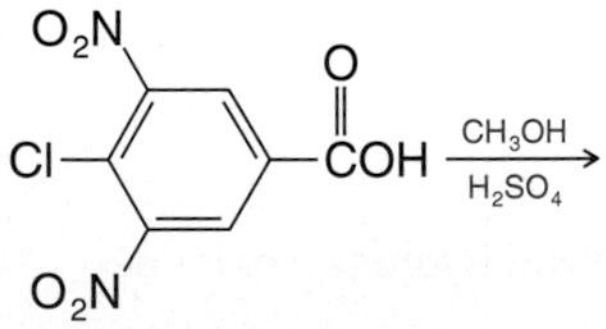

(*h*) (bicyclic alcohol, OH, H) + $CH_3\overset{O}{\overset{\|}{C}}O\overset{O}{\overset{\|}{C}}CH_3 \longrightarrow$

(*i*) Cl–$C_6H_2(NO_2)_2$–$\overset{O}{\overset{\|}{C}}OH \xrightarrow[H_2SO_4]{CH_3OH}$

(*j*) $CH_3\overset{O}{\overset{\|}{C}}O$, $\overset{O}{\overset{\|}{C}}OCH_3$, H_3C (substituted cyclohexene) $\xrightarrow[2.\ H_2O]{1.\ LiAlH_4}$

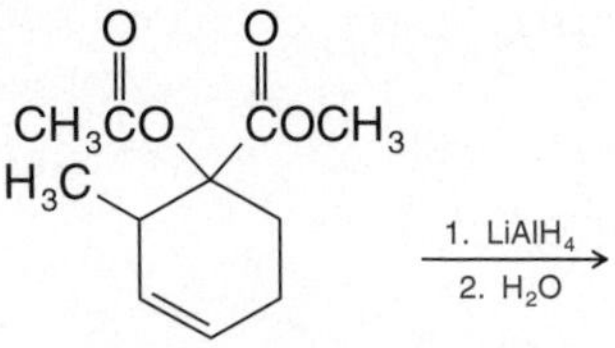

(*k*) Product of part (*j*) $\xrightarrow[CH_3OH,\ H_2O]{HIO_4}$

15.29 On heating 1,2,4-butanetriol in the presence of an acid catalyst, a cyclic ether of molecular formula $C_4H_8O_2$ was obtained in 81 to 88 percent yield. Suggest a reasonable structure for this product.

15.30 Give the Cahn-Ingold-Prelog *R* and *S* descriptors for the diol(s) formed from *cis*-2-pentene and *trans*-2-pentene on treatment with the osmium tetraoxide–*tert*-butyl hydroperoxide reagent.

15.31 Suggest reaction sequences and reagents suitable for carrying out each of the following conversions. Two synthetic operations are required in each case.

(*a*) 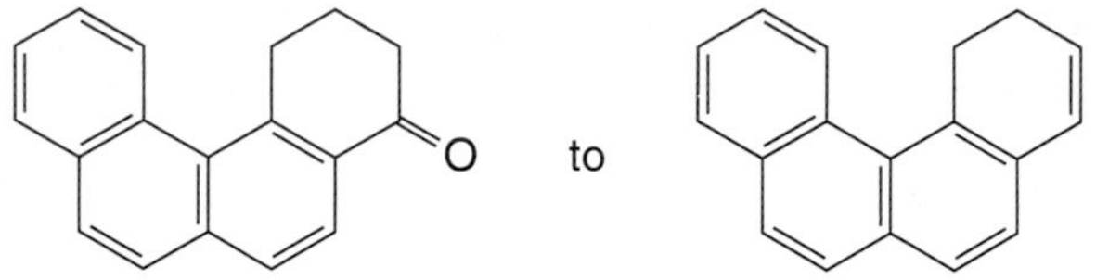 O to

(*b*) 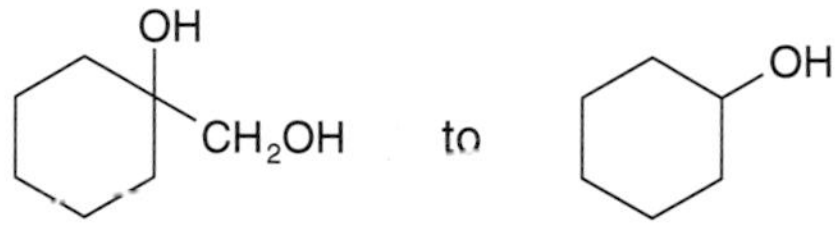 OH CH_2OH to OH

(*c*) OH C_6H_5 to OH C_6H_5 OH

15.32 The fungus responsible for Dutch elm disease is spread by European bark beetles when they burrow into the tree. Other beetles congregate at the site, attracted by the scent of a mixture of chemicals, some emitted by other beetles and some coming from the tree. One of the compounds given off by female bark beetles is 4-methyl-3-heptanol. Suggest an efficient synthesis of this pheromone from alcohols of five carbon atoms or fewer.

15.33 Show by a series of equations how you could prepare 3-methylpentane from ethanol and any necessary inorganic reagents.

15.34

(*a*) The cis isomer of 3-hexen-1-ol ($CH_3CH_2CH{=}CHCH_2CH_2OH$) has the characteristic odor of green leaves and grass. Suggest a synthesis for this compound from acetylene and any necessary organic or inorganic reagents.

(*b*) One of the compounds responsible for the characteristic odor of ripe tomatoes is the cis isomer of $CH_3CH_2CH{=}CHCH_2CH{=}O$. How could you prepare this compound?

15.35 R. B. Woodward was one of the leading organic chemists of the middle part of the twentieth century. Known primarily for his achievements in the synthesis of complex natural products, he was awarded the Nobel Prize in chemistry in 1965. He entered Massachusetts Institute of Technology as a 16-year-old freshman in 1933 and four years later was awarded the Ph.D. While a student there he carried out a synthesis of *estrone,* a female sex hormone. The early stages of Woodward's estrone synthesis required the conversion of *m*-methoxybenzaldehyde to *m*-methoxybenzyl cyanide, which was accomplished in three steps:

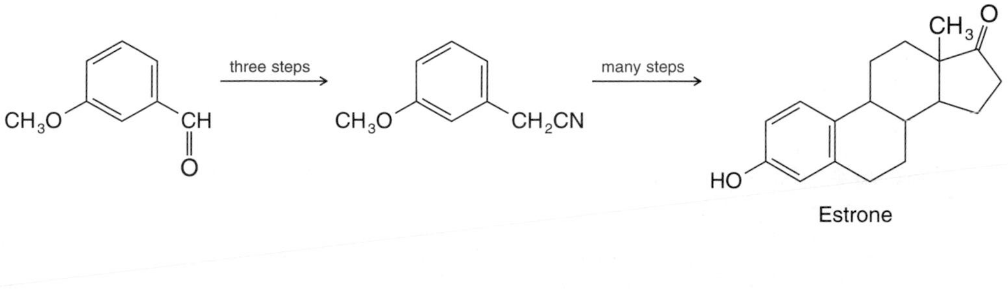

CH_3O CH=O —three steps→ CH_3O CH_2CN —many steps→ HO CH_3 O

Estrone

Suggest a reasonable three-step sequence, showing all necessary reagents, for the preparation of *m*-methoxybenzyl cyanide from *m*-methoxybenzaldehyde.

15.36 Complete the following series of equations by writing structural formulas for compounds A through I:

(*a*) cyclopentadiene $\xrightarrow{HCl}$ C_5H_7Cl (Compound A) $\xrightarrow[H_2O]{NaHCO_3}$ C_5H_8O (Compound B) $\xrightarrow[H_2SO_4,\ H_2O]{Na_2Cr_2O_7}$ C_5H_6O (Compound C)

(*b*) $CH_2{=}CHCH_2CH_2CH(OH)CH_3$ $\xrightarrow[pyridine]{SOCl_2}$ $C_6H_{11}Cl$ (Compound D) $\xrightarrow[2.\ reductive\ workup]{1.\ O_3}$ C_5H_9ClO (Compound E) $\xrightarrow{NaBH_4}$ $C_5H_{11}ClO$ (Compound F)

(*c*) 1-bromo-2-methylnaphthalene (Br, CH_3) $\xrightarrow[benzoyl\ peroxide,\ heat]{NBS}$ Compound G $\xrightarrow[heat]{H_2O,\ CaCO_3}$ Compound H $\xrightarrow[CH_2Cl_2]{PCC}$ $(C_{11}H_7BrO)$ Compound I

15.37 When 2-phenyl-2-butanol is allowed to stand in ethanol containing a few drops of sulfuric acid, the ether shown in the following is formed:

$$C_6H_5C(CH_3)(OH)CH_2CH_3 \xrightarrow[H_2SO_4]{CH_3CH_2OH} C_6H_5C(CH_3)(OCH_2CH_3)CH_2CH_3$$

Suggest a reasonable mechanism for this reaction based on the observation that the ether produced from optically active alcohol is racemic, and that alkenes can be shown not to be intermediates in the reaction.

15.38 Suggest a chemical test that would permit you to distinguish between the two glycerol monobenzyl ethers shown.

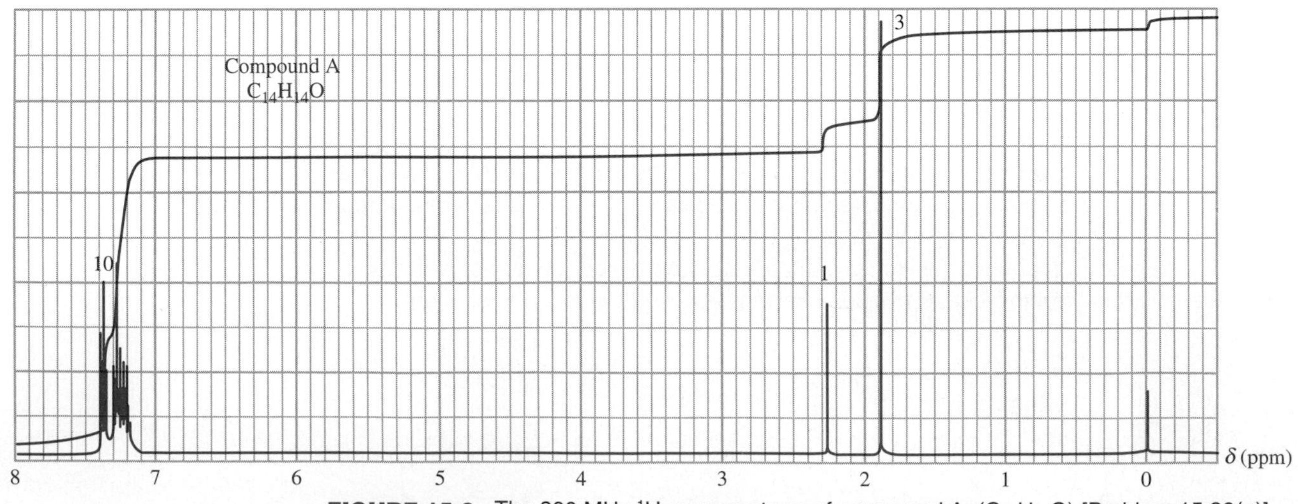

FIGURE 15.6 The 300 MHz ^1H nmr spectrum of compound A, ($C_{14}H_{14}O$) [Problem 15.39(*a*)].

$C_6H_5CH_2OCH_2\underset{\displaystyle OH}{\underset{|}{C}}HCH_2OH$ $HOCH_2\underset{\displaystyle OCH_2C_6H_5}{\underset{|}{C}}HCH_2OH$

1-*O*-Benzylglycerol   2-*O*-Benzylglycerol

15.39 Identify the following alcohols on the basis of their ^1H nmr spectra:

(*a*) Compound A, $C_{14}H_{14}O$ (Figure 15.6)
(*b*) Compound B, $C_9H_{11}BrO$ (Figure 15.7)

15.40 A diol ($C_8H_{18}O_2$) does not react with periodic acid. Its ^1H nmr spectrum contains 3 singlets at $\delta = 1.2$ (12 protons), 1.6 (4 protons), and 2.0 ppm (2 protons). What is the structure of this diol?

15.41 Identify each of the following $C_4H_{10}O$ isomers on the basis of their ^{13}C nmr spectra:

(*a*) 31.2 ppm: quartet
68.9 ppm: singlet
(*b*) 10.0 ppm: quartet
22.7 ppm: quartet
32.0 ppm: triplet
69.2 ppm: doublet
(*c*) 18.9 ppm: quartet, area 2
30.8 ppm: doublet, area 1
69.4 ppm: triplet, area 1

15.42 A compound $C_3H_7ClO_2$ exhibited three peaks in its ^{13}C nmr spectrum at 46.8 (triplet), 63.5 (triplet), and 72.0 ppm (doublet). What is the structure of this compound?

15.43 A compound $C_6H_{14}O$ has the ^{13}C nmr spectrum shown in Figure 15.8. Its mass spectrum has a prominent peak at *m/z* 31. Suggest a reasonable structure for this compound.

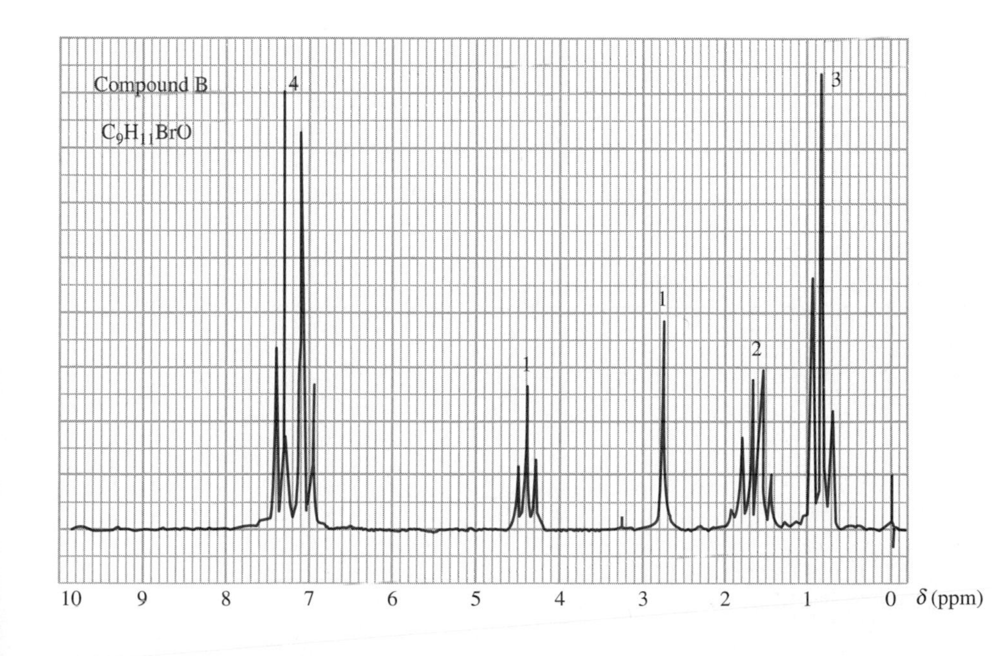

FIGURE 15.7 The 60 MHz ^1H nmr spectrum of compound B, ($C_9H_{11}BrO$) [Problem 15.39(*b*)].

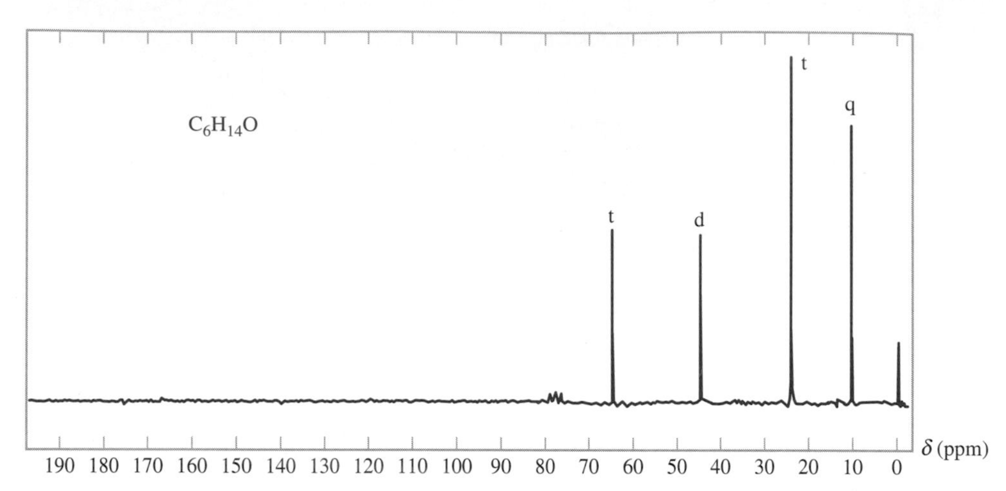

FIGURE 15.8 The ^{13}C nmr spectrum of the compound $C_6H_{14}O$ (Problem 15.43). (*The* ^{13}C *spectrum is taken from* Carbon-13 NMR Spectra: A Collection of Assigned, Coded, and Indexed Spectra, *by LeRoy F. Johnson and William C. Jankowski, Wiley-Interscience, New York, 1972. Reprinted by permission of John Wiley & Sons, Inc.*)

MOLECULAR MODELING EXERCISES

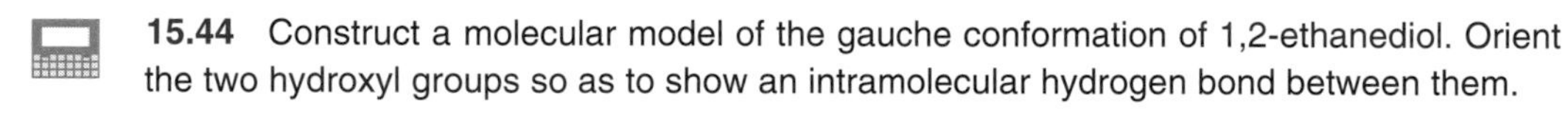

15.44 Construct a molecular model of the gauche conformation of 1,2-ethanediol. Orient the two hydroxyl groups so as to show an intramolecular hydrogen bond between them.

15.45 The extent of intramolecular hydrogen bonding between hydroxyl groups of vicinal diols has been studied using infrared spectroscopy. Such hydrogen bonding has been found to be very strong in the chiral diastereomer of $(CH_3)_3CCHCHC(CH_3)_3$ (with HO and OH on the two central CH carbons), and absent in the meso diastereomer. Construct a molecular model of each diastereomer and suggest a reason for the absence of an intramolecular hydrogen bond in the meso diol.

15.46 The major product in the lithium aluminum hydride reduction of 3,3,5-trimethylcyclohexanone corresponds to hydrogen transfer from the reagent to the less-hindered face of the carbonyl group. Construct a molecular model of the major product in its most stable conformation.

15.47 Two tertiary alcohols are formed in the addition of C_6H_5MgBr to 4-*tert*-butylcyclohexanone. Make a molecular model of each. The major product is the more stable diastereomer. Which one is this? (*Note:* This reaction, like most Grignard additions, is kinetically controlled. It just happens in this case that the diastereomer that is formed at the faster rate is also the more stable one.)

15.48 Construct a molecular model of *trans*-2-butene and the products formed on (*a*) syn hydroxylation, and (*b*) anti hydroxylation of the double bond. Repeat the exercise for *cis*-2-butene.

CHAPTER 16

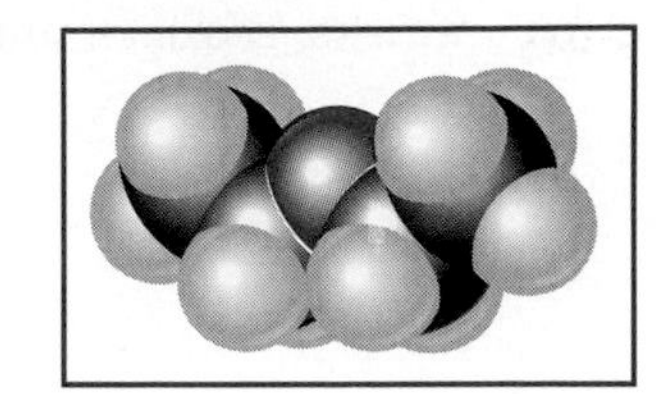

ETHERS, EPOXIDES, AND SULFIDES

In contrast to alcohols with their rich chemical reactivity, **ethers**—compounds containing a C—O—C unit—undergo relatively few chemical reactions. As you saw when we discussed Grignard reagents in Chapter 14 and lithium aluminum hydride reductions in Chapter 15, this lack of reactivity of ethers makes them valuable as solvents in a number of synthetically important transformations. In the present chapter you will learn of the conditions in which an ether linkage acts as a functional group, as well as the methods by which ethers are prepared.

Unlike most ethers, **epoxides**—compounds in which the C—O—C unit forms a three-membered ring—are very reactive substances. The principles of nucleophilic substitution reactions are important in understanding the preparation and properties of epoxides.

Sulfides (RSR′) are the sulfur analogs of ethers. Just as in the preceding chapter, where we saw that the properties of thiols (RSH) are different from those of alcohols, we will explore differences between sulfides and ethers in this chapter.

16.1 NOMENCLATURE OF ETHERS, EPOXIDES, AND SULFIDES

Ethers are named, in substitutive nomenclature, as *alkoxy* derivatives of alkanes. Radicofunctional names of ethers are derived by listing the two alkyl groups in the general structure ROR′ in alphabetical order as separate words, and then adding the word *ether* at the end. When both alkyl groups are the same the prefix *di-* precedes the name of the alkyl group.

	$CH_3CH_2OCH_2CH_3$	$CH_3CH_2OCH_3$	$CH_3CH_2OCH_2CH_2CH_2Cl$
Substitutive name:	Ethoxyethane	Methoxyethane	1-Chloro-3-ethoxypropane
Radicofunctional name:	Diethyl ether	Ethyl methyl ether	3-Chloropropyl ethyl ether

Ethers are described as **symmetrical** or **unsymmetrical** depending on whether the two groups bonded to oxygen are the same or different. Unsymmetrical ethers are also called **mixed ethers.** Diethyl ether is a symmetrical ether; ethyl methyl ether is an unsymmetrical ether.

Cyclic ethers have their oxygen as part of a ring—they are *heterocyclic compounds* (Section 3.16). Oxygen heterocycles of commonly encountered ring sizes have specific IUPAC names.

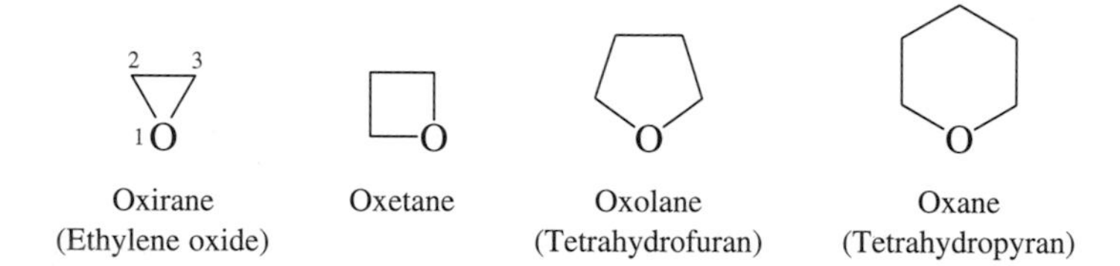

Recall from Section 6.18 that epoxides may be named as *-epoxy* derivatives of alkanes in substitutive IUPAC nomenclature.

In each case the ring is numbered starting at the oxygen. The IUPAC rules also permit oxirane (without substituents) to be called *ethylene oxide. Tetrahydrofuran* and *tetrahydropyran* are acceptable synonyms for oxolane and oxane, respectively.

PROBLEM 16.1 Each of the following ethers has been shown to be or is suspected to be a *mutagen,* which means it can induce mutations in test cells. Write the structure of each of these ethers.

(*a*) Chloromethyl methyl ether
(*b*) 2-(Chloromethyl)oxirane (also known as epichlorohydrin)
(*c*) 3,4-Epoxy-1-butene (2-vinyloxirane)

SAMPLE SOLUTION (*a*) Chloromethyl methyl ether has a chloromethyl group ($ClCH_2$—) and a methyl group (CH_3—) attached to oxygen. Its structure is $ClCH_2OCH_3$.

It is not unusual for substances to have more than one ether linkage. Two such compounds, often used as solvents, are the *diethers* 1,2-dimethoxyethane and 1,4-dioxane. Diglyme, also a commonly used solvent, is a *triether.*

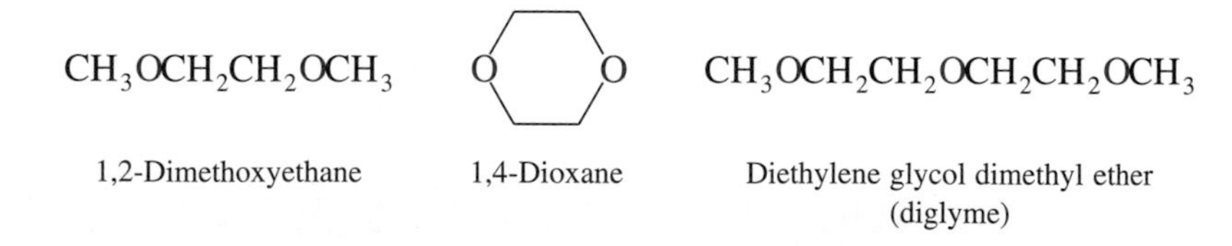

Molecules that contain several ether functions are referred to as *polyethers.* Polyethers have received much recent attention, and some examples of them will appear in Section 16.4.

Sulfides are sometimes informally referred to as *thioethers,* but this term is not part of systematic IUPAC nomenclature.

The sulfur analogs (RS—) of alkoxy groups are called *alkylthio* groups. The first two of the following examples illustrate the use of alkylthio prefixes in substitutive nomenclature of sulfides. Radicofunctional names of sulfides are derived in exactly the same way as those of ethers but end in the word *sulfide.* Sulfur heterocycles have names analogous to their oxygen relatives, except that *ox-* is replaced by *thi-*. Thus the sulfur heterocycles containing three-, four-, five-, and six-membered rings are named *thiirane, thietane, thiolane,* and *thiane,* respectively.

$CH_3CH_2SCH_2CH_3$ — Ethylthioethane, Diethyl sulfide

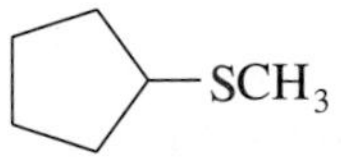

$C_5H_9-SCH_3$ — (Methylthio)cyclopentane, Cyclopentyl methyl sulfide

Thiirane

16.2 STRUCTURE AND BONDING IN ETHERS AND EPOXIDES

Bonding in ethers is readily understood by comparing ethers with water and alcohols. Van der Waals strain involving alkyl groups causes the bond angle at oxygen to be larger in ethers than alcohols, and larger in alcohols than in water. An extreme example is di-*tert*-butyl ether, where steric hindrance between the *tert*-butyl groups is responsible for a dramatic increase in the C—O—C bond angle.

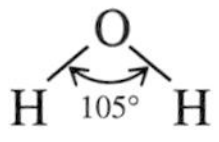

Water	Methanol	Dimethyl ether	Di-*tert*-butyl ether
H—O—H 105°	H—O—CH_3 108.5°	CH_3—O—CH_3 112°	$(CH_3)_3C$—O—$C(CH_3)_3$ 132°

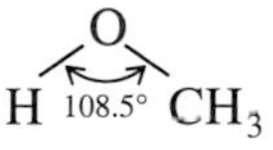

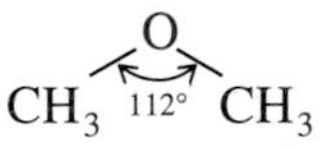

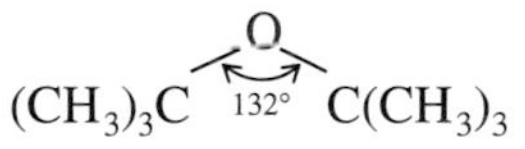

Carbon-oxygen bond distances are somewhat shorter than carbon-carbon bond distances. The C—O bond distances in dimethyl ether (141 pm) and methanol (142 pm) are similar to one another, and both are shorter than the C—C bond distance in ethane (153 pm).

An oxygen in a hydrocarbon chain affects the conformation of a molecule in much the same way that a CH_2 unit does. The most stable conformation of diethyl ether is the all-staggered anti conformation. Tetrahydropyran adopts the chair form in its most stable conformation—a fact that has an important bearing on the structures of many carbohydrates.

This chapter opened with a drawing of a space-filling model of diethyl ether.

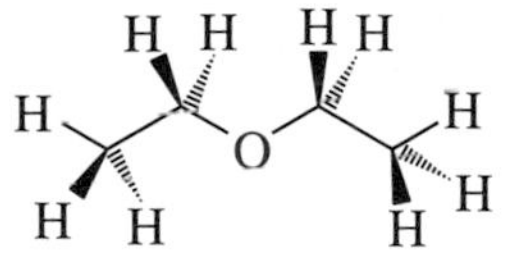

Anti conformation of diethyl ether

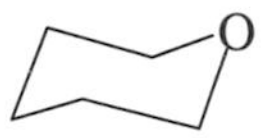

Chair conformation of tetrahydropyran

Incorporating an oxygen atom into a three-membered ring requires its bond angle to be seriously distorted from the normal tetrahedral value. In ethylene oxide, for example, the bond angle at oxygen is 61.5°.

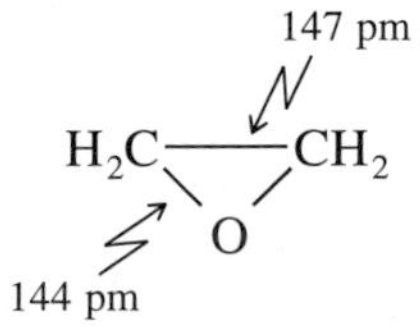

H_2C—CH_2 (147 pm), C—O 144 pm

C—O—C angle 61.5°

C—C—O angle 59.2°

Thus epoxides, like cyclopropanes, are strained. They tend to undergo reactions that open the three-membered ring by cleavage of one of the carbon-oxygen bonds.

PROBLEM 16.2 The heats of combustion of 1,2-epoxybutane (2-ethyloxirane) and tetrahydrofuran have been measured: one is 2499 kJ/mol (597.8 kcal/mol); the other is 2546 kJ/mol (609.1 kcal/mol). Match the heats of combustion with the respective compounds.

Ethers, like water and alcohols, are polar. Diethyl ether, for example, has a dipole moment of 1.2 D. Cyclic ethers have larger dipole moments; ethylene oxide and tetrahydrofuran have dipole moments in the 1.7- to 1.8-D range—about the same as that of water.

16.3 PHYSICAL PROPERTIES OF ETHERS

It is instructive to compare the physical properties of ethers with alkanes and alcohols. With respect to boiling point, ethers resemble alkanes more than alcohols; with respect to solubility in water, ethers resemble alcohols more than alkanes. Why?

	$CH_3CH_2OCH_2CH_3$	$CH_3CH_2CH_2CH_2CH_3$	$CH_3CH_2CH_2CH_2OH$
	Diethyl ether	Pentane	1-Butanol
Boiling point:	35°C	36°C	117°C
Solubility in water:	7.5 g/100 mL	Insoluble	9 g/100 mL

In general, the boiling points of alcohols are unusually high because of hydrogen bonding (Section 4.6). Attractive forces in the liquid phases of ethers and alkanes, which lack —OH groups and cannot form intermolecular hydrogen bonds, are much weaker, and their boiling points lower.

As shown in Figure 16.1, however, the presence of an oxygen atom permits ethers to participate in hydrogen bonds to water molecules. These attractive forces cause ethers to dissolve in water to approximately the same extent as comparably constituted alcohols. Alkanes cannot engage in hydrogen bonding to water.

PROBLEM 16.3 Ethers tend to dissolve in alcohols and vice versa. Represent the hydrogen bonding interaction between an alcohol molecule and an ether molecule.

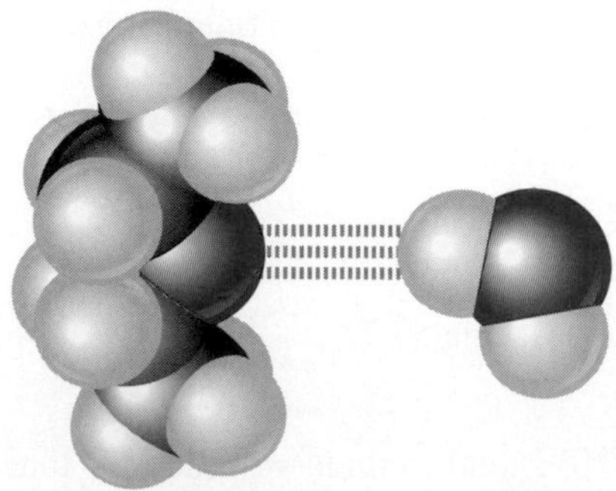

FIGURE 16.1 Molecular models depicting hydrogen bonding between diethyl ether and water. The dashed lines signify the attractive force between the negatively polarized oxygen of diethyl ether and one of the positively polarized hydrogens of water.

16.4 CROWN ETHERS

Their polar carbon-oxygen bonds and the presence of unshared electron pairs at oxygen contribute to the ability of ethers to form Lewis acid–Lewis base complexes with metal ions.

$$\underset{\text{(Lewis base)}}{\underset{\text{Ether}}{R_2\ddot{O}:}} \quad + \quad \underset{\text{(Lewis acid)}}{\underset{\text{Metal ion}}{M^+}} \quad \rightleftharpoons \quad \underset{\text{complex}}{\underset{\text{Ether–metal ion}}{R_2\overset{+}{\ddot{O}}—M}}$$

The strength of this bonding depends on the kind of ether. Simple ethers form relatively weak complexes with metal ions. A major advance in the area came in 1967 when Charles J. Pedersen of Du Pont described the preparation and properties of a class of *polyethers* that form much more stable complexes with metal ions than do simple ethers.

Pedersen was a corecipient of the 1987 Nobel Prize in chemistry.

Pedersen prepared a series of *macrocyclic polyethers,* cyclic compounds containing 4 or more ether linkages in a ring of 12 or more atoms. He called these compounds **crown ethers,** because their molecular models resemble crowns. Systematic nomenclature of crown ethers is somewhat cumbersome, and so Pedersen devised a shorthand description whereby the word *crown* is preceded by the total number of atoms in the ring and is followed by the number of oxygen atoms.

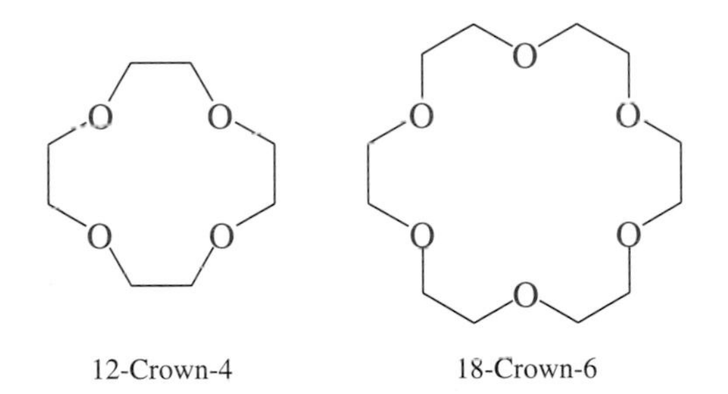

12-Crown-4 18-Crown-6

12-Crown-4 and 18-crown-6 are a cyclic tetramer and hexamer, respectively, of repeating $—OCH_2CH_2—$ units; they are polyethers based on ethylene glycol ($HOCH_2CH_2OH$) as the parent alcohol.

PROBLEM 16.4 What organic compound mentioned earlier in this chapter is a cyclic dimer of $—OCH_2CH_2—$ units?

The metal ion–complexing properties of crown ethers are illustrated by their effects on the solubility and reactivity of salts in nonpolar media. Potassium fluoride is ionic and practically insoluble in benzene, but 0.05 *M* solutions can be prepared when 18-crown-6 is present. This increased solubility of potassium fluoride in benzene is attributable to the formation of a stable complex between a potassium ion (K^+) and the crown ether. This complex, depicted with the aid of a space-filling model in Figure 16.2, is stabilized by ion-dipole forces between K^+ and the six oxygens of the polyether. Potassium ion, with an ionic radius of 266 pm, fits comfortably within the 260- to 320-pm internal cavity of 18-crown-6. Nonpolar CH_2 groups dominate the outer

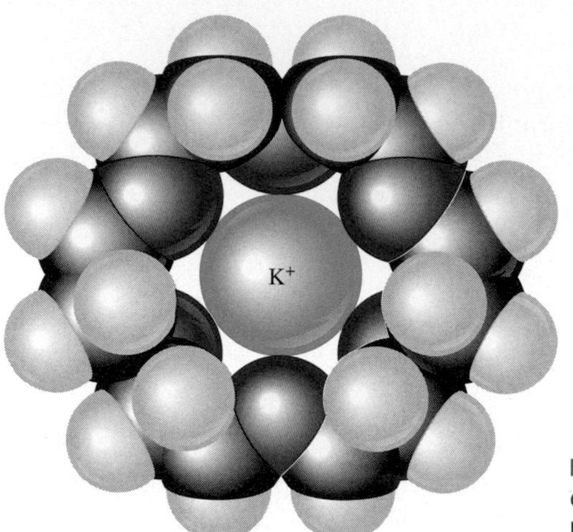

FIGURE 16.2 A space-filling model of the complex formed between 18-crown-6 and potassium ion.

surface of the complex, mask its polar interior, and permit the complex to dissolve in nonpolar solvents such as benzene. Every K^+ that is carried into benzene brings a fluoride ion (F^-) with it, resulting in a solution containing strongly complexed potassium ions and relatively unsolvated fluoride ions.

$$\text{18-Crown-6} + K^+F^- \xrightleftharpoons{\text{benzene}} \text{18-Crown-6-}K^+ \quad F^-$$

18-Crown-6 | Potassium fluoride (solid) | 18-Crown-6-potassium fluoride complex (in solution)

In media such as water and alcohols, fluoride ion is strongly solvated by hydrogen bonding and is neither very basic nor very nucleophilic. On the other hand, the poorly solvated, or ''naked,'' fluoride ions that are present when potassium fluoride dissolves in benzene in the presence of a crown ether are better able to express their anionic reactivity. Thus, alkyl halides react with potassium fluoride in benzene containing 18-crown-6, thereby providing a method for the preparation of otherwise difficultly accessible alkyl fluorides.

$$CH_3(CH_2)_6CH_2Br \xrightarrow[\text{18-crown-6}]{\text{KF, benzene, 90°C}} CH_3(CH_2)_6CH_2F$$

1-Bromooctane → 1-Fluorooctane (92%)

No reaction is observed when the process is carried out under comparable conditions but with the crown ether omitted.

Catalysis by crown ethers has been demonstrated in numerous organic reactions that involve anions as reactants.

POLYETHER ANTIBIOTICS

One way in which pharmaceutical companies search for new drugs is by growing colonies of microorganisms in nutrient broths and assaying the substances produced for their biological activity. This method has yielded thousands of antibiotic substances, of which hundreds have been developed into effective drugs. Antibiotics are, by definition, toxic (*anti* = "against"; *bios* = "life"), and the goal is to find substances that are more toxic to infectious organisms than to their human hosts.

Since 1950, a number of **polyether antibiotics** have been discovered using fermentation technology. They are characterized by the presence of several cyclic ether structural units, as illustrated for the case of *monensin* in Figure 16.3*a*. Monensin and other naturally occurring polyethers are similar to crown ethers in their ability to form stable complexes with metal ions. The structure of the monensin–sodium bromide complex is depicted in Figure 16.3*b*, where it can be seen that four ether oxygens and two hydroxyl groups surround a sodium ion. The alkyl groups are oriented toward the outside of the complex, while the polar oxygens and the metal ion are on the inside. The hydrocarbonlike surface of the complex permits it to carry its sodium ion through the hydrocarbonlike interior of a cell membrane. This disrupts the normal balance of sodium ions within the cell and interferes with important processes of cellular respiration. Small amounts of monensin are added to poultry feed in order to kill parasites that live in the intestines of chickens. Compounds such as monensin and the crown ethers that affect metal ion transport are referred to as **ionophores** ("ion carriers").

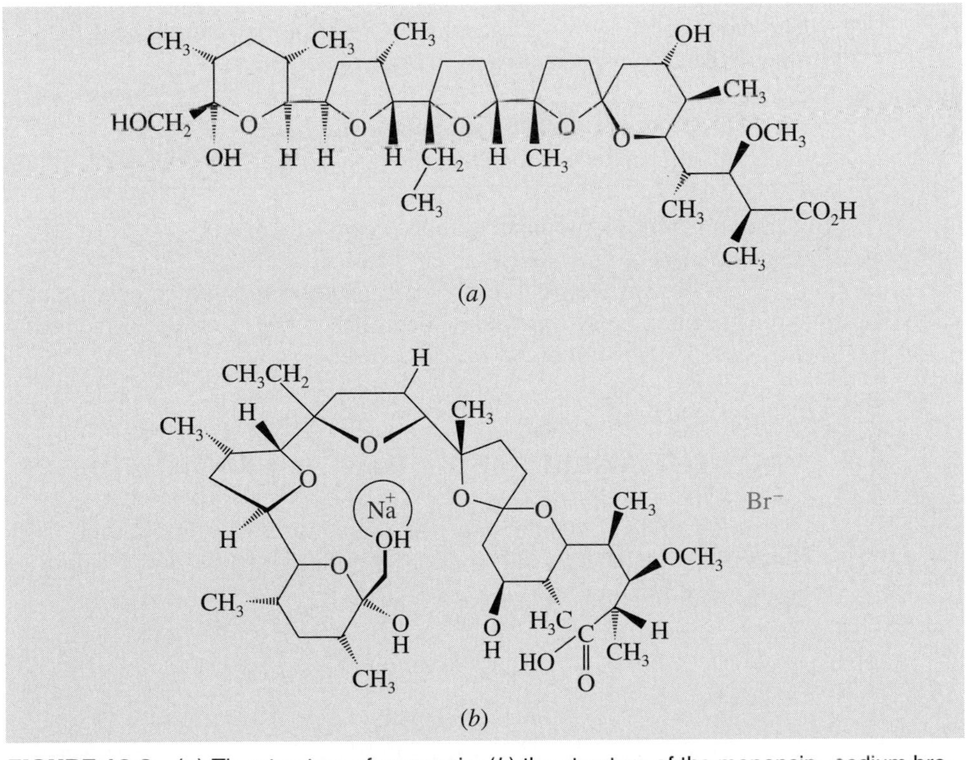

(*a*)

(*b*)

FIGURE 16.3 (*a*) The structure of monensin; (*b*) the structure of the monensin–sodium bromide complex showing coordination of sodium ion by oxygen atoms of monensin.

16.5 PREPARATION OF ETHERS

Because they are widely used as solvents, many simple dialkyl ethers are commercially available. Diethyl ether and dibutyl ether, for example, are prepared by acid-catalyzed condensation of the corresponding alcohols, as described earlier in Section 15.7.

$$\underset{\text{1-Butanol}}{2CH_3CH_2CH_2CH_2OH} \xrightarrow[130°C]{H_2SO_4} \underset{\text{Dibutyl ether (60\%)}}{CH_3CH_2CH_2CH_2OCH_2CH_2CH_2CH_3} + \underset{\text{Water}}{H_2O}$$

In general, this method is limited to the preparation of symmetrical ethers in which both alkyl groups are primary. Isopropyl alcohol, however, is readily available at low cost and gives yields of diisopropyl ether high enough to justify making $(CH_3)_2CHOCH(CH_3)_2$ by this method on an industrial scale.

Approximately 4×10^9 lb of *tert*-butyl methyl ether is prepared in the United States each year by the acid-catalyzed addition of methanol to 2-methylpropene:

$$\underset{\text{2-Methylpropene}}{(CH_3)_2C{=}CH_2} + \underset{\text{Methanol}}{CH_3OH} \xrightarrow{H^+} \underset{\textit{tert}\text{-Butyl methyl ether}}{(CH_3)_3COCH_3}$$

tert-Butyl methyl ether is often referred to as MTBE, standing for the incorrect name "methyl *tert*-butyl ether." Remember, italicized prefixes are ignored when alphabetizing, and *tert*-butyl precedes methyl.

Small amounts of *tert*-butyl methyl ether are added to gasoline as an octane enhancer. The daily consumption of gasoline is so high that the demand for *tert*-butyl methyl ether exceeds our present capacity to produce it.

PROBLEM 16.5 Outline a reasonable mechanism for the formation of *tert*-butyl methyl ether according to the preceding equation.

Solvomercuration-demercuration (Section 14.14) is sometimes used for the laboratory synthesis of ethers from alkenes and alcohols.

The following section describes a versatile method for the preparation of either symmetrical or unsymmetrical ethers that is based on the principles of bimolecular nucleophilic substitution.

16.6 THE WILLIAMSON ETHER SYNTHESIS

The reaction is named for Alexander Williamson, a British chemist who used it to prepare diethyl ether in 1850.

A method of long standing for the preparation of ethers is the **Williamson ether synthesis.** Nucleophilic substitution of an alkyl halide by an alkoxide gives the carbon-oxygen bond of an ether:

$$\underset{\text{Alkoxide ion}}{R\ddot{\underset{\cdot\cdot}{O}}:^-} + \underset{\text{Alkyl halide}}{R'{-}\ddot{\underset{\cdot\cdot}{X}}:} \longrightarrow \underset{\text{Ether}}{R\ddot{\underset{\cdot\cdot}{O}}R'} + \underset{\text{Halide ion}}{:\ddot{\underset{\cdot\cdot}{X}}:^-}$$

Preparation of ethers by the Williamson ether synthesis is most successful when the alkyl halide is one that is reactive toward S_N2 substitution. Methyl halides and primary alkyl halides are the best substrates.

$$\underset{\text{Sodium butoxide}}{CH_3CH_2CH_2CH_2ONa} + \underset{\text{Iodoethane}}{CH_3CH_2I} \longrightarrow \underset{\text{Butyl ethyl ether (71\%)}}{CH_3CH_2CH_2CH_2OCH_2CH_3} + \underset{\text{Sodium iodide}}{NaI}$$

PROBLEM 16.6 Write equations describing two different ways in which benzyl ethyl ether could be prepared by a Williamson ether synthesis.

Secondary and tertiary alkyl halides are not suitable substrates, because they tend to react with alkoxide bases by E2 elimination rather than by S_N2 substitution. Whether the alkoxide base is primary, secondary, or tertiary is much less important than the nature of the alkyl halide. Thus benzyl isopropyl ether is prepared in high yield from benzyl chloride, a primary chloride that is incapable of undergoing elimination, and sodium isopropoxide:

$$\underset{\text{Sodium isopropoxide}}{(CH_3)_2CHONa} + \underset{\text{Benzyl chloride}}{C_6H_5CH_2Cl} \longrightarrow \underset{\text{Benzyl isopropyl ether (84\%)}}{(CH_3)_2CHOCH_2C_6H_5} + \underset{\text{Sodium chloride}}{NaCl}$$

The alternative synthetic route using the sodium salt of benzyl alcohol and an isopropyl halide would be much less effective, because of increased competition from elimination as the alkyl halide becomes more sterically hindered.

PROBLEM 16.7 Only one combination of alkyl halide and alkoxide is appropriate for the preparation of each of the following ethers by the Williamson ether synthesis. What is the correct combination in each case?

(*a*) CH_3CH_2O–cyclopentyl

(*b*) $CH_2{=}CHCH_2OCH(CH_3)_2$

(*c*) $(CH_3)_3COCH_2C_6H_5$

SAMPLE SOLUTION (*a*) The ether linkage of cyclopentyl ethyl ether involves a primary carbon and a secondary one. Choose the alkyl halide corresponding to the primary alkyl group, leaving the secondary alkyl group to arise from the alkoxide nucleophile.

$$\underset{\text{Sodium cyclopentanolate}}{C_5H_9ONa} + \underset{\text{Ethyl bromide}}{CH_3CH_2Br} \xrightarrow{S_N2} \underset{\text{Cyclopentyl ethyl ether}}{C_5H_9OCH_2CH_3}$$

The alternative combination, cyclopentyl bromide and sodium ethoxide, is not appropriate, since elimination will be the major reaction:

$$\underset{\text{Sodium ethoxide}}{CH_3CH_2ONa} + \underset{\text{Bromocyclopentane}}{C_5H_9Br} \xrightarrow{E2} \underset{\text{Ethanol}}{CH_3CH_2OH} + \underset{\text{Cyclopentene}}{C_5H_8}$$

(major products)

Both reactants in the Williamson ether synthesis usually originate in alcohol precursors. Sodium and potassium alkoxides are prepared by reaction of an alcohol with the appropriate metal, and alkyl halides are most commonly made from alcohols by reaction with a hydrogen halide (Section 4.9), thionyl chloride (Section 4.15), or phosphorus tribromide (Section 4.15). Alternatively, alkyl *p*-toluenesulfonates may be used in place of alkyl halides; alkyl *p*-toluenesulfonates are also prepared from alcohols as their immediate precursors (Section 8.15).

16.7 REACTIONS OF ETHERS: A REVIEW AND A PREVIEW

Up to this point, no reactions of dialkyl ethers have been presented. Indeed, ethers are one of the least reactive of the functional groups we shall study. It is this low level of reactivity, along with an ability to dissolve nonpolar substances, that makes ethers so often used as solvents when carrying out organic reactions. Nevertheless, most ethers are hazardous materials, and precautions must be taken when using them. Diethyl ether is extremely flammable and because of its high volatility can form explosive mixtures in air relatively quickly. Open flames must never be present in laboratories where diethyl ether is being used. Other low-molecular-weight ethers must also be treated as fire hazards.

PROBLEM 16.8 Combustion in air is, of course, a chemical property of ethers that is shared by many other organic compounds. Write balanced chemical equations for the complete combustion (in air) of diethyl ether.

A second dangerous property of ethers is the ease with which they undergo oxidation in air to form explosive peroxides. Air oxidation of diethyl ether proceeds according to the equation

$$\underset{\text{Diethyl ether}}{CH_3CH_2OCH_2CH_3} + \underset{\text{Oxygen}}{O_2} \longrightarrow \underset{\text{1-Ethoxyethyl hydroperoxide}}{\underset{\displaystyle HOO}{CH_3\overset{}{C}HOCH_2CH_3}}$$

The reaction is a free-radical one, and oxidation occurs at the carbon that bears the ether oxygen to form a hydroperoxide, a compound of the type ROOH. Hydroperoxides tend to be unstable and shock-sensitive. On standing, they form related peroxidic derivatives, which are also prone to violent decomposition. Air oxidation leads to peroxides within a few days if ethers are even briefly exposed to atmospheric oxygen. For this reason, one should never use old bottles of dialkyl ethers, and extreme care must be exercised in their disposal.

16.8 ACID-CATALYZED CLEAVAGE OF ETHERS

Just as the carbon-oxygen bond of alcohols is cleaved on reaction with hydrogen halides (Section 4.9), so too is an ether linkage broken:

$$\underset{\text{Alcohol}}{ROH} + \underset{\text{Hydrogen halide}}{HX} \longrightarrow \underset{\text{Alkyl halide}}{RX} + \underset{\text{Water}}{H_2O}$$

$$\underset{\text{Ether}}{\text{ROR}'} + \underset{\text{Hydrogen halide}}{\text{HX}} \longrightarrow \underset{\text{Alkyl halide}}{\text{RX}} + \underset{\text{Alcohol}}{\text{R}'\text{OH}}$$

The cleavage of ethers is normally carried out under conditions (excess hydrogen halide, heat) such that the alcohol formed as one of the original products is subsequently converted to an alkyl halide. Thus, the reaction typically leads to two alkyl halide molecules:

$$\underset{\text{Ether}}{\text{ROR}'} + \underset{\text{Hydrogen halide}}{2\text{HX}} \xrightarrow{\text{heat}} \underset{\text{Two alkyl halides}}{\text{RX} + \text{R}'\text{X}} + \underset{\text{Water}}{H_2O}$$

$$\underset{\textit{sec}\text{-Butyl methyl ether}}{CH_3CH(OCH_3)CH_2CH_3} \xrightarrow[\text{heat}]{\text{HBr}} \underset{\text{2-Bromobutane (81\%)}}{CH_3CH(Br)CH_2CH_3} + \underset{\text{Bromomethane}}{CH_3Br}$$

Cyclic ethers yield one molecule of a dihalide:

$$\underset{\text{Tetrahydrofuran}}{\text{(cyclic } C_4H_8O)} \xrightarrow[150°C]{\text{HI}} \underset{\text{1,4-Diiodobutane (65\%)}}{ICH_2CH_2CH_2CH_2I}$$

The order of hydrogen halide reactivity is HI > HBr ≫ HCl. Hydrogen fluoride is not effective.

PROBLEM 16.9 A series of dialkyl ethers was allowed to react with excess hydrogen bromide, with the following results. Identify the ether in each case.

(*a*) One ether gave a mixture of bromocyclopentane and 1-bromobutane.
(*b*) Another ether gave only benzyl bromide.
(*c*) A third ether gave one mole of 1,5-dibromopentane per mole of ether.

SAMPLE SOLUTION (*a*) In the reaction of dialkyl ethers with excess hydrogen bromide, each alkyl group of the ether function is cleaved and forms an alkyl bromide. Since bromocyclopentane and 1-bromobutane are the products, the starting ether must be butyl cyclopentyl ether.

$$\underset{\text{Butyl cyclopentyl ether}}{C_5H_9\text{—}OCH_2CH_2CH_2CH_3} \xrightarrow[\text{heat}]{\text{HBr}} \underset{\text{Bromocyclopentane}}{C_5H_9\text{—}Br} + \underset{\text{1-Bromobutane}}{CH_3CH_2CH_2CH_2Br}$$

A mechanism for the cleavage of diethyl ether by hydrogen bromide is outlined in Figure 16.4. The key step is an S_N2-like attack on a dialkyloxonium ion by bromide (step 2). The following problem asks you to explore an alternative mechanism.

Overall Reaction:

$$CH_3CH_2OCH_2CH_3 + 2\,HBr \xrightarrow{\text{heat}} 2\,CH_3CH_2Br + H_2O$$

Diethyl ether | Hydrogen bromide | Ethyl bromide | Water

Mechanism:

Step 1: Proton transfer to the oxygen of the ether to give a dialkyloxonium ion.

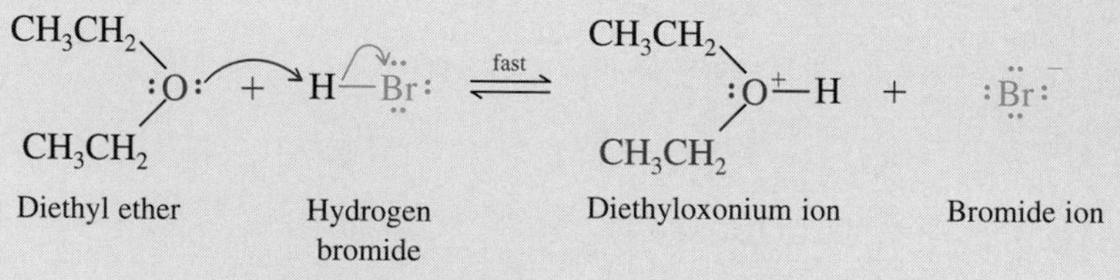

$$(CH_3CH_2)_2\ddot{O}: + H{-}\ddot{B}r: \overset{\text{fast}}{\rightleftharpoons} (CH_3CH_2)_2\overset{+}{O}{-}H + :\ddot{B}r:^-$$

Diethyl ether | Hydrogen bromide | Diethyloxonium ion | Bromide ion

Step 2: Nucleophilic attack of the halide anion on carbon of the dialkyloxonium ion. This step gives one molecule of an alkyl halide and one molecule of an alcohol.

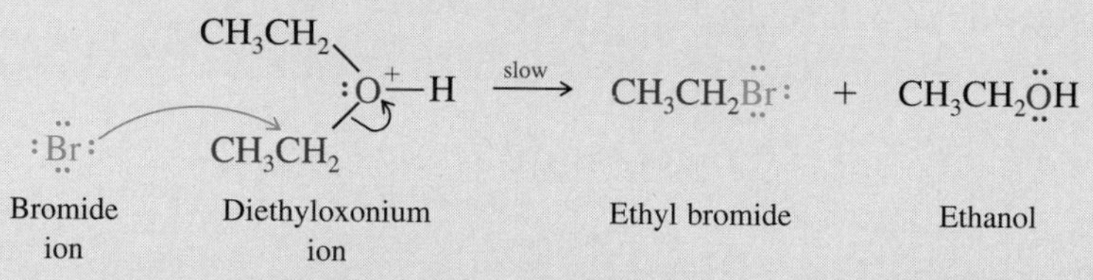

$$:\ddot{B}r:^- + (CH_3CH_2)_2\overset{+}{O}{-}H \xrightarrow{\text{slow}} CH_3CH_2\ddot{B}r: + CH_3CH_2\ddot{O}H$$

Bromide ion | Diethyloxonium ion | Ethyl bromide | Ethanol

Steps 3 and Step 4: These two steps do not involve an ether at all. They correspond to those in which an alcohol is converted to an alkyl halide (Sections 4.9-4.14).

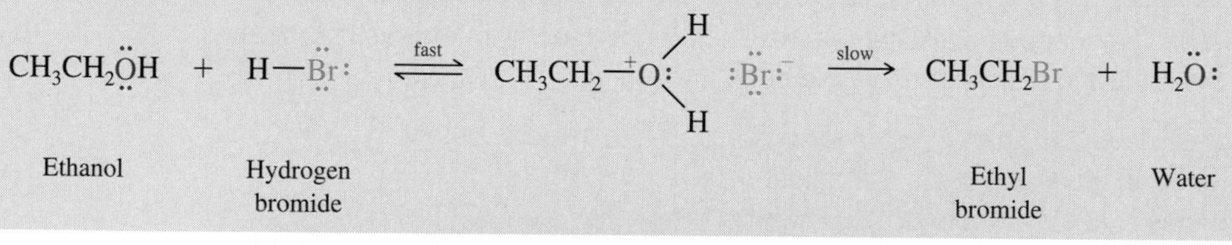

$$CH_3CH_2\ddot{O}H + H{-}\ddot{B}r: \overset{\text{fast}}{\rightleftharpoons} CH_3CH_2{-}\overset{+}{O}H_2 \;\; :\ddot{B}r:^- \xrightarrow{\text{slow}} CH_3CH_2Br + H_2\ddot{O}:$$

Ethanol | Hydrogen bromide | Ethyl bromide | Water

FIGURE 16.4 The mechanism for the cleavage of ethers by hydrogen halides, using the reaction of diethyl ether with hydrogen bromide as an example.

PROBLEM 16.10 Di-*tert*-butyl ether is rapidly cleaved, even by hydrogen chloride at room temperature:

$$(CH_3)_3COC(CH_3)_3 \xrightarrow{\text{HCl}} 2(CH_3)_3CCl$$

Di-*tert*-butyl ether | *tert*-Butyl chloride (95%)

The dialkyloxonium ion formed by protonation of di-*tert*-butyl ether is much too crowded to be attacked directly by chloride ion. Suggest an alternative process by which *tert*-butyl chloride is formed from di-*tert*-butyloxonium ion.

With mixed ethers of the type ROR′, the question arises which carbon-oxygen bond is broken first. While some studies have been carried out on this point of mechanistic detail, it is not one that we need examine at our level of study.

16.9 PREPARATION OF EPOXIDES: A REVIEW AND A PREVIEW

There are two main laboratory methods for the preparation of epoxides:

1. Epoxidation of alkenes by reaction with peroxy acids
2. Base-promoted ring closure of vicinal halohydrins

Epoxidation of alkenes was discussed in Section 6.18 and is represented by the general equation

$$\underset{\text{Alkene}}{R_2C{=}CR_2} + \underset{\text{Peroxy acid}}{R'\overset{O}{\overset{\|}{C}}OOH} \longrightarrow \underset{\text{Epoxide}}{R_2C\underset{O}{-}CR_2} + \underset{\text{Carboxylic acid}}{R'\overset{O}{\overset{\|}{C}}OH}$$

The reaction is easy to carry out, and yields are usually high. Epoxidation is a stereospecific syn addition.

$$\underset{(E)\text{-1,2-Diphenylethene}}{(C_6H_5)HC{=}CH(C_6H_5)} + \underset{\text{Peroxyacetic acid}}{CH_3\overset{O}{\overset{\|}{C}}OOH} \longrightarrow \underset{\substack{trans\text{-2,3-Diphenyloxirane} \\ (78\text{–}83\%)}}{(C_6H_5)HC\underset{O}{-}CH(C_6H_5)} + \underset{\text{Acetic acid}}{CH_3\overset{O}{\overset{\|}{C}}OH}$$

The following section describes the preparation of epoxides by the base-promoted ring closure of vicinal halohydrins. Since vicinal halohydrins are customarily prepared from alkenes (Section 6.17), both methods, epoxidation using peroxy acids and ring closure of halohydrins, are based on alkenes as the starting materials for preparing epoxides.

16.10 CONVERSION OF VICINAL HALOHYDRINS TO EPOXIDES

The formation of vicinal halohydrins from alkenes was described in Section 6.17. Halohydrins are readily converted to epoxides on treatment with base:

$$\underset{\text{Alkene}}{R_2C{=}CR_2} \xrightarrow[H_2O]{X_2} \underset{\text{Vicinal halohydrin}}{R_2\underset{HO}{\underset{|}{C}}-\underset{X}{\underset{|}{C}}R_2} \xrightarrow{HO^-} \underset{\text{Epoxide}}{R_2C\underset{O}{-}CR_2}$$

Reaction with base brings the alcohol function of the halohydrin into equilibrium with its corresponding alkoxide:

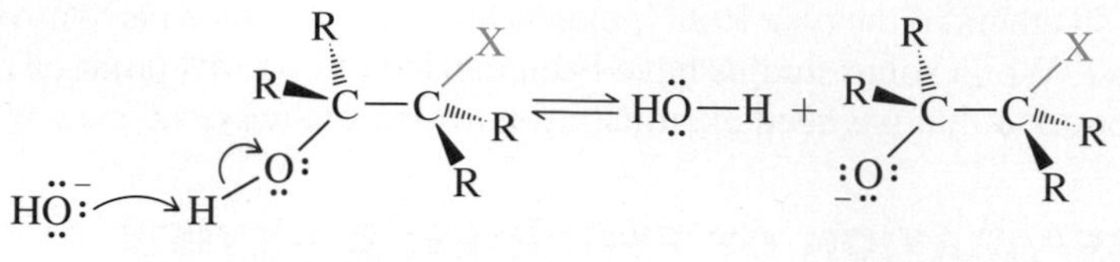

R X R X
R C—C R ⇌ HÖ—H + R C—C R
HÖ: H :O: R :O: R

Vicinal halohydrin

Next, in what amounts to an *intramolecular* Williamson ether synthesis, the alkoxide oxygen attacks the carbon that bears the halide leaving group, giving an epoxide. As in other nucleophilic substitutions, the nucleophile approaches carbon from the side opposite the bond to the leaving group:

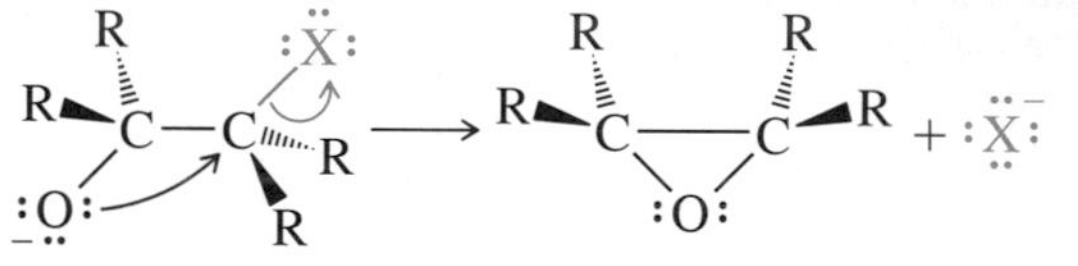

R :X: R R
R C—C R ⟶ R C——C R + :X:
:O: R :O:

Epoxide

H OH H
NaOH / H_2O ⟶ O
H Br H

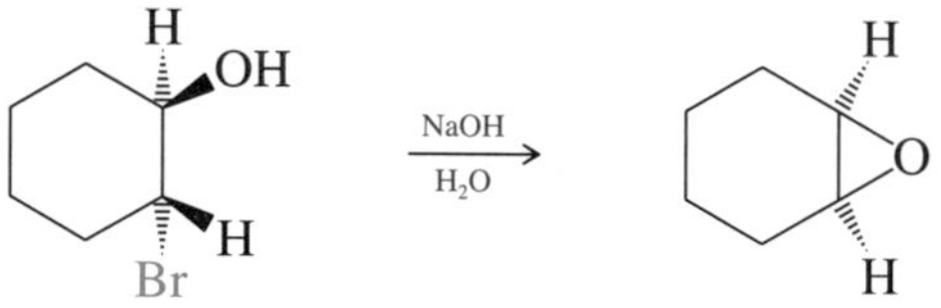

trans-2-Bromocyclohexanol 1,2-Epoxycyclohexane (81%)

Overall, the stereospecificity of this method for epoxide preparation is the same as that observed in peroxy acid oxidation of alkenes. Substituents that are cis to each other in the alkene remain cis in the epoxide. This is because formation of the bromohydrin involves anti addition and the ensuing intramolecular nucleophilic substitution reaction takes place with inversion of configuration at the carbon that bears the halide leaving group.

H_3C Br H_3C
H anti addition ⟶ H_3C H inversion of configuration ⟶ H_3C H
H_3C H_3C H H
H OH O

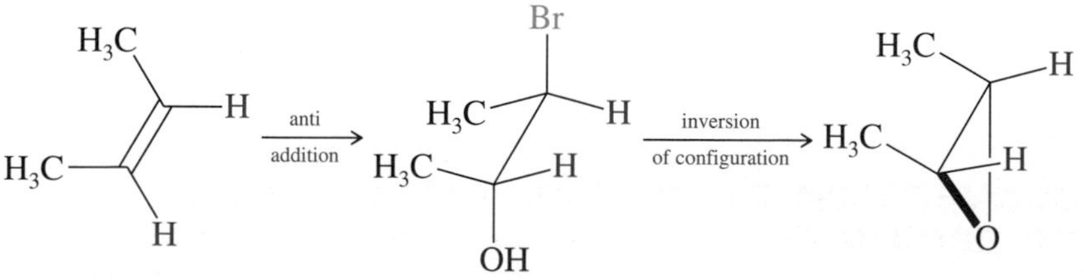

(*Z*)-2-Butene
(*cis*-2-butene)

cis-2,3-Epoxybutane

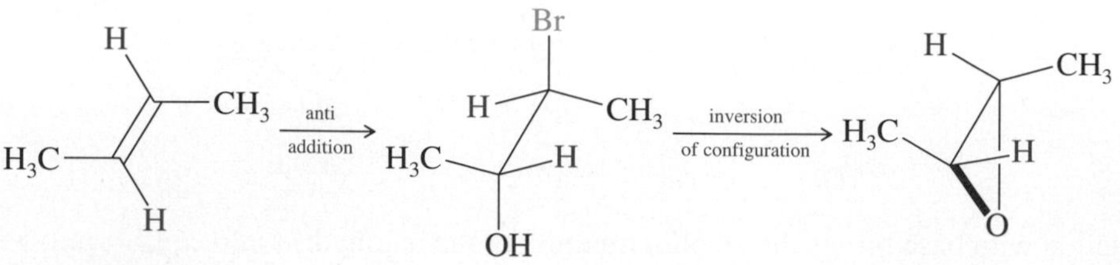

H Br H CH_3
CH_3 anti addition ⟶ H CH_3 inversion of configuration ⟶ H_3C H
H_3C H_3C H
H OH O

(*E*)-2-Butene
(*trans*-2-butene)

trans-2,3-Epoxybutane

PROBLEM 16.11 Is either of the epoxides formed in the preceding reactions chiral? Is either epoxide optically active when prepared from the alkene by this method?

About 2×10^9 lb/yr of 1,2-epoxypropane is produced in the United States as an intermediate in the preparation of various polymeric materials, including polyurethane plastics and foams and polyester resins. A large fraction of the 1,2-epoxypropane is made from propene through formation and base-promoted ring closure of the chlorohydrin.

16.11 REACTIONS OF EPOXIDES: A REVIEW AND A PREVIEW

The chemical property that most distinguishes epoxides is their far greater reactivity toward nucleophilic reagents compared with that exhibited by simple ethers. Epoxides react rapidly with nucleophiles under conditions in which other ethers are inert. This enhanced reactivity results from the ring strain of epoxides. Reactions that lead to ring opening relieve this strain.

Angle strain is the main source of strain in epoxides, but torsional strain that results from the eclipsing of bonds on adjacent carbons is also present. Both kinds of strain are relieved when a ring-opening reaction occurs.

We saw an example of nucleophilic ring opening of epoxides in Section 15.4, where the reaction of Grignard reagents with ethylene oxide was described as a synthetic route to primary alcohols:

$$RMgX + H_2C\underset{O}{\diagdown\!\diagup}CH_2 \xrightarrow[2.\ H_3O^+]{1.\ \text{diethyl ether}} RCH_2CH_2OH$$

Grignard reagent — Ethylene oxide — Primary alcohol

$$C_6H_5CH_2MgCl + H_2C\underset{O}{\diagdown\!\diagup}CH_2 \xrightarrow[2.\ H_3O^+]{1.\ \text{diethyl ether}} C_6H_5CH_2CH_2CH_2OH$$

Benzylmagnesium chloride — Ethylene oxide — 3-Phenyl-1-propanol (71%)

Nucleophiles other than Grignard reagents also lead to ring opening of epoxides. There are two fundamental ways in which these reactions are carried out. The first (Section 16.12) involves anionic nucleophiles in neutral or basic solution.

$$Y{:}^- + R_2C\underset{\ddot{O}}{\diagdown\!\diagup}CR_2 \longrightarrow R_2C(Y){-}CR_2({:}\ddot{O}{:}^-) \xrightarrow{H_2O} R_2C(Y){-}CR_2({:}\ddot{O}H)$$

Nucleophile — Epoxide — Alkoxide ion — Alcohol

These reactions are usually performed in water or alcohols as solvents, and the alkoxide ion intermediate is rapidly transformed to an alcohol by proton transfer.

Nucleophilic ring-opening reactions of epoxides may also be carried out under conditions of acid catalysis. Here the nucleophile is not an anion but rather a solvent molecule.

$$\text{HY:} + R_2C\underset{\ddot{\underset{..}{O}}}{\overline{\quad}}CR_2 \xrightarrow{H^+} R_2\overset{\text{Y:}}{\overset{|}{C}}-\underset{\underset{:\ddot{O}H}{|}}{CR_2}$$

Epoxide Alcohol

Acid-catalyzed ring opening of epoxides is discussed in Section 16.13.

There is an important difference in the regiochemistry of ring-opening reactions of epoxides depending on the reaction conditions. Unsymmetrically substituted epoxides tend to react with anionic nucleophiles at the less hindered carbon of the ring. Under conditions of acid catalysis, however, the more highly substituted carbon is attacked.

Nucleophiles attack here when reaction is catalyzed by acids. → RCH—CH_2 (O bridging) ← Anionic nucleophiles attack here.

The underlying reasons for this difference in site of nucleophilic attack will be explained in Section 16.13.

16.12 NUCLEOPHILIC RING-OPENING REACTIONS OF EPOXIDES

Ethylene oxide is a very reactive substance. It reacts rapidly and exothermically with anionic nucleophiles to yield 2-substituted derivatives of ethanol by cleaving the carbon-oxygen bond of the ring:

$$H_2C\underset{O}{\overline{\quad}}CH_2 \xrightarrow[\text{ethanol-water, 0°C}]{KSCH_2CH_2CH_2CH_3} CH_3CH_2CH_2CH_2SCH_2CH_2OH$$

Ethylene oxide (oxirane) 2-(Butylthio)ethanol (99%)

PROBLEM 16.12 What is the principal organic product formed in the reaction of ethylene oxide with each of the following?

(*a*) Sodium cyanide (NaCN) in aqueous ethanol
(*b*) Sodium azide (NaN_3) in aqueous ethanol
(*c*) Sodium hydroxide (NaOH) in water
(*d*) Phenyllithium (C_6H_5Li) in ether, followed by addition of dilute sulfuric acid
(*e*) 1-Butynylsodium ($CH_3CH_2C{\equiv}CNa$) in liquid ammonia

SAMPLE SOLUTION (*a*) Sodium cyanide is a source of the nucleophilic cyanide anion. Cyanide ion attacks ethylene oxide, opening the ring and forming 2-cyanoethanol:

$$H_2C\underset{O}{\overline{\quad}}CH_2 \xrightarrow[\text{ethanol-water}]{NaCN} NCCH_2CH_2OH$$

Ethylene oxide 2-Cyanoethanol

Nucleophilic ring opening of epoxides has many of the features of an S_N2 reaction. Inversion of configuration is observed at the carbon at which substitution occurs.

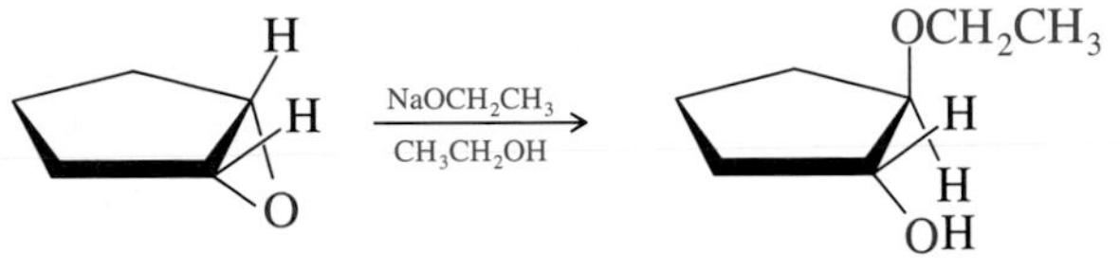

$$\text{1,2-Epoxycyclopentane} \xrightarrow[CH_3CH_2OH]{NaOCH_2CH_3} \textit{trans}\text{-2-Ethoxycyclopentanol (67\%)}$$

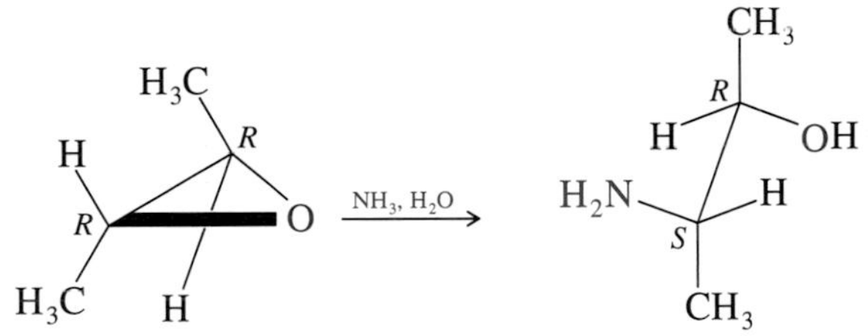

$$(2R,3R)\text{-2,3-Epoxybutane} \xrightarrow{NH_3,\ H_2O} (2R,3S)\text{-3-Amino-2-butanol (70\%)}$$

Unsymmetrical epoxides are preferentially attacked at the less substituted, less sterically hindered carbon of the ring:

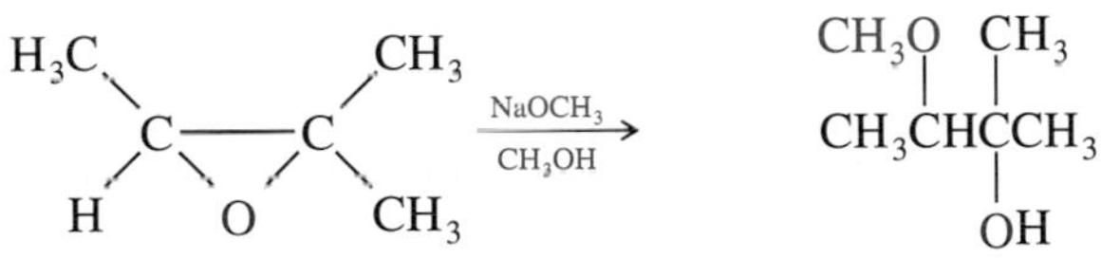

$$\text{2,2,3-Trimethyloxirane} \xrightarrow[CH_3OH]{NaOCH_3} \underset{\text{3-Methoxy-2-methyl-2-butanol (53\%)}}{CH_3CH(OCH_3)C(CH_3)(OH)CH_3}$$

PROBLEM 16.13 Given the starting material 1-methyl-1,2-epoxycyclopentane, of absolute configuration as shown, decide which one of the compounds A through C correctly represents the product of its reaction with sodium methoxide in methanol.

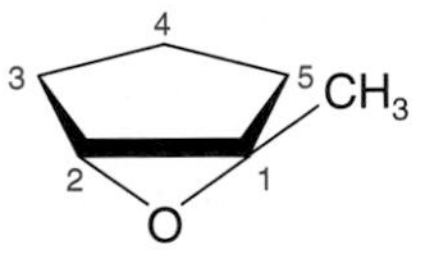

1-Methyl-1,2-epoxycyclopentane 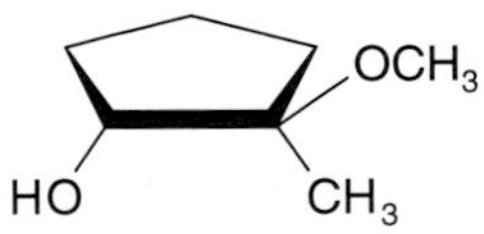   Compound A 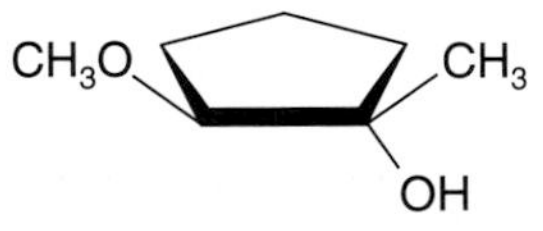   Compound B 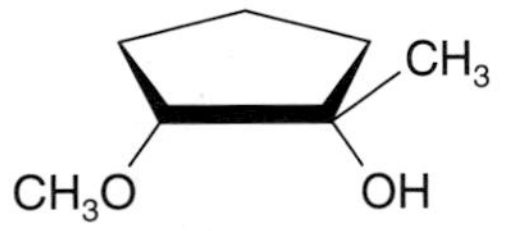   Compound C

The principles of nucleophilic substitution lead to the picture of epoxide ring opening shown in Figure 16.5. The nucleophile attacks the less crowded carbon from the side opposite the carbon-oxygen bond. Bond formation with the nucleophile accompanies carbon-oxygen bond breaking, and a substantial portion of the strain in the three-membered ring is relieved as it begins to open in the transition state. The initial product of nucleophilic substitution is an alkoxide anion, which rapidly abstracts a proton from the solvent to give a β-substituted alcohol as the isolated product.

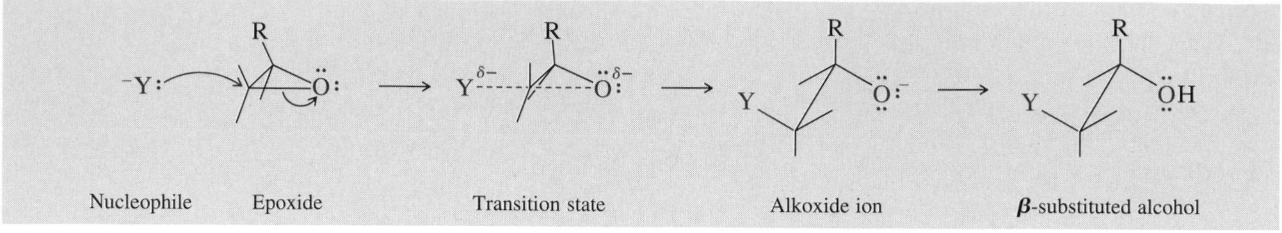

Nucleophile Epoxide Transition state Alkoxide ion β-substituted alcohol

FIGURE 16.5 Nucleophilic ring opening of an epoxide.

The reaction of Grignard reagents with epoxides is regioselective in the same sense, in that attack occurs at the less substituted ring carbon:

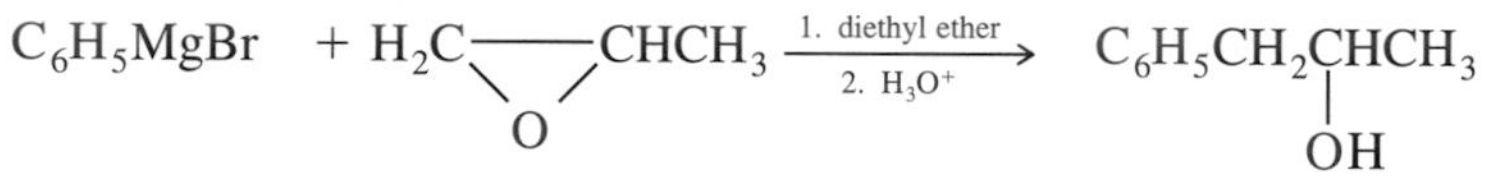

$$C_6H_5MgBr + H_2C\overset{O}{—}CHCH_3 \xrightarrow[2.\ H_3O^+]{1.\ \text{diethyl ether}} C_6H_5CH_2\underset{OH}{C}HCH_3$$

Phenylmagnesium bromide | 1,2-Epoxypropane | 1-Phenyl-2-propanol (60%)

Epoxides are reduced to alcohols on treatment with lithium aluminum hydride. Hydride is transferred to the less crowded carbon.

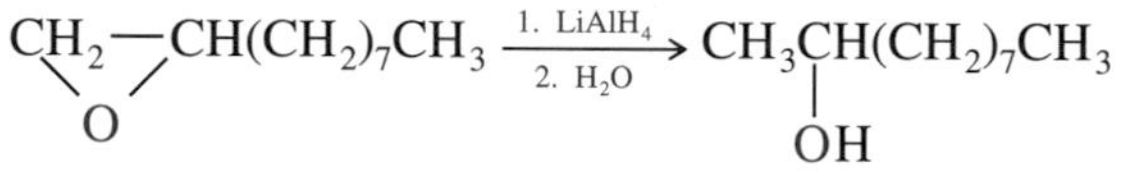

$$CH_2\overset{O}{—}CH(CH_2)_7CH_3 \xrightarrow[2.\ H_2O]{1.\ LiAlH_4} CH_3\underset{OH}{C}H(CH_2)_7CH_3$$

1,2-Epoxydecane | 2-Decanol (90%)

Epoxidation of an alkene, followed by lithium aluminum hydride reduction of the resulting epoxide, gives the same alcohol that would be obtained by acid-catalyzed hydration (Section 6.9) or oxymercuration-demercuration (Section 14.14) of the alkene.

16.13 ACID-CATALYZED RING-OPENING REACTIONS OF EPOXIDES

As we saw in the preceding section, nucleophilic ring opening of ethylene oxide yields 2-substituted derivatives of ethanol. The reactions previously described involved nucleophilic attack on the carbon of the ring under neutral or basic conditions. Other nucleophilic ring-opening reactions of epoxides likewise give 2-substituted derivatives of ethanol but either involve an acid as a reactant or occur under conditions of acid catalysis:

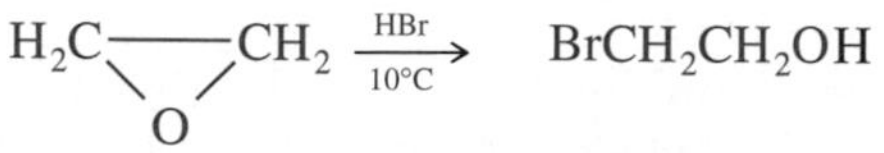

$$H_2C\overset{O}{—}CH_2 \xrightarrow[10°C]{HBr} BrCH_2CH_2OH$$

Ethylene oxide | 2-Bromoethanol (87–92%)

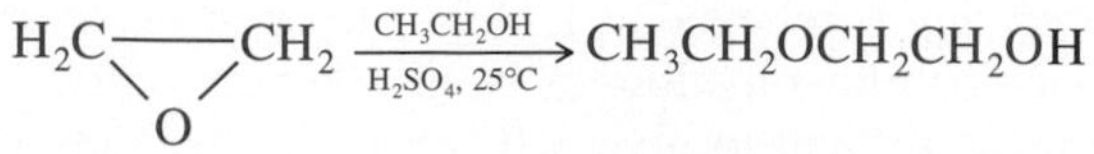

$$H_2C\overset{O}{—}CH_2 \xrightarrow[H_2SO_4,\ 25°C]{CH_3CH_2OH} CH_3CH_2OCH_2CH_2OH$$

Ethylene oxide | 2-Ethoxyethanol (85%)

A third example is the industrial preparation of ethylene glycol ($HOCH_2CH_2OH$) by hydrolysis of ethylene oxide in dilute sulfuric acid. This reaction and its mechanism are presented in Figure 16.6. As Figure 16.6 shows, there is a significant difference between the ring openings of epoxides discussed in the preceding section and the acid-catalyzed ones described here. Under conditions of acid catalysis, the species that is attacked by the nucleophile is not the epoxide itself, but rather its conjugate acid. The transition state for ring opening has a fair measure of carbocation character. Breaking of the ring carbon-oxygen bond is more advanced than formation of the bond to the nucleophile.

Transition state for attack by water on conjugate acid of ethylene oxide

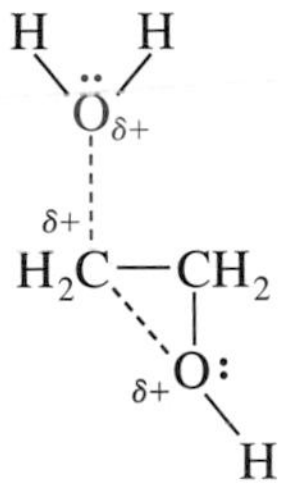

H H
O δ+
δ+
H_2C—CH_2
δ+ O:
H

Because *carbocation* character develops at the transition state, substitution is favored at the carbon that can better support a developing positive charge. Thus, in contrast to the reaction of epoxides with relatively basic nucleophiles, in which S_N2-like attack is faster at the less crowded carbon of the three-membered ring, acid catalysis promotes substitution at the position that bears the greater number of alkyl groups:

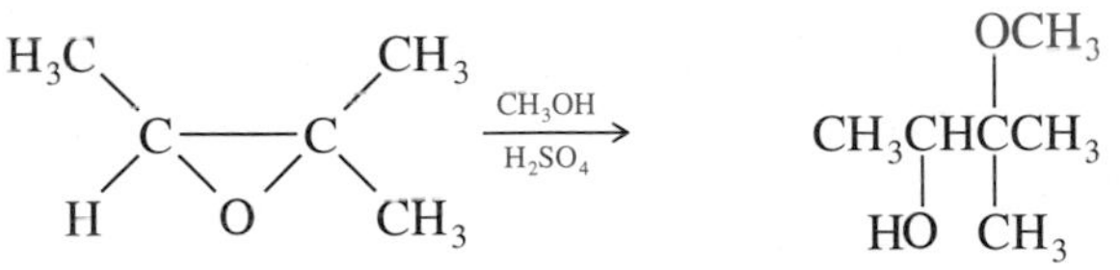

H_3C, H, CH_3, CH_3, O — $\xrightarrow[H_2SO_4]{CH_3OH}$ — $CH_3CHC(OCH_3)CH_3$ with HO and CH_3

2,2,3-Trimethyloxirane — 3-Methoxy-3-methyl-2-butanol (76%)

While nucleophilic participation at the transition state is slight, it is enough to ensure that substitution proceeds with inversion of configuration.

$\xrightarrow{HBr}$

1,2-Epoxycyclohexane — *trans*-2-Bromocyclohexanol (73%)

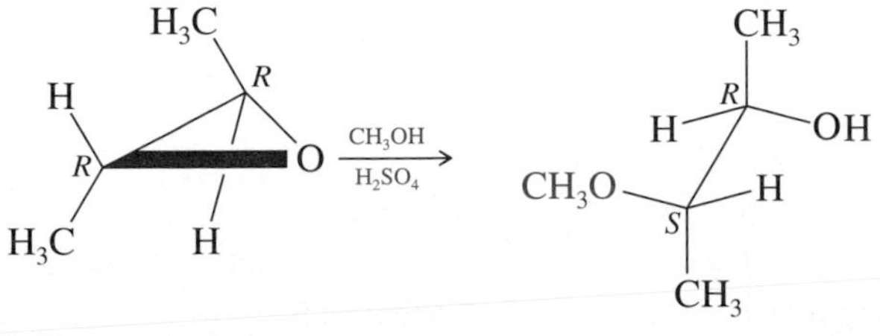

$\xrightarrow[H_2SO_4]{CH_3OH}$

(2*R*,3*R*)-2,3-Epoxybutane — (2*R*,3*S*)-3-Methoxy-2-butanol (57%)

Overall Reaction:

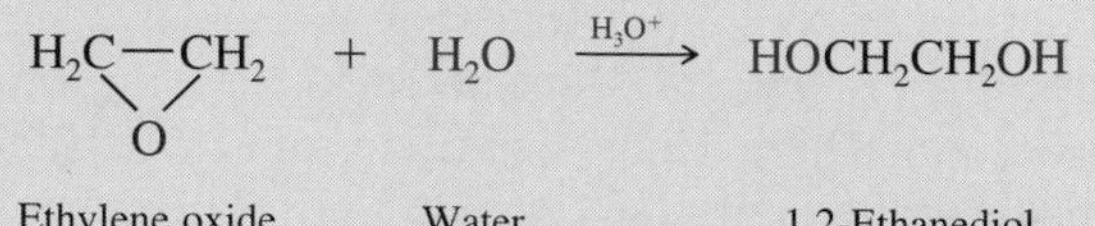

$$\underset{\text{Ethylene oxide}}{H_2C\overset{O}{—}CH_2} + \underset{\text{Water}}{H_2O} \xrightarrow{H_3O^+} \underset{\text{1,2-Ethanediol}}{HOCH_2CH_2OH}$$

Mechanism:

Step 1: Proton transfer to the oxygen of the epoxide to give an oxonium ion.

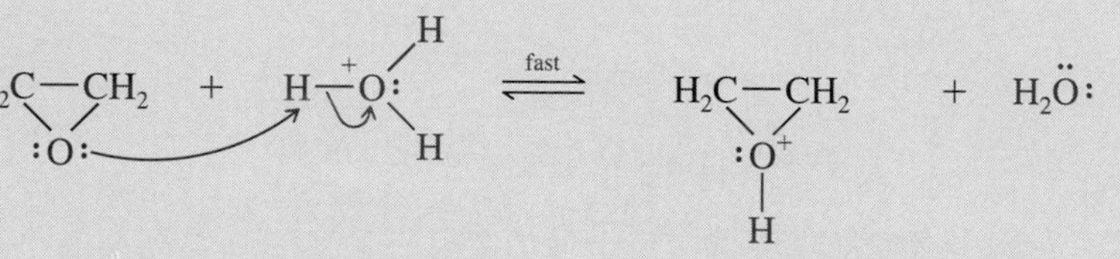

$$\underset{\text{Ethylene oxide}}{H_2C\overset{:O:}{—}CH_2} + \underset{\text{Hydronium ion}}{H—\overset{+}{O}H_2} \underset{}{\overset{\text{fast}}{\rightleftharpoons}} \underset{\text{Ethyleneoxonium ion}}{H_2C\overset{:O^+—H}{—}CH_2} + \underset{\text{Water}}{H_2\ddot{O}:}$$

Step 2: Nucleophilic attack by water on carbon of the oxonium ion. The carbon-oxygen bond of the ring is broken in this step and the ring opens.

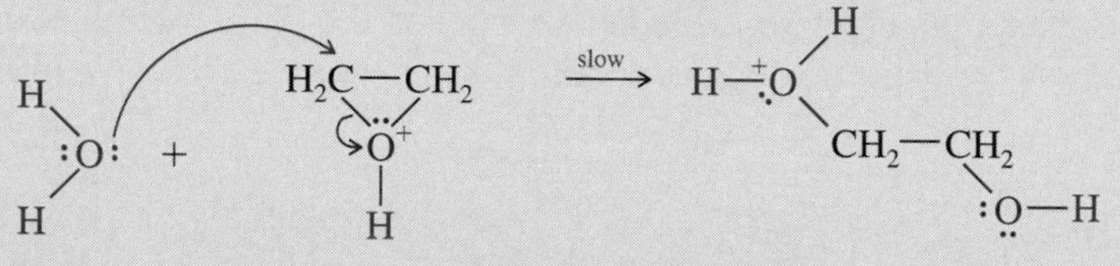

$$\underset{\text{Water}}{H_2\ddot{O}:} + \underset{\text{Ethyleneoxonium ion}}{H_2C\overset{O^+—H}{—}CH_2} \xrightarrow{\text{slow}} \underset{\text{2-Hydroxyethyloxonium ion}}{H_2\overset{+}{O}—CH_2—CH_2—\ddot{O}—H}$$

Step 3: Proton transfer to water completes the reaction and regenerates the acid catalyst.

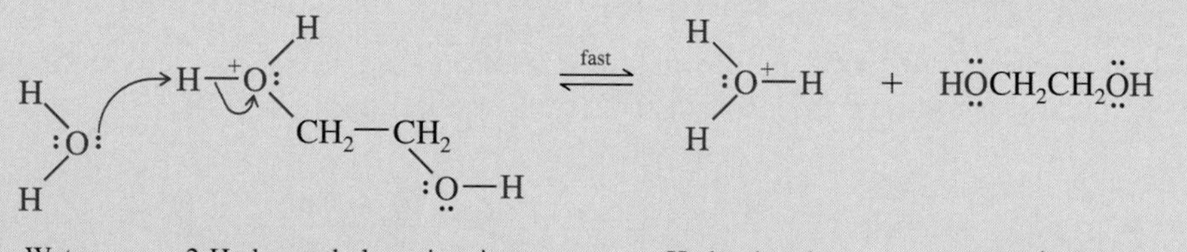

$$\underset{\text{Water}}{H_2\ddot{O}:} + \underset{\text{2-Hydroxyethyloxonium ion}}{H_2\overset{+}{O}—CH_2—CH_2—\ddot{O}—H} \overset{\text{fast}}{\rightleftharpoons} \underset{\text{Hydronium ion}}{H_2\overset{+}{O}—H} + \underset{\text{1,2-Ethanediol}}{H\ddot{O}CH_2CH_2\ddot{O}H}$$

FIGURE 16.6 The mechanism for the acid-catalyzed nucleophilic ring opening of ethylene oxide by water.

PROBLEM 16.14 Which product, compound A, B, or C, would you expect to be formed when 1-methyl-1,2-epoxycyclopentane of the absolute configuration shown is allowed to stand in methanol containing a few drops of sulfuric acid? Compare your answer with that given for Problem 16.13.

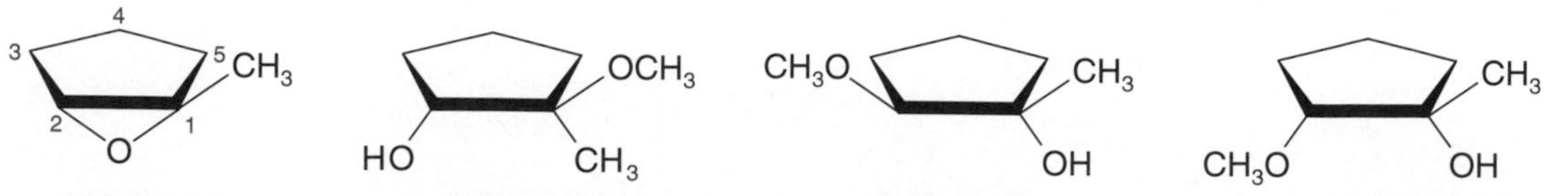

1-Methyl-1,2-epoxycyclopentane (ring carbons numbered 1–5, O bridging C-1 and C-2, CH_3 on C-1)

Compound A: OCH_3, CH_3, HO

Compound B: CH_3O, CH_3, OH

Compound C: CH_3, CH_3O, OH

A method for achieving net anti hydroxylation of alkenes combines two stereospecific processes: epoxidation of the double bond and hydrolysis of the derived epoxide.

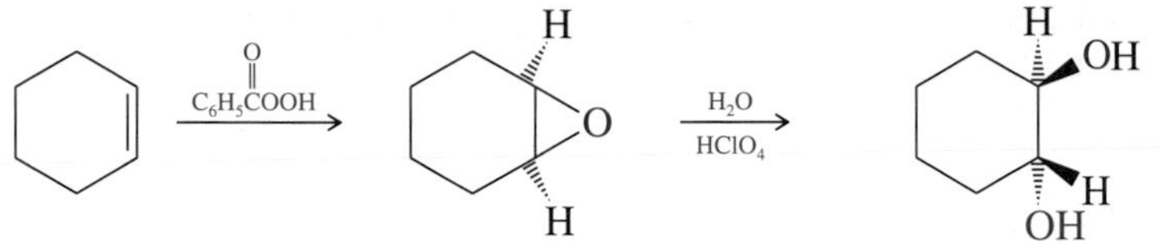

Cyclohexene $\xrightarrow{C_6H_5COOH}$ 1,2-Epoxycyclohexane $\xrightarrow[HClO_4]{H_2O}$ *trans*-1,2-Cyclohexanediol (80%)

PROBLEM 16.15 Which alkene, *cis*-2-butene or *trans*-2-butene, would you choose in order to prepare *meso*-2,3-butanediol by epoxidation followed by acid-catalyzed hydrolysis? Which alkene would yield *meso*-2,3,-butanediol by osmium tetraoxide hydroxylation?

16.14 EPOXIDES IN BIOLOGICAL PROCESSES

Many naturally occurring substances are epoxides. You have seen two examples of such compounds already in disparlure, the sex attractant of the gypsy moth (Section 6.18), and in the carcinogenic epoxydiol formed from benzo[a]pyrene (Section 11.10). In most cases, epoxides are biosynthesized by the enzyme-catalyzed transfer of one of the oxygen atoms of an O_2 molecule to an alkene. Since only one of the atoms of O_2 is transferred to the substrate, the enzymes that catalyze such transfers are classified as *monooxygenases.* A biological reducing agent, usually the coenzyme NADH (Section 15.11), is required as well.

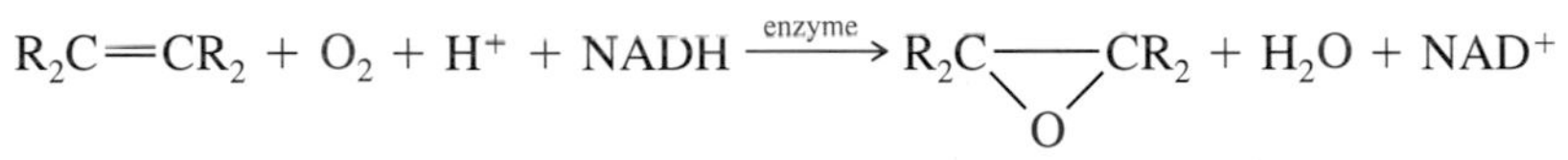

$$R_2C{=}CR_2 + O_2 + H^+ + NADH \xrightarrow{\text{enzyme}} R_2C\overset{O}{—}CR_2 + H_2O + NAD^+$$

A prominent example of such a reaction is the biological epoxidation of the polyene squalene.

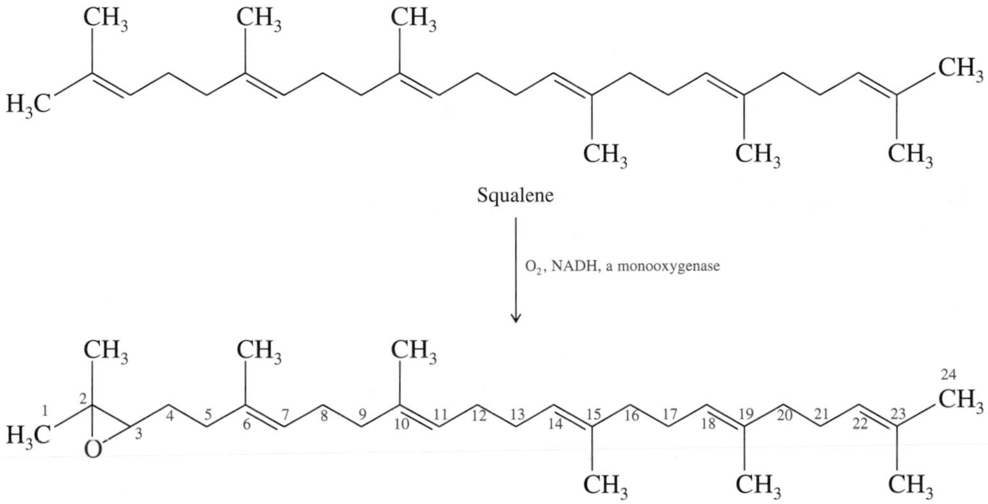

Squalene

$\downarrow$ O_2, NADH, a monooxygenase

Squalene 2,3-epoxide

The reactivity of epoxides toward nucleophilic ring opening is responsible for one of the biological roles they play. Squalene 2,3-epoxide, for example, is the biological precursor to cholesterol and the steroid hormones, including testosterone, progesterone, estrone, and cortisone. The pathway from squalene 2,3-epoxide to these compounds is triggered by epoxide ring opening and will be described in Chapter 26.

16.15 PREPARATION OF SULFIDES

Sulfides, compounds of the type RSR′, are prepared by nucleophilic substitution reactions. Treatment of a primary or secondary alkyl halide with an alkanethiolate ion (RS^-) gives a sulfide:

$$\underset{\text{Sodium alkanethiolate}}{R\ddot{S}:^- \ Na^+} + \underset{\text{Alkyl halide}}{R'—\ddot{X}:} \xrightarrow{S_N2} \underset{\text{Sulfide}}{R\ddot{S}R'} + \underset{\text{Sodium halide}}{Na^+ :\ddot{X}:^-}$$

$$\underset{\text{3-Chloro-1-butene}}{CH_3\underset{\displaystyle Cl}{\underset{|}{C}}HCH{=}CH_2} \xrightarrow[\text{methanol}]{NaSCH_3} \underset{\text{Methyl 1-methylallyl sulfide (62\%)}}{CH_3\underset{\displaystyle SCH_3}{\underset{|}{C}}HCH{=}CH_2}$$

K_a for CH_3SH is 1.8×10^{-11} ($pK_a = 10.7$).

It is not necessary to prepare and isolate the sodium alkanethiolate in a separate operation. Because thiols are more acidic than water, they are quantitatively converted to their alkanethiolate anions by sodium hydroxide. Thus, all that is normally done is to add a thiol to sodium hydroxide in a suitable solvent (water or an alcohol) followed by the alkyl halide.

PROBLEM 16.16 The *p*-toluenesulfonate derived from (*R*)-2-octanol and *p*-toluenesulfonyl chloride was allowed to react with sodium benzenethiolate (C_6H_5SNa). Give the structure, including stereochemistry and the appropriate *R* or *S* descriptor, of the product.

16.16 OXIDATION OF SULFIDES: SULFOXIDES AND SULFONES

We saw in Section 15.14 that thiols differ from alcohols in respect to their behavior toward oxidation. Similarly, sulfides differ from ethers in their behavior toward oxidizing agents. While ethers tend to undergo oxidation at carbon to give hydroperoxides (Section 16.7), sulfides are oxidized at sulfur to give **sulfoxides.** If the oxidizing agent is strong enough and present in excess, oxidation can proceed further to give **sulfones.**

Third-row elements such as sulfur can expand their valence shell beyond eight electrons, and so sulfur-oxygen bonds in sulfoxides and sulfones are sometimes represented as double bonds.

$$\underset{\text{Sulfide}}{R—\ddot{S}—R'} \xrightarrow{\text{oxidize}} \underset{\text{Sulfoxide}}{R—\overset{\displaystyle :\ddot{O}:^-}{\overset{|}{\ddot{S}^{+}}}—R'} \xrightarrow{\text{oxidize}} \underset{\text{Sulfone}}{R—\overset{\displaystyle :\ddot{O}:^-}{\overset{|}{\underset{\displaystyle :\ddot{O}:^-}{\underset{|}{S^{2+}}}}}—R'}$$

When the desired product is a sulfoxide, sodium metaperiodate ($NaIO_4$) is an ideal reagent. It oxidizes sulfides to sulfoxides in high yield but shows no tendency to oxidize sulfoxides to sulfones.

$$C_6H_5\ddot{S}CH_3 + NaIO_4 \xrightarrow{\text{water}} C_6H_5\overset{+}{S}(\ddot{O}^-)CH_3 + NaIO_3$$

Methyl phenyl sulfide | Sodium metaperiodate | Methyl phenyl sulfoxide (91%) | Sodium iodate

Peroxy acids, usually in dichloromethane as the solvent, are also reliable reagents for converting sulfides to sulfoxides.

One equivalent of a peroxy acid or of hydrogen peroxide converts sulfides to sulfoxides; two equivalents gives the corresponding sulfone.

$$C_6H_5\ddot{S}CH{=}CH_2 + 2H_2O_2 \xrightarrow{\text{acetic acid}} C_6H_5\overset{2+}{S}(\ddot{O}^-)_2CH{=}CH_2 + 2H_2O$$

Phenyl vinyl sulfide | Hydrogen peroxide | Phenyl vinyl sulfone (74–78%) | Water

---

PROBLEM 16.17 The bonds to sulfur are arranged in a trigonal pyramidal geometry in sulfoxides and in a tetrahedral geometry in sulfones. Is phenyl vinyl sulfoxide chiral? What about phenyl vinyl sulfone?

Oxidation of sulfides occurs in living systems as well. Among naturally occurring sulfoxides, one that has received recent attention is *sulforaphane,* which is present in broccoli and other vegetables. Sulforaphane holds promise as a potential anticancer agent because, unlike most anticancer drugs, which act by killing rapidly dividing tumor cells faster than they kill normal cells, sulforaphane is nontoxic and may simply inhibit the formation of tumors.

$$CH_3\overset{+}{S}(O^-)CH_2CH_2CH_2N{=}C{=}S$$

Sulforaphane

16.17 ALKYLATION OF SULFIDES: SULFONIUM SALTS

Sulfur is more nucleophilic than oxygen (Section 8.8), and sulfides react with alkyl halides much faster than do ethers. The products of these reactions, called **sulfonium salts,** are also more stable than the corresponding oxygen analogs.

$$RR'\ddot{S}: + R''{-}\ddot{X}: \xrightarrow{S_N2} RR'\overset{+}{S}{-}R'' \; :\ddot{X}:^-$$

Sulfide | Alkyl halide | Sulfonium salt

$$CH_3(CH_2)_{10}CH_2SCH_3 + CH_3I \longrightarrow CH_3(CH_2)_{10}CH_2\overset{+}{S}(CH_3)CH_3\ I^-$$

Dodecyl methyl sulfide — Methyl iodide — Dodecyldimethylsulfonium iodide

PROBLEM 16.18 What other combination of alkyl halide and sulfide will yield the same sulfonium salt shown in the preceding example? Predict which combination will yield the sulfonium salt at the faster rate.

The *S* in *S*-adenosylmethionine indicates that the adenosyl group is bonded to sulfur. It does not stand for the Cahn-Ingold-Prelog stereochemical descriptor.

A naturally occurring sulfonium salt, S-*adenosylmethionine* (*SAM*), is a key substance in certain biological processes. It is formed by a nucleophilic substitution in which the sulfur atom of methionine attacks the primary carbon of adenosine triphosphate, displacing the triphosphate leaving group as shown in Figure 16.7.

S-Adenosylmethionine acts as a biological methyl-transfer agent. Nucleophiles, particularly nitrogen atoms of amines, attack the methyl carbon of SAM, breaking the carbon-sulfur bond. The following equation represents the biological formation of *epinephrine* by methylation of *norepinephrine.* Only the methyl group and the sulfur of SAM are shown explicitly in the equation in order to draw attention to the similarity of this reaction, which occurs in living systems, to the more familiar S_N2 reactions we have studied.

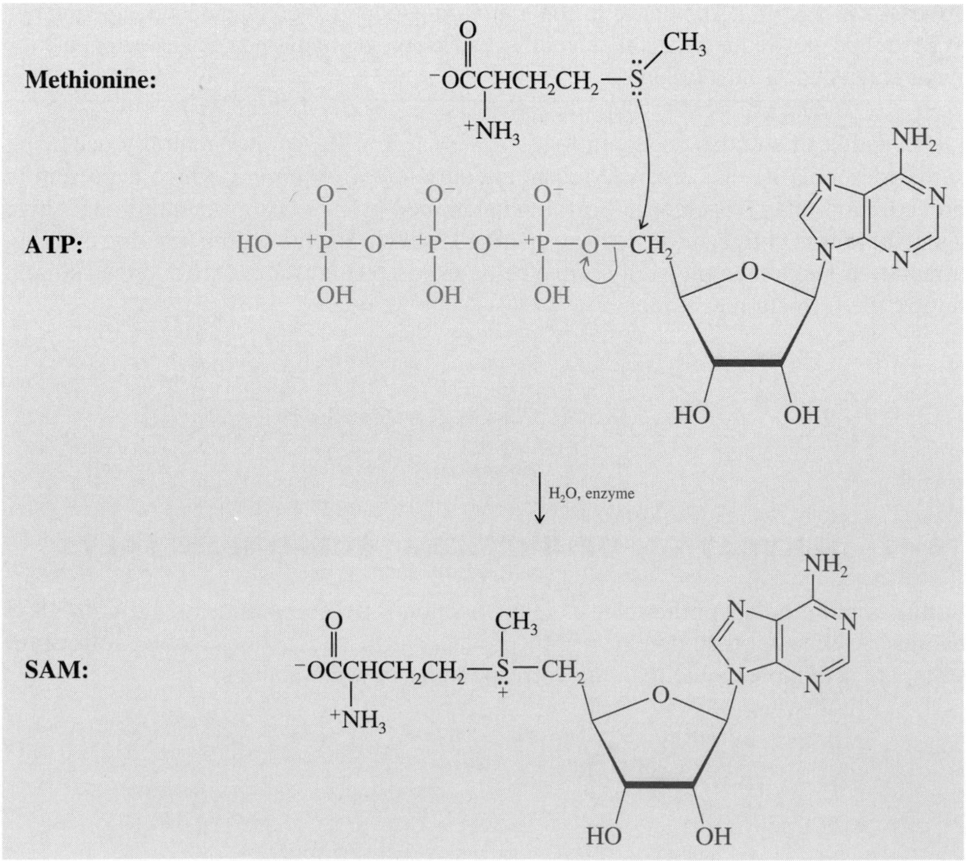

Methionine: $^-OCCHCH_2CH_2—\ddot{S}—CH_3$ (C=O; $^+NH_3$)

ATP: $HO—^+P(O^-)(OH)—O—^+P(O^-)(OH)—O—^+P(O^-)(OH)—O—CH_2$–ribose(HO, OH)–adenine($NH_2$)

↓ H_2O, enzyme

SAM: $^-OCCHCH_2CH_2—\overset{+}{S}(CH_3)—CH_2$–ribose(HO, OH)–adenine($NH_2$) (C=O; $^+NH_3$)

FIGURE 16.7 Nucleophilic substitution at the primary carbon of adenosine triphosphate (ATP) by the sulfur atom of methionine yields *S*-adenosylmethionine (SAM). The reaction is catalyzed by an enzyme.

HO, HO–C_6H_3–CHCH$_2$N:(H)(H)(OH) + CH_3–$\overset{+}{S}$: $\xrightarrow{\text{enzyme}}$

Norepinephrine SAM

HO, HO–C_6H_3–CHCH$_2\overset{+}{N}$(H)(H)–CH_3(OH) + :S: $\xrightarrow{-H^+}$ HO, HO–C_6H_3–CHCH$_2\ddot{N}$(H)–CH_3(OH)

Epinephrine

Epinephrine is also known as *adrenaline* and is a hormone with profound physiological effects designed to prepare the body for "fight or flight."

16.18 SPECTROSCOPIC ANALYSIS OF ETHERS

The infrared spectra of ethers are characterized by a strong, rather broad band due to C—O—C stretching between 1070 and 1150 cm^{-1}. Dialkyl ethers exhibit this band consistently at 1120 cm^{-1}, as illustrated in the infrared spectrum of dipropyl ether (Figure 16.8).

The chemical shift of the proton in the **H**—C—O—C unit of an ether is very similar to that of the proton in the **H**—C—OH unit of an alcohol. A range δ = 3.3 to 4.0 ppm is appropriate. In the ^{1}H nmr spectrum of dipropyl ether, shown in Figure 16.9, the assignment of signals to the various protons in the molecule is

δ = 0.9 ppm → CH_3; δ = 1.5 ppm → CH_2 (middle); δ = 3.4 ppm → OCH_2

$$CH_3CH_2CH_2OCH_2CH_2CH_3$$

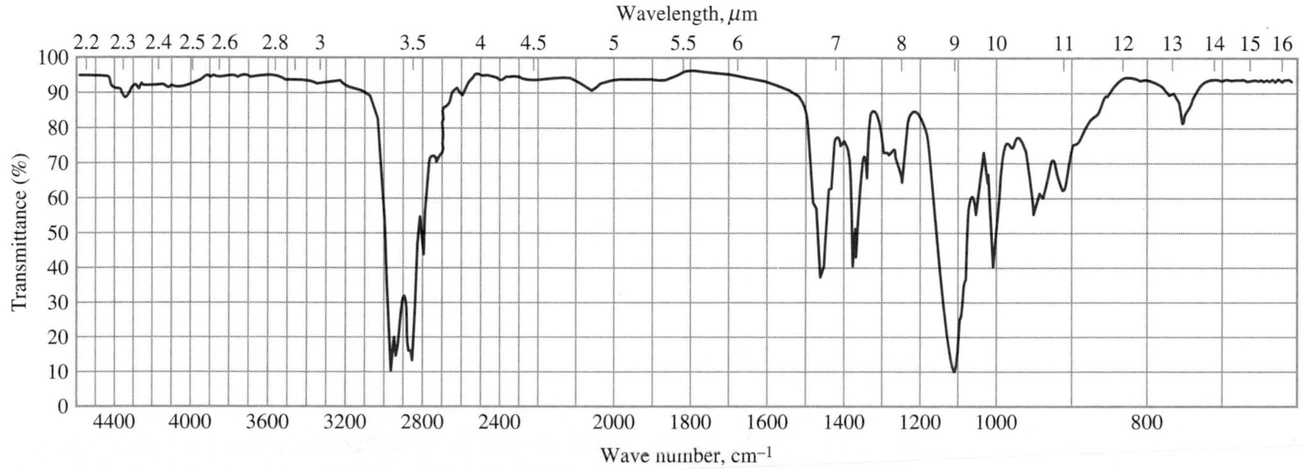

FIGURE 16.8 The infrared spectrum of dipropyl ether ($CH_3CH_2CH_2OCH_2CH_2CH_3$). The strong peak near 1100 cm^{-1} is due to C—O—C stretching.

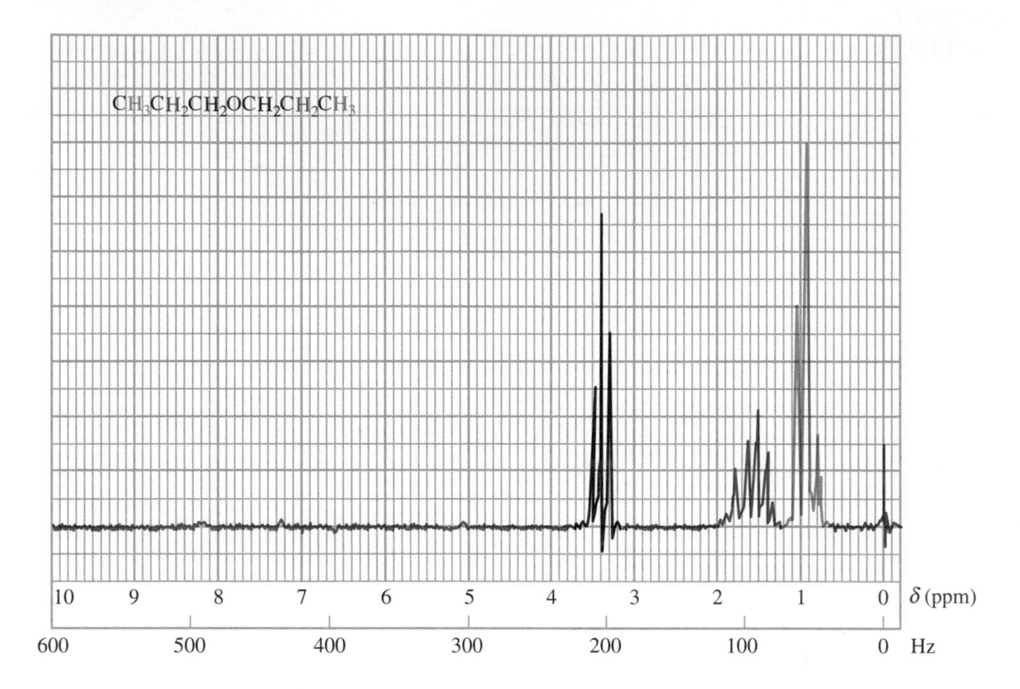

FIGURE 16.9 The 60-MHz ^1H nmr spectrum of dipropyl ether ($CH_3CH_2CH_2OCH_2CH_2CH_3$).

16.19 MASS SPECTROMETRY OF ETHERS

Ethers, like alcohols, lose an alkyl radical from their molecular ion to give an oxygen-stabilized cation. Thus, *m/z* 73 and *m/z* 87 are both more abundant than the molecular ion in the mass spectrum of *sec*-butyl ethyl ether.

$$CH_3CH_2\ddot{O}^{\cdot+}—\underset{\displaystyle CH_3}{\underset{|}{C}}HCH_2CH_3$$

m/z 102

$$CH_3CH_2\overset{+}{\underset{\cdot\cdot}{O}}=CHCH_3 + \cdot CH_2CH_3 \qquad CH_3CH_2\overset{+}{\underset{\cdot\cdot}{O}}=CHCH_2CH_3 + \cdot CH_3$$

m/z 73 *m/z* 87

PROBLEM 16.19 There is another oxygen-stabilized cation of *m/z* 87 capable of being formed by fragmentation of the molecular ion in the mass spectrum of *sec*-butyl ethyl ether. Suggest a reasonable structure for this ion.

16.20 SUMMARY

Ethers are compounds that contain a C—O—C linkage. In substitutive nomenclature they are named as *alkoxy* derivatives of alkanes. In radicofunctional nomenclature, one names each alkyl group as a separate word (in alphabetical order) and concludes with the word *ether*. **Epoxides** are normally named as *epoxy* derivatives of alkanes or as substituted *oxiranes.* **Sulfides** are sulfur analogs of ethers; they contain the C—S—C functional group. They are named as *alkylthio* derivatives of alkanes in substitutive nomenclature. The radicofunctional names of sulfides are derived in the same manner as those of ethers, but the concluding word is *sulfide* (Section 16.1).

Structurally, ethers resemble alkanes, and cyclic ethers resemble cycloalkanes. An sp^3 hybridized oxygen in a carbon chain affects its shape in much the same way as a $—CH_2—$ group does (Section 16.2).

The carbon-oxygen bond of ethers is a polar bond, and ethers can act as proton acceptors toward water and alcohols (Section 16.3). Since ethers lack —OH groups, they cannot act as the proton donor in forming hydrogen bonds to other molecules. The negatively polarized oxygen of an ether can coordinate with metal ions, such as the magnesium atom of a Grignard reagent. Certain cyclic polyethers, called **crown ethers,** are particularly effective in coordinating with Na^+ and K^+, and salts of these cations can be dissolved in nonpolar solvents when crown ethers are present (Section 16.4). Under these conditions the rates of many reactions involving anionic species are accelerated.

While ethers are commonplace substances in organic chemistry, there are only a few methods by which they are prepared. These are summarized in Table 16.1.

TABLE 16.1
Preparation of Ethers

Reaction (section) and comments	General equation and specific example
Acid-catalyzed condensation of alcohols (Sections 15.7 and 16.5) Two molecules of an alcohol condense in the presence of an acid catalyst to yield a dialkyl ether and water. The reaction is limited to the synthesis of symmetrical ethers from primary alcohols.	$2RCH_2OH \xrightarrow{H^+} RCH_2OCH_2R + H_2O$ Alcohol — Ether — Water $CH_3CH_2CH_2OH \xrightarrow[heat]{H_2SO_4} CH_3CH_2CH_2OCH_2CH_2CH_3$ Propyl alcohol — Dipropyl ether
The Williamson ether synthesis (Section 16.6) An alkoxide ion displaces a halide or similar leaving group in an S_N2 reaction. The alkyl halide cannot be one that is prone to elimination, and so this reaction is limited to primary alkyl halides. There is no limitation on the alkoxide ion that can be used.	$RO^- + R'CH_2X \longrightarrow ROCH_2R' + X^-$ Alkoxide ion — Primary alkyl halide — Ether — Halide ion $(CH_3)_2CHCH_2ONa + CH_3CH_2Br \longrightarrow (CH_3)_2CHCH_2OCH_2CH_3 + NaBr$ Sodium isobutoxide — Ethyl bromide — Ethyl isobutyl ether (66%) — Sodium bromide
Solvomercuration-demercuration (Section 14.14) Alkenes are converted to ethers by reaction with mercury(II) acetate in an alcohol solvent. The regioselectivity of addition follows Markovnikov's rule.	$R_2C{=}CHR \xrightarrow[R'OH]{Hg(O_2CCH_3)_2} R_2C(OR')—CHR(HgO_2CCH_3) \xrightarrow{NaBH_4} R_2C(OR')—CH_2R$ Alkene — Alkoxyalkylmercury(II) acetate — Ether $CH_3CH_2CH_2CH_2CH{=}CH_2 \xrightarrow[2.\ NaBH_4]{1.\ Hg(OAc)_2,\ CH_3CH_2OH} CH_3CH_2CH_2CH_2CH(OCH_2CH_3)CH_3$ 1-Hexene — 2-Ethoxyhexane (98%)

The only important reaction of ethers is their cleavage by hydrogen halides (Section 16.8):

$$ROR' + 2HX \longrightarrow RX + R'X + H_2O$$

Ether, Hydrogen halide, Alkyl halide, Alkyl halide, Water

The order of hydrogen halide reactivity is HI > HBr > HCl ≫ HF.

$$C_6H_5CH_2OCH_2CH_3 \xrightarrow[\text{heat}]{\text{HBr}} C_6H_5CH_2Br + CH_3CH_2Br$$

Benzyl ethyl ether, Benzyl bromide, Ethyl bromide

Epoxides are prepared by the methods listed in Table 16.2. Epoxides are much more reactive than ethers, especially in reactions that lead to cleavage of their three-membered ring (Section 16.11). Relief of ring strain provides the driving force for these reactions. Anionic nucleophiles usually attack the less substituted carbon of the epoxide in an S_N2-like fashion (Section 16.12).

TABLE 16.2
Preparation of Epoxides

Reaction (section) and comments	General equation and specific example
Peroxy acid oxidation of alkenes (Sections 6.17 and 16.9) Peroxy acids transfer oxygen to alkenes to yield epoxides. Stereospecific syn addition is observed.	$R_2C{=}CR_2 + R'COOH$ (peroxy acid, R'C(=O)OOH) $\longrightarrow$ $R_2C{-}CR_2$ (epoxide, bridged by O) $+ R'COH$ (R'C(=O)OH) Alkene, Peroxy acid, Epoxide, Carboxylic acid $(CH_3)_2C{=}C(CH_3)_2 \xrightarrow{CH_3CO_2OH}$ $(H_3C)_2C{-}C(CH_3)_2$ (epoxide, bridged by O) 2,3-Dimethyl-2-butene → 2,2,3,3-Tetramethyloxirane (70–80%)
Base-promoted cyclization of vicinal halohydrins (Section 16.10) This reaction is an intramolecular version of the Williamson ether synthesis. The alcohol function of a vicinal halohydrin is converted to its conjugate base, which then displaces halide from the adjacent carbon to give an epoxide.	$R_2C(OH){-}CR_2X \overset{HO^-}{\rightleftharpoons} R_2C(O^-){-}CR_2X \longrightarrow R_2C{-}CR_2$ (epoxide, bridged by O) Vicinal halohydrin, Epoxide $(CH_3)_2C(OH){-}CH(Br)CH_3 \xrightarrow[H_2O]{NaOH} (CH_3)_2C{-}CHCH_3$ (epoxide, bridged by O) 3-Bromo-2-methyl-2-butanol → 2,2,3-Trimethyloxirane (78%)

$$\underset{\text{Epoxide}}{\text{RCH}\overset{O}{-}\text{CR}_2} + \underset{\text{Nucleophile}}{Y^-} \longrightarrow \text{RCH(Y)}-\text{CR}_2\text{O}^- \longrightarrow \underset{\beta\text{-substituted alcohol}}{\text{RCH(Y)}-\text{CR}_2\text{OH}}$$

Nucleophile attacks this carbon.

$$\underset{\text{2,2,3-Trimethyloxirane}}{(H_3C)HC\overset{O}{-}C(CH_3)_2} \xrightarrow[CH_3OH]{NaOCH_3} \underset{\text{3-Methoxy-2-methyl-2-butanol (53\%)}}{CH_3CH(OCH_3)C(OH)(CH_3)CH_3}$$

Under conditions of acid catalysis, nucleophiles attack the carbon that can better support a positive charge. Carbocation character is developed in the transition state (Section 16.13).

$$\underset{\text{Epoxide}}{\text{RCH}\overset{O}{-}\text{CR}_2} + H^+ \rightleftharpoons \text{RCH}\overset{\overset{+}{O}H}{-}\text{CR}_2 \xrightarrow{HY} \text{RCH(OH)}-\text{CR}_2\overset{+}{Y}H \xrightarrow{-H^+} \underset{\beta\text{-substituted alcohol}}{\text{RCH(OH)CR}_2\text{Y}}$$

Nucleophile attacks this carbon.

$$\underset{\text{2,2,3-Trimethyloxirane}}{(H_3C)HC\overset{O}{-}C(CH_3)_2} \xrightarrow[H_2SO_4]{CH_3OH} \underset{\text{3-Methoxy-3-methyl-2-butanol (76\%)}}{CH_3CH(OH)C(OCH_3)(CH_3)CH_3}$$

Inversion of configuration is observed at the carbon that undergoes nucleophilic attack, irrespective of whether the reaction occurs in acid or in base.

The epoxide functional group occurs in a great many natural products, and epoxide ring opening is sometimes a key step in the biosynthesis of other substances (Section 16.14).

Sulfides are prepared by nucleophilic substitution (S_N2) in which an alkanethiolate ion attacks an alkyl halide (Section 16.15):

$$\underset{\text{Alkanethiolate}}{R\ddot{\underset{..}{S}}:^-} + \underset{\text{Alkyl halide}}{R'-\ddot{\underset{..}{X}}:} \longrightarrow \underset{\text{Sulfide}}{R\ddot{\underset{..}{S}}-R'} + \underset{\text{Halide}}{:\ddot{\underset{..}{X}}:^-}$$

$$\underset{\text{Benzenethiol}}{C_6H_5SH} \xrightarrow{NaOCH_2CH_3} \underset{\text{Sodium benzenethiolate}}{C_6H_5SNa} \xrightarrow{C_6H_5CH_2Cl} \underset{\text{Benzyl phenyl sulfide (60\%)}}{C_6H_5SCH_2C_6H_5}$$

Oxidation of sulfides yields sulfoxides, then sulfones (Section 16.16). Sodium metaperiodate is specific for oxidation of sulfides to sulfoxides, and no further. Hydrogen peroxide or peroxy acids can yield sulfoxides (1 mole of oxidant per mole of sulfide) or sulfones (2 moles of oxidant per mole of sulfide):

$$R{-}\ddot{S}{-}R' \xrightarrow{\text{oxidize}} R{-}\overset{+}{S}(O^-){-}R' \xrightarrow{\text{oxidize}} R{-}\overset{2+}{S}(O^-)_2{-}R'$$

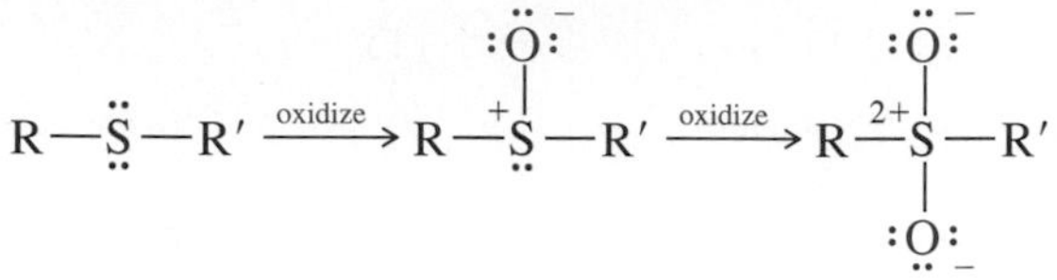

Sulfide Sulfoxide Sulfone

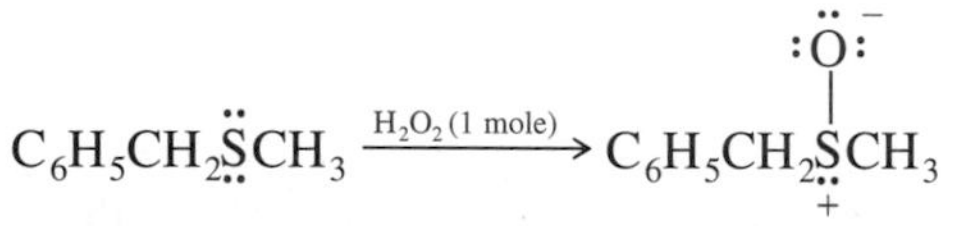

$$C_6H_5CH_2\ddot{S}CH_3 \xrightarrow{H_2O_2\,(1\text{ mole})} C_6H_5CH_2\overset{+}{S}(O^-)CH_3$$

Benzyl methyl sulfide Benzyl methyl sulfoxide (94%)

Sulfides react with alkyl halides to give sulfonium salts (Section 16.17):

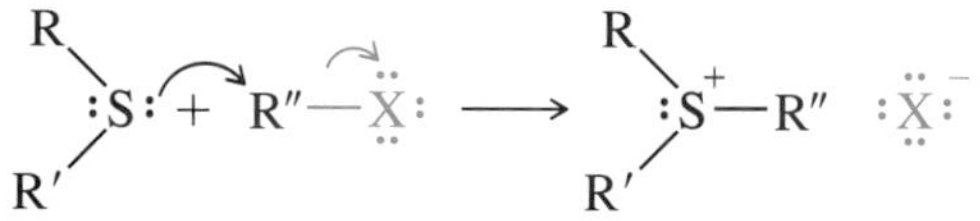

$$R_2S: + R''{-}X: \longrightarrow R_2\overset{+}{S}{-}R'' \quad :X:^-$$

Sulfide Alkyl halide Sulfonium salt

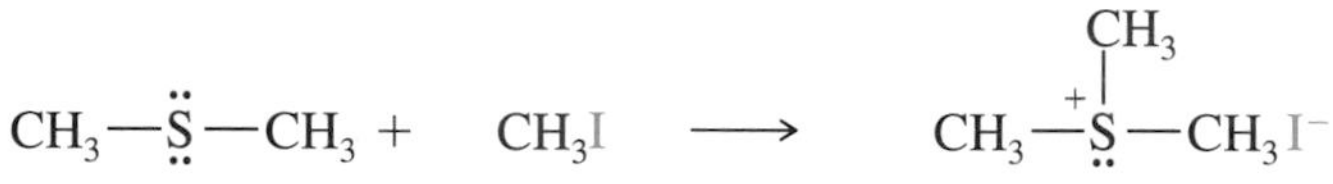

$$CH_3{-}\ddot{S}{-}CH_3 + CH_3I \longrightarrow CH_3{-}\overset{+}{S}(CH_3){-}CH_3\ I^-$$

Dimethyl sulfide Methyl iodide Trimethylsulfonium iodide (100%)

PROBLEMS

16.20 Write the structures of all the constitutionally isomeric ethers of molecular formula $C_5H_{12}O$ and give an acceptable name for each.

16.21 Many ethers, including diethyl ether, are effective as general anesthetics. Because simple ethers are quite flammable, their place in medical practice has been taken by highly halogenated nonflammable ethers. Two such general anesthetic agents are *isoflurane* and *enflurane.* These compounds are isomeric; isoflurane is 1-chloro-2,2,2-trifluoroethyl difluoromethyl ether, while enflurane is 2-chloro-1,1,2-trifluoroethyl difluoromethyl ether. Write the structural formulas of isoflurane and enflurane.

16.22 While epoxides are always considered to have their oxygen atom as part of a three-membered ring, the prefix *epoxy* in the IUPAC system of nomenclature can be used to denote a cyclic ether of various sizes. Thus

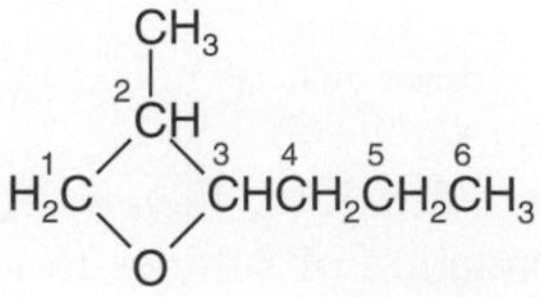

$$\underset{1}{H_2C}{-}\underset{2}{CH}(CH_3){-}\underset{3}{CH}(O\text{, bridging C1})\underset{4}{CH_2}\underset{5}{CH_2}\underset{6}{CH_3}$$

may be named 2-methyl-1,3-epoxyhexane. Using the epoxy prefix in this way, name each of the following compounds:

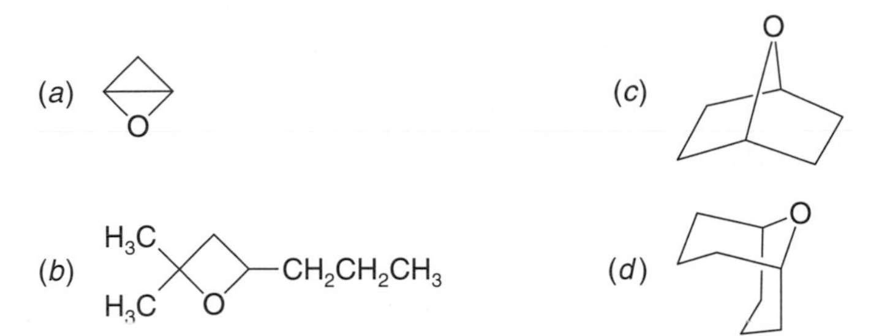

16.23 The name of the parent six-membered sulfur-containing heterocycle is *thiane.* It is numbered beginning at sulfur. Multiple incorporation of sulfur in the ring is indicated by the prefixes *di-, tri-,* and so on.

(*a*) How many methyl-substituted thianes are there? Which ones are chiral?
(*b*) Write structural formulas for 1,4-dithiane and 1,3,5-trithiane.
(*c*) Which dithiane isomer is a disulfide?
(*d*) Draw the two most stable conformations of the sulfoxide derived from thiane.

16.24 The most stable conformation of 1,3-dioxan-5-ol is the chair form that has its hydroxyl group in an axial orientation. Suggest a reasonable explanation for this fact.

OH, O, O — 1,3-Dioxan-5-ol

16.25 Outline the steps in the preparation of each of the constitutionally isomeric ethers of molecular formula $C_4H_{10}O$, starting with the appropriate alcohols. Use the Williamson ether synthesis as your key reaction.

16.26 Predict the principal organic product of each of the following reactions. Specify stereochemistry where appropriate.

(*a*) Cyclohexyl—Br + $CH_3CH_2CHCH_3$ (with ONa on CH) $\longrightarrow$

(*b*) CH_3CH_2I + CH_3CH_2, CH_3, H—C—ONa $\longrightarrow$

(*c*) $CH_3CH_2CHCH_2Br$ (with OH on CH) $\xrightarrow{NaOH}$

(*d*) C_6H_5(H)C=C(H)CH_3 + C_6H_5—$\overset{O}{\overset{\|}{C}}$OOH $\longrightarrow$

(*e*) [cyclohexene oxide] $\xrightarrow[\text{dioxane-water}]{NaN_3}$

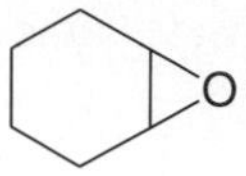

(*f*) [2-(2-bromophenyl)-2-methyloxirane] $\xrightarrow[\text{methanol}]{NH_3}$

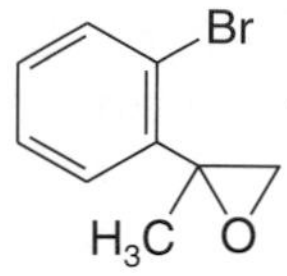

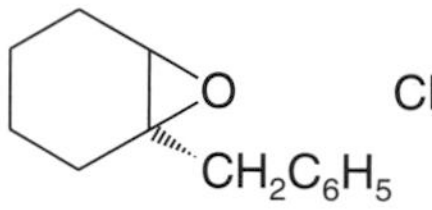

(*g*) [1-benzylcyclohexene oxide] $CH_3ONa \xrightarrow{CH_3OH}$

(*h*) $C_6H_5CH{-}CH_2$ (epoxide, O bridging) $\xrightarrow[CHCl_3]{HCl}$

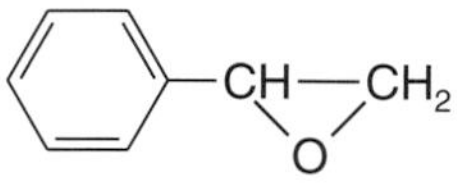

(*i*) $CH_3(CH_2)_{16}CH_2OTs + CH_3CH_2CH_2CH_2SNa \longrightarrow$

(*j*) [Fischer projection: C_6H_5 top; H—C—CH_3; Cl—C—H; C_6H_5 bottom] $\xrightarrow{C_6H_5SNa}$

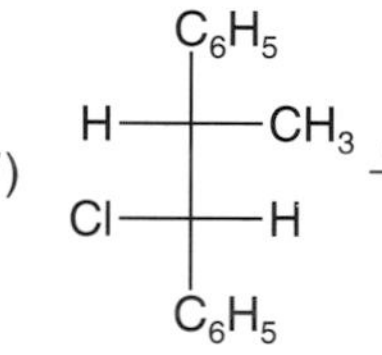

16.27 Oxidation of 4-*tert*-butylthiane (see Problem 16.23 for the structure of thiane) with sodium metaperiodate gives a mixture of two compounds of molecular formula $C_9H_{18}OS$. Both products give the same sulfone on further oxidation with hydrogen peroxide. What is the relationship between the two compounds?

16.28 When (*R*)-(+)-2-phenyl-2-butanol is allowed to stand in methanol containing a few drops of sulfuric acid, racemic 2-methoxy-2-phenylbutane is formed. Suggest a reasonable mechanism for this reaction.

16.29 Select reaction conditions that would allow you to carry out each of the following stereospecific transformations:

(*a*) [epoxide with H (hashed) and CH_3 (wedged)] $\longrightarrow$ (*R*)-1,2-propanediol

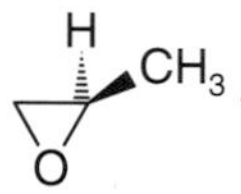

(*b*) [epoxide with H (hashed) and CH_3 (wedged)] $\longrightarrow$ (*S*)-1,2-propanediol

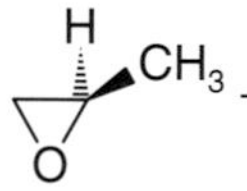

16.30 The last step in the synthesis of divinyl ether (used as an anesthetic under the name *Vinethene*) involves heating $ClCH_2CH_2OCH_2CH_2Cl$ with potassium hydroxide. Show how you could prepare the necessary starting material $ClCH_2CH_2OCH_2CH_2Cl$ from ethylene.

16.31 Suggest short, efficient reaction sequences suitable for preparing each of the following compounds from the given starting materials and any necessary organic or inorganic reagents:

(*a*) $C_6H_5CH_2OCH_3$ from $C_6H_5COCH_3$ (with C=O)

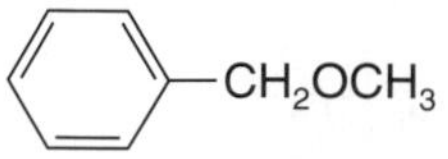

(*b*) [1-phenylcyclohexene oxide, C_6H_5 on epoxide carbon] from bromobenzene and cyclohexanol

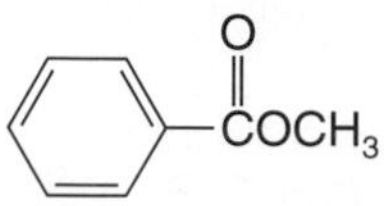

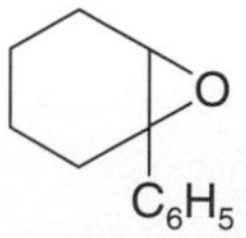

(*c*) $C_6H_5CH_2CH(OH)CH_3$ from bromobenzene and isopropyl alcohol (*d*) $C_6H_5CH_2CH_2CH_2OCH_2CH_3$ from benzyl alcohol and ethanol

(*e*) 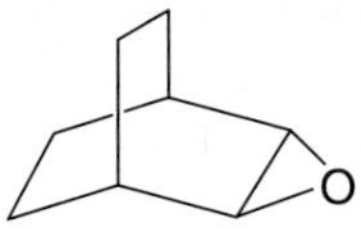 from 1,3-cyclohexadiene and ethanol

(*f*) $C_6H_5CH(OH)CH_2SCH_2CH_3$ from styrene and ethanol

16.32 Among the ways in which 1,4-dioxane may be prepared are the methods expressed in the equations shown:

(*a*) $2HOCH_2CH_2OH \xrightarrow[\text{heat}]{H_2SO_4}$ 1,4-Dioxane + $2H_2O$

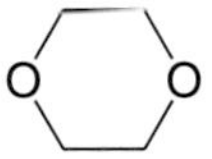

Ethylene glycol — 1,4-Dioxane — Water

(*b*) $ClCH_2CH_2OCH_2CH_2Cl \xrightarrow{NaOH}$ 1,4-Dioxane

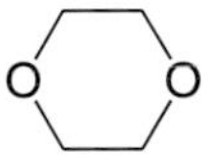

Bis(2-chloroethyl) ether — 1,4-Dioxane

Suggest reasonable mechanisms for each of these reactions.

16.33 Deduce the identity of the missing compounds in the following reaction sequences. Show stereochemistry in parts (*b*) through (*d*).

(*a*) $CH_2{=}CHCH_2Br \xrightarrow[\text{2. } CH_2{=}O \text{ 3. } H_3O^+]{\text{1. Mg}}$ compound A (C_4H_8O) $\xrightarrow{Br_2}$ compound B ($C_4H_8Br_2O$) $\xrightarrow{\text{KOH, 25°C}}$ compound C (C_4H_7BrO) $\xrightarrow[\text{heat}]{\text{KOH}}$  Compound D

(*b*) Cl–C(H)(CO$_2$H)(CH$_3$) $\xrightarrow[\text{2. } H_2O]{\text{1. } LiAlH_4}$ compound E (C_3H_7ClO) $\xrightarrow{KOH,\ H_2O}$ compound F (C_3H_6O)

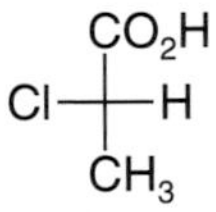

(*c*) CH$_3$, H–C–Cl, H–C–OH, CH$_3$ $\xrightarrow{NaOH}$ compound G (C_4H_8O) $\xrightarrow{NaSCH_3}$ compound H ($C_5H_{12}OS$)

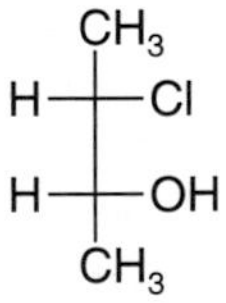

(*d*) Compound I (C_7H_{12}) $\xrightarrow[(CH_3)_3COH,\ HO^-]{OsO_4,\ (CH_3)_3COOH}$ compound J ($C_7H_{14}O_2$) (a liquid)

Compound I $\xrightarrow{C_6H_5CO_2OH}$ Compound K $\xrightarrow[H_2SO_4]{H_2O}$ compound L ($C_7H_{14}O_2$) (mp 99.5–101°C)

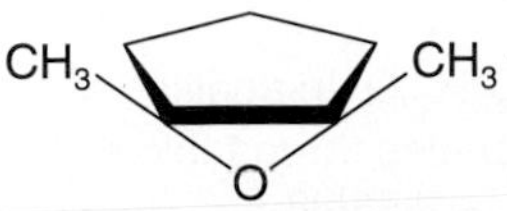

Compound K

16.34 Cineole is the chief component of eucalyptus oil; it has the molecular formula $C_{10}H_{18}O$ and contains no double or triple bonds. It reacts with hydrochloric acid to give the dichloride shown:

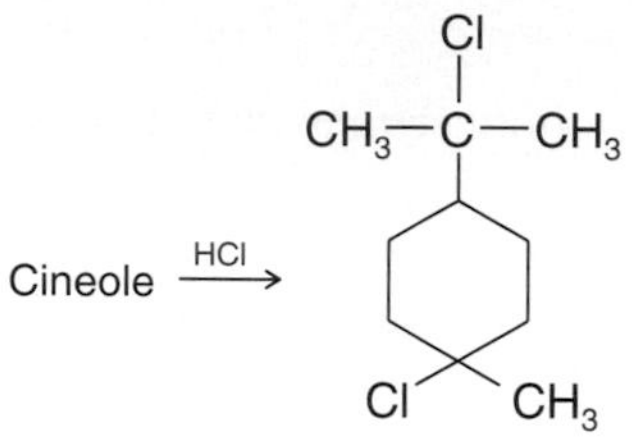

Cineole $\xrightarrow{HCl}$ 1-chloro-4-(1-chloro-1-methylethyl)-1-methylcyclohexane [structure: cyclohexane ring bearing CH_3—C(Cl)—CH_3 at top and Cl, CH_3 at bottom]

Deduce the structure of cineole.

16.35 The *p*-toluenesulfonate shown undergoes an intramolecular Williamson reaction on treatment with base to give a spirocyclic ether. Demonstrate your understanding of the terminology used in the preceding sentence by writing the structure, including stereochemistry, of the product.

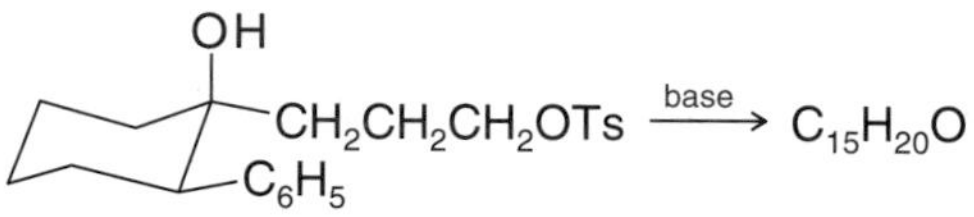

[cyclohexane ring bearing OH, $CH_2CH_2CH_2OTs$ and C_6H_5] $\xrightarrow{\text{base}}$ $C_{15}H_{20}O$

16.36 All the following questions pertain to ^1H nmr spectra of isomeric ethers having the molecular formula $C_5H_{12}O$.

(*a*) Which one has only singlets in its ^1H nmr spectrum?

(*b*) Along with other signals, this ether has a coupled doublet-heptet pattern. None of the protons responsible for this pattern are coupled to protons anywhere else in the molecule. Identify this ether.

(*c*) In addition to other signals in its ^1H nmr spectrum, this ether exhibits two signals at relatively low field. One is a singlet; the other is a doublet. What is the structure of this ether?

(*d*) In addition to other signals in its ^1H nmr spectrum, this ether exhibits two signals at relatively low field. One is a triplet; the other is a quartet. Which ether is this?

16.37 The ^1H nmr spectrum of compound A (C_8H_8O) consists of two singlets of equal area at $\delta = 5.1$ (sharp) and 7.2 ppm (broad). On treatment with excess hydrogen bromide, compound A is converted to a single dibromide ($C_8H_8Br_2$). The ^1H nmr spectrum of the dibromide is similar to that of A in that it exhibits two singlets of equal area at $\delta = 4.7$ (sharp) and 7.3 ppm (broad). Suggest reasonable structures for compound A and the dibromide derived from it.

16.38 The ^1H nmr spectrum of a compound ($C_{10}H_{13}BrO$) is shown in Figure 16.10. The compound gives benzyl bromide, along with a second compound $C_3H_6Br_2$, when heated with HBr. What is the first compound?

16.39 A compound is a cyclic ether of molecular formula $C_9H_{10}O$. In addition to signals for the six carbons of an aromatic ring, its ^{13}C nmr spectrum includes three triplets at $\delta = 28.3$, 65.3, and 67.9 ppm. Oxidation of the compound with sodium dichromate and sulfuric acid gave 1,2-benzenedicarboxylic acid. What is the compound?

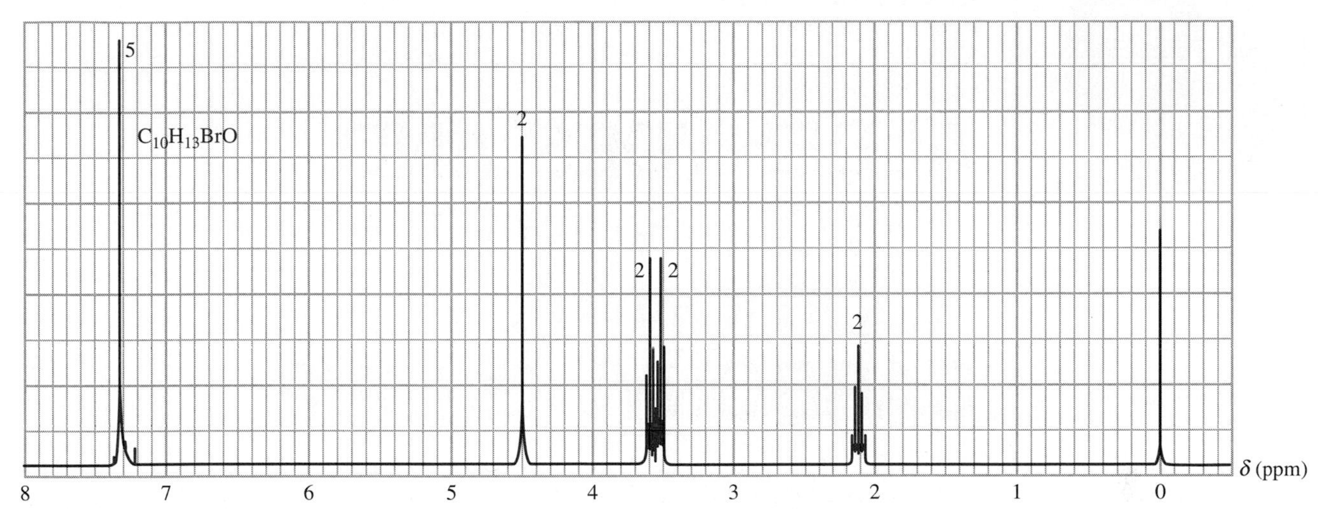

FIGURE 16.10 The 300-MHz ^{1}H nmr spectrum of a compound, $C_{10}H_{13}BrO$ (Problem 16.38). The integral ratios of the signals reading from left to right (low field to high field) are 5:2:2:2:2. The signals at 3.5 and 3.6 ppm are both triplets.

MOLECULAR MODELING EXERCISES

16.40 1,4-Butanediol undergoes acid-catalyzed cyclization to form a cyclic ether (tetrahydrofuran). Construct a molecular model of the oxonium ion formed from 1,4-butanediol and orient it in the appropriate geometry for cyclic ether formation.

16.41 Construct a molecular model of 18-crown-6 and its complex with K^+.

16.42 Construct a molecular model of *trans*-2-bromocyclohexanol in its most stable conformation. This conformation is ill-suited to undergo epoxide formation on treatment with base. Why? What must happen in order to produce 1,2-epoxycyclohexane from *trans*-2-bromocyclohexanol?

16.43 Construct a molecular model of *threo*-3-bromo-2-butanol. What is the stereochemistry (cis or trans) of the 2,3-epoxybutane formed on treatment of *threo*-3-bromo-2-butanol with base? Repeat the exercise for *erythro*-3-bromo-2-butanol.

16.44 The chair conformation having an axial oxygen is the most stable conformation of thiane 1-oxide. Construct a molecular model of this conformation. This conformation is believed to be stabilized by a dipole-induced dipole attractive force. Identify the atoms or groups involved.

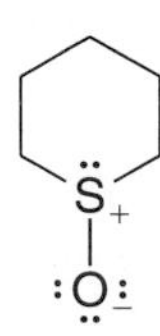

CHAPTER 17

ALDEHYDES AND KETONES. NUCLEOPHILIC ADDITION TO THE CARBONYL GROUP

Aldehydes and ketones are characterized by the presence of an acyl group $\overset{\text{O}}{\overset{\|}{\text{RC}}}$— bonded either to hydrogen or to another carbon.

$$\underset{\text{Formaldehyde}}{\overset{\text{O}}{\overset{\|}{\text{HCH}}}} \qquad \underset{\text{Aldehyde}}{\overset{\text{O}}{\overset{\|}{\text{RCH}}}} \qquad \underset{\text{Ketone}}{\overset{\text{O}}{\overset{\|}{\text{RCR}'}}}$$

While the present chapter includes the usual collection of topics designed to acquaint us with a particular class of compounds, its central theme is a fundamental reaction type, *nucleophilic addition to carbonyl groups.* The principles of nucleophilic addition to aldehydes and ketones developed here will be seen to have broad applicability in later chapters when transformations of various derivatives of carboxylic acids are discussed.

17.1 NOMENCLATURE

The -*e* ending of an alkane name is dropped before a suffix beginning with a vowel (-*al*) and retained before one beginning with a consonant (-*dial*).

The longest continuous chain that contains the —$\overset{\text{O}}{\overset{\|}{\text{C}}}$H group provides the base name for aldehydes. The -*e* ending of the corresponding alkane name is replaced by -*al,* and substituents are specified in the usual way. It is not necessary to specify the location of the —$\overset{\text{O}}{\overset{\|}{\text{C}}}$H group in the name, since the chain must be numbered by starting with this group as C-1. The suffix -*dial* is added to the appropriate alkane name when the compound contains two aldehyde functions.

$$\underset{\text{4,4-Dimethylpentanal}}{\mathrm{CH_3C(CH_3)_2CH_2CH_2CH{=}O}} \qquad \underset{\text{5-Hexenal}}{\mathrm{CH_2{=}CHCH_2CH_2CH_2CH{=}O}} \qquad \underset{\text{2-Phenylpropanedial}}{\mathrm{O{=}HCCH(C_6H_5)CH{=}O}}$$

4,4-Dimethylpentanal 5-Hexenal 2-Phenylpropanedial

When a formyl group (—CH═O) is attached to a ring, the ring name is followed by the suffix *-carbaldehyde.*

Cyclopentanecarbaldehyde 2-Naphthalenecarbaldehyde

Certain common names of familiar aldehydes are acceptable as IUPAC names. A few examples include

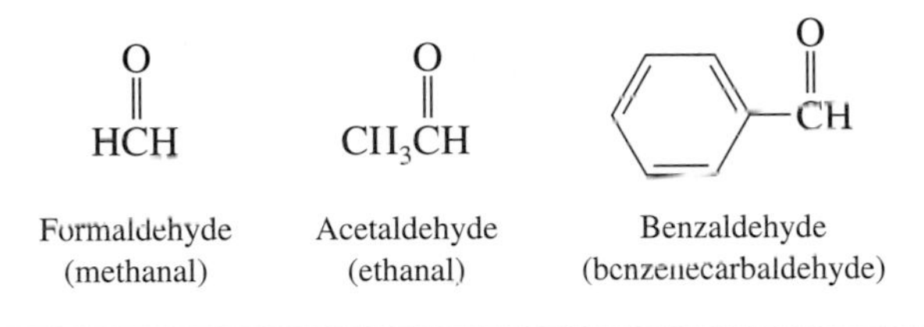

Formaldehyde (methanal) Acetaldehyde (ethanal) Benzaldehyde (benzenecarbaldehyde)

PROBLEM 17.1 The common names and structural formulas of a few aldehydes follow. Provide an alternative IUPAC name.

(*a*) $\mathrm{(CH_3)_2CHCH{=}O}$ (isobutyraldehyde)

(*b*) $\mathrm{O{=}HCCH_2CH_2CH_2CH{=}O}$ (glutaraldehyde)

(*c*) $\mathrm{C_6H_5CH{=}CHCH{=}O}$ (cinnamaldehyde)

(*d*) HO—(benzene ring with $\mathrm{CH_3O}$)—CH═O (vanillin)

SAMPLE SOLUTION (*a*) Don't be fooled by the fact that the common name is isobutyraldehyde. The longest continuous chain is three carbons, and so the base name is *propanal.* There is a methyl group at C-2; thus the compound is 2-methylpropanal.

$$\overset{3}{\mathrm{C}}\mathrm{H_3}\overset{2}{\mathrm{C}}\mathrm{H(CH_3)}\overset{1}{\mathrm{C}}\mathrm{H{=}O}$$

2-Methylpropanal (isobutyraldehyde)

With ketones, the *-e* ending of an alkane is replaced by *-one* in the longest continuous chain containing the carbonyl group. The location of the carbonyl is specified by numbering the chain in the direction that provides the lower number for this group.

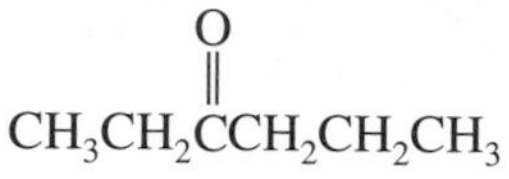

$CH_3CH_2\overset{O}{\overset{\|}{C}}CH_2CH_2CH_3$ — 3-Hexanone (*not* 4-hexanone)

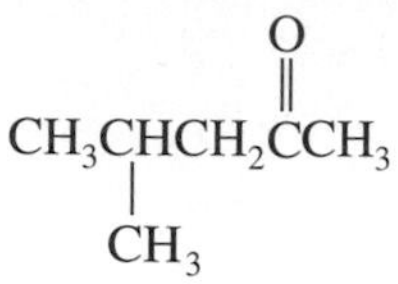

$CH_3\underset{CH_3}{\underset{|}{C}}HCH_2\overset{O}{\overset{\|}{C}}CH_3$ — 4-Methyl-2-pentanone (*not* 2-methyl-4-pentanone)

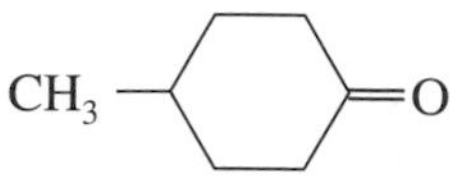

CH_3–(cyclohexane ring)=O — 4-Methylcyclohexanone

The substitutive names of ketones are developed in the same way as those of alcohols. The radicofunctional names of ketones are developed in the same way as those of ethers.

Although substitutive names of the type just described are preferred, the IUPAC rules also permit ketones to be named by radicofunctional nomenclature. The groups attached to the carbonyl group are named as separate words followed by the word *ketone.* The groups are listed alphabetically.

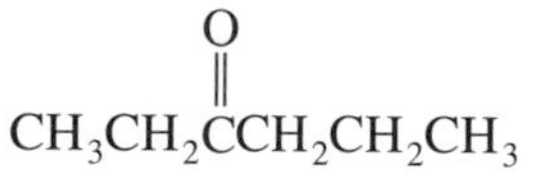

$CH_3CH_2\overset{O}{\overset{\|}{C}}CH_2CH_2CH_3$ — Ethyl propyl ketone

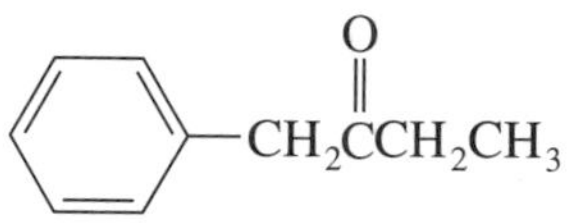

C_6H_5–$CH_2\overset{O}{\overset{\|}{C}}CH_2CH_3$ — Benzyl ethyl ketone

$CH_2{=}CH\overset{O}{\overset{\|}{C}}CH{=}CH_2$ — Divinyl ketone

PROBLEM 17.2 Convert each of the following radicofunctional names to a substitutive name.

(*a*) Dibenzyl ketone
(*b*) Ethyl isopropyl ketone
(*c*) Methyl 2,2-dimethylpropyl ketone
(*d*) Allyl methyl ketone

SAMPLE SOLUTION (*a*) First write the structure corresponding to the name. Dibenzyl ketone has two benzyl groups attached to a carbonyl.

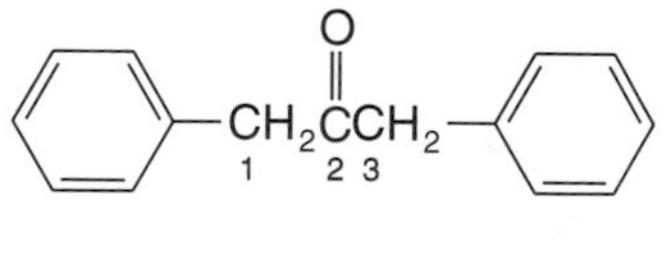

C_6H_5–$\underset{1}{CH_2}\underset{2}{\overset{O}{\overset{\|}{C}}}\underset{3}{CH_2}$–$C_6H_5$ — Dibenzyl ketone

The longest continuous chain contains three carbons, and C-2 is the carbon of the carbonyl group. The substitutive IUPAC name for this ketone is *1,3-diphenyl-2-propanone.*

A few of the common names acceptable for ketones in the IUPAC system are

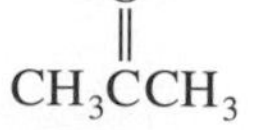

$CH_3\overset{O}{\overset{\|}{C}}CH_3$ — Acetone

C_6H_5–$\overset{O}{\overset{\|}{C}}CH_3$ — Acetophenone

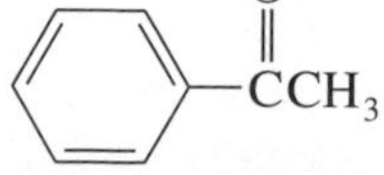

C_6H_5–$\overset{O}{\overset{\|}{C}}$–$C_6H_5$ — Benzophenone

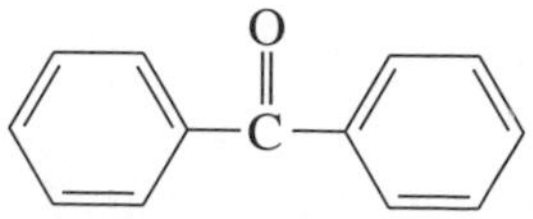

(The suffix *-phenone* indicates that the acyl group is attached to a benzene ring.)

17.2 STRUCTURE AND BONDING: THE CARBONYL GROUP

Two notable aspects of the carbonyl group are its geometry and its polarity. The carbonyl group and the atoms attached to it lie in the same plane. Formaldehyde, for example, is a planar molecule. The bond angles involving the carbonyl group are close to 120°.

Formaldehyde	Acetaldehyde	Acetone
H–C(=O)–H: O=C–H 121.7°, O=C–H 121.7°, H–C–H 116.5°	H_3C–C(=O)–H: O=C–CH_3 123.9°, O=C–H 118.6°, H_3C–C–H 117.5°	H_3C–C(=O)–CH_3: O=C–CH_3 121.4°, O=C–CH_3 121.4°, H_3C–C–CH_3 117.2°

The average carbon-oxygen double-bond distance of 122 pm in aldehydes and ketones is significantly shorter than the typical carbon-oxygen single-bond distance of 141 pm in alcohols and ethers.

The presence of a carbonyl group makes aldehydes and ketones rather polar. Their molecular dipole moments, for example, are substantially larger than those of comparable compounds that contain carbon-carbon double bonds.

$CH_3CH_2CH{=}CH_2$	$CH_3CH_2CH{=}O$
1-Butene	Propanal
Dipole moment: 0.3 D	Dipole moment: 2.5 D

Bonding in formaldehyde can be described according to an sp^2 hybridization model analogous to that of ethylene, as shown in Figure 17.1.

Because of the high electronegativity of oxygen, the electron density in both the σ and π components of the carbon-oxygen double bond is displaced toward oxygen. The carbonyl group is polarized so that carbon is partially positive and oxygen is partially negative.

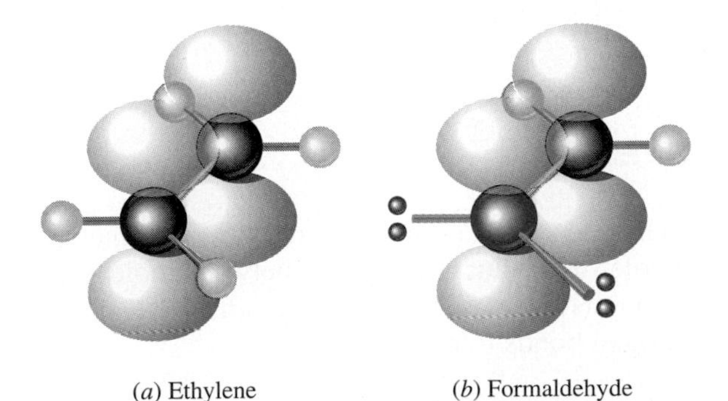

(*a*) Ethylene (*b*) Formaldehyde

FIGURE 17.1 Similarities between the orbital hybridization models of bonding in ethylene (*a*) and formaldehyde (*b*). Both molecules have the same number of electrons, and carbon is sp^2 hybridized in both. In formaldehyde, one of the carbons is replaced by an sp^2 hybridized oxygen (shown in red). Oxygen has two unshared electron pairs; each pair occupies an sp^2 hybridized orbital. Like the carbon-carbon double bond of ethylene, the carbon-oxygen double bond of formaldehyde is composed of a two-electron σ component and a two-electron π component.

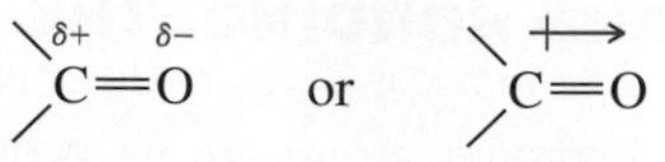

$$\overset{\delta+}{C}{=}\overset{\delta-}{O} \quad \text{or} \quad \overset{\longmapsto}{C{=}O}$$

In resonance terms, electron delocalization in the carbonyl group is represented by contributions from two principal resonance forms:

The chemistry of the carbonyl group is considerably simplified if you remember that carbon is partially positive (has carbocation character) and oxygen is partially negative (weakly basic).

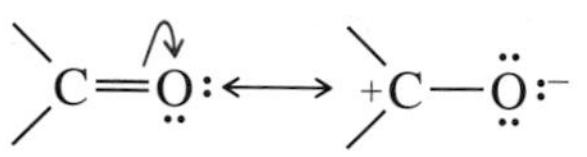

$$\underset{A}{C{=}\ddot{O}{:}} \longleftrightarrow \underset{B}{{}^{+}C{-}\ddot{O}{:}^{-}}$$

Of these two forms, A, having one more covalent bond and avoiding the separation of positive and negative charges that characterizes B, is the one that better approximates the bonding in a carbonyl group.

Alkyl substituents stabilize a carbonyl group in much the same way that they stabilize carbon-carbon double bonds and carbocations, namely, by releasing electrons to sp^2 hybridized carbon. Thus, as measured by their heats of combustion, the ketone 2-butanone is more stable than its aldehyde isomer butanal.

$$\underset{\text{Butanal}}{CH_3CH_2CH_2\overset{\overset{O}{\|}}{C}H} + \underset{\text{Oxygen}}{\tfrac{11}{2}O_2} \longrightarrow \underset{\text{Carbon dioxide}}{4CO_2} + \underset{\text{Water}}{4H_2O} \qquad \Delta H^\circ = -2475 \text{ kJ/mol } (-592.1 \text{ kcal/mol})$$

$$\underset{\text{2-Butanone}}{CH_3CH_2\overset{\overset{O}{\|}}{C}CH_3} + \underset{\text{Oxygen}}{\tfrac{11}{2}O_2} \longrightarrow \underset{\text{Carbon dioxide}}{4CO_2} + \underset{\text{Water}}{4H_2O} \qquad \Delta H^\circ = -2442 \text{ kJ/mol } (-584.2 \text{ kcal/mol})$$

Less heat is given off when 2-butanone is burned because it has less potential energy. A ketone carbonyl has two electron-releasing alkyl groups that contribute to its stabilization, while an aldehyde has only one. Structural effects on carbonyl group *stability* are an important factor in the *relative reactivities* of aldehydes and ketones, as will become evident later in this chapter.

17.3 PHYSICAL PROPERTIES

Physical constants such as melting point, boiling point, and solubility in water are collected for a variety of aldehydes and ketones in Appendix 1.

In general, aldehydes and ketones have higher boiling points than hydrocarbon analogs because they are more polar and the dipole-dipole attractive forces between molecules are stronger. They have lower boiling points than alcohols because, unlike alcohols, two carbonyl groups cannot form hydrogen bonds to each other.

	$CH_3CH_2CH{=}CH_2$ 1-Butene	$CH_3CH_2CH{=}O$ Propanal	$CH_3CH_2CH_2OH$ 1-Propanol
bp (1 atm)	−6°C	49°C	97°C
Solubility in water (g/100 mL)	Negligible	20	Miscible in all proportions

Aldehydes and ketones can form hydrogen bonds with the protons of water and are more soluble than hydrocarbons, but less soluble than alcohols in water.

17.4 SOURCES OF ALDEHYDES AND KETONES

Low-molecular-weight aldehydes and ketones are important industrial chemicals. While specialized procedures have been developed for making many of them, most can be prepared by oxidation (or dehydrogenation) of the corresponding alcohol. Formaldehyde, a starting material for a number of plastics, is prepared by oxidation of methanol over a silver or iron oxide–molybdenum oxide catalyst at elevated temperature.

The name *aldehyde* was invented to stand for *al*cohol *dehyd*rogenatum, indicating that aldehydes are related to alcohols by loss of hydrogen.

$$\underset{\text{Methanol}}{CH_3OH} + \underset{\text{Oxygen}}{\tfrac{1}{2}O_2} \xrightarrow[500°C]{\text{catalyst}} \underset{\text{Formaldehyde}}{HCH(=O)} + \underset{\text{Water}}{H_2O}$$

Similar processes are used to convert ethanol to acetaldehyde and isopropyl alcohol to acetone.

Acetaldehyde can be prepared by hydration of acetylene (Section 9.13), but a more economical procedure is air oxidation of ethylene in the presence of palladium chloride and copper(II) chloride as catalysts:

$$\underset{\text{Ethylene}}{CH_2{=}CH_2} + \underset{\text{Oxygen}}{\tfrac{1}{2}O_2} \xrightarrow[H_2O]{PdCl_2,\ CuCl_2} \underset{\text{Acetaldehyde}}{CH_3CH(=O)}$$

This is known as the *Wacker process.* An organopalladium compound is an intermediate.

Hydroformylation is a reaction in which alkenes are converted to aldehydes containing an additional carbon atom by reaction with carbon monoxide and hydrogen in the presence of a suitable cobalt- or rhodium-based catalyst.

$$\underset{\text{Alkene}}{RCH{=}CH_2} + \underset{\text{Carbon monoxide}}{CO} + \underset{\text{Hydrogen}}{H_2} \xrightarrow{Co_2(CO)_8} \underset{\text{Aldehyde}}{RCH_2CH_2CH(=O)}$$

Excess hydrogen brings about the hydrogenation of the aldehyde and allows the process to be adapted to the preparation of primary alcohols. Over 2×10^9 lb/yr of a variety of aldehydes and alcohols is prepared in the United States by hydroformylation. An organometallic compound based on cobalt or rhodium is an intermediate in the reaction.

Benzaldehyde is prepared industrially by hydrolysis of (dichloromethyl)benzene, also known as benzal chloride. Benzal chloride is made by free-radical chlorination of toluene:

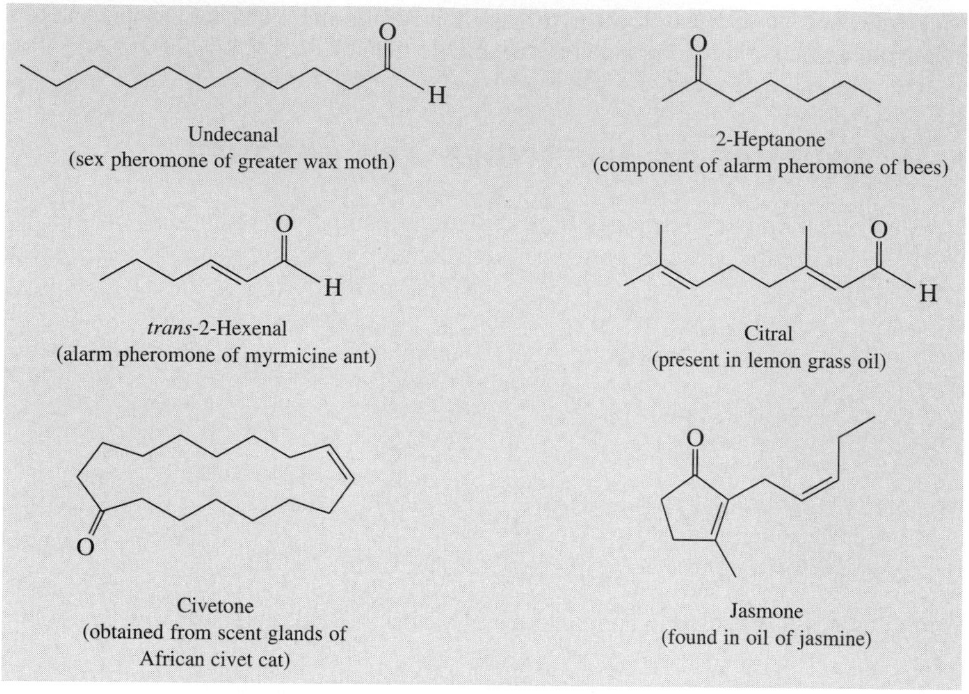

Undecanal
(sex pheromone of greater wax moth)

2-Heptanone
(component of alarm pheromone of bees)

trans-2-Hexenal
(alarm pheromone of myrmicine ant)

Citral
(present in lemon grass oil)

Civetone
(obtained from scent glands of African civet cat)

Jasmone
(found in oil of jasmine)

FIGURE 17.2 Some naturally occurring aldehydes and ketones.

$$C_6H_5-CH_3 \xrightarrow[\text{heat}]{Cl_2} C_6H_5-CHCl_2 \xrightarrow{H_2O} C_6H_5-\overset{O}{\overset{\|}{C}}H$$

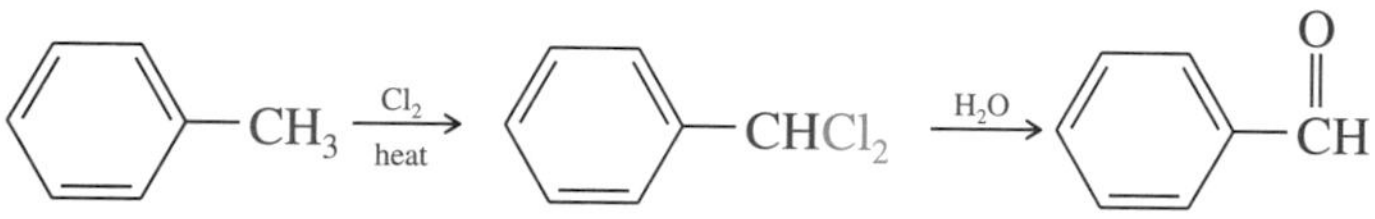

Toluene — (Dichloromethyl)benzene (benzal chloride) — Benzaldehyde

A number of aldehydes and ketones are prepared industrially and in the laboratory as well by a reaction known as the *aldol condensation.* This will be discussed in detail in Sections 18.9 through 18.12.

Many aldehydes and ketones occur naturally. Several of these are shown in Figure 17.2.

17.5 REACTIONS THAT LEAD TO ALDEHYDES AND KETONES: REVIEW AND APPLICATIONS

Reactions that yield aldehydes and ketones have appeared earlier among several functional group classes that we have already examined, namely, alkenes, alkynes, arenes, and alcohols. These reactions are summarized in Table 17.1. The most important of these reactions to the synthetic chemist are the last two in the table: the oxidation of primary alcohols to aldehydes and secondary alcohols to ketones. Indeed, when combined with reactions that yield alcohols, the oxidation methods are so versatile that *it will not be necessary to introduce any new methods for preparing aldehydes and ketones in this chapter.* A few examples will illustrate the preparation of aldehydes and ketones by way of alcohols.

First, consider the preparation of an aldehyde from a carboxylic acid. The most widely used method is an indirect one that combines the reduction of the carboxylic

TABLE 17.1
Summary of Reactions Discussed in Earlier Chapters That Yield Aldehydes and Ketones

Reaction (section) and comments	General equation and specific example
Ozonolysis of alkenes (Section 6.19) This cleavage reaction is more often seen in structural analysis than in synthesis. The substitution pattern around a double bond is revealed by identifying the carbonyl-containing compounds that make up the product. Hydrolysis of the ozonide intermediate in the presence of zinc (reductive workup) permits aldehyde products to be isolated without further oxidation.	$\underset{\text{Alkene}}{(R)(R')C{=}C(H)(R'')} \xrightarrow[\text{2. } H_2O,\ Zn]{\text{1. } O_3} \underset{\text{Two carbonyl compounds}}{R\overset{O}{\overset{\parallel}{C}}R' + R''\overset{O}{\overset{\parallel}{C}}H}$ $\underset{\text{2,6-Dimethyl-2-octene}}{(CH_3)_2C{=}CHCH_2CH_2CH(CH_3)CH_2CH_3} \xrightarrow[\text{2. } H_2O,\ Zn]{\text{1. } O_3} \underset{\text{Acetone}}{CH_3\overset{O}{\overset{\parallel}{C}}CH_3} + \underset{\text{4-Methylhexanal (91\%)}}{H\overset{O}{\overset{\parallel}{C}}CH_2CH_2\overset{CH_3}{\overset{\mid}{C}}HCH_2CH_3}$
Hydration of alkynes (Section 9.13) Reaction occurs by way of an enol intermediate formed by Markovnikov addition of water to the triple bond.	$\underset{\text{Alkyne}}{RC{\equiv}CR'} + H_2O \xrightarrow[HgSO_4]{H_2SO_4} \underset{\text{Ketone}}{R\overset{O}{\overset{\parallel}{C}}CH_2R'}$ $\underset{\text{1-Octyne}}{HC{\equiv}C(CH_2)_5CH_3} + H_2O \xrightarrow[HgSO_4]{H_2SO_4} \underset{\text{2-Octanone (91\%)}}{CH_3\overset{O}{\overset{\parallel}{C}}(CH_2)_5CH_3}$
Friedel-Crafts acylation of aromatic compounds (Section 12.7) Acyl chlorides and carboxylic acid anhydrides acylate aromatic rings in the presence of aluminum chloride. The reaction is one of electrophilic aromatic substitution in which acylium ions are generated and attack the ring.	$ArH + R\overset{O}{\overset{\parallel}{C}}Cl \xrightarrow{AlCl_3} Ar\overset{O}{\overset{\parallel}{C}}R + HCl$ or $ArH + R\overset{O}{\overset{\parallel}{C}}O\overset{O}{\overset{\parallel}{C}}R \xrightarrow{AlCl_3} Ar\overset{O}{\overset{\parallel}{C}}R + RCO_2H$ $\underset{\text{Anisole}}{CH_3O{-}C_6H_5} + \underset{\text{Acetic anhydride}}{CH_3\overset{O}{\overset{\parallel}{C}}O\overset{O}{\overset{\parallel}{C}}CH_3} \xrightarrow{AlCl_3} \underset{p\text{-Methoxyacetophenone (90–94\%)}}{CH_3O{-}C_6H_4{-}\overset{O}{\overset{\parallel}{C}}CH_3}$
Oxidation of primary alcohols to aldehydes (Section 15.10) Pyridinium dichromate (PDC) or pyridinium chlorochromate (PCC) in anhydrous media such as dichloromethane oxidizes primary alcohols to aldehydes while avoiding overoxidation to carboxylic acids.	$\underset{\text{Primary alcohol}}{RCH_2OH} \xrightarrow[CH_2Cl_2]{\text{PDC or PCC}} \underset{\text{Aldehyde}}{R\overset{O}{\overset{\parallel}{C}}H}$ $\underset{\text{1-Decanol}}{CH_3(CH_2)_8CH_2OH} \xrightarrow[CH_2Cl_2]{PDC} \underset{\text{Decanal (98\%)}}{CH_3(CH_2)_8\overset{O}{\overset{\parallel}{C}}H}$
Oxidation of secondary alcohols to ketones (Section 15.10) Many oxidizing agents are available for converting secondary alcohols to ketones. PDC or PCC may be used, as well as other Cr(VI)-based agents such as chromic acid or potassium dichromate and sulfuric acid.	$\underset{\text{Secondary alcohol}}{R\underset{OH}{\underset{\mid}{C}}HR'} \xrightarrow{Cr(VI)} \underset{\text{Ketone}}{R\overset{O}{\overset{\parallel}{C}}R'}$ $\underset{\text{1-Phenyl-1-pentanol}}{C_6H_5\underset{OH}{\underset{\mid}{C}}HCH_2CH_2CH_2CH_3} \xrightarrow[\text{acetic acid–water}]{CrO_3} \underset{\text{1-Phenyl-1-pentanone (93\%)}}{C_6H_5\overset{O}{\overset{\parallel}{C}}CH_2CH_2CH_2CH_3}$

acid to a primary alcohol with the subsequent oxidation of the primary alcohol to an aldehyde:

$$\underset{\text{Carboxylic acid}}{RCO_2H} \xrightarrow{\text{reduce}} \underset{\text{Primary alcohol}}{RCH_2OH} \xrightarrow{\text{oxidize}} \underset{\text{Aldehyde}}{RCH{=}O}$$

$$\underset{\text{Benzoic acid}}{C_6H_5COOH} \xrightarrow[\text{2. } H_2O]{\text{1. } LiAlH_4} \underset{\text{Benzyl alcohol (81\%)}}{C_6H_5CH_2OH} \xrightarrow[CH_2Cl_2]{\text{PDC}} \underset{\text{Benzaldehyde (83\%)}}{C_6H_5CH{=}O}$$

PROBLEM 17.3 Can catalytic hydrogenation be used to reduce a carboxylic acid to a primary alcohol in the first step of this sequence?

It is often necessary to prepare ketones by processes involving carbon-carbon bond formation. In such cases the standard method combines extension of the carbon chain by addition of a Grignard reagent to an aldehyde with oxidation of the resulting secondary alcohol to give the desired ketone:

$$\underset{\text{Aldehyde}}{RCH{=}O} \xrightarrow[\text{2. } H_3O^+]{\text{1. R'MgX, diethyl ether}} \underset{\text{Secondary alcohol}}{RCH(OH)R'} \xrightarrow{\text{oxidize}} \underset{\text{Ketone}}{RC({=}O)R'}$$

$$\underset{\text{Propanal}}{CH_3CH_2CH{=}O} \xrightarrow[\text{2. } H_3O^+]{\text{1. } CH_3(CH_2)_3MgBr\text{, diethyl ether}} \underset{\text{3-Heptanol}}{CH_3CH_2CH(OH)(CH_2)_3CH_3} \xrightarrow{H_2CrO_4} \underset{\text{3-Heptanone (57\% from propanal)}}{CH_3CH_2C({=}O)(CH_2)_3CH_3}$$

PROBLEM 17.4 Show how 2-butanone could be prepared by a procedure in which all of the carbons originate in acetic acid (CH_3CO_2H).

17.6 REACTIONS OF ALDEHYDES AND KETONES: A REVIEW AND A PREVIEW

A summary of our earlier encounters with the chemical reactions of aldehydes and ketones is presented in Table 17.2. All the transformations shown there are valuable tools of the synthetic organic chemist. Carbonyl groups provide access to hydrocarbons via Clemmensen or Wolff-Kishner reduction (Section 12.8), to alcohols of the same carbon skeleton by a variety of reduction methods (Section 15.2), and to alcohols of more complex structure by reaction with Grignard or organolithium reagents (Sections 14.6 and 14.7).

One of the characteristics of the carbonyl group is its tendency to undergo *nucleophilic addition* reactions of the type represented by the general equation

$$\overset{\delta+}{>}C\overset{\delta-}{=O} + \overset{\delta+}{X}-\overset{\delta-}{Y} \longrightarrow \text{C(O—X)(Y)}$$

Aldehyde or ketone — Product of nucleophilic addition

A negatively polarized atom or group attacks the positively polarized carbon of the carbonyl group in the rate-determining step of these reactions. Grignard reagents,

TABLE 17.2
Summary of Reactions of Aldehydes and Ketones Discussed in Earlier Chapters

Reaction (section) and comments	General equation and specific example
Reduction to hydrocarbons (Section 12.8) Two methods for converting carbonyl groups to methylene units are the Clemmensen reduction (zinc amalgam and concentrated hydrochloric acid) and the Wolff-Kishner reduction (heat with hydrazine and potassium hydroxide in a high-boiling alcohol).	$RCR' \text{ (C=O)} \longrightarrow RCH_2R'$ Aldehyde or ketone → Hydrocarbon Citronellal (CH=O) $\xrightarrow[\text{diethylene glycol, heat}]{H_2NNH_2,\ KOH}$ 2,6-Dimethyl-2-octene (80%)
Reduction to alcohols (Section 15.2) Aldehydes are reduced to primary alcohols and ketones are reduced to secondary alcohols by a variety of reducing agents. Catalytic hydrogenation over a metal catalyst and reduction with sodium borohydride or lithium aluminum hydride are general methods. Sodium in ethanol is useful for reduction of ketones.	$RCR' \text{ (C=O)} \longrightarrow RCHR' \text{ (OH)}$ Aldehyde or ketone → Alcohol $CH_3O-C_6H_4-CH{=}O \xrightarrow[CH_3OH]{NaBH_4} CH_3O-C_6H_4-CH_2OH$ p-Methoxybenzaldehyde → p-Methoxybenzyl alcohol (96%)
Addition of Grignard reagents and organolithium compounds (Sections 14.6–14.7) Aldehydes are converted to secondary alcohols and ketones to tertiary alcohols.	$RCR' \text{ (C=O)} + R''M \longrightarrow RC(O^-M^+)(R'')R' \xrightarrow{H_3O^+} RC(OH)(R'')R'$ Cyclohexanone + CH_3CH_2MgBr (Ethylmagnesium bromide) $\xrightarrow[\text{2. } H_3O^+]{\text{1. diethyl ether}}$ 1-Ethylcyclohexanol (HO, CH_2CH_3) (74%)

organolithium reagents, lithium aluminum hydride, and sodium borohydride all react with carbonyl compounds by nucleophilic addition.

The next section presents the mechanistic features of nucleophilic addition to aldehydes and ketones through a discussion of their *hydration,* the addition of a water molecule to the carbonyl group. Once we develop the principles of nucleophilic addition, we will survey a number of nucleophilic addition reactions of aldehydes and ketones; some of these are of synthetic interest, others are of mechanistic importance, and a few possess both qualities.

17.7 PRINCIPLES OF NUCLEOPHILIC ADDITION TO CARBONYL GROUPS: HYDRATION OF ALDEHYDES AND KETONES

Aldehydes and ketones react with water in a rapidly reversible equilibrium:

$$\underset{\substack{\text{Aldehyde}\\\text{or ketone}}}{RC(=O)R'} + \underset{\text{Water}}{H_2O} \overset{\text{fast}}{\rightleftharpoons} \underset{\substack{\text{Geminal diol}\\\text{(hydrate)}}}{RC(OH)_2R'} \qquad K_{hydr} = \frac{[\text{hydrate}]}{[\text{carbonyl compound}][\text{water}]}$$

Overall, the reaction is classified as an *addition.* The elements of water add to the carbonyl group. Hydrogen becomes bonded to the negatively polarized carbonyl oxygen, hydroxyl to the positively polarized carbon.

Table 17.3 compares the equilibrium constants K_{hydr} for hydration of some simple aldehydes and ketones. The position of equilibrium depends strongly on the nature of the carbonyl group and is influenced by a combination of *electronic* and *steric* effects.

TABLE 17.3
Equilibrium Constants (K_{hydr}) for Hydration of Some Aldehydes and Ketones

Carbonyl compound	Hydrate	K_{hydr}*	Percent conversion to hydrate†
$HC(=O)H$	$CH_2(OH)_2$	41	99.96
$CH_3C(=O)H$	$CH_3CH(OH)_2$	1.8×10^{-2}	50
$(CH_3)_3CC(=O)H$	$(CH_3)_3CCH(OH)_2$	4.1×10^{-3}	19
$CH_3C(=O)CH_3$	$(CH_3)_2C(OH)_2$	2.5×10^{-5}	0.14

* $K_{hydr} = \dfrac{[\text{hydrate}]}{[\text{carbonyl compound}][\text{water}]}$. Units of K_{hydr} are M^{-1}.

† Total concentration (hydrate plus carbonyl compound) assumed to be 1 *M*. Water concentration is 55.5 *M*.

Consider first the electronic effect of substituents on the stabilization of the carbonyl group. As the starting material becomes more stable, the smaller is its equilibrium constant for hydration. *Formaldehyde* has no alkyl substituents to stabilize its carbonyl group and is converted almost completely to its hydrate in aqueous solution. The carbonyl of *acetaldehyde* is stabilized by one alkyl substituent, the carbonyl of *acetone* by two. The proportion of hydrate present in an aqueous solution of a typical aldehyde is much less than that in an aqueous solution of formaldehyde, while ketones are converted to their hydrates to an even smaller extent.

A striking example of an electronic effect on carbonyl group stability and its relation to the equilibrium constant for hydration is seen in the case of hexafluoroacetone. In contrast to the almost negligible hydration of acetone, hexafluoroacetone is completely hydrated.

$$CF_3\overset{O}{\overset{\|}{C}}CF_3 + H_2O \rightleftharpoons CF_3\underset{OH}{\underset{|}{\overset{OH}{\overset{|}{C}}}}CF_3 \qquad K_{hydr} = 22{,}000$$

Hexafluoroacetone Water 1,1,1,3,3,3-Hexafluoro-2,2-propanediol

Instead of stabilizing the carbonyl group by electron donation as alkyl substituents do, trifluoromethyl groups destabilize it by withdrawing electrons. A less stabilized carbonyl group is associated with a greater equilibrium constant for addition.

PROBLEM 17.5 *Chloral* is one of the common names for trichloroethanal. A solution of chloral in water is called *chloral hydrate;* this material has featured prominently in countless detective stories as the notorious "Mickey Finn" knockout drops. Write a structural formula for chloral hydrate.

To understand the role played by steric effects, let us examine the geminal diol product (hydrate). The carbon that bears the two hydroxyl groups is sp^3 hybridized. Its substituents are more crowded than they are in the starting aldehyde or ketone. Increased crowding can be better tolerated when the substituents are small (hydrogen) than when they are large (alkyl groups).

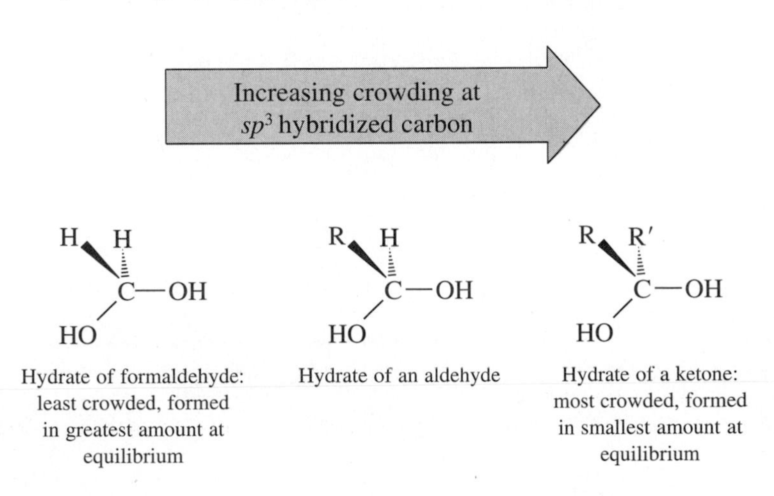

Hydrate of formaldehyde: least crowded, formed in greatest amount at equilibrium

Hydrate of an aldehyde

Hydrate of a ketone: most crowded, formed in smallest amount at equilibrium

Electronic and steric effects operate in the same direction to make the equilibrium constants for hydration of aldehydes more favorable than those of ketones.

Let us turn now to structural effects and the effects of catalysts on the *rate* of hydration. While the equilibrium for the hydration of aldehydes and ketones is rapidly established even under neutral conditions, it is markedly catalyzed by both acids and bases.

Figure 17.3 presents the two steps of the base-catalyzed mechanism. In the first step the nucleophile, a hydroxide ion, attacks the carbonyl group, forming a bond to the carbonyl carbon. The product of this step is an alkoxide anion. It abstracts a proton from water in the second step to yield the geminal diol product and regenerate hydroxide ion. The proton-transfer step, like other proton transfers between oxygens that we have seen, is fast. The first step is rate-determining.

The role of the basic catalyst (HO^-) is to increase the rate of the nucleophilic addition step. Hydroxide ion, the nucleophile in the base-catalyzed reaction, is much more reactive than a water molecule, the nucleophile in neutral media.

Aldehydes react faster than ketones for almost the same reasons that their equilibrium constants for hydration are more favorable. The $sp^2 \longrightarrow sp^3$ hybridization change that the carbonyl carbon undergoes on hydration is partially developed in the transition state for the rate-determining nucleophilic addition step (Figure 17.4). Alkyl groups at the reaction site increase the activation energy by simultaneously lowering the energy of the starting state (ketones have a more stabilized carbonyl group than aldehydes) and raising the energy of the transition state (a steric crowding effect).

Three steps are involved in the acid-catalyzed hydration reaction, as shown in Figure 17.5. The first and last are rapid proton-transfer processes. The second is the nucleophilic addition step. The role of the acid catalyst is to activate the carbonyl group toward attack by a weakly nucleophilic water molecule. Protonation of oxygen makes the carbonyl carbon of an aldehyde or a ketone much more electrophilic. Expressed in resonance terms, the protonated carbonyl has a greater degree of carbocation character than an unprotonated carbonyl.

$$\text{C}=\overset{+}{\text{O}}\text{–H} \longleftrightarrow \overset{+}{\text{C}}\text{–}\ddot{\text{O}}\text{–H}$$

Step 1: Nucleophilic addition of hydroxide ion to the carbonyl group

$$HO^- + RR'C=O \overset{\text{slow}}{\rightleftharpoons} RR'C(OH)–O^-$$

Hydroxide   Aldehyde or ketone

Step 2: Proton transfer from water to the intemediate formed in the first step

$$RR'C(OH)–O^- + H–OH \overset{\text{fast}}{\rightleftharpoons} RR'C(OH)–O–H + {}^-OH$$

Water   Geminal diol   Hydroxide ion

FIGURE 17.3 The mechanism of hydration of an aldehyde or a ketone under base-catalyzed conditions.

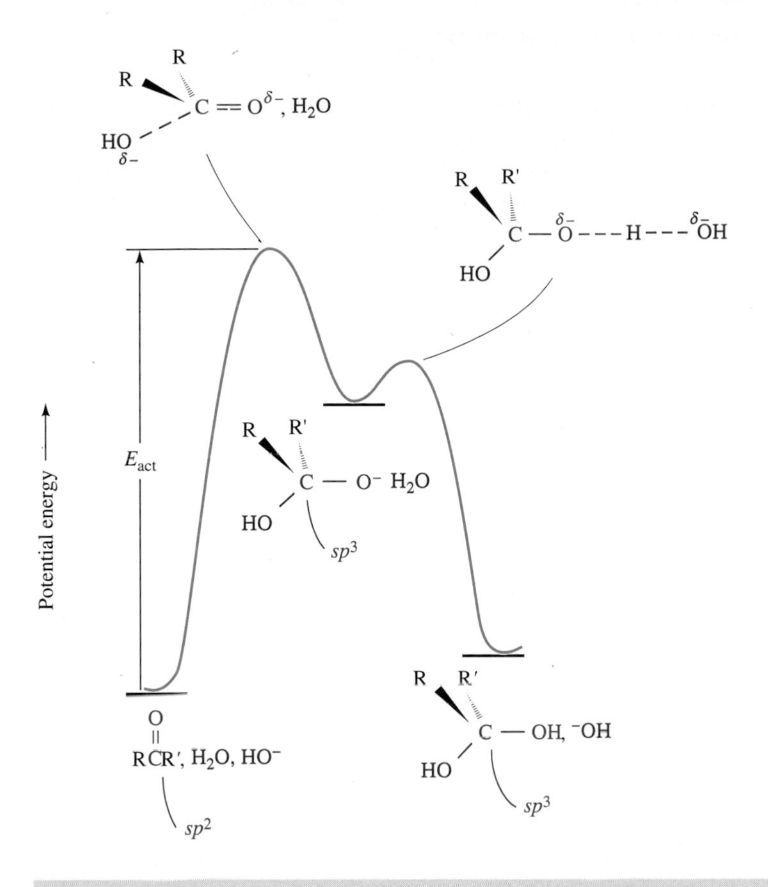

FIGURE 17.4 Potential energy diagram for base-catalyzed hydration of a carbonyl compound.

Step 1: Protonation of the carbonyl oxygen

$$R(R')C{=}\ddot{O}: + H{-}\overset{+}{\ddot{O}}H_2 \underset{}{\overset{\text{fast}}{\rightleftharpoons}} R(R')C{=}\overset{+}{\ddot{O}}{-}H + H\ddot{O}H$$

Aldehyde or ketone | Hydronium ion | Conjugate acid of carbonyl compound | Water

Step 2: Nucleophilic addition to the protonated aldehyde or ketone

$$H{-}\ddot{O}{-}H + R(R')C{=}\overset{+}{\ddot{O}}{-}H \overset{\text{slow}}{\rightleftharpoons} RR'C(\ddot{O}H)(\overset{+}{\ddot{O}}H_2)$$

Water | Conjugate acid of carbonyl compound | Conjugate acid of geminal diol

Step 3: Proton transfer from the conjugate acid of the geminal diol to a water molecule

$$RR'C(\ddot{O}H)(\overset{+}{\ddot{O}}H_2) + H_2\ddot{O}: \overset{\text{fast}}{\rightleftharpoons} RR'C(\ddot{O}H)(\ddot{O}H) + H_3\overset{+}{\ddot{O}}$$

Conjugate acid of geminal diol | Water | Geminal diol | Hydronium ion

FIGURE 17.5 The mechanism of hydration of an aldehyde or ketone under acid-catalyzed conditions.

Electron delocalization in the neutral carbonyl involves a charge-separated dipolar resonance form. Electron delocalization in the protonated carbonyl is more pronounced because there is no separation of opposite charges in either resonance form.

Steric and electronic effects influence the rate of nucleophilic addition to a protonated carbonyl group in much the same way as they do for the case of a neutral one, and protonated aldehydes react faster than protonated ketones.

With this as background, let us now examine how the principles of nucleophilic addition apply to the characteristic reactions of aldehydes and ketones. We shall begin with the addition of hydrogen cyanide.

17.8 CYANOHYDRIN FORMATION

The product of addition of hydrogen cyanide to an aldehyde or a ketone contains both a hydroxyl group and a cyano group bonded to the same carbon. Compounds of this type are called **cyanohydrins.**

$$\underset{\text{Aldehyde or ketone}}{RC(=O)R'} + \underset{\text{Hydrogen cyanide}}{HC\equiv N} \longrightarrow \underset{\text{Cyanohydrin}}{RC(OH)(C\equiv N)R'}$$

The mechanism of this reaction is outlined in Figure 17.6. It is analogous to the mechanism of base-catalyzed hydration in that the nucleophile (cyanide ion) attacks the carbonyl carbon in the first step of the reaction, after which proton transfer to the carbonyl oxygen takes place in the second step.

The addition of hydrogen cyanide is catalyzed by cyanide ion, but HCN is too weak an acid to provide enough $:\bar{C}\equiv N:$ for the reaction to proceed at a reasonable rate. Therefore cyanohydrins are normally prepared by adding an acid to a solution containing the carbonyl compound and sodium or potassium cyanide. This procedure ensures that free cyanide ion is always present in amounts sufficient to accelerate the rate of the reaction.

In substitutive IUPAC nomenclature, cyanohydrins are named as hydroxy derivatives of nitriles. Since nitrile nomenclature will not be discussed until Section 20.1, we will refer to cyanohydrins as derivatives of the parent aldehyde or ketone as shown in the examples. This conforms to the practice of most chemists.

Cyanohydrin formation is reversible, and the position of equilibrium depends on the steric and electronic factors governing nucleophilic addition to carbonyl groups described in the preceding section. Aldehydes and unhindered ketones give good yields of cyanohydrins.

$$\underset{\text{2,4-Dichlorobenzaldehyde}}{2,4\text{-}Cl_2C_6H_3CH(=O)} \xrightarrow[\text{then HCl}]{\text{NaCN, ether-water}} \underset{\text{2,4-Dichlorobenzaldehyde cyanohydrin (100\%)}}{2,4\text{-}Cl_2C_6H_3CH(OH)C\equiv N}$$

$$\underset{\text{Acetone}}{CH_3C(=O)CH_3} \xrightarrow[\text{then } H_2SO_4]{\text{NaCN, } H_2O} \underset{\text{Acetone cyanohydrin (77–78\%)}}{CH_3C(OH)(C\equiv N)CH_3}$$

The overall reaction:

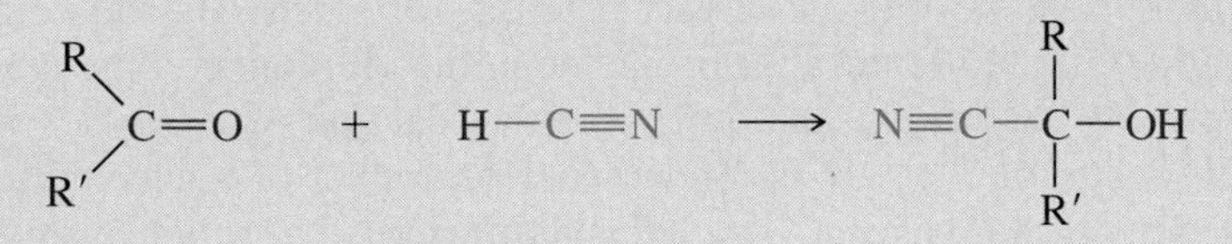

$$RR'C{=}O + H{-}C{\equiv}N \longrightarrow N{\equiv}C{-}CRR'{-}OH$$

Aldehyde or ketone   Hydrogen cyanide   Cyanohydrin

Step 1: Nucleophilic attack by the negatively charged carbon of cyanide ion at the carbonyl carbon of the aldehyde or ketone. Hydrogen cyanide itself is not very nucleophilic and does not ionize to form cyanide ion to a significant extent. Thus, a source of cyanide ion such as NaCN or KCN is used.

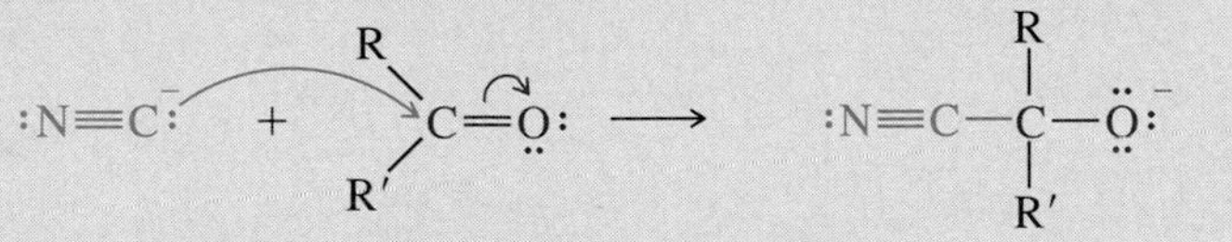

$$:N{\equiv}C:^- + RR'C{=}\ddot{O}: \longrightarrow :N{\equiv}C{-}CRR'{-}\ddot{O}:^-$$

Cyanide ion   Aldehyde or ketone   Conjugate base of cyanohydrin

Step 2: The alkoxide ion formed in the first step abstracts a proton from hydrogen cyanide. This step yields the cyanohydrin product and regenerates cyanide ion.

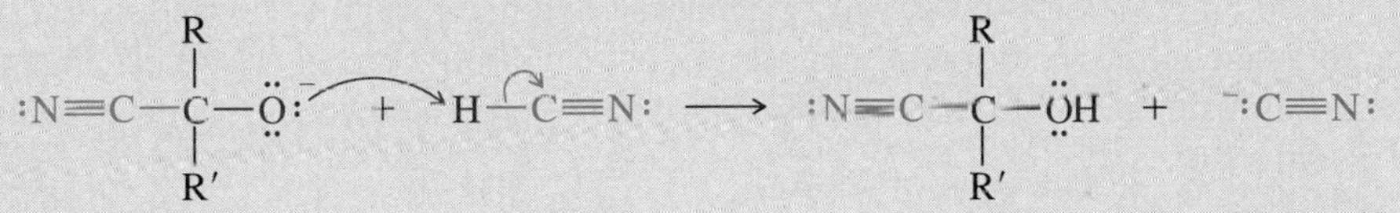

$$:N{\equiv}C{-}CRR'{-}\ddot{O}:^- + H{-}C{\equiv}N: \longrightarrow :N{\equiv}C{-}CRR'{-}\ddot{O}H + {}^-:C{\equiv}N:$$

Conjugate base of cyanohydrin   Hydrogen cyanide   Cyanohydrin   Cyanide ion

FIGURE 17.6 The mechanism of cyanohydrin formation from an aldehyde or a ketone.

Cyanohydrin formation is of synthetic value in that a new carbon-carbon bond is made by this process and a cyano group may be converted to a carboxylic acid function (by hydrolysis, to be discussed in Section 19.12) or to an amine (by reduction, to be discussed in Section 22.10).

PROBLEM 17.6 The hydroxyl group of a cyanohydrin is also a potentially reactive site. *Methacrylonitrile* is an industrial chemical used in the production of plastics and fibers. One method for its preparation is the acid-catalyzed dehydration of acetone cyanohydrin. Deduce the structure of *methacrylonitrile.*

A few cyanohydrins and ethers of cyanohydrins occur naturally. One species of millipede stores benzaldehyde cyanohydrin, along with an enzyme that catalyzes its cleavage to benzaldehyde and hydrogen cyanide, in separate compartments above its legs. When attacked, the insect ejects a mixture of the cyanohydrin and the enzyme, repelling the invader by spraying it with hydrogen cyanide.

17.9 ACETAL FORMATION

Many of the most interesting and useful reactions of aldehydes and ketones involve transformation of the initial product of nucleophilic addition to some other substance under the reaction conditions. An example is the reaction of aldehydes with alcohols under conditions of acid catalysis. The expected product of nucleophilic addition of the alcohol to the carbonyl group is called a **hemiacetal.** The product actually isolated, however, corresponds to reaction of one mole of the aldehyde with *two* moles of alcohol to give *geminal diethers* known as **acetals:**

$$\underset{\text{Aldehyde}}{\mathrm{RCH{=}O}} \xrightleftharpoons{\mathrm{R'OH,\ H^+}} \underset{\text{Hemiacetal}}{\mathrm{RCH(OH)(OR')}} \xrightleftharpoons{\mathrm{R'OH,\ H^+}} \underset{\text{Acetal}}{\mathrm{RCH(OR')_2}} + \underset{\text{Water}}{\mathrm{H_2O}}$$

$$\underset{\text{Benzaldehyde}}{\mathrm{C_6H_5CH{=}O}} + \underset{\text{Ethanol}}{\mathrm{2CH_3CH_2OH}} \xrightarrow{\mathrm{HCl}} \underset{\text{Benzaldehyde diethyl acetal (66\%)}}{\mathrm{C_6H_5CH(OCH_2CH_3)_2}}$$

The overall reaction proceeds in two stages. The hemiacetal is formed in the first stage by nucleophilic addition of the alcohol to the carbonyl group. The mechanism of hemiacetal formation is exactly analogous to that of acid-catalyzed hydration of aldehydes and ketones (Section 17.7):

$$\underset{\text{Aldehyde}}{\mathrm{RCH{=}\ddot{O}\!:}} \xrightleftharpoons{\mathrm{H^+}} \mathrm{RCH{=}\overset{+}{O}H} \xrightleftharpoons{\mathrm{R'\ddot{O}H}} \mathrm{RCH(\ddot{O}H)(\overset{+}{O}HR')} \xrightleftharpoons{\mathrm{-H^+}} \underset{\text{Hemiacetal}}{\mathrm{RCH(OH)(\ddot{O}R')}}$$

Under the acidic conditions of its formation, the hemiacetal is converted to an acetal by way of a carbocation intermediate:

$$\underset{\text{Hemiacetal}}{\mathrm{RCH(\ddot{O}H)(\ddot{O}R')}} \xrightleftharpoons{\mathrm{H^+,\ fast}} \mathrm{RCH(\overset{+}{O}H_2)(\ddot{O}R')} \xrightleftharpoons{\text{slow}} \underset{\text{Carbocation}}{\mathrm{R\overset{+}{C}H(\ddot{O}R')}} + \underset{\text{Water}}{\mathrm{H_2\ddot{O}\!:}}$$

This carbocation is stabilized by electron release from its oxygen substituent:

$$\underset{\curvearrowright :\ddot{O}R'}{\overset{R \quad H}{\overset{+}{C}}} \longleftrightarrow \underset{^{+}\ddot{O}R'}{\overset{R \quad H}{C}\Vert}$$

A particularly stable resonance form; both carbon and oxygen have octets of electrons.

Nucleophilic capture of the carbocation intermediate by an alcohol molecule leads to an acetal:

$$\underset{\text{Alcohol}}{R'\ddot{O}H} + \underset{^{+}\ddot{O}R'}{\overset{R \quad H}{C}\Vert} \underset{}{\overset{\text{fast}}{\rightleftharpoons}} \underset{:\ddot{O}R'}{\overset{H\overset{+}{\ddot{O}}R'}{RCH}} \underset{}{\overset{-H^{+},\ \text{fast}}{\rightleftharpoons}} \underset{\text{Acetal}}{\underset{:\ddot{O}R'}{\overset{:\ddot{O}R'}{RCH}}}$$

PROBLEM 17.7 Write a stepwise mechanism for the formation of benzaldehyde diethyl acetal from benzaldehyde and ethanol under conditions of acid catalysis.

Acetal formation is reversible in acid. An equilibrium is established between the reactants, i.e., the carbonyl compound and the alcohol, and the acetal product. The position of equilibrium is favorable for acetal formation from most aldehydes, especially when excess alcohol is present as the reaction solvent. For most ketones the position of equilibrium is unfavorable, and other methods must be used for the preparation of acetals from ketones.

At one time it was customary to designate the products of addition of alcohols to ketones as *ketals*. This term has been dropped from the IUPAC system of nomenclature, and the term *acetal* is now applied to the adducts of both aldehydes and ketones.

Diols that bear two hydroxyl groups in a 1,2 or 1,3 relationship to each other yield *cyclic acetals* on reaction with either aldehydes or ketones. The five-membered cyclic acetals derived from ethylene glycol are the most commonly encountered examples. Often the position of equilibrium is made more favorable by removing the water formed in the reaction by azeotropic distillation with benzene or toluene:

$$\underset{\text{Heptanal}}{CH_3(CH_2)_5\overset{O}{\overset{\Vert}{C}}H} + \underset{\substack{\text{Ethylene glycol} \\ \text{(1,2-ethanediol)}}}{HOCH_2CH_2OH} \xrightarrow[\text{benzene}]{p\text{-toluenesulfonic acid}} \underset{\substack{\text{2-Hexyl-1,3-dioxolane} \\ (81\%)}}{\text{cyclic acetal: } H,\ (CH_2)_5CH_3}$$

$$\underset{\substack{\text{Benzyl methyl} \\ \text{ketone}}}{C_6H_5CH_2\overset{O}{\overset{\Vert}{C}}CH_3} + \underset{\substack{\text{Ethylene glycol} \\ \text{(1,2-ethanediol)}}}{HOCH_2CH_2OH} \xrightarrow[\text{benzene}]{p\text{-toluenesulfonic acid}} \underset{\substack{\text{2-Benzyl-2-methyl-1,3-dioxolane} \\ (78\%)}}{\text{cyclic acetal: } C_6H_5CH_2,\ CH_3}$$

PROBLEM 17.8 Write the structures of the cyclic acetals derived from:

(*a*) Cyclohexanone and ethylene glycol
(*b*) Benzaldehyde and 1,3-propanediol
(*c*) Isobutyl methyl ketone and ethylene glycol
(*d*) Isobutyl methyl ketone and 2,2-dimethyl-1,3-propanediol.

SAMPLE SOLUTION (*a*) The cyclic acetals derived from ethylene glycol contain a five-membered 1,3-dioxolane ring.

$$\text{C}_6\text{H}_{10}\text{=O} + \text{HOCH}_2\text{CH}_2\text{OH} \xrightarrow{\text{H}^+} \text{(cyclic acetal)}$$

Cyclohexanone — Ethylene glycol — Acetal of cyclohexanone and ethylene glycol

Acetals are susceptible to hydrolysis in aqueous acid:

$$\text{RC(OR}'')_2\text{R}' + \text{H}_2\text{O} \overset{\text{H}^+}{\rightleftharpoons} \text{RC(=O)R}' + 2\text{R}''\text{OH}$$

Acetal — Aldehyde or ketone — Alcohol

This reaction is simply the reverse of the reaction by which acetals are formed—acetal formation is favored by excess alcohol, acetal hydrolysis by excess water. Acetal formation and acetal hydrolysis share the same mechanistic pathway but traverse that pathway in opposite directions. In the following section you will see how acetal formation and hydrolysis have been applied to synthetic organic chemistry as a means of carbonyl group protection.

PROBLEM 17.9 Problem 17.7 asked you to write a mechanism describing formation of benzaldehyde diethyl acetal from benzaldehyde and ethanol. Write a stepwise mechanism for the acid hydrolysis of this acetal.

17.10 ACETALS AS PROTECTING GROUPS

In the practice of organic synthesis, it frequently happens that one of the reactants contains a functional group that is incompatible with the reaction conditions. Consider, for example, the conversion

$$\text{CH}_3\text{C(=O)CH}_2\text{CH}_2\text{C}{\equiv}\text{CH} \longrightarrow \text{CH}_3\text{C(=O)CH}_2\text{CH}_2\text{C}{\equiv}\text{CCH}_3$$

5-Hexyn-2-one — 5-Heptyn-2-one

Seemingly straightforward enough, what is needed is to prepare the acetylenic anion, then alkylate it with methyl iodide (Section 9.7). There is a complication, however.

The carbonyl group in the starting alkyne will neither tolerate the strongly basic conditions required for anion formation nor survive in a solution containing carbanions. Acetylide ions add to carbonyl groups (Section 14.8). Thus, the necessary anion

$$CH_3\overset{\overset{\displaystyle O}{\|}}{C}CH_2CH_2C{\equiv}\overset{-}{C}{:}$$

is inaccessible.

The strategy that is routinely followed is to *protect* the carbonyl group during the reactions with which it is incompatible and then to *remove* the protecting group in a subsequent step. Acetals, especially those derived from ethylene glycol, are among the most useful groups for carbonyl protection, because they can be introduced and removed readily. A fact of vital importance is that acetals resemble ethers in being inert to many of the reagents, such as hydride reducing agents and organometallic compounds, that react readily with carbonyl groups. The sequence shown is the one actually used to bring about the desired transformation.

(*a*) Protection of carbonyl group

$$CH_3\overset{\overset{\displaystyle O}{\|}}{C}CH_2CH_2C{\equiv}CH \xrightarrow[\text{acid, benzene}]{HOCH_2CH_2OH,\ p\text{-toluenesulfonic}} \text{(1,3-dioxolane ring with } H_3C \text{ and } CH_2CH_2C{\equiv}CH)$$

5-Hexyn-2-one

2-(3′-Butynyl)-2-methyl-1,3-dioxolane (80%)

(*b*) Alkylation of alkyne

$$\text{(1,3-dioxolane: } H_3C,\ CH_2CH_2C{\equiv}CH) \xrightarrow[2.\ CH_3I]{1.\ NaNH_2,\ NH_3} \text{(1,3-dioxolane: } H_3C,\ CH_2CH_2C{\equiv}CCH_3)$$

2-Methyl-2-(3′-pentynyl)-1,3-dioxolane (78%)

(*c*) Unmasking of the carbonyl group by hydrolysis

$$\text{(1,3-dioxolane: } H_3C,\ CH_2CH_2C{\equiv}CCH_3) \xrightarrow[HCl]{H_2O} CH_3\overset{\overset{\displaystyle O}{\|}}{C}CH_2CH_2C{\equiv}CCH_3$$

2-Methyl-2-(3′-pentynyl)-1,3-dioxolane

5-Heptyn-2-one (96%)

Although protecting and unmasking the carbonyl group add two steps to the synthetic procedure, both steps are essential to its success. The tactic of functional group protection is frequently encountered in preparative organic chemistry, and considerable attention has been paid to the design of effective protecting groups for a variety of functionalities.

PROBLEM 17.10 Acetal formation is a characteristic reaction of aldehydes and ketones, but not of carboxylic acids. Show how you could advantageously use a cyclic acetal protecting group in the following synthesis:

Convert $CH_3\overset{O}{\overset{\|}{C}}-C_6H_4-\overset{O}{\overset{\|}{C}}OH$ to $CH_3\overset{O}{\overset{\|}{C}}-C_6H_4-CH_2OH$

17.11 REACTION WITH PRIMARY AMINES. NUCLEOPHILIC ADDITION-ELIMINATION

A second two-stage reaction that begins with nucleophilic addition to aldehydes and ketones is their reaction with primary amines, compounds of the type RNH_2 or $ArNH_2$. In the first stage of the reaction the amine adds to the carbonyl group to give a species known as a **carbinolamine.** Once formed, the carbinolamine undergoes dehydration to yield the product of the reaction, an *N*-alkyl- or *N*-aryl-substituted **imine:**

N-substituted imines are sometimes called **Schiff's bases,** after Hugo Schiff, a German chemist who described their formation in 1864.

$$\underset{\text{Aldehyde or ketone}}{R\overset{O}{\overset{\|}{C}}R'} + \underset{\text{Primary amine}}{R''\ddot{N}H_2} \overset{\text{addition}}{\rightleftharpoons} \underset{\text{Carbinolamine}}{R\overset{OH}{\overset{|}{\underset{H\ddot{N}R''}{\underset{|}{C}}}}R'} \overset{\text{elimination}}{\rightleftharpoons} \underset{N\text{-substituted imine}}{R\overset{:NR''}{\overset{\|}{C}}R'} + \underset{\text{Water}}{H_2O}$$

$$\underset{\text{Benzaldehyde}}{C_6H_5\overset{O}{\overset{\|}{C}}H} + \underset{\text{Methylamine}}{CH_3NH_2} \longrightarrow \underset{N\text{-Benzylidenemethylamine (70\%)}}{C_6H_5CH{=}NCH_3}$$

$$\underset{\text{Cyclohexanone}}{C_6H_{10}{=}O} + \underset{\text{Isobutylamine}}{(CH_3)_2CHCH_2NH_2} \longrightarrow \underset{N\text{-Cyclohexylideneisobutylamine (79\%)}}{C_6H_{10}{=}NCH_2CH(CH_3)_2}$$

Both the addition and the elimination phase of the reaction are accelerated by acid catalysis. Careful control of pH is essential, since sufficient acid must be present to give a reasonable equilibrium concentration of the protonated form of the aldehyde or ketone. However, too acidic a reaction medium converts the amine to its protonated form, a form that is not nucleophilic, and retards reaction.

PROBLEM 17.11 Write the structure of the carbinolamine intermediate and the imine product formed in the reaction of each of the following:

(*a*) Acetaldehyde and benzylamine, $C_6H_5CH_2NH_2$
(*b*) Benzaldehyde and butylamine, $CH_3CH_2CH_2CH_2NH_2$

(*c*) Cyclohexanone and *tert*-butylamine, $(CH_3)_3CNH_2$

(*d*) Acetophenone and cyclohexylamine, (cyclohexyl)$-NH_2$

SAMPLE SOLUTION The carbinolamine is formed by nucleophilic addition of the amine to the carbonyl group. Its dehydration gives the imine product.

$$\underset{\text{Acetaldehyde}}{CH_3\overset{O}{\overset{\|}{C}}H} + \underset{\text{Benzylamine}}{C_6H_5CH_2NH_2} \longrightarrow \underset{\text{Carbinolamine intermediate}}{CH_3\overset{OH}{\overset{|}{C}}H-\underset{H}{\underset{|}{N}}CH_2C_6H_5} \xrightarrow{-H_2O} \underset{\text{Imine product (N-ethylidenebenzylamine)}}{CH_3CH{=}NCH_2C_6H_5}$$

A number of compounds of the type H_2NZ, generally referred to as ''derivatives of ammonia,'' react with aldehydes and ketones in a manner analogous to that of primary amines. The carbonyl group (C=O) is converted to C=NZ, and a molecule of water is formed. Table 17.4 presents examples of some of these reactions. The mechanism by which each proceeds is similar to the nucleophilic addition-elimination mechanism described for the reaction of primary amines with aldehydes and ketones.

Imine formation, as well as the reactions listed in Table 17.4, are reversible and have been extensively studied from a mechanistic perspective because of their relevance to biological processes. Many biological reactions involve initial binding of a carbonyl compound to an enzyme or coenzyme via imine formation. The boxed essay that accompanies this section discusses this aspect of biological chemistry.

TABLE 17.4
Reaction of Aldehydes and Ketones with Derivatives of Ammonia: $R\overset{O}{\overset{\|}{C}}R' + H_2NZ \longrightarrow R\overset{NZ}{\overset{\|}{C}}R' + H_2O$

Reagent (H_2NZ)	Name of reagent	Type of product	Example
H_2NOH	Hydroxylamine	Oxime	$\underset{\text{Heptanal}}{CH_3(CH_2)_5\overset{O}{\overset{\|}{C}}H} \xrightarrow{H_2NOH} \underset{\text{Heptanal oxime (81–93\%)}}{CH_3(CH_2)_5\overset{NOH}{\overset{\|}{C}}H}$
$H_2NNHC_6H_5$*	Phenylhydrazine	Phenylhydrazone	$\underset{\text{Acetophenone}}{C_6H_5\overset{O}{\overset{\|}{C}}CH_3} \xrightarrow{H_2NNHC_6H_5} \underset{\text{Acetophenone phenylhydrazone (87–91\%)}}{C_6H_5\overset{NNHC_6H_5}{\overset{\|}{C}}CH_3}$
$H_2NNH\overset{O}{\overset{\|}{C}}NH_2$	Semicarbazide	Semicarbazone	$\underset{\text{2-Dodecanone}}{CH_3\overset{O}{\overset{\|}{C}}(CH_2)_9CH_3} \xrightarrow{H_2NNH\overset{O}{\overset{\|}{C}}NH_2} \underset{\text{2-Dodecanone semicarbazone (93\%)}}{CH_3\overset{NNH\overset{O}{\overset{\|}{C}}NH_2}{\overset{\|}{C}}(CH_2)_9CH_3}$

*Compounds related to phenylhydrazine react in an analogous way. *p*-Nitrophenylhydrazine yields *p*-nitrophenylhydrazones; 2,4-dinitrophenylhydrazine yields 2,4-dinitrophenylhydrazones.

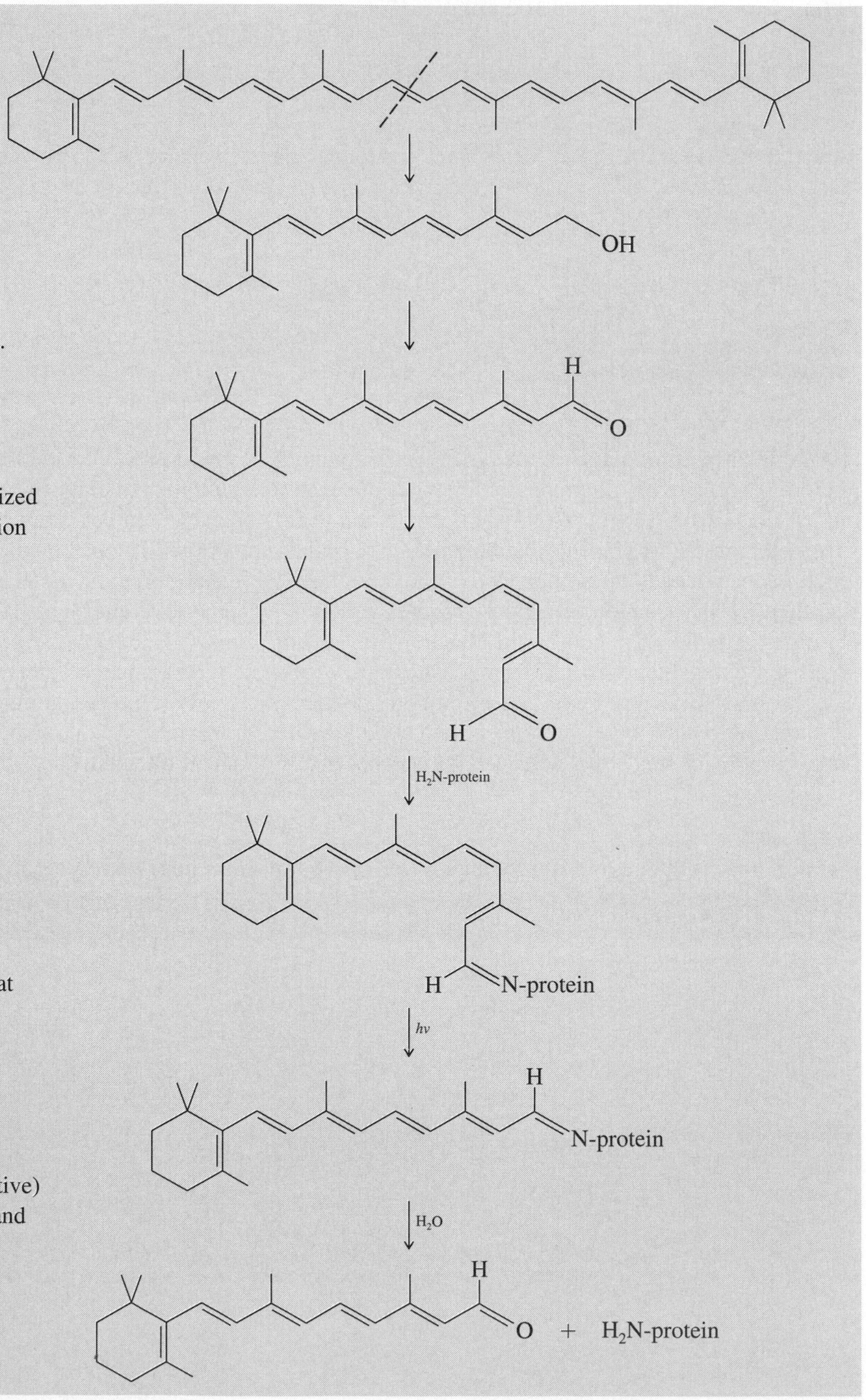

β-Carotene obtained from the diet is cleaved at its central carbon-carbon bond to give vitamin A (retinol)

Oxidation of retinol converts it to the corresponding aldehyde, retinal.

The double bond at C-11 is isomerized from the trans to the cis configuration

11-*cis*-Retinal is the biologically active stereoisomer and reacts with the protein opsin to form an imine. The covalently bound complex between 11-*cis*-retinal and ospin is called *rhodopsin*.

Rhodopsin absorbs a photon of light, causing the cis double-bond at C-11 to undergo a photochemical transformation to trans, which triggers a nerve impulse detected by the brain as a visual image.

Hydrolysis of the isomerized (inactive) form of rhodopsin liberates opsin and the all-trans isomer of retinal.

FIGURE 17.7 Summary diagram illustrating how imine formation between the aldehyde function of 11-*cis*-retinal and an amino group of a protein (opsin) is involved in the chemistry of vision. The numbering scheme used in retinal is based on one specifically developed for carotenes and compounds derived from them.

IMINES IN BIOLOGICAL CHEMISTRY

Many biological processes involve an "association" between two species in a step prior to some subsequent transformation. This association can take many forms. It can be a weak association of the attractive van der Waals type, or a stronger interaction such as a hydrogen bond. It can be an electrostatic attraction between a positively charged atom of one molecule and a negatively charged atom of another. Covalent bond formation between two species of complementary chemical reactivity represents an extreme kind of "association." It often occurs in biological processes in which aldehydes or ketones react with amines via imine intermediates.

An example of a biologically important aldehyde is *pyridoxal phosphate.* Pyridoxal phosphate is the active form of *vitamin* B_6 and is a coenzyme for many of the reactions of α-amino acids. In these reactions the amino acid binds to the coenzyme by reacting with it to form the imine shown below. Reactions then take place at the amino acid portion of the imine, modifying the amino acid. In the last step, enzyme-catalyzed hydrolysis cleaves the imine to pyridoxal and the modified amino acid.

A key step in the chemistry of vision is binding of an aldehyde to an enzyme via an imine. An outline of the steps involved is presented in Figure 17.7. It starts with *β-carotene,* a pigment that occurs naturally in several fruits and vegetables, including carrots. β-Carotene undergoes oxidative cleavage in the liver to give an alcohol known as *retinol* or *vitamin A.* Oxidation of vitamin A, followed by isomerization of one of its double bonds, gives the aldehyde *11-cis*-retinal. In the eye, the aldehyde function of 11-*cis*-retinal combines with an amino group of the protein *opsin* to form an imine called *rhodopsin.* When rhodopsin absorbs a photon of visible light, the cis double bond of the retinal unit undergoes a photochemical cis-to-trans isomerization, which is attended by a dramatic change in its shape and a change in the conformation of rhodopsin. This conformational change is translated into a nerve impulse perceived by the brain as a visual image. Enzyme-promoted hydrolysis of the photochemically isomerized rhodopsin regenerates opsin and a molecule of all-*trans*-retinal. Once all-*trans*-retinal has been enzymatically converted to its 11-cis isomer, it and opsin reenter the cycle shown in Figure 17.7.

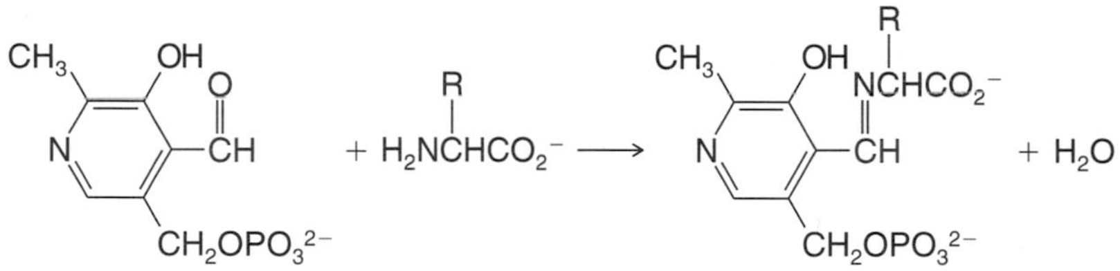

$$\text{Pyridoxal phosphate (pyridine ring with } CH_3\text{, OH, } CH{=}O\text{, } CH_2OPO_3^{2-}) + H_2N\overset{R}{\overset{|}{C}}HCO_2^- \longrightarrow \text{Imine (pyridine ring with } CH_3\text{, OH, } CH{=}N\overset{R}{\overset{|}{C}}HCO_2^-\text{, } CH_2OPO_3^{2-}) + H_2O$$

Pyridoxal phosphate — α-Amino acid — Imine

17.12 REACTION WITH SECONDARY AMINES: ENAMINES

Secondary amines are compounds of the type R_2NH. They add to aldehydes and ketones to form carbinolamines, but their carbinolamine intermediates can dehydrate to a stable product only in the direction that leads to a carbon-carbon double bond:

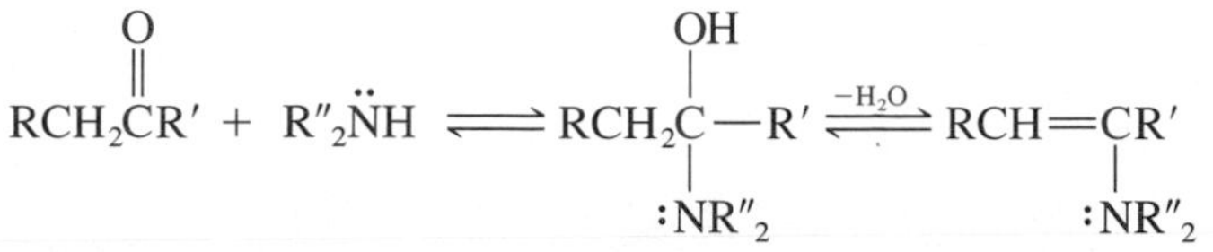

$$RCH_2\overset{O}{\overset{\|}{C}}R' + R''_2\ddot{N}H \rightleftharpoons RCH_2\underset{:NR''_2}{\underset{|}{\overset{OH}{\overset{|}{C}}}}-R' \xrightleftharpoons{-H_2O} RCH{=}\underset{:NR''_2}{\underset{|}{C}}R'$$

Aldehyde or ketone — Secondary amine — Carbinolamine — Enamine

The product of this dehydration is an alkenyl-substituted amine, or **enamine.**

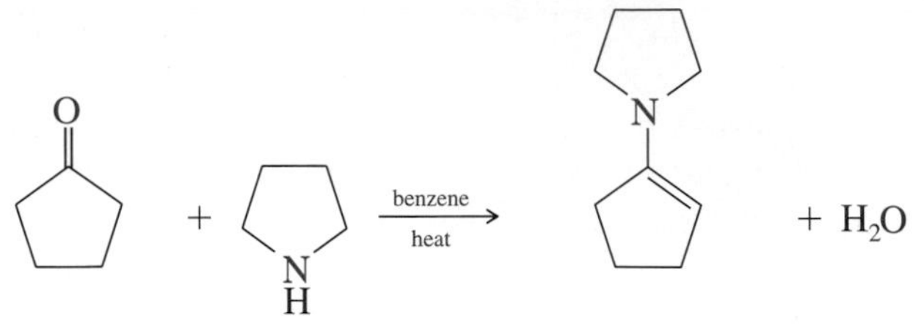

Cyclopentanone + Pyrrolidine $\xrightarrow[\text{heat}]{\text{benzene}}$ *N*-(1-Cyclopentenyl)pyrrolidine (80–90%) + H_2O (Water)

PROBLEM 17.12 Write the structure of the carbinolamine intermediate and the enamine product formed in the reaction of each of the following:

(*a*) Propanal and dimethylamine, CH_3NHCH_3

(*b*) 3-Pentanone and pyrrolidine

(*c*) Acetophenone and HN (piperidine ring)

SAMPLE SOLUTION (*a*) Nucleophilic addition of dimethylamine to the carbonyl group of propanal produces a carbinolamine:

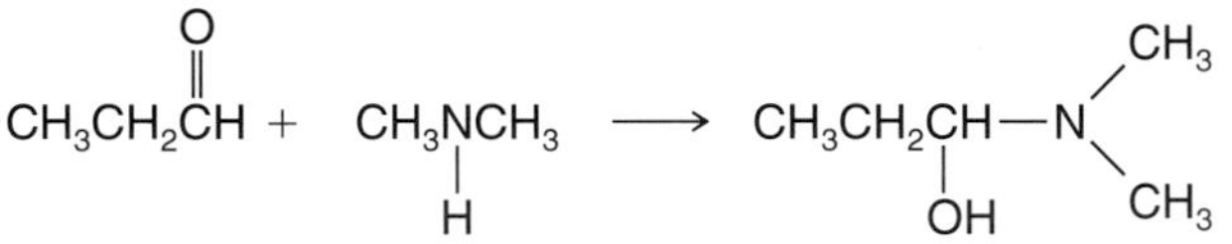

$$CH_3CH_2\overset{O}{\overset{\|}{C}}H + CH_3\underset{H}{\underset{|}{N}}CH_3 \longrightarrow CH_3CH_2\underset{OH}{\underset{|}{C}}H{-}N(CH_3)_2$$

Propanal — Dimethylamine — Carbinolamine intermediate

Dehydration of this carbinolamine yields the enamine:

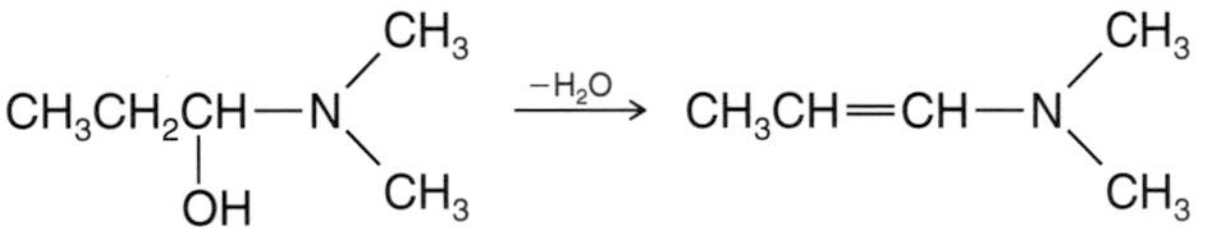

$$CH_3CH_2\underset{OH}{\underset{|}{C}}H{-}N(CH_3)_2 \xrightarrow{-H_2O} CH_3CH{=}CH{-}N(CH_3)_2$$

Carbinolamine intermediate — *N*-(1-Propenyl)dimethylamine

Enamines are used as reagents in synthetic organic chemistry and are involved in certain biochemical transformations.

17.13 THE WITTIG REACTION

A synthetic method of broad scope uses *phosphorus ylides* to convert aldehydes and ketones to alkenes:

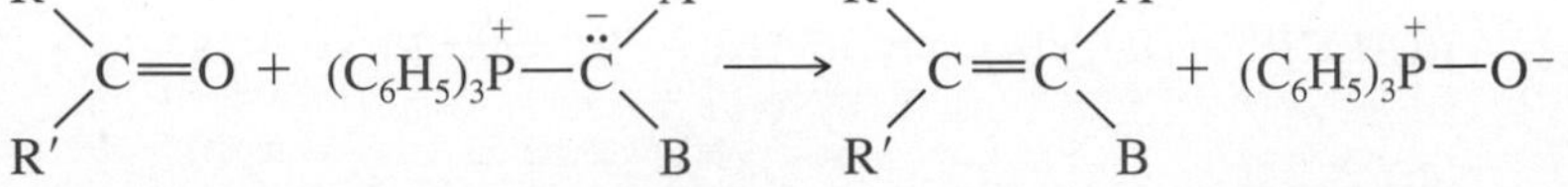

$$\underset{R'}{\overset{R}{>}}C{=}O + (C_6H_5)_3\overset{+}{P}{-}\overset{..}{\overset{-}{C}}\underset{B}{\overset{A}{<}} \longrightarrow \underset{R'}{\overset{R}{>}}C{=}C\underset{B}{\overset{A}{<}} + (C_6H_5)_3\overset{+}{P}{-}O^-$$

Aldehyde or ketone — Triphenylphosphonium ylide — Alkene — Triphenylphosphine oxide

The synthetic potential of this reaction was demonstrated by the German chemist Georg Wittig (who was awarded the Nobel Prize in chemistry for it in 1979). It is called the **Wittig reaction** and is a standard method for the preparation of alkenes.

Wittig reactions may be carried out in a number of different solvents; those chosen most often are tetrahydrofuran (THF) and dimethyl sulfoxide (DMSO).

$$\text{Cyclohexanone} + (C_6H_5)_3\overset{+}{P}-\overset{-}{\ddot{C}}H_2 \xrightarrow{\text{DMSO}} \text{Methylenecyclohexane} + (C_6H_5)_3\overset{+}{P}-O^-$$

Cyclohexanone — Methylenetriphenylphosphorane — Methylenecyclohexane (86%) — Triphenylphosphine oxide

One feature of the Wittig reaction that makes it a particularly attractive synthetic procedure is its regiospecificity. The position at which the double bond is introduced is never in doubt. The double bond is formed between the carbonyl carbon of the aldehyde or ketone and the negatively charged carbon of the ylide.

PROBLEM 17.13 Identify the alkene product in each of the following Wittig reactions:

(*a*) Benzaldehyde + $(C_6H_5)_3\overset{+}{P}-\overset{-}{\ddot{C}}$(cyclopentylidene)

(*b*) Butanal + $(C_6H_5)_3\overset{+}{P}-\overset{-}{\ddot{C}}HCH{=}CH_2$

(*c*) Cyclohexyl methyl ketone + $(C_6H_5)_3\overset{+}{P}-\overset{-}{\ddot{C}}H_2$

SAMPLE SOLUTION (*a*) In a Wittig reaction the negatively charged substituent attached to phosphorus is transferred to the aldehyde or ketone, replacing the carbonyl oxygen. The reaction shown has been used to prepare the indicated alkene in 65 percent yield.

$$C_6H_5\overset{O}{\overset{\|}{C}}H + (C_6H_5)_3\overset{+}{P}-\overset{-}{\ddot{C}}(\text{cyclopentylidene}) \longrightarrow C_6H_5CH{=}(\text{cyclopentylidene})$$

Benzaldehyde — Cyclopentylidenetriphenylphosphorane — Benzylidenecyclopentane (65%)

Before the mechanism of the Wittig reaction is described, a brief note about ylides is in order. **Ylides** are neutral molecules that have two oppositely charged atoms, each with an octet of electrons, directly bonded to each other. In the ylides used in the Wittig reaction, phosphorus has eight electrons and is positively charged; its attached carbon has eight electrons and is negatively charged. The negatively charged carbon of an ylide has carbanionic character and can act as a nucleophile toward carbonyl groups.

A mechanism for the Wittig reaction is outlined in Figure 17.8. The first stage is a cycloaddition in which the ylide reacts with the carbonyl group to give an intermediate containing a four-membered ring called an **oxaphosphetane.** This oxaphosphetane then dissociates to give an alkene and triphenylphosphine oxide. Presumably the direction of dissociation of the oxaphosphetane is dictated by the strong phosphorus-oxygen bond that results. The P—O bond strength in triphenylphosphine oxide has been estimated to be greater than 540 kJ/mol (130 kcal/mol).

The Wittig reaction is one that is still undergoing mechanistic investigation. Another possibility is that the oxaphosphetane intermediate is formed by a two-step process, rather than the one-step process shown in Figure 17.8.

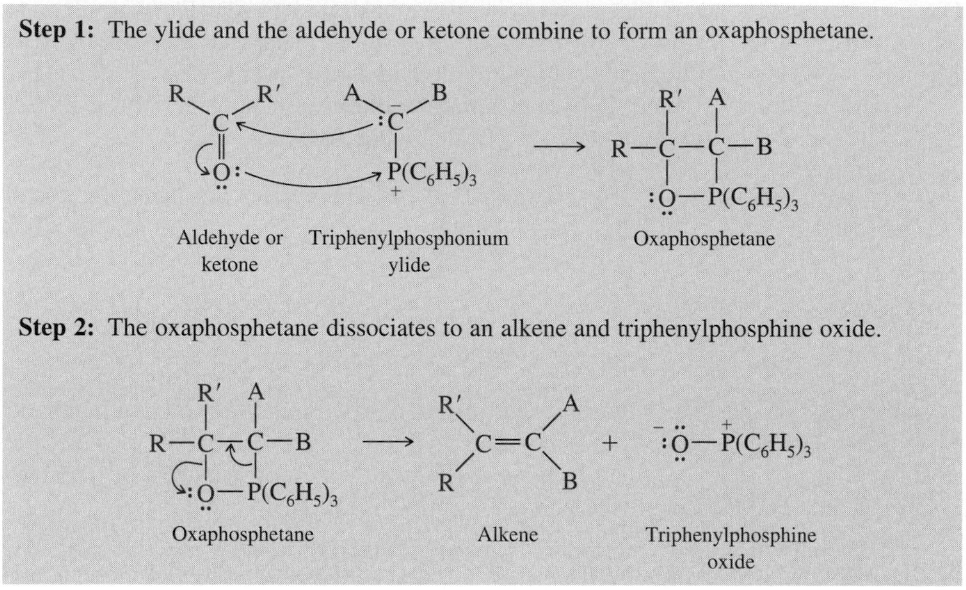

Step 1: The ylide and the aldehyde or ketone combine to form an oxaphosphetane.

$$R_2C{=}O \;(\text{Aldehyde or ketone}) + A(B)\bar{C}{-}\overset{+}{P}(C_6H_5)_3 \;(\text{Triphenylphosphonium ylide}) \longrightarrow \text{Oxaphosphetane}$$

Aldehyde or ketone — Triphenylphosphonium ylide — Oxaphosphetane

Step 2: The oxaphosphetane dissociates to an alkene and triphenylphosphine oxide.

$$\text{Oxaphosphetane} \longrightarrow RR'C{=}CAB + {}^{-}:\ddot{O}{-}\overset{+}{P}(C_6H_5)_3$$

Oxaphosphetane — Alkene — Triphenylphosphine oxide

FIGURE 17.8 The mechanism of the Wittig reaction.

17.14 PLANNING AN ALKENE SYNTHESIS VIA THE WITTIG REACTION

In order to identify the carbonyl compound and the ylide required to produce a given alkene, mentally disconnect the double bond so that one of its carbons is derived from a carbonyl group and the other is derived from an ylide. Taking styrene as a representative example, we see that two such disconnections are possible; either benzaldehyde or formaldehyde is an appropriate precursor.

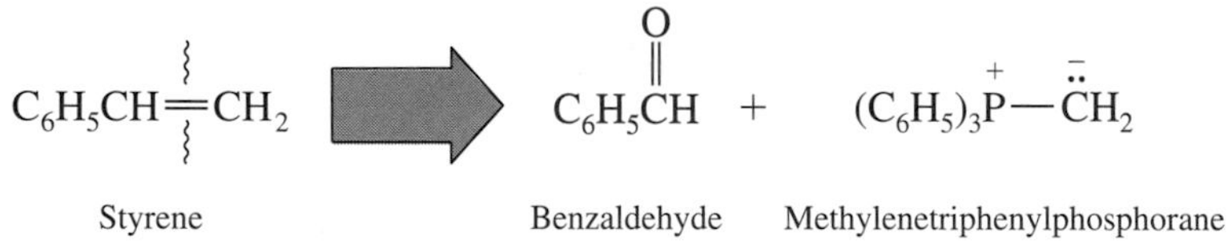

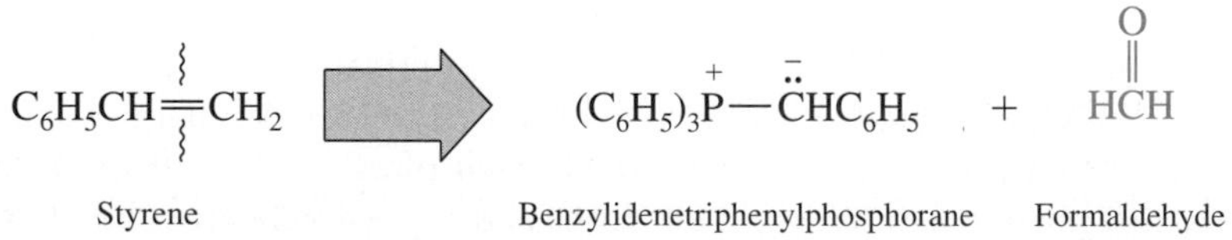

$$C_6H_5CH{=}CH_2 \Longrightarrow C_6H_5\overset{O}{\overset{\|}{C}}H + (C_6H_5)_3\overset{+}{P}{-}\overset{-}{\ddot{C}}H_2$$

Styrene — Benzaldehyde — Methylenetriphenylphosphorane

$$C_6H_5CH{=}CH_2 \Longrightarrow (C_6H_5)_3\overset{+}{P}{-}\overset{-}{\ddot{C}}HC_6H_5 + H\overset{O}{\overset{\|}{C}}H$$

Styrene — Benzylidenetriphenylphosphorane — Formaldehyde

Either route is a feasible one, and indeed styrene has been prepared from both combinations of reactants. Typically there will be two Wittig routes to an alkene, and any choice between them is made on the basis of availability of the particular starting materials.

PROBLEM 17.14 What combinations of carbonyl compound and ylide could you use to prepare each of the following alkenes?

(*a*) $CH_3CH_2CH_2CH{=}C(CH_3)CH_2CH_3$

(*b*) $CH_3CH_2CH_2CH{=}CH_2$

SAMPLE SOLUTION (*a*) There are two Wittig reaction routes that lead to the target molecule. One is represented by

$$\underset{\text{3-Methyl-3-heptene}}{CH_3CH_2CH_2CH{=}\underset{\substack{|\\CH_3}}{C}CH_2CH_3} \Longrightarrow \underset{\text{Butanal}}{CH_3CH_2CH_2\overset{\substack{O\\\|}}{C}H} + \underset{\text{1-Methylpropylidenetriphenylphosphorane}}{(C_6H_5)_3\overset{+}{P}{-}\overset{-}{\ddot{\underset{\substack{|\\CH_3}}{C}}}CH_2CH_3}$$

The other is

$$\underset{\text{3-Methyl-3-heptene}}{CH_3CH_2CH_2CH{=}\underset{\substack{|\\CH_3}}{C}CH_2CH_3} \Longrightarrow \underset{\text{Butylidenetriphenylphosphorane}}{CH_3CH_2CH_2\overset{-}{\ddot{C}}H{-}\overset{+}{P}(C_6H_5)_3} + \underset{\text{2-Butanone}}{CH_3\overset{\substack{O\\\|}}{C}CH_2CH_3}$$

Phosphorus ylides are prepared from alkyl halides by a two-step sequence. The first step is a nucleophilic substitution of the S_N2 type by triphenylphosphine on an alkyl halide to give an alkyltriphenylphosphonium salt:

$$\underset{\text{Triphenylphosphine}}{(C_6H_5)_3P:} \quad \underset{\text{Alkyl halide}}{\overset{A}{\underset{B}{>}}CH{-}X} \longrightarrow \underset{\text{Alkyltriphenylphosphonium halide}}{(C_6H_5)_3\overset{+}{P}{-}\overset{\substack{A\\|}}{C}H{-}B\ X^-}$$

Triphenylphosphine is a very powerful nucleophile, yet is not strongly basic. Consequently, methyl, primary, and secondary alkyl halides are all suitable substrates.

$$\underset{\text{Triphenylphosphine}}{(C_6H_5)_3P:} + \underset{\text{Bromomethane}}{CH_3Br} \xrightarrow{\text{benzene}} \underset{\text{Methyltriphenylphosphonium bromide (99\%)}}{(C_6H_5)_3\overset{+}{P}{-}CH_3\ Br^-}$$

The alkyltriphenylphosphonium salt products are ionic and crystallize in high yield from the nonpolar solvents in which they are prepared. After it is isolated, the alkyltriphenylphosphonium halide is converted to the desired ylide by deprotonation with a strong base:

$$\underset{\text{Alkyltriphenylphosphonium salt}}{(C_6H_5)_3\overset{+}{P}{-}\overset{\substack{A\\|}}{\underset{\substack{|\\H}}{C}}{-}B} + \underset{\text{Base}}{\ddot{Y}^-} \longrightarrow \underset{\text{Triphenylphosphonium ylide}}{(C_6H_5)_3\overset{+}{P}{-}\overset{-}{\ddot{C}}\overset{A}{\underset{B}{<}}} + \underset{\text{Conjugate acid of base used}}{HY}$$

Suitable strong bases include the sodium salt of dimethyl sulfoxide (in dimethyl sulfoxide as the solvent) and organolithium reagents (in diethyl ether or tetrahydrofuran).

$$(C_6H_5)_3\overset{+}{P}—CH_3\ Br^- + NaCH_2\overset{O}{\overset{\|}{S}}CH_3 \xrightarrow{DMSO} (C_6H_5)_3\overset{+}{P}—\overset{-}{\ddot{C}}H_2 + CH_3\overset{O}{\overset{\|}{S}}CH_3 + NaBr$$

Methyltriphenylphosphonium bromide | Sodiomethyl methyl sulfoxide | Methylenetriphenylphosphorane | Dimethyl sulfoxide | Sodium bromide

---

PROBLEM 17.15 The sample solution to Problem 17.14*a* showed the preparation of 3-methyl-3-heptene by a Wittig reaction involving the ylide shown. Write equations showing the formation of this ylide beginning with 2-bromobutane.

$$(C_6H_5)_3\overset{+}{P}—\overset{-}{\ddot{C}}(CH_3)CH_2CH_3$$

Normally the ylides are not isolated. Instead, the appropriate aldehyde or ketone is added directly to the solution in which the ylide was generated.

17.15 STEREOSELECTIVE ADDITION TO CARBONYL GROUPS

Nucleophilic addition to carbonyl groups sometimes leads to a mixture of stereoisomeric products. The preferred direction of attack is frequently controlled by steric factors, with the nucleophile approaching the carbonyl group at its less hindered face. Sodium borohydride reduction of 7,7-dimethylbicyclo[2.2.1]heptan-2-one illustrates this point:

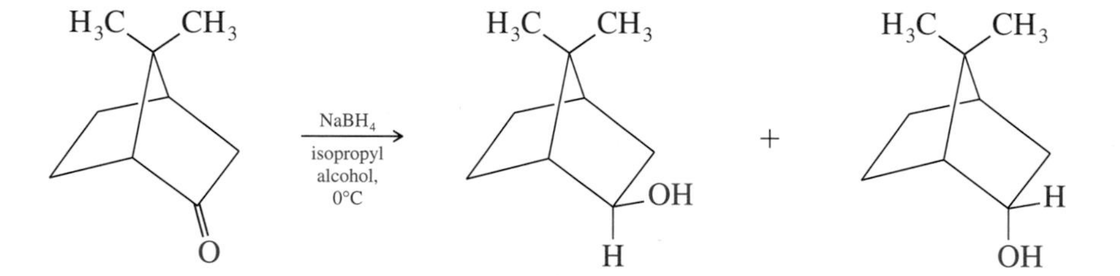

7,7-Dimethylbicyclo[2.2.1]-heptan-2-one $\xrightarrow[\text{isopropyl alcohol, 0°C}]{NaBH_4}$ *exo*-7,7-Dimethylbicyclo[2.2.1]-heptan-2-ol (80%) + *endo*-7,7-Dimethylbicyclo[2.2.1]-heptan-2-ol (20%)

Approach of borohydride to the upper face of the carbonyl group is sterically hindered by one of the methyl groups. The bottom face of the carbonyl group is less congested, and the major product is formed by hydride transfer from this direction.

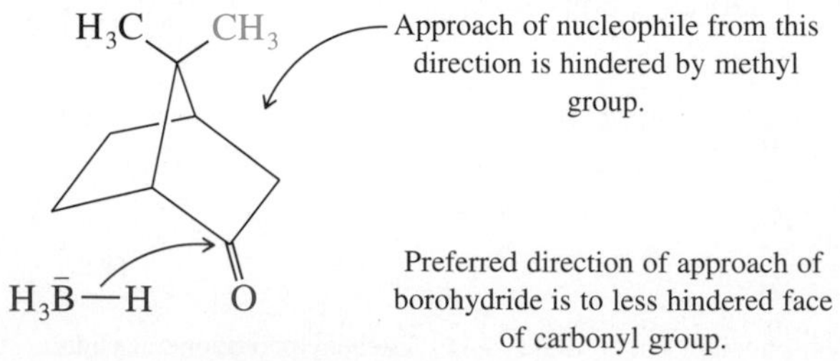

H_3C CH_3 $H_3\bar{B}—H$ O

Approach of nucleophile from this direction is hindered by methyl group.

Preferred direction of approach of borohydride is to less hindered face of carbonyl group.

The reduction reaction is *stereoselective.* A single starting material has the potential of forming two stereoisomeric forms of product but yields one isomer preferentially.

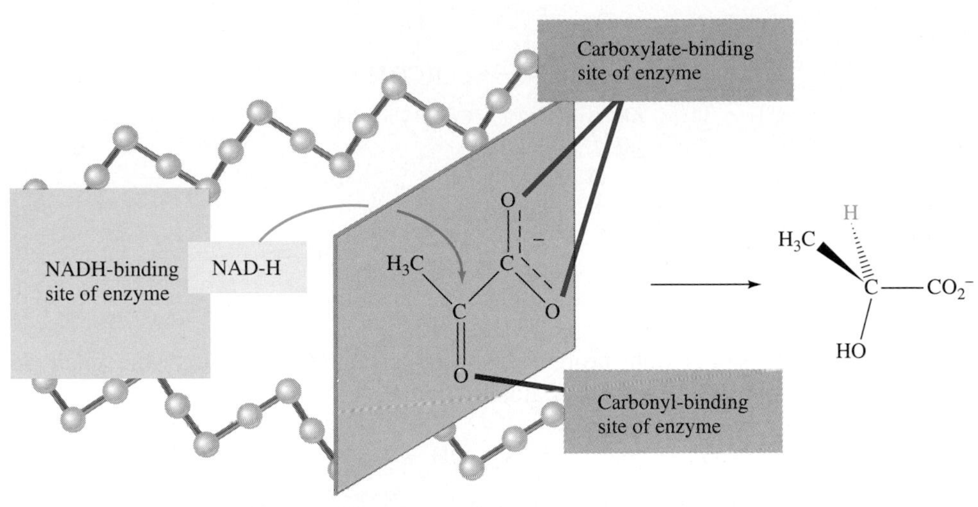

FIGURE 17.9 Schematic representation of enzyme-mediated reduction of pyruvate to (*S*)-(+)-lactate. A preferred orientation of binding of the substrate to the enzyme, coupled with a prescribed location of the reducing agent, the coenzyme NADH, leads to hydrogen transfer exclusively to a single face of the carbonyl group.

It is possible to predict the preferred stereochemical path of nucleophilic addition if one face of a carbonyl group is significantly more hindered to the approach of the reagent than the other. When no clear distinction between the two faces is evident, other, more subtle effects, which are still incompletely understood, come into play.

Enzyme-catalyzed reductions of carbonyl groups are, more often than not, completely stereoselective. Pyruvic acid is converted exclusively to (*S*)-(+)-lactic acid by the lactate dehydrogenase–NADH system (Section 15.11). The enantiomer (*R*)-(−)-lactic acid is not formed.

$$\underset{\text{Pyruvic acid}}{CH_3\overset{O}{\overset{\|}{C}}\overset{O}{\overset{\|}{C}}OH} + \underset{\text{Reduced form of coenzyme}}{NADH} + H^+ \xrightarrow{\text{lactate dehydrogenase}} \underset{\text{(S)-(+)-Lactic acid}}{(H_3C)(H)(HO)C{-}\overset{O}{\overset{\|}{C}}OH} + \underset{\text{Oxidized form of coenzyme}}{NAD^+}$$

Here the enzyme, a chiral molecule, binds the coenzyme and substrate in such a way that hydrogen is transferred exclusively to the face of the carbonyl group that leads to (*S*)-(+)-lactic acid (Figure 17.9).

The stereochemical outcome of enzyme-mediated reactions depends heavily on the way the protein chain is folded. Aspects of protein conformation will be discussed in Chapter 27.

17.16 OXIDATION OF ALDEHYDES

Aldehydes are readily oxidized to carboxylic acids by a number of reagents, including those based on Cr(VI) in aqueous media.

$$\mathrm{RCH}(=\mathrm{O}) \xrightarrow{\text{oxidize}} \mathrm{RCOH}(=\mathrm{O})$$

Aldehyde — Carboxylic acid

$$\text{Furfural (2-furyl–CH=O)} \xrightarrow[\mathrm{H_2SO_4,\ H_2O}]{\mathrm{K_2Cr_2O_7}} \text{2-furyl–}\mathrm{CO_2H}$$

Furfural — Furoic acid (75%)

Mechanistically these reactions probably proceed through the hydrate of the aldehyde and follow a course similar to that of alcohol oxidation.

$$\mathrm{RCH}(=\mathrm{O}) + \mathrm{H_2O} \rightleftharpoons \mathrm{RCH(OH)_2} \xrightarrow{\text{oxidize}} \mathrm{RCOH}(=\mathrm{O})$$

Aldehyde — Geminal diol (hydrate) — Carboxylic acid

17.17 BAEYER-VILLIGER OXIDATION OF KETONES

The reaction of ketones with peroxy acids is a novel and synthetically useful one. An oxygen from the peroxy acid is inserted between the carbonyl group and one of the attached carbons of the ketone to give an *ester*. Reactions of this type were first described by Adolf von Baeyer and Victor Villiger in 1899 and are known as **Baeyer-Villiger oxidations.**

Peroxy acids have been seen before as reagents for the epoxidation of alkenes (Section 6.18).

$$\mathrm{RCR'}(=\mathrm{O}) + \mathrm{R''COOH}(=\mathrm{O}) \longrightarrow \mathrm{RCOR'}(=\mathrm{O}) + \mathrm{R''COH}(=\mathrm{O})$$

Ketone — Peroxy acid — Ester — Carboxylic acid

Methyl ketones give esters of acetic acid; that is, oxygen insertion occurs between the carbonyl carbon and the larger of the two groups attached to it.

$$\text{C}_6\text{H}_{11}\text{–}\mathrm{CCH_3}(=\mathrm{O}) \xrightarrow[\mathrm{CHCl_3}]{\mathrm{C_6H_5CO_2OH}} \text{C}_6\text{H}_{11}\text{–}\mathrm{OCCH_3}(=\mathrm{O})$$

Cyclohexyl methyl ketone — Cyclohexyl acetate (67%)

The mechanism of the Baeyer-Villiger oxidation is shown in Figure 17.10. It begins with nucleophilic addition of the peroxy acid to the carbonyl group of the ketone, which is followed by migration of an alkyl group from the carbonyl group to oxygen. In general, it is the more substituted group that migrates. The *migratory aptitude* of the various alkyl groups is:

Tertiary alkyl > secondary alkyl > primary alkyl > methyl

The overall reaction:

$$\underset{\text{Ketone}}{\overset{O}{\overset{\|}{RCR'}}} + \underset{\text{Peroxy acid}}{\overset{O}{\overset{\|}{R''COOH}}} \longrightarrow \underset{\text{Ester}}{\overset{O}{\overset{\|}{ROCR'}}} + \underset{\text{Carboxylic acid}}{\overset{O}{\overset{\|}{R''COH}}}$$

Step 1: The peroxy acid adds to the carbonyl group of the ketone. This step is a nucleophilic addition analogous to *gem*-diol and hemiacetal formation.

$$\underset{\text{Ketone}}{\overset{O}{\overset{\|}{RCR'}}} + \underset{\text{Peroxy acid}}{\overset{O}{\overset{\|}{R''COOH}}} \longrightarrow RC(OH)R'(OOC(=O)R'')$$

(Peroxy monoester of *gem*-diol)

Step 2: The intermediate from step 1 undergoes rearrangement. Cleavage of the weak O—O bond of the peroxy ester is assisted by migration of one of the substituents from the carbonyl group to oxygen. The group R migrates with its pair of electrons in much the same way as alkyl groups migrate in carbocation rearrangements.

$$R(R')C(OH)\text{—O—OC(=O)R''} \longrightarrow \underset{\text{Ester}}{R'C(=O)OR} + \underset{\text{Carboxylic acid}}{\overset{O}{\overset{\|}{HOCR''}}}$$

FIGURE 17.10 Mechanism of the Baeyer-Villiger oxidation of a ketone.

PROBLEM 17.16 Using Figure 17.10 as a guide, write a mechanism for the Baeyer-Villiger oxidation of cyclohexyl methyl ketone by peroxybenzoic acid.

PROBLEM 17.17 Baeyer-Villiger oxidation of aldehydes yields carboxylic acids (for example, *m*-nitrobenzaldehyde yields *m*-nitrobenzoic acid). What group migrates to oxygen?

The reaction is stereospecific in the sense that the alkyl group migrates with retention of configuration.

$$\underset{cis\text{-1-Acetyl-2-methylcyclopentane}}{CH_3\text{—}C_5H_8\text{—}\overset{O}{\overset{\|}{C}}CH_3} \xrightarrow[CHCl_3]{C_6H_5CO_2OH} \underset{cis\text{-2-Methylcyclopentyl acetate}}{CH_3\text{—}C_5H_8\text{—}O\overset{O}{\overset{\|}{C}}CH_3}$$

cis-1-Acetyl-2-methylcyclopentane

cis-2-Methylcyclopentyl acetate (only product; 66% yield)

In the companion experiment carried out on the trans stereoisomer of the ketone, only the trans acetate was formed.

17.18 SPECTROSCOPIC ANALYSIS OF ALDEHYDES AND KETONES

Carbonyl groups are among the easiest functional groups to detect by infrared spectroscopy. The carbon-oxygen stretching mode of aldehydes and ketones gives rise to a very strong absorption in the region 1710 to 1750 cm^{-1}. Conjugation of the carbonyl group with an aromatic ring or with a double bond lowers this value to 1670 to 1700 cm^{-1}. In addition to strong absorption characteristic of the carbonyl group, aldehydes exhibit weak absorptions due to C—H stretching of the CH=O unit near 2720 and 2820 cm^{-1}, as the infrared spectrum of butanal in Figure 17.11 illustrates.

^1H nuclear magnetic resonance spectroscopy provides a very reliable way to identify aldehydes. The chemical shift of the proton of the CH=O unit occurs in a region (δ = 9 to 10 ppm) where few other proton signals appear. Figure 17.12 shows the ^1H nmr of 2-methylpropanal [$(CH_3)_2CHCH{=}O$], where the large chemical shift difference between the aldehyde proton and the other protons in the molecule is clearly evident. What is less apparent in the spectrum, but which can be discerned on expanding the scale, is that the aldehyde proton is a doublet, split by the proton at C-2. Coupling between the protons in HC—CH=O is much smaller than typical vicinal couplings, making the multiplicity of the aldehyde peak often difficult to see.

Methyl ketones, such as 2-butanone in Figure 17.13, are characterized by sharp singlets for the CH_3 group near δ = 2.1 ppm. Similarly, the deshielding effect of the carbonyl causes the CH_2 group attached to it to appear at lower field (δ = 2.4 ppm) than a CH_2 group in an alkane.

Aldehydes and ketones are also readily identified by their ^{13}C nmr spectra. Carbonyl carbons appear at very low field, some 190 to 220 ppm downfield from tetramethylsilane. Figure 17.14 illustrates this for 3-heptanone, where the chemical shift of the carbonyl carbon is 211 ppm. All seven carbons of 3-heptanone are clearly evident in this spectrum. Note that the intensity of the peak due to the carbonyl carbon is significantly less than that of the others, even though each peak in the spectrum corresponds to a single carbon atom. This decreased intensity is a characteristic of carbons that bear no hydrogen substituents.

Off-resonance decoupling of ^{13}C nmr spectra provides a ready means of distin-

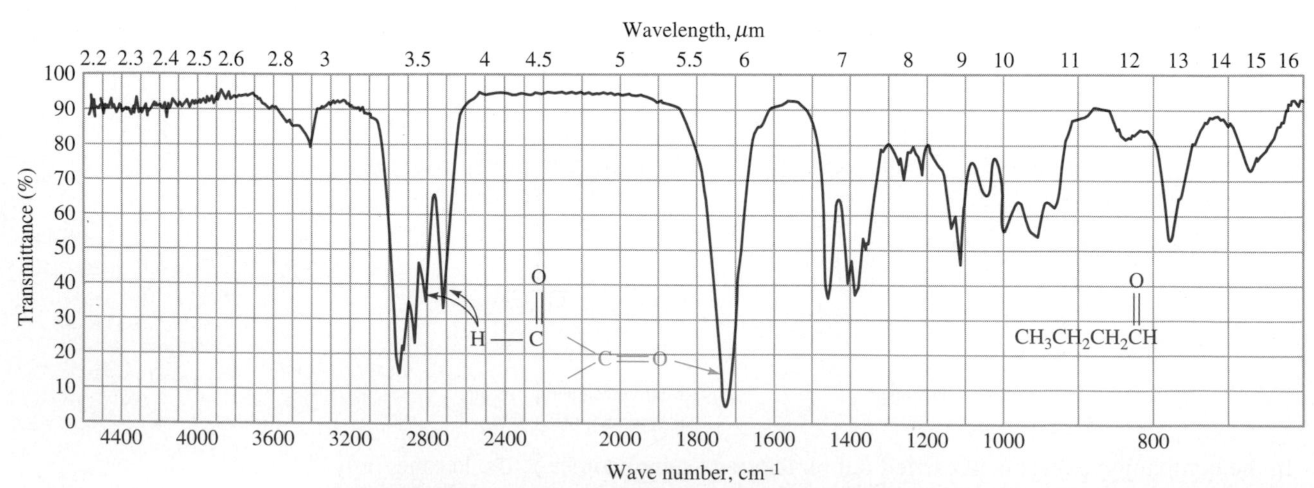

FIGURE 17.11 Infrared spectrum of butanal showing peaks characteristic of the CH=O unit: at 2700 and 2800 cm^{-1} (C—H) and at 1720 cm^{-1} (C=O).

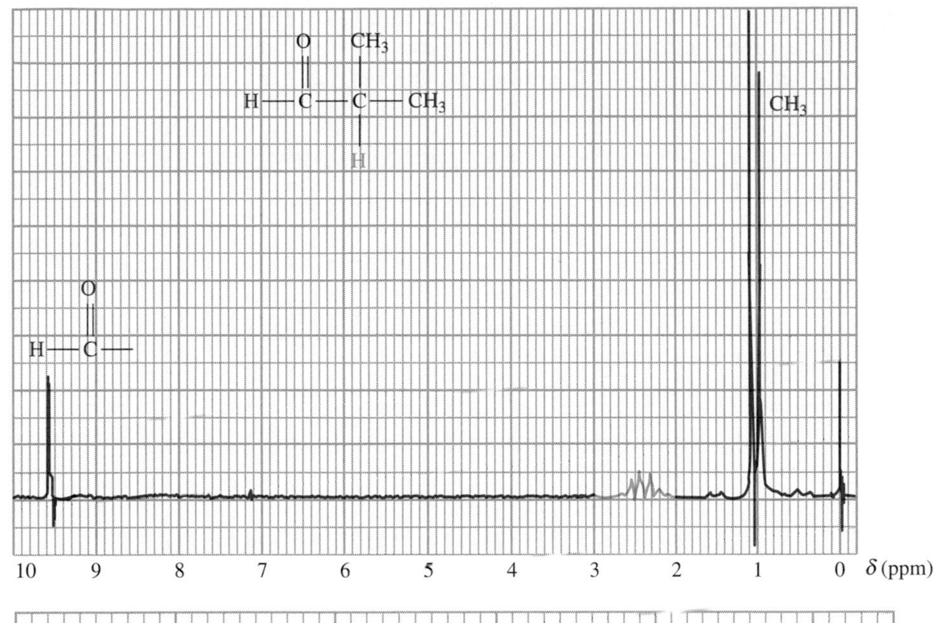

FIGURE 17.12 The 60 MHz ^{1}H nmr spectrum of 2-methylpropanal, showing the aldehyde proton as a doublet at low field (9.6 ppm).

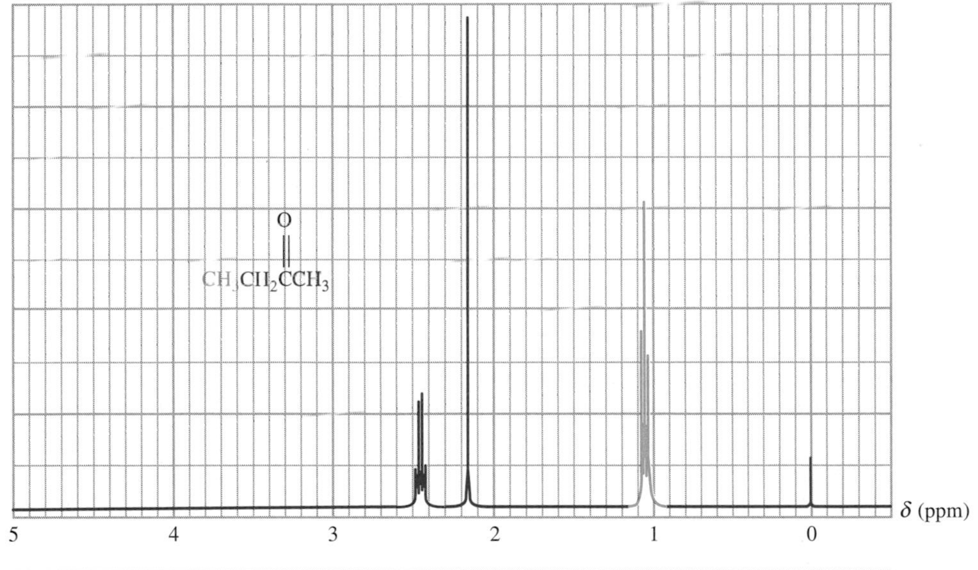

FIGURE 17.13 The 300 MHz ^{1}H nmr spectrum of 2-butanone.

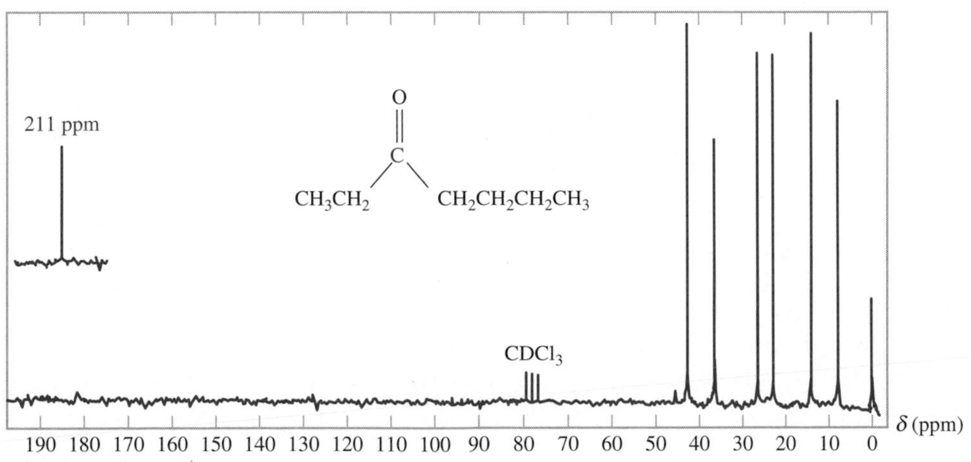

FIGURE 17.14 The ^{13}C nmr spectrum of 3-heptanone. *(Taken from LeRoy F. Johnson and William C. Jankowski, Carbon-13 NMR Spectra: A Collection of Assigned, Coded, and Indexed Spectra, Wiley-Interscience, New York, 1972. Reprinted by permission of John Wiley & Sons, Inc.)*

guishing aldehydes from ketones. The carbonyl carbon of an aldehyde appears as a doublet because of coupling to its attached proton, whereas a ketone carbonyl bears only carbon substituents and its ^{13}C nmr signal is a singlet.

17.19 MASS SPECTROMETRY OF ALDEHYDES AND KETONES

Aldehydes and ketones typically give a prominent molecular ion peak in their mass spectra. Aldehydes also exhibit an M-1 peak. A major fragmentation pathway for both aldehydes and ketones leads to formation of acyl cations (acylium ions) by cleavage of an alkyl group from the carbonyl. The most intense peak in the mass spectrum of diethyl ketone, for example, is *m/z* 57, corresponding to loss of ethyl radical from the molecular ion.

$$CH_3CH_2\overset{:O^{+\cdot}}{\overset{\|}{C}}CH_2CH_3 \longrightarrow CH_3CH_2C{\equiv}\ddot{O}^+ + \cdot CH_2CH_3$$

m/z 86   *m/z* 57

17.20 SUMMARY

The chemistry of the carbonyl group is probably the single most important aspect of organic chemical reactivity. Classes of compounds that contain the carbonyl group include many derived from carboxylic acids (acyl chlorides, acid anhydrides, esters, and amides) as well as the two related classes discussed in this chapter—*aldehydes* and *ketones.*

The substitutive names of aldehydes and ketones are developed by applying the principles of alkane nomenclature to the longest continuous chain that contains the carbonyl group and appending the suffixes *-al* for aldehydes and *-one* for ketones. The chain is numbered in the direction that gives the lowest locant to the carbon of the carbonyl group. Ketones are often named using radicofunctional nomenclature in which the two groups attached to the carbonyl group are cited in alphabetic order and *ketone* is appended as a separate word (Section 17.1).

The carbonyl carbon is sp^2 hybridized, and it and the atoms attached to it are coplanar (Section 17.2).

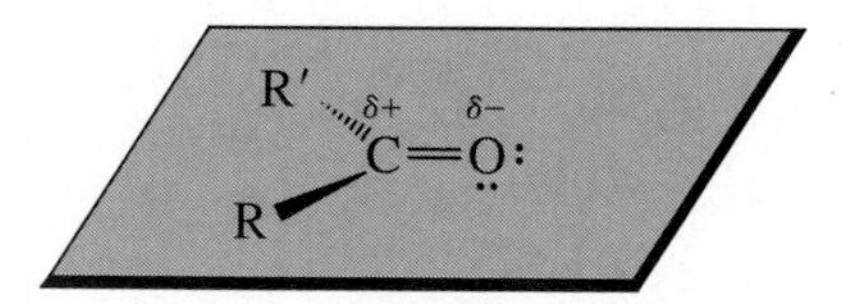

The carbon-oxygen double bond is rather polar. Carbon is positively polarized and oxygen negatively polarized. In addition to making aldehydes and ketones relatively polar molecules (Section 17.3), this polarization has an important effect on their chemical reactivity.

The characteristic reaction of aldehydes and ketones is *nucleophilic addition* to the carbonyl group (Section 17.6). Reagents of the type HY react according to the general equation

$$\overset{\delta+}{C}=\overset{\delta-}{O} + \overset{\delta+}{H}-\overset{\delta-}{Y} \rightleftharpoons Y-\underset{|}{\overset{|}{C}}-O-H$$

Aldehyde or ketone — Product of nucleophilic addition to carbonyl group

Aldehydes undergo nucleophilic addition more readily and have more favorable equilibrium constants for addition than do ketones. A summary of the nucleophilic addition reactions to aldehydes and ketones introduced in this chapter is presented in Table 17.5.

The step in which the nucleophile attacks the carbonyl carbon is rate-determining in both base-catalyzed and acid-catalyzed nucleophilic addition. In the base-catalyzed mechanism this is the first step.

$$\overset{-}{Y}: + \;>C=\ddot{O}: \overset{\text{slow}}{\rightleftharpoons} Y-\underset{|}{\overset{|}{C}}-\ddot{O}:^{-}$$

Nucleophile — Aldehyde or ketone

$$Y-\underset{|}{\overset{|}{C}}-\ddot{O}:^{-} + H-Y \overset{\text{fast}}{\rightleftharpoons} Y-\underset{|}{\overset{|}{C}}-\ddot{O}H + \overset{-}{Y}:$$

Product of nucleophilic addition

Under conditions of acid catalysis, the nucleophilic addition step follows protonation of the carbonyl oxygen. Protonation increases the carbocation character of a carbonyl group and makes it more electrophilic.

$$>C=\ddot{O}: + H-Y: \overset{\text{fast}}{\rightleftharpoons} >C=\overset{+}{\ddot{O}}H \longleftrightarrow >\overset{+}{C}-\ddot{O}H$$

Aldehyde or ketone — Resonance forms of protonated aldehyde or ketone

$$HY: + \;>C=\overset{+}{\ddot{O}}H \overset{\text{slow}}{\rightleftharpoons} H\overset{+}{Y}-\underset{|}{\overset{|}{C}}-\ddot{O}H \overset{-H^{+}}{\rightleftharpoons} Y-\underset{|}{\overset{|}{C}}-\ddot{O}H$$

Product of nucleophilic addition

Often the product of nucleophilic addition is not isolated but is an intermediate leading to the ultimate product. Most of the reactions in Table 17.5 are of this type.

Nucleophilic addition to the carbonyl group is *stereoselective*. When one direction of approach to the carbonyl group is less hindered than the other, the nucleophile normally attacks at the less hindered face (Section 17.15).

Aldehydes are easily oxidized to carboxylic acids (Section 17.16).

TABLE 17.5
Nucleophilic Addition to Aldehydes and Ketones

Reaction (section) and comments	General equation and typical example
Hydration (Section 17.7) Can be either acid- or base-catalyzed. Equilibrium constant is normally unfavorable for hydration of ketones unless R, R′, or both are strongly electron-withdrawing.	$RC(=O)R' + H_2O \rightleftharpoons RC(OH)_2R'$ Aldehyde or ketone; Water; Geminal diol $ClCH_2C(=O)CH_3 \xrightleftharpoons{H_2O} ClCH_2C(OH)_2CH_3$ Chloroacetone (90% at equilibrium); Chloroacetone hydrate (10% at equilibrium)
Cyanohydrin formation (Section 17.8) Reaction is catalyzed by cyanide ion. Cyanohydrins are useful synthetic intermediates; cyano group can be hydrolyzed to $-CO_2H$ or reduced to $-CH_2NH_2$.	$RC(=O)R' + HCN \rightleftharpoons RC(OH)(CN)R'$ Aldehyde or ketone; Hydrogen cyanide; Cyanohydrin $CH_3CH_2C(=O)CH_2CH_3 \xrightarrow[H^+]{KCN} CH_3CH_2C(OH)(CN)CH_2CH_3$ 3-Pentanone; 3-Pentanone cyanohydrin (75%)
Acetal formation (Sections 17.9–17.10) Reaction is acid-catalyzed. Equilibrium contant normally favorable for aldehydes, unfavorable for ketones. Cyclic acetals from vicinal diols form readily.	$RC(=O)R' + 2R''OH \xrightleftharpoons{H^+} RC(OR'')_2R' + H_2O$ Aldehyde or ketone; Alcohol; Acetal; Water $m\text{-}O_2NC_6H_4CH(=O) + CH_3OH \xrightarrow{HCl} m\text{-}O_2NC_6H_4CH(OCH_3)_2$ *m*-Nitrobenzaldehyde; Methanol; *m*-Nitrobenzaldehyde dimethyl acetal (76–85%)
Reaction with primary amines (Section 17.11) Isolated product is an imine (Schiff's base). A carbinolamine intermediate is formed, which undergoes dehydration to an imine.	$RC(=O)R' + R''NH_2 \rightleftharpoons RC(=NR'')R' + H_2O$ Aldehyde or ketone; Primary amine; Imine; Water $(CH_3)_2CHC(=O)H + (CH_3)_3CNH_2 \longrightarrow (CH_3)_2CHCH{=}NC(CH_3)_3$ 2-Methylpropanal; *tert*-Butylamine; *N*-(2-Methyl-1-propylidene)-*tert*-butylamine (50%)

TABLE 17.5
Nucleophilic Addition to Aldehydes and Ketones *(Continued)*

Reaction (section) and comments	General equation and typical example
Reaction with secondary amines (Section 17.12) Isolated product is an enamine. Carbinolamine intermediate cannot dehydrate to a stable imine.	$RCCH_2R'$ (C=O) + $(R'')_2NH$ ⇌ $RC{=}CHR'$ (with NR''_2 on C) + H_2O Aldehyde or ketone; Secondary amine; Enamine; Water Cyclohexanone + Morpholine —(benzene, heat)→ 1-Morpholinocyclohexene (85%)
The Wittig reaction (Sections 17.13–17.14) Reaction of a phosphorus ylide with aldehydes and ketones leads to the formation of an alkene. A very versatile method for the preparation of alkenes.	RCR' (C=O) + $(C_6H_5)_3\overset{+}{P}{-}\overset{-}{C}AB$ ⟶ $RR'C{=}CAB$ + $(C_6H_5)_3\overset{+}{P}{-}O^-$ Aldehyde or ketone; Wittig reagent (an ylide); Alkene; Triphenylphosphine oxide CH_3CCH_3 (C=O) + $(C_6H_5)_3\overset{+}{P}{-}\overset{-}{\ddot{C}}HCH_2CH_2CH_2CH_3$ —DMSO→ Acetone; 1-Pentylidenetriphenylphosphorane $(CH_3)_2C{=}CHCH_2CH_2CH_2CH_3$ + $(C_6H_5)_3\overset{+}{P}{-}O^-$ 2-Methyl-2-heptene (56%); Triphenylphosphine oxide

$$\underset{\text{Aldehyde}}{RCH} \xrightarrow[H_2O]{Cr(VI)} \underset{\text{Carboxylic acid}}{RCOH}$$

The oxidation of ketones with peroxy acids is called the *Baeyer-Villiger oxidation* and is a useful method for preparing esters (Section 17.17).

$$\underset{\text{Ketone}}{RCR'} \xrightarrow{R''COOH} \underset{\text{Ester}}{RCOR'}$$

With unsymmetrical ketones ($R \neq R'$), it is usually the more substituted carbon that becomes bonded to oxygen in the ester. Thus methyl ketones (CH_3CR) give acetate esters (CH_3CO_2R) on Baeyer-Villiger oxidation.

PROBLEMS

17.18

(*a*) Write structural formulas and provide IUPAC names for all the isomeric aldehydes and ketones that have the molecular formula $C_5H_{10}O$. Include stereoisomers.

(*b*) Which of the isomers in part (*a*) yield chiral alcohols on reaction with sodium borohydride?

(*c*) Which of the isomers in part (*a*) yield chiral alcohols on reaction with methylmagnesium iodide?

17.19 Each of the following aldehydes or ketones is known by a common name. Its IUPAC name is provided in parentheses. Write a structural formula for each one.

(*a*) Chloral (2,2,2-trichloroethanal)
(*b*) Pivaldehyde (2,2-dimethylpropanal)
(*c*) Acrolein (2-propenal)
(*d*) Crotonaldehyde [(*E*)-2-butenal]
(*e*) Citral [(*E*)-3,7-dimethyl-2,6-octadienal]
(*f*) Diacetone alcohol (4-hydroxy-4-methyl-2-pentanone)
(*g*) Carvone (5-isopropenyl-2-methyl-2-cyclohexenone)
(*h*) Biacetyl (2,3-butanedione)

17.20 Predict the product of reaction of propanal with each of the following:

(*a*) Lithium aluminum hydride
(*b*) Sodium borohydride
(*c*) Hydrogen (nickel catalyst)
(*d*) Methylmagnesium iodide, followed by dilute acid
(*e*) Sodium acetylide, followed by dilute acid
(*f*) Phenyllithium, followed by dilute acid
(*g*) Methanol containing dissolved hydrogen chloride
(*h*) Ethylene glycol, *p*-toluenesulfonic acid, benzene
(*i*) Aniline ($C_6H_5NH_2$)
(*j*) Dimethylamine, *p*-toluenesulfonic acid, benzene
(*k*) Hydroxylamine
(*l*) Hydrazine
(*m*) Product of part (*l*) heated in triethylene glycol with sodium hydroxide
(*n*) *p*-Nitrophenylhydrazine
(*o*) Semicarbazide
(*p*) Ethylidenetriphenylphosphorane [$(C_6H_5)_3\overset{+}{P}-\overset{-}{\ddot{C}}HCH_3$]
(*q*) Sodium cyanide with addition of sulfuric acid
(*r*) Chromic acid

17.21 Repeat the preceding problem for cyclopentanone instead of propanal.

17.22 Hydride reduction (with $LiAlH_4$ or $NaBH_4$) of each of the following ketones has been reported in the chemical literature and gives a mixture of two diastereomeric alcohols in each case. Give the structures of both alcohol products for each ketone.

(*a*) (*S*)-3-Phenyl-2-butanone
(*b*) 4-*tert*-Butylcyclohexanone

(*c*)

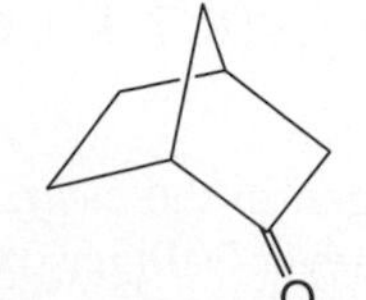

(*d*)

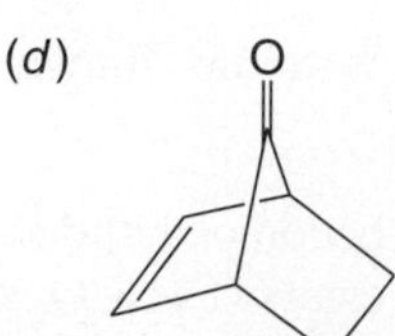

17.23 Choose which member in each of the following pairs reacts faster or has the more favorable equilibrium constant for reaction with the indicated reagent. Explain your reasoning.

(*a*) $C_6H_5\overset{O}{\overset{\|}{C}}H$ or $C_6H_5\overset{O}{\overset{\|}{C}}CH_3$ (rate of reduction with sodium borohydride)

(*b*) $Cl_3C\overset{O}{\overset{\|}{C}}H$ or $CH_3\overset{O}{\overset{\|}{C}}H$ (equilibrium constant for hydration)

(*c*) Acetone or 3,3-dimethyl-2-butanone (equilibrium constant for cyanohydrin formation)

(*d*) Acetone or 3,3-dimethyl-2-butanone (rate of reduction with sodium borohydride)

(*e*) $CH_2(OCH_2CH_3)_2$ or $(CH_3)_2C(OCH_2CH_3)_2$ (rate of acid-catalyzed hydrolysis)

17.24 Equilibrium constants for the dissociation (K_{diss}) of cyanohydrins according to the equation

$$R\underset{CN}{\underset{|}{\overset{OH}{\overset{|}{C}}}}R' \underset{}{\overset{K_{diss}}{\rightleftharpoons}} R\overset{O}{\overset{\|}{C}}R' + HCN$$

Cyanohydrin — Aldehyde or ketone — Hydrogen cyanide

have been measured for a number of cyanohydrins. Which cyanohydrin in each of the following pairs has the greater dissociation constant?

(*a*) $CH_3CH_2\overset{OH}{\overset{|}{C}}HCN$ or $(CH_3)_2\overset{OH}{\overset{|}{C}}CN$

(*b*) $C_6H_5\overset{OH}{\overset{|}{C}}HCN$ or $C_6H_5\underset{CH_3}{\underset{|}{\overset{OH}{\overset{|}{C}}}}CN$

17.25 Each of the following reactions has been reported in the chemical literature and gives a single organic product in good yield. What is the principal product in each reaction?

(*a*) 2-Bromo-3,4,5-trimethoxybenzaldehyde (benzene ring bearing CH_3O, CH_3O, CH_3O, Br, and $-\overset{O}{\overset{\|}{C}}H$) $+ HOCH_2CH_2CH_2OH \xrightarrow[\text{benzene, heat}]{p\text{-toluenesulfonic acid}}$

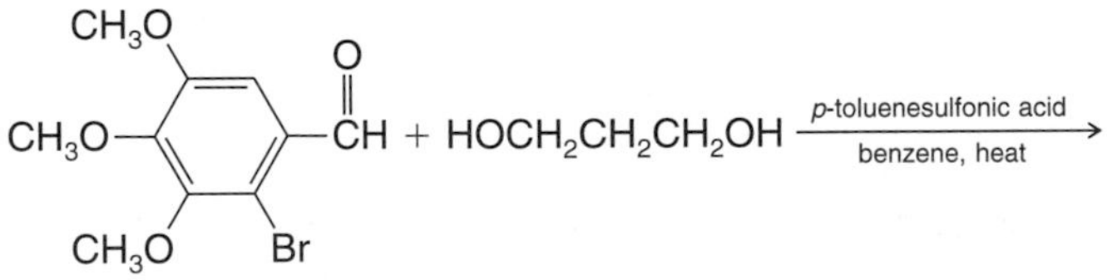

(*b*) 4-Hydroxy-2-methoxybenzaldehyde (benzene ring bearing HO, OCH_3, and $\overset{O}{\overset{\|}{C}}H$) $+ CH_3ONH_2 \longrightarrow$

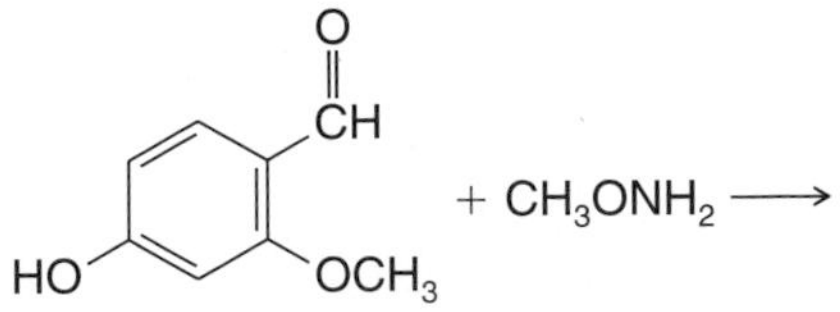

(*c*) $CH_3CH_2\overset{O}{\overset{\|}{C}}H + (CH_3)_2NNH_2 \longrightarrow$

(*d*) CH_3—C₆H₄—$\overset{CH_3}{|}{CHCH_2CH_2}$—(1,3-dioxolan-2-yl) $\xrightarrow[\text{heat}]{H_2O,\ HCl}$

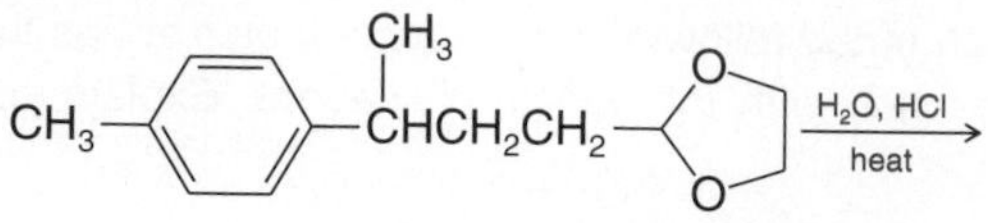

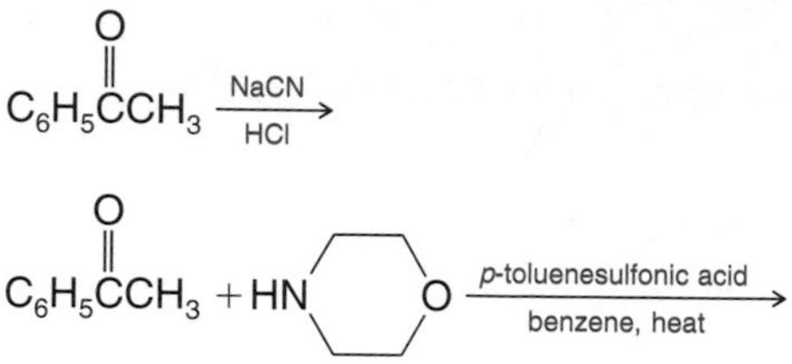

(*e*) $C_6H_5\overset{O}{\overset{\|}{C}}CH_3 \xrightarrow[HCl]{NaCN}$

(*f*) $C_6H_5\overset{O}{\overset{\|}{C}}CH_3$ + HN(morpholine)O $\xrightarrow[\text{benzene, heat}]{p\text{-toluenesulfonic acid}}$

(*g*) CH_3CH_2—C(CH_3)(C_6H_5)—$\overset{O}{\overset{\|}{C}}CH_3$ + $C_6H_5\overset{O}{\overset{\|}{C}}OOH \xrightarrow{CHCl_3}$

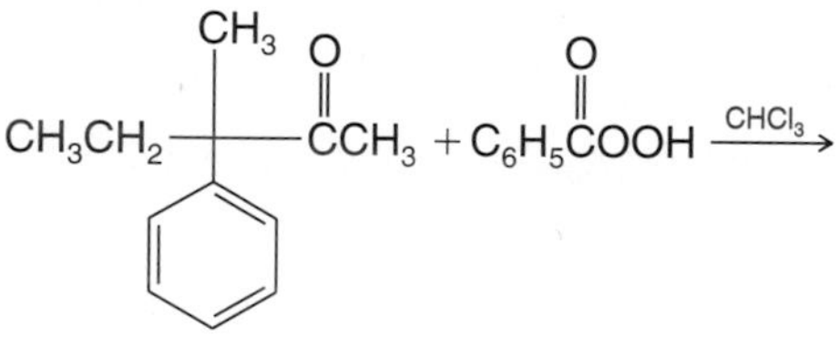

17.26 Wolff-Kishner reduction (hydrazine, KOH, ethylene glycol, 130°C) of the compound shown gave compound A. Treatment of compound A with *m*-chloroperoxybenzoic acid gave compound B, which on reduction with lithium aluminum hydride gave compound C. Oxidation of compound C with chromic acid gave compound D ($C_9H_{14}O$). Identify compounds A through D in this sequence.

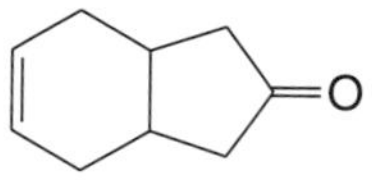

17.27 On standing in ^{17}O-labeled water, both formaldehyde and its hydrate are found to have incorporated the ^{17}O isotope of oxygen. Suggest a reasonable explanation for this observation.

17.28 Reaction of benzaldehyde with 1,2-octanediol in benzene containing a small amount of *p*-toluenesulfonic acid yields almost equal quantities of two products in a combined yield of 94 percent. Both products have the molecular formula $C_{15}H_{22}O_2$. Suggest reasonable structures for these products.

17.29 Compounds that contain both carbonyl and alcohol functional groups are often more stable as cyclic hemiacetals or cyclic acetals than as open-chain compounds. Examples of several of these are shown. Deduce the structure of the open-chain form of each.

(*a*) O OH

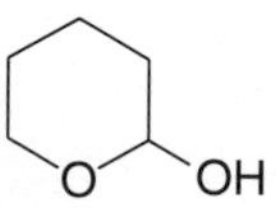

(*b*) HO O

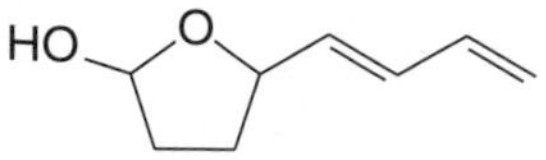

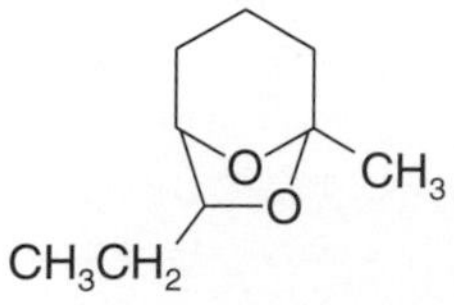

(*c*) O CH₃ O CH_3CH_2 — Brevicomin (sex attractant of Western pine beetle)

(*d*) [structure: spiroketal with CH_2CH_3, $HOCH_2$, OH substituents]

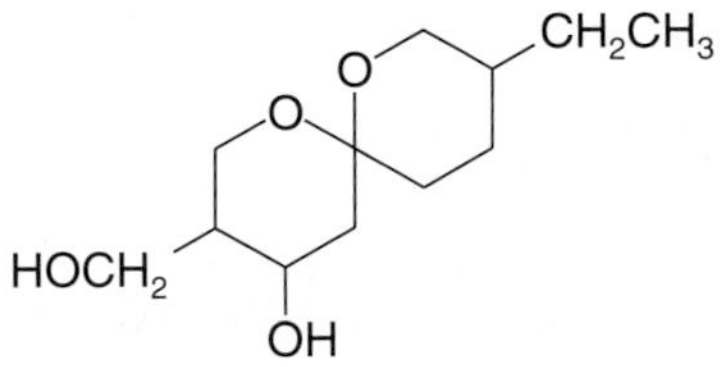

Talaromycin A (a toxic substance produced by a fungus that grows on poultry house litter)

17.30 Compounds that contain a carbon-nitrogen double bond are capable of stereoisomerism much like that seen in alkenes. The structures

$$\mathrm{RR'C{=}N(\ddot{\ })X} \quad \text{and} \quad \mathrm{RR'C{=}NX(\ddot{\ })}$$

are stereoisomeric. Specifying stereochemistry in these systems is best done by using *E-Z* descriptors and considering the nitrogen lone pair to be the lowest-priority group. Write the structures, clearly showing stereochemistry, of the following:

(*a*) (Z)-$CH_3CH{=}NCH_3$
(*b*) (*E*)-Acetaldehyde oxime
(*c*) (*Z*)-2-Butanone hydrazone
(*d*) (*E*)-Acetophenone semicarbazone

17.31 Compounds known as *nitrones* are formed when *N*-substituted derivatives of hydroxylamine react with aldehydes and ketones:

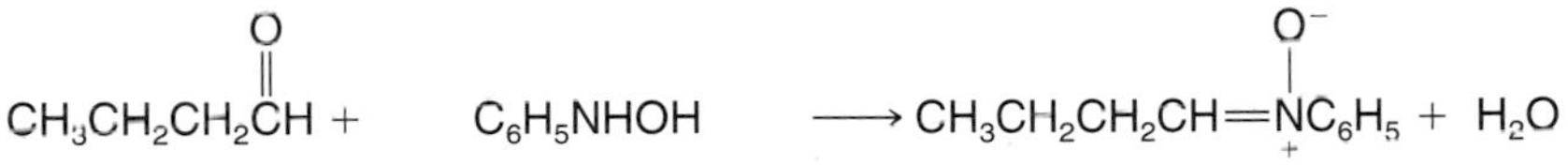

$$\underset{\text{Butanal}}{CH_3CH_2CH_2\overset{O}{\overset{\|}{C}}H} + \underset{N\text{-Phenylhydroxylamine}}{C_6H_5NHOH} \longrightarrow \underset{N\text{-(Butylidene)aniline } N\text{-oxide (a nitrone) (80\%)}}{CH_3CH_2CH_2CH{=}\overset{O^-}{\overset{|}{\underset{+}{N}}}C_6H_5} + \underset{\text{Water}}{H_2O}$$

Write a reasonable sequence of steps that describes the mechanism of this reaction.

17.32 Compounds known as *lactones,* which are cyclic esters, are formed on Baeyer-Villiger oxidation of cyclic ketones. Suggest a mechanism for the Baeyer-Villiger oxidation shown.

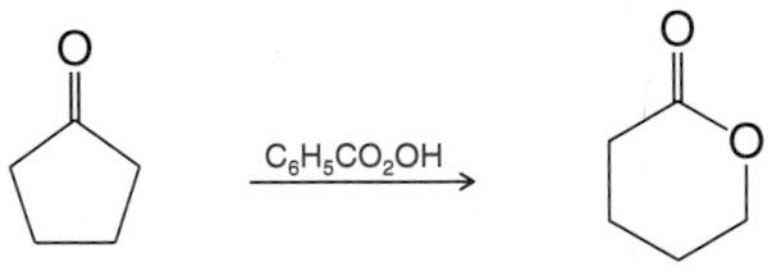

$$\text{Cyclopentanone} \xrightarrow{C_6H_5CO_2OH} \text{5-Pentanolide (78\%)}$$

17.33 Suggest reasonable mechanisms for each of the following reactions:

(*a*) $$(CH_3)_3C(Cl)\overset{O}{C{-}CH_2} \xrightarrow[CH_3OH]{NaOCH_3} (CH_3)_3C\overset{O}{\overset{\|}{C}}CH_2OCH_3 \quad (88\%)$$

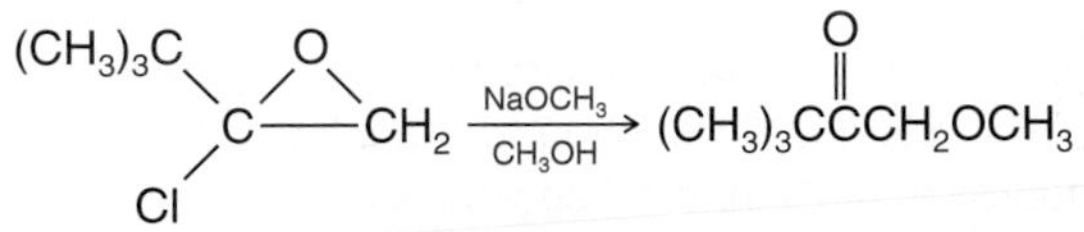

(*b*) $(CH_3)_3CCHCH(=O)$ with Cl on the central CH $\xrightarrow[CH_3OH]{NaOCH_3}$ $(CH_3)_3CCHCH(OCH_3)_2$ with OH on the central CH (72%)

17.34 *Amygdalin,* a substance present in peach, plum, and almond pits, is a derivative of the *R* enantiomer of benzaldehyde cyanohydrin. Give the structure of (*R*)-benzaldehyde cyanohydrin.

17.35 Using ethanol as the source of all the carbon atoms, describe efficient syntheses of each of the following, using any necessary organic or inorganic reagents:

(*a*) $CH_3CH(OCH_2CH_3)_2$

(*b*) 1,3-Dioxolane ring with H and CH_3 on C-2

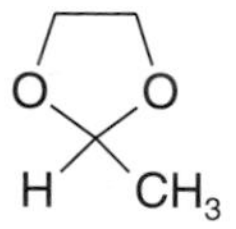

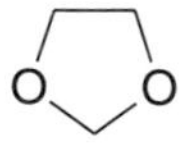

(*c*) 1,3-Dioxolane ring

(*d*) $CH_3CHC{\equiv}CH$ with OH on the CH carbon

(*e*) $HCCH_2C{\equiv}CH$ with =O on the first carbon ($HC(=O)CH_2C{\equiv}CH$)

(*f*) $CH_3CH_2CH_2CH_2OH$

17.36 Describe reasonable syntheses of benzophenone, $C_6H_5CC_6H_5$ (with =O on the central C), from each of the following starting materials and any necessary inorganic reagents.

(*a*) Benzoyl chloride and benzene
(*b*) Benzyl alcohol and bromobenzene
(*c*) Bromodiphenylmethane, $(C_6H_5)_2CHBr$
(*d*) Dimethoxydiphenylmethane, $(C_6H_5)_2C(OCH_3)_2$
(*e*) 1,1,2,2-Tetraphenylethene, $(C_6H_5)_2C{=}C(C_6H_5)_2$

17.37 The sex attractant of the female winter moth has been identified as $CH_3(CH_2)_8CH{=}CHCH_2CH{=}CHCH_2CH{=}CHCH{=}CH_2$. Devise a synthesis of this material from 3,6-hexadecadien-1-ol and allyl alcohol.

17.38 Suggest reasonable structures for compounds A and B:

$(C_6H_5)_2CHCHCCl$ (with C_6H_5 on the second CH and =O on the acyl carbon) $\xrightarrow{AlCl_3}$ compound A (mp 100–101°C)

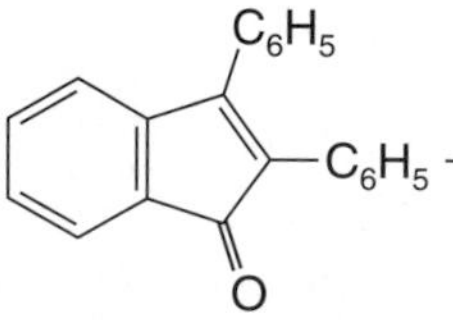

2,3-Diphenyl-1-indenone (inden-1-one with C_6H_5 at C-2 and C-3) $\xrightarrow{H_2,\ Pd}$ compound B (mp 151–154°C)

Both A and B have the molecular formula $C_{21}H_{16}O$ and exhibit a strong absorption peak in the infrared around 1680 cm^{-1}.

17.39 Hydrolysis of a compound A in dilute aqueous hydrochloric acid gave (along with methanol) a compound B, mp 164–165°C. Compound B had the molecular formula $C_{16}H_{16}O_4$; it exhibited hydroxyl absorption in its infrared spectrum at 3550 cm^{-1} but had no peaks in the carbonyl region. What is a reasonable structure for compound B?

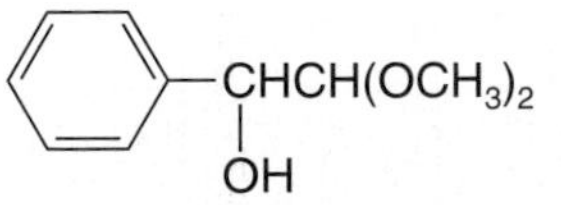

$C_6H_5CH(OH)CH(OCH_3)_2$ Compound A

17.40 Syntheses of each of the following compounds have been reported in the chemical literature. Using the indicated starting material and any necessary organic or inorganic reagents, describe short sequences of reactions that would be appropriate for each transformation.

(*a*) 1,1,5-Trimethylcyclononane from 5,5-dimethylcyclononanone

(*b*) $(C_6H_5)(H_3C)C{=}C_5H_8$ (cyclopentylidene) from $C_6H_5\overset{O}{\overset{\|}{C}}-C_5H_9$ (cyclopentyl)

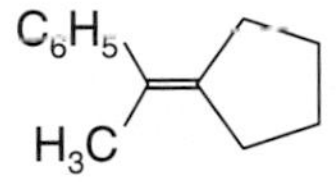

(*c*) o-$CH_3C_6H_4\overset{CH_2}{\overset{\|}{C}}CH_2CH_2CH_2CH{=}CH_2$ from *o*-bromotoluene and 5-hexenal

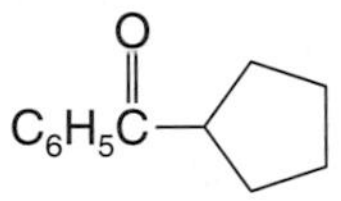

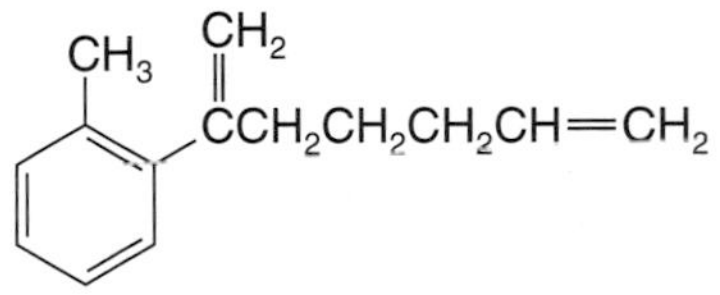

(*d*) $CH_3\overset{O}{\overset{\|}{C}}CH_2CH_2\overset{O}{\overset{\|}{C}}(CH_2)_5CH_3$ from $HC{\equiv}CCH_2CH_2CH_2OH$

(*e*) 3-Cl-2-$CH_3C_6H_3CH_2OCH_3$ from 3-chloro-2-methylbenzaldehyde

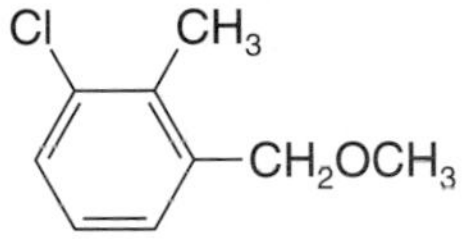

17.41 The following five-step synthesis has been reported in the chemical literature. Suggest reagents appropriate for each step.

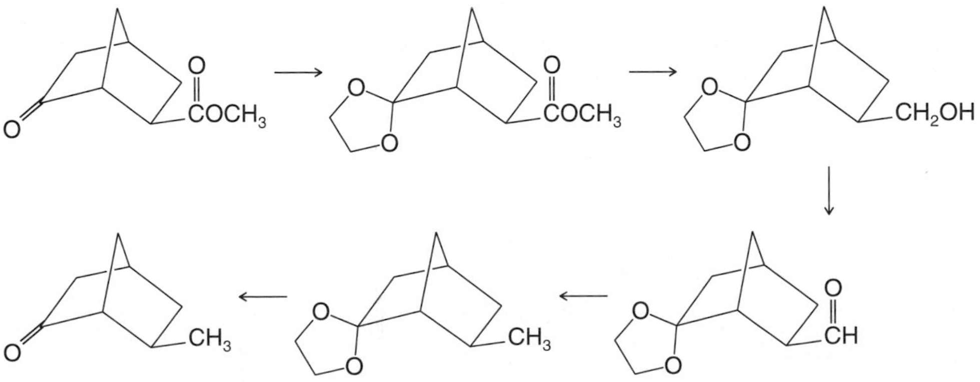

[Bicyclic keto ester with $\overset{O}{\overset{\|}{C}}OCH_3$ → ethylene ketal with $\overset{O}{\overset{\|}{C}}OCH_3$ → ethylene ketal with CH_2OH → ethylene ketal with $\overset{O}{\overset{\|}{C}}H$ → ethylene ketal with CH_3 → ketone with CH_3]

17.42 Increased "single-bond character" in a carbonyl group is associated with a decreased carbon-oxygen stretching frequency. Among the three compounds benzaldehyde, 2,4,6-trimethoxybenzaldehyde, and 2,4,6-trinitrobenzaldehyde, which one will have the lowest-frequency carbonyl absorption? Which one will have the highest?

17.43 A compound has the molecular formula C_4H_8O and contains a carbonyl group. Identify the compound on the basis of its ^1H nmr spectrum shown in Figure 17.15.

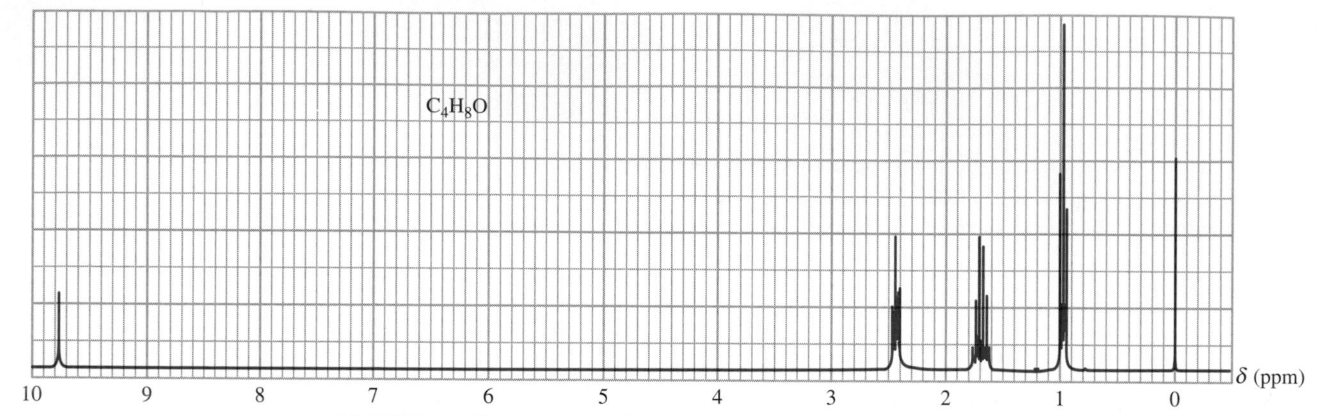

FIGURE 17.15 The 300 MHz ^{1}H nmr spectrum of a compound (C_4H_8O) (Problem 17.43). The signal at $\delta = 2.4$ ppm corresponds to two protons and can be shown on scale expansion to be a doublet of triplets. The signal at $\delta = 9.8$ ppm is a one-proton triplet.

17.44 A compound ($C_7H_{14}O$) has a strong peak in its infrared spectrum at 1710 cm^{-1}. Its ^{1}H nmr spectrum consists of three singlets in the ratio 9:3:2 at $\delta = 1.0$, 2.1, and 2.3 ppm, respectively. Identify the compound.

17.45 Compounds A and B are isomeric diketones of molecular formula $C_6H_{10}O_2$. The ^{1}H nmr spectrum of compound A contains two signals, both singlets, at $\delta = 2.2$ (6 protons) and 2.8 ppm (4 protons). The ^{1}H nmr spectrum of compound B contains two signals, one at $\delta = 1.3$ ppm (triplet, 6 protons) and the other at $\delta = 2.8$ ppm (quartet, 4 protons). What are the structures of compounds A and B?

17.46 A compound ($C_{11}H_{14}O$) has a strong peak in its infrared spectrum near 1700 cm^{-1}. Its 300 MHz ^{1}H nmr spectrum is shown in Figure 17.16. What is the structure of the compound?

17.47 A compound is a ketone of molecular formula $C_7H_{14}O$. Its ^{13}C nmr spectrum is shown in Figure 17.17. What is the structure of the compound?

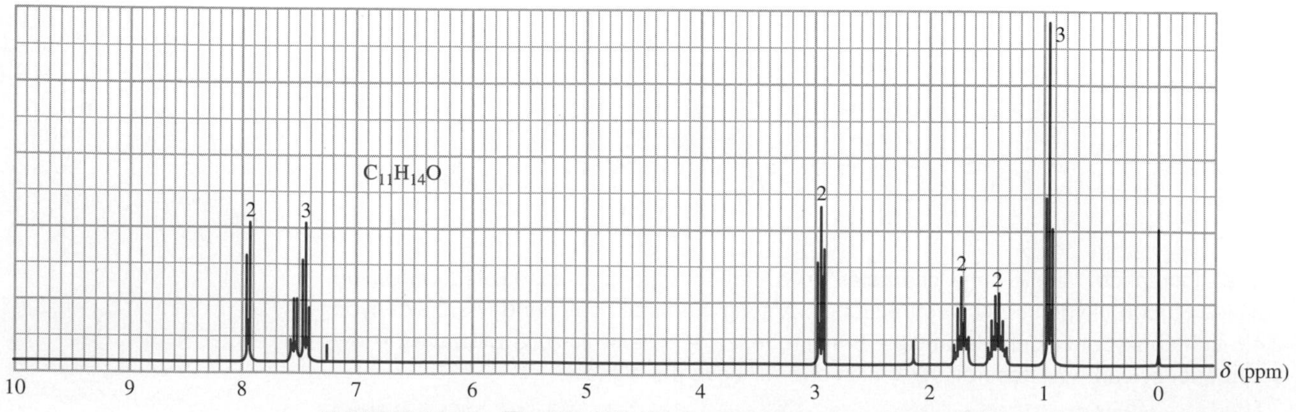

FIGURE 17.16 The 300 MHz ^{1}H nmr spectrum of a compound ($C_{11}H_{14}O$) (Problem 17.46).

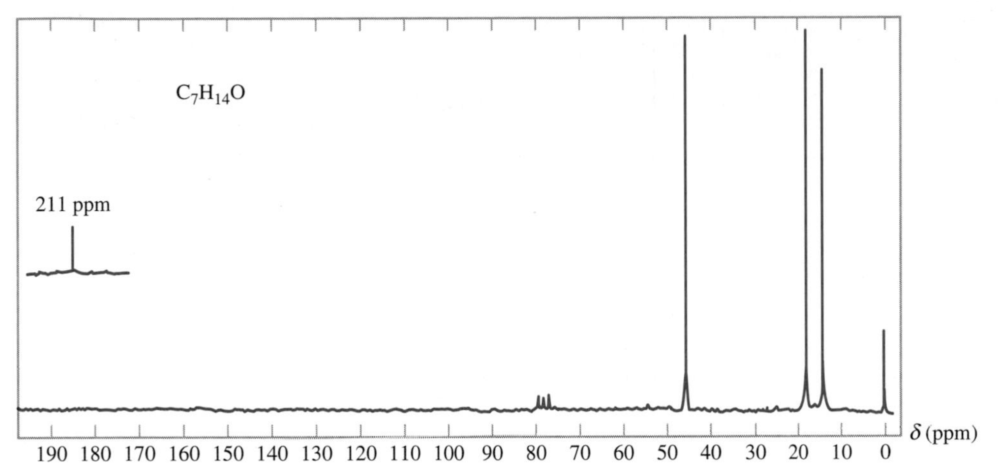

FIGURE 17.17 The ^{13}C nmr spectrum of a compound ($C_7H_{14}O$) (Problem 17.47). *(Taken from LeRoy F. Johnson and William C. Jankowski, Carbon-13 NMR Spectra: A Collection of Assigned, Coded, and Indexed Spectra, Wiley-Interscience, New York, 1972. Reprinted by permission of John Wiley & Sons, Inc.)*

17.48 Compound A and compound B are isomers having the molecular formula $C_{10}H_{12}O$. The mass spectrum of each compound contains an abundant peak at *m/z* 105. The ^{13}C nmr spectra of compound A (Figure 17.18) and compound B (Figure 17.19) are shown. Identify these two isomers.

MOLECULAR MODELING EXERCISES

17.49 The most stable conformation of acetone has one of the hydrogens of each methyl group eclipsed with the carbonyl oxygen. Construct a model of this conformation.

17.50 Construct a molecular model of cyclohexanone. Do either of the hydrogens of C-2 eclipse the carbonyl oxygen?

17.51 There are two stereoisomers of the oxime $CH_3CH{=}NOH$ (formed from acetaldehyde and hydroxylamine). Construct a molecular model of each stereoisomer. How many stereoisomers are formed in the reaction of acetone and hydroxylamine?

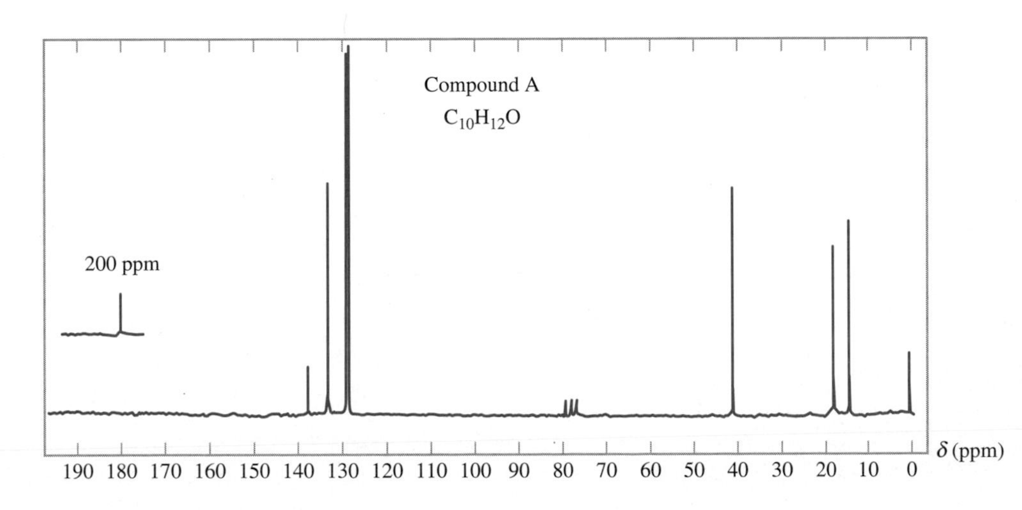

FIGURE 17.18 The ^{13}C nmr spectrum of compound A ($C_{10}H_{12}O$) (Problem 17.48). *(Taken from LeRoy F. Johnson and William C. Jankowski, Carbon-13 NMR Spectra: A Collection of Assigned, Coded, and Indexed Spectra, Wiley-Interscience, New York, 1972. Reprinted by permission of John Wiley & Sons, Inc.)*

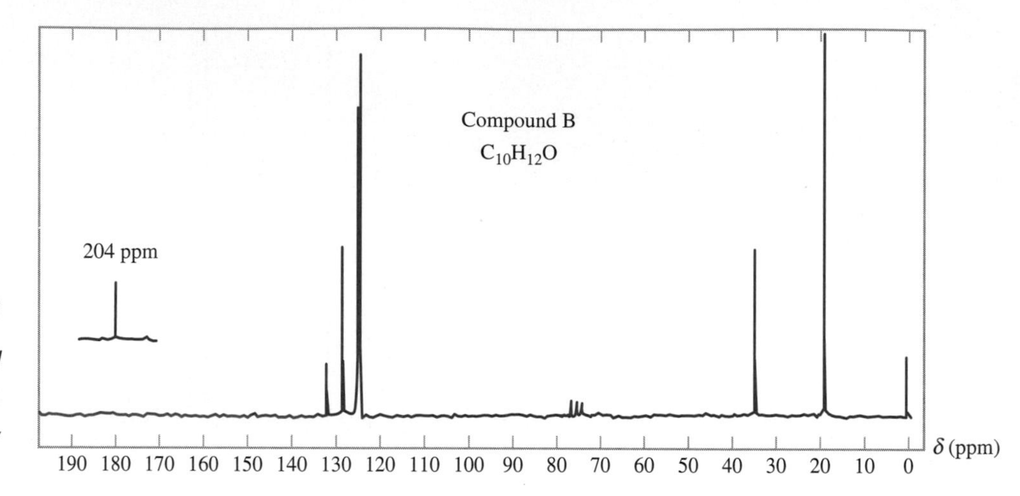

FIGURE 17.19 The ^{13}C nmr spectrum of compound B ($C_{10}H_{12}O$) (Problem 17.48). *(Taken from LeRoy F. Johnson and William C. Jankowski, Carbon-13 NMR Spectra: A Collection of Assigned, Coded, and Indexed Spectra, Wiley-Interscience, New York, 1972. Reprinted by permission of John Wiley & Sons, Inc.)*

17.52 The equilibrium constant for cyanohydrin formation is much greater for cyclohexanone than for 3,3-dimethylcyclohexanone. Construct a molecular model of each cyanohydrin and suggest a reason for this observation.

17.53 Construct a molecular model of the cyclic acetal shown. Be sure the six-membered ring is in a chair conformation and the *tert*-butyl group is equatorial. It is convenient to use wedges and dashes to distinguish among the protons designated H_w, H_x, H_y, and H_z. How misleading are the wedges and dashes in this case? What is the relationship between H_w and H_x? Are they enantiotopic or diastereotopic? Between H_w and H_y? Between H_w and H_z?

CHAPTER 18

ENOLS AND ENOLATES

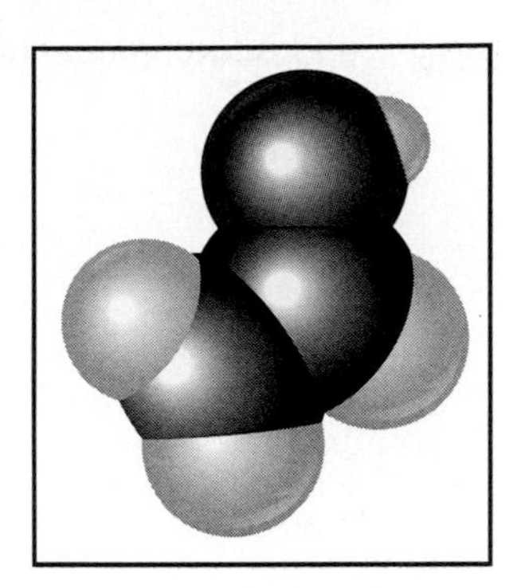

In the preceding chapter you learned that nucleophilic addition to the carbonyl group is one of the fundamental reaction types of organic chemistry. Aside from its own reactivity, a carbonyl group can affect the chemical properties of aldehydes and ketones in other ways. Aldehydes and ketones participate in an equilibrium with their **enol** isomers.

$$\underset{\text{Aldehyde or ketone}}{R_2CH\overset{\overset{\displaystyle O}{\|}}{C}R'} \rightleftharpoons \underset{\text{Enol}}{R_2C{=}\overset{\overset{\displaystyle OH}{|}}{C}R'}$$

In this chapter you will see a number of processes in which it is the enol form of an aldehyde or a ketone rather than the carbonyl compound itself which is the reactive species.

There is also an important group of reactions in which the carbonyl group acts as a powerful electron-withdrawing substituent, increasing the acidity of protons on the carbons to which it is attached:

$$R_2\underset{\underset{\displaystyle H}{|}}{C}\overset{\overset{\displaystyle O}{\|}}{C}R'$$

This proton (H) is far more acidic than a hydrogen in an alkane.

As an electron-withdrawing group on a carbon-carbon double bond, a carbonyl group renders the double bond susceptible to nucleophilic attack:

$$R_2C{=}CH\overset{O}{\overset{\|}{C}}R'$$

Normally, carbon-carbon double bonds are attacked by electrophilic reagents; a carbon-carbon double bond that is conjugated to a carbonyl group is attacked by nucleophilic reagents.

Substituent effects arising from the presence of a carbonyl group in a molecule make possible a number of chemical reactions that are of great synthetic and mechanistic importance. This chapter is complementary to the preceding one; the two chapters taken together demonstrate the extraordinary range of chemical reactions available to aldehydes and ketones.

18.1 THE α CARBON ATOM AND ITS HYDROGENS

It is convenient to identify various carbon atoms in aldehydes and ketones by using the Greek letters α, β, γ, and so forth, to specify their location in relation to the carbonyl group. The carbon atom adjacent to the carbonyl group is the α carbon atom, the next one down the chain is the β carbon, and so on. Butanal, for example, has an α carbon, a β carbon, and a γ carbon.

$$\underset{\gamma}{C}H_3\underset{\beta}{C}H_2\underset{\alpha}{C}H_2\overset{O}{\overset{\|}{C}}H$$

Carbonyl group is reference point; no Greek letter assigned to it.

Hydrogen substituents are identified by the same Greek letter as the carbon atom to which they are attached. A hydrogen connected to the α carbon atom is an α hydrogen. Butanal has two α protons, two β protons, and three γ protons. No Greek letter is assigned to the hydrogen attached directly to the carbonyl group of an aldehyde.

PROBLEM 18.1 How many α hydrogens are there in each of the following?

(*a*) 3,3-Dimethyl-2-butanone
(*b*) 2,2-Dimethylpropanal
(*c*) Benzyl methyl ketone
(*d*) Cyclohexanone

SAMPLE SOLUTION (*a*) This ketone has two different α carbons but only one of them has hydrogen substituents. There are three equivalent α hydrogens. The other nine hydrogens are attached to β carbon atoms.

$$\overset{\alpha}{C}H_3{-}\overset{O}{\overset{\|}{C}}{-}\overset{\alpha}{C}(\overset{\beta}{C}H_3)_2{-}\overset{\beta}{C}H_3$$

3,3-Dimethyl-2-butanone

Other than nucleophilic addition to the carbonyl group, the most important reactions of aldehydes and ketones involve substitution of an α hydrogen. Activation of the α position toward substitution is a direct result of its location with respect to the carbonyl group. A particularly well studied example is halogenation of aldehydes and ketones.

18.2 α HALOGENATION OF ALDEHYDES AND KETONES

Aldehydes and ketones react with halogens by *substitution* of halogen for one of the α hydrogens:

$$R_2\underset{H}{\underset{|}{C}}\overset{O}{\overset{\|}{C}}R' + X_2 \xrightarrow{H^+} R_2\underset{X}{\underset{|}{C}}\overset{O}{\overset{\|}{C}}R' + HX$$

Aldehyde or ketone — Halogen — α-Halo aldehyde or ketone — Hydrogen halide

The reaction is *regiospecific* for substitution of an α hydrogen. None of the hydrogens farther removed from the carbonyl group are affected.

$$\text{Cyclohexanone} + Cl_2 \xrightarrow{H_2O} \text{2-Chlorocyclohexanone} + HCl$$

Cyclohexanone — Chlorine — 2-Chlorocyclohexanone (61–66%) — Hydrogen chloride

Nor is the hydrogen directly attached to the carbonyl group in aldehydes affected. Only the α hydrogen is replaced.

$$\text{Cyclohexanecarbaldehyde} + Br_2 \xrightarrow{CHCl_3} \text{1-Bromocyclohexanecarbaldehyde} + HBr$$

Cyclohexanecarbaldehyde — Bromine — 1-Bromocyclohexanecarbaldehyde (80%) — Hydrogen bromide

---

PROBLEM 18.2 Chlorination of 2-butanone yields two isomeric products each having the molecular formula C_4H_7ClO. Identify these two compounds.

α Halogenation of aldehydes and ketones can be carried out in a variety of solvents (water and chloroform are shown in the examples, but acetic acid and diethyl ether are also often used). The reaction is catalyzed by acids. Since one of the reaction products, the hydrogen halide, is an acid and therefore a catalyst for the reaction, the process is said to be **autocatalytic.** Free radicals are *not* involved, and the reactions occur at room temperature in the absence of initiators. Mechanistically, acid-catalyzed halogenation of aldehydes and ketones is much different from free-radical halogenation of alkanes. While both processes lead to the replacement of a hydrogen by a halogen, they do so by pathways that are unrelated to each other.

18.3 MECHANISM OF α HALOGENATION OF ALDEHYDES AND KETONES

Lapworth was far ahead of his time in understanding how organic reactions occur. For an account of Lapworth's contributions to mechanistic organic chemistry, see the November 1972 issue of the *Journal of Chemical Education,* pp. 750–752.

In one of the earliest systematic investigations of a reaction mechanism in organic chemistry, Arthur Lapworth discovered in 1904 that the rates of chlorination and bromination of acetone were the same. Later he found that iodination of acetone proceeded at the same rate as chlorination and bromination. Moreover, the rates of all three halogenation reactions, while first-order in acetone, are independent of the halogen concentration. *Thus, the halogen does not participate in the reaction until after the rate-determining step.* These kinetic observations, coupled with the fact that substitution occurs exclusively at the α carbon atom, led Lapworth to propose that the rate-determining step is the conversion of acetone to a more reactive form, its enol isomer:

$$CH_3\overset{O}{\overset{\|}{C}}CH_3 \overset{slow}{\rightleftharpoons} CH_3\overset{OH}{\overset{|}{C}}{=}CH_2$$

Acetone — Propen-2-ol (enol form of acetone)

Once formed, this enol reacts rapidly with the halogen to form an α-halo ketone:

$$CH_3\overset{OH}{\overset{|}{C}}{=}CH_2 + X_2 \xrightarrow{fast} CH_3\overset{O}{\overset{\|}{C}}CH_2X + HX$$

Propen-2-ol (enol form of acetone) — Halogen — α-Halo derivative of acetone — Hydrogen halide

PROBLEM 18.3 Write the structures of the enol forms of 2-butanone that react with chlorine to give 1-chloro-2-butanone and 3-chloro-2-butanone.

Both phases of the Lapworth mechanism, enol formation and enol halogenation, are new to us. Let us examine them in reverse order. We can understand enol halogenation by analogy to halogen addition to alkenes. An enol is a very reactive kind of alkene. Its carbon-carbon double bond bears an electron-releasing hydroxyl group, which activates it toward attack by electrophiles.

$$CH_3\overset{:\ddot{O}H}{\overset{|}{C}}{=}CH_2 + :\ddot{\underset{..}{Br}}{-}\ddot{\underset{..}{Br}}: \xrightarrow{\text{very fast}} CH_3{-}\overset{:\ddot{O}H}{\overset{|}{\underset{+}{C}}}{-}CH_2\ddot{\underset{..}{Br}}: + :\ddot{\underset{..}{Br}}:^-$$

Propen-2-ol (enol form of acetone) — Bromine — Stabilized cationic intermediate — Bromide ion

The hydroxyl group stabilizes the cationic intermediate by delocalization of one of the unshared electron pairs of oxygen:

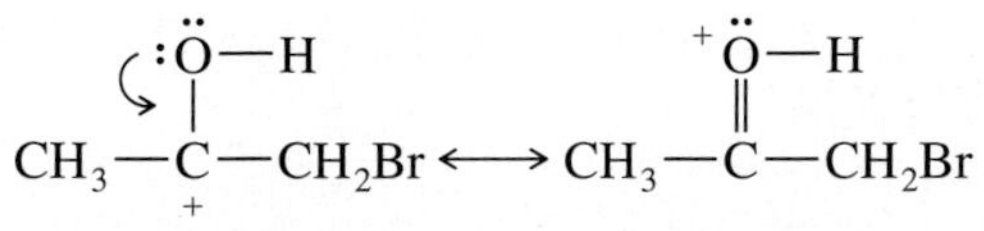

$$\mathrm{CH_3-\overset{\overset{\ddot{O}-H}{|}}{\underset{+}{C}}-CH_2Br} \longleftrightarrow \mathrm{CH_3-\overset{\overset{+\ddot{O}-H}{\|}}{C}-CH_2Br}$$

Less stable resonance form; six electrons on positively charged carbon.

More stable resonance form; all atoms (except hydrogen) have octets of electrons.

Participation by the enolic oxygen in the bromination step is responsible for the rapid attack on the carbon-carbon double bond of an enol. Organic chemists usually represent this participation explicitly:

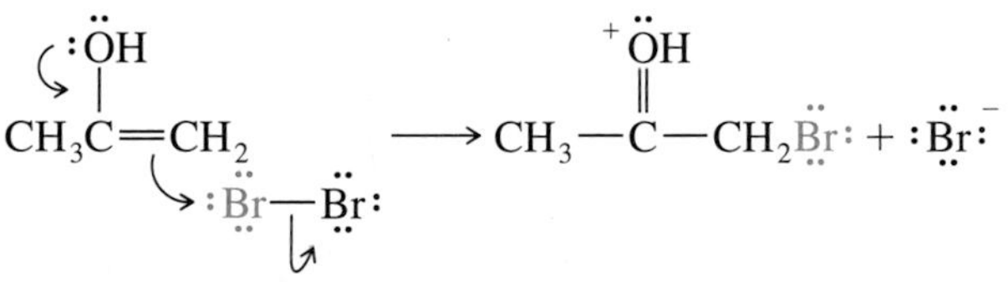

$$\mathrm{CH_3\overset{\overset{:\ddot{O}H}{|}}{C}{=}CH_2} \quad \mathrm{:\ddot{B}r-\ddot{B}r:} \longrightarrow \mathrm{CH_3-\overset{\overset{+\ddot{O}H}{\|}}{C}-CH_2\ddot{B}r:} + \mathrm{:\ddot{B}r:^-}$$

Representing the bromine addition step in this way emphasizes the increased nucleophilicity of the enol double bond and identifies the source of that increased nucleophilicity as the enolic oxygen.

PROBLEM 18.4 Represent the reaction of chlorine with each of the enol forms of 2-butanone (see Problem 18.3) according to the curved arrow formalism just described.

The cationic intermediate is simply the protonated form (conjugate acid) of the α-halo ketone. Deprotonation of the cationic intermediate gives the products.

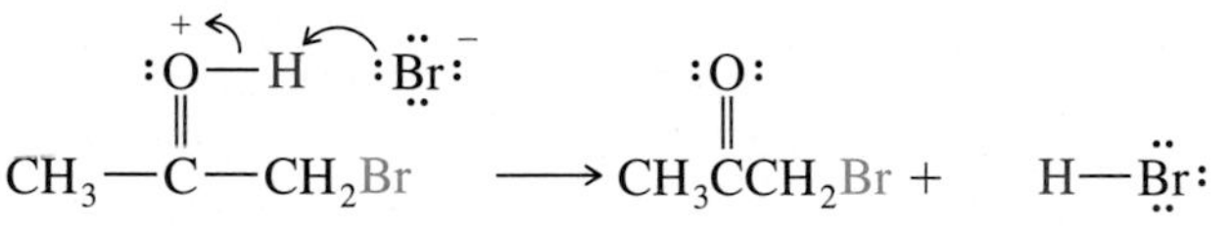

$$\mathrm{CH_3-\overset{\overset{+:O-H}{\|}}{C}-CH_2Br} \quad \mathrm{:\ddot{B}r:^-} \longrightarrow \mathrm{CH_3\overset{\overset{:O:}{\|}}{C}CH_2Br} + \mathrm{H-\ddot{B}r:}$$

Cationic intermediate — Bromoacetone — Hydrogen bromide

Having now seen how an enol, once formed, reacts with a halogen, let us consider the process of enolization itself.

18.4 ENOLIZATION AND ENOL CONTENT

Enols are related to an aldehyde or a ketone by a proton-transfer equilibrium known as **keto-enol tautomerism.** (*Tautomerism* refers to an interconversion between two structures that differ by the placement of an atom or a group.)

The keto and enol forms are constitutional isomers. In informal terminology they are referred to as *tautomers* of one another.

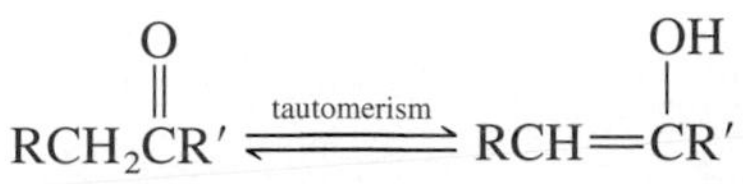

$$\mathrm{RCH_2\overset{\overset{O}{\|}}{C}R'} \underset{}{\overset{\text{tautomerism}}{\rightleftharpoons}} \mathrm{RCH{=}\overset{\overset{OH}{|}}{C}R'}$$

Keto form — Enol form

Overall reaction:

$$\underset{\text{Aldehyde or ketone}}{RCH_2\overset{\overset{\displaystyle O}{\|}}{C}R'} \xrightleftharpoons{H_3O^+} \underset{\text{Enol}}{RCH{=}\overset{\overset{\displaystyle OH}{|}}{C}R'}$$

Step 1: A proton is transferred from the acid catalyst to the carbonyl oxygen.

$$\underset{\text{Aldehyde or ketone}}{RCH_2\overset{\overset{\displaystyle :O:}{\|}}{C}R'} + \underset{\text{Hydronium ion}}{H{-}\overset{+}{O}H_2} \xrightleftharpoons{\text{fast}} \underset{\text{Conjugate acid of carbonyl compound}}{RCH_2\overset{\overset{\displaystyle :\overset{+}{O}{-}H}{\|}}{C}R'} + \underset{\text{Water}}{:OH_2}$$

Step 2: A water molecule acts as a Brønsted base to remove a proton from the α carbon atom of the protonated aldehyde or ketone.

$$\underset{\text{Conjugate acid of carbonyl compound}}{R\underset{\underset{\displaystyle H}{|}}{C}H{-}\overset{\overset{\displaystyle \overset{+}{O}{-}H}{\|}}{C}R'} + \underset{\text{Water}}{:OH_2} \xrightleftharpoons{\text{slow}} \underset{\text{Enol}}{RCH{=}\overset{\overset{\displaystyle :\ddot{O}{-}H}{|}}{C}R'} + \underset{\text{Hydronium ion}}{H{-}\overset{+}{O}H_2}$$

FIGURE 18.1 Sequence of steps that describes the acid-catalyzed enolization of an aldehyde or a ketone in aqueous solution.

The mechanism of enolization involves solvent-mediated proton transfer steps rather than a direct intramolecular ''proton jump'' from carbon to oxygen and is relatively slow in neutral media. The rate of enolization is catalyzed by acids as shown by the mechanism in Figure 18.1. In aqueous acid, a hydronium ion transfers a proton to the carbonyl oxygen in step 1, and a water molecule acts as a Brønsted base to remove a proton from the α carbon atom in step 2. The second step is slower than the first. The first step involves proton transfer between oxygens, while the second is a proton transfer from carbon to oxygen.

You have had earlier experience with enols in their role as intermediates in the hydration of alkynes (Section 9.13). The mechanism of enolization of aldehydes and ketones is precisely the reverse of the mechanism by which an enol is converted to a carbonyl compound.

The amount of enol present at equilibrium, the *enol content,* is quite small for simple aldehydes and ketones. The equilibrium constants for enolization, as shown by the following examples, are much less than unity:

$$\underset{\substack{\text{Acetaldehyde}\\ \text{(keto form)}}}{CH_3\overset{\overset{\displaystyle O}{\|}}{C}H} \rightleftharpoons \underset{\substack{\text{Vinyl alcohol}\\ \text{(enol form)}}}{CH_2{=}CHOH} \qquad K \cong 3 \times 10^{-7}$$

$$\underset{\text{Acetone (keto form)}}{CH_3\overset{O}{\overset{\|}{C}}CH_3} \rightleftharpoons \underset{\text{Propen-2-ol (enol form)}}{CH_2{=}\overset{OH}{\overset{|}{C}}CH_3} \qquad K \cong 6 \times 10^{-9}$$

In these and numerous other simple cases, the keto form is more stable than the enol by some 45 to 60 kJ/mol (11 to 14 kcal/mol). The chief reason for this difference is the greater resonance stabilization of a carbonyl group as compared with a carbon-carbon double bond.

With unsymmetrical ketones, enolization may occur in either of two directions:

$$\underset{\text{1-Buten-2-ol (enol form)}}{CH_2{=}\overset{OH}{\overset{|}{C}}CH_2CH_3} \rightleftharpoons \underset{\text{2-Butanone (keto form)}}{CH_3\overset{O}{\overset{\|}{C}}CH_2CH_3} \rightleftharpoons \underset{\text{2-Buten-2-ol (enol form)}}{CH_3\overset{OH}{\overset{|}{C}}{=}CHCH_3}$$

The ketone is by far the most abundant species present at equilibrium. Both enols are also present, but in very small concentrations. The enol with the more highly substituted double bond is the more stable of the two enols and is present in higher concentration than the other.

PROBLEM 18.5 Write structural formulas corresponding to

(*a*) The enol form of 2,4-dimethyl-3-pentanone
(*b*) The enol form of acetophenone
(*c*) The two enol forms of 2-methylcyclohexanone

SAMPLE SOLUTION (*a*) Remember that it is the α carbon atom that is deprotonated in the enolization process. The ketone 2,4-dimethyl-3-pentanone gives a single enol, since the two α carbons are equivalent.

$$\underset{\text{2,4-Dimethyl-3-pentanone (keto form)}}{(CH_3)_2CH\overset{O}{\overset{\|}{C}}CH(CH_3)_2} \rightleftharpoons \underset{\text{2,4-Dimethyl-2-penten-3-ol (enol form)}}{(CH_3)_2C{=}\overset{OH}{\overset{|}{C}}CH(CH_3)_2}$$

It is important to recognize that an enol is a real substance, capable of independent existence. An enol is *not* a resonance form of a carbonyl compound; the two are constitutional isomers of each other.

18.5 STABILIZED ENOLS

Certain structural features can make the keto $\rightleftharpoons$ enol equilibrium more favorable by stabilizing the enol form. Enolization of 2,4-cyclohexadienone is one such example:

O---H separation in intramolecular hydrogen bond is 166 pm

103 pm

133 pm

124 pm

134 pm

141 pm

FIGURE 18.2 Structure of the enol form of 2,4-pentanedione.

$$\text{2,4-Cyclohexadienone} \rightleftharpoons \text{Phenol}$$

2,4-Cyclohexadienone (keto form, not aromatic)

Phenol (enol form, aromatic)

K is too large to measure.

The enol is *phenol,* and the stabilization gained by forming an aromatic ring is more than sufficient to overcome the normal preference for the keto form. The amount of the keto form present in phenol at equilibrium is far too small to measure.

A 1,3 arrangement of two carbonyl groups (these compounds are called **β-diketones**) leads to a situation in which the keto and enol forms are of comparable stability.

$$CH_3\overset{O}{\overset{\|}{C}}CH_2\overset{O}{\overset{\|}{C}}CH_3 \rightleftharpoons CH_3\overset{OH}{\overset{|}{C}}{=}CH\overset{O}{\overset{\|}{C}}CH_3 \qquad K = 4$$

2,4-Pentanedione (20%) (keto form)

4-Hydroxy-3-penten-2-one (80%) (enol form)

The two most important structural features that stabilize the enol of a β-dicarbonyl compound are (1) conjugation of its double bond with the remaining carbonyl group and (2) the presence of a strong intramolecular hydrogen bond between the enolic hydroxyl group and the carbonyl oxygen (Figure 18.2).

In β-diketones it is the methylene group flanked by the two carbonyls that is involved in enolization. The alternative enol

$$CH_2{=}\overset{OH}{\overset{|}{C}}CH_2\overset{O}{\overset{\|}{C}}CH_3 \qquad \text{4-Hydroxy-4-penten-2-one}$$

does not have its carbon-carbon double bond conjugated with the carbonyl group, is not as stable, and is present in negligible amounts at equilibrium.

PROBLEM 18.6 Write structural formulas corresponding to

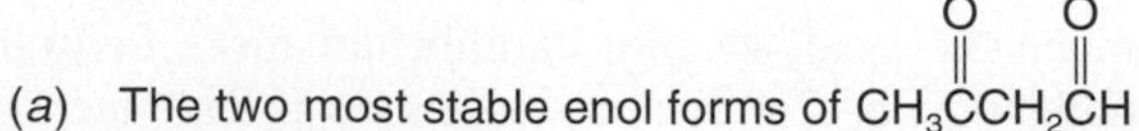

(*a*) The two most stable enol forms of $CH_3\overset{O}{\overset{\|}{C}}CH_2\overset{O}{\overset{\|}{C}}H$

(*b*) The two most stable enol forms of 1-phenyl-1,3-butanedione

SAMPLE SOLUTION (*a*) Enolization of this 1,3-dicarbonyl compound can involve either of the two carbonyl groups:

$$\text{CH}_3\overset{\text{O}\cdots\text{H}-\text{O}}{\text{C}=\text{CH}-\text{CH}} \rightleftharpoons \text{CH}_3\overset{\text{O}}{\overset{\|}{\text{C}}}\text{CH}_2\overset{\text{O}}{\overset{\|}{\text{C}}}\text{H} \rightleftharpoons \text{CH}_3\overset{\text{O}-\text{H}\cdots\text{O}}{\text{C}=\text{CH}-\text{CH}}$$

Both enols have their carbon-carbon double bonds conjugated to a carbonyl group and can form an intramolecular hydrogen bond. They are of comparable stability.

18.6 BASE-CATALYZED ENOLIZATION. ENOLATE ANIONS

The solvent-mediated proton-transfer equilibrium that interconverts a carbonyl compound and its enol can be catalyzed by bases as well as by acids. Figure 18.3 illustrates the roles of hydroxide ion and water in a base-catalyzed enolization occurring in aqueous medium. As in acid-catalyzed enolization, protons are transferred sequentially rather than in a single step. First (step 1), the base abstracts a proton from the α carbon atom to yield an anion. This anion is a resonance-stabilized species. Its negative charge is shared by the α carbon atom and the carbonyl oxygen.

Overall reaction:

$$\underset{\text{Aldehyde or ketone}}{\text{RCH}_2\overset{\text{O}}{\overset{\|}{\text{C}}}\text{R}'} \xrightleftharpoons{\text{HO}^-} \underset{\text{Enol}}{\text{RCH}=\overset{\text{OH}}{\overset{|}{\text{C}}}\text{R}'}$$

Step 1: A proton is abstracted by hydroxide ion from the α carbon atom of the carbonyl compound.

$$\underset{\text{Aldehyde or ketone}}{\text{RCH}(\text{H})-\overset{:\text{O}:}{\overset{\|}{\text{C}}}\text{R}'} + \underset{\text{Hydroxide ion}}{{}^-:\ddot{\text{O}}-\text{H}} \xrightleftharpoons{\text{slow}} \underset{\text{Conjugate base of carbonyl compound}}{\text{R}\ddot{\text{C}}^-\text{H}-\overset{:\text{O}:}{\overset{\|}{\text{C}}}\text{R}'} + \underset{\text{Water}}{\text{H}-\ddot{\text{O}}-\text{H}}$$

Step 2: A water molecule acts as a Brønsted acid to transfer a proton to the oxygen of the enolate ion.

$$\underset{\text{Conjugate base of carbonyl compound}}{\text{RCH}=\overset{:\ddot{\text{O}}:^-}{\overset{|}{\text{C}}}\text{R}'} + \underset{\text{Water}}{\text{H}-\ddot{\text{O}}-\text{H}} \xrightleftharpoons{\text{fast}} \underset{\text{Enol}}{\text{RCH}=\overset{:\ddot{\text{O}}-\text{H}}{\overset{|}{\text{C}}}\text{R}'} + \underset{\text{Hydroxide ion}}{{}^-:\ddot{\text{O}}-\text{H}}$$

FIGURE 18.3 Sequence of steps that describes the base-catalyzed enolization of an aldehyde or a ketone in aqueous solution.

$$R\bar{C}H-\overset{\overset{\displaystyle :\ddot{O}}{\|}}{C}R' \longleftrightarrow RCH=\overset{\overset{\displaystyle :\ddot{O}:^-}{|}}{C}R'$$

Resonance structures of
conjugate base

Protonation of this anion can occur either at the α carbon or at oxygen. Protonation of the α carbon simply returns the anion to the starting aldehyde or ketone. Protonation at oxygen, as shown in step 2 of Figure 18.3, produces the enol.

The key intermediate in this process, the conjugate base of the carbonyl compound, has carbanionic character. Organic chemists refer to it as an **enolate ion,** since it is at the same time the conjugate base of an enol. The term *enolate* is more descriptive of the electron distribution in this intermediate in that oxygen bears a greater share of the negative charge than does the α carbon atom.

The slow step in base-catalyzed enolization is formation of the enolate ion. The second step, proton transfer from water to the enolate oxygen, is very fast, as are almost all proton transfers from one oxygen atom to another.

Our experience to this point has been that C—H bonds are not very acidic. Compared with most hydrocarbons, however, aldehydes and ketones have relatively acidic protons on their α carbon atoms. Equilibrium constants for enolate formation from simple aldehydes and ketones are in the 10^{-16} to 10^{-20} range (pK_a = 16 to 20).

$$(CH_3)_2CH\overset{\overset{\displaystyle \ddot{O}:}{\|}}{C}H \rightleftharpoons H^+ + (CH_3)_2C=\overset{\overset{\displaystyle :\ddot{O}:^-}{|}}{C}H \qquad K_a = 3 \times 10^{-16} \; (pK_a = 15.5)$$

2-Methylpropanal

$$C_6H_5\overset{\overset{\displaystyle \ddot{O}:}{\|}}{C}CH_3 \rightleftharpoons H^+ + C_6H_5\overset{\overset{\displaystyle :\ddot{O}:^-}{|}}{C}=CH_2 \qquad K_a = 1.6 \times 10^{-16} \; (pK_a = 15.8)$$

Acetophenone

Delocalization of the negative charge onto the electronegative oxygen is responsible for the enhanced acidity of aldehydes and ketones. With K_a's in the 10^{-16} to 10^{-20} range, aldehydes and ketones are about as acidic as water and alcohols. Thus, hydroxide ion and alkoxide ions are sufficiently strong bases to produce solutions containing significant concentrations of enolate ions at equilibrium.

β-Diketones, such as 2,4-pentanedione, are even more acidic:

$$CH_3\overset{\overset{\displaystyle :\ddot{O}}{\|}}{C}CH_2\overset{\overset{\displaystyle \ddot{O}:}{\|}}{C}CH_3 \rightleftharpoons H^+ + CH_3\overset{\overset{\displaystyle ^-:\ddot{O}:}{|}}{C}=CH\overset{\overset{\displaystyle \ddot{O}:}{\|}}{C}CH_3 \qquad K_a = 10^{-9} \; (pK_a = 9)$$

In the presence of bases such as hydroxide, methoxide, ethoxide, these β-diketones are converted completely to their enolate ions. Notice that it is the methylene group flanked by the two carbonyl groups that is deprotonated. Both carbonyl groups participate in stabilizing the enolate by delocalizing its negative charge.

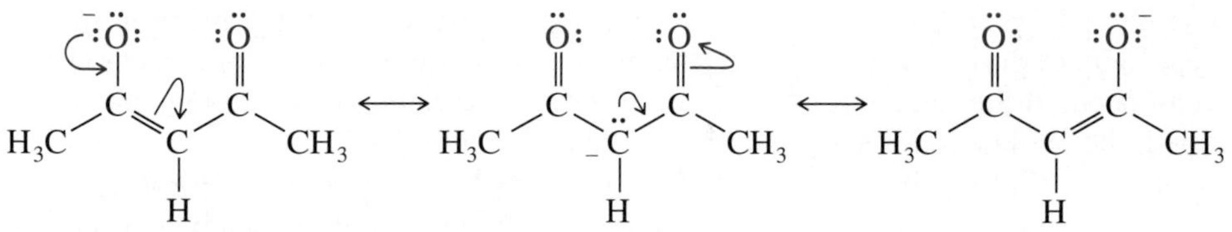

PROBLEM 18.7 Write the structure of the enolate ion derived from each of the following β-dicarbonyl compounds. Give the three most stable resonance forms of each.

(*a*) 2-Methyl-1,3-cyclopentanedione
(*b*) 1-Phenyl-1,3-butanedione
(*c*)

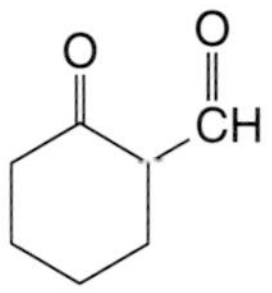

SAMPLE SOLUTION (*a*) First identify the proton that is removed by the base. It is on the carbon between the two carbonyl groups.

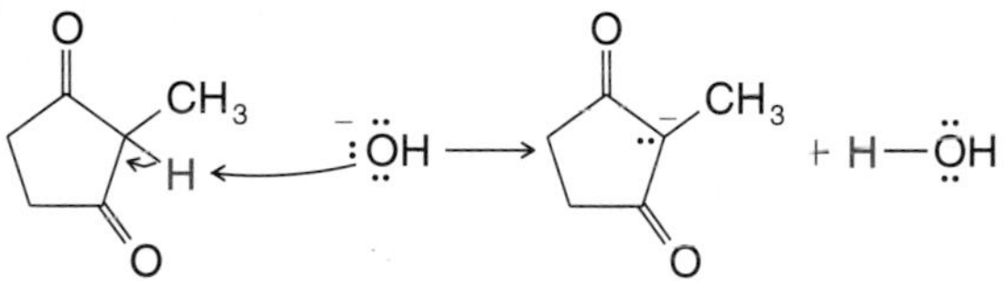

The three most stable resonance forms of this anion are

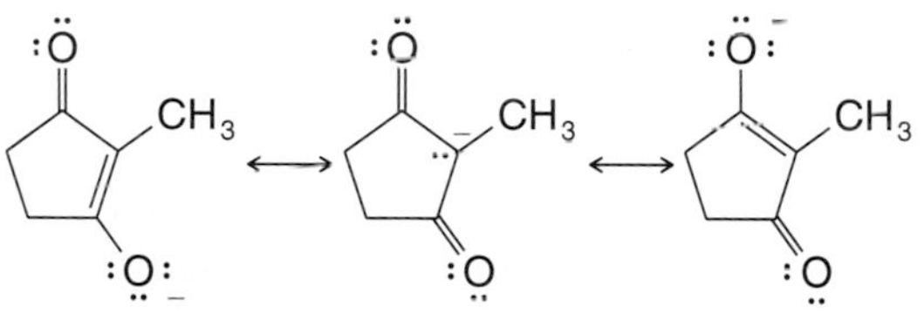

Enolate ions of β-dicarbonyl compounds are useful intermediates in organic synthesis. We shall see some examples of how they are employed in this way later in the chapter.

18.7 THE HALOFORM REACTION

Rapid halogenation of the α carbon atom takes place when an enolate ion is generated in the presence of chlorine, bromine, or iodine.

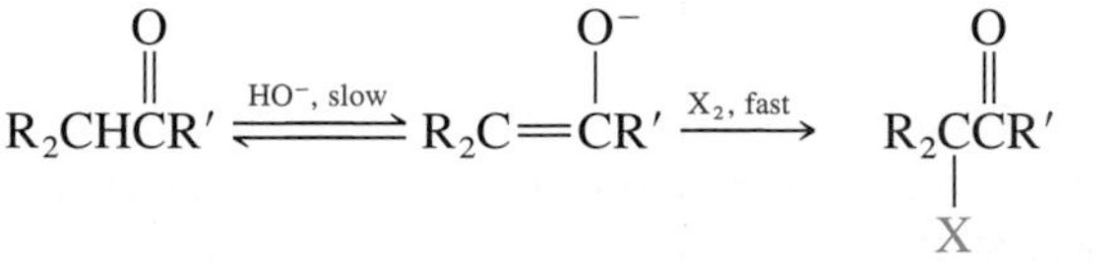

$$R_2CH\overset{O}{\overset{\|}{C}}R' \underset{}{\overset{HO^-,\ slow}{\rightleftharpoons}} R_2C{=}\overset{O^-}{\overset{|}{C}}R' \xrightarrow{X_2,\ fast} R_2\underset{X}{\underset{|}{C}}\overset{O}{\overset{\|}{C}}R'$$

Aldehyde or ketone — Enolate — α-Halo aldehyde or ketone

As in the acid-catalyzed halogenation of aldehydes and ketones, the reaction rate is independent of the concentration of the halogen; chlorination, bromination, and iodination all occur at the same rate. Formation of the enolate is rate-determining, and, once formed, the enolate ion reacts rapidly with the halogen.

Unlike its acid-catalyzed counterpart, α-halogenation in base cannot normally be limited to monohalogenation. Methyl ketones, for example, undergo a novel polyhalogenation and cleavage on treatment with a halogen in aqueous base.

$$\underset{\text{Methyl ketone}}{\overset{\displaystyle O \atop \|}{RCCH_3}} + \underset{\text{Halogen}}{3X_2} + \underset{\text{Hydroxide ion}}{4HO^-} \longrightarrow \underset{\text{Carboxylate ion}}{\overset{\displaystyle O \atop \|}{RCO^-}} + \underset{\text{Trihalomethane}}{CHX_3} + \underset{\text{Halide ion}}{3X^-} + \underset{\text{Water}}{3H_2O}$$

This is called the *haloform reaction* because the trihalomethane produced is chloroform, bromoform, or iodoform depending, of course, on the halogen used.

The mechanism of the haloform reaction begins with α halogenation via the enolate. The electron-attracting effect of an α halogen increases the acidity of the protons on the carbon to which it is bonded, making each subsequent halogenation *at that carbon* faster than the preceding one.

$$\overset{\displaystyle O \atop \|}{RCCH_3} \underset{\substack{\text{(slowest}\\ \text{halogenation}\\ \text{step)}}}{\xrightleftharpoons{X_2,\ HO^-}} \overset{\displaystyle O \atop \|}{RCCH_2X} \xrightarrow{X_2,\ HO^-} \overset{\displaystyle O \atop \|}{RCCHX_2} \underset{\substack{\text{(fastest}\\ \text{halogenation}\\ \text{step)}}}{\xrightarrow{X_2,\ HO^-}} \overset{\displaystyle O \atop \|}{RCCX_3}$$

The trihalomethyl ketone ($\overset{O \atop \|}{RCCX_3}$) so formed then undergoes nucleophilic addition of hydroxide ion to its carbonyl group, triggering its dissociation by cleavage of the bond to the CX_3 group.

$$\underset{\text{Trihalomethyl ketone}}{\overset{\displaystyle :\ddot{O}: \atop \|}{RCCX_3}} \xrightarrow{HO^-} RC(:\ddot{O}:^-)(:\ddot{O}H)\text{—}CX_3 \longrightarrow \overset{\displaystyle :\ddot{O}: \atop \|}{RC\text{—}\ddot{O}H} + :\bar{C}X_3$$

$$\overset{\displaystyle :\ddot{O}: \atop \|}{RC\text{—}\ddot{O}H} \xrightarrow{HO^-} \underset{\text{Carboxylate ion}}{\overset{\displaystyle :\ddot{O}: \atop \|}{RC\text{—}\ddot{O}:^-}} \qquad :\bar{C}X_3 \xrightarrow{H_2O} \underset{\text{Trihalomethane}}{HCX_3}$$

The three electron-withdrawing halogen substituents stabilize the negative charge of the trihalomethide ion ($^-{:}CX_3$) permitting it to act as a leaving group in the carbon-carbon bond-cleavage step.

The haloform reaction is sometimes used for the preparation of carboxylic acids from methyl ketones.

THE HALOFORM REACTION AND THE BIOSYNTHESIS OF TRIHALOMETHANES

Until scientists started looking specifically for them, it was widely believed that naturally occurring organohalogen compounds were rare. We now know that more than 2000 such compounds occur naturally, with the oceans being a particularly rich source.* Over 50 organohalogen compounds, including $CHBr_3$, CHBrClI, $BrCH_2CH_2I$, CH_2I_2, $Br_2CHCH{=}O$, I_2CHCO_2H, and $(Cl_3C)_2C{=}O$, have been found in a single species of Hawaiian red seaweed, for example. It is not surprising that organisms living in the oceans have adapted to their halide-rich environment by incorporating chlorine, bromine, and iodine into their metabolic processes. While chloromethane (CH_3Cl), bromomethane (CH_3Br), and iodomethane (CH_3I) are all produced by marine algae and kelp, land-based plants and fungi also contribute their share to the more than 5 million tons of the methyl halides formed each year by living systems. The ice plant, which grows in arid regions throughout the world and is cultivated as a ground cover along coastal highways in California, biosynthesizes CH_3Cl by a process in which nucleophilic attack by chloride ion (Cl^-) on the methyl group of *S*-adenosylmethionine is the key step (Section 16.17).

Interestingly, the trihalomethanes chloroform ($CHCl_3$), bromoform ($CHBr_3$), and iodoform (CHI_3) are biosynthesized by an entirely different process, one that is equivalent to the haloform reaction (Section 18.7) and begins with the formation of an α halo ketone. Unlike the biosynthesis of methyl halides, which requires attack by a halide nucleophile (X^-), α halogenation of a ketone requires attack by an electrophilic form of the halogen. For chlorination, the electrophilic form of the halogen is generated by oxidation of Cl^- in the presence of the enzyme *chloroperoxidase.* Thus, the overall equation for the enzyme-catalyzed chlorination of a methyl ketone may be written as

$$\underset{\text{Methyl ketone}}{CH_3\overset{O}{\overset{\|}{C}}R} + \underset{\text{Chloride}}{Cl^-} + \underset{\text{Oxygen}}{\tfrac{1}{2}O_2} \xrightarrow{\text{chloroperoxidase}} \underset{\text{Chloromethyl ketone}}{ClCH_2\overset{O}{\overset{\|}{C}}R} + \underset{\text{Hydroxide}}{HO^-}$$

(Compounds of the type $R\overset{O}{\overset{\|}{C}}CH_2\overset{O}{\overset{\|}{C}}OH$ also serve as biological precursors to $R\overset{O}{\overset{\|}{C}}CH_2Cl$ by a related process in which chlorination is accompanied by loss of a molecule of CO_2.) Further chlorination of the chloromethyl ketone gives the corresponding trichloromethyl ketone, which then undergoes hydrolysis to form chloroform.

$$\underset{\text{Chloromethyl ketone}}{ClCH_2\overset{O}{\overset{\|}{C}}R} \xrightarrow[Cl^-,\ O_2]{\text{chloroperoxidase}} \underset{\text{Dichloromethyl ketone}}{Cl_2CH\overset{O}{\overset{\|}{C}}R} \xrightarrow[Cl^-,\ O_2]{\text{chloroperoxidase}}$$

$$\underset{\text{Trichloromethyl ketone}}{Cl_3C\overset{O}{\overset{\|}{C}}R} \xrightarrow[HO^-]{H_2O} \underset{\text{Chloroform}}{Cl_3CH} + \underset{\text{Carboxylate}}{RCO_2^-}$$

Purification of drinking water, by adding Cl_2 to kill bacteria, is a source of electrophilic chlorine and contributes a nonenzymatic pathway for α chlorination and subsequent chloroform formation. While some of the odor associated with tap water may be due to chloroform, more of it probably results from chlorination of algae-produced organic compounds.

* The November 1994 edition of the *Journal of Chemical Education* contains as its cover story the article "Natural Organohalogens. Many More Than You Think!"

$$(CH_3)_3C\overset{O}{\overset{\|}{C}}CH_3 \xrightarrow[2.\ H^+]{1.\ Br_2,\ NaOH,\ H_2O} (CH_3)_3C\overset{O}{\overset{\|}{C}}OH + CHBr_3$$

3,3-Dimethyl-2-butanone — 2,2-Dimethylpropanoic acid (71–74%) — Tribromomethane (bromoform)

The methyl ketone shown in the example can enolize in only one direction and typifies the kind of reactant that can be converted to a carboxylic acid in synthetically acceptable yield by the haloform reaction. When C-3 of a methyl ketone bears enolizable hydrogens, as in $CH_3CH_2\overset{O}{\overset{\|}{C}}CH_3$, the first halogenation step is not very regioselective and the isolated yield of $CH_3CH_2CO_2H$ is on the order of only 50 percent.

The haloform reaction, using iodine, was once used as an analytical test in which the formation of a yellow precipitate of iodoform was taken as evidence that a substance was a methyl ketone. This application has been superseded by spectroscopic methods of structure determination. Interest in the haloform reaction has returned with the realization that chloroform and bromoform are natural products and are biosynthesized by an analogous process. (See the boxed essay ''The Haloform Reaction and the Biosynthesis of Trihalomethanes'' accompanying this section.)

18.8 SOME CHEMICAL AND STEREOCHEMICAL CONSEQUENCES OF ENOLIZATION

A number of novel reactions involving the α carbon atom of aldehydes and ketones involve enol and enolate anion intermediates.

Substitution of deuterium for hydrogen at the α carbon atom of an aldehyde or a ketone is a convenient way to introduce an isotopic label into a molecule and is readily carried out by treating the carbonyl compound with deuterium oxide (D_2O) and base.

$$\text{Cyclopentanone} + 4D_2O \xrightarrow[\text{reflux}]{\text{KOD}} \text{Cyclopentanone-2,2,5,5-}d_4 + 4DOH$$

Cyclopentanone — Cyclopentanone-2,2,5,5-d_4

Only the α hydrogens are replaced by deuterium in this reaction. The key intermediate is the enolate ion formed by proton abstraction from the α carbon atom of cyclopentanone. Transfer of deuterium from the solvent D_2O to the enolate gives cyclopentanone containing a deuterium atom in place of one of the hydrogens at the α carbon.

Formation of the enolate

$$\text{Cyclopentanone} + {}^{-}\!:\ddot{O}D \rightleftharpoons \text{Enolate of cyclopentanone} + H\ddot{O}D$$

Cyclopentanone — Enolate of cyclopentanone

Deuterium transfer to the enolate

$$\text{Enolate of cyclopentanone} + D{-}\ddot{O}{-}D \rightleftharpoons \text{Cyclopentanone-2-}d_1 + {}^{-}:\ddot{O}D$$

Enolate of cyclopentanone   Cyclopentanone-2-d_1

In excess D_2O the process continues until all four α protons are eventually replaced by deuterium.

If the α carbon atom of an aldehyde or a ketone is a stereogenic center, its stereochemical integrity is lost on enolization. Enolization of optically active *sec*-butyl phenyl ketone leads to its racemization by way of the achiral enol form with which it is in equilibrium.

$$C_6H_5C(=O){-}CH(CH_2CH_3)(CH_3) \xrightleftharpoons[\text{or } H_3O^+]{HO^-,\ H_2O,} C_6H_5C(OH){=}C(CH_2CH_3)(CH_3) \xrightleftharpoons[\text{or } H_3O^+]{HO^-,\ H_2O,} C_6H_5C(=O){-}CH(CH_2CH_3)(CH_3)$$

(*R*)-*sec*-Butyl phenyl ketone   Enol form [achiral; may be converted to either (*R*)- or (*S*)-*sec*-butyl phenyl ketone]   (*S*)-*sec*-Butyl phenyl ketone

Each act of proton abstraction from the α carbon atom converts a chiral molecule to an achiral enol or enolate anion. Careful kinetic studies have established that the rate of loss of optical activity of *sec*-butyl phenyl ketone is equal to its rate of hydrogen-deuterium exchange, its rate of bromination, and its rate of iodination. In each of the reactions, the rate-determining step is conversion of the starting ketone to the enol or enolate anion.

PROBLEM 18.8 Is the product from the α chlorination of (*R*)-*sec*-butyl phenyl ketone with Cl_2 in acetic acid chiral? Is it optically active?

18.9 THE ALDOL CONDENSATION: ALDEHYDES

As noted earlier, an aldehyde is partially converted to its enolate anion by bases such as hydroxide ion and alkoxide ions.

$$\underset{\text{Aldehyde}}{RCH_2\overset{O}{\overset{\|}{C}}H} + \underset{\text{Hydroxide}}{HO^-} \rightleftharpoons \underset{\text{Enolate}}{RCH{=}\overset{O^-}{\overset{|}{C}}H} + \underset{\text{Water}}{H_2O}$$

In a solution that contains both an aldehyde and its enolate ion, the enolate undergoes nucleophilic addition to the carbonyl group. This addition is analogous to the addition reactions of other nucleophilic reagents to aldehydes and ketones described in Chapter 17.

$$\mathrm{RCH_2\overset{\overset{\large :\ddot{O}:}{\|}}{C}H} \quad \mathrm{RCH{=}C(\ddot{O}:^-)H} \rightleftharpoons \mathrm{RCH_2\overset{\overset{-:\ddot{O}:}{|}}{C}H{-}\underset{\underset{R}{|}}{C}H\overset{\overset{:\ddot{O}}{\|}}{C}H} \xrightleftharpoons{H_2O} \mathrm{RCH_2\overset{\overset{:\ddot{O}H}{|}}{C}H{-}\underset{\underset{R}{|}}{C}H\overset{\overset{:\ddot{O}}{\|}}{C}H}$$

Product of aldol addition

Some of the earliest studies of the aldol reaction were carried out by Aleksandr Borodin. Though a physician by training and a chemist by profession, Borodin is remembered as the composer of some of the most familiar works in Russian music. See pp. 326–327 in the April 1987 issue of the *Journal of Chemical Education* for a biographical sketch of Borodin.

The alkoxide formed in the nucleophilic addition step then abstracts a proton from the solvent (usually water or ethanol) to yield the product of **aldol addition.** This product is known as an *aldol* because it contains both an aldehyde function and a hydroxyl group (*ald* + *ol* = *aldol*).

An important feature of aldol addition is that carbon-carbon bond formation occurs between the α carbon atom of one aldehyde and the carbonyl group of another. This is because carbanion (enolate) generation can involve proton abstraction *only* from the α carbon atom. The overall transformation can be represented schematically as shown in Figure 18.4.

Aldol addition occurs readily with aldehydes:

$$\mathrm{2CH_3\overset{\overset{O}{\|}}{C}H} \xrightarrow[4-5°C]{NaOH,\ H_2O} \mathrm{CH_3\underset{\underset{OH}{|}}{C}HCH_2\overset{\overset{O}{\|}}{C}H}$$

Acetaldehyde → 3-Hydroxybutanal (50%) (acetaldol)

$$\mathrm{2CH_3CH_2CH_2\overset{\overset{O}{\|}}{C}H} \xrightarrow[6-8°C]{KOH,\ H_2O} \mathrm{CH_3CH_2CH_2\underset{\underset{HO}{|}}{C}H\underset{\underset{CH_2CH_3}{|}}{C}H\overset{\overset{O}{\|}}{C}H}$$

Butanal → 2-Ethyl-3-hydroxyhexanal (75%)

$$\mathrm{RCH_2\overset{\overset{O}{\|}}{C}H} + \mathrm{\underset{\underset{R}{|}}{C}H_2\overset{\overset{O}{\|}}{C}H} \xrightarrow{base} \mathrm{RCH_2\overset{\overset{OH}{|}}{C}H{-}\underset{\underset{R}{|}}{C}H\overset{\overset{O}{\|}}{C}H}$$

FIGURE 18.4 The reactive sites in aldol addition are the carbonyl group of one aldehyde molecule and the α carbon atom of another.

PROBLEM 18.9 Write the structure of the aldol addition product of

(*a*) Pentanal, $CH_3CH_2CH_2CH_2\overset{O}{\overset{\|}{C}}H$

(*b*) 2-Methylbutanal, $CH_3CH_2\underset{CH_3}{\underset{|}{C}}H\overset{O}{\overset{\|}{C}}H$

(*c*) 3-Methylbutanal, $(CH_3)_2CHCH_2\overset{O}{\overset{\|}{C}}H$

SAMPLE SOLUTION (*a*) A good way to correctly identify the aldol addition product of any aldehyde is to work through the process mechanistically. Remember that the first step is enolate formation and that this *must* involve proton abstraction from the α carbon atom of the starting carbonyl compound.

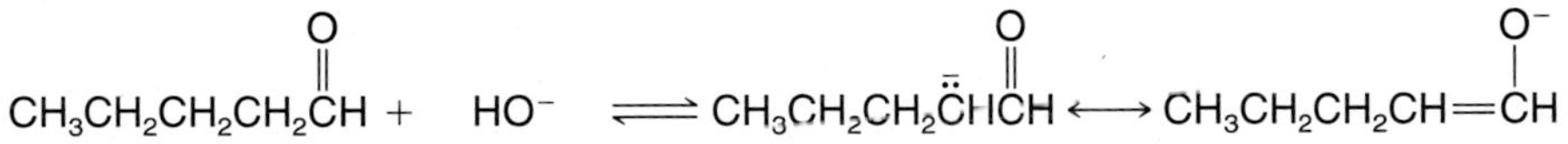

$$\underset{\text{Pentanal}}{CH_3CH_2CH_2CH_2\overset{O}{\overset{\|}{C}}H} + \underset{\text{Hydroxide}}{HO^-} \rightleftharpoons \underbrace{CH_3CH_2CH_2\ddot{C}H\overset{O}{\overset{\|}{C}}H \longleftrightarrow CH_3CH_2CH_2CH{=}\overset{O^-}{\overset{|}{C}}H}_{\text{Enolate of pentanal}}$$

Now use the negatively charged carbon of the enolate to form a new carbon-carbon bond to the carbonyl group. Proton transfer from the solvent to the alkoxide completes the process.

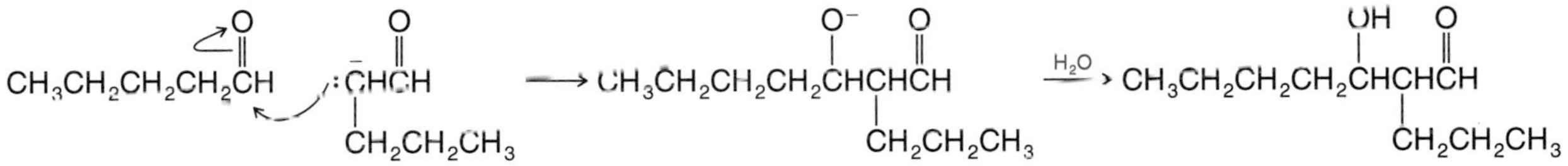

$$\underset{\text{Pentanal}}{CH_3CH_2CH_2CH_2\overset{O}{\overset{\|}{C}}H} \quad \underset{\text{Enolate of pentanal}}{:\overset{-}{C}H(CH_2CH_2CH_3)\overset{O}{\overset{\|}{C}}H} \longrightarrow CH_3CH_2CH_2CH_2\overset{O^-}{\overset{|}{C}}H\underset{CH_2CH_2CH_3}{\underset{|}{C}}H\overset{O}{\overset{\|}{C}}H \xrightarrow{H_2O} \underset{\text{3-Hydroxy-2-propylheptanal (aldol addition product of pentanal)}}{CH_3CH_2CH_2CH_2\overset{OH}{\overset{|}{C}}H\underset{CH_2CH_2CH_3}{\underset{|}{C}}H\overset{O}{\overset{\|}{C}}H}$$

The β-hydroxy aldehyde products of aldol addition undergo dehydration on heating, to yield *α,β-unsaturated aldehydes:*

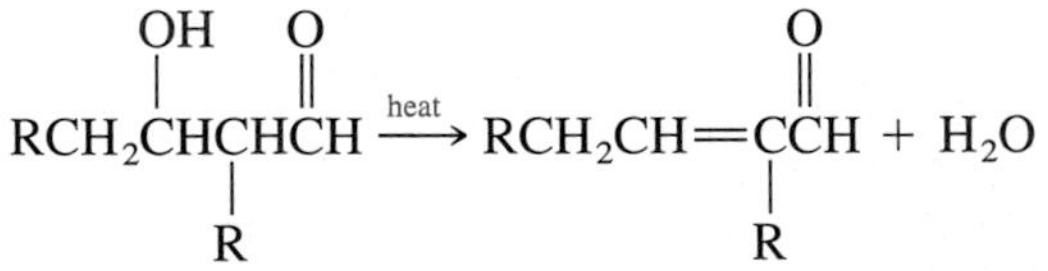

$$\underset{\beta\text{-Hydroxy aldehyde}}{RCH_2\overset{OH}{\overset{|}{C}}H\underset{R}{\underset{|}{C}}H\overset{O}{\overset{\|}{C}}H} \xrightarrow{\text{heat}} \underset{\alpha,\beta\text{-Unsaturated aldehyde}}{RCH_2CH{=}\underset{R}{\underset{|}{C}}\overset{O}{\overset{\|}{C}}H} + \underset{\text{Water}}{H_2O}$$

Conjugation of the newly formed double bond with the carbonyl group stabilizes the α,β-unsaturated aldehyde, provides the driving force for the dehydration process, and controls its regioselectivity. Dehydration can be effected by heating the aldol with acid or base. Normally, if the α,β-unsaturated aldehyde is the desired product, all that is done is to carry out the base-catalyzed aldol addition reaction at elevated temperature. Under these conditions, once the aldol addition product is formed, it rapidly loses water to form the α,β-unsaturated aldehyde.

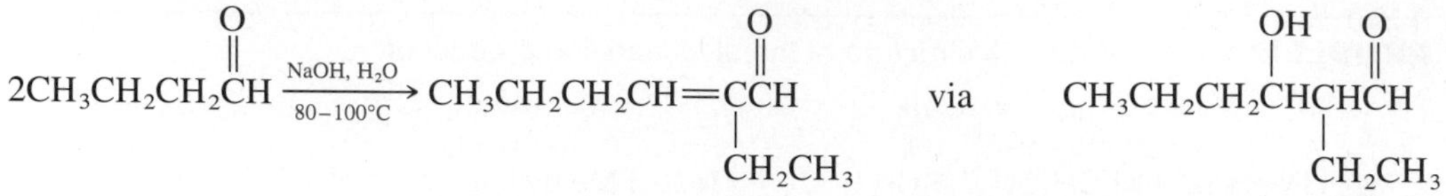

$$2CH_3CH_2CH_2\overset{\displaystyle O}{\overset{\|}{C}}H \xrightarrow[80-100°C]{NaOH,\ H_2O} CH_3CH_2CH_2CH{=}\underset{\displaystyle CH_2CH_3}{\underset{|}{\overset{\displaystyle O}{\overset{\|}{C}}}}H \quad \text{via} \quad CH_3CH_2CH_2\overset{\displaystyle OH}{\overset{|}{C}}H\underset{\displaystyle CH_2CH_3}{\underset{|}{C}}H\overset{\displaystyle O}{\overset{\|}{C}}H$$

Butanal — 2-Ethyl-2-hexenal (86%) — 2-Ethyl-3-hydroxyhexanal (not isolated; dehydrates under reaction conditions)

Recall from Section 15.7 that a condensation is a reaction in which two molecules combine to give a product along with some small (usually inorganic) molecule such as water.

Reactions in which two molecules of an aldehyde combine to form an α,β-unsaturated aldehyde and a molecule of water are called **aldol condensations.**

PROBLEM 18.10 Write the structure of the aldol condensation product of each of the aldehydes in Problem 18.9. One of these aldehydes can undergo aldol addition, but not aldol condensation. Which one? Why?

SAMPLE SOLUTION (*a*) Dehydration of the product of aldol addition of pentanal introduces the double bond between C-2 and C-3 to give an α,β-unsaturated aldehyde.

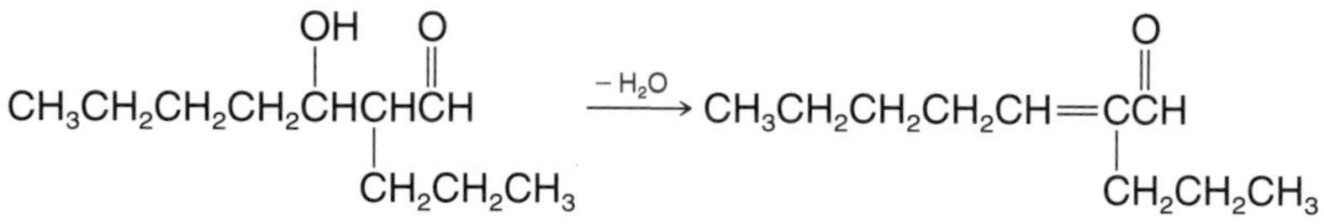

$$CH_3CH_2CH_2CH_2\overset{\displaystyle OH}{\overset{|}{C}}H\underset{\displaystyle CH_2CH_2CH_3}{\underset{|}{C}}H\overset{\displaystyle O}{\overset{\|}{C}}H \xrightarrow{-H_2O} CH_3CH_2CH_2CH_2CH{=}\underset{\displaystyle CH_2CH_2CH_3}{\underset{|}{\overset{\displaystyle O}{\overset{\|}{C}}}}H$$

Product of aldol addition of pentanal (3-hydroxy-2-propylheptanal) — Product of aldol condensation of pentanal (2-propyl-2-heptenal)

The point has been made earlier (Section 5.9) that alcohols require acid catalysis in order to undergo dehydration to alkenes. Thus, it may seem strange that aldol addition products can be dehydrated in base. This is another example of the way in which the enhanced acidity of protons at the α carbon atom affects the reactions of carbonyl compounds. Elimination may take place in a concerted E2 fashion or it may be stepwise and proceed through an enolate ion.

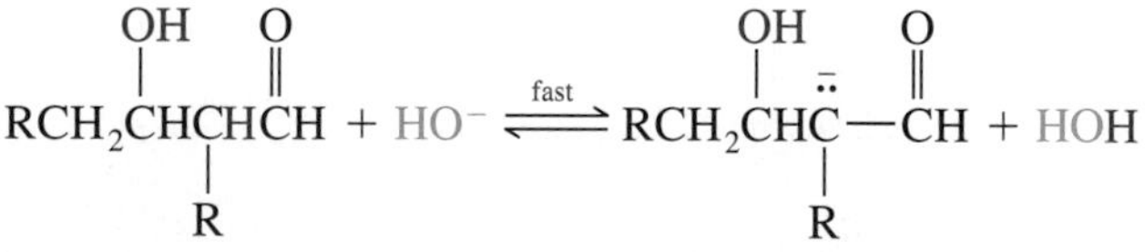

$$RCH_2\overset{\displaystyle OH}{\overset{|}{C}}H\underset{\displaystyle R}{\underset{|}{C}}H\overset{\displaystyle O}{\overset{\|}{C}}H + HO^- \underset{}{\overset{\text{fast}}{\rightleftharpoons}} RCH_2\overset{\displaystyle OH}{\overset{|}{C}}H\underset{\displaystyle R}{\underset{|}{\ddot{C}}}{-}\overset{\displaystyle O}{\overset{\|}{C}}H + HOH$$

β-Hydroxy aldehyde — Enolate ion of β-hydroxy aldehyde

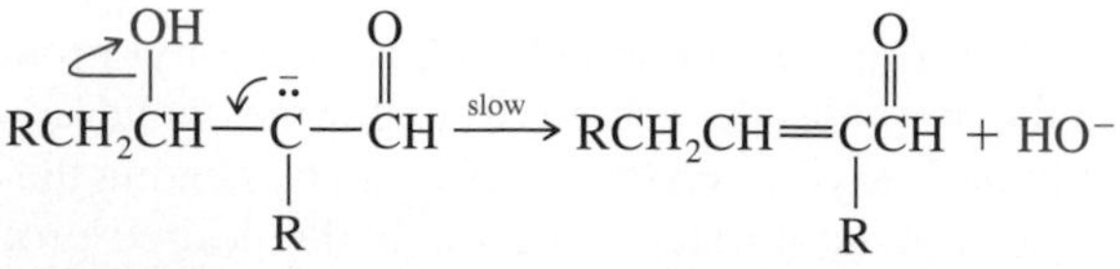

$$RCH_2\overset{\displaystyle OH}{\overset{|}{C}}H{-}\underset{\displaystyle R}{\underset{|}{\ddot{C}}}{-}\overset{\displaystyle O}{\overset{\|}{C}}H \xrightarrow{\text{slow}} RCH_2CH{=}\underset{\displaystyle R}{\underset{|}{\overset{\displaystyle O}{\overset{\|}{C}}}}H + HO^-$$

Enolate of β-Hydroxy aldehyde — α, β-Unsaturated aldehyde

18.10 THE ALDOL CONDENSATION: KETONES

As with other reversible nucleophilic addition reactions, the equilibria for aldol additions are less favorable for ketones than they are for aldehydes. For example, only 2 percent of the aldol addition product of acetone is present at equilibrium.

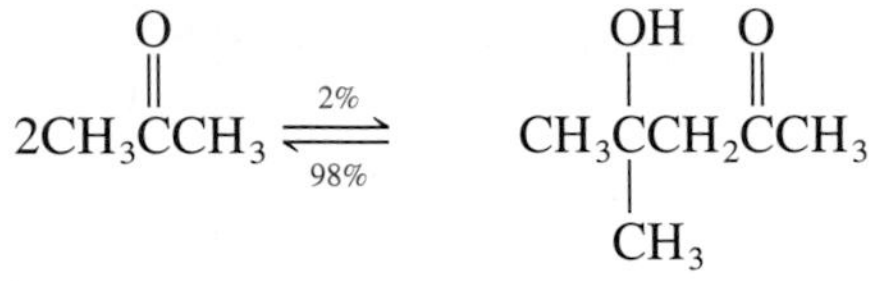

$$2CH_3\overset{O}{\overset{\|}{C}}CH_3 \underset{98\%}{\overset{2\%}{\rightleftharpoons}} CH_3\underset{CH_3}{\overset{OH}{C}}CH_2\overset{O}{\overset{\|}{C}}CH_3$$

Acetone 4-Hydroxy-4-methyl-2-pentanone

The situation is similar for aldol addition reactions of other ketones.

When conditions are chosen so as to favor dehydration of the aldol addition product, the position of equilibrium shifts to the right and α,β-unsaturated ketones result. One method uses aluminum tri-*tert*-butoxide at elevated temperature as a reagent to promote aldol condensation.

Aluminum tri-*tert*-butoxide acts as a catalyst in the aldol addition step and as a dehydrating agent in the elimination step.

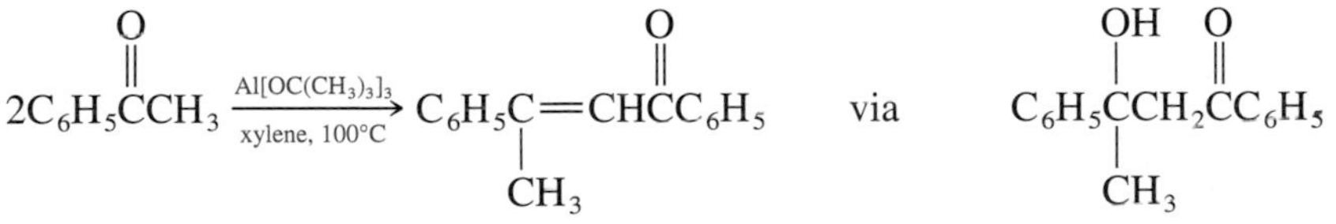

$$2C_6H_5\overset{O}{\overset{\|}{C}}CH_3 \xrightarrow[\text{xylene, 100°C}]{Al[OC(CH_3)_3]_3} C_6H_5\underset{CH_3}{C}{=}CH\overset{O}{\overset{\|}{C}}C_6H_5 \quad \text{via} \quad C_6H_5\underset{CH_3}{\overset{OH}{C}}CH_2\overset{O}{\overset{\|}{C}}C_6H_5$$

Acetophenone 1,3-Diphenyl-2-buten-1-one (77%) 3-Hydroxy-1,3-diphenyl-1-butanone (not isolated)

PROBLEM 18.11 *Mesityl oxide* is the common name of the α,β-unsaturated ketone formed by dehydration of the aldol addition product of acetone. Deduce the structural formula of mesityl oxide. What is its systematic IUPAC name?

PROBLEM 18.12 Cyclohexanone undergoes efficient aldol condensation on being heated with aluminum tri-*tert*-butoxide in xylene. Write the structure of the aldol condensation product along with that of its β-hydroxy ketone precursor.

Dicarbonyl compounds, that is, compounds having two carbonyl groups within the same molecule, undergo intramolecular aldol condensation reactions. Even bases as weak as sodium carbonate are adequate catalysts in these cases.

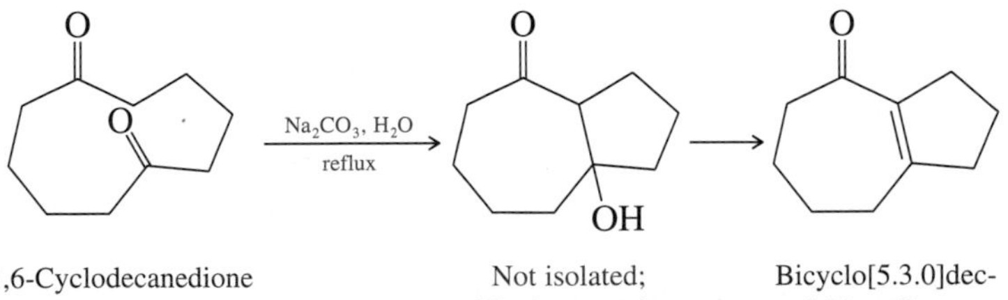

O O → (Na_2CO_3, H_2O, reflux) O, OH → O

1,6-Cyclodecanedione Not isolated; dehydrates under reaction conditions Bicyclo[5.3.0]dec-1(7)-en-2-one (96%)

Intramolecular aldol condensations proceed best when five- or six-membered rings result.

Since the aldol condensation leads to carbon-carbon bond formation, organic chemists have exploited it as a synthetic tool. Its usefulness has been extended beyond the *self-condensation* of aldehydes and ketones just described to include *mixed* (or *crossed*) *aldol condensations* between two different carbonyl compounds.

18.11 MIXED ALDOL CONDENSATIONS

Mixed aldol condensations are effective only when the number of reaction possibilities is limited. It would not be a useful procedure, for example, to treat a solution of acetaldehyde and propanal with base. A mixture of four aldol addition products forms under these conditions. Two of the products are those of self-addition:

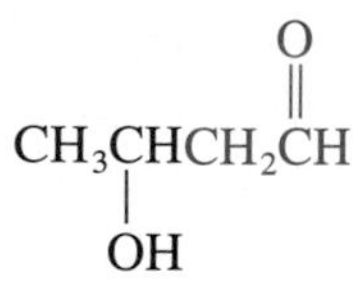

$$CH_3CH(OH)CH_2CH{=}O \qquad CH_3CH_2CH(OH)CH(CH_3)CH{=}O$$

3-Hydroxybutanal (from addition of enolate of acetaldehyde to acetaldehyde)

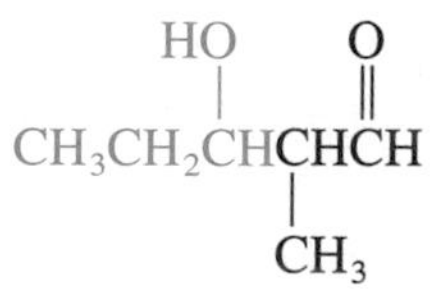

3-Hydroxy-2-methylpentanal (from addition of enolate of propanal to propanal)

Two are the products of mixed addition:

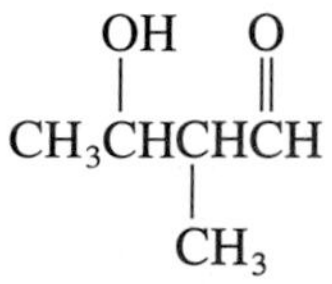

$$CH_3CH(OH)CH(CH_3)CH{=}O \qquad CH_3CH_2CH(OH)CH_2CH{=}O$$

3-Hydroxy-2-methylbutanal (from addition of enolate of propanal to acetaldehyde)

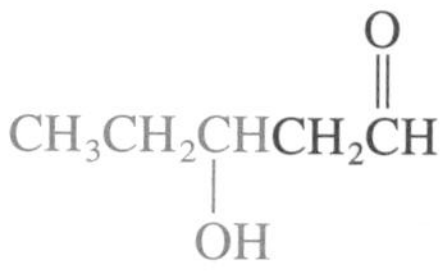

3-Hydroxypentanal (from addition of enolate of acetaldehyde to propanal)

The mixed aldol condensations that are the most synthetically useful are those in which one of the reactants is an aldehyde that cannot form an enolate. Formaldehyde, for example, has often been used successfully.

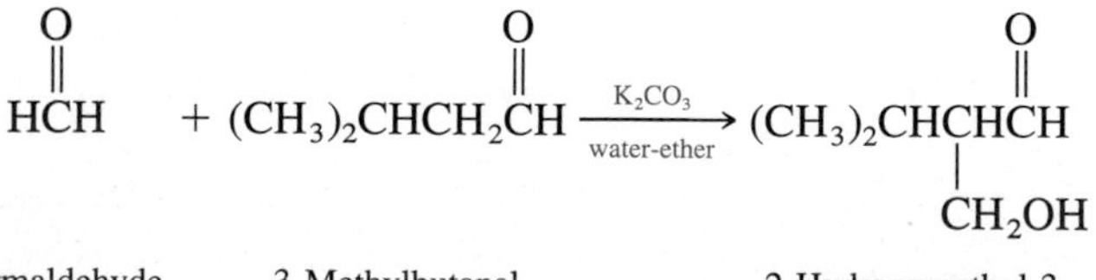

$$HCH{=}O + (CH_3)_2CHCH_2CH{=}O \xrightarrow[\text{water-ether}]{K_2CO_3} (CH_3)_2CHCH(CH_2OH)CH{=}O$$

Formaldehyde — 3-Methylbutanal — 2-Hydroxymethyl-3-methylbutanal (52%)

Not only is formaldehyde incapable of forming an enolate, but it is so reactive toward nucleophilic addition that it suppresses self-condensation of the other component by reacting rapidly with any enolate present.

Aromatic aldehydes cannot form enolates, and a large number of mixed aldol condensation reactions have been carried out in which an aromatic aldehyde reacts with an enolate.

$$CH_3O\text{-}C_6H_4\text{-}\overset{O}{\overset{\|}{C}}H + CH_3\overset{O}{\overset{\|}{C}}CH_3 \xrightarrow[30°C]{NaOH,\ H_2O} CH_3O\text{-}C_6H_4\text{-}CH{=}CH\overset{O}{\overset{\|}{C}}CH_3$$

p-Methoxybenzaldehyde | Acetone | 4-*p*-Methoxyphenyl-3-buten-2-one (83%)

Mixed aldol condensations using aromatic aldehydes always involve dehydration of the product of mixed addition and yield a product in which the double bond is conjugated to both the aromatic ring and the carbonyl group.

Mixed aldol condensations in which a ketone reacts with an aromatic aldehyde are known as *Claisen-Schmidt condensations.*

PROBLEM 18.13 Give the structure of the mixed aldol condensation product of benzaldehyde with

(*a*) Acetophenone, $C_6H_5\overset{O}{\overset{\|}{C}}CH_3$ (*b*) *tert*-Butyl methyl ketone, $(CH_3)_3C\overset{O}{\overset{\|}{C}}CH_3$
(*c*) Cyclohexanone

SAMPLE SOLUTION (*a*) The enolate of acetophenone reacts with benzaldehyde to yield the product of mixed addition. Dehydration of the intermediate occurs, giving the α,β-unsaturated ketone.

$$C_6H_5\overset{O}{\overset{\|}{C}}H + :\overset{-}{C}H_2\overset{O}{\overset{\|}{C}}C_6H_5 \longrightarrow C_6H_5\underset{OH}{\underset{|}{C}}HCH_2\overset{O}{\overset{\|}{C}}C_6H_5 \xrightarrow{-H_2O} C_6H_5CH{=}CH\overset{O}{\overset{\|}{C}}C_6H_5$$

Benzaldehyde | Enolate of acetophenone | 1,3-Diphenyl-2-propen-1-one

As actually carried out, the mixed aldol condensation product, 1,3-diphenyl-2-propen-1-one, has been isolated in 85 percent yield on treating benzaldehyde with acetophenone in an aqueous ethanol solution of sodium hydroxide at 15 to 30°C.

18.12 THE ALDOL CONDENSATION IN SYNTHETIC ORGANIC CHEMISTRY

Aldol condensations are one of the fundamental carbon-carbon bond–forming processes of synthetic organic chemistry. Furthermore, since the products of these aldol condensations contain functional groups capable of subsequent modification, access to a host of useful materials is gained.

To illustrate how aldol condensation may be coupled to functional group modification, consider the synthesis of 2-ethyl-1,3-hexanediol, a compound used as an insect repellent. This 1,3-diol is prepared by reduction of the aldol addition product of butanal:

$$CH_3CH_2CH_2\overset{O}{\overset{\|}{C}}H \xrightarrow{\text{aldol addition}} CH_3CH_2CH_2\overset{OH}{\overset{|}{C}}H\underset{CH_2CH_3}{\underset{|}{C}}H\overset{O}{\overset{\|}{C}}H \xrightarrow[Ni]{H_2} CH_3CH_2CH_2\overset{OH}{\overset{|}{C}}H\underset{CH_2CH_3}{\underset{|}{C}}HCH_2OH$$

Butanal | 2-Ethyl-3-hydroxyhexanal | 2-Ethyl-1,3-hexanediol

PROBLEM 18.14 Outline a synthesis of 2-ethyl-1-hexanol from butanal.

MIBK stands for *methyl isobutyl ketone,* which is, unfortunately, an incorrect name. Since group names are listed in alphabetical order, the correct name is *isobutyl methyl ketone.*

It is usually the carbon skeleton that provides a clue to the appropriate aldol route to a particular compound. Take, for example, 4-methyl-2-pentanone (MIBK, a common industrial solvent). It has the same carbon skeleton as the aldol condensation product of acetone.

$$CH_3\underset{\underset{CH_3}{|}}{C}HCH_2\overset{\overset{O}{\|}}{C}CH_3 \Longrightarrow CH_3\underset{\underset{CH_3}{|}}{C}{=}CH\overset{\overset{O}{\|}}{C}CH_3 \Longrightarrow 2CH_3\overset{\overset{O}{\|}}{C}CH_3$$

4-Methyl-2-pentanone (MIBK) — 4-Methyl-3-penten-2-one (mesityl oxide) — Acetone

Indeed, MIBK is prepared from acetone as its ultimate precursor by hydrogenation of the carbon-carbon double bond of the aldol condensation product of acetone.

18.13 EFFECTS OF CONJUGATION IN α,β-UNSATURATED ALDEHYDES AND KETONES

Aldol condensation offers an effective route to α,β-unsaturated aldehydes and ketones. These compounds have some interesting properties that result from conjugation of the carbon-carbon double bond with the carbonyl group. As shown in Figure 18.5, the π systems of the carbon-carbon and carbon-oxygen double bonds overlap to form an extended π system that permits increased electron delocalization.

This electron delocalization stabilizes a conjugated system. Under conditions chosen to bring about their interconversion, the equilibrium between a β,γ-unsaturated ketone and an α,β-unsaturated analog favors the conjugated isomer.

Figure 3.16 (page 103) shows how the composition of an equilibrium mixture of two components varies according to the free-energy difference between them. For the equilibrium shown in the accompanying equation, $\Delta G° = -4$ kJ/mol (-1 kcal/mol).

$$CH_3CH{=}CHCH_2\overset{\overset{O}{\|}}{C}CH_3 \underset{25°C}{\overset{K = 4.8}{\rightleftharpoons}} CH_3CH_2CH{=}CH\overset{\overset{O}{\|}}{C}CH_3$$

4-Hexen-2-one (17%) (β,γ-unsaturated ketone) — 3-Hexen-2-one (83%) (α,β-unsaturated ketone)

PROBLEM 18.15 Commercial mesityl oxide, $(CH_3)_2C{=}CH\overset{\overset{O}{\|}}{C}CH_3$, is often contaminated with about 10 percent of an isomer having the same carbon skeleton. What is a likely structure for this compound?

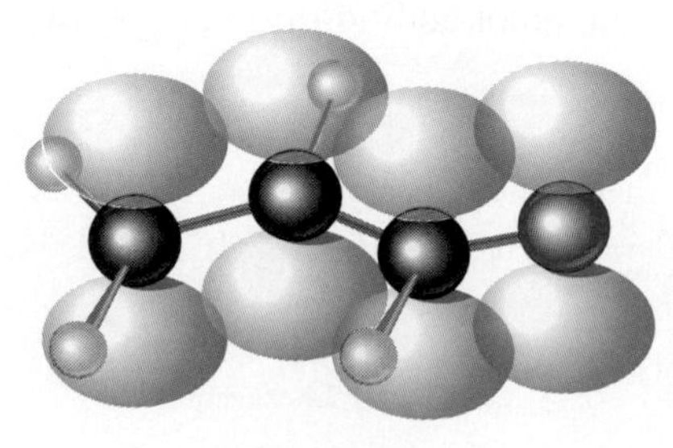

FIGURE 18.5 Acrolein ($CH_2{=}CHCH{=}O$) is a planar molecule. Oxygen and each carbon is sp^2 hybridized and each contributes one electron to a conjugated π electron system analogous to that of 1,3-butadiene.

In resonance terms, electron delocalization in α,β-unsaturated carbonyl compounds is represented by contributions from three principal resonance structures:

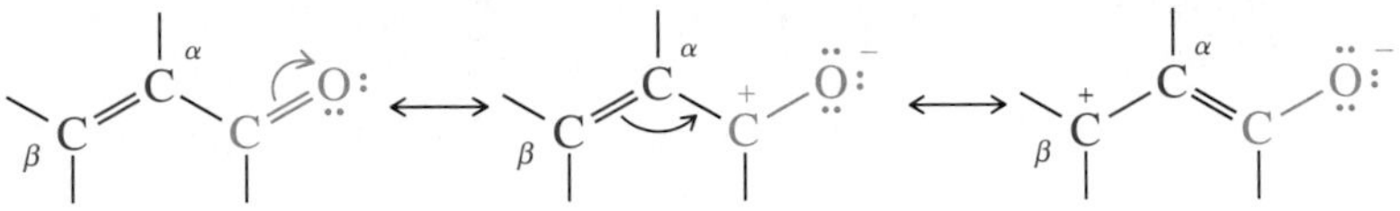

Most stable structure

The carbonyl group withdraws π electron density from the double bond, and both the carbonyl carbon and the β carbon are positively polarized. Their greater degree of charge separation makes the dipole moments of α,β unsaturated carbonyl compounds significantly larger than those of comparable aldehydes and ketones.

Butanal $\mu = 2.7$ D

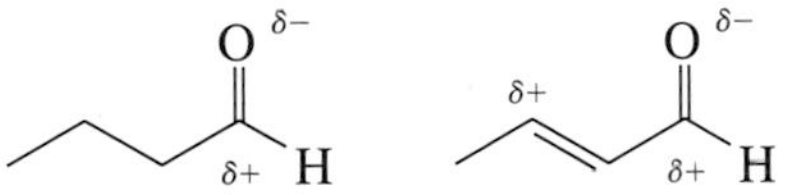

trans-2-Butenal $\mu = 3.7$ D

The diminished π electron density in the double bond makes α,β-unsaturated aldehydes and ketones less reactive than alkenes toward electrophilic addition. Electrophilic reagents—bromine and peroxy acids, for example—react more slowly with the carbon-carbon double bond of α,β-unsaturated carbonyl compounds than with simple alkenes.

On the other hand, the polarization of electron density in α,β-unsaturated carbonyl compounds makes their β carbon atoms rather electrophilic. Some chemical consequences of this enhanced electrophilicity are described in the following section.

18.14 CONJUGATE NUCLEOPHILIC ADDITION TO α,β-UNSATURATED CARBONYL COMPOUNDS

α,β-Unsaturated carbonyl compounds contain two electrophilic sites—the carbonyl carbon and the carbon atom that is β to it. Nucleophiles such as organolithium and Grignard reagents and lithium aluminum hydride tend to react by nucleophilic addition to the carbonyl group, as shown in the following example:

$$CH_3CH{=}CH\overset{O}{\overset{\|}{C}}H + HC{\equiv}CMgBr \xrightarrow[2.\ H_3O^+]{1.\ THF} CH_3CH{=}CH\overset{OH}{\overset{|}{C}}HC{\equiv}CH$$

2-Butenal | Ethynylmagnesium bromide | 4-Hexen-1-yn-3-ol (84%)

This is called *direct addition,* or *1,2 addition.* (The "1" and "2" do not refer to IUPAC locants but are used in a manner analogous to that employed in Section 10.10 to distinguish between direct and conjugate addition to conjugated dienes.)

With certain other nucleophiles, addition takes place at the carbon-carbon double bond rather than at the carbonyl group. Such reactions proceed via enol intermediates and are described as *conjugate addition,* or *1,4-addition,* reactions.

$$\underset{\alpha,\beta\text{-Unsaturated aldehyde or ketone}}{\overset{4}{\text{C}}=\overset{3}{\text{C}}-\overset{2}{\text{C}}(=\overset{1}{\text{O}})-} \xrightarrow{HY} \underset{\text{Enol formed by 1,4 addition}}{\text{Y}-\text{C}-\text{C}=\text{C}(\text{OH})-} \xrightarrow{fast} \underset{\text{Isolated product of 1,4-addition pathway}}{\text{Y}-\text{C}-\text{C}(\text{H})-\text{C}(=\text{O})-}$$

α, β-Unsaturated aldehyde or ketone — Enol formed by 1,4 addition — Isolated product of 1,4-addition pathway

The nucleophilic portion of the reagent (Y in HY) becomes bonded to the β carbon. For reactions carried out under conditions in which the attacking species is the anion $:Y^-$, an enolate ion precedes the enol.

$$^-\text{Y:} + \text{C}=\text{C}-\text{C}(=\ddot{\text{O}}:)- \longrightarrow \text{Y}-\text{C}-\overset{-}{\text{C}}:-\text{C}(=\ddot{\text{O}}:)- \longleftrightarrow \text{Y}-\text{C}-\text{C}=\text{C}(-\ddot{\text{O}}:^-)-$$

Enolate ion formed by nucleophilic addition of $:Y^-$ to β carbon

Ordinarily, nucleophilic addition to the carbon-carbon double bond of an alkene is very rare. It occurs with α,β-unsaturated carbonyl compounds because the carbanion that results is an enolate, which is more stable and formed with a lower activation energy than a simple alkyl anion.

Hydrogen cyanide and alkanethiols have K_a values in the 10^{-9} to 10^{-10} range ($pK_a = 9$ to 10), while K_a for acetylene is 10^{-26} ($pK_a = 26$).

Conjugate addition is most often observed when the nucleophile ($Y:^-$) is weakly basic. The nucleophiles in the two examples that follow are $^-:C{\equiv}N:$ and $C_6H_5CH_2\ddot{S}:^-$, respectively. Both are much weaker bases than acetylide ion, which was the nucleophile used in the example illustrating direct addition.

$$C_6H_5CH{=}CHCC_6H_5 \ (\text{C}{=}\text{O}) \xrightarrow[\text{ethanol–acetic acid}]{\text{KCN}} C_6H_5CH(CN)CH_2CC_6H_5 \ (\text{C}{=}\text{O})$$

1,3-Diphenyl-2-propen-1-one → 4-Oxo-2,4-diphenylbutanenitrile (93–96%)

$$\text{3-Methyl-2-cyclohexen-1-one} \xrightarrow[\text{HO}^-,\ H_2O]{C_6H_5CH_2SH} \text{3-Benzylthio-3-methylcyclohexanone} \ (CH_3,\ SCH_2C_6H_5)$$

3-Methyl-2-cyclohexen-1-one → 3-Benzylthio-3-methylcyclohexanone (58%)

One explanation for these observations is presented in Figure 18.6. Nucleophilic addition to α,β-unsaturated aldehydes and ketones may be governed either by *kinetic control* or by *thermodynamic control* (Section 10.10). 1,2 addition is faster than 1,4 addition and, under conditions in which the 1,2- and 1,4-addition products do not equilibrate, is the predominant pathway. Kinetic control operates with strongly basic nucleophiles to give the 1,2-addition product. A weakly basic nucleophile, however,

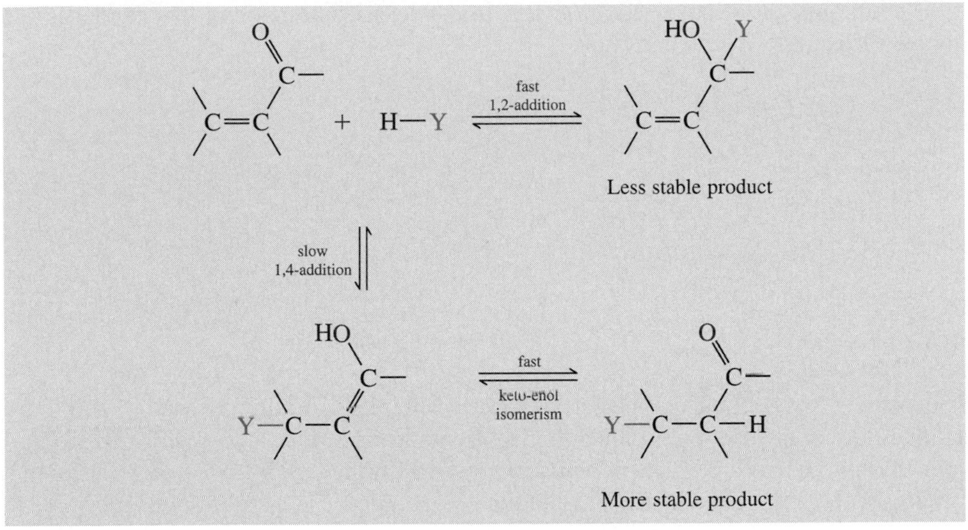

+ H—Y

fast 1,2-addition

Less stable product

slow 1,4-addition

fast keto-enol isomerism

More stable product

FIGURE 18.6 Nucleophilic addition to α,β-unsaturated aldehydes and ketones may take place either in a 1,2 or a 1,4 manner. Direct addition (1,2) occurs faster than conjugate addition (1,4) but gives a less stable product. The product of 1,4 addition retains the carbon-oxygen double bond, which is, in general, a more stable structural unit than a carbon-carbon double bond.

goes on and off the carbonyl carbon readily and permits the 1,2-addition product to equilibrate with the more slowly formed, but more stable, 1,4-addition product. Thermodynamic control is observed with weakly basic nucleophiles. The product of 1,4 addition, which retains the carbon-oxygen double bond, is more stable than the product of 1,2 addition, which retains the carbon-carbon double bond. In general, carbon-oxygen double bonds are more stable than carbon-carbon double bonds because the greater electronegativity of oxygen permits the π electrons to be bound more strongly.

PROBLEM 18.16 *Acrolein* ($CH_2{=}CHCH{=}O$) reacts exothermically with sodium azide (NaN_3) in aqueous acetic acid to form a compound, $C_3H_5N_3O$ in 71 percent yield. Propanal ($CH_3CH_2CH{=}O$), when subjected to the same reaction conditions, is recovered unchanged. Suggest a structure for the product formed from acrolein and offer an explanation for the difference in reactivity between acrolein and propanal.

18.15 ADDITION OF CARBANIONS TO α,β-UNSATURATED KETONES: THE MICHAEL REACTION

A synthetically useful reaction known as the **Michael reaction,** or **Michael addition,** involves nucleophilic addition of carbanions to α,β-unsaturated ketones. The most common types of carbanions used are enolate ions derived from β-diketones. These enolates are weak bases (Section 18.6) and react with α,β-unsaturated ketones by *conjugate addition.*

Arthur Michael, for whom the reaction is named, was an American chemist whose career spanned the period between the 1870s and the 1930s. He was independently wealthy and did much of his research in his own private laboratory.

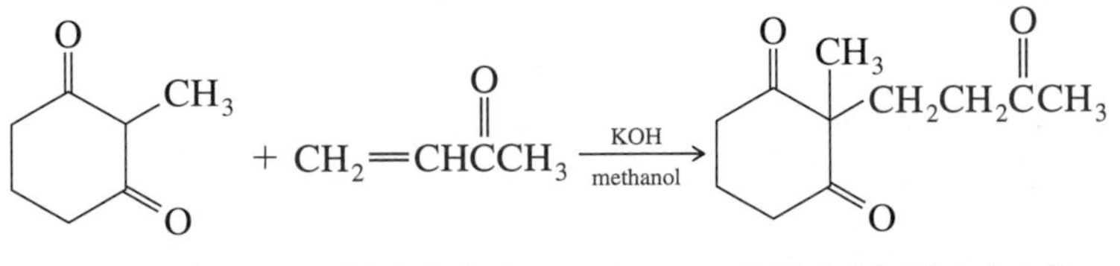

$$\text{2-Methyl-1,3-cyclohexanedione} + CH_2{=}CH\overset{O}{\overset{\|}{C}}CH_3 \xrightarrow[\text{methanol}]{\text{KOH}} \text{2-Methyl-2-(3'-oxobutyl)-1,3-cyclohexanedione (85\%)}$$

2-Methyl-1,3-cyclohexanedione | Methyl vinyl ketone | 2-Methyl-2-(3′-oxobutyl)-1,3-cyclohexanedione (85%)

The product of Michael addition has the necessary functionality to undergo an intramolecular aldol condensation:

$$\xrightarrow[\text{benzene, heat}]{Al[OC(CH_3)_3]_3} \qquad \xrightarrow{-H_2O}$$

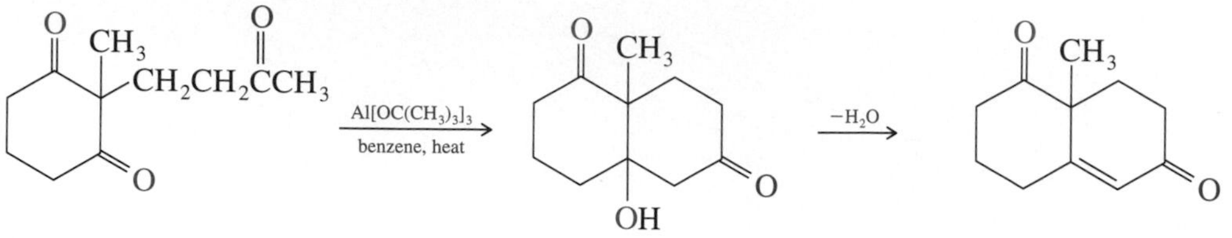

2-Methyl-2-(3′-oxobutyl)-1,3-cyclohexanedione — Intramolecular aldol addition product; not isolated — Δ^4-9-Methyloctalin-3,8-dione (40%)

The synthesis of cyclohexenone derivatives by Michael addition followed by intramolecular aldol condensation is called the **Robinson annulation,** after Sir Robert Robinson, who popularized its use. By *annulation* we mean the building of a ring onto some starting molecule. (The alternative spelling ''annelation'' is also often used.)

PROBLEM 18.17 Both the conjugate addition step and the intramolecular aldol condensation step can be carried out in one synthetic operation without isolating any of the intermediates along the way. For example, consider the reaction

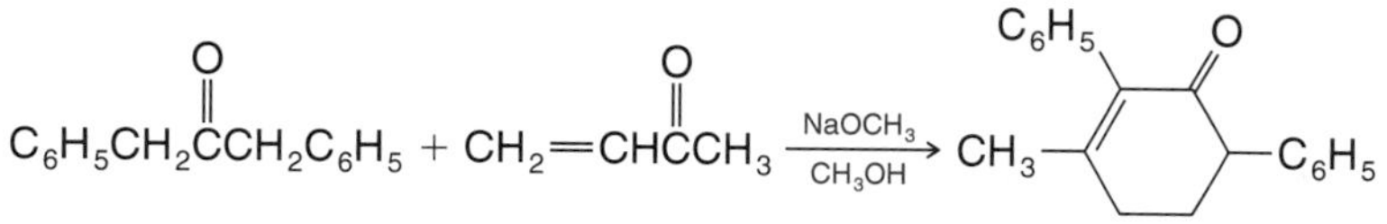

$$C_6H_5CH_2\overset{O}{\overset{\|}{C}}CH_2C_6H_5 + CH_2{=}CH\overset{O}{\overset{\|}{C}}CH_3 \xrightarrow[CH_3OH]{NaOCH_3}$$

Dibenzyl ketone — Methyl vinyl ketone — 3-Methyl-2,6-diphenyl-2-cyclohexen-1-one (55%)

Write structural formulas corresponding to the intermediates formed in the conjugate addition step and in the aldol addition step.

18.16 CONJUGATE ADDITION OF ORGANOCOPPER REAGENTS TO α,β-UNSATURATED CARBONYL COMPOUNDS

The preparation and some synthetic applications of lithium dialkylcuprates have been described earlier (Section 14.11). The most prominent feature of these reagents is their capacity to undergo conjugate addition to α,β-unsaturated aldehydes and ketones.

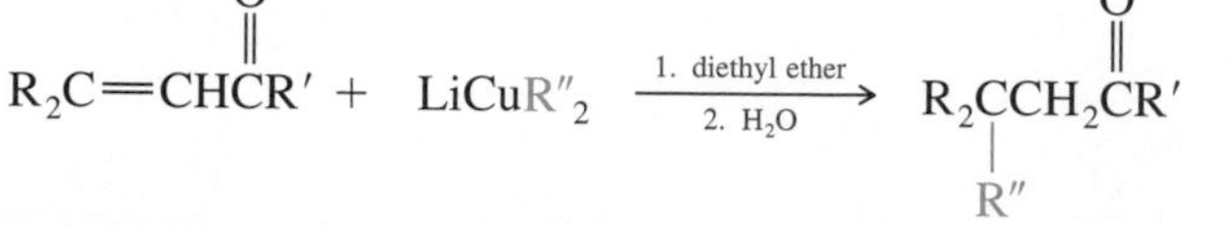

$$R_2C{=}CH\overset{O}{\overset{\|}{C}}R' + LiCuR''_2 \xrightarrow[2.\ H_2O]{1.\ \text{diethyl ether}} R_2\underset{R''}{\underset{|}{C}}CH_2\overset{O}{\overset{\|}{C}}R'$$

α,β-Unsaturated aldehyde or ketone — Lithium dialkylcuprate — Aldehyde or ketone alkylated at the β position

$$\text{3-Methyl-2-cyclohexen-1-one} + LiCu(CH_3)_2 \xrightarrow[\text{2. } H_2O]{\text{1. diethyl ether}} \text{3,3-Dimethylcyclohexanone (98\%)}$$

3-Methyl-2-cyclohexen-1-one | Lithium dimethylcuprate | 3,3-Dimethylcyclohexanone (98%)

PROBLEM 18.18 Outline two ways in which 4-methyl-2-octanone can be prepared by conjugate addition of an organocuprate to an α,β-unsaturated ketone.

SAMPLE SOLUTION Mentally disconnect one of the bonds to the β carbon so as to identify the group that comes from the lithium dialkylcuprate.

Disconnect this bond

$$CH_3CH_2CH_2CH_2 \text{—} \underset{CH_3}{\underset{|}{C}}HCH_2\overset{O}{\overset{\|}{C}}CH_3 \Longrightarrow CH_3CH_2CH_2\overset{-}{\underset{\cdot\cdot}{C}}H_2 + CH_3CH{=}CH\overset{O}{\overset{\|}{C}}CH_3$$

4-Methyl-2-octanone

According to this disconnection, the butyl group is derived from lithium dibutylcuprate. A suitable preparation is

$$CH_3CH{=}CH\overset{O}{\overset{\|}{C}}CH_3 + LiCu(CH_2CH_2CH_2CH_3)_2 \xrightarrow[\text{2. } H_2O]{\text{1. diethyl ether}} CH_3CH_2CH_2CH_2\underset{CH_3}{\underset{|}{C}}HCH_2\overset{O}{\overset{\|}{C}}CH_3$$

3-Penten-2-one | Lithium dibutylcuprate | 4-Methyl-2-octanone

Now see if you can identify the second possibility.

Like other carbon-carbon bond–forming reactions, organocuprate addition to enones is a powerful tool in organic synthesis.

18.17 ALKYLATION OF ENOLATE ANIONS

Since enolate anions are sources of nucleophilic carbon, one potential use in organic synthesis is their reaction with alkyl halides to give α-alkyl derivatives of aldehydes and ketones:

$$R_2CH\overset{O}{\overset{\|}{C}}R' \xrightarrow{\text{base}} R_2C{=}\overset{O^-}{\overset{|}{C}}R' \xrightarrow[S_N2]{R''X} R_2\underset{R''}{\underset{|}{C}}\text{—}\overset{O}{\overset{\|}{C}}R'$$

Aldehyde or ketone | Enolate anion | α-Alkyl derivative of an aldehyde or a ketone

Alkylation occurs by an S_N2 mechanism in which the enolate ion acts as a nucleophile toward the alkyl halide.

In practice, this reaction is difficult to carry out with simple aldehydes and ketones because aldol condensation competes with alkylation. Furthermore, it is not always possible to limit the reaction to the introduction of a single alkyl group. The most successful alkylation procedures use β-diketones as starting materials. Because they are relatively acidic, β-diketones can be converted quantitatively to their enolate ions by weak bases and do not self-condense. Ideally, the alkyl halide should be a methyl or primary alkyl halide.

$$CH_3\overset{O}{\overset{\|}{C}}CH_2\overset{O}{\overset{\|}{C}}CH_3 + CH_3I \xrightarrow{K_2CO_3} CH_3\overset{O}{\overset{\|}{C}}\underset{CH_3}{\underset{|}{C}H}\overset{O}{\overset{\|}{C}}CH_3$$

2,4-Pentanedione — Iodomethane — 3-Methyl-2,4-pentanedione (75–77%)

18.18 SUMMARY

Because aldehydes and ketones exist in equilibrium with their corresponding *enol* isomers, they can express a variety of different kinds of chemical reactivity.

$$R_2\underset{H}{\underset{|}{C}}-\overset{O}{\overset{\|}{C}}R' \rightleftharpoons R_2C=\overset{OH}{\overset{|}{C}}R'$$

α proton is relatively acidic; it can be removed by strong bases.

Carbonyl group is electrophilic; nucleophilic reagents add to carbonyl carbon.

α carbon atom of enol is nucleophilic; it attacks electrophilic reagents.

One reaction that proceeds via an enol is the α halogenation of carbonyl compounds (Sections 18.2 and 18.3). This reaction was used to introduce the idea of enol intermediates in reactions of aldehydes and ketones and is summarized in Table 18.1.

Ordinarily, the enol content of aldehydes and ketones is quite small. Typical aldehydes and ketones contain 10^{-5} to 10^{-7} percent enol at equilibrium (Section 18.4). Some representative data are included in Table 18.1.

A proton attached to the α carbon atom of an aldehyde or a ketone is about as acidic as an —OH proton of water or an alcohol and is capable of being removed by alkoxide bases (Section 18.6 and Table 18.1). The species formed by proton abstraction from an aldehyde or a ketone is called an *enolate ion.* Its negative charge is shared by the α carbon and the oxygen of the carbonyl group.

TABLE 18.1
Reactions of Aldehydes and Ketones That Involve Enol or Enolate Ion Intermediates

Reaction (section) and comments	General equation and typical example
Enolization (Sections 18.4 through 18.6) Aldehydes and ketones exist in equilibrium with their enol forms. The rate at which equilibrium is achieved is increased by acidic or basic catalysts. The enol content of simple aldehydes and ketones is quite small; β-diketones, however, are extensively enolized.	$R_2CH{-}\overset{O}{\overset{\Vert}{C}}R' \rightleftharpoons R_2C{=}\overset{OH}{\overset{\vert}{C}}R'$ Aldehyde or ketone; Enol Cyclopentanone $\overset{K}{\rightleftharpoons}$ Cyclopenten-1-ol $\quad K = 1 \times 10^{-8}$
α Halogenation (Sections 18.2 and 18.3) Halogens react with aldehydes and ketones by substitution; an α hydrogen is replaced by halogen. Reaction occurs by electrophilic attack of the halogen on the carbon-carbon double bond of the enol form of the aldehyde or ketone. An acid catalyst increases the rate of enolization, which is the rate-determining step.	$R_2CH\overset{O}{\overset{\Vert}{C}}R' + X_2 \longrightarrow R_2\underset{X}{\underset{\vert}{C}}\overset{O}{\overset{\Vert}{C}}R' + HX$ Aldehyde or ketone; Halogen; α-Halo aldehyde or ketone; Hydrogen halide $Br{-}C_6H_4{-}\overset{O}{\overset{\Vert}{C}}CH_3 + Br_2 \xrightarrow{\text{acetic acid}} Br{-}C_6H_4{-}\overset{O}{\overset{\Vert}{C}}CH_2Br + HBr$ *p*-Bromoacetophenone; Bromine; *p*-Bromophenacyl bromide (69–72%); Hydrogen bromide
Enolate ion formation (Section 18.6) An α proton of an aldehyde or a ketone is more acidic than most other protons bound to carbon. Aldehydes and ketones are weak acids, with K_a's in the range 10^{-16} to 10^{-20} (pK_a 16 to 20). Their enhanced acidity is due to the electron-withdrawing effect of the carbonyl group and the resonance stabilization of the enolate anion.	$R_2CH\overset{O}{\overset{\Vert}{C}}R' + HO^- \rightleftharpoons R_2C{=}\overset{:\ddot{O}:^-}{\overset{\vert}{C}}R' + H_2O$ Aldehyde or ketone; Hydroxide ion; Enolate anion; Water $CH_3CH_2\overset{:O:}{\overset{\Vert}{C}}CH_2CH_3 + HO^- \rightleftharpoons CH_3CH{=}\overset{:\ddot{O}:^-}{\overset{\vert}{C}}CH_2CH_3 + H_2O$ 3-Pentanone; Hydroxide ion; Enolate anion of 3-pentanone; Water
Haloform reaction (Section 18.7) Methyl ketones are cleaved on reactions with excess halogen in the presence of base. The products are a trihalomethane (haloform) and a carboxylate salt.	$R\overset{O}{\overset{\Vert}{C}}CH_3 + 3X_2 \xrightarrow{HO^-} R\overset{O}{\overset{\Vert}{C}}O^- + HCX_3$ Methyl ketone; Halogen; Carboxylate ion; Trihalomethane (haloform) $(CH_3)_3CCH_2\overset{O}{\overset{\Vert}{C}}CH_3 \xrightarrow[\text{2. } H^+]{\text{1. } Br_2,\ NaOH} (CH_3)_3CCH_2CO_2H + CHBr_3$ 4,4-Dimethyl-2-pentanone; 3,3-Dimethylbutanoic acid (89%); Bromoform

TABLE 18.1
Reactions of Aldehydes and Ketones That Involve Enol or Enolate Ion Intermediates (*Continued*)

Reaction (section) and comments	General equation and typical example
Aldol condensation (Sections 18.9 through 18.12) A reaction of great synthetic value for carbon-carbon bond formation. Nucleophilic addition of an enolate ion to a carbonyl group, followed by dehydration of the β-hydroxy aldehyde or ketone, yields an α,β unsaturated carbonyl compound.	$2RCH_2\overset{O}{\overset{\parallel}{C}}R' \xrightarrow{HO^-} RCH_2\underset{R'}{C}{=}\underset{R}{C}\overset{O}{\overset{\parallel}{C}}R' + H_2O$ Aldehyde or ketone; α,β-Unsaturated aldehyde or ketone; Water $CH_3(CH_2)_6\overset{O}{\overset{\parallel}{C}}H \xrightarrow[CH_3CH_2OH]{NaOCH_2CH_3} CH_3(CH_2)_6CH{=}\underset{HC=O}{C}(CH_2)_5CH_3$ Octanal; 2-Hexyl-2-decenal (79%)
Claisen-Schmidt reaction (Section 18.11) A mixed aldol condensation in which an aromatic aldehyde reacts with an enolizable aldehyde or ketone.	$Ar\overset{O}{\overset{\parallel}{C}}H + RCH_2\overset{O}{\overset{\parallel}{C}}R' \xrightarrow{HO^-} ArCH{=}\underset{R}{C}\overset{O}{\overset{\parallel}{C}}R' + H_2O$ Aromatic aldehyde; Aldehyde or ketone; α,β-Unsaturated carbonyl compound; Water $C_6H_5\overset{O}{\overset{\parallel}{C}}H + (CH_3)_3C\overset{O}{\overset{\parallel}{C}}CH_3 \xrightarrow[ethanol-water]{NaOH} C_6H_5CH{=}CH\overset{O}{\overset{\parallel}{C}}C(CH_3)_3$ Benzaldehyde; 3,3-Dimethyl-2-butanone; 4,4-Dimethyl-1-phenyl-1-penten-3-one (88–93%)
Conjugate addition to α,β-unsaturated carbonyl compounds (Sections 18.14 through 18.16) The β carbon atom of an α,β-unsaturated carbonyl compound is electrophilic; nucleophiles, especially weakly basic ones, yield the products of conjugate addition to α,β-unsaturated aldehydes and ketones.	$R_2C{=}CH\overset{O}{\overset{\parallel}{C}}R' + HY{:} \longrightarrow R_2\underset{Y}{C}CH_2\overset{O}{\overset{\parallel}{C}}R'$ α,β-Unsaturated aldehyde or ketone; Nucleophile; Product of conjugate addition $(CH_3)_2C{=}CH\overset{O}{\overset{\parallel}{C}}CH_3 \xrightarrow[H_2O]{NH_3} (CH_3)_2\underset{NH_2}{C}CH_2\overset{O}{\overset{\parallel}{C}}CH_3$ 4-Methyl-3-penten-2-one (mesityl oxide); 4-Amino-4-methyl-2-pentanone (63–70%)
Robinson annulation (Section 18.15) A combination of conjugate addition of an enolate anion to an α,β-unsaturated ketone with subsequent intramolecular aldol condensation.	2-Methylcyclohexanone + $CH_2{=}CH\overset{O}{\overset{\parallel}{C}}CH_3$ $\xrightarrow[2.\ KOH,\ heat]{1.\ NaOCH_2CH_3,\ CH_3CH_2OH}$ 6-Methylbicyclo[4.4.0]-1-decen-3-one 2-Methylcyclohexanone; Methyl vinyl ketone; 6-Methylbicyclo[4.4.0]-1-decen-3-one (46%)

TABLE 18.1
Reactions of Aldehydes and Ketones That Involve Enol or Enolate Ion Intermediates (*Continued*)

Reaction (section) and comments	General equation and typical example
Conjugate addition of organocopper compounds (Section 18.16) The principal synthetic application of lithium dialkylcuprate reagents is their reaction with α,β-unsaturated carbonyl compounds. Alkylation of the β carbon occurs.	$R_2C{=}CHCR'$ (C=O) + R''_2CuLi $\xrightarrow{\text{1. diethyl ether; 2. } H_2O}$ $R_2C(R'')-CH_2CR'$ (C=O) α,β-Unsaturated aldehyde or ketone; Lithium dialkylcuprate; β-Alkyl aldehyde or ketone 6-Methylcyclohept-2-enone $\xrightarrow{\text{1. } LiCu(CH_3)_2\text{; 2. } H_2O}$ 3,6-Dimethylcycloheptanone (85%)
α Alkylation of aldehydes and ketones (Sections 18.17) Alkylation of simple aldehydes and ketones via their enolates is difficult. β-Diketones can be converted quantitatively to their enolate anions, which react efficiently with primary alkyl halides.	$RCCH_2CR$ (two C=O) $\xrightarrow{R'CH_2X,\ HO^-}$ $RCCHCR$ with CH_2R' (two C=O) β-Diketone; α-Alkyl-β-diketone 2-Benzyl-1,3-cyclohexanedione + $C_6H_5CH_2Cl$ (Benzyl chloride) $\xrightarrow{KOCH_2CH_3,\ \text{ethanol}}$ 2,2-Dibenzyl-1,3-cyclohexanedione (69%)

$$R_2\overset{H}{C}-\overset{O}{\overset{\|}{C}}R' \underset{H_2O}{\overset{HO^-}{\rightleftharpoons}} \left[R_2\bar{C}-\overset{O}{\overset{\|}{C}}R' \leftrightarrow R_2C{=}\overset{O^-}{\overset{|}{C}}R'\right] \underset{HO^-}{\overset{H_2O}{\rightleftharpoons}} R_2C{=}\overset{OH}{\overset{|}{C}}R'$$

Aldehyde or ketone — Enolate ion — Enol

Enolate ions are nucleophilic and are key intermediates in base-promoted reactions of aldehydes and ketones. Table 18.1 includes examples of reactions of this type. In particular, the *aldol condensation* and variations of it provide a very useful method for creating carbon-carbon bonds (Sections 18.9 through 18.12).

Aldol condensation provides a ready route to *α,β-unsaturated aldehydes and ke-*

tones (Section 18.13). These compounds have the capacity to undergo either *1,2 addition (direct addition)* or *1,4 addition (conjugate addition)* on treatment with nucleophilic reagents (Section 18.14). Their reactions with strongly basic nucleophiles such as lithium aluminum hydride, Grignard reagents, and organolithium compounds are governed by kinetic control, and addition takes place at the carbonyl group just as with simple aldehydes and ketones. Thermodynamic control governs the reaction of weakly basic nucleophiles with α,β-unsaturated aldehydes and ketones, and these nucleophiles attack the β carbon. Table 18.1 contains an example illustrating 1,4 addition of ammonia to an α,β-unsaturated ketone.

The 1,4 addition of carbon nucleophiles to α,β-unsaturated ketones is called the *Michael reaction* (Section 18.15). The *Robinson annulation* (Table 18.1) is a synthetic method for ring formation that combines a Michael addition with an intramolecular aldol condensation.

Among compounds that undergo 1,4 addition to α,β-unsaturated carbonyl compounds, *lithium dialkylcuprates* are especially useful (Section 18.16 and Table 18.1).

Enolate ions are sometimes used as nucleophiles toward alkyl halides in order to introduce an alkyl group at the α carbon (Section 18.17 and Table 18.1).

PROBLEMS

18.19

(*a*) Write structural formulas for all the noncyclic aldehydes and ketones of molecular formula C_4H_6O.

(*b*) Are any of these compounds stereoisomeric?

(*c*) Are any of these compounds chiral?

(*d*) Which of these are α,β-unsaturated aldehydes or α,β-unsaturated ketones?

(*e*) Which of these can be prepared by a simple (i.e., not mixed) aldol condensation?

18.20 The simplest α,β-unsaturated aldehyde *acrolein* is prepared by heating glycerol with an acid catalyst. Suggest a mechanism for this reaction.

$$HOCH_2\underset{\displaystyle OH}{\underset{|}{C}}HCH_2OH \xrightarrow[\text{heat}]{KHSO_4} CH_2{=}CH\overset{\displaystyle O}{\overset{\|}{C}}H + H_2O$$

18.21 In each of the following pairs of compounds, choose the one that has the greater enol content and write the structure of its enol form:

(*a*) $(CH_3)_3\overset{O}{\overset{\|}{C}}H$ or $(CH_3)_2CH\overset{O}{\overset{\|}{C}}H$

(*b*) $C_6H_5\overset{O}{\overset{\|}{C}}C_6H_5$ or $C_6H_5CH_2\overset{O}{\overset{\|}{C}}CH_2C_6H_5$

(*c*) $C_6H_5\overset{O}{\overset{\|}{C}}CH_2\overset{O}{\overset{\|}{C}}C_6H_5$ or $C_6H_5CH_2\overset{O}{\overset{\|}{C}}CH_2C_6H_5$

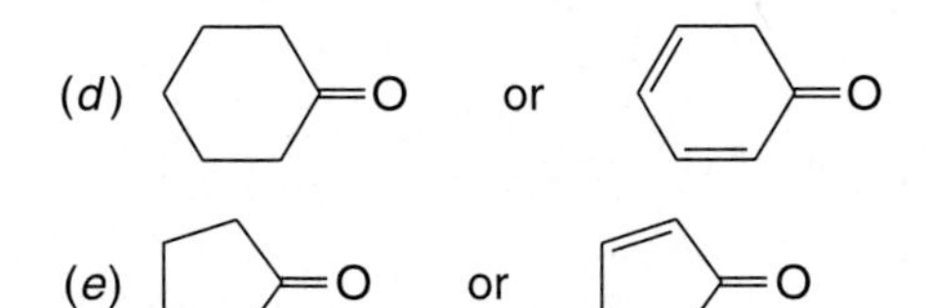

(e) O or O

(f) O O or O O

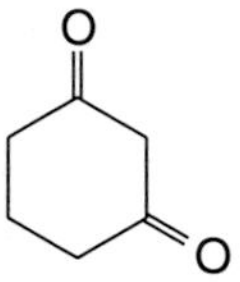

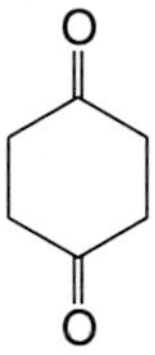

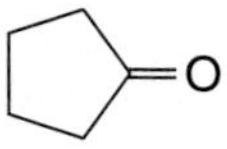

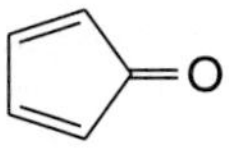

18.22 Give the structure of the expected organic product in the reaction of 3-phenylpropanal with each of the following:

(*a*) Chlorine in acetic acid
(*b*) Sodium hydroxide in ethanol, 10°C
(*c*) Sodium hydroxide in ethanol, 70°C
(*d*) Product of part (*c*) with lithium aluminum hydride; then H_2O
(*e*) Product of part (*c*) with sodium cyanide in acidic ethanol

18.23 Each of the following reactions has been reported in the chemical literature. Write the structure of the product(s) formed in each case

(*a*) $\overset{O}{\parallel}$ CCH_2CH_3 $\xrightarrow[CH_2Cl_2]{Cl_2}$ Cl

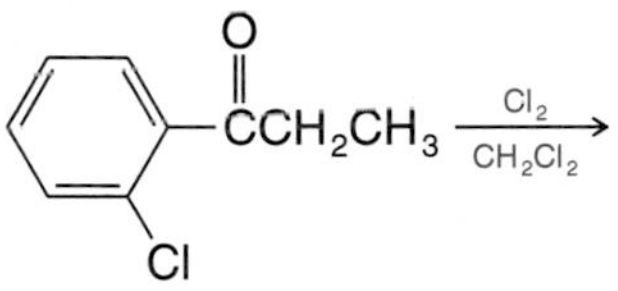

(*b*) CH_3 O $=C(CH_3)_2$ $\xrightarrow[NaOH,\ H_2O]{C_6H_5CH_2SH}$

(*c*) O C_6H_5 C_6H_5 $\xrightarrow[\text{diethyl ether}]{Br_2}$

(*d*) 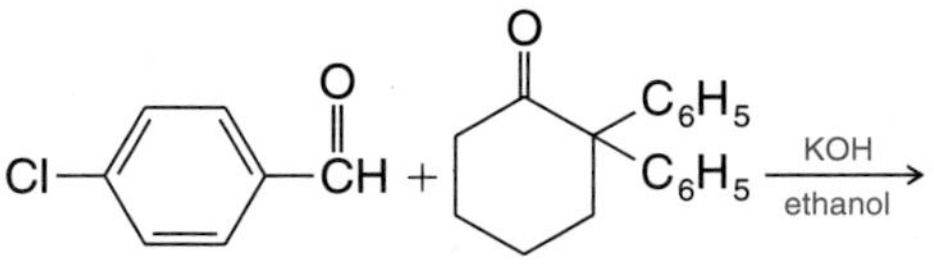 Cl $\overset{O}{\parallel}$ CH + O C_6H_5 C_6H_5 $\xrightarrow[\text{ethanol}]{KOH}$

(*e*) O $\overset{O}{\parallel}$ $CH + CH_3\overset{O}{\parallel}CCH_3$ $\xrightarrow[\text{water}]{NaOH}$

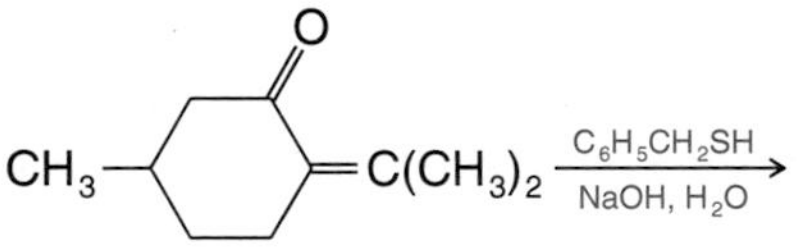

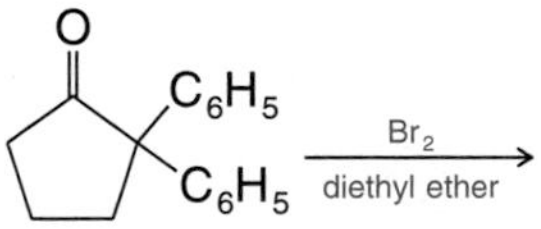

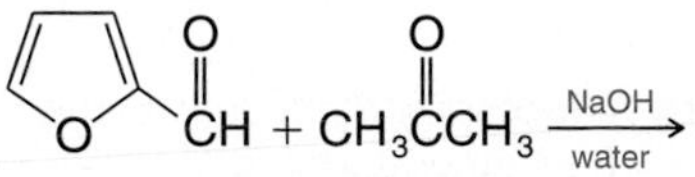

(*f*) [ring structure: O, CH_3, H_3C CH_3] $+ LiCu(CH_3)_2 \xrightarrow[2.\ H_2O]{1.\ \text{diethyl ether}}$

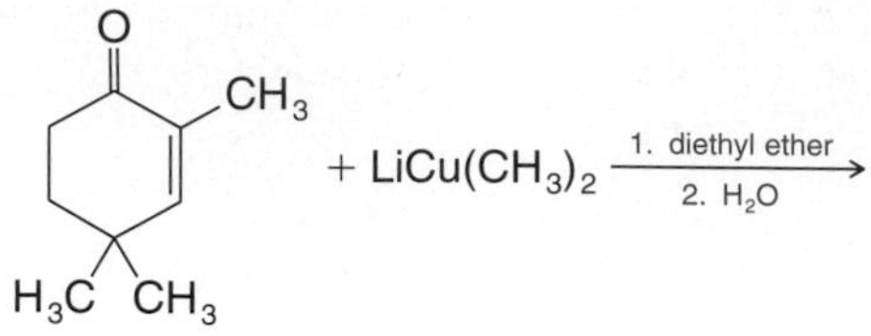

(*g*) [ring structure: O] $+ C_6H_5\overset{O}{\overset{\|}{C}}H \xrightarrow[\text{ethanol-water}]{\text{NaOH}}$

(*h*) [ring structure: O, O] $+ CH_2{=}CHCH_2Br \xrightarrow{\text{KOH}}$

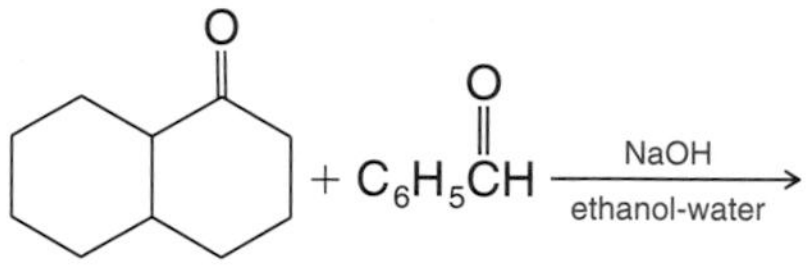

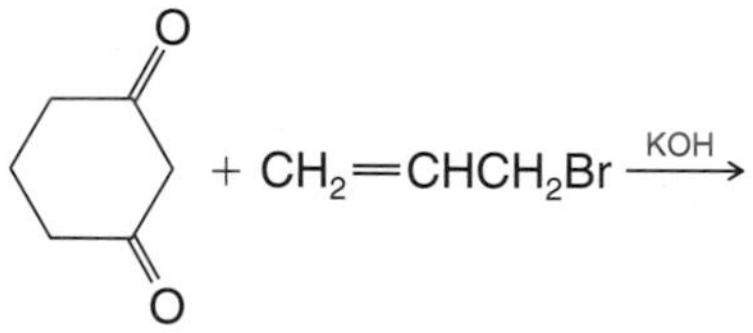

18.24 Show how each of the following compounds could be prepared from 3-pentanone. In most cases more than one synthetic transformation will be necessary.

(*a*) 2-Bromo-3-pentanone
(*b*) 1-Penten-3-one
(*c*) 1-Penten-3-ol
(*d*) 3-Hexanone
(*e*) 2-Methyl-1-phenyl-1-penten-3-one

18.25

(*a*) A synthesis that begins with 3,3-dimethyl-2-butanone gives the epoxide shown. Suggest reagents appropriate for each step in the synthesis.

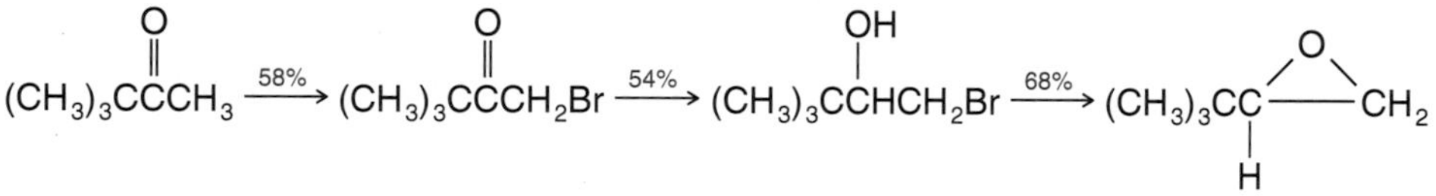

$$(CH_3)_3C\overset{O}{\overset{\|}{C}}CH_3 \xrightarrow{58\%} (CH_3)_3C\overset{O}{\overset{\|}{C}}CH_2Br \xrightarrow{54\%} (CH_3)_3C\overset{OH}{\overset{|}{C}}HCH_2Br \xrightarrow{68\%} (CH_3)_3C\underset{H}{\underset{|}{C}}\overset{O}{—}CH_2$$

(*b*) The yield for each step as actually carried out in the laboratory is given above each arrow. What is the overall yield for the three-step sequence?

18.26 Show how you could prepare each of the following compounds from cyclopentanone, D_2O, and any necessary organic or inorganic reagents.

(*a*) [structure: H, OH, D, D, D, D]

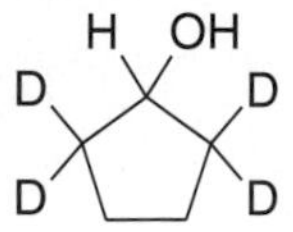

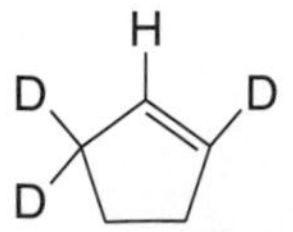

(*b*) [structure: H, D, D, D]

(*c*) [structure: D, D, D, H]

(*d*) [structure: D, D, D, D]

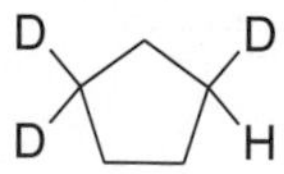

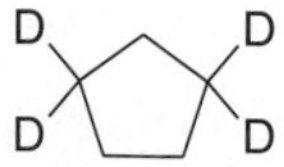

18.27

(*a*) At present, butanal is prepared industrially by hydroformylation of propene (Section 17.4). Write a chemical equation for this industrial synthesis.

(*b*) Before about 1970, the principal industrial preparation of butanal was from acetaldehyde. Outline a practical synthesis of butanal from acetaldehyde.

18.28 Identify the reagents appropriate for each step in the following syntheses:

(*a*) cyclohexanecarbaldehyde ⟶ 1-bromocyclohexanecarbaldehyde ⟶ cyclohex-1-enecarbaldehyde

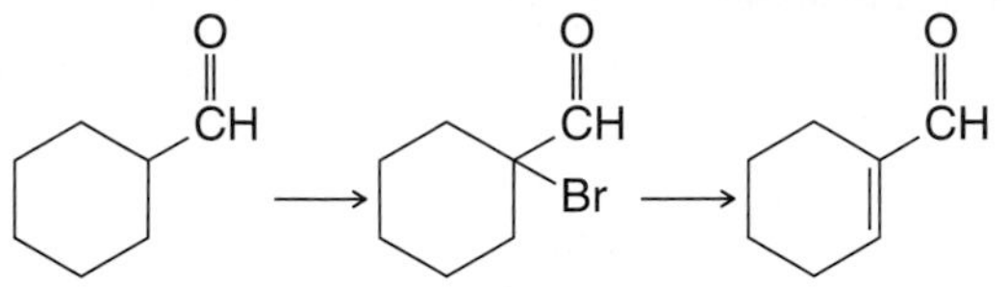

(*b*) trans-1,2-cyclohexanediol ⟶ $HCCH_2CH_2CH_2CH_2CH$ (with C=O at each end) ⟶ cyclopent-1-enecarbaldehyde

(*c*) 1-methyl-4-isopropylcyclohexene (CH_3, $CH(CH_3)_2$) ⟶ $CH_3\overset{O}{\overset{\|}{C}}CH_2CH_2\underset{CH(CH_3)_2}{CH}CH_2\overset{O}{\overset{\|}{C}}H$ ⟶ $(CH_3)_2CH$-cyclopentenyl-$\overset{O}{\overset{\|}{C}}CH_3$

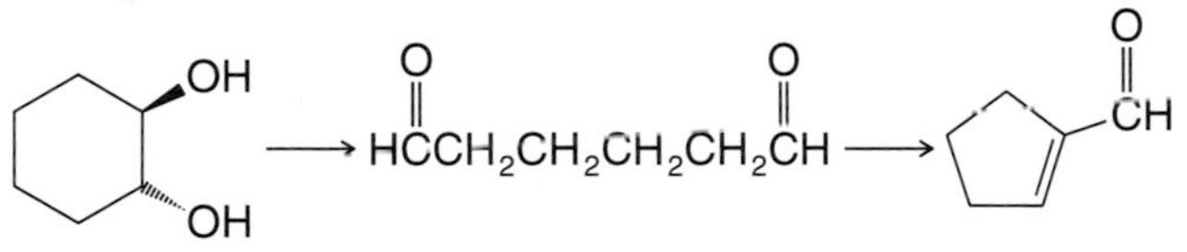

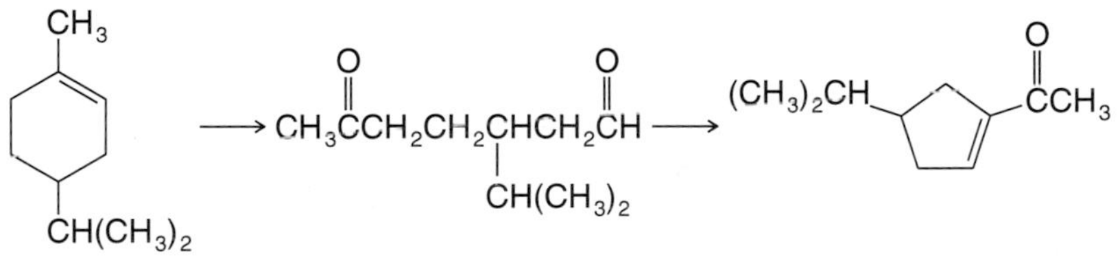

(*d*) $(CH_3)_2C{=}CHCH_2CH_2\overset{O}{\overset{\|}{C}}CH_3 \longrightarrow (CH_3)_2CH\underset{OH}{CH}CH_2CH_2\overset{O}{\overset{\|}{C}}CH_3$

$\downarrow$

3-isopropylcyclopent-2-enone ($(CH_3)_2CH$) $\longleftarrow (CH_3)_2CH\overset{O}{\overset{\|}{C}}CH_2CH_2\overset{O}{\overset{\|}{C}}CH_3$

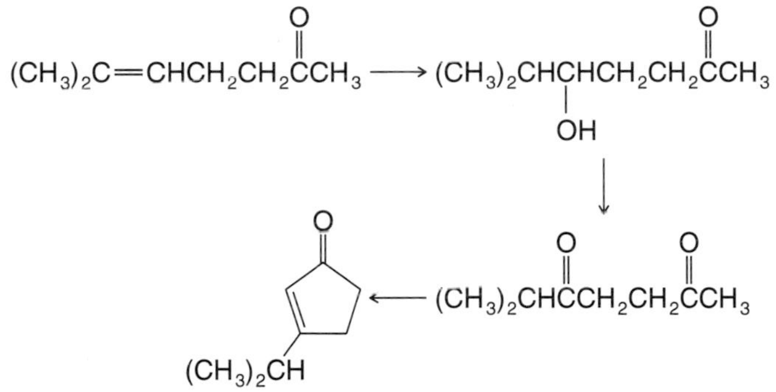

18.29 Give the structure of the product derived by intramolecular aldol condensation of the keto aldehyde shown:

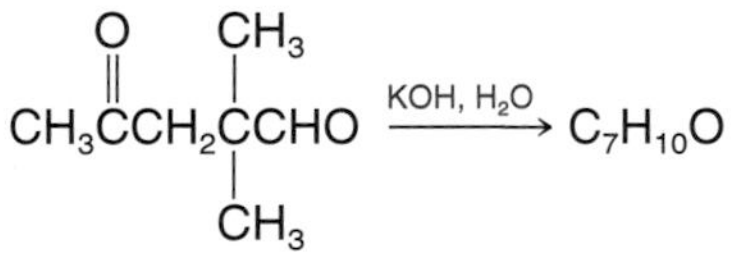

$$CH_3\overset{O}{\overset{\|}{C}}CH_2\underset{CH_3}{\overset{CH_3}{\overset{|}{\underset{|}{C}}}}CHO \xrightarrow{KOH,\ H_2O} C_7H_{10}O$$

18.30 Prepare each of the following compounds from the starting materials given and any necessary organic or inorganic reagents:

(*a*) $(CH_3)_2CH\underset{HO}{\underset{|}{C}H}\underset{CH_3}{\overset{CH_3}{\overset{|}{\underset{|}{C}}}}CH_2OH$ from $(CH_3)_2CHCH_2OH$

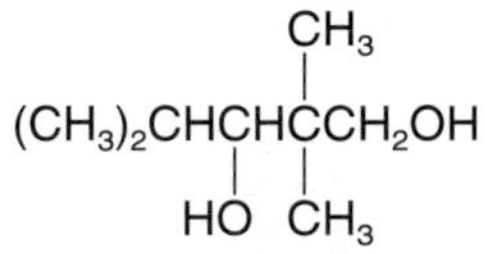

(*b*) $C_6H_5CH{=}\underset{CH_3}{\underset{|}{C}}CH_2OH$ from benzyl alcohol and 1-propanol

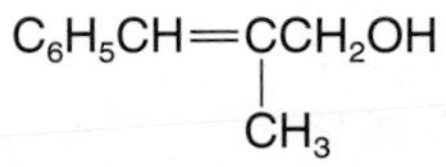

(c) 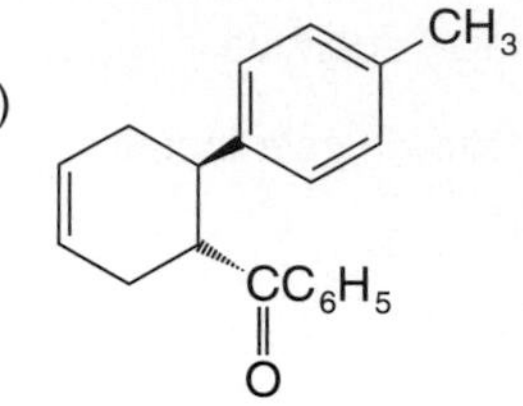 [structure: 4-methylphenyl-substituted cyclohexene bearing CC_6H_5 ketone] from acetophenone, 4-methylbenzyl alcohol, and 1,3-butadiene

18.31 *Terreic acid* is a naturally occurring antibiotic substance. Its actual structure is an enol isomer of the structure shown. Write the two most stable enol forms of terreic acid and choose which of those two is more stable.

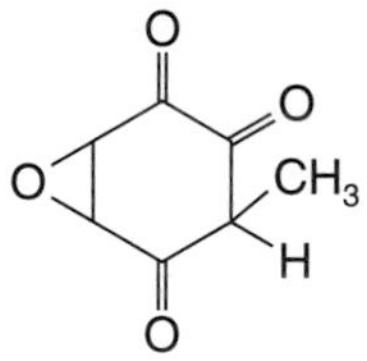

18.32 In each of the following the indicated observations were made before any of the starting material was transformed to aldol addition or condensation products:

(a) In aqueous acid, only 17 percent of $(C_6H_5)_2CHCH{=}O$ is present as the aldehyde; 2 percent of the enol is present. Some other species accounts for 81 percent of the material. What is it?

(b) In aqueous base, 97 percent of $(C_6H_5)_2CHCH{=}O$ is present as a species different from any of those in part (a). What is this species?

18.33

(a) For a long time attempts to prepare compound A were thwarted by its ready isomerization to compound B. The isomerization is efficiently catalyzed by traces of base. Write a reasonable mechanism for this isomerization.

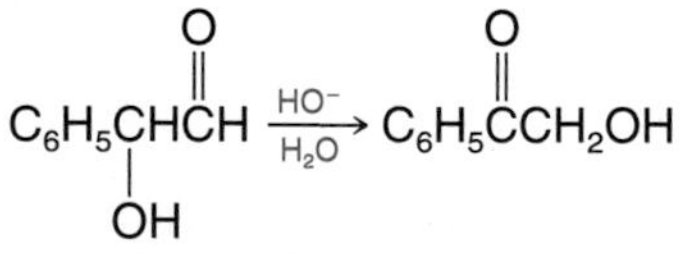

$$C_6H_5CH(OH)CH{=}O \xrightarrow[H_2O]{HO^-} C_6H_5C(=O)CH_2OH$$

Compound A Compound B

(b) Another attempt to prepare compound A by hydrolysis of its diethyl acetal gave only the 1,4-dioxane derivative C. How was compound C formed?

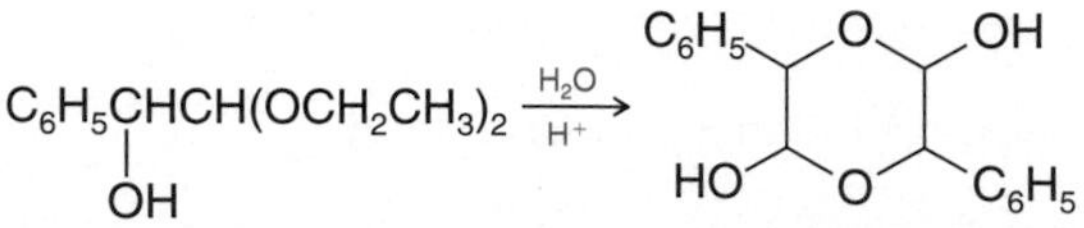

$$C_6H_5CH(OH)CH(OCH_2CH_3)_2 \xrightarrow[H^+]{H_2O}$$ [1,4-dioxane ring bearing C_6H_5, OH, HO, C_6H_5]

Compound C

18.34 Consider the ketones piperitone, menthone, and isomenthone.

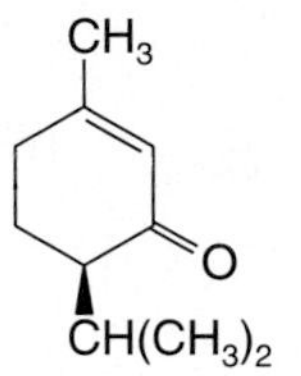

CH_3, $CH(CH_3)_2$ — (−)-Piperitone
CH_3, $CH(CH_3)_2$ — Menthone
CH_3, $CH(CH_3)_2$ — Isomenthone

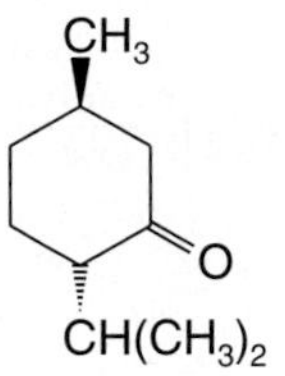

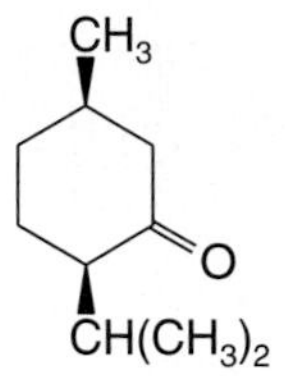

Suggest reasonable explanations for each of the following observations:

(*a*) Optically active piperitone (α_D − 32°) is converted to racemic piperitone on standing in a solution of sodium ethoxide in ethanol.

(*b*) Menthone is converted to a mixture of menthone and isomenthone on treatment with 90 percent sulfuric acid.

18.35 Many nitrogen-containing compounds engage in a proton-transfer equilibrium that is analogous to keto-enol tautomerism:

$$HX{-}N{=}Z \rightleftharpoons X{=}N{-}ZH$$

Each of the following compounds is the less stable partner of such a tautomeric pair. Write the structure of the more stable partner for each one.

(*a*) $CH_3CH_2N{=}O$

(*b*) $(CH_3)_2C{=}CHNHCH_3$

(*c*) $CH_3CH{=}\overset{+}{N}(O^-)OH$

(*d*) NH, NH, N

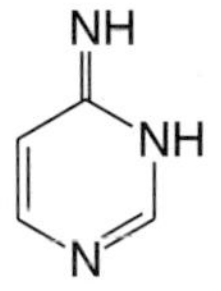

(*e*) $HN{=}C(OH)NH_2$

18.36 Outline reasonable mechanisms for each of the following reactions:

(*a*) $CH_2CH_2CH_2CH_2Br$ $\xrightarrow[\text{benzene}]{KOC(CH_3)_3}$ (76%)

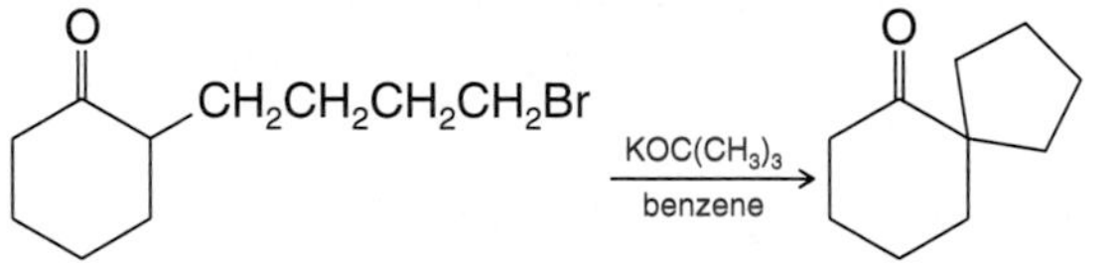

(*b*) $(CH_3)_2C{=}CHCH_2CH_2\underset{CH_3}{C}{=}CH\overset{O}{\overset{\|}{C}}H \xrightarrow[\text{heat}]{HO^-} (CH_3)_2C{=}CHCH_2CH_2\overset{O}{\overset{\|}{C}}CH_3 + CH_3\overset{O}{\overset{\|}{C}}H$
(96%)

(*c*) $H\overset{O}{\overset{\|}{C}}CH_2CH_2\underset{CH_3}{C}H\overset{O}{\overset{\|}{C}}CH_3 \xrightarrow[H_2O,\ CH_3OH]{KOH}$ CH_3, O (40%)

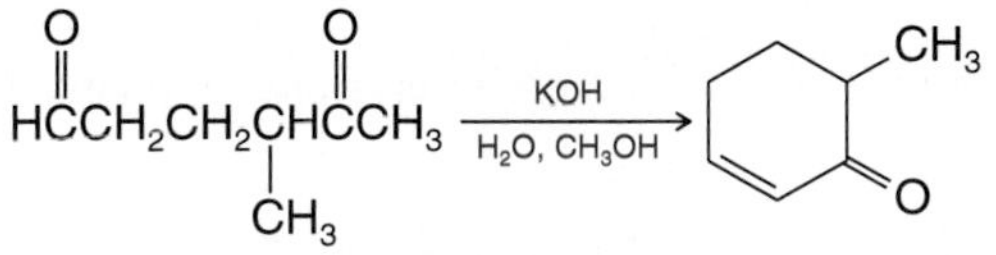

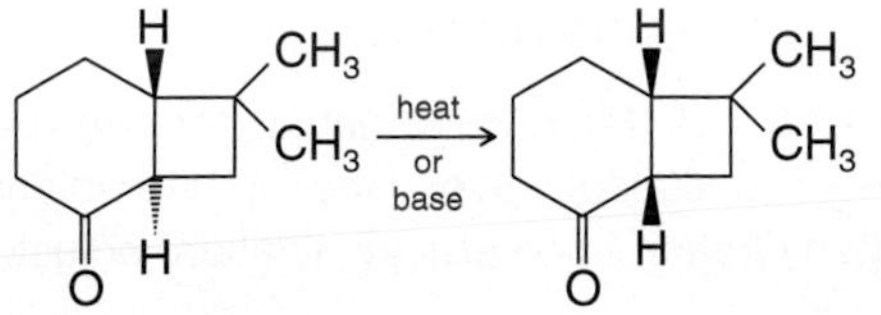

(*d*) H, CH_3, CH_3, H, O $\xrightarrow[\text{base}]{\text{heat or}}$ H, CH_3, CH_3, H, O (*e*) $C_6H_5\overset{O}{\overset{\|}{C}}\overset{O}{\overset{\|}{C}}C_6H_5 + C_6H_5CH_2\overset{O}{\overset{\|}{C}}CH_2C_6H_5 \xrightarrow[\text{ethanol}]{\text{KOH}}$ 2,3,4,5-tetraphenylcyclopentadienone (C_6H_5 groups at all four ring carbons, C=O) (91–96%)

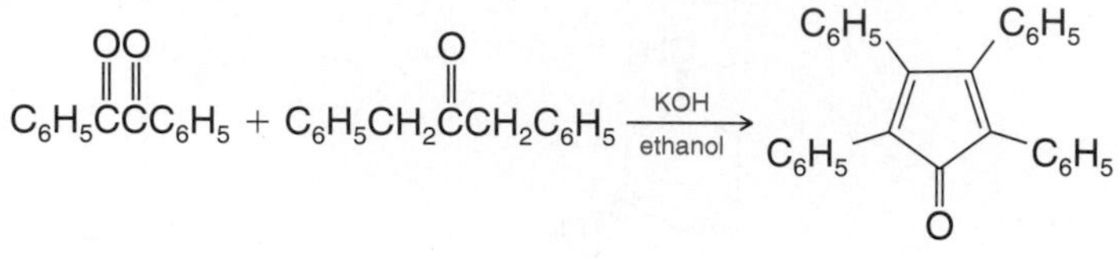

(*f*) $C_6H_5CH_2\overset{O}{\overset{\|}{C}}CH_2CH_3 + CH_2{=}\underset{C_6H_5}{\underset{|}{C}}\overset{O}{\overset{\|}{C}}C_6H_5 \xrightarrow[CH_3OH]{NaOCH_3}$ cyclohexenone product bearing C_6H_5, C_6H_5, C_6H_5, and CH_3 substituents (51%)

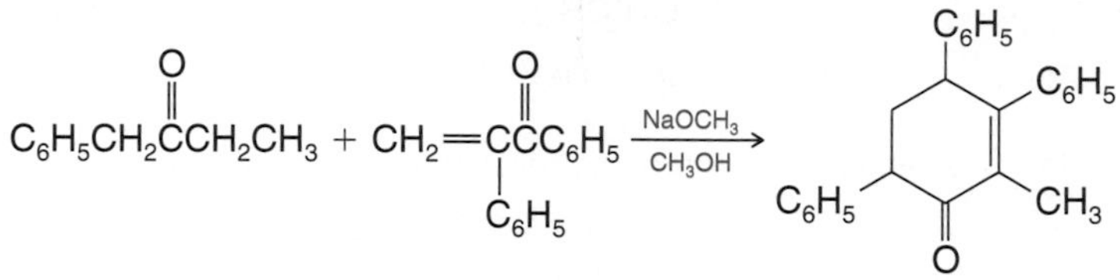

18.37 Suggest reasonable explanations for each of the following observations:

(*a*) The C=O stretching frequency of α,β-unsaturated ketones (about 1675 cm^{-1}) is less than that of typical dialkyl ketones (1710 to 1750 cm^{-1}).

(*b*) The C=O stretching frequency of cyclopropenone (1640 cm^{-1}) is lower than that of typical α,β-unsaturated ketones (1675 cm^{-1}).

(*c*) The dipole moment of diphenylcyclopropenone (μ = 5.1 D) is substantially larger than that of benzophenone (μ = 3.0 D)

(*d*) The β carbon of an α,β-unsaturated ketone is less shielded than the corresponding carbon of an alkene. Typical ^{13}C nmr chemical shift values are

$$CH_2{=}CH\overset{O}{\overset{\|}{C}}R \quad (\delta \cong 129 \text{ ppm}) \qquad CH_2{=}CHCH_2R \quad (\delta \cong 114 \text{ ppm})$$

18.38 Bromination of 3-methyl-2-butanone yielded two compounds, each having the molecular formula C_5H_9BrO, in a 95:5 ratio. The ^1H nmr spectrum of the major isomer A was characterized by a doublet at δ = 1.2 ppm (six protons), a heptet at δ = 3.0 ppm (one proton), and a singlet at δ = 4.1 ppm (two protons). The ^1H nmr spectrum of the minor isomer B exhibited two singlets, one at δ = 1.9 ppm and the other at δ = 2.5 ppm. The lower-field singlet had half the area of the higher-field singlet. Suggest reasonable structures for these two compounds.

18.39 Treatment of 2-butanone (1 mol) with Br_2 (2 mol) in aqueous HBr gave $C_4H_6Br_2O$. The ^1H nmr spectrum of the product was characterized by signals at δ = 1.9 ppm (doublet, three protons), 4.6 ppm (singlet, two protons), and 5.2 ppm (quartet, one proton). Identify this compound.

MOLECULAR MODELING EXERCISES

18.40 Construct molecular models of acetaldehyde and its enol isomer. Identify the possibilities for different conformations in each.

18.41 The most stable enolate of ethyl phenyl ketone has the *Z* configuration of its carbon-carbon double bond. Construct a molecular model of this species.

18.42 Construct a molecular model of acetaldol (3-hydroxybutanal; the aldol addition product of acetaldehyde). Dehydration of acetaldol gives the *E* stereoisomer of $CH_3CH{=}CHCH{=}O$. Assuming that the dehydration is an anti elimination, adjust the torsion angles in your model so as to show the geometry adopted by acetaldol in the transition state for its conversion to (*E*)-$CH_3CH{=}CHCH{=}O$. Why is this transition state lower in energy than the one for formation of (*Z*)-$CH_3CH{=}CHCH{=}O$ by anti elimination?

18.43 Construct a molecular model of (*E*)-$CH_3CH{=}CHCH{=}O$ in which the torsion angle between oxygen and C-3 is 180°. Describe the geometry of this α,β-unsaturated aldehyde.

18.44 Construct a molecular model of the enol form of $O{=}CHCH_2CH{=}O$. Choose a geometry that allows hydrogen bonding between the enolic —OH group and the carbonyl oxygen.

CHAPTER 19

CARBOXYLIC ACIDS

Carboxylic acids, compounds of the type $\mathrm{R\overset{\overset{O}{\|}}{C}OH}$, constitute one of the most frequently encountered classes of organic compounds. Countless natural products are carboxylic acids or are derived from them. Some carboxylic acids, such as acetic acid, have been known for centuries. Others, such as the prostaglandins, have been isolated only recently; scientists continue to discover more and more roles that these substances play as regulators of biological processes.

$$\mathrm{CH_3\overset{\overset{O}{\|}}{C}OH}$$

Acetic acid (present in vinegar)

O, $(CH_2)_6CO_2H$, $(CH_2)_4CH_3$, HO, OH

PGE_1 (a prostaglandin; a small amount of PGE_1 lowers blood pressure significantly)

The chemistry of carboxylic acids is the central theme of this chapter. The significance of carboxylic acids is magnified when one realizes that they are the parent compounds of a large group of derivatives that includes acyl chlorides, acid anhydrides, esters, and amides. Those classes of compounds will be discussed in the chapter following this one. Together, this chapter and the next tell the story of some of the most fundamental structural types and functional group transformations in organic and biological chemistry.

19.1 CARBOXYLIC ACID NOMENCLATURE

Nowhere in organic chemistry are common names as prevalent as they are among carboxylic acids. Many carboxylic acids are better known by common names than by their systematic names, and the framers of the IUPAC nomenclature rules have taken a liberal view toward accepting these common names as permissible alternatives to the systematic ones. Table 19.1 lists both the common and the systematic names of a number of important carboxylic acids.

Systematic names for carboxylic acids are derived by counting the number of carbons in the longest continuous chain that includes the carboxyl group and replacing the *-e* ending of the corresponding alkane by *-oic acid.* The first three acids in the table, methanoic (1 carbon), ethanoic (2 carbons), and octadecanoic acid (18 carbons), illustrate this point. When substituents are present, their locations are identified by number; numbering of the carbon chain always begins at the carboxyl group. This is illustrated in entries 4 and 5 in the table.

Notice that compounds 4 and 5 are named as hydroxy derivatives of carboxylic acids, rather than as carboxyl derivatives of alcohols. We have seen earlier that hy-

TABLE 19.1
Systematic and Common Names of Some Carboxylic Acids

	Structural formula	Systematic name	Common name
1.	HCO_2H	Methanoic acid	Formic acid
2.	CH_3CO_2H	Ethanoic acid	Acetic acid
3.	$CH_3(CH_2)_{16}CO_2H$	Octadecanoic acid	Stearic acid
4.	$CH_3CH(OH)CO_2H$	2-Hydroxypropanoic acid	Lactic acid
5.	$C_6H_5CH(OH)CO_2H$	2-Hydroxy-2-phenylethanoic acid	Mandelic acid
6.	$CH_2{=}CHCO_2H$	Propenoic acid	Acrylic acid
7.	$CH_3(CH_2)_7(H)C{=}C(H)(CH_2)_7CO_2H$ (cis)	(*Z*)-9-Octadecenoic acid	Oleic acid
8.	$C_6H_5CO_2H$	Benzenecarboxylic acid	Benzoic acid
9.	$o\text{-}HOC_6H_4CO_2H$	*o*-Hydroxybenzenecarboxylic acid	Salicylic acid
10.	$HO_2CCH_2CO_2H$	Propanedioic acid	Malonic acid
11.	$HO_2CCH_2CH_2CO_2H$	Butanedioic acid	Succinic acid
12.	$o\text{-}C_6H_4(CO_2H)_2$	1,2-Benzenedicarboxylic acid	Phthalic acid

droxyl groups take precedence over double bonds, and double bonds take precedence over halogens and alkyl groups, in naming compounds. Carboxylic acids outrank all the common groups we have encountered to this point.

Double bonds in the main chain are signaled by the ending *-enoic acid,* and their position is designated by a numerical prefix. Entries 6 and 7 are representative carboxylic acids that contain double bonds. Double-bond stereochemistry is specified by using either the cis-trans or the *E-Z* notation.

When a carboxyl group is attached to a ring, the parent ring is named (retaining the final *-e*) and the suffix *-carboxylic acid* is added, as shown in entries 8 and 9.

Compounds with two carboxyl groups, as illustrated by entries 10 through 12, are distinguished by the suffix *-dioic acid* or *-dicarboxylic acid* as appropriate. The final *-e* in the base name of the alkane is retained.

PROBLEM 19.1 The list of carboxylic acids in Table 19.1 is by no means exhaustive insofar as common names are concerned. Many others are known by their common names, a few of which follow. Give a systematic IUPAC name for each.

(*a*) $CH_2{=}C(CH_3)CO_2H$ (methacrylic acid)

(*b*) $H_3C(H)C{=}C(H)CO_2H$ (trans: H_3C and H on one carbon, H and CO_2H on the other) (crotonic acid)

(*c*) HO_2CCO_2H (oxalic acid)

(*d*) $CH_3-C_6H_4-CO_2H$ (*p*-toluic acid)

SAMPLE SOLUTION (*a*) Methacrylic acid is an industrial chemical used in the preparation of transparent plastics such as *Lucite* and *Plexiglas.* The carbon chain that includes both the carboxylic acid and the double bond is three carbon atoms in length. The compound is named as a derivative of *propenoic acid.* It is not necessary to locate the position of the double bond by number, as in "2-propenoic acid," because no other positions are structurally possible for it. The methyl group is at C-2, and so the correct systematic name for methacrylic acid is *2-methylpropenoic acid.*

19.2 STRUCTURE AND BONDING

The structural features characteristic of the carboxyl group ($—CO_2H$) are most easily seen by referring to the simplest carboxylic acid, formic acid. As shown in Figure 19.1, formic acid is planar, with one of its carbon-oxygen bonds significantly shorter than the other. The planar geometry and bond angles that are close to 120° suggest the sp^2 hybridization bonding model illustrated in Figure 19.2 and a $\sigma + \pi$ carbon-oxygen double bond analogous to that of aldehydes and ketones. Additionally, sp^2 hybridization of the hydroxyl oxygen allows one of its unshared electron pairs to be delocalized by orbital overlap with the π system of the carbonyl group.

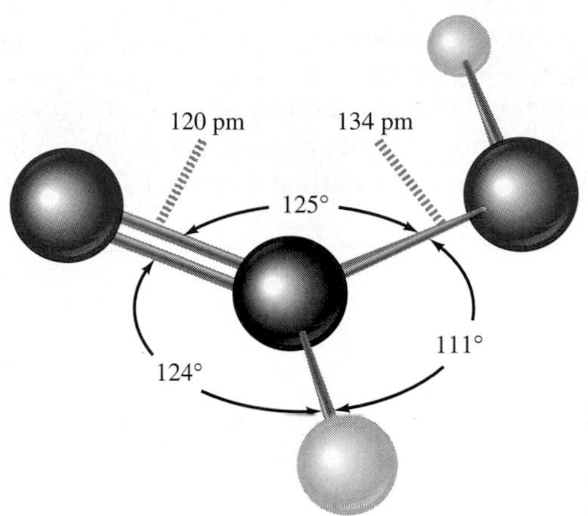

FIGURE 19.1 The structure of formic acid (HCO_2H), showing bond distances and bond angles involving the carboxyl group. All the atoms lie in the same plane.

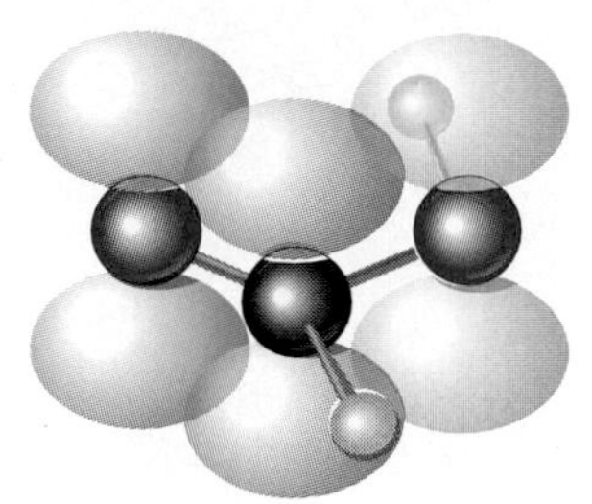

FIGURE 19.2 Carbon and both oxygens are sp^2 hybridized in formic acid. The π component of the C=O group (blue) and the *p* orbital of the OH oxygen (green) overlap to form an extended π system that includes carbon and the two oxygens.

In resonance terms, conjugation of the hydroxyl oxygen with the carbonyl group is represented as

$$\mathrm{H{-}C({=}\ddot{O}{:}){-}\ddot{O}H} \longleftrightarrow \mathrm{H{-}\overset{+}{C}({-}\ddot{O}{:}^{-}){-}\ddot{O}H} \longleftrightarrow \mathrm{H{-}C({-}\ddot{O}{:}^{-}){=}\overset{+}{O}H}$$

Lone-pair donation from the hydroxyl group makes the carbonyl group less electrophilic than that of an aldehyde or a ketone.

Carboxylic acids are fairly polar, and simple ones such as acetic acid, propanoic acid, and benzoic acid have dipole moments in the range 1.7 to 1.9 D.

19.3 PHYSICAL PROPERTIES

A summary of physical properties of some representative carboxylic acids is presented in Appendix 1.

The melting points and boiling points of carboxylic acids are higher than those of hydrocarbons and oxygen-containing organic compounds of comparable size and shape and indicate strong intermolecular attractive forces.

	2-Methyl-1-butene	2-Butanone	2-Butanol	Propanoic acid
bp (1 atm):	31°C	80°C	99°C	141°C

A unique hydrogen bonding arrangement, shown in Figure 19.3, contributes to these attractive forces. The hydroxyl group of one carboxylic acid molecule acts as a proton donor toward the carbonyl oxygen of a second. In a reciprocal fashion, the hydroxyl proton of the second carboxyl function interacts with the carbonyl oxygen of the first. The result is that the two carboxylic acid molecules are held together by *two* hydrogen bonds. So efficient is this hydrogen bonding that some carboxylic acids exist

RC(=O----H—O)(—O—H----O=)CR

FIGURE 19.3 Intermolecular hydrogen bonding between two carboxylic acid molecules.

as hydrogen-bonded dimers even in the gas phase. In the pure liquid a mixture of hydrogen-bonded dimers and higher aggregates is present.

In aqueous solution intermolecular association between carboxylic acid molecules is replaced by hydrogen bonding to water. The solubility properties of carboxylic acids are similar to those of alcohols. Carboxylic acids of four carbon atoms or fewer are miscible with water in all proportions.

19.4 ACIDITY OF CARBOXYLIC ACIDS

Carboxylic acids are the most acidic class of compounds that contain only carbon, hydrogen, and oxygen. With ionization constants K_a on the order of 10^{-5} (p$K_a \sim 5$), they are much stronger acids than water and alcohols. The case should not be overstated, however. Carboxylic acids are weak acids; a 0.1 *M* solution of acetic acid in water, for example, is only 1.3 percent ionized.

To understand the greater acidity of carboxylic acids compared with water and alcohols, compare the structural changes that accompany the ionization of a representative alcohol (ethanol) and a representative carboxylic acid (acetic acid). The equilibria that define K_a are

Ionization of ethanol

$$\underset{\text{Ethanol}}{CH_3CH_2OH} \rightleftharpoons H^+ + \underset{\text{Ethoxide ion}}{CH_3CH_2O^-} \qquad K_a = \frac{[H^+][CH_3CH_2O^-]}{[CH_3CH_2OH]} = 10^{-16}$$

Ionization of acetic acid

$$\underset{\text{Acetic acid}}{CH_3\overset{\overset{\displaystyle O}{\|}}{C}OH} \rightleftharpoons H^+ + \underset{\text{Acetate ion}}{CH_3\overset{\overset{\displaystyle O}{\|}}{C}O^-} \qquad K_a = \frac{[H^+][CH_3CO_2^-]}{[CH_3CO_2H]} = 1.8 \times 10^{-5}$$

Free energies of ionization are calculated from equilibrium constants according to the relationship

$$\Delta G° = -RT \ln K_a$$

From these K_a values, one calculates free energies of ionization ($\Delta G°$) of 91 kJ/mol (21.7 kcal/mol) for ethanol versus 27 kJ/mol (6.5 kcal/mol) for acetic acid. An energy diagram portraying these relationships is presented in Figure 19.4. Since it is *equilibria,* not *rates,* of ionization that are being compared, the diagram shows only the initial and final states. It is not necessary to be concerned about the energy of activation, since that affects only the rate of ionization, not the extent of ionization.

The large difference in the respective free energies of ionization between ethanol and acetic acid reflects a greater stabilization of acetate ion relative to ethoxide ion. Ionization of ethanol yields an alkoxide ion in which the negative charge is localized on oxygen. Solvation forces are the chief means by which ethoxide ion is stabilized. Acetate ion is also stabilized by solvation, but has two additional mechanisms for dispersing its negative charge that are not available to ethoxide ion:

1. *The inductive effect of the carbonyl group.* The carbonyl group is electron-withdrawing, and by attracting electrons away from the negatively charged oxygen, acetate anion is stabilized. This is an inductive effect, arising in the polarization of

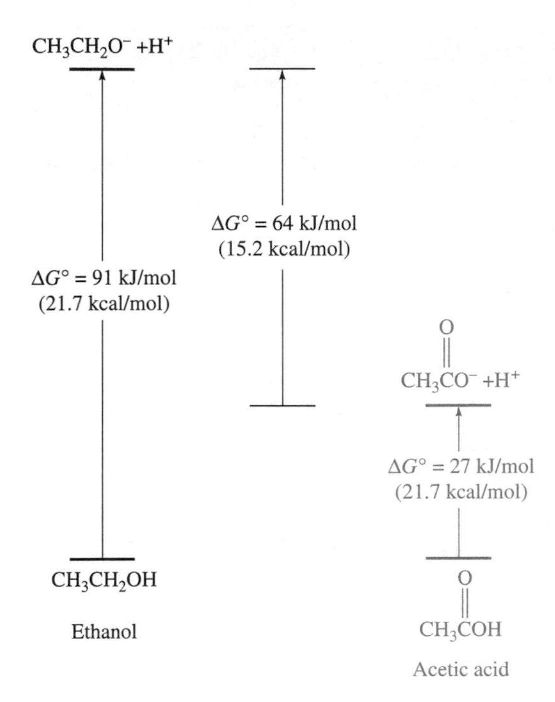

$CH_3\overset{O}{\overset{\|}{C}}O^- + H^+$

$\Delta G° = 27$ kJ/mol
(21.7 kcal/mol)

CH_3CH_2OH

Ethanol

$CH_3\overset{O}{\overset{\|}{C}}OH$

Acetic acid

FIGURE 19.4 Diagram comparing the free energies of ionization of ethanol and acetic acid in water.

the electron distribution in the σ bond between the carbonyl carbon and the negatively charged oxygen.

Positively polarized carbon attracts electrons from negatively charged oxygen.

$CH_3—\overset{\delta+}{C}(=\ddot{O}:^{\delta-})—\ddot{O}:^-$

CH_2 group has negligible effect on electron density at negatively charged oxygen.

$CH_3—CH_2—\ddot{O}:^-$

2. *The resonance effect of the carbonyl group.* Electron delocalization, expressed by resonance between the following Lewis structures, causes the negative charge in acetate to be shared equally by both oxygens. Electron delocalization of this type is not available to ethoxide ion.

$$CH_3C(=\ddot{O}:)\ddot{O}:^- \longleftrightarrow CH_3C(—\ddot{O}:^-)=\ddot{O}: \quad \text{or} \quad CH_3C(O^{-1/2})(O^{-1/2})$$

PROBLEM 19.2 Peroxyacetic acid ($CH_3\overset{O}{\overset{\|}{C}}OOH$) is a weaker acid than acetic acid; its K_a is 6.3×10^{-9} (pK_a 8.2) versus 1.8×10^{-5} for acetic acid (pK_a 4.7). Why are peroxy acids weaker than carboxylic acids?

Electron delocalization in acetate ion, as well as in carboxylate ions generally, is

QUANTITATIVE RELATIONSHIPS INVOLVING CARBOXYLIC ACIDS

Suppose you take two flasks, one containing pure water and the other a buffer solution maintained at a pH of 7.0. If you add 0.1 mol of acetic acid to each one and the final volume in each flask is 1 L, how much acetic acid is present at equilibrium? How much acetate ion? In other words, what is the extent of ionization of acetic acid in an unbuffered medium and in a buffered one?

The first case simply involves the ionization of a weak acid and is governed by the expression that defines K_a for acetic acid:

$$K_a = \frac{[H^+][CH_3CO_2^-]}{[CH_3CO_2H]} = 1.8 \times 10^{-5}$$

Since ionization of acetic acid gives one H^+ for each $CH_3CO_2^-$, the concentrations of the two ions are equal, and setting each one equal to x gives:

$$K_a = \frac{x^2}{0.1 - x} = 1.8 \times 10^{-5}$$

Solving for x gives the acetate ion concentration as:

$$x = 1.3 \times 10^{-3}$$

Thus when acetic acid is added to pure water, the ratio of acetate ion to acetic acid is

$$\frac{[CH_3CO_2^-]}{[CH_3CO_2H]} = \frac{1.3 \times 10^{-3}}{0.1} = 0.013$$

Only 1.3 percent of the acetic acid has ionized. Most of it (98.7 percent) remains unchanged.

Now consider what happens when the same amount of acetic acid is added to water that is buffered at pH = 7.0. Before doing the calculation, let us recognize that it is the $[CH_3CO_2^-]/[CH_3CO_2H]$ ratio in which we are interested and do a little algebraic manipulation. Since

$$K_a = \frac{[H^+][CH_3CO_2^-]}{[CH_3CO_2H]}$$

then

$$\frac{[CH_3CO_2^-]}{[CH_3CO_2H]} = \frac{K_a}{[H^+]}$$

This relationship is one form of the **Henderson-Hasselbalch equation.** It is a useful relationship in chemistry and biochemistry. One rarely needs to calculate the pH of a solution—pH is more often measured than calculated. It is much more common that one needs to know the degree of ionization of an acid at a particular pH, and the Henderson-Hasselbalch equation gives that ratio.

For the case at hand, the solution is buffered at pH = 7.0. Therefore,

$$\frac{[CH_3CO_2^-]}{[CH_3CO_2H]} = \frac{1.8 \times 10^{-5}}{10^{-7}} = 180$$

A very much different situation exists in an aqueous solution maintained at pH = 7.0 from the situation in pure water. We saw earlier that almost all the acetic acid in a 0.1 *M* solution in pure water was un-ionized. At pH 7.0, however, hardly any un-ionized acetic acid remains; it is almost completely converted to its carboxylate ion.

This difference in behavior for acetic acid in pure water versus water buffered at pH = 7.0 has some important practical consequences. Biochemists usually do not talk about acetic acid (or lactic acid, or salicylic acid, and so on). They talk about acetate (and lactate, and salicylate). Why? It is because biochemists are concerned with carboxylic acids as they exist in dilute aqueous solution at what is called *biological pH.* Biological fluids are naturally buffered. The pH of blood, for example, is maintained at 7.2, and at this pH carboxylic acids are almost entirely converted to their carboxylate anions.

An alternative form of the Henderson-Hasselbalch equation for acetic acid is

$$pH = pK_a + \log \frac{[CH_3CO_2^-]}{[CH_3CO_2H]}$$

From this equation it can be seen that when $[CH_3CO_2^-] = [CH_3CO_2H]$, then the second term is log 1 = 0, and pH = pK_a. This means that the pK_a of an acid is equal to the pH at which it is half-neutralized and is a relationship worth remembering.

supported by studies that reveal significant differences in the pattern of carbon-oxygen bond distances of carboxylic acids and their carboxylate anions.

$$CH_3C(=O)OH \quad (C{=}O: 121\ pm;\ C{-}OH: 136\ pm) \qquad CH_3C(O^{-1/2})_2\ \overset{+}{N}H_4 \quad (C{-}O: 125\ pm;\ 125\ pm)$$

As expected, the two carbon-oxygen bond distances in acetic acid are different from each other. The C═O double bond distance is shorter than the C—O single bond distance. In ammonium acetate, the two C—O bond distances are equal, as they should be according to both the resonance and molecular orbital pictures of bonding in carboxylate anions.

Organic chemists have, for many years, emphasized resonance in carboxylate ions in order to explain the acidity of carboxylic acids. That emphasis has recently been questioned and the view expressed that the inductive effect of the carbonyl group may be quantitatively more significant. It seems clear that, even though their relative contributions may be a matter of debate, both the inductive and the resonance effect of the carbonyl group are important in making carboxylic acids stronger acids than alcohols.

Similar principles apply to both the acidity of carboxylic acids and the α proton of aldehydes and ketones. In both cases a proton is lost from a species of the type $R\overset{O}{\overset{\|}{C}}{-}X{-}H$. Carboxylate ions and enolate ions are analogous to each other.

19.5 SALTS OF CARBOXYLIC ACIDS

In the presence of strong bases such as sodium hydroxide, carboxylic acids are neutralized rapidly and quantitatively:

$$R\overset{O}{\overset{\|}{C}}{-}\ddot{\underset{..}{O}}{-}H + {}^{-}{:}\ddot{\underset{..}{O}}H \xrightarrow{K = 10^{11}} R\overset{O}{\overset{\|}{C}}{-}\ddot{\underset{..}{O}}{:}^{-} + H{-}\ddot{\underset{..}{O}}H$$

Carboxylic acid (stronger acid)	Hydroxide ion (stronger base)	Carboxylate ion (weaker base)	Water (weaker acid)

PROBLEM 19.3 Write an ionic equation for the reaction of acetic acid with each of the following, and specify whether the equilibrium favors starting materials or products:

(*a*) Sodium ethoxide
(*b*) Potassium *tert*-butoxide
(*c*) Sodium bromide
(*d*) Sodium acetylide
(*e*) Potassium nitrate
(*f*) Lithium amide

SAMPLE SOLUTION (*a*) The reaction is an acid-base reaction; ethoxide ion is the base.

$$CH_3CO_2H + CH_3CH_2O^- \longrightarrow CH_3CO_2^- + CH_3CH_2OH$$

Acetic acid (stronger acid)	Ethoxide ion (stronger base)	Acetate ion (weaker base)	Ethanol (weaker acid)

The position of equilibrium lies well to the right. Ethanol, with a K_a of 10^{-16} (pK_a 16), is a much weaker acid than acetic acid.

The metal carboxylate salts formed on neutralization of carboxylic acids are named by first specifying the metal ion and then adding the name of the acid modified by replacing *-ic acid* by *-ate*. Monocarboxylate salts of diacids are designated by naming both the cation and hydrogen as substituents of carboxylate groups.

$CH_3\overset{O}{\overset{\|}{C}}OLi$ — Lithium acetate

$Cl-C_6H_4-\overset{O}{\overset{\|}{C}}ONa$ — Sodium *p*-chlorobenzoate

$HO\overset{O}{\overset{\|}{C}}(CH_2)_4\overset{O}{\overset{\|}{C}}OK$ — Potassium hydrogen hexanedioate

Metal carboxylate salts are ionic, and so long as the molecular weight is not too high, the sodium and potassium salts of carboxylic acids are soluble in water. Carboxylic acids therefore may be extracted from ether solutions into aqueous sodium or potassium hydroxide.

The solubility behavior of salts of carboxylic acids having 12 to 18 carbons is unusual and can be illustrated by considering sodium stearate:

$CH_3(CH_2)_{16}CO_2^-\ Na^+$

Sodium stearate
(sodium octadecanoate)

Sodium stearate has a polar carboxylate group at one end of a long hydrocarbon chain. The carboxylate group is **hydrophilic** (''water-loving'') and tends to confer water solubility on the molecule. The hydrocarbon chain is **lipophilic** (''fat-loving'') and tends to associate with other hydrocarbon chains. The compromise achieved by sodium stearate when it is placed in water is to form a colloidal dispersion of spherical aggregates called **micelles.** Each micelle is composed of 50 to 100 individual molecules. Micelles form spontaneously when the carboxylate concentration exceeds a certain minimum value called the **critical micelle concentration.** A representation of a micelle is shown in Figure 19.5.

Polar carboxylate groups dot the surface of the micelle. There they bind to water molecules and to sodium ions. The nonpolar hydrocarbon chains are directed toward the interior of the micelle, where individually weak but cumulatively significant induced dipole–induced dipole forces bind them together. Micelles are approximately spherical because a sphere encloses the maximum volume of material for a given surface area and disrupts the water structure least. Because their surfaces are negatively charged, two micelles repel each other rather than clustering to form higher aggregates.

It is the formation of micelles and their properties that are responsible for the cleansing action of soaps. Water that contains sodium stearate removes grease by enclosing it in the hydrocarbonlike interior of the micelles. The grease is washed away with the water, not because it dissolves in the water but because it dissolves in the micelles that are dispersed in the water. Sodium stearate is an example of a soap;

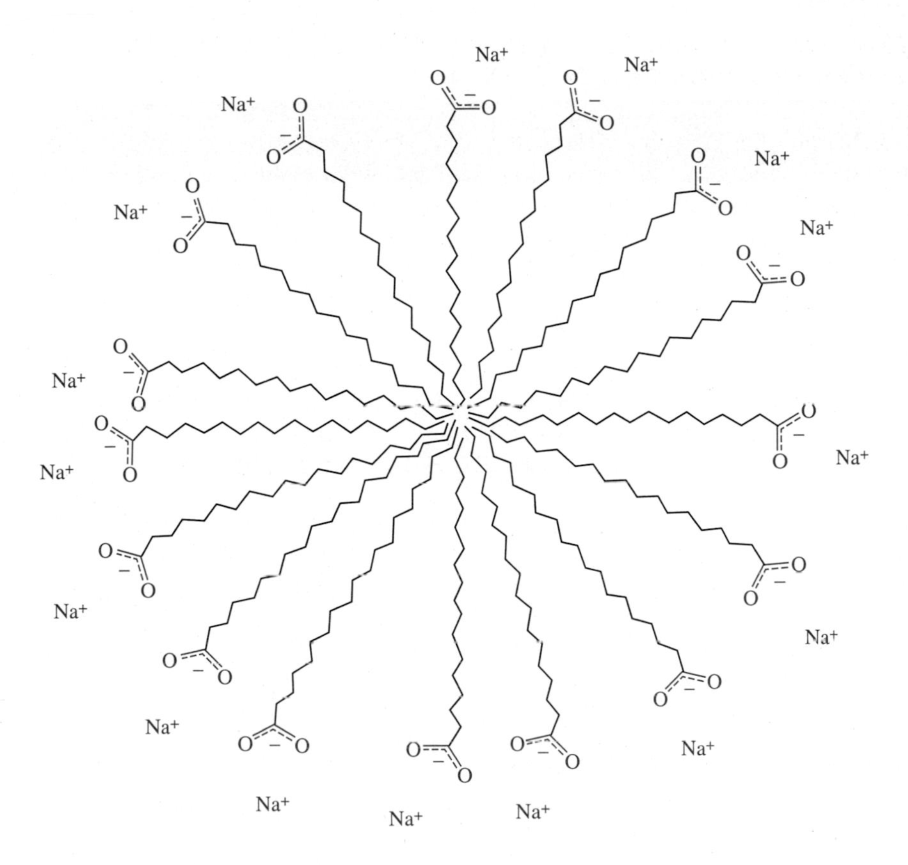

FIGURE 19.5 Idealized representation of the cross section of a sodium stearate micelle. A real micelle is more irregular. It contains voids and channels, and the carbon chains are not extended, but coiled back on themselves.

sodium and potassium salts of other C_{12} through C_{18} unbranched carboxylic acids possess similar properties.

Detergents are substances, including soaps, that cleanse by micellar action. A large number of synthetic detergents are known. One example is sodium lauryl sulfate. Sodium lauryl sulfate has a long hydrocarbon chain terminating in a polar sulfate ion and forms soaplike micelles in water.

^{-}O O^{-} S O ++ O^{-} Na^{+}

Sodium lauryl sulfate
(sodium dodecyl sulfate)

Detergents are designed to be effective in hard water, i.e., water containing calcium salts that form insoluble calcium carboxylates with soaps. These precipitates rob the soap of its cleansing power and form an unpleasant scum. The calcium salts of synthetic detergents such as sodium lauryl sulfate, however, are soluble and retain their micelle-forming ability in water.

19.6 SUBSTITUENTS AND ACID STRENGTH

Alkyl groups have little effect on the acidity of a carboxylic acid. The ionization constants of all acids that have the general formula $C_nH_{2n+1}CO_2H$ are very similar to

TABLE 19.2
Effect of Substituents on Acidity of Carboxylic Acids

Name of acid	Structure	Ionization constant K_a*	pK_a
Standard of comparison.			
Acetic acid	CH_3CO_2H	1.8×10^{-5}	4.7
Alkyl substituents have a negligible effect on acidity.			
Propanoic acid	$CH_3CH_2CO_2H$	1.3×10^{-5}	4.9
2-Methylpropanoic acid	$(CH_3)_2CHCO_2H$	1.6×10^{-5}	4.8
2,2-Dimethylpropanoic acid	$(CH_3)_3CCO_2H$	0.9×10^{-5}	5.1
Heptanoic acid	$CH_3(CH_2)_5CO_2H$	1.3×10^{-5}	4.9
α-Halogen substituents increase acidity.			
Fluoroacetic acid	FCH_2CO_2H	2.5×10^{-3}	2.6
Chloroacetic acid	$ClCH_2CO_2H$	1.4×10^{-3}	2.9
Bromoacetic acid	$BrCH_2CO_2H$	1.4×10^{-3}	2.9
Dichloroacetic acid	Cl_2CHCO_2H	5.0×10^{-2}	1.3
Trichloroacetic acid	Cl_3CCO_2H	1.3×10^{-1}	0.9
Electron-attracting groups increase acidity.			
Methoxyacetic acid	$CH_3OCH_2CO_2H$	2.7×10^{-4}	3.6
Cyanoacetic acid	$N{\equiv}CCH_2CO_2H$	3.4×10^{-3}	2.5
Nitroacetic acid	$O_2NCH_2CO_2H$	2.1×10^{-2}	1.7

* In water at 25°C.

one another and equal approximately 10^{-5} (pK_a 5). Table 19.2 lists some representative examples.

Electronegative substituents, particularly when they are attached to the α carbon, increase the acidity of a carboxylic acid. As the data in Table 19.2 show, all the monohaloacetic acids are about 100 times more acidic than acetic acid. Multiple halogen substitution increases the acidity even more; trichloroacetic acid is 7000 times more acidic than acetic acid!

Organic chemists have long attributed the acid-strengthening effect of electronegative atoms or groups to an inductive effect of the substituent transmitted through the σ bonds of the molecule. According to this model, the σ electrons in the carbon-chlorine bond of chloroacetate ion are drawn toward chlorine, leaving the α carbon atom with a slight positive charge. The α carbon, because of this positive character, attracts electrons from the negatively charged carboxylate, thus dispersing the charge and stabilizing the anion. The more stable the anion, the greater the equilibrium constant for its formation.

$$\overset{\delta-}{Cl}-\overset{\delta+}{C}H_2-C(=O)O^-$$

Chloroacetate anion is stabilized by electron-withdrawing effect of chlorine.

Inductive effects diminish rapidly as the number of σ bonds between the carboxyl group and the substituent increases. Consequently, the acid-strengthening effect of a halogen decreases as it becomes more remote from the carboxyl group:

$ClCH_2CO_2H$	$ClCH_2CH_2CO_2H$	$ClCH_2CH_2CH_2CO_2H$
Chloroacetic acid	3-Chloropropanoic acid	4-Chlorobutanoic acid
$K_a = 1.4 \times 10^{-3}$	$K_a = 1.0 \times 10^{-4}$	$K_a = 3.0 \times 10^{-5}$
$pK_a = 2.9$	$pK_a = 4.0$	$pK_a = 4.5$

PROBLEM 19.4 Which is the stronger acid in each of the following pairs?

(*a*) $(CH_3)_3CCH_2CO_2H$ or $(CH_3)_3\overset{+}{N}CH_2CO_2H$

(*b*) $CH_3CH_2CO_2H$ or $CH_3\underset{\displaystyle OH}{\underset{|}{C}}HCO_2H$

(*c*) $CH_3\overset{\displaystyle O}{\overset{\|}{C}}CO_2H$ or $CH_2{=}CHCO_2H$

(*d*) $CH_3CH_2CH_2CO_2H$ or $CH_3\underset{\displaystyle O}{\underset{\|}{\overset{\displaystyle O}{\overset{\|}{S}}}}CH_2CO_2H$

SAMPLE SOLUTION (*a*) The two compounds are viewed as substituted derivatives of acetic acid. A *tert*-butyl group is slightly electron-releasing and should have a modest effect on acidity. The compound $(CH_3)_3CCH_2CO_2H$ is expected to have an acid strength similar to that of acetic acid. A trimethylammonium substituent, on the other hand, is positively charged and is a powerful electron-withdrawing substituent. The compound $(CH_3)_3\overset{+}{N}CH_2CO_2H$ is expected to be a much stronger acid than $(CH_3)_3CCH_2CO_2H$. The measured ionization constants, shown as follows, confirm this prediction.

$(CH_3)_3CCH_2CO_2H$	$(CH_3)_3\overset{+}{N}CH_2CO_2H$
Weaker acid	Stronger acid
$K_a = 5 \times 10^{-6}$	$K_a = 1.5 \times 10^{-2}$
($pK_a = 5.3$)	($pK_a = 1.8$)

Another proposal that has been advanced to explain the acid-strengthening effect of polar substituents holds that the electron-withdrawing effect is transmitted through the water molecules that surround the carboxylate ion rather than through successive polarization of σ bonds. This is referred to as a **field effect.** Both field and inductive contributions to the polar effect tend to operate in the same direction, and it is believed that both are important.

It is a curious fact that substituents affect the entropy of ionization more than they do the enthalpy term in the expression

$$\Delta G^\circ = \Delta H^\circ - T\Delta S^\circ$$

The enthalpy term ΔH° is close to zero for the ionization of most carboxylic acids, regardless of their strength. The free energy of ionization ΔG° is dominated by the $-T\Delta S^\circ$ term. Ionization is accompanied by an increase in solvation forces, leading to a decrease in the entropy of the system; $\Delta S°$ is negative and $-T\,\Delta S°$ is positive. Anions that incorporate substituents capable of dispersing negative charge require less solvent reorganization for their stabilization, and less entropy is lost in their production than in the generation of anions that lack electron-withdrawing groups.

19.7 IONIZATION OF SUBSTITUTED BENZOIC ACIDS

A considerable body of data is available on the acidity of substituted benzoic acids. Benzoic acid itself is a somewhat stronger acid than acetic acid. Its carboxyl group is attached to an sp^2 hybridized carbon and ionizes to a greater extent than one that is attached to an sp^3 hybridized carbon. Remember, carbon becomes more electron-withdrawing as its *s* character increases.

CH_3CO_2H	$CH_2{=}CHCO_2H$	$C_6H_5CO_2H$
Acetic acid	Acrylic acid	Benzoic acid
$K_a = 1.8 \times 10^{-5}$	$K_a = 5.5 \times 10^{-5}$	$K_a = 6.3 \times 10^{-5}$
(pK_a 4.8)	(pK_a 4.3)	(pK_a 4.2)

PROBLEM 19.5 What is the most acidic neutral molecule characterized by the molecular formula $C_3H_xO_2$?

Table 19.3 lists the ionization constants of some substituted benzoic acids. The most pronounced effects are observed when strongly electron-withdrawing substituents are present at positions ortho to the carboxyl group. An *o*-nitro substituent, for example, increases the acidity of benzoic acid 100-fold. Substituent effects are small at positions meta and para to the carboxyl group. In those cases the pK_a values are clustered in the range 3.5 to 4.5.

TABLE 19.3
Acidity of Some Substituted Benzoic Acids

Substituent in $XC_6H_4CO_2H$	K_a (pK_a)* for different positions of substituent X		
	Ortho	Meta	Para
1. H	6.3×10^{-5} (4.2)	6.3×10^{-5} (4.2)	6.3×10^{-5} (4.2)
2. CH_3	1.2×10^{-4} (3.9)	5.3×10^{-5} (4.3)	4.2×10^{-5} (4.4)
3. F	5.4×10^{-4} (3.3)	1.4×10^{-4} (3.9)	7.2×10^{-5} (4.1)
4. Cl	1.2×10^{-3} (2.9)	1.5×10^{-4} (3.8)	1.0×10^{-4} (4.0)
5. Br	1.4×10^{-3} (2.8)	1.5×10^{-4} (3.8)	1.1×10^{-4} (4.0)
6. I	1.4×10^{-3} (2.9)	1.4×10^{-4} (3.9)	9.2×10^{-5} (4.0)
7. CH_3O	8.1×10^{-5} (4.1)	8.2×10^{-5} (4.1)	3.4×10^{-5} (4.5)
8. O_2N	6.7×10^{-3} (2.2)	3.2×10^{-4} (3.5)	3.8×10^{-4} (3.4)

* In water at 25°C.

19.8 DICARBOXYLIC ACIDS

Separate ionization constants, designated K_1 and K_2, respectively, characterize the two successive ionization steps of a dicarboxylic acid.

Oxalic acid is poisonous and occurs naturally in a number of plants including sorrel and begonia. It is a good idea to keep houseplants out of the reach of small children, who might be tempted to eat the leaves or berries.

$$\underset{\text{Oxalic acid}}{\mathrm{HO\overset{\displaystyle O}{\overset{\|}{C}}-\overset{\displaystyle O}{\overset{\|}{C}}OH}} \overset{K_1}{\rightleftharpoons} H^+ + \underset{\text{Hydrogen oxalate (monoanion)}}{\mathrm{HO\overset{\displaystyle O}{\overset{\|}{C}}-\overset{\displaystyle O}{\overset{\|}{C}}O^-}} \qquad K_1 = 6.5 \times 10^{-2} \quad pK_1 = 1.2$$

$$\underset{\text{Hydrogen oxalate (monoanion)}}{\mathrm{HO\overset{\displaystyle O}{\overset{\|}{C}}-\overset{\displaystyle O}{\overset{\|}{C}}O^-}} \overset{K_2}{\rightleftharpoons} H^+ + \underset{\text{Oxalate (Dianion)}}{\mathrm{{}^-O\overset{\displaystyle O}{\overset{\|}{C}}-\overset{\displaystyle O}{\overset{\|}{C}}O^-}} \qquad K_2 = 5.3 \times 10^{-5} \quad pK_2 = 4.3$$

The first ionization constant of dicarboxylic acids is larger than K_a for monocarboxylic analogs. One reason is statistical. There are two potential sites for ionization rather than one. Further, one carboxyl group acts as an electron-withdrawing group to facilitate dissociation of the other. This is particularly noticeable when the two carboxyl groups are separated by only a few bonds. Oxalic and malonic acid, for example, are several orders of magnitude stronger than simple alkyl derivatives of acetic acid. Heptanedioic acid, in which the carboxyl groups are well separated from each other, is only slightly stronger than acetic acid.

HO_2CCO_2H	$HO_2CCH_2CO_2H$	$HO_2C(CH_2)_5CO_2H$
Oxalic acid	Malonic acid	Heptanedioic acid
K_1 6.5×10^{-2}	K_1 1.4×10^{-3}	K_1 3.1×10^{-5}
(pK_1 1.2)	(pK_1 2.8)	(pK_1 4.3)

19.9 CARBONIC ACID

Through an accident of history, the simplest dicarboxylic acid, carbonic acid, $\mathrm{HO\overset{\displaystyle O}{\overset{\|}{C}}OH}$, is not even classified as an organic compound. Because many minerals are carbonate salts, nineteenth-century chemists placed carbonates, bicarbonates, and carbon dioxide in the inorganic realm. Nevertheless, the essential features of carbonic acid and its salts are easily understood on the basis of our knowledge of carboxylic acids.

Carbonic acid is formed when carbon dioxide reacts with water. Hydration of carbon dioxide is far from complete, however. Almost all the carbon dioxide that is dissolved in water exists as carbon dioxide; only 0.3 percent of it is converted to carbonic acid. Carbonic acid is a weak acid and ionizes to a small extent to bicarbonate ion.

The systematic name for bicarbonate ion is *hydrogen carbonate.* Thus, the systematic name for sodium bicarbonate ($NaHCO_3$) is *sodium hydrogen carbonate.*

$$\underset{\text{Carbon dioxide}}{CO_2} + \underset{\text{Water}}{H_2O} \rightleftharpoons \underset{\text{Carbonic acid}}{\mathrm{HO\overset{\displaystyle O}{\overset{\|}{C}}OH}} \rightleftharpoons H^+ + \underset{\text{Bicarbonate ion}}{\mathrm{HO\overset{\displaystyle O}{\overset{\|}{C}}O^-}}$$

The equilibrium constant for the overall reaction is related to an apparent equilibrium constant K_1 for carbonic acid ionization by the expression

$$K_1 = \frac{[H^+][HCO_3^-]}{[CO_2]} = 4.3 \times 10^{-7} \qquad pK_a = 6.4$$

These equations tell us that the reverse process, proton transfer from acids to bicarbonate to form carbon dioxide, will be favorable when K_a of the acid exceeds 4.3×10^{-7} ($pK_a < 6.4$). Among compounds containing carbon, hydrogen, and oxygen, only carboxylic acids are acidic enough to meet this requirement. They dissolve in aqueous sodium bicarbonate with the evolution of carbon dioxide. This behavior is the basis of a qualitative test for carboxylic acids.

PROBLEM 19.6 The value cited for the "apparent K_1" of carbonic acid, 4.3×10^{-7}, is the one normally given in reference books. It is determined by measuring the pH of water to which a known amount of carbon dioxide has been added. When we recall that only 0.3 percent of carbon dioxide is converted to carbonic acid in water, what is the "true K_1" of carbonic acid?

Carbonic anhydrase is an enzyme that catalyzes the hydration of carbon dioxide to bicarbonate. The uncatalyzed hydration of carbon dioxide is too slow to be effective in transporting carbon dioxide from the tissues to the lungs, and so mammals have developed catalysts to speed this process. The activity of carbonic anhydrase is remarkable; it has been estimated that one molecule of this enzyme can catalyze the hydration of 3.6×10^7 molecules of carbon dioxide per minute.

As with other dicarboxylic acids, the second ionization constant of carbonic acid is far smaller than the first.

$$\underset{\text{Bicarbonate ion}}{HO\overset{\overset{O}{\|}}{C}O^-} \overset{K_2}{\rightleftharpoons} H^+ + \underset{\text{Carbonate ion}}{{}^-O\overset{\overset{O}{\|}}{C}O^-}$$

The value of K_2 is 5.6×10^{-11} (pK_a 10.2). Bicarbonate is a weaker acid than carboxylic acids but a stronger acid than water and alcohols.

19.10 SOURCES OF CARBOXYLIC ACIDS

Many carboxylic acids were first isolated from natural sources and were given names based on their origin. Formic acid (Latin *formica,* "ant") was obtained by distilling ants. Since ancient times acetic acid (Latin *acetum,* "vinegar") has been known to be present in wine that has turned sour. Butyric acid (Latin *butyrum,* "butter") contributes to the odor of rancid butter, and lactic acid (Latin *lac,* "milk") has been isolated from sour milk.

While these humble origins make interesting historical notes, in most cases the large-scale preparation of carboxylic acids relies on chemical synthesis. Virtually none of the 3×10^9 lb of acetic acid produced in the United States each year is obtained from vinegar. Instead, most industrial acetic acid comes from the reaction of methanol with carbon monoxide.

$$\underset{\text{Methanol}}{CH_3OH} + \underset{\text{Carbon monoxide}}{CO} \xrightarrow[\text{heat, pressure}]{\text{cobalt or rhodium catalyst}} \underset{\text{Acetic acid}}{CH_3CO_2H}$$

The principal end use of acetic acid is in the production of vinyl acetate for paints and adhesives.

The carboxylic acid produced in the greatest amounts is 1,4-benzenedicarboxylic acid (terephthalic acid). About 5 × 10^9 lb/yr is produced in the United States as a

TABLE 19.4
Summary of Reactions Discussed in Earlier Chapters That Yield Carboxylic Acids

Reaction (section) and comments	General equation and specific example
Side-chain oxidation of alkylbenzenes (Section 11.15) A primary or secondary alkyl side chain on an aromatic ring is converted to a carboxyl group by reaction with a strong oxidizing agent such as potassium permanganate or chromic acid.	$\underset{\text{Alkylbenzene}}{ArCHR_2} \xrightarrow[K_2Cr_2O_7,\ H_2SO_4]{KMnO_4 \text{ or}} \underset{\text{Arenecarboxylic acid}}{ArCO_2H}$ 3-Methoxy-4-nitrotoluene $\xrightarrow[2.\ H^+]{1.\ KMnO_4,\ HO^-}$ 3-Methoxy-4-nitrobenzoic acid (100%)
Oxidation of primary alcohols (Section 15.10) Potassium permanganate and chromic acid convert primary alcohols to carboxylic acids by way of the corresponding aldehyde.	$\underset{\text{Primary alcohol}}{RCH_2OH} \xrightarrow[K_2Cr_2O_7,\ H_2SO_4]{KMnO_4 \text{ or}} \underset{\text{Carboxylic acid}}{RCO_2H}$ $(CH_3)_3CCH(CH_2OH)C(CH_3)_3 \xrightarrow[H_2O,\ H_2SO_4]{H_2CrO_4} (CH_3)_3CCH(CO_2H)C(CH_3)_3$ 2-*tert*-Butyl-3,3-dimethyl-1-butanol → 2-*tert*-Butyl-3,3-dimethylbutanoic acid (82%)
Oxidation of aldehydes (Section 17.16) Aldehydes are particularly sensitive to oxidation and are converted to carboxylic acids by a number of oxidizing agents, including potassium permanganate and chromic acid.	$\underset{\text{Aldehyde}}{RCHO} \xrightarrow{\text{oxidizing agent}} \underset{\text{Carboxylic acid}}{RCO_2H}$ Furan-2-carbaldehyde (furfural) $\xrightarrow[H_2SO_4,\ H_2O]{K_2Cr_2O_7}$ Furan-2-carboxylic acid (furoic acid) (75%)

starting material for the preparation of polyester fibers. One important process converts *p*-xylene to terephthalic acid by oxidation with nitric acid:

$$CH_3-C_6H_4-CH_3 \xrightarrow{HNO_3} HO_2C-C_6H_4-CO_2H$$

p-Xylene — 1,4-Benzenedicarboxylic acid (terephthalic acid)

You will recognize the side-chain oxidation of *p*-xylene to terephthalic acid as a reaction type discussed previously (Section 11.15). Examples of other reactions encountered earlier that can be applied to the synthesis of carboxylic acids are collected in Table 19.4.

The examples cited in the table give carboxylic acids that have the same number of carbon atoms as the starting material. The reactions to be described in the next two sections permit carboxylic acids to be prepared by extending a chain by one carbon atom and are of great value in laboratory syntheses of carboxylic acids.

19.11 SYNTHESIS OF CARBOXYLIC ACIDS BY THE CARBOXYLATION OF GRIGNARD REAGENTS

You have seen how Grignard reagents add to the carbonyl group of aldehydes, ketones, and esters. Grignard reagents react in much the same way with *carbon dioxide* to yield magnesium salts of carboxylic acids. Acidification converts these magnesium salts to the desired carboxylic acids.

$$\overset{\delta-}{R}-\overset{\delta+}{MgX} + O{=}C{=}O \longrightarrow RC(=O)OMgX \xrightarrow[H_2O]{H^+} RC(=O)OH$$

Grignard reagent acts as a nucleophile toward carbon dioxide — Halomagnesium carboxylate — Carboxylic acid

Overall, the carboxylation of Grignard reagents transforms an alkyl or aryl halide to a carboxylic acid in which the carbon skeleton has been extended by one carbon atom.

$$CH_3CH(Cl)CH_2CH_3 \xrightarrow[\text{2. } CO_2 \text{ 3. } H_3O^+]{\text{1. Mg, diethyl ether}} CH_3CH(CO_2H)CH_2CH_3$$

2-Chlorobutane — 2-Methylbutanoic acid (76–86%)

$$\text{9-Bromo-10-methylphenanthrene} \xrightarrow[\text{2. } CO_2 \text{ 3. } H_3O^+]{\text{1. Mg, diethyl ether}} \text{10-Methylphenanthrene-9-carboxylic acid}$$

9-Bromo-10-methylphenanthrene — 10-Methylphenanthrene-9-carboxylic acid (82%)

The major limitation to this procedure is that the alkyl or aryl halide must not bear substituents that are incompatible with Grignard reagents, such as OH, NH, SH, or C=O.

19.12 SYNTHESIS OF CARBOXYLIC ACIDS BY THE PREPARATION AND HYDROLYSIS OF NITRILES

Primary and secondary alkyl halides may be converted to the next higher carboxylic acid by a two-step synthetic sequence involving the preparation and hydrolysis of *nitriles*. Nitriles, also known as *alkyl cyanides*, are prepared by nucleophilic substitution.

$$:\ddot{X}-R + :\bar{C}\equiv N: \longrightarrow RC\equiv N + :\ddot{X}:^-$$

Primary or secondary alkyl halide; Cyanide ion; Nitrile (alkyl cyanide); Halide ion

The reaction is of the S_N2 type and works best with primary and secondary alkyl halides. Elimination is the only reaction observed with tertiary alkyl halides. Aryl and vinyl halides do not react. Dimethyl sulfoxide is the preferred solvent for this reaction, but alcohols and water-alcohol mixtures have also been used.

Once the cyano group has been introduced, the nitrile is subjected to hydrolysis. Usually this is carried out in aqueous acid at reflux.

$$RC\equiv N + 2H_2O + H^+ \xrightarrow{\text{heat}} RCOH + NH_4^+$$

(RCOH = carboxylic acid with C=O)

Nitrile; Water; Carboxylic acid; Ammonium ion

The mechanism of nitrile hydrolysis will be described in Section 20.19.

$$C_6H_5CH_2Cl \xrightarrow[\text{DMSO}]{\text{NaCN}} C_6H_5CH_2CN \xrightarrow[\text{heat}]{H_2O,\ H_2SO_4} C_6H_5CH_2COH$$

Benzyl chloride; Benzyl cyanide (92%); Phenylacetic acid (77%)

Dicarboxylic acids have been prepared from dihalides by this method:

$$BrCH_2CH_2CH_2Br \xrightarrow[H_2O]{\text{NaCN}} NCCH_2CH_2CH_2CN \xrightarrow[\text{heat}]{H_2O,\ HCl} HOCCH_2CH_2CH_2COH$$

1,3-Dibromopropane; 1,5-Pentanedinitrile (77–86%); 1,5-Pentanedioic acid (83–85%)

PROBLEM 19.7 Only one of the two procedures just described, preparation and carboxylation of a Grignard reagent or formation and hydrolysis of a nitrile, is appropriate to each of the following $RX \rightarrow RCO_2H$ conversions. Identify the correct procedure in each case, and specify why the other will fail.

(*a*) Bromobenzene → benzoic acid
(*b*) 2-Chloroethanol → 3-hydroxypropanoic acid
(*c*) *tert*-Butyl chloride → 2,2-dimethylpropanoic acid

SAMPLE SOLUTION (*a*) Bromobenzene is an aryl halide and is unreactive toward nucleophilic substitution by cyanide ion. The route $C_6H_5Br \rightarrow C_6H_5CN \rightarrow C_6H_5CO_2H$ fails because the first step fails. The route proceeding through the Grignard reagent is perfectly satisfactory and appears as an experiment in a number of introductory organic chemistry laboratory texts.

$$C_6H_5Br \xrightarrow[\text{diethyl ether}]{\text{Mg}} C_6H_5MgBr \xrightarrow[\text{2. } H_3O^+]{\text{1. } CO_2} C_6H_5CO_2H$$

Bromobenzene — Phenylmagnesium bromide — Benzoic acid

Nitrile groups in cyanohydrins (Section 17.8) are hydrolyzed under conditions similar to those of alkyl cyanides. Cyanohydrin formation followed by hydrolysis provides a route to the preparation of α-hydroxy carboxylic acids.

$$CH_3\overset{O}{\overset{\|}{C}}CH_2CH_2CH_3 \xrightarrow[\text{2. } H^+]{\text{1. NaCN}} CH_3\underset{CN}{\underset{|}{\overset{OH}{\overset{|}{C}}}}CH_2CH_2CH_3 \xrightarrow[\text{heat}]{H_2O,\ HCl} CH_3\underset{CO_2H}{\underset{|}{\overset{OH}{\overset{|}{C}}}}CH_2CH_2CH_3$$

2-Pentanone — 2-Pentanone cyanohydrin — 2-Hydroxy-2-methylpentanoic acid (60% from 2-pentanone)

19.13 REACTIONS OF CARBOXYLIC ACIDS: A REVIEW AND A PREVIEW

The most apparent chemical property of carboxylic acids, their acidity, has already been examined in earlier sections of this chapter. Three reactions of carboxylic acids —conversion to acyl chlorides, reduction, and esterification—have been encountered in previous chapters and are reviewed in Table 19.5. Acid-catalyzed esterification of carboxylic acids is one of the fundamental reactions of organic chemistry, and this portion of the chapter begins with an examination of the mechanism by which it occurs. Later, in Sections 19.16 and 19.17, two new reactions of carboxylic acids that are of synthetic value will be described.

19.14 MECHANISM OF ACID-CATALYZED ESTERIFICATION

An important question about the mechanism of acid-catalyzed esterification concerns the origin of the alkoxy oxygen. For example, does the methoxy oxygen in methyl benzoate come from methanol, or is it derived from benzoic acid?

$$C_6H_5\overset{O}{\overset{\|}{C}}OCH_3$$

Is this the oxygen originally present in benzoic acid, or is it the oxygen of methanol?

TABLE 19.5
Summary of Reactions of Carboxylic Acids Discussed in Earlier Chapters

Reaction (section) and comments	General equation and specific example
Formation of acyl chlorides (Section 12.7) Thionyl chloride reacts with carboxylic acids to yield acyl chlorides.	$RCO_2H + SOCl_2 \longrightarrow RCCl(=O) + SO_2 + HCl$ Carboxylic acid; Thionyl chloride; Acyl chloride; Sulfur dioxide; Hydrogen chloride m-$CH_3OC_6H_4CH_2CO_2H \xrightarrow[\text{heat}]{SOCl_2}$ m-$CH_3OC_6H_4CH_2CCl(=O)$ *m*-Methoxyphenylacetic acid → *m*-Methoxyphenylacetyl chloride (85%)
Lithium aluminum hydride reduction (Section 15.3) Carboxylic acids are reduced to primary alcohols by the powerful reducing agent lithium aluminum hydride.	$RCO_2H \xrightarrow[\text{2. } H_2O]{\text{1. } LiAlH_4\text{, diethyl ether}} RCH_2OH$ Carboxylic acid; Primary alcohol p-$F_3CC_6H_4CO_2H \xrightarrow[\text{2. } H_2O]{\text{1. } LiAlH_4\text{, diethyl ether}}$ p-$F_3CC_6H_4CH_2OH$ *p*-(Trifluoromethyl)benzoic acid → *p*-(Trifluoromethyl)benzyl alcohol (96%)
Esterification (Section 15.8) In the presence of an acid catalyst, carboxylic acids and alcohols react to form esters. The reaction is an equilibrium process but can be driven to favor the ester by removing the water that is formed.	$RCO_2H + R'OH \overset{H^+}{\rightleftharpoons} RCOR'(=O) + H_2O$ Carboxylic acid; Alcohol; Ester; Water $C_6H_5CO_2H + CH_3OH \xrightarrow{H_2SO_4} C_6H_5COCH_3(=O)$ Benzoic acid; Methanol; Methyl benzoate (70%)

The answer to this question is critical to mechanistic understanding, because it tells us whether it is the carbon-oxygen bond of the alcohol or a carbon-oxygen of the carboxylic acid that is broken during the esterification.

A clear-cut answer was provided by Irving Roberts and Harold C. Urey of Columbia University in 1938. They prepared methanol that had been enriched in the mass 18 isotope of oxygen. When this sample of methanol was esterified with benzoic acid, the methyl benzoate product contained all the ^{18}O label that was originally present in the methanol.

The overall reaction:

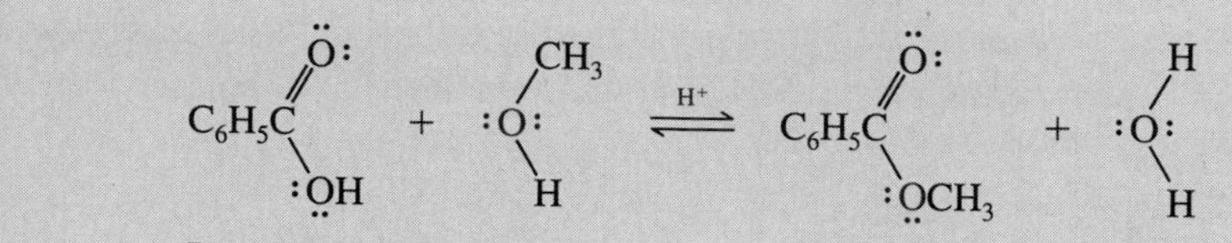

$$C_6H_5C(=\ddot{O}:)\ddot{O}H + CH_3\ddot{O}H \overset{H^+}{\rightleftharpoons} C_6H_5C(=\ddot{O}:)\ddot{O}CH_3 + H\ddot{O}H$$

Benzoic acid — Methanol — Methyl benzoate — Water

Step 1: The carboxylic acid is protonated on its carbonyl oxygen. The proton donor shown in the equation for this step is an alkyloxonium ion formed by proton transfer from the acid catalyst to the alcohol.

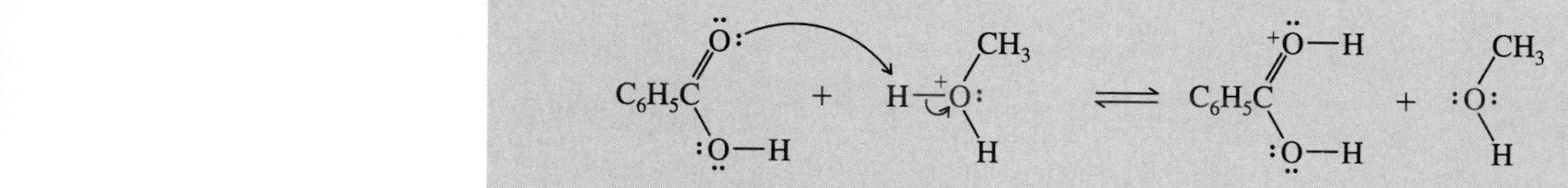

$$C_6H_5C(=\ddot{O}:)\ddot{O}{-}H + H{-}\overset{+}{O}(CH_3)H \rightleftharpoons C_6H_5C(=\overset{+}{\ddot{O}}{-}H)\ddot{O}{-}H + CH_3\ddot{O}H$$

Benzoic acid — Methyloxonium ion — Conjugate acid of benzoic acid — Methanol

Step 2: Protonation of the carboxylic acid increases the positive character of its carbonyl group. A molecule of the alcohol acts as a nucleophile and attacks the carbonyl carbon.

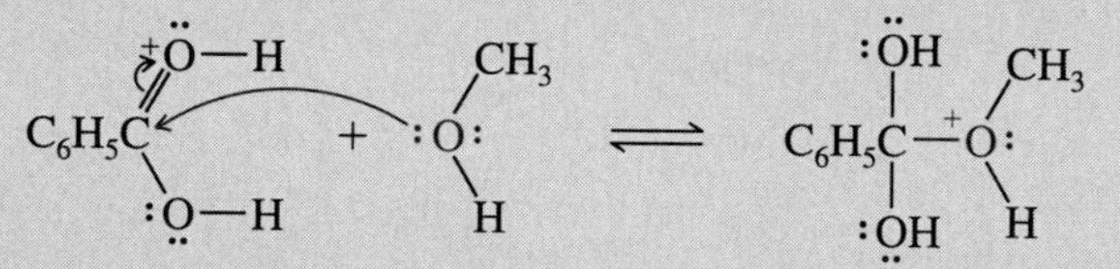

$$C_6H_5C(=\overset{+}{\ddot{O}}{-}H)\ddot{O}{-}H + CH_3\ddot{O}H \rightleftharpoons C_6H_5C(\ddot{O}H)_2{-}\overset{+}{\ddot{O}}(CH_3)H$$

Conjugate acid of benzoic acid — Methanol — Protonated form of tetrahedral intermediate

Step 3: The oxonium ion formed in step 2 loses a proton to give the tetrahedral intermediate in its neutral form. This step concludes the first stage in the mechanism.

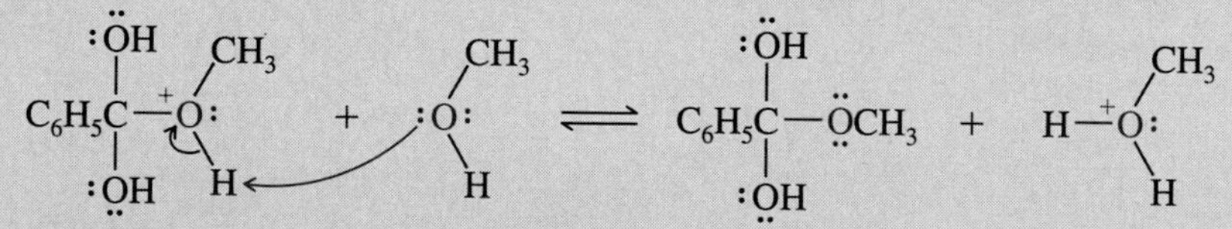

$$C_6H_5C(\ddot{O}H)_2{-}\overset{+}{\ddot{O}}(CH_3)H + CH_3\ddot{O}H \rightleftharpoons C_6H_5C(\ddot{O}H)_2{-}\ddot{O}CH_3 + H{-}\overset{+}{O}(CH_3)H$$

Protonated form of tetrahedral intermediate — Methanol — Tetrahedral intermediate — Methyloxonium ion

FIGURE 19.6 Sequence of steps that describes the mechanism of acid-catalyzed esterification.

In this equation, the red O signifies oxygen enriched in its mass 18 isotope; analysis of isotopic enrichment was performed by mass spectrometry.

$$C_6H_5\overset{O}{\overset{\|}{C}}OH + CH_3OH \xrightarrow{H^+} C_6H_5\overset{O}{\overset{\|}{C}}OCH_3 + H_2O$$

Benzoic acid — ^{18}O-enriched methanol — ^{18}O-enriched methyl benzoate — Water

The results of the Roberts-Urey experiment tell us that the C—O bond of the alcohol is preserved during esterification. The oxygen that is lost as a water molecule must come from the carboxylic acid. Any mechanism proposed for this reaction must be consistent with these observations.

Step 4: The second stage begins with protonation of the tetrahedral intermediate on one of its hydroxyl oxygens.

$$C_6H_5C(OH)_2-OCH_3 + H-\overset{+}{O}(CH_3)H \rightleftharpoons C_6H_5C(OH)(\overset{+}{O}H_2)-OCH_3 + CH_3OH$$

Tetrahedral intermediate | Methyloxonium ion | Hydroxyl-protonated tetrahedral intermediate | Methanol

Step 5: This intermediate loses a molecule of water to give the protonated form of the ester.

$$C_6H_5C(OH)(\overset{+}{O}H_2)-OCH_3 \rightleftharpoons C_6H_5C(=\overset{+}{O}H)OCH_3 + H_2O$$

Hydroxyl-protonated tetrahedral intermediate | Conjugate acid of methyl benzoate | Water

Step 6: Deprotonation of the species formed in step 5 gives the neutral form of the ester product.

$$C_6H_5C(=\overset{+}{O}H)OCH_3 + CH_3OH \rightleftharpoons C_6H_5C(=O)OCH_3 + H-\overset{+}{O}(CH_3)H$$

Conjugate acid of methyl benzoate | Methanol | Methyl benzoate | Methyloxonium ion

FIGURE 19.6 *(Continued)*

A mechanism consistent with these facts is presented in Figure 19.6. The six steps are best viewed as a combination of two distinct stages. *Formation* of a **tetrahedral intermediate** characterizes the first stage (steps 1 to 3), and *dissociation* of this tetrahedral intermediate characterizes the second (steps 4 to 6).

$$C_6H_5C(=O)OH + CH_3OH \underset{H^+}{\overset{\text{steps 1-3}}{\rightleftharpoons}} C_6H_5C(OH)_2-OCH_3 \underset{H^+}{\overset{\text{steps 4-6}}{\rightleftharpoons}} C_6H_5C(=O)OCH_3 + H_2O$$

Benzoic acid | Methanol | Tetrahedral intermediate | Methyl benzoate | Water

The species connecting the two stages is called a *tetrahedral intermediate* because the hybridization at carbon has changed from sp^2 in the carboxylic acid to sp^3 in the intermediate before returning to sp^2 in the ester product. *The tetrahedral intermediate is formed by nucleophilic addition of an alcohol to a carboxylic acid and is analogous to a hemiacetal formed by nucleophilic addition of an alcohol to an aldehyde or a ketone.* The three steps that lead to the tetrahedral intermediate in the first stage of esterification are analogous to those in the mechanism for acid-catalyzed nucleophilic addition of an alcohol to an aldehyde or a ketone. The tetrahedral intermediate cannot be isolated. It is unstable under the conditions of its formation and undergoes acid-catalyzed dehydration to form the ester.

Notice that the oxygen of methanol becomes incorporated into the methyl benzoate product according to the mechanism outlined in Figure 19.6, as the observations of the Roberts-Urey experiment require it to be.

Notice, too, that it is the carbonyl oxygen of the carboxylic acid that is protonated in the first step and not the hydroxyl oxygen. The species formed by protonation of the carbonyl oxygen is more stable, because it is stabilized by electron delocalization. The positive charge is shared equally by both oxygens.

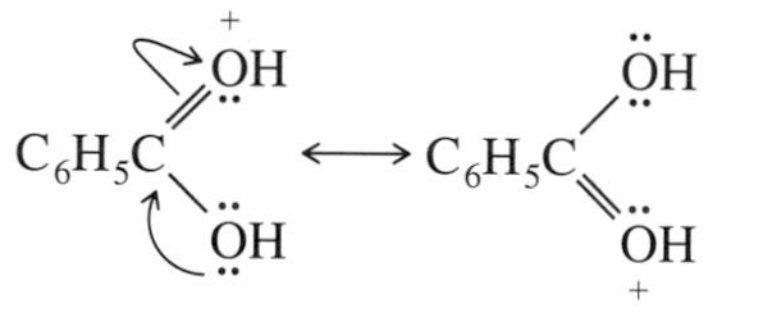

$$C_6H_5C(=\overset{+}{O}H)(\ddot{O}H) \longleftrightarrow C_6H_5C(\ddot{O}H)(=\overset{+}{O}H)$$

Electron delocalization in carbonyl-protonated benzoic acid

Protonation of the hydroxyl oxygen, on the other hand, yields a less stable cation:

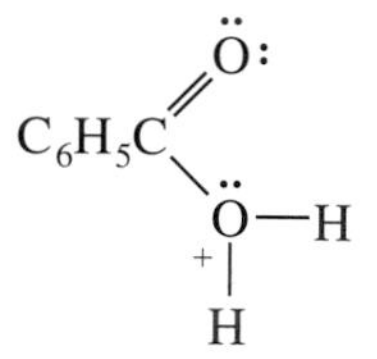

$$C_6H_5C(=\ddot{O}:)-\overset{+}{O}H_2$$

Localized positive charge in hydroxyl-protonated benzoic acid

The positive charge in this cation cannot be shared by the two oxygens; it is localized on one of them. Since protonation of the carbonyl oxygen gives a more stable cation, it is that cation that is formed preferentially.

PROBLEM 19.8 When benzoic acid is allowed to stand in water that is enriched in ^{18}O, the isotopic label becomes incorporated into the benzoic acid. The reaction is catalyzed by acids. Suggest an explanation for this observation.

In the next chapter the three elements of the mechanism just described will be seen again as part of the general theme that unites the chemistry of carboxylic acid derivatives. These elements are

1. Activation of the carbonyl group by protonation of the carbonyl oxygen
2. Nucleophilic addition to the protonated carbonyl to form a tetrahedral intermediate
3. Elimination from the tetrahedral intermediate to restore the carbonyl group

This sequence is one of the fundamental mechanistic patterns of organic chemistry.

19.15 INTRAMOLECULAR ESTER FORMATION: LACTONES

Hydroxy acids, i.e., compounds that contain both a hydroxyl and a carboxylic acid function within the same molecule, have the capacity to form cyclic esters called *lactones.* This intramolecular esterification takes place spontaneously and is especially favorable when the ring that is formed is five-membered or six-membered. Lactones that contain a five-membered cyclic ester are referred to as **γ-lactones** because they arise from γ-hydroxy carboxylic acids. Their six-membered analogs are known as **δ-lactones.**

$$HOCH_2CH_2CH_2\overset{O}{\overset{\|}{C}}OH \longrightarrow \text{(4-butanolide ring)}$$

4-Hydroxybutanoic acid — 4-Butanolide

$$HOCH_2CH_2CH_2CH_2\overset{O}{\overset{\|}{C}}OH \longrightarrow \text{(5-pentanolide ring)}$$

5-Hydroxypentanoic acid — 5-Pentanolide

A lactone is named by replacing the *-oic acid* ending of the parent carboxylic acid by *-olide* and identifying its oxygenated carbon by number. This system is illustrated in the lactones shown in the preceding equations. Both 4-butanolide and 5-pentanolide are better known by their common names, γ-butyrolactone and δ-valerolactone, respectively, and these two common names are permitted by the IUPAC rules.

Reactions that are expected to produce hydroxy acids often yield the derived lactones instead if a five- or six-membered ring can be formed.

$$CH_3\overset{O}{\overset{\|}{C}}CH_2CH_2CH_2CO_2H \xrightarrow[\text{2. } H_2O,\ H^+]{\text{1. } NaBH_4} \text{(5-hexanolide, } H_3C\text{)} \quad \text{via} \quad CH_3\underset{OH}{\underset{|}{C}}HCH_2CH_2CH_2CO_2H$$

5-Oxohexanoic acid — 5-Hexanolide (78%) — 5-Hydroxyhexanoic acid

Many natural products are lactones, and it is not unusual to find examples in which the ring size is rather large. A few naturally occurring lactones are shown in Figure 19.7. The *macrolide antibiotics,* of which erythromycin is one example, are macrocyclic (large-ring) lactones. The lactone ring of erythromycin is 14-membered.

PROBLEM 19.9 Write the structure of the hydroxy acid corresponding to each of the following lactones. The structure of each lactone is given in Figure 19.7.

(*a*) Mevalonolactone (*b*) Pentadecanolide (*c*) Vernolepin

SAMPLE SOLUTION (*a*) The ring oxygen of the lactone is derived from the hydroxyl group of the hydroxy acid, while the carbonyl group corresponds to that of the carboxyl function. To identify the hydroxy acid, disconnect the O—C(O) bond of the ester.

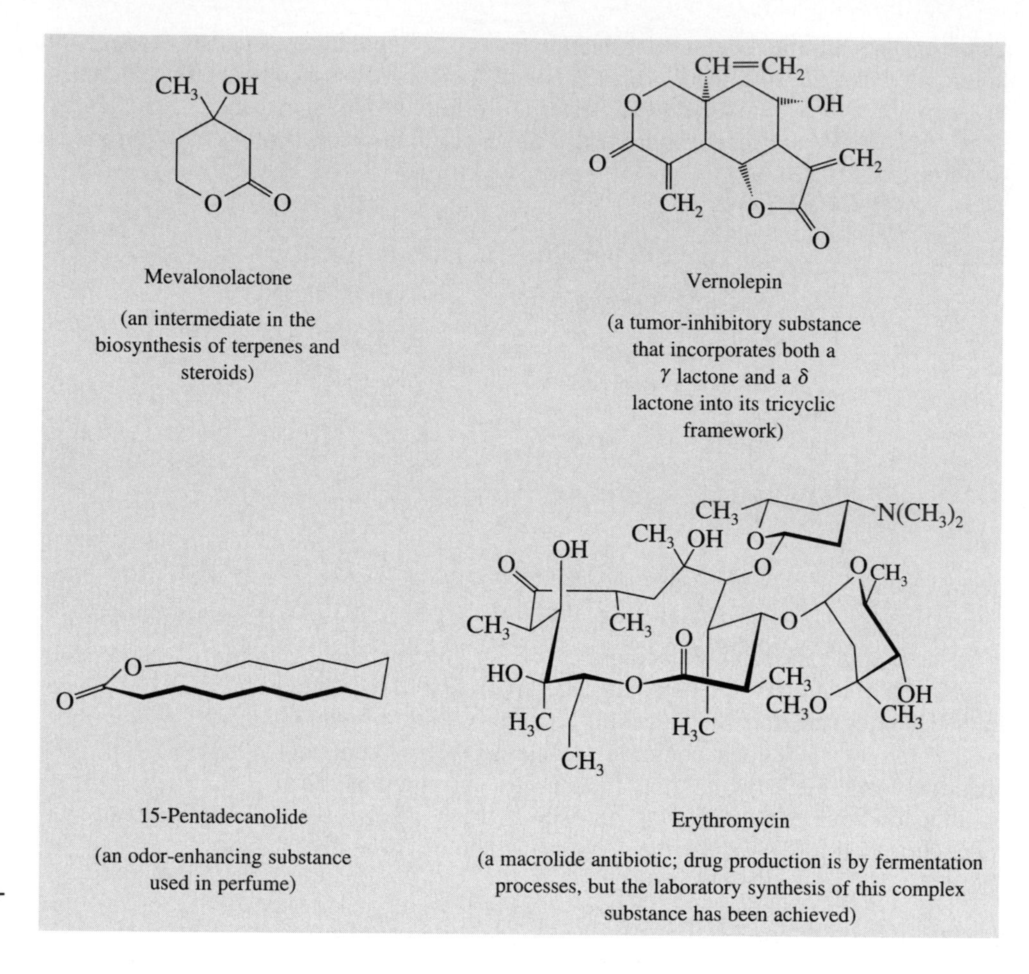

Mevalonolactone (an intermediate in the biosynthesis of terpenes and steroids)

Vernolepin (a tumor-inhibitory substance that incorporates both a γ lactone and a δ lactone into its tricyclic framework)

15-Pentadecanolide (an odor-enhancing substance used in perfume)

Erythromycin (a macrolide antibiotic; drug production is by fermentation processes, but the laboratory synthesis of this complex substance has been achieved)

FIGURE 19.7 Some naturally occurring lactones.

Mevalonolactone (disconnect bond indicated) ⟹ $HOCH_2CH_2C(CH_3)(OH)CH_2COOH$

Mevalonic acid

The compound *anisatin* is an example, perhaps the only example, of a naturally occurring β-lactone. Its isolation and structure determination were described in the journal *Tetrahedron Letters* (1982), p. 5111.

Lactones whose rings are three- or four-membered (α-lactones and β-lactones) are very reactive, making their isolation difficult. Special methods are normally required for the laboratory synthesis of small-ring lactones as well as those that contain rings larger than six-membered.

19.16 α HALOGENATION OF CARBOXYLIC ACIDS. THE HELL-VOLHARD-ZELINSKY REACTION

Esterification of carboxylic acids involves nucleophilic addition to the carbonyl group as a key step. In this respect the carbonyl group of a carboxylic acid resembles that of an aldehyde or a ketone. Do carboxylic acids resemble aldehydes and ketones in other ways? Do they, for example, form *enols,* and can they be halogenated at their α carbon atom via an enol in the way that aldehydes and ketones can?

The enol content of carboxylic acids is far less than that of aldehydes and ketones, and introduction of a halogen substituent at the α carbon atom requires a different set of reaction conditions. Bromination is the reaction that is normally carried out, and the usual procedure involves treatment of the carboxylic acid with bromine in the presence of a small amount of phosphorus trichloride as a catalyst for enolization.

$$\underset{\text{Carboxylic acid}}{R_2C(H)CO_2H} + \underset{\text{Bromine}}{Br_2} \xrightarrow{PCl_3} \underset{\alpha\text{-Bromo carboxylic acid}}{R_2C(Br)CO_2H} + \underset{\text{Hydrogen bromide}}{HBr}$$

$$\underset{\text{Phenylacetic acid}}{C_6H_5CH_2\overset{O}{\overset{\|}{C}}OH} \xrightarrow[\text{benzene, 80°C}]{Br_2,\ PCl_3} \underset{\alpha\text{-Bromophenylacetic acid (60–62\%)}}{C_6H_5CH(Br)\overset{O}{\overset{\|}{C}}OH}$$

This method of α bromination of carboxylic acids is called the **Hell-Volhard-Zelinsky** reaction. The Hell-Volhard-Zelinsky reaction is sometimes carried out by using a small amount of phosphorus instead of phosphorus trichloride as the catalyst. Phosphorus reacts with bromine to yield phosphorus tribromide as the active catalyst under these conditions.

The Hell-Volhard-Zelinsky reaction is of synthetic value in that the α halogen can be displaced by nucleophilic substitution:

$$\underset{\text{Butanoic acid}}{CH_3CH_2CH_2CO_2H} \xrightarrow[P]{Br_2} \underset{\text{2-Bromobutanoic acid (77\%)}}{CH_3CH_2CH(Br)CO_2H} \xrightarrow[H_2O,\ \text{heat}]{K_2CO_3} \underset{\text{2-Hydroxybutanoic acid (69\%)}}{CH_3CH_2CH(OH)CO_2H}$$

A standard method for the preparation of an α-amino acid uses α-bromo carboxylic acids as the substrate and aqueous ammonia as the nucleophile:

$$\underset{\text{3-Methylbutanoic acid}}{(CH_3)_2CHCH_2CO_2H} \xrightarrow[PCl_3]{Br_2} \underset{\text{2-Bromo-3-methylbutanoic acid (88\%)}}{(CH_3)_2CHCH(Br)CO_2H} \xrightarrow[H_2O]{NH_3} \underset{\text{2-Amino-3-methylbutanoic acid (48\%)}}{(CH_3)_2CHCH(NH_2)CO_2H}$$

PROBLEM 19.10 α-Iodo acids are not normally prepared by direct iodination of carboxylic acids under conditions of the Hell-Volhard-Zelinsky reaction. Show how you could convert octadecanoic acid to its 2-iodo derivative by an efficient sequence of reactions.

19.17 DECARBOXYLATION OF MALONIC ACID AND RELATED COMPOUNDS

The loss of a molecule of carbon dioxide from a carboxylic acid is known as a **decarboxylation** reaction.

$$\underset{\text{Carboxylic acid}}{RCO_2H} \longrightarrow \underset{\text{Alkane}}{RH} + \underset{\text{Carbon dioxide}}{CO_2}$$

Decarboxylation of simple carboxylic acids takes place with great difficulty and is rarely encountered.

Compounds that do undergo ready thermal decarboxylation include those related to malonic acid. On being heated above its melting point, malonic acid is converted to acetic acid and carbon dioxide.

$$\underset{\substack{\text{Malonic acid}\\\text{(propanedioic acid)}}}{HO_2CCH_2CO_2H} \xrightarrow{150°C} \underset{\substack{\text{Acetic acid}\\\text{(ethanoic acid)}}}{CH_3CO_2H} + \underset{\text{Carbon dioxide}}{CO_2}$$

It is important to recognize that only one carboxyl group is lost in this process. The second carboxyl group is retained. A mechanism recognizing the assistance that one carboxyl group gives to the departure of the other is represented by the equation

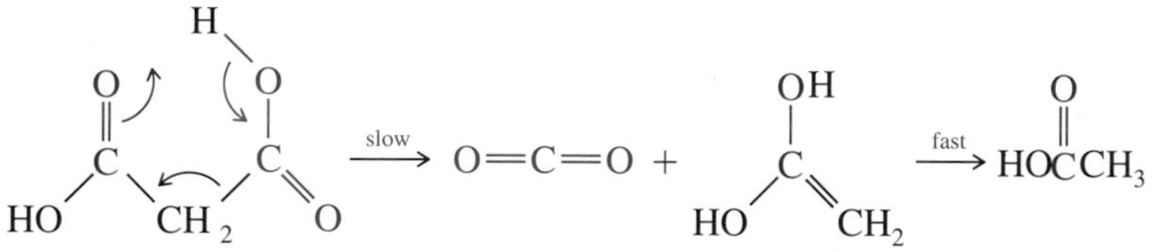

$$HOC(=O)CH_2C(=O)OH \xrightarrow{\text{slow}} \underset{\text{Carbon dioxide}}{O{=}C{=}O} + \underset{\substack{\text{Enol form of}\\\text{acetic acid}}}{HO(OH)C{=}CH_2} \xrightarrow{\text{fast}} \underset{\text{Acetic acid}}{HO\overset{O}{\overset{\|}{C}}CH_3}$$

The transition state involves the carbonyl oxygen of one carboxyl group—the one that stays behind—acting as a proton acceptor toward the hydroxyl group of the carboxyl that is lost. Carbon-carbon bond cleavage leads to the enol form of acetic acid, along with an equivalent amount of carbon dioxide.

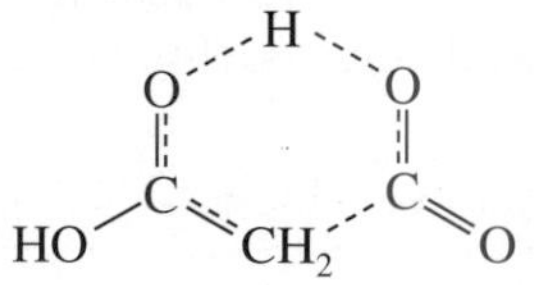

Representation of transition state in thermal decarboxylation of malonic acid

The enol intermediate is subsequently transformed to acetic acid by proton transfer processes.

The protons attached to C-2 of malonic acid are not directly involved in the process and so may be replaced by other substituents without much effect on the ease of decarboxylation. Analogs of malonic acid substituted at C-2 undergo efficient thermal decarboxylation.

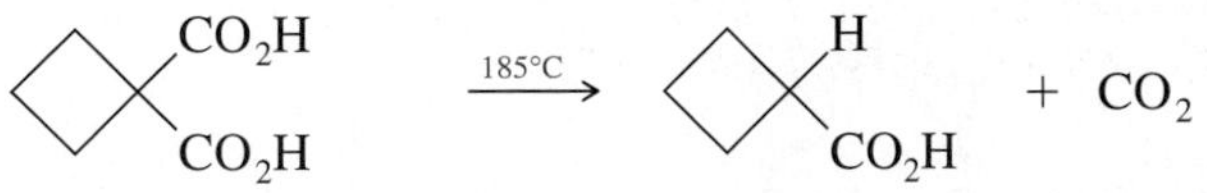

$$\text{1,1-Cyclobutanedicarboxylic acid} \xrightarrow{185°C} \text{Cyclobutanecarboxylic acid (74\%)} + CO_2 \text{ (Carbon dioxide)}$$

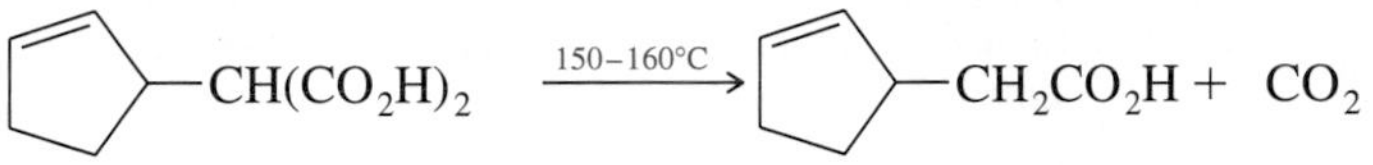

$$\text{2-(2-Cyclopentenyl)malonic acid } C_5H_7CH(CO_2H)_2 \xrightarrow{150-160°C} \text{(2-Cyclopentenyl)acetic acid (96–99\%) } C_5H_7CH_2CO_2H + CO_2 \text{ (Carbon dioxide)}$$

PROBLEM 19.11 What will be the product isolated after thermal decarboxylation of each of the following? Using curved arrows, represent the bond changes that take place at the transition state.

(*a*) $(CH_3)_2C(CO_2H)_2$

(*b*) $CH_3(CH_2)_6CHCO_2H$ with CO_2H on the CH carbon

(*c*) $C_6H_5C(CO_2H)_2CH_3$ (phenyl carbon bearing two CO_2H groups and a CH_3)

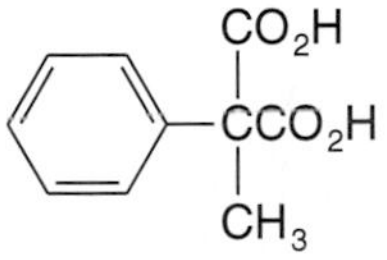

SAMPLE SOLUTION (*a*) Thermal decarboxylation of malonic acid derivatives leads to the replacement of one of the carboxyl groups by a hydrogen.

$$(CH_3)_2C(CO_2H)_2 \xrightarrow{\text{heat}} (CH_3)_2CHCO_2H + CO_2$$

2,2-Dimethylmalonic acid → 2-Methylpropanoic acid + Carbon dioxide

The transition state incorporates a cyclic array of six atoms:

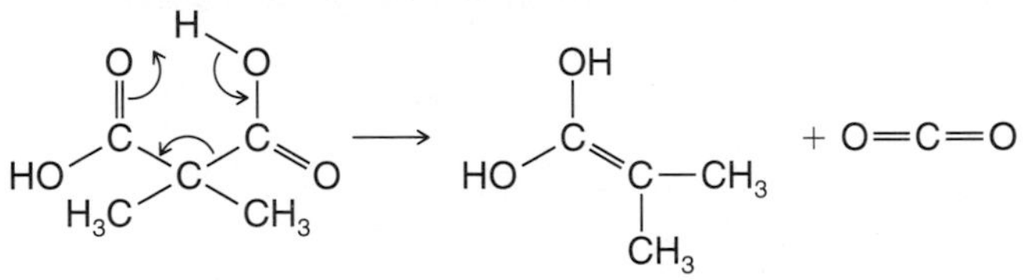

$$HO(O{=})C{-}C(CH_3)_2{-}C(=O)OH \longrightarrow HO(HO)C{=}C(CH_3){-}CH_3 + O{=}C{=}O$$

2,2-Dimethylmalonic acid → Enol form of 2-methylpropanoic acid + Carbon dioxide

Tautomerization of the enol form to 2-methylpropanoic acid completes the process.

The thermal decarboxylation of malonic acid derivatives is the last step in a multistep synthesis of carboxylic acids known as the *malonic ester synthesis*. This synthetic method will be described in Section 21.7.

Notice that the carboxyl group that stays behind during the decarboxylation of malonic acid has a hydroxyl function that is not directly involved in the process. Compounds that have substituents other than hydroxyl groups at this position undergo an analogous decarboxylation.

$$HO-C(=O)-CH_2-C(=O)-O-H \quad \text{(cyclic arrangement, curved arrows)}$$

Bonding changes during decarboxylation of malonic acid

$$R-C(=O)-CH_2-C(=O)-O-H \quad \text{(cyclic arrangement, curved arrows)}$$

Bonding changes during decarboxylation of a β-keto acid

The compounds most frequently encountered in this reaction are β-keto acids, that is, carboxylic acids in which the β carbon is a carbonyl function. Decarboxylation of β-keto acids leads to ketones.

$$\underset{\beta\text{-Keto acid}}{RCCH_2CO_2H}\ (\text{C=O on first C}) \xrightarrow{\text{heat}} \underset{\text{Carbon dioxide}}{CO_2} + \underset{\text{Enol form of ketone}}{RC(OH)=CH_2} \xrightarrow{\text{fast}} \underset{\text{Ketone}}{RCCH_3}\ (\text{C=O})$$

$$\underset{\text{2,2-Dimethylacetoacetic acid}}{CH_3C(=O)C(CH_3)_2CO_2H} \xrightarrow{25°C} \underset{\text{3-Methyl-2-butanone}}{CH_3CCH(CH_3)_2}\ (\text{C=O}) + \underset{\text{Carbon dioxide}}{CO_2}$$

PROBLEM 19.12 Show the bonding changes that occur and write the structure of the intermediate formed in the thermal decarboxylation of

(*a*) Benzoylacetic acid (*b*) 2,2-Dimethylacetoacetic acid

SAMPLE SOLUTION (*a*) By analogy to the thermal decarboxylation of malonic acid, we represent the corresponding reaction of benzoylacetic acid as

$$\underset{\text{Benzoylacetic acid}}{C_6H_5C(=O)-CH_2-C(=O)-O-H} \longrightarrow \underset{\text{Enol form of acetophenone}}{C_6H_5C(OH)=CH_2} + \underset{\text{Carbon dioxide}}{O=C=O}$$

Acetophenone is the isolated product; it is formed from its enol by proton-transfer reactions.

The thermal decarboxylation of β-keto acids is the last step in a ketone synthesis known as the *acetoacetic ester synthesis.* The acetoacetic ester synthesis is discussed in Section 21.6.

19.18 SPECTROSCOPIC ANALYSIS OF CARBOXYLIC ACIDS

The infrared and proton nuclear magnetic resonance spectra of a representative carboxylic acid, 4-phenylbutanoic acid, are shown in Figure 19.8. The hydroxyl absorp-

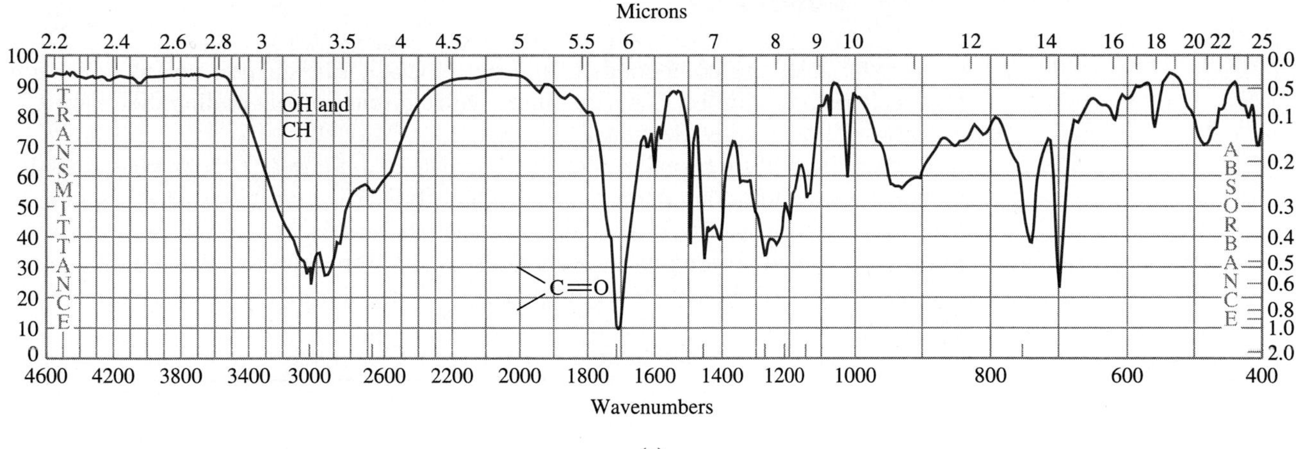

(*a*)

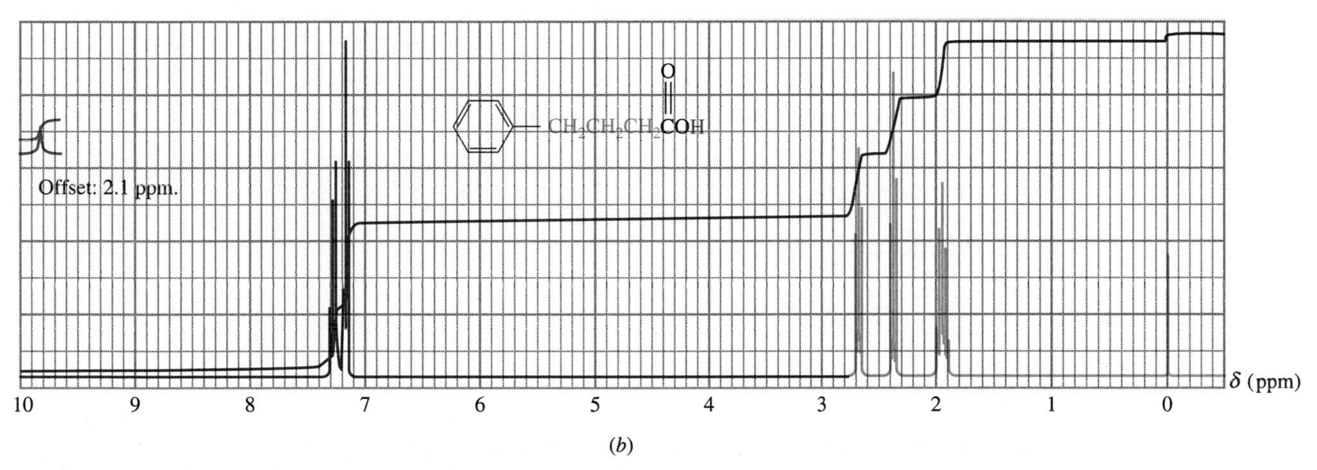

(*b*)

FIGURE 19.8 The infrared (top) and 300-MHz ^{1}H nmr (bottom) spectra of 4-phenylbutanoic acid.

tions overlap with the C—H stretching frequencies to produce a broad absorption in the 3500 to 2500 cm^{-1} region of the infrared spectrum. The carbonyl group gives rise to a strong band at 1700 cm^{-1}. In general, the position of the carbonyl absorption of carboxylic acids is found in the range 1650 to 1740 cm^{-1}.

The hydroxyl proton of a carboxyl group is normally the least shielded of all the protons of an nmr spectrum. Carboxyl protons appear 10 to 12 ppm downfield from tetramethylsilane (TMS), often as a broad peak. Since the customary sweep width of a ^{1}H nmr spectrum is 10 ppm, the spectrum must be offset as shown in Figure 19.8 in order to display this signal on the chart paper. The chemical shift of the carboxyl proton is 11.9 ppm and is equal to the sum of the offset (2.1 ppm) and the observed chart position (9.8 ppm). As with other acidic protons, the carboxyl proton can be identified by adding D_2O to the sample. Hydrogen-deuterium exchange converts —CO_2H to —CO_2D, and the signal corresponding to the carboxyl proton disappears.

The chemical shift of the carbonyl carbon of carboxylic acids is not as far downfield as that of an aldehyde or ketone carbonyl. While aldehyde and ketone carbonyl carbons appear between 190 and 215 ppm, carboxylic acid carbonyl carbons absorb between 160 and 185 ppm.

19.19 SUMMARY

Carboxylic acids are compounds of the type RCOH (with C=O). The chapter begins with their nomenclature (Section 19.1) and continues with a discussion of bonding (Section 19.2) and physical properties (Section 19.3).

Carboxylic acids are weak acids and dissociate to a modest extent in water, but are deprotonated almost completely at pH = 7 and above (Section 19.4). The dissociation constants K_a of most carboxylic acids are in the 10^{-4} to 10^{-5} range. The greater acidity of carboxylic acids compared with alcohols results from the inductive electron-withdrawing powers of the carbonyl group and electron delocalization in the carboxylate anion.

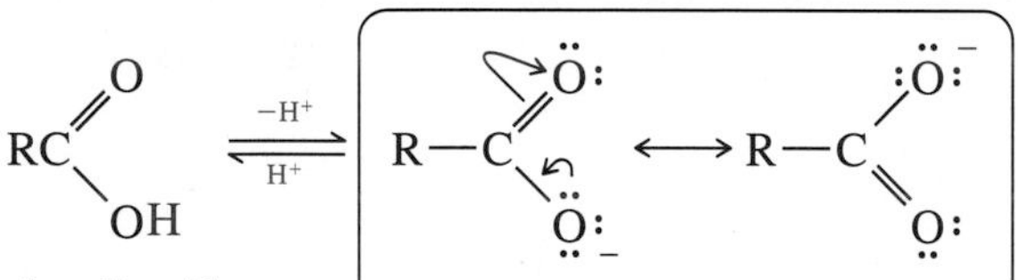

$$\mathrm{RC(=O)OH} \underset{\mathrm{H^+}}{\overset{-\mathrm{H^+}}{\rightleftharpoons}} \left[\mathrm{R{-}C(=\ddot{O}:)\ddot{O}:^-} \longleftrightarrow \mathrm{R{-}C(\ddot{O}:^-)(=\ddot{O}:)} \right]$$

Carboxylic acid

Resonance description of electron delocalization in carboxylate anion

Electron-withdrawing substituents, especially when located close to the carboxyl group, increase the acidity of carboxylic acids (Sections 19.6 through 19.9). Trifluoroacetic acid and 2,4,6-trinitrobenzoic acid are relatively strong acids—stronger than HF, for example.

CF_3CO_2H	$O_2N{-}C_6H_2(NO_2)_2{-}CO_2H$
Trifluoroacetic acid	2,4,6-Trinitrobenzoic acid
$K_a = 5.9 \times 10^{-1}$	$K_a = 2.2 \times 10^{-1}$
($pK_a = 0.2$)	($pK_a = 0.6$)

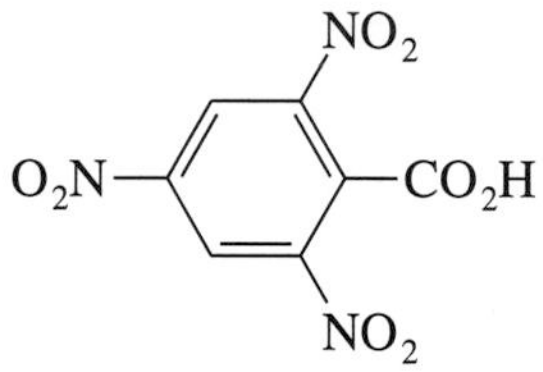

Carbon dioxide and carbonic acid are components of a hydration-dehydration equilibrium in water. The major species present is carbon dioxide (Section 19.9).

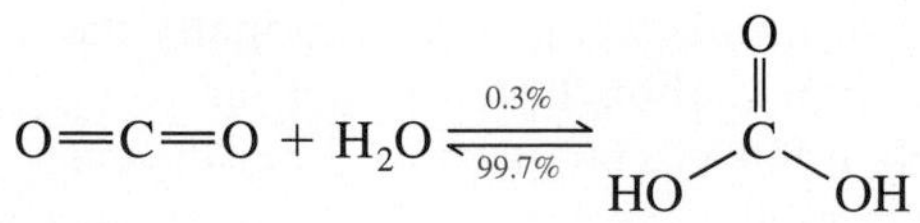

$$\mathrm{O{=}C{=}O} + \mathrm{H_2O} \underset{99.7\%}{\overset{0.3\%}{\rightleftharpoons}} \mathrm{HO{-}C(=O){-}OH}$$

Several reactions that have been encountered earlier lead to carboxylic acids and can be used for their synthesis. These are summarized in Table 19.4 (Section 19.10). Two new methods were introduced in this chapter and are particularly useful:

1. Carboxylation of Grignard reagents (Section 19.11)
2. Preparation and hydrolysis of nitriles (Section 19.12)

Grignard reagents are nucleophilic and add to carbon dioxide to yield carboxylic acids after acidification:

$$\underset{\text{Alkyl halide}}{RX} \xrightarrow[\text{diethyl ether}]{Mg} \underset{\text{Grignard reagent}}{RMgX} \xrightarrow{CO_2} \underset{\text{Halomagnesium carboxylate}}{R\overset{O}{\overset{\|}{C}}OMgX} \xrightarrow[H^+]{H_2O} \underset{\text{Carboxylic acid}}{RCO_2H}$$

Primary and secondary alkyl halides undergo nucleophilic substitution by cyanide ion to form nitriles. Nitriles are converted to carboxylic acids by hydrolysis:

$$\underset{\text{Primary or secondary alkyl halide}}{RX} \xrightarrow{CN^-} \underset{\text{Nitrile}}{RCN} \xrightarrow[H^+]{H_2O} \underset{\text{Carboxylic acid}}{RCO_2H}$$

Carboxylic acids can be transformed to acyl chlorides with thionyl chloride, to primary alcohols by reduction with lithium aluminum hydride, and to esters by acid-catalyzed condensation with alcohols (Section 19.13, Table 19.5). The mechanism of acid-catalyzed esterification involves some key elements that are important to understanding the chemistry of derivatives of carboxylic acids (Section 19.14):

$$RC(=O)OH + H^+ \rightleftharpoons RC(=\overset{+}{O}H)OH \overset{R'OH}{\rightleftharpoons} RC(OH)_2\text{—}\overset{+}{O}(R')H \rightleftharpoons \boxed{RC(OH)_2\text{—}OR'} + H^+$$

$$\boxed{RC(OH)_2\text{—}OR'} \rightleftharpoons$$

$$RC(=O)OR' + H^+ \rightleftharpoons R\overset{\overset{+}{O}\text{—}H}{\overset{\|}{C}}OR' \overset{-H_2O}{\rightleftharpoons} R\overset{OH}{\overset{|}{C}}OR'\text{—}\overset{+}{O}H_2$$

Protonation of the carbonyl oxygen activates the carbonyl group toward nucleophilic addition. Addition of an alcohol gives a tetrahedral intermediate (shown in the box in the preceding equation), which has the capacity to revert to starting materials or to undergo dehydration to yield an ester. It will be seen in the next chapter that esters can be hydrolyzed to carboxylic acids and alcohols under conditions of acid catalysis. The mechanism of that reaction is precisely the reverse of the one shown in the preceding equation.

An intramolecular esterification can occur when a molecule contains both a hydroxyl and a carboxyl group. Cyclic esters are called *lactones* and are most stable when the ring is five- or six-membered (Section 19.15).

Halogenation at the α carbon atom of carboxylic acids can be accomplished by the *Hell-Volhard-Zelinsky reaction* (Section 19.16). An acid is treated with chlorine or bromine in the presence of a catalytic quantity of phosphorus or a phosphorus trihalide:

$$\underset{\text{Carboxylic acid}}{R_2CHCO_2H} + \underset{\text{Halogen}}{X_2} \xrightarrow{\text{P or } PX_3} \underset{\alpha\text{-Halo acid}}{R_2C(X)CO_2H} + H{-}X$$

This reaction is of synthetic value in that α-halo acids are reactive substrates in nucleophilic substitution reactions.

1,1-Dicarboxylic acids and β-keto acids undergo thermal decarboxylation by a mechanism in which a β-carbonyl group assists the departure of carbon dioxide (Section 19.17).

$$XC(=O)CR_2CO_2H \xrightarrow{-CO_2} X(HO)C{=}CR_2 \longrightarrow XCCHR_2 \text{ (with C=O)}$$

X = OH: malonic acid derivative
X = alkyl or aryl: β-keto acid

Enol form of product

X = OH: carboxylic acid
X = alkyl or aryl: ketone

PROBLEMS

19.13 Many carboxylic acids are much better known by their common names than by their systematic names. Some of these follow. Provide a structural formula for each one on the basis of its systematic name.

(*a*) 2-Hydroxypropanoic acid (better known as *lactic acid,* it is found in sour milk and is formed in the muscles during exercise)

(*b*) 2-Hydroxy-2-phenylethanoic acid (also known as *mandelic acid,* it is obtained from plums, peaches, and other fruits)

(*c*) Tetradecanoic acid (also known as *myristic acid,* it can be obtained from a variety of fats)

(*d*) 10-Undecenoic acid (also called *undecylenic acid,* it is used, in combination with its zinc salt, to treat fungal infections such as athlete's foot)

(*e*) 3,5-Dihydroxy-3-methylpentanoic acid (also called *mevalonic acid,* it is an important intermediate in the biosynthesis of terpenes and steroids)

(*f*) *(E)*-2-Methyl-2-butenoic acid (also known as *tiglic acid,* it is a constituent of various natural oils)

(*g*) 2-Hydroxybutanedioic acid (also known as *malic acid,* it is found in apples and other fruits)

(*h*) 2-Hydroxy-1,2,3-propanetricarboxylic acid (better known as *citric acid,* it contributes to the tart taste of citrus fruits)

(*i*) 2-(*p*-Isobutylphenyl)propanoic acid (an anti-inflammatory drug better known as *ibuprofen*)

(*j*) *o*-Hydroxybenzenecarboxylic acid (better known as *salicylic acid,* it is obtained from willow bark)

19.14 Give an acceptable systematic IUPAC name for each of the following:

(*a*) $CH_3(CH_2)_6CO_2H$
(*b*) $CH_3(CH_2)_6CO_2K$
(*c*) $CH_2{=}CH(CH_2)_5CO_2H$
(*d*) H_3C, H / C=C / $(CH_2)_4CO_2H$, H
(*e*) $HO_2C(CH_2)_6CO_2H$
(*f*) $CH_3(CH_2)_4CH(CO_2H)_2$
(*g*) cycloheptane–CO_2H
(*h*) benzene–$CH(CH_2CH_3)(CH_2)_4CO_2H$

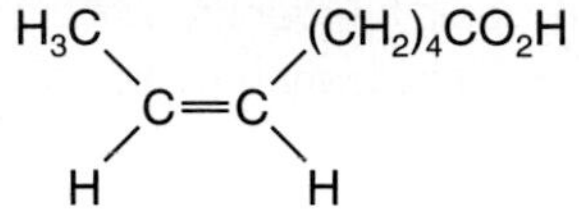

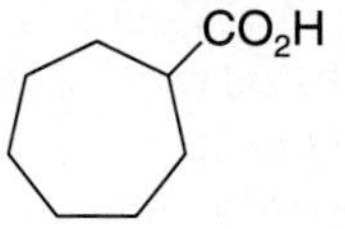

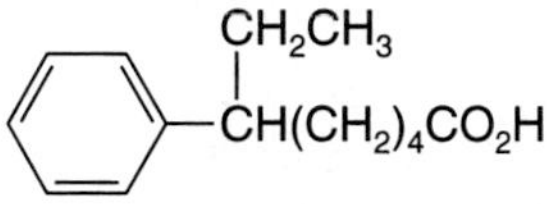

19.15 Rank the compounds in each of the following groups in order of decreasing acidity:

(*a*) Acetic acid, ethane, ethanol
(*b*) Benzene, benzoic acid, benzyl alcohol
(*c*) Propanedial, 1,3-propanediol, propanedioic acid, propanoic acid
(*d*) Acetic acid, ethanol, trifluoroacetic acid, 2,2,2-trifluoroethanol, trifluoromethanesulfonic acid (CF_3SO_2OH)
(*e*) Cyclopentanecarboxylic acid, 2,4-pentanedione, cyclopentanone, cyclopentene

19.16 In spite of the fact that entropy effects on the ionization of carboxylic acids are larger than enthalpy effects, reliable predictions concerning relative acidities can usually be made by considering how well substituents stabilize carboxylate anions by electron withdrawal. Using this procedure, identify the more acidic compound in each of the following pairs:

(*a*) $CF_3CH_2CO_2H$ or $CF_3CH_2CH_2CO_2H$
(*b*) $CH_3CH_2CH_2CO_2H$ or $CH_3C{\equiv}CCO_2H$
(*c*) cyclohexane–CO_2H or benzene–CO_2H
(*d*) F, F, F, F, F–benzene–CO_2H (pentafluorobenzoic acid) or benzene–CO_2H
(*e*) F, F, F, F, F–benzene–CO_2H or F, F, F, F, F–benzene–benzene–CO_2H
(*f*) furan (O)–2-CO_2H or furan (O)–3-CO_2H
(*g*) furan (O)–2-CO_2H or pyrrole (N–H)–2-CO_2H

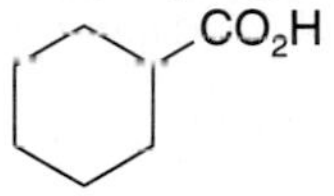

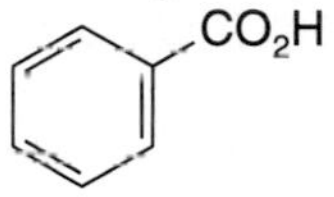

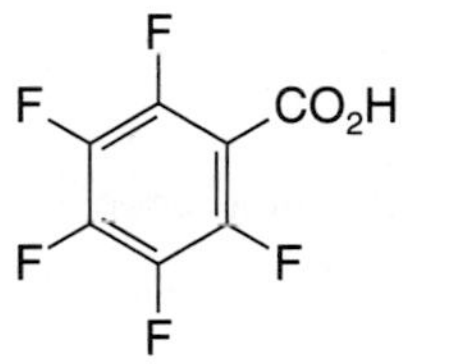

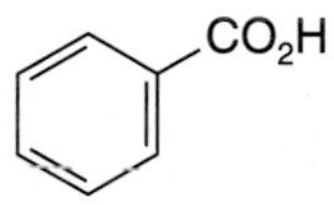

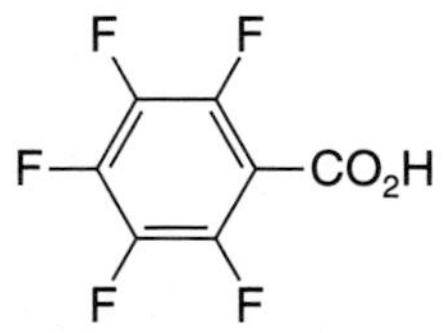

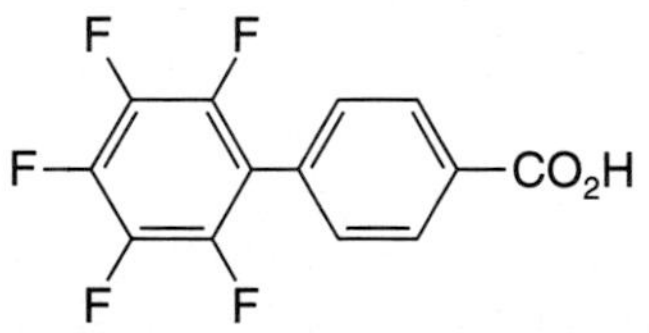

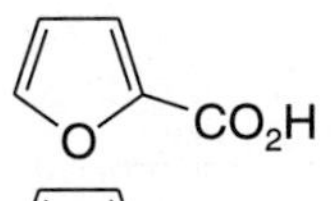

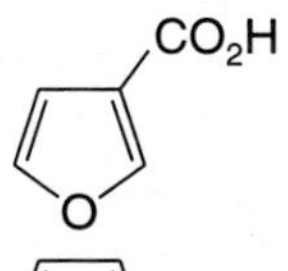

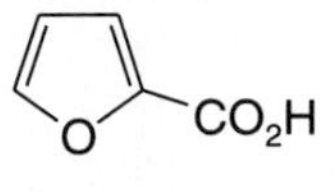

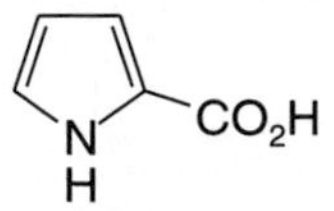

19.17 Propose methods for preparing butanoic acid from each of the following:

(*a*) 1-Butanol
(*b*) Butanal
(*c*) 1-Butene
(*d*) 1-Propanol
(*e*) 2-Propanol
(*f*) Acetaldehyde
(*g*) $CH_3CH_2CH(CO_2H)_2$

19.18 It is sometimes necessary to prepare isotopically labeled samples of organic substances for probing biological transformations and reaction mechanisms. Various sources of the radioactive mass 14 carbon isotope are available. Describe synthetic procedures by which benzoic acid, labeled with ^{14}C at its carbonyl carbon, could be prepared from benzene and the following ^{14}C-labeled precursors. You may use any necessary organic or inorganic reagents. (In the formulas shown, an asterisk indicates ^{14}C.)

(*a*) $\overset{*}{C}H_3Cl$ (*b*) $H\overset{*}{C}(=O)H$ (*c*) $\overset{*}{C}O_2$

19.19 Give the product of the reaction of pentanoic acid with each of the following reagents:

(*a*) Sodium hydroxide
(*b*) Sodium bicarbonate
(*c*) Thionyl chloride
(*d*) Phosphorus tribromide
(*e*) Benzyl alcohol, sulfuric acid (catalytic amount)
(*f*) Chlorine, phosphorus tribromide (catalytic amount)
(*g*) Bromine, phosphorus trichloride (catalytic amount)
(*h*) Product of part *(g)* treated with sodium iodide in acetone
(*i*) Product of part *(g)* treated with aqueous ammonia
(*j*) Lithium aluminum hydride, then hydrolysis
(*k*) Phenylmagnesium bromide

19.20 Show how butanoic acid may be converted to each of the following compounds:

(*a*) 1-Butanol
(*b*) Butanal
(*c*) 1-Chlorobutane
(*d*) Butanoyl chloride
(*e*) Phenyl propyl ketone
(*f*) 4-Octanone
(*g*) 2-Bromobutanoic acid
(*h*) 2-Butenoic acid

19.21 Show by a series of equations, using any necessary organic or inorganic reagents, how acetic acid can be converted to each of the following compounds:

(*a*) $H_2NCH_2CO_2H$
(*b*) $C_6H_5OCH_2CO_2H$
(*c*) $NCCH_2CO_2H$
(*d*) $HO_2CCH_2CO_2H$
(*e*) ICH_2CO_2H
(*f*) $BrCH_2CO_2CH_2CH_3$
(*g*) $(C_6H_5)_3\overset{+}{P}-\overset{-}{\ddot{C}}HCO_2CH_2CH_3$
(*h*) $C_6H_5CH{=}CHCO_2CH_2CH_3$

19.22 Each of the following reactions has been reported in the chemical literature and gives a single product in good yield. What is the product in each reaction?

(*a*) $(H_3C)(H)C{=}C(CH_3)(CO_2H) \xrightarrow{\text{ethanol, } H_2SO_4}$

(*b*) cyclopropyl$-CO_2H \xrightarrow[\text{2. } H_2O]{\text{1. } LiAlD_4}$

(*c*) cyclohexyl$-CO_2H \xrightarrow[P]{Br_2}$

(*d*) 3-bromo-1-(CF_3)benzene $\xrightarrow[\text{2. } CO_2 \text{ 3. } H_3O^+]{\text{1. Mg, diethyl ether}}$

(*e*) 3-chloro-1-(CH_2CN)benzene $\xrightarrow[H_2SO_4\text{, heat}]{H_2O\text{, acetic acid}}$

(*f*) $CH_2{=}CH(CH_2)_8CO_2H \xrightarrow[\text{benzoyl peroxide}]{HBr}$

19.23 Show by a series of equations how you could synthesize each of the following compounds from the indicated starting material and any necessary organic or inorganic reagents:

(*a*) 2-Methylpropanoic acid from *tert*-butyl alcohol
(*b*) 3-Methylbutanoic acid from *tert*-butyl alcohol
(*c*) 3,3-Dimethylbutanoic acid from *tert*-butyl alcohol
(*d*) $HO_2C(CH_2)_5CO_2H$ from $HO_2C(CH_2)_3CO_2H$
(*e*) 3-Phenyl-1-butanol from $CH_3CH(C_6H_5)CH_2CN$
(*f*) Br, CO_2H from cyclopentyl bromide
(*g*) Cl, CO_2H from (*E*)-$ClCH{=}CHCO_2H$
(*h*) 2,4-Dimethylbenzoic acid from *m*-xylene
(*i*) 4-Chloro-3-nitrobenzoic acid from *p*-chlorotoluene
(*j*) (*Z*)-$CH_3CH{=}CHCO_2H$ from propyne

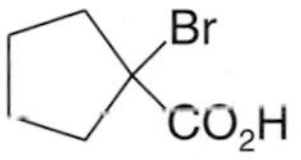

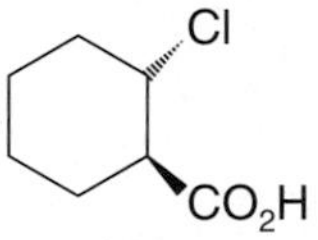

19.24 Suggest reasonable explanations for each of the following observations:

(*a*) Both hydrogens are anti to each other in the most stable conformation of formic acid.
(*b*) Oxalic acid has a dipole moment of zero in the gas phase.
(*c*) The dissociation constant of *o*-hydroxybenzoic acid is greater (by a factor of 12) than that of *o*-methoxybenzoic acid.
(*d*) One of the diastereomers of 3-hydroxycyclohexanecarboxylic acid readily forms a lactone while the other does not.
(*e*) Ascorbic acid (vitamin C), while not a carboxylic acid, is sufficiently acidic to cause carbon dioxide liberation on being dissolved in aqueous sodium bicarbonate.

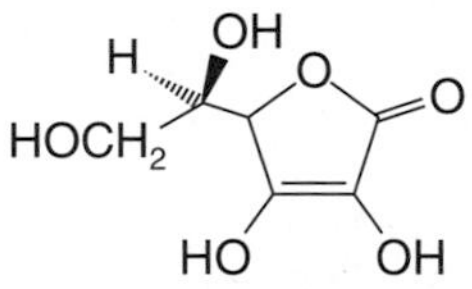

H, OH, $HOCH_2$, O, O, HO, OH

Ascorbic acid

19.25 When compound A is heated, two isomeric products are formed. What are these two products?

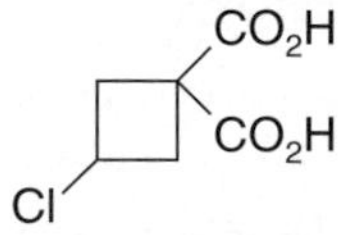

CO_2H, CO_2H, Cl

Compound A

19.26 A certain carboxylic acid ($C_{14}H_{26}O_2$), which can be isolated from whale blubber or sardine oil, yields nonanal and $O{=}CH(CH_2)_3CO_2H$ on ozonolysis. What is the structure of this acid?

19.27 When levulinic acid ($CH_3\overset{O}{\overset{\|}{C}}CH_2CH_2CO_2H$) was hydrogenated at high pressure over a nickel catalyst at 220°C, a single product, the compound $C_5H_8O_2$, was isolated in 94 percent yield. This compound lacks hydroxyl absorption in its infrared spectrum and does not immediately liberate carbon dioxide on being shaken with sodium bicarbonate. What is a reasonable structure for the compound?

19.28 On standing in dilute aqueous acid, compound A is smoothly converted to mevalonolactone.

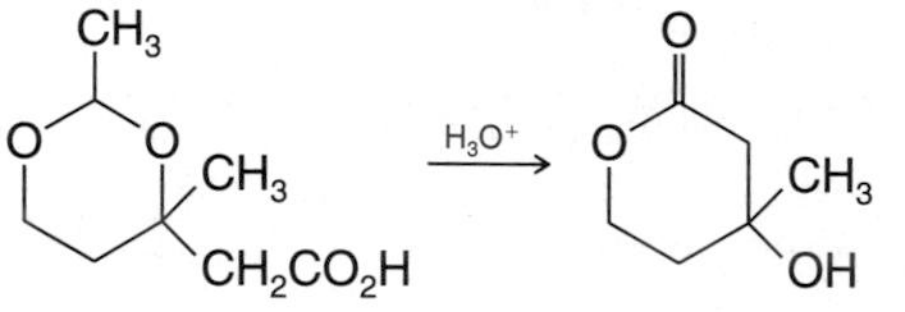

CH$_3$ ... O O ... CH$_3$, CH$_2$CO$_2$H $\xrightarrow{H_3O^+}$ O, O ... CH$_3$, OH

Compound A Mevalonolactone

Suggest a reasonable mechanism for this reaction. What other organic product is also formed?

19.29 Suggest reaction conditions suitable for the preparation of compound A from 5-hydroxy-2-hexynoic acid.

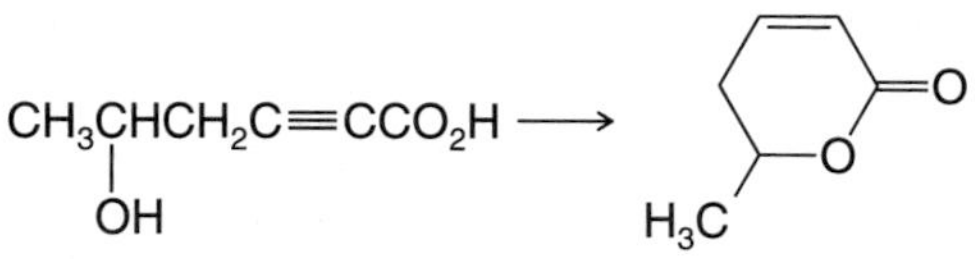

$CH_3\underset{OH}{\underset{|}{C}}HCH_2C{\equiv}CCO_2H \longrightarrow$ =O, O, H$_3$C

5-Hydroxy-2-hexynoic acid Compound A

19.30 In the presence of the enzyme *aconitase,* the double bond of aconitic acid undergoes hydration. The reaction is reversible, and the following equilibrium is established:

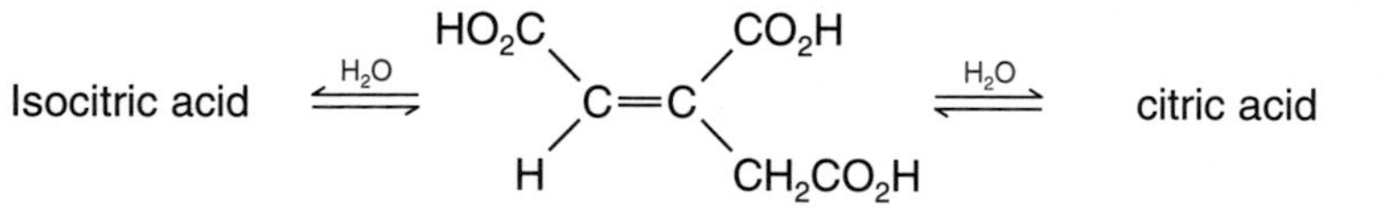

Isocitric acid $\underset{}{\overset{H_2O}{\rightleftharpoons}}$ (HO$_2$C)(H)C=C(CO$_2$H)(CH$_2$CO$_2$H) $\overset{H_2O}{\rightleftharpoons}$ citric acid

($C_6H_8O_7$) (6% at equilibrium) — Aconitic acid (4% at equilibrium) — ($C_6H_8O_7$) (90% at equilibrium)

(*a*) The major tricarboxylic acid present is *citric acid,* the substance responsible for the tart taste of citrus fruits. Citric acid is achiral. What is its structure?

(*b*) What must be the constitution of isocitric acid? (Assume that no rearrangements accompany hydration.) How many stereoisomers are possible for isocitric acid?

19.31 The ^{1}H nmr spectra of formic acid (HCO_2H), maleic acid (*cis*-$HO_2CCH{=}CHCO_2H$), and malonic acid ($HO_2CCH_2CO_2H$) are similar in that each is characterized by two singlets of equal intensity. Match these compounds with the designations A, B, and C on the basis of the appropriate ^{1}H nmr chemical shift data.

Compound A: signals at δ = 3.2 and 12.1 ppm
Compound B: signals at δ = 6.3 and 12.4 ppm
Compound C: signals at δ = 8.0 and 11.4 ppm

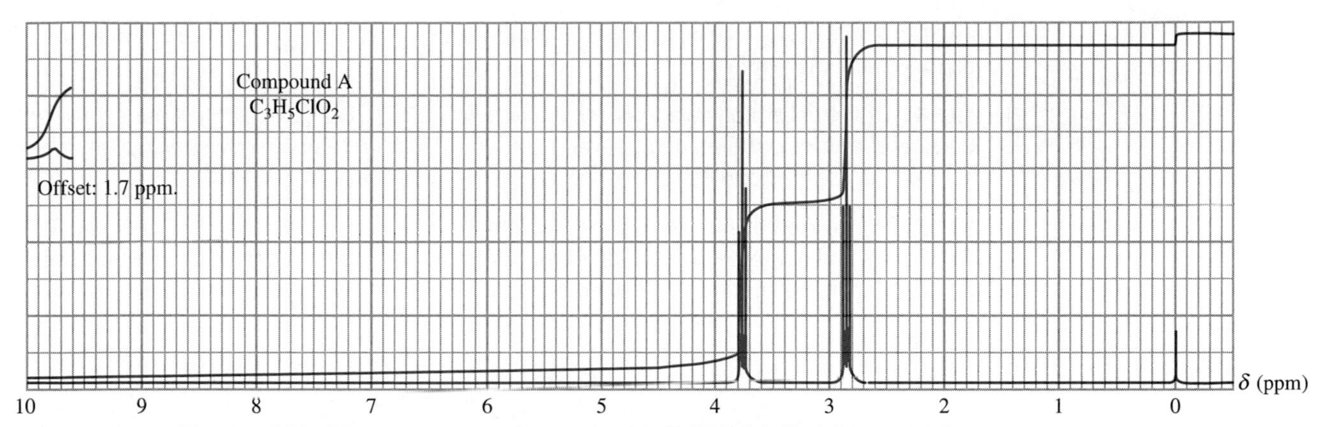

FIGURE 19.9 The 300-MHz ^{1}H nmr spectrum of compound A ($C_3H_5ClO_2$) (Problem 19.33*a*). The signals at $\delta = 2.9$ and 3.8 ppm are each triplets.

19.32 Compounds A and B are isomers having the molecular formula $C_4H_8O_3$. Identify A and B on the basis of their ^{1}H nmr spectra.

Compound A: $\delta = 1.3$ ppm (3H, triplet); 3.6 ppm (2H, quartet); 4.1 ppm (2H, singlet); 11.1 ppm (1H, broad singlet)

Compound B: $\delta = 2.6$ ppm (2H, triplet); 3.4 ppm (3H, singlet); 3.7 ppm (2H triplet); 11.3 ppm (1H, broad singlet)

19.33 Compounds A and B are carboxylic acids. Identify each one on the basis of its ^{1}H nmr spectrum.

(*a*) Compound A ($C_3H_5ClO_2$) (Figure 19.9).
(*b*) Compound B ($C_9H_9NO_4$) (Figure 19.10).

MOLECULAR MODELING EXERCISES

19.34 Formic acid is planar, and its two hydrogens are anti to one another. Construct a molecular model of formic acid.

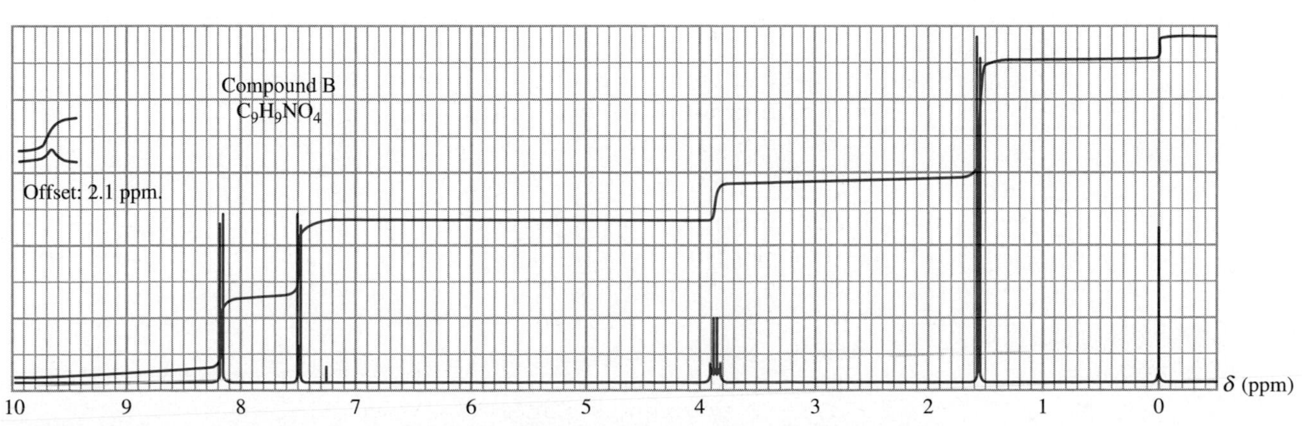

FIGURE 19.10 The 300-MHz ^{1}H nmr spectrum of compound B ($C_9H_9NO_4$) (Problem 19.33*b*).

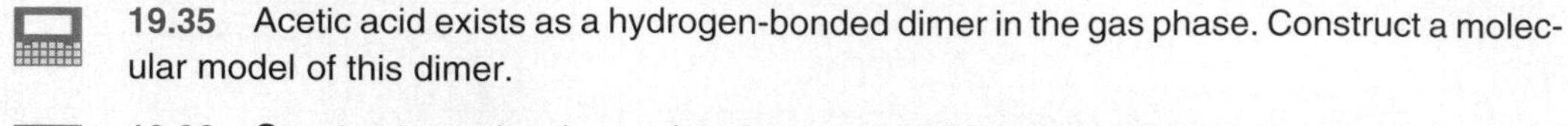

19.35 Acetic acid exists as a hydrogen-bonded dimer in the gas phase. Construct a molecular model of this dimer.

19.36 Construct a molecular model of stearic acid [$CH_3(CH_2)_{16}CO_2H$]. How does the introduction of a trans double bond between C-8 and C-9 affect the overall shape of the molecule? A cis double bond?

19.37 Which stereoisomer of 4-hydroxycyclohexanecarboxylic acid (cis or trans) can form a lactone? Make a molecular model of this lactone. What is the conformation of the cyclohexane ring in the starting hydroxy acid? In the lactone?

HO–C_6H_{10}–CO_2H

Aryl esters, that is, compounds of the type $R\overset{O}{\overset{\|}{C}}OAr$, are named in an analogous way.

The names of *amides* of the type $R\overset{O}{\overset{\|}{C}}NH_2$ are derived from carboxylic acids by replacing the suffix *-oic acid* or *-ic acid* by *-amide.*

$CH_3\overset{O}{\overset{\|}{C}}NH_2$	$C_6H_5\overset{O}{\overset{\|}{C}}NH_2$	$(CH_3)_2CHCH_2\overset{O}{\overset{\|}{C}}NH_2$
Acetamide	Benzamide	3-Methylbutanamide

We name compounds of the type $R\overset{O}{\overset{\|}{C}}NHR'$ and $R\overset{O}{\overset{\|}{C}}NR'_2$ as *N*-alkyl- and *N,N*-dialkyl-substituted derivatives of a parent amide.

$CH_3\overset{O}{\overset{\|}{C}}NHCH_3$	$C_6H_5\overset{O}{\overset{\|}{C}}N(CH_2CH_3)_2$	$CH_3CH_2CH_2\overset{O}{\overset{\|}{C}}\underset{\underset{CH_3}{\mid}}{N}CH(CH_3)_2$
N-Methylacetamide	*N,N*-Diethylbenzamide	*N*-Isopropyl-*N*-methylbutanamide

Substitutive names for *nitriles* add the suffix *-nitrile* to the name of the parent hydrocarbon chain that includes the carbon of the cyano group. Nitriles may also be named by replacing the *-ic acid* or *-oic acid* ending of the corresponding carboxylic acid by *-onitrile.* Alternatively, they are sometimes given radicofunctional names as alkyl cyanides.

$CH_3C{\equiv}N$	$C_6H_5C{\equiv}N$	$CH_3\underset{\underset{C\equiv N}{\mid}}{C}HCH_3$
Ethanenitrile (acetonitrile)	Benzonitrile	2-Methylpropanenitrile (isopropyl cyanide)

PROBLEM 20.1 Write a structural formula for each of the following compounds:

(*a*) 2-Phenylbutanoyl bromide
(*b*) 2-Phenylbutanoic anhydride
(*c*) Butyl 2-phenylbutanoate
(*d*) 2-Phenylbutyl butanoate
(*e*) 2-Phenylbutanamide
(*f*) *N*-Ethyl-2-phenylbutanamide
(*g*) 2-Phenylbutanenitrile

SAMPLE SOLUTION (*a*) A 2-phenylbutanoyl group is a four-carbon acyl unit that bears a phenyl substituent at C-2. When the name of an acyl group is followed by the name of a halide, it designates an *acyl halide.*

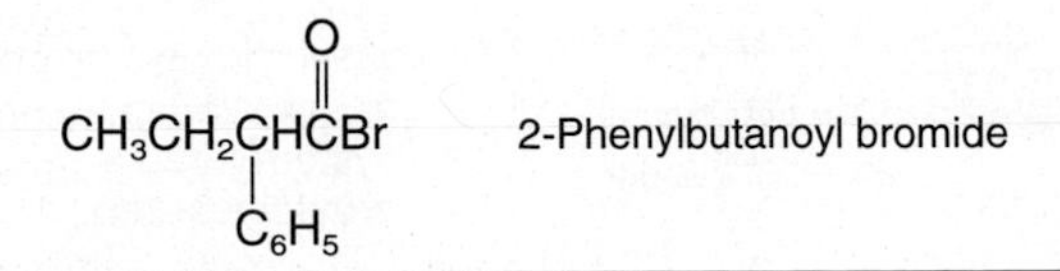

20.2 STRUCTURE OF CARBOXYLIC ACID DERIVATIVES

Like other carbonyl-containing compounds we have studied—aldehydes, ketones, and carboxylic acids—derivatives of carboxylic acids have a planar arrangement of bonds to their carbonyl group. An important structural feature of acyl chlorides, anhydrides, esters, and amides is that the atom attached to the acyl group bears an unshared pair of electrons that is capable of interacting with the carbonyl π system, as shown in Figure 20.1.

Electron delocalization in carboxylic acids and its derivatives is represented in resonance terms by contributions from the following structures:

$$R-C(=\ddot{O}:)-\ddot{X} \longleftrightarrow R-\overset{+}{C}(-\ddot{O}:^{-})-\ddot{X} \longleftrightarrow R-C(-\ddot{O}:^{-})=\overset{+}{X}$$

Electron release from the substituent stabilizes the carbonyl group and decreases its electrophilic character. The extent of this electron delocalization depends on the electron-donating properties of the substituent X. Generally, the less electronegative X is, the better it donates electrons to the carbonyl group and the greater its stabilizing effect.

Resonance stabilization in acyl chlorides is not nearly as pronounced as in other derivatives of carboxylic acids:

$$R-C(=\ddot{O}:)-\ddot{Cl}: \longleftrightarrow R-C(-\ddot{O}:^{-})=\overset{+}{Cl}:$$

Weak resonance stabilization

Because the carbon-chlorine bond is so long—typically on the order of 180 pm for acyl chlorides—overlap between the 3*p* orbitals of chlorine and the π orbital of the carbonyl group is poor. Consequently, there is little delocalization of the electron pairs of chlorine into the π system. The carbonyl group of an acyl chloride feels the normal electron-withdrawing inductive effect of a chlorine substituent without a significant compensating electron-releasing effect due to lone-pair donation by chlorine. This makes the carbonyl carbon of an acyl chloride more susceptible to attack by nucleophiles than that of other carboxylic acid derivatives.

Acid anhydrides are better stabilized by electron delocalization than are acyl chlorides. The lone-pair electrons of oxygen are delocalized more effectively into the carbonyl group. Resonance involves both carbonyl groups of an acid anhydride.

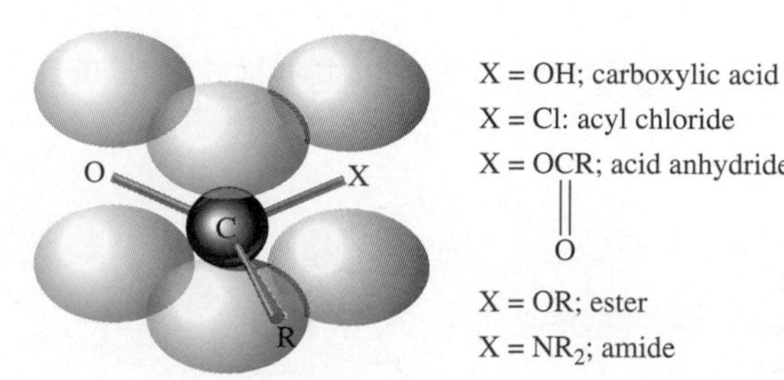

FIGURE 20.1 The three σ bonds originating at the carbonyl carbon are coplanar. The *p* orbitals of the carbonyl carbon, its oxygen, and the atom by which group X is attached to the acyl group overlap to form an extended π system through which the π electrons are delocalized.

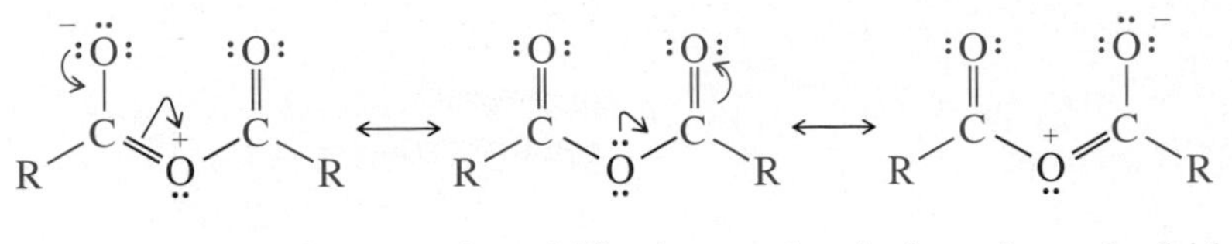

The carbonyl group of an ester is stabilized more than is that of an anhydride. Since both acyl groups of an anhydride compete for the oxygen lone pair, each carbonyl is stabilized less than the single carbonyl group of an ester.

is more effective than

Ester

Acid anhydride

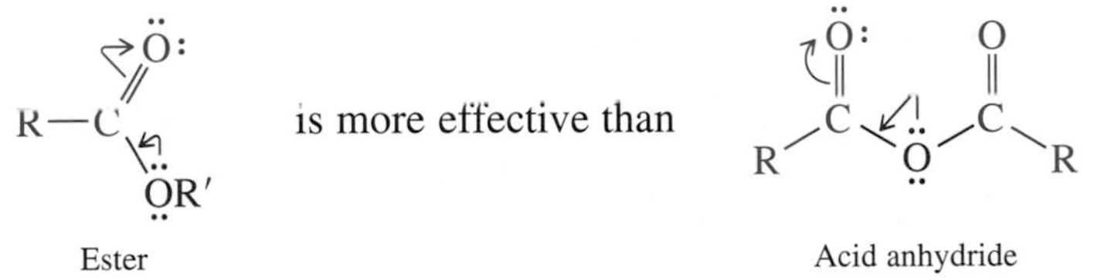

Esters are stabilized by resonance to about the same extent as carboxylic acids but not as much as amides. Nitrogen is less electronegative than oxygen and is a better electron-pair donor.

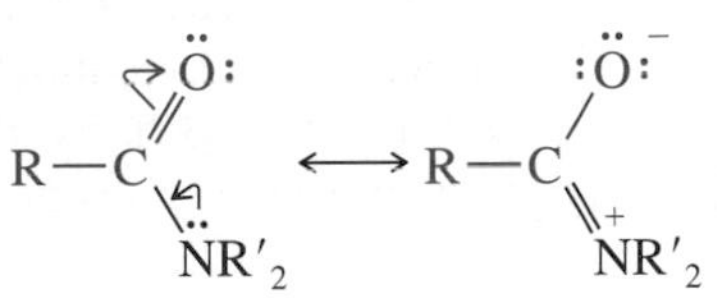

Very effective resonance stabilization

Amide resonance is a powerful stabilizing force and gives rise to a number of structural effects. Unlike the pyramidal arrangement of bonds in ammonia and amines, the bonds to nitrogen in amides lie in the same plane. Formamide, for example, as shown in Figure 20.2, is a planar molecule. The carbon-nitrogen bond has considerable double-bond character and, at 135 pm, is substantially shorter than the normal 147-pm carbon-nitrogen single-bond distance observed in amines.

The barrier to rotation about the carbon-nitrogen bond in amides is 75 to 85 kJ/mol (18 to 20 kcal/mol).

E_{act} = 75 to 85 kJ/mol (18 to 20 kcal/mol)

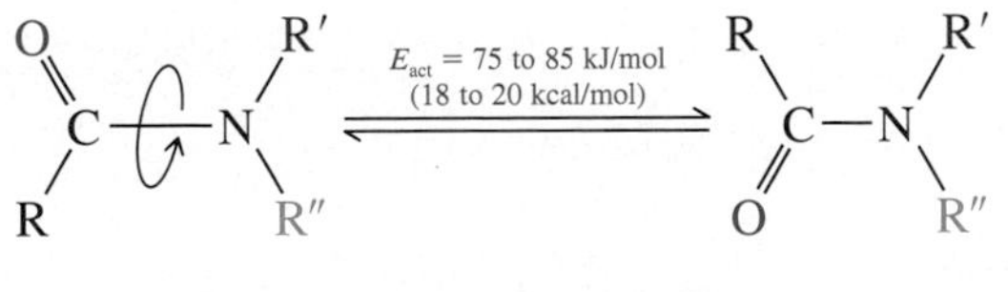

This is an unusually high rotational energy barrier for a single bond and indicates that the carbon-nitrogen bond has significant double-bond character, as the resonance picture suggests.

PROBLEM 20.2 The ^{1}H nuclear magnetic resonance spectrum of *N,N*-dimethylformamide shows a separate signal for each of the two methyl groups. Can you explain why?

Electron release from nitrogen stabilizes the carbonyl group of amides and decreases the rate at which nucleophiles attack the carbonyl carbon. Nucleophilic reagents attack electrophilic sites in a molecule; if electrons are donated to an electrophilic site in a molecule by a substituent, then the tendency of that molecule to react with external nucleophiles is moderated.

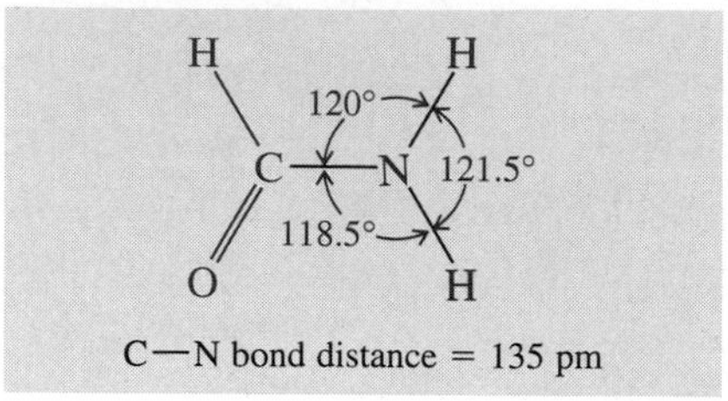

C—N bond distance = 135 pm

FIGURE 20.2 Structural features of formamide. The molecule is planar; all the atoms lie in the plane of the paper.

An extreme example of carbonyl group stabilization is seen in carboxylate anions:

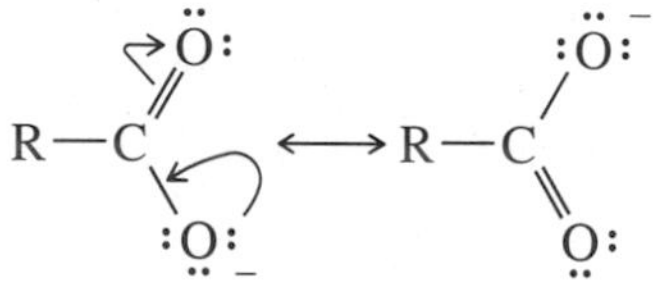

$$\mathrm{R{-}C({=}\ddot{O}{:})(\ddot{O}{:}^-) \longleftrightarrow R{-}C(\ddot{O}{:}^-)({=}\ddot{O}{:})}$$

The negatively charged oxygen substituent is a powerful electron donor to the carbonyl group. Resonance in carboxylate anions is more effective than resonance in carboxylic acids, acyl chlorides, anhydrides, esters, and amides.

Table 20.1 summarizes the stabilizing effects of substituents on carbonyl groups to which they are attached. In addition to a qualitative ranking, quantitative estimates of the relative rates of hydrolysis of the various classes of acyl derivatives are given. A weakly stabilized carboxylic acid derivative reacts with water faster than does a more stabilized one.

Most methods for their preparation convert one class of carboxylic acid derivative to another, and the order of carbonyl group stabilization given in Table 20.1 bears directly on the means by which these transformations may be achieved. A reaction that converts one carboxylic acid derivative to another that lies below it in the table is practical; a reaction that converts it to one that lies above it in the table is not. This is another way of saying that one carboxylic acid derivative can be converted to another if the reaction leads to a more stabilized carbonyl group. Numerous examples of

TABLE 20.1
Relative Stability and Reactivity of Carboxylic Acid Derivatives

Carboxylic acid derivative		Stabilization	Relative rate of hydrolysis*
Acyl chloride	$\mathrm{R\overset{\overset{O}{\|}}{C}Cl}$	Very small	10^{11}
Anhydride	$\mathrm{R\overset{\overset{O}{\|}}{C}O\overset{\overset{O}{\|}}{C}R}$	Small	10^{7}
Ester	$\mathrm{R\overset{\overset{O}{\|}}{C}OR'}$	Moderate	1.0
Amide	$\mathrm{R\overset{\overset{O}{\|}}{C}NR'_2}$	Large	$<10^{-2}$
Carboxylate anion	$\mathrm{R\overset{\overset{O}{\|}}{C}O^-}$	Very large	

* Rates are approximate and are relative to ester as standard substrate at pH 7.

reactions of this type will be presented in the sections that follow. We begin with reactions of acyl chlorides.

20.3 NUCLEOPHILIC ACYL SUBSTITUTION IN ACYL CHLORIDES

Acyl chlorides are readily prepared from carboxylic acids by reaction with thionyl chloride (Section 12.7).

One of the most useful reactions of acyl chlorides was presented in Section 12.7. Friedel-Crafts acylation of aromatic rings takes place when arenes are treated with acyl chlorides in the presence of aluminum chloride.

$$\underset{\text{Carboxylic acid}}{\mathrm{RCOH}} + \underset{\text{Thionyl chloride}}{\mathrm{SOCl_2}} \longrightarrow \underset{\text{Acyl chloride}}{\mathrm{RCCl}} + \underset{\text{Sulfur dioxide}}{\mathrm{SO_2}} + \underset{\text{Hydrogen chloride}}{\mathrm{HCl}}$$

(C=O in RCOH and RCCl)

$$\underset{\text{2-Methylpropanoic acid}}{\mathrm{(CH_3)_2CHCOH}} \xrightarrow[\text{heat}]{\mathrm{SOCl_2}} \underset{\text{2-Methylpropanoyl chloride (90\%)}}{\mathrm{(CH_3)_2CHCCl}}$$

On treatment with the appropriate nucleophile, an acyl chloride may be converted to an acid anhydride, an ester, an amide, or a carboxylic acid. Examples illustrating these transformations of acyl chlorides are presented in Table 20.2.

PROBLEM 20.3 Apply the knowledge gained by studying Table 20.2 to help you predict the major organic product obtained by reaction of benzoyl chloride with each of the following:

(*a*) Acetic acid
(*b*) Benzoic acid
(*c*) Ethanol
(*d*) Methylamine, CH_3NH_2
(*e*) Dimethylamine, $(CH_3)_2NH$
(*f*) Water

SAMPLE SOLUTION (*a*) As noted in Table 20.2, the reaction of an acyl chloride with a carboxylic acid yields an acid anhydride.

$$\underset{\text{Benzoyl chloride}}{\mathrm{C_6H_5CCl}} + \underset{\text{Acetic acid}}{\mathrm{CH_3COH}} \longrightarrow \underset{\text{Acetic benzoic anhydride}}{\mathrm{C_6H_5COCCH_3}}$$

The product is a mixed anhydride. Acetic acid acts as a nucleophile and substitutes for chloride on the benzoyl group.

The mechanisms of all the reactions cited in Table 20.2 are similar to the mechanism of hydrolysis of an acyl chloride.

$$\underset{\text{Acyl chloride}}{\mathrm{RCCl}} + \underset{\text{Water}}{\mathrm{H_2O}} \longrightarrow \underset{\text{Carboxylic acid}}{\mathrm{RCOH}} + \underset{\text{Hydrogen chloride}}{\mathrm{HCl}}$$

TABLE 20.2
Conversion of Acyl Chlorides to Other Carboxylic Acid Derivatives

Reaction (section) and comments	General equation and specific example
Reaction with carboxylic acids (Section 20.4) Acyl chlorides react with carboxylic acids to yield acid anhydrides. When this reaction is used for preparative purposes, a weak organic base such as pyridine is normally added. Pyridine is a catalyst for the reaction and also acts as a base to neutralize the hydrogen chloride that is formed.	$\underset{\text{Acyl chloride}}{R\overset{O}{\overset{\|}{C}}Cl} + \underset{\text{Carboxylic acid}}{R'\overset{O}{\overset{\|}{C}}OH} \longrightarrow \underset{\text{Acid anhydride}}{R\overset{O}{\overset{\|}{C}}O\overset{O}{\overset{\|}{C}}R'} + \underset{\text{Hydrogen chloride}}{HCl}$ $\underset{\text{Heptanoyl chloride}}{CH_3(CH_2)_5\overset{O}{\overset{\|}{C}}Cl} + \underset{\text{Heptanoic acid}}{CH_3(CH_2)_5\overset{O}{\overset{\|}{C}}OH} \xrightarrow{\text{pyridine}} \underset{\text{Heptanoic anhydride (78–83\%)}}{CH_3(CH_2)_5\overset{O}{\overset{\|}{C}}O\overset{O}{\overset{\|}{C}}(CH_2)_5CH_3}$
Reaction with alcohols (Section 15.8) Acyl chlorides react with alcohols to form esters. The reaction is typically carried out in the presence of pyridine.	$\underset{\text{Acyl chloride}}{R\overset{O}{\overset{\|}{C}}Cl} + \underset{\text{Alcohol}}{R'OH} \longrightarrow \underset{\text{Ester}}{R\overset{O}{\overset{\|}{C}}OR'} + \underset{\text{Hydrogen chloride}}{HCl}$ $\underset{\text{Benzoyl chloride}}{C_6H_5\overset{O}{\overset{\|}{C}}Cl} + \underset{\textit{tert-}\text{Butyl alcohol}}{(CH_3)_3COH} \xrightarrow{\text{pyridine}} \underset{\textit{tert-}\text{Butyl benzoate (80\%)}}{C_6H_5\overset{O}{\overset{\|}{C}}OC(CH_3)_3}$
Reaction with ammonia and amines (Section 20.13) Acyl chlorides react with ammonia and amines to form amides. A base such as sodium hydroxide is normally added to react with the hydrogen chloride produced.	$\underset{\text{Acyl chloride}}{R\overset{O}{\overset{\|}{C}}Cl} + \underset{\text{Ammonia or amine}}{R'_2NH} + \underset{\text{Hydroxide}}{HO^-} \longrightarrow \underset{\text{Amide}}{R\overset{O}{\overset{\|}{C}}NR'_2} + \underset{\text{Water}}{H_2O} + \underset{\text{Chloride ion}}{Cl^-}$ $\underset{\text{Benzoyl chloride}}{C_6H_5\overset{O}{\overset{\|}{C}}Cl} + \underset{\text{Piperidine}}{HN\bigcirc} \xrightarrow[H_2O]{NaOH} \underset{\textit{N-}\text{Benzoylpiperidine (87–91\%)}}{C_6H_5\overset{O}{\overset{\|}{C}}\text{—}N\bigcirc}$
Hydrolysis (Section 20.3) Acyl chlorides react with water to yield carboxylic acids. In base, the acid is converted to its carboxylate salt. The reaction has little preparative value because the acyl chloride is nearly always prepared from the carboxylic acid rather than vice versa.	$\underset{\text{Acyl chloride}}{R\overset{O}{\overset{\|}{C}}Cl} + \underset{\text{Water}}{H_2O} \longrightarrow \underset{\text{Carboxylic acid}}{R\overset{O}{\overset{\|}{C}}OH} + \underset{\text{Hydrogen chloride}}{HCl}$ $\underset{\text{Phenylacetyl chloride}}{C_6H_5CH_2\overset{O}{\overset{\|}{C}}Cl} + \underset{\text{Water}}{H_2O} \longrightarrow \underset{\text{Phenylacetic acid}}{C_6H_5CH_2\overset{O}{\overset{\|}{C}}OH} + \underset{\text{Hydrogen chloride}}{HCl}$

First stage: Formation of the tetrahedral intermediate by nucleophilic addition of water to the carbonyl group

$$H_2\ddot{O}: + RC(=\ddot{O}:)\ddot{Cl}: \xrightleftharpoons{\text{slow}} H_2\overset{+}{O}-C(R)(\ddot{Cl}:)-\ddot{O}:^{-} \xrightarrow{\text{fast}} H-\ddot{O}-C(R)(\ddot{Cl}:)-\ddot{O}-H$$

Water Acyl chloride Tetrahedral intermediate

Second stage: Dissociation of tetrahedral intermediate by dehydrohalogenation

$$H-\ddot{O}-C(R)(\ddot{Cl}:)-\ddot{O}-H + H_2\ddot{O}: \xrightarrow{\text{fast}} RC(=\ddot{O}:)\ddot{O}H + H_3\overset{+}{O}: + :\ddot{Cl}:^{-}$$

Tetrahedral intermediate Water Carboxylic acid Hydronium ion Chloride ion

FIGURE 20.3 Hydrolysis of an acyl chloride proceeds by way of a tetrahedral intermediate. Formation of the tetrahedral intermediate is rate-determining.

They differ with respect to the nucleophile that attacks the carbonyl group, which in hydrolysis is a water molecule. The accepted mechanism for hydrolysis of an acyl chloride is outlined in Figure 20.3.

In the first stage of the mechanism, water undergoes nucleophilic addition to the carbonyl group to form a tetrahedral intermediate. This stage of the process is analogous to the hydration of aldehydes and ketones discussed in Section 17.7.

The tetrahedral intermediate has three potential leaving groups on carbon: two hydroxyl groups and a chlorine. In the second stage of the reaction, the tetrahedral intermediate dissociates. Loss of chloride from the tetrahedral intermediate is faster than loss of hydroxide; chloride is less basic than hydroxide and is a better leaving group. The tetrahedral intermediate dissociates because this dissociation restores the resonance-stabilized carbonyl group.

PROBLEM 20.4 Write the structure of the tetrahedral intermediate formed in each of the reactions given in Problem 20.3. Using curved arrows, show how each tetrahedral intermediate dissociates to the appropriate products.

SAMPLE SOLUTION (*a*) The tetrahedral intermediate arises by nucleophilic addition of acetic acid to benzoyl chloride.

$$C_6H_5\overset{O}{\overset{\|}{C}}Cl + CH_3\overset{O}{\overset{\|}{C}}OH \longrightarrow C_6H_5\underset{Cl}{\underset{|}{\overset{HO}{\overset{|}{C}}}}O\overset{O}{\overset{\|}{C}}CH_3$$

Benzoyl chloride Acetic acid Tetrahedral intermediate

Loss of a proton and of chloride ion from the tetrahedral intermediate yields the mixed anhydride.

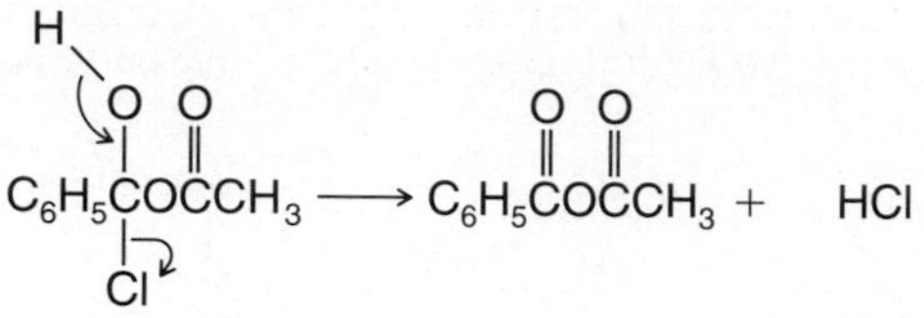

$$C_6H_5\underset{Cl}{\overset{OH}{C}}O\overset{O}{\overset{\|}{C}}CH_3 \longrightarrow C_6H_5\overset{O}{\overset{\|}{C}}O\overset{O}{\overset{\|}{C}}CH_3 + HCl$$

Tetrahedral intermediate | Acetic benzoic anhydride | Hydrogen chloride

Nucleophilic substitution in *acyl* chlorides occurs much more readily than nucleophilic substitution in *alkyl* chlorides. Benzoyl chloride ($C_6H_5\overset{O}{\overset{\|}{C}}Cl$), for example, is approximately 1000 times as reactive as benzyl chloride ($C_6H_5CH_2Cl$) toward solvolysis (80% ethanol : 20% water, 25°C). Nucleophilic acyl substitution does not involve carbocation intermediates such as those formed in S_N1 reactions of alkyl halides, nor does it proceed by way of the crowded pentacoordinate transition states that characterize the S_N2 mechanism of nucleophilic alkyl substitution. The transition state for nucleophilic acyl substitution leads to an intermediate that has a tetrahedral arrangement of bonds. This transition state can be achieved with a relatively modest expenditure of energy, especially when the carbonyl carbon is fairly electrophilic, as it is in an acyl halide.

20.4 PREPARATION OF CARBOXYLIC ACID ANHYDRIDES

After acyl halides, the next most reactive class of carboxylic acid derivatives is acid anhydrides. The most readily available acid anhydride is acetic anhydride. Millions of tons of acetic anhydride are prepared annually by the reaction of acetic acid with a substance known as *ketene.*

The bonding in ketene is similar to that in allene ($CH_2{=}C{=}CH_2$). The carbon that bears two double bonds is *sp* hybridized.

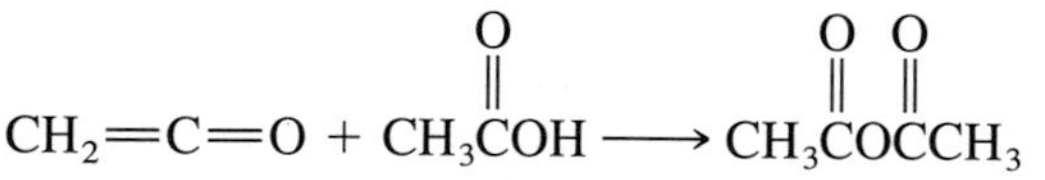

$$CH_2{=}C{=}O + CH_3\overset{O}{\overset{\|}{C}}OH \longrightarrow CH_3\overset{O}{\overset{\|}{C}}O\overset{O}{\overset{\|}{C}}CH_3$$

Ketene | Acetic acid | Acetic anhydride

Acetic anhydride has several commercial applications, including the synthesis of aspirin (Section 24.10) and the preparation of cellulose acetate for use in plastics and fibers. Two cyclic anhydrides, phthalic anhydride and maleic anhydride, are industrial chemicals.

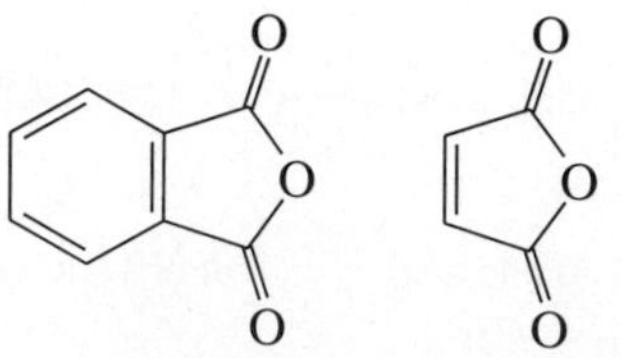

Phthalic anhydride | Maleic anhydride

The customary method for the laboratory synthesis of acid anhydrides is the reaction of acyl chlorides with carboxylic acids (Table 20.2).

Acid anhydrides rarely occur naturally. One example is the putative aphrodisiac *cantharidin,* obtained from a species of beetle.

$$\mathrm{RCCl} + \mathrm{R'COH} + \text{Pyridine} \longrightarrow \mathrm{RCOCR'} + \text{Pyridinium } \mathrm{Cl^-}$$

Acyl chloride | Carboxylic acid | Pyridine | Carboxylic acid anhydride | Pyridinium chloride

This procedure is applicable to the preparation of both symmetrical anhydrides (R and R′ the same) and mixed anhydrides (R and R′ different).

PROBLEM 20.5 Benzoic anhydride has been prepared in excellent yield by adding one molar equivalent of water to two molar equivalents of benzoyl chloride. How do you suppose this reaction takes place?

Cyclic anhydrides in which the ring is five-membered or six-membered are sometimes prepared by heating the corresponding dicarboxylic acids in an inert solvent:

$$\text{Maleic acid} \xrightarrow[130^\circ\mathrm{C}]{\text{tetrachloroethane}} \text{Maleic anhydride (89\%)} + H_2O$$

Maleic acid | Maleic anhydride (89%) | Water

20.5 REACTIONS OF CARBOXYLIC ACID ANHYDRIDES

The most important reactions of acid anhydrides involve cleavage of a bond between oxygen and one of the carbonyl groups. One acyl group is transferred to an attacking nucleophile; the other retains its single bond to oxygen and becomes the acyl group of a carboxylic acid.

$$\mathrm{RC{-}OCR} + \mathrm{HY{:}} \longrightarrow \mathrm{RC{-}\ddot{Y}} + \mathrm{HOCR}$$

Bond cleavage occurs here in an acid anhydride. | Nucleophile | Product of nucleophilic acyl substitution | Carboxylic acid

One reaction of this type is familiar to you, namely, Friedel-Crafts acylation (Section 12.7):

$$\underset{\text{Acid anhydride}}{\overset{\text{O O}}{\text{RCOCR}}} + \underset{\text{Arene}}{\text{ArH}} \xrightarrow{AlCl_3} \underset{\text{Ketone}}{\overset{\text{O}}{\text{RCAr}}} + \underset{\text{Carboxylic acid}}{\overset{\text{O}}{\text{RCOH}}}$$

$$\underset{\text{Acetic anhydride}}{CH_3\overset{O}{C}O\overset{O}{C}CH_3} + \underset{o\text{-Fluoroanisole}}{\text{(2-F-C}_6\text{H}_4\text{)}-OCH_3} \xrightarrow{AlCl_3} \underset{\text{3-Fluoro-4-methoxyacetophenone (70–80\%)}}{CH_3\overset{O}{C}-\text{(3-F-C}_6\text{H}_3\text{)}-OCH_3} + \underset{\text{Acetic acid}}{CH_3CO_2H}$$

An acyl cation is an intermediate in Friedel-Crafts acylation reactions.

PROBLEM 20.6 Write a structural formula for the acyl cation intermediate in the preceding reaction.

Conversions of acid anhydrides to other carboxylic acid derivatives are illustrated in Table 20.3. Since a more highly stabilized carbonyl group must result in order for acyl transfer to be effective, acid anhydrides are readily converted to carboxylic acids, esters, and amides but not to acyl chlorides.

PROBLEM 20.7 Apply the knowledge gained by studying Table 20.3 to help you predict the major organic product of each of the following reactions:

(*a*) Benzoic anhydride + methanol $\xrightarrow{H^+}$
(*b*) Acetic anhydride + ammonia (2 mol) $\longrightarrow$
(*c*) Phthalic anhydride + $(CH_3)_2NH$ (2 mol) $\longrightarrow$
(*d*) Phthalic anhydride + sodium hydroxide (2 mol) $\longrightarrow$

SAMPLE SOLUTION (*a*) Nucleophilic acyl substitution by an alcohol on an acid anhydride yields an ester.

$$\underset{\text{Benzoic anhydride}}{C_6H_5\overset{O}{C}O\overset{O}{C}C_6H_5} + \underset{\text{Methanol}}{CH_3OH} \xrightarrow{H^+} \underset{\text{Methyl benzoate}}{C_6H_5\overset{O}{C}OCH_3} + \underset{\text{Benzoic acid}}{C_6H_5\overset{O}{C}OH}$$

The first example cited in Table 20.3 introduces a new aspect of nucleophilic acyl substitution that is relevant not only to the reactions of acid anhydrides but also to those of acyl chlorides, esters, and amides as well. Nucleophilic acyl substitutions are subject to catalysis by acids.

We can describe how an acid catalyst increases the rate of nucleophilic acyl substitution by considering the hydrolysis of an acid anhydride. Formation of the tetrahedral intermediate is rate-determining, and it is this step that is accelerated by the catalyst. The acid anhydride is activated toward nucleophilic addition by protonation of one of its carbonyl groups:

$$\mathrm{RC(=\ddot{O})O C(=\ddot{O})R} + \mathrm{H^+} \xrightarrow{\text{fast}} \mathrm{RC(=\overset{+}{O}H)OC(=\ddot{O})R}$$

Acid anhydride — Proton — Protonated form of acid anhydride

TABLE 20.3
Conversion of Acid Anhydrides to Other Carboxylic Acid Derivatives

Reaction (section) and comments	General equation and specific example
Reaction with alcohols (Section 15.8) Acid anhydrides react with alcohols to form esters. The reaction may be carried out in the presence of pyridine or it may be catalyzed by acids. In the example shown, only one acetyl group of acetic anhydride becomes incorporated into the ester; the other becomes the acetyl group of an acetic acid molecule.	$\mathrm{RCOCR} + \mathrm{R'OH} \longrightarrow \mathrm{RCOR'} + \mathrm{RCOH}$ (Acid anhydride, Alcohol, Ester, Carboxylic acid) $\mathrm{CH_3COCCH_3} + \mathrm{HOCH(CH_3)CH_2CH_3} \xrightarrow{H_2SO_4} \mathrm{CH_3COCH(CH_3)CH_2CH_3}$ (Acetic anhydride, *sec*-Butyl alcohol, *sec*-Butyl acetate (60%))
Reaction with ammonia and amines (Section 20.13) Acid anhydrides react with ammonia and amines to form amides. Two molar equivalents of amine are required. In the example shown, only one acetyl group of acetic anhydride becomes incorporated into the amide; the other becomes the acetyl group of the amine salt of acetic acid.	$\mathrm{RCOCR} + 2\mathrm{R'_2NH} \longrightarrow \mathrm{RCNR'_2} + \mathrm{RCO^-\ H_2\overset{+}{N}R'_2}$ (Acid anhydride, Amine, Amide, Ammonium carboxylate salt) $\mathrm{CH_3COCCH_3} + \mathrm{H_2N{-}C_6H_4{-}CH(CH_3)_2} \longrightarrow \mathrm{CH_3CNH{-}C_6H_4{-}CH(CH_3)_2}$ (Acetic anhydride, *p*-Isopropylaniline, *p*-Isopropylacetanilide (98%))
Hydrolysis (Section 20.5) Acid anhydrides react with water to yield two carboxylic acid functions. Cyclic anhydrides yield dicarboxylic acids.	$\mathrm{RCOCR'} + \mathrm{H_2O} \longrightarrow 2\mathrm{RCOH}$ (Acid anhydride, Water, Carboxylic acid) Phthalic anhydride + $\mathrm{H_2O}$ $\longrightarrow$ Phthalic acid ($\mathrm{C_6H_4(COOH)_2}$) (Phthalic anhydride, Water, Phthalic acid)

The protonated form of the acid anhydride is present to only a very small extent, but it is quite electrophilic. Water (and other nucleophiles) add to a protonated carbonyl group much faster than they do to a neutral one. Thus, the rate-determining nucleophilic addition of water to form a tetrahedral intermediate takes place more rapidly in the presence of an acid than in its absence.

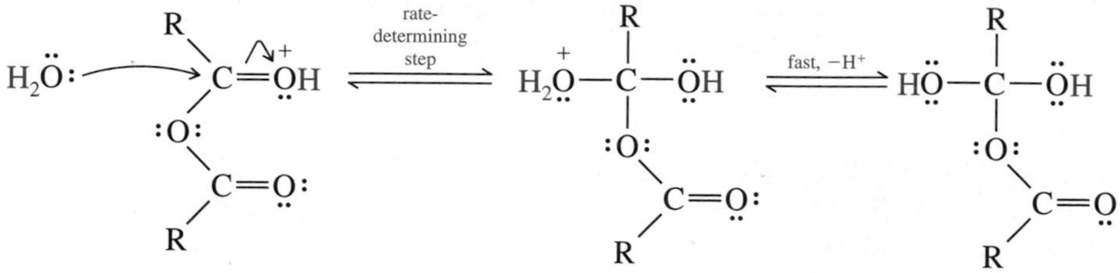

rate-determining step; fast, $-H^+$

Water — Protonated form of an acid anhydride — Tetrahedral intermediate

An acid catalyst also facilitates the dissociation of the tetrahedral intermediate. Protonation of its carbonyl permits the leaving group to depart as a neutral carboxylic acid molecule, which is a less basic leaving group than a carboxylate anion.

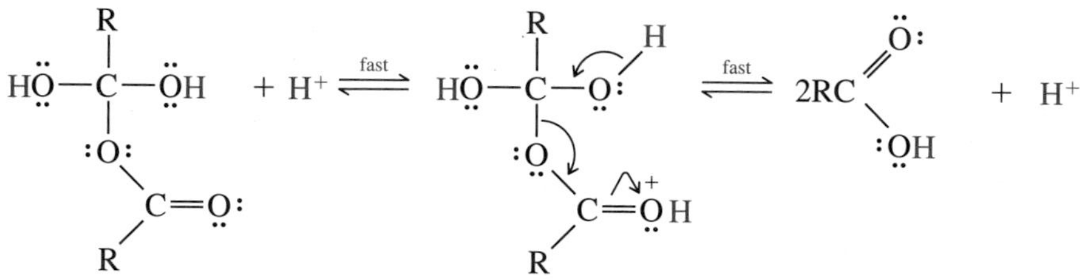

Tetrahedral intermediate $+ H^+$ (fast) ⇌ (fast) ⇌ $2RC(=O)OH + H^+$

Tetrahedral intermediate — Two carboxylic acid molecules — Proton

This pattern of increased reactivity resulting from carbonyl group protonation has been seen before in nucleophilic additions to aldehydes and ketones (Section 17.7) and in the mechanism of the acid-catalyzed esterification of carboxylic acids (Section 19.14). Many biological reactions involve nucleophilic acyl substitution and are catalyzed by enzymes that act by donating a proton to the carbonyl oxygen, the leaving group, or both.

Section 27.19 gives an example of protein hydrolysis in which an enzyme acts as a very efficient acid catalyst.

PROBLEM 20.8 Write the structure of the tetrahedral intermediate formed in each of the reactions given in Problem 20.7. Using curved arrows, show how each tetrahedral intermediate dissociates to the appropriate products.

SAMPLE SOLUTION (*a*) The reaction given is the acid-catalyzed esterification of methanol by benzoic anhydride. The first step is the activation of the anhydride toward nucleophilic addition by protonation.

$$C_6H_5COCC_6H_5 + H^+ \rightleftharpoons C_6H_5COCC_6H_5 \text{ (protonated)}$$

Benzoic anhydride — Proton — Protonated form of benzoic anhydride

The tetrahedral intermediate is formed by nucleophilic addition of methanol to the protonated carbonyl group.

$$CH_3\ddot{O}H + (C_6H_5)C{=}\overset{+}{O}H\text{-}O\text{-}C(=O)C_6H_5 \rightleftharpoons CH_3\overset{+}{O}(H)\text{-}C(C_6H_5)(OH)\text{-}O\text{-}C(=O)C_6H_5 \rightleftharpoons CH_3O\text{-}C(C_6H_5)(OH)\text{-}O\text{-}C(=O)C_6H_5 + H^+$$

Methanol | Protonated form of benzoic anhydride | Tetrahedral intermediate

$$CH_3O\text{-}C(C_6H_5)(OH)\text{-}O\text{-}C(=\ddot{O})C_6H_5 + H^+ \rightleftharpoons CH_3O\text{-}C(C_6H_5)(OH)\text{-}O\text{-}C(=\overset{+}{O}H)C_6H_5 \longrightarrow C_6H_5\overset{O}{\overset{\|}{C}}OCH_3 + C_6H_5\overset{O}{\overset{\|}{C}}OH + H^+$$

Tetrahedral intermediate | Proton | Methyl benzoate | Benzoic acid | Proton

---

Acid anhydrides are more stable and less reactive than acyl chlorides. Acetyl chloride, for example, undergoes hydrolysis about 10^5 times more rapidly than acetic anhydride at 25°C.

20.6 SOURCES OF ESTERS

Many esters occur naturally. Those of low molecular weight are fairly volatile, and many have pleasing odors. Esters often form a significant fraction of the fragrant oil of fruits and flowers.

$$CH_3\overset{O}{\overset{\|}{C}}OCH_2CH_2CH(CH_3)_2$$

3-Methylbutyl acetate (contributes to characteristic odor of bananas)

$$o\text{-}HOC_6H_4\overset{O}{\overset{\|}{C}}OCH_3$$

Methyl salicylate (principal component of oil of wintergreen)

Among the chemicals used by insects to communicate with one another, esters occur frequently.

Notice that (*Z*)-5-tetradecen-4-olide is a cyclic ester. Recall from Section 19.15 that cyclic esters are called *lactones* and that the suffix *-olide* is characteristic of IUPAC names for lactones.

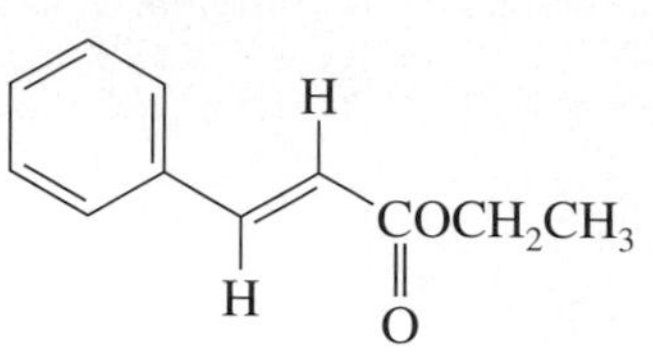

$C_6H_5CH{=}CHCOCH_2CH_3$ (with C=O)

Ethyl cinnamate
(one of the constituents of the sex pheromone of the male oriental fruit moth)

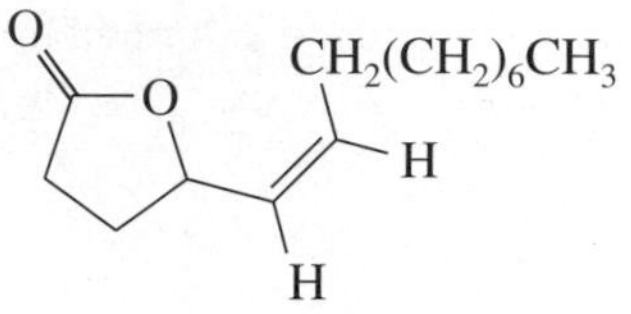

$CH_2(CH_2)_6CH_3$

(*Z*)-5-Tetradecen-4-olide
(sex pheromone of female Japanese beetle)

Esters of glycerol, called *glycerol triesters, triacylglycerols,* or *triglycerides,* are abundant natural products. The most important group of glycerol triesters includes those in which each acyl group is unbranched and has 14 or more carbon atoms.

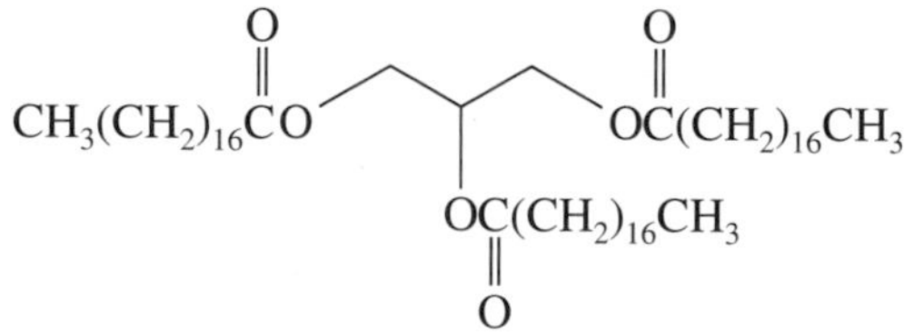

$$CH_3(CH_2)_{16}\overset{O}{\overset{\|}{C}}O\text{–}CH_2\text{–}CH(O\overset{O}{\overset{\|}{C}}(CH_2)_{16}CH_3)\text{–}CH_2\text{–}O\overset{O}{\overset{\|}{C}}(CH_2)_{16}CH_3$$

Tristearin, a trioctadecanoyl ester of glycerol found in many animal and vegetable fats (the three carbons and three oxygens of glycerol are shown in red)

Fats and **oils** are naturally occurring mixtures of glycerol triesters. Fats are mixtures that are solids at room temperature; oils are liquids. The long-chain carboxylic acids obtained from fats and oils by hydrolysis are known as **fatty acids.**

The chief methods used to prepare esters in the laboratory have all been described earlier, and are summarized in Table 20.4.

20.7 PHYSICAL PROPERTIES OF ESTERS

Esters are moderately polar, with dipole moments in the 1.5- to 2.0-D range. Dipole-dipole interactions contribute to the intermolecular attractive forces that cause esters to have higher boiling points than hydrocarbons of similar shape and molecular weight. Because they lack hydroxyl groups, ester molecules cannot form hydrogen bonds to each other; consequently, esters have lower boiling points than alcohols of comparable molecular weight.

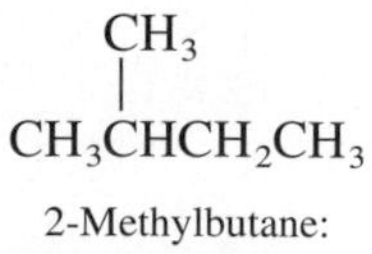

$$CH_3\overset{CH_3}{\overset{|}{C}}HCH_2CH_3 \qquad CH_3\overset{O}{\overset{\|}{C}}OCH_3 \qquad CH_3\overset{OH}{\overset{|}{C}}HCH_2CH_3$$

2-Methylbutane: mol wt 72, bp 28°C | Methyl acetate: mol wt 74, bp 57°C | 2-Butanol: mol wt 74, bp 99°C

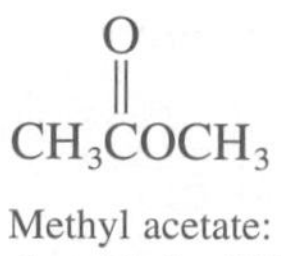

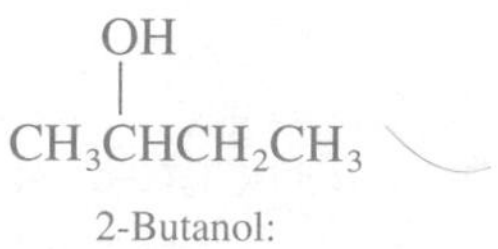

Esters can participate in hydrogen bonds with substances that contain hydroxyl groups (water, alcohols, carboxylic acids). This confers some measure of water solubility on low-molecular-weight esters; methyl acetate, for example, dissolves in water to the extent of 33 g/100 mL. Water solubility decreases as the carbon content of the ester increases. Fats and oils, the glycerol esters of long-chain carboxylic acids, are practically insoluble in water.

TABLE 20.4
Preparation of Esters

Reaction (section) and comments	General equation and specific example
From carboxylic acids (Sections 15.8 and 19.14) In the presence of an acid catalyst, alcohols and carboxylic acids react to form an ester and water. This is the Fischer esterification.	$\underset{\text{Carboxylic acid}}{RCOH} + \underset{\text{Alcohol}}{R'OH} \overset{H^+}{\rightleftharpoons} \underset{\text{Ester}}{RCOR'} + \underset{\text{Water}}{H_2O}$ (each C=O shown) $\underset{\text{Propanoic acid}}{CH_3CH_2COH} + \underset{\text{1-Butanol}}{CH_3CH_2CH_2CH_2OH} \xrightarrow{H_2SO_4} \underset{\text{Butyl propanoate (85\%)}}{CH_3CH_2COCH_2CH_2CH_2CH_3} + \underset{\text{Water}}{H_2O}$
From acyl chlorides (Sections 15.8 and 20.3) Alcohols react with acyl chlorides by nucleophilic acyl substitution to yield esters. These reactions are typically performed in the presence of a weak base such as pyridine.	$\underset{\text{Acyl chloride}}{RCCl} + \underset{\text{Alcohol}}{R'OH} + \underset{\text{Pyridine}}{C_5H_5N} \longrightarrow \underset{\text{Ester}}{RCOR'} + \underset{\text{Pyridinium chloride}}{C_5H_5NH^+\ Cl^-}$ $\underset{\text{3,5-Dinitrobenzoyl chloride}}{3,5\text{-}(O_2N)_2C_6H_3CCl} + \underset{\text{Isobutyl alcohol}}{(CH_3)_2CHCH_2OH} \xrightarrow{\text{pyridine}} \underset{\text{Isobutyl 3,5-dinitrobenzoate (85\%)}}{3,5\text{-}(O_2N)_2C_6H_3COCH_2CH(CH_3)_2}$
From carboxylic acid anhydrides (Sections 15.8 and 20.5) Acyl transfer from an acid anhydride to an alcohol is a standard method for the preparation of esters. The reaction is subject to catalysis by either acids (H_2SO_4) or bases (pyridine).	$\underset{\text{Acid anhydride}}{RCOCR} + \underset{\text{Alcohol}}{R'OH} \longrightarrow \underset{\text{Ester}}{RCOR'} + \underset{\text{Carboxylic acid}}{RCOH}$ $\underset{\text{Acetic anhydride}}{CH_3COCCH_3} + \underset{m\text{-Methoxybenzyl alcohol}}{m\text{-}CH_3OC_6H_4CH_2OH} \xrightarrow{\text{pyridine}} \underset{m\text{-Methoxybenzyl acetate (99\%)}}{m\text{-}CH_3OC_6H_4CH_2OCCH_3}$
Baeyer-Villiger oxidation of ketones (Section 17.18) Ketones are converted to esters on treatment with peroxy acids. The reaction proceeds by migration of the group R′ from carbon to oxygen. It is the more highly substituted group that migrates. Methyl ketones give acetate esters.	$\underset{\text{Ketone}}{RCR'} + \underset{\text{Peroxy acid}}{R''COOH} \longrightarrow \underset{\text{Ester}}{RCOR'} + \underset{\text{Carboxylic acid}}{R''COH}$ $\underset{\text{Cyclopropyl methyl ketone}}{CH_3C\text{-}C_3H_5} \xrightarrow{CF_3CO_2OH} \underset{\text{Cyclopropyl acetate (53\%)}}{CH_3CO\text{-}C_3H_5}$

20.8 REACTIONS OF ESTERS: A REVIEW AND A PREVIEW

The reaction of esters with Grignard reagents and with lithium aluminum hydride, both useful in the synthesis of alcohols, have been described earlier. They are reviewed in Table 20.5.

Nucleophilic acyl substitutions at the ester carbonyl group are summarized in Table 20.6. Esters are less reactive than acyl chlorides and acid anhydrides. Nucleophilic acyl substitution in esters, especially ester hydrolysis, has been extensively investigated from a mechanistic perspective. Indeed, much of what we know concerning the general topic of nucleophilic acyl substitution derives from studies carried out on esters. The following sections describe those mechanistic studies.

20.9 ACID-CATALYZED ESTER HYDROLYSIS

Ester hydrolysis is the most studied and best understood of all nucleophilic acyl substitutions. Esters are fairly stable in neutral aqueous media but are cleaved when heated

TABLE 20.5
Summary of Reactions of Esters Discussed in Earlier Chapters

Reaction (section) and comments	General equation and specific example
Reaction with Grignard reagents (Section 14.10) Esters react with two equivalents of a Grignard reagent to produce tertiary alcohols. Two of the groups bonded to the carbon that bears the hydroxyl group in the tertiary alcohol are derived from the Grignard reagent.	$RC(=O)OR' + 2R''MgX \xrightarrow[2.\ H_3O^+]{1.\ \text{diethyl ether}} RC(OH)R''_2 + R'OH$ Ester, Grignard reagent, Tertiary alcohol, Alcohol $\text{cyclopropyl-}C(=O)OCH_2CH_3 + 2CH_3MgI \xrightarrow[2.\ H_3O^+]{1.\ \text{diethyl ether}} \text{cyclopropyl-}C(OH)(CH_3)CH_3 + CH_3CH_2OH$ Ethyl cyclopropanecarboxylate, Methylmagnesium iodide, 2-Cyclopropyl-2-propanol (93%), Ethanol
Reduction with lithium aluminum hydride (Section 15.3) Lithium aluminum hydride cleaves esters to yield two alcohols.	$RC(=O)OR' \xrightarrow[2.\ H_2O]{1.\ LiAlH_4} RCH_2OH + R'OH$ Ester, Primary alcohol, Alcohol $C_6H_5C(=O)OCH_2CH_3 \xrightarrow[2.\ H_2O]{1.\ LiAlH_4} C_6H_5CH_2OH + CH_3CH_2OH$ Ethyl benzoate, Benzyl alcohol (90%), Ethyl alcohol

TABLE 20.6
Conversion of Esters to Other Carboxylic Acid Derivatives

Reaction (section) and comments	General equation and specific example
Reaction with ammonia and amines (Section 20.13) Esters react with ammonia and amines to form amides. Methyl and ethyl esters are the most reactive.	$RCOR' + R''_2NH \longrightarrow RCNR''_2 + R'OH$ (C=O on each carbonyl carbon) Ester, Amine, Amide, Alcohol $FCH_2COCH_2CH_3 + NH_3 \xrightarrow{H_2O} FCH_2CNH_2 + CH_3CH_2OH$ Ethyl fluoroacetate, Ammonia, Fluoroacetamide (90%), Ethanol
Hydrolysis (Sections 20.9 and 20.10) Ester hydrolysis may be catalyzed either by acids or by bases. Acid-catalyzed hydrolysis is an equilibrium-controlled process, the reverse of the Fischer esterification. Hydrolysis in base is irreversible and is the method usually chosen for preparative purposes.	$RCOR' + H_2O \longrightarrow RCOH + R'OH$ Ester, Water, Carboxylic acid, Alcohol $m\text{-}O_2NC_6H_4COCH_3 \xrightarrow[2.\ H^+]{1.\ H_2O,\ NaOH} m\text{-}O_2NC_6H_4COH + CH_3OH$ Methyl *m*-nitrobenzoate, *m*-Nitrobenzoic acid (90–96%), Methanol

with water in the presence of strong acids or bases. The hydrolysis of esters in dilute aqueous acid is the reverse of the Fischer esterification (Sections 15.8 and 19.14):

$$RCOR' + H_2O \overset{H^+}{\rightleftharpoons} RCOH + R'OH$$

Ester, Water, Carboxylic acid, Alcohol

When esterification is the objective, water is removed from the reaction mixture to encourage ester formation. When ester hydrolysis is the objective, the reaction is carried out in the presence of a generous excess of water.

$$C_6H_5CHCOCH_2CH_3(Cl) + H_2O \xrightarrow[heat]{HCl} C_6H_5CHCOH(Cl) + CH_3CH_2OH$$

Ethyl 2-chloro-2-phenylacetate, Water, 2-Chloro-2-phenylacetic acid (80–82%), Ethyl alcohol

PROBLEM 20.9 The compound having the structure shown was heated with dilute sulfuric acid to give a product having the molecular formula $C_5H_{12}O_3$ in 63 to 71 percent yield. Propose a reasonable structure for this product. What other organic compound is formed in this reaction?

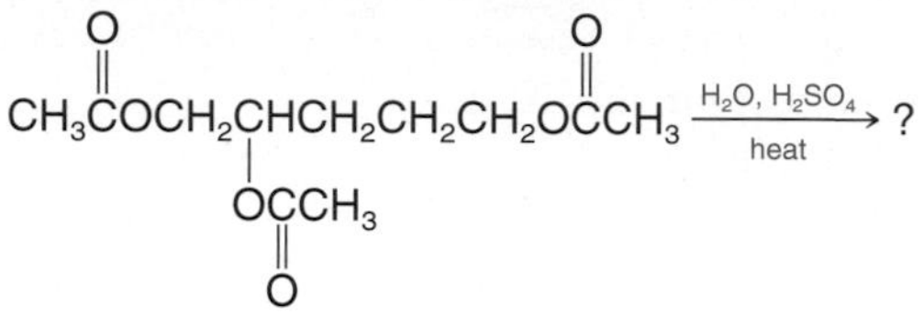

$$CH_3\overset{O}{\overset{\|}{C}}OCH_2\underset{O\underset{\underset{O}{\|}}{C}CH_3}{\underset{|}{C}}HCH_2CH_2CH_2O\overset{O}{\overset{\|}{C}}CH_3 \xrightarrow[\text{heat}]{H_2O,\ H_2SO_4} ?$$

The mechanism of acid-catalyzed ester hydrolysis is presented in Figure 20.4. It is precisely the reverse of the mechanism given for acid-catalyzed ester formation in Section 19.14. Like other nucleophilic acyl substitutions, it proceeds in two stages. A tetrahedral intermediate is formed in the first stage, and this tetrahedral intermediate dissociates to products in the second stage.

A key feature of the first stage is the site at which the starting ester is protonated. Protonation of the carbonyl oxygen, as shown in step 1 of Figure 20.4, gives a cation that is stabilized by electron delocalization. The alternative site of protonation, the alkoxy oxygen, gives rise to a much less stable cation.

Protonation of carbonyl oxygen **Protonation of alkoxy oxygen**

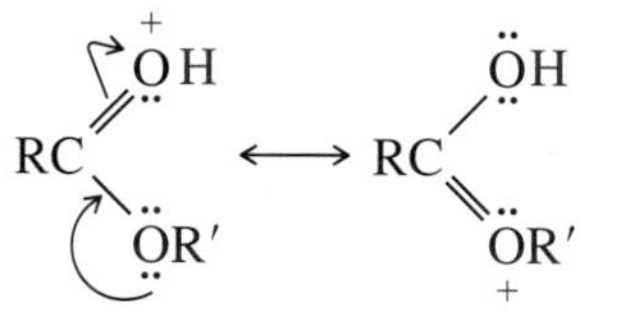

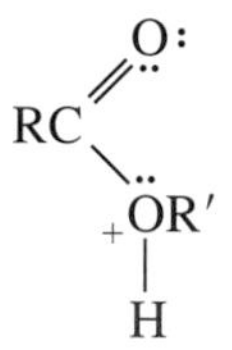

$$RC(=\overset{+}{O}H)-\ddot{O}R' \longleftrightarrow RC(-\ddot{O}H)=\overset{+}{O}R' \qquad\qquad RC(=\ddot{O}:)-\overset{+}{O}(H)R'$$

Positive charge is delocalized. Positive charge is localized on a single oxygen.

Protonation of the carbonyl oxygen, as noted earlier in the reactions of aldehydes and ketones, makes the carbonyl group more susceptible to nucleophilic attack (Section 17.7). A water molecule adds to the carbonyl group of the protonated ester in step 2. Loss of a proton from the resulting oxonium ion gives the neutral form of the tetrahedral intermediate in step 3 and completes the first stage of the mechanism.

Once formed, the tetrahedral intermediate can revert to starting materials by merely reversing the reactions that formed it, or it can continue onward to products. In the second stage of the reaction, the tetrahedral intermediate dissociates to an alcohol and a carboxylic acid. In step 4 of Figure 20.4, protonation of the tetrahedral intermediate at its alkoxy oxygen gives a new oxonium ion, which loses a molecule of alcohol in step 5. Along with the alcohol, the protonated form of the carboxylic acid arises by dissociation of the tetrahedral intermediate. Its deprotonation in step 6 completes the process.

PROBLEM 20.10 On the basis of the general mechanism for acid-catalyzed ester hydrolysis shown in Figure 20.4, write an analogous sequence of steps for the specific case of ethyl benzoate hydrolysis.

The most important facet of the mechanism outlined in Figure 20.4 is the tetrahedral intermediate. Evidence in support of the existence of such a species was developed by

Step 1: Protonation of the carbonyl oxygen of the ester

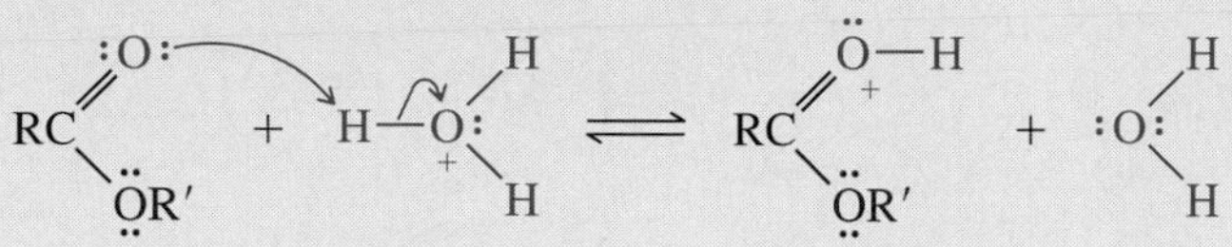

Ester + Hydronium ion ⇌ Protonated form of ester + Water

Step 2: Nucleophilic addition of water to protonated form of ester

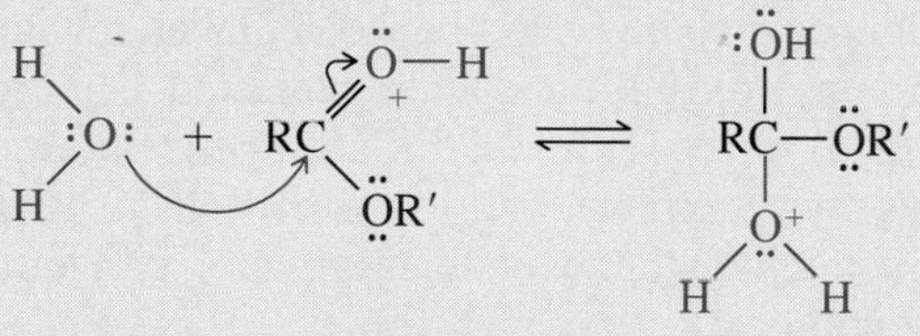

Water + Protonated form of ester ⇌ Oxonium ion

Step 3: Deprotonation of the oxonium ion to give the neutral form of the tetrahedral intermediate

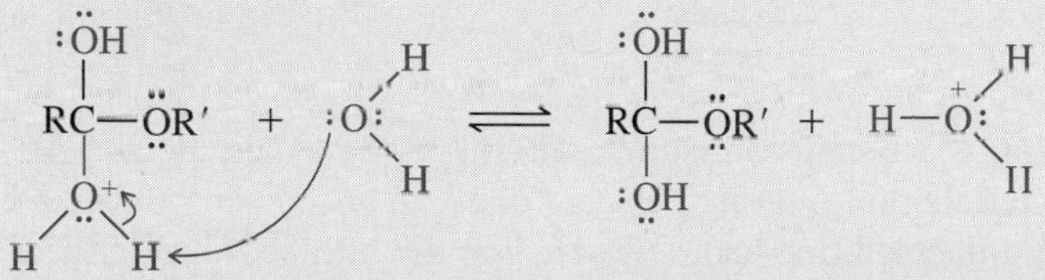

Oxonium ion + Water ⇌ Tetrahedral intermediate + Hydronium ion

Step 4: Protonation of the tetrahedral intermediate at its alkoxy oxygen

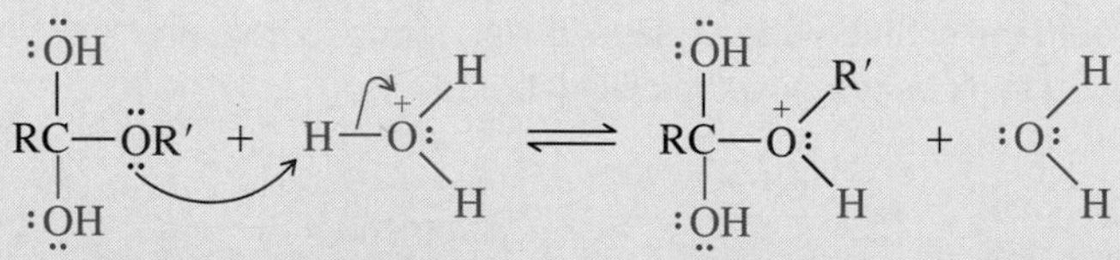

Tetrahedral intermediate + Hydronium ion ⇌ Oxonium ion + Water

Step 5: Dissociation of the protonated form of the tetrahedral intermediate to an alcohol and the protonated form of the carboxylic acid

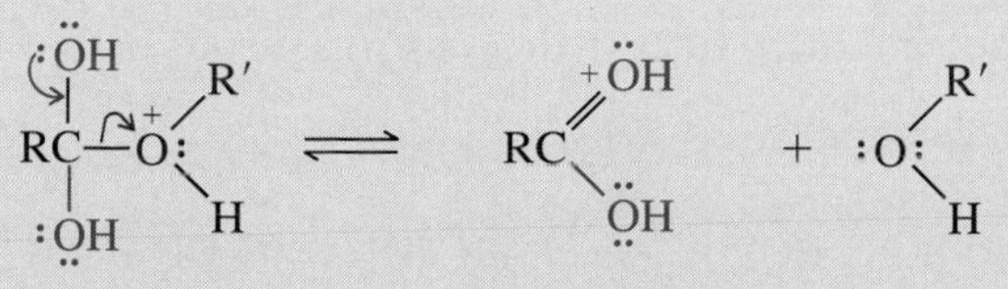

Oxonium ion ⇌ Protonated form of carboxylic acid + Alcohol

FIGURE 20.4 Sequence of steps that describes the mechanism of acid-catalyzed ester hydrolysis. Steps 1 through 3 show the formation of the tetrahedral intermediate. Dissociation of the tetrahedral intermediate is shown in steps 4 through 6.

Step 6: Deprotonation of the protonated carboxylic acid

$$RC(=\overset{+}{O}H)OH + H_2O \rightleftharpoons RC(=O)OH + H_3O^+$$

Protonated form of carboxylic acid | Water | Carboxylic acid | Hydronium ion

FIGURE 20.4 *(Continued)*

Professor Myron Bender on the basis of isotopic labeling experiments he carried out at the University of Chicago. Bender prepared ethyl benzoate, labeled with the mass 18 isotope of oxygen at the carbonyl oxygen, and then subjected it to acid-catalyzed hydrolysis in ordinary (unlabeled) water. He found that ethyl benzoate, recovered from the reaction before hydrolysis was complete, had lost a portion of its isotopic label. This observation is consistent only with the reversible formation of a tetrahedral intermediate under the reaction conditions:

$$C_6H_5C(=O)OCH_2CH_3 + H_2O \overset{H^+}{\rightleftharpoons} C_6H_5C(OH)(OH)OCH_2CH_3 \overset{H^+}{\rightleftharpoons} C_6H_5C(=O)OCH_2CH_3 + H_2O$$

Ethyl benzoate (labeled with ^{18}O) | Water | Tetrahedral intermediate | Ethyl benzoate | Water (labeled with ^{18}O)

In this equation, O indicates an isotopically labeled oxygen atom. The two hydroxyl groups in the tetrahedral intermediate are equivalent to each other, and so either the labeled or the unlabeled one can be lost when the tetrahedral intermediate reverts to ethyl benzoate. Both are retained when the tetrahedral intermediate goes on to form benzoic acid.

PROBLEM 20.11 In a similar experiment, unlabeled 4-butanolide was allowed to stand in an acidic solution in which the water had been labeled with ^{18}O. When the lactone was extracted from the solution after 4 days, it was found to contain ^{18}O. Which oxygen of the lactone do you think became isotopically labeled?

4-Butanolide

20.10 BASE-PROMOTED ESTER HYDROLYSIS: SAPONIFICATION

Unlike its acid-catalyzed counterpart, ester hydrolysis in aqueous base is *irreversible.*

$$RCOR' + HO^- \longrightarrow RCO^- + R'OH$$

(RCOR′ and RCO⁻ each with C=O)

Ester | Hydroxide ion | Carboxylate ion | Alcohol

This is because carboxylic acids are converted to their corresponding carboxylate anions under these conditions, and these anions are incapable of acyl transfer to alcohols. Base is consumed in the reaction, and so we speak of it as *base-promoted* rather than base-catalyzed.

In cases in which ester hydrolysis is carried out for preparative purposes, basic conditions are normally chosen.

$$o\text{-}CH_3C_6H_4CH_2OCCH_3\ (C{=}O) + NaOH \xrightarrow[\text{heat}]{\text{water-methanol}} NaOCCH_3\ (C{=}O) + o\text{-}CH_3C_6H_4CH_2OH$$

o-Methylbenzyl acetate — Sodium hydroxide — Sodium acetate — *o*-Methylbenzyl alcohol (95–97%)

When one wishes to isolate the carboxylic acid product, a separate acidification step following hydrolysis is necessary. Acidification converts the carboxylate salt to the free acid.

$$CH_2{=}C(CH_3)COCH_3\ (C{=}O) \xrightarrow[\text{2. } H_2SO_4]{\text{1. NaOH, } H_2O\text{, heat}} CH_2{=}C(CH_3)COH\ (C{=}O) + CH_3OH$$

Methyl 2-methylpropenoate (methyl methacrylate) — 2-Methylpropenoic acid (87%) (methacrylic acid) — Methyl alcohol

Base-promoted ester hydrolysis is called **saponification,** which means ''soap making.'' Over 2000 years ago, the Phoenicians made soap by heating animal fat with wood ashes. Animal fat is rich in glycerol triesters, and wood ashes are a source of potassium carbonate. Base-promoted cleavage of the fats produced a mixture of long-chain carboxylic acids as their potassium salts.

$$CH_3(CH_2)_xCO{-}CH_2CH(OC(CH_2)_yCH_3)CH_2{-}OC(CH_2)_zCH_3 \xrightarrow[\text{heat}]{K_2CO_3,\ H_2O} HOCH_2CHCH_2OH\ (OH) +$$

Glycerol

$$KOC(CH_2)_xCH_3 + KOC(CH_2)_yCH_3 + KOC(CH_2)_zCH_3$$

Potassium carboxylate salts

Potassium and sodium salts of long-chain carboxylic acids form micelles that dissolve grease (Section 19.5) and have cleansing properties. The carboxylic acids obtained by saponification of fats are called *fatty acids.*

PROBLEM 20.12 *Trimyristin* is obtained from coconut oil and has the molecular formula $C_{45}H_{86}O_6$. On being heated with aqueous sodium hydroxide followed by acidification, trimyristin was converted to glycerol and tetradecanoic acid as the only products. What is the structure of trimyristin?

In one of the earliest kinetic studies of an organic reaction, carried out over a century ago, the rate of hydrolysis of ethyl acetate in aqueous sodium hydroxide was shown to be first-order in ester and first-order in base.

$$\underset{\text{Ethyl acetate}}{CH_3\overset{O}{\overset{\|}{C}}OCH_2CH_3} + \underset{\text{Sodium hydroxide}}{NaOH} \longrightarrow \underset{\text{Sodium acetate}}{CH_3\overset{O}{\overset{\|}{C}}ONa} + \underset{\text{Ethanol}}{CH_3CH_2OH}$$

$$\text{Rate} = k[CH_3\overset{O}{\overset{\|}{C}}OCH_2CH_3][NaOH]$$

Overall, the reaction exhibits second-order kinetics. Both the ester and the base are involved in the rate-determining step or in a rapid step that precedes it.

Two processes that are consistent with second-order kinetics both involve hydroxide ion as a nucleophile but differ in the site of nucleophilic attack. One of these processes is simply an S_N2 reaction in which hydroxide displaces carboxylate from the alkyl group of the ester. We say that this pathway involves *alkyl-oxygen cleavage,* because it is the bond between oxygen and the alkyl group of the ester that breaks. The other process involves *acyl-oxygen cleavage,* with hydroxide attacking the carbonyl group.

Alkyl-oxygen cleavage

$$\underset{\text{Ester}}{R\overset{:O:}{\overset{\|}{C}}-\ddot{O}-R'} \quad \underset{\text{Hydroxide ion}}{:\ddot{O}H^-} \longrightarrow \underset{\text{Carboxylate ion}}{R\overset{:O:}{\overset{\|}{C}}\ddot{O}:^-} + \underset{\text{Alcohol}}{R'-\ddot{O}H}$$

Acyl-oxygen cleavage

$$\underset{\text{Hydroxide ion}}{H\ddot{O}:^-} + \underset{\text{Ester}}{R\overset{:O:}{\overset{\|}{C}}-\ddot{O}R'} \xrightarrow{\text{slow}} R\overset{:O:}{\overset{\|}{C}}\ddot{O}H + R'\ddot{O}:^-$$

$$\downarrow \text{fast}$$

$$\underset{\text{Carboxylate ion}}{R\overset{:O:}{\overset{\|}{C}}\ddot{O}:^-} + \underset{\text{Alcohol}}{R'\ddot{O}H}$$

Convincing evidence that base-promoted ester hydrolysis proceeds by the second of these two paths, namely, acyl-oxygen cleavage, has been obtained from several sources. In one experiment, ethyl propanoate labeled with ^{18}O in the ethoxy group was hydrolyzed in base. On isolating the products, all the ^{18}O was found in the ethyl alcohol; there was no ^{18}O enrichment in the sodium propanoate (as shown in the following equation, in which O indicates ^{18}O).

$$\underset{\text{}^{18}\text{O-labeled ethyl propanoate}}{CH_3CH_2\overset{O}{\overset{\|}{C}}OCH_2CH_3} + \underset{\text{Sodium hydroxide}}{NaOH} \longrightarrow \underset{\text{Sodium propanoate}}{CH_3CH_2\overset{O}{\overset{\|}{C}}ONa} + \underset{\text{}^{18}\text{O-labeled ethyl alcohol}}{CH_3CH_2OH}$$

Therefore, the carbon-oxygen bond broken in the process is the one between oxygen and the propanoyl group. The bond between oxygen and the ethyl group remains intact.

PROBLEM 20.13 In a similar experiment, pentyl acetate was subjected to saponification with ^{18}O-labeled hydroxide in ^{18}O-labeled water. What product do you think became isotopically labeled here, acetate ion or 1-pentanol?

Identical conclusions in support of acyl-oxygen cleavage have been obtained from stereochemical studies. Saponification of esters of optically active alcohols proceeds with *retention of configuration.*

$$\underset{(R)\text{-}(+)\text{-1-Phenylethyl acetate}}{CH_3\overset{O}{\overset{\|}{C}}{-}O{-}C(H)(C_6H_5)(CH_3)} \xrightarrow[\text{ethanol-water}]{\text{KOH}} \underset{\text{Potassium acetate}}{CH_3\overset{O}{\overset{\|}{C}}OK} + \underset{\substack{(R)\text{-}(+)\text{-1-Phenylethyl alcohol} \\ \text{(80\% yield; same optical purity as ester)}}}{HO{-}C(H)(C_6H_5)(CH_3)}$$

None of the bonds to the stereogenic center are broken when acyl-oxygen cleavage occurs. Had alkyl-oxygen cleavage occurred instead, it would have been accompanied by inversion of configuration at the stereogenic center to give (*S*)-(−)-1-phenylethyl alcohol.

Once it was established that hydroxide ion attacks the carbonyl group in base-promoted ester hydrolysis, the next question to be addressed concerned whether the reaction is concerted or involves an intermediate. In a concerted reaction acyl-oxygen cleavage occurs at the same time that hydroxide ion attacks the carbonyl group.

$$\underset{\text{Hydroxide ion}}{HO^-} + \underset{\text{Ester}}{R\overset{O}{\overset{\|}{C}}OR'} \longrightarrow \underset{\text{Representation of transition state for concerted displacement}}{\overset{\delta-}{HO}\text{---}\underset{\underset{O}{\|}}{\overset{\overset{R}{|}}{C}}\text{---}\overset{\delta-}{OR'}} \longrightarrow \underset{\text{Carboxylic acid}}{R\overset{O}{\overset{\|}{C}}OH} + \underset{\text{Alkoxide ion}}{R'O^-}$$

In an extension of the work described in the preceding section, Bender showed that base-promoted ester hydrolysis was *not* concerted and, like acid hydrolysis, took place by way of a tetrahedral intermediate. The nature of the experiment was the same, and the results were similar to those observed in the acid-catalyzed reaction. Ethyl benzoate enriched in ^{18}O at the carbonyl oxygen was subjected to base-promoted hydrolysis, and samples were isolated before saponification was complete. The recovered

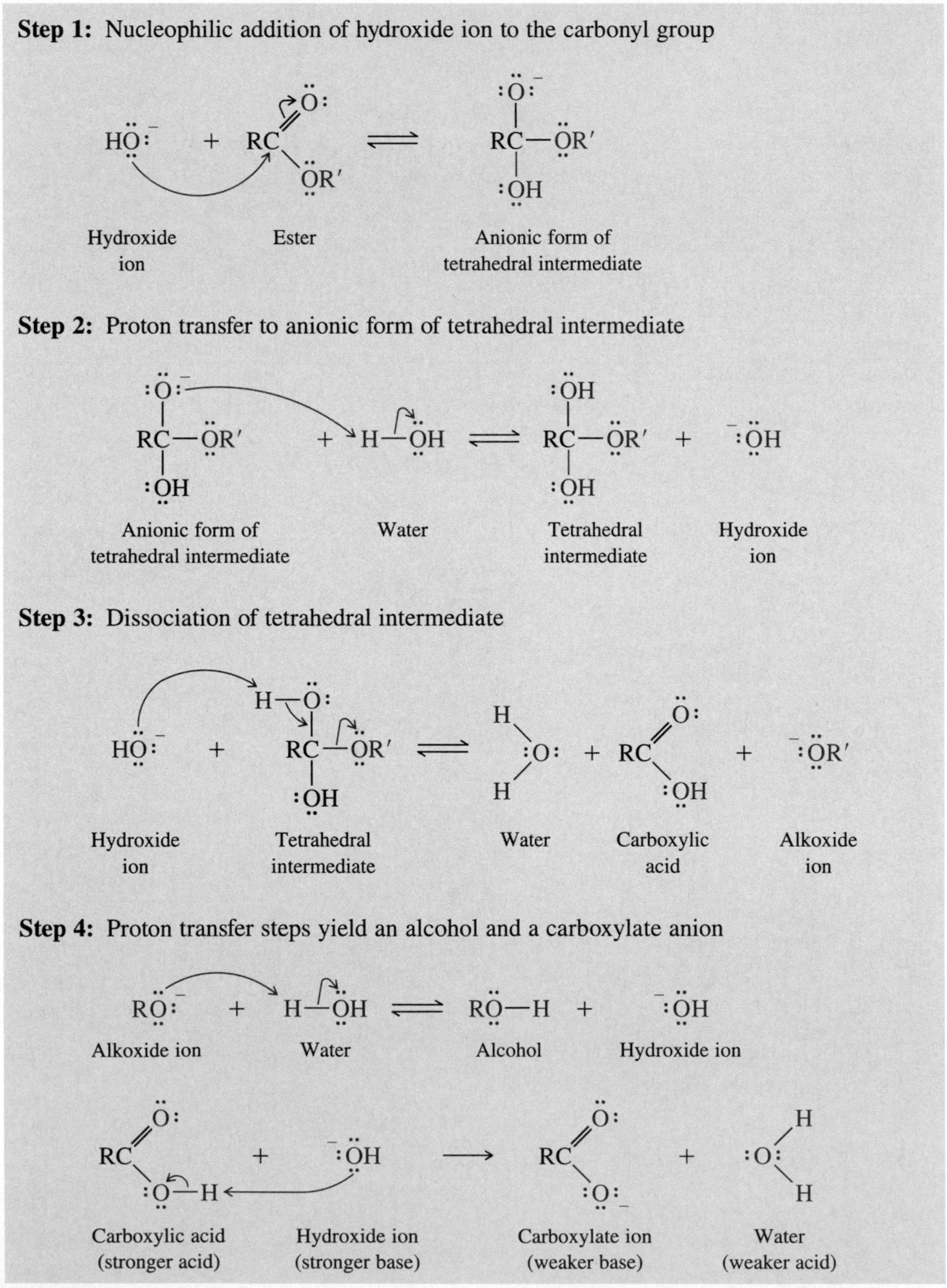

Step 1: Nucleophilic addition of hydroxide ion to the carbonyl group

$$HO^- + RC(=O)OR' \rightleftharpoons RC(O^-)(OR')(OH)$$

Hydroxide ion | Ester | Anionic form of tetrahedral intermediate

Step 2: Proton transfer to anionic form of tetrahedral intermediate

$$RC(O^-)(OR')(OH) + H{-}OH \rightleftharpoons RC(OH)(OR')(OH) + {}^-OH$$

Anionic form of tetrahedral intermediate | Water | Tetrahedral intermediate | Hydroxide ion

Step 3: Dissociation of tetrahedral intermediate

$$HO^- + RC(OH)(OR')(OH) \rightleftharpoons H_2O + RC(=O)OH + {}^-OR'$$

Hydroxide ion | Tetrahedral intermediate | Water | Carboxylic acid | Alkoxide ion

Step 4: Proton transfer steps yield an alcohol and a carboxylate anion

$$RO^- + H{-}OH \rightleftharpoons RO{-}H + {}^-OH$$

Alkoxide ion | Water | Alcohol | Hydroxide ion

$$RC(=O)O{-}H + {}^-OH \longrightarrow RC(=O)O^- + H_2O$$

Carboxylic acid (stronger acid) | Hydroxide ion (stronger base) | Carboxylate ion (weaker base) | Water (weaker acid)

FIGURE 20.5 Sequence of steps that describes the mechanism of base-promoted ester hydrolysis.

ethyl benzoate had lost a portion of its isotopic label, an observation consistent with the formation of a tetrahedral intermediate:

$$C_6H_5C(=O)OCH_2CH_3 + H_2O \underset{}{\overset{HO^-}{\rightleftharpoons}} C_6H_5C(OH)(OH)OCH_2CH_3 \underset{}{\overset{HO^-}{\rightleftharpoons}} C_6H_5C(=O)OCH_2CH_3 + H_2O$$

Ethyl benzoate (labeled with ^{18}O) — Water — Tetrahedral intermediate — Ethyl benzoate — Water (labeled with ^{18}O)

All these facts—the observation of second-order kinetics, acyl-oxygen cleavage, and the involvement of a tetrahedral intermediate—are accommodated by the reaction mechanism shown in Figure 20.5. Like the acid-catalyzed mechanism, it has two distinct stages, namely, formation of the tetrahedral intermediate and its subsequent dissociation. All the steps are reversible except the last one. The equilibrium constant for proton abstraction from the carboxylic acid by hydroxide is so large that step 4 is, for all intents and purposes, irreversible, and this makes the overall reaction irreversible.

Steps 2 and 4 are proton-transfer reactions and are very fast. Nucleophilic addition to the carbonyl group has a higher activation energy than dissociation of the tetrahedral intermediate; step 1 is rate-determining.

PROBLEM 20.14 On the basis of the general mechanism for base-promoted ester hydrolysis shown in Figure 20.5, write an analogous sequence of steps for the saponification of ethyl benzoate.

20.11 REACTION OF ESTERS WITH AMMONIA AND AMINES

Esters react with ammonia to form amides.

$$\mathrm{RCOR'} + NH_3 \longrightarrow \mathrm{RCNH_2} + \mathrm{R'OH}$$

(with C=O in RCOR′ and RCNH₂)

Ester — Ammonia — Amide — Alcohol

Ammonia is more nucleophilic than water, and thus this reaction can be carried out in aqueous solution.

$$CH_2{=}C(CH_3)COCH_3 + NH_3 \xrightarrow{H_2O} CH_2{=}C(CH_3)CNH_2 + CH_3OH$$

Methyl 2-methylpropenoate — Ammonia — 2-Methylpropenamide (75%) — Methyl alcohol

Amines, which are substituted derivatives of ammonia, react similarly:

$$FCH_2\overset{O}{\overset{\|}{C}}OCH_2CH_3 + C_6H_{11}NH_2 \xrightarrow{\text{heat}} FCH_2\overset{O}{\overset{\|}{C}}NH{-}C_6H_{11} + CH_3CH_2OH$$

Ethyl fluoroacetate | Cyclohexylamine | N-Cyclohexylfluoroacetamide (61%) | Ethyl alcohol

The amine must be primary (RNH_2) or secondary (R_2NH). Tertiary amines (R_3N) cannot form amides, because they have no proton on nitrogen that can be replaced by an acyl group.

PROBLEM 20.15 Give the structure of the expected product of the following reaction:

(γ-lactone with CH_3 substituent) $+ CH_3NH_2 \longrightarrow$

The reaction of ammonia and amines with esters follows the same general mechanistic course as other nucleophilic acyl substitution reactions. A tetrahedral intermediate is formed in the first stage of the process and dissociates in the second stage.

Formation of tetrahedral intermediate

$$R\overset{O}{\overset{\|}{C}}OR' + :NH_3 \rightleftharpoons R\overset{O^-}{\overset{|}{\underset{^+NH_3}{\underset{|}{C}}}}OR' \rightleftharpoons R\overset{OH}{\overset{|}{\underset{NH_2}{\underset{|}{C}}}}OR'$$

Ester | Ammonia | | Tetrahedral intermediate

Dissociation of tetrahedral intermediate

$$R\overset{O-H}{\overset{|}{\underset{NH_2}{\underset{|}{C}}}}{-}OR' \rightleftharpoons R\overset{O}{\overset{\|}{C}}NH_2 + R'OH$$

Tetrahedral intermediate | Amide | Alcohol

While both stages are written as equilibrium processes, the position of equilibrium of the overall reaction lies far to the right because the amide carbonyl is stabilized to a much greater extent than the ester carbonyl.

20.12 THIOESTERS

Thioesters, compounds of the type $R\overset{O}{\overset{\|}{C}}SR'$, undergo the same kinds of reactions as esters and by similar mechanisms. Nucleophilic acyl substitution of a thioester gives a *thiol* along with the product of acyl transfer. For example:

$$\underset{\text{S-2-Phenoxyethyl ethanethioate}}{CH_3\overset{O}{\overset{\|}{C}}SCH_2CH_2OC_6H_5} + \underset{\text{Methanol}}{CH_3OH} \xrightarrow{HCl} \underset{\text{Methyl acetate}}{CH_3\overset{O}{\overset{\|}{C}}OCH_3} + \underset{\text{2-Phenoxyethanethiol (90\%)}}{HSCH_2CH_2OC_6H_5}$$

In this reaction an acetyl group is transferred from sulfur to the oxygen of methanol.

PROBLEM 20.16 Write the structure of the tetrahedral intermediate formed in the reaction just described.

The carbon-sulfur bond of a thioester is rather long—typically on the order of 180 pm—and delocalization of the sulfur lone-pair electrons into the π orbital of the carbonyl group is not as pronounced as in esters. Nucleophilic acyl substitution reactions of thioesters occur faster than those of simple esters. A number of important biological processes involve thioesters; several of these are described in Chapter 26.

20.13 PREPARATION OF AMIDES

Amides are readily prepared by acylation of ammonia and amines with acyl chlorides, anhydrides, or esters.

Acylation of *ammonia* (NH_3) yields an amide ($R'\overset{O}{\overset{\|}{C}}NH_2$).

Primary amines (RNH_2) yield *N*-substituted amides ($R'\overset{O}{\overset{\|}{C}}NHR$).

Secondary amines (R_2NH) yield *N,N*-disubstituted amides ($R'\overset{O}{\overset{\|}{C}}NR_2$).

Examples illustrating these reactions may be found in Tables 20.2, 20.3, and 20.6.

Two molar equivalents of amine are required in the reaction with acyl chlorides and acid anhydrides; one molecule of amine acts as a nucleophile, the second as a Brønsted base.

$$\underset{\text{Amine}}{2R_2NH} + \underset{\text{Acyl chloride}}{R'\overset{O}{\overset{\|}{C}}Cl} \longrightarrow \underset{\text{Amide}}{R'\overset{O}{\overset{\|}{C}}NR_2} + \underset{\text{Hydrochloride salt of amine}}{R_2\overset{+}{N}H_2 \quad Cl^-}$$

$$\underset{\text{Amine}}{2R_2NH} + \underset{\text{Acid anhydride}}{R'\overset{O}{\overset{\|}{C}}O\overset{O}{\overset{\|}{C}}R'} \longrightarrow \underset{\text{Amide}}{R'\overset{O}{\overset{\|}{C}}NR_2} + \underset{\text{Carboxylate salt of amine}}{R_2\overset{+}{N}H_2 \quad {}^-O\overset{O}{\overset{\|}{C}}R'}$$

It is possible to use only one molar equivalent of amine in these reactions if some other base, such as sodium hydroxide, is present in the reaction mixture to react with the hydrogen chloride or carboxylic acid that is formed. This is a useful procedure in those cases in which the amine is a valuable one or is available only in small quantities.

Esters and amines react in a 1 : 1 molar ratio to give amides. No acidic product is formed from the acylating agent, and so no additional base is required.

$$\underset{\text{Amine}}{R_2NH} + \underset{\text{Methyl ester}}{R'\overset{O}{\overset{\|}{C}}OCH_3} \longrightarrow \underset{\text{Amide}}{R'\overset{O}{\overset{\|}{C}}NR_2} + \underset{\text{Methanol}}{CH_3OH}$$

PROBLEM 20.17 Write an equation showing the preparation of the following amides from the indicated carboxylic acid derivative:

(*a*) $(CH_3)_2CH\overset{O}{\overset{\|}{C}}NH_2$ from an acyl chloride

(*b*) $CH_3\overset{O}{\overset{\|}{C}}NHCH_3$ from an acid anhydride

(*c*) $H\overset{O}{\overset{\|}{C}}N(CH_3)_2$ from a methyl ester

SAMPLE SOLUTION (*a*) Amides of the type $R\overset{O}{\overset{\|}{C}}NH_2$ are derived by acylation of ammonia.

$$\underset{\text{2-Methylpropanoyl chloride}}{(CH_3)_2CH\overset{O}{\overset{\|}{C}}Cl} + \underset{\text{Ammonia}}{2NH_3} \longrightarrow \underset{\text{2-Methylpropanamide}}{(CH_3)_2CH\overset{O}{\overset{\|}{C}}NH_2} + \underset{\text{Ammonium chloride}}{NH_4Cl}$$

Two molecules of ammonia are needed because its acylation produces, in addition to the desired amide, a molecule of hydrogen chloride. Hydrogen chloride (an acid) reacts with ammonia (a base) to give ammonium chloride.

All these reactions proceed by nucleophilic addition of the amine to the carbonyl group. Dissociation of the tetrahedral intermediate proceeds in the direction that leads to an amide.

$$\underset{\text{Acylating agent}}{R\overset{:O:}{\overset{\|}{C}}X} + \underset{\text{Amine}}{R'_2\ddot{N}H} \rightleftharpoons \underset{\text{Tetrahedral intermediate}}{R\overset{:\ddot{O}-H}{\underset{:NR'_2}{C}}-X} \rightleftharpoons \underset{\text{Amide}}{R\overset{:O:}{\overset{\|}{C}}\ddot{N}R'_2} + \underset{\text{Conjugate acid of leaving group}}{HX}$$

The carbonyl group of an amide is stabilized to a greater extent than that of an acyl chloride, anhydride, or ester; amides are formed rapidly and in high yield from each of these carboxylic acid derivatives.

Amides are sometimes prepared directly from carboxylic acids and amines by a two-step process. The first step is an acid-base reaction in which the acid and the amine combine to form an ammonium carboxylate salt. On heating, the ammonium carboxylate salt loses water to form an amide.

$$\underset{\text{Carboxylic acid}}{\mathrm{R\overset{O}{\overset{\|}{C}}OH}} + \underset{\text{Amine}}{\mathrm{R'_2NH}} \longrightarrow \underset{\text{Ammonium carboxylate salt}}{\mathrm{R\overset{O}{\overset{\|}{C}}O^- \quad R'_2\overset{+}{N}H_2}} \xrightarrow{\text{heat}} \underset{\text{Amide}}{\mathrm{R\overset{O}{\overset{\|}{C}}NR'_2}} + \underset{\text{Water}}{\mathrm{H_2O}}$$

In practice, both steps may be combined in a single operation by simply heating a carboxylic acid and an amine together:

$$\underset{\text{Benzoic acid}}{\mathrm{C_6H_5\overset{O}{\overset{\|}{C}}OH}} + \underset{\text{Aniline}}{\mathrm{C_6H_5NH_2}} \xrightarrow{225^\circ\mathrm{C}} \underset{\substack{N\text{-Phenylbenzamide} \\ (80-84\%)}}{\mathrm{C_6H_5\overset{O}{\overset{\|}{C}}NHC_6H_5}} + \underset{\text{Water}}{\mathrm{H_2O}}$$

A similar reaction in which ammonia and carbon dioxide are heated under pressure is the basis of the industrial synthesis of *urea*. Here, the reactants first combine, yielding a salt called *ammonium carbamate:*

$$\underset{\text{Ammonia}}{\mathrm{H_3N:}} + \underset{\text{Carbon dioxide}}{\mathrm{:\ddot{O}{=}C{=}\ddot{O}:}} \longrightarrow \mathrm{H_3\overset{+}{N}{-}C(=\ddot{O}:)(\ddot{O}:^-)} \xrightarrow{\mathrm{NH_3}} \underset{\text{Ammonium carbamate}}{\mathrm{H_2\ddot{N}{-}C(=\ddot{O}:)(\ddot{O}:^-\ \overset{+}{N}H_4)}}$$

On being heated, ammonium carbamate undergoes dehydration to form urea:

$$\underset{\text{Ammonium carbamate}}{\mathrm{H_2\ddot{N}{-}C(=\ddot{O}:)(\ddot{O}:^-\ \overset{+}{N}H_4)}} \xrightarrow{\text{heat}} \underset{\text{Urea}}{\mathrm{H_2\ddot{N}\overset{\ddot{O}:}{\overset{\|}{C}}\ddot{N}H_2}} + \underset{\text{Water}}{\mathrm{H_2\ddot{O}:}}$$

Over 10^{10} lb of urea—most of it used as fertilizer—is produced annually in the United States by this method.

These thermal methods for preparing amides are limited in their generality. Most often amides are prepared in the laboratory from acyl chlorides, acid anhydrides, or esters, and these are the methods that you should apply to solving synthetic problems.

20.14 LACTAMS

Lactams are cyclic amides and are analogous to lactones, which are cyclic esters. Most lactams are known by their common names, as the examples shown illustrate.

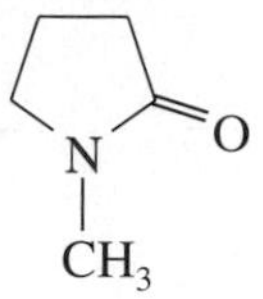

N-Methylpyrrolidone (a polar aprotic solvent)

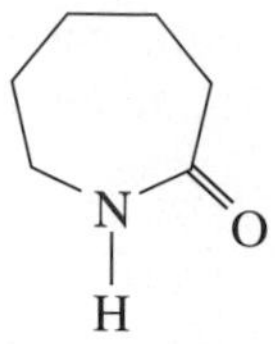

ε-Caprolactam (industrial chemical used to prepare a type of nylon)

Just as amides are more stable than esters, lactams are more stable than lactones. Thus, while β-lactones are difficultly accessible (Section 19.15), β-lactams are among the best-known products of the pharmaceutical industry. The penicillins and cephalosporins, which are so useful in treating bacterial infections, are β-lactams and are customarily referred to as *β-lactam antibiotics.*

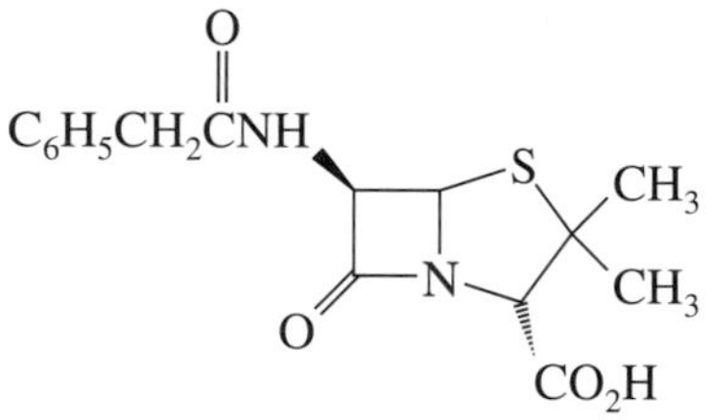

Penicillin G

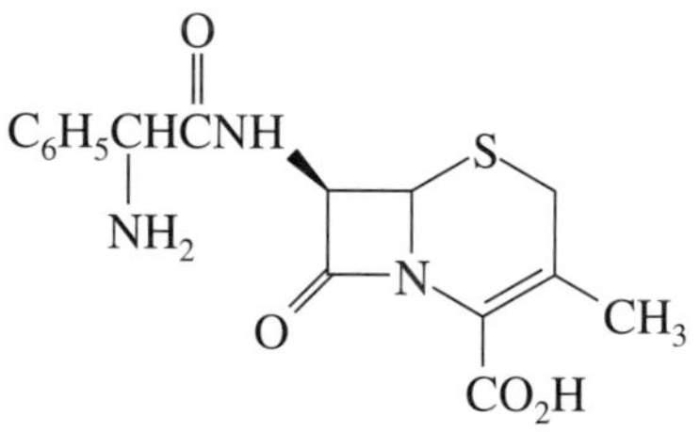

Cephalexin

These antibiotics act by inhibiting an enzyme that is essential for bacterial cell wall formation. A nucleophilic site on the enzyme reacts with the carbonyl group in the four-membered ring, and the ring opens to acylate the enzyme. Once its nucleophilic site is acylated, the enzyme is no longer active and the bacteria die. The β-lactam rings of the penicillins and cephalosporins combine just the right level of stability in aqueous media with reactivity toward nucleophilic substitution to be effective acylating agents toward this critical bacterial enzyme.

20.15 IMIDES

Compounds that have two acyl groups bonded to a single nitrogen are known as **imides.** The most common imides are cyclic ones:

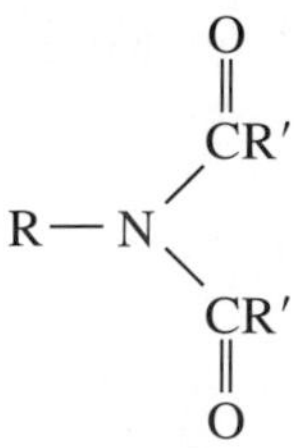

Imide

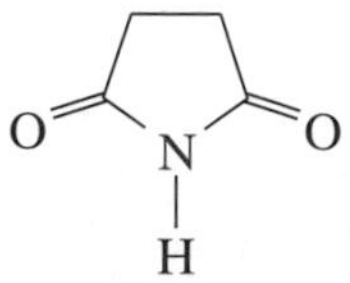

Succinimide

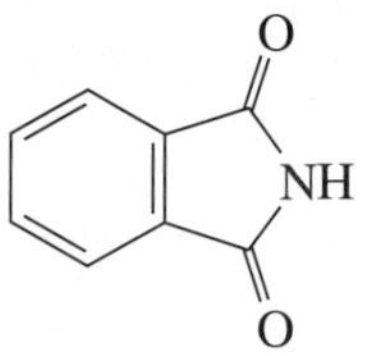

Phthalimide

Cyclic imides can be prepared by heating the ammonium salts of dicarboxylic acids:

$$\underset{\text{Succinic acid}}{\mathrm{HO\overset{O}{\overset{\|}{C}}CH_2CH_2\overset{O}{\overset{\|}{C}}OH}} + \underset{\text{Ammonia}}{2NH_3} \longrightarrow \underset{\text{Ammonium succinate}}{\overset{+}{N}H_4\ {}^{-}O\overset{O}{\overset{\|}{C}}CH_2CH_2\overset{O}{\overset{\|}{C}}O^{-}\ \overset{+}{N}H_4} \xrightarrow{\text{heat}} \underset{\substack{\text{Succinimide}\\(82-83\%)}}{\text{(cyclic imide: NH)}}$$

Replacement of the proton on nitrogen in succinimide by bromine gives *N*-bromosuccinimide, a reagent used for allylic and benzylic brominations (Sections 10.4 and 11.14).

PROBLEM 20.18 Phthalimide has been prepared in 95 percent yield by heating the compound formed on reaction of phthalic anhydride (Section 20.4) with excess ammonia. This compound has the molecular formula $C_8H_{10}N_2O_3$. What is its structure?

20.16 HYDROLYSIS OF AMIDES

The only nucleophilic acyl substitution reaction that amides undergo is hydrolysis. Amides are fairly stable in water, but the amide bond is cleaved on heating in the presence of strong acids or bases. Nominally, this cleavage produces an amine and a carboxylic acid.

$$\underset{\text{Amide}}{\mathrm{R\overset{O}{\overset{\|}{C}}\ddot{N}R'_2}} + \underset{\text{Water}}{H_2O} \longrightarrow \underset{\text{Carboxylic acid}}{\mathrm{R\overset{O}{\overset{\|}{C}}OH}} + \underset{\text{Amine}}{\mathrm{H{-}\ddot{N}R'_2}}$$

In acid, however, the amine is protonated, giving an ammonium ion, $R'_2\overset{+}{N}H_2$:

$$\underset{\text{Amide}}{\mathrm{R\overset{O}{\overset{\|}{C}}\ddot{N}R'_2}} + \underset{\text{Hydronium ion}}{H_3O^+} \longrightarrow \underset{\text{Carboxylic acid}}{\mathrm{R\overset{O}{\overset{\|}{C}}OH}} + \underset{\text{Ammonium ion}}{\mathrm{R'{-}\overset{+}{N}H_2{-}R'}}$$

In base the carboxylic acid is deprotonated, giving a carboxylate ion:

$$\underset{\text{Amide}}{\mathrm{R\overset{O}{\overset{\|}{C}}\ddot{N}R'_2}} + \underset{\text{Hydroxide ion}}{HO^-} \longrightarrow \underset{\text{Carboxylate ion}}{\mathrm{R\overset{O}{\overset{\|}{C}}O^-}} + \underset{\text{Amine}}{\mathrm{R'{-}\ddot{N}HR'}}$$

The acid-base reactions that occur after the amide bond is broken render the overall hydrolysis irreversible in both cases. The amine product is protonated in acid; the carboxylic acid is deprotonated in base.

$$\text{CH}_3\text{CH}_2\text{CH(C}_6\text{H}_5\text{)CNH}_2\ (\text{C=O}) \xrightarrow[\text{heat}]{\text{H}_2\text{O, H}_2\text{SO}_4} \text{CH}_3\text{CH}_2\text{CH(C}_6\text{H}_5\text{)COH}\ (\text{C=O}) + \overset{+}{\text{N}}\text{H}_4\ \ \text{HSO}_4^-$$

2-Phenylbutanamide — 2-Phenylbutanoic acid (88–90%) — Ammonium hydrogen sulfate

$$\text{CH}_3\text{CNH}\text{—C}_6\text{H}_4\text{—Br}\ (\text{C=O}) \xrightarrow[\text{ethanol-water, heat}]{\text{KOH}} \text{CH}_3\text{CO}^-\ \text{K}^+\ (\text{C=O}) + \text{H}_2\text{N—C}_6\text{H}_4\text{—Br}$$

N-(4-Bromophenyl)acetamide (p-bromoacetanilide) — Potassium acetate — p-Bromoaniline (95%)

In 1989 a controversy arose about *Alar* residues on apples. Alar is a growth regulator, applied to apples to keep them on the trees longer so that fruit of higher quality is produced. Alar itself is not harmful, but a substance formed when Alar undergoes hydrolysis, called *UDMH,* which stands for *u*nsymmetrical *dim*ethyl*h*ydrazine, appears to be a carcinogen. Alar is an amide, and the equation for its hydrolysis is

$$\underset{\text{Alar}}{\text{HO}\overset{\text{O}}{\overset{\|}{\text{C}}}\text{CH}_2\text{CH}_2\overset{\text{O}}{\overset{\|}{\text{C}}}\text{NHN(CH}_3)_2} + \underset{\text{Water}}{\text{H}_2\text{O}} \longrightarrow \underset{\text{Succinic acid}}{\text{HO}\overset{\text{O}}{\overset{\|}{\text{C}}}\text{CH}_2\text{CH}_2\overset{\text{O}}{\overset{\|}{\text{C}}}\text{OH}} + \underset{\text{UDMH}}{\text{H}_2\text{NN(CH}_3)_2}$$

Much of the concern centers on UDMH levels in apple juice, for two reasons. First, heating is involved in the processing of apples into juice, and the rate of conversion of Alar to UDMH, as with all amide hydrolyses, increases with temperature. Second, young children consume apple juice in disproportionately high amounts and so are at greater risk than adults. Shortly after the issue arose, the manufacturer of Alar voluntarily halted its sale in the United States.

Mechanistically, amide hydrolysis is similar to the hydrolysis of other carboxylic acid derivatives. The mechanism of the acid-promoted hydrolysis is presented in Figure 20.6. It proceeds in two stages; a tetrahedral intermediate is formed in the first stage and dissociates in the second.

The amide is activated toward nucleophilic attack by protonation of its carbonyl oxygen. The cation produced in this step is stabilized by resonance involving the nitrogen lone pair and is more stable than the intermediate in which the amide nitrogen is protonated.

Protonation of carbonyl oxygen

$$\text{RC}(=\overset{+}{\text{O}}\text{H})\ddot{\text{N}}\text{H}_2 \longleftrightarrow \text{RC}(\text{—}\ddot{\text{O}}\text{H})=\overset{+}{\text{N}}\text{H}_2$$

Most stable resonance forms of an O-protonated amide

Protonation of amide nitrogen

$$\text{RC}(=\ddot{\text{O}}\text{:})\text{—}\overset{+}{\text{N}}\text{H}_3$$

An acylammonium ion; the positive charge is localized on nitrogen

Step 1: Protonation of the carbonyl oxygen of the amide

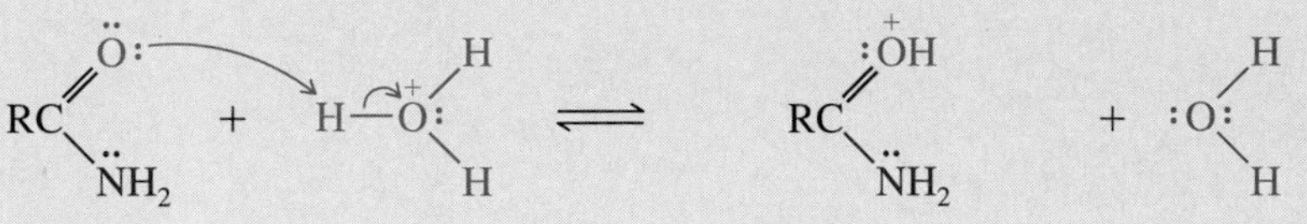

Amide + Hydronium ion ⇌ Protonated form of amide + Water

Step 2: Nucleophilic addition of water to the protonated form of the amide

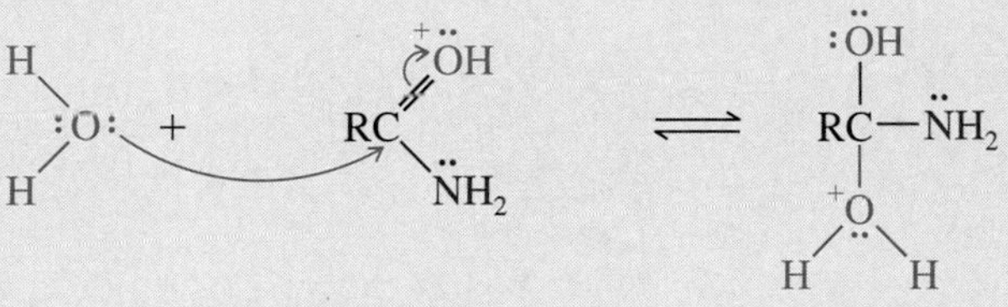

Water + Protonated form of amide ⇌ Oxonium ion

Step 3: Deprotonation of the oxonium ion to give the neutral form of the tetrahedral intermediate

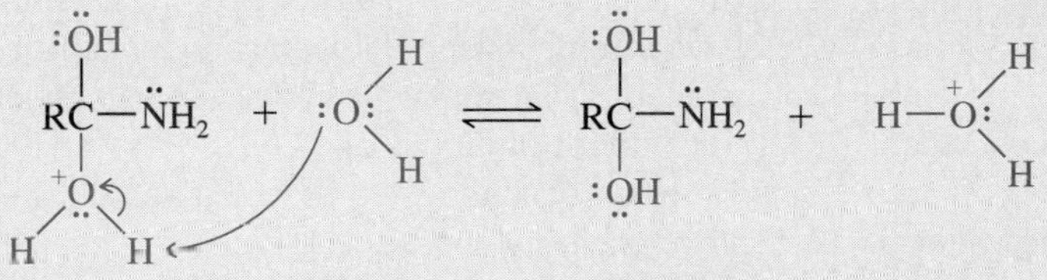

Oxonium ion + Water ⇌ Tetrahedral intermediate + Hydronium ion

Step 4: Protonation of the tetrahedral intermediate at its amino nitrogen

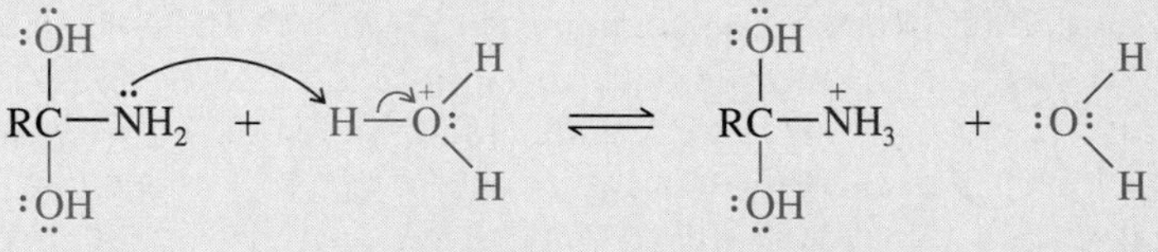

Tetrahedral intermediate + Hydronium ion ⇌ Ammonium ion + Water

Step 5: Dissociation of the *N*-protonated form of the tetrahedral intermediate to ammonia and the protonated form of the carboxylic acid

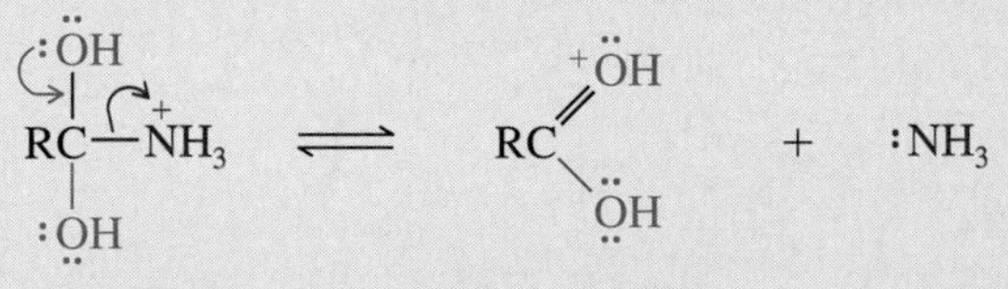

Ammonium ion ⇌ Protonated form of carboxylic acid + Ammonia

FIGURE 20.6 Sequence of steps that describes the mechanism of acid-promoted amide hydrolysis. Steps 1 through 3 show the formation of the tetrahedral intermediate. Dissociation of the tetrahedral intermediate is shown in steps 4 through 6.

Step 6: Proton transfer processes yielding ammonium ion and the carboxylic acid

$$H_2\overset{+}{O}{-}H + :NH_3 \longrightarrow H_2O: + \overset{+}{N}H_4$$

Hydronium ion — Ammonia — Water — Ammonium ion

$$RC(=\overset{+}{O}H)OH + H_2O: \longrightarrow RC(=O)OH + H{-}\overset{+}{O}H_2$$

Protonated form of carboxylic acid — Water — Carboxylic acid — Hydronium ion

FIGURE 20.6 (*Continued*)

CONDENSATION POLYMERS. POLYAMIDES AND POLYESTERS

All fibers are polymers of one kind or another. Cotton, for example, is cellulose, and cellulose is a naturally occurring polymer of glucose. Silk and wool are naturally occurring polymers of amino acids. An early goal of inventors and entrepreneurs was to produce fibers from other naturally occurring polymers. Their earliest efforts consisted of chemically modifying the short cellulose fibers obtained from wood so that they could be processed into longer fibers more like cotton and silk. These efforts were successful, and the resulting fibers of modified cellulose, known generically as *rayon,* have been produced by a variety of techniques since the late nineteenth century.

A second approach involved direct chemical synthesis of polymers by connecting appropriately chosen small molecules together into a long chain. In 1938 E. I. Du Pont de Nemours and Company announced the development of *nylon,* the first synthetic polymer fiber.

The leader of Du Pont's effort was Wallace H. Carothers,* who reasoned that he could reproduce the properties of silk by constructing a polymer chain held together, as is silk, by amide bonds. The necessary amide bonds were formed by heating a dicarboxylic acid with a diamine. The combination of hexanedioic acid (*adipic acid*) and 1,6-hexanediamine (*hexamethylenediamine*) reacts to give a salt that, when heated, gives a **polyamide** called *Nylon 66.* The amide bonds form by a condensation reaction, and Nylon 66 is an example of a **condensation polymer.**

$$HO\overset{O}{\overset{\|}{C}}(CH_2)_4\overset{O}{\overset{\|}{C}}OH + H_2N(CH_2)_6NH_2 \longrightarrow {}^-O\overset{O}{\overset{\|}{C}}(CH_2)_4\overset{O}{\overset{\|}{C}}O^- \quad H_3\overset{+}{N}(CH_2)_6\overset{+}{N}H_3$$

Adipic acid — Hexamethylenediamine

heat, $-H_2O$

$${}^-O\overset{O}{\overset{\|}{C}}(CH_2)_4\overset{O}{\overset{\|}{C}}{-}\left[NH(CH_2)_6NH\overset{O}{\overset{\|}{C}}(CH_2)_4\overset{O}{\overset{\|}{C}}{-}\right]_n NH(CH_2)_6\overset{+}{N}H_3$$

Nylon 66

The first "6" in Nylon 66 stands for the number of carbons in the diamine, the second for the number of carbons in the dicarboxylic acid. Nylon 66 was an immediate success and fostered the development of a large number of related polyamides, many of which have also found their niche in the marketplace.

A slightly different class of polyamides is the *aramids* (*aromatic polyamides*). Like the nylons, the aramids are prepared from a dicarboxylic acid and a diamine, but the functional groups are anchored to benzene rings. An example of an aramid is *Kevlar,* which is a polyamide derived from 1,4-benzenedicarboxylic acid (*terephthalic acid*) and 1,4-benzenediamine (p-*phenylenediamine*):

$$^{-}OC(=O)-C_6H_4-C(=O)-\left[NH-C_6H_4-NHC(=O)-C_6H_4-C(=O)-\right]_n NH-C_6H_4-\overset{+}{N}H_3$$

Kevlar (a polyamide of the aramid class)

Kevlar fibers are very strong, and this property makes Kevlar an increasingly popular choice in applications where the ratio of strength to weight is important. For example, a cable made from Kevlar weighs only one-fifth as much as a steel cable but is just as strong. Kevlar is also used to make lightweight bulletproof vests.

Nomex is another aramid fiber. Kevlar and Nomex differ only in that the substitution pattern in the aromatic rings is para in Kevlar but meta in Nomex. Nomex is best known for its fire-resistant properties and is used in protective clothing for firefighters, astronauts, and race-car drivers.

Polyesters represent a second class of condensation polymers, and the principles behind their synthesis parallel those of polyamides. Ester formation between the functional groups of a dicarboxylic acid and a diol serve to connect small molecules together into a long polyester. The most familiar example of a polyester is *Dacron,* which is prepared from 1,4-benzenedicarboxylic acid and 1,2-ethanediol (*ethylene glycol*):

$$HOC(=O)-C_6H_4-C(=O)-\left[OCH_2CH_2OC(=O)-C_6H_4-C(=O)-\right]_n OCH_2CH_2OH$$

Dacron (a polyester)

The production of polyester fibers leads that of all other types. Annual United States production of polyester fibers is 1.6 million tons versus 1.4 million tons for cotton and 1.0 million tons for nylon. Wool and silk trail far behind at 0.04 and 0.01 million tons, respectively.

Not all synthetic polymers are used as fibers. *Mylar,* for example, is chemically the same as Dacron, but is prepared in the form of a thin film instead of a fiber. *Lexan* is a polyester which, because of its impact resistance, is used as a shatterproof substitute for glass. It is a **polycarbonate** having the structure shown:

$$HO-C_6H_4-C(CH_3)_2-C_6H_4-O-\left[C(=O)-O-C_6H_4-C(CH_3)_2-C_6H_4-O-\right]_n H$$

Lexan (a polycarbonate)

In terms of the number of scientists and engineers involved, research and development in polymer chemistry has been the principal activity of the chemical industry in the United States for most of this century. The initial goal of making synthetic materials that are the equal of natural fibers has been more than met; it has been far exceeded. What is also important is that all of this did not begin with a chance discovery. It began with a management decision to do basic research in a specific area, and to support it in the absence of any guarantee that success would be quickly achieved.†

* For an account of Carothers' role in the creation of nylon, see the September 1988 issue of the *Journal of Chemical Education* (pp. 803–808).

† The April 1988 issue of the *Journal of Chemical Education* contains a number of articles on polymers, including a historical review entitled "Polymers Are Everywhere" (pp. 327–334) and a glossary of terms (pp. 314–319).

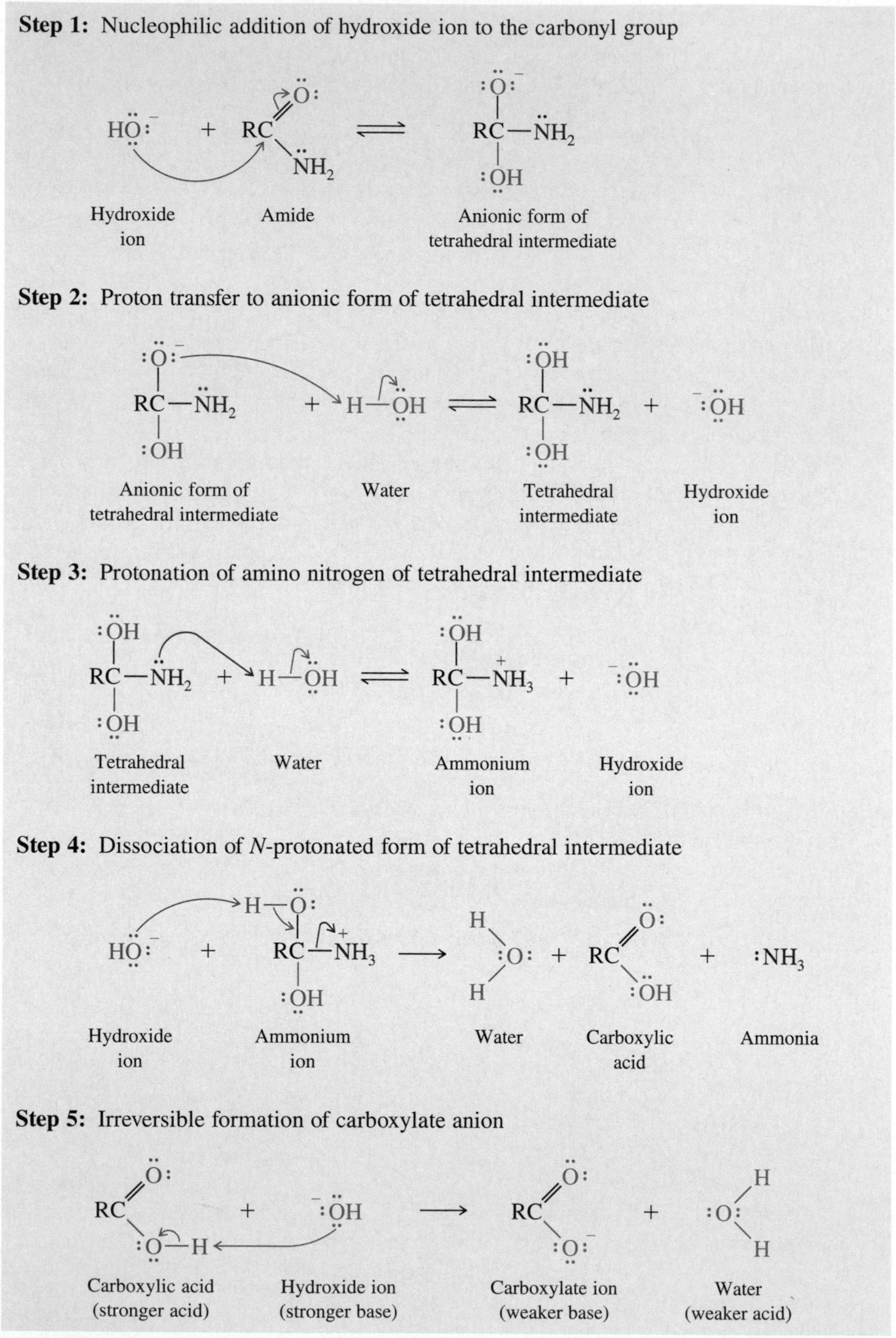

Step 1: Nucleophilic addition of hydroxide ion to the carbonyl group

$$HO^- + RC(=O)NH_2 \rightleftharpoons RC(O^-)(OH)NH_2$$

Hydroxide ion | Amide | Anionic form of tetrahedral intermediate

Step 2: Proton transfer to anionic form of tetrahedral intermediate

$$RC(O^-)(OH)NH_2 + H-OH \rightleftharpoons RC(OH)_2NH_2 + {}^-OH$$

Anionic form of tetrahedral intermediate | Water | Tetrahedral intermediate | Hydroxide ion

Step 3: Protonation of amino nitrogen of tetrahedral intermediate

$$RC(OH)_2NH_2 + H-OH \rightleftharpoons RC(OH)_2\overset{+}{N}H_3 + {}^-OH$$

Tetrahedral intermediate | Water | Ammonium ion | Hydroxide ion

Step 4: Dissociation of *N*-protonated form of tetrahedral intermediate

$$HO^- + RC(OH)_2\overset{+}{N}H_3 \longrightarrow H_2O + RC(=O)OH + :NH_3$$

Hydroxide ion | Ammonium ion | Water | Carboxylic acid | Ammonia

Step 5: Irreversible formation of carboxylate anion

$$RC(=O)OH + {}^-OH \longrightarrow RC(=O)O^- + H_2O$$

Carboxylic acid (stronger acid) | Hydroxide ion (stronger base) | Carboxylate ion (weaker base) | Water (weaker acid)

FIGURE 20.7 Sequence of steps that describes the mechanism of base-promoted amide hydrolysis.

Once formed, the *O*-protonated intermediate is attacked by a water molecule in step 2. The intermediate formed in this step loses a proton in step 3 to give the neutral form of the tetrahedral intermediate. The tetrahedral intermediate has its amino group ($—NH_2$) attached to sp^3 hybridized carbon, and this amino group is the site at which protonation occurs in step 4. Cleavage of the carbon-nitrogen bond in step 5 yields the protonated form of the carboxylic acid, along with a molecule of ammonia. In acid solution ammonia is immediately protonated to give ammonium ion, as shown in step 6. This protonation step has such a large equilibrium constant that it makes the overall reaction irreversible.

PROBLEM 20.19 On the basis of the general mechanism for acid-promoted amide hydrolysis shown in Figure 20.6, write an analogous sequence of steps for the hydrolysis of acetanilide, $CH_3\overset{O}{\overset{\|}{C}}NHC_6H_5$.

In base the tetrahedral intermediate is formed in a manner analogous to that proposed for ester saponification. Steps 1 and 2 in Figure 20.7 show the formation of the tetrahedral intermediate in the base-promoted hydrolysis of amides. In step 3 the basic amino group of the tetrahedral intermediate abstracts a proton from water, and in step 4 the derived ammonium ion undergoes base-promoted dissociation. Conversion of the carboxylic acid to its corresponding carboxylate anion in step 5 completes the process and renders the overall reaction irreversible.

PROBLEM 20.20 On the basis of the general mechanism for base-promoted hydrolysis shown in Figure 20.7, write an analogous sequence for the hydrolysis of *N,N*-dimethylformamide, $H\overset{O}{\overset{\|}{C}}N(CH_3)_2$.

20.17 THE HOFMANN REARRANGEMENT OF *N*-BROMO AMIDES

On treatment with bromine in basic solution, amides of the type $R\overset{O}{\overset{\|}{C}}NH_2$ undergo an interesting reaction that leads to amines. This reaction was discovered by the German chemist August W. Hofmann over 100 years ago and is called the **Hofmann rearrangement.**

$$\underset{\text{Amide}}{R\overset{O}{\overset{\|}{C}}NH_2} + \underset{\text{Bromine}}{Br_2} + \underset{\text{Hydroxide ion}}{4HO^-} \longrightarrow \underset{\text{Amine}}{RNH_2} + \underset{\text{Bromide ion}}{2Br^-} + \underset{\text{Carbonate ion}}{CO_3^{2-}} + \underset{\text{Water}}{2H_2O}$$

The group R attached to the carboxamide function may be alkyl or aryl.

The relationship of the amine product to the amide reactant is rather remarkable. The overall reaction appears as if the carbonyl group had been plucked out of the amide, leaving behind a primary amine having one less carbon atom than the amide.

$$(CH_3)_3CCH_2\overset{O}{\overset{\|}{C}}NH_2 \xrightarrow[H_2O]{Br_2,\ NaOH} (CH_3)_3CCH_2NH_2$$

3,3-Dimethylbutanamide → 2,2-Dimethylpropanamine (94%)

$$m\text{-}BrC_6H_4\overset{O}{\overset{\|}{C}}NH_2 \xrightarrow[H_2O]{Br_2,\ KOH} m\text{-}BrC_6H_4NH_2$$

m-Bromobenzamide → *m*-Bromoaniline (87%)

PROBLEM 20.21 Outline an efficient synthesis of 1-propanamine ($CH_3CH_2CH_2NH_2$) from butanoic acid.

The mechanism of the Hofmann rearrangement is presented in Figure 20.8. It involves three stages:

1. Formation of an *N*-bromo amide intermediate (steps 1 and 2)
2. Rearrangement of the *N*-bromo amide to an isocyanate (steps 3 and 4)
3. Hydrolysis of the isocyanate (steps 5 and 6)

Formation of the *N*-bromo amide intermediate is relatively straightforward. The base converts the amide to its corresponding anion (step 1), which acts as a nucleophile toward bromine (step 2).

Conversion of the *N*-bromo amide to its conjugate base in step 3 is also easy to understand. It is an acid-base reaction exactly analogous to that of step 1. The anion produced in step 3 is a key intermediate, which rearranges in step 4 by migration of the alkyl (or aryl) group from carbon to nitrogen, with loss of bromide from nitrogen. The product of this rearrangement is an isocyanate. The isocyanate formed in the rearrangement step then undergoes base-promoted hydrolysis in steps 5 and 6 to give the observed amine.

Among the experimental observations that contributed to elaboration of the mechanism shown in Figure 20.8 are the following:

1. Only amides of the type $R\overset{O}{\overset{\|}{C}}NH_2$ undergo the Hofmann rearrangement. The amide nitrogen must have *two* protons attached to it, of which one is replaced by bromine to give the *N*-bromo amide, while abstraction of the second by base is necessary to trigger the rearrangement. Amides of the type $R\overset{O}{\overset{\|}{C}}NHR'$ form *N*-bromo amides under the reaction conditions, but these *N*-bromo amides do not rearrange.
2. Rearrangement proceeds with *retention of configuration* at the migrating group.

$$C_6H_5CH_2\text{–}\overset{H}{C}(CH_3)\text{–}\overset{O}{\overset{\|}{C}}NH_2 \xrightarrow[H_2O]{Br_2,\ NaOH} C_6H_5CH_2\text{–}\overset{H}{C}(CH_3)\text{–}NH_2$$

(*S*)-(+)-2-Methyl-3-phenylpropanamide → (*S*)-(+)-1-Phenyl-2-propanamine

Overall Reaction

$$\underset{\text{Amide}}{RCNH_2} + \underset{\text{Bromine}}{Br_2} + \underset{\text{Hydroxide ion}}{4HO^-} \longrightarrow \underset{\text{Amine}}{RNH_2} \quad \underset{\text{Bromide ion}}{2Br^-} + \underset{\text{Carbonate ion}}{CO_3^{2-}} + \underset{\text{Water}}{2H_2O}$$

Step 1: Deprotonation of the amide. Amides of the type $RC(=O)NH_2$ are about as acidic as water, so appreciable quantities of the conjugate base are present at equilibrium in aqueous base. The conjugate base of an amide is stabilized by electron delocalization in much the same way that an enolate anion is.

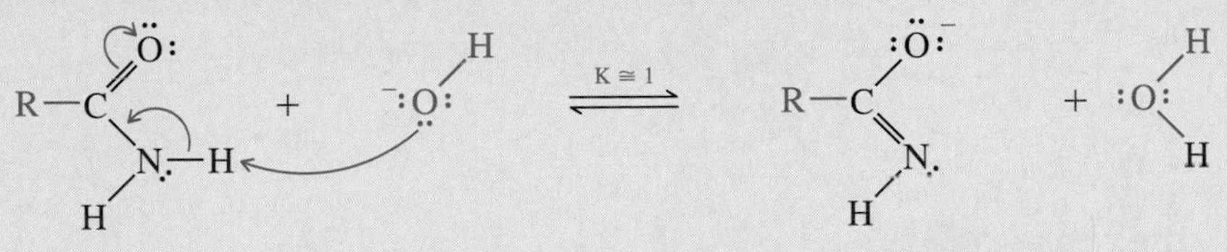

Amide + Hydroxide ion $\underset{}{\overset{K \cong 1}{\rightleftharpoons}}$ Conjugate base of amide + Water

Step 2: Reaction of the conjugate base of the amide with bromine. The product of this step is an *N*-bromo amide.

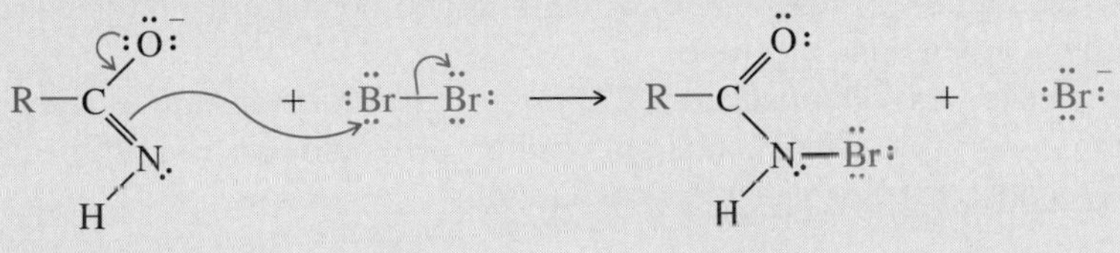

Conjugate base of amide + Bromine $\longrightarrow$ *N*-Bromo amide + Bromide ion

Step 3: Deprotonation of the *N*-bromo amide. The electron-withdrawing effect of the bromine substituent reinforces that of the carbonyl group and makes the *N*-bromo amide even more acidic than the starting amide.

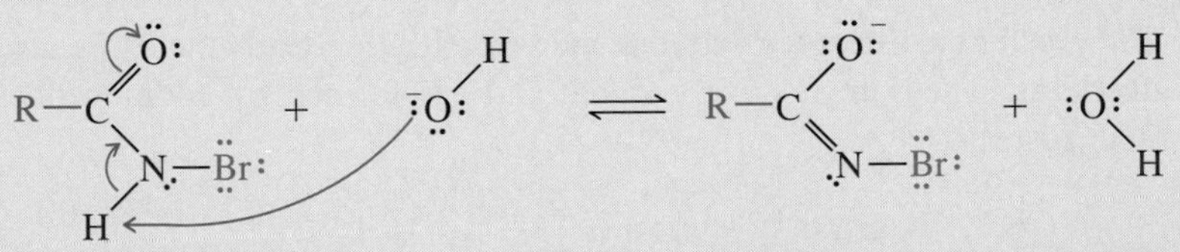

N-bromo amide + Hydroxide ion $\rightleftharpoons$ Conjugate base of *N*-bromo amide + Water

Step 4: Rearrangement of the conjugate base of the *N*-bromo amide. The group R migrates from carbon to nitrogen, and bromide is lost as a leaving group from nitrogen. The product of this rearrangement is an *N*-alkyl isocyanate.

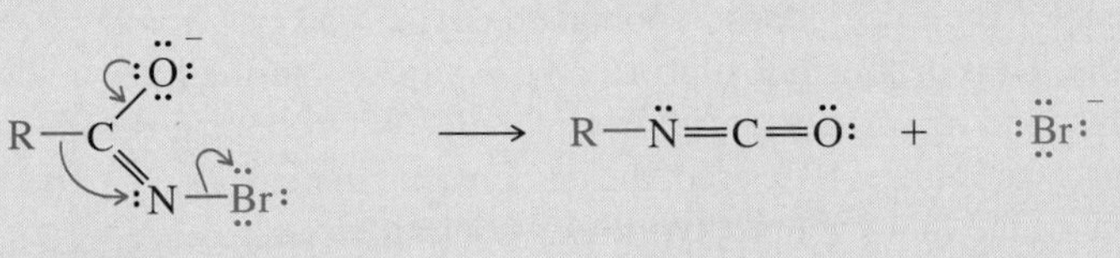

Conjugate base of *N*-bromo amide $\longrightarrow$ $R—N=C=O$ + Br^-

N-Alkyl isocyanate; Bromide ion

FIGURE 20.8 Sequence of steps that describes the mechanism of the Hofmann rearrangement.

Step 5: Hydrolysis of the isocyanate begins by base-catalyzed addition of water to form an *N*-alkylcarbamic acid.

$$R-\ddot{N}=C=\ddot{O}: + H_2O \longrightarrow (R)(H)\ddot{N}-C(=\ddot{O}:)OH$$

N-Alkyl isocyanate — *N*-Alkylcarbamic acid

Step 6: The *N*-alkycarbamic acid is unstable and dissociates to an amine and carbon dioxide. Carbon dioxide is converted to carbonate ion in base. (Several steps are actually involved; in the interests of brevity, they are summarized as shown.)

$$(R)(H)N-C(=O)OH + 2HO^- \longrightarrow RNH_2 + CO_3^{2-} + H_2O$$

N-Alkylcarbamic acid — Hydroxide ion — Amine — Carbonate ion — Water

FIGURE 20.8 (*Continued*)

The new carbon-nitrogen bond is formed at the same face of the migrating carbon as the bond that is broken. The rearrangement step depicted in Figure 20.8 satisfies this requirement. Presumably, carbon-nitrogen bond formation is concerted with carbon-carbon bond cleavage.

3. Isocyanates are intermediates. When the reaction of an amide with bromine is carried out in methanol containing sodium methoxide instead of in aqueous base, the product that is isolated is a **carbamate.**

$$CH_3(CH_2)_{14}\overset{O}{\overset{\|}{C}}NH_2 \xrightarrow[CH_3OH]{Br_2,\ NaOCH_3} CH_3(CH_2)_{14}NH\overset{O}{\overset{\|}{C}}OCH_3$$

Hexadecanamide — Methyl *N*-pentadecylcarbamate (84–94%)

Carbamates are esters of **carbamic acid** ($H_2N\overset{O}{\overset{\|}{C}}OH$). Carbamates are also known as **urethans.** They are relatively stable and are formed by addition of alcohols to isocyanates.

$$RN=C=O + CH_3OH \longrightarrow RNH\overset{O}{\overset{\|}{C}}OCH_3$$

Isocyanate — Methanol — Methyl *N*-alkylcarbamate

Carbamic acid itself ($H_2N\overset{O}{\overset{\|}{C}}OH$) and *N*-substituted derivatives of carbamic acid are unstable; they decompose spontaneously to carbon dioxide and ammonia or an amine. Thus in aqueous solution, an isocyanate intermediate yields an amine via the corresponding carbamic acid; in methanol, an isocyanate is converted to an isolable methyl carbamate. If desired, the carbamate can be isolated, purified, and converted to an amine in a separate hydrolysis operation.

While the Hofmann rearrangement is complicated with respect to mechanism, it is easy to carry out and gives amines that are sometimes difficult to prepare by other methods.

20.18 PREPARATION OF NITRILES

Nitriles are organic compounds that contain the $-C\equiv N$ functional group. We have already discussed the two main procedures by which alkyl cyanides are prepared, namely, the nucleophilic substitution of alkyl halides by cyanide and the conversion of aldehydes and ketones to cyanohydrins. Table 20.7 reviews aspects of these reactions. Neither of the reactions in Table 20.7 is suitable for aryl cyanides ($ArC\equiv N$); these compounds are readily prepared by a reaction to be discussed in Chapter 22.

Both alkyl and aryl cyanides are accessible by dehydration of amides.

$$\underset{\substack{\text{Amide}\\\text{(R may be alkyl}\\\text{or aryl)}}}{RCNH_2}\ (\text{C}{=}\text{O}) \longrightarrow \underset{\substack{\text{Nitrile}\\\text{(R may be alkyl}\\\text{or aryl)}}}{RC\equiv N} + \underset{\text{Water}}{H_2O}$$

Among the reagents used to effect the dehydration of amides is the compound P_4O_{10}, known by the common name *phosphorus pentoxide* because it was once thought to have the molecular formula P_2O_5. Phosphorus pentoxide is the anhydride of phosphoric acid and is used in a number of reactions requiring dehydrating agents.

$$\underset{\text{2-Methylpropanamide}}{(CH_3)_2CH\overset{O}{\overset{\|}{C}}NH_2} \xrightarrow[200°C]{P_4O_{10}} \underset{\substack{\text{2-Methylpropanenitrile}\\(69-86\%)}}{(CH_3)_2CHC\equiv N}$$

TABLE 20.7
Preparation of Nitriles

Reaction (section) and comments	General equation and specific example
Nucleophilic substitution by cyanide ion (Sections 8.1, 8.14) Cyanide ion is a good nucleophile and reacts with alkyl halides to give alkyl cyanides. The reaction is of the S_N2 type and is limited to primary and secondary alkyl halides. Tertiary alkyl halides undergo elimination; aryl and vinyl halides do not react.	$\underset{\text{Cyanide ion}}{:N\equiv\bar{C}:} + \underset{\text{Alkyl halide}}{R-X} \longrightarrow \underset{\text{Nitrile}}{RC\equiv N} + \underset{\text{Halide ion}}{X^-}$ $\underset{\text{1-Chlorodecane}}{CH_3(CH_2)_8CH_2Cl} \xrightarrow[\text{ethanol-water}]{KCN} \underset{\text{Undecanenitrile (95\%)}}{CH_3(CH_2)_8CH_2CN}$
Cyanohydrin formation (Section 17.8) Hydrogen cyanide adds to the carbonyl group of aldehydes and ketones.	$\underset{\text{Aldehyde or ketone}}{R\overset{O}{\overset{\|}{C}}R'} + \underset{\text{Hydrogen cyanide}}{HCN} \longrightarrow \underset{\text{Cyanohydrin}}{R\overset{OH}{\overset{\mid}{\underset{C\equiv N}{\underset{\mid}{C}}}}R'}$ $\underset{\text{3-Pentanone}}{CH_3CH_2\overset{O}{\overset{\|}{C}}CH_2CH_3} \xrightarrow[H^+]{KCN} \underset{\text{3-Pentanone cyanohydrin (75\%)}}{CH_3CH_2\overset{OH}{\overset{\mid}{\underset{CN}{\underset{\mid}{C}}}}CH_2CH_3}$

PROBLEM 20.22 Show how ethyl alcohol could be used to prepare (*a*) CH_3CN and (*b*) CH_3CH_2CN. Along with ethyl alcohol you may use any necessary inorganic reagents.

An important nitrile is *acrylonitrile,* $CH_2{=}CHCN$. It is prepared industrially from propene, ammonia, and oxygen in the presence of a special catalyst. Polymers of acrylonitrile have many applications, the most prominent being their use in the preparation of acrylic fibers.

20.19 HYDROLYSIS OF NITRILES

Nitriles are classified as carboxylic acid derivatives because they are converted to carboxylic acids on hydrolysis. The conditions required are similar to those for the hydrolysis of amides, namely, heating in aqueous acid or base for several hours. Like the hydrolysis of amides, nitrile hydrolysis is irreversible in the presence of acids or bases. Acid hydrolysis yields ammonium ion and a carboxylic acid.

$$\underset{\text{Nitrile}}{RC{\equiv}N} + \underset{\text{Water}}{H_2O} + \underset{\text{Hydronium ion}}{H_3O^+} \longrightarrow \underset{\text{Carboxylic acid}}{R\overset{O}{\overset{\|}{C}}OH} + \underset{\text{Ammonium ion}}{\overset{+}{N}H_4}$$

$$\underset{p\text{-Nitrobenzyl cyanide}}{O_2N{-}C_6H_4{-}CH_2CN} \xrightarrow[\text{heat}]{H_2O,\ H_2SO_4} \underset{p\text{-Nitrophenylacetic acid (92–95\%)}}{O_2N{-}C_6H_4{-}CH_2\overset{O}{\overset{\|}{C}}OH}$$

In aqueous base, hydroxide ion abstracts a proton from the carboxylic acid. In order to isolate the acid a subsequent acidification step is required.

$$\underset{\text{Nitrile}}{RC{\equiv}N} + \underset{\text{Water}}{H_2O} + \underset{\text{Hydroxide ion}}{HO^-} \longrightarrow \underset{\text{Carboxylate ion}}{R\overset{O}{\overset{\|}{C}}O^-} + \underset{\text{Ammonia}}{NH_3}$$

$$\underset{\text{Undecanenitrile}}{CH_3(CH_2)_9CN} \xrightarrow[2.\ H^+]{1.\ KOH,\ H_2O,\ heat} \underset{\text{Undecanoic acid (80\%)}}{CH_3(CH_2)_9\overset{O}{\overset{\|}{C}}OH}$$

Nitriles are susceptible to nucleophilic addition. In their hydrolysis, water adds across the carbon-nitrogen triple bond. In a series of proton-transfer steps, an amide is produced:

$$\text{RC}\equiv\text{N:} + \text{H}_2\text{O} \rightleftharpoons \text{RC}\begin{smallmatrix}\text{OH}\\ \text{NH}\end{smallmatrix} \rightleftharpoons \text{RC}\begin{smallmatrix}\text{O}\\ \text{NH}_2\end{smallmatrix}$$

Nitrile Water Imino acid Amide

We have already discussed both the acid- and base-promoted hydrolysis of amides (Section 20.16). All that remains to complete the mechanistic picture of nitrile hydrolysis is to examine the conversion of the nitrile to the corresponding amide.

Nucleophilic addition to the nitrile may be either acid- or base-catalyzed. In aqueous base, hydroxide adds to the carbon-nitrogen triple bond:

$$\text{HO:}^- + \text{RC}\equiv\text{N:} \rightleftharpoons \text{RC}\begin{smallmatrix}\text{OH}\\ \text{N:}^-\end{smallmatrix} \xrightarrow[\text{OH}^-]{\text{H}_2\text{O}} \text{RC}\begin{smallmatrix}\text{OH}\\ \text{NH}\end{smallmatrix}$$

Hydroxide ion Nitrile Imino acid

The imino acid is transformed to the amide by the sequence

$$\text{RC}\begin{smallmatrix}\text{O—H}\\ \text{NH}\end{smallmatrix} + \text{:OH}^- \rightleftharpoons \text{RC}\begin{smallmatrix}\text{O}\\ \text{NH}^-\end{smallmatrix} + \text{H—OH} \rightleftharpoons \text{RC}\begin{smallmatrix}\text{O}\\ \text{NH}_2\end{smallmatrix} + \text{:OH}^-$$

Imino acid Hydroxide ion Amide anion Water Amide Hydroxide ion

---

PROBLEM 20.23 Suggest a reasonable mechanism for the conversion of a nitrile (RCN) to the corresponding amide in aqueous acid.

Nucleophiles other than water can also add to the carbon-nitrogen triple bond of nitriles. In the following section we will see a synthetic application of such a nucleophilic addition.

20.20 ADDITION OF GRIGNARD REAGENTS TO NITRILES

The carbon-nitrogen triple bond of nitriles is much less reactive toward nucleophilic addition than is the carbon-oxygen double bond of aldehydes and ketones. Strongly basic nucleophiles such as Grignard reagents, however, do react with nitriles in a reaction that is of synthetic value:

$$\text{RC}\equiv\text{N} + \text{R}'\text{MgX} \xrightarrow[\text{2. H}_2\text{O}]{\text{1. diethyl ether}} \text{R}\overset{\text{NH}}{\overset{\|}{\text{C}}}\text{R}' \xrightarrow[\text{heat}]{\text{H}_2\text{O, H}^+} \text{R}\overset{\text{O}}{\overset{\|}{\text{C}}}\text{R}'$$

Nitrile Grignard reagent Imine Ketone

The imine formed by nucleophilic addition of the Grignard reagent to the nitrile is normally not isolated but is hydrolyzed directly to a ketone. The overall sequence is used as a means of preparing ketones.

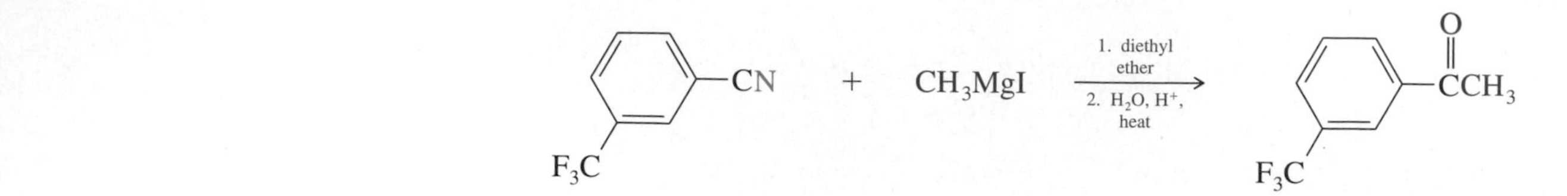

$$m\text{-}CF_3C_6H_4CN + CH_3MgI \xrightarrow[\text{2. } H_2O,\ H^+,\ \text{heat}]{\text{1. diethyl ether}} m\text{-}CF_3C_6H_4\overset{O}{\overset{\|}{C}}CH_3$$

m-(Trifluoromethyl)benzonitrile — Methylmagnesium iodide — m-(Trifluoromethyl)acetophenone (79%)

PROBLEM 20.24 Write an equation showing how you could prepare ethyl phenyl ketone from propanenitrile and a Grignard reagent. What is the structure of the imine intermediate?

Organolithium reagents react in the same way and are often used instead of Grignard reagents.

20.21 SPECTROSCOPIC ANALYSIS OF CARBOXYLIC ACID DERIVATIVES

Infrared spectroscopy is quite useful in identifying carboxylic acid derivatives. The carbonyl stretching vibration is very strong, and its position is sensitive to the nature of the carbonyl group. In general, electron donation from the substituent decreases the double bond character of the bond between carbon and oxygen and decreases the stretching frequency. Two distinct absorptions are observed for the symmetric and antisymmetric stretching vibrations of the anhydride function.

$CH_3\overset{O}{\overset{\|}{C}}Cl$	$CH_3\overset{O}{\overset{\|}{C}}O\overset{O}{\overset{\|}{C}}CH_3$	$CH_3\overset{O}{\overset{\|}{C}}OCH_3$	$CH_3\overset{O}{\overset{\|}{C}}NH_2$
Acetyl chloride	Acetic anhydride	Methyl acetate	Acetamide
$\nu_{C=O} = 1822\ cm^{-1}$	$\nu_{C=O} = 1748\ cm^{-1}$ and $1815\ cm^{-1}$	$\nu_{C=O} = 1736\ cm^{-1}$	$\nu_{C=O} = 1694\ cm^{-1}$

Nitriles are readily identified by absorption due to —C≡N stretching in the 2210 to 2260 cm^{-1} region.

Even at 60 MHz, chemical-shift differences in their ^1H nmr spectra aid the structure determination of esters. Consider the two isomeric esters ethyl acetate and methyl propanoate. As Figure 20.9 shows, the number of signals and their multiplicities are the same for both esters. Both have a methyl singlet and a triplet-quartet pattern for their ethyl group.

$$CH_3\overset{O}{\overset{\|}{C}}OCH_2CH_3$$

Ethyl acetate: Singlet $\delta = 2.0$ ppm; Quartet $\delta = 4.1$ ppm; Triplet $\delta = 1.3$ ppm

$$CH_3O\overset{O}{\overset{\|}{C}}CH_2CH_3$$

Methyl propanoate: Singlet $\delta = 3.6$ ppm; Quartet $\delta = 2.3$ ppm; Trip[illegible] $\delta = 1.2$[illegible]

Notice, however, that there is a significant difference in the chemical shifts of the corresponding signals in the two spectra. The methyl singlet is more shielded (δ = 2.0 ppm) when it is bonded to the carbonyl group of ethyl acetate than when it is

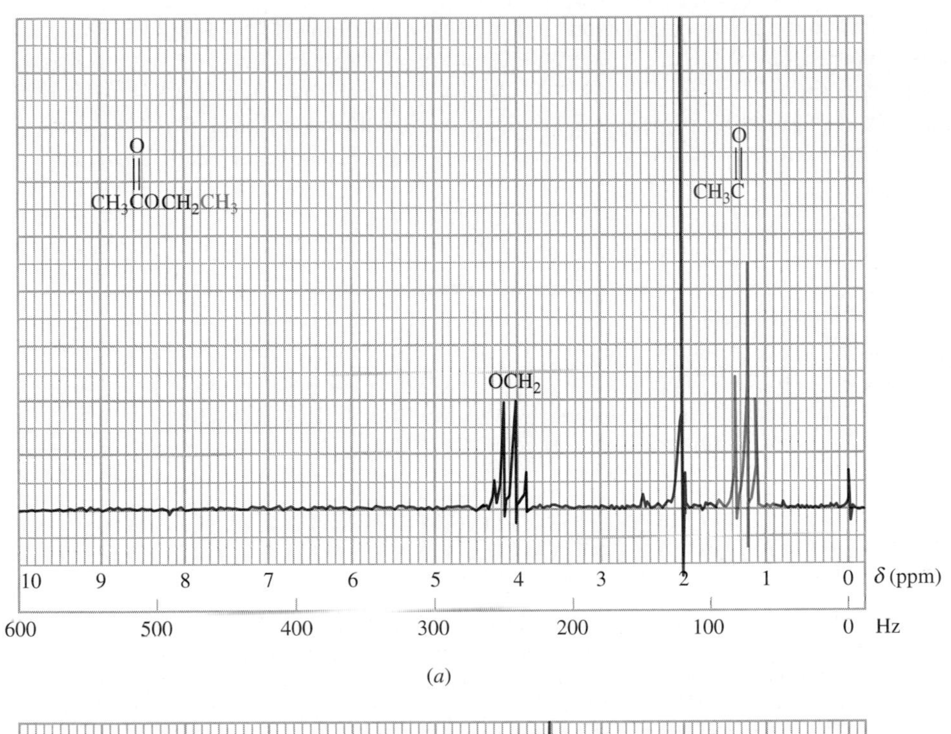

(*a*)

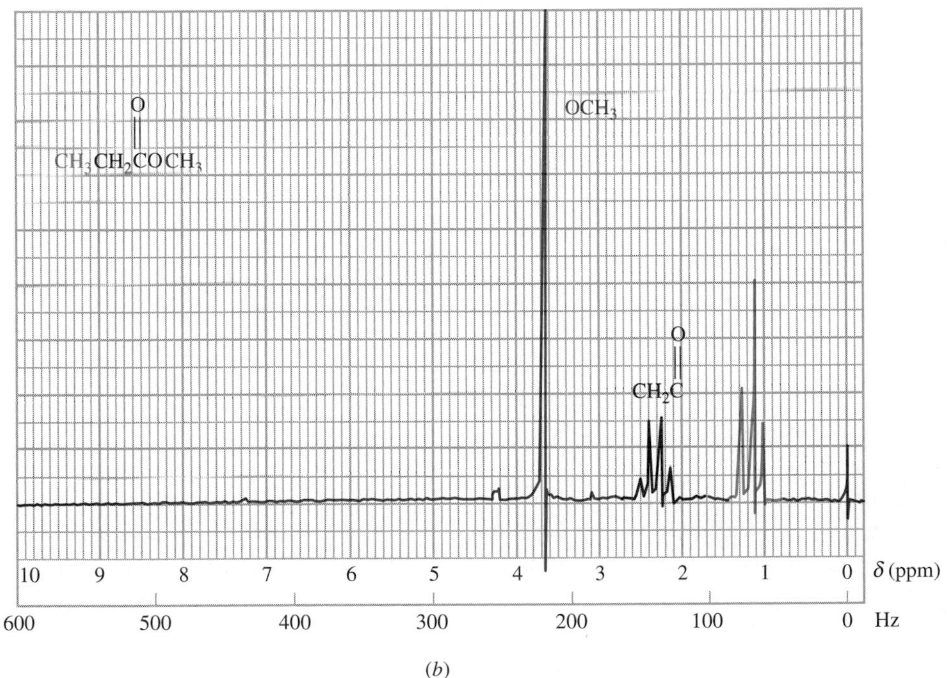

(*b*)

FIGURE 20.9 The 60-MHz ^{1}H nmr spectra of (*a*) ethyl acetate and (*b*) methyl propanoate.

bonded to the oxygen of methyl propanoate ($\delta = 3.6$ ppm). The methylene quartet is more shielded ($\delta = 2.3$ ppm) when it is bonded to the carbonyl group of methyl propanoate than when it is bonded to the oxygen of ethyl acetate ($\delta = 4.1$ ppm). Analysis of the number of peaks and their splitting patterns will not provide an unambiguous answer to structure assignment in esters; chemical-shift data must also be considered.

The chemical shift of the N—H proton of amides appears in the range $\delta = 5$ to 8 ppm. It is often a very broad peak; sometimes it is so broad that it does not rise much over the baseline and can be lost in the background noise.

The ^{13}C nmr spectra of carboxylic acid derivatives, like the spectra of carboxylic acids themselves, are characterized by a low-field resonance for the carbonyl carbon in the range $\delta = 160$ to 180 ppm. The carbonyl carbons of carboxylic acid derivatives are more shielded than those of aldehydes and ketones, but less shielded than the sp^2 hybridized carbons of alkenes and arenes.

20.22 MASS SPECTROMETRY OF CARBOXYLIC ACID DERIVATIVES

A prominent peak in the mass spectra of most carboxylic acid derivatives corresponds to an acylium ion derived by cleavage of the bond to the carbonyl group:

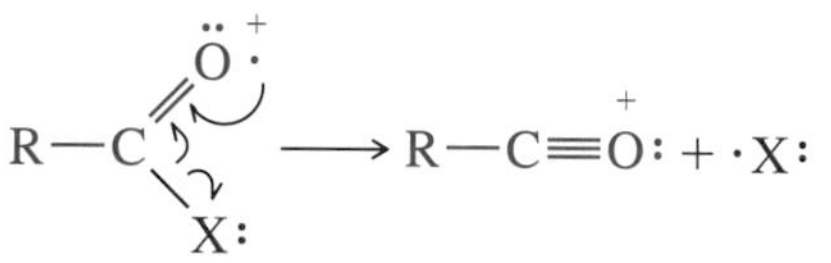

$$\mathrm{R{-}C(=\ddot{O}^{+\cdot}){-}X{:}} \longrightarrow \mathrm{R{-}C{\equiv}\overset{+}{O}{:}} + \cdot\mathrm{X{:}}$$

Amides, however, tend to cleave in the opposite direction to produce a nitrogen-stabilized acylium ion:

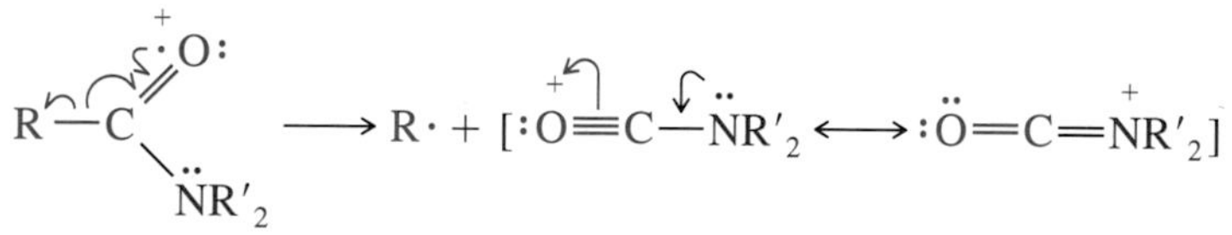

$$\mathrm{R{-}C(=\overset{+}{O}{:}\cdot){-}\ddot{N}R'_2} \longrightarrow \mathrm{R}\cdot + [\mathrm{{:}\overset{+}{O}{\equiv}C{-}\ddot{N}R'_2} \longleftrightarrow \mathrm{{:}\ddot{O}{=}C{=}\overset{+}{N}R'_2}]$$

20.23 SUMMARY

This chapter concerns the preparation and reactions of *acyl chlorides, acid anhydrides, esters, amides,* and *nitriles.* These compounds are generally classified as carboxylic acid derivatives, and their nomenclature is based on that of carboxylic acids (Section 20.1).

$\mathrm{RC(=O)Cl}$	$\mathrm{RC(=O)OC(=O)R}$	$\mathrm{RC(=O)OR'}$	$\mathrm{RC(=O)NR'_2}$	$\mathrm{RC{\equiv}N}$
Acyl chloride	Carboxylic acid anhydride	Ester	Amide	Nitrile

The characteristic reaction of acyl chlorides, acid anhydrides, esters, and amides is **nucleophilic acyl substitution.** Addition of a nucleophilic reagent HY: to the carbonyl group leads to a tetrahedral intermediate that dissociates to give the product of substitution:

$$\mathrm{R\overset{O}{\overset{\|}{C}}{-}X} + \mathrm{HY{:}} \rightleftharpoons \mathrm{R\underset{Y}{\underset{|}{\overset{OH}{\overset{|}{C}}}}{-}X} \rightleftharpoons \mathrm{R\overset{O}{\overset{\|}{C}}{-}Y} + \mathrm{HX{:}}$$

Carboxylic acid derivative | Nucleophile | Tetrahedral intermediate | Product of nucleophilic acyl substitution | Conjugate acid of leaving group

The order of reactivity of carboxylic acid derivatives toward nucleophilic acyl substitution is

Most reactive | Least reactive

$$\mathrm{R\overset{O}{\overset{\|}{C}}Cl} > \mathrm{R\overset{O}{\overset{\|}{C}}O\overset{O}{\overset{\|}{C}}R} > \mathrm{R\overset{O}{\overset{\|}{C}}OR'} > \mathrm{R\overset{O}{\overset{\|}{C}}NR'_2}$$

Least stabilized carbonyl group | Most stabilized carbonyl group

The more stable the carbonyl group, the less reactive it is toward attack by a nucleophile. Nitrogen is a better electron pair donor than oxygen, and amides have a more stabilized carbonyl than esters and anhydrides. Chlorine is the poorest electron pair donor, and acyl chlorides have the least stabilized carbonyl group and are the most reactive (Section 20.2).

Acyl chlorides are powerful acylating agents and can be converted to anhydrides, esters, and amides by nucleophilic acyl substitution (Section 20.3):

$$\mathrm{R\overset{O}{\overset{\|}{C}}Cl} + \mathrm{R'\overset{O}{\overset{\|}{C}}OH} \longrightarrow \mathrm{R\overset{O}{\overset{\|}{C}}O\overset{O}{\overset{\|}{C}}R'} + \mathrm{HCl}$$

Acyl chloride | Carboxylic acid | Acid anhydride | Hydrogen chloride

$$\mathrm{R\overset{O}{\overset{\|}{C}}Cl} + \mathrm{R'OH} \longrightarrow \mathrm{R\overset{O}{\overset{\|}{C}}OR'} + \mathrm{HCl}$$

Acyl chloride | Alcohol | Ester | Hydrogen chloride

$$\mathrm{R\overset{O}{\overset{\|}{C}}Cl} + \mathrm{2R'_2NH} \longrightarrow \mathrm{R\overset{O}{\overset{\|}{C}}NR'_2} + \mathrm{R'_2\overset{+}{N}H_2\ Cl^-}$$

Acyl chloride | Amine | Amide | Ammonium chloride salt

Examples of each of these reactions may be found in Table 20.2.

Acid anhydrides may be prepared from acyl chlorides in the laboratory, but the most commonly encountered anhydrides are industrial chemicals prepared by specialized methods (Section 20.4). Although they are less reactive acylating agents than acyl chlorides, acid anhydrides are useful reagents for the preparation of esters and amides (Section 20.5):

$$\underset{\text{Acid anhydride}}{\overset{\text{O}\ \ \text{O}}{\overset{\|\ \ \|}{\text{RCOCR}}}} + \underset{\text{Alcohol}}{\text{R}'\text{OH}} \longrightarrow \underset{\text{Ester}}{\overset{\text{O}}{\overset{\|}{\text{RCOR}'}}} + \underset{\text{Carboxylic acid}}{\overset{\text{O}}{\overset{\|}{\text{RCOH}}}}$$

$$\underset{\text{Acid anhydride}}{\overset{\text{O}\ \ \text{O}}{\overset{\|\ \ \|}{\text{RCOCR}}}} + \underset{\text{Amine}}{2\text{R}'_2\text{NH}} \longrightarrow \underset{\text{Amide}}{\overset{\text{O}}{\overset{\|}{\text{RCNR}'_2}}} + \underset{\text{Ammonium carboxylate salt}}{\text{R}'_2\overset{+}{\text{N}}\text{H}_2 \quad {}^-\overset{\text{O}}{\overset{\|}{\text{OCR}}}}$$

Table 20.3 presents examples of these reactions.

Esters occur naturally or are prepared from alcohols by the Fischer esterification procedure or by acylation with acyl chlorides or acid anhydrides (Section 20.6; Table 20.4). Section 20.7 compared the properties of esters with those of alkanes and alcohols to assess the influence of polarity and opportunities for hydrogen bonding on boiling point and solubility in water. Previously encountered reactions of esters (with Grignard reagents and with lithium aluminum hydride) were recalled in Section 20.8.

Sections 20.9 and 20.10 compared the mechanisms of ester hydrolysis in acidic and in basic media. In both cases hydrolysis proceeds through the same tetrahedral intermediate. This intermediate is analogous to the hydrates formed by nucleophilic addition of water to aldehydes and ketones.

$$\text{R}-\underset{\text{OH}}{\underset{|}{\overset{\text{OH}}{\overset{|}{\text{C}}}}}-\text{OR}' \qquad \text{Tetrahedral intermediate in ester hydrolysis}$$

Ester hydrolysis in acid is the reverse of the Fischer esterification. Acid-catalyzed ester hydrolysis and the Fischer esterification proceed through the same series of intermediates, but their order of appearance is reversed.

Ester hydrolysis in base is called **saponification** and is irreversible because the carboxylic acid is converted to the corresponding carboxylate anion under these conditions.

$$\underset{\text{Ester}}{\overset{\text{O}}{\overset{\|}{\text{RCOR}'}}} + \underset{\text{Hydroxide ion}}{\text{HO}^-} \longrightarrow \underset{\text{Carboxylate ion}}{\overset{\text{O}}{\overset{\|}{\text{RCO}^-}}} + \underset{\text{Alcohol}}{\text{R}'\text{OH}}$$

Esters react with amines to give amides (Section 20.11):

$$\underset{\text{Ester}}{\overset{\text{O}}{\overset{\|}{\text{RCOR}'}}} + \underset{\text{Amine}}{\text{R}''_2\text{NH}} \longrightarrow \underset{\text{Amide}}{\overset{\text{O}}{\overset{\|}{\text{RCNR}''_2}}} + \underset{\text{Alcohol}}{\text{R}'\text{OH}}$$

Thioesters (Section 20.12) undergo reactions analogous to those of esters, but at faster rates. A sulfur atom stabilizes the carbonyl group less effectively than an oxygen.

Amides are prepared from amines by reactions in which acyl chlorides, anhydrides, or esters act as acylating agents. In a limited number of cases, amides are prepared by heating ammonium carboxylate salts (Section 20.13). Lactams are cyclic amides (Section 20.14). Imides are characterized by the attachment of two acyl groups to the same nitrogen (Section 20.15).

Like ester hydrolysis, amide hydrolysis can be achieved in either aqueous acid or aqueous base (Section 20.16). It is irreversible in both media. In base, the carboxylic acid is converted to the carboxylate anion; in acid, the amine is protonated to an ammonium ion:

$$\underset{\text{Amide}}{\overset{\overset{\large O}{\|}}{RCNR'_2}} + \underset{\text{Water}}{H_2O} \begin{cases} \xrightarrow{H_3O^+} \underset{\text{Carboxylic acid}}{\overset{\overset{\large O}{\|}}{RCOH}} + \underset{\text{Ammonium ion}}{R'_2\overset{+}{N}H_2} \\ \xrightarrow{HO^-} \underset{\text{Carboxylate ion}}{\overset{\overset{\large O}{\|}}{RCO^-}} + \underset{\text{Amine}}{R'_2NH} \end{cases}$$

The **Hofmann rearrangement** converts amides of the type $\overset{\overset{O}{\|}}{RCNH_2}$ to primary amines (RNH_2; Section 20.17). The carbon chain is shortened by one carbon with loss of the carbonyl group:

$$\underset{\text{Amide}}{\overset{\overset{\large O}{\|}}{RCNH_2}} \xrightarrow[\text{NaOH}]{Br_2} \underset{\text{Amine}}{RNH_2}$$

Nitriles are prepared by nucleophilic substitution of alkyl halides with cyanide ion, by converting aldehydes or ketones to cyanohydrins, or by dehydration of amides (Section 20.18). Nitriles are useful starting materials for the preparation of carboxylic acids by hydrolysis (Section 20.19) or ketones by reaction with Grignard reagents (Section 20.20).

$$\underset{\text{Nitrile}}{RC{\equiv}N} \xrightarrow[\text{or} \; 1.\ H_2O,\ HO^- \; 2.\ H^+]{H_2O,\ H^+} \underset{\text{Carboxylic acid}}{\overset{\overset{\large O}{\|}}{RCOH}}$$

$$\underset{\text{Nitrile}}{RC{\equiv}N} + \underset{\text{Grignard reagent}}{R'MgX} \xrightarrow[2.\ H_2O,\ H^+]{1.\ \text{diethyl ether}} \underset{\text{Ketone}}{\overset{\overset{\large O}{\|}}{RCR'}}$$

PROBLEMS

20.25 Write a structural formula for each of the following compounds:

(*a*) *m*-Chlorobenzoyl bromide

(*b*) Trifluoroacetic anhydride

(*c*) *cis*-1,2-Cyclopropanedicarboxylic anhydride

(*d*) Ethyl cycloheptanecarboxylate

(*e*) 1-Phenylethyl acetate

(*f*) 2-Phenylethyl acetate

(*g*) *p*-Ethylbenzamide
(*h*) *N*-Ethylbenzamide
(*i*) 2-Methylhexanenitrile

20.26 Give an acceptable IUPAC name for each of the following compounds:

(*a*) $CH_3\underset{\displaystyle Cl}{\underset{|}{C}}HCH_2\overset{\displaystyle O}{\overset{\|}{C}}Br$

(*b*) $CH_3\overset{\displaystyle O}{\overset{\|}{C}}OCH_2—C_6H_5$

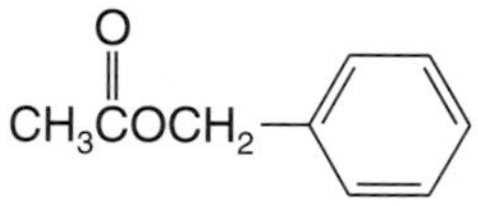

(*c*) $CH_3O\overset{\displaystyle O}{\overset{\|}{C}}CH_2—C_6H_5$

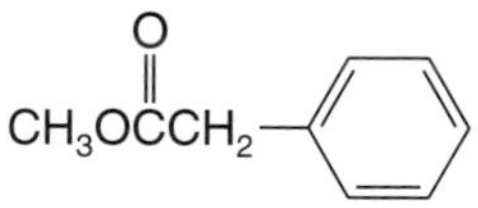

(*d*) $ClCH_2CH_2\overset{\displaystyle O}{\overset{\|}{C}}O\overset{\displaystyle O}{\overset{\|}{C}}CH_2CH_2Cl$

(*e*) H_3C, H_3C; O, O, O

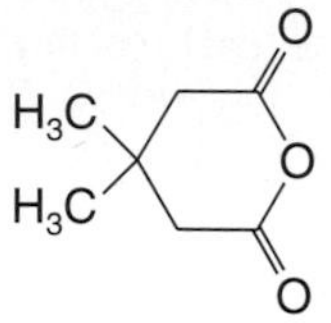

(*f*) $(CH_3)_2CHCH_2CH_2C≡N$

(*g*) $(CH_3)_2CHCH_2CH_2\overset{\displaystyle O}{\overset{\|}{C}}NH_2$

(*h*) $(CH_3)_2CHCH_2CH_2\overset{\displaystyle O}{\overset{\|}{C}}NHCH_3$

(*i*) $(CH_3)_2CHCH_2CH_2\overset{\displaystyle O}{\overset{\|}{C}}N(CH_3)_2$

20.27 Write a structural formula for the principal organic product or products of each of the following reactions:

(*a*) Acetyl chloride and bromobenzene, $AlCl_3$
(*b*) Acetyl chloride and butanethiol
(*c*) Propanoyl chloride and sodium propanoate
(*d*) Butanoyl chloride and benzyl alcohol
(*e*) *p*-Chlorobenzoyl chloride and ammonia

(*f*) O O O and water

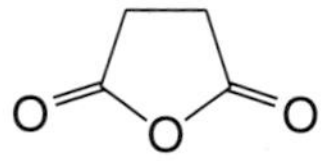

(*g*) O O O and aqueous sodium hydroxide

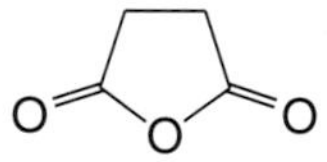

(*h*) O O O and aqueous ammonia

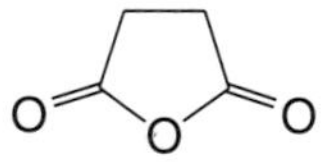

(*i*) O O O and benzene, $AlCl_3$

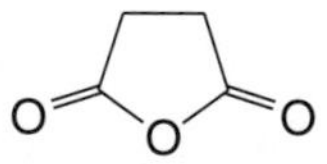

(*j*) O O O and 1,3-pentadiene

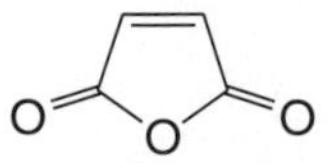

(*k*) Acetic anhydride and 3-pentanol

(*l*) O O and aqueous sodium hydroxide

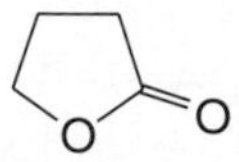

(*m*) O O and aqueous ammonia

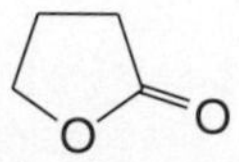

(*n*) O O and lithium aluminum hydride, then H_2O

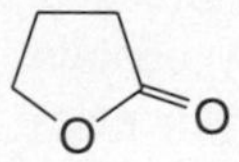

(*o*) (structure) and excess methylmagnesium bromide, then H_3O^+

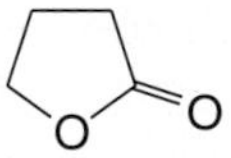

(*p*) Ethyl phenylacetate and methylamine (CH_3NH_2)

(*q*) (structure, N–CH_3) and aqueous sodium hydroxide

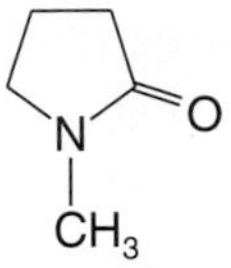

(*r*) (structure, N–CH_3) and aqueous hydrochloric acid, heat

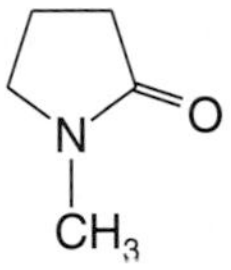

(*s*) (structure, N–CH_3) and aqueous sodium hydroxide

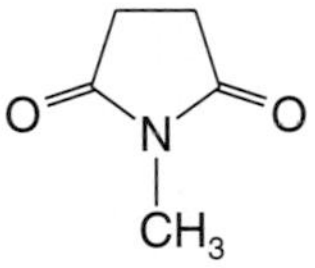

(*t*) (structure, N–CH_3) and aqueous hydrochloric acid, heat

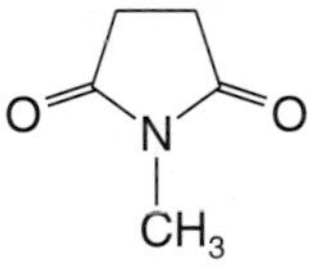

(*u*) $C_6H_5NHCCH_3$ (C=O) and aqueous hydrochloric acid, heat

(*v*) $C_6H_5CNHCH_3$ (C=O) and aqueous sulfuric acid, heat

(*w*) (cyclopentyl)–CNH_2 (C=O) and P_4O_{10}

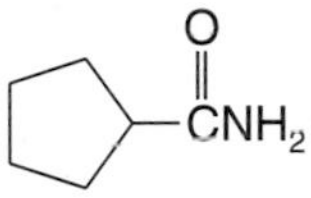

(*x*) $(CH_3)_2CHCH_2C{\equiv}N$ and aqueous hydrochloric acid, heat

(*y*) *p*-Methoxybenzonitrile and aqueous sodium hydroxide, heat

(*z*) Propanenitrile and methylmagnesium bromide, then H_3O^+

(*aa*) (structure: H_3C, CH_3, C, O, NH_2) $+ Br_2 \xrightarrow[CH_3OH]{NaOCH_3}$

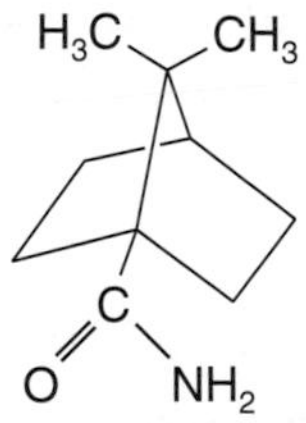

(*bb*) Product of (*aa*) $\xrightarrow{KOH,\ H_2O}$

20.28 Using ethanol as the ultimate source of all the carbon atoms, along with any necessary inorganic reagents, show how you could prepare each of the following:

(*a*) Acetyl chloride
(*b*) Acetic anhydride
(*c*) Ethyl acetate
(*d*) Ethyl bromoacetate
(*e*) 2-Bromoethyl acetate
(*f*) Ethyl cyanoacetate
(*g*) Acetamide
(*h*) Methylamine (CH_3NH_2)
(*i*) 2-Hydroxypropanoic acid

20.29 Using toluene as the ultimate source of all the carbon atoms, along with any necessary inorganic reagents, show how you could prepare each of the following:

(*a*) Benzoyl chloride
(*b*) Benzoic anhydride
(*c*) Benzyl benzoate
(*d*) Benzamide
(*e*) Benzonitrile
(*f*) Benzyl cyanide
(*g*) Phenylacetic acid
(*h*) *p*-Nitrobenzoyl chloride
(*i*) *m*-Nitrobenzoyl chloride
(*j*) Aniline

20.30 The saponification of ^{18}O-labeled ethyl propanoate has been described in Section 20.10 as one of the significant experiments that demonstrated acyl-oxygen cleavage in ester hydrolysis. The ^{18}O-labeled ethyl propanoate used in this experiment was prepared from ^{18}O-labeled ethyl alcohol, which in turn was obtained from acetaldehyde and ^{18}O-enriched water. Write a series of equations showing the preparation of $CH_3CH_2\overset{\overset{\displaystyle O}{\|}}{C}OCH_2CH_3$ (where $O = {}^{18}O$) from these starting materials.

20.31 Suggest a reasonable explanation for each of the following observations:

(*a*) The second-order rate constant *k* for saponification of ethyl trifluoroacetate is over 1 million times as great as that for ethyl acetate (25°C).
(*b*) The second-order rate constant for saponification of ethyl 2,2-dimethylpropanoate, $(CH_3)_3CCO_2CH_2CH_3$, is almost 100 times as small as that for ethyl acetate (30°C).
(*c*) The second-order rate constant *k* for saponification of methyl acetate is 100 times as great as that for *tert*-butyl acetate (25°C).
(*d*) The second-order rate constant *k* for saponification of methyl *m*-nitrobenzoate is 40 times as great as that for methyl benzoate (25°C).
(*e*) The second-order rate constant *k* for saponification of 5-pentanolide is over 20 times as great as that for 4-butanolide (25°C).

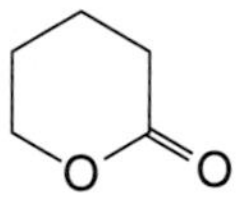

5-Pentanolide 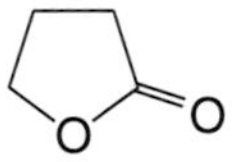   4-Butanolide

(*f*) The second-order rate constant *k* for saponification of ethyl *trans*-4-*tert*-butylcyclohexanecarboxylate is 20 times as great as that for its cis diastereomer (25°C).

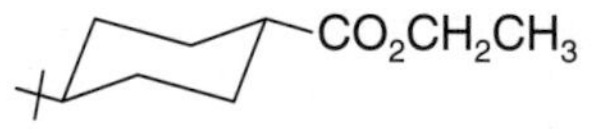

$CO_2CH_2CH_3$ $CO_2CH_2CH_3$

Ethyl *trans*-4-*tert*-butylcyclohexanecarboxylate 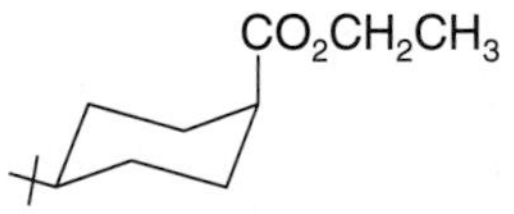   Ethyl *cis*-4-*tert*-butylcyclohexanecarboxylate

20.32 The preparation of *cis*-4-*tert*-butylcyclohexanol from its trans stereoisomer was carried out by the following sequence of steps. Write structural formulas, including stereochemistry, for compounds A and B.

Step 1: (OH) + CH_3–C_6H_4–SO_2Cl $\xrightarrow{\text{pyridine}}$ compound A ($C_{17}H_{26}O_3S$)

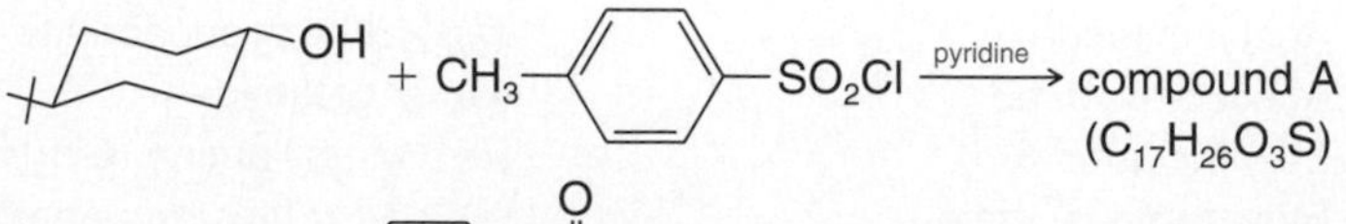

Step 2: Compound A + $C_6H_5\overset{\overset{\displaystyle O}{\|}}{C}ONa$ $\xrightarrow[\text{heat}]{N,N\text{-dimethylformamide}}$ compound B ($C_{17}H_{24}O_2$)

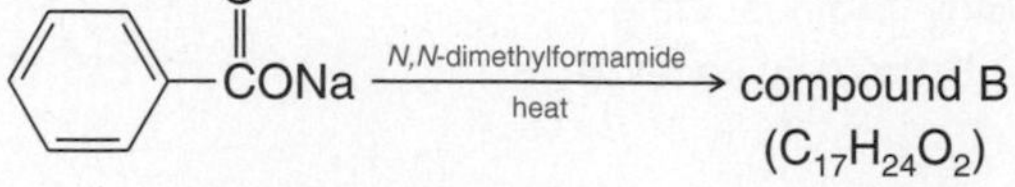

Step 3: Compound B $\xrightarrow[H_2O]{NaOH}$ (*trans*-4-*tert*-butylcyclohexanol)

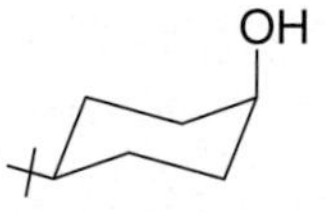

20.33 The ketone shown was prepared in a three-step sequence from ethyl trifluoroacetate. The first step in the sequence involved treating ethyl trifluoroacetate with ammonia to give a compound A. Compound A was in turn converted to the desired ketone by way of a compound B. Fill in the missing reagents in the sequence shown and give the structures of compounds A and B.

$$CF_3\overset{O}{\overset{\|}{C}}OCH_2CH_3 \xrightarrow{NH_3} \text{compound A} \longrightarrow \text{compound B} \longrightarrow CF_3\overset{O}{\overset{\|}{C}}C(CH_3)_3$$

20.34 *Ambrettolide* is obtained from hibiscus and has a musklike odor. Its preparation from a compound A is outlined in the table that follows. Write structural formulas, ignoring stereochemistry, for compounds B through G in this synthesis. (*Hint:* Zinc, as used in Step 4, converts vicinal dibromides to alkenes.)

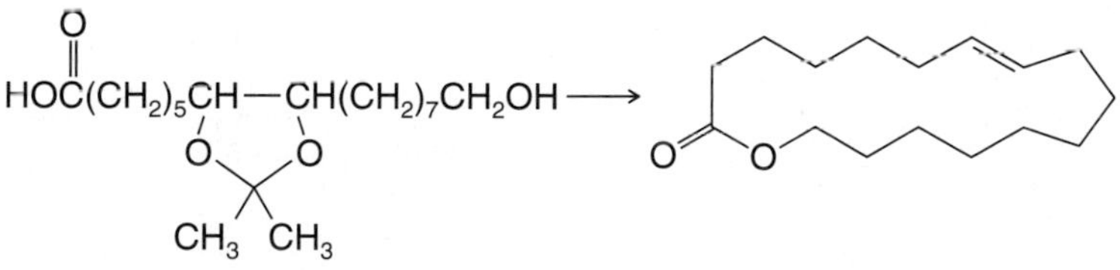

$$HO\overset{O}{\overset{\|}{C}}(CH_2)_5CH—CH(CH_2)_7CH_2OH \longrightarrow \text{Ambrettolide}$$

(In compound A, the two CH carbons are joined through O atoms to a $C(CH_3)_2$ group, forming a five-membered ring.)

Compound A — Ambrettolide

Step	Reactant	Reagents	Product
1.	Compound A	H_2O, H^+, heat	Compound B ($C_{16}H_{32}O_5$)
2.	Compound B	HBr	Compound C ($C_{16}H_{29}Br_3O_2$)
3.	Compound C	Ethanol, H_2SO_4	Compound D ($C_{18}H_{33}Br_3O_2$)
4.	Compound D	Zinc, ethanol	Compound E ($C_{18}H_{33}BrO_2$)
5.	Compound E	Sodium acetate, acetic acid	Compound F ($C_{20}H_{36}O_4$)
6.	Compound F	KOH, ethanol, then H^+	Compound G ($C_{16}H_{30}O_3$)
7.	Compound G	Heat	Ambrettolide ($C_{16}H_{28}O_2$)

20.35 The preparation of the sex pheromone of the bollworm moth, (*E*)-9,11-dodecadien-1-yl acetate, from a compound A has been described. Suggest suitable reagents for each step in this sequence.

(*a*) $$HOCH_2CH{=}CH(CH_2)_7CO_2CH_3 \longrightarrow H\overset{O}{\overset{\|}{C}}CH{=}CH(CH_2)_7CO_2CH_3$$

Compound A (*E* isomer) — Compound B

(*b*) Compound B $\longrightarrow$ $CH_2{=}CHCH{=}CH(CH_2)_7CO_2CH_3$
Compound C

(*c*) Compound C $\longrightarrow$ $CH_2{=}CHCH{=}CH(CH_2)_7CH_2OH$
Compound D

(*d*) Compound D $\longrightarrow$ $CH_2{=}CHCH{=}CH(CH_2)_7CH_2O\overset{O}{\overset{\|}{C}}CH_3$
(*E*)-9,11-Dodecadien-1-yl acetate

20.36 Outline reasonable mechanisms for each of the following reactions:

(*a*) γ-butyrolactone + $BrMgCH_2CH_2CH_2CH_2MgBr$ $\xrightarrow[2.\ H_3O^+]{1.\ THF}$ 1-(3-hydroxypropyl)cyclopentanol (HO, $CH_2CH_2CH_2OH$)

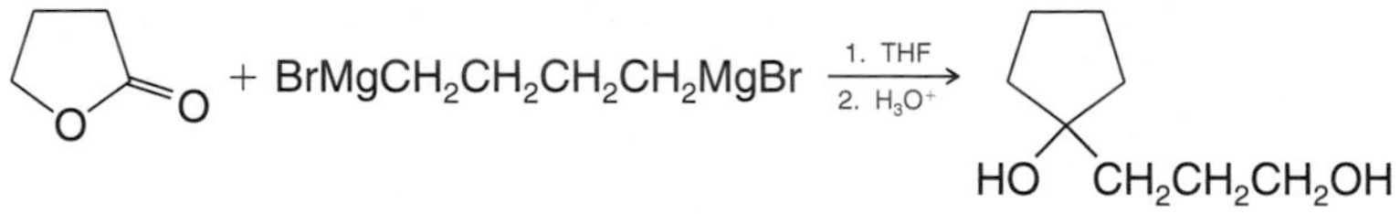

(*b*) $H_2NCH_2CH_2$-substituted thiane-2-one (δ-thiolactone) $\xrightarrow{\text{spontaneous}}$ eight-membered lactam bearing HS (N–H)

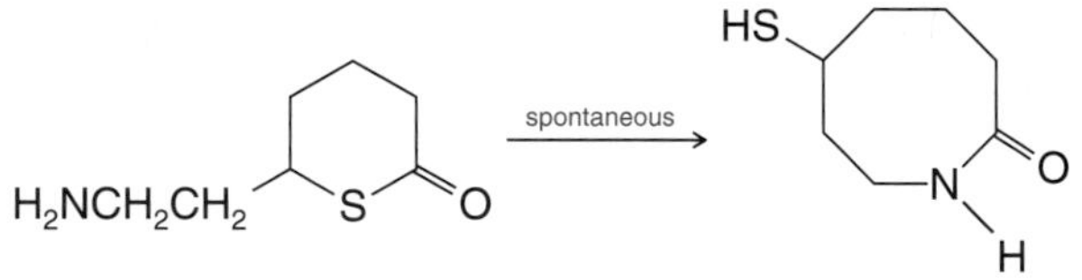

20.37 Identify compounds A through C in the following equations:

(*a*) $CH_3O-C_6H_4-\overset{O}{\overset{\|}{C}}Cl$ + $C_6H_5-\overset{O}{\overset{\|}{C}}-\underset{OH}{\underset{|}{C}H}-C_6H_5$ $\xrightarrow{\text{pyridine}}$ compound A ($C_{22}H_{18}O_4$)

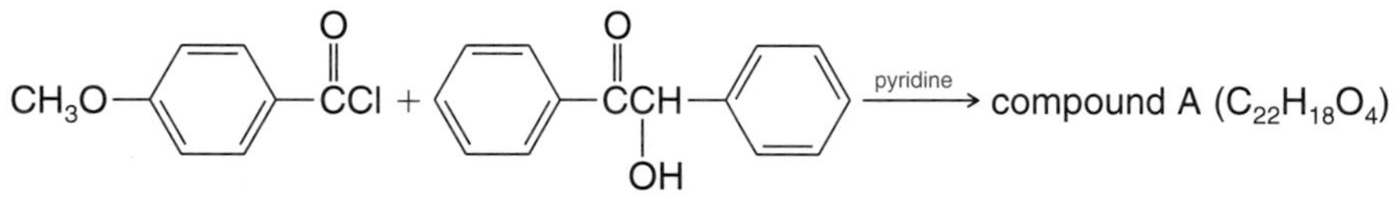

(*b*) $CH_3\overset{O}{\overset{\|}{C}}CH_2CH_2\overset{O}{\overset{\|}{C}}OCH_2CH_3$ $\xrightarrow[2.\ H_3O^+]{1.\ CH_3MgI\ (1\ equiv),\ diethyl\ ether}$ compound B (a lactone, $C_6H_{10}O_2$)

(*c*) 1,2,4-benzenetricarboxylic acid (three $\overset{O}{\overset{\|}{C}}OH$ groups) $\xrightarrow{\text{heat}}$ compound C ($C_9H_4O_5$) + H_2O

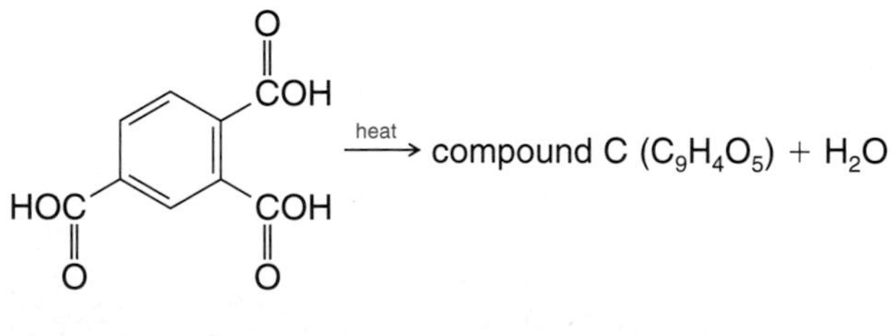

20.38 When compounds of the type represented by A are allowed to stand in pentane, they are converted to a constitutional isomer.

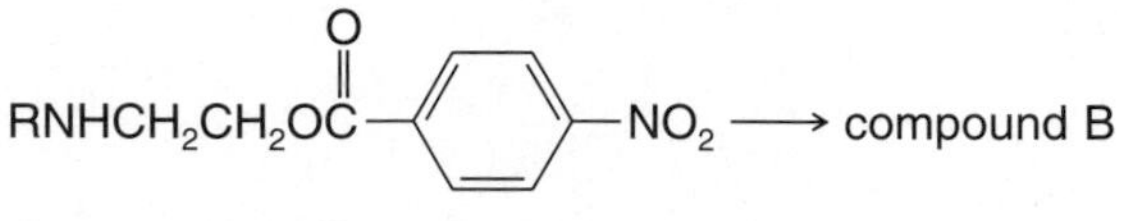

$RNHCH_2CH_2O\overset{O}{\overset{\|}{C}}-C_6H_4-NO_2$ $\longrightarrow$ compound B
Compound A

Hydrolysis of either A or B yields $RNHCH_2CH_2OH$ and *p*-nitrobenzoic acid. Suggest a reasonable structure for compound B and demonstrate your understanding of the mechanism of this reaction by writing the structure of the key intermediate in the conversion of compound A to compound B.

20.39

(*a*) In the presence of dilute hydrochloric acid, compound A is converted to a constitutional isomer, compound B.

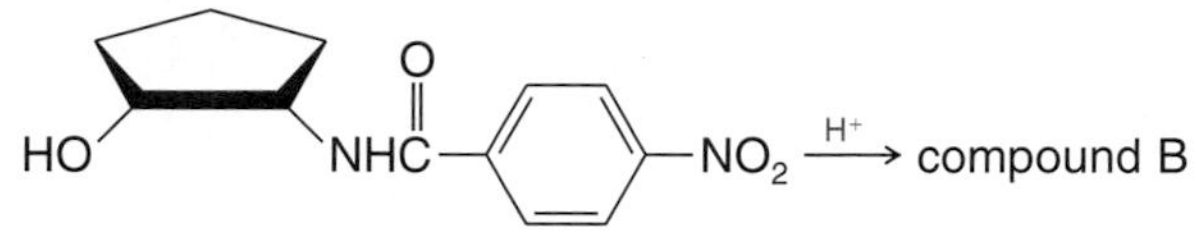

$$\text{HO}\!-\!\text{(cyclopentane)}\!-\!\text{NHC(=O)}\!-\!\text{C}_6\text{H}_4\!-\!\text{NO}_2 \xrightarrow{\text{H}^+} \text{compound B}$$

Compound A

Suggest a reasonable structure for compound B.

(*b*) The trans stereoisomer of compound A is stable under the reaction conditions. Why does it not rearrange?

20.40 Poly(vinyl alcohol) is a useful water-soluble polymer. It cannot be prepared directly from vinyl alcohol, because of the rapidity with which vinyl alcohol ($CH_2{=}CHOH$) isomerizes to acetaldehyde. Vinyl acetate, however, does not rearrange and can be polymerized to poly(vinyl acetate). How could you make use of this fact to prepare poly(vinyl alcohol)?

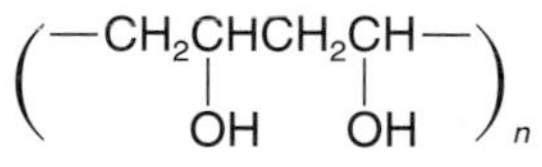

$$\left(\!-CH_2\underset{\large OH}{CH}CH_2\underset{\large OH}{CH}-\!\right)_n \qquad \left(\!-CH_2\underset{\large CH_3\overset{}{C}(=O)O}{CH}CH_2\underset{\large O\overset{}{C}(=O)CH_3}{CH}-\!\right)_n$$

Poly(vinyl alcohol) 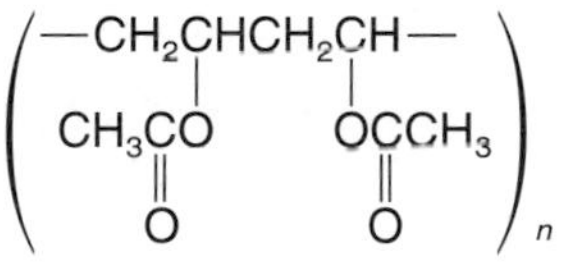 Poly(vinyl acetate)

20.41 *Lucite* is a polymer of methyl methacrylate.

(*a*) Assuming the first step in the polymerization of methyl methacrylate is as shown,

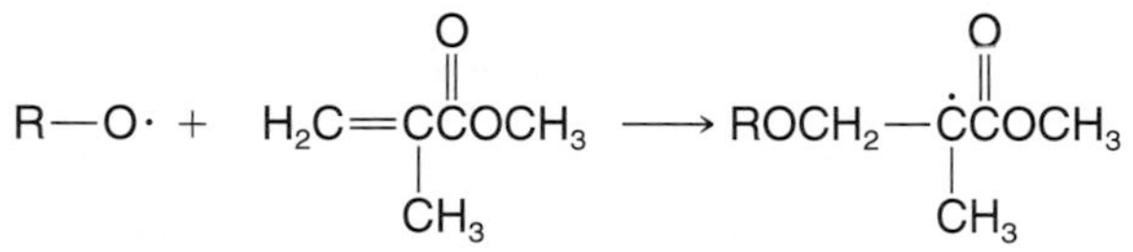

$$R{-}O\cdot + H_2C{=}\underset{\large CH_3}{C}\overset{\large O}{\overset{\|}{C}}OCH_3 \longrightarrow ROCH_2{-}\underset{\large CH_3}{\dot{C}}\overset{\large O}{\overset{\|}{C}}OCH_3$$

Methyl methacrylate

write a structural formula for the free radical produced after the next two propagation steps.

(*b*) Outline a synthesis of methyl methacrylate from acetone, sodium cyanide, and any necessary organic or inorganic reagents.

20.42 A compound ($C_4H_6O_2$) has a strong band in the infrared at 1760 cm^{-1}. Its ^{13}C nmr spectrum exhibits signals at $\delta = 20.2$ (quartet), 96.8 (triplet), 141.8 (doublet), and 167.6 ppm (singlet). The ^1H nmr spectrum of the compound has a three-proton singlet at $\delta = 2.1$ ppm along with three other signals, each of which is a doublet of doublets, at $\delta = 4.7$, 4.9, and 7.3 ppm. What is the structure of the compound?

20.43 A compound has a molecular weight of 83 and contains nitrogen. Its infrared spectrum contains a moderately strong peak at 2270 cm^{-1}. Its 300-MHz ^1H nmr spectrum is shown in Figure 20.10. What is the structure of the compound?

20.44 A compound has a molecular formula of $C_8H_{14}O_4$, and its infrared spectrum contains

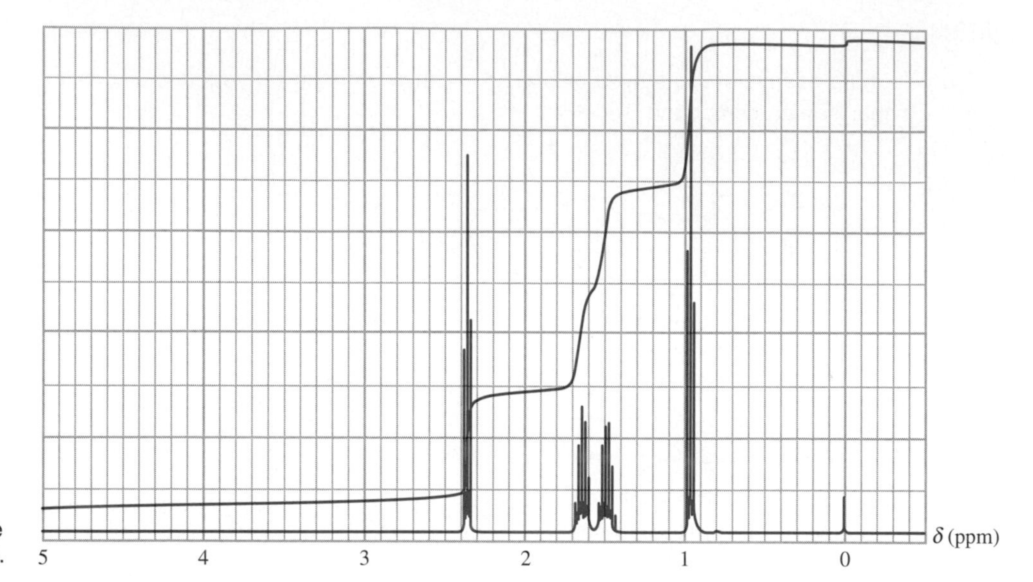

FIGURE 20.10 The 300-MHz ^1H nmr spectrum of the compound in Problem 20.43.

an intense peak at 1730 cm^{-1}. The 300-MHz ^1H nmr spectrum of the compound is shown in Figure 20.11. What is its structure?

20.45 The infrared spectrum of a compound (C_3H_6ClNO) has an intense peak at 1680 cm^{-1}. Its ^1H nmr spectrum is shown in Figure 20.12. What is the structure of the compound? How could you prepare the compound from propanoic acid?

MOLECULAR MODELING EXERCISES

20.46 Formamide ($H_2NCH{=}O$) is planar. What hybridization state of nitrogen should you choose in order to construct a molecular model of formamide?

20.47 Make a molecular model of ketene ($CH_2{=}C{=}O$), an intermediate in the industrial preparation of acetic anhydride. Describe its geometry.

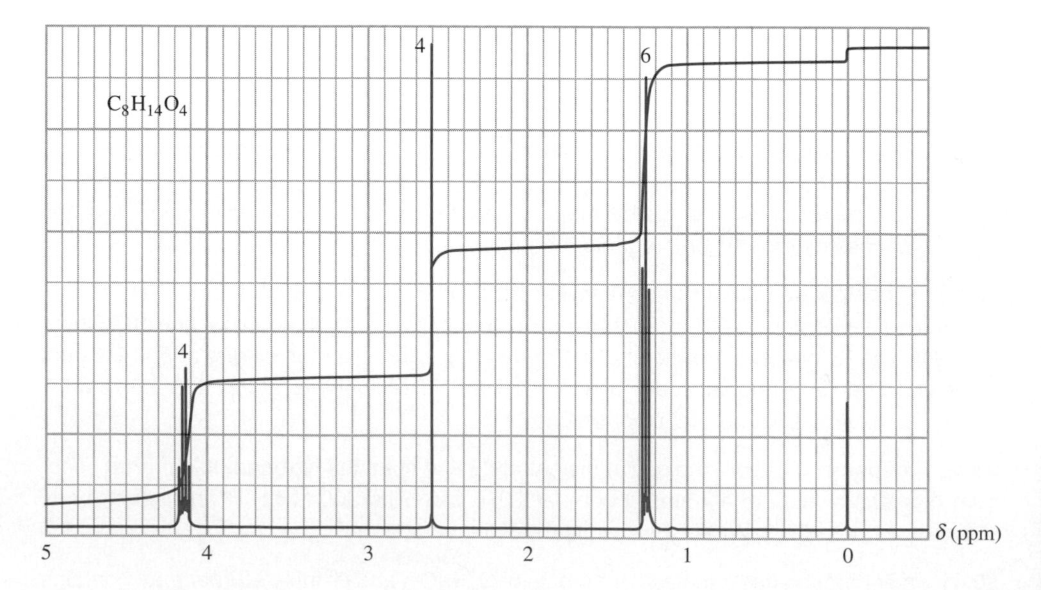

FIGURE 20.11 The 300-MHz ^1H nmr spectrum of compound $C_8H_{14}O_4$ in Problem 20.44.

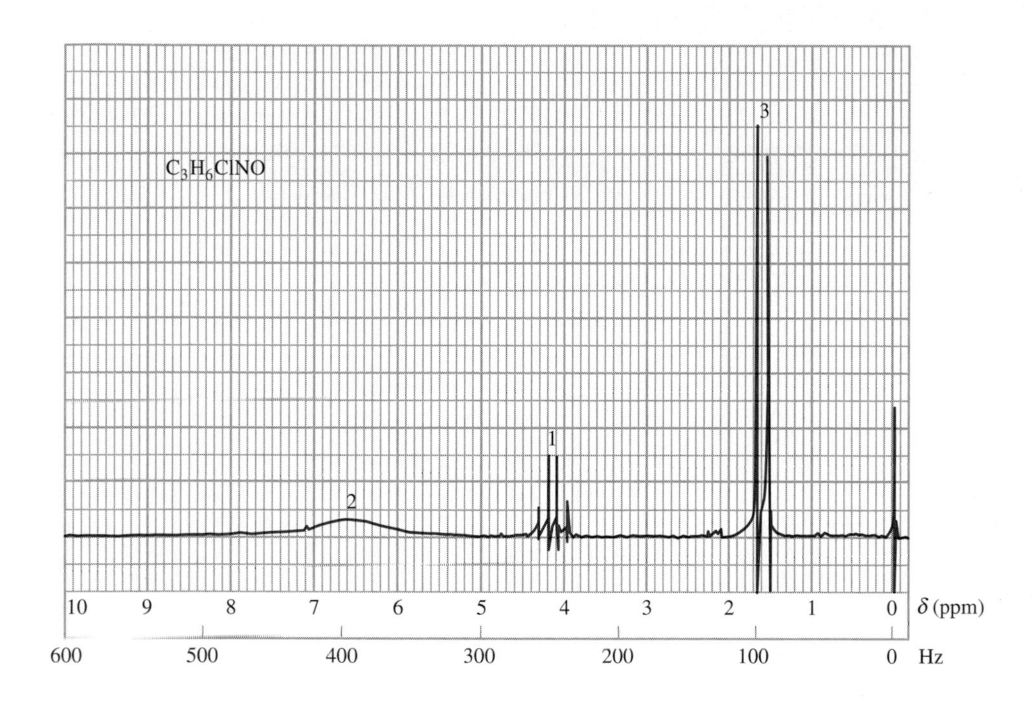

FIGURE 20.12 The 60-MHz ^1H nmr spectrum of compound C_3H_6ClNO in Problem 20.45.

20.48 Excluding enantiomers, there are three isomeric cyclopropanedicarboxylic acids. Two of them, A and B, are consitutional isomers of each other and each forms a cyclic anhydride on being heated. The third diacid, C, does not form a cyclic anhydride. C is a constitutional isomer of A and a stereoisomer of B. Identify A, B, and C. Construct molecular models of the cyclic anhydrides formed on heating A and B. Why doesn't C form a cyclic anhydride?

20.49 Construct a molecular model of the tetrahedral intermediate formed in the reaction of methyl formate with ammonia.

CHAPTER 21

ESTER ENOLATES

You have already had considerable experience with carbanionic compounds and their applications in synthetic organic chemistry. The first carbanion you studied in detail was acetylide ion in Chapter 9. This was followed in Chapter 14 by an examination of organometallic compounds, where you saw that species of this type—Grignard reagents, for example—act as sources of negatively polarized carbon. In Chapter 18 you learned that enolate ions—reactive intermediates generated from aldehydes and ketones—are nucleophilic and that this property can be used to advantage in organic synthesis as a method for carbon-carbon bond formation.

The present chapter extends our study of carbanions to the enolate ions derived from esters. **Ester enolates** are important reagents in synthetic organic chemistry. The stabilized enolates derived from **β-keto esters** are particularly useful.

$$\mathrm{R{-}\overset{O}{\overset{\|}{C}}{-}CH_2{-}\overset{O}{\overset{\|}{C}}{-}OR'}$$

β-Keto ester: a ketone carbonyl is β to the carbonyl group of the ester.

A proton attached to the α carbon atom of a β-keto ester is relatively acidic. Typical acid dissociation constants K_a for β-keto esters are $\sim 10^{-11}$ (pK_a 11). Because the α carbon atom is flanked by two electron-withdrawing carbonyl groups, a carbanion formed at this site is highly stabilized. The electron delocalization in the anion of a β-keto ester is represented by the resonance structures

$$\mathrm{R{-}\overset{^-:\ddot{O}:}{\overset{|}{C}}{=}\underset{H}{\underset{|}{C}}{-}\overset{:\ddot{O}:}{\overset{\|}{C}}{-}OR'} \longleftrightarrow \mathrm{R{-}\overset{:\ddot{O}}{\overset{\|}{C}}{-}\underset{H}{\underset{|}{\ddot{C}^-}}{-}\overset{:\ddot{O}:}{\overset{\|}{C}}{-}OR'} \longleftrightarrow \mathrm{R{-}\overset{:\ddot{O}}{\overset{\|}{C}}{-}\underset{H}{\underset{|}{C}}{=}\overset{:\ddot{O}:^-}{\overset{|}{C}}{-}OR'}$$

Principal resonance structures of the anion of a β-keto ester

This chapter begins by describing the preparation and properties of β-keto esters, proceeds to a discussion of their synthetic applications, continues to an examination of related species, and concludes by exploring some recent developments in the active field of synthetic carbanion chemistry.

21.1 BASE-PROMOTED ESTER CONDENSATION: THE CLAISEN CONDENSATION

Ludwig Claisen was a German chemist who worked during the last two decades of the nineteenth century and the first two decades of the twentieth. His name is associated with three reactions. The *Claisen-Schmidt reaction* was presented in Section 18.11, the *Claisen condensation* is discussed in this section, and the *Claisen rearrangement* will be introduced in Section 24.13.

Before describing how β-keto esters are used as reagents for organic synthesis, we need to see how these compounds themselves are prepared. The principal method for the preparation of β-keto esters is a reaction known as the **Claisen condensation:**

$$2RCH_2\overset{O}{\overset{\|}{C}}OR' \xrightarrow[2.\ H_3O^+]{1.\ NaOR'} RCH_2\overset{O}{\overset{\|}{C}}\underset{R}{\underset{|}{C}}H\overset{O}{\overset{\|}{C}}OR' + R'OH$$

Ester β-Keto ester Alcohol

On treatment with alkoxide bases, esters undergo self-condensation to give a β-keto ester and an alcohol. Ethyl acetate, for example, undergoes a Claisen condensation on treatment with sodium ethoxide to give a β-keto ester known by its common name *ethyl acetoacetate* (also called *acetoacetic ester*):

$$2CH_3\overset{O}{\overset{\|}{C}}OCH_2CH_3 \xrightarrow[2.\ H_3O^+]{1.\ NaOCH_2CH_3} CH_3\overset{O}{\overset{\|}{C}}CH_2\overset{O}{\overset{\|}{C}}OCH_2CH_3 + CH_3CH_2OH$$

Ethyl acetate — Ethyl acetoacetate (75%) (acetoacetic ester) — Ethanol

The systematic name of ethyl acetoacetate is *ethyl 3-oxobutanoate.* The presence of a ketone carbonyl group is indicated by the designation "*oxo*" along with the appropriate locant. Thus, there are four carbon atoms in the acyl group of ethyl 3-oxobutanoate, C-3 being the carbonyl carbon of the ketone function.

The mechanism of the Claisen condensation of ethyl acetate is presented in Figure 21.1. The first two steps of the mechanism are analogous to those of aldol addition (Section 18.9). An enolate ion is generated in step 1, which undergoes nucleophilic addition to the carbonyl group of a second ester molecule in step 2. The species formed in this step is a tetrahedral intermediate of the same type that we encountered in our discussion of nucleophilic acyl substitution of esters in Section 20.10. It dissociates by expelling an ethoxide ion, as shown in step 3, which restores the carbonyl group to give the β-keto ester. Steps 1 to 3 show two different types of ester reactivity: one molecule of the ester gives rise to an enolate; the second molecule acts as an acylating agent.

Claisen condensations involve two distinct experimental operations. The first stage concludes in step 4 of Figure 21.1, where the base removes a proton from C-2 of the β-keto ester. Because this proton is relatively acidic, the position of equilibrium for step 4 lies far to the right.

In general, the equilibrium represented by the sum of steps 1 to 3 is not favorable for

Overall reaction:

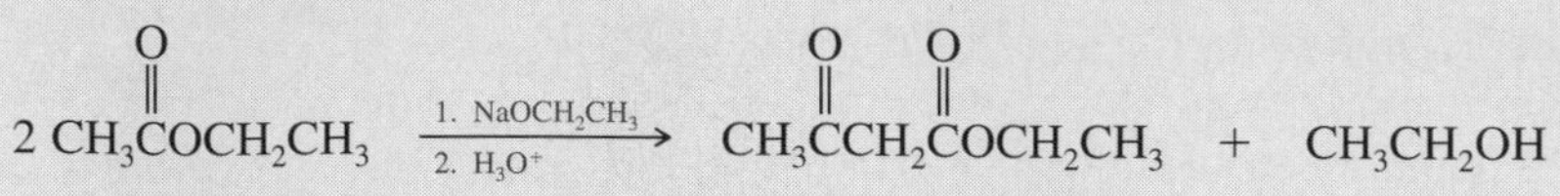

$$2\ CH_3\overset{O}{\overset{\|}{C}}OCH_2CH_3 \xrightarrow[2.\ H_3O^+]{1.\ NaOCH_2CH_3} CH_3\overset{O}{\overset{\|}{C}}CH_2\overset{O}{\overset{\|}{C}}OCH_2CH_3 + CH_3CH_2OH$$

Ethyl acetate → Ethyl 3-oxobutanoate (ethyl acetoacetate) + Ethanol

Step 1: Proton abstraction from the α carbon atom of ethyl acetate to give the corresponding enolate.

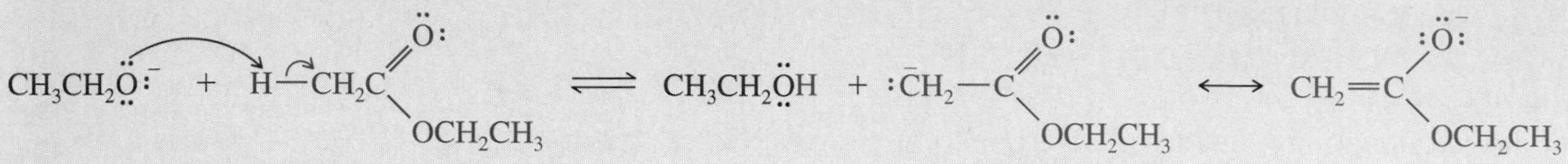

$$CH_3CH_2\ddot{O}:^- + H-CH_2C(=\ddot{O}:)OCH_2CH_3 \rightleftharpoons CH_3CH_2\ddot{O}H + :\bar{C}H_2-C(=\ddot{O}:)OCH_2CH_3 \longleftrightarrow CH_2=C(-\ddot{O}:^-)OCH_2CH_3$$

Ethoxide | Ethyl acetate | Ethanol | Enolate of ethyl acetate

Step 2: Nucleophilic addition of the ester enolate to the carbonyl group of the neutral ester. The product is the anionic form of the tetrahedral intermediate.

$$CH_3\overset{\ddot{O}:}{\overset{\|}{C}}OCH_2CH_3 + CH_2=C(-\ddot{O}:^-)OCH_2CH_3 \rightleftharpoons CH_3\overset{:\ddot{O}:^-}{\overset{|}{C}}(OCH_2CH_3)CH_2\overset{\ddot{O}:}{\overset{\|}{C}}OCH_2CH_3$$

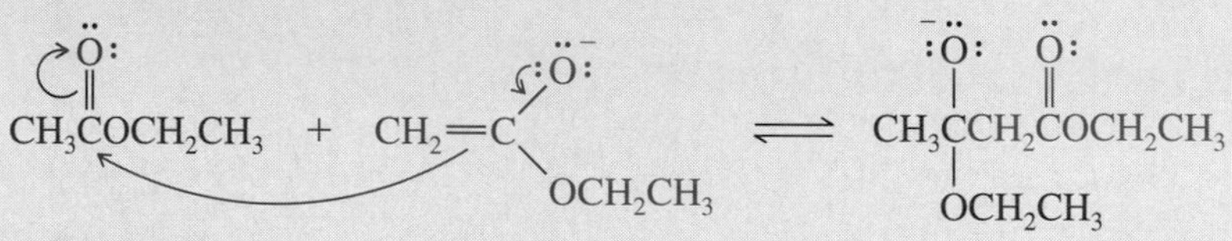

Ethyl acetate | Enolate of ethyl acetate | Anionic form of tetrahedral intermediate

Step 3: Dissociation of the tetrahedral intermediate.

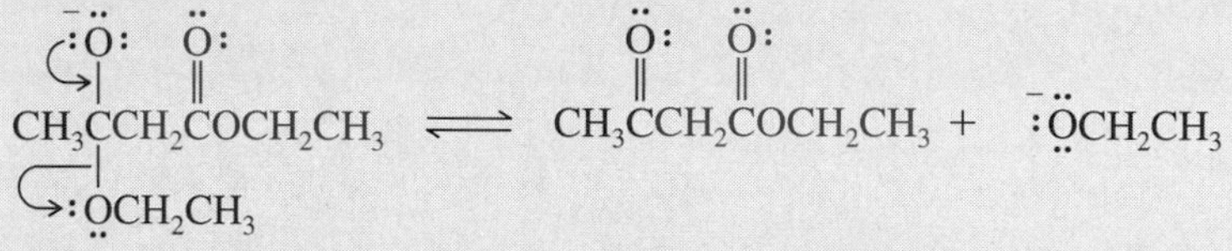

$$CH_3\overset{:\ddot{O}:^-}{\overset{|}{C}}(:\ddot{O}CH_2CH_3)CH_2\overset{\ddot{O}:}{\overset{\|}{C}}OCH_2CH_3 \rightleftharpoons CH_3\overset{\ddot{O}:}{\overset{\|}{C}}CH_2\overset{\ddot{O}:}{\overset{\|}{C}}OCH_2CH_3 + :\ddot{O}CH_2CH_3^-$$

Anionic form of tetrahedral intermediate | Ethyl 3-oxobutanoate | Ethoxide ion

Step 4: Deprotonation of the β-keto ester product.

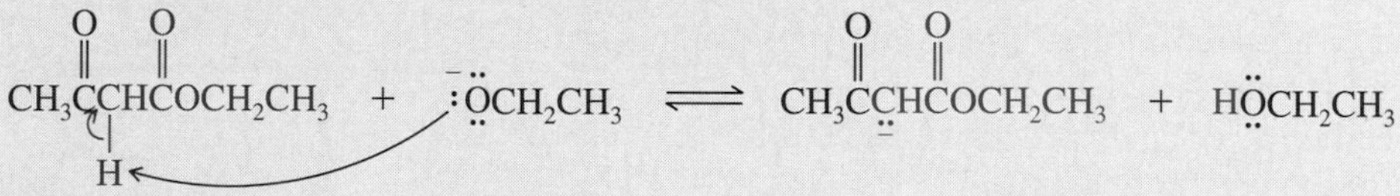

$$CH_3\overset{O}{\overset{\|}{C}}\underset{H}{\underset{|}{C}}H\overset{O}{\overset{\|}{C}}OCH_2CH_3 + :\ddot{O}CH_2CH_3^- \rightleftharpoons CH_3\overset{O}{\overset{\|}{C}}\underset{..}{\overset{-}{C}}H\overset{O}{\overset{\|}{C}}OCH_2CH_3 + H\ddot{O}CH_2CH_3$$

Ethyl 3-oxobutanoate (stronger acid) | Ethoxide ion (stronger base) | Conjugate base of ethyl 3-oxobutanoate (weaker base) | Ethanol (weaker acid)

Step 5: Acidification of the reaction mixture. This is performed in a separate synthetic operation to give the product in its neutral form for eventual isolation.

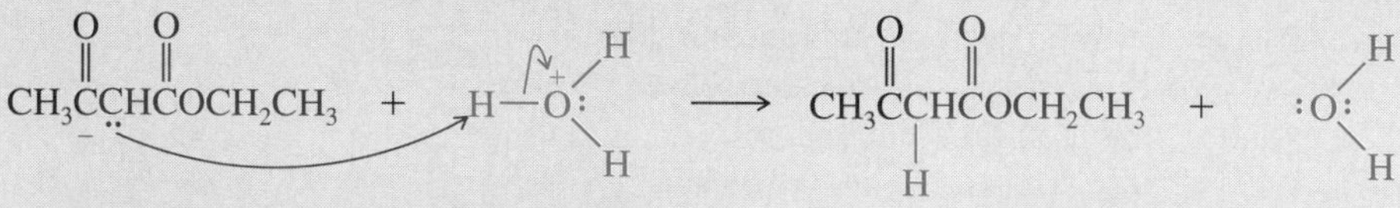

$$CH_3\overset{O}{\overset{\|}{C}}\underset{..}{\overset{-}{C}}H\overset{O}{\overset{\|}{C}}OCH_2CH_3 + H-\overset{+}{O}H_2 \longrightarrow CH_3\overset{O}{\overset{\|}{C}}\underset{H}{\underset{|}{C}}H\overset{O}{\overset{\|}{C}}OCH_2CH_3 + :\ddot{O}H_2$$

Conjugate base of ethyl 3-oxobutanoate (stronger base) | Hydronium ion (stronger acid) | Ethyl 3-oxobutanoate (weaker acid) | Water (weaker base)

FIGURE 21.1 Sequence of steps that describes the mechanism of the Claisen condensation of ethyl acetate.

condensation of two ester molecules to a β-keto ester. (Two ester carbonyl groups are more stable than one ester plus one ketone carbonyl.) However, because the β-keto ester is deprotonated under the reaction conditions, the equilibrium represented by the sum of steps 1 to 4 does lie to the side of products. On subsequent acidification (step 5), the anion of the β-keto ester is converted to its neutral form and isolated.

Organic chemists sometimes write equations for the Claisen condensation in a form that shows both stages explicitly:

$$2CH_3\overset{O}{\overset{\|}{C}}OCH_2CH_3 \xrightarrow{NaOCH_2CH_3} CH_3\overset{O}{\overset{\|}{C}}\underset{Na^+}{\underset{\ddot{-}}{C}}H\overset{O}{\overset{\|}{C}}OCH_2CH_3 \xrightarrow{H_3O^+} CH_3\overset{O}{\overset{\|}{C}}CH_2\overset{O}{\overset{\|}{C}}OCH_2CH_3$$

Ethyl acetate — Sodium salt of ethyl acetoacetate — Ethyl acetoacetate

Like aldol condensations, Claisen condensations always involve bond formation between the α carbon atom of one molecule and the carbonyl carbon of another:

$$2CH_3CH_2\overset{O}{\overset{\|}{C}}OCH_2CH_3 \xrightarrow[2.\ H_3O^+]{1.\ NaOCH_2CH_3} CH_3CH_2\overset{O}{\overset{\|}{C}}\underset{CH_3}{\underset{|}{C}}H\overset{O}{\overset{\|}{C}}OCH_2CH_3 + CH_3CH_2OH$$

Ethyl propanoate — Ethyl 2-methyl-3-oxopentanoate (81%) — Ethanol

---

PROBLEM 21.1 One of the following esters cannot undergo the Claisen condensation. Which one? Write structural formulas for the Claisen condensation products of the other two.

$CH_3CH_2CH_2CH_2CO_2CH_2CH_3$	$C_6H_5CO_2CH_2CH_3$	$C_6H_5CH_2CO_2CH_2CH_3$
Ethyl pentanoate	Ethyl benzoate	Ethyl phenylacetate

Unless the β-keto ester can form a stable anion by deprotonation as in Step 4 of Figure 21.1, the Claisen condensation product is present in only trace amounts at equilibrium. Ethyl 2-methylpropanoate, for example, does not give any of its condensation product under the customary conditions of the Claisen condensation.

$$2(CH_3)_2CH\overset{O}{\overset{\|}{C}}OCH_2CH_3 \overset{NaOCH_2CH_3}{\rightleftharpoons} (CH_3)_2CH\overset{O}{\overset{\|}{C}}\underset{H_3C\ \ CH_3}{C}\overset{O}{\overset{\|}{C}}OCH_2CH_3$$

Ethyl 2-methylpropanoate — Ethyl 2,2,4-trimethyl-3-oxopentanoate (cannot form a stable anion; formed in no more than trace amounts)

At least two protons must be present at the α carbon in order for the equilibrium to favor product formation. Claisen condensation is possible for esters of the type $R\mathbf{CH_2}CO_2R'$, but not for $R_2\mathbf{CH}CO_2R'$.

21.2 INTRAMOLECULAR CLAISEN CONDENSATION: THE DIECKMANN REACTION

Esters of *dicarboxylic acids* undergo an intramolecular version of the Claisen condensation when a five- or six-membered ring can be formed.

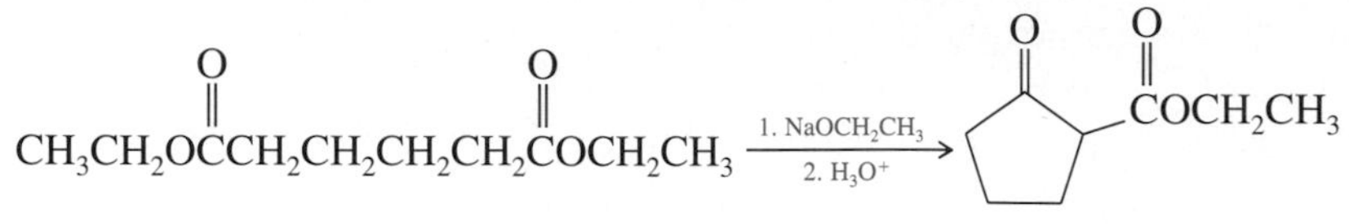

$$CH_3CH_2OCCH_2CH_2CH_2CH_2COCH_2CH_3 \xrightarrow[2.\ H_3O^+]{1.\ NaOCH_2CH_3}$$ Ethyl (2-oxocyclopentane)carboxylate

Diethyl hexanedioate

Ethyl (2-oxocyclopentane)-carboxylate (74–81%)

Walter Dieckmann was a German chemist and a contemporary of Claisen.

This reaction is an example of a **Dieckmann cyclization.** The anion formed by proton abstraction at the carbon α to one carbonyl group attacks the other carbonyl to form a five-membered ring.

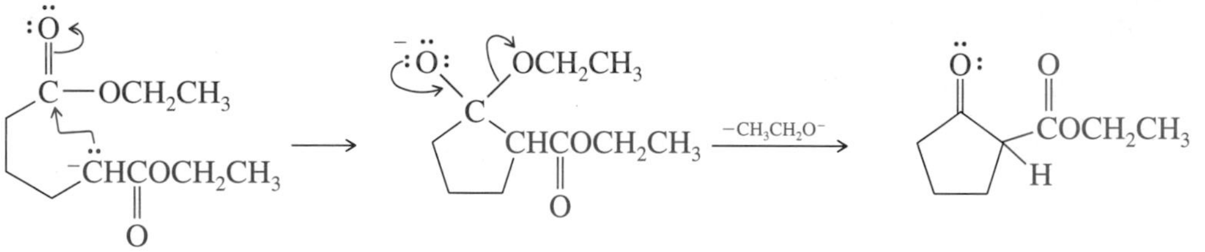

$\longrightarrow$ $\xrightarrow{-CH_3CH_2O^-}$

Enolate of diethyl hexanedioate

Ethyl (2-oxocyclopentane)carboxylate

PROBLEM 21.2 Write the structure of the Dieckmann cyclization product formed on treatment of each of the following diesters with sodium ethoxide, followed by acidification.

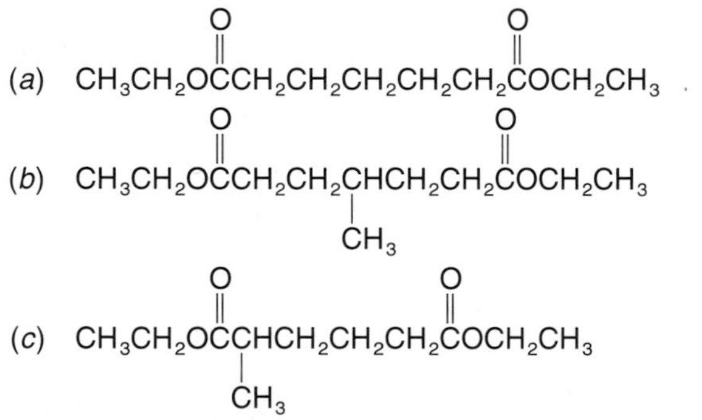

(*a*) $CH_3CH_2OCCH_2CH_2CH_2CH_2CH_2COCH_2CH_3$

(*b*) $CH_3CH_2OCCH_2CH_2CHCH_2CH_2COCH_2CH_3$ (with CH_3 on CH)

(*c*) $CH_3CH_2OCCHCH_2CH_2CH_2COCH_2CH_3$ (with CH_3 on CH)

SAMPLE SOLUTION (*a*) Diethyl heptanedioate has one more methylene group in its chain than the diester cited in the example (diethyl hexanedioate). Its Dieckmann cyclization product contains a six-membered ring instead of the five-membered ring formed from diethyl hexanedioate.

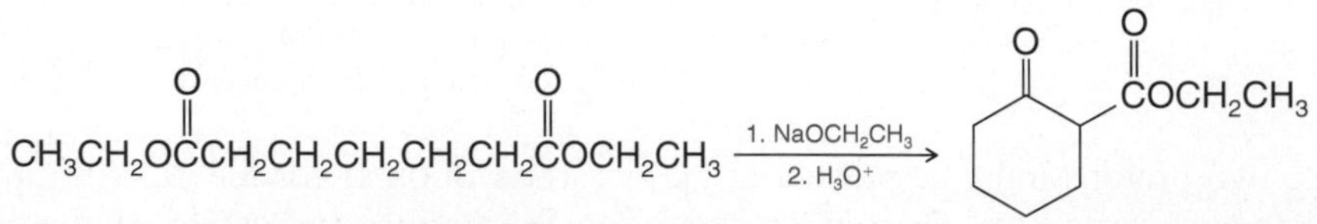

$$CH_3CH_2OCCH_2CH_2CH_2CH_2CH_2COCH_2CH_3 \xrightarrow[2.\ H_3O^+]{1.\ NaOCH_2CH_3}$$

Diethyl heptanedioate

Ethyl (2-oxocyclohexane)-carboxylate

21.3 MIXED CLAISEN CONDENSATIONS

Analogous to the mixed aldol condensations discussed in Section 18.10, mixed Claisen condensations involve carbon-carbon bond formation between the α carbon atom of one ester and the carbonyl carbon of another.

$$\underset{\text{Ester}}{\mathrm{R\overset{O}{\overset{\|}{C}}OCH_2CH_3}} + \underset{\text{Another ester}}{\mathrm{R'CH_2\overset{O}{\overset{\|}{C}}OCH_2CH_3}} \xrightarrow[\text{2. } H_3O^+]{\text{1. NaOCH}_2\text{CH}_3} \underset{\beta\text{-Keto ester}}{\mathrm{R\overset{O}{\overset{\|}{C}}\underset{R'}{\underset{|}{C}}H\overset{O}{\overset{\|}{C}}OCH_2CH_3}}$$

The best results are obtained in those mixed Claisen condensations in which one of the ester components, the acylating agent, is incapable of forming an enolate. Esters of this type include the following:

$$\underset{\text{Formate esters}}{\mathrm{H\overset{O}{\overset{\|}{C}}OR}} \qquad \underset{\text{Carbonate esters}}{\mathrm{RO\overset{O}{\overset{\|}{C}}OR}} \qquad \underset{\text{Oxalate esters}}{\mathrm{RO\overset{O}{\overset{\|}{C}}\overset{O}{\overset{\|}{C}}OR}} \qquad \underset{\text{Benzoate esters}}{\mathrm{C_6H_5{-}\overset{O}{\overset{\|}{C}}OR}}$$

The following equation shows an example of a mixed Claisen condensation in which a benzoate ester is used as the nonenolizable component:

$$\underset{\substack{\text{Methyl benzoate}\\ \text{(cannot form an enolate)}}}{\mathrm{C_6H_5{-}\overset{O}{\overset{\|}{C}}OCH_3}} + \underset{\text{Methyl propanoate}}{\mathrm{CH_3CH_2\overset{O}{\overset{\|}{C}}OCH_3}} \xrightarrow[\text{2. } H_3O^+]{\text{1. NaOCH}_3} \underset{\substack{\text{Methyl 2-methyl-3-oxo-}\\ \text{3-phenylpropanoate (60\%)}}}{\mathrm{C_6H_5{-}\overset{O}{\overset{\|}{C}}\underset{CH_3}{\underset{|}{C}}H\overset{O}{\overset{\|}{C}}OCH_3}}$$

PROBLEM 21.3 Give the structure of the product obtained when ethyl phenylacetate ($C_6H_5CH_2CO_2CH_2CH_3$) is treated with each of the following esters under conditions of the mixed Claisen condensation:

(*a*) Diethyl carbonate (*b*) Diethyl oxalate (*c*) Ethyl formate

SAMPLE SOLUTION (*a*) Diethyl carbonate cannot form an enolate, but ethyl phenylacetate can. Nucleophilic acyl substitution on diethyl carbonate by the enolate of ethyl phenylacetate yields a *diester.*

$$\mathrm{CH_3CH_2O{-}\overset{O}{\overset{\|}{C}}{-}OCH_2CH_3} + \mathrm{C_6H_5\overset{..}{\overset{-}{C}}H{-}\overset{O}{\overset{\|}{C}}OCH_2CH_3} \longrightarrow \mathrm{C_6H_5CH(\overset{O}{\overset{\|}{C}}OCH_2CH_3)_2}$$

Diethyl 2-phenylpropanedioate
(diethyl phenylmalonate)

The reaction proceeds in good yield (86 percent), and the product is a useful one in further synthetic transformations of the type to be described in Section 21.7.

21.4 ACYLATION OF KETONES WITH ESTERS

In a reaction that is related to the mixed Claisen condensation, nonenolizable esters are used as acylating agents for ketone enolates. Ketones (via their enolates) are converted to β-keto esters by reaction with diethyl carbonate.

Sodium hydride was used as the base in this example. It is often used instead of sodium ethoxide in these reactions.

$$CH_3CH_2O\overset{O}{\overset{\|}{C}}OCH_2CH_3 + \text{Cycloheptanone} \xrightarrow[2.\ H_3O^+]{1.\ NaH} \text{Ethyl (2-oxocycloheptane)carboxylate}$$

Diethyl carbonate | Cycloheptanone | Ethyl (2-oxocycloheptane)-carboxylate (91–94%)

Esters of nonenolizable monocarboxylic acids such as ethyl benzoate give β-diketones on reaction with ketone enolates:

$$C_6H_5\overset{O}{\overset{\|}{C}}OCH_2CH_3 + CH_3\overset{O}{\overset{\|}{C}}C_6H_5 \xrightarrow[2.\ H_3O^+]{1.\ NaOCH_2CH_3} C_6H_5\overset{O}{\overset{\|}{C}}CH_2\overset{O}{\overset{\|}{C}}C_6H_5$$

Ethyl benzoate | Acetophenone | 1,3-Diphenyl-1,3-propanedione (62–71%)

Intramolecular acylation of ketones yields cyclic β-diketones when the ring that is formed is five- or six-membered.

$$CH_3CH_2\overset{O}{\overset{\|}{C}}CH_2CH_2\overset{O}{\overset{\|}{C}}OCH_2CH_3 \xrightarrow[2.\ H_3O^+]{1.\ NaOCH_3} \text{2-Methyl-1,3-cyclopentanedione}$$

Ethyl 4-oxohexanoate | 2-Methyl-1,3-cyclopentanedione (70–71%)

PROBLEM 21.4 Write an equation for the carbon-carbon bond–forming step in the cyclization reaction just cited. Show clearly the structure of the enolate ion and use curved arrows to represent its nucleophilic addition to the appropriate carbonyl group. Write a second equation showing dissociation of the tetrahedral intermediate formed in the carbon-carbon bond–forming step.

Even though ketones have the potential to react with themselves by aldol addition, recall that the position of equilibrium for such reactions lies to the side of the starting materials (Section 18.9). On the other hand, acylation of ketone enolates gives products (β-keto esters or β-diketones) that are converted to delocalized anions under the reaction conditions. Consequently, ketone acylation is observed to the exclusion of aldol addition when ketones are treated with base in the presence of esters.

21.5 KETONE SYNTHESIS VIA β-KETO ESTERS

The carbon-carbon bond–forming potential inherent in the Claisen and Dieckmann reactions has been extensively exploited in organic synthesis. Subsequent transformations of the β-keto ester products permit the synthesis of other functional groups. One of these transformations converts β-keto esters to ketones; it is based on the fact that β-keto *acids* (not esters!) undergo decarboxylation readily (Section 19.17). Indeed, β-keto acids, and their corresponding carboxylate anions as well, lose carbon dioxide so easily that they tend to decarboxylate under the conditions of their formation.

$$RC(=O)CH(R')C(=O)OH \xrightarrow[-CO_2]{\text{heat}} RC(OH){=}CHR' \longrightarrow RC(=O)CH_2R'$$

β-Keto acid — Enol form of ketone — Ketone

Thus, 5-nonanone has been prepared from ethyl pentanoate by the sequence

$$CH_3CH_2CH_2CH_2\overset{O}{\overset{\|}{C}}OCH_2CH_3 \xrightarrow[2.\ H_3O^+]{1.\ NaOCH_2CH_3} CH_3CH_2CH_2CH_2\overset{O}{\overset{\|}{C}}\underset{CH_2CH_2CH_3}{\underset{|}{C}H}\overset{O}{\overset{\|}{C}}OCH_2CH_3$$

Ethyl pentanoate — Ethyl 3-oxo-2-propylheptanoate (80%)

1. KOH, H_2O, 70–80°C
2. H_3O^+

$$CH_3CH_2CH_2CH_2\overset{O}{\overset{\|}{C}}CH_2CH_2CH_2CH_3 \xleftarrow[-CO_2]{70-80°C} CH_3CH_2CH_2CH_2\overset{O}{\overset{\|}{C}}\underset{CH_2CH_2CH_3}{\underset{|}{C}H}\overset{O}{\overset{\|}{C}}OH$$

5-Nonanone (81%) — 3-Oxo-2-propylheptanoic acid (not isolated; decarboxylates under conditions of its formation)

The sequence begins with a Claisen condensation of ethyl pentanoate to give a β-keto ester. The ester is hydrolyzed, and the resulting β-keto acid decarboxylates to yield the desired ketone.

PROBLEM 21.5 Write appropriate chemical equations showing how you could prepare cyclopentanone from diethyl hexanedioate.

The most prominent application of β-keto esters to organic synthesis employs a similar pattern of ester saponification and decarboxylation as its final stage, as described in the following section.

21.6 THE ACETOACETIC ESTER SYNTHESIS

Ethyl acetoacetate (acetoacetic ester), available by the Claisen condensation of ethyl acetate, possesses properties that make it a useful starting material for the preparation of ketones. These properties are

1. The acidity of the α proton
2. The ease with which acetoacetic acid undergoes thermal decarboxylation

Ethyl acetoacetate is a stronger acid than ethanol and is quantitatively converted to its anion on treatment with sodium ethoxide in ethanol.

$$CH_3C(=O)CH_2C(=O)OCH_2CH_3 + NaOCH_2CH_3 \longrightarrow [CH_3C(=O)CHC(=O)OCH_2CH_3]^- Na^+ + CH_3CH_2OH$$

Ethyl acetoacetate (stronger acid) K_a 10^{-11} (pK_a 11)

Sodium ethoxide (stronger base)

Sodium salt of ethyl acetoacetate (weaker base)

Ethanol (weaker acid) K_a 10^{-16} (pK_a 16)

The anion produced by proton abstraction from ethyl acetoacetate is nucleophilic. Adding an alkyl halide to a solution of the sodium salt of ethyl acetoacetate leads to alkylation of the α carbon.

$$Na^+ \; CH_3C(=O)\bar{C}HC(=O)OCH_2CH_3 + R{-}X \longrightarrow CH_3C(=O)CH(R)C(=O)OCH_2CH_3 + NaX$$

Sodium salt of ethyl acetoacetate; alkyl halide

2-Alkyl derivative of ethyl acetoacetate

Sodium halide

The new carbon-carbon bond is formed by an S_N2-type reaction, and so the alkyl halide must be one that is not sterically hindered. Methyl and primary alkyl halides work best, while secondary alkyl halides give lower yields. Tertiary alkyl halides react only by elimination, not substitution.

Saponification and decarboxylation of the alkylated derivative of ethyl acetoacetate yields a ketone.

$$H_3C\overset{O}{\overset{\|}{C}}\underset{H\;\;R}{C}\overset{O}{\overset{\|}{C}}OCH_2CH_3 \xrightarrow[2.\ H^+]{1.\ HO^-,\ H_2O} H_3C\overset{O}{\overset{\|}{C}}\underset{H\;\;R}{C}\overset{O}{\overset{\|}{C}}OH \xrightarrow[-CO_2]{heat} CH_3\overset{O}{\overset{\|}{C}}CH_2R$$

2-Alkyl derivative of ethyl acetoacetate — 2-Alkyl derivative of acetoacetic acid — Ketone

This reaction sequence is called the **acetoacetic ester synthesis.** It is a standard procedure for the preparation of ketones from alkyl halides, as the conversion of 1-bromobutane to 2-heptanone illustrates.

$$CH_3\overset{O}{\overset{\|}{C}}CH_2\overset{O}{\overset{\|}{C}}OCH_2CH_3 \xrightarrow[2.\ CH_3CH_2CH_2CH_2Br]{1.\ NaOCH_2CH_3,\ ethanol} CH_3\overset{O}{\overset{\|}{C}}\underset{CH_2CH_2CH_2CH_3}{\underset{|}{C}H}\overset{O}{\overset{\|}{C}}OCH_2CH_3 \xrightarrow[3.\ heat\ -CO_2]{1.\ NaOH,\ H_2O\ 2.\ H^+} CH_3\overset{O}{\overset{\|}{C}}CH_2CH_2CH_2CH_2CH_3$$

Ethyl acetoacetate — Ethyl 2-butyl-3-oxobutanoate (70%) — 2-Heptanone (60%)

The acetoacetic ester synthesis accomplishes the overall transformation

$$R{-}X \longrightarrow R{-}CH_2\overset{O}{\overset{\|}{C}}CH_3$$

Primary or secondary alkyl halide — α-Alkylated derivative of acetone

Thus an alkyl halide is converted to an alkyl derivative of acetone.

We call a structural unit in a molecule that is related to a synthetic operation a **synthon.** The three-carbon unit $-CH_2\overset{O}{\overset{\|}{C}}CH_3$ is a synthon that alerts us to the possibility that a particular molecule may be accessible by the acetoacetic ester synthesis.

E. J. Corey (page 577) invented the word *synthon* in connection with his efforts to formalize synthetic planning.

PROBLEM 21.6 Show how you could prepare each of the following ketones from ethyl acetoacetate and any necessary organic or inorganic reagents:

(*a*) 1-Phenyl-1,4-pentanedione

(*b*) 4-Phenyl-2-butanone
(*c*) 5-Hexen-2-one

SAMPLE SOLUTION (*a*) Approach these syntheses in a retrosynthetic way. Identify the synthon $-CH_2\overset{O}{\overset{\|}{C}}CH_3$ and mentally disconnect the bond to the α carbon atom. The $-CH_2\overset{O}{\overset{\|}{C}}CH_3$ synthon is derived from ethyl acetoacetate; the remainder of the molecule originates in the alkyl halide.

$$C_6H_5\overset{O}{\overset{\|}{C}}CH_2 \vdots \overset{\alpha}{} CH_2\overset{O}{\overset{\|}{C}}CH_3 \Longrightarrow C_6H_5\overset{O}{\overset{\|}{C}}CH_2X + \overset{-}{:}CH_2\overset{O}{\overset{\|}{C}}CH_3$$

Disconnect here

1-Phenyl-1,4-pentanedione — Required alkyl halide — Derived from ethyl acetoacetate

Analyzing the target molecule in this way reveals that the required alkyl halide is an α-halo ketone. Thus, a suitable starting material would be bromomethyl phenyl ketone.

Can you think of how bromomethyl phenyl ketone might be prepared?

$$C_6H_5\overset{O}{\overset{\|}{C}}CH_2Br + CH_3\overset{O}{\overset{\|}{C}}CH_2\overset{O}{\overset{\|}{C}}OCH_2CH_3 \xrightarrow[\text{2. NaOH, } H_2O \text{ 3. } H^+ \text{ 4. heat}]{\text{1. } NaOCH_2CH_3\text{, ethanol}} C_6H_5\overset{O}{\overset{\|}{C}}CH_2CH_2\overset{O}{\overset{\|}{C}}CH_3$$

Bromomethyl phenyl ketone — Ethyl acetoacetate — 1-Phenyl-1,4-pentanedione

Dialkylation of ethyl acetoacetate can also be accomplished, opening the way to ketones with two alkyl substituents at the α carbon:

$$CH_3\overset{O}{\overset{\|}{C}}-\underset{}{\overset{CH_2CH=CH_2}{\overset{|}{C}H}}-CO_2CH_2CH_3 \xrightarrow[\text{2. } CH_3CH_2I]{\text{1. } NaOCH_2CH_3} CH_3\overset{O}{\overset{\|}{C}}-\underset{CH_2CH_3}{\underset{|}{\overset{CH_2CH=CH_2}{\overset{|}{C}}}}-CO_2CH_2CH_3 \xrightarrow[\text{2. } H^+ \text{ 3. heat}]{\text{1. NaOH, } H_2O} CH_3\overset{O}{\overset{\|}{C}}\underset{CH_2CH_3}{\underset{|}{C}H}CH_2CH=CH_2$$

Ethyl 2-allylacetoacetate — Ethyl 2-allyl-2-ethylacetoacetate (75%) — 3-Ethyl-5-hexen-2-one (48%)

The starting material in the example is obtained by alkylation of ethyl acetoacetate with allyl bromide.

Recognize, too, that the reaction sequence is one that is characteristic of β-keto esters in general and not limited to just ethyl acetoacetate and its derivatives. Thus,

Ethyl 2-oxocyclohexanecarboxylate $\xrightarrow[2.\ CH_2{=}CHCH_2Br]{1.\ NaOCH_2CH_3}$ Ethyl 1-allyl-2-oxocyclohexanecarboxylate (89%) $\xrightarrow[2.\ H^+ \quad 3.\ heat]{1.\ KOH,\ H_2O}$ 2-Allylcyclohexanone (66%)

The starting material in this example is the Dieckmann cyclization product of diethyl heptanedioate (see Problem 21.2*a*).

It is reasonable to ask why one would prepare a ketone by way of a keto ester (ethyl acetoacetate, for example) rather than by direct alkylation of the enolate of a ketone. One reason is that the monoalkylation of ketones via their enolates is a difficult reaction to carry out in good yield. (Remember, however, that *acylation* of ketone enolates as described in Section 21.4 is achieved readily.) A second reason is that the delocalized enolates of β-keto esters, being far less basic than ketone enolates, give a higher substitution/elimination ratio when they react with alkyl halides. This can be quite important in those syntheses in which the alkyl halide is expensive or difficult to obtain.

Anions of β-keto esters are said to be *synthetically equivalent* to the enolates of ketones. The anion of ethyl acetoacetate is synthetically equivalent to the enolate of acetone, for example. The use of synthetically equivalent groups is a common tactic in synthetic organic chemistry. One of the skills that characterize the most creative practitioners of organic synthesis is an ability to recognize situations in which otherwise difficult transformations can be achieved through the use of synthetically equivalent reagents.

21.7 THE MALONIC ESTER SYNTHESIS

The **malonic ester synthesis** is a method for the preparation of carboxylic acids and is represented by the general equation

$$RX + CH_2(COOCH_2CH_3)_2 \xrightarrow[ethanol]{NaOCH_2CH_3} RCH(COOCH_2CH_3)_2 \xrightarrow[3.\ heat]{1.\ HO^-,\ H_2O \quad 2.\ H^+} RCH_2COOH$$

Alkyl halide; Diethyl malonate (malonic ester); α-Alkylated derivative of diethyl malonate; Carboxylic acid

The malonic ester synthesis is conceptually analogous to the acetoacetic ester synthesis. The overall transformation is

$$\underset{\text{Primary or secondary alkyl halide}}{R{-}X} \longrightarrow \underset{\alpha\text{-Alkylated derivative of acetic acid}}{R{-}CH_2\overset{O}{\overset{\|}{C}}OH}$$

Diethyl malonate (also known as malonic ester) serves as a source of the synthon $-CH_2\overset{O}{\overset{\|}{C}}OH$ in the same way that the ethyl acetoacetate serves as a source of the synthon $-CH_2\overset{O}{\overset{\|}{C}}CH_3$.

The properties of diethyl malonate that make the malonic ester synthesis a useful procedure are the same as those responsible for the synthetic value of ethyl acetoacetate. The protons at C-2 of diethyl malonate are relatively acidic and one is readily removed on treatment with sodium ethoxide.

$$CH_3CH_2O\overset{O}{\overset{\|}{C}}CH_2\overset{O}{\overset{\|}{C}}OCH_2CH_3 + NaOCH_2CH_3 \longrightarrow$$

Diethyl malonate (stronger acid) K_a 10^{-13} (pK_a 13) — Sodium ethoxide (stronger base)

A space-filling model of the sodium salt of diethyl malonate appears at the top of the first page of this chapter.

$$\left[CH_3CH_2O\overset{O}{\overset{\|}{C}}\overset{H}{C}\overset{O}{\overset{\|}{C}}OCH_2CH_3\right]^- Na^+ + CH_3CH_2OH$$

Sodium salt of diethyl malonate (weaker base) — Ethanol (weaker acid) K_a 10^{-16} (pK_a 16)

Treatment of the anion of diethyl malonate with alkyl halides leads to alkylation at C-2.

$$Na^+\ CH_3CH_2O\overset{O}{\overset{\|}{C}}\overset{-}{\underset{H}{C}}\overset{O}{\overset{\|}{C}}OCH_2CH_3 \quad R{-}X \longrightarrow CH_3CH_2O\overset{O}{\overset{\|}{C}}\underset{H\ \ R}{C}\overset{O}{\overset{\|}{C}}OCH_2CH_3 + NaX$$

Sodium salt of diethyl malonate; alkyl halide — 2-Alkyl derivative of diethyl malonate — Sodium halide

Converting the C-2 alkylated derivative to the corresponding malonic acid derivative by ester hydrolysis gives a compound susceptible to thermal decarboxylation. Temperatures of approximately 180°C are normally required.

$$CH_3CH_2O\overset{O}{\overset{\|}{C}}\underset{H\ \ R}{C}\overset{O}{\overset{\|}{C}}OCH_2CH_3 \xrightarrow[2.\ H^+]{1.\ HO^-,\ H_2O} HO\overset{O}{\overset{\|}{C}}\underset{H\ \ R}{C}\overset{O}{\overset{\|}{C}}OH \xrightarrow[-CO_2]{heat} RCH_2\overset{O}{\overset{\|}{C}}OH$$

2-Alkyl derivative of diethyl malonate — 2-Alkyl derivative of malonic acid — Carboxylic acid

In a typical example of the malonic ester synthesis, 6-heptenoic acid has been prepared from 5-bromo-1-pentene:

$$CH_2{=}CHCH_2CH_2CH_2Br + CH_2(COOCH_2CH_3)_2 \xrightarrow[ethanol]{NaOCH_2CH_3}$$

5-Bromo-1-pentene — Diethyl malonate

$$CH_2{=}CHCH_2CH_2CH_2CH(COOCH_2CH_3)_2$$

Diethyl 2-(4-pentenyl)malonate (85%)

$$CH_2{=}CHCH_2CH_2CH_2CH(\overset{O}{\overset{\|}{C}}OCH_2CH_3)_2 \xrightarrow[3.\ heat]{1.\ HO^-,\ H_2O \quad 2.\ H^+} CH_2{=}CHCH_2CH_2CH_2CH_2\overset{O}{\overset{\|}{C}}OH$$

Diethyl 2-(4-pentenyl)malonate — 6-Heptenoic acid (75%)

PROBLEM 21.7 Show how you could prepare each of the following carboxylic acids from diethyl malonate and any necessary organic or inorganic reagents:

(*a*) 3-Methylpentanoic acid
(*b*) Nonanoic acid
(*c*) 4-Methylhexanoic acid
(*d*) 3-Phenylpropanoic acid

SAMPLE SOLUTION (*a*) Analyze the target molecule retrosynthetically by mentally disconnecting a bond to the α carbon atom.

$$CH_3CH_2\underset{CH_3}{\underset{|}{CH}}\ \vdots\ \overset{\alpha}{C}H_2\overset{O}{\overset{\|}{C}}OH \Rightarrow CH_3CH_2\underset{CH_3}{\underset{|}{CH}}{-}X + \ :\bar{C}H_2\overset{O}{\overset{\|}{C}}OH$$

Disconnect here

3-Methylpentanoic acid — Required alkyl halide — Derived from diethyl malonate

We see that a secondary alkyl halide is needed as the alkylating agent. The anion of diethyl malonate is a weaker base than ethoxide ion and reacts with secondary alkyl halides by substitution rather than elimination. Thus, the synthesis of 3-methylpentanoic acid begins with the alkylation of the anion of diethyl malonate by 2-bromobutane.

$$CH_3CH_2CHBr(CH_3) + CH_2(COOCH_2CH_3)_2 \xrightarrow[\text{3. } H^+ \text{ 4. heat}]{\text{1. } NaOCH_2CH_3\text{, ethanol 2. NaOH, } H_2O} CH_3CH_2CH(CH_3)CH_2COH(=O)$$

2-Bromobutane — Diethyl malonate — 3-Methylpentanoic acid

As actually carried out and reported in the chemical literature, diethyl malonate has been alkylated with 2-bromobutane in 83 to 84 percent yield and the product of that reaction converted to 3-methylpentanoic acid by saponification, acidification, and decarboxylation in 62 to 65 percent yield.

By performing two successive alkylation steps, the malonic ester synthesis can be applied to the synthesis of α,α-disubstituted derivatives of acetic acid:

$$CH_2(COOCH_2CH_3)_2 \xrightarrow[\text{2. } CH_3Br]{\text{1. } NaOCH_2CH_3\text{, ethanol}} CH_3CH(COOCH_2CH_3)_2$$

Diethyl malonate — Diethyl 2-methyl-1,3-propanedioate (79–83%)

$$\xrightarrow[\text{2. } CH_3(CH_2)_8CH_2Br]{\text{1. } NaOCH_2CH_3\text{, ethanol}} (H_3C)(CH_3(CH_2)_8CH_2)C(COOCH_2CH_3)_2$$

Diethyl 2-decyl-2-methyl-1,3-propanedioate

$$\xrightarrow[\text{3. heat}]{\text{1. KOH, ethanol-water 2. } H^+} (H_3C)(H)C(CH_3(CH_2)_8CH_2)(COOH)$$

2-Methyldodecanoic acid (61–74%)

PROBLEM 21.8 Ethyl acetoacetate may also be subjected to double alkylation. Show how you could prepare 3-methyl-2-butanone by double alkylation of ethyl acetoacetate.

The malonic ester synthesis has been adapted to the preparation of cycloalkanecarboxylic acids from dihaloalkanes:

$$CH_2(COOCH_2CH_3)_2 \xrightarrow[\text{2. } BrCH_2CH_2CH_2Br]{\text{1. } NaOCH_2CH_3\text{, ethanol}} BrCH_2CH_2CH_2CH(COOCH_2CH_3)_2$$

Diethyl malonate — (Not isolated; cyclizes in the presence of sodium ethoxide)

$$\longrightarrow \text{cyclo-}(CH_2CH_2CH_2)C(COOCH_2CH_3)_2$$

Diethyl 1,1-cyclobutanedicarboxylate (60–65%)

$$\xrightarrow[\text{2. heat}]{\text{1. } H_3O^+} \text{cyclobutane-}CH(COOH)$$

Cyclobutanecarboxylic acid (80% from diester)

The cyclization step is limited to the formation of rings of seven carbons or fewer.

PROBLEM 21.9 Cyclopentyl methyl ketone has been prepared from 1,4-dibromobutane and ethyl acetoacetate. Outline the steps in this synthesis by writing a series of equations showing starting materials, reagents, and isolated intermediates.

21.8 BARBITURATES

Diethyl malonate has uses other than in the synthesis of carboxylic acids. One particularly valuable application lies in the preparation of *barbituric acid* by nucleophilic acyl substitution with urea:

Barbituric acid was first prepared in 1864 by Adolph von Baeyer (page 93). A historical account of his work and the later development of barbiturates as sedative-hypnotics appears in the October 1951 issue of the *Journal of Chemical Education* (pp. 524–526).

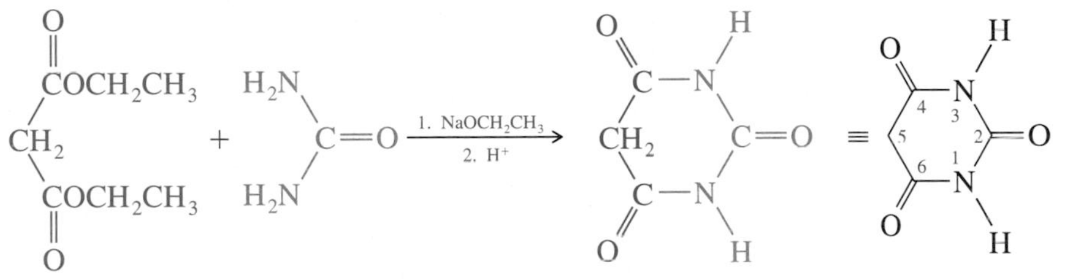

$$CH_2(COOCH_2CH_3)_2 + (H_2N)_2C{=}O \xrightarrow[2.\ H^+]{1.\ NaOCH_2CH_3} \text{Barbituric acid}$$

Diethyl malonate   Urea   Barbituric acid (72–78%)

Barbituric acid is the parent of a group of compounds known as **barbiturates.** The barbiturates are classified as *sedative-hypnotic agents,* meaning that they decrease the responsiveness of the central nervous system and promote sleep. Thousands of derivatives of the parent ring system of barbituric acid have been tested for sedative-hypnotic activity; the most useful are the 5,5-disubstituted derivatives.

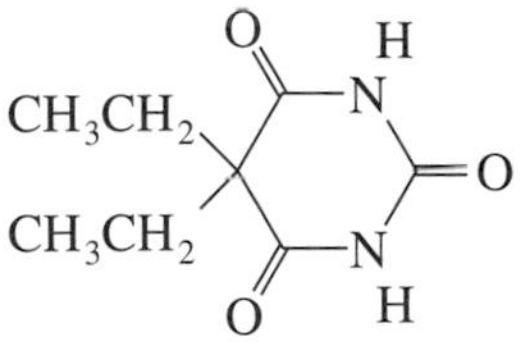

5,5-Diethylbarbituric acid (barbital; Veronal)

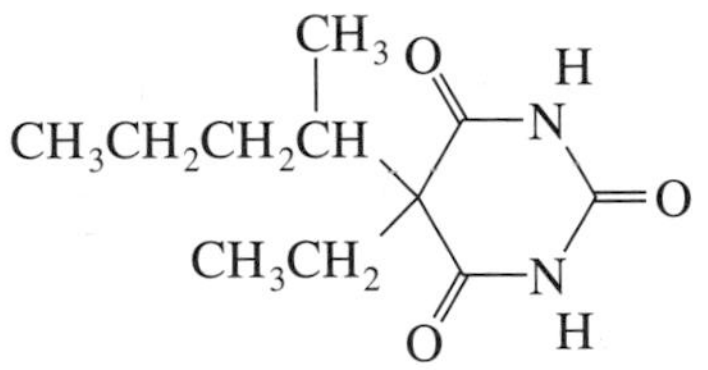

5-Ethyl-5-(1-methylbutyl)-barbituric acid (pentobarbital; Nembutal)

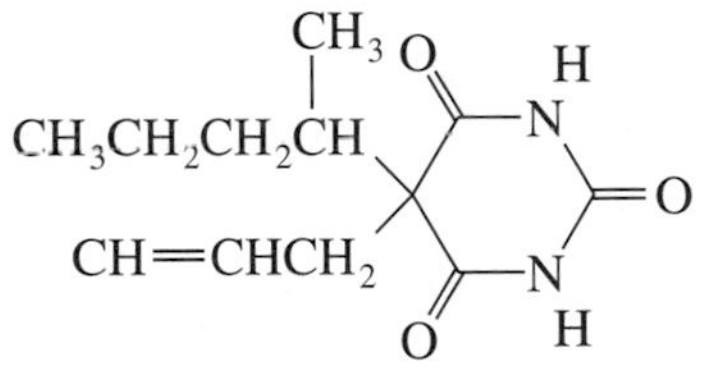

5-Allyl-5-(1-methylbutyl)-barbituric acid (secobarbital; Seconal)

These compounds are prepared in a manner analogous to that of barbituric acid itself. Diethyl malonate is alkylated twice, then treated with urea.

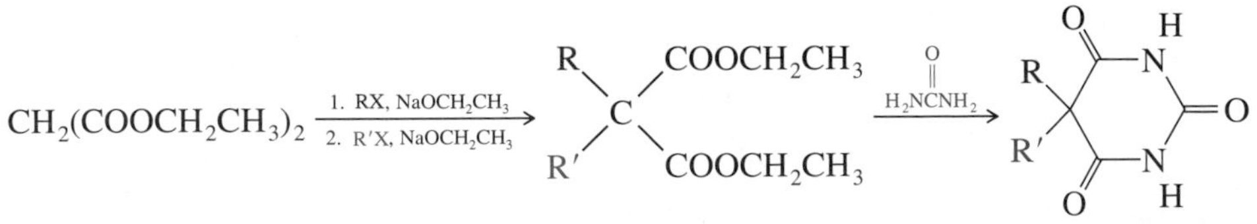

$$CH_2(COOCH_2CH_3)_2 \xrightarrow[2.\ R'X,\ NaOCH_2CH_3]{1.\ RX,\ NaOCH_2CH_3} RR'C(COOCH_2CH_3)_2 \xrightarrow{H_2NCNH_2 (C{=}O)} \text{5,5-disubstituted barbituric acid}$$

Diethyl malonate   Dialkylated derivative of diethyl malonate   5,5-Disubstituted derivative of barbituric acid

PROBLEM 21.10 Show, by writing a suitable sequence of reactions, how you could prepare pentobarbital from diethyl malonate. (The structure of pentobarbital has been shown in this section.)

Barbituric acids, as their name implies, are weakly acidic and are converted to their sodium salts (sodium barbiturates) in aqueous sodium hydroxide. Sometimes the drug is dispensed in its neutral form; sometimes the sodium salt is used. The salt is designated by appending the word *sodium* to the name of the barbituric acid—*pentobarbital sodium,* for example.

PROBLEM 21.11 Thiourea ($H_2NC(=S)NH_2$) reacts with diethyl malonate and its alkyl derivatives in the same way that urea does. Give the structure of the product obtained when thiourea is used instead of urea in the synthesis of pentobarbital. The anesthetic *thiopental* (*Pentothal*) *sodium* is the sodium salt of this product. What is the structure of this compound?

PROBLEM 21.12 Aryl halides react too slowly to undergo substitution by the S_N2 mechanism with the sodium salt of diethyl malonate, and so the phenyl substituent of *phenobarbital* cannot be introduced in the way that alkyl substituents can.

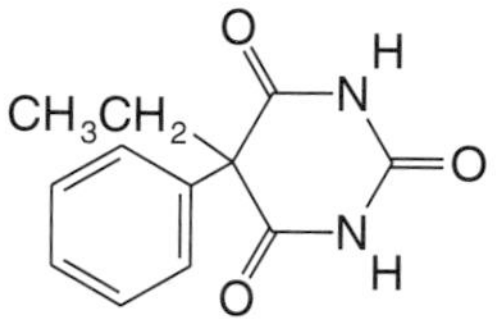

5-Ethyl-5-phenylbarbituric acid (phenobarbital)

One synthesis of phenobarbital begins with ethyl phenylacetate and diethyl carbonate. Using these starting materials and any necessary organic or inorganic reagents, devise a synthesis of phenobarbital. (*Hint:* See the sample solution to Problem 21.3*a*.)

The various barbiturates differ in the time required for the onset of sleep and in the duration of their effects. All the barbiturates must be used only in strict accordance with instructions to avoid potentially lethal overdoses. Drug dependence in some individuals is also a problem.

21.9 MICHAEL ADDITIONS OF STABILIZED ANIONS

Stabilized anions exhibit a pronounced tendency to undergo conjugate addition to α,β-unsaturated carbonyl compounds. This reaction, called the *Michael reaction,* has been described for anions derived from β-diketones in Section 18.14. The enolates of ethyl acetoacetate and diethyl malonate also undergo Michael addition to the β carbon atom of α,β-unsaturated aldehydes, ketones, and esters. For example,

$$CH_3\overset{O}{\overset{\|}{C}}CH{=}CH_2 + CH_2(COOCH_2CH_3)_2 \xrightarrow[\text{ethanol}]{\text{KOH}} CH_3\overset{O}{\overset{\|}{C}}CH_2CH_2CH(COOCH_2CH_3)_2$$

Methyl vinyl ketone — Diethyl malonate — Ethyl 2-carboethoxy-5-oxohexanoate (83%)

In this reaction the enolate of diethyl malonate adds to the β carbon of methyl vinyl ketone.

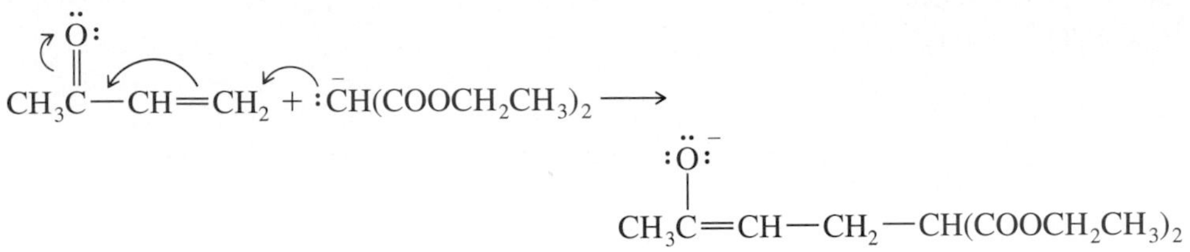

$$CH_3\overset{\overset{\ddot{O}:}{\|}}{C}-CH=CH_2 + :\overset{-}{C}H(COOCH_2CH_3)_2 \longrightarrow CH_3\overset{\overset{:\ddot{O}:^-}{|}}{C}=CH-CH_2-CH(COOCH_2CH_3)_2$$

The intermediate formed in the nucleophilic addition step abstracts a proton from the solvent to give the observed product.

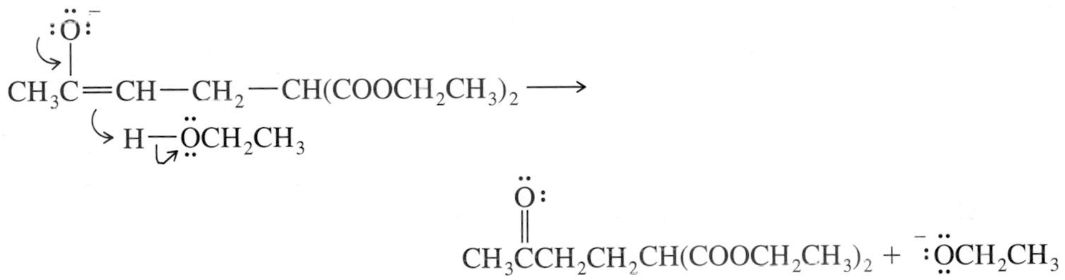

$$CH_3\overset{\overset{:\ddot{O}:^-}{|}}{C}=CH-CH_2-CH(COOCH_2CH_3)_2 \quad H-\ddot{O}CH_2CH_3 \longrightarrow CH_3\overset{\overset{\ddot{O}:}{\|}}{C}CH_2CH_2CH(COOCH_2CH_3)_2 + \,^{-}:\ddot{O}CH_2CH_3$$

After isolation, the Michael adduct may be subjected to ester hydrolysis and decarboxylation. When α,β-unsaturated ketones are carried through this sequence, the final products are 5-keto acids (δ-keto acids).

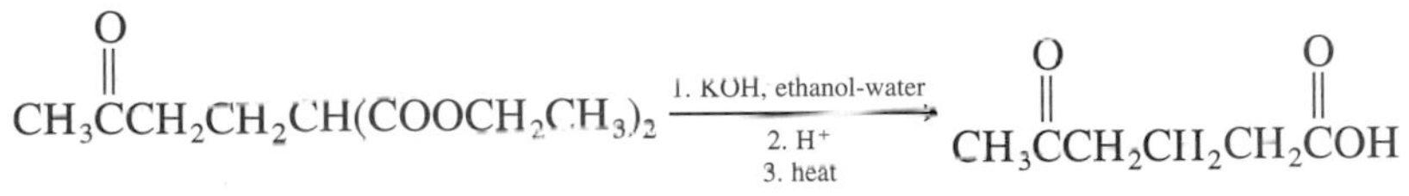

$$CH_3\overset{\overset{O}{\|}}{C}CH_2CH_2CH(COOCH_2CH_3)_2 \xrightarrow[\text{2. H}^+ \text{ 3. heat}]{\text{1. KOH, ethanol-water}} CH_3\overset{\overset{O}{\|}}{C}CH_2CH_2CH_2\overset{\overset{O}{\|}}{C}OH$$

Ethyl 2-carboethoxy-5-oxohexanoate (from diethyl malonate and methyl vinyl ketone)

5-Oxohexanoic acid (42%)

PROBLEM 21.13 Ethyl acetoacetate behaves similarly to diethyl malonate in its reactivity toward α,β-unsaturated carbonyl compounds. Give the structure of the product of the following reaction sequence:

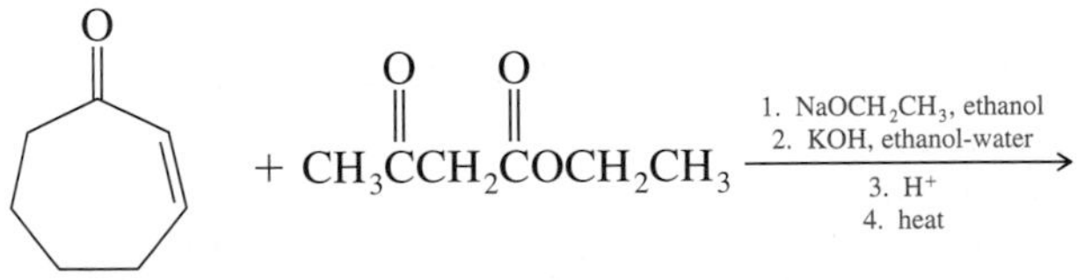

$$\text{2-Cycloheptenone} + CH_3\overset{\overset{O}{\|}}{C}CH_2\overset{\overset{O}{\|}}{C}OCH_2CH_3 \xrightarrow[\text{3. H}^+ \text{ 4. heat}]{\text{1. NaOCH}_2\text{CH}_3\text{, ethanol; 2. KOH, ethanol-water}}$$

2-Cycloheptenone Ethyl acetoacetate

21.10 α DEPROTONATION OF CARBONYL COMPOUNDS BY LITHIUM DIALKYLAMIDES

Most of the reactions of ester enolates described so far have centered on stabilized enolates derived from 1,3-dicarbonyl compounds such as diethyl malonate and ethyl acetoacetate. While the synthetic value of these and related stabilized enolates is clear, chemists have long been interested in extending the usefulness of nonstabilized enolates derived from simple esters. Consider the deprotonation of an ester as represented by the acid-base reaction

$$\underset{\text{Ester}}{\mathrm{RCH(H)C(=O)OR'}} + \underset{\text{Base}}{:\mathrm{B}^-} \rightleftharpoons \underset{\text{Ester enolate}}{\mathrm{RCH{=}C(O^-)(OR')}} + \underset{\text{Conjugate acid of base}}{\mathrm{H{-}B}}$$

We already know what happens when simple esters are treated with alkoxide bases—they undergo the Claisen condensation (Section 21.1). Simple esters have acid dissociation constants K_a of approximately 10^{-22} (pK_a 22) and are incompletely converted to their enolates with alkoxide bases. The small amount of enolate that is formed reacts by nucleophilic addition to the carbonyl group of the ester.

What happens if the base is much stronger than an alkoxide ion? If the base is strong enough, it will convert the ester completely to its enolate. Under these conditions the Claisen condensation is suppressed because there is no neutral ester present for the enolate to add to. A very strong base is one that is derived from a very weak acid. Referring to the table of acidities (Table 4.2, page 131), we see that ammonia is quite a weak acid; its K_a is 10^{-36} (pK_a 36). Therefore, amide ion ($H_2\ddot{N}{:}^-$) is a very strong base—more than strong enough to deprotonate an ester quantitatively. However, amide ion also tends to add to the carbonyl group of esters; to avoid this complication, highly hindered analogs of $H_2\ddot{N}{:}^-$ are used instead. The most frequently used base for ester enolate formation is *lithium diisopropylamide* (LDA):

Lithium diisopropylamide is commercially available. Alternatively, it may be prepared by the reaction of butyllithium with $[(CH_3)_2CH]_2NH$ (see Problem 14.4*a* for a related reaction).

$$\mathrm{Li^+\ (CH_3)_2CH{-}\ddot{\underset{..}{N}}{-}CH(CH_3)_2} \qquad \text{Lithium diisopropylamide}$$

Lithium diisopropylamide is a strong enough base to abstract a proton from the α carbon atom of an ester, but because it is so sterically hindered, it does not add readily to the carbonyl group. To illustrate,

$$\mathrm{CH_3CH_2CH_2C(=O)OCH_3} + \mathrm{[(CH_3)_2CH]_2NLi} \longrightarrow \mathrm{CH_3CH_2CH{=}C(OLi)(OCH_3)} + \mathrm{[(CH_3)_2CH]_2NH}$$

Methyl butanoate (stronger acid) K_a 10^{-22} (pK_a 22); Lithium diisopropylamide (stronger base); Lithium enolate of methyl butanoate (weaker base); Diisopropylamine (weaker acid) K_a 10^{-36} (pK_a 36)

Direct alkylation of esters has been achieved by formation of the ester enolate with LDA followed by addition of an alkyl halide. Tetrahydrofuran (THF) is the solvent most often used in these reactions.

$$\underset{\text{Methyl butanoate}}{\mathrm{CH_3CH_2CH_2C(=O)OCH_3}} \xrightarrow[\text{THF}]{\text{LDA}} \underset{\text{Lithium enolate of methyl butanoate}}{\mathrm{CH_3CH_2CH{=}C(OLi)(OCH_3)}} \xrightarrow{\mathrm{CH_3CH_2I}} \underset{\text{Methyl 2-ethylbutanoate (92\%)}}{\mathrm{CH_3CH_2CH(CH_2CH_3)C(=O)OCH_3}}$$

Ester enolates generated by proton abstraction with dialkylamide bases add to aldehydes and ketones to give β-hydroxy esters.

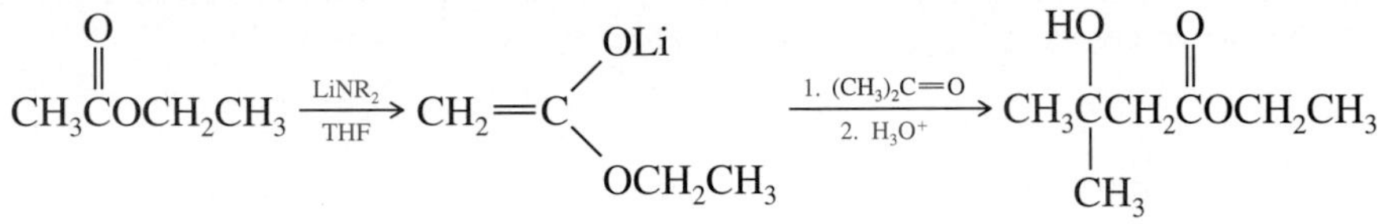

$$CH_3\overset{O}{\overset{\|}{C}}OCH_2CH_3 \xrightarrow[THF]{LiNR_2} CH_2{=}C(OLi)(OCH_2CH_3) \xrightarrow[2.\ H_3O^+]{1.\ (CH_3)_2C{=}O} CH_3\underset{CH_3}{\overset{HO}{C}}CH_2\overset{O}{\overset{\|}{C}}OCH_2CH_3$$

Ethyl acetate — Lithium enolate of ethyl acetate — Ethyl 3-hydroxy-3-methylbutanoate (90%)

Lithium dialkylamides are excellent bases for the preparation of ketone enolates as well. Ketone enolates generated in this way can be alkylated with alkyl halides or, as illustrated in the following equation, treated with an aldehyde or a ketone.

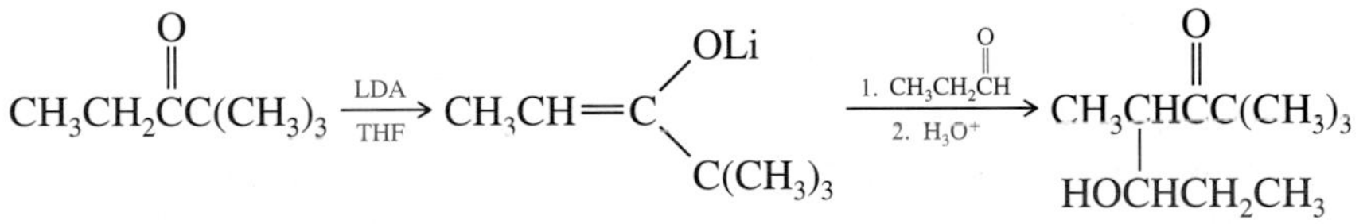

$$CH_3CH_2\overset{O}{\overset{\|}{C}}C(CH_3)_3 \xrightarrow[THF]{LDA} CH_3CH{=}C(OLi)(C(CH_3)_3) \xrightarrow[2.\ H_3O^+]{1.\ CH_3CH_2\overset{O}{\overset{\|}{C}}H} CH_3\underset{HOCHCH_2CH_3}{CH}\overset{O}{\overset{\|}{C}}C(CH_3)_3$$

2,2-Dimethyl-3-pentanone — Lithium enolate of 2,2-dimethyl-3-pentanone — 5-Hydroxy-2,2,4-trimethyl-3-heptanone (81%)

Thus, mixed aldol additions can be achieved by the tactic of quantitative enolate formation using LDA followed by addition of a different aldehyde or ketone.

PROBLEM 21.14 Outline efficient syntheses of each of the following compounds from readily available aldehydes, ketones, esters, and alkyl halides according to the methods described in this section:

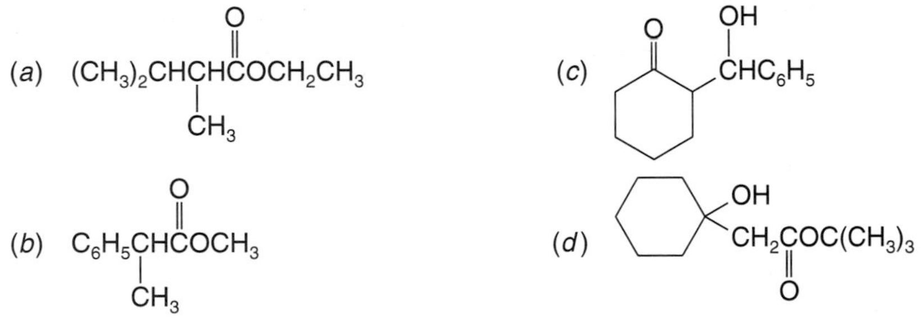

(*a*) $(CH_3)_2CH\underset{CH_3}{CH}\overset{O}{\overset{\|}{C}}OCH_2CH_3$

(*b*) $C_6H_5\underset{CH_3}{CH}\overset{O}{\overset{\|}{C}}OCH_3$

(*c*) 2-(CHC_6H_5 bearing OH)cyclohexanone: $\overset{OH}{CH}C_6H_5$ attached to the carbon α to the C=O of cyclohexanone

(*d*) Cyclohexane carbon bearing OH and $CH_2\overset{O}{\overset{\|}{C}}OC(CH_3)_3$

SAMPLE SOLUTION (*a*) The α carbon atom of the ester has two different alkyl groups attached to it.

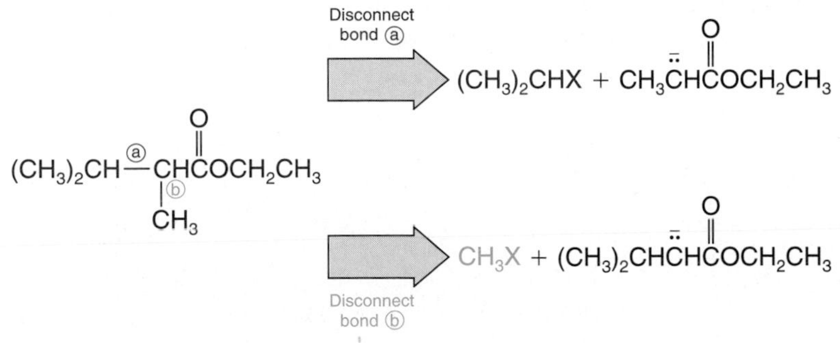

$$(CH_3)_2CH \overset{ⓐ}{—} \underset{ⓑ\ CH_3}{CH}\overset{O}{\overset{\|}{C}}OCH_2CH_3$$

Disconnect bond ⓐ ⟹ $(CH_3)_2CHX + CH_3\ddot{C}H\overset{O}{\overset{\|}{C}}OCH_2CH_3$

Disconnect bond ⓑ ⟹ $CH_3X + (CH_3)_2CH\ddot{C}H\overset{O}{\overset{\|}{C}}OCH_2CH_3$

The critical carbon-carbon bond–forming step requires nucleophilic substitution on an alkyl halide by an ester enolate. Methyl halides are more reactive than isopropyl halides in S_N2 reactions and cannot undergo elimination as a competing process. Therefore, choose the synthesis in which bond ⓑ is formed by alkylation.

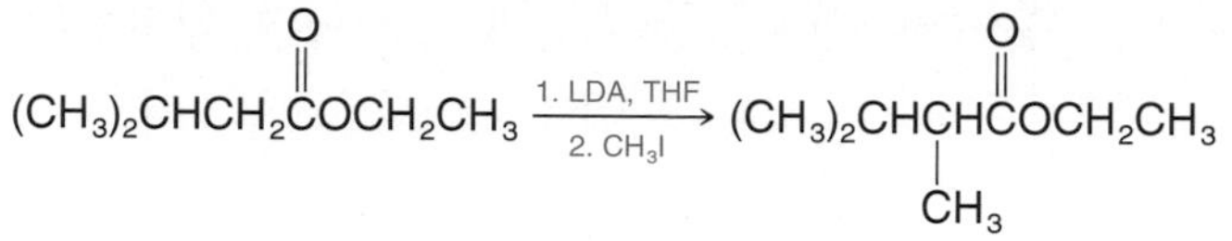

$$(CH_3)_2CHCH_2\overset{O}{\overset{\|}{C}}OCH_2CH_3 \xrightarrow[\text{2. } CH_3I]{\text{1. LDA, THF}} (CH_3)_2CH\underset{CH_3}{\underset{|}{C}H}\overset{O}{\overset{\|}{C}}OCH_2CH_3$$

Ethyl 3-methylbutanoate — Ethyl 2,3-dimethylbutanoate

(This synthesis has been reported in the chemical literature and gives the desired product in 95 percent yield.)

21.11 SUMMARY

β-Keto esters are useful reagents for a number of carbon-carbon bond–forming reactions. They are prepared by the procedures summarized in Table 21.1.

Hydrolysis of β-keto esters, such as those shown in Table 21.1, gives β-keto acids which undergo rapid decarboxylation, forming ketones (Section 21.5).

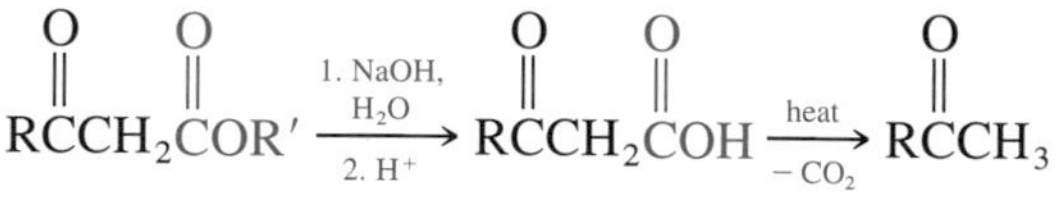

$$R\overset{O}{\overset{\|}{C}}CH_2\overset{O}{\overset{\|}{C}}OR' \xrightarrow[\text{2. } H^+]{\text{1. NaOH, } H_2O} R\overset{O}{\overset{\|}{C}}CH_2\overset{O}{\overset{\|}{C}}OH \xrightarrow[-CO_2]{\text{heat}} R\overset{O}{\overset{\|}{C}}CH_3$$

β-Keto ester — β-Keto acid — Ketone

β-Keto esters are characterized by K_a's of about 10^{-11} (pK_a 11) and are quantitatively converted to their enolates on treatment with alkoxide bases.

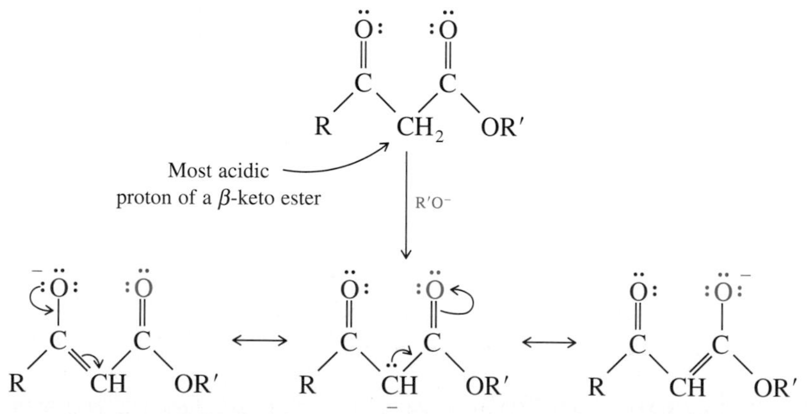

$$R\overset{O}{\overset{\|}{C}}CH_2\overset{O}{\overset{\|}{C}}OR' \xrightarrow{R'O^-} \text{enolate}$$

Most acidic proton of a β-keto ester

Resonance forms illustrating charge delocalization in enolate of a β-keto ester

$$R\overset{O^-}{\overset{|}{C}}{=}CH\overset{O}{\overset{\|}{C}}OR' \longleftrightarrow R\overset{O}{\overset{\|}{C}}\overset{-}{C}H\overset{O}{\overset{\|}{C}}OR' \longleftrightarrow R\overset{O}{\overset{\|}{C}}CH{=}\overset{O^-}{\overset{|}{C}}OR'$$

The anion of a β-keto ester may be alkylated at carbon with an alkyl halide and the product of this reaction subjected to ester hydrolysis and decarboxylation to give a ketone.

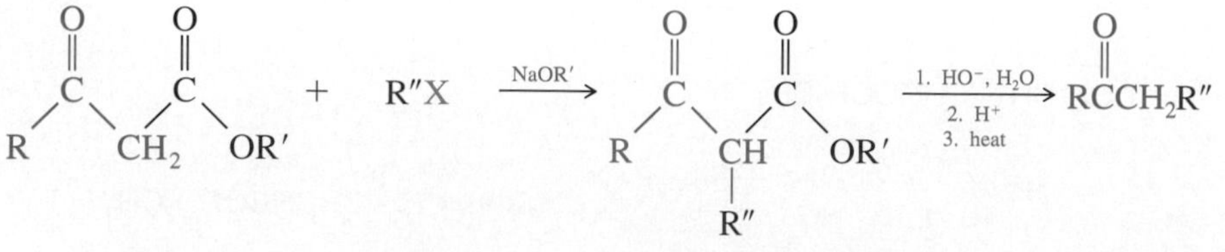

$$R\overset{O}{\overset{\|}{C}}CH_2\overset{O}{\overset{\|}{C}}OR' + R''X \xrightarrow{NaOR'} R\overset{O}{\overset{\|}{C}}\underset{R''}{\underset{|}{C}H}\overset{O}{\overset{\|}{C}}OR' \xrightarrow[\text{2. } H^+ \text{ 3. heat}]{\text{1. } HO^-, H_2O} R\overset{O}{\overset{\|}{C}}CH_2R''$$

β-Keto ester — Alkyl halide — Alkylated β-keto ester — Ketone

TABLE 21.1
Preparation of β-Keto Esters

Reaction (section) and comments	General equation and specific example
Claisen condensation (Section 21.1) Esters of the type $RCH_2\overset{O}{\overset{\|}{C}}OR'$ are converted to β-keto esters on treatment wtih alkoxide bases. One molecule of an ester is converted to its enolate; a second molecule of ester acts as an acylating agent toward the enolate.	$2RCH_2\overset{O}{\overset{\|}{C}}OR' \xrightarrow[2.\ H^+]{1.\ NaOR'} RCH_2\overset{O}{\overset{\|}{C}}\underset{R}{\underset{\|}{C}}H\overset{O}{\overset{\|}{C}}OR' + R'OH$ Ester β-Keto ester Alcohol $2CH_3CH_2CH_2\overset{O}{\overset{\|}{C}}OCH_2CH_3 \xrightarrow[2.\ H^+]{1.\ NaOCH_2CH_3} CH_3CH_2CH_2\overset{O}{\overset{\|}{C}}\underset{CH_2CH_3}{\underset{\|}{C}}H\overset{O}{\overset{\|}{C}}OCH_2CH_3$ Ethyl butanoate Ethyl 2-ethyl-3-oxohexanoate (76%)
Dieckmann cyclization (Section 21.2) An intramolecular analog of the Claisen condensation. Cyclic β-keto esters in which the ring is five- to seven-membered may be formed by using this reaction.	$C_6H_4(CH_2\overset{O}{\overset{\|}{C}}OCH_2CH_3)_2 \xrightarrow[2.\ H^+]{1.\ NaOCH_2CH_3}$ (indan-2-one bearing $\overset{O}{\overset{\|}{C}}OCH_2CH_3$ at C-1) Diethyl 1,2-benzenediacetate Ethyl indan-2-one-1-carboxylate (70%)
Mixed Claisen condensations (Section 21.3) Diethyl carbonate, diethyl oxalate, ethyl formate, and benzoate esters cannot form ester enolates but can act as acylating agents toward other ester enolates.	$R\overset{O}{\overset{\|}{C}}OCH_2CH_3 + R'CH_2\overset{O}{\overset{\|}{C}}OCH_2CH_3 \xrightarrow[2.\ H^+]{1.\ NaOCH_2CH_3} R\overset{O}{\overset{\|}{C}}\underset{R'}{\underset{\|}{C}}H\overset{O}{\overset{\|}{C}}OCH_2CH_3$ Ester Another ester β-Keto ester $CH_3CH_2\overset{O}{\overset{\|}{C}}OCH_2CH_3 + CH_3CH_2O\overset{O}{\overset{\|}{C}}\overset{O}{\overset{\|}{C}}OCH_2CH_3 \xrightarrow[2.\ H^+]{1.\ NaOCH_2CH_3} CH_3\underset{\underset{O}{\underset{\|}{C}}-\underset{O}{\underset{\|}{C}}OCH_2CH_3}{\underset{\|}{C}}H\overset{O}{\overset{\|}{C}}OCH_2CH_3$ Ethyl propanoate Diethyl oxalate Diethyl 3-methyl-2-oxobutanedioate (60–70%)
Acylation of ketones (Section 21.4) Diethyl carbonate and diethyl oxalate can be used to acylate ketone enolates to give β-keto esters.	$RCH_2\overset{O}{\overset{\|}{C}}R' + CH_3CH_2O\overset{O}{\overset{\|}{C}}OCH_2CH_3 \xrightarrow[2.\ H^+]{1.\ NaOCH_2CH_3} R\underset{\overset{O}{\overset{\|}{C}}OCH_2CH_3}{\underset{\|}{C}}H\overset{O}{\overset{\|}{C}}R'$ Ketone Diethyl carbonate β-Keto ester $(CH_3)_3CCH_2\overset{O}{\overset{\|}{C}}CH_3 + CH_3CH_2O\overset{O}{\overset{\|}{C}}OCH_2CH_3 \xrightarrow[2.\ H^+]{1.\ NaOCH_2CH_3} (CH_3)_3CCH_2\overset{O}{\overset{\|}{C}}CH_2\overset{O}{\overset{\|}{C}}OCH_2CH_3$ 4,4-Dimethyl-2-pentanone Diethyl carbonate Ethyl 5,5-dimethyl-3-oxohexanoate (66%)

The **acetoacetic ester synthesis** (Section 21.6) is a procedure in which ethyl acetoacetate is alkylated with an alkyl halide as the first step in the preparation of ketones of the type $CH_3\overset{O}{\overset{\|}{C}}CH_2R$.

$$\underset{\text{Ethyl acetoacetate}}{CH_3\overset{O}{\overset{\|}{C}}CH_2\overset{O}{\overset{\|}{C}}OCH_2CH_3} \xrightarrow[CH_3CH{=}CHCH_2Br]{NaOCH_2CH_3} CH_3\overset{O}{\overset{\|}{C}}\underset{CH_2CH=CHCH_3}{\underset{|}{C}H}\overset{O}{\overset{\|}{C}}OCH_2CH_3 \xrightarrow[\text{2. } H^+ \text{ 3. heat}]{\text{1. } HO^-, H_2O} \underset{\text{5-Hepten-2-one (81\%)}}{CH_3\overset{O}{\overset{\|}{C}}CH_2CH_2CH=CHCH_3}$$

The **malonic ester synthesis** (Section 21.7) is related to the acetoacetic ester synthesis. Alkyl halides (RX) are converted to carboxylic acids of the type RCH_2COOH by reaction with the enolate ion derived from diethyl malonate, followed by saponification and decarboxylation.

$$\underset{\text{Diethyl malonate}}{CH_2(COOCH_2CH_3)_2} \xrightarrow[\text{3-chlorocyclopentene}]{NaOCH_2CH_3} \text{(2-cyclopentenyl)}{-}CH(COOCH_2CH_3)_2 \xrightarrow[\text{2. } H^+ \text{ 3. heat}]{\text{1. } HO^-, H_2O} \underset{\text{(2-Cyclopentenyl)acetic acid (66\%)}}{\text{(2-cyclopentenyl)}{-}CH_2\overset{O}{\overset{\|}{C}}OH}$$

Alkylation of diethyl malonate, followed by reaction with urea, gives derivatives of barbituric acid, called **barbiturates,** which are useful sleep-promoting drugs (Section 21.8).

$$\underset{\text{Diethyl malonate}}{CH_2(COOCH_2CH_3)_2} \xrightarrow{RX,\ NaOCH_2CH_3} \underset{\text{Alkylated derivative of diethyl malonate}}{RCH(COOCH_2CH_3)_2} \xrightarrow{H_2N\overset{O}{\overset{\|}{C}}NH_2} \underset{\text{Alkylated derivative of barbituric acid}}{\text{5-R-barbituric acid}}$$

Michael addition of the enolate ions derived from ethyl acetoacetate and diethyl malonate provides an alternative method for preparing their α-alkyl derivatives (Section 21.9).

$$\underset{\text{Diethyl malonate}}{CH_2(COOCH_2CH_3)_2} \quad \underset{\text{Ethyl 2-butenoate}}{CH_3CH=CH\overset{O}{\overset{\|}{C}}OCH_2CH_3} \xrightarrow[CH_3CH_2OH]{NaOCH_2CH_3} \underset{\text{Triethyl 2-methylpropane-1,1,3-tricarboxylate (95\%)}}{CH_3\underset{CH(COOCH_2CH_3)_2}{\underset{|}{C}H}CH_2\overset{O}{\overset{\|}{C}}OCH_2CH_3}$$

It is possible to generate ester enolates by direct deprotonation provided that the base used is very strong. Lithium diisopropylamide (LDA) is often used for this purpose. It also converts ketones quantitatively to their enolates (Section 21.10).

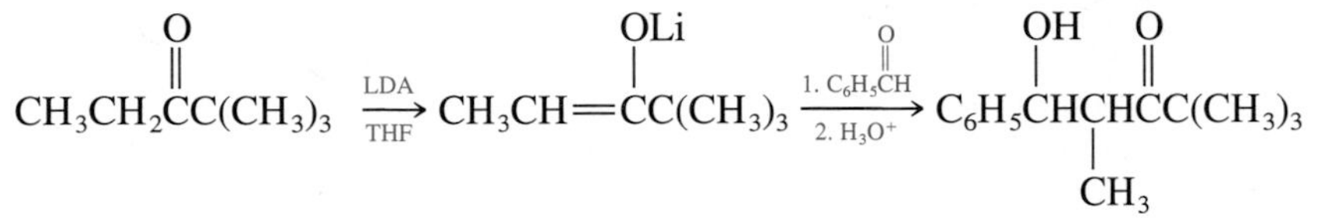

$$\underset{\text{2,2-Dimethyl-3-pentanone}}{CH_3CH_2\overset{\overset{\displaystyle O}{\|}}{C}C(CH_3)_3} \xrightarrow[\text{THF}]{\text{LDA}} CH_3CH{=}\overset{\overset{\displaystyle OLi}{|}}{C}C(CH_3)_3 \xrightarrow[\text{2. } H_3O^+]{\text{1. } C_6H_5\overset{\overset{O}{\|}}{C}H} \underset{\text{1-Hydroxy-2,4,4-trimethyl-1-phenyl-3-pentanone (78\%)}}{C_6H_5\overset{\overset{\displaystyle OH}{|}}{C}H\underset{\underset{\displaystyle CH_3}{|}}{C}H\overset{\overset{\displaystyle O}{\|}}{C}C(CH_3)_3}$$

PROBLEMS

21.15 The following questions pertain to the esters shown and their behavior under conditions of the Claisen condensation.

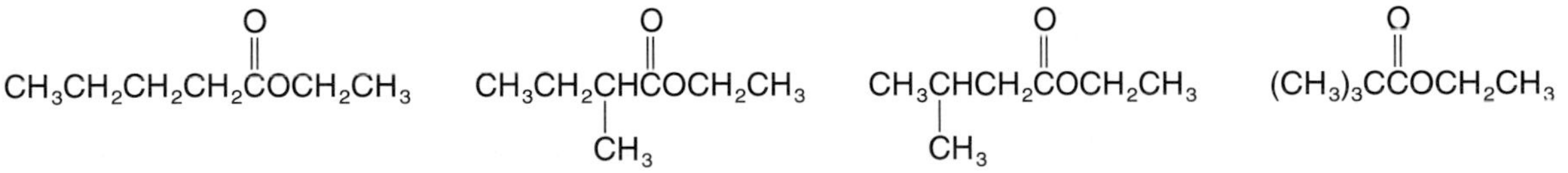

$$\underset{\text{Ethyl pentanoate}}{CH_3CH_2CH_2CH_2\overset{\overset{\displaystyle O}{\|}}{C}OCH_2CH_3} \qquad \underset{\text{Ethyl 2-methylbutanoate}}{CH_3CH_2\underset{\underset{\displaystyle CH_3}{|}}{C}H\overset{\overset{\displaystyle O}{\|}}{C}OCH_2CH_3} \qquad \underset{\text{Ethyl 3-methylbutanoate}}{CH_3\underset{\underset{\displaystyle CH_3}{|}}{C}HCH_2\overset{\overset{\displaystyle O}{\|}}{C}OCH_2CH_3} \qquad \underset{\text{Ethyl 2,2-dimethylpropanoate}}{(CH_3)_3C\overset{\overset{\displaystyle O}{\|}}{C}OCH_2CH_3}$$

(*a*) Two of these esters are converted to β-keto esters in good yield on treatment with sodium ethoxide and subsequent acidification of the reaction mixture. Which two are these? Write the structure of the Claisen condensation product of each one.

(*b*) One ester is capable of being converted to a β-keto ester on treatment with sodium ethoxide, but the amount of β-keto ester that can be isolated after acidification of the reaction mixture is quite small. Which ester is this?

(*c*) One ester is incapable of reaction under conditions of the Claisen condensation. Which one? Why?

21.16

(*a*) Give the structure of the Claisen condensation product of ethyl phenylacetate ($C_6H_5CH_2COOCH_2CH_3$).

(*b*) What ketone would you isolate after saponification and decarboxylation of this Claisen condensation product?

(*c*) What ketone would you isolate after treatment of the Claisen condensation product of ethyl phenylacetate with sodium ethoxide and allyl bromide, followed by saponification and decarboxylation?

(*d*) Give the structure of the mixed Claisen condensation product of ethyl phenylacetate and ethyl benzoate.

(*e*) What ketone would you isolate after saponification and decarboxylation of the product in part (*d*)?

(*f*) What ketone would you isolate after treatment of the product in part (*d*) with sodium ethoxide and allyl bromide, followed by saponification and decarboxylation?

21.17 All the following questions concern ethyl (2-oxocyclohexane)carboxylate.

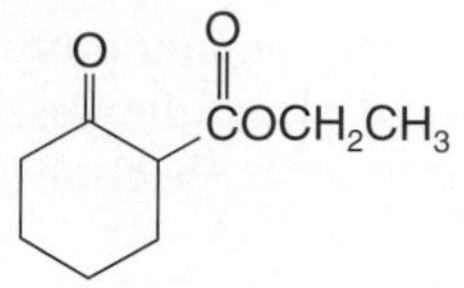

Ethyl (2-oxocyclohexane)carboxylate

(*a*) Write a chemical equation showing how you could prepare ethyl (2-oxocyclohexane)carboxylate by a Dieckmann reaction.
(*b*) Write a chemical equation showing how you could prepare ethyl (2-oxocyclohexane)carboxylate by acylation of a ketone.
(*c*) Write structural formulas for the two most stable enol forms of ethyl (2-oxocyclohexane)carboxylate.
(*d*) Write the three most stable resonance forms for the most stable enolate derived from ethyl (2-oxocyclohexane)carboxylate.
(*e*) Show how you could use ethyl (2-oxocyclohexane)carboxylate to prepare 2-methylcyclohexanone.
(*f*) Give the structure of the product formed on treatment of ethyl (2-oxocyclohexane)carboxylate with acrolein ($CH_2{=}CH\overset{O}{\overset{\|}{C}}H$) in ethanol in the presence of sodium ethoxide.

21.18 Give the structure of the product formed on reaction of ethyl acetoacetate with each of the following:

(*a*) 1-Bromopentane and sodium ethoxide
(*b*) Saponification and decarboxylation of the product in part (*a*)
(*c*) Methyl iodide and the product in part (*a*) treated with sodium ethoxide
(*d*) Saponification and decarboxylation of the product in part (*c*)
(*e*) 1-Bromo-3-chloropropane and one equivalent of sodium ethoxide
(*f*) Product in part (*e*) treated with a second equivalent of sodium ethoxide
(*g*) Saponification and decarboxylation of the product in part (*f*)
(*h*) Phenyl vinyl ketone and sodium ethoxide
(*i*) Saponification and decarboxylation of the product in part (*h*)

21.19 Repeat the preceding problem for diethyl malonate.

21.20

(*a*) Only a small amount (less than 0.01 percent) of the enol form of diethyl malonate is present at equilibrium. Write a structural formula for this enol.
(*b*) Enol forms are present to the extent of about 8 percent in ethyl acetoacetate. There are three constitutionally isomeric enols possible. Write structural formulas for these three enols. Which one do you think is the most stable? The least stable? Why?
(*c*) Bromine reacts rapidly with both diethyl malonate and ethyl acetoacetate. The reaction is acid-catalyzed and liberates hydrogen bromide. What is the product formed in each reaction?

21.21

(*a*) On addition of one equivalent of methylmagnesium iodide to ethyl acetoacetate, the Grignard reagent is consumed, but the only organic product obtained after working up the reaction mixture is ethyl acetoacetate. Why? What happens to the Grignard reagent?

(*b*) On repeating the reaction but using D_2O and DCl to work up the reaction mixture, it is found that the recovered ethyl acetoacetate contains deuterium. Where is this deuterium located?

21.22 Give the structure of the principal organic product of each of the following reactions:

(*a*) Ethyl octanoate $\xrightarrow[\text{2. H}^+]{\text{1. NaOCH}_2\text{CH}_3}$

(*b*) Product of part (*a*) $\xrightarrow[\text{2. H}^+ \text{ 3. heat}]{\text{1. NaOH, H}_2\text{O}}$

(*c*) Ethyl acetoacetate + 1-bromobutane $\xrightarrow{\text{NaOCH}_2\text{CH}_3\text{, ethanol}}$

(*d*) Product of part (*c*) $\xrightarrow[\text{2. H}^+ \text{ 3. heat}]{\text{1. NaOH, H}_2\text{O}}$

(*e*) Product of part (*c*) + 1-iodobutane $\xrightarrow{\text{NaOCH}_2\text{CH}_3\text{, ethanol}}$

(*f*) Product of part (*e*) $\xrightarrow[\text{2. H}^+ \text{ 3. heat}]{\text{1. NaOH, H}_2\text{O}}$

(*g*) Acetophenone + diethyl carbonate $\xrightarrow[\text{2. H}^+]{\text{1. NaOCH}_2\text{CH}_3}$

(*h*) Acetone + diethyl oxalate $\xrightarrow[\text{2. H}^+]{\text{1. NaOCH}_2\text{CH}_3}$

(*i*) Diethyl malonate + 1-bromo-2-methylbutane $\xrightarrow{\text{NaOCH}_2\text{CH}_3\text{, ethanol}}$

(*j*) Product of part (*i*) $\xrightarrow[\text{2. H}^+ \text{ 3. heat}]{\text{1. NaOH, H}_2\text{O}}$

(*k*) Diethyl malonate + 6-methyl-2-cyclohexenone $\xrightarrow{\text{NaOCH}_2\text{CH}_3\text{, ethanol}}$

(*l*) Product of part (*k*) $\xrightarrow{\text{H}_2\text{O, HCl, heat}}$

(*m*) *tert*-Butyl acetate $\xrightarrow[\text{2. benzaldehyde 3. H}^+]{\text{1. [(CH}_3)_2\text{CH]}_2\text{NLi, THF}}$

21.23 Give the structure of the principal organic product of each of the following reactions:

(*a*) [cyclopentanone bearing CH_3CH_2 and two $COOCH_2CH_3$ groups] $\xrightarrow[\text{heat}]{\text{H}_2\text{O, H}_2\text{SO}_4}$ $C_7H_{12}O$

(*b*) [triester with three $COOCH_2CH_3$ groups] $\xrightarrow[\text{2. H}^+]{\text{1. NaOCH}_2\text{CH}_3}$ $C_{12}H_{18}O_5$

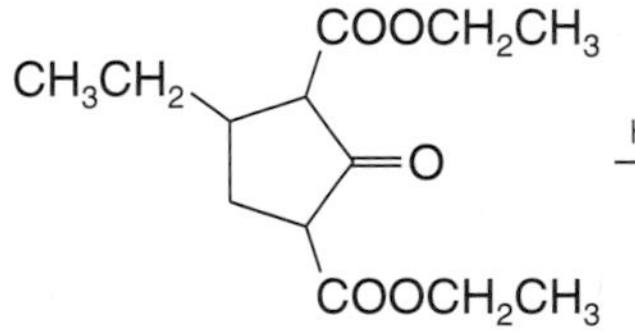

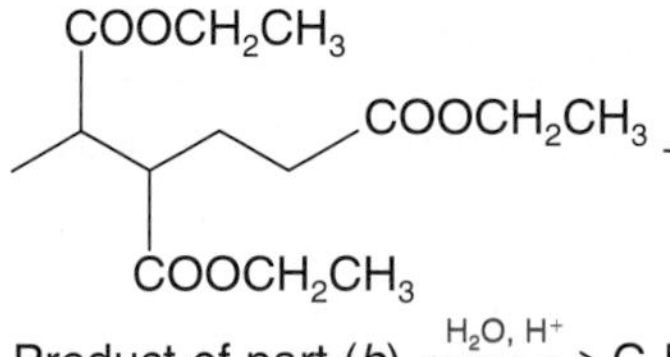

(*c*) Product of part (*b*) $\xrightarrow[\text{heat}]{\text{H}_2\text{O, H}^+}$ $C_7H_{10}O_3$

(*d*) [cis-1,2-cyclopropane with H, H, $CH_2COOCH_2CH_3$, $CH_2COOCH_2CH_3$] $\xrightarrow[\text{2. H}^+]{\text{1. NaOCH}_2\text{CH}_3}$ $C_9H_{12}O_3$

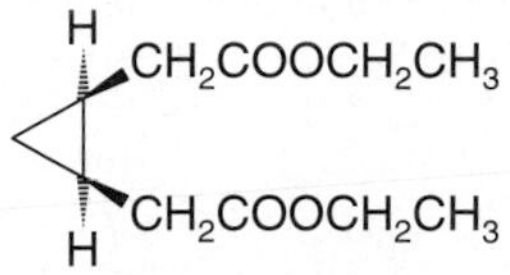

(*e*) Product of part (*d*) $\xrightarrow[\text{2. H}^+ \text{ 3. heat}]{\text{1. HO}^-,\ \text{H}_2\text{O}}$ C_6H_8O

21.24 Show how you could prepare each of the following compounds. Use the starting material indicated along with ethyl acetoacetate or diethyl malonate and any necessary inorganic reagents. Assume also that the customary organic solvents are freely available.

(*a*) 4-Phenyl-2-butanone from benzyl alcohol
(*b*) 3-Phenylpropanoic acid from benzyl alcohol
(*c*) 2-Allyl-1,3-propanediol from propene
(*d*) 4-Penten-1-ol from propene
(*e*) 5-Hexen-2-ol from propene
(*f*) Cyclopropanecarboxylic acid from 1,2-dibromoethane
(*g*) 1,1-cyclopropane with two $\overset{O}{\overset{\|}{C}}NH_2$ groups from 1,2-dibromoethane
(*h*) $HO_2C(CH_2)_{10}CO_2H$ from $HO_2C(CH_2)_6CO_2H$

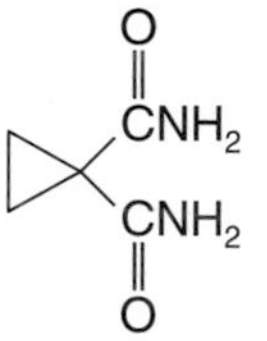

21.25 *Diphenadione* inhibits the clotting of blood; i.e., it is an *anticoagulant.* It is used to control vampire bat populations in South America by a "Trojan horse" strategy. A few bats are trapped, smeared with diphenadione, and then released back into their normal environment. Other bats, in the course of grooming these diphenadione-coated bats, ingest the anticoagulant and bleed to death, either internally or through accidental bites and scratches.

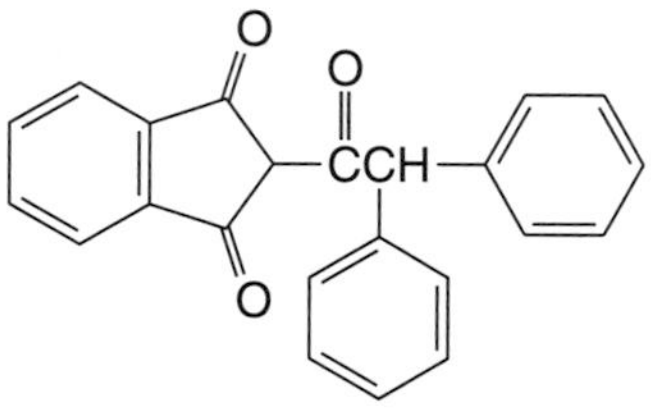

Diphenadione

Suggest a synthesis of diphenadione from 1,1-diphenylacetone and dimethyl 1,2-benzenedicarboxylate.

21.26 *Phenylbutazone* is a frequently prescribed anti-inflammatory drug. It is prepared by the reaction shown.

$$\underset{\text{Diethyl butylmalonate}}{CH_3CH_2CH_2CH_2CH(COOCH_2CH_3)_2} + \underset{\text{1,2-Diphenylhydrazine}}{C_6H_5NHNHC_6H_5} \longrightarrow \underset{\text{Phenylbutazone}}{C_{19}H_{20}N_2O_2}$$

What is the structure of phenylbutazone?

21.27 The use of epoxides as alkylating agents for diethyl malonate provides a useful route to γ-lactones. Write equations illustrating such a sequence for styrene oxide as the starting epoxide. Is the lactone formed by this reaction 3-phenylbutanolide, or is it 4-phenylbutanolide?

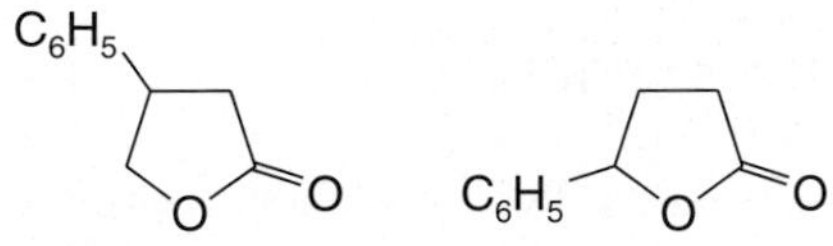

C_6H_5 (3-substituted lactone ring) $\qquad$ C_6H_5 (4-substituted lactone ring)

3-Phenylbutanolide $\qquad$ 4-Phenylbutanolide

21.28 Diethyl malonate is prepared commercially by hydrolysis and esterification of ethyl cyanoacetate.

$$N{\equiv}CCH_2\overset{O}{\overset{\|}{C}}OCH_2CH_3 \qquad \text{Ethyl cyanoacetate}$$

The preparation of ethyl cyanoacetate proceeds via ethyl chloroacetate and begins with acetic acid. Write a sequence of reactions describing this synthesis.

21.29 The tranquilizing drug *meprobamate* has the structure shown.

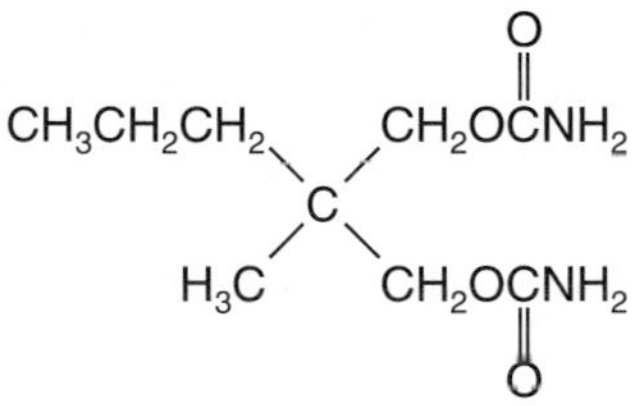

$$(CH_3CH_2CH_2)(H_3C)C(CH_2O\overset{O}{\overset{\|}{C}}NH_2)_2 \qquad \text{Meprobamate}$$

Devise a synthesis of meprobamate from diethyl malonate and any necessary organic or inorganic reagents. *Hint: Carbamate esters,* i.e., compounds of the type $RO\overset{O}{\overset{\|}{C}}NH_2$, are prepared from alcohols by the sequence of reactions

$$\underset{\text{Alcohol}}{ROH} + \underset{\text{Phosgene}}{Cl\overset{O}{\overset{\|}{C}}Cl} \longrightarrow \underset{\text{Chlorocarbonate ester}}{RO\overset{O}{\overset{\|}{C}}Cl} \xrightarrow{NH_3,\ H_2O} \underset{\text{Carbamate ester}}{RO\overset{O}{\overset{\|}{C}}NH_2}$$

21.30 When the compound shown was heated in refluxing hydrochloric acid for 60 hours, a product with the molecular formula $C_5H_6O_3$ was isolated in 97 percent yield. Identify this product. Along with this product, three other carbon-containing substances are formed. What are they?

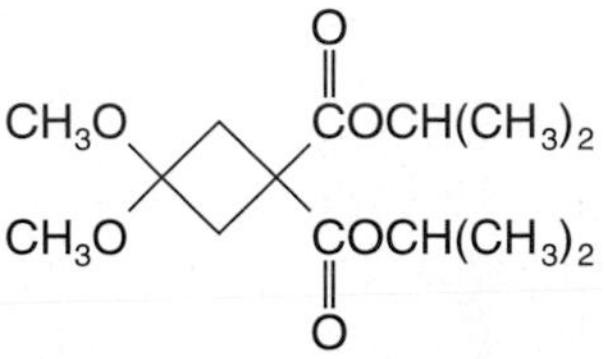

$$(CH_3O)_2C_4H_4(\overset{O}{\overset{\|}{C}}OCH(CH_3)_2)_2$$

(cyclobutane ring: two CH_3O groups on one carbon, two $\overset{O}{\overset{\|}{C}}OCH(CH_3)_2$ groups on the opposite carbon)

MOLECULAR MODELING EXERCISES

21.31 Construct molecular models of two constitutionally isomeric enol forms of ethyl acetoacetate.

21.32 Construct a molecular model of the *Z* stereoisomer of the lithium enolate of ethyl propanoate.

21.33 Construct a molecular model of acetoacetic acid (3-oxobutanoic acid) in the geometry in which it undergoes decarboxylation.

21.34 The compound shown does not undergo thermal decarboxylation under conditions that suffice for most other β-keto acids. Why? (*Hint:* Examine a molecular model of the enol intermediate formed during decarboxylation for sources of strain.)

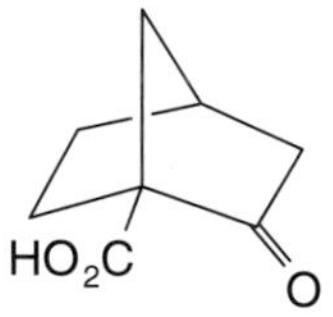

HO_2C O

21.35 Construct a molecular model showing the transition state geometry for the following reaction.

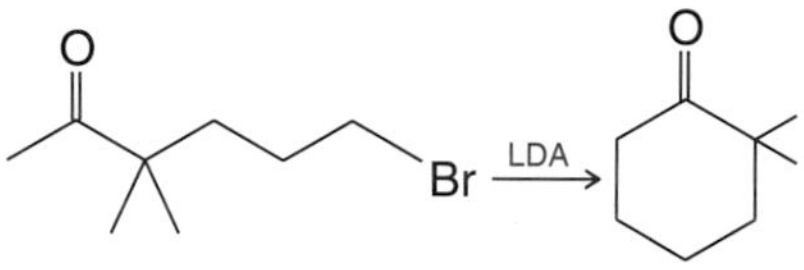

O Br $\xrightarrow{\text{LDA}}$ O

CHAPTER 22

AMINES

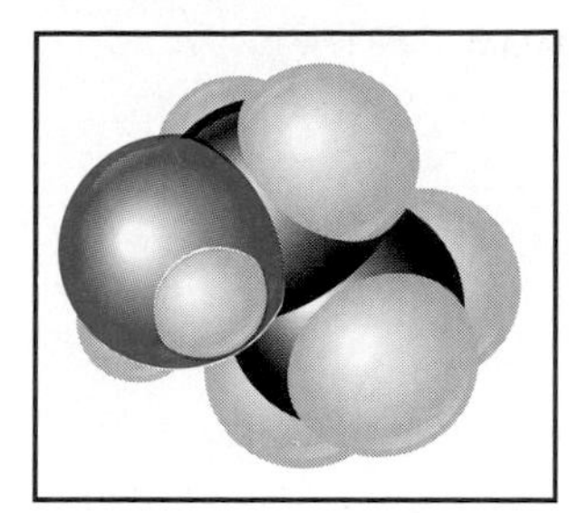

Nitrogen-containing compounds are essential to life and are ultimately derived from atmospheric nitrogen. By a process known as *nitrogen fixation,* atmospheric nitrogen is reduced to ammonia, then converted to organic nitrogen compounds. This chapter describes the chemistry of **amines,** organic derivatives of ammonia. Amines are classified either as *alkylamines* or *arylamines.* **Alkylamines** have their nitrogen attached to sp^3 hybridized carbon; **arylamines** have their nitrogen attached to an sp^2 hybridized carbon of a benzene or benzenelike ring.

$$\mathrm{R{-}\ddot{N}\langle} \qquad \mathrm{Ar{-}\ddot{N}\langle}$$

R = alkyl group: alkylamine   Ar = aryl group: arylamine

Amines, like ammonia, are weak bases. They are, however, the strongest bases that are found in significant quantities under physiological conditions. Amines are usually the bases involved in biological acid-base reactions; they are often the nucleophiles in biological nucleophilic substitutions.

Our word *vitamin* was coined in 1912 in the belief that the substances present in the diet that prevented scurvy, pellagra, beriberi, rickets, and other diseases were ''vital amines.'' In many cases, that belief was confirmed; certain vitamins did prove to be amines. In many other cases, however, vitamins were not amines. Nevertheless, the name *vitamin* entered our language and stands as a reminder that early chemists were acutely aware of the crucial place occupied by amines in biological processes.

22.1 AMINE NOMENCLATURE

Unlike alcohols and alkyl halides, which are classified as primary, secondary, or tertiary according to the degree of substitution at the carbon that bears the functional group, amines are classified according to their *degree of substitution at nitrogen.* An amine with one carbon attached to nitrogen is a *primary amine,* an amine with two is a *secondary amine,* and an amine in which all three bonds from nitrogen are to carbon is a *tertiary amine.*

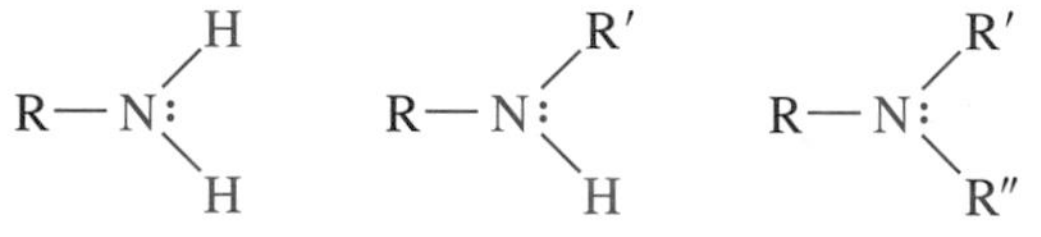

$R{-}\ddot{N}(H)H$	$R{-}\ddot{N}(R')H$	$R{-}\ddot{N}(R')R''$
Primary amine	Secondary amine	Tertiary amine

The groups attached to nitrogen may be any combination of alkyl or aryl groups.

Amines are named in two main ways, either as *alkylamines* or as *alkanamines.* When primary amines are named as alkylamines, the ending *-amine* is added to the name of the alkyl group that bears the nitrogen. When named as alkanamines, the alkyl group is named as an alkane and the *-e* ending replaced by *-amine.*

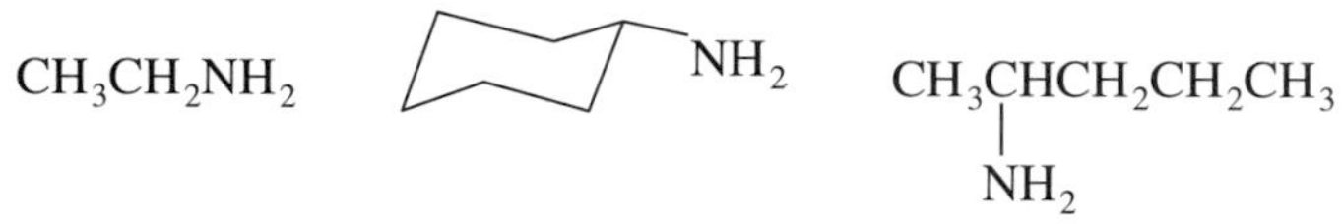

$CH_3CH_2NH_2$	(cyclohexane ring)$-NH_2$	$CH_3CH(NH_2)CH_2CH_2CH_3$
Ethylamine (ethanamine)	Cyclohexylamine (cyclohexanamine)	1-Methylbutylamine (2-pentanamine)

The alkanamine naming system was introduced by *Chemical Abstracts* and is more versatile and easier to use than the older IUPAC system of alkylamine names. The latest version of the IUPAC rules accepts both.

PROBLEM 22.1 Give an acceptable alkylamine or alkanamine name for each of the following amines:

(*a*) $C_6H_5CH_2CH_2NH_2$ (*b*) $C_6H_5CH(CH_3)NH_2$ (*c*) $CH_2{=}CHCH_2NH_2$

SAMPLE SOLUTION (*a*) The amino substituent is bonded to an ethyl group that bears a phenyl substituent at C-2. The compound $C_6H_5CH_2CH_2NH_2$ may be named as either 2-phenylethylamine or 2-phenylethanamine.

Aniline was first isolated in 1826 as a degradation product of indigo, a dark-blue dye obtained from the West Indian plant *Indigofera anil,* from which the name *aniline* is derived.

Aniline is the parent IUPAC name for amino-substituted derivatives of benzene. Substituted derivatives of aniline are numbered beginning at the carbon that bears the amino group. Substituents are listed in alphabetical order, and the direction of numbering is governed by the usual "first point of difference" rule.

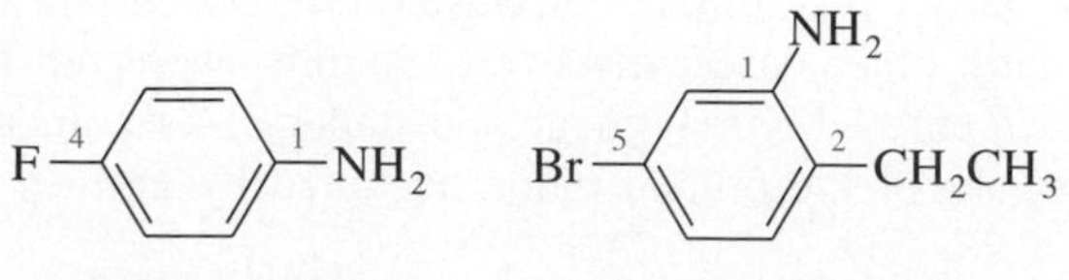

$F{-}C_6H_4{-}NH_2$ (F at C-4, NH_2 at C-1)	NH_2 at C-1, CH_2CH_3 at C-2, Br at C-5
p-Fluoroaniline	5-Bromo-2-ethylaniline

Arylamines may also be named as *arenamines.* Thus, *benzenamine* becomes an alternative, but rarely used, name for aniline.

Compounds with two amino groups are named by adding the suffix *-diamine* to the name of the corresponding alkane or arene. The final *-e* of the parent hydrocarbon is retained.

$H_2NCH_2CHCH_3$ (with NH_2 on C-2) — 1,2-Propanediamine

$H_2NCH_2CH_2CH_2CH_2CH_2CH_2NH_2$ — 1,6-Hexanediamine

$H_2N{-}C_6H_4{-}NH_2$ — 1,4-Benzenediamine

Amino groups rank rather low in seniority when the parent compound is identified for naming purposes. Hydroxyl groups and carbonyl groups outrank amino groups. In these cases, the amino group is named as a substituent.

$HOCH_2CH_2NH_2$ — 2-Aminoethanol

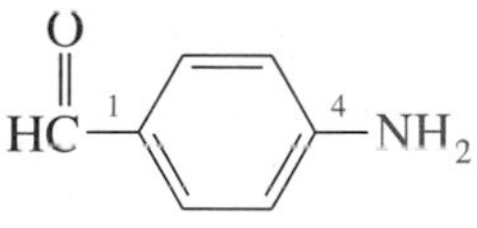

$HC(=O){-}C_6H_4{-}NH_2$ (CHO at C-1, NH_2 at C-4) — *p*-Aminobenzaldehyde (4-Aminobenzenecarbaldehyde)

Secondary and tertiary amines are named as *N*-substituted derivatives of primary amines. The parent primary amine is taken to be the one with the longest carbon chain. The prefix *N*- is added as a locant to identify substituents on the amino nitrogen as needed.

$CH_3NHCH_2CH_3$ — *N*-Methylethylamine (a secondary amine)

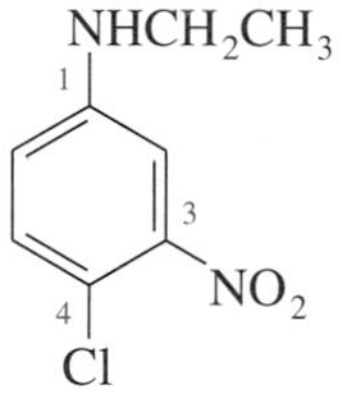

Benzene ring with $NHCH_2CH_3$ at C-1, NO_2 at C-3, Cl at C-4 — 4-Chloro-*N*-ethyl-3-nitroaniline (a secondary amine)

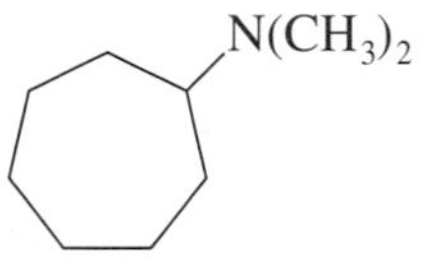

Cycloheptane ring with $N(CH_3)_2$ — *N,N*-Dimethylcycloheptylamine (a tertiary amine)

PROBLEM 22.2 Assign alkanamine names to *N*-methylethylamine and to *N,N*-dimethylcycloheptylamine.

SAMPLE SOLUTION *N*-Methylethylamine (given as $CH_3NHCH_2CH_3$ in the preceding example) is an *N*-substituted derivative of ethanamine; it is *N*-methylethanamine.

PROBLEM 22.3 Classify the following amine as primary, secondary, or tertiary, and give it an acceptable IUPAC name.

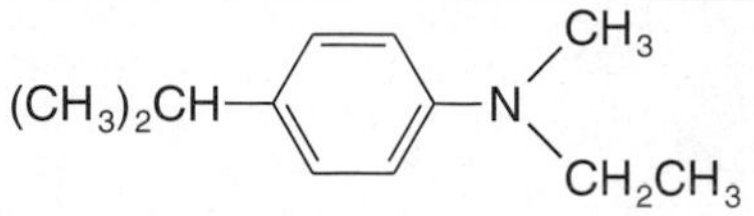

$(CH_3)_2CH{-}C_6H_4{-}N(CH_3)(CH_2CH_3)$

A nitrogen that bears four substituents is positively charged and is named as an *ammonium* ion. The anion that is associated with it is also identified in the name:

$CH_3\overset{+}{N}H_3\ \ Cl^-$	$C_5H_9\overset{+}{N}(CH_3)(H)CH_2CH_3\ \ CF_3CO_2^-$	$C_6H_5CH_2\overset{+}{N}(CH_3)_3\ \ I^-$
Methylammonium chloride	*N*-Ethyl-*N*-methylcyclopentylammonium trifluoroacetate	Benzyltrimethylammonium iodide (a quaternary ammonium salt)

Ammonium salts that have four alkyl groups bonded to nitrogen are called **quaternary ammonium salts.**

22.2 STRUCTURE AND BONDING

Alkylamines. The structure of alkylamines can be illustrated by reference to methylamine. As shown in Figure 22.1 methylamine, like ammonia, has a pyramidal arrangement of substituents at nitrogen. Its H—N—H angles (106°) are slightly smaller than the tetrahedral value of 109.5°, while the C—N—H angle (112°) is slightly larger. The C—N bond distance of 147 pm lies between typical C—C bond distances in alkanes (153 pm) and C—O bond distances in alcohols (143 pm).

An orbital hybridization description of bonding in methylamine is shown in Figure 22.2. Nitrogen and carbon are both sp^3 hybridized and are joined by a σ bond. The unshared electron pair on nitrogen occupies an sp^3 hybridized orbital. It is this lone pair that is involved in reactions in which amines act as bases or nucleophiles.

Arylamines. Aniline, like alkylamines, has a pyramidal arrangement of bonds around nitrogen, but its pyramid is somewhat shallower. One measure of the extent of this flattening is given by the angle between the carbon-nitrogen bond and the bisector of the H—N—H angle.

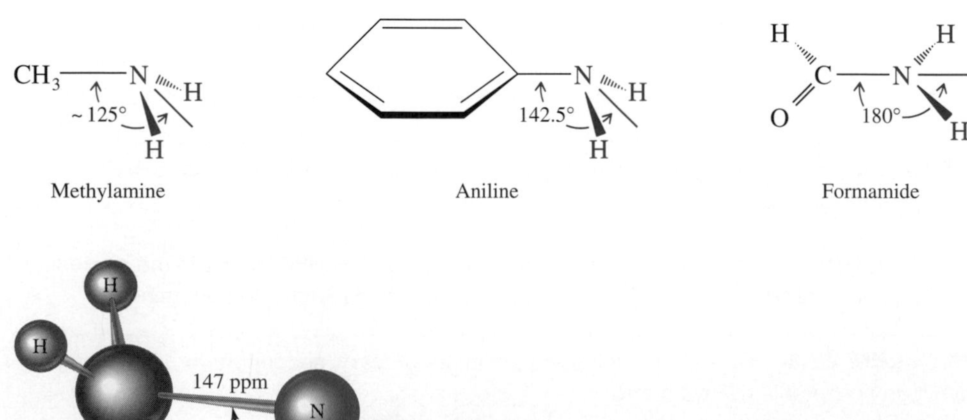

FIGURE 22.1 A ball-and-stick model of methylamine showing the trigonal pyramidal arrangement of bonds to nitrogen. The most stable conformation has the staggered arrangement of bonds shown. Other alkylamines have similar geometries.

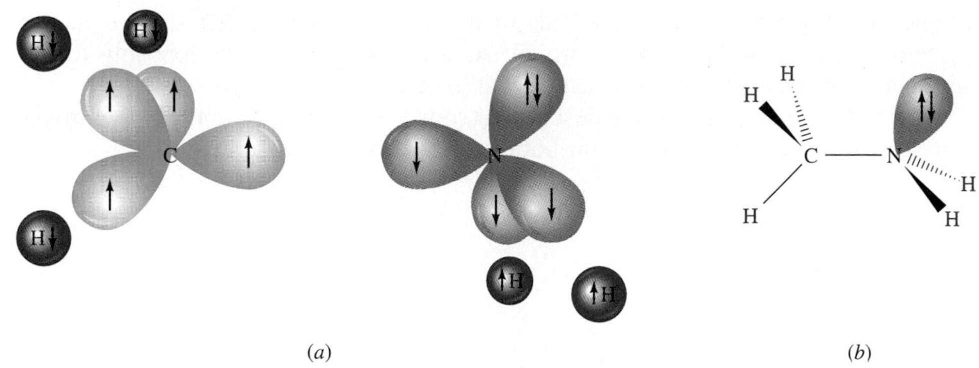

FIGURE 22.2 Orbital hybridization description of bonding in methylamine. (*a*) Carbon has four valence electrons; each of four equivalent sp^3 hybridized orbitals contains one electron. Nitrogen has five valence electrons. Three of its sp^3 hybrid orbitals contain one electron each; the fourth sp^3 hybrid orbital contains two electrons. (*b*) Nitrogen and carbon are connected by a σ bond in methylamine. This σ bond is formed by overlap of an sp^3 hybrid orbital on each atom. The five hydrogen atoms of methylamine are jointed to carbon and nitrogen by σ bonds. The two remaining electrons of nitrogen occupy an sp^3 hybridized orbital.

For sp^3 hybridized nitrogen, this angle (not the same as the C—N—H bond angle) is 125°, and the measured angles in simple alkylamines are close to that. The corresponding angle for sp^2 hybridization at nitrogen with a planar arrangement of bonds, as in amides, for example, is 180°. The measured value for this angle in aniline is 142.5°, indicative of a hybridization somewhat closer to sp^3 than to sp^2.

The structure of aniline reflects a compromise between two modes of binding the nitrogen lone pair. These electrons are bound both by the attractive force exerted by the nitrogen nucleus and by their delocalization into the aromatic ring (Figure 22.3). The electrons are more strongly attracted to nitrogen when they are in an orbital with some *s* character—an sp^3 hybridized orbital, for example—than when they are in a *p* orbital. Increasing the amount of *s* character in an orbital brings the electrons closer to the nucleus. On the other hand, delocalization of these electrons into the aromatic π system is better achieved if they occupy a *p* orbital. A *p* orbital of nitrogen is better

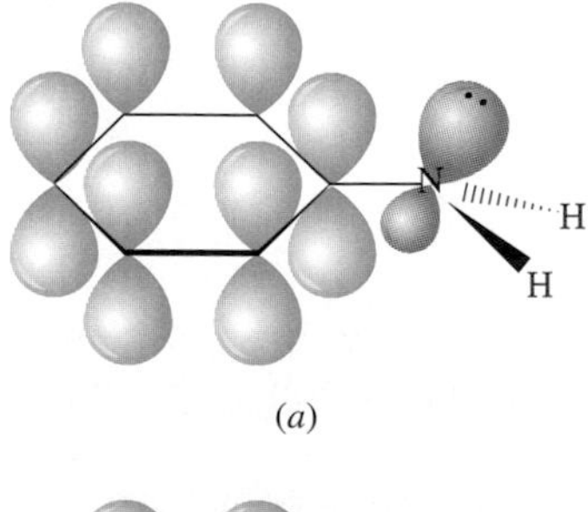

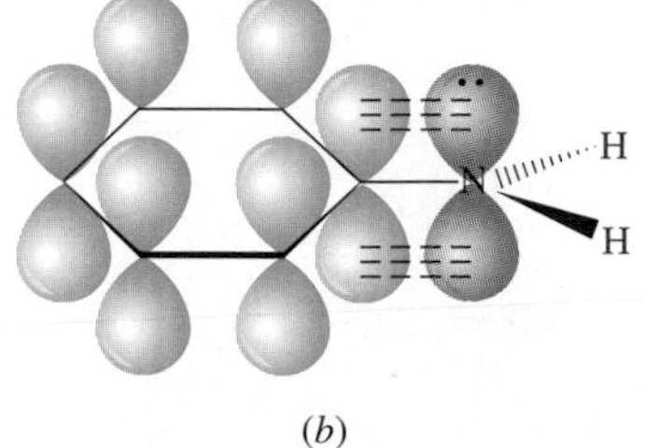

FIGURE 22.3 Orbital hybridization and lone-pair binding in aniline. (*a*) The lone pair is strongly bound by nitrogen when it occupies an sp^3 hybridized orbital. (*b*) The lone pair is delocalized best into the ring when it occupies a *p* orbital that is aligned for overlap with the aromatic π system. The actual structure combines features of both *a* and *b*; nitrogen adopts an orbital hybridization that is between sp^3 and sp^2.

aligned for overlap with the p orbitals of the benzene ring to form an extended π system than is an sp^3 hybridized orbital. As a result of these two opposing forces, nitrogen adopts an orbital hybridization that is between sp^3 and sp^2.

The corresponding resonance description shows the delocalization of the nitrogen lone-pair electrons in terms of contributions from dipolar structures.

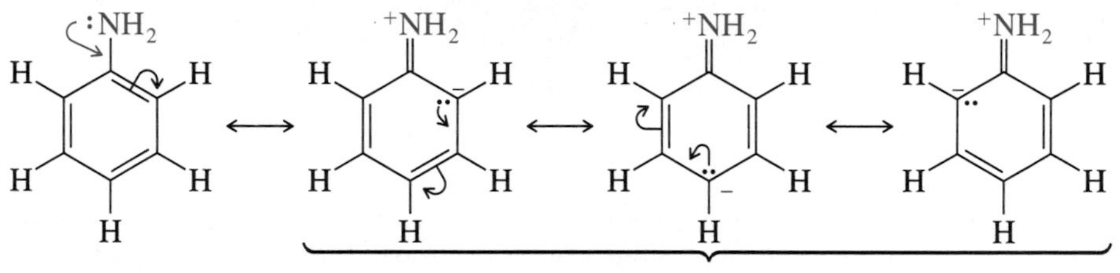

Most stable Lewis structure for aniline

Dipolar resonance forms of aniline

The orbital and resonance models for bonding in arylamines are simply alternative ways of describing the same phenomenon. Delocalization of the nitrogen lone pair decreases the electron density at nitrogen while increasing it in the π system of the aromatic ring. We have already seen one chemical consequence of this in the high level of reactivity of aniline in electrophilic aromatic substitution reactions (Section 12.12). Other ways in which electron delocalization affects the properties of arylamines are described in later sections of this chapter.

22.3 PHYSICAL PROPERTIES

A collection of physical properties of some representative amines is given in Appendix 1. Most commonly encountered alkylamines are liquids with unpleasant, "fishy" odors.

We have seen on a number of occasions that the polar nature of a substance can affect physical properties such as boiling point. This is true for amines, which are more polar than alkanes but less polar than alcohols. For similarly constituted compounds, alkylamines have boiling points which are higher than those of alkanes but lower than those of alcohols.

$CH_3CH_2CH_3$	$CH_3CH_2NH_2$	CH_3CH_2OH
Propane	Ethylamine	Ethanol
$\mu = 0$ D	$\mu = 1.2$ D	$\mu = 1.7$ D
bp −42°C	bp 17°C	bp 78°C

Dipole-dipole interactions, especially hydrogen bonding, are stronger in amines than in alkanes. The less polar nature of amines as compared with alcohols, however, makes these intermolecular forces weaker in amines than in alcohols.

Among isomeric amines, primary amines have the highest boiling points, and tertiary amines the lowest.

$CH_3CH_2CH_2NH_2$	$CH_3CH_2NHCH_3$	$(CH_3)_3N$
Propylamine	*N*-Methylethylamine	Trimethylamine
(a primary amine)	(a secondary amine)	(a tertiary amine)
bp 50°C	bp 34°C	bp 3°C

Primary and secondary amines can participate in intermolecular hydrogen bonding, while tertiary amines cannot.

Amines that have fewer than six or seven carbon atoms are soluble in water. All amines, even tertiary amines, can act as proton acceptors in hydrogen bonding to water molecules.

The simplest arylamine, aniline, is a liquid at room temperature and has a boiling point of 184°C. Almost all other arylamines have higher boiling points. Aniline is only slightly soluble in water (3 g/100 mL). Substituted derivatives of aniline tend to be even less water-soluble.

22.4 MEASURES OF AMINE BASICITY

There are two conventions used to measure the basicity of amines. One of them defines a **basicity constant K_b** for the amine acting as a proton acceptor from water:

$$R_3N\colon + H-\ddot{O}H \rightleftharpoons R_3\overset{+}{N}-H + \colon\ddot{\overset{-}{O}}H$$

$$K_b = \frac{[R_3NH^+][HO^-]}{[R_3N]} \quad \text{and} \quad pK_b = -\log K_b$$

For ammonia, $K_b = 1.8 \times 10^{-5}$ ($pK_b = 4.7$). A typical amine such as methylamine (CH_3NH_2) is a stronger base than ammonia and has $K_b = 4.4 \times 10^{-4}$ ($pK_b = 3.3$).

The other convention relates the basicity of an amine (R_3N) to the *acid dissociation constant K_a* of its conjugate acid (R_3NH^+):

$$R_3\overset{+}{N}-H \rightleftharpoons H^+ + R_3N\colon$$

where K_a and pK_a have their usual meaning:

$$K_a = \frac{[H^+][R_3N]}{[R_3NH^+]} \quad \text{and} \quad pK_a = -\log K_a$$

The conjugate acid of ammonia is ammonium ion (NH_4^+), which has $K_a = 5.6 \times 10^{-10}$ ($pK_a = 9.3$). The conjugate acid of methylamine is methylammonium ion ($CH_3NH_3^+$), which has $K_a = 2 \times 10^{-11}$ ($pK_a = 10.7$). *The more basic the amine, the weaker is its conjugate acid.* Methylamine is a stronger base than ammonia; methylammonium ion is a weaker acid than ammonium ion.

The relationship between the equilibrium constant K_b for an amine (R_3N) and K_a for its conjugate acid (R_3NH^+) is:

$$K_aK_b = 10^{-14} \quad \text{and} \quad pK_a + pK_b = 14$$

PROBLEM 22.4 A chemistry handbook lists K_b for *quinine* as 1×10^{-6}. What is pK_b for quinine? What are the values of K_a and pK_a for the conjugate acid of quinine?

Citing amine basicity according to the acidity of the conjugate acid permits acid-base reactions involving amines to be analyzed according to the usual Brønsted relationships. By comparing the acidity of an acid with the conjugate acid of an amine, for example, we see that amines are converted to ammonium ions by acids even as weak as acetic acid:

Recall from Section 4.6 that acid-base reactions are characterized by equilibrium constants greater than unity when the stronger acid is on the left side of the equation and the weaker acid on the right.

$$CH_3\ddot{N}H_2 + H{-}\ddot{O}CCH_3 \longrightarrow CH_3\overset{+}{N}H_3 + {}^{-}\!:\!\ddot{O}CCH_3$$

Methylamine; Acetic acid (stronger acid; $pK_a = 4.7$); Methylammonium ion (weaker acid; $pK_a = 10.7$); Acetate ion

Conversely, adding sodium hydroxide to an ammonium salt converts it to the free amine:

$$CH_3\overset{+}{N}H_3 + {}^{-}\!:\!\ddot{O}H \longrightarrow CH_3\ddot{N}H_2 + H{-}\ddot{O}H$$

Methylammonium ion (stronger acid; $pK_a = 10.7$); Hydroxide ion; Methylamine; Water (weaker acid; $pK_a = 15.7$)

PROBLEM 22.5 Apply the Henderson-Hasselbalch equation (see "Quantitative Relationships Involving Carboxylic Acids," the box accompanying Section 19.4) to calculate the $CH_3NH_3^+/CH_3NH_2$ ratio in water at pH 7.

Their basicity provides a means by which amines may be separated from neutral organic compounds. A mixture containing an amine is dissolved in diethyl ether and shaken with dilute hydrochloric acid to convert the amine to an ammonium salt. The ammonium salt, being ionic, dissolves in the aqueous phase, which is separated from the ether layer. Adding sodium hydroxide to the aqueous layer converts the ammonium salt back to the free amine, which is then removed from the aqueous phase by extraction with a fresh portion of ether.

22.5 BASICITY OF AMINES

Amines are weak bases, but as a class *amines are the strongest bases of all neutral molecules.* Table 22.1 lists basicity data for a number of amines. The most important relationships to be drawn from the data are

1. Alkylamines are slightly stronger bases than ammonia.
2. Alkylamines differ very little among themselves in basicity. Their basicities cover a range of less than 10 in equilibrium constant (one pK unit).
3. Arylamines are much weaker bases than ammonia and alkylamines. Their basicity constants are on the order of 10^6 smaller than those of alkylamines (six pK units).

Alkylamines. While most alkylamines are very similar in basicity, it is generally true that their basicities increase in the order

$$NH_3 < RNH_2 \sim R_3N < R_2NH$$

Ammonia (least basic); Primary amine; Tertiary amine; Secondary amine (most basic)

TABLE 22.1
Base Strength of Amines As Measured by Their Basicity Constants and the Dissociation Constants of Their Conjugate Acids*

		Basicity		Acidity of conjugate acid	
Compound	**Structure**	K_b	pK_b	K_a	pK_a
Ammonia	NH_3	1.8×10^{-5}	4.7	5.5×10^{-10}	9.3
Primary amines					
Methylamine	CH_3NH_2	4.4×10^{-4}	3.4	2.3×10^{-11}	10.6
Ethylamine	$CH_3CH_2NH_2$	5.6×10^{-4}	3.2	1.8×10^{-11}	10.8
Isopropylamine	$(CH_3)_2CHNH_2$	4.3×10^{-4}	3.4	2.3×10^{-11}	10.6
tert-Butylamine	$(CH_3)_3CNH_2$	2.8×10^{-4}	3.6	3.6×10^{-11}	10.4
Aniline	$C_6H_5NH_2$	3.8×10^{-10}	9.4	2.6×10^{-5}	4.6
Secondary amines					
Dimethylamine	$(CH_3)_2NH$	5.1×10^{-4}	3.3	2.0×10^{-11}	10.7
Diethylamine	$(CH_3CH_2)_2NH$	1.3×10^{-3}	2.9	7.7×10^{-12}	11.1
N-Methylaniline	$C_6H_5NHCH_3$	6.1×10^{-10}	9.2	1.6×10^{-5}	4.8
Tertiary amines					
Trimethylamine	$(CH_3)_3N$	5.3×10^{-5}	4.3	1.9×10^{-10}	9.7
Triethylamine	$(CH_3CH_2)_3N$	5.6×10^{-4}	3.2	1.8×10^{-11}	10.8
N,N-Dimethylaniline	$C_6H_5N(CH_3)_2$	1.2×10^{-9}	8.9	8.3×10^{-6}	5.1

* In water at 25°C.

Diethylamine, for example, is more basic than either ethylamine or triethylamine, and all these compounds are more basic than ammonia, as measured in *aqueous solution.*

Basicity of amines in aqueous solution

$$NH_3 < CH_3CH_2NH_2 \sim (CH_3CH_2)_3N < (CH_3CH_2)_2NH$$

Ammonia	Ethylamine	Triethylamine	Diethylamine
K_b 1.8×10^{-5}	K_b 5.6×10^{-4}	K_b 5.6×10^{-4}	K_b 1.3×10^{-3}
(pK_b 4.7)	(pK_b 3.2)	(pK_b 3.2)	(pK_b 2.9)

The discontinuity in basicity among the various classes of amines suggests that there are at least two substituent effects involved and that they operate in opposite directions.

An alkyl group can increase the base strength of an amine by releasing electrons to nitrogen. The positive charge of an ammonium ion is dispersed better by having alkyl groups instead of hydrogen as substituents on nitrogen.

Ethyl group releases electrons to nitrogen; positive charge is dispersed and ion is stabilized.

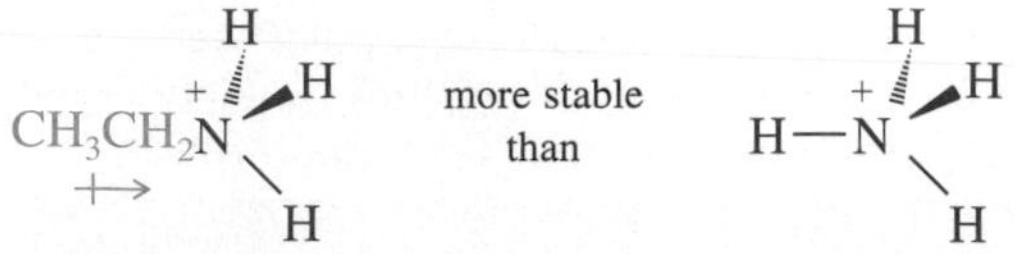

$CH_3CH_2\overset{+}{N}H_3$ (Ethylammonium ion) more stable than $H-\overset{+}{N}H_3$ (Ammonium ion)

By stabilizing the ammonium ion, alkyl groups increase the equilibrium constant for amine protonation.

Were this the only effect of alkyl groups, the basicity of amines would increase with increasing alkyl substitution. Indeed, this is precisely what is observed for proton transfer to amines in the *gas phase*.

Gas-phase basicity of amines

$$NH_3 < CH_3CH_2NH_2 < (CH_3CH_2)_2NH < (CH_3CH_2)_3N$$

Ammonia (least basic) — Ethylamine — Diethylamine — Triethylamine (most basic)

Electron release from alkyl groups provides the principal mechanism by which the conjugate acid of an amine is stabilized in the gas phase. The more alkyl groups that are attached to the positively charged nitrogen, the more stable the alkylammonium ion becomes.

Basicity as measured by K_b, however, refers to equilibrium measurements made in dilute aqueous solution. The altered order of amine basicities in solution, as compared with those in the gas phase, must arise from *solvation effects.*

While alkyl substituents increase the ability of an ammonium ion to disperse its positive charge, they decrease its ability to form hydrogen bonds to water molecules. Dialkylammonium ions, formed by protonation of secondary amines, have two hydrogen substituents on nitrogen that can participate in hydrogen bonding. Trialkylammonium ions have only one and are therefore less stabilized by solvation than are their dialkyl counterparts.

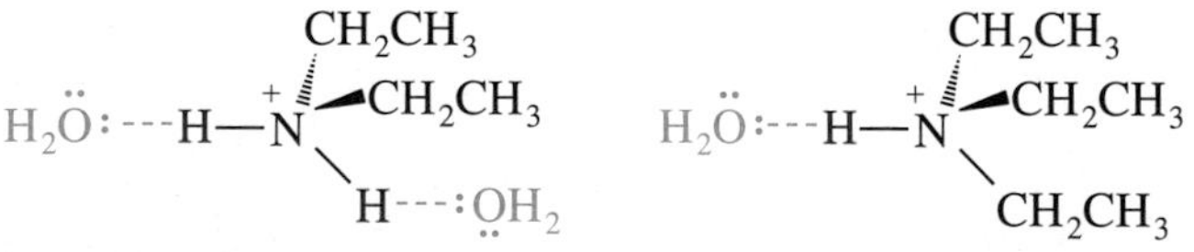

$H_2\ddot{O}\text{:}\cdots H-\overset{+}{N}(CH_2CH_3)_2-H\cdots\text{:}\ddot{O}H_2$

Diethylammonium ion (two protons on nitrogen available for hydrogen bonding)

$H_2\ddot{O}\text{:}\cdots H-\overset{+}{N}(CH_2CH_3)_3$

Triethylammonium ion (only one proton on nitrogen available for hydrogen bonding)

Dialkylamines are slightly more basic than either primary or tertiary amines because their conjugate acids possess the best combination of alkyl and hydrogen substituents to permit stabilization both by electron release from alkyl groups and by solvation due to hydrogen bonding.

PROBLEM 22.6 Identify the stronger base in each of the following pairs. Explain your reasoning.

(*a*) $FCH_2CH_2NH_2$ or $CH_3CH_2NH_2$

(*b*) $F_2CHCH_2NH_2$ or $F_3CCH_2NH_2$

SAMPLE SOLUTION (*a*) The strongly electronegative fluorine substituent diminishes the ability of the substituent to stabilize the protonated form of the amine.

$$F\overleftarrow{CH_2CH_2}\overset{+}{N}H_3 \quad \text{is less stabilized than} \quad CH_3\overrightarrow{CH_2}\overset{+}{N}H_3$$

As measured by their respective K_b's, ethylamine is 90 times more basic than its 2-fluoro derivative.

$CH_3CH_2NH_2$	$FCH_2CH_2NH_2$
Ethylamine	2-Fluoroethylamine
stronger base: K_b 5.6×10^{-4}	weaker base: K_b 6.3×10^{-6}
(pK_b 3.2)	(pK_b 5.2)

Arylamines. Arylamines are several orders of magnitude less basic than alkylamines. While K_b for most alkylamines is on the order of 10^{-4} (pK_b 4), arylamines have K_b's in the 10^{-10} range. The sharply decreased basicity of arylamines arises because the stabilizing effect of lone-pair electron delocalization is sacrificed on protonation.

$$C_6H_5\ddot{N}RR' + H_2O \rightleftharpoons C_6H_5\overset{+}{N}HRR' + HO^-$$

Amine is stabilized by delocalization of lone pair into π system of ring, decreasing the electron density at nitrogen.

Lone-pair electrons transformed to N—H bonded pair

The aromatic ring does very little to disperse the positive charge in the ammonium ion. Indeed, since the ring carbon attached to nitrogen is sp^2 hybridized, it is electron-withdrawing and destabilizes the ammonium ion. Stabilization of the amine and destabilization of the ammonium ion combine to make the equilibrium constant for amine protonation much smaller for arylamines than for alkylamines. This relationship is depicted in Figure 22.4, where the free energies of protonation of cyclohexylamine and aniline are compared. As measured by their respective K_b's, cyclohexylamine is almost 1 million times more basic than aniline.

$$C_6H_5NH_2 + H_2O \rightleftharpoons C_6H_5\overset{+}{N}H_3 + HO^- \qquad (K_b\ 3.8 \times 10^{-10};\ pK_b\ 9.4)$$

Aniline — Water — Anilinium ion — Hydroxide ion

$$C_6H_{11}NH_2 + H_2O \rightleftharpoons C_6H_{11}\overset{+}{N}H_3 + HO^- \qquad (K_b\ 4.4 \times 10^{-4};\ pK_b\ 3.4)$$

Cyclohexylamine — Water — Cyclohexylammonium ion — Hydroxide ion

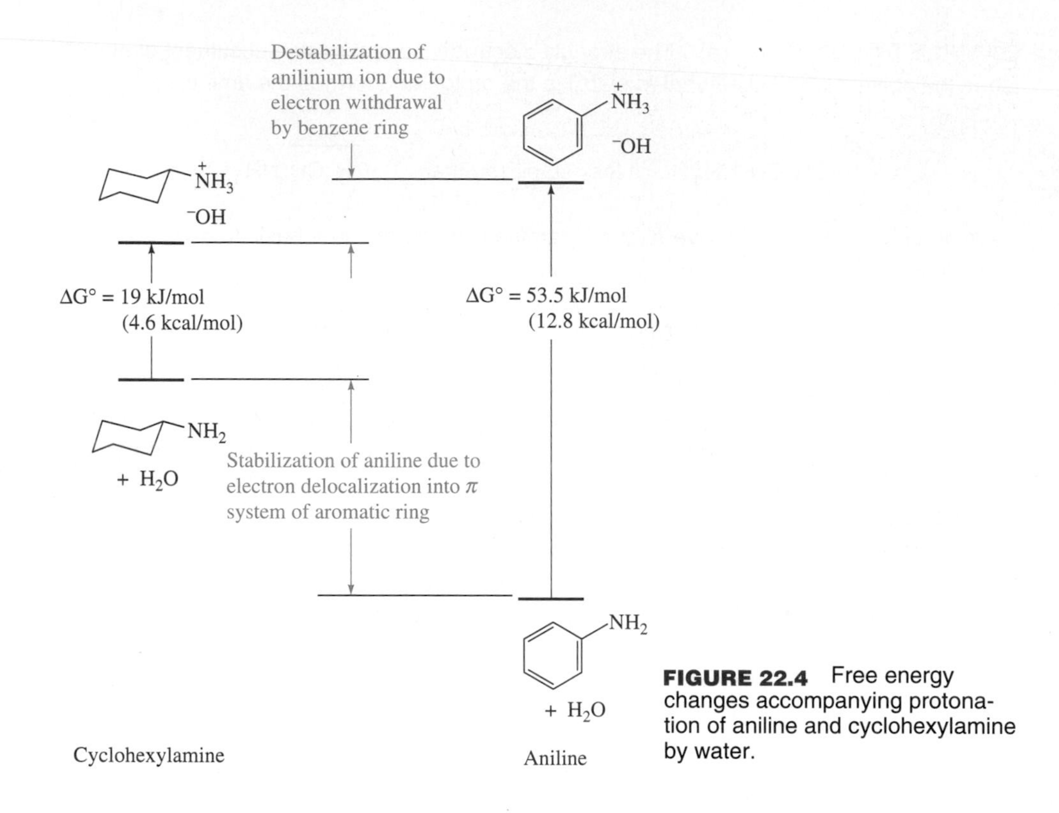

FIGURE 22.4 Free energy changes accompanying protonation of aniline and cyclohexylamine by water.

When the proton donor is a strong acid, arylamines can be completely protonated. Aniline is extracted from an ether solution into 1 *M* hydrochloric acid because it is converted to a water-soluble anilinium ion salt under these conditions.

PROBLEM 22.7 The two amines shown differ by a factor of 40,000 in their K_b values. Which is the stronger base? Why?

NH

N H

Tetrahydroquinoline Tetrahydroisoquinoline

Conjugation of the amino group of an arylamine with a second aromatic ring, then a third, reduces its basicity even further. Diphenylamine is 6300 times less basic than aniline, while triphenylamine is scarcely a base at all, being estimated as 10^8 times less basic than aniline and 10^{14} times less basic than ammonia.

$C_6H_5NH_2$	$(C_6H_5)_2NH$	$(C_6H_5)_3N$
Aniline	Diphenylamine	Triphenylamine
(K_b 3.8×10^{-10};	(K_b 6×10^{-14};	(K_b $\sim 10^{-19}$;
pK_b 9.4)	pK_b 13.2)	pK_b $\sim$ 19)

TABLE 22.2
Effect of Substituents on the Basicity of Aniline

	X	K_b	pK_b
X—C₆H₄—NH₂	H	4×10^{-10}	9.4
	CH_3	2×10^{-9}	8.7
	CF_3	2×10^{-12}	11.5
	O_2N	1×10^{-13}	13.0

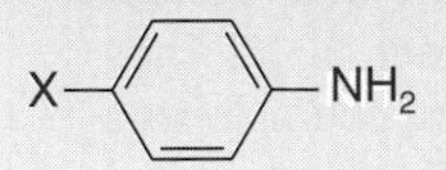

In general, electron-donating substituents on the aromatic ring increase the basicity of arylamines slightly. Thus, as shown in Table 22.2, an electron-donating methyl group in the para position *increases* the basicity of aniline by a factor of only 5 to 6 (less than 1 p*K* unit). Electron-withdrawing groups are base-weakening and exert larger effects. A *p*-trifluoromethyl group *decreases* the basicity of aniline by a factor of 200 and a *p*-nitro group by a factor of 3800. In the case of *p*-nitroaniline a resonance interaction of the type shown provides for extensive delocalization of the unshared electron pair of the amine group.

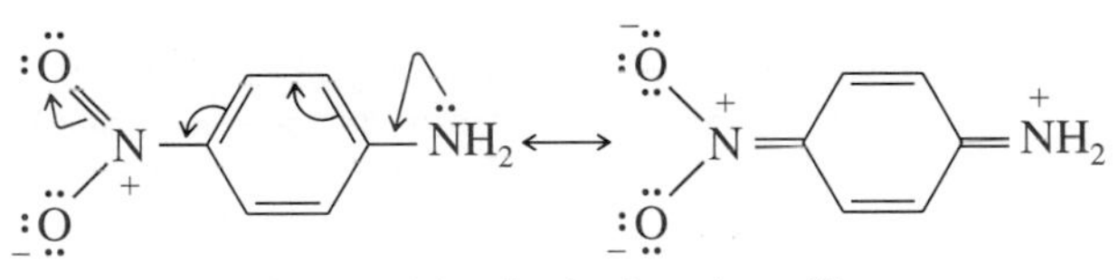

Electron delocalization in *p*-nitroaniline

Just as aniline is much less basic than alkylamines because the unshared electron pair of nitrogen is delocalized into the π system of the ring, *p*-nitroaniline is even less basic because the extent of this delocalization is greater and involves the oxygens of the nitro group.

PROBLEM 22.8 Each of the following is a much weaker base than aniline. Present a resonance argument to explain the effect of the substituent in each case.

(*a*) *o*-Cyanoaniline (*b*) $C_6H_5NHCCH_3$ (with C=O) (*c*) *p*-Aminoacetophenone

SAMPLE SOLUTION (*a*) A cyano substituent is strongly electron-withdrawing. When present at a position ortho to an amino group on an aromatic ring, a cyano substituent increases the delocalization of the amine lone-pair electrons by a direct resonance interaction.

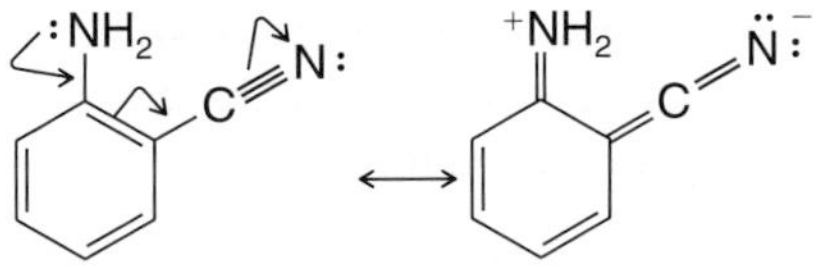

This resonance stabilization is lost when the amine group becomes protonated, and *o*-cyanoaniline is therefore a weaker base than aniline.

AMINES AS NATURAL PRODUCTS

The ease with which amines are extracted into aqueous acid, combined with their regeneration on treatment with base, makes it a simple matter to separate amines from other plant materials, and nitrogen-containing natural products were among the earliest organic compounds to be studied. Their basic properties led amines obtained from plants to be called **alkaloids.** The number of known alkaloids exceeds 5000. They are of special interest because most are characterized by a high level of biological activity. Some examples include *cocaine, coniine,* and *morphine.*

Cocaine

(A central nervous system stimulant obtained from the leaves of the coca plant.)

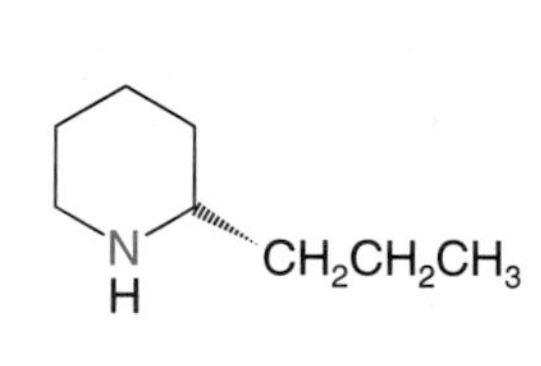

Coniine

(Present along with other alkaloids in the hemlock extract used to poison Socrates.)

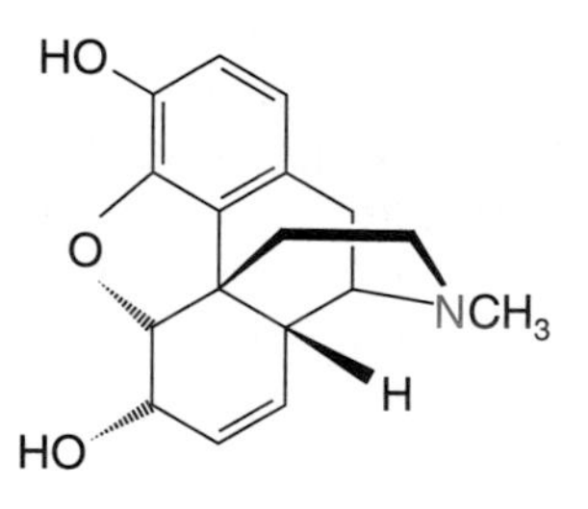

Morphine

(An opium alkaloid. While it is an excellent analgesic, its use is restricted because of potential for addiction. Heroin is the diacetate ester of morphine.)

Many alkaloids, such as *nicotine* and *quinine,* contain two (or more) nitrogen atoms. The nitrogens shown in red in quinine and nicotine are part of a substituted quinoline and pyridine ring, respectively.

Quinine

(Alkaloid of cinchona bark used to treat malaria)

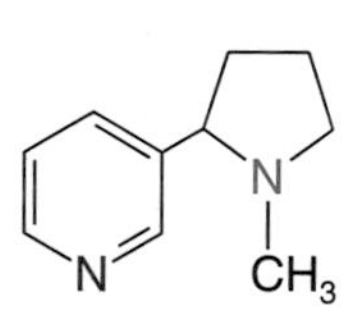

Nicotine

(An alkaloid present in tobacco; a very toxic compound sometimes used as an insecticide)

Several naturally occurring amines mediate the transmission of nerve impulses and are referred to as **neurotransmitters.** Two examples are *epinephrine* and *serotonin.* (Strictly speaking, these compounds are not classified as alkaloids, because they are not isolated from plants.)

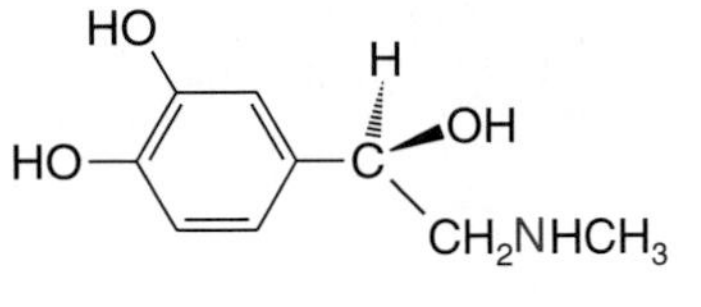

Epinephrine

(Also called *adrenaline*; a hormone secreted by the adrenal gland that prepares the organism for "flight or fight.")

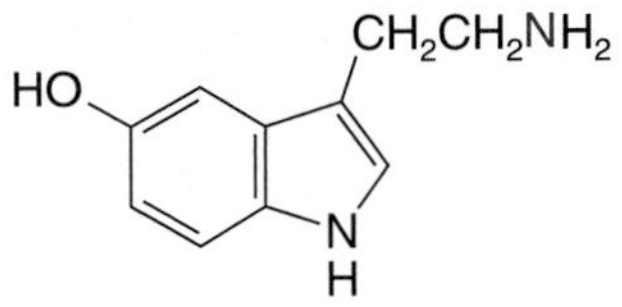

Serotonin

(A hormone synthesized in the pineal gland. Certain mental disorders are believed to be related to serotonin levels in the brain.)

Among the more important amine derivatives found in the body are a group of compounds known as **polyamines,** which contain two to four nitrogen atoms separated by several methylene units:

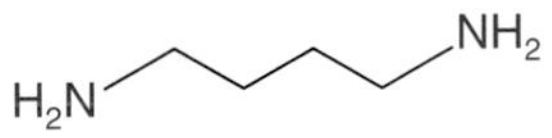

Putrescine

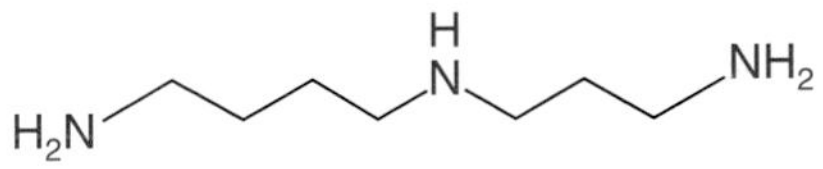

Spermidine

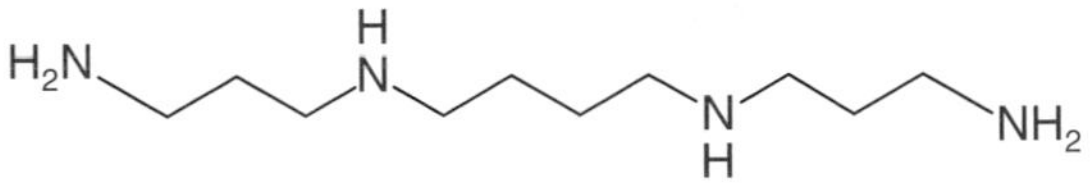

Spermine

These compounds are present in almost all mammalian cells, where they are believed to be involved in cell differentiation and proliferation. Because each nitrogen of a polyamine is protonated at physiological pH (7.4), putrescine, spermidine, and spermine exist as cations with a charge of +2, +3, and +4, respectively, in body fluids. Structural studies suggest that these polyammonium ions affect the conformation of biological macromolecules by electrostatic binding to specific anionic sites—the negatively charged phosphate groups of DNA, for example.

Multiple substitution by strongly electron-withdrawing groups diminishes the basicity of arylamines still more. As just noted, aniline is 3800 times as strong a base as *p*-nitroaniline; however, it is 10^9 times more basic than 2,4-dinitroaniline. A practical consequence of this is that arylamines that bear two or more strongly electron-withdrawing groups are often not capable of being extracted from ether solution into dilute aqueous acid.

Heterocyclic Amines. Nitrogen in nonaromatic heterocyclic compounds, piperidine, for example, is similar in basicity to nitrogen in alkylamines. When nitrogen is part of an aromatic ring, however, its basicity decreases markedly. Pyridine, for example, resembles arylamines in being almost 1 million times less basic than piperidine.

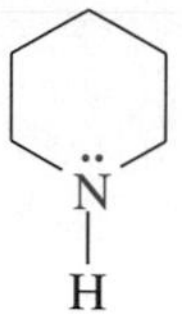

is more basic than

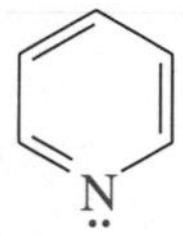

Piperidine
($K_b = 1.6 \times 10^{-3}$; $pK_b = 2.8$)

Pyridine
($K_b = 1.4 \times 10^{-9}$; $pK_b = 8.8$)

Pyridine and imidazole were two of the heterocyclic aromatic compounds described in Section 11.23.

Imidazole and its derivatives form an interesting and important class of heterocyclic aromatic amines. Imidazole is approximately 100 times more basic than pyridine. Protonation of imidazole yields an ion that is stabilized by the electron delocalization represented in the resonance structures shown:

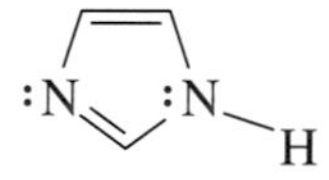

Imidazole
($K_b = 1 \times 10^{-7}$; $pK_b = 7$)

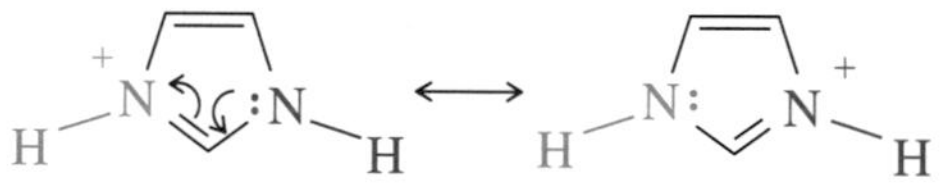

Imidazolium ion

An imidazole ring is a structural unit in the amino acid *histidine* (Section 27.1) and is involved in a large number of biological processes as a base and as a nucleophile.

22.6 TETRAALKYLAMMONIUM SALTS AS PHASE-TRANSFER CATALYSTS

In spite of being ionic compounds, many quaternary ammonium salts dissolve in nonpolar media. The four alkyl groups attached to nitrogen shield its positive charge and impart *lipophilic* character to the tetraalkylammonium ion. The following two quaternary ammonium salts, for example, are soluble in solvents of low polarity such as benzene, decane, and halogenated hydrocarbons:

$CH_3\overset{+}{N}(CH_2CH_2CH_2CH_2CH_2CH_2CH_2CH_3)_3\ Cl^-$

Methyltrioctylammonium chloride

$C_6H_5-CH_2\overset{+}{N}(CH_2CH_3)_3\ Cl^-$

Benzyltriethylammonium chloride

This property of quaternary ammonium salts is used to advantage in an experimental technique known as **phase-transfer catalysis.** Imagine that you wish to carry out the reaction

$$CH_3CH_2CH_2CH_2Br + NaCN \longrightarrow CH_3CH_2CH_2CH_2CN + NaBr$$

Butyl bromide — Sodium cyanide — Pentanenitrile — Sodium bromide

Sodium cyanide does not dissolve in butyl bromide. The two reactants contact each other only at the surface of the solid sodium cyanide, and the rate of reaction under these conditions is too slow to be of synthetic value. Dissolving the sodium cyanide in water is of little help, since butyl bromide is not soluble in water and reaction can occur only at the interface between the two phases. Adding a small amount of benzyltrimethylammonium chloride, however, causes pentanenitrile to form rapidly even at room temperature. The quaternary ammonium salt is acting as a *catalyst;* it increases the reaction rate. How does it do this?

Quaternary ammonium salts catalyze the reaction between an anion and an organic substrate by transferring the anion from the aqueous phase, where it cannot contact the substrate, to the organic phase. In the example just cited, the first step occurs in the aqueous phase and is an exchange of the anionic partner of the quaternary ammonium salt for cyanide ion:

$$C_6H_5CH_2\overset{+}{N}(CH_3)_3\ Cl^- + CN^- \overset{\text{fast}}{\rightleftharpoons} C_6H_5CH_2\overset{+}{N}(CH_3)_3\ CN^- + Cl^-$$

Benzyltrimethylammonium chloride (aqueous) | Cyanide ion (aqueous) | Benzyltrimethylammonium cyanide (aqueous) | Chloride ion (aqueous)

The benzyltrimethylammonium ion migrates to the butyl bromide phase, carrying a cyanide ion along with it.

$$C_6H_5CH_2\overset{+}{N}(CH_3)_3\ CN^- \overset{\text{fast}}{\rightleftharpoons} C_6H_5CH_2\overset{+}{N}(CH_3)_3\ CN^-$$

Benzyltrimethylammonium cyanide (aqueous) | Benzyltrimethylammonium cyanide (in butyl bromide)

Once in the organic phase, cyanide ion is only weakly solvated and is far more reactive than it is in water or ethanol, where it is strongly solvated by hydrogen bonding. Nucleophilic substitution takes place rapidly.

$$CH_3CH_2CH_2CH_2Br + C_6H_5CH_2\overset{+}{N}(CH_3)_3\ CN^- \overset{\text{fast}}{\rightleftharpoons}$$

Butyl bromide | Benzyltrimethylammonium cyanide (in butyl bromide)

$$CH_3CH_2CH_2CH_2CN + C_6H_5CH_2\overset{+}{N}(CH_3)_3\ Br^-$$

Pentanenitrile (in butyl bromide) | Benzyltrimethylammonium bromide (in butyl bromide)

The benzyltrimethylammonium bromide formed in this step returns to the aqueous phase, where it can repeat the cycle.

Phase-transfer catalysis succeeds for two reasons. First, it provides a mechanism for introducing an anion into the medium that contains the reactive substrate. More important, the anion is introduced in a weakly solvated, highly reactive state. You have already encountered phase-transfer catalysis in another form in Section 16.4, where the metal-complexing properties of crown ethers were described. Crown ethers permit metal salts to dissolve in nonpolar solvents by surrounding the cation with a lipophilic cloak, leaving the anion free to react without the encumbrance of strong solvation forces.

Phase-transfer catalysis is the subject of an article in the April 1978 issue of the *Journal of Chemical Education* (pp. 235–238). This article includes examples of a variety of reactions carried out under phase-transfer conditions.

22.7 REACTIONS THAT LEAD TO AMINES: A REVIEW AND A PREVIEW

Methods for preparing amines address either or both of the following questions:

1. How is the required carbon-nitrogen bond to be formed?
2. Given a nitrogen-containing organic compound such as an amide, a nitrile, or a nitro compound, how is the correct oxidation state of the desired amine to be achieved?

A number of reactions that lead to carbon-nitrogen bond formation have been presented in earlier chapters and are summarized in Table 22.3. Among the reactions in the table, the nucleophilic ring opening of epoxides, reaction of α-halo acids with ammonia, and the Hofmann rearrangement give amines directly and need no further elaboration. The other reactions in Table 22.3 yield products that are converted to amines by some subsequent procedure. As these procedures are described in the following sections, you will see that they are largely applications of principles that you have already learned. You will encounter some new reagents and some new uses for familiar reagents, but very little in the way of new reaction types is involved.

22.8 PREPARATION OF AMINES BY ALKYLATION OF AMMONIA

Alkylamines are, in principle, capable of being prepared by nucleophilic substitution reactions of alkyl halides with ammonia.

$$\underset{\text{Alkyl halide}}{RX} + \underset{\text{Ammonia}}{2NH_3} \longrightarrow \underset{\text{Primary amine}}{RNH_2} + \underset{\text{Ammonium halide salt}}{\overset{+}{N}H_4\,X^-}$$

While this reaction is useful for preparing α-amino acids (Table 22.3, fifth entry), it is not a general method for the synthesis of amines. Its major limitation is that the expected primary amine product is itself a nucleophile and competes with ammonia for the alkyl halide.

$$\underset{\text{Alkyl halide}}{RX} + \underset{\text{Primary amine}}{RNH_2} + \underset{\text{Ammonia}}{NH_3} \longrightarrow \underset{\text{Secondary amine}}{RNHR} + \underset{\text{Ammonium halide salt}}{\overset{+}{N}H_4\,X^-}$$

When 1-bromooctane, for example, is allowed to react with ammonia, both the primary amine and the secondary amine are isolated in comparable amounts.

$$\underset{\substack{\text{1-Bromooctane}\\ \text{(1 mol)}}}{CH_3(CH_2)_6CH_2Br} \xrightarrow{NH_3\ (2\ mol)} \underset{\substack{\text{Octylamine}\\ \text{(45\%)}}}{CH_3(CH_2)_6CH_2NH_2} + \underset{\substack{N,N\text{-Dioctylamine}\\ \text{(43\%)}}}{[CH_3(CH_2)_6CH_2]_2NH}$$

In a similar manner, competitive alkylation may continue, resulting in formation of a trialkylamine.

TABLE 22.3
Methods for Carbon-Nitrogen Bond Formation Discussed in Earlier Chapters

Reaction (section) and comments	General equation and specific example
Nucleophilic substitution by azide ion on an alkyl halide (Sections 8.1, 8.14) Azide ion is a very good nucleophile and reacts with primary and secondary alkyl halides to give alkyl azides. Phase-transfer catalysts accelerate the rate of reaction.	$:\ddot{N}^{-}{=}\overset{+}{N}{=}\ddot{N}:^{-} + R{-}X \longrightarrow :\ddot{N}^{-}{=}\overset{+}{N}{=}\ddot{N}{-}R + X^-$ Azide ion; Alkyl halide; Alkyl azide; Halide ion $CH_3CH_2CH_2CH_2CH_2Br \xrightarrow[\text{phase-transfer catalyst}]{NaN_3} CH_3CH_2CH_2CH_2CH_2N_3$ Pentyl bromide (1-bromopentane); Pentyl azide (89%) (1-azidopentane)
Nitration of arenes (Section 12.3) The standard method for introducing a nitrogen atom as a substituent on an aromatic ring is nitration with a mixture of nitric acid and sulfuric acid. The reaction proceeds by electrophilic aromatic substitution.	$ArH + HNO_3 \xrightarrow{H_2SO_4} ArNO_2 + H_2O$ Arene; Nitric acid; Nitroarene; Water $C_6H_5CH{=}O \xrightarrow[H_2SO_4]{HNO_3}$ m-$O_2NC_6H_4CH{=}O$ Benzaldehyde; *m*-Nitrobenzaldehyde (75–84%)
Nucleophilic ring opening of epoxides by ammonia (Section 16.12) The strained ring of an epoxide is opened on nucleophilic attack by ammonia and amines to give β-amino alcohols. Azide ion also reacts with epoxides; the products are β-azido alcohols.	$H_3N: + R_2C{-}CR_2$ (epoxide, O bridging) $\longrightarrow H_2\ddot{N}{-}CR_2{-}CR_2{-}OH$ Ammonia; Epoxide; β-Amino alcohol (2*R*,3*R*)-2,3-Epoxybutane $\xrightarrow[\text{water}]{NH_3}$ (2*R*,3*S*)-3-Amino-2-butanol (70%)
Nucleophilic addition of amines to aldehydes and ketones (Sections 17.11, 17.12) Primary amines undergo nucleophilic addition to the carbonyl group of aldehydes and ketones to form carbinolamines. These carbinolamines dehydrate under the conditions of their formation to give *N*-substituted imines. Secondary amines yield enamines.	$R\ddot{N}H_2 + R'\overset{O}{\overset{\Vert}{C}}R'' \longrightarrow R'\overset{:NR}{\overset{\Vert}{C}}R'' + H_2O$ Primary amine; Aldehyde or ketone; Imine; Water $CH_3NH_2 + C_6H_5\overset{O}{\overset{\Vert}{C}}H \longrightarrow C_6H_5CH{=}NCH_3$ Methylamine; Benzaldehyde; *N*-Benzylidenemethylamine (70%)

TABLE 22.3
Methods for Carbon-Nitrogen Bond Formation Discussed in Earlier Chapters *(Continued)*

Reaction (section) and comments	General equation and specific example
Nucleophilic substitution by ammonia on α-halo acids (Section 19.16) The α-halo acids obtained by halogenation of carboxylic acids under conditions of the Hell-Volhard-Zelinsky reaction are reactive substrates in nucleophilic substitution processes. A standard method for the preparation of α-amino acids is displacement of halide from α-halo acids by nucleophilic substitution using excess aqueous ammonia.	$H_3N{:} + RCH(X)CO_2H \longrightarrow RCH(\overset{+}{N}H_3)CO_2^- + NH_4X$ Ammonia (excess); α-Halo carboxylic acid; α-Amino acid; Ammonium halide $(CH_3)_2CHCH(Br)CO_2H \xrightarrow[H_2O]{NH_3} (CH_3)_2CHCH(\overset{+}{N}H_3)CO_2^-$ 2-Bromo-3-methylbutanoic acid; 2-Amino-3-methylbutanoic acid (47–48%)
Nucleophilic acyl substitution (Section 20.11) Acylation of ammonia and amines by an acyl chloride, acid anhydride, or ester is an exceptionally effective method for the formation of carbon-nitrogen bonds.	$R_2\ddot{N}H + R'C(=O)X \longrightarrow R_2\ddot{N}\overset{O}{\overset{\Vert}{C}}R' + HX$ Primary or secondary amine, or ammonia; Acyl chloride, acid anhydride, or ester; Amide 2 Pyrrolidine + $CH_3\overset{O}{\overset{\Vert}{C}}Cl \longrightarrow$ N-Acetylpyrrolidine ($\overset{O}{\overset{\Vert}{NC}}CH_3$) + Pyrrolidine hydrochloride ($\overset{+}{N}H_2$ Cl^-) Pyrrolidine; Acetyl chloride; *N*-Acetylpyrrolidine (79%); Pyrrolidine hydrochloride
The Hofmann rearrangement (Section 20.17) Amides are converted to amines by reaction with bromine in basic media. An *N*-bromo amide is an intermediate; it rearranges to an isocyanate. Hydrolysis of the isocyanate yields an amine.	$R\overset{O}{\overset{\Vert}{C}}NH_2 \xrightarrow[H_2O]{Br_2,\ HO^-} RNH_2$ Amide; Amine $(CH_3)_3C\overset{O}{\overset{\Vert}{C}}NH_2 \xrightarrow[H_2O]{Br_2,\ HO^-} (CH_3)_3CNH_2$ 2,2-Dimethylpropanamide; *tert*-Butylamine (64%)

$$RX + R_2NH + NH_3 \longrightarrow R_3N + \overset{+}{N}H_4\,X^-$$

Alkyl halide; Secondary amine; Ammonia; Tertiary amine; Ammonium halide salt

Even the tertiary amine competes with ammonia for the alkylating agent. The product is a quaternary ammonium salt.

$$RX + R_3N \longrightarrow R_4\overset{+}{N}\,X^-$$

Alkyl halide; Tertiary amine; Quaternary ammonium salt

Because alkylation of ammonia can lead to a complex mixture of products, it is used to prepare primary amines only when the starting alkyl halide is not particularly expensive and the desired amine can be easily separated from the other components of the reaction mixture.

PROBLEM 22.9 Alkylation of ammonia is sometimes employed in industrial processes; the resulting mixture of amines is separated by distillation. The ultimate starting materials for the industrial preparation of allylamine are propene, chlorine, and ammonia. Write a series of equations showing the industrial preparation of allylamine from these starting materials. (Allylamine has a number of uses, including the preparation of the diuretic drugs *meralluride* and *mercaptomerin.*)

Aryl halides do not normally react with ammonia under these conditions. The few exceptions are special cases and will be described in Section 23.5.

22.9 THE GABRIEL SYNTHESIS OF PRIMARY ALKYLAMINES

A method that achieves the same end result as that desired by alkylation of ammonia but which avoids the formation of secondary and tertiary amines as by-products is the **Gabriel synthesis.** Alkyl halides are converted to primary alkylamines without contamination by secondary or tertiary amines. The key reagent is the potassium salt of phthalimide, prepared by the reaction

The Gabriel synthesis is based on work carried out by Siegmund Gabriel at the University of Berlin in the 1880s. A detailed discussion of each step in the Gabriel synthesis of benzylamine can be found in the October 1975 *Journal of Chemical Education* (pp. 670–671).

$$\text{Phthalimide (:NH)} + KOH \longrightarrow N\text{-Potassiophthalimide } (:\overset{-}{N}:K^+) + H_2O$$

Phthalimide   *N*-Potassiophthalimide   Water

Phthalimide, with a K_a of 5×10^{-9} (pK_a 8.3), can be quantitatively converted to its potassium salt by an acid-base reaction with potassium hydroxide. The potassium salt of phthalimide has a negatively charged nitrogen atom, which acts as a nucleophile toward primary alkyl halides in a bimolecular nucleophilic substitution (S_N2) process.

DMF is an abbreviation for *N,N*-dimethylformamide, $HC(=O)N(CH_3)_2$. DMF is a polar aprotic solvent (Section 8.13) and an excellent medium for S_N2 reactions.

$$:\overset{-}{N}:K^+ + C_6H_5CH_2Cl \xrightarrow{DMF} :NCH_2C_6H_5 + KCl$$

N-Potassiophthalimide   Benzyl chloride   *N*-Benzylphthalimide (74%)   Potassium chloride

The product of this reaction is an imide (Section 20.15), a diacyl derivative of an amine. Either aqueous acid or aqueous base can be used to hydrolyze its two amide bonds and liberate the desired primary amine. A more effective method of cleaving the two amide bonds is by acyl transfer to hydrazine:

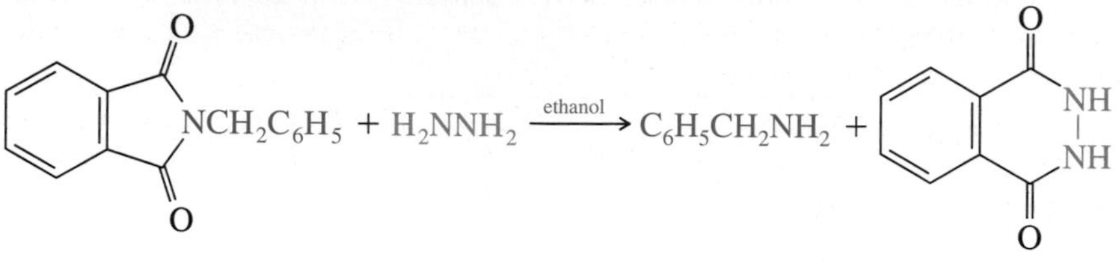

$$\text{N-Benzylphthalimide} + H_2NNH_2 \xrightarrow{\text{ethanol}} C_6H_5CH_2NH_2 + \text{Phthalhydrazide}$$

N-Benzylphthalimide (phthalimide–$NCH_2C_6H_5$) + Hydrazine (H_2NNH_2) → Benzylamine ($C_6H_5CH_2NH_2$) (97%) + Phthalhydrazide

Aryl halides cannot be converted to arylamines by the Gabriel synthesis, because they do not undergo nucleophilic substitution with *N*-potassiophthalimide in the first step of the procedure.

Among compounds other than simple alkyl halides, α-halo ketones and α-halo esters have been employed as substrates in the Gabriel synthesis. Alkyl *p*-toluenesulfonate esters have also been used. Because phthalimide can undergo only a single alkylation, the formation of secondary and tertiary amines does not occur, and the Gabriel synthesis is a valuable procedure for the laboratory preparation of primary amines.

PROBLEM 22.10 Which of the following amines can be prepared by the Gabriel synthesis? Which ones cannot? Write equations showing the successful applications of this method.

(*a*) Butylamine
(*b*) Isobutylamine
(*c*) *tert*-Butylamine
(*d*) 2-Phenylethylamine
(*e*) *N*-Methylbenzylamine
(*f*) Aniline

SAMPLE SOLUTION (*a*) The Gabriel synthesis is limited to preparation of amines of the type RCH_2NH_2, that is, primary alkylamines in which the amino group is bonded to a primary carbon. Butylamine may be prepared from butyl bromide by this method.

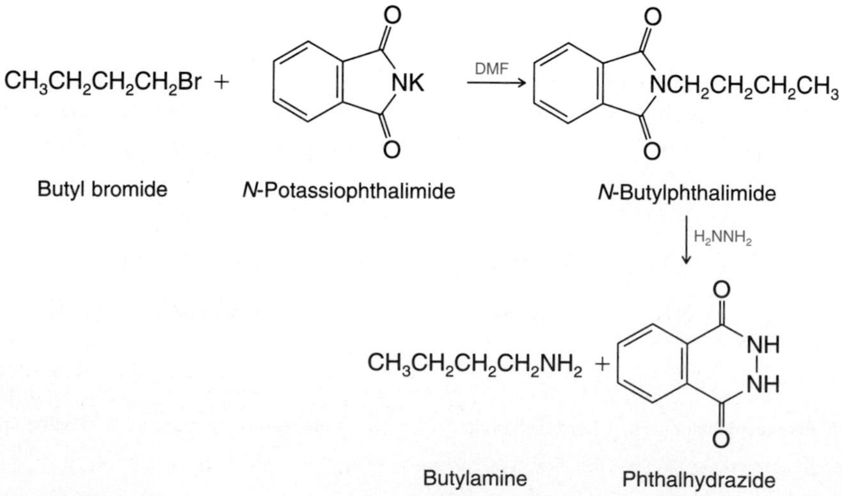

$$CH_3CH_2CH_2CH_2Br + \text{N-Potassiophthalimide (NK)} \xrightarrow{\text{DMF}} \text{N-Butylphthalimide } (NCH_2CH_2CH_2CH_3)$$

Butyl bromide + N-Potassiophthalimide → N-Butylphthalimide

$$\text{N-Butylphthalimide} \xrightarrow{H_2NNH_2} CH_3CH_2CH_2CH_2NH_2 + \text{Phthalhydrazide}$$

Butylamine + Phthalhydrazide

22.10 PREPARATION OF AMINES BY REDUCTION

Almost any nitrogen-containing organic compound can be reduced to an amine. The synthesis of amines then becomes a question of the availability of suitable precursors and the choice of an appropriate reducing agent.

Alkyl *azides,* prepared by nucleophilic substitution of alkyl halides by sodium azide, as shown in the first entry of Table 22.3, are reduced to alkylamines by a variety of reagents, including lithium aluminum hydride.

$$R{-}\ddot{N}{=}\overset{+}{N}{=}\ddot{N}{:}^{-} \xrightarrow{\text{reduce}} R\ddot{N}H_2$$

Alkyl azide → Primary amine

$$C_6H_5CH_2CH_2N_3 \xrightarrow[\text{2. } H_2O]{\text{1. LiAlH}_4\text{, diethyl ether}} C_6H_5CH_2CH_2NH_2$$

2-Phenylethyl azide → 2-Phenylethylamine (89%)

Catalytic hydrogenation is also effective:

1,2-Epoxycyclohexane $\xrightarrow[\text{dioxane-water}]{NaN_3}$ *trans*-2-Azidocyclohexanol (61%) $\xrightarrow{H_2,\ Pt}$ *trans*-2-Aminocyclohexanol (81%)

In its overall design, this procedure is similar to the Gabriel synthesis; a nitrogen nucleophile is used in a carbon-nitrogen bond–forming operation and then converted to an amino group in a subsequent transformation.

The same reduction methods may be applied to the conversion of *nitriles* to primary amines.

$$RC{\equiv}N \xrightarrow[H_2\text{, catalyst}]{\text{LiAlH}_4\text{ or}} RCH_2NH_2$$

Nitrile → Primary amine

$$F_3C{-}C_6H_4{-}CH_2CN \xrightarrow[\text{2. } H_2O]{\text{1. LiAlH}_4\text{, diethyl ether}} F_3C{-}C_6H_4{-}CH_2CH_2NH_2$$

p-(Trifluoromethyl)benzyl cyanide → 2-(*p*-Trifluoromethyl)phenylethylamine (53%)

$$CH_3CH_2CH_2CH_2CN \xrightarrow[\text{diethyl ether}]{H_2\text{ (100 atm), Ni}} CH_3CH_2CH_2CH_2CH_2NH_2$$

Pentanenitrile → 1-Pentanamine (56%)

Since nitriles can be prepared from alkyl halides by a nucleophilic substitution reaction with cyanide ion, the overall process $RX \rightarrow RC{\equiv}N \rightarrow RCH_2NH_2$ leads to primary amines that have one more carbon atom than the starting alkyl halide.

The preparation of pentanenitrile under phase-transfer conditions was described in Section 22.6.

Cyano groups in *cyanohydrins* (Section 17.8) are reduced under the same reaction conditions.

Nitro groups are readily reduced to primary amine functions by a variety of methods. Catalytic hydrogenation over platinum, palladium, or nickel is often used, as is reduction by iron or tin in hydrochloric acid. The ease with which nitro groups are reduced is especially useful in the preparation of arylamines, where the sequence $ArH \rightarrow ArNO_2 \rightarrow ArNH_2$ is the standard route to these compounds.

$CH(CH_3)_2$, NO_2 $\xrightarrow[\text{methanol}]{H_2,\ Ni}$ $CH(CH_3)_2$, NH_2

o-Isopropylnitrobenzene *o*-Isopropylaniline (92%)

For reductions carried out in acidic media, a pH adjustment with sodium hydroxide is required in the last step in order to convert $ArNH_3^+$ to $ArNH_2$.

Cl—C₆H₄—NO_2 $\xrightarrow[\text{2. NaOH}]{\text{1. Fe, HCl}}$ Cl—C₆H₄—NH_2

p-Chloronitrobenzene *p*-Chloroaniline (95%)

O_2N, O, $\overset{\|}{C}CH_3$ $\xrightarrow[\text{2. NaOH}]{\text{1. Sn, HCl}}$ H_2N, O, $\overset{\|}{C}CH_3$

m-Nitroacetophenone *m*-Aminoacetophenone (82%)

PROBLEM 22.11 Outline syntheses of each of the following arylamines from benzene:

(*a*) *o*-Isopropylaniline
(*b*) *p*-Isopropylaniline
(*c*) 4-Isopropyl-1,3-benzenediamine
(*d*) *p*-Chloroaniline
(*e*) m-Aminoacetophenone

SAMPLE SOLUTION (*a*) The last step in the synthesis of *o*-isopropylaniline, the reduction of the corresponding nitro compound by catalytic hydrogenation, is given as one of the three preceding examples. The necessary nitroarene is obtained by fractional distillation of the ortho-para mixture formed during nitration of isopropylbenzene.

$CH(CH_3)_2$ $\xrightarrow{HNO_3}$ $CH(CH_3)_2$, NO_2 + $CH(CH_3)_2$, NO_2

Isopropylbenzene *o*-Isopropylnitrobenzene (bp 110°C) *p*-Isopropylnitrobenzene (bp 131°C)

As actually performed, a 62 percent yield of a mixture of ortho and para nitration products has been obtained with an ortho-para ratio of about 1:3.

Isopropylbenzene is prepared by the Friedel-Crafts alkylation of benzene using isopropyl chloride and aluminum chloride, or propene in the presence of an acid catalyst (Section 12.6).

Reduction of an azide, a nitrile, or a nitro compound furnishes a primary amine. A method that provides access to primary, secondary, or tertiary amines is reduction of the carbonyl group of an amide by lithium aluminum hydride.

$$\underset{\text{Amide}}{\mathrm{R\overset{\overset{\displaystyle O}{\|}}{C}NR'_2}} \xrightarrow[\text{2. } H_2O]{\text{1. } LiAlH_4} \underset{\text{Amine}}{\mathrm{RCH_2NR'_2}}$$

In this general equation, R and R′ may be either alkyl or aryl groups. Primary amides yield primary amines:

$$\underset{\text{3-Phenylbutanamide}}{\mathrm{C_6H_5\underset{\underset{\displaystyle CH_3}{|}}{C}HCH_2\overset{\overset{\displaystyle O}{\|}}{C}NH_2}} \xrightarrow[\text{2. } H_2O]{\text{1. } LiAlH_4\text{, diethyl ether}} \underset{\text{3-Phenyl-1-butanamine (59\%)}}{\mathrm{C_6H_5\underset{\underset{\displaystyle CH_3}{|}}{C}HCH_2CH_2NH_2}}$$

N-Substituted amides yield secondary amines:

Acetanilide is an acceptable IUPAC synonym for *N*-phenylethanamide.

$$\underset{\text{Acetanilide}}{\mathrm{C_6H_5{-}NH\overset{\overset{\displaystyle O}{\|}}{C}CH_3}} \xrightarrow[\text{2. } H_2O]{\text{1. } LiAlH_4\text{, diethyl ether}} \underset{N\text{-Ethylaniline (92\%)}}{\mathrm{C_6H_5{-}NHCH_2CH_3}}$$

N,N-Disubstituted amides yield tertiary amines:

$$\underset{N,N\text{-Dimethylcyclohexanecarboxamide}}{\mathrm{C_6H_{11}{-}\overset{\overset{\displaystyle O}{\|}}{C}N(CH_3)_2}} \xrightarrow[\text{2. } H_2O]{\text{1. } LiAlH_4\text{, diethyl ether}} \underset{N,N\text{-Dimethyl(cyclohexylmethyl)amine (88\%)}}{\mathrm{C_6H_{11}{-}CH_2N(CH_3)_2}}$$

Because amides are so easy to prepare, this is a versatile method for the preparation of amines.

The preparation of amines by the methods described in this section involves the prior synthesis and isolation of some reducible material that has a carbon-nitrogen bond—an azide, a nitrile, a nitro-substituted arene, or an amide. The following section describes a method that combines the two steps of carbon-nitrogen bond formation and reduction into a single operation. Like the reduction of amides, it offers the possibility of preparing primary, secondary, or tertiary amines by proper choice of starting materials.

22.11 REDUCTIVE AMINATION

A class of nitrogen-containing compounds that was omitted from the section just discussed includes *imines* and their derivatives. Imines are formed by the reaction of aldehydes and ketones with ammonia. Imines can be reduced to primary amines by catalytic hydrogenation.

$$\underset{\text{Aldehyde or ketone}}{RC(=O)R'} + \underset{\text{Ammonia}}{NH_3} \longrightarrow \underset{\text{Imine}}{RC(=NH)R'} \xrightarrow[\text{catalyst}]{H_2} \underset{\text{Primary amine}}{RCH(NH_2)R'}$$

The overall reaction converts a carbonyl compound to an amine by carbon-nitrogen bond formation and reduction; it is commonly known as **reductive amination.** What makes it a particularly valuable synthetic procedure is that it can be carried out in a single operation by hydrogenation of a solution containing both ammonia and the carbonyl compound along with a hydrogenation catalyst. The intermediate imine is not isolated but undergoes reduction under the conditions of its formation. Also, the reaction is broader in scope than implied by the preceding equation. All classes of amines—primary, secondary, and tertiary—may be prepared by reductive amination.

When primary amines are desired, the reaction is carried out as just described:

$$\underset{\text{Cyclohexanone}}{C_6H_{10}{=}O} + \underset{\text{Ammonia}}{NH_3} \xrightarrow[\text{ethanol}]{H_2,\ Ni} \underset{\text{Cyclohexylamine (80\%)}}{C_6H_{11}NH_2} \quad \text{via} \quad C_6H_{10}{=}NH$$

Secondary amines are prepared by hydrogenation of a carbonyl compound in the presence of a primary amine. An N-substituted imine, or *Schiff's base,* is an intermediate:

$$\underset{\text{Heptanal}}{CH_3(CH_2)_5CH(=O)} + \underset{\text{Aniline}}{H_2N{-}C_6H_5} \xrightarrow[\text{ethanol}]{H_2,\ Ni} \underset{N\text{-Heptylaniline (65\%)}}{CH_3(CH_2)_5CH_2NH{-}C_6H_5}$$

$$\text{via} \quad CH_3(CH_2)_5CH{=}N{-}C_6H_5$$

Reductive amination has been successfully applied to the preparation of tertiary amines from carbonyl compounds and secondary amines even though a neutral imine is not possible in this case.

$$\underset{\text{Butanal}}{CH_3CH_2CH_2CH(=O)} + \underset{\text{Piperidine}}{C_5H_{10}NH} \xrightarrow[\text{ethanol}]{H_2,\ Ni} \underset{N\text{-Butylpiperidine (93\%)}}{CH_3CH_2CH_2CH_2{-}NC_5H_{10}}$$

Presumably, the species that undergoes reduction here is a carbinolamine or an iminium ion derived from it.

$$CH_3CH_2CH_2\underset{}{\overset{OH}{|}}{CH}-N(\text{piperidine}) \rightleftharpoons CH_3CH_2CH_2CH{=}\overset{+}{N}(\text{piperidine}) + HO^-$$

Carbinolamine — Iminium ion

PROBLEM 22.12 Show how you could prepare each of the following amines from benzaldehyde by reductive amination:

(*a*) Benzylamine
(*b*) Dibenzylamine
(*c*) *N,N*-Dimethylbenzylamine
(*d*) *N*-Benzylpiperidine

SAMPLE SOLUTION (*a*) Since benzylamine is a primary amine, it is derived from ammonia and benzaldehyde.

$$C_6H_5\overset{O}{\overset{\|}{C}}H + NH_3 + H_2 \xrightarrow{Ni} C_6H_5CH_2NH_2 + H_2O$$

Benzaldehyde, Ammonia, Hydrogen, Benzylamine (89%), Water

The reaction proceeds by initial formation of the imine $C_6H_5CH{=}NH$, followed by its hydrogenation.

A variation of the classical reductive amination procedure uses sodium cyanoborohydride ($NaBH_3CN$) instead of hydrogen as the reducing agent and is better suited to amine syntheses in which only a few grams of material are needed. All that is required is to add sodium cyanoborohydride to an alcohol solution of the carbonyl compound and an amine.

$$C_6H_5\overset{O}{\overset{\|}{C}}H + CH_3CH_2NH_2 \xrightarrow[\text{methanol}]{NaBH_3CN} C_6H_5CH_2NHCH_2CH_3$$

Benzaldehyde, Ethylamine, *N*-Ethylbenzylamine (91%)

22.12 REACTIONS OF AMINES: A REVIEW AND A PREVIEW

The noteworthy properties of amines are their *basicity* and their *nucleophilicity*. The basicity of amines has been discussed in Section 22.5. Several reactions in which amines act as nucleophiles have already been encountered in earlier chapters. These are summarized in Table 22.4.

Both the basicity and the nucleophilicity of amines originate in the unshared electron pair of nitrogen. When an amine acts as a base, this electron pair abstracts a proton from a Brønsted acid. When an amine undergoes the reactions summarized in Table 22.4, the first step in each case is the attack of the unshared electron pair on the positively polarized carbon of a carbonyl group.

TABLE 22.4
Reactions of Amines Discussed in Previous Chapters*

Reaction (section) and comments	General equation and specific example
Reaction of primary amines with aldehydes and ketones (Section 17.11) Imines are formed by nucleophilic addition of a primary amine to the carbonyl group of an aldehyde or a ketone. The key step is formation of a carbinolamine intermediate, which then dehydrates to the imine.	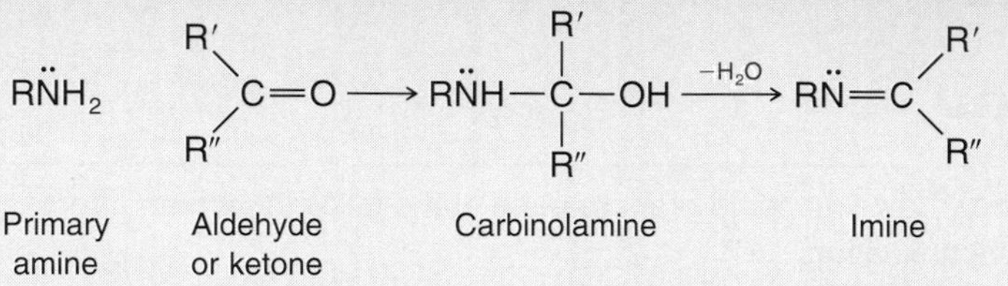 $R\ddot{N}H_2 + R'R''C{=}O \longrightarrow R\ddot{N}H{-}C(R')(R''){-}OH \xrightarrow{-H_2O} R\ddot{N}{=}C(R')R''$ Primary amine; Aldehyde or ketone; Carbinolamine; Imine $CH_3NH_2 + C_6H_5CH{=}O \longrightarrow C_6H_5CH{=}NCH_3 + H_2O$ Methylamine; Benzaldehyde; *N*-Benzylidenemethylamine (70%); Water
Reaction of secondary amines with aldehydes and ketones (Section 17.12) Enamines are formed in the corresponding reaction of secondary amines with aldehydes and ketones.	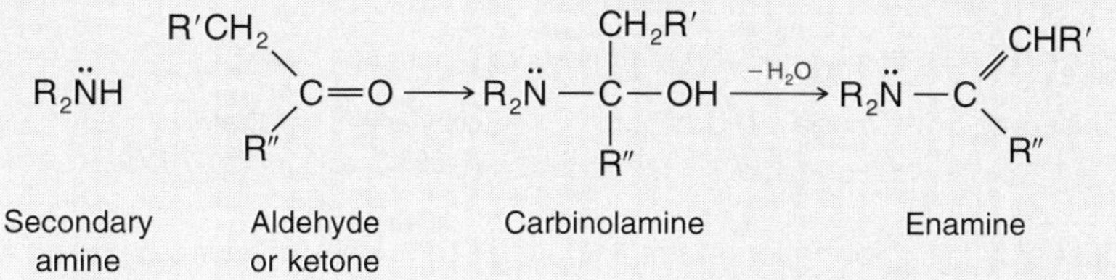 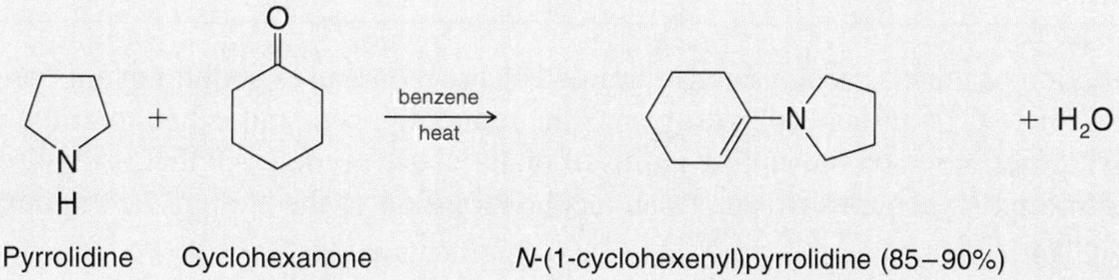 $R_2\ddot{N}H + R'CH_2(R'')C{=}O \longrightarrow R_2\ddot{N}{-}C(CH_2R')(R''){-}OH \xrightarrow{-H_2O} R_2\ddot{N}{-}C(R''){=}CHR'$ Secondary amine; Aldehyde or ketone; Carbinolamine; Enamine Pyrrolidine + Cyclohexanone $\xrightarrow{\text{benzene, heat}}$ *N*-(1-cyclohexenyl)pyrrolidine (85–90%) + H_2O
Reaction of amines with acyl chlorides (Section 20.3) Amines are converted to amides on reaction with acyl chlorides. Other acylating agents, such as carboxylic acid anhydrides and esters, may also be used but are less reactive.	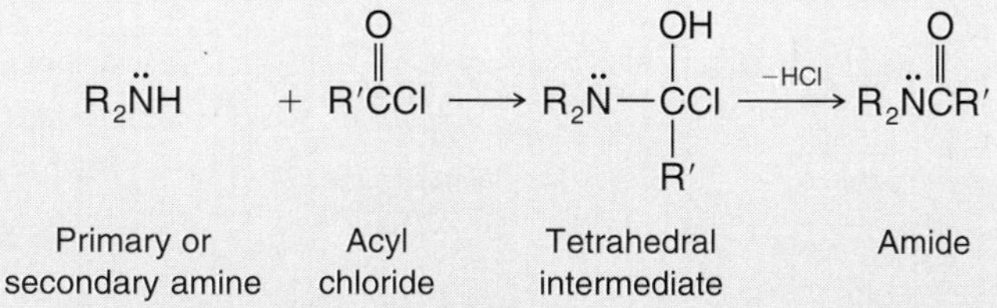 $R_2\ddot{N}H + R'\overset{O}{\overset{\parallel}{C}}Cl \longrightarrow R_2\ddot{N}{-}C(OH)(R')Cl \xrightarrow{-HCl} R_2\ddot{N}\overset{O}{\overset{\parallel}{C}}R'$ Primary or secondary amine; Acyl chloride; Tetrahedral intermediate; Amide $CH_3CH_2CH_2CH_2NH_2 + CH_3CH_2CH_2CH_2\overset{O}{\overset{\parallel}{C}}Cl \longrightarrow CH_3CH_2CH_2CH_2\overset{O}{\overset{\parallel}{C}}NHCH_2CH_2CH_2CH_3$ Butylamine; Pentanoyl chloride; *N*-Butylpentanamide (81%)

* Both alkylamines and arylamines undergo these reactions.

Amine acting as a base

Amine acting as a nucleophile

In addition to being more basic than arylamines, alkylamines are also more nucleophilic. All the reactions in Table 22.4 take place faster with alkylamines than with arylamines.

The sections that follow introduce some additional reactions of amines. In all cases our understanding of how these reactions take place starts with a consideration of the role of the unshared electron pair of nitrogen.

We will begin with an examination of the reactivity of amines as nucleophiles in S_N2 reactions.

22.13 REACTION OF AMINES WITH ALKYL HALIDES

Nucleophilic substitution results when primary alkyl halides are treated with amines.

$$\underset{\text{Primary amine}}{R\ddot{N}H_2} + \underset{\text{Primary alkyl halide}}{R'CH_2X} \longrightarrow \underset{\text{Ammonium halide salt}}{R\overset{+}{N}H_2-CH_2R'\ X^-} \longrightarrow \underset{\text{Secondary amine}}{R\ddot{N}H-CH_2R'} + \underset{\text{Hydrogen halide}}{HX}$$

$$\underset{\text{Aniline (4 mol)}}{C_6H_5NH_2} + \underset{\text{Benzyl chloride (1 mol)}}{C_6H_5CH_2Cl} \xrightarrow[90°C]{NaHCO_3} \underset{N\text{-Benzylaniline (85–87\%)}}{C_6H_5NHCH_2C_6H_5}$$

A second alkylation may follow, converting the secondary amine to a tertiary amine. Alkylation need not stop there; the tertiary amine may itself be alkylated, giving a quaternary ammonium salt.

$$\underset{\text{Primary amine}}{R\ddot{N}H_2} \xrightarrow{R'CH_2X} \underset{\text{Secondary amine}}{R\ddot{N}HCH_2R'} \xrightarrow{R'CH_2X} \underset{\text{Tertiary amine}}{R\ddot{N}(CH_2R')_2} \xrightarrow{R'CH_2X} \underset{\text{Quaternary ammonium salt}}{R\overset{+}{N}(CH_2R')_3\ X^-}$$

Because of its high reactivity toward nucleophilic substitution, methyl iodide is the alkyl halide most frequently encountered in amine alkylations designed to proceed to the quaternary ammonium salt stage.

$$\underset{\text{(Cyclohexylmethyl)amine}}{C_6H_{11}-CH_2NH_2} + \underset{\text{Methyl iodide}}{3CH_3I} \xrightarrow[heat]{methanol} \underset{\text{(Cyclohexylmethyl)trimethylammonium iodide (99\%)}}{C_6H_{11}-CH_2\overset{+}{N}(CH_3)_3\ I^-}$$

Quaternary ammonium salts, as we have seen, are useful in synthetic organic chemistry as phase-transfer catalysts. In another, more direct application, quaternary ammonium *hydroxides* are used as substrates in an elimination reaction to form alkenes.

22.14 THE HOFMANN ELIMINATION

The halide anion of quaternary ammonium iodides may be replaced by hydroxide by treatment with an aqueous slurry of silver oxide. Silver iodide precipitates, and a solution of the quaternary ammonium hydroxide is formed.

$$2(R_4\overset{+}{N}\ I^-) + Ag_2O + H_2O \longrightarrow 2(R_4\overset{+}{N}\ \overset{-}{O}H) + 2AgI$$

Quaternary ammonium iodide — Silver oxide — Water — Quaternary ammonium hydroxide — Silver iodide

$$C_6H_{11}{-}CH_2\overset{+}{N}(CH_3)_3\ I^- \xrightarrow[H_2O,\ CH_3OH]{Ag_2O} C_6H_{11}{-}CH_2\overset{+}{N}(CH_3)_3\ HO^-$$

(Cyclohexylmethyl)trimethylammonium iodide — (Cyclohexylmethyl)trimethylammonium hydroxide

When quaternary ammonium hydroxides are heated, they undergo β-elimination to form an alkene and an amine.

$$C_6H_{10}(H){-}CH_2{-}\overset{+}{N}(CH_3)_3\ \ {:}\ddot{\underset{..}{O}}H^- \xrightarrow{160°C} C_6H_{10}{=}CH_2 + (CH_3)_3N{:} + H_2O$$

(Cyclohexylmethyl)trimethylammonium hydroxide — Methylenecyclohexane (69%) — Trimethylamine — Water

This reaction is known as the **Hofmann elimination;** it was developed by August W. Hofmann in the mid-nineteenth century and is both a synthetic method to prepare alkenes and an analytical tool for structure determination.

A novel aspect of the Hofmann elimination is its regioselectivity. Elimination in alkyltrimethylammonium hydroxides proceeds in the direction that gives the less substituted alkene.

$$CH_3\underset{^+N(CH_3)_3}{\underset{|}{C}H}CH_2CH_3\ HO^- \xrightarrow[-H_2O \\ -(CH_3)_3N]{heat} CH_2{=}CHCH_2CH_3 + CH_3CH{=}CHCH_3$$

sec-Butyltrimethylammonium hydroxide — 1-Butene (95%) — 2-Butene (5%) (cis and trans)

It is the less sterically hindered β hydrogen that is removed by the base in Hofmann elimination reactions. Methyl groups are deprotonated in preference to methylene groups, and methylene groups are deprotonated in preference to methines. The regioselectivity of Hofmann elimination is opposite to that predicted by the Zaitsev rule (Section 5.10). Base-promoted elimination reactions of alkyltrimethylammonium salts are said to obey the **Hofmann rule;** they yield the less substituted alkene.

PROBLEM 22.13 Give the structure of the alkene formed in major amounts when the hydroxide of each of the following quaternary ammonium ions is heated.

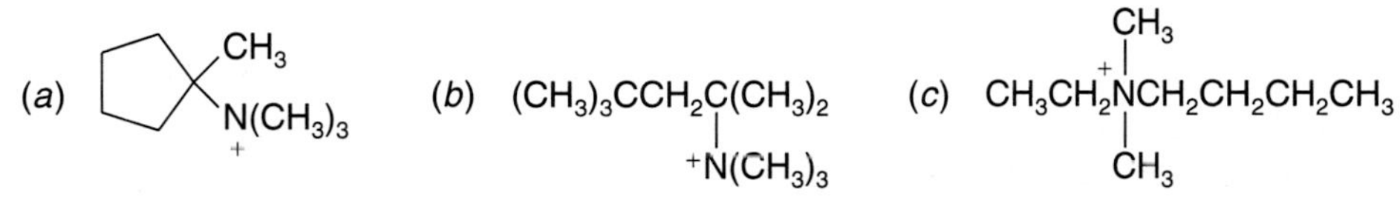

(*a*) 1-methylcyclopentyl with CH_3 and $\overset{+}{N}(CH_3)_3$

(*b*) $(CH_3)_3CCH_2C(CH_3)_2$ with $^{+}N(CH_3)_3$

(*c*) $CH_3CH_2\overset{+}{N}(CH_3)_2CH_2CH_2CH_2CH_3$

SAMPLE SOLUTION (*a*) Two alkenes are capable of being formed by β elimination, methylenecyclopentane and 1-methylcyclopentene.

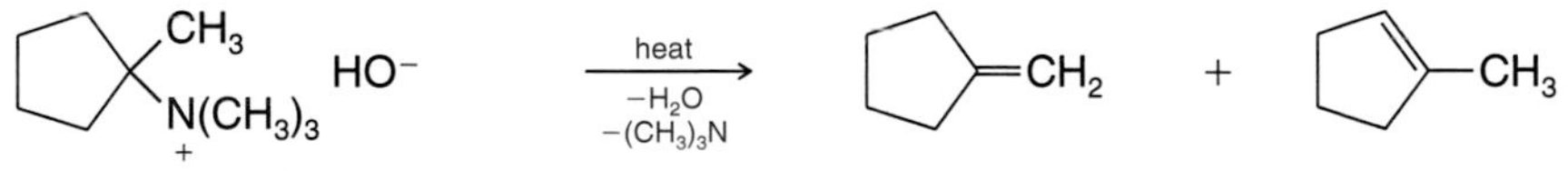

$$\text{(1-Methylcyclopentyl)trimethylammonium hydroxide} \xrightarrow[-H_2O,\ -(CH_3)_3N]{heat} \text{Methylenecyclopentane} + \text{1-Methylcyclopentene}$$

(1-Methylcyclopentyl)trimethylammonium hydroxide — Methylenecyclopentane — 1-Methylcyclopentene

Methylenecyclopentane has the less substituted double bond and is the major product. The reported isomer distribution is 91 percent methylenecyclopentane and 9 percent 1 mothylcyclopentene.

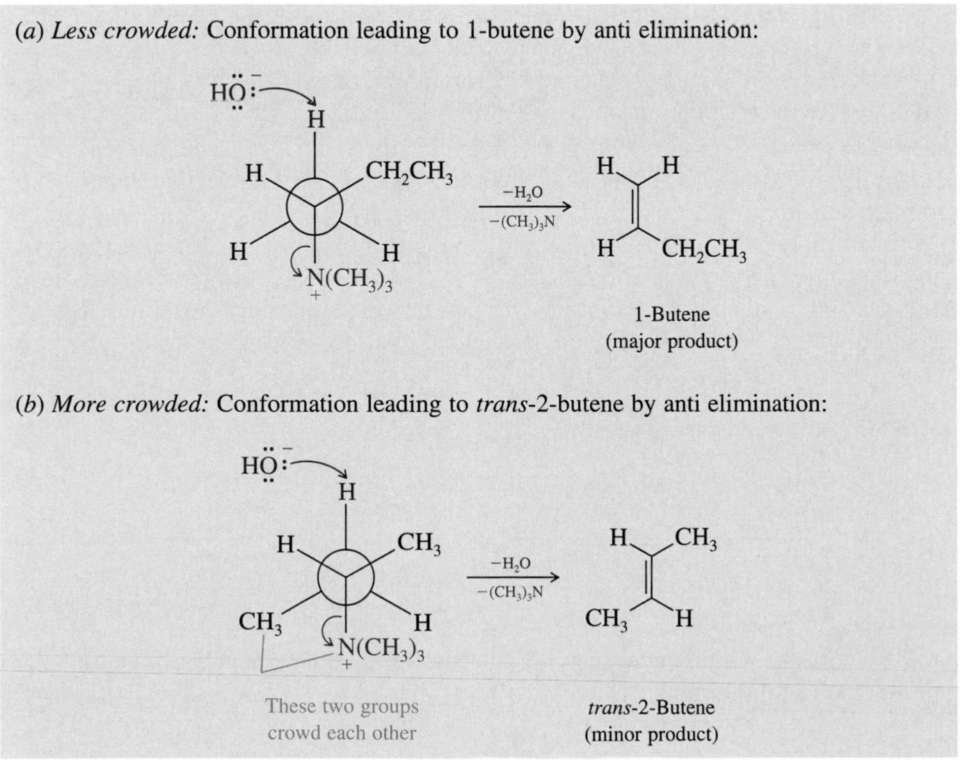

(*a*) *Less crowded:* Conformation leading to 1-butene by anti elimination:

HO:⁻ → H; Newman projection with H, CH_2CH_3, H, H, $\overset{+}{N}(CH_3)_3$ $\xrightarrow[-(CH_3)_3N]{-H_2O}$ $H_2C{=}CHCH_2CH_3$

1-Butene
(major product)

(*b*) *More crowded:* Conformation leading to *trans*-2-butene by anti elimination:

HO:⁻ → H; Newman projection with H, CH_3, CH_3, H, $\overset{+}{N}(CH_3)_3$ $\xrightarrow[-(CH_3)_3N]{-H_2O}$ $CH_3CH{=}CHCH_3$

These two groups crowd each other

trans-2-Butene
(minor product)

FIGURE 22.5 Newman projections showing the conformations leading to (*a*) 1-butene and (*b*) *trans*-2-butene by Hofmann elimination of *sec*-butyltrimethylammonium hydroxide. The major product is 1-butene.

We can understand the regioselectivity of the Hofmann elimination by comparing steric effects in the E2 transition states for formation of 1-butene and *trans*-2-butene from *sec*-butyltrimethylammonium hydroxide. In terms of its size, $(CH_3)_3\overset{+}{N}—$ (trimethylammonio) is comparable to $(CH_3)_3C—$ (*tert*-butyl). As Figure 22.5 illustrates, the E2 transition state requires an anti relationship between the proton that is removed and the trimethylammonio group. No serious van der Waals repulsions are evident in the transition-state geometry for formation of 1-butene. The conformation leading to *trans*-2-butene, however, is destabilized by van der Waals strain between the trimethylammonio group and a methyl group gauche to it. Thus, the activation energy for formation of *trans*-2-butene exceeds that of 1-butene, which becomes the major product because it is formed faster.

With a regioselectivity opposite to that of the Zaitsev rule, the Hofmann elimination is sometimes used in synthesis to prepare alkenes not accessible by dehydrohalogenation of alkyl halides. This application has decreased in importance since the Wittig reaction (Section 17.14) became established as a synthetic method beginning in the 1950s. Similarly, most of the analytical applications of Hofmann elimination have been replaced by spectroscopic methods.

22.15 ELECTROPHILIC AROMATIC SUBSTITUTION IN ARYLAMINES

Arylamines contain two functional groups, the amine group and the aromatic ring; they are **difunctional compounds.** The reactivity of the amine group is affected by its aryl substituent, and the reactivity of the ring is affected by its amine substituent. The same electron delocalization that reduces the basicity and the nucleophilicity of an arylamine nitrogen increases the electron density in the aromatic ring and makes arylamines extremely reactive toward electrophilic aromatic substitution.

The reactivity of arylamines was noted in Section 12.12, where it was pointed out that $—\ddot{N}H_2$, $—\ddot{N}HR$, and $—\ddot{N}R_2$ are ortho, para–directing and exceedingly powerful activating groups. These substituents are such powerful activators that electrophilic aromatic substitution is only rarely performed directly on arylamines.

Direct nitration of aniline and other arylamines, for example, is difficult to carry out and is accompanied by oxidation that leads to the formation of dark-colored "tars." As a solution to this problem it is standard practice to first protect the amino group by acylation with either acetyl chloride or acetic anhydride.

$$\underset{\text{Arylamine}}{ArNH_2} \xrightarrow[\text{or } CH_3\overset{O}{\overset{\|}{C}}O\overset{O}{\overset{\|}{C}}CH_3]{CH_3\overset{O}{\overset{\|}{C}}Cl} \underset{N\text{-Acetylarylamine}}{ArNH\overset{O}{\overset{\|}{C}}CH_3}$$

Amide resonance within the *N*-acetyl group competes with delocalization of the nitrogen lone pair into the ring.

Amide resonance in acetanilide

Protecting the amino group of an arylamine in this way moderates its reactivity and permits nitration of the ring to be achieved. The acetamido group is activating toward electrophilic aromatic substitution and is ortho, para–directing.

$$p\text{-Isopropylaniline} \xrightarrow[\text{(protection step)}]{CH_3COCCH_3} p\text{-Isopropylacetanilide (98\%)} \xrightarrow[\text{(nitration step)}]{HNO_3,\ 20°C} \text{4-Isopropyl-2-nitroacetanilide (94\%)}$$

Another feature of the *N*-acetyl protecting group is that after it has served its purpose, it may be removed by hydrolysis, liberating a free amino group.

$$\underset{N\text{-Acetylarylamine}}{ArNHCCH_3} \xrightarrow[\text{or 1. } H_3O^+ \text{ 2. } HO^-]{H_2O,\ HO^-} \underset{\text{Arylamine}}{ArNH_2}$$

$$\text{4-Isopropyl-2-nitroacetanilide} \xrightarrow[\text{heat (``deprotection'' step)}]{\text{KOH, ethanol}} \text{4-Isopropyl-2-nitroaniline (100\%)}$$

The net effect of the sequence *protect-nitrate-deprotect* is the same as if the substrate had been nitrated directly. Because direct nitration is impossible, however, the indirect route is the only practical method.

PROBLEM 22.14 Outline syntheses of each of the following from aniline and any necessary organic or inorganic reagents:

(*a*) *p*-Nitroaniline (*b*) 2,4-Dinitroaniline (*c*) p-Aminoacetanilide

SAMPLE SOLUTION (*a*) It has already been stated that direct nitration of aniline is not a practical reaction. The amino group must first be protected as its *N*-acetyl derivative.

$$\text{Aniline} \xrightarrow{CH_3COCCH_3} \text{Acetanilide} \xrightarrow[H_2SO_4]{HNO_3} o\text{-Nitroacetanilide} + p\text{-Nitroacetanilide}$$

Nitration of acetanilide yields a mixture of ortho and para substitution products. The para isomer is separated, then subjected to hydrolysis to give *p*-nitroaniline.

$$p\text{-Nitroacetanilide} \xrightarrow[\text{or 1. } H_3O^+ \text{ 2. } HO^-]{H_2O,\ HO^-} p\text{-Nitroaniline}$$

Unprotected arylamines are so reactive toward halogenation that it is difficult to limit the reaction to monosubstitution. Generally, halogenation proceeds rapidly to replace all the available hydrogens that are ortho or para to the amino group.

$$p\text{-Aminobenzoic acid} \xrightarrow[\text{acetic acid}]{Br_2} \text{4-Amino-3,5-dibromobenzoic acid (82\%)}$$

Decreasing the electron-donating ability of an amino group by acylation makes it possible to limit halogenation to monosubstitution.

$$\text{2-Methylacetanilide} \xrightarrow[\text{acetic acid}]{Cl_2} \text{4-Chloro-2-methylacetanilide (74\%)}$$

Friedel-Crafts reactions are normally not successful when attempted on an arylamine, but can be carried out readily once the amino group is protected.

$$\text{2-Ethylacetanilide} + CH_3\overset{O}{\overset{\|}{C}}Cl \xrightarrow{AlCl_3} \text{4-Acetamido-3-ethylacetophenone (57\%)}$$

2-Ethylacetanilide 4-Acetamido-3-ethylacetophenone (57%)

22.16 NITROSATION OF ALKYLAMINES

When solutions of sodium nitrite ($NaNO_2$) are acidified, a number of species are formed that act as **nitrosating agents.** That is, they react as sources of nitrosyl cation, $:\overset{+}{N}{=}\ddot{O}:$. In order to simplify discussion, organic chemists group all these species together and speak of the chemistry of one of them, *nitrous acid,* as a generalized precursor to nitrosyl cation.

Nitrosyl cation is also called *nitrosonium* ion. It can be represented by the two resonance structures

$$:\overset{+}{N}{=}\ddot{O}: \longleftrightarrow :N{\equiv}\overset{+}{O}:$$

$$:\ddot{O}^{-}{-}\ddot{N}{=}\ddot{O}: \xrightarrow{H^+} H{-}\ddot{O}{-}\ddot{N}{=}\ddot{O}: \xrightarrow{H^+} H{-}\overset{+}{O}(H){-}\ddot{N}{=}\ddot{O}: \xrightarrow{-H_2O} :\overset{+}{N}{=}\ddot{O}:$$

Nitrite ion (from sodium nitrite) Nitrous acid Nitrosyl cation

Nitrosation of amines is best illustrated by examining what happens when a secondary amine "reacts with nitrous acid." The amine acts as a nucleophile, attacking the nitrogen of nitrosyl cation.

$$R_2\ddot{N}H + :\overset{+}{N}{=}\ddot{O}: \longrightarrow R_2\overset{+}{N}(H){-}\ddot{N}{=}\ddot{O}: \xrightarrow{-H^+} R_2\ddot{N}{-}\ddot{N}{=}\ddot{O}:$$

Secondary alkylamine Nitrosyl cation *N*-Nitroso amine

The intermediate that is formed in the first step loses a proton to give an *N*-nitroso amine as the isolated product.

$$(CH_3)_2\ddot{N}H \xrightarrow[H_2O]{NaNO_2,\ HCl} (CH_3)_2\ddot{N}{-}\ddot{N}{=}\ddot{O}:$$

Dimethylamine *N*-Nitrosodimethylamine (88–90%)

PROBLEM 22.15 *N*-Nitroso amines are stabilized by electron delocalization. Write the two most stable resonance forms of *N*-nitrosodimethylamine, $(CH_3)_2NNO$.

N-Nitroso amines are more often called *nitrosamines,* and because many of them are potent carcinogens, they have been the object of much recent investigation. We encounter nitrosamines in the environment on a daily basis. A few of these, all of which are known carcinogens, are:

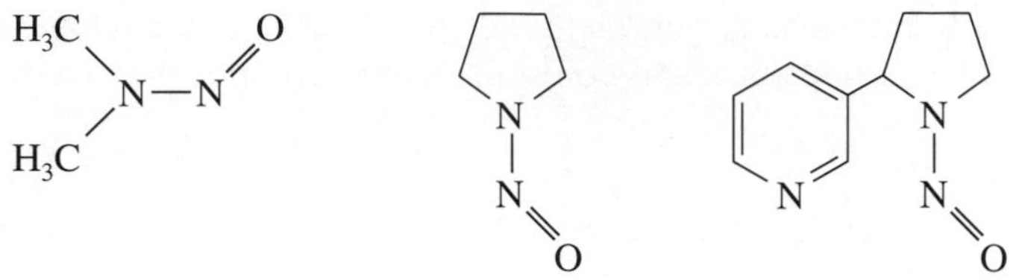

$$\underset{\text{(formed during tanning of leather; also found in beer and herbicides)}}{\underset{N\text{-Nitrosodimethylamine}}{(H_3C)_2N-N=O}}$$

N-Nitrosodimethylamine (formed during tanning of leather; also found in beer and herbicides)

N-Nitrosopyrrolidine (formed when bacon that has been cured with sodium nitrite is fried)

N-Nitrosonornicotine (present in tobacco smoke)

The July 1977 issue of the *Journal of Chemical Education* contains an article entitled "Formation of Nitrosamines in Food and in the Digestive System."

Nitrosamines are formed whenever nitrosating agents come in contact with secondary amines. Indeed, more nitrosamines are probably synthesized within our body than enter it by environmental contamination. Enzyme-catalyzed reduction of nitrate (NO_3^-) produces nitrite (NO_2^-), which combines with amines present in the body to form *N*-nitroso amines.

When primary amines are nitrosated, the *N*-nitroso compounds produced can react further.

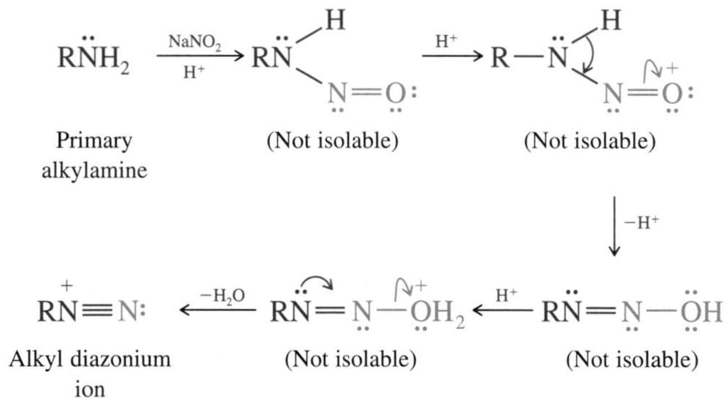

$$R\ddot{N}H_2 \xrightarrow[H^+]{NaNO_2} R\ddot{N}(H)-\ddot{N}=\ddot{O}: \xrightarrow{H^+} R-\ddot{N}(H)-\ddot{N}=\overset{+}{\ddot{O}}H$$

Primary alkylamine (Not isolable) (Not isolable)

$$\downarrow -H^+$$

$$R\overset{+}{N}\equiv N: \xleftarrow{-H_2O} R\ddot{N}=\ddot{N}-\overset{+}{O}H_2 \xleftarrow{H^+} R\ddot{N}=\ddot{N}-\ddot{O}H$$

Alkyl diazonium ion (Not isolable) (Not isolable)

The product of this sequence of steps is an alkyl diazonium ion. The amine, in being converted to a diazonium ion, is said to have been **diazotized.** Alkyl diazonium ions are not very stable, decomposing rapidly under the conditions of their formation. Molecular nitrogen is a leaving group par excellence, and the reaction products arise by solvolysis of the diazonium ion. Usually, a carbocation intermediate is involved.

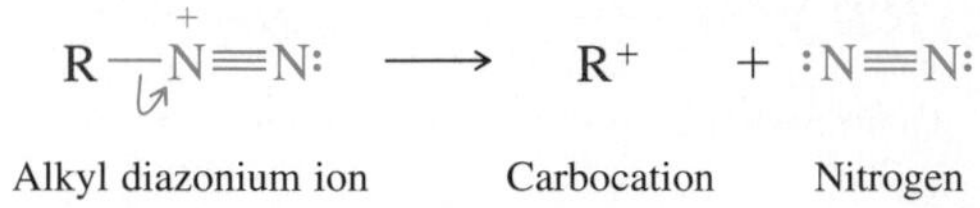

Recall from Section 8.15 that decreasing basicity is associated with increasing leaving-group ability. Molecular nitrogen is an exceedingly weak base and an excellent leaving group.

$$R-\overset{+}{N}\equiv N: \longrightarrow R^+ + :N\equiv N:$$

Alkyl diazonium ion Carbocation Nitrogen

Figure 22.6 depicts the sequence of events that occur when a typical primary alkylamine reacts with nitrous acid.

Since nitrogen-free products result from the formation and decomposition of diazonium ions, these reactions are often referred to as **deamination reactions.** Alkyl diazonium ions are rarely used in synthetic work but have been studied extensively to probe the behavior of carbocations generated under conditions in which the leaving group is lost rapidly and irreversibly.

$$\underset{\text{1,1-Dimethylpropylamine}}{CH_3CH_2C(CH_3)_2NH_2} \xrightarrow{HONO} \underset{\text{1,1-Dimethylpropyl diazonium ion}}{CH_3CH_2C(CH_3)_2\overset{+}{N}\equiv N:} \longrightarrow \underset{\text{1,1-Dimethylpropyl cation}}{CH_3CH_2\overset{+}{C}(CH_3)_2} + \underset{\text{Nitrogen}}{:N\equiv N:}$$

$-H^+$ / H_2O

$$\underset{\substack{\text{2-Methyl-2-butene}\\(2\%)}}{CH_3CH{=}C(CH_3)_2} + \underset{\substack{\text{2-Methyl-1-butene}\\(3\%)}}{CH_3CH_2C(CH_3){=}CH_2} + \underset{\substack{\text{2-Methyl-2-butanol}\\(80\%)}}{CH_3CH_2C(CH_3)_2OH}$$

FIGURE 22.6 The diazonium ion generated by treatment of a primary alkylamine with nitrous acid loses nitrogen to give a carbocation. The isolated products are derived from the carbocation and include, in this example, alkenes (by loss of a proton) and an alcohol (nucleophilic capture by water).

PROBLEM 22.16 Nitrous acid deamination of 2,2-dimethylpropylamine, $(CH_3)_3CCH_2NH_2$, gives the same products as were indicated as being formed from 1,1-dimethylpropylamine in Figure 22.6. Suggest a mechanistic explanation for the formation of these compounds from 2,2-dimethylpropylamine.

Aryl diazonium ions, prepared by nitrous acid diazotization of primary arylamines, are substantially more stable than alkyl diazonium ions and are of enormous synthetic value. Their use in the synthesis of substituted aromatic compounds is described in the following two sections.

The nitrosation of tertiary alkylamines is rather complicated, and no generally useful chemistry is associated with reactions of this type.

22.17 NITROSATION OF ARYLAMINES

We learned in the preceding section that different reactions are observed when the various classes of alkylamines—primary, secondary, and tertiary—react with nitrosating agents. While no useful chemistry attends the nitrosation of tertiary alkylamines, electrophilic aromatic substitution by nitrosyl cation ($:\overset{+}{N}\equiv O:$) takes place with *N,N*-dialkylarylamines.

$$\underset{\textit{N,N}\text{-Diethylaniline}}{C_6H_5N(CH_2CH_3)_2} \xrightarrow[\text{2. HO}^-]{\text{1. NaNO}_2\text{, HCl, H}_2\text{O, 8°C}} \underset{\textit{N,N}\text{-Diethyl-}\textit{p}\text{-nitrosoaniline (95\%)}}{p\text{-ON}{-}C_6H_4N(CH_2CH_3)_2}$$

Nitrosyl cation is a relatively weak electrophile and attacks only very strongly activated aromatic rings.

N-Alkylarylamines resemble secondary alkylamines in that they form *N*-nitroso compounds on reaction with nitrous acid.

$$C_6H_5NHCH_3 \xrightarrow[H_2O,\ 10°C]{NaNO_2,\ HCl} C_6H_5N(CH_3)-N=O$$

N-Methylaniline → *N*-Methyl-*N*-nitrosoaniline (87–93%)

Primary arylamines, like primary alkylamines, form diazonium ion salts on nitrosation. Aryl diazonium ions are considerably more stable than their alkyl counterparts. Whereas alkyl diazonium ions decompose under the conditions of their formation, aryl diazonium salts are stable enough to be stored in aqueous solution at 0 to 5°C for reasonable periods of time. Loss of nitrogen from an aryl diazonium ion generates an unstable aryl cation and is much slower than loss of nitrogen from an alkyl diazonium ion.

$$C_6H_5NH_2 \xrightarrow[H_2O,\ 0-5°C]{NaNO_2,\ HCl} C_6H_5\overset{+}{N}\equiv N:\ Cl^-$$

Aniline → Benzenediazonium chloride

$$(CH_3)_2CH-C_6H_4-NH_2 \xrightarrow[H_2O,\ 0-5°C]{NaNO_2,\ H_2SO_4} (CH_3)_2CH-C_6H_4-\overset{+}{N}\equiv N:\ HSO_4^-$$

p-Isopropylaniline → *p*-Isopropylbenzenediazonium hydrogen sulfate

Aryl diazonium ions undergo a variety of reactions that make them versatile intermediates for the preparation of a host of ring-substituted aromatic compounds. In these reactions, summarized in Figure 22.7 and discussed individually in the following

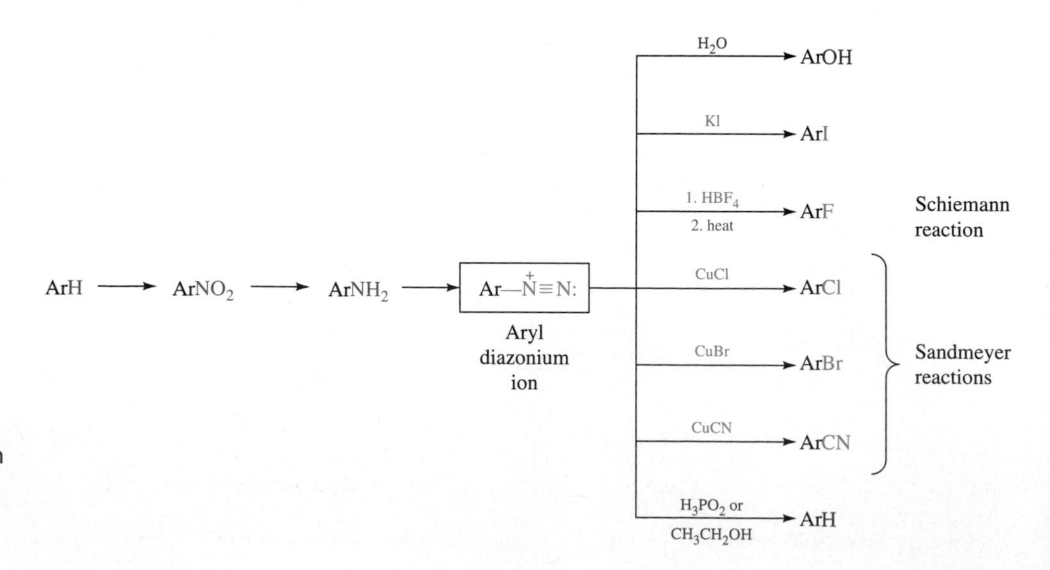

FIGURE 22.7 Flowchart showing the synthetic origin of aryl diazonium ions and their most useful transformations.

section, molecular nitrogen acts as a leaving group and is replaced by another atom or group. All the reactions are regiospecific; the entering group becomes bonded to precisely the ring position from which nitrogen departs.

22.18 SYNTHETIC TRANSFORMATIONS OF ARYL DIAZONIUM SALTS

An important reaction of aryl diazonium ions is their conversion to *phenols* by hydrolysis:

$$\underset{\text{Aryl diazonium ion}}{\mathrm{Ar\overset{+}{N}{\equiv}N:}} + \underset{\text{Water}}{\mathrm{H_2O}} \longrightarrow \underset{\text{A phenol}}{\mathrm{ArOH}} + \mathrm{H^+} + \underset{\text{Nitrogen}}{\mathrm{:N{\equiv}N:}}$$

This is the most general method for the preparation of phenols. It is easily performed; the aqueous acidic solution in which the diazonium salt is prepared is heated and gives the phenol directly. An aryl cation is probably generated, which is then captured by water acting as a nucleophile.

$$\underset{p\text{-Isopropylaniline}}{\mathrm{(CH_3)_2CH{-}C_6H_4{-}NH_2}} \xrightarrow[\text{2. } \mathrm{H_2O}\text{, heat}]{\text{1. } \mathrm{NaNO_2, H_2SO_4, H_2O}} \underset{p\text{-Isopropylphenol (73\%)}}{\mathrm{(CH_3)_2CH{-}C_6H_4{-}OH}}$$

Sulfuric acid is normally used instead of hydrochloric acid in the diazotization step so as to minimize the competition with water for capture of the cationic intermediate. Hydrogen sulfate anion (HSO_4^-) is less nucleophilic than chloride.

PROBLEM 22.17 Design a synthesis of *m*-bromophenol from benzene.

The reaction of an aryl diazonium salt with potassium iodide is the standard method for the preparation of *aryl iodides*. The diazonium salt is prepared from a primary aromatic amine in the usual way, a solution of potassium iodide is then added, and the reaction mixture is brought to room temperature or heated to accelerate the reaction.

$$\underset{\text{Aryl diazonium ion}}{\mathrm{Ar{-}\overset{+}{N}{\equiv}N:}} + \underset{\text{Iodide ion}}{\mathrm{I^-}} \longrightarrow \underset{\text{Aryl iodide}}{\mathrm{ArI}} + \underset{\text{Nitrogen}}{\mathrm{:N{\equiv}N:}}$$

$$\underset{o\text{-Bromoaniline}}{o\text{-}\mathrm{BrC_6H_4NH_2}} \xrightarrow[\text{KI, room temperature}]{\mathrm{NaNO_2, HCl, H_2O}, 0\text{–}5^\circ\mathrm{C}} \underset{o\text{-Bromoiodobenzene (72–83\%)}}{o\text{-}\mathrm{BrC_6H_4I}}$$

PROBLEM 22.18 Show by a series of equations how you could prepare *m*-bromoiodobenzene from benzene.

Diazonium salt chemistry provides the principal synthetic method for the preparation of *aryl fluorides* through a process known as the **Schiemann reaction.** In this procedure the aryl diazonium ion is isolated as its fluoroborate salt, which then yields the desired aryl fluoride on being heated.

$$\underset{\text{Aryl diazonium fluoroborate}}{Ar-\overset{+}{N}\equiv N: \quad \overset{-}{BF_4}} \xrightarrow{\text{heat}} \underset{\text{Aryl fluoride}}{ArF} + \underset{\text{Boron trifluoride}}{BF_3} + \underset{\text{Nitrogen}}{:N\equiv N:}$$

A standard way to form the aryl diazonium fluoroborate salt is to add fluoroboric acid (HBF_4) or a fluoroborate salt to the diazotization medium.

$$\underset{m\text{-Aminophenyl ethyl ketone}}{m\text{-}H_2NC_6H_4CCH_2CH_3(=O)} \xrightarrow[\text{3. heat}]{\text{1. NaNO}_2\text{, H}_2\text{O, HCl; 2. HBF}_4} \underset{\text{Ethyl } m\text{-fluorophenyl ketone (68\%)}}{m\text{-}FC_6H_4CCH_2CH_3(=O)}$$

PROBLEM 22.19 Show the proper sequence of synthetic transformations in the conversion of benzene to ethyl *m*-fluorophenyl ketone.

While *aryl chlorides* and *aryl bromides* are capable of being prepared by electrophilic aromatic substitution, it is often necessary to prepare these compounds from an aromatic amine. The amine is converted to the corresponding diazonium salt and then treated with copper(I) chloride or copper(I) bromide as appropriate.

$$\underset{\text{Aryl diazonium ion}}{Ar-\overset{+}{N}\equiv N:} \xrightarrow{\text{CuX}} \underset{\text{Aryl chloride or bromide}}{ArX} + \underset{\text{Nitrogen}}{:N\equiv N:}$$

$$\underset{m\text{-Nitroaniline}}{m\text{-}H_2NC_6H_4NO_2} \xrightarrow[\text{2. CuCl, heat}]{\text{1. NaNO}_2\text{, HCl, H}_2\text{O, 0–5°C}} \underset{m\text{-Chloronitrobenzene (68–71\%)}}{m\text{-}ClC_6H_4NO_2}$$

$$\underset{o\text{-Chloroaniline}}{o\text{-}ClC_6H_4NH_2} \xrightarrow[\text{2. CuBr, heat}]{\text{1. NaNO}_2\text{, HBr, H}_2\text{O, 0–10°C}} \underset{o\text{-Bromochlorobenzene (89–95\%)}}{o\text{-}ClC_6H_4Br}$$

Reactions that employ copper(I) salts as reagents for replacement of nitrogen in diazonium salts are called **Sandmeyer reactions.** The Sandmeyer reaction using copper(I) cyanide is a good method for the preparation of aromatic *nitriles:*

$$Ar-\overset{+}{N}\equiv N: \xrightarrow{CuCN} ArCN + :N\equiv N:$$

Aryl diazonium ion — Aryl nitrile — Nitrogen

$$o\text{-Toluidine} \xrightarrow[\text{2. CuCN, heat}]{\text{1. NaNO}_2\text{, HCl, H}_2\text{O, 0°C}} o\text{-Methylbenzonitrile (64–70\%)}$$

CH_3, NH_2 — CH_3, CN

o-Toluidine — *o*-Methylbenzonitrile (64–70%)

Since cyano groups may be hydrolyzed to carboxylic acids (Section 20.19), the Sandmeyer preparation of aryl nitriles is a key step in the conversion of arylamines to substituted benzoic acids. In the example just cited, the *o*-methylbenzonitrile that was formed was subsequently subjected to acid-catalyzed hydrolysis and gave *o*-methylbenzoic acid in 80 to 89 percent yield.

The preparation of aryl chlorides, bromides, and cyanides by the Sandmeyer reaction is mechanistically complicated and may involve arylcopper intermediates.

It is possible to replace amino substituents on an aromatic nucleus by hydrogen by reducing a diazonium salt with hypophosphorous acid (H_3PO_2) or with ethanol. These reductions are free-radical reactions in which ethanol or hypophosphorous acid acts as a hydrogen atom donor:

$$Ar-\overset{+}{N}\equiv N: \xrightarrow[CH_3CH_2OH]{H_3PO_2 \text{ or}} ArH + :N\equiv N:$$

Aryl diazonium ion — Arene — Nitrogen

Reactions of this type are known as **reductive deaminations.**

$$o\text{-Toluidine} \xrightarrow[H_3PO_2]{NaNO_2,\ H_2SO_4,\ H_2O} \text{Toluene (70–75\%)}$$

CH_3, NH_2 — CH_3

o-Toluidine — Toluene (70–75%)

$$\text{4-Isopropyl-2-nitroaniline} \xrightarrow[CH_3CH_2OH]{NaNO_2,\ HCl,\ H_2O} m\text{-Isopropylnitrobenzene (59\%)}$$

$CH(CH_3)_2$, NO_2, NH_2 — $CH(CH_3)_2$, NO_2

4-Isopropyl-2-nitroaniline — *m*-Isopropylnitrobenzene (59%)

Sodium borohydride has also been used to reduce aryl diazonium salts in reductive deamination reactions.

PROBLEM 22.20 Cumene (isopropylbenzene) is a relatively inexpensive commercially available starting material. Show how you could prepare *m*-isopropylnitrobenzene from cumene.

The value of diazonium salts in synthetic organic chemistry rests on two main points. Through the use of diazonium salt chemistry:

1. Substituents that are otherwise difficultly accessible, such as fluoro, iodo, cyano, and hydroxyl, may be introduced onto a benzene ring.
2. Compounds that have substitution patterns not directly available by electrophilic aromatic substitution can be prepared.

The first of these two features is readily apparent and is illustrated by Problems 22.17 to 22.19. If you have not done these problems yet, you are strongly encouraged to attempt them now.

The second point is somewhat less obvious but is readily illustrated by the synthesis of 1,3,5-tribromobenzene. This particular substitution pattern cannot be obtained by direct bromination of benzene, because bromine is an ortho, para director. Instead, advantage is taken of the powerful activating and ortho, para–directing effects of the amino group in aniline. Bromination of aniline yields 2,4,6-tribromoaniline in quantitative yield. Diazotization of the resulting 2,4,6-tribromoaniline and reduction of the diazonium salt gives the desired 1,3,5-tribromobenzene.

NH_2 Br Br Br NH_2

$\xrightarrow[H_2O]{Br_2}$ $\xrightarrow[CH_3CH_2OH]{NaNO_2,\ H_2SO_4,\ H_2O}$

Aniline 2,4,6-Tribromoaniline (100%) 1,3,5-Tribromobenzene (74–77%)

In order to exploit the versatility inherent in the transformations available to aryl diazonium salts, be prepared to reason backward. When you see a fluorine substituent in a synthetic target, for example, realize that it probably will have to be introduced by a Schiemann reaction of an arylamine; realize that the required arylamine is derived from a nitroarene, and that the nitro group is introduced by nitration. Be aware that an unsubstituted position of an aromatic ring need not have always been that way. It might once have borne an amino group that was used to control the orientation of electrophilic aromatic substitution reactions before being removed by reductive deamination. The strategy of synthesis is intellectually demanding, and a considerable sharpening of your reasoning power can be gained by attacking the synthesis problems at the end of each chapter. Remember, plan your sequence of accessible intermediates by reasoning backward from the target; then fill in the details on how each transformation is to be carried out.

22.19 AZO COUPLING

A reaction of aryl diazonium salts that does not involve loss of nitrogen takes place when they react with phenols and arylamines. Aryl diazonium ions are relatively weak

FROM DYES TO SULFA DRUGS

The medicine cabinet was virtually bare of antibacterial agents until **sulfa drugs** burst upon the scene in the 1930s. Before sulfa drugs became available, bacterial infection might transform a small cut or puncture wound to a life-threatening event. The story of how sulfa drugs were developed is an interesting example of being right for the wrong reasons. It was known that many bacteria absorbed dyes, and staining was a standard method for making bacteria more visible under the microscope. Might there not be some dye that is both absorbed by bacteria and toxic to them? Acting on this hypothesis, scientists at the German dyestuff manufacturer I. G. Farbenindustrie undertook a program to test the thousands of compounds in their collection for their antibacterial properties.

In general, *in vitro* testing of drugs precedes *in vivo* testing. The two terms mean, respectively, "in glass" and "in life." In vitro testing of antibiotics is carried out using bacterial cultures in test tubes or petri dishes. Drugs that are found to be active in vitro progress to the stage of in vivo testing. In vivo testing is carried out in living organisms: laboratory animals or human volunteers. The I. G. Farben scientists found that some dyes did possess antibacterial properties, both in vitro and in vivo. Others were active in vitro but were converted to inactive substances in vivo and therefore of no use as drugs. Unexpectedly, an azo dye called *Prontosil* was inactive in vitro but active in vivo. In 1932 a member of the I. G. Farben research group, Gerhard Domagk, used Prontosil to treat a 10-month-old boy suffering from a serious, potentially fatal staphylococcal infection and observed a dramatic recovery. Domagk was awarded the Nobel Prize in medicine or physiology in 1939.

In spite of the rationale upon which the testing of dyestuffs as antibiotics rested, subsequent research revealed that the antibacterial properties of Prontosil had nothing at all to do with its being a dye! In the body, Prontosil undergoes a reductive cleavage of its azo linkage to form *sulfanilamide,* which is the substance actually responsible for the observed biological activity. This is why Prontosil is active in vivo, but not in vitro.

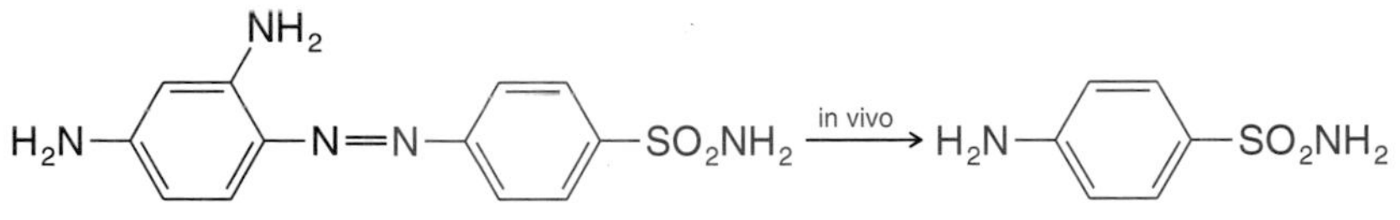

$$H_2N-C_6H_3(NH_2)-N{=}N-C_6H_4-SO_2NH_2 \xrightarrow{\text{in vivo}} H_2N-C_6H_4-SO_2NH_2$$

Prontosil Sulfanilamide

Bacteria require *p*-aminobenzoic acid in order to biosynthesize *folic acid,* a growth factor. Structurally, sulfanilamide resembles *p*-aminobenzoic acid and is mistaken for it by the bacteria. Folic acid biosynthesis is inhibited and bacterial growth is slowed sufficiently to allow the body's natural defenses to effect a cure. Since animals do not biosynthesize folic acid but obtain it in their food, sulfanilamide halts the growth of bacteria without harm to the host.

Identification of the mechanism by which Prontosil combats bacterial infections was an early triumph of **pharmacology,** a branch of science at the interface of physiology and biochemistry that studies the mechanism of drug action. By recognizing that sulfanilamide was the active agent, the task of preparing structurally modified analogs with potentially superior properties was considerably simplified. Instead of preparing Prontosil analogs, chemists synthesized sulfanilamide analogs. They did this with a vengeance; over 5000 compounds related to sulfanilamide were prepared during the period 1935–1946. Two of the most widely used sulfa drugs are *sulfathiazole* and *sulfadiazine.*

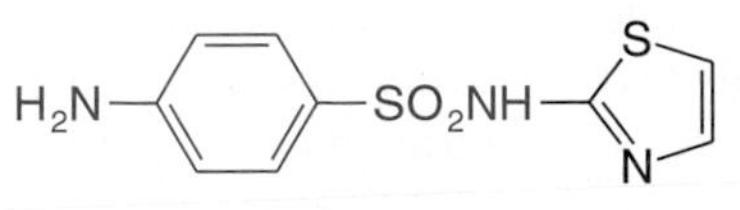

$$H_2N-C_6H_4-SO_2NH-(2\text{-thiazolyl}) \qquad H_2N-C_6H_4-SO_2NH-(2\text{-pyrimidinyl})$$

Sulfathiazole 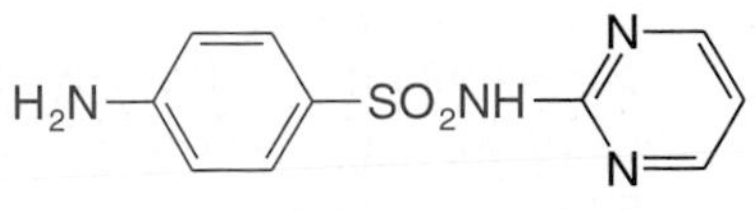 Sulfadiazine

We tend to take the efficacy of modern drugs for granted. One comparison with the not-too-distant past might put this view into better perspective. Once sulfa drugs were introduced in the United States, the number of pneumonia deaths alone decreased by an estimated 25,000 per year. Most present-day applications of sulfa drugs, however, are in the area of veterinary medicine. Although sulfa drugs were widely used in human medicine during the 1940s, they have been superseded by even more effective antibiotics such as the penicillins and the tetracyclines.

electrophiles but have sufficient reactivity to attack strongly activated aromatic rings. The reaction is known as *azo coupling;* two aryl groups are joined together by an azo (—N=N—) function.

ERG — H + $:\overset{+}{N}{\equiv}NAr$ ⟶ ERG — (H)(N=NAr) $\xrightarrow{-H^+}$ ERG — N=NAr

(ERG is a powerful electron-releasing group such as —OH or —NR_2)

Aryl diazonium ion

Intermediate in electrophilic aromatic substitution

Azo compound

Azo compounds are often highly colored, and many of them are used as dyes.

1-Naphthol + $C_6H_5\overset{+}{N}{\equiv}N:\ Cl^-$ ⟶ $N{=}NC_6H_5$ (OH)

1-Naphthol

Benzenediazonium chloride

2-(Phenylazo)-1-naphthol

A number of pH indicators, methyl red, for example, are azo compounds.

The colors of azo compounds vary with the nature of the aryl group, with its substituents, and with pH. Substituents also affect the water-solubility of azo dyes and how well they bind to a particular fabric. Countless combinations of diazonium salts and aromatic substrates have been examined with a view toward obtaining azo dyes suitable for a particular application.

22.20 SPECTROSCOPIC ANALYSIS OF AMINES

The absorptions of interest in the infrared spectra of amines are those associated with N—H vibrations. Primary alkyl- and arylamines exhibit two peaks in the range 3000 to 3500 cm^{-1}, which are due to symmetric and antisymmetric N—H stretching modes.

Symmetric N—H stretching mode of a primary amine

R—N(H)(H)   R—N(H)(H)

Antisymmetric N—H stretching mode of a primary amine

These two vibrations are clearly visible at 3270 and 3380 cm^{-1} in the infrared spectrum of butylamine, shown in Figure 22.8*a*. Secondary amines such as diethylamine, shown in Figure 22.8*b*, exhibit only one peak, which is due to N—H stretching, at 3280 cm^{-1}. Tertiary amines, of course, are transparent in this region, since they have no N—H bonds.

Characteristics of the nuclear magnetic resonance spectra of amines may be illustrated by comparing 4-methylbenzylamine (Figure 22.9*a*) with 4-methylbenzyl alcohol (Figure 22.9*b*). Nitrogen is less electronegative than oxygen and so shields neighboring nuclei to a greater extent. The benzylic methylene group attached to nitrogen in 4-methylbenzylamine appears at higher field (δ 3.7 ppm) than the benzylic methylene of 4-methylbenzyl alcohol (δ 4.6 ppm). The N—H protons are somewhat more shielded than the O—H protons of an alcohol. In 4-methylbenzylamine the protons of the amino group correspond to the signal at δ 1.4 ppm, while the hydroxyl proton signal of 4-methylbenzyl alcohol is found at δ 1.9 ppm. The chemical shifts of amino group protons, like those of hydroxyl protons, are variable and are sensitive to solvent, concentration, and temperature.

Similarly, carbons that are bonded to nitrogen are more shielded than those bonded to oxygen, as revealed by comparing the ^{13}C chemical shifts of methylamine and methanol.

26.9 ppm CH_3NH_2 — Methylamine

48.0 ppm CH_3OH — Methanol

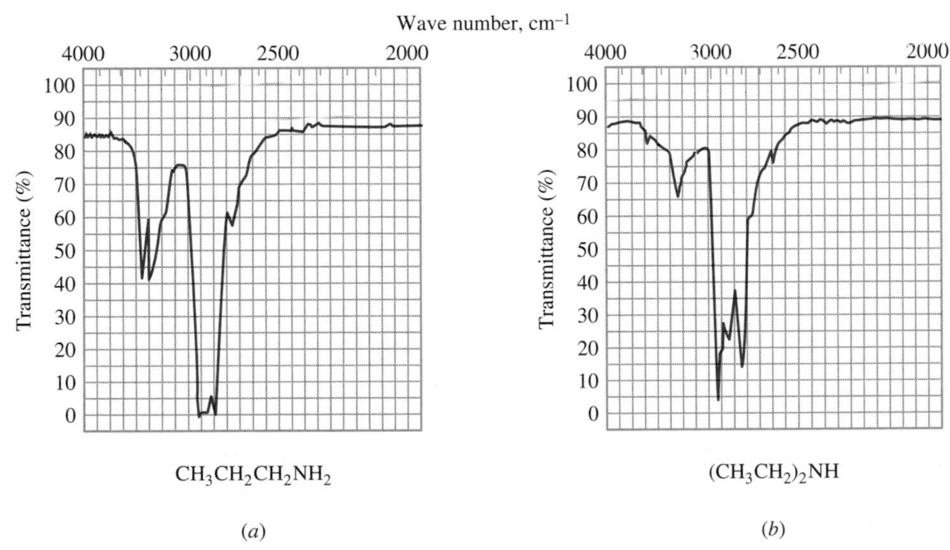

FIGURE 22.8 Portions of the infrared spectrum of (*a*) butylamine and (*b*) diethylamine. Primary amines exhibit two peaks due to N—H stretching in the 3270 to 3380 cm^{-1} region, while secondary amines show only one.

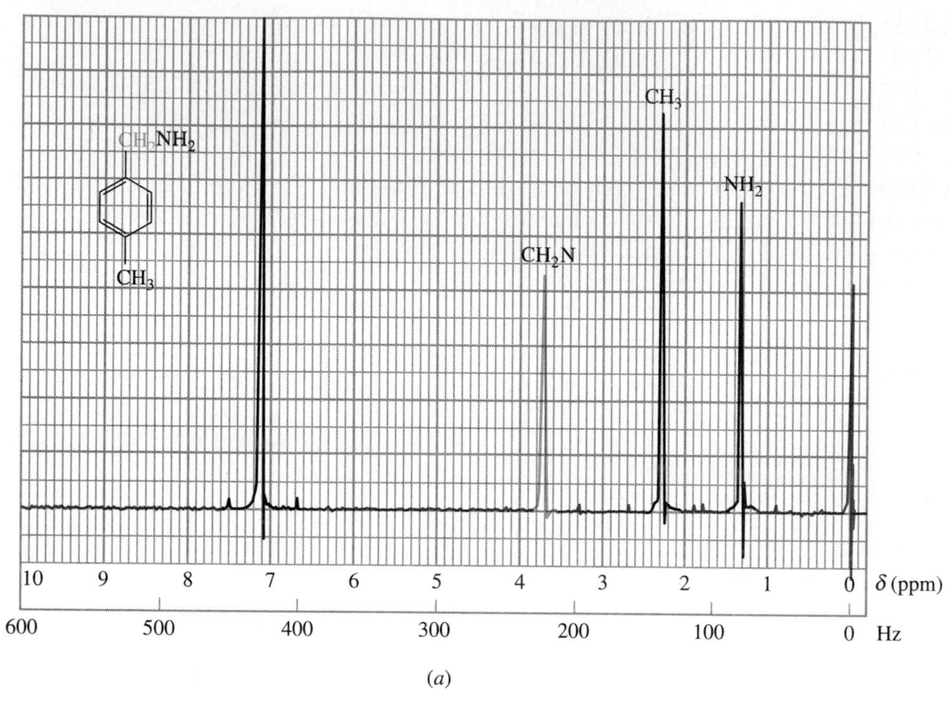

(*a*)

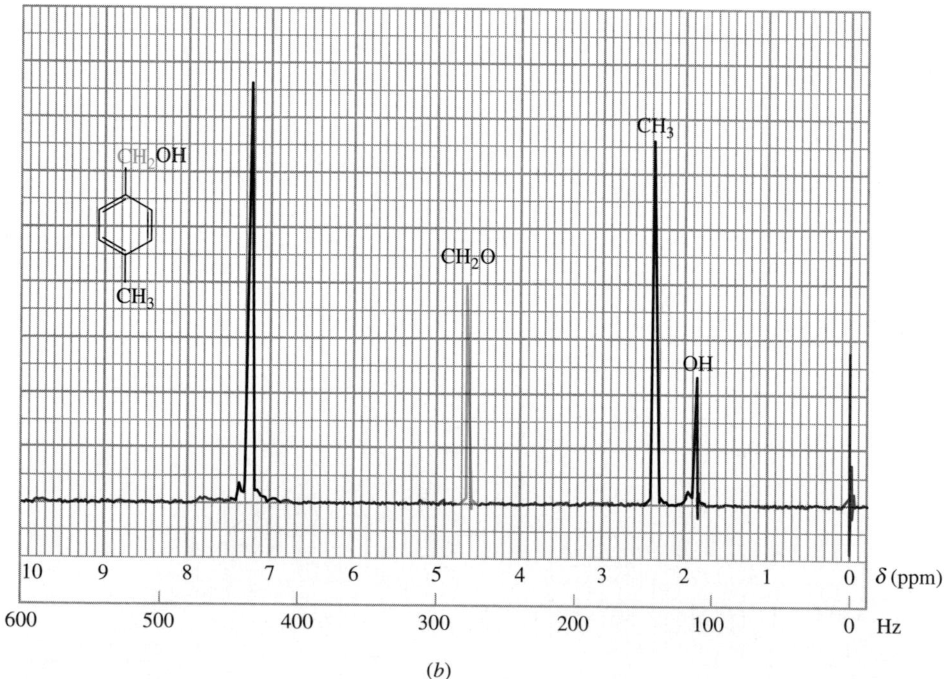

(*b*)

FIGURE 22.9 The 60-MHz ^{1}H nmr spectra of (*a*) 4-methylbenzylamine and (*b*) 4-methylbenzyl alcohol. The singlet corresponding to CH_2N in (*a*) is more shielded than that of CH_2O in (*b*).

22.21 MASS SPECTROMETRY OF AMINES

A number of features combine to make amines easily identifiable by mass spectrometry.

First, the peak for the molecular ion M^+ for all compounds that contain only carbon, hydrogen, and oxygen has an *m/z* value that is an even number. The presence of a nitrogen atom in the molecule requires that the *m/z* value for the molecular ion be odd. An odd number of nitrogens corresponds to an odd value of the molecular weight; an even number of nitrogens corresponds to an even molecular weight.

Second, nitrogen is exceptionally good at stabilizing adjacent carbocation sites. The fragmentation pattern seen in the mass spectra of amines is dominated by cleavage of groups from the carbon atom attached to the nitrogen, as the data for the following pair of constitutionally isomeric amines illustrate:

$$(CH_3)_2\ddot{N}CH_2CH_2CH_2CH_3 \xrightarrow{e^-} (CH_3)_2\overset{+\cdot}{N}-CH_2-CH_2CH_2CH_3 \longrightarrow (CH_3)_2\overset{+}{N}=CH_2 + \cdot CH_2CH_2CH_3$$

N,N-Dimethyl-1-butanamine; M^+ (*m/z* 101); (*m/z* 58) (most intense peak)

$$CH_3\ddot{N}HCH_2CH_2CH(CH_3)_2 \xrightarrow{e^-} CH_3\overset{+\cdot}{N}H-CH_2-CH_2CH(CH_3)_2 \longrightarrow CH_3\overset{+}{N}H=CH_2 + \cdot CH_2CH(CH_3)_2$$

N,3-Dimethyl-1-butanamine; M^+ (*m/z* 101); (*m/z* 44) (most intense peak)

22.22 SUMMARY

Alkylamines are compounds of the type shown, where R, R′, and R″ are alkyl groups. One or more of these groups is an aryl group in arylamines.

RNH_2	$RR'NH$	$RR'R''N$
Primary amine	Secondary amine	Tertiary amine

Alkylamines are named in two ways (Section 22.1). An older method appends the ending *-amine* to the name of the alkyl group. A newer method applies the principles of substitutive nomenclature by replacing the *-e* ending of an alkane name by *-amine* and uses appropriate locants to identify the position of the amino group. Arylamines are named as derivatives of aniline.

The unshared electron pair on nitrogen is of paramount importance in understanding the properties of amines (Section 22.2). Alkylamines have a pyramidal arrangement of bonds to nitrogen, and the unshared electron pair resides in an sp^3 hybridized orbital. The geometry at nitrogen in arylamines is somewhat flatter than in alkylamines, and the unshared electron pair is delocalized into the π system of the ring. Delocalization binds the electron pair more strongly in arylamines than in alkylamines. Arylamines are less basic and less nucleophilic than alkylamines.

Basicity of amines is expressed either as a basicity constant K_b (pK_b) of the amine or as a dissociation constant K_a (pK_a) of its conjugate acid (Section 22.4).

$$R_3N\!: + H_2O \rightleftharpoons R_3\overset{+}{N}H + HO^- \qquad K_b = \frac{[R_3\overset{+}{N}H][HO^-]}{[R_3N]}$$

The basicity constants of alkylamines lie in the range 10^{-3} to 10^{-5}. Arylamines are much weaker bases, with K_b values in the range 10^{-9} to 10^{-11} (Section 22.5).

Quaternary ammonium salts, compounds of the type $R_4N^+ X^-$, find application in a technique called **phase-transfer catalysis.** A small amount of a quaternary ammonium salt promotes the transfer of an anion from aqueous solution, where it is highly solvated, to an organic solvent, where it is much less solvated and much more reactive (Section 22.6).

Methods for the preparation of amines are summarized in Table 22.5.

The reactions of amines are summarized in Table 22.6. A particularly important amine reaction is the formation of aryl diazonium salts from primary arylamines. These diazonium salts are valuable synthetic intermediates in reactions summarized in Table 22.7.

PROBLEMS

22.21 Write structural formulas for all the amines of molecular formula $C_4H_{11}N$. Give an acceptable name for each one and classify it as a primary, secondary, or tertiary amine.

22.22 Provide a structural formula for each of the following compounds:

(*a*) 2-Ethyl-1-butanamine
(*b*) *N*-Ethyl-1-butanamine
(*c*) Dibenzylamine
(*d*) Tribenzylamine
(*e*) Tetraethylammonium hydroxide
(*f*) *N*-Allylcyclohexylamine
(*g*) *N*-Allylpiperidine
(*h*) Benzyl 2-aminopropanoate
(*i*) 4-(*N,N*-Dimethylamino)cyclohexanone
(*j*) 2,2-Dimethyl-1,3-propanediamine

22.23 Many naturally occurring nitrogen compounds and many nitrogen-containing drugs are better known by common names than by their systematic names. A few of these follow. Write a structural formula for each one.

(*a*) *trans*-2-Phenylcyclopropylamine, better known as *tranylcypromine:* an antidepressant drug
(*b*) *N*-Benzyl-*N*-methyl-2-propynylamine, better known as *pargyline:* a drug used to treat high blood pressure
(*c*) 1-Phenyl-2-propanamine, better known as *amphetamine:* a stimulant
(*d*) 1-(*m*-Hydroxyphenyl)-2-(methylamino)ethanol: better known as *phenylephrine:* a nasal decongestant

22.24

(*a*) Give the structures and an acceptable name for all the isomers of molecular formula C_7H_9N that contain a benzene ring.
(*b*) Which one of these isomers is the strongest base?

TABLE 22.5
Preparation of Amines

Reaction (section) and comments	General equation and specific example
Alkylation methods	
Alkylation of ammonia (Section 22.8) Ammonia can act as a nucleophile toward primary and some secondary alkyl halides to give primary alkylamines. Yields tend to be modest because the primary amine is itself a nucleophile and undergoes alkylation. Alkylation of ammonia can lead to a mixture containing a primary amine, a secondary amine, a tertiary amine, and a quaternary ammonium salt.	$RX + 2NH_3 \longrightarrow RNH_2 + NH_4X$ Alkyl halide, Ammonia, Alkylamine, Ammonium halide $C_6H_5CH_2Cl \xrightarrow{NH_3\ (8\ mol)} C_6H_5CH_2NH_2 + (C_6H_5CH_2)_2NH$ Benzyl chloride (1 mol), Benzylamine (53%), Dibenzylamine (39%)
Alkylation of phthalimide. The Gabriel synthesis (Section 22.9) The potassium salt of phthalimide reacts with alkyl halides to give *N*-alkylphthalimide derivatives. Hydrolysis or hydrazinolysis of this derivative yields a primary alkylamine.	RX + (phthalimide $N^- K^+$) $\longrightarrow$ (phthalimide NR) Alkyl halide, *N*-Potassiophthalimide, *N*-Alkylphthalimide (phthalimide NR) + $H_2NNH_2 \longrightarrow RNH_2$ + (phthalhydrazide) *N*-Alkylphthalimide, Hydrazine, Primary amine, Phthalhydrazide $CH_3CH{=}CHCH_2Cl \xrightarrow[2.\ H_2NNH_2,\ ethanol]{1.\ N\text{-potassiophthalimide, DMF}} CH_3CH{=}CHCH_2NH_2$ 1-Chloro-2-butene, 2-Buten-1-amine (95%)
Reduction methods	
Reduction of alkyl azides (Section 22.10) Alkyl azides, prepared by nucleophilic substitution by azide ion in primary or secondary alkyl halides, are reduced to primary alkylamines by lithium aluminum hydride or by catalytic hydrogenation.	$R\ddot{N}{=}\overset{+}{N}{=}\ddot{N}{:}^- \xrightarrow{reduce} R\ddot{N}H_2$ Alkyl azide, Primary amine $CF_3CH_2CH(N_3)CO_2CH_2CH_3 \xrightarrow{H_2,\ Pd} CF_3CH_2CH(NH_2)CO_2CH_2CH_3$ Ethyl 2-azido-4,4,4-trifluorobutanoate, Ethyl 2-amino-4,4,4-trifluorobutanoate (96%)
Reduction of nitriles (Section 22.10) Nitriles are reduced to primary amines by lithium aluminum hydride or by catalytic hydrogenation.	$RC{\equiv}N \xrightarrow{reduce} RCH_2NH_2$ Nitrile, Primary amine (cyclopropyl)–CN $\xrightarrow[2.\ H_2O]{1.\ LiAlH_4}$ (cyclopropyl)–CH_2NH_2 Cyclopropyl cyanide, Cyclopropylmethanamine (75%)

TABLE 22.5
Preparation of Amines *(Continued)*

Reaction (section) and comments	General equation and specific example
Reduction of aryl nitro compounds (Section 22.10) The standard method for the preparation of an arylamine is by nitration of an aromatic ring, followed by reduction of the nitro group. Typical reducing agents include iron or tin in hydrochloric acid or catalytic hydrogenation.	$ArNO_2 \xrightarrow{\text{reduce}} ArNH_2$ Nitroarene → Arylamine $C_6H_5NO_2 \xrightarrow[\text{2. HO}^-]{\text{1. Fe, HCl}} C_6H_5NH_2$ Nitrobenzene → Aniline (97%)
Reduction of amides (Section 22.10) Lithium aluminum hydride reduces the carbonyl group of an amide to a methylene group. Primary, secondary, or tertiary amines may be prepared by proper choice of the starting amide. R and R′ may be either alkyl or aryl.	$RCNR'_2$ (C=O) $\xrightarrow{\text{reduce}} RCH_2NR'_2$ Amide → Amine $CH_3CNHC(CH_3)_3$ (C=O) $\xrightarrow[\text{2. H}_2\text{O}]{\text{1. LiAlH}_4} CH_3CH_2NHC(CH_3)_3$ *N-tert*-Butylacetamide → *N*-Ethyl-*tert*-butylamine (60%)
Reductive amination (Section 22.11) Reaction of ammonia or an amine with an aldehyde or a ketone in the presence of a reducing agent is an effective method for the preparation of primary, secondary, or tertiary amines. The reducing agent may be either hydrogen in the presence of a metal catalyst or sodium cyanoborohydride. R, R′, and R″ may be either alkyl or aryl.	RCR' (C=O) $+ R''_2NH \xrightarrow{\text{reducing agent}} RCR'$ (with NR''_2 and H on C) Aldehyde or ketone + Ammonia or an amine → Amine CH_3CCH_3 (C=O) + Cyclohexylamine (NH_2) $\xrightarrow{H_2,\ Pt}$ $HNCH(CH_3)_2$-cyclohexane Acetone + Cyclohexylamine → *N*-Isopropylcyclohexylamine (79%)

(*c*) Which, if any, of these isomers yield an *N*-nitroso amine on treatment with sodium nitrite and hydrochloric acid?

(*d*) Which, if any, of these isomers undergo nitrosation of their benzene ring on treatment with sodium nitrite and hydrochloric acid?

22.25 Arrange the following compounds and/or anions in each group in order of decreasing basicity:

(*a*) H_3C^-, H_2N^-, HO^-, F^-

(*b*) H_2O, NH_3, HO^-, H_2N^-

(*c*) HO^-, H_2N^-, $:\overline{C}{\equiv}N:$, NO_3^-

(*d*) [phthalimide anion, N⁻], [isoindolinone anion, N⁻], [isoindoline anion, N⁻]

TABLE 22.6
Reactions of Amines Discussed in This Chapter

Reaction (section) and comments	General equation and specific example
Alkylation (Section 22.13) Amines act as nucleophiles toward alkyl halides. Primary amines yield secondary amines, secondary amines yield tertiary amines, and tertiary amines yield quaternary ammonium salts.	$R\ddot{N}H_2 \xrightarrow{R'CH_2X} R\ddot{N}HCH_2R'$ (Primary amine → Secondary amine) $\xrightarrow{R'CH_2X} R\ddot{N}(CH_2R')_2$ (Tertiary amine) $\xrightarrow{R'CH_2X} R\overset{+}{N}(CH_2R')_3\,X^-$ (Quaternary ammonium salt) 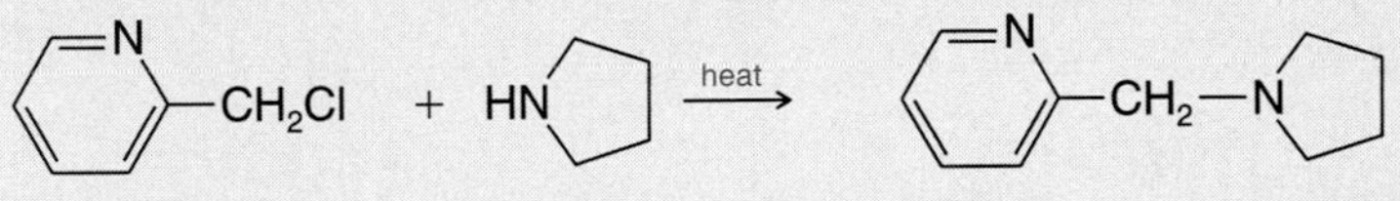 2-Chloromethylpyridine + Pyrrolidine $\xrightarrow{heat}$ 2-(Pyrrolidinylmethyl)pyridine (93%)
Hofmann elimination (Section 22.14) Quaternary ammonium hydroxides undergo elimination on being heated. It is an anti elimination of the E2 type. The regioselectivity of the Hofmann elimination is opposite to that of the Zaitsev rule and leads to the less highly substituted alkene.	$RCH_2CH(\overset{+}{N}(CH_3)_3)R'\ HO^- \xrightarrow{heat} RCH{=}CHR' + :N(CH_3)_3 + H_2O$ Alkyltrimethylammonium hydroxide → Alkene + Trimethylamine + Water 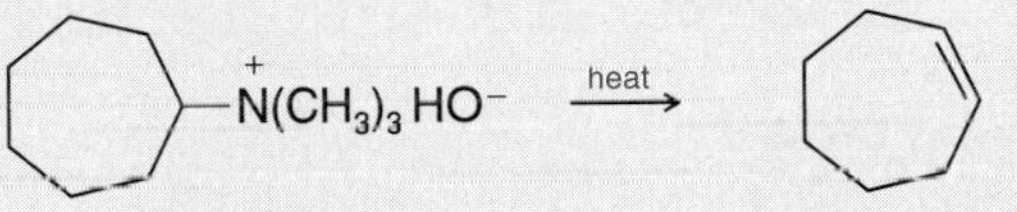 Cycloheptyltrimethylammonium hydroxide ($\overset{+}{N}(CH_3)_3\ HO^-$) $\xrightarrow{heat}$ Cycloheptene (87%)
Electrophilic aromatic substitution (Section 22.15) Arylamines are very reactive toward electrophilic aromatic substitution. It is customary to protect arylamines as their *N*-acyl derivatives before carrying out ring nitration, chlorination, bromination, sulfonation, or Friedel-Crafts reactions.	$ArH + E^+ \longrightarrow ArE + H^+$ Arylamine + Electrophile → Product of electrophilic aromatic substitution + Proton 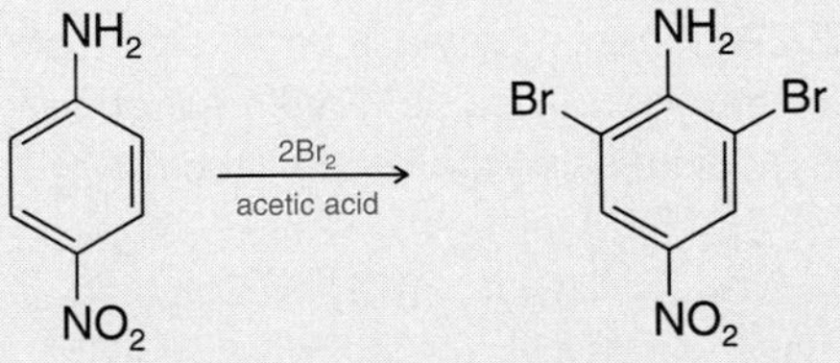 *p*-Nitroaniline $\xrightarrow[\text{acetic acid}]{2Br_2}$ 2,6-Dibromo-4-nitroaniline (95%)

22.26 Arrange the members of each group in order of decreasing basicity:

(*a*) Ammonia, aniline, methylamine
(*b*) Acetanilide, aniline, *N*-methylaniline
(*c*) 2,4-Dichloroaniline, 2,4-dimethylaniline, 2,4-dinitroaniline
(*d*) 3,4-Dichloroaniline, 4-chloro-2-nitroaniline, 4-chloro-3-nitroaniline
(*e*) Dimethylamine, diphenylamine, *N*-methylaniline

TABLE 22.6
Reactions of Amines Discussed in This Chapter *(Continued)*

Reaction (section) and comments	General equation and specific example
Nitrosation (Section 22.16) Nitrosation of amines occurs when sodium nitrite is added to a solution containing an amine and an acid. *Primary amines* yield alkyl diazonium salts. Alkyl diazonium salts are very unstable and yield carbocation-derived products. Aryl diazonium salts are exceedingly useful synthetic intermediates. Their reactions are described in Table 22.7.	$RNH_2 \xrightarrow[H^+,\ H_2O]{NaNO_2} R\overset{+}{N}{\equiv}N:$ Primary amine → Diazonium ion 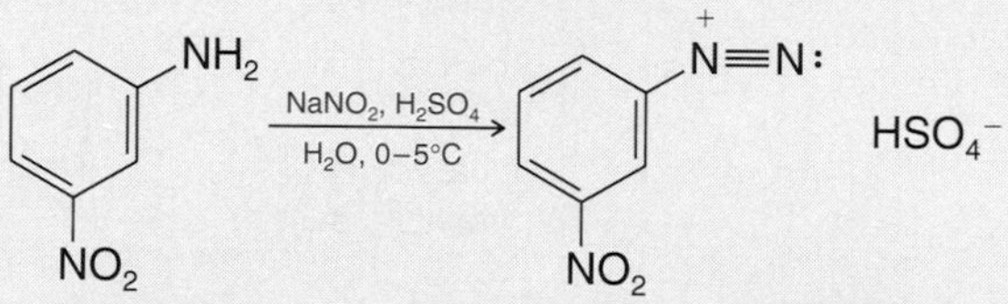 *m*-Nitroaniline → *m*-Nitrobenzenediazonium hydrogen sulfate
Secondary alkylamines and secondary arylamines yield *N*-nitroso amines.	$R_2NH \xrightarrow[H_2O]{NaNO_2,\ H^+} R_2N{-}N{=}O$ Secondary amine → *N*-Nitroso amine 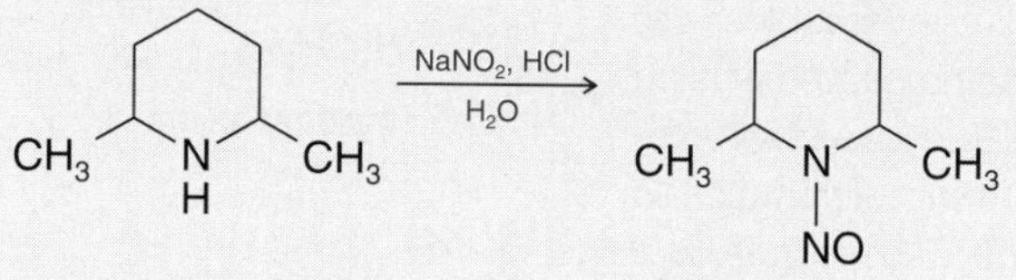 2,6-Dimethylpiperidine → 2,6-Dimethyl-*N*-nitrosopiperidine (72%)
Tertiary alkylamines illustrate no useful chemistry on nitrosation. *Tertiary arylamines* undergo nitrosation of the ring by electrophilic aromatic substitution.	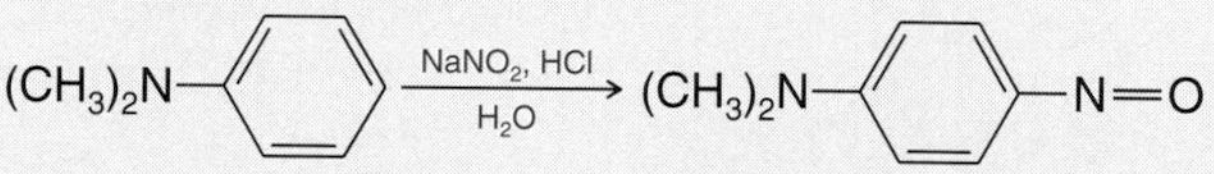 *N,N*-Dimethylaniline → *N,N*-Dimethyl-4-nitrosoaniline (80–89%)

22.27 *Physostigmine* is an alkaloid obtained from a west African plant; it is used in the treatment of glaucoma. Treatment of physostigmine with methyl iodide gives a quaternary ammonium salt. What is the structure of this salt?

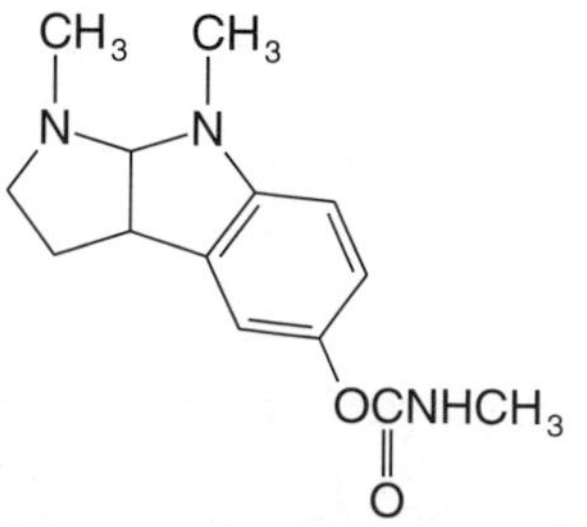

Physostigmine

22.28 Describe procedures for preparing each of the following compounds, using ethanol as the source of all their carbon atoms. Once you prepare a compound, you need not repeat its synthesis in a subsequent part of this problem.

TABLE 22.7
Synthetically Useful Transformations Involving Aryl Diazonium Ions

Reaction and comments	General equation and specific example
Preparation of phenols Heating its aqueous acidic solution converts a diazonium salt to a phenol. This is the most general method for the synthesis of phenols.	$ArNH_2 \xrightarrow[2.\ H_2O,\ heat]{1.\ NaNO_2,\ H_2SO_4,\ H_2O} ArOH$ Primary arylamine → Phenol *m*-Nitroaniline $\xrightarrow[2.\ H_2O,\ heat]{1.\ NaNO_2,\ H_2SO_4,\ H_2O}$ *m*-Nitrophenol (81–86%)
Preparation of aryl fluorides Addition of fluoroboric acid to a solution of a diazonium salt causes the precipitation of an aryl diazonium fluoroborate. When the dry aryl diazonium fluoroborate is heated, an aryl fluoride results. This is the Schiemann reaction; it is the most general method for the preparation of aryl fluorides.	$ArNH_2 \xrightarrow[2.\ HBF_4]{1.\ NaNO_2,\ H^+,\ H_2O} Ar\overset{+}{N}{\equiv}N:\ \overset{-}{B}F_4 \xrightarrow{heat} ArF$ Primary arylamine → Aryl diazonium fluoroborate → Aryl fluoride *m*-Toluidine $\xrightarrow[2.\ HBF_4]{1.\ NaNO_2,\ HCl,\ H_2O}$ *m*-Methylbenzenediazonium fluoroborate (76–84%) *m*-Methylbenzenediazonium fluoroborate $\xrightarrow{heat}$ *m*-Fluorotoluene (89%)
Preparation of aryl iodides Aryl diazonium salts react with sodium or potassium iodide to form aryl iodides. This is the most general method for the synthesis of aryl iodides.	$ArNH_2 \xrightarrow[2.\ NaI\ or\ KI]{1.\ NaNO_2,\ H^+,\ H_2O} ArI$ Primary arylamine → Aryl iodide 2,6-Dibromo-4-nitroaniline $\xrightarrow[2.\ NaI]{1.\ NaNO_2,\ H_2SO_4,\ H_2O}$ 1,3-Dibromo-2-iodo-5-nitrobenzene (84–88%)

TABLE 22.7

Synthetically Useful Transformations Involving Aryl Diazonium Ions *(Continued)*

Reaction and comments	General equation and specific example
Preparation of aryl chlorides In the Sandmeyer reaction a solution containing an aryl diazonium salt is treated with copper (I) chloride to give an aryl chloride.	$ArNH_2 \xrightarrow[2.\ CuCl]{1.\ NaNO_2,\ HCl,\ H_2O} ArCl$ Primary arylamine → Aryl chloride o-Toluidine (CH_3, NH_2) $\xrightarrow[2.\ CuCl]{1.\ NaNO_2,\ HCl,\ H_2O}$ *o*-Chlorotoluene (CH_3, Cl) (74–79%) *o*-Toluidine → *o*-Chlorotoluene (74–79%)
Preparation of aryl bromides The Sandmeyer reaction using copper (I) bromide is applicable to the conversion of primary arylamines to aryl bromides.	$ArNH_2 \xrightarrow[2.\ CuBr]{1.\ NaNO_2,\ HBr,\ H_2O} ArBr$ Primary arylamine → Aryl bromide *m*-Bromoaniline (NH_2, Br) $\xrightarrow[2.\ CuBr]{1.\ NaNO_2,\ HBr,\ H_2O}$ *m*-Dibromobenzene (Br, Br) (80–87%)
Preparation of aryl cyanides Copper (I) cyanide converts aryl diazonium salts to aryl cyanides.	$ArNH_2 \xrightarrow[2.\ CuCN]{1.\ NaNO_2,\ H_2O} ArCN$ Primary arylamine → Aryl cyanide *o*-Nitroaniline (NH_2, NO_2) $\xrightarrow[2.\ CuCN]{1.\ NaNO_2,\ HCl,\ H_2O}$ *o*-Nitrobenzonitrile (CN, NO_2) (87%)
Reductive deamination of primary arylamines The amino substituent of an arylamine can be replaced by hydrogen by treatment of its derived diazonium salt with ethanol or with hypophosphorous acid.	$ArNH_2 \xrightarrow[2.\ CH_3CH_2OH\ or\ H_3PO_2]{1.\ NaNO_2,\ H^+,\ H_2O} ArH$ 4-Methyl-2-nitroaniline (CH_3, NO_2, NH_2) $\xrightarrow[2.\ H_3PO_2]{1.\ NaNO_2,\ HCl,\ H_2O}$ *m*-Nitrotoluene (CH_3, NO_2) (80%)

(*a*) Ethylamine
(*b*) *N*-Ethylacetamide
(*c*) Diethylamine
(*d*) *N,N*-Diethylacetamide
(*e*) Triethylamine
(*f*) Tetraethylammonium bromide

22.29 Show by writing the appropriate sequence of equations how you could carry out each of the following transformations:

(*a*) 1-Butanol to 1-pentanamine
(*b*) *tert*-Butyl chloride to 2,2-dimethyl-1-propanamine
(*c*) Cyclohexanol to *N*-methylcyclohexylamine
(*d*) Isopropyl alcohol to 1-amino-2-methyl-2-propanol
(*e*) Isopropyl alcohol to 1-amino-2-propanol
(*f*) Isopropyl alcohol to 1-(*N,N*-dimethylamino)-2-propanol
(*g*)

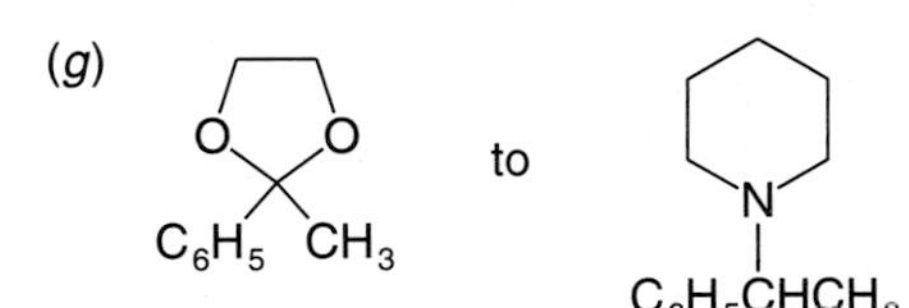

N
|
$C_6H_5CHCH_3$

22.30 Each of the following dihaloalkanes gives an *N*-(haloalkyl)phthalimide on reaction with one equivalent of the potassium salt of phthalimide. Write the structure of the phthalimide derivative formed in each case and explain the basis for your answer.

(*a*) FCH_2CH_2Br
(*b*) $BrCH_2CH_2CH_2CHCH_3$ (with Br on the CH)
(*c*) $BrCH_2CCH_2CH_2Br$ (with CH_3 above and CH_3 below the C)

22.31 Give the structure of the expected product formed when benzylamine reacts with each of the following reagents:

(*a*) Hydrogen bromide
(*b*) Sulfuric acid
(*c*) Acetic acid
(*d*) Acetyl chloride
(*e*) Acetic anhydride
(*f*) Acetone
(*g*) Acetone and hydrogen (nickel catalyst)
(*h*) Ethylene oxide
(*i*) 1,2-Epoxypropane
(*j*) Excess methyl iodide
(*k*) Sodium nitrite in dilute hydrochloric acid

22.32 Write the structure of the product formed on reaction of aniline with each of the following:

(*a*) Hydrogen bromide
(*b*) Excess methyl iodide
(*c*) Acetaldehyde
(*d*) Acetaldehyde and hydrogen (nickel catalyst)
(*e*) Acetic anhydride
(*f*) Benzoyl chloride
(*g*) Sodium nitrite, aqueous sulfuric acid, 0–5°C
(*h*) Product of part (*g*), heated in aqueous acid
(*i*) Product of part (*g*), treated with copper(I) chloride
(*j*) Product of part (*g*), treated with copper(I) bromide
(*k*) Product of part (*g*), treated with copper(I) cyanide
(*l*) Product of part (*g*), treated with hypophosphorous acid
(*m*) Product of part (*g*), treated with potassium iodide

(*n*) Product of part (*g*), treated with fluoroboric acid, then heated
(*o*) Product of part (*g*), treated with phenol
(*p*) Product of part (*g*), treated with *N,N*-dimethylaniline

22.33 Write the structure of the product formed on reaction of acetanilide with each of the following:

(*a*) Lithium aluminum hydride
(*b*) Nitric acid and sulfuric acid
(*c*) Sulfur trioxide and sulfuric acid
(*d*) Bromine in acetic acid
(*e*) *tert*-Butyl chloride, aluminum chloride
(*f*) Acetyl chloride, aluminum chloride
(*g*) 6 *M* hydrochloric acid, reflux
(*h*) Aqueous sodium hydroxide, reflux

22.34 Identify the principal organic products of each of the following reactions:

(*a*) Cyclohexanone + cyclohexylamine $\xrightarrow{H_2,\ Ni}$

(*b*) O NCH$_2$CH$_3$ $\xrightarrow[2.\ H_2O,\ HO^-]{1.\ LiAlH_4}$

(*c*) $C_6H_5CH_2CH_2CH_2OH \xrightarrow[2.\ (CH_3)_2NH\ (excess)]{1.\ p\text{-toluenesulfonyl chloride, pyridine}}$

(*d*) $(CH_3)_2CHNH_2$ + CH$_3$O, OCH$_3$, —CH—CH$_2$ (O) $\longrightarrow$

(*e*) $(C_6H_5CH_2)_2NH + CH_3\overset{O}{\overset{\|}{C}}CH_2Cl \xrightarrow[THF]{triethylamine}$

(*f*) H$_3$C, H$_3$C, CH$_3$, $\overset{+}{N}(CH_3)_3$ HO$^-$ $\xrightarrow{heat}$

(*g*) $(CH_3)_2CHNHCH(CH_3)_2 \xrightarrow[HCl,\ H_2O]{NaNO_2}$

22.35 Each of the following reactions has been reported in the chemical literature and proceeds in good yield. Identify the principal organic product of each reaction.

(*a*) 1,2-Diethyl-4-nitrobenzene $\xrightarrow[ethanol]{H_2,\ Pt}$

(*b*) 1,3-Dimethyl-2-nitrobenzene $\xrightarrow[2.\ HO^-]{1.\ SnCl_2,\ HCl}$

(*c*) Product of part (*b*) + $ClCH_2\overset{O}{\overset{\|}{C}}Cl \longrightarrow$

(*d*) Product of part (*c*) + $(CH_3CH_2)_2NH \longrightarrow$

(*e*) Product of part (*d*) + HCl $\longrightarrow$

(*f*) $C_6H_5NH\overset{O}{\overset{\|}{C}}CH_2CH_2CH_3 \xrightarrow[2.\ HO^-]{1.\ LiAlH_4}$

(*g*) Aniline + heptanal $\xrightarrow{H_2,\ Ni}$

(*h*) Acetanilide + $ClCH_2\overset{O}{\overset{\|}{C}}Cl \xrightarrow{AlCl_3}$

(*i*) Br—(C₆H₄)—(C₆H₄)—$NO_2 \xrightarrow[2.\ HO^-]{1.\ Fe,\ HCl}$

(*j*) Product of part (*i*) $\xrightarrow[2.\ H_2O,\ heat]{1.\ NaNO_2,\ H_2SO_4,\ H_2O}$

(*k*) 2,6-Dinitroaniline $\xrightarrow[2.\ CuCl]{1.\ NaNO_2,\ H_2SO_4,\ H_2O}$

(*l*) *m*-Bromoaniline $\xrightarrow[2.\ CuBr]{1.\ NaNO_2,\ HBr,\ H_2O}$

(*m*) *o*-Nitroaniline $\xrightarrow[2.\ CuCN]{1.\ NaNO_2,\ HCl,\ H_2O}$

(*n*) 2,6-Diiodo-4-nitroaniline $\xrightarrow[2.\ KI]{1.\ NaNO_2,\ H_2SO_4,\ H_2O}$

(*o*) $:N\equiv\overset{+}{N}$—(C₆H₄)—(C₆H₄)—$\overset{+}{N}\equiv N:\ \ 2\overset{-}{B}F_4 \xrightarrow{heat}$

(*p*) 2,4,6-Trinitroaniline $\xrightarrow[H_2O,\ H_3PO_2]{NaNO_2,\ H_2SO_4}$

(*q*) 2-Amino-5-iodobenzoic acid $\xrightarrow[2.\ CH_3CH_2OH]{1.\ NaNO_2,\ HCl,\ H_2O}$

(*r*) Aniline $\xrightarrow[2.\ 2,3,6\text{-trimethylphenol}]{1.\ NaNO_2,\ H_2SO_4,\ H_2O}$

(*s*) $(CH_3)_2N$—(C₆H₄, *m*-CH_3) $\xrightarrow[2.\ HO^-]{1.\ NaNO_2,\ HCl,\ H_2O}$

22.36 Provide a reasonable explanation for each of the following observations:

(*a*) 4-Methylpiperidine has a higher boiling point than *N*-methylpiperidine.

4-Methylpiperidine (bp 129°C)   *N*-Methylpiperidine (bp 106°C)

(*b*) Two isomeric quaternary ammonium salts are formed in comparable amounts when 4-*tert*-butyl-*N*-methylpiperidine is treated with benzyl chloride.

CH_3N(piperidine ring)—$C(CH_3)_3$ 4-*tert*-Butyl-*N*-methylpiperidine

(*c*) When tetramethylammonium hydroxide is heated at 130°C, trimethylamine and methanol are formed.

(*d*) The major product formed on treatment of 1-propanamine with sodium nitrite in dilute hydrochloric acid is 2-propanol.

22.37 Give the structures, including stereochemistry, of compounds A through C.

(*S*)-2-Octanol + CH_3–C_6H_4–SO_2Cl $\xrightarrow{\text{pyridine}}$ compound A $\xrightarrow{NaN_3,\ \text{methanol-water}}$ compound B $\xrightarrow{1.\ LiAlH_4;\ 2.\ HO^-}$ compound C

22.38 Devise efficient syntheses of each of the following compounds from the designated starting materials. You may also use any necessary organic or inorganic reagents.

(*a*) 3,3-Dimethyl-1-butanamine from 1-bromo-2,2-dimethylpropane

(*b*) $CH_2{=}CH(CH_2)_8CH_2{-}N$(pyrrolidine ring) from 10-undecenoic acid and pyrrolidine

(*c*) 2-(C_6H_5O)cyclopentyl–NH_2 from 2-(C_6H_5O)cyclopentanol (OH)

(*d*) $C_6H_5CH_2N(CH_3)CH_2CH_2CH_2CH_2NH_2$ from $C_6H_5CH_2NHCH_3$ and $BrCH_2CH_2CH_2CN$

(*e*) NC–C_6H_4–$CH_2N(CH_3)_2$ from NC–C_6H_4–CH_3

22.39 Each of the following compounds has been prepared from *p*-nitroaniline. Outline a reasonable series of steps leading to each one.

(*a*) *p*-Nitrobenzonitrile
(*b*) 3,4,5-Trichloroaniline
(*c*) 1,3-Dibromo-5-nitrobenzene
(*d*) 3,5-Dibromoaniline
(*e*) *p*-Acetamidophenol *(acetaminophen)*

22.40 Each of the following compounds has been prepared from *o*-anisidine (*o*-methoxyaniline). Outline a series of steps leading to each one.

(*a*) o-Bromoanisole
(*b*) o-Fluoroanisole
(*c*) 3-Fluoro-4-methoxyacetophenone
(*d*) 3-Fluoro-4-methoxybenzonitrile
(*e*) 3-Fluoro-4-methoxyphenol

22.41 Design syntheses of each of the following compounds from the indicated starting material and any necessary organic or inorganic reagents:

(*a*) *p*-Aminobenzoic acid from *p*-methylaniline

(*b*) *p*-$FC_6H_4\overset{O}{\overset{\|}{C}}CH_2CH_3$ from benzene

(*c*) 1-Bromo-2-fluoro-3,5-dimethylbenzene from *m*-xylene

(*d*) 1-Bromo-4-fluoro-2-methylnaphthalene (Br, CH_3, F) from 2-methyl-4-nitro-1-naphthylamine (NH_2, CH_3, NO_2)

(*e*) *o*-$BrC_6H_4C(CH_3)_3$ from *p*-$O_2NC_6H_4C(CH_3)_3$
(*f*) *m*-$ClC_6H_4C(CH_3)_3$ from *p*-$O_2NH_6H_4C(CH_3)_3$
(*g*) 1-Bromo-3,5-diethylbenzene from *m*-diethylbenzene

(*h*) CF_3, I, Br from CF_3, $NHCCH_3$ (O), H_2N, Br

(*i*) NH, CH_3O, CH_3O from CH_2COCH_3 (O), CH_3O, CH_3O, O_2N

22.42 Ammonia and amines undergo conjugate addition to α,β-unsaturated carbonyl compounds (Section 18.13). On the basis of this information, predict the principal organic product of each of the following reactions:

(*a*) $(CH_3)_2C{=}CH\overset{O}{\overset{\|}{C}}CH_3 + NH_3 \longrightarrow$

(*b*) $=O$ + HN $\longrightarrow$

(*c*) $C_6H_5\overset{O}{\overset{\|}{C}}CH{=}CHC_6H_5$ + HN O $\longrightarrow$

(*d*) O, $(CH_2)_3CH(CH_2)_4CH_3$, NH_2 $\xrightarrow{\text{spontaneous}}$

22.43 A number of compounds of the type represented by compound A were prepared for evaluation as potential analgesic drugs. Their preparation is described in a retrosynthetic format as shown.

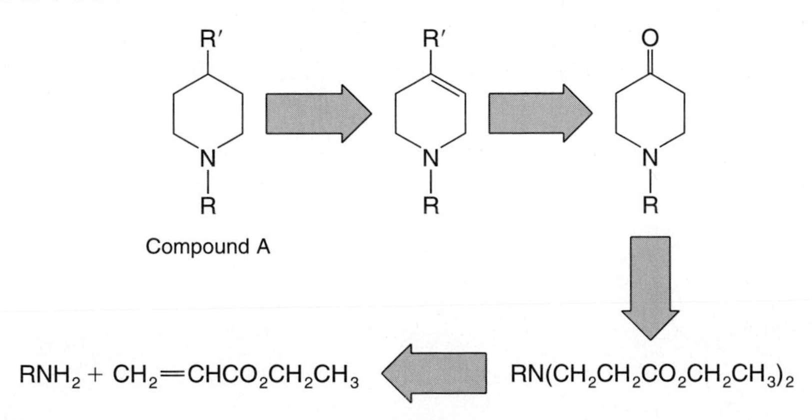

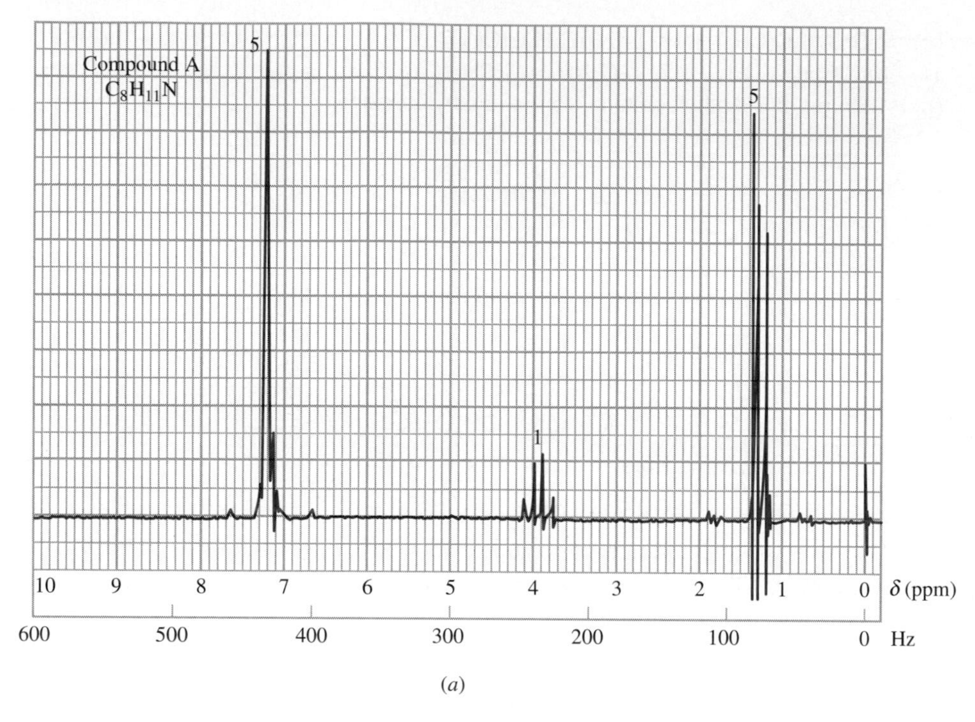

(*a*)

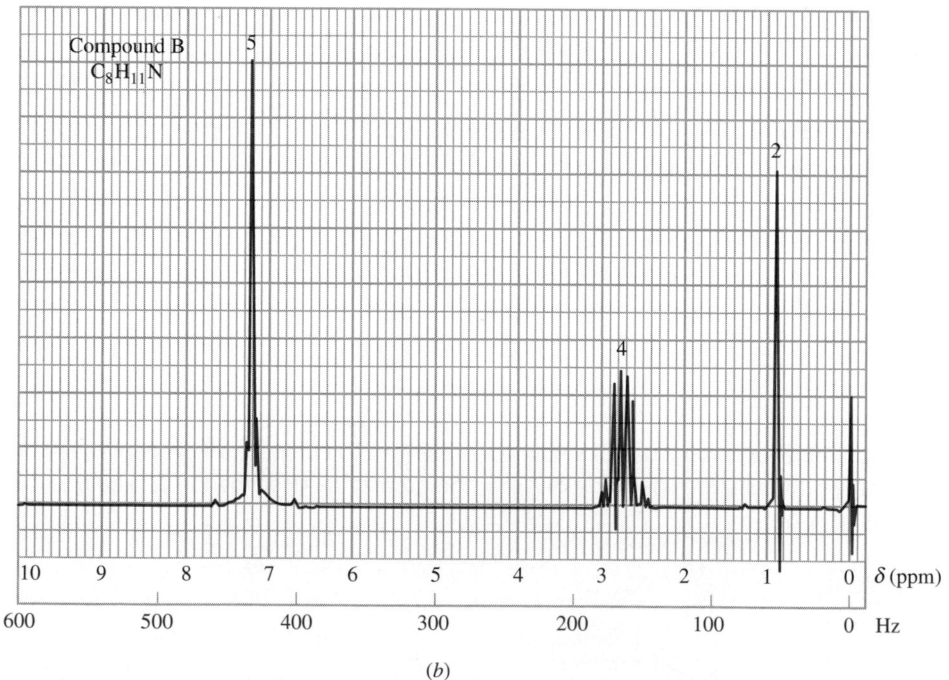

(*b*)

FIGURE 22.10 The 60-MHz ^1H nmr spectra of (*a*) compound A and (*b*) compound B (Problem 22.48).

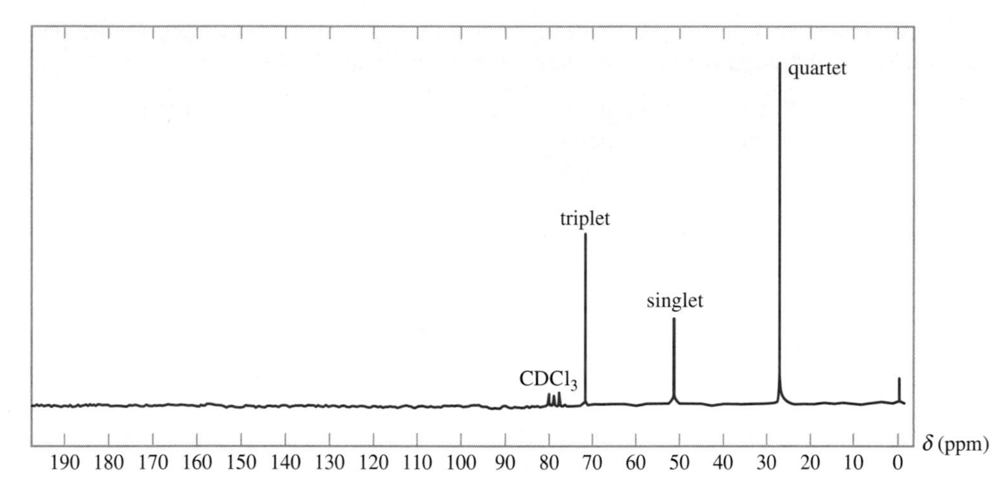

FIGURE 22.11 The ^{13}C nmr spectrum of the compound described in Problem 22.49. The peak multiplicities were determined by off-resonance decoupling in a separate experiment. *(Taken from Carbon-13 NMR Spectra: A Collection of Assigned, Coded, and Indexed Spectra, by LeRoy F. Johnson and William C. Jankowski, Wiley-Interscience, New York, 1972. Reprinted by permission of John Wiley & Sons, Inc.)*

On the basis of this retrosynthetic analysis, design a synthesis of *N*-methyl-4-phenylpiperidine (compound A, where R = CH_3, R′ = C_6H_5). Present your answer as a series of equations, showing all necessary reagents and isolated intermediates.

22.44 *Mescaline,* a hallucinogenic amine obtained from the peyote cactus, has been synthesized in two steps from 3,4,5-trimethoxybenzyl bromide. The first step is nucleophilic substitution by sodium cyanide. The second step is a lithium aluminum hydride reduction. What is the structure of mescaline?

22.45 *Methamphetamine* is a notorious street drug. One synthesis involves reductive amination of benzyl methyl ketone with methylamine. What is the structure of methamphetamine?

22.46 The basicity constants of *N,N*-dimethylaniline and pyridine are almost the same, while 4-(*N,N*-dimethylamino)pyridine is considerably more basic than either.

$N(CH_3)_2$ N $N(CH_3)_2$ N

N,N-Dimethylaniline	Pyridine	4-(*N,N*-Dimethylamino)pyridine
K_b 1.3 × 10^{-9}	K_b 2 × 10^{-9}	K_b = 5 × 10^{-5}
pK_b 8.9	pK_b 8.7	pK_b 4.3

Identify the more basic of the two nitrogens of 4-(*N,N*-dimethylamino)pyridine and suggest an explanation for its enhanced basicity as compared with pyridine and *N,N*-dimethylaniline.

22.47 The compound shown is a somewhat stronger base than ammonia. Which nitrogen do you think is protonated when it is treated with an acid? Write a structural formula for the species that results.

CH_3 N N

5-Methyl-γ-carboline (pK_b = 3.5)

22.48 Compounds A and B are isomeric amines of molecular formula $C_8H_{11}N$. Identify each isomer on the basis of the ^1H nmr spectra given in Figure 22.10.

22.49 Does the ^{13}C nmr spectrum shown in Figure 22.11 correspond to that of 1-amino-2-methyl-2-propanol or to 2-amino-2-methyl-1-propanol? Could this compound be prepared by reaction of an epoxide with ammonia?

MOLECULAR MODELING EXERCISES

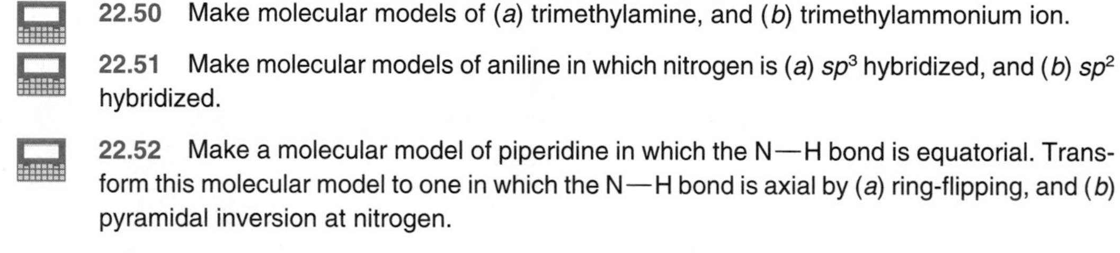

22.50 Make molecular models of (*a*) trimethylamine, and (*b*) trimethylammonium ion.

22.51 Make molecular models of aniline in which nitrogen is (*a*) sp^3 hybridized, and (*b*) sp^2 hybridized.

22.52 Make a molecular model of piperidine in which the N—H bond is equatorial. Transform this molecular model to one in which the N—H bond is axial by (*a*) ring-flipping, and (*b*) pyramidal inversion at nitrogen.

22.53 Make a molecular model of benzenediazonium ion ($C_6H_5N_2^+$). Is it planar? What is the geometry at the nitrogen attached to the ring? Identify the vacant orbital that results when benzenediazonium ion loses N_2 to form $C_6H_5^+$.

22.54 Consider the following derivatives of aniline with regard to their basicity.

Compound	R	R′	pK_b	pK_a of conjugate acid
Aniline	H	H	9.40	4.60
N-Methylaniline	H	CH_3	9.15	4.85
N-Ethylaniline	H	CH_2CH_3	8.89	5.11
N,N-Dimethylaniline	CH_3	CH_3	8.85	5.15
N,N-Diethylaniline	CH_2CH_3	CH_2CH_3	7.44	6.56

There is something unusual about *N,N*-diethylaniline that makes it more than 1 p*K* unit more basic than expected. See if you can deduce the steric effect responsible for this increased basicity by inspecting a molecular model of *N,N*-diethylaniline.

CHAPTER 23

ARYL HALIDES

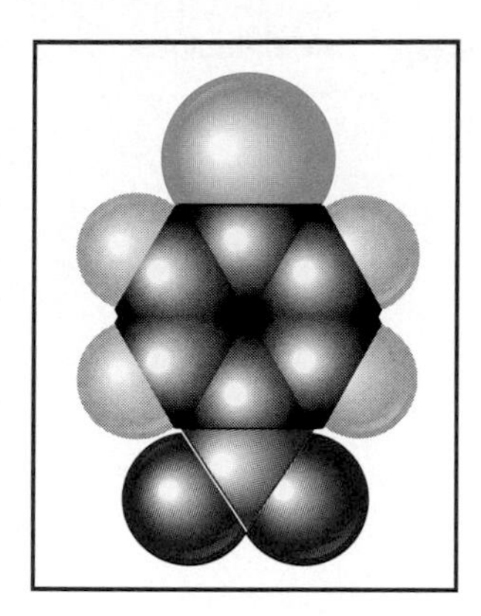

The value of *alkyl halides* as starting materials for the preparation of a variety of organic functional groups has been stressed many times. In those earlier discussions, we noted that *aryl halides* are normally much less reactive than alkyl halides in reactions that involve carbon-halogen bond cleavage. In the present chapter you will see that aryl halides can exhibit their own patterns of chemical reactivity, and that these reactions are novel, useful, and mechanistically interesting.

23.1 BONDING IN ARYL HALIDES

Aryl halides are compounds in which a halogen substituent is attached directly to an aromatic ring. Representative aryl halides include

Fluorobenzene	1-Chloro-2-nitrobenzene	1-Bromonaphthalene	*p*-Iodobenzyl alcohol

Halogen-containing organic compounds in which the halogen substituent is not directly bonded to an aromatic ring, even though an aromatic ring may be present, are not aryl halides. Benzyl chloride ($C_6H_5CH_2Cl$), for example, is not an aryl halide.

The carbon-halogen bonds of aryl halides are both shorter and stronger than the carbon-halogen bonds of alkyl halides, and in this respect as well as in their chemical behavior, they resemble vinyl halides more than alkyl halides. A hybridization effect

TABLE 23.1
Carbon-Hydrogen and Carbon-Chlorine Bond Dissociation Energies of Selected Compounds

Compound	Hybridization of carbon to which X is attached	Bond energy, kJ/mol (kcal/mol) X = H	X = Cl
CH_3CH_2X	sp^3	410 (98)	339 (81)
$CH_2{=}CHX$	sp^2	452 (108)	368 (88)
C_6H_5—X	sp^2	469 (112)	406 (97)

seems to be responsible because, as the data in Table 23.1 indicate, similar patterns are seen for both carbon-hydrogen bonds and carbon-halogen bonds. An increase in *s* character from 25 percent (sp^3 hybridization) to 33.3 percent *s* character (sp^2 hybridization) increases the tendency of carbon to attract electrons and to bind substituents more strongly.

PROBLEM 23.1 Consider all the isomers of C_7H_7Cl containing a benzene ring and write the structure of the one that has the weakest carbon-chlorine bond as measured by its bond dissociation energy.

The strength of their carbon-halogen bonds causes aryl halides to react very slowly in reactions in which carbon-halogen bond cleavage is rate-determining, nucleophilic substitution, for example. Later in this chapter we will see examples of such reactions that do take place at reasonable rates but proceed by mechanisms distinctly different from the classical S_N1 and S_N2 pathways.

23.2 SOURCES OF ARYL HALIDES

The two main methods for the preparation of aryl halides—direct halogenation of arenes by electrophilic aromatic substitution and preparation by way of aryl diazonium salts—have been described earlier and are reviewed in Table 23.2. A number of aryl halides occur naturally, some of which are shown in Figure 23.1.

23.3 PHYSICAL PROPERTIES OF ARYL HALIDES

Melting points and boiling points for some representative aryl halides are listed in Appendix 1.

Aryl halides resemble alkyl halides in many of their physical properties. All are practically insoluble in water and most are denser than water.

Aryl halides are polar molecules but are less polar than alkyl halides.

Chlorocyclohexane
μ 2.2 D

Chlorobenzene
μ 1.7 D

TABLE 23.2
Summary of Reactions Discussed in Earlier Chapters That Yield Aryl Halides

Reaction (section) and comments	General equation and specific example
Halogenation of arenes (Section 12.5) Aryl chlorides and bromides are conveniently prepared by electrophilic aromatic substitution. The reaction is limited to chlorination and bromination. Fluorination is difficult to control; iodination is too slow to be useful.	$ArH + X_2 \xrightarrow[FeX_3]{Fe \text{ or}} ArX + HX$ Arene, Halogen, Aryl halide, Hydrogen halide $O_2N-C_6H_5 + Br_2 \xrightarrow{Fe} m\text{-}BrC_6H_4NO_2$ Nitrobenzene, Bromine, *m*-Bromonitrobenzene (85%)
The Sandmeyer reaction (Section 22.18) Diazotization of a primary arylamine followed by treatment of the diazonium salt with copper(I) bromide or copper(I) chloride yields the corresponding aryl bromide or aryl chloride.	$ArNH_2 \xrightarrow[2.\ CuX]{1.\ NaNO_2,\ H_3O^+} ArX$ Primary arylamine, Aryl halide 1-Amino-8-chloronaphthalene $\xrightarrow[2.\ CuBr]{1.\ NaNO_2,\ HBr}$ 1-Bromo-8-chloronaphthalene (62%)
The Schiemann reaction (Section 22.18) Diazotization of an arylamine followed by treatment with fluoroboric acid gives an aryl diazonium fluoroborate salt. Heating this salt converts it to an aryl fluoride.	$ArNH_2 \xrightarrow[2.\ HBF_4]{1.\ NaNO_2,\ H_3O^+} Ar\overset{+}{N}{\equiv}N\!:\ \overset{-}{B}F_4 \xrightarrow{heat} ArF$ Primary arylamine, Aryl diazonium fluoroborate, Aryl fluoride $C_6H_5NH_2 \xrightarrow[2.\ HBF_4;\ 3.\ heat]{1.\ NaNO_2,\ H_2O,\ HCl} C_6H_5F$ Aniline, Fluorobenzene (51–57%)
Reaction of aryl diazonium salts with iodide ion (Section 22.18) Adding potassium iodide to a solution of an aryl diazonium ion leads to the formation of an aryl iodide.	$ArNH_2 \xrightarrow[2.\ KI]{1.\ NaNO_2,\ H_3O^+} ArI$ Primary arylamine, Aryl iodide $C_6H_5NH_2 \xrightarrow[2.\ KI]{1.\ NaNO_2,\ HCl,\ H_2O} C_6H_5I$ Aniline, Iodobenzene (74–76%)

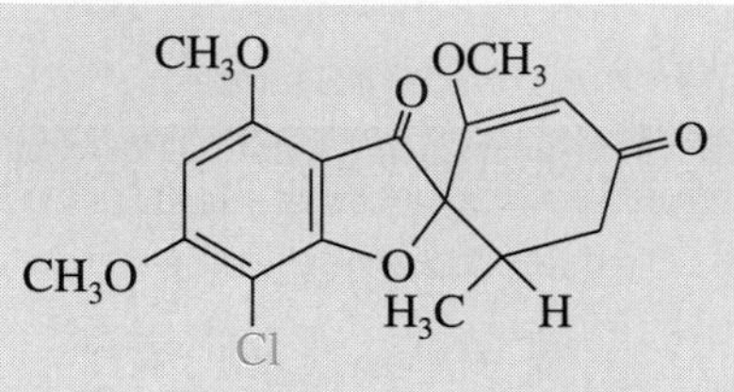

Griseofulvin: biosynthetic product of a particular microorganism, used as an orally administered antifungal agent.

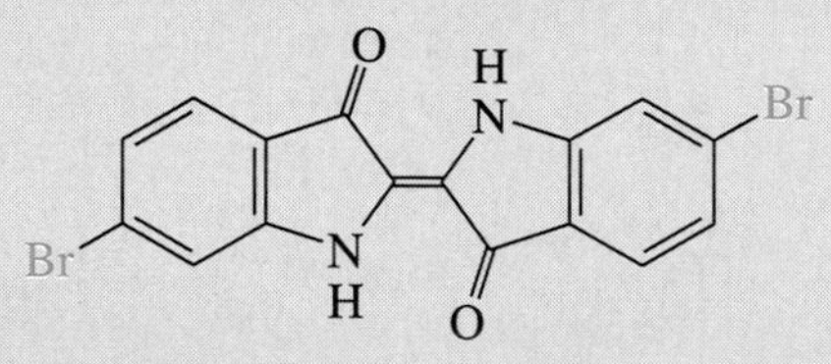

Dibromoindigo: principal constituent of a dye known as Tyrian purple, which is isolated from a species of Mediterranean sea snail and was much prized by the ancients for its vivid color.

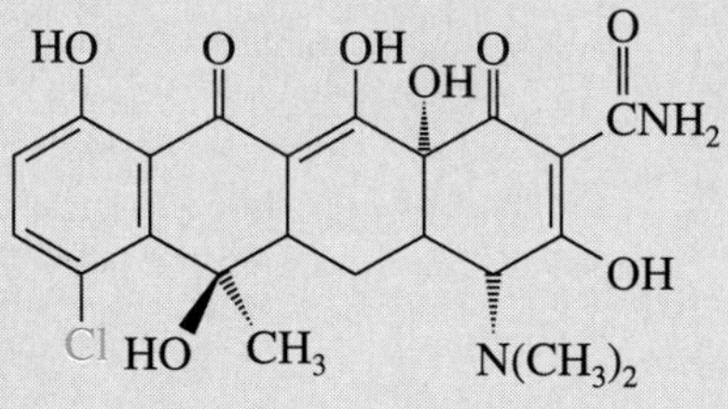

Chlortetracycline: an antibiotic.

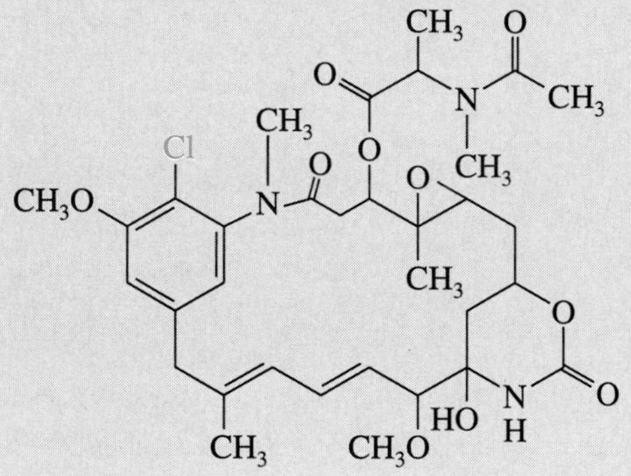

Maytansine: a potent antitumor agent isolated from a bush native to Kenya; 10 tons of plant yielded 6 g of maytansine.

FIGURE 23.1 Some naturally occurring aryl halides.

Since carbon is sp^2 hybridized in chlorobenzene, it is more electronegative than the sp^3 hybridized carbon of chlorocyclohexane. Consequently, the withdrawal of electron density away from carbon by chlorine is less pronounced in aryl halides than in alkyl halides, and the molecular dipole moment is smaller.

23.4 REACTIONS OF ARYL HALIDES: A REVIEW AND A PREVIEW

Table 23.3 summarizes the reactions of aryl halides that we have encountered to this point.

Noticeably absent from Table 23.3 are nucleophilic substitutions. We have, to this point, seen no nucleophilic substitution reactions of aryl halides in this text. Chlorobenzene, for example, is essentially inert to aqueous sodium hydroxide at room tem-

TABLE 23.3
Summary of Reactions of Aryl Halides Discussed in Earlier Chapters

Reaction (section) and comments	General equation and specific example
Electrophilic aromatic substitution (Section 12.14) Halogen substituents are slightly deactivating and ortho, para-directing.	Bromobenzene $\xrightarrow[AlCl_3]{CH_3COCCH_3}$ p-Bromoacetophenone (69–79%)
Formation of aryl Grignard reagents (Section 14.4) Aryl halides react with magnesium to form the corresponding arylmagnesium halide. Aryl iodides are the most reactive, aryl fluorides the least. A similar reaction occurs with lithium to give aryllithium reagents (Section 14.3).	ArX (Aryl halide) + Mg (Magnesium) $\xrightarrow{\text{diethyl ether}}$ ArMgX (Arylmagnesium halide) Bromobenzene + Mg (Magnesium) $\xrightarrow{\text{diethyl ether}}$ Phenylmagnesium bromide (95%)

perature. Reaction temperatures over 300°C are required for nucleophilic substitution to proceed at a reasonable rate.

The mechanism of this reaction is discussed in Section 23.8.

$$\text{Chlorobenzene} \xrightarrow[2.\ H^+]{1.\ NaOH,\ H_2O,\ 370°C} \text{Phenol (97\%)}$$

Aryl halides are much less reactive than alkyl halides in nucleophilic substitution reactions. The carbon-halogen bonds of aryl halides are too strong, and aryl cations are too high in energy, to permit aryl halides to ionize readily in S_N1-type processes. Further, as Figure 23.2 depicts, the optimal transition-state geometry required for S_N2 processes cannot be achieved. Nucleophilic attack from the side opposite the carbon-halogen bond is blocked by the aromatic ring.

FIGURE 23.2 Nucleophilic substitution, with inversion of configuration, is blocked by the benzene ring of an aryl halide. (*a*) *Alkyl halide:* The new bond is formed by attack of the nucleophile at carbon from the side opposite the bond to the leaving group. Inversion of configuration is observed. (*b*) *Aryl halide:* The aromatic ring blocks approach of the nucleophile to carbon at the side opposite the bond to the leaving group. Inversion of configuration is impossible.

23.5 NUCLEOPHILIC SUBSTITUTION IN NITRO-SUBSTITUTED ARYL HALIDES

One group of aryl halides that do undergo nucleophilic substitution with reasonable ease consists of those that bear a nitro group ortho or para to the halogen.

$$p\text{-}ClC_6H_4NO_2 + NaOCH_3 \xrightarrow[85°C]{CH_3OH} p\text{-}CH_3OC_6H_4NO_2 + NaCl$$

p-Chloronitrobenzene | Sodium methoxide | *p*-Nitroanisole (92%) | Sodium chloride

An *ortho*-nitro group exerts a comparable rate-enhancing effect. *m*-Chloronitrobenzene, while much more reactive than chlorobenzene itself, is thousands of times less reactive than either *o*- or *p*-chloronitrobenzene.

The effect of *o*- and *p*-nitro substituents is cumulative, as the rate data for substitution in a series of nitro-substituted chlorobenzene derivatives demonstrate:

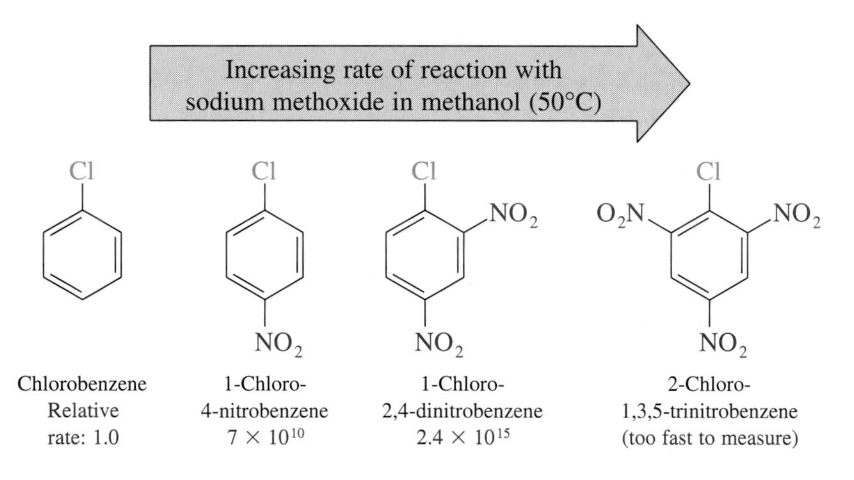

PROBLEM 23.2 Write the structure of the expected product from the reaction of 1-chloro-2,4-dinitrobenzene with each of the following reagents:

(*a*) CH_3CH_2ONa (*c*) NH_3
(*b*) $C_6H_5CH_2SNa$ (*d*) CH_3NH_2

SAMPLE SOLUTION (*a*) Sodium ethoxide is a source of the nucleophile $CH_3CH_2O^-$, which displaces chloride from 1-chloro-2,4-dinitrobenzene.

$$ClC_6H_3(NO_2)_2 + CH_3CH_2O^- \longrightarrow CH_3CH_2OC_6H_3(NO_2)_2 + Cl^-$$

1-Chloro-2,4-dinitrobenzene | Ethoxide anion | 1-Ethoxy-2,4-dinitrobenzene

---

In contrast to nucleophilic substitution in alkyl halides, where *alkyl fluorides* are exceedingly unreactive, *aryl fluorides* undergo nucleophilic substitution readily when the ring bears an *o*- or a *p*-nitro group.

The compound 1-fluoro-2,4-dinitrobenzene is exceedingly reactive toward nucleophilic aromatic substitution and was used in an imaginative way by Frederick Sanger (Section 27.10) in his determination of the structure of insulin.

$$p\text{-}FC_6H_4NO_2 + KOCH_3 \xrightarrow[85°C]{CH_3OH} p\text{-}CH_3OC_6H_4NO_2 + KF$$

p-Fluoronitrobenzene + Potassium methoxide → *p*-Nitroanisole (93%) + Potassium fluoride

Indeed, the order of leaving-group reactivity in nucleophilic aromatic substitution is the opposite of that seen in aliphatic substitution. *Fluoride is the most reactive leaving group in nucleophilic aromatic substitution, iodide the least reactive.*

$p\text{-}XC_6H_4NO_2$

Relative reactivity toward sodium methoxide in methanol (50°C):

X = F	312
X = Cl	1.0
X = Br	0.8
X = I	0.4

Kinetic studies of many of the reactions described in this section have demonstrated that they follow a second-order rate law:

$$\text{Rate} = k[\text{aryl halide}][\text{nucleophile}]$$

Second-order kinetics is usually interpreted in terms of a bimolecular rate-determining step. In this case, then, we look for a mechanism in which both the aryl halide and the nucleophile are involved in the slowest step of the sequence. Such a mechanism is described in the following section.

23.6 THE ADDITION-ELIMINATION MECHANISM OF NUCLEOPHILIC AROMATIC SUBSTITUTION

The generally accepted mechanism for nucleophilic aromatic substitution in nitro-substituted aryl halides, illustrated for the reaction of *p*-fluoronitrobenzene with sodium methoxide, is outlined in Figure 23.3. It is a two-step **addition-elimination mechanism,** in which addition of the nucleophile to the aryl halide is followed by elimination of the halide leaving group. The mechanism is consistent with the following experimental observations:

1. *Kinetics:* As the observation of second-order kinetics requires, the rate-determining step (step 1) involves both the aryl halide and the nucleophile.
2. *Rate-enhancing effect of the nitro group:* The nucleophilic addition step is slow because the aromatic character of the ring must be sacrificed in order to form the cyclohexadienyl anion intermediate. Only when the anionic intermediate is stabilized by the presence of a strong electron-withdrawing substituent ortho or para to the leaving group will the activation energy for its formation be low enough to

Overall reaction:

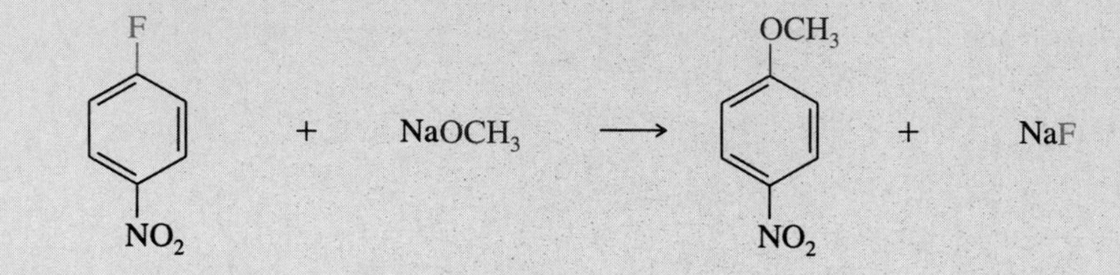

p-Fluoronitrobenzene + $NaOCH_3$ ⟶ p-Nitroanisole + NaF

p-Fluoronitrobenzene | Sodium methoxide | p-Nitroanisole | Sodium fluoride

Step 1: Addition stage. The nucleophile, in this case methoxide ion, adds to the carbon atom that bears the leaving group to give a cyclohexadienyl anion intermediate.

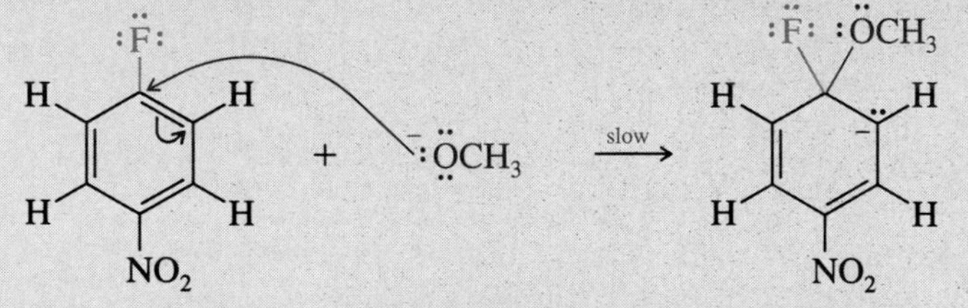

p-Fluoronitrobenzene + Methoxide ion $\xrightarrow{\text{slow}}$ Cyclohexadienyl anion intermediate

Step 2: Elimination stage. Loss of halide from the cyclohexadienyl intermediate restores the aromaticity of the ring and gives the product of nucleophilic aromatic substitution.

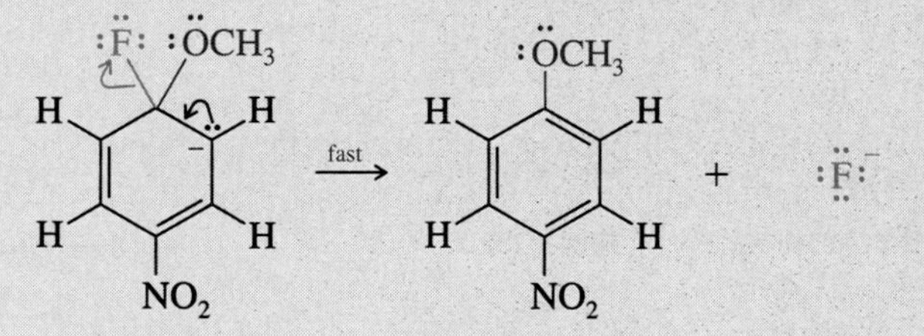

Cyclohexadienyl anion intermediate $\xrightarrow{\text{fast}}$ p-Nitroanisole + Fluoride ion

FIGURE 23.3 Sequence of steps that describes the addition-elimination mechanism of nucleophilic aromatic substitution.

provide a reasonable reaction rate. We can illustrate the stabilization that a *p*-nitro group provides by examining the resonance structures for the cyclohexadienyl anion formed from methoxide and *p*-fluoronitrobenzene:

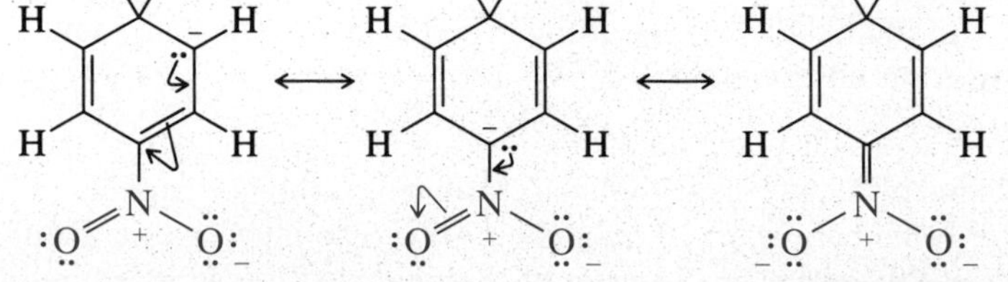

Most stable resonance structure; negative charge is on oxygen

PROBLEM 23.3 Write the most stable resonance structure for the cyclohexadienyl anion formed by reaction of methoxide ion with *o*-fluoronitrobenzene.

m-Fluoronitrobenzene reacts with sodium methoxide 10^5 times more slowly than do its ortho and para isomers. According to the resonance description, direct conjugation of the negatively charged carbon with the nitro group is not possible in the cyclohexadienyl anion intermediate from *m*-fluoronitrobenzene, and the decreased reaction rate reflects the decreased stabilization afforded this intermediate.

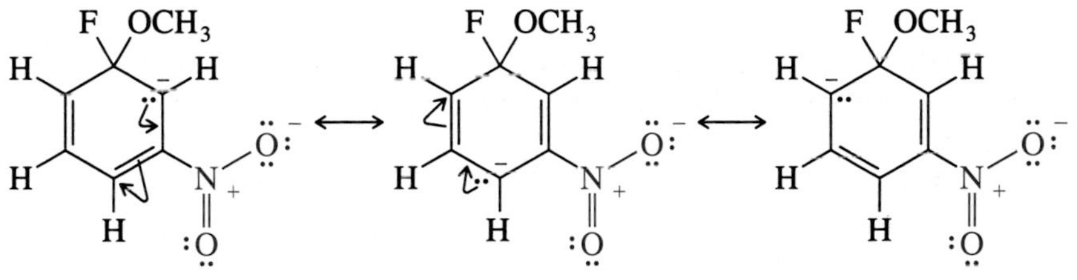

(Negative charge is restricted to carbon in all resonance forms)

PROBLEM 23.4 Reaction of 1,2,3-tribromo-5-nitrobenzene with sodium ethoxide in ethanol gave a single product, $C_8H_7Br_2NO_3$, in quantitative yield. Suggest a reasonable structure for this compound.

3. *Leaving-group effects:* Since aryl fluorides have the strongest carbon-halogen bond and react fastest, the rate-determining step cannot involve carbon-halogen bond cleavage. According to the mechanism in Figure 23.3 the carbon-halogen bond breaks in the rapid elimination step that follows the rate-determining addition step. The unusually high reactivity of aryl fluorides arises because fluorine is the most electronegative of the halogens, and its greater ability to attract electrons increases the rate of formation of the cyclohexadienyl anion intermediate in the first step of the mechanism.

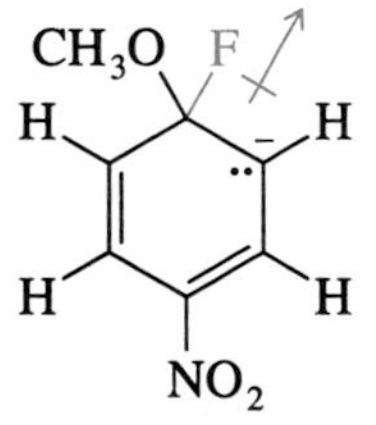

is more stable than

Fluorine stabilizes cyclohexadienyl anion by withdrawing electrons.

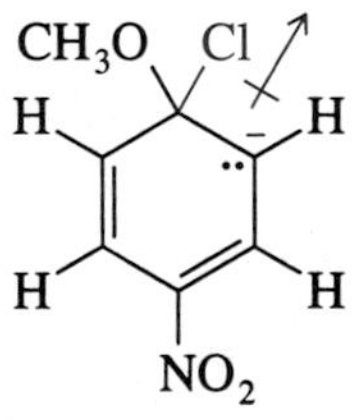

Chlorine is less electronegative than fluorine and does not stabilize cyclohexadienyl anion to as great an extent.

Before leaving this mechanistic discussion, we should mention that the addition-elimination mechanism for nucleophilic aromatic substitution illustrates a principle you should remember. The words *activating* and *deactivating* as applied to substituent effects in organic chemistry are without meaning when they stand alone. When we say that a group is activating or deactivating, we need to specify the reaction type that is being considered. A nitro group is a strongly *deactivating* substituent in *electrophilic* aromatic substitution, where it markedly destabilizes the key cyclohexadienyl cation intermediate:

Nitrobenzene and an electrophile

very slow

Cyclohexadienyl cation intermediate; nitro group is destabilizing

$-H^+$ fast

Product of electrophilic aromatic substitution

A nitro group is a strongly *activating* substituent in *nucleophilic aromatic substitution,* where it stabilizes the key cyclohexadienyl anion intermediate:

o-Halonitrobenzene ($X = F$, Cl, Br, or I) and a nucleophile

slow addition

Cyclohexadienyl anion intermediate; nitro group is stabilizing

fast elimination

Product of nucleophilic aromatic substitution

$+ :X^-$

A nitro group behaves the same way in both reactions: it attracts electrons. Reaction is retarded when electrons flow from the aromatic ring to the attacking species (electrophilic aromatic substitution). Reaction is facilitated when electrons flow from the attacking species to the aromatic ring (nucleophilic aromatic substitution). By being aware of the connection between reactivity and substituent effects, you will sharpen your appreciation of how chemical reactions occur.

23.7 RELATED NUCLEOPHILIC AROMATIC SUBSTITUTION REACTIONS

The most common types of aryl halides in nucleophilic aromatic substitutions are those that bear *o*- or *p*-nitro substituents. Among other classes of reactive aryl halides, a few merit special consideration. One class includes highly fluorinated aromatic compounds such as hexafluorobenzene, which undergoes substitution of one of its fluorines on reaction with nucleophiles such as sodium methoxide.

$$C_6F_6 \xrightarrow[CH_3OH,\ 65°C]{NaOCH_3} C_6F_5OCH_3$$

Hexafluorobenzene

2,3,4,5,6-Pentafluoroanisole (72%)

Here it is the combined electron-attracting effects of the six fluorine substituents that combine to stabilize the cyclohexadienyl anion intermediate and permit the reaction to proceed so readily.

Halides derived from certain heterocyclic aromatic compounds are often quite reactive toward nucleophiles. 2-Chloropyridine, for example, reacts with sodium methoxide some 230 million times faster than chlorobenzene at 50°C.

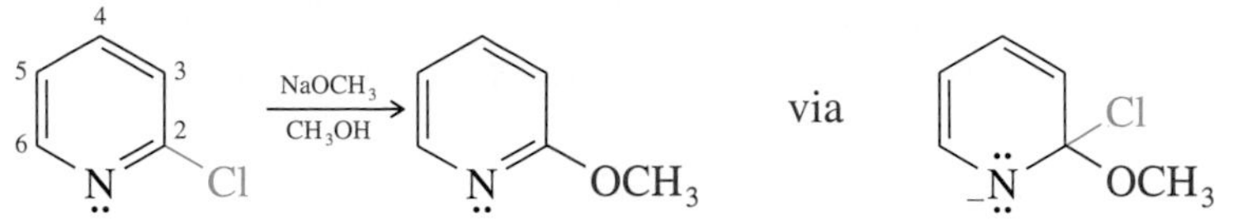

$\xrightarrow[CH_3OH]{NaOCH_3}$ via

2-Chloropyridine 2-Methoxypyridine Anionic intermediate

Again, rapid reaction is attributed to the stability of the intermediate formed in the addition step. In contrast to chlorobenzene, where the negative charge of the intermediate must be borne by carbon, the anionic intermediate in the case of 2-chloropyridine has its negative charge on nitrogen. Since nitrogen is more electronegative than carbon, the intermediate is more stable and is formed faster than the one from chlorobenzene.

PROBLEM 23.5 Offer an explanation for the observation that 4-chloropyridine is more reactive toward nucleophiles than 3-chloropyridine.

There is another type of nucleophilic aromatic substitution which occurs under quite different reaction conditions from those discussed to this point and which proceeds by a different and rather surprising mechanism. It is described in the following section.

23.8 THE ELIMINATION-ADDITION MECHANISM OF NUCLEOPHILIC AROMATIC SUBSTITUTION. BENZYNE

Very strong bases such as sodium or potassium amide react readily with aryl halides, even those without electron-withdrawing substituents, to give products corresponding to nucleophilic substitution of halide by the base.

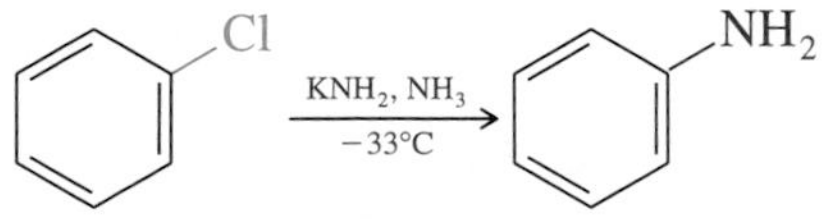

$\xrightarrow[-33°C]{KNH_2,\ NH_3}$

Chlorobenzene Aniline (52%)

For a long time, observations concerning the regiochemistry of these reactions presented organic chemists with a puzzle. Substitution did not occur exclusively at the carbon from which the halide leaving group departed. Rather, a mixture of regioisomers was obtained in which the amine group was either on the carbon that originally bore the leaving group or on one of the carbons adjacent to it. Thus *o*-bromotoluene gave a mixture of *o*-methylaniline and *m*-methylaniline; *p*-bromotoluene gave *m*-methylaniline and *p*-methylaniline.

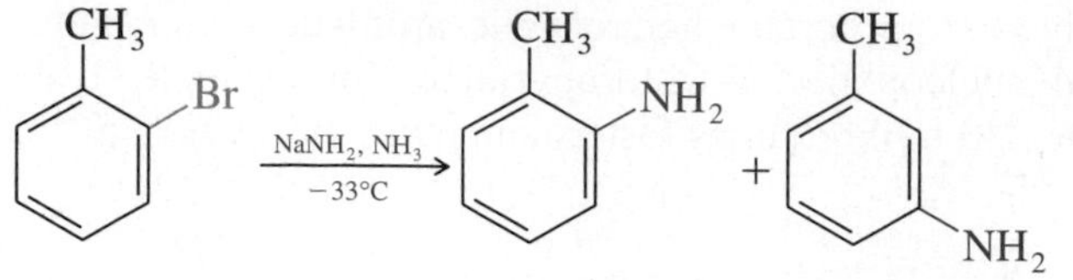

o-Bromotoluene $\xrightarrow[-33°C]{NaNH_2, NH_3}$ o-Methylaniline + m-Methylaniline

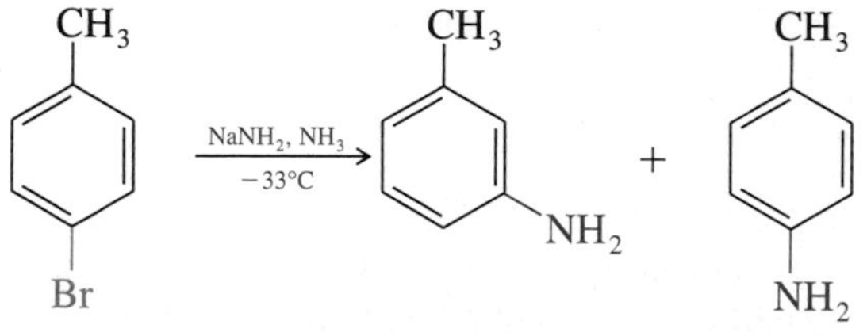

p-Bromotoluene $\xrightarrow[-33°C]{NaNH_2, NH_3}$ m-Methylaniline + p-Methylaniline

Three regioisomers (*o*-, *m*-, and *p*-methylaniline) were formed from *m*-bromotoluene.

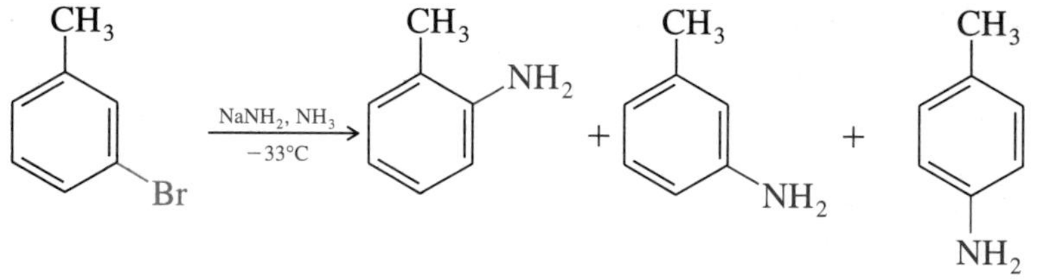

m-Bromotoluene $\xrightarrow[-33°C]{NaNH_2, NH_3}$ o-Methylaniline + m-Methylaniline + p-Methylaniline

These results rule out nucleophilic aromatic substitution by the addition-elimination mechanism, since that mechanism requires the attacking nucleophile to attach itself to the carbon from which the leaving group departs.

A solution to the question of the mechanism of these reactions was provided by John D. Roberts in 1953 on the basis of an imaginative experiment. Roberts prepared a sample of chlorobenzene in which one of the carbons, the one bearing the chlorine, was the radioactive mass 14 isotope of carbon. Reaction with potassium amide in liquid ammonia yielded aniline containing almost exactly half of its ^{14}C label at C-1 and half at C-2:

This work was done while Roberts was at MIT. He later moved to the California Institute of Technology, where he became a leader in applying nmr spectroscopy to nuclei other than protons, especially ^{13}C and ^{15}N.

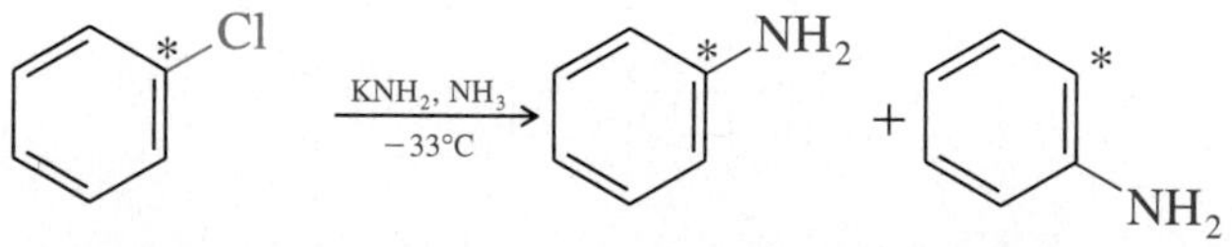

Chlorobenzene-1-^{14}C (* = ^{14}C) $\xrightarrow[-33°C]{KNH_2, NH_3}$ Aniline-1-^{14}C (48%) + Aniline-2-^{14}C (52%)

The mechanism most consistent with the observations of this isotopic labeling experiment is the **elimination-addition mechanism** outlined in Figure 23.4. The first stage in this mechanism is a base-promoted dehydrohalogenation of chlorobenzene. The intermediate formed in this step contains a triple bond in an aromatic ring and is called **benzyne.** Aromatic compounds related to benzyne are known as **arynes.** The triple bond in benzyne is somewhat different from the usual triple bond of an alkyne, however. In benzyne one of the π components of the triple bond is part of the delocal-

Overall reaction:

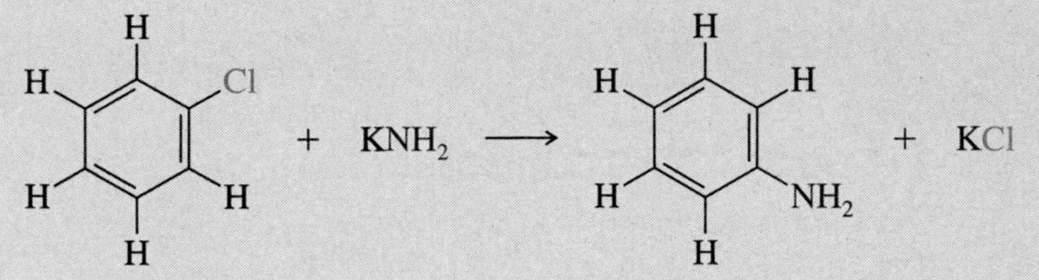

Chlorobenzene + KNH_2 ⟶ Aniline + KCl

Step 1: Elimination stage. Amide ion is a very strong base and brings about the dehydrohalogenation of chlorobenzene by abstracting a proton from the carbon adjacent to the one that bears the leaving group. The product of this step is an unstable intermediate called *benzyne.*

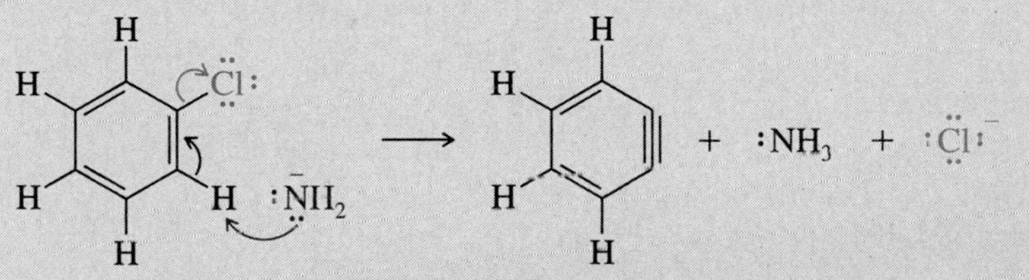

Chlorobenzene + $:\ddot{N}H_2^-$ ⟶ Benzyne + $:NH_3$ + $:\ddot{Cl}:^-$

Step 2: Beginning of addition phase. Amide ion acts as a nucleophile and adds to one of the carbons of the triple bond. The product of this step is a carbanion.

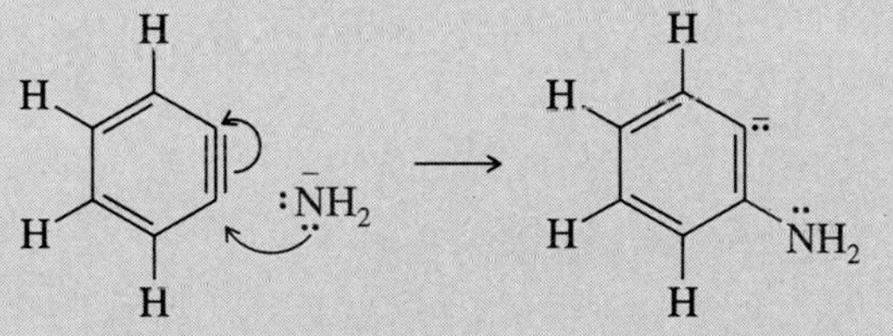

Benzyne + $:\ddot{N}H_2^-$ ⟶ Aryl anion

Step 3: Completion of addition phase. The aryl anion abstracts a proton from the ammonia used as the solvent in the reaction.

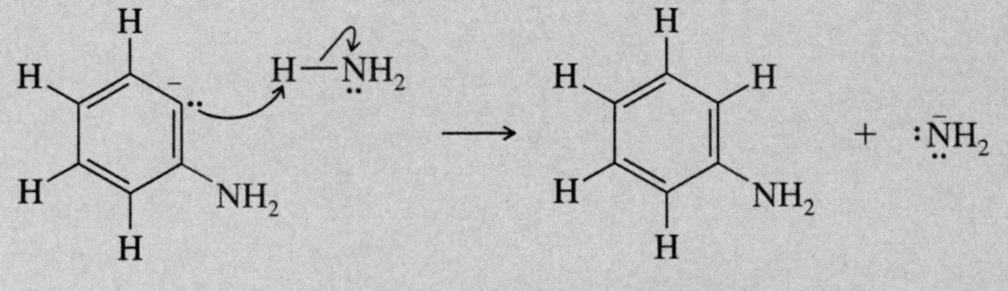

Aryl anion + $H—\ddot{N}H_2$ ⟶ Aniline + $:\ddot{N}H_2^-$

FIGURE 23.4 Sequence of steps that describes the elimination-addition mechanism of nucleophilic aromatic substitution.

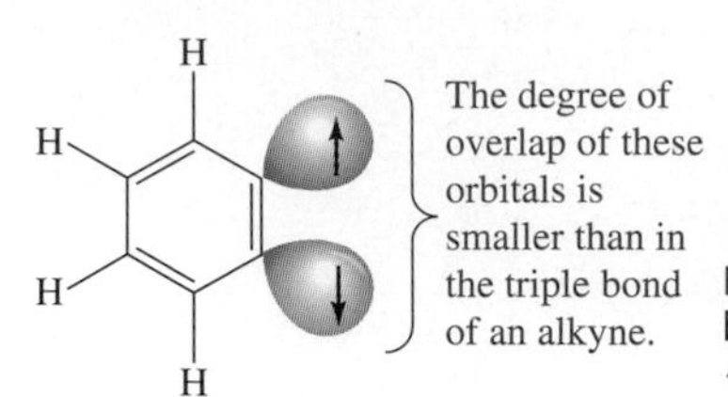

FIGURE 23.5 The sp^2 orbitals in the plane of the ring in benzyne are not properly aligned for good overlap, and π bonding is weak.

ized π system of the aromatic ring. The second π component results from overlapping sp^2 hybridized orbitals (*not p-p* overlap), lies in the plane of the ring, and does not interact with the aromatic π system. This π bond is relatively weak, since, as illustrated in Figure 23.5, its contributing sp^2 orbitals are not oriented properly for effective overlap.

Because the ring prevents linearity of the C—C≡C—C unit and because the π bonding in that unit is weak, benzyne is strained and highly reactive. This enhanced reactivity is evident in the second stage of the elimination-addition mechanism as shown in steps 2 and 3 of Figure 23.4. In this stage the base acts as a nucleophile and adds to the strained bond of benzyne to form a carbanion. The carbanion, an *aryl anion,* then abstracts a proton from ammonia to yield the observed product.

The carbon that bears the leaving group and a carbon ortho to it become equivalent in the benzyne intermediate. Thus when chlorobenzene-1-^{14}C is the substrate, the amino group may be introduced with equal likelihood at either position.

PROBLEM 23.6 2-Bromo-1,3-dimethylbenzene is inert to nucleophilic aromatic substitution on treatment with sodium amide in liquid ammonia. It is recovered unchanged even after extended contact with the reagent. Suggest an explanation for this lack of reactivity.

Once the intermediacy of an aryne intermediate was established, the reason for the observed regioselectivity of substitution in *o*-, *m*-, and *p*-chlorotoluene became evident. Only a single aryne intermediate may be formed from *o*-chlorotoluene, but this aryne yields a mixture containing comparable amounts of *o*- and *m*-methylaniline.

o-Chlorotoluene $\xrightarrow[NH_3]{KNH_2}$ 3-Methylbenzyne $\xrightarrow[NH_3]{KNH_2}$ *o*-Methylaniline + *m*-Methylaniline

Similarly, *p*-chlorotoluene gives a single aryne, and this aryne gives a mixture of *m*- and *p*-methylaniline.

p-Chlorotoluene $\xrightarrow[NH_3]{KNH_2}$ 4-Methylbenzyne $\xrightarrow[NH_3]{KNH_2}$ *m*-Methylaniline + *p*-Methylaniline

Two isomeric arynes give the three isomeric substitution products formed from *m*-chlorotoluene:

CH_3 KNH_2 / NH_3 CH_3 KNH_2 / NH_3 CH_3 NH_2 + CH_3 NH_2

Cl

m-Chlorotoluene

3-Methylbenzyne *o*-Methylaniline *m*-Methylaniline

CH_3 KNH_2 / NH_3 CH_3 NH_2 + CH_3 NH_2

4-Methylbenzyne *m*-Methylaniline *p*-Methylaniline

While nucleophilic aromatic substitution by the elimination-addition mechanism is most commonly seen with very strong amide bases, it also occurs with bases such as hydroxide ion at high temperatures. A ^{14}C-labeling study revealed that hydrolysis of chlorobenzene proceeds by way of a benzyne intermediate.

* Cl NaOH, H_2O / 395°C * OH + * OH

Chlorobenzene-1-^{14}C Phenol-1-^{14}C (54%) Phenol-2-^{14}C (43%)

PROBLEM 23.7 Two isomeric phenols are obtained in comparable amounts on hydrolysis of *p*-iodotoluene with 1 *M* sodium hydroxide at 300°C. Suggest reasonable structures for these two products.

23.9 DIELS-ALDER REACTIONS OF BENZYNE

Alternative methods for its generation have made it possible to use benzyne as an intermediate in a number of synthetic applications. One such method involves treating *o*-bromofluorobenzene with magnesium, usually in tetrahydrofuran as the solvent.

Br F Mg, THF / heat

o-Bromofluorobenzene Benzyne

The reaction proceeds by formation of the Grignard reagent from *o*-bromofluorobenzene. Since the order of reactivity of magnesium with aryl halides is ArI > ArBr > ArCl > ArF, the Grignard reagent has the structure shown and forms benzyne by loss of the salt FMgBr:

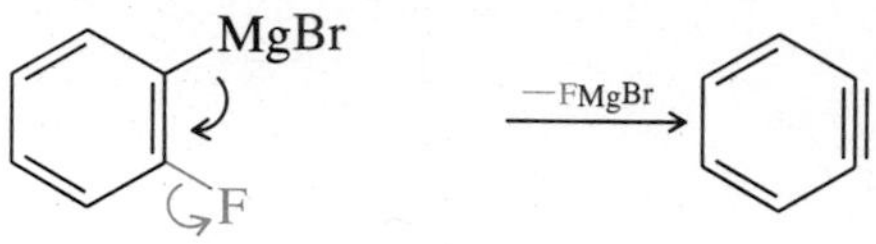

MgBr, F — FMgBr →

o-Fluorophenylmagnesium bromide   Benzyne

Its strained triple bond makes benzyne a relatively good dienophile, and when benzyne is generated in the presence of a conjugated diene, Diels-Alder cycloaddition occurs.

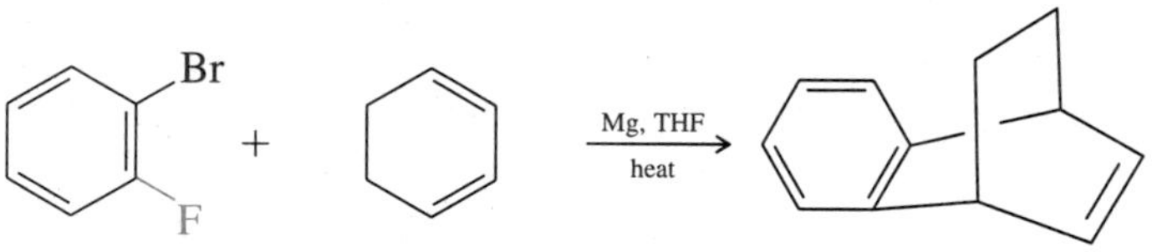

Br, F + → Mg, THF / heat →

o-Bromofluorobenzene   1,3-Cyclohexadiene   5,6-Benzobicyclo[2.2.2]octa-2,5-diene (46%)

via

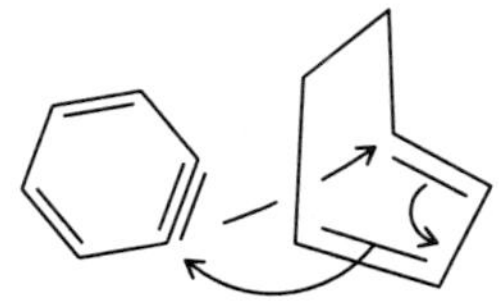

PROBLEM 23.8 Give the structure of the cycloaddition product formed when benzyne is generated in the presence of furan. (See Section 11.23, if necessary, to remind yourself of the structure of furan.)

Benzyne may also be generated by treating *o*-bromofluorobenzene with lithium. In this case, *o*-fluorophenyllithium is formed, which then loses lithium fluoride to form benzyne.

23.10 SUMMARY

Aryl halides have stronger carbon-halogen bonds (Section 23.1) and undergo nucleophilic substitution much more slowly (Section 23.4) than alkyl halides. Only when certain requirements are met do nucleophilic substitution reactions of aryl halides occur readily.

Aryl halides that bear a nitro group ortho or para to the halogen constitute the major class of compounds that react readily with nucleophiles (Section 23.5). *o*- or *p*-nitro-substituted aryl halides react with nucleophiles by an **addition-elimination mechanism** (Section 23.6).

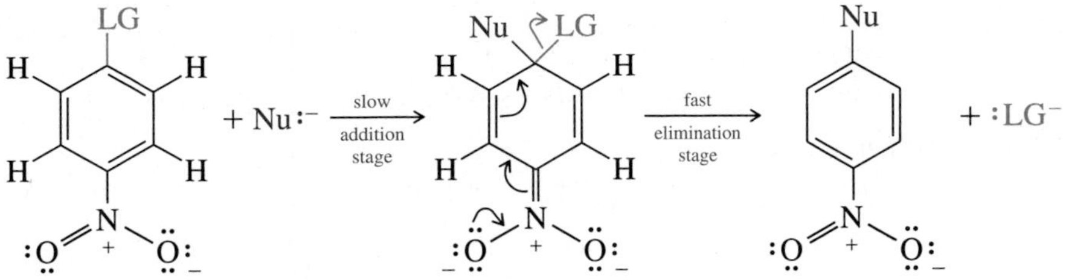

LG, H, H, H, H, + Nu:⁻ slow addition stage → Nu LG, H, H, H, H → fast elimination stage → Nu + :LG⁻

Nitro-substituted aryl halide

Cyclohexadienyl anion intermediate

Product of nucleophilic aromatic substitution

The rate-determining intermediate is a cyclohexadienyl anion and is stabilized by electron-withdrawing substituents. Other aryl halides that give stabilized anions can also undergo nucleophilic aromatic substitution at synthetically useful rates (Section 23.7).

Nucleophilic aromatic substitution can also occur by an **elimination-addition mechanism** (Section 23.8). This pathway is followed when the nucleophile is an exceptionally strong base such as amide ion in the form of sodium amide ($NaNH_2$) or potassium amide (KNH_2). **Benzyne** and related **arynes** are intermediates in nucleophilic aromatic substitutions that proceed by the elimination-addition mechanism.

LG, H + B:⁻ → slow elimination stage → B:⁻, BH fast addition stage → B, H

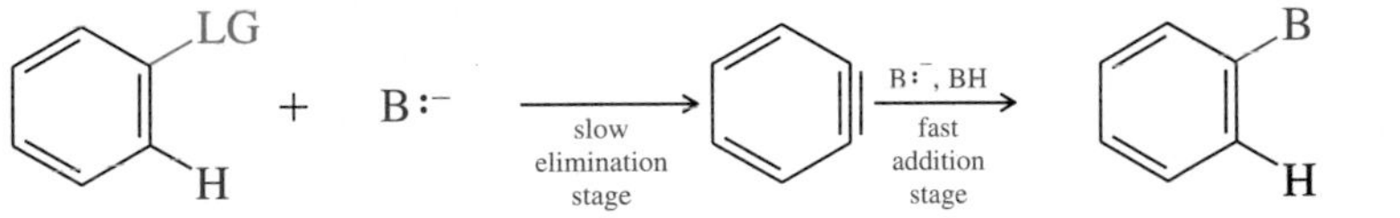

Aryl halide

Strong base

Benzyne

Product of nucleophilic aromatic substitution

Nucleophilic aromatic substitution by the elimination-addition mechanism can lead to substitution on the same carbon that bore the leaving group or on an adjacent carbon.

Benzyne is a reactive dienophile and gives Diels-Alder products when generated in the presence of dienes (Section 23.9). In these cases it is convenient to form benzyne by dissociation of the Grignard reagent of *o*-bromofluorobenzene.

PROBLEMS

23.9 Write a structural formula for each of the following:

(*a*) *m*-Chlorotoluene
(*b*) 2,6-Dibromoanisole
(*c*) *p*-Fluorostyrene
(*d*) 4,4′-Diiodobiphenyl
(*e*) 2-Bromo-1-chloro-4-nitrobenzene
(*f*) 1-Chloro-1-phenylethane
(*g*) *p*-Bromobenzyl chloride
(*h*) 2-Chloronaphthalene
(*i*) 1,8-Dichloronaphthalene
(*j*) 9-Fluorophenanthrene

23.10 Identify the principal organic product of each of the following reactions. If two regioisomers are formed in appreciable amounts, show them both.

(*a*) Chlorobenzene + acetyl chloride $\xrightarrow{AlCl_3}$

(*b*) Bromobenzene + magnesium $\xrightarrow{\text{diethyl ether}}$
(*c*) Product of part (*b*) + dilute hydrochloric acid ⟶
(*d*) Iodobenzene + lithium $\xrightarrow{\text{diethyl ether}}$
(*e*) Bromobenzene + sodium amide $\xrightarrow{\text{liquid ammonia, } -33°C}$
(*f*) *p*-Bromotoluene + sodium amide $\xrightarrow{\text{liquid ammonia, } -33°C}$
(*g*) 1-Bromo-4-nitrobenzene + ammonia ⟶
(*h*) *p*-Bromobenzyl bromide + sodium cyanide ⟶
(*i*) *p*-Chlorobenzenediazonium chloride + *N,N*-dimethylaniline ⟶
(*j*) Hexafluorobenzene + sodium hydrogen sulfide ⟶

23.11 Potassium *tert*-butoxide reacts with halobenzenes on heating in dimethyl sulfoxide to give *tert*-butyl phenyl ether.

(*a*) *o*-Fluorotoluene yields *tert*-butyl *o*-methylphenyl ether almost exclusively under these conditions. By which mechanism (addition-elimination or elimination-addition) do aryl fluorides react with potassium *tert*-butoxide in dimethyl sulfoxide?
(*b*) At 100°C, bromobenzene reacts over 20 times as fast as fluorobenzene. By which mechanism do aryl bromides react?

23.12 Predict the products formed when each of the following isotopically substituted derivatives of chlorobenzene is treated with sodium amide in liquid ammonia. Estimate as quantitatively as possible the composition of the product mixture. The asterisk in part (*a*) designates ^{14}C, and D in part (*b*) is ^{2}H.

(*a*) *—Cl (chlorobenzene labeled with * at C-1)

(*b*) D, D —Cl (2,6-dideuterochlorobenzene)

23.13 Choose the compound in each of the following pairs that reacts faster with sodium methoxide in methanol at 50°C:

(*a*) Chlorobenzene or *o*-chloronitrobenzene
(*b*) *o*-Chloronitrobenzene or *m*-chloronitrobenzene
(*c*) 4-Chloro-3-nitroacetophenone or 4-chloro-3-nitrotoluene
(*d*) 2-Fluoro-1,3-dinitrobenzene or 1-fluoro-3,5-dinitrobenzene
(*e*) 1,4-Dibromo-2-nitrobenzene or 1-bromo-2,4-dinitrobenzene

23.14 In each of the following reactions, an amine or a lithium amide derivative reacts with an aryl halide. Give the structure of the expected product and specify the mechanism by which it is formed.

(*a*) Br (bromobenzene) + LiN(pyrrolidine) ⟶

(*b*) Br, NO_2, NO_2 (1-bromo-2,4-dinitrobenzene) + piperidine (N—H) ⟶

(*c*) Br, NO_2, Br (1,4-dibromo-2-nitrobenzene) + piperidine (N—H) ⟶

23.15 Piperidine, the amine reactant in parts (*b*) and (*c*) of the preceding problem, reacts with 1-bromonaphthalene on heating at 230°C to give a single product, compound A ($C_{15}H_{17}N$), as a noncrystallizable liquid. The same reaction using 2-bromonaphthalene yielded an isomeric product, compound B, a solid melting at 50 to 53°C. Mixtures of A and B were formed when either 1- or 2-bromonaphthalene was allowed to react with sodium piperidide in piperidine. Suggest reasonable structures for compounds A and B and offer an explanation for their formation under each set of reaction conditions.

23.16 1,2,3,4,5-Pentafluoro-6-nitrobenzene reacts readily with sodium methoxide in methanol at room temperature to yield two major products, each having the molecular formula $C_7H_3F_4NO_3$. Suggest reasonable structures for these two compounds.

23.17 Predict the principal organic product in each of the following reactions:

(*a*) Cl, NO_2, CH_3 (benzene ring) + $C_6H_5CH_2SK \longrightarrow$

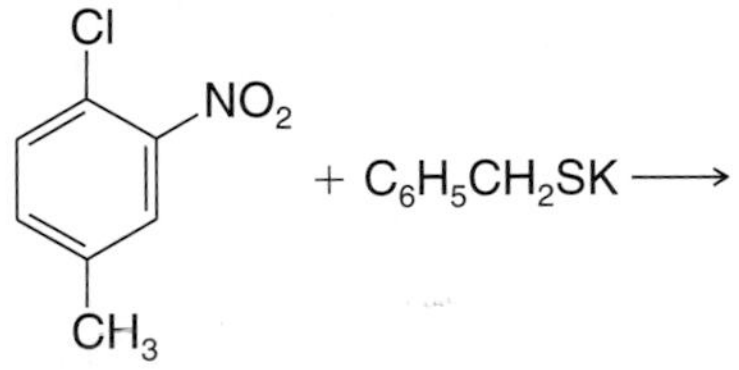

(*b*) Cl, NO_2, NO_2 (benzene ring) $\xrightarrow[\text{triethylene glycol}]{H_2NNH_2}$ $C_6H_6N_4O_4$

(*c*) Cl, Cl (benzene ring) $\xrightarrow[\text{2. } NH_3\text{, ethylene glycol, 140°C}]{\text{1. } HNO_3\text{, } H_2SO_4\text{, 120°C}}$ $C_6H_6N_4O_4$

(*d*) CF_3, Cl (benzene ring) $\xrightarrow[\text{2. } NaOCH_3\text{, } CH_3OH]{\text{1. } HNO_3\text{, } H_2SO_4}$ $C_8H_6F_3NO_3$

(*e*) I–(benzene ring)–CH_2Br + $(C_6H_5)_3P \longrightarrow$

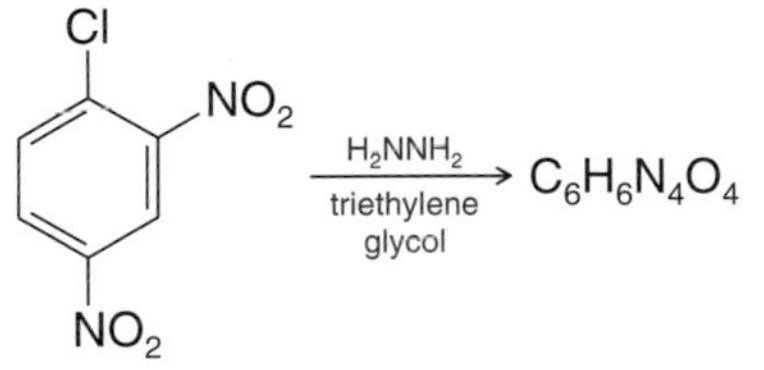

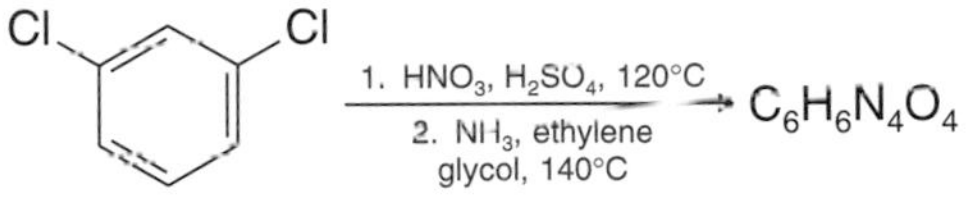

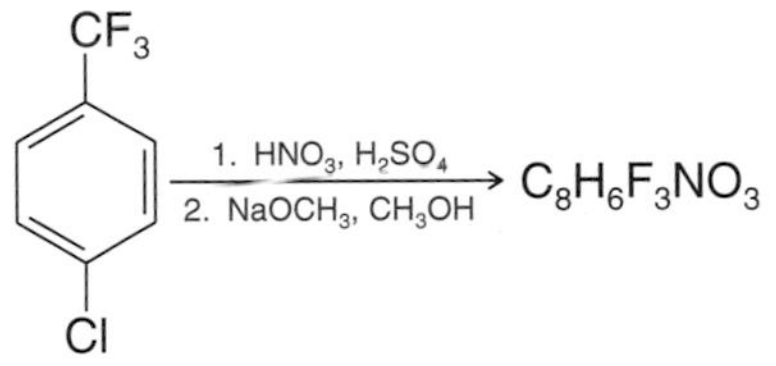

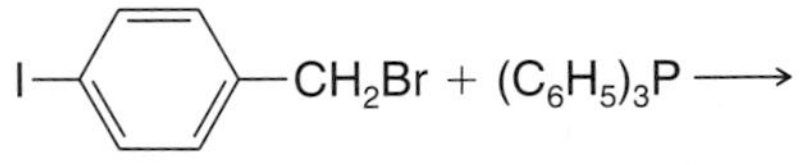

(*f*) Br–(benzene ring, CH_3)–OCH_3 $\xrightarrow[\text{2. } NaSCH_3]{\text{1. NBS, benzoyl peroxide, } CCl_4\text{, heat}}$ $C_9H_{11}BrOS$

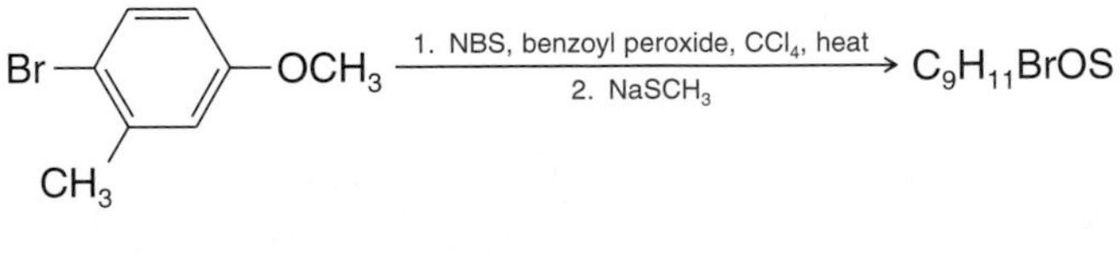

23.18 Hydrolysis of *p*-bromotoluene with aqueous sodium hydroxide at 300°C yields *m*-methylphenol and *p*-methylphenol in a 5:4 ratio. What is the meta-para ratio for the same reaction carried out on *p*-chlorotoluene?

23.19 The herbicide *trifluralin* is prepared by the following sequence of reactions. Identify compound A and deduce the structure of trifluralin.

CF_3 / Cl (benzene) $\xrightarrow[\text{heat}]{HNO_3,\ H_2SO_4}$ compound A ($C_7H_2ClF_3N_2O_4$) $\xrightarrow{(CH_3CH_2CH_2)_2NH}$ trifluralin

23.20 *Chlorbenside* is a pesticide used to control red spider mites. It is prepared by the sequence shown. Identify compounds A and B in this sequence. What is the structure of Chlorbenside?

O_2N–C₆H₄–CH_2Cl + NaS–C₆H₄–Cl ⟶ compound A

compound A $\xrightarrow{\text{1. Fe, HCl; 2. NaOH}}$ compound B

Chlorbenside $\xleftarrow[\text{2. CuCl}]{\text{1. } NaNO_2,\ HCl}$ compound B

23.21 A method for the generation of benzyne involves heating the diazonium salt from *o*-aminobenzoic acid (benzenediazonium-2-carboxylate). Using curved arrows, show how this substance forms benzyne. What two inorganic compounds are formed in this reaction?

$\overset{+}{N}{\equiv}N{:}$ / CO_2^- Benzenediazonium-2-carboxylate

23.22 The compound *triptycene* may be prepared as shown. What is the structure of compound A?

F / Br (benzene) + Compound A ($C_{14}H_{10}$) $\xrightarrow[\text{heat}]{\text{Mg, THF}}$

Triptycene

23.23 Nitro-substituted aromatic compounds that do not bear halide leaving groups react with nucleophiles according to the equation

NO_2, NO_2, X + :Y^- ⟶ NO_2, Y, X, NO_2

The product of this reaction, as its sodium salt, is called a *Meisenheimer complex* after the German chemist Jacob Meisenheimer, who reported on their formation and reactions in

1902. A Meisenheimer complex corresponds to the product of the nucleophilic addition stage in the addition-elimination mechanism for nucleophilic aromatic substitution.

(*a*) Give the structure of the Meisenheimer complex formed by addition of sodium ethoxide to 2,4,6-trinitroanisole.
(*b*) What other combination of reactants yields the same Meisenheimer complex as that of part (*a*)?

23.24 A careful study of the reaction of 2,4,6-trinitroanisole with sodium methoxide revealed that two different Meisenheimer complexes were present. Suggest reasonable structures for these two complexes.

23.25 Suggest a reasonable mechanism for each of the following reactions:

(*a*) $C_6H_5Br + CH_2(COOCH_2CH_3)_2 \xrightarrow[\text{2. } H_3O^+]{\text{1. excess } NaNH_2,\ NH_3} C_6H_5CH(COOCH_2CH_3)_2$

(*b*) $CH_2CH_2CH_2CH_2\overset{O}{\overset{\|}{C}}OCH_2CH_3$; Cl $\xrightarrow[\text{2. } H_3O^+]{\text{1. excess } NaNH_2,\ NH_3}$ $COOCH_2CH_3$

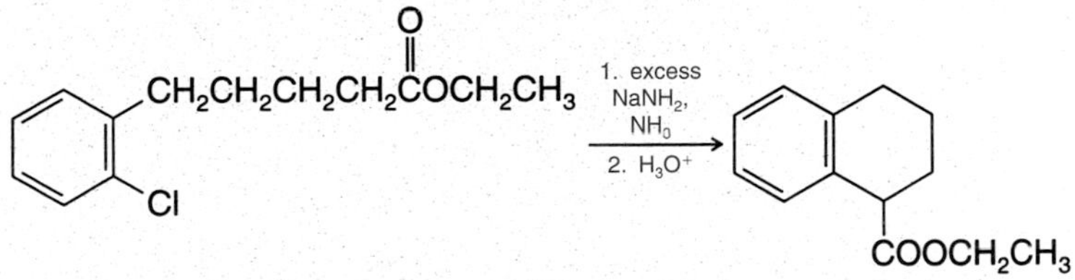

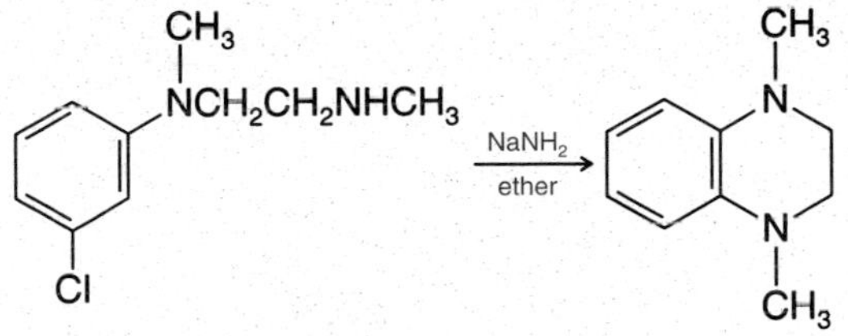

(*c*) CH_3; $NCH_2CH_2NHCH_3$; Cl $\xrightarrow[\text{ether}]{NaNH_2}$ CH_3; N; N; CH_3

(*d*) F; F; F; F; F; OCH_2CH_2OH $\xrightarrow[\text{heat}]{K_2CO_3}$ F; F; F; F; O; O

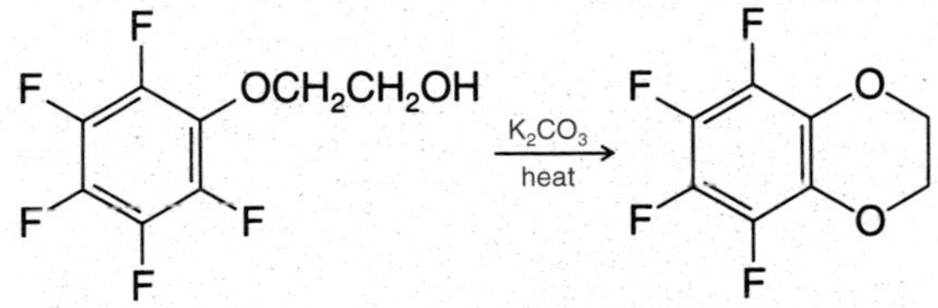

23.26 Mixtures of chlorinated derivatives of biphenyl, called *polychlorinated biphenyls,* or *PCBs,* were once prepared industrially on a large scale as insulating materials in electrical equipment. As equipment containing PCBs was discarded, the PCBs entered the environment at a rate that reached an estimated 25,000 lb/yr. PCBs are very stable and accumulate in the fatty tissue of fish, birds, and mammals. They have been shown to be *teratogenic,* meaning that they induce mutations in the offspring of affected individuals. Some countries have banned the use of PCBs. A large number of chlorinated biphenyls are possible, and the commercially produced material is a mixture of many compounds.

(*a*) How many monochloro derivatives of biphenyl are possible?
(*b*) How many dichloro derivatives are possible?
(*c*) How many octachloro derivatives are possible?
(*d*) How many nonachloro derivatives are possible?

23.27 DDT-resistant insects have the ability to convert DDT to a less toxic substance called DDE. The mass spectrum of DDE shows a cluster of peaks for the molecular ion at *m/z* 316, 318, 320, 322, and 324. Suggest a reasonable structure for DDE.

$Cl-C_6H_4-CH(CCl_3)-C_6H_4-Cl$ DDT (*d*ichloro*d*iphenyl*t*richloroethane)

MOLECULAR MODELING EXERCISES

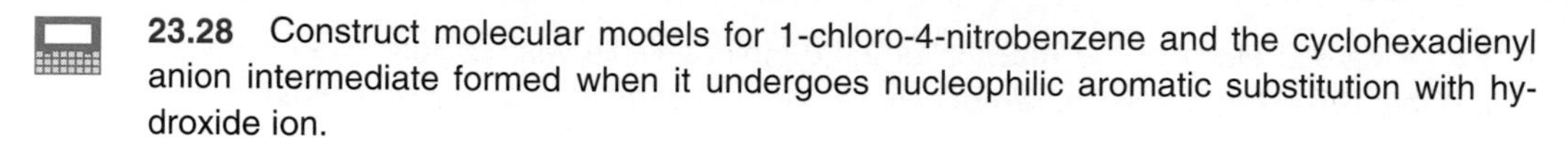

23.28 Construct molecular models for 1-chloro-4-nitrobenzene and the cyclohexadienyl anion intermediate formed when it undergoes nucleophilic aromatic substitution with hydroxide ion.

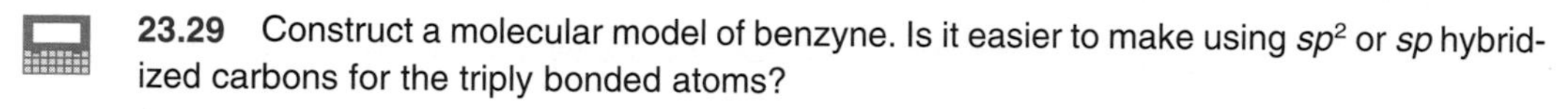

23.29 Construct a molecular model of benzyne. Is it easier to make using sp^2 or sp hybridized carbons for the triply bonded atoms?

CHAPTER 24

PHENOLS

P*henols* are compounds that have a hydroxyl group bonded directly to a benzene or benzenoid ring. The parent compound of this group, C_6H_5OH, called simply *phenol,* is an important industrial chemical. Many of the properties of phenols are analogous to those of alcohols, but this similarity is something of an oversimplification. Like arylamines, phenols are difunctional compounds; the hydroxyl group and the aromatic ring interact strongly, affecting each other's reactivity. This interaction leads to some novel and useful properties of phenols. A key step in the synthesis of aspirin, for example, is without parallel in the reactions of either alcohols or arenes. With periodic reminders of the ways in which phenols resemble alcohols and arenes, this chapter emphasizes the ways in which phenols are unique.

24.1 NOMENCLATURE

An old name for benzene was *phene,* and its hydroxyl derivative came to be called *phenol.* This, like many other entrenched common names, is an acceptable IUPAC name. Likewise, *o*-, *m*-, and *p*-cresol are acceptable names for the various ring-substituted hydroxyl derivatives of toluene. More highly substituted compounds are named as derivatives of phenol. Numbering of the ring begins at the hydroxyl-substituted carbon and proceeds in the direction that gives the lower number to the next substituted carbon. Substituents are cited in alphabetical order.

The systematic name for phenol is *benzenol.*

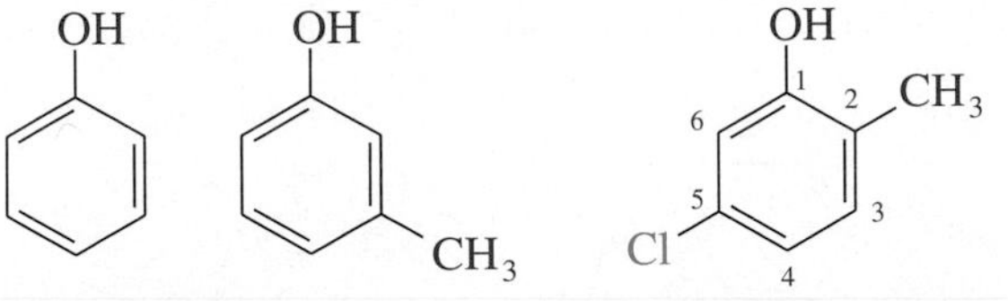

OH OH OH
1 2 CH_3
6
5 3
CH_3 Cl 4

Phenol *m*-Cresol 5-Chloro-2-methylphenol

Pyrocatechol is often called *catechol.*

The three dihydroxy derivatives of benzene may be named as 1,2-, 1,3-, and 1,4-benzenediol, respectively, but each is more familiarly known by the common name indicated in parentheses below. These common names are permissible IUPAC names.

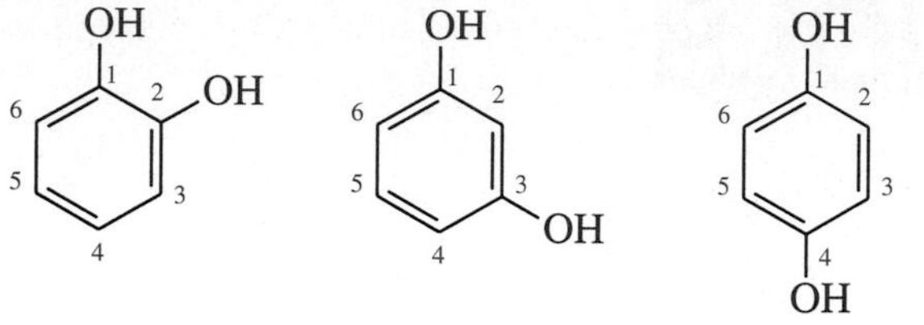

1,2-Benzenediol (pyrocatechol)

1,3-Benzenediol (resorcinol)

1,4-Benzenediol (hydroquinone)

The common names for the two hydroxy derivatives of naphthalene are 1-naphthol and 2-naphthol. These are also acceptable IUPAC names.

PROBLEM 24.1 Write structural formulas for each of the following compounds:

(*a*) Pyrogallol (1,2,3-benzenetriol)
(*b*) *o*-Benzylphenol
(*c*) 3-Nitro-1-naphthol
(*d*) 4-Chlororesorcinol

SAMPLE SOLUTION (*a*) Like the dihydroxybenzenes, the isomeric trihydroxybenzenes have unique names. Pyrogallol, used as a developer of photographic film, is 1,2,3-benzenetriol. The three hydroxyl groups occupy adjacent positions on a benzene ring.

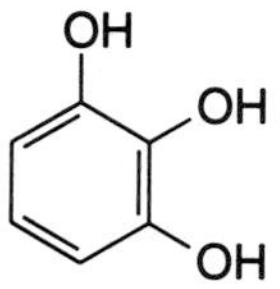

Pyrogallol (1,2,3-benzenetriol)

Carboxyl and acyl groups take precedence over the phenolic hydroxyl in determining the base name. The hydroxyl is treated as a substituent in these cases.

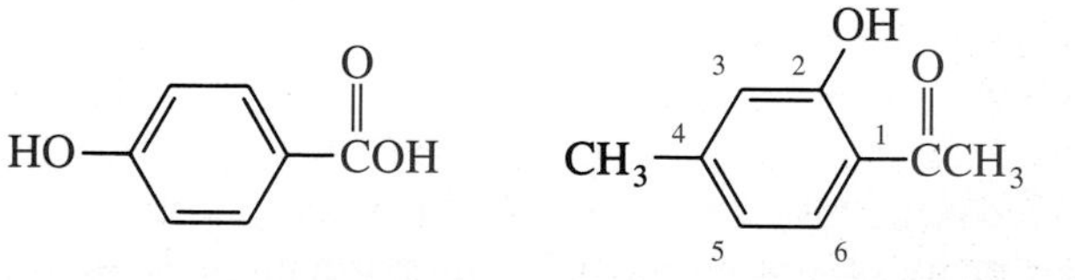

p-Hydroxybenzoic acid

2-Hydroxy-4-methylacetophenone

24.2 STRUCTURE AND BONDING

Phenol is planar, with a C—O—H angle of 109°, almost the same as the tetrahedral angle and not much different from the 108.5° C—O—H angle of methanol:

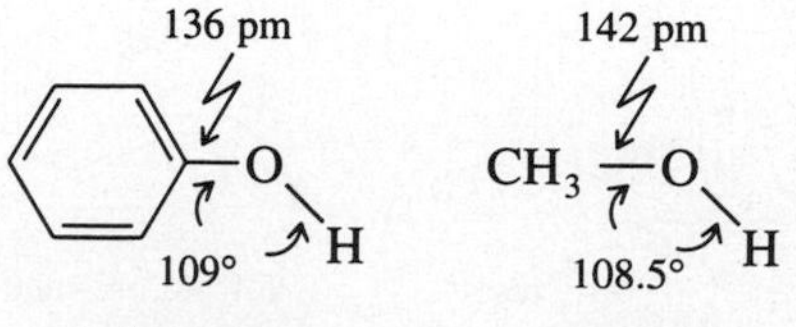

136 pm, 109°
Phenol

142 pm, 108.5°
Methanol

As we have noted on a number of occasions, bonds to sp^2 hybridized carbon are shorter than those to sp^3 hybridized carbon, and the case of phenols is no exception. The carbon-oxygen bond distance in phenol is slightly less than that in methanol.

In resonance terms, the shorter carbon-oxygen bond distance in phenol is attributed to the partial double-bond character that results from conjugation of the unshared electron pair of oxygen with the aromatic ring.

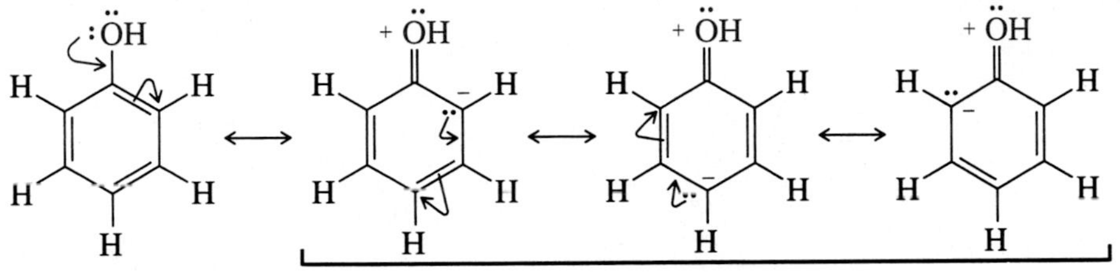

Most stable Lewis structure for phenol

Dipolar resonance forms of phenol

Many of the properties of phenols reflect the polarization implied by the resonance description. The hydroxyl oxygen is less basic, and the hydroxyl proton more acidic, in phenols than in alcohols. Electrophiles attack the aromatic ring of phenols much faster than they attack benzene, indicating that the ring, especially the positions ortho and para to the hydroxyl group, is relatively ''electron-rich.''

24.3 PHYSICAL PROPERTIES

The physical properties of phenols are strongly influenced by the hydroxyl group, which permits phenols to form hydrogen bonds with other phenol molecules (Figure 24.1*a*) and with water (Figure 24.1*b*). Thus, phenols have higher melting points and boiling points and are more soluble in water than arenes and aryl halides of comparable molecular weight. Table 24.1 compares phenol, toluene, and fluorobenzene with regard to these physical properties.

The physical properties of some representative phenols are collected in Appendix 1.

Some ortho-substituted phenols, such as *o*-nitrophenol, have boiling points that are significantly lower than those of the meta and para isomers. This is because the *intramolecular* hydrogen bond that forms between the hydroxyl group and the substituent partially compensates for the energy required to go from the liquid state to the vapor.

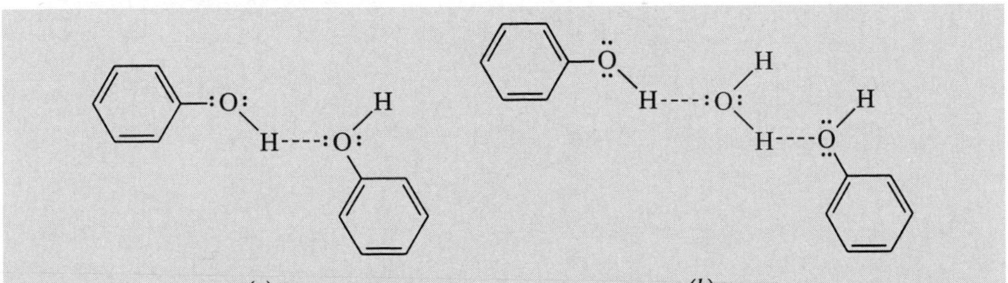

(*a*) (*b*)

FIGURE 24.1 (*a*) A hydrogen bond between two phenol molecules; (*b*) hydrogen bonds between water and phenol molecules.

TABLE 24.1
Comparison of Physical Properties of an Arene, a Phenol, and an Aryl Halide

	Compound		
Physical property	Toluene, $C_6H_5CH_3$	Phenol, C_6H_5OH	Fluorobenzene, C_6H_5F
Molecular weight	92	94	96
Melting point	−95°C	43°C	−41°C
Boiling point (1 atm)	111°C	132°C	85°C
Solubility in water (25°C)	0.05 g/100 mL	8.2 g/100 mL	0.2 g/100 mL

Intramolecular hydrogen bond in *o*-nitrophenol

PROBLEM 24.2 One of the hydroxybenzoic acids is known by the common name *salicylic acid.* Its methyl ester, methyl salicylate, occurs in oil of wintergreen. Methyl salicylate boils over 50°C lower than either of the other two methyl hydroxybenzoates. What is the structure of methyl salicylate? Why is its boiling point so much lower than that of either of its regioisomers?

24.4 ACIDITY OF PHENOLS

Because of its acidity, phenol was known as *carbolic acid* when Joseph Lister introduced it as an antiseptic in 1865 to prevent postoperative bacterial infections that were then a life-threatening hazard in even minor surgical procedures.

The most characteristic property of phenols is their acidity. Phenols are more acidic than alcohols but less acidic than carboxylic acids. Recall that carboxylic acids have ionization constants K_a of approximately 10^{-5} (pK_a 5), while the K_a's of alcohols are in the 10^{-16} to 10^{-20} range (pK_a 16 to 20). The K_a for most phenols is about 10^{-10} (pK_a 10).

To help understand why phenols are more acidic than alcohols, let us compare the ionization equilibria for phenol and ethanol. In particular, consider the differences in charge delocalization in ethoxide ion and in phenoxide ion. The negative charge in ethoxide ion is localized on oxygen and is stabilized only by solvation forces.

$$CH_3CH_2\ddot{O}{-}H \rightleftharpoons H^+ + CH_3CH_2\ddot{O}{:}^- \qquad K_a = 10^{-16}\ (pK_a = 16)$$

Ethanol   Proton   Ethoxide ion

The negative charge in phenoxide ion is stabilized both by solvation and by electron delocalization into the ring.

$$C_6H_5\ddot{O}{-}H \rightleftharpoons H^+ + C_6H_5\ddot{O}{:}^- \qquad K_a = 10^{-10}\ (pK_a = 10)$$

Phenol   Proton   Phenoxide ion

Electron delocalization in phenoxide is represented by resonance between the structures:

The negative charge in phenoxide ion is shared by the oxygen and the carbons that are ortho and para to it. Delocalization of its negative charge strongly stabilizes phenoxide ion.

A free energy diagram comparing the ionization of phenol with that of ethanol is shown in Figure 24.2. The energy difference between the two product states is quite large because of electron delocalization in phenoxide ion. The energy difference between the two starting states is smaller because electron delocalization in phenol, as described in Section 24.2, is accompanied by charge separation. Overall, the free energy change for ionization of phenol is considerably less than for ionization of ethanol, and the equilibrium constant is larger.

To place the acidity of phenol in perspective, note that while phenol is more than 1 million times more acidic than ethanol, it is over 100,000 times weaker than acetic acid. Thus, phenols can be separated from alcohols because they are more acidic, and from carboxylic acids because they are less acidic. On shaking an ether solution containing both an alcohol and a phenol with dilute sodium hydroxide, the phenol is

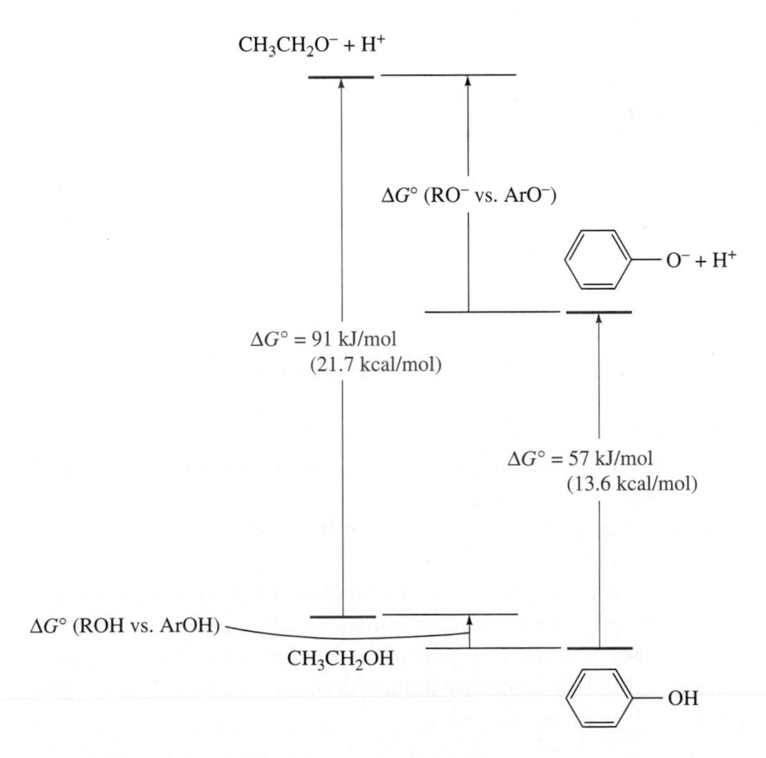

FIGURE 24.2 Free energies of ionization of ethanol and phenol in water. Most of the difference between the two is due to the large difference in stabilization energy of the phenoxide anion relative to the ethoxide anion.

converted quantitatively to its sodium salt, which is extracted into the aqueous phase. The alcohol remains in the ether phase.

$$C_6H_5{-}OH + HO^- \underset{}{\overset{K > 1}{\rightleftharpoons}} C_6H_5{-}O^- + H_2O$$

Phenol (stronger acid) | Hydroxide ion (stronger base) | Phenoxide ion (weaker base) | Water (weaker acid)

On shaking an ether solution of a phenol and a carboxylic acid with dilute sodium bicarbonate, the carboxylic acid is converted quantitatively to its sodium salt and extracted into the aqueous phase. The phenol remains in the ether phase.

$$C_6H_5{-}OH + HCO_3^- \overset{K < 1}{\rightleftharpoons} C_6H_5{-}O^- + H_2CO_3$$

Phenol (weaker acid) | Bicarbonate ion (weaker base) | Phenoxide ion (stronger base) | Carbonic acid (stronger acid)

It is necessary to keep the acidity of phenols in mind when we discuss preparation and reactions. Reactions that produce phenols, when carried out in basic solution, require an acidification step in order to convert the phenoxide ion to the neutral form of the phenol.

$$C_6H_5{-}O^- + H_3O^+ \overset{K > 1}{\rightleftharpoons} C_6H_5{-}OH + H_2O$$

Phenoxide ion (stronger base) | Hydronium ion (stronger acid) | Phenol (weaker acid) | Water (weaker base)

Many synthetic reactions involving phenols as nucleophiles are carried out in the presence of sodium or potassium hydroxide. Under these conditions the phenol is converted to the phenoxide ion, which is a far better nucleophile.

24.5 SUBSTITUENT EFFECTS ON THE ACIDITY OF PHENOLS

As Table 24.2 shows, most phenols have ionization constants that are similar to that of phenol itself. Substituent effects, in general, are small.

Alkyl substitution produces negligible changes in acidities, as do weakly electronegative groups attached to the ring.

Only when the substituent is strongly electron-withdrawing, as is a nitro group, is a substantial change in acidity noted. The ionization constants of *o*- and *p*-nitrophenol are several hundred times greater than that of phenol. An ortho- or para-nitro group greatly stabilizes the phenoxide ion by permitting a portion of the negative charge to be borne by its own oxygens.

TABLE 24.2
Acidities of Some Phenols

Compound name	Ionization constant K_a	pK_a
Monosubstituted phenols		
Phenol	1.0×10^{-10}	10.0
o-Cresol	4.7×10^{-11}	10.3
m-Cresol	8.0×10^{-11}	10.1
p-Cresol	5.2×10^{-11}	10.3
o-Chlorophenol	2.7×10^{-9}	8.6
m-Chlorophenol	7.6×10^{-9}	9.1
p-Chlorophenol	3.9×10^{-9}	9.4
o-Methoxyphenol	1.0×10^{-10}	10.0
m-Methoxyphenol	2.2×10^{-10}	9.6
p-Methoxyphenol	6.3×10^{-11}	10.2
o-Nitrophenol	5.9×10^{-8}	7.2
m-Nitrophenol	4.4×10^{-9}	8.4
p-Nitrophonol	6.9×10^{-8}	7.2
Di- and trinitrophenols		
2,4-Dinitrophenol	1.1×10^{-4}	4.0
3,5-Dinitrophenol	2.0×10^{-7}	6.7
2,4,6-Trinitrophenol	4.2×10^{-1}	0.4
Naphthols		
1-Naphthol	5.9×10^{-10}	9.2
2-Naphthol	3.5×10^{-10}	9.5

Recall from Section 24.1 that cresols are methyl-substituted derivatives of phenol.

Electron delocalization in *o*-nitrophenoxide ion

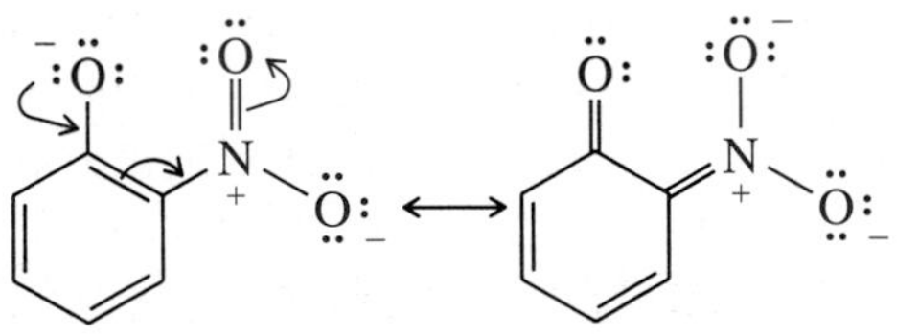

Electron delocalization in *p*-nitrophenoxide ion

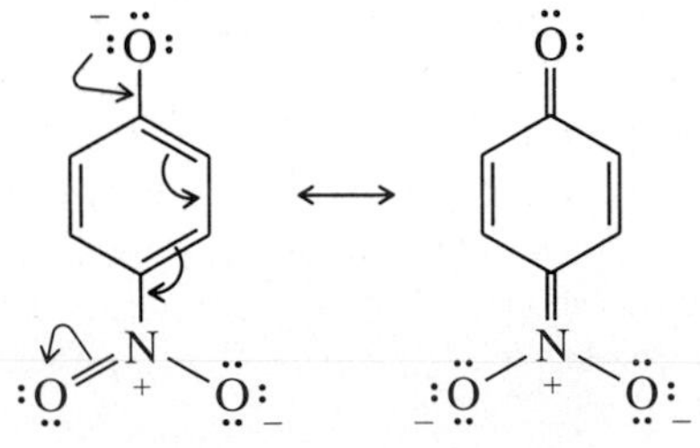

A meta-nitro group is not directly conjugated to the phenoxide oxygen and thus stabilizes a phenoxide ion to a smaller extent. *m*-Nitrophenol is more acidic than phenol but less acidic than either *o*- or *p*-nitrophenol.

PROBLEM 24.3 Which is the stronger acid in each of the following pairs? Explain your reasoning.

(*a*) Phenol or *p*-hydroxybenzaldehyde
(*b*) *m*-Cyanophenol or *p*-cyanophenol
(*c*) *o*-Fluorophenol or *p*-fluorophenol

SAMPLE SOLUTION (*a*) The best approach when comparing the acidities of different phenols is to assess opportunities for stabilization of negative charge in their anions. Electron delocalization in the anion of *p*-hydroxybenzaldehyde is very effective because of conjugation with the formyl group.

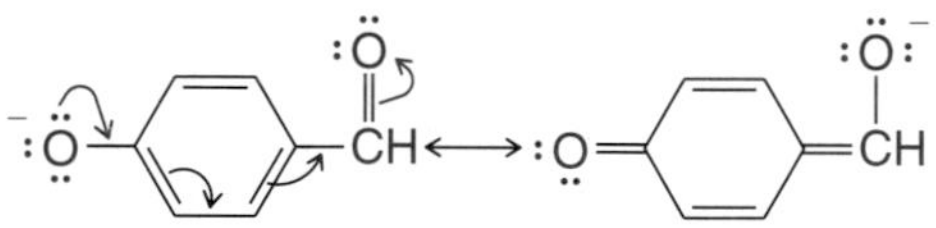

$$^{-}:\ddot{O}-C_6H_4-CH{=}\ddot{O}: \longleftrightarrow :\ddot{O}{=}C_6H_4{=}CH-\ddot{O}:^{-}$$

A formyl substituent, like a nitro group, is strongly electron-withdrawing and acid-strengthening, especially when ortho or para to the hydroxyl group. *p*-Hydroxybenzaldehyde, with a K_a of 2.4×10^{-8}, is a stronger acid than phenol.

Multiple substitution by strongly electron-withdrawing groups greatly increases the acidity of phenols, as the K_a values for 2,4-dinitrophenol (K_a 1.1×10^{-4}) and 2,4,6-trinitrophenol (K_a 4.2×10^{-1}) in Table 24.2 attest.

24.6 SOURCES OF PHENOLS

Phenol was first isolated in the early nineteenth century from coal tar, and a small portion of the more than 4 billion lb of phenol produced in the United States each year comes from this source. While significant quantities of phenol are used to prepare aspirin and dyes, most of it is converted to phenolic resins used in adhesives and plastics. Almost all the phenol produced commercially is synthetic, with several different processes in current use. These are summarized in Table 24.3.

The reaction of benzenesulfonic acid with sodium hydroxide (first entry in Table 24.3) proceeds by the addition-elimination mechanism of nucleophilic aromatic substitution (Section 23.6). Hydroxide replaces sulfite ion (SO_3^{2-}) at the carbon atom that bears the leaving group. Thus, *p*-toluenesulfonic acid is converted exclusively to *p*-cresol by an analogous reaction:

Can you recall how to prepare *p*-toluenesulfonic acid?

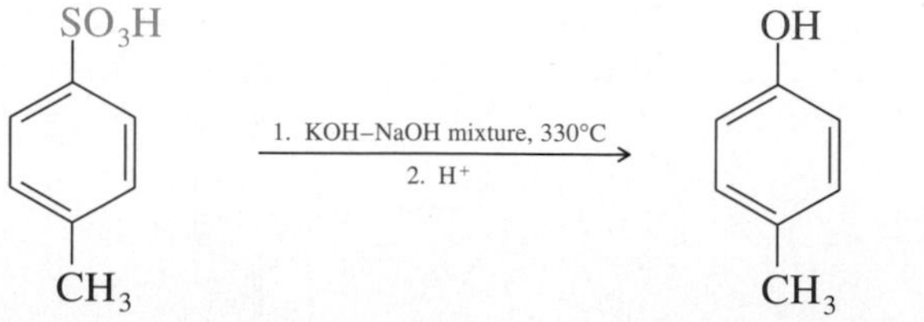

$$p\text{-}CH_3C_6H_4SO_3H \xrightarrow[\text{2. } H^+]{\text{1. KOH–NaOH mixture, 330°C}} p\text{-}CH_3C_6H_4OH$$

p-Toluenesulfonic acid → *p*-Cresol (63–72%)

TABLE 24.3
Industrial Syntheses of Phenol

Reaction and comments	Chemical equation
Reaction of benzenesulfonic acid with sodium hydroxide This is the oldest method for the preparation of phenol. Benzene is sulfonated and the benzenesulfonic acid heated with molten sodium hydroxide. Acidification of the reaction mixture gives phenol.	Benzenesulfonic acid ($C_6H_5SO_3H$) $\xrightarrow[2.\ H^+]{1.\ NaOH,\ 300-350°C}$ Phenol (C_6H_5OH)
Hydrolysis of chlorobenzene Heating chlorobenzene with aqueous sodium hydroxide at high pressure gives phenol after acidification.	Chlorobenzene (C_6H_5Cl) $\xrightarrow[2.\ H^+]{1.\ NaOH,\ H_2O,\ 370°C}$ Phenol (C_6H_5OH)
From cumene Almost all the phenol produced in the United States is prepared by this method. Oxidation of cumene takes place at the benzylic position to give a hydroperoxide. On treatment with dilute sulfuric acid, this hydroperoxide is converted to phenol and acetone.	Isopropylbenzene (cumene) ($C_6H_5CH(CH_3)_2$) $\xrightarrow{O_2}$ 1-Methyl-1-phenylethyl hydroperoxide ($C_6H_5C(CH_3)_2OOH$) 1-Methyl-1-phenylethyl hydroperoxide ($C_6H_5C(CH_3)_2OOH$) $\xrightarrow[H_2SO_4]{H_2O}$ Phenol (C_6H_5OH) + Acetone ($(CH_3)_2C{=}O$)

PROBLEM 24.4 Write a stepwise mechanism for the conversion of *p*-toluenesulfonic acid to *p*-cresol under the conditions shown in the preceding equation.

On the other hand, ^{14}C-labeling studies have shown that the base-promoted hydrolysis of chlorobenzene (second entry in Table 24.3) proceeds by the elimination-addition mechanism and involves benzyne as an intermediate.

Can you recall how to prepare chlorobenzene?

PROBLEM 24.5 Write a stepwise mechanism for the hydrolysis of chlorobenzene under the conditions shown in Table 24.3.

The most commonly used industrial synthesis of phenol is based on isopropylbenzene (cumene) as the starting material and is shown in the third entry of Table 24.3. The economically attractive features of this process are its use of cheap reagents (oxygen and sulfuric acid) and the fact that it yields two high-volume industrial chemicals, phenol and acetone. The mechanism of this novel synthesis forms the basis of Problem 24.31 at the end of this chapter.

Can you recall how to prepare isopropylbenzene?

The most important synthesis of phenols in the laboratory is from amines by hydrolysis of their corresponding diazonium salts, as described in Section 22.18:

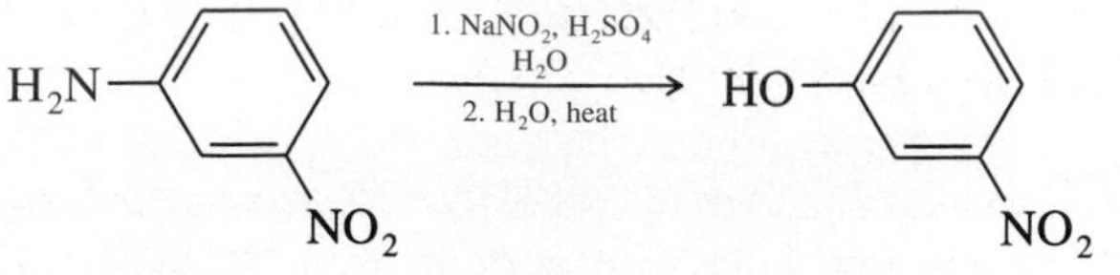

$$\text{H}_2\text{N-C}_6\text{H}_4\text{-NO}_2 \xrightarrow[\text{2. H}_2\text{O, heat}]{\text{1. NaNO}_2\text{, H}_2\text{SO}_4\text{, H}_2\text{O}} \text{HO-C}_6\text{H}_4\text{-NO}_2$$

m-Nitroaniline → *m*-Nitrophenol (81–86%)

24.7 NATURALLY OCCURRING PHENOLS

Phenolic compounds are commonplace natural products. Figure 24.3 presents a sampling of some naturally occurring phenols. Phenolic natural products can arise by a number of different biosynthetic pathways. In mammals aromatic rings are hydroxylated by way of arene oxide intermediates formed by the enzyme-catalyzed reaction between an aromatic ring and molecular oxygen:

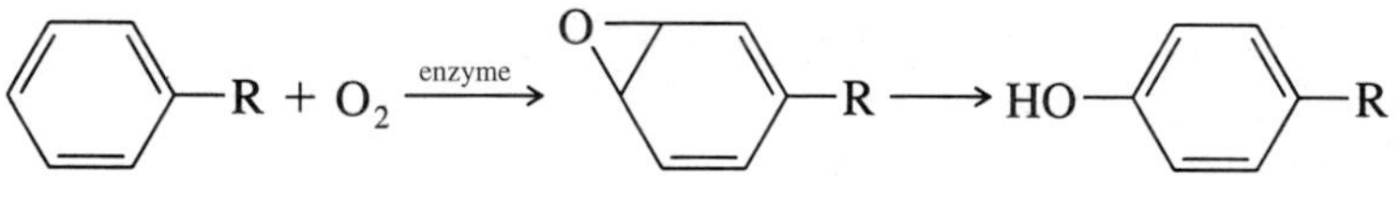

$$\text{Ar-R} + \text{O}_2 \xrightarrow{\text{enzyme}} \text{Arene oxide} \longrightarrow \text{HO-C}_6\text{H}_4\text{-R}$$

Arene → Arene oxide → Phenol

In plants phenol biosynthesis proceeds by building the aromatic ring from carbohydrate precursors that already contain the required hydroxyl group.

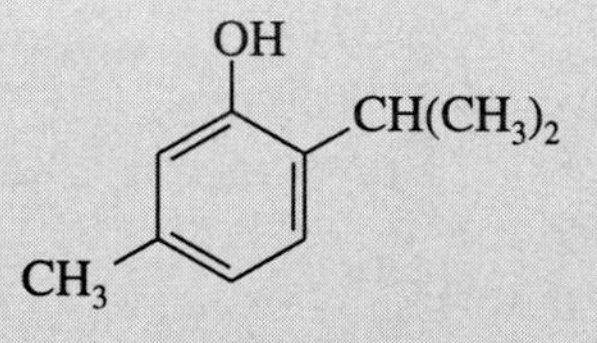

Thymol
(major constituent of oil of thyme)

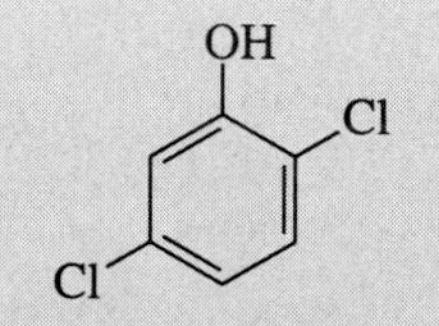

2,5-Dichlorophenol
(isolated from defensive secretion of a species of grasshopper)

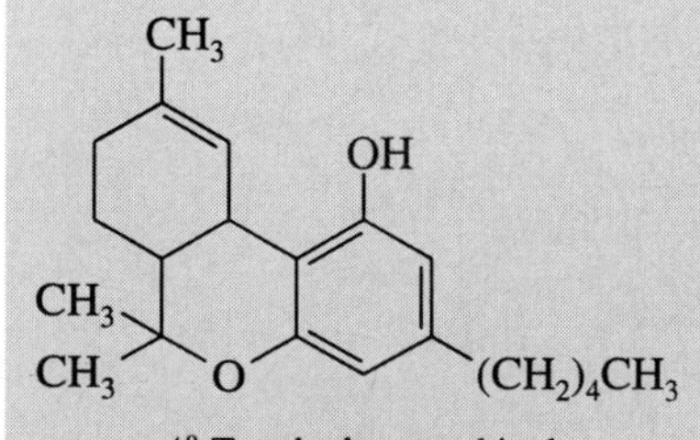

Δ^9-Tetrahydrocannabinol
(active component of marijuana)

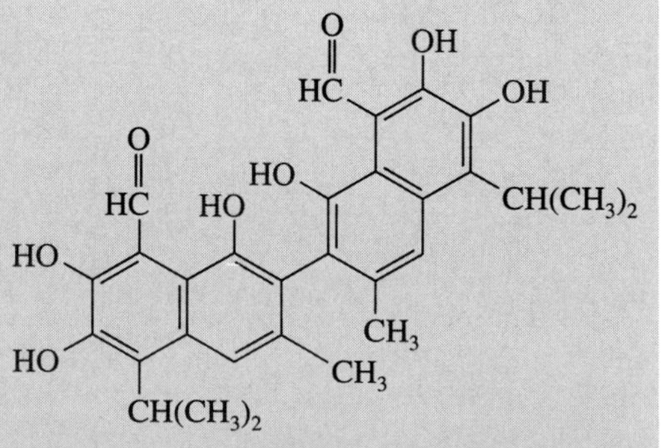

Gossypol
(About 10^9 lb of this material is obtained each year in the United States as a by product of cotton-oil production.)

FIGURE 24.3 Some naturally occurring phenols.

24.8 REACTIONS OF PHENOLS. ELECTROPHILIC AROMATIC SUBSTITUTION

In most of their reactions phenols behave as nucleophiles, and the reagents that act upon them are electrophiles. Either the hydroxyl oxygen or the aromatic ring may be the site of nucleophilic reactivity in a phenol. Reactions that take place on the ring lead to electrophilic aromatic substitution; Table 24.4 summarizes the behavior of phenols in reactions of this type.

A hydroxyl group is a very powerful activating substituent, and electrophilic aromatic substitution in phenols occurs far faster, and under milder conditions, than in benzene. The first entry in Table 24.4, for example, depicts the monobromination of phenol in high yield at low temperature and in the absence of any catalyst. In this case, the reaction was carried out in the nonpolar solvent 1,2-dichloroethane. In polar solvents such as water it is difficult to limit the bromination of phenols to monosubstitution. In the following example, all three positions that are ortho or para to the hydroxyl undergo rapid substitution:

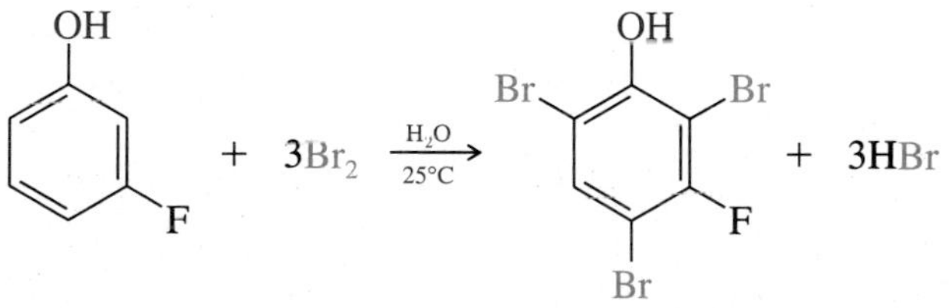

$$m\text{-Fluorophenol} + 3Br_2 \xrightarrow[25°C]{H_2O} \text{2,4,6-Tribromo-3-fluorophenol (95\%)} + 3HBr$$

m-Fluorophenol   Bromine   2,4,6-Tribromo-3-fluorophenol (95%)   Hydrogen bromide

Other typical electrophilic aromatic substitution reactions—nitration (second entry), sulfonation (fourth entry), and Friedel-Crafts alkylation and acylation (fifth and sixth entries)—take place readily and are synthetically useful. Phenols also undergo electrophilic substitution reactions that are limited to only the most active aromatic compounds; these include nitrosation (third entry) and coupling with diazonium salts (seventh entry).

PROBLEM 24.6 Each of the following reactions has been reported in the chemical literature and gives a single organic product in high yield. Identify the product in each case.

(*a*) 3-Benzyl-2,6-dimethylphenol treated with bromine in chloroform
(*b*) 4-Bromo-2-methylphenol treated with 2-methylpropene and sulfuric acid
(*c*) 2-Isopropyl-5-methylphenol (thymol) treated with sodium nitrite and dilute hydrochloric acid
(*d*) *p*-Cresol treated with propanoyl chloride and aluminum chloride

SAMPLE SOLUTION (*a*) The ring that bears the hydroxyl group is much more reactive than the other. In electrophilic aromatic substitution reactions of rings that bear several substituents, it is the most activating substituent that controls the orientation. Bromination occurs para to the hydroxyl group.

TABLE 24.4
Electrophilic Aromatic Substitution Reactions of Phenols

Reaction and comments	Specific example
Halogenation Bromination and chlorination of phenols occur readily even in the absence of a catalyst. Substitution occurs primarily at the position para to the hydroxyl group. When the para position is blocked, ortho substitution is observed.	Phenol + Bromine (Br_2) $\xrightarrow[0°C]{ClCH_2CH_2Cl}$ *p*-Bromophenol (93%) + Hydrogen bromide (HBr)
Nitration Phenols are nitrated on treatment with a dilute solution of nitric acid in either water or acetic acid. It is not necessary to use mixtures of nitric and sulfuric acids, because of the high reactivity of phenols.	*p*-Cresol $\xrightarrow[\text{acetic acid, 5°C}]{HNO_3}$ 4-Methyl-2-nitrophenol (73–77%)
Nitrosation On acidification of aqueous solutions of sodium nitrite, the nitrosonium ion ($:N{\equiv}\overset{+}{O}:$) is formed, which is a weak electrophile and attacks the strongly activated ring of a phenol. The product is a nitroso phenol.	2-Naphthol $\xrightarrow[H_2SO_4,\ H_2O,\ 0°C]{NaNO_2}$ 1-Nitroso-2-naphthol (99%)
Sulfonation Heating a phenol with concentrated sulfuric acid causes sulfonation of the ring.	2,6-Dimethylphenol $\xrightarrow[100°C]{H_2SO_4}$ 4-Hydroxyl-3,5-dimethylbenzenesulfonic acid (69%)
Friedel-Crafts alkylation Alcohols in combination with acids serve as sources of carbocations. Attack of a carbocation on the electron-rich ring of a phenol brings about its alkylation.	*o*-Cresol + *tert*-Butyl alcohol ($(CH_3)_3COH$) $\xrightarrow[60°C]{H_3PO_4}$ 4-*tert*-Butyl-2-methylphenol (63%)

TABLE 24.4
Electrophilic Aromatic Substitution Reactions of Phenols *(Continued)*

Reaction and comments	Specific example
Friedel-Crafts acylation In the presence of aluminum chloride, acyl chlorides and carboxylic acid anhydrides acylate the aromatic ring of phenols.	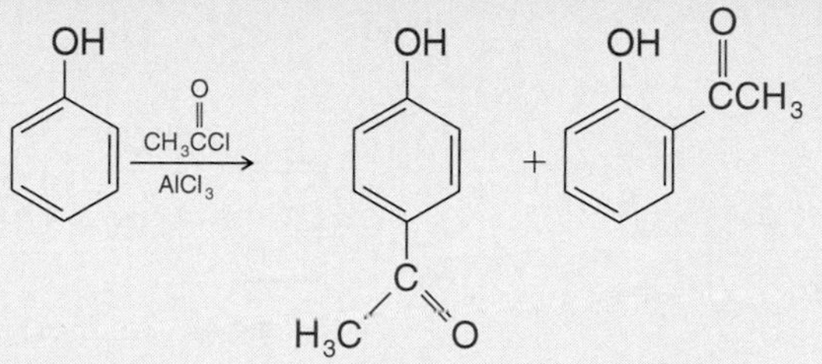 Phenol $\xrightarrow[AlCl_3]{CH_3CCl\ (=O)}$ *p*-Hydroxyacetophenone (74%) + *o*-Hydroxyacetophenone (16%)
Reaction with arenediazonium salts Addition of a phenol to a solution of a diazonium salt formed from a primary aromatic amine leads to formation of an azo compound. The reaction is carried out at a pH such that a significant portion of the phenol is present as its phenoxide ion. The diazonium ion acts as an electrophile toward the strongly activated ring of the phenoxide ion.	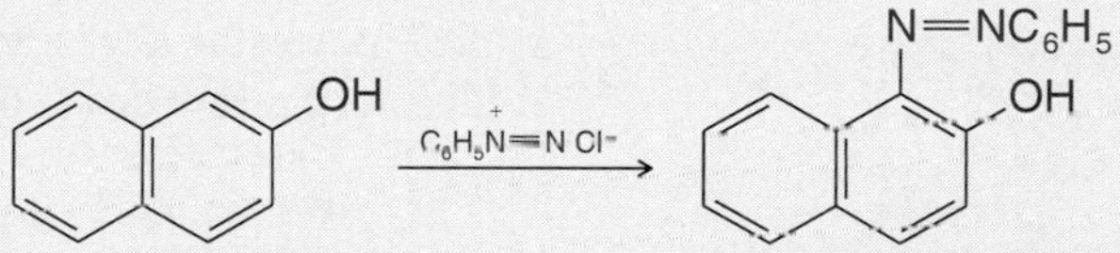 2-Naphthol $\xrightarrow{C_6H_5\overset{+}{N}{\equiv}N\ Cl^-}$ 1-Phenylazo-2-naphthol (48%)

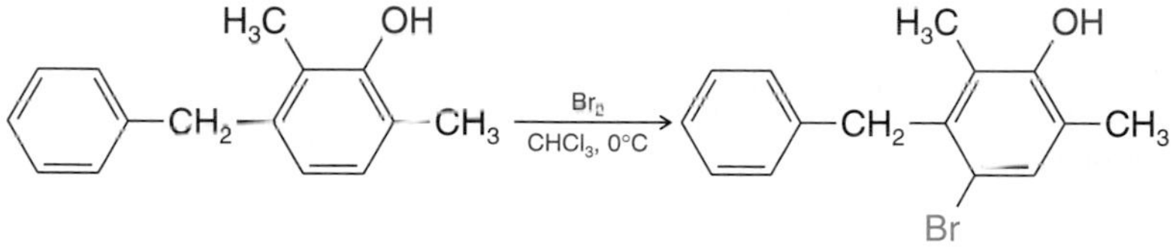

3-Benzyl-2,6-dimethylphenol $\xrightarrow[CHCl_3,\ 0°C]{Br_2}$ 3-Benzyl-4-bromo-2,6-dimethylphenol (isolated in 100% yield)

The aromatic ring of a phenol, like that of an arylamine, is regarded as an electron-rich functional unit and is capable of a variety of reactions. In some cases, however, it is the hydroxyl oxygen that reacts instead. An example of this kind of chemical reactivity is described in the following section.

24.9 ACYLATION OF PHENOLS

Acylating agents, such as acyl chlorides and carboxylic acid anhydrides, can react with phenols either at the aromatic ring (*C* acylation) or at the hydroxyl oxygen (*O* acylation):

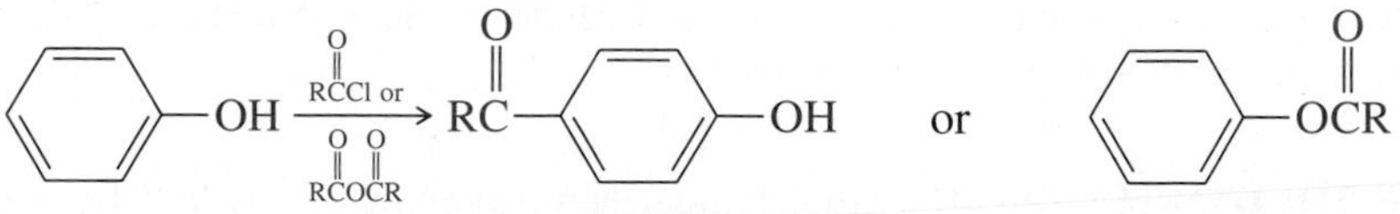

Phenol $\xrightarrow{RCCl\ \text{or}\ RCOCR}$ Aryl ketone (product of *C* acylation) or Aryl ester (product of *O* acylation)

As shown in the sixth entry of Table 24.4 in the preceding section, *C* acylation of phenols is observed under the customary conditions of the Friedel-Crafts reaction (treatment with an acyl chloride or acid anhydride in the presence of aluminum chloride). In the absence of aluminum chloride, however, *O* acylation occurs instead.

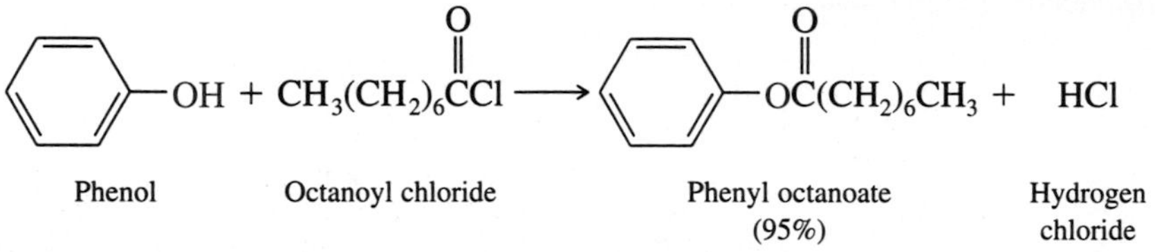

$$C_6H_5OH + CH_3(CH_2)_6\overset{O}{\overset{\|}{C}}Cl \longrightarrow C_6H_5O\overset{O}{\overset{\|}{C}}(CH_2)_6CH_3 + HCl$$

Phenol — Octanoyl chloride — Phenyl octanoate (95%) — Hydrogen chloride

The *O* acylation of phenols with carboxylic acid anhydrides can be conveniently catalyzed in either of two ways. One method involves converting the acid anhydride to a more powerful acyl transfer agent by protonation of one of its carbonyl oxygens. Addition of a few drops of sulfuric acid is usually sufficient.

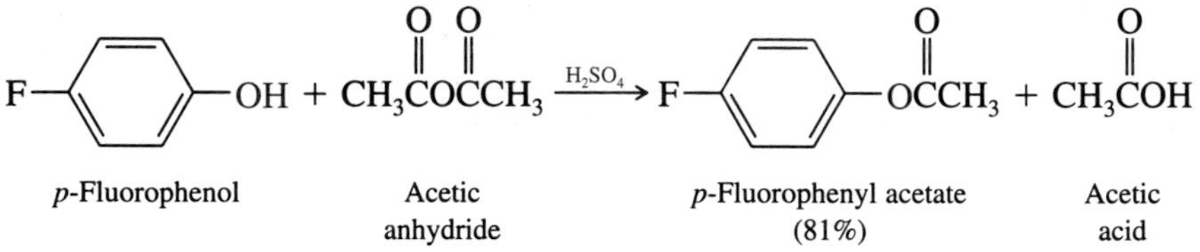

$$F{-}C_6H_4{-}OH + CH_3\overset{O}{\overset{\|}{C}}O\overset{O}{\overset{\|}{C}}CH_3 \xrightarrow{H_2SO_4} F{-}C_6H_4{-}O\overset{O}{\overset{\|}{C}}CH_3 + CH_3\overset{O}{\overset{\|}{C}}OH$$

p-Fluorophenol — Acetic anhydride — *p*-Fluorophenyl acetate (81%) — Acetic acid

An alternative approach is to increase the nucleophilicity of the phenol by converting it to its phenoxide anion in basic solution:

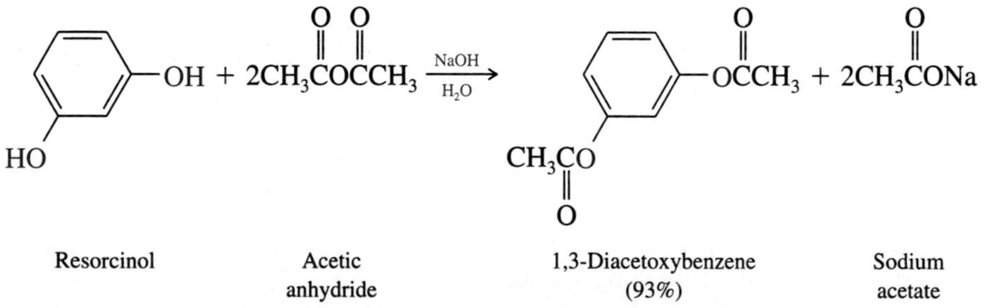

$$HO{-}C_6H_4{-}OH + 2CH_3\overset{O}{\overset{\|}{C}}O\overset{O}{\overset{\|}{C}}CH_3 \xrightarrow[H_2O]{NaOH} CH_3\overset{O}{\overset{\|}{C}}O{-}C_6H_4{-}O\overset{O}{\overset{\|}{C}}CH_3 + 2CH_3\overset{O}{\overset{\|}{C}}ONa$$

Resorcinol — Acetic anhydride — 1,3-Diacetoxybenzene (93%) — Sodium acetate

---

PROBLEM 24.7 Write chemical equations expressing each of the following transformations:

(*a*) Preparation of *o*-nitrophenyl acetate by sulfuric acid catalysis of the reaction between a phenol and a carboxylic acid anhydride.
(*b*) Esterification of 2-naphthol with acetic anhydride in aqueous sodium hydroxide
(*c*) Reaction of phenol with benzoyl chloride

SAMPLE SOLUTION (*a*) The problem specifies that an acid anhydride be used. Therefore, use acetic anhydride to prepare the acetate ester of *o*-nitrophenol:

$$o\text{-}O_2NC_6H_4OH + CH_3COCCH_3 \xrightarrow{H_2SO_4} o\text{-}O_2NC_6H_4OCCH_3 + CH_3COH$$

o-Nitrophenol — Acetic anhydride — *o*-Nitrophenyl acetate (isolated in 93% yield by this method) — Acetic acid

The preference for *O* acylation of phenols arises because these reactions are *kinetically controlled.* *O* acylation is faster than *C* acylation. The *C*-acyl isomers are more stable, however, and it is known that aluminum chloride is a very effective catalyst for the conversion of aryl esters to aryl ketones. (This isomerization is called the **Fries rearrangement.**)

$$C_6H_5OCC_6H_5 \xrightarrow{AlCl_3} o\text{-}HOC_6H_4CC_6H_5 + C_6H_5C\text{-}C_6H_4\text{-}OH\ (para)$$

Phenyl benzoate — *o*-Hydroxybenzophenone (9%) — *p*-Hydroxybenzophenone (64%)

Thus, ring acylation of phenols is observed under Friedel-Crafts conditions because the presence of aluminum chloride causes that reaction to be subject to *thermodynamic (equilibrium) control* (Section 10.10).

Fischer esterification, in which a phenol and a carboxylic acid condense in the presence of an acid catalyst, is not used to prepare aryl esters.

24.10 CARBOXYLATION OF PHENOLS. ASPIRIN AND THE KOLBE-SCHMITT REACTION

The best-known aryl ester is *O*-acetylsalicylic acid, better known as *aspirin.* It is prepared by acetylation of the phenolic hydroxyl group of salicylic acid:

$$o\text{-}HOC_6H_4CO_2H + CH_3COCCH_3 \xrightarrow{H_2SO_4} o\text{-}CH_3COC_6H_4CO_2H + CH_3COH$$

Salicylic acid (*o*-hydroxybenzoic acid) — Acetic anhydride — *O*-Acetylsalicylic acid (aspirin) — Acetic acid

Aspirin possesses a number of properties that make it the most often recommended drug. It is an analgesic, effective in relieving headache pain. It is also an anti-inflammatory agent, providing some relief from the swelling associated with arthritis and minor injuries. Aspirin is an antipyretic compound; that is, it reduces fever. Each year,

An entertaining account of the history of aspirin can be found in the 1991 book *The Aspirin Wars. Money, Medicine, and 100 Years of Rampant Competition*, by Charles C. Mann.

more than 40 million lb of aspirin is produced in the United States, a rate that translates to 300 tablets per year for every man, woman, and child.

The key compound in the synthesis of aspirin, salicylic acid, is prepared from phenol by a process discovered over 100 years ago by the German chemist Hermann Kolbe. In the Kolbe synthesis, also known as the **Kolbe-Schmitt reaction,** sodium phenoxide is heated with carbon dioxide under pressure, and the reaction mixture is subsequently acidified to yield salicylic acid:

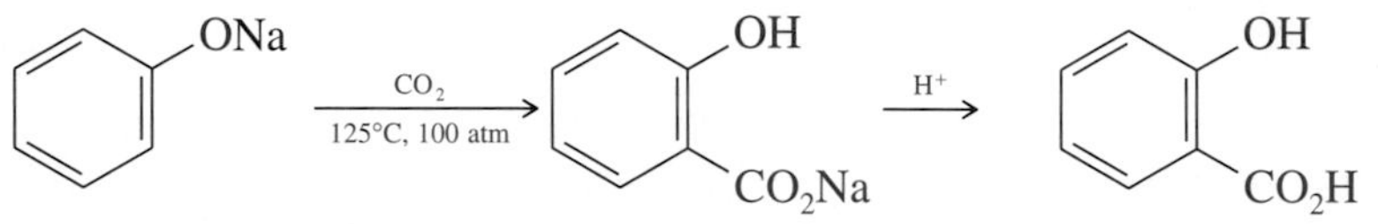

ONa $\xrightarrow[125°C,\ 100\ atm]{CO_2}$ OH, CO_2Na $\xrightarrow{H^+}$ OH, CO_2H

Sodium phenoxide — Sodium salicylate — Salicylic acid (79%)

While a hydroxyl group strongly activates an aromatic ring toward electrophilic attack, an oxyanion substituent is an even more powerful activator. Electron delocalization in phenoxide anion leads to increased electron density at the positions ortho and para to oxygen.

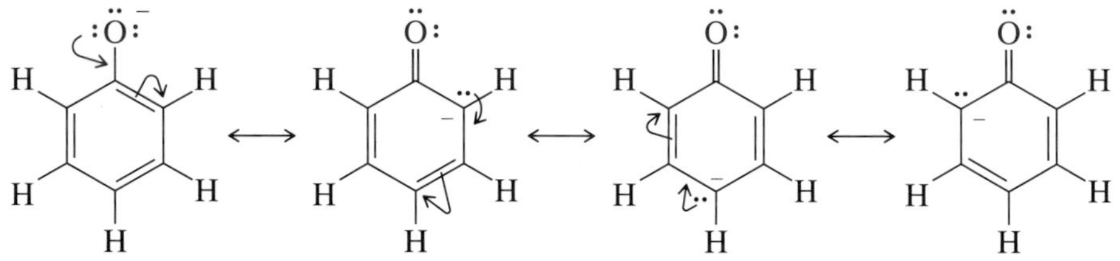

The increased nucleophilicity of the ring permits it to react with carbon dioxide. An intermediate is formed that is simply the keto form of salicylate anion:

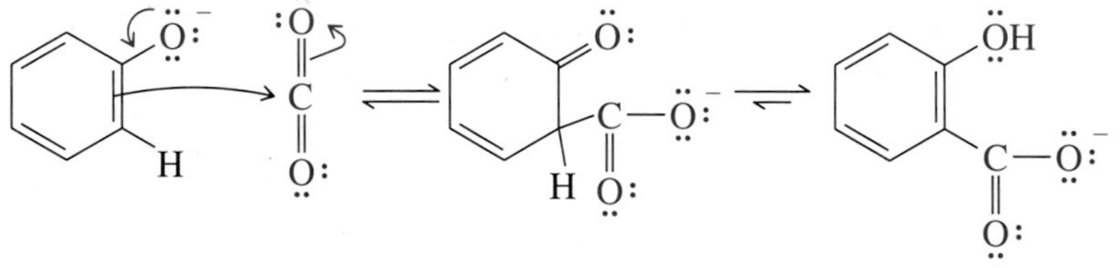

Phenoxide anion (stronger base) — Carbon dioxide — Cyclohexadienone intermediate — Salicylate anion (weaker base)

The Kolbe-Schmitt reaction is an equilibrium process governed by thermodynamic control. The position of equilibrium favors formation of the weaker base (salicylate ion) at the expense of the stronger one (phenoxide ion). Thermodynamic control is also responsible for the pronounced bias toward ortho over para substitution. Salicylate anion is a weaker base than *p*-hydroxybenzoate and so is the predominant species at equilibrium.

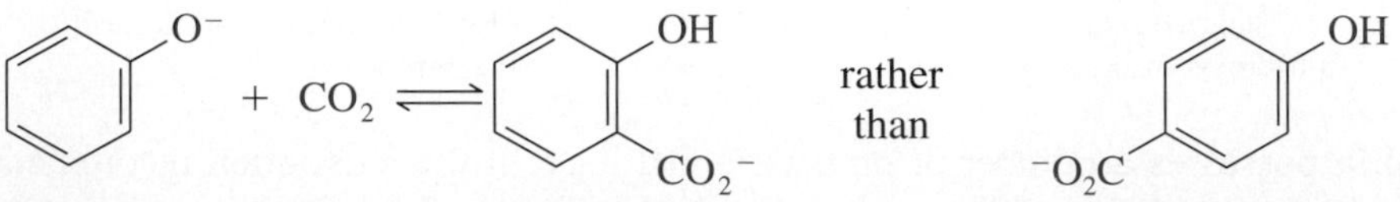

O^- + CO_2 ⇌ OH, CO_2^- rather than ^-O_2C, OH

Phenoxide ion (strongest base; K_a of conjugate acid, 10^{-10}) — Carbon dioxide — Salicylate anion (weakest base; K_a of conjugate acid, 1.06×10^{-3}) — *p*-Hydroxybenzoate anion (K_a of conjugate acid, 3.3×10^{-5})

Salicylate anion is a weaker base than *p*-hydroxybenzoate because it is stabilized by intramolecular hydrogen bonding.

Intramolecular hydrogen bonding in salicylate anion

The Kolbe-Schmitt reaction has been applied to the preparation of other *o*-hydroxybenzoic acids. Alkyl derivatives of phenol behave very much like phenol itself.

p-Cresol $\xrightarrow[\text{3. } H^+]{\text{1. NaOH} \quad \text{2. } CO_2\text{, 125°C, 100 atm}}$ 2-Hydroxy-5-methylbenzoic acid (78%)

Phenols that bear strongly electron-withdrawing substituents usually give low yields of carboxylated products; their derived phenoxide anions are less basic, and the equilibrium constants for their carboxylation are smaller.

24.11 PREPARATION OF ARYL ETHERS

Aryl ethers are best prepared by the Williamson method (Section 16.6). Alkylation of the hydroxyl oxygen of a phenol takes place readily when a phenoxide anion reacts with an alkyl halide.

$$ArO^{-} + R{-}X \xrightarrow{S_N2} ArOR + X^{-}$$

Phenoxide anion + Alkyl halide → Alkyl aryl ether + Halide anion

$$C_6H_5ONa + CH_3I \xrightarrow[\text{heat}]{\text{acetone}} C_6H_5OCH_3 + NaI$$

Sodium phenoxide + Iodomethane → Anisole (95%) + Sodium iodide

As the synthesis is normally performed, a solution of the phenol and alkyl halide is simply heated in the presence of a suitable base such as potassium carbonate:

$$C_6H_5OH + CH_2{=}CHCH_2Br \xrightarrow[\text{acetone, heat}]{K_2CO_3} C_6H_5OCH_2CH{=}CH_2$$

Phenol + Allyl bromide → Allyl phenyl ether (86%)

This is an example of an S_N2 reaction in a polar aprotic solvent.

The alkyl halide must be one that reacts readily in an S_N2 process. Thus, methyl and primary alkyl halides are the most effective alkylating agents. Elimination becomes competitive with substitution when secondary alkyl halides are used and is the only reaction observed with tertiary alkyl halides.

PROBLEM 24.8 Reaction of phenol with 1,2-epoxypropane in aqueous sodium hydroxide at 150°C gives a single product, $C_9H_{12}O_2$, in 90 percent yield. Suggest a reasonable structure for this compound.

AGENT ORANGE AND DIOXIN

The once widely used herbicide 2,4,5-trichlorophenoxyacetic acid (2,4,5-T) is prepared by reaction of the sodium salt of 2,4,5-trichlorophenol with chloroacetic acid:

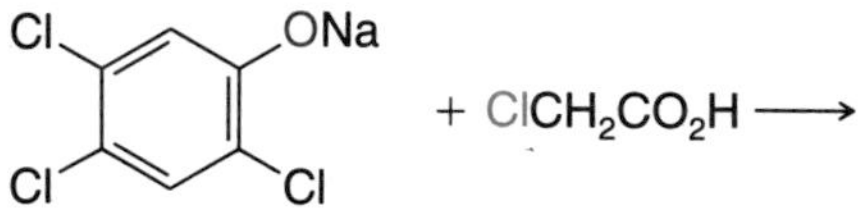

Sodium 2,4,5-trichlorophenolate + $ClCH_2CO_2H$ ⟶ 2,4,5-Trichlorophenoxyacetic acid (2,4,5-T) + NaCl

Cl, ONa, Cl, Cl — Sodium 2,4,5-trichlorophenolate; Chloroacetic acid; Cl, OCH_2CO_2H, Cl, Cl — 2,4,5-Trichlorophenoxyacetic acid (2,4,5-T)

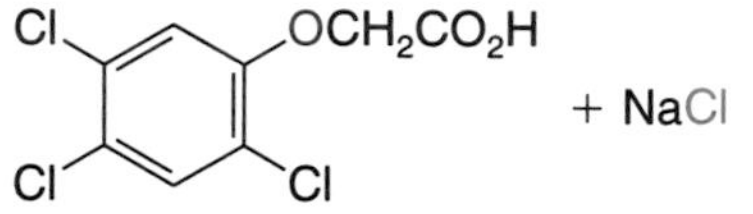

The starting material for this process, 2,4,5-trichlorophenol, is made by treating 1,2,4,5-tetrachlorobenzene with aqueous base. Nucleophilic aromatic substitution of one of the chlorines by an addition-elimination mechanism yields 2,4,5-trichlorophenol:

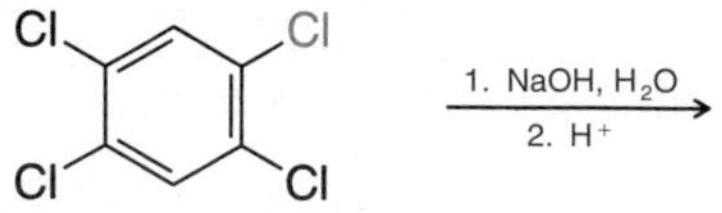

1,2,4,5-Tetrachlorobenzene $\xrightarrow[2.\ H^+]{1.\ NaOH,\ H_2O}$ 2,4,5-Trichlorophenol

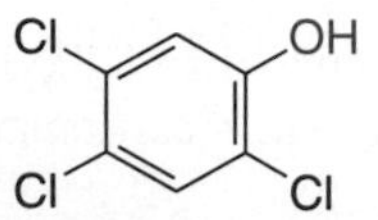

In the course of making 2,4,5-trichlorophenol, it almost always becomes contaminated with small amounts of 2,3,7,8-tetrachlorodibenzo-*p*-dioxin, better known as *dioxin.*

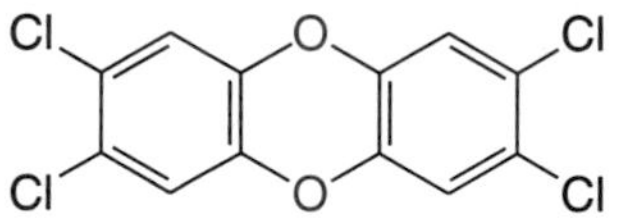

2,3,7,8-Tetrachlorodibenzo-*p*-dioxin (dioxin)

Dioxin is carried along when 2,4,5-trichlorophenol is converted to 2,4,5-T, and enters the environment when 2,4,5-T is sprayed on vegetation. Typically, the amount of dioxin present in 2,4,5-T is very small. *Agent Orange,* a 2,4,5-T–based defoliant used on a large scale in Vietnam, contained about 2 ppm of dioxin.

Tests with animals have revealed that dioxin is one of the most toxic substances known. Toward mice it is about 2000 times more toxic than strychnine and about 150,000 times more toxic than sodium cyanide. Fortunately, however, available evidence indicates that humans are far more resistant to dioxin than are test animals, and so far there have been no human fatalities directly attributable to dioxin. The most prominent short-term symptom seen so far has been a severe skin disorder known as *chloracne.* Yet to be determined is the answer to the question of long-term effects. A 1991 study of the health records of over 5000 workers who were exposed to dioxin-contaminated chemicals indicated a 15 percent increase in incidences of cancer compared with those of a control group. Workers who were exposed to higher dioxin levels for prolonged periods exhibited a 50 percent increase in their risk of dying from cancer, especially soft-tissue sarcomas, compared with the control group.

Since 1979 the use of 2,4,5-T has been regulated in the United States.

The reaction between an alkoxide ion and an aryl halide can be used to prepare alkyl aryl ethers only when the aryl halide is one that reacts rapidly by the addition-elimination mechanism of nucleophilic aromatic substitution (Section 23.6).

$$p\text{-}FC_6H_4NO_2 \xrightarrow[CH_3OH,\ 25^\circ C]{KOCH_3} p\text{-}CH_3OC_6H_4NO_2$$

p-Fluoronitrobenzene → *p*-Nitroanisole (93%)

PROBLEM 24.9 Which of the following two combinations of reactants is more appropriate for the preparation of *p*-nitrophenyl phenyl ether?

(*a*) Fluorobenzene and *p*-nitrophenol (*b*) *p*-Fluoronitrobenzene and phenol

---

24.12 CLEAVAGE OF ARYL ETHERS BY HYDROGEN HALIDES

The cleavage of *dialkyl ethers* by hydrogen halides has been discussed in Section 16.8, where it was noted that the same pair of alkyl halides results irrespective of the order in which the carbon-oxygen bonds of the ether are broken.

$$ROR' + 2HX \longrightarrow RX + R'X + H_2O$$

Dialkyl ether + Hydrogen halide → Two alkyl halides + Water

Cleavage of *alkyl aryl ethers* by hydrogen halides always proceeds so that the alkyl-oxygen bond is broken and yields an alkyl halide and a phenol:

$$ArOR + HX \longrightarrow ArOH + RX$$

Alkyl aryl ether + Hydrogen halide → Phenol + Alkyl halide

Since phenols are not converted to aryl halides by reaction with hydrogen halides, reaction proceeds no further.

$$o\text{-}CH_3OC_6H_4OH \xrightarrow[heat]{HBr} o\text{-}C_6H_4(OH)_2 + CH_3Br$$

Guaiacol → Pyrocatechol (85–87%) + Methyl bromide (57–72%)

Guaiacol is obtained by chemical treatment of *lignum vitae,* the wood from a species of tree that grows in warm climates. It is sometimes used as an expectorant to help relieve bronchial congestion.

The first step in the reaction of an alkyl aryl ether with a hydrogen halide is protonation of oxygen to form an alkylaryloxonium ion:

$$Ar\ddot{O}R + H-\ddot{X}: \overset{fast}{\rightleftharpoons} Ar\overset{+}{\ddot{O}}(R)H + :\ddot{X}:^-$$

Alkyl aryl ether + Hydrogen halide ⇌ Alkylaryloxonium ion + Halide ion

This is followed by a nucleophilic substitution step:

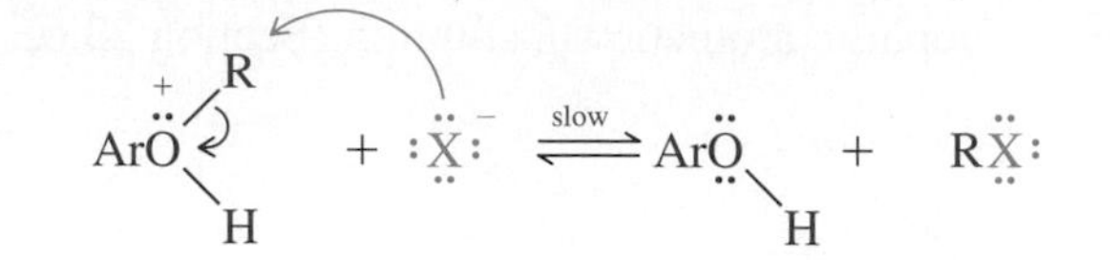

$$\mathrm{Ar\ddot{O}^{+}(R)H} + :\ddot{X}:^{-} \xrightleftharpoons{\text{slow}} \mathrm{Ar\ddot{O}H} + \mathrm{R\ddot{X}:}$$

Alkylaryloxonium ion — Halide ion — Phenol — Alkyl halide

Attack by the halide nucleophile at the sp^3 hybridized carbon of the alkyl group is analogous to what takes place in the cleavage of dialkyl ethers. Attack at the sp^2 hybridized carbon of the aromatic ring is much slower. Indeed, nucleophilic aromatic substitution does not occur at all under these conditions.

24.13 CLAISEN REARRANGEMENT OF ALLYL ARYL ETHERS

Allyl aryl ethers undergo an interesting reaction, called the **Claisen rearrangement,** on being heated. The allyl group migrates from oxygen to the ring carbon ortho to it.

Allyl phenyl ether is prepared by the reaction of phenol with allyl bromide, as described in Section 24.11

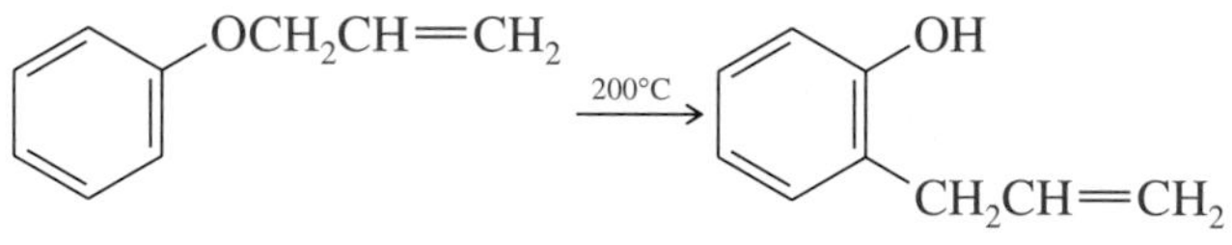

$$C_6H_5OCH_2CH{=}CH_2 \xrightarrow{200^\circ C} o\text{-}HOC_6H_4CH_2CH{=}CH_2$$

Allyl phenyl ether — o-Allylphenol (73%)

Carbon-14 labeling of the allyl group revealed that the terminal carbon of the allyl group is the one that becomes bonded to the ring and suggests a mechanism involving a concerted electron reorganization in the first step. This step is followed by enolization of the resulting cyclohexadienone to regenerate the aromatic ring.

● = ^{14}C

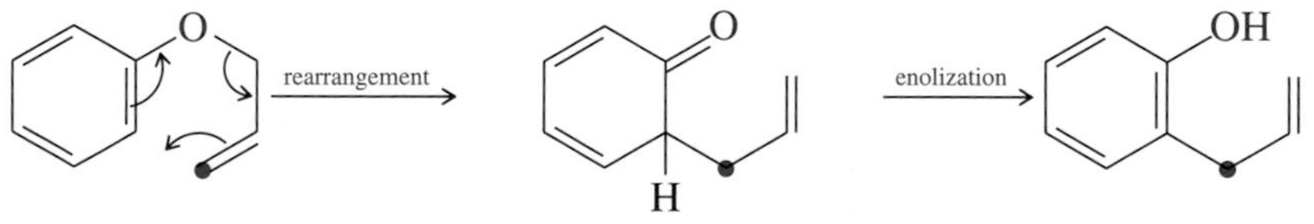

Allyl phenyl ether —rearrangement→ 6-Allyl-2,4-cyclohexadienone —enolization→ o-Allylphenol

PROBLEM 24.10 The mechanism of the Claisen rearrangement of other allylic ethers of phenol is analogous to that of allyl phenyl ether. What is the product of the Claisen rearrangement of $C_6H_5OCH_2CH{=}CHCH_3$?

The transition state for the first step of the Claisen rearrangement bears much in common with the transition state for the Diels-Alder cycloaddition. Both involve a concerted six-electron reorganization.

via

via

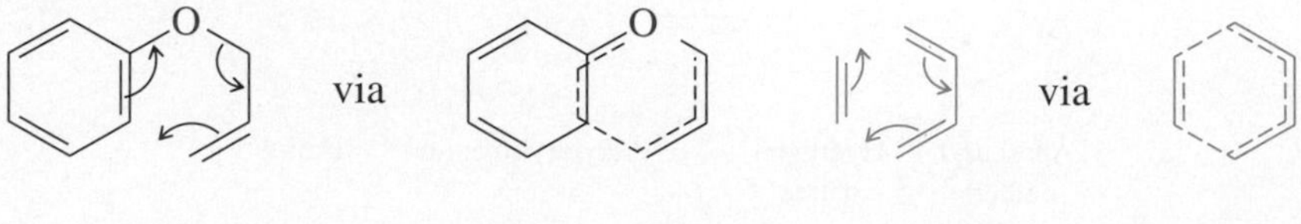

Recall from Section 10.12 that the Diels-Alder reaction, along with electrocyclic reactions (Section 10.14), belongs to a class of processes called *pericyclic* reactions. The Claisen rearrangement is an example of a third type of pericyclic reaction known as **sigmatropic rearrangements.** A sigmatropic rearrangement is characterized by a transition state in which a σ bond migrates from one end of a conjugated π electron system to the other. In this case the σ bond to oxygen at one end of an allyl unit is broken and replaced by a σ bond to the ring carbon at the other end.

24.14 OXIDATION OF PHENOLS. QUINONES

Phenols are more easily oxidized than alcohols, and a large number of inorganic oxidizing agents have been used for this purpose. The phenol oxidations that are of the most use to the organic chemist are those involving derivatives of 1,2-benzenediol (pyrocatechol) and 1,4-benzenediol (hydroquinone). Oxidation of compounds of this type with silver oxide or with chromic acid yields conjugated dicarbonyl compounds called **quinones.**

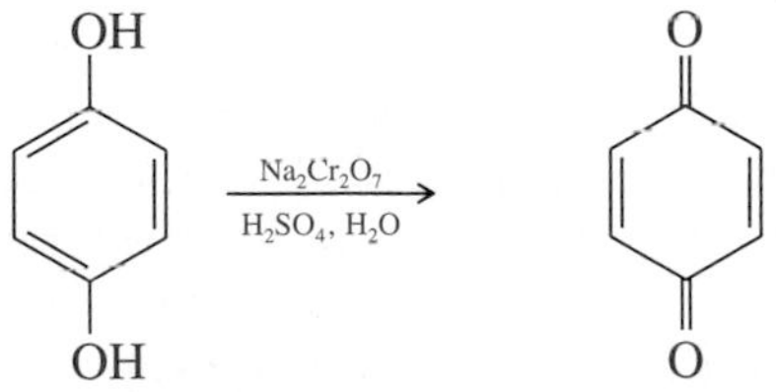

Hydroquinone $\xrightarrow[H_2SO_4,\ H_2O]{Na_2Cr_2O_7}$ *p*-Benzoquinone (76–81%)

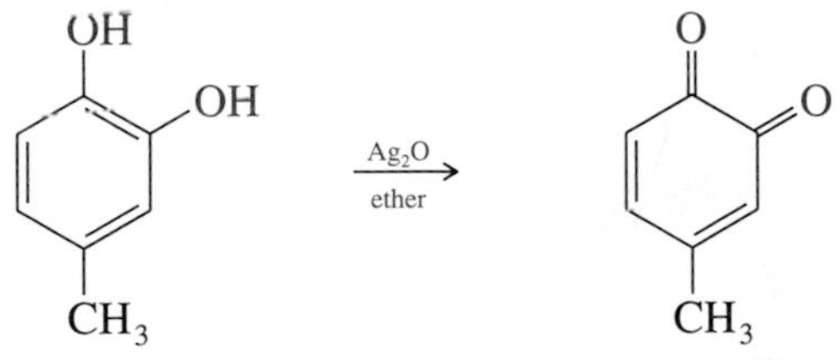

4-Methylpyrocatechol (4-methyl-1,2-benzenediol) $\xrightarrow[ether]{Ag_2O}$ 4-Methyl-1,2-benzoquinone (68%)

Silver oxide is a weak oxidizing agent.

Quinones are colored; *p*-benzoquinone, for example, is yellow. Many occur naturally and have been used as dyes. *Alizarin* is a red pigment extracted from the roots of the madder plant. Its preparation from anthracene, a coal tar derivative, in 1868 was a significant step in the development of the synthetic dyestuff industry.

Alizarin

Quinones that are based on the anthracene ring system are called *anthraquinones.* Alizarin is one example of an *anthraquinone dye.*

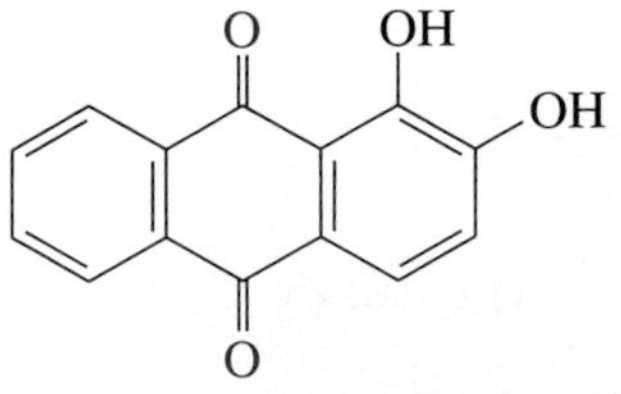

The oxidation-reduction process that connects hydroquinone and benzoquinone involves two 1-electron transfers:

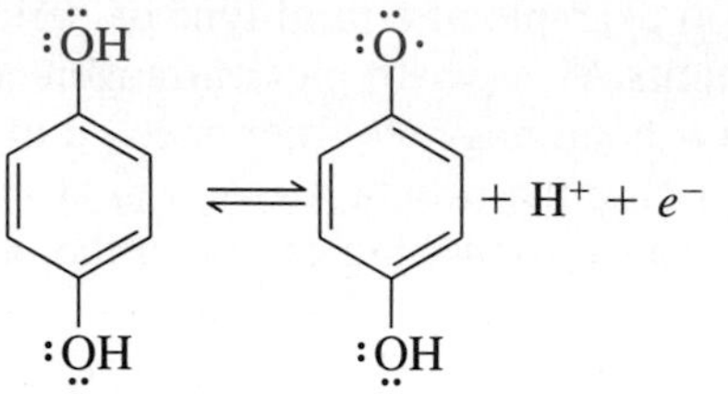

$$\text{Hydroquinone} \rightleftharpoons \text{semiquinone radical} + H^+ + e^-$$

Hydroquinone

$$\text{semiquinone radical} \rightleftharpoons \text{Benzoquinone} + H^+ + e^-$$

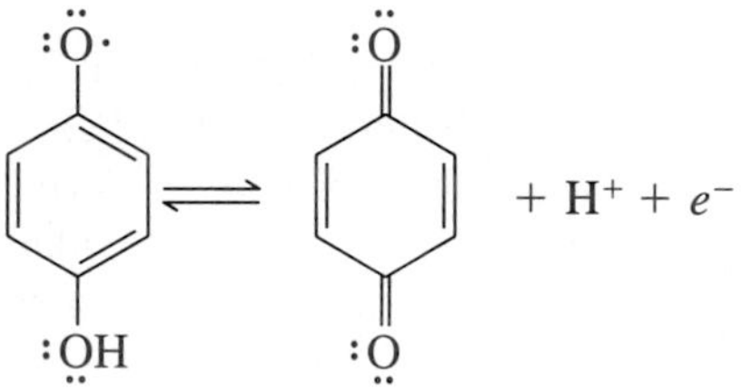

Benzoquinone

The ready reversibility of this reaction is essential to the role that quinones play in cellular respiration, the process by which an organism uses molecular oxygen to convert its food to carbon dioxide, water, and energy. Electrons are not transferred directly from the substrate molecule to oxygen but instead are transferred by way of an *electron transport chain* involving a succession of oxidation-reduction reactions. A key component of this electron transport chain is the substance known as *ubiquinone,* or coenzyme Q:

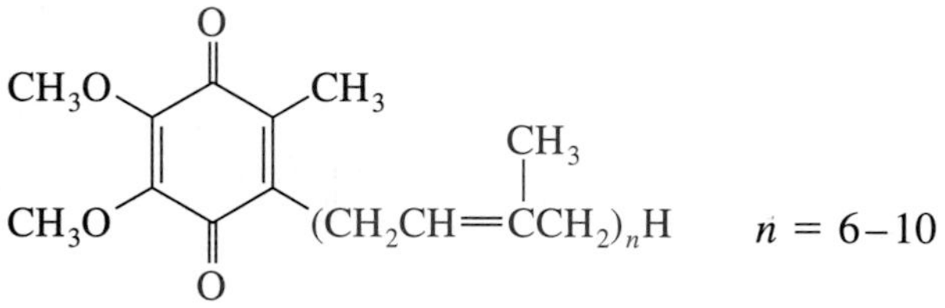

$$CH_3O,\ CH_3O,\ CH_3,\ (CH_2CH{=}C(CH_3)CH_2)_nH \qquad n = 6\text{–}10$$

Ubiquinone (coenzyme Q)

The name *ubiquinone* is a shortened form of *ubiquitous quinone,* a term coined to describe the observation that this substance can be found in all cells. The length of its side chain varies among different organisms; the most common form in vertebrates has $n = 10$, while ubiquinones in which $n = 6$ to 9 are found in yeasts and plants.

Another physiologically important quinone is vitamin K. Here "K" stands for *koagulation* (Danish), since this substance was first identified as essential for the normal clotting of blood.

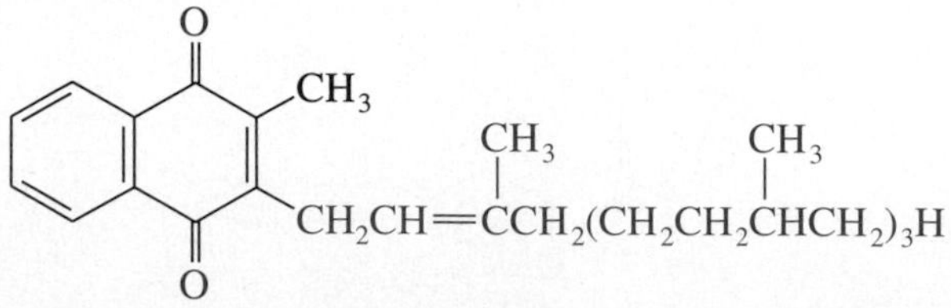

$$CH_3,\ CH_2CH{=}C(CH_3)CH_2(CH_2CH_2CH(CH_3)CH_2)_3H$$

Vitamin K

Some vitamin K is provided in the normal diet, but a large proportion of that required by humans is produced by their intestinal flora.

"Intestinal flora" is a general term for the bacteria, yeast, and fungi that live in the large intestine.

24.15 SPECTROSCOPIC ANALYSIS OF PHENOLS

The infrared spectra of phenols combine features of those of alcohols and aromatic compounds. Hydroxyl absorbances resulting from O—H stretching are found in the 3600 cm^{-1} region, and the peak due to C—O stretching appears around 1200 to 1250 cm^{-1}. These features can be seen in the infrared spectrum of *p*-cresol, shown in Figure 24.4.

The ^1H nmr signals for the hydroxyl protons of phenols generally appear at lower field than those of alcohols. Chemical shifts of protons in phenolic hydroxyl groups are in the range $\delta = 4$ to 12 ppm. Often the signal is very broad. Figure 24.5 is the ^1H nmr spectrum of *p*-cresol.

A peak for the molecular ion is usually quite prominent in the mass spectra of phenols. It is, for example, the most intense peak in phenol.

24.16 SUMMARY

Phenol is both an important industrial chemical and the parent of a large class of compounds widely distributed as natural products. Although *benzenol* is the systematic name for C_6H_5OH, the IUPAC rules permit *phenol* to be used instead. Substituted derivatives are named on the basis of phenol as the parent compound (Section 24.1).

Phenols are polar compounds, resembling arylamines in having an electron-rich aromatic ring (Section 24.2). Their polar nature and the prospect of involving the —OH group in hydrogen bonding make phenolic compounds more soluble in water and give them higher boiling points than arenes and aryl halides (Section 24.3).

With K_a's of approximately 10^{-10} ($pK_a = 10$), phenols are stronger acids than alcohols, but weaker than carboxylic acids (Section 24.4). They are converted quantitatively to phenoxide anions on treatment with aqueous sodium hydroxide.

$$ArOH + NaOH \longrightarrow ArONa + H_2O$$

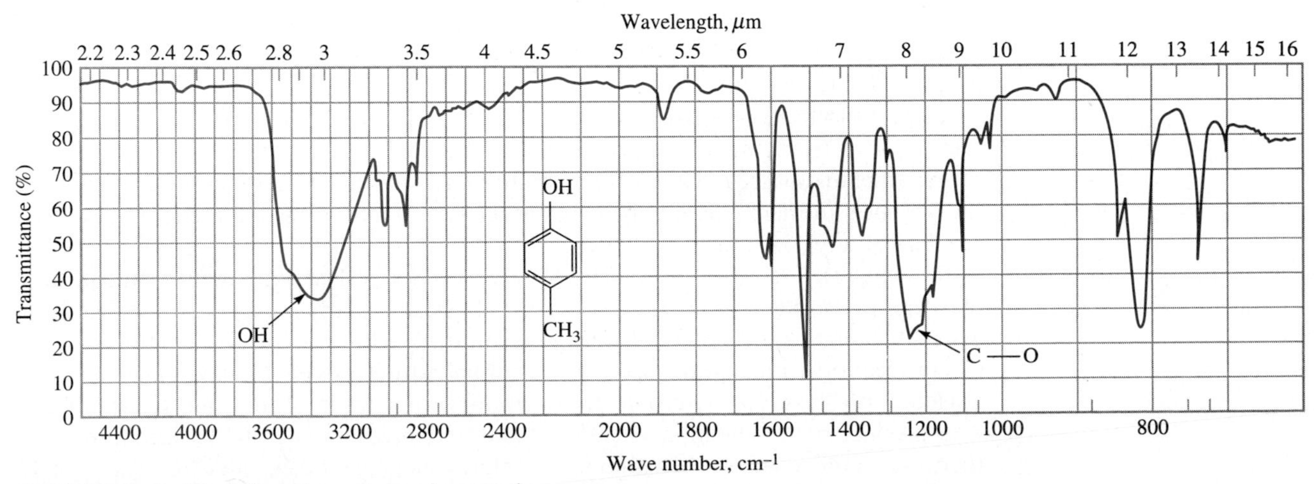

FIGURE 24.4 The infrared spectrum of *p*-cresol.

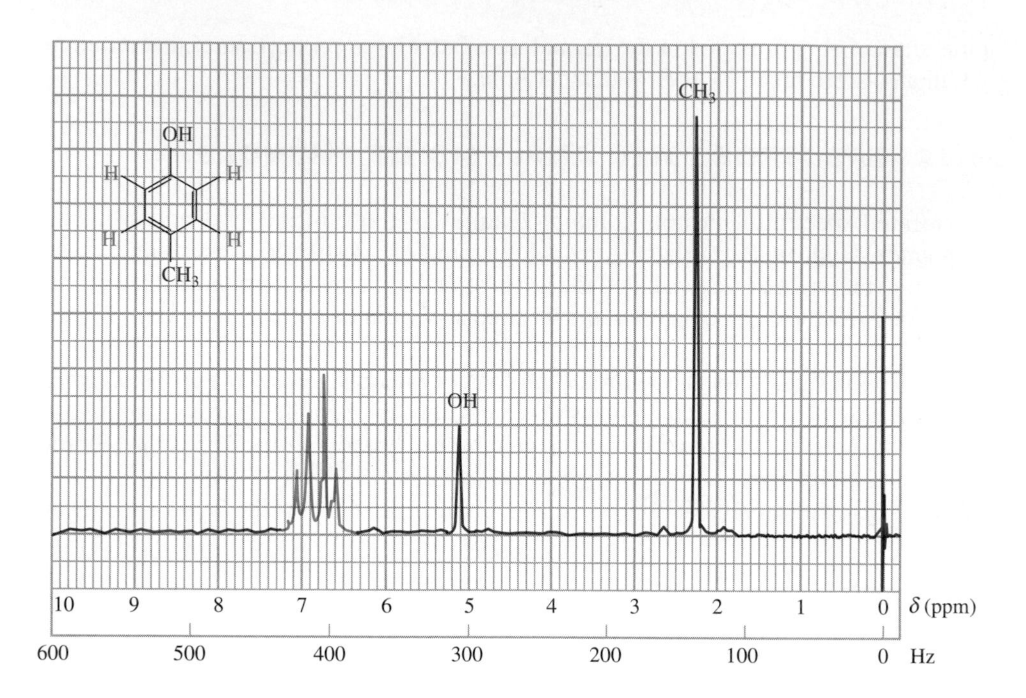

FIGURE 24.5 The 60-MHz ^{1}H nmr spectrum of *p*-cresol.

Strongly electron-withdrawing substituents on the ring increase the acidity of phenols. Electron-releasing substituents have a negligible effect (Section 24.5).

A number of industrial syntheses are available for the preparation of phenol (Table 24.3, Section 24.6). Phenols are normally prepared in the laboratory by hydrolysis of aryl diazonium salts.

$$\underset{\text{Arylamine}}{ArNH_2} \xrightarrow{NaNO_2,\ H^+} \underset{\text{Aryl diazonium ion}}{Ar\overset{+}{N}{\equiv}N:} \xrightarrow[\text{heat}]{H_2O} \underset{\text{A phenol}}{ArOH}$$

$$\underset{\text{3-Fluoro-4-methoxyaniline}}{CH_3O\text{-}C_6H_3(F)\text{-}NH_2} \xrightarrow[\text{2. } H_2O,\ \text{heat}]{\text{1. } NaNO_2,\ H_2SO_4,\ H_2O} \underset{\text{3-Fluoro-4-methoxyphenol (70\%)}}{CH_3O\text{-}C_6H_3(F)\text{-}OH}$$

Many phenols occur naturally, in both plants and animals (Section 24.7). Their biosynthesis in mammals involves oxidation of aromatic rings via arene oxide intermediates.

The hydroxyl group of a phenol is a very powerful activating substituent, and electrophilic aromatic substitution occurs with great ease in phenol and its derivatives. Typical examples of these reactions were summarized in Table 24.4 (Section 24.8).

On reaction with acyl chlorides and acid anhydrides, phenols may undergo either acylation of the hydroxyl group (*O* acylation) or acylation of the ring (*C* acylation). The product of *C* acylation is more stable and predominates under conditions of thermodynamic control when aluminum chloride is present (see entry 6 in Table 24.4, Section 24.8). *O* acylation is faster than *C* acylation, and aryl esters are formed under conditions of kinetic control (Section 24.9).

$$\text{ArOH} + \text{RCX (O)} \longrightarrow \text{ArOCR (O)} + \text{HX}$$

A phenol — Acylating agent — Aryl ester

$$o\text{-}O_2NC_6H_4OH \xrightarrow[H_2SO_4]{CH_3COCCH_3 \text{ (O O)}} o\text{-}O_2NC_6H_4OCCH_3 \text{ (O)}$$

o-Nitrophenol — *o*-Nitrophenyl acetate (93%)

The **Kolbe-Schmitt synthesis** of salicylic acid is a vital step in the preparation of aspirin (Section 24.10). Phenols, as their sodium salts, undergo highly regioselective ortho carboxylation on treatment with carbon dioxide at elevated temperature and pressure.

$$p\text{-}(CH_3)_3CC_6H_4ONa \xrightarrow[2.\ H^+]{1.\ CO_2,\ \text{heat, pressure}} \text{5-}(CH_3)_3C\text{-2-}HOC_6H_3CO_2H$$

Sodium *p-tert*-butylphenoxide — 5-*tert*-Butyl-2-hydroxybenzoic acid (74%)

Phenoxide anions are nucleophilic toward alkyl halides, and the preparation of alkyl aryl ethers is easily achieved under S_N2 conditions (Section 24.11).

$$\text{ArO}^- + \text{RX} \longrightarrow \text{ArOR} + \text{X}^-$$

Phenoxide anion — Alkyl halide — Alkyl aryl ether — Halide anion

$$o\text{-}O_2NC_6H_4OH \xrightarrow[K_2CO_3]{CH_3CH_2CH_2CH_2Br} o\text{-}O_2NC_6H_4OCH_2CH_2CH_2CH_3$$

o-Nitrophenol — Butyl *o*-nitrophenyl ether (75/80%)

The cleavage of alkyl aryl ethers by hydrogen halides yields a phenol and an alkyl halide (Section 24.12).

$$\text{ArOR} + \text{HX} \xrightarrow{\text{heat}} \text{ArOH} + \text{RX}$$

Alkyl aryl ether — Hydrogen halide — A phenol — Alkyl halide

$$m\text{-}CH_3OC_6H_4CH_2CO_2H \xrightarrow[\text{heat}]{HI} m\text{-}HOC_6H_4CH_2CO_2H + CH_3I$$

m-Methoxyphenylacetic acid — *m*-Hydroxyphenylacetic acid (72%) — Methyl iodide

On being heated, allyl aryl ethers undergo a **Claisen rearrangement** to form *o*-allylphenols. A cyclohexadienone, formed by a concerted six-π-electron reorganization, is an intermediate (Section 24.13).

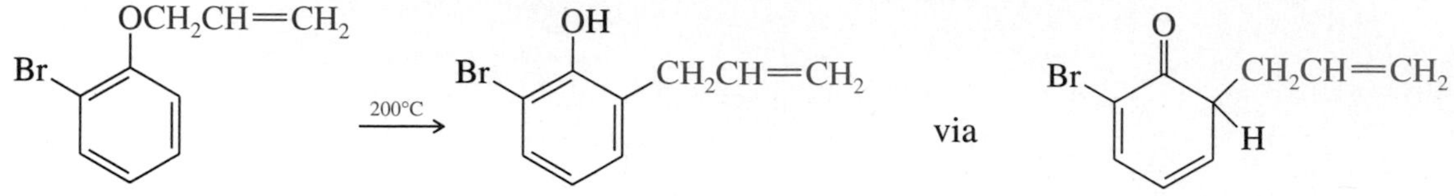

Br, $OCH_2CH{=}CH_2$ $\xrightarrow{200°C}$ OH, Br, $CH_2CH{=}CH_2$ via O, Br, $CH_2CH{=}CH_2$, H

Allyl *o*-bromophenyl ether

2-Allyl-6-bromophenol (82%)

6-Allyl-2-bromo-2,4-cyclohexadienone

Oxidation of 1,2- and 1,4-benzenediols gives colored compounds known as **quinones** (Section 24.14).

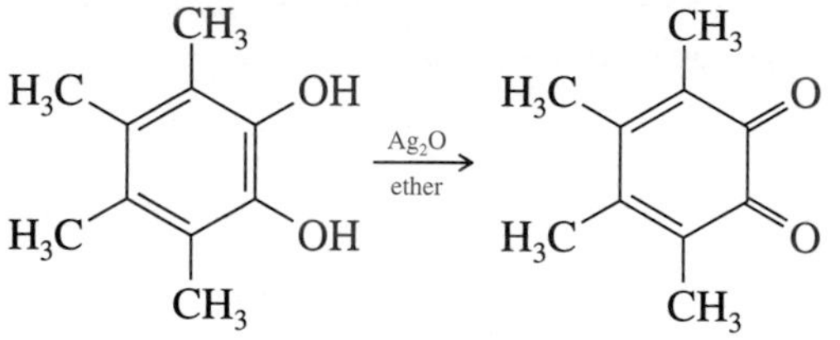

CH_3, H_3C, OH, H_3C, OH, CH_3 $\xrightarrow[\text{ether}]{Ag_2O}$ CH_3, H_3C, O, H_3C, O, CH_3

3,4,5,6-Tetramethyl-1,2-benzenediol

3,4,5,6-Tetramethyl-1,2-benzoquinone (81%)

PROBLEMS

24.11 The IUPAC rules permit the use of common names for a number of familiar phenols and aryl ethers. These common names are listed here along with their systematic names. Write the structure of each compound.

(*a*) *Vanillin* (4-hydroxy-3-methoxybenzaldehyde): a component of vanilla bean oil, which contributes to its characteristic flavor

(*b*) *Thymol* (2-isopropyl-5-methylphenol): obtained from oil of thyme

(*c*) *Carvacrol* (5-isopropyl-2-methylphenol): present in oil of thyme and marjoram

(*d*) *Eugenol* (4-allyl-2-methoxyphenol): obtained from oil of cloves

(*e*) *Gallic acid* (3,4,5-trihydroxybenzoic acid): prepared by hydrolysis of tannins derived from plants

(*f*) *Salicyl alcohol* (*o*-hydroxybenzyl alcohol): obtained from bark of poplar and willow trees

24.12 Name each of the following compounds:

(*a*) OH, CH_2CH_3, NO_2

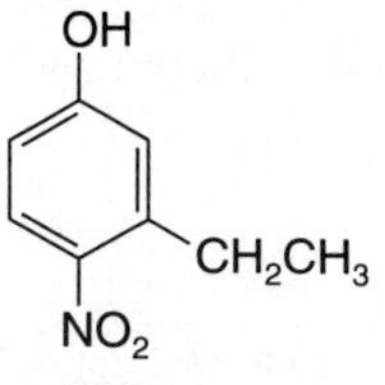

(*b*) OH, NO_2, CH_2CH_3

(*c*) Cl, HO, CH_2

(*d*) OCH_3, H_3C, CH_3, $CH(CH_3)_2$

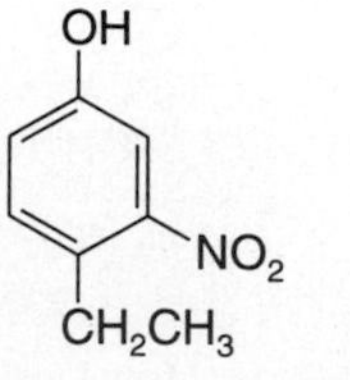

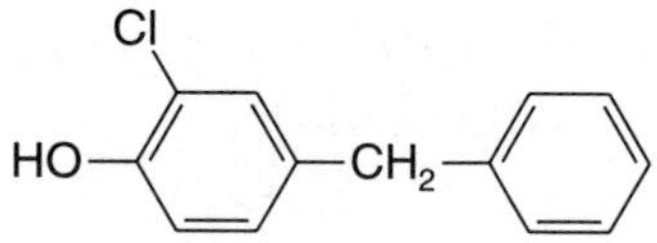

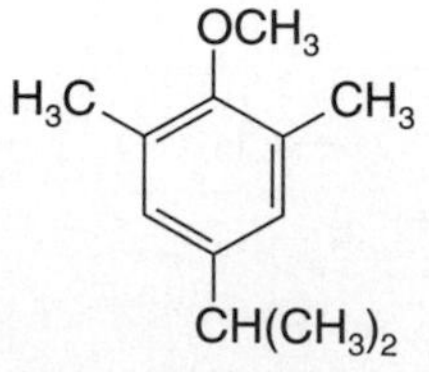

(*e*) [structure: 2,5-dichlorophenyl ester, with Cl substituents on ring] $-OCCCl_3$ (with C=O)

Cl

Cl

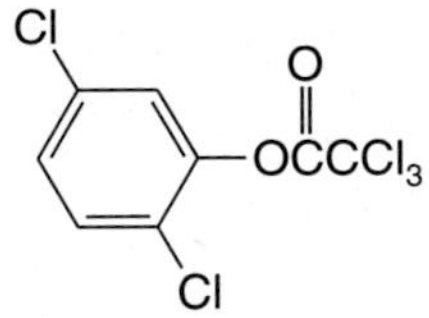

24.13 Write a balanced chemical equation for each of the following reactions:

(*a*) Phenol + sodium hydroxide
(*b*) Product of part (*a*) + ethyl bromide
(*c*) Product of part (*a*) + butyl *p*-toluenesulfonate
(*d*) Product of part (*a*) + acetic anhydride
(*e*) *o*-Cresol + benzoyl chloride
(*f*) *m*-Cresol + ethylene oxide
(*g*) 2,6-Dichlorophenol + bromine
(*h*) *p*-Cresol + excess aqueous bromine
(*i*) Isopropyl phenyl ether + excess hydrogen bromide + heat

24.14 Outline four different syntheses of phenol, beginning with benzene and using any necessary organic or inorganic reagents.

24.15 Which phenol in each of the following pairs is more acidic? Justify your choice.

(*a*) 2,4,6-Trimethylphenol or 2,4,6-trinitrophenol
(*b*) 2,6-Dichlorophenol or 3,5-dichlorophenol
(*c*) 3-Nitrophenol or 4-nitrophenol
(*d*) Phenol or 4-cyanophenol
(*e*) 2,5-Dinitrophenol or 2,6-dinitrophenol

24.16 Choose the reaction in each of the following pairs that proceeds at the faster rate. Explain your reasoning.

(*a*) Base-promoted hydrolysis of phenyl acetate or *m*-nitrophenyl acetate
(*b*) Base-promoted hydrolysis of *m*-nitrophenyl acetate or *p*-nitrophenyl acetate
(*c*) Reaction of ethyl bromide with phenol or with the sodium salt of phenol
(*d*) Reaction of ethylene oxide with the sodium salt of phenol or with the sodium salt of *p*-nitrophenol
(*e*) Bromination of phenol or phenyl acetate

24.17 Pentafluorophenol is readily prepared by heating hexafluorobenzene with potassium hydroxide in *tert*-butyl alcohol:

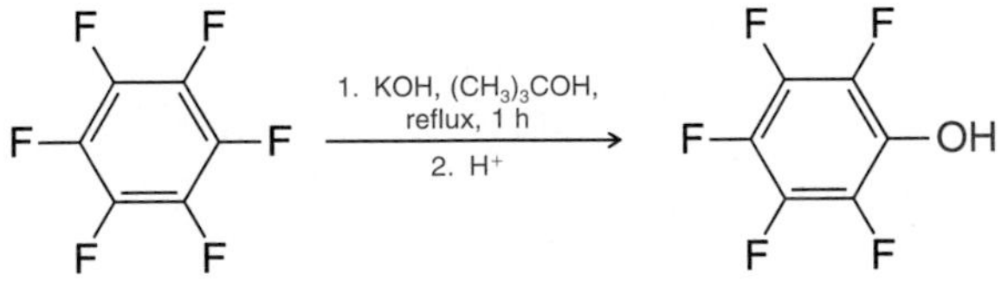

$$C_6F_6 \xrightarrow[2.\ H^+]{1.\ KOH,\ (CH_3)_3COH,\ \text{reflux, 1 h}} C_6F_5OH$$

Hexafluorobenzene — Pentafluorophenol (71%)

What is the most reasonable mechanism for this reaction? Comment on the comparative ease with which this conversion occurs.

24.18 Each of the following reactions has been reported in the chemical literature and proceeds cleanly in good yield. Identify the principal organic product in each case.

(*a*) 2-methoxyphenol (ring with OH and OCH_3) + $CH_2{=}CHCH_2Br$ $\xrightarrow[\text{acetone}]{K_2CO_3}$

(*b*) C_6H_5ONa + $ClCH_2CH(OH)CH_2OH$ $\longrightarrow$

(*c*) 4-hydroxy-3-methoxybenzaldehyde (HO, OCH_3, $\overset{O}{\parallel}CH$) $\xrightarrow[\text{acetic acid, heat}]{HNO_3}$

(*d*) $CH_3\overset{O}{\overset{\parallel}{C}}NH-C_6H_4-OCH_2CH{=}CH_2$ $\xrightarrow{\text{heat}}$

(*e*) ring with OH, OCH_2CH_3, NO_2 + Br_2 $\xrightarrow{\text{acetic acid}}$

(*f*) ring with OH, Cl, OH $\xrightarrow[H_2SO_4]{K_2Cr_2O_7}$

(*g*) ring with CH_3, $O\overset{O}{\overset{\parallel}{C}}CH_3$, $CH(CH_3)_2$ $\xrightarrow{AlCl_3}$

(*h*) ring with OH, H_3C, CH_3 + ring with Cl, NO_2 $\xrightarrow[\text{dimethyl sulfoxide, 90°C}]{NaOH}$

(*i*) ring with Cl, OH, Cl + $2Cl_2$ $\xrightarrow{\text{acetic acid}}$

(*j*) [o-toluidine (CH_3, NH_2)] + [phenyl salicylate: OH, C(=O)–O–C_6H_5] $\xrightarrow{\text{heat}}$

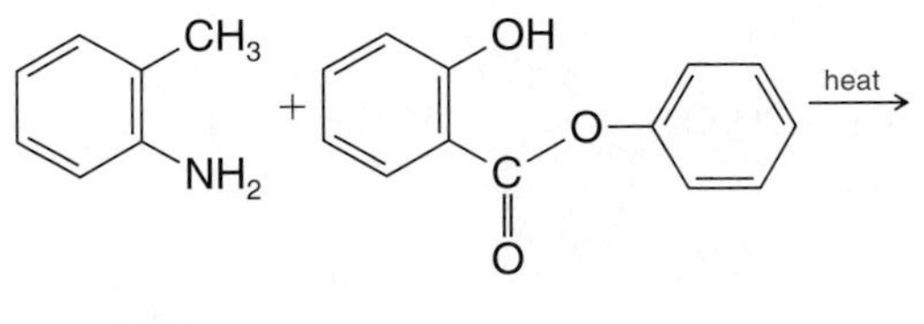

(*k*) [2,4,5-trichlorophenol: OH, Cl, Cl, Cl] + $C_6H_5\overset{+}{N}{\equiv}N\ Cl^-$ $\longrightarrow$

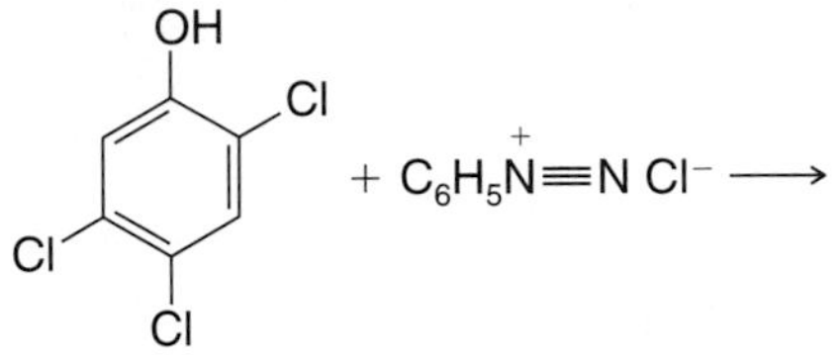

24.19 A synthesis of the analgesic substance *phenacetin* is outlined in the following equation. What is the structure of phenacetin?

$$p\text{-Nitrophenol} \xrightarrow[\substack{2.\ \text{Fe, HCl; then HO}^-\\ 3.\ CH_3\overset{O}{\overset{\|}{C}}O\overset{O}{\overset{\|}{C}}CH_3}]{1.\ CH_3CH_2Br,\ NaOH} \text{phenacetin}$$

24.20 Identify compounds A through C in the synthetic sequence represented by equations (*a*) through (*c*).

(*a*) Phenol + $H_2SO_4 \xrightarrow{\text{heat}}$ compound A ($C_6H_6O_7S_2$)

(*b*) Compound A + $Br_2 \xrightarrow[2.\ H^+]{1.\ HO^-}$ compound B ($C_6H_5BrO_7S_2$)

(*c*) Compound B + $H_2O \xrightarrow[\text{heat}]{H^+}$ compound C (C_6H_5BrO)

24.21 Treatment of 3,5-dimethylphenol with dilute nitric acid, followed by steam distillation of the reaction mixture, gave a compound A ($C_8H_9NO_3$, mp 66°C) in 36 percent yield. The nonvolatile residue from the steam distillation gave a compound B ($C_8H_9NO_3$, mp 108°C) in 25 percent yield on extraction with chloroform. Identify compounds A and B.

24.22 Outline a reasonable synthesis of 4-nitrophenyl phenyl ether from chlorobenzene and phenol.

24.23 As an allergen for testing purposes, synthetic 3-pentadecylcatechol is more useful than natural poison ivy extracts (of which it is one component). A stable crystalline solid, it is efficiently prepared in pure form from readily available starting materials. Outline a reasonable synthesis of this compound from 2,3-dimethoxybenzaldehyde and any necessary organic or inorganic reagents.

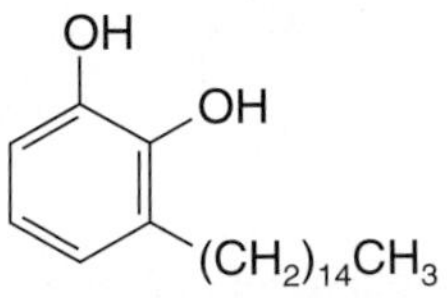

[Structure: benzene ring with OH, OH, $(CH_2)_{14}CH_3$]

3-Pentadecylcatechol

24.24 Describe a scheme for carrying out the following synthesis. (In the synthesis reported in the literature, four separate operations were required.)

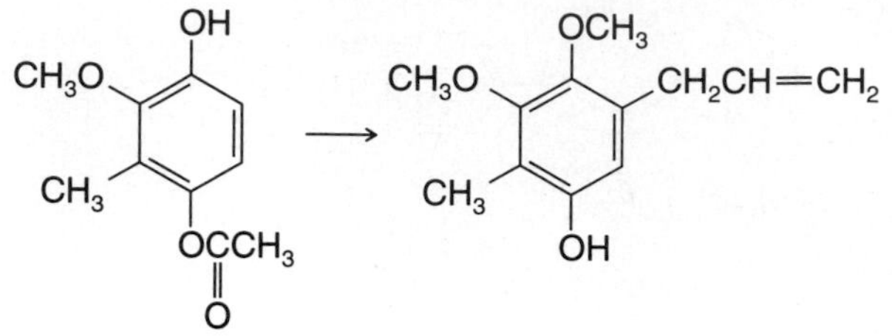

24.25 In a general reaction known as the *cyclohexadienone-phenol rearrangement,* cyclohexadienones are converted to phenols under conditions of acid catalysis. An example is

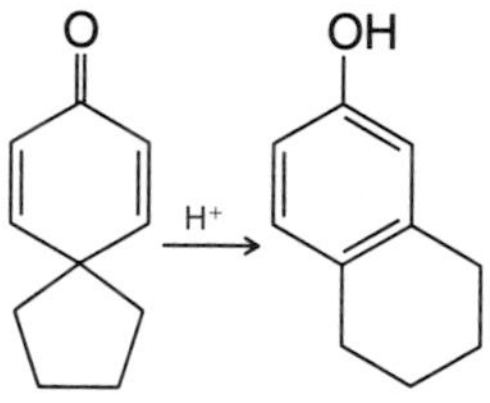

(100%)

Write a reasonable mechanism for this reaction.

24.26 Treatment of *p*-hydroxybenzoic acid with aqueous bromine leads to the evolution of carbon dioxide and the formation of 2,4,6-tribromophenol. Explain.

24.27 Treatment of phenol with excess aqueous bromine is actually more complicated than expected. A white precipitate forms rapidly, which on closer examination is not 2,4,6-tribromophenol but is instead 2,4,4,6-tetrabromocyclohexadienone. Explain the formation of this product.

24.28 Treatment of 2,4,6-tri-*tert*-butylphenol with bromine in cold acetic acid gives the compound $C_{18}H_{29}BrO$ in quantitative yield. The infrared spectrum of this compound contains absorptions at 1630 and 1655 cm^{-1}. Its ^1H nmr spectrum shows only three peaks (all singlets), at $\delta = 1.2$, 1.3, and 6.9 ppm, in the ratio 9:18:2. What is a reasonable structure for the compound?

24.29 A compound A undergoes hydrolysis of its acetal function in dilute sulfuric acid to yield 1,2-ethanediol and compound B ($C_6H_6O_2$), mp 54°C. Compound B exhibits a carbonyl stretching band in the infrared at 1690 cm^{-1} and has two singlets in its ^1H nmr spectrum, at $\delta = 2.9$ and 6.7 ppm, in the ratio 2:1. On standing in water or ethanol, compound B is converted cleanly to an isomeric substance, compound C, mp 172 to 173°C. Compound C has no peaks attributable to carbonyl groups in its infrared spectrum. Identify compounds B and C.

Compound A

24.30 Fomecin A ($C_8H_8O_5$) is an antibacterial fungal metabolite. It is soluble in dilute sodium hydroxide solution, and reacts with 2,4-dinitrophenylhydrazine. Its ^1H nmr spectrum includes singlets at $\delta = 4.6$ ppm (2 H), 6.5 ppm (1 H), and 10.2 ppm (1 H) as well as broad signals for four O—H protons. Treatment of fomecin A with acetic anhydride gives the product shown in the following equation. Suggest a structure for fomecin A.

Fomecin A $\xrightarrow[\text{pyridine}]{\text{acetic anhydride}}$ (tetra-acetylated product: $CH_3\overset{O}{\overset{\|}{C}}O$ groups on the benzene ring and $O\overset{O}{\overset{\|}{C}}CH_3$ on the fused five-membered ring containing O)

24.31 One of the industrial processes for the preparation of phenol, discussed in Section 24.6, includes an acid-catalyzed rearrangement of cumene hydroperoxide as a key step. This reaction proceeds by way of an intermediate hemiacetal:

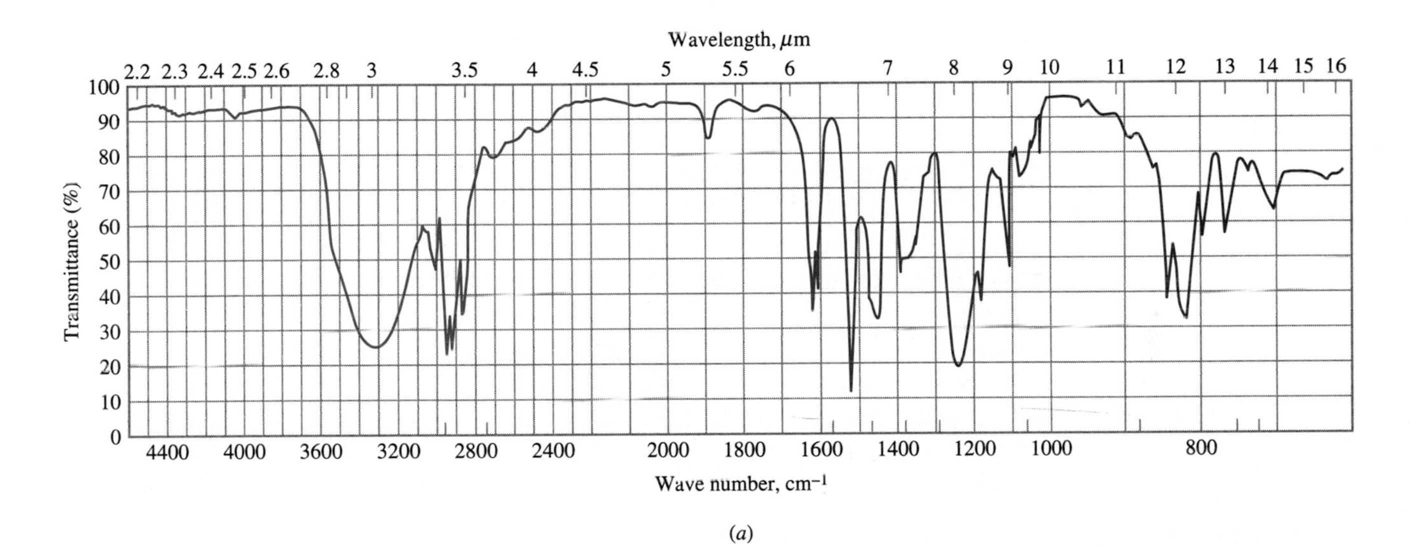

(*a*)

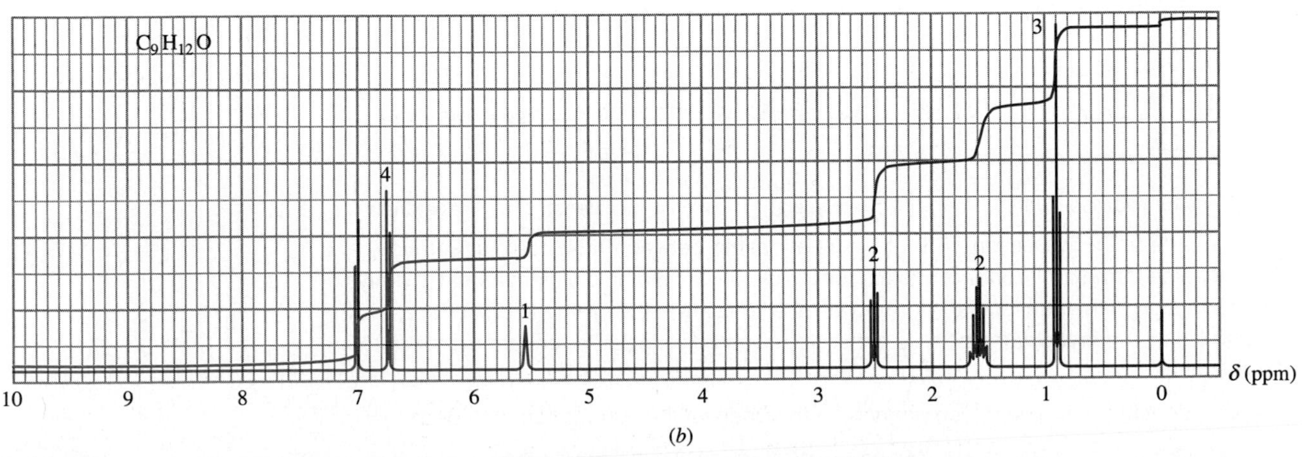

(*b*)

FIGURE 24.6 (*a*) Infrared and (*b*) 300-MHz ^1H nmr spectra of compound $C_9H_{12}O$ (Problem 24.32*a*).

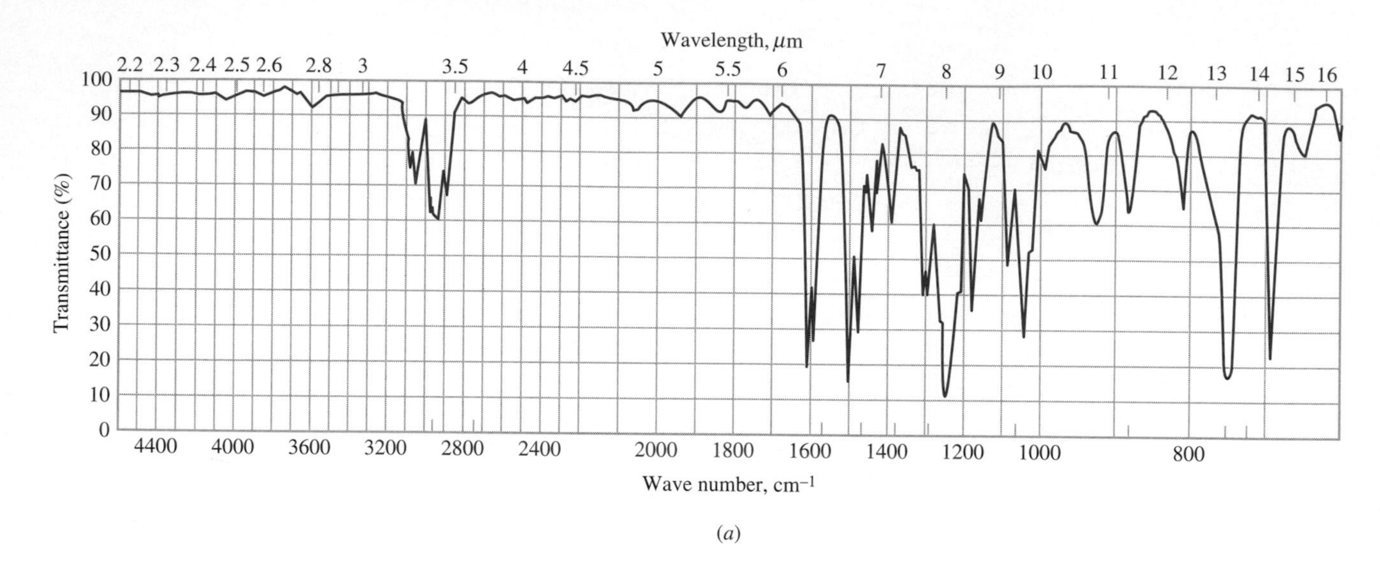

(*a*)

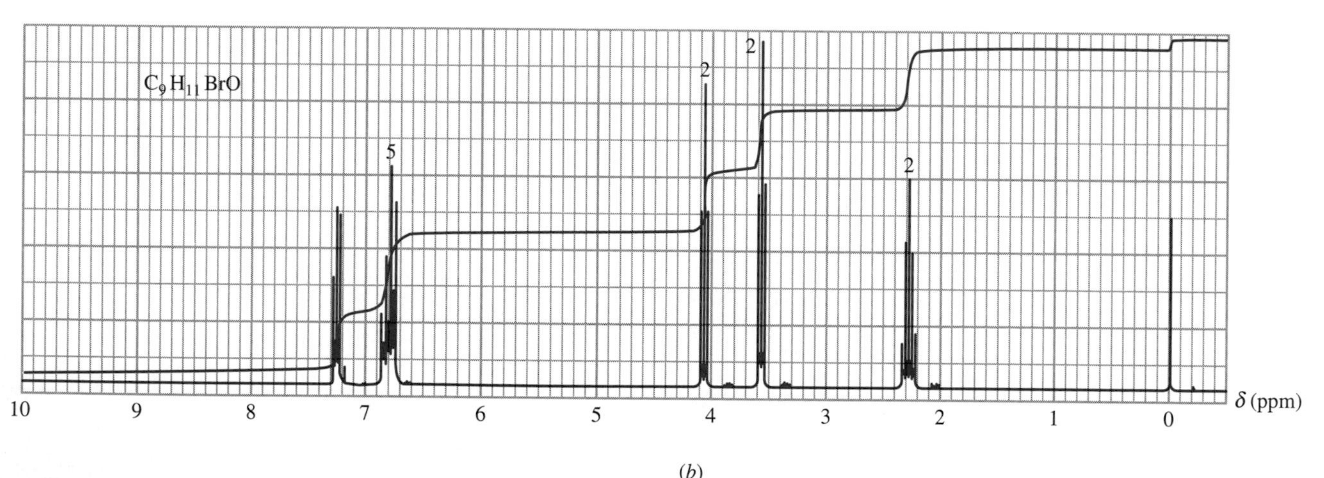

(*b*)

FIGURE 24.7 (*a*) Infrared and (*b*) 300-MHz ^{1}H nmr spectra of compound $C_9H_{11}BrO$ (Problem 24.32*b*).

$$C_6H_5C(CH_3)_2OOH \xrightarrow[H_2O]{H_2SO_4} C_6H_5OC(CH_3)_2OH \xrightarrow{H_2O} C_6H_5OH + CH_3\overset{O}{\overset{\|}{C}}CH_3$$

Cumene hydroperoxide — Hemiacetal — Phenol — Acetone

You have learned in Section 17.9 of the relationship between hemiacetals, ketones, and alcohols; the formation of phenol and acetone is simply an example of hemiacetal hydrolysis. The formation of the hemiacetal intermediate is a key step in the synthetic procedure; it is the step in which the aryl-oxygen bond is generated. Can you suggest a reasonable mechanism for this step?

24.32 Identify the following compounds on the basis of the information provided:

(*a*) $C_9H_{12}O$: Its infrared and ^{1}H nmr spectra are shown in Figure 24.6.
(*b*) $C_9H_{11}BrO$: Its infrared and ^{1}H nmr spectra are shown in Figure 24.7.
(*c*) $C_{10}H_{11}ClO_2$: Its infrared and ^{1}H nmr spectra are shown in Figure 24.8.

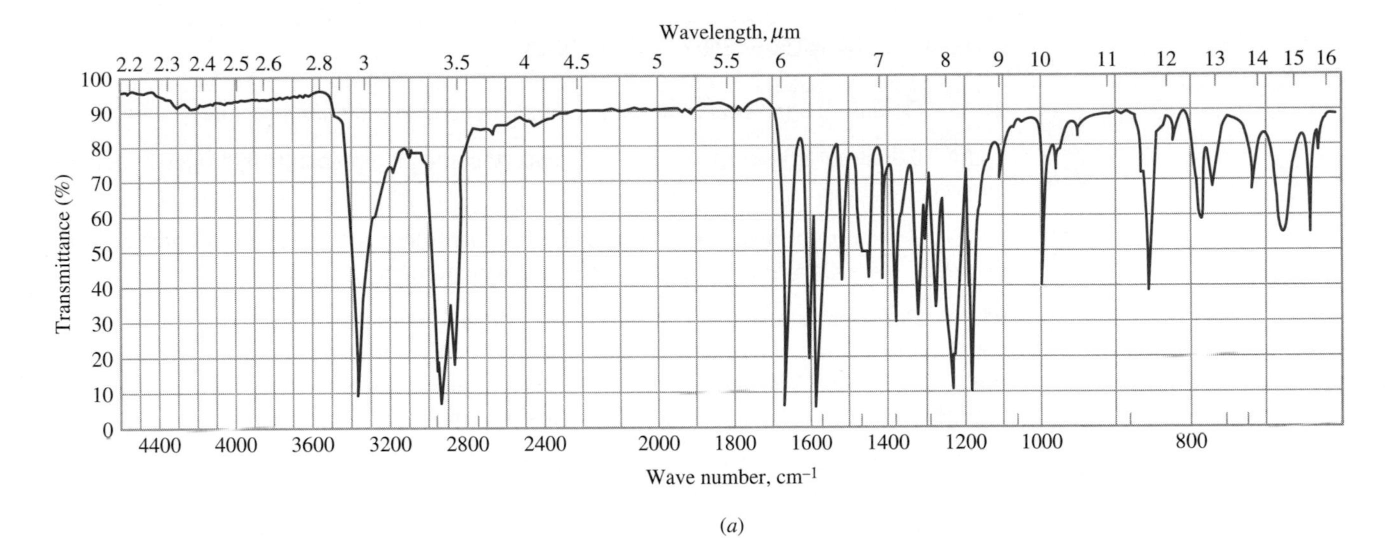

(*a*)

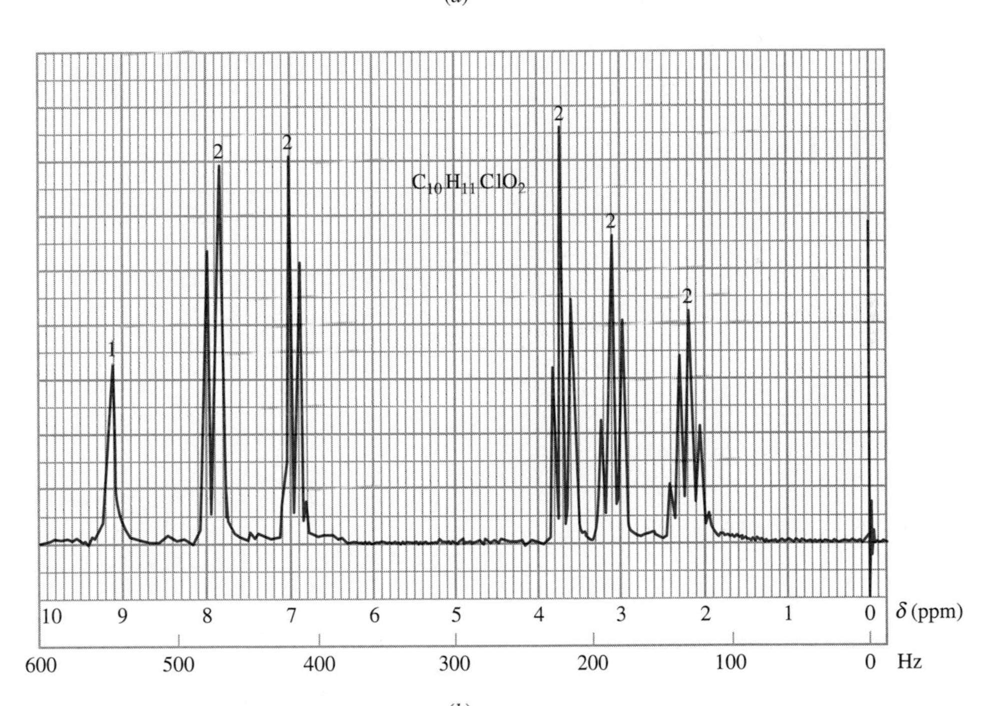

(*b*)

FIGURE 24.8 (*a*) Infrared and (*b*) 60-MHz ^{1}H nmr spectra of compound $C_{10}H_{11}ClO_2$ (Problem 24.32*c*).

MOLECULAR MODELING EXERCISES

24.33 Construct a molecular model of phenol, assuming a planar structure.

24.34 Construct a molecular model of aspirin. Assume intramolecular hydrogen bonding is present.

24.35 Construct a molecular model of allyl phenyl ether and arrange the atoms in the geometry necessary to undergo the Claisen rearrangement.

24.36 Construct a molecular model of 3,5-dimethyl-4-nitrophenol and suggest an explanation for the decreased acidity of this substance ($pK_a = 8.2$) compared to *p*-nitrophenol ($pK_a = 7.2$). (*Note*: The decreased acidity is too large to be explained on the basis of methyl groups acting only as electron-releasing substituents.)

24.37 The biological properties of dioxin include an ability to bind to a protein known as the AH (aromatic hydrocarbon) receptor. Dioxin is not a hydrocarbon, but it shares a certain structural property with aromatic hydrocarbons. Construct molecular models of dioxin and anthracene to reveal their similarities.

CHAPTER 25

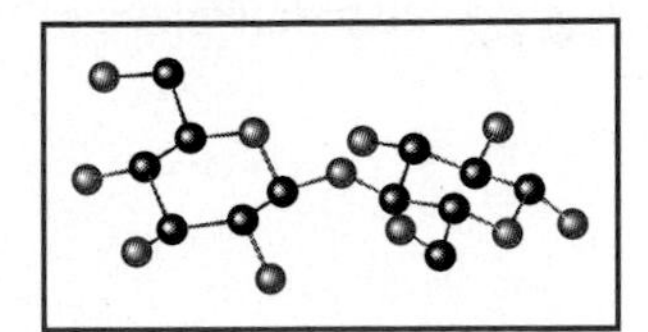

CARBOHYDRATES

The major classes of organic compounds common to living systems are *lipids, proteins, nucleic acids,* and *carbohydrates.* Carbohydrates are very familiar to us—we call many of them "sugars." They constitute a substantial portion of the food we eat and provide most of the energy that keeps the human engine running. Carbohydrates are structural components of the walls of plant cells and the wood of trees. Genetic information is stored and transferred by way of nucleic acids, specialized derivatives of carbohydrates, which will be examined in more detail in Chapter 27.

Historically, carbohydrates were once considered to be "hydrates of carbon" because their molecular formulas in many (but not all) cases correspond to $C_n(H_2O)_m$. It is more realistic to define a carbohydrate as a *polyhydroxy aldehyde or polyhydroxy ketone,* a point of view closer to structural reality and more suggestive of chemical reactivity.

This chapter is divided into two parts. The first, and major, portion is devoted to carbohydrate structure. You will see how the principles of stereochemistry and conformational analysis combine to aid our understanding of this complex subject. The remainder of the chapter describes chemical reactions of carbohydrates. Most of these reactions are simply extensions of what you have already learned concerning alcohols, aldehydes, ketones, and acetals.

25.1 CLASSIFICATION OF CARBOHYDRATES

Sugar is a combination of the Sanskrit words *su* (sweet) and *gar* (sand). Thus, its literal meaning is "sweet sand."

The Latin word for *sugar* is *saccharum,* and the derived term *saccharide* is the basis of a system of carbohydrate classification. A **monosaccharide** is a simple carbohydrate, one that on attempted hydrolysis is not cleaved to smaller carbohydrates. *Glucose* ($C_6H_{12}O_6$), for example, is a monosaccharide. A **disaccharide** on hydrolysis is cleaved

TABLE 25.1
Some Classes of Monosaccharides

Number of carbon atoms	Aldose	Ketose
Four	Aldotetrose	Ketotetrose
Five	Aldopentose	Ketopentose
Six	Aldohexose	Ketohexose
Seven	Aldoheptose	Ketoheptose
Eight	Aldooctose	Ketooctose

to two monosaccharides, which may be the same or different. *Sucrose*—common table sugar—is a disaccharide that yields one molecule of glucose and one of fructose on hydrolysis.

$$\text{Sucrose } (C_{12}H_{22}O_{11}) + H_2O \longrightarrow \text{glucose } (C_6H_{12}O_6) + \text{fructose } (C_6H_{12}O_6)$$

An **oligosaccharide** (*oligos* is a Greek word that in its plural form means "few") yields 3 to 10 monosaccharide units on hydrolysis. **Polysaccharides** are hydrolyzed to more than 10 monosaccharide units. *Cellulose* is a polysaccharide molecule that gives thousands of glucose molecules when completely hydrolyzed.

Over 200 different monosaccharides are known. They can be grouped according to the number of carbon atoms they contain and whether they are polyhydroxy aldehydes or polyhydroxy ketones. Monosaccharides that are polyhydroxy aldehydes are called **aldoses;** those that are polyhydroxy ketones are **ketoses.** Aldoses and ketoses are further classified according to the number of carbon atoms in the main chain. Table 25.1 lists the terms applied to monosaccharides having four to eight carbon atoms.

25.2 FISCHER PROJECTIONS AND THE D-L NOTATIONAL SYSTEM

Stereochemistry is the key to understanding carbohydrate structure, a fact that was clearly appreciated by the German chemist Emil Fischer. The projection formulas used by Fischer to represent stereochemistry in chiral molecules that were described in Section 7.7 are particularly well-suited to carbohydrates. Figure 25.1 illustrates their application to the enantiomers of *glyceraldehyde* (2,3-dihydroxypropanal), a fundamental molecule in carbohydrate stereochemistry. When the Fischer projection is oriented as shown in the figure, with the carbon chain vertical and the aldehyde carbon at the top, the C-2 hydroxyl group points to the right in (+)-glyceraldehyde and to the left in (−)-glyceraldehyde.

Adopting the enantiomers of glyceraldehyde as stereochemical reference compounds originated with proposals made in 1906 by M. A. Rosanoff, a chemist at New York University.

Techniques for determining the absolute configuration of chiral molecules were not developed until the 1950s, and so it was not possible for Fischer and his contemporaries to relate the sign of rotation of any substance to its absolute configuration. A system evolved based on the arbitrary assumption, later shown to be correct, that the enantiomers of glyceraldehyde have the signs of rotation and absolute configurations shown in Figure 25.1. Two stereochemical descriptors were defined, D and L. The absolute configuration of (+)-glyceraldehyde, as depicted in the figure, was said to be

CH═O, H–C–OH, CH_2OH (three-dimensional representation) CH═O, H–C–OH, CH_2OH (Fischer projection)

(*R*)-(+)-Glyceraldehyde

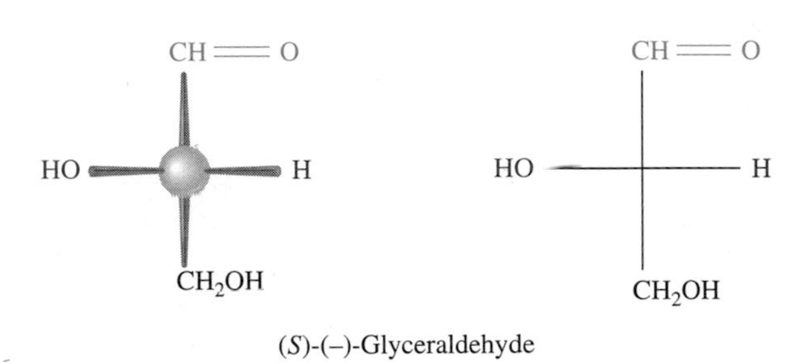

(*S*)-(−)-Glyceraldehyde

FIGURE 25.1 Three-dimensional representations and Fischer projections of the enantiomers of glyceraldehyde.

D and that of its enantiomer, (−)-glyceraldehyde, L. Compounds that had a spatial arrangement of substituents analogous to those of D-(+)- and L-(−)-glyceraldehyde were said to have the D and L configurations, respectively.

PROBLEM 25.1 Identify each of the following Fischer projections as either D- or L-glyceraldehyde:

(*a*) CH_2OH / HO–+–H / CHO (*b*) H / $HOCH_2$–+–CHO / OH (*c*) CHO / $HOCH_2$–+–H / OH

SAMPLE SOLUTION (*a*) Redraw the Fischer projection so as to more clearly show the true spatial orientation of the groups. Next, reorient the molecule so that its relationship to the glyceraldehyde enantiomers in Figure 25.1 is apparent.

CH_2OH / HO–+–H / CHO is equivalent to CH_2OH / HO–C–H / CHO $\xrightarrow{\text{turn } 180^\circ}$ CHO / H–C–OH / CH_2OH

The structure is the same as that of (+)-glyceraldehyde in the figure. It is D-glyceraldehyde.

Both Fischer projection formulas and the D-L system of stereochemical notation proved to be so helpful in representing carbohydrate stereochemistry that the chemical and biochemical literature is replete with their use. To read that literature you need to be acquainted with both of these devices, in addition to the more modern Cahn-Ingold-Prelog system.

25.3 THE ALDOTETROSES

Glyceraldehyde is considered to be the simplest chiral carbohydrate. It is an **aldotriose** and, since it contains one stereogenic center, exists in two stereoisomeric forms, the D and L enantiomers. Moving up the scale in complexity, next come the **aldotetroses.** Examination of their structures illustrates the application of the Fischer system to compounds that contain more than one stereogenic center.

The aldotetroses are the four stereoisomers of 2,3,4-trihydroxybutanal. Fischer projection formulas are constructed by orienting the molecule in an eclipsed conformation with the aldehyde group at the top. The four carbon atoms define the main chain of the Fischer projection and are arranged vertically. Horizontal bonds are directed outward, vertical bonds back.

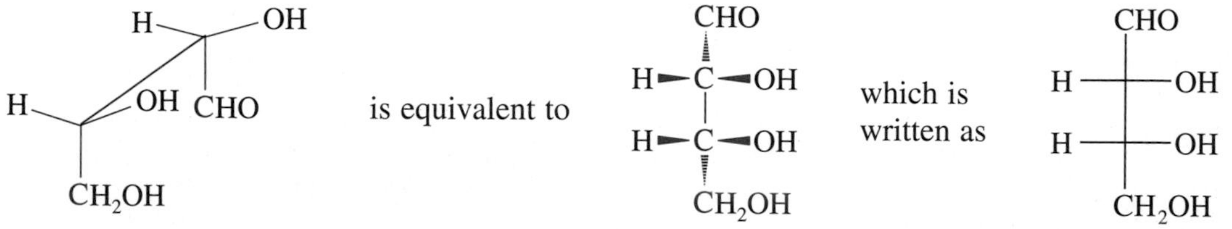

H OH H OH CHO CH_2OH is equivalent to CHO H—C—OH H—C—OH CH_2OH which is written as CHO H—OH H—OH CH_2OH

Eclipsed conformation of a tetrose

Fischer projection formula of a tetrose

The particular aldotetrose just shown is called D-*erythrose.* The prefix D tells us that the configuration at the *highest-numbered stereogenic center* is analogous to that of D-(+)-glyceraldehyde. Its mirror image is L-erythrose.

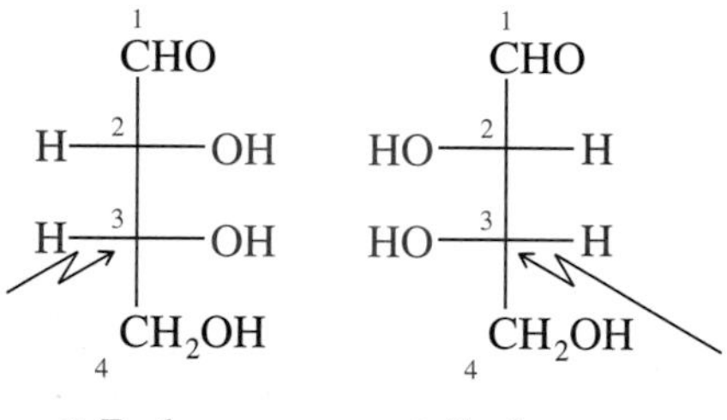

Highest-numbered stereogenic center has configuration analogous to that of D-glyceraldehyde

D-Erythrose

L-Erythrose

Highest-numbered stereogenic center has configuration analogous to that of L-glyceraldehyde

Relative to each other, both hydroxyl groups are on the same side in Fischer projections of the erythrose enantiomers. The remaining two stereoisomers have their hydroxyl groups on opposite sides in Fischer projection formulas. They are diastereomers of D- and L-erythrose and are called D- and L-*threose.* The D and L prefixes again specify the configuration of the highest-numbered stereogenic center. D-Threose and L-threose are enantiomers of each other:

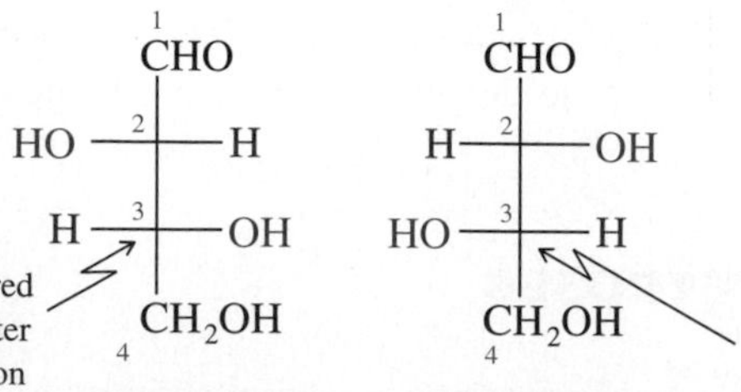

Highest-numbered stereogenic center has configuration analogous to that of D-glyceraldehyde

D-Threose

L-Threose

Highest-numbered stereogenic center has configuration analogous to that of L-glyceraldehyde

PROBLEM 25.2 Which of the four aldotetroses just discussed is the following?

H OH O
CH
$HOCH_2$ H
OH

As shown for the aldotetroses, an aldose belongs to the D or the L series according to the configuration of the stereogenic center farthest removed from the aldehyde function. Individual names, such as erythrose and threose, specify the particular arrangement of stereogenic centers within the molecule relative to each other. Optical activities cannot be determined directly from the D and L prefixes. As it turns out, both D-erythrose and D-threose are levorotatory, while D-glyceraldehyde is dextrorotatory.

25.4 ALDOPENTOSES AND ALDOHEXOSES

Aldopentoses have three stereogenic centers. The eight stereoisomers are divided into a set of four D-aldopentoses and an enantiomeric set of four L-aldopentoses. The aldopentoses are named *ribose, arabinose, xylose,* and *lyxose.* Fischer projection formulas of the D stereoisomers of the aldopentoses are given in Figure 25.2. Notice that all these diastereomers have the same configuration at C-4 and that this configuration is analogous to that of D-(+)-glyceraldehyde.

PROBLEM 25.3 L-(+)-Arabinose is a naturally occurring L sugar. It is obtained by acid hydrolysis of the polysaccharide present in mesquite gum. Write a Fischer projection formula for L-(+)-arabinose.

Among the aldopentoses, D-ribose is a component of many biologically important substances, most notably the ribonucleic acids, while D-xylose is very abundant and is isolated by hydrolysis of the polysaccharides present in corncobs and the wood of trees.

The aldohexoses include some of the most familiar of the monosaccharides, as well as the most abundant organic compound on earth, D-(+)-glucose. With four stereogenic centers, 16 stereoisomeric aldohexoses are possible; 8 belong to the D series and 8 to the L series. All are known, either as naturally occurring substances or as the products of synthesis. The eight D-aldohexoses are given in Figure 25.2; it is the spatial arrangement at C-5, hydrogen to the left in a Fischer projection and hydroxyl to the right, that identifies them as carbohydrates of the D series.

Strictly speaking cellulose is the most abundant organic compound, but each cellulose molecule is a polysaccharide composed of thousands of glucose units (Section 25.15).

PROBLEM 25.4 Name the following sugar:

CHO
H—OH
H—OH
H—OH
HO—H
CH_2OH

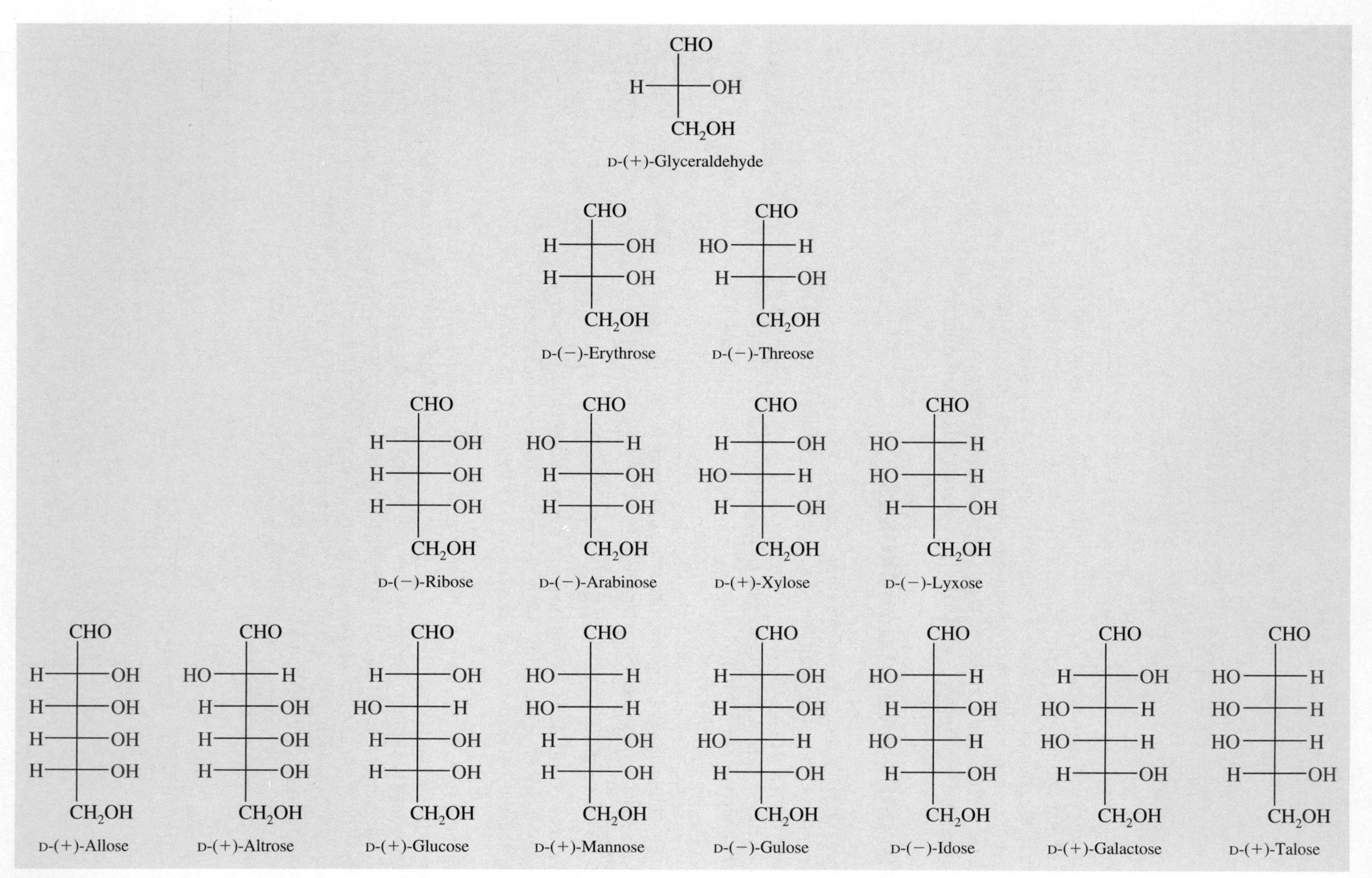

FIGURE 25.2 Configurations of the D series of aldoses containing three through six carbon atoms.

Of all the monosaccharides, D-(+)-*glucose* is the best known, most important, and most abundant. Its formation from carbon dioxide, water, and sunlight is the central theme of photosynthesis. Carbohydrate formation by photosynthesis is estimated to be on the order of 10^{11} tons per year, a source of stored energy utilized, directly or indirectly, by all higher forms of life on the planet. Glucose was isolated from raisins in 1747 and by hydrolysis of starch in 1811. Its structure was determined, in work culminating in 1900, by Emil Fischer.

D-(+)-*Galactose* is a constituent of numerous polysaccharides. It is best obtained by acid hydrolysis of lactose (milk sugar), a disaccharide of D-glucose and D-galactose. L(−)-Galactose also occurs naturally and can be prepared by hydrolysis of flaxseed gum and agar. The principal source of D-(+)-*mannose* is hydrolysis of the polysaccharide of the ivory nut, a large, nutlike seed obtained from a South American palm.

25.5 A MNEMONIC FOR CARBOHYDRATE CONFIGURATIONS

The task of relating carbohydrate configurations to names requires either a world-class memory or an easily recalled mnemonic. The mnemonic device that serves us well here was popularized by the husband-wife team of Louis F. Fieser and Mary Fieser of Harvard University in their 1956 textbook, *Organic Chemistry.* As with many mnemonic devices, it is not clear who actually invented it, and references to this particular one appeared in the chemical education literature before publication of the Fiesers' text. The mnemonic has two features: (1) a system for setting down all the stereoisomeric D-aldohexoses in a logical order; and (2) a way to assign the correct name to each one.

See, for example, the November 1955 issue of the *Journal of Chemical Education* (p. 584). An article giving references to a variety of chemistry mnemonics appears in the July 1960 issue of the *Journal of Chemical Education* (p. 366).

A systematic way to set down all the D-hexoses (as in Fig. 25.2) is to draw skeletons of the necessary eight Fischer projection formulas, placing the hydroxyl group at C-5 to the right in each so as to guarantee that they all belong to the D series. Working up the carbon chain, place the hydroxyl group at C-4 to the right in the first four structures, and to the left in the next four. In each of these two sets of four, place the C-3 hydroxyl group to the right in the first two and to the left in the next two; in each of the resulting four sets of two, place the C-2 hydroxyl group to the right in the first one and to the left in the second.

Once the eight Fischer projections have been written, they are named in order with the aid of the sentence: All altruists gladly make gum in gallon tanks. The words of the sentence stand for *allose, altrose, glucose, mannose, gulose, idose, galactose, talose.*

An analogous pattern of configurations can be seen in the aldopentoses when they are arranged in the order *ribose, arabinose, xylose, lyxose.* (RAXL is an easily remembered nonsense word that gives the correct sequence.) This pattern is discernible even in the aldotetroses erythrose and threose.

25.6 CYCLIC FORMS OF CARBOHYDRATES: FURANOSE FORMS

Aldoses incorporate two functional groups, C=O and OH, which are capable of reacting with each other. We saw in Section 17.9 that nucleophilic addition of an alcohol function to a carbonyl group gives a hemiacetal. When the hydroxyl and

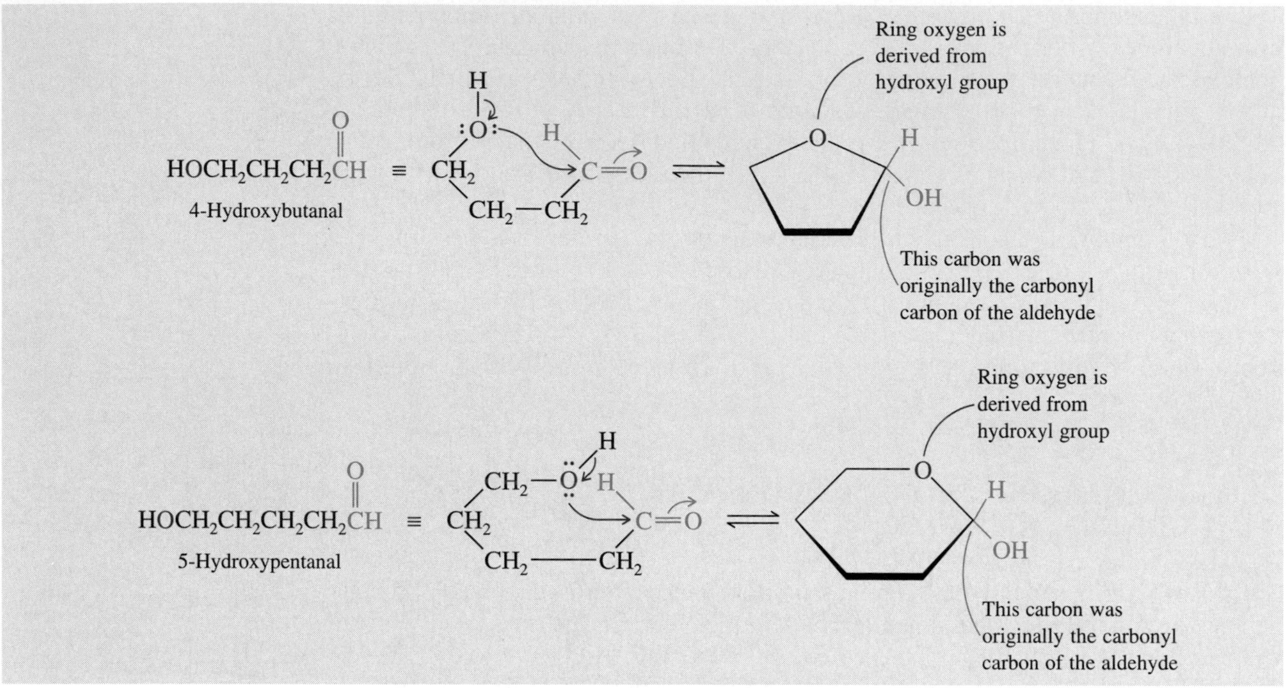

$HOCH_2CH_2CH_2CH{=}O$ ≡ (ring-closing arrows) ⇌ five-membered ring

4-Hydroxybutanal

Ring oxygen is derived from hydroxyl group

This carbon was originally the carbonyl carbon of the aldehyde

$HOCH_2CH_2CH_2CH_2CH{=}O$ ≡ (ring-closing arrows) ⇌ six-membered ring

5-Hydroxypentanal

Ring oxygen is derived from hydroxyl group

This carbon was originally the carbonyl carbon of the aldehyde

FIGURE 25.3 Cyclic hemiacetal formation in 4-hydroxybutanal and 5-hydroxypentanal.

carbonyl groups are part of the same molecule, a *cyclic hemiacetal* results, as illustrated in Figure 25.3.

Cyclic hemiacetal formation is most common when the ring that results is five-membered or six-membered. Five-membered cyclic hemiacetals of carbohydrates are called **furanose** forms; six-membered ones are called **pyranose** forms. The ring carbon that is derived from the carbonyl group, the one that bears two oxygen substituents, is called the **anomeric** carbon.

Aldoses exist almost exclusively as their cyclic hemiacetals; very little of the open-chain form is present at equilibrium. In order to understand their structures and chemical reactions, we need to be able to translate Fischer projection formulas of carbohydrates into their cyclic hemiacetal forms. Consider first cyclic hemiacetal formation in D-erythrose. So as to visualize furanose ring formation more clearly, redraw the Fischer projection in a form more suited to cyclization, being careful to maintain the stereochemistry at each stereogenic center.

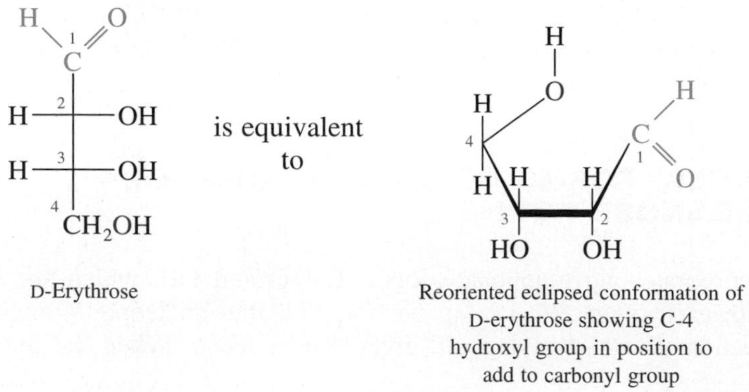

D-Erythrose (C-1 CHO; C-2 H—OH; C-3 H—OH; C-4 CH_2OH) is equivalent to

Reoriented eclipsed conformation of D-erythrose showing C-4 hydroxyl group in position to add to carbonyl group

Hemiacetal formation between the carbonyl group and the terminal hydroxyl yields the five-membered furanose ring form. The anomeric carbon becomes a new stereogenic center; its hydroxyl group can be either cis or trans to the other hydroxyl groups of the molecule.

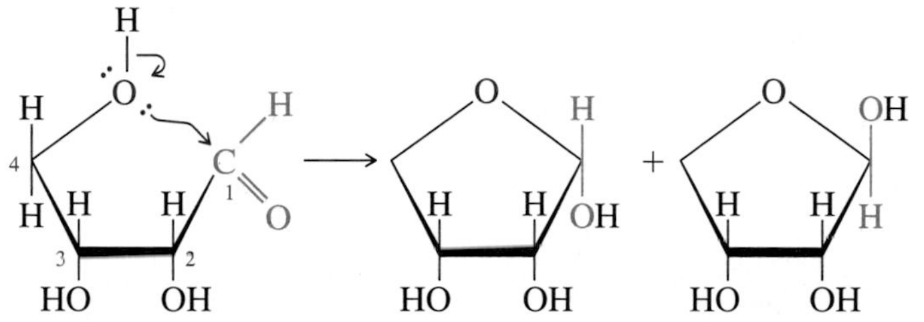

D-Erythrose

α-D-Erythrofuranose (hydroxyl group at anomeric carbon is down)

β-D-Erythrofuranose (hydroxyl group at anomeric carbon is up)

Representations of carbohydrates constructed in this manner are called **Haworth formulas,** after the British carbohydrate chemist Sir Norman Haworth (St. Andrew's University and the University of Birmingham). Early in his career Haworth contributed to the discovery that sugars exist as cyclic hemiacetals rather than in open-chain forms. Later he collaborated on an efficient synthesis of vitamin C from carbohydrate precursors. This was the first chemical synthesis of a vitamin and provided an inexpensive route to its preparation on a commercial scale. Haworth was a corecipient of the Nobel Prize for chemistry in 1937.

The two stereoisomeric furanose forms of D-erythrose are named α-D-erythrofuranose and β-D-erythrofuranose. The prefixes α and β describe *relative configuration.* The configuration of the anomeric carbon is α when its hydroxyl group is on the same side of a Fischer projection as the hydroxyl group at the highest numbered stereogenic center. When the hydroxyl groups at the anomeric carbon and the highest numbered stereogenic center are on opposite sides of a Fischer projection, the configuration at the anomeric carbon is β.

Substituents that are to the right in a Fischer projection are "down" in the corresponding Haworth formula.

Generating Haworth formulas to show stereochemistry in furanose forms of higher aldoses is slightly more complicated and requires an additional operation. Furanose forms of D-ribose are frequently encountered building blocks in biologically important organic molecules. They result from hemiacetal formation between the aldehyde group and the hydroxyl at C-4:

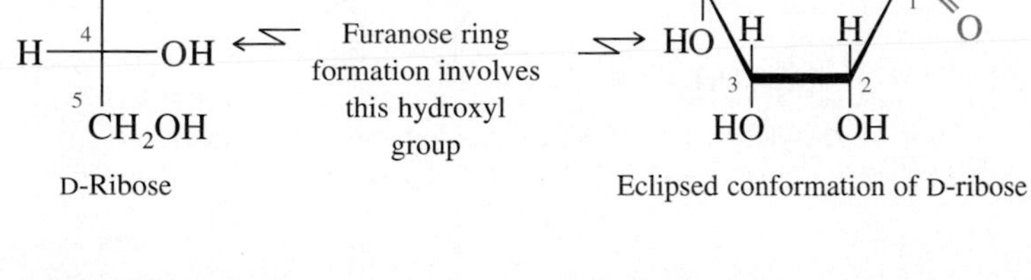

Furanose ring formation involves this hydroxyl group

D-Ribose

Eclipsed conformation of D-ribose

Notice that the eclipsed conformation of D-ribose derived directly from the Fischer projection does not have its C-4 hydroxyl group properly oriented for furanose ring formation. We must redraw it in a conformation that permits the five-membered cyclic hemiacetal to form. This is accomplished by rotation about the C(3)—C(4) bond, taking care that the configuration at C-4 is not changed.

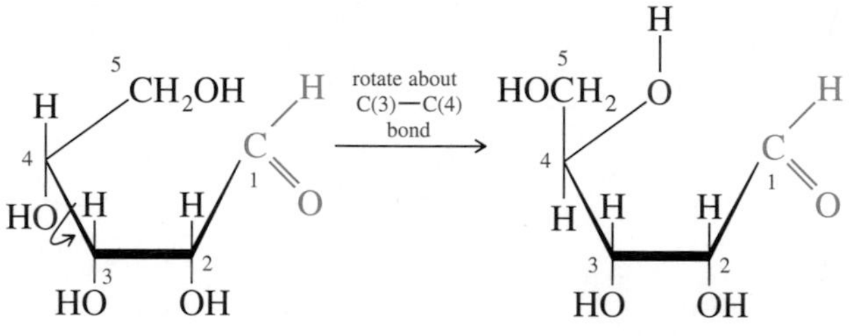

rotate about C(3)—C(4) bond

Conformation of D-ribose suitable for furanose ring formation

As viewed in the drawing, a 120° anticlockwise rotation of C-4 places its hydroxyl group in the proper position. At the same time, this rotation moves the CH_2OH group to a position such that it will become a substituent that is ‘‘up’’ on the five-membered ring. The hydrogen at C-4 then will be ‘‘down’’ in the furanose form.

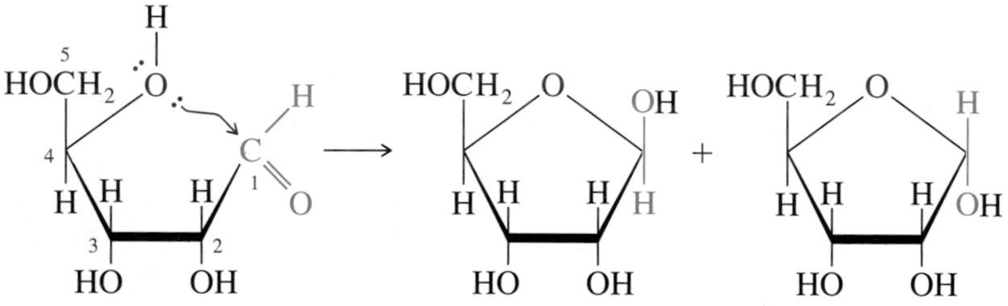

β-D-Ribofuranose

α-D-Ribofuranose

PROBLEM 25.5 Write Haworth formulas corresponding to the furanose forms of each of the following carbohydrates:

(*a*) D-Xylose
(*b*) D-Arabinose
(*c*) L-Arabinose
(*d*) D-Threose

SAMPLE SOLUTION (*a*) The Fischer projection formula of D-xylose is given in Figure 25.2.

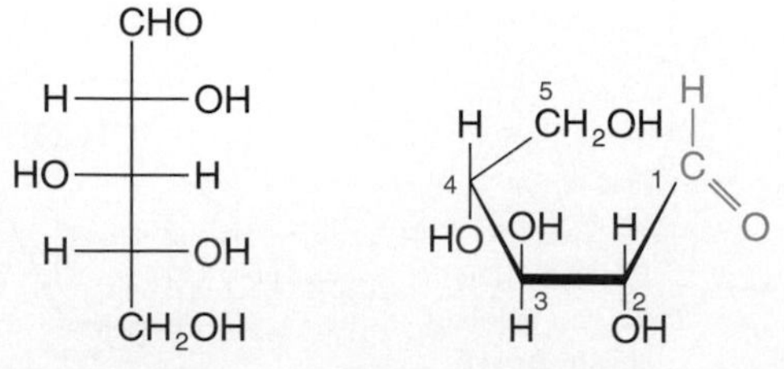

D-Xylose

Eclipsed conformation of D-xylose

Carbon-4 of D-xylose must be rotated in an anticlockwise sense in order to bring its hydroxyl group into the proper orientation for furanose ring formation.

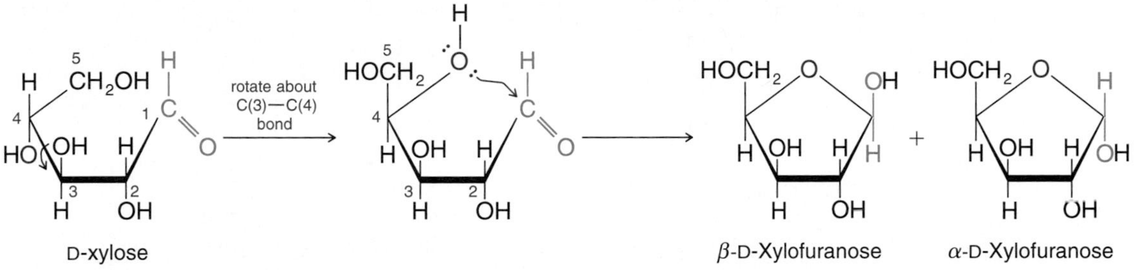

rotate about C(3)—C(4) bond

D-xylose

β-D-Xylofuranose + α-D-Xylofuranose

25.7 CYCLIC FORMS OF CARBOHYDRATES: PYRANOSE FORMS

During the discussion of hemiacetal formation in D-ribose in the preceding section, you may have noticed that aldopentoses have the potential of forming a six-membered cyclic hemiacetal via addition of the C-5 hydroxyl to the carbonyl group. This mode of ring closure leads to α- and β-*pyranose* forms:

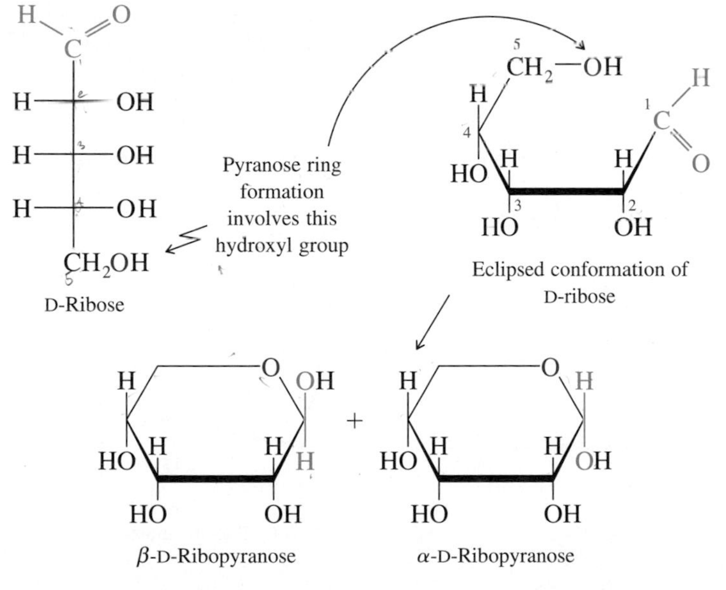

D-Ribose

Pyranose ring formation involves this hydroxyl group

Eclipsed conformation of D-ribose

β-D-Ribopyranose + α-D-Ribopyranose

Like aldopentoses, aldohexoses such as D-glucose are capable of forming two furanose forms (α and β) and two pyranose forms (α and β). The Haworth representations of the pyranose forms of D-glucose are constructed as shown in Figure 25.4; each has a CH_2OH group as a substituent on the six-membered ring.

Haworth formulas are satisfactory for representing *configurational* relationships in pyranose forms but are uninformative as to carbohydrate *conformations.* X-ray crystallographic studies of a large number of carbohydrates reveal that the six-membered pyranose ring of D-glucose adopts a chair conformation:

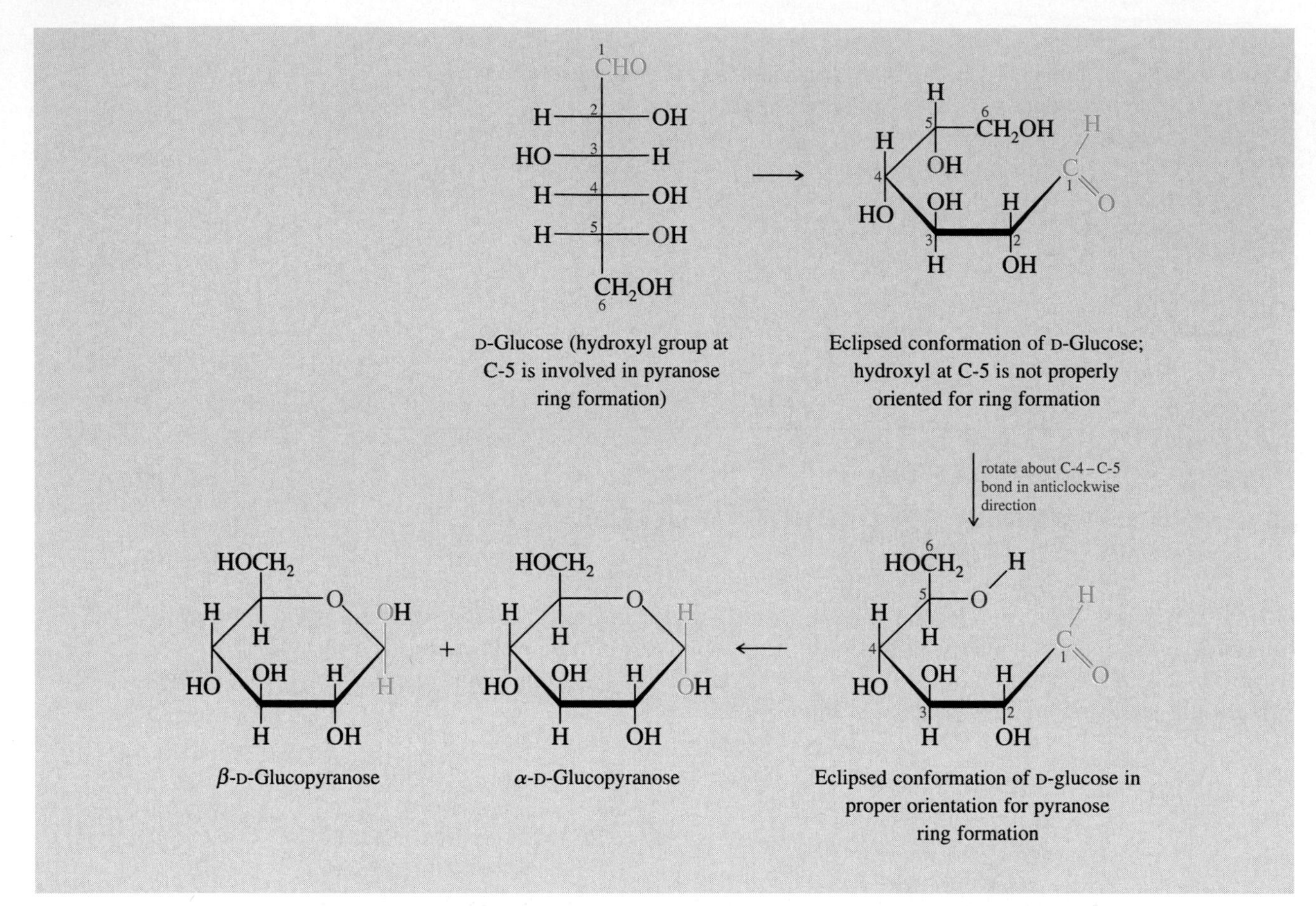

FIGURE 25.4 Haworth formulas for α- and β-pyranose forms of D-glucose.

β-D-Glucopyranose

α-D-Glucopyranose

All the ring substituents other than hydrogen in β-D-glucopyranose are equatorial in the most stable chair conformation. Only the anomeric hydroxyl group is axial in the α isomer; all the other substituents are equatorial.

Other aldohexoses behave similarly in adopting chair conformations that permit the CH_2OH substituent to occupy an equatorial orientation. Normally the CH_2OH group is the bulkiest, most conformationally demanding substituent in the pyranose form of a hexose.

PROBLEM 25.6 Clearly represent the most stable conformation of the β-pyranose form of each of the following sugars:

(*a*) D-Galactose (*b*) D-Mannose (*c*) L-Mannose (*d*) L-Ribose

SAMPLE SOLUTION (*a*) By analogy with the procedure outlined for D-glucose in Figure 25.4, first generate a Haworth formula for β-D-galactopyranose:

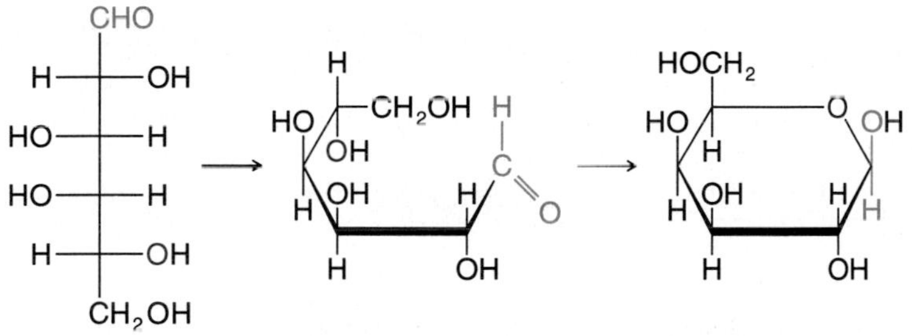

D-Galactose

β-D-Galactopyranose
(Haworth formula)

Next, redraw the planar Haworth formula more realistically as a chair conformation, choosing the one that has the CH_2OH group equatorial.

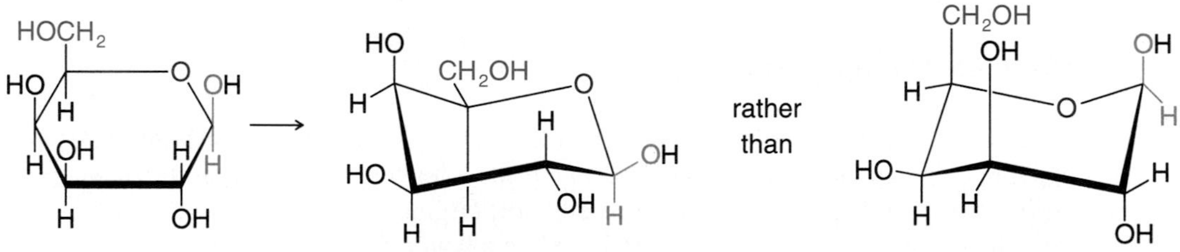

rather than

Most stable chair conformation of β-D-galactopyranose

Less stable chair; CH_2OH group is axial

Galactose differs from glucose in configuration at C-4. The C-4 hydroxyl is axial in β-D-galactopyranose, while it is equatorial in β-D-glucopyranose.

Since six-membered rings are normally less strained than five-membered ones, pyranose forms are usually present in greater amounts than furanose forms at equilibrium, and the concentration of the open-chain form is quite small. The distribution of carbohydrates among their various hemiacetal forms has been examined by using ^1H and ^{13}C nmr spectroscopy. In aqueous solution, for example, D-ribose is found to contain the various α- and β-furanose and pyranose forms in the amounts shown in

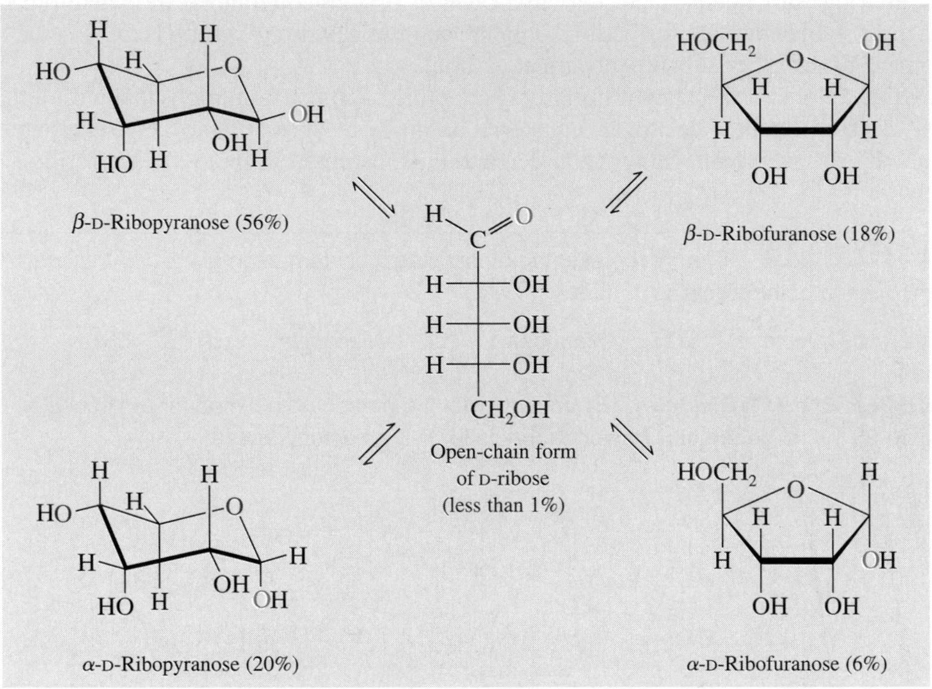

β-D-Ribopyranose (56%)

β-D-Ribofuranose (18%)

Open-chain form of D-ribose (less than 1%)

α-D-Ribopyranose (20%)

α-D-Ribofuranose (6%)

FIGURE 25.5 Distribution of furanose, pyranose, and open-chain forms of D-ribose in aqueous solution as measured by ^1H and ^{13}C nmr spectroscopy.

Figure 25.5. The concentration of the open-chain form at equilibrium is too small to measure directly. Nevertheless, it occupies a central position, in that interconversions of α and β anomers and furanose and pyranose forms take place by way of the open-chain form as an intermediate. As will be seen later, certain chemical reactions also proceed by way of the open-chain form.

25.8 MUTAROTATION

In spite of their easy interconversion in solution, α and β forms of carbohydrates are capable of independent existence, and many have been isolated in pure form as crystalline solids. When crystallized from ethanol, D-glucose yields α-D-glucopyranose, mp 146°C, $[\alpha]_D$ +112.2°. Crystallization from a water-ethanol mixture produces β-D-glucopyranose, mp 148 to 155°C, $[\alpha]_D$ +18.7°. In the solid state the two forms do not interconvert and are stable indefinitely. Their structures have been unambiguously confirmed by x-ray crystallography.

The optical rotations just cited for each isomer are those measured immediately after each one is dissolved in water. On standing, the rotation of the solution containing the α isomer decreases from +112.2° to +52.5°; the rotation of the solution of the β isomer increases from +18.7° to the same value of +52.5°. This phenomenon is called **mutarotation.** What is happening is that each solution, initially containing only one anomeric form, undergoes equilibration to the same mixture of α- and β-pyranose forms. The open-chain form is an intermediate in the process.

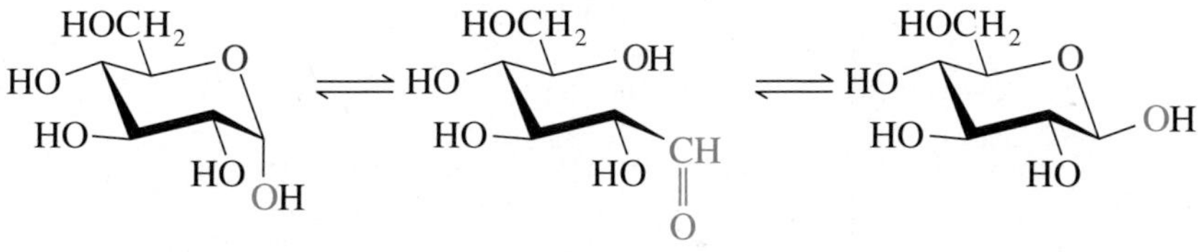

α-D-Glucopyranose (mp 146°C; $[\alpha]_D$ +112.2°) ⇌ Open-chain form of D-glucose ⇌ β-D-Glucopyranose (mp 148–155°C; $[\alpha]_D$ +18.7°)

The distribution between the α and β anomeric forms at equilibrium is readily calculated from the optical rotations of the pure isomers and the final optical rotation of the solution and is determined to be 36 percent α to 64 percent β. Independent measurements have established that only the pyranose forms of D-glucose are present in significant quantities at equilibrium.

A 1987 carbon-13 nmr study of D-glucose in water detected five species: the α-pyranose (38.8%), β-pyranose (60.9%), α-furanose (0.14%), and β-furanose (0.15%) forms, and the hydrate of the open-chain form (0.0045%).

PROBLEM 25.7 The specific optical rotations of pure α- and β-D-mannopyranose are +29.3° and −17.0°, respectively. When either form is dissolved in water, mutarotation occurs and the observed rotation of the solution changes until a final rotation of +14.2° is observed. Assuming that only α- and β-pyranose forms are present, calculate the percent of each isomer at equilibrium.

It is not possible to tell by inspection whether the α- or β-pyranose form of a particular carbohydrate predominates at equilibrium. As just described, the β-pyranose form is the major species present in an aqueous solution of D-glucose, while the α-pyranose form predominates in a solution of D-mannose (Problem 25.7). The relative abundance of α- and β-pyranose forms in solution is a complicated issue and depends on several factors. One is solvation of the anomeric hydroxyl group. An equatorial OH is less crowded and better solvated by water than an axial one. This effect stabilizes the β-pyranose form in aqueous solution. A second factor, called the **anomeric effect,** involves an electronic interaction between the ring oxygen and the anomeric substituent and preferentially stabilizes the axial OH of the α-pyranose form. Because the two effects operate in different directions but are comparable in magnitude in aqueous solution, the α-pyranose form is more abundant for some carbohydrates and the β-pyranose form for others.

The anomeric effect is best explained by a molecular orbital analysis that is beyond the scope of this text.

25.9 KETOSES

Up to this point all our attention has been directed toward aldoses, carbohydrates having an aldehyde function in their open-chain form. Aldoses are more common than ketoses, and their role in biological processes has been more thoroughly studied. Nevertheless, a large number of ketoses are known, and several of them are pivotal intermediates in carbohydrate biosynthesis and metabolism. Examples of some ketoses include D-*ribulose,* L-*xylulose,* and D-*fructose:*

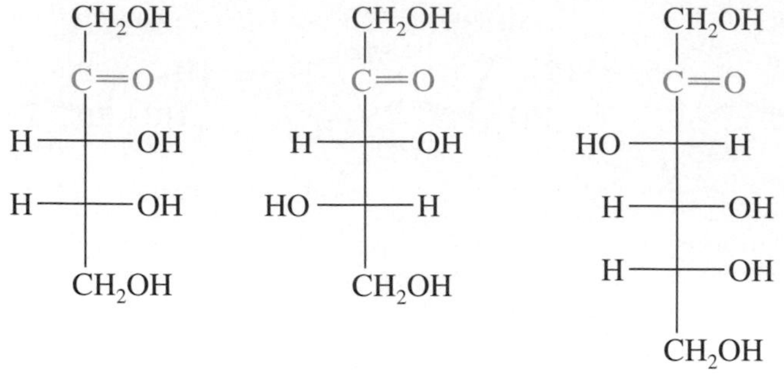

D-Ribulose (a 2-ketopentose that is a key compound in photosynthesis)

L-Xylulose (a 2-ketopentose excreted in excessive amounts in the urine of persons afflicted with the mild genetic disorder pentosuria)

D-Fructose (a 2-ketohexose also known as *levulose;* it is found in honey and is significantly sweeter than table sugar)

In these three examples the carbonyl group is located at C-2, which is the most common location for the carbonyl function in naturally occurring ketoses.

PROBLEM 25.8 How many ketotetroses are possible? Write Fischer projection formulas for each.

Ketoses, like aldoses, exist mainly as cyclic hemiacetals. In the case of D-ribulose, furanose forms result from addition of the C-5 hydroxyl to the carbonyl group.

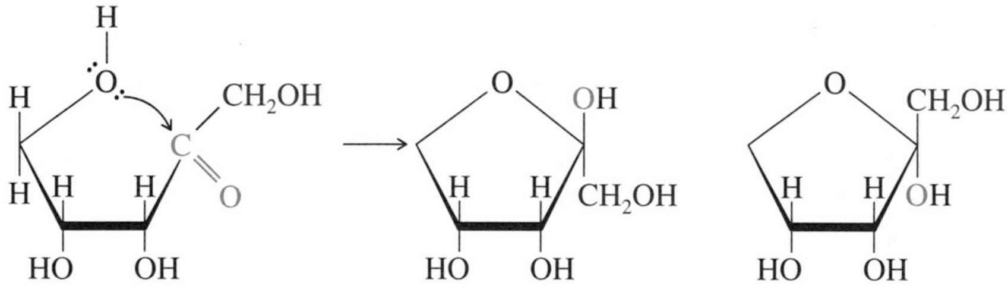

Eclipsed conformation of D-ribulose

β-D-Ribulofuranose

α-D-Ribulofuranose

The anomeric carbon of a furanose or pyranose form of a ketose bears both a hydroxyl group and a carbon substituent. In the case of 2-ketoses, this substituent is a CH_2OH group. As with aldoses, the anomeric carbon of a cyclic hemiacetal is readily identifiable because it is bonded to two oxygens.

25.10 DEOXY SUGARS

A commonplace variation on the general pattern seen in carbohydrate structure is the replacement of one or more of the hydroxyl substituents by some other atom or group.

In **deoxy sugars** the hydroxyl group is replaced by hydrogen. Two examples of deoxy sugars are 2-deoxy-D-ribose and L-rhamnose:

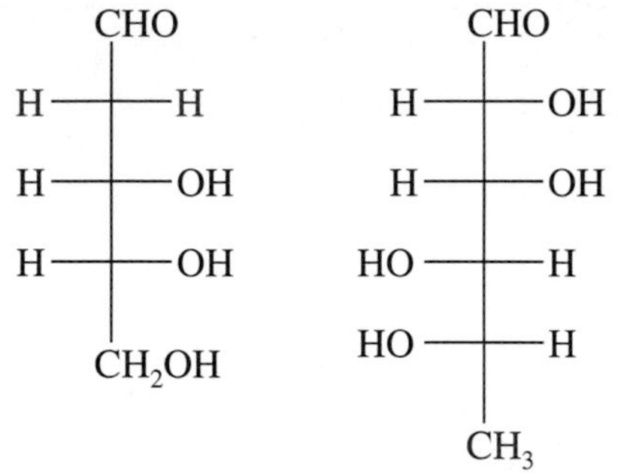

```
       CHO                    CHO
  H ────┼──── H          H ────┼──── OH
  H ────┼──── OH         H ────┼──── OH
  H ────┼──── OH        HO ────┼──── H
       CH2OH            HO ────┼──── H
                              CH3
```

2-Deoxy-D-ribose

L-Rhamnose (6-deoxy-L-mannose)

The hydroxyl at C-2 in D-ribose is absent in 2-deoxy-D-ribose. In Chapter 27 we shall see how derivatives of 2-deoxy-D-ribose, called *deoxyribonucleotides,* are the fundamental building blocks of deoxyribonucleic acid (DNA), the material responsible for storing genetic information. L-Rhamnose is a compound isolated from a number of plants. Its carbon chain terminates in a methyl rather than a CH_2OH group.

PROBLEM 25.9 Write Fischer projection formulas for

(*a*) *Cordycepose* (3-deoxy-D-ribose): a deoxy sugar isolated by hydrolysis of the antibiotic substance *cordycepin*

(*b*) L-*Fucose* (6-deoxy-L-galactose): obtained from seaweed

SAMPLE SOLUTION (*a*) The hydroxyl group at C-3 in D-ribose is replaced by hydrogen in 3-deoxy-D-ribose.

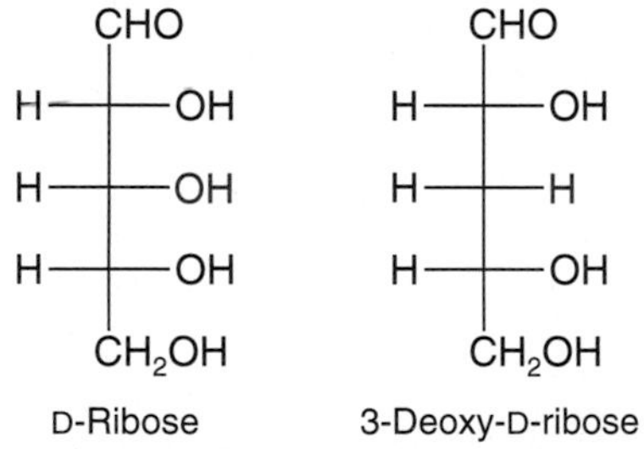

```
       CHO                    CHO
  H ────┼──── OH         H ────┼──── OH
  H ────┼──── OH         H ────┼──── H
  H ────┼──── OH         H ────┼──── OH
       CH2OH                  CH2OH
```

D-Ribose (from Figure 25.2)

3-Deoxy-D-ribose (cordycepose)

25.11 AMINO SUGARS

Another structural variation is the replacement of a hydroxyl group in a carbohydrate by an amine group. Compounds characterized by such a structure are called **amino sugars.** More than 60 amino sugars are known, many of them having been isolated and identified only recently as components of antibiotic substances.

For a review of the isolation of chitin from natural sources and some of its uses, see the November 1990 issue of the *Journal of Chemical Education* (pp. 938–942).

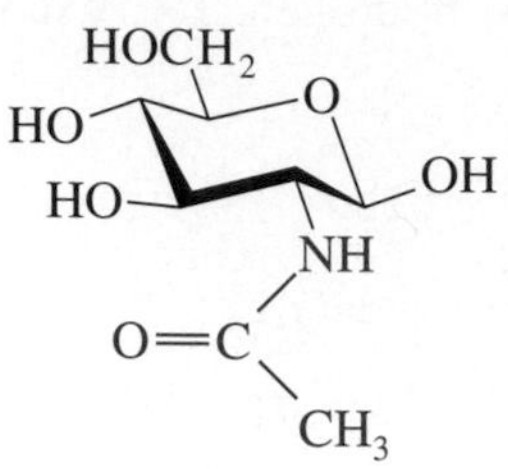

N-Acetyl-D-glucosamine (shown in its β-pyranose form; principal component of the polysaccharide that is major constituent of *chitin,* the substance that makes up the tough outer skeleton of arthropods and insects)

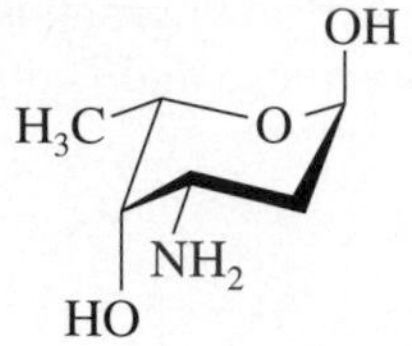

L-Daunosamine (shown in its α-pyranose form; obtained from daunomycin and Adriamycin, two powerful drugs used in cancer chemotherapy)

25.12 BRANCHED-CHAIN CARBOHYDRATES

Carbohydrates that have a carbon substituent attached to the main chain are said to have a **branched chain.** D-Apiose and L-vancosamine are representative branched-chain carbohydrates:

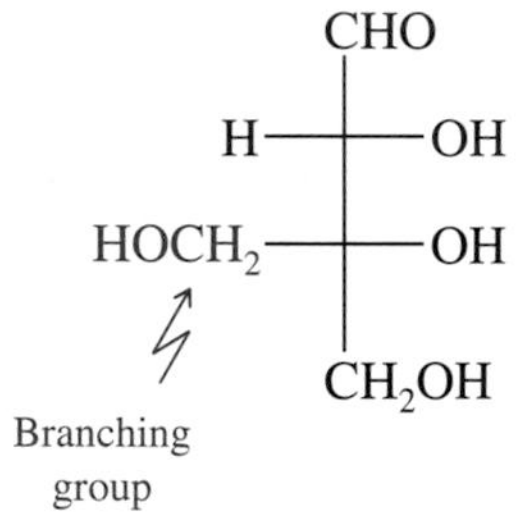

D-Apiose (shown in its open-chain form; isolated from parsley and from the cell wall polysaccharide of various marine plants)

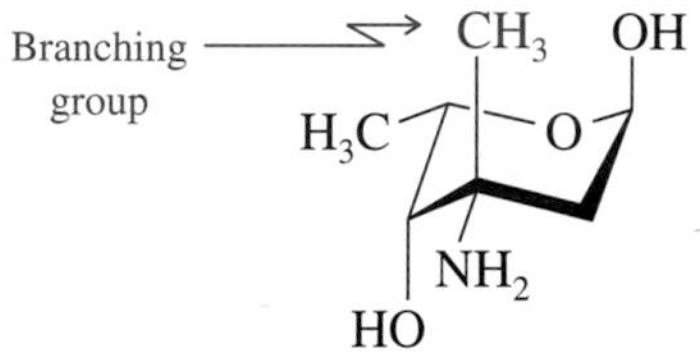

L-Vancosamine (shown in its α-pyranose form; a component of the antibiotic compound vancomycin)

25.13 GLYCOSIDES

Glycosides are a large and very important class of carbohydrate derivatives characterized by the replacement of the anomeric hydroxyl group by some other substituent. Glycosides are termed *O*-glycosides, *N*-glycosides, *S*-glycosides, and so on, according to the atom attached to the anomeric carbon.

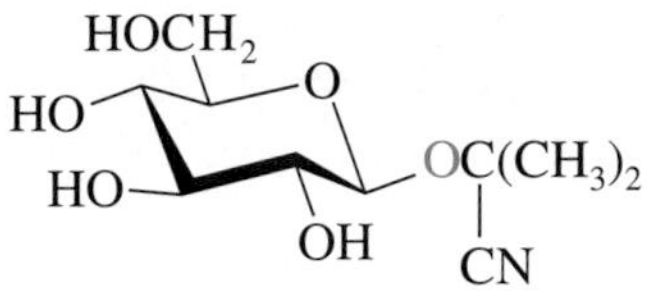

Linamarin
(an *O*-glycoside:
obtained from manioc,
a type of yam widely
distributed in
southeast Asia)

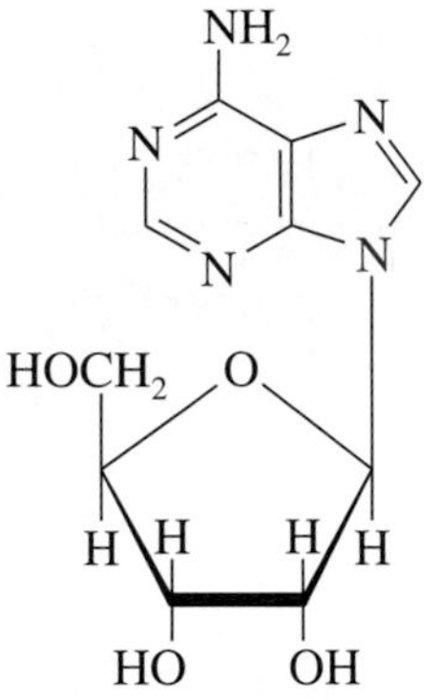

Adenosine
(an *N*-glycoside: also
known as a nucleoside;
adenosine is one of the
fundamental molecules
of biochemistry)

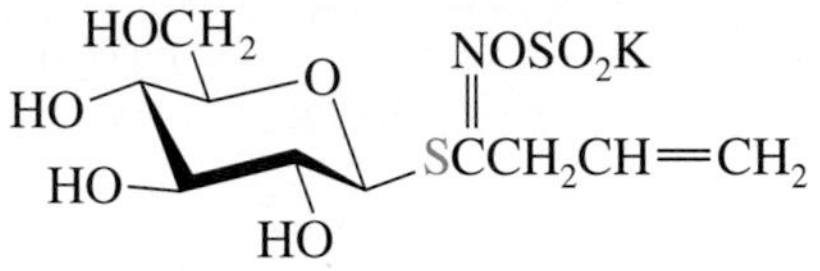

Sinigrin
(an *S*-glycoside:
contributes to the
characteristic flavor
of mustard and
horseradish)

Usually, the term *glycoside* without a prefix is taken to mean an *O*-glycoside and will be used that way in this chapter. Glycosides are classified as α or β in the customary way, according to the configuration at the anomeric carbon. All three of the glycosides just shown are β-glycosides. Linamarin and sinigrin are glycosides of D-glucose; adenosine is a glycoside of D-ribose.

Structurally, *O*-glycosides are mixed acetals that involve the anomeric position of furanose and pyranose forms of carbohydrates. Recall the sequence of intermediates in acetal formation (Section 17.9):

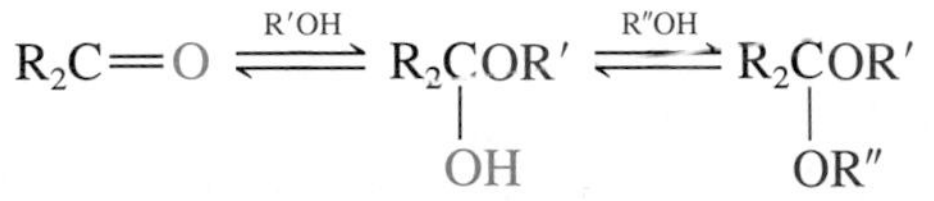

$$\underset{\text{Aldehyde or ketone}}{R_2C{=}O} \xrightleftharpoons{R'OH} \underset{\text{Hemiacetal}}{R_2C(OH)OR'} \xrightleftharpoons{R''OH} \underset{\text{Acetal}}{R_2C(OR'')OR'}$$

When this sequence is applied to carbohydrates, the first step takes place *intramolecularly* and spontaneously to yield a cyclic hemiacetal. The second step is *intermolecular,* requires an alcohol R″OH as a reactant, and proceeds readily only in the presence of an acid catalyst. An oxygen-stabilized carbocation is an intermediate.

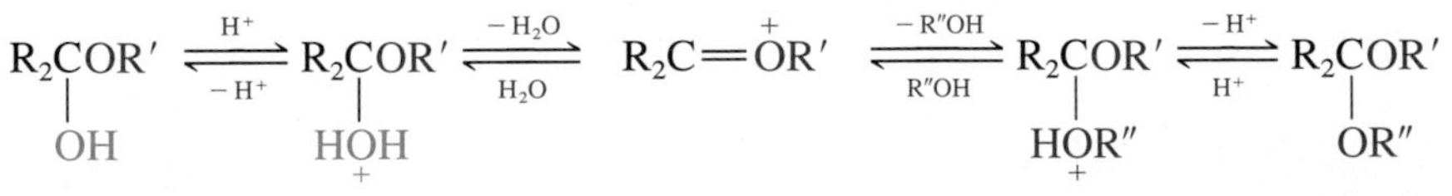

$$\underset{\text{Hemiacetal}}{R_2C(OH)OR'} \underset{-H^+}{\overset{H^+}{\rightleftharpoons}} R_2C(\overset{+}{O}H_2)OR' \underset{H_2O}{\overset{-H_2O}{\rightleftharpoons}} \underset{\text{Oxygen-stabilized carbocation}}{R_2C{=}\overset{+}{O}R'} \underset{R''OH}{\overset{-R''OH}{\rightleftharpoons}} R_2C(\overset{+}{O}HR'')OR' \underset{H^+}{\overset{-H^+}{\rightleftharpoons}} \underset{\text{Mixed acetal}}{R_2C(OR'')OR'}$$

The preparation of glycosides in the laboratory is carried out by simply allowing a carbohydrate to react with an alcohol in the presence of an acid catalyst:

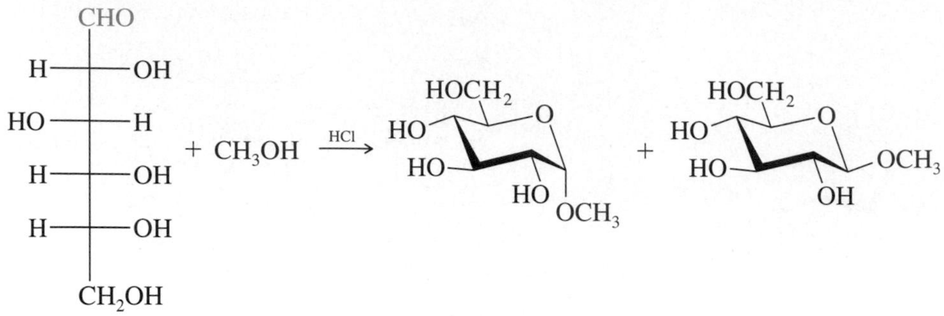

D-Glucose + Methanol $\xrightarrow{HCl}$ Methyl α-D-glucopyranoside (major product; isolated in 49% yield) + Methyl β-D-glucopyranoside (minor product)

A point to be emphasized about glycoside formation is that, despite the presence of a number of other hydroxyl groups in the carbohydrate, it is *only the anomeric hydroxyl group that is replaced.* This is because a carbocation at the anomeric position is stabilized by the ring oxygen and is the only one capable of being formed under the reaction conditions.

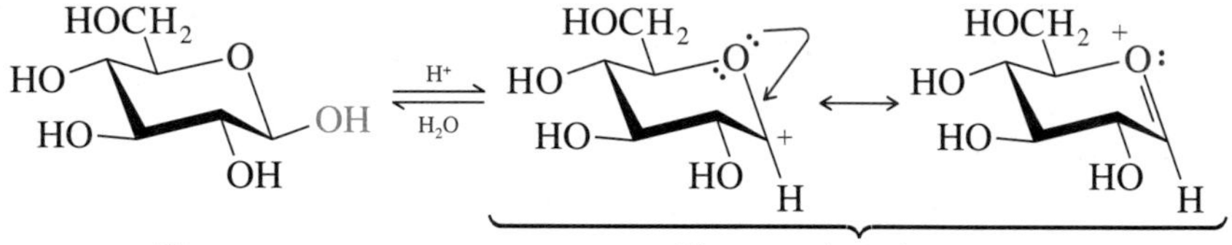

D-Glucose (shown in β-pyranose form)

Electron pair on ring oxygen can stabilize carbocation at anomeric position only.

Once the carbocation is formed, it is captured by the alcohol acting as a nucleophile. Attack can occur at either the β face or the α face of the carbocation.

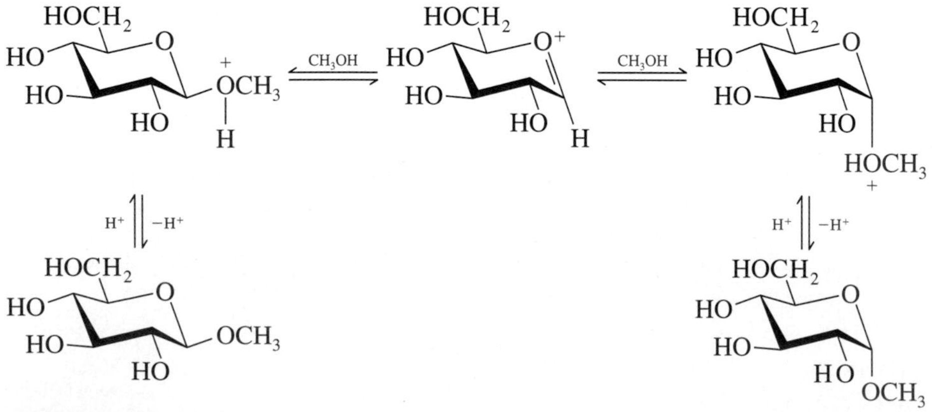

Methyl β-D-glucopyranoside

Methyl α-D-glucopyranoside

Equilibration of the glycosides usually takes place under the conditions of this reaction. It is an example of a reaction that is subject to *thermodynamic* (or *equilibrium*) *control.* Glucopyranosides are more stable than glucofuranosides and predominate at equilibrium.

PROBLEM 25.10 Methyl glycosides of 2-deoxy sugars have been prepared by the acid-catalyzed addition of methanol to unsaturated carbohydrates known as *glycals.*

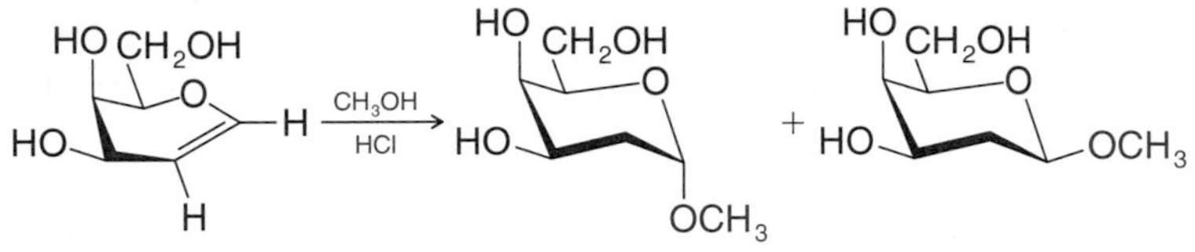

$\xrightarrow[\text{HCl}]{CH_3OH}$

D-Galactal

Methyl 2-deoxy-α-D-lyxohexopyranoside (38%)

Methyl 2-deoxy-β-D-lyxohexopyranoside (36%)

Suggest a reasonable mechanism for the acid-catalyzed formation of glycosides from glycals.

Under neutral or basic conditions glycosides are configurationally stable; unlike the free sugars from which they are derived, glycosides do not exhibit mutarotation. Converting the anomeric hydroxyl group to an ether function (hemiacetal → acetal) prevents its reversion to the open-chain form in neutral or basic media. In aqueous acid, acetal formation can be reversed and the glycoside hydrolyzed to an alcohol and the free sugar.

25.14 DISACCHARIDES

Disaccharides are carbohydrates that yield two monosaccharide molecules on hydrolysis. Structurally, disaccharides are *glycosides* in which the alkoxy group attached to the anomeric carbon is derived from a second sugar molecule.

Maltose, obtained by the hydrolysis of starch, and *cellobiose,* by the hydrolysis of cellulose, are isomeric disaccharides. In both maltose and cellobiose two D-glucopyranose units are joined by a glycosidic bond between C-1 of one unit and C-4 of the other. The two are diastereomers, differing only in the stereochemistry at the anomeric carbon of the glycoside bond; maltose is an α-glycoside, cellobiose is a β-glycoside.

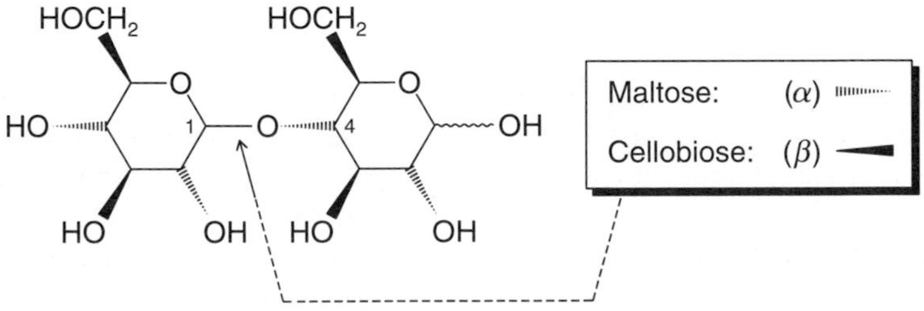

Maltose: (α)

Cellobiose: (β)

The stereochemistry and points of connection of glycosidic bonds are commonly designated by symbols such as $\alpha(1,4)$ for maltose and $\beta(1,4)$ for cellobiose; α and β designate the stereochemistry at the anomeric position, while the numerals specify the ring carbons involved.

The free anomeric hydroxyl group is the one shown at the far right of the preceding structural formula. The symbol ~~~ is used to represent a bond of variable stereochemistry.

Both maltose and cellobiose have a free anomeric hydroxyl group that is not involved in a glycoside bond. The configuration at the free anomeric center is variable and may be either α or β. Indeed, two stereoisomeric forms of maltose have been isolated; one has its anomeric hydroxyl group in an equatorial orientation, while the other has an axial anomeric hydroxyl.

PROBLEM 25.11 The two stereoisomeric forms of maltose just mentioned undergo mutarotation when dissolved in water. What is the structure of the key intermediate in this process?

The single difference in their structures, the stereochemistry of the glycosidic bond, causes maltose and cellobiose to differ significantly in their three-dimensional shape, as the molecular models of Figure 25.6 illustrate. This difference in shape affects the way in which maltose and cellobiose interact with other chiral molecules such as proteins, and they behave much differently toward enzyme-catalyzed hydrolysis. An enzyme known as *maltase* catalyzes the hydrolytic cleavage of the α-glycosidic bond of maltose but is without effect in promoting the hydrolysis of the β-glycosidic bond of cellobiose. A different enzyme, *emulsin,* produces the opposite result: emulsin catalyzes the hydrolysis of cellobiose but not of maltose. The behavior of each enzyme is general for glucosides. Maltase catalyzes the hydrolysis of α-glucosides and is also known as *α-glucosidase,* whereas emulsin catalyzes the hydrolysis of β-glucosides and is known as *β-glucosidase.* The specificity of these enzymes offers a useful tool for structure determination in that it allows the stereochemistry of glycosidic linkages to be assigned.

Lactose is a disaccharide constituting 2 to 6 percent of milk and is known as *milk sugar.* It differs from maltose and cellobiose in that only one of its monosaccharide units is D-glucose. The other monosaccharide unit, the one that contributes its anomeric carbon to the glycoside bond, is D-galactose. Like cellobiose, lactose is a β-glycoside.

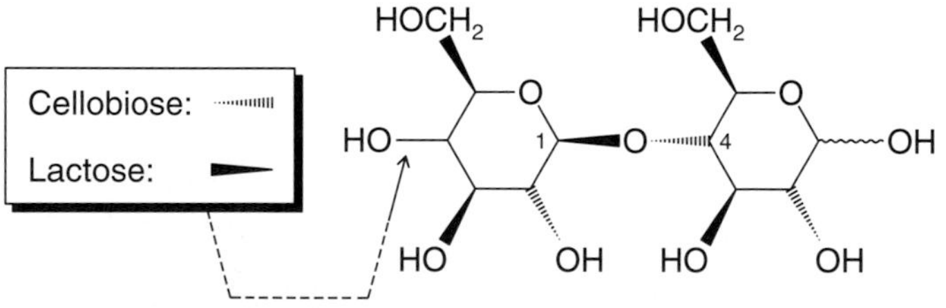

Digestion of lactose is facilitated by the β-glycosidase *lactase.* A deficiency of this enzyme makes it difficult to digest lactose and causes abdominal discomfort. Lactose intolerance is a genetic trait; it is treatable through over-the-counter formulations of lactase and restrictions on the amount of milk in the diet.

The most familiar of all the carbohydrates is *sucrose*—common table sugar. Sucrose is a disaccharide in which D-glucose and D-fructose are joined at their anomeric carbons by a glycosidic bond (Figure 25.7). Its chemical composition is the same irrespective of its source; sucrose from cane and sucrose from sugar beets are chemically identical. Since sucrose does not have a free anomeric hydroxyl group, it does not undergo mutarotation.

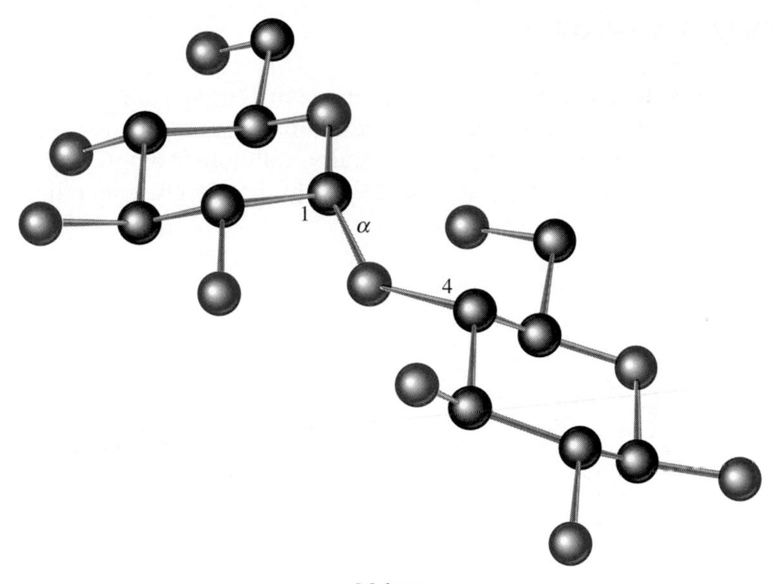

Maltose

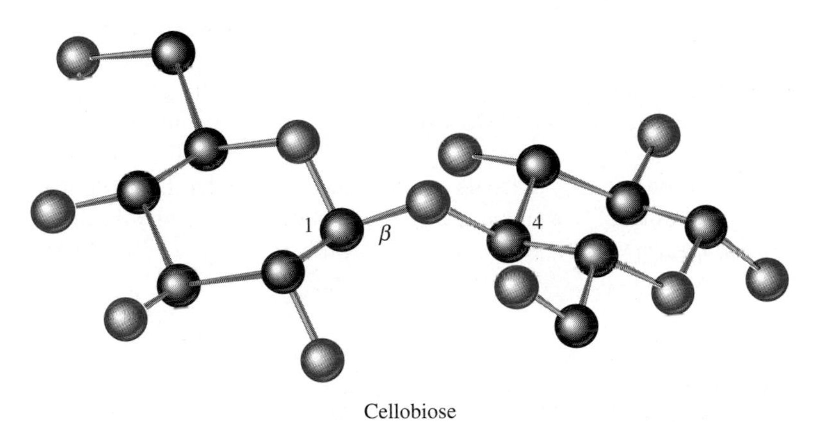

Cellobiose

FIGURE 25.6 Molecular models of the isomeric disaccharides maltose and cellobiose. Each is a disaccharide in which the anomeric carbon of one D-glucopyranose unit is connected to the C-4 equatorial oxygen of a second D-glucopyranose unit. The glycosidic bond has the α orientation in maltose and a β orientation in cellobiose.

D-Glucose portion of molecule

$HOCH_2$

O

HO

HO

HO

O

HO

OH

O

CH_2OH

CH_2OH

D-Fructose portion of molecule

α-Glycoside bond to anomeric position of D-glucose

β-Glycoside bond to anomeric position of D-fructose

FIGURE 25.7 The structure of sucrose.

25.15 POLYSACCHARIDES

Cellulose is the principal structural component of vegetable matter. Wood is 30 to 40 percent cellulose, cotton over 90 percent. Photosynthesis in plants is responsible for the formation of 10^9 tons per year of cellulose. Structurally, cellulose is a polysaccharide composed of several thousand D-glucose units joined by $\beta(1,4)$ glycosidic linkages:

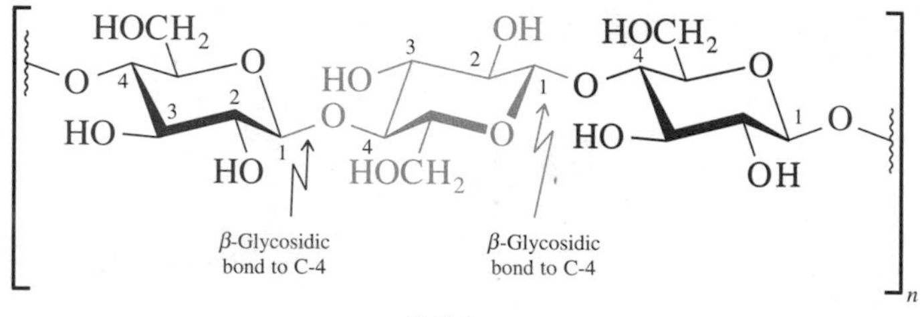

Cellulose

Complete hydrolysis of all the glycosidic bonds of cellulose yields D-glucose. The disaccharide fraction that results from partial hydrolysis is cellobiose.

Animals do not possess the enzymes necessary to catalyze the hydrolysis of cellulose and so cannot digest it. Cattle and other ruminants can use cellulose as a food source in an indirect way. Colonies of microorganisms that live in the digestive tract of these animals consume cellulose and in the process convert it to other substances that are digestible by the host.

A more direct source of energy for animals is provided by the starches found in many foods. Starch is a mixture of a water-dispersible fraction called *amylose* and a second component, *amylopectin.* Amylose is a polysaccharide made up of about 100 to several thousand D-glucose units joined by $\alpha(1,4)$ glycosidic bonds:

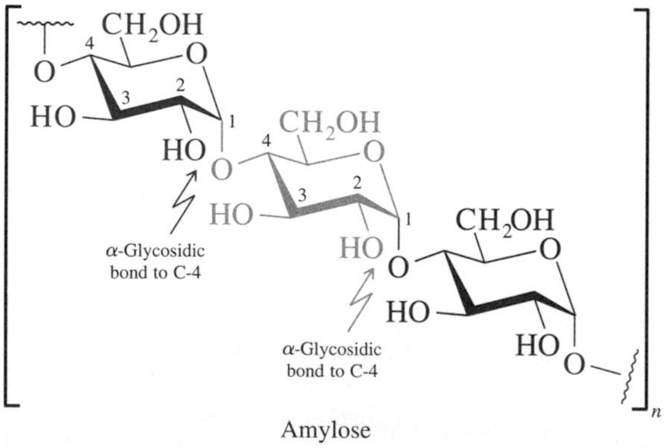

Amylose

PROBLEM 25.12 Complete the analogy: Cellulose is to cellobiose as amylose is to __________________.

Like amylose, amylopectin is a polysaccharide of $\alpha(1,4)$-linked D-glucose units. Instead of being a continuous length of $\alpha(1,4)$ units, however, amylopectin is branched. Attached to C-6 at various points on the main chain are short polysaccharide branches of 24 to 30 glucose units joined by $\alpha(1,4)$ glycosidic bonds. A portion of an amylopectin molecule is represented in Figure 25.8.

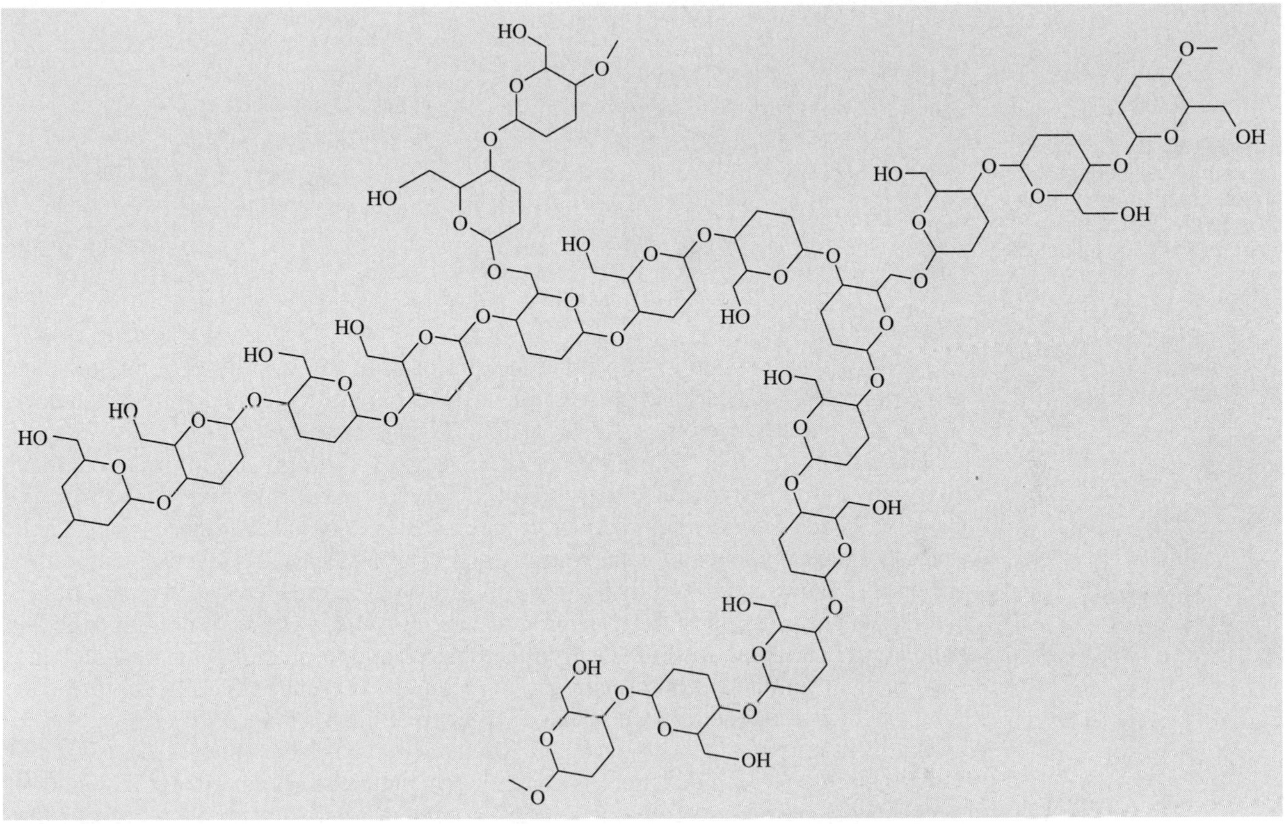

FIGURE 25.8 A portion of the amylopectin structure. α-Glycosidic bonds link C-1 and C-4 of adjacent glucose units. Glycoside bonds from C-6 to the polysaccharide chains are designated by red oxygen atoms. Hydroxyl groups attached to individual rings have been omitted for clarity.

Starch is a plant's way of storing glucose to meet its energy needs. Animals can tap that source by eating starchy foods and, with the aid of their α-glycosidase enzymes, hydrolyze the starch to glucose. When more glucose is available than is needed as fuel, animals store it as glycogen. *Glycogen* is similar to amylopectin in that it is a branched polysaccharide of $\alpha(1,4)$-linked D-glucose units with subunits connected to C-6 of the main chain.

25.16 CELL-SURFACE GLYCOPROTEINS

That carbohydrates play an informational role in biological interactions is a recent revelation of great importance. *Glycoproteins,* protein molecules covalently bound to carbohydrates, are often the principal species involved. When a cell is attacked by a virus or bacterium or when it interacts with another cell, the drama begins when the foreign particle attaches itself to the surface of the host cell. The invader recognizes the host by the glycoproteins on the cell surface. More specifically, it recognizes particular carbohydrate sequences at the end of the glycoprotein. For example, the receptor on the cell surface to which an influenza virus attaches itself has been identified as a glycoprotein terminating in a disaccharide of *N*-acetylgalactosamine and *N*-acetyl-neuraminic acid (Figure 25.9). Since attachment of the invader to the surface of the

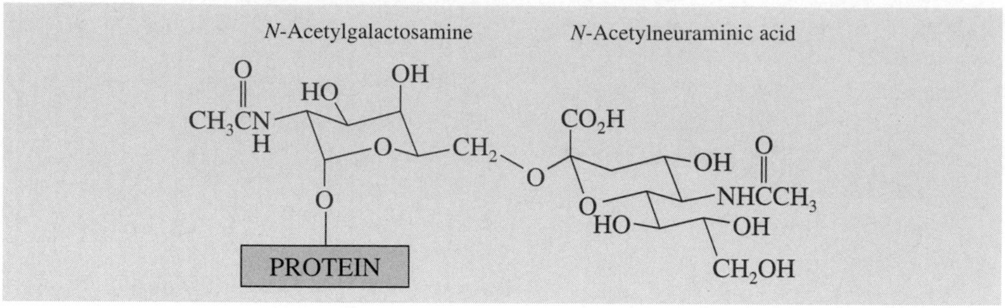

FIGURE 25.9 Schematic diagram of cell-surface glycoprotein, showing disaccharide unit that is recognized by invading influenza virus.

host cell is the first step in infection, one approach to disease prevention is to selectively inhibit this "host-guest" interaction. Identifying the precise nature of the interaction is the first step in the rational design of drugs that prevent it.

Human blood group substances offer another example of the informational role played by carbohydrates. The structure of the glycoproteins attached to the surface of blood cells determines whether blood is type A, B, AB, or O. Differences between the carbohydrate components of the various glycoproteins have been identified and are shown in Figure 25.10. Compatibility of blood types is dictated by *antigen-antibody* interactions. The cell-surface glycoproteins are *antigens. Antibodies* present in certain blood types can cause the blood cells of certain other types to clump together, and thus set practical limitations on transfusion procedures. The antibodies "recognize" the antigens they act upon by their terminal saccharide units.

Antigen-antibody interactions are the fundamental basis by which the immune system functions. These interactions are chemical in nature and often involve associations between glycoproteins of an antigen and complementary glycoproteins of the antibody. The precise chemical nature of antigen-antibody association is an area of active investigation, with significant implications for chemistry, biochemistry, and physiology.

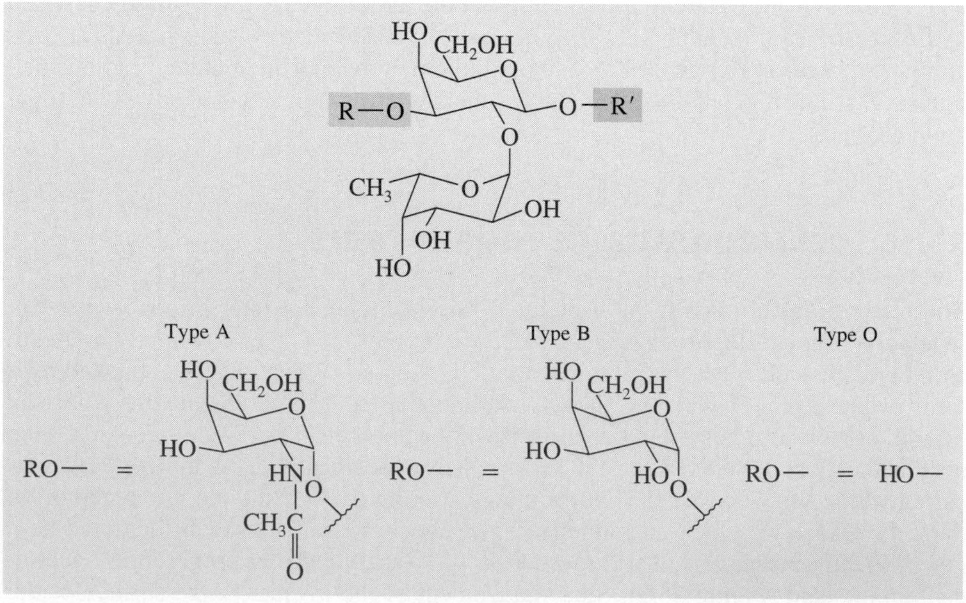

FIGURE 25.10 Terminal carbohydrate units of human blood group glycoproteins. The group R′ is a polymer of *N*-acetylgalactosamine units attached to a protein. The structural difference between the type A, type B, and type O glycoproteins lies in the group designated R.

25.17 CARBOHYDRATE STRUCTURE DETERMINATION

Present-day techniques for structure determination in carbohydrate chemistry are substantially the same as those for any other type of compound. The full range of modern instrumental methods, including mass spectrometry and infrared and nuclear magnetic resonance spectroscopy, is brought to bear on the problem. If the unknown substance is crystalline, x-ray diffraction can provide precise structural information that in the best cases is equivalent to taking a photograph of the molecule.

Before the widespread availability of instrumental methods, the major approach to structure determination relied on a battery of chemical reactions and tests. The response of an unknown substance to various reagents and procedures provided a body of data from which the structure could be deduced. Some of these procedures are still used to supplement the information obtained by instrumental methods. To better understand the scope and limitations of these tests, a brief survey of the chemical reactions of carbohydrates is in order. In many cases these reactions are simply applications of chemistry you have already learned. Certain of the transformations, however, are virtually unique to carbohydrates.

The classical approach to structure determination in carbohydrate chemistry is best exemplified by Fischer's work with D-glucose. A detailed account of this study appears in the August 1941 issue of the *Journal of Chemical Education* (pp. 353–357).

25.18 REDUCTION OF CARBOHYDRATES

While carbohydrates exist almost entirely as cyclic hemiacetals in aqueous solution, they are in rapid equilibrium with their open-chain forms, and most of the reagents that react with simple aldehydes and ketones react in an analogous way with the carbonyl functional groups of carbohydrates.

The carbonyl group of carbohydrates can be reduced to an alcohol function. Typical procedures include catalytic hydrogenation and sodium borohydride reduction. Lithium aluminum hydride is not suitable, because it is not compatible with the solvents (water, alcohols) that are required to dissolve carbohydrates. The products of carbohydrate reduction are called **alditols.** Since these alditols lack a carbonyl group, they are, of course, incapable of forming cyclic hemiacetals and exist exclusively in noncyclic forms.

α-D-Galactofuranose, or
β-D-Galactofuranose, or
α-D-Galactopyranose, or
β-D-Galactopyranose
⇌

	CHO				CH_2OH	
H	—	OH		H	—	OH
HO	—	H	$\xrightarrow[H_2O]{NaBH_4}$	HO	—	H
HO	—	H		HO	—	H
H	—	OH		H	—	OH
	CH_2OH				CH_2OH	

D-Galactose → D-Galactitol (90%)

PROBLEM 25.13 Does sodium borohydride reduction of D-ribose yield an optically active product? Explain.

Another name for glucitol, obtained by reduction of D-glucose, is *sorbitol;* it is used as a sweetener, especially in special diets required to be low in sugar. Reduction of

D-fructose yields a mixture of glucitol and mannitol, corresponding to the two possible configurations at the newly generated stereogenic center at C-2.

25.19 OXIDATION OF CARBOHYDRATES

A characteristic property of an aldehyde function is its sensitivity to oxidation. A solution of copper(II) sulfate as its citrate complex **(Benedict's reagent)** is capable of oxidizing aliphatic aldehydes to the corresponding carboxylic acid.

Benedict's reagent is the key material in a test kit available from drugstores that permits individuals to monitor the glucose levels in their urine.

$$RCH{=}O + 2Cu^{2+} + 5HO^- \longrightarrow RCO_2^- + Cu_2O + 3H_2O$$

Aldehyde; From copper(II) sulfate; Hydroxide ion; Carboxylate anion; Copper(I) oxide; Water

The formation of a red precipitate of copper(I) oxide by reduction of Cu(II) is taken as a positive test for an aldehyde. Carbohydrates that give positive tests with Benedict's reagent are termed **reducing sugars.**

Aldoses are, of course, reducing sugars, since they possess an aldehyde function in their open-chain form. Ketoses are also reducing sugars. Under the conditions of the test, ketoses equilibrate with aldoses by way of *enediol intermediates,* and the aldoses are oxidized by the reagent.

$$\underset{\text{Ketose}}{R{-}C({=}O){-}CH_2OH} \rightleftharpoons \underset{\text{Enediol}}{R{-}C(OH){=}CHOH} \rightleftharpoons \underset{\text{Aldose}}{R{-}CHOH{-}CH{=}O} \xrightarrow{Cu^{2+}} \text{positive test } (Cu_2O \text{ formed})$$

The same kind of equilibrium is available to α-hydroxy ketones generally; such compounds give a positive test with Benedict's reagent. Any carbohydrate that contains a free hemiacetal function is a reducing sugar. The free hemiacetal is in equilibrium with the open-chain form and through it is susceptible to oxidation. Maltose, for example, gives a positive test with Benedict's reagent.

Maltose $\rightleftharpoons$ Open-chain form of maltose $\xrightarrow{Cu^{2+}}$ positive test (Cu_2O formed)

Glycosides, in which the anomeric carbon is part of an acetal function, do not give a positive test and are not reducing sugars.

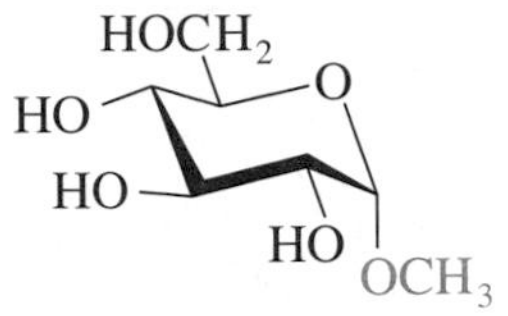

Methyl α-D-glucopyranoside: not a reducing sugar

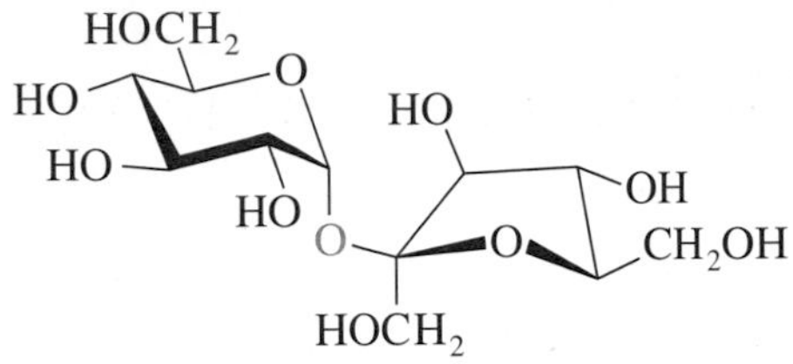

Sucrose: not a reducing sugar

PROBLEM 25.14 Which of the following would be expected to give a positive test with Benedict's reagent? Why?

(*a*) D-Galactitol
(*b*) L-Arabinose
(*c*) 1,3-Dihydroxyacetone
(*d*) D-Fructose
(*e*) Lactose
(*f*) Amylose

SAMPLE SOLUTION

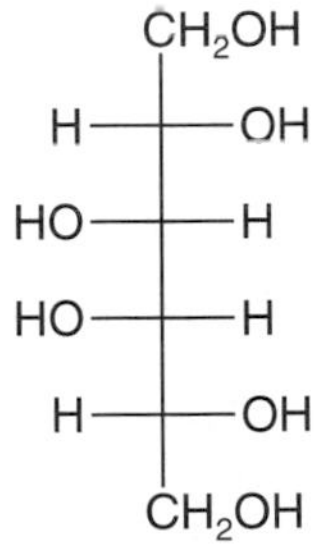

D-Galactitol

(*a*) Recall from the preceding section that D-galactitol has the structure shown. Lacking an aldehyde, an α-hydroxy ketone, or a hemiacetal function, it cannot be oxidized by Cu^{2+} and will not give a positive test with Benedict's reagent.

Fehling's solution, a tartrate complex of copper(II) sulfate, has also been used as a test for reducing sugars.

Derivatives of aldoses in which the terminal aldehyde function is oxidized to a carboxylic acid are called **aldonic acids.** Aldonic acids are named by replacing the *-ose* ending of the aldose by *-onic acid.* Oxidation of aldoses with bromine is the most commonly used method for the preparation of aldonic acids and involves the furanose or pyranose form of the carbohydrate.

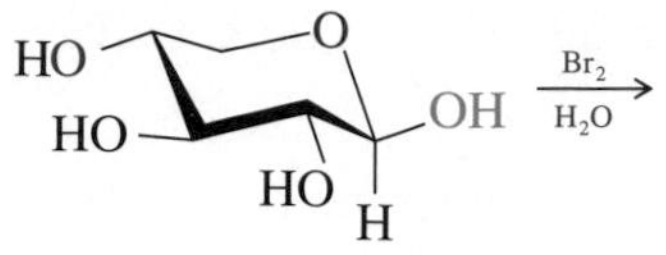

HO, HO, HO, OH, H, O $\xrightarrow[H_2O]{Br_2}$

β-D-Xylopyranose

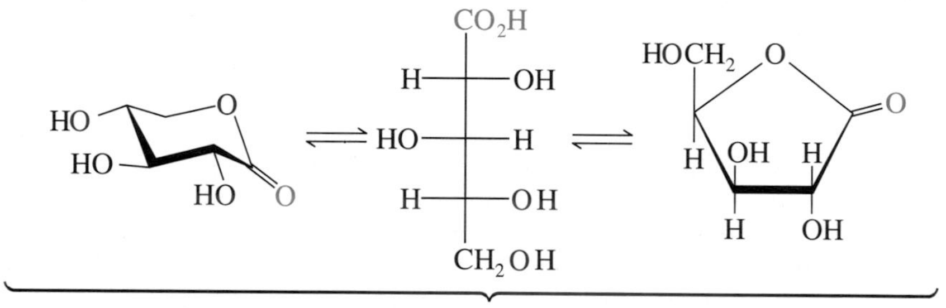

$\rightleftharpoons$ $\rightleftharpoons$

D-Xylonic acid
(90%)

Aldonic acids exist in equilibrium with their five- or six-membered lactones. They can be isolated as carboxylate salts of their open-chain forms on treatment with base.

The reaction of aldoses with nitric acid leads to the formation of **aldaric acids** by oxidation of both the aldehyde and the terminal primary alcohol function to carboxylic acid groups. Aldaric acids are also known as *saccharic acids* and are named by substituting *-aric acid* for the *-ose* ending of the corresponding carbohydrate.

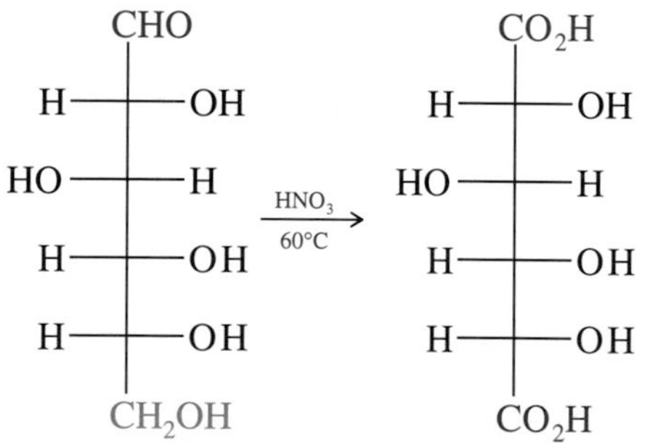

$\xrightarrow[60°C]{HNO_3}$

D-Glucose D-Glucaric acid (41%)

Like aldonic acids, aldaric acids exist mainly as lactones.

PROBLEM 25.15 Another hexose gives the same aldaric acid on oxidation as does D-glucose. Which one?

Uronic acids occupy an oxidation state between aldonic and aldaric acids. They have an aldehyde function at one end of their carbon chain and a carboxylic acid group at the other.

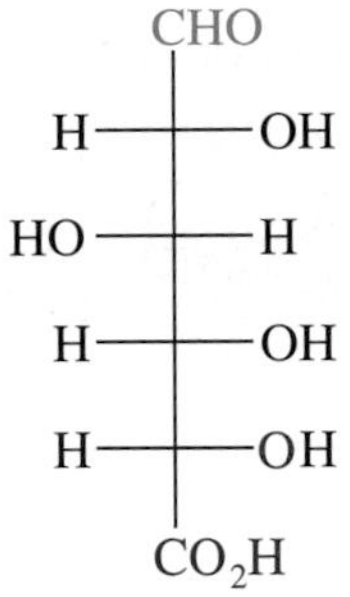

CHO
H—OH
HO—H
H—OH
H—OH
CO_2H

Fischer projection formula of D-glucuronic acid

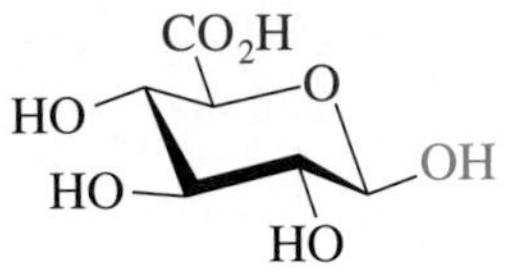

CO_2H, O, HO, HO, OH, HO

β-Pyranose form of D-glucuronic acid

Uronic acids are biosynthetic intermediates in various metabolic processes; ascorbic acid (vitamin C), for example, is biosynthesized by way of glucuronic acid. Many metabolic waste products are excreted in the urine as their glucuronate salts.

25.20 CYANOHYDRIN FORMATION AND CARBOHYDRATE CHAIN EXTENSION

The presence of an aldehyde function in their open-chain forms makes aldoses reactive toward addition of hydrogen cyanide. Addition yields a mixture of diastereomeric cyanohydrins.

α-L-Arabinofuranose, or
β-L-Arabinofuranose, or
α-L-Arabinopyranose, or
β-L-Arabinopyranose ⇌

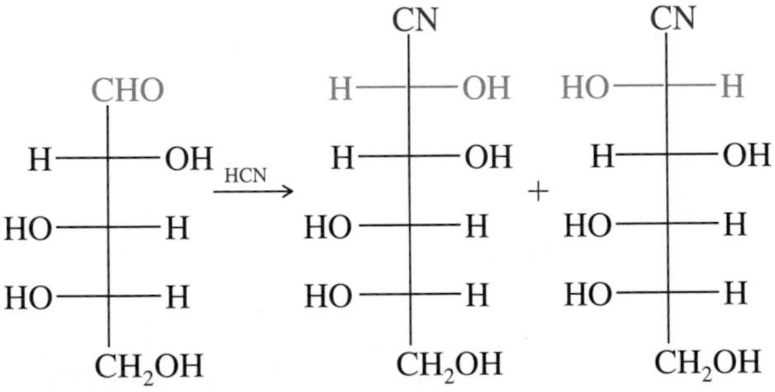

L-Arabinose		L-Mannononitrile		L-Glucononitrile
CHO		CN		CN
		H—OH		HO—H
H—OH	$\xrightarrow{HCN}$	H—OH	+	H—OH
HO—H		HO—H		HO—H
HO—H		HO—H		HO—H
CH_2OH		CH_2OH		CH_2OH

The reaction is used for the chain extension of aldoses in the synthesis of new or unusual sugars. In this case, L-arabinose is an abundant natural product and possesses the correct configurations at its three stereogenic centers for elaboration to the relatively rare L-enantiomers of glucose and mannose. After cyanohydrin formation, the cyano groups are converted to aldehyde functions by hydrogenation in aqueous solution. Under these conditions, —C≡N is reduced to —CH=NH and hydrolyzes rapidly to —CH=O. Use of a poisoned palladium-on-barium sulfate catalyst prevents further reduction to the alditols.

```
          CN                              CHO
   H ──┼── OH                       H ──┼── OH
   H ──┼── OH     H2, H2O           H ──┼── OH
  HO ──┼── H    ───────────→       HO ──┼── H
  HO ──┼── H      Pd/BaSO4         HO ──┼── H
        CH2OH                           CH2OH
```

L-Mannononitrile

L-Mannose (56% yield from L-arabinose)

(Similarly, L-gluconpremonitrile has been reduced to L-glucose; its yield was 26 percent from L-arabinose.)

An older version of this sequence is called the **Kiliani-Fischer** synthesis. It, too, proceeds through a cyanohydrin, but it uses a less efficient method for converting the cyano group to the required aldehyde.

25.21 EPIMERIZATION, ISOMERIZATION, AND RETRO-ALDOL CLEAVAGE REACTIONS OF CARBOHYDRATES

Carbohydrates undergo a number of isomerization and degradation reactions under both laboratory and physiological conditions. For example, a mixture of glucose, fructose, and mannose results when any one of the individual components is treated with aqueous base. The nature of this reaction can be understood by examining the consequences of base-catalyzed enolization of glucose:

```
        CHO                         CHOH                          CHO
         |                           ||                            |
    H—C—OH                         C—OH                      HO—C—H
   HO ──┼── H    HO⁻, H2O     HO ──┼── H     HO⁻, H2O     HO ──┼── H
    H ──┼── OH   ⇌             H ──┼── OH    ⇌             H ──┼── OH
    H ──┼── OH                 H ──┼── OH                  H ──┼── OH
        CH2OH                      CH2OH                       CH2OH
```

D-Glucose (*R* configuration at C-2)

Enediol (C-2 not a stereogenic center)

D-Mannose (*S* configuration at C-2)

Because the configuration at C-2 is lost on enolization, the enediol intermediate can revert either to D-glucose or to D-mannose. Two stereoisomers that have multiple stereogenic centers but differ in configuration at only one of them are referred to as **epimers.** Glucose and mannose are epimeric at C-2. Under these conditions epimerization occurs only at C-2 because it alone is α to the carbonyl group.

There is another reaction available to the enediol intermediate. Proton transfer from water to C-1 converts the enediol not to an aldose but to the ketose D-fructose:

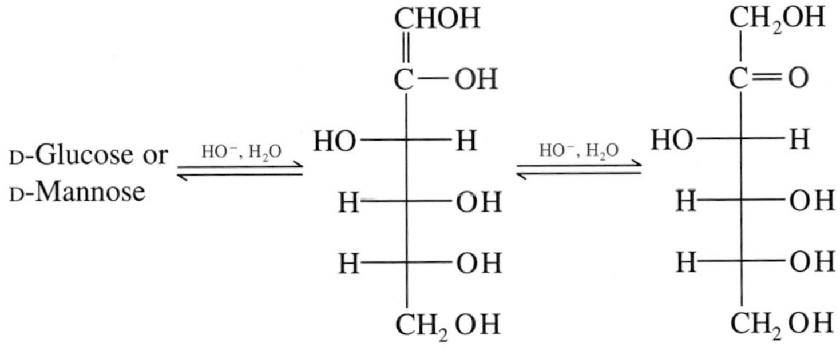

D-Glucose or D-Mannose $\xrightleftharpoons{HO^-, H_2O}$ Enediol $\xrightleftharpoons{HO^-, H_2O}$ D-Fructose

Enediol

D-Fructose

Fructose is sweeter than sucrose, and the "high-fructose corn syrup" commonly used in beverages is prepared by enzyme-catalyzed isomerization of glucose.

The isomerization of D-glucose to D-fructose by way of an enediol intermediate is an important step in **glycolysis,** a complex process (11 steps) by which an organism converts glucose to chemical energy in the absence of oxygen. The substrate is not glucose itself but its 6-phosphate ester. The enzyme that catalyzes the isomerization is called *phosphoglucose isomerase.*

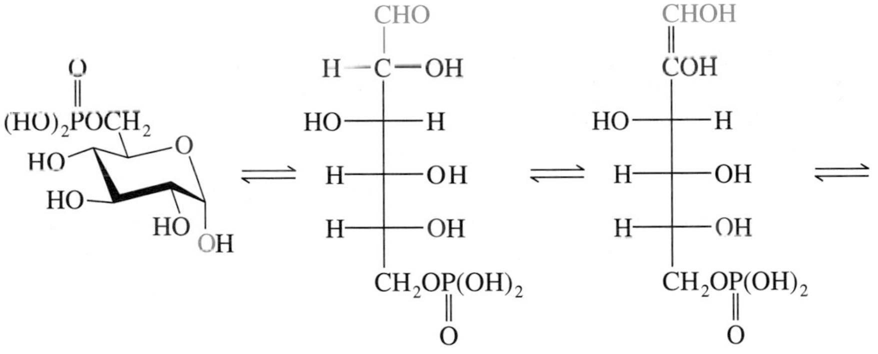

D-Glucose 6-phosphate

Open-chain form of D-glucose 6-phosphate

Enediol

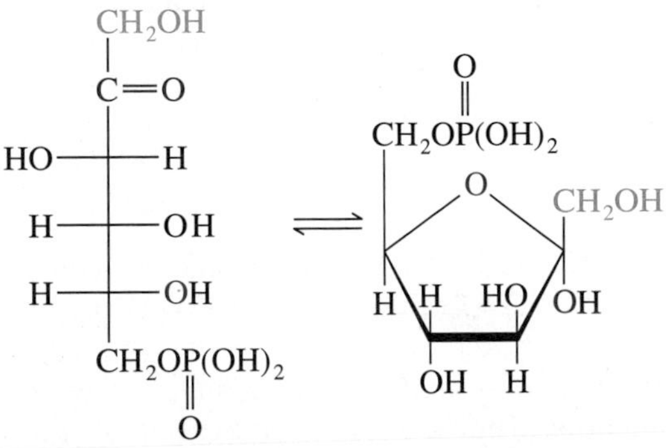

Open-chain form of D-fructose 6-phosphate

D-Fructose 6-phosphate

Following its formation, D-fructose 6-phosphate is converted to its corresponding 1,6-phosphate diester, which is then cleaved to two 3-carbon fragments under the influence of the enzyme *aldolase:*

$$\begin{array}{c} CH_2OP(=O)(OH)_2 \\ | \\ C{=}O \\ | \\ HO{-}C{-}H \\ | \\ H{-}C{-}OH \\ | \\ H{-}C{-}OH \\ | \\ CH_2OP(=O)(OH)_2 \end{array} \underset{}{\overset{\text{aldolase}}{\rightleftharpoons}} \begin{array}{c} CH_2OP(=O)(OH)_2 \\ | \\ C{=}O \\ | \\ CH_2OH \end{array} + \begin{array}{c} CH{=}O \\ | \\ H{-}C{-}OH \\ | \\ CH_2OP(=O)(OH)_2 \end{array}$$

D-Fructose 1,6-diphosphate; Dihydroxyacetone phosphate; D-Glyceraldehyde 3-phosphate

This cleavage is a *retro-aldol* reaction. It is the reverse of the process by which D-fructose 1,6-diphosphate would be formed by addition of the enolate of dihydroxyacetone phosphate to D-glyceraldehyde 3-phosphate. The enzyme aldolase catalyzes both the aldol condensation of the two components and, in glycolysis, the retro-aldol cleavage of D-fructose 1,6-diphosphate.

Further steps in glycolysis use the D-glyceraldehyde 3-phosphate formed in the aldolase-catalyzed cleavage reaction as a substrate. Its coproduct, dihydroxyacetone phosphate, is not wasted, however. The enzyme *triose phosphate isomerase* converts dihydroxyacetone phosphate to D-glyceraldehyde 3-phosphate, which enters the glycolysis pathway for further transformations.

PROBLEM 25.16 Suggest a reasonable structure for the intermediate in the conversion of dihydroxyacetone phosphate to D-glyceraldehyde 3-phosphate.

Cleavage reactions of carbohydrates also occur on treatment with aqueous base for prolonged periods as a consequence of base-catalyzed retro-aldol reactions. As pointed out in Section 18.9, aldol addition is a reversible process, and β-hydroxy carbonyl compounds can be cleaved to an enolate and either an aldehyde or a ketone.

25.22 ACYLATION AND ALKYLATION OF HYDROXYL GROUPS IN CARBOHYDRATES

The alcohol groups of carbohydrates undergo chemical reactions typical of hydroxyl functions. They are converted to esters by reaction with acyl chlorides and carboxylic acid anhydrides.

HOCH$_2$, HO, HO, HO, OH (ring O) + $5CH_3\overset{O}{\overset{\|}{C}}O\overset{O}{\overset{\|}{C}}CH_3$ $\xrightarrow{\text{pyridine}}$ $CH_2O\overset{O}{\overset{\|}{C}}CH_3$, $CH_3\overset{O}{\overset{\|}{C}}O$, $CH_3\underset{O}{\underset{\|}{C}}O$, $CH_3\underset{O}{\underset{\|}{C}}O$, $O\underset{O}{\underset{\|}{C}}CH_3$ (ring O)

α-D-Glucopyranose

Acetic anhydride

1,2,3,4,6-Penta-*O*-acetyl-α-D-glucopyranose (88%)

Ethers are formed under conditions of the Williamson ether synthesis. Methyl ethers of carbohydrates are efficiently prepared by alkylation with methyl iodide in the presence of silver oxide.

HOCH$_2$, HO, HO, HO, OCH$_3$ (ring O) + $4CH_3I$ $\xrightarrow[CH_3OH]{Ag_2O}$ CH$_3$OCH$_2$, CH$_3$O, CH$_3$O, CH$_3$O, OCH$_3$ (ring O)

Methyl α-D-glucopyranoside

Methyl iodide

Methyl 2,3,4,6-tetra-*O*-methyl-α-D-glucopyranoside (97%)

This reaction has been used in an imaginative way to determine the ring size of glycosides. Once all the free hydroxyl groups of a glycoside have been methylated, the glycoside is subjected to acid-catalyzed hydrolysis. Only the anomeric methoxy group is hydrolyzed under these conditions—another example of the ease of carbocation formation at the anomeric position.

CH$_3$OCH$_2$, CH$_3$O, CH$_3$O, CH$_3$O, OCH$_3$ (ring O) $\xrightarrow[H^+]{H_2O}$ CH$_3$OCH$_2$, CH$_3$O, CH$_3$O, CH$_3$O, OH (ring O) $\rightleftharpoons$ CHO; H—OCH$_3$; CH$_3$O—H; H—OCH$_3$; H—OH; CH$_2$OCH$_3$

Methyl 2,3,4,6-tetra-*O*-methyl-α-D-glucopyranoside

2,3,4,6-Tetra-*O*-methyl-D-glucose

Notice that all the hydroxyl groups in the free sugar except C-5 are methylated. Carbon-5 is not methylated, because it was originally the site of the ring oxygen in the methyl glycoside. Once the position of the hydroxyl group in the free sugar has been determined, either by spectroscopy or by converting the sugar to a known compound, the ring size stands revealed.

25.23 PERIODIC ACID OXIDATION OF CARBOHYDRATES

Periodic acid oxidation (Section 15.12) finds extensive use as an analytical method in carbohydrate chemistry. Structural information is obtained by measuring the number of equivalents of periodic acid that react with a given compound and by identifying the reaction products. A vicinal diol consumes one equivalent of periodate and is cleaved to two carbonyl compounds:

$$\underset{\text{Vicinal diol}}{R_2C(OH)\text{—}C(OH)R'_2} + \underset{\text{Periodic acid}}{HIO_4} \longrightarrow \underset{\text{Two carbonyl compounds}}{R_2C{=}O + R'_2C{=}O} + \underset{\text{Iodic acid}}{HIO_3} + \underset{\text{Water}}{H_2O}$$

α-Hydroxy carbonyl compounds are cleaved to a carboxylic acid and a carbonyl compound:

$$\underset{\alpha\text{-Hydroxy carbonyl compound}}{RC({=}O)C(OH)R'_2} + \underset{\text{Periodic acid}}{HIO_4} \longrightarrow \underset{\text{Carboxylic acid}}{RC({=}O)OH} + \underset{\text{Carbonyl compound}}{R'_2C{=}O} + \underset{\text{Iodic acid}}{HIO_3}$$

When three contiguous carbons bear hydroxyl groups, two moles of periodate are consumed per mole of carbohydrate and the central carbon is oxidized to a molecule of formic acid:

$$R_2C(OH)\text{—}CH(OH)\text{—}C(OH)R'_2 + \underset{\text{Periodic acid}}{2HIO_4} \longrightarrow \underset{\text{Carbonyl compound}}{R_2C{=}O} + \underset{\text{Formic acid}}{HC({=}O)OH} + \underset{\text{Carbonyl compound}}{R'_2C{=}O} + \underset{\text{Iodic acid}}{2HIO_3}$$

Points at which cleavage occurs

Ether and acetal functions are not affected by the reagent.

An example of the use of periodic acid oxidation in structure determination can be found in a case in which a previously unknown methyl glycoside was obtained by the reaction of D-arabinose with methanol and hydrogen chloride. The size of the ring was identified as five-membered because only one mole of periodic acid was consumed per mole of glycoside and no formic acid was produced. Were the ring six-membered, two moles of periodic acid would be required per mole of glycoside and one mole of formic acid would be produced.

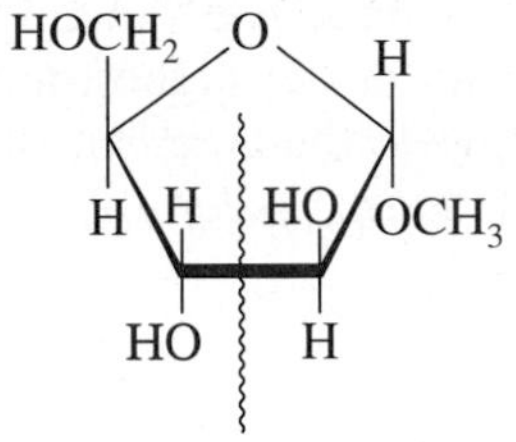

Only one site for periodic acid cleavage in methyl α-D-arabinofuranoside

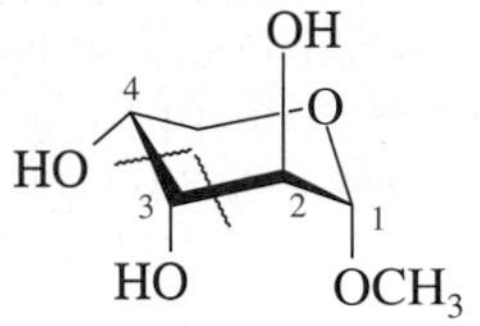

Two sites of periodic acid cleavage in methyl α-D-arabinopyranoside, C-3 lost as formic acid

---

PROBLEM 25.17 Give the products of periodic acid oxidation of each of the following. How many moles of reagent will be consumed per mole of substrate in each case?

(*a*) D-Arabinose
(*b*) D-Ribose
(*c*) Methyl β-D-glucopyranoside
(*d*)

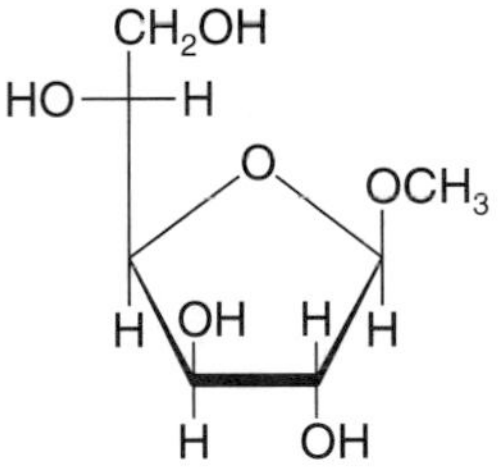

SAMPLE SOLUTION (*a*) The α-hydroxy aldehyde unit at the end of the sugar chain is cleaved, as well as all the vicinal diol functions. Four moles of periodic acid are required per mole of D-arabinose. Four moles of formic acid and one mole of formaldehyde are produced.

D-Arabinose, showing points of cleavage by periodic acid; each cleavage requires one equivalent of HIO_4.

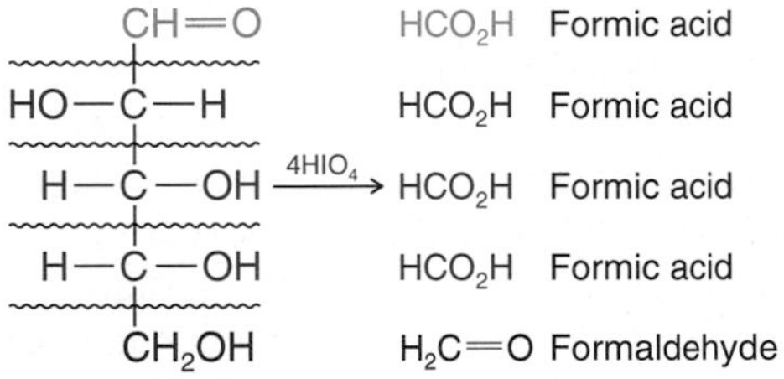

$$\begin{array}{c} CH{=}O \\ HO-C-H \\ H-C-OH \\ H-C-OH \\ CH_2OH \end{array} \xrightarrow{4HIO_4} \begin{array}{ll} HCO_2H & \text{Formic acid} \\ HCO_2H & \text{Formic acid} \\ HCO_2H & \text{Formic acid} \\ HCO_2H & \text{Formic acid} \\ H_2C{=}O & \text{Formaldehyde} \end{array}$$

25.24 SUMMARY

Carbohydrates are marvelous molecules! In most of them, every carbon bears a functional group, and the nature of the functional groups changes as the molecule interconverts between open-chain and cyclic hemiacetal forms. Any approach to understanding carbohydrates must begin with structure.

Carbohydrates are polyhydroxylic aldehydes and ketones. Those derived from aldehydes are classified as **aldoses;** those derived from ketones are **ketoses** (Section 25.1).

Fischer projection formulas offer a convenient way to represent configurational relationships in carbohydrates. When the hydroxyl group at the highest-numbered stereogenic center is to the right, the carbohydrate belongs to the D series; when this hydroxyl is to the left, the carbohydrate belongs to the L series (Section 25.2).

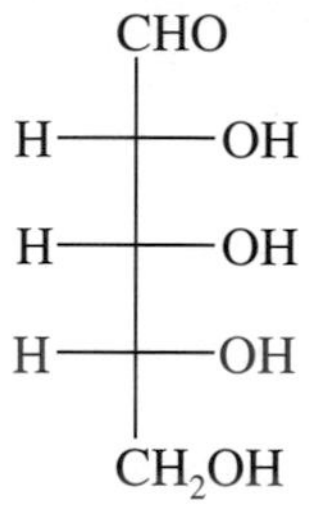

D-Ribose 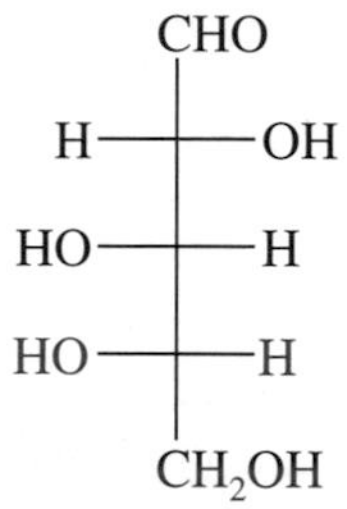   L-Arabinose

D-Ribose and L-arabinose are aldoses that have five carbons and so are classified as **aldopentoses** (Section 25.4). **Aldohexoses** are aldoses that have six carbons. Figure 25.2 (Section 25.4) gives the Fischer projections of aldoses having six or fewer carbons.

Most carbohydrates exist as cyclic hemiacetals. Five-membered cyclic hemiacetals are called **furanose** forms (Section 25.6), while six-membered cyclic hemiacetals are called **pyranose** forms (Section 25.7).

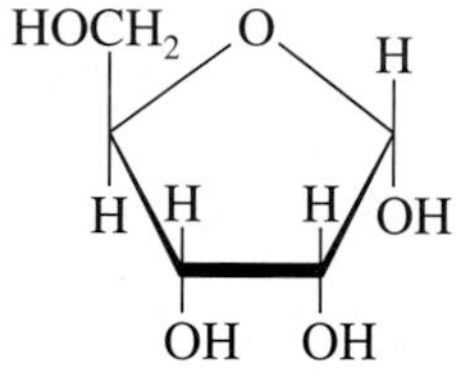

α-D-Ribofuranose 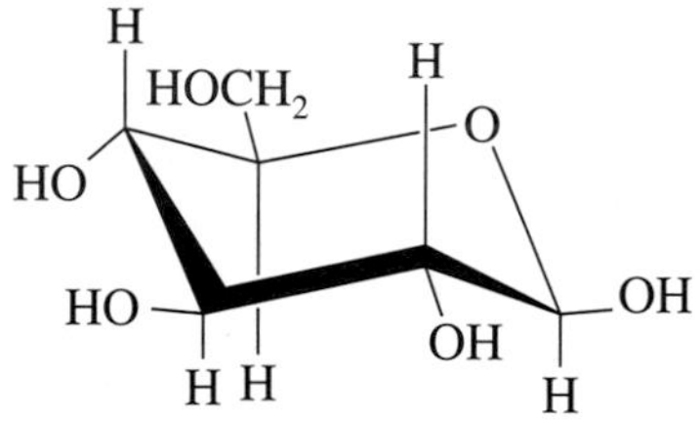 β-D-Glucopyranose

The **anomeric carbon** in a cyclic acetal is the one attached to *two* oxygens. It is the carbon that corresponds to the carbonyl carbon in the open-chain form. The symbols α and β refer to the configuration at the anomeric carbon.

A particular carbohydrate can interconvert between furanose and pyranose forms and between the α and β configuration of each form. The change from one form to an equilibrium mixture of all the possible hemiacetals causes a change in optical rotation which is called **mutarotation** (Section 25.8).

Ketoses are characterized by the ending *-ulose* in their name. Most naturally occurring ketoses have their carbonyl group located at C-2. Like aldoses, ketoses cyclize to hemiacetals and exist as furanose or pyranose forms (Section 25.9).

Structurally modified carbohydrates include **deoxy sugars** (Section 25.10), **amino sugars** (Section 25.11), and **branched-chain carbohydrates** (Section 25.12).

Glycosides are acetals, compounds in which the anomeric hydroxyl group has been replaced by an alkoxy group (Section 25.13). Glycosides are easily prepared in the laboratory by allowing a carbohydrate and an alcohol to stand in the presence of an acid catalyst.

TABLE 25.2

Summary of Reactions of Carbohydrates

Reaction (section) and comments	Example
Transformations of the carbonyl group	
Reduction (Section 25.18) The carbonyl group of aldoses and ketoses is reduced by sodium borohydride or by catalytic hydrogenation. The products are called *alditols.*	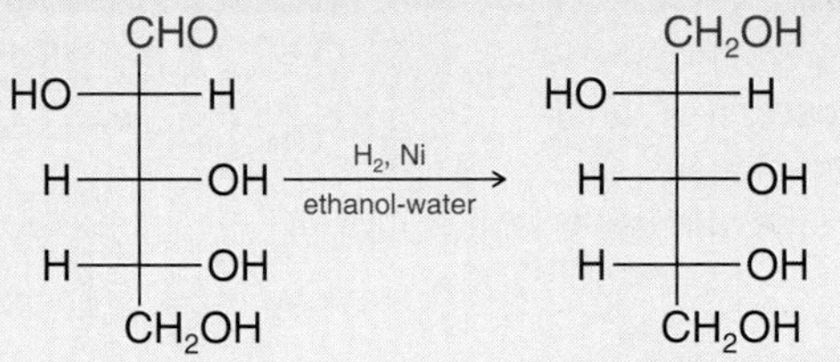 D-Arabinose →(H_2, Ni; ethanol-water) D-Arabinitol (80%)
Oxidation with Benedict's reagent (Section 25.19) Sugars that contain a free hemiacetal function are called reducing sugars. They react with copper(II) sulfate in a sodium citrate/sodium carbonate buffer (Benedict's reagent) to form a red precipitate of copper(I) oxide. Used as a qualitative test for reducing sugars.	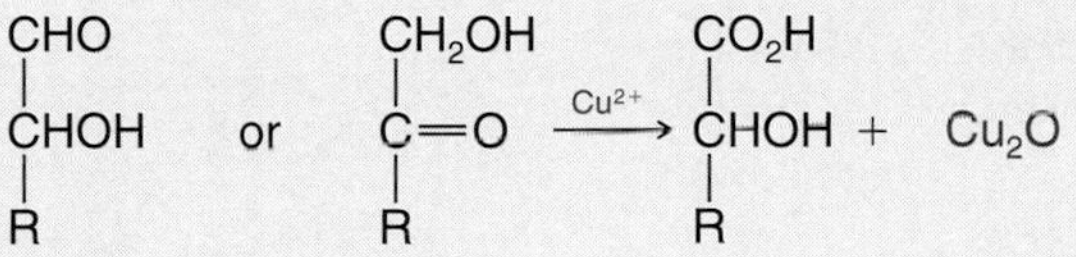 CHO–CHOH–R (Aldose) or CH_2OH–C=O–R (Ketose) →(Cu^{2+}) CO_2H–CHOH–R (Aldonic acid) + Cu_2O (Copper(I) oxide)
Oxidation with bromine (Section 25.19) When a preparative method for an aldonic acid is required, bromine oxidation is used. The aldonic acid is formed as its lactone. More properly described as a reaction of the anomeric hydroxyl group than of a free aldehyde.	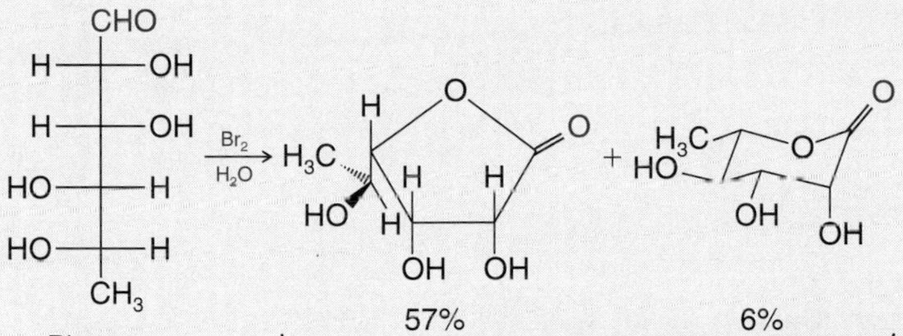 L-Rhamnose →(Br_2, H_2O) L-Rhamnonolactone (57% + 6%)
Chain extension by way of cyanohydrin formation (Section 25.20) The Kiliani-Fischer synthesis proceeds by nucleophilic addition of HCN to an aldose, followed by conversion of the cyano group to an aldehyde. A mixture of stereoisomers results; the two aldoses are epimeric at C-2. Section 25.20 describes the modern version of the Kiliani-Fischer synthesis. The example at the right illustrates the classical version.	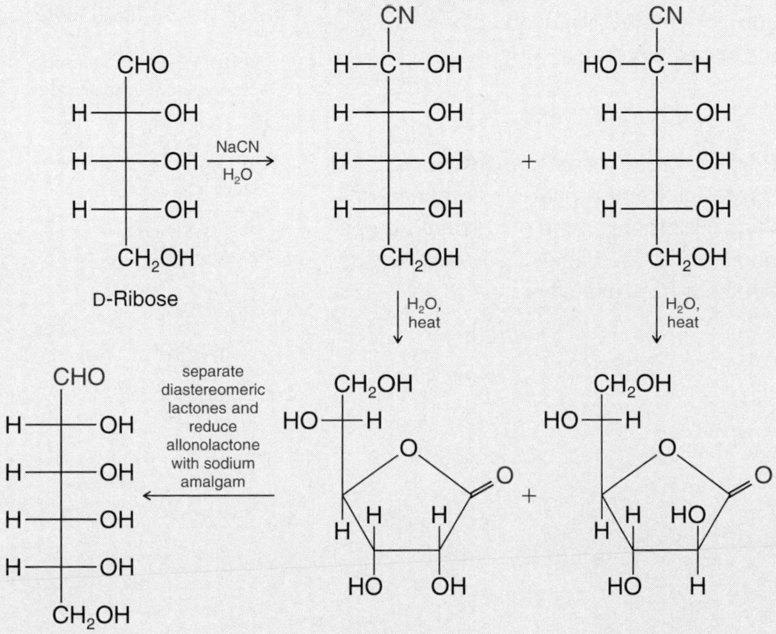 D-Ribose →(NaCN, H_2O) two cyanohydrins →(H_2O, heat) Allonolactone (35–40%) + Altronolactone (about 45%); separate diastereomeric lactones and reduce allonolactone with sodium amalgam → D-Allose (34%)

TABLE 25.2
Summary of Reactions of Carbohydrates *(Continued)*

Reaction (section) and comments	Example
Enediol formation (Section 25.21) Enolization of an aldose or a ketose gives an enediol. Enediols can revert to aldoses or ketoses with loss of stereochemical integrity at the α carbon atom.	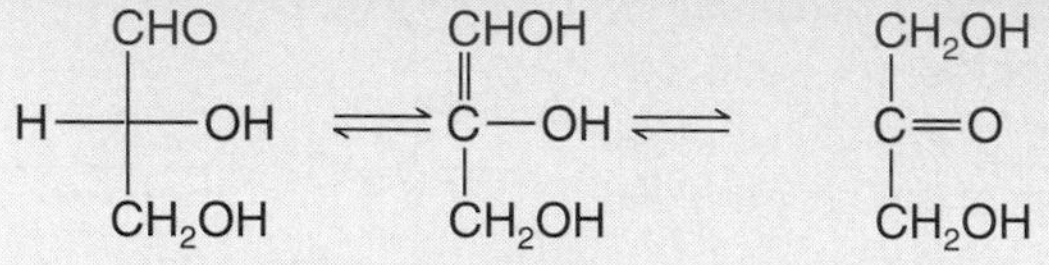 D-Glyceraldehyde ⇌ Enediol ⇌ 1,3-Dihydroxyacetone
Reactions of the hydroxyl group	
Acylation (Section 25.22) Esterification of the available hydroxyl groups occurs when carbohydrates are treated with acylating agents.	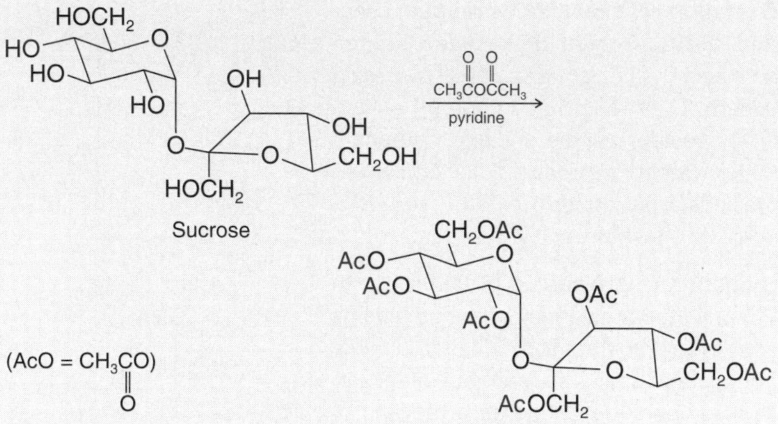 Sucrose $\xrightarrow[\text{pyridine}]{CH_3COCCH_3}$ Sucrose octaacetate (66%) $(AcO = CH_3\overset{O}{\overset{\parallel}{C}}O)$
Alkylation (Section 25.22) Alkyl halides react with carbohydrates to form ethers at the available hydroxyl groups. An application of the Williamson ether synthesis to carbohydrates.	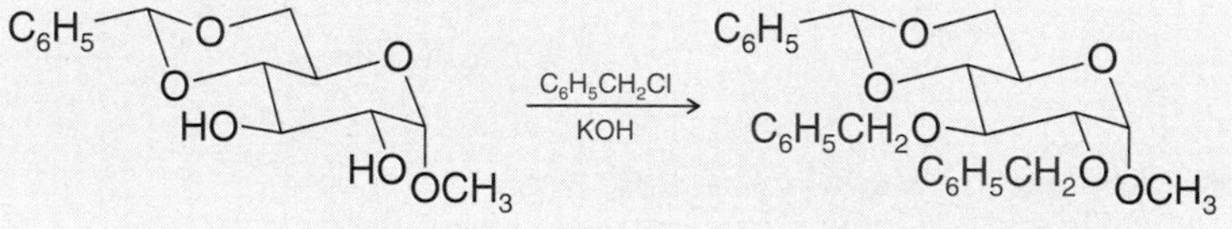 Methyl 4,6-*O*-benzylidene-α-D-glucopyranoside $\xrightarrow[KOH]{C_6H_5CH_2Cl}$ Methyl 2,3-di-*O*-benzyl-4,6-*O*-benzylidene-α-D-glucopyranoside (92%)
Periodic acid oxidation (Section 25.23) Vicinal diol and α-hydroxy carbonyl functions in carbohydrates are cleaved by periodic acid. Used analytically as a tool for structure determination.	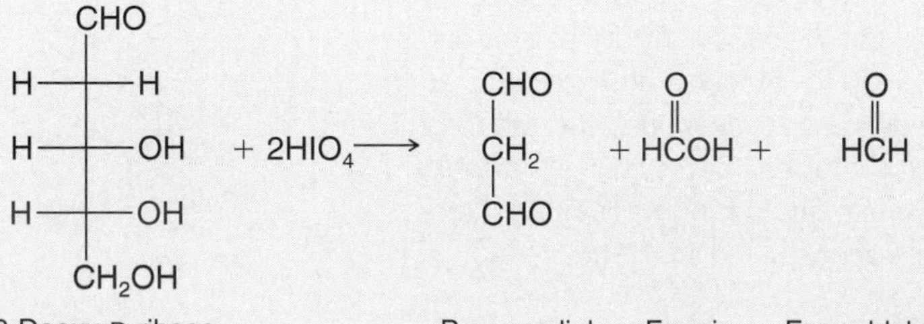 2-Deoxy-D-ribose $+ 2HIO_4 \longrightarrow$ Propanedial + Formic acid ($HCOOH$) + Formaldehyde ($HCHO$)

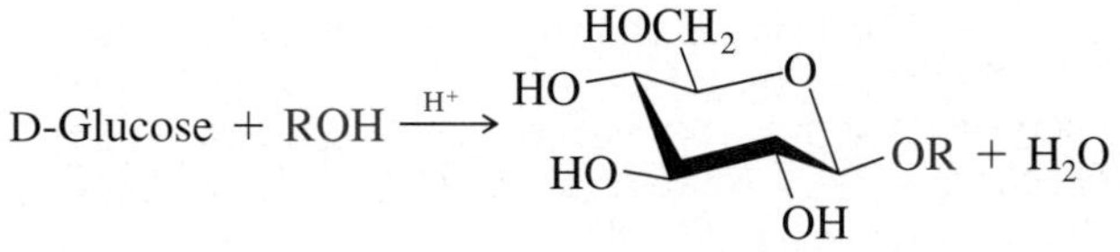

$$\text{D-Glucose} + \text{ROH} \xrightarrow{H^+} \text{Glycoside} + H_2O$$

Glycoside

Disaccharides are carbohydrates in which two monosaccharides are joined by a glycoside bond (Section 25.14). **Polysaccharides** have many monosaccharide units connected through glycosidic linkages (Section 25.15). Complete hydrolysis of disaccharides and polysaccharides cleaves the glycoside bonds, yielding the free monosaccharide components.

Carbohydrates undergo chemical reactions characteristic of aldehydes and ketones, alcohols, diols, and other classes of compounds, depending on their structure. A review of the reactions described in this chapter is presented in Table 25.2. While some of the reactions have synthetic value, many of them are used in analysis and structure determination.

PROBLEMS

25.18 Refer to the Fischer projection formula of D-(+)-xylose in Figure 25.2 (Section 25.4) and give structural formulas for

(*a*) (−)-Xylose (Fischer projection)
(*b*) D-Xylitol
(*c*) β-D-Xylopyranose
(*d*) α-L-Xylofuranose
(*e*) Methyl α-L-xylofuranoside
(*f*) D-Xylonic acid (open-chain Fischer projection)
(*g*) δ-Lactone of D-xylonic acid
(*h*) γ-Lactone of D-xylonic acid
(*i*) D-Xylaric acid (open-chain Fischer projection)

25.19 From among the carbohydrates shown in Figure 25.2, choose the D-aldohexoses that yield

(*a*) An optically inactive product on reduction with sodium borohydride
(*b*) An optically inactive product on oxidation with bromine
(*c*) An optically inactive product on oxidation with nitric acid
(*d*) The same enediol

25.20 Write the Fischer projection formula of the open-chain form of each of the following:

(*a*)

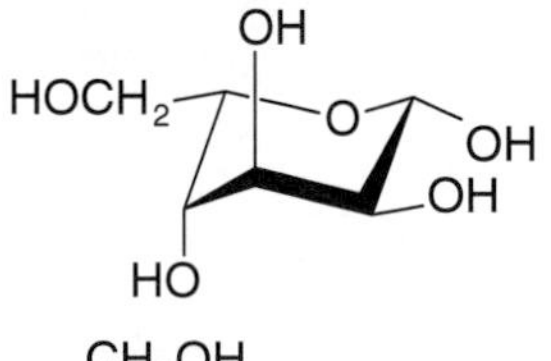

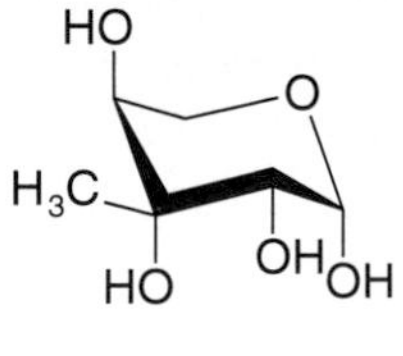

(*b*)

(*c*)

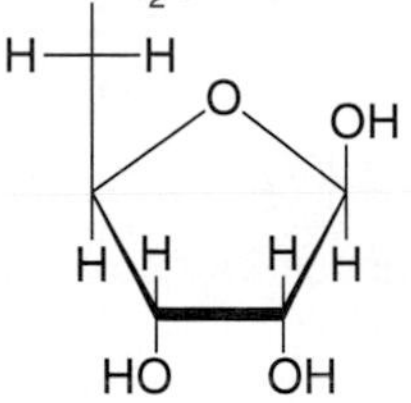

(*d*)

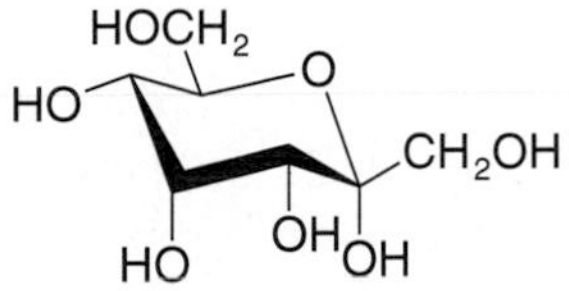

25.21 From among the carbohydrates shown in Problem 25.20 choose the one(s) that

(*a*) Belong to the L series
(*b*) Are deoxy sugars
(*c*) Are branched-chain sugars
(*d*) Are ketoses
(*e*) Are furanose forms
(*f*) Have the α configuration at their anomeric carbon

25.22 How many pentuloses are possible? Write their Fischer projection formulas.

25.23 The Fischer projection formula of the branched-chain carbohydrate D-apiose has been presented in Section 25.12.

(*a*) How many stereogenic centers are there in the open-chain form of D-apiose?
(*b*) Does D-apiose form an optically active alditol on reduction?
(*c*) How many stereogenic centers are there in the furanose forms of D-apiose?
(*d*) How many stereoisomeric furanose forms of D-apiose are possible? Write their Haworth formulas.

25.24 Treatment of D-mannose with methanol in the presence of an acid catalyst yields four isomeric products having the molecular formula $C_7H_{14}O_6$. What are these four products?

25.25 Maltose and cellobiose (Section 25.14) are examples of disaccharides derived from D-glucopyranosyl units.

(*a*) How many other disaccharides are possible that meet this structural requirement?
(*b*) How many of these are reducing sugars?

25.26 Gentiobiose has the molecular formula $C_{12}H_{22}O_{11}$ and has been isolated from gentian root and by hydrolysis of amygdalin. Gentiobiose exists in two different forms, one melting at 86°C and the other at 190°C. The lower-melting form is dextrorotatory ($[\alpha]_D^{22}$ +16°), while the higher-melting one is levorotatory ($[\alpha]_D^{22}$ −6°). The rotation of an aqueous solution of either form, however, gradually changes until a final value of $[\alpha]_D^{22}$ + 9.6° is observed. Hydrolysis of gentiobiose is efficiently catalyzed by emulsin and produces two moles of D-glucose per mole of gentiobiose. Gentiobiose forms an octamethyl ether, which on hydrolysis in dilute acid yields 2,3,4,6-tetra-*O*-methyl-D-glucose and 2,3,4-tri-*O*-methyl-D-glucose. What is the structure of gentiobiose?

25.27 *Cyanogenic glycosides* are potentially toxic because they liberate hydrogen cyanide on enzyme-catalyzed or acidic hydrolysis. Give a mechanistic explanation for this behavior for the specific cases of

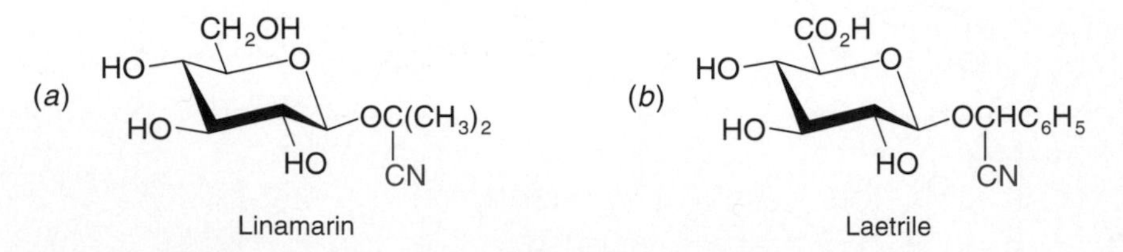

25.28 The following are the more stable anomers of the pyranose forms of D-glucose, D-mannose, and D-galactose:

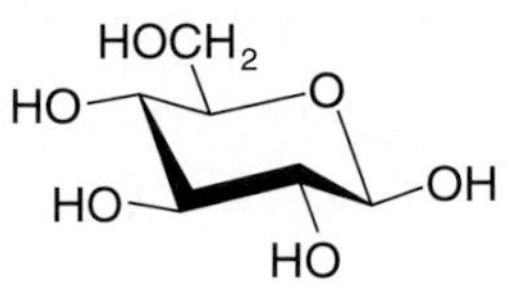

β-D-Glucopyranose (64% at equilibrium)

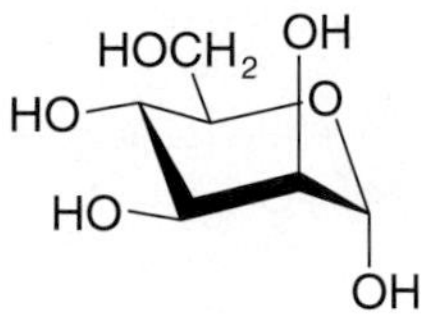

α-D-Mannopyranose (68% at equilibrium)

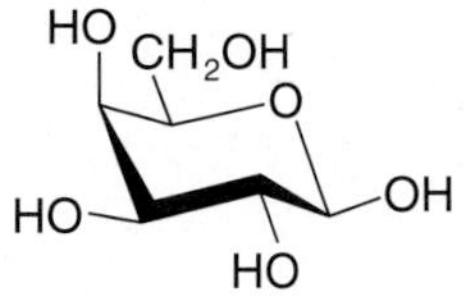

β-D-Galactopyranose (64% at equilibrium)

On the basis of these empirical observations and your own knowledge of steric effects in six-membered rings, predict the preferred form (α- or β-pyranose) at equilibrium in aqueous solution for each of the following:

(*a*) D-Gulose
(*b*) D-Talose
(*c*) D-Xylose
(*d*) D-Lyxose

25.29 Basing your answers on the general mechanism for the first stage of acid-catalyzed acetal hydrolysis

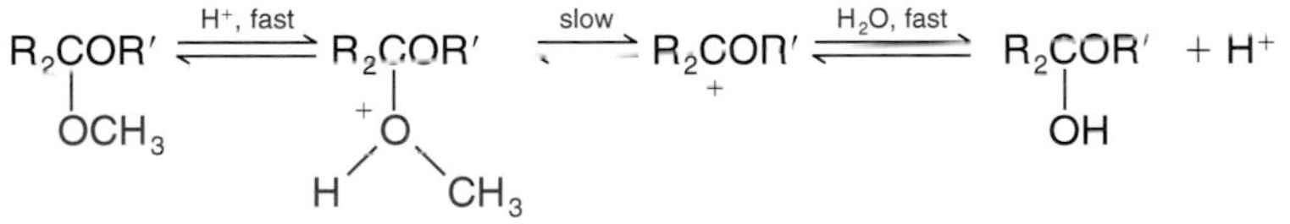

$$R_2C(OCH_3)OR' \xrightleftharpoons{H^+,\ \text{fast}} R_2C(\overset{+}{O}HCH_3)OR' \xrightarrow{\text{slow}} R_2\overset{+}{C}OR' \xrightleftharpoons{H_2O,\ \text{fast}} R_2C(OH)OR' + H^+$$

Acetal → Hemiacetal

suggest reasonable explanations for the following observations:

(*a*) Methyl α-D-fructofuranoside (A) undergoes acid-catalyzed hydrolysis some 10^5 times faster than methyl α-D-glucofuranoside (B).

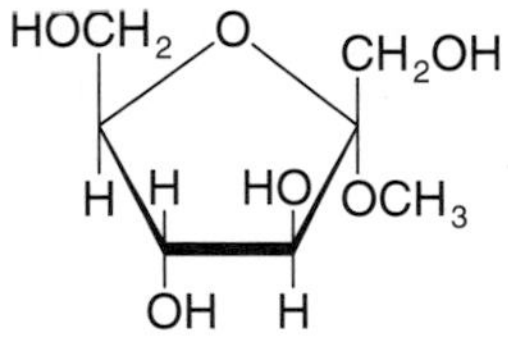

Compound A

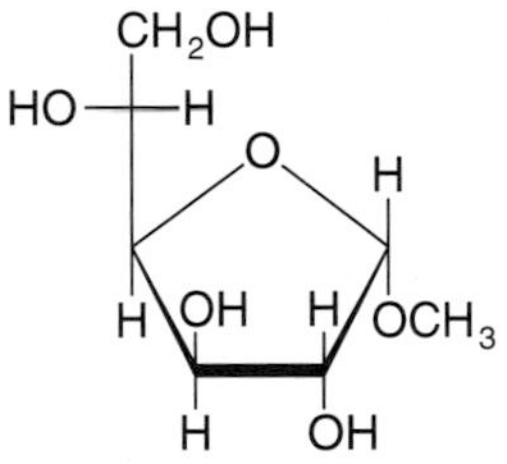

Compound B

(*b*) The β-methyl glucopyranoside of 2-deoxy-D-glucose (C) undergoes hydrolysis several thousand times faster than that of D-glucose (D).

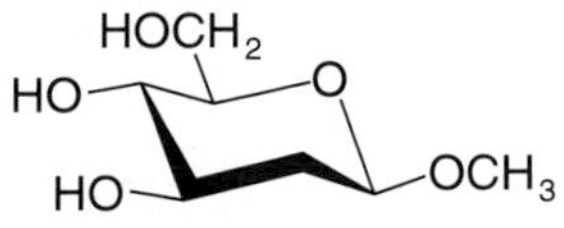

Compound C

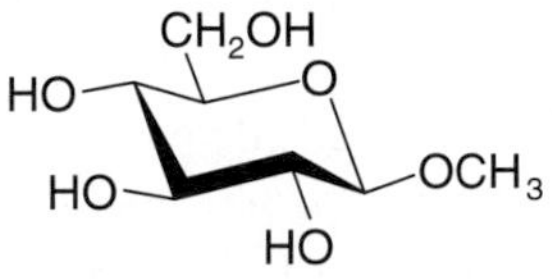

Compound D

25.30 D-Altrosan is converted to D-altrose by dilute aqueous acid. Suggest a reasonable mechanism for this reaction.

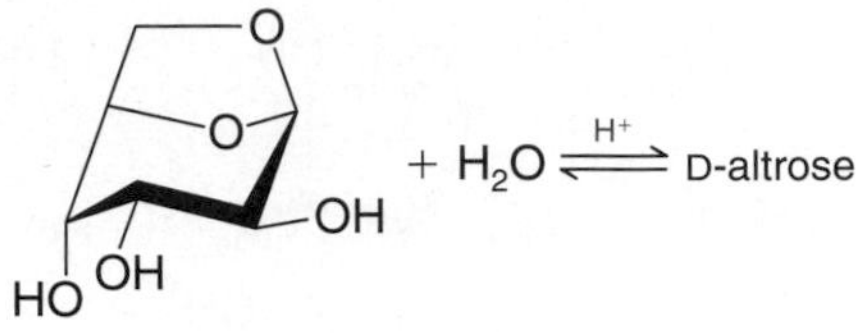

$+ H_2O \overset{H^+}{\rightleftharpoons}$ D-altrose

D-Altrosan

25.31 When D-galactose was heated at 165°C, a small amount of compound A was isolated:

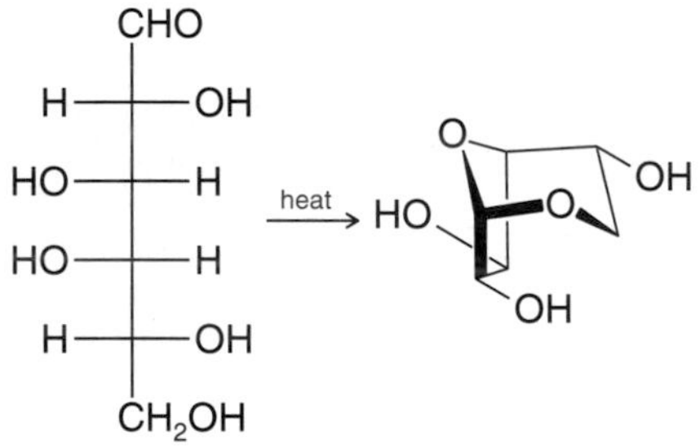

heat

D-Galactose Compound A

The structure of compound A was established, in part, by converting it to known compounds. Treatment of A with excess methyl iodide in the presence of silver oxide, followed by hydrolysis with dilute hydrochloric acid, gave a trimethyl ether of D-galactose. Comparing this trimethyl ether with known trimethyl ethers of D-galactose allowed the structure of compound A to be deduced.

How many trimethyl ethers of D-galactose are there? Which one is the same as the product derived from A?

25.32 Phlorizin is obtained from the root bark of apple, pear, cherry, and plum trees. It has the molecular formula $C_{21}H_{24}O_{10}$ and yields a compound A and D-glucose on hydrolysis in the presence of emulsin. When phlorizin was treated with excess methyl iodide in the presence of potassium carbonate and then subjected to acid-catalyzed hydrolysis, a compound B was obtained. Deduce the structure of phlorizin from this information.

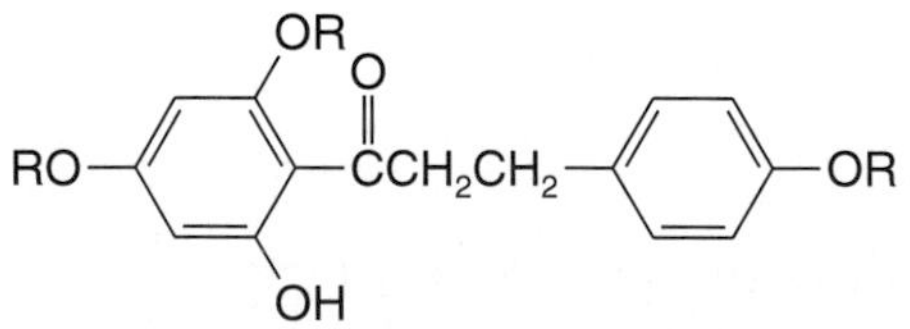

Compound A: R = H
Compound B: $R = CH_3$

25.33 Emil Fischer's determination of the structure of glucose was carried out as the nineteenth century ended and the twentieth began. The structure of no other sugar was known at that time, and none of the spectroscopic techniques that aid organic analysis were then available. All Fischer had was information from chemical transformations, polarimetry, and his own intellect. Fischer realized that (+)-glucose could be represented by 16 possible stereostructures. By arbitrarily assigning a particular configuration to the stereogenic center at C-5, the configurations of C-2, C-3, and C-4 could be determined relative to it, reducing the number of structural possibilities to eight. Thus, he started with a structural representation shown as follows, where C-5 of (+)-glucose has what is now known as the D configuration.

$$
\begin{array}{c}
\text{CHO} \\
| \\
\text{CHOH} \\
| \\
\text{CHOH} \\
| \\
\text{CHOH} \\
| \\
\text{H}-\text{C}-\text{OH} \\
| \\
\text{CH}_2\text{OH}
\end{array}
$$

Eventually, Fischer's arbitrary assumption proved to be correct, and the structure he proposed for (+)-glucose is correct in an absolute as well as a relative sense. The following exercise uses information available to Fischer and leads you through a reasoning process similar to that employed in his determination of the structure of (+)-glucose. See if you can work out the configuration of (+)-glucose from the information provided, assuming the configuration of C-5 as just shown.

1. Chain extension of the aldopentose (−)-arabinose by way of the derived cyanohydrin gave a mixture of (+)-glucose and (+)-mannose.
2. Oxidation of (−)-arabinose with warm nitric acid gave an optically active aldaric acid.
3. Both (+)-glucose and (+)-mannose were oxidized to optically active aldaric acids with nitric acid.
4. There is another sugar, (+)-gulose, that gives the same aldaric acid on oxidation as does (+)-glucose.

MOLECULAR MODELING EXERCISES

25.34 Construct a molecular model of D-glyceraldehyde. Arrange the molecular model so that its projection onto a flat surface appears the same as a Fischer projection.

25.35 Write a Fischer projection of D-ribose and transform it into a molecular model. What are the *R*,*S* configurations of its three stereogenic centers?

25.36 Convert the molecular model of D-ribose constructed in the preceding exercise to (*a*) α-D-ribofuranose, and (*b*) β-D-ribopyranose.

25.37 Construct a molecular model of β-D-glucopyranose in its most stable conformation. Alter the configurations at the appropriate carbons so as to transform this molecular model into (*a*) β-D-mannopyranose, and (*b*) α-D-allopyranose.

25.38 Construct molecular models of (*a*) cellobiose, and (*b*) maltose.

CHAPTER 26

ACETATE-DERIVED NATURAL PRODUCTS

Photochemical energy is stored as chemical potential energy by the process of *photosynthesis* (Section 25.4). Carbon dioxide and water combine in the presence of visible light and chlorophyll to form carbohydrates. The energy that is stored in the carbohydrate is released during carbohydrate *metabolism:*

$$\underset{\text{Glucose}}{C_6H_{12}O_6} + \underset{\text{Oxygen}}{6O_2} \longrightarrow \underset{\text{Carbon dioxide}}{6CO_2} + \underset{\text{Water}}{H_2O} + \text{energy}$$

During one stage of carbohydrate metabolism, called *glycolysis,* glucose is converted to lactic acid. Pyruvic acid is an intermediate.

$$\underset{\text{Glucose}}{C_6H_{12}O_6} \longrightarrow \underset{\text{Pyruvic acid}}{CH_3\overset{\overset{\displaystyle O}{\|}}{C}CO_2H} \longrightarrow \underset{\text{Lactic acid}}{CH_3\overset{\overset{\displaystyle OH}{|}}{C}HCO_2H}$$

In most biochemical reactions the pH of the medium is close to 7. At this pH carboxylic acids are nearly completely converted to their conjugate bases. Thus, it is common practice in biological chemistry to specify the derived carboxylate anion rather than the carboxylic acid itself. For example, we say that glycolysis leads to *lactate* by way of *pyruvate.*

Pyruvate is utilized by living systems in a number of different ways. One pathway, the one leading to lactate and beyond, is concerned with energy storage and production. This is not the only pathway available to pyruvate, however. A significant fraction of it is converted to acetate for use as a starting material in the *biosynthesis* of more complex substances. These *acetate-derived natural products* play key roles in the chemistry of life. Their diverse structural types and the biosynthetic thread that connects all of them to acetate are the principal concerns of this chapter.

We begin with a brief description of the biosynthetic origin of acetate.

26.1 ACETYL COENZYME A

The form in which acetate is used in most of its important biochemical reactions is **acetyl coenzyme A** (Figure 26.1*a*). Acetyl coenzyme A is a *thioester* (Section 20.12). Its formation from pyruvate involves several steps and is summarized in the overall equation:

$$\underset{\text{Pyruvic acid}}{CH_3\overset{O}{\overset{\|}{C}}\overset{O}{\overset{\|}{C}}OH} + \underset{\text{Coenzyme A}}{CoASH} + \underset{\text{Oxidized form of nicotinamide adenine dinucleotide}}{NAD^+} \longrightarrow \underset{\text{Acetyl coenzyme A}}{CH_3\overset{O}{\overset{\|}{C}}SCoA} + \underset{\text{Reduced form of nicotinamide adenine dinucleotide}}{NADH} + \underset{\text{Carbon dioxide}}{CO_2} + \underset{\text{Proton}}{H^+}$$

All the individual steps are catalyzed by enzymes. NAD^+ (Section 15.11) is required as an oxidizing agent, and coenzyme A (Figure 26.1*b*) is the acetyl group acceptor. Coenzyme A is a *thiol;* its chain terminates in a *sulfhydryl* (—SH) group. Acetylation of the sulfhydryl group of coenzyme A gives acetyl coenzyme A.

Because sulfur does not donate electrons to an attached carbonyl group as well as oxygen does, compounds of the type $R\overset{O}{\overset{\|}{C}}SR'$ are better acyl transfer agents than is $R\overset{O}{\overset{\|}{C}}OR'$. They also contain a greater proportion of enol at equilibrium. Both properties are apparent in the properties of acetyl coenzyme A. In some of its reactions acetyl coenzyme A acts as an acetyl transfer agent, whereas in others the α carbon atom of the acetyl group is the reactive site.

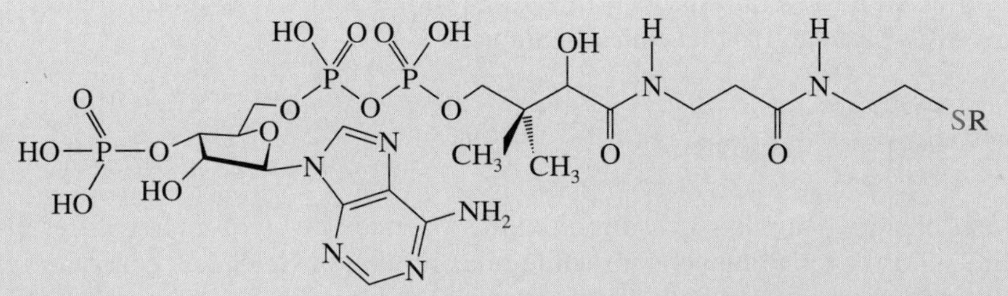

(*a*) $R = \overset{O}{\overset{\|}{C}}CH_3$ Acetyl coenzyme A (abbreviation: $CH_3\overset{O}{\overset{\|}{C}}SCoA$)

(*b*) $R = H$ Coenzyme A (abbreviation: CoASH)

FIGURE 26.1 Structures of (*a*) acetyl coenzyme A and (*b*) coenzyme A.

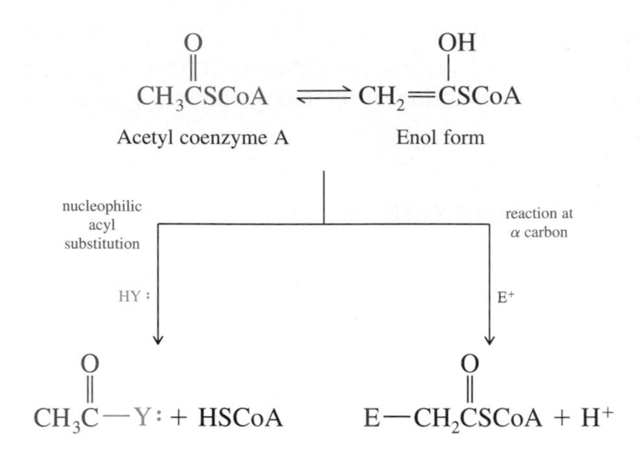

You will see numerous examples of both reaction types in the following sections. Keep in mind that in vivo reactions (reactions in living systems) are enzyme-catalyzed and occur at rates that are far greater than when the same transformations are carried out in vitro (''in glass'') in the absence of enzymes. In spite of the rapidity with which enzyme-catalyzed reactions take place, the nature of these transformations is essentially the same as the fundamental processes of organic chemistry described throughout this text.

The September 1986 issue of the *Journal of Chemical Education* (pp. 772–775) contains the last of a series of three papers on metabolism. This paper reviews lipids.

Among the natural products that are biosynthesized from acetate are a number of members of the **lipid** class. Lipids are naturally occurring substances that are soluble in nonpolar solvents. This is an operational, rather than a structural, distinction. Material isolated from some natural source is shaken with a polar solvent (water or an alcohol-water mixture) and a nonpolar one (diethyl ether, hexane, or dichloromethane). Carbohydrates, proteins, nucleic acids, and related compounds are polar and do not dissolve in the nonpolar solvent—they either dissolve in the aqueous phase or remain behind undissolved. The portion of the natural material that dissolves in the nonpolar solvent is the lipid fraction.

Fats are one type of lipid. They have a number of functions in living systems, including that of energy storage. While carbohydrates serve as a source of readily available energy, an equal weight of fat delivers over twice the amount of energy. It is more efficient for an organism to store energy in the form of fat because it requires less mass than storing the same amount of energy in carbohydrate molecules.

How living systems convert acetate to fats is an exceedingly complex story, one that is well understood in broad outline and becoming increasingly clear in detail as well. We will examine several aspects of this topic in the next few sections, focusing primarily on its structural and chemical features.

26.2 FATTY ACIDS

An experiment describing the analysis of the triglyceride composition of several vegetable oils is described in the May 1988 issue of the *Journal of Chemical Education* (pp. 464–466).

Fats and oils are naturally occurring mixtures of triacylglycerols. They differ in that fats are solids at room temperature while oils are liquids. Scientists generally ignore this distinction and refer to both groups of compounds as fats. Two typical triacylglycerols are shown in Figure 26.2. All three acyl groups in a triacylglycerol may be the same, all three may be different, or one may be different from the other two. The triacylglycerol shown in Figure 26.2*a*, 2-oleyl-1,3-distearylglycerol, occurs naturally and accounts for about 20 percent of the triacylglycerol mixture of cocoa butter. The

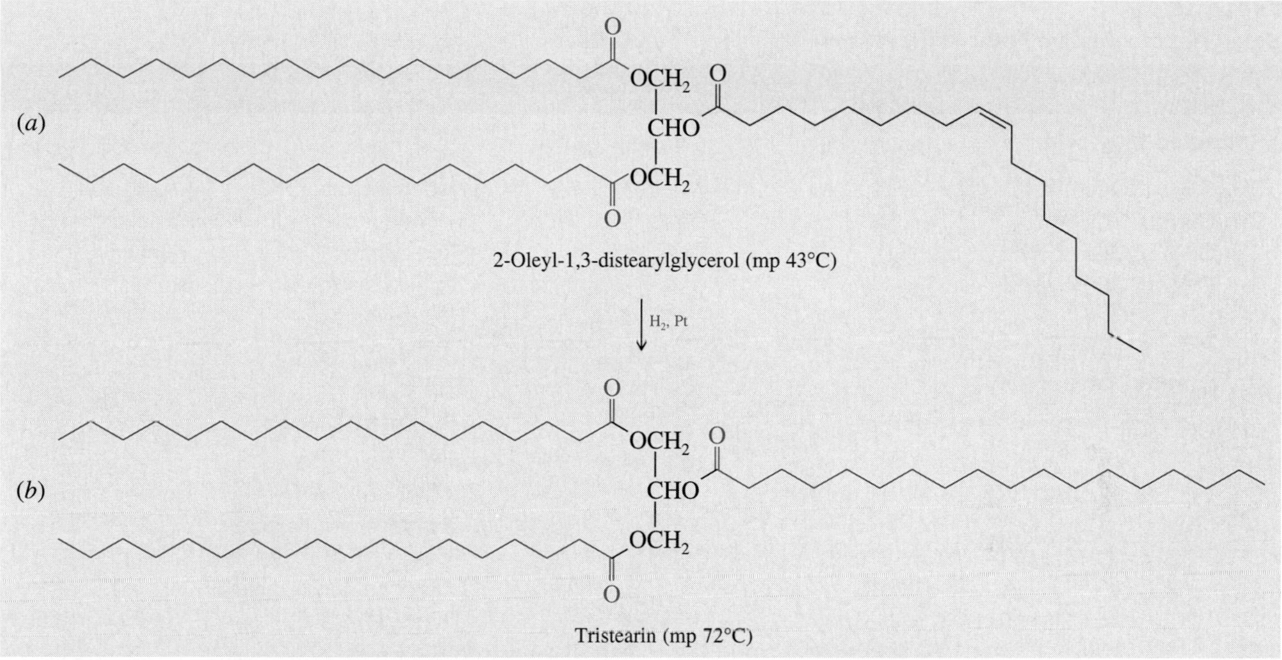

(a)

2-Oleyl-1,3-distearylglycerol (mp 43°C)

H_2, Pt

(b)

Tristearin (mp 72°C)

FIGURE 26.2 The structures of two typical triacylglycerols. (*a*) 2-Oleyl-1,3-distearylglycerol is a naturally occurring triacylglycerol found in cocoa butter. The cis double bond of its oleyl group gives the molecule a shape that interferes with efficient crystal packing. (*b*) Catalytic hydrogenation converts 2-oleyl-1,3-distearylglycerol to tristearin. Tristearin has a higher melting point than 2-oleyl-1,3-distearylglycerol.

presence of a cis double bond in the oleyl side chain interferes with the ability of neighboring triacylglycerol molecules to pack together into a crystal lattice, and this substance is a relatively low-melting solid (mp 43°C). Hydrogenation of the double bond converts 2-oleyl-1,3-distearylglycerol to *tristearin* (Figure 26.2*b*). Tristearin has a higher melting point (mp 72°C) than 2-oleyl-1,3-distearylglycerol because all three of its acyl groups can adopt the extended zigzag conformation that permits efficient crystal packing. Catalytic hydrogenation is commonly used in the food industry to transform readily available liquid vegetable oils to solid "shortenings."

Hydrolysis of fats yields glycerol and long-chain **fatty acids.** Thus, tristearin gives glycerol and three molecules of stearic acid (octadecanoic acid) on hydrolysis. Table 26.1 lists a few representative fatty acids. As the examples in Table 26.1 indicate, most naturally occurring fatty acids possess an even number of carbon atoms and an unbranched carbon chain. The carbon chain may be completely saturated or may incorporate one or more multiple bonds. When double bonds are present, they almost always have the cis (or *Z*) configuration. Acyl groups containing 14 to 20 carbon atoms are the most abundant in triacylglycerols.

Strictly speaking, the term *fatty acid* is restricted to those carboxylic acids that occur naturally in triacylglycerols. Many chemists and biochemists, however, refer to all unbranched carboxylic acids, irrespective of their origin and chain length, as *fatty acids.*

PROBLEM 26.1 What fatty acids are produced on hydrolysis of 2-oleyl-1,3-distearylglycerol? What other triacylglycerol gives the same fatty acids and in the same proportions as 2-oleyl-1,3-distearylglycerol?

TABLE 26.1
Some Representative Fatty Acids

Structural formula	Systematic name	Common name
Saturated fatty acids		
$CH_3(CH_2)_{10}COOH$	Dodecanoic acid	Lauric acid
$CH_3(CH_2)_{12}COOH$	Tetradecanoic acid	Myristic acid
$CH_3(CH_2)_{14}COOH$	Hexadecanoic acid	Palmitic acid
$CH_3(CH_2)_{16}COOH$	Octadecanoic acid	Stearic acid
$CH_3(CH_2)_{18}COOH$	Icosanoic acid	Arachidic acid
Unsaturated fatty acids		
$CH_3(CH_2)_7CH{=}CH(CH_2)_7COOH$	(*Z*)-9-Octadecenoic acid	Oleic acid
$CH_3(CH_2)_4CH{=}CHCH_2CH{=}CH(CH_2)_7COOH$	(9*Z*,12*Z*)-9,12-Octadecadienoic acid	Linoleic acid
$CH_3CH_2CH{=}CHCH_2CH{=}CHCH_2CH{=}CH(CH_2)_7COOH$	(9*Z*,12*Z*,15*Z*)-9,12,15-Octadecatrienoic acid	Linolenic acid
$CH_3(CH_2)_4CH{=}CHCH_2CH{=}CHCH_2CH{=}CHCH_2CH{=}CH(CH_2)_3COOH$	(5*Z*,8*Z*,11*Z*,14*Z*)-5,8,11,14-Icosatetraenoic acid	Arachidonic acid

In the form of triacylglycerols, fatty acids are found in both plants and animals, where they are biosynthesized from acetate by way of acetyl coenzyme A. The following section outlines the mechanism of fatty acid biosynthesis.

26.3 FATTY ACID BIOSYNTHESIS

The major elements of fatty acid biosynthesis may be described by considering the formation of butanoic acid from two molecules of acetyl coenzyme A. The ''machinery'' responsible for accomplishing this conversion is a complex of enzymes known as **fatty acid synthetase.** Certain portions of this complex, referred to as **acyl carrier protein (ACP),** bear a side chain that is structurally similar to coenzyme A. An important early step in fatty acid biosynthesis is the transfer of the acetyl group from a molecule of acetyl coenzyme A to the sulfhydryl group of acyl carrier protein.

$$\underset{\text{Acetyl coenzyme A}}{CH_3\overset{O}{\overset{\|}{C}}SCoA} + \underset{\text{Acyl carrier protein}}{HS{-}ACP} \longrightarrow \underset{S\text{-Acetyl acyl carrier protein}}{CH_3\overset{O}{\overset{\|}{C}}S{-}ACP} + \underset{\text{Coenzyme A}}{HSCoA}$$

PROBLEM 26.2 Using HSCoA and HS—ACP as abbreviations for coenzyme A and acyl carrier protein, respectively, write a structural formula for the tetrahedral intermediate in the preceding reaction.

A second molecule of acetyl coenzyme A reacts with carbon dioxide (actually bicarbonate ion at biological pH) to give malonyl coenzyme A:

$$\underset{\text{Acetyl coenzyme A}}{CH_3\overset{O}{\overset{\|}{C}}SCoA} + \underset{\text{Bicarbonate}}{HCO_3^-} \longrightarrow \underset{\text{Malonyl coenzyme A}}{{}^-O\overset{O}{\overset{\|}{C}}CH_2\overset{O}{\overset{\|}{C}}SCoA} + \underset{\text{Water}}{H_2O}$$

Formation of malonyl coenzyme A is followed by a nucleophilic acyl substitution, which transfers the malonyl group to the acyl carrier protein as a thioester.

$$\underset{\text{Malonyl coenzyme A}}{{}^-O\overset{O}{\overset{\|}{C}}CH_2\overset{O}{\overset{\|}{C}}SCoA} + \underset{\text{Acyl carrier protein}}{HS—ACP} \longrightarrow \underset{\text{S-Malonyl acyl carrier protein}}{{}^-O\overset{O}{\overset{\|}{C}}CH_2\overset{O}{\overset{\|}{C}}S—ACP} + \underset{\text{Coenzyme A}}{HSCoA}$$

When both building block units are in place on the acyl carrier protein, carbon-carbon bond formation occurs between the α carbon atom of the malonyl group and the carbonyl carbon of the acetyl group. This is shown in step 1 of Figure 26.3. Carbon-carbon bond formation is accompanied by decarboxylation and produces a four-carbon acetoacetyl (3-oxobutanoyl) group bound to acyl carrier protein.

The acetoacetyl group is then transformed to a butanoyl group by the reaction sequence illustrated in steps 2 to 4 of Figure 26.3.

The four carbon atoms of the butanoyl group originate in two molecules of acetyl coenzyme A. Carbon dioxide assists the reaction but is not incorporated into the product. The same carbon dioxide that is used to convert one molecule of acetyl coenzyme A to malonyl coenzyme A is regenerated in the decarboxylation step that accompanies carbon-carbon bond formation.

Successive repetitions of the steps shown in Figure 26.3 give unbranched acyl groups having 6, 8, 10, 12, 14, and 16 carbon atoms. In each case, chain extension occurs by reaction with a malonyl group bound to the acyl carrier protein. Thus, the biosynthesis of the 16-carbon acyl group of hexadecanoic (palmitic) acid can be represented by the equation:

$$\underset{\text{S-Acetyl acyl carrier protein}}{CH_3\overset{O}{\overset{\|}{C}}S—ACP} + \underset{\text{S-Malonyl acyl carrier protein}}{7HO\overset{O}{\overset{\|}{C}}CH_2\overset{O}{\overset{\|}{C}}S—ACP} + \underset{\text{Reduced form of coenzyme}}{14\ NADPH} + \underset{\text{Hydronium ion}}{14\ H_3O^+} \longrightarrow$$

$$\underset{\text{S-Hexadecanoyl acyl carrier protein}}{CH_3(CH_2)_{14}\overset{O}{\overset{\|}{C}}S—ACP} + \underset{\text{Carbon dioxide}}{7CO_2} + \underset{\text{Acyl carrier protein}}{7HS—ACP} + \underset{\text{Oxidized form of coenzyme}}{14\ NADP^+} + \underset{\text{Water}}{21\ H_2O}$$

Step 1: An acetyl group is transferred to the α carbon atom of the malonyl group with evolution of the carbon dioxide. Presumably decarboxylation gives an enol, which attacks the acetyl group.

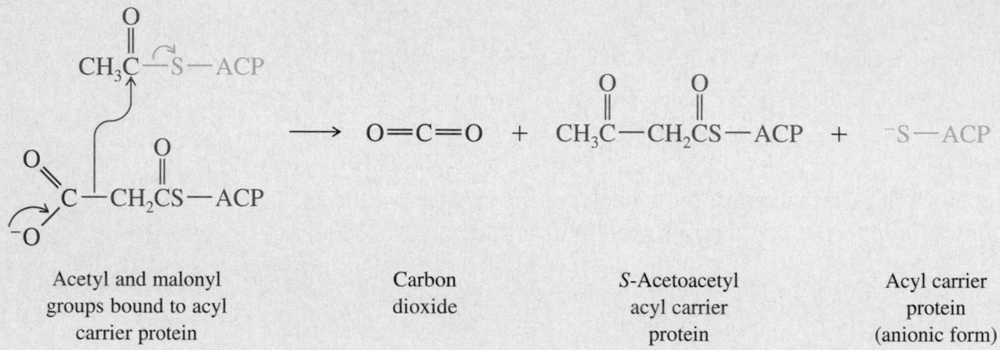

$$\underset{\text{Acetyl and malonyl groups bound to acyl carrier protein}}{CH_3\overset{O}{\overset{\|}{C}}-S-ACP \quad {}^-O(O=)C-CH_2\overset{O}{\overset{\|}{C}}S-ACP} \longrightarrow \underset{\text{Carbon dioxide}}{O=C=O} + \underset{\text{S-Acetoacetyl acyl carrier protein}}{CH_3\overset{O}{\overset{\|}{C}}-CH_2\overset{O}{\overset{\|}{C}}S-ACP} + \underset{\text{Acyl carrier protein (anionic form)}}{{}^-S-ACP}$$

Step 2: The ketone carbonyl of the acetoacetyl group is reduced to an alcohol function. This reduction requires NADPH as a coenzyme. (NADPH is the phosphate ester of NADH and reacts similarly to it.)

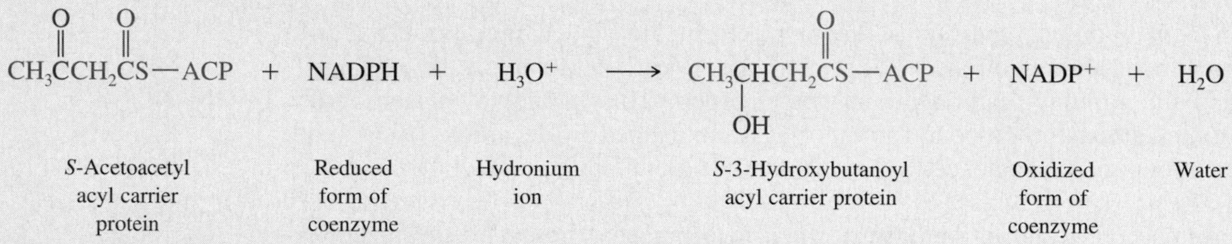

$$\underset{\text{S-Acetoacetyl acyl carrier protein}}{CH_3\overset{O}{\overset{\|}{C}}CH_2\overset{O}{\overset{\|}{C}}S-ACP} + \underset{\text{Reduced form of coenzyme}}{NADPH} + \underset{\text{Hydronium ion}}{H_3O^+} \longrightarrow \underset{\text{S-3-Hydroxybutanoyl acyl carrier protein}}{CH_3\underset{OH}{\underset{|}{C}}HCH_2\overset{O}{\overset{\|}{C}}S-ACP} + \underset{\text{Oxidized form of coenzyme}}{NADP^+} + \underset{\text{Water}}{H_2O}$$

Step 3: Dehydration of the β-hydroxy acyl group.

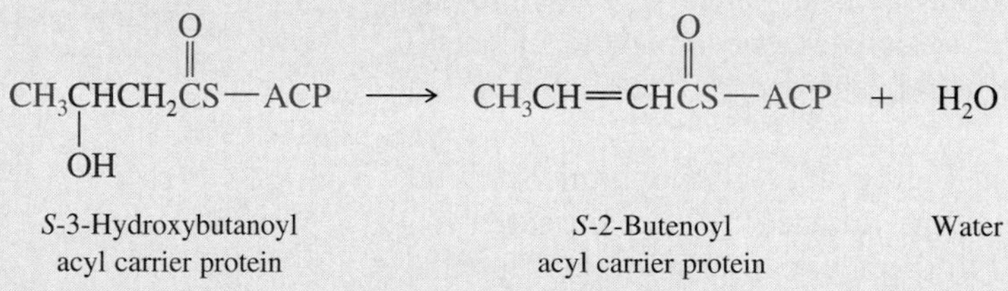

$$\underset{\text{S-3-Hydroxybutanoyl acyl carrier protein}}{CH_3\underset{OH}{\underset{|}{C}}HCH_2\overset{O}{\overset{\|}{C}}S-ACP} \longrightarrow \underset{\text{S-2-Butenoyl acyl carrier protein}}{CH_3CH=CH\overset{O}{\overset{\|}{C}}S-ACP} + \underset{\text{Water}}{H_2O}$$

Step 4: Reduction of the double bond of the α,β-unsaturated acyl group. This step requires NADPH as a coenzyme.

$$\underset{\text{S-2-Butenoyl acyl carrier protein}}{CH_3CH=CH\overset{O}{\overset{\|}{C}}S-ACP} + \underset{\text{Reduced form of coenzyme}}{NADPH} + \underset{\text{Hydronium ion}}{H_3O^+} \longrightarrow \underset{\text{S-Butanoyl acyl carrier protein}}{CH_3CH_2CH_2\overset{O}{\overset{\|}{C}}S-ACP} + \underset{\text{Oxidized form of coenzyme}}{NADP^+} + \underset{\text{Water}}{H_2O}$$

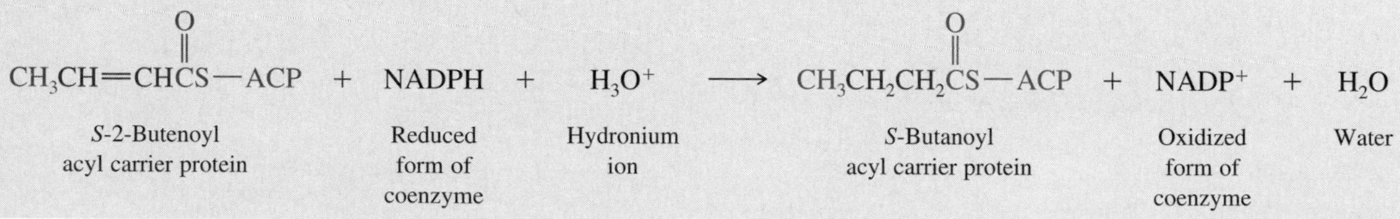

FIGURE 26.3 Steps in the formation of a butanoyl group from acetyl and malonyl building blocks.

PROBLEM 26.3 By analogy to the intermediates given in steps 1 to 4 of Figure 26.3, write the sequence of acyl groups that are attached to the acyl carrier protein in the conversion of $CH_3(CH_2)_{12}\overset{O}{\overset{\|}{C}}S{-}ACP$ to $CH_3(CH_2)_{14}\overset{O}{\overset{\|}{C}}S{-}ACP$.

This phase of fatty acid biosynthesis concludes with the transfer of the acyl group from acyl carrier protein to coenzyme A. The resulting acyl coenzyme A molecules can then undergo a number of subsequent biological transformations. One such transformation is chain extension, leading to acyl groups with more than 16 carbons. Another is the introduction of one or more carbon-carbon double bonds. A third is acyl transfer from sulfur to oxygen to form esters such as triacylglycerols. The process by which acyl coenzyme A molecules are converted to triacylglycerols involves a type of intermediate called a *phospholipid* and is discussed in the following section.

26.4 PHOSPHOLIPIDS

Triacylglycerols arise, not by acyl transfer to glycerol itself, but by a sequence of steps in which the first stage is acyl transfer to L-glycerol 3-phosphate (from reduction of dihydroxyacetone 3-phosphate, formed as described in Section 25.21). The product of this stage is called a **phosphatidic acid.**

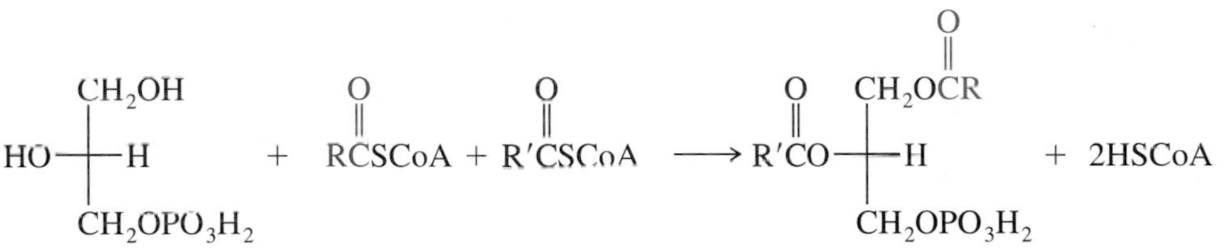

$$\begin{array}{c}CH_2OH\\ HO{-}\overset{|}{\underset{|}{C}}{-}H\\ CH_2OPO_3H_2\end{array} + R\overset{O}{\overset{\|}{C}}SCoA + R'\overset{O}{\overset{\|}{C}}SCoA \longrightarrow \begin{array}{c}CH_2O\overset{O}{\overset{\|}{C}}R\\ R'\overset{O}{\overset{\|}{C}}O{-}\overset{|}{\underset{|}{C}}{-}H\\ CH_2OPO_3H_2\end{array} + 2HSCoA$$

L-Glycerol 3-phosphate; Two acyl coenzyme A molecules (R and R′ may be the same or they may be different); Phosphatidic acid; Coenzyme A

PROBLEM 26.4 What is the absolute configuration (*R* or *S*) of L-glycerol 3-phosphate? What must be the absolute configuration of the naturally occurring phosphatidic acids biosynthesized from it?

Hydrolysis of the phosphate ester function of the phosphatidic acid gives a diacylglycerol, which then reacts with a third acyl coenzyme A molecule to produce a triacylglycerol.

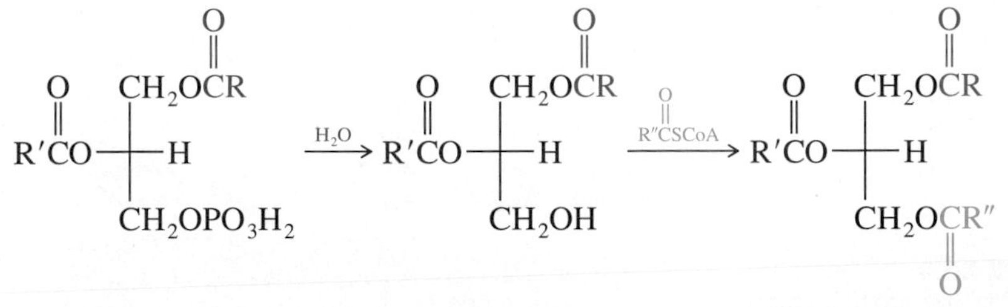

$$\begin{array}{c}CH_2O\overset{O}{\overset{\|}{C}}R\\ R'\overset{O}{\overset{\|}{C}}O{-}\overset{|}{\underset{|}{C}}{-}H\\ CH_2OPO_3H_2\end{array} \xrightarrow{H_2O} \begin{array}{c}CH_2O\overset{O}{\overset{\|}{C}}R\\ R'\overset{O}{\overset{\|}{C}}O{-}\overset{|}{\underset{|}{C}}{-}H\\ CH_2OH\end{array} \xrightarrow{R''\overset{O}{\overset{\|}{C}}SCoA} \begin{array}{c}CH_2O\overset{O}{\overset{\|}{C}}R\\ R'\overset{O}{\overset{\|}{C}}O{-}\overset{|}{\underset{|}{C}}{-}H\\ CH_2O\underset{\underset{O}{\|}}{C}R''\end{array}$$

Phosphatidic acid; Diacylglycerol; Triacylglycerol

Phosphatidic acids not only are intermediates in the biosynthesis of triacylglycerols but also are biosynthetic precursors of other members of a group of compounds called **phosphoglycerides** or **glycerol phosphatides.** Phosphorus-containing derivatives of lipids are known as **phospholipids,** and phosphoglycerides are one type of phospholipid.

Lecithin is added to foods such as mayonnaise as an emulsifying agent to prevent the fat and water from separating into two layers.

One important phospholipid is **phosphatidylcholine,** also called *lecithin.* Phosphatidylcholine is a mixture of diesters of phosphoric acid. One ester function is derived from a diacylglycerol, whereas the other is a choline [$-OCH_2CH_2\overset{+}{N}(CH_3)_3$] unit.

$$\begin{array}{l} \qquad\qquad CH_2OC(=O)R \\ R'C(=O)O-C-H \\ \qquad\qquad CH_2OPO_2^- \\ \qquad\qquad\qquad OCH_2CH_2\overset{+}{N}(CH_3)_3 \end{array}$$

Phosphatidylcholine
(R and R′ are usually different)

Phosphatidylcholine possesses a polar "head group" (the positively charged choline and negatively charged phosphate units) and two nonpolar "tails" (the acyl groups). Under certain conditions, such as at the interface of two aqueous phases, phosphatidylcholine forms what is called a *lipid bilayer,* as shown in Figure 26.4. Because there are two long-chain acyl groups in each molecule, the most stable assembly has the polar groups solvated by water molecules at the top and bottom surfaces and the lipophilic acyl groups directed toward the interior of the bilayer.

Phosphatidylcholine is one of the principal components of cell membranes. It is

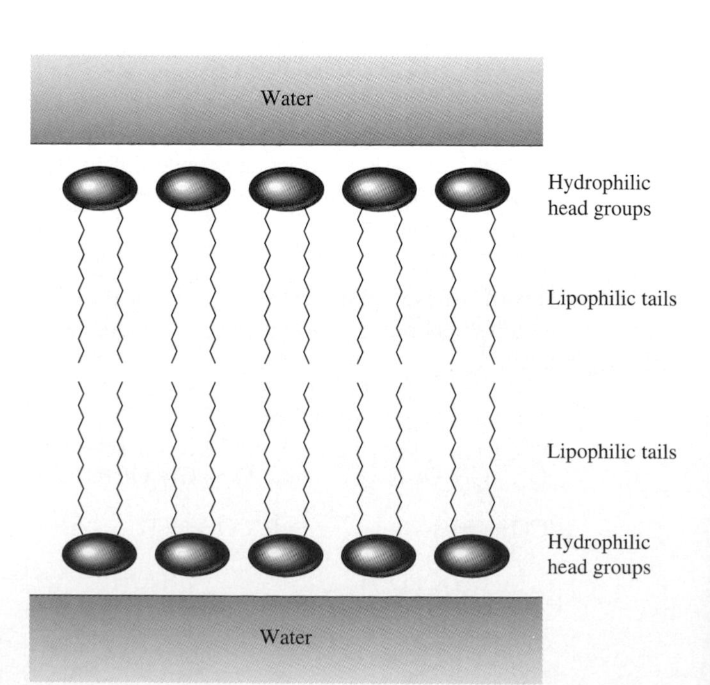

FIGURE 26.4 Schematic drawing of the cross section of a phospholipid bilayer.

believed that these membranes are composed of lipid bilayers analogous to those of Figure 26.4. Nonpolar materials can diffuse through the bilayer from one side to the other relatively easily; polar materials, particularly metal ions such as Na^+, K^+, and Ca^{2+}, cannot. The transport of metal ions through a membrane is usually assisted by certain proteins present in the lipid bilayer, which contain a metal ion binding site surrounded by a lipophilic exterior. The metal ion is picked up at one side of the lipid bilayer and delivered at the other, surrounded at all times by a polar environment on its passage through the hydrocarbonlike interior of the membrane. Ionophore antibiotics such as monensin (Section 16.4) disrupt the normal functioning of cells by facilitating metal ion transport across cell membranes.

26.5 WAXES

Waxes are solid materials that make up part of the protective coatings of a number of living things, including the leaves of plants, the fur of animals, and the feathers of birds. They are usually mixtures of esters in which both the alkyl and acyl group are unbranched and contain a dozen or more carbon atoms. Beeswax, for example, contains the ester triacontyl hexadecanoate as one component of a complex mixture of hydrocarbons, alcohols, and esters.

$$CH_3(CH_2)_{14}\overset{O}{\overset{\|}{C}}OCH_2(CH_2)_{28}CH_3$$

Triacontyl hexadecanoate

PROBLEM 26.5 Spermaceti is a wax obtained from the sperm whale. It contains, among other materials, an ester known as *cetyl palmitate,* which is used as an emollient in a number of soaps and cosmetics. The systematic name for cetyl palmitate is *hexadecyl hexadecanoate.* Write a structural formula for this substance.

Fatty acids normally occur naturally as esters; fats, oils, phospholipids, and waxes all are unique types of fatty acid esters. There is, however, an important class of fatty acid derivatives that exists and carries out its biological role in the form of the free acid. This class of fatty acid derivatives is described in the following section.

26.6 PROSTAGLANDINS

Research in physiology carried out in the 1930s established that the lipid fraction of semen contains small amounts of substances that exert powerful effects on smooth muscle. Sheep prostate glands proved to be a convenient source of this material and yielded a mixture of structurally related substances referred to collectively as **prostaglandins.** We now know that prostaglandins are present in almost all animal tissues, where they carry out a variety of regulatory functions.

Prostaglandins are extremely potent substances and exert their physiological effects at very small concentrations. Because of this, their isolation was difficult, and it was not until 1960 that the first members of this class, designated PGE_1 and $PGF_{1\alpha}$ (Figure 26.5), were obtained as pure compounds. More than a dozen structurally related prostaglandins have since been isolated and identified. All the prostaglandins are 20-car-

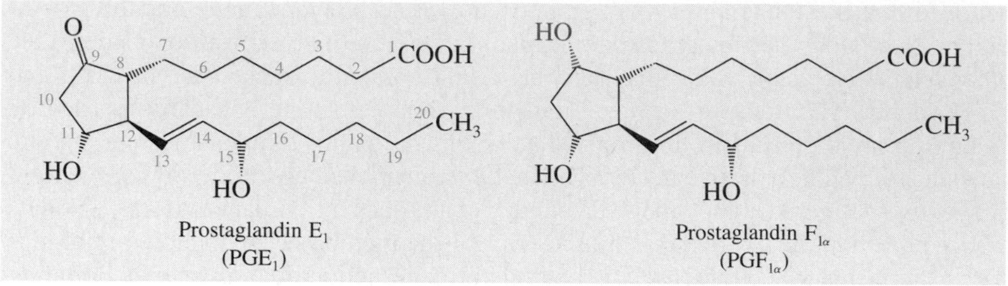

Prostaglandin E_1 (PGE_1)

Prostaglandin $F_{1\alpha}$ ($PGF_{1\alpha}$)

FIGURE 26.5 Structures of two representative prostaglandins. The numbering scheme is illustrated in the structure of PGE_1.

bon carboxylic acids and contain a cyclopentane ring. All have hydroxyl groups at C-11 and C-15 (for the numbering of the positions in prostaglandins, see Figure 26.5). Prostaglandins belonging to the F series have an additional hydroxyl group at C-9, while a carbonyl function is present at this position in the various PGEs. The subscript numerals in their abbreviated names indicate the number of double bonds.

Arachidonic acid gets its name from *arachidic acid,* the saturated C_{20} fatty acid isolated from peanut *(Arachis hypogaea)* oil.

Prostaglandins are believed to arise from unsaturated C_{20}-carboxylic acids such as arachidonic acid (Table 26.1). Mammals cannot biosynthesize arachidonic acid directly. They obtain linoleic acid (Table 26.1) from vegetable oils in their diet and extend the carbon chain of linoleic acid from 18 to 20 carbons while introducing two more double bonds. Linoleic acid is said to be an **essential fatty acid,** forming part of the dietary requirement of mammals. Animals fed on diets that are deficient in linoleic acid grow poorly and suffer a number of other disorders, some of which are reversed on feeding them vegetable oils rich in linoleic acid and other *polyunsaturated fatty acids.* One function of these substances is to provide the raw materials for prostaglandin biosynthesis.

PROBLEM 26.6 Arachidonic acid is the biosynthetic precursor to PGE_2. The structures of PGE_1 (Figure 26.5) and PGE_2 are identical except that PGE_2 has one more double bond than PGE_1. Suggest a reasonable structure for PGE_2.

Physiological responses to prostaglandins encompass a variety of effects. Some prostaglandins relax bronchial muscle, others contract it. Some stimulate uterine contractions and have been used to induce therapeutic abortions. PGE_1 dilates blood vessels and lowers blood pressure; it inhibits the aggregation of platelets and offers promise as a drug to reduce the formation of blood clots.

The long-standing question of the mode of action of aspirin has been addressed in terms of its effects on prostaglandin biosynthesis. Prostaglandin biosynthesis is the body's response to tissue damage and is manifested by pain and inflammation at the affected site. Aspirin has been shown to inhibit the activity of an enzyme required for prostaglandin formation. Aspirin reduces pain and inflammation—and probably fever as well—by reducing prostaglandin levels in the body.

Much of the fundamental work on prostaglandins and related compounds was carried out by Sune Bergstrom and Bengt Samuelsson of the Karolinska Institute (Sweden) and by John Vane of the Wellcome Foundation (Great Britain). These three shared the Nobel Prize for physiology or medicine in 1982. Bergstrom began his research on prostaglandins because he was interested in the oxidation of fatty acids. That research led to the identification of a whole new class of biochemical mediators. Prostaglandin research has now revealed that other derivatives of oxidized polyunsaturated fatty acids, structurally distinct from the prostaglandins, are also physiologically important. These fatty acid derivatives include, for example, a group of substances known as the **leukotrienes,** which have been implicated as mediators in immunological processes.

26.7 TERPENES. THE ISOPRENE RULE

The word *essential* as applied to naturally occurring organic substances can have two different meanings. For example, as used in the previous section with respect to fatty acids, *essential* means "necessary." Linoleic acid is an "essential" fatty acid; it must be included in the diet in order for animals to grow properly because they lack the ability to biosynthesize it directly.

Essential is also used as the adjective form of the noun *essence.* The mixtures of substances that make up the fragrant material of plants are called *essential oils* because they contain the essence, i.e., the odor, of the plant. The study of the composition of essential oils ranks as one of the oldest areas of organic chemical research. Very often, the principal volatile component of an essential oil belongs to a class of chemical substances called the **terpenes.**

Myrcene, a hydrocarbon isolated from bayberry oil, is a typical terpene:

$$(CH_3)_2C{=}CHCH_2CH_2C({=}CH_2)CH{=}CH_2$$

Myrcene

The structural feature that distinguishes terpenes from other natural products is the **isoprene unit.** The carbon skeleton of myrcene (exclusive of its double bonds) corresponds to the head-to-tail union of two isoprene units.

$$CH_2{=}C(CH_3){-}CH{=}CH_2$$

tail

head

Isoprene
(2-methyl-1,3-butadiene)

Two isoprene units
linked head to tail

Terpenes are often referred to as *isoprenoid* compounds. They are classified according to the number of carbon atoms they contain, as summarized in Table 26.2.

TABLE 26.2
Classification of Terpenes

Class	Number of carbon atoms
Monoterpene	10
Sesquiterpene	15
Diterpene	20
Sesterpene	25
Triterpene	30
Tetraterpene	40

Monoterpenes

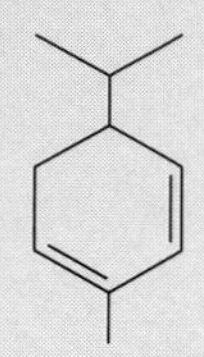

α-Phellandrene
(eucalyptus)

Menthol
(peppermint)

Citral
(lemon grass)

Sesquiterpenes

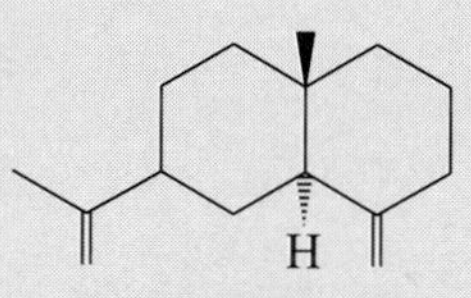

α-Selinene
(celery)

Farnesol
(ambrette)

Abscisic acid
(a plant homone)

Diterpenes

Cembrene
(pine)

Vitamin A
(present in mammalian tissue and fish oil;
important substance in the chemistry of vision)

Triterpenes

Squalene
(shark liver oil)

Tetraterpenes

β-Carotene
(present in carrots and other vegetables;
enzymes in the body cleave β-carotene to vitamin A)

FIGURE 26.6 Some representative terpenes and related natural products. Structures are customarily depicted as carbon skeleton formulas when describing compounds of isoprenoid origin.

While the term *terpene* once referred only to hydrocarbons, current usage includes functionally substituted derivatives as well. Figure 26.6 presents the structural formulas for a number of representative terpenes. The isoprene units in some of these are relatively easy to identify. The three isoprene units in the sesquiterpene **farnesol,** for example, are indicated as follows in color. They are joined in a head-to-tail fashion.

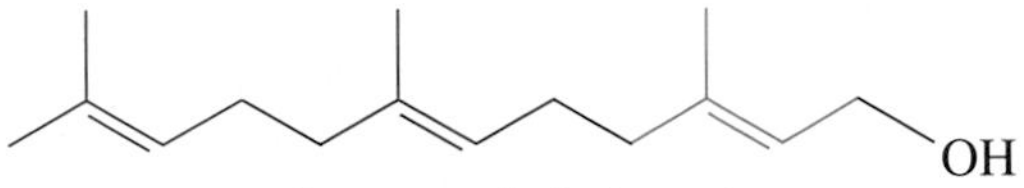

OH

Isoprene units in farnesol

Many terpenes contain one or more rings, but these also can be viewed as collections of isoprene units. An example is ***α*-selinene.** Like farnesol, it is made up of three isoprene units linked head to tail.

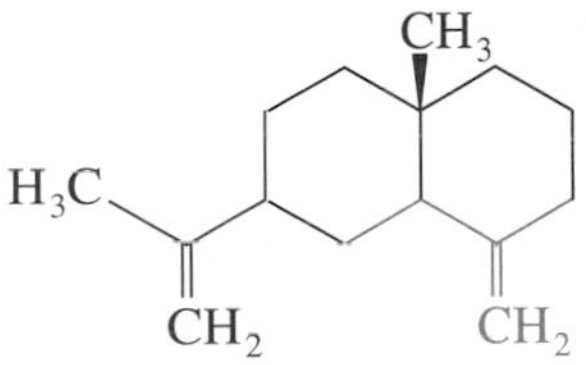

CH_3 H_3C CH_2 CH_2

Isoprene units in α-selinene

PROBLEM 26.7 Locate the isoprene units in each of the monoterpenes, sesquiterpenes, and diterpenes shown in Figure 26.6. (In some cases there are two equally correct arrangements.)

Tail-to-tail linkages of isoprene units sometimes occur, especially in the higher terpenes. The C(12)—C(13) bond of squalene unites two C_{15} units in a tail-to-tail manner. (Notice, however, that isoprene units are joined head to tail within each C_{15} unit of squalene.)

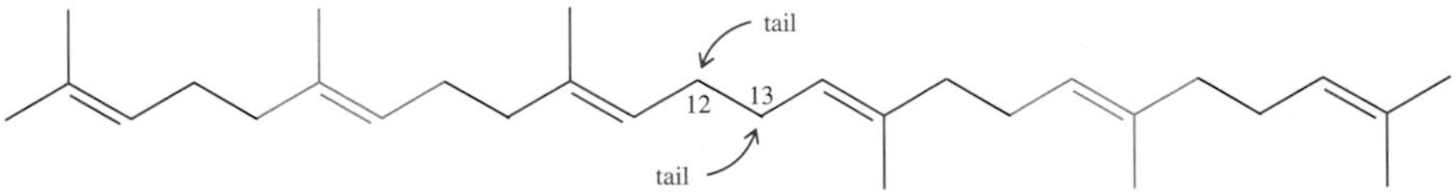

tail 12 13 tail

Isoprene units in squalene

PROBLEM 26.8 Identify the isoprene units in β-carotene (Figure 26.6). Which carbons are joined by a tail-to-tail link between isoprene units?

The German chemist Otto Wallach (Nobel Prize in chemistry, 1910) established the structures of many monoterpenes and is credited with recognizing that they can be viewed as collections of isoprene units. Leopold Ruzicka of the Swiss Federal Institute of Technology (Zürich), in his studies of sesquiterpenes and higher terpenes, extended and refined what we now know as the **isoprene rule.** He was a corecipient of the Nobel Prize in chemistry in 1939. While exceptions to it are known, the isoprene rule is a useful guide to terpene structures and has stimulated research in the biosynthetic origin of these compounds. It is a curious fact that terpenes contain isoprene units but isoprene does not occur naturally. What is the *biological isoprene unit,* how is it biosynthesized, and how do individual isoprene units combine to give terpenes?

26.8 ISOPENTENYL PYROPHOSPHATE: THE BIOLOGICAL ISOPRENE UNIT

Isoprenoid compounds are biosynthesized from acetate by a process that involves several stages. The first stage is the formation of *mevalonic acid* from three molecules of acetic acid:

$$3CH_3\overset{O}{\overset{\|}{C}}OH \xrightarrow{\text{several steps}} HO\overset{O}{\overset{\|}{C}}CH_2\underset{OH}{\underset{|}{\overset{CH_3}{\overset{|}{C}}}}CH_2CH_2OH$$

Acetic acid — Mevalonic acid

In the second stage, mevalonic acid is converted to *3-methyl-3-butenyl pyrophosphate* (*isopentenyl pyrophosphate*):

It is convenient to use the symbol —OPP to represent the pyrophosphate group

$$HO\overset{O}{\overset{\|}{C}}CH_2\underset{OH}{\underset{|}{\overset{CH_3}{\overset{|}{C}}}}CH_2CH_2OH \xrightarrow{\text{several steps}} CH_2{=}\overset{CH_3}{\overset{|}{C}}CH_2CH_2O\underset{HO}{\underset{|}{\overset{O}{\overset{\|}{P}}}}O\underset{OH}{\underset{|}{\overset{O}{\overset{\|}{P}}}}OH \equiv \text{(skeletal)}\ OPP$$

Mevalonic acid — Isopentenyl pyrophosphate

Isopentenyl pyrophosphate is the biological isoprene unit; it contains five carbon atoms connected in the same order as in isoprene.

Isopentenyl pyrophosphate undergoes an enzyme-catalyzed reaction that converts it, in an equilibrium process, to *3-methyl-2-butenyl pyrophosphate* (*dimethylallyl pyrophosphate*):

$$\text{OPP} \underset{-H^+}{\overset{H^+}{\rightleftharpoons}} \left[\overset{+}{\ }\ \text{OPP}\right] \underset{H^+}{\overset{-H^+}{\rightleftharpoons}} \text{OPP}$$

Isopentenyl pyrophosphate — Carbocation intermediate — Dimethylallyl pyrophosphate

Isopentenyl pyrophosphate and dimethylallyl pyrophosphate are structurally similar—both contain a double bond and a pyrophosphate ester unit—but the chemical reactivity expressed by each is different. The principal site of reaction in dimethylallyl pyrophosphate is the carbon that bears the pyrophosphate group. Pyrophosphate is a reasonably good leaving group in nucleophilic substitution reactions, especially when, as in dimethylallyl pyrophosphate, it is located at an allylic carbon.

Isopentenyl pyrophosphate, on the other hand, does not have its leaving group attached to an allylic carbon and is far less reactive than dimethylallyl pyrophosphate toward nucleophilic reagents. The principal site of reaction in isopentenyl pyrophosphate is the carbon-carbon double bond, which, like the double bonds of simple alkenes, is reactive toward electrophiles.

26.9 CARBON-CARBON BOND FORMATION IN TERPENE BIOSYNTHESIS

The chemical properties of isopentenyl pyrophosphate and dimethylallyl pyrophosphate are complementary in a way that permits them to react with each other to form a carbon-carbon bond that unites two isoprene units. Using the π electrons of its double bond, isopentenyl pyrophosphate acts as a nucleophile and displaces pyrophosphate from dimethylallyl pyrophosphate.

OPP OPP $\xrightarrow{-(^{-}OPP)}$ OPP

Dimethylallyl pyrophosphate Isopentenyl pyrophosphate Ten-carbon carbocation

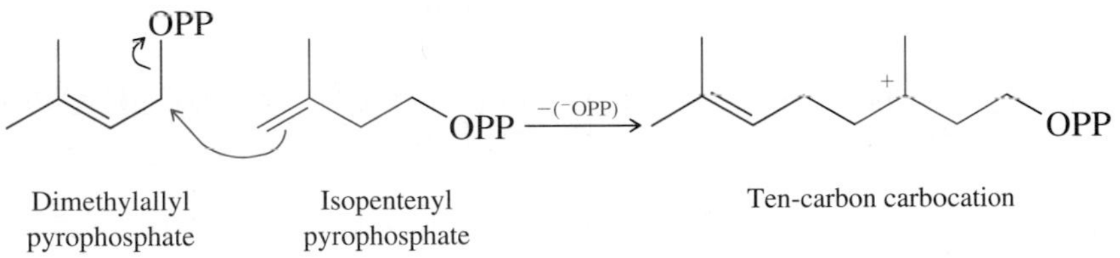

The tertiary carbocation formed in this step can react according to any of the various reaction pathways available to carbocations. One of these is loss of a proton to give a double bond.

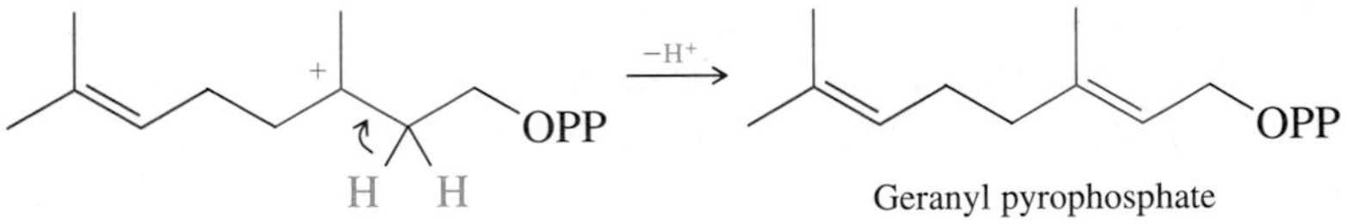

OPP H H $\xrightarrow{-H^+}$ OPP

Geranyl pyrophosphate

The product of this reaction is *geranyl pyrophosphate.* Hydrolysis of the pyrophosphate ester group gives *geraniol,* a naturally occurring monoterpene found in rose oil.

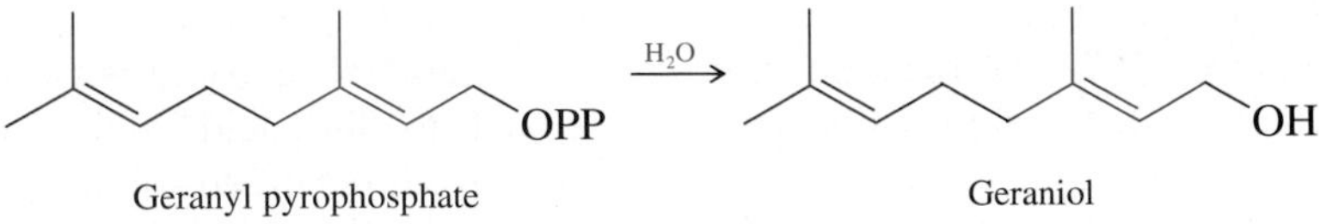

OPP $\xrightarrow{H_2O}$ OH

Geranyl pyrophosphate Geraniol

Geranyl pyrophosphate is an allylic pyrophosphate and, like dimethylallyl pyrophosphate, can act as an alkylating agent toward a molecule of isopentenyl pyrophosphate. A 15-carbon carbocation is formed, which, on deprotonation, gives *farnesyl pyrophosphate.*

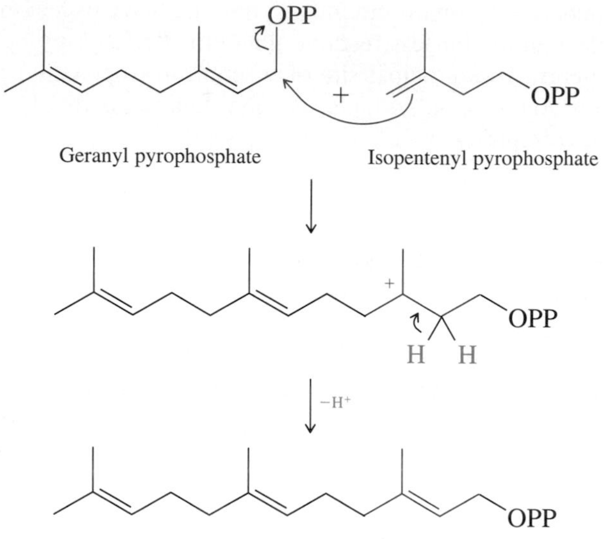

Geranyl pyrophosphate Isopentenyl pyrophosphate

Farnesyl pyrophosphate

Hydrolysis of the pyrophosphate ester group converts farnesyl pyrophosphate to the corresponding alcohol *farnesol* (see Figure 26.6 for the structure of farnesol).

A repetition of the process just shown produces the diterpene geranylgeraniol from farnesyl pyrophosphate.

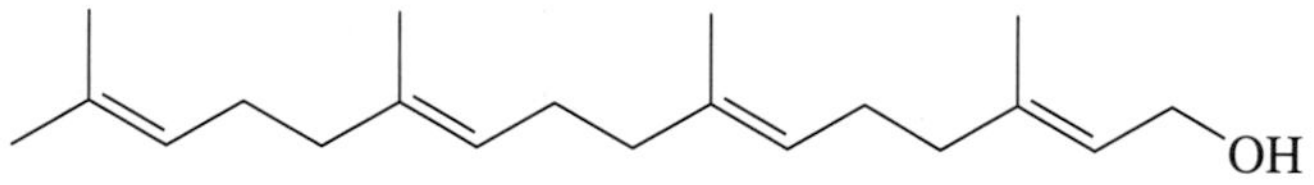

Geranylgeraniol

PROBLEM 26.9 Write a sequence of reactions that describes the formation of geranylgeraniol from farnesyl pyrophosphate.

The higher terpenes are formed not by successive addition of C_5 units but by the coupling of simpler terpenes. Thus, the triterpenes (C_{30}) are derived from two molecules of farnesyl pyrophosphate, and the tetraterpenes (C_{40}) from two molecules of geranylgeranyl pyrophosphate. These carbon-carbon bond–forming processes involve tail-to-tail couplings and proceed by a more complicated mechanism than that just described.

The enzyme-catalyzed reactions that lead to geraniol and farnesol (as their pyrophosphate esters) are mechanistically related to the acid-catalyzed dimerization of alkenes discussed in Section 6.21. The reaction of an allylic pyrophosphate or a carbocation with a source of π electrons is a recurring theme in terpene biosynthesis and is invoked to explain the origin of more complicated structural types. Consider, for example, the formation of cyclic monoterpenes. *Neryl pyrophosphate,* formed by an enzyme-catalyzed isomerization of the *E* double bond in geranyl pyrophosphate, has the proper geometry to form a six-membered ring via intramolecular attack of the double bond on the allylic pyrophosphate unit.

OPP → OPP → + $^{-}$OPP

Geranyl pyrophosphate Neryl pyrophosphate Tertiary carbocation

Loss of a proton from the tertiary carbocation formed in this step gives *limonene,* an abundant natural product found in many citrus fruits. Capture of the carbocation by water gives *α-terpineol,* also a known natural product.

$-H^+$ → Limonene

H_2O → HO α-Terpineol

The same tertiary carbocation serves as the precursor to numerous bicyclic monoterpenes. A carbocation having a bicyclic skeleton is formed by intramolecular attack of the π electrons of the double bond on the positively charged carbon.

Bicyclic carbocation

This bicyclic carbocation then undergoes many reactions typical of carbocation intermediates to provide a variety of bicyclic monoterpenes, as outlined in Figure 26.7.

PROBLEM 26.10 The structure of the bicyclic monoterpene borneol is shown in Figure 26.7. Isoborneol, a stereoisomer of borneol, can be prepared in the laboratory by a two-step sequence. In the first step, borneol is oxidized to camphor by treatment with chromic acid. In the second step, camphor is reduced with sodium borohydride to a mixture of 85% isoborneol and 15% borneol. On the basis of these transformations, deduce structural formulas for isoborneol and camphor.

Analogous processes involving cyclizations and rearrangements of carbocations derived from farnesyl pyrophosphate produce a rich variety of structural types in the sesquiterpene series. We will have more to say about the chemistry of higher terpenes, especially the triterpenes, later in this chapter. For the moment, however, let us return to smaller molecules in order to complete the picture of how isoprenoid compounds arise from acetate.

A. Loss of a proton from the bicyclic carbocation yields α-pinene and β-pinene. The pinenes are the most abundant of the monoterpenes. They are the main constituents of turpentine.

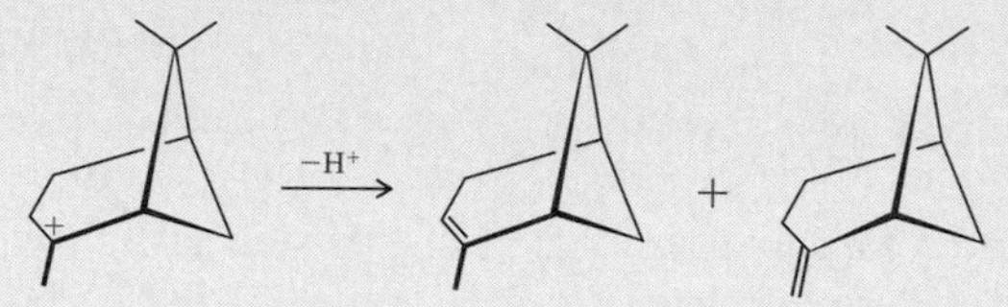

$-H^+$

α-Pinene β-Pinene

B. Capture of the carbocation by water, accompanied by rearrangement of the bicyclo[3.1.1] carbon skeleton to a bicyclo[2.2.1] unit, yields borneol. Borneol is found in the essential oil of certain trees that grow in Indonesia.

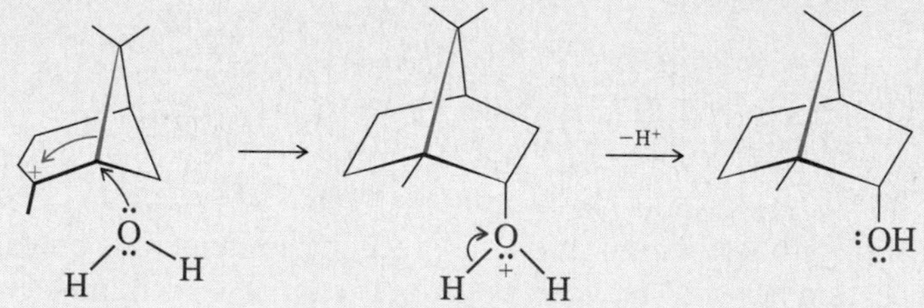

$-H^+$

Borneol

FIGURE 26.7 Two of the reaction pathways available to the C_{10} bicyclic carbocation formed from neryl pyrophosphate. The same carbocation can lead to monoterpenes based on either the bicyclo[3.1.1] or the bicyclo[2.2.1] carbon skeleton.

26.10 THE PATHWAY FROM ACETATE TO ISOPENTENYL PYROPHOSPHATE

The introduction to Section 26.8 pointed out that mevalonic acid is the biosynthetic precursor of isopentenyl pyrophosphate. The early steps in the biosynthesis of mevalonate from three molecules of acetic acid are analogous to those in fatty acid biosynthesis (Section 26.3) except that they do not involve acyl carrier protein. Thus, the reaction of acetyl coenzyme A with malonyl coenzyme A yields a molecule of acetoacetyl coenzyme A.

$$CH_3\overset{O}{\overset{\|}{C}}SCoA + {}^-O_2CCH_2\overset{O}{\overset{\|}{C}}SCoA \longrightarrow CH_3\overset{O}{\overset{\|}{C}}CH_2\overset{O}{\overset{\|}{C}}SCoA + CO_2$$

Acetyl coenzyme A; Malonyl coenzyme A; Acetoacetyl coenzyme A; Carbon dioxide

Carbon-carbon bond formation then occurs between the ketone carbonyl of acetoacetyl coenzyme A and the α carbon of a molecule of acetyl coenzyme A.

$$CH_3\overset{O}{\overset{\|}{C}}CH_2\overset{O}{\overset{\|}{C}}SCoA + CH_3\overset{O}{\overset{\|}{C}}SCoA \longrightarrow CH_3\overset{HO}{\overset{|}{\underset{CH_2COH}{\underset{|}{C}}}}CH_2\overset{O}{\overset{\|}{C}}SCoA + CoASH$$

(in the product, the CH_2COH group bears a C=O: $CH_2\overset{\|}{C}OH$ with O)

Acetoacetyl coenzyme A; Acetyl coenzyme A; β-Hydroxy-β-methylglutaryl coenzyme A (HMG CoA); Coenzyme A

The product of this reaction, known by the common name β-hydroxy-β-methylglutaryl coenzyme A (HMG CoA), has the carbon skeleton of mevalonic acid and is converted to it by enzymatic reduction.

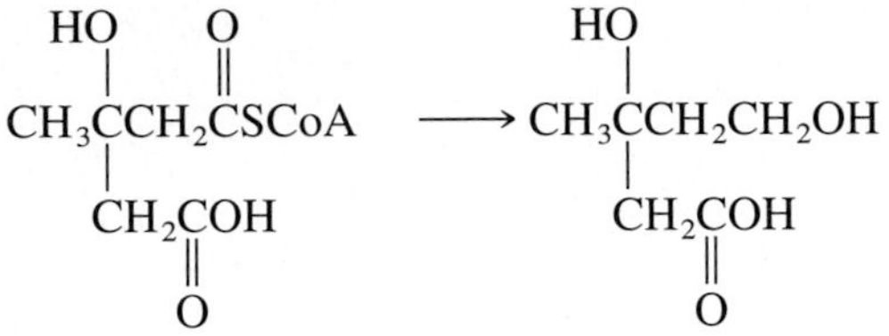

$$\mathrm{CH_3C(OH)(CH_2COOH)CH_2CSCoA} \longrightarrow \mathrm{CH_3C(OH)(CH_2COOH)CH_2CH_2OH}$$

β-Hydroxy-β-methylglutaryl coenzyme A (HMG CoA) — Mevalonic acid

In keeping with its biogenetic origin in three molecules of acetic acid, mevalonic acid has six carbon atoms. The conversion of mevalonate to isopentenyl pyrophosphate involves loss of the ''extra'' carbon as carbon dioxide. First, the alcohol hydroxyl groups of mevalonate are converted to phosphate ester functions—they are enzymatically *phosphorylated,* with introduction of a simple phosphate at the tertiary site and a pyrophosphate at the primary site. Decarboxylation, in concert with loss of the tertiary phosphate, introduces a carbon-carbon double bond and gives isopentenyl pyrophosphate, the fundamental building block for formation of isoprenoid natural products.

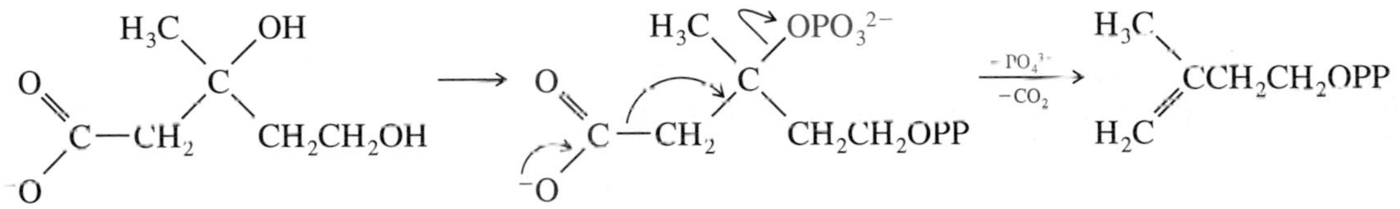

$$\mathrm{^-O_2CCH_2C(CH_3)(OH)CH_2CH_2OH} \longrightarrow \mathrm{^-O_2CCH_2C(CH_3)(OPO_3^{2-})CH_2CH_2OPP} \xrightarrow[-CO_2]{-PO_4^{3-}} \mathrm{H_2C{=}C(CH_3)CH_2CH_2OPP}$$

Mevalonate — (Unstable; undergoes rapid decarboxylation with loss of phosphate) — Isopentenyl pyrophosphate

Much of what we know concerning the pathway from acetate to mevalonate to isopentenyl pyrophosphate to terpenes comes from ''feeding'' experiments, in which plants are grown in the presence of radioactively labeled organic substances and the distribution of the radioactive label is determined in the products of biosynthesis. To illustrate, eucalyptus plants were allowed to grow in a medium containing acetic acid enriched with ^{14}C in its methyl group. *Citronellal* was isolated from the mixture of monoterpenes produced by the plants and shown, by a series of chemical degradations, to contain the radioactive ^{14}C label at carbons 2, 4, 6, and 8, as well as at the carbons of both branching methyl groups.

Citronellal occurs naturally as the principal component of citronella oil and is used as an insect repellent.

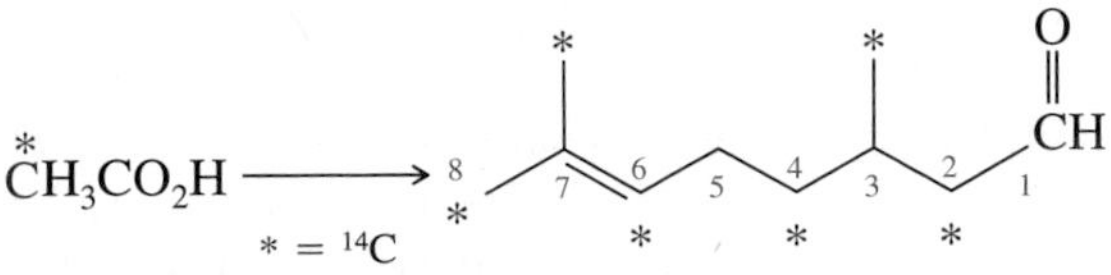

$$\overset{*}{\mathrm{C}}\mathrm{H_3CO_2H} \longrightarrow \text{Citronellal (labeled at C-2, C-4, C-6, C-8 and both branching methyls)}$$

$* = {}^{14}C$

Figure 26.8 traces the ^{14}C label from its origin in acetic acid to its experimentally determined distribution in citronellal.

$$*CH_3CO_2H \longrightarrow *CH_3\overset{O}{\overset{\|}{C}}\overset{*}{C}H_2\overset{O}{\overset{\|}{C}}SCoA \longrightarrow *CH_3\overset{OH}{\overset{|}{C}}\overset{*}{C}H_2\overset{O}{\overset{\|}{C}}SCoA\,(*CH_2CO_2H) \longrightarrow *CH_3\overset{OPO_3^{2-}}{\overset{|}{C}}\overset{*}{C}H_2CH_2OPP\,(*CH_2CO_2^-) \longrightarrow$$

FIGURE 26.8 Diagram showing the distribution of the ^{14}C label in citronellal biosynthesized from acetate in which the methyl carbon was isotopically enriched with ^{14}C.

PROBLEM 26.11 How many carbon atoms of citronellal would be radioactively labeled if the acetic acid used in the experiment were enriched with ^{14}C at C-1 instead of at C-2? Identify these carbon atoms.

A more recent experimental technique employs ^{13}C as the isotopic label. Instead of locating the position of a ^{14}C label by a laborious degradation procedure, the ^{13}C nmr spectrum of the natural product is recorded. The signals for the carbons that are enriched in ^{13}C are far more intense than those corresponding to carbons in which ^{13}C is present only at the natural abundance level.

Isotope incorporation experiments have demonstrated the essential correctness of the scheme presented in this and preceding sections for terpene biosynthesis. Considerable effort has been expended toward its detailed elaboration because of the common biosynthetic origin of terpenes and another class of acetate-derived natural products, the steroids.

26.11 STEROIDS. CHOLESTEROL

Cholesterol is the central compound in any discussion of steroids. Its name is a combination of the Greek words for ''bile'' (*chole*) and ''solid'' (*stereos*) preceding the characteristic alcohol suffix *-ol*. It is the most abundant steroid present in humans and the most important one as well, since all other steroids arise from it. An average adult has over 200 g of cholesterol; it is found in almost all body tissues, with relatively large amounts present in the brain and spinal cord and in gallstones. Cholesterol is the chief constituent of the plaque that builds up on the walls of arteries and restricts the flow of blood in the circulatory disorder known as *atherosclerosis*.

Cholesterol was isolated in the eighteenth century, but its structure is so complex that its correct constitution was not determined until 1932 and its stereochemistry not verified until 1955. Steroids are characterized by the tetracyclic ring system shown in Figure 26.9*a*. As shown in Figure 26.9*b*, cholesterol contains this tetracyclic skeleton modified to include an alcohol function at C-3, a double bond at C-5, methyl groups at C-10 and C-13, and a C_8H_{17} side chain at C-17. Isoprene units may be discerned in various portions of the cholesterol molecule, but the overall correspondence with the isoprene rule is far from perfect. Indeed, cholesterol has only 27 carbon atoms, three too few for it to be classed as a triterpene.

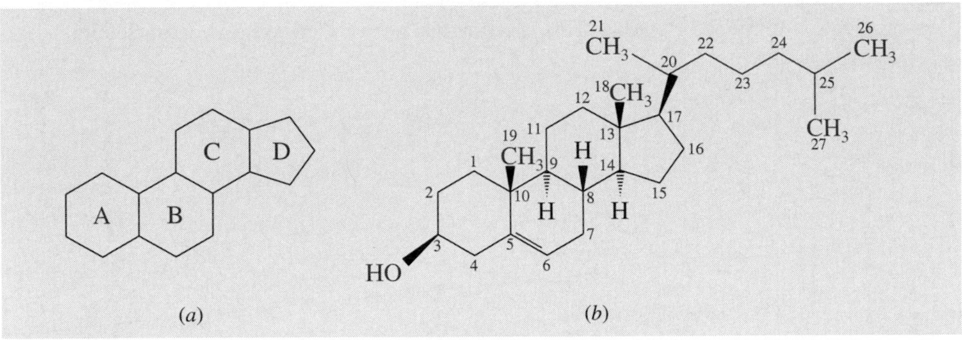

FIGURE 26.9 (*a*) The tetracyclic ring system characteristic of steroids. The rings are designated A, B, C, and D as shown. (*b*) The structure of cholesterol. A unique numbering system is used for steroids and is indicated in the structural formula.

Mammals accumulate cholesterol from their diet, but are also able to biosynthesize it from acetate. The pioneering work that identified the key intermediates in the complicated pathway of cholesterol biosynthesis was carried out by Konrad Bloch (Harvard) and Feodor Lynen (Munich), corecipients of the 1962 Nobel Prize for physiology or medicine. An important discovery was that the triterpene *squalene* (Figure 26.6) is an intermediate in the formation of cholesterol from acetate. Thus, *the early stages of cholesterol biosynthesis are the same as those of terpene biosynthesis* described in Sections 26.8 through 26.10. (In fact, a significant fraction of our knowledge of terpene biosynthesis is a direct result of experiments carried out in the area of steroid biosynthesis.)

How does the tetracyclic steroid cholesterol arise from the acyclic triterpene squalene? Figure 26.10 outlines the stages involved. It has been shown that the first step is oxidation of squalene to the corresponding 2,3-epoxide. Enzyme-catalyzed ring opening of this epoxide in step 2 is accompanied by a cyclization reaction, in which the electrons of four of the five double bonds of squalene 2,3-epoxide are used to close the A, B, C, and D rings of the potential steroid skeleton. The carbocation that results from the cyclization reaction of step 2 is then converted to a triterpene known as *lanosterol* by the rearrangement shown in step 3. Step 4 of Figure 26.10 simply indicates the structural changes that remain to be accomplished in the transformation of lanosterol to cholesterol.

Lanosterol is one component of *lanolin,* a mixture of many substances that coats the wool of sheep.

PROBLEM 26.12 The biosynthesis of cholesterol as outlined in Figure 26.10 is admittedly quite complicated. It will aid your understanding of the process if you consider the following questions:

(*a*) Which carbon atoms of squalene 2,3-epoxide correspond to the doubly bonded carbons of cholesterol?

(*b*) Which two hydrogen atoms of squalene 2,3-epoxide are the ones that migrate in step 3?

(*c*) Which methyl group of squalene 2,3-epoxide becomes the methyl group at the C, D ring junction of cholesterol?

(*d*) What three methyl groups of squalene 2,3-epoxide are lost during the conversion of lanosterol to cholesterol?

SAMPLE SOLUTION (*a*) As the structural formula in step 4 of Figure 26.10 indicates, the double bond of cholesterol unites C-5 and C-6 (steroid numbering). The corresponding carbons in the cyclization reaction of step 2 in the figure may be identified as C-7 and C-8 of squalene 2,3-epoxide (systematic IUPAC numbering).

Step 1: Squalene undergoes enzymic oxidation to the 2,3-epoxide. This reaction has been described earlier, in Section 16.14.

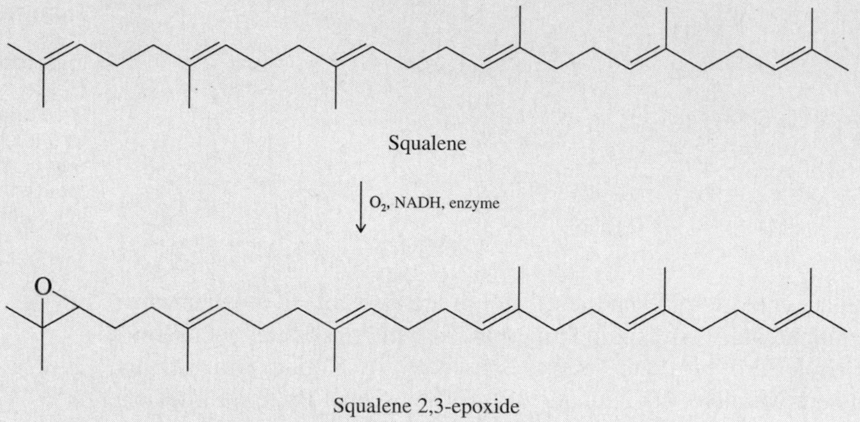

Squalene

O_2, NADH, enzyme

Squalene 2,3-epoxide

Step 2: Cyclization of squalene 2,3-epoxide, shown in its coiled form, is triggered by ring opening of the epoxide. Cleavage of the carbon-oxygen bond is assisted by protonation of oxygen and by nucleophilic participation of the π electrons of the neighboring double bond. A series of ring closures leads to the tetracyclic carbocation shown.

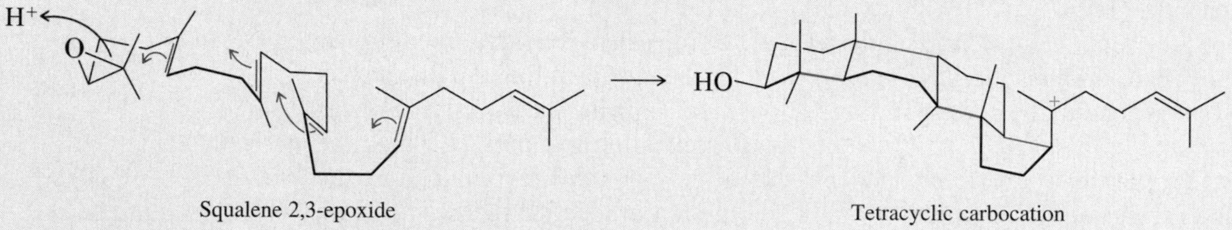

Squalene 2,3-epoxide

Tetracyclic carbocation

Step 3: Rearrangement of the tertiary carbocation formed by cyclization produces lanosterol. Two hydride shifts, from C-17 to C-20 and from C-13 to C-17, are accompanied by methyl shifts from C-14 to C-13 and from C-8 to C-14. A double bond is formed at C-8 by loss of the proton at C-9.

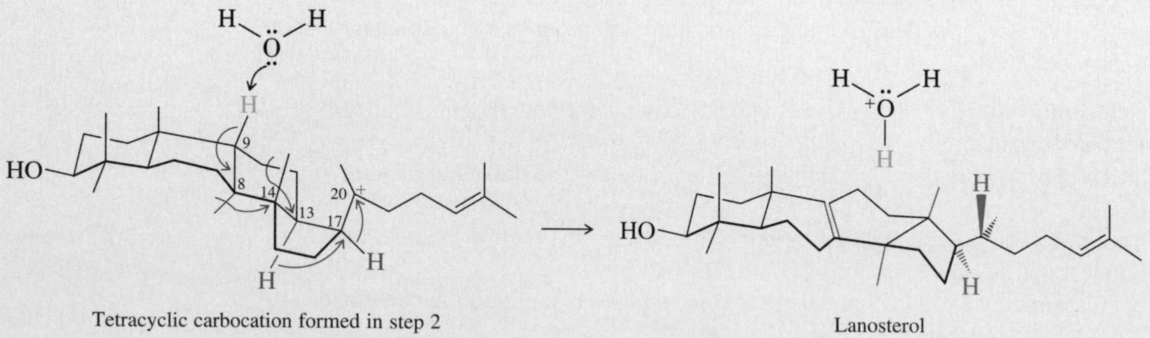

Tetracyclic carbocation formed in step 2

Lanosterol

FIGURE 26.10 The biosynthetic conversion of squalene to cholesterol proceeds through lanosterol. Lanosterol is formed by a cyclization reaction of squalene-2,3-epoxide.

Step 4: A series of enzyme-catalyzed reactions converts lanosterol to cholesterol. The three methyl groups shown in blue in the structural formula of lanosterol are lost via separate multistep operations, the C-8 and C-24 double bonds are reduced, and a new double bond is introduced at C-5.

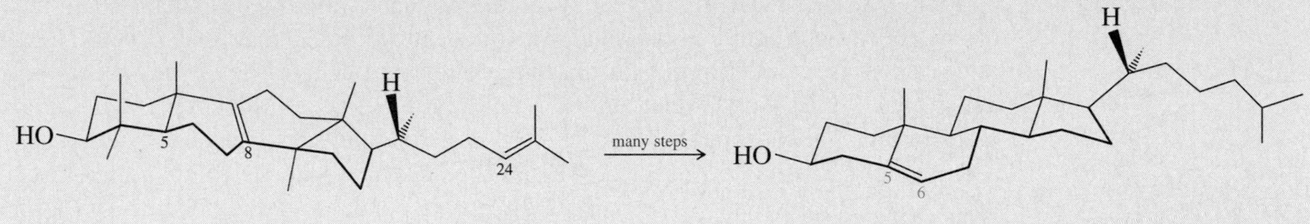

Lanosterol → (many steps) → Cholesterol

FIGURE 26.10 *(Continued)*

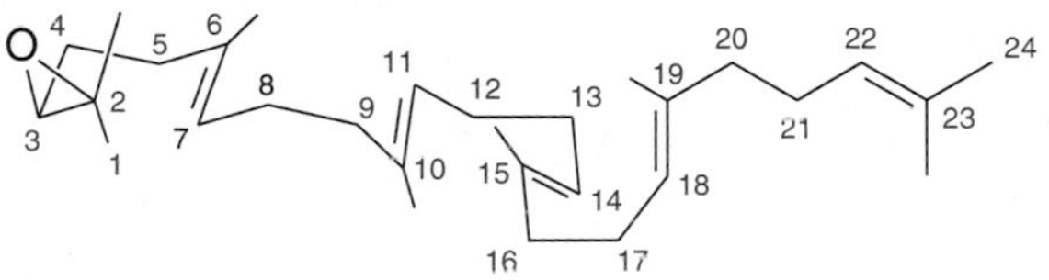

Coiled form of squalene 2,3-epoxide

PROBLEM 26.13 The biosynthetic pathway shown in Figure 26.10 was developed with the aid of isotopic labeling experiments. Which carbon atoms of cholesterol would you expect to be labeled when acetate enriched with ^{14}C in its methyl group ($^{14}CH_3COOH$) is used as the carbon source?

Once formed in the body, cholesterol can undergo a number of transformations. A very common one is acylation of its C-3 hydroxyl group by reaction with coenzyme A derivatives of fatty acids. Other processes convert cholesterol to the biologically important steroids described in the following sections.

26.12 VITAMIN D

A steroid very closely related structurally to cholesterol is its 7-dehydro derivative. 7-Dehydrocholesterol is formed by enzymic oxidation of cholesterol and has a conjugated diene unit in its B ring. 7-Dehydrocholesterol is present in the tissues of the skin, where it is transformed to vitamin D_3 by a sunlight-induced photochemical reaction.

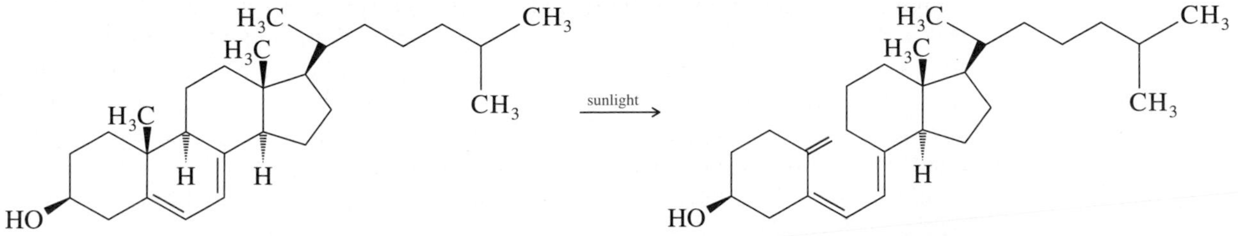

7-Dehydrocholesterol → (sunlight) → Vitamin D_3

Vitamin D_3 is a key compound in the process by which Ca^{2+} is absorbed from the intestine. Low levels of vitamin D_3 lead to Ca^{2+} concentrations in the body that are insufficient to support proper bone growth, resulting in the disease called *rickets.*

Rickets was once a serious health problem. It was thought to be a dietary deficiency disease because it could be prevented in children by feeding them fish liver oil. Actually, rickets is an environmental disease brought about by a deficiency of sunlight. Where the winter sun is weak, children may not be exposed to enough of its light to convert the 7-dehydrocholesterol in their skin to vitamin D_3 at levels sufficient to promote the growth of strong bones. Fish have adapted to an environment that screens them from sunlight, and so they are not directly dependent on photochemistry for their vitamin D_3 and accumulate it by a different process. While fish liver oil is a good source of vitamin D_3, it is not very palatable. Synthetic vitamin D_3, prepared from cholesterol, is often added to milk and other foods to ensure that children receive enough of the vitamin for their bones to develop properly. *Irradiated ergosterol* is another dietary supplement added to milk and other foods for the same purpose. Ergosterol, a steroid obtained from yeast, is structurally similar to 7-dehydrocholesterol and, on irradiation with sunlight or artificial light, is converted to vitamin D_2, a substance analogous to vitamin D_3 and comparable with it in antirachitic activity.

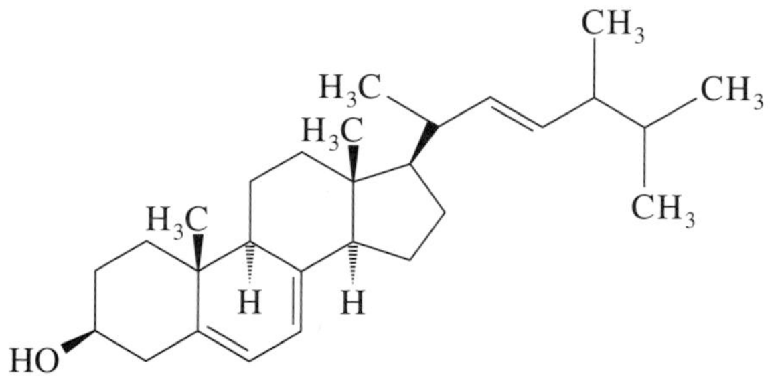

Ergosterol

PROBLEM 26.14 Suggest a reasonable structure for vitamin D_2.

26.13 BILE ACIDS

A significant fraction of the body's cholesterol is used to form **bile acids.** Oxidation in the liver removes a portion of the C_8H_{17} side chain, and additional hydroxyl groups are introduced at various positions on the steroid nucleus. *Cholic acid* is the most abundant of the bile acids. In the form of certain amide derivatives called **bile salts,** of which *sodium taurocholate* is one example, bile acids act as emulsifying agents to aid the digestion of fats. Bile salts have detergent properties similar to those of salts of long-chain fatty acids and promote the transport of lipids through aqueous media.

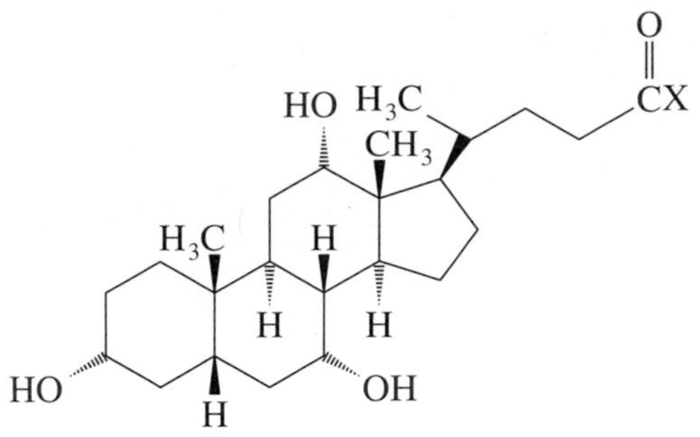

X = OH: cholic acid
X = $NHCH_2CH_2SO_3Na$: sodium taurocholate

26.14 CORTICOSTEROIDS

The outer layer, or *cortex,* of the adrenal gland is the source of a large group of substances known as **corticosteroids.** Like the bile acids, they are derived from cholesterol by oxidation, with cleavage of a portion of the alkyl substituent on the D ring. *Cortisol* is the most abundant of the corticosteroids, while *cortisone* is probably the best known. Cortisone is commonly prescribed as an anti-inflammatory drug, especially in the treatment of rheumatoid arthritis.

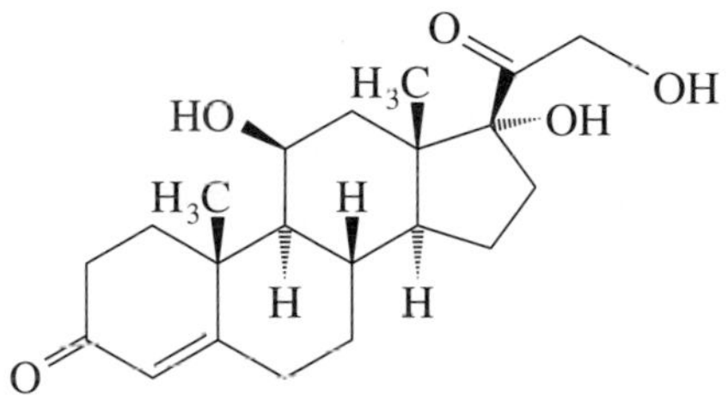

Cortisol

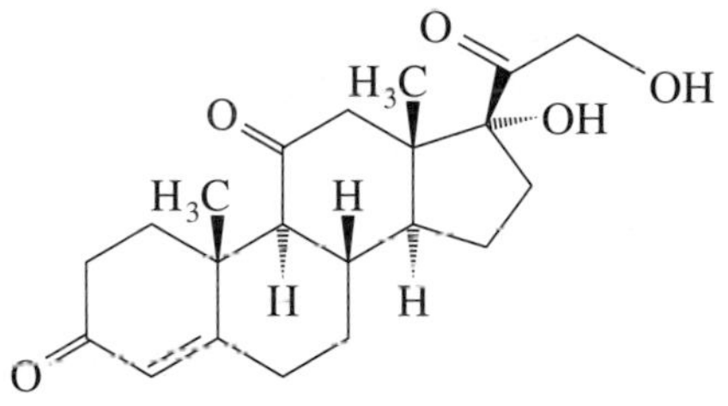

Cortisone

Corticosteroids exhibit a wide range of physiological effects. One important function is to assist in maintaining the proper electrolyte balance in body fluids. They also play a vital regulatory role in the metabolism of carbohydrates and in mediating the allergic response.

26.15 SEX HORMONES

Hormones are the chemical messengers of the body; they are secreted by the endocrine glands and regulate biological processes. Corticosteroids, described in the preceding section, are hormones produced by the adrenal glands. The sex glands—testes in males, ovaries in females—secrete a number of hormones that are involved in sexual development and reproduction. *Testosterone* is the principal male sex hormone; it is an **androgen.** Testosterone promotes muscle growth, deepening of the voice, the growth of body hair, and other male secondary sex characteristics. Testosterone is formed from cholesterol and is the biosynthetic precursor of estradiol, the principal female sex hormone, or **estrogen.** *Estradiol* is a key substance in the regulation of the menstrual cycle and the reproductive process. It is the hormone most responsible for the development of female secondary sex characteristics.

ANABOLIC STEROIDS

As we have seen in this chapter, steroids have a number of functions in human physiology. Cholesterol is a component part of cell membranes and is found in large amounts in the brain. Derivatives of cholic acid assist the digestion of fats in the small intestine. Cortisone and its derivatives are involved in maintaining the electrolyte balance in body fluids. The sex hormones responsible for masculine and feminine characteristics as well as numerous aspects of pregnancy from conception to birth are steroids.

In addition to being an androgen, the principal male sex hormone testosterone promotes muscle growth and is classified as an **anabolic** steroid hormone. Biological chemists distinguish between two major classes of metabolism, **catabolic** and **anabolic** processes. Catabolic processes are degradative pathways in which larger molecules are broken down to smaller ones. Anabolic processes are the reverse; larger molecules are synthesized from smaller ones. While the body mainly stores energy from food in the form of fat, a portion of that energy goes toward producing muscle from protein. An increase in the amount of testosterone, accompanied by an increase in the amount of food consumed, will cause an increase in the body's muscle mass.

The pharmaceutical industry has developed and studied a number of anabolic steroids for use in veterinary medicine and in rehabilitation from injuries that are accompanied by deterioration of muscles. The ideal agent would be one that possessed the anabolic properties of testosterone without its androgenic (masculinizing) effects. *Dianabol* and *stanozolol* are among the many synthetic anabolic steroids available by prescription.

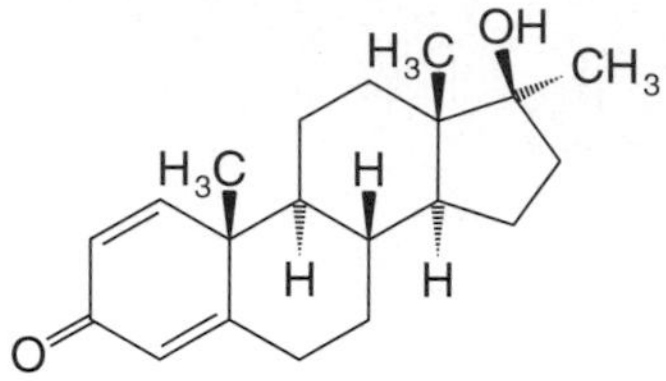

The sprinter Ben Johnson lost the gold medal he won in the 100-meter dash at the 1988 Olympic Games when tests revealed the presence of stanozolol in his urine. Abuse of anabolic steroids probably exists in all sports but appears to be most prevalent in competitions that place a premium on size and strength such as football, weight lifting, the weight events in track and field (discus and shot put), and bodybuilding.

H$_3$C OH CH$_3$ H$_3$C H H H O

Dianabol

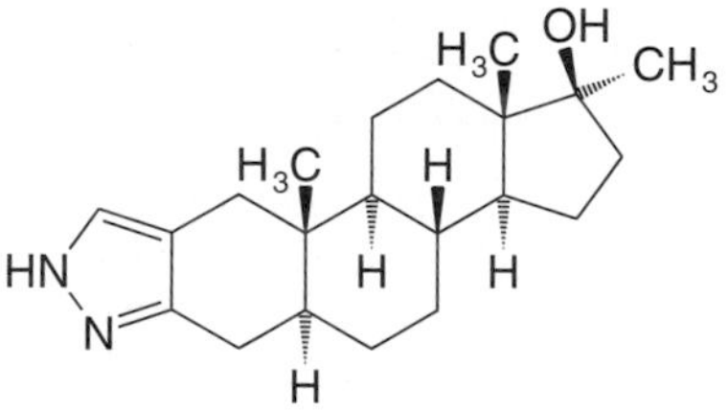

H$_3$C OH CH$_3$ H$_3$C H HN N H H H

Stanozolol

Some scientific studies indicate that the gain in performance obtained through the use of anabolic steroids is small. This may be a case, however, where the anecdotal evidence of the athletes is more accurate than the scientific studies. The scientific studies are done under ethical conditions where patients are treated with "prescription-level" doses of steroids. A 240-pound offensive tackle ("too small" by today's standards) may take several anabolic steroids at a time at 10 to 20 times their prescribed doses in order to weigh the 280 pounds he (or his coach) feels is necessary. The price athletes pay for gains in size and strength can be enormous. This price includes emotional costs (friendships lost because of heightened aggressiveness), sterility, testicular atrophy (the testes cease to function once the body starts to obtain a sufficient supply of testosterone-like steroids from outside), and increased risk of premature death from liver cancer or heart disease.

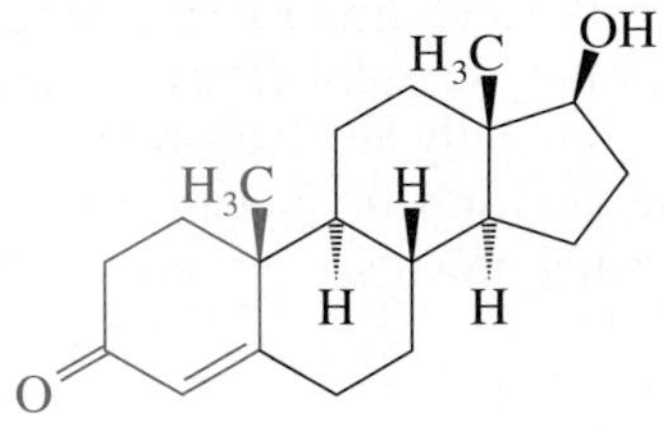

Testosterone

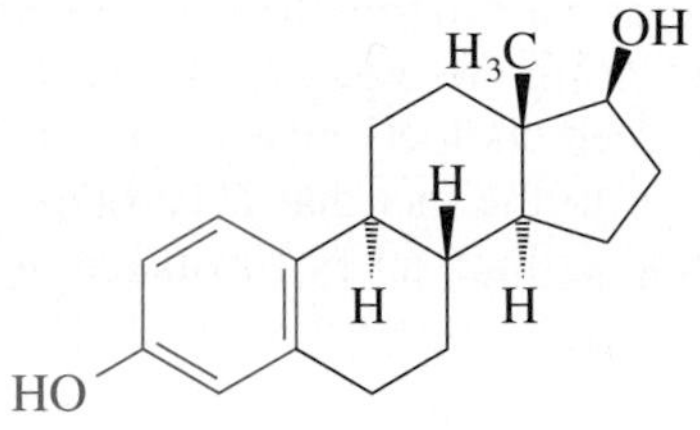

Estradiol

Testosterone and estradiol are present in the body in only minute amounts, and their isolation and identification required heroic efforts. In order to obtain 0.012 g of estradiol for study, for example, 4 tons of sow ovaries had to be extracted!

A separate biosynthetic pathway leads from cholesterol to *progesterone,* a female sex hormone. One function of progesterone is to suppress ovulation at certain stages of the menstrual cycle and during pregnancy. Synthetic substances, such as *norethindrone,* have been developed that are superior to progesterone when taken orally to ''turn off'' ovulation. By inducing temporary infertility, they form the basis of most oral contraceptive agents.

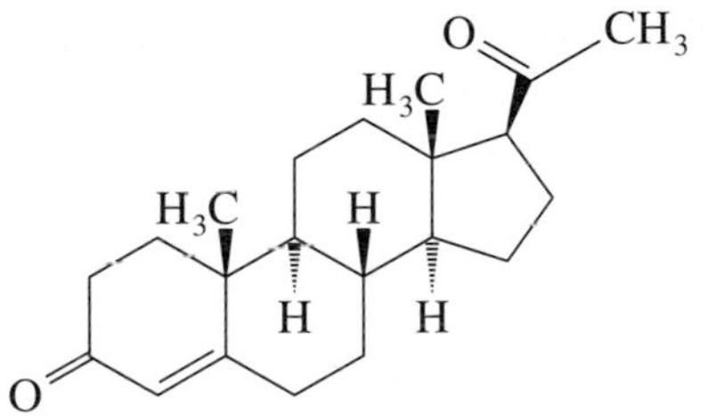

Progesterone

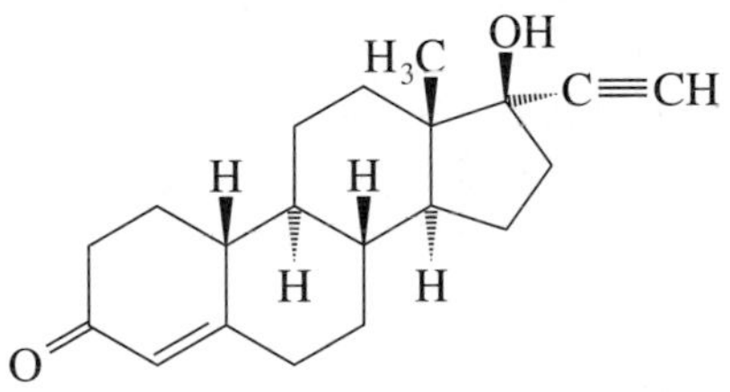

Norethindrone

26.16 CAROTENOIDS

Carotenoids are natural pigments characterized by a tail-to-tail linkage between two C_{20} units and an extended conjugated system of double bonds. They are the most widely distributed of the substances that give color to our world and occur in flowers, fruits, plants, insects, and animals. It has been estimated that biosynthesis from acetate produces approximately 100 million tons of carotenoids per year. The most familiar carotenoids are lycopene and β-carotene, pigments found in numerous plants and easily isolable from ripe tomatoes and carrots, respectively.

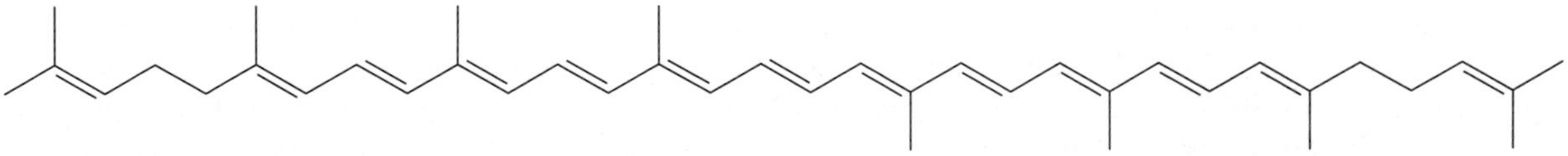

Lycopene

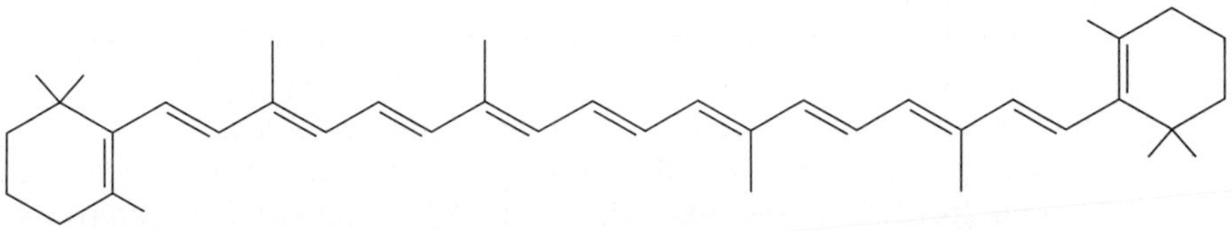

β-Carotene

Carotenoids absorb visible light (Section 13.19) and dissipate its energy as heat, thereby protecting the organism from any potentially harmful effects associated with sunlight-induced photochemistry. They are also indirectly involved in the chemistry of vision, owing to the fact that β-carotene is the biosynthetic precursor of vitamin A, also known as retinol, a key substance in the visual process.

The structural chemistry of the visual process, beginning with β-carotene, was described in the boxed essay entitled *Imines in Biological Chemistry* that accompanied Section 17.11.

26.17 SUMMARY

Chemists and biochemists find it convenient to divide the principal organic substances present in cells into four main groups: *carbohydrates, proteins, nucleic acids,* and **lipids.** Structural differences separate carbohydrates from proteins, and both of these are structurally distinct from nucleic acids. Lipids, on the other hand, are characterized by a *physical property,* their solubility in nonpolar solvents, rather than by their structure. In this chapter we have examined lipid molecules that share a common biosynthetic origin in that all their carbons are derived from acetic acid (acetate).

The form in which acetate acts out its biosynthetic role is as **acetyl coenzyme A** (Figure 26.1, Section 26.1). Acetyl coenzyme A is the biosynthetic precursor to the **fatty acids** (Sections 26.2 and 26.3). Fatty acids most often occur naturally as esters. **Fats** (and oils) are glycerol esters of long-chain carboxylic acids. Typically, these chains are unbranched and contain even numbers of carbon atoms.

```
      O
      ‖
   RCOCH2 O
        |  ‖
       CHOCR′
        |
   R″COCH2
      ‖
      O
```

Triacylglycerol
(R, R′, and R″ may be the same or different)

Phospholipids (Section 26.4) are intermediates in the biosynthesis of triacylglycerols from fatty acids and are the principal constituents of cell membranes.

Waxes (Section 26.5) are mixtures of substances that usually contain esters of fatty acids and long-chain alcohols.

A group of compounds called **prostaglandins** (Section 26.6) are potent regulators of biochemical processes. The prostaglandins are biosynthesized from C_{20} fatty acids by a combination of reactions that lead to oxygen incorporation and five-membered ring formation.

Terpenes and related *isoprenoid* compounds (Section 26.7) are biosynthesized from acetate by way of *mevalonate* and *isopentenyl pyrophosphate* (Section 26.8). The processes that form carbon-carbon bonds between isoprene units can be understood on the basis of nucleophilic attack of the π electrons of a double bond on a carbocation or carbocation precursor (Section 26.9).

The pathway from acetate to **steroids** (Section 26.11) proceeds by way of the triterpene *squalene.* Enzymatic cyclization of 2,3-epoxysqualene gives *lanosterol,* which in turn is converted to *cholesterol.* Cholesterol is an abundant natural product and is the precursor to other physiologically important substances including *vitamin* D_3 (Section 26.12), the **bile acids** (Section 26.13), the **corticosteroids** (Section 26.14), and the **sex hormones** (Section 26.15).

Carotenoids (Section 26.16) are *tetraterpenes* that most commonly occur as plant pigments.

PROBLEMS

26.15 Identify the carbon atoms expected to be labeled with ^{14}C when each of the following substances is biosynthesized from acetate enriched with ^{14}C in its methyl group:

(*a*) $CH_3(CH_2)_{14}CO_2H$ Palmitic acid

(*b*) PGE$_2$ (structure: O, COOH, CH_3, HO, OH)

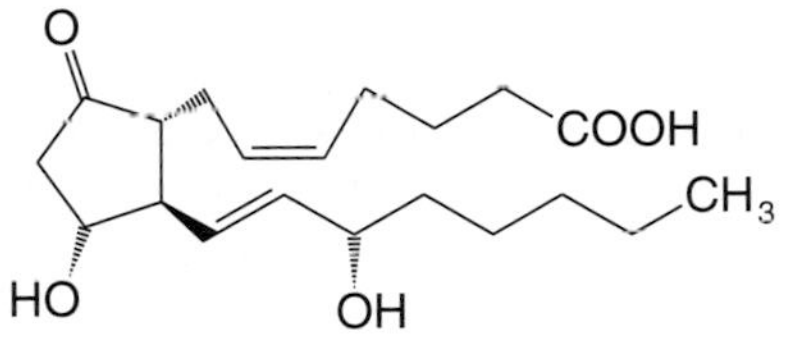

(*c*) Limonene

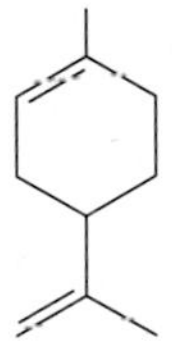

(*d*) β-Carotene

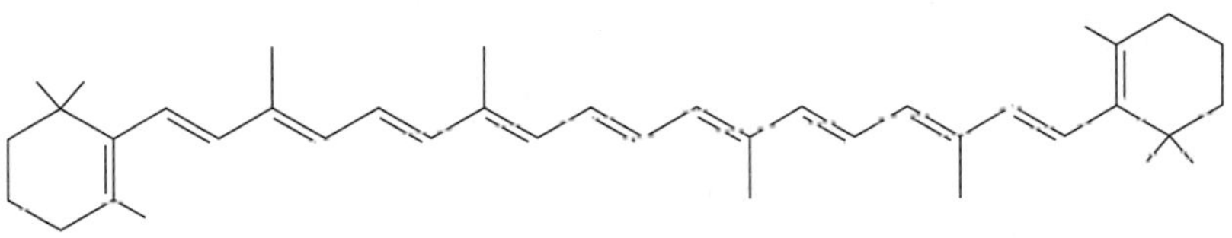

26.16 The biosynthetic pathway to prostaglandins leads also to a class of physiologically potent substances known as *prostacyclins.* Which carbon atoms of the prostacyclin shown here would you expect to be enriched in ^{14}C if it were biosynthesized from acetate labeled with ^{14}C in its methyl group?

(structure: COOH, O, CH_3, HO, OH)

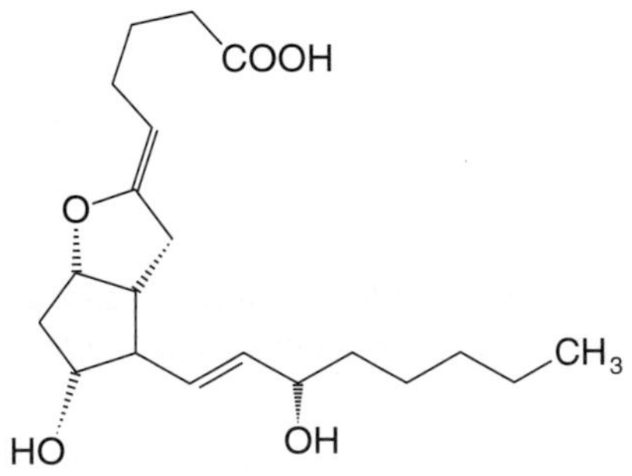

26.17 Identify the isoprene units in each of the following naturally occurring substances:

(*a*) *Ascaridole,* a naturally occurring peroxide present in chenopodium oil:

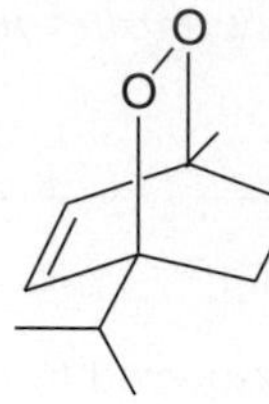

(*b*) *Dendrolasin,* a constituent of the defense secretion of a species of ant:

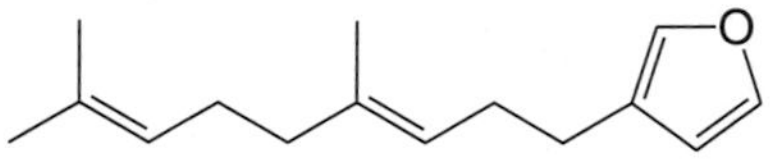

(*c*) γ-*Bisabolene,* a sesquiterpene found in the essential oils of a large number of plants:

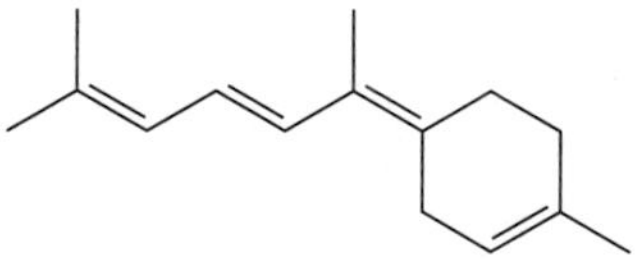

(*d*) α-*Santonin,* an anthelmintic substance isolated from artemisia flowers:

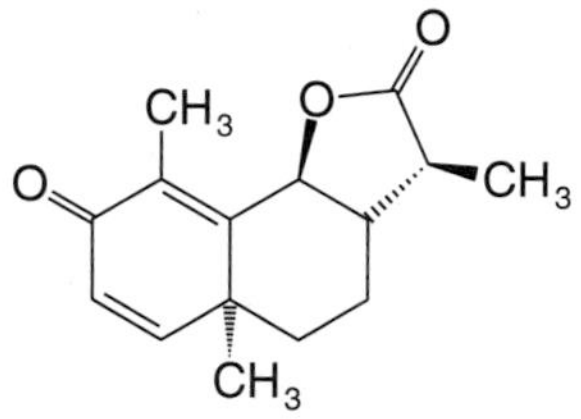

(*e*) *Tetrahymanol,* a pentacyclic triterpene isolated from a species of protozoans:

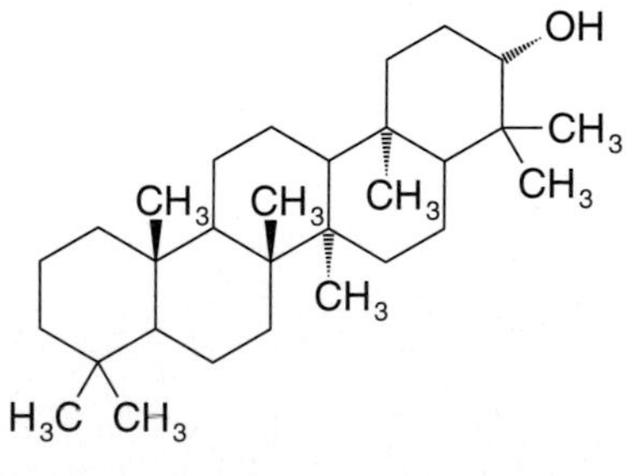

26.18 *Cubitene* is a diterpene present in the defense secretion of a species of African termite. What unusual feature characterizes the joining of isoprene units in cubitene?

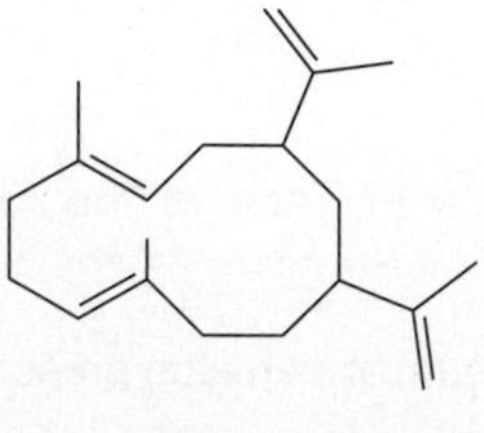

26.19 *Pyrethrins* are a group of naturally occurring insecticidal substances found in the flowers of various plants of the chrysanthemum family. The following is the structure of a typical pyrethrin, *cinerin I* (exclusive of stereochemistry):

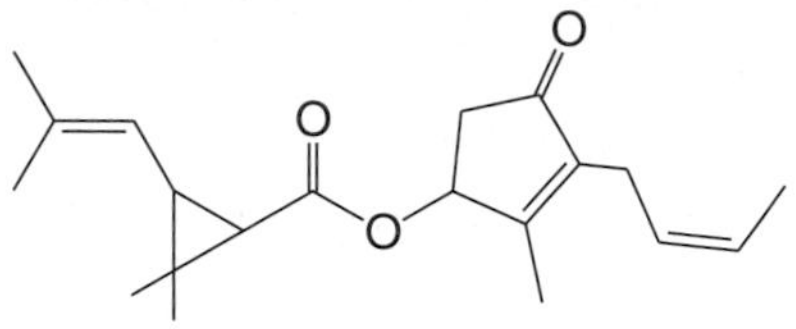

(*a*) Locate any isoprene units present in cinerin I.

(*b*) Hydrolysis of cinerin I gives an optically active carboxylic acid, (+)-chrysanthemic acid. Ozonolysis of (+)-chrysanthemic acid, followed by oxidation, gives acetone and an optically active dicarboxylic acid, (−)-caronic acid ($C_7H_{10}O_4$). What is the structure of (−)-caronic acid? Are the two carboxyl groups cis or trans to each other? What does this information tell you about the structure of (+)-chrysanthemic acid?

26.20 *Cerebrosides* are found in the brain and in the myelin sheath of nerve tissue. The structure of the cerebroside *phrenosine* is

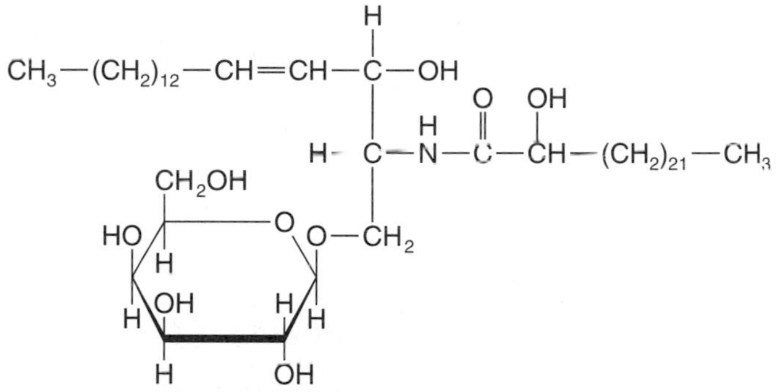

(*a*) What hexose is formed on hydrolysis of the glycoside bond of phrenosine? Is phrenosine an α- or a β-glycoside?

(*b*) Hydrolysis of phrenosine gives, in addition to the hexose in part (*a*), a fatty acid called *cerebronic acid,* along with a third substance called *sphingosine.* Write structural formulas for both cerebronic acid and sphingosine.

26.21 Each of the following reactions has been reported in the chemical literature and proceeds in good yield. What are the principal organic products of each reaction? (In some of the exercises more than one diastereomer may be theoretically possible, but in such instances one diastereomer is either the major product or the only product. For those reactions in which one diastereomer is formed preferentially, indicate its expected stereochemistry.)

(*a*) $CH_3(CH_2)_7C{\equiv}C(CH_2)_7COOH + H_2 \xrightarrow{\text{Lindlar Pd}}$

(*b*) $CH_3(CH_2)_7C{\equiv}C(CH_2)_7COOH \xrightarrow[\text{2. } H^+]{\text{1. Li, } NH_3}$

(*c*) $(Z)\text{-}CH_3(CH_2)_7CH{=}CH(CH_2)_7\overset{O}{\overset{\|}{C}}OCH_2CH_3 + H_2 \xrightarrow{\text{Pt}}$

(*d*) (*Z*)-$CH_3(CH_2)_5CHCH_2CH{=}CH(CH_2)_7\overset{O}{\overset{\|}{C}}OCH_3$ $\xrightarrow[\text{2. } H_2O]{\text{1. } LiAlH_4}$
OH

(*e*) (*Z*)-$CH_3(CH_2)_7CH{=}CH(CH_2)_7COOH + C_6H_5CO_2OH \longrightarrow$

(*f*) Product of part (*e*) + $H_3O^+ \longrightarrow$

(*g*) (*Z*)-$CH_3(CH_2)_7CH{=}CH(CH_2)_7COOH \xrightarrow[\text{2. } H^+]{\text{1. } OsO_4,\ (CH_3)_3COOH,\ HO^-}$

(*h*) H_3C CH_3 CH_3 $\xrightarrow[\text{2. } H_2O_2,\ HO^-]{\text{1. } B_2H_6,\ \text{diglyme}}$

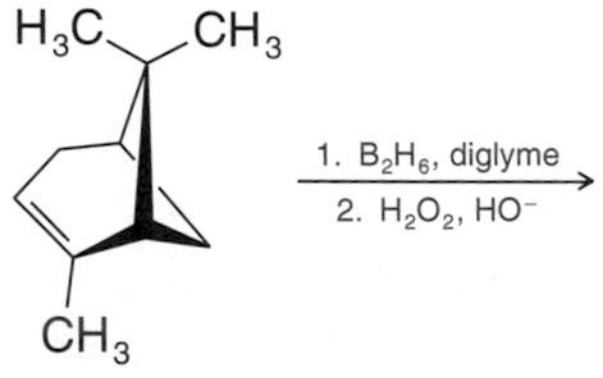

(*i*) H_3C CH_3 CH_2 $\xrightarrow[\text{2. } H_2O_2,\ HO^-]{\text{1. } B_2H_6,\ \text{diglyme}}$

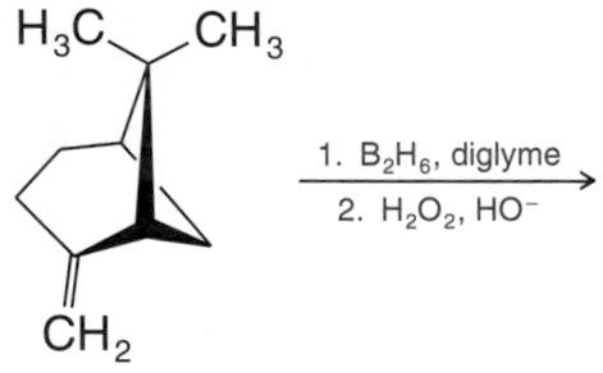

(*j*) CH_3 OH CH_3 CH_3 H H H H CH_3O CH_3O $\xrightarrow{HCl,\ H_2O} C_{21}H_{34}O_2$

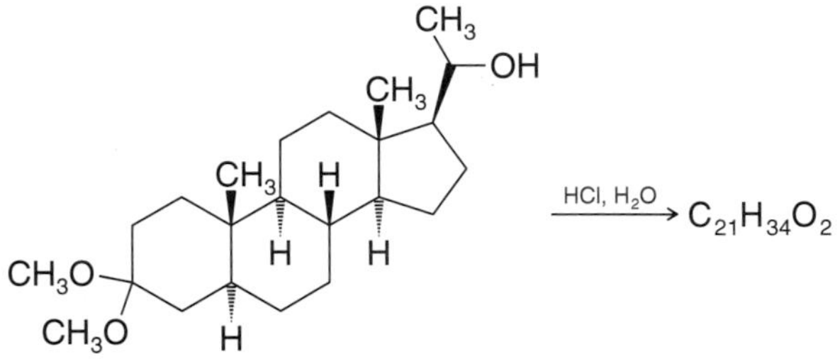

26.22 Describe an efficient synthesis of each of the following compounds from octadecanoic (stearic) acid using any necessary organic or inorganic reagents:

(*a*) Octadecane
(*b*) 1-Phenyloctadecane
(*c*) 3-Ethylicosane
(*d*) Icosanoic acid
(*e*) 1-Heptadecanamine
(*f*) 1-Octadecanamine
(*g*) 1-Nonadecanamine

26.23 A synthesis of triacylglycerols has been described that begins with the substance shown.

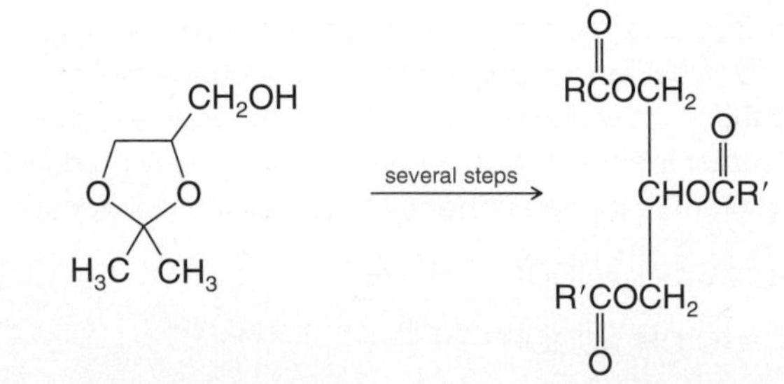

CH_2OH, O, O, H_3C CH_3 $\xrightarrow{\text{several steps}}$ $R\overset{O}{\overset{\|}{C}}OCH_2$, $CHO\overset{O}{\overset{\|}{C}}R'$, $R'\underset{O}{\underset{\|}{C}}OCH_2$

4-(Hydroxymethyl)-2,2-dimethyl-1,3-dioxolane

Triacylglycerol

Outline a series of reactions suitable for the preparation of a triacylglycerol of the type illustrated in the equation, where R and R′ are different.

26.24 The isoprenoid compound shown is a scent marker present in the urine of the red fox. Suggest a reasonable synthesis for this substance from 3-methyl-3-buten-1-ol and any necessary organic or inorganic reagents.

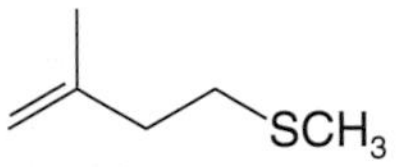

SCH_3

26.25 *Sabinene* is a monoterpene found in the oil of citrus fruits and plants. It has been synthesized from 6-methyl-2,5-heptanedione by the sequence that follows. Suggest reagents suitable for carrying out each of the indicated transformations.

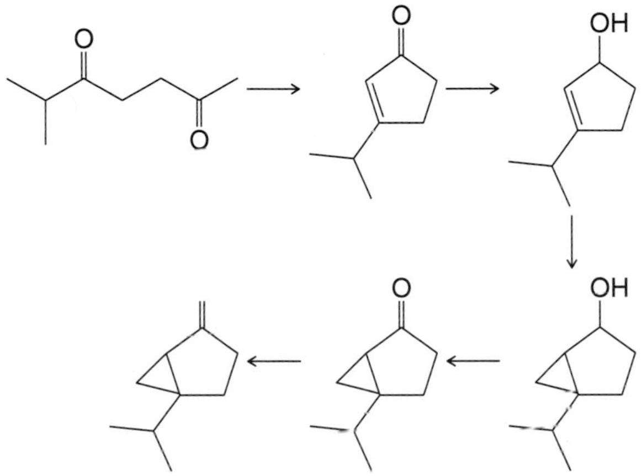

O O OH OH O

Sabinene

26.26 Isoprene has sometimes been used as a starting material in the laboratory synthesis of terpenes. In one such synthesis, the first step is the electrophilic addition of 2 moles of hydrogen bromide to isoprene to give 1,3-dibromo-3-methylbutane.

$$CH_2{=}C(CH_3)CH{=}CH_2 + 2HBr \longrightarrow (CH_3)_2CBrCH_2CH_2Br$$

2-Methyl-1,3-butadiene (isoprene) + Hydrogen bromide → 1,3-Dibromo-3-methylbutane

Write a series of equations describing the mechanism of this reaction.

26.27 The ionones are fragrant substances present in the scent of iris and are used in perfume. A mixture of α- and β-ionone can be prepared by treatment of pseudoionone with sulfuric acid.

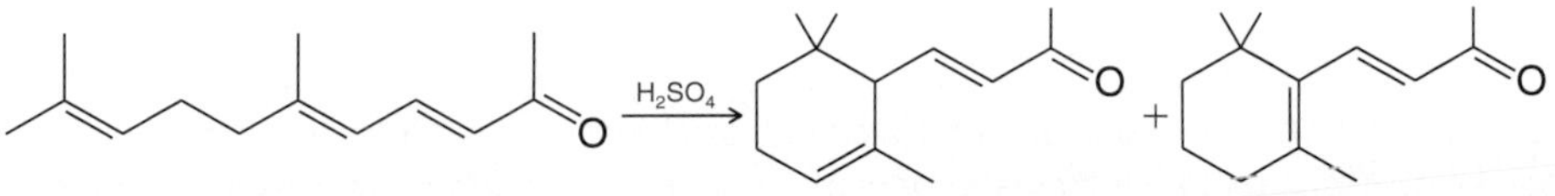

Pseudoionone $\xrightarrow{H_2SO_4}$ α-Ionone + β-Ionone

Write a stepwise mechanism for this reaction.

26.28 β,γ-Unsaturated steroidal ketones represented by the partial structure shown below are readily converted in acid to their α,β-unsaturated isomers. Write a stepwise mechanism for this reaction.

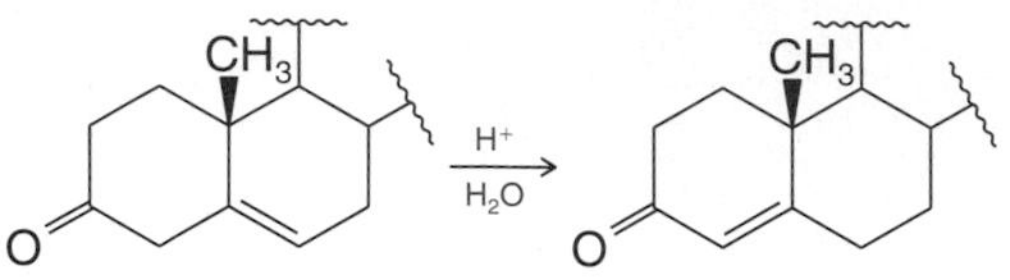

26.29

(*a*) Suggest a mechanism for the following reaction.

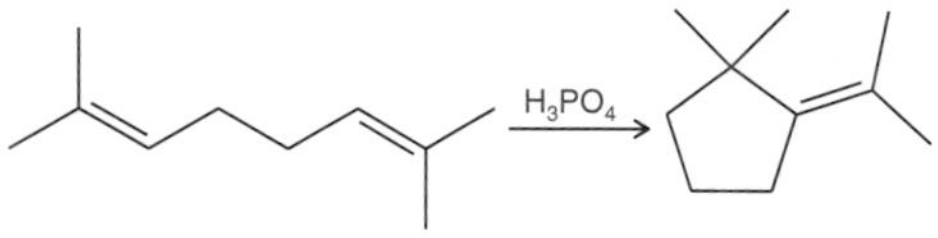

(*b*) The following two compounds are also formed in the reaction given in part *a*. How are these two products formed?

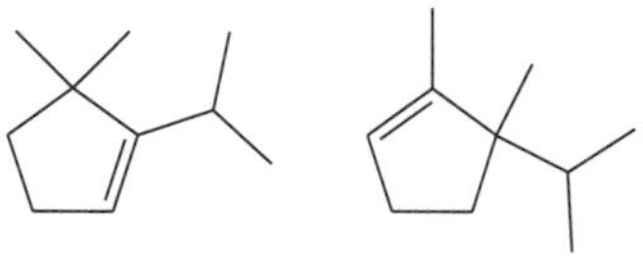

(*Note:* The solution to this problem is not given in the *Solutions Manual and Study Guide.* It is discussed in detail, however, in a very interesting article on pages 541–542 of the June 1995 issue of the *Journal of Chemical Education.*)

MOLECULAR MODELING EXERCISES

26.30 Compare their molecular shapes by building molecular models of the triglycerides shown in Figure 26.2.

26.31 The sesquiterpene shown below is called β-cadinene and is obtained from the essential oil of juniper. Construct a molecular model of β-cadinene and disconnect the bonds necessary to reveal its three isoprene units.

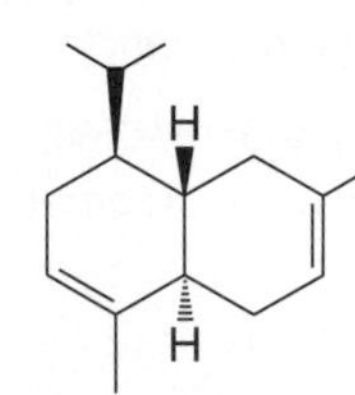

26.32 Certain phenols, such as orsellinic acid, are biosynthesized from acetate via "tetraacetic acid." Construct a molecular model of tetraacetic acid and show how it can be transformed to orsellinic acid.

$$CH_3\overset{O}{\overset{\|}{C}}CH_2\overset{O}{\overset{\|}{C}}CH_2\overset{O}{\overset{\|}{C}}CH_2CO_2H \longrightarrow$$

Tetraacetic acid

Orsellinic acid

26.33 The compound shown is *diethyl stilbestrol* (DES); it has a number of therapeutic uses in estrogen-replacement therapy. DES is not a steroid, but can adopt a shape that allows it to mimic estrogens such as estradiol (p 1083) and bind to the same receptor sites. Construct molecular models of DES and estradiol that illustrate this similarity in molecular size, shape, and location of polar groups.

26.34 As described in Section 26.8, mevalonic acid is a key early intermediate in the biosynthesis of terpenes and steroids. Its Fischer projection is shown below. Construct a molecular model of mevalonic acid showing the correct configuration at C-3. Is the configuration *R* or *S*? Mevalonic acid is in equilibrium with its δ lactone called mevalonolactone. Convert your molecular model to mevalonolactone.

CH_2CO_2H

H_3C — — OH

CH_2CH_2OH

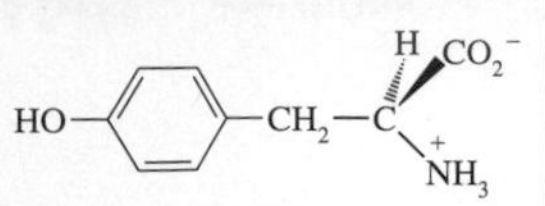

$HO-C_6H_4-CH_2-C(H)(CO_2^-)(\overset{+}{N}H_3)$

CHAPTER 27

AMINO ACIDS, PEPTIDES, AND PROTEINS. NUCLEIC ACIDS

The relationship between structure and function reaches its ultimate expression in the chemistry of amino acids, peptides, and proteins.

Amino acids are carboxylic acids that contain an amine function. Under certain conditions the amine group of one molecule and the carboxyl group of a second can react, uniting the two amino acids by an amide bond.

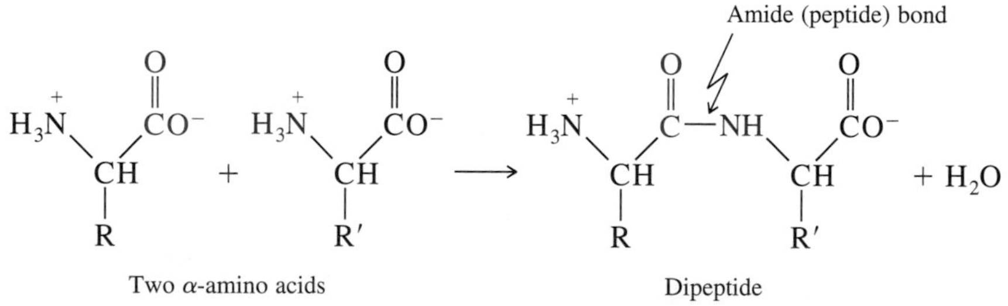

$$H_3\overset{+}{N}CH(R)CO^- (C{=}O) + H_3\overset{+}{N}CH(R')CO^- (C{=}O) \longrightarrow H_3\overset{+}{N}CH(R)C(=O)\text{—}NHCH(R')CO^- (C{=}O) + H_2O$$

Two α-amino acids — Dipeptide — Amide (peptide) bond

Amide linkages between amino acids are known as **peptide bonds,** and the product of peptide bond formation between two amino acids is called a **dipeptide.** The peptide chain may be extended to incorporate three amino acids in a **tripeptide,** four in a **tetrapeptide,** and so on. **Polypeptides** contain many amino acid units. **Proteins** are naturally occurring polypeptides that contain more than 50 amino acid units—most proteins are polymers of 100 to 300 amino acids.

The most striking thing about proteins is the diversity of their roles in living systems: silk, hair, skin, muscle, and connective tissue are proteins, and almost all enzymes are proteins. As in most aspects of chemistry and biochemistry, structure is the key to function. We will explore the structure of proteins by first concentrating on their fundamental building block units, the α-amino acids. Then, after developing the principles of peptide structure and conformation, we will see how the insights gained from these smaller molecules aid our understanding of proteins.

The chapter concludes with a discussion of the **nucleic acids,** which are the genetic material of living systems and direct the biosynthesis of proteins. These two types of biopolymers, nucleic acids and proteins, are the organic chemicals of life.

27.1 CLASSIFICATION OF AMINO ACIDS

Amino acids are classified as α, β, γ, and so on, according to the location of the amine group on the carbon chain that contains the carboxylic acid function.

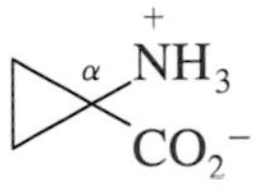

$\overset{+}{N}H_3$, CO_2^- (cyclopropane ring, α carbon) — 1-Aminocyclopropanecarboxylic acid: an α-amino acid that is the biological precursor to ethylene in plants

$H_3\overset{+}{N}\underset{\beta}{C}H_2\underset{\alpha}{C}H_2CO_2^-$ — 3-Aminopropanoic acid: known as β-alanine, it is a β-amino acid that makes up one of the structural units of coenzyme A

$H_3\overset{+}{N}\underset{\gamma}{C}H_2\underset{\beta}{C}H_2\underset{\alpha}{C}H_2CO_2^-$ — 4-Aminobutanoic acid: known as γ-aminobutyric acid (GABA), it is a γ-amino acid and is involved in the transmission of nerve impulses

While more than 700 different amino acids are known to occur naturally, a group of 20 of them commands special attention. These 20 are the amino acids that are normally present in proteins and are listed in Table 27.1. All the amino acids from which proteins are derived are α-amino acids, and all but one of these contain a primary amino function and conform to the general structure

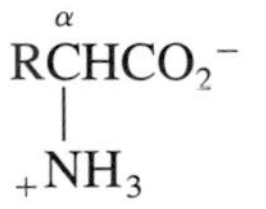

$$R\overset{\alpha}{C}HCO_2^- \quad (\text{with } {}^+NH_3 \text{ on the } \alpha \text{ carbon})$$

The one exception is proline, a secondary amine in which the amino nitrogen is incorporated into a five-membered ring.

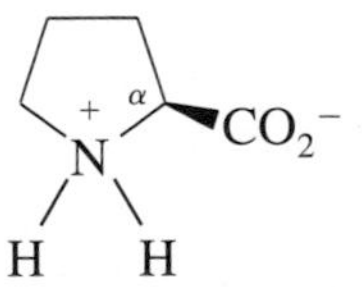

$$\text{Proline: five-membered ring with } \overset{+}{N}H_2 \text{ and } \alpha\text{-carbon bearing } CO_2^-$$

Table 27.1 includes the customary three-letter abbreviations for the common amino acids as well as their newer one-letter abbreviations.

While humans possess the capacity to biosynthesize some of the amino acids shown in the table, they must obtain certain of the others from their diet. Those that must be included in our dietary requirements are called **essential amino acids** and are identified as such in the table.

TABLE 27.1
α-Amino Acids Found in Proteins

Name	Abbreviation	Structural formula*
Amino acids with nonpolar side chains		
Glycine	Gly (G)	$H{-}\overset{\overset{+}{N}H_3}{\overset{\vert}{C}}HCO_2^-$
Alanine	Ala (A)	$CH_3{-}\overset{\overset{+}{N}H_3}{\overset{\vert}{C}}HCO_2^-$
Valine†	Val (V)	$(CH_3)_2CH{-}\overset{\overset{+}{N}H_3}{\overset{\vert}{C}}HCO_2^-$
Leucine†	Leu (L)	$(CH_3)_2CHCH_2{-}\overset{\overset{+}{N}H_3}{\overset{\vert}{C}}HCO_2^-$
Isoleucine†	Ile (I)	$CH_3CH_2\overset{CH_3}{\overset{\vert}{C}}H{-}\overset{\overset{+}{N}H_3}{\overset{\vert}{C}}HCO_2^-$
Methionine†	Met (M)	$CH_3SCH_2CH_2{-}\overset{\overset{+}{N}H_3}{\overset{\vert}{C}}HCO_2^-$
Proline	Pro (P)	$H_2C{-}\overset{+}{N}H_2$, H_2C, $H_2C{-}CHCO_2^-$ (five-membered ring)
Phenylalanine†	Phe (F)	$C_6H_5{-}CH_2{-}\overset{\overset{+}{N}H_3}{\overset{\vert}{C}}HCO_2^-$
Tryptophan†	Trp (W)	indol-3-yl$-CH_2{-}\overset{\overset{+}{N}H_3}{\overset{\vert}{C}}HCO_2^-$ (ring N–H)
Amino acids with polar but nonionized side chains		
Asparagine	Asn (N)	$H_2N\overset{O}{\overset{\Vert}{C}}CH_2{-}\overset{\overset{+}{N}H_3}{\overset{\vert}{C}}HCO_2^-$
Glutamine	Gln (Q)	$H_2N\overset{O}{\overset{\Vert}{C}}CH_2CH_2{-}\overset{\overset{+}{N}H_3}{\overset{\vert}{C}}HCO_2^-$

TABLE 27.1
α-Amino Acids Found in Proteins (*Continued*)

Serine	Ser (S)	$HOCH_2{-}CH(NH_3^+)CO_2^-$
Threonine†	Thr (T)	$CH_3CH(OH){-}CH(NH_3^+)CO_2^-$
Amino acids with acidic side chains		
Aspartic acid	Asp (D)	$^-OC(=O)CH_2{-}CH(NH_3^+)CO_2^-$
Glutamic acid	Glu (E)	$^-OC(=O)CH_2CH_2{-}CH(NH_3^+)CO_2^-$
Tyrosine	Tyr (Y)	$HO{-}C_6H_4{-}CH_2{-}CH(NH_3^+)CO_2^-$
Cysteine	Cys (C)	$HSCH_2{-}CH(NH_3^+)CO_2^-$
Amino acids with basic side chains		
Lysine†	Lys (K)	$H_3\overset{+}{N}CH_2CH_2CH_2CH_2{-}CH(NH_3^+)CO_2^-$
Arginine†	Arg (R)	$H_2NC(=\overset{+}{N}H_2)NHCH_2CH_2CH_2{-}CH(NH_3^+)CO_2^-$
Histidine†	His (H)	(imidazol-4-yl)$-CH_2{-}CH(NH_3^+)CO_2^-$

* All amino acids are shown in the form present in greatest concentration at pH 7.

† An essential amino acid, which must be present in the diet of mammals to ensure normal growth.

27.2 STEREOCHEMISTRY OF AMINO ACIDS

Glycine (aminoacetic acid) is the simplest amino acid and the only one in Table 27.1 that is achiral. The α carbon atom is a stereogenic center in all the others. Configurations in amino acids are normally specified by the D, L notational system. All the chiral amino acids obtained from proteins have the L configuration at their α carbon atom.

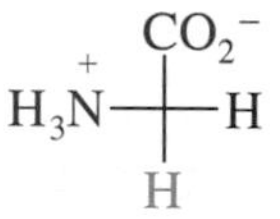

CO_2^- / $H_3\overset{+}{N}$—C—H / H

Glycine (achiral)

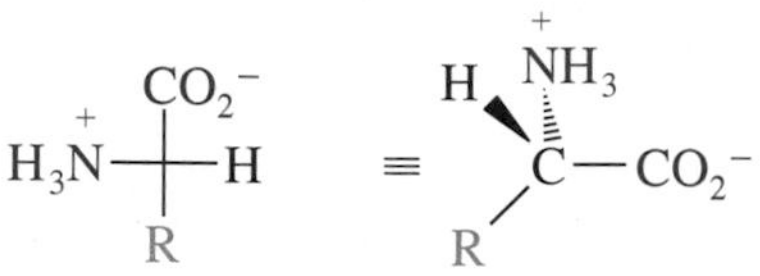

CO_2^- / $H_3\overset{+}{N}$—C—H / R $\equiv$ H, $\overset{+}{N}H_3$, C—CO_2^-, R

Fischer projection formula of an L amino acid

PROBLEM 27.1 What is the absolute configuration (*R* or *S*) at the α carbon atom in each of the following L amino acids?

(*a*) CO_2^- / $H_3\overset{+}{N}$—C—H / CH_2OH

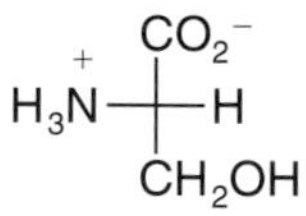

L-Serine

(*b*) CO_2^- / $H_3\overset{+}{N}$—C—H / CH_2SH

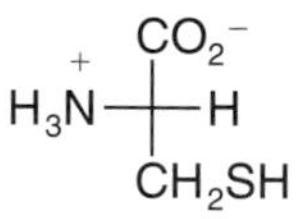

L-Cysteine

(*c*) CO_2^- / $H_3\overset{+}{N}$—C—H / $CH_2CH_2SCH_3$

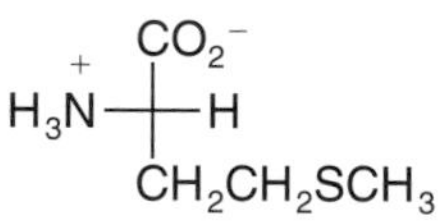

L-Methionine

SAMPLE SOLUTION (*a*) First identify the four substituents attached directly to the stereogenic center and rank them in order of decreasing sequence rule precedence. For L-serine the substituents are

$$H_3\overset{+}{N}— > —CO_2^- > —CH_2OH > H$$

Highest-ranked Lowest-ranked

Next, translate the Fischer projection formula of L-serine to a three-dimensional representation and orient it so that the lowest-ranked substituent at the stereogenic center is directed away from you.

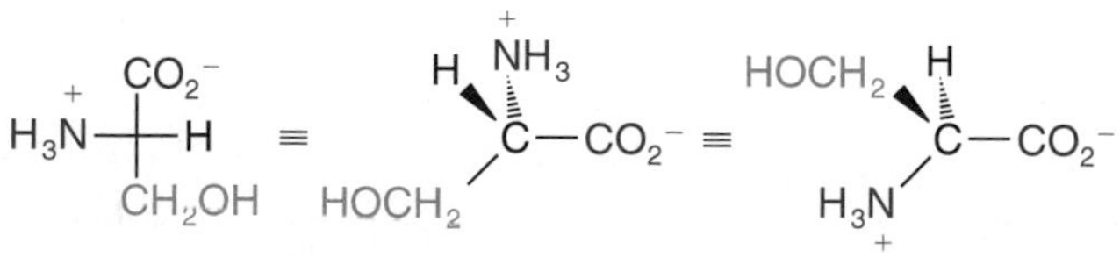

CO_2^- / $H_3\overset{+}{N}$—C—H / CH_2OH $\equiv$ H, $\overset{+}{N}H_3$, C—CO_2^-, $HOCH_2$ $\equiv$ $HOCH_2$, H, C—CO_2^-, $H_3\overset{+}{N}$

The order of decreasing sequence rule precedence of the three highest-ranked substituents traces an anticlockwise path.

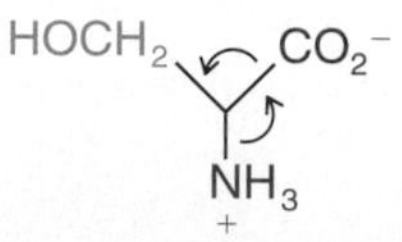

$HOCH_2$, CO_2^-, $\overset{+}{N}H_3$

The absolute configuration of L-serine is *S*.

PROBLEM 27.2 Which of the amino acids in Table 27.1 have more than one stereogenic center?

While all the chiral amino acids obtained from proteins have the L configuration at their α carbon, that should not be taken to mean that D amino acids are unknown. There are, in fact, quite a number of naturally occurring D amino acids. D-Alanine, for example, is a constituent of bacterial cell walls. The significant point is that D amino acids are not constituents of proteins.

27.3 ACID-BASE BEHAVIOR OF AMINO ACIDS

The physical properties of a typical amino acid such as glycine suggest that it is a very polar substance, much more polar than would be expected on the basis of its formulation as $H_2NCH_2CO_2H$. Glycine is a crystalline solid; it does not melt, but on being heated it eventually decomposes at 233°C. It is very soluble in water but practically insoluble in nonpolar organic solvents. These properties are attributed to the fact that the stable form of glycine is a **zwitterion,** or **inner salt.**

The zwitterion is also often referred to as a *dipolar ion.* Note, however, that it is not an ion, but is a neutral molecule.

$$H_2NCH_2C(=O)OH \rightleftharpoons H_3\overset{+}{N}CH_2C(=O)O^-$$

Zwitterionic form of glycine

The equilibrium expressed by the equation lies overwhelmingly to the side of the zwitterion.

Glycine, as well as other amino acids, is *amphoteric,* containing an acidic functional group and a basic functional group. The acidic functional group is the ammonium ion $H_3\overset{+}{N}—$; the basic functional group is the carboxylate ion $—CO_2^-$. How do we know this? Aside from the physical properties cited in the preceding paragraph, the acid-base properties of glycine, as illustrated by the titration curve in Figure 27.1, require it. In a strongly acidic medium the species present is $H_3\overset{+}{N}CH_2CO_2H$. As the pH is raised, a proton is removed from this species. Is the proton removed from the positively charged nitrogen or from the carboxyl group? We know what to expect for the relative acid strengths of $R\overset{+}{N}H_3$ and RCO_2H. A typical ammonium ion has $pK_a \cong 9$, and a typical carboxylic acid has $pK_a \cong 5$. The measured pK_a for the conjugate acid of glycine is 2.35, a value closer to that expected for deprotonation of the carboxyl group. As the pH is raised, a second deprotonation step, corresponding to removal of a proton from nitrogen of the zwitterion, is observed. The pK_a associated with this step is 9.78, much like that of typical alkylammonium ions.

$$H_3\overset{+}{N}CH_2C(=O)OH \underset{+H^+}{\overset{-H^+}{\rightleftharpoons}} H_3\overset{+}{N}CH_2C(=O)O^- \underset{+H^+}{\overset{-H^+}{\rightleftharpoons}} H_2NCH_2C(=O)O^-$$

Species present in strong acid; Zwitterion; predominant species in solutions near neutrality; Species present in strong base

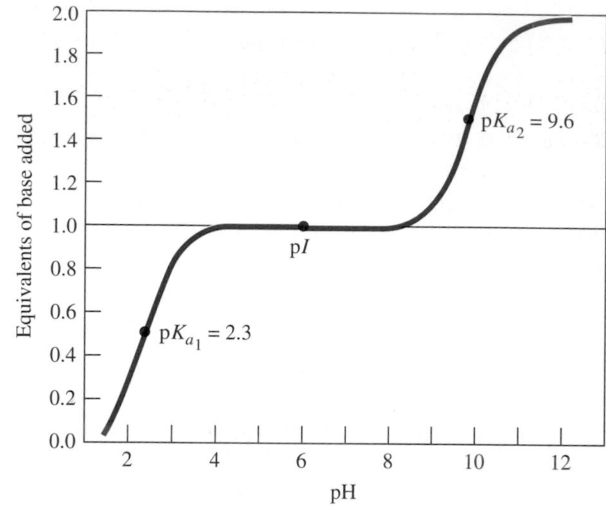

FIGURE 27.1 The titration curve of glycine. At pH values less than pK_{a_1}, $H_3\overset{+}{N}CH_2CO_2H$ is the major species present. At pH values between pK_{a_1} and pK_{a_2}, the principal species is the zwitterion $H_3\overset{+}{N}CH_2CO_2^-$. The concentration of the zwitterion is a maximum at the isoelectric point pI. At pH values greater than pK_{a_2}, $H_2NCH_2CO_2^-$ is the species present in greatest concentration.

Thus, glycine is characterized by two pK_a values: the one corresponding to the more acidic site is designated pK_{a_1}, the one corresponding to the less acidic site is designated pK_{a_2}. Table 27.2 lists pK_{a_1} and pK_{a_2} values for the α-amino acids that have neutral side chains, which are the first two groups of amino acids given in Table 27.1. In all cases their pK_a values are similar to those of glycine.

Table 27.2 includes a column labeled pI, which gives **isoelectric point** values. The isoelectric point is the pH at which the amino acid bears no net charge; it corresponds to the pH at which the concentration of the zwitterion is a maximum. For the amino acids in Table 27.2 this is the average of pK_{a_1} and pK_{a_2}. It lies slightly to the acid side of neutrality because amino acids are somewhat stronger acids than bases.

TABLE 27.2
Acid-Base Properties of Amino Acids with Neutral Side Chains

Amino acid	pK_{a_1}*	pK_{a_2}*	pI
Glycine	2.34	9.60	5.97
Alanine	2.34	9.69	6.00
Valine	2.32	9.62	5.96
Leucine	2.36	9.60	5.98
Isoleucine	2.36	9.60	6.02
Methionine	2.28	9.21	5.74
Proline	1.99	10.60	6.30
Phenylalanine	1.83	9.13	5.48
Tryptophan	2.83	9.39	5.89
Asparagine	2.02	8.80	5.41
Glutamine	2.17	9.13	5.65
Serine	2.21	9.15	5.68
Threonine	2.09	9.10	5.60

* In all cases pK_{a_1} corresponds to ionization of the carboxyl group; pK_{a_2} corresponds to deprotonation of the ammonium ion.

TABLE 27.3
Acid-Base Properties of Amino Acids with Ionizable Side Chains

Amino acid	pK_{a_1}*	pK_{a_2}	pK_{a_3}	pI
Aspartic acid	1.88	3.65	9.60	2.77
Glutamic acid	2.19	4.25	9.67	3.22
Tyrosine	2.20	9.11	10.07	5.66
Cysteine	1.96	8.18	10.28	5.07
Lysine	2.18	8.95	10.53	9.74
Arginine	2.17	9.04	12.48	10.76
Histidine	1.82	6.00	9.17	7.59

* In all cases pK_{a_1} corresponds to ionization of the carboxyl group of $RCH(\overset{+}{N}H_3)CO_2H$.

Some amino acids, including those listed in the last two sections of Table 27.1, have side chains that bear acidic or basic groups. As Table 27.3 indicates, these amino acids are characterized by three pK_a values. The "extra" pK_a value (it can be either pK_{a_2} or pK_{a_3}) reflects the nature of the function present in the side chain. The isoelectric points of the amino acids in Table 27.3 are midway between the pK_a values of the monocation and monoanion and are well removed from neutrality when the side chain bears a carboxyl group (aspartic acid, for example) or a basic amine function (lysine, for example).

PROBLEM 27.3 Write the most stable structural formula for tyrosine:

(*a*) In its cationic form
(*b*) In its zwitterionic form
(*c*) As a monoanion
(*d*) As a dianion

SAMPLE SOLUTION (*a*) The cationic form of tyrosine is the one present at low pH. The positive charge is on nitrogen, and the species present is an ammonium ion.

$$HO-C_6H_4-CH_2CH(\overset{+}{N}H_3)CO_2H$$

PROBLEM 27.4 Write structural formulas for the principal species present when the pH of a solution containing lysine is raised from 1 to 9 and again to 13.

The acid-base properties of their side chains are one way in which individual amino acids differ. This is important in peptides and proteins, where the properties of the substance depend on its amino acid constituents, especially on the nature of the side chains. It is also important in analyses in which a complex mixture of amino acids is separated into its components by taking advantage of the differences in their proton-donating and proton-accepting abilities.

ELECTROPHORESIS

Electrophoresis is a method for separation and purification that relies on the movement of charged particles in an electric field. Its principles can be introduced by considering the electrophoretic behavior of some representative amino acids. The medium is a cellulose acetate strip that is moistened with an aqueous solution buffered at a particular pH. The opposite ends of the strip are placed in separate compartments containing the buffer, and each compartment is connected to a source of direct electric current by an electrode (Figure 27.2*a*). If the buffer solution is more acidic than the isoelectric point (pI) of the amino acid, the amino acid has a net positive charge and migrates toward the negatively charged electrode. Conversely, when the buffer is more basic than the pI of the amino acid, the amino acid has a net negative charge and migrates toward the positively charged electrode. When the pH of the buffer corresponds to the pI, the amino acid has no net charge and does not migrate from the origin.

Thus if a mixture containing alanine, aspartic acid, and lysine is subjected to electrophoresis in a buffer that matches the isoelectric point of alanine (pH 6.0), aspartic acid (pI = 2.8) migrates toward the positive electrode, alanine remains at the origin, and lysine (pI = 9.7) migrates toward the negative electrode (Figure 27.2*b*).

$^-O_2CCH_2CHCO_2^-$ (with $^+NH_3$ on CH)	$CH_3CHCO_2^-$ (with $^+NH_3$ on CH)	$H_3\overset{+}{N}(CH_2)_4CHCO_2^-$ (with $^+NH_3$ on CH)
Aspartic acid (monoanion)	Alanine (neutral)	Lysine (monocation)

A mixture of amino acids

$^-O_2CCH_2CHCO_2^-$ | $^+NH_3$ $\quad$ $CH_3CHCO_2^-$ | $^+NH_3$ $\quad$ $H_3\overset{+}{N}(CH_2)_4CHCO_2^-$ | $^+NH_3$

is placed at the center of a sheet of cellulose acetate. The sheet is soaked with an aqueous solution buffered at a pH of 6.0. At this pH aspartic acid exists as its –1 anion, alanine as its zwitterion, and lysine as its +1 ion.

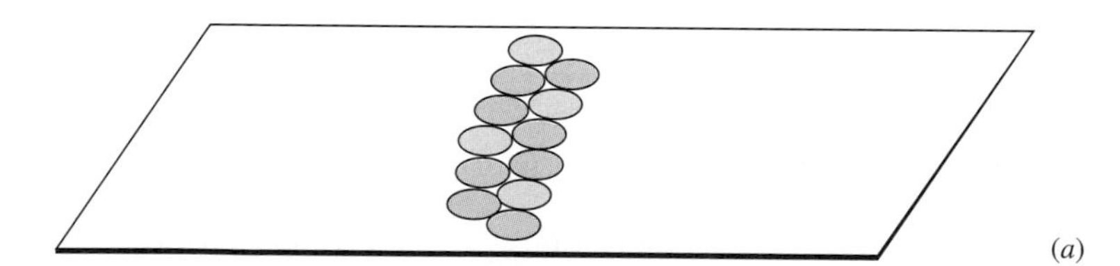

Application of an electric current causes the negatively charged ions to migrate to the + electrode, and the positively charged ions to migrate to the – electrode. The zwitterion, with a net charge of zero, remains at its original position.

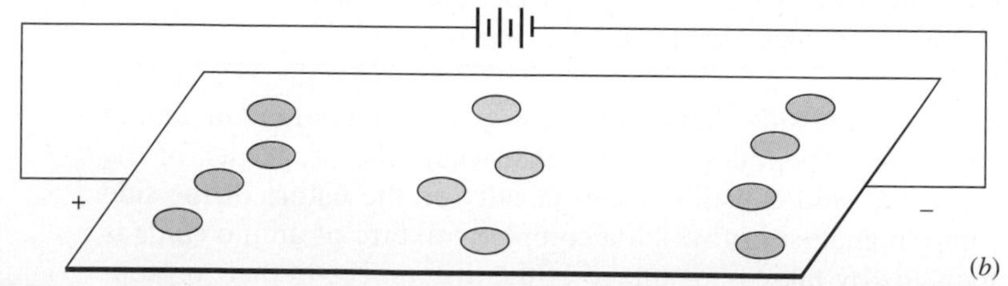

FIGURE 27.2 Application of electrophoresis to the separation of aspartic acid, alanine, and lysine according to their charge type at a pH corresponding to the isoelectric point (pI) of alanine.

Electrophoresis is used primarily to analyze mixtures of peptides and proteins, rather than individual amino acids, but analogous principles apply. Because they incorporate different numbers of amino acids and because their side chains are different, two peptides will have slightly different acid-base properties and slightly different net charges at a particular pH. Thus, their mobilities in an electric field will be different, and electrophoresis can be used to separate them. The medium used to separate peptides and proteins is typically a polyacrylamide gel, leading to the term **gel electrophoresis** for this technique.

A second factor that governs the rate of migration during electrophoresis is the size (length and shape) of the peptide or protein. Larger molecules move through the polyacrylamide gel more slowly than smaller ones. In current practice, the experiment is modified to exploit differences in size more than differences in net charge, especially in the **SDS gel electrophoresis** of proteins. Approximately 1.5 g of the detergent *sodium dodecyl sulfate* (SDS, page 775) per gram of protein is added to the aqueous buffer. SDS binds to the protein, causing the protein to unfold so that it is roughly rod-shaped with the $CH_3(CH_2)_{10}CH_2$—groups of SDS associated with the lipophilic portions of the protein. The negatively charged sulfate groups are exposed to the water. The SDS molecules that they carry ensure that all the protein molecules are negatively charged and migrate toward the positive electrode. Further, all the proteins in the mixture now have similar shapes and tend to travel at rates proportional to their chain length. Thus, when carried out on a preparative scale, SDS gel electrophoresis permits proteins in a mixture to be separated according to their molecular weight. On an analytical scale, it is used to estimate the molecular weight of a protein by comparing its electrophoretic mobility with that of proteins of known molecular weight.

Later, in Section 27.28, we will see how gel electrophoresis is used in nucleic acid chemistry.

27.4 SYNTHESIS OF AMINO ACIDS

Of the various methods available for the laboratory synthesis of amino acids, one of the oldest dates back over 100 years and is simply a nucleophilic substitution in which ammonia reacts with an α-halo carboxylic acid.

$$\underset{\text{2-Bromopropanoic acid}}{CH_3\underset{\displaystyle Br}{\underset{|}{C}}HCO_2H} + \underset{\text{Ammonia}}{2NH_3} \xrightarrow{H_2O} \underset{\text{Alanine (65–70\%)}}{CH_3\underset{\displaystyle \overset{+}{N}H_3}{\underset{|}{C}}HCO_2^-} + \underset{\text{Ammonium bromide}}{NH_4Br}$$

The α-halo acid is normally prepared by the Hell-Volhard-Zelinsky reaction (Section 19.16).

PROBLEM 27.5 Outline the steps in a synthesis of valine from 3-methylbutanoic acid.

In the **Strecker synthesis** an aldehyde is converted to an α-amino acid with one more carbon atom by a two-stage procedure in which an α-amino nitrile is an intermediate. The α-amino nitrile is formed by reaction of the aldehyde with ammonia or an ammonium salt and a source of cyanide ion. Hydrolysis of the nitrile group to a carboxylic acid function completes the synthesis.

$$\underset{\text{Acetaldehyde}}{CH_3\overset{O}{\overset{\|}{C}}H} \xrightarrow[NaCN]{NH_4Cl} \underset{\text{2-Aminopropanenitrile}}{CH_3\underset{\displaystyle NH_2}{\underset{|}{C}}HC{\equiv}N} \xrightarrow[\text{2. }HO^-]{\text{1. }H_2O,\ HCl,\ \text{heat}} \underset{\text{Alanine (52–60\%)}}{CH_3\underset{\displaystyle \overset{+}{N}H_3}{\underset{|}{C}}HCO_2^-}$$

The synthesis of alanine was described by Adolf Strecker of the University of Würzburg (Germany) in a paper published in 1850.

PROBLEM 27.6 Outline the steps in the preparation of valine by the Strecker synthesis.

The most widely used method for the laboratory synthesis of α-amino acids is a modification of the malonic ester synthesis (Section 21.7). The key reagent is *diethyl acetamidomalonate,* a derivative of malonic ester that already has the critical nitrogen substituent in place at the α carbon atom. The side chain is introduced by alkylating the anion of diethyl acetamidomalonate with an alkyl halide in exactly the same way as diethyl malonate itself is alkylated.

$$\underset{\text{Diethyl acetamidomalonate}}{CH_3\overset{O}{\overset{\|}{C}}NHCH(CO_2CH_2CH_3)_2} \xrightarrow[CH_3CH_2OH]{NaOCH_2CH_3} \underset{\text{Sodium salt of diethyl acetamidomalonate}}{CH_3\overset{O}{\overset{\|}{C}}NH\underset{Na^+}{\ddot{C}}(CO_2CH_2CH_3)_2} \xrightarrow{C_6H_5CH_2Cl} \underset{\text{Diethyl acetamidobenzylmalonate (90\%)}}{CH_3\overset{O}{\overset{\|}{C}}NH\underset{CH_2C_6H_5}{\underset{|}{C}}(CO_2CH_2CH_3)_2}$$

Hydrolysis of the alkylated derivative of diethyl acetamidomalonate removes the acetyl group from nitrogen and converts the two ester functions to carboxyl groups. Decarboxylation of the diacid gives the desired product.

$$\underset{\text{Diethyl acetamidobenzylmalonate}}{CH_3\overset{O}{\overset{\|}{C}}NH\underset{CH_2C_6H_5}{\underset{|}{C}}(CO_2CH_2CH_3)_2} \xrightarrow[H_2O,\ heat]{HBr} \underset{\text{(not isolated)}}{H_3\overset{+}{N}\underset{CH_2C_6H_5}{\underset{|}{C}}(CO_2H)_2} \xrightarrow[-\,CO_2]{heat} \underset{\text{Phenylalanine (65\%)}}{C_6H_5CH_2\underset{\overset{+}{N}H_3}{\underset{|}{C}}HCO_2^-}$$

PROBLEM 27.7 Outline the steps in the synthesis of valine from diethyl acetamidomalonate. The overall yield of valine by this method is reported to be rather low (31 percent). Can you think of a reason why this synthesis is not very efficient?

Unless a resolution step is introduced into the reaction scheme, the α-amino acids prepared by the synthetic methods just described are racemic. Optically active amino acids, when desired, may be obtained by resolving a racemic mixture or by **enantioselective synthesis.** A synthesis is described as enantioselective if it produces one enantiomer of a chiral compound in an amount greater than its mirror image. Recall from Section 7.9 that optically inactive reactants cannot give optically active products. Therefore enantioselective syntheses of amino acids require an enantiomerically enriched chiral reagent or catalyst at some point in the process. If the chiral reagent or catalyst is enantiomerically homogeneous and if the reaction sequence is completely enantioselective, an optically pure amino acid is obtained. Chemists have succeeded in preparing α-amino acids by techniques that are more than 95 percent enantioselective. While this is an impressive feat, we must not lose sight of the fact that the reactions that produce amino acids in living systems do so with 100 percent enantioselectivity.

27.5 REACTIONS OF AMINO ACIDS

Amino acids undergo reactions characteristic of both their amine and carboxylic acid functional groups. Acylation is a typical reaction of the amino group.

$$\underset{\text{Glycine}}{H_3\overset{+}{N}CH_2CO_2^-} + \underset{\text{Acetic anhydride}}{CH_3\overset{O}{\overset{\|}{C}}O\overset{O}{\overset{\|}{C}}CH_3} \longrightarrow \underset{N\text{-Acetylglycine (89–92\%)}}{CH_3\overset{O}{\overset{\|}{C}}NHCH_2CO_2H} + \underset{\text{Acetic acid}}{CH_3CO_2H}$$

Ester formation is a typical reaction of the carboxyl group.

$$\underset{\text{Alanine}}{CH_3\underset{\overset{|}{\underset{+}{NH_3}}}{C}HCO_2^-} + \underset{\text{Ethanol}}{CH_3CH_2OH} \xrightarrow{HCl} \underset{\text{Hydrochloride salt of alanine ethyl ester (90–95\%)}}{CH_3\underset{\overset{|}{\underset{+}{NH_3}}}{C}H\overset{O}{\overset{\|}{C}}OCH_2CH_3\ Cl^-}$$

A reaction of amino acids that is used to detect their presence is the formation of a purple color on treatment with *ninhydrin.* The same compound responsible for the purple color is formed from all amino acids in which the α-amino group is primary.

$$2\ \text{Ninhydrin} + H_3\overset{+}{N}\underset{\underset{R}{|}}{C}HCO_2^- + HO^- \longrightarrow \text{Violet dye} + \boxed{R\overset{O}{\overset{\|}{C}}H,\ CO_2,\ 4\ H_2O}$$

Ninhydrin

Violet dye

(Formed, but not normally isolated)

Proline, in which the α-amino group is secondary, gives an orange compound on reaction with ninhydrin.

PROBLEM 27.8 Suggest a reasonable mechanism for the reaction of an α-amino acid with ninhydrin.

27.6 SOME BIOCHEMICAL REACTIONS INVOLVING AMINO ACIDS

The 20 amino acids listed in Table 27.1 are biosynthesized by a number of different pathways, and we will touch on only a few of them in an introductory way. We will examine the biosynthesis of glutamic acid first, since it illustrates a biochemical process analogous to a reaction we have discussed earlier in the context of amine synthesis, *reductive amination* (Section 22.11).

Glutamic acid is formed in most organisms from ammonia and α-ketoglutaric acid. α-Ketoglutaric acid is one of the intermediates in the **tricarboxylic acid cycle** (also called the **Krebs cycle**) and arises via metabolic breakdown of food sources—carbohydrates, fats, and proteins.

The August 1986 issue of the *Journal of Chemical Education* (pp. 673–677) contains a review of the Krebs cycle.

$$\underset{\alpha\text{-Ketoglutaric acid}}{HO_2CCH_2CH_2\overset{O}{\overset{\|}{C}}CO_2H} + \underset{\text{Ammonia}}{NH_3} \xrightarrow[\text{reducing agents}]{\text{enzymes}} \underset{\text{L-Glutamic acid}}{HO_2CCH_2CH_2\underset{\overset{+}{N}H_3}{\underset{|}{C}}HCO_2^-}$$

Ammonia reacts with the ketone carbonyl group to give an imine ($>C{=}NH$), which is then reduced to the amine function of the α-amino acid. Both imine formation and imine reduction are enzyme-catalyzed reactions. The reduced form of nicotinamide diphosphonucleotide (NADPH) is a coenzyme and acts as a reducing agent. The step in which the imine is reduced is the one in which the stereogenic center is introduced and gives only L-glutamic acid.

L-Glutamic acid is not an essential amino acid. It need not be present in the diet, since mammals possess the ability to biosynthesize it from sources of α-ketoglutaric acid. It is, however, a key intermediate in the biosynthesis of other amino acids by a process known as **transamination.** L-Alanine, for example, is formed from pyruvic acid by transamination from L-glutamic acid.

$$\underset{\text{Pyruvic acid}}{CH_3\overset{O}{\overset{\|}{C}}CO_2H} + \underset{\text{L-Glutamic acid}}{HO_2CCH_2CH_2\underset{\overset{+}{N}H_3}{\underset{|}{C}}HCO_2^-} \xrightarrow{\text{enzymes}} \underset{\text{L-Alanine}}{CH_3\underset{\overset{+}{N}H_3}{\underset{|}{C}}HCO_2^-} + \underset{\alpha\text{-Ketoglutaric acid}}{HO_2CCH_2CH_2\overset{O}{\overset{\|}{C}}CO_2H}$$

In transamination an amine group is transferred from L-glutamic acid to pyruvic acid. An outline of the mechanism of transamination is presented in Figure 27.3.

One amino acid often serves as the biological precursor to another. L-Phenylalanine is classified as an essential amino acid, whereas its *p*-hydroxy derivative, L-tyrosine, is not. This is because mammals can convert L-phenylalanine to L-tyrosine by hydroxylation of the aromatic ring. An *arene oxide* (Section 24.7) is an intermediate.

$$\underset{\text{L-Phenylalanine}}{C_6H_5{-}CH_2\underset{\overset{+}{N}H_3}{\underset{|}{C}}HCO_2^-} \xrightarrow[\text{enzyme}]{O_2} \underset{\text{Arene oxide intermediate}}{(\text{arene oxide}){-}CH_2\underset{\overset{+}{N}H_3}{\underset{|}{C}}HCO_2^-} \xrightarrow{\text{enzyme}} \underset{\text{L-Tyrosine}}{HO{-}C_6H_4{-}CH_2\underset{\overset{+}{N}H_3}{\underset{|}{C}}HCO_2^-}$$

Step 1: The amine function of L-glutamate reacts with the ketone carbonyl of pyruvate to form an imine.

$$\underset{\text{L-Glutamate}}{(^-O_2C)(^-O_2CCH_2CH_2)CH{-}\overset{+}{N}H_3} + \underset{\text{Pyruvate}}{O{=}C(CO_2^-)CH_3} \longrightarrow \underset{\text{Imine}}{(^-O_2C)(^-O_2CCH_2CH_2)CH{-}N{=}C(CO_2^-)CH_3}$$

Step 2: Enzyme-catalyzed proton-transfer steps cause migration of the double bond, converting the imine formed in step 1 to an isomeric imine.

base: H—acid

$$\underset{\text{Imine from step 1}}{(^-O_2C)(^-O_2CCH_2CH_2)C(H){-}N{=}C(CO_2^-)CH_3} \longrightarrow \underset{\text{Rearranged imine}}{(^-O_2C)(^-O_2CCH_2CH_2)C{=}N{-}C(H)(CO_2^-)CH_3}$$

Step 3: Hydrolysis of the rearranged imine gives L-alanine and α-ketoglutarate.

$$\underset{\text{Rearranged imine}}{(^-O_2C)(^-O_2CCH_2CH_2)C{=}N{-}CH(CO_2^-)CH_3} + \underset{\text{Water}}{H_2O} \longrightarrow \underset{\alpha\text{-Ketoglutarate}}{(^-O_2C)(^-O_2CCH_2CH_2)C{=}O} + \underset{\text{L-Alanine}}{H_3\overset{+}{N}{-}CH(CO_2^-)CH_3}$$

FIGURE 27.3 The mechanism of transamination. All the steps are enzyme-catalyzed.

Some people lack the enzymes necessary to convert L-phenylalanine to L-tyrosine. The L-phenylalanine that they obtain from their diet is instead diverted along an alternative metabolic pathway which produces phenylpyruvic acid:

$$\underset{\text{L-Phenylalanine}}{C_6H_5{-}CH_2CH(\overset{+}{N}H_3)CO_2^-} \xrightarrow{\text{enzymes}} \underset{\text{Phenylpyruvic acid}}{C_6H_5{-}CH_2\overset{O}{\overset{\|}{C}}CO_2H}$$

Phenylpyruvic acid can cause mental retardation in infants who are deficient in the enzymes necessary to convert L-phenylalanine to L-tyrosine. They are said to suffer from **phenylketonuria,** or **PKU disease.** PKU disease can be detected by a simple test routinely administered to newborns. It cannot be cured, but is controlled by restricting the dietary intake of L-phenylalanine. In practice this means avoiding foods such as meat that are rich in L-phenylalanine.

Among the biochemical reactions that amino acids undergo is *decarboxylation* to amines. Decarboxylation of histidine, for example, gives histamine, a powerful vaso-

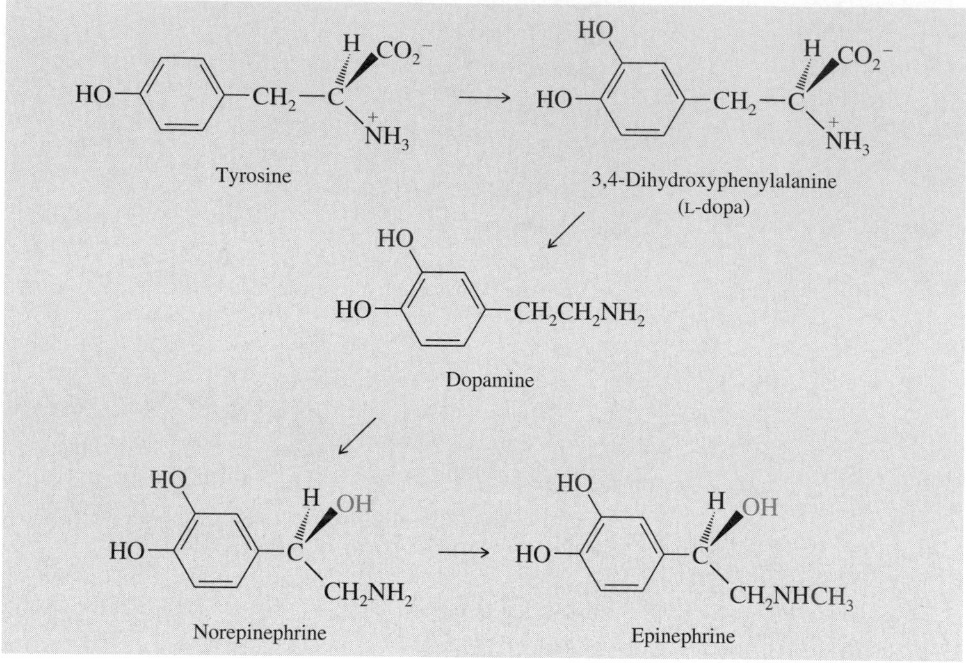

FIGURE 27.4 Tyrosine is the biosynthetic precursor to a number of neurotransmitters. Each transformation is enzyme-catalyzed. Hydroxylation of the aromatic ring of tyrosine converts it to 3,4-dihydroxyphenylalanine (L-dopa), decarboxylation of which gives dopamine. Hydroxylation of the benzylic carbon of dopamine converts it to norepinephrine (noradrenaline), and methylation of the amino group of norepinephrine yields epinephrine (adrenaline).

Tyrosine → 3,4-Dihydroxyphenylalanine (L-dopa) → Dopamine → Norepinephrine → Epinephrine

dilator normally present in tissue and formed in excessive amounts under conditions of traumatic shock.

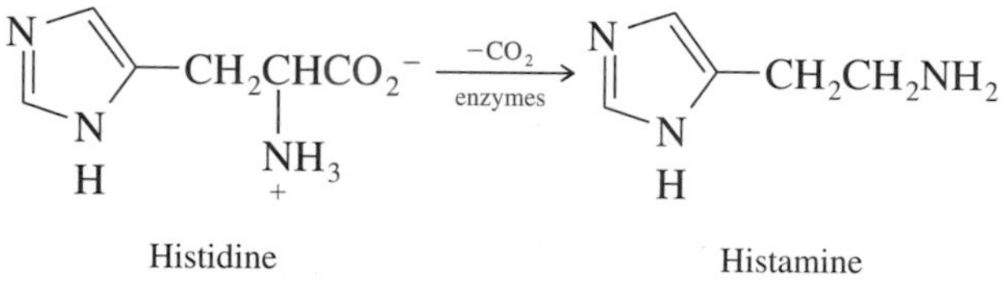

Histidine $\xrightarrow[\text{enzymes}]{-CO_2}$ Histamine

Histamine is responsible for many of the symptoms associated with hay fever and other allergies. An antihistamine relieves these symptoms by blocking the action of histamine.

PROBLEM 27.9 One of the amino acids in Table 27.1 is the biological precursor to γ-aminobutyric acid (4-aminobutanoic acid), which it forms by a decarboxylation reaction. Which amino acid is this?

For a review of neurotransmitters, see the February 1988 issue of the *Journal of Chemical Education* (pp. 108–111).

The chemistry of the brain and central nervous system is affected by a group of substances called **neurotransmitters.** Several of these neurotransmitters arise from L-tyrosine by structural modification and decarboxylation, as outlined in Figure 27.4.

27.7 PEPTIDES

A key biochemical reaction of amino acids is their conversion to peptides, polypeptides, and proteins. In all these substances amino acids are linked together by amide bonds. The amide bond between the amino group of one amino acid and the carboxyl of another is called a **peptide bond.** Alanylglycine is a representative dipeptide.

N-terminal amino acid $\overset{+}{H_3N}CHC(=O)—NHCH_2CO_2^-$ C-terminal amino acid

$$\overset{+}{H_3N}\underset{\displaystyle CH_3}{CH}\overset{\displaystyle O}{\overset{\|}{C}}—NHCH_2CO_2^-$$

Alanylglycine
(Ala-Gly)

By agreement, peptide structures are written so that the amino group (as $H_3\overset{+}{N}$— or H_2N—) is at the left and the carboxyl group (as CO_2^- or CO_2H) is at the right. The left and right ends of the peptide are referred to as the **N terminus** (or amino terminus) and the **C terminus** (or carboxyl terminus), respectively. Alanine is the N-terminal amino acid in alanylglycine while glycine is the C-terminal amino acid. A dipeptide is named as an acyl derivative of the C-terminal amino acid. We call the precise order of bonding in a peptide its amino acid **sequence.** The amino acid sequence is conveniently specified by using the three-letter amino acid abbreviations for the respective amino acids and connecting them by hyphens. Individual amino acid components of peptides are often referred to as amino acid **residues.**

It is understood that α-amino acids occur as their L stereoisomers unless otherwise indicated. The D notation is explicitly shown when a D amino acid is present, and a racemic amino acid is identified by the prefix DL.

PROBLEM 27.10 Write structural formulas showing the constitution of each of the following dipeptides:

(*a*) Gly-Ala
(*b*) Ala-Phe
(*c*) Phe-Ala
(*d*) Gly-Glu
(*e*) Lys-Gly
(*f*) D-Ala-D-Ala

SAMPLE SOLUTION (*a*) Gly-Ala is a constitutional isomer of Ala-Gly. Glycine is the N-terminal amino acid in Gly-Ala; alanine is the C-terminal amino acid.

N-terminal amino acid C-terminal amino acid

$$\overset{+}{H_3N}CH_2\overset{\displaystyle O}{\overset{\|}{C}}—NH\underset{\displaystyle CH_3}{CH}CO_2^-$$

Glycylalanine
(Gly-Ala)

Figure 27.5 shows the structure of Ala-Gly as determined by x-ray crystallography. An important feature is the planar geometry associated with the peptide bond, and the most stable conformation with respect to this bond has the two α carbon atoms anti to each other. Rotation about the amide linkage is slow because delocalization of the unshared electron pair of nitrogen into the carbonyl group gives partial double-bond character to the carbon-nitrogen bond.

PROBLEM 27.11 Expand your answer to Problem 27.10 by showing the structural formula for each dipeptide in a manner that reveals the stereochemistry at the α carbon atom.

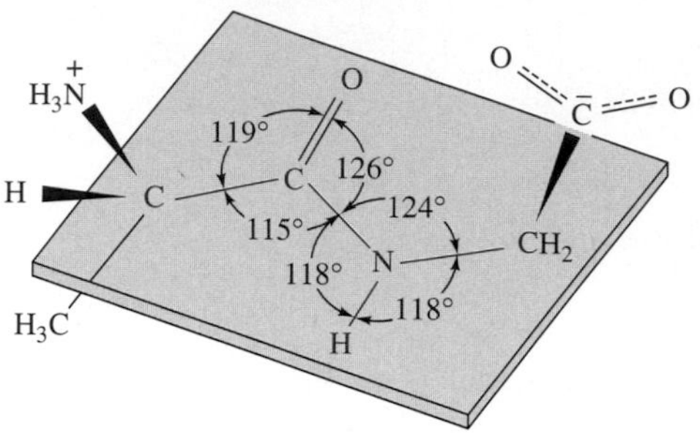

FIGURE 27.5 The structure of the dipeptide L-alanylglycine as revealed by x-ray crystallography. The six atoms shown in red all lie in the same plane, and the α carbon atoms of the two amino acids are anti to each other.

SAMPLE SOLUTION (*a*) Glycine is achiral, and so Gly-Ala has only one stereogenic center, the α carbon atom of the L-alanine residue. When the carbon chain is drawn in an extended zigzag fashion and L-alanine is the C terminus, its structure is as shown:

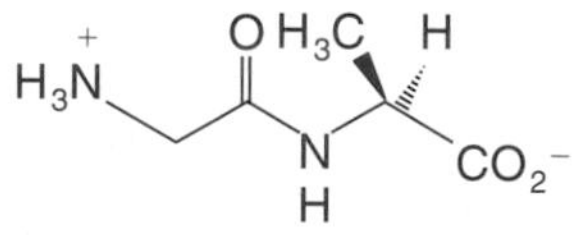

Glycyl-L-alanine (Gly-Ala)

Higher peptides are treated in an analogous fashion. Figure 27.6 gives the structural formula and amino acid sequence of a naturally occurring pentapeptide known as *leucine enkephalin.* Enkephalins are pentapeptide components of **endorphins,** polypeptides present in the brain that act as the body's own painkillers. A second substance, known as *methionine enkephalin,* is also present in endorphins. Methionine enkephalin differs from leucine enkephalin only in having methionine instead of leucine as its C-terminal amino acid.

PROBLEM 27.12 What is the amino acid sequence (using three-letter abbreviations) of methionine enkephalin?

Peptides having structures slightly different from those described to this point are known. One such variation is seen in the nonapeptide *oxytocin,* shown in Figure 27.7. Oxytocin is a hormone secreted by the pituitary gland that stimulates uterine contractions during childbirth. Rather than terminating in a carboxyl group, the terminal glycine residue in oxytocin has been modified so that it exists as the corresponding amide. Two cysteine units, one of them the N-terminal amino acid, are joined by the

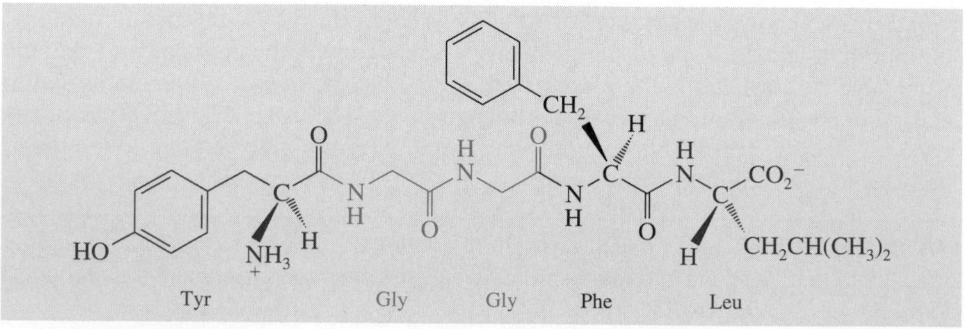

FIGURE 27.6 The structure of the pentapeptide leucine enkephalin. Its amino acid sequence is Tyr-Gly-Gly-Phe-Leu, tyrosine being the N-terminal and leucine the C-terminal amino acid.

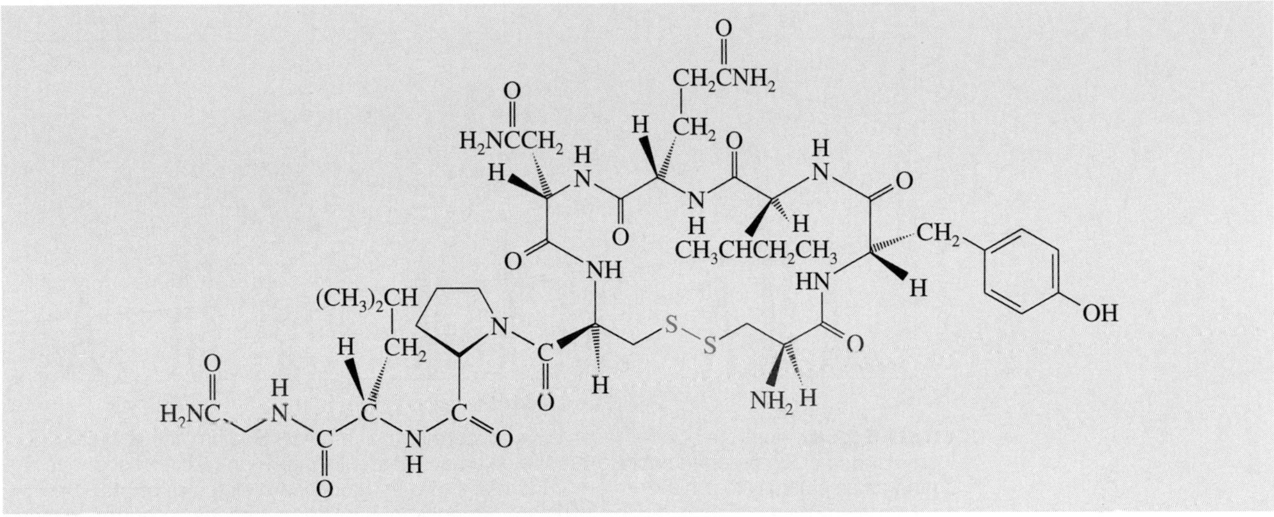

FIGURE 27.7 The structure of oxytocin, a nonapeptide containing a disulfide bond between two cysteine residues. One of these cysteines is the N-terminal amino acid; it is shown in blue. The C-terminal amino acid is the amide of glycine and is shown in red. There are no free carboxyl groups in the molecule; all exist in the form of carboxamides.

sulfur-sulfur bond of a large-ring cyclic disulfide unit. This is a common structural modification in polypeptides and proteins that contain cysteine residues. It provides a covalent bond between regions of peptide chains that may be many amino acid residues removed from each other.

Recall from Section 15.14 that compounds of the type RSH are readily oxidized to RSSR.

27.8 PEPTIDE STRUCTURE DETERMINATION: AMINO ACID ANALYSIS

Chemists and biochemists distinguish between several levels of peptide structure. The **primary structure** of a peptide is its constitution and corresponds to its amino acid sequence plus any disulfide links. Determining the primary structure can be a truly formidable task, since the 20 amino acids of Table 27.1 provide a number of molecular building blocks large enough to produce a staggering number of dipeptides, tripeptides, tetrapeptides, and so forth. Given a peptide of unknown structure, how does one go about determining its amino acid sequence?

The first step is to identify which amino acids are present and the relative amounts of each one. The unknown peptide is subjected to acid-catalyzed hydrolysis by heating it in 6 *M* hydrochloric acid for about 24 h. Under these conditions the amide bonds are cleaved, and a solution is obtained that contains all the amino acids originally present in the peptide. This mixture is then separated by a technique known as **ion-exchange chromatography.** Ion-exchange chromatography separates amino acids according to their acid-base properties and, to a lesser extent, depends on differences in the lipophilic properties of their side chains. After adsorption of the mixture of amino acids on a polymeric ion-exchange resin, the resin is washed with aqueous buffers of increasing pH. Individual amino acids pass through the column at different rates and appear in different fractions of the aqueous effluent. The exit of amino acids from the column is determined by mixing the effluent with ninhydrin. The intensity of the ninhydrin color

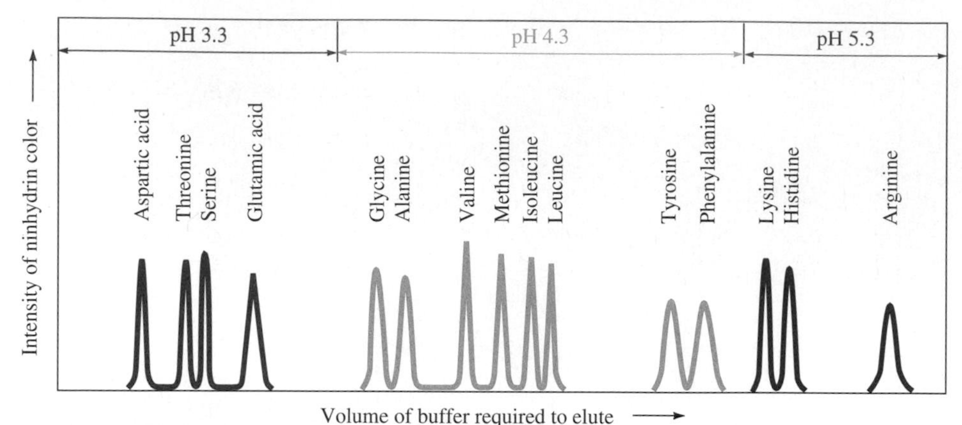

FIGURE 27.8 Analysis of a representative mixture of amino acids. Sodium citrate buffers were used to elute the amino acids from the column. The peaks shown in red correspond to amino acids eluted with a buffer of pH 3.3; those shown in green were eluted at pH 4.3. Because amino acids with basic side chains are strongly held by the adsorbent, a shorter column was used for their analysis. These were eluted with a buffer of pH 5.3 and are shown in blue at the right. The horizontal scale is different for these three amino acids from that for the others.

is monitored electronically and the appearance of an amino acid recorded as a peak on a strip chart.

A typical amino acid analysis is illustrated in Figure 27.8. The volume of buffer of specified pH required to remove the various amino acids from the ion-exchange column is compared with that required for authentic samples of known amino acids and serves to identify each component in the mixture. Since peak areas are proportional to the amount of amino acid present, the molar ratios of amino acids are also determined. The entire operation is carried out automatically on an apparatus called an **amino acid analyzer.** The analytical procedure is extremely sensitive and can be performed routinely with 10^{-5} to 10^{-7} g of peptide.

An amino acid analysis identifies the component amino acids in a peptide and their ratios. Additional information typically includes molecular weight, which for small peptides can be determined exactly by mass spectrometry. Molecular weights of large peptides and proteins are estimated by a variety of specialized physical methods designed for this purpose, such as SDS gel electrophoresis.

27.9 PEPTIDE STRUCTURE DETERMINATION: SEQUENCE ANALYSIS

The principles of peptide sequencing may be illustrated by considering a representative example. Suppose we have a tetrapeptide of unknown sequence but which, on amino acid analysis, was shown to contain Ala, Gly, Phe, and Val in equimolar ratios. What additional information do we need in order to determine the sequence of this tetrapeptide?

While there are 24 possible tetrapeptides that may be derived from Ala, Gly, Phe, and Val, a relatively modest amount of information is required to distinguish among them. The following three experimentally determined facts are sufficient to define the primary structure of this tetrapeptide:

1. Valine is the N-terminal amino acid.

2. Hydrolysis of the tetrapeptide yielded a number of fragments, one of which was a tripeptide composed of Gly, Phe, and Val.
3. Also present in the hydrolysis mixture was a dipeptide containing Ala and Gly.

We interpret these data as follows. Since Val is the N-terminal amino acid and the tripeptide obtained on partial hydrolysis contains Val but not Ala, Ala must be the C-terminal amino acid. Since Ala and Gly are found together in the same dipeptide fragment from a partial hydrolysis, Gly must be the third amino acid residue, counting from the N terminus. Therefore, the primary structure of the tetrapeptide is

Val-Phe-Gly-Ala

Partial hydrolysis of the tetrapeptide can give dipeptide and tripeptide fragments other than those that were cited, and so more information is normally available than the three facts that yielded the solution to the problem. In practice, as many fragments as possible are isolated in pure form, and their amino acid composition is determined. Thus, the internal consistency of the entire pattern of results serves as a check on the proposed sequence.

PROBLEM 27.13 List all the substances that might be present in a hydrolysis mixture obtained from Val-Phe-Gly-Ala.

Amino acid sequencing of higher peptides is a challenging task, one that requires experimental skill and a systematic analysis of the data. While the approach just outlined is, in principle, capable of being applied to a peptide of any length, peptide sequencing has been simplified by the introduction of a number of techniques that provide additional information. Some of these techniques are described in the following sections. We will begin with **end group analysis.** A very important piece of information in our example of peptide sequencing was the identification of valine as the N-terminal amino acid. How was the identity of this amino acid determined?

27.10 END GROUP ANALYSIS: THE N TERMINUS

Identification of the N-terminal amino acid rests on the fact that the α-amino groups of all the amino acids in a peptide chain, except the N-terminal amino acid, are incorporated into amide bonds. The α-amino group of the N-terminal amino acid is free and capable of acting as a nucleophile. One method for N-terminal amino acid identification involves treatment of a peptide with 1-fluoro-2,4-dinitrobenzene, which is very reactive in nucleophilic aromatic substitutions that proceed by the addition-elimination mechanism (Section 23.6).

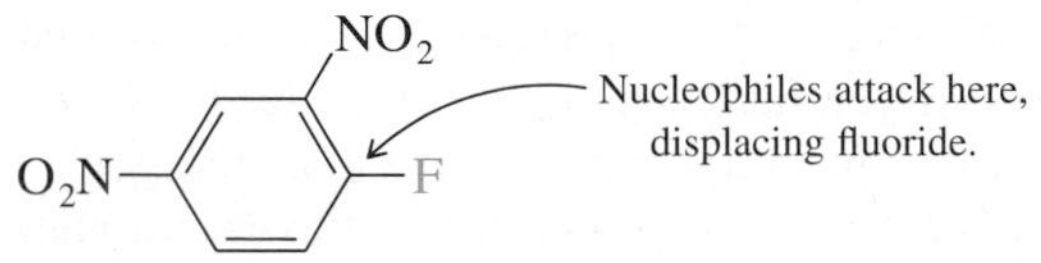

NO_2

O_2N—(benzene ring)—F

Nucleophiles attack here, displacing fluoride.

1-Fluoro-2,4-dinitrobenzene

The amino group of the N-terminal amino acid displaces fluoride from 1-fluoro-2,4-dinitrobenzene and gives a peptide in which the N-terminal nitrogen is labeled with a

The reaction is carried out by mixing the peptide and 1-fluoro-2,4-dinitrobenzene in the presence of a weak base such as sodium carbonate. In the first step the base abstracts a proton from the terminal $H_3\overset{+}{N}$ group to give a free amino function. The nucleophilic amino group attacks 1-fluoro-2,4-dinitrobenzene, displacing fluoride.

$$O_2N\text{-}C_6H_3(NO_2)\text{-}F \; + \; H_2NCH[(CH_3)_2CH]C(=O)\text{—}NHCH(CH_2C_6H_5)C(=O)\text{—}NHCH_2C(=O)\text{—}NHCH(CH_3)CO_2^-$$

1-Fluoro-2,4-dinitrobenzene   Val-Phe-Gly-Ala

↓

$$O_2N\text{-}C_6H_3(NO_2)\text{—}NHCH[CH(CH_3)_2]C(=O)\text{—}NHCH(CH_2C_6H_5)C(=O)\text{—}NHCH_2C(=O)\text{—}NHCH(CH_3)CO_2^-$$

DNP-Val-Phe-Gly-Ala

Acid hydrolysis cleaves the amide bonds of the 2,4-dinitrophenyl-labeled peptide, giving the 2,4-dinitrophenyl-labeled N-terminal amino acid and a mixture of unlabeled amino acids.

↓ H_3O^+

$$O_2N\text{-}C_6H_3(NO_2)\text{—}NHCH[CH(CH_3)_2]CO_2H \; + \; H_3\overset{+}{N}CH(CH_2C_6H_5)CO_2H \; + \; H_3\overset{+}{N}CH_2CO_2H \; + \; H_3\overset{+}{N}CH(CH_3)CO_2H$$

DNP-Val   Phe   Gly   Ala

FIGURE 27.9 Use of 1-fluoro-2,4-dinitrobenzene as a reagent to identify the N-terminal amino acid of a peptide.

2,4-dinitrophenyl (DNP) group. This is shown for the case of Val-Phe-Gly-Ala in Figure 27.9. The 2,4-dinitrophenyl-labeled peptide DNP-Val-Phe-Gly-Ala is isolated and subjected to hydrolysis, after which the 2,4-dinitrophenyl derivative of the N-terminal amino acid is isolated and identified as DNP-Val by comparing its chromatographic behavior with that of standard samples of 2,4-dinitrophenyl-labeled amino acids. None of the other amino acid residues bear a 2,4-dinitrophenyl group; they appear in the hydrolysis product as the free amino acids.

The application of 1-fluoro-2,4-dinitrobenzene to N-terminal amino acid identification was introduced by Frederick Sanger of Cambridge University (England), and chemists often refer to 1-fluoro-2,4-dinitrobenzene as **Sanger's reagent.** Sanger used this technique extensively in research that led to the determination of the amino acid sequence of insulin. He and his coworkers began this work in 1944 and completed it 10 years later, and in 1958 Sanger was awarded the Nobel Prize in chemistry for this pioneering achievement.

Sanger was a corecipient of a second Nobel Prize in 1980 for devising methods for sequencing nucleic acids. Sanger's strategy for nucleic acid sequencing will be described in Section 27.28.

A related method employs 5-(dimethylamino)naphthalene-1-sulfonyl chloride as the reagent that reacts with the amino group of the N-terminal amino acid:

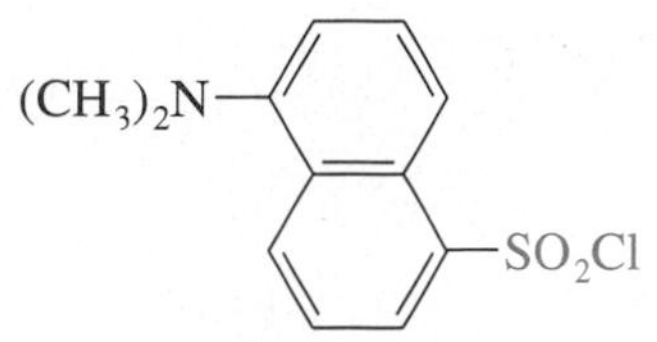

$(CH_3)_2N$—[naphthalene]—SO_2Cl

5-(*D*imethyl*a*mino)*n*aphthalene-1-*s*ulfon*yl* chloride (dansyl chloride)

This compound is known as *dansyl* chloride (dansyl is a combination of the letters indicated in italics in its systematic name), and the procedure is called **dansylation.** Identification of the N terminus by dansylation is carried out in exactly the same way as in Sanger's method. Dansyl chloride labels the peptide at the α-amino group of the N-terminal amino acid as a sulfonamide derivative, which is isolated and identified after hydrolysis.

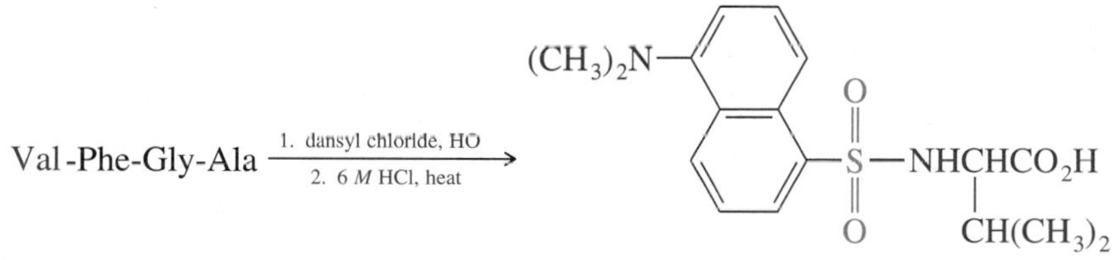

Val-Phe-Gly-Ala $\xrightarrow[\text{2. 6 } M \text{ HCl, heat}]{\text{1. dansyl chloride, HO}}$ $(CH_3)_2N$—[naphthalene]—SO_2—$NHCHCO_2H$ (with $CH(CH_3)_2$ on the α carbon)

Dansyl derivative of valine

The amount of peptide required for dansylation is very small. The (dimethyl-amino)naphthyl group imparts a strong fluorescence to dansyl derivatives, and they can be located on paper or thin-layer chromatography plates in minute amounts, much smaller than those required to detect 2,4-dinitrophenyl derivatives.

The **Edman degradation,** developed by Pehr Edman (University of Lund, Sweden), is the most frequently used method for determining the N-terminal amino acid. It has an advantage over other methods in that it permits successive identification of amino acids, one by one, beginning at the N terminus. The Edman degradation is based on the chemistry shown in Figure 27.10. An amino acid reacts with phenyl isothiocyanate to give a *phenylthiocarbamoyl* (PTC) derivative, as shown in the first step. This PTC derivative is then treated with an acid in an *anhydrous* medium (Edman used nitromethane saturated with hydrogen chloride) to cleave the amide bond between the N-terminal amino acid and the remainder of the peptide. No other peptide bonds are cleaved in this step. Amide bond hydrolysis requires water. When the PTC derivative is treated with acid in an anhydrous medium, the sulfur atom of the $\rangle C{=}S$ unit acts as an internal nucleophile, and the only amide bond cleaved under these conditions is the one to the N-terminal amino acid. The product of this cleavage, called a *thiazolone,* is unstable under the conditions of its formation and rearranges to a *phenylthiohydantoin* (PTH), which is isolated and identified by comparing it with standard samples of PTH derivatives of known amino acids. This is normally done by chromatographic methods, but mass spectrometry has also been used.

Notice that only the N-terminal amide bond is broken in the Edman degradation; the rest of the peptide chain remains intact. It can be isolated and subjected to a second Edman procedure to determine its new N terminus. Thus one can proceed along a peptide chain by beginning with the N terminus and determining each amino acid in order. The sequence is given directly by the structure of the PTH derivative formed in each successive degradation.

Step 1: A peptide is treated with phenyl isothiocyanate to give a phenylthiocarbamoyl (PTC) derivitive.

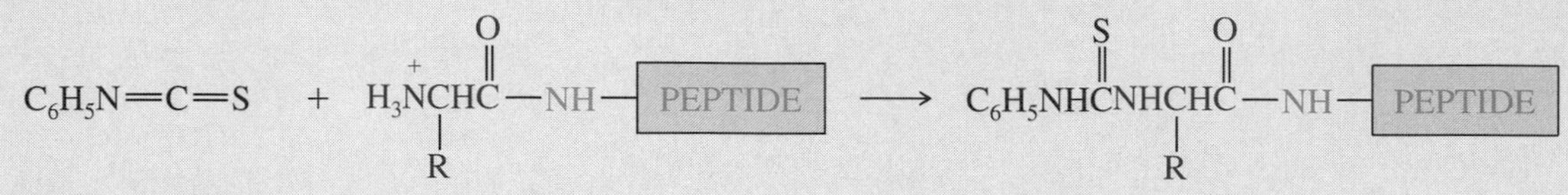

$$C_6H_5N{=}C{=}S \;+\; H_3\overset{+}{N}CH(R)C({=}O){-}NH{-}\boxed{\text{PEPTIDE}} \longrightarrow C_6H_5NHC({=}S)NHCH(R)C({=}O){-}NH{-}\boxed{\text{PEPTIDE}}$$

Phenyl isothiocyanate | PTC derivative

Step 2: On reaction with hydrogen chloride in an anhydrous solvent, the thiocarbonyl sulfur of the PTC derivative attacks the carbonyl carbon of the N-terminal amino acid. The N-terminal amino acid is cleaved as a thiazolone derivative from the remainder of the peptide.

$$\text{PTC derivative} \xrightarrow{HCl} \text{Thiazolone } (C_6H_5NH{-}C, S, C{=}O, N{-}CH{-}R) \;+\; H_3\overset{+}{N}{-}\boxed{\text{PEPTIDE}}$$

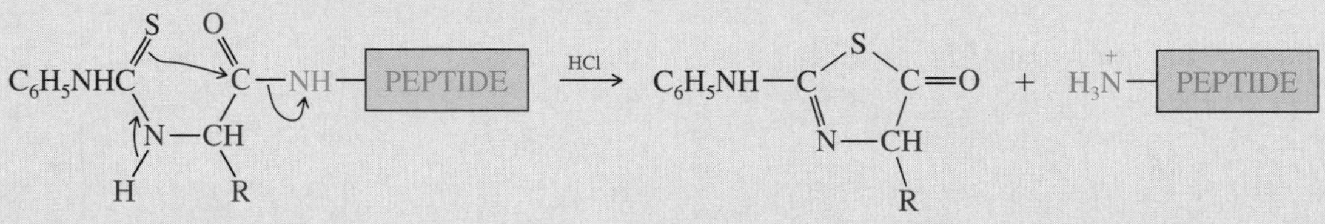

PTC derivative | Thiazolone | Remainder of peptide

Step 3: Once formed, the thiazolone derivative isomerizes to a more stable phenylthiohydantoin (PTH) derivative, which is isolated and characterized, thereby providing identification of the N-terminal amino acid. The remainder of the peptide (formed in step 2) can be isolated and subjected to a second Edman degradation.

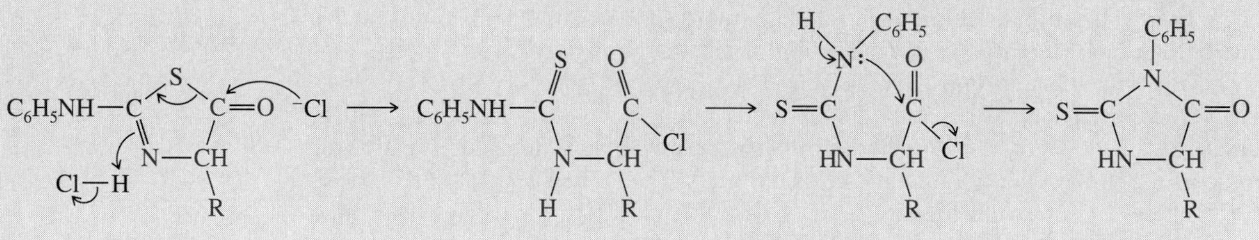

$$\text{Thiazolone} \longrightarrow C_6H_5NH{-}C({=}S){-}NH{-}CH(R){-}C({=}O)Cl \longrightarrow S{=}C(N(H)C_6H_5)(HN{-}CH(R){-}C({=}O)Cl) \longrightarrow \text{PTH derivative}$$

Thiazolone | PTH derivative

FIGURE 27.10 The chemical reactions involved in the identification of the N-terminal amino acid of a peptide by Edman degradation.

PROBLEM 27.14 Give the structure of the PTH derivative isolated in the second Edman cycle of our representative tetrapeptide Val-Phe-Gly-Ala.

Ideally, one could determine the primary structure of even the largest protein by repeating the Edman procedure. Because anything less than 100 percent conversion in any single Edman degradation gives a mixture containing some of the original peptide along with the degraded one, two different PTH derivatives are formed in the next Edman cycle, and the ideal is not realized in practice. Nevertheless, some impressive results have been achieved. It is a fairly routine matter to sequence the first 20 amino acids from the N terminus by repetitive Edman cycles, and even 60 residues have been determined on a single sample of the protein myoglobin. The entire procedure has been automated and incorporated into a device called an **Edman sequenator,** which carries out all the operations under computer control.

27.11 END GROUP ANALYSIS: THE C TERMINUS

While there are a few purely chemical techniques for determining the amino acid at the carboxyl, or C, terminus of a peptide, the most widely used method is one based on enzymatic hydrolysis. Enzymes that catalyze the hydrolysis of peptide bonds are called **peptidases, proteases,** or **proteolytic enzymes.** Many are quite selective with respect to the type of peptide bonds that they cleave. One group of pancreatic enzymes, known as **carboxypeptidases,** catalyzes only the hydrolysis of the peptide bond involving the C-terminal amino acid (Figure 27.11). Once the C-terminal amino acid has been cleaved, yielding a peptide shorter by one amino acid, the carboxypeptidase then catalyzes the cleavage of the peptide bond to the new C-terminal amino acid. This property is used to advantage in sequence analysis. A peptide is incubated with a carboxypeptidase, and the concentration of free amino acids is monitored as a function of time. The order of their decreasing concentration indicates the position of each amino acid with respect to the C terminus of the original peptide. Normally, the sequence of only a few amino acids from the C terminus can be determined in this way.

Papain, the active component of most meat tenderizers, is a proteolytic enzyme.

27.12 SELECTIVE HYDROLYSIS OF PEPTIDES

Carboxypeptidase-catalyzed hydrolysis is an example of one way in which biochemical degradation of a peptide can be used in sequence determination. Living organisms contain a number of different peptidases, several of which find use in peptide sequencing because they permit the controlled hydrolysis of peptides by cleavage of certain specific amide bonds.

Trypsin, a digestive enzyme present in the intestine, catalyzes only the hydrolysis of peptide bonds involving the carboxyl group of a lysine or arginine residue. *Chymotrypsin,* another digestive enzyme, is selective for the hydrolysis of peptide bonds involving the carboxyl group of amino acids with aromatic groups in their side chains, namely, phenylalanine, tyrosine, and tryptophan. The catalytic activity of *pepsin* is much like that of chymotrypsin, but pepsin also catalyzes the hydrolysis of peptide bonds to the carbonyl groups of methionine and leucine.

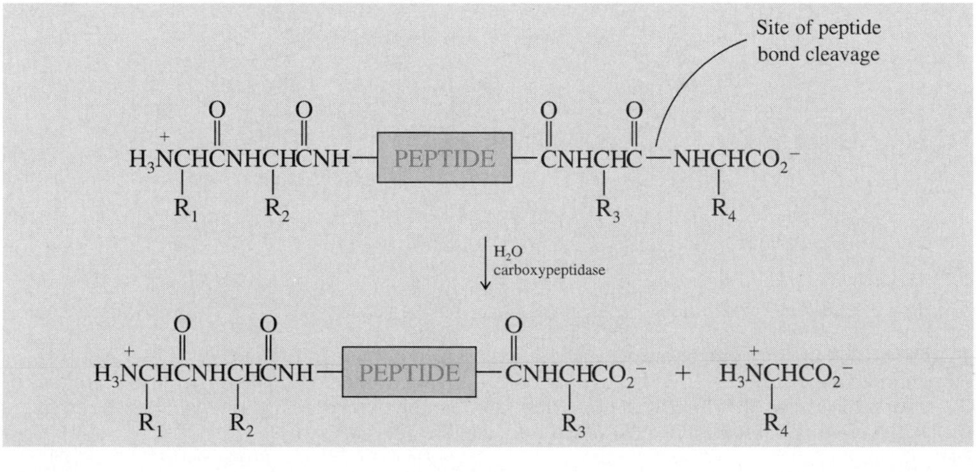

Site of peptide bond cleavage

$H_3\overset{+}{N}CH(R_1)C(=O)NHCH(R_2)C(=O)NH$—PEPTIDE—$C(=O)NHCH(R_3)C(=O)$—$NHCH(R_4)CO_2^-$

$\xrightarrow{H_2O,\ \text{carboxypeptidase}}$

$H_3\overset{+}{N}CH(R_1)C(=O)NHCH(R_2)C(=O)NH$—PEPTIDE—$C(=O)NHCH(R_3)CO_2^-$ + $H_3\overset{+}{N}CH(R_4)CO_2^-$

FIGURE 27.11 Diagram showing the site of carboxypeptidase-catalyzed hydrolysis. Carboxypeptidase is selective for cleavage of the peptide bond to the C-terminal amino acid. Once this peptide bond has been cleaved, carboxypeptidase proceeds to catalyze the hydrolytic cleavage of the new C-terminal amino acid.

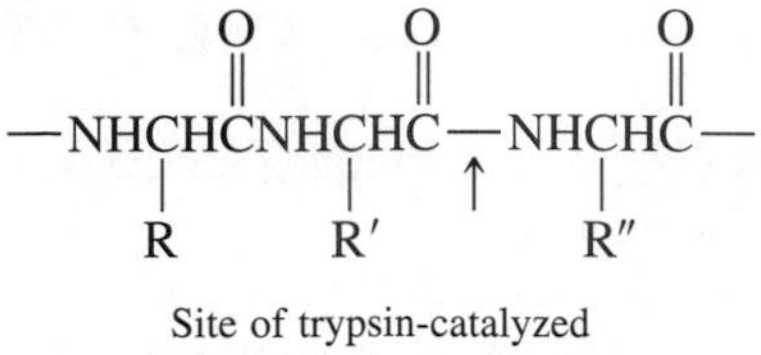

$$-\mathrm{NH\underset{\underset{R}{|}}{C}H\overset{\overset{O}{\|}}{C}NH\underset{\underset{R'}{|}}{C}H\overset{\overset{O}{\|}}{C}\underset{\uparrow}{-}NH\underset{\underset{R''}{|}}{C}H\overset{\overset{O}{\|}}{C}}-$$

Site of trypsin-catalyzed hydrolysis when red amino acid residue is lysine or arginine

Unlike hydrolysis in 6 *M* hydrochloric acid, which gives a complicated mixture of peptide fragments and free amino acids, selective enzyme-catalyzed hydrolysis produces only a limited number of fragments. After these fragments have been isolated and sequenced, their primary structures are examined for points of overlap in order to deduce the structure of the peptide that produced them. An example of this is shown in Figure 27.12, which depicts some of the work that led to Sanger's determination of the structure of insulin. Insulin, with 51 amino acids, has two structural units, a 21–amino acid peptide (the A chain) and a 30–amino acid peptide (the B chain). The A chain and the B chain are joined by disulfide bonds between two pairs of cysteine residues (CysS-SCys). Reaction with 1-fluoro-2,4-dinitrobenzene established that phenylalanine is the N terminus of the B chain. The four peptides shown in blue in Figure 27.12 were isolated by pepsin-catalyzed hydrolysis of the B chain, and the amino acid sequence of each was determined independently. Together these four peptides contain 27 of the 30 amino acids of the insulin B chain, but there are no points of overlap between them. Since, however, the sequences of two tetrapeptides isolated by acid hydrolysis

1 2 3 4 5 6 7 8 9 10 11
Phe-Val-Asn-Gln-His-Leu-Cys-Gly-Ser-His-Leu

Ser-His-Leu-Val

Leu-Val-Glu-Ala

12 13 14 15
Val-Glu-Ala-Leu

Ala-Leu-Tyr

16 17
Tyr-Leu-Val-Cys

18 19 20 21 22 23 24
Val-Cys-Gly-Glu-Arg-Gly-Phe

25
Gly-Phe-Phe-Tyr-Thr-Pro-Lys

26 27 28 29 30
Tyr-Thr-Pro-Lys-Ala

1 5 10 15 20 25 30
Phe-Val-Asn-Gln-His-Leu-Cys-Gly-Ser-His-Leu-Val-Glu-Ala-Leu-Tyr-Leu-Val-Cys-Gly-Glu-Arg-Gly-Phe-Phe-Tyr-Thr-Pro-Lys-Ala

FIGURE 27.12 Diagram showing how the amino acid sequence of the B chain of bovine insulin can be determined by overlap of peptide fragments. Pepsin-catalyzed hydrolysis produced the fragments shown in blue, trypsin produced the one shown in green, and acid-catalyzed hydrolysis gave many fragments, including the four shown in red.

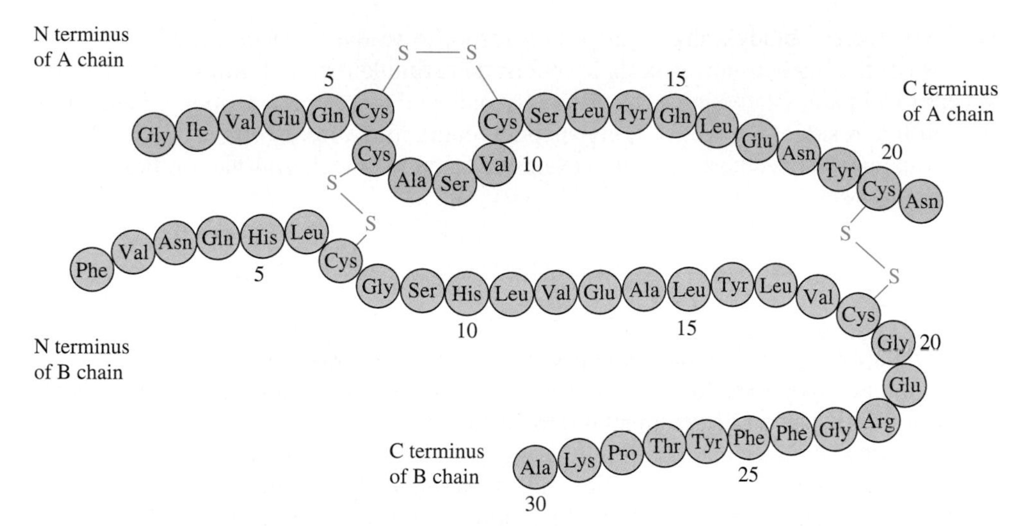

FIGURE 27.13 Depiction of the amino acid sequence in bovine insulin. The A chain is shown in red and the B chain in blue. The A chain is joined to the B chain by two disulfide units (shown in green). There is also a disulfide bond linking cysteines 6 and 11 in the A chain. Human insulin has threonine and isoleucine at residues 8 and 10, respectively, in the A chain and threonine as the C-terminal amino acid in the B chain.

(shown in red) overlap with both the N-terminal fragment and a second peptide from the pepsin-catalyzed hydrolysis, it is possible to establish the first 15 residues beginning with the N terminus. Similarly, overlaps with two other fragments from the acid hydrolysis (red) reveals that the sequence Tyr-Leu joins the first 15 residues to the third peptide obtained from pepsin-catalyzed hydrolysis. Thus, the first 24 residues are identified. The peptide shown in green in Figure 27.12 was isolated by trypsin-catalyzed hydrolysis; its amino acid sequence overlaps with two of the peptides from pepsin-catalyzed hydrolysis, providing the necessary information to complete the entire sequence of 30 amino acids.

Sanger also determined the sequence of the A chain and identified the cysteine residues involved in disulfide bonds between the A and B chains as well as in the disulfide linkage within the A chain. The complete insulin structure is shown in Figure 27.13. The structure shown is that of bovine insulin (from cattle). The A chains of human insulin and bovine insulin differ in only two amino acid residues; their B chains are identical except for the amino acid at the C terminus.

Once a peptide has been isolated and purified, a whole battery of techniques is brought to bear on the question of its amino acid sequence. These include Edman degradation and carboxypeptidase determination of the C terminus, along with partial hydrolysis and sequence analysis of smaller fragments. Since Sanger's successful determination of the primary structure of insulin in 1954, the amino acid sequences of hundreds of proteins, some with more than 300 amino acids in their chains, have been determined. In a recent innovation, it has become possible to determine the nucleotide sequence of the gene that directs the synthesis of a peptide and, on the basis of the genetic code (Section 27.27), to deduce the sequence of the amino acids in the peptide.

27.13 THE STRATEGY OF PEPTIDE SYNTHESIS

One way to confirm the structure proposed for a peptide is to synthesize a peptide having a specific sequence of amino acids and compare the two. This was done, for example, in the case of *bradykinin,* a peptide present in blood that acts to lower blood pressure. Excess bradykinin, formed as a response to the sting of wasps and other insects containing substances in their venom that stimulate bradykinin release, causes severe local pain. Bradykinin was originally believed to be an octapeptide containing two proline residues. However, a nonapeptide containing three prolines in the following sequence was synthesized and determined to be identical with natural bradykinin in every respect, including biological activity:

Arg-Pro-Pro-Gly-Phe-Ser-Pro-Phe-Arg Bradykinin

A reevaluation of the original sequence data established that natural bradykinin was indeed the nonapeptide shown. Here the synthesis of a peptide did more than confirm structure; synthesis was instrumental in determining structure.

Chemists and biochemists also synthesize peptides in order to promote understanding of the mechanism of their action. By systematically altering the sequence, researchers can sometimes determine which amino acids are intimately involved in the reactions that involve a particular peptide. Many synthetic peptides have been prepared in searching for new drugs.

PROBLEM 27.15 The C-terminal methyl ester of the dipeptide Asp-Phe was prepared in connection with research on digestive enzymes known as *gastrins.* The chemist who synthesized it noted that it had a sweet taste. This synthetic dipeptide ester is now known as *aspartame;* it is 200 times sweeter than sucrose and is widely used as an artificial sweetener. Write a structural formula for aspartame.

The objective in peptide synthesis may be simply stated: to connect amino acids in a prescribed sequence by amide bond formation between them. A number of very effective methods and reagents have been designed for peptide bond formation, so that the joining together of amino acids by amide linkages is not difficult. (Some of these methods are described in Section 27.16.) The real difficulty lies in ensuring that the correct sequence is obtained. This can be illustrated by considering the synthesis of a representative dipeptide, Phe-Gly. Random peptide bond formation in a mixture containing phenylalanine and glycine would be expected to lead to four dipeptides:

$$\underset{\text{Phenylalanine}}{H_3\overset{+}{N}\underset{\underset{CH_2C_6H_5}{|}}{C}HCO_2^-} + \underset{\text{Glycine}}{H_3\overset{+}{N}CH_2CO_2^-} \longrightarrow \text{Phe-Gly} + \text{Phe-Phe} + \text{Gly-Phe} + \text{Gly-Gly}$$

The amino groups of both phenylalanine and glycine can be involved in peptide bond formation. Similarly, the carboxyl groups of both amino acids can be involved.

In order to direct the synthesis so that only Phe-Gly is formed, the amino group of phenylalanine and the carboxyl group of glycine must be protected so that they cannot react under the conditions of peptide bond formation. We can represent the peptide bond formation step by the following equation, where X and Y are amine- and carboxyl-protecting groups, respectively:

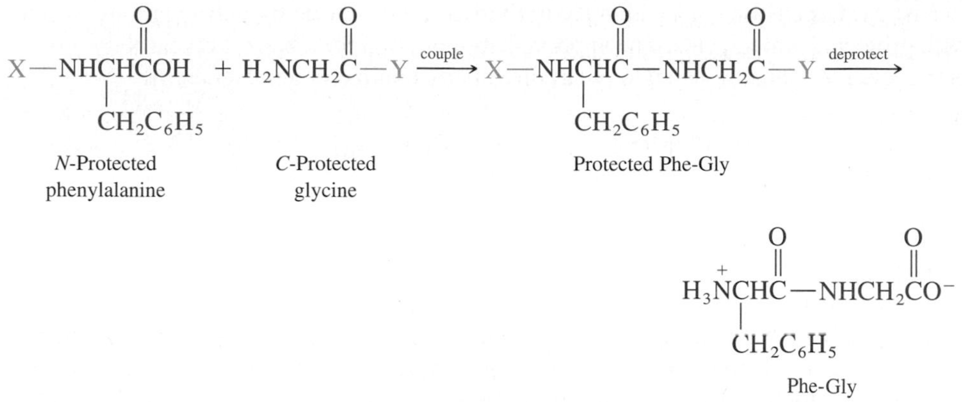

$$\underset{\substack{N\text{-Protected}\\ \text{phenylalanine}}}{X{-}NH\underset{\substack{|\\ CH_2C_6H_5}}{C}HCOH} \; + \; \underset{\substack{C\text{-Protected}\\ \text{glycine}}}{H_2NCH_2\overset{\overset{O}{\|}}{C}{-}Y} \xrightarrow{\text{couple}} \underset{\text{Protected Phe-Gly}}{X{-}NH\underset{\substack{|\\ CH_2C_6H_5}}{C}H\overset{\overset{O}{\|}}{C}{-}NHCH_2\overset{\overset{O}{\|}}{C}{-}Y} \xrightarrow{\text{deprotect}}$$

$$\underset{\text{Phe-Gly}}{H_3\overset{+}{N}\underset{\substack{|\\ CH_2C_6H_5}}{C}H\overset{\overset{O}{\|}}{C}{-}NHCH_2\overset{\overset{O}{\|}}{C}O^-}$$

Thus, the synthesis of a dipeptide of prescribed sequence requires at least three operations:

1. *Protect* the amino group of the N-terminal amino acid and the carboxyl group of the C-terminal amino acid.
2. *Couple* the two protected amino acids by amide bond formation between them.
3. *Deprotect* the amino group at the N terminus and the carboxyl group at the C terminus.

Higher peptides are prepared in an analogous way by a direct extension of the logic just outlined for the synthesis of dipeptides.

Sections 27.14 through 27.17 elaborate on these features of peptide synthesis by describing the chemistry associated with the protection and deprotection of amino and carboxyl functions, along with methods for peptide bond formation.

27.14 AMINO GROUP PROTECTION

The reactivity of an amino group is suppressed by converting it to an amide function, and amino groups are most often protected by acylation. The benzyloxycarbonyl group ($C_6H_5CH_2O\overset{\overset{O}{\|}}{C}{-}$) is one of the most often used amino-protecting groups. It is attached by acylation of an amino acid with benzyloxycarbonyl chloride.

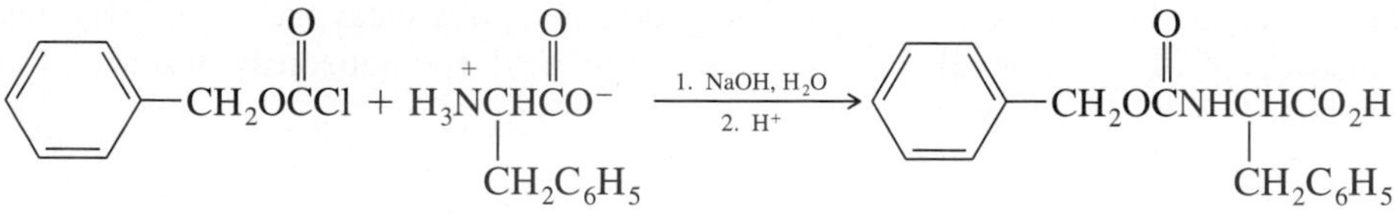

$$\underset{\substack{\text{Benzyloxycarbonyl}\\ \text{chloride}}}{C_6H_5CH_2O\overset{\overset{O}{\|}}{C}Cl} \; + \; \underset{\text{Phenylalanine}}{H_3\overset{+}{N}\underset{\substack{|\\ CH_2C_6H_5}}{C}H\overset{\overset{O}{\|}}{C}O^-} \xrightarrow[\text{2. } H^+]{\text{1. NaOH, } H_2O} \underset{\substack{N\text{-Benzyloxycarbonylphenylalanine}\\ (82\text{–}87\%)}}{C_6H_5CH_2O\overset{\overset{O}{\|}}{C}NH\underset{\substack{|\\ CH_2C_6H_5}}{C}HCO_2H}$$

Another name for the benzyloxycarbonyl group is *carbobenzoxy.* This name, and its abbreviation *Cbz,* while often found in the older literature, is no longer acceptable in IUPAC nomenclature.

PROBLEM 27.16 Lysine reacts with two equivalents of benzyloxycarbonyl chloride to give a derivative containing two benzyloxycarbonyl groups. What is the structure of this compound?

Just as it is customary to identify individual amino acids by abbreviations, so too with protected amino acids. The approved abbreviation for a benzyloxycarbonyl group is the letter Z. Thus, *N*-benzyloxycarbonylphenylalanine is represented as

$$\underset{\displaystyle CH_2C_6H_5}{ZNH\underset{|}{C}HCO_2H} \quad \text{or more simply as} \quad \text{Z-Phe}$$

The value of the benzyloxycarbonyl protecting group is that it is easily removed by reactions other than hydrolysis. In peptide synthesis, amide bonds are formed. We protect the N terminus as an amide but need to remove the protecting group without cleaving the very amide bonds we labored so hard to construct. Removing the protecting group by hydrolysis would surely bring about cleavage of peptide bonds as well. One advantage that the benzyloxycarbonyl protecting group enjoys over more familiar acyl groups such as acetyl is that it can be removed by *hydrogenolysis* in the presence of palladium. The following equation illustrates this for the removal of the benzyloxycarbonyl protecting group from the ethyl ester of Z-Phe-Gly:

The term *hydrogenolysis* refers to the cleavage of a molecule under conditions of catalytic hydrogenation.

$$C_6H_5CH_2O\overset{\overset{\displaystyle O}{\|}}{C}NH\underset{\underset{\displaystyle CH_2C_6H_5}{|}}{C}H\overset{\overset{\displaystyle O}{\|}}{C}NHCH_2CO_2CH_2CH_3 \xrightarrow[\text{Pd}]{H_2}$$

N-Benzyloxycarbonylphenylalanylglycine ethyl ester

$$\underset{\text{Toluene}}{C_6H_5CH_3} + \underset{\text{Carbon dioxide}}{CO_2} + \underset{\text{Phenylalanylglycine ethyl ester (100\%)}}{H_2N\underset{\underset{\displaystyle CH_2C_6H_5}{|}}{C}H\overset{\overset{\displaystyle O}{\|}}{C}NHCH_2CO_2CH_2CH_3}$$

The products derived from the protecting group, toluene and carbon dioxide, are easily removed, which facilitates purification. The resulting dipeptide has its carboxyl group protected and its amino group free. It can be coupled with an *N*-protected amino acid to give a tripeptide or subjected to ester hydrolysis to yield the dipeptide Phe-Gly.

Hydrogenolysis of the benzyloxycarbonyl protecting group involves initial cleavage of the benzyl-oxygen bond. The products of this cleavage are toluene and a carbamic acid (Section 20.17). The carbamic acid spontaneously loses carbon dioxide.

$$\underset{\text{Z-protected peptide}}{C_6H_5CH_2{-}O\overset{\overset{\displaystyle O}{\|}}{C}{-}NHR} + \underset{\text{Hydrogen}}{H_2} \xrightarrow{\text{Pd}} \underset{\text{Toluene}}{C_6H_5CH_3} + \underset{\text{Carbamic acid (not isolated)}}{[HO\overset{\overset{\displaystyle O}{\|}}{C}{-}NHR]} \longrightarrow \underset{\text{Carbon dioxide}}{CO_2} + \underset{\text{Peptide}}{H_2NR}$$

Alternatively, the benzyloxycarbonyl protecting group may be removed by treatment with hydrogen bromide in acetic acid:

$$C_6H_5CH_2OC(=O)NHCH(CH_2C_6H_5)C(=O)NHCH_2CO_2CH_2CH_3 \xrightarrow{HBr}$$

N-Benzyloxycarbonylphenylalanylglycine ethyl ester

$$C_6H_5CH_2Br + CO_2 + H_3\overset{+}{N}CH(CH_2C_6H_5)C(=O)NHCH_2CO_2CH_2CH_3\ Br^-$$

Benzyl bromide | Carbon dioxide | Phenylalanylglycine ethyl ester hydrobromide (82%)

Deprotection by this method rests on the ease with which benzyl esters are cleaved by nucleophilic attack at the benzylic carbon in the presence of strong acids. Bromide ion is the nucleophile, and a carbamic acid is an intermediate.

A related N-terminal-protecting group is *tert*-butoxycarbonyl, abbreviated *Boc:*

$$(CH_3)_3COC(=O)—$$

tert-Butoxycarbonyl (Boc-)

$$(CH_3)_3COC(=O)—NHCH(CH_2C_6H_5)CO_2H \quad \text{also written as} \quad BocNHCH(CH_2C_6H_5)CO_2H$$

N-tert-Butoxycarbonylphenylalanine | Boc-Phe

Like the benzyloxycarbonyl protecting group, the Boc group may be removed by treatment with hydrogen bromide (it is stable to hydrogenolysis, however):

$$(CH_3)_3COC(=O)NHCH(CH_2C_6H_5)C(=O)NHCH_2CO_2CH_2CH_3 \xrightarrow{HBr}$$

N-tert-Butoxycarbonylphenylalanylglycine ethyl ester

$$(CH_3)_2C{=}CH_2 + CO_2 + H_3\overset{+}{N}CH(CH_2C_6H_5)C(=O)NHCH_2CO_2CH_2CH_3\ Br^-$$

2-Methylpropene | Carbon dioxide | Phenylalanylglycine ethyl ester hydrobromide (86%)

The *tert*-butyl group is cleaved as the corresponding carbocation, leaving a carbamic acid, which loses carbon dioxide to give the unprotected amino group. Loss of a proton from *tert*-butyl cation converts it to 2-methylpropene. Because of the ease with which a *tert*-butyl group is cleaved as a carbocation, other acidic reagents, such as trifluoroacetic acid, may also be used.

27.15 CARBOXYL GROUP PROTECTION

Carboxyl groups of amino acids and peptides are normally protected as esters. Methyl and ethyl esters are prepared by Fischer esterification, as the example cited in Section 27.5 illustrates. Deprotection of methyl and ethyl esters is accomplished by hydrolysis in dilute base. Benzyl esters are a popular choice because they can be removed by hydrogenolysis. Thus a synthetic peptide, protected at both its N terminus with a Z group and at its C terminus as a benzyl ester, can be completely deprotected in a single operation.

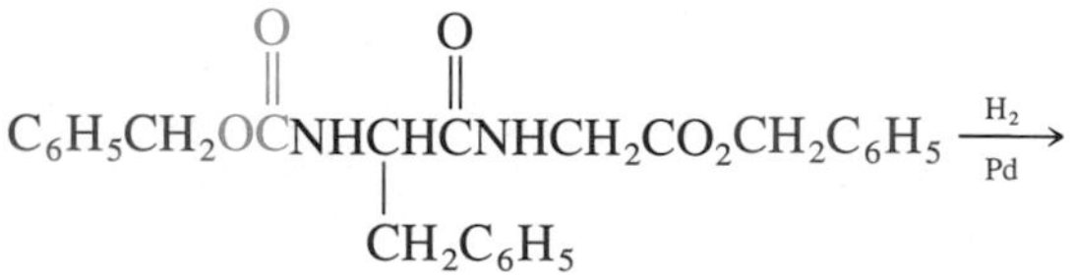

$$C_6H_5CH_2OC(=O)NHCH(CH_2C_6H_5)C(=O)NHCH_2CO_2CH_2C_6H_5 \xrightarrow[Pd]{H_2}$$

N-Benzyloxycarbonylphenylalanylglycine benzyl ester

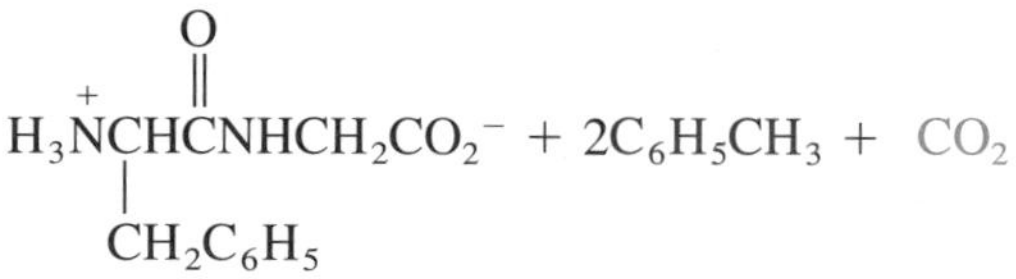

$$H_3\overset{+}{N}CH(CH_2C_6H_5)C(=O)NHCH_2CO_2^- + 2C_6H_5CH_3 + CO_2$$

Phenylalanylglycine (87%) — Toluene — Carbon dioxide

Several of the amino acids listed in Table 27.1 bear side-chain functional groups, which must also be protected during peptide synthesis. In most cases, protecting groups are available that can be removed by hydrogenolysis.

27.16 PEPTIDE BOND FORMATION

In order to form a peptide bond between two suitably protected amino acids, the free carboxyl group of one of them must be *activated* so that it is a reactive acylating agent. The most familiar acylating agents are acyl chlorides, and they were once extensively used to couple amino acids. Certain drawbacks to this approach, however, led chemists to seek alternative methods.

In one method, treatment with *N,N'*-dicyclohexylcarbodiimide (DCCI) of a solution containing the *N*-protected and the *C*-protected amino acids leads directly to peptide bond formation:

$$ZNHCH(CH_2C_6H_5)C(=O)OH + H_2NCH_2C(=O)OCH_2CH_3 \xrightarrow[\text{chloroform}]{\text{DCCI}} ZNHCH(CH_2C_6H_5)C(=O){-}NHCH_2C(=O)OCH_2CH_3$$

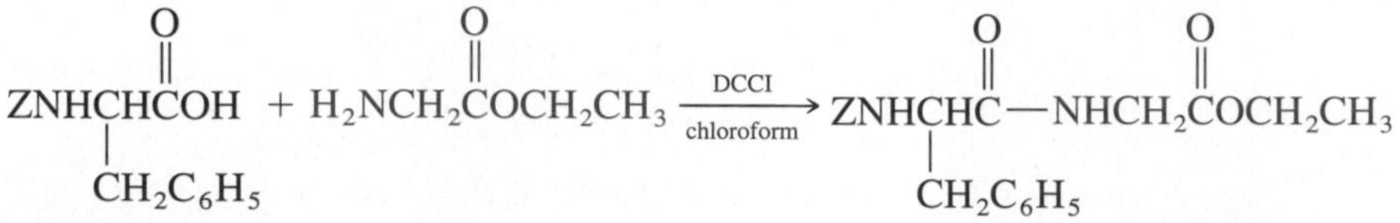

Z-Protected phenylalanine — Glycine ethyl ester — Z-Protected Phe-Gly ethyl ester (83%)

N,N'-Dicyclohexylcarbodiimide has the structure shown:

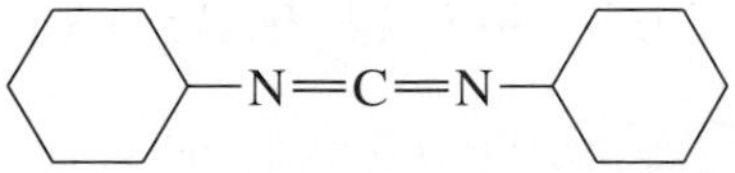

$$C_6H_{11}-N=C=N-C_6H_{11}$$

N,N'-Dicyclohexylcarbodiimide (DCCI)

The mechanism by which DCCI promotes the condensation of an amine and a carboxylic acid to give an amide is outlined in Figure 27.14.

PROBLEM 27.17 Show the steps involved in the synthesis of Ala-Leu from alanine and leucine using benzyloxycarbonyl and benzyl ester protecting groups and DCCI-promoted peptide bond formation.

In the second major method of peptide synthesis the carboxyl group is activated by converting it to an *active ester,* usually a *p*-nitrophenyl ester. Recall from Section 20.11 that esters react with ammonia and amines to give amides. *p*-Nitrophenyl esters are much more reactive than methyl and ethyl esters in these reactions because *p*-nitrophenoxide is a better (less basic) leaving group than methoxide and ethoxide. Simply allowing the active ester and a *C*-protected amino acid to stand in a suitable solvent is sufficient to bring about peptide bond formation by nucleophilic acyl substitution.

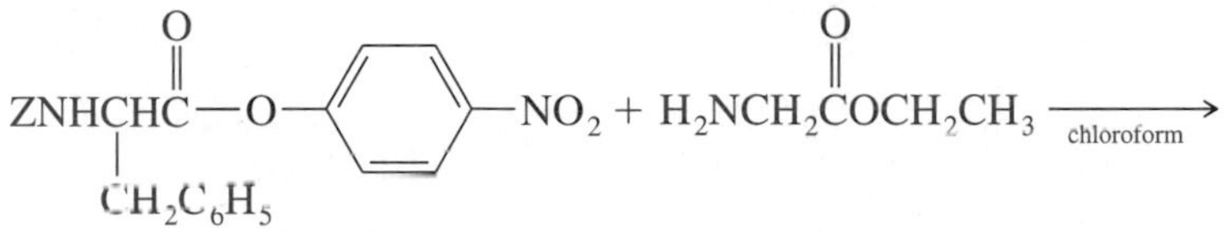

$$\underset{\text{Z-Protected phenylalanine } p\text{-nitrophenyl ester}}{\text{ZNHCH(CH}_2\text{C}_6\text{H}_5)\text{C(=O)}-\text{O}-\text{C}_6\text{H}_4-\text{NO}_2} + \underset{\text{Glycine ethyl ester}}{H_2NCH_2C(=O)OCH_2CH_3} \xrightarrow{\text{chloroform}}$$

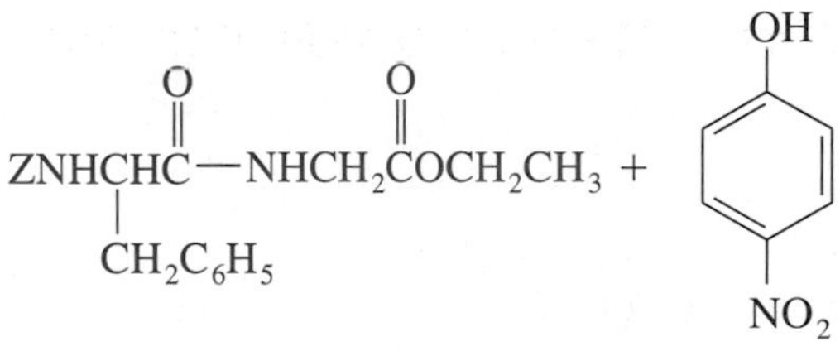

$$\underset{\text{Z-Protected Phe-Gly ethyl ester (78\%)}}{\text{ZNHCH(CH}_2\text{C}_6\text{H}_5)\text{C(=O)}-\text{NHCH}_2\text{C(=O)OCH}_2\text{CH}_3} + \underset{p\text{-Nitrophenol}}{HO-C_6H_4-NO_2}$$

The *p*-nitrophenol formed as a by-product in this reaction is easily removed by extraction with dilute aqueous base. Unlike free amino acids and peptides, protected peptides are not zwitterionic and are more soluble in organic solvents than in water.

PROBLEM 27.18 *p*-Nitrophenyl esters are made from Z-protected amino acids by reaction with *p*-nitrophenol in the presence of *N,N'*-dicyclohexylcarbodiimide. Suggest a reasonable mechanism for this reaction.

PROBLEM 27.19 Show how you could convert the ethyl ester of Z-Phe-Gly to Leu-Phe-Gly (as its ethyl ester) by the active ester method.

Overall reaction:

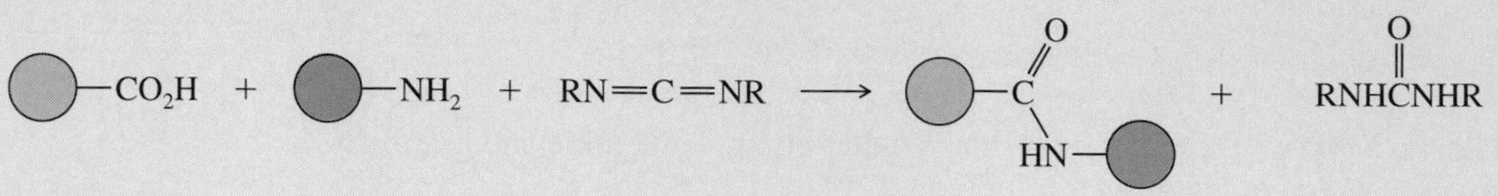

Carboxylic acid + Amine + DCCI ⟶ Amide + *N*,*N*′-Dicyclohexylurea

$$\text{—CO}_2\text{H} + \text{—NH}_2 + \text{RN=C=NR} \longrightarrow \text{—C(=O)—HN—} + \text{RNHC(=O)NHR}$$

DCCI = *N*,*N*′-dicyclohexylcarbodiimide; R = cyclohexyl

Mechanism:

Step 1: In the first stage of the reaction, the carboxylic acid adds to one of the double bonds of DCCI to give an *O*-acylisourea.

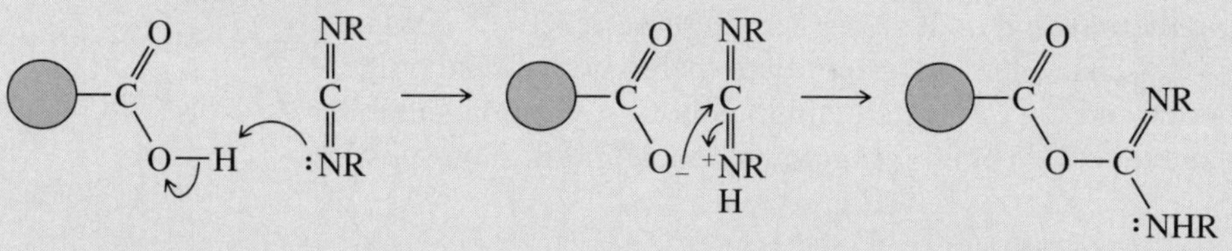

Carboxylic acid + DCCI ⟶ ⟶ *O*-Acylisourea

Step 2: Structurally, *O*-acylisoureas resemble carboxylic acid anhydrides and are powerful acylating agents. In the reaction's second stage the amine adds to the carbonyl group of the *O*-acylisourea to give a tetrahedral intermediate.

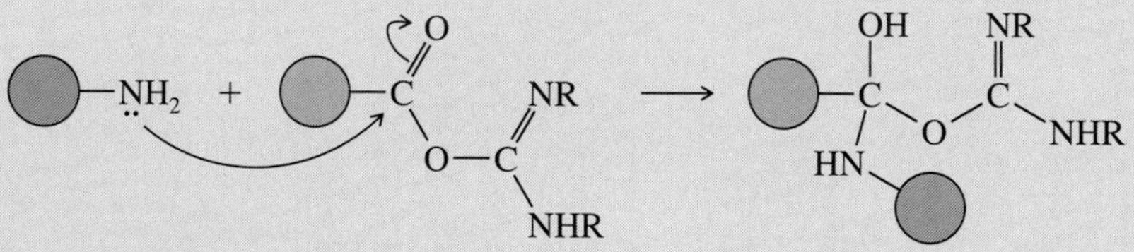

Amine + *O*-Acylisourea ⟶ Tetrahedral intermediate

Step 3: The tetrahedral intermediate dissociates to an amide and *N*,*N*′-dicyclohexylurea.

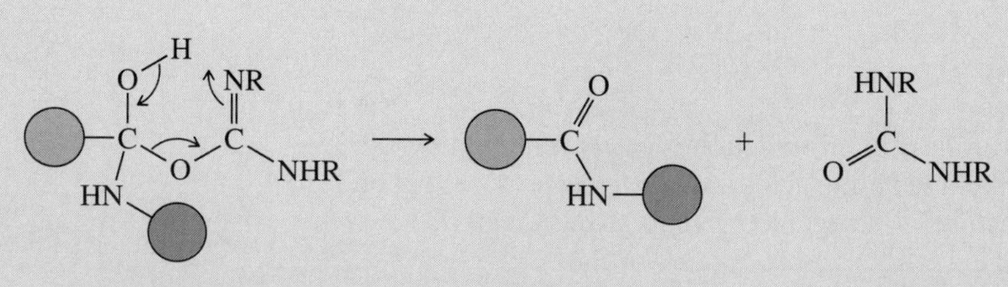

Tetrahedral intermediate ⟶ Amide + *N*,*N*′-Dicyclohexylurea

FIGURE 27.14 The mechanism of amide bond formation by *N*,*N*′-dicyclohexylcarbodiimide-promoted condensation of a carboxylic acid and an amine.

Higher peptides are prepared either by stepwise extension of peptide chains, one amino acid residue at a time, or by coupling of fragments containing several residues (the **fragment condensation** approach). Human pituitary adrenocorticotropic hormone (ACTH), for example, has 39 amino acids and was synthesized by coupling of smaller peptides containing residues 1 through 10, 11 through 16, 17 through 24, and 25 through 39. An attractive feature of this approach is that the various protected peptide fragments may be individually purified, thereby simplifying the purification of the final product. Included among the substances that have been synthesized by fragment condensation are insulin (51 amino acids) and the protein ribonuclease A (124 amino acids). In the stepwise extension approach, the starting peptide in a particular step differs from the coupling product by only one amino acid residue and the properties of the two peptides may be so similar as to preclude purification by conventional techniques. The following section describes a method by which many of the difficulties involved in the purification of intermediates have been surmounted.

27.17 SOLID-PHASE PEPTIDE SYNTHESIS. THE MERRIFIELD METHOD

In 1962 R. Bruce Merrifield of Rockefeller University reported the synthesis of the nonapeptide bradykinin (see Section 27.13) by a novel method. In Merrifield's method, peptide coupling and deprotection are carried out not in homogeneous solution but at the surface of an insoluble polymer, or *solid support.* Beads of a copolymer prepared from styrene containing about 2% divinylbenzene are treated with chloromethyl methyl ether and tin(IV) chloride to give a resin in which about 10 percent of the aromatic rings bear —CH_2Cl groups (Figure 27.15). The growing peptide is anchored to this polymer, and excess reagents, impurities, and by-products are removed by thorough washing after each operation. This greatly simplifies the purification of intermediates.

Merrifield was awarded the 1984 Nobel Prize in chemistry for developing the solid-phase method of peptide synthesis.

The actual process of solid-phase peptide synthesis, outlined in Figure 27.16, begins with the attachment of the C-terminal amino acid to the chloromethylated polymer in step 1. Nucleophilic substitution by the carboxylate anion of an *N*-Boc-protected C-terminal amino acid displaces chloride from the chloromethyl group of the polymer to form an ester, protecting the C terminus while anchoring it to a solid support. Next, the Boc group is removed by treatment with acid (step 2), and the polymer containing the unmasked N terminus is washed with a series of organic

CH_2 CH CH_2 CH CH_2 CH CH_2 CH

CH_2Cl

FIGURE 27.15 A section of polystyrene showing one of the benzene rings modified by chloromethylation. Individual polystyrene chains in the resin used in solid-phase peptide synthesis are connected to one another at various points (cross-linked) by adding a small amount of *p*-divinylbenzene to the styrene monomer. The chloromethylation step is carried out under conditions such that only about 10 percent of the benzene rings bear —CH_2Cl groups.

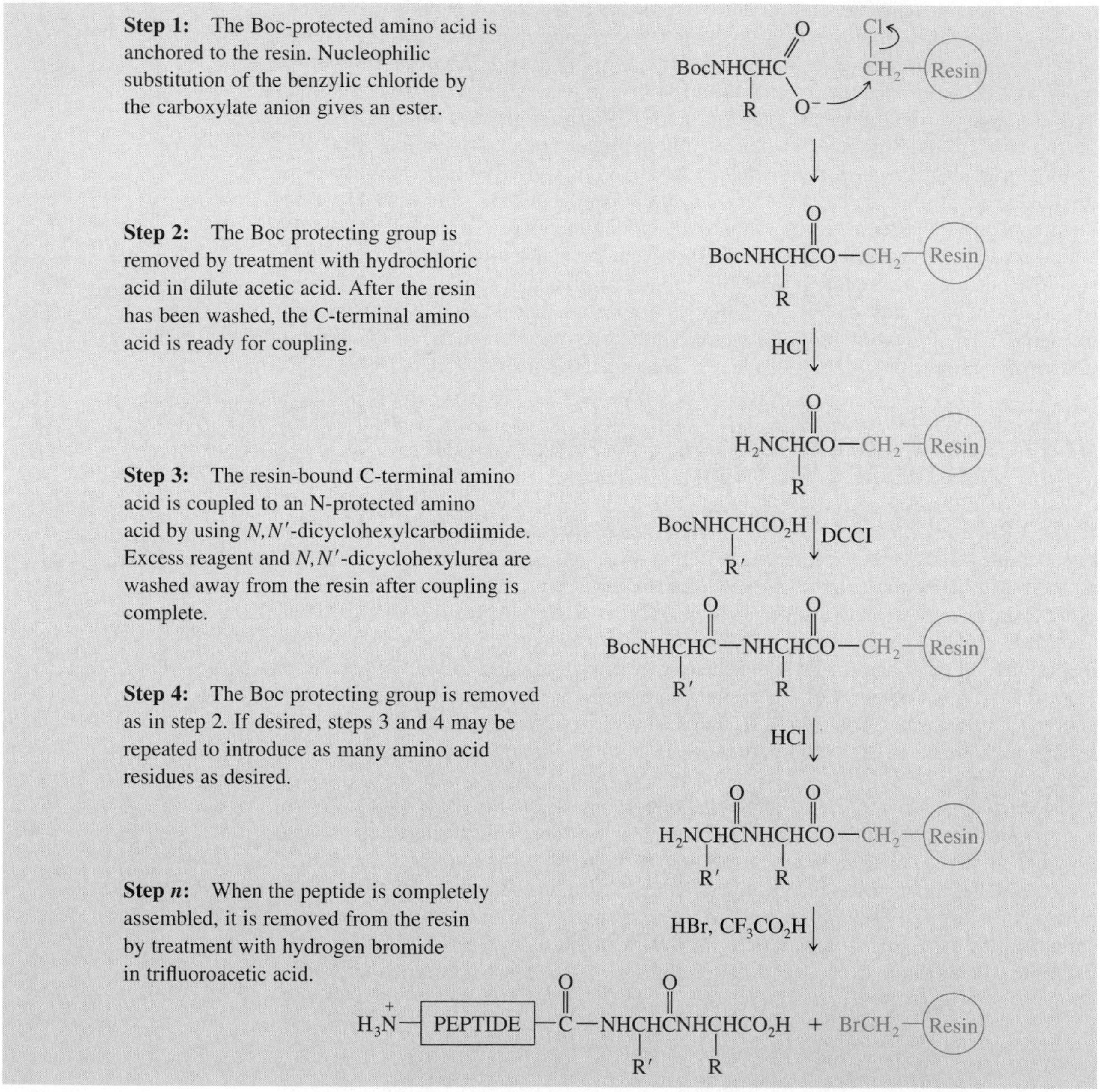
Step 1: The Boc-protected amino acid is anchored to the resin. Nucleophilic substitution of the benzylic chloride by the carboxylate anion gives an ester.
Step 2: The Boc protecting group is removed by treatment with hydrochloric acid in dilute acetic acid. After the resin has been washed, the C-terminal amino acid is ready for coupling.
Step 3: The resin-bound C-terminal amino acid is coupled to an N-protected amino acid by using *N,N'*-dicyclohexylcarbodiimide. Excess reagent and *N,N'*-dicyclohexylurea are washed away from the resin after coupling is complete.
Step 4: The Boc protecting group is removed as in step 2. If desired, steps 3 and 4 may be repeated to introduce as many amino acid residues as desired.
Step *n*: When the peptide is completely assembled, it is removed from the resin by treatment with hydrogen bromide in trifluoroacetic acid.
HCl
DCCI
HCl
HBr, CF_3CO_2H
PEPTIDE
Resin

FIGURE 27.16 Diagram illustrating peptide synthesis by the solid-phase method of Merrifield. Amino acid residues are attached sequentially beginning at the C terminus.

solvents. By-products are removed, and only the polymer and its attached C-terminal amino acid residue remain. Next (step 3), a peptide bond to an *N*-Boc-protected amino acid is formed by condensation in the presence of *N,N'*-dicyclohexylcarbodiimide. Again, the polymer is washed thoroughly. The Boc protecting group is then removed by acid treatment (step 4), and after washing, the polymer is now ready for the addition of another amino acid residue by a repetition of the cycle. When all the amino acids have been added, the synthetic peptide is removed from the polymeric support by treatment with hydrogen bromide in trifluoroacetic acid.

By successively adding amino acid residues to the C-terminal amino acid, it took Merrifield only 8 days to synthesize bradykinin in 68 percent yield. The biological activity of synthetic bradykinin was identical with that of natural material.

PROBLEM 27.20 Starting with phenylalanine and glycine, outline the steps in the preparation of Phe-Gly by the Merrifield method.

Merrifield successfully automated all the steps in solid-phase peptide synthesis, and computer-controlled equipment is now commercially available. Using an early version of his "peptide synthesizer," in collaboration with coworker Bernd Gutte, Merrifield reported the synthesis of the enzyme ribonuclease in 1969. It took them only 6 weeks to perform the 369 reactions and 11,391 steps necessary to assemble the sequence of 124 amino acids of ribonuclease.

Solid-phase peptide synthesis does not solve all purification problems, however. Even if every coupling step in the ribonuclease synthesis proceeded in 99 percent yield, the product would be contaminated with many different peptides containing 123 amino acids, 122 amino acids, and so on. Thus, Merrifield and Gutte's 6 weeks of synthesis was followed by 4 months spent in purifying the final product. The technique has since been refined to the point that yields at the 99 percent level and greater are achieved with current instrumentation, and thousands of peptides and peptide analogs have been prepared by the solid-phase method.

27.18 SECONDARY STRUCTURES OF PEPTIDES AND PROTEINS

The primary structure of a peptide is its constitution, and its constitution is dictated by the amino acid sequence. We also speak of the **secondary structure** of a peptide, that is, the conformational relationship of nearest neighbor amino acids with respect to each other. We apply unique names to certain arrangements of peptide units, just as we identify certain conformations of small molecules by descriptive names such as gauche, anti, chair, boat, and so on. On the basis of x-ray crystallographic studies and careful examination of molecular models, Linus Pauling and Robert B. Corey of the California Institute of Technology showed that certain peptide conformations were more stable than others. Two arrangements, the ***α* helix** and the **pleated *β* sheet,** stand out as secondary structural units that are both particularly stable and commonly encountered. Both of these incorporate two features that are especially important:

1. The geometry of the peptide bond is planar and the main chain is arranged in an anti conformation (Section 27.7).
2. Hydrogen bonding can occur when the N—H group of one amino acid unit and the C═O group of another are close in space; conformations that maximize the number of these hydrogen bonds are stabilized by them.

Figure 27.17 depicts one type of secondary structure, a flat sheet characterized by C═O · · · H—N hydrogen bonds between two peptide chains. This is not a particularly stable structure, because of van der Waals repulsions between the substituents on the α carbon atom of one chain and the corresponding substituents on the neighboring chain. Rotation about the single bond between the α carbon and nitrogen permits

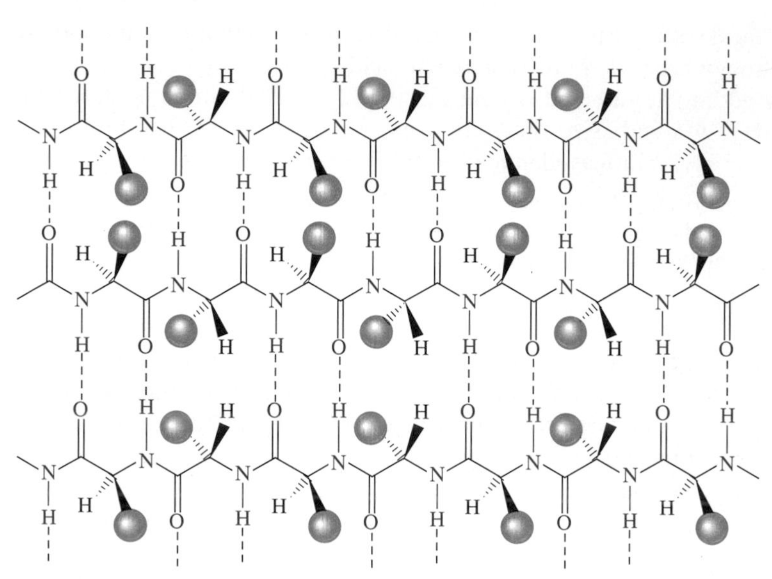

FIGURE 27.17 Diagram of the β-sheet secondary structure of a protein, showing hydrogen bonds between amide protons (blue) and carbonyl oxygens (red) of adjacent chains. Van der Waals repulsions between substituents (green) at the α carbon atoms are relieved by a conformational change within each chain, which has the effect of introducing vertical creases in the sheet. This motion places the substituents almost perpendicular to the plane of the page, alternating above and below the plane as one proceeds along the chain. The secondary structure that results is known as the *pleated β sheet.*

substituents on adjacent chains to move away from each other. Within each chain, then, the carbon atoms alternate with respect to their orientation—one is above the mean plane of the sheet, the next is below this mean plane, and so on. The peptide chains are not flat but are rippled and give the three-dimensional surface the appearance of a *pleated sheet.* The pleated β sheet is an important secondary structure, especially in proteins that are rich in amino acids with small substituents on their α carbon atoms. *Fibroin,* the major protein of most silk fibers, is almost entirely pleated β sheet, and over 80 percent of fibroin is a repeating sequence of the six-residue unit

-Gly-Ser-Gly-Ala-Gly-Ala-

The pleated β sheet is flexible, but since the peptide chains are nearly in an extended conformation, it resists stretching.

Unlike the pleated β sheet, in which hydrogen bonds are formed between two chains, the α *helix* is stabilized by hydrogen bonds within a single chain. Figure 27.18 illustrates a section of peptide α helix constructed from L amino acids. A right-handed helical conformation with about 3.6 amino acids per turn permits each carbonyl oxygen to be hydrogen-bonded to an amide proton and vice versa. The α helix is found in many proteins; the principal protein components of muscle (*myosin*) and wool (*α-keratin*), for example, contain high percentages of α helix. When wool fibers are stretched, these helical regions are elongated by the breaking of hydrogen bonds. Disulfide bonds between cysteine residues of neighboring α-keratin chains are too

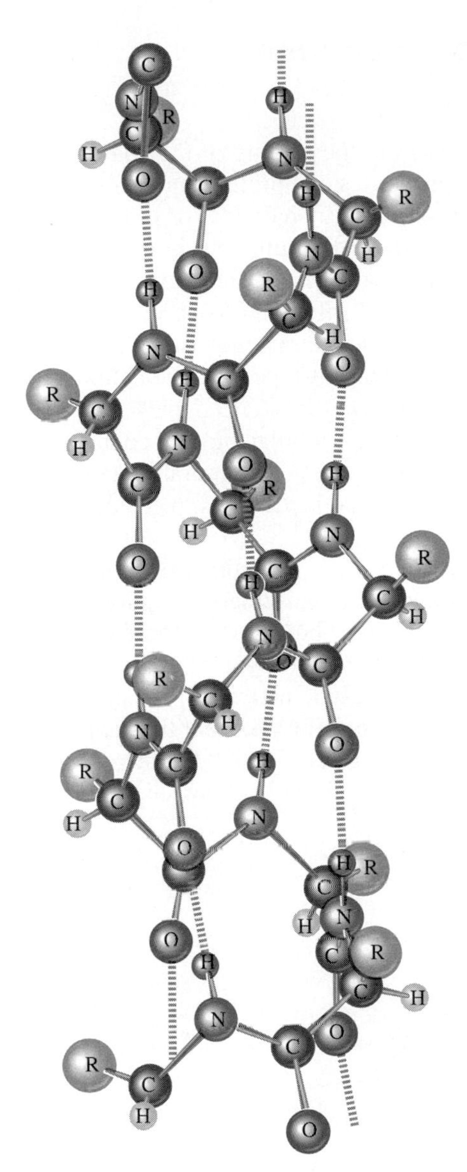

FIGURE 27.18 A protein α helix. The α helix is stabilized by hydrogen bonds within the chain between amide protons (blue) and carbonyl oxygens (red). The substituents at the α carbon are shown in green. When viewed along the helical axis, the chain turns in a clockwise direction and is designated as a right-handed helix.

strong to be broken during stretching, however, and they limit the extent of distortion. After the stretching force is removed, the hydrogen bonds re-form spontaneously and the wool fiber returns to its original shape. Wool has properties that are different from those of silk because the secondary structures of the two fibers are different, and their secondary structures are different because their primary structures are different.

Proline is the only amino acid in Table 27.1 that is a secondary amine, and its presence in a peptide chain introduces an amide nitrogen that has no hydrogen substituent available for hydrogen bonding. This disrupts the network of hydrogen bonds and divides the peptide into two separate regions of α helix. The presence of proline is often associated with a bend in the peptide chain.

Proteins, or sections of proteins, sometimes exist as **random coils,** an arrangement that lacks the regularity of the α helix or pleated β sheet.

27.19 TERTIARY STRUCTURE OF PEPTIDES AND PROTEINS

The **tertiary structure** of a peptide or protein refers to the folding of the chain. The way the chain is folded affects both the physical properties of a protein and its biological function. Structural proteins, such as those present in skin, hair, tendons, wool, and silk, may have either helical or pleated-sheet secondary structures but in general are elongated in shape, with a chain length many times the chain diameter. They are classed as *fibrous* proteins and, as befits their structural role, tend to be insoluble in water. Many other proteins, including most enzymes, operate in aqueous media; some are soluble but most are dispersed as colloids. Proteins of this type are called **globular** proteins. Globular proteins are approximately spherical. Figure 27.19 depicts the peptide backbone of carboxypeptidase A (Section 27.11), a globular protein containing 307 amino acids. A typical protein such as carboxypeptidase A incorporates elements of a number of secondary structures: some segments are helical; others, pleated sheet; and still others correspond to no simple description.

The shape of a large protein is influenced by many factors, including, of course, its primary and secondary structure. The disulfide bond shown in green in Figure 27.19 links Cys-138 of carboxypeptidase A to Cys-161 and contributes to the tertiary structure. Carboxypeptidase A contains a Zn^{2+} ion (shown in red in Figure 27.19), making it a **metalloenzyme.** The Zn^{2+} ion is essential to the catalytic activity of the enzyme, and its presence influences the tertiary structure. The Zn^{2+} ion lies near the center of the enzyme, where it is coordinated to the imidazole nitrogens of two histidine residues (His-69, His-196) and to the carboxylate side chain of Glu-72. (These ligands are shown in blue in the figure.)

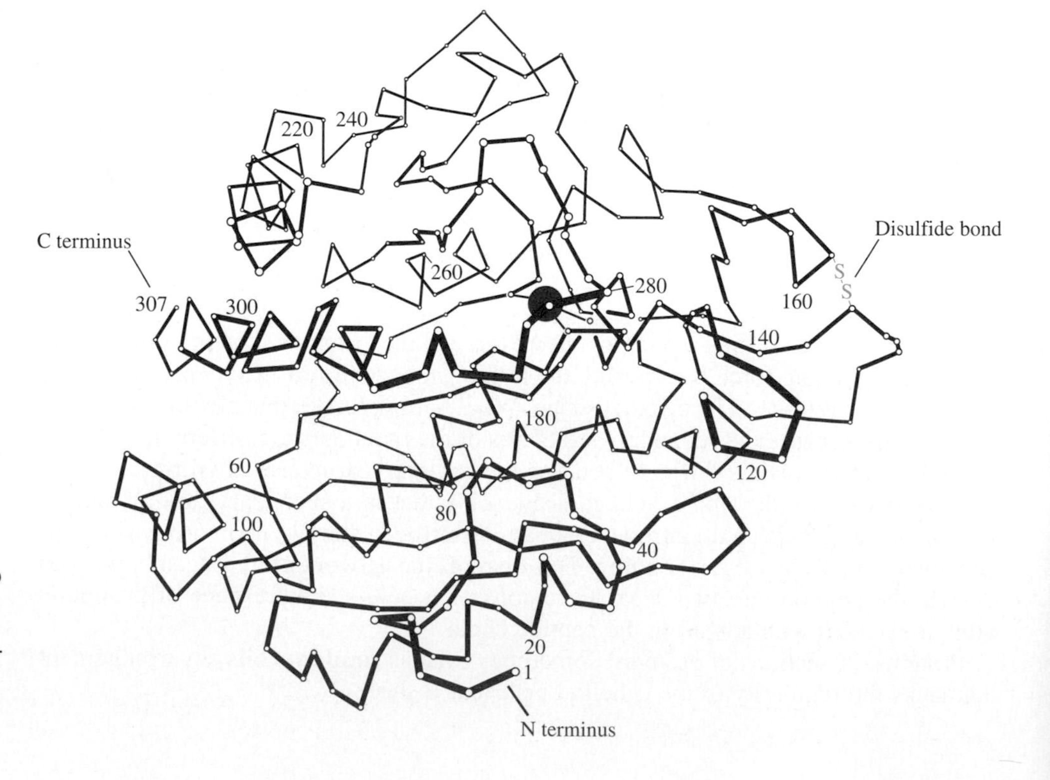

FIGURE 27.19 The peptide backbone of carboxypeptidase A. The substituents at the α carbon atom of each of the amino acids have been omitted so that the tertiary structure of the enzyme can be more easily seen. Amino acids are numbered beginning at the N terminus (bottom center). The C-terminal amino acid is residue 307 at the left. A disulfide bridge between two cysteines is shown in green, and Zn^{2+} is shown in red. The amino acids that are coordinated to Zn^{2+} (His-69, His-196, and Glu-72) are indicated in blue.

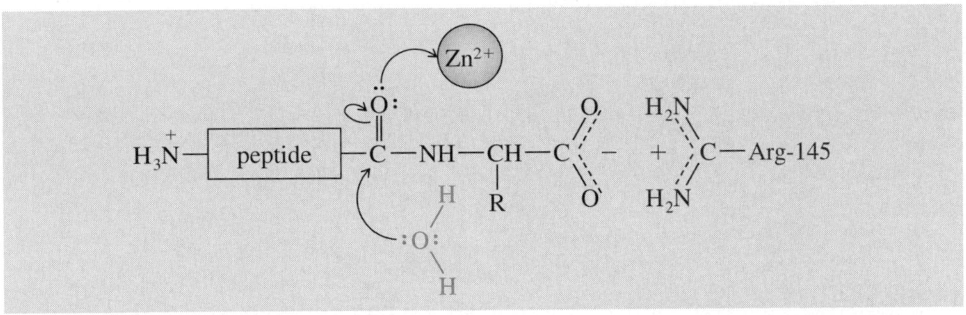

FIGURE 27.20 Proposed mechanism of hydrolysis of a peptide catalyzed by carboxypeptidase A. The peptide is bound at the active site by an ionic bond between its C-terminal amino acid and the positively charged side chain of arginine-145. Coordination of Zn^{2+} to oxygen makes the carbon of the carbonyl group more positive and increases the rate of nucleophilic attack by water.

Protein tertiary structure is also influenced by its environment. In water a globular protein usually adopts a shape that places its lipophilic groups toward the interior, with its polar groups on the surface, where they are solvated by water molecules. About 65 percent of the mass of most cells is water, and the proteins present in cells are said to be in their *native state*—the tertiary structure in which they express their biological activity. When the tertiary structure of a protein is disrupted by adding substances that cause the protein chain to unfold, the protein becomes *denatured* and loses most of, if not all, its activity. Evidence that supports the view that the tertiary structure is dictated by the primary structure includes experiments in which proteins are denatured and allowed to stand, whereupon they are observed to spontaneously readopt their native-state conformation with full recovery of biological activity.

Tertiary structures of proteins are determined by x-ray crystallography. Myoglobin, the oxygen storage protein of muscle, was the first protein to have its tertiary structure revealed (1957). For their work on myoglobin and hemoglobin, respectively, John C. Kendrew and Max F. Perutz were awarded the 1962 Nobel Prize in chemistry. A knowledge of how the protein chain is folded is an indispensable step toward understanding the mechanism of enzyme action.

Consider, for example, carboxypeptidase A, which catalyzes the hydrolysis of the peptide bond at the C terminus of a peptide (Section 27.11). It is believed that an ionic bond between the positively charged side chain of an arginine residue (Arg-145) of the enzyme and the negatively charged carboxylate group of the substrate's terminal amino acid binds the peptide at the **active site,** the region of the enzyme's interior where the catalytically important functional groups are located. There, the Zn^{2+} ion acts as a Lewis acid toward the carbonyl oxygen of the peptide substrate, increasing its susceptibility toward attack by a water molecule (Figure 27.20).

A proposed mechanism for enzyme action is like any other reaction mechanism in that it represents our best present interpretation of the facts and must be refined or, in some cases, abandoned, as additional experimental evidence becomes available.

Living systems contain thousands of different enzymes. As we have seen, all are structurally quite complex, and there are no sweeping generalizations that can be made to include all aspects of enzymic catalysis. The case of carboxypeptidase A illustrates one mode of enzyme action, the bringing together of reactants and catalytically active functions at the active site.

27.20 COENZYMES

The number of chemical processes that protein side chains can engage in is rather limited. Most prominent among them are proton donation, proton abstraction, and

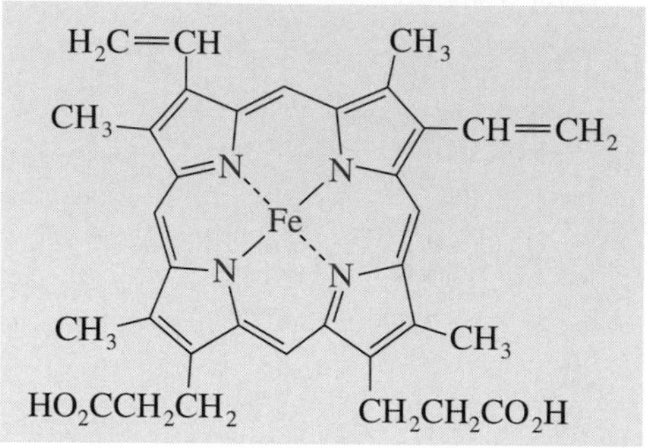

FIGURE 27.21 The structure of heme.

nucleophilic addition to carbonyl groups. In many biological processes a richer variety of reactivity is required, and proteins often act in combination with nonprotein organic molecules to bring about the necessary chemistry. These ''helper molecules,'' referred to as **coenzymes, cofactors,** or **prosthetic groups,** interact with both the enzyme and the substrate to produce the necessary chemical change. Acting alone, for example, proteins lack the necssary functionality to be effective oxidizing or reducing agents. They can catalyze biological oxidations and reductions, however, in the presence of a suitable coenzyme. In earlier sections we have seen numerous examples of these reactions in which the coenzyme NAD^+ acted as an oxidizing agent, and others in which NADH acted as a reducing agent.

Heme (Figure 27.21) is an important prosthetic group in which iron(II) is coordinated with the four nitrogen atoms of a type of tetracyclic aromatic substance known as a *porphyrin.* The oxygen-storing protein of muscle, myoglobin, represented schematically in Figure 27.22, consists of a heme group surrounded by a protein of 153 amino

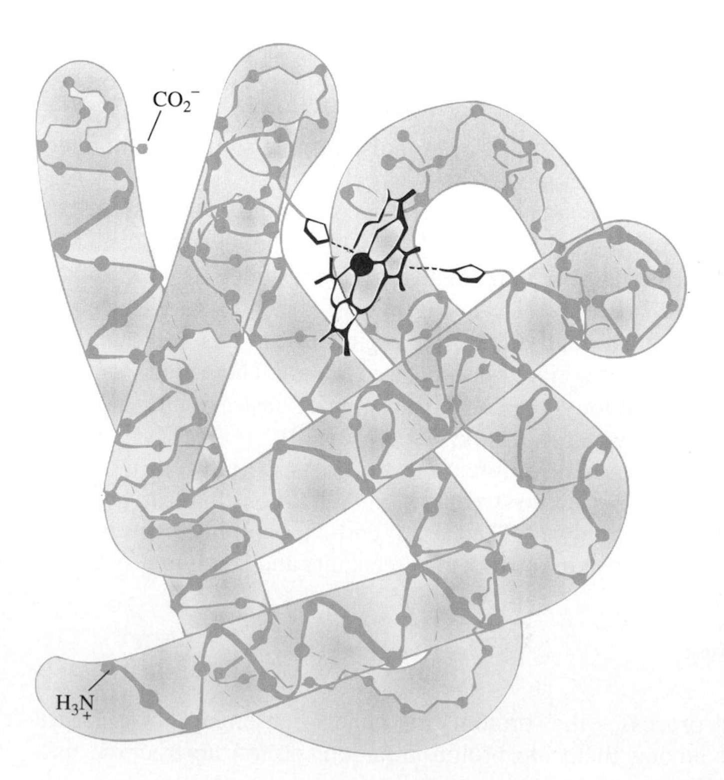

FIGURE 27.22 Diagram showing the folding of the protein chain of myoglobin. The heme portion is shown in red. The blue rings represent the side chains of histidine residues.

acids. Four of the six available coordination sites of Fe^{2+} are taken up by the nitrogens of the porphyrin, one by a histidine residue of the protein, and the last by a water molecule. Myoglobin stores oxygen obtained from the blood by formation of an Fe-O_2 complex. The oxygen displaces water as the sixth ligand on iron and is held there until needed. The protein serves as a container for the heme and prevents oxidation of Fe^{2+} to Fe^{3+}, an oxidation state in which iron lacks the ability to bind oxygen. Separately, neither heme nor the protein binds oxygen in aqueous solution; together, they do it very well.

27.21 PROTEIN QUATERNARY STRUCTURE. HEMOGLOBIN

Rather than existing as a single polypeptide chain, some proteins are assemblies of two or more chains. The manner in which these subunits are organized is called the **quaternary structure** of the protein.

Hemoglobin is the oxygen-carrying protein of blood. It binds oxygen at the lungs and transports it to the muscles, where it is stored by myoglobin. Hemoglobin binds oxygen in very much the same way as myoglobin, using heme as the prosthetic group. Hemoglobin is much larger than myoglobin, however, having a molecular weight of 64,500, whereas that of myoglobin is 17,500; hemoglobin contains four heme units, myoglobin only one. Hemoglobin is an assembly of four hemes and four protein chains, including two identical chains called the *alpha chains* and two identical chains called the *beta chains.*

An article entitled "Hemoglobin: Its Occurrence, Structure, and Adaptation" appeared in the March 1982 issue of the *Journal of Chemical Education* (pp. 173–178).

Some substances, such as CO, form strong bonds to the iron of heme, strong enough to displace O_2 from it. Carbon monoxide binds 30 to 50 times more effectively than oxygen to myoglobin and hundreds of times better than oxygen to hemoglobin. Strong binding of CO at the active site interferes with the ability of heme to perform its biological task of transporting and storing oxygen, with potentially lethal results.

How function depends on structure can be seen in the case of the genetic disorder *sickle cell anemia.* This is a debilitating, sometimes fatal, disease in which red blood cells become distorted ("sickle-shaped") and interfere with the flow of blood through the capillaries. This condition results from the presence of an abnormal hemoglobin in affected people. The primary structures of the beta chain of normal and sickle cell hemoglobin differ by a single amino acid out of 149; sickle cell hemoglobin has valine in place of glutamic acid as the sixth residue from the N terminus. A tiny change in amino acid sequence can produce a life-threatening result! This modification is genetically controlled and probably became established in the gene pool because bearers of the trait have an increased resistance to malaria.

27.22 PYRIMIDINES AND PURINES

One of the major scientific achievements of the twentieth century has been the identification, at the molecular level, of the chemical interactions that are involved in the transfer of genetic information and the control of protein biosynthesis. The substances involved are biological macromolecules called **nucleic acids.** Nucleic acids were isolated over 100 years ago, and, as their name implies, they are acidic substances present in the nuclei of cells. There are two major kinds of nucleic acids, ribonucleic acid (RNA) and deoxyribonucleic acid (DNA). In order to understand the complex structure of nucleic acids, we first need to examine some simpler substances, nitrogen-con-

Recall that heterocyclic aromatic compounds were introduced in Section 11.23.

taining aromatic heterocycles called *pyrimidines* and *purines*. The parent substance of each class and the numbering system used are shown:

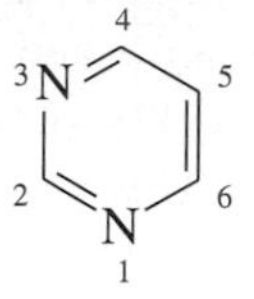

Pyrimidine 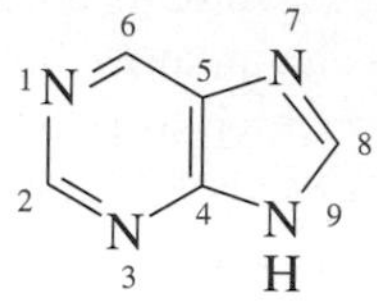 Purine

The pyrimidines that occur in DNA are cytosine and thymine. Cytosine is also a structural unit in RNA, which, however, contains uracil instead of thymine. Other pyrimidine derivatives are sometimes present but in small amounts.

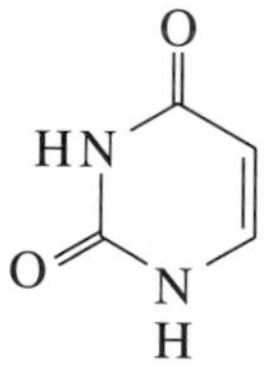

Uracil (occurs in RNA) Thymine (occurs in DNA) Cytosine (occurs in both RNA and DNA)

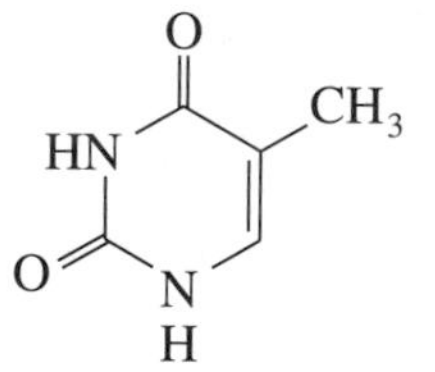

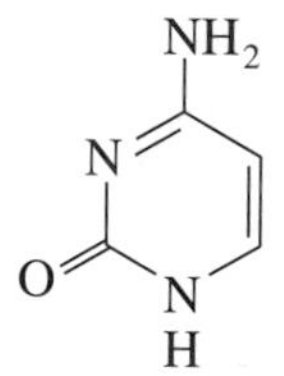

PROBLEM 27.21 5-Fluorouracil is a drug used in cancer chemotherapy. What is its structure?

Adenine and guanine are the principal purines of both DNA and RNA.

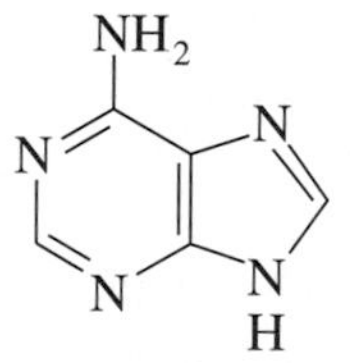

Adenine 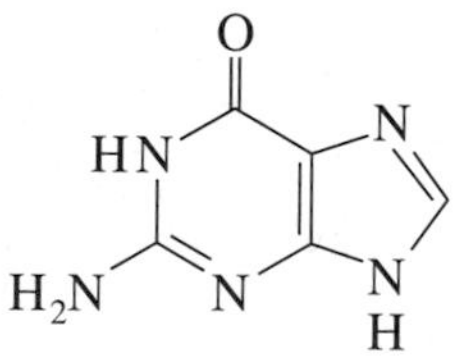 Guanine

The rings of purines and pyrimidines are aromatic and planar. You will see how important this flat shape is when we consider the structure of nucleic acids.

Pyrimidines and purines occur naturally in substances other than nucleic acids. Coffee, for example, is a familiar source of caffeine. Tea contains both caffeine and theobromine.

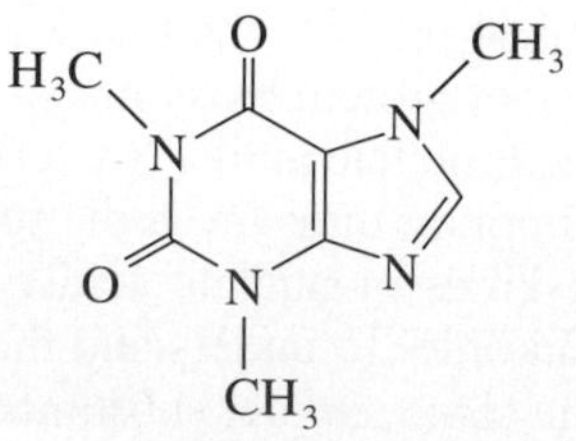

Caffeine 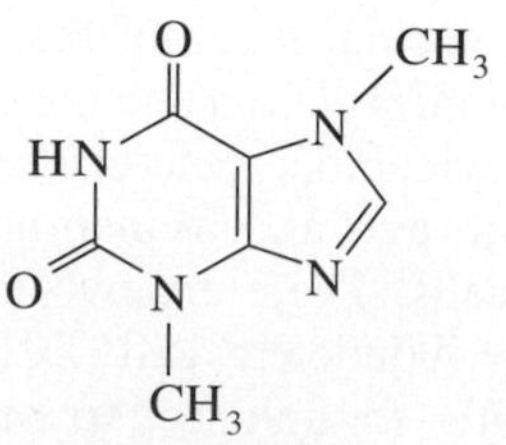 Theobromine

27.23 NUCLEOSIDES

The term **nucleoside** was once restricted to pyrimidine and purine *N*-glycosides of D-ribofuranose and 2-deoxy-D-ribofuranose, because these are the substances present in nucleic acids. The term is used more liberally now with respect to the carbohydrate portion, but is still usually limited to pyrimidine and purine substituents at the anomeric carbon. *Uridine* is a representative pyrimidine nucleoside; it bears a D-ribofuranose group at N-1. *Adenosine* is a representative purine nucleoside; its carbohydrate unit is attached at N-9.

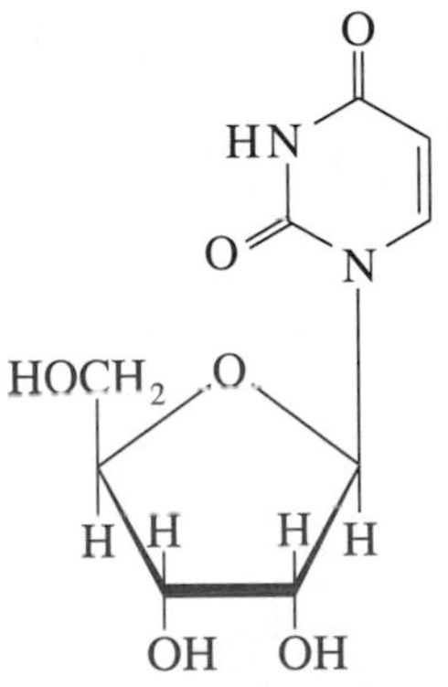

Uridine
(1-β-D-ribofuranosyluracil)

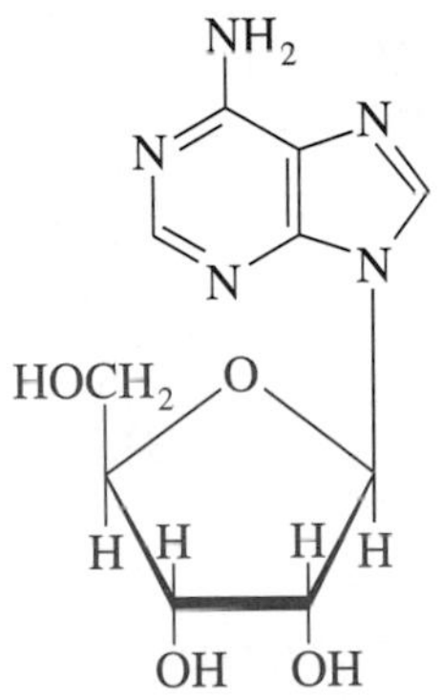

Adenosine
(9-β-D-ribofuranosyladenine)

It is customary to refer to the noncarbohydrate portion of a nucleoside as a purine or pyrimidine *base*.

PROBLEM 27.22 The names of the principal nucleosides obtained from RNA and DNA are listed. Write a structural formula for each one.

(*a*) Thymidine (thymine-derived nucleoside in DNA)
(*b*) Cytidine (cytosine-derived nucleoside in RNA)
(*c*) Guanosine (guanine-derived nucleoside in RNA)

SAMPLE SOLUTION (*a*) Thymine is a pyrimidine base present in DNA; its carbohydrate substituent is 2-deoxyribofuranose, which is attached to N-1 of thymine.

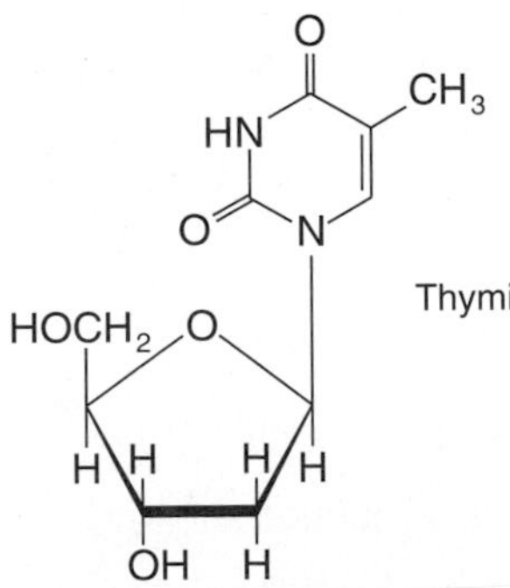

Thymidine

Nucleosides of 2-deoxyribose are named in the same way. Carbons in the carbohydrate portion of the molecule are identified as 1′, 2′, 3′, 4′, and 5′ to distinguish them from atoms in the purine or pyrimidine base. Thus, the adenine nucleoside of 2-deoxyribose is called 2′-deoxyadenosine or 9-β-2′-deoxyribofuranosyladenine.

27.24 NUCLEOTIDES

Nucleotides are phosphoric acid esters of nucleosides. The 5′-monophosphate of adenosine is called *5′-adenylic acid* or *adenosine 5′-monophosphate* (AMP).

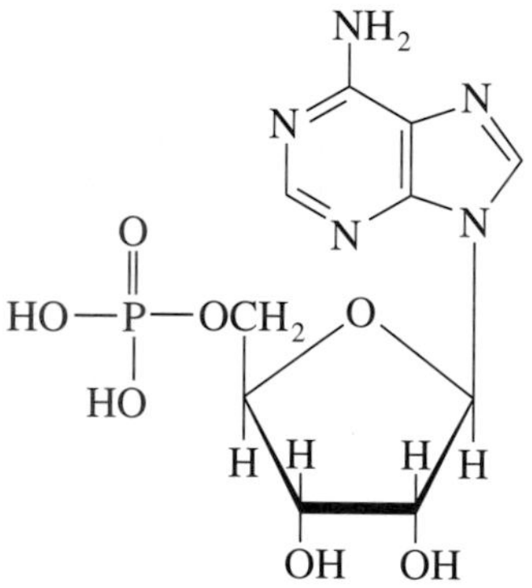

5′-Adenylic acid (AMP)

As its name implies, 5′-adenylic acid is an acidic substance; it is a dibasic acid with pK_a's for ionization of 3.8 and 6.2, respectively. In aqueous solution at pH 7, both OH groups of the $P(O)(OH)_2$ unit are ionized.

The analogous D-ribonucleotides of the other purines and pyrimidines are *uridylic acid, guanylic acid,* and *cytidylic acid. Thymidylic acid* is the 5′-monophosphate of thymidine (the carbohydrate is 2-deoxyribose in this case).

Other important 5′-nucleotides of adenosine include *adenosine diphosphate* (ADP) and *adenosine triphosphate* (ATP):

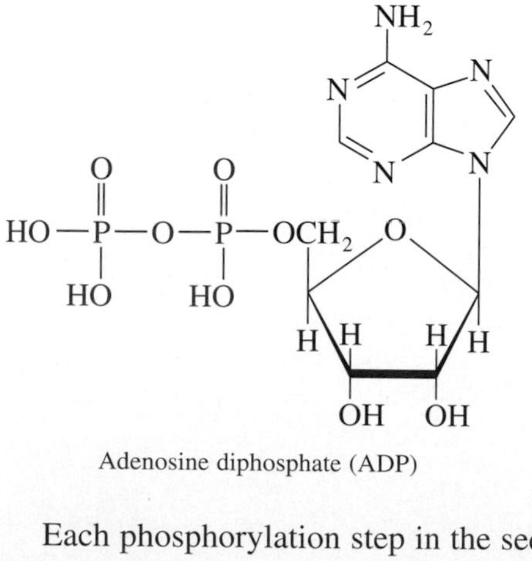

Adenosine diphosphate (ADP)

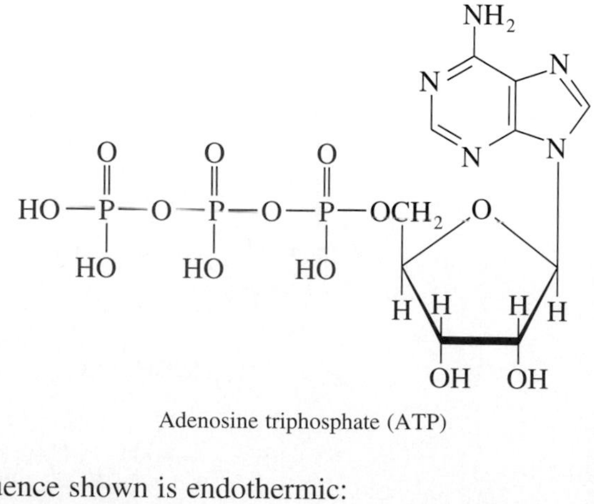

Adenosine triphosphate (ATP)

Each phosphorylation step in the sequence shown is endothermic:

$$\text{Adenosine} \xrightarrow[\text{enzymes}]{PO_4^{3-}} \text{AMP} \xrightarrow[\text{enzymes}]{PO_4^{3-}} \text{ADP} \xrightarrow[\text{enzymes}]{PO_4^{3-}} \text{ATP}$$

The energy to drive each step comes from carbohydrates by the process of glycolysis (Section 25.21). It is convenient to view ATP as the storage vessel for the energy released during conversion of carbohydrates to carbon dioxide and water. That energy becomes available to the cells when ATP undergoes hydrolysis. The hydrolysis of ATP to ADP and phosphate has a $\Delta G°$ value of -35 kJ/mol (-8.4 kcal/mol).

For a discussion of glycolysis, see the July 1986 issue of the *Journal of Chemical Education* (pp. 566–570).

Adenosine 3′-5′-cyclic monophosphate *(cyclic AMP)* is an important regulator of a large number of biological processes. It is a cyclic ester of phosphoric acid and adenosine involving the hydroxyl groups at C-3′ and C-5′.

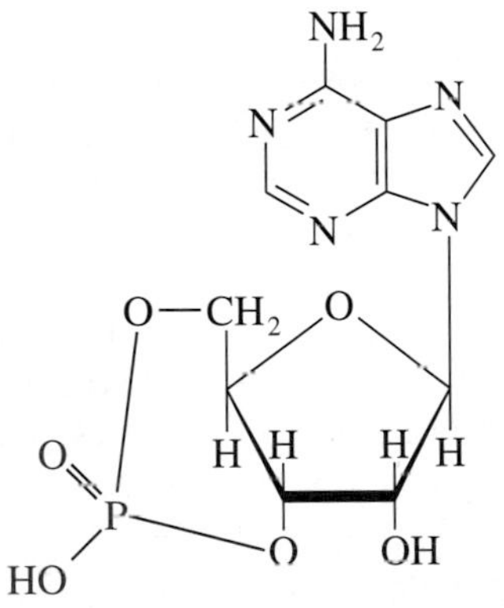

Adenosine 3′-5′-cyclic monophosphate (*cyclic AMP*)

27.25 NUCLEIC ACIDS

Nucleic acids are **polynucleotides** in which a phosphate ester unit links the 5′ oxygen of one nucleotide to the 3′ oxygen of another. Figure 27.23 is a generalized depiction

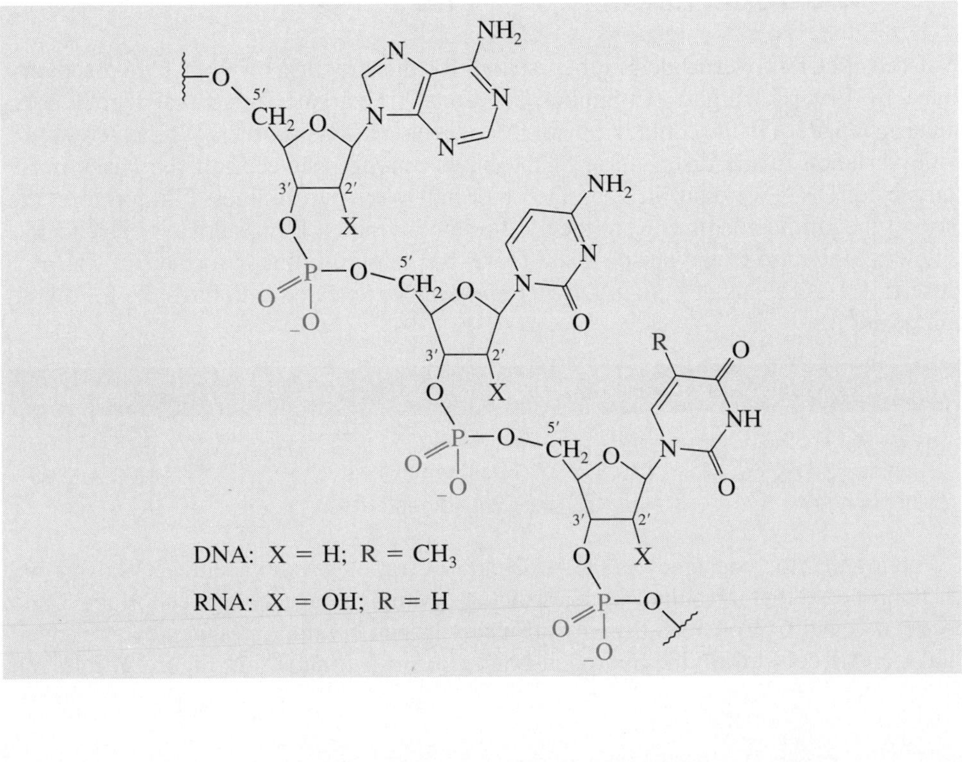

DNA: X = H; R = CH_3

RNA: X = OH; R = H

FIGURE 27.23 A portion of a polynucleotide chain.

of the structure of a nucleic acid. Nucleic acids are classified as ribonucleic acids (RNA) or deoxyribonucleic acids (DNA) depending on the carbohydrate present.

Research on nucleic acids progressed slowly until it became evident during the 1940s that they played a role in the transfer of genetic information. It was known that the genetic information of an organism resides in the chromosomes present in each of its cells and that individual chromosomes are made up of smaller units called *genes.* When it became apparent that genes are DNA, interest in nucleic acids intensified. There was a feeling that once the structure of DNA was established, the precise way in which it carried out its designated role would become more evident. In some respects the problems are similar to those of protein chemistry. Knowing that DNA is a polynucleotide is comparable with knowing that proteins are polyamides. What is the nucleotide sequence (primary structure)? What is the precise shape of the polynucleotide chain (secondary and tertiary structure)? Is the genetic material a single strand of DNA, or is it an assembly of two or more strands? The complexity of the problem can be indicated by noting that a typical strand of human DNA contains approximately 10^8 nucleotides; if uncoiled it would be several centimeters long, yet it and many others like it reside in cells too small to see with the naked eye.

Watson and Crick have each written accounts of their work, and both are well worth reading. Watson's is entitled *The Double Helix.* Crick's is *What Mad Pursuit: A Personal View of Scientific Discovery.*

In 1953 James D. Watson and Francis H. C. Crick pulled together data from biology, biochemistry, chemistry, and x-ray crystallography, along with the insight they gained from molecular models, to propose a structure for DNA and a mechanism for its replication. Their two brief papers paved the way for an explosive growth in our understanding of life processes at the molecular level, the field we now call *molecular biology.* Along with Maurice Wilkins, who was responsible for the x-ray crystallographic work, Watson and Crick shared the 1962 Nobel Prize in physiology or medicine.

27.26 STRUCTURE AND REPLICATION OF DNA. THE DOUBLE HELIX

Watson and Crick were aided in their search for the structure of DNA by a discovery made by Erwin Chargaff (Columbia University). Chargaff found that there was a consistent pattern in the composition of DNAs from various sources. While there was a wide variation in the distribution of the bases among species, half the bases in all samples of DNA were purines and the other half were pyrimidines. Furthermore, the ratio of the purine adenine (A) to the pyrimidine thymine (T) was always close to 1:1. Likewise, the ratio of the purine guanine (G) to the pyrimidine cytosine (C) was also close to 1:1. Analysis of human DNA, for example, revealed it to have the following composition:

Purine	Pyrimidine	Base ratio
Adenine (A) 30.3%	Thymine (T) 30.3%	A/T = 1.00
Guanine (G) 19.5%	Cytosine (C) 19.9%	G/C = 0.98
Total purines 49.8%	Total pyrimidines 50.1%	

Feeling that the constancy in the A/T and G/C ratios was no accident, Watson and Crick proposed that it resulted from a structural complementarity between A and T and between G and C. Consideration of various hydrogen bonding arrangements revealed that A and T could form the hydrogen-bonded *base pair* shown in Figure 27.24*a* and

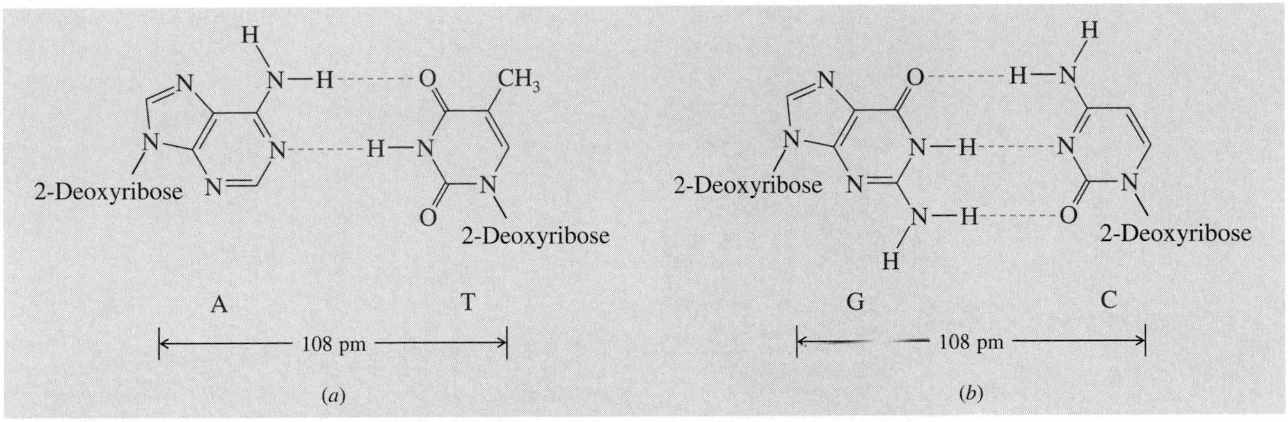

FIGURE 27.24 Base pairing between (*a*) adenine and thymine and (*b*) guanine and cytosine.

that G and C could associate as in Figure 27.24*b*. Specific base pairing of A to T and of G to C by hydrogen bonds is a key element in the Watson-Crick model for the structure of DNA. We shall see that it is also a key element in the replication of DNA.

Because each hydrogen-bonded base pair contains one purine and one pyrimidine, A---T and G---C are approximately the same size. Thus, two nucleic acid chains may be aligned side by side with their bases in the middle, as illustrated in Figure 27.25.

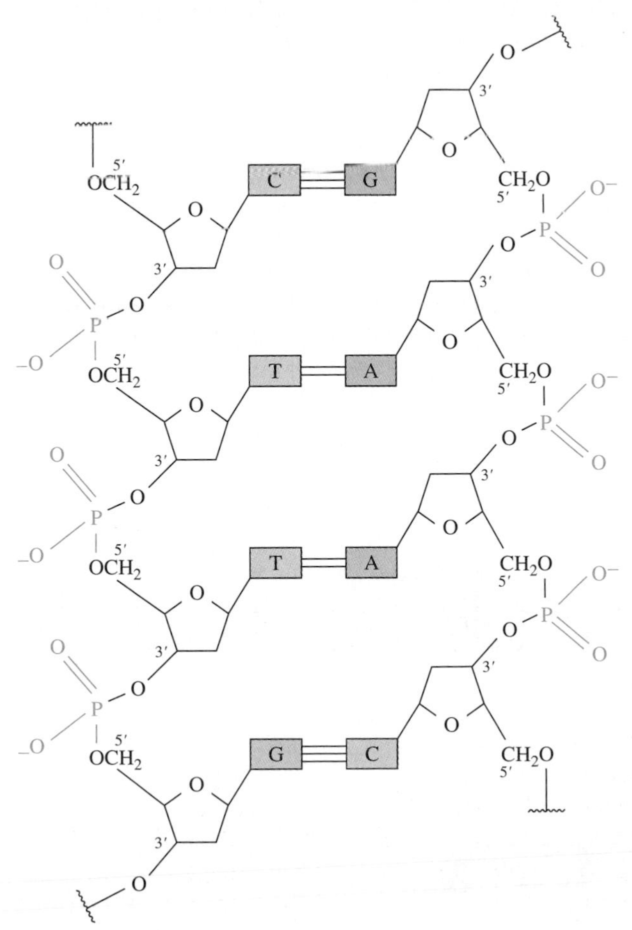

FIGURE 27.25 Hydrogen bonds between complementary bases (A and T, G and C) permit pairing of two DNA strands. The strands are antiparallel; the 5′ end of the left strand is at the top, while the 5′ end of the right strand is at the bottom.

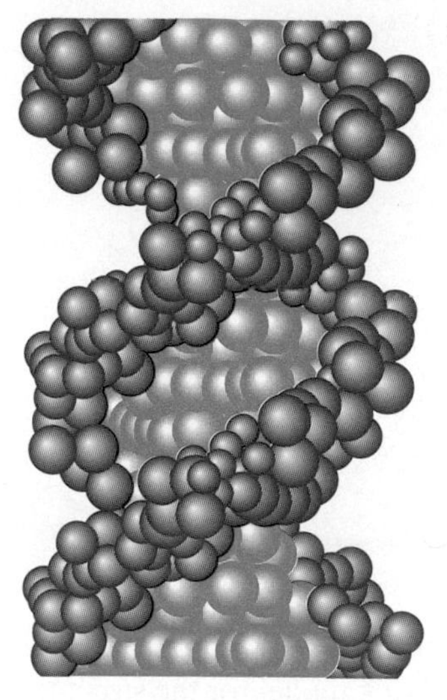

FIGURE 27.26 A model of a portion of a DNA double helix. The red atoms on the outside of the double helix belong to the carbohydrate-phosphate "backbone" of the molecule. The green atoms belong to the purine and pyrimidine bases and are stacked as steps on a spiral staircase on the inside.

The two chains are joined by the network of hydrogen bonds between the paired bases A---T and G---C. Since x-ray crystallographic data indicated a helical structure, Watson and Crick proposed that the two strands are intertwined as a **double helix** (Figure 27.26).

The Watson-Crick base pairing model for DNA structure holds the key to understanding the process of DNA **replication.** During cell division a cell's DNA is duplicated, that in the new cell being identical with that in the original cell. At one stage of cell division the DNA double helix begins to unwind, separating the two chains. As portrayed in Figure 27.27, each strand serves as the template upon which a new DNA strand is constructed. Each new strand is exactly like the original partner because the A---T, G---C base pairing requirement ensures that the new strand is the precise complement of the template, just as the old strand was. As the double helix unravels, each strand becomes one half of a new and identical DNA double helix.

The structural requirements for the pairing of nucleic acid bases are also critical for utilizing genetic information, and in living systems this means protein biosynthesis.

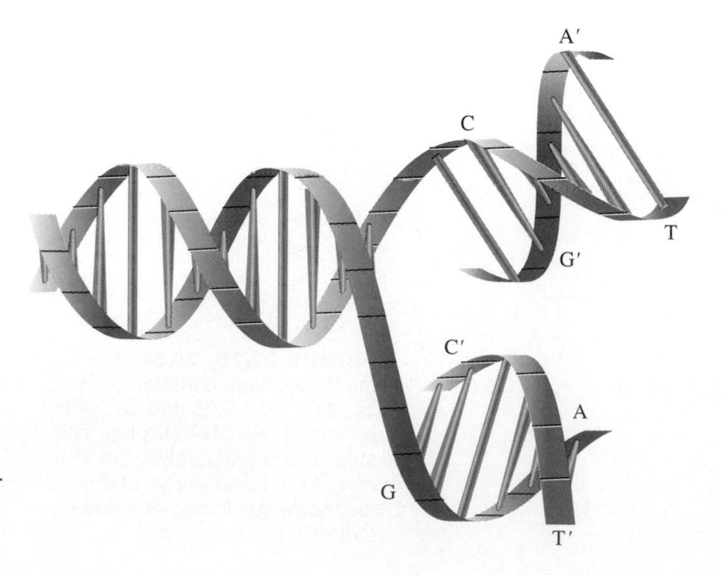

FIGURE 27.27 During DNA replication the double helix unwinds, and each of the original strands serves as a template for the synthesis of its complementary strand.

27.27 DNA-DIRECTED PROTEIN BIOSYNTHESIS

Protein biosynthesis is directed by DNA through the agency of several types of ribonucleic acid called *messenger RNA (mRNA)*, *transfer RNA (tRNA)*, and *ribosomal RNA (rRNA)*. There are two main stages in protein biosynthesis, **transcription** and **translation.**

In the transcription stage a molecule of mRNA having a nucleotide sequence complementary to one of the strands of a DNA double helix is constructed. A diagram illustrating transcription is presented in Figure 27.28. Transcription begins at the 5′ end of the DNA molecule, and ribonucleotides with bases complementary to the DNA bases are polymerized with the aid of the enzyme *RNA polymerase.* Thymine does not occur in RNA; the base that pairs with adenine in RNA is uracil. Unlike DNA, RNA is single-stranded.

In the translation stage, the nucleotide sequence of the mRNA is decoded and ''read'' as an amino acid sequence to be constructed. Since there are only four different bases in mRNA and 20 amino acids to be coded for, codes using either one nucleotide to one amino acid or two nucleotides to one amino acid are inadequate. If nucleotides are read in sets of three, however, the four mRNA bases (A, U, C, G) generate 64 possible ''words,'' more than sufficient to code for 20 amino acids. It has been established that the *genetic code* is indeed made up of triplets of adjacent nucleotides called *codons.* The amino acids corresponding to each of the 64 possible codons of mRNA have been determined (Table 27.4).

PROBLEM 27.23 It was pointed out in Section 27.21 that sickle cell hemoglobin has valine in place of glutamic acid at one point in its protein chain. Compare the codons for valine and glutamic acid. How do they differ?

The mechanism of translation makes use of the same complementary base pairing principle used in replication and transcription. Each amino acid is associated with a particular tRNA. Transfer RNA is much smaller than DNA and mRNA. It is single-

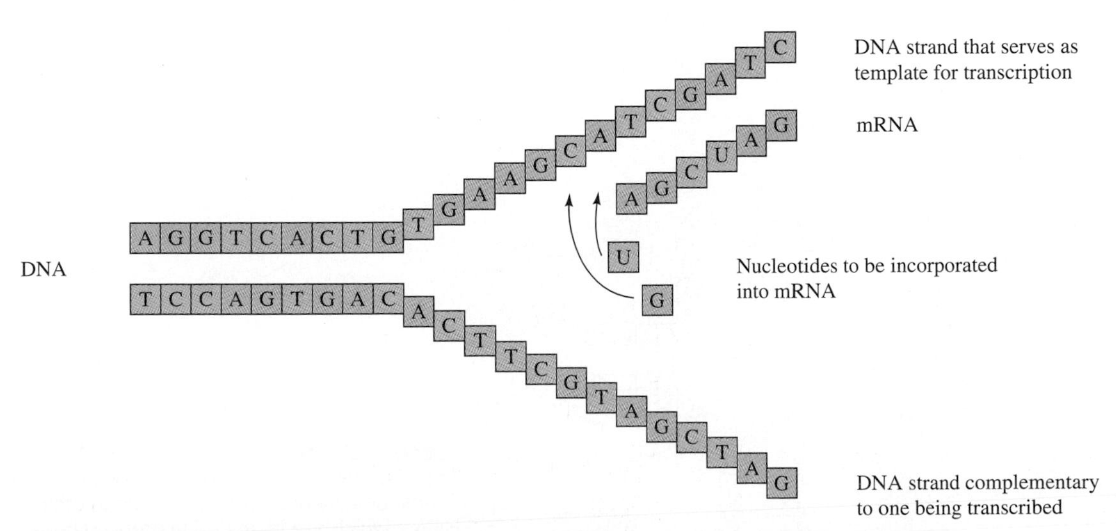

FIGURE 27.28 During transcription, a molecule of mRNA is assembled by using DNA as a template.

TABLE 27.4
The Genetic Code (Messenger RNA Codons)*

Alanine	Arginine	Asparagine	Aspartic acid	Cysteine
GCU GCA	CGU CGA AGA	AAU	GAU	UGU
GCC GCG	CGC CGG AGG	AAC	GAC	UGC
Glutamic acid	Glutamine	Glycine	Histidine	Isoleucine
GAA	CAA	GGU GGA	CAU	AUU AUA
GAG	CAG	GGC GGG	CAC	AUC
Leucine	Lysine	Methionine	Phenylalanine	Proline
UUA CUU CUA	AAA	AUG	UUU	CCU CCA
UUG CUC CUG	AAG		UUC	CCC CCG
Serine	Threonine	Tryptophan	Tyrosine	Valine
UCU UCA AGU	ACU ACA	UGG	UAU	GUU GUA
UCC UCG AGC	ACC ACG		UAC	GUC GUG

* The first letter of each triplet corresponds to the nucleotide nearer the 5′ terminus, the last letter to the nucleotide nearer the 3′ terminus. UAA, UGA, and UAG are not included in the table; they are chain-terminating codons.

stranded and contains 70 to 90 ribonucleotides arranged in a "cloverleaf" pattern (Figure 27.29). Its characteristic shape results from the presence of paired bases in some regions and their absence in others. All tRNAs have a CCA triplet at their 3′ terminus, to which is attached, by an ester linkage, an amino acid unique to that particular tRNA. At one of the loops of tRNA there is a nucleotide triplet called the *anticodon,* which is complementary to a codon of mRNA. The codons of mRNA are read by the anticodons of tRNA, and the proper amino acids are transferred in sequence to the growing protein.

According to Crick, the so-called central dogma of molecular biology is "DNA makes RNA makes protein."

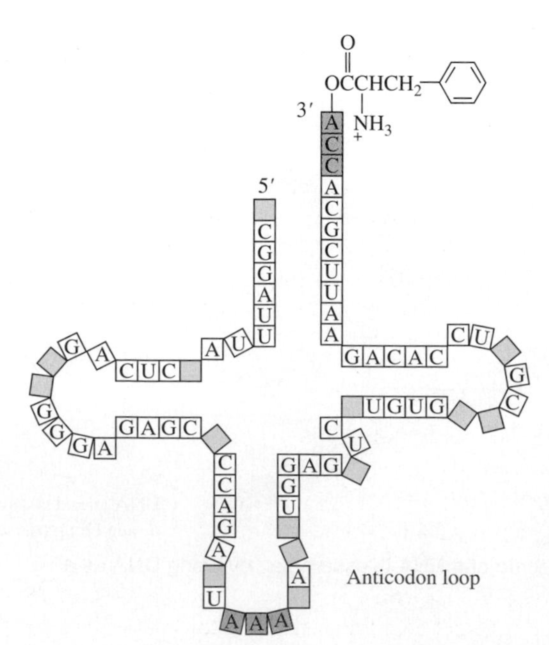

FIGURE 27.29 Phenylalanine tRNA. Transfer RNAs usually contain modified bases, slightly different from those in other RNAs, which are indicated in green. The anticodon for phenylalanine is shown in red, and the CCA triplet, which bears the phenylalanine, is in blue.

AIDS

The explosive growth of our knowledge of nucleic acid chemistry and its role in molecular biology in the 1980s happened to coincide with a challenge to human health that would have defied understanding a generation ago. That challenge is *acquired immune deficiency syndrome,* or **AIDS.** AIDS is a condition in which the body's immune system is devastated by a viral infection to the extent that it can no longer perform its vital function of identifying and destroying invading organisms. AIDS victims die from "opportunistic" infections—diseases that are normally held in check by a healthy immune system but which can become deadly when the immune system is compromised. In the short time since its discovery, AIDS has claimed the lives of over 200,000 victims in the United States, and the World Health Organization estimates that as many as 10 million people throughout the world will have died by the year 2000.

The virus responsible for almost all the AIDS cases in the United States was identified by scientists at the Louis Pasteur Institute in Paris in 1983 and is known as *human immunodeficiency virus 1* (HIV-1). HIV-1 is believed to have originated in Africa, where a related virus, HIV-2, was discovered in 1986 by the Pasteur Institute group. Both HIV-1 and HIV-2 are classed as **retroviruses,** because their genetic material is RNA rather than DNA. HIVs require a host cell to reproduce, and the hosts in humans are the so-called T4 lymphocytes, which are the cells primarily responsible for inducing the immune system to respond when provoked. The HIV penetrates the cell wall of a T4 lymphocyte and deposits both its RNA and an enzyme called *reverse transcriptase* inside the T4 cell, where the reverse transcriptase catalyzes the formation of a DNA strand that is complementary to the viral RNA. The transcribed DNA then serves as the template for formation of double-helical DNA, which, with the information it carries for reproduction of the HIV, becomes incorporated into the T4 cell's own genetic material. The viral DNA induces the host lymphocyte to begin producing copies of the virus, which then leave the host to infect other T4 cells. In the course of HIV reproduction, the ability of the T4 lymphocyte to reproduce itself is hampered. As the number of T4 cell decreases, so does the body's ability to combat infections.

There is no present known cure for AIDS. The drug most used to treat AIDS patients in the United States is *zidovudine,* also known as *azidothymine,* or *AZT.* AZT does not kill the AIDS virus but rather interferes with its ability to reproduce. Thus AZT fights an AIDS infection but does not cure it.

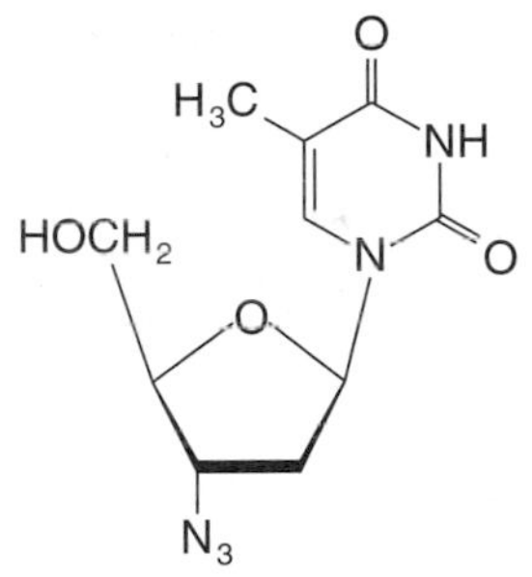

Zidovudine (AZT)

Recent evidence even suggests that, while AZT delays the onset of AIDS symptoms, it may not significantly extend the life expectancy of those infected with HIV.

The AIDS outbreak has been and continues to be a tragedy on a massive scale. Until a cure is discovered, or a vaccine developed, sustained efforts at preventing its transmission offer our best weapon against the spread of AIDS.

27.28 DNA SEQUENCING

In 1988 the United States Congress authorized the first allocation of funds in what may be a \$3 billion project dedicated to determining the sequence of bases that make up the human **genome.** (The genome is the aggregate of all the genes that determine what an organism becomes.) Given that the human genome contains approximately 3×10^9 base pairs, this expenditure amounts to \$1 per base pair—a strikingly small cost when one considers both the complexity of the project and the increased understanding of human biology that is sure to result. DNA sequencing, which lies at the heart of the human genome project, is a relatively new technique but one that has seen dramatic advances in efficiency in a very short time.

In order to explain how DNA sequencing works, we must first mention **restriction enzymes.** Like all organisms, bacteria are subject to infection by external invaders (viruses and other bacteria, for example) and possess defenses in the form of restriction enzymes that destroy the intruder by cleaving its DNA. About 200 different restriction enzymes are known. They differ in respect to the nucleotide sequence they recognize, and each restriction enzyme cleaves DNA at a specific nucleotide site. Thus, one can take a large piece of DNA and, with the aid of restriction enzymes, cleave it into units small enough to be sequenced conveniently. These smaller DNA fragments are separated and purified by gel electrophoresis. At a pH of 7.4, each phosphate link between adjacent nucleotides is ionized, giving the DNA fragments a negative charge and causing them to migrate to the positively charged electrode. Separation is size-dependent. Larger polynucleotides move more slowly through the polyacrylamide gel than smaller ones. The technique is so sensitive that two polynucleotides differing in length by only a single nucleotide can be separated from each other on polyacrylamide gels.

Gel electrophoresis of proteins was described in the boxed essay accompanying Section 27.3.

Once the DNA is separated into smaller fragments, each fragment is sequenced independently. Again, gel electrophoresis is used, this time as an analytical tool. In the technique devised by Frederick Sanger, the two strands of a sample of a small fragment of DNA, 100 to 200 base pairs in length, are separated and one strand is used as a template to create complements of itself. The single-stranded sample is divided among four test tubes, each of which contains the materials necessary for DNA synthesis. These materials include the four nucleosides present in DNA, 2′-deoxyadenosine (dA), 2′-deoxythymidine (dT), 2′-deoxyguanosine (dG), and 2′-deoxycytidine (dC) as their triphosphates dATP, dTTP, dGTP, and dCTP.

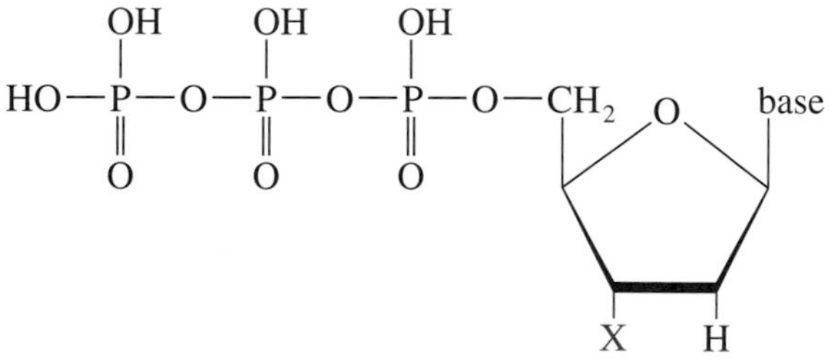

OH OH OH
HO—P—O—P—O—P—O—CH_2 O base
O O O
X H

X = OH	X = H
dATP	ddATP
dTTP	ddTTP
dGTP	ddGTP
dCTP	ddCTP

Also present in the first test tube is a synthetic analog of adenosine triphosphate in which both the 2′ and 3′ hydroxyl groups have been replaced by hydrogen substituents. This compound is called 2′,3′-dideoxyadenosine triphosphate (ddATP). Similarly, ddTTP is added to the second tube, ddGTP to the third, and ddCTP to the fourth. Each tube also contains a "primer." The primer is a short section of the complementary DNA strand, which has been labeled with an isotope of phosphorus (^{32}P) that emits α particles. When the electrophoresis gel is examined at the end of the experiment, the positions of the DNAs formed by chain extension of the primer are located by detecting their α emission by a technique called *autoradiography.*

As DNA synthesis proceeds, nucleotides from the solution are added to the growing polynucleotide chain. Chain extension takes place without complication so long as the incorporated nucleotides are derived from dATP, dTTP, dGTP, and dCTP. If, however, the incorporated species is derived from a dideoxy analog, chain extension stops. Because the dideoxy species ddA, ddT, ddG, and ddC lack hydroxyl groups at 3′, they cannot engage in the 3′ → 5′ phosphodiester linkage necessary for chain extension. Thus, the first tube—the one containing ddATP—contains a mixture of DNA fragments of different length, *all of which terminate in ddA*. Similarly, all the polynucleotides in the second tube terminate in ddT, those in the third tube terminate in ddG, and those in the fourth terminate in ddC.

The contents of each tube are then subjected to electrophoresis in separate lanes on the same sheet of polyacrylamide gel and the DNAs located by autoradiography. A typical electrophoresis gel of a DNA fragment containing 50 nucleotides will exhibit a pattern of 50 bands distributed among the four lanes with no overlaps. Each band corresponds to a polynucleotide that is one nucleotide longer than the one that precedes it (which may be in a different lane). One then simply ''reads'' the nucleotide sequence according to the lane in which each succeeding band appears.

The Sanger method for DNA sequencing is summarized in Figure 27.30.

This work produced a second Nobel Prize for Sanger. (His first was for protein sequencing in 1958.) Sanger shared the 1980 chemistry prize with Walter Gilbert of Harvard University, who developed a chemical method for DNA sequencing (the Maxam-Gilbert method), and with Paul Berg of Stanford University, who was responsible for many of the most important techniques in nucleic acid chemistry and biology.

A recent modification of Sanger's method has resulted in the commercial availabil-

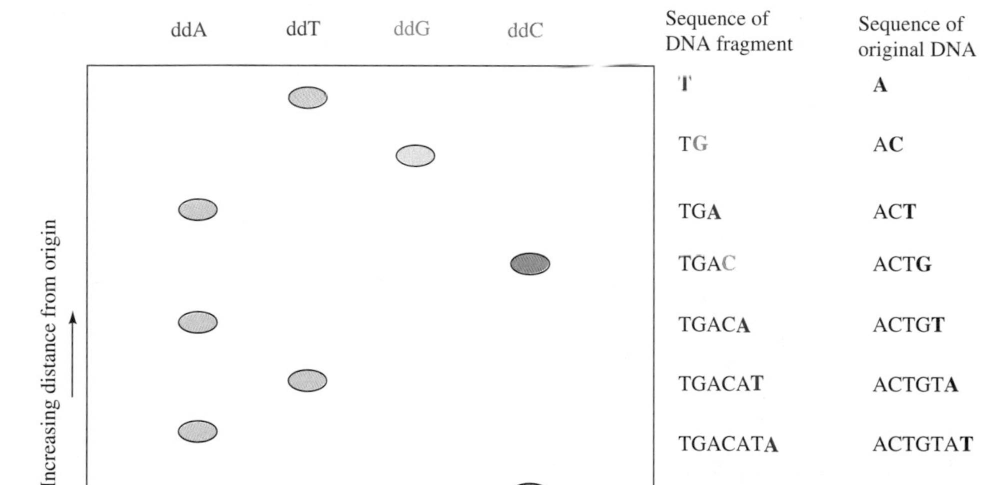

FIGURE 27.30 Sequencing of a short strand of DNA (10 bases) by Sanger's method using dideoxynucleotides to halt polynucleotide chain extension. Double-stranded DNA is separated and one of the strands used to produce complements of itself in four different tubes. All the tubes contain a primer tagged with ^{32}P, dATP, dTTP, dGTP, and dCTP (see text for abbreviations). The first tube also contains ddATP, the second ddTTP, the third ddGTP, and the fourth ddCTP. All the DNA fragments in the first tube terminate in A, those in the second terminate in T, those in the third terminate in G, and those in the fourth terminate in C. Location of the zones by autoradiographic detection of ^{32}P identifies the terminal nucleoside. The original DNA strand is its complement.

In 1995 a team of U.S. scientists announced the complete sequencing of the 1.8 million base genome of a species of influenza bacterium.

ity of a **DNA sequenator.** An instrument developed by Du Pont is based on Sanger's use of dideoxy analogs of nucleotides. Instead, however, of tagging a primer with ^{32}P, the purine and pyrimidine base portions of the dideoxynucleotides are each modified to contain a side chain that bears a different fluorescent dye, and all the dideoxy analogs are present in the same reaction. After electrophoretic separation of the products in a single lane, the gel is read by argon-laser irradiation at four different wavelengths. One wavelength causes the modified ddA-containing polynucleotides to fluoresce, another causes modified-ddT fluorescence, and so on. The data are stored and analyzed in a computer and printed out as the DNA sequence. It is claimed that a single instrument can sequence 10,000 nucleotides per day, making the hope of sequencing the 3 billion base pairs in the human genome a not-impossible goal.

27.29 SUMMARY

This chapter revolves around **proteins.** The first third describes the building blocks of proteins, progressing through **amino acids** and **peptides.** The middle third deals with proteins themselves. The last third discusses **nucleic acids** and their role in the biosynthesis of proteins.

A group of 20 amino acids, listed in Table 27.1, regularly appears as the hydrolysis products of proteins (Section 27.1). All are α-amino acids, and with the exception of glycine, all are chiral and have the L configuration (Section 27.2). The most stable structure of a neutral amino acid is a **zwitterion** (Section 27.3). The pH of an aqueous solution at which the concentration of the zwitterion is a maximum is called the **isoelectric point** (pI).

CO_2^-

$H_3\overset{+}{N}$—C—H

$CH(CH_3)_2$

Fischer projection formula of L-valine in its zwitterionic form

Amino acids are synthesized in the laboratory from

1. α-Halo acids by reaction with ammonia
2. Aldehydes by reaction with ammonia and cyanide ion (the Strecker synthesis)
3. Alkyl halides by reaction with the enolate anion derived from diethyl acetamidomalonate

The amino acids prepared by these methods are formed as racemic mixtures and are optically inactive (Section 27.4).

Amino acids undergo reactions characteristic of the amino group (such as amide formation) and the carboxyl group (such as esterification). Amino acid side chains undergo reactions characteristic of the functional groups they contain (Section 27.5).

The reactions that amino acids undergo in living systems include **transamination** and **decarboxylation** (Section 27.6).

An amide linkage between two α-amino acids is called a **peptide bond** (Section 27.7). The **primary structure** of a peptide is given by its amino acid sequence plus any disulfide bonds between two cysteine residues. By convention, peptides are named and written beginning at the N terminus.

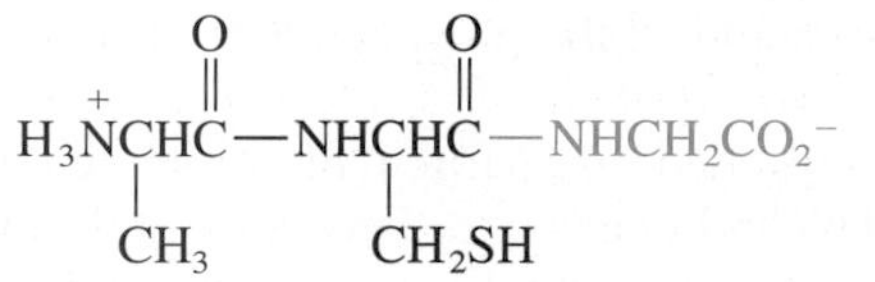

$$\overset{+}{H_3N}\underset{\displaystyle CH_3}{CH}\overset{O}{\overset{\|}{C}}-NH\underset{\displaystyle CH_2SH}{CH}\overset{O}{\overset{\|}{C}}-NHCH_2CO_2^- \qquad \text{Ala-Cys-Gly}$$

Alanylcysteinylglycine

The first step in establishing the primary structure involves acid hydrolysis of all the peptide bonds and determining what amino acids are present and their relative amounts by **ion-exchange chromatography** using an **amino acid analyzer** (Section 27.8).

Sequencing follows a logical strategy (Section 27.9) and is facilitated by **end group analysis.** The N terminus (Section 27.10) may be identified by reaction of the peptide with

1. 1-Fluoro-2,4-dinitrobenzene (Sanger's reagent)
2. Phenyl isothiocyanate (Edman's reagent)
3. 5-(Dimethylamino)naphthalene-1-sulfonyl chloride (dansyl chloride)

The C terminus is identified by hydrolysis with *carboxypeptidase,* an enzyme that selectively catalyzes the hydrolysis of the peptide bond to the C-terminal amino acid (Section 27.11). Enzymes such as *trypsin, chymotrypsin,* and *pepsin* are also used in peptide sequencing. They promote selective hydrolysis of the peptide into a limited number of smaller fragments, each of which may be individually sequenced to provide data on the overall sequence (Section 27.12).

Peptide synthesis requires that the number of possible reactions be limited by the judicious use of protecting groups (Section 27.13). Amino protecting groups include *benzyloxycarbonyl* (Z) and tert-*butoxycarbonyl* (Boc).

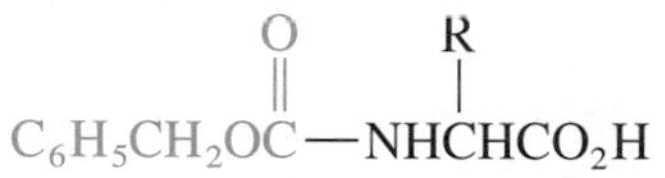

$$C_6H_5CH_2O\overset{O}{\overset{\|}{C}}-NH\overset{R}{\overset{|}{C}}HCO_2H \qquad (CH_3)_3CO\overset{O}{\overset{\|}{C}}-NH\overset{R}{\overset{|}{C}}HCO_2H$$

Benzyloxycarbonyl-protected amino acid

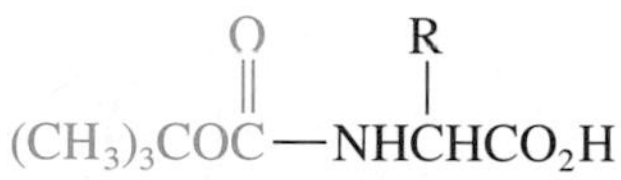

tert-Butoxycarbonyl-protected amino acid

Hydrogen bromide may be used to remove either the benzyloxycarbonyl or *tert*-butoxycarbonyl protecting group. The benzyloxycarbonyl protecting group may also be removed by catalytic hydrogenolysis.

Carboxyl groups are normally protected as benzyl, methyl, or ethyl esters (Section 27.15). Hydrolysis in dilute base is normally used to deprotect methyl and ethyl esters. Benzyl protecting groups are removed by hydrogenolysis.

Peptide bond formation (Section 27.16) between a protected amino acid having a free carboxyl group and a protected amino acid having a free amino group can be accomplished with the aid of *N,N'*-dicyclohexylcarbodiimide (DCCI).

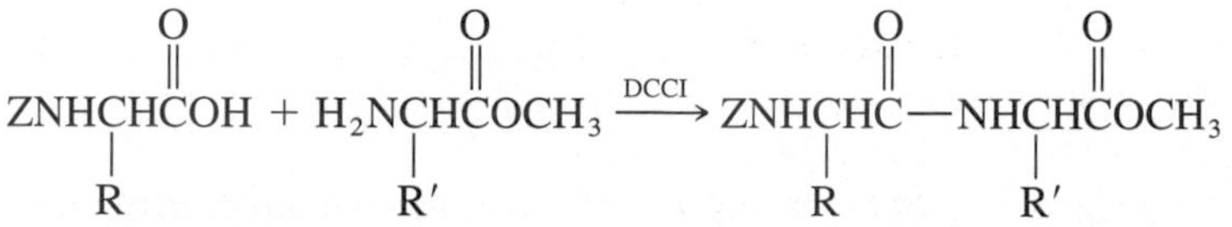

$$ZNH\underset{\displaystyle R}{CH}\overset{O}{\overset{\|}{C}}OH + H_2N\underset{\displaystyle R'}{CH}\overset{O}{\overset{\|}{C}}OCH_3 \xrightarrow{\text{DCCI}} ZNH\underset{\displaystyle R}{CH}\overset{O}{\overset{\|}{C}}-NH\underset{\displaystyle R'}{CH}\overset{O}{\overset{\|}{C}}OCH_3$$

In the **Merrifield method** (Section 27.17) the carboxyl group of an amino acid is anchored to a solid support and the chain extended one amino acid at a time. When all the amino acid residues have been added, the polypeptide is removed from the solid support.

Two **secondary structures** of proteins are particularly prominent (Section 27.18). The *pleated β sheet* is stabilized by hydrogen bonds between N—H and C=O groups of adjacent chains. The *α helix* is stabilized by hydrogen bonds within a single polypeptide chain.

The folding of a peptide chain is its **tertiary structure** (Section 27.19). Many proteins consist of two or more chains, and the way in which the various units are assembled into the protein in its native state is its **quaternary structure** (Section 27.21).

Almost, but not all, enzymes are proteins. For identifying certain RNA-catalyzed biological processes Sidney Altman (Yale University) and Thomas R. Cech (University of Colorado) shared the 1989 Nobel Prize in chemistry.

Enzymes are catalysts. They accelerate the rates of chemical reactions in biological systems, but the kinds of reactions that take place are the basic reactions of organic chemistry. One way in which enzymes accelerate these reactions is by bringing reactive functions together in the presence of catalytically active functions of the protein. Often those catalytically active functions are nothing more than proton donors and proton acceptors. In many cases a protein acts in cooperation with a **coenzyme,** a small molecule having the proper functionality to carry out a chemical change not otherwise available to the protein itself (Sections 27.20 and 27.21).

Carbohydrate derivatives of purine and pyrimidine (Section 27.22) are among the most important compounds of biological chemistry. *N*-Glycosides of D-ribose and 2-deoxy-D-ribose in which the substituent at the anomeric position is a derivative of purine or pyrimidine are called **nucleosides** (Section 27.23). **Nucleotides** are phosphate esters of nucleosides (Section 27.24). **Nucleic acids** are polymers of nucleotides (Section 27.25).

Nucleic acids derived from 2-deoxy-D-ribose **(DNA)** are responsible for storing and transmitting genetic information (Section 27.26). DNA exists as a double-stranded pair of helices in which hydrogen bonds are responsible for complementary base pairing between adenine (A) and thymine (T), and between guanine (G) and cytosine (C). During cell division the two strands of DNA unwind and are duplicated. Each strand acts as a template on which its complement is constructed.

In the **transcription** stage of protein biosynthesis (Section 27.27), a molecule of **messenger RNA** (mRNA) having a nucleotide sequence complementary to that of DNA is assembled. Transcription is followed by **translation,** in which triplets of nucleotides of mRNA called **codons** are recognized by **transfer RNA** (tRNA) for a particular amino acid, and that amino acid is added to the growing peptide chain.

The nucleotide sequence of DNA can be determined by a technique in which a short section of single-stranded DNA is allowed to produce its complement in the presence of dideoxy analogs of ATP, TTP, GTP, and CTP. DNA formation terminates when a dideoxy analog is incorporated into the growing polynucleotide chain. A mixture of polynucleotides differing from one another by an incremental nucleoside is produced and analyzed by electrophoresis. From the observed sequence of the complementary chain, the sequence of the original DNA is deduced (Section 27.28).

PROBLEMS

27.24 The imidazole ring of the histidine side chain acts as a proton acceptor in certain enzyme-catalyzed reactions. Which is the more stable protonated form of the histidine residue, A or B? Why?

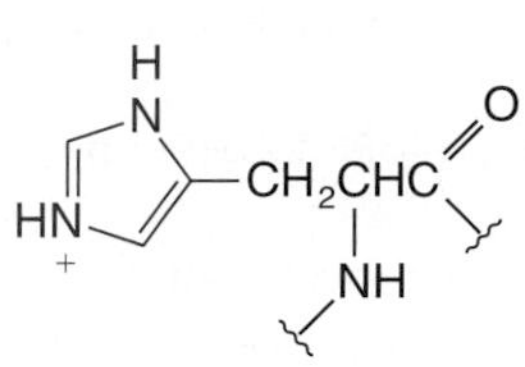

A 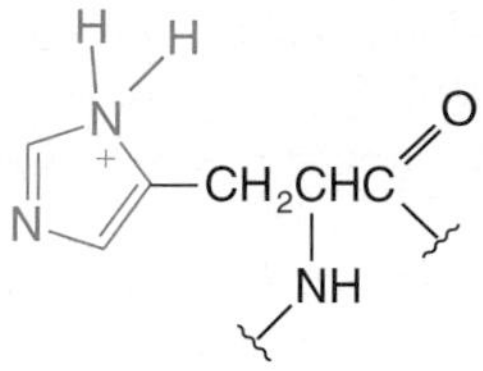 B

27.25 Acrylonitrile ($CH_2{=}CHC{\equiv}N$) readily undergoes conjugate addition when treated with nucleophilic reagents. Describe a synthesis of β-alanine ($H_3\overset{+}{N}CH_2CH_2CO_2^-$) that takes advantage of this fact.

27.26 (*a*) Isoleucine has been prepared by the following sequence of reactions. Give the structure of compounds A through D isolated as intermediates in this synthesis.

$$CH_3CH_2\underset{\displaystyle Br}{\underset{|}{C}}HCH_3 \xrightarrow[\text{sodium ethoxide}]{\text{diethyl malonate}} A \xrightarrow[\text{2. HCl}]{\text{1. KOH}} B\ (C_7H_{12}O_4)$$

$$B \xrightarrow{Br_2} C\ (C_7H_{11}BrO_4) \xrightarrow{\text{heat}} D \xrightarrow[H_2O]{NH_3} \text{isoleucine (racemic)}$$

(*b*) An analogous procedure has been used to prepare phenylalanine. What alkyl halide would you choose as the starting material for this synthesis?

27.27 Hydrolysis of the following compound in concentrated hydrochloric acid for several hours at 100°C gives one of the amino acids in Table 27.1. Which one? Is it optically active?

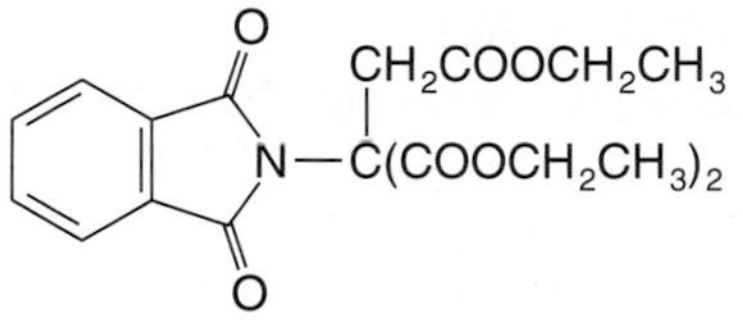

$CH_2COOCH_2CH_3$

N—$C(COOCH_2CH_3)_2$

27.28 If you synthesized the tripeptide Leu-Phe-Ser from amino acids prepared by the Strecker synthesis, how many stereoisomers would you expect to be formed?

27.29 How many peaks would you expect to see on the strip chart after amino acid analysis of bradykinin?

Arg-Pro-Pro-Gly-Phe-Ser-Pro-Phe-Arg Bradykinin

27.30 Automated amino acid analysis of peptides containing asparagine (Asn) and glutamine (Gln) residues gives a peak corresponding to ammonia. Why?

27.31 What are the products of each of the following reactions? Your answer should account for all the amino acid residues in the starting peptides.

(*a*) Reaction of Leu-Gly-Ser with 1-fluoro-2,4-dinitrobenzene
(*b*) Hydrolysis of the compound in part (*a*) in concentrated hydrochloric acid (100°C)

(*c*) Treatment of Met-Val-Pro with dansyl chloride, followed by hydrolysis in concentrated hydrochloric acid (100°C)
(*d*) Treatment of Ile-Glu-Phe with $C_6H_5N{=}C{=}S$, followed by hydrogen bromide in nitromethane
(*e*) Reaction of Asn-Ser-Ala with benzyloxycarbonyl chloride
(*f*) Reaction of the product of part (*e*) with *p*-nitrophenol and *N,N'*-dicyclohexylcarbodiimide
(*g*) Reaction of the product of part (*f*) with the ethyl ester of valine
(*h*) Hydrogenolysis of the product of part (*g*) over palladium

27.32 Hydrazine cleaves amide bonds to form *acylhydrazides* according to the general mechanism of nucleophilic acyl substitution discussed in Chapter 20:

$$\underset{\text{Amide}}{R\overset{O}{\overset{\|}{C}}NHR'} + \underset{\text{Hydrazine}}{H_2NNH_2} \longrightarrow \underset{\text{Acylhydrazide}}{R\overset{O}{\overset{\|}{C}}NHNH_2} + \underset{\text{Amine}}{R'NH_2}$$

This reaction forms the basis of one method of terminal residue analysis. A peptide is treated with excess hydrazine in order to cleave all the peptide linkages. One of the terminal amino acids is cleaved as the free amino acid and identified, while all the other amino acid residues are converted to acylhydrazides. Which amino acid is identified by *hydrazinolysis,* the N terminus or the C terminus?

27.33 *Somatostatin* is a tetradecapeptide of the hypothalamus that inhibits the release of pituitary growth hormone. Its amino acid sequence has been determined by a combination of Edman degradations and enzymic hydrolysis experiments. On the basis of the following data, deduce the primary structure of somatostatin:

1. Edman degradation gave PTH-Ala.
2. Selective hydrolysis gave peptides having the following indicated sequences:
   Phe-Trp
   Thr-Ser-Cys
   Lys-Thr-Phe
   Thr-Phe-Thr-Ser-Cys
   Asn-Phe-Phe-Trp-Lys
   Ala-Gly-Cys-Lys-Asn-Phe
3. Somatostatin has a disulfide bridge.

27.34 What protected amino acid would you anchor to the solid support in the first step of a synthesis of oxytocin (Figure 27.7) by the Merrifield method?

27.35 *Nebularine* is a toxic nucleoside isolated from a species of mushroom. Its systematic name is 9-β-D-ribofuranosylpurine. Write a structural formula for nebularine.

27.36 The nucleoside *vidarabine* (ara-A) shows promise as an antiviral agent. Its structure is identical with that of adenosine (Section 27.23) except that D-arabinose replaces D-ribose as the carbohydrate component. Write a structural formula for this substance.

27.37 When 6-chloropurine is heated with aqueous sodium hydroxide, it is quantitatively converted to *hypoxanthine.* Suggest a reasonable mechanism for this reaction.

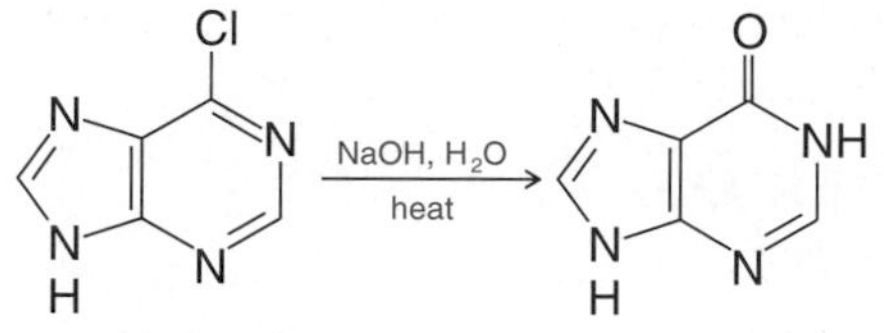

6-Chloropurine $\xrightarrow[\text{heat}]{NaOH,\ H_2O}$ Hypoxanthine

27.38 Treatment of adenosine with nitrous acid gives a nucleoside known as *inosine:*

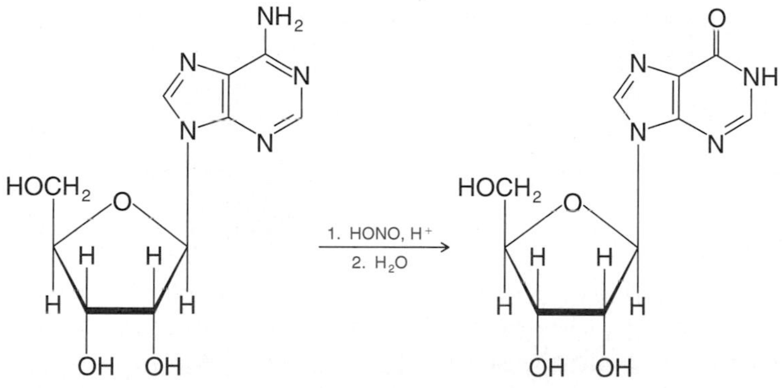

Adenosine $\xrightarrow[2.\ H_2O]{1.\ HONO,\ H^+}$ Inosine

Suggest a reasonable mechanism for this reaction.

27.39 (*a*) The 5′-nucleotide of inosine, *inosinic acid* ($C_{10}H_{13}N_4O_8P$), is added to foods as a flavor enhancer. What is the structure of inosinic acid? (The structure of inosine is given in Problem 27.38.)

(*b*) The compound *2′,3′-dideoxyinosine (DDI)* holds promise as a drug for the treatment of AIDS. What is the structure of DDI?

27.40 In one of the early experiments designed to elucidate the genetic code, Marshall Nirenberg of the U.S. National Institutes of Health (Nobel Prize in physiology or medicine, 1968) prepared a synthetic mRNA in which all the bases were uracil. He added this poly(U) to a cell-free system containing all the necessary materials for protein biosynthesis. A polymer of a single amino acid was obtained. What amino acid was polymerized?

MOLECULAR MODELING EXERCISES

27.41 Construct a molecular model of L-alanine in its zwitterionic form.

27.42 Connect a glycine unit to your model of L-alanine by a peptide bond so as to give Ala-Gly.

27.43 Connect a glycine unit to your model of L-alanine by a peptide bond so as to give Gly-Ala.

27.44 Construct molecular models of adenine and thymine assuming all ring carbons and all nitrogens are sp^2 hybridized. Use the models of these purine and pyrimidine bases to illustrate the hydrogen bonds between A and T in DNA.

APPENDIX 1

A. SELECTED PHYSICAL PROPERTIES OF REPRESENTATIVE HYDROCARBONS

Compound name	Molecular formula	Structural formula	Melting point, °C	Boiling point, °C (1 atm)
Alkanes				
Methane	CH_4	CH_4	−182.5	−160
Ethane	C_2H_6	CH_3CH_3	−183.6	−88.7
Propane	C_3H_8	$CH_3CH_2CH_3$	−187.6	−42.2
Butane	C_4H_{10}	$CH_3CH_2CH_2CH_3$	−139.0	−0.4
2-Methylpropane	C_4H_{10}	$(CH_3)_3CH$	−160.9	−10.2
Pentane	C_5H_{12}	$CH_3(CH_2)_3CH_3$	−129.9	36.0
2-Methylbutane	C_5H_{12}	$(CH_3)_2CHCH_2CH_3$	−160.5	27.9
2,2-Dimethylpropane	C_5H_{12}	$(CH_3)_4C$	−16.6	9.6
Hexane	C_6H_{14}	$CH_3(CH_2)_4CH_3$	−94.5	68.8
Heptane	C_7H_{16}	$CH_3(CH_2)_5CH_3$	−90.6	98.4
Octane	C_8H_{18}	$CH_3(CH_2)_6CH_3$	−56.9	125.6
Nonane	C_9H_{20}	$CH_3(CH_2)_7CH_3$	−53.6	150.7
Decane	$C_{10}H_{22}$	$CH_3(CH_2)_8CH_3$	−29.7	174.0
Dodecane	$C_{12}H_{26}$	$CH_3(CH_2)_{10}CH_3$	−9.7	216.2
Pentadecane	$C_{15}H_{32}$	$CH_3(CH_2)_{13}CH_3$	10.0	272.7
Icosane	$C_{20}H_{42}$	$CH_3(CH_2)_{18}CH_3$	36.7	205 (15 mm)
Hectane	$C_{100}H_{202}$	$CH_3(CH_2)_{98}CH_3$	115.1	
Cycloalkanes				
Cyclopropane	C_3H_6		−127.0	−32.9
Cyclobutane	C_4H_8			13.0
Cyclopentane	C_5H_{10}		−94.0	49.5
Cyclohexane	C_6H_{12}		6.5	80.8
Cycloheptane	C_7H_{14}		−13.0	119.0
Cyclooctane	C_8H_{16}		13.5	149.0
Cyclononane	C_9H_{18}			171
Cyclodecane	$C_{10}H_{20}$		9.6	201
Cyclopentadecane	$C_{15}H_{30}$		60.5	112.5 (1 mm)
Alkenes and cycloalkenes				
Ethene (ethylene)	C_2H_4	$CH_2{=}CH_2$	−169.1	−103.7
Propene	C_3H_6	$CH_3CH{=}CH_2$	−185.0	−47.6
1-Butene	C_4H_8	$CH_3CH_2CH{=}CH_2$	−185	−6.1
2-Methylpropene	C_4H_8	$(CH_3)_2C{=}CH_2$	−140	−6.6
Cyclopentene	C_5H_8		−98.3	44.1

Compound name	Molecular formula	Structural formula	Melting point, °C	Boiling point, °C (1 atm)
1-Pentene	C_5H_{10}	$CH_3CH_2CH_2CH{=}CH_2$	−138.0	30.2
2-Methyl-2-butene	C_5H_{10}	$(CH_3)_2C{=}CHCH_3$	−134.1	38.4
Cyclohexene	C_6H_{10}		−104.0	83.1
1-Hexene	C_6H_{12}	$CH_3CH_2CH_2CH_2CH{=}CH_2$	−138.0	63.5
2,3-Dimethyl-2-butene	C_6H_{12}	$(CH_3)_2C{=}C(CH_3)_2$	−74.6	73.5
1-Heptene	C_7H_{14}	$CH_3(CH_2)_4CH{=}CH_2$	−119.7	94.9
1-Octene	C_8H_{16}	$CH_3(CH_2)_5CH{=}CH_2$	−104	119.2
1-Decene	$C_{10}H_{20}$	$CH_3(CH_2)_7CH{=}CH_2$	−80.0	172.0
Alkynes				
Ethyne (acetylene)	C_2H_2	$HC{\equiv}CH$	−81.8	−84.0
Propyne	C_3H_4	$CH_3C{\equiv}CH$	−101.5	−23.2
1-Butyne	C_4H_6	$CH_3CH_2C{\equiv}CH$	−125.9	8.1
2-Butyne	C_4H_6	$CH_3C{\equiv}CCH_3$	−32.3	27.0
1-Hexyne	C_6H_{10}	$CH_3(CH_2)_3C{\equiv}CH$	−132.4	71.4
3,3-Dimethyl-1-butyne	C_6H_{10}	$(CH_3)_3CC{\equiv}CH$	−78.2	37.7
1-Octyne	C_8H_{14}	$CH_3(CH_2)_5C{\equiv}CH$	−79.6	126.2
1-Nonyne	C_9H_{16}	$CH_3(CH_2)_6C{\equiv}CH$	−36.0	160.6
1-Decyne	$C_{10}H_{18}$	$CH_3(CH_2)_7C{\equiv}CH$	−40.0	182.2
Arenes				
Benzene	C_6H_6		5.5	80.1
Toluene	C_7H_8	$-CH_3$	−95	110.6
Styrene	C_8H_8	$-CH{=}CH_2$	−33	145
p-Xylene	C_8H_{10}	H_3C- $-CH_3$	−13	138
Ethylbenzene	C_8H_{10}	$-CH_2CH_3$	−94	136.2
Naphthalene	$C_{10}H_8$		80.3	218
Diphenylmethane	$C_{13}H_{12}$	$(C_6H_5)_2CH_2$	26	261
Triphenylmethane	$C_{19}H_{16}$	$(C_6H_5)_3CH$	94	

B. SELECTED PHYSICAL PROPERTIES OF REPRESENTATIVE ORGANIC HALOGEN COMPOUNDS

Alkyl halides

Compound name	Structural formula	Boiling point, °C (1 atm)				Density, g/mL (20°C)		
		Fluoride	Chloride	Bromide	Iodide	Chloride	Bromide	Iodide
Halomethane	CH_3X	−78	−24	3	42			2.279
Haloethane	CH_3CH_2X	−32	12	38	72	0.903	1.460	1.933
1-Halopropane	$CH_3CH_2CH_2X$	−3	47	71	103	0.890	1.353	1.739
2-Halopropane	$(CH_3)_2CHX$	−11	35	59	90	0.859	1.310	1.714
1-Halobutane	$CH_3CH_2CH_2CH_2X$		78	102	130	0.887	1.276	1.615
2-Halobutane	$CH_3\underset{\substack{\vert \\ X}}{C}HCH_2CH_3$		68	91	120	0.873	1.261	1.597
1-Halo-2-methylpropane	$(CH_3)_2CHCH_2X$	16	68	91	121	0.878	1.264	1.603
2-Halo-2-methylpropane	$(CH_3)_3CX$		51	73	99	0.847	1.220	1.570
1-Halopentane	$CH_3(CH_2)_3CH_2X$	65	108	129	157	0.884	1.216	1.516
1-Halohexane	$CH_3(CH_2)_4CH_2X$	92	134	155	180	0.879	1.175	1.439
1-Halooctane	$CH_3(CH_2)_6CH_2X$	143	183	202	226	0.892	1.118	1.336
Halocyclopentane	–X		114	138	166	1.005	1.388	1.694
Halocyclohexane	–X		142	167	192	0.977	1.324	1.626

Aryl halides

Compound	Halogen substituent (X)							
	Fluorine		Chlorine		Bromine		Iodine	
	mp	bp	mp	bp	mp	bp	mp	bp
C_6H_5X	−41	85	−45	132	−31	156	−31	188
o-$C_6H_4X_2$	−34	91	−17	180	7	225	27	286
m-$C_6H_4X_2$	−59	83	−25	173	−7	218	35	285
p-$C_6H_4X_2$	−13	89	53	174	87	218	129	285
1,3,5-$C_6H_3X_3$	−5	76	63	208	121	271	184	
C_6X_6	5	80	230	322	327		350	

* All boiling points and melting points cited are in °C.

C. SELECTED PHYSICAL PROPERTIES OF REPRESENTATIVE ALCOHOLS, ETHERS, AND PHENOLS

Alcohols

Compound name	Structural formula	Melting point, °C	Boiling point, °C (1 atm)	Solubility, g/100 mL H_2O
Methanol	CH_3OH	−94	65	∞
Ethanol	CH_3CH_2OH	−117	78	∞
1-Propanol	$CH_3CH_2CH_2OH$	−127	97	∞
2-Propanol	$(CH_3)_2CHOH$	−90	82	∞
1-Butanol	$CH_3CH_2CH_2CH_2OH$	−90	117	9
2-Butanol	$CH_3CH(OH)CH_2CH_3$	−115	100	26
2-Methyl-1-propanol	$(CH_3)_2CHCH_2OH$	−108	108	10
2-methyl-2-propanol	$(CH_3)_3COH$	26	83	∞
1-Pentanol	$CH_3(CH_2)_3CH_2OH$	79	138	
1-Hexanol	$CH_3(CH_2)_4CH_2OH$	−52	157	0.6
1-Dodecanol	$CH_3(CH_2)_{10}CH_2OH$	26	259	Insoluble
Cyclohexanol	(cyclohexane ring)–OH	25	161	3.6

Ethers

Compound name	Structural formula	Melting point, °C	Boiling point, °C (1 atm)	Solubility, g/100 mL H_2O
Dimethyl ether	CH_3OCH_3	−138.5	−24	Very soluble
Diethyl ether	$CH_3CH_2OCH_2CH_3$	−116.3	34.6	7.5
Dipropyl ether	$CH_3CH_2CH_2OCH_2CH_2CH_3$	−122	90.1	Slight
Diisopropyl ether	$(CH_3)_2CHOCH(CH_3)_2$	−60	68.5	0.2
1,2-Dimethoxyethane	$CH_3OCH_2CH_2OCH_3$		83	∞
Diethylene glycol dimethyl ether (diglyme)	$CH_3OCH_2CH_2OCH_2CH_2OCH_3$		161	∞
Ethylene oxide	(three-membered ring with O)	−111.7	10.7	∞
Tetrahydrofuran	(five-membered ring with O)	−108.5	65	∞

Phenols

Compound name	Melting point, °C	Boiling point, °C	Solubility, g/100 mL H_2O
Phenol	43	182	8.2
o-Cresol	31	191	2.5
m-Cresol	12	203	0.5
p-Cresol	35	202	1.8
o-Chlorophenol	7	175	2.8
m-Chlorophenol	32	214	2.6
p-Chlorophenol	42	217	2.7
o-Nitrophenol	45	217	0.2
m-Nitrophenol	96		1.3
p-Nitrophenol	114	279	1.6
1-Naphthol	96	279	Slight
2-Naphthol	122	285	0.1
Pyrocatechol	105	246	45.1
Resorcinol	110	276	147.3
Hydroquinone	170	285	6

D. SELECTED PHYSICAL PROPERTIES OF REPRESENTATIVE ALDEHYDES AND KETONES

Compound name	Structural formula	Melting point, °C	Boiling point, °C (1 atm)	Solubility g/100 mL H_2O
Aldehydes				
Formaldehyde	$HC(=O)H$	−92	−21	Very soluble
Acetaldehyde	$CH_3C(=O)H$	−123.5	20.2	∞
Propanal	$CH_3CH_2C(=O)H$	−81	49.5	20
Butanal	$CH_3CH_2CH_2C(=O)H$	−99	75.7	4
Benzaldehyde	$C_6H_5C(=O)H$	−26	178	0.3

Compound name	Structural formula	Melting point, °C	Boiling point, °C (1 atm)	Solubility g/100 mL H_2O
Ketones				
Acetone	$CH_3C(=O)CH_3$	−94.8	56.2	∞
2-Butanone	$CH_3C(=O)CH_2CH_3$	−86.9	79.6	37
2-Pentanone	$CH_3C(=O)CH_2CH_2CH_3$	−77.8	102.4	Slight
3-Pentanone	$CH_3CH_2C(=O)CH_2CH_3$	−39.9	102.0	4.7
Cyclopentanone	(cyclopentane ring)=O	−51.3	130.7	43.3
Cyclohexanone	(cyclohexane ring)=O	−45	155	
Acetophenone	$C_6H_5C(=O)CH_3$	21	202	Insoluble
Benzophenone	$C_6H_5C(=O)C_6H_5$	48	306	Insoluble

E. SELECTED PHYSICAL PROPERTIES OF REPRESENTATIVE CARBOXYLIC ACIDS AND DICARBOXYLIC ACIDS

Compound name	Structural formula	Melting point, °C	Boiling point, °C (1 atm)	Solubility, g/100 mL H_2O
Carboxylic acids				
Formic acid	HCO_2H	8.4	101	∞
Acetic acid	CH_3CO_2H	16.6	118	∞
Propanoic acid	$CH_3CH_2CO_2H$	−20.8	141	∞
Butanoic acid	$CH_3CH_2CH_2CO_2H$	−5.5	164	∞
Pentanoic acid	$CH_3(CH_2)_3CO_2H$	−34.5	186	3.3 (16°C)
Decanoic acid	$CH_3(CH_2)_8CO_2H$	31.4	269	0.003 (15°C)
Benzoic acid	$C_6H_5CO_2H$	122.4	250	0.21 (17°C)
Dicarboxylic acids				
Oxalic acid	HO_2CCO_2H	186	Sublimes	10 (20°C)
Malonic acid	$HO_2CCH_2CO_2H$	130–135	Decomposes	138 (16°C)
Succinic acid	$HO_2CCH_2CH_2CO_2H$	189	235	6.8 (20°C)
Glutaric acid	$HO_2CCH_2CH_2CH_2CO_2H$	97.5		63.9 (20°C)

F. SELECTED PHYSICAL PROPERTIES OF REPRESENTATIVE AMINES

Alkylamines

Compound name	Structural formula	Melting point, °C	Boiling point, °C	Solubility, g/100 mL H_2O
Primary amines				
Methylamine	CH_3NH_2	−92.5	−6.7	Very high
Ethylamine	$CH_3CH_2NH_2$	−80.6	16.6	∞
Butylamine	$CH_3CH_2CH_2CH_2NH_2$	−50	77.8	∞
Isobutylamine	$(CH_3)_2CHCH_2NH_2$	−85	68	∞
sec-Butylamine	$CH_3CH_2\underset{\underset{CH_3}{\mid}}{C}HNH_2$	−104	66	∞
tert-Butylamine	$(CH_3)_3CNH_2$	−67.5	45.2	
Hexylamine	$CH_3(CH_2)_5NH_2$	−19	129	Slightly soluble
Cyclohexylamine	$C_6H_{11}-NH_2$	−18	134.5	∞
Benzylamine	$C_6H_5CH_2NH_2$	10	184.5	∞
Secondary amines				
Dimethylamine	$(CH_3)_2NH$	−92.2	6.9	Very soluble
Diethylamine	$(CH_3CH_2)_2NH$	−50	55.5	Very soluble
N-Methylpropylamine	$CH_3NHCH_2CH_2CH_3$		62.4	Soluble
Piperidine	(ring) N–H	−10.5	106.4	∞
Tertiary amines				
Trimethylamine	$(CH_3)_3N$	−117.1	2.9	41
Triethylamine	$(CH_3CH_2)_3N$	−114.7	89.4	∞
N-Methylpiperidine	(ring) N–CH_3	3	107	

Arylamines

Compound name	Melting point, °C	Boiling point, °C
Primary amines		
Aniline	−6.3	184
o-Toluidine	−14.7	200
m-Toluidine	−30.4	203
p-Toluidine	44	200
o-Chloroaniline	−14	209
m-Chloroaniline	−10	230
p-Chloroaniline	72.5	232
o-Nitroaniline	71.5	284
m-Nitroaniline	114	306
p-Nitroaniline	148	332
Secondary amines		
N-Methylaniline	−57	196
N-Ethylaniline	−63	205
Tertiary amines		
N,N-Dimethylaniline	2.4	194
Triphenylamine	127	365

APPENDIX 2

ANSWERS TO IN-TEXT PROBLEMS

Problems are of two types: in-text problems that appear within the body of each chapter, and end-of-chapter problems. This appendix gives brief answers to all the in-text problems. More detailed discussions of in-text problems as well as detailed solutions to all the end-of-chapter problems are provided in a separate *Study Guide and Student Solutions Manual.* Answers to part (*a*) of those in-text problems with multiple parts have been provided in the form of a sample solution within each chapter and are not repeated here.

CHAPTER 1

1.1 4

1.2 All the third-row elements have a neon core containing 10 electrons ($1s^22s^22p^6$). The elements in the third row, their atomic numbers Z, and their electron configurations beyond the neon core are Na (Z = 11) $3s^1$; Mg (Z = 12) $3s^2$; Al (Z = 13) $3s^23p_x^{\,1}$; Si (Z = 14) $3s^23p_x^{\,1}\cdot 3p_y^{\,1}$; P ($Z$ = 15) $3s^23p_x^{\,1}3p_y^{\,1}3p_z^{\,1}$; S ($Z$ = 16) $3s^23p_x^{\,2}3p_y^{\,1}3p_z^{\,1}$; Cl ($Z$ = 17) $3s^23p_x^{\,2}3p_y^{\,2}3p_z^{\,1}$; Ar ($Z$ = 18) $3s^23p_x^{\,2}3p_y^{\,2}3p_z^{\,2}$.

1.3 Those ions that possess a noble gas electron configuration are (*a*) K^+; (*c*) H^-; (*e*) F^-; and (*f*) Ca^{2+}.

1.4 Electron configuration of C^+ is $1s^22s^22p^1$; electron configuration of C^- is $1s^22s^22p^3$. Neither C^+ nor C^- possesses a noble gas electron configuration.

1.5 H:F̤̈:

1.6

```
    H H
  H:C̤̈:C̤̈:H
    H H
```

1.7 (*b*)

```
  :F̈         F̈:
     \      /
      C == C         ;
     /      \
  :F̈         F̈:
```

(*c*)

```
  H         H
    \      /
     C == C
    /      \
  H         C≡N:
```

1.8 Fluorine is *more* electronegative than chlorine, and so fluorine is negative end of dipole in FCl. Iodine is *less* electronegative than chlorine, and so chlorine is negative end of dipole in ICl.

1.9 (*b*) Neither phosphorus nor bromine has a formal charge in PBr_3; (*c*) sulfur has a formal charge of +2 in the Lewis structure given for sulfuric acid, the two oxygens bonded only to sulfur each have a formal charge of −1, and the oxygens and hydrogens of the two OH groups have no formal charge; (*d*) none of the atoms have a formal charge in the Lewis structure given for nitrous acid.

1.10 The electron counts of nitrogen in ammonium ion and boron in borohydride ion are both 4 (half of 8 electrons in covalent bonds). Since a neutral nitrogen has 5 electrons in its valence shell, an electron count of 4 gives it a formal charge of +1. A neutral boron has 3 valence electrons, so that an electron count of 4 in borohydride ion corresponds to a formal charge of −1.

1.11 (*b*)

```
            H
            |
     H  H—C—H  H
     |      |     |
  H—C——C——C—H ;
     |      |     |
     H      H     H
```

(*c*)

```
       H   H
       |   |
  :C̈l—C—C—C̈l: ;
       |   |
       H   H
```

(*d*)

```
     H   H
     |   |
  H—C—C—C̈l: ;
     |   |
     H  :C̈l:
```

(*e*)

```
     H       H   H
     |   ..  |   |
  H—C—N—C—C—H ;
     |   |   |   |
     H   H   H   H
```

(*f*)

```
               H
               |
             H—C—H
     H         |
     |         |
  H—C————C—C=Ö:
     |         |   |
     H         H   H
```

1.12 (*b*) $(CH_3)_2CHCH(CH_3)_2$; (*c*)

```
HOCH2CHCH(CH3)2;
      |
      CH3
```

(*d*)

```
   CH2—CH2
  /       \
CH2        CH—C(CH3)3
  \       /
   CH2—CH2
```

1.13

```
          :O:
           ||
  H—N̈—C—Ö—H
     |
     H
```

1.14 (*b*) $CH_3CH_2CH_2OH$, $(CH_3)_2CHOH$, and $CH_3CH_2OCH_3$. (*c*) There are seven isomers of $C_4H_{10}O$. Four have —OH groups: $CH_3CH_2CH_2CH_2OH$, $(CH_3)_2CHCH_2OH$, $(CH_3)_3COH$, and

```
CH3CHCH2CH3.
   |
   OH
```

Three have C—O—C units: $CH_3OCH_2CH_2CH_3$, $CH_3CH_2OCH_2CH_3$, and $(CH_3)_2CHOCH_3$.

1.15

(*b*) $^{-}:\ddot{O}-C(=\ddot{O}:)-\ddot{O}-H \longleftrightarrow :\ddot{O}=C(-\ddot{O}:^{-})-\ddot{O}-H$

(*c*) $^{-}:\ddot{O}-C(=\ddot{O}:)-\ddot{O}:^{-} \longleftrightarrow :\ddot{O}=C(-\ddot{O}:^{-})-\ddot{O}:^{-}$

and $^{-}:\ddot{O}-C(=\ddot{O}:)-\ddot{O}:^{-} \longleftrightarrow {}^{-}:\ddot{O}-C(-\ddot{O}:^{-})=\ddot{O}:$

(*d*) $^{-}:\ddot{O}-B(=\ddot{O}:^{-})-\ddot{O}:^{-} \longleftrightarrow :\ddot{O}=B(-\ddot{O}:^{-})-\ddot{O}:^{-}$

and $^{-}:\ddot{O}-B(-\ddot{O}:^{-})=\ddot{O}: \longleftrightarrow {}^{-}:\ddot{O}-B(=\ddot{O}:)-\ddot{O}:^{-}$

1.16 The H—B—H angles in BH_4^- are 109.5° (tetrahedral).

1.17 (*b*) Tetrahedral; (*c*) linear; (*d*) trigonal planar.

1.18 (*b*) Oxygen is negative end of dipole moment directed along bisector of H—O—H angle; (*c*) no dipole moment; (*d*) dipole moment directed along axis of C—Cl bond, with chlorine at negative end, and carbon and hydrogens partially positive; (*e*) dipole moment directed along bisector of H—C—H angle, with oxygen at negative end; (*f*) dipole moment aligned with axis of linear molecule, with nitrogen at negative end.

1.19 (*b*) sp^2; (*c*) carbon of CH_2 group is sp^2, while carbon of C=O is *sp*; (*d*) two doubly bonded carbons are each sp^2, while carbon of CH_3 group is sp^3; (*e*) carbon of C=O is sp^2, while carbons of CH_3 group are sp^3; (*f*) two doubly bonded carbons are each sp^2, while carbon bonded to nitrogen is *sp.*

CHAPTER 2

2.1 All 10 of the bonds in propane $CH_3CH_2CH_3$ are σ bonds. The eight C—H σ bonds arise from C(sp^3)—H(1*s*) overlap. The two C—C bonds arise from C(sp^3)—C(sp^3) overlap.

2.2 $CH_3(CH_2)_{26}CH_3$

2.3 The molecular formula is $C_{11}H_{24}$; the condensed structural formula is $CH_3(CH_2)_9CH_3$.

2.4 $CH_3CH_2CH(CH_3)CH_2CH_3$ or [skeletal structure];

$CH_3CH(CH_3)CH(CH_3)CH_3$ or [skeletal structure]; and $CH_3CH_2C(CH_3)_2CH_3$ or [skeletal structure]

2.5 (*b*) $CH_3(CH_2)_{26}CH_3$; (*c*) undecane

2.6 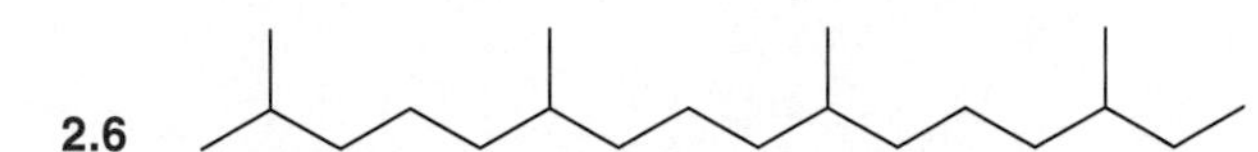

2.7 (*b*) $CH_3CH_2CH_2CH_2CH_3$ (pentane), $(CH_3)_2CHCH_2CH_3$ (2-methylbutane), $(CH_3)_4C$ (2,2-dimethylpropane); (*c*) 2,2,4-trimethylpentane; (*d*) 2,2,3,3-tetramethylbutane

2.8 $CH_3CH_2CH_2CH_2CH_2—$ (pentyl, primary); $CH_3CH_2CH_2CHCH_3$ (1-methylbutyl, secondary); $CH_3CH_2CHCH_2CH_3$ (1-ethylpropyl, secondary); $(CH_3)_2CHCH_2CH_2—$ (3-methylbutyl, primary); $CH_3CH_2CH(CH_3)CH_2—$ (2-methylbutyl, primary); $(CH_3)_2CCH_2CH_3$ (1,1-dimethylpropyl, tertiary); and $(CH_3)_2CHCHCH_3$ (1,2-dimethylpropyl, secondary)

2.9 (*b*) 4-Ethyl-2-methylhexane; (*c*) 8-ethyl-4-isopropyl-2,6-dimethyldecane

2.10 (*b*) 4-Isopropyl-1,1-dimethylcyclodecane; (*c*) cyclohexylcyclohexane

2.11 2,2,3,3-Tetramethylbutane (106°C); 2-methylheptane (116°C); octane (126°C); nonane (151°C)

2.12 $+ 9O_2 \longrightarrow 6CO_2 + 6H_2O$

2.13 13,313 kJ/mol

2.14 Hexane ($CH_3CH_2CH_2CH_2CH_2CH_3$) > pentane ($CH_3CH_2CH_2CH_2CH_3$) > isopentane [$(CH_3)_2CHCH_2CH_3$] > neopentane [$(CH_3)_4C$]

2.15 (*b*) Oxidation of carbon; (*c*) reduction of carbon

CHAPTER 3

3.1 (*b*) Butane; (*c*) 2-methylbutane; (*d*) 3-methylhexane

3.2 Shape of potential energy diagram is identical with that for ethane (Figure 3.4). Activation energy for internal rotation is higher than that of ethane, lower than that of butane.

3.3 150°

3.4 (*b*) (*c*) (*d*)

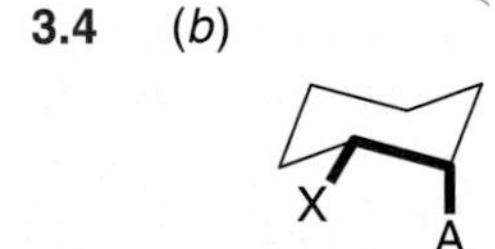

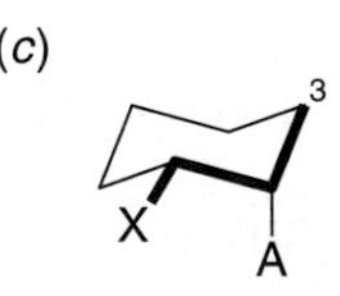

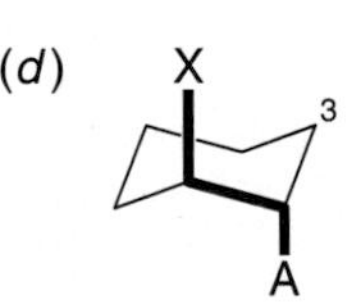

3.5 (*b*) Less stable; (*c*) methyl is equatorial and down

3.6 CH_3 $C(CH_3)_3$

3.7 Ethylcyclopropane: 3384 kJ/mol (808.8 kcal/mol); methylcyclobutane: 3352 kJ/mol (801.2 kcal/mol)

3.8 1,1-Dimethylcyclopropane, ethylcyclopropane, methylcyclobutane, and cyclopentane

3.9 *cis*-1,3-5-Trimethylcyclohexane is more stable.

3.10 (*b*) H_3C H H $C(CH_3)_3$ (*c*) H H_3C $C(CH_3)_3$ H

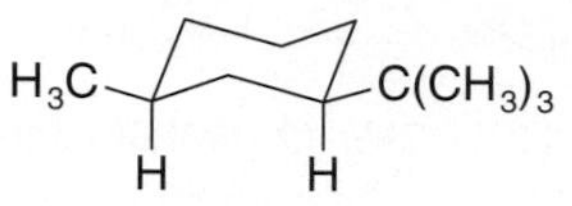

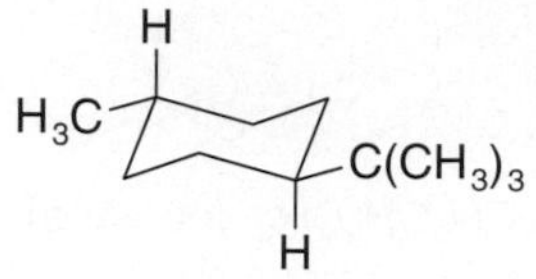

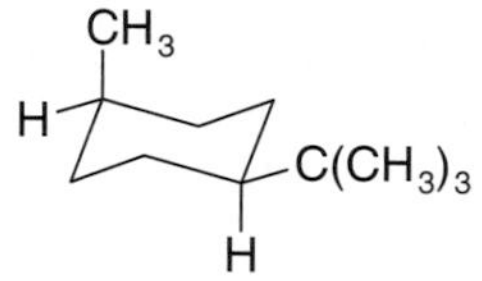

(*d*) CH_3 H $C(CH_3)_3$ H

3.11 CH_3 CH_3 CH_2 ⟶ CH_3 CH_3 CH_2 ⟶ CH_3 CH_3 CH_2

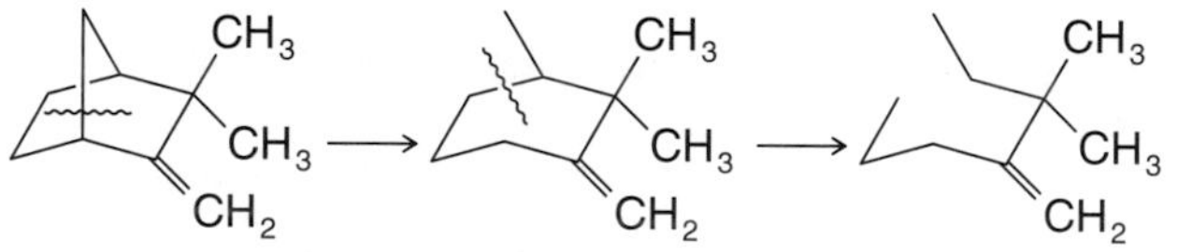

Other pairs of bond cleavages are also possible.

3.12 (*b*)

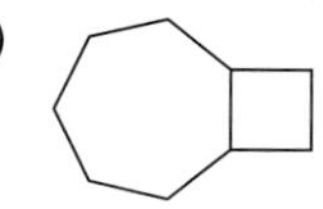

(*c*) 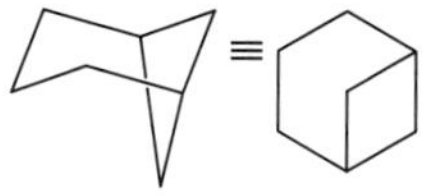 ≡ (*d*)

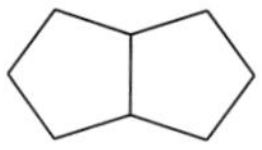

3.13 N—CH_3

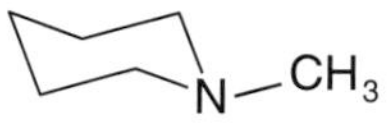

CHAPTER 4

4.1

	$CH_3CH_2CH_2CH_2Cl$	$CH_3CHCH_2CH_3$ (Cl on C-2)
Substitutive name:	1-Chlorobutane	2-Chlorobutane
Radicofunctional names:	*n*-Butyl chloride or butyl chloride	*sec*-Butyl chloride or 1-methylpropyl chloride
	$(CH_3)_2CHCH_2Cl$	$(CH_3)_3CCl$
	1-Chloro-2-methylpropane	2-Chloro-2-methylpropane
	Isobutyl chloride or 2-methylpropyl chloride	*tert*-Butyl chloride or 1,1-dimethylethyl chloride

4.2

	$CH_3CH_2CH_2CH_2OH$	$CH_3CHCH_2CH_3$ (OH on C-2)
Substitutive name:	1-Butanol	2-Butanol
Radicofunctional names:	*n*-Butyl alcohol or butyl alcohol	*sec*-Butyl alcohol or 1-methylpropyl alcohol
	$(CH_3)_2CHCH_2OH$	$(CH_3)_3COH$
	2-Methyl-1-propanol	2-Methyl-2-propanol
	Isobutyl alcohol or 2-methylpropyl alcohol	*tert*-Butyl alcohol or 1,1-dimethylethyl alcohol

4.3 $CH_3CH_2CH_2CH_2OH$ $CH_3CH(OH)CH_2CH_3$ $(CH_3)_2CHCH_2OH$ $(CH_3)_3COH$

Primary   Secondary   Primary   Tertiary

4.4 The carbon-bromine bond is longer than the carbon-chlorine bond; therefore, while the charge e in the dipole moment expression $\mu = e \cdot d$ is smaller for the bromine than for the chlorine compound, the distance d is greater.

4.5 Hydrogen bonding in ethanol (CH_3CH_2OH) makes its boiling point higher than that of dimethyl ether (CH_3OCH_3), for which hydrogen bonding is absent.

4.6 $H_3N\!: + H\text{—}\ddot{Cl}\!: \rightleftharpoons H_3\overset{+}{N}\text{—}H + :\ddot{Cl}:^-$

Base   Acid   Conjugate acid   Conjugate base

4.7 $K_a = 8 \times 10^{-10}$; hydrogen cyanide is a weak acid.

4.8 Hydrogen cyanide is a stronger acid than water; its conjugate base (CN^-) is a weaker base than hydroxide (HO^-).

4.9 $(CH_3)_3C\text{—}\ddot{O}\text{—}H + H\text{—}\ddot{Cl}: \rightleftharpoons (CH_3)_3C\text{—}\overset{+}{O}H_2 + :\ddot{Cl}:^-$

Base   Acid   Conjugate acid   Conjugate base

4.10 Greater than 1

4.11 $(CH_3)_3C\text{—}\overset{\delta+}{O}(H)\cdots H\cdots \overset{\delta-}{Cl}$

4.12

(b) $(CH_3CH_2)_3COH + HCl \longrightarrow (CH_3CH_2)_3CCl + H_2O$
(c) $CH_3(CH_2)_{12}CH_2OH + HBr \longrightarrow CH_3(CH_2)_{12}CH_2Br + H_2O$

4.13 $(CH_3)_2\overset{+}{C}CH_2CH_3$

4.14 **1-Butanol:** Rate-determining step is bimolecular; therefore S_N2.

1. $CH_3CH_2CH_2CH_2\ddot{O}H + H\text{—}\ddot{Br}: \xrightarrow{\text{fast}} CH_3CH_2CH_2CH_2\overset{+}{O}H_2 + :\ddot{Br}:^-$

2. $:\ddot{Br}:^- + CH_3CH_2CH_2\text{—}CH_2\text{—}\overset{+}{O}H_2 \xrightarrow{\text{slow}} CH_3CH_2CH_2CH_2Br + :\ddot{O}H_2$

2-Butanol: Rate-determining step is unimolecular; therefore S_N1.

1. $CH_3CH_2CH(\ddot{O}H)CH_3 + H\text{—}\ddot{Br}: \xrightarrow{\text{fast}} CH_3CH_2CH(\overset{+}{O}H_2)CH_3 + :\ddot{Br}:^-$

2. 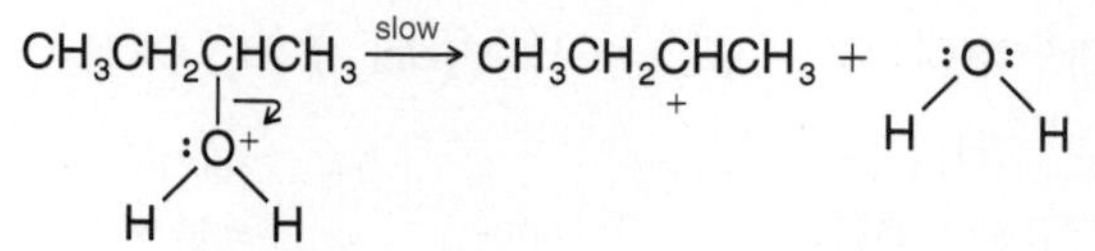 $CH_3CH_2CHCH_3$ (with $:\overset{+}{O}H_2$ on C-2, H H) $\xrightarrow{slow}$ $CH_3CH_2\overset{+}{C}HCH_3$ + $H\ddot{O}H$

3. $:\ddot{Br}:^-$ + $CH_3CH_2\overset{+}{C}HCH_3$ $\xrightarrow{fast}$ $CH_3CH_2CHCH_3$ (Br on C-2)

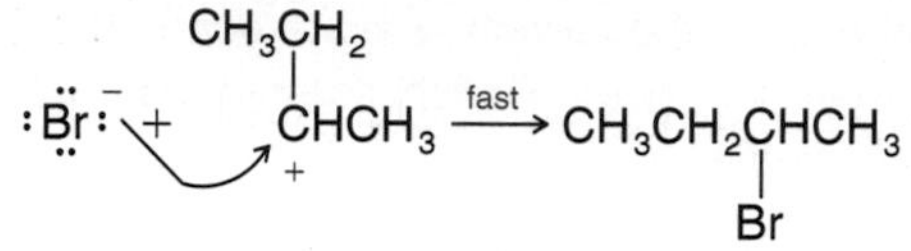

4.15 $(CH_3)_2\dot{C}CH_2CH_3$

4.16 (*b*) The carbon-carbon bond dissociation energy is lower for 2-methylpropane because it yields a more stable (secondary) radical; propane yields a primary radical. (*c*) The carbon-carbon bond dissociation energy is lower for 2,2-dimethylpropane because it yields a still more stable tertiary radical.

4.17 CH_3CHCl_2 and $ClCH_2CH_2Cl$

4.18 1-Chloropropane (43%); 2-chloropropane (57%)

4.19 (*b*) Br, $C(CH_3)_2$, CH_3; (*c*) Br

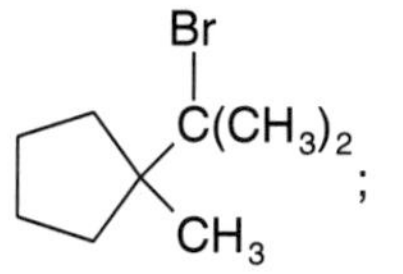

CHAPTER 5

5.1 (*b*) 3,3-Dimethyl-1-butene; (*c*) 2-methyl-2-hexene; (*d*) 4-chloro-1-pentene; (*e*) 4-penten-2-ol

5.2 Cl 1 5 4 2 3 — Cl 3 2 4 1 5 — 1 5 Cl 4 2 3

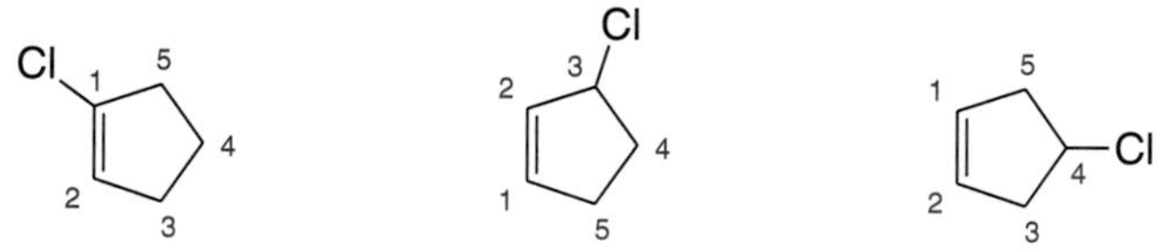

1-Chlorocyclopentene 3-Chlorocyclopentene 4-Chlorocyclopentene

5.3 (*b*) 3-Ethyl-3-hexene; (*c*) two carbons are sp^2 hybridized, six are sp^3 hybridized; (*d*) there are three sp^2-sp^3 σ bonds and three sp^3-sp^3 σ bonds.

5.4

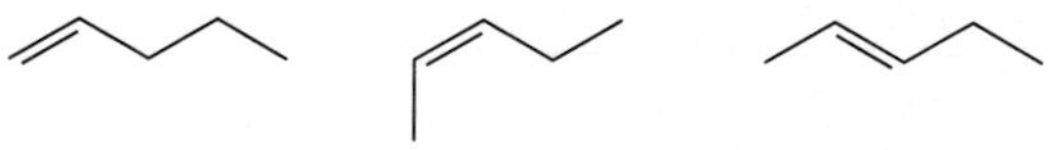

1-Pentene *cis*-2-Pentene *trans*-2-Pentene

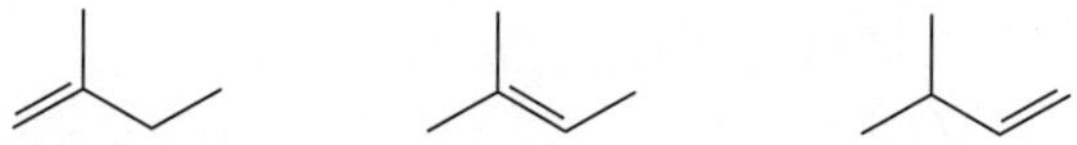

2-Methyl-1-butene 2-Methyl-2-butene 3-Methyl-1-butene

5.5 $CH_3(CH_2)_7$, $(CH_2)_{12}CH_3$, C=C, H, H

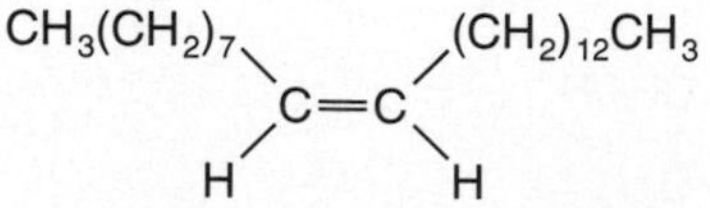

5.6 (*b*) *Z*; (*c*) *E*; (*d*) *E*

5.7

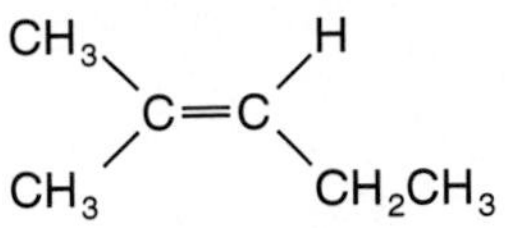

CH_3, CH_3 C=C H, CH_2CH_3
2-Methyl-2-pentene

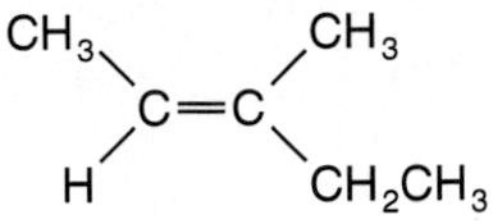

CH_3, H C=C CH_3, CH_2CH_3
(*E*)-3-Methyl-2-pentene

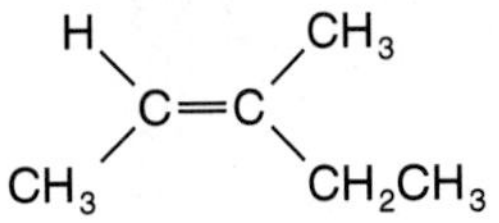

H, CH_3 C=C CH_3, CH_2CH_3
(*Z*)-3-Methyl-2-pentene

5.8 $(CH_3)_2C{=}C(CH_3)_2$

5.9 2-Methyl-2-butene (most stable) > (*E*)-2-pentene > (*Z*)-2-pentene > 1-pentene (least stable)

5.10

(*c*) H 1, 2 H, 3, CH_3

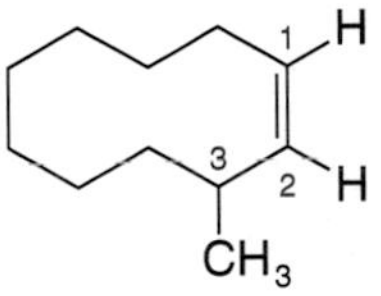

(*e*) H 1, H 2, 3, 4, 5, CH_3

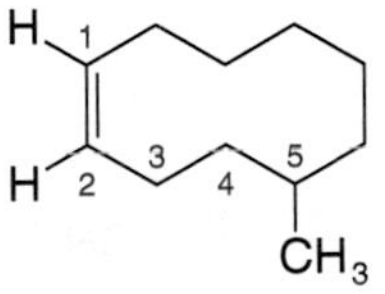

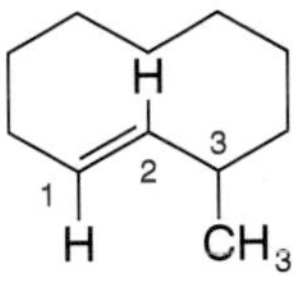

(*d*) H, 1, 2, 3, H, CH_3

(*f*) H 2, 1, 3, H, 4, 5, CH_3

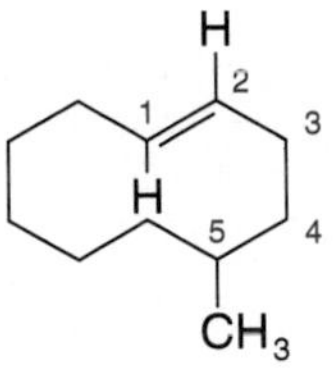

5.11 (*b*) Propene; (*c*) propene; (*d*) 2,3,3-trimethyl-1-butene

5.12

(*b*) CH_3 and CH_2

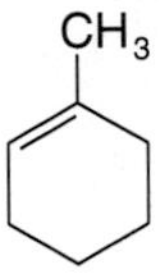

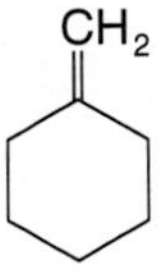

Major Minor

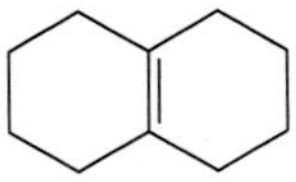

(*c*) and H

Major Minor

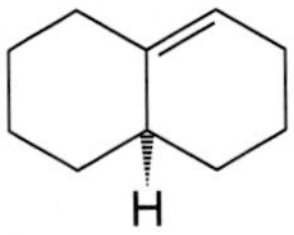

5.13 1-Pentene, *cis*-2-pentene, and *trans*-2-pentene

5.14

(*b*) H_3C OH → CH_3 + + H_2O

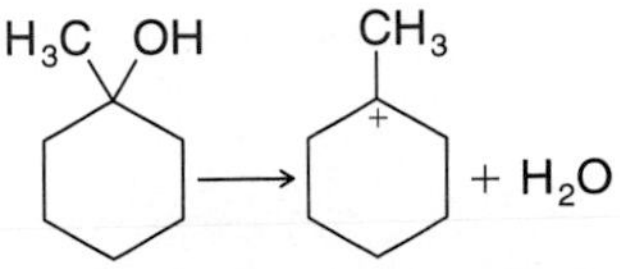

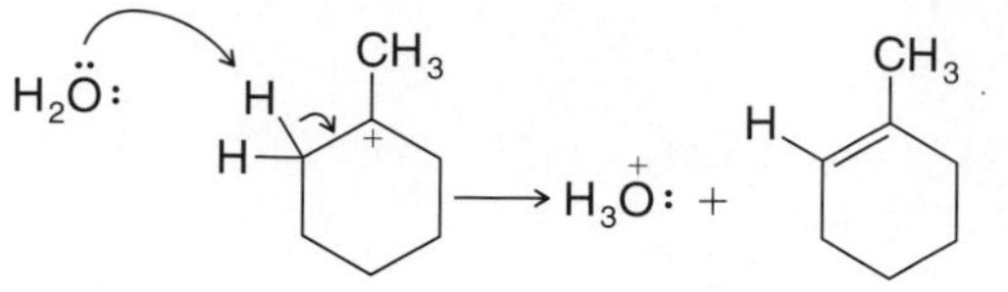

$H_2\ddot{O}$: H H–(cyclohexyl cation with CH_3) ⟶ $H_3\overset{+}{O}$: + 1-methylcyclohexene

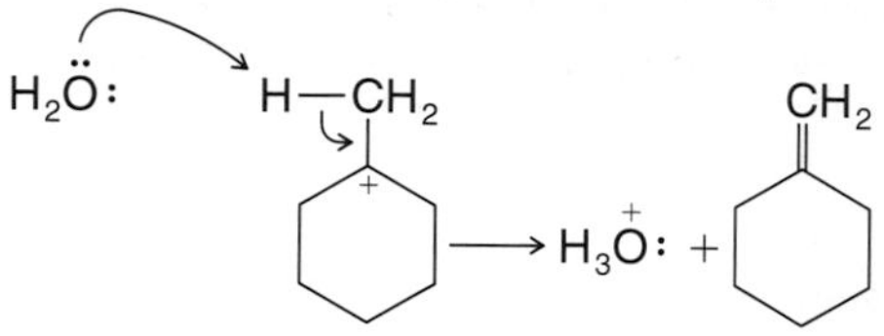

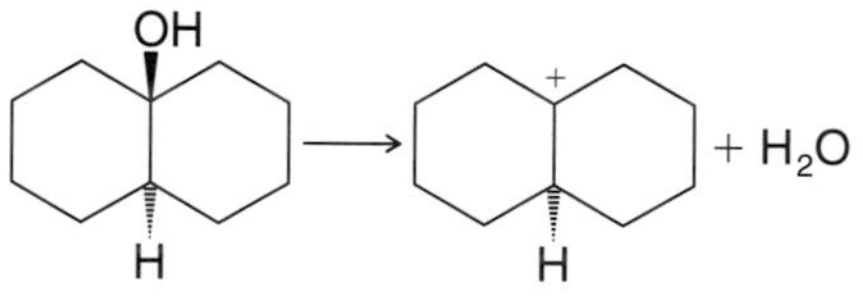

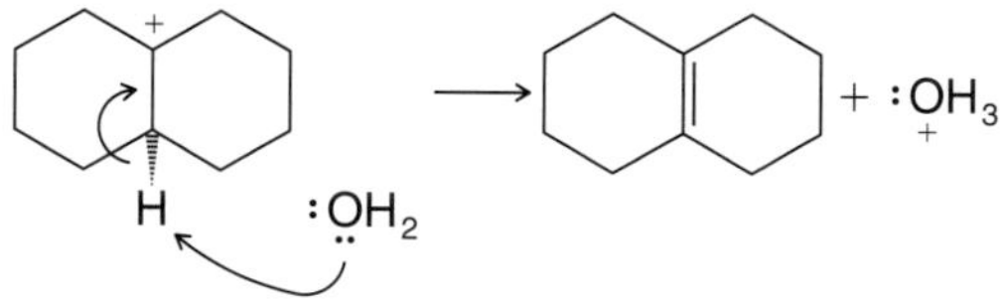

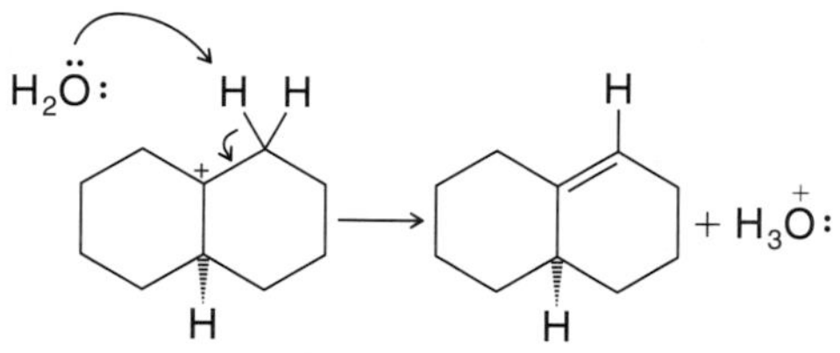

and $H_2\ddot{O}$: H–CH_2–(cyclohexyl cation) ⟶ $H_3\overset{+}{O}$: + methylenecyclohexane

(*c*) OH, H ⟶ (+), H + H_2O

(+), H, :OH_2 ⟶ + :$\overset{+}{O}H_3$

and $H_2\ddot{O}$: H H, (+), H ⟶ H, H + $H_3\overset{+}{O}$:

5.15 CH_3, CH_3, H, OH $\xrightarrow[-H_2O]{H^+}$ CH_3, CH_3, H (+) ⟶ CH_3, H, CH_3 (+) $\xrightarrow{-H^+}$ CH_3, CH_3

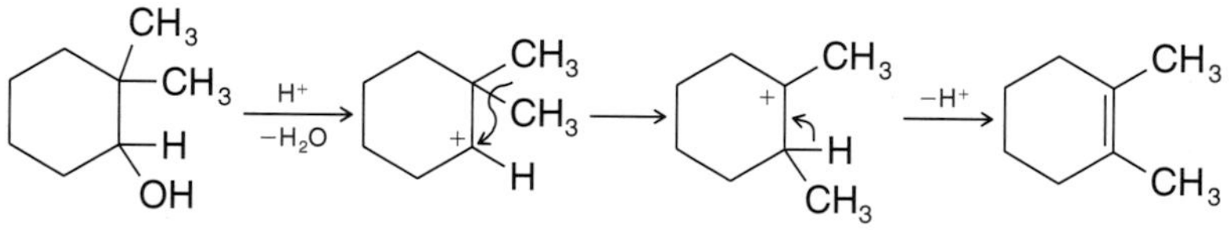

5.16 (*b*) $(CH_3)_2C{=}CH_2$; (*c*) $CH_3CH{=}C(CH_2CH_3)_2$; (*d*) $CH_3CH{=}C(CH_3)_2$ (major) and $CH_2{=}CHCH(CH_3)_2$ (minor); (*e*) $CH_2{=}CHCH(CH_3)_2$; (*f*) 1-methylcyclohexene (major) and methylenecyclohexane (minor)

5.17 $CH_2{=}CHCH_2CH_3$, *cis*-$CH_3CH{=}CHCH_3$, and *trans*-$CH_3CH{=}CHCH_3$.

5.18 $CH_3\ddot{O}:^-$ H–C(H)(H)–C(CH_3)(CH_3)(:Cl:) ⟶ $CH_3\ddot{O}$–H + $H_2C{=}C(CH_3)_2$ + $:\ddot{C}l:^-$

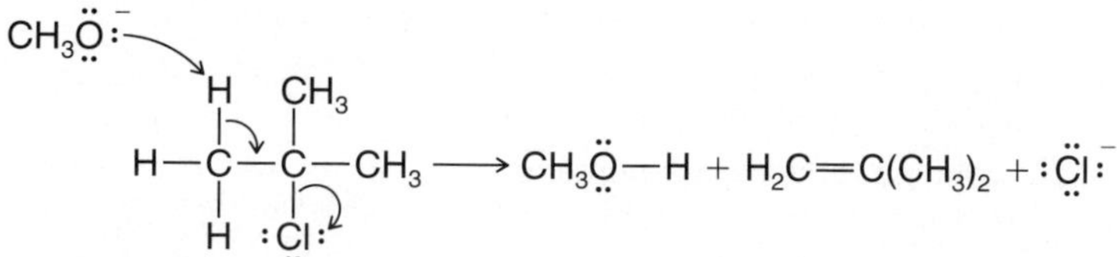

5.19 :Br:$^-$, $(CH_3)_3C$–, H, $(CH_3)_3C$–$\ddot{O}:^-$

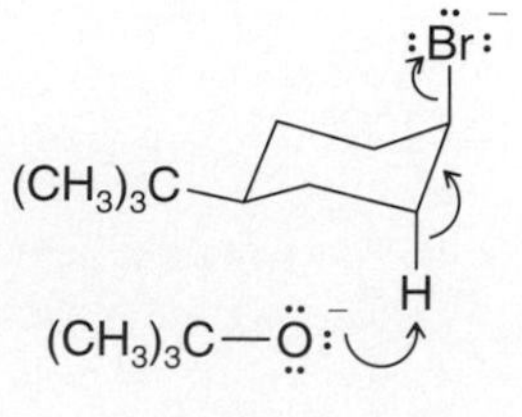

CHAPTER 6

6.1 2-Methyl-1-butene, 2-methyl-2-butene, and 3-methyl-1-butene

6.2 2-Methyl-2-butene (112 kJ/mol, 26.7 kcal/mol), 2-methyl-1-butene (118 kJ/mol, 28.2 kcal/mol), and 3-methyl-1-butene (126 kJ/mol, 30.2 kcal/mol)

6.3 (*b*) $(CH_3)_2C(Cl)CH_2CH_3$; (*c*) $CH_3CH(Cl)CH_2CH_3$; (*d*) 1-chloro-1-ethylcyclohexane (CH_3CH_2 and Cl on the same ring carbon)

6.4 (*b*) $(CH_3)_2\overset{+}{C}CH_2CH_3$; (*c*) $CH_3\overset{+}{C}HCH_2CH_3$; (*d*) CH_3CH_2–cyclohexyl cation (+ on ring carbon)

6.5 $CH_3C(CH_3)_2-CH{=}CH_2 \xrightarrow{HCl} CH_3C(CH_3)_2-\overset{+}{C}HCH_3 \xrightarrow{CH_3\ shift} CH_3\overset{+}{C}(CH_3)-CH(CH_3)CH_3$

$\downarrow Cl^-$: $CH_3C(CH_3)_2-CH(Cl)CH_3$

$\downarrow Cl^-$: $CH_3C(Cl)(CH_3)-CH(CH_3)CH_3$

6.6 Addition in accordance with Markovnikov's rule gives 1,2-dibromopropane. Addition opposite to Markovnikov's rule gives 1,3-dibromopropane.

6.7 Absence of peroxides: (*b*) 2-bromo-2-methylbutane; (*c*) 2-bromobutane; (*d*) 1-bromo-1-ethylcyclohexane. Presence of peroxides: (*b*) 1-bromo-2-methylbutane; (*c*) 2-bromobutane; (*d*) 1-(bromoethyl)cyclohexane.

6.8 Cyclohexene $\xrightarrow{H_2SO_4}$ Cyclohexyl hydrogen sulfate (cyclohexyl–OSO_2OH)

Cyclohexene Cyclohexyl hydrogen sulfate

6.9 The concentration of hydroxide ion is too small in acid solution to be chemically significant.

6.10 cyclopropyl–$C(CH_3){=}CH_2$ is more reactive, because it gives a tertiary carbocation cyclopropyl–$\overset{+}{C}(CH_3)_2$ when it is protonated in acid solution.

6.11 E1

6.12 (*b*) $CH_3CH(OH)CH_2CH_3$ (*c*) cyclobutyl with H and CH_2OH (*d*) cyclopentyl–OH

(*e*) $CH_3CH(OH)CH(CH_2CH_3)_2$ (*f*) $HOCH_2CH_2CH(CH_2CH_3)_2$

6.13 CH_3 CH_3 H HO CH_3 H

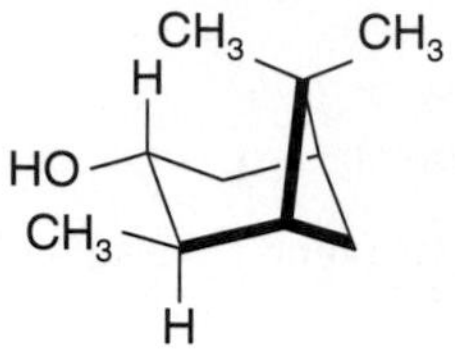

6.14 ^{82}Br Br H ^{82}Br

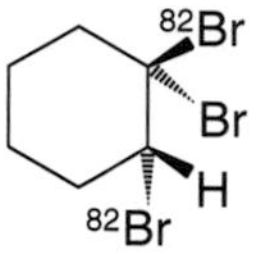

6.15 2-Methyl-2-butene (most reactive) > 2-methyl-1-butene > 3-methyl-1-butene (least reactive)

6.16 (*b*) $(CH_3)_2C(OH)-CHBrCH_3$; (*c*) $BrCH_2CH(OH)CH(CH_3)_2$; (*d*) Br CH_3 OH

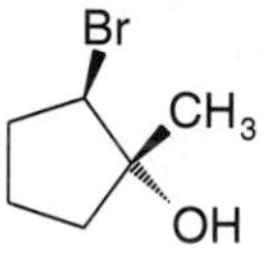

6.17 *cis*-2-Methyl-7,8-epoxyoctadecane

6.18 *cis*-$(CH_3)_2CHCH_2CH_2CH_2CH_2CH{=}CH(CH_2)_9CH_3$

6.19 2,4,4-Trimethyl-1-pentene

6.20 (*b*) 3; (*c*) 2; (*d*) 3; (*e*) 2; (*f*) 2

6.21 Hydrogenation over a metal catalyst such as platinum, palladium, or nickel

6.22 $(CH_3)_3CBr \xrightarrow[\text{heat}]{NaOCH_2CH_3} (CH_3)_2C{=}CH_2 \xrightarrow[H_2O]{Br_2} (CH_3)_2C(OH){-}CH_2Br$

CHAPTER 7

7.1 (*c*) C-2 is a stereogenic center; (*d*) no stereogenic centers.

7.2 (*c*) C-2 is a stereogenic center; (*d*) no stereogenic centers.

7.3

(*b*) (*Z*)-1,2-Dichloroethene is achiral. The plane of the molecule is a plane of symmetry. A second plane of symmetry is perpendicular to the plane of the molecule and bisects the carbon-carbon bond.

(*c*) *cis*-1,2-Dichlorocyclopropane is achiral. It has a plane of symmetry that bisects the C-1—C-2 bond and passes through C-3.

(*d*) *trans*-1,2-Dichlorocyclopropane is chiral. It has neither a plane of symmetry nor a center of symmetry.

7.4 $[\alpha]_D - 39°$

7.5 Two-thirds (66.7 percent)

7.6 (*b*) *R*; (*c*) *S*; (*d*) *S*

7.7 (*b*) H_3C, F, H, F on cyclopropane ring (cis-difluoro structure)

7.8 (*b*) Fischer projection: CH_3 top; FCH_2—C—H; CH_2CH_3 bottom (*c*) CH_3 top; H—C—CH_2Br; CH_2CH_3 bottom (*d*) CH_3 top; H—C—OH; $CH{=}CH_2$ bottom

7.9 *S*

7.10

Erythro	Erythro	Threo	Threo
CH_3	CH_3	CH_3	CH_3
H—C—OH	HO—C—H	H—C—OH	HO—C—H
H—C—NH_2	H_2N—C—H	H_2N—C—H	H—C—NH_2
CH_3	CH_3	CH_3	CH_3

7.11 2*S*,3*R*

7.12 2,4-Dibromopentane

7.13 *cis*-1,3-Dimethylcyclohexane

7.14 *RRR* *RRS* *RSR* *SRR* *SSS* *SSR* *SRS* *RSS*

7.15 Eight

7.16 Epoxidation of *cis*-2-butene gives *meso*-2,3-epoxybutane; *trans*-2-butene gives a racemic mixture of (2*R*,3*R*) and (2*S*,3*S*)-2,3-epoxybutane.

7.17 No. The major product *cis*-1,2-dimethylcyclohexane is less stable than the minor product *trans*-1,2-dimethylcyclohexane.

7.18 CO_2H top; HO—C(*S*)—H; H—C(*S*)—OH; CO_2H bottom and CO_2H top; H—C(*R*)—OH; H—C(*S*)—OH; CO_2H bottom

7.19 No

7.20 (*S*)-1-Phenylethylammonium (*S*)-malate

CHAPTER 8

8.1 (*b*) $CH_3OCH_2CH_3$; (*c*) $CH_3OC(=O)C_6H_5$; (*d*) $CH_3\ddot{N}{=}\overset{+}{N}{=}\overset{-}{\ddot{N}}:$; (*e*) $CH_3C{\equiv}N$;

(*f*) CH_3SH

8.2 $ClCH_2CH_2CH_2C{\equiv}N$

8.3 No

8.4

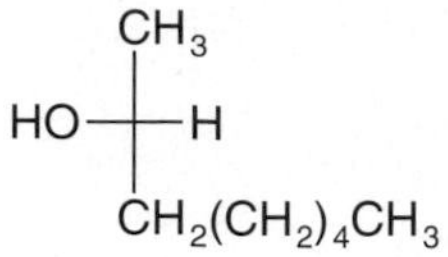

$$HO-\overset{CH_3}{\underset{CH_2(CH_2)_4CH_3}{C}}-H$$

8.5 Hydrolysis of (*R*)-(−)-2-bromooctane by the S_N2 mechanism yields optically active (*S*)-(+)-2-octanol. The 2-octanol obtained by hydrolysis of racemic 2-bromooctane is not optically active.

8.6 (*b*) 1-Bromopentane; (*c*) 2-chloropentane; (*d*) 2-bromo-5-methylhexane; (*e*) 1-bromodecane

8.7 $CH_3\underset{NO_2}{CH}(CH_2)_5CH_3$ and $CH_3\underset{ONO}{CH}(CH_2)_5CH_3$

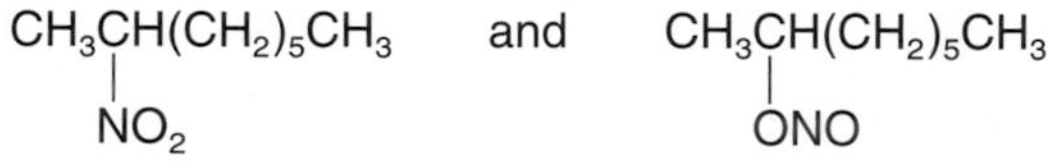

8.8 Product is $(CH_3)_3COCH_3$. The mechanism of solvolysis is S_N1.

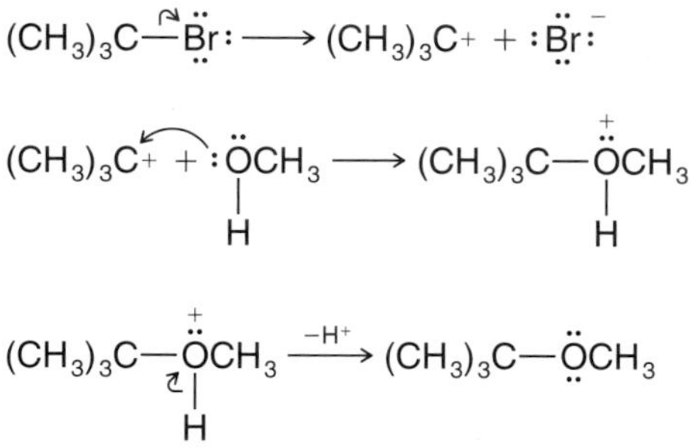

$$(CH_3)_3C-\ddot{Br}: \longrightarrow (CH_3)_3C^+ + :\ddot{Br}:^-$$

$$(CH_3)_3C^+ + :\ddot{O}(H)CH_3 \longrightarrow (CH_3)_3C-\overset{+}{\ddot{O}}(H)CH_3$$

$$(CH_3)_3C-\overset{+}{\ddot{O}}(H)CH_3 \xrightarrow{-H^+} (CH_3)_3C-\ddot{O}CH_3$$

8.9 (*b*) 1-Methylcyclopentyl iodide; (*c*) cyclopentyl bromide; (*d*) *tert*-butyl iodide

8.10 Both *cis*- and *trans*-1,4-dimethylcyclohexanol are formed in the hydrolysis of either *cis*- or *trans*-1,4-dimethylcyclohexyl bromide.

8.11 A hydride shift produces a tertiary carbocation; a methyl shift produces a secondary carbocation.

8.12 (*b*) cyclohexyl$-OCH_2CH_3$; (*c*) $CH_3\underset{OCH_3}{CH}CH_2CH_3$;

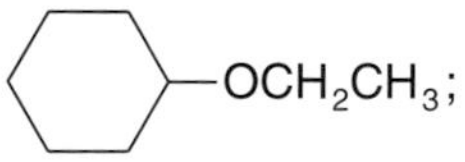

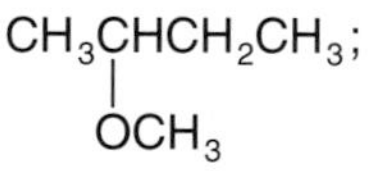

(*d*) *cis*- and *trans*-$CH_3CH{=}CHCH_3$ and $CH_2{=}CHCH_2CH_3$

8.13 $CH_3(CH_2)_{16}CH_2OH + CH_3-C_6H_4-SO_2Cl \xrightarrow{\text{pyridine}}$

$$CH_3(CH_2)_{16}CH_2OSO_2-C_6H_4-CH_3 + HCl$$

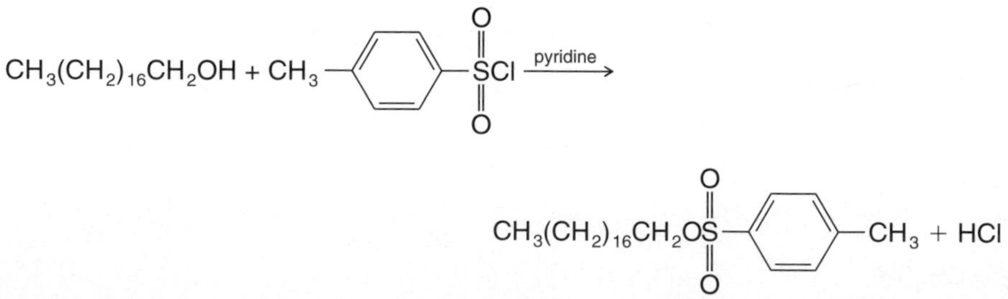

8.14 (*b*) $CH_3(CH_2)_{16}CH_2I$; (*c*) $CH_3(CH_2)_{16}CH_2C{\equiv}N$; (*d*) $CH_3(CH_2)_{16}CH_2SH$; (*e*) $CH_3(CH_2)_{16}CH_2SCH_2CH_2CH_2CH_3$

8.15 The product has the R configuration and a specific rotation $[\alpha]_D$ of $-9.9°$.

$$CH_3(CH_2)_5(H)(H_3C)C\text{—}OTs \xrightarrow{H_2O} HO\text{—}C(H)((CH_2)_5CH_3)(CH_3)$$

8.16 $CH_3CH_2C(Cl)(CH_3)_2$

CHAPTER 9

9.1

$$:\bar{C}{\equiv}\bar{C}: + H\text{—}\ddot{O}\text{—}H \longrightarrow :\bar{C}{\equiv}C\text{—}H + :\ddot{O}^{-}\text{—}H$$

Carbide ion, Water, Acetylide ion, Hydroxide ion

$$H\text{—}\ddot{O}\text{—}H + :\bar{C}{\equiv}C\text{—}H \longrightarrow H\text{—}\ddot{O}:^{-} + H\text{—}C{\equiv}C\text{—}H$$

Water, Acetylide ion, Hydroxide ion, Acetylene

9.2 $CH_3CH_2CH_2C{\equiv}CH$ (1-pentyne), $CH_3CH_2C{\equiv}CCH_3$ (2-pentyne), $(CH_3)_2CHC{\equiv}CH$ (3-methyl-1-butyne)

9.3 The bond from the methyl group in 1-butyne is to an sp^3 hybridized carbon and so is longer than the bond from the methyl group in 2-butyne, which is to an sp hybridized carbon.

9.4

(*b*) $HC{\equiv}C\text{—}H + :\bar{C}H_2CH_3 \underset{}{\overset{K \gg 1}{\rightleftharpoons}} HC{\equiv}\bar{C}: + CH_3CH_3$

Acetylene (stronger acid), Ethyl anion (stronger base), Acetylide ion (weaker base), Ethane (weaker acid)

(*c*) $CH_2{=}CH\text{—}H + :\bar{\ddot{N}}H_2 \overset{K \ll 1}{\rightleftharpoons} CH_2{=}\ddot{\bar{C}}H + :NH_3$

Ethylene (weaker acid), Amide ion (weaker base), Vinyl anion (stronger base), Ammonia (stronger acid)

(*d*) $CH_3C{\equiv}CCH_2\ddot{O}\text{—}H + :\bar{\ddot{N}}H_2 \overset{K \gg 1}{\rightleftharpoons} CH_3C{\equiv}CCH_2\ddot{O}:^{-} + :NH_3$

2-Butyn-1-ol (stronger acid), Amide ion (stronger base), 2-Butyn-1-olate anion (weaker base), Ammonia (weaker acid)

9.5

(*b*) $HC{\equiv}CH \xrightarrow[\text{2. } CH_3Br]{\text{1. } NaNH_2,\ NH_3} CH_3C{\equiv}CH \xrightarrow[\text{2. } CH_3CH_2CH_2CH_2Br]{\text{1. } NaNH_2,\ NH_3} CH_3C{\equiv}CCH_2CH_2CH_2CH_3$

(or reverse order of alkylations)

(*c*) $HC{\equiv}CH \xrightarrow[\text{2. } CH_3CH_2CH_2Br]{\text{1. } NaNH_2,\ NH_3} CH_3CH_2CH_2C{\equiv}CH \xrightarrow[\text{2. } CH_3CH_2Br]{\text{1. } NaNH_2,\ NH_3} CH_3CH_2CH_2C{\equiv}CCH_2CH_3$

(or reverse order of alkylations)

9.6 Both $CH_3CH_2CH_2C{\equiv}CH$ and $CH_3CH_2C{\equiv}CCH_3$ can be prepared by alkylation of acetylene. The alkyne $(CH_3)_2CHC{\equiv}CH$ cannot be prepared by alkylation of acetylene, because the required alkyl halide, $(CH_3)_2CHBr$, is secondary and will react with the strongly basic acetylide ion by elimination.

9.7 $(CH_3)_3C\overset{Br}{\underset{Br}{C}}CH_3$ or $(CH_3)_3CCH_2CHBr_2$ or $(CH_3)_3C\underset{Br}{C}HCH_2Br$

9.8

(*b*) $CH_3CH_2CH_2OH \xrightarrow[\text{heat}]{H_2SO_4} CH_3CH{=}CH_2 \xrightarrow{Br_2} CH_3\overset{Br}{C}HCH_2Br \xrightarrow[2.\ H^+]{1.\ NaNH_2} CH_3C{\equiv}CH$

(*c*) $(CH_3)_2CHBr \xrightarrow{NaOCH_2CH_3} CH_3CH{=}CH_2$; then proceed as in (*a*) and (*b*).

(*d*) $CH_3CHCl_2 \xrightarrow[2.\ H_2O]{1.\ NaNH_2} HC{\equiv}CH \xrightarrow[2.\ CH_3Br]{1.\ NaNH_2} CH_3C{\equiv}CH$

(*e*) $CH_3CH_2OH \xrightarrow[\text{heat}]{H_2SO_4} CH_2{=}CH_2 \xrightarrow{Br_2} BrCH_2CH_2Br \xrightarrow[2.\ H_2O]{1.\ NaNH_2} HC{\equiv}CH$; then proceed as in (*d*).

9.9 $HC{\equiv}CH \xrightarrow[2.\ CH_3CH_2CH_2Br]{1.\ NaNH_2,\ NH_3} CH_3CH_2CH_2C{\equiv}CH \xrightarrow[2.\ CH_3CH_2CH_2Br]{1.\ NaNH_2,\ NH_3}$

$CH_3CH_2CH_2C{\equiv}CCH_2CH_2CH_3 \xrightarrow[Pt]{H_2} CH_3(CH_2)_6CH_3$

or $HC{\equiv}CH \xrightarrow[2.\ CH_3(CH_2)_5Br]{1.\ NaNH_2,\ NH_3} CH_3(CH_2)_5C{\equiv}CH \xrightarrow[Pt]{H_2} CH_3(CH_2)_6CH_3$

9.10 Oleic acid is *cis*-$CH_3(CH_2)_7CH{=}CH(CH_2)_7CO_2H$. Stearic acid is $CH_3(CH_2)_{16}CO_2H$.

9.11 Elaidic acid is *trans*-$CH_3(CH_2)_7CH{=}CH(CH_2)_7CO_2H$.

9.12

$CH_3C{\equiv}CH \xrightarrow[2.\ CH_3CH_2CH_2CH_2Br]{1.\ NaNH_2,\ NH_3} CH_3C{\equiv}CCH_2CH_2CH_2CH_3$

$\xrightarrow{Li,\ NH_3}$ (H₃C)(H)C=C(H)(CH₂CH₂CH₂CH₃) [trans]

$\xrightarrow[\text{Lindlar Pd}]{H_2}$ (H₃C)(H)C=C(CH₂CH₂CH₂CH₃)(H) [cis]

9.13

(*b*) $CH_2{=}CHCl \xrightarrow{HCl} CH_3CHCl_2$

(*c*) $CH_2{=}CHBr \xrightarrow[2.\ H_2O]{1.\ NaNH_2,\ NH_3} HC{\equiv}CH \xrightarrow{2HCl} CH_3CHCl_2$

(*d*) $CH_3CHBr_2 \xrightarrow[2.\ H_2O]{1.\ NaNH_2,\ NH_3} HC{\equiv}CH \xrightarrow{2HCl} CH_3CHCl_2$

9.14 $CH_3C{\equiv}CCH_3 \xrightarrow[H_2SO_4]{H_2O,\ Hg^{2+}} \left[CH_3\underset{OH}{C}{=}CHCH_3\right] \longrightarrow CH_3\overset{O}{\overset{\|}{C}}CH_2CH_3$

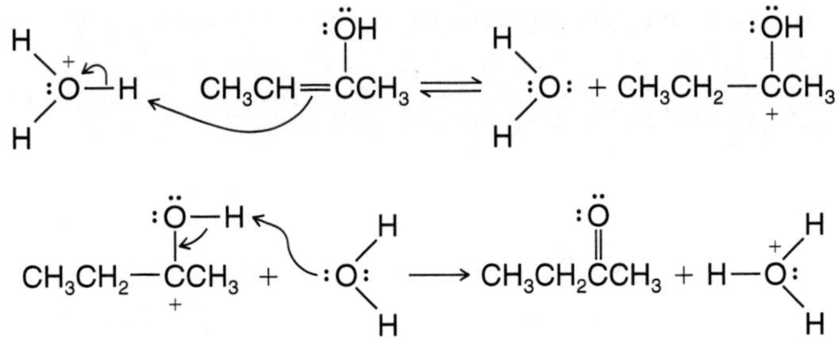

$$H_2\overset{+}{O}{-}H + CH_3CH{=}C(OH)CH_3 \rightleftharpoons H_2O: + CH_3CH_2{-}\overset{+}{C}(OH)CH_3$$

$$CH_3CH_2{-}\overset{+}{C}(OH)CH_3 + :OH_2 \longrightarrow CH_3CH_2\overset{O}{\overset{\|}{C}}CH_3 + H{-}\overset{+}{O}H_2$$

9.15 2-Octanone is prepared as shown:

$$HC{\equiv}CH \xrightarrow[\text{2. } CH_3(CH_2)_4CH_2Br]{\text{1. } NaNH_2,\ NH_3} CH_3(CH_2)_4CH_2C{\equiv}CH \xrightarrow[HgSO_4]{H_2O,\ H_2SO_4} CH_3(CH_2)_4CH_2\overset{O}{\overset{\|}{C}}CH_3$$

4-Octyne is prepared as described in Problem 9.9 and converted to 4-octanone by hydration with H_2O, H_2SO_4, and $HgSO_4$.

9.16 $CH_3(CH_2)_4C{\equiv}CCH_2CH_2C{\equiv}C(CH_2)_4CH_3$

CHAPTER 10

10.1

(*b*) $CH_2{=}C(CH_3){-}\overset{+}{C}H_2 \longleftrightarrow \overset{+}{C}H_2{-}C(CH_3){=}CH_2$

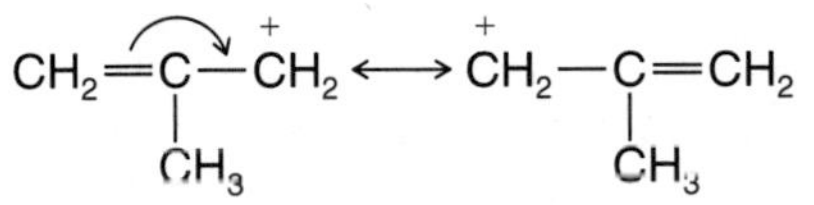

(*c*) [cyclohexyl cation]$=C(CH_3)_2 \longleftrightarrow$ [cyclohexenyl]$-\overset{+}{C}(CH_3)_2$

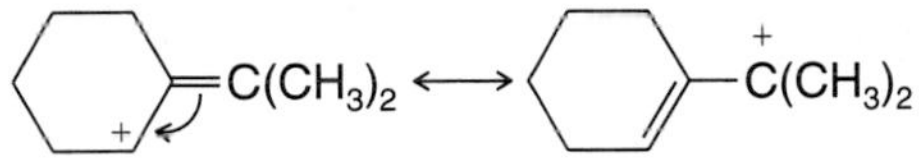

10.2 3-Bromo-1-methylcyclohexene and 3-chloro-3-methylcyclohexene

10.3

(*b*) 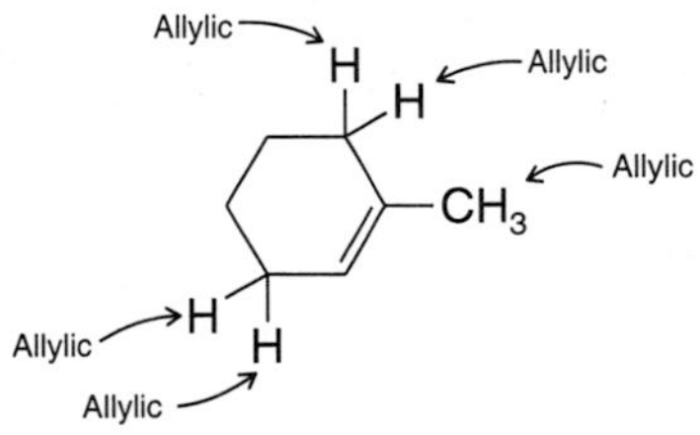 Allylic, Allylic, Allylic, Allylic, Allylic (H, H, CH_3, H, H)

(*c*) CH_3 ← Allylic

(*d*) Allylic → H H ← Allylic

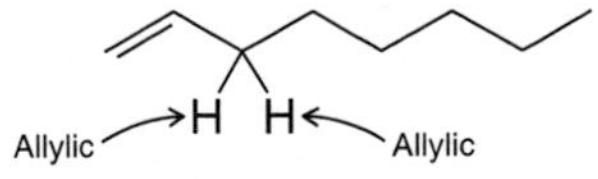

10.4 (Propagation step 1)

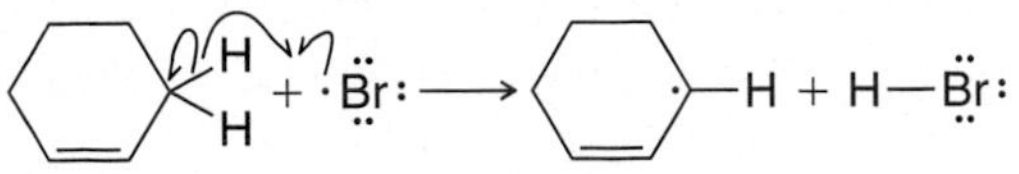

$$\text{cyclohexene}{-}H_2 + \cdot\ddot{Br}: \longrightarrow \text{cyclohexenyl}\cdot{-}H + H{-}\ddot{Br}:$$

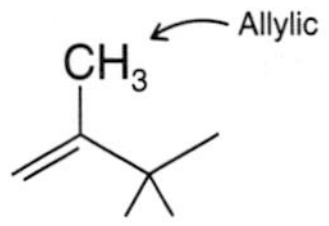

(Propagation step 2)

$$\text{cyclohexenyl}\cdot{-}H + :\ddot{Br}:\ddot{Br}: \longrightarrow \text{cyclohexenyl}(Br)(H) + \cdot\ddot{Br}:$$

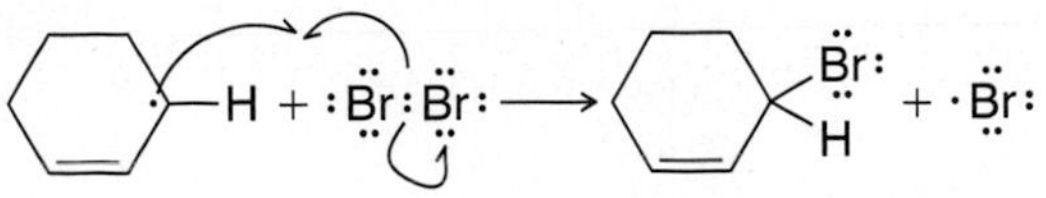

10.5 2,3,3-Trimethyl-1-butene gives only $(CH_3)_3CC(CH_2Br){=}CH_2$. 1-Octene gives a mixture of $CH_2{=}CHCH(Br)(CH_2)_4CH_3$ as well as the cis and trans stereoisomers of $BrCH_2CH{=}CH(CH_2)_4CH_3$.

10.6 (*b*) All the double bonds in humulene are isolated. (*c*) Two of the double bonds in cembrene are conjugated to each other but isolated from the remaining double bonds in the molecule. (*d*) The CH=C=CH unit is a cumulated double bond; it is conjugated to the double bond at C-2.

10.7 1,2-Pentadiene (3251 kJ/mol, 777.1 kcal/mol); (*E*)-1,3-pentadiene (3186 kJ/mol, 761.6 kcal/mol); 1,4-pentadiene (3217 kJ/mol, 768.9 kcal/mol)

10.8 $CH_2{=}CHCH_2C(CH_3){=}CHCH_3$ (*cis* + *trans*) and $CH_2{=}CHCH_2C({=}CH_2)CH_2CH_3$

10.9 $(CH_3)_2C(Cl)CH{=}CH_2$

10.10 3,4-Dibromo-3-methyl-1-butene, 3,4-dibromo-2-methyl-1-butene, and 1,4-dibromo-2-methyl-2-butene

10.11

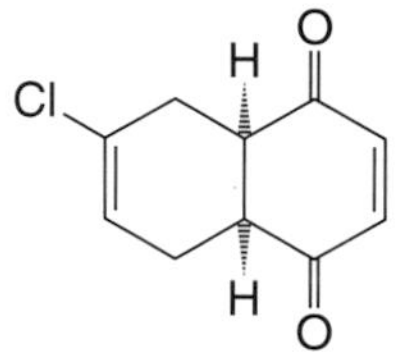

Cl H O H O

10.12

(*b*) $CH_2{=}CHCH{=}CH_2$ + *cis*-$N{\equiv}CCH{=}CHC{\equiv}N$

(*c*) $CH_3CH{=}CHCH{=}CH_2$ + O O O

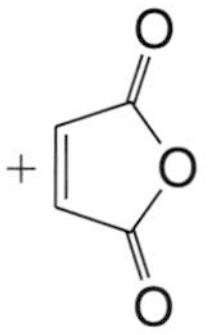

10.13

CO_2CH_3 CO_2CH_3

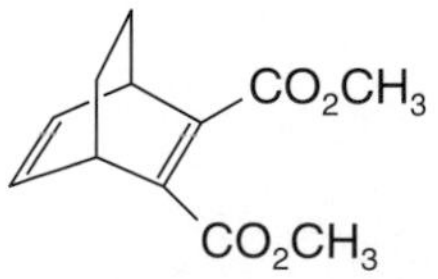

10.14 Reaction of 1,3-cyclopentadiene and dimethyl fumarate

10.15 If the electrocyclic ring opening of *cis*-3,4-dimethylcyclobutene is conrotatory but the ring closure of the resulting diene is disrotatory, the stereochemical integrity of the substances involved will be lost. Both reactions must have the same stereochemistry (conrotatory).

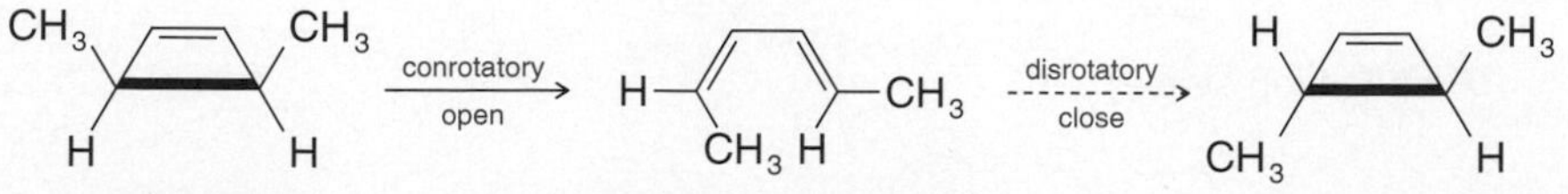

CH_3 CH_3 H H — conrotatory open → H — CH_3 CH_3 H — disrotatory close --→ H CH_3 CH_3 H

cis-3,4-Dimethylcyclobutene *cis,trans*-2,4-Hexadiene *trans*-3,4-Dimethylcyclobutene

Identical reasoning applies to the 1,3,5-hexatriene to 1,3-cyclohexadiene transformation. The ring-opening and ring-closing reactions have the same stereochemistry (disrotatory).

10.16 The two stereoisomers that undergo conrotatory closure to give 1,3-5-cyclooctatrienes in which methyl groups are trans to each other on the ring are the ones shown.

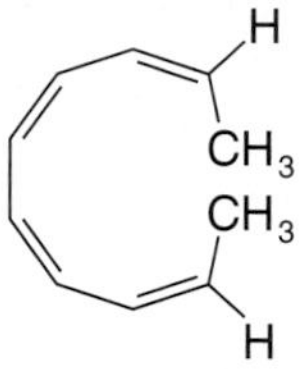

H, CH_3, CH_3, H

cis,cis,cis,cis-2,4,6,8-Decatetraene

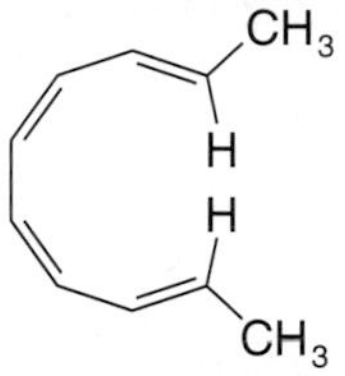

CH_3, H, H, CH_3

trans,cis,cis,trans-2,4,6,8-Decatetraene

CHAPTER 11

11.1

(*a*) CH_3 ⟷ CH_3 CO_2H ⟷ CO_2H

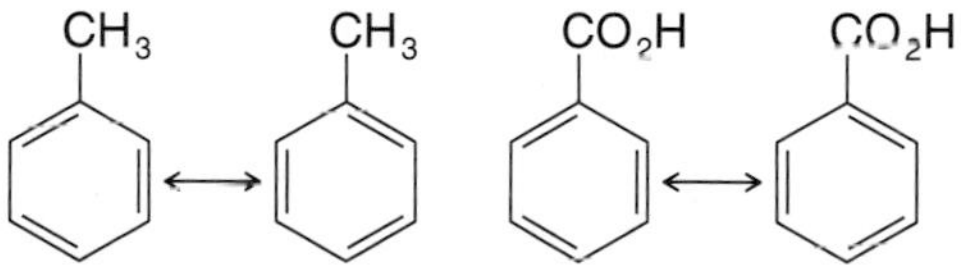

(*b*) CH_3 CO_2H

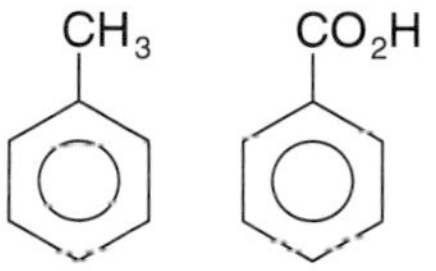

11.2 (*b*) $CH{=}CH_2$, Cl (*c*) NH_2, NO_2

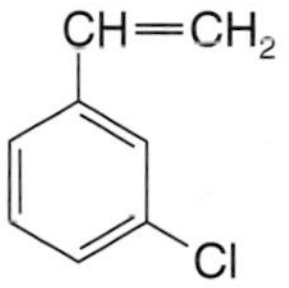

11.3

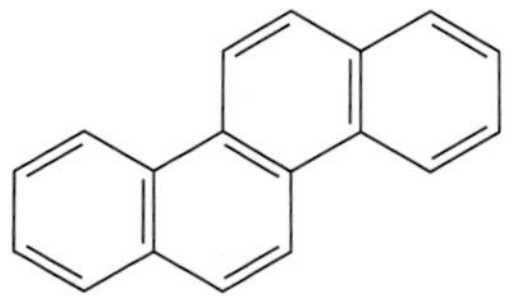

11.4 CH_3—⟨⟩—CH_3

11.5 (*b*) $BrCH_2$—⟨⟩—OCH_3, O_2N

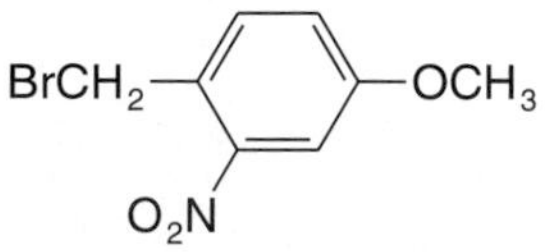

11.6 $(CH_3)_3C$—⟨⟩—CO_2H, CO_2H

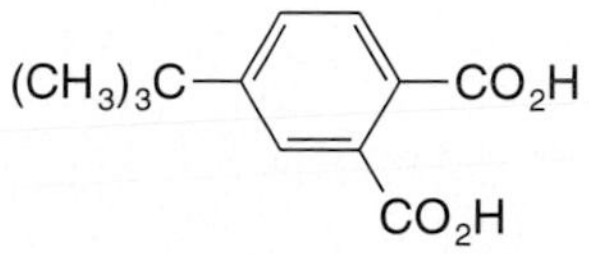

11.7 (*b*) $C_6H_5CH_2OC(CH_3)_3$; (*c*) $C_6H_5CH_2\ddot{N}{=}\overset{+}{N}{=}\ddot{N}:^{-}$; (*d*) $C_6H_5CH_2SH$; (*e*) $C_6H_5CH_2I$

11.8 1,2-Dihydronaphthalene, 101 kJ/mol (24.1 kcal/mol); 1,4-dihydronaphthalene, 113 kJ/mol (27.1 kcal/mol)

11.9 (*b*) $C_6H_5CH(CH_3)CH_2OH$; (*c*) $C_6H_5CH(OH)CH_2Br$; (*d*) $C_6H_5CH{-}CH_2$ (epoxide, O bridging) $+ C_6H_5CO_2H$

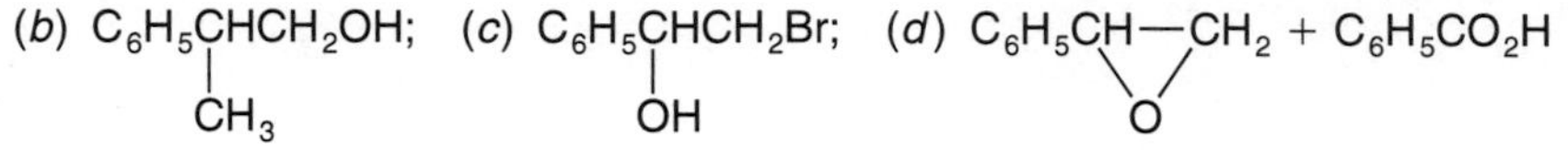

11.10 Styrene, 4393 kJ/mol (1050 kcal/mol); cyclooctatetraene, 4543 kJ/mol (1086 kcal/mol)

11.11 Diels-Alder reaction

11.12 (*b*) Five doubly occupied bonding orbitals plus two half-filled nonbonding orbitals plus five vacant antibonding orbitals

11.13 Divide the heats of combustion by the number of carbons. The two aromatic hydrocarbons (benzene and [18]-annulene) have heats of combustion per carbon that are less than those of the nonaromatic hydrocarbons (cyclooctatetraene and [16]-annulene). On a per carbon basis, the aromatic hydrocarbons have lower potential energy (are more stable) than the nonaromatic hydrocarbons.

11.14

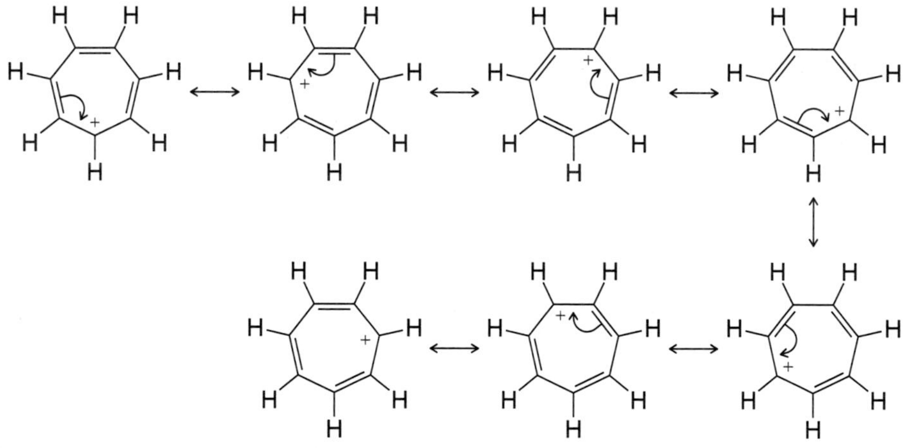

11.15

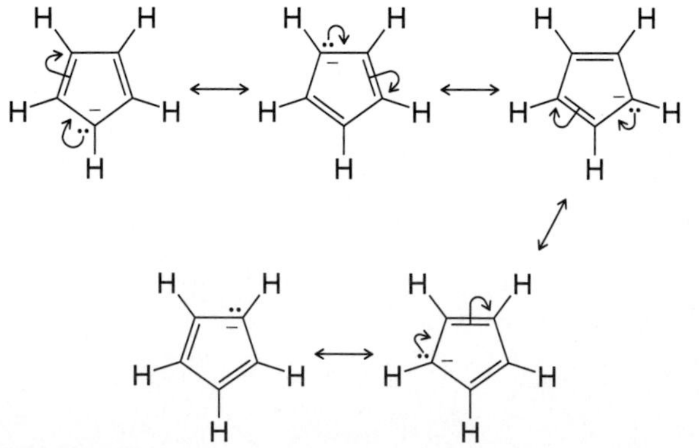

11.16 $+ :\ddot{N}H_2^{-} \longrightarrow$ $+ \ddot{N}H_3$

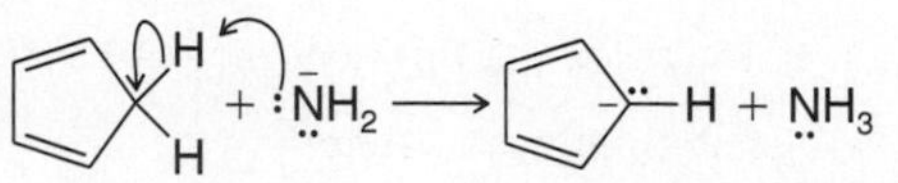

11.17 (*b*) Cyclononatetraenide anion is aromatic.

11.18 Indole is more stable than isoindole.

Six-membered ring corresponds to benzene.

Indole: more stable

Isoindole: less stable

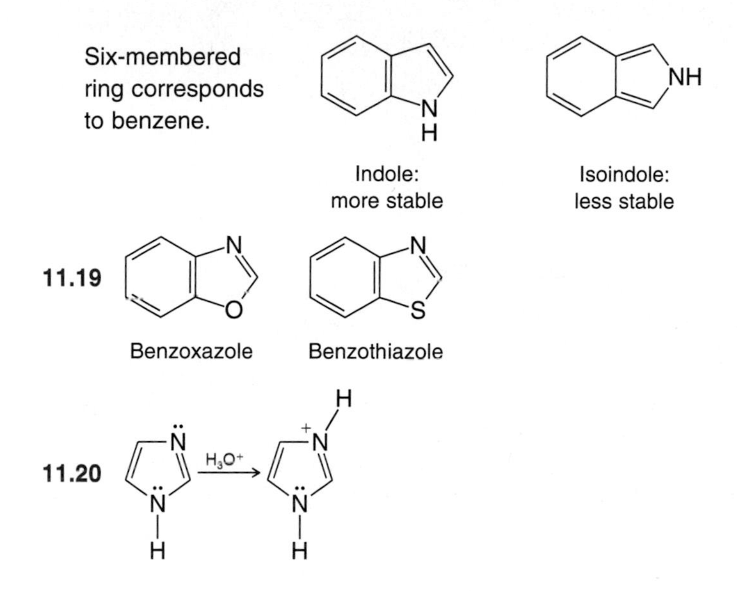

Six-membered ring does not have same pattern of bonds as benzene.

11.19

Benzoxazole Benzothiazole

11.20

CHAPTER 12

12.1 The positive charge is shared by the three carbons indicated in the three most stable resonance structures:

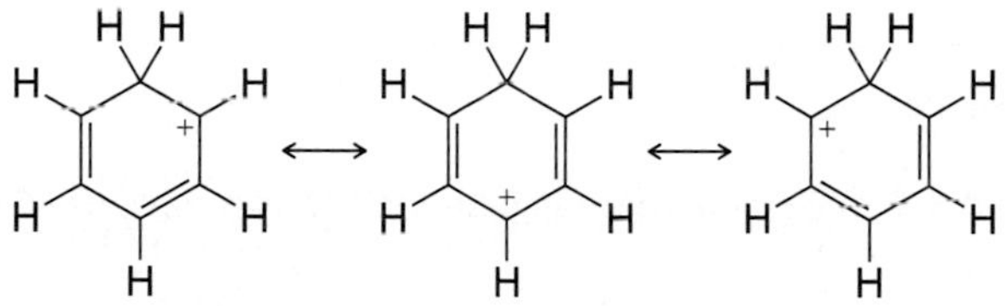

Provided that these structures contribute equally, the resonance picture coincides with the MO treatment in assigning one-third of a positive charge (+0.33) to each of the indicated carbons.

12.2

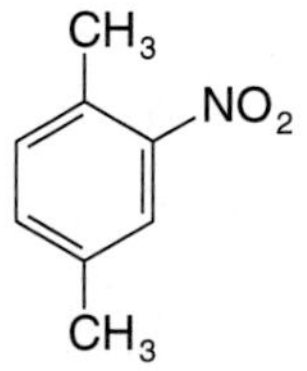

12.3

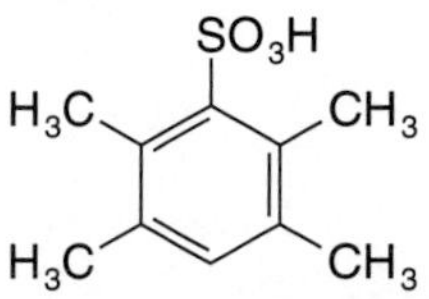

12.4 The major product is isopropylbenzene. Ionization of 1-chloropropane is accompanied by a hydride shift to give $CH_3\overset{+}{C}HCH_3$, which then attacks benzene.

12.5 $CH_3CH{=}CH_2 + H{-}F \longrightarrow CH_3\overset{+}{C}HCH_3 + F^-$;

then $CH_3\overset{+}{C}HCH_3$ + benzene ⟶ cyclohexadienyl cation (H, $CH(CH_3)_2$) $\xrightarrow{-H^+}$ $C_6H_5{-}CH(CH_3)_2$

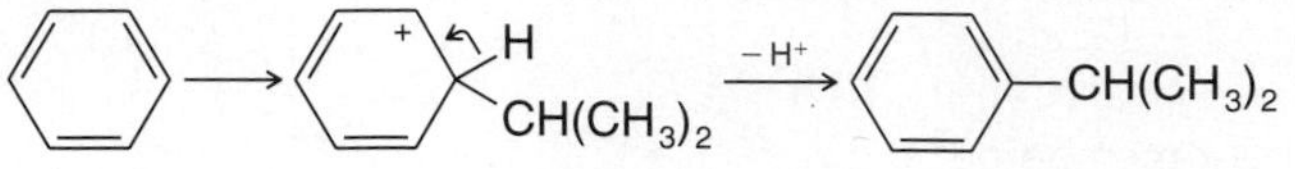

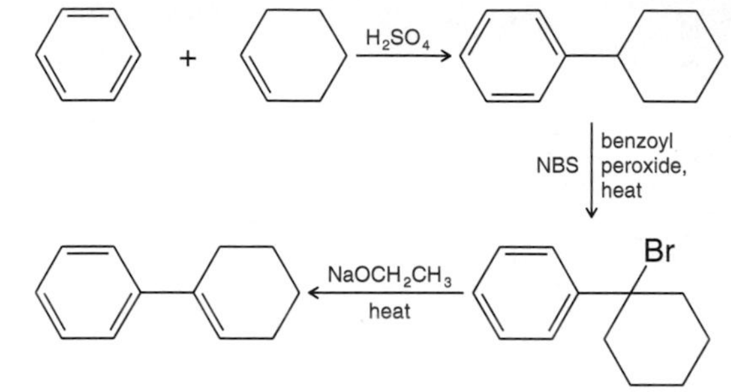

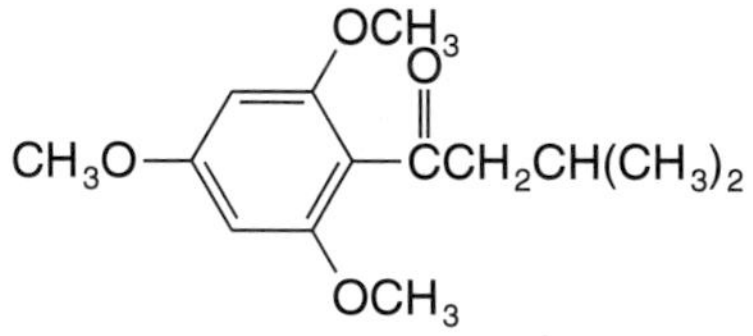

12.6 benzene + cyclohexene $\xrightarrow{H_2SO_4}$ cyclohexylbenzene $\xrightarrow{\text{NBS, benzoyl peroxide, heat}}$ 1-bromo-1-phenylcyclohexane (Br) $\xrightarrow{NaOCH_2CH_3,\ \text{heat}}$ 1-phenylcyclohexene

12.7 $CH_3O{-}C_6H_2(OCH_3)_2{-}\overset{O}{\overset{\|}{C}}CH_2CH(CH_3)_2$

12.8 $C_6H_5{-}\overset{O}{\overset{\|}{C}}CH_2CH_2\overset{O}{\overset{\|}{C}}OH$

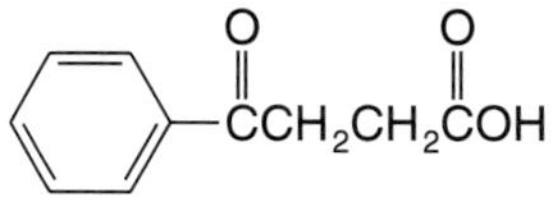

12.9 (*b*) Friedel-Crafts acylation of benzene with $(CH_3)_3C\overset{O}{\overset{\|}{C}}Cl$, followed by reduction with Zn(Hg) and hydrochloric acid

12.10 (*b*) Toluene is 1.7 times more reactive than *tert*-butylbenzene. (*c*) Ortho (10%), meta (6.7%), para (83.3%)

12.11

$-CH_2Cl$	$-CHCl_2$	$-CCl_3$
Deactivating	Deactivating	Deactivating
ortho-para directing	ortho-para directing	meta directing

12.12 (*b*) Resonance structures of the cyclohexadienyl cation (:NH_2, Br, H) ⟷ ⟷

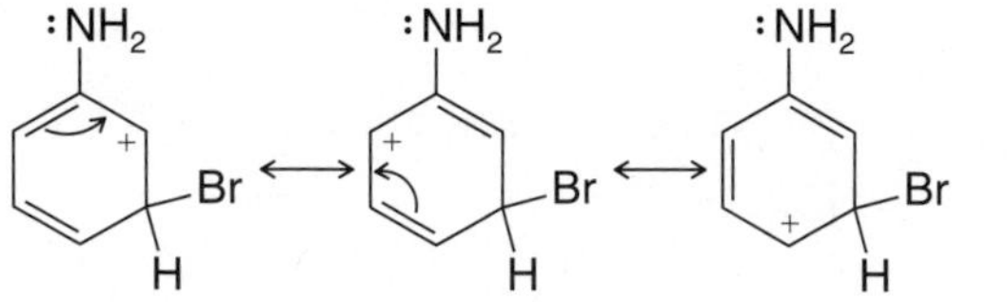

(*c*) Resonance structures of the cyclohexadienyl cation (:NH_2, H, Br) ⟷ ⟷ ($^+NH_2$) ⟷

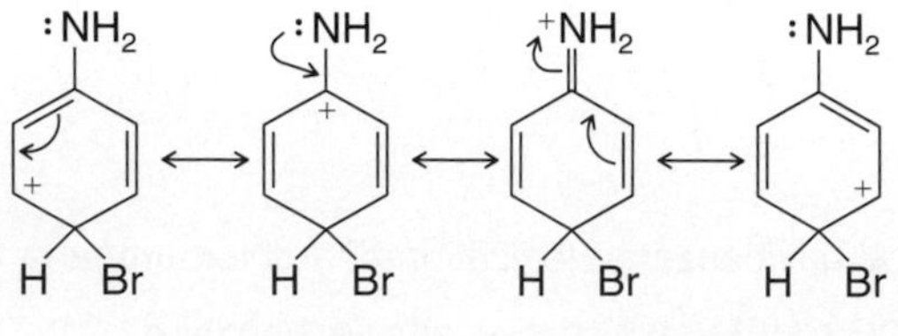

12.13 [2-nitrobiphenyl] and [4-nitrobiphenyl, $-NO_2$]

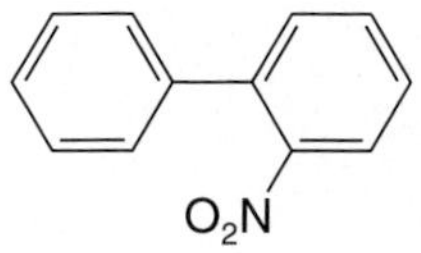

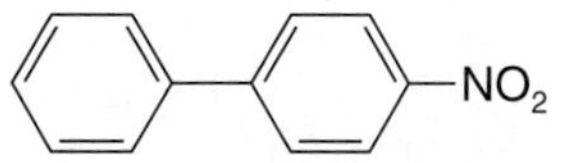

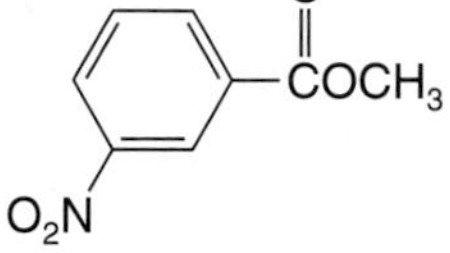

12.14 (*b*) [m-O_2N–C_6H_4–$\overset{O}{\|}{C}OCH_3$] (*c*) [m-O_2N–C_6H_4–$\overset{O}{\|}{C}CH_2CH_3$]

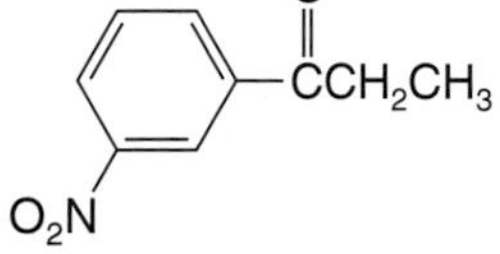

12.15 The group $-\overset{+}{N}(CH_3)_3$ is strongly deactivating and meta-directing. Its positively charged nitrogen makes it a powerful electron-withdrawing substituent. It resembles a nitro group.

12.16 [Cl (ortho)–C_6H_4–CH_2–C_6H_4–Cl] and [Cl–C_6H_4–CH_2–C_6H_4–Cl]

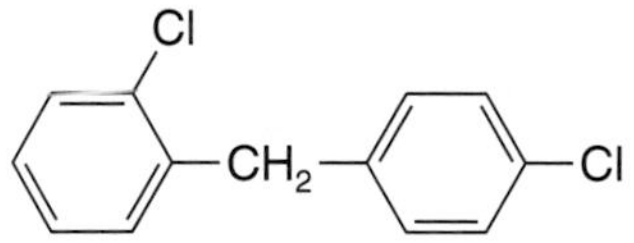

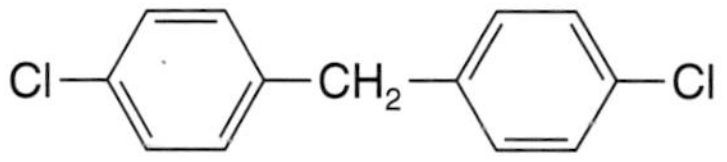

12.17 (*b*) [Cl, Cl, NO_2 substituted benzene] (*c*) [NO_2, O_2N, NO_2 substituted benzene] (*d*) $CH_3\overset{O}{\|}{C}$–[benzene with NO_2]–OCH_3

(*e*) CH_3–[benzene with NO_2]–OCH_3 (*f*) [benzene with OCH_3, Br, Br, NO_2]

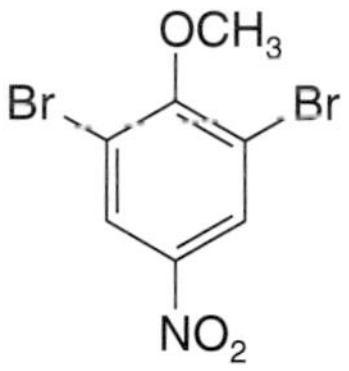

12.18 *m*-Bromonitrobenzene:

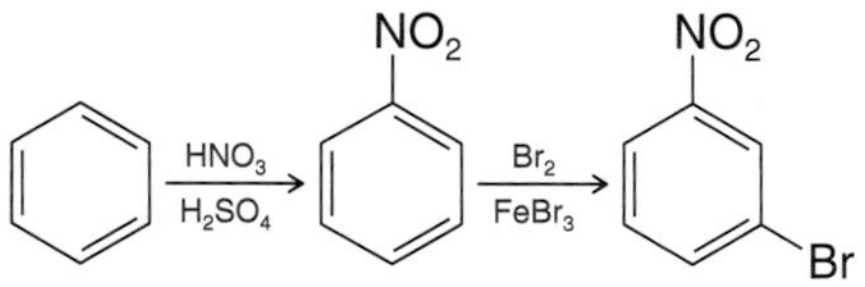

$$C_6H_6 \xrightarrow[H_2SO_4]{HNO_3} C_6H_5NO_2 \xrightarrow[FeBr_3]{Br_2} m\text{-}BrC_6H_4NO_2$$

p-Bromonitrobenzene:

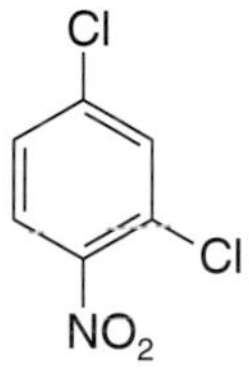

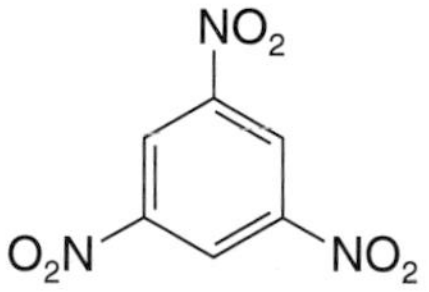

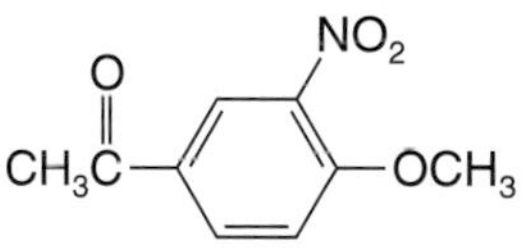

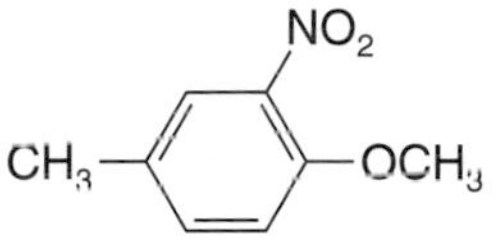

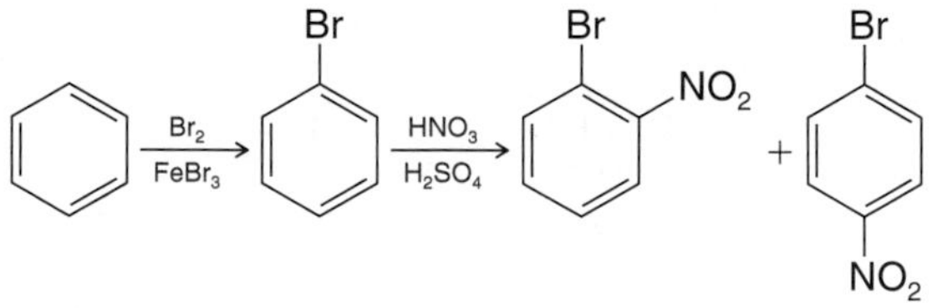

$$C_6H_6 \xrightarrow[FeBr_3]{Br_2} C_6H_5Br \xrightarrow[H_2SO_4]{HNO_3} o\text{-}BrC_6H_4NO_2 + p\text{-}BrC_6H_4NO_2$$

12.19

$$C_6H_5CH_3 \xrightarrow[H_2SO_4,\ H_2O,\ heat]{Na_2Cr_2O_7} C_6H_5CO_2H \xrightarrow[H_2SO_4]{HNO_3} m\text{-}O_2NC_6H_4CO_2H$$

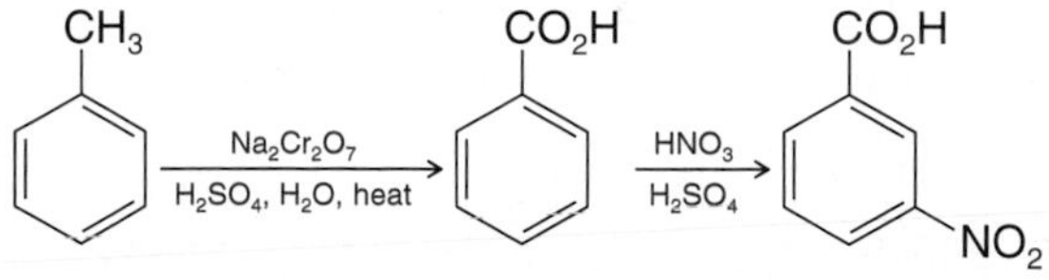

12.20 [structures: naphthalenesulfonic acids; H, SO_3H — Formed faster + H, SO_3H — More stable]

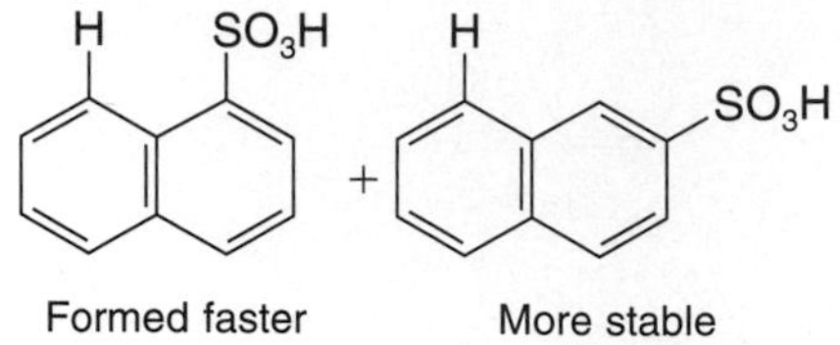

Formed faster More stable

The hydrogen at C-8 (the one shown in the structural formulas) crowds the —SO_3H group in the less stable isomer.

12.21 [structure: thiophene ring, S, SO_3H]

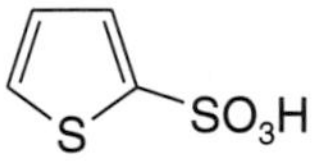

CHAPTER 13

13.1 (*b*) 5.37 ppm; (*c*) 3.07 ppm

13.2 (*b*) Five; (*c*) two; (*d*) two; (*e*) three; (*f*) one; (*g*) four; (*h*) three

13.3 (*b*) One; (*c*) one; (*d*) one; (*e*) four; (*f*) four

13.4 (*b*) One signal (singlet); (*c*) two signals (doublet and triplet); (*d*) two signals (both singlets); (*e*) two signals (doublet and quartet)

13.5 (*b*) Three signals (singlet, triplet, and quartet); (*c*) two signals (triplet and quartet); (*d*) three signals (singlet, triplet, and quartet); (*e*) four signals (three triplets and quartet)

13.6 $Cl_2CHCH(OCH_2CH_3)_2$

13.7 Both H_b and H_c appear as doublets of doublets:

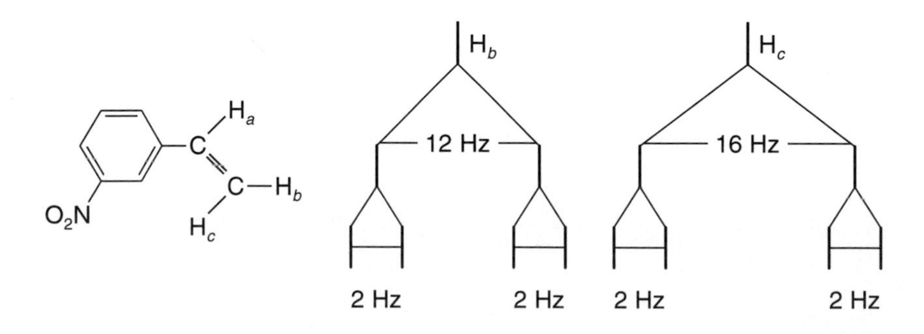

13.8 (*b*) The signal for the proton at C-2 is split into a quartet by the methyl protons, and each line of this quartet is split into a doublet by the aldehyde proton. It appears as a doublet of quartets.

13.9 (*b*) Six; (*c*) six; (*d*) nine; (*e*) three

13.10

(*b*) 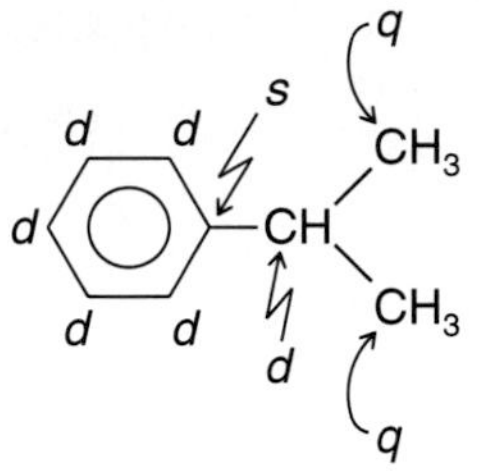 (*d*) 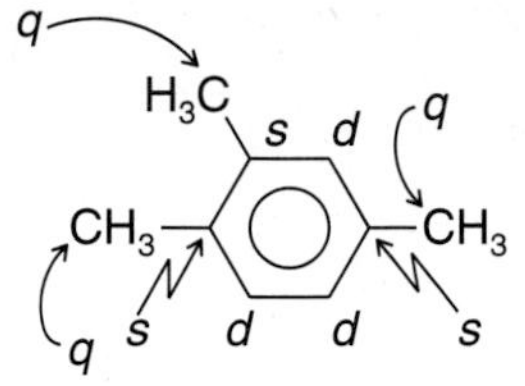 (*c*) 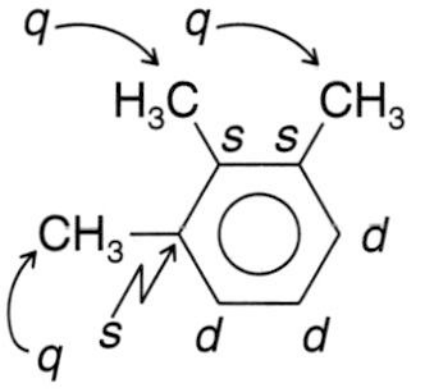 (*e*)

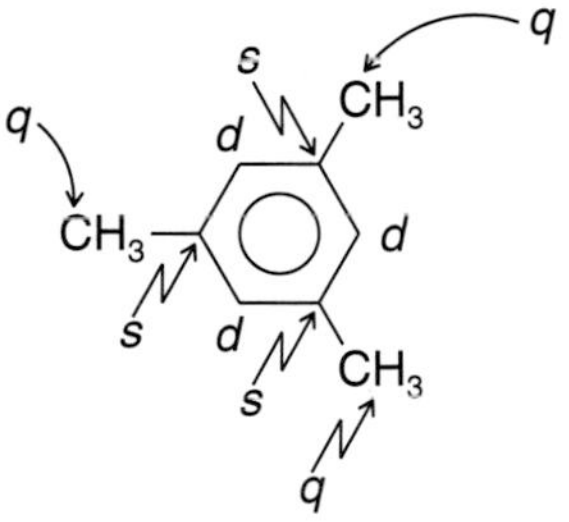

13.11 1,2,4-Trimethylbenzene

13.12 Benzyl alcohol. Infrared spectrum has peaks for O—H and sp^3 C—H; lacks peak for C=O.

13.13 (*b*) Three peaks (*m/z* 146, 148, and 150); (*c*) three peaks (*m/z* 234, 236, and 238); (*d*) three peaks (*m/z* 190, 192, and 194)

13.14

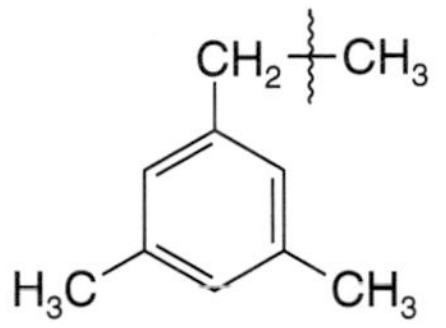

Base peak $C_9H_{11}^+$ (*m/z* 119)

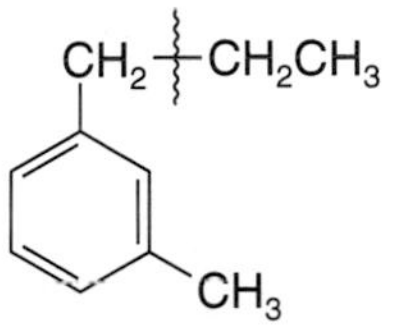

Base peak $C_8H_9^+$ (*m/z* 105)

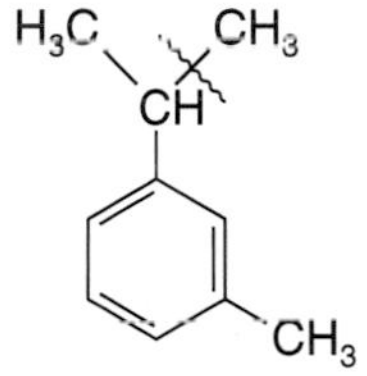

Base peak $C_9H_{11}^+$ (*m/z* 119)

CHAPTER 14

14.1 (*b*) Cyclohexylmagnesium chloride; (*c*) iodomethylzinc iodide

14.2

(*b*) $CH_3CH(Br)CH_2CH_3 + 2Li \longrightarrow CH_3CH(Li)CH_2CH_3 + LiBr$

(*c*) $C_6H_5CH_2Br + 2Na \longrightarrow C_6H_5CH_2Na + NaBr$

14.3 (*b*) $CH_2{=}CHCH_2MgCl$; (*c*) cyclobutyl—MgI; (*d*) cyclohexenyl—MgBr

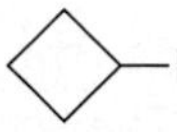

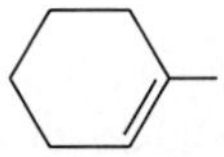

14.4 (*b*) $CH_3(CH_2)_4CH_2OH + CH_3CH_2CH_2CH_2Li \longrightarrow CH_3CH_2CH_2CH_3 + CH_3(CH_2)_4CH_2OLi$; (*c*) $C_6H_5SH + CH_3CH_2CH_2CH_2Li \longrightarrow CH_3CH_2CH_2CH_3 + C_6H_5SLi$

14.5 (*b*) $C_6H_5CHCH_2CH_2CH_3$ (OH on C-1); (*c*) 1-propylcyclohexanol ($CH_2CH_2CH_3$ and OH on the same ring carbon); (*d*) $CH_3CH_2CH_2C(OH)(CH_3)CH_2CH_3$

14.6

(*b*) $CH_3MgI + C_6H_5\overset{O}{\overset{\|}{C}}CH_3 \xrightarrow[2.\ H_3O^+]{1.\ \text{diethyl ether}} C_6H_5C(OH)(CH_3)CH_3$

and $C_6H_5MgBr + CH_3\overset{O}{\overset{\|}{C}}CH_3 \xrightarrow[2.\ H_3O^+]{1.\ \text{diethyl ether}} C_6H_5C(OH)(CH_3)CH_3$

14.7 (*b*) $2C_6H_5MgBr$ + cyclopropyl$-\overset{O}{\overset{\|}{C}}OCH_2CH_3$

14.8 (*b*) $LiCu(CH_3)_2$ + 1-bromo-3,3-dimethylcyclopentene (Br on the double-bond carbon; two CH_3 groups on C-3)

14.9 (*b*) methylenecyclobutane (cyclobutane$=CH_2$)

14.10 *cis*-2-Butene ⟶ dibromocyclopropane with the two CH_3 groups cis (H, H; Br, Br on the third carbon)   *trans*-2-Butene ⟶ dibromocyclopropane with the two CH_3 groups trans (Br, Br on the third carbon)

14.11 2-Pentanol and 3-pentanol

14.12 (*b*) methylenecyclopentane (CH_2) and 1-methylcyclopentene (CH_3)   (*c*) *cis*-3-Hexene and *trans*-3-hexene

14.13

(*b*) $CH_2{=}CH_2 \xrightarrow[2.\ NaBH_4,\ HO^-]{1.\ Hg(OAc)_2,\ CH_3CH(OH)CH_2CH_2CH_2CH_3} CH_3CH(OCH_2CH_3)CH_2CH_2CH_2CH_3$

and $CH_2{=}CHCH_2CH_2CH_2CH_3 \xrightarrow[2.\ NaBH_4,\ HO^-]{1.\ Hg(OAc)_2,\ CH_3CH_2OH} CH_3CH(OCH_2CH_3)CH_2CH_2CH_2CH_3$

CHAPTER 15

15.1 The primary alcohols $CH_3CH_2CH_2CH_2OH$ and $(CH_3)_2CHCH_2OH$ can each be prepared by hydrogenation of an aldehyde. The secondary alcohol $CH_3CH(OH)CH_2CH_3$ can be prepared by hydrogenation of a ketone. The tertiary alcohol $(CH_3)_3COH$ cannot be prepared by hydrogenation.

15.2 (*b*) $CH_3C(D)(OD)CH_3$; (*c*) $C_6H_5C(D)(H)OH$; (*d*) DCH_2OD

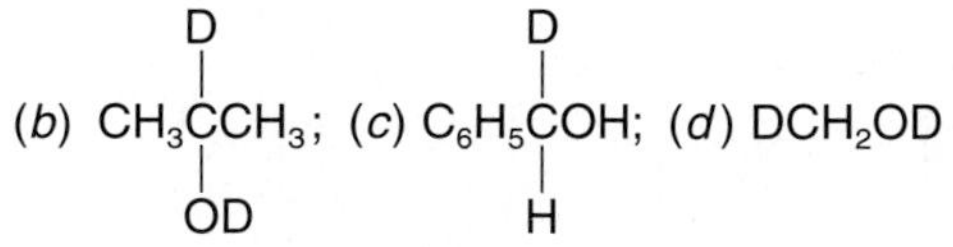

15.3 $CH_3CH_2\overset{O}{\overset{\|}{C}}OCH(CH_3)_2$

15.4 (*b*) (cyclohexyl)MgBr

15.5 $HO\overset{O}{\overset{\|}{C}}CH_2\underset{CH_3}{\underset{|}{C}}HCH_2\overset{O}{\overset{\|}{C}}OH \xrightarrow[2.\ H_2O]{1.\ LiAlH_4} HOCH_2CH_2\underset{CH_3}{\underset{|}{C}}HCH_2CH_2OH$

$CH_3O\overset{O}{\overset{\|}{C}}CH_2\underset{CH_3}{\underset{|}{C}}HCH_2\overset{O}{\overset{\|}{C}}OCH_3 \xrightarrow[2.\ H_2O]{1.\ LiAlH_4} HOCH_2CH_2\underset{CH_3}{\underset{|}{C}}HCH_2CH_2OH + 2CH_3OH$

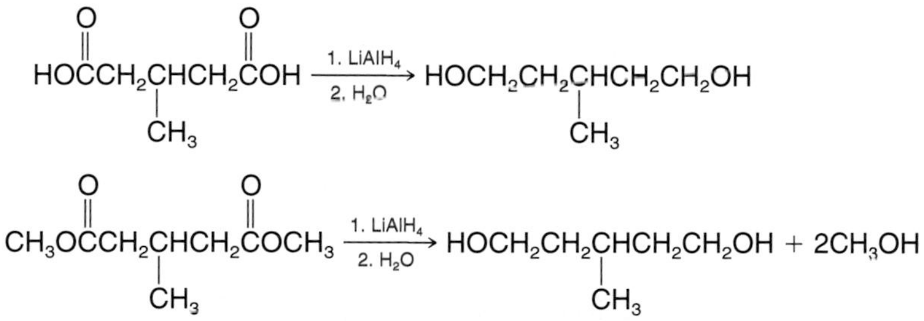

15.6 *cis*-2-Butene yields the meso stereoisomer of 2,3-butanediol:

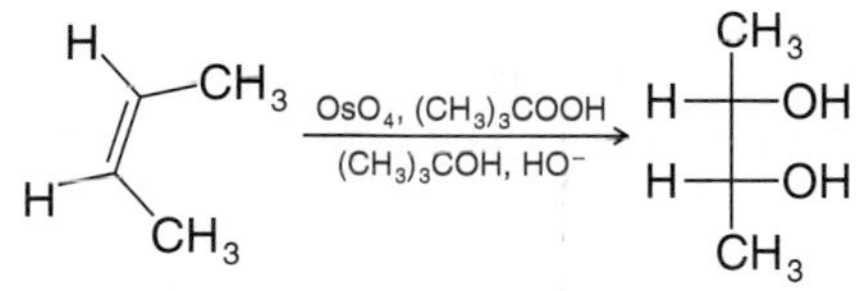

cis-2-Butene $\xrightarrow[(CH_3)_3COH,\ HO^-]{OsO_4,\ (CH_3)_3COOH}$ Fischer projection: CH_3 top; H—C—OH; H—C—OH; CH_3 bottom

trans-2-Butene gives equal quantities of the two enantiomers of the chiral diol:

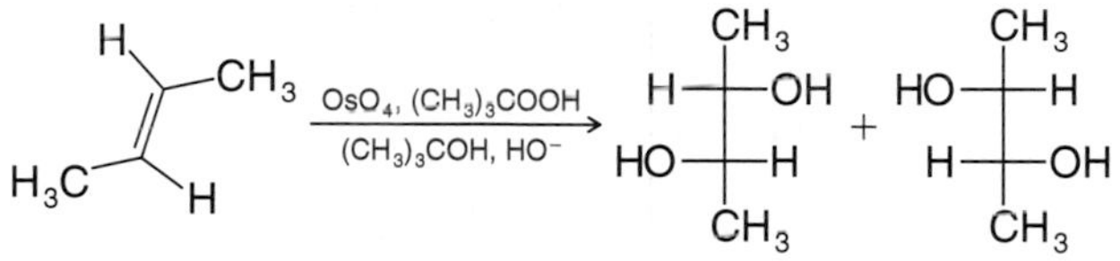

trans-2-Butene $\xrightarrow[(CH_3)_3COH,\ HO^-]{OsO_4,\ (CH_3)_3COOH}$ (CH_3; H—C—OH; HO—C—H; CH_3) + (CH_3; HO—C—H; H—C—OH; CH_3)

15.7

Step 1: $H\ddot{O}CH_2CH_2CH_2CH_2CH_2—\ddot{O}H + H—\ddot{O}SO_2OH \longrightarrow$

$H\ddot{O}CH_2CH_2CH_2CH_2CH_2—\overset{+}{O}H_2 + {}^{-}:\ddot{O}SO_2OH$

Step 2: (cyclization; ring closure with loss of water) $\longrightarrow$ (protonated oxane, $\overset{+}{O}$—H) $+ H_2O$

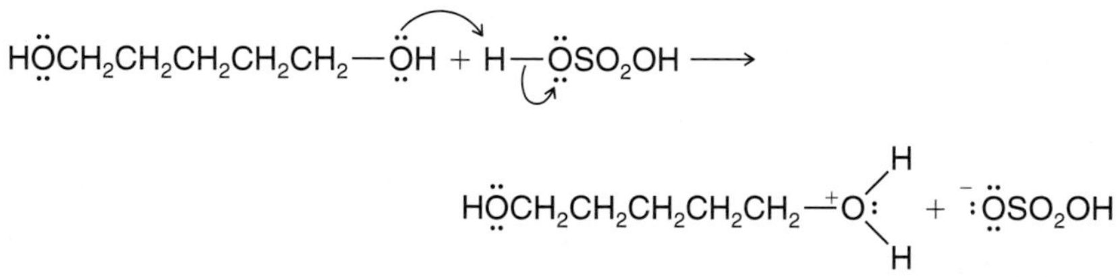

Step 3: (protonated oxane) $+ {}^{-}:\ddot{O}SO_2OH \longrightarrow$ (oxane) $+ H—\ddot{O}SO_2OH$

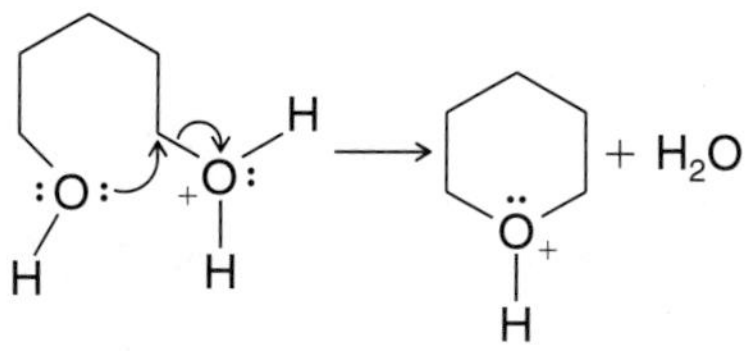

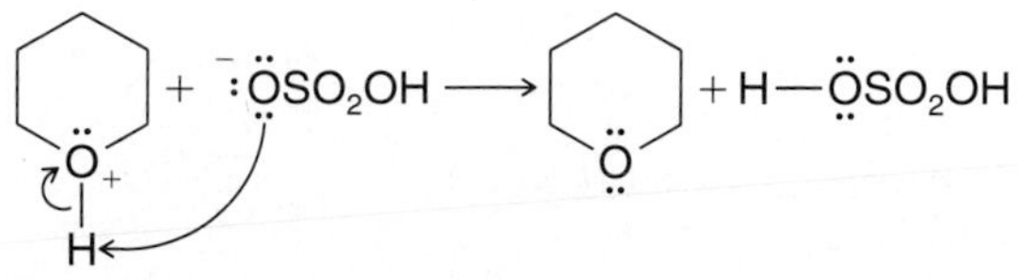

15.8 (*b*) $CH_3OC(=O)-C_6H_4-C(=O)OCH_3$

15.9 $(CH_3)_3C$–cyclohexyl–OH (axial) $\xrightarrow{\text{acetic anhydride}}$ $(CH_3)_3C$–cyclohexyl–$OC(=O)CH_3$ (axial)

$(CH_3)_3C$–cyclohexyl–OH (equatorial) $\xrightarrow{\text{acetic anhydride}}$ $(CH_3)_3C$–cyclohexyl–$OC(=O)CH_3$ (equatorial)

15.10 $O_2NOCH_2CH(ONO_2)CH_2ONO_2$

15.11 (*b*) $CH_3C(=O)(CH_2)_5CH_3$; (*c*) $CH_3(CH_2)_5CH(=O)$

15.12 (*b*) One; (*c*) none

15.13 (*b*) $(CH_3)_2CHCH_2CH(=O) + C_6H_5CH_2CH(=O)$; (*c*) cyclopentanone (cyclopentyl=O) + $HCH(=O)$

15.14 (*S*)-2-Butanol gives (*S*)-*sec*-butyl *p*-toluenesulfonate, which gives (*R*)-2-butanethiol.

15.15 The peak at *m/z* 70 corresponds to loss of water from the molecular ion. The peaks at *m/z* 59 and 73 correspond to the cleavages indicated:

$CH_3 \nmid C(\overset{+\cdot}{\ddot{O}H})(CH_3) \nmid CH_2CH_3$ (cleavage of the $C-CH_2CH_3$ bond gives 59; cleavage of the CH_3-C bond gives 73)

CHAPTER 16

16.1 (*b*) CH_2-CHCH_2Cl (epoxide, ring O); (*c*) $CH_2-CHCH=CH_2$ (epoxide, ring O)

16.2 1,2-Epoxybutane, 2546 kJ/mol (609.1 kcal/mol); tetrahydrofuran, 2499 kJ/mol (597.8 kcal/mol)

16.3 $R_2\ddot{O}\text{:}\cdots\cdots H-\ddot{\underset{\cdot\cdot}{O}}-R$

16.4 1,4-Dioxane

16.5 $(CH_3)_2C{=}CH_2 \xrightarrow{H^+} (CH_3)_2\overset{+}{C}{-}CH_3 \xrightarrow{H\ddot{O}CH_3} (CH_3)_3C{-}\overset{+}{\ddot{O}}CH_3(H) \xrightarrow{-H^+} (CH_3)_3C\ddot{O}CH_3$

16.6 $C_6H_5CH_2ONa + CH_3CH_2Br \longrightarrow C_6H_5CH_2OCH_2CH_3 + NaBr$
and $CH_3CH_2ONa + C_6H_5CH_2Br \longrightarrow C_6H_5CH_2OCH_2CH_3 + NaBr$

16.7

(*b*) $(CH_3)_2CHONa + CH_2{=}CHCH_2Br \longrightarrow CH_2{=}CHCH_2OCH(CH_3)_2 + NaBr$
(*c*) $(CH_3)_3COK + C_6H_5CH_2Br \longrightarrow (CH_3)_3COCH_2C_6H_5 + KBr$

16.8 $CH_3CH_2OCH_2CH_3 + 6O_2 \longrightarrow 4CO_2 + 5H_2O$

16.9 (*b*) $C_6H_5CH_2OCH_2C_6H_5$; (*c*) [tetrahydropyran ring with O]

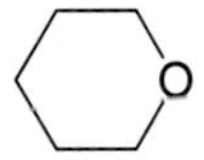

16.10 $(CH_3)_3COC(CH_3)_3 \xrightarrow{HCl} (CH_3)_3C\overset{+}{O}(H)C(CH_3)_3$

$$(CH_3)_3C{-}\overset{+}{O}(H)C(CH_3)_3 \longrightarrow (CH_3)_3C^+ + HOC(CH_3)_3$$

$$(CH_3)_3C^+ + Cl^- \longrightarrow (CH_3)_3CCl$$

$$(CH_3)_3COH + HCl \longrightarrow (CH_3)_3CCl + H_2O$$

16.11 Only the trans epoxide is chiral. As formed in this reaction, neither product is optically active.

16.12 (*b*) $N_3CH_2CH_2OH$; (*c*) $HOCH_2CH_2OH$; (*d*) $C_6H_5CH_2CH_2OH$;
(*e*) $CH_3CH_2C{\equiv}CCH_2CH_2OH$

16.13 Compound B

16.14 Compound A

16.15 *trans*-2-Butene gives *meso*-2,3-butanediol on epoxidation followed by acid-catalyzed hydrolysis. *cis*-2-Butene gives *meso*-2,3-butanediol on osmium tetraoxide hydroxylation.

16.16 The product has the *S* configuration.

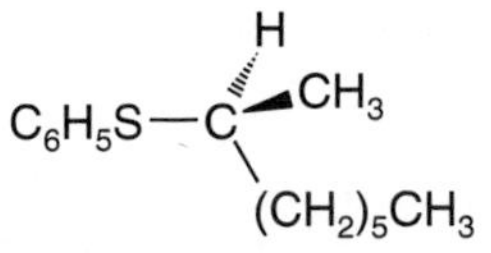

$C_6H_5S{-}C(H)(CH_3)(CH_2)_5CH_3$

16.17 Phenyl vinyl sulfoxide is chiral. Phenyl vinyl sulfone is achiral.

16.18 $CH_3SCH_3 + CH_3(CH_2)_{10}CH_2I$ will yield the same sulfonium salt. This combination is not as effective as $CH_3I + CH_3(CH_2)_{10}CH_2SCH_3$, because the reaction mechanism is S_N2 and CH_3I is more reactive than $CH_3(CH_2)_{10}CH_2I$ in reactions of this type because it is less crowded.

16.19 $CH_2{=}\overset{+}{\ddot{O}}CH(CH_3)CH_2CH_3$

CHAPTER 17

17.1 (*b*) Pentanedial; (*c*) 3-phenyl-2-propenal; (*d*) 4-hydroxy-3-methoxybenzaldehyde

17.2 (*b*) 2-Methyl-3-pentanone; (*c*) 4,4-dimethyl-2-pentanone; (*d*) 4-penten-2-one

17.3 No. Carboxylic acids are inert to catalytic hydrogenation.

17.4

$$CH_3\overset{O}{\overset{\|}{C}}OH \xrightarrow[2.\ H_2O]{1.\ LiAlH_4} CH_3CH_2OH \xrightarrow[CH_2Cl_2]{PCC} CH_3\overset{O}{\overset{\|}{C}}H$$

$$CH_3CH_2OH \xrightarrow[heat]{HBr} CH_3CH_2Br \xrightarrow[diethyl\ ether]{Mg} CH_3CH_2MgBr$$

$$CH_3CH_2MgBr + CH_3\overset{O}{\overset{\|}{C}}H \xrightarrow[2.\ H_3O^+]{1.\ diethyl\ ether} CH_3\overset{OH}{\overset{|}{C}}HCH_2CH_3 \xrightarrow[CH_2Cl_2]{PCC} CH_3\overset{O}{\overset{\|}{C}}CH_2CH_3$$

17.5 $Cl_3CCH(OH)_2$

17.6 $CH_2{=}C(CH_3)C{\equiv}N$

17.7

Step 1: $C_6H_5CH{=}\ddot{O}: + H{-}\overset{+}{O}(H)CH_2CH_3 \rightleftharpoons C_6H_5CH{=}\overset{+}{\ddot{O}}{-}H + :\ddot{O}(H)CH_2CH_3$

Step 2: $C_6H_5CH{=}\overset{+}{\ddot{O}}{-}H + :\ddot{O}(H)CH_2CH_3 \rightleftharpoons C_6H_5\overset{:\ddot{O}H}{\overset{|}{C}}H{-}\overset{+}{O}(H)CH_2CH_3$

Step 3: $C_6H_5\overset{:\ddot{O}H}{\overset{|}{C}}H{-}\overset{+}{O}(H)CH_2CH_3 + :\ddot{O}(H)CH_2CH_3 \rightleftharpoons$

$$C_6H_5\overset{:\ddot{O}H}{\overset{|}{C}}H{-}\ddot{O}CH_2CH_3 + H{-}\overset{+}{O}(H)CH_2CH_3$$

Formation of the hemiacetal is followed by loss of water to give a carbocation.

Step 4: $C_6H_5\underset{H}{\overset{H\ddot{O}:}{C}}-\ddot{O}CH_2CH_3 + H-\overset{+}{O}(H)CH_2CH_3 \rightleftharpoons C_6H_5\underset{H}{\overset{H\overset{+}{\ddot{O}}H}{C}}-\ddot{O}CH_2CH_3 + :\ddot{O}(H)CH_2CH_3$

Step 5: $C_6H_5\overset{H_2\overset{+}{\ddot{O}}}{C}H-\ddot{O}CH_2CH_3 \rightleftharpoons C_6H_5\overset{+}{C}H-\ddot{O}CH_2CH_3 + H-\ddot{O}-H$

Step 6: $C_6H_5\overset{+}{C}H-\ddot{O}CH_2CH_3 + :\ddot{O}(H)CH_2CH_3 \rightleftharpoons C_6H_5CH(:\overset{+}{O}(H)CH_2CH_3)-\ddot{O}CH_2CH_3$

Step 7: $C_6H_5CH(:\overset{+}{O}(H)CH_2CH_3)-\ddot{O}CH_2CH_3 + :\ddot{O}(H)CH_2CH_3 \rightleftharpoons C_6H_5CH(:\ddot{O}CH_2CH_3)-\ddot{O}CH_2CH_3 + H-\overset{+}{\ddot{O}}(H)CH_2CH_3$

17.8 (*b*) O, O (six-membered ring); C_6H_5, H (*c*) O, O (five-membered ring); $(CH_3)_2CHCH_2$, CH_3 (*d*) H_3C, CH_3; O, O (six-membered ring); $(CH_3)_2CHCH_2$, CH_3

17.9

Step 1: $C_6H_5CH(:\ddot{O}CH_2CH_3)-\ddot{O}CH_2CH_3 + H-\overset{+}{\ddot{O}}(H)CH_2CH_3 \rightleftharpoons C_6H_5CH(:\overset{+}{O}(H)CH_2CH_3)-\ddot{O}CH_2CH_3 + :\ddot{O}(H)CH_2CH_3$

Step 2: $C_6H_5CH(\overset{+}{\ddot{O}}(H)CH_2CH_3)-\ddot{O}CH_2CH_3 \rightleftharpoons C_6H_5\overset{+}{C}H-\ddot{O}CH_2CH_3 + :\ddot{O}(H)CH_2CH_3$

Step 3: $C_6H_5\overset{+}{C}H-\ddot{O}CH_2CH_3 + :\ddot{O}H_2 \rightleftharpoons C_6H_5CH(\overset{+}{O}H_2)-\ddot{O}CH_2CH_3$

Step 4: $C_6H_5CH(\overset{+}{O}H_2)-\ddot{O}CH_2CH_3 + :\ddot{O}(H)CH_2CH_3 \rightleftharpoons C_6H_5CH(\ddot{O}H)-\ddot{O}CH_2CH_3 + H-\overset{+}{O}(H)CH_2CH_3$

Step 5: $C_6H_5CH(\ddot{O}H)-\ddot{O}CH_2CH_3 + H-\overset{+}{O}(H)CH_2CH_3 \rightleftharpoons C_6H_5CH(\ddot{O}H)-\overset{+}{O}(H)CH_2CH_3 + :\ddot{O}(H)CH_2CH_3$

Step 6: $C_6H_5CH(\ddot{O}H)-\overset{+}{O}(H)CH_2CH_3 \rightleftharpoons C_6H_5CH{=}\overset{+}{O}-H + :\ddot{O}(H)CH_2CH_3$

Step 7: $C_6H_5CH{=}\overset{+}{O}-H + :\ddot{O}(H)CH_2CH_3 \rightleftharpoons C_6H_5CH{=}\ddot{O}: + H-\overset{+}{O}(H)CH_2CH_3$

17.10

$CH_3C(O)-C_6H_4-COOH \xrightarrow{HOCH_2CH_2OH,\ H^+,\ heat}$ (2-methyl-1,3-dioxolan-2-yl)$-C_6H_4-COOH \xrightarrow{1.\ LiAlH_4;\ 2.\ H_2O}$ (2-methyl-1,3-dioxolan-2-yl)$-C_6H_4-CH_2OH \xrightarrow{H_2O,\ H^+,\ heat} CH_3C(O)-C_6H_4-CH_2OH$

17.11

(*b*) $C_6H_5CH(OH)NHCH_2CH_2CH_2CH_3 \longrightarrow C_6H_5CH{=}NCH_2CH_2CH_2CH_3$

(*c*) 1-(tert-butylamino)cyclohexanol [$C_6H_{10}(OH)NHC(CH_3)_3$] $\longrightarrow$ $C_6H_{10}{=}NC(CH_3)_3$

(*d*) $C_6H_5C(OH)(CH_3)-NH-C_6H_{11} \longrightarrow C_6H_5C(CH_3){=}N-C_6H_{11}$

17.12

(*b*) $CH_3CH_2C(N(CH_2)_4)(OH)CH_2CH_3 \longrightarrow CH_3CH{=}C(N(CH_2)_4)CH_2CH_3$

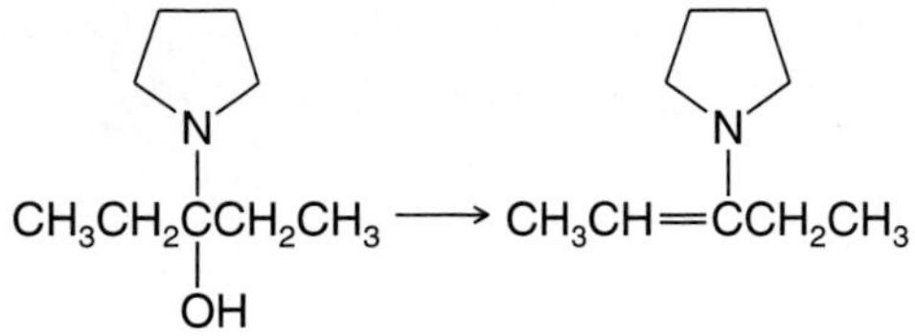

(*c*) $C_6H_5C(N(CH_2)_5)(OH)CH_3 \longrightarrow C_6H_5C(N(CH_2)_5){=}CH_2$

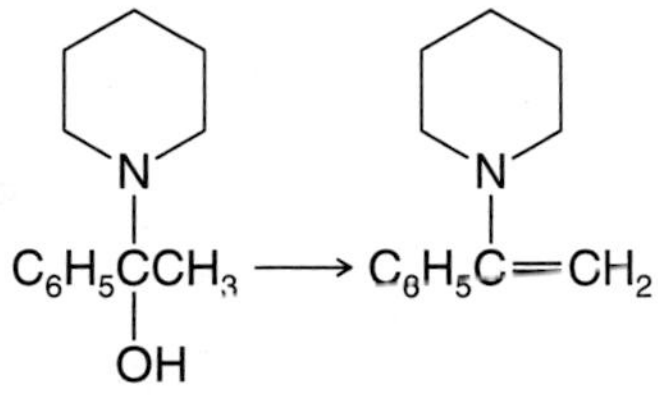

17.13 (*b*) $CH_3CH_2CH_2CH{=}CHCH{=}CH_2$; (*c*) $C_6H_{11}C(CH_3){=}CH_2$

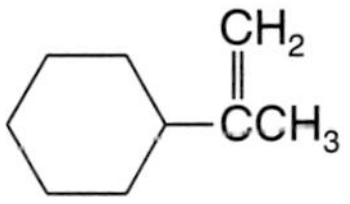

17.14

(*b*) $CH_3CH_2CH_2CH{=}O + (C_6H_5)_3\overset{+}{P}{-}\ddot{C}H_2^-$ or $HCH{=}O + CH_3CH_2CH_2\ddot{C}H^-{-}\overset{+}{P}(C_6H_5)_3$

17.15 $CH_3CH(Br)CH_2CH_3 + (C_6H_5)_3P{:} \longrightarrow CH_3CH(\overset{+}{P}(C_6H_5)_3)CH_2CH_3\ Br^-$

$CH_3CH(\overset{+}{P}(C_6H_5)_3)CH_2CH_3\ Br^- \xrightarrow[DMSO]{NaCH_2\overset{O}{\overset{\|}{S}}CH_3} CH_3\ddot{C}(\overset{+}{P}(C_6H_5)_3)CH_2CH_3$

17.16 $C_6H_{11}COCH_3 + C_6H_5COOOH \longrightarrow C_6H_{11}C(OH)(CH_3)OOCC_6H_5$

$C_6H_{11}C(CH_3)(OH){-}O{-}OCC_6H_5 \longrightarrow CH_3C(=O){-}O{-}C_6H_{11} + HOC(=O)C_6H_5$

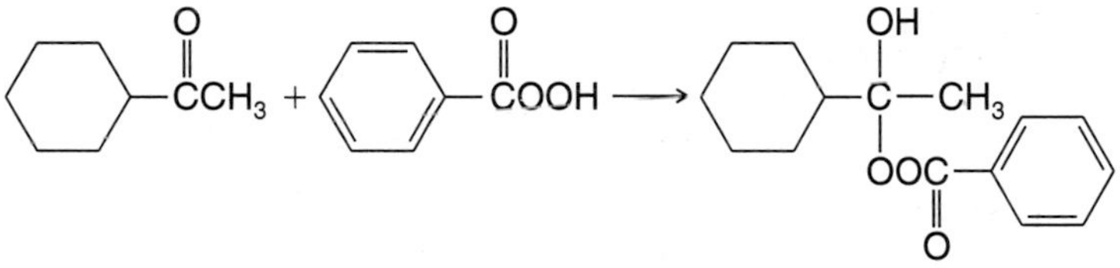

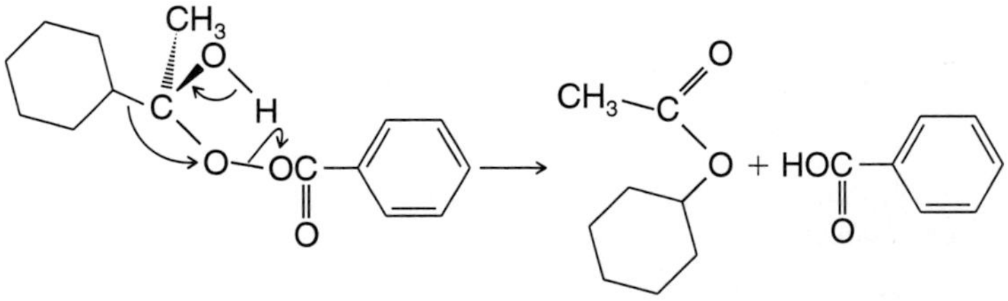

17.17 Hydrogen migrates to oxygen (analogous to a hydride shift in a carbocation).

CHAPTER 18

18.1 (*b*) Zero; (*c*) five; (*d*) four

18.2 $ClCH_2\overset{O}{\overset{\|}{C}}CH_2CH_3$ and $CH_3\overset{O}{\overset{\|}{C}}CH(Cl)CH_3$

18.3 $CH_2{=}\overset{OH}{C}CH_2CH_3 \xrightarrow{Cl_2} ClCH_2\overset{O}{\overset{\|}{C}}CH_2CH_3 \qquad CH_3\overset{OH}{C}{=}CHCH_3 \xrightarrow{Cl_2} CH_3\overset{O}{\overset{\|}{C}}\underset{Cl}{C}HCH_3$

18.4 $CH_2{=}\overset{:\ddot{O}H}{C}CH_2CH_3 \;\; :\ddot{Cl}{-}\ddot{Cl}: \longrightarrow :\ddot{Cl}CH_2\overset{{}^{+}\ddot{O}H}{\overset{\|}{C}}CH_2CH_3 \;\; :\ddot{Cl}:^- \qquad CH_3\overset{:\ddot{O}H}{C}{=}CHCH_3 \;\; :\ddot{Cl}{-}\ddot{Cl}: \longrightarrow CH_3\overset{{}^{+}\ddot{O}H}{\overset{\|}{C}}\underset{:\ddot{Cl}:}{C}HCH_3 \;\; :\ddot{Cl}:$

18.5 (*b*) $C_6H_5\overset{OH}{C}{=}CH_2$; (*c*) [cyclohexene ring]—OH, CH_3 and [cyclohexene ring]—OH, CH_3

18.6 (*b*) $C_6H_5\overset{O}{\overset{\|}{C}}CH{=}\overset{HO}{C}CH_3$ and $C_6H_5\overset{OH}{C}{=}CH\overset{O}{\overset{\|}{C}}CH_3$

18.7 (*b*) $C_6H_5\overset{O}{\overset{\|}{C}}CH{=}\overset{O^-}{C}CH_3 \longleftrightarrow C_6H_5\overset{O}{\overset{\|}{C}}\underset{..}{C}H\overset{O}{\overset{\|}{C}}CH_3 \longleftrightarrow C_6H_5\overset{O^-}{C}{=}CH\overset{O}{\overset{\|}{C}}CH_3$

(*c*) [cyclohexanone]$={\overset{O^-}{C}H}$ $\longleftrightarrow$ [cyclohexanone carbanion]$-\overset{O}{\overset{\|}{C}}H$ $\longleftrightarrow$ [cyclohexene]$(O^-)-\overset{O}{\overset{\|}{C}}H$

18.8 Product is chiral, but is formed as a racemic mixture because it arises from an achiral intermediate (the enol). Therefore it is not optically active.

18.9 (*b*) $CH_3CH_2\underset{CH_3}{C}H\overset{HO}{C}H{-}\overset{CH_3}{\underset{HC=O}{C}}CH_2CH_3$ (*c*) $(CH_3)_2CHCH_2\overset{OH}{C}H{-}\underset{HC=O}{C}HCH(CH_3)_2$

18.10 (*b*) $CH_3CH_2\underset{CH_3}{C}H\overset{HO}{C}H{-}\overset{CH_3}{\underset{HC=O}{C}}CH_2CH_3$ (α) (*c*) $(CH_3)_2CHCH_2CH{=}\underset{HC=O}{C}CH(CH_3)_2$

Cannot dehydrate; no protons on α carbon atom

18.11 $(CH_3)_2C{=}CH\overset{O}{\overset{\|}{C}}CH_3$ (4-Methyl-3-penten-2-one)

18.12 [structure: OH, O] $\xrightarrow{-H_2O}$ [structure: O]

18.13 (*b*) $C_6H_5CH{=}CHC(=O)C(CH_3)_3$ (*c*) $C_6H_5CH{=}$(2-oxocyclohexylidene)

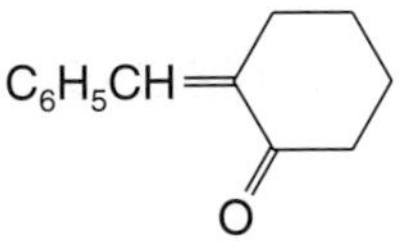

18.14 $CH_3CH_2CH_2CH(=O) \xrightarrow[H_2O,\ heat]{NaOH} CH_3CH_2CH_2CH{=}C(CH_2CH_3)CH(=O) \xrightarrow[Pt]{H_2} CH_3CH_2CH_2CH_2CH(CH_2CH_3)CH_2OH$

18.15 $CH_3C(=CH_2)CH_2C(=O)CH_3$

18.16 Acrolein ($CH_2{=}CHCH{=}O$) undergoes conjugate addition with sodium azide in aqueous solution to give $N_3CH_2CH_2CH{=}O$. Propanal is not an α,β-unsaturated carbonyl compound and cannot undergo conjugate addition.

18.17 $C_6H_5CH_2C(=O)CH(C_6H_5)CH_2CH_2C(=O)CH_3$ and 3-hydroxy-3-methyl-2,6-diphenylcyclohexanone (H, C_6H_5, O, H_3C, HO, C_6H_5)

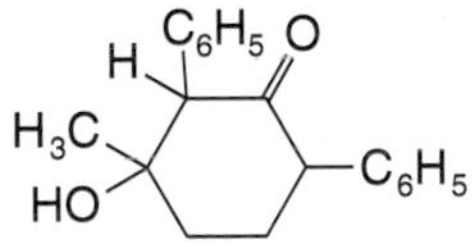

18.18 $CH_3CH_2CH_2CH_2CH{=}CHC(=O)CH_3 + LiCu(CH_3)_2$

CHAPTER 19

19.1 (*b*) (*E*)-2-butenoic acid; (*c*) ethanedioic acid; (*d*) *p*-methylbenzoic acid or 4-methylbenzoic acid.

19.2 The negative charge in $CH_3C(=O)OO^-$ cannot be delocalized into the carbonyl group.

19.3

(*b*) $CH_3CO_2H + (CH_3)_3CO^- \rightleftharpoons CH_3CO_2^- + (CH_3)_3COH$
(The position of equilibrium lies to the right.)

(*c*) $CH_3CO_2H + Br^- \rightleftharpoons CH_3CO_2^- + HBr$
(The position of equilibrium lies to the left.)

(*d*) $CH_3CO_2H + HC{\equiv}C{:}^- \rightleftharpoons CH_3CO_2^- + HC{\equiv}CH$
(The position of equilibrium lies to the right.)

(*e*) $CH_3CO_2H + NO_3^- \rightleftharpoons CH_3CO_2^- + HNO_3$
(The position of equilibrium lies to the left.)

(*f*) $CH_3CO_2H + H_2N^- \rightleftharpoons CH_3CO_2^- + NH_3$
(The position of equilibrium lies to the right.)

19.4 (*b*) $CH_3CH(OH)CO_2H$; (*c*) $CH_3C(=O)CO_2H$; (*d*) $CH_3S(=O)_2CH_2CO_2H$

19.5 $HC{\equiv}CCO_2H$

19.6 The "true K_1" for carbonic acid is 1.4×10^{-4}.

19.7

(*b*) The conversion proceeding by way of the nitrile is satisfactory.

$$HOCH_2CH_2Cl \xrightarrow{NaCN} HOCH_2CH_2CN \xrightarrow{\text{hydrolysis}} HOCH_2CH_2CO_2H$$

Since 2-chloroethanol has a proton bonded to oxygen, it is not an appropriate substrate for conversion to a stable Grignard reagent.

(*c*) The procedure involving a Grignard reagent is satisfactory.

$$(CH_3)_3CCl \xrightarrow{Mg} (CH_3)_3CMgCl \xrightarrow[2.\ H_3O^+]{1.\ CO_2} (CH_3)_3CCO_2H$$

The reaction of *tert*-butyl chloride with cyanide ion proceeds by elimination rather than substitution.

19.8 Water labeled with ^{18}O adds to benzoic acid to give the tetrahedral intermediate shown. This intermediate can lose unlabeled H_2O to give benzoic acid containing ^{18}O.

$$C_6H_5C(={}^{18}O)OH \xleftarrow{-H_2O} C_6H_5C(OH)_2{-}^{18}OH \xrightarrow{-H_2O} C_6H_5C(=O){-}^{18}OH$$

19.9 (*b*) $HOCH_2(CH_2)_{13}CO_2H$; (*c*) [structure: cyclohexane ring bearing $HOCH_2$ and $CH{=}CH_2$ on one carbon, OH groups, H, and two $C(=CH_2)CO_2H$ substituents (HO_2C, H_2C; CH_2, CO_2H)]

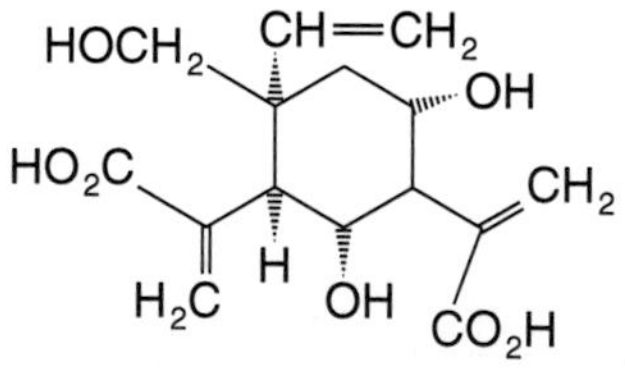

19.10 $CH_3(CH_2)_{15}CH_2CO_2H \xrightarrow[PCl_3]{Br_2} CH_3(CH_2)_{15}CH(Br)CO_2H \xrightarrow[\text{acetone}]{NaI} CH_3(CH_2)_{15}CH(I)CO_2H$

19.11

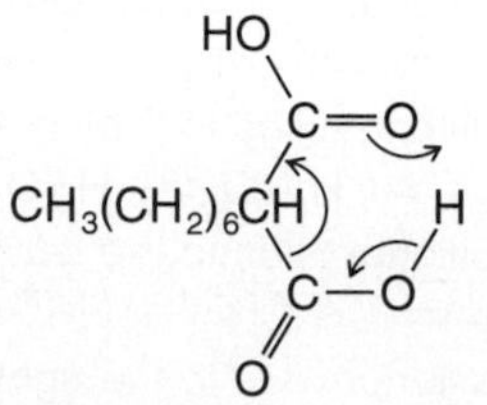

(*b*) $CH_3(CH_2)_6CH_2CO_2H$ via $CH_3(CH_2)_6CH(CO_2H)_2$ [cyclic transition state: HO–C=O and C(=O)–O–H with curved arrows] (*c*) $C_6H_5\underset{\underset{CH_3}{|}}{C}HCO_2H$ via $\text{O=C(–O–H)–C(C}_6\text{H}_5\text{)(CH}_3\text{)–C(=O)OH}$

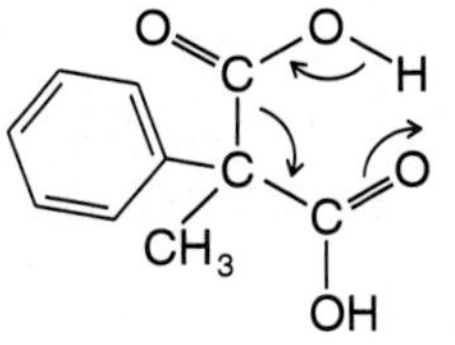

19.12 (*b*) $CH_3\overset{\overset{O}{\|}}{C}–C(CH_3)_2–\overset{\overset{O}{\|}}{C}–O–H \xrightarrow{-CO_2} CH_3\overset{\overset{OH}{|}}{C}=\underset{\underset{CH_3}{|}}{C}CH_3 \longrightarrow CH_3\overset{\overset{O}{\|}}{C}CH(CH_3)_2$

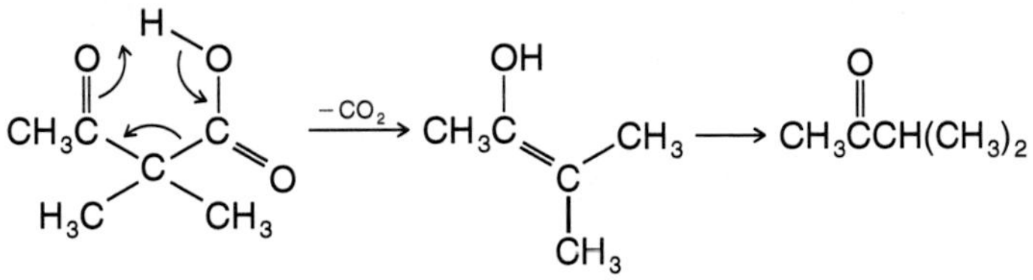

CHAPTER 20

20.1

(*b*) $CH_3CH_2\underset{\underset{C_6H_5}{|}}{C}H\overset{\overset{O}{\|}}{C}O\overset{\overset{O}{\|}}{C}\underset{\underset{C_6H_5}{|}}{C}HCH_2CH_3$; (*c*) $CH_3CH_2\underset{\underset{C_6H_5}{|}}{C}H\overset{\overset{O}{\|}}{C}OCH_2CH_2CH_2CH_3$;

(*d*) $CH_3CH_2CH_2\overset{\overset{O}{\|}}{C}OCH_2\underset{\underset{C_6H_5}{|}}{C}HCH_2CH_3$; (*e*) $CH_3CH_2\underset{\underset{C_6H_5}{|}}{C}H\overset{\overset{O}{\|}}{C}NH_2$;

(*f*) $CH_3CH_2\underset{\underset{C_6H_5}{|}}{C}H\overset{\overset{O}{\|}}{C}NHCH_2CH_3$; (*g*) $CH_3CH_2\underset{\underset{C_6H_5}{|}}{C}HC\equiv N$

20.2 Rotation about the carbon-nitrogen bond is slow in amides. The methyl groups of *N,N*-dimethylformamide are nonequivalent because one is cis to oxygen, the other cis to hydrogen.

20.3 (*b*) $C_6H_5\overset{\overset{O}{\|}}{C}O\overset{\overset{O}{\|}}{C}C_6H_5$; (*c*) $C_6H_5\overset{\overset{O}{\|}}{C}OCH_2CH_3$; (*d*) $C_6H_5\overset{\overset{O}{\|}}{C}NHCH_3$; (*e*) $C_6H_5\overset{\overset{O}{\|}}{C}N(CH_3)_2$;
(*f*) $C_6H_5\overset{\overset{O}{\|}}{C}OH$

20.4

(*b*) $C_6H_5\overset{\overset{H–O}{|}}{\underset{\underset{Cl}{|}}{C}}O\overset{\overset{O}{\|}}{C}C_6H_5 \longrightarrow C_6H_5\overset{\overset{O}{\|}}{C}O\overset{\overset{O}{\|}}{C}C_6H_5 + HCl$

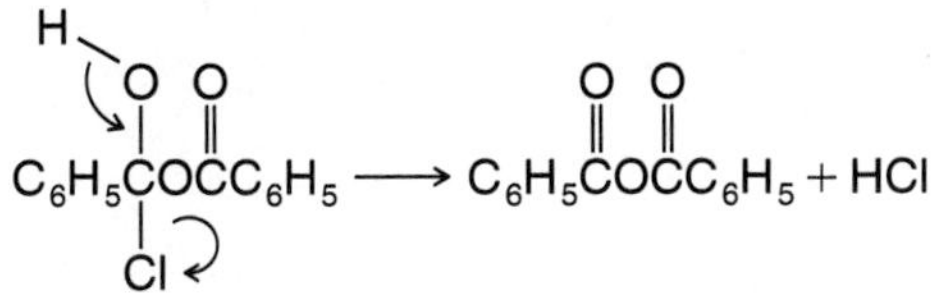

(*c*) $C_6H_5\overset{\overset{H–O}{|}}{\underset{\underset{Cl}{|}}{C}}OCH_2CH_3 \longrightarrow C_6H_5\overset{\overset{O}{\|}}{C}OCH_2CH_3 + HCl$

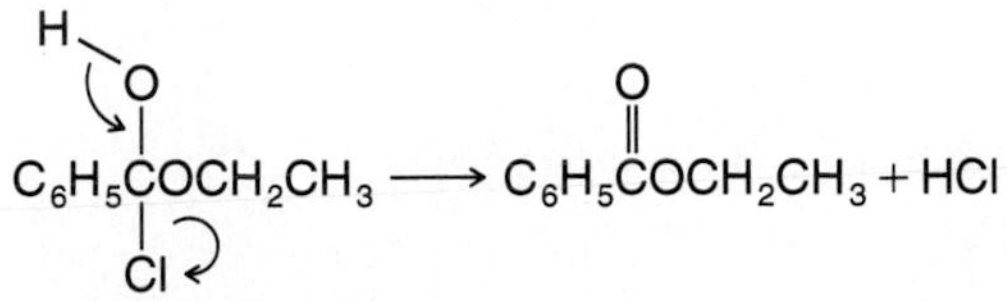

(*d*) $C_6H_5C(OH)(Cl)NHCH_3 \longrightarrow C_6H_5C(=O)NHCH_3 + CH_3\overset{+}{N}H_3\ Cl^-$ (base: $CH_3\ddot{N}H_2$)

(*e*) $C_6H_5C(OH)(Cl)N(CH_3)_2 \longrightarrow C_6H_5C(=O)N(CH_3)_2 + (CH_3)_2\overset{+}{N}H_2\ Cl^-$ (base: $(CH_3)_2\ddot{N}H$)

(*f*) $C_6H_5C(OH)(Cl)OH \longrightarrow C_6H_5C(=O)OH + HCl$

20.5 $C_6H_5C(=O)Cl + H_2O \longrightarrow C_6H_5C(=O)OH + HCl$

$C_6H_5C(=O)Cl + C_6H_5C(=O)OH \longrightarrow C_6H_5C(=O)OC(=O)C_6H_5 + HCl$

20.6 $CH_3\overset{+}{C}{=}\ddot{O}: \longleftrightarrow CH_3C{\equiv}\overset{+}{O}:$

20.7 (*b*) $CH_3C(=O)NH_2 + CH_3CO_2^-\ \overset{+}{N}H_4$; (*c*) $C_6H_4(C(=O)N(CH_3)_2)(CO_2^-)\ \ H_2\overset{+}{N}(CH_3)_2$;

(*d*) $C_6H_4(CO_2^-\ Na^+)_2$

20.8

(*b*) $CH_3C(OH)(NH_2){-}OC(=O)CH_3 \longrightarrow CH_3C(=O)NH_2 + H_4\overset{+}{N}\overset{-}{O}C(=O)CH_3$ (base: $H_3N:$)

(*c*)

$(CH_3)_2\ddot{N}H$ ⟶ H–O (tetrahedral carbon bearing $N(CH_3)_2$, ring O, C=O on benzene ring)

⟶ C_6H_4 with $\overset{O}{\parallel}CN(CH_3)_2$ and $\underset{O}{\parallel}CO^-$ $H_2\overset{+}{N}(CH_3)_2$

(*d*)

HO^- ⟶ H–O (tetrahedral carbon bearing OH, ring O, C=O on benzene ring)

⟶ C_6H_4 with $\overset{O}{\parallel}COH$ and $\underset{O}{\parallel}CO^-$ $+ H_2O$

20.9 $HOCH_2\underset{OH}{CH}CH_2CH_2CH_2OH$ ($C_5H_{12}O_3$) and CH_3CO_2H

20.10

Step 1: Protonation of the carbonyl oxygen

$$C_6H_5C(=\ddot{O}:)\ddot{O}CH_2CH_3 + H{-}\overset{+}{\ddot{O}}(H)H \rightleftharpoons C_6H_5C(=\overset{+}{\ddot{O}}H)\ddot{O}CH_2CH_3 + :\ddot{O}(H)H$$

Step 2: Nucleophilic addition of water

$$H_2\ddot{O}: + C_6H_5C(=\overset{+}{\ddot{O}}H)\ddot{O}CH_2CH_3 \rightleftharpoons C_6H_5\underset{+\ddot{O}H_2}{\overset{:\ddot{O}H}{C}}{-}\ddot{O}CH_2CH_3$$

Step 3: Deprotonation of oxonium ion to give neutral form of tetrahedral intermediate

$$C_6H_5\underset{+\ddot{O}H_2}{\overset{:\ddot{O}H}{C}}{-}\ddot{O}CH_2CH_3 + :\ddot{O}(H)H \rightleftharpoons C_6H_5\underset{H\ddot{O}:}{\overset{:\ddot{O}H}{C}}{-}\ddot{O}CH_2CH_3 + H{-}\overset{+}{\ddot{O}}(H)H$$

Step 4: Protonation of ethoxy oxygen

$$C_6H_5\underset{H\ddot{O}:}{\overset{:\ddot{O}H}{C}}{-}\ddot{O}CH_2CH_3 + H{-}\overset{+}{\ddot{O}}(H)H \longrightarrow C_6H_5\underset{H\ddot{O}:}{\overset{:\ddot{O}H}{C}}{-}\underset{H}{\overset{+}{\ddot{O}}}CH_2CH_3 + :\ddot{O}(H)H$$

Step 5: Dissociation of protonated form of tetrahedral intermediate

$$C_6H_5C(OH)_2-\overset{+}{O}(H)CH_2CH_3 \rightleftharpoons C_6H_5C(=\overset{+}{O}H)OH + HOCH_2CH_3$$

Step 6: Deprotonation of protonated form of benzoic acid

$$C_6H_5C(=\overset{+}{O}-H)OH + H_2O \rightleftharpoons C_6H_5C(=O)OH + H-\overset{+}{O}H_2$$

20.11 The carbonyl oxygen of the lactone became labeled with ^{18}O.

20.12 $CH_3(CH_2)_{12}COOCH_2CH(OC(O)(CH_2)_{12}CH_3)CH_2OC(O)(CH_2)_{12}CH_3$

20.13 The isotopic label appeared in the acetate ion.

20.14

Step 1: Nucleophilic addition of hydroxide ion to the carbonyl group

$$HO^- + C_6H_5C(=O)OCH_2CH_3 \rightleftharpoons C_6H_5C(O^-)(OH)-OCH_2CH_3$$

Step 2: Proton transfer from water to give neutral form of tetrahedral intermediate

$$C_6H_5C(O^-)(OH)-OCH_2CH_3 + H-OH \rightleftharpoons C_6H_5C(OH)_2-OCH_2CH_3 + {}^-OH$$

Step 3: Hydroxide ion–promoted dissociation of tetrahedral intermediate

$$HO^- + C_6H_5C(OH)_2-OCH_2CH_3 \rightleftharpoons HOH + C_6H_5C(=O)OH + {}^-OCH_2CH_3$$

Step 4: Proton abstraction from benzoic acid

$$C_6H_5C(=O)O-H + {}^-OH \longrightarrow C_6H_5C(=O)O^- + HOH$$

20.15 $CH_3NHC(=O)CH_2CH_2CH(OH)CH_3$

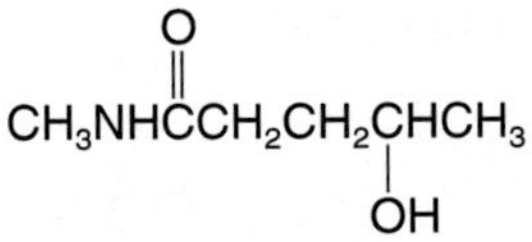

20.16 $CH_3C(OH)(OCH_3)SCH_2CH_2OC_6H_5$

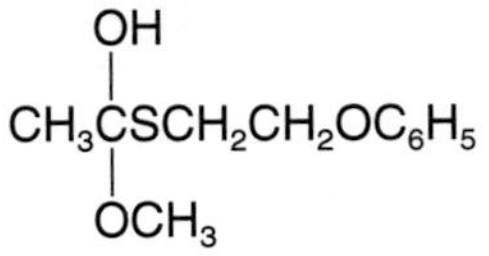

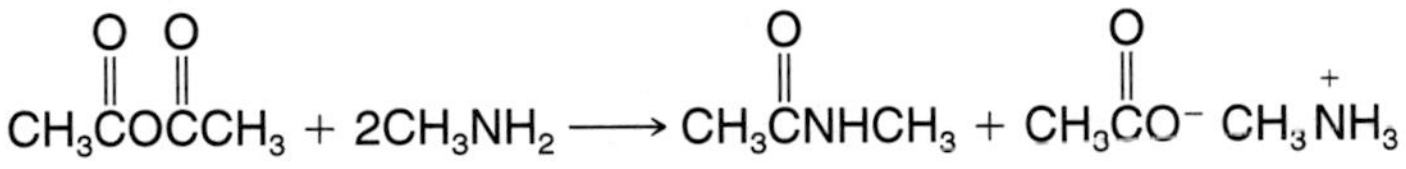

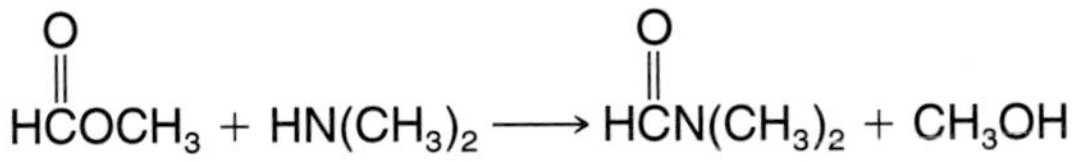

20.17

(*b*) $CH_3C(=O)OC(=O)CH_3 + 2CH_3NH_2 \longrightarrow CH_3C(=O)NHCH_3 + CH_3C(=O)O^- \; CH_3\overset{+}{N}H_3$

(*c*) $HC(=O)OCH_3 + HN(CH_3)_2 \longrightarrow HC(=O)N(CH_3)_2 + CH_3OH$

20.18 $o\text{-}C_6H_4(CNH_2{=}O)(CO_2^-) \; \overset{+}{N}H_4$

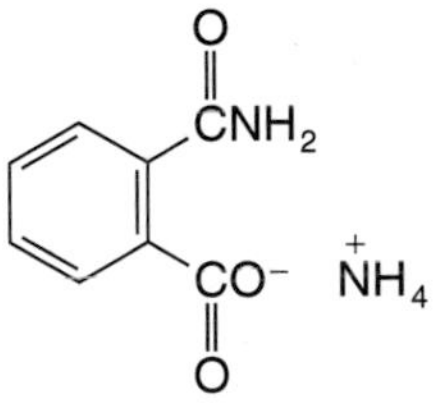

20.19

Step 1: Protonation of the carbonyl oxygen

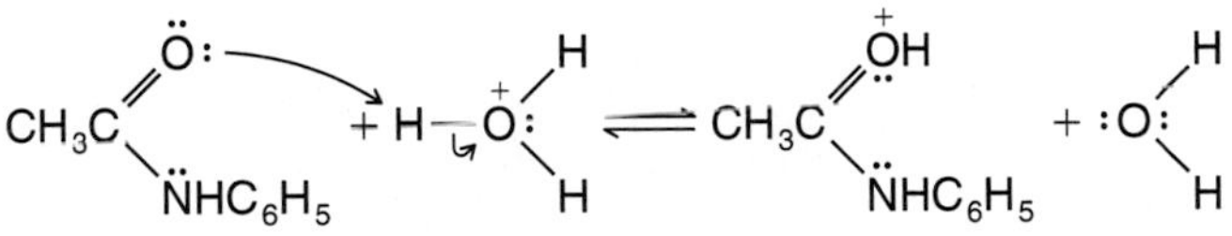

$CH_3C(=\ddot{O}:)\ddot{N}HC_6H_5 + H{-}\overset{+}{O}H_2 \rightleftharpoons CH_3C(=\overset{+}{O}H)\ddot{N}HC_6H_5 + :\ddot{O}H_2$

Step 2: Nucleophilic addition of water

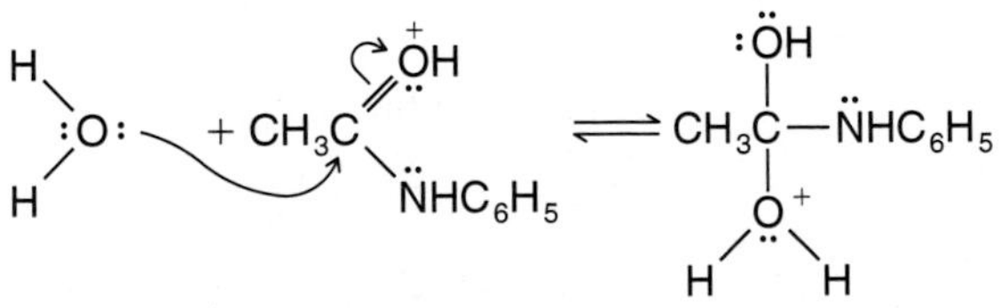

$:\ddot{O}H_2 + CH_3C(=\overset{+}{O}H)\ddot{N}HC_6H_5 \rightleftharpoons CH_3C(:\ddot{O}H)(\ddot{N}HC_6H_5)(\overset{+}{O}H_2)$

Step 3: Deprotonation of oxonium ion to give neutral form of tetrahedral intermediate

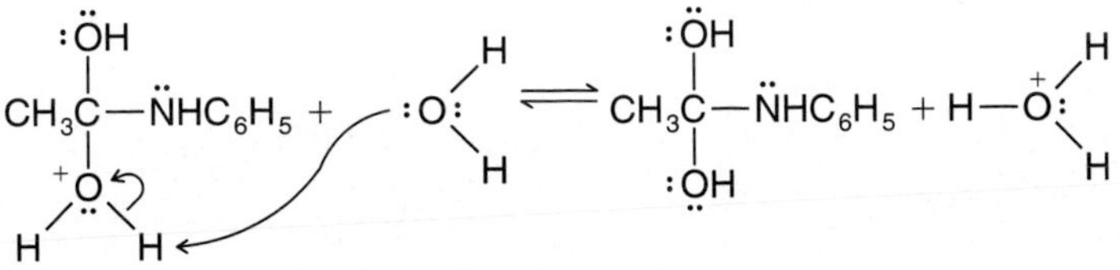

$CH_3C(:\ddot{O}H)(\ddot{N}HC_6H_5)(\overset{+}{O}H_2) + :\ddot{O}H_2 \rightleftharpoons CH_3C(:\ddot{O}H)_2{-}\ddot{N}HC_6H_5 + H{-}\overset{+}{O}H_2$

Step 4: Protonation of amino group of tetrahedral intermediate

$$CH_3C(OH)_2-\ddot{N}HC_6H_5 + H-\overset{+}{O}H_2 \rightleftharpoons CH_3C(OH)_2-\overset{+}{N}H_2C_6H_5 + :OH_2$$

Step 5: Dissociation of *N*-protonated form of tetrahedral intermediate

$$CH_3C(OH)_2-\overset{+}{N}H_2C_6H_5 \rightleftharpoons CH_3C(=\overset{+}{O}H)OH + H_2\ddot{N}C_6H_5$$

Step 6: Proton-transfer processes

$$H_2\overset{+}{O}-H + H_2\ddot{N}C_6H_5 \rightleftharpoons H_2O: + H_2\overset{+}{N}HC_6H_5$$

$$CH_3C(=\overset{+}{O}-H)OH + :OH_2 \rightleftharpoons CH_3C(=O)OH + H-\overset{+}{O}H_2$$

20.20

Step 1: Nucleophilic addition of hydroxide ion to the carbonyl group

$$H\ddot{O}:^- + HC(=O)\ddot{N}(CH_3)_2 \longrightarrow HC(-O^-)(OH)-\ddot{N}(CH_3)_2$$

Step 2: Proton transfer to give neutral form of tetrahedral intermediate

$$HC(-O^-)(OH)-\ddot{N}(CH_3)_2 + H-\ddot{O}H \longrightarrow HC(OH)_2-\ddot{N}(CH_3)_2 + {}^-:\ddot{O}H$$

Step 3: Proton transfer from water to nitrogen of tetrahedral intermediate

$$HC(OH)_2-\ddot{N}(CH_3)_2 + H-\ddot{O}H \rightleftharpoons HC(OH)_2-\overset{+}{N}H(CH_3)_2 + {}^-:\ddot{O}H$$

Step 4: Dissociation of *N*-protonated form of tetrahedral intermediate

$$HO:^- + HC(OH)(OH)-\overset{+}{N}H(CH_3)_2 \rightleftharpoons H_2O: + HC(=O)OH + HN(CH_3)_2$$

Step 5: Irreversible formation of formate ion

$$HC(=O)-O-H + :OH^- \longrightarrow HC(=O)O^- + HOH$$

20.21 $CH_3CH_2CH_2CO_2H \xrightarrow[2.\ NH_3]{1.\ SOCl_2} CH_3CH_2CH_2\overset{O}{\overset{\|}{C}}NH_2 \xrightarrow[H_2O,\ NaOH]{Br_2} CH_3CH_2CH_2NH_2$

20.22

$$CH_3CH_2OH \xrightarrow[H_2SO_4,\ heat]{Na_2Cr_2O_7,\ H_2O} CH_3\overset{O}{\overset{\|}{C}}OH \xrightarrow[2.\ NH_3]{1.\ SOCl_2} CH_3\overset{O}{\overset{\|}{C}}NH_2 \xrightarrow{P_4O_{10}} CH_3C{\equiv}N$$

$$CH_3CH_2OH \xrightarrow[or\ HBr]{PBr_3} CH_3CH_2Br \xrightarrow{NaCN} CH_3CH_2CN$$

20.23 In acid, the nitrile is protonated on nitrogen. Nucleophilic addition of water yields an imino acid.

$$H_2\ddot{O} + RC{\equiv}\overset{+}{N}-H \rightleftharpoons RC(\overset{+}{O}H_2)(=NH) \underset{H_3O^+}{\overset{H_2O}{\rightleftharpoons}} RC(OH)(=NH)$$

A series of proton transfers converts the imino acid to an amide.

$$RC(OH)(=NH) + H-\overset{+}{O}H_2 \rightleftharpoons RC(OH)(=\overset{+}{N}H_2) + :OH_2 \rightleftharpoons RC(=O)NH_2 + H-\overset{+}{O}H_2$$

20.24 $CH_3CH_2CN + C_6H_5MgBr \xrightarrow[2.\ H_2O,\ H^+,\ heat]{1.\ diethyl\ ether} C_6H_5\overset{O}{\overset{\|}{C}}CH_2CH_3$

The imine intermediate is $C_6H_5\overset{NH}{\overset{\|}{C}}CH_2CH_3$.

CHAPTER 21

21.1 Ethyl benzoate cannot undergo the Claisen condensation.

Claisen condensation product of ethyl pentanoate:

$$CH_3CH_2CH_2CH_2\overset{O}{\overset{\|}{C}}\underset{CH_2CH_2CH_3}{\underset{|}{C}H}\overset{O}{\overset{\|}{C}}OCH_2CH_3$$

Claisen condensation product of ethyl phenylacetate:

$$C_6H_5CH_2\overset{O}{\overset{\|}{C}}\underset{C_6H_5}{\underset{|}{C}H}\overset{O}{\overset{\|}{C}}OCH_2CH_3$$

21.2 (*b*) O, C, OCH_2CH_3, O, CH_3 (c) O, C, OCH_2CH_3, O, CH_3

21.3 (*b*) $C_6H_5\underset{\underset{O}{\overset{\|}{C}}OCH_2CH_3}{\underset{|}{C}H}\overset{O}{\overset{\|}{C}}\overset{O}{\overset{\|}{C}}OCH_2CH_3$ (*c*) $C_6H_5\underset{\underset{O}{\overset{\|}{C}}OCH_2CH_3}{\underset{|}{C}H}\overset{O}{\overset{\|}{C}}H$

21.4 CH_3—$\bar{C}H$, C, O, CH_2, CH_3CH_2O, CH_2, C, O $\longrightarrow$ CH_3, O, CH_3CH_2O O^-

CH_3, O, CH_3CH_2O O^- $\longrightarrow$ CH_3, O, O + $^-OCH_2CH_3$

21.5 $CH_3CH_2O\overset{O}{\overset{\|}{C}}CH_2CH_2CH_2CH_2\overset{O}{\overset{\|}{C}}OCH_2CH_3 \xrightarrow[\text{2. H}^+]{\text{1. NaOCH}_2\text{CH}_3}$

O, O, $\overset{\|}{C}OCH_2CH_3$ $\xrightarrow[\text{2. H}^+ \text{ 3. heat}]{\text{1. HO}^-\text{, H}_2\text{O}}$ O

21.6

(*b*) $C_6H_5CH_2Br + CH_3\overset{O}{\overset{\|}{C}}CH_2\overset{O}{\overset{\|}{C}}OCH_2CH_3 \xrightarrow[\text{3. H}^+ \text{ 4. heat}]{\text{1. NaOCH}_2\text{CH}_3 \text{ 2. HO}^-\text{, H}_2\text{O}} C_6H_5CH_2CH_2\overset{O}{\overset{\|}{C}}CH_3$

(*c*) $CH_2{=}CHCH_2Br + CH_3\overset{O}{\overset{\|}{C}}CH_2\overset{O}{\overset{\|}{C}}OCH_2CH_3 \xrightarrow[\text{3. H}^+\text{ 4. heat}]{\text{1. NaOCH}_2\text{CH}_3\text{ 2. HO}^-\text{, H}_2\text{O}} CH_2{=}CHCH_2CH_2\overset{O}{\overset{\|}{C}}CH_3$

21.7

(*b*) $CH_3(CH_2)_5CH_2Br + CH_2(COOCH_2CH_3)_2 \xrightarrow[\text{ethanol}]{\text{NaOCH}_2\text{CH}_3} CH_3(CH_2)_5CH_2CH(COOCH_2CH_3)_2$

$\xrightarrow{\text{1. HO}^-\text{, H}_2\text{O; 2. H}^+\text{; 3. heat}} CH_3(CH_2)_5CH_2CH_2\overset{O}{\overset{\|}{C}}OH$

(*c*) $CH_3CH_2\underset{CH_3}{\underset{|}{C}}HCH_2Br + CH_2(COOCH_2CH_3)_2 \xrightarrow[\text{ethanol}]{\text{NaOCH}_2\text{CH}_3} CH_3CH_2\underset{CH_3}{\underset{|}{C}}HCH_2CH(COOCH_2CH_3)_2$

$\xrightarrow{\text{1. HO}^-\text{, H}_2\text{O; 2. H}^+\text{; 3. heat}} CH_3CH_2\underset{CH_3}{\underset{|}{C}}HCH_2CH_2\overset{O}{\overset{\|}{C}}OH$

(*d*) $C_6H_5CH_2Br + CH_2(COOCH_2CH_3)_2 \xrightarrow[\text{ethanol}]{\text{NaOCH}_2\text{CH}_3} C_6H_5CH_2CH(COOCH_2CH_3)_2$

$\xrightarrow{\text{1. HO}^-\text{, H}_2\text{O; 2. H}^+\text{; 3. heat}} C_6H_5CH_2CH_2\overset{O}{\overset{\|}{C}}OH$

21.8 $CH_3\overset{O}{\overset{\|}{C}}CH_2\overset{O}{\overset{\|}{C}}OCH_2CH_3 \xrightarrow[\text{CH}_3\text{Br}]{\text{NaOCH}_2\text{CH}_3} CH_3\overset{O}{\overset{\|}{C}}\underset{CH_3}{\underset{|}{C}}H\overset{O}{\overset{\|}{C}}OCH_2CH_3 \xrightarrow[\text{CH}_3\text{Br}]{\text{NaOCH}_2\text{CH}_3}$

$CH_3\overset{O}{\overset{\|}{C}}\underset{H_3C\ \ CH_3}{C}\overset{O}{\overset{\|}{C}}OCH_2CH_3 \xrightarrow[\text{2. H}^+\text{ 3. heat}]{\text{1. HO}^-\text{, H}_2\text{O}} CH_3\overset{O}{\overset{\|}{C}}CH(CH_3)_2$

21.9 $CH_3\overset{O}{\overset{\|}{C}}CH_2\overset{O}{\overset{\|}{C}}OCH_2CH_3 + BrCH_2CH_2CH_2CH_2Br \xrightarrow{\text{NaOCH}_2\text{CH}_3}$

1-acetyl-1-(ethoxycarbonyl)cyclopentane (cyclopentane ring bearing $\overset{O}{\overset{\|}{C}}CH_3$ and $CO_2CH_2CH_3$) $\xrightarrow[\text{2. H}^+\text{, heat}]{\text{1. HO}^-\text{, H}_2\text{O}}$ cyclopentane ring bearing $\overset{O}{\overset{\|}{C}}CH_3$ and H

21.10 $CH_2(COOCH_2CH_3)_2 \xrightarrow[\text{2. } NaOCH_2CH_3,\ CH_3CH_2Br]{\text{1. } NaOCH_2CH_3,\ CH_3CH_2CH_2CH(Br)CH_3}$ $CH_3CH_2CH_2CH(CH_3)$ / CH_3CH_2 $>C(COOCH_2CH_3)_2$

$CH_3CH_2CH_2CH(CH_3)$ / CH_3CH_2 $>C<$ $COCH_2CH_3$ (C=O) / $COCH_2CH_3$ (C=O) $+ H_2NC(=O)NH_2 \longrightarrow$ $CH_3CH_2CH_2CH(CH_3)$ / CH_3CH_2 at C-5 of a pyrimidine ring: C(=O)–N(H)–C(=O)–N(H)–C(=O)

21.11 $CH_3CH_2CH_2CH(CH_3)$ / CH_3CH_2 at C-5 of ring: C(=O)–$\bar{N}$–C(=S)–N(H)–C(=O), Na^+ $\longleftrightarrow$ $CH_3CH_2CH_2CH(CH_3)$ / CH_3CH_2 at C-5 of ring: C(=O)–N=C(S^-)–N(H)–C(=O), Na^+

21.12 $C_6H_5CH_2C(=O)OCH_2CH_3 + CH_3CH_2OC(=O)OCH_2CH_3 \xrightarrow{NaOCH_2CH_3} C_6H_5CH(COOCH_2CH_3)_2$

$C_6H_5CH(COOCH_2CH_3)_2 \xrightarrow[CH_3CH_2Br]{NaOCH_2CH_3} C_6H_5C(CH_2CH_3)(COOCH_2CH_3)_2 \xrightarrow{H_2NC(=O)NH_2}$ C_6H_5 / CH_3CH_2 at C-5 of ring: C(=O)–N(H)–C(=O)–N(H)–C(=O)

21.13

Cycloheptanone with $CH_2C(=O)CH_3$ at C-3

21.14

(*b*) $C_6H_5CH_2CO_2CH_3 \xrightarrow[\text{2. } CH_3I]{\text{1. LDA, THF}} C_6H_5CH(CH_3)CO_2CH_3$

(*c*) Cyclohexanone $\xrightarrow[\text{2. } C_6H_5CHO \quad \text{3. } H_2O]{\text{1. LDA, THF}}$ 2-[$CH(OH)C_6H_5$]cyclohexanone

(*d*) $CH_3CO_2C(CH_3)_3 \xrightarrow[\text{2. cyclohexanone} \quad \text{3. } H_2O]{\text{1. LDA, THF}}$ cyclohexane with OH and $CH_2CO_2C(CH_3)_3$ on C-1

CHAPTER 22

22.1 (*b*) 1-Phenylethanamine or 1-phenylethylamine; (*c*) 2-propen-1-amine or allylamine

22.2 *N,N*-Dimethylcycloheptanamine

22.3 Tertiary amine; *N*-ethyl-4-isopropyl-*N*-methylaniline

22.4 $pK_b = 6$; K_a of conjugate acid $= 1 \times 10^{-8}$; pK_a of conjugate acid $= 8$

22.5 $\log (CH_3NH_3^+/CH_3NH_2) = 10.7 - 7 = 3.7$; $(CH_3NH_3^+/CH_3NH_2) = 10^{3.7} = 5000$

22.6 (*b*) $F_2CHCH_2NH_2$

22.7 Tetrahydroisoquinoline is a stronger base than tetrahydroquinoline. The unshared electron pair of tetrahydroquinoline is delocalized into the aromatic ring, and this substance resembles aniline in its basicity, whereas tetrahydroisoquinoline resembles an alkylamine.

22.8 (*b*) The lone pair of nitrogen is delocalized into the carbonyl group by amide resonance.

$$C_6H_5\ddot{N}H{-}C(={:}\ddot{O}{:})CH_3 \longleftrightarrow C_6H_5\overset{+}{N}H{=}C({:}\ddot{O}{:}^-)CH_3$$

(*c*) The amino group is conjugated to the carbonyl group through the aromatic ring.

$$H_2\ddot{N}{-}C_6H_4{-}C(={:}\ddot{O}{:})CH_3 \longleftrightarrow H_2\overset{+}{N}{=}C_6H_4{=}C({:}\ddot{O}{:}^-)CH_3$$

22.9 $CH_2{=}CHCH_3 \xrightarrow[400°C]{Cl_2} CH_2{=}CHCH_2Cl \xrightarrow{NH_3} CH_2{=}CHCH_2NH_2$.

22.10 Isobutylamine and 2-phenylethylamine can be prepared by the Gabriel synthesis. *tert*-Butylamine, *N*-methylbenzylamine, and aniline cannot.

(*b*) $(CH_3)_2CHCH_2Br$ + potassium phthalimide (NK) $\longrightarrow$ $NCH_2CH(CH_3)_2$ (N-isobutylphthalimide) $\xrightarrow{H_2NNH_2}$ $(CH_3)_2CHCH_2NH_2$ + phthalhydrazide (NH–NH) (*d*) $C_6H_5CH_2CH_2Br$ + (potassium phthalimide, NK) ⟶ (N-substituted phthalimide, $NCH_2CH_2C_6H_5$)

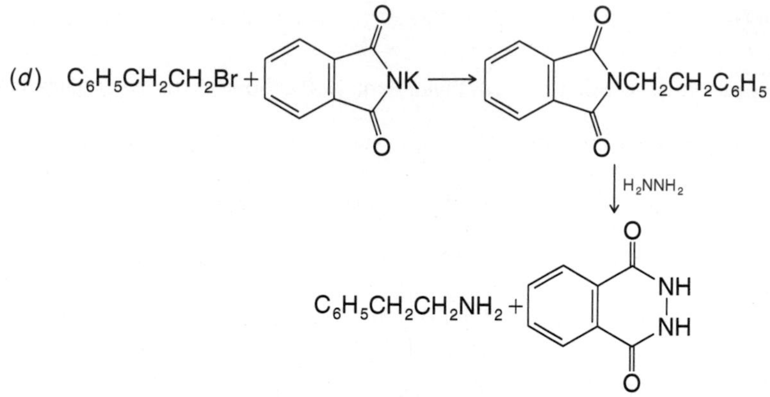

$\downarrow H_2NNH_2$

$C_6H_5CH_2CH_2NH_2$ + (phthalhydrazide, NH–NH)

22.11 (*b*) Prepare *p*-isopropylnitrobenzene as in (*a*); then reduce with H_2, Ni (or Fe + HCl or Sn + HCl, followed by base). (*c*) Prepare isopropylbenzene as in (*a*); then dinitrate with $HNO_3 + H_2SO_4$; then reduce both nitro groups. (*d*) Chlorinate benzene with $Cl_2 + FeCl_3$; then nitrate (HNO_3, H_2SO_4), separate the desired para isomer from the unwanted ortho isomer, and reduce. (*e*) Acetylate benzene by a Friedel-Crafts reaction (acetyl chloride + $AlCl_3$); then nitrate (HNO_3, H_2SO_4); then reduce the nitro group.

22.12

(*b*) $C_6H_5\overset{O}{\overset{\|}{C}}H + C_6H_5CH_2NH_2 \xrightarrow{H_2,\ Ni} C_6H_5CH_2NHCH_2C_6H_5$

(*c*) $C_6H_5\overset{O}{\overset{\|}{C}}H + (CH_3)_2NH \xrightarrow{H_2,\ Ni} C_6H_5CH_2N(CH_3)_2$

(*d*) $C_6H_5\overset{O}{\overset{\|}{C}}H$ + HN(piperidine) $\xrightarrow{H_2,\ Ni}$ $C_6H_5CH_2$—N(piperidine)

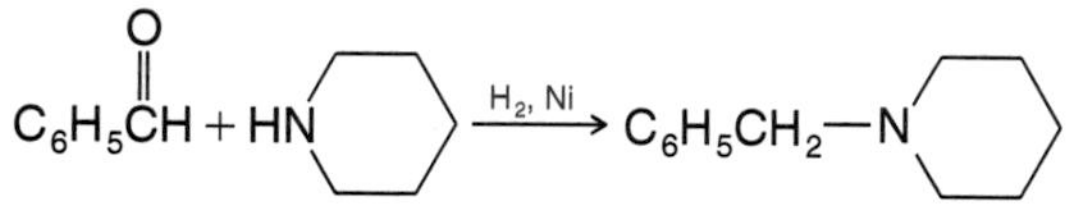

22.13 (*b*) $(CH_3)_3CCH_2\underset{CH_3}{\underset{|}{C}}{=}CH_2$; (*c*) $CH_2{=}CH_2$

22.14 (*b*) Prepare acetanilide as in (*a*); dinitrate (HNO_3, H_2SO_4); then hydrolyze the amide in either acid or base. (*c*) Prepare *p*-nitroacetanilide as in (*a*); then reduce the nitro group with H_2 (or Fe + HCl or Sn + HCl, followed by base).

22.15 $(H_3C)_2\ddot{N}{-}\ddot{N}{=}\ddot{O}: \longleftrightarrow (H_3C)_2\overset{+}{N}{=}\ddot{N}{-}\ddot{O}:^-$

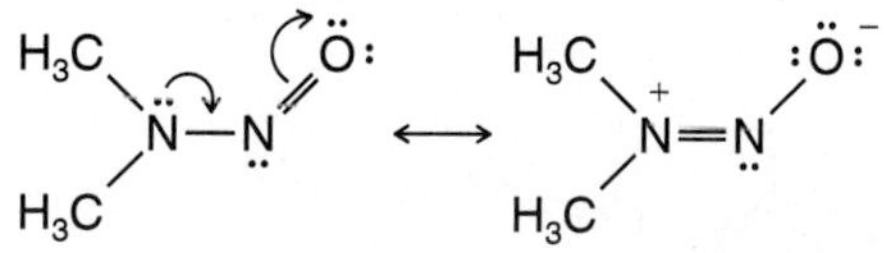

22.16 The diazonium ion from 2,2-dimethylpropylamine rearranges via a methyl shift on loss of nitrogen to give 1,1-dimethylpropyl cation.

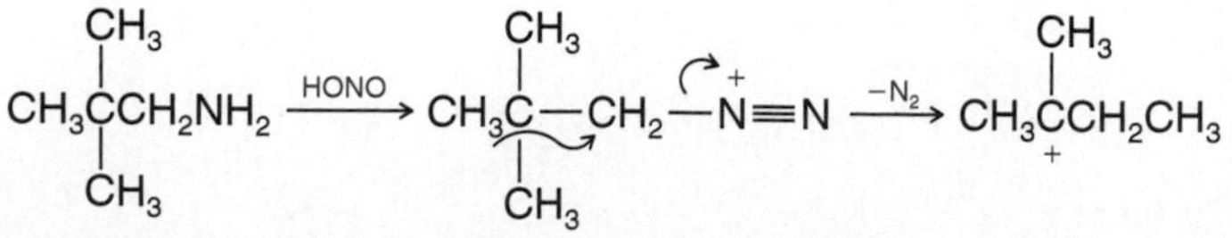

$$CH_3\underset{CH_3}{\overset{CH_3}{C}}CH_2NH_2 \xrightarrow{HONO} CH_3\underset{CH_3}{\overset{CH_3}{C}}{-}CH_2{-}\overset{+}{N}{\equiv}N \xrightarrow{-N_2} CH_3\overset{CH_3}{\underset{+}{C}}CH_2CH_3$$

22.17 Intermediates: benzene to nitrobenzene to *m*-bromonitrobenzene to *m*-bromoaniline to *m*-bromophenol. Reagents: HNO_3, H_2SO_4; Br_2, $FeBr_3$; Fe, HCl then HO^-; $NaNO_2$, H_2SO_4, H_2O, then heat in H_2O.

22.18 Prepare *m*-bromoaniline as in Problem 22.17; then $NaNO_2$, HCl, H_2O followed by KI.

22.19 Intermediates: benzene to ethyl phenyl ketone to ethyl *m*-nitrophenyl ketone to *m*-aminophenyl ethyl ketone to ethyl *m*-fluorophenyl ketone. Reagents: propanoyl chloride, $AlCl_3$; HNO_3, H_2SO_4; Fe, HCl, then HO^-; $NaNO_2$, H_2O, HCl, then HBF_4, then heat.

22.20 Intermediates: isopropylbenzene to *p*-isopropylnitrobenzene to *p*-isopropylaniline to *p*-isopropylacetanilide to 4-isopropyl-2-nitroacetanilide to 4-isopropyl-2-nitroaniline to *m*-isopropylnitrobenzene. Reagents: HNO_3, H_2SO_4; Fe, HCl, then HO^-; acetyl chloride; HNO_3, H_2SO_4; acid or base hydrolysis; $NaNO_2$, HCl, H_2O, and CH_3CH_2OH or H_3PO_2.

CHAPTER 23

23.1 $C_6H_5CH_2Cl$

23.2 (*b*) $SCH_2C_6H_5$, NO_2, NO_2 (*c*) NH_2, NO_2, NO_2 (*d*) $NHCH_3$, NO_2, NO_2

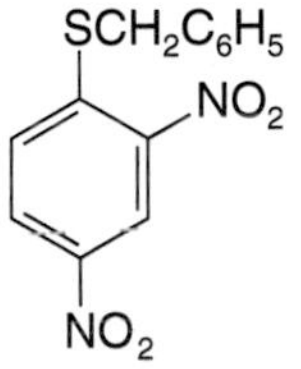

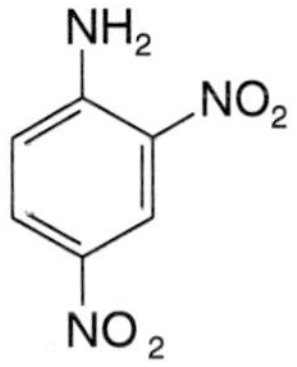

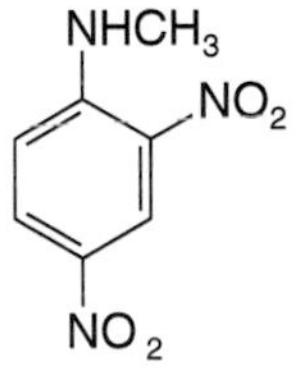

23.3 CH_3O, F, :Ö:⁻, N⁺, Ö:⁻

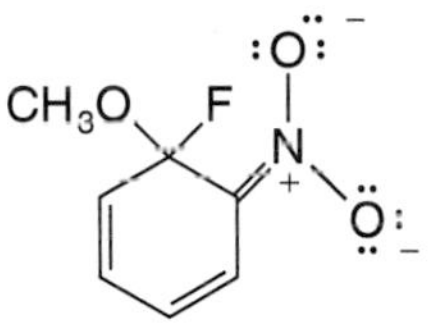

23.4 OCH_2CH_3, Br, Br, NO_2

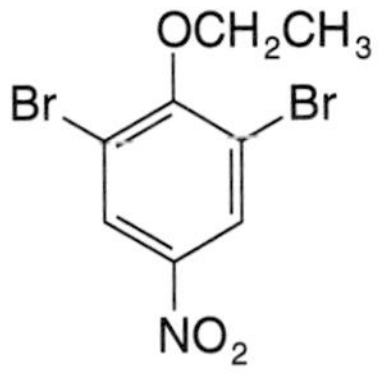

23.5 Nitrogen bears a portion of the negative charge in the anionic intermediate formed in the nucleophilic addition step in 4-chloropyridine, but not in 3-chloropyridine.

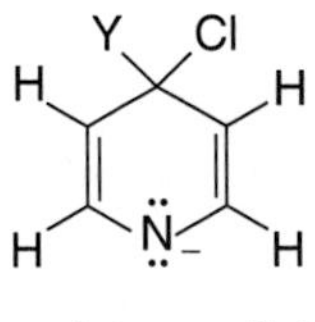

Y Cl H H H N H is more stable and formed faster than H H Y Cl H N H

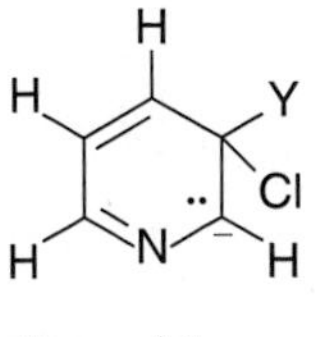

23.6 A benzyne intermediate is impossible because neither of the carbons ortho to the intended leaving group bears a proton.

23.7 3-Methylphenol and 4-methylphenol (*m*-cresol and *p*-cresol)

23.8 O

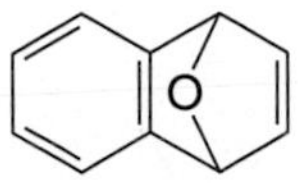

CHAPTER 24

24.1 (*b*) 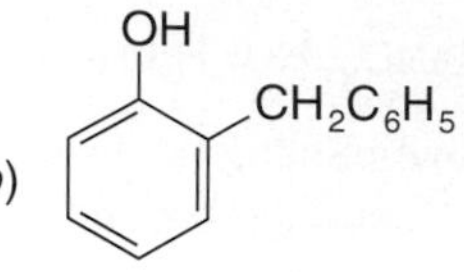 (*c*) 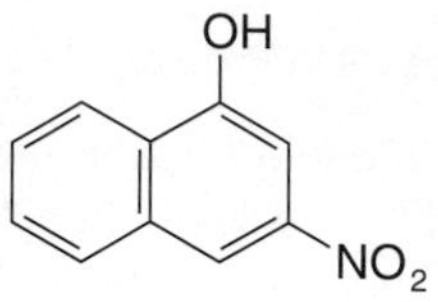 (*d*)

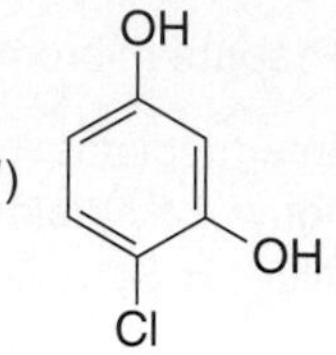

OH $CH_2C_6H_5$ NO_2 OH OH Cl

24.2 Methyl salicylate is the methyl ester of *o*-hydroxybenzoic acid. Intramolecular (rather than intermolecular) hydrogen bonding is responsible for its relatively low boiling point.

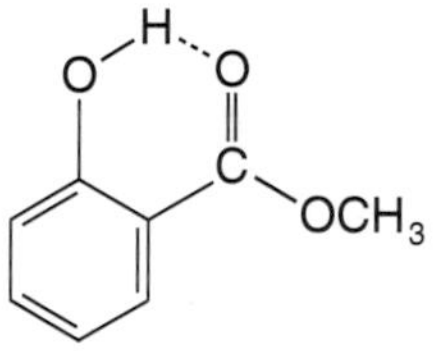

OCH_3

24.3 (*b*) *p*-Cyanophenol is stronger acid because of conjugation of cyano group with phenoxide oxygen. (*c*) *o*-Fluorophenol is stronger acid because electronegative fluorine substituent can stabilize negative charge better when fewer bonds intervene between it and the phenoxide oxygen.

24.4 CH_3—C_6H_4—SO_3^- + ^-OH ⟶ CH_3 OH SO_3^- $\xrightarrow{-SO_3^{2-}}$ CH_3—C_6H_4—OH

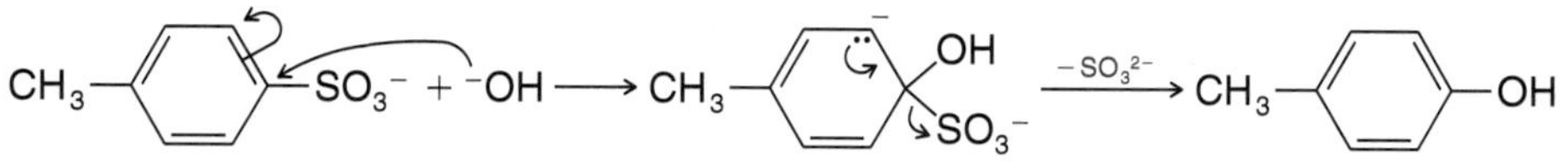

24.5 Cl + ^-OH ⟶ + H_2O + Cl^-;

then + ^-OH ⟶ OH $\xrightarrow{H_2O}$ OH

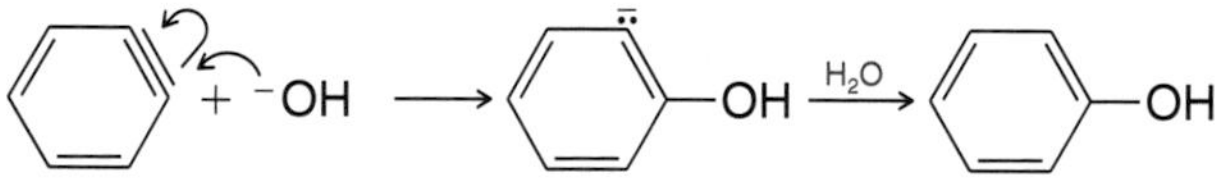

24.6 (*b*) $(CH_3)_3C$ OH CH_3 Br (*c*) OH $CH(CH_3)_2$ H_3C N O (*d*) OH $\overset{O}{\|}$ CCH_2CH_3 CH_3

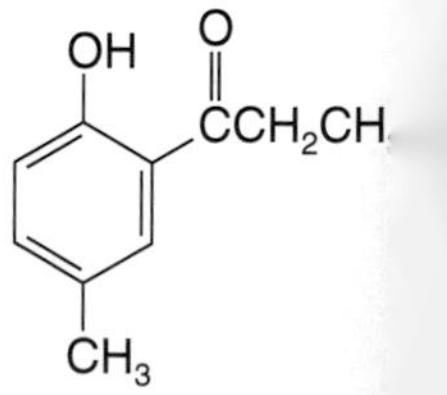

24.7

(*b*) OH + $CH_3\overset{O}{\overset{\|}{C}}O\overset{O}{\overset{\|}{C}}CH_3$ $\xrightarrow{NaOH}$ $O\overset{O}{\overset{\|}{C}}CH_3$ + $CH_3\overset{O}{\overset{\|}{C}}ONa$

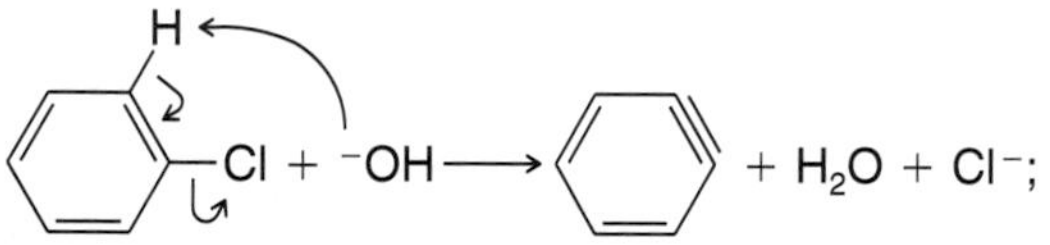

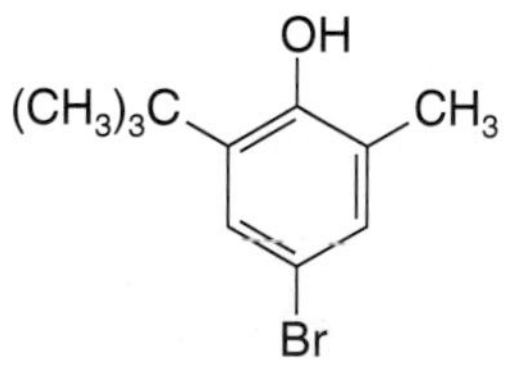

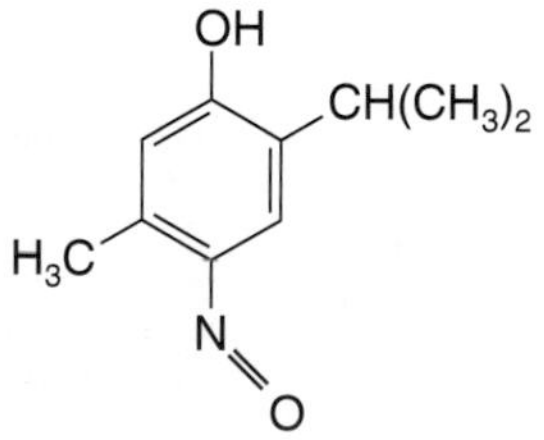

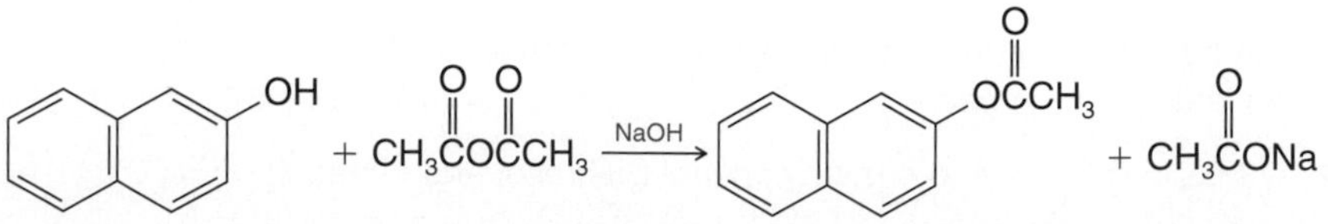

(*c*) $C_6H_5OH + C_6H_5\overset{O}{\overset{\|}{C}}Cl \longrightarrow C_6H_5O\overset{O}{\overset{\|}{C}}C_6H_5 + HCl$

24.8 $C_6H_5OCH_2CHCH_3$ (with OH on the CH)

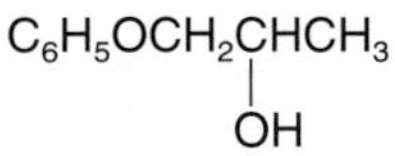

24.9 *p*-Fluoronitrobenzene and phenol (as its sodium or potassium salt)

24.10 OH, $CHCH{=}CH_2$, CH_3

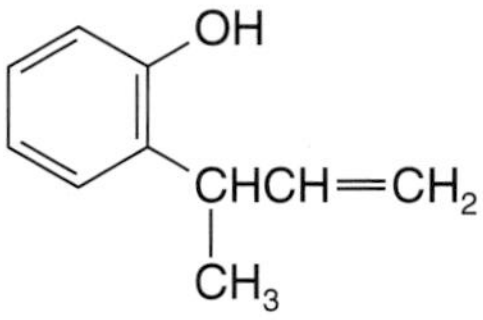

CHAPTER 25

25.1 (*b*) L-Glyceraldehyde; (*c*) D-glyceraldehyde

25.2 L-Erythrose

25.3 CHO; H—OH; HO—H; HO—H; CH_2OH

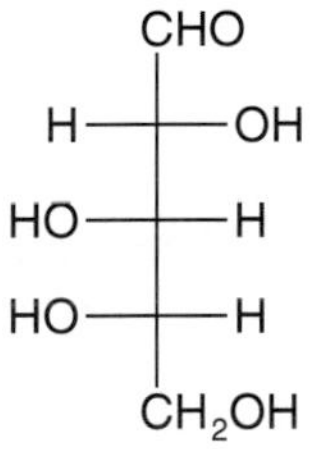

25.4 L-Talose

25.5

(*b*) $HOCH_2$, O, OH, H, H, HO, H, HO, H and $HOCH_2$, O, H, H, H, HO, OH, HO, H

(*c*) H, O, OH, $HOCH_2$, OH, H, H, H, OH and H, O, H, $HOCH_2$, OH, H, OH, H, OH

(*d*) O, OH, H, HO, H, HO, H and O, H, H, HO, OH, HO, H

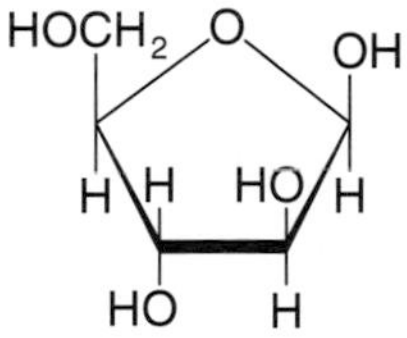

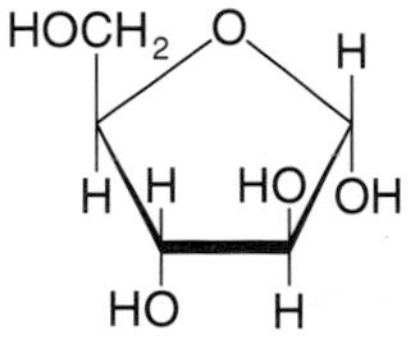

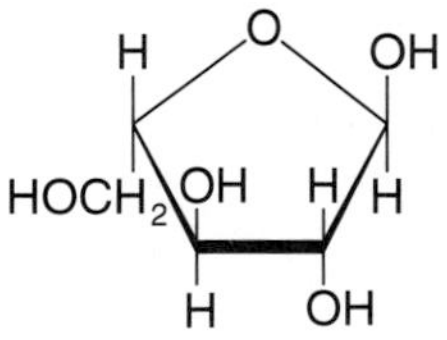

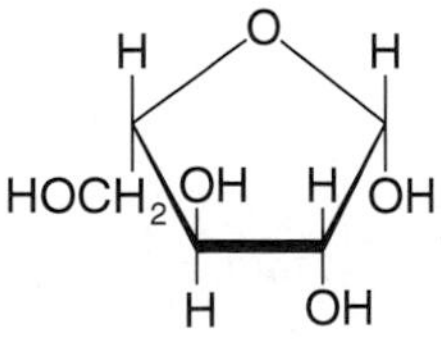

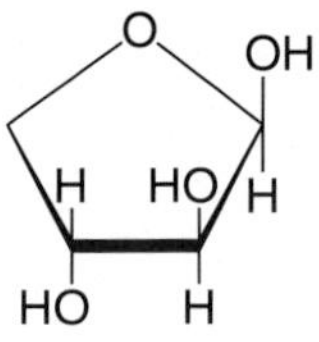

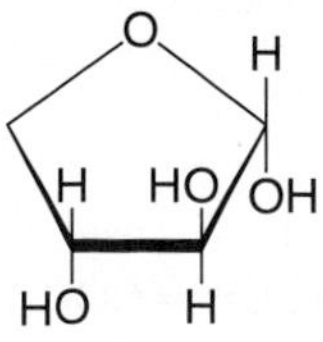

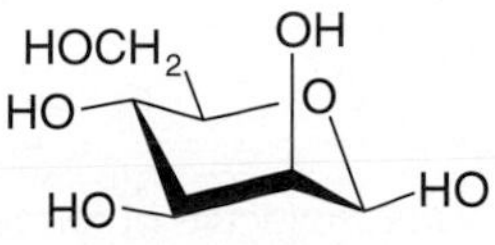

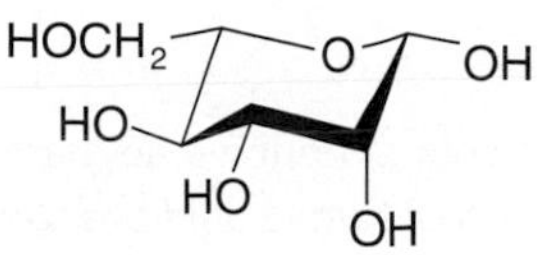

25.6 (*b*) $HOCH_2$, OH, HO, O, HO, HO (*c*) $HOCH_2$, O, OH, HO, HO, OH

(*d*) HO, HO, O, OH, OH

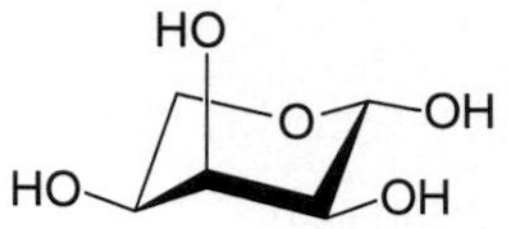

25.7 67% α, 33% β

25.8

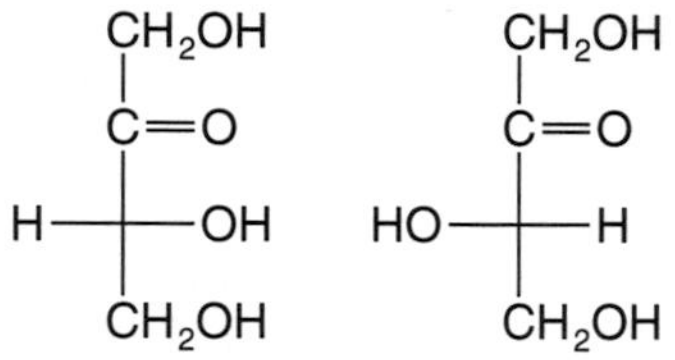

CH_2OH / C=O / H—C—OH / CH_2OH $\qquad$ CH_2OH / C=O / HO—C—H / CH_2OH

25.9 (*b*)

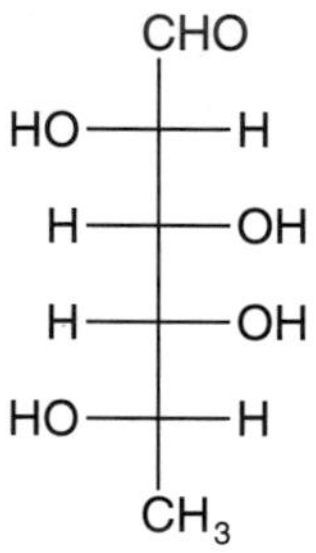

CHO / HO—C—H / H—C—OH / H—C—OH / HO—C—H / CH_3

25.10

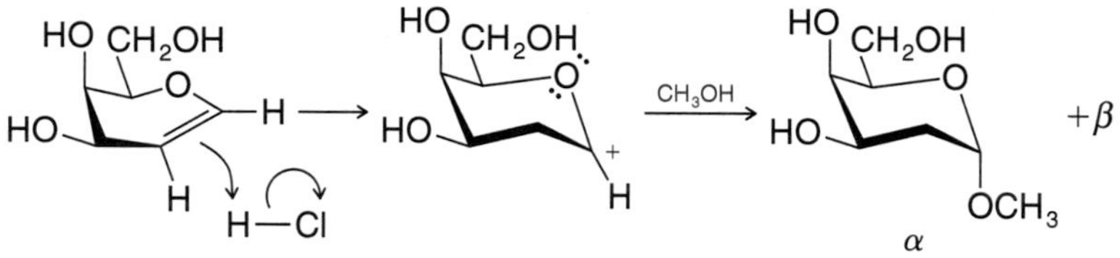

HO CH_2OH, O, HO, H, H, H—Cl $\longrightarrow$ HO CH_2OH, O, HO, +, H $\xrightarrow{CH_3OH}$ HO CH_2OH, O, HO, OCH_3 $\quad +\beta$

α

25.11

CH_2OH, O, HO, HO, HO, O, CH_2OH, OH, HO, HO, C=O, H

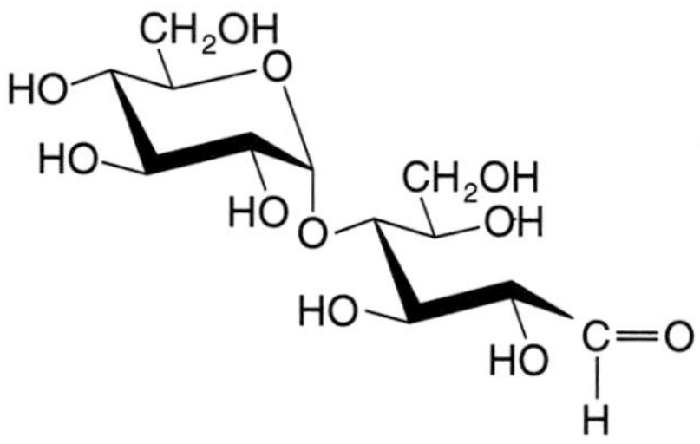

25.12 Maltose

25.13 No. The product is a meso form.

25.14 All (*b*) through (*f*) will give positive tests.

25.15 L-Gulose

25.16 The intermediate is an enediol, $HOCH{=}C(OH)CH_2OP(O)(OH)_2$

25.17

(*b*) Four equivalents of periodic acid are required. One molecule of formaldehyde and four molecules of formic acid are formed from each molecule of D-ribose.

(*c*) Two equivalents

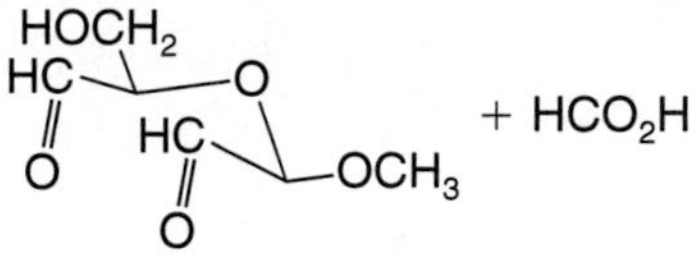

$$\text{HOCH}_2\text{-CH(CHO)-O-CH(OCH}_3\text{)-CHO} + HCO_2H$$

(*d*) Two equivalents

$$\text{furanoside with two CHO groups, CH(O) and OCH}_3 + \overset{O}{\overset{\|}{HCH}}$$

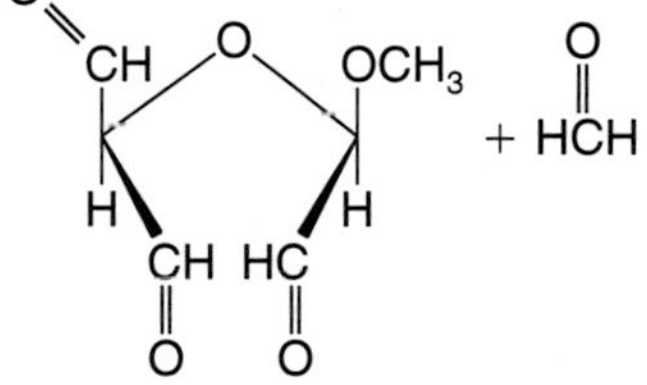

CHAPTER 26

26.1 Hydrolysis gives $CH_3(CH_2)_{16}CO_2H$ (two moles) and (Z)-$CH_3(CH_2)_7CH{=}CH(CH_2)_7$-$CO_2H$ (one mole). The same mixture of products is formed from 1-oleyl-2,3-distearylglycerol.

26.2 $CH_3C(OH)(SCoA)(S{-}ACP)$

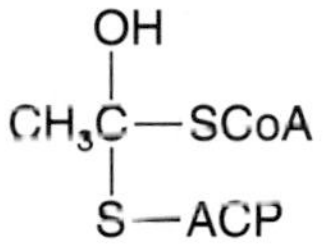

26.3 $CH_3(CH_2)_{12}\overset{O}{\overset{\|}{C}}S{-}ACP \longrightarrow CH_3(CH_2)_{12}\overset{O}{\overset{\|}{C}}CH_2\overset{O}{\overset{\|}{C}}{-}ACP \longrightarrow$

$CH_3(CH_2)_{12}\underset{OH}{\underset{|}{C}}HCH_2\overset{O}{\overset{\|}{C}}S{-}ACP \longrightarrow CH_3(CH_2)_{12}CH{=}CH\overset{O}{\overset{\|}{C}}S{-}ACP \longrightarrow$

$CH_3(CH_2)_{12}CH_2CH_2\overset{O}{\overset{\|}{C}}S{-}ACP$

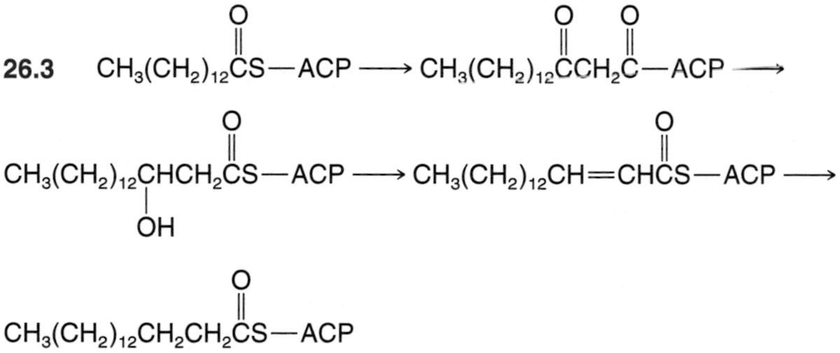

26.4 *R* in both cases

26.5 $CH_3(CH_2)_{14}\overset{O}{\overset{\|}{C}}O(CH_2)_{15}CH_3$

26.6 [structure: cyclopentanone with HO, CO_2H side chain, and OH side chain]

HO OH CO_2H

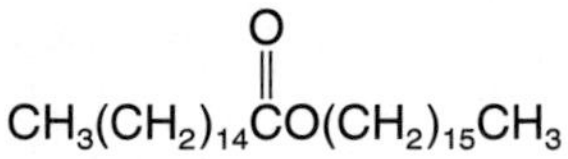

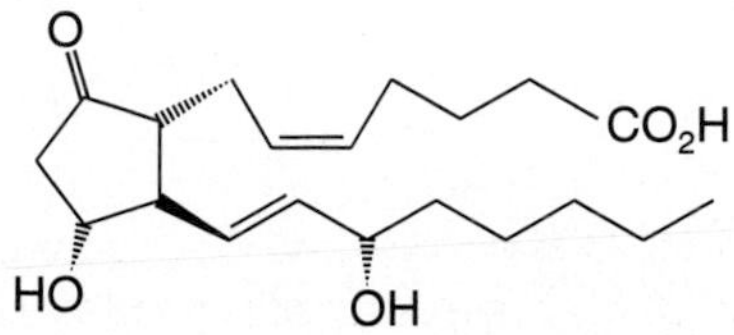

26.7

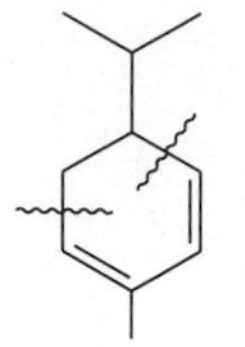

O
CH
OH

α-Phellandrene 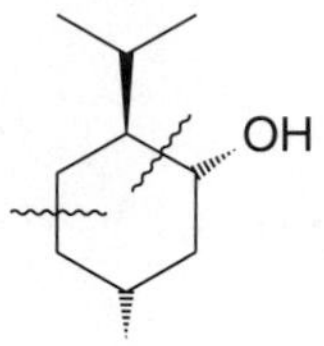   Menthol 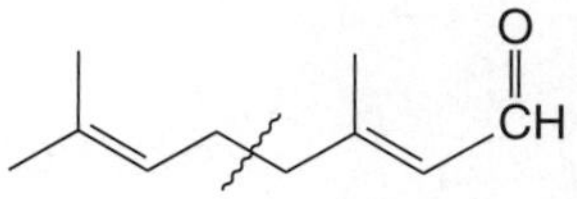   Citral

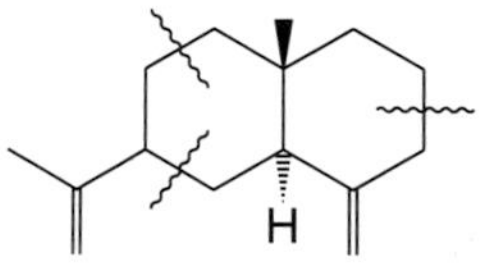

H
OH
OH
O
CO$_2$H

α-Selinene 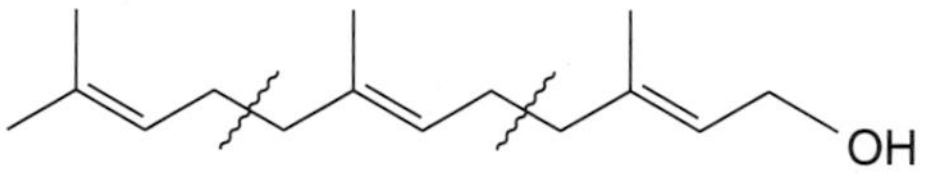   Farnesol 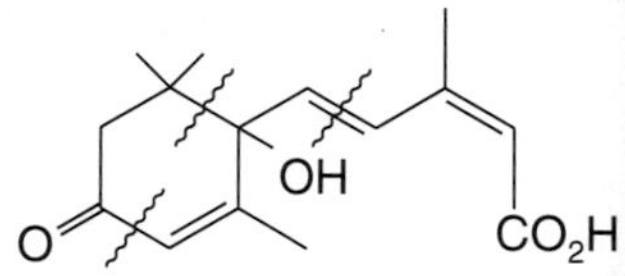   Abscisic acid

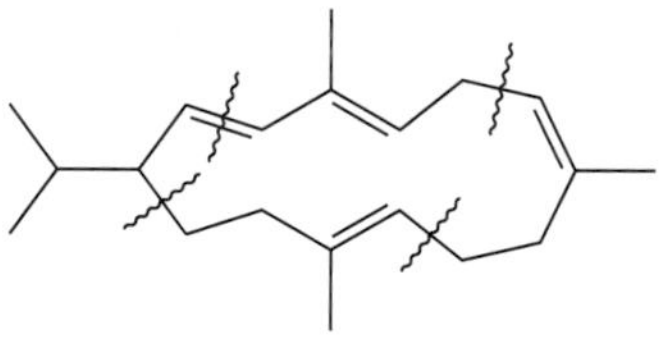

OH

Cembrene 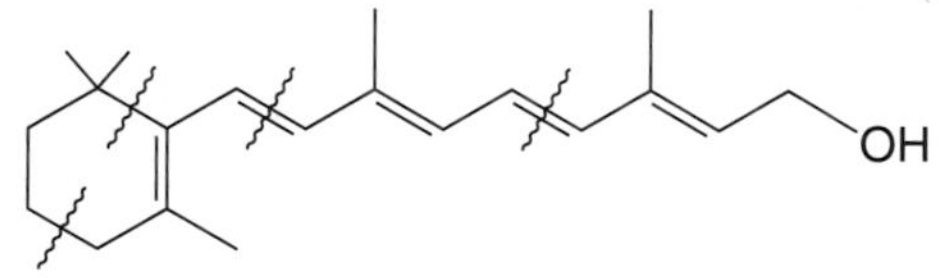   Vitamin A

26.8

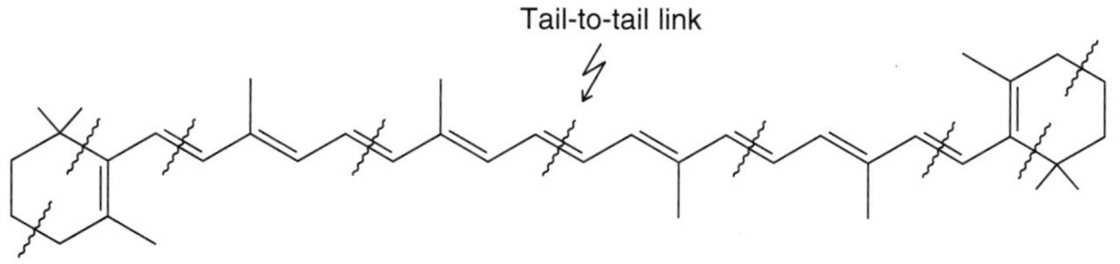

Tail-to-tail link

26.9

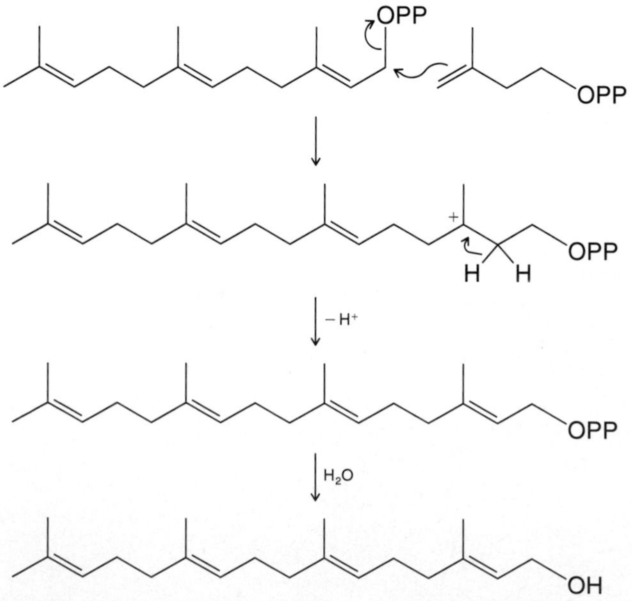

OPP
OPP

+
H H
OPP

– H$^+$

OPP

H_2O

OH

26.10

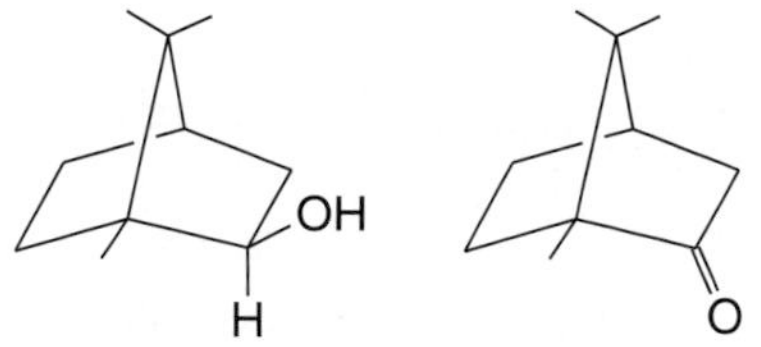

Isoborneol　　Camphor

26.11 Four carbons would be labeled with ^{14}C; they are C-1, C-3, C-5, and C-7.

26.12 (*b*) Hydrogens that migrate are those originally attached to C-13 and C-17 (steroid numbering); (*c*) the methyl group attached to C-15 of squalene 2,3-epoxide; (*d*) the methyl groups at C-2 and C-10 plus the terminal methyl group of squalene 2,3-epoxide.

26.13 All the methyl groups are labeled, plus C-1, C-3, C-5, C-7, C-9, C-13, C-15, C-17, C-20, and C-24 (steroid numbering).

26.14 The structure of vitamin D_2 is the same as that of vitamin D_3 except that vitamin D_2 has a double bond between C-22 and C-23 and a methyl substituent at C-24.

CHAPTER 27

27.1 (*b*) *R*; (*c*) *S*

27.2 Isoleucine and threonine

27.3

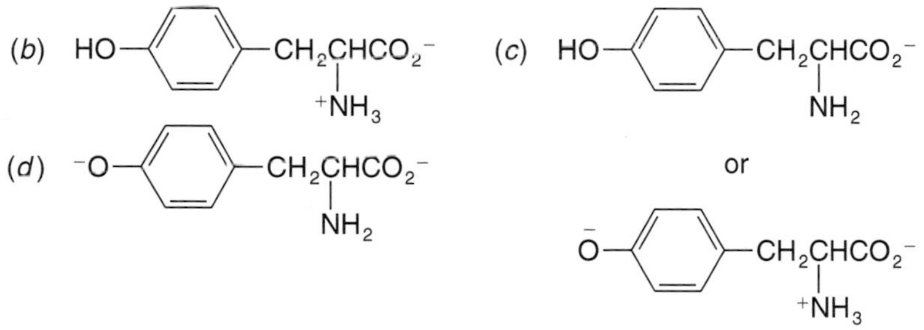

(*b*) $HO-C_6H_4-CH_2CH(\overset{+}{N}H_3)CO_2^-$　(*c*) $HO-C_6H_4-CH_2CH(NH_2)CO_2^-$

(*d*) $^-O-C_6H_4-CH_2CH(NH_2)CO_2^-$　or　$^-O-C_6H_4-CH_2CH(\overset{+}{N}H_3)CO_2^-$

27.4 At pH 1:

$H_3\overset{+}{N}CH_2CH_2CH_2CH_2CH(\overset{+}{N}H_3)CO_2H$

At pH 9:

$H_3\overset{+}{N}CH_2CH_2CH_2CH_2CH(NH_2)CO_2^-$

At pH 13:

$H_2NCH_2CH_2CH_2CH_2CH(NH_2)CO_2^-$

27.5 $(CH_3)_2CHCH_2CO_2H \xrightarrow[P]{Br_2} (CH_3)_2CHCH(Br)CO_2H \xrightarrow{NH_3} (CH_3)_2CHCH(\overset{+}{N}H_3)CO_2^-$

27.6 $(CH_3)_2CHCH{=}O \xrightarrow[NaCN]{NH_4Cl} (CH_3)_2CHCH(NH_2)CN \xrightarrow[2.\ HO^-]{1.\ H_2O,\ HCl,\ heat} (CH_3)_2CHCH(\overset{+}{N}H_3)CO_2^-$

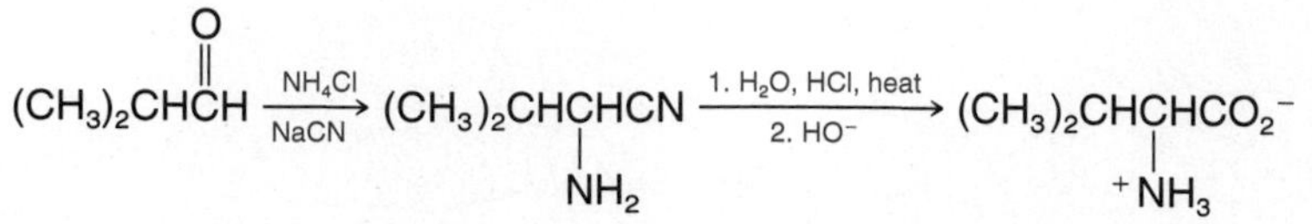

27.7 Treat the sodium salt of diethyl acetamidomalonate with isopropyl bromide. Remove the amide and ester functions by hydrolysis in aqueous acid; then heat to cause $(CH_3)_2CHC(\overset{+}{N}H_3)(CO_2H)_2$ to decarboxylate to give valine. The yield is low because isopropyl bromide is a secondary alkyl halide, because it is sterically hindered to nucleophilic attack, and because elimination competes with substitution.

27.8

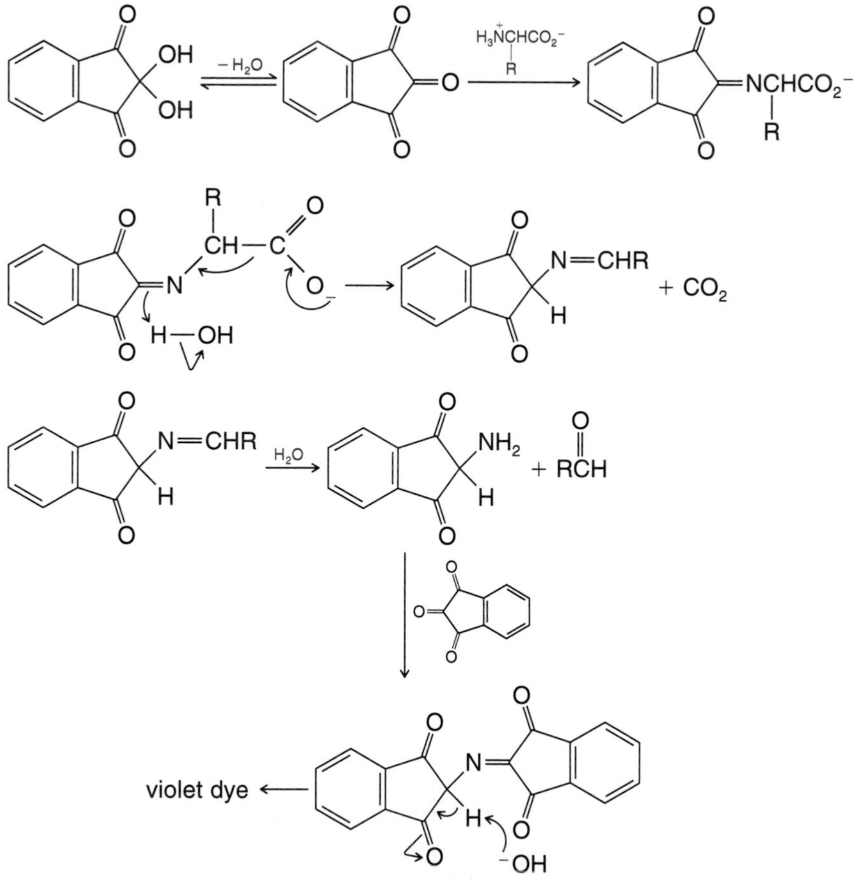

OH / OH (ninhydrin) $\underset{}{\overset{-H_2O}{\rightleftharpoons}}$ =O $\xrightarrow{H_3\overset{+}{N}CHRCO_2^-}$ =NCHRCO$_2^-$

=N–CHR–C(=O)O$^-$, H—OH → N=CHR, H + CO_2

N=CHR, H $\xrightarrow{H_2O}$ NH$_2$, H + RCH(=O)

(with ninhydrin, O=, O, O) ↓

violet dye ← N=, H, $^-$OH

27.9 Glutamic acid

27.10 (*b*) $H_3\overset{+}{N}CH(CH_3)C(=O)NHCH(CH_2C_6H_5)CO_2^-$ (*c*) $H_3\overset{+}{N}CH(CH_2C_6H_5)C(=O)NHCH(CH_3)CO_2^-$ (*d*) $H_3\overset{+}{N}CH_2C(=O)NHCH(CH_2CH_2CO_2^-)CO_2^-$

(*e*) $H_3\overset{+}{N}CH(CH_2CH_2CH_2CH_2\overset{+}{N}H_3)C(=O)NHCH_2CO_2^-$ (*f*) $H_3\overset{+}{N}CH(CH_3)C(=O)NHCH(CH_3)CO_2^-$

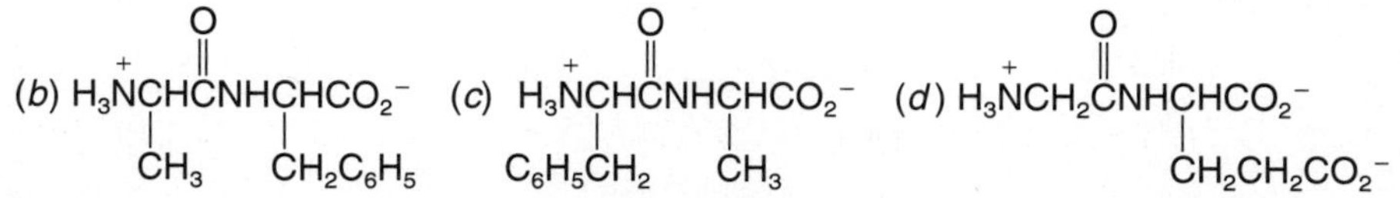

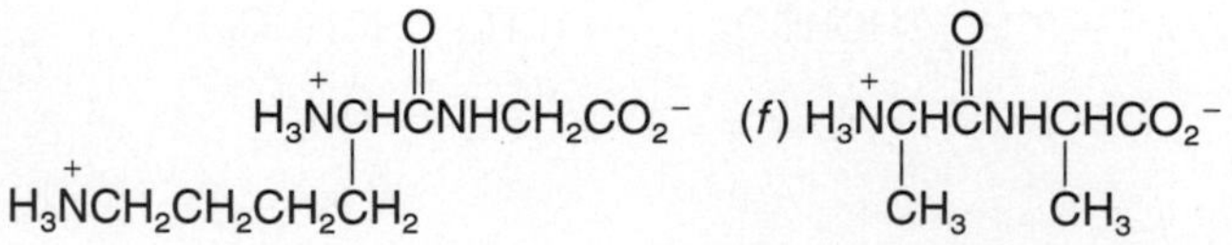

27.11 (*b*) $H_3\overset{+}{N}$, H_3C, H, O, N, H, $CH_2C_6H_5$, H, CO_2^-

(*c*) $H_3\overset{+}{N}$, $C_6H_5CH_2$, H, O, N, H, CH_3, H, CO_2^-

(*d*) $H_3\overset{+}{N}$, O, N, H, $CH_2CH_2CO_2^-$, H, CO_2^-

(*e*) $H_3\overset{+}{N}$, $H_3\overset{+}{N}CH_2CH_2CH_2CH_2$, H, O, N, H, CO_2^-

(*f*) $H_3\overset{+}{N}$, H, CH_3, O, N, H, H, CH_3, CO_2^-

27.12 Tyr-Gly-Gly-Phe-Met

27.13 Val; Phe; Gly; Ala; Val-Phe; Phe-Gly; Gly-Ala; Val-Phe-Gly; Phe-Gly-Ala

27.14 C_6H_5, N, S, O, HN, $CH_2C_6H_5$

27.15 $H_3\overset{+}{N}CH(CH_2CO_2^-)CNHCH(CH_2C_6H_5)COCH_3$ (C=O on both carbonyl carbons)

27.16 $C_6H_5CH_2OCNHCHCO_2H$ (C=O), with $CH_2CH_2CH_2CH_2NHCOCH_2C_6H_5$ (C=O) on the α-carbon

27.17 $H_3\overset{+}{N}CH(CH_3)CO_2^- + C_6H_5CH_2OCCl \longrightarrow C_6H_5CH_2OCNHCH(CH_3)CO_2H$

$H_3\overset{+}{N}CH(CH_2CH(CH_3)_2)CO_2^- + C_6H_5CH_2OH \xrightarrow[\text{2. HO}^-]{\text{1. H}^+\text{, heat}} H_2NCH(CH_2CH(CH_3)_2)CO_2CH_2C_6H_5$

$C_6H_5CH_2OCNHCH(CH_3)CO_2H + H_2NCH(CH_2CH(CH_3)_2)COCH_2C_6H_5 \xrightarrow{\text{DCCI}} C_6H_5CH_2OCNHCH(CH_3)CNHCH(CH_2CH(CH_3)_2)COCH_2C_6H_5$

$C_6H_5CH_2OCNHCH(CH_3)CNHCH(CH_2CH(CH_3)_2)COCH_2C_6H_5 \xrightarrow[\text{Pd}]{H_2}$ Ala-Leu

27.18 An *O*-acylisourea is formed by addition of the Z-protected amino acid to *N,N'*-dicyclohexylcarbodiimide, as shown in Figure 27.14. This *O*-acylisourea is attacked by *p*-nitrophenol.

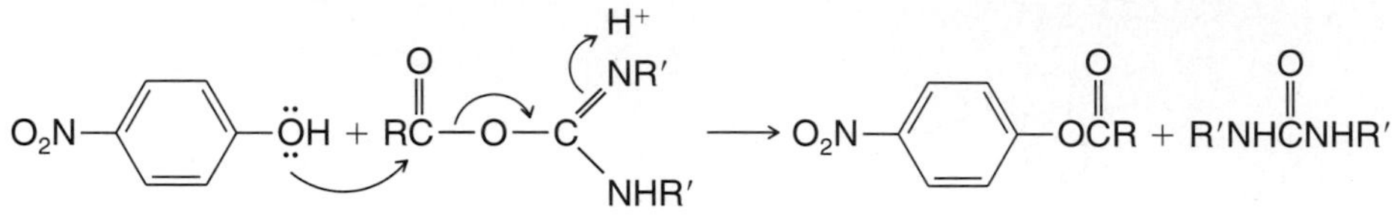

$$O_2N-C_6H_4-\ddot{O}H + RC(=O)-O-C(=NR')(NHR') \xrightarrow{H^+} O_2N-C_6H_4-OC(=O)R + R'NHC(=O)NHR'$$

27.19 Remove the Z protecting group from the ethyl ester of Z-Phe-Gly by hydrogenolysis. Couple with the *p*-nitrophenyl ester of Z-Leu; then remove the Z group of the ethyl ester of Z-Leu-Phe-Gly.

27.20 Protect glycine as its Boc derivative and anchor this to the solid support. Remove the protecting group and treat with Boc-protected phenylalanine and DCCI. Remove the Boc group with HCl; then treat with HBr in trifluoroacetic acid to cleave Phe-Gly from the solid support.

27.21 [5-fluorouracil structure: HN, O, F, N, H]

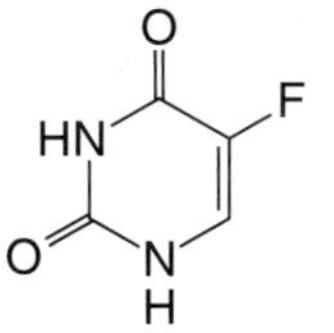

27.22 (*b*) Cytidine (*c*) Guanosine

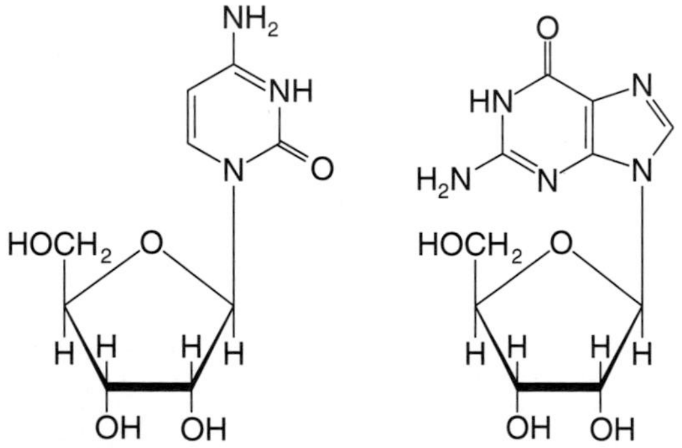

[Structures: cytidine — NH_2, NH, N, O, $HOCH_2$, O, H, H, H, H, OH, OH; guanosine — O, HN, N, H_2N, N, N, $HOCH_2$, O, H, H, H, H, OH, OH]

27.23 The codons for glutamic acid (GAA and GAG) differ by only one base from two of the codons for valine (GUA and GUG).

GLOSSARY

Absolute configuration (Section 7.5): The three-dimensional arrangement of atoms or groups at a stereogenic center.

Acetal (Section 17.9): Product of the reaction of an aldehyde or a ketone with two moles of an alcohol according to the equation

$$\mathrm{R\overset{O}{\overset{\|}{C}}R' + 2R''OH \xrightarrow{H^+} R\overset{OR''}{\underset{OR''}{C}}R' + H_2O}$$

Acetoacetic ester synthesis (Section 21.6): A synthetic method for the preparation of ketones in which alkylation of the enolate of ethyl acetoacetate ($\mathrm{CH_3\overset{O}{\overset{\|}{C}}CH_2\overset{O}{\overset{\|}{C}}OCH_2CH_3}$) is the key carbon-carbon bond–forming step.

Acetyl coenzyme A (Section 26.1): A thiol ester (abbreviated as $\mathrm{CH_3\overset{O}{\overset{\|}{C}}SCoA}$) that acts as the source of acetyl groups in biosynthetic processes involving acetate.

Acetylene (Sections 1.16 and 9.1): The simplest alkyne, HC≡CH.

Achiral (Section 7.1): Opposite of *chiral*. An achiral object is superimposable on its mirror image.

Acid (Section 4.6): According to the Arrhenius definition, a substance that ionizes in water to produce protons. According to the Brønsted-Lowry definition, a substance that donates a proton to some other substance. According to the Lewis definition, an electron pair acceptor.

Acid anhydride (Section 20.1): Compound of the type $\mathrm{R\overset{O}{\overset{\|}{C}}O\overset{O}{\overset{\|}{C}}R}$. Both R groups are usually the same, although they need not always be.

Acid dissociation constant K_a (Section 4.6): Equilibrium constant for dissociation of an acid:

$$K_a = \frac{[\mathrm{H^+}][\mathrm{A^-}]}{[\mathrm{HA}]}$$

Activating substituent (Sections 12.10 and 12.12): A group that when present in place of a hydrogen substituent causes a particular reaction to occur faster. Term is most often applied to substituents that increase the rate of electrophilic aromatic substitution.

Active site (Section 27.19): The region of an enzyme at which the substrate is bound.

Acylation (Section 12.7 and Chapter 20): Reaction in which an acyl group becomes attached to some structural unit in a molecule. Examples include the Friedel-Crafts acylation and the conversion of amines to amides.

Acyl chloride (Section 20.1): Compound of the type $\mathrm{R\overset{O}{\overset{\|}{C}}Cl}$. R may be alkyl or aryl.

Acyl group (Sections 12.7 and 20.1): The group $\mathrm{R\overset{O}{\overset{\|}{C}}—}$. R may be alkyl or aryl.

Acylium ion (Section 12.7): The cation $\mathrm{R—C{\equiv}\overset{+}{O}:}$.

Acyl transfer (Section 20.3): A nucleophilic acyl substitution. A reaction in which one type of carboxylic acid derivative is converted to another.

Addition (Section 6.1): Reaction in which a reagent X—Y adds to a multiple bond so that X becomes attached to one of the carbons of the multiple bond and Y to the other.

1,2 addition (Section 10.10): Addition of reagents of the type X—Y to conjugated dienes in which X and Y add to adjacent doubly bonded carbons:

$$\mathrm{R_2C{=}CH{-}CH{=}CR_2 \xrightarrow{X-Y} R_2\underset{X}{C}{-}\underset{Y}{C}H{-}CH{=}CR_2}$$

1,4 addition (Section 10.10): Addition of reagents of the type X—Y to conjugated dienes in which X and Y add to the termini of the diene system:

$$\mathrm{R_2C{=}CH{-}CH{=}CR_2 \xrightarrow{X-Y} R_2\underset{X}{C}{-}CH{=}CH{-}\underset{Y}{C}R_2}$$

Addition-elimination mechanism (Section 23.6): Two-stage mechanism for nucleophilic aromatic substitution. In the addition stage, the nucleophile adds to the carbon that bears the leaving group. In the elimination stage, the leaving group is expelled.

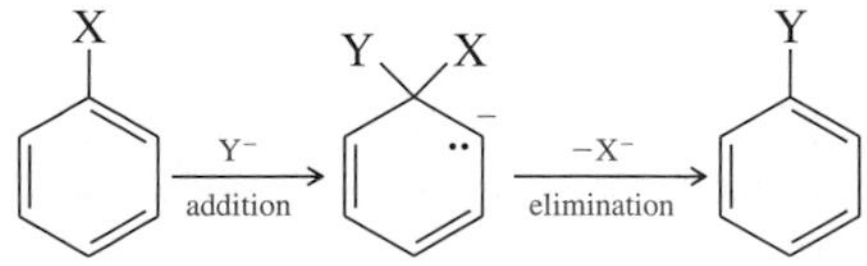

Alcohol (Section 4.2): Compound of the type ROH.

Alcohol dehydrogenase (Section 15.11): Enzyme in the liver that catalyzes the oxidation of alcohols to aldehydes and ketones.

Aldaric acid (Section 25.19): Carbohydrate in which carboxylic acid functions are present at both ends of the chain. Aldaric acids are typically prepared by oxidation of aldoses with nitric acid.

Aldehyde (Section 17.1): Compound of the type $RCH(=O)$ or $ArCH(=O)$.

Alder rule (Section 10.13): The major stereoisomer formed in a Diels-Alder reaction is derived from the transition state in which the multiple bonds are closest to one another.

Alditol (Section 25.18): The polyol obtained on reduction of the carbonyl group of a carbohydrate.

Aldol addition (Section 18.9): Nucleophilic addition of an aldehyde or ketone enolate to the carbonyl group of an aldehyde or a ketone. The most typical case involves two molecules of an aldehyde, and is usually catalyzed by bases.

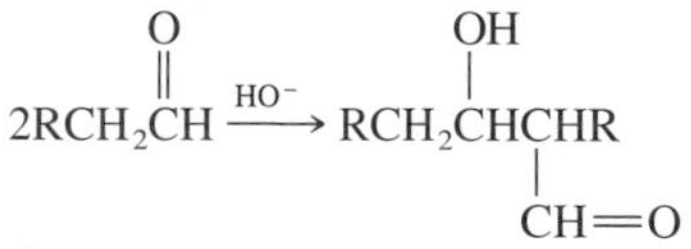

$$2RCH_2CH(=O) \xrightarrow{HO^-} RCH_2CH(OH)CH(CH{=}O)R$$

Aldol condensation (Sections 18.9–18.12): When an aldol addition is carried out so that the *β*-hydroxy aldehyde or ketone dehydrates under the conditions of its formation, the product is described as arising by an aldol condensation.

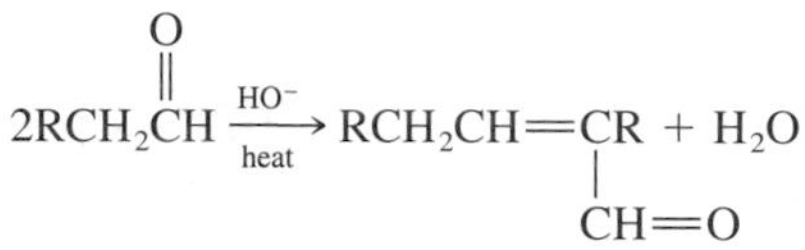

$$2RCH_2CH(=O) \xrightarrow[\text{heat}]{HO^-} RCH_2CH{=}C(CH{=}O)R + H_2O$$

Aldonic acid (Section 25.19): Carboxylic acid obtained by oxidation of the aldehyde function of an aldose.

Aldose (Section 25.1): Carbohydrate that contains an aldehyde carbonyl group in its open-chain form.

Alicyclic (Section 2.14): Term describing an *ali*phatic *cyclic* structural unit.

Aliphatic (Section 2.1): Term applied to compounds that do not contain benzene or benzenelike rings as structural units. (Historically, *aliphatic* was used to describe compounds derived from fats and oils.)

Alkadiene (Section 10.5): Hydrocarbon that contains two carbon-carbon double bonds; commonly referred to as a *diene.*

Alkaloid (Section 22.5): Amine that occurs naturally in plants. The name derives from the fact that such compounds are weak bases.

Alkane (Section 2.1): Hydrocarbon in which all the bonds are single bonds. Alkanes have the general formula C_nH_{2n+2}.

Alkene (Section 2.1): Hydrocarbon that contains a carbon-carbon double bond (C=C); also known by the older name *olefin.*

Alkoxide ion (Section 4.8): Conjugate base of an alcohol; a species of the type $R{-}\ddot{\underset{..}{O}}{:}^-$.

Alkylamine (Section 22.1): Amine in which the organic groups attached to nitrogen are alkyl groups.

Alkylation (Section 9.7): Reaction in which an alkyl group is attached to some structural unit in a molecule.

Alkyl group (Section 2.12): Structural unit related to an alkane by replacing one of the hydrogens by a potential point of attachment to some other atom or group. The general symbol for an alkyl group is R—.

Alkyl halide (Section 4.1): Compound of the type RX, in which X is a halogen substituent (F, Cl, Br, I).

Alkyloxonium ion (Section 4.6): Positive ion of the type ROH_2^+.

Alkyne (Section 2.1): Hydrocarbon that contains a carbon-carbon triple bond.

Allene (Section 10.5): The compound $CH_2{=}C{=}CH_2$.

Allyl cation (Section 10.2): The carbocation $CH_2{=}CHCH_2^+$. The carbocation is stabilized by delocalization of the *π* electrons of the double bond, and the positive charge is shared by the two CH_2 groups. Substituted analogs of allyl cation are called *allylic carbocations.*

Allyl group (Sections 5.1, 10.1): The group $CH_2{=}CHCH_2{-}$.

Allylic rearrangement (Section 10.2): Functional group transformation in which double-bond migration has converted one allylic structural unit to another, as in:

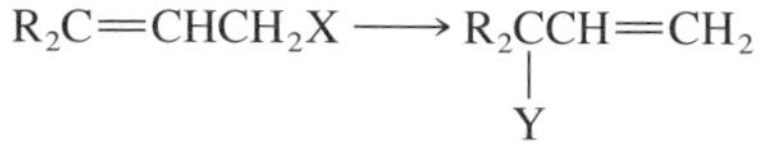

$$R_2C{=}CHCH_2X \longrightarrow R_2C(Y)CH{=}CH_2$$

Amide (Section 20.1): Compound of the type $RC(=O)NR'_2$.

Amine (Chapter 22): Molecule in which a nitrogen-containing group of the type $-NH_2$, $-NHR$, or $-NR_2$ is attached to an alkyl or aryl group.

***α*-Amino acid** (Section 27.1): A carboxylic acid that contains an amino group at the *α* carbon atom. *α*-Amino acids are the building blocks of peptides and proteins. An *α*-amino acid normally exists as a *zwitterion.*

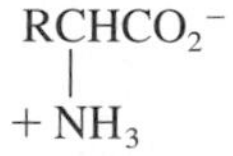

$$RCH(\overset{+}{N}H_3)CO_2^-$$

L-Amino acid (Section 27.2): A description of the stereochemistry at the *α* carbon atom of a chiral amino acid. The Fischer projection of an *α*-amino acid has the amino group on the left when the carbon chain is vertical with the carboxyl group at the top.

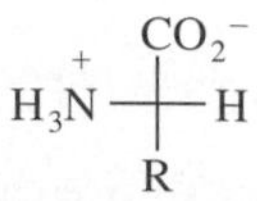

$$\begin{array}{c} CO_2^- \\ H_3\overset{+}{N}{-}\!\!\!\!+\!\!\!\!{-}H \\ R \end{array}$$

Amino acid residues (Section 27.7): Individual amino acid components of a peptide or protein.

Amino sugar (Section 25.11): Carbohydrate in which one of the hydroxyl groups has been replaced by an amino group.

Amylopectin (Section 25.15): A polysaccharide present in starch. Amylopectin is a polymer of α(1,4)-linked glucose units, as is amylose (see *amylose*). Unlike amylose, amylopectin contains branches of 24 to 30 glucose units connected to the main chain by an α(1,6) linkage.

Amylose (Section 25.15): The water-dispersible component of starch. It is a polymer of α(1,4)-linked glucose units.

Anabolic steroid (Section 26.15): A steroid that promotes muscle growth.

Androgen (Section 26.15): A male sex hormone.

Angle strain (Section 3.5): The strain a molecule possesses because its bond angles are distorted from their normal values.

Anion (Section 1.2): Negatively charged ion.

Annulene (Section 11.21): Monocyclic hydrocarbon characterized by a completely conjugated system of double bonds. Annulenes may or may not be aromatic.

Anomeric carbon (Section 25.6): The carbon atom in a furanose or pyranose form that is derived from the carbonyl carbon of the open-chain form. It is the ring carbon that is bonded to two oxygens.

Anomeric effect (Section 25.8): The preference for an electronegative substitutent, especially a hydroxyl group, to occupy an axial orientation when bonded to the anomeric carbon in the pyranose form of a carbohydrate.

Anti (Section 3.1): Term describing relative position of two substituents on adjacent atoms when the angle between their bonds is on the order of 180°. Atoms X and Y in the structure shown are anti to each other.

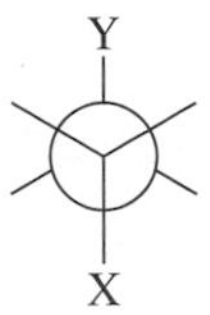

Anti addition (Section 6.3): Addition reaction in which the two portions of the attacking reagent X—Y add to opposite faces of the double bond.

Antibonding orbital (Section 1.12): An orbital in a molecule in which an electron is less stable than when localized on an isolated atom.

Anticodon (Section 27.27): Sequence of three bases in a molecule of tRNA that is complementary to the codon of mRNA for a particular amino acid.

Anti-Markovnikov addition (Sections 6.8, 6.11): Addition reaction for which the regioselectivity is opposite to that predicted on the basis of Markovnikov's rule.

Aprotic solvent (Section 8.13): A solvent that does not have easily exchangeable protons such as those bonded to oxygen of hydroxyl groups.

Ar— (Section 2.2): Symbol for an aryl group.

Arene (Section 2.1): Aromatic hydrocarbon. Often abbreviated ArH.

Arenium ion (Section 12.2): The carbocation intermediate formed by attack of an electrophile on an aromatic substrate in electrophilic aromatic substitution. See *cyclohexadienyl cation.*

Aromatic compound (Section 11.3): An electron-delocalized species that is much more stable than any structure written for it in which all the electrons are localized either in covalent bonds or as unshared electron pairs.

Aromaticity (Section 11.3): Special stability associated with aromatic compounds.

Arylamine (Section 22.1): An amine that has an aryl group attached to the amine nitrogen.

Aryne (Section 23.8): A species that contains a triple bond within an aromatic ring (see *benzyne*).

Asymmetric (Section 7.1): Lacking all significant symmetry elements; an asymmetric object does not have a plane, axis, or center of symmetry.

Asymmetric center (Section 7.2): Obsolete name for a *stereogenic center.*

Atactic polymer (Section 7.15): Polymer characterized by random stereochemistry at its stereogenic centers. An atactic polymer, unlike an isotactic or a syndiotactic polymer, is not a stereoregular polymer.

Atomic number (Section 1.1): The number of protons in the nucleus of a particular atom. The symbol for atomic number is Z, and each element has a unique atomic number.

Axial bond (Section 3.7): A bond to a carbon in the chair conformation of cyclohexane oriented like the six "up-and-down" bonds in the following:

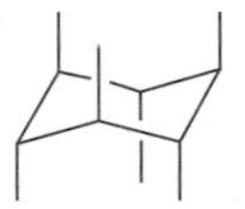

Azo coupling (Section 22.19): Formation of a compound of the type ArN=NAr′ by reaction of an aryl diazonium salt with an arene. The arene must be strongly activated toward electrophilic aromatic substitution; i.e., it must bear a powerful electron-releasing substituent such as —OH or $—NR_2$.

Baeyer strain theory (Section 3.5): Incorrect nineteenth-century theory that considered the rings of cycloalkanes to be planar, and assessed their stabilities according to how much the angles of a corresponding regular polygon deviated from the tetrahedral value of 109.5°.

Baeyer-Villiger oxidation (Section 17.17): Oxidation of an aldehyde or, more commonly, a ketone with a peroxy acid. The product of Baeyer-Villiger oxidation of a ketone is an ester.

$$\mathrm{R}\overset{\overset{\mathrm{O}}{\|}}{\mathrm{C}}\mathrm{R'} \xrightarrow{\mathrm{R''}\overset{\overset{\mathrm{O}}{\|}}{\mathrm{C}}\mathrm{OH}} \mathrm{R}\overset{\overset{\mathrm{O}}{\|}}{\mathrm{C}}\mathrm{OR'}$$

Ball-and-stick model (Section 1.10): Type of molecular model in which balls representing atoms are connected by sticks representing bonds.

Base (Section 4.6): According to the Arrhenius definition, a substance that ionizes in water to produce hydroxide ions. According to the Brønsted-Lowry definition, a substance that accepts a proton from some suitable donor. According to the Lewis definition, an electron pair donor.

Base pair (Section 27.26): Term given to the purine of a nucleotide and its complementary pyrimidine. Adenine (A) is complementary to thymine (T), and guanine (G) is complementary to cytosine (C).

Base peak (Section 13.20): The most intense peak in a mass spectrum. The base peak is assigned a relative intensity of 100, and the intensities of all other peaks are cited as a percentage of the base peak.

Basicity constant K_b (Section 22.4): A measure of base strength, especially of amines.

$$K_b = \frac{[R_3NH^+][HO^-]}{[R_3N]}$$

Bending vibration (Section 13.18): The regular, repetitive motion of an atom or a group along an arc the radius of which is the bond connecting the atom or group to the rest of the molecule. Bending vibrations are one type of molecular motion that gives rise to a peak in the infrared spectrum.

Benedict's reagent (Section 25.19): A solution containing the citrate complex of $CuSO_4$. It is used to test for the presence of reducing sugars.

Benzene (Section 11.1): The most typical aromatic hydrocarbon:

Benzyl group (Section 11.9): The group $C_6H_5CH_2$—.

Benzylic carbon (Section 11.12): A carbon directly attached to a benzene ring. A hydrogen attached to a benzylic carbon is a benzylic hydrogen. A carbocation in which the benzylic carbon is positively charged is a benzylic carbocation. A free radical in which the benzylic carbon bears the unpaired electron is a benzylic radical.

Benzyne (Section 23.8): The compound

Benzyne is formed as a reactive intermediate in the reaction of aryl halides with very strong bases such as potassium amide.

Bile acids (Section 26.13): Steroid derivatives biosynthesized in the liver that aid digestion by emulsifying fats.

Bimolecular (Section 4.7): A process in which two particles react in the same elementary step.

Biological isoprene unit (Section 26.8): Isopentenyl pyrophosphate, the biological precursor to terpenes and steroids:

Birch reduction (Section 11.13): Reduction of an aromatic ring to a 1,4-cyclohexadiene on treatment with a group I metal (Li, Na, K) and an alcohol in liquid ammonia.

Boat conformation (Section 3.6): An unstable conformation of cyclohexane, depicted as

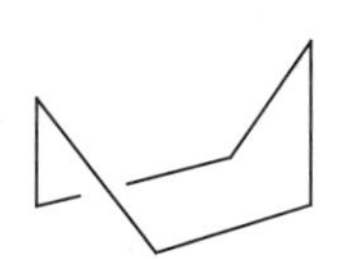

π bond (Section 1.15): In alkenes, a bond formed by overlap of *p* orbitals in a side-by-side manner. A π bond is weaker than a σ bond. The carbon-carbon double bond in alkenes consists of two sp^2 hybridized carbons joined by a σ bond and a π bond.

σ bond (Section 1.12): A connection between two atoms in which the electron probability distribution has rotational symmetry along the internuclear axis. A cross section perpendicular to the internuclear axis is a circle.

Bond dissociation energy (Section 1.3): For a substance A:B, the energy required to break the bond between A and B so that each retains one of the electrons in the bond. Table 4.3 (Section 4.18) gives bond dissociation energies for some representative compounds.

Bonding orbital (Section 1.12): An orbital in a molecule in which an electron is more stable than when localized on an isolated atom. All the bonding orbitals are normally doubly occupied in stable neutral molecules.

Bond-line formula (Section 1.7): Formula in which connections between carbons are shown but individual carbons and hydrogens are not. The bond-line formula represents the compound $(CH_3)_2CHCH_2CH_3$.

Boundary surface (Section 1.1): The surface that encloses the region of an orbital where the probability of finding an electron is high (90 to 95 percent).

Branched-chain carbohydrate (Section 25.12): Carbohydrate in which the main carbon chain bears a carbon substituent in place of a hydrogen or hydroxyl group.

Bromohydrin (Section 6.17): A halohydrin in which the halogen is bromine (see *halohydrin*).

Bromonium ion (Section 6.16): A halonium ion in which the halogen is bromine (see *halonium ion*).

Brønsted acid See *acid.*

Brønsted base See *base.*

Buckminsterfullerene (Chapter 11, box, ''Carbon Clusters and Fullerenes''): Name given to the C_{60} cluster with structure resembling the geodesic domes of R. Buckminster Fuller; see front cover.

***n*-Butane** (Section 2.7): Common name for butane $CH_3CH_2CH_2CH_3$.

***n*-Butyl group** (Section 2.12): The group $CH_3CH_2CH_2CH_2—$.

***sec*-Butyl group** (Section 2.12): The group $CH_3CH_2\underset{|}{C}HCH_3$.

***tert*-Butyl group** (Section 2.12): The group $(CH_3)_3C—$.

Cahn-Ingold-Prelog notation (Section 7.6): System for specifying absolute configuration as *R* or *S* on the basis of the order in which atoms or groups are attached to a stereogenic center. Groups are ranked in order of precedence according to rules based on atomic number.

Carbamate (Section 20.17): An ester of carbamic acid (H_2NCO_2H); a compound of the type H_2NCO_2R.

Carbanion (Section 9.6): Anion in which the negative charge is borne by carbon. An example is acetylide ion.

Carbene (Section 14.13): A neutral molecule in which one of the carbon atoms is associated with six valence electrons.

Carbinolamine (Section 17.11): Compound of the type $HO—\overset{|}{\underset{|}{C}}—NR_2$. Carbinolamines are formed by nucleophilic addition of an amine to a carbonyl group and are intermediates in the formation of imines and enamines.

Carbocation (Section 4.10): Positive ion in which the charge resides on carbon. An example is *tert*-butyl cation, $(CH_3)_3C^+$. Carbocations are unstable species that, though they cannot normally be isolated, are believed to be intermediates in certain reactions.

Carboxylate ion (Section 19.5): The conjugate base of a carboxylic acid, an ion of the type RCO_2^-.

Carboxylation (Section 19.11): In the preparation of a carboxylic acid, the reaction of a carbanion with carbon dioxide. Typically, the carbanion source is a Grignard reagent.

$$RMgX \xrightarrow[2.\ H_3O^+]{1.\ CO_2} RCO_2H$$

Carboxylic acid (Section 19.1): Compound of the type $R\overset{O}{\overset{\|}{C}}OH$, also written as RCO_2H.

Carboxylic acid derivative (Section 20.1): Compound that yields a carboxylic acid on hydrolysis. Carboxylic acid derivatives include acyl chlorides, anhydrides, esters, and amides.

Carotenoids (Section 26.16): Naturally occurring tetraterpenoid plant pigments.

Cation (Section 1.2): Positively charged ion.

Cellobiose (Section 25.14): A disaccharide in which two glucose units are joined by a $\beta(1,4)$ linkage. Cellobiose is obtained by the hydrolysis of cellulose.

Cellulose (Section 25.15): A polysaccharide in which thousands of glucose units are joined by $\beta(1,4)$ linkages.

Center of symmetry (Section 7.3): A point in the center of a structure located so that a line drawn from it to any element of the structure, when extended an equal distance in the opposite direction, encounters an identical element. Benzene, for example, has a center of symmetry.

Chain reaction (Section 4.19): Reaction mechanism in which a sequence of individual steps repeats itself many times, usually because a reactive intermediate consumed in one step is regenerated in a subsequent step. The halogenation of alkanes is a chain reaction proceeding via free-radical intermediates.

Chair conformation (Section 3.6): The most stable conformation of cyclohexane:

Chemical shift (Section 13.4): A measure of how shielded the nucleus of a particular atom is. Nuclei of different atoms have different chemical shifts, and nuclei of the same atom have chemical shifts that are sensitive to their molecular environment. In proton and carbon-13 nmr, chemical shifts are cited as δ, or parts per million (ppm), from the hydrogens or carbons, respectively, of tetramethylsilane.

Chiral (Section 7.1): Term describing an object that is not superimposable on its mirror image.

Chiral carbon atom (Section 7.2): A carbon that is bonded to four groups, all of which are different from one another. Also called an *asymmetric carbon atom.* A more modern term is *stereogenic center.*

Chiral center (Section 7.2): See *stereogenic center.*

Chlorohydrin (Section 6.17): A halohydrin in which the halogen is chlorine (see *halohydrin*).

Chloronium ion (Section 6.16): A halonium ion in which the halogen is chlorine (see *halonium ion*).

Cholesterol (Section 26.11): The most abundant steroid in animals and the biological precursor to other naturally occurring steroids, including the bile acids, sex hormones, and corticosteroids.

Chromatography (Section 13.20): A method for separation and analysis of mixtures based on the different rates at which different compounds are removed from a stationary phase by a moving phase.

Chromophore (Section 13.19): The structural unit of a molecule principally responsible for absorption of radiation of a particular frequency; a term usually applied to ultraviolet-visible spectroscopy.

Chymotrypsin (Section 27.12): A digestive enzyme that catalyzes the hydrolysis of proteins. Chymotrypsin selectively catalyzes the cleavage of the peptide bond between the carboxyl group of phenylalanine, tyrosine, or tryptophan and some other amino acid.

***cis*-** (Section 3.13): Stereochemical prefix indicating that two substituents are on the same side of a ring or double bond. (Contrast with the prefix *trans*-.)

Claisen condensation (Section 21.1): Reaction in which a β-keto ester is formed by condensation of two moles of an ester in base:

$$RCH_2\overset{O}{\overset{\|}{C}}OR' \xrightarrow[2.\ H^+]{1.\ NaOR'} RCH_2\overset{O}{\overset{\|}{C}}\underset{R}{\underset{|}{C}H}\overset{O}{\overset{\|}{C}}OR' + R'OH$$

Claisen rearrangement (Section 24.13): Thermal conversion of an allyl phenyl ether to an *o*-allyl phenol. The rearrangement proceeds via a cyclohexadienone intermediate.

$$C_6H_5OCH_2CH{=}CH_2 \xrightarrow{heat} o\text{-}HOC_6H_4CH_2CH{=}CH_2$$

Claisen-Schmidt condensation (Section 18.11): A mixed aldol condensation involving a ketone enolate and an aromatic aldehyde or ketone.

Clemmensen reduction (Section 12.8): Method for reducing the carbonyl group of aldehydes and ketones to a methylene group ($C{=}O \rightarrow CH_2$) by treatment with zinc amalgam [Zn(Hg)] in concentrated hydrochloric acid.

Closed-shell electron configuration (Sections 1.1 and 11.8): Stable electron configuration in which all the lowest-energy orbitals of an atom (in the case of the noble gases), an ion (e.g., Na^+), or a molecule (e.g., benzene) are filled.

^{13}C nmr (Section 13.16): Nuclear magnetic resonance spectroscopy in which the environments of individual carbon atoms are examined via their mass 13 isotope.

Codon (Section 27.27): Set of three successive nucleotides in mRNA that is unique for a particular amino acid. The 64 codons possible from combinations of A, T, G, and C code for the 20 amino acids from which proteins are constructed.

Coenzyme (Section 27.20): Molecule that acts in combination with an enzyme to bring about a reaction.

Coenzyme Q (Section 24.14): Naturally occurring group of related quinones involved in the chemistry of cellular respiration. Also known as *ubiquinone*.

Combustion (Section 2.17): Burning of a substance in the presence of oxygen. All hydrocarbons yield carbon dioxide and water when they undergo combustion.

Common nomenclature (Section 2.10): Names given to compounds on some basis other than a comprehensive, systematic set of rules.

Concerted reaction (Section 4.7): Reaction that occurs in a single step.

Condensation polymer (Section 20.16): Polymer in which the bonds that connect the monomers are formed by condensation reactions. Typical condensation polymers include polyesters and polyamides.

Condensation reaction (Section 15.7): Reaction in which two molecules combine to give a product accompanied by the expulsion of some small stable molecule (such as water). An example is acid-catalyzed ether formation:

$$2ROH \xrightarrow{H_2SO_4} ROR + H_2O$$

Condensed structural formula (Section 1.7): A standard way of representing structural formulas in which subscripts are used to indicate replicated atoms or groups, as in $(CH_3)_2CHCH_2CH_3$.

Conformational analysis (Section 3.1): Study of the conformations available to a molecule, their relative stability, and the role they play in defining the properties of the molecule.

Conformations (Section 3.1): Nonidentical representations of a molecule generated by rotation about single bonds.

Conformers (Section 3.1): Different conformations of a single molecule.

Conjugate acid (Section 4.6): The species formed from a Brønsted base after it has accepted a proton.

Conjugate addition (Sections 10.10 and 18.13): Addition reaction in which the reagent adds to the termini of the conjugated system with migration of the double bond; synonymous with 1,4 addition. The most common examples include conjugate addition to 1,3-dienes and to α,β-unsaturated carbonyl compounds.

Conjugate base (Section 4.6): The species formed from a Brønsted acid after it has donated a proton.

Conjugated diene (Section 10.5): System of the type $C{=}C{-}C{=}C$, in which two pairs of doubly bonded carbons are joined by a single bond. The π electrons are delocalized over the unit of four consecutive sp^2 hybridized carbons.

Connectivity (Section 1.7): Order in which a molecule's atoms are connected. Synonymous with *constitution.*

Conrotatory (Section 10.14): Describing the stereochemical pathway in which substituents at the ends of the π electron system rotate in the same sense (both clockwise or both counterclockwise) during an electrocyclic reaction. The opposite of *conrotatory* is *disrotatory.*

Constitution (Section 1.7): Order of atomic connections that defines a molecule.

Constitutional isomers (Section 1.8): Isomers that differ in respect to the order in which the atoms are connected. Butane ($CH_3CH_2CH_2CH_3$) and isobutane [$(CH_3)_3CH$] are constitutional isomers.

Copolymer (Section 10.11): Polymer formed from two or more different monomers.

Coupling constant *J* (Section 13.8): A measure of the extent to which two nuclear spins are coupled. In the simplest cases, it is equal to the distance between adjacent peaks in a split nmr signal.

Covalent bond (Section 1.3): Chemical bond between two atoms that results from their sharing of two electrons.

Cracking (Section 2.15): A key step in petroleum refining in which high-molecular-weight hydrocarbons are converted to lower-molecular-weight ones by thermal or catalytic carbon-carbon bond cleavage.

Critical micelle concentration (Section 19.5): Concentration above which substances such as salts of fatty acids aggregate to form micelles in aqueous solution.

Crown ether (Section 16.4): A cyclic polyether that, via ion-dipole attractive forces, forms stable complexes with metal ions. Such complexes, along with their accompanying anion, are soluble in nonpolar solvents.

C terminus (Section 27.7): The amino acid at the end of a peptide or protein chain that has its carboxyl group intact—i.e., in which the carboxyl group is not part of a peptide bond.

Cumulated diene (Section 10.5): Diene of the type C=C=C, in which a single carbon atom participates in double bonds with two others.

Cyanohydrin (Section 17.8): Compound of the type

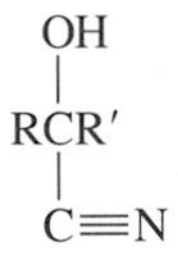

```
 OH
 |
RCR′
 |
 C≡N
```

Cyanohydrins are formed by nucleophilic addition of HCN to the carbonyl group of an aldehyde or a ketone.

Cycloaddition (Section 10.12): Addition, such as the Diels-Alder reaction, in which a ring is formed via a cyclic transition state.

Cycloalkane (Section 2.14): An alkane in which a ring of carbon atoms is present.

Cycloalkene (Section 5.1): A cyclic hydrocarbon characterized by a double bond between two of the ring carbons.

Cycloalkyne (Section 9.4): A cyclic hydrocarbon characterized by a triple bond between two of the ring carbons.

Cyclohexadienyl anion (Section 23.6): The key intermediate in nucleophilic aromatic substitution by the addition-elimination mechanism. It is represented by the general structure shown, where Y is the nucleophile and X is the leaving group.

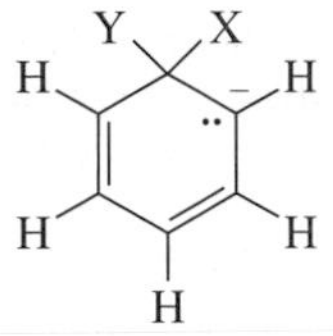

Cyclohexadienyl cation (Section 12.2): The key intermediate in electrophilic aromatic substitution reactions. It is represented by the general structure

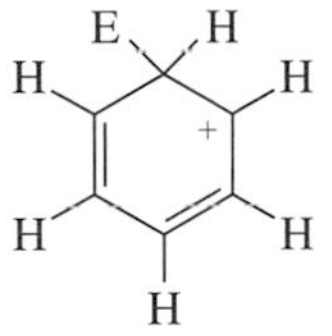

where E is derived from the electrophile that attacks the ring.

Deactivating substituent (Sections 12.10 and 12.13): A group that when present in place of a hydrogen substituent causes a particular reaction to occur more slowly. The term is most often applied to the effect of substituents on the rate of electrophilic aromatic substitution.

Debye unit (D) (Section 1.5): Unit customarily used for measuring dipole moments:

$$1\ \text{D} = 1 \times 10^{-18}\ \text{esu} \cdot \text{cm}$$

Decarboxylation (Section 19.17): Reaction of the type $RCO_2H \rightarrow RH + CO_2$, in which carbon dioxide is lost from a carboxylic acid. Decarboxylation normally occurs readily only when the carboxylic acid is a 1,3-dicarboxylic acid or a β-keto acid.

Decoupling (Section 13.17): In nmr spectroscopy, any process that destroys the coupling of nuclear spins between two nuclei. Two types of decoupling are employed in ^{13}C nmr spectroscopy. *Broadband decoupling* removes all the $^{1}H-^{13}C$ couplings; *off-resonance decoupling* removes all of $^{1}H-^{13}C$ couplings except those between directly bonded atoms.

Dehydration (Section 5.9): Removal of H and OH from adjacent atoms. The term is most commonly employed in the preparation of alkenes by heating alcohols in the presence of an acid catalyst.

Dehydrogenation (Section 5.1): Removal of the elements of H_2 from adjacent atoms. The term is most commonly encountered in the industrial preparation of ethylene from ethane, propene from propane, 1,3-butadiene from butane, and styrene from ethylbenzene.

Dehydrohalogenation (Section 5.14): Reaction in which an alkyl halide, on being treated with a base such as sodium ethoxide, is converted to an alkene by loss of a proton from one carbon and the halogen from the adjacent carbon.

Delocalization (Section 1.9): Association of an electron with more than one atom. The simplest example is the shared electron pair (covalent) bond. Delocalization is important in conjugated π electron systems, where an electron may be associated with several carbon atoms.

Deoxy sugar (Section 25.10): A carbohydrate in which one of the hydroxyl groups has been replaced by a hydrogen.

Detergents (Section 19.5): Substances that clean by micellar action. While the term usually refers to a synthetic detergent, soaps are also detergents.

Diastereomers (Section 7.10): Stereoisomers that are not enantiomers—stereoisomers that are not mirror images of one another.

Diastereotopic (Section 13.7): Describing two atoms or groups in a molecule that are attached to the same atom but are in stereochemically different environments that are not mirror images of each other. The two protons shown in bold in $\mathbf{CH_2}$=CHCl, for example, are diastereotopic. One is cis to chlorine, the other is trans.

1,3-Diaxial repulsion (Section 3.9): Repulsive forces between axial substituents on the same side of a cyclohexane ring.

Diazonium ion (Sections 22.16–22.17): Ion of the type R—$\overset{+}{N}$≡N:. Aryl diazonium ions are formed by treatment of primary aromatic amines with nitrous acid. They are extremely useful in the preparation of aryl halides, phenols, and aryl cyanides.

Diazotization (Section 22.16): The reaction by which a primary arylamine is converted to the corresponding diazonium ion by nitrosation.

Dieckmann reaction (Section 21.2): An intramolecular version of the Claisen condensation.

Dielectric constant (Section 8.13): A measure of the ability of a material to disperse the force of attraction between oppositely charged particles. The symbol for dielectric constant is ϵ.

Diels-Alder reaction (Section 10.12): Conjugate addition of an alkene to a conjugated diene to give a cyclohexene derivative. Diels-Alder reactions are extremely useful in synthesis.

Dienophile (Section 10.12): The alkene that adds to the diene in a Diels-Alder reaction.

β-Diketone (Section 18.5): Compound of the type RC(=O)CH$_2$C(=O)R′; also referred to as a 1,3-diketone.

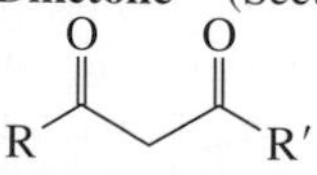

Dimer (Section 6.21): Molecule formed by the combination of two identical molecules.

Dipeptide (Section 27.7): A compound in which two α-amino acids are linked by an amide bond between the amino group of one and the carboxyl group of the other:

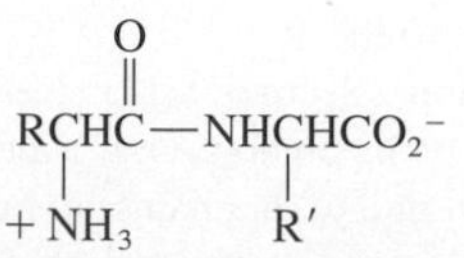

$$\underset{\overset{|}{+NH_3}}{RCH}\overset{O}{\overset{\|}{C}}—NH\underset{\overset{|}{R'}}{CH}CO_2^-$$

Dipole-dipole attraction (Section 2.16): A force of attraction between oppositely polarized atoms.

Dipole–induced dipole attraction (Section 4.5): A force of attraction that results when a species with a permanent dipole induces a complementary dipole in a second species.

Dipole moment (Section 1.5): Product of the attractive force between two opposite charges and the distance between them. Dipole moment has the symbol μ and is measured in Debye units (D).

Disaccharide (Sections 25.1 and 25.14): A carbohydrate that yields two monosaccharide units (which may be the same or different) on hydrolysis.

Dispersion force (Section 2.16): See *induced dipole–induced dipole attraction.*

Disrotatory (Section 10.14): Describing the stereochemical pathway in which substituents at the ends of the π electron system rotate in opposite senses during an electrocyclic reaction. The opposite of *disrotatory* is *conrotatory.*

Dissymmetric (Section 7.1): Describing an object that is not superposable on its mirror image. In referring to molecules, the term *chiral* is used.

Disubstituted alkene (Section 5.6): Alkene of the type R_2C=CH_2 or RCH=CHR. The groups R may be the same or different, they may be any length, and they may be branched or unbranched. The significant point is that there are two carbons *directly* bonded to the carbons of the double bond.

Disulfide bridge (Section 27.7): An S—S bond between the sulfur atoms of two cysteine residues in a peptide or protein.

DNA (deoxyribonucleic acid) (Section 27.25): A polynucleotide of 2′-deoxyribose present in the nuclei of cells that serves to store and replicate genetic information. Genes are DNA.

Double bond (Section 1.4): Bond formed by the sharing of four electrons between two atoms.

Double dehydrohalogenation (Section 9.8): Reaction in which a geminal dihalide or vicinal dihalide, on being treated with a very strong base such as sodium amide, is converted to an alkyne by loss of two protons and the two halogen substituents.

Double helix (Section 27.26): The form in which DNA normally occurs in living systems. Two complementary strands of DNA are associated with each other by hydrogen bonds between their base pairs, and each DNA strand adopts a helical shape.

Downfield (Section 13.4): The low-field region of an nmr spectrum. A signal that is downfield with respect to another lies to its left on the spectrum.

Eclipsed conformation (Section 3.1): Conformation in which bonds on adjacent atoms are aligned with one another. For example, the C—H bonds indicated in the structure shown are eclipsed.

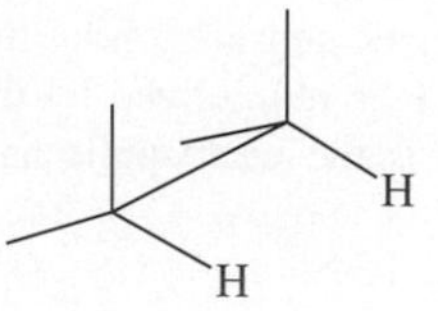

H
H

Edman degradation (Section 27.10): Method for determining the N-terminal amino acid of a peptide or protein. It involves treating the material with phenyl isothiocyanate ($C_6H_5N{=}C{=}S$), cleaving with acid, and then identifying the phenylthiohydantoin (PTH derivative) produced.

Elastomer (Section 10.11): A synthetic polymer that possesses elasticity.

Electrocyclic reaction (Section 10.14): A pericyclic reaction involving the formation of a σ bond between the ends of a conjugated π system or its reverse.

Electromagnetic radiation (Section 13.1): Various forms of radiation propagated at the speed of light. Electromagnetic radiation includes (among others) visible light; infrared, ultraviolet, and microwave radiation; and radio waves, cosmic rays, and x-rays.

Electron affinity (Section 1.2): Energy change associated with the capture of an electron by an atom.

Electronegativity (Section 1.5): A measure of the ability of an atom to attract the electrons in a covalent bond toward itself. Fluorine is the most electronegative element.

Electronic effect (Section 5.6): An effect on structure or reactivity that is attributed to the change in electron distribution that a substituent causes in a molecule.

Electron impact (Section 13.20): Method for producing positive ions in mass spectrometry whereby a molecule is bombarded by high-energy electrons.

Electrophile (Section 4.11): A species (ion or compound) that can act as a Lewis acid, or electron pair acceptor; an "electron seeker." Carbocations are one type of electrophile.

Electrophilic addition (Section 6.4): Mechanism of addition in which the species that first attacks the multiple bond is an electrophile ("electron seeker").

Electrophilic aromatic substitution (Section 12.1): Fundamental reaction type exhibited by aromatic compounds. An electrophilic species (E^+) attacks an aromatic ring and replaces one of the hydrogen substituents.

$$\text{Ar—H} + \text{E—Y} \longrightarrow \text{Ar—E} + \text{H—Y}$$

Electrophoresis (Section 27.3): Method for separating substances on the basis of their tendency to migrate to a positively or negatively charged electrode at a particular pH.

Electrostatic attraction (Section 1.2): Force of attraction between oppositely charged particles.

Elementary step (Section 4.7): A step in a reaction mechanism in which each species shown in the equation for this step participates in the same transition state. An elementary step is characterized by a single transition state.

Elements of unsaturation: See *SODAR.*

β-Elimination (Section 5.8): Reaction in which a double or triple bond is formed by loss of atoms or groups from adjacent atoms. (See *dehydration, dehydrogenation, dehydrohalogenation,* and *double dehydrohalogenation.*)

Elimination-addition mechanism (Section 23.8): Two-stage mechanism for nucleophilic aromatic substitution. In the first stage, an aryl halide undergoes elimination to form an aryne intermediate. In the second stage, nucleophilic addition to the aryne yields the product of the reaction.

Elimination bimolecular (E2) mechanism (Section 5.15): Mechanism for elimination of alkyl halides characterized by a transition state in which the attacking base removes a proton at the same time that the bond to the halide leaving group is broken.

Elimination unimolecular (E1) mechanism (Section 5.17): Mechanism for elimination characterized by the slow formation of a carbocation intermediate followed by rapid loss of a proton from the carbocation to form the alkene.

Enamine (Section 17.12): Product of the reaction of a secondary amine and an aldehyde or a ketone. Enamines are characterized by the general structure $R_2C{=}\underset{\displaystyle NR'_2}{\overset{}{C}}R$.

Enantiomeric excess (Section 7.4): Difference between the percentage of the major enantiomer present in a mixture and the percentage of its mirror image. An optically pure material has an enantiomeric excess of 100 percent. A racemic mixture has an enantiomeric excess of zero.

Enantiomers (Section 7.1): Stereoisomers that are related as an object and its nonsuperimposable mirror image.

Enantioselective synthesis (Section 27.4): Reaction that converts an achiral or racemic starting material to a chiral product in which one enantiomer is present in excess of the other.

Enantiotopic (Section 13.7): Describing two atoms or groups in a molecule whose environments are nonsuperposable mirror images of each other. The two protons shown in bold in $CH_3\mathbf{CH_2}Cl$, for example, are enantiotopic. Replacement of first one, then the other, by some arbitrary test group yields compounds that are enantiomers of each other.

Endothermic (Section 1.2): Term describing a process or reaction that absorbs heat.

Enediyne antibiotics (Section 9.4): A family of tumor-inhibiting substances characterized by the presence of a $C{\equiv}C{-}C{=}C{-}C{\equiv}C$ unit as part of a 9- or 10-membered ring.

Energy of activation (Section 3.2): Minimum energy that a reacting system must possess above its most stable state in order to undergo a chemical or structural change.

Enol (Section 9.13): Compound of the type $R\overset{\displaystyle OH}{\overset{|}{C}}{=}CR_2$. Enols are in equilibrium with an isomeric aldehyde or ketone, but are normally much less stable than aldehydes and ketones.

Enolate ion (Section 18.6): The conjugate base of an enol. Enolate ions are stabilized by electron delocalization.

$$R\overset{O^-}{\overset{|}{C}}{=}CR_2 \longleftrightarrow R\overset{O}{\overset{\|}{C}}{-}\ddot{C}R_2^{\,-}$$

Enthalpy (Section 2.17): The heat content of a substance; symbol, *H*.

Envelope (Section 3.11): One of the two most stable conformations of cyclopentane. Four of the carbons in the envelope conformation are coplanar; the fifth carbon lies above or below this plane.

Enzyme (Section 27.19): A protein that catalyzes a chemical reaction in a living system.

Epimers (Section 25.22): Diastereomers that differ in configuration at only one of their stereogenic centers.

Epoxidation (Section 6.18): Conversion of an alkene to an epoxide by treatment with a peroxy acid.

Epoxide (Section 6.18): Compound of the type

$$R_2C\overset{\quad}{\underset{O}{\diagdown\!\diagup}}CR_2.$$

Equatorial bond (Section 3.7): A bond to a carbon in the chair conformation of cyclohexane oriented approximately along the equator of the molecule.

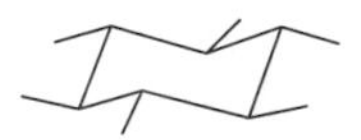

Erythro (Section 7.11): Term applied to the relative configuration of two stereogenic centers within a molecule. The erythro stereoisomer has like substituents on the same side of a Fischer projection.

Essential amino acids (Section 27.1): Amino acids that must be present in the diet for normal growth and good health.

Essential fatty acids (Section 26.6): Fatty acids that must be present in the diet for normal growth and good health.

Essential oils (Section 26.7): Pleasant-smelling oils of plants consisting of mixtures of terpenes, esters, alcohols, and other volatile organic substances.

Ester (Section 20.1): Compound of the type $R\overset{O}{\overset{\|}{C}}OR'$.

Estrogen (Section 26.15): A female sex hormone.

Ethene (Section 5.1): IUPAC name for $CH_2{=}CH_2$. The common name *ethylene,* however, is used far more often, and the IUPAC rules permit its use.

Ether (Section 16.1): Molecule that contains a C—O—C unit such as ROR′, ROAr, or ArOAr. When the two groups bonded to oxygen are the same, the ether is described as a *symmetrical ether.* When the groups are different, it is called a *mixed ether.*

Ethylene (Section 5.1): $CH_2{=}CH_2$, the simplest alkene and the most important industrial organic chemical.

Ethyl group (Section 2.12): The group CH_3CH_2—.

Exothermic (Section 1.2): Term describing a reaction or process that gives off heat.

Extinction coefficient: See *molar absorptivity.*

***E-Z* notation for alkenes** (Section 5.4): System for specifying double-bond configuration that is an alternative to cis-trans notation. When higher-ranked substituents are on the same side of the double bond, the configuration is *Z.* When higher-ranked substituents are on opposite sides, the configuration is *E.* Rank is determined by the Cahn-Ingold-Prelog system.

Fats and oils (Section 20.6): Triesters of glycerol. Fats are solids at room temperature, oils are liquids.

Fatty acid (Section 26.2): Carboxylic acids obtained by hydrolysis of fats and oils. Fatty acids typically have unbranched chains and contain an even number of carbon atoms in the range of 12 to 20 carbons. They may include one or more double bonds.

Fatty acid synthetase (Section 26.3): Complex of enzymes that catalyzes the biosynthesis of fatty acids from acetate.

Field effect (Section 19.5): An electronic effect in a molecule that is transmitted from a substituent to a reaction site via the medium (e.g., solvent).

Fingerprint region (Section 13.18): The region 1400–625 cm^{-1} of an infrared spectrum. This region is less characteristic of functional groups than others, but varies so much from one molecule to another that it can be used to determine whether two substances are identical or not.

Fischer esterification (Sections 15.8 and 19.14): Acid-catalyzed ester formation between an alcohol and a carboxylic acid:

$$R\overset{O}{\overset{\|}{C}}OH + R'OH \xrightarrow{H^+} R\overset{O}{\overset{\|}{C}}OR' + H_2O$$

Fischer projection (Section 7.7): Method for representing stereochemical relationships. The four bonds to a stereogenic carbon are represented by a cross. The horizontal bonds are understood to project toward the viewer and the vertical bonds away from the viewer.

w, x, y, z on C (wedge/dash) is represented in a Fischer projection as w—+—y with x at top and z at bottom.

Formal charge (Section 1.6): The charge, either positive or negative, on an atom calculated by subtracting from the number of valence electrons in the neutral atom a number equal to the sum of its unshared electrons plus half the electrons in its covalent bonds.

Fragmentation pattern (Section 13.20): In mass spectrometry, the ions produced by dissociation of the molecular ion.

Free energy (Section 3.9): The available energy of a system; symbol, *G.*

Free radical (Section 4.18): Neutral species in which one of the electrons in the valence shell of carbon is unpaired. An example is methyl radical, $\cdot CH_3$.

Frequency (Section 13.1): Number of waves per unit time. Although often expressed in hertz (Hz), or cycles per second, the SI unit for frequency is s^{-1}.

Friedel-Crafts acylation (Section 12.7): An electrophilic aromatic substitution in which an aromatic compound reacts with an acyl chloride or a carboxylic acid anhydride in the presence of aluminum chloride. An acyl group becomes bonded to the ring.

$$Ar{-}H + RC({=}O){-}Cl \xrightarrow{AlCl_3} Ar{-}C({=}O)R$$

Friedel-Crafts alkylation (Section 12.6): An electrophilic aromatic substitution in which an aromatic compound reacts with an alkyl halide in the presence of aluminum chloride. An alkyl group becomes bonded to the ring.

$$Ar{-}H + R{-}X \xrightarrow{AlCl_3} Ar{-}R$$

Fries rearrangement (Section 24.9): Aluminum chloride-promoted rearrangement of an aryl ester to a ring-acylated derivative of phenol.

$$C_6H_5{-}OC({=}O)R \xrightarrow{AlCl_3} RC({=}O){-}C_6H_4{-}OH$$

Frontier orbitals (Section 10.15): Orbitals involved in a chemical reaction, usually the highest-occupied molecular orbital of one reactant and the lowest-unoccupied molecular orbital of the other.

Functional group (Section 2.2): An atom or a group of atoms in a molecule responsible for its reactivity under a given set of conditions.

Furanose form (Section 25.6): Five-membered ring arising via cyclic hemiacetal formation between the carbonyl group and a hydroxyl group of a carbohydrate.

Gabriel synthesis (Section 22.9): Method for the synthesis of primary alkylamines in which a key step is the formation of a carbon-nitrogen bond by alkylation of the potassium salt of phthalimide.

$$\text{(phthalimide)}NK \xrightarrow{RX} \text{(phthalimide)}N{-}R \longrightarrow RNH_2$$

Gauche (Section 3.1): Term describing the position relative to each other of two substituents on adjacent atoms when the angle between their bonds is on the order of 60°. Atoms X and Y in the structure shown are gauche to each other.

(Newman projection with X and Y gauche)

Geminal dihalide (Section 9.8): A dihalide of the form R_2CX_2, in which the two halogen substituents are located on the same carbon.

Geminal diol (Section 17.7): The hydrate $R_2C(OH)_2$ of an aldehyde or a ketone.

Genome (Section 27.28): The aggregate of all the genes that determine what an organism becomes.

Glycogen (Section 25.15): A polysaccharide present in animals that is derived from glucose. Similar in structure to amylopectin.

Glycoside (Section 25.13): A carbohydrate derivative in which the hydroxyl group at the anomeric position has been replaced by some other group. An *O*-glycoside is an ether of a carbohydrate in which the anomeric position bears an alkoxy group.

Grain alcohol (Section 4.2): A nonsystematic name for ethanol (CH_3CH_2OH).

Grignard reagent (Section 14.4): An organomagnesium compound of the type RMgX formed by the reaction of magnesium with an alkyl or aryl halide.

Half-chair (Section 3.11): One of the two most stable conformations of cyclopentane. Three consecutive carbons in the half-chair conformation are coplanar. The fourth and fifth carbon lie, respectively, above and below the plane.

Haloform reaction (Section 18.7): The formation of CHX_3 (X = Br, Cl, or I) brought about by cleavage of a methyl ketone on treatment with Br_2, Cl_2, or I_2 in aqueous base.

$$RC({=}O)CH_3 \xrightarrow[HO^-]{X_2} RC({=}O)O^- + CHX_3$$

Halogenation (Sections 4.16 and 12.5): Replacement of a hydrogen substituent by a halogen. The most frequently encountered examples are the free-radical halogenation of alkanes and the halogenation of arenes by electrophilic aromatic substitution.

Halohydrin (Section 6.17): A compound that contains both a halogen atom and a hydroxyl group. The term is most often used for compounds in which the halogen and the hydroxyl are on adjacent atoms (*vicinal halohydrins*). The most commonly encountered halohydrins are *chlorohydrins* and *bromohydrins*.

Halonium ion (Section 6.16): A species that incorporates a positively charged halogen. Bridged halonium ions are intermediates in the addition of halogens to the double bond of an alkene.

Hammond's postulate (Section 4.13): Principle used to deduce the approximate structure of a transition state. If two states, such as a transition state and an unstable intermediate derived from it, are similar in energy, they are believed to be similar in structure.

Haworth formulas (Section 25.6): Planar representations of furanose and pyranose forms of carbohydrates.

Heat of combustion (Section 2.17): Heat evolved on combustion of a substance. It is the value of $-\Delta H°$ for the combustion reaction.

Heat of formation (Section 2.17): The value of $\Delta H°$ for formation of a substance from its elements.

Heat of hydrogenation (Section 6.1): Heat evolved on hydrogenation of a substance. It is the value of $-\Delta H°$ for the addition of H_2 to a multiple bond.

α helix (Section 27.18): One type of protein secondary structure. It is a right-handed helix characterized by hydrogen bonds between NH and C=O groups. It contains approximately 3.6 amino acids per turn.

Hell-Volhard-Zelinsky reaction (Section 19.16): The phosphorus trihalide–catalyzed α halogenation of a carboxylic acid:

$$R_2CHCO_2H + X_2 \xrightarrow[\text{or } PX_3]{P} R_2C(X)CO_2H + HX$$

Hemiacetal (Section 17.9): Product of nucleophilic addition of one molecule of an alcohol to an aldehyde or a ketone. Hemiacetals are compounds of the type $R_2C(OH){-}OR'$.

Hemiketal (Section 17.9): An old name for a hemiacetal derived from a ketone.

Henderson-Hasselbalch equation (Section 19.4): An equation that relates degree of dissociation of an acid at a particular pH to its pK_a.

$$pH = pK_a + \log \frac{[\text{conjugate base}]}{[\text{acid}]}$$

Heteroatom (Section 1.7): An atom in an organic molecule that is neither carbon nor hydrogen.

Heterocyclic compound (Section 3.16): Cyclic compound in which one or more of the atoms in the ring are elements other than carbon. Heterocyclic compounds may or may not be aromatic.

Heterogeneous reaction (Section 6.1): A reaction involving two or more substances present in different phases. Hydrogenation of alkenes is a heterogeneous reaction that takes place on the surface of an insoluble metal catalyst.

Heterolytic cleavage (Section 4.18): Dissociation of a two-electron covalent bond in such a way that both electrons are retained by one of the initially bonded atoms.

Hexose (Section 25.4): A carbohydrate with six carbon atoms.

Hofmann elimination (Section 22.14): Conversion of a quaternary ammonium hydroxide, especially an alkyltrimethylammonium hydroxide, to an alkene on heating. Elimination occurs in the direction that gives the less substituted double bond.

$$R_2CH{-}CR'_2(\overset{+}{N}(CH_3)_3)\ HO^- \xrightarrow{\text{heat}} R_2C{=}CR'_2 + N(CH_3)_3 + H_2O$$

Hofmann rearrangement (Section 20.17): Reaction in which an amide reacts with bromine in basic solution to give a primary amine having one less carbon atom than the amide.

$$RC(=O)NH_2 \xrightarrow[\text{NaOH, } H_2O]{Br_2} RNH_2$$

HOMO (Section 10.15): Highest occupied molecular orbital (the orbital of highest energy that contains at least one of a molecule's electrons).

Homologous series (Section 2.8): Group of structurally related substances in which successive members differ by a CH_2 group.

Homolytic cleavage (Section 4.18): Dissociation of a two-electron covalent bond in such a way that one electron is retained by each of the initially bonded atoms.

Hückel's rule (Section 11.21): Completely conjugated planar monocyclic hydrocarbons possess special stability when the number of their π electrons $= 4n + 2$, where n is an integer.

Hund's rule (Section 1.1): When two orbitals are of equal energy, they are populated by electrons so that each is half-filled before either one is doubly occupied.

Hydration (Section 6.10): Addition of the elements of water (H, OH) to a multiple bond.

Hydride shift (Section 5.13): Migration of a hydrogen with a pair of electrons (H:) from one atom to another. Hydride shifts are most commonly seen in carbocation rearrangements.

Hydroboration-oxidation (Section 6.11): Reaction sequence involving a separate hydroboration stage and oxidation stage. In the hydroboration stage, diborane adds to an alkene to give an alkylborane. In the oxidation stage, the alkylborane is oxidized with hydrogen peroxide to give an alcohol. The reaction product is an alcohol corresponding to the anti-Markovnikov, syn hydration of an alkene.

Hydrocarbon (Section 2.1): A compound that contains only carbon and hydrogen.

Hydrogenation (Section 6.1): Addition of H_2 to a multiple bond.

Hydrogen bonding (Section 4.5): Type of dipole-dipole attractive force in which a positively polarized hydrogen of one molecule is weakly bonded to a negatively polarized atom of an adjacent molecule. Hydrogen bonds typically involve the hydrogen of one —OH group and the oxygen of another.

Hydrolysis (Section 6.9): Water-induced cleavage of a bond.

Hydronium ion (Section 4.6): The species H_3O^+.

Hydrophilic (Section 19.5): Literally, "water-loving"; a term applied to substances that are soluble in water, usually because of their ability to form hydrogen bonds with water.

Hydrophobic (Section 19.5): Literally, "water-hating"; a term applied to substances that are not soluble in water, but are soluble in nonpolar, hydrocarbonlike media.

Hydroxylation (Section 15.5): Reaction or sequence of reactions in which an alkene is converted to a vicinal diol.

Imide (Section 20.15): Compound of the type $RN(\overset{O}{\overset{\|}{C}}R')_2$, in which two acyl groups are bonded to the same nitrogen.

Imine (Section 17.11): Compound of the type $R_2C{=}NR'$ formed by the reaction of an aldehyde or a ketone with a primary amine ($R'NH_2$). Imines are sometimes called *Schiff's bases.*

Index of hydrogen deficiency: See *SODAR.*

Induced dipole–induced dipole attraction (Section 2.16): Force of attraction resulting from a mutual and complementary polarization of one molecule by another. Also referred to as *London forces* or *dispersion forces.*

Inductive effect (Section 4.11): An electronic effect transmitted by successive polarization of the σ bonds within a molecule or an ion.

Infrared (ir) spectroscopy (Section 13.18): Analytical technique based on energy absorbed by a molecule as it vibrates by stretching and bending bonds. Infrared spectroscopy is useful for analyzing the functional groups in a molecule.

Initiation step (Section 4.19): A process which causes a reaction, usually a free-radical reaction, to begin but which by itself is not the principal source of products. The initiation step in the halogenation of an alkane is the dissociation of a halogen molecule to two halogen atoms.

Integrated area (Section 13.7): The relative area of a signal in an nmr spectrum. Areas are proportional to the number of equivalent protons responsible for the peak.

Intermediate (Section 3.8): Transient species formed during a chemical reaction. Typically, an intermediate is not stable under the conditions of its formation and proceeds further to form the product. Unlike a transition state, which corresponds to a maximum along a potential energy surface, an intermediate lies at a potential energy minimum.

Intermolecular forces (Section 2.16): Forces, either attractive or repulsive, between two atoms or groups in *separate* molecules.

Intramolecular forces (Section 2.17): Forces, either attractive or repulsive, between two atoms or groups *within* the same molecule.

Inversion of configuration (Section 8.5): Reversal of the three-dimensional arrangement of the four bonds to sp^3 hybridized carbon. The representation shown illustrates inversion of configuration in a nucleophilic substitution where LG is the leaving group and Nu is the nucleophile.

$$(x)(w)(y)C{-}LG \longrightarrow Nu{-}C(x)(w)(y)$$

Ionic bond (Section 1.2): Chemical bond between oppositely charged particles that results from the electrostatic attraction between them.

Ionization energy (Section 1.2): Amount of energy required to remove an electron from some species.

Isobutane (Section 2.7): The common name for 2-methylpropane, $(CH_3)_3CH$.

Isobutyl group (Section 2.12): The group $(CH_3)_2CHCH_2{-}$.

Isoelectric point (Section 27.3): pH at which the concentration of the zwitterionic form of an amino acid is a maximum. At a pH below the isoelectric point the dominant species is a cation. At higher pH, an anion predominates. At the isoelectric point the amino acid has no net charge.

Isolated diene (Section 10.5): Diene of the type $C{=}C{-}(C)_x{-}C{=}C$, in which the two double bonds are separated by one or more sp^3 hybridized carbons. Isolated dienes are slightly less stable than isomeric conjugated dienes.

Isomers (Section 1.8): Different compounds that have the same molecular formula. Isomers may be either constitutional isomers or stereoisomers.

Isopentane (Section 2.9): The common name for 2-methylbutane, $(CH_3)_2CHCH_2CH_3$.

Isoprene unit (Section 26.7): The characteristic five-carbon structural unit found in terpenes:

Isopropyl group (Section 2.12): The group $(CH_3)_2CH{-}$.

Isotactic polymer (Section 7.15): A stereoregular polymer in which the substituent at each successive stereogenic center is on the same side of the zigzag carbon chain.

Isotopic cluster (Section 13.20): In mass spectrometry, a group of peaks that differ in m/z because they incorporate different isotopes of their component elements.

IUPAC nomenclature (Section 2.10): The most widely used method of naming organic compounds. It uses a set of rules proposed and periodically revised by the International Union of Pure and Applied Chemistry.

Kekulé structure (Section 11.3): Structural formula for an aromatic compound that satisfies the customary rules of bonding and is usually characterized by a pattern of alternating single and double bonds. There are two Kekulé formulations for benzene:

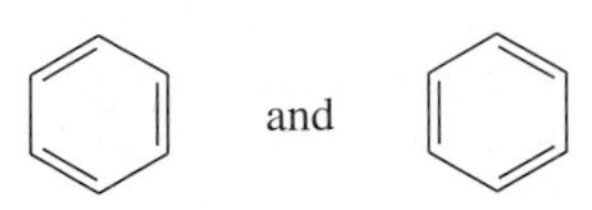

A single Kekulé structure does not completely describe the actual bonding in the molecule.

Ketal (Section 17.9): An old name for an acetal derived from a ketone.

Keto-enol tautomerism (Section 18.4): Process by which an aldehyde or a ketone and its enol equilibrate:

$$\mathrm{R\overset{O}{\overset{\|}{C}}{-}CHR_2 \rightleftharpoons R\overset{OH}{\overset{|}{C}}{=}CR_2}$$

Ketone (Section 17.1): A member of the family of compounds in which both atoms attached to a carbonyl group (C=O) are carbon, as in $\mathrm{R\overset{O}{\overset{\|}{C}}R}$, $\mathrm{R\overset{O}{\overset{\|}{C}}Ar}$, $\mathrm{Ar\overset{O}{\overset{\|}{C}}Ar}$.

Ketose (Section 25.1): A carbohydrate that contains a ketone carbonyl group in its open-chain form.

Kiliani-Fischer synthesis (Section 25.20): A synthetic method for carbohydrate chain extension. The new carbon-carbon bond is formed by converting an aldose to its cyanohydrin. Reduction of the cyano group to an aldehyde function completes the synthesis.

Kinetically controlled reaction (Section 10.10): Reaction in which the major product is the one that is formed at the fastest rate.

Kolbe-Schmitt reaction (Section 24.10): The high-pressure reaction of the sodium salt of a phenol with carbon dioxide to give an *o*-hydroxybenzoic acid. The Kolbe-Schmitt reaction is used to prepare salicylic acid in the synthesis of aspirin.

Lactam (Section 20.14): A cyclic amide.

Lactone (Section 19.15): A cyclic ester.

Lactose (Section 25.14): Milk sugar; a disaccharide formed by a β glycosidic linkage between C-4 of glucose and C-1 of galactose.

LDA (Section 21.10): Abbreviation for lithium diisopropylamide $LiN[CH(CH_3)_2]$. LDA is a strong, sterically hindered base, used to convert esters to their enolates.

Leaving group (Section 5.15): The group, normally a halide ion, that is lost from carbon in a nucleophilic substitution or elimination.

Le Châtelier's principle (Section 6.10): A reaction at equilibrium responds to any stress imposed upon it by shifting the equilibrium in the direction that minimizes the stress.

Lewis acid: See *acid.*

Lewis base: See *base.*

Lewis structure (Section 1.3): A chemical formula in which electrons are represented by dots. Two dots (or a line) between two atoms represent a covalent bond in a Lewis structure. Unshared electrons are explicitly shown, and stable Lewis structures are those in which the octet rule is satisfied.

Lipid bilayer (Section 26.4): Arrangement of two layers of phospholipids that constitutes cell membranes. The polar termini are located at the inner and outer membrane-water interfaces, and the lipophilic hydrocarbon tails cluster on the inside.

Lipids (Section 26.1): Biologically important natural products characterized by high solubility in nonpolar organic solvents.

Lipophilic (Section 19.5): Literally, ''fat-loving''; synonymous in practice with *hydrophobic.*

London force (Section 2.16): See *induced dipole–induced dipole attraction.*

LUMO (Section 10.15): The orbital of lowest energy that contains none of a molecule's electrons; the lowest unoccupied molecular orbital.

Magnetic resonance imaging (MRI) (Section 13.17): A diagnostic method in medicine in which tissues are examined by nmr.

Malonic ester synthesis (Section 21.7): Synthetic method for the preparation of carboxylic acids involving alkylation of the enolate of diethyl malonate ($\mathrm{CH_3CH_2O\overset{O}{\overset{\|}{C}}CH_2\overset{O}{\overset{\|}{C}}OCH_2CH_3}$) as the key carbon-carbon bond–forming step.

Maltose (Section 25.14): A disaccharide obtained from starch in which two glucose units are joined by an $\alpha(1,4)$ glycosidic link.

Markovnikov's rule (Section 6.5): An unsymmetrical reagent adds to an unsymmetrical double bond in the direction that places the positive part of the reagent on the carbon of the double bond that has the greater number of hydrogen substituents.

Mass spectrometry (Section 13.20): Analytical method in which a molecule is ionized and the various ions are examined on the basis of their mass-to-charge ratio.

Mechanism (Section 4.7): The sequence of steps that describes how a chemical reaction occurs; a description of the intermediates and transition states that are involved during the transformation of reactants to products.

Mercaptan (Section 15.13): An old name for the class of compounds now known as *thiols.*

Merrifield method: See *solid-phase peptide synthesis.*

Meso stereoisomer (Section 7.11): An achiral molecule that has stereogenic centers. The most common kind of *meso* compound is a molecule with two stereogenic centers and a plane of symmetry.

Messenger RNA (mRNA) (Section 27.27): A polynucleotide of ribose that ''reads'' the sequence of bases in DNA and interacts with tRNAs in the ribosomes to promote protein biosynthesis.

Meta (Section 11.9): Term describing a 1,3 relationship between substituents on a benzene ring.

Meta director (Section 12.9): A group that when present on a benzene ring directs an incoming electrophile to a position meta to itself.

Metalloenzyme (Section 27.19): An enzyme in which a metal ion at the active site contributes in a chemically significant way to the catalytic activity.

Methine group (Section 2.6): The group CH.

Methylene group (Section 2.6): The group $—CH_2—$.

Methyl group (Section 1.14): The group $CH_3—$.

Mevalonic acid (Section 26.10): An intermediate in the biosynthesis of steroids from acetyl coenzyme A.

Micelle (Section 19.5): A spherical aggregate of species such as carboxylate salts of fatty acids that contain a lipophilic end and a hydrophilic end. Micelles containing 50 to 100 carboxylate salts of fatty acids are soaps.

Michael addition (Sections 18.15 and 21.9): The conjugate addition of a carbanion (usually an enolate) to an α,β-unsaturated carbonyl compound.

Microscopic reversibility (Section 6.10): The principle that the intermediates and transition states in the forward and backward stages of a reversible reaction are identical, but are encountered in the reverse order.

Molar absorptivity (Section 13.19): A measure of the intensity of a peak, usually in uv-vis spectroscopy.

Molecular dipole moment (Section 1.11): The overall measured dipole moment of a molecule. It can be calculated as the resultant (or vector sum) of all the individual bond dipole moments.

Molecular formula (Section 1.7): Chemical formula in which subscripts are used to indicate the number of atoms of each element present in one molecule. In organic compounds, carbon is cited first, hydrogen second, and the remaining elements in alphabetical order.

Molecular ion (Section 13.20): In mass spectrometry, the species formed by loss of an electron from a molecule.

Molecular orbital theory (Section 1.12): Theory of chemical bonding in which electrons are assumed to occupy orbitals in molecules much as they occupy orbitals in atoms. The molecular orbitals are described as combinations of the orbitals of all of the atoms that make up the molecule.

Monomer (Section 6.21): The simplest stable molecule from which a particular polymer may be prepared.

Monosaccharide (Section 25.1): A carbohydrate that cannot be hydrolyzed further to yield a simpler carbohydrate.

Monosubstituted alkene (Section 5.6): An alkene of the type $RCH{=}CH_2$, in which there is only one carbon *directly* bonded to the carbons of the double bond.

Multiplicity (Section 13.8): The number of peaks into which a signal is split in nuclear magnetic resonance spectroscopy. Signals are described as *singlets, doublets, triplets,* and so on, according to the number of peaks into which they are split.

Mutarotation (Section 25.8): The change in optical rotation that occurs when a single form of a carbohydrate is allowed to equilibrate to a mixture of isomeric hemiacetals.

Neopentane (Section 2.9): The common name for 2,2-dimethylpropane, $(CH_3)_4C$.

Neurotransmitter (Section 22.5): Substance, usually a naturally occurring amine, that mediates the transmission of nerve impulses.

Newman projection (Section 3.1): Method for depicting conformations in which one sights down a carbon-carbon bond and represents the front carbon by a point and the back carbon by a circle.

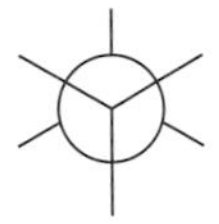

Nitration (Section 12.3): Replacement of a hydrogen substituent by an $—NO_2$ group. The term is usually used in connection with electrophilic aromatic substitution.

$$Ar—H \xrightarrow[H_2SO_4]{HNO_3} Ar—NO_2$$

Nitrile (Section 20.1): A compound of the type $RC{\equiv}N$. R may be alkyl or aryl. Also known as *alkyl* or *aryl cyanides.*

Nitrosamine See *N-nitrosamine.*

***N*-Nitrosamine** (Section 22.16): A compound of the type $R_2N—N{=}O$. R may be alkyl or aryl groups, which may be the same or different. *N*-Nitrosamines are formed by nitrosation of secondary amines.

Nitrosation (Section 22.16): The reaction of a substance, usually an amine, with nitrous acid. Primary amines yield diazonium ions; secondary amines yield *N*-nitrosamines. Tertiary aromatic amines undergo nitrosation of their aromatic ring.

Noble gases (Section 1.1): The elements in group VIIIA of the periodic table (helium, neon, argon, krypton, xenon, radon). Also known as the *rare gases,* they are, with few exceptions, chemically inert.

Nodal surface (Section 1.1): A plane drawn through an orbital where the algebraic sign of a wave function changes. The probability of finding an electron at a node is close to zero.

N terminus (Section 27.7): The amino acid at the end of a peptide or protein chain that has its α-amino group intact; i.e., the α-amino group is not part of a peptide bond.

Nuclear magnetic resonance (nmr) spectroscopy (Section 13.3): A method for structure determination based on the effect of molecular environment on the energy required to promote a given nucleus from a lower-energy spin state to a higher-energy state.

Nucleic acid (Section 27.25): A polynucleotide present in the nuclei of cells.

Nucleophile (Section 4.11): An atom that has an unshared electron pair which can be used to form a bond to carbon. Nucleophiles are Lewis bases.

Nucleophilic acyl substitution (Section 20.3): Nucleophilic substitution at the carbon atom of an acyl group.

Nucleophilic addition (Section 17.6): The characteristic reaction of an aldehyde or a ketone. An atom possessing an unshared electron pair bonds to the carbon of the C=O group, and some other species (normally hydrogen) bonds to the oxygen.

$$\mathrm{R\overset{\overset{\displaystyle O}{\|}}{C}R' + H{-}Y{:} \longrightarrow R\overset{\overset{\displaystyle OH}{|}}{\underset{\underset{\displaystyle R'}{|}}{C}}{-}Y{:}}$$

Nucleophilic aliphatic substitution (Chapter 8): Reaction in which a nucleophile replaces a leaving group, usually a halide ion, from sp^3 hybridized carbon. Nucleophilic aliphatic substitution may proceed by either an S_N1 or an S_N2 mechanism.

Nucleophilic aromatic substitution (Chapter 23): A reaction in which a nucleophile replaces a leaving group as a substituent on an aromatic ring. Substitution may proceed by an addition-elimination mechanism or an elimination-addition mechanism.

Nucleophilicity (Section 8.8): A measure of the reactivity of a Lewis base in a nucleophilic substitution reaction.

Nucleoside (Section 27.23): The combination of a purine or pyrimidine base and a carbohydrate, usually ribose or 2-deoxyribose.

Nucleotide (Section 27.24): The phosphate ester of a nucleoside.

Octane rating (Section 2.15): The capacity of a sample of gasoline to resist "knocking," expressed as a number equal to the percentage of 2,2,4-trimethylpentane ("isooctane") in an isooctane-heptane mixture that has the same knocking characteristics.

Octet rule (Section 1.3): When forming compounds, atoms gain, lose, or share electrons so that the number of their valence electrons is the same as that of the nearest noble gas. For the elements carbon, nitrogen, oxygen, and the halogens, this number is 8.

Oligosaccharide (Section 25.1): A carbohydrate that gives 3 to 10 monosaccharides on hydrolysis.

Optical activity (Section 7.4): Ability of a substance to rotate the plane of polarized light. In order to be optically active, a substance must be chiral, and one enantiomer must be present in excess of the other.

Optically pure (Section 7.4): Describing a chiral substance in which only a single enantiomer is present.

Orbital (Section 1.1): A region in space near an atom where there is a high probability (90 to 95 percent) of finding an electron.

σ Orbital (Section 1.12): A bonding orbital characterized by rotational symmetry.

σ* Orbital (Section 1.12): An antibonding orbital characterized by rotational symmetry.

Organometallic compound (Section 14.1): A compound that contains a carbon-to-metal bond.

Ortho (Section 11.9): Term describing a 1,2 relationship between substituents on a benzene ring.

Ortho-para director (Section 12.9): A group that when present on a benzene ring directs an incoming electrophile to the positions ortho and para to itself.

Oxidation (Section 2.18): A decrease in the number of electrons associated with an atom. In organic chemistry, oxidation of carbon occurs when a bond between carbon and an atom that is less electronegative than carbon is replaced by a bond to an atom that is more electronegative than carbon.

Oxime (Section 17.11): A compound of the type $R_2C{=}NOH$, formed by the reaction of hydroxylamine (NH_2OH) with an aldehyde or a ketone.

Oxonium ion (Section 4.6): Specific name for the species H_3O^+ (also called *hydronium ion*). General name for species such as alkyloxonium ions ROH_2^+ analogous to H_3O^+.

Oxymercuration-demercuration (Section 14.14): Synthetic method for converting alkenes to alcohols. An alkene is treated with a solution of mercury(II) acetate in aqueous tetrahydrofuran, followed by reduction with sodium borohydride. Hydration of the double bond occurs with a regiochemistry consistent with Markovnikov's rule.

Ozonolysis (Section 6.19): Ozone-induced cleavage of a carbon-carbon double or triple bond.

Para (Section 11.9): Term describing a 1,4 relationship between substituents on a benzene ring.

Paraffin hydrocarbons (Section 2.17): An old name for alkanes and cycloalkanes.

Partial rate factor (Section 12.10): In electrophilic aromatic substitution, a number that compares the rate of attack at a particular ring carbon with the rate of attack at a single position of benzene.

Pauli exclusion principle (Section 1.1): No two electrons can have the same set of four quantum numbers. An equivalent expression is that only two electrons can occupy the same orbital, and then only when they have opposite spins.

PCC (Section 15.10): Abbreviation for pyridinium chlorochromate $C_5H_5NH^+ \ ClCrO_3^-$. When used in an anhydrous medium, PCC oxidizes primary alcohols to aldehydes and secondary alcohols to ketones.

PDC (Section 15.10): Abbreviation for pyridinium dichromate $(C_5H_5NH)_2^{2+} \ Cr_2O_7^{2-}$. Used in same manner and for same purposes as PCC (see preceding entry).

***n*-Pentane** (Section 2.8): The common name for pentane, $CH_3CH_2CH_2CH_2CH_3$.

Pentose (Section 25.4): A carbohydrate with five carbon atoms.

Peptide (Section 27.7): Structurally, a molecule composed of two or more α-amino acids joined by peptide bonds.

Peptide bond (Section 27.7): An amide bond between the carboxyl group of one α-amino acid and the amino group of another:

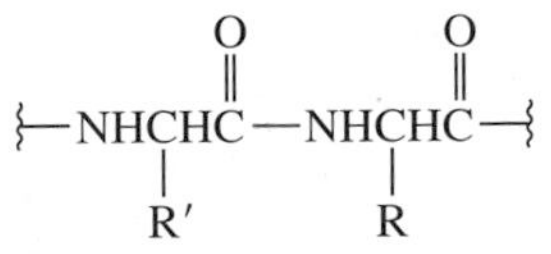

$$\text{—NHCHC(=O)—NHCHC(=O)—}$$
(R′ on the first CH, R on the second CH)

(The blue bond is the peptide bond.)

Pericyclic reaction (Section 10.12): A reaction that proceeds through a cyclic transition state.

Period (Section 1.1): A horizontal row of the periodic table.

Peroxide (Section 6.8): A compound of the type ROOR.

Peroxide effect (Section 6.8): Reversal of regioselectivity observed in the addition of hydrogen bromide to alkenes brought about by the presence of peroxides in the reaction mixture.

Phase-transfer catalysis (Section 22.6): Method for increasing the rate of a chemical reaction by transporting an ionic reactant from an aqueous phase where it is solvated and less reactive to an organic phase where it is not solvated and is more reactive. Typically, the reactant is an anion that is carried to the organic phase as its quaternary ammonium salt.

Phenols (Section 24.1): Family of compounds characterized by a hydroxyl substituent on an aromatic ring as in ArOH. *Phenol* is also the name of the parent compound, C_6H_5OH.

Phenyl group (Section 11.9): The group

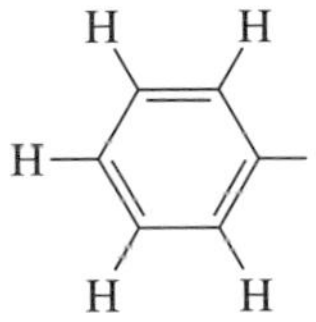

(benzene ring bearing five H atoms and one open bond)

It is often abbreviated C_6H_5—.

Phospholipid (Section 26.4): A diacylglycerol bearing a choline-phosphate "head group." Also known as *phosphatidylcholine.*

Photochemical reaction (Section 4.20): A chemical reaction that occurs when light is absorbed by a substance.

Photon (Section 13.1): Term for an individual "bundle" of energy, or particle, of electromagnetic radiation.

pK_a (Section 4.6): A measure of acid strength defined as $-\log K_a$. The stronger the acid, the smaller the value of pK_a.

Planck's constant (Section 13.1): Constant of proportionality (h) in the equation $E = h\nu$, which relates the energy (E) to the frequency (ν) of electromagnetic radiation.

Plane of symmetry (Section 7.3): A plane that bisects an object, such as a molecule, into two mirror-image halves; also called a *mirror plane.* When a line is drawn from any element in the object perpendicular to such a plane and extended an equal distance in the opposite direction, a duplicate of the element is encountered.

Pleated β sheet (Section 27.18): Type of protein secondary structure characterized by hydrogen bonds between NH and C=O groups of adjacent parallel peptide chains. The individual chains are in an extended zigzag conformation.

Polar covalent bond (Section 1.5): A shared electron pair bond in which the electrons are drawn more closely to one of the bonded atoms than the other.

Polarimeter (Section 7.4): An instrument used to measure optical activity.

Polarizability (Section 4.5): A measure of the ease of distortion of the electric field associated with an atom or a group. A fluorine atom in a molecule, for example, holds its electrons tightly and is very nonpolarizable. Iodine is very polarizable.

Polarized light (Section 7.4): Light in which the electric field vectors vibrate in a single plane. Polarized light is used in measuring optical activity.

Polyamide (Section 20.16): A polymer in which individual structural units are joined by amide bonds. Nylon is a synthetic polyamide; proteins are naturally occurring polyamides.

Polyamine (Section 22.5): A compound that contains many amino groups. The term is usually applied to a group of naturally occurring substances, including spermine, spermidine, and putrescine, that are believed to be involved in cell differentiation and proliferation.

Polycyclic aromatic hydrocarbon (Section 11.10): An aromatic hydrocarbon characterized by the presence of two or more benzenelike rings.

Polycyclic hydrocarbon (Section 3.15): A hydrocarbon in which two carbons are common to two or more rings.

Polyester (Section 20.16): A polymer in which individual structural units are joined by ester bonds.

Polyether (Section 16.4): A molecule that contains many ether linkages. Polyethers occur naturally in a number of antibiotic substances.

Polyethylene (Section 6.21): A polymer of ethylene.

Polymer (Section 6.21): Large molecule formed by the repetitive combination of many smaller molecules (monomers).

Polymerization (Section 6.21): Process by which a polymer is prepared. The principal processes include *free-radical, cationic, coordination,* and *condensation polymerization.*

Polypeptide (Section 27.1): A polymer made up of "many" (more than 8 to 10) amino acid residues.

Polypropylene (Section 6.21): A polymer of propene.

Polysaccharide (Sections 25.1 and 25.15): A carbohydrate that yields "many" monosaccharide units on hydrolysis.

Potential energy (Section 2.17): The energy a system has exclusive of its kinetic energy.

Potential energy diagram (Section 2.17): Plot of potential energy versus some arbitrary measure of the degree to which a reaction has proceeded (the reaction coordinate). The point of maximum potential energy is the transition state.

Primary alkyl group (Section 2.12): Structural unit of the type RCH_2—, in which the point of attachment is to a primary carbon.

Primary amine (Section 22.1): An amine with a single alkyl or aryl substituent and two hydrogens: an amine of the type RNH_2 (primary alkylamine) or $ArNH_2$ (primary arylamine).

Primary carbon (Section 2.12): A carbon that is directly attached to only one other carbon.

Primary structure (Section 27.8): The sequence of amino acids in a peptide or protein.

Principal quantum number (Section 1.1): The quantum number (n) of an electron that describes its energy level. An electron with $n = 1$ must be an s electron; one with $n = 2$ has s and p states available.

Propagation steps (Section 4.19): Fundamental steps that repeat over and over again in a chain reaction. Almost all of the products in a chain reaction arise from the propagation steps.

Protecting group (Section 17.10): A temporary alteration in the nature of a functional group so that it is rendered inert under the conditions in which reaction occurs somewhere else in the molecule. In order to be synthetically useful, a protecting group must be stable under a prescribed set of reaction conditions, yet be easily introduced and removed.

Protein (Chapter 27): A naturally occurring polymer that typically contains 100 to 300 amino acid residues.

Protic solvent (Section 8.13): A solvent that has easily exchangeable protons, especially protons bonded to oxygen as in hydroxyl groups.

Purine (Section 27.22): The heterocyclic aromatic compound.

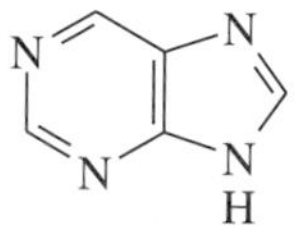

Pyranose form (Section 25.7): Six-membered ring arising via cyclic hemiacetal formation between the carbonyl group and a hydroxyl group of a carbohydrate.

Pyrimidine (Section 27.22): The heterocyclic aromatic compound.

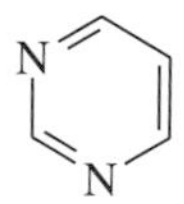

Quantum (Section 13.1): The energy associated with a photon.

Quaternary ammonium salt (Section 22.1): Salt of the type $R_4N^+ X^-$. The positively charged ion contains a nitrogen with a total of four organic substituents (any combination of alkyl and aryl groups).

Quaternary carbon (Section 2.12): A carbon that is directly attached to four other carbons.

Quaternary structure (Section 27.21): Description of the way in which two or more protein chains, not connected by chemical bonds, are organized in a larger protein.

Quinone (Section 24.14): The product of oxidation of an ortho or para dihydroxybenzene derivative. Examples of quinones include

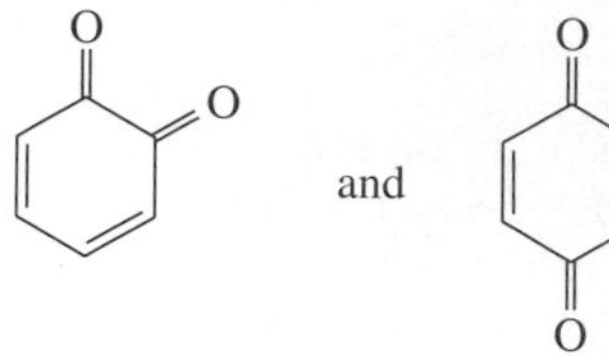

and

R (Section 2.2): Symbol for an alkyl group.

Racemic mixture (Section 7.4): Mixture containing equal quantities of enantiomers.

Radicofunctional nomenclature (Section 4.1): Type of IUPAC nomenclature in which compounds are named according to functional group families. The last word in the name identifies the functional group; the first word designates the alkyl or aryl group that bears the functional group. *Methyl bromide, ethyl alcohol,* and *diethyl ether* are examples of radicofunctional names.

Rate-determining step (Section 4.12): Slowest step of a multistep reaction mechanism. The overall rate of a reaction can be no faster than its slowest step.

Rearrangement (Section 5.13): Intramolecular migration of an atom, a group, or a bond from one atom to another.

Reducing sugar (Section 25.19): A carbohydrate that can be oxidized with substances such as Benedict's reagent. In general, a carbohydrate with a free hydroxyl group at the anomeric position.

Reduction (Section 2.18): Gain in the number of electrons associated with an atom. In organic chemistry, reduction of carbon occurs when a bond between carbon and an atom which is more electronegative than carbon is replaced by a bond to an atom which is less electronegative than carbon.

Reductive amination (Section 22.11): Method for the preparation of amines in which an aldehyde or a ketone is treated with ammonia or an amine under conditions of catalytic hydrogenation.

Refining (Section 2.15): Conversion of crude oil to useful materials, especially gasoline.

Reforming (Section 2.15): Step in oil refining in which the proportion of aromatic and branched-chain hydrocarbons in petroleum is increased so as to improve the octane rating of gasoline.

Regioselective (Section 4.20): Term describing a reaction that can produce two (or more) constitutional isomers but gives one of them in greater amounts than the other. A reaction that is 100 percent regioselective is termed regiospecific.

Relative configuration (Section 7.5): Stereochemical configuration on a comparative, rather than an absolute, basis. Terms such as D, L, erythro, threo, α, and β describe relative configuration.

Resolution (Section 7.14): Separation of a racemic mixture into its enantiomeric constituents.

Resonance (Section 1.9): Method by which electron delocalization may be shown using Lewis structures. The true electron distribution in a molecule is regarded as a hybrid of the various Lewis structures that can be written for a molecule.

Resonance energy (Section 10.6): Extent to which a substance is stabilized by electron delocalization. It is the difference in energy between the substance and a hypothetical model in which the electrons are localized.

Restriction enzymes (Section 27.28): Enzymes that catalyze the cleavage of DNA at specific sites.

Retention of configuration (Section 6.13): Stereochemical pathway observed when a new bond is made that has the same spatial orientation as the bond that was broken.

Retrosynthetic analysis (Section 14.9): Technique for synthetic planning based on reasoning backward from the target molecule to appropriate starting materials. An arrow of the type ⇨ designates a retrosynthetic step.

Ring inversion (Section 3.8): Process by which a chair conformation of cyclohexane is converted to a mirror-image chair. All of the equatorial substituents become axial, and vice versa. Also called *ring flipping,* or *chair-chair interconversion.*

RNA (ribonucleic acid) (Section 27.25): A polynucleotide of ribose.

Robinson annulation (Section 18.15): The combination of a Michael addition and an intramolecular aldol condensation used as a synthetic method for ring formation.

Rotamer (Section 3.1): Synonymous with *conformer.*

Sandmeyer reaction (Section 22.18): Reaction of an aryl diazonium ion with CuCl, CuBr, or CuCN to give respectively an aryl chloride, aryl bromide, or aryl cyanide (nitrile).

Sanger's reagent (Section 27.10): The compound 1-fluoro-2,4-dinitrobenzene, used in N-terminal amino acid identification.

Saponification (Section 20.10): Hydrolysis of esters in basic solution. The products are an alcohol and a carboxylate salt. The term means "soap making" and derives from the process whereby animal fats were converted to soap by heating with wood ashes.

Saturated hydrocarbon (Section 6.1): A hydrocarbon in which there are no multiple bonds.

Schiemann reaction (Section 22.18): Preparation of an aryl fluoride by heating the diazonium fluoroborate formed by addition of tetrafluoroboric acid (HBF_4) to a diazonium ion.

Schiff's base (Section 17.11): Another name for an imine; a compound of the type $R_2C{=}NR'$.

Secondary alkyl group (Section 2.12): Structural unit of the type $R_2CH—$, in which the point of attachment is to a secondary carbon.

Secondary amine (Section 22.1): An amine with any combination of two alkyl or aryl substituents and one hydrogen on nitrogen; an amine of the type RNHR′, RNHAr, or ArNHAr′.

Secondary carbon (Section 2.12): A carbon that is directly attached to two other carbons.

Secondary structure (Section 27.18): The conformation with respect to nearest-neighbor amino acids in a peptide or protein. The α helix and the pleated β sheet are examples of protein secondary structures.

Sequence rule (Section 7.6): Foundation of the Cahn-Ingold-Prelog system. It is a procedure for ranking substituents on the basis of atomic number.

Shielding (Section 13.4): Effect of a molecule's electrons that decreases the strength of an external magnetic field felt by a proton or another nucleus.

Sigmatropic rearrangement (Section 24.13): Migration of a σ bond from one end of a conjugated π electron system to the other. The Claisen rearrangement is an example.

Simmons-Smith reaction (Section 14.12): Reaction of an alkene with iodomethylzinc iodide to form a cyclopropane derivative.

Skew boat (Section 3.6): An unstable conformation of cyclohexane. It is slightly more stable than the boat conformation.

Soaps (Section 19.5): Cleansing substances obtained by the hydrolysis of fats in aqueous base. Soaps are sodium or potassium salts of unbranched carboxylic acids having 12 to 18 carbon atoms.

SODAR (Section 6.20): Acronym for *sum of double bonds and rings.* Examination of the molecular formula can provide information about a substance if one remembers that each double bond or ring in a molecule causes its molecular formula to contain two fewer hydrogens than the corresponding alkane.

$$\text{SODAR} = \tfrac{1}{2}(C_nH_{2n+2} - C_nH_x)$$

Solid-phase peptide synthesis (Section 27.17): Method for peptide synthesis in which the C-terminal amino acid is covalently attached to an inert solid support and successive amino acids are attached via peptide bond formation. At the completion of the synthesis the polypeptide is removed from the support.

Solvolysis reaction (Section 8.8): Nucleophilic substitution in a medium in which the only nucleophiles present are the solvent and its conjugate base.

Space-filling model (Section 1.10): A type of molecular model that attempts to represent the volume occupied by the atoms.

Specific rotation (Section 7.4): Optical activity of a substance per unit concentration per unit path length:

$$[\alpha] = \frac{100\alpha}{cl}$$

where α is the observed rotation in degrees, c is the concentration in g/100 mL, and l is the path length in decimeters.

Spectrometer (Section 13.1): Device designed to measure absorption of electromagnetic radiation by a sample.

Spectrum (Section 13.2): Output, usually in chart form, of a spectrometer. Analysis of a spectrum provides information about molecular structure.

***sp* hybridization** (Section 1.16): Hybridization state adopted by carbon when it bonds to two other atoms as, for example, in alkynes. The *s* orbital and one of the 2*p* orbitals mix to form two equivalent *sp* hybridized orbitals. A linear geometry is characteristic of *sp* hybridization.

sp^2 hybridization (Section 1.15): A model to describe the bonding of a carbon attached to three other atoms or groups. The carbon 2*s* orbital and the two 2*p* orbitals are combined to give a set of three equivalent sp^2 orbitals having 33.3 percent *s* character and 66.7 percent *p* character. One *p* orbital remains unhybridized. A trigonal planar geometry is characteristic of sp^2 hybridization.

sp^3 hybridization (Section 1.13): A model to describe the bonding of a carbon attached to four other atoms or groups. The carbon 2*s* orbital and the three 2*p* orbitals are combined to give a set of four equivalent orbitals having 25 percent *s* character and 75 percent *p* character. These orbitals are directed toward the corners of a tetrahedron.

Spin quantum number (Section 1.1): One of the four quantum numbers that describe an electron. An electron may have either of two different spin quantum numbers, $+\frac{1}{2}$ or $-\frac{1}{2}$.

Spin-spin coupling (Section 13.8): The communication of nuclear spin information between two nuclei.

Spin-spin splitting (Section 13.8): The splitting of nmr signals caused by the coupling of nuclear spins. Only nonequivalent nuclei (such as protons with different chemical shifts) can split one another's signals.

Spirocyclic hydrocarbon (Section 3.15): A hydrocarbon in which a single carbon is common to two rings.

Squalene (Section 26.11): A naturally occurring triterpene from which steroids are biosynthesized.

Staggered conformation (Section 3.1): Conformation of the type shown, in which the bonds on adjacent carbons are as far away from one another as possible.

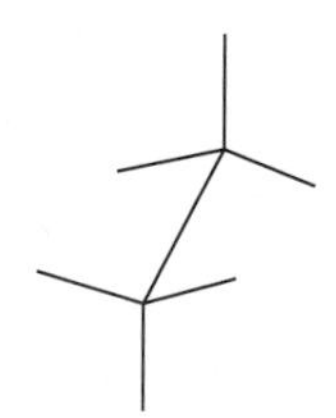

Stereochemistry (Chapter 7): Chemistry in three dimensions; the relationship of physical and chemical properties to the spatial arrangement of the atoms in a molecule.

Stereoelectronic effect (Section 5.15): An electronic effect that depends on the spatial arrangement between the orbitals of the electron donor and acceptor.

Stereogenic axis (Section 10.8): Line drawn through a molecule that is analogous to the long axis of a right-handed or left-handed screw or helix.

Stereogenic center (Section 7.2): An atom that has four nonequivalent atoms or groups attached to it. At various times stereogenic centers have been called *asymmetric centers* or *chiral centers.*

Stereoisomers (Section 3.13): Isomers which have the same constitution but which differ in respect to the arrangement of their atoms in space. Stereoisomers may be either *enantiomers* or *diastereomers.*

Stereoregular polymer (Section 7.15): Polymer containing stereogenic centers according to a regular repeating pattern. Syndiotactic and isotactic polymers are stereoregular.

Stereoselective reaction (Sections 5.13 and 6.3): Reaction in which a single starting material has the capacity to form two or more stereoisomeric products but forms one of them in greater amounts than any of its stereoisomers. Terms such as *addition to the less hindered side* describe stereoselectivity.

Stereospecific reaction (Section 7.13): Reaction in which stereoisomeric starting materials give stereoisomeric products. Terms such as *syn addition, anti elimination,* and *inversion of configuration* describe stereospecific reactions.

Steric hindrance (Section 7.7): In nucleophilic substitution, resistance to nucleophilic attack caused by the presence of large (or numerous) atoms or groups near the carbon that bears the leaving group.

Steric strain (Section 3.3): Synonymous with *van der Waals strain.*

Steroid (Section 26.11): Type of lipid present in both plants and animals characterized by a nucleus of four fused rings (three are six-membered, one is five-membered). Cholesterol is the most abundant steroid in animals.

Strecker synthesis (Section 27.4): Method for preparing amino acids in which the first step is reaction of an aldehyde with ammonia and hydrogen cyanide to give an amino nitrile, which is then hydrolyzed.

$$\underset{}{\overset{\displaystyle O}{\overset{\|}{R\text{CH}}}} \xrightarrow[\text{HCN}]{\text{NH}_3} \underset{\displaystyle \text{NH}_2}{\text{RCHC}{\equiv}\text{N}} \xrightarrow{\text{hydrolysis}} \underset{\displaystyle {}^{+}\text{NH}_3}{\text{RCHCO}_2^-}$$

Stretching vibration (Section 13.18): A regular, repetitive motion of two atoms or groups along the bond that connects them.

Structural isomer (Section 1.8): Synonymous with *constitutional isomer.*

Substitution nucleophilic bimolecular (S_N2) mechanism (Section 8.4): Concerted mechanism for nucleophilic substitution in which the nucleophile attacks carbon from the side opposite the bond to the leaving group and assists the departure of the leaving group.

Substitution nucleophilic unimolecular (S_N1) mechanism (Section 8.9): Mechanism for nucleophilic substitution characterized by a two-step process. The first step is rate-determining and is the ionization of an alkyl halide to a carbocation and a halide ion.

Substitution reaction (Section 4.9): Chemical reaction in which an atom or a group of a molecule is replaced by a different atom or group.

Substitutive nomenclature (Section 4.1): Type of IUPAC nomenclature in which a substance is identified by a name ending in a suffix characteristic of the type of compound. 2-Methylbutanol, 3-pentanone, and 2-phenylpropanoic acid are examples of substitutive names.

Sucrose (Section 25.14): A disaccharide of glucose and fructose in which the two monosaccharides are joined at their anomeric positions.

Sulfide (Section 16.1): A compound of the type RSR′. Sulfides are the sulfur analogs of ethers.

Sulfonation (Section 12.4): Replacement of a hydrogen substituent by an $—SO_3H$ group. The term is usually used in connection with electrophilic aromatic substitution.

$$\text{Ar—H} \xrightarrow[\text{H}_2\text{SO}_4]{\text{SO}_3} \text{Ar—SO}_3\text{H}$$

Sulfone (Section 16.16): Compound of the type $R—\overset{\ddot{O}:^-}{\underset{\ddot{O}:^-}{S^{2+}}}—R$.

Sulfoxide (Section 16.16): Compound of the type $R—\overset{\ddot{O}:^-}{\ddot{S}^+}—R$.

Symmetry-allowed reaction (Section 10.16): Concerted reaction in which the orbitals involved overlap in phase at all stages of the process. The conrotatory ring opening of cyclobutene to 1,3-butadiene is a symmetry-allowed reaction.

Symmetry-forbidden reaction (Section 10.16): Concerted reaction in which the orbitals involved do not overlap in phase at all stages of the process. The disrotatory ring opening of cyclobutene to 1,3-butadiene is a symmetry-forbidden reaction.

Syn addition (Section 6.3): Addition reaction in which the two portions of the reagent which add to a multiple bond add from the same side.

Syndiotactic polymer (Section 7.15): Stereoregular polymer in which the configuration of successive stereogenic centers alternates along the chain.

Synthon (Section 21.6): A structural unit in a molecule that is related to a synthetic operation.

Systematic nomenclature (Section 2.10): Names for chemical compounds that are developed on the basis of a prescribed set of rules. Usually the IUPAC system is meant when the term *systematic nomenclature* is used.

Tautomerism (Sections 9.13 and 18.4): Process by which two isomers are interconverted by an actual or formal movement of an atom or a group. Enolization is a form of tautomerism.

$$R\overset{O}{\overset{\|}{C}}—CHR_2 \rightleftharpoons R\overset{OH}{\overset{|}{C}}=CR_2$$

Terminal alkyne (Section 9.1): Alkyne of the type $RC\equiv CH$, in which the triple bond appears at the end of the chain.

Termination steps (Section 4.19): Reactions that halt a chain reaction. In a free-radical chain reaction, termination steps consume free radicals without generating new radicals to continue the chain.

Terpenes (Section 26.7): Compounds that can be analyzed as clusters of isoprene units. Terpenes with 10 carbons are classified as *monoterpenes,* those with 15 are *sesquiterpenes,* those with 20 are *diterpenes,* and those with 30 are *triterpenes.*

Tertiary alkyl group (Section 2.12): Structural unit of the type $R_3C—$, in which the point of attachment is to a tertiary carbon.

Tertiary amine (Section 22.1): Amine of the type R_3N with any combination of three alkyl or aryl substituents on nitrogen.

Tertiary carbon (Section 2.12): A carbon that is directly attached to three other carbons.

Tertiary structure (Section 27.19): A description of how a protein chain is folded.

Tetrahedral intermediate (Section 19.14 and Chapter 20): The key intermediate in nucleophilic acyl substitution. Formed by nucleophilic addition to the carbonyl group of a carboxylic acid derivative.

Tetramethylsilane (TMS) (Section 13.5): The molecule $(CH_3)_4Si$, used as a standard to calibrate proton and carbon-13 nmr spectra.

Tetrasubstituted alkene (Section 5.6): Alkene of the type $R_2C=CR_2$, in which there are four carbons *directly* bonded to the carbons of the double bond. (The R groups may be the same or different.)

Tetrose (Section 25.3): A carbohydrate with four carbon atoms.

Thermochemistry (Section 2.17): The study of heat changes that accompany chemical processes.

Thermodynamically controlled reaction (Section 10.10): Reaction in which the reaction conditions permit two or more products to equilibrate, giving a predominance of the most stable product.

Thioester (Section 20.12): An *S*-acyl derivative of a thiol; a compound of the type $R\overset{O}{\overset{\|}{C}}SR'$.

Thiol (Section 15.13): Compound of the type RSH or ArSH.

Threo (Section 7.11): Term applied to the relative configuration of two stereogenic centers within a molecule. The threo stereoisomer has like substituents on opposite sides of a Fischer projection.

Torsional strain (Section 3.2): Decreased stability of a molecule that results from the eclipsing of bonds.

trans- (Section 3.13): Stereochemical prefix indicating that two substituents are on opposite sides of a ring or a double bond. (Contrast with the prefix *cis*-).

Transcription (Section 27.27): Construction of a strand of mRNA complementary to a DNA template.

Transfer RNA (tRNA) (Section 27.27): A polynucleotide of ribose that is bound at one end to a unique amino acid. This amino acid is incorporated into a growing peptide chain.

Transition state (Section 3.2): The point of maximum energy in an elementary step of a reaction mechanism.

Translation (Section 27.27): The "reading" of mRNA by various tRNAs, each one of which is unique for a particular amino acid.

Triacylglycerol (Section 26.2): A derivative of glycerol (1,2,3-propanetriol) in which the three oxygens bear acyl groups derived from fatty acids.

Tripeptide (Section 27.1): A compound in which three α-amino acids are linked by peptide bonds.

Triple bond (Section 1.4): Bond formed by the sharing of six electrons between two atoms.

Trisubstituted alkene (Section 5.6): Alkene of the type $R_2C{=}CHR$, in which there are three carbons *directly* bonded to the carbons of the double bond. (The R groups may be the same or different.)

Trivial nomenclature (Section 2.10): Term synonymous with *common nomenclature.*

Trypsin (Section 27.12): A digestive enzyme that catalyzes the hydrolysis of proteins. Trypsin selectively catalyzes the cleavage of the peptide bond between the carboxyl group of lysine or arginine and some other amino acid.

Twist boat (Section 3.6): Synonymous with *skew boat.*

Ultraviolet-visible (uv-vis) spectroscopy (Section 13.19): Analytical method based on transitions between electronic energy states in molecules. Useful in studying conjugated systems such as polyenes.

Unimolecular (Section 4.12): Describing a step in a reaction mechanism in which only one particle undergoes a chemical change at the transition state.

α,β-Unsaturated aldehyde or ketone (Section 18.13): Aldehyde or ketone that bears a double bond between its α and β carbons as in $R_2C{=}CH\overset{O}{\overset{\|}{C}}R'$.

Unsaturated hydrocarbon (Section 6.1): A hydrocarbon that can undergo addition reactions; i.e., one which contains multiple bonds.

Upfield (Section 13.4): The high-field region of an nmr spectrum. A signal that is upfield with respect to another lies to its right on the spectrum.

Urethan (Section 20.17): Another name for a carbamate ester; a compound of the type (H_2NCO_2R).

Uronic acids (Section 25.19): Carbohydrates that have an aldehyde function at one end of their carbon chain and a carboxylic acid group at the other.

Valence-bond theory (Section 1.13): Theory of chemical bonding based on overlap of half-filled atomic orbitals between two atoms. Orbital hybridization is an important element of valence-bond theory.

Valence electrons (Section 1.1): The outermost electrons of an atom. For second-row elements these are the 2*s* and 2*p* electrons.

Valence-shell electron pair repulsion model (Section 1.10): Method for predicting the shape of a molecule based on the notion that electron pairs surrounding a central atom repel each other. Four electron pairs will arrange themselves in a tetrahedral geometry, three will assume a trigonal planar geometry, and two electron pairs will adopt a linear arrangement.

Van der Waals force (Section 2.16): Force of attraction or repulsion between two atoms or groups that are not bonded to each other.

Van der Waals radius (Section 2.16): A measure of the effective size of an atom or a group. The repulsive force between two atoms increases rapidly when they approach each other at distances less than the sum of their van der Waals radii.

Van der Waals strain (Section 3.3): Destabilization that results when two atoms or groups approach each other too closely. Also known as *van der Waals repulsion.*

Vicinal (Section 6.14): Describing two substituents that are located on adjacent atoms.

Vicinal coupling (Section 13.8): Coupling of the nuclear spins of atoms X and Y when they are substituents on adjacent atoms as in X—A—B—Y. Vicinal coupling is the most common cause of spin-spin splitting in ^1H nmr spectroscopy.

Vicinal diol (Section 15.5): Compound that has two hydroxyl (—OH) groups which are on adjacent sp^3 hybridized carbons.

Vinyl group (Section 5.1): The group $CH_2{=}CH{-}$.

Vitalism (Introduction): A nineteenth-century theory that divided chemical substances into two main classes, organic and inorganic, according to whether they originated in living (animal or vegetable) or nonliving (mineral) matter, respectively. Vitalist doctrine held that the conversion of inorganic substances to organic ones could be accomplished only through the action of some "vital force."

Wacker process (Section 17.4): Industrial process for the preparation of acetaldehyde by air oxidation of ethylene in the presence of palladium chloride.

Walden inversion (Section 8.5): Originally, a reaction sequence developed by Paul Walden whereby a chiral starting material was transformed to its enantiomer by a series of stereospecific reactions. Current usage is more general and refers to the inversion of configuration that attends any bimolecular nucleophilic substitution.

Wave functions (Section 1.1): The solutions to arithmetic expressions that express the energy of an electron in an atom.

Wavelength (Section 13.1): Distance between two successive maxima (peaks) or two successive minima (troughs) of a wave.

Wave numbers (Section 13.18): Conventional units in infrared spectroscopy that are proportional to frequency. Wave numbers are in reciprocal centimeters (cm^{-1}).

Wax (Section 26.5): A mixture of water-repellent substances that form a protective coating on the leaves of plants, the fur of animals, and the feathers of birds, among other things. A principal component of a wax is often an ester in which both the acyl portion and the alkyl portion are characterized by long carbon chains.

Williamson ether synthesis (Section 16.6): Method for the preparation of ethers involving an S_N2 reaction between an alkoxide ion and a primary alkyl halide:

$$\text{RONa} + \text{R'CH}_2\text{Br} \longrightarrow \text{R'CH}_2\text{OR} + \text{NaBr}$$

Wittig reaction (Section 17.13): Method for the synthesis of alkenes by the reaction of an aldehyde or a ketone with a phosphorus ylide.

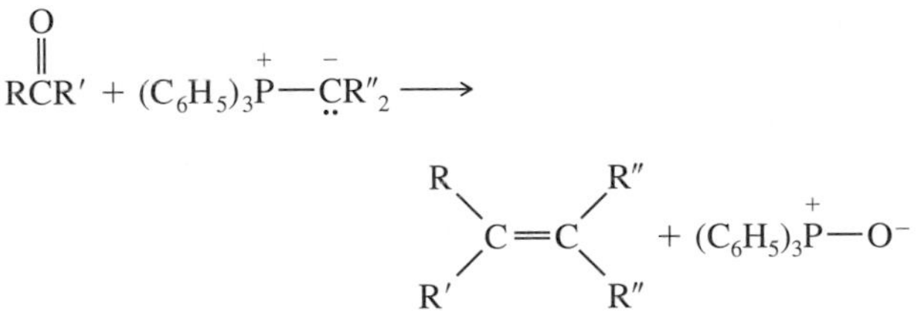

$$\text{R}\overset{\text{O}}{\overset{\|}{\text{C}}}\text{R'} + (\text{C}_6\text{H}_5)_3\overset{+}{\text{P}}-\overset{-}{\underset{\cdot\cdot}{\text{C}}}\text{R''}_2 \longrightarrow \text{RR'C}=\text{CR''}_2 + (\text{C}_6\text{H}_5)_3\overset{+}{\text{P}}-\text{O}^-$$

Wolff-Kishner reduction (Section 12.8): Method for reducing the carbonyl group of aldehydes and ketones to a methylene group ($C{=}O \rightarrow CH_2$) by treatment with hydrazine (H_2NNH_2) and base (KOH) in a high-boiling alcohol solvent.

Wood alcohol (Section 4.2): A nonsystematic name for methanol, CH_3OH.

Ylide (Section 17.13): A neutral molecule in which two oppositely charged atoms, each with an octet of electrons, are directly bonded to each other. The compound $(C_6H_5)_3\overset{+}{P}-\overset{-}{\underset{\cdot\cdot}{C}}H_2$ is an example of an ylide.

Zaitsev's rule (Section 5.10): When two or more alkenes are capable of being formed by an elimination reaction, the one with the more highly substituted double bond (the more stable alkene) is the major product.

Zwitterion (Section 27.3): The form in which neutral amino acids actually exist. The amino group is in its protonated form and the carboxyl group is present as a carboxylate ion.

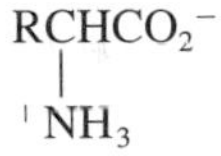

$$\text{R}\underset{\overset{|}{\overset{+}{\text{N}}\text{H}_3}}{\text{C}}\text{HCO}_2^-$$

INDEX